Proterozoic tectonic evolution of the Grenville orogen in North America

Edited by

Richard P. Tollo
Department of Earth and Environmental Sciences
George Washington University
Washington, D.C. 20052
USA

Louise Corriveau
Geological Survey of Canada
GSC-Québec, 880 Chemin Sainte-Foy
Québec, Québec G1S 2L2
Canada

James M. McLelland
Department of Geosciences
Skidmore College
Saratoga Springs, New York 12866
USA

Mervin J. Bartholomew
Department of Earth Sciences
University of Memphis
Memphis, Tennessee 38152
USA

THE
GEOLOGICAL
SOCIETY
OF AMERICA

Memoir 197

3300 Penrose Place, P.O. Box 9140 ▪ Boulder, Colorado 80301-9140, USA

2004

Published by The Geological Society of America, Inc.
3300 Penrose Place, P.O. Box 9140, Boulder, Colorado 80301-9140, USA
www.geosociety.org

Printed in U.S.A.

GSA Books Science Editor: Abhijit Basu

Library of Congress Cataloging-in-Publication Data

Proterozoic tectonic evolution of the Grenville orogen in North America / edited by Richard P. Tollo . . . [et al.].
p. cm.—(Memoir ; 197)
Includes bibliographical references, index.
ISBN 0-8137-1197-5
1. Geology, Stratigraphic—Proterozoic. 2. Geology—East (U.S.) 3. Geology—Canada, Eastern. I. Tollo, Richard P. II. Memoir (Geological Society of America) ; 197.

QE653.5.P78 2004
51.7′15′0974—dc22

2003071036

Cover: Strongly foliated garnetiferous gneiss of Grenvillian age with abundant light-colored leucosomes oriented parallel to foliation, photographed within Shenandoah National Park in the Blue Ridge province of Virginia.

10 9 8 7 6 5 4 3 2 1

Contents

Texas and Mexico

Acknowledgments

This memoir benefited greatly from thoughtful and careful reviews by the following individuals: John N. Aleinikoff, J. Lawford Anderson, Graham B. Baird, Suzanne Baldwin, Robert L. Bauer, James Beard, Pat Bickford, Andy Bobyarchick, John Bursnall, Sharon Carr, Sébastien Castonguay, Ray A. Coish, Marc Constantin, David Cornell, David Corrigan, Peter S. Dahl, Tony Davidson, Alan Dickin, Richard Diecchio, Craig Dietsch, Michael Easton, G. Nelson Eby, Mark A. Evans, Stewart S. Farrar, Ron Fodor, Paul D. Fullagar, Jack Garrihan, Charles Gilbert, Charles F. Gower, Lyal B. Harris, James Hibbard, John P. Hogan, Craig Johnson, Eric L. Johnson, Michael Higgins, Paul Karabinos, John Ketchum, Stephen A. Kish, Gunther Kletetschka, Z.X. Li, Wolf Maier, Nathalie Marchildon, Jacques Martignole, Brent Miller, David P. Moecher, Renato de Moraes, Paul Mueller, Peter Muller, Grant Alan Osborne, Victor Owen, Brent Owens, Yuanning Pan, William H. Peck, Virginia Peterson, John H. Puffer, Douglas W. Rankin, Toby Rivers, Scott Samson, Shelia J. Seaman, Donald T. Secor, Steve Sheppard, Phil Simony, Diane R. Smith, Luigi Solari, Frank S. Spear, Mark G. Steltenpohl, Kevin G. Stewart, Skip Stoddard, Margaret M. Streepey, Javier Fernandez Suarez, Paul Thayer, Eric Tohver, James F. Tull, David Valentino, Randy van Schmus, Gregory J. Walsh, Richard Wardle, Hardolf Wasteneys, Bodo Weber, Anthony Williams-Jones, Robert P. Wintsch, and Philip R. Whitney.

Preface

The compilation of papers included in this volume presents a systemwide perspective on Mesoproterozoic rocks and processes of the Grenville orogen and Appalachian inliers, and is the first such volume to include numerous contributions concerning the Grenvillian geology of Canada, the United States, and Mexico. This volume originated with the topical sessions entitled "Proterozoic tectonic evolution of the Grenville orogen in eastern North America," held at the annual meeting of the Geological Society of America on November 5, 2001, in Boston, Massachusetts. The goal of these sessions was to provide a forum for new data and ideas resulting from recent and ongoing studies conducted throughout the orogen, from southeastern Canada through the Adirondacks and Appalachian inliers to Texas and Mexico. The success of the Boston sessions provided motivation to compile a volume summarizing the state of multidisciplinary geological knowledge regarding the Grenville orogen. Compilation of this volume comes at an especially propitious time in the evolution of ideas concerning this widespread tectonic entity, because significant advances in understanding of the geology of the Grenville orogen in North America have materialized in recent years as a result of:

1. Multidimensional lithospheric mapping, which integrates process-oriented field studies, xenolith petrology, geophysical investigations, and remote-sensing analysis;
2. Breakthroughs in understanding intrusive processes, crustal rheology, protolith identification, and structural geology;
3. Application of improved analytical techniques, resulting in more accurate dating, characterization of source compositions, and terrane analysis;
4. Increased worldwide research focus on paleogeographic reconstructions and plate tectonic processes during the late Mesoproterozoic to late Neoproterozoic; and
5. Recent publication of geological syntheses for important sections of the orogen in Canada, including the volumes by Hynes and Ludden (2000) and Wardle and Hall (2002).

Publication of this volume commemorates the twentieth anniversary of "The Grenville Event in the Appalachians and related topics" (Bartholomew, 1984), which was the first major synthesis devoted to developing an integrated understanding of Grenvillian rocks in the United States. In the 20 years that have elapsed since publication of the Bartholomew (1984) volume, the application of new techniques and fresh geological perspectives has resulted in development of a more systemwide scientific approach toward the Grenville orogen that we hope is accurately reflected in this volume.

REFERENCES CITED

Bartholomew, M.J., ed., 1984, The Grenville Event in the Appalachians and related topics: Boulder, Colorado, Geological Society of America Special Paper 194, 287 p.

Hynes, A., and Ludden, J.N., eds., 2000, Special issue: The Lithoprobe Abitibi-Grenville transect: Canadian Journal of Earth Sciences, v. 37, no. 2 and 3.

Wardle, R.J., and Hall, J., eds., 2002, Special issue: Proterozoic evolution of the northeastern Canadian Shield: Lithoprobe eastern Canadian Shield onshore-offshore transect: Canadian Journal of Earth Sciences, v. 39, no. 5.

Geological Society of America
Memoir 197
2004

Proterozoic tectonic evolution of the Grenville orogen in North America: An introduction

Richard P. Tollo
Department of Earth Sciences, George Washington University, Washington, D.C. 20052, USA
Louise Corriveau
Geological Survey of Canada, GSC-Québec, 880 Chemin Sainte-Foy, Québec G1S 2L2, Canada
James McLelland
Department of Geosciences, Skidmore College, Saratoga Springs, New York 12866, USA
Mervin J. Bartholomew
Department of Earth Sciences, University of Memphis, Memphis, Tennessee 38152, USA

As illustrated by many papers in this volume, the fundamental tools used in geologic analysis of the complex igneous and metamorphic rocks constituting the Grenville Province and its various outliers include many traditional pursuits such as detailed field mapping, petrologic and geochemical analysis, and geochronologic investigation. However, continuous refinement of these tools and application of new strategies in protolith studies have provided novel avenues to improve understanding of the pre-orogenic nature of many of the gneiss- and granitoid-dominated terranes. Moreover, the techniques involved in such analysis have benefited significantly from recent technical advances, including the high spatial resolution available for isotopic analyses utilizing ion microprobe techniques, elemental mapping obtained through electron microprobe analysis, and the increased resolution of crustal seismic profiling. The numerous multiauthored papers in this volume underscore the importance of interdisciplinary studies in developing an improved understanding of Grenvillian and related rocks, and the geologic processes involved in their genesis.

In this volume, the term *Grenville orogen* is used to refer to all areas affected by dominantly convergent-style orogenesis during the interval ca. 1.3–1.0 Ga. From the outset, this project was designed to include contributions concerned with the entire historical span of the Grenville orogen, including those (1) identifying geological precursors to terrane accretion, (2) deciphering the sequence of tectonic events involved in creation of the orogen, and (3) documenting its eventual geologic destruction. Most papers focus on aspects of the ca. 1.3–1.0 Ga time frame, which represents an important period of convergent tectonics along the southeastern margin of Laurentia. Because of the protracted nature of this convergence and the various geologic entities involved, these papers cover a broad range of topics that collectively illustrate the complexity of the orogen and the processes involved in its creation. Such papers present results from studies located throughout the geographic extent of the Grenville orogen and related outliers in North America (Fig. 1). A subset of papers concerns the Late Neoproterozoic rifting of Rodinia and the subsequent breakup of the Grenville orogen. The geographic focus of these latter papers is centered on the southern Appalachians and the Blue Ridge Province of Virginia, where the geologic record of Neoproterozoic extension is well preserved. Individually, these papers involve topics related to the three main stages in the regional Neoproterozoic extensional regime, including (1) failed rifting at ca. 750 Ma, (2) successful rifting at ca. 570 Ma, and (3) subsequent establishment of a passive tectonic margin.

TEMPORAL SUBDIVISIONS AND NOMENCLATURE

Many of the studies concerned with Grenvillian (*sensu lato*) orogenesis utilize results from detailed mapping and U-Pb isotopic analysis to calibrate the chronology of geologic events on both local and regional scales. Such temporal calibrations and broad correlations are elements critical to developing a firm understanding of local geological evolution, which is, in turn, an essential requirement for establishing regional plate tectonic models. In these pursuits, clear communication of geologically constrained time intervals and potentially correlatable events is extremely important. Nevertheless, recent advances in geological knowledge and analytical technology have resulted in sev-

Tollo, R.P., Corriveau, L., McLelland, J., and Bartholomew, M.J., 2004, Proterozoic tectonic evolution of the Grenville orogen in North America: An introduction, *in* Tollo, R.P., Corriveau, L., McLelland, J., and Bartholomew, M.J., eds., Proterozoic tectonic evolution of the Grenville orogen in North America: Boulder, Colorado, Geological Society of America Memoir 197, p. 1–18. For permission to copy, contact editing@geosociety.org.

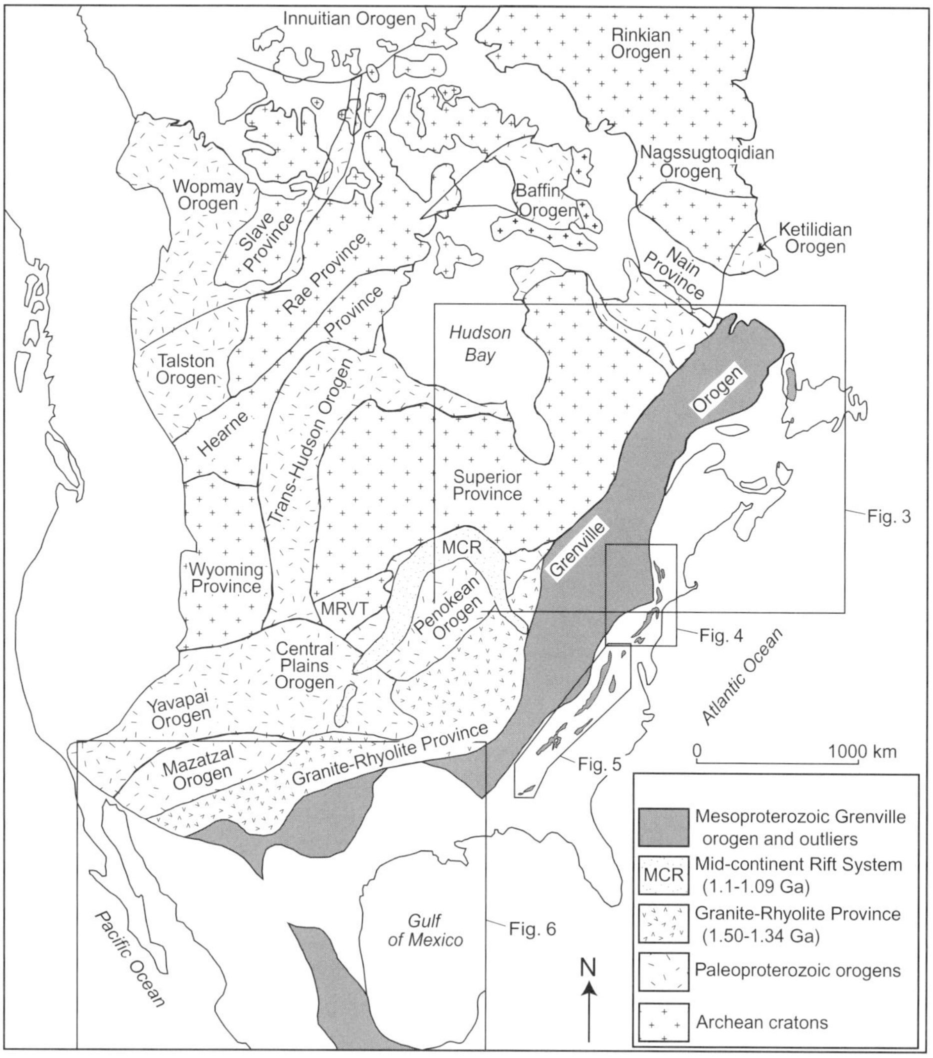

Figure 1. Generalized map showing the location of the Grenville orogen, including both its subsurface extent and the location of exposed outliers occurring within the Appalachians and its continuation in Texas and Mexico, in relation to other principal Precambrian lithotectonic elements of North America. Areas of younger orogens and rocks are unpatterned. Abbreviations include MCR—Mid-continent Rift System; MRVT—Minnesota River Valley terrane. Map modified from Rankin et al. (1990), Wheeler et al. (1996), Rivers (1997), Card and Poulsen (1998), and Davidson (1998a).

eral competing time scales and nomenclature schemes for tectonic events in the Grenville Province of Canada and the Adirondacks (Fig. 2). Some of the key contributions in this regard are summarized below.

The nature and timing of geological events that resulted in the development of the Canadian Grenville Province serve as a benchmark for interpretation of similar rocks throughout North America, and as a linchpin for reconstruction of plate tectonic configurations during the Mesoproterozoic. Identification and interpretation of the timing and sequence of events in the Canadian Grenville have evolved over time as (1) new field-based information has become available, (2) geophysical crustal mapping techniques have improved, (3) modeling of plate tectonic processes has become more sophisticated, and (4) new, high-precision isotopic age information for both igneous and metamorphic and/or deformational events has been obtained. The following brief synopsis of published syntheses regarding the timing and nature of Grenvillian orogenesis is presented on the following pages to provide readers with a reference frame in which to better understand the geologic context and nomenclature used to define orogenies in the papers included within this volume. Davidson (1998a) and Gower and Krogh (2002) provide more detailed historical perspectives and additional background on the evolution of geological knowledge in this province. Considering the present uncertainty with regard to the merits of various models, we did not request that contributors of papers to this volume (references noted in italics) conform to any particular scheme.

Wynne-Edwards (1972) and Moore and Thompson (1980) were among the first researchers to recognize multiple episodes of orogenesis within the Grenville Province. Moore and Thompson (1980) noted that deposition of the Flinton Group, which they constrained to the interval 1080–1050 Ma, occurred after arc-related magmatism, uplift, and erosion, and prior to major regional metamorphism. On this basis, they proposed the term "Elzevirian orogeny" to denote the period of activity that pre-

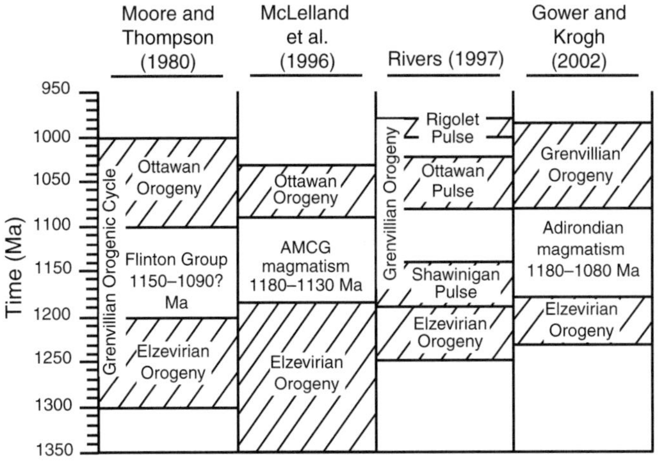

Figure 2. Comparison of calibrated time intervals and tectonic nomenclature for the 1350–980 Ma time interval of orogenic activity within the Grenville Province of Canada and the Adirondacks. The diagonal-lined pattern indicates orogenic episodes. AMCG—anorthosite-mangerite-charnockite-granite.

dated the Flinton Group, and "Ottawan orogeny" for the period that postdated its deposition (Fig. 2). By analogy with the Paleozoic Appalachian orogenic cycle of Rogers (1967), which includes at least three major regional pulses of activity, Moore and Thompson (1980) further suggested that the Elzevirian and Ottawan orogenies collectively constitute a corresponding Grenvillian orogenic cycle.

McLelland et al. (1996) defined similar periods of ca. 1350–1185 Ma arc-related Elzevirian and 1090–1035 Ma continent-continent–style Ottawan orogenic activity in the Adirondacks (Fig. 2). They further recognized a period of widespread magmatism involving primarily anorthosite-mangerite-charnockite-granite (AMCG) compositions, beginning shortly after cessation of Elzevirian activity, and extending through the interval 1180–1130 Ma that they inferred to be a result of lithospheric delamination associated with orogenic collapse. In the Adirondacks, AMCG magmatism was followed by the 1100–1090 Ma emplacement of the mildly alkaline Hawkeye suite, completing the local cycle of igneous activity occurring between the two major orogenic periods. Farther north in the western Grenville province of Canada, AMCG magmatism occurred in a time frame similar to that in the Adirondacks (e.g., 1165–1135 Ma for the Morin AMCG suites in the Morin terrane; Doig, 1991, and Martignole et al., 2000), partly coeval with granite-monzonite-gabbro intrusions such as the Frontenac and Chevreuil mafic-felsic suites in the Central Metasedimentary Belt (the Composite Arc Belt of Carr et al., 2000) (Davidson and van Breemen, 2000; Corriveau and van Breemen, 2000). Subsequent potassic alkaline magmatism occurred across most of the Composite Arc Belt between 1090 and 1075 Ma (Corriveau et al., 1990) providing further linkage with geology in the Adirondack segment of the orogen (Corrigan and Hanmer, 1997).

Rivers (1997) suggested that including both the Andean-style Elzevirian and Himalayan-style Ottawan orogenic episodes (Corrigan, 1995) within a single Grenvillian orogenic cycle (Moore and Thompson, 1980; Davidson, 1998a) obscured differences in the tectonic character of each episode of crustal activity. Following Corrigan and Hanmer (1997), Rivers (1997) proposed that accretionary events associated with closure of the Elzevir back-arc and coeval continental back-arc basins between ca. 1250 and 1190 Ma constituted the Elzevirian orogeny, whereas the largely compressional tectonics associated with terminal continent-continent collision defined the Grenvillian orogeny, which temporally overlapped the Ottawan orogeny originally defined by Moore and Thompson (1980) (Fig. 2). Rivers (1997) further suggested that the Grenvillian orogenic episode included at least three pulses of thrusting and high-grade metamorphism in the Grenville Province for which he suggested the names Shawinigan (ca. 1.19–1.14 Ga), Ottawan (1.08–1.02 Ga), and Rigolet (1.01–0.98 Ga) (Fig. 2). In a modification of this terminology, Rivers et al. (2002) subsequently upgraded the pulses to orogenies.

Considering both Grenvillian and pre-Grenvillian events in the development of the orogen, Gower and Krogh (2002) separated the geological evolution of the eastern Grenville Province into three stages: (1) Stage 1, which included pre-Labradorian (>1710 Ma) and Labradorian (1710–1600 Ma) events in which an outboard arc was created and, after short-lived subduction, subsequently accreted to pre-Labradorian proto–North America, leading to the establishment of a post-Labradorian passive margin; (2) Stage 2, in which a continental-margin arc was created during Pinwarian (1520–1460 Ma) orogenesis followed by flat-subduction-related Elsonian (1460–1230 Ma) magmatism; and (3) Stage 3, which included far-field effects of Elzevirian orogenesis (1230–1180 Ma), back-arc tectonism inboard of a post-Elzevirian continental margin (1180–1080 Ma), and Grenvillian continent-continent collision (1080–980 Ma). Gower and Krogh (2002) suggested that Grenvillian orogenic activity is best regarded as including the culminating collisional events between 1080 and 980 Ma (Fig. 2), arguing that only tectonism that occurred during this period is pan-Grenville in nature. In doing so, Gower and Krogh (2002) proposed a refinement in the use of the original terminology of Moore and Thompson (1980). These authors also suggested that orogenesis peaked at different times in different places in the eastern Grenville Province as the locus of crustal thrusting migrated throughout the area, resulting in periodic shifting of the "pressure point" responsible for compressional tectonism. Gower and Krogh (2002) referred to the extended post-Elzevirian, pre-Grenvillian period (1180–1080 Ma) of largely AMCG magmatism, which is widespread in the eastern Grenville Province, as "Adirondian" (Fig. 2) in acknowledgment of the many examples preserved in the Adirondacks.

It is apparent from the foregoing discussion that, as new data have become available, tectonic interpretations and temporal nomenclature for rocks of the Canadian Grenville Province and beyond have evolved considerably, and that the resolution of competing interpretations is still in progress. For this reason, readers of this volume are encouraged to consider and evaluate the nomenclatural framework used by individual authors in order to maximize understanding of the data and interpretations presented. Differences between the proposed time intervals for orogenic activity within Canada, and between Canada and the Adirondacks, reflect differences in location within the orogen, as well as contrasts in scientific approaches and interpretation of data that are beyond the scope of this introduction. Nevertheless, the preponderance of evidence available demonstrates the following:

1. A major series of tectonic events occurred throughout the Grenville Province of Canada and the Adirondacks at ca. 1080 to 980 Ma—the "Grenvillian Orogeny" of Gower and Krogh (2002), the "Ottawan Orogeny" of Moore and Thompson (1980) and McLelland et al. (1996), and the "Ottawan" and "Rigolet" pulses of Rivers (1997)—reflecting a period of strong continent-continent collision. Tectonism that occurred during this interval resulted in the creation of the Grenville Province in its present configuration (Gower and Krogh, 2002), and is well documented throughout the geographic area addressed by the papers in this volume.
2. An earlier period of largely accretionary tectonic activity ended at ca. 1180 Ma. However, effects of Elzevirian orogenesis are largely absent south of the Green Mountains and the Adirondacks (McLelland et al., 1996) and, in Canada, are limited to the southwest portions of the Grenville Province (Gower and Krogh, 2002).
3. An extended period of dominantly AMCG magmatism occurred during the interval between the two orogenies (McLelland et al., 1996), and was also widespread farther east in the Grenville Province, where evidence for the accretionary Elzirian orogeny is lacking. Results from papers included in this volume indicate that such AMCG magmatism also occurred within the Appalachian inliers (*Tollo et al.*), and possibly in Texas (*Mosher et al.*).

Thus, with our present knowledge, only the 1080–980 Ma period of orogeny and the preceding episode of AMCG-dominated magmatism are recognized throughout most areas of Grenvillian rocks in North America. The pan-Grenvillian orogenesis that ended at ca. 980 Ma, and is the subject of many papers in this volume, resulted in the development of an orogenic belt of considerable width and lateral extent. These characteristics, and the extensive crustal thickening that resulted in widespread attainment of high ambient temperatures of metamorphism, bear witness to the major convergence and crustal shortening that attended Grenvillian orogenesis, and that mark the tectonic amalgamation of cratons that ultimately resulted in the creation of Rodinia (Dalziel et al., 2000).

CANADIAN GRENVILLE PROVINCE

The Grenville Province of Canada, for which the toponym is the village of Grenville in Québec (e.g., Logan, 1863), represents the classic location for and the longest continuous segment of Late Mesoproterozoic orogenic belts in the world (Wynne-Edwards, 1972; Davidson, 1995). The province extends over a broad area of eastern Canada located southeast of the Grenville Front (Fig. 3) and constitutes the youngest orogenic belt in the Canadian Shield. The Grenville Province consists of a mosaic of geological terranes that collectively record more than one billion years of Paleoproterozoic through Neoproterozoic crustal growth on an Archean core along and outboard of the southeastern margin of proto–North America (Wynne-Edwards, 1972; Gower et al., 1990; Rivers, 1997; Davidson, 1998a; Rivers and Corrigan, 2000; Gower and Krogh, 2003). Because the region did not experience widespread recrystallization resulting from subsequent tectonic events that occurred along the Laurentian margin, the Grenville Province in Canada is a key laboratory in which to study pre-Grenvillian- and Grenvillian-age tectonic evolution, and the processes that led to the crustal accretion of Laurentia and the assembly of the Mesoproterozoic supercontinent Rodinia (e.g., Dalziel et al., 2000).

Past landmarks in the study of the Canadian Grenville Province were recorded and brought into perspective in the volumes entitled *The Grenville Problem* (Thompson, 1956) and *The Grenville Province* (Moore, 1986). These volumes, together with the paper of Wynne-Edwards (1972), discussions at a major Grenville symposium in the 1970s (Baer, 1974), and demonstrations of significant thrust stacking along ductile shear zones in western Labrador by Rivers (1983) and in the Central Gneiss Belt by Davidson (1984), framed the currently prevailing view of the Grenville Province as a Himalayan-type orogen (e.g., Windley, 1986). During the last decade, new tectonic and metallogenic models arose from government surveys, academic research, mapping programs, the Lithoprobe Abitibi-Grenville Transect and Eastern Canadian Shield Onshore-Offshore Transect (ESCOOT) projects, and the International Geological Correlation Programme (IGCP) Project No. 440, the Assembly and Breakup of Rodinia (e.g., Easton, 1992, 2000; Sangster et al., 1992; Davidson, 1998a, 1998b; Berman et al., 2000; Martignole et al., 2000; Ludden and Hynes, 2000a, 2000b; White et al., 2000; Meert and Powell, 2001; Wardle and Hall, 2002). Such research also resulted in a wide variety of lithotectonic and timing nomenclature (e.g., Wynne-Edwards, 1972; Hoffman, 1989; Rivers et al., 1989; Easton, 1992; Davidson, 1995, 1998a; Gower, 1996; Rivers, 1997; Carr et al., 2000; Ludden and Hynes, 2000b; Gower and Krogh, 2003; Rivers et al., 2002), as discussed earlier.

Acquiring the ability to distinguish orogenwide events from those that are local or terrane-restricted represents an important emphasis in current research and a key for future reconciliation of ideas and models (e.g., Gower and Krogh, 2002). The papers

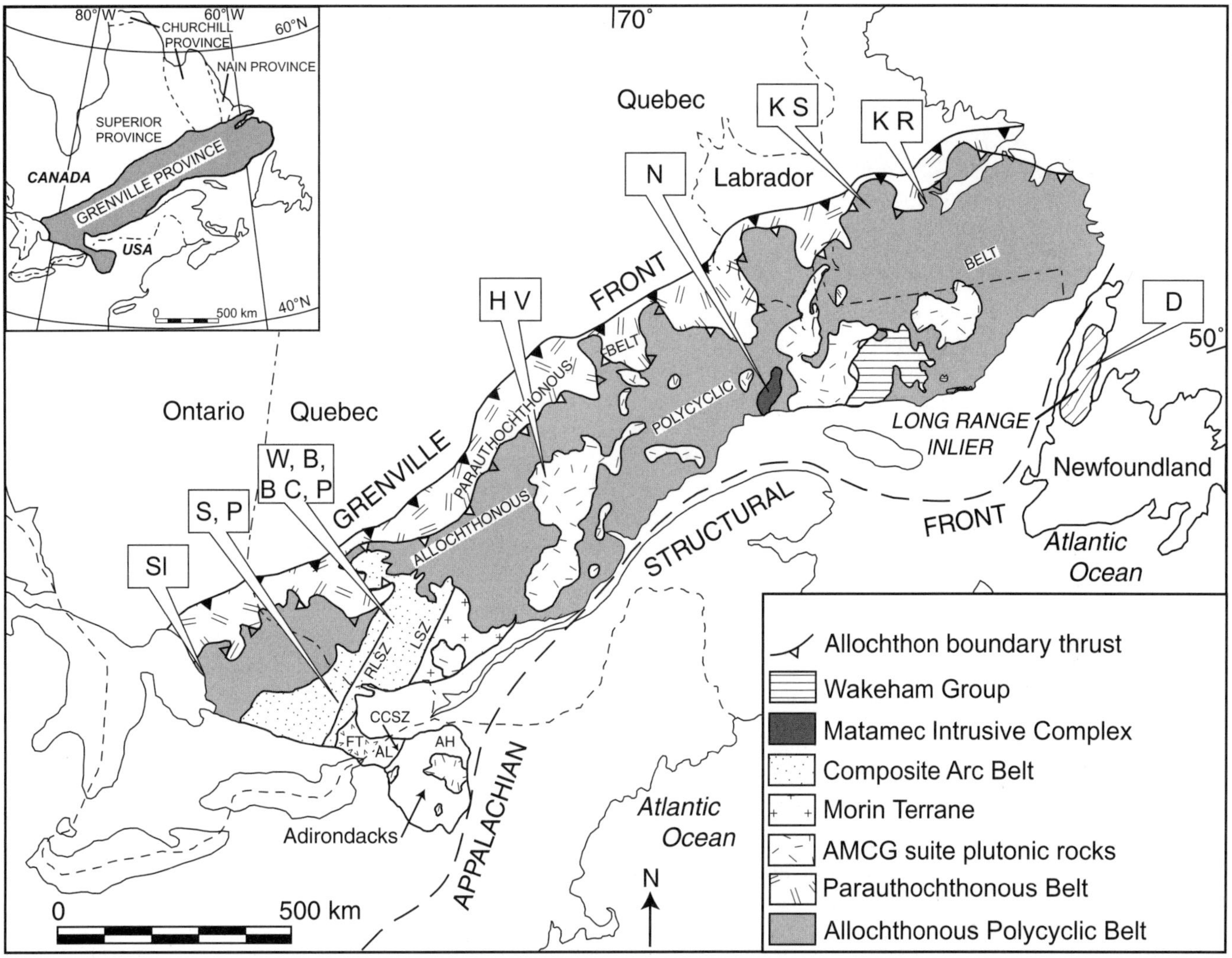

Figure 3. Map of the Grenville Province in Canada showing generalized locations of the study areas pertaining to papers included in this volume. Modified and simplified from Davidson (1998a, Figure 3.6). The eastern extension of the Wakeham Group and the Matamec intrusive complex are from Corriveau et al. (2003b) and Gobeil et al. (1999), respectively. Abbreviations include RLSZ—Robertson Lake shear zone; LSZ—Labelle shear zone; CCSZ—Carthage-Colton shear zone; FT—Frontenac terrane; AL—Adirondack Lowlands; AH—Adirondack Highlands. Nomenclature for the principal lithotectonic belts of the Canadian Grenville Province adopted from Rivers et al. (1989) and Carr et al. (2000). Labeled study areas correspond to the following: K R—Krauss and Rivers; K S—Korhonen and Stout; N—Nabil et al.; H V—Hébert and van Breemen; W—Wodicka et al.; B—Blein et al.; B C—Boggs and Corriveau; P—Peck et al.; S—Schwerdtner; Sl—Slagstad et al.; D—Dickin.

in this volume by *Blein et al., Boggs and Corriveau,* and *Wodicka et al.* on the Bondy Gneiss Complex, and those by *Peck et al.* and *Schwerdtner et al.* on its host, the Composite Arc Belt, combined with the contribution of *Slagstad et al.* on the structurally underlying Allochthonous Polycyclic Belt farther to the west, provide examples of the types of studies currently undertaken in the western Grenville Province. These studies, along with the papers by *Hébert and van Breemen* and by *Nabil et al.* on the Allochthonous Polycyclic Belt in the central and eastern Grenville Province, as well as those by *Korhonen and Stout* and by *Krauss and Rivers* on the eastern Grenville segment of the Parautochthonous and Allochthonous Polycyclic Belts, are contributions that bridge the knowledge gap between the eastern and the western Grenville Province in Canada (Fig. 3). These new findings from Canada provide a geological framework that can be compared to results of studies in the Adirondacks, the Appalachian inliers, and the occurrences in Texas and Mexico. Therefore, this collection of papers provides an orogenwide perspective that can serve as a foundation for a renewed and more comprehensive vision of the Grenvillian orogen in North America.

Although great strides have been made during the past twenty years, a large part of the Grenville Province of Canada remains simplistically mapped as undifferentiated pink or gray gneiss complexes. These high-grade metamorphic gneiss terranes were formerly considered by many as geologically intractable and economically sterile and thus too risky for mineral exploration. However, several papers in this volume, such as those by *Nabil et al.* and by *Blein et al.* on the central Grenville Province and by *Selleck et al.* on the Adirondacks, capitalize on leading-edge research on magma emplacement mechanisms in large mafic-felsic intrusions (see also Wiebe and Collins, 1998; Saint-Germain and Corriveau, 2003) and on new field strategies (see also Corriveau et al., 2003a, 2003b) to reassess the origin and economic potential of such terranes.

Several orogenic events contributed to the formation of the Grenville Province. With the exception of the Archean and Early to Middle Paleoprotorozoic orogenesis that built precursors of the Parautochthonous Belt (e.g., the 1.89–1.80 Ga Makkovikian orogen in Labrador), the main crust-building events were the Late Paleoproterozoic Labradorian (1.71–1.60 Ga) and the Mesoproterozoic Pinwarian (1.52–1.46 Ga), Elzevirian (1.23–1.18 Ga), and Grenvillian (1.08–0.98 Ga) orogenies. Within the Canadian Grenville province, the protolith ages of rocks are generally younger toward the southeast, with exceptions occurring in the Labradorian crust of northern Labrador and in the Composite Arc Belt (Rivers, 1997; Gower and Krogh, 2002). Rivers (1997) interpreted this polarity in crustal ages as evidence for the accretionary growth of the southeast margin of proto–North America from ca. 1.75–1.0 Ga.

The Labradorian and Pinwarian events are best documented in the eastern Grenville of Labrador, where U-Pb isotopic dating and Sm-Nd isotopic mapping have played a key role in elucidating crustal relations (Rivers and Corrigan, 2000; Gower and Krogh, 2003). Assuming that Sm-Nd model ages do not reflect mixtures of Archean and Mesoproterozoic rocks, such isotopic mapping can be used to determine the distribution of crust-formation ages in high-grade terranes. Sm-Nd isotopic mapping of meta-igneous rocks in this region extends the distribution of concealed Labradorian crust as far east as the Long Range inlier in the Appalachian orogen of Newfoundland (Fig. 3), where it is in contact with a major juvenile crustal block that, in central Québec, has ca. 1575–1450 Ma Sm-Nd model ages (Dickin and Higgins, 1992; Dickin, 2000). The contribution of *Dickin* indicates that the younger 1.55 Ga block in the Long Range is correlated with the Mesoproterozoic gneisses of the Blair River inlier in Nova Scotia, which contains gneiss that predates Grenvillian plutons (Miller and Barr, 2000). Dickin's interpretation and the systematic presence of Labradorian-age xenocrysts in Pinwarian volcanics and intrusive rocks in Labrador and along the north shore of the St. Lawrence River, east of the Wakeham Group (van Breemen and Corriveau, 2001; Corriveau et al., 2003b), give further support for a pre-Grenvillian Andean-type continental margin arc setting along the southeastern margin of proto–North America (e.g., Gower and Tucker, 1980; van Breemen and Davidson, 1988; Rivers, 1997). *Hébert and van Breemen* document a widespread Pinwarian event in the Saguenay area of the central Grenville Province and tightly constrain its evolution through the Mesoproteorozoic. They demonstrate that supracrustal rocks of the Saguenay Gneiss Complex surround the Cap de la Mer amphibolite unit, dated at 1506 ± 13 Ma, and were intruded 100 m.y. later by plutonic rocks forming the Cap à l'Est Gneiss Complex (1391 +7/–8 Ma) and the Cyriac rapakivi granite (1383 ± 16 Ma).

Recent studies of the Wakeham Group to the east (newly recognized as 1.5 Ga in age and coeval with emplacement of subvolcanic granitoids), the 1375 Ma Matamec intrusive complex, and adjacent gneiss terranes led to mineral exploration in the eastern Grenville Province of Québec (e.g., Verpaelst et al., 1997; Corriveau et al., 2003a, 2003b; Clark, 2003; Gobeil et al., 2003; Wodicka et al., 2003). Discovery of the spectacular Lac Volant Ni-Cu prospect on the heels of the Voisey's Bay discovery north of the Grenville front in Labrador provoked an unprecedented staking rush in Québec's North Shore region (Perreault et al., 1997). Although results from drilling eventually indicated that the massive sulfide bodies are limited in extent (Raymond et al., 1997), *Nabil et al.* demonstrate that the gabbronorite dike and its magmatic Ni-Cu sulfides share similar compositional features, emplacement style, magma type, and age with the Norwegian Ertelien and Flåt deposits in the Sveconorwegian Province and with Voisey's Bay in Labrador. Therefore, the Lac Volant find constitutes key evidence that magmatic sulfides can be an important exploration target in the Grenville Province.

Writing on the pervasively deformed Allochthonous Polycyclic Belt of the southwestern Grenville Province in Ontario, *Slagstad et al.* offer a contribution toward unraveling the early history of high-grade ortho- and paragneisses with 1500–1350 Ma protolith ages. Like *Wodicka et al.* and *Blein et al.,* they demonstrate that, in addition to the zircon U-Pb isotopic system, whole-rock geochemical compositions remain robust and amenable to detailed protolith studies. Their geochemical and geochronological data point to the presence of ca. 1500 and 1450 Ma arc and back-arc environments where, following cessation of magmatic activity, high-grade metamorphism occurred between ca. 1450 and 1430 Ma in the back-arc regions. The ca. 1360 Ma bimodal rhyolitic and basaltic magmatism they document suggests a return to an extensional tectonic regime along the arc, an event also documented in the Composite Arc Belt to the east by *Wodicka et al.* Their detailed study raises both similarities to and differences from previous "Andean"-type models (e.g., Rivers and Corrigan, 2000) and offers further insight to the evolution of the southeastern margin of Laurentia.

Although the Composite Arc Belt (Fig. 3) is interpreted as a composite terrane of post-1300 Ma magmatic arcs and marginal basins juxtaposed with the Allochthonous Polycyclic Belt (Easton, 1992; Davidson, 1995; Carr et al., 2000), *Wodicka et al.* provide new U-Pb zircon ages obtained through sensitive high-resolution ion microprobe (SHRIMP) techniques that reveal the presence of older 1.4–1.3 Ga volcano-plutonic components likely built on proto–North American crust. The samples studied were selected from metavolcanic and metaplutonic gneisses of the Bondy Gneiss Complex either within or near a Cu-Au hydrothermal system metamorphosed at granulite facies. The U-Pb data, in conjunction with geochemical and Nd data (Blein et al., 2003), support the interpretation that the Bondy Gneiss Complex represents a ca. 1.39 Ga volcano-plutonic edifice formed in a juvenile arc and back-arc setting. As such, the complex provides a key link to the arc setting documented farther west by *Slagstad et al.* as well as to other ≤1.45–1.3 Ga juvenile rocks from the Central Granulite terrane in Québec (*Hébert and van Breemen*) and in the Adirondack Highlands (McLelland et al., 1996).

Recent field work combined with geochemical studies has resulted in identification of volcanic rocks among the largely uncharted felsic gneiss terranes of the Grenville Province in Canada (e.g., Gower, 1996; Blein et al., 2003; Slagstad et al., this volume; Wodicka et al., this volume). Associated volcanic-plutonic settings have, to date, only rarely been identified in the Grenville Province even though they form an intrinsic component of many other orogenic belts and serve as sites for many ore deposits (e.g., Ohmoto, 1996; Corbett and Leach, 1998; Hitzman, 2000; Large et al., 2001). *Blein et al.* illustrate that the mineral potential of the Grenville Province will be significantly underestimated until volcanic settings and associated intrusions become routinely recognized during regional mapping and their geochemistry studied in detail (see also Corriveau et al., 2003a, 2003b). The identification, by *Blein et al.,* of cordierite-orthopyroxene white gneiss as the granulite-facies equivalent of a chlorite alteration zone played a key role in finding a Cu-Au-Fe oxide hydrothermal system in the Bondy Gneiss Complex. Supporting evidence for a hydrothermal origin for these gneisses is given by mass-balance calculations that record a loss in Ca and a gain in Mg, K, and Si with respect to the adjacent nonaltered rhyolitic protolith, compatible with feldspar destruction, sericitization, and chloritization. These lithofacies display pronounced birdwing-shaped rare-earth-element (REE) profiles and thus differ from known, Cl-driven, hydrothermal systems with cordierite-orthopyroxene or -orthoamphibole rocks that are associated with metamorphosed, volcanic-hosted-massive-sulfide deposits, but share similarities with F-driven hydrothermal signatures. Two episodes of metamorphism affected these rocks, as recorded by mantles on zoned zircons and whole new zircon crystals dated by *Wodicka et al.* High-pressure and -temperature (*P-T*) granulite-facies metamorphism and partial melting occurred at 1.21–1.18 Ga, whereas a younger, albeit localized, metamorphic overprint occurred at ca. 1.15–1.13 Ga. *Boggs and Corriveau* demonstrate that the peak pressure of 10 kilobars was reached at 1.20 Ga, with crystallization of peak-temperature (950 °C) anatectic melt at 1.18 Ga followed by isothermal decompression. The conditions recorded in the Bondy Gneiss Complex are 150–200 °C and 2 kilobars higher than those previously documented in the region, a feature that *Boggs and Corriveau,* in conjunction with Corriveau et al. (1998), demonstrate to result from regional-scale strain partitioning during 1.17–1.16 Ga magma emplacement and deformation, and from the coarse grain size. Their paper reinforces the efficacy of the long-standing field strategy indicating that, in high-grade, polyphase orogenic belts, the early orogenic record can sometimes be deciphered by careful study of nonreactivated areas in otherwise largely overprinted granulite-facies terranes. In this case, fine-scale delineation of the early event was made possible by identification of particularly responsive aluminous assemblages of the fossil hydrothermal system in the gneiss complex.

Peck et al. make use of the refractory nature of $\delta^{18}O$ values of zircon to characterize the source signature of a series of granitic rocks related to 1.18–1.13 Ga AMCG plutonism that stitched the Frontenac and Elzevir terranes in Ontario with the Composite Arc Belt of Québec, the Adirondack Highlands, and the Morin terrane. By calculating the original $\delta^{18}O$ of the magma, they provide constraints on terrane boundaries and discontinuities in the lower crust of this polyphase orogen as well as on the origin of the granitoids themselves. The documented variability in $\delta^{18}O$ (zircon) for these granitoids corresponds to geographic location, with high values of $\delta^{18}O$ (magma) in the central Frontenac terrane in Ontario and lower values in the other terranes of New York, Ontario, and Québec. In fact, the high, 12.4 to 14.3‰, $\delta^{18}O$ (magma) values of plutons in the Frontenac terrane are some of the highest magmatic oxygen-isotope ratios recognized worldwide in rocks without unusual whole-rock geochemical or radiogenic isotope compositions. Building upon results of previous oxygen-isotope studies by Shieh (1985) and Shieh and Schwarcz (1978), the authors attribute the high $\delta^{18}O$ values to subduction or underthrusting of hydrothermally altered basalts and sediments at the base of the Frontenac terrane at or prior to 1.2 Ga during closure of an ocean basin between the Frontenac terrane and the Adirondack Highlands.

During the past two decades, northwest-directed thrusting has been increasingly recognized as playing an important role in the structural evolution of the Grenville Province (e.g., Wynne-Edwards, 1972; Davidson et al., 1982; Culshaw et al., 1983, 1997; Rivers, 1983; Davidson, 1984; Green et al., 1988; Hanmer, 1988; Rivers et al., 1989; Gower et al., 1990; Easton, 1992; Martignole and Calvert, 1996). Lithoprobe Lines 32 and 33 and their reflection-seismic profiles across the Composite Arc Belt in southeastern Ontario outline shallow-crustal reflectors interpreted as gently dipping thrusts (Carr et al., 2000; White et al., 2000). However, *Schwerdtner et al.* point out that this crustal view contrasts with the pervasively steeply inclined to vertical planar fabrics observed in the field. They document north- to east-northeast-striking, longitudinal zones of high strain that are linked tentatively to ductile thrusting at the sole of fold nappes. Modeling of the prevailing steeply dipping planar fabrics combined with shear-strain indicators is difficult to reconcile with shallow-crustal, ductile-thrusting models, but is compatible with midcrustal stretching thrusts. Their study suggests that the high-strain zones and their walls were stretched or shortened progressively in the direction of simultaneous tangential shearing and/or irrotational slip, indicating that the longitudinal dislocations qualify as stretching thrusts active in a midcrustal regime.

Papers by *Korhonen and Stout* and by *Krauss and Rivers* address the formation of mid- to high-pressure granulite and retrogade eclogite in the eastern Grenville Province. With their study of low variance, sapphirine-bearing assemblages in the aluminous gneisses of the Wilson Lake allochthon (Nunn et al., 1985; Thomas, 1993), *Korhonen and Stout* demonstrate that, by searching for low-variance metamorphic assemblages in high-grade metamorphic terranes, it is possible to document equilibrium conditions of regional metamorphism even where retrograde and nonequilibrium textures prevail. Their phase-equilibrium arguments, supported by electron microprobe analyses, allow characterization of high-temperature silicate-oxide equilibria, which, in turn, permit improved understanding of the petrologic origins of the pronounced aeromagnetic high that extends over much of the Red Wine Mountains segment in the Wilson Lake allochthon. As the search for Cu-Au-Fe oxide deposits intensify in high-grade metamorphic terranes (e.g., *Selleck et al.*), understanding the petrologic significance of aeromagnetic highs will be critically important. A recently recognized high-pressure belt, with eclogite-facies and high-pressure granulite-facies assemblages, occurs discontinuously along the orogen within the structurally lowest segments of the Allochthonous Polycyclic Belt (Ludden and Hynes, 2000b; Rivers et al., 2002), where it is tectonically overlain by a low-pressure belt. *Krauss and Rivers* provide precise geothermobarometric determinations of the metamorphic conditions experienced by the Grand Lake thrust system at the base of the thick-skinned Cape Caribou River allochthon in the high-pressure belt. Their estimate of P-T conditions of ~14 kilobars at 875 °C is interpreted as the pressure at the maximum temperature experienced by the rocks in the shear zone. This thrust system is interpreted as the northeasterly extension of the high-pressure belt along the Allochthon Boundary thrust system of the Grenville Province.

The tectonic record of the Grenville Province of Canada preserves many clues bearing on assembly of the supercontinent Rodinia between 1.2 and 1.0 Ga (Hoffman, 1991; Dalziel et al., 2000) but little evidence for the ocean that surrounded it at ca. 1.0 to 0.75 Ga. *Murphy et al.* review worldwide evidence for ca. 0.9 to 0.8 Ga ophiolites and ensimatic arc complexes and ca. 0.8 to 0.6 Ga recycled mafic to felsic arc complexes accreted to the leading edges of Rodinia following its breakup beginning at ca. 0.75 Ga. Using the Brasiliano and Trans-Saharan orogens as case examples, *Murphy et al.* illustrate that, where subduction culminated in Late Neoproterozoic continental collision, vestiges of peri-Rodinian crust became cratonized between the colliding cratons. In contrast, the Arabian Shield and peri-Gondwana provide examples of accretionary orogens in which subduction was not terminated by continental collision. In these cases, peri-Rodinian crust subsequently became involved in Paleozoic orogenesis and recycled. Such recycling is exemplified by the Avalonian belt of the Canadian Appalachians, as recorded by Sm-Nd depleted mantle (T_{DM}) model ages that overlap with the life span of the supercontinent. The geodynamic linkage that *Murphy et al.* present provides important constraints on the Rodinian orogenic cycle, specifically on its configuration and the timing of major tectonothermal events.

ADIRONDACKS

The slightly elliptical Adirondack dome connects with the main Grenville Province (Figs. 3 and 4) via the Thousand Islands region, where the Frontenac arch crosses the St. Lawrence River and brings Proterozoic rocks to the surface through their Paleozoic cover. Both the Elzevirian and the Ottawan orogenies of Moore and Thompson's (1980) Grenvillian orogenic cycle affected the region and resulted in high-grade metamorphism together with northwest-verging, ductile nappes. Early investigations of Adirondack geology, which focused upon field mapping and structural and petrologic problems (e.g., Buddington, 1939, deWaard, 1965, McLelland and Isachsen, 1986, among others), represent important studies that continue to form the geological foundation for current research. Similarly, the U-Pb zircon studies of Silver (1969) that indicated the considerable potential of geochronological data in high-grade terranes opened the way to well-constrained, contemporary models of Adirondack evolution. Coupled with modern petrological investigations (e.g., Bohlen et al., 1985; Valley et al., 1990; Spear and Markusen, 1997), geochronological data have brought increasing clarity to the evolution of the Adirondacks and its relationship to the Grenville occurrences to the north and the south (e.g., McLelland et al., 1996; Wasteneys et al., 1999). The contributions summarized in the following paragraphs are representative of the impact that modern geochronology and petrology have had on our understanding of Adirondack geology and reflect the increasing agreement among most Adirondack researchers concerning the evolution of this important sector of the Grenville Province.

The Adirondack region consists of two major subdivisions, referred to as the Highlands (granulite facies) to the southeast and the Lowlands (upper amphibolite facies) to the northwest, separated by a northwest-dipping (~45°) high-strain zone known as the Carthage-Colton shear zone (Fig. 4). The last movement along this zone was clearly down to the west, but it had a protracted and enigmatic history whose clarification is crucial to unraveling the evolution of the region. Recently, a great deal of attention has been focused on the Carthage-Colton shear zone, and this volume contains four papers concerning this feature. Two of these (*Johnson et al.* and *Streepey et al.*) utilize U-Pb titanite ages and $^{40}Ar/^{39}Ar$ hornblende and biotite ages to fix the timing of local and regional events associated with the Carthage-Colton shear zone. Both sets of investigators agree that, during the ca. 1090–1035 Ma Ottawan orogeny, the Carthage-Colton shear zone acted as a steeply dipping, right lateral oblique shear, or transpressional, boundary that juxtaposed the Lowlands and the Highlands, and thrust the former over the

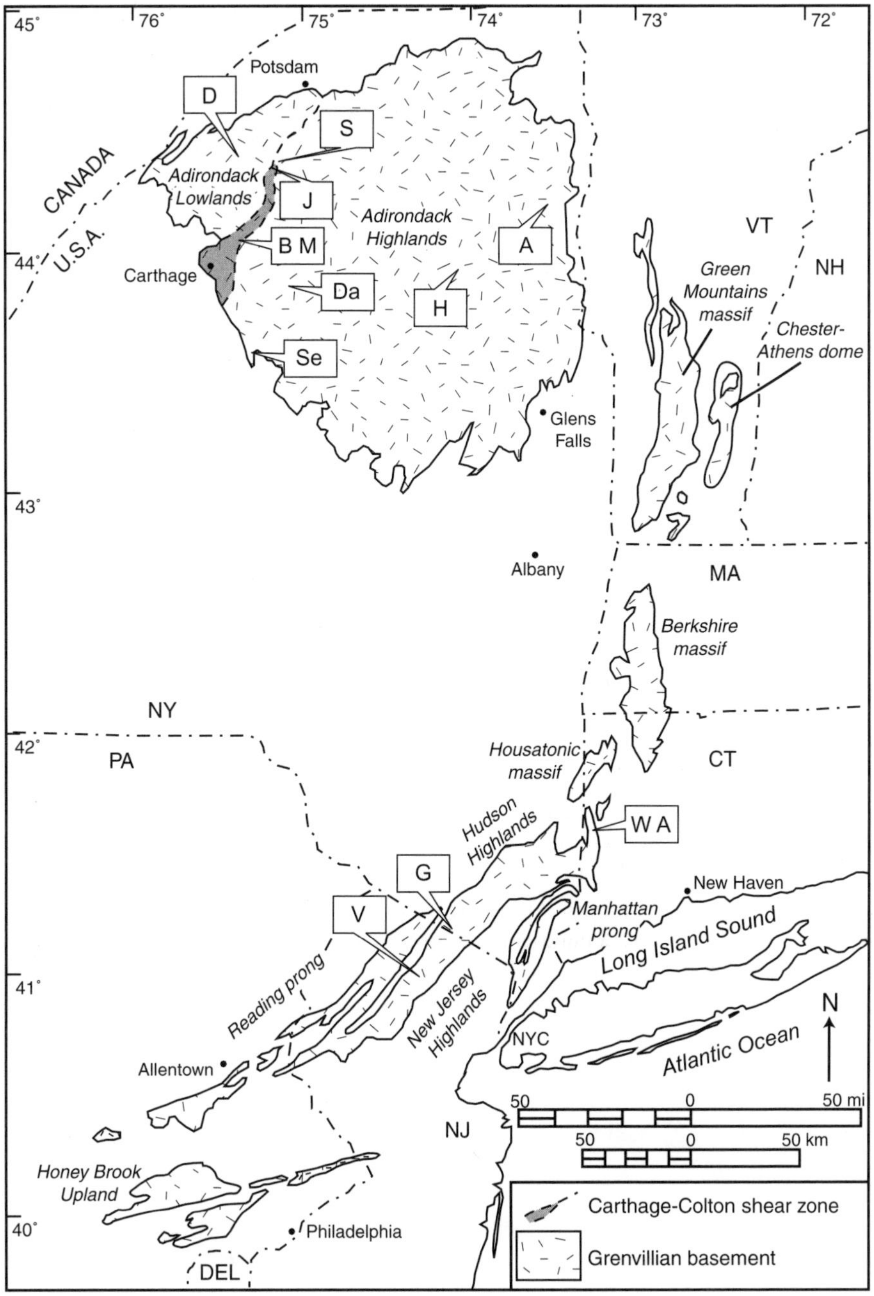

Figure 4. Map showing the Adirondacks and locations of major exposures of Grenvillian basement rocks in the northern Appalachians with generalized locations of study areas pertaining to papers included in this volume. Map modified from King and Beikman (1974), Rankin et al. (1990), and Rankin (1993). Labeled study areas correspond to the following: D—Dahl et al.; S—Streepey et al.; J—Johnson et al.; B M—Baird and MacDonald; Da—Darling et al.; H—Hamilton et al.; A—Alcock et al.; Se—Selleck et al.; W A—Walsh and Aleinikoff; G—Gorring et al.; V—Volkert.

latter for an unspecified distance. By ca. 1050 Ma, the region began to undergo uplift, during which the amphibolite-facies Lowlands moved down to the northwest along the Carthage-Colton shear zone relative to the granulite-facies Highlands, reaching its current position by ca. 1020 Ma. This sequence of events provides a geologically reasonable Ottawan history for the Carthage-Colton shear zone. Until these authors presented their transpressional-collapse model, it had been very difficult to account for the differences in metamorphic grade and U-Pb titanite and $^{40}Ar/^{39}Ar$ hornblende ages across the Carthage-Colton shear zone. For example, the Lowlands showed U-Pb titanite ages of 1150–1100 Ma, whereas Highland titanites yielded ages of ca. 1030 Ma (Mezger et al., 1991). The transpressional model resolves this apparent inconsistency by making it possible for the Lowlands to have been laterally and vertically separated from the Highlands during much of the Ottawan orogeny, and does so without requiring the closing of an intervening ocean basin, for which there is no evidence. This model has important implications for the Labelle shear zone, which occurs along strike of the Carthage-Colton shear zone in Québec and which, together with the latter, separates the Central Granulite terrane from the Frontenac terrane.

In addition to the Ottawan evolution of the Carthage-Colton shear zone described previously, a post-Ottawan history is documented by $^{40}Ar/^{39}Ar$ data for hornblende, biotite, and alkali feldspar (*Streepey et al.*). These data demonstrate that regional extension took place along crustal-scale shear zones, including the Carthage-Colton shear zone, from ca. 950–780 Ma. $^{40}Ar/^{39}Ar$ dating along the east-dipping Robertson Lake shear zone (Fig. 3) in southeastern Ontario and southwestern Québec also documents extensional displacement within the same time interval and suggests that the intervening Frontenac terrane represents a large, down-dropped block within the Composite Arc Belt. Finite-element modeling of regional extension is consistent with uplift and extensional collapse possibly associated with ascent of a large mantle plume coupled with far-field extension related to the impending breakup of Rodinia. $^{40}Ar/^{39}Ar$ ages for hornblende and biotite in the Lowlands were studied by *Dahl et al.,* whose preferred interpretation of the data is that regional younging from northwest to southeast reflects cooling (1100–900 Ma) followed by regional tilting of ~9 ± 3° NW (900–520 Ma), a result that is in general agreement with *Streepey et al.* As a starting point, the *Dahl et al.* model also places the Lowlands above the western Highlands prior to extensional collapse along the Carthage-Colton shear zone at 1050–1040 Ma, thus juxtaposing the cooler Lowlands against the warmer Highlands. Reactivation of regional extensional collapse along the Carthage-Colton shear zone resumed by ca. 945 Ma and may have resulted in 4–6 km of displacement.

Significantly, the investigations of all three groups discussed earlier concur on the transpressional nature of the Carthage-Colton shear zone during the Ottawan orogeny followed by protracted periods of extensional collapse along the zone. This model is able to accommodate the observation that titanite ages in the Lowlands range from 1150 to 1100 Ma, implying that this region did not experience the granulite-facies metamorphism of the Highlands, where titanite ages of 1030 Ma are common. However, ca. 1155 Ma rocks in the Lowlands are isoclinally folded, and the only currently known orogenic event that could have been responsible for post-1155 Ma regional deformation is the Ottawan orogeny. Thus, it is significant that *Baird and MacDonald* utilized petrofabrics and magnetofabrics to show that a previously unrecognized foliation and lineation exists within the ca. 1155 Ma Diana Syenite that straddles the Carthage-Colton shear zone. This fabric crosscuts a magmatic flow foliation in the syenite, as well as a presyenite foliation in the country rocks, and is therefore younger than 1155 Ma. The authors interpret the fabric as Ottawan in age and note that it has the same character and orientation as a fabric that is common in the Lowlands, and that it also predates two other fabrics and fold sets in the Lowlands. This result suggests that Ottawan structures are present in the Lowlands, although temperatures remained well below those in the Highlands, and is consistent with the transpressive model of the Carthage-Colton shear zone, in which there is sufficient lateral displacement between the two terranes to account for observed thermal differences.

The pressure-temperature-time (*P-T-t*) path of Ottawan metamorphism was investigated by Bohlen (1987), Spear and Markusen (1997), and McLelland et al. (2001), all of whom concluded that the path was characterized by initial heating followed by loading during collision. *Darling et al.* shed new light on this issue through their study of a spectacular prismatine-bearing metapelite exposed in the Moose River near McKeever. In addition to prismatine, a B-rich end member of the kornerupine group, the assemblages in these rocks contain minerals such as garnet, orthopyroxene, spinel, cordierite, biotite, plagioclase, and quartz. Various combinations of these phases produce low-variance assemblages that are amenable to net transfer and exchange thermobarometry and/or analysis on *P-T* grids. Deduced pressures and temperatures, together with textures, indicate approximately isobaric cooling from 850 ± 20 °C and 6.6 ± 0.6 kilobars to 675 ± 50 °C and 5 ± 0.6 kilobars consistent with the results of Spear and Markusen (1997). These results are indicative of Ottawan heating followed by loading and orogenic collapse at ca. 1050 Ma.

The timing and origin of the distinctive AMCG suite are issues of long-standing interest within the Adirondacks and other Proterozoic terranes. Accordingly, McLelland and Chiarenzelli (1990) dated the Adirondack AMCG suite by obtaining conventional multigrain ages on the granitoid members of the suite. These results indicated an age range of 1157 to 1125 Ma for six granitoids and, by extension, a similar age range for associated, and mutually crosscutting, anorthositic rocks. *Hamilton et al.* report additional U-Pb isotopic ages obtained using SHRIMP techniques on zircons from four AMCG granitoids and a ferrodiorite dike that is inferred to be a late differentiate of the anorthosite it crosscuts. Ages for the granitoids fall into the age interval 1175–1155 Ma, and the ferrodiorite yields an age of 1156 ± 7 Ma. These results amplify the previous data and indicate that the AMCG suite was emplaced within the interval 1175–1125 Ma, shortly after the Elzevirian orogeny, suggesting that this plutonic activity is probably related to delamination of an overthickened orogen and subsequent magma ponding near the base of the crust (McLelland et al., 1996). However, as indicated by studies in some of the Appalachian inliers (e.g., *Tollo et al.* and *Hughes et al.*), AMCG magmatism occurred in some areas that apparently did not undergo Elsevirian orogenesis, and thus a causal mechanism related to the orogeny is difficult to establish universally for this suite.

Delineation of the timing of closely spaced geologic events is especially difficult in terranes such as the Adirondacks that have undergone multiple episodes of metamorphism, magmatism, and deformation. *Alcock et al.* illustrate a practical method for obtaining high-precision U-Pb isotopic ages from zircons in rocks produced by high-temperature anatexis, utilizing a combination of imaging techniques, chemical analyses, and isotope-dilution thermal ionization mass spectrometry methods to obtain precise ages for geologic events responsible for producing separate age domains in complex zircons from leucocratic segregations in charnockitic gneiss. Their results indicate that melts produced by anatexis in the northeastern Adirondacks crystallized at 1054–1040 Ma, following peak metamorphic conditions associated with Ottawan orogenesis by 20–30 m.y. (McLelland et al., 2001). Their success in elucidating the timing of geologic events in rocks of the northeastern Adirondacks suggests that their methodology offers a means of obtaining high-precision, geologically meaningful, geochronologic information from isotopically complex zircons in situations where the timing of closely spaced events might not be resolvable by ion microprobe techniques.

Selleck et al. describe a large quartz-sillimanite vein complex that is interpreted to be the partial result of hydrothermal leaching of alkali cations from feldspar in a 1035 ± 4 Ma leucogranite. Further leaching removed quartz and produced residual diaspore lenses that contain significant quantities of zircon. SHRIMP analysis of these zircons yields ages that overlap the age of the granite and suggest that the zircons were derived from the granite as it underwent extreme leaching shortly after emplacement. Similar hydrothermal leaching is common within young Adirondack leucogranites (Foose and McLelland, 1995) and is responsible for abundant quartz-sillimanite nodules and albitization associated with these rocks. These hydrothermally altered rocks are associated with historically important Kiruna-type, rare-earth-element-enriched, Fe oxide deposits similar to those that occur elsewhere within the Mesoproterozoic of North America and support a hydrothermal origin for these important ore concentrations.

APPALACHIAN INLIERS

Rocks of Late Mesoproterozoic age (1.2–1.0 Ga) occur near the western margin of the Appalachian metamorphic core, forming a series of inliers within the Appalachian orogenic belt (Rankin et al., 1989). These basement rocks are exposed in the cores of a chain of fault-bounded anticlinoria collectively described, using Alpine terminology, as external massifs by Hatcher (1983). Rankin (1976) referred to this linear chain as defining the Appalachian Blue-Green-Long axis, referring to the Blue Ridge and Green Mountains and the Long Range, respectively. Basement rocks of similar age are exposed in a series of antiforms and internal massifs to the east within the Appalachian Piedmont, and include the Chester and Athens domes, Manhattan prong, Baltimore and Goochland terranes, and Pine Mountain window, some of which are the locations of studies included within this volume (Fig. 5). Unlike the rocks constituting the Grenville Province of Canada and the Adirondack massif, those constituting the external and internal Appalachian massifs underwent recrystallization and local deformation associated with Paleozoic orogenesis. Such effects locally complicate interpretation of Mesoproterozic events.

Correlation of geologic events recorded in the Grenvillian rocks of the Appalachian inliers with those in the Adirondacks and the main Grenville Province is an important focus for researchers concerned with deciphering regional-scale tectonic processes, and correspondingly constitutes a major theme of many papers included in this volume. South of the Adirondacks, Grenvillian rocks occur within the Reading prong–New Jersey Highlands–Hudson Highlands region (Fig. 4), and are the subject of three papers in the volume. *Walsh and Aleinikoff* report an age of 1311 ± 7 Ma for a granite gneiss from western Connecticut, establishing it as one of the oldest rocks in the region and suggesting that emplacement occurred during early stages of the Elzevirian orogeny, as redefined by McLelland et al. (1996) (Fig. 1). *Volkert* presents a summary of the geologic history of the New Jersey Highlands, integrated with results of recent mapping and geochemical studies, and establishes a basis for comparison with neighboring Grenvillian terranes. Within the Highlands, the undated Losee metamorphic suite is interpreted to have been derived from a calc-alkaline plutonic-volcanic assemblage that was likely generated in a continental-margin magmatic arc prior to the collisional Ottawan orogeny, and was subsequently juxtaposed with a supracrustal sequence developed inboard of the arc. Available evidence in the New Jersey Highlands indicates that, following emplacement of the ca. 1100 Ma Byram and Lake Hopatcong intrusive suites, the area experienced a major episode of tectonothermal activity between 1090 and 1030 Ma that was coeval with the Ottawan orogeny recorded in the Adirondacks and the main Grenville Province. *Walsh and Aleinikoff* report igneous activity associated with Ottawan orogenesis, including syntectonic migmatite, granitoids, and augen gneiss with ages ranging from 1057 ± 10 Ma to 1045 ± 8 Ma. The undeformed, ca. 1020 Ma Mount Eve granite suite, which *Gorring et al.* demonstrate to have compositional features similar to A-type granitoids, appears to be representative of late to post-Ottawan magmatism in the region. Together, these papers illustrate the episodic and interrelated nature of magmatism, metamorphism, and deformation associated with Grenvillian orogenesis, and establish a basis for correlation with nearby areas such as the Adirondack massif.

Grenvillian basement in the Virginia Blue Ridge (Fig. 5) is dominated by rocks of igneous and metaigneous origin (Bartholomew and Lewis, 1984; Rankin et al., 1989). As a result, recent geologic investigations typically involve some combination of (1) detailed field mapping to determine field relations, (2) geochemical analysis to decipher petrologic characteristics, and (3) U-Pb isotopic analysis to elucidate temporal relations. *Tollo et al.* present geologic and geochronologic evidence for an episode of deformation and metamorphism that occurred in the northern Blue Ridge at 1078–1050 Ma and that may represent local effects of regional Ottawan orogenesis, as documented to the north (McLelland et al., 1996). These data, and results from other recent investigations (Aleinikoff et al., 2000), indicate that Grenvillian magmatism in the Blue Ridge defines three pulses of activity that occurred at ca. 1160–1140 Ma, ca. 1112 Ma, and ca. 1080–1050 Ma, correlative with three episodes of post-Elzevirian magmatism in the Adirondacks. *Hughes et al.* present geochemical data indicating compositional similarities between major basement units occurring in

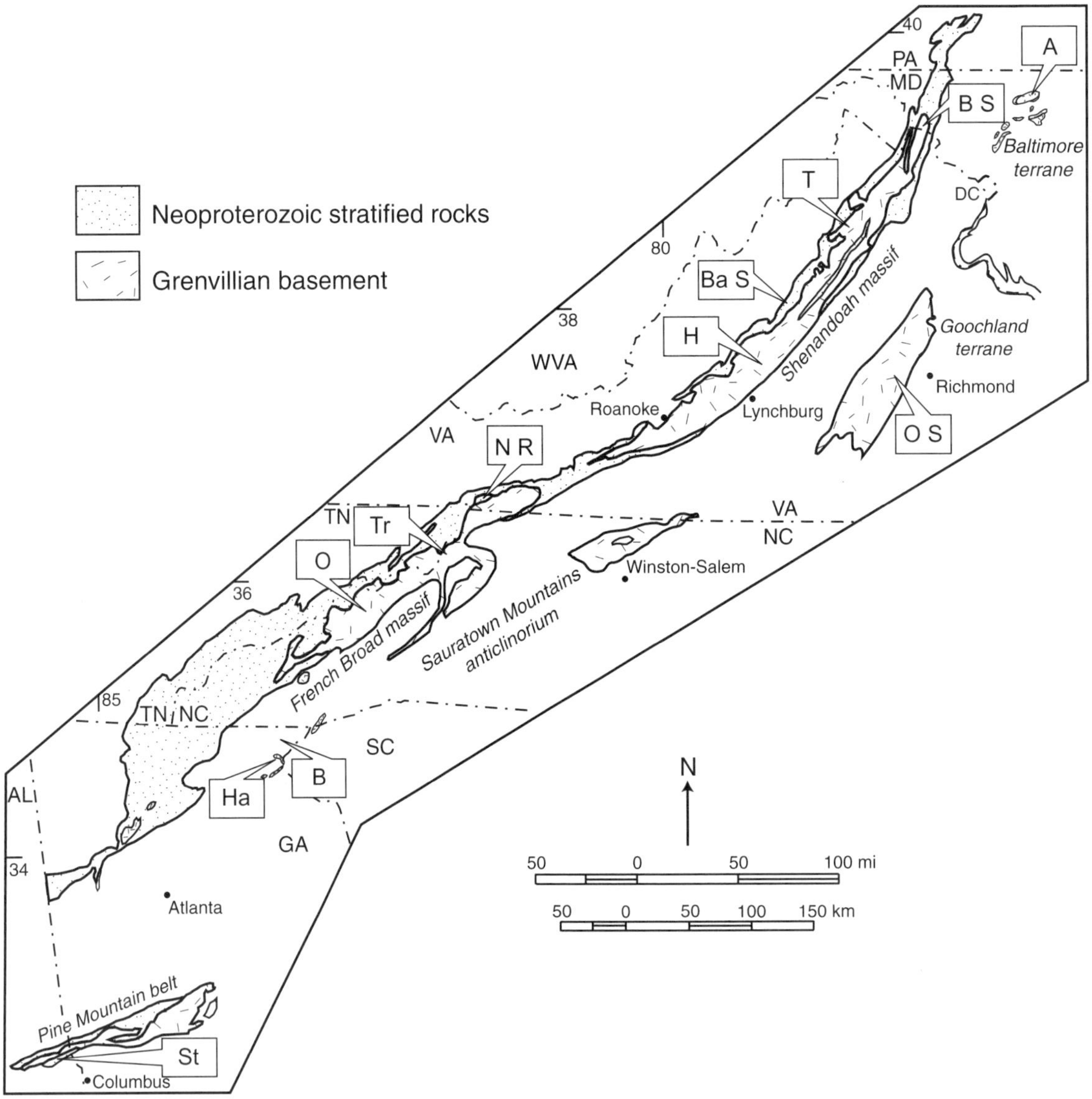

Figure 5. Map showing locations of major exposures of Grenvillian basement rocks in the central and southern Appalachians with generalized locations of study areas pertaining to papers included in this volume. Map modified from King and Beikman (1974), Rankin et al. (1990), Rankin (1993), and Owens and Samson (this volume). Labeled study areas correspond to the following: A—Aleinikoff et al.; B—Bream et al.; B S—Burton and Southworth; T—Tollo et al.; Ba S—Badger and Sinha; H—Hughes et al.; O S—Owens and Samson; N R—Novak and Rankin; Tr—Trupe et al.; O—Ownby et al.; Ha—Hatcher et al.; St—Steltenpohl et al.

the eastern and western portions of the central Blue Ridge (the Lovingston and Pedlar massifs, respectively, of Bartholomew and Lewis, 1984), and note lithologic similarities to comparable anorthosite-bearing suites in the main Grenville Province. Although as yet uncalibrated by high-precision isotopic dating, the age relationships described by *Hughes et al.* for the central Blue Ridge are similar to those defined in the northern Blue Ridge by *Tollo et al.* and by Aleinikoff et al. (2000). The structural significance of the regional Rockfish Valley fault zone (Bartholomew et al., 1981) in the central Blue Ridge has been a controversial issue, with arguments centered on the role played by the fault zone in controlling the observed distribution of Grenvillian basement rocks (c.f. Sinha and Bartholomew, 1984, and Evans, 1991). *Burton and Southworth* describe field-based observations pertaining to the Short Hill fault zone in the northern Blue Ridge and propose a common ancestry with the Rockfish Valley fault zone, using a combination of detailed mapping and structural analysis to suggest an extended evolutionary history for structural development of the northern Blue Ridge. Collectively, these papers represent an initial attempt to calibrate the temporal sequence of igneous, metamorphic, and deformational events responsible for constructing the Mesoproterozoic geologic framework of the Virginia Blue Ridge.

Internal basement massifs occurring within the Appalachian crystalline core have posed a dilemma for geologists because of the possible exotic nature of these crustal blocks with respect to Laurentia during the Appalachian orogenic cycle. A subset of papers in this volume concerns two of the most enigmatic of these terranes: the Baltimore terrane located west of Baltimore, and the Goochland terrane located near Richmond (Fig. 5). Another crustal block of uncertain origin, the Mars Hill terrane of western North Carolina and adjacent Tennessee, is the subject of a third paper. *Aleinikoff et al.* present new geochronologic data indicating ca. 1250 Ma emplacement ages for granitic rocks of the Baltimore terrane, demonstrating that these basement lithologies are significantly older than most other dated Mesoproterozoic rocks occurring in Grenvillian massifs of the central and southern Appalachians, but are similar in age to rocks in the Adirondacks, Vermont, and southeastern Canada. These data suggest that the Baltimore terrane has a geologic history that is different from the histories of neighboring external massifs and that it may be exotic with respect to them. Nd-isotopic data obtained for ca. 1045 Ma rocks of the Goochland terrane in Virginia by *Owens and Samson* indicate similarities to Mesoproterozoic basement of the Blue Ridge, with which it shares some similar lithologic assemblages. These data collectively suggest that the Goochland terrane is not exotic with respect to Laurentia, but may represent a crustal block that was translated laterally during oblique convergence, possibly from the north (Bartholomew and Lewis, 1992). *Ownby et al.* present new geochronologic data for rocks of the Mars Hill terrane in western North Carolina and adjacent Tennessee, indicating that rocks of this lithologically diverse terrane are distinct in age, protolith composition, and metamorphic history from rocks of the structurally overlying Eastern Blue Ridge and underlying Western Blue Ridge. These authors document a magmatic event at 1.2 Ga, but also present U-Pb geochronologic data and Sm-Nd isotopic features supportive of an extensive pre-Grenvillian history. Such characteristics suggest that the Mars Hill terrane may have an exotic origin with respect to nearby Grenvillian crust, and could have been derived from Paleoproterozoic crust located either in a distant portion of Laurentia or on another continent.

Lithologic assemblages constituting Grenvillian basement in the southern Appalachians share many characteristics with rocks of the Shenandoah massif located to the north, except for the absence of anorthosite and a lesser abundance of orthopyroxene-bearing granitoids (Rankin et al., 1989). Blocks of Grenvillian rocks south of Virginia include both external massifs, such as the large French Broad massif, and smaller internal massifs exposed in structural windows (Fig. 5). The latter include the Sauratown Mountains and Pine Mountain windows, as well as several other small isolated blocks. Throughout much of this area, Grenvillian fabrics and mineral assemblages are variably overprinted by the effects of Paleozoic metamorphism and deformation, locally hindering delineation of Mesoproterozic features. Throughout many parts of this region, ongoing investigations involve combinations of detailed mapping and structural study, petrologic characterization, and isotopic analysis, which, in the case of the internal massifs especially, is typically applied toward elucidating the tectonic origin of possibly exotic crustal blocks. *Hatcher et al.* discuss the structure and lithologic characteristics of basement rocks within three internal massifs in the central and eastern Blue Ridge in the Carolinas and in northeastern Georgia (Fig. 5) with similar Grenvillian tectonic histories but contrasting Paleozoic structural sequences. A variety of geologic data suggest that these three terranes were all derived as small blocks rifted from Laurentia during Late Neoproterozoic extension, eventually forming a crustal collage that was reaccreted to Laurentia during episodes of Paleozoic convergence. *Trupe et al.* demonstrate the usefulness of working in areas that were relatively unaffected by Appalachian deformation and recrystallization and located away from major thrust faults, utilizing new and existing mapping and structural studies to compile a tectonic map of rocks located in the western Blue Ridge along the North Carolina–Tennessee border. These authors also construct a restoration of the Grenvillian belts to their pre-Paleozoic positions, placing the Mars Hill terrane on the outboard edge of the Grenville orogen, a position consistent with accretion to the proto–North American margin during Grenville convergence.

The Late Neoproterozoic rifting of Rodinia and the subsequent dismemberment of the extensive Grenville orogenic belt represent one of the most significant periods of continental breakup in the geological evolution of Earth. An exceptionally detailed and complete geologic record of these events is preserved in the Blue Ridge (the Shenandoah and French Broad massifs, Fig. 5) and elsewhere in the southern Appalachians as rift-related plutonic rocks and stratified metasedimentary and metavolcanic units that locally overlie and intrude Grenvillian-age basement. Papers concerning this activity discuss the three major stages in the extensional process: (1) unsuccessful rifting at ca. 750–700 Ma, (2) successful rifting at ca. 570 Ma, and (3) Late Neoproterozoic to Early Paleozoic transition to a passive continental margin. *Novak and Rankin* present field and geochemical data characterizing a 760 Ma compositionally zoned welded ash-flow sheet that constitutes part of the rift-related Mount Rogers Formation in southwestern Virginia. This mildly peralkaline to metaluminous volcanic center is closely associated in both space and time with numerous similarly A-type plutons in the northwestern North Carolina–southwestern Virginia Blue Ridge. Collectively these rocks appear to represent products of an initial pulse of magmatism developed in the early stages of a long-term process of crustal thinning that ultimately resulted in the rifting of Rodinia. *Badger and Sinha* demonstrate the utility of integrating field evidence with trace-element geochemical data to define compositional and physical stratigraphy in metamorphosed flood basalts of the 570 Ma Catoctin Formation in central Virginia, and use this framework to model the petrogenetic development of the voluminous Catoctin Volcanic Province. They illustrate similarities between the Catoctin and other flood basalt provinces, and suggest derivation from a fertile, possibly plume-related mantle source associated with terminal stages of Late Neoproterozoic rifting. Catoctin volcanism was followed by deposition of the regional Chilhowee Group, which reflects stabilization of the Laurentian passive margin, thus bringing to an end the tectonic cycle that began 700 m.y. earlier with initial development of the Grenville orogen in the crustal segment currently represented by the Blue Ridge.

Geochronological studies of these postrift sedimentary deposits provide important insight into the nature and paleogeographic location of source terranes, and thus serve to constrain the timing relations of regional tectonic models. The Pine Mountain belt in Georgia and Alabama (Fig. 5) represents the southernmost occurrence of Grenvillian crust in the Appalachian region, and, like other internal massifs, the paleogeographic origin of these rocks has significant implications for regional tectonics. *Steltenpohl et al.* report results from a U-Pb isotopic study of detrital zircons from a Late Neoproterozoic–Early Cambrian quartzite overlying Grenvillian basement, and these data identify a bimodal population that was likely derived from Laurentian crust of 1.2–1.0 Ga and ca. 1.4 Ga age. However, the presence of an additional population of much older zircons (2.3–2.4 Ga) may indicate proximity of a different source terrane, and suggests that the Pine Mountain basement might have originated as a crustal fragment located between Laurentia and Gondwana (source of the older zircons), and was subsequently thrust to its present position during Paleozoic orogenesis. *Bream et al.* also report U-Pb and Nd isotopic data bearing on the nature of the Laurentian margin during Late Neproterozoic rifting, presenting results obtained from sedimentary sequences occurring across a broad area of the southern Appalachian Blue Ridge and Inner Piedmont (Fig. 5). Based on their results and other corroborating information, these authors suggest that extensive sedimentary packages in the southern Appalachian Blue Ridge and western Inner Piedmont were derived from Laurentia. However, the presence of detrital zircons of Neoproterozoic and Ordovician age in paragneisses from the eastern Inner Piedmont indicate that these rocks were deposited much later and could have been derived from either a Panafrican source or possibly a mixture of Panafrican and recycled Laurentian margin sources.

TEXAS AND MEXICO

Grenvillian rocks of the Pine Mountain window disappear southwestward, along with the other rocks of the Paleozoic Appalachian orogen, beneath the Coastal Plain sediments of the Gulf of Mexico. Twelve hundred kilometers to the west, Mesoproterozoic rocks, which are now considered part of the Grenville orogen (Mosher, 1998), are exposed in the Llano Uplift of central Texas (Fig. 6), where they lie west of the structural front of the Paleozoic Ouachita fold and thrust belt. Thus, as in the Adirondacks and the Grenville Province of Canada, these Mesoproterozoic rocks of central and western Texas still occupy approximately the same relative positions that they acquired as

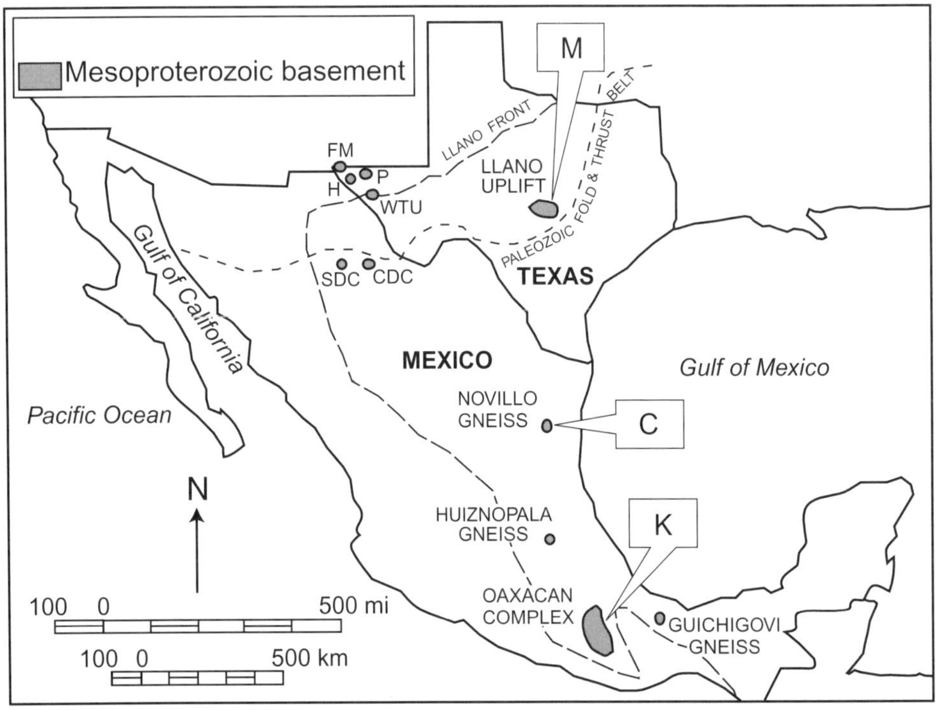

Figure 6. Map showing locations of major exposures of Mesoproterozoic basement rocks in Texas and Mexico with generalized locations of study areas pertaining to papers included in this volume. Map modified from de Cserna (1989) and Mosher (1998). Labeled study areas correspond to the following: M—Mosher et al.; C—Cameron et al.; K—Keppie et al. Abbreviations include FM—Franklin Mountains; H—Hueco Mountains; P—Pump Station Hills; WTU—West Texas Uplift; SDC—Sierra Del Cuervo, CDC—Cerro Del Carrizalillo.

a result of Grenvillian orogenesis associated with the formation of Laurentia. The same Mesoproterozoic rocks occur on both sides of the Llano Front in West Texas, but the front is interpreted to represent a structural equivalent of the Grenville Front in Canada because rocks in the Franklin Mountains, the Hueco Mountains, and the Pump Station Hills (Fig. 6) lack the characteristic signatures of Grenvillian deformation (Mosher, 1998). Mosher's earlier work concerned the western part of the Llano Uplift, the west Texas rocks, and rocks exposed at Sierra Del Cuervo and Cerro Del Carrizalillo in Mexico. In their paper included in this volume, *Mosher et al.* document the same type of relationships in the eastern Llano Uplift as those observed previously in central and west Texas, thus supporting the interpretation of a Mesoproterozoic suture separating rocks with Laurentian affinities to the north from rocks of an accreted ensimatic arc to the south. Where the Llano Front extends westward into northern Mexico is currently unknown. Moreover, exactly how the structural front of the Paleozoic fold and thrust belt is related to Mesoproterozoic rocks across northern Mexico is also poorly understood because of limited geologic research and the effects of extensive deformation of Mesozoic and Cenozoic age in this area.

Previous work in eastern and southern Mexico by Patchett and Ruiz (1987), Ortega-Gutiérrez et al. (1995), and Weber and Köhler (1999), among others, documented a protracted Grenvillian history within a large part of Mexico that lies east of the accreted terranes of Mesozoic and Cenozoic age in Baja California and southwestern Mexico. The papers by *Cameron et al.* and *Keppie et al.* continue work on selected aspects of this relatively new and little-known frontier of the Grenville orogen and support the earlier interpretations that much of the Grenvillian rocks of Mexico were part of an arc terrane accreted to Laurentia late in the development of the Grenville orogen. These workers in Mexico attribute the present location of the extensive Mexican portion of the Grenville orogen to Paleozoic accretion of this terrane after the breakup of Rodinia. Thus, like the relationships of Grenvillian inliers of the Appalachians relative to the Adirondacks (e.g., Bartholomew and Lewis, 1988, 1992), the relationships between Grenvillian rocks in Mexico and Texas may remain obscure until more research is done. Indeed, nowhere is it more obvious that new research is needed than in the vicinity of Sierra Del Cuervo and Cerro Del Carrizalillo (Fig. 6), which the researchers based in Texas consider compatible with interpretations based on the nearby rocks of central and western Texas. In contrast, Mexican researchers generally include these rocks within the Mexican terrane but south of a Paleozoic suture that includes the Oaxacan Complex of southern Mexico. Clearly, research on the Grenvillian rocks of Mexico is in its nascency and promises to be very rewarding.

ACKNOWLEDGMENTS

The authors acknowledge and express their appreciation to Pat Bickford, Tony Davidson, Charles Gower, Douglas Rankin, and Toby Rivers for the detailed reviews each provided for this chapter. Our sincere gratitude is also extended to Elizabeth Anne Borduas for her technical assistance in the production of the final version.

REFERENCES CITED

Aleinikoff, J.N., Burton, W.C., Lyttle, P.T., Nelson, A.E., and Southworth, C.S., 2000, U-Pb geochronology of zircon and monazite from Mesoproterozoic granitic gneisses of the northern Blue Ridge, Virginia and Maryland, USA: Precambrian Research, v. 99, p. 113–146.

Baer, A.J., 1974, Grenville geology and plate tectonics: Geoscience Canada, v. 1, p. 54–61.

Bartholomew, M.J., and Lewis, S.E., 1984, Evolution of Grenville massifs in the Blue Ridge geologic province, southern and central Appalachians, *in* Bartholomew, M.J., et al., eds., The Grenville event in the Appalachians and related topics: Boulder, Colorado, Geological Society of America Special Paper 194, p. 229–254.

Bartholomew, M.J., and Lewis, S.E., 1988, Peregrination of Middle Proterozoic massifs and terranes within the Appalachian orogen, eastern U.S.A.: Trabajos de Geología, Universidad de Oviedo, v. 17, p. 155–165.

Bartholomew, M.J., and Lewis, S.E., 1992, Appalachian Grenville massifs: Pre-Appalachian translational tectonics, *in* Mason, R., ed., Basement Tectonics 7: Amsterdam, Kluwer Academic Publishers, p. 363–374.

Bartholomew, M.J., Gathright, T.M., II, and Henika, W.S., 1981, A tectonic model for the Blue Ridge in central Virginia: American Journal of Science, v. 281, no. 9, p. 1164–1183.

Berman, R., Easton, R.M., and Nadeau, L., 2000, A new metamorphic map of the Canadian Shield, introduction: Canadian Mineralogist, v. 38, p. 277–285.

Blein, O., LaFlèche, M.R., and Corriveau, L., 2003, Geochemistry of the granulitic Bondy Gneiss Complex: A 1.4 Ga arc in the Central Metasedimentary Belt, Grenville Province, Canada: Precambrian Research, v. 120, p. 193–218.

Bohlen, S.R., 1987, Pressure-temperature-time paths and a tectonic model for the origin of granulites: Journal of Geology, v. 95, p. 617–632.

Bohlen, S.R., Valley, J., Essene, E., 1985, Metamorphism in the Adirondacks, I: Petrology, pressure, and temperature: Journal of Petrology v. 26, p. 971–992.

Buddington, A.F., 1939, Adirondack igneous rocks and their metamorphism: Boulder, Colorado, Geological Society of America Memoir 7, 354 p.

Card, K.D., and Poulsen, K.H., 1998, Geology and mineral deposits of the Superior Province of the Canadian Shield, *in* Lucas, S.B., and St-Onge, M.R., coordinators, Geology of the Precambrian Superior and Grenville provinces and Precambrian fossils in North America: Boulder, Colorado, Geological Society of America, The Geology of North America, v. C-1, p. 15–204.

Carr, S.D., Easton, R.M., Jamieson, R.A., and Culshaw, N.G., 2000, Geologic transect across the Grenville orogen of Ontario and New York: Canadian Journal of Earth Sciences, v. 37, p. 193–216.

Clark, T., 2003, Métallogénie des métaux usuels et précieux, des éléments radioactifs et des éléments des terres rares, région de la moyenne Côte-Nord, *in* Brisebois, D., and Clark, T., coords., Géologie et minérales de la partie est de la Province de Grenville: Québec, Ministère des Ressources naturelles, DV 2002–03, p. 269–326.

Corbett, G.J., and Leach, T.M., 1998, Southwest Pacific rim gold-copper systems: Structure, alteration, and mineralisation: Littleton, Colorado, Society of Economic Geologists Special Publication 6, 236 p.

Corrigan, D., 1995, Mesoproterozoic evolution of the south-central Grenville orogen: Structural, metamorphic and geochronologic constraints from the Maurice transect [Ph.D. dissertation]: Ottawa, Ontario, Carleton University, 282 p.

Corrigan, D., and Hanmer, S., 1997, Anorthosites and related granitoids in the

Grenville orogen: A product of convective thinning of the lithosphere? Geology, v. 25, p. 61–64.
Corriveau, L., and van Breemen, O., 2000, Docking of the Central Metasedimentary Belt to Laurentia in geon 12: Evidence from the 1.17–1.16 Ga Chevreuil intrusive suite and host gneisses, Québec: Canadian Journal of Earth Sciences, v. 37, p. 253–269.
Corriveau, L., Heaman, L.M., Marcantonio, F., and van Breemen, O., 1990, 1.1 Ga K-rich alkaline plutonism in the SW Grenville Province: U-Pb constraints for the timing of subduction-related magmatism: Contributions to Mineralogy and Petrology, v. 105, p. 473–485.
Corriveau, L., Rivard, B., and van Breemen, O., 1998, Rheological controls on Grenvillian intrusive suites: Implications for tectonic analysis: Journal of Structural Geology, v. 20, p. 1191–1204.
Corriveau, L., Blein, O., Bonnet, A.-L., Fu, W., Pilote, P., and van Breemen, O., 2003a, Field identification of Cu-Au-Fe-oxides bearing hydrothermal systems in undifferentiated gneiss complexes of the Grenville Province: Canadian Institute of Mining (CIM) Mining Industry Conference and Exhibition, Montreal, 2003, CIM Technical Paper, CD-ROM.
Corriveau, L., Bonnet, A.-L., van Breemen, O., and Pilote, P., 2003b, Tracking the Wakeham Group volcanics and associated Cu-Fe-oxydes hydrothermal activity from La Romaine eastward, Eastern Grenville Province, Québec: Ottawa, Ontario, Geological Survey of Canada Paper 2003-C12, 11 p.
Culshaw, N.G., Davidson, A., and Nadeau, L., 1983, Structural subdivisions of the Grenville Province in the Parry Sound–Algonquin region, Ontario: Ottawa, Ontario, Geological Survey of Canada Paper 83–1B, p. 243–252.
Culshaw, N.G., Jamieson, R.A., Ketchum, J.W.F., Wodicka, N., Corrigan, D., and Reynolds, P.H., 1997, Transect across the northwestern Grenville orogen, Georgian Bay, Ontario: Polystage convergence and extension in the lower orogenic crust: Tectonics, v. 16, p. 966–982.
Dalziel, I.W.D., Mosher, S., and Gahagan, L.M., 2000, Laurentia-Kalahari collision and the assembly of Rodinia: Journal of Geology, v. 108, p. 499–513.
Davidson, A., 1984, Identification of ductile shear zones in the southwestern Grenville Province of the Canadian Shield, *in* Kröner, A., and Greiling, E., eds., Precambrian tectonics illustrated: Stuttgart, Schweizerbart'sche Verlagsbuchhandlung, p. 263–279.
Davidson, A., 1995, A review of the Grenville orogen in its North American type area: Australian Geological Survey Organisation Journal of Australian Geology and Geophysics, v. 16, p. 3–24.
Davidson, A., 1998a, An overview of Grenville Province geology, Canadian Shield, Chapter 3, *in* Lucas, S.B., and St-Onge, M.R., coordinators, Geology of the Precambrian Superior and Grenville provinces and Precambrian fossils in North America: Ottawa, Ontario, Geological Survey of Canada, Geology of Canada, no. 7, p. 205–270 (also Boulder, Colorado, Geological Society of America, The Geology of North America, v. C-1).
Davidson, A., 1998b, Geological map of the Grenville Province, Canada and adjacent parts of the United States of America: Ottawa, Ontario, Geological Survey of Canada Map 1947A, scale 1:2,000,000.
Davidson, A., and van Breemen, O., 2000, Age and extent of the Frontenac plutonic suite in the Central Metasedimentary Belt, Grenville Province, southeastern Ontario: Current Research: Ottawa, Ontario, Geological Survey of Canada Paper 2000-F4, 7 p.
Davidson, A., Culshaw, N.G., and Nadeau, L., 1982, A tectono-metamorphic framework for part of the Grenville Province, Parry Sound region, Ontario, *in* Current Research, Part A: Ottawa, Ontario, Geological Survey of Canada Paper 82–1A, p. 175–190.
de Cserna, Z., 1989, An outline of the geology of Mexico, *in* Bally, A.W., and Palmer, A.R., eds., The Geology of North America—An overview: Boulder, Colorado, Geological Society of America, The Geology of North America, v. A, p.233–264.
deWaard, D., 1965, The occurrence of garnet in the granulite facies terrane of the Adirondack Highlands: Journal of Petrology, v. 6, p. 165–191.
Dickin, A.P., 2000, Crustal formation in the Grenville Province: Nd-isotope evidence: Canadian Journal of Earth Sciences, v. 37, p. 165–181.
Dickin, A.P., and Higgins, M.H., 1992, Sm/Nd evidence for a major 1.5 Ga crust-forming event in the central Grenville Province: Geology, v. 20, p. 137–140.
Doig, R., 1991, U-Pb zircon dates of the Morin anorthosite suite rocks, Grenville Province, Québec: Journal of Geology, v. 99, p. 729–738.
Easton, R.M., 1992, The Grenville Province and the Proterozoic history of central and southern Ontario, *in* Thurston, P.C., et al., eds., Geology of Ontario: Toronto, Ontario Geological Survey Special Volume 4, Part 2, p. 715–904.
Easton, R.M., 2000, Metamorphism of the Canadian shield, Ontario, Canada, II: Proterozoic metamorphic history: Canadian Mineralogist, v. 38, p. 319–344.
Evans, N., 1991, Latest Precambrian to Ordovician metamorphism in the Virginia Blue Ridge: Origin of the contrasting Lovingston and Pedlar basement massifs: American Journal of Science, v. 291, p. 425–452.
Foose, M.P., and McLelland, J.M., 1995, Proterozoic low-Ti iron-oxide deposits in New York and New Jersey: Relation to Fe-oxide (Cu-U-Au-rare earth element) deposits and tectonic implications: Geology, v. 23, p. 665–668.
Gobeil, A., Chevé, S., Clark, T., Corriveau, L., Perreault, S., Dion, D.J., and Nabil, H., 1999, Géologie de la région du lac Nipisso (SNRC 22I/13): Québec, Ministère des Ressources naturelles, RG 98-19, 60 p.
Gobeil, A., Brisebois, D., Clark, T., Verpaelst, P., Madore, L., Wodicka, N., and Chevé, S., 2003, Géologie de la moyenne Côte-Nord, *in* Brisebois, D., and Clark, T., coords., Synthèse géologique et métallogénique de la partie est de la Province de Grenville: Québec, Ministère des Ressources naturelles, DV 2002–03, p. 9–58.
Gower, C.F., 1996, The evolution of the Grenville Province in eastern Labrador, Canada, *in* Brewer, T.S., ed., Precambrian crustal evolution in the North Atlantic Region: Geological Society [London] Special Publication 112, p. 197–218.
Gower, C.F., and Krogh, T.E., 2002, A U-Pb geochronological review of the Proterozoic history of the eastern Grenville Province: Canadian Journal of Earth Sciences, v. 39, p. 795–829.
Gower, C.F., and Krogh, T.E., 2003, A U-Pb geochronological review of the Pre-Labradorian and Labradorian geological history of the eastern Grenville Province, *in* Brisebois, D., and Clark, T., coords., Géologie et ressources minérales de la partie est de la Province de Grenville: Québec, Ministère des Ressources naturelles, DV 2002-03, 142–172.
Gower, C.F., and Tucker, R.D., 1980, Distribution of pre-1400-Ma crust in the Grenville Province: Implications for rifting in Laurentia-Baltica during geon 14: Geology, v. 22, p. 827–830.
Gower, C.F., Ryan, A.B., and Rivers, T., 1990, Mid-Proterozoic Laurentia-Baltica: An overview and a summary of the contributions made by this volume: St. John's, Newfoundland, Geological Association of Canada Special Paper 38, p. 1–20.
Green, A.G., Milkereit, B., Davidson, A., Spencer, C., Hutchinson, D.R., Cannon, W.F., Lee, M.W., Agena, W.F., Behrendt, J.C., and Hinze, W.J., 1988, Crustal structure of the Grenville Front and adjacent terranes: Geology, v. 16, p. 788–792.
Hanmer, S., 1988, Ductile thrusting at mid-crustal level, southwestern Grenville Province: Canadian Journal of Earth Sciences, v. 25, p. 1049–1059.
Hatcher, R.D., 1983, Basement massifs in the Appalachians: Their role in deformation during the Appalachian orogenies: Geological Journal, v. 18, p. 255–265.
Hitzman, M.W., 2000, Iron Oxide-Cu-Au deposits: What, where, when and why, *in* Porter, T.M., ed., Hydrothermal iron oxide copper-gold and related deposits: A global perspective, v. 1: Adelaide, PGC Publishing, p. 9–25.
Hoffman, P.F., 1989, Precambrian geology and tectonic history of North America, *in* Bally, A.W., and Palmer, A.R., eds., The geology of North America: An overview: Boulder, Colorado, Geological Society of America, The Geology of North America, v. A, p. 447–511.
Hoffman, P.F., 1991, Did the breakout of Laurentia turn Gondwana inside out? Science, v. 252, p. 1409–1412.
King, P.B., and Beikman, H.M., 1974, Geologic map of the United States: Reston, Virginia, U.S. Geological Survey Professional Paper 901, map and explanatory text.

Large, R.R., McPhie, J., Gemmel, J.B., Herrmann, W., and Davidson, J., 2001, The spectrum of ore deposit types, volcanic environments, alteration halos, and related exploration vectors in submarine volcanic successions: Some examples from Australia: Economic Geology, v. 96, p. 913–938.

Logan, W.E., 1863, Report on the geology of Canada: Ottawa, Ontario, Geological Survey of Canada, Report of Progress from its Commencement to 1863, 983 p.

Ludden, J., and Hynes, A., 2000a, The Abitibi-Grenville Lithoprobe transect, part III: Introduction: Canadian Journal of Earth Sciences, v. 37, p. 115–116.

Ludden, J., and Hynes, A., 2000b, The Lithoprobe Abitibi-Grenville transect: Two billion years of crust formation and recycling in the Precambrian Shield of Canada: Canadian Journal of Earth Sciences, v. 37, p. 459–476.

Martignole, J., and Calvert, A.J., 1996, Crustal-scale shortening and extension across the Grenville Province of western Québec: Tectonics, v. 15, p. 376–386.

Martignole, J., Calvert, A.J., Friedman, R., and Reynolds, P., 2000, Crustal evolution along a seismic section across the Grenville Province (western Québec): Canadian Journal of Earth Sciences, v. 37, p. 291–306.

McLelland, J., and Chiarenzelli, J., 1990, Isotopic constraints on the emplacement age of the Marcy anorthosite massif, Adirondack Mountains, New York: Journal of Geology, v. 98, p. 19–41.

McLelland, J., and Isachsen, Y., 1986, Geological synthesis of the Adirondack Mountains and their setting within the Grenville Province of Canada, *in* Moore, J., et al., eds., The Grenville Province: St. John's, Newfoundland, Geological Association of Canada Special Paper 31, p. 75–95.

McLelland, J., Daly, S., and McLelland, J.M., 1996, The Grenville Orogenic Cycle (ca. 1350–1000 Ma): An Adirondack perspective: Tectonophysics, v. 265, p. 1–28.

McLelland, J., Hamilton, M., Selleck, B., McLelland, J.M., Orrell, S., and Walker, D., 2001, Zircon U-Pb geochronology of the Ottawan Orogeny, Adirondack Highlands, New York: Regional and tectonic implications: Precambrian Research, v. 109, p. 39–72.

Meert, J.G., and Powell, C.M., 2001, Assembly and break-up of Rodinia: Introduction to the special volume: Precambrian Research, v. 110, p 1–8.

Mezger, K., Rawnsley, C.M., Bohlen, S.R., and Hanson, G.N., 1991, U-Pb garnet, sphene, monazite and rutile ages: Implications for the duration of metamorphism and cooling histories, Adirondack Mountains, New York: Journal of Geology, v. 99, p. 415–428.

Miller, B.V., and Barr, S.M., 2000, Petrology and isotopic composition of a Grenvillian basement fragment in the northern Appalachian orogen: Blair River inlier, Nova Scotia, Canada: Journal of Petrology, v. 41, p. 1777–1804.

Moore, J.M., 1986, Introduction: The "Grenville Problem" then and now, *in* Moore, J.M., et al., eds., The Grenville Province: St. John's, Newfoundland, Geological Association of Canada Special Paper 31, p. 107–117.

Moore, J.M., and Thompson, P., 1980, The Flinton group: A late Precambrian metasedimentary sequence in the Grenville Province of eastern Ontario: Canadian Journal of Earth Sciences, v. 17, p. 1685–1707.

Mosher, S., 1998, Tectonic evolution of the southern Laurentian Grenville orogenic belt: Geological Society of America Bulletin, v. 110, p. 1357–1375.

Nunn, G.A.G., Thomas, A., and Krogh, T.E., 1985, The Labradorian orogeny: Geochronological database, *in* Current Research: St. John's, Newfoundland Department of Mines and Energy, Mineral Development Division, Reprint 85-1, p. 43–54.

Ohmoto, H., 1996, Formation of volcanogenic massive sulfide deposits: The Kuroko perspective: Ore Geology Reviews, v. 10, p. 135–177.

Ortega-Gutiérrez, F., Ruiz, J., and Centano-Garcia, E., 1995, Oaxaquia, a Proterozoic microcontinent accreted to North America during the late Paleozoic: Geology, v. 23, p. 1127–1130.

Patchett, P.J., and Ruiz, J., 1987, Nd isotopic ages of crust formation and metamorphism in the Precambrian of eastern and southern Mexico: Contributions to Mineralogy and Petrology, v. 96, p. 523–528.

Perreault, S., Clark, T., Gobeil, A., Chevé, S., Dion, D.-J., Corriveau, L., Nabil, H., and Lortie, P., 1997, The Cu-Ni-Co potential of the Sept-Îles region: The lac Volant showing: Québec, Ministère des Ressources naturelles, PRO 97-03, 10 p.

Rankin, D.W., 1976, Appalachian salients and recesses: Late Precambrian continental breakup and the opening of the Iapetus ocean: Journal of Geophysical Research, v. 81, p. 5605–5619.

Rankin, D.W., 1993, Distribution and geologic setting of Precambrian (Proterozoic) rocks east and southeast of the Grenville front in eastern United States and adjacent Canada, *in* Reed, J.C., Jr., et al., eds., Precambrian: Conterminous U.S.: Boulder, Colorado, Geological Society of America, The Geology of North America, v. C-2, plate 5.

Rankin, D.W., Drake, A.A., Jr., Glover, L., III, Goldsmith, R., Hall, L.M., Murray, D.P., Ratcliffe, N.M., Read, J.F., Secor, D.T., Jr., and Stanley, R.S., 1989, Pre-orogenic terranes, *in* Hatcher, R.D., Jr., et al., eds., The Appalachian-Ouachita orogen in the United States: Boulder, Colorado, Geological Society of America, The Geology of North America, v. F-2, p. 7–100.

Rankin, D.W., Drake, A.A., Jr., and Ratcliffe, N.M., 1990, Geologic map of the U.S. Appalachians showing the Laurentian margin and Taconic orogen, *in* Hatcher, R.D., Jr., et al., eds., The Appalachian-Ouachita orogen in the United States: Boulder, Colorado, Geological Society of America, The Geology of North America, v. F-2, plate 2.

Raymond, D., Perry, C., and Roy, I., 1997, Le projet de SOQUEM: Une première année d'exploration: Québec, Ministère des Ressources naturelles, DV 97-03, p. 15.

Rivers, T., 1983, The northern margin of the Grenville Province in western Labrador—Anatomy of an ancient orogenic front: Precambrian Research, v. 22, p. 41–73.

Rivers, T., 1997, Lithotectonic elements of the Grenville Province: Review and tectonic implications: Precambrian Research, v. 86, p. 117–154.

Rivers, T., and Corrigan, D., 2000, Convergent margin on southeastern Laurentia during the Mesoproterozoic: Tectonic implications: Canadian Journal of Earth Sciences, v. 37, p. 359–383.

Rivers, T., Martignole, J., Gower, C.F., and Davidson, A., 1989, New tectonic divisions of the Grenville Province, southeast Canadian Shield: Tectonics, v. 8, p. 63–84.

Rivers, T., Ketchum, J., Indares, A., and Hynes, A., 2002, The High Pressure belt in the Grenville Province: Architecture, timing, and exhumation: Canadian Journal of Earth Sciences, v. 39, p. 867–893.

Rogers, J., 1967, Chronology of tectonic movements in the Appalachian region of eastern North America: American Journal of Science, v. 265, p. 408–427.

Saint-Germain, P., and Corriveau, L., 2003, Évolution magmatique et géochimique des gabbronorites et des monzonites du Complexe de Matamec, région de Sept-Îles, *in* Brisebois, D., and Clark, T., coords., Géologie et ressources minérales de la partie est de la Province de Grenville: Québec, Ministère des Ressources naturelles, DV 2002-03, p. 179–212.

Sangster, A., Gauthier, M., and Gower, C.F., 1992, Metallogeny of structural zones, Grenville Province, northeastern North America: Precambrian Research, v. 58, p. 410–426.

Shieh, Y., 1985, High ^{18}-O grantic plutons from the Frontenac Axis, Grenville Province of Ontario, Canada: Geochimica et Cosmochimica Acta, v. 49, p. 117–123.

Shieh, Y., and Schwarcz, H.P., 1978, The oxygen isotope composition of the surface of the Canadian Shield: Canadian Journal of Earth Sciences, v. 15, p. 1773–1782.

Silver, L., 1969, A geochronological investigation of the anorthosite complex, Adirondack Mountains, New York, *in* Isachsen, Y.W., ed., Origin of Adirondack anorthosites and related rocks: Albany, New York State Museum Memoir 18, p. 233–252.

Sinha, A.K., and Bartholomew, M.J., 1984, Evolution of the Grenville terrane in the central Virginia Appalachians, *in* Bartholomew, M.J., et al., eds., The Grenville event in the Appalachians and related topics: Boulder, Colorado, Geological Society of America Special Paper 194, p. 175–186.

Spear, F.S., and Markusen, J.C., 1997, Mineral zoning, *P-T*-X-M phase relations and metamorphic evolution of some Adirondack granulites, New York: Journal of Petrology, v. 38, p. 757–783.

Thomas, A., 1993, Geology of the Winokapau Lake area, Grenville Province, Central Labrador: Ottawa, Ontario, Geological Survey of Canada Paper 89-18, p. 52.

Thompson, J.E., 1956, Preface, *in* Thompson, J.E., ed., The Grenville problem: Ottawa, Ontario, Royal Society of Canada Special Publication 1, p. 3–13.

Valley, J., Bohlen, S.R., Essene, E.J., and Lamb, W., 1990, Metamorphism in the Adirondacks, II: The role of fluids: Journal of Petrology, v. 31, p. 555–596.

van Breemen, O., and Corriveau, L., 2001, 1.5 Ga pyroclastic volcanism in the Wakeham Group, eastern Grenville Province, Québec: Boulder, Colorado, Geological Society of America Abstracts with Programs, v. 33, no. 6, p. A-89.

van Breemen, O., and Davidson, A., 1988, Northeast extension of Proterozoic terranes of mid-continental North America: Geological Society of America Bulletin, v. 100, p. 630–638.

Verpaelst, P., Madore, L., Brisebois, D., Choinière, J., Dion, D.J., and David, J., 1997, Géologie de la région du lac Bohier (12N/03): Québec, Ministère des Ressources naturelles, RG 97-07, 29 p.

Wardle, R.J., and Hall, J., 2002, Proterozoic evolution of the northeastern Canadian Shield: Lithoprobe Eastern Canadian Shield Onshore-Offshore Transect (ESCOOT), introduction and summary: Canadian Journal of Earth Sciences, v. 39, p. 563–567.

Wasteneys, H., McLelland, J., and Lumbers, S., 1999, Precise zircon geochronology in the Adirondack Lowlands and implications for revising plate tectonic models of the Central Metasedimentary Belt and Adirondack Mountains, Grenville Province, Ontario and New York: Canadian Journal of Earth Sciences, v. 36, p. 967–984.

Weber, B., and Köhler, H., 1999, Sm/Nd, Rb/Sr, and U-Pb geochronology of a Grenville terrane in southern Mexico: Origin and geologic history of the Guichicovi complex: Precambrian Research, v. 96, p. 245–262.

Wheeler, J.O., Hoffman, P.F., Card, K.D., Davidson, A., Sanford, B.V., Okulitch, A.V., and Roest, W.R., 1996, Geological Map of Canada: Ottawa, Ontario, Geological Survey of Canada Map 1860A, scale 1:5,000,000.

White, D.J., Forsyth, D.A., Asudeh, I., Carr, S.D., Wu, H., Easton, R.M., and Mereu, R.F., 2000, A seismic-based cross-section of the Grenville Orogen in southern Ontario and western Québec: Canadian Journal of Earth Sciences, v. 37, p. 183–192.

Wiebe, R.A., and Collins, W.J., 1998, Depositional features and stratigraphic sections in granitic plutons: Implications for the emplacement and crystallization of granitic magma: Journal of Structural Geology, v. 20, p. 1273–1289.

Windley, B.F., 1986, Comparative tectonics of the western Grenville and the western Himalaya, *in* Moore, J.M., et al., eds., The Grenville Province: St. John's, Newfoundland, Geological Association of Canada Special Paper 31, p. 341–348.

Wodicka, N., David, J., Parent, M., Gobeil, A., and Verpaelst, P., 2003, Géochronologie U-Pb et Pb-Pb de la région de Sept-Îles–Natashquan, Province de Grenville, Moyenne-Côte-Nord, *in* Brisebois, D., and Clark, T., coords., Géologie et ressources minérales de la partie est de la Province de Grenville: Québec, Ministère des Ressources naturelles, DV 2002–03. p. 59–118.

Wynne-Edwards, H.R., 1972, The Grenville Province, *in* Price, R.A., and Douglas, R.J.W., eds., Variations in tectonic styles in Canada: St. John's, Newfoundland, Geological Association of Canada Special Paper 11, p. 263–334.

Manuscript Accepted by the Society August 25, 2003

Printed in the USA

Geological Society of America
Memoir 197
2004

Cordierite-orthopyroxene white gneiss: A key to unveiling premetamorphic hydrothermal activity in the Bondy gneiss complex, Grenville Province, Québec

Olivier Blein*
Université du Québec, Institut National de la Recherche Scientifiques–Eau Terre Environment (INRS-ETE), P.O. Box 7500, Sainte-Foy, Québec, G1V 4C7, Canada
Louise Corriveau
Geological Survey of Canada, GSC-Québec, 880 Chemin Ste-Foy, Room 840, Québec, G1S 2L2, Canada
Marc R.-LaFlèche
Université du Québec, INRS-ETE, P.O. Box 7500, Sainte-Foy, Québec G1V 4C7, Canada

ABSTRACT

The 1.4 Ga volcanic-plutonic arc-related Bondy gneiss complex in the western Grenville Province of Québec, Canada, comprises a Cu-Au-Fe oxides hydrothermal system characterized by a series of showings among aluminous and magnesian lithofacies, such as tourmalinite, aluminous sillimanite-bearing gneiss, garnetite, and cordierite-orthopyroxene white gneiss. The latter has a mineral assemblage and a major element chemistry comparable to those of cordierite-orthoamphibole gneiss found in alteration pipes associated with volcanogenic massive sulfide (VMS) deposits. Supporting evidence for a hydrothermal origin for the Bondy cordierite-orthopyroxene gneiss is given by mass-balance calculations that record a loss in Ca and gain in Mg, K, and Si with respect to the adjacent nonaltered rhyolitic protolith, compatible with feldspar destruction, sericitization, and chloritization. These lithofacies present birdwing-shaped rare earth element (REE) profiles, reflecting a mass change of REE and high field strength elements (HFSE) involving relative leaching of light rare earth elements (LREE) and enrichment in heavy rare earth elements (HREE), with a general calculated mass increase from Gd to Yb. These trace element characteristics differ from those of known, Cl-driven, VMS-associated hydrothermal cordierite-orthopyroxene or orthoamphibole rocks, but share similarities with F-driven hydrothermal signatures. The decoupling between major element and trace element signatures suggest that the type of fluids driving the postvolcanic or premetamorphic hydrothermal alteration and mineralization event in the Bondy gneiss complex evolved through time from seawater related to orthomagmatic.

Keywords: cordierite, mobility, rare earth elements

*E-mail: oblein@libertysurf.fr.

Blein, O., Corriveau, L., and LaFlèche, M.R., 2004, Cordierite-orthopyroxene white gneiss: A key to unveiling premetamorphic hydrothermal activity in the Bondy gneiss complex, Grenville Province, Québec, *in* Tollo, R.P., Corriveau, L., McLelland, J., and Bartholomew, M.J., eds., Proterozoic tectonic evolution of the Grenville orogen in North America: Boulder, Colorado, Geological Society of America Memoir 197, p. 19–33. For permission to copy, contact editing@geosociety.org.

INTRODUCTION

In exploration for volcanogenic massive sulfide (VMS) deposits, geochemistry is used to characterize the distribution and variation of hydrothermal alteration assemblages in the footwalls of deposits. Alteration in such settings is driven by Cl-rich fluids and characterized by the breakdown of sodic plagioclase and volcanic glass, and their replacement by sericite and chlorite. Sericitization is typical of the outer parts of the alteration system (e.g., Date et al., 1983; Eastoe et al., 1987), and involves a loss of Na_2O and CaO and a gain of K_2O. Chloritization prevails in footwall pipe zones (e.g., Sangster, 1972; Lydon, 1988; Large, 1992; Lentz, 1999; Schardt et al., 2001), and the pervasive replacement of sericite by chlorite involves a loss of K_2O and a gain of FeO and MgO. Mg-Fe chlorite alteration is typically developed close to the deposits themselves where hydrothermal temperatures and water/rock ratios are at their peak (e.g., Franklin et al., 1981; Lydon, 1988; Schardt et al., 2001).

In the last two decades, a large number of field studies and laboratory experiments have investigated interactions between volcanic rocks and seawater, including petrological processes and geochemical characteristics (e.g., Humphries and Thompson, 1978a, 1978b; Menzies et al., 1979; Ludden et al., 1982; Mottl, 1983; Seyfried, 1987). These studies demonstrated that the mobility of trace elements, particularly the high field strength elements (HFSE) and rare earth elements (REE), is highly dependent on the nature and physicochemical conditions of the fluids, the mineralogy of original and alteration assemblages, and the transport of REE by ligand complexes (Lottermoser, 1992; Wood and Williams-Jones, 1994). Convincing examples of REE mobility under hydrothermal conditions have been observed in chlorite-rich rocks from alteration pipes and haloes associated with several VMS deposits (e.g., Finlow-Bates and Stumpfl, 1981; Campbell et al., 1984; MacLean, 1988; Whitford et al., 1988; Lottermoser, 1992; Pan et al., 1994).

In medium-grade metamorphic terrains, cordierite-orthoamphibole rocks associated with VMS deposits represent metamorphosed equivalents of such chlorite-rich rocks (Vallance, 1967; Franklin et al., 1981), though other genetic models apply to some specific occurrences (e.g., metamorphosed Mg- and Fe-rich sedimentary rocks; Reinhardt, 1987; Pan et al., 1991). Described early by Eskola (1914), cordierite-orthoamphibole rocks, characterized by the presence of orthoamphiboles in association with Al-rich phases, particularly cordierite, Al_2SiO_5 phases, and garnet and staurolite, have attracted much attention in petrological studies (Robinson et al., 1982; Schumacher and Robinson, 1987; Schneiderman and Tracy, 1991; Humphreys, 1993). Such mineral assemblages reflect the unusual bulk compositions of these rocks. Cordierite-orthopyroxene (Crd-Opx) gneiss is a high-grade equivalent of cordierite-orthoamphibole rocks (Schreurs and Westra, 1985; Guiraud et al., 1996; symbols for rock-forming minerals after Kretz, 1983).

Located in the Central Metasedimentary Belt of Québec, in the western Grenville Province, Canada (Fig. 1), the Bondy gneiss complex is a volcano-plutonic edifice of 1.4 Ga metamorphosed at granulite facies. It comprises an anomalous zone rich in magnetite and hosts a series of Cu showings among aluminous and magnesian lithofacies, including tourmalinite, garnetite and sillimanite-quartz, sillimanite-biotite-garnet, sillimanite-orthopyroxene-cordierite, and cordierite-kornerupine-orthopyroxene gneisses. The present study focuses on the geo-

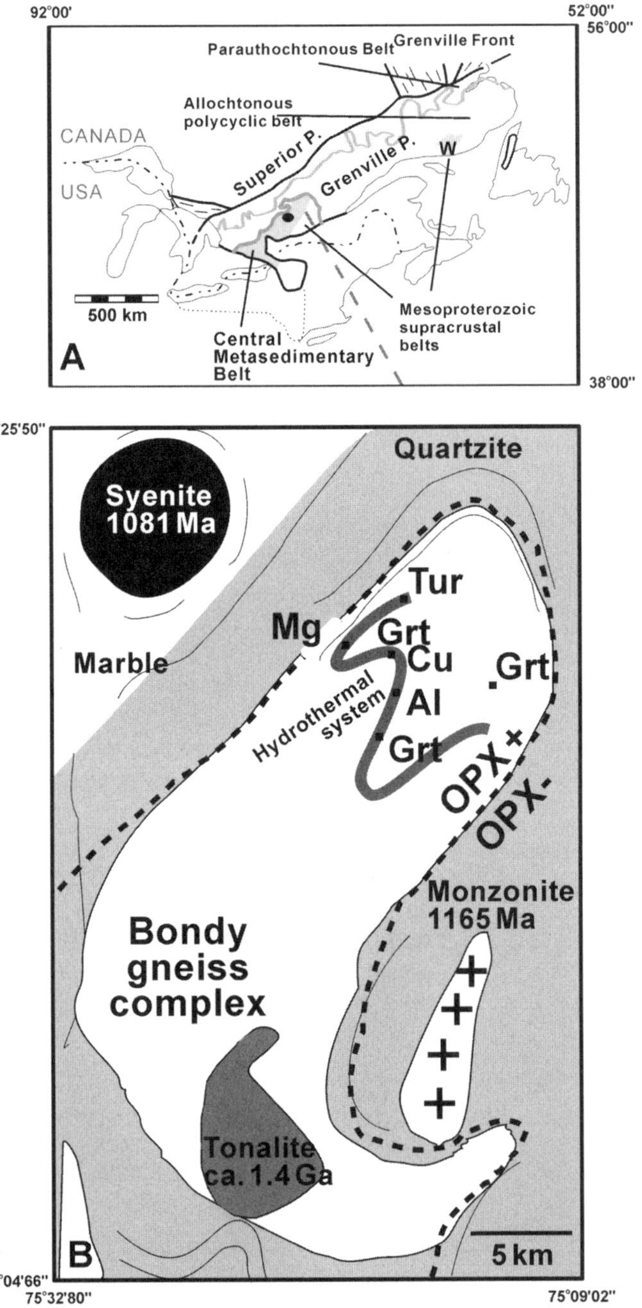

Figure 1. Distribution of principal tectonic elements in the Grenville Province, modified from Rivers et al. (1989). Black dot: Bondy gneiss complex. Bondy gneiss complex geologic context: Al—aluminous lithofacies; Cu—copper showing; Grt—garnetite; Mg—magnesian lithofacies; Opx—orthopyroxene; Tur—tourmalinite.

chemistry of these rocks, and attempts to (a) determine protoliths and origin of the Crd-Opx gneisses, and (b) examine the behavior of major elements, REE, and HFSE during mineralization and hydrothermal processes. Exploration strategies have generally considered large quartzofeldspathic gneiss complexes to be monotonous and sterile and have ignored them in the past. We fill part of the information gap on such terrains through research on the Crd-Opx gneiss of the Bondy gneiss complex. Such lithofacies are key to mineral exploration of volcanogenic massive sulfide deposits worldwide, and may provide a tool that can be used to understand the seemingly intractable gneiss complexes of the Grenville Province.

THE BONDY GNEISS COMPLEX

The Bondy gneiss complex is a volcano-plutonic complex composed mostly of leucocratic granitic and tonalitic orthogneiss with sporadic, meter- to kilometer-thick mafic amphibolite and mafic granulite units, and with a metatonalitic pluton in its southern part. In its northern part, the Bondy gneiss complex comprises aluminous lithofacies, which can be traced along gneissosity for more than 12 km following mostly a 100 m wide folded band (Fig. 2). From north to south, these aluminous lithofacies are (1) a thick magnesian tourmalinite unit; (2) a Crd-Opx white gneiss unit; (3) a variety of sillimanite- and/or cordierite-

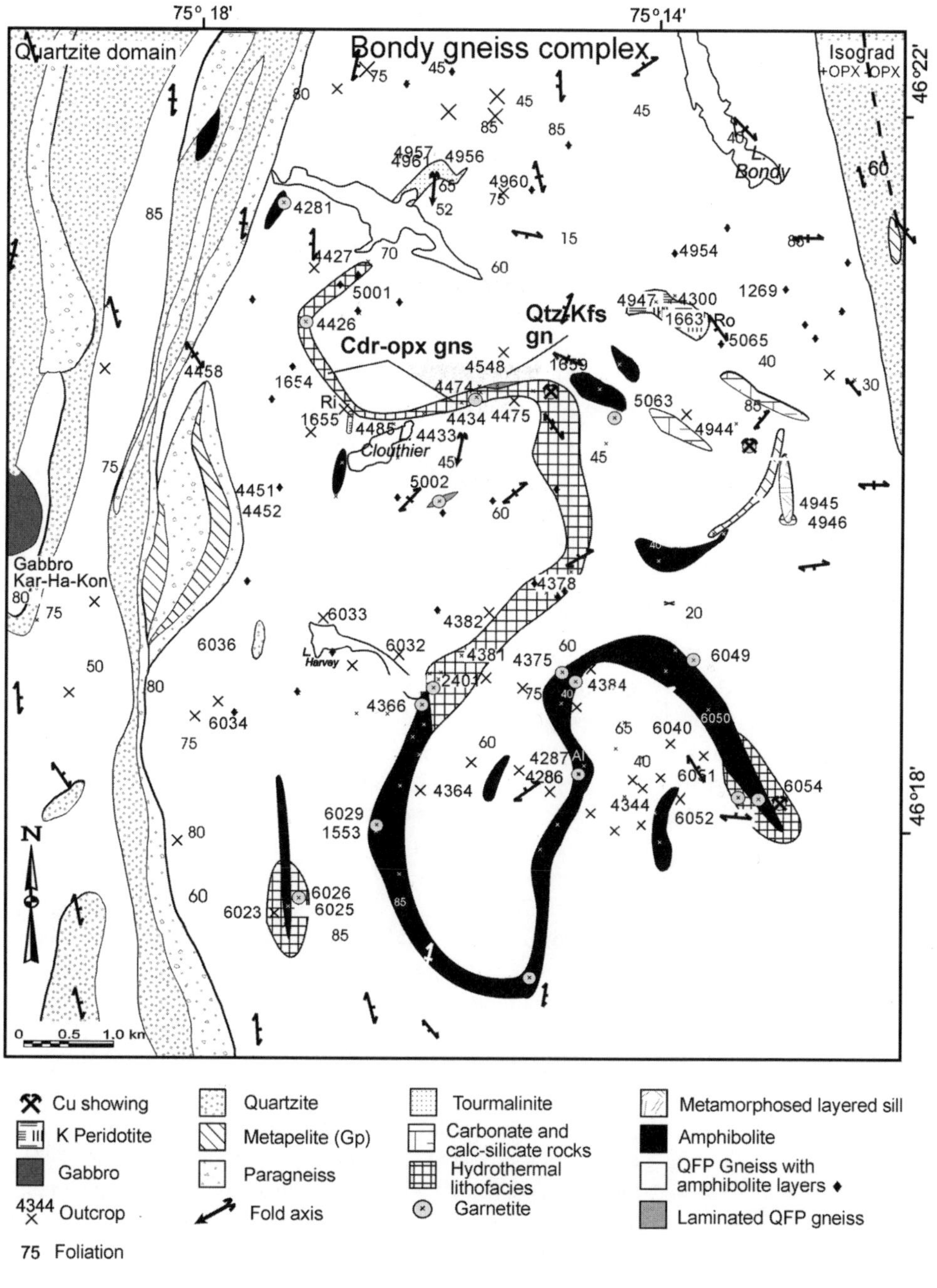

Figure 2. Geology of the hydrothermal system in the Bondy gneiss complex (National Topographical System map 31J/6). Numbers in figure are sample numbers. Crd—cordierite; gns—gneiss; Kfs—K feldspar; Opx—orthopyroxene; QFP—quartzofeldspathic; Qtz—quartz.

bearing gneisses adjacent to a laminated quartzofeldspathic gneiss (resembling a tuff or a flow-banded rhyolite); (4) a series of magnetite-rich garnetite, biotite-orthopyroxene-garnet gneiss, and amphibolite, locally mineralized in chalcopyrite; (5) a hyperaluminous sillimanite and quartz gneiss; (6) a biotite-rich garnetite among biotite-garnet gneisses; and (7) a layered garnet-bearing amphibolite unit with hornblende-bearing garnetite. This band can be subdivided into two parts: a northern part composed mainly of gneissic facies, and a southern part characterized by its enrichment in amphibolite (Fig. 2). Mafic granulites are commonly found among, and in, the vicinity of the aluminous lithofacies but prevail only in the southern zone. Six Cu showings with up to 6000 to 8980 ppm Cu on representative samples (up to 10 kg) have been discovered during regional mapping (Corriveau et al., 1996), mainly in magnetite-rich garnetite and biotite-orthopyroxene-garnet gneiss. Anomalous gold values of 644 and 906 ppb were detected in sillimanite-rich gneiss, while Au values of 50 ppb are common in associated orthopyroxene-biotite gneiss; this is a value significantly higher than the few ppb normally present in the paragneiss units of the Central Metasedimentary Belt of Québec (Corriveau et al., 1996).

The chemical compositions of mafic granulites, tonalitic gneisses, tonalites, and laminated quartzofeldspathic gneisses are summarized in Table 1. Their protoliths are the equivalent of mafic to felsic volcanic and plutonic rocks with tholeiitic to calc-alkaline affinities derived from a depleted mantle source within a subduction zone and a back-arc setting (Blein et al., 2003). According to major element and trace element chemistry, mafic granulites have been subdivided into three groups. Geochemical data of mafic granulites in Groups I and II are transitional between normal mid-oceanic ridge basalts (N-MORB) and intraoceanic island-arc tholeiites. Group I mafic granulites, with flat REE patterns (Fig. 3, A), are similar to N-MORB tholeiites, whereas Group II mafic granulites, with slight enrichment in light rare earth elements (LREE) (Fig. 3, B), are closer to island-arc tholeiites. Such an association is observed in back-arc environments, and Group II mafic granulites may be related to an early stage of basin opening when the influence of fluids derived from dehydration of subducted slabs was still important. Finally, the protolith of the laminated quartzofeldspathic gneiss, with enrichment in LREE, large Eu negative anomaly, and flat heavy rare earth elements (HREE) patterns (Fig. 3, C), was probably a high-silica rhyolite, such as those commonly observed in back-arc settings associated with basalts (Takagi et al., 1999; Piercey et al., 2002). Group III mafic granulites and tonalitic gneisses have a calc-alkaline affinity with enrichment in LREE (Figs. 3, D and E), and tonalites share similarities with adakites characterized by steep REE patterns (Fig. 3, F). New sensitive high-resolution ion microprobe (SHRIMP) ages suggest that the tonalites crystallized at 1.4 Ga (Wodicka et al., this volume). These rocks were metamorphosed to granulite facies at 1.20 Ga (Corriveau and van Breemen, 2000; Boggs and Corriveau, this volume).

LITHOLOGIES AND PETROGRAPHY

Geological indicators of mineralization in the Bondy gneiss complex are mainly associated with the northern part of the complex, where Crd-Opx gneiss occurrences are associated with tourmalinite, garnetite, aluminous gneiss, and laminated quartzofeldspathic gneiss.

Cordierite-Orthopyroxene Gneiss

In the field, the cordierite-orthopyroxene gneisses are characterized by a very unusual white color. The metamorphic assemblage of these gneisses is a high-grade equivalent to that of cordierite-anthophyllite schists (Schreurs and Westra, 1985). The transition from the amphibolite to the granulite facies is characterized by the breakdown of hornblende and biotite to orthopyroxene (± clinopyroxene).

The Crd-Opx gneiss consists of white, leucocratic, decimeter- to meter-thick layers with plagioclase (45%), quartz (25–30%) and phlogopite (10%). These layers are intercalated with brownish melanocratic, millimeter- to centimeter-thick layers of tourmalinite containing millimetric sea-green kornerupine (15–35%), millimetric black tourmaline (5–20%), honey-colored orthopyroxene (10–30%), striking blue cordierite (20–25%), quartz (30–35%), and variable amounts of plagioclase, with rutile, garnet, and zircon as accessory minerals. Some centimeter- to decimeter-thick layers are rusty and include trace amounts of pyrite, pyrrhotite, and chalcopyrite.

In thin section, orthopyroxene has a pale green, yellow to light pink pleochroism indicative of its magnesian character; kornerupine is sea-green and up to 2 × 15 mm in size; tourmaline is fine-grained, granoblastic, and pleochroic, with a yellow color, and it displays diffuse zoning. Zircon is commonly round and composite. Sulfides are generally disseminated, but are locally associated with garnet and/or phlogopite. Tourmaline and orthopyroxene are oriented parallel to the gneissosity. Kornerupine and cordierite are randomly oriented.

At other sites, white gneiss with orthopyroxene-cordierite-kornerupine-phlogopite layers either prevails (e.g., outcrop 4434 in Fig. 2), or is associated with orthopyroxene-kornerupine gneiss (e.g., 4475 in Fig. 2) and tourmalinite layers with kornerupine (4656, 4657, Fig. 2). Kornerupine at 4475 forms up to 20% of the rock. It is dark green macroscopically, and pleochroic, with a striking blue color under the microscope. Based on the color intensity, the kornerupine in this outcrop is more Fe-rich than the translucent kornerupine in 1654.

Associated Rocks

Tourmalinite and quartz-tourmaline gneiss occur as a 200-meter-thick unit (Fig. 2), and also as sporadic millimeter-thick laminae concordant with the gneissosity in Crd-Opx gneiss. Tourmalinite is made up of quartz, tourmaline, and traces of apatite, biotite, monazite, rutile, and sulfide. Locally, centi-

TABLE 1. AVERAGE WHOLE-ROCK GEOCHEMISTRIES OF MAFIC GRANULITES, TONALITIC GNEISSES, TONALITES, AND LAMINATED QUARTZOFELDSPATHIC GNEISSES OF THE BONDY GNEISS COMPLEX

	Mafic granulites											
	Group I		Group II		Group III		Tonalitic gneisses		Tonalites		Laminated Quartz-feldspar gneisses	
Rock	12		11		6		6		7		13	
n	av	sd	av	sd	av	sd	av	sd	av	sd	av	sd
SiO_2 (wt%)	47.71	1.50	48.08	2.93	49.79	2.95	57.17	6.69	65.60	2.64	69.81	4.55
TiO_2	1.63	0.53	1.62	0.49	1.05	0.18	1.11	0.35	0.45	0.24	0.39	0.15
Al_2O_3	15.18	1.48	16.30	1.60	17.10	1.95	16.25	2.76	17.09	1.64	14.05	2.12
$Fe2O_3^t$	14.38	2.08	14.10	1.71	10.37	1.37	9.09	2.56	3.89	1.24	2.78	1.42
MnO	0.22	0.03	0.30	0.17	0.23	0.12	0.26	0.14	0.09	0.02	0.03	0.03
MgO	7.84	0.75	6.31	1.15	5.68	2.18	2.90	0.48	1.13	0.72	1.30	1.42
CaO	9.94	1.55	8.81	2.08	8.49	1.30	6.13	1.49	3.69	0.95	0.82	0.84
Na_2O	2.54	0.63	2.89	0.74	3.55	1.00	4.17	0.91	4.85	0.76	2.36	1.52
K_2O	0.52	0.38	1.13	0.84	2.31	1.09	1.49	0.75	2.26	0.90	7.38	2.53
P_2O_5	0.15	0.07	0.23	0.09	0.34	0.14	0.41	0.18	0.12	0.05	0.05	0.05
LOI	0.40	0.36	0.35	0.40	1.25	0.36	0.91	0.57	0.60	0.32	0.91	0.25
Total	100.52	0.83	100.12	0.76	100.16	1.22	99.90	0.38	99.76	0.69	99.88	0.21
Pb (ppm)	4.5	2.1	6.5	6.1	8.6	1.7	7.7	4.2	8.1	4.1	4.0	1.4
Rb	2.9	1.7	19.5	21.6	67.0	77.9	19.2	12.8	27.2	18.4	110.6	23.9
Cs	0.030	0.023	0.275	0.348	2.103	3.558	0.298	0.333	0.109	0.078	0.217	0.060
Ba	39	29	310	355	571	312	559	209	873	251	1268	380
Sr	145	26	380	170	541	343	344	125	536	157	65	78
Zr	89	25	90	22	134	90	194	45	191	70	331	56
Y	31	7	32	8	25	3	37	6	14	4	38	18
La	4.30	1.15	10.17	1.99	18.79	7.02	19.22	2.40	16.25	7.00	28.75	10.62
Ce	13.25	3.89	29.49	5.99	48.18	18.98	48.03	8.19	31.93	15.55	64.52	22.17
Pr	2.09	0.61	4.34	0.94	6.63	2.54	6.37	1.09	3.72	1.79	7.06	2.32
Nd	10.88	2.66	20.06	4.65	28.42	11.07	26.79	4.02	13.95	6.53	26.44	7.99
Sm	3.62	0.81	5.11	1.25	5.96	2.19	5.61	0.64	2.43	1.03	4.78	1.24
Eu	1.33	0.32	1.65	0.30	1.63	0.37	1.45	0.36	0.87	0.41	0.75	0.18
Gd	4.01	0.83	5.07	1.12	4.86	1.66	5.59	0.89	2.10	0.65	5.73	1.58
Tb	0.70	0.15	0.83	0.18	0.66	0.15	0.81	0.12	0.28	0.09	0.78	0.20
Dy	4.49	0.84	5.00	1.09	3.92	0.86	5.08	0.76	1.67	0.53	4.80	1.48
Ho	0.91	0.17	1.00	0.22	0.74	0.14	1.06	0.17	0.34	0.11	1.03	0.35
Er	2.65	0.52	2.86	0.64	2.13	0.38	3.06	0.47	1.00	0.36	3.17	1.12
Tm	0.37	0.07	0.40	0.09	0.29	0.05	0.45	0.07	0.15	0.06	0.50	0.18
Yb	2.42	0.50	2.52	0.56	1.87	0.19	2.78	0.48	0.97	0.43	3.14	1.13
Lu	0.36	0.07	0.39	0.09	0.29	0.04	0.44	0.07	0.16	0.07	0.53	0.18
Ta	0.135	0.048	0.218	0.158	0.236	0.118	0.383	0.191	0.252	0.328	0.495	0.227
Nb	2.18	0.76	4.38	2.18	6.57	4.16	10.45	3.05	5.95	4.10	13.48	3.82
Hf	1.56	0.33	1.49	0.38	2.47	1.55	2.56	1.27	1.89	1.60	7.60	1.13
Th	0.42	0.16	0.43	0.17	1.65	2.47	0.96	1.12	1.32	2.48	9.59	3.47
U	0.34	0.24	0.19	0.09	0.80	1.06	0.78	1.14	0.62	1.11	2.28	0.92

Source: Blein et al. (2003).

meter-thick layers contain ~90% tourmaline. Among aluminous gneiss, tourmalinite locally contains cordierite and pyrite, with disseminated pyrrhotite in traces.

Garnetite with magnetite occurs as decimeter-thick layers with garnet, magnetite, biotite or orthopyroxene, and feldspars. They are locally mineralized in chalcopyrite, in association with Fe sulfides (pyrite). Garnetite with quartz is made up of garnet, quartz, biotite, feldspars, and magnetite.

Aluminous gneiss is varied and includes, among others, sillimanite-cordierite-orthopyroxene, sillimanite-biotite-garnet, biotite-garnet-orthopyroxene, and cordierite-garnet-orthopyroxene. Aluminous gneiss associated with tourmalinite is made up mostly of quartz, with biotite or phlogopite, cordierite, kornerupine, orthopyroxene, and sillimanite. The biotite-rich aluminous gneiss comprises biotite-garnet, biotite-pyrite, biotite-garnet-sillimanite, or orthopyroxene-biotite-garnet assemblages.

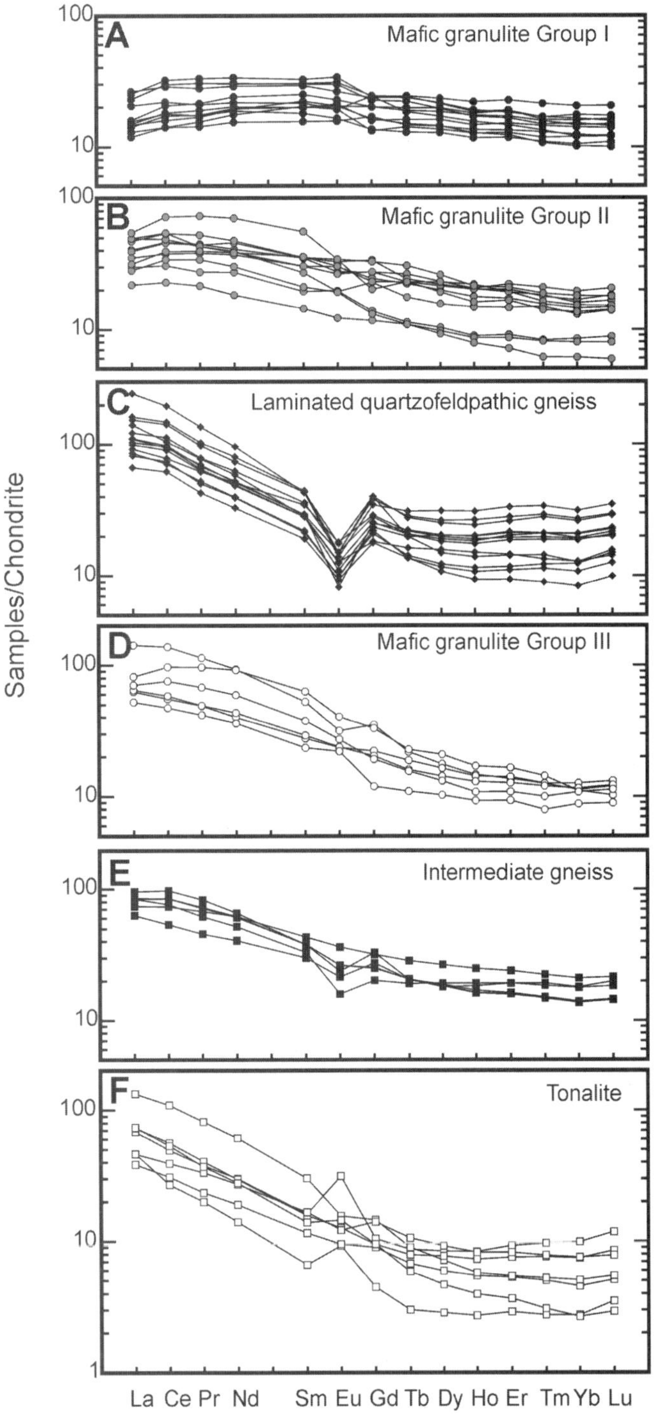

Figure 3. Chondrite-normalized rare-earth element (REE) patterns for the mafic granulites, tonalitic gneisses, tonalites, and laminated quartzofeldspathic gneisses of the Bondy gneiss complex (normalizing values from Sun and McDonough, 1989).

These gneisses show minor disseminated sulfides as pyrite, pyrrhotite, and chalcopyrite, found in nearly equal proportion. Magnetite and spinel are common accessory minerals. Cordierite occurs locally. Zircon is common and prismatic. The biotite-poor aluminous gneiss comprises garnet-orthopyroxene siliceous gneiss and garnet-cordierite gneiss. This lithology is in close association with sillimanite-biotite rusty gneiss or garnetite. These layers are composed of orthopyroxene, garnet, biotite, quartz, feldspars, and cordierite. With these antiferromagnetic topologies, these gneisses were first mapped as metapelites. The area was then remapped thoroughly in detail, and mineral showings were found (Corriveau et al., 1996).

Laminated quartzofeldspathic gneisses are characterized by leucocratic, cyclic, asymmetric, centimeter-scale compositional layers locally repeated in a very systematic order starting with (1) a coarse-grained K-feldspar-rich layer, (2) a medium-grained granitic layer with K-feldspar and quartz, and (3) a biotite-rich layer with K-feldspar, quartz, and orthopyroxene. The contact between K-feldspar-rich and biotite-rich layers is abrupt. In contrast to layers in straight gneiss, these layers have a cyclic and graded grain size, sharp contacts, and no porphyroclasts. The cyclic and asymmetric nature of the layering is inferred to be primary rather than an artifact of deformation or segregation during partial melting (Blein et al., 2003). The layering has been transposed but was not created by metamorphism and associated deformation. The lack of quartz-rich units among the laminated quartzofeldspathic gneiss, combined with the nature and the texture of the layering, is best reconciled with a rhyolitic tuff protolith.

SAMPLES AND METHOD

Thirteen Crd-Opx gneiss samples were selected for whole-rock geochemical analyses (Table 2); their locations are identified in Figure 2. The major elements and two trace elements (Zr and Y) were determined by X-ray fluorescence at the laboratories of the Centre de Recherche Minérale (Québec). The analytical precision is better than 1% for the major elements and better than 5% for most trace elements. Rare-earth and other trace elements (Ba, Rb, Sr, Cs, Th, U, Hf, Nb, Ta, Pb) were determined by inductively coupled plasma mass spectroscopy (ICP-MS) at the laboratories of Institut National de la Recherche Scientifiques–Eau Terre Environment (Québec) with analytical precision better than 3 to 5%. For ICP-MS, we used Parr acid digestion bombs in order to dissolve all accessory minerals. One hundred mg of rock powder were dissolved with 3 ml of HF and 1 ml of HNO_3 in Teflon cups enclosed in steel jackets, and placed on a hot plate (at 150 °C) for four days. After evaporation to dryness, the residues were dissolved with concentrated 6N HCl solution, and evaporated to dryness. The results are listed in Table 2, where samples are grouped according to metamorphic assemblages.

TABLE 2. WHOLE-ROCK GEOHEMISTRIES OF CORDIERITE-ORTHOPYROXENE GNEISSES

	Cordierite-orthopyroxene gneiss												
	Group I	Group I	Group I	Group I	Group I	Group I	Group I	Group II	Group II	Group III	Group III	Group III	Group III
Samples	1654F2	1654A2	1659U1	1654H2	4475A2	1654J2	1654I	4434A2	1659X	1654G2	1654L	1654K	1654G3
SiO_2 (wt%)	58.70	68.30	69.70	71.09	71.40	73.00	75.5	69.50	74.40	61.70	64.20	66.00	71.60
TiO_2	0.82	0.34	0.53	0.66	0.31	0.26	0.37	0.31	0.03	0.40	0.45	0.42	0.31
Al_2O_3	13.50	12.50	14.90	13.11	13.20	14.20	10.50	13.40	14.10	18.80	14.00	17.80	13.30
$Fe_2O_3^t$	7.20	2.70	3.09	2.50	3.56	0.23	0.73	4.09	1.28	2.64	4.42	3.08	1.09
MnO	0.04	0.04	0.06	0.02	0.03	0.01	0.01	0.04	0.05	0.02	0.08	0.05	0.01
MgO	7.08	7.63	2.25	5.07	5.24	2.63	5.53	5.80	0.53	0.78	10.40	8.91	4.70
CaO	5.85	3.84	3.90	1.23	0.54	1.26	0.95	0.75	2.48	0.73	0.76	0.44	0.82
Na_2O	2.80	3.84	3.69	4.55	3.85	5.17	4.06	4.59	4.19	2.45	3.54	2.26	5.86
K_2O	2.08	0.18	0.52	1.53	0.80	0.98	0.94	0.59	0.87	11.00	0.43	0.38	1.27
P_2O_5	0.09	0.08	0.04	0.17	0.06	0.57	0.32	0.15	0.03	0.06	0.01	0.01	0.02
LOI	1.76	0.67	0.47	1.59	0.97	1.51	1.14	0.90	0.86	1.25	1.33	0.84	1.09
Total	99.92	100.12	99.15	101.52	99.96	99.82	100.05	100.12	98.82	99.83	99.62	100.19	100.07
Zr (ppm)	116	339	348	280	343	95	279	356	414	392	440	411	335
Y	91	39	52	62	40	23	58	28	55	8	7	6	6
Rb	1.9	3.5	3.3	37.9	19.5	23.2	20.8	14.6	5.6	33.4	10.7	9.8	33.5
Sr	15.8	75.5	80.6	89.6	20.9	65.4	56.5	28.2	308.5	96.9	52.1	36.9	105.9
Nb	17.27	8.69	13.30	28.42	7.56	3.16	8.48	9.61	<0.01	169.52	30.34	30.89	11.47
Cs	0.072	0.070	0.024	0.143	0.145	0.252	0.230	0.115	0.045	0.424	0.104	0.107	421
Ba	9.4	29.7	13.6	114.9	387.2	44.7	31.7	260.0		87.9	55.4	42.0	84.4
La	23.59	10.87	11.80	12.22	17.46	4.98	5.59	6.50	19.47	0.921	0.911	1.091	0.765
Ce	50.15	21.08	26.36	27.37	39.47	12.91	14.95	12.52	33.98	1.808	1.807	1.849	1.357
Pr	6.61	2.77	3.37	3.43	4.88	1.73	2.26	1.30	3.46	0.219	0.166	0.197	0.168
Nd	28.87	11.85	13.59	15.33	18.24	7.47	10.82	4.58	11.30	1.010	0.631	0.639	0.738
Sm	7.29	3.18	3.37	3.86	3.38	1.95	3.01	0.82	1.74	0.261	0.088	0.158	0.260
Eu	0.90	0.41	0.93	0.39	0.36	0.22	0.23	0.13	1.11	0.049	0.052	0.042	0.049
Gd	7.82	3.91	4.98	4.91	4.04	2.30	4.16	1.12	1.97	0.270	0.110	0.153	0.341
Tb	1.59	0.65	1.05	0.76	0.62	0.44	0.78	0.22	0.50	0.066	0.027	0.033	0.064
Dy	11.33	4.60	7.41	5.22	4.15	2.84	5.52	2.29	5.03	0.659	0.301	0.385	0.606
Ho	2.64	1.12	1.61	1.20	0.97	0.66	1.34	0.79	1.44	0.169	0.122	0.116	0.147
Er	8.64	3.67	5.14	3.69	3.36	2.06	4.38	2.69	5.53	0.604	0.472	0.538	0.573
Tm	1.44	0.63	0.76	0.59	0.57	0.34	0.67	0.51	1.00	0.122	0.127	0.135	0.110
Yb	9.99	4.82	5.30	4.02	4.17	2.39	4.27	4.18	7.18	1.122	1.239	1.286	0.973
Lu	1.78	0.86	0.84	0.64	0.72	0.41	0.72	0.81	1.24	0.216	0.251	0.289	0.221
Hf	3.26	8.70	10.42	6.81	8.52	2.69	7.63	10.25	8.48	9.76	8.22	10.97	8.56
Ta	0.582	0.414	0.635	1.399	0.614	0.123	0.317	0.439	0.010	2.269	0.642	0.601	0.741
Pb	9.57	3.71	18.20	4.36	2.26	2.57	3.02	2.71	5.40	1.94	1.75	1.21	2.54
Th	104.01	16.66	6.53	8.36	6.78	10.53	9.40	5.74	0.38	4.81	0.29	1.45	4.36
U	4.70	2.58	2.92	12.10	2.74	13.16	5.44	2.94	0.65	0.92	0.69	1.79	1.23

GEOCHEMISTRY

Major and Trace Element Geochemical Data

The general chemical characteristics of the Crd-Opx gneisses are highly variable SiO_2 and Al_2O_3 contents, high $Fe_2O_3^t$ and MgO contents, and quite low alkali and CaO contents. The contents of Al_2O_3 (14.1 ± 2.1), $Fe_2O_3^t$ (2.8 ± 1.8), MgO (5.1 ± 2.9), and SiO_2 (69 ± 5) represent 90 wt% of the total content of major elements. The TiO_2 content (0.40 ± 0.19) is low, and the contents of MnO and P_2O_5 are below 0.08 and 0.6 wt%, respectively. The content of K_2O ranges from 0.18 up to 11 wt%. Locally, a lower MgO/FeO and higher K_2O content results in the additional phase phlogopite. Most Crd-Opx gneisses fall close to the trend defined by the laminated quartzofeldspathic gneisses in a TiO_2-Zr plot (Fig. 4, A), supporting the possibility that they were derived from that unit; however, five samples are scattered, on the right or the left of the regression line of laminated quartzofeldspathic gneisses, suggesting the mobility of TiO_2. Moreover, in the Al_2O_3-TiO_2 and Y-Zr plots the Crd-Opx gneisses are scattered (Fig. 4, B and C), confirming the mobility of TiO_2 and Y.

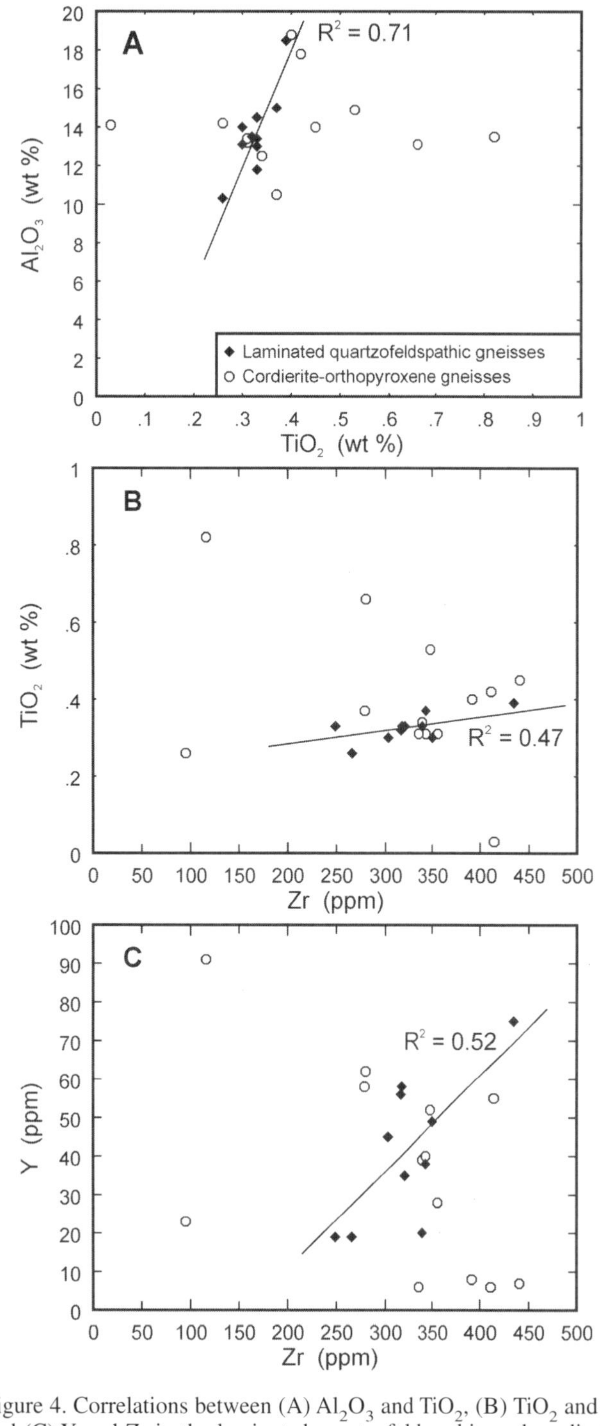

Figure 4. Correlations between (A) Al_2O_3 and TiO_2, (B) TiO_2 and Zr, and (C) Y and Zr in the laminated quartzofeldspathic and cordierite-orthopyroxene gneisses (Blein et al., 2003).

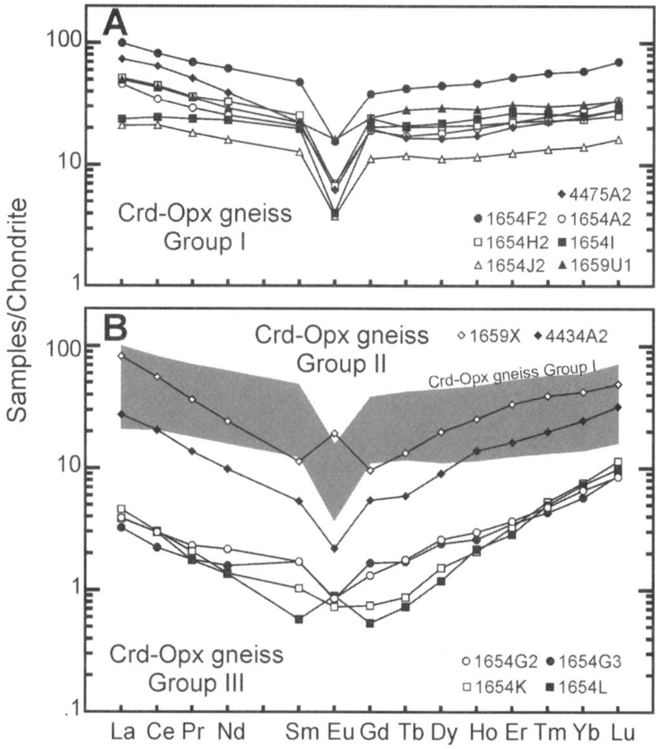

Figure 5. Chondrite-normalized rare-earth element (REE) patterns for the cordierite-orthopyroxene gneisses (normalizing values from Sun and McDonough, 1989). The shaded area is the field of Group I Cordierite-Orthopyroxene gneisses. Numbers in figure are sample numbers. Crd—cordierite; Opx—orthopyroxene.

Chondrite-normalized REE patterns of Crd-Opx gneisses are plotted in Figure 5. The patterns can be separated into three groups. Groups I and II show distinctive birdwing-shaped profiles, with negative LREE, positive HREE slopes, and negative Eu anomalies (Fig. 5, A and B). Group II samples show higher LREE and HREE fractionation than those of Group I (Fig. 5, B). In Group III, the total amount of REE in the samples is lower than in other groups. As in other groups, the samples show birdwing-shaped profiles characterized by higher contents of HREE than LREE (Fig. 5, B). All patterns, except those seen in 1659X and 1654L, show negative Eu anomalies, with chrondrite-normalized ratios of measured Eu divided by the hypothetical Eu concentration required to produce REE pattern with no Eu anomaly (Eu/Eu*) in the range 0.21–0.77. The amplitude of the negative Eu anomaly is not related to the total amount of REE in the samples. These REE patterns are unlike those of common rock types, and we interpret this feature as a legacy of the processes that generated the unusual major element chemistry of these rocks.

With a large range of values (6–91 ppm), Y, as expected, correlates well with the HREE (r = 0.97). All samples have high Zr contents (95–440 ppm) and display superchondritic $[Zr/Sm]_n$ ratios (Group I, 3.1 ± 1.2; Group II, 13.3 ± 6.3; Group III, 102 ± 58). Zr does not correlate with the HREE, and Group III, with the lower HREE contents, has the high Zr content (395 ± 38). However, the $[Zr/Sm]_n$ ratios correlate negatively with the $[Tb/Yb]_n$ ratios (r = –0.73; Fig. 6).

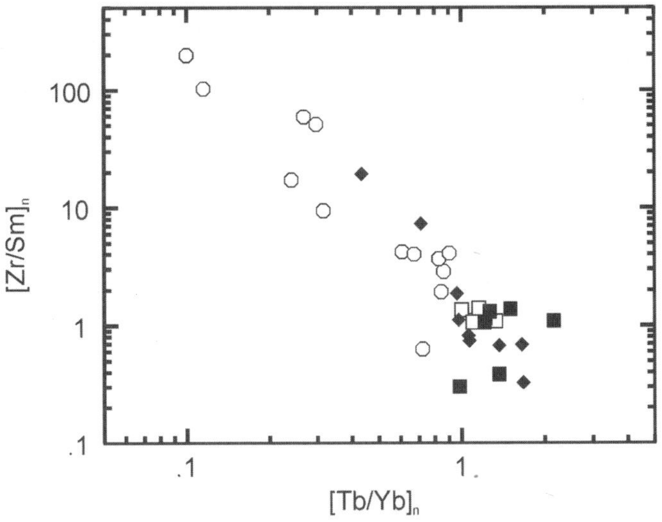

Figure 6. $[Zr/Sm]_n$ versus $[Tb/Yb]_n$ (chondrite normalizing values from Sun and McDonough, 1989). In this diagram, we observed a negative correlation between the increase of $[Zr/Sm]_n$ ratios and the decrease of ratios $[Tb/Yb]_n$ for the cordierite-orthopyroxene gneisses. Open circle—cordierite-orthopyroxene gneisses; black diamond—Central Australian sapphirine granulites (Windrim et al., 1984); open square—cordierite-orthoamphibole rocks from Orijärvi, Finland (Smith et al., 1992); black square—cordierite-orthoamphibole gneiss from Manitouwadge mining camp, Ontario (Pan and Fleet, 1995).

Results of Mass-Balance Calculations

Spatial association and geochemical data suggest that the Crd-Opx gneisses were most likely derived from the laminated quartzofeldspathic gneisses. Major and trace element geochemistry, as well as field observations, suggest that the laminated quartzofeldspathic gneisses were probably rhyolitic or high-silica rhyolitic tuffs characterized by variable degrees of potassic metasomatism prior to metamorphism (Blein et al., 2003). Less altered laminated quartzofeldspathic gneiss samples have been used for mass-balance calculations. The mass-balance calculations, following the method of MacLean and Kranidiotis (1987) and MacLean and Barrett (1993), are based on an assumption of Al immobility (e.g., Humphries and Thompson, 1978b; Pan et al., 1994).

The calculated mass changes indicate addition of SiO_2 and MgO and depletion of TiO_2, $Fe_2O_3^t$, MnO, and CaO for the Crd-Opx gneiss with respect to the inferred metarhyolite protolith (Fig. 7). These changes are similar to those commonly recorded in the footwalls of modern and ancient synvolcanic base metal sulfide systems (Franklin et al., 1981; Lydon, 1984). In low-grade metamorphic rocks, major element mobility is usually due to feldspar destruction (loss of Ca), chloritization (gain of Mg), and silicification. The Mg-rich character of the alteration facies is indicative of a seawater-dominated hydrothermal fluid (Trägårdh, 1991; Shriver and MacLean, 1993; Barrett and MacLean, 1994). This hydrothermal fluid probably developed a slightly acidic pH due to the release of H^+ in the chlorite-forming reaction (Reed, 1997), and therefore sericite rather than feldspar was the stable K-bearing phase.

Moreover, the Crd-Opx gneiss, with different absolute concentrations of REE, different interelement ratios, and positive slope for HREE, is indicative of REE mobility. This is supported by mass-balance calculations, which indicate losses or gains of up to 317% for Lu (Table 3). Europium is lower, or higher for two samples, in calculated mass change than the neighboring REE (Table 3), indicating preferential mobility of Eu. The

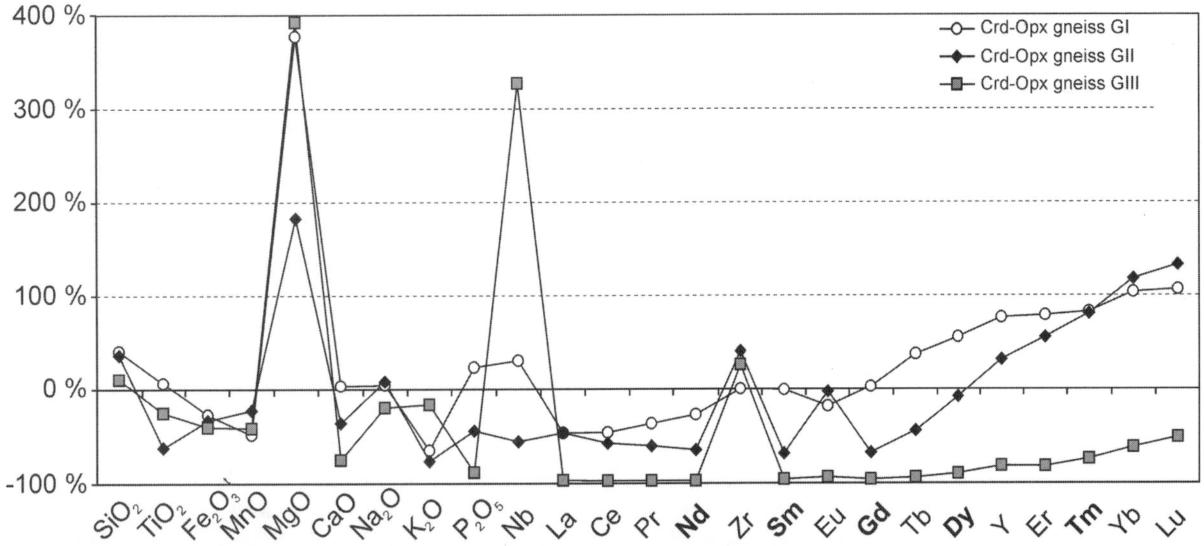

Figure 7. Mass-balance diagrams showing the calculated mass changes (percent) of major elements, rare earth elements (REE), and high field strength elements (HFSE) in the cordierite-orthopyroxene gneiss. Crd—cordierite; Opx—orthopyroxene.

TABLE 3. MASS CHANGES (PERCENT) FOR THE FORMATION OF CORDIERITE–ORTHOPYROXENE GNEISSES

	Cordierite–bronzite gneiss												
Samples	Group I 1654A2	Group I 1654F2	Group I 1659H2	Group I 1654I	Group I 1654J2	Group I 1659U1	Group I 4475A2	Group II 1659X	Group II 4434A2	Group III 1654G2	Group III 1654G3	Group III 1654K	Group III 1654L
SiO_2	42	13	41	87	34	22	41	38	35	–14	40	–3.3	20
TiO_2	–46	21	0.08	–30	–64	–29	–53	–96	–54	–58	–54	–53	–36
$Fe_2O_3^t$	–35	61	–42	–79	–95	–37	–18	–73	–7.8	–58	–75	–48	–4.6
MnO	–47	–51	–75	–84	–94	–34	–63	–41	–51	–82	–94	–54	–5.7
MgO	360	295	191	297	40	14	199	–72	226	–69	166	277	460
CaO	112	199	–35	–38	–39	81	–72	21	–61	–73	–57	–83	–63
Na_2O	41	–4.7	60	78	67	14	34	37	57	–40	103	–42	16
K_2O	–95	–51	–63	–71	–78	–89	–81	–80	–86	88	–69	–93	–90
P_2O_5	76	83	257	738	1004	–26	25	–41	208	–12	–59	–85	–90
Nb	–15	57	259	–1.2	–73	9.2	–30	–98	–12	1004	5.6	112	165
La	–50	0.29	–47	–69	–80	–55	–24	–21	–72	–97	–97	–96	–96
Ce	–57	–5.0	–47	–64	–77	–55	–24	–38	–76	–98	–97	–97	–97
Pr	–48	14	–39	–50	–72	–47	–14	–43	–77	–97	–97	–97	–97
Nd	–41	33	–27	–36	–67	–43	–14	–50	–79	–97	–97	–98	–97
Zr	35	–57	6.5	32	–67	16	30	46	32	3.9	26	15	57
Sm	–12	86	1.7	–0.91	–53	–22	–12	–58	–79	–95	–93	–97	–98
Eu	–29	47	–34	–51	–66	37	–40	74	–79	–94	–92	–95	–92
Gd	–9.9	67	7.9	14	–53	–3.8	–12	–60	–76	–96	–93	–98	–98
Tb	9.2	149	23	57	–34	50	– 0.96	–25	–65	–93	–90	–96	–96
Dy	27	189	37	81	–31	71	8.0	23	–41	–88	–84	–93	–93
Y	35	193	105	140	–30	52	32	69	–9.3	–82	–80	–85	–78
Er	53	233	47	117	–25	80	32	104	4.4	–83	–78	–84	–82
Tm	67	253	47	110	–21	69	42	133	25	–79	–73	–75	–70
Yb	103	289	61	114	–12	87	66	168	64	–69	–62	–62	–53
Lu	115	312	52	113	–10	76	70	173	88	–64	–48	–49	–44

LREE systematically decrease, and the HREE increase, except in the third group of Crd-Opx gneiss. In addition, mass-balance calculations show that there is a general increase in the size of calculated mass increase from La to Yb (Fig. 7). Such behavior is characteristic of high fluid/rocks ratios with fluoride and/or carbonate complexing agents (Taylor and Fryer, 1983).

The mass-balance calculations reveal that HFSE also were mobilized during the formation of the Crd-Opx gneiss, decreasing or increasing for Zr (–66 to 57) and Nb (–98 to 1017; Fig. 7; Table 3).

DISCUSSION

Overview

Mineral replacement related to hydrothermal alteration by heated seawater provides an attractive explanation of the origin of the Bondy gneiss complex Crd-Opx gneiss protoliths. The chemical imprint of seawater hydrothermal alteration is well documented empirically though experiments and observations of natural occurrences as well as on theoretical grounds (e.g., Vallance, 1967; Bischoff and Dickson, 1975; Humphries and Thompson, 1978a, 1978b; Ludden and Thompson, 1979; Seyfried and Mottl, 1982; Baker and de Groot, 1983; Mottl, 1983; Hajash and Chandler, 1987). These studies show that the alteration process proceeds mainly through loss of Ca and gain of Mg in starting materials ranging in composition from basalt to rhyolite. The exact nature of the alteration products is highly dependent on a number of factors (temperature, initial composition of altered material, water/rock ratio, etc.), but the result generally corresponds to chlorite schist's containing variable amounts of quartz and sericite.

Origin of Cordierite-Orthopyroxene Gneiss

Since ~95–98% of the bulk compositions of most rocks can be specified by the four components MgO, FeO* (i.e., total Fe as FeO), Al_2O_3, and SiO_2, the postulated primary protolith can be assessed in a molar SiO_2-Al_2O_3-(MgO + FeO) diagram (Fig. 8, A). The Crd-Opx gneisses plot in an area described by a quartz–K-feldspar–garnet triangle, and eight samples fall in the quartz-chlorite field. This supports the assumption that many Crd-Opx gneisses have originated from rocks containing chlorite or Mg-Al clay minerals such as smectite. Additionally, the Al_2O_3- (MgO + FeO*)-K_2O diagram (Fig. 8, B) suggests that chlorite could be a dominant component, with minor amounts of serpentine, illite, and/or muscovite.

The alteration trends of Crd-Opx gneisses can be summarized by viewing compositions of least altered and altered rocks in a ($Fe_2O_3^t$ + MgO)-Al_2O_3/2-K_2O-(Na_2O + CaO) tetrahedron

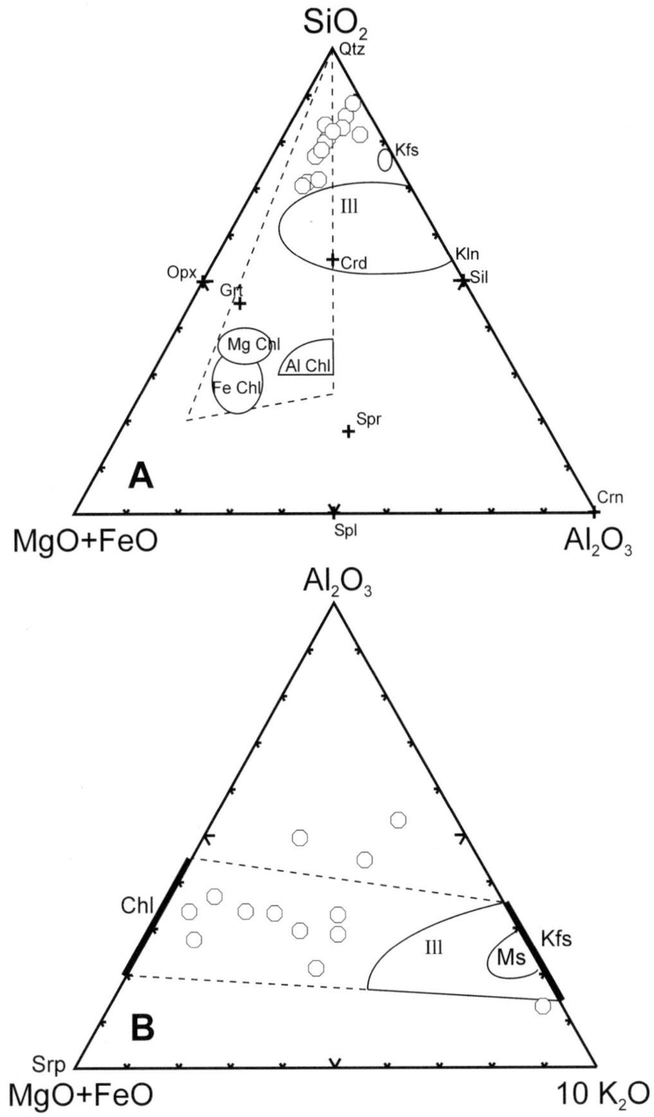

Figure 8. Diagrams comparing whole-rock samples with the compositional fields of various minerals: (A) SiO_2-Al_2O_3-(FeO + MgO) diagram; (B) Al_2O_3-K_2O-(FeO + MgO) diagram. Chl—chlorite; Crd—cordierite; Crn—corundum; Grt—garnet; Ill—illite; Kfs—K feldspar; Kln—kaolinite; Ms—muscovite; Opx—orthopyroxene; Qtz—quartz; Sil—sillimanite; Spl—spinel; Spr—sapphirine; Srp—serpentine.

(Fig. 9; Riverin and Hodgson, 1980; Luff et al., 1992). Crd-Opx gneisses define a clear trend from low to high ($Fe_2O_3^t$ + MgO) in a ($Fe_2O_3^t$ + MgO)-(Na_2O + CaO)-Al_2O_3/2 projection. Starting with assemblages of the composition of rhyolite, the alteration proceeds mainly through loss of Ca and gain of Mg, as suggested by mass-balance changes. Such chemical zoning, with loss of Ca and gain of Mg, is the basis of lithogeochemical prospecting around VMS deposits (e.g., Hodges and Manojlovic, 1993).

With the exception of some unusual rocks, such as boninites, negative or flat HREE slopes and highly variable LREE fractionation characterize normal igneous and sedimentary rocks. Birdwing-shaped REE profiles suggest selective REE mobility related to fluid-rock interactions. There are numerous studies that suggest that REE, particularly the LREE, can be mobilized during hydrothermal alteration below VMS deposits (Taylor and Fryer, 1980; Baker and de Groot, 1983; Campbell et al., 1984; Bence and Taylor, 1985; Vocke et al., 1987; MacLean, 1988; Morogan, 1989; Oreskes and Einaudi, 1990; MacLean and Hoy, 1991). The mobility of REE is controlled by pH, temperature, the stability of parent minerals, and the complexing agents present in the solutions (Lottermoser 1992; Wood and Williams-Jones, 1994). Factors favoring REE mobility during fluid-rock interactions appear to be long fluid residence times, high fluid/rock ratios, and abundant REE complexing agents in the fluids (Taylor and Fryer, 1980; Staudigel and Hart, 1983). Destabilization of LREE and HREE complexes by change in temperature, pressure, pH, alkalinity, or ligand activity will lead to the fractionation and deposition of REE (Kubach et al., 1981). However, the calculations presented by Wood and Williams-Jones (1994) suggest that ancient analogues of modern seafloor hydrothermal vent fluids could not have transported REE in the quantities required to explain the extent of REE mobility documented for alteration zones associated with massive sulfide deposits.

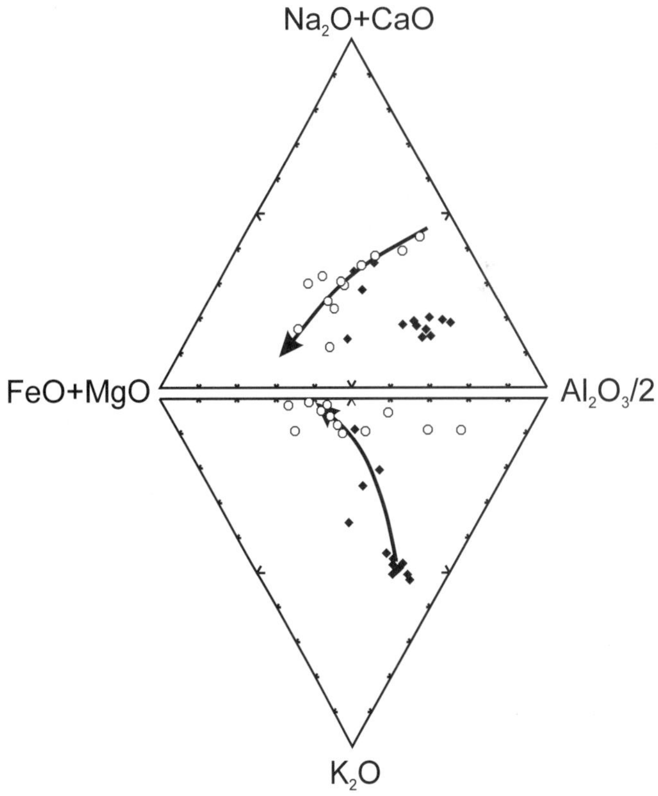

Figure 9. Tetrahedron showing compositional trends associated with cordierite-orthopyroxene gneiss, garnetite, tourmalinite, and siliceous gneiss. The form of the plot is based on Riverin and Hodgson (1980) and Luff et al. (1992). Open circles—cordierite-orthopyroxene gneisses, filled diamonds—laminated quartzofeldspathic gneisses.

Comparison with Other Al-Mg Gneiss Occurrences

The questions concerning the origin of Crd-Opx gneisses are not restricted to the Grenville Province. For that reason, we compare our results to the results obtained for other occurrences, the first from the Early Proterozoic Svecofennian supracrustal rocks of Orijärvi in southwest Finland (Smith et al., 1992), the second from the Archean VMS deposits of the Manitouwadge mining camp in Ontario, Canada (Pan and Fleet, 1995), and the third from the sapphirine granulites of central Australia (Windrim et al., 1984). Although the major element compositions of these other occurrences of Crd-Opx rocks are similar to those of the Bondy gneiss complex, their trace element characteristics are different.

Cordierite-orthoamphibole (anthophyllite-gedrite) rocks occur in the Early Proterozoic Svecofennian supracrustal rocks at Orijärvi, Finland (Smith et al., 1992), in association with Cu-Zn sulfide deposits (Colley and Westra, 1987; Beeson, 1988). They are easily recognized by the presence of cordierite knots and anthophyllite rosettes on weathered surfaces. SiO_2 (57.3–75.7 wt%), MgO (3.4–7.5 wt%), $Fe_2O_3^t$ (4.9–11.3 wt%), and Al_2O_3 (8.8–18.7 wt%) represent between 90 and 98% of the total content of major elements, and CaO (0.1–1.7 wt%) and Na_2O (0.5–1.9 wt%) are low. The protoliths of these facies were chlorite-quartz-sericite schists produced from rhyolitic precursors. The REE patterns for the Orijärvi cordierite-orthoamphibole rocks, which are enriched in LREE with negative Eu anomalies (Fig. 10, A), strongly resemble those of stratigraphically equivalent Svecofennian metarhyolites from central Sweden related to arc environments (Baker and de Groot, 1983). This reveals that there has been very little redistribution of REE. The alteration process is coeval with the Cu-Zn mineralization; however, geochemical data do not indicate particular enrichment of REE in cordierite-orthoamphibole rocks.

The Manitouwadge mining camp, Ontario, Canada, represents one of the classic localities of cordierite-orthoamphibole (anthophyllite-gedrite) rocks associated with Archean VMS deposits (Pan and Fleet, 1995; Schandl et al., 1995; Zaleski and Peterson, 1995). The contents of major elements, particularly SiO_2 (37.6–51.2 wt%), MgO (6.5–12.3 wt%), $Fe_2O_3^t$ (18.3–29.1 wt%), and Al_2O_3 (13.2–18.3 wt%), vary widely in the cordierite-orthoamphibole gneiss, and CaO (1.5–3.2 wt%) and Na_2O (0.25–2.4 wt%) are low. The chondrite-normalized REE patterns are slightly enriched in LREE ($[La/Yb]_n$ = 1.8–7.5), with high concentrations in HREE ($[Yb]_n$ = 7–80) and a Eu anomaly that becomes more negative with increasing abundances of REE (Fig. 10, B; Pan and Fleet, 1995). Mass-balance calculations, assuming tholeiitic basalt as the protolith, indicate that HFSE and REE were mobilized and that there was a general increase in the size of calculated mass increases from La to Yb. This is consistent with the increased stability of REE complexes (carbonate, fluoride, etc.) with increase in atomic number (Wood, 1991). The trace element behavior supports field relationships showing that most of the cordierite-orthoamphi-

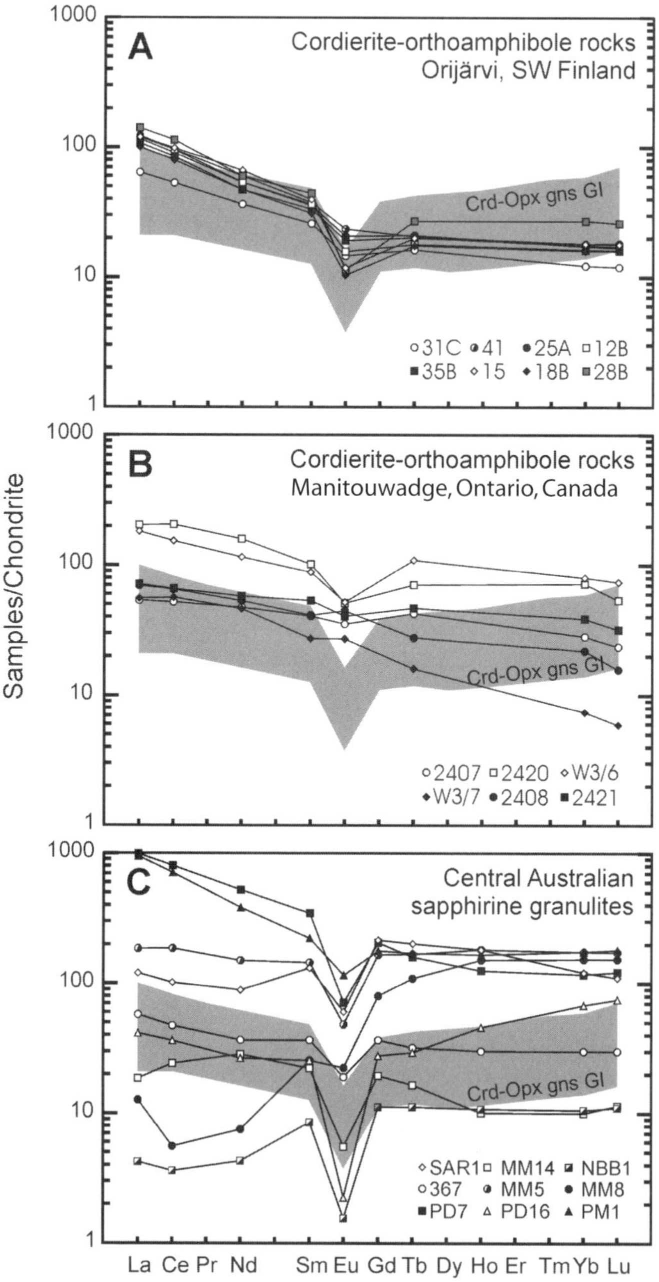

Figure 10. Chondrite-normalized rare earth element (REE) patterns for the cordierite-orthoamphibole rocks from Orijärvi, SW Finland (Smith et al., 1992); the cordierite-orthoamphibole gneisses from the Manitouwadge camp, Ontario, Canada (Pan and Fleet, 1995); and the Central Australian sapphirine granulites from Windrim et al. 1984 (normalizing values from Sun and McDonough, 1989). Numbers in figure are sample numbers. Crd—cordierite; gns—gneiss; Opx—orthopyroxene.

bole gneisses represent metamorphosed equivalents of chlorite-rich rocks that were derived hydrothermally from tholeiitic basalts. In such facies, HFSE and REE cannot be used for discrimination tectonic setting because of their mobility.

Unusual REE profiles have been reported in central Australian sapphirine granulites (Windrim et al., 1984). Field relationships, petrography, and the chemistry of the sapphirine granulites suggest that their protoliths comprised chlorite-rich rocks that were generated by hydrothermal alteration of a range of rock types prior to metamorphism. MgO (7–27%), Al_2O_3 (7–38%), and $Fe_2O_3^t$ (3.5–21%) are generally very high but variable, with low SiO_2 (35–60%). Central Australian sapphirine granulites are characterized by large variations in $[La/Sm]_n$ (0.5–4.3) and $[Tb/Yb]_n$ ratios (0.4–1.7), the development of birdwing-shaped profiles, and the presence of negative Eu anomalies in all samples (Fig. 10, C). In these samples, there is a correspondence between modal abundance and relative proportions of the minor minerals (zircon, monazite, titanite, and apatite) and abundance of REE and HFSE. Moreover, central Australian sapphirine granulites display chondritic to superchondritic $[Zr/Sm]_n$ ratios, and these ratios correlate negatively with the $[Tb/Yb]_n$ ratios as in the Crd-Opx gneisses of the Bondy hydrothermal system (Fig. 6). The Nd isotopic data and REE profiles of central Australian sapphirine granulites demonstrate mobility of REE and HFSE during the formation of stratiform Cu-Pb-Zn sulfide deposits, with which they are closely associated.

Essentially 90–98% of the bulk composition of most cordierite-orthoamphibole or orthopyroxene rocks can be specified by the four components MgO, $Fe_2O_3^t$, Al_2O_3, and SiO_2, and these express the silicification, chloritization, and/or sericitization of the protoliths associated with volcanogenic-exhalative hydrothermal systems. It is clear from this that the protoliths must have been chlorite-dominated mixtures such as chlorite + quartz and chlorite + sericite + quartz. The presence of a hydrothermal alteration zone is confirmed by the presence of tourmalinites and garnetites associated with Crd-Opx gneiss, since tourmalinites and garnetites or coticules are known guides to ore associated with VMS hydrothermal system (Spry et al., 2000).

The mobility of REE and HFSE is not systematic in VMS systems. The mobility of REE, particularly the LREE, has been observed in various VMS deposits, and related to hydrothermal alteration. Wood and Williams-Jones (1994) concluded that seafloor hydrothermal vent fluids could not have transported REE in the quantities required to explain the documented extent of REE mobility. In VMS hydrothermal systems, metals leached from the volcanic pile are commonly proposed as the source of ore concentrations (e.g., Lydon, 1988; Galley, 1993), although some workers argue that magmatically derived metals are significant or even dominant (Stanton, 1990; Sawkins, 1994; Yang and Scott, 1996). Such mobility of REE and HFSE, e.g., Zr, is common in F-rich hydrothermal systems, characterized by the presence of complexing agents, and related to alkalic F-rich igneous suites (Rubin et al., 1993).

CONCLUSIONS

The Crd-Opx gneisses of the Bondy gneiss complex were derived from rhyolites that had experienced hydrothermal alteration prior to metamorphism. They represent metamorphosed equivalents of chlorite-rich rocks derived from interactions between volcanic rocks and seawater on the seafloor associated with VMS deposits (Vallance, 1967; Franklin et al., 1981; Schandl et al., 1995; Zaleski and Peterson, 1995). The alteration related to VMS deposits is confirmed by the presence of minerals that are highly aluminous and magnesian, such as cordierite, tourmaline, kornerupine, orthopyroxene, and garnet (Boggs and Corriveau, this volume). Concentrations of Al_2O_3, $Fe_2O_3^t$, MgO, and SiO_2 that represent 90 wt% of the total contents of major elements are compatible with the composition of seafloor alteration products. Moreover, mass-balance calculations, assuming a metarhyolite prehydrothermal protolith, indicate the addition of SiO_2, K_2O, and MgO due to silicification, sericitization, and chloritization, and depletion of CaO as a result of plagioclase destruction.

The worldwide cordierite-orthoamphibole gneiss occurrences are systematically characterized by the presence of Al-rich phases, and a chemical composition enriched in MgO, FeO, and Al_2O_3 and depleted in alkalis as compared to rock types that constitute precursors. However, the mobility of REE and HFSE is not systematic in cordierite-orthoamphibole rocks. In the Bondy gneiss complex, the development of birdwing-shaped REE profiles is the result of the leaching of LREE, from La to Eu, and the addition of HREE, from Gd to Lu, to the rocks. Moreover, the addition of HREE increases from Gd to Lu. Such mobility is generally observed in hydrothermal systems characterized by high fluid/rock ratios with fluoride and/or carbonate complexing agents (Taylor and Fryer, 1983).

These results suggest that trace element chemistry is decoupled from the signature outlined by mineralogy and major element chemistry. It appears that the Crd-Opx gneisses record (1) a broad alteration characterized by destruction of primary feldspar and formation of hydrothermal phyllosilicates such as chlorite and (2) a focused flux of hydrothermal fluids, transporting REE, and especially HREE, in the Bondy gneiss complex.

ACKNOWLEDGMENTS

This paper is part of project 920002QN of the Geological Survey of Canada, and was funded through Institut National de la Recherche Scientifiques (INRS) project 15267-689, contracted out by STREX of the St. Geneviève Group. We express our thanks to our collaborators L.B. Harris, B. Rivard, and N. Wodicka, and to INRS geochemical laboratory and administration staff. Thoughtful and constructive reviews by Yumming Pan, Suzanne Paradis, and Anthony Williams-Jones helped clarify the ideas in the manuscript. The Papineau-Labelle Wildlife Reserve provided lodging during field work.

REFERENCES CITED

Baker, J.H., and de Groot, P.A., 1983, Proterozoic seawater-felsic volcanics interaction W. Bergslagen, Sweden: Evidence for high REE mobility and implications for 1.8 Ga seawater compositions: Contributions to Mineralogy and Petrology, v. 82, p. 119–130.

Barrett, T.J., and MacLean, W.H., 1994, Mass changes in hydrothermal alteration zones associated with VMS deposits of the Noranda area: Exploration Mining Geology, v. 3, p. 131–160.

Beeson, R., 1988, Identification of cordierite-anthophyllite rock types associated with sulphide deposits of copper, lead, and zinc: Transactions Institute of Mineralogy and Metallogeny (Section B: Applied Earth Sciences), v. 97, p. 108–115.

Bence, A.E., and Taylor, B.E., 1985, Rare earth element systematics of West Shata metavolcanic rocks: Petrogenesis and hydrothermal alteration: Economic Geology, v. 80, p. 2164–2176.

Bischoff, J.L., and Dickson, F.W., 1975, Seawater-basalt interaction at 2000 °C and 500 kbars: Implications for origin of sea-floor heavy metal deposits and regulation of seawater chemistry: Earth and Planetary Sciences Letters, v. 25, p. 385–397.

Blein, O., LaFlèche, M., and Corriveau, L., 2003, Geochemistry of the granulitic Bondy gneiss complex: 1.4 Ga arc in the Central Metasedimentary Belt, Grenville Province, Canada: Precambrian Research, v. 120, p. 193–218.

Campbell, I.H., Lesher, C.M., Coad, P., Franklin, J.M., Gorton, M.P., and Thurston, P.C., 1984, Rare-earth element mobility in alteration pipes below massive Cu-Zn sulphide deposits: Chemical Geology, v. 45, p. 181–202.

Colley, H., and Westra, L., 1987, The volcano-tectonic setting and mineralization of the Early Proterozoic Kemiö-Orijärvi-Lohja belt, SW Finland, *in* Pharaoh, T.C., et al., eds., Geochemistry and mineralization of Proterozoic volcanic suites, Geological Society [London] Special Publication, v. 33, p. 95–107.

Corriveau, L., and van Breemen, O., 2000, Docking of the Central Metasedimentary Belt to Laurentia: Evidence from the 1.17–1.16 Ga Chevreuil intrusive suite and host gneisses, Quebec: Canadian Journal of Earth Sciences, v. 37, p. 253–269.

Corriveau L., Tellier, M.L., and Morin, D., 1996, Le dyke de minette de Rivard et le complexe gneissique cuprifère de Bondy: Implications tectoniques et métallogéniques pour la région de Mont-Laurier, Québec: Ottawa, Ontario, Geological Survey of Canada, Open File Report, v. 3078, 70 p.

Date, J., Watanabe, Y., and Saeki, Y., 1983, Zonal alteration around the Fukazawa Kuroko deposits, Akita Prefecture, northern Japan: Economic Geology Monography, v. 5, p. 365–386.

Eastoe, C.J., Solomon, M., and Walshe, J.L., 1987, District-scale alteration associated with massive sulphide deposits in the Mount Rad Volcanics, western Tasmania: Economic Geology, v. 82, p. 1239–1258.

Eskola, P., 1914, On the petrology of the Orijävi region in south-western Finland: Helsinki, Finland, Bulletin Commission of Geology, 40 p.

Finlow-Bates, T., and Stumpfl, E.F., 1981, The behaviour of so-called immobile elements in hydrothermally altered rocks associated with volcanogenic submarine-exhalative ore deposits: Mineralium Deposita, v. 16, p. 319–328.

Franklin, J.M., Lydon, J.W., and Sangster, D.F., 1981, Volcanic-associated massive sulphide deposits: Economic Geology, 75th Anniversary Volume, p. 486–627.

Galley, A.G., 1993, Characteristics of semi-conformable alteration zones associated with volcanogenic massive sulphide districts: Journal of Geochemical Exploration, v. 48, p. 175–200.

Guiraud, M., Kienast, J.-R., and Rahmani, A., 1996, Petrology study of high-temperature granulites from In Ouzzal, Algeria: Some implications on the phase relationships in the FMASTOCr system: European Journal of Mineralogy, v. 8, p. 1375–1390.

Hajash, A., and Chandler, G.W., 1987, An experimental investigation of high-temperature interactions between seawater and rhyolite, andesite, basalt, and peridotite: Contributions to Mineralogy and Petrology, v. 78, p. 240–254.

Hodges, D.J, and Manojlovic, P.M., 1993, Application of lithogeochemistry to exploration for deep VMS deposits in high-grade metamorphic rocks, Snow Lake, Manitoba: Journal of Geochemical Exploration, v. 48, p. 201–224.

Humphreys, H.C., 1993, Metamorphic evolution in amphibole-bearing aluminous gneiss, South Africa: American Mineralogist, v. 78, p. 1041–1055.

Humphries, S.E., and Thompson, G., 1978a, Hydrothermal alteration of oceanic basalts by seawater: Geochimica et Cosmochimica Acta, v. 42, p. 107–125.

Humphries, S.E., and Thompson, G., 1978b, Trace element mobility during hydrothermal alteration of oceanic basalt: Geochimica et Cosmochimica Acta, v. 42, p. 127–136.

Kretz, R., 1983, Symbols for rock-forming minerals: American Mineralogist, v. 68, p. 277–279.

Kubach, I., Schubert, P., and Töpper, W., 1981, Y, La und Lanthanide: Berlin, Geochemie Gesamterde: Magmatische Abfolge, *in* Gmelin Handbuch der Anorganischen Geochemie, A5: Berlin, Springer Verlag, 475 p.

Large, R.R., 1992, Australian volcanic-hosted massive sulphide deposits: Features, styles, and genetic models: Economic Geology, v. 87, p. 471–510.

Lentz, D.R., 1999, Petrology, geochemistry and oxygen isotope interpretation of felsic volcanic and related rocks hosting the Brunswick 6 and 12 massive sulphide deposits (Brunswick belt) Bathurst mining camp, New Brunswick, Canada: Economic Geology, v. 94, p. 57–86.

Lottermoser, B.G., 1992, Rare earth elements and hydrothermal ore formation processes: Ore Geology Reviews, v. 7, p. 25–41.

Ludden, J.N., and Thompson, G., 1979, An evaluation of the behavior of the rare-earth elements during the weathering of sea-floor basalt: Earth and Planetary Sciences Letters, v. 43, p. 85–92.

Ludden, J.N., Gélinas, L., and Trudel, P., 1982, Archean metavolcanics from the Noranda district, Abitibi greenstone belt, Quebec, 2: Mobility of trace elements and petrogenetic constraints: Canadian Journal of Earth Sciences, v. 19, p. 2276–2287.

Luff, W.M., Goodfellow, W.D., and Juras, S.J., 1992, Evidence for a feeder pipe and associated alteration at the Brunswick n°12 massive-sulphide deposit: Exploration Mining Geology, v. 1, p. 167–185.

Lydon, J.W., 1984, Some observations on the mineralogical and chemical zonation patterns of volcanogenic sulphide deposits of Cyprus: Ottawa, Ontario, Geological Survey of Canada, Paper 84-1A, p. 611–616.

Lydon, J.W., 1988, Volcanogenic massive sulphide deposits, Part 2: Genetic models: Geoscience Canada Reprint Ser. 3, p. 155–182.

MacLean, W.H., 1988, Rare earth element mobility at constant-REE ratios in the alteration zone at the Phelps Dodge massive sulphide deposit, Matagami, Quebec: Mineralium Deposita, v. 23, p. 231–238.

MacLean, W.H., and Barrett, L.D., 1993, Lithogeochemical techniques using immobile elements: Journal of Geochemical Exploration, v. 48, p. 109–133.

MacLean, W.H., and Hoy, L.D., 1991, Geochemistry of hydrothermally altered rocks at the Horne mine, Noranda, Quebec: Economic Geology, v. 86, p. 506–528.

MacLean, W.H., and Kranidiotis, P., 1987, Immobile elements as monitors of mass transfer in hydrothermal alteration: Phelps Dodge massive sulphide deposit, Matagami, Quebec: Economic Geology, v. 82, p. 951–962.

Menzies, M., Seyfried, W.E., and Blanchard, D., 1979, Experimental evidence of rare earth element mobility in greenstone: Nature, v. 282, p. 39–399.

Morogan, K., 1989, Mass transfer and REE mobility during fenitization at Alnö, Sweden: Contributions to Mineralogy and Petrology, v. 103, p. 25–84.

Mottl, M.J., 1983, Metabasalts, axial hot springs and the structure of hydrothermal systems at mid-ocean ridges: Geological Society of America Bulletin, v. 94, p. 161–180.

Oreskes, N., and Einaudi, M.T., 1990, Origin of rare earth element-enriched hematite breccias at the Olympic Dam Cu-U-Au-Ag deposit, Roxby Downs, south Australia: Economic Geology, v. 85, p. 1–28.

Pan, Y., and Fleet, M.E., 1995, Geochemistry and origin of cordierite-orthoamphibole gneiss and associated rocks at an Archean volcanogenic massive sulphide camp: Manitouwadge, Ontario, Canada: Precambrian Research, v. 74, p. 73–89.

Pan, Y., Fleet, M.E., and Stone, W.E., 1991, Geochemistry of metasedimentary rocks in the late Archean Hemlo-Heron Bay greenstone belt, Superior Province, Ontario: Implications for provenance and tectonic setting: Precambrian Research, v. 52, p. 53–69.

Pan, Y., Fleet, M.E., and Barnett, R.L., 1994 Rare earth mineralogy and geochemistry at the Mattagami Lake volcanogenic massive sulphide deposits, Quebec: Canadian Mineralogist, v. 32, p. 133–147.

Piercey, S.J., Paradis, S., Peter, J.M., and Tucker, T.L., 2002, Geochemistry of basalts from the Wolverine volcanic-hosted massive-sulphide deposit, Finlayson Lake district, Yukon Territory: Ottawa, Ontario, Geological Survey of Canada, Current Research 2002-A03, 11 p.

Reed, M.H., 1997, Hydrothermal alteration and its relationship to ore fluid composition, *in* Barnes, H.L., ed., Geochemistry of hydrothermal ore deposits: New York, Wiley, p. 303–358.

Reinhardt, J., 1987, Cordierite-anthophyllite rocks from north-west Queensland, Australia: Metamorphosed magnesian pelites: Journal of Metamorphic Geology, v. 4, p. 451–472.

Riverin, G., and Hodgson, C.J., 1980, Wall-rock alteration at the Millenbach Cu-Zn mine, Noranda, Quebec: Economic Geology, v. 75, p. 424–444.

Rivers, T., Martignole, J., Gower, C.F., and Davidson, A., 1989, New tectonic divisions of the Grenville Province, southeast Canadian Shield: Tectonics, v. 8, p. 63–84.

Robinson, P., Spear, F.S., and Schumacher, J.C., 1982, Phase relations of metamorphic amphiboles: Natural occurrences and theory, *in* Veblen, D.R., and Ribbe, P.H., eds., Amphiboles: Petrology and experimental phase relations: Review in Mineralogy: Chelsea, Michigan, Book Crafters, p. 1–228.

Rubin, J.N., Henry, C.D., and Price, J.G., 1993, The mobility of zirconium and other "'immobile'" elements during hydrothermal alteration: Chemical Geology, v. 110, p. 29–47.

Sangster, D.F., 1972, Precambrian volcanogenic massive sulphide deposits in Canada: A review: Ottawa, Ontario, Geological Survey of Canada, Paper 72-22, 44 p.

Sawkins, F.J., 1994, Integrated tectonic-genetic model for volcanic-hosted massive sulphide deposits: Geology, v. 18, p. 1061–1064.

Schandl, E.S., Gorton, M.P., and Wasteneys, H.A., 1995, Rare-earth element geochemistry of the metamorphosed volcanogenic massive sulphide deposits of the Manitouwadge mining camp, Superior Province, Canada: A potential exploration tool?: Economic Geology, v. 90, p. 1217–1236.

Schardt, C., Cooke, D.R., Gemmell, J.B., and Large, R.R., 2001, Geochemical modeling of the zoned footwall alteration pipe, Hellyer volcanic-hosted massive sulphide deposit, western Tasmania, Australia: Economic Geology, v. 96, p. 1037–1054.

Schneiderman, J.S., and Tracy, R.J., 1991, Petrology of orthoamphibole-cordierite gneisses from the Orijarvi area, southwest Finland: American Mineralogist, v. 76, p. 942–955.

Schreurs, J., and Westra, L., 1985, Cordierite-orthopyroxene rocks: The granulite equivalents of the Orijarvi cordierite-anthophyllite rocks in West Uusimaa, southwest Finland: Lithos, v. 18, p. 215–228.

Schumacher, J.C., and Robinson, P., 1987, Mineral chemistry and metasomatic growth of aluminous enclaves in gedrite-cordierite-gneiss from southwestern New Hampshire, USA: Journal of Petrology, v. 28, p. 1033–1073.

Seyfried, W.E., 1987, Experimental and theoretical constraints on hydrothermal alteration processes at mid-ocean ridges: Annual Review of Earth and Planetary Sciences, v. 15, p. 317–335.

Seyfried, W.E., and Mottl, M.J., 1982, Hydrothermal alteration of basalts by seawater under seawater-dominated conditions: Geochimica et Cosmochimica Acta, v. 46, p. 985–1002.

Shriver, N.A., and MacLean, W.H., 1993, Mass, volume and chemical changes in the alteration zone at the Norbec mine, Noranda, Quebec: Mineralium Deposita, v. 28, p. 157–166.

Smith, M.S., Dumek, R., and Schneiderman, J.S., 1992, Implications of trace-element geochemistry for the origin of cordierite-orthoamphibole rock from Orijarvi, SW Finland: Journal of Geology, v. 100, p. 545–559.

Spry, P.G., Peter, J.M., and Slack, J.F., 2000, Meta-exhalites as exploration guides to ore, *in* Spry, P.G., et al., eds., Metamorphic and metamorphogenic ore deposits: Reviews in Economic Geology, v. 11, p. 163–201.

Stanton, R.L., 1990, Magmatic evolution and the ore type–lava type affiliation of volcanogenic exhalative ores, *in* Hughes, F.E., ed., Geology of the mineral deposits of Australia and Papua New Guinea: Melbourne, Australian Institute of Mineralogy and Metallogeny, p. 101–107.

Staudigel, H., and Hart, S.R., 1983, Alteration of basaltic glass: Mechanisms and significance for the oceanic crust–seawater budget: Geochimica et Cosmochimica Acta, v. 47, p. 337–350.

Sun, S.S., and McDonough, W.F., 1989, Chemical and isotopic systematics of oceanic basalts: Implications for mantle composition and processes, *in* Saunders, A.D., and Norry, M.J., eds., Magmatism in the ocean basins: Geological Society [London] Special Publication 42, p. 313–345.

Takagi, T., Orihashi, Y., Naito, K., Watanabe, Y., 1999, Petrology of a mantle-derived rhyolite, Hokkaido, Japan: Chemical Geology, v. 160, p. 425–445.

Taylor, R.P., and Fryer, B.J., 1980, Multi-stage hydrothermal alteration in porphyry copper systems in northern Turkey: The temporal interplay of potassic, propylitic and phyllic fluids: Canadian Journal of Earth Sciences, v. 17, p. 901–926.

Taylor, R.P., and Fryer, B.J., 1983, Rare earth element litho-geochemistry of granitoid mineral deposits: Canadian Institute of Mineralogy and Metallogeny Bulletin, v. 76, p. 901–926.

Trägårdh, J., 1991, Metamorphism of magnesium-altered felsic volcanic rocks from Bergslagen, central Sweden: A transition from Mg-chlorite- to cordierite-rich rocks: Ore Geology Review, v. 6, p. 485–497.

Vallance, T.G., 1967, Mafic rock alteration and isochemical development of some cordierite-anthophyllite rocks: Journal of Petrology, v. 8, p. 84–96.

Vocke, R.D., Jr., Hanson, G.N., and Grünenfelder, M., 1987, Rare earth element mobility in the Roffna Gneiss, Switzerland: Contributions to Mineralogy and Petrology, v. 95, p. 145–154.

Whitford, D.J., Korsch, M.J., Porritt, P.M., and Craven, S.J., 1988, Rare earth element mobility around the volcanogenic polymetallic massive sulphide deposit at Que River, Tasmania, Australia: Chemical Geology, v. 68, p. 105–119.

Windrim, D.P., McCulloch, M.T., Chappell, B.W., and Cameron, W.E., 1984, Nd isotopic systematics and chemistry of Central Australian sapphirine granulites: An example of rare earth element mobility: Earth and Planetary Sciences Letters, v. 70, p. 27–39.

Wood, S.A., 1991, The aqueous geochemistry of the rare earth elements and yttrium 2: Theoretical predictions of speciation in hydrothermal solutions to 350 °C at saturated water vapour pressure: Chemical Geology, v. 88, p. 99–125.

Wood, S.A., and Williams-Jones, A.E., 1994, The aqueous geochemistry of the rare earth elements and yttrium 4: Monazite solubility and REE mobility in exhalative massive sulphide-depositing environments: Chemical Geology, v. 115, p. 47–60.

Yang, K.H., and Scott, S.D., 1996, Possible contribution of a metal-rich magmatic fluid to a sea-floor hydrothermal system: Nature, v. 383, p. 420–423.

Zaleski, E., and Peterson, V.L., 1995, Depositional setting and deformation of massive sulphide deposits, Iron-Formation, and associated alteration in the Manitouwadge Greenstone Belt, Superior Province, Ontario: Economic Geology, v. 90, p. 2244–2261.

Manuscript Accepted by the Society August 25, 2003

Printed in the USA

Geological Society of America
Memoir 197
2004

Granulite-facies P-T-t *paths and the influence of retrograde cation diffusion during polyphase orogenesis, western Grenville Province, Québec*

Katherine J.E. Boggs*
Département des Sciences appliquées, Université du Québec à Chicoutimi, 555 Blvd de l'Université, Chicoutimi, Québec G7H 2B1, Canada
Louise Corriveau
Geological Survey of Canada, GSC-Québec, 880 Chemin Sainte-Foy, rm. 840, Québec G1S 2L2, Canada

ABSTRACT

The Bondy gneiss complex preserves evidence of an early 1.20–1.18 Ga granulite-facies tectonometamorphic event within the Central Metasedimentary Belt of Québec, western Grenville Province. Peak metamorphic conditions, estimated with garnet-biotite, garnet-orthopyroxene, garnet-aluminosilicate-quartz-plagioclase, and Al-in-orthopyroxene thermobarometers, reached 950 °C and 10 kilobars, compatible with the presence of sillimanite with orthopyroxene in aluminous gneiss. Peak pressure occurred at 1.20 Ga and was followed by crystallization of peak temperature anatectic melt at 1.18 Ga and isothermal decompression as recorded by garnet-cordierite-orthopyroxene-bearing leucosomes and garnet zonation profiles and textures. These conditions are 150–200 °C and 2 kilobars higher than those previously documented in the region. Peak assemblages were overprinted in the Nominingue-Chénéville deformation zone, to the east of the Bondy gneiss complex, during deformation at amphibolite facies and intrusion of monzonite-diorite-gabbro sheets at 1.17–1.16 Ga. High concentrations of ferromagnesian inclusions in large garnet grains reduced the effective diffusional domain size, which, when combined with retrograde cation diffusion, locally obliterated evidence of peak metamorphic conditions in the gneiss complex. Retrograde cooling paths in the gneiss complex and the deformation zone are distinct, that of the gneiss complex being at least 0.5 kilobar lower at any given temperature.

Keywords: Grenville, granulite, *P-T-t* trajectories, retrograde cation diffusion

INTRODUCTION

During high-grade polyphase orogenesis, tectonic reactivation may reset the chemical and textural record of peak metamorphic assemblages, and thus affect our ability to decipher early orogenic events (Spear and Florence, 1992; Erambert and Austrheim, 1993). However, by carefully studying nonreactivated areas, the early orogenic record can be deciphered in otherwise largely overprinted granulite-facies terrains (Tommasi and Vauchez, 1997; Corriveau et al., 1998). In the Central Metasedimentary Belt of Québec (Fig. 1), in the western Grenville Province (Wynne-Edwards, 1972), a strong orogenic and

*E-mail: kboggs@mtroyal.ca. Current address: Department of Earth Sciences, Mount Royal College, 4828 Richard Road SW, Calgary, Alberta T3E 6K6, Canada.

Boggs, K.J.E., and Corriveau, L., 2004, Granulite-facies *P-T-t* paths and the influence of retrograde cation diffusion during polyphase orogenesis, western Grenville Province, Québec, *in* Tollo, R.P., Corriveau, L., McLelland, J., and Bartholomew, M.J., eds., Proterozoic tectonic evolution of the Grenville orogen in North America: Boulder, Colorado, Geological Society of America Memoir 197, p. 35–64. For permission to copy, contact editing@geosociety.org.

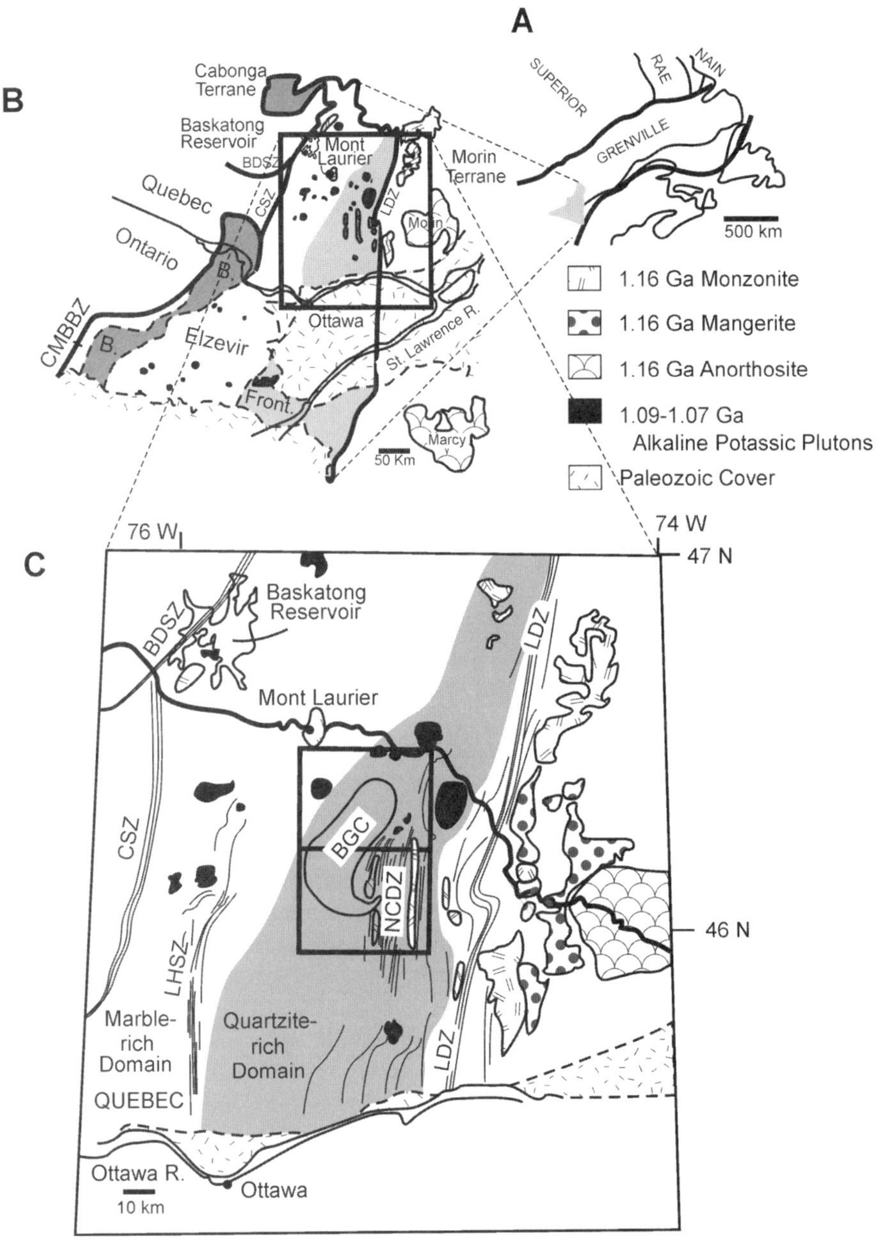

Figure 1. Location of study area (modified after Corriveau and van Breemen, 2000). (A) Canadian Grenville Province, with the Central Metasedimentary Belt in pale gray; (B) divisions of the Central Metasedimentary Belt; (C) detail of the Central Metasedimentary Belt in Québec, with the study area outlined. B—Bancroft terrane; BGC—Bondy gneiss complex; BDSZ—Baskatong-Dozois shear zone; CMBBZ—Central Metasedimentary Belt Boundary Zone (Davidson, 1995); CSZ—Cayamant shear zone; Front.—Frontenac terrane; HSZ—Heney shear zone; LDZ—Labelle deformation zone; NCDZ—Nominingue-Chénéville deformation zone.

magmatic pulse was recorded between 1.17 and 1.14 Ga, coeval with the onset of 1165–1135 Ma Morin anorthosite-mangerite-charnockite-granite magmatism in the adjacent Morin terrane (Corriveau and van Breemen, 2000; Martignole et al., 2000). The Bondy gneiss complex (hereafter referred to as the complex) escaped this strong metamorphic overprint and preserves a record of an earlier 1.20–1.18 Ga tectonothermal event (Figs. 1, 2, and 3; Corriveau et al., 1998; Corriveau and van Breemen, 2000; Harris et al., 2001; Wodicka et al., this volume). Detection of this early record first occurred upon the realization that an undeformed mafic dike swarm in this gneiss complex was coeval with deformed mafic dikes in adjacent areas (Corriveau et al., 1998). Fine monitoring of this early event is permitted by particularly responsive aluminous gneiss assemblages of a fossil hydrothermal system in the complex (Blein et al., this volume; Wodicka et al., this volume). The complex is flanked to the east by the Nominingue-Chénéville deformation zone (hereafter referred to as the deformation zone), a site of syntectonic emplacement of 1.17–1.16 Ga monzonite-diorite-gabbro sheet intrusions and dike swarms (Corriveau and van Breemen, 2000).

In this paper we describe the pressure-temperature-time (*P-T-t*) path from the complex, which was unaffected by late-stage deformation events and only mildly affected by mineral reequilibration during 1.17–1.16 Ga deformation and retrogression (Harris et al., 2001; Wodicka et al., this volume), and contrast it

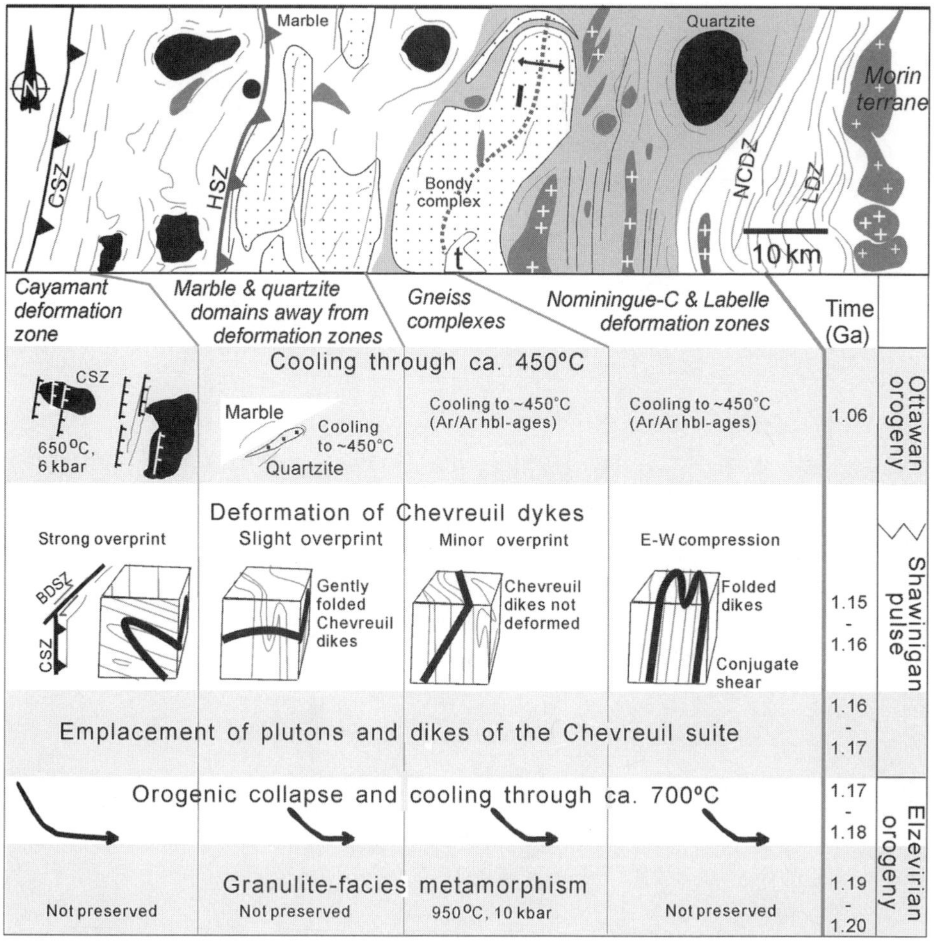

Figure 2. Schematic model for tectonomagmatic events in the Central Metasedimentary Belt of Québec. BDSZ—Baskatong-Dozois shear zone; CSZ—Cayamant shear zone; HSZ—Heney shear zone; LDZ—Labelle deformation zone; NCDZ—Nominingue-Chénéville deformation zone.

with that of the strongly reactivated deformation zone (Figs. 1, 2, and 3). These *P-T-t* paths diverge after 1.17–1.16 Ga magma emplacement and deformation. To explain why certain mineral parageneses in the complex preserved evidence of true peak metamorphic conditions, we examine the effects of retrograde cation diffusion and high concentrations of ferromagnesian mineral inclusions in garnet. We show that with careful selection of minerals for geothermometry and geobarometry scant evidence of peak conditions can be extracted from largely retrograded granulite-facies terrains.

GEOLOGICAL SETTING

The Québec segment of the Central Metasedimentary Belt consists of Mesoproterozoic supracrustal and intrusive rocks, structurally juxtaposed on Laurentian Paleoproterozoic and Archean gneisses that were metamorphosed during the Grenvillian orogeny (Guo and Dickin, 1996; Corriveau and Morin, 2000; Martignole et al., 2000). The belt is bounded on the east by the Morin terrane, which is typified by granulite-facies supracrustal rocks intruded by the large 1165–1135 Ma Morin anorthosite-mangerite-charnockite-granite complex (Martignole and Schrijver, 1970; Emslie and Hunt, 1990; Doig, 1991). The belt comprises at least three discrete crustal domains, from west to east: a marble-rich domain, a quartzite-rich domain, and a domain of gneiss complexes, such as the Bondy gneiss complex (Figs. 1 and 3). These domains are bounded by north-south trending deformation zones that acted as preferential sites for emplacement of the 1.17–1.16 Ga monzonite-diorite-gabbro Chevreuil suite and coeval deformation (Figs. 1, 2, and 3; Corriveau et al., 1998; Corriveau and van Breemen, 2000).

The belt extends southwesterly into Ontario and New York, where it comprises the Bancroft, Elzevir, and Frontenac terranes (Fig. 1; Davidson, 1995). Metamorphic pressures and temperatures in those terranes were compiled by Streepey et al. (1997). Calculated pressures and temperatures increase outward from the south central portion of the Central Metasedimentary Belt in Ontario, reaching ~800 °C and 6–8 kilobars near the Ottawa River (e.g., Anovitz and Essene, 1990). In the Central Metasedimentary Belt of Québec, granulite-facies rocks are mostly restricted to gneiss complexes, whereas the marble and quartzite-rich domains, as well as the deformation zones, are mostly at amphibolite facies (Figs. 2 and 3). Indares and Martignole (1984, 1990a, 1990b) estimated metamorphic temperatures and

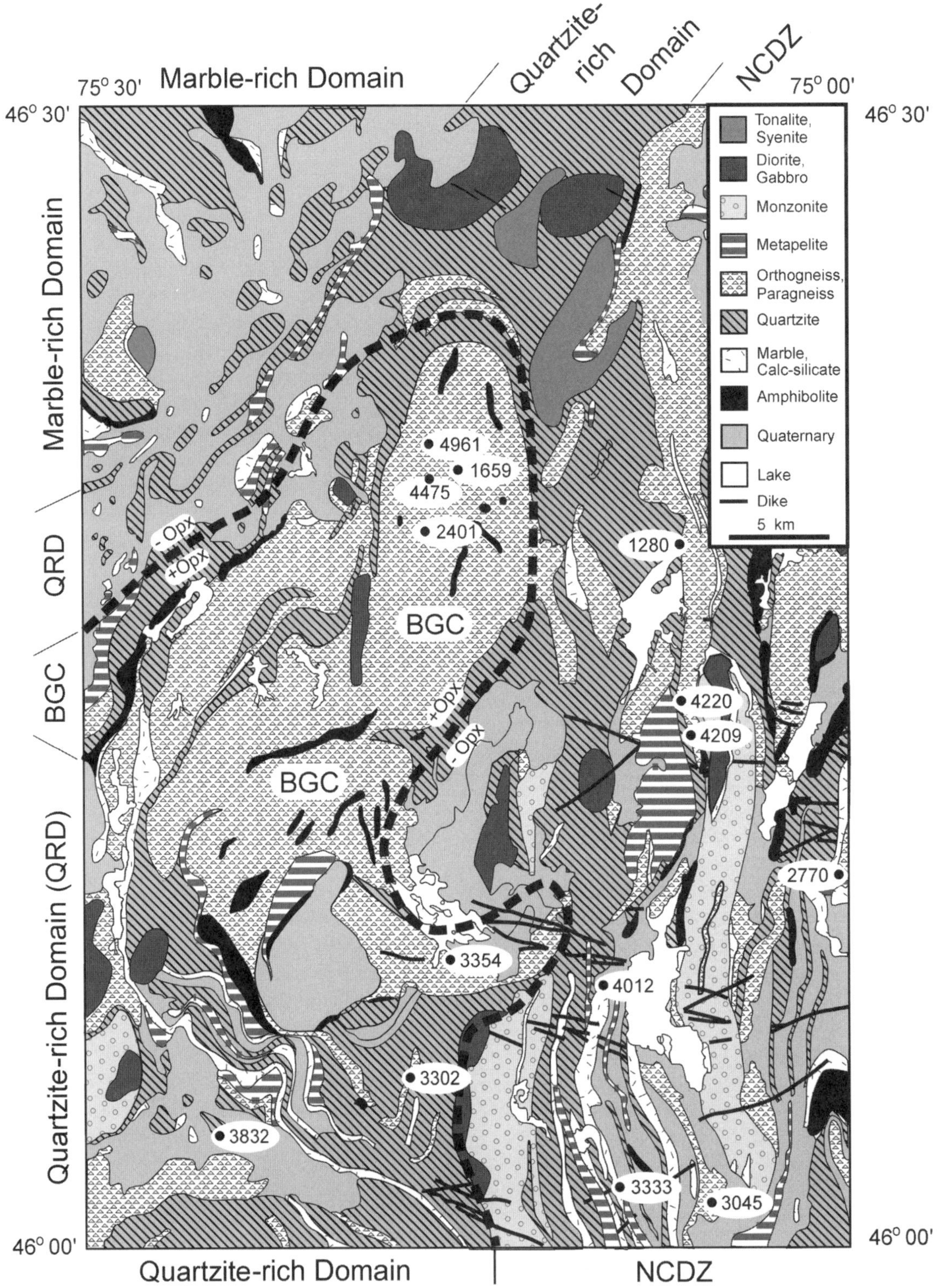

Figure 3. Geological map of study area with sample localities (modified after Corriveau et al., 1996). BGC—Bondy gneiss complex; NCDZ—Nomininque-Chénéville deformation zone; Opx—orthopyroxene; QRD—quartzite-rich domain.

pressures along a transect across the belt (Figs. 1, C, and 4). From garnet zonation patterns in metapelites and from garnet amphibolites they inferred a series of cooling and decompression paths and maximum peak metamorphic conditions of 800 °C and 8.8 kilobars.

The distinctive 1165–1156 Ma Chevreuil igneous suite represents a regional marker that constrains timing of deformation (Fig. 2; Corriveau and van Breemen, 2000; Harris et al., 2001). Concordance of Chevreuil pluton ages within the margin of error suggests a burst of magmatic activity at 1165 Ma followed by a minor influx of mafic and felsic dike swarms for ~10 m.y. By mapping the extent of Chevreuil mafic dike deformation, Corriveau et al. (1998) documented that, once metamorphosed, the lithotectonic domains of the Central Metasedimentary Belt had contrasting mechanical properties, which range from nonreactivated to completely overprinted orogenic segments (Fig. 2). The Bondy gneiss complex is a key rheological window where Chevreuil mafic dikes are not deformed and where granulite-facies mineral paragenesis prevail in the gneisses.

Conventional U-Pb geochronology of metamorphic zircons from the complex (Corriveau and van Breemen, 2000) constrained the timing of gneiss formation and peak metamorphism as prior to the 1.17–1.16 Ga overprint. Zircon fractions from an aluminous gneiss without significant anatectic segregation gave an upper intercept age of 1195 ± 7 Ma consistent with the timing of the peak pressure. Concordant zircon fractions from a tonalitic gneiss with granitic leucosomes gave $^{207}Pb/^{206}Pb$ ages of 1177.1 ± 2.9 Ma and 1178.9 ± 1.5 Ma. The zircon morphology and younger ages are consistent with an anatectic origin. Metamorphic monazite separated from an aluminous gneiss in the complex and from a metapelite in the quartzite domain yielded $^{207}Pb/^{206}Pb$ model ages of 1182 ± 2 Ma and 1174 ± 2 Ma, respectively. These ages are interpreted as cooling ages below ~800 ± 50 °C (Bingen and van Breemen, 1998). Sensitive high-resolution ion microprobe (SHRIMP) data on metamorphic zircons from the complex cluster around two main events at 1.20–1.18 and 1.15–1.13 Ga (Wodicka et al., this volume). Large errors in this SHRIMP data permit correlation of these results with the 1.20–1.18 and 1.17–1.15 Ga ID-TIMS (isotope dilution–thermal ionization mass spectrometry) ages (Bingen and van Breemen, 1998; Corriveau and van Breemen, 2000; Wodicka et al., this volume).

Features of fabric-forming events in the Central Metasedimentary Belt of Quèbec are summarized in Table 1 (Harris et al.,

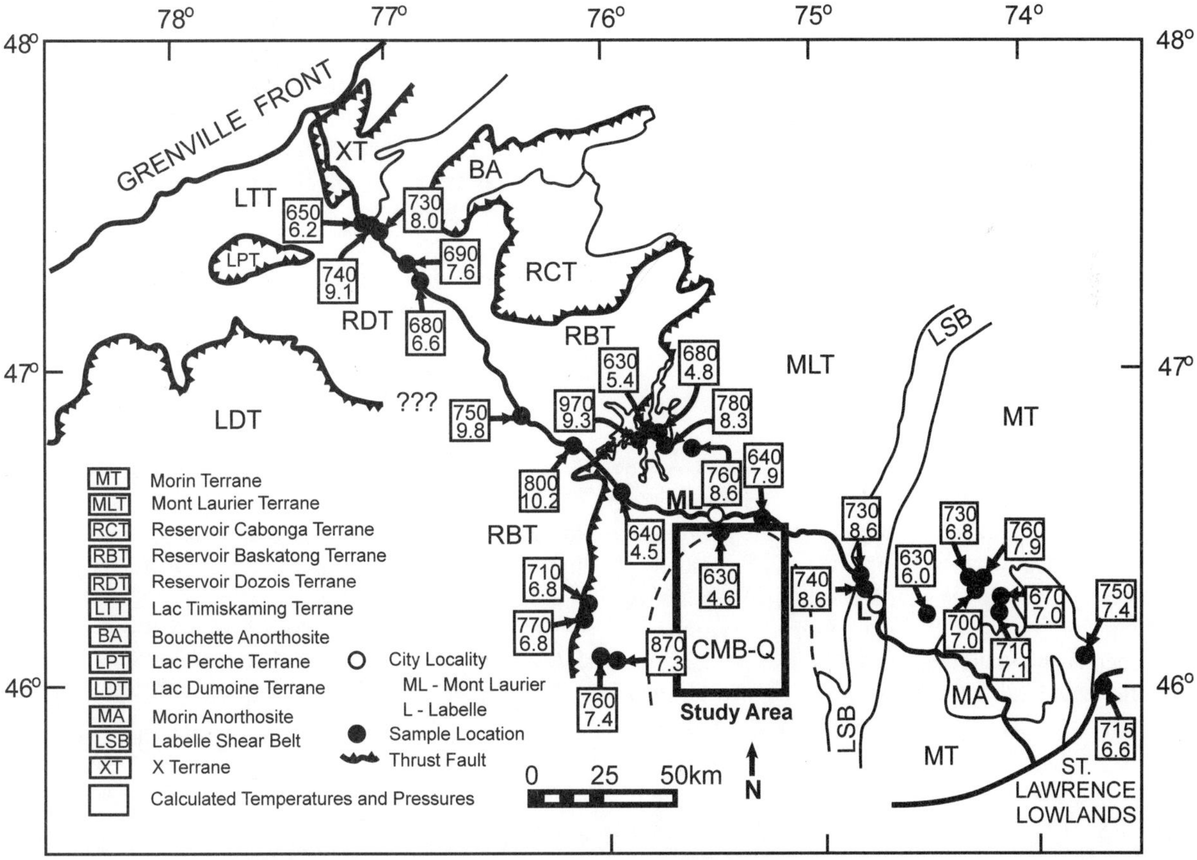

Figure 4. Metamorphic pressures and temperatures in the southern Grenville Province of Québec (modified after Indares and Martignole, 1990a, 1990b, with additions from Kretz et al., 1989). CMB-Q—Central Metasedimentary Belt in Québec; ???—regions where pressures and temperatures are unknown.

TABLE 1. SUMMARY OF STRUCTURAL EVENTS IN STUDY AREA

Event	Characteristic features	Timing (Ga)
D1	Metamorphic foliation, transposed layering parallel to S1 Massive leucosomes parallel to S1 First and main phase of crustal thickening	1.2–1.9 (Corriveau and van Breemen, 2000)
F2/F3	Isoclinal folds, fold S1 and layer-parallel leucosomes Axial planar foliation crosscut by ameboidal leucosomes (Fig. 5, A and B)	1.18 (leucosomes)
D4	Ductile shear zones crosscut and boudinage D3 features	
D5	Regional-scale open upright NS folds, folds D4 shear zones	
D6	Later shallowly dipping axial surfaces in quartzite-rich domain Crosscut by Chevreuil suite intrusives	1.17–1.16 (Chevreuil suite)
D7	Compressive event focused along NS deformation zones such as the Nominingue-Chénéville deformation zone (NCDZ) Formed NE dextral and SSE sinistral shear zones and NNE-striking thrusts in NCDZ (Rivard et al., 1999) Started during Chevreuil suite emplacement	1.17–1.156 (Corriveau and van Breemen, 2000)
D8	Late open folds with ENE axial surfaces Produced dome and basin interference patterns	

Source: After Harris et al. (1998, 2001).

1998, 2001). An orthopyroxene-in isograd (Fig. 3) cuts obliquely across F3 folds, indicating that retrogression from 1.20–1.18 Ga granulite-facies metamorphism to amphibolite facies postdates D3. Monazite ages, which are 1.18 Ga or slightly younger, record cooling through 850–750 °C, likely after D3 (Boggs, 1996; Corriveau and van Breemen, 2000). All D3 and older fabrics are cut by 1.17–1.16 Ga Chevreuil suite intrusive bodies (Fig. 2; Harris et al., 2001).

MINERAL ASSEMBLAGES AND TEXTURES

Textural information, discussed later, is supportive of reactions used for geothermobarometry and of calculated geothermobarometric results. Samples were chosen from outcrops within the complex and the deformation zone (Fig. 3) that contained peak or retrograde mineral assemblages suitable for geothermobarometric analyses. Outcrops 2401, 1659, and 4475 (complex) preserved evidence of peak mineral assemblages such as Spl-Grt-Sil (outcrop 2401) and Sil-Opx (outcrops 1659 and 4475; mineral abbreviations are in Appendix 1). Rocks from the deformation zone did not preserve evidence of peak metamorphic conditions because they were strongly affected by the 1.17 Ga Chevreuil intrusions and coeval deformation.

The compositionally diverse aluminous gneiss of the complex contains between three and five phase (in AFM projection) peak and retrograde assemblages that denote the early 1.20–1.18 Ga Grenvillian tectonometamorphic event (Corriveau et al., 1998). Some of these assemblages were both easily identified in coarse-grained rocks in the field and confirmed by petrographic studies—e.g., Crd-Grt-Opx in leucosomes (Fig. 5, A and B), an assemblage equivalent to Bt in gneisses (Fig. 5, C) and Crd-Qtz symplectite (Fig. 5, D). Other assemblages, such as prograde and retrograde Bt-Crd-Opx-Sil (Fig. 5, E and F) and retrograde Bt + Sil at the expense of Crd (Fig. 5, G and H), were observed only petrographically.

Constraints on Prograde Conditions in the Bondy Gneiss Complex

Evidence of peak conditions is contained within armored cores of porphyroblasts (e.g., garnet 2401D) and as coarse-grained idioblastic porphyroblasts (e.g., Fig. 6, A, B, and C). Lowest-variance assemblages include (1) Bt + Crd + Grt + Sil, (2) Bt + Crd + Grt, (3) Bt + Grt + Opx, (4) Bt + Crd + Opx, or (5) Crd + Opx + Sil (Figs. 5 and 6); all contain alkali feldspar, ilmenite, plagioclase, and quartz. Following the petrogenetic

→

Figure 5. Outcrop and thin-section photographs of rocks from the Bondy gneiss complex (see Fig. 3 for localities; for mineral abbreviations see Appendix 1). (A) Bt-Grt-Cpx-Opx gneiss with ameboidal leucosomes (sample 4475); (B) Crd-Grt-Kfs-Opx-Pl-Qtz assemblage observable on the outcrop scale only due to its very coarse grains (4475); (C) Bt-Crd-Grt-Opx assemblage (4475, plane polarized light); (D) aluminous Crd-Kfs-Opx-Pl-Qtz gneiss with intergrown Crd-Qtz (1659); (E) aluminous Bt-Crd-Opx-Sil gneiss (4961, plane light); (F) photomicrograph of an aluminous gneiss with Sil forming at the contact between Crd and Opx (1659); (G) Crd and retrograde green Bt forming in Sil clots (1659); (H) local retrograde green Bt-Sil in late shear zones (4475).

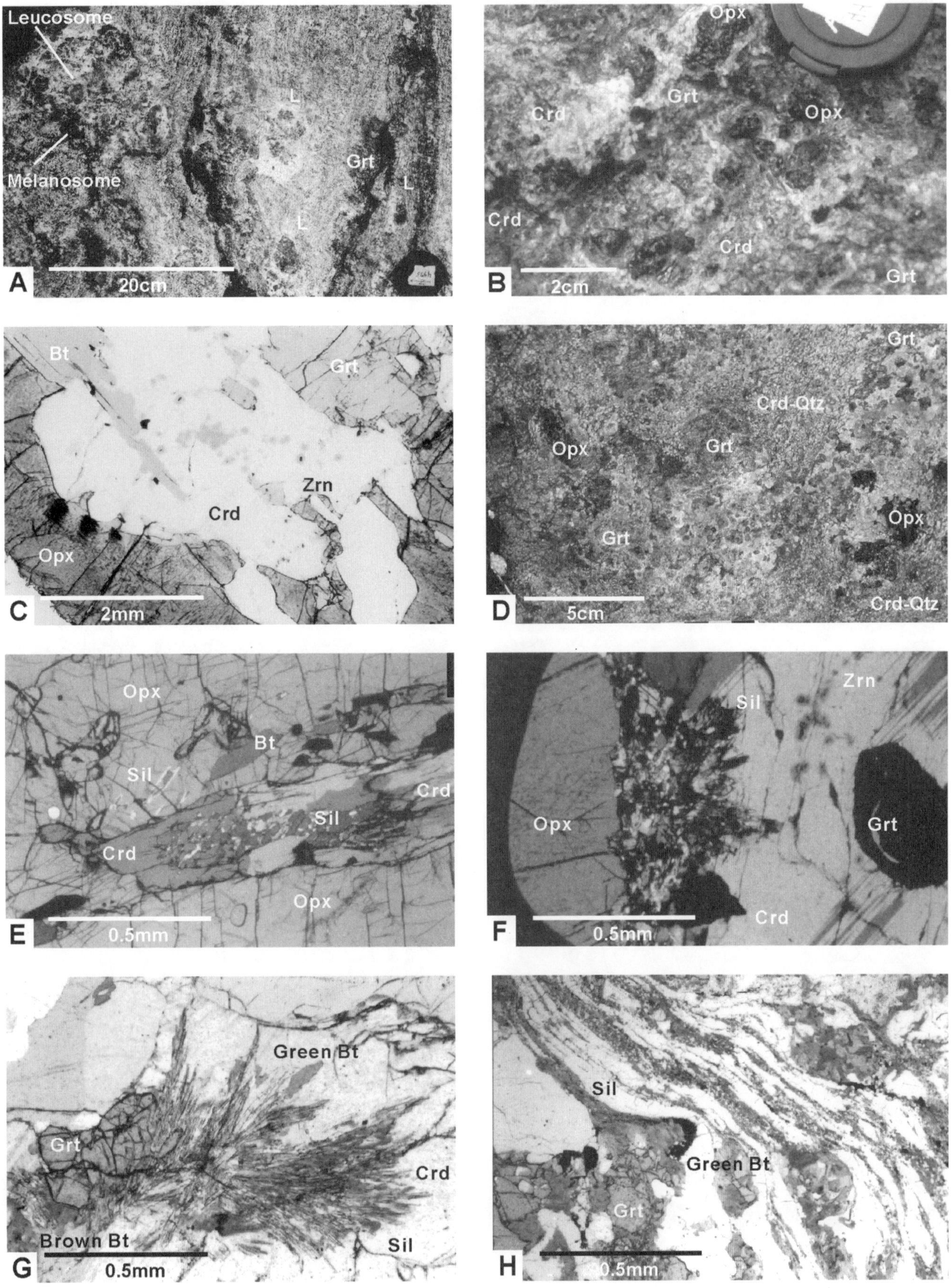
Leucosome
L
Mélanosome
Grt
L
L
A
20cm
Opx
Crd
Grt
Opx
Crd
Crd
Grt
B
2cm
Bt
Grt
Zrn
Crd
Opx
C
2mm
Grt
Crd-Qtz
Opx
Grt
Opx
Grt
Crd-Qtz
D
5cm
Opx
Bt
Sil
Crd
Sil
Crd
Opx
E
0.5mm
Sil
Zrn
Opx
Grt
Crd
F
0.5mm
Green Bt
Grt
Crd
Brown Bt
Sil
G
0.5mm
Sil
Green Bt
Grt
H
0.5mm

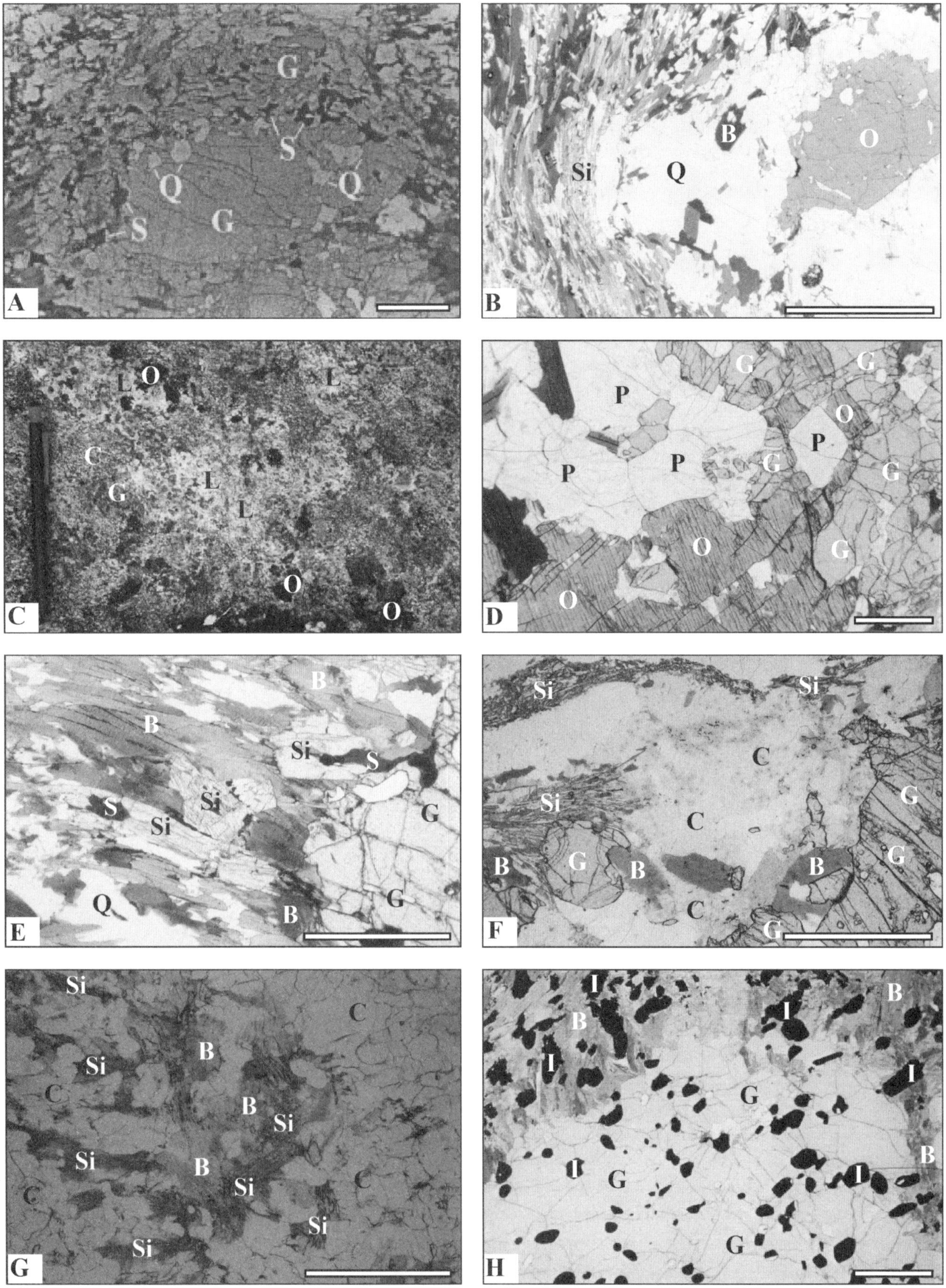

A
G
S
Q
Q
S
G
B
Si
Q
B
O
C
L
O
L
C
G
L
L
O
O
D
P
P
P
P
G
G
O
G
G
O
G
O
E
B
B
Si
S
Si
S
Si
G
Q
B
G
F
Si
Si
C
Si
C
G
G
B
G
B
B
C
G
G
Si
C
Si
B
C
B
Si
Si
B
Si
C
C
Si
Si
H
I
B
I
B
I
I
G
B
I
G
I
G

grid of Carrington and Harley (1995, their Fig. 11) for biotite melting in the KFMASH (K_2O-FeO-MgO-Al_2O_3-SiO_2-H_2O) metapelite system, these assemblages depict three changes of AFM (Al_2O_3-FeO-MgO) topology, which are accounted for by osumilite-absent univariant reactions under 8 kilobars:

Bt + Sil + Kfs + Qtz = Grt + Crd + melt (A)

Bt + Grt + Kfs + Qtz = Crd + Opx + melt (B)

and for pressure equal to or greater than 8 kilobars:

Bt + Sil + Kfs + Qtz = Opx + Crd + melt (C)

Sillimanite prisms in contact with cordierite and biotite enclosed in an orthopyroxene grain that lies in the main gneissosity (S1 gneiss fabric at outcrop 4961, Fig. 5, E) provide evidence of a Crd + Opx + Sil assemblage, which indicates pressures greater than 8.5 kilobars and temperatures lower than 900 °C (Carrington and Harley, 1995). This texture reveals that peak pressure (1.2 Ga) was achieved at an early stage of foliation development. Lack of osumilite in magnesian rocks of the complex implies that peak metamorphic temperatures did not exceed 975 °C at 10 kilobars (Carrington and Harley, 1995). The coarse-grained Crd-Grt-Opx assemblage, which is diagnostic of peak temperatures higher than 850 °C (Carrington and Harley, 1995), occurs in massive leucosomes that cut F3 folds of the S1 gneissosity in biotite-sillimanite gneiss (Fig. 5, A and B). This postfabric assemblage is inferred to be coeval with the 1.18 Ga anatectic melt-laden gneiss (Corriveau and van Breemen, 2000).

Outcrop 2401 consists of a migmatitic aluminous gneiss with undulose irregular granitic leucosomes (Fig. 6, C). Leucosomes contain minor amounts of garnet and locally a greenish alteration product (possibly pinite after cordierite). Melanosomes contain three separate assemblages: Grt + Spl + Sil + Pl, Opx + Grt + Bt, and Grt + Bt + Kfs + Sil + Ap + Pl + Qtz + Bt. Intergrown orthopyroxene and garnet in melanosomes (Fig. 6, D) suggest the reaction

Bt + Pl + Qtz = Opx + Grt (+Kfs) + melt (D)

with melt migrating to form leucosomes, leaving behind Opx + Grt. One garnet contains a relict inclusion-free core surrounded by an inclusion-rich rim (Figs. 6, A; 7, F; and 8). The location of spinel (as inclusions in garnet with inclusions of quartz and/or surrounded by sillimanite; Fig. 6, E) suggests the reaction

Grt + Als = Spl + Qtz + melt (E)

producing Spl + Qtz (Fig. 9, III).

Constraints on Retrograde Conditions in the Bondy Gneiss Complex

Migmatized aluminous gneiss of outcrops 1659 and 4961 (Figs. 5, E and F, and 6, B) contain radiating fine-grained to coarsely crystalline sillimanite at the contacts between porphyroblasts of cordierite and orthopyroxene in a melanosome. This sillimanite-forming texture indicates that pressure was still high after the main fabric-forming event.

During retrogression from the peak metamorphic temperature, the *P-T* trajectory is inferred to have followed reaction (E) closely because spinel production is closely followed by spinel consumption. Evidence of this process is present in garnet from outcrop 2401, where a rim of spinel inclusions surrounds an inclusion-free core (Figs. 6, A and 9, III and IV). Absence of the assemblage Spl + Crd in quartz-bearing rocks implies that the reaction

V + Qtz + Spl = Crd (Fig. 9) (F)

did not occur. Garnet porphyroblasts from sample 1659 commonly have partial coronas of cordierite, suggesting the following reaction:

Grt + Sil + Qtz + V = Crd (Fig. 9, V) (G)

Cordierite coronas surrounding garnet in sample 1659 contain retrograde brown biotite, suggestive of the following reaction (Fig. 6, F):

L + Grt = Bt + Crd + Qtz + V (H)

Brown biotite + cordierite aggregates also have sillimanite associated with them (Fig. 6, G) implying the crossing of reaction:

L + Crd = Bt + Als + Qtz + V (Fig. 9, VII) (I)

These retrograde reactions involving cordierite require that the retrograde *P-T* path passed below ~5.5 kilobars at 770 °C.

A minor amount of the distinctive intergrowth of light green biotite and radially or randomly oriented prismatic sillimanite occurs within a cordierite host (outcrop 1659; Fig. 5, G). This biotite differs from fabric-forming brown biotite and the Sil + brown Bt intergrowth seen among most cordierite grains (Fig. 6, E). In the model system KFMASH, the reaction

Crd + Kfs + H_2O = Bt + Sil + Qtz (J)

Figure 6. Photomicrographs of samples from the Bondy gneiss complex (see Fig. 3 for localities). All scale bars are 1 mm. B—biotite; C—cordierite; G—garnet; I—ilmenite; L—liquid; O—orthopyroxene; S—spinel; Si—sillimanite; Q—quartz. All views in plane polarized light. (A) Garnet (sample 2401D); (B) sillimanite and orthopyroxene (1659); (C) flecky leucosome texture (2401); (D) intergrown orthopyroxene and garnet (2401F); (E) spinel surrounded by sillimanite (edge of garnet 2401D); (F) cordierite, biotite, and sillimanite surrounding garnet (1659); (G) retrograde biotite and sillimanite in cordierite (1659); (H) poikiloblastic garnet (4475D3) with ilmenite inclusions.

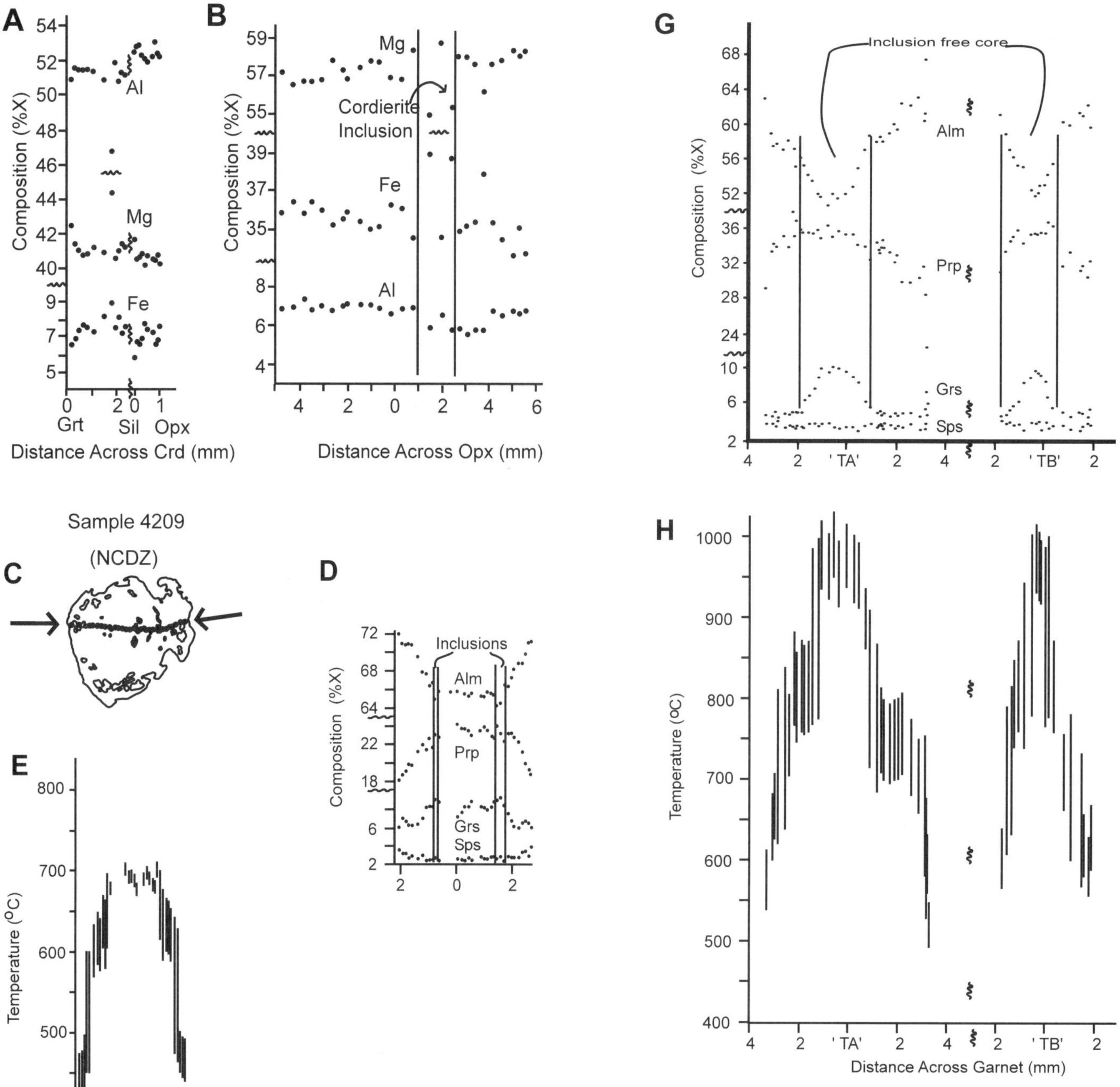

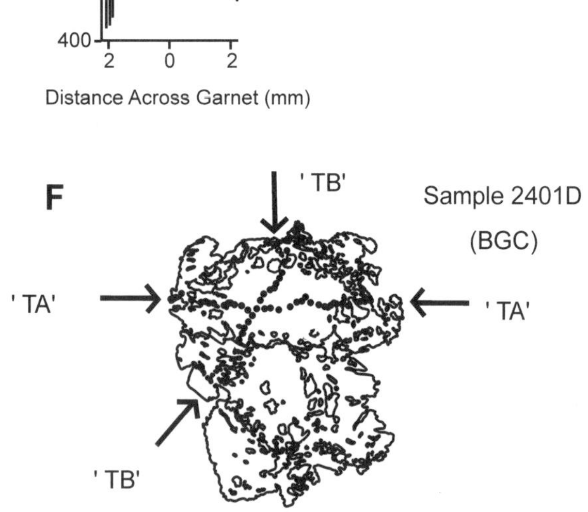

Figure 7. Sketches of garnet grains, compositional trends across minerals, and the calculated range in temperatures for each position analyzed. All are from the Bondy gneiss complex except 3832 (quartzite-rich domain). (A) Transect across cordierite (Crd) porphyroblast (sample 1659). Grt—garnet; Sil—sillimanite; Opx—orthopyroxene. (B) Transect across orthopyroxene porphyroblast (1659). (C), (F), (I), and (L) are sketches of analyzed garnet grains. Arrows point toward transects, black dots are spots analyzed with the electron microprobe, and gray irregular shapes are inclusions (mostly spinel and magnetite, with minor amounts of biotite, quartz, and feldspar). Scale of grains is the same as the plots below the sketches. (D), (G), (J), and (M) are plots of analyzed composition—% almandine (Alm), pyrope (Prp), grossular (Grs), and spessartine (Sps)—across the grains in sketches C, F, I, and L. (E), (H), (K), and (N) are the range of calculated temperatures from the Ferry and Spear (1978) biotite-garnet thermometer. The lines represent the range in temperatures calculated for the composition in the corresponding analyses from the graphs D, G, J, and M. BGC—Bondy gneiss complex; NCDZ—Nomininque-Chénéville deformation zone (Davidson, 1995); QD—quartzite-rich domain 'TA'—transect A; 'TB'—transect B.

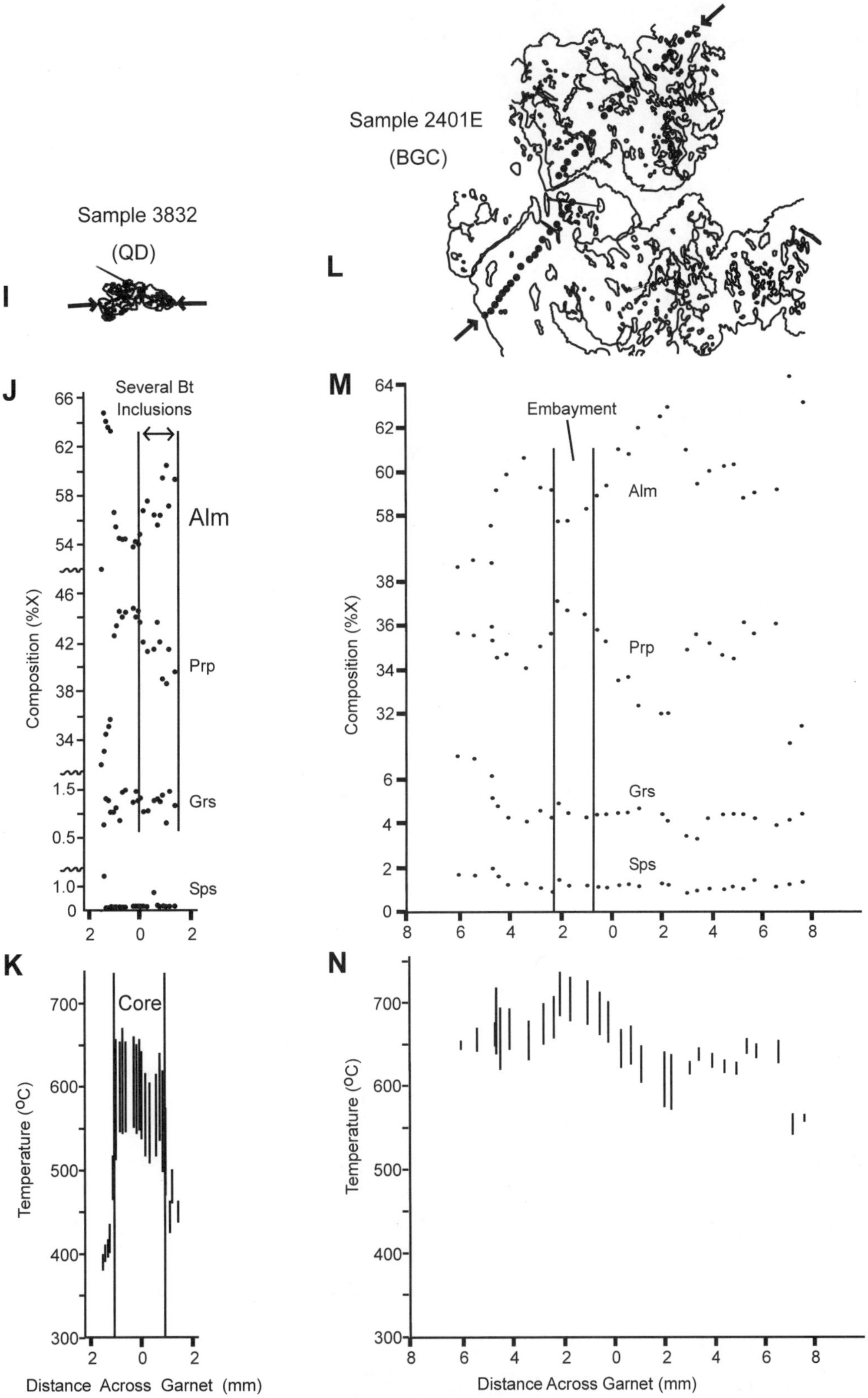
Sample 3832
(QD)
I
Sample 2401E
(BGC)
L
J
Several Bt Inclusions
Alm
Prp
Grs
Sps
Composition (%X)
M
Embayment
K
Core
Temperature (°C)
N
Distance Across Garnet (mm)

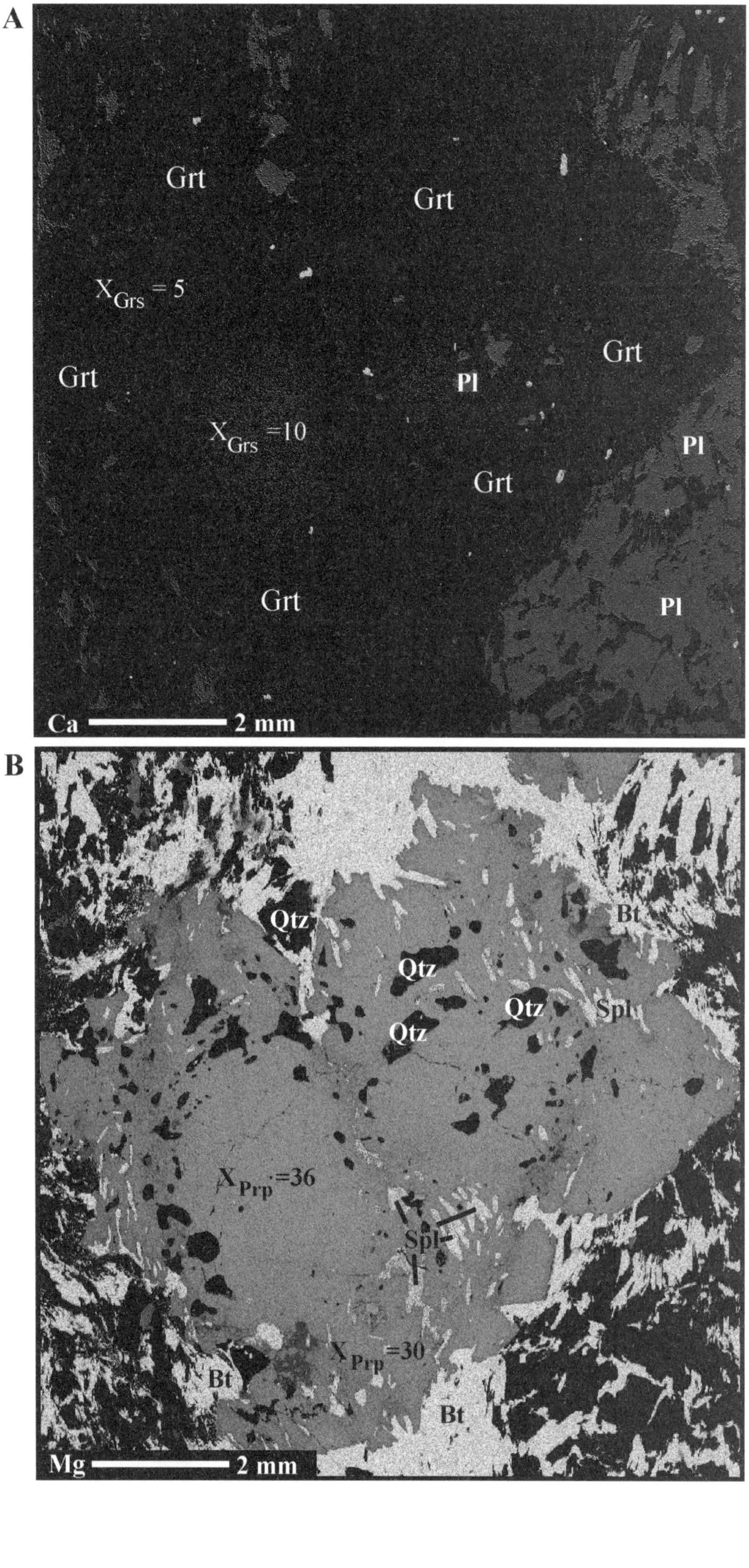

Figure 8. Calcium (A) and magnesium (B) compositional maps of garnet sample 2401D (same garnet as in Fig. 7 F, G, and H). The brighter, warmer, colors correspond to higher concentrations of calcium and magnesium. For mineral abbreviations see Appendix 1.

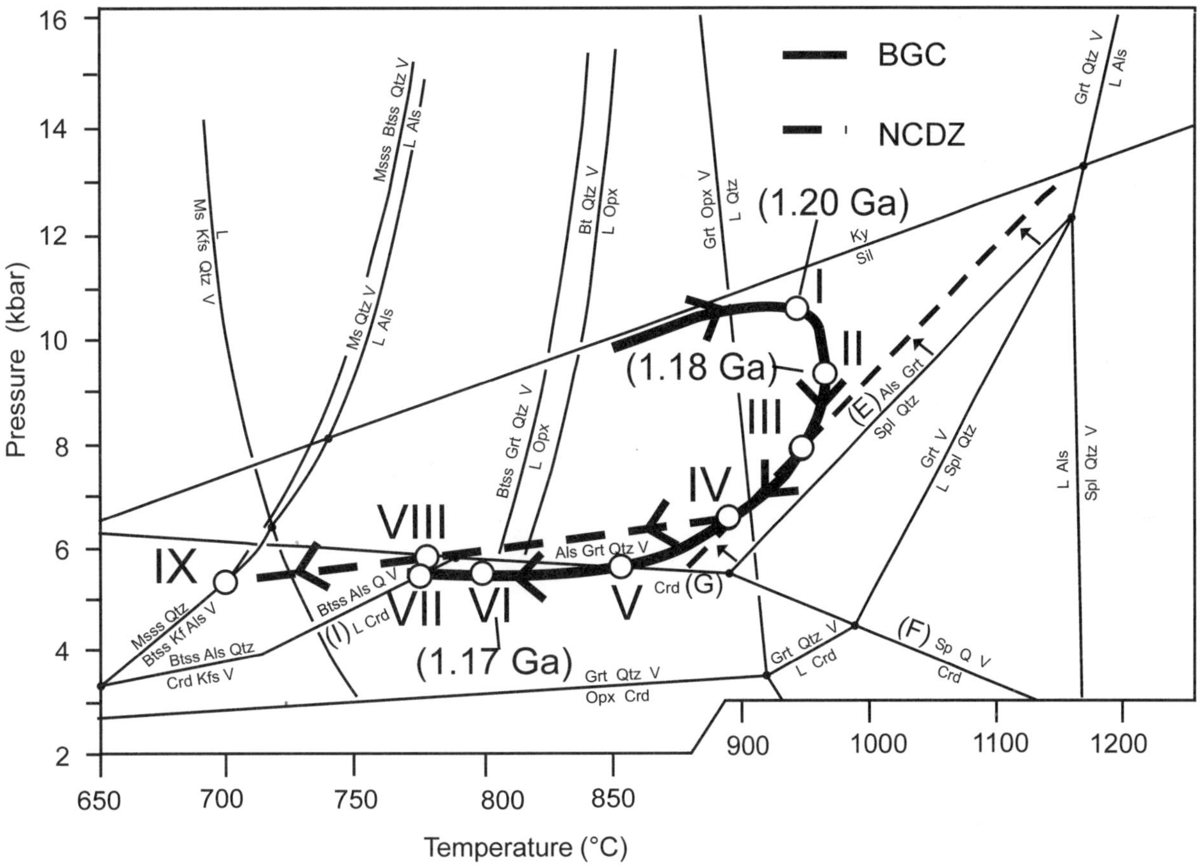

Figure 9. *P-T-t* trajectories using both mineral assemblages and geothermobarometric results for the Bondy gneiss complex (BGC) and the Nominingue-Chénéville deformation zone (NCDZ). Petrogenetic grid simplified after Vielzeuf and Holloway, 1988; age data from Bingen and van Breemen, 1998; Corriveau and van Breemen, 2000). Bold Roman numbers represent pressures and temperatures calculated from geothermobarometers and supported by mineral assemblages. Finer bold dashed line represents recalculated location for Als + Grt = Spl + Qtz reaction in *P-T* space due to zinc content in spinel, using the Nichols et al. (1992) barometer. Capitalized bold letters correspond to reactions quoted in the text. For abbreviations see Appendix 1.

would be compatible with the observed texture (Corriveau, 1982). This reaction is trivarient, with a gentle positive slope in *P-T* space and cordierite on the low-pressure side of the reaction curve (Holdaway and Lee, 1977). It contrasts with the negatively sloped reaction by which green biotite is formed at the expense of cordierite and brown biotite (Vernon, 1978). During retrograde metamorphism, a decrease in temperature combined with an increase in pressure or P_{H_2O} and an increase in water activity or a_{H_2O} would be required to form green biotite and sillimanite at the expense of cordierite. Compatible with the observation that Chevreuil suite dikes have minor late-stage deformation in the complex, this retrograde texture is not usually deformed. One exception is a series of discrete shear zones expressed by the development of a local fabric within this retrograde assemblage (outcrop 4475; Fig. 5, H) The relatively rare occurrence of this distinctive green biotite suggests a weaker chemical equilibration for the green biotite and sillimanite reaction products. This is compatible with this assemblage's having grown during a younger event, i.e., likely the 1.17–1.16 Ga reequilibration event (Wodicka et al., this volume).

Constraints on Retrograde Conditions in the Nominingue-Chénéville Deformation Zone

Cordierite was not observed in the deformation zone, suggesting that retrograde conditions were nearly isobaric and above reaction (G), which constrains the retrograde conditions to a path that is greater than 5.8 kilobars at 780 °C (Fig. 9, IV–VIII). K-feldspar was also not observed, forcing retrogression to proceed below 5.5 kilobars at 700 °C (Fig. 9, IX). The observed mineral assemblages and textures thus indicate that retrograde pressures in the deformation zone were up to 0.5 kilobars higher than in the complex (Fig. 9, IV, VIII, and IX). Alternatively, reaction (J), where cordierite is replaced by biotite and sillimanite, might have proceeded to completion in the strongly reactivated deformation zone, where early gneissosity is largely transposed parallel to the pervasive north-south fabric within amphibolite-facies conditions. The presence of pegmatite dikes with ubiquitous muscovite provides ample evidence that the deformation zone experienced an important influx of fluids, which could have facilitated the breakdown of cordierite.

MINERAL CHEMISTRY

Mineral Analyses

In this study minerals from two main rock types were analyzed: (1) Sil + Bt + Grt, Sil + Opx + Crd, Crd + Krn + Opx, Qtz + Pl + Crd + Opx + Krn + Tur + Phl aluminous gneiss from the complex and (2) a series of metapelite and garnet-biotite mixed paragneisses from the deformation zone. The analytical conditions, sources of error, and criteria for accepting microprobe analyses are summarized in Appendix 2, sample localities are shown in Figure 3, and representative mineral analyses are listed in Tables 2–9.

Plagioclase (Table 2) is mostly labradorite, except for two samples with An_{25} and An_{88}, both from the complex. No consistent zonation patterns were observed in plagioclase. However, the plagioclase in sample 3045 varied nonsystematically in composition from An_{50} to An_{54}.

The composition of alkali feldspar ranges from $Or_{70}Ab_{17}An_{13}$ to $Or_{89}Ab_{11}An_{0}$ (Table 3), with both extremes from the complex. Metapelite from the deformation zone has

TABLE 2. REPRESENTATIVE PLAGIOCLASE ANALYSES

Sample	Weight percent oxides					
number	SiO_2	Al_2O_3	CaO	Na_2O	Total	%An
1280-7	57.0	27.4	9.53	6.40	100.3	62
1280-8	57.7	27.0	8.66	6.65	100.0	59
1280-34	58.9	25.5	7.90	7.20	99.6	55
1280-44	57.5	27.1	9.00	6.48	100.0	61
1280-45	57.5	27.1	9.00	6.60	100.2	60
1280-46	57.3	27.2	9.06	6.44	100.0	61
1280-15	56.6	27.8	9.56	6.12	100.0	63
2039-41	57.3	27.3	8.70	6.50	99.8	60
2039-42	56.4	27.9	9.69	5.98	100.0	64
2039-53	56.4	27.9	9.61	5.93	99.8	64
2039-53	56.4	27.8	9.65	5.96	99.8	64
2039-54	56.5	27.8	9.60	6.01	99.9	64
3045-23	55.6	28.4	10.43	5.69	100.1	50
3045-28	53.2	29.9	12.23	4.61	99.9	59
3045-25	55.1	28.8	10.71	5.44	100.0	52
3045-26	55.6	28.4	10.40	5.60	100.0	51
3045-27	53.9	29.4	11.60	5.00	99.9	56
3045-59	53.7	26.7	11.80	4.90	100.1	57
3045-61	53.4	29.8	11.87	4.90	100.0	57
3302-64	61.8	24.0	5.28	8.69	99.7	25
4012-74	46.1	34.9	18.04	1.31	100.4	88
4209-25	58.9	26.2	7.88	7.19	100.1	55
4209-26	58.6	26.1	7.87	7.24	99.8	55
4209-27	58.9	26.2	7.80	7.20	100.1	55
4209-29	58.7	26.4	8.00	7.10	100.2	56
4209-30	58.9	26.2	8.02	7.24	100.4	55
4209-31	59.1	26.0	7.68	7.29	100.1	54

TABLE 3. REPRESENTATIVE K-FELDSPAR ANALYSES

Sample	Weight percent oxides								
number	SiO_2	Al_2O_3	CaO	Na_2O	K_2O	Total	%Or	%Ab	%An
1280-40	64.1	19.0	0.26	1.23	14.7	99.3	86	11	3
1280-53	64.1	18.6	0.11	1.32	14.6	98.7	87	12	1
1280-54	64.3	18.7	0.25	1.27	14.8	99.4	86	11	2
1280-55	64.3	18.8	0.21	1.10	14.9	99.4	88	10	2
2401D-38	64.9	18.8	0.19	1.65	14.9	100.4	84	14	2
2401D-39	65.0	18.5	0.36	1.76	14.5	100.1	82	15	3
2401D-41	64.7	18.9	0.14	1.66	14.7	100.1	84	14	1
2401D-42	64.7	19.1	0.24	1.06	15.4	100.5	88	9	2
2401D-43	64.6	18.7	0.19	1.37	15.0	99.8	86	12	2
2401D-44	64.1	19.0	0.23	1.39	15.0	99.7	86	12	2
2401D-44	64.0	20.1	0.22	1.39	15.0	100.8	86	12	2
2401D-45	65.2	18.8	0.25	1.72	14.6	100.6	83	15	2
2401D-46	64.7	18.9	0.39	1.98	13.9	100.0	79	17	4
2401D-47	64.6	18.8	0.25	1.43	15.1	100.1	85	12	2
2401D-48	64.3	18.9	0.15	1.52	14.9	99.7	85	13	1
2401D-49	65.1	18.7	0.08	1.36	15.3	100.5	87	12	1
2401D-50	64.5	19.0	bdl	1.58	15.2	100.3	86	14	0
2401D-51	65.0	18.9	0.34	1.43	14.9	100.6	84	12	3
2401D-52	63.7	20.0	1.48	2.07	13.0	100.1	70	17	13
2401D-55	64.4	19.1	0.10	1.30	15.1	100.1	88	11	1
3302-65	63.9	19.5	bdl	1.14	14.7	99.2	89	11	0
3302-66	63.7	19.3	0.41	1.16	14.7	99.3	86	10	4
3302-68	63.8	19.4	0.32	1.25	14.4	99.3	86	11	3
4220-5	64.7	18.2	0.30	0.94	15.2	99.4	89	8	3

Note: bdl—below detection limit.

more consistent K-feldspar composition, with a range from $Or_{86}Ab_{11}An_3$ to $Or_{89}Ab_8An_3$ (including perthite lamellae too small to be separately analyzed). Analyses from sample 2401 range from $Or_{70}Ab_{17}An_{13}$ to $Or_{88}Ab_{11}An_1$. The difference of 18 mol% Or is likely due to the exsolution of a sodic plagioclase from the initially homogenous ternary feldspar.

Cordierite from outcrop 1659 in the complex varies in Fe/(Fe + Mg) ratio from 0.12 to 0.17 (Table 4). No regular zonation pattern was detected. Toward contacts with garnet, the mol% Mg in cordierite increases slightly (Fig. 7, A).

Orthopyroxene analyses are listed in Table 5. Figure 7, B, illustrates a profile across an orthopyroxene grain from aluminous gneiss 1659 in the complex. The Fe/(Fe + Mg) ratios vary from 0.33 (adjacent to garnet inclusions) to 0.41 (porphyroblast core), suggesting exchange with adjacent ferromagnesian minerals.

Spinel, observed only in thin sections from outcrop 2401 in the complex, is a hercynite-spinel solid solution (X_{Fe} 0.51 to 0.56, X_{Mg} 0.29 to 0.39) with a minor gahnite component (X_{Zn} 0.06 to 0.15; Table 6). Addition of zinc increases the stability field of spinel in the KFMASH system (Nichols et al., 1992). The magnitude of this effect was calculated, and the temperature of the spinel-producing reaction was reduced accordingly by ~70 °C (Fig. 9).

Biotite chemistry differs significantly between aluminous rocks of the complex and metapelites of the deformation zone, probably reflecting host-rock chemistry. In pelitic rocks, biotite plots above Fe/(Fe + Mg) = 0.38. In contrast, biotite from hydrothermally leached aluminous gneisses plot dominantly in the phlogopite field below Fe/(Fe + Mg) = 0.35 (Table 7, Fig. 10). No zonation was observed in biotite.

Garnet Zonation Profiles

Garnet (Tables 8 and 9) is predominantly an almandine solid solution (X_{Fe} 0.50–0.78) with large amounts of pyrope (X_{Mg} 0.14–0.48) and smaller fractions of grossular (X_{Ca} 0.00–0.16) and spessartine (X_{Mn} < 0.05) components. Therefore, garnet is considered essentially ternary (Alm, Prp, and Grs). Five representative transects, from both the gneiss complex and deformation zone, conducted to examine effects of post-peak metamorphic diffusion in garnet, revealed U-shaped almandine profiles, bell-shaped pyrope profiles, and flat spessartine profiles, suggesting the influence of retrograde cation exchange (Fig. 7, D and G; Boggs, 1996).

Grossular profiles are less consistent (Fig. 7, D and G). Samples 4209 and 2401D have an X_{Grs} profile that increases inward from the rims up to 5%. In 4209, this increase is grad-

TABLE 4. REPRESENTATIVE CORDIERITE ANALYSES

Sample	Weight percent oxides						
number	SiO_2	Al_2O_3	FeO	MgO	Total	X_{Fe}	X_{Mg}
1659-30	48.1	33.3	3.18	10.6	95.2	0.14	0.86
1659-31	48.3	33.0	3.36	10.4	95.1	0.15	0.85
1659-33	48.9	33.0	3.44	10.4	95.7	0.16	0.84
1659-34	48.4	33.4	3.38	10.6	95.8	0.15	0.85
1659-36	48.4	33.2	3.78	10.6	96.0	0.17	0.83
1659-38	48.8	32.9	3.40	10.2	95.4	0.16	0.84
1659-39	48.4	32.8	3.73	10.5	95.4	0.17	0.83
1659-40	48.6	33.1	3.31	10.6	95.5	0.15	0.85
1659-41	48.2	33.1	3.49	10.6	95.4	0.16	0.84
1659-72	49.1	33.7	3.34	10.7	96.8	0.15	0.85
1659-73	49.1	33.3	3.45	10.7	96.5	0.15	0.85
1659-74	49.3	33.6	3.39	10.3	96.6	0.16	0.84
1659-77	49.4	33.5	3.30	10.5	96.7	0.15	0.85
1659-31	49.6	33.4	2.93	10.6	96.6	0.14	0.86
1659-7	49.3	33.3	2.29	10.7	95.7	0.11	0.89
1659-13	49.7	33.5	2.83	10.5	96.5	0.13	0.87
1659-15	49.3	33.8	2.67	10.6	96.4	0.12	0.88
1659-10	49.2	33.5	3.02	10.2	95.9	0.14	0.86
1659-22	49.3	33.4	2.94	10.2	95.8	0.14	0.86
1659-23	49.1	33.3	3.12	10.3	95.9	0.15	0.85
1659-24	49.7	33.4	3.52	10.2	96.8	0.16	0.84
1659-25	49.2	33.6	3.38	10.5	96.6	0.15	0.85
1659-27	49.1	33.3	3.27	10.2	95.8	0.15	0.85
1659-28	49.3	33.6	2.95	10.1	96.0	0.14	0.86
1659-29	48.9	33.3	3.08	10.2	95.5	0.14	0.86
1659-30	49.0	33.3	3.45	10.2	95.9	0.16	0.84
1659-66	48.7	33.5	2.71	11.1	95.9	0.12	0.88

TABLE 5. REPRESENTATIVE ORTHOPYROXENE ANALYSES

Sample number	Weight percent oxides							
	SiO_2	Al_2O_3	TiO_2	FeO	MgO	Total	X_{Fe}	X_{Mg}
1659-9	48.3	6.55	0.11	23.6	20.6	99.2	0.39	0.61
1659-12	47.5	6.29	0.13	24.4	20.6	98.8	0.40	0.60
1659-22	49.1	7.33	bdl	20.8	22.5	99.8	0.34	0.66
1659-23	48.8	7.66	bdl	20.4	22.3	99.2	0.34	0.66
1659-24	49.2	7.50	0.11	21.4	22.2	100.3	0.35	0.65
1659-25	48.8	7.59	bdl	21.0	22.3	99.8	0.35	0.65
1659-26	49.0	6.78	0.12	20.7	22.3	99.0	0.34	0.66
1659-27	49.9	6.15	bdl	20.6	23.3	99.9	0.33	0.67
1659-17	47.6	6.21	bdl	24.1	21.2	99.0	0.39	0.61
1659-8	47.9	6.62	bdl	23.5	20.8	98.9	0.39	0.61
1659-17	46.7	5.59	0.65	25.9	20.8	99.6	0.41	0.59
1659-34	49.1	5.71	bdl	22.6	21.8	99.2	0.37	0.63
1659-37	49.3	5.66	bdl	21.6	22.3	98.9	0.35	0.65
1659-38	49.4	5.50	0.08	21.8	22.0	98.8	0.36	0.64
1659-76	48.6	6.39	bdl	23.5	20.7	99.2	0.39	0.61
1659-78	49.2	6.38	bdl	23.3	21.0	99.9	0.38	0.62

Note: bdl—below detection limit.

ual. In sample 2401D, the increase is abrupt, corresponding to the transition from an inclusion-rich rim to an inclusion-free core. Two distinct growth episodes are inferred for garnet in sample 2401D (Figs. 6, A; 7, F and G; and 8).

Indares and Martignole (1990a) concluded that the garnet zonation profiles they documented in the Central Metasedimentary Belt are suggestive of isothermal decompression after peak metamorphic conditions. Their observed outward decrease in Ca, without a significant variation in Mg, is similar to the zonation pattern observed in the inclusion-free core of garnet from outcrop 2401 (Fig. 7, G). This proposed isothermal decompression is compatible with the orogenic collapse event proposed by Harris et al. (2001), which is supported by rapid cooling through 800 ± 50 °C at 1.18–1.17 Ga (Bingen and van Breemen, 1998). The outer spinel inclusion-rich zone of garnet 2401D could reflect the end of the rapid cooling and the start of the 1.17–1.16 Ga overprint (Corriveau and van Breemen, 2000).

PETROGENETIC GRIDS, MELT-DERIVED ASSEMBLAGES, AND EXPERIMENTAL STUDIES

Migmatites are formed from multiple processes involving metasomatism, melt injection, metamorphic differentiation, or partial melting, or a combination of those processes (Olsen, 1983). For most rocks of pelitic composition the progression through fluid-absent melting reactions occurs within a fairly narrow temperature interval. Due to the slow diffusion rates of H_2O in H_2O-undersaturated melts and the endothermic nature of these reactions, evidence of initial stages of melting is preserved in many migmatites (Le Breton and Thompson, 1988).

Melting of pelites in the middle and deep crust involves the breakdown of mica. Fluid-absent melting reactions were studied experimentally in great detail (e.g., Vielzeuf and Holloway, 1988; Puziewicz and Johannes, 1990; Patiño-Douce and

TABLE 6. REPRESENTATIVE SPINEL ANALYSES

Sample number	Weight percent oxides									
	SiO_2	Al_2O_3	FeO	MgO	CoO	ZnO	Total	X_{Fe}	X_{Mg}	X_{Zn}
2401D-59	bdl	62.3	22.6	9.66	bdl	5.04	99.6	0.51	0.39	0.10
2401D-65	bdl	62.0	24.6	8.59	bdl	4.44	99.7	0.56	0.35	0.09
2401D-67	0.67	61.3	22.6	7.68	0.17	6.63	99.0	0.54	0.33	0.14
2401D-71	bdl	62.1	25.9	9.84	bdl	2.96	101.0	0.56	0.38	0.06
2401D-50	bdl	62.9	25.1	8.84	bdl	4.31	101.0	0.56	0.35	0.09
2401D-32	0.14	59.2	26.0	7.38	bdl	7.91	101.0	0.56	0.29	0.15

Note: bdl—below detection limit.

TABLE 7. REPRESENTATIVE BIOTITE ANALYSES

Sample	Weight percent oxides											
number	SiO_2	Al_2O_3	TiO_2	FeO	MgO	MnO	CaO	Na_2O	K_2O	F	Cl	Total
1280-25	37.8	17.1	1.93	10.6	17.8	bdl	0.08	0.28	9.83	na	na	95.5
1280-36	36.1	16.5	4.14	18.0	11.3	bdl	0.07	0.16	9.61	na	na	96.0
1280-48	36.3	16.6	4.20	17.5	11.3	bdl	0.06	0.29	9.73	na	na	95.9
1280-56	36.5	16.6	4.12	17.4	11.5	0.14	0.16	0.29	9.37	na	na	96.0
1280-59	36.5	16.3	4.02	17.4	11.1	0.11	0.12	0.24	9.73	na	na	95.5
1280-1*	35.2	15.8	4.13	18.2	10.9	0.03	0.03	0.11	11.50	0.143	0.009	96.0
1280-2*	35.0	16.0	4.07	18.2	10.8	0.05	0.02	0.06	11.36	0.134	0.008	95.7
1280-3*	35.4	16.0	3.97	18.0	10.9	0.02	0.02	0.08	11.67	0.135	0.017	96.1
1659-50	40.8	17.5	0.65	7.0	21.4	bdl	0.27	bdl	8.50	na	na	96.2
1659-43	39.9	16.1	1.78	9.0	20.4	bdl	0.27	bdl	8.75	na	na	96.1
1659-41	39.7	16.7	1.73	9.2	20.5	bdl	0.23	bdl	8.40	na	na	96.4
1659-39	39.8	16.3	1.58	9.1	20.8	bdl	0.28	bdl	8.82	na	na	96.6
1659-1*	38.7	15.7	1.50	9.2	19.8	0.02	bdl	0.41	10.51	1.289	bdl	97.1
1659-2*	38.6	15.8	1.71	9.3	19.7	0.013	0.02	0.40	10.37	1.442	0.009	97.4
1659-3*	38.2	15.9	1.70	9.3	19.3	bdl	0.05	0.35	10.30	1.366	0.009	96.4
1659-4*	38.3	15.9	1.70	9.1	19.6	bdl	0.05	0.42	10.32	1.646	0.007	96.9
1659-5*	38.8	16.2	1.84	9.1	19.8	bdl	0.03	0.36	10.27	1.322	0.005	97.6
1659-6*	39.0	16.0	1.48	8.4	20.5	0.008	bdl	0.42	10.56	1.376	bdl	97.7
2039-1*	36.2	15.9	2.82	16.8	12.7	0.17	0.02	0.16	11.35	0.275	bdl	96.4
2039-2*	36.1	15.6	2.75	17.0	12.7	0.15	0.01	0.17	11.25	0.30	bdl	95.9
2039-4*	36.1	15.8	2.82	16.7	12.6	0.19	0.03	0.14	11.46	0.336	bdl	96.2
2401-18	37.0	18.6	4.16	12.4	14.0	0.17	bdl	0.26	10.10	na	na	96.7
2401-101	36.3	16.9	4.96	11.4	16.8	bdl	bdl	bdl	9.62	na	na	96.0
2401-134	36.6	18.6	3.99	10.1	17.2	bdl	bdl	bdl	10.20	na	na	96.6
2401-135	36.5	17.9	4.42	10.1	17.1	bdl	bdl	bdl	10.10	na	na	96.1
2401-1*	35.6	16.6	3.95	13.7	14.3	bdl	bdl	0.13	12.03	0.469	0.015	96.8
2401-2*	35.5	16.6	3.96	13.7	14.4	0.01	0.01	0.13	12.03	0.40	0.019	96.8
2401-3*	36.1	16.4	5.22	12.7	14.9	0.02	0.03	0.18	12.06	0.372	0.021	98.0
2770-7	35.8	19.8	3.59	17.3	10.0	bdl	0.11	0.38	9.70	na	na	96.7
2770-17	35.8	20.0	3.59	17.4	9.9	bdl	0.10	0.29	9.50	na	na	96.5
2770-20	34.9	19.3	3.11	18.6	9.6	bdl	0.07	0.34	9.14	na	na	95.1
2770-22	35.6	20.8	2.78	18.0	9.9	bdl	0.26	0.27	9.22	na	na	96.9
3302-44	38.8	17.0	3.63	7.2	19.4	bdl	0.29	bdl	9.24	na	na	95.6
3302-32	38.6	17.6	4.16	7.9	18.6	bdl	bdl	bdl	9.61	na	na	96.5
3302-32	38.2	17.4	4.09	7.9	18.3	bdl	bdl	bdl	9.94	na	na	95.8
3302-32	37.6	17.7	3.82	7.8	18.6	bdl	bdl	bdl	9.74	na	na	95.3
4209-5	36.5	18.0	4.40	16.9	11.0	bdl	bdl	bdl	9.93	na	na	96.7
4209-6	36.6	18.1	4.56	16.2	10.9	bdl	bdl	bdl	9.49	na	na	95.8
4209-10	37.0	17.9	4.42	16.3	11.6	bdl	bdl	bdl	9.82	na	na	97.0
4209-7	36.3	18.0	4.54	16.2	11.1	bdl	bdl	bdl	9.51	na	na	95.7
4209-8	36.7	18.2	4.41	16.1	11.1	bdl	bdl	bdl	9.49	na	na	96.0

Note: bdl—below detection limit; na—not analyzed; *—second set of analyses.

Johnston, 1991; Vielzeuf and Montel, 1994; Clemens, 1995). Determination of the series of reactions that take place provides precise limits on pressures and temperatures of melting and verification of results obtained from conventional geothermobarometers.

Experimental results at 10 kilobars from Patiño-Douce and Johnston (1991) suggest progressive melting through three vapor-absent reactions (Bt_2 is less hydrous than Bt_1):

$$Ms + Bt_1 + Pl + Qtz = Bt_2 + Als + melt \qquad (K)$$

$$Bt + Als + Qtz \pm Ilm = Grt \pm Rt + melt \qquad (L)$$

$$Grt + Als = Spl + Qtz + melt \qquad (M)$$

The muscovite dehydration-melting reaction (K) is complete between 800 and 820 °C. Between 825 and 975 °C biotite reacts out via the biotite-dehydration-melting reaction (L). In the complex, all muscovite has reacted out, but biotite remains, implying a peak temperature of between 820 and 975 °C (Patiño-Douce and Johnston, 1991).

TABLE 8. REPRESENTATIVE GARNET ANALYSES

Sample number	Weight percent oxides							
	SiO_2	Al_2O_3	TiO_2	FeO	MgO	MnO	CaO	Total
1280-1	37.1	21.5	0.08	32.1	0.71	5.35	3.29	100.1
1280-2	37.4	21.4	bdl	31.4	0.98	5.67	3.34	100.2
1280-3	37.3	21.4	0.07	31.2	0.94	5.58	3.49	99.9
1280-4	37.2	21.8	bdl	31.3	0.97	5.78	3.61	100.6
1280-5	37.5	21.4	bdl	31.1	0.87	5.40	3.62	100.0
1659-29	38.6	21.6	bdl	27.2	11.90	bdl	0.41	99.8
1659-30	38.6	21.8	bdl	26.8	11.80	bdl	0.48	99.4
1659-31	38.9	21.9	bdl	26.7	11.60	0.45	0.34	99.9
1659-28	38.7	21.9	bdl	26.2	11.61	0.31	0.45	99.2
2401D-1	38.1	23.0	bdl	26.3	9.33	0.97	3.07	100.8
2401D-3	38.3	23.0	bdl	25.2	9.26	1.17	3.70	100.6
2401D-5	38.0	22.8	bdl	27.5	9.11	1.16	2.27	100.8
2401D-10	37.9	22.5	bdl	28.7	9.04	0.92	1.86	100.9
2401D-11	37.9	22.8	bdl	28.0	9.12	1.06	1.67	100.6
2770-4	36.5	21.8	bdl	35.4	3.79	1.15	1.48	100.1
2770-1	36.4	21.7	bdl	34.9	4.48	0.92	1.48	99.8
2770-2	37.0	21.9	bdl	33.8	5.07	0.86	1.67	100.4
2770-3	37.0	21.8	bdl	34.1	5.10	0.97	1.53	100.5
2770-30	36.5	21.9	bdl	35.6	3.90	1.14	1.47	100.5
3302-1	38.6	22.4	0.14	25.9	10.54	0.43	1.66	99.7
3302-2	38.5	22.3	bdl	26.3	10.70	bdl	1.81	99.7
3302-3	38.0	22.3	bdl	26.3	10.29	bdl	2.12	98.9
3302-4	38.4	22.0	bdl	26.7	10.35	0.41	2.33	100.2
3302-5	27.8	22.6	bdl	26.5	10.10	0.49	2.41	99.8
4209-2	38.0	21.7	0.15	29.9	5.96	1.26	3.25	100.2
4209-3	38.1	22.0	bdl	30.4	5.57	1.16	3.14	100.4
4209-4	38.0	21.9	bdl	30.4	5.84	1.14	2.97	100.2
4209-5	38.0	22.0	bdl	30.9	5.50	1.11	2.95	100.5
4209-6	37.9	21.9	bdl	31.1	5.67	1.36	2.50	100.4

Note: bdl—below detection limit.

TABLE 9. REPRESENTATIVE END MEMBER COMPONENTS OF GARNET

Sample number	Core analyses				Rim analyses			
	X_{Alm}	X_{Prp}	X_{Sps}	X_{Grs}	X_{Alm}	X_{Prp}	X_{Sps}	X_{Grs}
1659-1	0.50	0.48	0.02	bdl	0.58	0.40	0.01	bdl
1659-2	0.51	0.48	0.01	bdl	0.57	0.43	bdl	bdl
1659-3	0.53	0.44	0.02	bdl	0.51	0.47	0.01	bdl
2401D-1	0.56	0.36	0.01	0.08	0.65	0.26	0.01	0.05
2401D-2	0.51	0.37	0.01	0.10	0.67	0.23	0.01	0.04
2401E	0.63	0.32	0.01	0.04	0.56	0.36	0.02	0.07
2770-1	0.74	0.19	0.02	0.05	0.77	0.15	0.03	0.05
2770-2	0.74	0.20	0.02	0.04	0.78	0.14	0.03	0.04
3045	0.60	0.25	0.03	0.13	0.55	0.31	0.02	0.11
3302-AB	0.57	0.35	0.01	0.07	0.62	0.35	0.02	0.05
3302-CD	0.58	0.33	0.01	0.08	0.64	0.30	0.03	0.05
3302-E	0.62	0.32	0.01	0.04	0.62	0.32	0.02	0.05
3302-Leuc	0.55	0.37	0.01	0.07	0.56	0.39	0.01	0.05
3333-36	0.67	0.25	0.05	0.02	0.72	0.19	0.04	0.05
3333-35	0.68	0.24	0.03	0.05	0.72	0.18	0.04	0.05
3354	0.54	0.41	0.01	0.03	0.64	0.32	bdl	0.03
3832-23	0.54	0.44	bdl	0.02	0.64	0.34	0.01	0.01
3382-17	0.55	0.04	bdl	0.01	0.57	0.41	0.01	0.02
3832-16	0.56	0.43	bdl	0.01	0.60	0.35	0.01	0.01
4209-13	0.64	0.25	0.03	0.10	0.72	0.17	0.04	0.04
4209-16	0.71	0.20	0.03	0.06	0.76	0.15	0.04	0.06
4209-17	0.68	0.22	0.02	0.04	0.73	0.16	0.04	0.06
4220-19	0.65	0.20	0.02	0.13	0.70	0.15	0.01	0.16
4220-20	0.69	0.19	0.03	0.10	0.67	0.15	0.01	0.16
4220-21	0.65	0.21	0.02	0.13	0.70	0.15	0.02	0.15
4475D	0.56	0.43	bdl	0.01	0.69	0.28	0.02	0.02

Note: For mineral abbreviations see Appendix 1; bdl—below detection limit; Leuc—leucosome analysis.

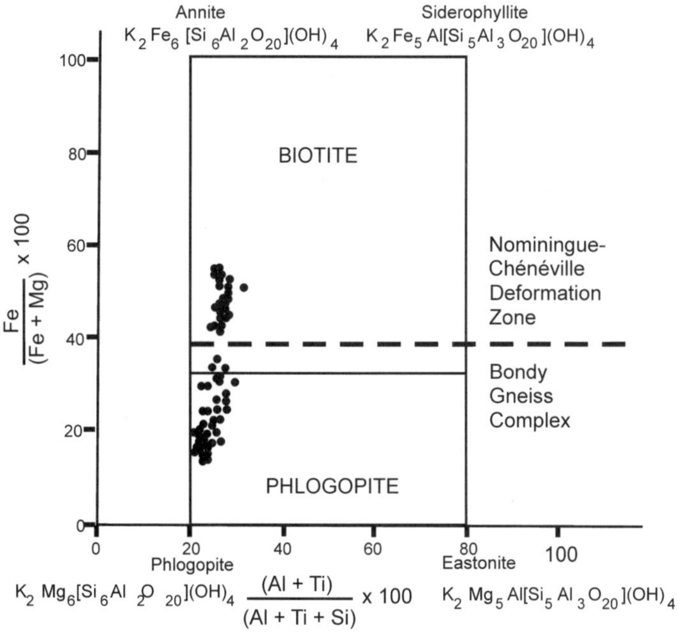

Figure 10. Plot of biotite composition (modified after Deer et al., 1966).

GEOTHERMOBAROMETRIC RESULTS

Calculations using both the biotite-garnet Fe-Mg exchange thermometer (Fig. 7E, H, K, and N; Table 10; Ferry and Spear, 1978; Perchuk and Lavrent'eva, 1983) and the garnet-orthopyroxene thermometer (Table 11; Harley, 1984a; Sen and Bhattacharya, 1984) resulted in a calculated peak metamorphic temperature of 950 °C in the complex (Fig. 9, I). Above 530 °C, calculated temperatures from the Ferry and Spear thermometer were greater than those of the Perchuk and Lavrent'eva thermometer. Below 530 °C, the reverse was true.

Results (Fig. 7) from the Ferry and Spear thermometer for four garnets (thin sections of 4209 in the deformation zone, 2401D2 and 2401E in the complex, and 3832 in the quartzite-rich domain) were chosen because they occur in samples with 15–50% biotite. These higher concentrations of biotite provide an "infinite reservoir" for influx of the Fe/Mg ratio into garnet (Spear and Florence, 1992). The biotites of this study are not zoned, which supports the presence of this infinite reservoir. Ranges of calculated temperatures occur because each garnet analysis was paired with several biotite analyses from the same thin section (e.g., garnet analyses from rims of garnet porphyroblasts were paired with all biotite analyses from rims of biotite porphyroblasts). Where biotites are included in garnet, all analyses of these biotite inclusions were paired with garnet analyses near these inclusions (i.e., garnet 3832 and the inclusion-rich rim of garnet 2401D). Garnet analyses from the inclusion-free core of garnet 2401D and the inclusion-free portions of other garnets were paired with all biotite analyses from cores of biotite porphyroblasts. All appropriate biotite analyses were used in these pairings because of the possibility that Fe/Mg diffusion continued in some biotite cores after diffusion had ceased in the garnet cores (Spear and Florence, 1992).

TABLE 10. BIOTITE-GARNET GEOTHERMOMETRY RESULTS

Thin section (garnet)	Analysis location (point on garnet)	Geothermometer calculation results: FS range (°C)	Average (°C)	Peak *T* subtract Av *T*	PL range (°C)	Average (°C)	Δ (FS-PL)
2039	C-6	763–788	778	172	608–619	614	164
(10 mm,	C-7	731–755	746	204	291–602	598	148
small garnet)	C-8	697–720	711	239	579–589	585	126
	M-4	727–751	741	209	591–602	598	143
	M-9	671–192	684	266	565–575	571	113
	R-1	654–675	667	283	557–567	563	104
	R-2	709–732	723	227	583–593	588	135
	R-3	716–740	730	220	585–595	591	139
	R-13	684–706	697	253	572–582	578	119
2770	C-27	572–744	682	268	551–637	606	76
(10 mm,	C-3	589–769	704	246	559–646	615	89
small garnet)	C-4	593–777	710	240	561–659	618	92
	C-5	616–809	739	211	573–665	632	107
	M-2	588–767	702	248	559–646	615	87
	M-29	524–676	621	329	523–604	575	46
	M-2	616–810	739	211	575–666	634	105
	M-21	550–714	655	295	538–621	591	64
	R-1	528–607	613	337	520–599	571	42
	R-6	514–661	608	342	517–596	568	40
	R-34	581–758	694	256	539–622	593	101
	R-22	545–706	648	302	536–620	590	58

(continued)

TABLE 10. *Continued*

Thin section (garnet)	Analysis location (point on garnet)	Geothermometer calculation results					
		FS range (°C)	Average (°C)	Peak *T* subtract Av *T*	PL range (°C)	Average (°C)	Δ (FS-PL)
3045	C-22	657–741	703	247	587–626	610	93
(7 mm,	C-6	629–707	671	279	572–610	594	77
small garnet)	C-32	707–801	758	192	615–656	638	120
	C-30	653–736	698	252	587–626	610	88
	C-9	603–678	644	306	559–595	580	64
	C-19	658–742	703	247	590–629	613	90
	R-1	649–731	694	256	584–623	606	88
	R-2	670–756	717	233	592–631	614	103
	R-20	662–747	708	242	593–632	616	92
	R-18	656–739	701	249	590–629	613	88
	R-16	661–746	707	243	592–631	614	93
3302	LC-6	480–757	649	301	500–635	589	60
(10 mm,	LC-7	490–773	662	288	509–647	599	63
small garnet)	LC-8	494–780	668	282	511–650	603	65
	LC-21	491–775	664	286	505–642	595	69
	LC-22	492–777	665	285	507–644	597	68
	LM-10	494–781	669	281	511–651	603	66
	LM-57	492–778	666	284	507–645	598	68
	LM-10	435–674	581	369	472–597	554	27
	LM-13	457–713	613	337	484–613	569	44
	LM-3	452–703	605	345	482–610	566	39
	LR-1	495–784	671	279	509–648	600	71
	LR-14	481–757	650	300	504–641	594	56
	LR-17	448–697	600	350	479–607	563	37
	LR-11	448–697	600	350	479–606	562	38
	LR-1	449–698	601	349	479–607	563	38
	MC-22	400–611	529	421	469–592	550	–21
	MC-21	415–637	551	399	459–579	538	13
	MC-49	415–638	552	398	460–580	539	13
	MC-50	416–640	553	397	461–581	540	13
	MC-51	418–642	555	395	461–581	540	15
	MM-46	418–643	556	394	461–582	540	16
	MM-47	424–654	565	385	465–587	545	20
	MM-53	420–647	559	391	464–585	543	16
	MM-54	420–646	558	392	462–584	542	16
	MM-62	388–591	513	437	443–557	517	–4
	MR-63	391–595	516	434	444–558	519	–3
	MR-65	347–522	455	495	415–519	483	–28
	MR-68	403–616	533	417	454–573	532	1
	MR-18	400–611	529	421	450–567	527	2
	MR-33	414–636	550	400	459–579	538	12
3333A	R-53	529–698	660	290	526–612	594	66
(20 mm,	R-46	545–721	682	268	535–624	605	77
intermed.	R-47	562–746	705	245	545–636	617	88
garnet,	R-48	543–719	679	271	534–622	603	76
deeply	R-49	537–709	671	279	530–617	598	73
embayed)	R-40	556–738	697	253	540–630	611	86
	R-46	565–751	709	241	545–636	617	92
	R-48	567–753	711	239	546–638	618	93
	R-53	578–771	727	223	552–644	624	103
	R-24	575–766	723	227	552–644	624	99
3354	C-10	565–742	653	297	552–640	597	56
(10 mm,	C-11	549–718	632	318	539–624	583	49
small garnet)	C-12	572–752	661	289	552–640	597	64

TABLE 10. *Continued*

Thin section (garnet)	Analysis location (point on garnet)	Geothermometer calculation results: FS range (°C)	Average (°C)	Peak *T* subtract Av *T*	PL range (°C)	Average (°C)	Δ (FS-PL)
	C-31	523–681	601	349	528–610	570	31
	C-32	530–690	609	341	529–611	571	38
	M-7	539–704	621	329	534–618	577	44
	M-17	552–723	637	313	541–627	585	52
	M-18	521–678	599	351	527–610	569	30
	M-38	493–638	565	385	511–590	551	14
	M-28	498–645	571	379	514–593	554	17
	M-46	518–673	595	355	525–607	567	28
	R-1	458–588	523	427	490–564	528	–5
	R-22	521–678	599	351	514–593	554	45
	R-43	498–645	571	379	511–590	551	20
	R-55	510–662	586	364	518–598	559	27
	R-41	496–642	569	381	513–592	553	16
4012	C-17	537–708	636	314	526–613	578	58
(20 mm,	C-19	520–683	615	335	516–600	566	49
intermed.	M-93	506–664	598	352	509–591	558	40
garnet,	M-92	494–646	583	367	502–583	551	32
deeply	M-90	491–642	579	371	502–583	551	28
embayed)	M-88	533–702	631	319	522–608	574	57
	M-22	541–716	643	307	528–615	580	63
	R-96	523–688	619	331	518–602	568	51
	R-84	496–649	585	365	502–584	552	33
	R-1	537–709	637	313	526–613	578	59
	R-79	523–688	619	331	518–602	568	51
	R-80	527–693	624	326	520–605	571	53
	R-82	502–658	593	357	506–588	555	38
	R-2	472–614	555	395	489–567	536	19
	R-72	471–613	554	396	488–566	535	19
4475A	C-3	560–650	618	332	545–593	576	42
(110 mm,	C-33	576–670	636	314	555–603	586	50
large garnet,	C-4	549–636	605	345	540–586	570	35
full Ilm incls.)	M-26	614–716	680	270	575–626	608	72
	M-13	611–713	677	273	573–624	606	71
	M-22	490–564	537	413	504–547	531	6
	M-7	499–575	548	402	511–554	538	10
	M-35	590–686	652	298	561–610	593	59
	R-37	436–500	477	473	471–509	495	–18
	R-9	428–490	468	482	465–503	489	–21
	R-8	530–613	583	367	529–574	558	25
	R-10	431–494	472	478	468–507	493	–21
4475D2	C-16	448–597	505	445	481–565	514	–9
(120 mm,	C-61	456–608	514	436	488–574	522	–8
large garnet,	C-62	453–604	511	439	486–572	520	–9
full Ilm incls.)	C-21	451–600	508	442	485–570	519	–11
	C-22	450–599	507	443	484–569	518	–11
	M-4	430–571	484	466	472–554	505	–21
	M-5	429–570	483	467	471–553	504	–21
	M-69	436–580	492	458	476–559	509	–17
	M-71	422–559	475	475	467–548	499	–24
	M-27	435–578	490	460	475–558	508	–18
	R-78	385–506	432	518	441–516	470	–38
	R-79	394–519	442	508	445–521	476	–34
	R-47	357–467	400	550	421–491	449	–49
	R-38	370–489	415	535	429–502	458	–43
	R-39	366–480	410	540	427–499	456	–46

(continued)

TABLE 10. *Continued*

Thin section (garnet)	Analysis location (point on garnet)	Geothermometer calculation results FS range (°C)	Average (°C)	Peak *T* subtract Av *T*	PL range (°C)	Average (°C)	Δ (FS-PL)
4475D3	C-3	524–670	605	345	528–606	572	33
(135 mm,	C-4	532–681	615	335	534–612	578	37
large garnet,	C-5	523–668	603	347	528–605	571	32
full Ilm incls.)	C-21	524–669	604	346	528–606	572	32
	C-22	522–667	602	348	528–605	571	31
	M-9	527–674	609	341	531–609	575	34
	M-11	523–668	604	346	528–606	572	32
	M-13	518–661	598	352	526–602	569	29
	M-15	504–642	581	369	517–593	560	21
	M-72	479–607	550	400	502–574	542	8
	M-73	463–586	532	418	493–563	532	0
	R-22	430–542	493	457	470–536	507	–14
	R-38	443–559	508	442	477–544	515	–7
	R-65	407–510	464	486	454–518	490	–26
	R-1	384–480	437	513	439–499	473	–36

Note: L—leucosome; M—melanosome; C—core; M (by itself or second)—mid; R—rim.
? (FS–PL)—difference between calculated values from two geothermometers.
Thin section (garnet)—description of garnet; mm is estimation of 2D size of garnet; Ilm—ilmenite.
T—temperature calculated from geothermometer; Av—average.
FS—Ferry and Spear (1978); PL—Perchuk and Lavrent'eva (1983).

TABLE 11. GARNET-OPRTHOPYROXENE GEOTHERMOMETER RESULTS

Garnet	Harley C	S & B C	Lal C
Core #40	1009	1010	795
	941	942	751
	963	964	768
Core #41	992	991	785
	926	925	743
	947	946	759
Rim #35	932	930	745
	872	870	707
	891	890	722
Rim #36	947	947	756
	885	886	717
	905	906	732
Rim #34	979	978	774
	915	913	733
	936	934	749

Note: Analyses from sample 1659.
C—degrees Celsius, calculated from geothermometer; Harley—Harley (1984b); S & B—Sen and Bhattacharya (1984); Lal—Lal (1983).

Garnet 2401E is poikiloblastic, with ilmenite and spinel inclusions throughout (Figs. 6, H, and 7, L); garnet grains 4209 and 3832 are small (<10 mm^2; Fig. 7, C and I). The inclusion-free core of garnet 2401D2 (Fig. 8) preserved evidence of a calculated peak metamorphic temperature of 950 °C, 150 °C greater than that of the adjacent inclusion-rich zone of the same garnet (Fig. 7, F and H). This peak metamorphic temperature was also 300–350 °C greater than that of the cores of smaller garnets 4209 (Fig. 7, E) and 3832 (Fig. 7, K), and 200–250 °C greater than that of the large poikiloblastic garnet 2401E (Fig. 7, N), which is full of spinel and ilmenite inclusions.

The major proportion of garnet was formed in the presence of plagioclase (as in Fig. 6, D), during the prograde portion of the *P-T* trajectory. Indeed, the highest-temperature melt-forming reaction generated garnet as a residual phase. This suggests that the garnet-aluminosilicate-quartz-plagioclase barometer would record the maximum pressure. Koziol and Newton's (1988) calibration, used because it requires the least extrapolation, resulted in a calculated maximum pressure of ~10 kilobars (9.9 ± 0.6 kilobars; see sample 2401; Fig. 9, II) for the maximum temperature of 950 °C (Table 12).

Harley's (1984b) calibration of the barometer based on Al solubility in orthopyroxene was used because it is the most recent and covers the widest compositional system. The maximum temperature of 950 °C was used to determine a lower limit for the maximum pressure of 9.3 kilobars (Table 13) for the beginning of the decompression episode in both the deformation zone and the complex (Fig. 9, between I and II).

The calibration by Nichols et al. (1992) of the univariant pressure-sensitive reaction

$$\text{Als} + \text{Grt} = \text{Spl} + \text{Qtz} \quad \text{(N)}$$

was used with Berman's (1990) garnet model and the spinel model of Nichols et al. (1992) to calculate 8.4 kilobars at 950 °C, between points II and III in Figure 9 (Table 14).

TABLE 12. GARNET-ALUMINOSILCATE-SILICA-PLAGIOCLASE GEOBAROMETER RESULTS

Calc. *P* sample	750 °C (kbar)	850 °C (kbar)	950 °C (kbar)	Calc. *P* sample	750 °C (kbar)	850 °C (kbar)	950 °C (kbar)
2401D2	5.39	8.48	10.95	4012	6.18	9.23	10.42
	5.42	7.66	10.12		6.21	9.26	10.46
	5.37	7.52	9.98		6.23	9.08	10.47
	5.36	7.42	9.89		5.93	8.67	10.27
	5.40	7.50	9.97		5.89	7.50	10.24
					6.19	9.32	10.44
2401E	4.49	7.20	9.43		6.19	9.02	10.45
	4.50	7.73	9.45		5.92	8.31	10.26
	4.48	7.43	9.43		7.86	10.22	11.61
	4.47	8.00	9.42		5.90	8.71	10.24
3302				4209	6.60	9.46	11.93
Leuc.	10.60	13.25	15.33		6.65	9.54	11.97
	10.35	13.01	15.33		6.63	9.41	11.95
	10.73	13.38	15.35		6.72	10.29	12.01
	10.60	13.09	15.98		6.78	10.86	12.05
	10.92	13.43	15.99		6.79	11.31	12.06
					6.86	9.69	12.11
Mel.	11.19	13.87	15.23		6.64	9.30	11.96
	10.85	13.51	15.28		6.64	9.90	11.96
	78.91	10.55	15.31		6.62	9.86	11.95
	8.55	11.21	15.29		6.73	11.78	12.02
	8.27	10.93	15.30		6.75	10.73	12.04

Note: Leuc.—leucosome; Mel. —melanosome; C—degrees Celsius; Calc. *P*—pressure, calculated from Koziol and Newton (1988) geobarometer.

One cordierite breakdown reaction, based on the Fe-end members and suitable to the peraluminous assemblages in this study, was reformulated by Bhattacharya (1986):

$$\text{Fe-Crd} = \text{Alm} + \text{Als} + \text{Qtz} \qquad (O)$$

Geobarometers based on the assemblage Crd + Grt + Sil + Qtz (Hutcheson et al., 1974; Thompson, 1976a, 1976b; Holdaway and Lee, 1977; Newton and Wood, 1979; Lonker, 1980; Martignole and Sisi, 1981; Bhattacharya and Sen, 1985) result in calculated pressures that disagree greatly (Bhattacharya, 1986). For simplicity, the Bhattacharya (1986) barometer was used because the results from this barometer are of the correct magnitude. Bhattacharya (1986) attributed these more realistic results to (1) more refined modeling of cordierite hydration at variable P-T-a_{H_2O} conditions, (2) improved activity-composition relations used to account for mixing in garnet, and (3) the use of the internally consistent thermodynamic database of pure phases for calibrating this geobarometer. Free water would have been consumed during the prograde path; therefore, the X_{H_2O} (molar water content) of cordierite would have been low (0.3–0.1). The resultant pressures are 5.3–6.3 kilobars at 800 °C (a close approximation to Fig. 9, point VI (sample 2401); Table 15).

The only evidence of retrograde conditions in the deformation zone is contained in mineral assemblages and textures (discussed previously), which were not suitable for geothermobarometric analysis. The feldspar thermometer was not used in this study because exsolution was highly variable.

DISCUSSION

Pressure-Temperature-Time Constraints

The proposed peak metamorphic temperature of 950 °C (Fig. 9, II) calculated for the complex is supported by the assemblage Spl + Grt + Sil (outcrop 2401), which occurs at temperatures greater than 850 °C (Vielzeuf and Holloway, 1988), and by the assemblage Opx + Sil (outcrops 1659 and 4956), which occurs at temperatures greater than 900 °C (Carrington and Harley, 1995).

Field observations, garnet zonation profiles, and garnet textures support isothermal decompression through the peak metamorphic temperature (Fig. 9, between II and III), which was originally proposed by Indares and Martignole (1984, 1990a). Therefore, the peak metamorphic temperature of 950 °C was used to calculate the peak metamorphic pressure of 9.9 ± 0.6 kilobars (Fig. 9, I). This high pressure is supported by the observed Opx-Sil assemblage (outcrops 1659 and 4956), which was constrained to greater than 8.5 kilobars by Carrington and Harley (1995). These peak conditions would correspond to the 1.20–1.18 Ga metamorphic zircon U-Pb ages (Boggs, 1996; Corriveau and van Breemen, 2000).

TABLE 13. ALUMINA IN ORTHOPYROXENE GEOBAROMETER RESULTS

Calc. *P* sample	750 °C (kbar)	850 °C (kbar)	950 °C (kbar)	Calc. *P* sample	750 °C (kbar)	850 °C (kbar)	950 °C (kbar)
Opx #34	4.86	6.16	7.57	Opx #25	5.41	6.78	8.27
	5.36	6.70	8.16		4.76	9.08	7.51
	5.83	7.22	8.72		5.34	9.71	8.19
	5.56	6.93	8.40		5.72	7.13	8.65
	5.76	7.15	8.64		6.22	7.67	9.23
	6.15	7.57	9.10		5.77	7.18	8.69
	5.64	7.02	8.50		5.63	7.03	8.54
Opx #37	4.21	5.47	6.84		6.85	8.37	10.00
	4.71	6.01	7.43		6.22	7.68	9.25
	5.18	6.53	7.99		5.97	7.41	8.96
	4.91	6.23	7.67				
	5.11	6.45	7.91	Opx #26	5.35	6.73	8.24
	5.49	6.88	8.38		4.71	6.03	7.48
	4.99	6.32	7.77		5.28	6.66	8.16
Opx #38	4.71	6.00	7.40		5.66	7.07	8.61
	5.20	6.54	7.99		6.16	7.62	9.19
	5.67	7.06	8.56		5.72	7.13	8.66
	5.41	6.76	8.23		5.57	6.98	8.50
	5.61	6.99	8.48		6.79	8.32	9.97
	5.99	7.41	8.94		6.16	7.63	9.21
	5.49	6.86	8.33		5.92	7.36	8.92
Opx #39	5.01	6.32	7.72				
	5.51	6.86	8.31	Opx #27	6.01	7.43	8.96
	5.98	7.37	8.87		5.37	6.73	8.19
	5.71	7.08	8.55		6.95	7.36	8.88
	5.91	7.30	8.79		6.32	7.77	9.33
	6.29	7.72	9.25		6.83	8.32	9.91
	5.79	7.17	8.64		6.38	7.82	9.37
Opx #17	7.21	8.66	10.20		6.24	7.68	9.22
	7.70	9.20	10.80		7.45	9.01	10.68
	8.17	9.72	11.36		6.82	8.32	9.93
	7.91	9.43	11.04		6.58	8.06	9.64
	8.11	9.65	11.28				
	8.49	10.07	11.74				
	7.99	9.52	11.13				

Note: Calc. *P*—calculated pressure, from Harley (1984b) and Harley and Green (1982) alumina in orthopyroxene barometers; Opx—orthopyroxene.

The position of spinel in the inclusion-rich garnet rims of outcrop 2401 implies that late stages of isothermal decompression likely occurred once the Als + Grt = Spl + Qtz univariant curve was reached (8 kilobars at 950 °C; Fig. 9, III). In sample 2401D2, spinel is surrounded by garnet and sillimanite and not in contact with quartz, suggesting that the Als + Grt = Spl + Qtz univariant curve was crossed in reverse (Fig. 9, IV). Cordierite surrounds garnet and sillimanite in outcrop 1659, suggesting that the retrograde portion of the *P-T-t* trajectory passed between the calculated 5.3 and 6.3 kilobars at 800 °C (Fig. 9, VI). Cooling below ~800 °C (monazite U-Pb, ID-TIMS ages) occurred at 1.18–1.17 Ga (Boggs, 1996; Bingen and van Breemen, 1998).

The absence of cordierite suggests that the retrograde pressures within the deformation zone were up to 0.5 kilobars higher than in the complex (Fig. 9, IV and VIII), possibly due to emplacement of monzonite-diorite complexes. The absence of observed muscovite with K-feldspar in the paragneisses of the Nominingue-Chénéville deformation zone places the retrograde portion of the *P-T-t* trajectory below the univariant curve that consumes K-feldspar (Fig. 9, IX).

Retrograde Cation Diffusion and Peak Metamorphic Conditions

The calculated peak metamorphic temperature of 950 °C for the complex is 150–200 °C above those documented by previous workers who examined regions strongly reactivated at 1.17–1.16 Ga (Anovitz and Essene, 1990; Indares and Martignole, 1990a; Streepey et al., 1997). This discrepancy can be attributed, at least in part, to retrograde cation exchange, a process that may have been particularly efficient in zones of severe amphibolite-facies reactivation such as observed in the deformation zone (Boggs, 1996; Corriveau et al., 1998). The U-

TABLE 14. GARNET-SPINEL GEOBAROMETER RESULTS

Garnet core calculations						Garnet rim calculations					
Spl-Prp 750 °C (kbar)	Hc-Alm 750 °C (kbar)	Spl-Prp 850 °C (kbar)	Hc-Alm 850 °C (kbar)	Spl-Prp 950 °C (kbar)	Hc-Alm 950 °C (kbar)	Spl-Prp 750 °C (kbar)	Hc-Alm 750 °C (kbar)	Spl-Prp 850 °C (kbar)	Hc-Alm 850 °C (kbar)	Spl-Prp 950 °C (kbar)	Hc-Alm 950 °C (kbar)
3.98	4.73	5.91	7.28	7.83	9.82	4.04	4.42	5.98	6.94	7.91	9.45
4.02	4.60	5.95	7.13	7.89	9.66	4.04	4.38	5.99	6.89	7.93	9.39
3.91	4.83	5.82	7.39	7.73	9.95	3.90	4.74	5.82	7.29	7.74	9.83
4.03	4.60	5.95	7.14	7.87	9.68	3.90	4.71	5.82	7.25	7.74	9.79
4.12	4.41	6.05	6.94	7.97	9.47	3.87	4.83	5.80	7.38	7.72	9.92
4.08	4.48	6.02	7.01	7.94	9.54	4.08	4.38	6.02	6.90	7.95	9.42
3.91	4.87	5.83	7.43	7.75	9.98	4.18	4.18	6.09	6.71	7.99	9.24
4.12	4.37	6.06	6.89	7.99	9.41	3.56	5.25	5.40	7.85	7.25	10.50
4.12	4.27	6.10	6.78	8.04	9.29	4.18	3.77	6.15	6.21	8.11	8.66
4.12	4.07	6.08	6.55	8.03	9.02	4.18	3.72	6.16	6.16	8.13	8.59
4.16	3.94	6.12	6.40	8.08	8.86	4.04	4.09	5.99	6.56	7.94	9.03
4.05	4.17	5.99	6.66	7.93	9.15	4.04	4.05	5.99	6.52	7.93	8.99
4.17	3.93	6.12	6.41	8.06	8.88	4.02	4.17	5.97	6.65	7.92	9.12
4.26	3.75	6.21	6.21	8.17	8.67	4.22	3.72	6.19	6.17	8.14	8.62
4.22	3.82	6.19	6.29	8.14	8.74	4.32	3.53	6.26	5.98	8.19	8.44
4.05	4.21	6.00	6.70	7.95	9.18	3.70	4.59	5.57	7.12	7.44	9.65
4.26	3.71	6.23	6.17	8.19	8.61	4.37	4.20	6.36	6.67	8.34	9.15
4.30	3.61	6.28	6.05	8.24	8.49	4.38	4.15	6.37	6.62	8.36	9.19
4.32	4.50	6.29	7.01	8.26	9.52	4.24	4.52	6.20	7.03	8.17	9.53
4.35	4.37	6.34	6.86	8.31	9.36	4.23	4.48	6.20	6.99	8.17	9.49
4.24	4.60	6.20	7.13	8.16	9.64	4.21	4.60	6.18	7.11	8.15	9.62

Note: For mineral abbreviations see Appendix 1. Calculated pressures after Nichols et al. (1992).

shaped almandine, bell-shaped pyrope, and flat spessartine profiles (Fig. 7, D and G) suggest the influence of retrograde cation diffusion (Boggs, 1996). Spear (1991), with numerical models, demonstrated that retrograde cation diffusion could cause preservation of chemical conditions representative of temperatures up to 350 °C below peak metamorphic conditions.

Plots of calculated temperatures versus distance across garnet showed that the calculated temperatures of garnet rims are significantly lower than those of garnet cores (Fig. 7, E, H, K, and N). Garnet grains under 3.5 mm^2 were completely rehomogenized, recording evidence of temperatures up to 350 °C below peak metamorphic temperatures (Fig. 7, K; Table 10). Most garnets in the deformation zone were larger than or equal to 20 mm^2 (other than the one in Fig. 7C), whereas many garnets in the gneiss complex were 40–140 mm^2 (Fig. 7, F and L).

Garnet grains larger than 50 mm^2 (from the gneiss complex) recorded evidence of temperatures 100–550 °C below the 950 °C peak metamorphic temperature (Fig. 7, H and N; Table 10). These larger garnet grains have high concentrations of ferromagnesian and feldspar inclusions (Figs. 6, H, and 8). Biotite inclusions in garnets are typically affected by an Fe-Mg exchange reaction with the surrounding host during retrograde metamorphism from middle- to high-grade metamorphic conditions (Spear and Florence, 1982; Spear, 1991). Whitney (1991) also demonstrated the possibility of postentrapment changes in plagioclase inclusions within garnets related to fractures in the host garnet. Compositional modification of both the host garnet and inclusions may be driven by development of microcracks at metamorphic temperatures (Whitney, 1991; Erambert and Austrheim, 1993; Hames and Menard, 1993; Whitney, 1996; Whitney and Dilek, 1998). Thus, the presence of these inclusions provided not only a focus for local Fe-Mg exchange but also a focus for the development of fractures (Whitney et al., 2000), thus opening these garnet cores to increased effects of retrograde cation diffusion (e.g., Fig. 7, G and H, versus 7, M and N).

The influence of inclusions on exposing inclusion-rich portions of garnet to the influence of retrograde cation exchange is visible in calcium and magnesium X-ray maps of garnet 2401D (Figs. 7, G, and 8). In the inclusion-rich rim of this garnet, molar concentrations of calcium and magnesium decreased. In the inclusion-free core of this garnet, molar concentrations of calcium and magnesium were higher.

In the gneiss complex, the end result of retrograde cation diffusion combined with reduced diffusion domain size is to locally decrease the temperatures calculated from the geothermometers (Boggs, 1996). Retrograde cation diffusion enhanced by grain size reduction and fluid infiltration associated with the 1.17–1.16 Ga deformation decrease the temperatures calculated from the geothermometers in the deformation zone. Careful selection of medium-size (here 15–50 mm^2), inclusion-free garnets is thus vital for determination of true peak metamorphic conditions when constructing *P-T-t* trajectories in granulite-facies terranes.

The calculated peak metamorphic pressure of 9.9 ± 0.6 kilobars in the complex is ~2 kilobars above those documented in

TABLE 15. GARNET-CORDIERITE GEOBAROMETER RESULTS

T a_{H_2O}	750 °C 0.1 (kbar)	750 °C 0.3 (kbar)	800 °C 0.1 (kbar)	800 °C 0.3 (kbar)	850 °C 0.1 (kbar)	850 °C 0.3 (kbar)
	4.43	5.27	4.36	5.25	4.31	5.23
	4.63	5.47	4.58	5.46	4.53	5.45
	4.59	5.43	4.54	5.42	4.48	5.40
	4.62	5.46	4.57	5.45	4.52	5.44
	4.64	5.48	4.59	5.47	4.54	5.46
	5.20	6.04	5.18	6.06	5.16	6.28
	5.41	6.25	5.40	6.27	5.38	6.30
	5.36	6.20	5.35	6.23	5.33	6.25
	5.40	6.24	5.39	6.26	5.37	6.29
	5.42	6.25	5.40	6.28	5.39	6.31
	4.53	5.37	4.47	5.35	4.42	5.34
	4.73	5.57	4.69	5.57	4.64	5.56
	4.69	5.53	4.64	5.52	4.59	5.51
	4.73	5.56	4.68	5.56	4.63	5.55
	4.74	5.58	4.70	5.57	4.65	5.57
	4.74	5.58	4.70	5.57	4.65	5.57
	4.74	5.78	4.91	5.79	4.87	5.79
	4.90	5.74	4.86	5.74	4.82	5.74
	4.94	5.77	4.90	5.78	4.86	5.78
	4.95	5.79	4.92	5.79	4.88	5.80
	7.71	8.63	7.62	8.50	7.53	8.37
	7.68	8.60	7.59	8.47	7.50	8.34
	7.70	8.62	7.61	8.49	7.52	8.36
	7.68	8.60	7.59	8.47	7.50	8.33
	7.62	8.54	7.53	8.40	7.44	8.28
	7.58	8.50	7.50	8.38	7.41	8.25
	7.61	8.53	7.52	8.40	7.44	8.27
	7.58	8.50	7.50	8.38	7.41	8.25
	7.57	8.49	7.49	8.36	7.40	8.24
	7.54	8.46	7.45	8.33	7.37	8.21
	7.56	8.48	7.47	8.35	7.39	8.23
	7.54	8.46	7.45	8.33	7.37	8.21

Note: T—temperature; a_{H_2O}—activity of water. Calculated pressures after Bhattacharya (1986).

previous studies (Fig. 4; Anovitz and Essene, 1990; Indares and Martignole, 1990b; Streepey et al., 1997). One possible interpretation is that the complex contains a deeper portion of the crust than those areas previously studied. However, because most of the area was reactivated after peak metamorphism (Corriveau et al., 1998), we attribute this discrepancy to the influence of retrograde cation diffusion assisted by deformation in large parts of the Central Metasedimentary Belt of Québec. Approximation of true metamorphic conditions is attained only in areas such as the complex that were shielded by high competency from pervasive reactivation. In previous studies, peak metamorphic temperatures were underestimated; these lower temperatures were then used to estimate the peak metamorphic pressure. When a peak metamorphic temperature of 950 °C is used to calculate peak pressure, the result increases by 1–2 kilobars (e.g., Tables 12, 13, 14, and 15).

Tectonic Implications

High pressures documented in the complex record substantial thickening of the crust (D1) attributed to continental collision at 1.20 Ga (Corriveau and van Breemen, 2000), whereas the 1.18 Ga age is linked to anatectic melting and the peak temperature mineral assemblages. Harris et al. (2001) suggested that the fabric hosting the peak pressure assemblages developed from transposition of S1 during rebound of a lithospheric root thickened during terrane assembly. The high temperature achieved, following crustal thickening, could thus be associated with the D2–D5 orogenic collapse described by Harris et al. (2001). A series of mafic and ultramafic xenoliths in a minette dike provides evidence of voluminous mafic magma underplating at the base of the crust directly below the complex, possibly associated with this orogenic collapse (Corriveau and Morin,

2000). This high temperature event would be compatible with the pervasive 1.18 Ga metamorphic ID-TIMS and SHRIMP ages obtained in the complex and with the sporadic oldest ages obtained in the rest of the Central Metasedimentary Belt of Québec (Corriveau and van Breemen, 2000; Wodicka et al., this volume). The monazite age (1182 ± 2 Ma) of the complex is evidence of rapid cooling between the calculated peak temperature of 950 °C and the closure temperature of monazite (800 °C), because the monazite age is essentially identical to the age of anatectic melting. This rapid cooling supports isothermal decompression in the complex.

Indares and Martignole (1990a) concluded that the garnet zonation profiles they documented in the Central Metasedimentary Belt of Québec, which are similar to the garnet core profiles of this study, are suggestive of isothermal decompression after attainment of peak metamorphic conditions. They concluded that this decompression resulted from tectonic unroofing after crustal thickening by thrusting. This proposal is compatible with the observed decompression reaction rim of plagioclase around some garnet (Boggs, 1996) in the complex, as well as with the orogenic collapse event described by Harris et al. (2001) for the 1.19–1.17 Ga period. The composition of outer garnet zones could reflect the 1.16 Ga overprint. The proposed divergence of cooling paths in the complex and in the deformation zone appears to require greater uplift of the rigid gneiss complex (Corriveau et al., 1998; Harris et al., 2001) than of the bordering high-strain zone by the time at which mineral assemblages were quenched, which could relate to greater buoyancy of the complex during extensional unroofing.

Other alternatives are possible. A crust thickened in an Andean-type setting would be compatible with the models of Rivers and Corrigan (2000), while the doming structure and the high calculated pressure and temperature could have resulted from channel flow (Grujic et al., 2002). While all these avenues are of interest in deciphering the orogenic evolution of the western Grenville Province, the assessment of which tectonic model best takes into account all observed textures and data is beyond the scope of this paper.

CONCLUSIONS

Peak metamorphic conditions in the Bondy gneiss complex are estimated at 950 °C and 10 kilobars. These values are 150–200 °C and 2 kilobars higher than those documented previously in the Central Metasedimentary Belt of Québec. Initial decompression, inferred to result from tectonic unroofing during extension, was isothermal. Subsequent retrograde cooling paths in the gneiss complex and the deformation zone are distinct; that of the gneiss complex is at least 0.5 kilobar lower at any given temperature.

Retrograde cation diffusion, enhanced by fluid infiltration and grain size reduction associated with the 1.17–1.16 Ga deformation event, decreased calculated temperatures from the geothermometers in the deformation zone. In the gneiss complex, which was only locally affected by the 1.17–1.16 Ga deformation event, the presence of ferromagnesian inclusions that reduced the size of diffusional domains, combined with the influence of retrograde cation diffusion, masks evidence of peak metamorphic conditions in most garnet grains. Careful selection of appropriate minerals for geothermometry is again shown as vital for constructing *P-T-t* trajectories in granulite terrains.

APPENDIX 1—LIST OF ABBREVIATIONS

The following mineral and other abbreviations are after Kretz (1983).

Ab	albite	Ilm	ilmenite	Pl	plagioclase
Alm	almandine	Kfs	K-feldspar	Prp	pyrope
Als	aluminosilicate	Krn	kornerupine	Rt	rutile
An	anorthite	Ky	kyanite	Qtz	quartz
Ap	apatite	L	liquid	Sil	sillimanite
Bt	biotite	Ms	muscovite	Spl	spinel
Btss	biotite, solid state	Msss	muscovite, solid state	Sps	spessartine
				Tur	tourmaline
Crd	cordierite	Opx	orthopyroxene	V	vapor
Grs	grossular	Or	orthoclase		
Grt	garnet	Phl	phlogopite		

APPENDIX 2—ELECTRON MICROPROBE ANALYSES

Analytical Conditions

The ARL-SEMQ II electron microprobe at the Université du Québec à Chicoutimi was used to obtain mineral compositions. This machine is equipped with a NORAN energy detector with a beryllium window that gives a resolution of 143 eV. This microprobe is controlled by the quantitative microanalysis system TN-5500: Series II from NORAN. The analytical results of silicates were corrected using Bence-Albee matrix corrections.

The diameter of the beam is ~5 microns, which restrains the analyzed surface to a grain size of ~8 microns. An accelerating voltage of 15 kV was used, with an electron beam current of 10 nA. Counts for each element were made for 100 seconds. The program MICROQ was used to reduce the analyses to weight percent oxides.

Data Consistency and Sources of Error

The errors due to day-to-day variations and drift during use were assessed by analyzing one spot on a garnet both once a day, and ten times in one day. The means, standard deviations, and coefficients of variation were calculated for each oxide. For the five main oxides in this garnet (SiO_2, Al_2O_3, FeO, MgO, CaO), the coefficients of variation were less than 3.6% of the oxide concentration for both the day-to-day consistency and the consistency during one day. For iron, the coefficient of variation was

0.7% over one day and 1.8% over the ten days of analyses. The coefficients of variation for magnesium were slightly higher, at 1.8% for one day's analyses and 3.3% for the ten days of microprobe analyses. The 17.7% for manganese is disproportionately high due to the low abundance of manganese present at the test site. For an oxide concentration under 2%, any small variation would result in high standard deviation and coefficient of variation values that were not representative of the overall accuracy obtained for the oxide analyses.

Corrections of Systematic Errors

Plagioclase analyses and repetitive standard analyses were used to correct for the two sources of error discussed earlier. When present, plagioclase was analyzed to provide a control for the calibration of silicon and aluminum during microprobe standardization. Because of the coupled substitution between silicon-aluminum and calcium-sodium, it is possible to predict the silicon and aluminum content from the mole fraction of anorthite. These correction values were applied to the weight percent oxides analyzed by the microprobe before the stoichiometric values were calculated.

Repetitive analyses of known garnet standards were used to correct the measured results without readjusting the microprobe standardization file. The known standard values were divided by the measured values to obtain a correction factor for each oxide measurement. These correction factors were then applied to the mineral analyses that were made closest in time to that particular standard.

Criteria for Accepting Microprobe Analyses

The choice of reliable mineral analysis is important for the determination of pressure and temperature from geothermobarometers. First, the totals were examined. For anhydrous minerals, if the totals were not between 99.0% and 101.0%, the analysis was rejected. For biotites and cordierites, if the totals were not between 95.0 and 97.0%, the analyses were rejected.

Furthermore, for anhydrous minerals, the site occupancy was calculated to determine the analysis reliability. For example, for garnet the acceptable relative error values would be equivalent to the coefficient of variance: ± 2.0% for silicon, ± 2.0% for aluminum, ± 2.1% for iron, ± 3.6% for magnesium, and ± 3.3% for calcium. If the analyses were not within these ranges, they were rejected.

ACKNOWLEDGMENTS

Careful reviews by Nathalie Marchildon and Lawford Anderson were very beneficial. Constructive comments by N. Culshaw, M. Easton, E. Essene, and E. Sawyer greatly improved an earlier version of this paper. Further comments by J. Moore and N. Wodicka are greatly appreciated. This paper is composed of a portion of the senior author's M.S. thesis, written at Université du Québec à Chicoutimi (UQAC). Academic supervision and support by E. Sawyer were greatly appreciated and very constructive.

This study was financed by a National Science and Engineering Research Council (NSERC) grant to E. Sawyer, in particular the microprobe analyses at UQAC and a UQAC fellowship granted to KB. Maintenance of the microprobe was financed by a NSERC infrastructure grant to S.-J. Barnes. Field work was funded by the Geological Survey of Canada, project 920002QN.

The assistance of R.C. Newton with the garnet-aluminosilicate-silica-plagioclase barometer was greatly appreciated. G. Couture, L. Madore, M. Mainville, N. Mohan, and D. Morin are thanked for their assistance in the field. The friendly assistance of the staff at the Papineau Labelle Wildlife Preserve was greatly appreciated.

This is Geological Survey of Canada contribution number 2003006. This paper is dedicated to the memory of Alain Demers.

REFERENCES CITED

Anovitz, L.M., and Essene, E.J., 1990, Thermobarometry and pressure-temperature paths in the Grenville Province of Ontario: Journal of Petrology, v. 31, p. 197–241.

Berman, R.G., 1990, Mixing properties of Ca-Mg-Fe-Mn garnets: American Mineralogist, v. 75, p. 328–344.

Bhattacharya, A., 1986, Some geobarometers involving cordierite in the FeO-Al_2O_3-SiO_2 (H_2O) system: Refinements, thermodynamic calibration, and applicability in granulite facies rocks: Contributions to Mineralogy and Petrology, v. 94, p. 387–394.

Bhattacharya, A., and Sen, S.K., 1985, Energetics of hydration of cordierite and water barometry in cordierite granulites: Contributions to Mineralogy and Petrology, v. 89, p. 370–378.

Bingen, B., and van Breemen, O., 1998, U-Pb monazite ages in amphibolite- to granulite-facies orthogneiss reflect hydrous mineral breakdown reactions: Sveconorwegian Province of SW Norway: Contributions to Mineralogy and Petrology, v. 132, p. 336–353.

Boggs, K.J.E., 1996, Retrograde cation exchange in garnets during slow cooling of mid crustal granulites and the *P-T-t* trajectories from the Mont Laurier region, Grenville Province, Quebec [M.S. thesis]: Chicoutimi, Université du Québec à Chicoutimi, 352 p.

Carrington, D.P., and Harley, S.L., 1995, Partial melting and phase relations in high-grade metapelites: An experimental petrogenetic grid in the KMASH system: Contributions to Mineralogy and Petrology, v. 120, p. 270–291.

Clemens, J.D., 1995, Phlogopite stability in the silica-saturated portion of the system K_2O-Al_2O_3-MgO-SiO_2-H_2O: New data and a reappraisal of phase relations to 1.5 GPa: American Mineralogist, v. 80, p. 982–997.

Corriveau L., 1982, Physical conditions of the regional and retrograde metamorphism of the Chicoutimi area, Québec [M.S. thesis]: Kingston, Ontario, Queen's University, 264 p.

Corriveau, L., and Morin, D., 2000, Modeling 3D architecture of western Grenville from xenoliths, styles of magma emplacement and Lithoprobe reflectors: Canadian Journal of Earth Sciences, v. 37, p. 235–251.

Corriveau, L., and van Breemen, O., 2000, Docking of the Central Metasedimentary Belt to Laurentia in geon 12: Evidence from the 1.17–1.16 Ga Chevreuil intrusive suite and host gneisses, Québec: Canadian Journal of Earth Sciences, v. 37, p. 253–269.

Corriveau, L., Tellier, M.L., and Morin, D., 1996, Le dyke de minette de Rivard et le complexe gneissique cuprifère de Bondy: Implications tectoniques et métallogéniques pour la région de Mont-Laurier, Québec: Ottawa, Ontario, Geological Survey of Canada Open File Report, v. 3078, 70 p.

Corriveau, L., Rivard, B., and van Breemen, O., 1998, Rheological controls on Grenvillian intrusive suites: Implications for tectonic analysis: Journal of Structural Geology, v. 20, p. 1191–1204.

Davidson, A., 1995, A review of the Grenville orogen in its North American type area: AGSO [Australian Geological Survey Organisation] Journal of Australian Geology and Geophysics, v. 16, p. 3–24.

Deer, W.A., Howie, R.A., and Zussman, J., 1966, An introduction to the rock-forming minerals: New York, Longman Scientific and Technical, 528 p.

Doig, R., 1991, U-Pb zircon dates of the Morin anorthosite suite rocks, Grenville Province, Québec: Journal of Geology, v. 99, p. 729–738.

Emslie, R.F., and Hunt, P.A., 1990, Ages and petrogenetic significance of igneous mangerite-charnockite suites associated with massif anorthosites, Grenville Province: Journal of Geology, v. 98, p. 213–231.

Erambert, M., and Austrheim, H., 1993, The effect of fluid and deformation on zoning and inclusion patterns in polymetamorphic garnets: Contributions to Mineralogy and Petrology, v. 115, p. 204–214.

Ferry, J.M., and Spear, F.S., 1978, Experimental calibration of the partitioning of Fe and Mg between biotite and garnet: Contributions to Mineralogy and Petrology, v. 66, p. 113–117.

Grujic, D., Hollister, L.S., and Parrish, R.R., 2002, Himalayan metamorphic sequence as an orogenic channel: Insight from Bhutan: Earth and Planetary Science Letters, v. 198, p. 177–191.

Guo, A., and Dickin, A.P., 1996, Tectonic significance of Nd model age mapping in the Grenville Province of western Québec: Precambrian Research, v. 77, p. 231–241.

Hames, W.E., and Menard, T., 1993, Fluid-assisted modification of garnet composition along rims, cracks, and mineral inclusion boundaries in samples of amphibolite facies schists: American Mineralogist, v. 78, p. 338–344.

Harley, S.L., 1984a, An experimental study of the partitioning of Fe and Mg between garnet and orthopyroxene: Contributions to Mineralogy and Petrology, v. 86, 359–373.

Harley, S.L., 1984b, The solubility of alumina in orthopyroxene coexisting with garnet in FeO-MgO-Al_2O_3-SiO_2 and CaO-FeO-MgO-Al_2O_3-SiO_2: Journal of Petrology, v. 25, p. 665–696.

Harley, S.L., and Green, D.H., 1982, Garnet-orthopyroxene barometry for granulites and peridotites: Nature, v. 300, p. 697–701.

Harris, L., Rivard, B., and Corriveau, L., 1998, Crustal-scale extensional collapse of the Elzevir orogen, Grenville Province, SW Québec imaged by Lithoprobe: Geological Society of Australia Abstracts, v. 50, p. 39.

Harris, L., Rivard, B., and Corriveau, L., 2001, Structure of the Lac Nominingue-Chénéville zone in the Mont-Laurier region, Central Metasedimentary Belt, Québec, Grenville Province: Canadian Journal of Earth Sciences, v. 38, p. 787–802.

Holdaway, M.J., and Lee, S.M., 1977, Fe-Mg cordierite stability in high grade pelitic rocks based on experimental, theoretical and natural observations: Contributions to Mineralogy and Petrology, v. 63, p. 175–198.

Hutcheson, I., Froese, E., and Gordon, T.M., 1974, The assemblage quartz-sillimanite-garnet-cordierite as an indicator of metamorphic conditions in the Daly Bay complex, N.W.T.: Contributions to Mineralogy and Petrology, v. 44, p. 29–34.

Indares, A., and Martignole, J., 1984, Evolution of *P-T* conditions during a high-grade metamorphic event in the Maniwaki area (Grenville Province): Canadian Journal of Earth Sciences, v. 21, p. 853–886.

Indares, A., and Martignole, J., 1990a, Metamorphic constraints on the tectonic evolution of the allochthonous monocyclic belt of the Grenville Province, western Quebec: Canadian Journal of Earth Sciences, v. 27, p. 371–386.

Indares, A., and Martignole, J., 1990b, Metamorphic constraints on the evolution of the gneisses from the parautochthonous and allochthonous polycyclic belts, Grenville Province, western Quebec: Canadian Journal of Earth Sciences, v. 27, p. 357–370.

Koziol, A.M., and Newton, R.C., 1988, Redetermination of the anorthite breakdown reaction and improvement of the plagioclase-garnet-Al_2SiO_5-quartz geobarometer: American Mineralogist, v. 73, p. 216–223.

Kretz, R., 1983, Symbols for rock-forming minerals: American Mineralogist, v. 68, p. 277–279.

Lal, R.K., 1993, Internally consistent recalibrations of mineral equilibria for geothermobarometry involving garnet-orthopyroxene-plagioclase-quartz assemblages and their application to the South Indian granulites: Journal of Metamorphic Geology, v. 95, p. 855–866.

Le Breton, N., and Thompson, A.B., 1988, Fluid-absent (dehydration) melting of biotite in metapelites in the early stages of crustal anatexis: Contributions to Mineralogy and Petrology, v. 99, p. 226–237.

Lonker, S.W., 1980, Conditions of metamorphism in high-grade pelitic gneisses from the Frontenac Axis, Ontario, Canada: Canadian Journal of Earth Sciences, v. 17, p. 1666–1686.

Martignole, J., and Schrijver, K., 1970, Tectonic setting and evolution of the Morin anorthosite, Grenville Province, Quebec: Bulletin of the Geological Society of Finland, v. 42, p. 165–209.

Martignole, J., and Sisi, S.-C., 1981, Cordierite-garnet-H_2O equilibrium: A geological thermometer, barometer and water fugacity indicator: Contributions to Mineralogy and Petrology, v. 77, p. 38–46.

Martignole, J., Calvert, A.J., Friedman, R., and Reynolds, P., 2000, Crustal evolution along a seismic section across the Grenville Province (western Québec): Canadian Journal of Earth Sciences, v. 37, p.291–306.

Newton, R.C., and Wood, B.J., 1979, Thermodynamics of water in cordierite and petrologic consequences of cordierite as a hydrous phase: Contributions to Mineralogy and Petrology, v. 68, p. 391–405.

Nichols, G.T., Berry, R.F., and Green, D.H., 1992, Internally consistent gahnitic spinel-cordierite-garnet equilibria in the FMASHZn system: Geothermobarometry and applications: Contributions to Mineralogy and Petrology, v. 111, p. 362–377.

Olsen, S.N., 1983, A quantitative approach to local mass balance in migmatites, *in* Arton, M.P., and Gribble, C.D., eds., Migmatites, melting and metamorphism: Nantwich, England, Shiva, p. 201–233.

Patiño-Douce, A.E., and Johnston, A.D., 1991, Phase equilibria and melt productivity in the pelitic system: Implications for the origin of peraluminous granitoids and aluminous granulites: Contributions to Mineralogy and Petrology, v. 107, p. 202–218.

Perchuk, L.L., and Lavrent'eva, I.V., 1983, Experimental investigation of exchange equilibria in the system cordierite-garnet-biotite, in Saxena, S.K., ed., Kinetics and equilibrium in mineral reactions: New York, Springer-Verlag, p. 199–239.

Puziewicz, J., and Johannes, W., 1990, Experimental study of a biotite-bearing granitic system under water-saturated and water-undersaturated conditions: Contributions to Mineralogy and Petrology, v. 104, p. 397–406.

Rivard, B., Corriveau, L., and Harris, L., 1999, Structural reconnaissance of a deep crustal orogen using satellite imagery and airborne geophysics: Canadian Journal of Remote-Sensing, v. 25, no. 3, p. 258–267.

Rivers, T., and Corrigan, D., 2000, Convergent margin on southeastern Laurentia during the Mesoproterozoic: Tectonic implications: Canadian Journal of Earth Sciences, v. 37, p. 359–383.

Sen, S.K., and Bhattacharya, A., 1984, An orthopyroxene-garnet thermometer and its application to the Madras charnockites: Contributions to Mineralogy and Petrology, v. 88, p. 64–71.

Spear, F., 1991, On the interpretation of peak metamorphic temperatures in the light of garnet diffusion during cooling: Journal of Metamorphic Geology, v. 9, p. 279–388.

Spear, F., and Florence, F.F., 1992, Thermobarometry in granulites: Pitfalls and new approaches: Precambrian Research, v. 55, p. 209–241.

Streepey, M.M., Essene, E.J., and van der Pluijm, B.A., 1997, A compilation of thermobarometric data from the metasedimentary belt of the Grenville Province, Ontario, and New York State: The Canadian Mineralogist, v. 35, p. 1237–1247.

Thompson, A.B., 1976a, Mineral reactions in pelitic rocks, I: Prediction of *P-T*-X (Fe-Mg) phase relations: American Journal of Science, v. 276, p. 401–424.

Thompson, A.B., 1976b, Mineral reactions in pelitic rocks, II: Calculation of some *P-T*-X (Fe-Mg) phase relations: American Journal of Science, v. 276, p. 425–454.

Tommasi, A., and Vauchez, A., 1997, Continental-scale rheological heterogeneities and complex intraplate tectono-metamorphic patterns: Insights from a case-study and numerical models: Tectonophysics, v. 279, p. 327–350.

Vernon, R.H., 1978, Pseudomorphous replacement of cordierite by symplectic intergrowths of andalusite, biotite and quartz: Lithos, v. 11, p. 283–289.

Vielzeuf, D., and Holloway, J.R., 1988, Experimental determination of the fluid-absent melting relations in the pelitic system: Contributions to Mineralogy and Petrology, v. 98, p. 257–276.

Vielzeuf, D., and Montel, J.M., 1994, Partial melting of metagreywackes, Part I: Fluid-absent experiments and phase relationships: Contributions to Mineralogy and Petrology, v. 117, p. 375–393.

Whitney, D.L., 1991, Calcium depletion halos and Fe-Mn-Mg zoning around faceted plagioclase inclusions in garnet from a high-grade pelitic gneiss: American Mineralogist, v. 76, p. 493–501.

Whitney, D.L., 1996, Garnets as open systems during regional metamorphism: Geology, v. 24, p. 147–150.

Whitney, D.L., and Dilek, Y., 1998, Metamorphism during crustal thickening and extension in central Anatolia: The Niode metamorphic core complex: Journal of Petrology, v. 39, p. 1385–1403.

Whitney, D.L., Cooke, M.L., and Du Frane, S.A., 2000, Modeling of radial microcracks at corners of inclusions in garnet using fracture mechanics: Journal of Geophysical Research, v. 105, p. 2843–2853.

Wynne-Edwards, H.R., 1972, The Grenville Province: St. John's, Newfoundland, Geological Association of Canada Special Paper, v. 11, p. 263–334.

Manuscript Accepted by the Society August 25, 2003

Printed in the USA

Geological Society of America
Memoir 197
2004

Mesoproterozoic basement of the Lac St. Jean Anorthosite Suite and younger Grenvillian intrusions in the Saguenay region, Québec: Structural relationships and U-Pb geochronology

Claude Hébert*
Ressources naturelles, Québec City, Québec G1H 6R1, Canada
Otto van Breemen
Geological Survey of Canada, Ottawa, Ontario K1A 0E8, Canada

ABSTRACT

The Chicoutimi Gneiss Complex was previously considered to host the huge 1160–1140 Ma Lac St. Jean anorthosite suite (20,000 km^2), the 1082 ± 3 Ma Chicoutimi Mangerite, and the 1067 ± 3 Ma La Baie Granite. New geological mapping and geochronological data from the Chicoutimi Gneiss Complex now demonstrate clear correlations among some gneissic units and the partially deformed Chicoutimi Mangerite and La Baie Granite. The remaining gneissic units in the Chicoutimi Gneiss Complex have been grouped into six new lithodems. The three oldest were generated during two distinct events. The Saguenay Gneiss Complex, which is one of these lithodems and includes the Cap de la Mer Amphibolite unit dated at 1506 ± 13 Ma, is related to a widespread Pinwarian event. It was followed, 100 Ma later, by the 1391 +7/–8 Ma Cap à l'Est Gneiss Complex and the 1383 ± 16 Ma Cyriac Rapakivi Granite. Two younger lithodems are the 1155–1135 Ma Kénogami Charnockite and the 1150 ± 3 Ma Baie à Cadie Mafic-Ultramafic Suite, which are genetically related to the Lac St. Jean Anorthosite Suite. Finally, the 1045 ± 5 Ma Simoncouche Gabbro is the youngest Grenvillian igneous unit dated so far in the Saguenay region. The Lac St. Jean Anorthosite Suite, the Chicoutimi Mangerite, the La Baie Granite, and their host rocks (the Saguenay Gneiss Complex, the Cap à l'Est Gneiss Complex, and the Cyriac Rapakivi Granite) are everywhere in tectonic contact with each other. These contacts are part of the St. Fulgence shear zone, which is over 400 km long and several kilometers wide. Since the Simoncouche Gabbro outcrops within the St. Fulgence shear zone and is weakly deformed, it can be concluded that 1045 ± 5 Ma marked the end of the final movement along the St. Fulgence shear zone in the Saguenay region.

Keywords: anorthosite, geochronology, Grenville, basement, structure

* E-mail: claude.hebert@mrn.gouv.qc.ca.

Hébert, C., and van Breemen, O., 2004, Mesoproterozoic basement of the Lac St. Jean Anorthosite Suite and younger Grenvillian intrusions in the Saguenay region, Québec: Structural relationships and U-Pb geochronology, *in* Tollo, R.P., Corriveau, L., McLelland, J., and Bartholomew, M.J., eds., Proterozoic tectonic evolution of the Grenville orogen in North America: Boulder, Colorado, Geological Society of America Memoir 197, p. 65–79. For permission to copy, contact editing@geosociety.org.

INTRODUCTION

During the 1970s it was recognized that the geology of the Saguenay region (Fig. 1) is characterized by anorthosite-mangerite-charnockite plutons emplaced into migmatitic gneisses, both intruded by younger felsic plutons (Vallée and Dubuc, 1970; Laurin and Sharma, 1975; Woussen et al., 1986). It was inferred that the migmatitic gneisses were derived from supracrustal rocks, granitic orthogneiss, and amphibolite. Owen et al. (1980) and Owen (1981) introduced the term "Old Gneiss Complex" for the migmatitic gneisses, and this term was employed by Higgins and van Breemen (1992) and by Hervet et al. (1994). Woussen et al. (1981) used the term "basement gneiss" for the Old Gneiss Complex, whereas Dimroth et al. (1981) used the term "Gneiss Complex of Chicoutimi." The most recent name for these gneisses is "Chicoutimi Gneiss Complex," coined by Higgins and van Breemen (1996), who considered the gneiss complex the basement of the Saguenay region. Only the 1393 +22/−10 Ma U-Pb age of the Ruisseau à Jean-Guy Mafic Intrusion (Hervet et al., 1994) and the Rb/Sr isochron age 1482 ± 72 Ma of a paragneiss (Frith and Doig, 1973) from the Saguenay region have been geochronologically demonstrated to be older than the Lac St. Jean Anorthosite Suite (or AMCG, from anorthosite-mangerite-charnockite-granite). A 1.55 Ga Sm/Nd isochron age (Dickin and Higgins, 1992) has been obtained from tonalitic gneiss that occurs within the host rocks near the southern margin of the Lac St. Jean Anorthosite Suite.

A regional geological mapping program was carried out from 1994 to 1996 in order to enhance the knowledge of the Saguenay region. Previously known felsic plutons of the Chicoutimi Mangerite, the La Baie Granite, the Laurent Syenite, the Lac des Îlets Granite, and the Anse à Phillipe Layered Intrusion were mapped in greater detail. Furthermore, six new lithodemic assemblages were defined within the gneiss complex. These are the Saguenay Gneiss Complex, which include the Cap de la Mer Amphibolite unit, the Cap à l'Est Gneiss Complex, the Cyriac Rapakivi Granite, the Kénogami Charnockite, the Baie à Cadie Mafic-Ultramafic Suite, and the Simoncouche Gabbro (Fig. 2). This study summarizes the geological mapping and also presents new geochronological data.

Figure 2. Geology and geochronology of the Saguenay region. Locations of U-Pb mineral ages from different sources are indicated. The code following each point indicates the age (in Ma), the rock type (Am—amphibolite; An—anorthosite; C–charnockite; Di—diorite; FD—ferrodiorite; G—granite; Gb—gabbro; Gn—gabbronorite; Gs—gneiss; Lg—leucogabbro; Lt—leucotroctolite; Ma—mangerite; Mo—monzonite), and the source of the data: 1—Higgins and van Breemen, 1992; 2—Emslie and Hunt, 1990; 3—Hervet et al., 1994; 4—Higgins and van Breemen 1996; 5—This work. *Use of the lithodemic codes follows the North American Commission on Stratigraphic Nomenclature (NACSN, 1983).

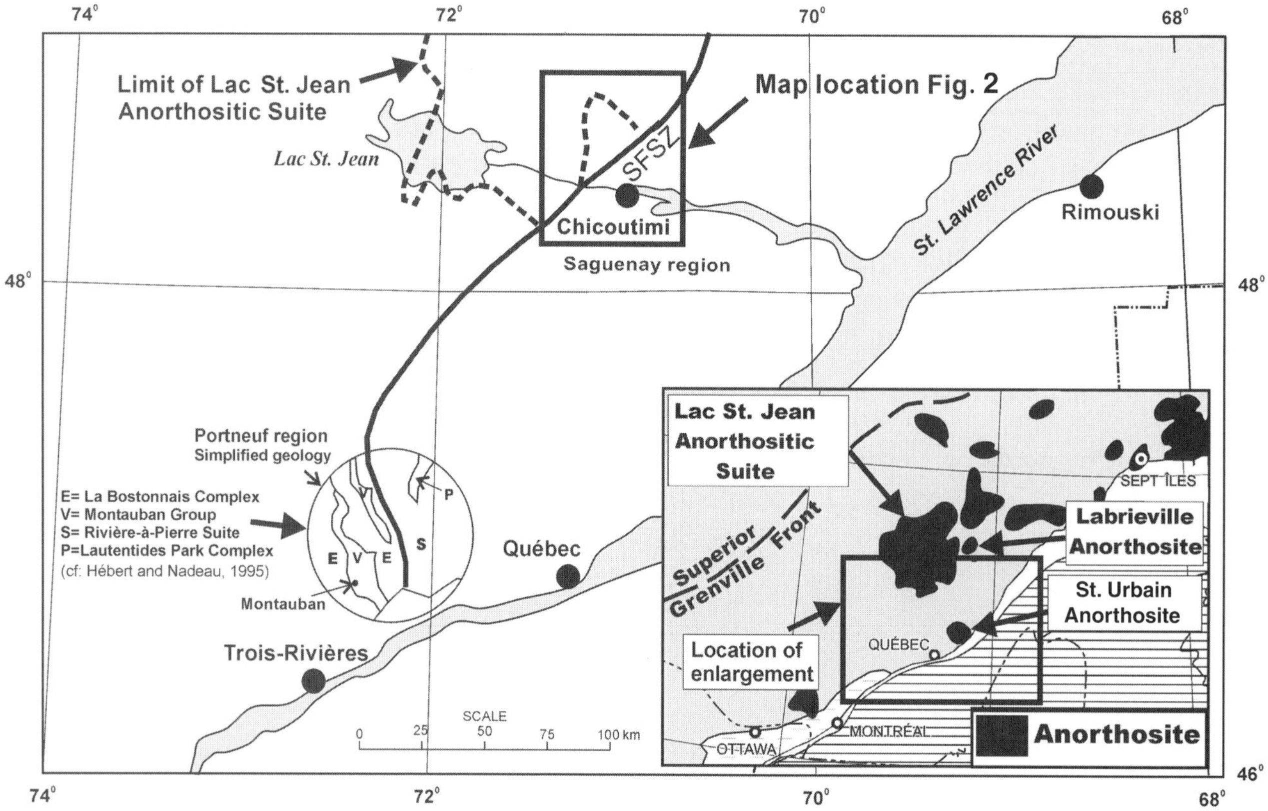

Figure 1. Location map of the area discussed in this paper. SFSZ—St. Fulgence shear zone.

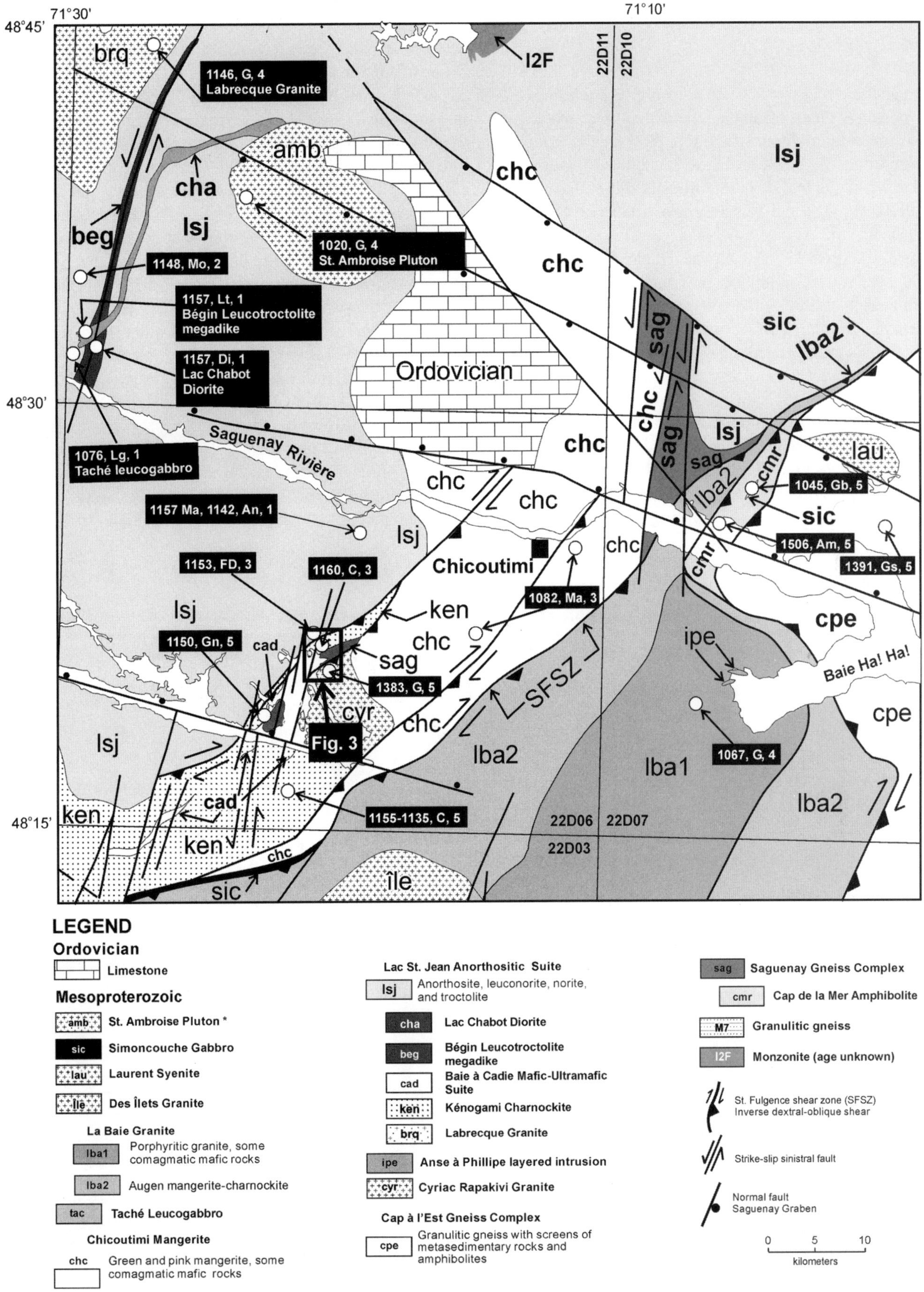

71°30'
71°10'
48°45'
48°30'
48°15'
22D11
22D10
22D06
22D07
22D03
brq
1146, G, 4
Labrecque Granite
I2F
amb
cha
beg
lsj
1020, G, 4
St. Ambroise Pluton
1148, Mo, 2
1157, Lt, 1
Bégin Leucotroctolite
megadike
1157, Di, 1
Lac Chabot
Diorite
chc
Ordovician
sag
sic
lba2
lau
Saguenay Rivière
1076, Lg, 1
Taché leucogabbro
1157 Ma, 1142, An, 1
1045, Gb, 5
cmr
1506, Am, 5
1391, Gs, 5
Chicoutimi
1153, FD, 3
1160, C, 3
1082, Ma, 3
ken
cpe
ipe
Baie Ha! Ha!
1150, Gn, 5
cad
1383, G, 5
SFSZ
cyr
Fig. 3
1067, G, 4
lba1
1155-1135, C, 5
île
LEGEND
Ordovician
Limestone
Mesoproterozoic
amb St. Ambroise Pluton *
sic Simoncouche Gabbro
lau Laurent Syenite
île Des Îlets Granite
La Baie Granite
lba1 Porphyritic granite, some comagmatic mafic rocks
lba2 Augen mangerite-charnockite
tac Taché Leucogabbro
Chicoutimi Mangerite
chc Green and pink mangerite, some comagmatic mafic rocks
Lac St. Jean Anorthositic Suite
lsj Anorthosite, leuconorite, norite, and troctolite
cha Lac Chabot Diorite
beg Bégin Leucotroctolite megadike
cad Baie à Cadie Mafic-Ultramafic Suite
ken Kénogami Charnockite
brq Labrecque Granite
ipe Anse à Phillipe layered intrusion
cyr Cyriac Rapakivi Granite
Cap à l'Est Gneiss Complex
cpe Granulitic gneiss with screens of metasedimentary rocks and amphibolites
sag Saguenay Gneiss Complex
cmr Cap de la Mer Amphibolite
M7 Granulitic gneiss
I2F Monzonite (age unknown)
St. Fulgence shear zone (SFSZ)
Inverse dextral-oblique shear
Strike-slip sinistral fault
Normal fault
Saguenay Graben
0 5 10
kilometers

GENERAL GEOLOGY

The gneissic rocks of the Saguenay region are represented by the Saguenay Gneiss Complex, which includes the Cap de la Mer Amphibolite unit and the Cap à l'Est Gneiss Complex. The most extensive lithodem of the region is, however, the 1160–1140 Ma Lac St. Jean Anorthosite Suite (Emslie and Hunt, 1990; Higgins and van Breemen, 1992, 1996; Hébert, 1997; Hébert and Lacoste, 1998a) (Figs. 1 and 2). The Kénogami Charnockite, the Cyriac Rapakivi Granite, and the 1082 ± 3 Ma Chicoutimi Mangerite are intrusive along the southern margin of this anorthositic mass. The 1067 ± 3 Ma La Baie Granite (Higgins and van Breemen, 1996) occurs on the east side of the Chicoutimi Mangerite. The Baie à Cadie Mafic-Ultramafic Intrusion is located near the contact of the anorthosite within the Kénogami Charnockite. The Simoncouche Gabbro outcrops within the Cap de la Mer Amphibolite and along the contact between the Kénogami Charnockite, the Chicoutimi Mangerite, and the La Baie Granite. Finally, the Laurent Syenite is intrusive within the Cap à l'Est Gneiss Complex, on the north shore of the Rivière Saguenay, while the Lac des Îlets Granite and the Anse à Phillipe Layered Intrusion occur within the La Baie Granite.

More recently, Higgins and van Breemen (1996) proposed that the intrusive anorthosite-mangerite-charnockite plutons of the Saguenay region belong to three AMCG generations. The oldest of these is the Lac St. Jean Anorthosite Suite (1160–1140 Ma) (Fig. 2; Table 1). The second generation is the St. Urbain Anorthosite Suite (AMCG; 1082–1050 Ma), an anorthositic member of which outcrops farther south outside of the Saguenay area (Fig. 1). Within the studied area, the St. Urbain Anorthosite Suite is represented by the 1082 ± 3 Ma Chicoutimi Mangerite (Hervet et al., 1994; Higgins and van Breemen, 1996) and the 1067 ± 4 Ma La Baie Granite (Higgins and van Breemen, 1996). The youngest AMCG generation, termed the 1020–1010 Ma Labrieville Anorthosite Suite (Fig. 1), includes the Labrieville Anorthosite (Owens et al., 1992) and the 1020 +4/–3 Ma St. Ambroise Pluton (Higgins and van Breemen, 1996), located within the Lac St. Jean Anorthosite Suite (Fig. 2; Table 1). Higgins and van Breemen (1996) also suggested that the previous structural or stratigraphic notion of late to postkinematic plutons should be abandoned, as in the Lac St.

TABLE 1. U-PB AGES IN THE SAGUENAY REGION

1506 Ma event; part of the structural basement	1506 ± 13 Ma§§	Cap de la Mer Amphibolite (Saguenay Gneiss Complex)
1393–1383 Ma event; part of the structural basement	1393 +22/–10 Ma*	Ruisseau à Jean-Guy Mafic Intrusion
	1391 +8/–7 Ma§§	Cap à l'Est Gneiss Complex
	1383 ± 16 Ma§§	Cyriac Rapakivi Granite
1160–1140 Ma event: Lac St. Jean Anorthosite Suite (AMCG)	1160–1140 Ma**	Anorthosite samples from Saguenay region
	~1160 Ma*	Lac St. Jean farsundite (same as Kénogami Charnockite)
	1157 ± 2 Ma#	Lac Chabot Diorite
	1154 ± 2 Ma#	Granophyric segregation, Jonquière
	1157 ± 2 Ma#	Bégin Leucotroctolite Megadyke
	1150 ± 3 Ma§§	Baie à Cadie Mafic-Ultramafic Suite
	1146 ± 3 Ma**	Labrecque Granite
	1155–1135 Ma§§	Kénogami Charnockite
	1148 ± 4 Ma§	Pyroxene Monzonite
1082–1067 Ma event; related to St. Urbain AMCG Suite	1082 ± 3 Ma* **	Chicoutimi Mangerite: Green and pink mangerite
	1076 ± 2 Ma#	Taché Leucogabbro
	1067 ± 4 Ma#	La Baie Granite
1050–1045 Ma event	1050 ± 10 Ma*	Metabasite dike III-B (Lac Kénogami)
	1045 ± 5 Ma§§	Simoncouche Gabbro
1020 Ma event; related to Labrieville AMGC Suite	1020 +4/–3 Ma**	St. Ambroise Pluton

Note: AMCG—anorthosite-mangerite-charnockite-granite.
*Hervet et al. (1994).
§Emslie and Hunt (1990).
#Higgins and van Breemen (1992).
**Higgins and van Breemen (1996).
§§This paper.

Jean AMCG suite less deformed rocks are not always younger than more deformed rocks.

LITHOSTRATIGRAPHY

Assemblages Observed within the Limits of the Chicoutimi Gneiss Complex

Gneissic Rocks

The Saguenay Gneiss Complex (Fig. 2) contains the oldest known Precambrian rocks in the area. Two main facies are recognized, namely supracrustal and plutonic rocks (Hébert and Lacoste 1998a, 1998b, 1998c). Sillimanite-garnet-cordierite-bearing paragneiss (Dagenais, 1983), cordierite-bearing quartzite, calc-silicate rocks, marble, and amphibolite dominate the supracrustal series. The Cap de la Mer Amphibolite unit (Fig. 2) is composed principally of amphibolite, although primary gabbroic texture and primary orthopyroxene are clearly visible in less deformed patches (Hébert and Lacoste, 1998b). This unit has been included within the Saguenay Gneiss Complex. Peak regional granulite-facies metamorphism in the supracrustal rocks reached 780 °C and 6.2 kb (Corriveau, 1982), and local retrograde metamorphism at 690 °C and 4.7 kb related to the emplacement of the Lac St. Jean Anorthosite Suite has also been observed in the paragneiss (Corriveau, 1982). Frith and Doig (1973) obtained a Rb/Sr isochron age of 1482 ± 72 Ma for a paragneiss collected within the new lithodem, defined and named as the Saguenay Gneiss Complex. Although this age was interpreted as a metamorphic age, no evidence for isotopic homogenization was presented, and we prefer to interpret this age in terms of an averaged provenance. The plutonic facies consist mostly of intrusive rocks such as orthopyroxene gabbro, diorite, and tonalite but include screens of supracrustal rocks. The intrusive rocks are generally foliated, gneissic, or migmatitic. On the northeast shore of Lac Kénogami, a gabbroic unit termed the Ruisseau à Jean-Guy Mafic Intrusion was mapped and dated as 1393 +22/–10 Ma (Hervet et al., 1994; Fig. 3). On the basis of our field observations we include this unit within the plutonic facies of the Saguenay Gneiss Complex. It is likely that pre-Grenvillian migmatization and tectonism affected the Saguenay Gneiss Complex because straight gneiss structures are involved in inferred Grenvillian folding.

The Cap de la Mer Amphibolite (Fig. 2) is composed principally of amphibolite, although primary gabbroic texture and primary orthopyroxene are clearly visible in less deformed lenses (Hébert and Lacoste, 1998b). This unit is interpreted as a distinct lithodem, although it is surrounded by supracrustal facies of the Saguenay Gneiss Complex. The Cap à l'Est Gneiss Complex (Fig. 2) is principally composed of granulite-facies monzonitic, granitic, and syenitic gneiss (Hébert, 1997; Hébert and Lacoste, 1998b, 1998c, 1998d, 1998e, 1998f). Ubiquitous amphibolite boudins are also components of the complex, but account for less than 5% of its volume. Amphibolite is the most abundant lithology (40%) within five to six kilometers of the contact with the Cap de la Mer Amphibolite, suggesting a genetic link between the boudins and the Cap de la Mer Amphibolite. The Cap à l'Est Gneiss Complex also contains screens of paragneiss, quartzite, and marble, as well as fragments or enclaves of massive to gneissic tonalite, gabbro, pyroxenite, norite, and anorthosite. The field relationships described earlier suggest that the Cap à l'Est Gneiss Complex is younger than the Cap de la Mer Amphibolite.

Intrusive Rocks

The Cyriac Rapakivi Granite (Fig. 2) corresponds to the adamellite described by Hervet et al. (1994; Fig. 3). Enclaves of mafic rocks observed within the Cyriac Rapakivi Granite are similar to the already described 1393 +22/–10 Ma Ruisseau à Jean-Guy Mafic Intrusion.

The Kénogami Charnockite (Fig. 2) comprises essentially coarse-grained augen charnockite and some mangerite. The Baie à Cadie Mafic-Ultramafic Suite (Fig. 2) contains many distinct facies, such as harzburgite, dunite, olivine gabbronorite, and gabbronorite (Vaillancourt, 2001). Toward the southwest, the extension of the Baie à Cadie Mafic-Ultramafic Suite can be followed for several kilometers (Fig. 2).

The Simoncouche Gabbro (Fig. 2) is a coarse- to medium-grained black to dark gray rock. The texture is subophitic, with hornblende the most abundant mafic mineral, although pyroxene is rare. This unit is undeformed except for its weakly foliated margins (Lacoste and Hébert, 1998).

Igneous Rocks Observed Outside of the Chicoutimi Gneiss Complex

Lac St. Jean Anorthosite Suite

The Lac St. Jean Anorthosite Suite (Emslie and Hunt, 1990; Hébert, 1997; Hébert and Lacoste, 1998a) is composed of multiple intrusions dated between 1160 and 1140 Ma (Higgins and van Breemen, 1996; Table 1). Mafic and anorthositic rocks comprise anorthosite, leuconorite, norite, gabbro, olivine gabbro (troctolite), diorite, pyroxenite, peridotite, dunite, and magnetitite-ilmeninite (Hébert et al., 1998b). The anorthositic suite hosts the 1157 ± 3 Ma Bégin Leucotroctolite megadike, the 1157 ± 3 Ma Lac Chabot Diorite (Higgins and van Breemen, 1992; Fig. 2; Table 1), and the 1148 ± 4 Ma Pyroxene Monzonite (Emslie and Hunt 1990; Fig. 2; Table 1). The 1146 ± 3 Ma Labrecque Granite (Fig. 2; Table 1; Higgins and van Breemen, 1996), is the only felsic unit related to the Lac St. Jean Anorthosite Suite (AMCG).

Younger Intrusions

The Chicoutimi Mangerite, the La Baie Granite, and the Taché Leucogabbro represent the St. Urbain Anorthosite Suite in this study area.

The Chicoutimi Mangerite comprises mainly porphyritic pink or green mangerite, with homogeneous prismatic K-feldspar

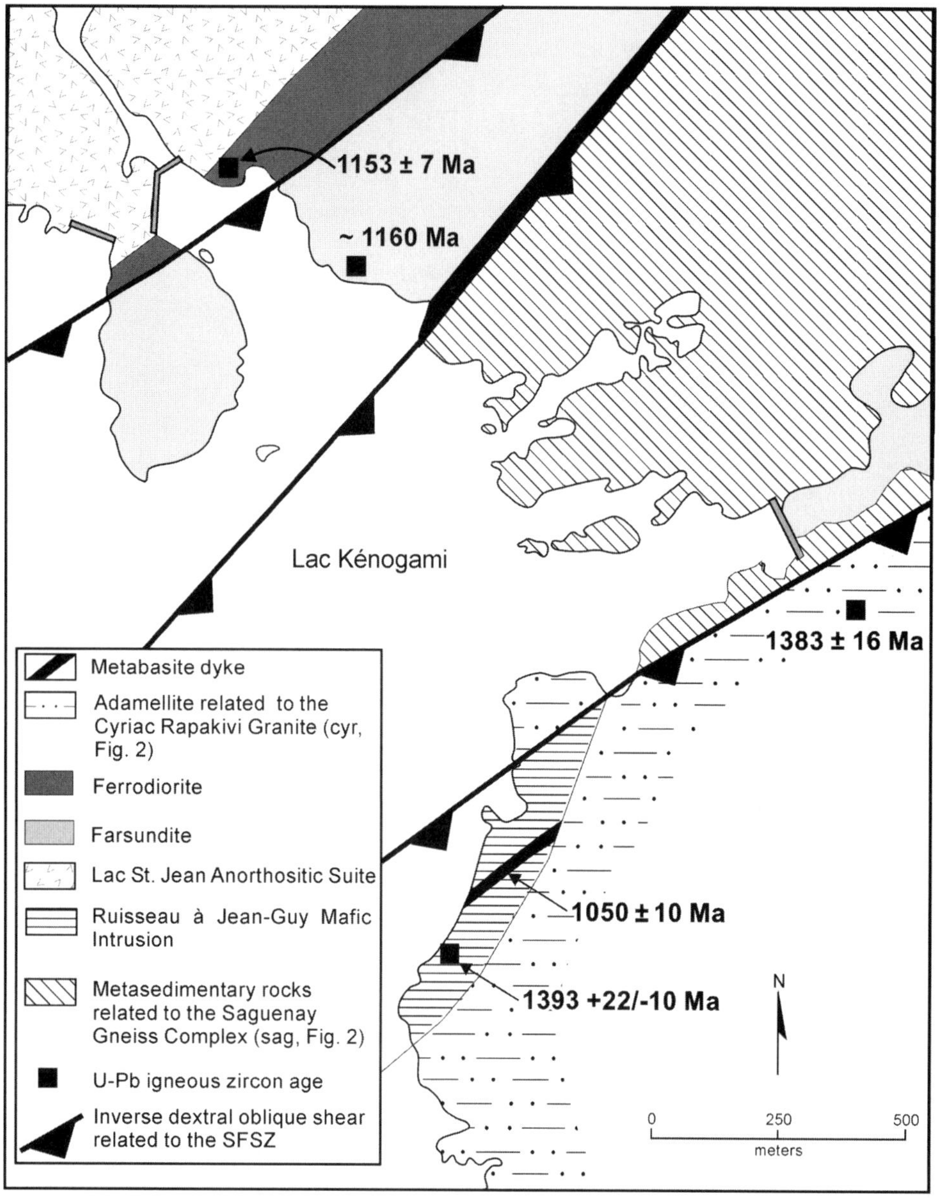

Figure 3. Geology of the margin of the Lac St. Jean Anorthosite Suite at Lac Kénogami, modified from Hervet et al. (1994).

phenocrysts and orthopyroxene and hornblende. It also contains some augen to gneissic equigranular monzonitic rocks and gneissic comagmatic mafic rocks (amphibolite and diorite) that locally contain orthopyroxene (Hébert and Lacoste, 1998a). The green and pink facies yielded an age of 1082 ± 3 Ma (Higgins and van Breemen, 1992; Hervet et al., 1994; Fig. 2; Table 1).

The La Baie Granite contains principally a porphyritic rapakivi granite (Fig. 2, lba1) and an augen-textured mangerite-charnockite (Fig. 2, lba2). It also contains some outcrops of comagmatic mafic rocks (Hébert and Lacoste, 1998b). The porphyritic rapakivi granite phase was dated at 1067 ± 4 Ma (Higgins and van Breemen, 1996; Table 1; Fig. 2). Taché Leucogabbro, represented by small intrusions that occur within the Lac St. Jean Anorthosite Suite (Fig. 2), yields an age of 1076 ± 2 Ma (Higgins and van Breemen, 1992). Higgins and van Breemen (1996) linked these three units to the Sm/Nd isochron age of 1079 ± 22 Ma of the St. Urbain Anorthosite (AMCG; Ashwal and Wooden, 1983).

Unassigned Intrusions

Several undated, and hence unassigned, intrusions have also been mapped. The Anse à Philippe Layered Intrusion occurs as large enclaves of a mafic to ultramafic layered sequence within the La Baie Granite along the shoreline of Baie Ha! Ha! (Fig. 2). Lithologies include gabbro, pyroxenite, peridotite, and dunite. These rocks can be grouped into two units (Lavoie, 1991). The first consists of olivine mafic and ultramafic rocks enclosing pockets of gabbro near the top, and then pockets of layered hornblende gabbronorite and pyroxenite. The second unit is composed of olivine mafic rocks and olivine orthopyroxenite. The noritic composition of these rocks suggests a possible correlation with the Lac St. Jean Anorthosite Suite.

The Des Îlets Granite (Fig. 2) is a pink monzogranite, granite, and quartz monzonite. Biotite is accompanied by minor amphibole. Mortar and granoblastic textures are ubiquitous. This granite could be a late phase of the La Baie Granite.

The Laurent Syenite (Fig. 2) is partly truncated in its southern part by one of the Saguenay Graben faults. The syenite is pink to red, medium-grained, and massive. The margin of the intrusion is monzonitic and strongly foliated along its southern contact.

STRUCTURE

The Grenvillian deformational history of the region can be separated into three events. The first event (D_1) is related to a major period of thrusting that produced foliation or gneissosity generally oriented from east-west to east-southeast (S_1). Dip directions are variable, although a moderate northerly dip clearly predominates. The stretching lineation has a low rake in the S_1 plane, and plunges mainly toward the north. The S_1 fabric and the thrust faults were affected by a D_2 event, which makes it difficult to reconstitute the initial geometry of the nappes.

Evidence for thrusting is often observed within the Cap à l'Est Gneiss Complex and the anorthositic suite. Undoing the effect of syn- to post-tectonic thrusting events would result in the anorthositic suite's covering a surface of 20,000 km^2, and would make this assemblage the world's largest documented anorthositic suite. The Chicoutimi Mangerite also shows an early east-west gneissic foliation but no evidence for thrusting.

The D_2 event was a period of shortening and strike-slip movement, and is responsible for the dominant northeast-southwest tectonic fabric. An S_2 foliation with a strike direction varying from 025° to 060° is recognized throughout the region and is generally the dominant foliation. Open to tight P_2 folds affect the S_1 foliation where the axial plane (S_2) has a subvertical dip. Fold axes have low rakes that plunge toward the north. When the deformation is more intense, the S_2 fabric is represented by subvertical straight gneiss where the rake of stretching lineations varies from 45° toward north to subvertical. The most important straight gneiss zone is a kilometer-wide deformation corridor named the St. Fulgence shear zone (Hébert and Lacoste, 1994; Hébert et al, 1998a). The available shear-sense indicators systematically show dextral displacement. The St. Fulgence shear zone cuts the D_1 thrusts and has evolved into dextral strike-slip faults during postcollisional readjustments. This structural combination generated an anastomosing tectonic pattern in which shear zones isolate immense, elongate blocks tens of kilometers in size, retaining a well-preserved east-west to east-southeast S_1 foliation as in some units.

The St. Fulgence shear zone is a major structure in this region and is present at the contact between all the lithodems observed. This structure can be traced from the Saguenay region for more than 400 km south to the Portneuf region, where it was named the St. Maurice lithotectonic zone (Nadeau and Corrigan, 1991; Hébert and Nadeau, 1995; Fig. 1).

The D_3 deformation produced numerous brittle-ductile subvertical faults with northerly to north-northeast trends. Stretching lineations associated with these faults are subhorizontal. Depending on the lithology affected, spectacular mylonite zones contain a variety of shear-sense indicators. Shear-sense indicator fabric and dragging of older fabrics indicate systematic sinistral displacement. The offsets observed along these structures are typically tens of meters in length, but rarely exceed several hundred meters. These phenomena are particularly well exposed in the Lac Kénogami area (Fig. 2; Hébert and Lacoste, 1998a), where the anorthosite contact is offset in an échelon pattern.

Relating the timing of these three deformations to the various AMCG emplacement events is the subject of an ongoing study.

GEOCHRONOLOGY

Six samples (Fig. 2; Table 2) have been collected for U-Pb thermal ionization mass spectrometry (TIMS) geochronology that may span much of the duration of the tectonic history of the Saguenay region. Field relationships indicate that two assemblages were older than the AMCG suite: namely, the Saguenay Gneiss Complex and the Cap à l'Est Gneiss Complex. The Cap de la Mer Amphibolite unit that represents a supracrustal component of the Saguenay Gneiss Complex and granulitic gneiss from the Cap à l'Est Gneiss Complex were sampled for geochronology (Fig. 2). A third sample was collected from the Cyriac Rapakivi Granite in order to test whether this granite is correlative with the porphyritic facies of the La Baie Granite and also to determine if the rapakivi and the augen facies are parts of the same suite. The fourth sample was selected from a charnockite of the augen Kénogami Charnockite, because this facies was inferred to be related to the augen charnockite-mangerite facies of the La Baie Granite. This sample was also analyzed using the U-Pb-SHRIMP (sensitive high-resolution ionization microprobe) technique. The fifth sample was collected from the Baie à Cadie Mafic-Ultramafic Suite, because some facies appeared to be related to the Lac St. Jean Anorthosite Suite. The sixth comes from the Simoncouche Gabbro that outcrops along the St. Fulgence shear zone within the Cap de la Mer Amphibolite. A weak foliation suggests that this unit was emplaced late in the tectonic history of this region.

Analytical Techniques

The isotope dilution and thermal ionization techniques (TIMS) used for measuring U-Pb isotopes in zircon at the Geological Survey of Canada (GSC) are summarized by Parrish et al. (1987), and are based on Krogh (1973). Prior to TIMS analysis, all zircon fractions were strongly abraded until the

TABLE 2. U-PB TIMS[†] ISOTOPIC DATA FOR ZIRCON

Fraction[§]	Weight[#] (mg)	U (ppm)	Pb** (ppm)	^{206}Pb[§§]/^{204}Pb	Pb[##] (pg)	^{208}Pb***/^{206}Pb	^{207}Pb***/^{235}U	^{206}Pb***/^{238}U	^{207}Pb***/^{206}Pb	Age (Ma) ^{207}Pb[§§§]/^{206}Pb	Disc[###]
Cap de la Mer Amphibolite. Location: NTS 22D/07, zone 19, 359393E, 5367344N (NAD 83)											
E, 50*75	34	397	79	43458	4	0.057	0.2027 ± 0.08	2.285 ± 0.09	0.08175 ± 0.03	1240 ± 1	4.40
D, 50*100	39	401	82	45324	4	0.058	0.2074 ± 0.08	2.367 ± 0.10	0.08274 ± 0.03	1263 ± 1	4.17
C, 75*150	44	313	68	17573	11	0.068	0.2179 ± 0.10	2.559 ± 0.12	0.08519 ± 0.07	1320 ± 3	4.10
B, 85*125	63	125	30	15072	7	0.141	0.2241 ± 0.08	2.678 ± 0.10	0.08668 ± 0.04	1354 ± 1	4.09
A, 100*175	43	254	60	1322	120	0.096	0.2300 ± 0.09	2.786 ± 0.14	0.08786 ± 0.09	1379 ± 4	3.62
Cap à l'Est Gneiss Complex. Location: NTS 22D/07, zone 19, 363778E, 5367017N (NAD 83)											
A, 50*150	32	307	71	40921	3	0.186	0.2128 ± 0.08	2.448 ± 0.09	0.08344 ± 0.03	1279 ± 1	3.05
D, 75*100	17	359	83	14224	6	0.154	0.2166 ± 0.09	2.512 ± 0.10	0.08412 ± 0.03	1295 ± 1	2.68
B, 50*125	54	331	82	75504	3	0.194	0.2253 ± 0.08	2.665 ± 0.10	0.08579 ± 0.03	1334 ± 1	1.97
C, 100*200	22	352	88	45769	2	0.195	0.2278 ± 0.08	2.709 ± 0.09	0.08623 ± 0.03	1343 ± 1	1.66
Cyriac Rapakivi Granite. Location: NTS 22D/06, zone 19, 332815E, 5357885N (NAD 83)											
A1, 50*125	14	236	57	14431	2	0.109	0.2326 ± 0.08	2.800 ± 0.10	0.08728 ± 0.03	1367 ± 1	1.49
B1, 75*175	31	297	72	39495	2	0.107	0.2374 ± 0.08	2.896 ± 0.10	0.08845 ± 0.03	1392 ± 1	1.51
A2, 50*125	22	268	66	21694	4	0.107	0.2389 ± 0.08	3.006 ± 0.09	0.09129 ± 0.03	1453 ± 1	5.49
A3, 50*125	25	280	70	25040	4	0.111	0.2405 ± 0.08	3.055 ± 0.09	0.09211 ± 0.03	1470 ± 1	6.06
C1, 35*95	4	113	28	455	14	0.131	0.2391 ± 0.13	2.935 ± 0.30	0.08903 ± 0.24	1405 ± 9	1.82
C2, 40*100	5	163	40	1486	9	0.116	0.2372 ± 0.09	2.880 ± 0.13	0.08804 ± 0.08	1384 ± 3	0.90
C3, 40*105	4	112	28	1468	5	0.114	0.2384 ± 0.12	2.921 ± 0.16	0.08888 ± 0.10	1402 ± 4	1.86
Kénogami Charnockite. Location: NTS 22D/06, zone 19, 327066E, 5348795N (NAD 83)											
Z1, 50*200	15	55	11	2984	3	0.119	2.046 ± 0.18	0.1921 ± 0.16	0.07722 ± 0.10	1127 ± 4	–0.60
Z2, 50*200	21	105	20	1603	16	0.103	2.000 ± 0.17	0.1887 ± 0.11	0.07687 ± 0.10	1118 ± 4	–0.30
Z3, 50*200	20	72	14	7427	2	0.117	2.062 ± 0.14	0.1924 ± 0.13	0.07772 ± 0.08	1140 ± 3	0.50
Z6, 50*200	28	154	30	20130	2	0.1033	2.016 ±0.13	0.1892 ± 0.11	0.07728 ± 0.04	1128 ± 2	1.1
Baie à Cadie Mafic-Ultramafic Suite. Location: NTS 22D/06, zone 19, 329427E, 5354256N (NAD 83)											
A1, 75*125	12	195	43	1872	15	0.241	0.1949 ± 0.10	2.099 ± 0.13	0.07810 ± 0.08	1149 ± 3	0.15
A2, 75*125	10	99	21	3850	1	0.169	0.1953 ± 0.12	2.106 ± 0.13	0.07819 ± 0.08	1152 ± 3	0.14
B2, 75*125	21	434	99	5450	21	0.271	0.1966 ± 0.08	2.131 ± 0.10	0.07862 ± 0.04	1163 ± 2	0.54
B1, 75*125	19	252	56	2928	21	0.184	0.2032 ± 0.09	2.246 ± 0.11	0.08017 ± 0.05	1201 ± 2	0.81
Simoncouche Gabbro. Location: NTS 22D/07, zone 19, 362098E, 5369345N (NAD 83)											
C, 50*60	7	142	25	3196	4	0.103	0.1755 ± 0.11	1.796 ± 0.14	0.07424 ± 0.09	1048 ± 4	0.61
D, 50*100	8	221	39	4183	2	0.104	0.1748 ± 0.09	1.793 ± 0.11	0.07441 ± 0.06	1053 ± 2	1.44
B, 75*80	8	305	55	6708	0	0.107	0.1762 ± 0.09	1.810 ± 0.10	0.07446 ± 0.04	1054 ± 2	0.79
A, 40*70	9	279	50	2726	10	0.107	0.1766 ± 0.09	1.818 ± 0.11	0.07465 ± 0.06	1059 ± 2	1.07

[†]Thermal ionization mass spectrometry.
*Sizes in μm before abrasion; all fractions are nonmagnetic at a side slope of –0.5° on a Frantz isodynamic magnetic separator operating at 1.8 amps.
[#]Error on weight = ±1 μg.
**Radiogenic Pb.
[§§]Measured ratio corrected for spike and Pb fractionation of 0.09 ± 0.045% per AMU.
[##]Total common Pb on analysis corrected for fractionation and spike.
***Corrected for blank Pb and U, common Pb; errors quoted are 1 sigma in percent.
[§§§]Age errors quoted are 2 SE in Ma.
[###]Discordance in % along a discordia to origin.

crystals assumed a well-rounded shape (Krogh, 1982). Abrasion was used to minimize the effects of peripheral lead loss and/or to remove metamorphic rims. Mass spectrometry, data reduction, and propagation of analytical uncertainties of the relevant components in the calculation of isotopic ratios and ages were performed following the numerical procedure of Roddick et al. (1987). A modified form of York's (1969) method for linear regression analysis was used (see Parrish et al., 1987). The isotopic data are presented in Table 2. All age uncertainties are given at the 95% confidence level.

For U-Pb analysis using the GSC SHRIMP II facility, ~150 grains were selected at random from the least magnetic fraction, with an average size of ca. 100 microns. Zircon grains were mounted in a 2.5 cm epoxy disk (#IP206) along with fragments of the GSC zircon standard 6266. Bias in the measured Pb/U ratio was corrected relative to 6266 zircon using the $^{206}Pb^{*}/^{270}[UO_2]$ versus $^{254}[UO]/^{238}[U]$ technique described by Stern and Amelin (2002, personal commun.). The polished zircons were imaged in cathodoluminescence in order to guide the placement of the analysis site. Selected areas of the zircons were sputtered using a mass-filtered O^- primary beam operating in Kohler mode. The primary beam averaged 13 nA, generating elliptical pits of 20 microns in diameter. Pits were less than 1 micron deep after completion of 20 minute analyses. Further details of analytical procedures using the SHRIMP II ion microprobe at the GSC are available in Stern (1996 and 1997). Age estimates and error calculations were performed using Isoplot/Ex version 2.2 (Ludwig, 2000). Since all the SHRIMP analyses in this study were concordant to slightly discordant, the weighted mean $^{207}Pb/^{206}Pb$ ages were considered adequate for estimating the ages of zircon growth. The SHRIMP U-Pb data are listed in Table 3 and shown in Figure 4, D, with age uncertainties presented at the 68% confidence level.

Results

Sample 1: Cap de la Mer Amphibolite

The sample is a medium- to fine-grained, foliated gabbroic amphibolite with a saccharoïdal and equigranular texture. It mainly contains 1–2% secondary quartz, 45–48% plagioclase, and nearly 40% amphibole. Zircons range from small, well-formed but rounded, clear and colorless prisms with a length/breadth ratio of 2:1/3:1 to small, equidimensional, clear and colorless grains. The prismatic zircons have lower U concentrations than the equidimensional grains, and contain fluid inclusions. On a concordia plot (Fig. 4, A), data points corresponding to five analyses are aligned along a discordia with upper and lower intercept ages of 1493 ± 23 Ma and 1061 ± 18 Ma. Fractions A, B, and C, which were closer to the upper intercept, consisted of prismatic zircons, whereas fractions D and E, closer to the lower intercept, were comprised of small equidimensional crystals. The mean square of weighted deviates (MSWD) from the regression calculation is 5.0, which indicates that there is considerable "geological" scatter among the data points that is beyond the analytical uncertainty. If the youngest data point, E, is removed from the regression, an MSWD of 0.83 is obtained where all the scatter is accounted for by the analytical uncertainty. This regression is considered to yield better estimates of the upper and lower intercept ages, 1506 ± 13 Ma and 1077 ± 12 Ma respectively, where the upper intercept age is interpreted as that of the igneous precursor and the lower intercept age that of new zircon growth during Grenville metamorphism. There may also have been a component of Grenville lead loss from the igneous zircon.

Sample 2: Cap à l'Est Gneiss Complex

The sample is a green gneissic charnockite. It is coarse grained, containing 34% quartz, 17% plagioclase (albite), 38% perthitic potash feldspar, and 2% clinopyroxene. Zircons are clear and colorless irregular fragments with no inclusions. These zircons are represented by fractions B, C, and D. Fraction A consists of slim, irregular prisms. The four fractions are linearly aligned, with the coarse fractions C and B closer to the upper intercept of 1391 +8/–7 Ma and the smaller fractions D and A plotting closer to the lower intercept of 1027 ± 18 Ma (Fig. 4, B). An MSWD of 0.67 indicated that the scatter of data points falls within analytical uncertainty. The upper intercept age is attributed to late magmatic crystallization. The lower intercept age is interpreted in terms of Pb loss during Grenville metamorphism.

Sample 3: Cyriac Rapakivi Granite

The sample is a pink, megacrystic, hornblende-biotite granite with rapakivi texture. Quartz (15%), plagioclase (15%), perthitic potash feldspar (~50%), hornblende (6–7%), and biotite (7%) are the main constituents. Zircons are subhedral prismatic. Cathodoluminescence images of polished zircons show locally small rounded overgrowths. TIMS results for this sample can be found in Table 2. On a concordia plot (Fig. 4, C), four fractions (B1, C1–C3) are clustered near the concordia ca. 1380 Ma, with discordances of 0.9 to 1.9% and $^{207}Pb/^{206}Pb$ ages in the range from 1405 Ma to 1384 Ma. Fraction A1 yields a younger $^{207}Pb/^{206}Pb$ age of 1367 Ma, while more discordant fractions A2 and A3 yield older $^{207}Pb/^{206}Pb$ ages of 1470 Ma and 1453 Ma. Although an age in the range from 1410 Ma to 1360 Ma is indicated, the data are open to alternative interpretations in that the cluster of data do not form a linear array and suggest a significant inherited component.

In order to clarify the ambiguity in the TIMS results, a further SHRIMP study was undertaken (see Table 3). Sixteen U-Pb analyses performed using the SHRIMP yield ages in three groupings (Fig. 4, D): four Grenville $^{207}Pb/^{206}Pb$ ages in the range from 1152 Ma to 1054 Ma, a cluster of eight ages ca. 1380 Ma, and a single analysis at 1715 Ma. The weighted mean of the central cluster yields an age of 1391 ± 21 Ma, with a zero probability of fit (MSWD = 6.6). By excluding the youngest age of 1349 Ma and the ages between 1432 Ma and 1457 Ma from the middle grouping, a weighted mean age of 1383 ± 16 Ma is

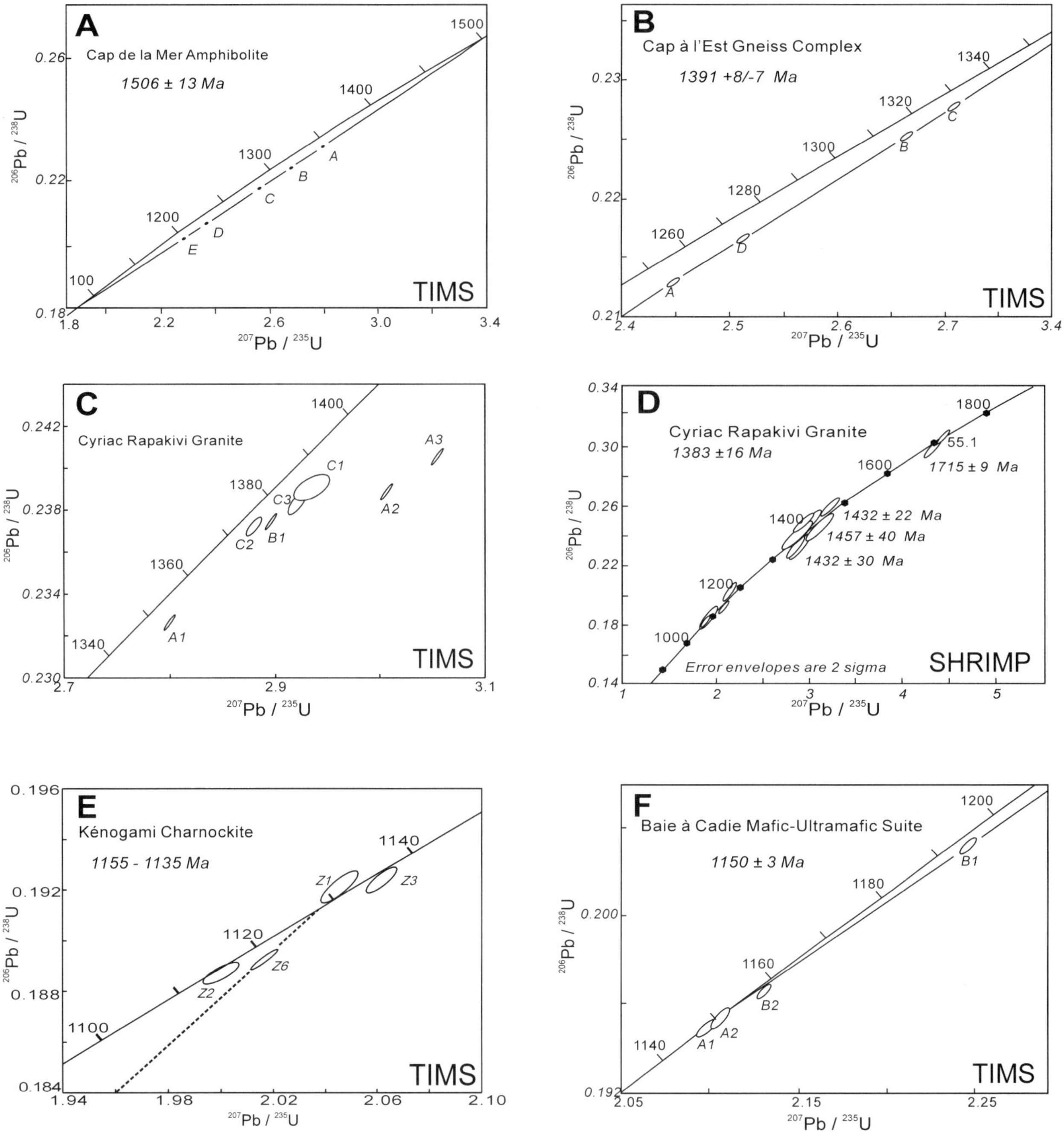

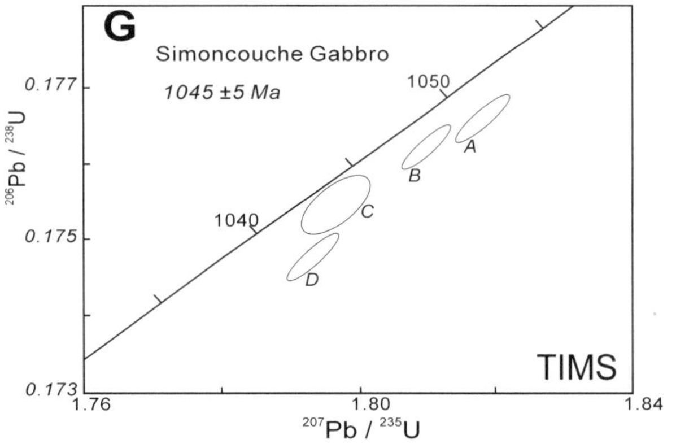

Figure 4. Uranium-lead concordia diagrams for (A) thermal ionization mass spectrometry (TIMS) data for zircons from the Cap de la Mer Amphibolite; (B) TIMS data for zircons from the Cap à l'Est Gneiss Complex; (C) TIMS data for zircons from the Cyriac Rapakivi Granite; (D) sensitive high-resolution ionization microprobe (SHRIMP) data for zircons from the Cyriac Rapakivi Granite; (E) TIMS for zircons from the Kénogami Charnockite; (F) data for zircons from the Baie à Cadie Mafic-Ultramafic Suite; and (G) TIMS data for zircons from the Simoncouche Gabbro. Error envelopes are at the 2 sigma level for both TIMS and SHRIMP data.

TABLE 3. U-PB SHRIMP* ISOTOPIC DATA FOR ZIRCON FROM CYRIAC RAPAKIVI GRANITE

																Age (Ma)				
Spot name§	U (ppm)	Th (ppm)	Th/U	Pb (ppm)	^{204}Pb (ppb)	^{204}Pb/^{206}Pb	±^{204}Pb/^{206}Pb	f206#	^{208}Pb/^{206}Pb	±^{208}Pb/^{206}Pb	^{206}Pb/^{238}U	±^{206}Pb/^{238}U	^{207}Pb/^{235}U	±^{207}Pb/^{235}U	^{207}Pb/^{206}Pb	±^{207}Pb/^{206}Pb	^{206}Pb/^{238}U	±^{206}Pb/^{238}U	^{207}Pb/^{206}Pb	±^{207}Pb/^{206}Pb	Conc. (%)**
Metamorphic zircon																					
Cyriac-70.1	1350	21	0.0162	228	2	1.00E-05	6.40E-06	0.00017	0.0061	0.00026	0.18190	0.00233	1.8674	0.02632	0.07445	0.00033	1077.3	12.72	1053.8	9.07	102.2
Cyriac-87.1	3188	51	0.0165	553	1	2.57E-06	2.39E-06	4.00E-05	0.005	1.00E-04	0.18699	0.00248	1.9342	0.03134	0.07502	0.00058	1105.1	13.5	1069.0	15.69	103.4
Cyriac-46.1	317	43	0.1402	62	1	1.00E-05	1.00E-05	0.00017	0.03935	0.00078	0.20237	0.00264	2.137	0.0322	0.07659	0.00046	1188.1	14.19	1110.4	12.14	107
Cyriac-50.1	637	14	0.0235	115	2	1.93E-05	9.95E-06	0.00033	0.00592	0.00045	0.19417	0.00276	2.0931	0.03239	0.07818	0.00035	1143.9	14.94	1151.6	9.04	99.3
Igneous zircon																					
Cyriac-33.1	464	160	0.3555	114	0	3.02E-06	6.46E-06	5.00E-05	0.1053	0.001	0.23900	0.00316	2.8504	0.0425	0.08650	0.00047	1381.5	16.47	1349.3*	10.5	102.4
Cyriac-87.2	256	117	0.4696	67	2	2.98E-05	1.05E-05	0.00052	0.14063	0.00098	0.24820	0.00331	2.9722	0.04863	0.08685	0.00069	1429.2	17.12	1357.2	15.42	105.3
Cyriac-85.1	314	61	0.2018	74	1	1.91E-05	2.32E-05	0.00033	0.06033	0.00203	0.23866	0.00394	2.8651	0.06712	0.08707	0.00128	1379.7	20.52	1362.0	28.58	101.3
Cyriac-70.2	318	79	0.2561	79	0	5.41E-06	1.24E-05	9.00E-05	0.07667	0.00099	0.24733	0.00318	2.9699	0.06421	0.08709	0.00138	1424.7	16.45	1362.4	30.85	104.6
Cyriac-46.2	343	135	0.4061	88	1	1.00E-05	1.00E-05	0.00017	0.11609	0.0017	0.24665	0.00327	2.9691	0.04437	0.08731	0.00047	1421.2	16.95	1367.2	10.46	103.9
Cyriac-39.1	420	142	0.3487	107	0	3.22E-06	1.31E-05	6.00E-05	0.10412	0.00132	0.24849	0.00333	3.0075	0.04484	0.08778	0.00045	1430.7	17.22	1377.7	9.79	103.8
Cyriac-87.3	354	130	0.3796	87	1	1.00E-05	1.00E-05	0.00017	0.11519	0.00054	0.23644	0.00304	2.8849	0.04216	0.08849	0.00049	1368.2	15.88	1393.2	10.6	98.2
Cyriac-84.1	380	72	0.1946	87	1	1.00E-05	1.00E-05	0.00017	0.05842	0.00096	0.23254	0.00304	2.8503	0.04081	0.08890	0.00039	1347.8	15.93	1402.0	8.49	96.1
Inherited zircon																					
Cyriac-22.1	260	90	0.3574	63	3	5.11E-05	1.78E-05	0.00088	0.11162	0.00129	0.23301	0.00349	2.9014	0.0511	0.09031	0.00068	1350.2	18.3	1432.1	14.47	94.3
Cyriac-60.1	145	45	0.3233	38	0	1.00E-05	1.00E-05	0.00017	0.09481	0.00086	0.25587	0.00352	3.1864	0.04915	0.09032	0.00049	1468.7	18.09	1432.3	10.47	102.5
Cyriac-77.1	153	111	0.7457	43	2	5.17E-05	2.77E- 05	0.0009	0.22138	0.00181	0.24601	0.00344	3.104	0.05762	0.09151	0.00097	1417.8	17.83	1457.3	20.24	97.3
Cyriac-55.1	319	136	0.4404	102	2	1.91E-05	9.05E-06	0.00033	0.12893	0.00102	0.30100	0.00383	4.3588	0.05788	0.10503	0.00026	1696.3	19.02	1714.8	4.58	98.9

Note: Uncertainties reported at 1 sigma and calculated by numerical propagation of all known sources of error (Stern, 1997).

*Sensitive high-resolution ionization microprobe.

§Excluded from calculation of weighted mean; isotopic system likely disturbed during metamorphism.

#f206 refers to the mole fraction of total ^{206}Pb that is due to common Pb; data reflect readings for common Pb corrected according to procedures outlined in Stern (1997).

**Conc. = 100 × (^{206}Pb/^{238}U age)/(^{207}Pb/^{206}Pb age).

obtained, with a probability of fit of 5.6% (MSWD = 2.0). This age is taken as the best estimate of the time of igneous crystallization of the zircons and the Cyriac Rapakivi Granite. The three oldest ages from the middle grouping, 1432 Ma, 1432 Ma, and 1457 Ma, are interpreted as inherited zircon, as is the oldest of all ages, 1715 Ma. The 1349 Ma age likely represents igneous zircon that suffered minor Pb loss during the Grenvillian orogeny. The four youngest ages are interpreted in terms of local zircon growth during metamorphic events of the Grenville orogeny.

Sample 4: Kénogami Charnockite

The sample is charnockite, where rapikivi texture is easily recognized within the augen potassic feldspar. Quartz (25%), feldspar (25%), and plagioclase (35%) are the main constituents. Zircons are clear and colorless, with no visible cores. Prisms have length/breadth ratios of 3:1 to 6:1, and the tips of prisms are slightly rounded. Four U-Pb zircon data points are concordant to slightly discordant and show $^{207}Pb/^{206}Pb$ ages ranging from 1140 Ma to 1127 Ma. Because of the slight rounding of the zircon crystals, the data are interpreted in terms of an igneous age that has been slightly disturbed by subsequent Grenville metamorphic effects. Following this interpretation, 1140 Ma is a minimum age for igneous crystallization (Fig. 4, E). In view of the grouping of the data points and the almost euhedral nature of the zircon crystals, it is unlikely that the maximum age greatly exceeds this age. Accordingly, a maximum age is assigned at 1155 Ma, with a minimum at 1135 Ma, where the minimum takes into account the analytical uncertainty of the oldest analysis.

Sample 5: Baie à Cadie Mafic-Ultramafic Suite

This sample is a pegmatitic gabbronorite showing a cumulate plagioclase and coronitic clinopyroxene surrounded by hornblende. Zircons are anhedral, consistent with late growth from a crystallizing magma. They are mostly clear. The best estimate of age is taken to be the average $^{207}Pb/^{206}Pb$ age of fractions A1 and A2, which is 1150 ± 3 Ma (Fig. 4, F). A line through this age and data point B1 yields an upper intercept age of 1368 ± 67 Ma, which is the same as the age of other igneous units in the area and likely reflects an inherited component.

Sample 6: Simoncouche Gabbro

The sample is a subophitic to foliated gabbro where plagioclase (>50%), hornblende (30%), and biotite (15%) are the main constituents. Zircons are anhedral fractions somewhat rounded in appearance. It is not certain whether the rounded aspect is the result of metamorphic overgrowth or resorbtion. Resorbtion is considered more likely, however, as the crystals are quite platy, often with one flat side. Crystals are very clear and colorless, and there is no obvious crystal morphology, internal or external. Four data points are clustered near the concordia (Fig. 4, G), between 1050 Ma and 1040 Ma. $^{207}Pb/^{206}Pb$ ages range from 1048 ± 4 Ma to 1059 ± 2 Ma. It is considered unlikely that a significant component of the age distribution results from metamorphic zircon growth after 1048 Ma; hence the bulk of the scatter is attributed to an inherited component. The age and uncertainty of the time of igneous crystallization are assigned at 1045 ± 5 Ma, approximated by the most concordant point, C (Fig. 4, G).

DISCUSSION

Within the Saguenay region of the Central Grenville Province, the new and previous U-Pb geochronological data show six igneous age groupings in the interval between 1506 and 1020 Ma (Table 1). Interpretations of these igneous events are given in this section, from the oldest to the youngest.

Lithodems Older than the Lac St. Jean Anorthosite Suite

Some strongly foliated rocks that were previously included within the Chicoutimi Gneiss Complex, or "basement" of the Saguenay region (Higgins and van Breemen, 1996), have been shown to belong to younger felsic intrusives groups that include the Chicoutimi Mangerite and the La Baie Granite (Fig. 2). The gneissic appearance of these younger intrusions was generated during the Grenvillian deformation. On the basis of new mapping and radiometric data presented, we redefine the basement of the Lac St. Jean Anorthosite Suite (AMCG) and younger intrusions of the Saguenay region, including the following units (Fig. 2).

1506 Ma Event

The Cap de la Mer Amphibolite (Fig. 2) yielded an age of 1506 ± 13 Ma (Fig. 4, A). This unit is assigned to the Saguenay Gneiss Complex, which outcrops along its western margin, and is interpreted to represent a part of its supracrustal component. It is interesting to note that Frith and Doig (1973) obtained a similar Rb/Sr isochron age of 1482 ± 72 Ma on a paragneiss collected in this area.

The 1506 ± 3 Ma age suggests that deposition of these supracrustal rocks may be associated with the Pinwarian event, a 1510–1450 Ma (Wasteneys et al., 1997) plutonic event found throughout the Grenville Province from Labrador-Québec to the mid-continental United States (Gower and Krogh, 2002) and attributed to an Andean-style continental-margin magmatic arc on the southeastern Laurentian margin (Tucker and Gower, 1994; Gower, 1996; Wasteneys et al., 1997). Felsic plutonic rocks generally characterize this event. In the eastern Grenville, Gower and Krogh (2002) interpreted the Pinwarian magmatism in terms of a continental arc built on pre-Pinwarian crust as demonstrated by the Sm-Nd isotopic data of Dickin (2000). However, in central Québec, from Sept Iles to Trois-Rivières and north of Lac St. Jean (Fig. 1), Dickin and Higgins (1992) and Dickin (2000) identified a broad terrane with ca. 1575–1450 Ma Sm-Nd model ages. They interpret this accreted "Quebecia" terrane in terms of a juvenile island arc. Such an interpretation is consistent with the apparent scarcity of felsic Pinwarian ages in this region and the mafic supracrustal nature of the Cap de la Mer Amphibolite.

1393–1383 Ma Event

A U-Pb zircon age of 1391 +8/–7 Ma (Fig. 4, B) was determined for a granulitic gneiss (charnockitic) sample from the Cap à l'Est Gneiss Complex (Fig. 2). These rocks could be derived from a large porphyritic, rapakivi-textured, felsic mass that intruded the Saguenay Gneiss Complex. Many enclaves (meters to kilometers long) of the latter (paragneiss, quartzite, amphibolite) have been mapped within the Cap à l'Est Gneiss Complex and could be ca. 1500 Ma or older.

The Cyriac Rapakivi Granite (Fig. 2) yielded a U-Pb zircon age of 1383 ± 16 Ma (Fig. 4, C and D), thus precluding correlation with the porphyritic facies of the La Baie Granite (1067 ± 3 Ma). A slightly older U-Pb age of 1393 +22/–10 Ma was obtained for the Ruisseau à Jean-Guy Mafic Intrusion. The latter also occurs as gabbro enclaves within the intrusive adamellite of Hervet et al. (1994), which corresponds to the Cyriac Rapakivi Granite.

The 1506 Ma and the 1390–1380 Ma lithodems are the host rocks of the Lac St. Jean Anorthosite Suite, the Chicoutimi Mangerite, and the La Baie Granite. As already mentioned, the relations between these host rocks and the younger intrusions are clearly structural and are represented by the St. Fulgence shear zone. So these host rocks have to be considered as structural basement to the Grenvillian intrusions.

On a more regional scale, the 1391 +8/–7 Ma Cap à l'Est Gneiss Complex, the 1383 ± 16 Ma Cyriac Rapakivi Granite, and the previously dated 1393 +22/–10 Ma Ruisseau à Jean-Guy Mafic Intrusion (Hervet et al., 1994) in the Saguenay area are coeval with the La Bostonnais Complex in the Portneuf area (Fig. 1), with ages in the 1410–1370 Ma range (Nadeau and van Breemen, 1994; Corrigan and van Breemen, 1997). The La Bostonnais Complex (Rondot, 1978; Hébert and Nadeau, 1990; Nadeau et al., 1992) occurs along the St. Maurice lithotectonic zone, which corresponds to the St. Fulgence shear zone in the Saguenay region. Dickin (2000) interpreted the La Bostonnais Complex in terms of an ensialic arc that reworked part of the Quebecia terrane, perhaps during accretion to Laurentia.

No age equivalent was found in this study for the 1.45 Ma Montauban Group, which contains mafic and felsic volcanics in the Portneuf area (Fig. 1; Nadeau and van Breemen, 1994).

Lac St. Jean Anorthosite Suite

1160–1140 Ma Event

The Baie à Cadie Mafic-Ultramafic Suite (Fig. 2) is one of the few major ultramafic bodies documented near the southern limit of the Lac St. Jean Anorthosite Suite. The U-Pb zircon age of 1150 ± 3 Ma (Fig. 5, F) obtained from a sample of a pegmatitic gabbro-norite of the Baie à Cadie Mafic-Ultramafic Suite shows that this assemblage is coeval with the Lac St. Jean Anorthosite Suite (1160–1140 Ma). Assuming that it is also cogenetic, these mafic-ultramafic rocks may represent a cumulate magma related to the anorthositic suite. Their position, separated from the main anorthositic mass, is interpreted to have resulted from deeper-level thrusting along the St. Fulgence shear zone within the Kénogami Charnockite (1140 ± 3 Ma). The 1155–1135 Ma Kénogami Charnockite is now recognized as a felsic component of the Lac St. Jean Anorthosite Suite (AMCG), as is the 1146 ± 3 Ma Labrecque Granite (Fig. 2).

The Baie à Cadie Mafic-Ultramafic Suite is economically important because it contains Ni-Cu showings that could eventually be comparable to other Ni-Cu mineralization found in the main mass of the Lac St. Jean Anorthosite Suite (Cimon and Hébert, 1998; Clark and Hébert, 1998; Hébert and Beaumier, 2000).

Lithologies Post–Lac St. Jean Anorthosite Suite

1082–1067 Ma Event

Higgins and van Breemen (1996) have proposed that the 1082 ± 3 Ma Chicoutimi Mangerite and the 1067 ± 3 Ma La Baie Granite are correlative to the Sm/Nd isochron age 1079 ± 22 Ma St. Urbain Anorthosite (AMCG). No new lithodems have been recognized to belong to this event, but geological mapping suggests a greater areal extent of both the Chicoutimi Mangerite and the La Baie Granite beyond previously accepted limits.

1050–1045 Ma Event

Numerous mafic dikes have been studied in the Saguenay area, and an intrusive chronology was proposed by Woussen et al. (1981). The chronology of these dike swarms was used to establish the regional lithostratigraphy. However, only one dike has been dated, a metabasite dike on the northeast shore of Lac Kénogami (Fig. 3) that yielded a U-Pb age of 1050 ± 10 Ma (Hervet et al., 1994). This dike cuts the Ruisseau à Jean-Guy Mafic Intrusion.

The Simoncouche Gabbro was collected for dating because it outcrops along the St. Fulgence shear zone at the contact between the Kénogami Charcockite and the La Baie Granite. The same gabbro outcrops within the Cap de la Mer Amphibolite. The agreement in age of this 1045 ± 5 Ma gabbro (Fig. 4, G) with the 1050 ± 10 Ma metabasite dike in the Lac Kénogami area (Hervet et al., 1994) is consistent with the interpretation that the Simoncouche Gabbro is also a dike. While these dikes could represent a late mafic phase (dike swarm) related to the emplacement of the Chicoutimi Mangerite and the La Baie Granite, they are more likely to represent a later distinct magmatic phase. They appear to be late to post–St. Fulgence shear zone because of their position relative to the shear zone. This is especially true of the long dike near the southern end of Lac Kénogami, which is weakly foliated along its margins (Fig. 2) and therefore dates the last movements along this regionally important deformation zone.

1020–1010 Ma Event

The 1020 +4/–3 Ma St. Ambroise Pluton (Higgins and van Breemen, 1996) remains the only igneous unit in the Saguenay region that likely belongs to the 1020–1010 Ma Labrieville AMCG Suite.

ACKNOWLEDGMENTS

The authors acknowledge Diane Bellerive, Carole Lafontaine, and Natalie Morisset for analytical help. In the field, special thanks are addressed to Pierre Lacoste, Johanne Nadeau, and Christine Vaillancourt, for their help in choosing and collecting samples for geochronology. The manuscript was critically read by Charlie Gower, Michael Higgins, Nicole Rayner, and Natasha Wodicka.

REFERENCES CITED

Ashwal, L.D., and Wooden, J.L., 1983, Sr and Nd isotope geochronology, geologic history and the origin of the Adirondack anorthosite: Geochimica Cosmochimica Acta, v. 47, p. 1875–1885.

Cimon, J., and Hébert, C., 1998, Processus magmatique à l'origine d'une séquence jotunitique différenciée dans l'anorthosite du Lac-Saint-Jean: Québec, Ressources naturelles du Québec, DV 98–05, p. 41.

Clark, T., and Hébert, C., 1998, Étude du gîte de Cu-Ni-Co de McNickel: Suite anorthositique de Lac-Saint-Jean: Québec, Ressources naturelles du Québec, ET 98-02, 52 p.

Corrigan, D., and van Breemen, O., 1997, U-Pb age constraints for the lithotectonic evolution of the Grenville province along the Mauricie transect, Quebec: Canadian Journal of Earth Sciences, v. 34, p. 299–316.

Corriveau, L., 1982, Physical conditions of the regional and the retrograde metamorphism in the pelitic gneiss of the Chicoutimi area, Quebec [M.S. thesis]: Kingston, Ontario, Queen's University, 264 p.

Dagenais, S., 1983, Pétrographie et stratigraphie de la séquence des paragneiss de Saint-Fulgence, Région du Haut-Saguenay, Québec [M.S. thesis]: Chicoutimi, Université du Québec à Chicoutimi, 165 p.

Dickin, A.P., 2000, Crustal formation in the Grenville Province; Nd-isotope evidence: Canadian Journal of Earth Sciences, v. 37, p. 165–181.

Dickin, A.P., and Higgins, M.H., 1992, Sm/Nd evidence for a major 1.5 Ga crust-forming event in the central Grenville province: Geology, v. 20, p. 137–140.

Dimroth, E., Woussen, G., and Roy, D.W., 1981, Geologic history of the Saguenay region, Quebec (Central Granulite Terrain of the Grenville Province): A working hypothesis: Canadian Journal of Earth Sciences, v. 8, p. 1506–1522.

Emslie, R.F., and Hunt, P.A., 1990, Age and petrogenic significance of igneous mangerite-charnockite suites associated with massif anorthosites, Grenville Province: Journal of Geology, v. 98, p. 213–232.

Frith, R., and Doig, R., 1973, Rb-Sr isotopic ages and petrologic studies of the rocks of the Lac-Saint-Jean area, Quebec: Canadian Journal of Earth Sciences, v. 10, p. 881–889.

Gower, C.F., 1996, The evolution of the Grenville Province in the eastern Labrador, Canada, *in* Brewer, T.S., ed., Precambrian Crustal Evolution in the North Atlantic Region: Geological Society [London] Special Publication 112, p. 197–218.

Gower, C.F., and Krogh, T.E., 2002, A U-Pb geochronological review of the Proterozoic history of the eastern Grenville Province: Canadian Journal of Earth Sciences, v. 39, p. 795–829.

Hébert, C., 1997, Géologie et compilation de la région de Chicoutimi (22D): Québec, Ministère de Ressources Naturelles du Québec, carte SI-22D-G2P-97K.

Hébert, C., and Beaumier, M., 2000, Géologie de la région du lac à Paul (SNRC 22E/15): Québec, Ministère de Ressources Naturelles du Québec, RG 99–05, 31 p.

Hébert, C., and Lacoste, P., 1994, Linéament de Saint-Fulgence-Poulin-de-Courval, *in* Séminaire d'information sur la recherche géologique: Programme et résumés 1994: Québec, Ministère de Ressources Naturelles du Québec, DV-94–09, p. 56.

Hébert, C., and Lacoste, P., 1998a, Géologie de la région de Jonquière-Chicoutimi (22D/06): Québec, Ministère de Ressources Naturelles du Québec, RG 97–08, 31 p.

Hébert, C., and Lacoste, P., 1998b, Géologie de région de Bagotville (22D/07): Québec, Ministère de Ressources Naturelles du Québec, RG 97–06, 21 p.

Hébert, C., and Lacoste, P., 1998c, Géologie de la région de Lac Jalobert (22D/10): Québec, Ministère de Ressources Naturelles du Québec, RG 97–05, 15 p.

Hébert, C., and Lacoste, P., 1998d, Géologie de la région du lac Poulin-de-Courval (22D16): Québec, Ministère de Ressources Naturelles du Québec, RG-97–03, 13 p.

Hébert, C., and Lacoste, P., 1998e, Géologie de la région du lac Moncouche (22D/15): Québec, Ministère de Ressources Naturelles du Québec, carte SI-22D15-C3G-98X.

Hébert, C., and Lacoste, P., 1998f, Géologie de la région de lac des Savanes (22D/09): Québec, Ministère de Ressources Naturelles du Québec, carte SI-22D09-C3G-98X.

Hébert, C., and Nadeau, L., 1990, Déformation et extension de l'assemblage métasédimentaite de Montauban dans la réserve de Portneuf: Résumé des conférences: Québec, Ministère de Ressources Naturelles du Québec, DV 90–40, p. 11–14.

Hébert, C., and Nadeau, L., 1995, Géologie de la région de Talbot (Portneuf) (31P/01): Québec, Ministère de Ressources Naturelles du Québec, ET 95–01, 16 p.

Hébert, C., Chown, E.H., and Daigneault, R., 1998a, Tectono-magmatic history of the Saguenay region: Guidebook of the pre-congress field trip no. A-06: St. John's, Newfoundland–Ottawa, Ontario, Geological Association of Canada–Mineralogical Association of Canada.

Hébert, C., van Breemen, O., and Lacoste, P., 1998b, Tectonic setting and U-Pb zircon age of the Poulin-de-Courval Mangerite, Saguenay–Lac-Saint-Jean area, Grenville Province, Quebec, *in* Radiogenic Age and Isotopic Studies: Report 11: Ottawa, Ontario, Geological Survey of Canada, Current Research 1998-F, p. 69–76.

Hervet, M.D., van Breemen, O., and Higgins, M., 1994, U-Pb igneous crystallisation age of intrusive rocks near the southeastern margin of the Lac-Saint-Jean Anorthosite Complex, Grenville Province, Québec, *in* Radiogenic Age and Isotopic Studies: Report 8: Ottawa, Ontario, Geological Survey of Canada, Current Research 1994-F, p. 115–124.

Higgins, M.D., and van Breemen, O., 1992, The age of the Lac-Saint-Jean Anorthosite Suite intrusion and associated mafic rocks, Grenville Province, Canada: Canadian Journal of Earth Sciences, v. 29, p.1412–1423.

Higgins, M.D., and van Breemen, O., 1996, Three generations of AMCG magmatism, contact metamorphism and tectonism in the Saguenay–Lac-Saint-Jean region, Grenville Province, Canada: Precambrian Research, v. 79, p. 327–346.

Krogh T.E., 1973, A low contamination method for hydrothermal decomposition of zircon and extraction of U and Pb for isotopic age determinations: Geochimica et Cosmochimica Acta, v. 377, p. 485–494.

Krogh T.E., 1982, Improved accuracy of U-Pb ages by the creation of more concordant systems using an air abrasion technique: Geochimica et Cosmochimica Acta, v. 46, p. 637–649.

Lacoste, P., and Hébert, C., 1998, Carte géologie du feuillet Rivière Pikauba, 22D/03: Québec, Ministère de Ressources naturelles du Québec, RG 98–10, 18 p.

Laurin, A.F., and Sharma, K.N.M., 1975, Région des rivières Mistassini, Péribonka et Saguenay (Grenville 1965–67): Québec, Ressources naturelles du Québec, RG 161, 89 p.

Lavoie, J., 1991, Description des principales unités du complex stratiforme de

La Baie: Chicoutimi, Université du Québec à Chicoutimi, Projet de fin d'étude.

Ludwig, K.R., 2000, Isoplot/Ex, version 2.2: A geochronological toolkit for Microsoft Excel: Berkeley, California, Berkeley Geochronology Center Special Publication 1a, 53 p.

Nadeau, L., and Corrigan, D., 1991, Preliminary notes on the geology of the Saint-Maurice tectonic zone, Grenville orogen, Quebec: Ottawa, Ontario, Geological Survey of Canada, Current Research, Paper 91-1E, p. 245–255.

Nadeau, L., and van Breemen, O., 1994. Do the 1.45–1.39 Ga Montauban group and the La Bostonnais complex constitute a Grenvillian accreted terrane? Toronto-Ottawa, Ontario, Geological Association of Canada–Mineralogical Association of Canada Program with Abstracts, v. 19, p. A81.

Nadeau, L., van Breemen, O., and Hébert, C., 1992, Géologie, âge et extension géographique du Groupe de Montauban et du complex de la Bostonnais: Québec, Ministère de l'Énergie et des Ressources du Québec, DV 92–03, p. 35–39.

North American Commission on Stratigraphic Nomenclature (NACSN), 1983, North American Stratigraphic Code: American Association of Petroleum Geologists Bulletin, v. 67, no. 5, p. 841–875.

Owen, S.V., 1981, Petrography of leucocratic segregation in the migmatitic Old Gneiss Complex east of Chicoutimi, Québec [M.S. thesis]: Chicoutimi, Université du Québec à Chicoutimi, 172 p.

Owen, S.V., Dimroth, E., and Woussen, G., 1980, The Old Gneiss Complex east of Chicoutimi, Québec: Ottawa, Ontario, Geological Survey of Canada, Current Research, Paper 80-1A, p. 137–146.

Owens, B.E., Dymek, R.F., Tucker, R.D., Brannon, J.C., and Podosek, F.A., 1992, Age and radiogenic isotopic composition of a late- to post-tectonic anorthosite in the Grenville Province, Labrieville massif, Quebec: Lithos, v. 31, p. 186–206.

Parrish, R.R., Roddick, J.C., Loveridge, W.D., and Sullivan, R.W., 1987, Uranium-lead analytical techniques at the Geochronology Laboratory, Geological Survey of Canada, *in* Radiogenic Age and Isotopic Studies: Report 1: Ottawa, Ontario, Geological Survey of Canada, Paper 87-2, p. 3–7.

Rivers, T., Martignole, J., Gower, C., and Davidson, T., 1989, New tectonic divisions of the Grenville province, southeast Canadian shield: Tectonics, v. 8, p. 63–84.

Roddick, J. C., 1987. Generalized numerical error analysis with application to geochronology and thermodynamics. Geochemica et Cosmochemica Acta, v. 51, pp. 359–362.

Rondot, J., 1978, Région de Saint-Maurice: Québec, Ministère de Ressources naturelles, Québec, DPV-594.

Stern, R.A., 1996, A SHRIMP II ion microprobe at the Geological Survey of Canada: Geosciences Canada, v. 23, p. 73–76.

Stern, R.A., 1997, The GSC Sensitive High Resolution Ion Microprobe (SHRIMP): Analytical techniques of zircon U-Th-Pb age determinations and performance evaluation, *in* Radiogenic age and isotopic studies, Report 10: Ottawa, Ontario, Geological Survey of Canada, Current Research, 1997-F, p. 1–31.

Tucker, R.D., and Gower, C.F., 1994, A U-Pb geochronological framework for the Pinware terrane, Grenville Province, southeast Labrador: Journal of Geology, v. 102, p. 67–78.

Vaillancourt, C., 2001, Étude géochimique et économique de la Suite mafique-ultramafique de la Baie à Cadie au lac Kénogami, Saguenay-Lac-Saint-Jean: Chicoutimi, Université du Québec à Chicoutimi, Québec Mémoire de maîtrise.

Vallée, M., and Dubuc, F., 1970, Le Complexe de Carbonatite de St-Honoré, Québec: Bulletin Canadian Mines Institute Transactions, v. 73, p. 246–357.

Wasteneys, H.A., Kamo, S.L., Moser, D., Krogh, T.E., Gower, C.F., and Owen, J.V., 1997, U-Pb geochronological constraints on the geological evolution of the Pinware terrane and adjacent areas, Grenville Province, southeast Labrador, Canada: Precambrian Research, v. 81, p. 101–128.

Woussen, G., Dimroth, E., and Roy, D.W., 1978, Évolution chronologique des roches précambriennes du haut Saguenay: GEOS, Energy, Mines and Resources, Canada, summer 1979.

Woussen, G., Dimroth, E., Corriveau, L., and Archer, P., 1981, Crystallisation and emplacement of the Lac-Saint-Jean Anorthosite Suite massif (Québec, Canada): Contributions to Mineralogy and Petrology, v. 76, p. 343–350.

Woussen, G., Roy, D.W., Dimroth, E., and Chown, E.H., 1986, Mid-Proterozoic extensional tectonics in the core zone of the Grenville Province, *in* Moore, J.M., et al., eds., The Grenville Province: Toronto, Ontario, Geological Association of Canada, Special Paper 31, p. 297–312.

York, D., 1969, Least squares fitting of a straight line with correlated errors: Earth and Planetary Science Letters, v. 5, p. 320.

MANUSCRIPT ACCEPTED BY THE SOCIETY AUGUST 25, 2003

Printed in the USA

Geological Society of America
Memoir 197
2004

Low-variance sapphirine-bearing assemblages from Wilson Lake, Grenville Province of Labrador

Fawna J. Korhonen*
James H. Stout
Department of Geology and Geophysics, University of Minnesota, 310 Pillsbury Drive SE, Minneapolis, Minnesota 55455-0219, USA

ABSTRACT

Sapphirine + quartz-bearing pelitic gneisses from Wilson Lake, in the Grenville Province of central Labrador, and from other granulite-facies terranes, are well known for their reaction rims and nonequilibrium textures. However, some corundum-bearing gneisses from the Red Wine Mountains massif in the Wilson Lake area have a low variance assemblage that appears to record equilibrium conditions of regional metamorphism. The silica-undersaturated assemblage sapphirine (Spr) + orthopyroxene (Opx) + sillimanite (Sil) + garnet (Grt) + spinel (Spl) + corundum (Crn) + magnetite + titanhematite (+ plagioclase + biotite) approaches invariance in the six-component system FeO-MgO-Al_2O_3-SiO_2-Fe_2O_3-TiO_2. The resulting cordierite [Crd]-absent invariant point appears to be stable near 900 °C and 1000 MPa, and at an oxygen fugacity (fO_2) defined by coexisting pure magnetite and titanhematite (after integration of exsolved ferrian ilmenite). Further evidence of high fO_2 is exsolved titanhematite ($X_{Fe_2O_3}$ = 0.73) in orthopyroxene, exsolved hematite in sillimanite, and high dissolved Fe^{3+} in sillimanite and corundum. Systematic partitioning of Fe^{3+} in the coexisting silicate and oxide phases documents their mutual equilibrium.

The *P-T*-fO_2 conditions represented by the [Crd]-absent invariant point in corundum-bearing rocks are consistent with the stability of spinel + quartz. This would not be the case if X_{Mg}(Grt) < X_{Mg}(Spl). The phase relationships presented here and the absence of garnet + cordierite assemblages in this area further suggests that the [Crd]-absent invariant point is stable in the field of sapphirine + quartz. The partitioning sequence $X_{Fe^{3+}}$(Spr) > $X_{Fe^{3+}}$(Spl) > $X_{Fe^{3+}}$(Opx) > $X_{Fe^{3+}}$(Grt) we estimate from microprobe analyses extends the stabilities of sapphirine and spinel relative to orthopyroxene and garnet, shifting the [Crd]-absent invariant point to a lower temperature.

Keywords: oxygen fugacity, sapphirine, corundum, granulite facies, Labrador

INTRODUCTION

The Red Wine Mountains of central Labrador host continental crust that has been metamorphosed to the granulite facies. These rocks are part of the Wilson Lake terrane (Fig. 1), one of several allochthonous, thrust-bound crustal segments that characterize the Grenville Province in Labrador (Rivers et al., 1989; Gower, 1996). The allochthons include the Lac Joseph terrane to the west and the Mealy Mountains and Lake Melville terranes to the east of the Wilson Lake terrane (Fig. 1). These terranes are

* E-mail, corresponding author: korh0011@umn.edu.

Korhonen, F.J., and Stout, J.H., 2004, Low-variance sapphirine-bearing assemblages from Wilson Lake, Grenville Province of Labrador, *in* Tollo, R.P., Corriveau, L., McLelland, J., and Bartholomew, M.J., eds., Proterozoic tectonic evolution of the Grenville orogen in North America: Boulder, Colorado, Geological Society of America Memoir 197, p. 81–103. For permission to copy, contact editing@geosociety.org.

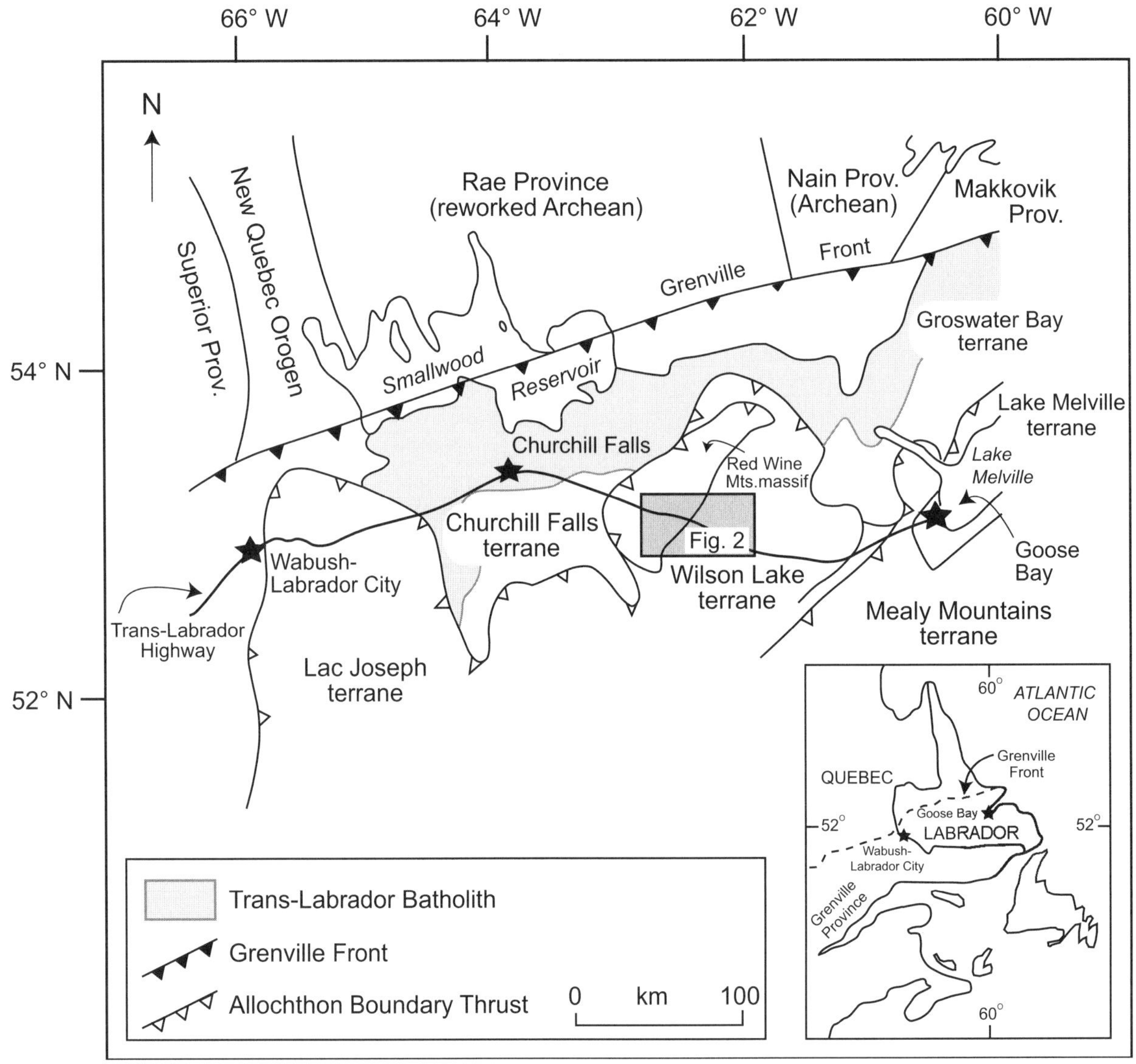

Figure 1. General tectonic map of central Labrador, after Corrigan et al. (1997), and location of the study area (Fig. 2).

generally distributed parallel to the Grenville front and are separated from it by the Trans-Labrador batholith for a distance of nearly 600 km. The Allochthon Boundary Thrust separates the allochthonous terranes from the underlying rocks to the north (Nunn et al., 1985; Rivers et al., 1989; Corrigan et al., 1997).

In broad terms, the protolith ages of the terranes in the Grenville Province are generally younger to the southeast, which is interpreted by Rivers (1997) as evidence for the accretionary growth of the southeast margin of Laurentia from ca. 1750 Ma to ca. 1000 Ma. Two pieces of this history are recorded in the Wilson Lake area. The earlier is the deformation and metamorphism associated with the Labradorian orogeny, dated at 1710–1600 Ma (Nunn et al., 1985; Thomas et al., 1986; Gower, 1996). Rocks affected by this event are exposed in the Red Wine Mountains massif (Thomas, 1993, p. 5) and are characterized by sapphirine-bearing aluminous paragneisses (Arima et al., 1986; Thomas et al., 2000). The calc-alkaline rocks of the Trans-Labrador batholith north of the Wilson Lake terrane (Fig. 1) have been interpreted to represent an Andean-style magmatic arc that formed during subduction beneath the Laurentian margin at ca. 1650 Ma (Wardle et al., 1990; Gower et al., 1992). Labradorian metamorphism persisted until ca. 1636 Ma in the Lac Joseph terrane to the west (Fig. 1) (Connelly et al., 1995) and until ca. 1620–1610 Ma in the Lake Melville terrane to the east (Corrigan et al., 2000).

The second piece of accretionary history recorded in the Wilson Lake area is attributed to renewed convergence during the ca. 1080–985 Ma Grenvillian orogeny (Currie and Loveridge, 1985; Corrigan et al., 2000; Gower and Krogh, 2002). During this time, thrusting toward the northwest resulted in tectonic stacking on a crustal scale. The Red Wine Mountains massif (Nunn et al., 1985; Rivers and Nunn, 1985; Thomas, 1993; Fig. 1) is one example where quartzofeldspathic gneisses with various sapphirine-bearing assemblages are situated on top of similar gneisses that lack sapphirine. The extent to which the earlier metamorphic record has been affected by Grenvillian metamorphism is unclear in the Wilson Lake area.

The Red Wine Mountains massif (Fig. 2) is characterized by a pronounced regional aeromagnetic high that extends over much of the Red Wine Mountains. The anomaly is caused by a strong, stable remanence carried by grains of titanhematite closely associated with sapphirine and other granulite-facies mineral assemblages (Kletetschka and Stout, 1998). In contrast to the high-grade rocks of the massif, the underlying lower-grade rocks have a relatively low magnetic expression. The natural remanent magnetization (NRM) can potentially be useful in unraveling the complicated metamorphic and tectonic history of this region, and understanding the petrologic origins of the remanence is an important aspect of this study.

Previous petrologic studies in the Wilson Lake area (Morse and Talley, 1971; Arima et al., 1986; Currie and Gittens, 1988) have focused on the phase equilibria among coexisting silicate phases. In this paper, we combine electron microprobe analyses and phase equilibrium arguments to evaluate the high-temperature silicate-oxide equilibria in this area for the first time.

REGIONAL AND LOCAL GEOLOGY

The predominant rock unit in the Wilson Lake area is known as the Disappointment Lake paragneiss (Thomas et al., 2000; Fig. 2, unit Awph), a migmatitic quartzofeldspathic gneiss that underlies much of the highland areas of the Red Wine Mountains. The paragneisses and associated gabbronorites (discussed later) make up the entirety of the Red Wine Mountains

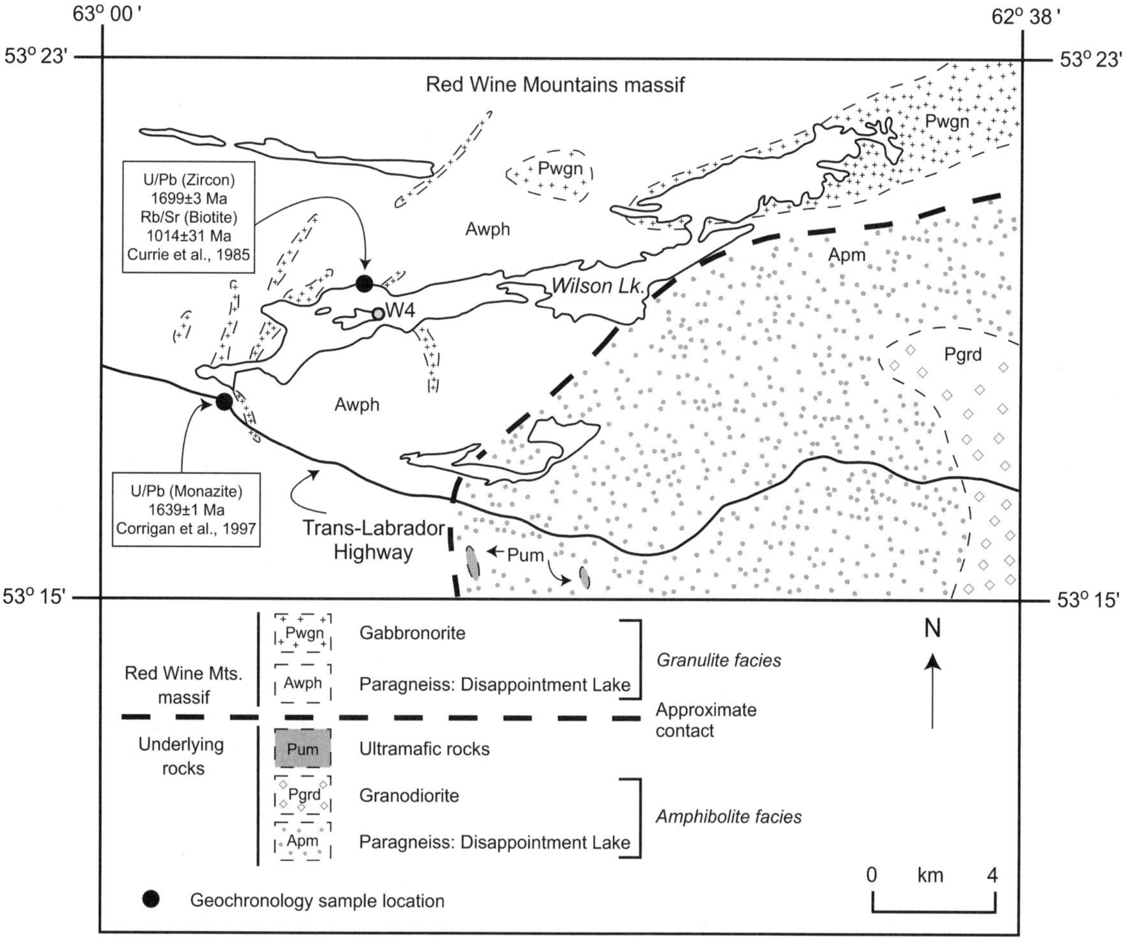

Figure 2. Geologic map of the Wilson Lake area, after Thomas et al. (2000), showing the distribution of rocks in the Red Wine Mountains massif and the underlying units. The rocks in the massif are quartzofeldspathic paragneisses (unit Awph) with sapphirine-bearing assemblages and associated gabbronorites (unit Pwgn). The underlying rocks are quartzofeldspathic paragneisses (unit Apm) lacking sapphirine, as well as granodiorite (unit Pgrd) and a few isolated outcrops of ultramafic rocks (unit Pum).

massif. The hard, dense aspect of the paragneiss, its resistance to breakage across foliation, and the absence of hydrous minerals in most parts of it are characteristic of this unit of the massif. The paragneisses are highly deformed and exhibit at least two periods of isoclinal folding.

Bodies of deformed medium-grained gabbronorite (Fig. 2, unit Pwgn) are also present in the Red Wine Mountains massif. Many of them cut across the paragneiss foliation and are clearly derived from mafic dikes and sills. Spectacular outcrops along a new section of the Trans-Labrador Highway (Fig. 2) reveal at least two generations of these dikes. The rocks are generally massive but become foliated toward their margins. Plagioclase, orthopyroxene, augite, hornblende, biotite, and magnetite-ilmenite are the main constituents, with rare garnet and spinel.

The most distinctive mineral assemblages within the Red Wine Mountains massif are orthopyroxene + sillimanite + quartz and sapphirine + quartz. These assemblages are found within the restitic layers of the paragneiss and are widespread in the massif (Morse and Talley, 1971; Arima et al., 1986; Currie and Gittens, 1988). Experimental studies (Hensen and Green, 1971; Bertrand et al., 1991) in the ideal FeO-MgO-Al_2O_3-SiO_2 (FMAS) system indicate extremely high temperatures (>1000 °C) and moderate pressures (>1000 MPa) for these assemblages. These results suggest that much of the massif was subject to the extremely high-temperature conditions of metamorphism prior to its tectonic emplacement over the lower–grade rocks. These conditions and tectonic history contrast with those inferred for other terranes in the Grenville Province, and in particular with the high-pressure belt of Grenvillian eclogite-facies mafic rocks (Rivers et al., 2002) that are part of a thrust slice below the Lac Joseph terrane (Fig. 1) to the west. Pressure-temperature estimates for the eclogitic rocks are 1800 MPa and 850 °C (Rivers et al., 2002).

The underlying rocks (Fig. 2, units Apm and Pgrd) are middle to upper amphibolite-facies paragneisses. Unit Apm is part of the Disappointment Lake paragneiss, but differs from unit Awph in the ubiquitous presence of muscovite and/or biotite and the absence of sapphirine (Thomas et al., 2000). These rocks split readily along foliation. Sillimanite and/or kyanite, garnet, and cordierite are also common. Herd et al. (1987), Currie and Gittens (1988), and Thomas (1993) interpreted the cordierite and hydrous phases as indicative of an amphibolite-facies event inferred to be of Grenvillian age. This metamorphic event was pervasive within the underlying rocks but localized only within the Red Wine Mountains massif. Unit Pgrd is a coarse-grained, strongly foliated granodiorite that is interpreted by Thomas et al. (2000) as intrusive into unit Awph.

Kletetschka and Stout (1998) demonstrated that coarse-grained magnetite is present in the amphibolite-facies Disappointment Lake paragneiss (Fig. 2, unit Apm) and in unit Pgrd in about the same abundance as in unit Awph of the Red Wine Mountains massif. This observation accounts for the similar bulk magnetic susceptibilities of all major rock units shown in Figure 2. These authors also showed that lesser amounts of titanhematite in units of the underlying rocks accounts for its low magnetic field strength. The occurrence (Fig. 2, unit Pum) of isolated outcrops of ultramafic rocks on or near the boundary between the Red Wine Mountains massif and the underlying rocks in the Wilson Lake area is consistent with the inference that this is an important tectonic boundary.

Constraints on the protolith ages and timing of metamorphism of the Red Wine Mountains massif and the underlying rocks are sparse. Currie and Loveridge (1985) obtained a U-Pb upper concordia intercept age of 1699 ± 3 Ma on zircon separated from a 40 kg specimen of paragneiss collected along the north shore of Wilson Lake (Fig. 2). They interpret this age as that of the granulite-facies metamorphism. Biotite from the same specimen yielded a Rb-Sr date of 1014 ± 31 Ma, which they interpret as a retrograde Grenvillian thermal event. Three monazite grains from a sapphirine-bearing gneiss in a roadcut on the Trans-Labrador Highway south of Wilson Lake (Fig. 2) yield a concordant ^{207}Pb-^{206}Pb age of 1639 ± 1 Ma and 2 discordant ages of 1635 ± 3 Ma and 1640 ± 5 Ma (Corrigan et al., 1997). These authors attribute these ages to the waning stages of Labradorian metamorphism. U-Pb dates on zircon from a foliated orthogneiss that intrudes paragneiss in the Wilson Lake terrane south of the study area give an igneous emplacement age of 1650 Ma (James et al., 2002). The lower intercept age based on these data is poorly constrained, but suggests a Grenvillian overprinting event between 1080 and 1000 Ma. Rb-Sr dates on biotite from the Red Wine Mountains massif and the underlying rocks are interpreted to date a retrograde event associated with Grenville emplacement (1100–900 Ma) and uplift (Currie and Loveridge, 1985; Thomas, 1993).

Some additional age constraints are provided in the Lac Joseph terrane, located west of the Wilson Lake terrane (Fig. 1). Both the Lac Joseph terrane and the Wilson Lake terrane are part of the allochthonous belt of Rivers et al. (1989). The rocks in this region are mainly amphibolite- to granulite-facies migmatites that exhibit two distinct generations of leucosomes associated with a restite assemblage of sillimanite + garnet + biotite + magnetite (Connelly and Heaman, 1993). Monazite from these leucosomes yields U-Pb ages of 1660 and 1636 Ma, which are interpreted by Connelly and Heaman (1993) as ages of crystallization of partial melting during a protracted Labradorian orogeny.

U-Pb dating on zircon and monazite from a ductile shear zone in the same area suggests that deformation and metamorphism were pre-Grenville in the Lac Joseph terrane (Connelly and Heaman, 1993; Connelly et al., 1995).

DISAPPOINTMENT LAKE PARAGNEISS

General Mineral Assemblages

Several petrologic studies (Morse and Talley, 1971; Leong and Moore, 1972; Jackson and Finn, 1982; Currie and Gittens, 1988) of high-grade paragneisses near Wilson Lake have focused on the silicate assemblages orthopyroxene + sillimanite

+ quartz and sapphirine + quartz. With one exception, sillimanite is the only Al_2SiO_5 polymorph described in these studies, which is consistent with our own observations both in the field and in about 150 thin sections. Morse and Talley (1971) describe kyanite with sapphirine + quartz from a ridge 2 km north of the eastern part of Wilson Lake. This observation was recently checked and confirmed (Morse, 2002, personal commun.).

Most rocks in the massif appear to have preserved these primary high-grade assemblages, whereas some assemblages show evidence for disequilibrium and variable overprinting. All of the primary assemblages include variable amounts of titanhematite and magnetite. As discussed in more detail later, the primary assemblages show evidence of having formed under relatively high oxidizing conditions. Oxygen fugacity (fO_2) at near-peak conditions was sufficient to stabilize titanhematite in the presence of magnetite. Evidence that the silicate phases were also subjected to high fO_2 is exsolved titanhematite in orthopyroxene (Fig. 3, A). Grains that appear to be part of the primary assemblages occur as euhedral to subhedral grains with strong pink to pale green pleochroism. Aligned titanhematite lamellae that range in size from ~1 μm to submicron follow rational planes in the structure. Their internal red reflections impart a

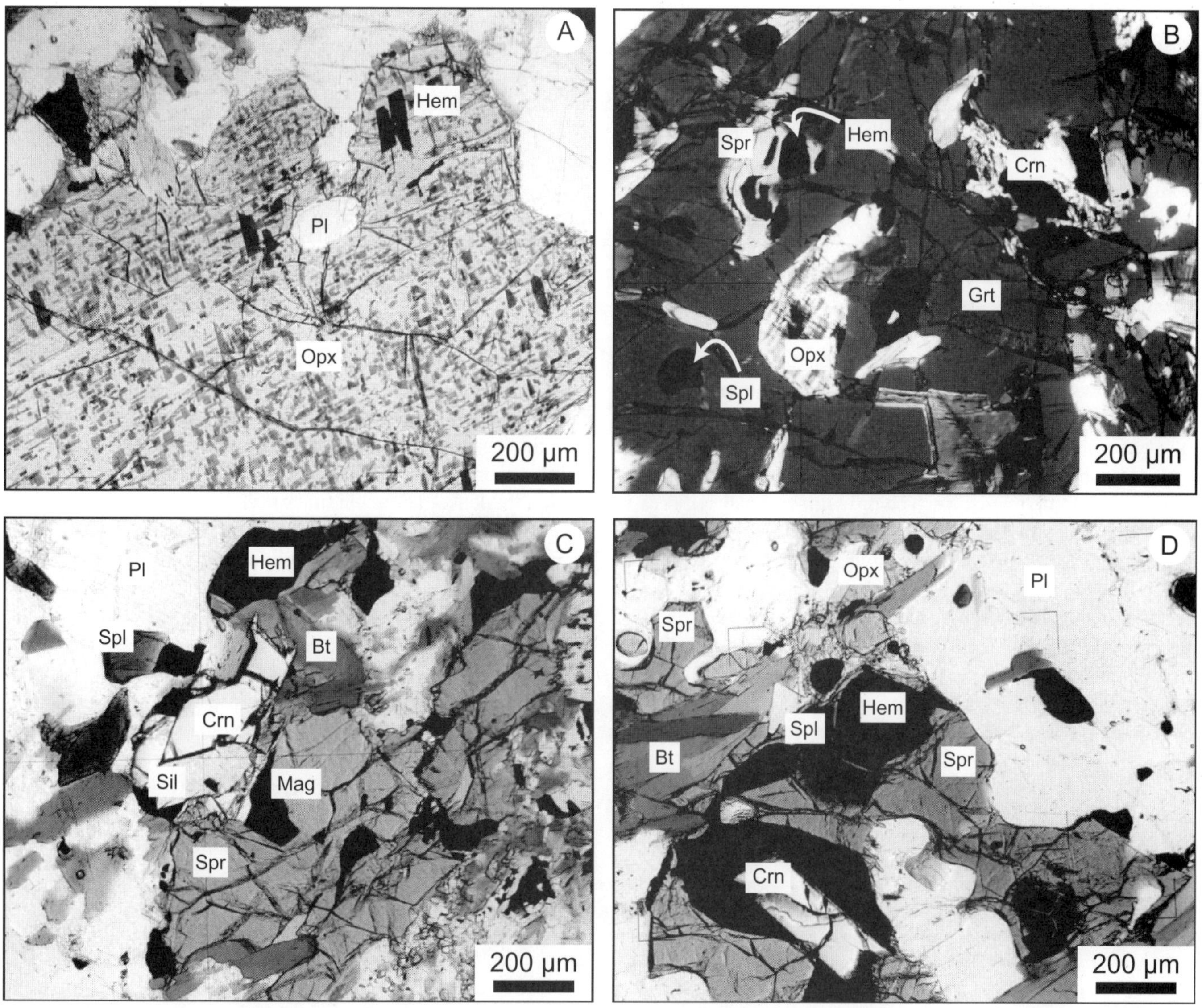

Figure 3. Photomicrographs showing critical contact relationships observed in the low-variance assemblage from sample W4. (A) Titanhematite exsolution in orthopyroxene. (B) Coarse-grained garnet with inclusions of sapphirine, spinel with magnetite exsolution, corundum, titanhematite, and orthopyroxene with titanhematite exsolution. (C) Close association of corundum and sillimanite. (D) Sapphirine and orthopyroxene coexisting with corundum, titanhematite, spinel, and coarse-grained biotite. Bt—biotite; Crn—corundum; Grt—garnet; Hem—titanhematite; Mag—magnetite; Opx—orthopyroxene; Pl—plagioclase; Sil—sillimanite; Spl—spinel; Spr—sapphirine.

distinctive pink color to the grains. Currie and Gittens's (1988) general description (p. 611) of orthopyroxene in this area as "riddled with aligned biotite inclusions" almost certainly refers to this exsolved oxide phase. Further evidence for oxide-silicate equilibria is exsolved hematite in sillimanite, and high dissolved Fe^{3+} in sillimanite and corundum. All of these features appear to be unique to primary assemblages and serve to distinguish them from secondary assemblages, which, on textural grounds, have formed at a later time.

Equilibrium Textures

In most cases, distinction between rocks that are a product of the primary metamorphism and those that have been subsequently recrystallized and/or deformed can be made only petrographically. Quartz-bearing assemblages with orthopyroxene, sillimanite, nearly pure magnetite, and titanhematite occur and appear to have achieved textural equilibrium. Corundum-bearing, quartz-absent assemblages with orthopyroxene, sillimanite, spinel, titanhematite, and magnetite also occur and have similar textures. A small group of rocks in this study from a single locality (Fig. 2, sample W4) contains the unusual association of sapphirine (Spr) + orthopyroxene (Opx) + sillimanite (Sil) + corundum (Crn) + spinel (Spl) + garnet (Grt) + titanhematite (Hem) + magnetite (Mag) + plagioclase (Pl) + biotite (Bt), all within single thin sections (Fig 3; abbreviations of minerals are after Kretz, 1983). There are variations in the modal abundances of phases within a thin section, and at no single point can all phases be found in mutual contact. Table 1 lists the observed phases in mutual contact in the form of subassemblages in one specific polished thin section. Figure 3 illustrates some of the critical contact relationships. The only mineral pair not seen in direct physical contact is garnet and sillimanite. There is no textural evidence for disequilibrium within or between the subassemblages.

Although these samples seem to preserve intergranular textural equilibrium, several phases exhibit intragranular exsolution. Titanhematite grains have ferrian ilmenite lamellae (Fig. 4), spinel grains have exsolved magnetite, orthopyroxene has titanhematite lamellae, and sillimanite has hematite lamellae.

TABLE 1. SUBASSEMBLAGES OBSERVED IN SAMPLE W4

Grt	Spr	Sil	Opx	Spl	Crn	Mag	Hem	Pl	Bt
X	X		X	X	X		X	X	X
X	X		X	X	X	X		X	X
	X	X	X	X	X	X	X	X	X
	X	X	X	X	X	X	X	X	X
	X	X		X	X	X	X	X	X

Note: Abbreviations, after Kretz (1983): Grt—garnet; Spr—sapphirine; Sil—sillimanite; Opx—orthopyroxene; Spl—spinel; Crn—corundum; Mag—magnetite; Hem—titanhematite; Pl—plagioclase; Bt—biotite.

OXIDE LENSES

Titanhematite and magnetite are present in variable amounts and proportions in most of the sapphirine-bearing gneisses. In addition, the oxide phases are concentrated as deformed lenses and boudin-shaped masses up to tens of meters in length and a few meters in thickness. Veins consist of mostly magnetite and titanhematite but can have up to 20% silicates, commonly sapphirine, orthopyroxene, and spinel. Both sapphirine and orthopyroxene display abundant intracrystalline exsolution of opaque oxides, which is similar to observations in the paragneiss as discussed earlier. The oxide lenses have been observed in a few places cutting across the foliation of the surrounding gneisses, attesting to some degree of mobilization after formation of the paragneiss foliation.

The titanhematite grains in the oxide lenses contain ferrian ilmenite exsolution. Some lamellae have minor amounts of additional oxide phases, with variable amounts of Al_2O_3, MgO, and MnO. The exsolution lamellae in titanhematite grains from oxide lenses vary in scale, with the largest lamellae occurring in Ti-rich layers, suggesting a bulk compositional control. Additionally, these samples often contain several generations of lamellae within a single grain.

The oxide lenses are extensive enough in the central parts of Wilson Lake to have attracted mild economic interest (Currie and Gittens, 1988). Their origin and age are uncertain. Leong and Moore (1972) interpreted the lenses as concordant with the surrounding gneisses and originally part of an aluminous, ferruginous protolith. However, Gittens and Currie (1979) believed that the lenses postdated the foliation and were of metasomatic origin (Currie and Gittens, 1988). The silicate assemblages within the lenses and their similar relationship to oxide phases, as we document in the paragneiss later, suggests that the lenses have fully participated in the high-temperature metamorphism.

EVIDENCE FOR MULTIPLE METAMORPHIC EVENTS

Many of the paragneisses in the Red Wine Mountains massif provide evidence for more than a single metamorphic event, whereas other closely associated layers seem to preserve primary assemblages. These differences are not easily distinguished in the field. In some areas, mylonites are interlayered with coarser-grained granulites that appear not to have been affected by deformation. Petrographic observations of some of the latter rocks reveal undeformed coronas of secondary orthopyroxene, cordierite, and sillimanite separating sapphirine and quartz. Such nonequilibrium textures and reaction rims have been well described in rocks from Wilson Lake (e.g., Morse and Talley, 1971; Leong and Moore, 1972; Grew, 1980; Arima et al., 1986; Currie and Gittens, 1988). Figure 5 shows an example of this texture, which is indicative of the retrograde reaction sapphirine + quartz = orthopyroxene + sillimanite + cordierite. As

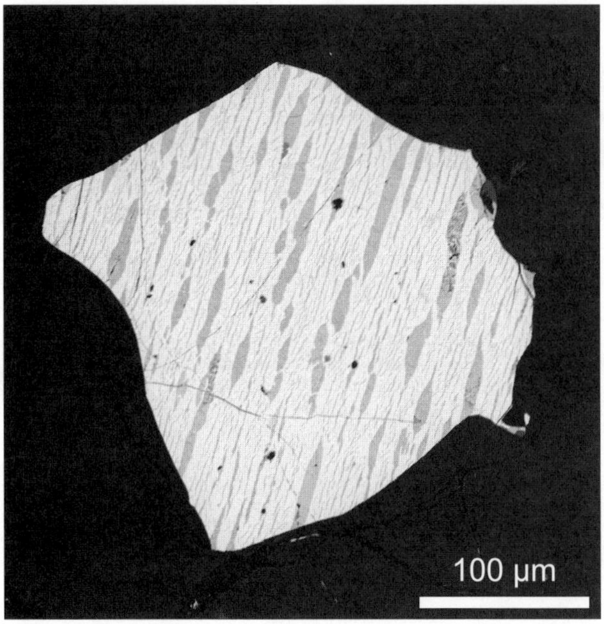

Figure 4. Backscattered electron image of a titanhematite grain with ferrian ilmenite exsolution. The titanhematite host appears as the lighter phase, and the ferrian ilmenite exsolution is gray. Compositions of exsolution lamellae were integrated into the host compositions using computer image analysis of such grains. The integration procedure is discussed in the text, and the results for titanhematite are given in Table 4.

discussed in the following section, the orthopyroxene rims in these coronas have a composition consistent with crystallization at a lower fO_2 than the orthopyroxene that appears to be part of the primary assemblages.

Mylonitic zones are characterized by quartz ribbons wrapping around porphyroclasts of primary orthopyroxene and sillimanite (Fig. 6). Parallel necklaces of secondary orthopyroxene, sillimanite, quartz, plagioclase, and titanhematite define the mylonitic fabric and suggest that the *P-T* conditions of deformation were not far removed from that of the primary metamorphism. The recrystallized orthopyroxene has a different composition than both the primary orthopyroxene and the orthopyroxene rims in the sapphirine + quartz coronas (Fig. 8). The titanhematites are small, elongate grains that are clearly part of the mylonitic fabric. These grains contain exsolution lamellae of ferrian ilmenite 1 to 2 μm in thickness that are undeformed (Fig. 7). Coarse plagioclase grains, however, do have deformed polysynthetic twinning (Fig. 6).

MINERAL CHEMISTRY

Analytical Methods

Electron microprobe

Mineral compositions were obtained with a JEOL JXA-8900R wavelength-dispersive electron microprobe at the University of Minnesota. Quantitative point analyses for most phases were collected with a focused 1 μm beam at 15 kV and a beam current of 20 nA. A defocused beam was used for titan-

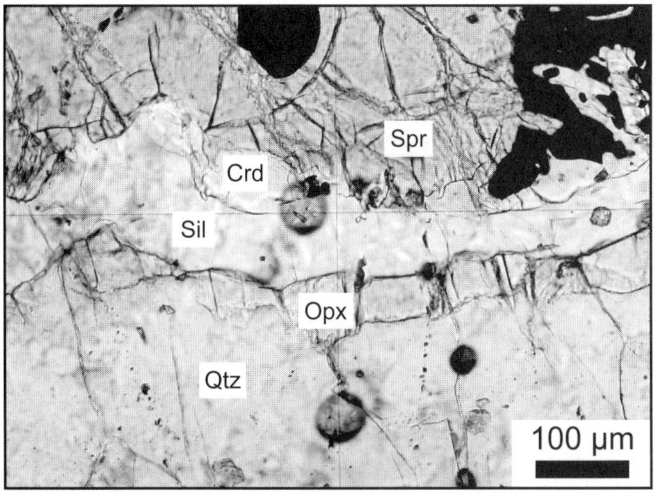

Figure 5. Photomicrograph of retrograde reaction sapphirine + quartz = orthopyroxene + sillimanite + cordierite. The composition of the orthopyroxene rims in these coronas and the absence of exsolved hematite are consistent with crystallization at a lower fO_2 than the orthopyroxene that appears to be part of the primary assemblages (see text for discussion). Crd—cordierite; Opx—orthopyroxene; Qtz—quartz; Sil—sillimanite; Spr—sapphirine.

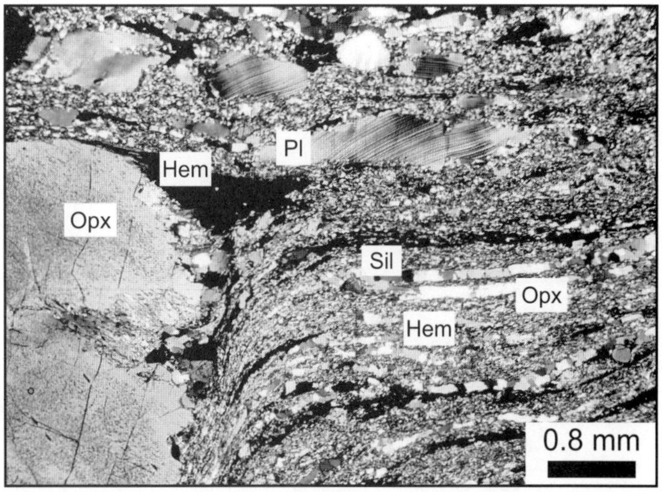

Figure 6. Photomicrograph of mylonitic fabric wrapping around large porphyroclast of primary orthopyroxene. Secondary orthopyroxene, sillimanite, and titanhematite suggest that the *P-T* conditions of deformation were not far removed from that of the primary metamorphism. Hem—titanhematite; Opx—orthopyroxene; Pl—plagioclase; Sil—sillimanite.

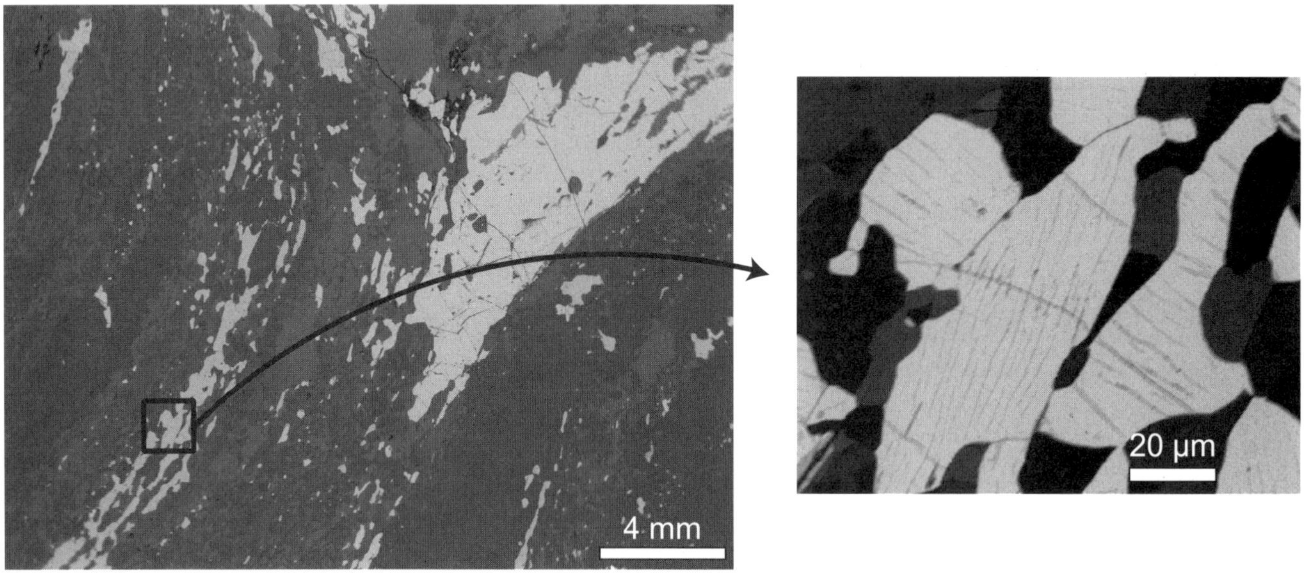

Figure 7. Backscattered electron image of titanhematite with ferrian ilmenite exsolution in mylonitic fabric. The small, elongated grains are clearly part of the mylonitic fabric. The undeformed lamellae suggest that exsolution likely postdates the deformation.

hematite grains with submicron exsolution lamellae. Natural mineral standards were used that most closely resembled the unknown phases. Most point analyses were based on 20 second peak counting times and 10 second background times.

The analyses in Table 2 are representative point analyses rather than averages. The total numbers of point analyses on which the representative ones are based are listed in parentheses.

Ferric Iron Recalculation

Ferric iron was calculated for all relevant phases with an oxygen-based formula and a cation normalization factor using the stoichiometric method outlined by Schumacher (1991). Our results (Table 2) compare well with the recalculated sapphirine analyses discussed by Higgins et al. (1979), for which independent Mössbauer determinations were made for ferrous and ferric iron. Nonetheless, ferric iron calculated from microprobe analyses is highly sensitive to error accumulated in analyzing the other oxide components, and that error may not be the same for the different phases. The values given in Table 2 are thus considered a reliable indication of the relative Fe^{3+} contents of the phases, but the absolute abundances may not be accurate.

Integration Procedure

In order to better characterize the equilibrium compositions of the exsolved spinel and titanhematite grains, the compositions of the lamellae were integrated back into the compositions of the host grains. Areal proportions of host and lamellae were calculated by computer analysis of backscattered electron images. In order to minimize the edge effects on the grains, several internal areas for each grain were analyzed and averaged together. Very little areal variation within individual grains was observed by this method, so the calculated areal modes were assumed to be representative of the volumetric modes. Average point analyses for lamellae and host for each grain were added in proportion to the areal proportion as volume proxies to obtain integrated compositions (Tables 3 and 4). In a few cases, lamellae in titanhematite grains were so small that reliable point analyses could not be obtained. A defocused electron microprobe beam (20 to 30 μm) was used for these grains.

Results

Orthopyroxene

The composition of primary orthopyroxene in the rocks studied is relatively consistent, although minor zoning is observed between the cores and the rims (Table 2). The cores are enriched in Al (up to 0.5 cations/6 oxygens), and are slightly depleted in Mg and Si relative to the rims. This variation is attributed to the Tschermak substitution 2Al = Mg + Si, as shown in Figure 8. There is also a positive correlation of Al with calculated ferric iron. Based on stoichiometric recalculations, ~17% of the total iron is Fe^{3+}.

Most of the oxide lamellae in orthopyroxene are less than 1 μm wide, making it difficult to obtain their composition even with the smallest possible beam diameter. In order to calculate the oxide composition, a 1 μm beam was stepped across a single lamella in submicron intervals such that varying proportions of lamella and host orthopyroxene were analyzed at each step. These point analyses were extrapolated to the orthopyroxene-free composition in Figure 9, where the Fe^{3+} presumed to be in the lamellae as hematite-ilmenite binary solution is plotted against Si as a proxy for the relative amount

TABLE 2. REPRESENTATIVE MICROPROBE ANALYSES

Phase	Orthopyroxene (180 analyses)						
Sample	W4 rim	W4 rim	W4 core	W4 core	W3 mylonite recrystallized	W3 mylonite recrystallized	W23 reaction rim
(wt%)							
TiO_2	0.05	0.01	0.05	0.06	0.03	0.03	0.00
Cr_2O_3	0.02	0.01	0.04	0.04	0.00	0.00	0.02
SiO_2	47.81	48.30	47.30	47.18	50.40	50.24	53.50
Al_2O_3	9.87	9.37	11.15	11.34	5.98	6.16	3.14
CaO	0.08	0.07	0.08	0.09	0.01	0.08	0.07
MgO	23.25	24.18	23.24	22.97	24.81	24.82	30.38
FeO	18.09	17.21	17.73	17.87	16.66	16.59	10.98
MnO	0.67	0.51	0.68	0.70	1.33	1.29	1.24
Na_2O	0.01	0.01	0.02	0.00	0.02	0.01	0.00
K_2O	n.d.	n.d.	n.d.	n.d.	n.d.	n.d.	n.d.
Total	99.84	99.66	100.31	100.24	99.33	99.24	99.33
(Cation atoms)							
Fe^{3+}	0.09	0.09	0.10	0.09	0.06	0.06	0.07
Fe^{2+}	0.46	0.43	0.43	0.45	0.44	0.44	0.25
Ti	0.00	0.00	0.00	0.00	0.00	0.00	0.00
Cr	0.00	0.00	0.00	0.00	0.00	0.00	0.00
Si	1.74	1.75	1.71	1.71	1.84	1.84	1.90
Al	0.42	0.40	0.48	0.48	0.26	0.27	0.13
Ca	0.00	0.00	0.00	0.00	0.00	0.00	0.00
Mg	1.26	1.31	1.25	1.24	1.35	1.35	1.61
Mn	0.02	0.02	0.02	0.02	0.04	0.04	0.04
Na	0.00	0.00	0.00	0.00	0.00	0.00	0.00
K	n.d.	n.d.	n.d.	n.d.	n.d.	n.d.	n.d.
Total cations	4	4	4	4	4	4	4
Total oxygens	6	6	6	6	6	6	6
(Moles oxide)							
Fe_2O_3	0.04	0.05	0.05	0.04	0.03	0.03	0.04
FeO	0.46	0.43	0.43	0.45	0.44	0.44	0.25
TiO_2	0.00	0.00	0.00	0.00	0.00	0.00	0.00
Cr_2O_3	0.00	0.00	0.00	0.00	0.00	0.00	0.00
SiO_2	1.74	1.75	1.71	1.71	1.84	1.84	1.90
Al_2O_3	0.21	0.20	0.24	0.24	0.13	0.13	0.07
CaO	0.00	0.00	0.00	0.00	0.00	0.00	0.00
MgO	1.26	1.31	1.25	1.24	1.35	1.35	1.61
MnO	0.02	0.02	0.02	0.02	0.04	0.04	0.04
Na_2O	0.00	0.00	0.00	0.00	0.00	0.00	0.00
K_2O	n.d.	n.d.	n.d.	n.d.	n.d.	n.d.	n.d.
X_{Mg}	0.74	0.75	0.74	0.74	0.75	0.75	0.86

Phase	Garnet (90 analyses)		Sillimanite (50 analyses)			Corundum (70 analyses)	
Sample	W4 rim	W4 core	W4	W4	W23 with exsolved titanhematite	W4	W4
(Wt%)							
TiO_2	0.04	0.00	0.05	0.01	0.00	0.00	0.02
Cr_2O_3	n.d.	n.d.	n.d.	n.d.	n.d.	0.00	0.05
SiO_2	38.99	39.49	36.65	36.72	36.80	n.d.	n.d.
Al_2O_3	23.05	23.63	61.93	61.83	61.11	99.11	98.92
CaO	2.23	1.52	n.d.	n.d.	n.d.	n.d.	n.d.
MgO	9.96	11.74	n.d.	n.d.	n.d.	n.d.	n.d.
FeO	23.15	21.21	1.24	1.04	1.69	1.22	1.13
MnO	2.27	2.08	0.00	0.00	0.00	n.d.	n.d.
Na_2O	n.d.	n.d.	n.d.	n.d.	n.d.	n.d.	n.d.
K_2O	n.d.	n.d.	n.d.	n.d.	n.d.	n.d.	n.d.
Total	99.69	99.68	99.86	99.59	99.60	100.33	100.12

(continued)

TABLE 2. *Continued*

Phase	Garnet (90 analyses)		Sillimanite (50 analyses)			Corundum (70 analyses)	
Sample	W4 rim	W4 core	W4	W4	W23 with exsolved titanhematite	W4	W4
(Cation atoms)							
Fe^{3+}	0.00	0.00	0.03	0.03	0.04	0.02	0.02
Fe^{2+}	1.33	1.33	0.00	0.00	0.00	0.00	0.00
Ti	0.00	0.00	0.00	0.00	0.00	0.00	0.00
Cr	n.d.	n.d.	n.d.	n.d.	n.d.	0.00	0.00
Si	2.99	2.99	0.99	1.00	1.00	n.d.	n.d.
Al	2.08	2.11	1.98	1.98	1.96	1.98	1.98
Ca	0.18	0.12	n.d.	n.d.	n.d.	n.d.	n.d.
Mg	1.14	1.32	n.d.	n.d.	n.d.	n.d.	n.d.
Mn	0.15	0.13	0.00	0.00	0.00	n.d.	n.d.
Na	n.d.	n.d.	n.d.	n.d.	n.d.	n.d.	n.d.
K	n.d.	n.d.	n.d.	n.d.	n.d.	n.d.	n.d.
Total cations	8	8	3	3	3	2	2
Total oxygens	12	12	5	5	5	3	3
(Moles oxide)							
Fe_2O_3	0.00	0.00	0.02	0.01	0.02	0.01	0.01
FeO	1.33	1.33	0.00	0.00	0.00	0.00	0.00
TiO_2	0.00	0.00	0.00	0.00	0.00	0.00	0.00
Cr_2O_3	n.d.	n.d.	n.d.	n.d.	n.d.	0.00	0.00
SiO_2	2.99	2.99	0.99	1.00	1.00	n.d.	n.d.
Al_2O_3	1.04	1.05	0.99	0.99	0.98	0.99	0.99
CaO	0.18	0.12	n.d.	n.d.	n.d.	n.d.	n.d.
MgO	1.14	1.32	n.d.	n.d.	n.d.	n.d.	n.d.
MnO	0.15	0.13	0.00	0.00	0.00	n.d.	n.d.
Na_2O	n.d.	n.d.	n.d.	n.d.	n.d.	n.d.	n.d.
K_2O	n.d.	n.d.	n.d.	n.d.	n.d.	n.d.	n.d.
X_{Mg}	0.46	0.50	N.A.	N.A.	N.A.	N.A.	N.A.

Phase	Sapphirine (115 analyses)				Magnetite (80 analyses)		Biotite (55 analyses)
Sample	W4 core	W4 core	W4 core	W4 core	W4	W4	W4
(Wt%)							
TiO_2	0.02	0.01	0.03	0.01	0.02	0.00	3.81
Cr_2O_3	n.d.	n.d.	n.d.	n.d.	0.23	0.34	n.d.
SiO_2	11.74	13.74	11.79	13.74	0.00	0.00	36.85
Al_2O_3	63.09	58.75	62.24	58.66	0.21	0.24	17.33
CaO	0.00	0.02	0.00	0.02	0.06	0.00	0.05
MgO	15.00	15.55	15.11	15.46	0.01	0.01	16.70
FeO	9.94	11.14	9.69	11.24	90.90	91.00	10.27
MnO	0.22	0.29	0.24	0.32	0.00	0.00	0.09
Na_2O	n.d.	n.d.	n.d.	n.d.	0.00	0.03	0.05
K_2O	n.d.	n.d.	n.d.	n.d.	n.d.	n.d.	9.26
Total	100.01	99.49	99.11	99.45	91.43	91.62	94.41
Fe^{3+}	0.28	0.29	0.28	0.29	1.98	1.98	n.d.
Fe^{2+}	0.71	0.83	0.68	0.84	1.00	1.00	1.28
Ti	0.00	0.00	0.00	0.00	0.00	0.00	0.43
Cr	n.d.	n.d.	n.d.	n.d.	0.01	0.01	n.d.
Si	1.41	1.66	1.42	1.67	0.00	0.00	5.59
Al	8.91	8.38	8.86	8.38	0.01	0.01	3.10
Ca	0.00	0.00	0.00	0.00	0.00	0.00	0.01
Mg	2.68	2.81	2.72	2.79	0.00	0.00	3.77
Mn	0.02	0.03	0.02	0.03	0.00	0.00	0.01

TABLE 2. *Continued*

Phase	Sapphirine (115 analyses)				Magnetite (80 analyses)		Biotite (55 analyses)
Sample	W4 core	W4 core	W4 core	W4 core	W4	W4	W4
Na	n.d.	n.d.	n.d.	n.d.	0.00	0.00	0.01
K	n.d.	n.d.	n.d.	n.d.	n.d.	n.d.	1.79
Total cations	14	14	14	14	3	3	16
Total oxygens	20	20	20	20	4	4	22
Fe_2O_3	0.14	0.14	0.14	0.14	0.99	0.99	n.d.
FeO	0.71	0.83	0.68	0.84	1.00	1.00	1.28
TiO_2	0.00	0.00	0.00	0.00	0.00	0.00	0.43
Cr_2O_3	n.d.	n.d.	n.d.	n.d.	0.00	0.01	n.d.
SiO_2	1.41	1.66	1.42	1.67	0.00	0.00	5.59
Al_2O_3	4.45	4.19	4.43	4.19	0.00	0.01	1.55
CaO	0.00	0.00	0.00	0.00	0.00	0.00	0.01
MgO	2.68	2.81	2.72	2.79	0.00	0.00	3.77
MnO	0.02	0.03	0.02	0.03	0.00	0.00	0.01
Na_2O	n.d.	n.d.	n.d.	n.d.	0.00	0.00	0.01
K_2O	n.d.	n.d.	n.d.	n.d.	n.d.	n.d.	0.90
X_{Mg}	0.79	0.77	0.80	0.77	N.A.	N.A.	N.A.

Phase	Biotite (55 analyses)	Plagioclase (230 analyses)					Cordierite (3 analyses)
Sample	W4	W4	W4 garnet-bearing subassemblage	W4 sillimanite-bearing	W3 mylonite	W3 mylonite	W23 reaction rim
(Wt%)							
TiO_2	3.68	n.d.	n.d.	n.d.	n.d.	n.d.	0.00
Cr_2O_3	n.d.	n.d.	n.d.	n.d.	n.d.	n.d.	0.00
SiO_2	36.82	55.29	56.38	54.99	58.49	58.98	49.02
Al_2O_3	17.77	28.86	28.34	29.09	26.68	25.76	32.89
CaO	0.01	10.38	9.56	10.68	8.15	7.42	0.03
MgO	17.44	n.d.	n.d.	n.d.	n.d.	n.d.	11.70
FeO	9.46	0.26	0.29	0.08	0.14	0.08	1.47
MnO	0.09	n.d.	n.d.	n.d.	n.d.	n.d.	0.17
Na_2O	0.11	5.36	5.68	5.03	6.92	7.09	0.12
K_2O	8.84	0.20	0.28	0.17	0.14	0.37	0.00
Total	94.25	100.35	100.53	100.03	100.52	99.69	95.40
Fe^{3+}	n.d.	0.00	0.00	0.00	0.00	0.00	0.00
Fe^{2+}	1.18	0.01	0.01	0.00	0.00	0.00	0.13
Ti	0.42	n.d.	n.d.	n.d.	n.d.	n.d.	0.00
Cr	n.d.	n.d.	n.d.	n.d.	n.d.	n.d.	0.00
Si	5.56	2.48	2.53	2.48	2.60	2.64	5.05
Al	3.16	1.53	1.50	1.55	1.40	1.36	3.99
Ca	0.00	0.50	0.46	0.52	0.39	0.36	0.00
Mg	3.93	n.d.	n.d.	n.d.	n.d.	n.d.	1.80
Mn	0.01	n.d.	n.d.	n.d.	n.d.	n.d.	0.01
Na	0.03	0.47	0.49	0.44	0.60	0.62	0.02
K	1.70	0.01	0.02	0.01	0.01	0.02	0.00
Total cations	16	5	5	5	5	5	11
Total oxygens	22	8	8	8	8	8	18
Fe_2O_3	n.d.	0.00	0.00	0.00	0.00	0.00	0.00
FeO	1.18	0.01	0.01	0.00	0.00	0.00	0.13
TiO_2	0.42	n.d.	n.d.	n.d.	n.d.	n.d.	0.00
Cr_2O_3	n.d.	n.d.	n.d.	n.d.	n.d.	n.d.	0.00
SiO_2	5.56	2.48	2.53	2.48	2.60	2.64	5.05

(continued)

TABLE 2. *Continued*

Phase	Biotite (55 analyses)	Plagioclase (230 analyses)					Cordierite (3 analyses)
Sample	W4	W4	W4 garnet-bearing subassemblage	W4 sillimanite-bearing	W3 mylonite	W3 mylonite	W23 reaction rim
Al_2O_3	1.58	0.76	0.75	0.77	0.70	0.68	2.00
CaO	0.00	0.50	0.46	0.52	0.39	0.36	0.00
MgO	3.93	n.d.	n.d.	n.d.	n.d.	n.d.	1.80
MnO	0.01	n.d.	n.d.	n.d.	n.d.	n.d.	0.01
Na_2O	0.02	0.23	0.25	0.22	0.30	0.31	0.01
K_2O	0.85	0.01	0.01	0.00	0.00	0.01	0.00
X_{Mg}	n.d.	N.A.	N.A.	N.A.	N.A.	N.A.	0.93
Anorthite content	N.A.	0.51	0.47	0.53	0.39	0.36	N.A.

Note: n.d.—no data (not analyzed); N.A.—not applicable; $X_{Mg} = Mg/(Mg + Fe^{2+})$.

TABLE 3. REINTEGRATED COMPOSITIONS OF SPINEL

Analysis	W4B-L4	W4B-L5	W4B-L6	W4B-L8	W4B-L12	W4B-L9	W4B-L-L9	W4B-L11
Spinel host								
Areal proportion (%)	94.54	96.75	94.30	96.71	96.75	94.70	94.70	95.15
Mass	378.16	387.00	377.20	386.84	387.00	378.80	378.80	380.60
Moles	2.38	2.45	2.33	2.41	2.40	2.39	2.35	2.39
Fe (cations)	1.10	1.11	1.28	1.14	1.15	1.04	1.29	1.13
Mg (cations)	1.24	1.41	1.09	1.19	1.12	1.28	1.12	1.21
Magnetite lamellae								
Areal proportion (%)	5.46	3.25	5.70	3.29	3.25	5.30	5.30	4.85
Mass	28.39	16.90	29.64	17.11	16.90	27.56	27.56	25.22
Moles	0.12	0.07	0.13	0.07	0.07	0.12	0.12	0.11
Fe (cations)	0.37	0.22	0.38	0.22	0.22	0.36	0.36	0.33
Integrated composition (cations)								
Fe total	1.47	1.33	1.67	1.36	1.37	1.39	1.65	1.46
Mg total	1.24	1.41	1.09	1.19	1.12	1.28	1.12	1.21
Al total	4.67	4.74	4.46	4.68	4.67	4.70	4.51	4.67
Cr total	0.03	0.02	0.01	0.01	0.02	0.01	0.01	0.01
Mn total	0.01	0.00	0.02	0.03	0.02	0.02	0.02	0.03
Zn total	0.10	0.08	0.12	0.18	0.21	0.12	0.09	0.11
Integrated cations normalized to 4 oxygens								
Fe^{3+}	0.12	0.12	0.18	0.11	0.10	0.17	0.12	0.12
Fe^{2+}	0.46	0.41	0.50	0.44	0.45	0.50	0.44	0.46
Ti	0.00	0.00	0.00	0.00	0.00	0.00	0.00	0.00
Cr	0.01	0.01	0.01	0.00	0.01	0.00	0.01	0.00
Si	0.00	0.00	0.00	0.00	0.00	0.00	0.00	0.00
Al	1.87	1.88	1.82	1.88	1.89	1.87	1.83	1.87
Ca	0.00	0.00	0.00	0.00	0.00	0.00	0.00	0.00
Mg	0.49	0.56	0.44	0.48	0.45	0.51	0.45	0.48
Mn	0.00	0.00	0.01	0.01	0.01	0.01	0.01	0.01
Na	0.00	0.00	0.00	0.00	0.00	0.00	0.00	0.00
Zn	0.04	0.03	0.05	0.07	0.09	0.05	0.04	0.04
Total	3	3	3	3	3	3	3	3

Note: Areal proportions calculated from backscattered electron image analysis.
Density of spinel assumed to be 4.0 g/cm³, mass = 4.0 * volume.
Density of magnetite assumed to be = 5.2 g/cm³.

TABLE 4. REINTEGRATED COMPOSITIONS OF TITANHEMATITE

Analysis	W4-L8	W4-L4	W4-L2	W4-L3	W4-LL9	W4-RL12	W4-L7	W4-L6
Hematite host								
Areal proportion (%)	62.74	60.40	59.96	59.80	66.63	63.54	59.38	60.20
Mass	330.01	317.70	315.39	314.55	350.47	334.22	312.34	316.65
Moles	2.09	2.02	2.01	1.99	2.22	2.12	1.98	2.00
Ti (cations)	0.45	0.47	0.50	0.39	0.41	0.49	0.34	0.36
Fe^{3+} (cations)	3.28	3.09	3.00	3.19	3.59	3.26	3.26	3.28
Fe^{2+} (cations)	0.43	0.44	0.48	0.38	0.40	0.47	0.33	0.34
Ilmenite lamellae								
Areal proportion (%)	37.26	39.60	40.04	40.20	33.37	36.46	40.62	39.80
Mass	175.87	186.91	188.99	189.74	157.51	172.09	191.73	187.86
Moles	1.16	1.25	1.24	1.25	1.04	1.13	1.26	1.24
Ti (cations)	1.05	1.15	1.19	1.06	0.96	1.03	1.21	1.16
Fe^{3+} (cations)	0.22	0.19	0.11	0.37	0.15	0.21	0.11	0.15
Fe^{2+} (cations)	0.98	1.04	0.98	1.00	0.92	0.93	0.95	0.97
Integrated composition (cations)								
Ti	1.50	1.62	1.69	1.45	1.38	1.51	1.55	1.52
Fe^{3+}	3.50	3.28	3.10	3.56	3.74	3.46	3.37	3.43
Fe^{2+}	1.41	1.48	1.46	1.38	1.32	1.40	1.28	1.31
Integrated cations normalized to 3 oxygens								
Ti	0.47	0.50	0.53	0.45	0.43	0.47	0.49	0.48
Fe^{3+}	1.09	1.02	0.98	1.11	1.16	1.08	1.07	1.08
Fe^{2+}	0.44	0.46	0.46	0.43	0.41	0.44	0.41	0.41
Total	2	2	2	2	2	2	2	2
X_{hem}	0.55	0.51	0.50	0.56	0.58	0.54	0.54	0.55

Note: Areal proportions calculated from backscattered electron image analysis.
Density of hematite assumed to be 5.26 g/cm³.
Density of ilmenite assumed to be 4.72 g/cm³.

of orthopyroxene. The resulting $Y_{Fe^{3+}} = 0.84$, where $Y_{Fe^{3+}} = Fe^{3+}/(Fe^{3+} + Fe^{2+} + Ti)$, converts to a titanhematite composition of 0.73 Fe_2O_3 + 0.27 $FeTiO_3$. The mass fraction of oxide is so small that its contribution to the primary orthopyroxene chemistry is negligible.

Secondary orthopyroxene occurs as small, recrystallized grains within mylonitic bands, and as coronas involved in the reaction sapphirine + quartz = orthopyroxene + sillimanite + cordierite. The compositions of these secondary orthopyroxene grains are different from one another, and are consistently less

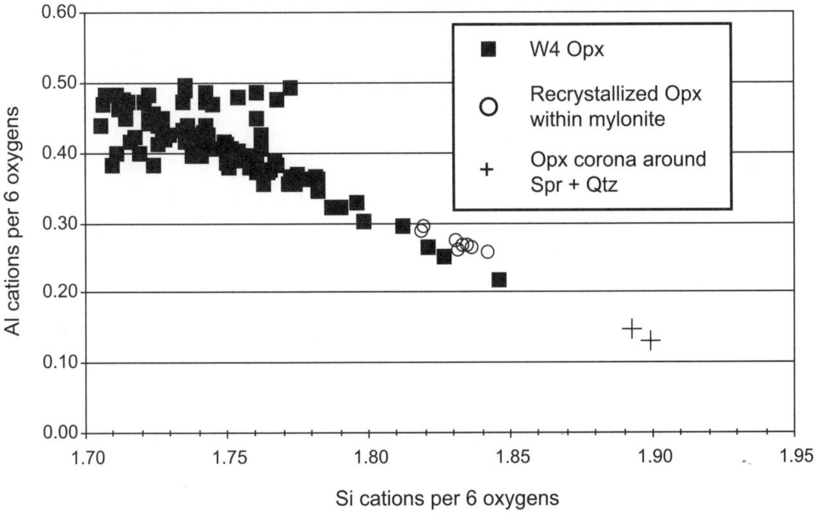

Figure 8. Aluminum cations versus silicon cations for orthopyroxene. Samples from W4 generally show a linear decrease in aluminum with increasing silicon and magnesium. This variation is attributed to the Tschermak substitution 2Al = Mg + Si. The recrystallized orthopyroxene grains from the mylonites are less aluminous than W4, and the orthopyroxene reaction rims surrounding sapphirine + quartz are the least aluminous orthopyroxene of any analyzed. Opx—orthopyroxene; Qtz—quartz; Spr—sapphirine.

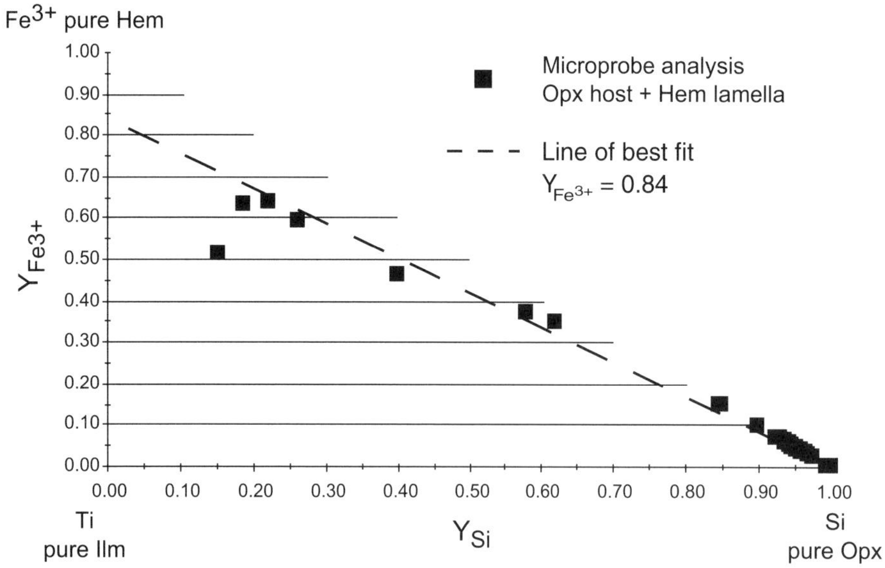

Figure 9. Point analyses of variable proportions of orthopyroxene and oxide exsolution lamellae are plotted as cation fractions in the ternary system Fe^{3+}-Ti-Si, where $Y_{Fe^{3+}} = Fe^{3+}/(Fe^{3+} + Ti + Si)$ and $Y_{Si} = Si/(Fe^{3+} + Ti + Si)$. Si is used as a proxy for the relative amount of orthopyroxene, and the Fe^{3+} + Ti join is used to characterize the hematite-ilmenite binary solution of the lamellae. The dashed line is a best fit extrapolated from a pure orthopyroxene host to an orthopyroxene-free composition of $Y_{Fe^{3+}} = 0.84$, which converts to a lamellae composition of 0.73 Fe_2O_3 + 0.27 $FeTiO_3$. Hem—titanhematite; Ilm—ilmenite; Opx—orthopyroxene.

aluminous and more magnesian-rich than grains considered primary (Fig. 8). Orthopyroxene coronas between sapphirine and quartz have a much different composition than the other grains, with lesser amounts of Fe (total) and Al, and greater amounts of Si and Mg. The orthopyroxene in these coronas is more magnesian than the sapphirine from which it formed.

Sapphirine

Sapphirine generally occurs as coarse, euhedral to subhedral grains in close association with all of the phases listed in Table 1. It is consistently more magnesian than primary orthopyroxene. Approximately 25% of the total iron calculates as Fe^{3+}. There appears to be some variation in sapphirine composition among the subassemblages listed in Table 1; the extremes of the range are listed in Table 2. The compositions lie between the ideal $7MgO \cdot 9Al_2O_3 \cdot 3SiO_2$ (= 7:9:3) and $3MgO \cdot 5Al_2O_3 \cdot 1SiO_2$ (= 3:5:1) compositions (Higgins et al., 1979).

Sillimanite

Sillimanite generally occurs as euhedral prismatic grains and is commonly distinctive in the field because of its light brown color, imparted by high dissolved ferric iron and/or finely exsolved hematite. Fleet and Arima (1985) described oriented, lath-shaped hematite inclusions in sillimanite from Wilson Lake and attributed their formation to intracrystalline precipitation on rational planes. The dimensions of the laths are typically a few tenths of microns but occasionally are large enough to enable microprobe analyses. Our analyses confirm that the laths are indeed pure hematite and that dissolved Fe^{3+} ranges up to 0.034 cations/5 oxygens (Table 2). Sillimanite also is observed as microfibrolitic overgrowths in some rocks and as part of various reaction rims in other rocks that show evidence for disequilibrium. Exploratory microprobe analyses show them to be low in dissolved Fe^{3+}.

Spinel

Spinel occurs as small green grains so peppered with magnetite exsolution as to appear nearly opaque in transmitted light (Fig. 3, D). Our analyses show that the magnetite is pure Fe_3O_4. Analyses of the host spinel reveal a small but possibly significant ZnO concentration. Similar analyses are reported by Currie and Gittens (1988).

In order to characterize the composition of the exsolved spinel grains at equilibrium, the magnetite lamellae were integrated into the composition of the spinel host (Table 3).

Garnet

Garnet is not commonly observed in sapphirine-bearing assemblages (Currie and Gittens, 1988), although in this study garnet was observed in direct contact with sapphirine (Fig. 3, B), along with orthopyroxene, spinel, corundum, titanhematite, plagioclase, and minor biotite (Table 1). The garnet is coarse-grained, with inclusions of several other phases, including exsolved orthopyroxene, spinel, and exsolved titanhematite. The garnet composition is relatively homogeneous across a single grain, as well as throughout the subassemblages listed in Table 1. Minor amounts of MnO and CaO are also present. Recalculated analyses reveal little or no Fe^{3+} (Table 2). These results are also consistent with analyses compiled by Currie and Gittens (1988).

Corundum

Corundum is common in many of the melanocratic gneisses and in the thicker oxide lenses as euhedral grains, commonly associated with intergrown magnetite. Analyses of corundum in association with other silicates and oxides having relatively high ferric iron indicate a total dissolved Fe^{3+} of up to 0.017 cations/3 oxygen formula units (Table 2).

Titanhematite

Titanhematite occurs as coarse, equant grains both in the mafic layers of the aluminous paragneisses and in the oxide-rich lenses. Ferrian ilmenite lamellae exsolved from the titanhematites are always observed, although their size and chemistry may vary between rocks. Titanhematite associated with the subassemblages of sample W4 listed in Table 1 exhibit several generations of lamellae (Fig. 4). The first-generation lamella is commonly greater than 1 μm in diameter. Subsequent generations of lamellae are much finer.

In order to characterize the composition of the titanhematite grains at equilibrium, the ferrian ilmenite lamellae were integrated into the composition of the host (Table 4). Although the various lamellae appeared to have similar compositions based on backscattered electron images, it was not possible to obtain quantitative analyses of the finest-scale lamellae, so we assumed that the coarse and fine lamellae had the same composition. The representative integrated composition was calculated to be $X_{Fe_2O_3} = 0.56$ (Fig. 10), where $X_{Fe_2O_3}$ is the mole fraction of hematite in the hematite-ilmenite binary. These grains were also analyzed with a defocused beam to use as a comparison, which yielded a composition of $X_{Fe_2O_3} \approx 0.64$. Regardless of the integration technique, the grains in these assemblages have the most Ti of any of the titanhematite analyzed. Titanhematite from oxide-rich lenses has very fine-scale lamellae. A defocused beam was used to obtain an integrated composition of $X_{Fe_2O_3} \approx 0.76$.

Titanhematite within mylonitic layers occurs as small and elongate grains that are clearly part of the mylonitic fabric. These grains contain fine exsolution lamellae that are undeformed. A defocused beam was used to obtain an integrated composition of $X_{Fe_2O_3} \approx 0.70$.

Magnetite

Magnetite occurs primarily as coarse anhedral grains, although small secondary magnetite is locally present along grain boundaries. The analyzed magnetite is nearly pure except for small amounts of Cr_2O_3 and Al_2O_3 (Table 2). No titaniferous magnetite was observed in this study, although it has been observed by others (e.g., Currie and Gittens, 1988). Pure magnetite coexisting with a hematite-rich solid solution, as observed in Wilson Lake, is consistent with Fe-Ti oxide phase equilibria (Burton, 1991). Titaniferous magnetite that coexists with hematite-ilmenite solid solutions is possible only for rocks with more Ti-rich bulk compositions than those studied here.

Plagioclase

The plagioclase associated with the subassemblages listed in Table 1 typically range between An 47 in contact with garnet in garnet-bearing subassemblages and An 53 in contact with sillimanite in sillimanite-bearing subassemblages. Plagioclase within the mylonitic layers have a lower An content (An 37).

Biotite

Secondary biotite is common as seams and aggregates throughout many samples. However, coarse euhedral biotite is locally observed that texturally appears to be primary. Analyses (Table 2) of this biotite show significant TiO_2 concentrations. We have made no attempt to calculate the ferric iron in biotite based on stoichiometry because of the uncertainties in hydroxyl content.

Cordierite

Cordierite was observed as a corona phase between sapphirine and quartz, and as rare individual grains in some samples. The cordierite analysis in Table 2 is from the sapphirine + quartz reaction rim. This analysis is essentially identical with the analyses reported by Currie and Gittens (1988), who also interpret the cordierite as a retrograde phase. It is the most magnesian of the Fe-Mg silicates.

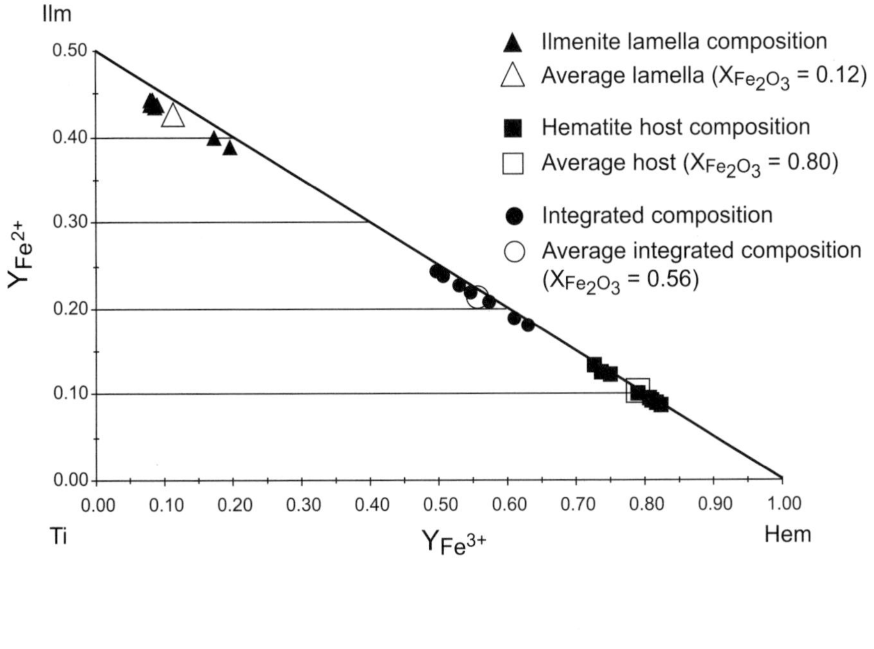

Figure 10. Hematite-ilmenite binary solution in ternary system Fe^{3+}-Fe^{2+}-Ti. Point analyses are plotted as cation fractions, where $Y_{Fe^{3+}} = Fe^{3+}/(Fe^{3+} + Fe^{2+} + Ti)$ and $Y_{Fe^{2+}} = Fe^{2+}/(Fe^{3+} + Fe^{2+} + Ti)$. The compositions of the ferrian ilmenite lamellae were integrated into the composition of the host. The representative integrated composition was calculated to be $X_{Fe_2O_3} = 0.56$, where $X_{Fe_2O_3}$ is the mole fraction of hematite in the hematite-ilmenite binary. These results are generally consistent with results acquired with a defocused beam. Hem—titanhematite; Ilm—ilmenite.

ELEMENT PARTITIONING

The phase chemistry previously described focuses on the subassemblages listed in Table 1. The relative uniformity of the analyses for each phase from layer to layer provides good evidence for chemical equilibria among the silicate and oxide phases. As there can be no variations in pressure and temperature at the scale of a single thin section, possible variations in the chemical potentials of components as determined by variations in mineral compositions would be the only criteria for supposing that equilibria were not achieved.

A predictable consequence of the equilibrium assumption is a systematic partitioning of elements among the phases involved. As partitioning can vary with pressure and temperature, it is imperative that only those analyses from coexisting phases be considered. Analyses of phases from specimen W4 (Table 1) show that X_{Mg}(Spr) ≥ X_{Mg}(Opx) > X_{Mg}(Grt) > X_{Mg}(integrated Spl) (Fig. 11), where $X_{Mg} = Mg/(Mg + Fe^{2+})$. We infer that cordierite would be the most magnesian phase if the bulk composition were more magnesian. However, these low-variance assemblages from Wilson Lake are sufficiently rich in Fe so that primary cordierite is not present. This partitioning sequence is consistent with a few other studies (Harris, 1981; Waters, 1991), in which the assemblages can be documented as having formed at relatively high fO_2 conditions. Most studies, however, invoke the reversed sequence X_{Mg}(Spl) > X_{Mg}(Grt), which is applicable to rocks formed at lower fO_2 conditions (Ellis et al., 1980; Hensen, 1987). We note that the partitioning sequence of Hensen (1987) for low fO_2 conditions would hold if the magnetite exsolution had not been integrated back into the spinel composition (Fig. 11).

Additional evidence for silicate-oxide equilibration is the regular increase in the Fe_2SiO_5 content of sillimanite with increasing Fe_2O_3 content of coexisting ilmenite (reintegrated) as shown in Figure 12. The Wilson Lake data and Grew's (1980) data and compilation provide a nearly complete saturation curve for the Fe_2SiO_5 content of sillimanite and the Fe_2O_3 content of coexisting ilmenite. Grew's data and compilation (1980) show that the increases of Fe_2O_3 in ilmenite and Fe_2SiO_5 in sillimanite are linear at low concentrations of Fe^{3+}, thus behaving as ideal dilute solutions. The Wilson Lake data show a continuing increase in dissolved Fe_2O_3 and Fe_2SiO_5, but the departure of the data from the linear trend established by Grew (1980) suggests nonideal behavior at higher Fe^{3+} concentrations. The endpoint for the overall trend of the data in Figure 12 should be considered a minimum in dissolved Fe_2SiO_5 because that par-

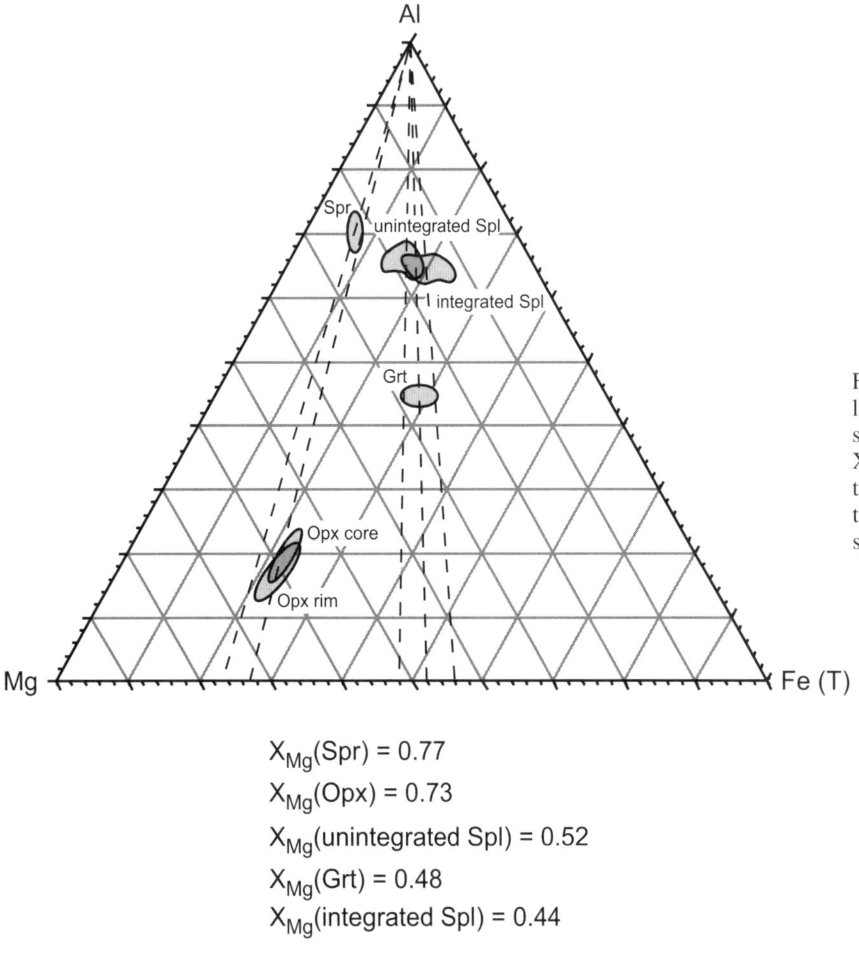

Figure 11. Compositions of the Fe-Mg phases for the low variance assemblage W4 exhibit the partitioning sequence X_{Mg}(sapphirine) ≥ X_{Mg}(orthopyroxene) > X_{Mg}(garnet) > X_{Mg}(integrated spinel). This partitioning is consistent with relatively high fO_2 conditions. Grt—garnet; Opx—orthopyroxene; Spl—spinel; Spr—sapphirine.

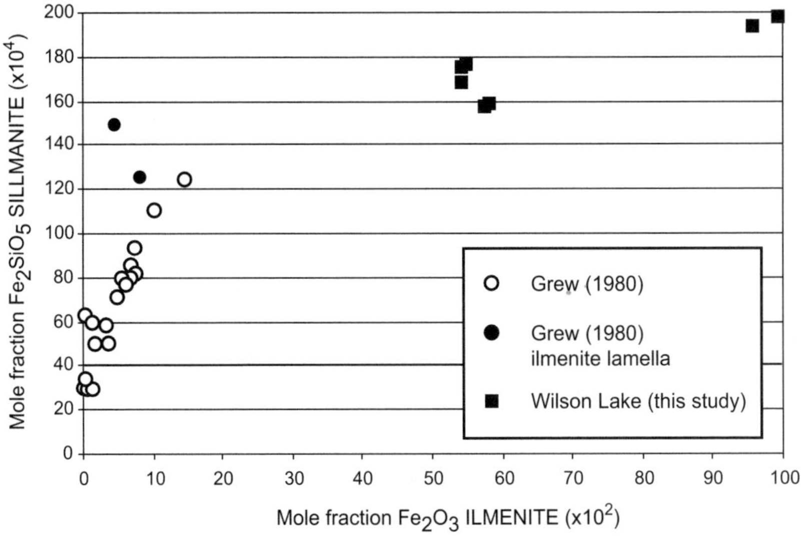

Figure 12. A regular increase in the Fe_2SiO_5 content of sillimanite with the increasing Fe_2O_3 content of coexisting ilmenite suggests that these two phases crystallized in equilibrium (after Grew, 1980). Sillimanite from Wilson Lake contains significant dissolved Fe^{3+}, with the highest values associated with exsolved hematite.

ticular sillimanite contains exsolved hematite that presumably formed at temperatures below those of peak metamorphism. It should be noted that the two points from Grew's (1980) data with elevated Fe_2SiO_5 contents are plotted against ilmenite lamellae analyses in magnetite and hematite rather than against an integrated composition, emphasizing the importance of integrating host and lamellae compositions.

It is to be expected that Fe-Mg silicate phases that equilibrated with Fe^{3+}-rich oxide phases under relatively high fO_2 conditions would have higher dissolved Fe^{3+} than those that equilibrated under more reducing conditions. Such is the case for the sapphirine and orthopyroxene coexisting with magnetite and titanhematite. Sapphirine ($X_{Fe^{3+}} = Fe^{3+}/(Fe^{3+} + Fe^{2+}) = 0.27$) partitions Fe^{3+} over orthopyroxene ($X_{Fe^{3+}} = 0.17$), whereas in the same assemblage there is no evidence from recalculated formulae for Fe^{3+} in garnet. Integrated spinel compositions have $X_{Fe^{3+}} = 0.22$, whereas the nonintegrated spinel host has $X_{Fe^{3+}} = 0.10$.

The overall partitioning sequence $X_{Fe^{3+}}$(Spr) > $X_{Fe^{3+}}$(Spl) > $X_{Fe^{3+}}$(Opx) > $X_{Fe^{3+}}$(Grt) observed in the Wilson Lake assemblages is appropriate for the relatively high fO_2 conditions defined by coexisting titanhematite and magnetite. Other natural occurrences (Caporuscio and Morse, 1978; Lal et al., 1984) in which $X_{Fe^{3+}}$(Spr) > $X_{Fe^{3+}}$(Spl) are also characterized by relatively high fO_2 conditions of metamorphism. The reversed sequence $X_{Fe^{3+}}$(Spl) > $X_{Fe^{3+}}$(Spr) is more commonly invoked (e.g., Hensen, 1986) for lower fO_2 conditions where hematite and magnetite are not stable.

PHASE EQUILIBRIA

The broad phase relationships among the coexisting silicate and oxide phases in the Wilson Lake area can be represented in the system FeO-MgO-Al_2O_3-SiO_2-TiO_2-Fe_2O_3. The relevant phases are orthopyroxene, sapphirine, garnet, sillimanite, spinel, corundum, magnetite, and titanhematite. For simplicity at this point, we exclude biotite as the only K-bearing phase and plagioclase as the only Na-bearing phase. Assuming that the phases listed in Table 1 and their compositions as given in Tables 2–4 approach equilibrium, the thermodynamic variance approaches zero.

Figure 13, A, represents the phase relationships graphically in the system FeO-MgO-Al_2O_3-SiO_2 (FMAS) in the presence of magnetite and titanhematite. This is not a rigorous projection through magnetite and titanhematite because of our desire to show qualitatively with some exaggeration the relative partitioning of Fe^{2+} and Mg among the six remaining phases. The six-phase assemblage is shown as an eight-sided volume in Figure 13, A. Any bulk composition within this volume would be expressed as the invariant assemblage (plus magnetite and titanhematite) under the *P, T,* and fO_2 conditions of the invariant point. The fO_2 in this assemblage, defined by the coexistence of magnetite and titanhematite with the integrated composition given in Table 4, is near but below that defined by pure magnetite and hematite at the *P* and *T* conditions of the invariant equilibrium. The compositions of the six phases, including their ferric iron contents, are completely determined.

Bulk compositions that would produce quartz-bearing assemblages lie outside the eight-sided volume. More SiO_2-rich compositions would likely produce the divariant assemblage Opx + Sil + Grt + Qtz as depicted in Figure 13, B, rather than assemblages involving cordierite and garnet. Nowhere in the Wilson Lake literature are cordierite and garnet described as an equilibrium pair. The assemblage Sil + Opx + Qtz is common in the Wilson Lake area, as is Spr + Qtz ± Sil ± Opx for more Mg-rich compositions, depicted qualitatively in Figure 13, B. Cordierite (Crd) may be stable under the conditions of the invariant point as suggested in Figure 13, B, but only for bulk compositions more magnesian than those typically observed in the Wilson Lake area.

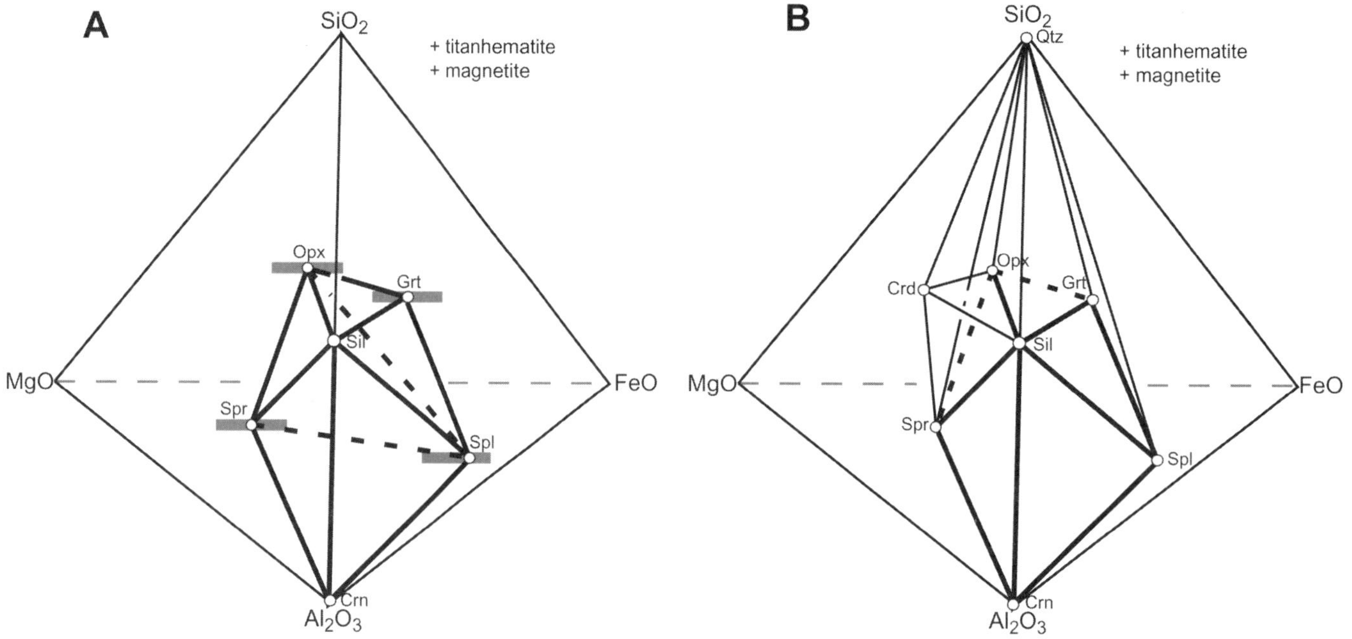

Figure 13. (A) $FeO-MgO-Al_2O_3-SiO_2$ (FMAS) tetrahedron for the invariant assemblage in sample W4 in the presence of magnetite + titanhematite. The thick black lines outline the eight-sided volume defined by the six phases (orthopyroxene-garnet-sillimanite-sapphirine-spinel-corundum). Any bulk composition within this volume would be expressed as the invariant assemblage (+ Mag + Hem) under the *P, T,* and fO_2 conditions of the invariant point. Gray bars indicate a schematic solid solution for the Fe-Mg phases. (B) FMAS invariant assemblage (+ Mag + Hem) extended to include cordierite and quartz. The original six-phase volume is still outlined, but other assemblages are also possible and are depicted in the FMAS tetrahedron as subvolumes. Cordierite may be stable under the conditions of the invariant point, but only for more magnesian bulk compositions. The quartz-bearing assemblages depicted above are dependent on Fe^{2+}-Mg partitioning, and hence on fO_2 (see text for further discussion). Crd—cordierite; Crn—corundum; Grt—garnet; Opx—orthopyroxene; Qtz—quartz; Sil—sillimanite; Spl—spinel; Spr—sapphirine.

The existence of the quartz-bearing assemblages we infer in Figure 13, B, critically depends on the observed sequence of Fe^{2+}-Mg partitioning. For example, if the Fe^{2+}-Mg partitioning between sapphirine and orthopyroxene were to be reversed (i.e., $X_{Mg}(Opx) > X_{Mg}(Spr)$) from our observations, a stable sapphirine + quartz tieline would pierce the Opx-Grt-Sil triangle in Figure 13, B, and thus be incompatible with the invariant point assemblage (Fig. 13, A). Similar reasoning applies to the spinel + quartz tieline which we infer to be stable in Figure 13, B. If the Fe^{2+}-Mg partitioning between garnet and spinel were to be reversed (i.e., $X_{Mg}(Spl) > X_{Mg}(Grt)$) from our observations, the spinel + quartz tieline would pierce the Opx-Grt-Sil triangle and thus be incompatible with the invariant point assemblage. We note that the assemblage spinel (with up to 2.5 wt% ZnO) + quartz has been described by Currie and Gittens (1988) in leucogranite veins within paragneiss at Wilson Lake.

Biotite is an important phase in the W4 assemblage (Table 1) because of its TiO_2 content and likely Fe^{3+} content. Titanhematite is the only other TiO_2-bearing phase in the assemblage, and thus biotite must be included in the overall equilibrium for there to be a viable cation exchange (e.g., $2Fe^{3+} = Ti + Fe^{2+}$) involving titanhematite. The addition of biotite as a phase requires the addition of K_2O as a component. However, the variance of the system does not change, and assemblage W4 remains invariant.

An important consequence of this line of reasoning is the need to invoke another potassic phase in the W4 assemblage in order to provide mass balance with biotite. In the absence of K-feldspar or muscovite, a K_2O-bearing melt phase is the most likely candidate. As there is no direct evidence for this hypothetical phase in assemblage W4, we can only infer its former presence as leucosome in the surrounding migmatites.

It is worth noting that under relatively oxidizing conditions in which Fe^{3+} is preferred over Fe^{2+}, iron partitions into the oxide phases and enriches the remaining silicate phases in Mg. The fact that garnet is the most Fe^{2+}-rich of the silicates and is more likely to be found in more Fe^{2+}-rich bulk compositions likely accounts for its relative scarcity in the Wilson Lake area.

PARTIAL PETROGENETIC GRID

The phase associations in sample W4 (Table 1) are exceptional because to our knowledge they are the first direct evidence that these complex solid solutions have mutually equilibrated. The compositions of Fe-Mg solutions in higher-variance systems change with *P* and *T* and are fixed only at the specific *P, T,* and other intensive variables (e.g., fO_2) that are unique to each invariant point. The partial petrogenetic grid of Figure 14 represents a framework of reactions and invariant points in the FMAS system that is relevant to the Wilson Lake assemblages because it represents both corundum-bearing (left half) and quartz-bearing (right half) bulk compositions, joined at the [Qtz, Crn]-absent invariant point (Hensen, 1987; Hensen and Harley, 1990). The invariant assemblage W4 (Table 1) corresponds to the stable [Crd]-absent invariant point, which is positioned in this grid at ~900 °C and 1000 MPa based on experimental constraints in the simpler MAS system and presumed stabilities of other invariant points in the system (Hensen, 1987). This estimate is consistent with and independent of those determined by cation exchange thermobarometry for quartz-bearing assemblages in the Wilson Lake area (Arima et al., 1986; Currie and Gittens, 1988).

The stability fields of sapphirine + quartz and spinel + quartz are outlined in the high-temperature half of Figure 14, which is relevant to silica-saturated bulk compositions. As discussed in the following section, the apparent inconsistency of the *P-T* coordinates of the [Crd]-absent invariant point and the fields of Spr + Qtz and Spl + Qtz is likely due to the effects of ferric iron in sapphirine and spinel, and possibly of ZnO in spinel.

The petrogenetic grid in Figure 14 is not entirely satisfactory for interpreting metamorphic rocks that have formed under relatively high fO_2 conditions. A high-fO_2 alternative (Hensen, 1987) to the silica-saturated part of Figure 14 inverts the stabilities of the three stable invariant points shown and renders stable the four metastable invariant points not shown in Figure 14. Application of this result to quartz-bearing assemblages in granulite-facies rocks that include magnetite and hematite has been met with some success (e.g., Goscombe 1992; Dasgupta et al., 1995). The same inversion cannot be applied to the silica-undersaturated part of the grid because of the presumed stability of the [Crd]-absent invariant point at both low fO_2 and high fO_2. A complete analysis of both quartz-bearing and corundum-bearing

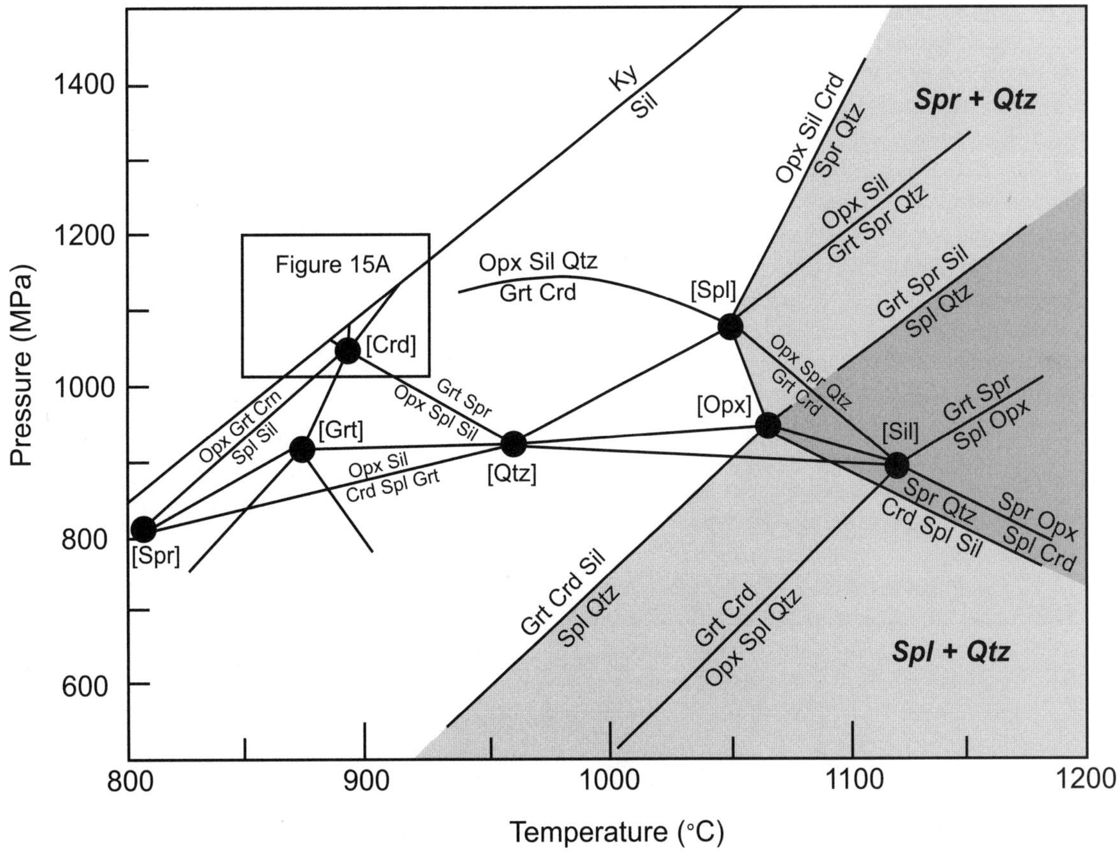

Figure 14. Partial petrogenetic grid in the FeO-MgO-Al_2O_3-SiO_2 (FMAS) system linking corundum-bearing bulk compositions (lower temperatures) and quartz-bearing bulk compositions (higher temperatures) through the [Qtz]-absent invariant point. The [Qtz]-absent invariant point in the Si-undersaturated system is equivalent to the [Crn]-absent invariant point in Si-saturated system. The stability fields for sapphirine + quartz and spinel + quartz are shaded. Modified after Hensen (1987); Hensen and Harley (1990); Spear (1995, Figs. 10–24). Location of kyanite-sillimanite reaction after Hensen and Harley (1990). The invariant assemblage W4 corresponds to the stable cordierite [Crd]-absent invariant point shown in Figure 15, A and B. Crd—cordierite; Crn—corundum; Grt—garnet; Ky—kyanite; Opx—orthopyroxene; Qtz—quartz; Sil—sillimanite; Spl—spinel; Spr—sapphirine.

assemblages at high fO_2 conditions requires an evaluation of the many possible n + 4 grids (Hensen, 1987), which is beyond the scope of the present study.

DISCUSSION

The departure of the phase compositions described in this paper from the ideal FMAS system at relatively low fO_2 leads to several important modifications of Figure 14. The most important of these is the shift of reaction curves in the *P-T* projection, and hence shifts in the locations of invariant points, due to ferric iron. The partitioning sequence $X_{Fe^{3+}}$(Spr) > $X_{Fe^{3+}}$(Spl) > $X_{Fe^{3+}}$(Opx) > $X_{Fe^{3+}}$(Grt) observed for the [Crd]-absent invariant point assemblage will extend the stabilities of sapphirine and spinel relative to orthopyroxene and garnet everywhere along the (Spr)-, (Grt)-, and (Spl)-absent reactions shown in Figure 15, A, which shows the details of the inset box from Figure 14. The [Crd]-absent invariant point shifts to lower temperature. The simultaneous effect of the gahnite ($ZnAl_2O_4$) component in spinel shifts the invariant point to higher pressure along the (Spl)-absent reaction. In the course of this shift toward higher fO_2 conditions, magnetite + titanhematite stabilizes in the presence of TiO_2, and the Fe^{2+}-Mg partitioning in silicates reverses such that spinel becomes less magnesian than coexisting garnet as inferred in Figure 13, B. The reversal of partitioning between garnet and spinel changes the mass balance coefficients of reactions in the immediate vicinity of the [Crd]-absent invariant point now at high fO_2 (Fig. 15, B) such that Opx ((Spr)-absent reaction) and Spr ((Opx)-absent reaction) become product phases at high temperature rather than reactant phases as shown in Figure 15, A. Additional Fe^{3+} in sapphirine and spinel, and Zn in spinel, further shifts the [Crd]-absent invariant point in Figure 15, B, to lower *T* and higher *P*. The magnitude of this shift must be such that the invariant point remains within the stability field of sillimanite. We note, however, that additional Fe^{3+} in the form of a Fe_2SiO_5 component will expand the stability of sillimanite relative to kyanite and shift the sillimanite-kyanite equilibrium (Fig. 14) to higher pressure (Grew, 1980).

Reaction curves and invariant points in the silica-saturated part of Figure 14 will also be displaced due to additional components. Ferric iron stabilizes the assemblage Spr + Qtz, extending its stability to lower temperature along the (Spr)-absent reaction. Both Fe^{3+} and Zn solubility in spinel will extend the stability of the assemblage Spl + Qtz to lower *T* and higher *P* as the [Opx]-absent invariant point shifts to higher *P* along the reaction joining the [Opx]- and [Spl]-absent invariant points (Harley, 1986). We have not yet made a full evaluation of the quartz-bearing assemblages in the Wilson Lake area, so we can only speculate that the stability fields of Spr + Qtz and Spl + Qtz in the presence of magnetite and titanhematite will expand to include the *P-T* coordinates of the [Crd]-absent invariant point.

The relative timing and conditions of formation of secondary orthopyroxene, cordierite, and sillimanite at the expense of sapphirine and quartz are unclear. This reaction affects only bulk compositions more SiO_2- and MgO-rich than the invariant assemblage volume shown in Figure 13, B. Products of this reaction break the sapphirine + quartz tieline and stabilize the assemblages Opx + Sil + Crd + Qtz and Opx + Sil + Crd + Spr. This reaction lies on the low temperature side of the Spr + Qtz sta-

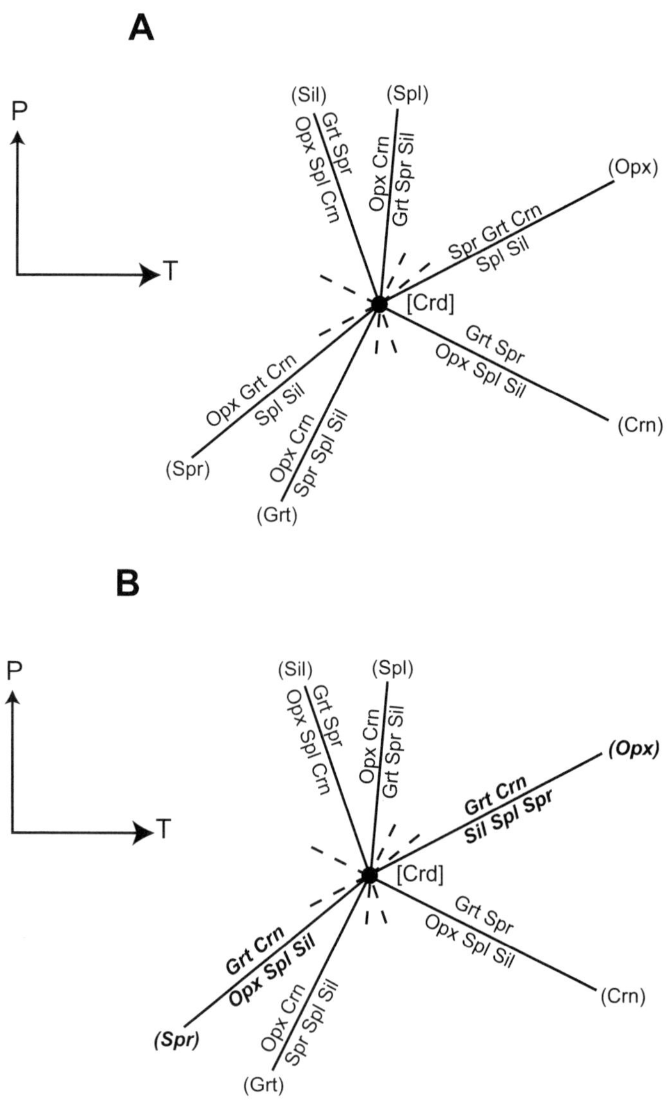

Figure 15. (A) Cordierite [Crd]-absent invariant point in FeO-MgO-Al_2O_3-SiO_2 for corundum-bearing bulk compositions at low fO_2 conditions (from Hensen and Harley, 1990). The addition of Fe_2O_3 and ZnO as components shifts the invariant point to lower *T* and higher *P* (see text for further discussion). (B) A schematic representation of Figure 15, A, more appropriate for high fO_2 conditions. The reversal of partitioning between garnet and spinel at high fO_2 changes the mass balance coefficients such that Opx ((Spr)-absent reaction) and Spr ((Opx)-absent reaction) become product phases at high temperature rather than reactant phases as shown in Figure 15, A. These reactions that change under high fO_2 are shown in italics. The addition of Fe_2O_3 and ZnO will shift the invariant point to lower *T* and higher *P*. Crd—cordierite; Crn—corundum; Grt—garnet; Opx—orthopyroxene; Sil—sillimanite; Spl—spinel; Spr—sapphirine.

bility field in Figure 14 and is interpreted by Morse and Talley (1971) and by Currie and Gittens (1988) as evidence for isobaric cooling. The absence of exsolved hematite in secondary orthopyroxene suggests reaction under less oxidizing conditions than those of the primary assemblages.

Further evidence for the lower temperature history of the Wilson Lake granulites is recorded in the exsolution of ferrian ilmenite from reintegrated titanhematite, exsolved magnetite in spinel, exsolved titanhematite in orthopyroxene, and exsolved hematite in sillimanite. Our integrated titanhematite composition represents the composition of this primary phase prior to exsolution, which must have been at temperatures above the solvus shown in Figure 16. The minimum temperature estimate is on the order of 650 °C, which suggests that the magnetic remanence, which is due to host titanhematite in exsolved primary grains (Kletetschka and Stout, 1998), is most likely due to cooling from above the Curie temperature. The host and lamellae compositions suggest that exsolution would occur at a minimum temperature of 450 °C.

One further constraint on the relative timing of Fe-Ti oxide exsolution is the observation of undeformed exsolution lamellae in mylonitic titanhematite that clearly postdates porphyroclastic, oxidized orthopyroxene (Fig. 7). The sillimanite + quartz fabric with reequilibrated orthopyroxene (Fig. 8) in the mylonite is still within the granulite facies. Had the exsolution in the titanhematite occurred prior to the mylonitic deformation, the lamellae would likely be deformed. Thus, it appears that the exsolution in the titanhematites postdates this mylonitic deformation. However, the age of the mylonites is problematic. U-Pb dating on zircon in mylonites from the Lac Joseph terrane yield Labradorian ages (Connelly and Heaman, 1993), which are slightly younger than the Labradorian ages these authors report for the primary metamorphism. On the other hand, both the Wilson Lake and the Lac Joseph terranes are floored with Grenvillian shear zones, and we cannot at this time eliminate a Grenvillian age for the mylonites.

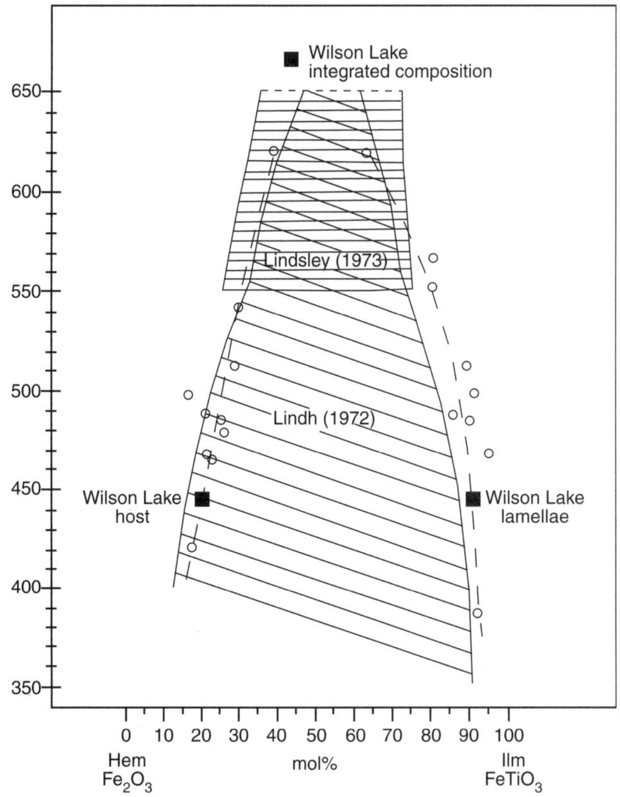

Figure 16. A temperature-composition diagram for the Fe-Ti oxides, after Braun and Raith (1985). The integrated titanhematite composition represents the composition prior to exsolution. The host and lamellae compositions suggest that exsolution would occur at a minimum temperature of 450 °C. Hem—titanhematite; Ilm—ilmenite.

CONCLUSIONS

The predominant mineral assemblages in the restitic parts of paragneisses in the Red Wine Mountains massif are Opx + Sil + Qtz, Spr + Qtz, and possibly Spl + Qtz. Closely associated with these assemblages is the low-variance, silica-undersaturated assemblage Spr + Opx + Sil + Grt + Spl + Crn + Mag + titanhematite (+ Pl + Bt), which approaches invariance in the six-component system FeO-MgO-Al_2O_3-SiO_2-Fe_2O_3-TiO_2. The resulting [Crd]-absent invariant point appears to be stable near 900 °C and near 1000 MPa, and at an fO_2 defined by coexisting pure magnetite and titanhematite (after integration of exsolved ferrian ilmenite). Further evidence of high fO_2 is exsolved titanhematite ($X_{Fe_2O_3} = 0.73$) in orthopyroxene, exsolved hematite in sillimanite, and high dissolved Fe^{3+} in sillimanite and corundum. Systematic partitioning of Fe^{3+} in the coexisting silicate and oxide phases documents their mutual equilibrium.

The observed sequence of Fe^{2+}-Mg partitioning after appropriate integration of exsolved phases is X_{Mg}(Spr) ≥ X_{Mg}(Opx) > X_{Mg}(Grt) > X_{Mg}(Spl). The Fe^{2+}-Mg partitioning between garnet and spinel in assemblage W4 is reversed from that normally invoked for FMAS petrogenetic grids applicable to low fO_2 conditions. This reversal is critical to the stability of Spl + Qtz under high fO_2 conditions.

The partitioning sequence $X_{Fe^{3+}}$(Spr) > $X_{Fe^{3+}}$(Spl) > $X_{Fe^{3+}}$(Opx) > $X_{Fe^{3+}}$(Grt) observed for the [Crd]-absent invariant point assemblage extends the stabilities of sapphirine and spinel relative to orthopyroxene and garnet, shifting the [Crd]-absent invariant point to lower temperature. The simultaneous effect of the gahnite ($ZnAl_2O_4$) component in spinel shifts the invariant point to higher pressure. Fe^{3+} stabilizes the assemblage Spr + Qtz, extending its stability to lower temperature. Both Fe^{3+} and Zn solubility in spinel extend the stability of the assemblage Spl + Qtz to lower *T* and higher *P*. Previous estimates of peak metamorphic temperatures in the Red Wine Mountains massif are likely too high.

Evidence from our study and the literature supports a multiphase history for the migmatitic paragneisses that comprise the

Red Wine Mountains massif. The remanence-dominated aeromagnetic high associated with the massif is carried by coarse-grained host titanhematite in exsolved primary grains. The remanence acquisition likely postdates the primary granulite-facies metamorphism and a later granulite-facies mylonitic deformation.

ACKNOWLEDGMENTS

We acknowledge Drs. Gunther Kletetschka and Toby Rivers for their thoughtful and careful reviews. We also acknowledge Dr. Peter McSwiggen for his assistance with the electron microprobe, and Kristina Gross for her assistance with some of the figures. We acknowledge the Graduate School of the University of Minnesota for support of the field and analytical work.

REFERENCES CITED

Arima, M., Kerrich, R., and Thomas, A., 1986, Sapphirine-bearing paragneiss from the northern Grenville province in Labrador, Canada: Protolith composition and metamorphic P-T conditions: Geology, v. 14, p. 844–847.

Bertrand, P., Ellis, D.J., and Green, D.H., 1991, The stability of sapphirine-quartz and hypersthene-sillimanite-quartz assemblages: An experimental investigation in the system FeO-MgO-Al_2O_3-SiO_2 under H_2O and CO_2 conditions: Contributions to Mineralogy and Petrology, v. 108, p. 55–71.

Braun, E., and Raith, M., 1985, Fe-Ti-oxides in metamorphic basites from the Eastern Alps, Austria: A contribution to the formation of solid solutions of natural Fe-Ti-oxide assemblages: Contributions to Mineralogy and Petrology, v. 90, p. 199–213.

Burton, B.P., 1991, Interplay of chemical and magnetic ordering, *in* Lindsley, D.H., ed., Oxide Minerals: Petrologic and magnetic significance: Washington, D.C., Mineralogical Society of America, v. 25, p. 303–319.

Caporuscio, F.A., and Morse, S.A., 1978, Occurrence of sapphirine plus quartz at Peekskill, New York: American Journal of Science, v. 278, p. 1334–1342.

Connelly, J.N., and Heaman, L.M., 1993, U-Pb geochronological constraints on the tectonic evolution of the Grenville Province, western Labrador: Precambrian Research, v. 63, p. 123–142.

Connelly, J.N., Rivers, T., and James, D.T., 1995, Thermotectonic evolution of the Grenville Province of Western Labrador: Tectonics, v. 14, p. 202–217.

Corrigan, D., Rivers, T., and Dunning, G.R., 1997, Preliminary report on the evolution of the Allochthon Boundary Thrust in Eastern Labrador, Mechin River to Goose Bay: [Eastern Canadian Shield Onshore-Offshore Transect] Meeting, 14–15 April, Ottawa, Ontario, Lithoprobe Report No. 61, p. 45–55.

Corrigan, D., Rivers, T., and Dunning, G., 2000, U-Pb constraints for the plutonic and tectonometamorphic evolution of Lake Melville terrane, Labrador, and implications for basement reworking in the northeastern Grenville Province: Precambrian Research, v. 99, p. 65–90.

Currie, K.L., and Gittens, J., 1988, Contrasting sapphirine parageneses from Wilson Lake, Labrador, and their tectonic significance: Metamorphic Geology, v. 6, p. 603–622.

Currie, K.L., and Loveridge, W.D., 1985, Geochronology of retrogressed granulites from Wilson Lake, Labrador, *in* Current Research, Part B: Ottawa, Ontario, Geological Survey of Canada, Paper 85-1B, p. 191–197.

Dasgupta, S., Sengupta, P., Ehl, J., Raith, M., and Bardhan, S., 1995, Reaction textures in a suite of spinel granulites from the Eastern Ghats Belt, India: Evidence for polymetamorphism, a partial petrogenetic grid in the system KFMASH and the roles of ZnO and Fe_2O_3: Journal of Petrology, v. 36, p. 435–461.

Ellis, D.J., Sheraton, J.W., England, R.N., Dallwitz, W.B., 1980, Osumilite-sapphirine-quartz granulites from Enderby Land, Antarctica—Mineral assemblages and reactions: Contributions to Mineralogy and Petrology, v. 72, p. 353–367.

Fleet, M.E., and Arima, M., 1985, Oriented hematite inclusions in sillimanite: American Mineralogist, v. 70, p. 1232–1237.

Gittens, J., and Currie, K.L., 1979, Petrologic studies of sapphirine-bearing granulites around Wilson Lake, Labrador, *in* Current Research, Part A: Ottawa, Ontario, Geological Survey of Canada, Paper 79-1A, p. 77–82.

Goscombe, B., 1992, Silica-undersaturated sapphirine, spinel and kornerupine granulite facies rocks, NE Strangways Range, Central Australia: Journal of Metamorphic Geology, v. 10, p. 181–201.

Gower, C.F., 1996, The evolution of the Grenville Province in eastern Labrador, Canada, *in* Brewer, T.S., ed., Precambrian crustal evolution in the North Atlantic region: Québec, Geological Society of Canada Special Publication 112, p. 197–218.

Gower, C.F., and Krogh, T.E., 2002, A U-Pb geochronological review of the Proterozoic history of the eastern Grenville Province: Canadian Journal of Earth Sciences, v. 39, p. 795–829.

Gower, C.F., Schärer, U., and Heaman, L.M., 1992, The Labradorian Orogeny in the Grenville Province, eastern Labrador: Canadian Journal of Earth Sciences, v. 29, p. 1944–1957.

Grew, E.S., 1980, Sapphirine and quartz association from Archean rocks in Enderby Land, Antarctica: American Mineralogist, v. 65, p. 821–836.

Harley, S.L., 1986, A sapphirine-cordierite-garnet-sillimanite granulite from Enderby Land, Antarctica: Implications for FMAS petrogenetic grids in the granulite facies: Contributions to Mineralogy and Petrology, v. 94, p. 452–460.

Harris, N., 1981, The application of spinel-bearing metapelites to P-T determinations: An example from South India: Contributions to Mineralogy and Petrology, v. 76, p. 229–233.

Hensen, B.J., 1986, Theoretical phase relations involving cordierite and garnet revisited: The influence of oxygen fugacity on the stability of sapphirine and spinel in the system Mg-Fe-Al-Si-O: Contributions to Mineralogy and Petrology, v. 92, p. 362–367.

Hensen, B.J., 1987, P-T paths for silica-undersaturated granulites in the systems MAS (n + 4) and FMAS (n + 3)—Tools for the derivation of P-T paths of metamorphism: Journal of Metamorphic Geology, v. 5, p. 255–271.

Hensen, B.J., and Green, D.H., 1971, Experimental study of the stability of cordierite and garnet in pelitic compositions at high pressures and temperatures, I: Compositions with excess alumino-silicate: Contributions to Mineralogy and Petrology, v. 33, p. 309–330.

Hensen, B.J., and Harley, S.L., 1990, Graphical analysis of P-T-X relations in granulite facies metapelites, *in* Ashworth, J.R., and Brown, M., eds., High temperature metamorphism and crustal anatexis: London, Mineralogical Society of Great Britain, Unwin-Hyman, p. 19–56.

Herd, R.K., Ackermand, D., Thomas, A., and Windley, B.F., 1987, Oxygen fugacity variations and mineral reactions in sapphirine-bearing paragneisses, E. Grenville province, Canada: Mineralogical Magazine, v. 51, p. 203–206.

Higgins, J.B., Ribbe, P.H., Herd, R.K., 1979, Sapphirine, I: Contributions to Mineralogy and Petrology, v. 68, p. 349–356.

Jackson, V., and Finn, G., 1982, Geology, petrography and petrochemistry of granulite rocks from Wilson Lake, Labrador: St. John's, Newfoundland, Government of Newfoundland and Labrador Open-File, Labrador 13E/T(40), 22 p.

James, D.T., Kamo, S., Krogh, T., and Nadeau, L., 2002, Preliminary report on U-Pb ages for intrusive rocks from the Western Mealy Mountains and Wilson Lake terranes, Grenville Province, southern Labrador, *in* Current Research: St. John's, Newfoundland Department of Mines and Energy Geological Survey, Report 02–1, p. 67–77.

Kletetschka, G., and Stout, J.H., 1998, The origin of magnetic anomalies in lower crustal rocks, Labrador: Geophysical Research Letters, v. 25, p. 199–202.

Kretz, R., 1983, Symbols for rock forming minerals: American Mineralogist, v. 68, p. 277–279.

Lal, R.K., Ackermand, D., Raith, M., Raase, P., and Seifert, F., 1984, Sapphirine bearing assemblages from Kiranur, Southern India: A study of chemographic relationships in the Na_2O-FeO-MgO-Al_2O_3-SiO_2-H_2O system: Neues Jahrbuch Mineralogische Abhandlung, v. 150, p. 121–152.

Leong, K.M., and Moore, J.M., 1972, Sapphirine-bearing rocks from Wilson Lake, Labrador: Canadian Mineralogist, v. 11, p. 777–790.

Lindh, A., 1972, A hydrothermal investigation of the system FeO, Fe_2O_3, TiO_2: Lithos, v. 5, p. 325–343.

Lindsley, D.H., 1973, Delimitation of the hematite-ilmenite miscibility gap: Geological Society of America Bulletin, v. 84, p. 657–661.

Morse, S.A., and Talley, H., 1971, Sapphirine reactions in deep-seated granulites near Wilson Lake, Central Labrador, Canada: Earth and Planetary Science Letters, v. 10, p. 325–328.

Nunn, G.A.G., Thomas, A., and Krogh, T.E., 1985, The Labradorian Orogeny: Geochronological database, *in* Current Research: St. John's, Newfoundland Department of Mines and Energy, Mineral Development Division, Reprint 85-1, p. 43–54.

Rivers, T., 1997, Lithotectonic elements of the Grenville Province: Review and tectonic implications: Precambrian Research, v. 86, p. 117–154.

Rivers, T., and Nunn, G.A.G., 1985, A reassessment of the Grenvillian orogeny in western Labrador, *in* Tobi, A.C., and Touret, J.L.R., eds., The deep proterozoic crust in the North Atlantic provinces: Dordrecht, The Netherlands, Reidel Publishing Company, NATO Advanced Science Institute Series, Ser. C, v. 158, p. 163–174.

Rivers, T., Martignole, J., Gower, C.F., and Davidson, A., 1989, New tectonic divisions of the Grenville Province, southeast Canadian Shield: Tectonics, v. 8, p. 63–84.

Rivers, T., Ketchum, J., Andares, A., Hynes, A., 2002, The High Pressure belt in the Grenville Province: Architecture, timing, and exhumation: Canadian Journal of Earth Sciences, v. 39, p. 867–893.

Schumacher, J.C., 1991, Empirical ferric iron corrections: Necessity, assumptions, and effects on selected geothermobarometers: Mineralogical Magazine, v. 55, p. 3–18.

Spear, F.S., 1995, Metamorphic phase equilibria and pressure-temperature-time paths: Washington, D.C., Mineralogical Society of America, 799 p.

Thomas, A., 1993, Geology of the Winokapau Lake Area, Grenville Province, Central Labrador: Ottawa, Ontario, Geological Survey of Canada, Paper 89-18, 52 p.

Thomas, A., Nunn, G.A.G., and Krogh, T.E., 1986, The Labradorian Orogeny: Evidence for a newly defined 1600 to 1700 Ma orogenic event in Grenville Province crystalline rocks from Central Labrador, *in* Moore, J.M., Davidson, A., and Baer, A.J., eds., The Grenville Province: St. John's, Newfoundland, Geological Association of Canada, Special Paper 31, p. 175–189.

Thomas, A., Blomberg, P., Jackson, V., and Finn, G., 2000, Bedrock geology map of the Wilson Lake area (13E/SE), Labrador: St. John's, Newfoundland, Newfoundland and Labrador Department of Mines and Energy, Geological Survey, Map 2000–01, scale 1:100,000, Open File 013E/0061.

Wardle, R.J., Gower, C.F., and Kerr, A., 1990, The southeast margin of Laurentia ca. 1.7 Ga: The case of the missing crust, *in* Geological Association of Canada–Mineralogical Association of Canada Joint Annual Meeting, 16–18 May, Vancouver, British Columbia, Programs with Abstracts, v. 15, p. A137.

Waters, D.J., 1991, Hercynite-quartz granulites: Phase relations, and implications for crustal processes: European Journal of Mineralogy, v. 3, p. 367–386.

Manuscript Accepted by the Society August 25, 2003

Printed in the USA

Geological Society of America
Memoir 197
2004

High-pressure granulites in the Grenvillian Grand Lake thrust system, Labrador: Pressure-temperature conditions and tectonic evolution

Jason B. Krauss
Toby Rivers*
Department of Earth Sciences, Memorial University of Newfoundland, St. John's, Newfoundland A1B 3X5, Canada

ABSTRACT

The Cape Caribou River allochthon is a thick Grenvillian thrust sheet composed of Paleoproterozoic orthogneiss and Mesoproterozoic mafic dikes in which penetrative syn-thrusting deformation and recrystallization are largely restricted to the 1- to 2-km-wide basal shear zone, the Grand Lake thrust system. Grenvillian mylonitic fabrics in orthogneiss in the Grand Lake thrust system are characterized by the peak high-pressure (H*P*) granulite-facies assemblages garnet-clinopyroxene-plagioclase-clinoamphibole-quartz (Grt-Cpx-Pl-Cam-Qtz) and orthopyroxene-clinopyroxene-plagioclase-garnet-clinoamphibole-quartz (Opx-Cpx-Pl-Grt-Cam-Qtz) in mafic and intermediate lithologies, respectively. Evidence for the prograde reactions Cam + Pl = Cpx + Grt + Qtz + H_2O and Opx + Pl = Grt + Cpx + Qtz is preserved in some samples. Partial retrograde replacement of the subassemblage Grt-Cpx by Cam-Pl is widespread, but retrogression was domainal on both outcrop and thin-section scales. Thermobarometry applied to the syntectonic H*P* granulite-facies assemblage Grt-Cpx-Pl-Qtz-Cam ± Opx in the Grand Lake thrust system shear zone and adjacent hangingwall and footwall has yielded apparent peak pressure-temperature (*P-T*) estimates of ~14 kilobars/875 °C that are considered to closely approximate the *P* at maximum *T* experienced by rocks, despite evidence in some samples for postpeak resetting of thermometers. The rocks are inferred to have followed a clockwise *P-T* path that involved quasi-isothermal decompression following peak *T* conditions, although details of the path remain only qualitatively constrained because of the unknown extent of postpeak reequilibration. The *P-T* results and tectonic setting suggest that the Cape Caribou River allochthon is part of the H*P* belt of the Grenville Province, and that the localized Grenvillian granulite-facies recrystallization in the Grand Lake thrust system was possibly partly a result of frictional heating.

Keywords: high-pressure granulite, Grenville Province, geothermobarometry, thick-skinned thrust systems, shear heating, granulite uncertainty principle

*E-mail, corresponding author: trivers@esd.mun.ca.

Krauss, J.B., and Rivers, T., 2004, High-pressure granulites in the Grenvillian Grand Lake thrust system, Labrador: Pressure-temperature conditions and tectonic evolution, *in* Tollo, R.P., Corriveau, L., McLelland, J., and Bartholomew, M.J., eds., Proterozoic tectonic evolution of the Grenville orogen in North America: Boulder, Colorado, Geological Society of America Memoir 197, p. 105–133.

INTRODUCTION

Precise geothermobarometric determinations of the metamorphic conditions experienced in and adjacent to major crustal-scale shear zones are important because they provide constraints on the depth and temperature of shearing, thereby enabling a better understanding of the orogenic processes undergone by deeply buried crustal segments. In thick-skinned thrust systems, penetrative deformation is localized in the shear zones at the base of each thrust slice, and internal deformation within the thick (commonly >10 km) slices is less intense (e.g., Culshaw et al., 1997; Indares et al., 1998). This distribution of strain may be preserved within the thrust slices comprising an orogenic wedge if the latter maintains a critical taper during emplacement, thereby minimizing subsequent internal deformation (e.g., Platt, 1986; Jamieson and Beaumont, 1989). As a result, recrystallization and the development of syntectonic metamorphic mineral assemblages associated with thick-skinned thrusting are typically localized in and adjacent to the basal shear zone, leaving the interiors of the thick thrust slices relatively unaffected.

One such crustal-scale, thick-skinned thrust system, exposed in the eastern Grenville Province near Goose Bay, Labrador, is the Cape Caribou River allochthon and its basal shear zone known as the Grand Lake thrust system (Ryan et al., 1982; Wardle and Ash, 1984, 1986; Wardle et al., 1990). The Grand Lake thrust system is a 1- to 2-km-thick ductile shear zone that separates the Cape Caribou River allochthon in its hangingwall from underlying rocks of the Lake Melville and Groswater Bay terranes in its footwall (Wardle et al., 1990). The Grand Lake thrust system is part of the crustal-scale Allochthon Boundary Thrust system separating the Parautochthonous and Allochthonous Belts of the Grenville Province, which can be traced along the length of the Grenville orogen (Fig. 1) (Rivers et al., 1989, 2002). The Grand Lake thrust system developed during north-northwest–directed Grenvillian thrusting of the Cape Caribou River allochthon (Fig. 2) (Wardle et al., 1990), with available geochronological data indicating that displacement on the thrust system occurred during both the Ottawan (ca. 1080–1020 Ma) and the Rigolet (ca. 1010–990 Ma) orogenies (Krogh, 1986; Krogh and Heaman, 1988; Philippe et al., 1993; Bussy et al., 1995; Corrigan et al., 1997).

The shear zone is inferred to have formed under granulite-facies conditions based on the presence of orthopyroxene, which, with quartz and plagioclase, define the widespread stretching lineation in the granitoid rocks in the shear zone (Ryan et al., 1982; Wardle and Ash, 1984, 1986; Krauss, 2002). As in other thick-skinned thrust systems, field and petrographic evidence suggests that the effects of penetrative strain diminish over a short distance (meters to tens or hundreds of meters) away from the ductile mylonitic fabrics of the Grand Lake thrust system (Krauss, 2002). Based on these observations, it is inferred that the shear zone was developed at the base of a thick orogenic wedge that was not pervasively internally deformed during its emplacement. The Grand Lake thrust system offers an opportunity to study the tectonic processes associated with deep-level, thick-skinned thrusting in a crustal-scale shear zone and in adjacent footwall and hangingwall rocks. The aim of this study is to establish the pressure-temperature (*P-T*) conditions during thrusting in the shear zone at the base of the Cape Caribou River

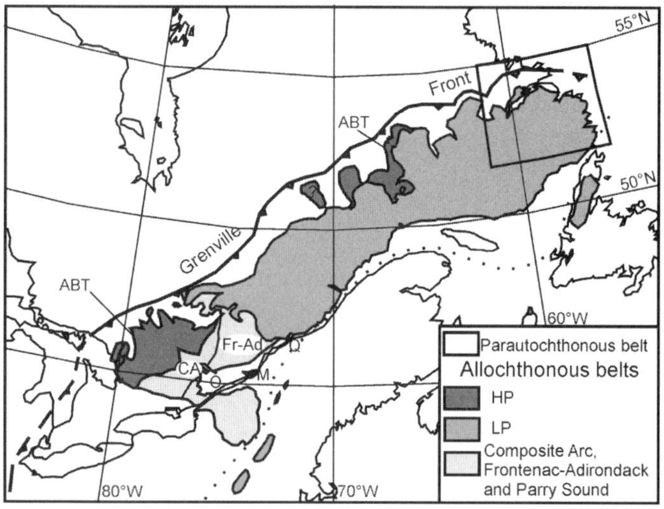

Figure 1. Tectonic elements of the Grenville Province (after Rivers et al., 2002), showing the Parautochthonous Belt and the overlying Allochthonous high-pressure (*HP*) and low-pressure (*LP*) belts. ABT—Allochthon Boundary Thrust; CA—Composite arc belt; Fr-Ad—Frontenac-Adirondack belt. Box indicates the location of Figure 2.

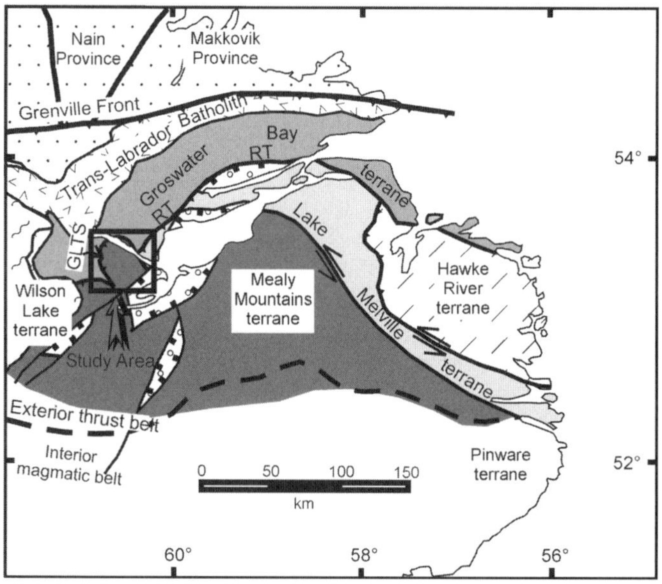

Figure 2. Major structural divisions and Grenvillian terranes in the eastern Grenville Province (after Wardle and Ryan, 1996). Location of the Goose Bay region (Fig. 3) is indicated by the box. GLTS—Grand Lake thrust system; RT—Rigolet thrust; Neoproterozoic Lake Melville rift system indicated by normal faults, and rift clastics by open-circle fill pattern.

allochthon and in the adjacent footwall and hangingwall rocks. Details of the time of formation of the syntectonic mineral assemblages in the shear zone will be discussed in a future paper. Based on the presence of syntectonic metamorphic orthopyroxene in the Grand Lake thrust system (Ryan et al., 1982; Wardle and Ash, 1984, 1986; Wardle et al., 1990; Krauss, 2002), it is likely that the rocks in the shear zone attained maximum temperatures in excess of 800 °C, implying that the minerals would have been susceptible to cationic exchange of Fe and Mg (and possibly other elements) during cooling (e.g., Spear, 1991). Such retrograde exchange renders estimation of "peak" *P-T* conditions (i.e., the pressure at maximum temperature in a clockwise *P-T* path) difficult in many granulites, a problem that Frost and Chacko (1989) referred to as the "granulite uncertainty principle." In addition, grain size reduction during shearing was important in the Grand Lake thrust system (Krauss, 2002), thereby increasing the possibility of diffusional postpeak resetting of thermometers and barometers (Spear, 1991). Implications of the granulite uncertainty principle with respect to the *P-T* conditions determined in this study and the *P-T* path for the Grand Lake thrust system are discussed after presentation of the results.

REGIONAL SETTING

Tectonic Elements of the Northern Grenville Province in Central Labrador

The major Grenvillian tectonic elements of the northern Grenville orogen are shown in Figure 1. The orogen has been subdivided into a series of (semi-) continuous subparallel stacked belts known as the Parautochthonous Belt and the overlying Allochthonous Belt (Rivers et al., 1989, 2002). The recently recognized high-pressure (H*P*) belt (Ludden and Hynes, 2000; Rivers et al., 2002), comprising the structurally lowest of the allochthonous belts, is characterized by relict high-pressure (eclogite-facies and H*P* granulite-facies) assemblages. It is tectonically overlain by a low-pressure (L*P*) belt in which high-pressure assemblages are lacking. Each of the belts has previously been subdivided into a series of terranes; the term is used in the Grenvillian literature in the sense of a metamorphic terrane, defined as a tectonically bounded segment of orogenic crust distinguished from adjacent terranes on the basis of, for example, its metamorphic, structural, or geochronologic character (e.g., Rivers et al., 1989; Rivers, 1997). There is no connotation of an exotic provenance during Grenvillian orogenesis. In the northern Grenville Province of central Labrador (Fig. 2), the Groswater Bay terrane (part of the Parautochthonous Belt), and the Lake Melville and Mealy Mountains terranes (part of the Allochthonous Belt) are principally composed of Late Paleoproterozoic para- and orthogneisses that were variably reworked during Grenvillian orogenesis. They are separated by major Grenvillian shear zones, of which at least one was a reactivated Paleoproterozoic thrust (Corrigan et al., 2000). The Cape Caribou River allochthon has been linked on lithological grounds to the Mealy Mountains terrane, from which it is separated by the Neoproterozoic Lake Melville rift system (Wardle et al., 1990) (see discussion later). The Paleoproterozoic tectonic evolution of these terranes is discussed briefly in the next section, before a review of the effects of Grenvillian reworking, the subject of this study.

Pre-Grenvillian Tectonic Evolution

The Late Paleoproterozoic rocks in the Grenville Province of central Labrador comprise juvenile arc terranes that were first assembled and accreted to Laurentia during the ca. 1680–1660 Ma Labradorian orogeny (Gower, 1996). The ca. 1650 Ma post-tectonic calc-alkaline Trans-Labrador batholith is interpreted as a continental-margin arc (Gower et al., 1992; Connelly et al., 1995; Gower, 1996) that intruded along the approximate location of the suture between the Laurentian crust and the accreted Labradorian arc assemblage (Wardle et al., 1986). Other post-tectonic magmatism includes the anorthosite, mangerite, charnockite, and granite (AMCG) of the voluminous ca. 1640–1620 Ma Mealy Mountains intrusive suite. There is evidence for a second regional deformation event during the latest Paleoproterozoic (ca. 1620–1610 Ma), based on ages of 1622 ± 6 Ma and 1613 ± 40 Ma for syntectonic pegmatite in the Cape Caribou River allochthon and the Lake Melville terrane (Krogh and Heaman, 1988) and ca. 1610 Ma for early displacement on the Rigolet shear zone east of the study area (Corrigan et al., 2000). These ages have been associated with thrusting of the Lake Melville terrane over the Groswater Bay terrane in the Goose Bay and Rigolet areas (Corrigan et al., 2000). After this event, there was a period of relative crustal stability until emplacement of the voluminous Michikamau, Harp Lake, and Mistastin AMCG suites between ca. 1450 and 1420 Ma (Krogh and Davis 1973; Emslie and Stirling, 1993) and the mafic dikes of the Michael and Shabogamo suites between 1472 and 1426 Ma (Schärer et al., 1986; Connelly et al., 1995; Corrigan et al., 2000), possibly in an ensialic back-arc setting (Rivers and Corrigan, 2000). Other magmatism in the study area during the Middle to Late Mesoproterozoic includes the 1307 ± 26 Ma composite gabbronorite-monzonite Arrowhead Lake pluton in the Groswater Bay terrane (Rb-Sr isochron; Fryer, 1983), the 1250 ± 2 Ma Mealy dikes in the Mealy Mountains terrane (Hamilton and Emslie, 1997), and the possibly correlative Northwest River dikes in the Lake Melville terrane and the Cape Caribou River allochthon (discussed later).

Grenvillian Tectonics in Central Labrador

Grenvillian tectonics in central Labrador principally involved thick-skinned imbrication of Labradorian terranes with variable reworking of the rocks within individual terranes. The Lake Melville terrane was thrust toward the northwest over the Groswater Bay terrane along the reactivated Rigolet thrust at 1047 Ma (Corrigan et al., 2000), and the Cape Caribou River allochthon was thrust toward the north-northwest over the assembled Groswater Bay and Lake Melville terranes such that the basal shear zone, the Grand Lake thrust system, truncates the Rigolet thrust (Fig. 3), implying that the Grand Lake thrust system originated as a result of out-of-sequence thrusting (Wardle et al., 1990). The degree of Grenvillian reworking probably varies significantly among and within the terranes, but in general is difficult to assess except where Mesoproterozoic marker units, such as the Northwest River dikes, are present, allowing discrimination between Labradorian and Grenvillian fabrics. The U-Pb ages of metamorphic zircon from the Lake Melville and Groswater Bay terranes and the Cape Caribou River allochthon are predominantly Labradorian, but in all three units Grenvillian U-Pb ages in the range from 1060 to 1010 Ma have been determined in the vicinity of the Grand Lake thrust system shear zone (Philippe et al., 1993; Bussy et al., 1995; Corrigan et al., 1997; Corrigan et al., 2000). This distribution is consistent with localized high-temperature Grenvillian recrystallization and mineral growth in the shear zone associated with thick-skinned Grenvillian thrusting as described above. Grenvillian reworking of the Mealy Mountains terrane appears to have been very minor since no evidence for growth of Grenvillian metamorphic zircon has been found (Gower, 1996), and regional heating was insufficient to reset the $^{40}Ar/^{39}Ar$ system in hornblende in the Mealy dikes (Reynolds, 1989), implying that temperatures were less than ~550 °C (e.g., Hanes, 1991).

The Cape Caribou River Allochthon and the Grand Lake Thrust System

The Grand Lake thrust system comprises a 1- to 2-km-thick intensely deformed zone of ductile mylonite and straight gneiss (e.g., Hanmer, 1988) at the base of the Cape Caribou River allochthon that is largely developed in dioritic to granodioritic orthogneiss, amphibolite and mafic granulite, and minor metasedimentary gneiss (metapelite and calc-silicate) derived from the footwall adjacent to the shear zone (Fig. 3). The overlying Cape Caribou River allochthon is principally composed of a Paleoproterozoic AMCG suite. The structurally lowest, and highest-strain, part of the Cape Caribou River allochthon comprises a locally orthopyroxene-bearing basal mylonite, gneissic granitoids, and mafic straight gneisses. A layered mafic gneiss unit is also present in some sections. These rocks are overlain by a recrystallized gabbro-diorite unit in the middle of the allochthon, with the top composed of massive leuco-gabbronorite and the Northwest River anorthosite. Pyroxene-bearing granitoids of the Dome Mountain suite are present throughout the allochthon except the basal mylonite, and Northwest River dikes occur in all units of the allochthon. The intensity of strain diminishes abruptly above the basal mylonite and straight gneisses, and the upper part of the allochthon, although displaying widespread evidence for static recrystallization, is not penetratively deformed and many original igneous contact relationships and textures are well preserved (Wardle et al., 1990; Krauss, 2002).

Lithological similarities between units in the Mealy Mountains terrane and the Cape Caribou River allochthon suggest a pre-Grenvillian lithological linkage between them (Wardle et al., 1990). For instance, the Paleoproterozoic ca. 1660–1620 Ma (Emslie and Hunt, 1990; James et al., 2000) Mealy Mountains intrusive suite in the Mealy Mountains terrane has been correlated with the ca. 1630–1625 Ma Northwest River anorthosite and the associated Dome Mountain monzonite suite in the Cape Caribou River allochthon (Bussy et al., 1995). A possible Mesoproterozoic correlation, noted earlier, is the mafic dikes of the Mealy Mountains and Northwest River suites, both of which are northeast-trending (Emslie, 1976; Krauss, 2002). The Northwest River dikes, which occur in both the Lake Melville terrane and the Cape Caribou River allochthon (Corrigan et al., 2000; Krauss, 2002), are undated, but are chemically comparable to the Mealy dikes; both are subalkaline tholeiites with similar trace element compositions (Krauss, 2002).

STRUCTURE OF THE CAPE CARIBOU RIVER ALLOCHTHON

Figure 4, A, is a structural map of the Cape Caribou River allochthon showing the orientations of the mylonitic fabrics (S_m) and stretching lineations (L_s) in the Grand Lake thrust system and adjacent terranes. The stretching lineation displays a maximum value of 25° → 165° determined from contoured data, a value compatible with limited kinematic evidence of northwest-directed thrusting (Wardle et al., 1990; Krauss 2002). Cross-sections through the northeast margin of the allochthon and perpendicular to L_s are shown in Figure 4, B. The present arcuate shape of the Cape Caribou River allochthon in map view is inferred to be largely a function of frontal and lateral ramping of the thrust sheet toward the northwest, possibly enhanced by Late Grenvillian, northwest-trending, large-wavelength and -amplitude crossfolds.

GRENVILLIAN FABRICS, MINERAL ASSEMBLAGES, AND REACTIONS

Footwall Fabrics, Mineral Assemblages, and Reactions

Grenvillian fabrics in footwall rocks away from the Grand Lake thrust system are commonly difficult to detect in the field and in thin section. The para- and orthogneisses in the Groswater Bay terrane generally lack evidence for crosscutting relationships with respect to older (Labradorian) fabrics, but on the basis of the mapped reorientation of the L_s toward the overlying Grand Lake thrust system over a distance of 1000 m in the field (Fig. 4, C), it is inferred that the Labradorian fabrics in the

Figure 3. Geology of the Goose Bay region (after Ryan et al., 1982 [northwest part], and Wardle and Ash, 1984, 1986 [northeast and south parts]), with minor modifications by Krauss (2002).

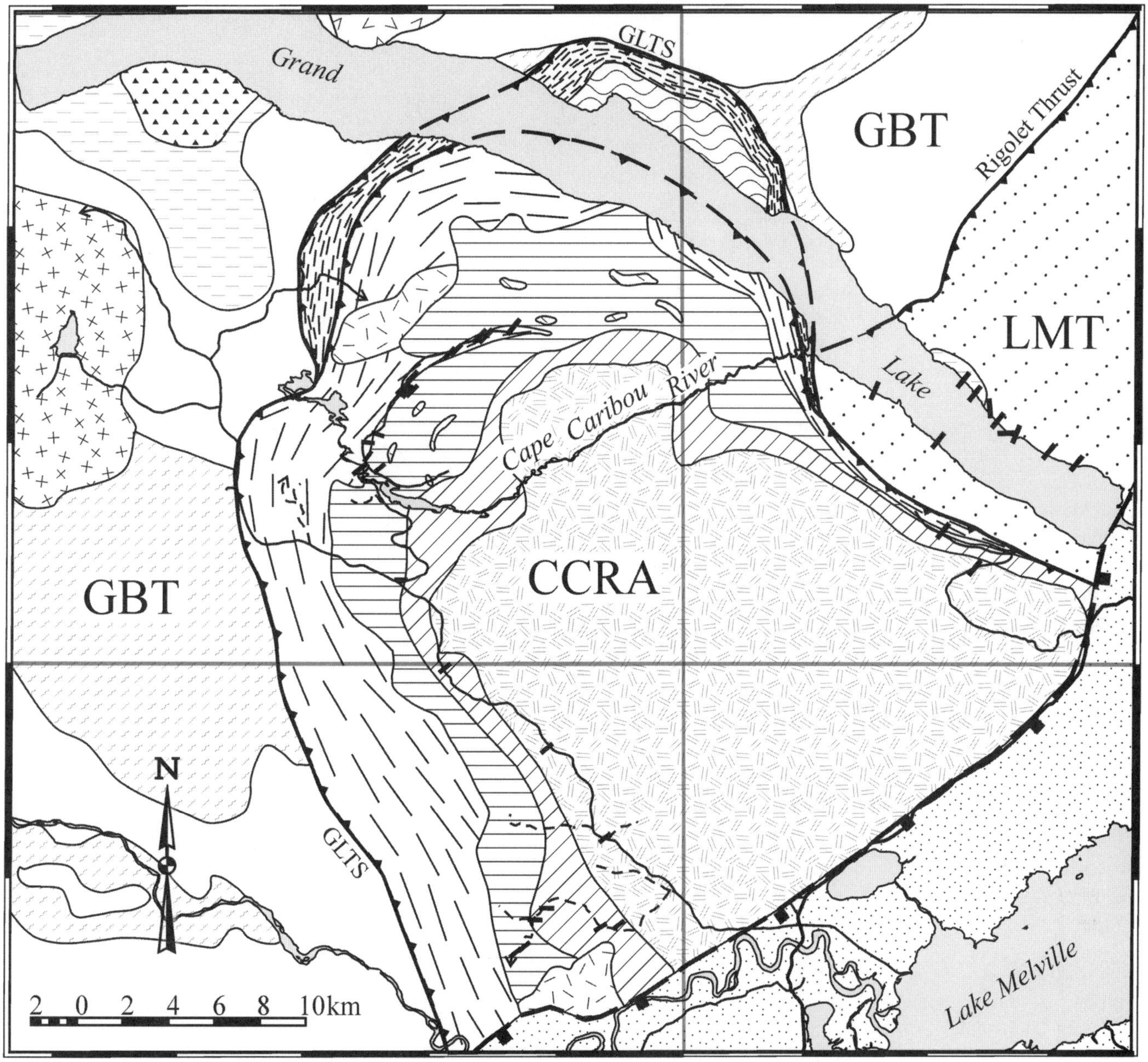

Cape Caribou River Allochthon (CCRA)
- Northwest River Anorthosite
- Layered Gabbro Monzonite
- Recrystallized Gabbro-Diorite
- Granitoid Gneiss
- Dome Mountain Suite
- Mafic Gneiss unit
- Mylonite Zone

Lake Melville terrane (LMT)
- Lake Melville Orthogneiss
- Susan River Quartz Diorite

Groswater Bay terrane (GBT)
- Paragneiss
- Orthogneiss
- Trans-Labrador Batholith
- Unnamed Metadiorite
- Susan River Quartz Diorite
- Arrowhead Lake pluton

- Northwest River dikes
- Lake Melville Rift System

Symbols
- Thrust/ductile shear zone
- Geological Contact: known, inferred
- Normal Fault
- Roads: Two lanes (paved)
- Logging road (dirt)
- Not maintained

Groswater Bay terrane were progressively rotated in a ductile manner into subparallelism with the Grenvillian strain field as a result of displacement on the Grand Lake thrust system. In contrast, in the Lake Melville terrane two distinct fabrics can be observed locally in the field, with a Grenvillian mylonitic fabric crosscutting an older (presumably Labradorian) gneissic layering (Fig. 5, A). Northwest River dikes are observed to crosscut the Labradorian gneissic layering in some localities of low Grenvillian strain, implying post-Labradorian emplacement; elsewhere dikes correlated with the Northwest River swarm have a well developed foliation that is attributed to Grenvillian tectonism (Wardle and Ryan, 1996; Krauss, 2002). As such, the Northwest River dikes are a useful marker of Grenvillian metamorphic and deformational effects and were specifically sought out for this purpose in this study.

Metamorphic assemblages in dioritic orthogneiss in the Groswater Bay terrane consist of Grt-Cam-Pl-Kfs-Qtz ± Bt ± Scp ± Cpx, with the subassemblage Grt-Cpx occurring only in mylonitic rocks adjacent to the shear zone (Cam—clinoamphibole; other mineral abbreviations after Kretz, 1983; see Appendix 2 for a complete listing of all mineral and component abbreviations). No evidence of leucosome formation was observed in this lithology. We relate the formation of the Grt-Cpx subassemblage to the generalized prograde, compressional reaction

$$\text{Cam} + \text{Pl} + \text{Qtz} = \text{Grt} + \text{Cpx} + H_2O \qquad (1a)$$

(e.g., Wells, 1979, Pattison, 2003), which we infer, on the basis of field relations and published geochronology, took place during Grenvillian thrust emplacement of the Cape Caribou River allochthon and associated recrystallization in the Grand Lake thrust system. Grt porphyroblasts are anhedral in shape and contain inclusions of Cam and Qtz. Generally, the SPO (shape-preferred orientation) of Cam, Bt, Qtz, and Cpx, where present, defines the mineral L_s in the orthogneiss. Scp occurs in layers parallel to the foliation, suggesting that CO_2-bearing fluids were present during deformation.

The paragneiss unit in the Groswater Bay terrane footwall is a pelitic migmatite composed of granitic leucosomes and a melanosome assemblage of Qtz-Pl-Kfs-Grt-Bt-Ky, locally with postpeak Sil and Ms. The leucosomes are inferred to have developed by dehydration melting of Ms following the generalized reaction

$$\text{Ms} + \text{Pl} = \text{Kfs} + \text{Ky} + \text{L} \qquad (2)$$

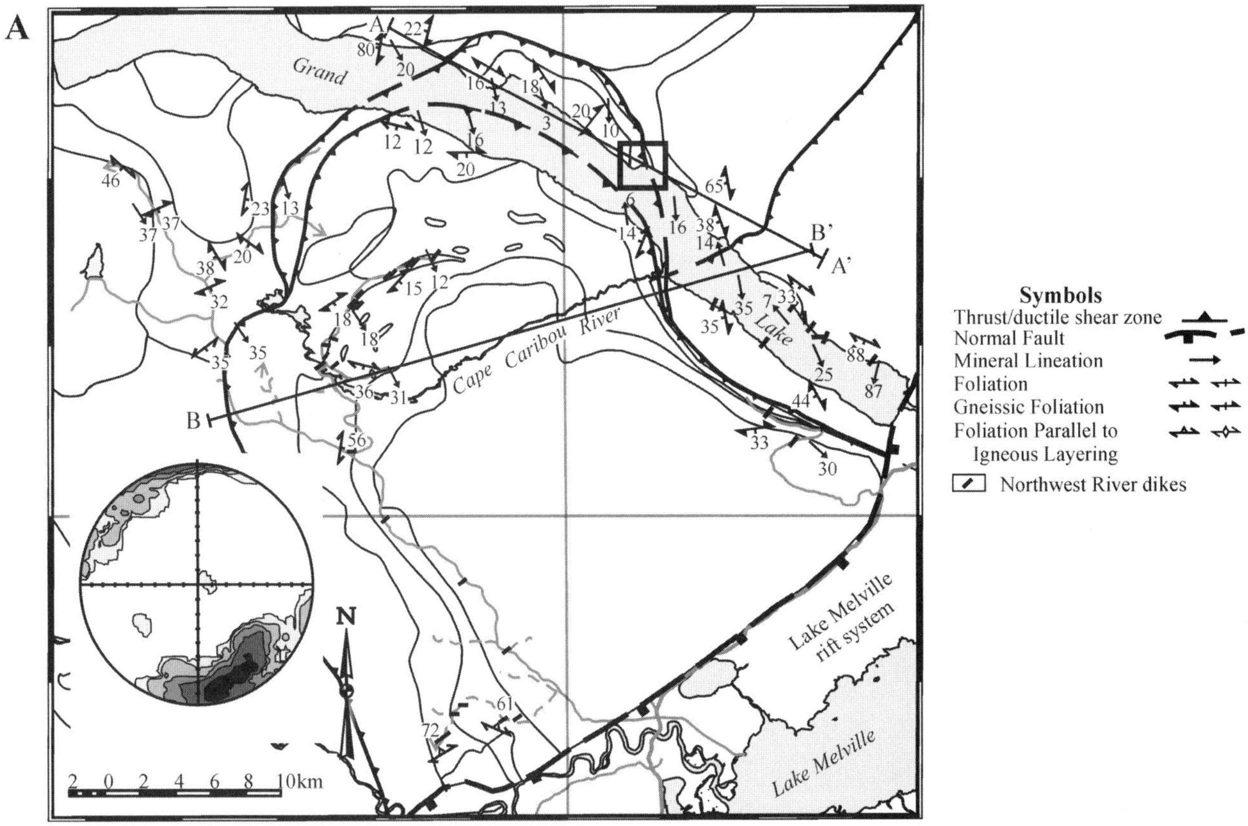

Figure 4. (A)

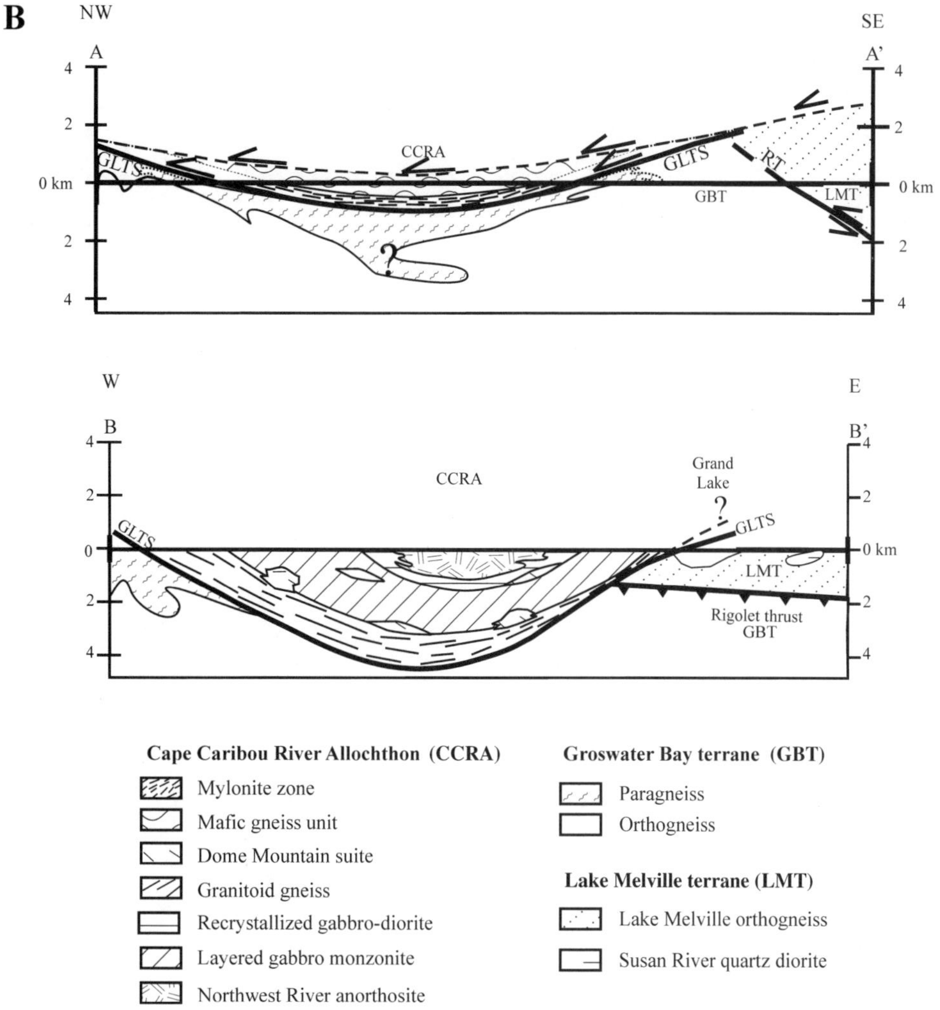

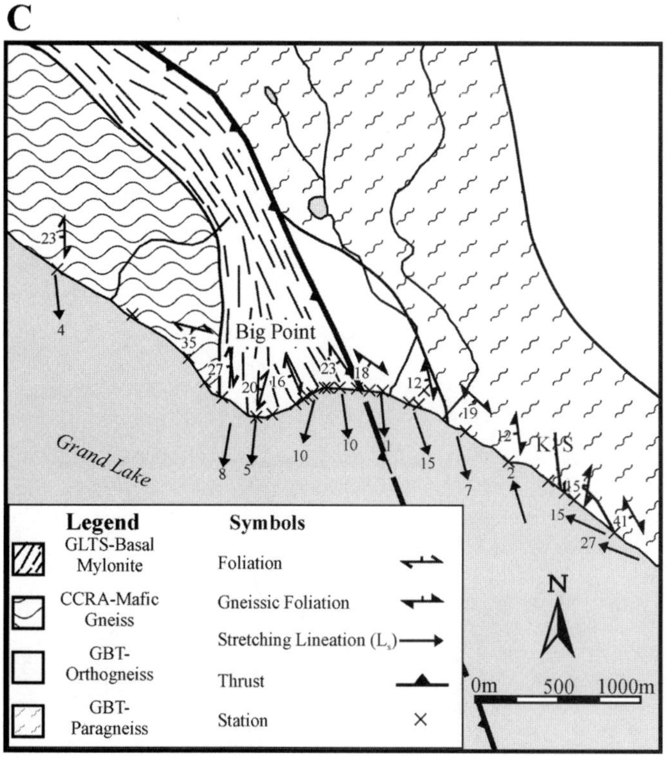

Figure 4 (*on this and previous page*). (A) Structural sketch map of the Goose Bay region showing representative orientations of foliations and stretching lineations related to the formation of the Grand Lake thrust system. Inset diagram shows contoured Grenvillian stretching lineation (L_s) data (lower hemisphere equal area projection). Box on Grand Lake shows the location of Figure 4, C. Lines A–A′ and B–B′ are locations of cross-sections shown in Figure 4, B. (B) Cross-sections of the Cape Caribou River allochthon along lines A–A′ and B–B′ (approximately parallel and perpendicular to the transport direction, respectively). GLTS—Grand Lake thrust system; RT—Rigolet thrust. (C) Structural sketch map of footwall rocks of the Groswater Bay terrane, the Grand Lake thrust system, and the lower parts of the Cape Caribou River allochthon showing the reorientation of the regional footwall foliation and L_s into the Grenvillian Grand Lake thrust system.

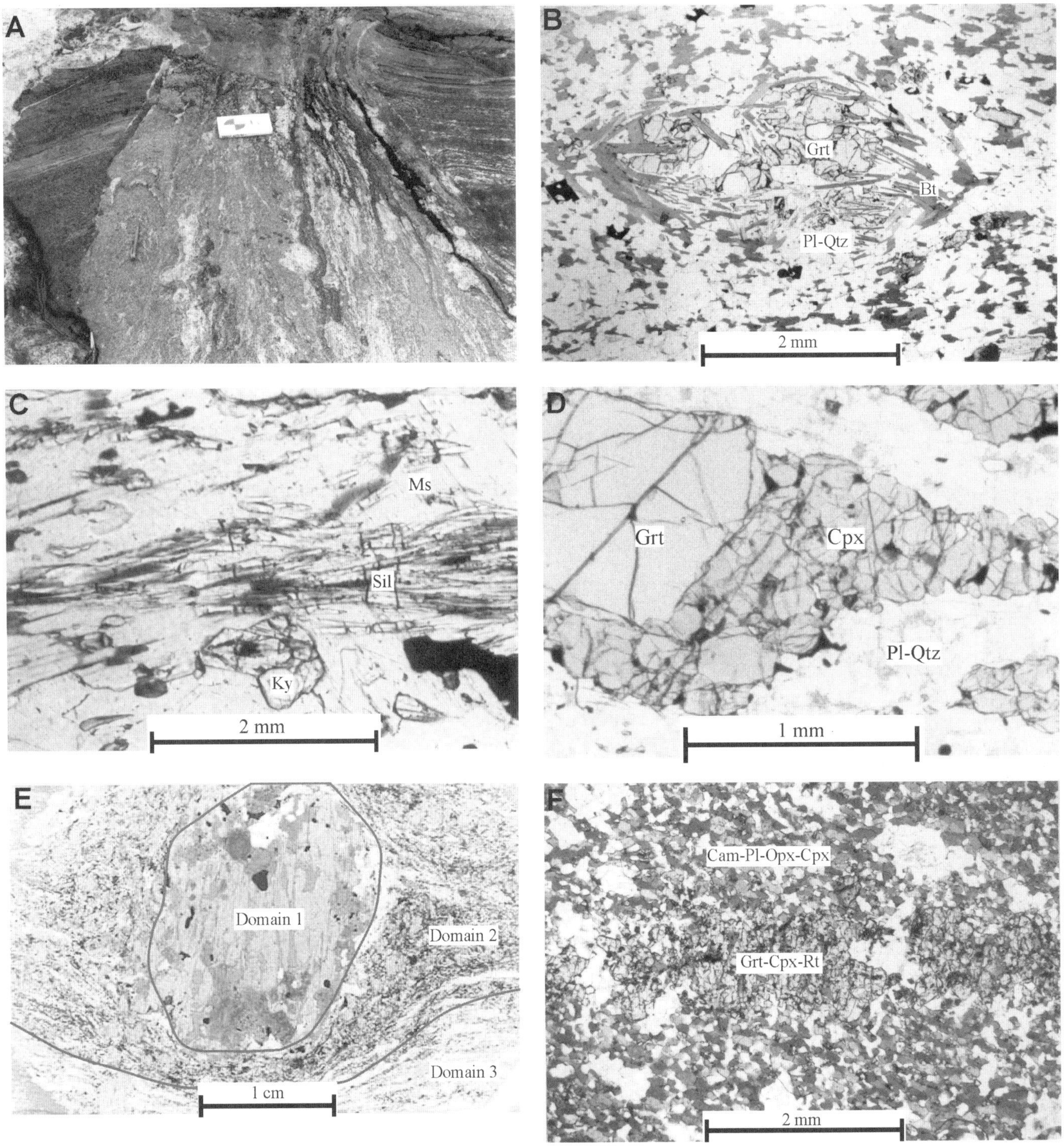

Figure 5. Field photographs and photomicrographs of important relationships observed in the study area. See Appendix 2 for a complete listing of all mineral and component abbreviations. (A) Crosscutting shear zone fabrics in the Lake Melville terrane. East-west fabric is inferred to be Labradorian and crosscutting and north-south fabric is inferred to be Grenvillian and has been dated at ca.1044 Ma (Krogh, 1986). Note also the presence of porphyroclasts of disaggregated pegmatite parallel to the Grenvillian fabric. North shore of Grand Lake in southeast part of map area. (B) Resorbed Grt partially replaced by secondary Bt, Pl, and Qtz in paragneiss from the Groswater Bay terrane. Plane polarized light (PPL). Sample JK-99-001. (C) Resorbed Ky and retrograde Ms and Sil in foliation, pelitic paragneiss. PPL. Sample JK-99-024. (D) Syntectonic Grt-Cpx-Pl-Qtz assemblage in basal mylonite of the Cape Caribou River allochthon, with granoblastic Cpx in the pressure shadow of Grt. PPL. Sample JK-99-029. (E) Photomicrograph of straight gneiss, Grand Lake thrust system. This sample displays three textural and mineralogical domains: (1) the large Grt porphyroblast, with Cpx, Pl, and Qtz inclusions; (2) a fine-grained Cpx-Grt-Cam-Pl-Qtz assemblage defining the foliation in the upper part of of the figure; and (3) a fine-grained Grt-Cpx-Opx-Cam-Pl-Qtz assemblage defining the foliation at the bottom of the figure. Sam-

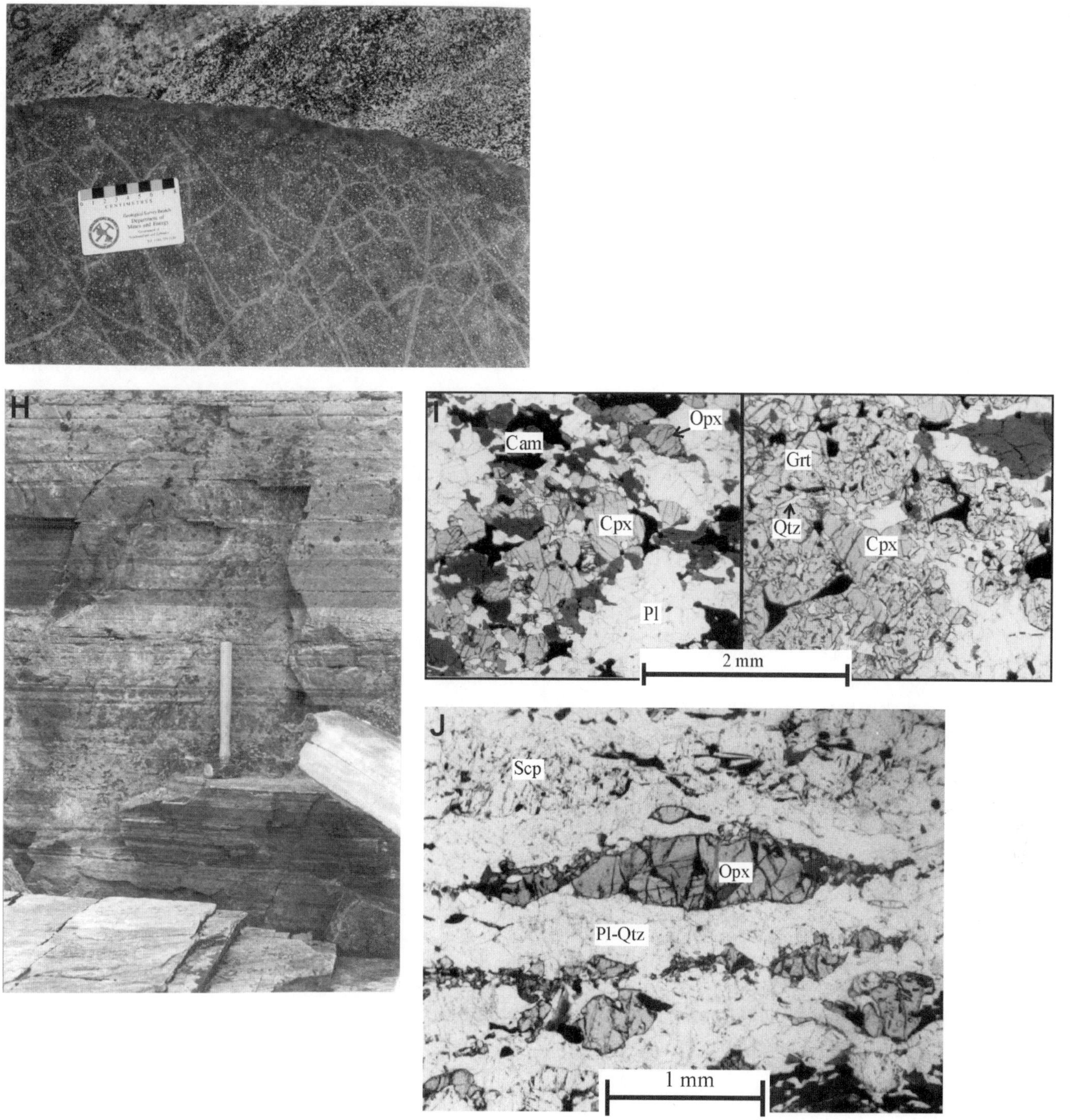

ple JK-99-012. (F) Northwest River dike with assemblage Cam-Pl-Opx-Cpx, except in vein in which assemblage is Grt-Cpx-Cam-Cal-Rt. Sample JK-99-122–1. (G) Northwest River dike in the Cape Caribou River allochthon showing preserved chilled margin and relict Pl phenocrysts. The host rock is a two-pyroxene Grt granulite, and the dike is a two-pyroxene and Amp-bearing granulite. The veins in the dike are principally composed of Grt and Cpx. Cape Caribou forestry access road, same location as U-Pb sample C-050 AM of Bussy et al. (1995). (H) Straight gneiss in basal shear zone of the Cape Caribou River allochthon showing typical alternating mafic and intermediate-felsic layering. North shore of Grand Lake. Hammer handle is 40 cm long. (I) Photomicrographs of two-pyroxene–Cam granulite and Grt granulite from within the same thin section (sample JK99-065a). Example of domainal equilibrium, with assemblages representing both sides of the reaction Opx + Pl = Grt + Cpx + Qtz in different parts of the thin section. Sample JK-99-065A from Cape Caribou. (J) Photomicrograph of shape-preferred orientation (SPO) of Opx and Scp defining the mylonitic foliation in the GLTS. PPL. Sample JK-99-028.

On the basis of available geochronological data, obtained by the conventional thermal ionization mass spectrometry (TIMS) on zircon, most migmatization is inferred to be Labradorian (Philippe et al., 1993, Bussy et al., 1995), but unpublished Grenvillian Zrn and Mnz ages from this unit indicate that significant Grenvillian recrystallization and possibly melting also occurred. Adjacent to the Grand Lake thrust system shear zone, leucosomes are progressively stretched out into thin Kfs-Pl-Qtz–rich aggregates, with elongate Qtz grains displaying lattice-preferred orientation (LPO; Krauss 2002). Carbonate and calc-silicate layers associated with the paragneiss unit consist of Cal ± Dol and Grt-Cpx-Ep-Qtz-Ttn ± Ol, and contacts between calc-silicate and metapelitic rocks are sites of strain inhomogeneity commonly intruded by pegmatite.

Most samples from the Groswater Bay terrane footwall show evidence for partial retrogression following formation of the peak Grenvillian assemblage. In the orthogneiss unit, common retrograde textures include secondary Cam replacing Cpx, and secondary Bt, Pl, and Qtz after Grt and Kfs (Fig. 5, B). The inferred reactions that led to these assemblages are

$$\text{Grt} + \text{Cpx} + \text{H}_2\text{O} = \text{Cam} + \text{Pl} + \text{Qtz} \quad (1b)$$

and

$$\text{Grt} + \text{Kfs} + \text{H}_2\text{O} = \text{Bt} + \text{Pl} + \text{Qtz} \quad (3)$$

respectively. Reaction 1b is reaction 1a in reverse. In the paragneiss unit, retrogression and decompression involved the formation of retrograde Ms and/or Sil (Fig. 5, C) by the generalized reactions

$$\text{Ky} + \text{Kfs} + \text{H}_2\text{O} = \text{Sil} + \text{Ms} + \text{Qtz} \quad (4)$$

or

$$\text{Grt} + \text{Kfs} + \text{H}_2\text{O} = \text{Sil} + \text{Bt} + \text{Pl} + \text{Qtz} \quad (5)$$

The water involved in these reactions may have been derived from crystallization of leucosomes formed as a result of reaction 2 if partial melting was Grenvillian. Reaction 1b suggests that retrogression was accompanied by decompression (e.g., Pattison, 2003), an interpretation that is also compatible with reactions 3, 4, and 5.

Many orthogneiss samples from the Lake Melville terrane contain the assemblage Opx-Cpx-Cam-Bt-Pl-Qtz ± Grt ± Kfs, with the two-pyroxene-plagioclase subassemblage providing mineralogical evidence for granulite-facies metamorphism. The age of this metamorphism is not dated, but on the basis of data from elsewhere in the Lake Melville terrane (Schärer et al., 1986; Gower, 1996) it is probably Labradorian. Grt is relatively rare, but has been observed adjacent to the Grand Lake thrust system with Opx, Cpx, Pl, and Cam, suggesting the general reaction

$$\text{Opx} + \text{Pl} = \text{Cpx} + \text{Grt} + \text{Qtz} \quad (6)$$

The spatial association of the product assemblage with the Grand Lake thrust system suggests that this reaction occurred during Grenvillian thrusting. As in the Groswater Bay terrane, many rocks in the Lake Melville terrane show evidence for a retrograde amphibolite-facies overprint that resulted in variable replacement of pyroxene by amphibole. As noted earlier, mafic bodies correlated with the Northwest River dikes provide a useful marker for Grenvillian metamorphism in the Lake Melville terrane. They are recrystallized, with some preserving evidence for a two–pyroxene-plagioclase metamorphic assemblage, whereas others are completely amphibolitized. This suggests that both the granulite-facies metamorphism and the amphibolite-facies overprint in the dikes are Grenvillian effects in these bodies, which is compatible with the U-Pb zircon age of 1011 ± 3 Ma obtained from a metamorphosed Northwest River dike by Bussy et al. (1995).

Hangingwall Fabrics, Mineral Assemblages, and Reactions

The base of the Cape Caribou River allochthon comprises strongly lineated mylonitic and straight gneiss of granodioritic, granitic, and tonalitic compositions, which is locally Opx-bearing. The straight gneiss and mylonite are composed of the assemblage Pl-Kfs-Qtz-Cam-Bt-Grt ± Opx ± Cpx ± Scp and have yielded Paleoproterozoic upper intercept ages of ca. 1.6 Ga and Grenvillian lower intercept U-Pb zircon ages of 1016 and 1008 Ma, which have been interpreted as the times of emplacement and deformation, respectively (Bussy et al., 1995). Syntectonic textures in the mylonite include discrete pods of Grt-Cpx-Pl ± Qtz ± Opx wrapped by Qtz ribbons and granoblastic feldspar domains, with Cpx and/or Opx forming elongate grains that define the stretching lineation, as in the Grand Lake thrust system. Cpx also occurs in pressure shadows adjacent to Grt and as granoblastic lens-shaped aggregates wrapped by Qtz and Pl ribbons (Fig. 5, D), implying that all these phases were stable during mylonitization. Opx and Cpx commonly show evidence for partial replacement by Bt and Cam, respectively. Two distinct textural domains were observed in outcrop and thin section in the basal mylonite of the Cape Caribou River allochthon: large porphyroblasts of Grt with Cpx and Pl inclusions, and the surrounding fine-grained fabric composed of Grt and elongate Cpx and Pl grains (Fig. 5, E).

Higher up in the hangingwall, Grenvillian strain is lower. In the granitoid gneiss unit, textures are gneissic, with a matrix foliation wrapping around Grt porphyroblasts. The peak assemblage in this unit is Grt-Opx-Cpx-Pl-Cam, with textures suggesting operation of the prograde compression reactions 1a and 6. Near the top of the allochthon, Grenvillian strain was minor. For instance, plutons of the Dome Mountain suite are commonly only weakly foliated, and igneous textures are well preserved in much of the Northwest River anorthosite. Grenvillian recrystallization in much of the Northwest River anorthosite was limited to the formation of coronas of Oam, Cam, and Bt between igneous Opx and Pl (Butt, 2000).

As in the Lake Melville terrane in the footwall of the Grand Lake thrust system, the Northwest River dikes are also an important marker unit for Grenvillian metamorphism in the Cape Caribou River allochthon in its hangingwall. Their metamorphic mineralogy varies with location in the hangingwall. Near the base, rare relict Pl phenocrysts contain abundant inclusions of spinel, and the recrystallized matrix consists of the granulite-facies assemblage Opx-Cpx-Cam-Pl ± Grt ± Qtz ± Fe-Ti oxide. Some of these dikes show evidence for formation of the subassemblage Grt-Cpx-Cal-Rt in veins (Fig. 5, F and G). Formation of this subassemblage is attributed to the reaction

$$\mathrm{Cam + Pl + Qtz + CO_2 = Grt + Cpx + Cal + Rt + H_2O} \quad (1c),$$

which is a modified version of reaction 1a wherein the breakdown of Cam was promoted by reduced activity of H_2O in the vein fluid. Formation of rutile in the product assemblage is compatible with measured elevated TiO_2 in reactant Cam (discussed later). Despite their high-grade metamorphic history, many of the dikes are not penetratively deformed, and intrusive relations and relict igneous textures are widespread (Fig. 5, G).

Dikes higher up the allochthon are also not penetratively deformed and show only minor evidence of recrystallization to amphibolite- and greenschist-facies assemblages, and some of those present in the uppermost levels of the Cape Caribou River allochthon are apparently unaffected by Grenvillian metamorphism. Metamorphism in these dikes, where present, was static (i.e., not accompanied by the high strain deformation characteristic of the base of the allochthon) and driven by the influx of H_2O. We interpret the present arrangement of metamorphic assemblages in the Grand Lake thrust system and the Cape Caribou River allochthon to be the combined result of two separate processes; i.e., the high to intermediate strain, granulite-facies assemblages in the Grand Lake thrust system and the base of the Cape Caribou River allochthon are related to Grenvillian thrust emplacement of the Cape Caribou River allochthon, whereas the patchily distributed static amphibolite- and greenschist-facies assemblages in the upper parts of the Cape Caribou River allochthon are a result of reaction with hydrous fluids. There is no evidence to suggest that thrusting at the base of the allochthon was related to fluid ingress at the top, so the two processes were probably not coeval, and thus the distribution of granulite- to greenschist-facies assemblages in the Cape Caribou River allochthon should not be interpreted in terms of a single metamorphic field gradient.

Shear Zone Fabrics, Mineral Assemblages, and Reactions

Excellent exposures of the Grand Lake thrust system shear zone on the northern shore of Grand Lake consist of alternating layers of mafic and intermediate-felsic mylonite with abundant garnet porphyroblasts (Fig. 5, H). The mafic layers are principally composed of Grt-Cpx-Pl-Cam ± Opx ± Qtz, with Opx-Pl commonly partially replaced by the subassemblage Grt-Cpx (Fig. 5, I), implying operation of reaction 6. The intermediate-felsic layers consist of Pl-Qtz-Grt-Opx-Cam ± Cpx, in which Opx is a stable phase and defines the mineral stretching lineation together with Pl-Kfs and Qtz (Fig. 5, J). We infer that the stability of the subassemblage Opx-Pl in the intermediate-felsic mylonite versus its apparent instability and replacement by Grt-Cpx in the mafic mylonite is a function of bulk composition, possibly reflecting the more Ca-rich nature of the mafic unit (c.f. Pattison, 2003). As in the mylonites of the hangingwall granitoid gneiss unit, the shear zone mylonites display syntectonic fabrics, with Grt porphyroblasts and their Cpx pressure shadows wrapped by matrix granoblastic Pl (see Fig. 5, D). Textural evidence in the form of 0.5–1.0 cm, top-to-the-north, σ-shaped winged porphyroclasts of amphibole suggests that this phase was also stable during mylonitization, defining a peak Grt-Cpx-Pl-Cam assemblage in the mafic layers. There is also widespread evidence for later growth of Cam at the expense of Cpx and Grt (reaction 1b) that resulted in complete amphibolitization of some mafic layers. U-Pb zircon ages of 1038 ± 2 Ma (Philippe et al., 1993) and 1007 +3/–2 Ma (Corrigan et al., 1997) determined from the Grand Lake thrust system indicate that the mylonitization was Grenvillian.

Summary

In summary, based on the mapping and petrography of Ryan et al. (1982), Wardle and Ash (1984, 1986), and Krauss (2002) and the U-Pb dating studies noted earlier, it can be stated that penetrative Grenvillian deformation was principally restricted to the Grand Lake thrust system shear zone at the base of the Cape Caribou River allochthon, with Grenvillian strain in the footwall away from the Grand Lake thrust system generally difficult to distinguish from older Labradorian deformation and Grenvillian strain in much of the hangingwall low. Grenvillian U-Pb zircon metamorphic ages in the Grand Lake thrust system shear zone and in Northwest River dikes imply that significant new zircon growth occurred in the shear zone, and formation of the synkinematic subassemblages Opx-Pl and Grt-Cpx imply temperatures of >800 °C during formation of the Grand Lake thrust system shear zone.

Figure 6 is a schematic $CaO\text{-}FeO\text{-}MgO\text{-}Al_2O_3\text{-}SiO_2\text{-}H_2O$ (CFMASH) grid (after Pattison 2003) showing the distribution of reactions involving Cam-Pl-Grt-Cpx-Opx in Qtz-bearing rocks relevant to this study. The inference from reaction textures that the [Opx] and [Cam] reactions (reactions 1a and 6, respectively) occurred in both shear zone and dike samples in the study area suggests that *P-T* conditions were in the vicinity of the invariant point in the model $CaO\text{-}MgO\text{-}Al_2O_3\text{-}SiO_2\text{-}H_2O$ (CMASH) or $CaO\text{-}FeO\text{-}Al_2O_3\text{-}SiO_2\text{-}H_2O$ (CFASH) system,

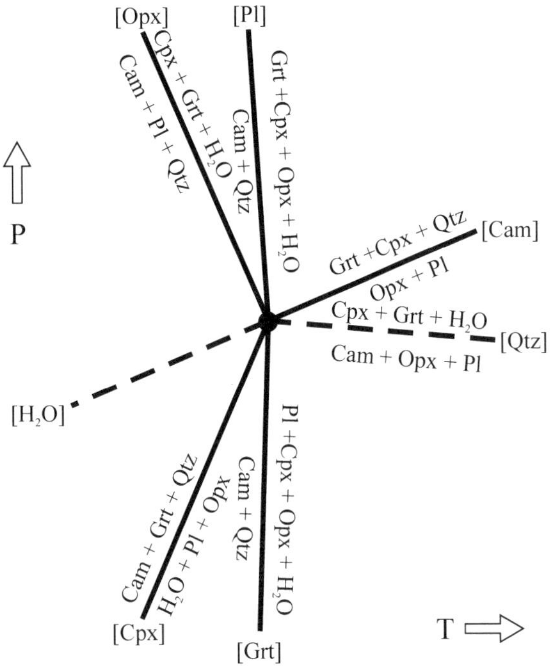

Figure 6. Phase equilibria in the CaO-MgO-FeO-Al_2O_3-SiO_2-H_2O (CMASH or CFASH) system assuming the presence of quartz and the absence of melt, after Pattison (2003). See Appendix 2 for a complete listing of all mineral and component abbreviations.

and that part of the compressional segment of the *P-T* path has been recorded locally. Textural evidence for reaction 1b indicates that the retrograde decompression [Opx] reaction is also recorded by the shear zone rocks, reaction 1c demonstrates the role of local differences in fluid composition in controlling the mineralogy, and both reactions imply the occurrence of post-peak domainal equilibrium in these rocks. Other evidence of postpeak retrograde decompression is inferred from reactions 3, 4, and 5 in pelitic rocks. We have made use of these interpretations in choosing assemblages and grains for analysis in the geothermobarometry study described in the next section.

GEOTHERMOBAROMETRY

Sampling and Analytical Methods

The choice of samples for geothermobarometry involved three principal criteria, that is, an appropriate assemblage, an inference of equilibrium among participating phases, and textural and/or geochronological evidence that the mineralogy was Grenvillian. Samples of dioritic orthogneiss with the assemblage Grt-Cpx-Pl-Qtz ± Opx ± Cam were chosen for geothermobarometry, as orthogneiss with this assemblage occurs in the footwall, hangingwall, and shear zone and was considered most likely to retain a record of peak *P-T* conditions. Metapelites from the adjacent footwall, although providing useful qualitative constraints on the *P-T* path, were not used, as they were considered likely to have reequilibrated following peak *P-T* conditions. The presence of Qtz is not always obvious optically in the fine-grained orthogneiss assemblages, and in some samples it was necessary to perform BSE (backscattered electron) imaging and qualitative ED (energy dispersive) analyses to confirm its presence. In other samples, Qtz, although readily observed, was distributed inhomogeneously throughout the thin section, and it was not clear a priori whether it was in equilibrium with the other analysed phases. In such cases, following mineral analyses of all phases in the assemblage, comparison of *P-T* results with those from other nearby samples was employed to make an a posteriori determination concerning its participation in the equilibrium (see discussion later). Appropriate assemblages were carefully selected on the basis of inferred domainal equilibrium, as discussed earlier, with special attention paid in the first instance to syntectonic fabrics in mafic rocks with the assemblage Grt-Cpx-Pl-Qtz. Assemblages with Grt-Cam-Pl-Qtz were also utilized if the Cam was interpreted to have crystallized during formation of the syntectonic fabric. In some cases both assemblages occurred in a single sample, allowing comparisons between the two calibrations. In a few samples in which all phases appeared to be in textural equilibrium, the assemblage Grt-Cpx-Opx-Cam-Pl-Qtz was used. Typically more than one textural domain in a single thin section was chosen to test the reproducibility and accuracy of the *P-T* estimates and/or to test for domainal equilibrium within the sample. In other cases, different textural domains in a single sample were chosen to enable estimation of peak and retrograde conditions. In all, a total of eighteen samples fulfilled these criteria and were used for geothermobarometry. Their locations are shown in Appendix 1, and details of the lithologies and assemblages analysed are given in Appendix 2.

Mineral analyses for geothermobarometry were performed on a Cameca SX-50 electron microprobe equipped with a LINK energy-dispersive spectrometer located in the Department of Earth Sciences at Memorial University of Newfoundland. Analytical conditions were a 15 kV accelerating voltage and a 20 nA beam current for all silicates except plagioclase feldspar, for which a 10 nA beam current was used to minimize the volatilization of Na during analyses. Analysis time was 100 s. A range of natural and pure standards was used to calibrate the instrument, and, based on repeated analyses of the standard materials, accuracy is estimated to be ± 5% relative or better for the major elements. The results were corrected using ZAF software. Fe^{3+} and site distributions were determined following the method of Droop (1987) for pyroxene and garnet, and that of Leake et al. (1997) for amphibole. However, the Fe^{3+} estimates displayed considerable variation, in some cases even among analyses of single grains, presumably because of minor errors in the determination of other cations, especially Si, and so thermobarometry was carried out under the assumption that $Fe^{tot} = Fe^{2+}$.

The zoning patterns of Ca, Fe, Mg, and Mn (as appropriate) were measured in Grt, Cpx, Opx, and Cam to determine the nature of the zoning and enable selection of appropriate sites for analysis. Geothermobarometry was performed on compositions of coexisting phases in low-variance assemblages, with equilibrium *P-T* conditions determined using the TWEEQU (thermobarometry with estimation of equilibrium state) software of Berman (1991). Version 2.02 was used for Grt-Cpx-Pl-Opx assemblages, whereas version 1.01 was used for assemblages with Cam. The calculations involved the *alm, prp,* and *grs* end-members of Grt, the *di* and *hd* end-members of Cpx, the *en* and *fs* end-members of Opx, the *an* and *ab* end-members of Pl, and the *prg, feprg, ts,* and *fets* end-members of Cam (see Appendix 2 for an explanation of the abbreviations).

In order to estimate the "peak" *P-T* conditions (i.e., the *P* at peak *T*) preserved in the rocks, the following geobarometers were used: Grt-Cpx-Pl-Qtz (GCPQ), Grt-Opx-Pl-Qtz (GOPQ), and Grt-Cam-Pl-Qtz (GHPQ) in association with the Grt-Cpx, Grt-Opx, or Grt-Cam thermometer, as appropriate. On the basis of textural criteria, the GHPQ barometer and Grt-Cam thermometer were used to constrain both prograde (peak) and retrograde metamorphic conditions. Standard errors in the *P-T* estimates resulting from analytical errors and uncertainties associated with thermochemical input parameters and mixing models are considered to be ± 1 kilobars and ± 50 °C (e.g., Spear, 1993).

The term *apparent peak* is used to describe the *P-T* results determined from the cores of minerals in syntectonic assemblages in view of the possibility of postpeak resetting as indicated by the granulite uncertainty principle. An empirical evaluation of the importance of postpeak resetting follows presentation of the *P-T* results.

Mineral Chemistry

Representative analyses are presented in Table 1, and Figure 7 shows typical zoning patterns present in Grt, Cpx, and Pl from a selected shear zone sample that reflects the majority of the zoning patterns observed. Except for cases in which retrogression is visible as optically distinct rims (e.g., on Cam), the minerals typically show very flat core-to-rim zoning patterns (Fig. 7), indicating that their compositions were homogenized during Grenvillian metamorphism. Figure 8, A, B, C, and D, shows the compositions of Grt, Cpx and Opx, Cam, and Pl in terms of their most abundant end-members. X_{sps} is low (~0.002) in Grt, which consists mostly of *alm-prp-grs* mixtures with X_{alm} = 0.47–0.62, X_{prp} = 0.10–0.34, and X_{grs} = 0.15–0.28. Cpx and Opx are essentially quadrilateral mixtures, with X_{jd} in Cpx ~0.01 and X_{di} = 0.57–0.81, and X_{en} in Opx = 0.65–0.69. Cam is *prg* to *feprg* according to the classification scheme of Leake et al. (1997), in which Fe^{3+} is low and TiO_2 is around 2 wt%. Recrystallized Pl compositions range from X_{an} = 12% to X_{an} =

TABLE 1. REPRESENTATIVE MINERAL ANALYSES
Representative garnet (Grt) analyses

Sample	029-1		071		086-1		056		142	
Mineral	Grt1-4	Grt1-8	Grt2-1	Grt2-5	Grt2-4	Grt1-4	Grt1-1	Grt1-6	Grt1-1	Grt1-6
Point	Core	Rim	Core	Rim	Core	Rim	Core	Rim	Core	Rim
SiO_2	38.78	38.69	38.97	38.73	38.58	38.64	38.70	38.68	38.48	38.58
Al_2O_3	21.04	21.13	21.58	21.52	20.92	21.09	21.28	21.27	20.96	21.05
TiO_2	0.05	0.12	0.11	0.01	0.23	0.06	0.09	0.09	0.06	0.06
Cr_2O_3	0.06	0.05	0.04	0.06	0.00	0.15	0.00	0.00	0.02	0.04
FeO	27.11	27.46	26.14	25.91	27.94	25.74	28.03	27.87	27.75	27.34
MgO	5.72	5.37	7.30	7.00	5.21	3.05	5.56	4.71	6.26	5.89
MnO	0.74	0.77	0.81	0.86	1.45	3.91	2.52	4.01	1.11	1.10
CaO	7.60	7.89	6.72	6.76	7.55	10.73	5.60	5.66	6.56	6.89
Total	101.09	101.48	101.67	100.85	101.89	103.37	101.78	102.28	101.20	100.95
Si apfu	3.003	2.990	2.979	2.983	2.985	2.977	2.995	2.993	2.983	2.993
Al	1.921	1.925	1.944	1.953	1.907	1.915	1.942	1.940	1.915	1.925
Ti	0.003	0.007	0.006	0.001	0.013	0.003	0.005	0.005	0.004	0.004
Cr	0.004	0.003	0.002	0.004	0.000	0.009	0.000	0.000	0.001	0.002
Fe (total)	1.755	1.774	1.671	1.669	1.808	1.658	1.814	1.803	1.799	1.774
Mg	0.660	0.618	0.832	0.804	0.601	0.350	0.642	0.543	0.723	0.682
Mn	0.049	0.050	0.053	0.056	0.095	0.255	0.165	0.263	0.073	0.072
Ca	0.631	0.653	0.550	0.558	0.626	0.886	0.464	0.469	0.544	0.573
Cation total	8.026	8.020	8.037	8.028	8.035	8.053	8.027	8.016	8.042	8.025
X_{Prp}	0.213	0.200	0.268	0.260	0.192	0.111	0.208	0.177	0.230	0.220
X_{Alm}	0.567	0.573	0.538	0.541	0.578	0.526	0.588	0.586	0.573	0.572
X_{Grs}	0.204	0.211	0.177	0.181	0.200	0.281	0.150	0.152	0.173	0.185
X_{Sps}	0.016	0.016	0.017	0.018	0.030	0.081	0.054	0.085	0.023	0.023

Note: Mineral formulas based on 12 oxygens; apfu—atoms per formula unit.

(continued)

TABLE 1. *Continued*

Representative clinopyroxene (Cpx) analyses

Sample	**029-1**		**071**		**086-1**		**142**	
Mineral	Cpx1-3	Cpx1-2	Cpx2-4	Cpx2-1	Cpx1-5	Cpx1-1	Cpx1-4	Cpx1-1
Point	Core	Rim	Core	Rim	Core	Rim	Core	Rim
SiO_2	50.86	50.94	51.18	51.11	49.94	52.13	50.89	51.50
Al_2O_3	3.89	3.63	4.34	4.00	4.60	3.57	3.33	3.00
TiO_2	0.37	0.34	0.50	0.44	0.18	0.19	0.26	0.28
Cr_2O_3	0.16	0.16	0.01	0.17	0.13	0.08	0.08	0.06
FeO	11.73	10.91	10.32	9.87	13.11	12.53	10.90	11.03
MgO	11.02	11.23	12.16	11.84	9.67	10.27	11.85	12.18
MnO	0.15	0.05	0.16	0.09	0.47	0.24	0.20	0.15
CaO	21.64	22.34	20.65	20.79	21.79	22.71	20.97	21.03
Na_2O	1.07	1.16	1.56	1.60	0.97	1.14	1.54	1.35
Total	100.90	100.76	100.90	99.90	100.88	102.93	100.04	100.59
Si apfu	1.89	1.89	1.88	1.90	1.85	1.91	1.89	1.91
Al	0.17	0.16	0.19	0.18	0.20	0.15	0.15	0.13
Ti	0.01	0.01	0.01	0.01	0.01	0.01	0.01	0.01
Cr	0.00	0.00	0.00	0.00	0.00	0.00	0.00	0.00
Fe^{3+}	0.10	0.12	0.13	0.11	0.11	0.09	0.16	0.13
Fe^{2+}	0.27	0.23	0.19	0.20	0.31	0.29	0.18	0.21
Mg	0.61	0.62	0.67	0.66	0.54	0.56	0.66	0.67
Mn	0.00	0.00	0.00	0.00	0.01	0.01	0.01	0.00
Ca	0.86	0.89	0.81	0.83	0.88	0.89	0.84	0.83
Na	0.08	0.08	0.11	0.12	0.07	0.08	0.11	0.10
Cation total	3.99	4.00	3.99	4.01	3.98	3.99	4.01	3.99
X_{Di}	0.635	0.688	0.683	0.688	0.582	0.620	0.714	0.689
X_{Hd}	0.283	0.250	0.196	0.208	0.333	0.325	0.199	0.232
X_{Jd}	0.006	0.005	0.011	0.012	0.007	0.070	0.006	0.005
X_{Ae}	0.010	0.013	0.020	0.017	0.010	0.010	0.025	0.017

Note: Ferric iron contents were calculated using the procedure outlined by Droop (1987).
Mineral formulas based on 6 oxygens; apfu—atoms per formula unit.

Representative amphibole (Cam) analyses

Sample	**086-1**		**056**	
Mineral	Amp1-3	Amp2-1	Amp1-1	Amp1-1
Point	Core	Rim	Core	Rim
SiO_2	41.46	41.16	42.29	42.29
Al_2O_3	12.34	12.94	12.25	12.25
TiO_2	1.72	1.91	1.88	1.88
Cr_2O_3	0.00	0.15	0.00	0.00
FeO	19.88	19.91	16.65	16.65
MgO	8.04	8.01	10.25	10.25
MnO	0.16	0.22	0.21	0.21
CaO	11.81	11.80	11.55	11.55
Na_2O	1.47	1.66	1.77	1.77
K_2O	1.97	1.99	1.31	1.31
Total	98.86	99.76	98.16	98.16
Si apfu	6.307	6.213	6.329	6.399
Al	2.212	2.302	2.162	2.134
Ti	0.197	0.217	0.211	0.160
Cr	0.000	0.018	0.000	0.011
Fe^{3+}	0.000	0.000	0.000	0.011
Fe^{2+}	2.597	2.567	1.991	1.809
Mg	1.824	1.803	2.286	2.446
Mn	0.021	0.028	0.027	0.026
Ca	1.841	1.851	1.852	1.900
Na	0.435	0.485	0.515	0.448
K	0.383	0.384	0.250	0.206
Cation total	15.817	15.868	15.623	15.550

Note: Ferric iron estimated using the procedure of Leake et al. (1997).
Mineral formulas based on 23 oxygens; apfu—atoms per formula unit.

(continued)

TABLE 1. *Continued*

Representative plagioclase (Pl) analyses

Sample	029-1		071		086-1		056		142	
Mineral	Pl1-3	Pl1-1	Pl2-4	Pl2-2	Pl4-2	Pl1-2	Pl1-3	Pl1-1	Pl1-3	Pl1-2
Point	Core	Rim	Core	Rim	Core	Rim	Core	Rim	Core	Rim
SiO_2	60.90	60.49	60.28	60.52	60.93	60.62	62.62	61.45	60.52	60.27
Al_2O_3	24.84	24.14	23.93	23.98	24.48	25.69	24.97	25.48	23.99	24.50
FeO	0.09	0.00	0.00	0.25	0.07	0.22	0.21	0.36	0.12	0.27
CaO	6.44	6.43	5.62	5.80	6.93	7.06	5.51	6.56	5.82	6.22
Na_2O	7.94	7.93	7.17	7.94	7.71	6.95	8.59	8.00	8.26	8.47
K_2O	0.37	0.36	1.68	0.47	0.24	0.51	0.17	0.24	0.29	0.31
Total	100.58	99.35	98.67	98.96	100.36	101.06	102.07	102.09	99.01	100.04
Si apfu	2.695	2.707	2.724	2.720	2.703	2.667	2.719	2.681	2.719	2.689
Al	1.296	1.273	1.275	1.270	1.280	1.332	1.278	1.310	1.270	1.288
Fe (total)	0.003	0.000	0.000	0.009	0.003	0.008	0.008	0.013	0.005	0.010
Ca	0.305	0.308	0.272	0.279	0.329	0.333	0.257	0.307	0.280	0.297
Na	0.682	0.688	0.628	0.692	0.663	0.593	0.723	0.677	0.719	0.732
K	0.021	0.021	0.097	0.027	0.014	0.028	0.009	0.013	0.017	0.018
Cation total	5.002	4.997	4.996	4.997	4.992	4.961	4.994	5.001	5.010	5.034
X_{An}	0.303	0.303	0.273	0.280	0.327	0.349	0.259	0.308	0.276	0.284
X_{Ab}	0.677	0.676	0.630	0.693	0.659	0.622	0.731	0.679	0.708	0.699
X_{Or}	0.020	0.020	0.097	0.027	0.014	0.030	0.010	0.013	0.017	0.017

Note: Mineral formulas based on 8 oxygens; apfu—atoms per formula unit.

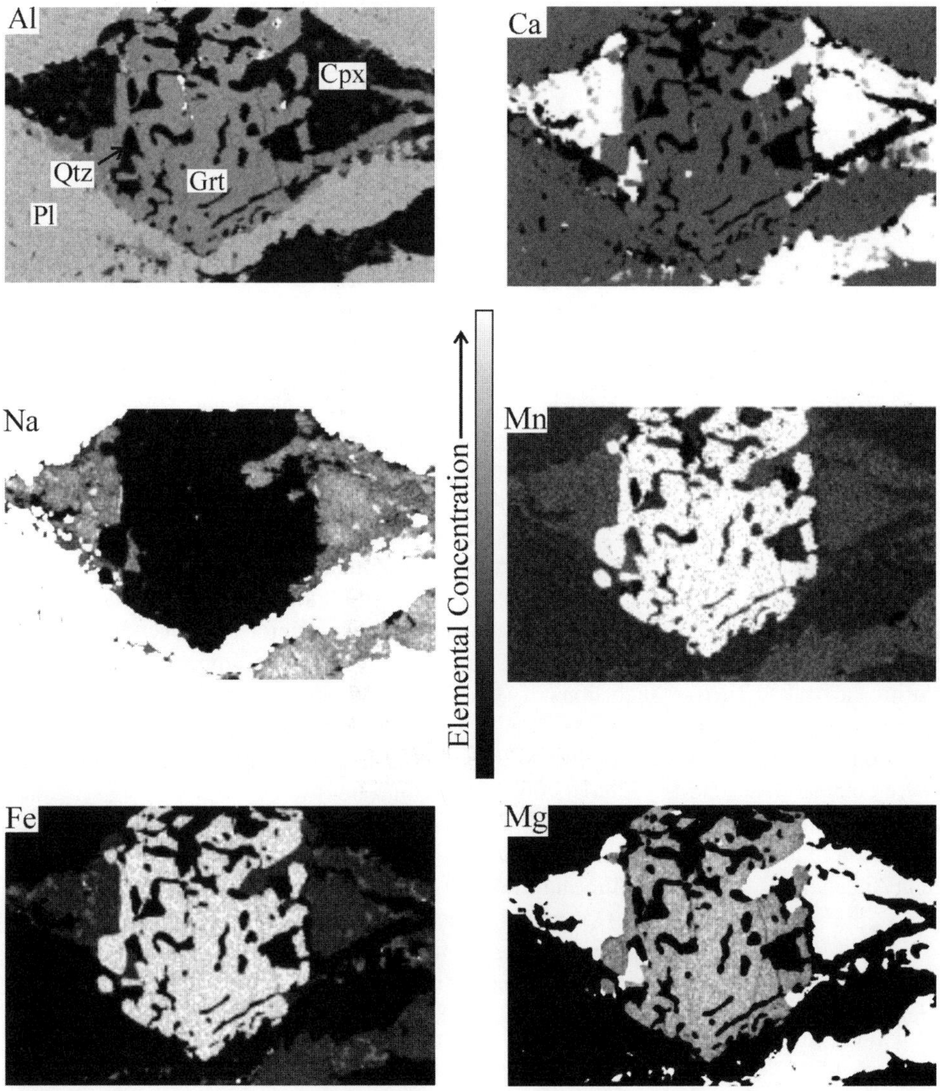

Figure 7. X-ray maps of Grt, Cpx, and Pl in a representative syntectonic assemblage composed of Grt-Cpx-Pl-Qtz. See Appendix 2 for a complete listing of all mineral and component abbreviations. Note absence of zoning. Sample JK-99-071.

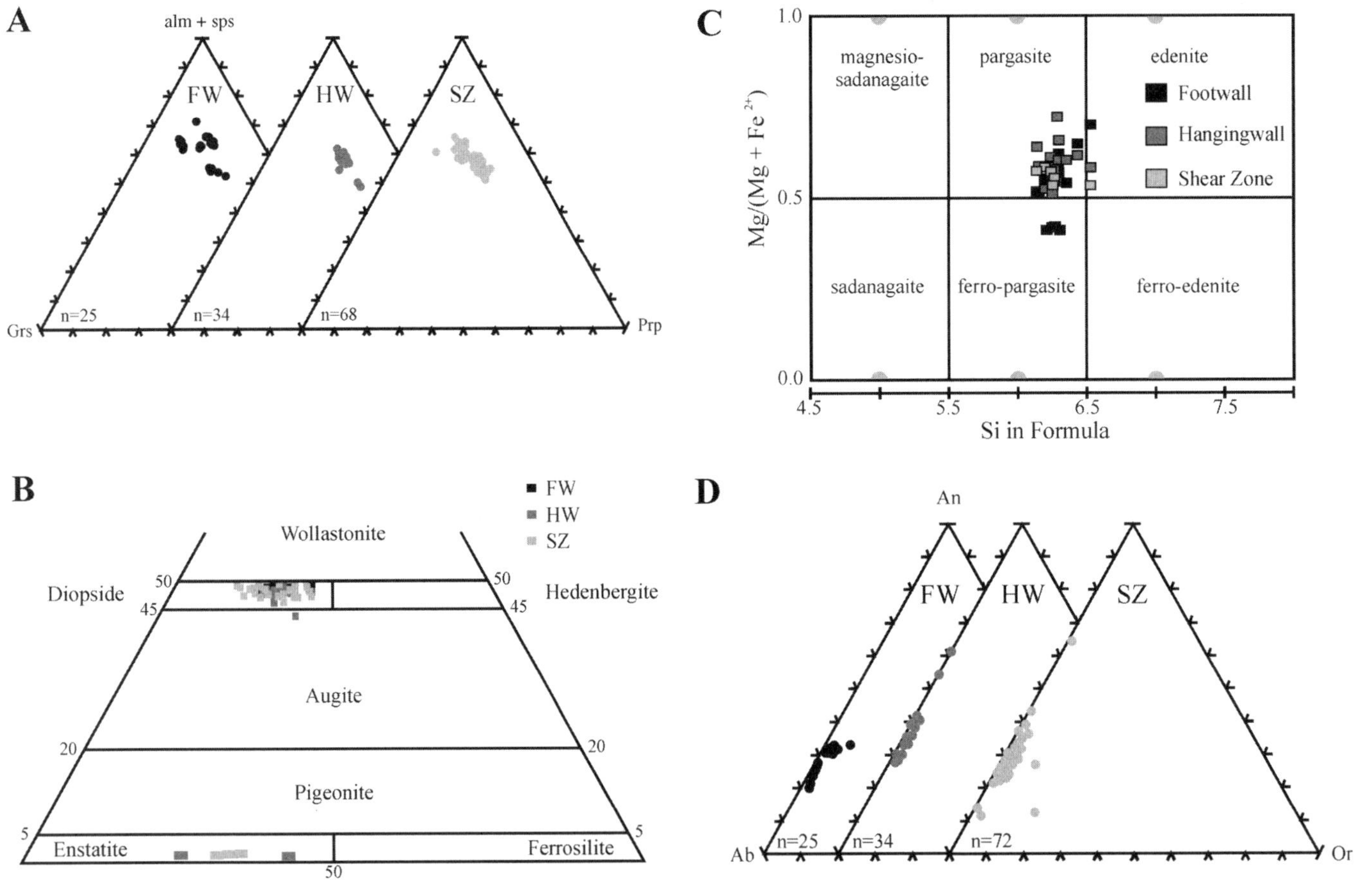

Figure 8. Chemical compositions of analyzed phases plotted in terms of their principal end-member components: (A) garnet, (B) pyroxene, (C) amphibole, and (D) plagioclase. FW—footwall; HW—hangingwall; SZ—shear zone. See Appendix 2 for a complete listing of all mineral and component abbreviations.

40% and are strongly dependent on bulk composition, with the more *an*-rich compositions obtained from the Northwest River dikes (Krauss, 2002).

P-T Results

P-T estimates made using the TWEEQU software are presented in Tables 2 and 3. Results from the footwall, hangingwall, and Grand Lake thrust system shear zone are discussed separately. In light of the inference that the compositions of the major silicate phases were diffusionally reset during the peak Grenvillian granulite-facies metamorphism, apparent peak *P-T* estimates were obtained from core compositions of coexisting minerals and retrograde *P-T* estimates were obtained from rim compositions. Brief descriptions of the textural settings of the grains analyzed are included with the discussion of the results, and more detailed descriptions are given in Appendix 2.

Footwall Samples

Four footwall samples were chosen for geothermobarometry, two each from the Lake Melville terrane and the Groswater Bay terrane. The samples from the Lake Melville terrane are from the mafic gneiss unit and a Northwest River dike, both close to the shear zone and therefore expected to yield Grenvillian *P-T* conditions. The mafic gneiss (sample 049) preserves petrographic evidence for reaction 1a and yields apparent peak conditions of about 12 kilobars/825 °C with the GCPQ barometer and the Grt-Cpx thermometer. Retrograde effects are minimal, with *P-T* conditions of 11.7 kilobars/800 °C obtained from rim analyses. Sample 056, from a Northwest River dike, is composed of Grt-Cam-Pl-Qtz. Grt has an amoeboid texture with inclusions of Qtz, Pl, and Bt, with the assemblage and texture inferred to imply rapid growth. Apparent peak *P-T* conditions of 8.9 kilobars/740 °C were calculated, with retrograde conditions 7 kilobars/660 °C.

Sample 074, dioritic orthogneiss from the Groswater Bay terrane, contains Grt-Cpx-Opx-Pl-Qtz. The assemblage displays equilibrium textures among most phases, but the lack of Grt-Opx associations implies that Opx was not stable in the Grt-Cpx-Pl-Qtz assemblage. Apparent peak metamorphic conditions determined for this sample using the GCPQ barometer are 12.9 kilobars/725 °C, with retrograde conditions of 11.3 kilo-

TABLE 2. RESULTS OF AVERAGED GEOTHERMOBAROMETRIC ESTIMATES BASED ON GCPQ GEOTHERMOBAROMETRY

	Sample	Geothermobarometer	a_{SiO_2}	Apparent peak	Retrograde
FW			Unity unless stated	*P* (kbar), *T* (°C)	*P* (kbar), *T* (°C)
	049	GCPQ		12.0, 825	11.7, 800
	056	GHPQ		8.9, 740	7.0, 660
	074	GCPQ		12.9, 725	11.3, 630
	084	GCPQ	0.63 (2 domains)	13.3, 890	12.9, 840
HW	122-3	GCPQ		11.4, 800	9.2, 780
	142	GCPQ		12.2, 840	11.5, 775
	196a	GCPQ		11.2, 800	7.7, 690
SZ	008	GCPQ	0.62 (1 domain)	13.9, 875	11.8, 725
	012	GCPQ	0.71 (1domain)	14.2, 850	11.8, 775
	017	GCPQ	0.96 (3 domains)	14.0, 855	n.d.
	028-1	GCPQ		11.5, 875	8.1, 785
	029-1	GCPQ		14.4, 870	13.7, 820
	086-1	GCPQ		13.2, 940	11.2, 775
	059a	GCPQ		13.0, 880	11.7, 790
	065a	GCPQ		13.7, 850	9.3, 614
	071	GCPQ		14.3, 880	14.0, 860
	072	GCPQ		14.0, 875	13.4, 840
	090	GCPQ		13.9, 850	9.7, 785

Note: Values for the activity of silica (a_{SiO_2}) indicate the calculated activity (from Krauss, 2002) and the number of domains in thin section in which applicable.
GCPQ—garnet-clinopyroxene-plagioclase-quartz; FW—footwall; HW—hangingwall; n.d.—not determined; SZ—shear zone.

bars/630 °C. The presence of Opx in this sample is not compatible with the temperature obtained by Grt-Cpx thermometry, suggesting that postpeak resetting has occurred. Groswater Bay terrane sample 084, also dioritic orthogneiss, comes from directly beneath the shear zone. The assemblage is Grt-Cpx-Cam-Pl-Qtz, with Cpx inclusions in Grt implying that the two minerals crystallized together, so GCPQ geothermobarometry was used to determine peak *P-T* conditions. Minor retrogression resulted in secondary Cam and Pl. Apparent peak metamorphic conditions are 13.3 kilobars/890 °C, and retrograde conditions are 12.9 kilobars/840 °C, implying that retrogression occurred near apparent peak *P-T* conditions.

Hangingwall Samples

Sample 122-3 is a mafic orthogneiss, with the inferred peak assemblage Grt-Cpx-Opx-Cam-Pl-Qtz, and numerous examples of equilibrium textures between Grt-Opx, Grt-Cpx, and Grt-Cam pairs can be observed. The GCPQ calibration yields a

TABLE 3. RESULTS OF AVERAGE *P-T* ESTIMATES BASED ON GOPQ AND GHPQ GEOTHERMOBAROMETRY

	Sample	Geothermobarometer	Apparent peak	Retrograde
HW			*P* (kbar), *T* (°C)	*P* (kbar), *T* (°C)
	122-3	GOPQ	9.7, 781	9.6, 739
		GHPQ	9.1, 774	9.2, 786
	142	GHPQ	11.8, 839	n.d.
	196a	GHPQ	8.9, 751	7.6, 692
SZ	008	GHPQ	11.6, 751	n.d.
	012	GHPQ	14.4, 815	n.d.
	028-1	GOPQ	11.8, 845	9.0, 735
		GHPQ	N/A	8.1, 786
	029-1	GHPQ	N/A	12.2, 765
	059a	GHPQ	12.6, 874	11.5, 776
	086-1	GHPQ	12.7, 967	10.7, 781
	090	GHPQ	N/A	9.7, 785

Note: GOPQ—garnet-orthopyroxene-plagioclase-quartz; GHPQ—garnet-clinoamphibole-plagioclase-quartz; FW—footwall; HW—hangingwall; N/A—not available; n.d.—not determined; SZ—shear zone.

wide range of apparent peak *P-T* estimates from 14.1 kilobars/850 °C to 10.7 kilobars/800 °C. Excluding a single Cpx analysis that plots in the augite field and may be an igneous relic, the remaining results yield a mean of 11.4 kilobars/800 °C. GOPQ and GHPQ calibrations both yield similar apparent peak *P-T* conditions of ~11.4 kilobars/800 °C.

Northwest River dikes were sampled from two localities in the Cape Caribou River allochthon close to the basal shear zone. As noted earlier, the dikes generally lack penetrative fabrics, have granoblastic textures, and are cut by late carbonate-bearing veins. The matrix assemblage in sample 142 is Pl-Cam, and in 196a the matrix assemblage is Pl-Cam-Cpx-Opx; the vein assemblage in both samples is Grt-Cpx-Pl-Cal-Qtz, implying operation of reactions 1c and 6. Apparent peak *P-T* conditions for the two samples estimated by the GCPQ calibration are 12.6 kilobars/840 °C and 11.2 kilobars/800 °C. Application of the GHPQ calibration to sample 142 yielded apparent peak *P-T* conditions of 11.8 kilobars/840 °C, within error of the estimate made using the GCPQ calibration, supporting textural evidence that Cam was part of the peak assemblage in this rock. Retrograde conditions determined by GCPQ vary widely, from 11.5 to 7.7 kilobars and 775 to 690 °C. Implications of these results are discussed in a later section.

Shear Zone Samples

Rocks from the shear zone are interlayered mafic and intermediate-felsic mylonites, where the mafic layers contain Cpx-Opx-Grt-Cam-Bt-Pl ± Scp ± Qtz and the intermediate-felsic layers contain Pl-Qtz-Kfs-Opx-Grt ± Cpx ± Cam. Of the eleven shear zone samples, six were from the mylonitic unit (samples 008, 012, 017, 028-1, 029-1, and 086-1), and five were from structurally higher levels of the shear zone in the granitoid gneiss and the overlying mafic gneiss unit (059a, 065a, 071, 072, and 090). In sample 012, granoblastic Cpx occurs in the pressure shadows of Grt porphyroclasts (see Fig. 5, E), and three texturally distinct Grt-Cpx associations can be recognized. However, apparent peak *P-T* estimates from all three sites are identical within error, with a mean of 14.2 kilobars/850 °C. Other results from the shear zone are given in Tables 2 and 3. In almost all cases the GCPQ geothermobarometer was used to estimate the apparent peak and retrograde metamorphic conditions. The average apparent peak *P-T* conditions in the shear zone are very consistent, with a mean of 13.9 ± 0.5 kilobar/873 ± 40 °C. The retrograde conditions calculated with GCPQ are less consistent (Table 2), suggesting heterogeneous retrogression, as discussed later. Application of the GHPQ calibration to shear zone samples yielded several estimates around 12.5 kilobars/875 °C (i.e., within error of the GCPQ estimates), with others as low as 8.1 kilobars/785 °C.

SiO_2 Activity

In most cases, several *P-T* determinations from a single sample yielded very similar estimates, providing confidence in the results. However, in four of the eighteen samples (numbers 008, 012, 017, and 084), one or more of the *P-T* estimates was significantly different from the remainder, indicating a local source of error in the calculations. Examination of the anomalous data indicated that the temperature estimates were similar, but the pressure estimates were systematically higher (by 0.2 to 2.2 kilobars). Assuming that equilibrium was achieved between the minerals analyzed, as implied by the textures, it is likely that the anomalous samples crystallized under a reduced activity of SiO_2. To test this hypothesis, samples that yielded anomalously high-pressure estimates were carefully examined using BSE imaging to search for Qtz, as noted earlier. This showed that Qtz is inhomogeneously distributed in these samples, commonly restricted to particular layers, leading to the conclusion that the a_{SiO_2} (activity of silica) may have varied on a domainal scale that is smaller than the thin section. Where Qtz is abundant (>5% by area), the *P-T* estimates are consistent. However, in domains in which Qtz is absent, shielded in another mineral, or low in abundance (<5% by area), the *P-T* estimates differ from those made elsewhere in the section. Assuming that the average apparent peak *P-T* derived from the majority of the determinations in these samples is correct, it is possible to quantify the reduced a_{SiO_2} that caused the anomalous pressure estimates. These calculations were carried out by Krauss (2002), who showed that the true a_{SiO_2} was 0.62, 0.71, 0.96, and 0.63, respectively, in certain domains of the four samples. Inputting these values of a_{SiO_2} into the calculations, in addition to "correcting" the *P* estimate, also resulted in small changes to the estimated apparent peak and retrograde temperatures. *P-T* results incorporating these a_{SiO_2} corrections are discussed later.

Interpretation of *P-T* Estimates

Table 2 summarizes the calculated apparent peak (core) and retrograde (rim) conditions. All *shear zone* samples, except 028-1, yield consistent apparent peak metamorphic conditions, estimated by GCPQ, of 13.9 ± 0.5 kilobar and 873 ± 40 °C (028-1 yields 11.5 kilobars and 875 °C). This may be taken as empirical confirmation that pervasive recrystallization and diffusional homogenization took place in the Grand Lake thrust system during shearing and thrust emplacement of the allochthon, as postulated earlier. We infer from the lower apparent peak metamorphic conditions obtained from 028-1 that this sample reequilibrated on the retrograde path. Estimated retrograde conditions in the shear zone samples are much more variable than those for the apparent peak, ranging from 14.0 to 8.1 kilobars and 860 to 614 °C, with overlapping estimates calculated by GCPQ and GHPQ implying that domainal equilibrium and fluid ingress were important. Several samples that exhibit evidence for retrograde reaction 1b yield *P-T* estimates in the range 12 to 11 kilobars and 800 to 750 °C, suggesting that hydrous retrogression was initiated shortly after apparent peak *P-T* conditions.

Hangingwall samples yield rather consistent apparent peak metamorphic conditions of 11.6 ± 0.6 kilobar/810 ± 25 °C. All samples analyzed exhibited evidence for prograde reaction 1c (due to CO_2-rich vein fluids) and/or 6. Comparison of this result with the calculated retrograde *P-T* conditions in the shear zone (see earlier) suggests that prograde reactions in the hangingwall took place under *P-T* conditions similar to the retrograde reactions in the subjacent shear zone. *Footwall* samples show a wide range of apparent peak *P-T* estimates, from 13.3 to 8.9 kilobars and 890 to 740 °C. The significance of these results is discussed later after consideration of the effects of the granulite uncertainty principle on the shear zone samples.

Granulite Uncertainty Principle

The estimated range of the apparent peak *P-T* conditions in the shear zone is consistent with the mineral assemblages. For instance, all samples plot in the Ky field, consistent with the inference that this mineral forms part of the Grenvillian mineral assemblage in pelites in the Groswater Bay terrane footwall. Furthermore, with respect to the apparent peak *T,* there is no evidence for Grenvillian partial melting in dioritic orthogneiss, which might imply *T* conditions in excess of 900 °C, but the presence of syntectonic migmatitic patches and veins in pelitic rocks confirms that temperatures sufficient for partial melting of pelite were attained. With respect to *P,* the presence of Pl and the lack of a significant *jd* component in Cpx in mafic orthogneiss indicate that eclogite-facies conditions were not attained. These rocks are therefore high-pressure granulites, and there appears to be no significant discrepancy between the estimated apparent peak *P-T* conditions and the mineral assemblages in the shear zone.

However, given the granulite uncertainty principle, the important question arises: Are the apparent peak *P-T* estimates in the shear zone samples accurate estimates of the *P* at peak *T?* Since diffusional resetting is grain-size dependent, it could be inferred from the consistency of the apparent peak *P-T* estimates (determined from grain cores) in the Grand Lake thrust system that all grains analyzed must have had a similar grain size. However, the analyzed phases (especially Grt) exhibit a range of grain size spanning at least an order of magnitude, so we infer that all grains, regardless of size, underwent diffusional homogenization at the metamorphic peak, perhaps in part because of increased rates of intragranular diffusion due to elevated defect densities in these high-strain rocks. As a result, we are forced to conclude that, although yielding reasonable apparent peak *P-T* conditions, the core analyses may not reflect the true peak *P-T;* instead, either (1) they may be points on the retrograde path or (2) they may be meaningless artifacts if there was nonsynchronous closure of the thermometer and barometer.

It therefore becomes important to try to determine the magnitude of the absolute difference between the true and the apparent peak *P-T* conditions. There are at least two lines of qualitative petrological evidence suggesting that the difference between the two may not be large: (1) There is no evidence for ultra-high-temperature (>900 °C) mineralogy in the shear zone, nor, as pointed out earlier, for partial melting in rocks of dioritic composition, suggesting that the average apparent peak *T* estimate of 870 ± 40 °C does not significantly underestimate the true peak *T.* (2) Although formation of Grenvillian leucosomes in pelitic gneiss by dehydration melting of muscovite is likely (see discussion earlier), there is no evidence for widespread dehydration melting of Bt. Bt is an abundant phase in the restite assemblage in pelitic gneiss, and much of it appears to have been part of the peak *P-T* assemblage, suggesting that temperatures did not greatly exceed ~850 °C. We therefore tentatively conclude that the apparent peak *T* conditions are close to, but less than, the true peak *T* conditions by a small but undetermined amount. Since geothermometers are known to reset more readily than geobarometers (e.g., Spear, 1993), we therefore further assume that the peak *P* conditions are not significantly underestimated.

P-T ***Paths***

Having considered the significance of the apparent peak *P-T* estimates, we now turn to the interpretation of the apparent *P-T* paths. Frost and Chacko (1989) and Selverstone and Chamberlain (1990) have pointed out that nonsynchronous closure of the geothermometer and geobarometer (i.e., closure of the net transfer geobarometer at a higher *T* than the exchange geothermometer) yields apparent *P-T* paths with slopes similar to the isopleths of the barometric reaction. Such apparent *P-T* paths may significantly underestimate the slope of the true *P-T* path; they are an artifact of the method and do not reflect the true instantaneous *P-T* conditions at any stage of the uplift of the high-grade rocks. Mengel and Rivers (1991) pointed out an empirical criterion for discrimination between real and apparent *P-T* paths in samples exhibiting domainal equilibrium; that is, in real *P-T* paths, the *P-T* vectors determined from individual samples are parallel to and lie on the overall path derived from a larger suite of samples, whereas in apparent *P-T* paths the vectors for individual samples and the suite as a whole are oblique to each other. We now examine the results from this study using these criteria.

Figure 9 shows the core and rim (apparent peak and retrograde) conditions for the eighteen samples analyzed. It is immediately evident that the apparent *P-T* paths followed by some of the samples are parallel to or nearly parallel to the slopes of the geobarometers, suggesting that nonsynchronous closure of the geobarometer and the geothermometer has occurred. The implications of this for the footwall, hangingwall, and shear zone samples are considered separately.

For the four samples from the *footwall,* all the apparent *P-T* paths can be interpreted in terms of nonsynchronous closure of the geobarometer and the geothermometer to some degree. For samples 084 and 049, in particular, the apparent *P-T* paths closely parallel isopleths of the GCPQ calibration (Fig. 9, A). The effects of nonsynchronous closure are minimal in these samples, however, due to the short *P-T* vectors. Sample 074 involves the same barometer, but the *P-T* vector is steeper and longer than in 084 and 049. This suggests that nonsynchronous

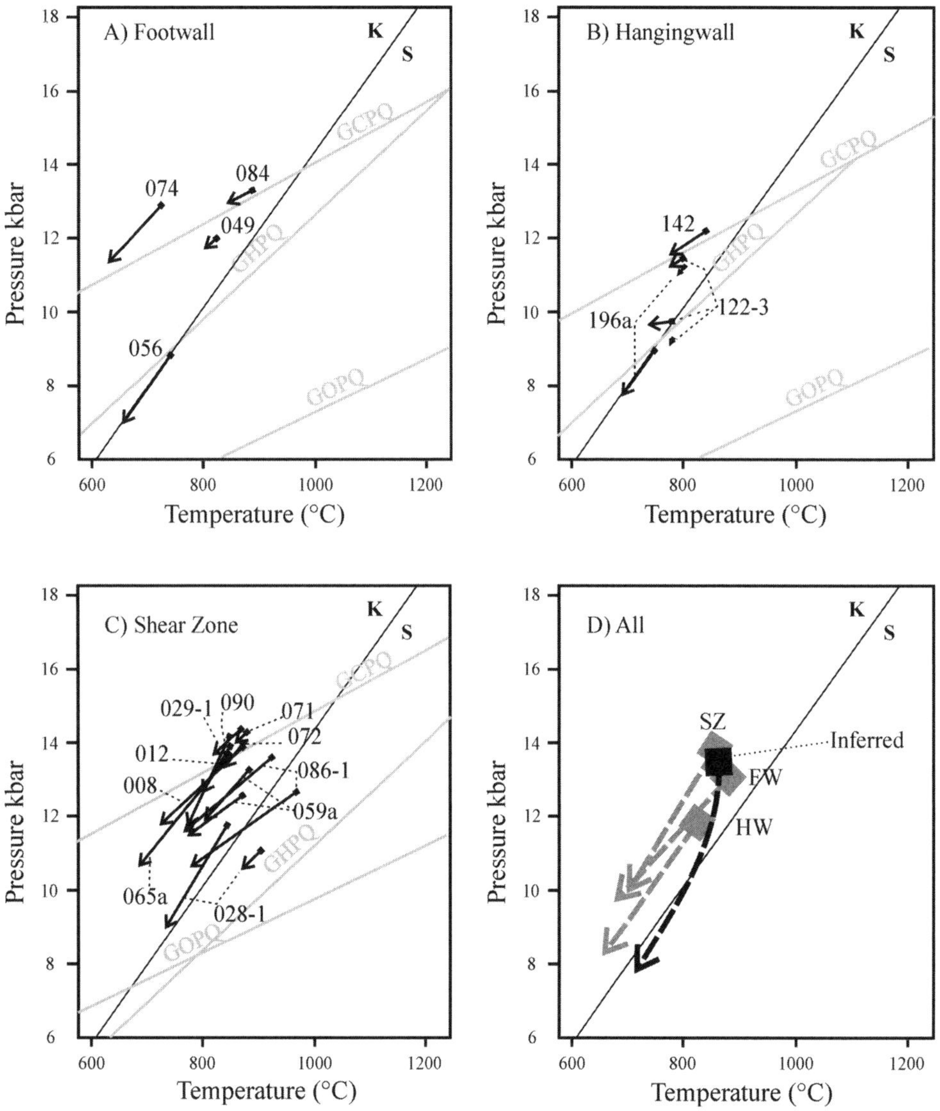

Figure 9. Segments of apparent pressure-temperature *(P-T)* paths (arrows) for (A) footwall, (B) hangingwall, and (C) shear zone rocks of the Grand Lake thrust system. The black line labeled K-S represents the boundary between Ky and Sil fields. See Appendix 2 for a complete listing of all mineral and component abbreviations; gray lines in A, B, and C indicate the slopes of the three geobarometers used in *P-T* determinations; GCPQ—Grt-Cpx-Pl-Qtz; GHPQ—Grt-Cam-Pl-Qtz; GOPQ—Grt-Opx-Pl-Qtz. Closure of the geobarometer at higher temperature than the geothermometer would cause the *P-T* paths to rotate into parallelism with isopleths of the geobarometers (see text for discussion). (D) Extrapolated *P-T* paths from the results in A, B, and C and the inferred shear zone and footwall *P-T* path incorporating observed reaction relationships (i.e., Ky ↔ Sil). See text for discussion.

closure was less severe in this sample, but since the apparent peak *T* is only 725 °C (>100 °C lower than in the other samples), it is unlikely that the apparent *P-T* path closely approximates the true path of the rock. Sample 056 involves the GHPQ calibration that has steeper isopleths than the GCPQ calibration. The calculated *P-T* vector is even steeper than isopleths for the GHPQ reaction, and the mineralogy is compatible with the *T* estimate of <700 °C, so it permits the interpretation that this sample records a segment of the real retrograde *P-T* path. However, petrographic evidence in pelitic gneiss in the Groswater Bay terrane for the growth of Sil after Ky (both inferred to be Grenvillian), noted earlier, implies that the footwall postpeak *P-T* path crossed from the Ky field to the Sil field (i.e., was steeper than the calculated *P-T* vector for sample 056). Assuming that the apparent peak estimates for samples 084 and 049 are close to the real peak *P-T* conditions, as discussed earlier, we conclude that the real *P-T* path must have exhibited isothermal decompression into the Sil field, that is, that the slope is underestimated by all the calculated segments shown in Figure 9, A.

Samples 142 and 122-3 from the *hangingwall* also yield apparent *P-T* vectors that are subparallel to the GCPQ isopleths, indicating that nonsynchronous closure in the GCPQ calibration occurred after peak metamorphism (Fig. 9, B). In both cases, the *P-T* vectors are short, indicating that the thermometer closed shortly after apparent peak conditions were locked in. The same is true of sample 196a, although in this case the *P-T* vector is steeper, suggesting that decompression was occurring during cooling. Apparent *P-T* vectors obtained with the GHPQ calibration for sample 196a also indicate decompression during cooling. As with the footwall samples, it is concluded that the slope of the steepest *P-T* vectors is the minimum possible for the hangingwall rocks. However, in this case there is no independent mineralogical evidence to support a steeper *P-T* path than that shown in Figure 9, B.

With respect to the *shear zone* samples, the apparent peak *P-T* estimates exhibit limited scatter, as discussed earlier, but the apparent retrograde paths of individual samples differ considerably in length and slope (Fig. 9, C). Four of the apparent retrograde paths follow slopes that are parallel or subparallel to the isopleths of the barometer used for the *P-T* calibration (e.g., samples 029-1, 059a, 071, and 086-1), implying closure of the thermometer after the barometer, as discussed earlier. However, the remaining seven samples yield *P-T* vectors that are steeper than the slope of the geobarometer, indicating that cooling was accompanied by decompression. The magnitudes of the *P-T* vectors vary from sample to sample, compatible with domainal equilibrium during retrogression in the shear zone rocks and indicating that recrystallization occurred over a range of *P-T* conditions, and thus at different times in samples from different parts of the shear zone. The small size of some of the domains in which this has occurred is indicated by the different lengths of vectors derived from a single sample, e.g., sample 028-1 (Fig. 9, C). The *P-T* vectors for these seven samples are subparallel to each other and together define a longer apparent *P-T* path for the shear zone, which is within the Ky field and subparallel to the Ky-Sil phase boundary. Once again, prudence dictates that independent evidence be sought to establish whether the slope of the *P-T* path is underestimated due to systematic nonsynchronous closure of the thermometer and barometer, but unfortunately in this case no such evidence was found.

Comparison of P-T *Paths*

Table 2 indicates that the average apparent peak *P-T* estimate from the shear zone (13.9 ± 0.5 kilobar/873 ± 40 °C) is greater than those from the hangingwall and the footwall. Consideration of the mechanics of shear zone propagation suggests that this result may be real. Models of ductile shear zones indicate that they may undergo progressive broadening during displacements across them as the wallrocks become mechanically weakened by strain accumulation (i.e., strain softening; Passchier and Trouw, 1998). Thus, the wallrocks may be incorporated into a long-lived growing shear zone at different times during its evolution. The lower state of strain and the lower apparent peak *P* estimates in the hangingwall and the footwall are thus compatible with progressive incorporation of the adjacent hangingwall and footwall into the shear zone after significant uplift had already occurred. As a result, the lower apparent peak *P* estimates in the hangingwall and the footwall may be real and reflect their later incorporation into the shear zone. This pattern is especially well displayed by the Northwest River dikes in the hangingwall. These rocks are monocyclic, recording only the imprint of the Grenvillian metamorphism. *P-T* estimates from the dikes (e.g., samples 142 and 196a) record apparent peak *P-T* conditions of ~11.5 kilobars/820 °C, suggesting that recrystallization of the hangingwall rocks occurred under these conditions, that is, ~2 kilobars (~5–6 km vertical height) above the base of the shear zone. Farther away from the Grand Lake thrust system shear zone, equilibration during Grenvillian shearing would have been progressively less penetrative as a result of the lower input of strain energy, and only partial recrystallization of the Labradorian assemblages is likely, but these rocks were not sampled in this study.

Figure 9, D, shows the extrapolated composite *P-T* paths based on the pooled results from the shear zone, footwall, and hangingwall samples, but omitting those that show clear evidence for nonsynchronous closure of the thermometer and the barometer (i.e., using only the samples with the steepest *P-T* vectors). With these restrictions, the slopes of the *P-T* vectors for individual samples are subparallel to the *P-T* path of the overall population, compatible with the criterion of Mengel and Rivers (1991). However, these are still the minimal slopes, and the true slopes of the *P-T* paths are inferred to have been steeper than the calculated slopes on the basis of the succession of Al silicate polymorphs in the footwall assemblage as noted earlier. Thus, in Figure 9, D, we have indicated a plausible qualitative *P-T* path for the footwall, shear zone, and hangingwall based on *P-T* and mineralogical evidence. This suggests that the rocks initially experienced a steep decompression path with minor cooling in the kyanite field, with the lower-pressure *P-T* evolution in the Sil field, possibly approximately subparallel to the Ky-Sil boundary. Comparison of these qualitative *P-T* paths with the calculated *P-T* vectors suggests that the thermometers continued to reset at temperatures up to 100 °C or so lower than the barometers, compatible with diffusion theory, thereby resulting in apparent *P-T* paths that are less steep than the true paths. As a result, the difference between the estimated and actual *P* conditions became progressively greater during retrograde cooling down to ~650 °C, as predicted by the granulite uncertainty principle.

TECTONIC MODEL FOR THE CAPE CARIBOU RIVER ALLOCHTHON

We now attempt to integrate the available *P-T* and structural data into a tectonic model for Grenvillian evolution of the Cape Caribou River allochthon and adjacent terranes. Details of the timing of the tectonic events described here are discussed elsewhere, although the reader is reminded that both Ottawan and Rigolet ages have been obtained in the vicinity of the Grand Lake thrust system shear zone (Philippe et al., 1993; Bussy et al., 1995; Corrigan et al., 1997, 2000).

Figure 10, A, shows the inferred pre-Grenvillian setting, with crustal stacking and inactive thrust faults (e.g., the Rigolet fault) inherited from the Paleoproterozoic Labradorian orogeny (Corrigan et al., 2000) and negligible topographic relief. Initiation of Grenvillian crustal thickening in the interior of the orogen (Fig. 10, B) resulted in the reactivation of the Rigolet thrust

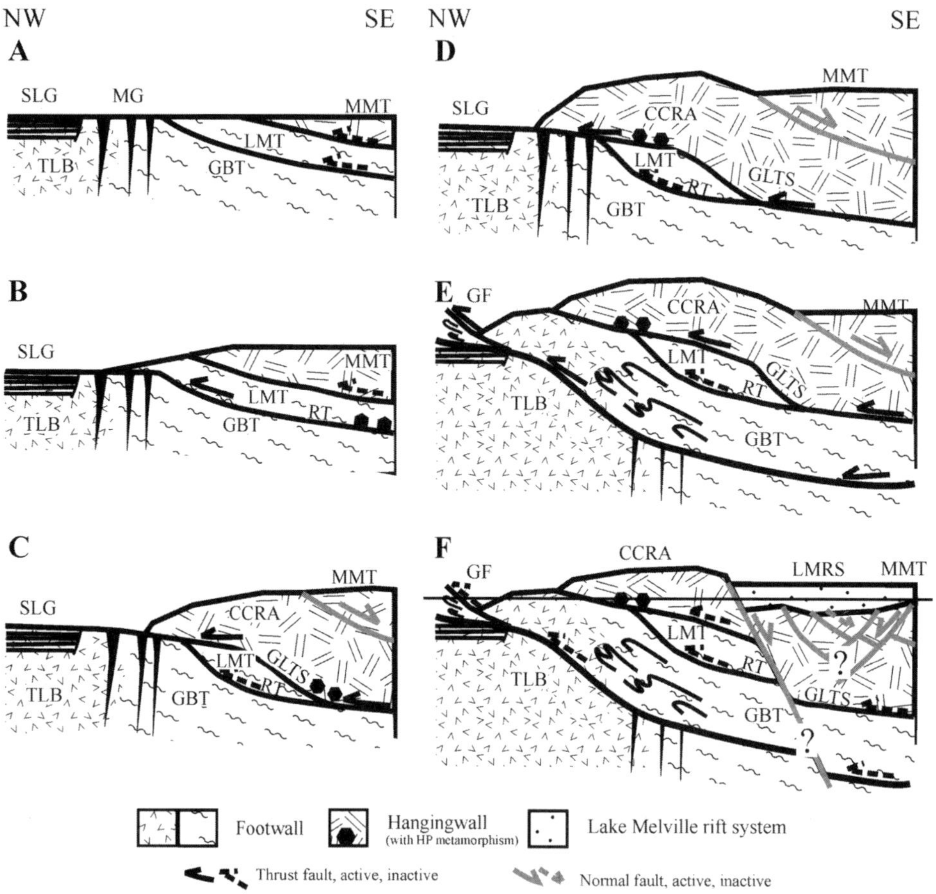

Figure 10. Schematic tectonic model showing the evolution of the Cape Caribou River allochthon (CCRA). GBT—Groswater Bay terrane; GF—Grenville Front; GLTS—Grand Lake thrust system; *HP*—high-pressure; LMRS—Lake Melville rift system; LMT—Lake Melville terrane; MG—Michael gabbro; MMT—Mealy Mountains terrane; RT—Rigolet thrust; SLG—Seal Lake Group; TLB—Trans-Labrador batholith; ? indicates the uncertainty in extrapolation of structures and lithologies at depth beneath the LMRS. (A) Late Mesoproterozoic–pre-Grenvillian situation: eroded Labradorian terranes separated by inactive thrust faults, cut by dikes of Michael gabbro and unconformably overlain by the Seal Lake Group. (B) Beginning of Grenvillian orogenesis (Early Ottawan orogeny, ca. 1080–1060 Ma): reactivation of the Rigolet thrust, with the Lake Melville terrane and the Mealy Mountains terrane together forming a crustal-scale orogenic wedge that was thrust toward the northwest over the Groswater Bay terrane. (C) Ottawan orogeny (continued): out-of-sequence thrusting, formation of the Grand Lake thrust system with inferred ramp-flat geometry. The Cape Caribou River allochthon was thrust northward over the Lake Melville terrane, and the Rigolet thrust was cut off, becoming inactive. Formation of the Grand Lake thrust system comprised the first stage of uplift of *HP* granulites up a crustal-scale ramp. Inferred initiation of normal displacement of the Mealy Mountains terrane with respect to the Cape Caribou River allochthon due to collapse of the upper part of the orogenic wedge, as the wedge exceeded critical taper. (D) Ottawan orogeny (continued): emplacement of *HP* granulites onto the flat as the Mealy Mountains terrane and Lake Melville terrane were thrust further north-northwest; continued normal faulting beneath the Mealy Mountains terrane at the top of the wedge. (E) Rigolet orogeny (ca. 1010–1000 Ma): advance of the orogen into its foreland, formation of the Grenville Front, and incorporation of the Groswater Bay terrane into the orogen; deformation of the Michael gabbro and its host rocks; continued normal faulting (inferred) in the upper parts of the thrust wedge. (F) Neoproterozoic (ca. 600 Ma): formation of the Lake Melville rift system; normal faults associated with LMRS further separated the Cape Caribou River allochthon from the Mealy Mountains terrane, obscuring evidence of inferred Grenvillian extensional faults. Horizontal line indicates the approximate location of the present erosion level exposing the *HP* rocks at the base of the Cape Caribou River allochthon.

and the formation of an orogenic wedge comprising the Lake Melville terrane and the Cape Caribou River allochthon–Mealy Mountains terrane as the latter were thrust onto the foreland to the northwest. Peak pressure estimates of ~14 kilobars (this study) determined from the high-pressure granulites in the shear zone and the adjacent hangingwall of the Rigolet thrust imply that the orogenic wedge was at least 40 km thick, and thus that in its Grenvillian manifestation the southeast-dipping Rigolet thrust was a crustal-scale structure. Based on the lack of penetrative Grenvillian fabrics and metamorphic overprint in many units in the Mealy Mountains terrane (e.g., the Mealy Mountains anorthosite; Gower, 1996), and the preservation of pre-Grenvillian $^{40}Ar/^{39}Ar$ ages in the Mealy dikes (Reynolds, 1989), it is inferred that the Mealy Mountains terrane comprised the upper, relatively cold and little deformed, part of this crustal-scale thrust wedge, whereas the Cape Caribou River allochthon formed the base.

Following early (Ottawan) displacement on the Rigolet thrust (at the base of the Lake Melville terrane), the basal thrust of the orogenic wedge cut structurally up section to form the Grand Lake thrust system (at the base of the Cape Caribou River allochthon), thereby truncating the Rigolet thrust (Fig. 10, C). Thrusting thus stepped back into the thrust wedge (out-of-sequence thrusting or overstacking), and the Rigolet thrust became inactive. That part of the Grand Lake thrust system exposed around Grand Lake represents a flat, which is inferred to give way to a crustal-scale ramp (not observed) toward the southeast. Uplift of the Cape Caribou River allochthon with the high-pressure granulites in its basal shear zone (the Grand Lake thrust system), and associated exhumation and cooling, are indicated in Figure 10, D. Cooling to below the closure temperatures of the geothermometers (~725–500 °C, Frost and Chacko, 1989) must have been rapid in order to preserve the mineral compositions and assemblages of the high-temperature granulites (875 ± 50 °C, this study). Uplift would have been enhanced by the inferred ramp geometry, and rapid exhumation and cooling may have been accomplished by southeast-directed normal faulting at the top of the orogenic wedge that is inferred to have occurred in response to overthickening. We infer that the Mealy Mountains terrane, with evidence for only low-*T* Grenvillian metamorphism, represents (part of) this downthrown hangingwall.

On the basis of independently acquired data, it is known that the orogenic wedge cut down into its footwall and advanced into the foreland during the ensuing Rigolet orogeny, leading to the formation of the Grenville Front about 100 km north of the Grand Lake thrust system, and incorporation of the Groswater Bay terrane into the orogen as part of the Parautochthonous Belt (Fig. 10, E). Deformation along the Grenville Front has been dated at ca. 1005 Ma (Krogh, 1994). Lithoprobe crustal-scale seismic sections in the western and central Grenville Province (Ludden and Hynes, 2000) show that the Grenville Front is the lowest thrust in the Grenville orogen, and so it is inferred to underlie the Grand Lake thrust system at depth (Fig. 10, E). Thrust displacement on the Grenville Front shear zone would have resulted in the strain concentration at the base of the enlarged orogenic wedge, with the Mealy Mountains terrane and the Cape Caribou River allochthon essentially passively transported at this time. However, uplift of the orogenic wedge along the Grenville Front shear zone may have promoted renewed normal faulting in the upper part of the wedge, thereby further separating the Mealy Mountains terrane from the Cape Caribou River allochthon (Fig. 10, E), and contributing to uplift and exhumation of the Cape Caribou River allochthon.

We infer from the thrust-loading model described earlier that formation of the high-pressure granulite-facies assemblages occurred in a classical clockwise *P-T* path, as shown in Figure 11. "Apparent peak" *P-T* conditions (i.e., the *P* at the apparent peak *T*) were followed by decompression and cooling, with the inferred retrograde *P-T* path crossing from the Ky to the Sil field in accord with petrographic observations. Decompression and ingress of H_2O in the Grand Lake thrust system caused the subassemblage Grt-Cpx to react to Cam-Pl-Qtz (reaction 1b), although ingress of H_2O was domainal on all scales and both subassemblages are present in many samples. As discussed earlier, peak *P-T* conditions in the hangingwall rocks adjacent to the shear zone may have been achieved during uplift, after apparent peak conditions in the shear zone itself.

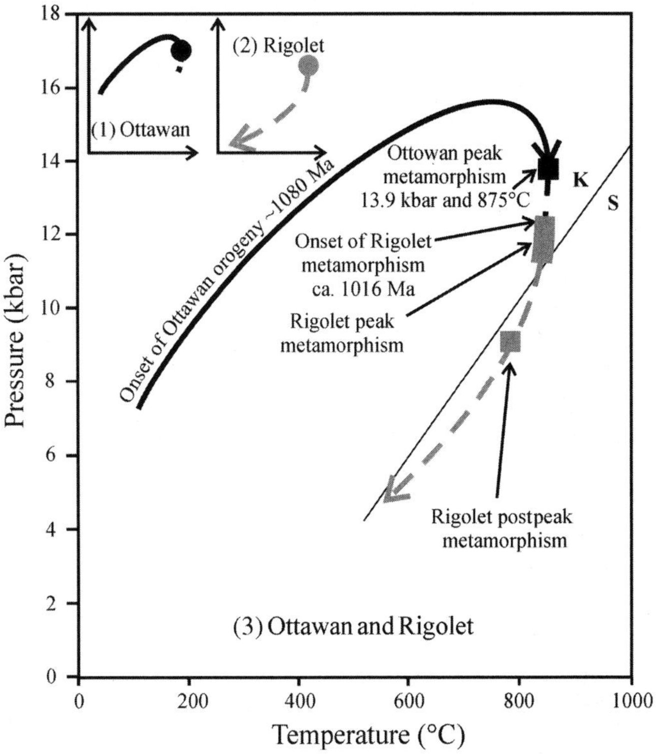

Figure 11. Schematic pressure-temperature-time (*P-T-t*) path for the Grand Lake thrust system and the Cape Caribou River allochthon. (1) The *P-T* path taken during the Ottawan orogeny (tectonic burial and early uplift or decompression). (2) The *P-T* path taken during the Rigolet orogeny (decompression and cooling). (3) The overall *P-T-t* path during the Grenvillian orogenesis, showing portion preserved in analyzed mineral assemblages. The black line labeled K-S represents the boundary between kyanite and sillimanite fields.

The final stage in the tectonic evolution, illustrated in Figure 10, F, was formation of the Lake Melville rift in the Late Neoproterozoic. This graben structure is a failed rift associated with the opening of the Iapetus Ocean and the initiation of the Appalachian "Wilson cycle" (Gower, 1986). From the perspective of this study, one important result of the formation of the Lake Melville rift is that the Grenvillian relationship between the Cape Caribou River allochthon and the Mealy Mountains terrane has been obscured. In the model presented in Figure 10,

F, the two terranes are inferred to be separated by a major Grenvillian extensional shear zone, as discussed earlier, that was possibly also reworked during the Neoproterozoic. However, this shear zone has not been observed in the field and hence is a putative structure. Its existence is supported by the convincing lithological correlations between the Mealy Mountains terrane and the Cape Caribou River allochthon (Wardle et al., 1990) and their strongly contrasting Grenvillian metamorphic signatures as documented in this study.

DISCUSSION

Relationship to the H*P* Belt

The Grand Lake thrust system occurs at the base of a thick-skinned Grenvillian thrust wedge composed of the Cape Caribou River allochthon and the Mealy Mountains terrane, defining the boundary between the thrust wedge and the underlying footwall rocks of the Parautochthonous Belt (Fig. 1). The Grenvillian high-pressure granulite-facies conditions (~14 kilobars/875 °C) experienced by the rocks in the shear zone have not been previously recorded elsewhere in the eastern Grenville Province. However, Grenvillian high-pressure rocks (eclogite and high-pressure granulite and their retrogressed equivalents) are known in both the western and the central Grenville Province, and in both locations also directly overlie the Parautochthonous Belt. They comprise the high-pressure (H*P*) belt, which is in turn overlain by a low-pressure (L*P*) belt that lacks evidence for Grenvillian high-pressure metamorphism (Fig. 1; Ludden and Hynes, 2000; Rivers et al. 2002). Thus, it is possible that the high-pressure granulites in the Grand Lake thrust system and the base of the Cape Caribou River allochthon form the northeasterly continuation of the H*P* belt into the eastern Grenville Province, a proposition we now evaluate.

In terms of its tectonic setting, the Cape Caribou River allochthon and the overlying Mealy Mountains terrane occupy identical positions to the Molson Lake terrane and the overlying Lac Joseph terrane, respectively, of the central Grenville Province (Fig. 1). The Molson Lake terrane, with eclogite-facies and high-pressure granulite-facies assemblages in mafic rocks, is part of the H*P* belt, whereas the Lac Joseph terrane, which shows evidence for negligible Grenvillian metamorphism, is part of the overlying *LP* belt (Rivers et al., 2002). Thus, given their metamorphic character, there is a compelling first-order logic for inclusion of the Cape Caribou River allochthon within the H*P* belt and the Mealy Mountains terrane within the L*P* belt. However, although the tectonic setting of the Cape Caribou River allochthon is similar to other H*P* terranes, it exhibits several differences from the H*P* terranes in the central and western Grenville Province that merit discussion. In the first place, Grenvillian deformation and recrystallization in the more westerly H*P* terranes were penetrative, with evidence for very high strains recorded in the interiors of some of the H*P* thrust sheets (e.g., Indares et al., 2000; Rivers et al., 2002). Thus, although these H*P* terranes were several kilometers thick, thrusting was not thick-skinned in the sense in which the term has been applied herein to the Cape Caribou River allochthon, with negligible deformation away from the basal shear zone. Second, although the estimated apparent peak temperatures in the Grand Lake thrust system are similar to those estimated for the H*P* terranes farther west, the maximum apparent peak pressures in the Grand Lake thrust system are significantly lower (i.e., ~14 kilobars in the Grand Lake thrust system versus ~18 kilobars in the Molson Lake and Lelukuau terranes in the central Grenville Province; Rivers et al., 2002). This difference could reflect a real change in the architecture of the orogenic wedge along strike, with the base of the wedge cutting down deeper in the central Grenville Province. Alternatively, the Cape Caribou River allochthon could represent the leading northwest edge of a partially exhumed H*P* terrane that experienced progressively higher pressures toward the southeast, as has been demonstrated for the Molson Lake terrane, in which estimated pressures range from ~14 kilobars in the northwest to ~18 kilobars in the southeast (Indares and Rivers, 1995). We are unable to distinguish between these two possibilities with the data available. A third possibility, that the peak pressures in the Grand Lake thrust system are significantly underestimated, has been discussed earlier, and although it cannot be completely ruled out, it has been argued on petrological grounds that the calculated maximum apparent peak pressures probably do not significantly underestimate the actual peak pressures. In summary, we conclude that the Cape Caribou River allochthon represents the northeasterly continuation of the H*P* belt into central Labrador, but acknowledge that the architecture and strain of the belt differ significantly between the two areas.

Role of Shear Heating

We now turn to the question of whether shear or frictional heating could have been important in generating the high temperatures in the Grand Lake thrust system at the base of the Cape Caribou River allochthon. The general problem of quantifying the magnitude of shear heating in the vicinity of large thrusts has been considered by several authors (e.g., Brun and Cobbold, 1980; Pavlis, 1986; Peacock, 1992). Given the many uncertainties in input parameters for the calculations, we seek a qualitative answer here. Pavlis (1986) has suggested that large continental thrust systems can be subdivided into two segments, a lower crustal–upper mantle segment where dry rocks are in contact on either side of the fault, and an upper crustal segment in which wet rocks from the hangingwall are in contact with dry crystalline rocks in the footwall. Our interest here lies in the former setting, as both footwall and hangingwall terranes adjacent to the Grand Lake thrust system can be characterized as dry. In such settings, virtually all the mechanical energy involved in rock deformation is converted into heat, and under plate tectonic rates of deformation of 5–10 cm per year, moderate shear stresses of a few tens of megapascals, and no partial melting, calculations show that temperature increases of >100 °C are possible in the vicinity of narrow shear zones of up to a few kilometers width (Pavlis, 1986). The magnitude of the temperature increase is strongly dependent on the width of the shear zone, the magnitude of the shear stress, the displacement velocity, and the rheology of the material involved (Pavlis, 1986). Given the thick-skinned nature of the thrusting of the Cape Caribou River allochthon, the absence of widespread partial melting, and the maximum width of the Grand Lake thrust system of 1 to 2 km, these results suggest that heating of significantly more than 100 °C could have occurred during displacements on the Grand Lake thrust system. Such local heating in the shear zone, in addition to regional thermal relaxation in response to tectonic burial, could therefore account for the unusual prograde character of the Grand Lake thrust system shear zone. Unfortunately, given the large number of unknowns, we are unable to further quantify this conclusion.

Returning to the general setting of the Cape Caribou River allochthon in the H*P* belt, we can conclude that shear heating may have been greater in the Grand Lake thrust system than at the base of the other H*P* terranes farther west if Grenvillian deformation was more strongly partitioned into the Grand Lake thrust system. If this was the case, the similar maximum temperatures achieved throughout the H*P* belt may be a coincidence rather than reflecting the underlying thermal regime.

CONCLUSIONS

The Cape Caribou River allochthon is an example of a thick-skinned thrust sheet in which penetrative deformation and recrystallization are restricted to the basal shear zone, the Grand Lake thrust system. In the footwall and hangingwall away from the shear zone, Labradorian igneous and metamorphic assemblages are variably preserved. Mylonitic orthogneiss in the Grand Lake thrust system is characterized by the peak Grenvillian assemblages Grt-Cpx-Pl-Cam-Qtz and Opx-Cpx-Pl-Cam-Qtz in mafic and intermediate-felsic lithologies, respectively, which implies high-pressure granulite-facies conditions. These assemblages have yielded apparent peak *P-T* estimates of ~14 kilobars/875 °C that are inferred to closely approximate the *P* at maximum *T* experienced by rocks in the shear zone, despite evidence in some samples for postpeak resetting. This distribution of syntectonic high-pressure granulite-facies assemblages suggests that recrystallization in the Grand Lake thrust system was driven by input of strain energy, with a possibly significant but unquantified component of shear heating leading to prograde metamorphism in the shear zone. The role of fluids in promoting recrystallization in the Grand Lake thrust system shear zone is unclear. Although hydrous minerals, e.g., Cam, are present in orthogneiss in the footwall, hangingwall, and shear zone, there is widespread textural evidence for Cam-breakdown reactions in peak orthogneiss assemblages. Thus the shear zone formed in rocks that were dehydrating, so shearing was not initiated as a result of hydraulic weakening driven by fluid ingress. Some of the fluid produced by the Cam-breakdown reactions in mafic orthogneiss may have been incorporated in syntectonic leucosomes in pelitic rocks in the adjacent footwall, and some may have entered the low-strain Northwest River dikes in the hangingwall, which display evidence for both hydration and carbonation reactions during Grenvillian metamorphism.

Following peak *P-T* conditions, the rocks are inferred to have followed a clockwise *P-T* path that involved quasi-isothermal decompression from the kyanite to the sillimanite field, although details of the path remain only qualitatively constrained because of the unknown degree of postpeak reequilibration in the mineral assemblages. Ingress of H_2O-rich fluids during decompression, possibly derived in part from crystallizing leucosomes in adjacent pelitic rocks, led to domainal retrogression of the peak high-pressure granulite-facies assemblage to an amphibolite-facies assemblage.

On the basis of its tectonic setting and the estimated *P-T* conditions in the Grand Lake thrust system, it is inferred that the Cape Caribou River allochthon represents the northeasterly continuation of the H*P* belt in the eastern Grenville Province. However, in detail it differs significantly in tectonic character from the H*P* terranes farther west.

ACKNOWLEDGMENTS

This study forms part of the M.S. thesis of JBK at Memorial University of Newfoundland. Jennifer Butt assisted JBK in the field, and we thank the Geological Survey of Newfoundland for the loan of field equipment and Bill Michelin of Northwest River for the use of his cabin while mapping on Grand Lake. JBK acknowledges receipt of a graduate fellowship from the School of Graduate Studies, Memorial University. The study was funded through a National Sciences and Engineering Council (Canada) grant to TR. We thank R.J. Wardle and D. Corrigan for their insightful reviews of the manuscript. This is Lithoprobe contribution 1354.

APPENDIX 1. SAMPLE LOCATIONS

Map of field area showing the locations of samples chosen for geothermobarometry.

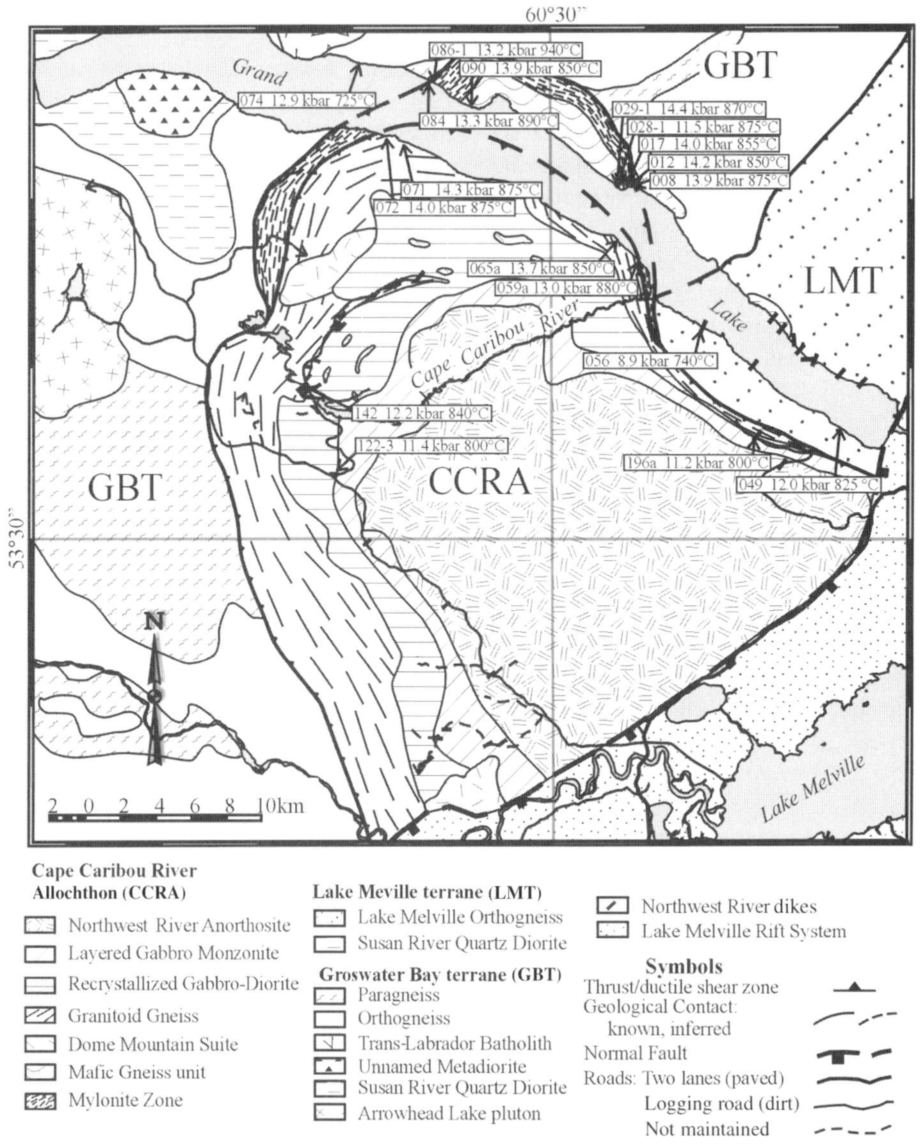

APPENDIX 2. SAMPLE DESCRIPTIONS

The following are brief petrographic descriptions of the samples chosen for geothermobarometry. Full descriptions may be found in Krauss (2002). Sample locations are shown in Appendix 1. Mineral and end-member component abbreviations are as follows (component abbreviations in italics): *ab*—albite; *ae*—aegirine; *alm*—almandine; Amp—amphibole; *an*—anorthite; Ap—apatite; Bt—biotite; Cal—calcite; Cam—clinoamphibole; Cpx—clinopyroxene; *di*—diopside; Dol—dolomite; *en*—enstatite; Ep—epidote; *feprg*—ferropargasite; *fets*—Fe-tschermakite; *fs*—ferrosilite; *grs*—grossular; Grt—garnet; *hd*—hedenbergite; *jd*—jadeite; Kfs—K feldspar; Ky—kyanite; L—granitic liquid; Mnz—monazite; Ms—muscovite; Oam—orthoamphibole; Ol—olivine; Opx—orthopyroxene; Or—orthoclase; Pl—plagioclase; *prg*—pargasite; *prp*—pyrope; Qtz—quartz; Rt—rutile; Scp—scapolite; Sil—sillimanite; *sps*—spessartine; *ts*—tschermakite; Ttn—titanite; Zrn—zircon.

Footwall Samples

JK-99-049, dioritic orthogneiss, Lake Melville terrane. The inferred peak metamorphic assemblage is Grt-Cpx-Cam-Pl-Qtz, with accessory Fe-Ti-oxides and Rt. The sample consists of two compositional domains, Grt-Cpx-Cam-Pl-Qtz with evidence for reaction 1a, and Cam-Cpx-Pl. Grt occurs as skeletal grains with abundant inclusions of Pl-Qtz-Cpx ± Cam. Retrogression (partial replacement of Cpx by Cam) is minimal.

JK-99-056, Northwest River dike, Lake Melville terrane. The peak metamorphic assemblage is Grt-Cam-Bt-Pl-Qtz. Grt occurs as skeletal grains with inclusions of Qtz-Bt-Pl.

JK-99-074, dioritic gneiss, Groswater Bay terrane. The assemblage is Grt-Opx-Cpx-Bt-Pl-Qtz, but the lack of Grt-Opx associations suggests that Opx is metastable. The texture is fine-grained and equigranular, with partial replacement of Opx and Grt by Bt.

JK-99-084, mafic gneiss, Groswater Bay terrane, adjacent to the basal mylonite zone of the Grand Lake thrust system. The peak assemblage is Grt-Cam-Cpx-Bt-Pl-Qtz, with accessory Ap-Zrn-Fe-Ti oxides-Rt. Grt porphyroblasts up to ~1 cm in diameter contain inclusions of Cpx-Fe-Ti oxide-Pl-Rt ± Qtz, suggesting that Grt, Cpx, Pl, and Qtz were in equilibrium during shearing. Retrogression occurs as secondary Cam and Pl via reaction 1b.

Hangingwall Samples

JK-99–122-3, recrystallized gabbro-diorite unit, Cape Caribou River allochthon. [Gabbro-diorite is host to a Northwest River dike; same locality as the U-Pb sample (C-050AM) collected by Bussy et al. (1995).] The peak metamorphic assemblage is Pl-Cam-Opx-Cpx-Grt-Bt ± Qtz. There is no penetrative fabric; Opx and Cpx occur as inclusions in Grt; Qtz is commonly associated with Grt-Cpx and less commonly with Grt-Opx; equilibrium textures among Grt, Opx, Cpx, and Cam suggest that they were all stable at the metamorphic peak; there is no mineralogical evidence for retrogression.

JK-99-142, Northwest River dike, Cape Caribou River allochthon. [Same location as U-Pb sample C-050D of Bussy et al. (1995).] The dike contains a matrix assemblage of Cam-Pl with veins composed of Grt-Cpx-Pl-Qtz-Cal, suggesting that CO_2-bearing fluids facilitated the production of the Grt-Cpx-Cal assemblage via reaction 1c.

JK-99-196a, Northwest River dike, Cape Caribou River allochthon. This dike also preserves Grt-Cpx veins in a Pl-Cam-Opx matrix (reaction 1c); in addition, Opx occurs as relatively small grains inferred to be growing at the expense of Cam. The vein assemblage is Grt-Cpx-Cam-Pl-Qtz, and Grt porphyroblasts contain inclusions of Cpx-Cam-Pl-Qtz.

Shear Zone Samples

JK-99-008, layered straight gneiss, Grand Lake thrust system. The mafic layers contain the peak assemblage Grt-Cpx-Cam-Bt-Pl ± Qtz, whereas the intermediate-felsic layers contain Pl-Qtz-Kfs ± Grt. Porphyroblasts of Cpx with SPO, partially replaced by secondary Cam, define the mineral stretching lineation together with Cam-Pl and Qtz. Qtz is generally restricted to the intermediate-felsic layers, but also occurs as inclusions in Grt.

JK-99-012, layered mafic-felsic straight gneiss with abundant large (up to 3cm) Grt porphyroblasts, Grand Lake thrust system. The peak metamorphic assemblage is Grt-Cpx-Opx-Cam-Bt-Pl-Qtz. Three textural domains are present: (1) large (~1 cm) Grt porphyroblasts with inclusions of Cpx-Pl wrapped by the mylonitic fabric, (2) equigranular Grt porphyroblasts with SPO grains of Cpx-Cam-Pl and stringers of Qtz, and (3) Opx in a mylonitic foliation composed of Grt-Cam-Pl-Qtz.

JK-99-017, mafic layer of layered mafic and intermediate-felsic straight gneiss, Grand Lake thrust system. The peak assemblage is Grt-Cpx-Pl-Qtz ± Cam, with Cpx occurring in the pressure shadows of Grt and Pl-Qtz occurring as granoblastic grains around Grt-Cpx. Cpx is partially replaced by secondary Cam.

JK-99-028, dioritic straight gneiss, Grand Lake thrust system. The peak assemblage is Grt-Opx-Cpx-Cam-Pl-Qtz, where SPO grains of Opx-Cpx and Cam define the mineral stretching lineation. Opx is inferred to have formed by breakdown of Grt during decompression via reaction 6 in reverse.

JK-99-029 contact of mylonitic straight gneiss and mafic gneiss units, Grand Lake thrust system. The peak assemblage is Grt-Cpx-Cam-Bt-Pl ± Qtz. Grt occurs as subhedral crystals with inclusions of Cpx-Cam-Pl, and as inclusion-free syntectonic porphyroblasts with granoblastic Cpx in pressure shadows wrapped by granoblastic matrix Pl.

JK-99-059a, straight gneiss, Cape Caribou, Grand Lake thrust system. The sample contains two compositional domains, characterized by the assemblages Cam-Opx-Cpx-Pl ± Qtz and Grt-Cpx-Pl ± Qtz ± Cam, suggesting the operation of reaction 6. Grt occurs as subhedral grains, with inclusions of Cpx-Pl ± Qtz inferred to represent the equilibrium assemblage of the rock.

JK-99-065a, mafic gneiss with large (up to 5 cm) porphyroblasts of Grt, Cape Caribou, Grand Lake thrust system. This sample contains two compositional domains composed of the assemblages Opx-Cpx-Cam-Pl and Grt-Cpx-Pl-Qtz. Grt occurs as collars around Fe-Ti oxides, where it may contain inclusions of Pl-Qtz, and also in association with Cpx, the texture inferred to be a result of reaction 6. There is no evidence for retrogression in this sample.

JK-99-071, dioritic gneiss, Grand Lake thrust system. The peak metamorphic assemblage is Grt-Cpx-Cam-Pl-Qtz-Scp, where SPO grains of Cpx and Cam define the mineral stretching lineation. Grt occurs as euhedral to anhedral grains that may contain inclusions of Cpx-Pl-Qtz inferred to have been in equilibrium during shearing.

JK-99-072, dioritic gneiss, Grand Lake thrust system. The peak assemblage is Grt-Cpx-Pl-Qtz-Scp-Cam ± Opx, with SPO grains of Cpx defining the mineral stretching lineation. Grt occurs as subhedral to anhedral porphyroblasts with inclusions of Qtz-Pl-Cpx. Cpx and Pl grains have undergone subgrain formation, producing granoblastic Cpx domains in the pressure shadows of Grt and granoblastic Pl associated with Qtz stringers, respectively. Opx is generally retrogressed to Cam and does not occur within the Grt-Cpx-Pl-Qtz domains.

JK-99-086, straight gneiss, base of Grand Lake thrust system. The peak assemblage in the sampled mafic layers is Grt-Cam-Cpx-Pl-Bt ± Qtz. The foliation is defined by Cam, and Grt occurs as heavily resorbed grains with collars of partially sericitized Pl and Cam separating it from Cpx, inferred to be the result of reaction 1b.

JK-99-090, mylonitic straight gneiss, Grand Lake thrust system. The sampled mafic layer contains the peak assemblage Grt-Cpx-Cam-Pl-Qtz ± Bt, where SPO grains of Cpx and Cam define the mineral stretching lineation. Grt occurs as flattened anhedral grains with inclusions of Cpx-Pl-Qtz. Partial replacement of Cpx by Cam is evidence for reaction 1b.

REFERENCES CITED

Berman, R.G., 1991, Thermobarometry using multi-equilibrium calculations: A new technique, with petrological applications: Canadian Mineralogist, v. 29, p. 833–855.

Brun, J.P., and Cobbold, P.R., 1980, Strain heating and thermal softening in continental shear zones: A review: Journal of Structural Geology, v. 2, p. 149–158.

Bussy, F., Krogh, T.E., and Wardle, R.J., 1995, Late Labradorian metamorphism and anorthosite-granitoid intrusion, Cape Caribou River Allochthon, Grenville Province, Labrador: Evidence from U-Pb geochronology: Canadian Journal of Earth Sciences, v. 32, p. 1411–1425.

Butt, J.M., 2000, Mineral chemistry and mineral reactions in a meta-anorthosite complex, Cape Caribou River Allochthon, Grenville Province, Labrador [B.S. honors dissertation]: St. John's, Memorial University of Newfoundland, 111 p.

Connelly, J.N., Rivers, T., and James, D.T., 1995, Thermotectonic evolution of the Grenville Province, western Labrador: Tectonics, v. 14, p. 202–217.

Corrigan, D., Rivers, T., and Dunning, G.R., 1997, Preliminary report on the evolution of the Allochthon Boundary Thrust in eastern Labrador, Mechin River to Goose Bay, *in* East Coast Seismic Offshore-Onshore Transect Meeting, 14–15 April, Ottawa, Ontario, Lithoprobe Report No. 61, p. 45–56.

Corrigan, D., Rivers, T., and Dunning, G.R., 2000, U-Pb constraints for the plutonic and tectonometamorphic evolution of the Lake Melville terrane, Labrador, and implications for basement reworking in the northeastern Grenville Province: Precambrian Research, v. 99, p. 65–90.

Culshaw, N.G., Jamieson, R.A., Ketchum, J.W.F., Wodicka, N., Corrigan, D., and Reynolds, P.H., 1997, Transect across the northwestern Grenville orogen, Georgian Bay, Ontario: Polystage convergence and extension in the lower orogenic crust: Tectonics, v. 16, p. 966–982.

Droop, G.T.R., 1987, A general equation for estimating Fe^{3+} concentrations in ferromagnesian silicates and oxides for microprobe analyses, using stoichiometric criteria: Mineralogical Magazine, v. 51, p. 431–435.

Emslie, R.F., 1976, Mealy Mountains Complex, Grenville Province, southern Labrador, *in* Report of Activities, Part A: Ottawa, Ontario, Geological Survey of Canada, Paper 76-1A, p. 165–170.

Emslie, R.F., and Hunt, P.A., 1990, Ages and petrogenetic significance of igneous mangerite-charnockite suites associated with massif anorthosites, Grenville Province: Journal of Geology, v. 98, p. 213–231.

Emslie, R.F., and Stirling, J.A.R, 1993, Rapakivi and related granitoids of the Nain Plutonic Suite: Geochemistry, mineral assemblages and fluid equilibria: Canadian Mineralogist, v. 31, p. 821–847.

Frost, B.R., and Chacko, T., 1989, The granulite uncertainty principle: Limitations on thermobarometry in granulites: Journal of Geology, v. 97, p. 435–450.

Fryer, B.J., 1983, Report of geochronology—Labrador mapping [unpublished report]: St. John's, Newfoundland Department of Mines and Energy, Open File Lab-617.

Gower, C.F., 1986, Geology of the Double Mer White Hills and surrounding region, Grenville Province, eastern Labrador: Ottawa, Ontario, Geological Survey of Canada, Paper 86–15.

Gower, C.F., 1996, The evolution of the Grenville Province in eastern Labrador, Canada, *in* Brewer, T.S., ed., Precambrian crustal evolution in the North Atlantic region: Geological Society [London] Special Publication 112, p. 197–218.

Gower, C.F., Schärer, U., and Heaman, L.M, 1992, The Labradorian orogeny in the Grenville Province, eastern Labrador, Canada: Canadian Journal of Earth Sciences, v. 29, p. 1944–1957.

Hamilton, M.A., and Emslie, R.F., 1997, Mealy dykes, Labrador, revisited: U-Pb baddeleyite age and implications for the eastern Grenville Province [abs]: Geological Association of Canada–Mineralogical Association of Canada Annual Meeting, 19–21 May, Ottawa, Ontario, v. 22, p. A-62.

Hanes, J.A., 1991, K-Ar and $^{40}Ar/^{39}Ar$ geochronology: Methods and applications, *in* Heaman, L., and Ludden, J.N., eds., Applications of radiogenic isotope systems to problems in geology: Short course handbook: Ottawa, Ontario, Mineralogical Association of Canada, Short Course Handbook, v. 19, p. 27–57.

Hanmer, S., 1988, Ductile thrusting at mid-crustal level, southwestern Grenville Province: Canadian Journal of Earth Sciences, v. 25, p. 1049–1059.

Indares, A., and Rivers, T., 1995, Textures, metamorphic reactions and thermobarometry of eclogitized metagabbros: A Proterozoic example: European Journal of Mineralogy, v. 7, p. 43–56.

Indares, A., Dunning, G., Cox, R., Gale, D., and Connelly, J., 1998, High-pressure, high-temperature rocks from the base of thick continental crust: Geology and age constraints from the Manicouagan Imbricate Zone, eastern Grenville Province: Tectonics, v. 17, p. 426–440.

Indares, A., Dunning, G., and Cox, R., 2000, Tectono-thermal evolution of deep crust in a Mesoproterozoic continental collision setting: The Manicouagan example: Canadian Journal of Earth Sciences, v. 37, p. 325–340.

James, D.T., Kamo, S., and Krogh, T.E., 2000, Preliminary U-Pb geochronological data from the Mealy Mountains terrane, Grenville Province, southern Labrador: St. John's, Geological Survey of Newfoundland, Current Research Report 2000-1, p. 169–178.

Jamieson, R.A., and Beaumont, C., 1989, Deformation and metamorphism in convergent margins: A model for uplift and exhumation of metamorphic terrains, *in* Daly, J.S., et al., eds., Evolution of metamorphic belts: Geological Society [London] Special Publication 43, p. 117–129.

Krauss, J.B, 2002, High-pressure (HP), granulite-facies thrusting in a thick-skinned thrust system in the eastern Grenville Province, central Labrador [M.S. thesis]: St. John's, Memorial University of Newfoundland, 335 p.

Kretz, R., 1983, Symbols for rock-forming minerals: American Mineralogist, v. 68, p. 277–279.

Krogh, T.E., 1986, Report to Newfoundland Department of Mines and Energy on U-Pb isotopic dating results from the 1985–86 contract [unpublished report]: St. John's, Geological Survey Branch, Newfoundland Department of Mines and Energy, Open File LAB (707), 106 p.

Krogh, T.E., 1994, Precise U-Pb ages for Grenvillian and pre-Grenvillian thrusting of Proterozoic and Archean metamorphic assemblages in the Grenville Front tectonic zone, Canada: Tectonics, v. 13, p. 963–982.

Krogh, T.E., and Davis, G.L., 1973, The significance of inherited zircons on the age and origin of igneous rocks: An investigation of the ages of Labradorian adamellites: Carnegie Institution of Washington Year Book, v. 72, p. 610–613.

Krogh, T.E., and Heaman, L., 1988, Report on 1987–1988 contract geochronology, Labrador [unpublished report]: St. John's, Mineral Development Division, Newfoundland Department of Mines, Open File, LAB (765), 102 p.

Leake, B.R., Wooley, A.R., Arps, C.E.S., Birch, W.D., Gilbert, M.C., Grice, J.D., Hawthorne, F.C., Kato, A., Kisch, H.J., Krivovichev, V.G., Linthout, K., Laird, J., Mandarino, J.A., Maresch, W.V., Nickel, E.H., Rock, N.M.S., Schumacher, J.C., Smith, D.C., Stephenson, N.C.N., Ungaretti, L., Whittaker, E.J.W., and Youzhi, G., 1997, Nomenclature of amphiboles: Report of the subcommittee on amphiboles of the International Mineralogical Association. Commission on new minerals and mineral names: American Mineralogist, v. 82, p. 1019–1037.

Ludden, J., and Hynes, A., 2000, The Lithoprobe Abitibi-Grenville transect: Two billion years of crust formation and recycling in the Precambrian Shield of Canada: Canadian Journal of Earth Sciences, v. 37, p. 459–476.

Mengel, F., and Rivers, T., 1991, Decompression reactions and *P-T* conditions in high-grade rocks, northern Labrador: *P-T-t* paths from individual samples and implications for early Proterozoic tectonic evolution: Journal of Petrology, v. 32, p. 139–167.

Passchier, C.W., and Trouw, R.A.J., 1998, Micro-tectonics: Berlin, Springer-Verlag, 289 p.

Pattison, D.R.M., 2003, Petrogenetic significance of orthopyroxene-free garnet + clinopyroxene + plagioclase-bearing metabasites with respect to the amphibolite and granulite facies: Journal of Metamorphic Geology, v. 21, p. 21–34.

Pavlis, T.L., 1986, The role of strain heating in the evolution of megathrusts: Journal of Geophysical Research, v. 91, no. B12, p. 12407–12422.

Peacock, S.M., 1992, Blueschist-facies metamorphism, shear heating, and P-T-t paths in subduction shear zones: Journal of Geophysical Research, B, Solid Earth and Planets, v. 97, 12, p. 693–707.

Philippe, S., Wardle, R.J., and Schärer, U., 1993, Labradorian and Grenvillian crustal evolution of the Goose Bay region, Labrador: New U-Pb geochronological constraints: Canadian Journal of Earth Sciences, v. 30, p. 2315–2327.

Platt, J.P., 1986, Dynamics of orogenic wedges and the uplift of high-pressure metamorphic rocks: Geological Society of America Bulletin, v. 97, p. 1037–1053.

Reynolds, P.H., 1989, $^{40}Ar/^{39}Ar$ dating of the Mealy Dykes of Labrador: Paleomagnetic implications: Canadian Journal of Earth Sciences, v. 26, p. 1567–1573.

Rivers, T., 1997, Lithotectonic elements of the Grenville Province: Review and tectonic implications: Precambrian Research, v. 86, p. 117–154.

Rivers, T., and Corrigan, D., 2000, Convergent margin on southeastern Laurentia during the Mesoproterozoic: Tectonic implications: Canadian Journal of Earth Sciences, v. 37, p. 359–383.

Rivers, T., Martignole, J., Gower, C.F., and Davidson, A., 1989, New tectonic divisions of the Grenville Province, southeast Canadian Shield: Tectonics, v. 8, p. 63–84.

Rivers, T., Ketchum, J., Indares, A., and Hynes, A., 2002, The High Pressure belt in the Grenville Province: Architecture, timing and exhumation models: Canadian Journal of Earth Sciences, v. 39, p. 867–893.

Ryan, A.B., Neale, T., and McGuire, J., 1982, Descriptive notes to accompany geological maps of the Grand Lake area, Labrador 13F/10, 11, 14, 15: St. John's, Newfoundland, Newfoundland and Labrador Department of Mines and Energy, Mineral Development Division, Maps 82-64 to 82-67.

Schärer, U., Krogh, T.E., and Gower, C.F., 1986, Age and evolution of the Grenville Province in eastern Labrador from U-Pb systematics in accessory minerals: Contributions to Mineralogy and Petrology, v. 94, p. 438–451.

Selverstone, J., and Chamberlain, C.P., 1990, Apparent isobaric cooling paths from granulites: Two counter examples from British Columbia and New Hampshire: Geology, v. 18, p. 307–310.

Spear, F.S., 1991, On the interpretation of peak metamorphic temperatures in light of garnet diffusion during cooling: Journal of Metamorphic Geology, v. 9, p. 379–388.

Spear, F.S., 1993, Metamorphic phase equilibria and Pressure-Temperature-time paths: Washington, D.C., Mineralogical Society of America Monograph, 799 p.

Wardle, R.J., and Ash, C., 1984, Geology of the Northwest River area, *in* Current Research: St. John's, Newfoundland, Mineral Development Division, Newfoundland and Labrador Department of Mines and Energy, Report 84–1, p. 53–67.

Wardle, R.J., and Ash, C., 1986, Geology of the Goose Bay–Goose River area, *in* Current Research: St. John's, Newfoundland, Mineral Development Division, Newfoundland and Labrador Department of Mines and Energy, Report 86–1, p. 113–123.

Wardle, R.J., and Ryan, A.B., 1996, Field Excursion: The Grenville Province in the Goose Bay area: Mid-conference field excursion: Proterozoic Evolution in the North Atlantic Realm: Correlation of PreCambrian of Europe and North America. Eastern Canadian Seismic Onshore-Offshore Transect–International Basement Tectonics Association conference, 29 July–2 August, Goose Bay, Labrador, Field Excursion Guide No. 3, 40 p.

Wardle, R.J., Rivers, T., Gower, C.F., Nunn, G.A.G., and Thomas, A., 1986, The northeastern Grenville Province: New insights, *in* Moore, J.M., et al., eds., The Grenville Province: St. John's, Newfoundland, Geological Association of Canada, Special Paper No. 31, p. 13–29.

Wardle, R.J., Ryan, A.B., Philippe, S., and Schärer, U., 1990, Proterozoic crustal development, Goose Bay region, Grenville Province, Labrador, Canada, *in* Gower, C.F., et al., eds., Mid-Proterozoic Laurentia-Baltica: St. John's, Newfoundland, Geological Association of Canada, Special Paper No. 38, p. 197–214.

Wells, P.R.A., 1979, Chemical and thermal evolution of Archaean sialic crust, southern West Greenland: Journal of Petrology, v. 20, p. 253–265.

Manuscript Accepted by the Society August 25, 2003

Printed in the USA

Geological Society of America
Memoir 197
2004

Neoproterozoic juvenile crust development in the peri-Rodinian ocean: Implications for Grenvillian orogenesis

J. Brendan Murphy*
Department of Earth Sciences, St. Francis Xavier University, P.O. Box 5000, Antigonish, Nova Scotia B2G 2W5, Canada
Jaroslav Dostal
Department of Geology, St. Mary's University, Halifax, Nova Scotia B3H 3C3, Canada
R. Damian Nance
Department of Geological Sciences, Ohio University, 316 Clippinger Laboratories, Athens, Ohio 45701, USA
J. Duncan Keppie
Instituto de Geología, Universidad Nacional Autónoma de México, 04510 México D.F., México

ABSTRACT

The supercontinent Rodinia is thought to have been formed by 1.2 to 1.0 Ga continent-continent collisions and to have dispersed between 0.75 and 0.6 Ga. The existence of Rodinia implies the presence of a Panthalassalike peri-Rodinian ocean between ca. 1.0 and 0.75 Ga within which juvenile crust developed. Although the vast majority of this crust was later subducted, vestiges are preserved in terranes that accreted to the leading edges of the dispersing continents following the breakup of Rodinia. These terranes are recognized by their ca. 1.2 to 0.75 Ga Sm-Nd depleted mantle (T_{DM}) model ages, coeval with the life of Rodinia. They include ca. 0.9 to 0.8 Ga ophiolites and ensimatic arc complexes, and ca. 0.8 to 0.6 Ga recycled mafic to felsic arc complexes. Formed within the peri-Rodinian ocean, the terranes were accreted to their respective continental margins in the Late Neoproterozoic. For orogens in which subduction culminated in Late Neoproterozoic continental collision (e.g., the Southern Yangste margin, Brasiliano, and Trans-Saharan orogens), vestiges of peri-Rodinian crust became cratonized within the suture zones between the colliding cratons. In accretionary orogens, in which subduction was not terminated by continental collision (e.g., the Arabian Shield and peri-Gondwana orogens), the terranes were subsequently involved in Paleozoic orogenesis. More generally, crustal formation in Panthalassa-type oceans and the subsequent recycling of this crust can be recognized by Sm-Nd T_{DM} model ages that overlap the life span of the supercontinent.

Keywords: Grenville, Gondwana, Rodinia, Sm-Nd isotopes, supercontinent

*E-mail: bmurphy@stfx.ca.

Murphy, J.B., Dostal, J., Nance, R.D., and Keppie, J.D., 2004, Neoproterozoic juvenile crust development in the peri-Rodinian ocean: Implications for Grenvillian orogenesis, *in* Tollo, R.P., Corriveau, L., McLelland, J., and Bartholomew, M.J., eds., Proterozoic tectonic evolution of the Grenville orogen in North America: Boulder, Colorado, Geological Society of America Memoir 197, p. 135–144. For permission to copy, contact editing@geosociety.org.

INTRODUCTION

The origin of the Grenville belt of eastern North America and other collisional orogenic belts of similar age is attributed to the assembly of the supercontinent Rodinia between 1.2 and 1.1 Ga by a series of arc-continent, followed by continent-continent, collisions (e.g., Hoffman, 1991; Dalziel, 1992; Dalziel et al., 2000; Fig. 1). In most belts, these collisions recycled ancient crust, as demonstrated by the Sm-Nd isotopic signature of gneissic and igneous bodies in the Grenville Province, which reflect crustal sources with time-integrated enrichment in light rare earth elements (LREE) and yield depleted mantle "model ages" considerably older than that of the orogenic activity itself (Dickin and McNutt, 1989, 1990; Daly and McLelland, 1991; Dickin, 2000). Although the genetic significance of the model ages is unclear, taken together they suggest that an important component of recycled ancient crust was involved in the genesis of the igneous rocks.

One of the corollaries of the existence of a supercontinent is the coeval presence of a Panthalassalike "peri-Rodinian" ocean. By comparison with Panthalassa and the modern Pacific, such an ocean would have contained ocean ridges, plume-related hot spots, and ensimatic subduction zones, each generating juvenile crust broadly coeval with the life span of Rodinia. Although most of this crust was subsequently destroyed by subduction, vestiges have been preserved in terranes that accreted to the leading edges of Rodinia following its breakup beginning at ~0.75 Ma (Wingate and Giddins, 2000). These terranes are characterized by having U-Pb crystallization ages and calculated Sm-Nd depleted mantle model ages that are coeval with the 1.2 to 0.75 Ga life span of Rodinia. Examples occur in Neoproterozoic collisional orogenic belts in southeast China, West Africa (e.g., the Trans-Saharan orogenic belt), and Brazil (e.g., the Tocantins and Boborema Provinces). Peri-Rodinian crust is also preserved in accretionary orogens such as the Arabian Shield (Kröner et al., 1987, 1992; Blasband et al., 2000). Some terranes are prone to crustal recycling after accretion if subduction zones become reestablished along continental margins after accretion. Such recycling is exemplified by the Avalonian belt of Atlantic Canada, which is dominated by 0.63–0.57 Ga arc-related igneous activity, but has ca. 1.0 Ga Sm-Nd depleted mantle model ages consistent with initial crustal formation within the peri-Rodinian ocean followed by recycling during Neoproterozoic subduction (Murphy et al., 2000).

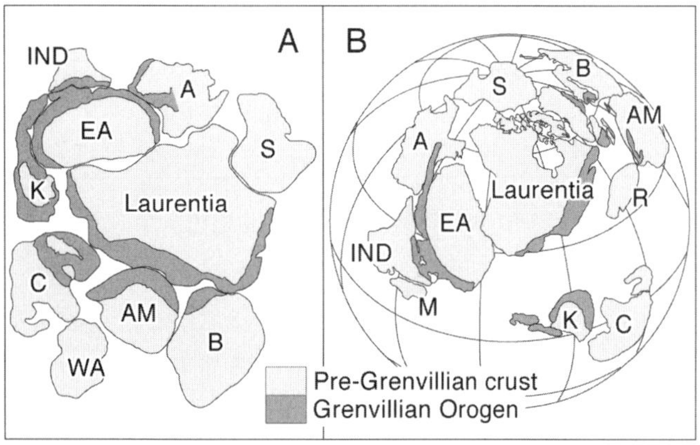

Figure 1. Two examples of the proposed configuration of Rodinia: (A) after Hoffman (1991); (B) after Dalziel et al. (2000). A—Australia; AM—Amazonia; B—Baltica; C—Congo; EA—East Antarctica; IND—India; K—Kalahari; M—Madagascar; R—Rio de la Plata; S—Siberia; WA—West Africa.

Sm-Nd ISOTOPIC SYSTEMATICS

Samarium and neodymium are light rare earth parent-daughter elements that behave coherently in the crust. As a result, the Sm/Nd ratio is rarely affected by intracrustal processes such as anatexis, fractionation, or regional metamorphism. Instead, variations in Sm/Nd in crustal rocks are largely inherited from the depleted mantle, which preferentially retains samarium over neodymium (DePaolo and Wasserburg, 1976; DePaolo, 1981, 1988).

Since the Sm/Nd ratio in typical crustal rocks is ~40% lower than that of the depleted mantle, the decay of ^{147}Sm to ^{143}Nd over geologic time produces $^{143}Nd/^{144}Nd$ ratios that are significantly higher in the depleted mantle than in the bulk earth, represented by the "chondrite uniform reservoir" (CHUR), the ratio of which is, in turn, higher than that in the crust. Upon melting, magmas acquire the $^{143}Nd/^{144}Nd$ ratio of their source, which is known as its initial ratio $(^{143}Nd/^{144}Nd)_0$.

The contrasting values of $(^{143}Nd/^{144}Nd)_0$ can be used to distinguish igneous rocks derived from juvenile crust (such as midoceanic ridges, plumes, oceanic arcs, and back-arcs) from those derived from the recycling of ancient crust provided that (1) the isotopic evolution of the depleted mantle is known, (2) the sample's Sm/Nd ratio has not been modified by coeval or subsequent intracrustal processes (e.g., assimilation of host rock, fractionation of rare earth element (REE)–bearing phases, hydrothermal activity), (3) a crustal Sm/Nd ratio was acquired during or very soon after the sample was emplaced in the crust, and (4) all the material in the sample was derived from a single event in the mantle (see Arndt and Goldstein, 1987). Since these conditions are generally met (Murphy and Nance, 2003), the Sm-Nd systematics of Neoproterozoic accreted terranes provides a powerful tool with which to examine juvenile crustal development associated with the amalgamation of Rodinia.

PERI-RODINIAN CRUST IN NEOPROTEROZOIC OROGENS

Although most of the peri-Rodinian ocean was consumed by subsequent subduction, vestiges of this oceanic crust are preserved as terranes that accreted to continental margins during Neoproterozoic orogenesis (Fig. 2) and as recycled products of this crust produced by subduction beneath the previously

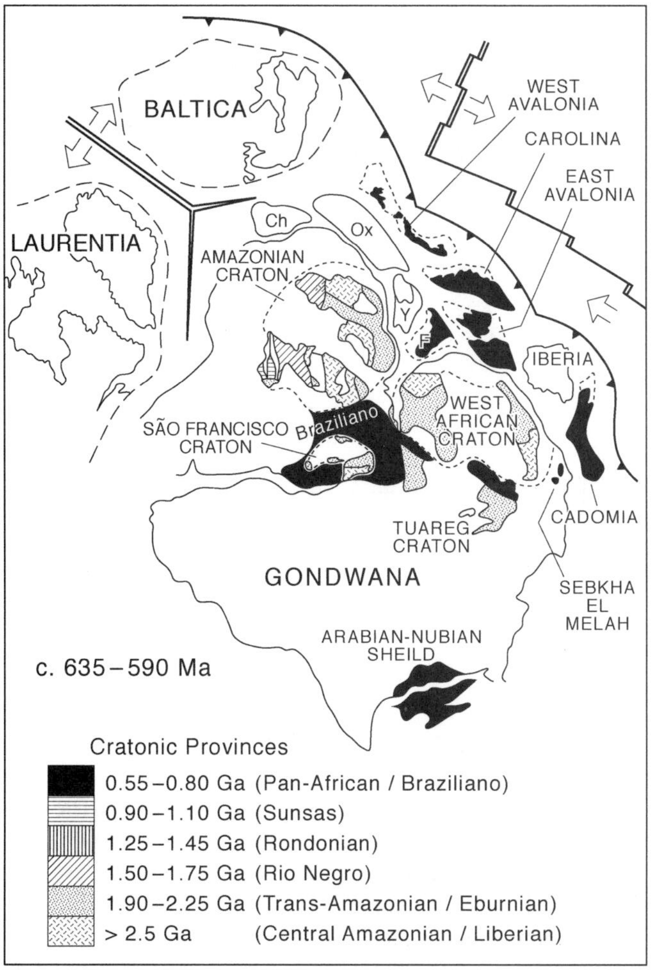

Figure 2. Late Neoproterozoic reconstruction (modified from Nance et al., 2002) relative to the continental reconstruction of Dalziel (1997) showing Late Neoproterozoic orogens and adjacent cratonic provinces, except for Baltica (from Hartz and Torsvik, 2002). Collisional belts shown include the Brasiliano fold belts (which contain the Tocantins and Borborema Provinces), and the Trans-Saharan belt of West Africa. Accretionary belts include the Arabian-Nubian shield. Peri-Gondwanan terranes occur along the northern periphery of Gondwana. Ch—Chortis block; Ox—Oaxaquia; Y—Yucatan block; F—Florida.

accreted terranes. There are two predominant end-member styles of Neoproterozoic orogenic belts (e.g., Murphy and Nance, 1991): those that culminate in continental collision, such as in northeastern Brazil and West Africa, and those that represent ongoing subduction and terrane accretion that terminated without continental collision, such as in Avalonia and the Arabian Shield.

Brazil

Neoproterozoic fold belts in Brazil are exemplified by the Borborema Province in northeastern Brazil and the Tocantins Province in central Brazil (Brito Neves, 1983; Tiexiera et al., 1989). On pre-Mesozoic continental reconstructions, the Borborema Province lies between the West African–Sao Luis and Sao Francisco–Congo-Kasai cratons (e.g., de Wit et al., 1988). The province was highly deformed and metamorphosed by collisions between these cratons during the 500 to 600 Ma Pan-African Brasiliano orogeny. However, several fold belts (including the Sergipano and Oros belts) predate the Brasiliano orogeny. Archean and Paleoproterozoic basement gneisses occur in the southern part of the province and are thought to represent tectonic slivers of the Sao Francisco craton. Pre-Brasiliano supracrustal sequences typically have ages of 2.0–1.9 Ga, 1.8–1.7 Ga, 1.1–1.0 Ga, and ~0.6 Ga. Metavolcanic and granitoid units with crystallization ages of ca. 1.0 Ga have depleted mantle (T_{DM}) ages that range from 1.17 to 1.77 Ga (van Schmus et al., 1995). These Sm-Nd isotopic characteristics are interpreted to reflect mixing between ancient crust (>2.1 Ga) and a juvenile (ca. 1.0 Ga) component (Fig. 3, B) that is thought to reflect a major crust-forming event (van Schmus et al., 1995). The Tocantins Province lies between the Amazon, the Sao Francisco, and possibly the Parana cratons and similarly consists of a number of fold belts (Tiexiera et al., 1989; Pimental et al., 1997). The Paraguai belt along the Amazonian margin predominantly comprises Neoproterozoic and Eocambrian platformal metasedimentary rocks. Along the western flank of the Sao Francisco craton, the Uruaça and Brasilia belts expose ca. 800 Ma polydeformed supracrustals that contain terranes with ophiolitic affinities (Pimental et al., 1991). The Goias massif in the central portion of the province is characterized by metavolcanic and metasedimentary assemblages that are juxtaposed against orthogneisses along steeply dipping mylonite zones (Pimental and Fuck, 1987). The massif is dominated by ca. 950–850 Ma mafic igneous protoliths that evolved within a primitive ensimatic arc setting, and younger (ca. 760–600 Ma) arc-related rocks of more felsic composition (Pimental et al., 1991, 1997; Pimental and Fuck, 1992). Sm-Nd isotopic constraints for this province suggest derivation from juvenile sources. Initial ε_{Nd} values (calculated for the age of crystallization) range from +0.2 to +6.9, while T_{DM} model ages lie between 0.9 and 1.2 Ga (Pimental and Fuck, 1992; Fig. 3, A).

West Africa

The Trans-Saharan orogenic belt (Cahen et al., 1984) occurs along the eastern margin of the West African craton and is thought by most authors to record the collision between the passive margin of this craton and the Tuareg Shield (Caby et al., 1989; Black et al., 1994; Dostal et al., 1994). The Tuareg Shield consists of a number of displaced arc-related terranes with highly variable tectonostratigraphic, metamorphic, and magmatic characteristics. Separated by subvertical shear zones, the terranes were accreted to the East Saharan craton at various times between 750 and 550 Ma (Black et al., 1994). Thus, the Trans-Saharan belt sequentially underwent arc-arc, arc-continent, and continent-continent collisional orogenesis in the Neoproterozoic (Caby et al., 1989; Black et al., 1994).

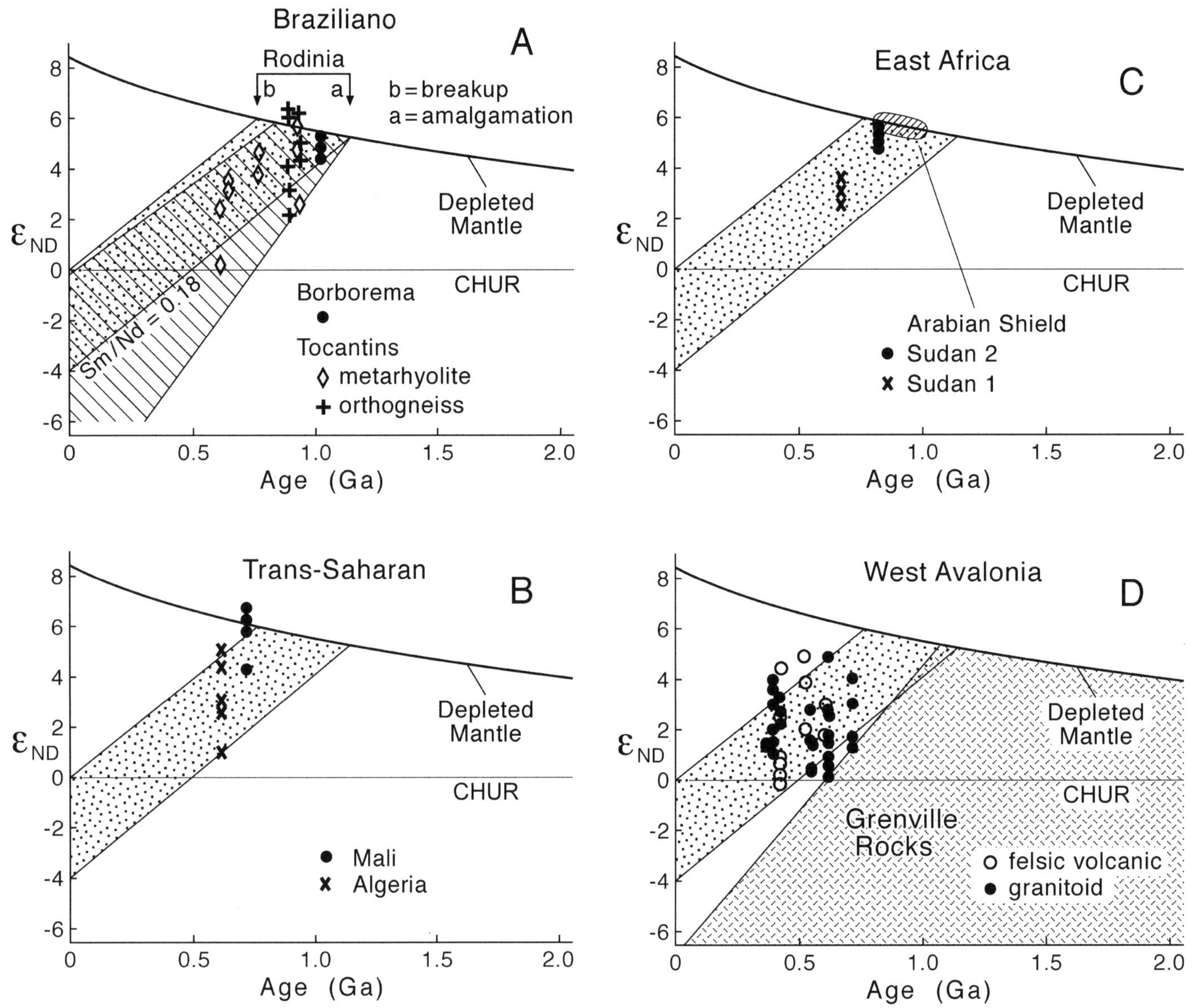

Figure 3. Summary diagrams $(\varepsilon_{Nd})_t$ versus time (Ga) showing examples of inferred peri-Rodinian crust preserved in Neoproterozoic orogens: (A) the Borborema and Tocantins Provinces of the Brasiliano orogenic belt, Brazil (Pimental and Fuck, 1992; van Schmus et al., 1995); (B) the Trans-Saharan belt of West Africa (Mali from Caby et al., 1989; Algeria from Dostal et al., 2002); (C) East African orogens such as the Arabian Shield (Stern et al., 1991; Kröner et al., 1992; Kuster and Liégeois, 2001; Stern 2002); (D) West Avalonia (Murphy et al., 1996, 2000; Murphy and Nance 2002). In each diagram, the evolution of ε_{Nd} with time for peri-Rodinian oceanic lithosphere is shown in stipples, and is defined by assuming a depleted mantle composition for the oceanic lithosphere formed between the time of the amalgamation (a) and the breakup (b) of Rodinia, at ca. 1.0 Ga and 0.75 Ga, respectively. The evolution assumes a typical Sm/Nd crustal ratio of 0.18. In A, the stippled region is the Sm-Nd isotopic envelope for the Brasiliano orogens. In D, the field for Grenvillian rocks (shaded) is compiled from Dickin and McNutt (1989), Patchett and Ruiz (1989), Marcantonio et al. (1990), Daly and McLelland (1991), and Dickin (2000). CHUR—chondrite uniform reservoir.

In contrast to the West African craton, the Tuareg Shield contains Neoproterozoic igneous rocks, indicating that convergence between these cratons was accompanied by easterly directed subduction (Caby, 1994). In Mali, a ca. 720 Ma complex of intraoceanic arcs known as the Tilemsi belt formed above such an east-dipping subduction zone and was accreted to the continental margin during Pan-African collisional orogenesis. The arcs are characterized by mafic to intermediate volcanic and plutonic rocks with calc-alkalic and island arc tholeiitic affinities. The volcanics are interbedded with metagreywackes (Dostal et al., 1994). The Sm-Nd isotopic data available indicate that the plutonic rocks and the metagreywackes have high ε_{Nd} values (+6.3 to +6.6 and +4.4 to +5.8 at t = 730 Ma), with model ages of 710 to 760 Ma and 840 to 940 Ma, respectively. These data indicate that the plutonic rocks were derived from the depleted mantle, whereas the metagreywackes, although predominantly derived from juvenile crust, had a limited contribution from sialic crust (Caby et al., 1989). To the north, in southwestern Algeria and southern Morocco, inliers of Trans-Saharan mafic rocks include shoshonites with arc-related geochemical features and high ε_{Nd} values (+1.0 to +5.0) that yield model ages of 0.95 to 1.20 Ga (Dostal et al., 2002; Fig. 3, B).

Arabian-Nubian Shield

Stretching for 5000 km along the East African margin, a succession of Neoproterozoic orogenic belts, ranging from ca. 950 to 550 Ma in age, form a classic example of the varying styles of Pan-African orogenesis (Kröner, 1984). To the north, the Arabian-Nubian Shield consists of a collage of ca. 850–650 Ma arcs and Paleoproterozoic–Late Archean continental microplates that accreted at various times during the Pan-African orogeny (Kröner, 1985; Blasband et al., 2000). Although models vary in detail, accretion appears to have occurred in a number of phases, including closure of fore-arc (Greenwood et al., 1980) or back-arc (Kröner et al., 1987) basins followed by the docking of younger arcs to the shield's eastern margin (present coordinates). Arc terranes are separated by ophiolitic complexes that represent sutures between the oceanic microplates (Kröner, 1985; Pallister et al., 1988; Kröner et al., 1992).

To the south, the Mozambique belt is a zone of highly deformed and metamorphosed rocks. These rocks are either vestiges of juvenile crust, or reworked Archean crust that could form the basement to the Arabian-Nubian Shield (Burke and Şengör, 1986; Shackleton, 1986). However, the presence of highly dismembered ophiolites in suture zones that can be traced from the Arabian-Nubian Shield (e.g., Behre, 1990) suggests that the Mozambique belt was generated by collisional tectonics (Stern and Dawoud, 1991).

A transition zone between the Arabian-Nubian Shield and the Mozambique belt occurs in Yemen (Windley et al., 1996; Whitehouse et al., 2001), Somalia (Lenoir et al., 1994), and Sudan (Stern and Dawoud, 1991; Kuster and Liégeois, 2001). Yemen comprises microcontinental blocks consisting of Late Archean and Proterozoic gneisses separated by obducted island arc complexes. To the south, in Somalia, comparable island arc terranes consist of Neoproterozoic and Early Paleozoic granitoids, some of which are derived from juvenile crust (Lenoir et al., 1994).

Geochemical and isotopic data from voluminous Neoproterozoic, predominantly mafic volcanic and interbedded metasedimentary rocks in the Bayuda Desert of Sudan (Kuster and Liégeois, 2001) record an oceanic island arc or back-arc environment. These rocks were metamorphosed at ca. 700 Ma during an event attributed to collision with the East Saharan craton. The Bayuda sequence has two components, a juvenile ca. 800 Ma mafic and sedimentary sequence characterized by high ε_{Nd} (+3.6 to +5.2 at $t = 800$ Ma) with T_{DM} ages of 0.78 to 0.90 Ga (Fig. 3, C), and a metasedimentary sequence derived from an older continental source with T_{DM} ages of ~2.1 Ga. To the south, ca. 740 Ma granulites and charnockites are interpreted to have formed as a result of collisional orogenesis followed by exhumation (Behre, 1990). The ε_{Nd} data available for these high-grade rocks (Stern and Dawoud, 1991; Stern 2002) range from +2.9 to +3.4 ($t = 740$ Ma) and yield T_{DM} ages of 0.96 to 0.98 Ga, suggesting derivation from a relatively juvenile mantle source.

Avalonia and Related Terranes

Avalonia is one of a group of terranes collectively referred to as peri-Gondwanan terranes. On the basis of faunal, lithostratigraphic, geochemical, and paleomagnetic data, these terranes are traditionally interpreted to represent segments of the northern (Amazonian and West African) Gondwanan margin in the Neoproterozoic and Early Paleozoic (Theokritoff, 1979; Keppie, 1985; van der Voo, 1988; Murphy and Nance, 1989; Cocks and Fortey, 1990; McNamara et al., 2001; Murphy et al., 2002; Mac Niocaill et al., 2002). Each of these terranes is characterized by Neoproterozoic magmatism that records a history of subduction beneath the Gondwanan continental margin (e.g., O'Brien et al., 1983; Keppie, 1985; Murphy et al., 1990; Nance et al., 1991; Egal et al., 1996; Linnemann et al., 2000; von Raumer et al., 2002).

The terranes were rifted from Gondwana in the Early Paleozoic and subsequently involved in Paleozoic and younger orogenesis so that they are preserved as a collection of suspect terranes in younger orogenic belts. They are now distributed (1) along the southeastern margin of the Appalachian belt, including the Florida subsurface (Heatherington et al., 1996), the Carolinas (Shervais et al., 1996; Dennis and Wright, 1997; Hibbard, 2000; Hibbard et al., 2002), and "West Avalonia," which stretches from New England to southeastern Newfoundland (O'Brien et al., 1983; Murphy and Nance, 1989, 1991; Keppie et al., 1991), and (2) along the southern margin of the Caledonide-Hercynian belt in "East Avalonia," which includes southeastern Ireland (Max and Roddick, 1989) and southern Britain (Tucker and Pharoah, 1991; Gibbons and Horak, 1996), the Armorican massif (Cadomia) of northwestern France (Egal et al., 1996; Strachan et al., 1996a), the Iberian peninsula (Quesada, 1990; Eguíluz et al., 2000; Fernández-Suárez et al., 2000), isolated inliers in Germany and the Czech Republic (e.g., Bohemian massif, Linnemann et al., 2000; Zulauf et al., 1999), and recently recognized vestiges in the Alpine belt (Neubauer, 2002; von Raumer et al., 2002).

Neoproterozoic tectonothermal activity begins with an early phase of ca. 760–660 Ma arc-related magmatism and is followed by ca. 660–650 Ma medium- to high-grade metamorphism in West Avalonia (e.g., Krogh et al., 1988; Swinden and Hunt, 1991; Bevier et al., 1993; Doig et al., 1993; O'Brien et al., 1996), East Avalonia (Tucker and Pharoah, 1991; Gibbons and Horak, 1996; Strachan et al., 1996b), and Cadomia (Egal et al., 1996). The early arc-related complexes of Avalonia probably formed outboard of the Gondwanan margin, while the metamorphic event is interpreted to record their accretion to this margin (Murphy et al., 2000; Nance et al., 2002).

At ca. 635 Ma, the main phase of arc-related activity commenced broadly synchronously along the Gondwanan margin. However, the termination of arc magmatism was diachronous, varying from 590 to 540 Ma. Arc-related rocks typically include voluminous calc alkaline mafic to felsic volcanics, coeval plutons, and pull-apart sedimentary basin deposits that contain detritus derived from the arc. With the diachronous termination of subduction came a transition to an intracontinental wrench regime that is interpreted to record ridge-trench collision in a manner analogous to the Oligocene collision between western North America and the East Pacific Rise and the diachronous initiation of the San Andreas transform margin (Murphy et al., 1999; Nance et al., 2002).

Sm-Nd isotopic data from crustally derived rocks in both the early arc and the main arc phases of West and East Avalonia yield similar ranges in initial ε_{Nd} values (–2.5 and +5.0), and T_{DM} model ages of 0.8 to 1.2 Ga in Atlantic Canada and 1.0 to 1.3 Ga in the British Midlands (Murphy et al., 1996, 2000; Fig. 3, D). Initial ε_{Nd} values of +0.5 to +5.9 and T_{DM} model ages of 0.7 to 1.1 Ga have also been reported from the main phase Virgilina sequence of Carolina (Samson et al., 1995; Wortman et al., 2000).

There is no record of tectonothermal activity in Avalonia or Carolina between the oldest T_{DM} ages of ca. 1.2 Ga and ca. 760 Ma, suggesting that the isotopic signature of Avalonia and Carolina represents a relatively simple evolution involving the recycling of ca. 1.0 to 1.2 Ga crust. If so, the depleted mantle model ages represent a genuine tectonothermal event that produced juvenile crust that was recycled by subsequent Neoproterozoic tectonothermal activity (Murphy et al., 2000). These data are therefore interpreted to reflect the evolution of the primitive Avalonian and Carolinian crust (proto-Avalonia-Carolina) at 1.2–1.0 Ga in the peri-Rodinian ocean, followed by the development of mature oceanic arcs at 750–650 Ma (early Avalonian magmatism). Their accretion to Gondwana at ca. 650 Ma was then followed by Andean-style arc development at 635–570 Ma (main Avalonian-Carolinian magmatism), which was predominantly generated by recycling preexisting ca. 1.2–1.0 Ga crust that was enriched in LREE.

Sm-Nd isotopic data for Cadomia, however, suggest that the relationship of this terrane to Gondwana was different. ε_{Nd} values from main phase arc rocks in northern France and the Channel Islands are predominantly negative, ranging from +1.9 to –9.9, with T_{DM} model ages ranging from 1.0 to 1.9 Ga. These data are interpreted to reflect the mixing of material derived from the mantle at ca. 600 Ma with continental basement represented by the 2.1 Ga Icartian Gneiss (D'Lemos and Brown, 1993; Nance and Murphy, 1996). In contrast to Avalonia and Carolina, there is no evidence of the significant influence of a juvenile crustal component in Cadomia. Instead, Cadomia appears to have originated as an ensialic arc built upon crust like that of the West African craton.

In addition to these terranes, several crustal blocks in Middle America contain assemblages with Early Paleozoic fauna that suggest an origin along the northern Gondwanan margin (Keppie and Ramos, 1999). However, these terranes do not preserve evidence of Neoproterozoic arc activity, suggesting that they were located inboard of the magmatic arc. These terranes expose basement of Pan-African (Yucatan block) and Grenville (Oaxaquia and Chortis block; Keppie and Ortega-Gutiérrez, 1995, 1999) age. The Yucatan block is thought to have been contiguous with the Florida basement until the opening of the Gulf of Mexico in the Mesozoic. The Grenville-age basement of Oaxaquia and the Chortis block is isotopically distinct from that of the Grenville belt in Laurentia, and has been correlated with basement massifs of Grenville age in Columbia (Ruiz et al., 1999).

SUMMARY AND DISCUSSION

Just as the amalgamation and dispersal of Pangea exerted a first-order influence on tectonothermal events in the Phanerozoic, Middle to Late Proterozoic global-scale tectonics were profoundly influenced by the amalgamation of the supercontinent Rodinia at ca. 1.2–1.0 Ga and its subsequent breakup beginning at 0.75 Ga. Most tectonic studies on the evolution of Rodinia concentrate on the collisional orogenies that resulted in its amalgamation, and the related rift-to-drift lithologies formed during its breakup. Since magmatism associated with Grenvillian collisional orogenesis is predominantly derived from recycled crust, T_{DM} model ages are generally significantly older than the orogenic event itself (e.g., Daly and McLelland, 1991; Dickin, 2000). However, during the life span of Rodinia, ocean ridge processes, ensimatic subduction, oceanic plateau generation, and plume activity resulted in the formation in the peri-Rodinian ocean of juvenile oceanic crust with ca. 1.2–0.75 Ga Sm-Nd depleted mantle (T_{DM}) model ages. Vestiges of this crust are preserved in Late Neoproterozoic orogenic belts as terranes that accreted to the leading margins of continents following the breakup and dispersal of Rodinia. Taken together, the formation and Neoproterozoic evolution of peri-Rodinian crust is geodynamically linked to the Rodinian cycle and so provides constraints on its configuration and the timing of its major tectonothermal events (e.g., Murphy et al., 2000).

A schematic paleocontinental reconstruction of Rodinia at ca. 0.8 Ga showing the inferred location of 1.0–0.8 Ga juvenile crust generated in the peri-Rodinian ocean is shown in Figure 4. For collisional orogens, vestiges of juvenile crust are shown between the converging cratons. For accretionary orogens, the juvenile crust is shown adjacent to the craton to which it accreted.

Vestiges of the peri-Rodinian oceanic crust can be identified by (1) crystallization ages and T_{DM} model ages that coincide with Rodinia's existence and (2) strongly positive (i.e., mantle-derived) ε_{Nd} isotopic signatures. Examples of such crust are preserved in Neoproterozoic collisional (interior) orogens such as in the Borborema and Tocantins Provinces of Brazil and in West Africa, in which arc accretion generally preceded ter-

Figure 4. Schematic paleocontinental reconstruction of Rodinia at ca. 0.8 Ga showing the inferred location of 1.0–0.8 Ga juvenile continental crust generated in the peri-Rodinian ocean. The diagram is modified from Moores (2002). The Sao Francisco–Congo–West Africa-Amazonia–Rio de la Plata relationship is modified from Alkmim et al. (2001) and the Laurentia-Baltica relationship from Hartz and Torsvik (2002). The location of India and the Rayner craton (too small to be shown) is assumed to be near the North Pole (Fitzsimons, 2000; Powell and Pisarevsky, 2002). The Mawson craton (Ma) is from Fitzsimons (2000). The locations of the belts of juvenile crust are not known with any certainty, but they are positioned outboard from the respective cratons to which they had become accreted by the end of the Neoproterozoic.

minal collision. This crust would have been stabilized and cratonized by terminal collision. Examples of peri-Rodinian crust are also preserved in accretionary (or peripheral) orogens such as in the Arabian Shield.

During the Late Neoproterozoic, portions of the northern Gondwanan margin evolved from western to eastern Pacific-style subduction (e.g., Avalonia, Carolina). Although this crust was recycled by subsequent orogenic activity, its existence can be identified from the ε_{Nd} signatures in younger crustally derived igneous rocks that indicate derivation predominantly from a crustal source that was itself extracted from the mantle while Rodinia existed as a coherent supercontinent.

ACKNOWLEDGMENTS

The constructive reviews of Alan Dickin and X.L. Li resulted in significant improvements in the manuscript. We are grateful to Louise Corriveau for editorial comments and to Sergei Pisarevksy for discussions. JBM and JD acknowledge the continuing support of the Natural Sciences and Engineering Research Council (Canada), RDN acknowledges the support of the National Science Foundation (EAR-0308105), and JDK acknowledges the support of the Programa de Apoyo a Proyectos de Investigación e Innovación Tecnológica (grant IN103003). Additional funding was derived from the University Council for Research at St. Francis Xavier University and a Gledden Senior Research Fellowship at the University of Western Australia (awarded to JBM). Contribution to International Geological Correlation Programme Project 453.

REFERENCES CITED

Alkmim, F.F., Marshak, S., and Fonseca, M.A., 2001, Assembling West Gondwana in the Neoproterozoic: Clues from the Sao Francisco region, Brazil: Geology, v. 29, p. 319–322.

Arndt, N.T., and Goldstein, S.L., 1987, Use and abuse of crust formation ages: Geology v. 15, 893–895.

Behre, S.M., 1990, Ophiolites in northeast and East Africa: Implications for Proterozoic crustal growth: Journal of the Geological Society of London, v. 147, p. 41–57.

Bevier, M.L., Barr, S.M., White, C.E., and Macdonald, A.S., 1993, U-Pb geochronologic constraints of the volcanic evolution of the Mira (Avalon) terrane, southeastern Cape Breton Island, Nova Scotia: Canadian Journal of Earth Sciences, v. 30, p. 1–10.

Black, R., Latouche, L., Liégeois, J.P., Caby, R., and Bertrand, J.M., 1994, Pan-African displaced terranes of the Tuareg shield (central Sahara): Geology, v. 22, p. 641–644.

Blasband, B., White, S., Brooijmans, P., De Boorder, H., and Visser, W., 2000, Late Proterozoic extensional collapse in the Arabian-Nubian Shield: Journal of the Geological Society of London, v. 157, p. 615–628.

Brito Neves, B.B., 1983, O mapa geológico de Nordeste oriental do Brasil, escala 1/1000000 [livre docencia thesis]: Sao Paulo, Brazil, Instituto de Geosciencias, Universidad de Sao Paulo, 177 p.

Burke, K., and Şengör, A.M.C., 1986, Tectonic escape in the evolution of the continental crust: American Geophysical Union Geodynamic Series, v. 14, p. 41–53.

Caby, R., 1994, First record of Precambrian coesite from northern Mali: Implications for Late Proterozoic plate tectonics around the West African craton: European Journal of Mineralogy, v. 6, p. 235–244.

Caby, R., Andreopoulos-Renaud, U., and Pin, C., 1989, Late Proterozoic arc-continent and continent-continent collision in the trans-Saharan belt: Canadian Journal of Earth Sciences, v. 26, p. 1136–1146.

Cahen, L., Snelling, N.J., Delhal, J., and Vail, J.R., 1984, The geochronology and evolution of Africa: Oxford, England, Clarendon Press, 512 p.

Cocks, L.R.M., and Fortey, R.A., 1990, Biogeography of Ordovician and Silurian faunas, *in* McKerrow, W.S., and Scotese, C.R., eds., Paleozoic paleogeography and biogeography: Geological Society [London] Memoir 12, p. 97–104.

Daly, J.S., and McLelland, J.M., 1991, Juvenile Middle Proterozoic crust in the Adirondack Highlands, Grenville Province, northeastern North America: Geology, v. 19, p. 119–122.

Dalziel, I., 1992, On the organization of American plates in the Neoproterozoic and the breakout of Laurentia: GSA Today, v. 2, no. 11, p. 238–241.

Dalziel, I.W.D., 1997, Overview: Neoproterozoic-Paleozoic geography and tectonics: Review, hypotheses and environmental speculations: Boulder, Colorado, Geological Society of America Bulletin, v. 109, p. 16–42.

Dalziel, I.W.D., Mosher, S., and Gahagan, L.M., 2000, Laurentia-Kalahari collision and the assembly of Rodinia: Journal of Geology, v. 108, p. 499–513.

Dennis, A., and Wright, J.E., 1997, The Carolina terrane in northwestern South Carolina: Age of deformation and metamorphism in an exotic arc: Tectonics, v. 16, p. 460–473.

DePaolo, D.J., 1981, Neodymium isotopes in the Colorado Front Range and crust-mantle evolution in the Proterozoic: Nature, v. 29, p. 193–196.

DePaolo, D.J., 1988, Neodymium isotope geochemistry: An introduction: New York, Springer-Verlag, 187 p.

DePaolo, D.J., and Wasserburg, G.J., 1976, Nd isotopic variations and petrogenetic models: Geophysical Research Letters, v. 3, p. 249–252.

de Wit, M., Jeffrey, M., Bergh, H., and Nicolaysen, L., 1988, Geological map of sectors of Gondwana reconstructed to their disposition ca. 150 Ma: Tulsa, Oklahoma, American Association of Petroleum Geologists, 2 sheets.

Dickin, A.P., 2000, Crustal formation in the Grenville Province: Nd isotopic evidence: Canadian Journal of Earth Sciences, v. 37, p. 165–181.

Dickin, A.P., and McNutt, R.H., 1989, Nd model age mapping of the southeast margin of the Archean foreland in the Grenville Province of Ontario: Geology, v. 17, p. 299–302.

Dickin, A.P., and McNutt, R.H., 1990, Nd model-age mapping of Grenville lithotectonic domains: Mid-Proterozoic crustal evolution in Ontario, *in* Gower, G.F., et al., eds., Mid-Proterozoic Laurentia-Baltica: St. John's, Newfoundland, Geological Association of Canada, Special Paper 38, p. 79–94.

D'Lemos, R.S., and Brown, M., 1993, Sm-Nd isotope characteristics of late Cadomian granite magmatism in northern France and the Channel Islands: Geological Magazine, v. 130, p. 797–804.

Doig, R., Murphy, J.B., and Nance, R.D., 1993, Tectonic significance of the Late Proterozoic Economy River Gneiss, Cobequid Highlands, Avalon composite terrane, Nova Scotia: Canadian Journal of Earth Sciences, v. 30, p. 474–479.

Dostal, J., Dupuy, C., and Caby, R., 1994, Geochemistry of the Neoproterozoic Tilemsi belt of Iforas (Mali, Sahara): A crustal section beneath an oceanic island arc: Precambrian Research, v. 65, p. 55–69.

Dostal, J., Caby, R., Keppie, J.D., and Maza, M., 2002, Neoproterozoic magmatism in Southwestern Algeria (Sebkha el Melah inlier): A northerly extension of the Trans-Saharan orogen: Journal of African Earth Sciences, v. 35, p. 213–225.

Egal, E., Guerrot, C., Le Goff, D., Thieblemont, D., and Chantraine, J., 1996, The Cadomian orogeny revisited in northern France, *in* Nance, R.D., and Thompson, M.D., eds., Avalonian and related peri-Gondwanan terranes of the circum North Atlantic: Boulder, Colorado, Geological Society of America Special Paper 304, p. 281–318.

Eguíluz, L., Gil Ibarguchi, J.I., Ábalos, B., and Apraiz, A., 2000, Superposed Hercynian and Cadomian orogenic cycles in the Ossa-Morena zone and related areas of the Iberian Massif: Geological Society of America Bulletin, v. 112, p. 1398–1413.

Fernández-Suárez, J., Gutiérrez-Alonso, G., Jenner, G.A., and Tubrett, M.N., 2000, New ideas on the Proterozoic-early Palaeozoic evolution of NW Iberia: Insights from U-Pb detrital zircon ages: Precambrian Research, v. 102, p. 185–206.

Fitzsimons, I.C.W., 2000, Grenville-age basement provinces in East Antarctica: Evidence for three separate collisional orogens: Geology, v. 28, p. 879–882.

Gibbons, W., and Horak, J., 1996, The evolution of the Neoproterozoic Avalonian subduction system: Evidence from the British Isles, *in* Nance, R.D., and Thompson, M.D., eds., Avalonian and related peri-Gondwanan terranes of the circum North Atlantic: Boulder, Colorado, Geological Society of America Special Paper 304, p. 269–280.

Greenwood, W.R., Hadley, D.G., Anderson, R.E., Fleck, R.J., and Roberts, R.J., 1980, Precambrian geologic history and plate tectonic evolution of the Arabian shield: Saudi Arabian Directorate General of Mineral Resources Bulletin, v. 24, 35 p.

Hartz, E.H., and Torsvik, T.H., 2002, Baltica upside down: A new plate tectonic model for Rodinia and the Iapetus Ocean: Geology, v. 30, p. 255–258.

Heatherington, A.L., Mueller, P.A., and Nutman, A.P., 1996, Neoproterozoic magmatism in the Suwannee terrane: Implications for terrane correlation, *in* Nance, R.D., and Thompson, M.D., eds., Avalonian and related peri-Gondwanan terranes of the circum–North Atlantic: Boulder, Colorado, Geological Society of America Special Paper 304, p. 257–68.

Hibbard, J., 2000, Docking Carolina: Mid-Paleozoic accretion in the southern Appalachians: Geology, v. 28, p. 127–130.

Hibbard, J.P., Stoddard, E.F., Secor, D.T., and Dennis, A.J., 2002, The Carolina Zone: Overview of Neoproterozoic to Early Paleozoic peri-Gondwana terranes along the eastern flank of the southern Appalachians: Earth Sciences Reviews, v. 57, p. 299–339.

Hoffman, P.F., 1991, Did the breakout of Laurentia turn Gondwana inside out? Science, v. 252, p. 1409–1412.

Keppie, J.D., 1985, The Appalachian Collage, *in* Gee, D.G., and Sturt, B.A., eds., The Caledonide orogen: Scandinavia and related areas: New York, Wiley, p. 1217–1226.

Keppie, J.D., and Ortega-Gutiérrez, F., 1995, Provenance of Mexican terranes: Isotopic constraints: International Geology Review, v. 37, p. 813–824.

Keppie, J.D., and Ortega-Gutiérrez, F., 1999, Middle American Precambrian basement: A missing piece of the reconstructed 1 Ga orogen, *in* Ramos, V.S., and Keppie, J.D., eds., Laurentia-Gondwana connections before Pangea: Boulder, Colorado, Geological Society of America Special Paper 336, p. 199–210.

Keppie, J.D., and Ramos, V.S., 1999, Odyssey of terranes in the Iapetus and Rheic Oceans during the Paleozoic, *in* Ramos, V.S., and Keppie, J.D., eds., Laurentia-Gondwana connections before Pangea: Boulder, Colorado, Geological Society of America Special Paper 336, p. 267–276.

Keppie, J.D., Nance, R.D., Murphy, J.B., and Dostal, J., 1991, The Avalon terrane, *in* Dallmeyer, R.D., and Lecorché, J.P., eds., The western African orogens and circum Atlantic correlatives: New York, Springer-Verlag, p. 315–333.

Krogh, T.E., Strong, D.F., O'Brien, S.J., and Papezik, V.S., 1988, Precise U-Pb zircon dates from the Avalon Terrane in Newfoundland: Canadian Journal of Earth Sciences, v. 25, p. 442–453.

Kröner, A., 1984, Late Precambrian tectonics and orogeny: A need to re-define the term Pan-African, in Klerkx, J., and Michot, J., eds., African Geology: Tervuren, Belgium, Musée Royal de l'Afrique Centrale, p. 23–28.

Kröner, A., 1985, Ophiolites and the evolution of tectonic boundaries in the Late Proterozoic Arabian-Nubian Shield of northeast Africa and Arabia: Precambrian Research, v. 27, p. 277–300.

Kröner, A., Greiling, R., Reischman, T., Hussein, I.M., Stern, R.J., Durr, S., Kruger, J., and Zimmer, M., 1987, Pan-African crustal evolution in the Nubian segment of northeast Africa, *in* Kröner, A., ed., Proterozoic lithospheric evolution: American Geophysical Union Geodynamic Series, v. 15, p. 235–257.

Kröner, A., Pallister, J.S., and Fleck, R.J., 1992, Age of initial oceanic magmatism in the Late Proterozoic Arabian Shield: Geology, v. 20, p. 803–806.

Kuster, D., and Liégeois, J.-P., 2001, Sr, Nd isotopes and geochemistry of the Bayuda Desert high grade metamorphic basement (Sudan): An early Pan-African oceanic convergent margin, not the edge of the East Saharan ghost craton: Precambrian Research, v. 109, p. 1–23.

Lenoir, J.-L., Kuster, D., Liégeois, J.-P., Utke, A., Haider, A, and Matheis, G., 1994, Origin and regional significance of late Precambrian and early Paleozoic granitoids in the Pan-African belt of Somalia: Geologische Rundschau, v. 83, p. 624–641.

Linnemann, U., Gehmlich, M., Tichomirowa, M., Buschmann, B., Nasdala, L., Jonas, P., Lützner, H., and Bombach, K., 2000, From Cadomian subduction to Early Palaeozoic rifting: The evolution of Saxo-Thuringia at the margin of Gondwana in the light of single zircon geochronology and basin development (Central European Variscides, Germany), *in* Franke, W., et al., eds., Orogenic processes: Quantification and modelling in the Variscan Belt: Geological Society of London Special Publication 179, p. 131–153.

Mac Niocaill, C., van der Pluijm, B.A., and Van der Voo, R., and McNamara, A.K., 2002, West African proximity of the Avalon terrane in the latest Precambrian: Reply: Geological Society of America Bulletin, v. 114, p. 1051–1052.

Marcantonio, F., Dickin, A.P., McNutt, R.H., and Heaman, L.M., 1990, Isotopic evidence for the crustal evolution of the Frontenac Arch in the Grenville Province of Ontario, Canada: Chemical Geology, v. 83, p. 297–314.

Max, M.D., and Roddick, J.C., 1989, Age of metamorphism in the Rosslare Complex, SE Ireland: Proceedings of the Geological Association, v. 100, p. 113–121.

McNamara, A.K., Mac Niocaill, C., van der Pluijm, B.A., and Van der Voo, R., 2001, West African proximity of the Avalon terrane in the latest Precambrian: Geological Society of America Bulletin, v. 113, p. 1161–1170.

Moores, E.M., 2002, Pre-1 Ga (pre-Rodinian ophiolites): Their tectonic and environmental implications: Geological Society of America Bulletin, v. 114, p. 80–95.

Murphy, J.B., and Nance, R.D., 1989, Model for the evolution of the Avalonian-Cadomian belt: Geology, v. 17, p. 735–738.

Murphy, J.B., and Nance, R.D., 1991, Supercontinent model for the contrasting character of Late Proterozoic orogenic belts: Geology, v. 19, p. 469–472.

Murphy, J.B., and Nance, R.D., 2002. Nd-Sm isotopic systematics as tectonic tracers: An example from West Avalonia, Canadian Appalachians: Earth Science Reviews, v. 59, p. 77–100.

Murphy, J.B., and Nance, R.D., 2003, Do supercontinents introvert or extrovert? Sm-Nd isotope evidence: Geologh, v. 31, p. 873–876.

Murphy, J.B., Keppie, J.D., Dostal, J., and Hynes, A.J., 1990, Late Precambrian Georgeville group: A volcanic arc rift succession in the Avalon Terrane of Nova Scotia: Geological Society of London Special Publication 51, p. 383–393.

Murphy, J.B., Keppie, J.D., Dostal, J., and Cousens, B.L., 1996, Repeated lower crustal melting beneath the Antigonish Highlands, Avalon Composite Terrane, Nova Scotia: Nd isotopic evidence and tectonic implications, *in* Nance, R.D., and Thompson, M.D., eds., Avalonian and related peri-Gondwanan terranes of the circum North Atlantic: Boulder, Colorado, Geological Society of America Special Paper 304, p. 109–120.

Murphy, J.B., Keppie, J.D., Dostal, J., and Nance, R.D., 1999, Neoproterozoic-early Paleozoic evolution of Avalonia and Cenozoic analogues, *in* Ramos, V., and Keppie, J.D., eds., Laurentia-Gondwana connections before Pangea: Boulder, Colorado, Geological Society of America Special Paper 336, p. 253–266.

Murphy, J.B., Strachan, R.A., Nance, R.D., Parker, K.D., and Fowler, M.B., 2000, Proto-Avalonia: A 1.2–1.0 Ga tectonothermal event and constraints for the evolution of Rodinia: Geology, v. 28, p. 1071–1074.

Murphy, J.B., Nance, R.D., and Keppie, J.D., 2002, West African proximity of the Avalon terrane in the latest Precambrian: Discussion: Geological Society of America Bulletin, v. 114, p. 1049–1050.

Nance, R.D., and Murphy, J.B., 1996, Basement isotopic signatures and Neoproterozoic paleogeography of Avalonian-Cadomian and related terranes in the circum North Atlantic, *in* Nance, R.D., and Thompson, M.D., eds., Avalonian and related peri-Gondwanan terranes of the circum North Atlantic: Boulder, Colorado, Geological Society of America Special Paper 304, p. 333–346.

Nance, R.D., Murphy, J.B., Strachan, R.A., D'Lemos, R.D., and Taylor, G.K., 1991, Tectonostratigraphic evolution of the Avalonian-Cadomian belt and the breakup of a late Precambrian Supercontinent: An interpretive review, *in* Stern, R.J., and Kroner, A., eds., Tectonic regimes and crustal evolution in the Late Proterozoic: Precambrian Research, v. 53, p. 41–78.

Nance, R.D, Murphy, J.B., and Keppie, J.D., 2002, A Cordilleran model for the evolution of Avalonia: Tectonophysics, v. 352, p. 11–31.

Neubauer, F., 2002, Evolution of late Neoproterozoic to early Paleozoic tectonic elements in Central and Southeast European Alpine mountain belts: Review and synthesis: Tectonophysics, v. 352, p. 87–103.

O'Brien, S.J., Wardle, R.J., and King, A.F., 1983, The Avalon zone: A Pan-African terrane in the Appalachian orogen of Canada: Geological Journal, v. 18, p. 195–222.

O'Brien, S.J., O'Brien, B.H., Dunning, G.R., and Tucker, R.D., 1996, Late Proterozoic Avalonian and related peri-Gondwanan rocks of the Newfoundland Appalachians, *in* Nance, R.D., and Thompson, M.D., eds., Avalonian and related peri-Gondwanan terranes of the circum North Atlantic: Boulder, Colorado, Geological Society of America Special Paper 304, p. 9–28.

Pallister, J.S., Stacey, J.S., Fischer, L.B., and Premo, W.R., 1988, Precambrian ophiolites of Arabia: Geologic settings, U-Pb geochronology, Pb isotopic characteristics and implications for continental accretion: Precambrian Research, v. 38, p. 1–54.

Patchett, P.J., and Ruiz, J., 1989, Nd isotopes and the origin of the Grenville-aged rocks in Texas: Implications for Proterozoic evolution of the United States mid-continent region: Journal of Geology, v. 97, p. 685–695.

Pimental, M.M, and Fuck, R.A., 1987, Late Proterozoic granitic magmatism in southwestern Goias, Brazil: Revisita Brasiliera Geosciocias, v. 17, p. 415–425.

Pimental, M.M, and Fuck, R.A., 1992, Neoproterozoic crustal accretion in central Brazil: Geology, v. 20, p. 375–379.

Pimental, M.M., Heaman, L.M., Fuck, R.A., and Marini, O.J., 1991, U-Pb zircon geochronology of Precambrian tin-bearing continental-type acid magmatism in central Brazil: Precambrian Research, v. 52, p. 321–335.

Pimental, M.M., Whitehouse, M.J., Viana, M. das G., Fuck, R.A., and Machado, N., 1997, The Mara Rosa arc in the Tocantins province: Further evidence for Neoproterozoic crustal accretion in central Brazil: Precambrian Research, v. 81, p. 299–310.

Powell, C. McA., and Pisarevsky, S.A., 2002, Late Neoproterozoic assembly of East Gondwana: Geology, v. 30, p. 3–6.

Quesada, C., 1990, Precambrian terranes in the Iberian Variscan foldbelt, *in* Strachan, R.A., and Taylor, G.K., eds., Avalonian and Cadomian geology of the North Atlantic: London, Blackie, p. 109–133.

Ruiz, J., Tosdal, R.M., Restrepo, P.A., and Murillo-Muñetón, G., 1999, Pb isotope evidence for Columbia–southern Mexico connections in the Proterozoic, *in* Ramos, V., and Keppie, J.D., eds., Laurentia-Gondwana connections before Pangea: Boulder, Colorado, Geological Society of America Special Paper 336, p. 183–197.

Samson, S.D., Hibbard, J.P., and Wortman, G.L., 1995, Nd isotopic evidence for juvenile crust in the Carolina terrane, southern Appalachians: Contribution to Mineralogy and Petrology, v. 121, p. 171–184.

Shackleton, R.M., 1986, Precambrian collision tectonics in Africa, *in* Coward, M.C., and Ries, A.C., eds., Collision tectonics: Geological Society [London] Special Publication 19, p. 329–349.

Shervais, J.W., Shelley, S.A., and Secor, D.T., Jr., 1996, Geochemistry of volcanic rocks of the Carolina and Augusta terranes in central South Carolina: An exotic rifted arc? *in* Nance R.D., and Thompson, M.D., eds., Avalonian and related peri-Gondwanan terranes of the circum North Atlantic: Boulder, Colorado, Geological Society of America Special Paper 304, p. 219–236.

Stern, R.J., 2002, Crustal evolution in the East African Orogen: A neodymium isotopic perspective: Journal of African Earth Sciences, v. 34, p. 109–117.

Stern, R.J., and Dawoud, A.M., 1991, Late Precambrian (740 Ma) charnockite, enderbite and granite from Moya, Sudan: A link between the Mozambique belt and the Arabian-Nubian shield: Journal of Geology, v. 99, p. 648–659.

Strachan, R.A., D'Lemos, R.S., and Dallmeyer, R.D., 1996a, Late Precambrian evolution of an active plate margin: North Armorican Massif, France, *in* Nance R.D., and Thompson, M.D., eds., Avalonian and related peri-Gondwanan terranes of the circum North Atlantic: Boulder, Colorado, Geological Society of America Special Paper 304, p. 319–332.

Strachan, R.A., Nance, R.D., Dallmeyer, R.D., D'Lemos, R.S., and Murphy, J.B., 1996b, Late Precambrian tectonothermal evolution of the Malverns Complex, U.K: Geological Society [London] Journal, v. 153, p. 589–600.

Swinden, H.S., and Hunt, P.A., 1991, A U-Pb zircon age from the Connaigre Bay Group, southwestern Avalon Zone, Newfoundland: Implications for regional correlations and metallogenesis: Ottawa, Ontario, Geological Survey of Canada Paper 90-2, p. 3–10.

Theokritoff, G., 1979, Early Cambrian provincialism and biogeographic boundaries in the North Atlantic region: Lethaia, v. 12, p. 281–295.

Tiexiera, W., Tassinari, C.C.G., Cordani, U.G., and Kawashita, K., 1989, A review of the geochronology of the Amazon craton: Tectonic implications: Precambrian Research, v. 42, p. 213–227.

Tucker, R.D., and Pharoah, T.C., 1991, U-Pb zircon ages for Late Precambrian igneous rocks in southern Britain: Geological Society [London] Journal, v. 148, p. 435–443.

van der Voo, R., 1988, Paleozoic paleogeography of North America, Gondwana, and intervening displaced terranes: Comparisons of paleomagnetism with paleoclimatology and biogeographical patterns: Geological Society of America Bulletin, v. 100, p. 311–324.

van Schmus, W.R., de Brito Neves, B.B., Hackspacher, P., and Babinski, M., 1995, U/Pb and Sm/Nd geochronologic studies of the eastern Borborema Province, Northeastern Brazil: Initial conclusions: Journal of South American Sciences, v. 8, p. 267–288.

von Raumer, J.F., Stampfli, G.M., Borel, G., and Bussy, F, 2002, Organization of pre-Variscan basement areas at the north-Gondwanan margin: International Journal of Earth Sciences, v. 91, p. 35–52.

Whitehouse, M.J., Windley, B.F., Stoeser, D.B., Al-Khirbash, S., Ba-Bttat, M.A.O., and Haider, A., 2001, Precambrian basement character of Yemen and correlations with Saudi Arabia and Somalia: Precambrian Research, v. 105, p. 357–369.

Windley, B.F., Whitehouse, M.J., and Ba-Btatt, M.A.O., 1996, Early Precambrian gneiss terranes and Pan-African island arcs in Yemen: Crustal accretion of the eastern Arabian shield: Geology, v. 24, p. 131–134.

Wingate, M.T.D., and Giddins, J.W., 2000, Age and paleomagnetism of the Mundine Well dyke system, Western Australia: Implications for an Australia-Laurentia connection at 755 Ma: Precambrian Research, v. 100, p. 335–357.

Wortman, G.L., Samson, S.D., and Hibbard, S.D., 2000, Precise U-Pb zircon constraints on the earliest magmatic history of the Carolina terrane: Journal of Geology, v. 108, p. 321–338.

Manuscript Accepted by the Society August 25, 2003

Printed in the USA

Geological Society of America
Memoir 197
2004

A Ni-Cu-Co-PGE massive sulfide prospect in a gabbronorite dike at Lac Volant, eastern Grenville Province, Québec

Hassan Nabil*
Sciences de la Terre, Université du Québec à Chicoutimi, 555 Bv. de l'Université, Chicoutimi, Québec G7H 2B1, Canada
Thomas Clark
Géologie Québec, Ministère des Ressources naturelles, 5700 4e Avenue Ouest, Bureau A-210, Charlesbourg, Québec G1H 6R1, Canada
Sarah-Jane Barnes
Sciences de la Terre, Université du Québec à Chicoutimi, 555 Bv. de l'Université, Chicoutimi, Québec G7H 2B1, Canada

ABSTRACT

The Lac Volant Ni-Cu-Co-PGE prospect, located 75 km northeast of Sept-Iles in the Grenville Province of Québec, is an example of magmatic sulfide mineralization associated with mafic magmas in a high-grade metamorphic terrain. Disseminated and massive sulfides occur within a 20- to 25-m-thick zone in a gabbronorite dike intruding the gabbronorite of the host Matamec Complex. The average composition (*n* = 29) of the sulfides is estimated to be 2.0% Cu, 1.5% Ni, 0.12% Co, 67 ppb Pt, and 256 ppb Pd. The age of the dike is 1351 ± 6 Ma (U-Pb zircon age), which is identical to the age of the Rivière-Pentecôte anorthosite (1354 ± 3 Ma) located 130 km to the southwest. The dike originated by means of multiple injections of sulfide-bearing, tholeiitic magmas derived from a depleted, N-type mid-ocean ridge basalt (N-MORB)-type source. The dike is chemically similar to the gabbronoritic host, suggesting similar parental magmas. Geochemical variation within the dike was caused by fractional crystallization of silicates (orthopyroxene, clinopyroxene, and plagioclase) and by ingestion of crustal rocks. Assimilation is suggested by the presence of xenoliths of granite and metasediment within the dike or narrow branching dikes; enrichment of the dike in Rb, Th, Ba, and light rare earth elements; and a negative Ta anomaly with respect to Th. The observed composition of the dike can be explained by 15% crustal contamination, which would have occurred at depth before the magma began to crystallize mafic minerals. Massive, matrix, and disseminated sulfides, interpreted to be of magmatic origin, are composed of pyrrhotite (the principal species, 75%), pentlandite (altered to bravoite and violarite), chalcopyrite, and pyrite. Magmatic breccia structures are common in gabbronorite containing disseminated sulfides. The presence of metasedimentary xenoliths and a high S/Se ratio (9,000–16,000) suggest that sulfide saturation was caused by contamination of the magma. The sulfide liquid interacted with a small volume of magma (R = silicate liquid/sulfide liquid = 200). Impoverishment of the sulfides in platinum-group elements (PGE) relative to Ni and Cu may have been caused by the loss of PGE during an earlier and deeper episode of sulfide separation. Limited sulfide fractionation is suggested by the chemical similarity of the three types of sulfide. Meta-

*E-mail: Hassan_Nabil@uqac.uquebec.ca.

Nabil, H., Clark, T., and Barnes, S.-J., 2004, A Ni-Cu-Co-PGE massive sulfide prospect in a gabbronorite dike at Lac Volant, eastern Grenville province, Québec, *in* Tollo, R.P., Corriveau, L., McLelland, J., and Bartholomew, M.J., eds., Proterozoic tectonic evolution of the Grenville orogen in North America: Boulder, Colorado, Geological Society of America Memoir 197, p. 145–161. For permission to copy, contact editing@geosociety.org.

morphic recrystallization produced xenomorphic pyrrhotite and idiomorphic pyrite crystals up to 25 cm and 4 cm in size, respectively. Alteration by meteoric water caused the transformation of pentlandite into bravoite and violarite. The Lac Volant prospect is similar in terms of composition, style of emplacement, magma type, and age to certain Norwegian deposits (the Ertelian and Flåt deposits in the Sveconorwegian Province) and to the Voisey's Bay deposit in Labrador.

Keywords: Grenville, gabbronorite, dike, magmatic, sulfides, Ni-Cu-PGE

INTRODUCTION

The discovery, in 1992, of the Voisey's Bay deposit in Labrador, evaluated at 32 million tonnes grading 2.83% Ni, 1.68% Cu, and 0.12% Co and located in a dike associated with a troctolitic-gabbroic complex (Ryan et al., 1995; Naldrett, 1997), has spurred exploration of the anorthosite and troctolite intrusions of the Grenville Province, with the hope of finding similar deposits.

In August 1996, the Lac Volant prospect was discovered during a regional mapping survey of the Ministère des Ressources naturelles du Québec (MRNQ), provoking an unprecedented staking rush in Québec's North Shore region (Perreault et al., 1996, 1997). The Lac Volant prospect was extensively explored, using airborne electromagnetic (EM) and magnetic surveys, ground-based gravity and EM surveys, and drilling. The massive sulfide bodies seen at surface were found to have limited depth extension, although disseminated sulfides were discovered to the northeast of the initial discovery (Raymond et al., 1997).

GEOLOGY OF THE MANITOU-NIPISSO AREA

The Manitou-Nipisso area is part of the Allochthonous Polycyclic Belt of the Grenville Province (Rivers et al., 1989). Integration of recent geological survey work, previous surveys (Jenkins, 1957; Hogan and Grenier, 1971), and aeromagnetic data has resulted in the recognition of four main geological entities (Gobeil et al., 1996, 1999; Chevé et al., 1999): the Manitou Gneiss Complex, the Matamec Igneous Complex, and the Tortue and Havre-Saint-Pierre anorthositic suites (Fig. 1).

GEOLOGY OF THE LAC VOLANT PROPERTY

In the vicinity of the prospect, 1:50,000-scale geological mapping covering NTS (National Topographical System, Canada) sheets 22I/13 (east half) and 22I/14 (east half) has revealed that the area is underlain mainly by the gabbronorite of the Matamec Complex (Gobeil et al., 1996, 1999; Saint-Germain, 2002; Saint-Germain and Corriveau, 2003). These fine-grained, granoblastic rocks with a salt-and-pepper appearance display a strong aeromagnetic pattern oriented north-northeast. The gabbronorites are mostly massive and are intercalated with sheetlike intrusions of porphyritic, pyroxene-bearing monzonite, quartz monzonite, and granite and with a few bands of ortho- and paragneiss. These intrusions are weakly to strongly deformed, and their orientation defines the structural grain within the complex. Field observations of such things as magma mingling textures show that the gabbronorite

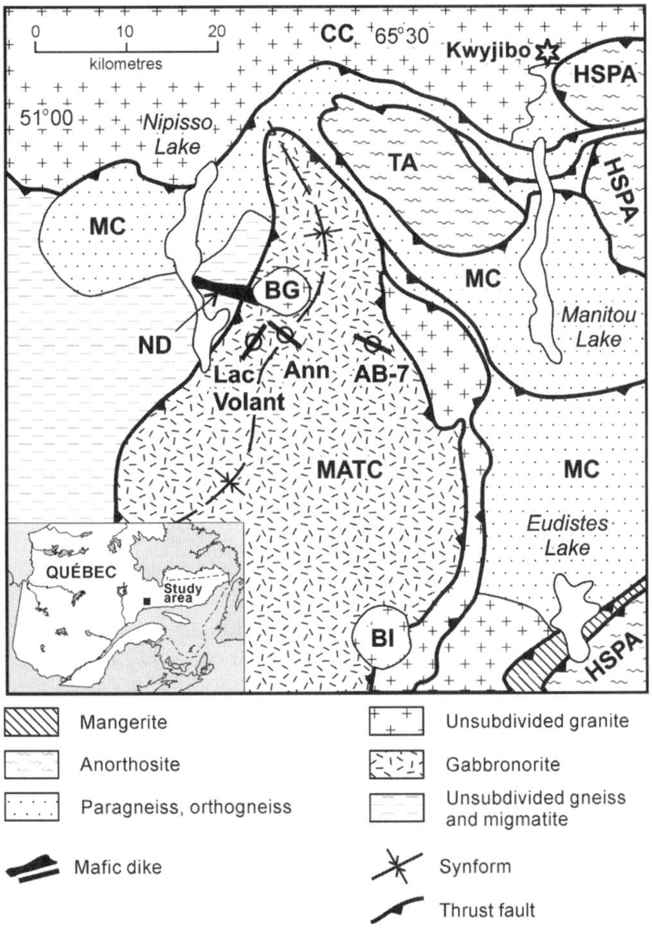

Figure 1. Regional geology of the Nipisso and Manitou Lakes area, showing the locations of magmatic Cu-Ni prospects, including Lac Volant. In the Matamec Complex (MATC), the orientation of the symbols indicates the strike of the mineralized dikes (Lac Volant, Ann, and Ab-7). Other abbreviations: HSPA—Havre-Saint-Pierre Anorthosite; TA—Tortue Anorthosite; CC—Canatiche Complex, host to the Kwyjibo Fe-Cu-REE (rare earth element) deposit (Chartrand et al., 2003; Clark, 2003); MC—Manitou Complex; ND—Nipisso Dike; BI—Boutereau Intrusion.

and the monzonite intrusions formed contemporaneously, and they have been interpreted as formed within a mafic-silicic layered intrusion (MASLI) by Saint-Germain and Corriveau (2003). The magnetic pattern suggests the presence of several faults, most of which correspond to regional lineaments oriented north-northeast–south-southwest, north-south, and northwest-southeast. Certain north-northeast–south-southwest and north-south lineaments contain fine- to medium-grained, gabbro and gabbronorite dikes. One of these contains the Lac Volant prospect (Perreault et al., 1996, 1997; Nabil, 1999) (Fig. 2), which U-Pb zircon dating gave an age of 1351 ± 6 Ma (Gobeil et al., 1999).

The Lac Volant dike strikes roughly parallel to the foliation of its host rocks. However, it is oblique to them locally (Fig. 2), and dips more steeply than the host rocks, cutting them at an approximate angle of 15 degrees (Perry and Roy, 1997). The dike is located on the western flank of a regional synform, a fold structure that originated either during the thrust emplacement of the Matamec during the Grenville orogeny (Gobeil et al., 1999), our preferred hypothesis, or during collapse of a MASLI intru-

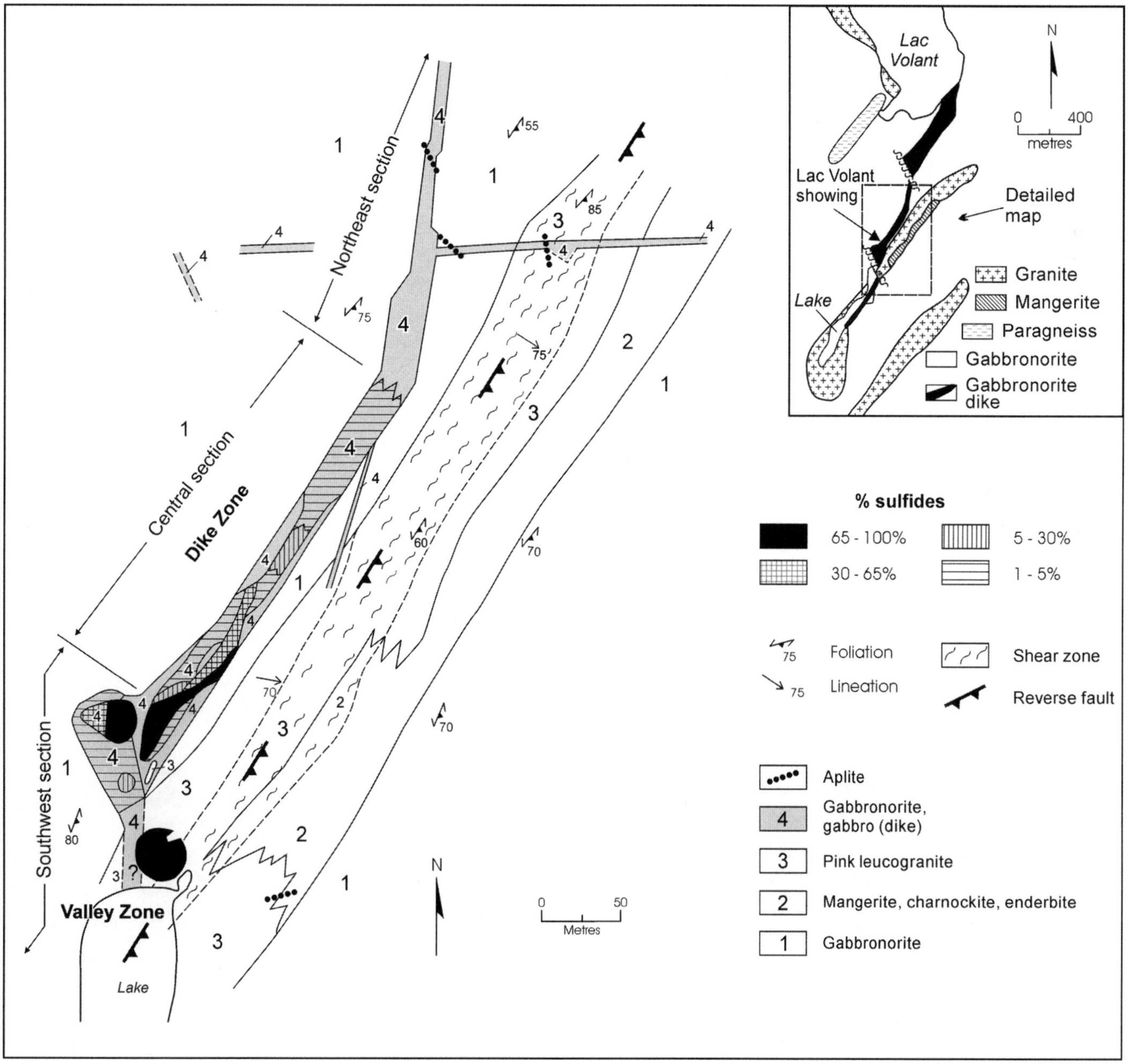

Figure 2. Detailed geology of the Lac Volant Cu-Ni-PGE prospect within the gabbronorite host of the Matamec Complex, modified from Perreault et al. (1996).

sion (Saint-Germain, 2002; Saint-Germain and Corriveau, 2003). The present orientation of the dike is interpreted as the result of this folding. If the regional synform is unfolded, the dike and its host rocks return to a subhorizontal position, which was likely the attitude of the dike when it was emplaced (cf. Saint-Germain and Corriveau, 2003, for an alternative).

The Lac Volant gabbro has a granular texture (Fig. 3) and is composed of plagioclase (30 to 35%), pyroxene (orthopyroxene and clinopyroxene, 35 to 40%), green hornblende (5 to 10%), biotite (<5%), and quartz (<5%). The dike locally contains large (centimeter- to decimeter-size), undeformed phenocrysts of plagioclase. Ilmenite is the most common oxide mineral. Even though the rocks have been affected by one or more episodes of deformation preceding the Grenville orogeny, they still display igneous textures. The dike is deformed at its contacts, but appears to be little deformed internally.

The dike may be subdivided into three parts (Fig. 2): (1) a central section, 20 to 25 m thick and 300 m long, oriented N35°E parallel to the local structural grain in the Matamec Complex; (2) a northeast section, in which the dike bifurcates toward the north (N5°E to N10°E), cutting the local foliation, and then again toward the northeast; and (3) a southwest section, characterized by a protuberance ~50 m thick located where the dike appears to adopt a north-south orientation, and farther southwest, by a north-northeast–south-southwest orientation where it is partly hidden beneath a small lake. The thickness of the dike varies from ~25 m near the mineralized zone to ~180 m farther to the northeast.

The mineralization consists of massive, semimassive, and disseminated sulfides occurring in two zones, namely the Valley Zone and the Dike Zone (Fig. 2). The main prospect of massive and disseminated mineralization extends over a distance of ~300 m; minor disseminated sulfides occur farther to the northeast. The Valley Zone is located in the southwest section and contains (at surface) only massive sulfides. It forms a low, circular hill measuring 23 m by 28 m and ~2 m high. The Dike Zone occurs in the central section and is possibly composed of two distinct bodies of massive sulfides, one of which is elongate, bordered by rocks containing semimassive (matrix) and disseminated sulfides. It is not clear whether the three bodies of massive sulfides shown in Figure 2 are in fact connected below surface. The elongate body of massive sulfides is ~75 m long and 4 to 10 m wide, and is oriented subparallel to the gabbronorite dike. Drilling has indicated that its vertical thickness is restricted to a few meters (Perry and Roy, 1997). Thus, the massive sulfide body, as preserved, has a cigar shape in a horizontal position. The protuberance near this body may represent a depression in the floor or a concavity in the wall of the dike. It was perhaps formed through magmatic erosion, and its presence may have been important for the accumulation of the massive sulfides.

The dike and mineralization consist of a magmatic breccia characterized by angular to rounded gabbronorite fragments, generally from a few centimeters to decimeters in diameter, enclosed within a gabbronorite matrix that is itself more finely brecciated. Both the fragments and the matrix may contain sulfides, but the latter commonly is richer in sulfides (Fig. 4). Barren breccia also occurs ~200 m south of the mineralized zone. These relations suggest at least two mineralizing events, of which the second appears to have been more important. In addition, the border zones of the dike are largely barren of sulfides (Fig. 2), a fact that suggests the earliest injection of magma was devoid of sulfides.

A xenolith of laminated quartzite, two decimeters across, occurs in the central section of the dike. It is bordered by a thin (mm) film of pyrite. Angular to rounded, decimeter-size xenoliths of granite, several of which are mylonitic, occur in a narrow, subsidiary, mafic dike (Fig. 5). In another subsidiary dike, a few small (several mm) xenoliths of pyroxenite and anorthosite have been observed.

Figure 3. Photomicrograph showing the granular texture of the gabbronorite dike and a phenocryst of plagioclase (doubly polarized light).

Figure 4. Breccia structure consisting of small, barren, or mineralized fragments in a sulfide-rich matrix.

Figure 5. Enclaves of granite of various sizes and shapes in a gabbroic matrix; some fragments are mylonitic (the scale is in centimeters).

PETROGRAPHY

Silicate Minerals

Plagioclase has a gray to violet-gray color to the naked eye. In thin section, plagioclase grains are equidimensional and are 1 to 7 mm in size. They form an aggregate of twinned, partially sutured grains with undulatory extinction. In a narrow, subsidiary, mafic dike, the texture is ophitic, and plagioclases form elongate prisms outlining the spaces occupied by the mafic silicates. Late deformation has caused some plagioclase prisms to be either broken or bent. The texture of plagioclase in the dike system contrasts with the recrystallized, granoblastic texture of plagioclase in the host gabbronorite. Hypersthene (25%) in the dike forms 1 to 4 mm, subautomorphic to xenomorphic crystals and is uralitized to varying degrees. Clinopyroxene includes twinned augite, pigeonite, and diopside (10–15%). Pigeonite crystals, which are subautomorphic to xenomorphic, reach 1 cm in diameter. The size and shape of pigeonite grains suggest that they crystallized slowly. Hornblende, constituting up to 5% of the rock, is subidiomorphic to xenomorphic and generally altered. It partially to completely replaces pyroxene. Biotite (<5%) is subidiomorphic to xenomorphic and generally develops around opaque oxides. In places, it surrounds ilmenite, and may be associated with trace amounts of apatite, quartz, and potassic feldspar representing late-stage residual liquid.

Massive Sulfide Bodies

The three bodies of massive sulfides (Fig. 2) have a similar mineralogy. The sulfides are composed essentially of pyrrhotite as the main constituent, along with pentlandite (generally altered to bravoite and violarite), chalcopyrite, and pyrite. Pyrite typically forms centimeter-scale cubes surrounded by chalcopyrite, a texture that is thought to be due to metamorphism. Magnetite forms 5 to 10% of the opaque phases. Galena occurs in trace amounts in the massive sulfides.

Pyrrhotite ($Fe_{1-x}S$)

Pyrrhotite is the principal mineral (75%–80%) forming the massive sulfides. Crystals are commonly a few centimeters across, but can reach 25 cm. Pyrrhotite crystals poikilitically include chalcopyrite, pyrite, magnetite, and pentlandite grains. Crystals are generally fractured and exhibit bent cleavage planes and, locally, undulatory extinction. Electron microprobe analysis shows that pyrrhotite corresponds to the monoclinic variety (Fe_7S_8) and contains traces of Ni, Co, and Cu (Nabil, 1999). In other Ni-Cu deposits—e.g., those at Selebi Phikwe, Strathcona (Naldrett, 1981), Voisey's Bay, and Duluth (Thériault et al., 1997)—pyrrhotite is typically of two crystallographic types: hexagonal (Fe_9S_{10}) or monoclinic (Fe_7S_8), the latter of which is relatively poor in Ni compared to hexagonal pyrrhotite.

Chalcopyrite

Chalcopyrite represents between 5 and 10% of the massive sulfides. It forms irregularly shaped masses ranging from millimeters to several centimeters in size. Grains are generally xenomorphic, enclosing idiomorphic crystals of pyrite, magnetite, and pyrrhotite. These chalcopyrite masses usually have irregular contacts and locally surround pyrrhotite (Fig. 6). In places, chalcopyrite also forms straight veins filling fractures in massive sulfides. Chalcopyrite has a stoichiometric composition (Nabil, 1999), and contains low concentrations of Ni and Zn.

Pentlandite

Pentlandite, which constitutes 5 to 8% of the massive sulfides, forms chains of millimeter- to centimeter-scale grains that are generally altered to bravoite and violarite. Pentlandite grains fill fractures developed in pyrrhotite (Fig. 7) that may have formed as a result of deformation and thermal contraction. In

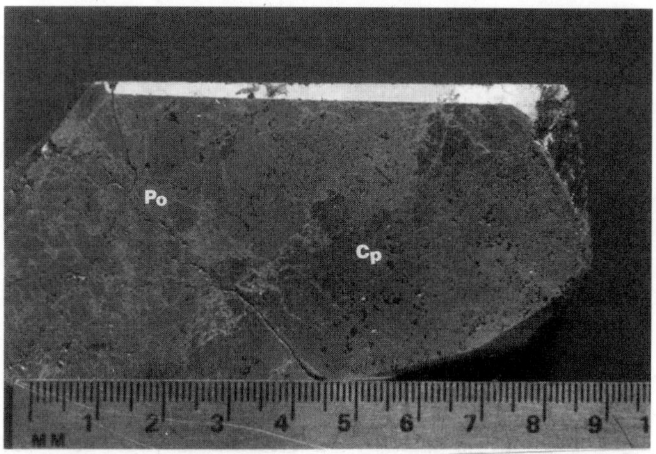

Figure 6. Irregular contact between a mass of chalcopyrite (Cp) and pyrrhotite (Po). The black spots are magnetite grains intergrown with pyrrhotite.

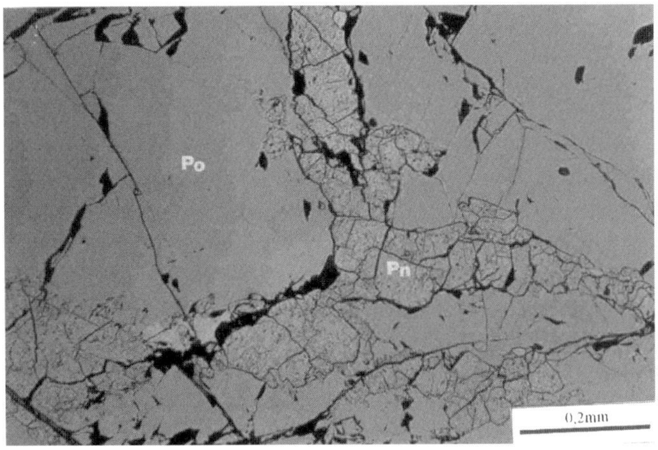

Figure 7. Reflected light photomicrograph showing pentlandite (Pn), altered to bravoite, forming chains in pyrrhotite (Po).

places, idiomorphic to subidiomorphic crystals of bravoite occur. The development of bravoite at Lac Volant may be the result of the loss of nickel and iron from pentlandite due to meteoric alteration (Nabil, 1999). In many other Ni-Cu deposits, such as those at Sudbury (Naldrett, 1984) and Noril'sk (Genkin et al., 1973), pentlandite occurs as flames in pyrrhotite formed as a result of exsolution. Flame texture makes nickel extraction difficult. Electron microprobe analysis of Lac Volant pentlandite reveals a rather variable composition, with Ni grades varying between 23 and 38% (Nabil, 1999).

Magnetite

Magnetite forms between 5 and 10% of the massive sulfides. It forms small (submillimetric to millimetric), rounded, amoeboid grains disseminated in pyrrhotite and chalcopyrite (Fig. 6). Some magnetite crystals contain spherical sulfide inclusions composed of pyrrhotite, chalcopyrite, and pentlandite. This texture can be explained by crystallization of magnetite from a sulfide liquid (Lightfoot et al., 1984).

Pyrite

Pyrite represents less than 5% of the massive sulfides. It typically forms idiomorphic to subidiomorphic crystals varying in size from 1 mm to several centimeters (Fig. 8). Pyrite generally occurs completely or partly surrounded by chalcopyrite. Three mechanisms might be responsible for initial pyrite formation: (1) oxidation of pyrrhotite to form magnetite and pyrite; (2) reaction of pyrrhotite with externally derived sulfur, perhaps introduced after the beginning of pentlandite crystallization; or (3) low-temperature reaction of pyrrhotite with a sulfur-rich liquid (Naldrett and Gasparrini, 1971). The close association of pyrite with chalcopyrite suggests addition of sulfur and copper and, consequently, formation of pyrite by means of one of the first two mechanisms described earlier. Thereafter, pyrite may have recrystallized during regional metamorphism, as indicated by the presence of the same texture in showings elsewhere in the region.

Galena

Galena occurs in trace amounts in the massive sulfides, as disseminated grains associated with pyrite or as subidiomorphic to xenomorphic grains in late-stage veins of chalcopyrite.

Disseminated and Semimassive Sulfide Bodies

Disseminated and semimassive sulfides represent more than 70% of the sulfides in the Lac Volant prospect. The sulfides are interstitial to the silicate minerals (Fig. 9). Pyrrhotite is the principal sulfide species, making up more than 85% of the sulfide mass. It occurs as xenomorphic grains and is generally fractured. Chalcopyrite makes up 5% of the sulfide mass and is xenomorphic; locally, it contains pyrite crystals. Pentlandite (<5%), which occurs in fractures in pyrrhotite, is generally altered to bravoite and violarite. Magnetite and ilmenite constitute 3 to 5% of the opaque minerals. In places, disseminated sulfides are cut by veinlets of pyrrhotite and chalcopyrite. Globular masses of sulfide (1–10 mm), mainly pyrrhotite, may occur along with disseminated sulfides. They may represent droplets of liquid sulfide trapped during the crystallization of the silicates (Naldrett, 1981). Such droplets could imply immiscible relations between a silicate liquid and a sulfide liquid.

SUMMARY OF CONCLUSIONS BASED ON FIELD RELATIONS AND PETROGRAPHY

The petrographic characteristics of the gabbronorite dike and the mafic and felsic country rocks of the Lac Volant prospect point to the following conclusions:

1. The mineralogy of the dike, in particular the absence of olivine, indicates that the magma was relatively evolved (not directly derived from the mantle source). Instead, the parent

Figure 8. Pyrite (Py) surrounded by chalcopyrite (Cp) in a mass of pyrrhotite (Po). The small black spots are magnetite crystals. Inset is a centimeter-scale cube of pyrite.

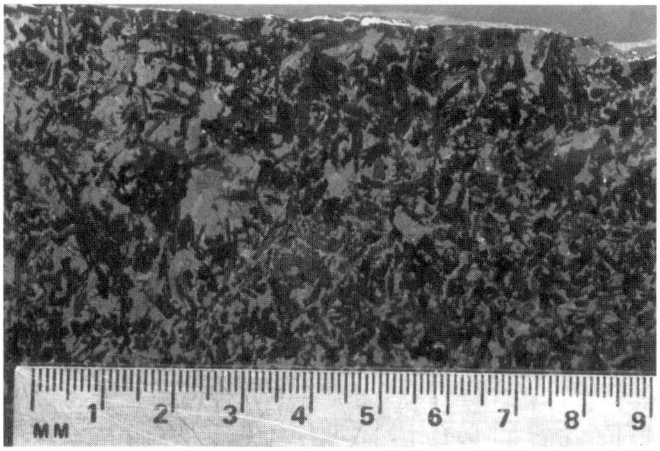

Figure 9. Matrix texture of semimassive sulfides in Lac Volant gabbronorite. Plagioclase forms randomly oriented laths.

magma must have undergone fractional crystallization prior to emplacement.

2. The presence of magmatic breccias suggests a dynamic magmatic system.

3. The presence of large plagioclase crystals (up to 8 cm) (Nabil, 1999, plate 11) and xenoliths of pyroxenite and anorthosite indicate slow crystallization and differentiation of the magma at depth.

4. The local occurrence of quartzitic metasedimentary xenoliths in the Lac Volant dike and the occurrence of granite xenoliths in subsidiary dikes suggests that the magma may have been contaminated by crustal material.

5. The slight deformation experienced by the dike is of brittle character, as suggested by broken plagioclase crystals and the angular form of the xenoliths.

GEOCHEMISTRY OF HOST GABBRONORITE

Three samples of gabbronorite from the Matamec Complex host have been analyzed for major oxides and trace elements, including platinum-group elements (PGE). The gabbronorite is tholeiitic, and its average Mg # is 0.59 (three samples). With the exception of higher concentrations of K_2O, Ba, and Rb, the gabbronorite host has a composition similar to that of the gabbronorite dike. This compositional similarity suggests that the respective magmas had similar origins and histories.

GEOCHEMISTRY OF THE DIKE

Table 1 shows the average composition of gabbronorite in the dike. SiO_2 and Al_2O_3 concentrations are from 45.6 to 52.6% and from 12.0 to 16.5%, respectively. MgO varies considerably, between 3.92 and 9.49%. TiO_2 varies between 2.01 and 4.81%. Na_2O and P_2O_5 vary between 2.01 and 4.81% and between 0.14 and 0.61%, respectively. A sample from a subsidiary dike containing xenoliths of leucogranite has a distinct composition. It is poorer in MgO, CaO, and MnO, and is richer in Fe_2O_3, TiO_2, Na_2O, K_2O, and P_2O_5. These compositional characteristics suggest a more differentiated composition for the magma and/or contamination of the magma by assimilation of country rock granite. Alkali and silica concentrations indicate that the dike magma was subalkalic (Nabil, 1999). In an AFM diagram (Irvine and Baragar, 1971; Fig. 10), the samples follow an iron-enrichment trend typical of tholeiitic rocks. The fact that some samples lie in the calc-alkalic field may be due to the mobile nature of Na and K.

The Lac Volant gabbronorite has moderately fractionated rare earth element (REE) patterns (Fig. 11), similar to those of E-type mid-ocean ridge basalt (E-MORB) and continental flood basalt but strongly enriched in light REE compared to island arc tholeiites (Fig. 11). However, the Lac Volant gabbronorite differs from these rock types with respect to other trace elements (Fig. 12). The plot proposed by Shervais (1982; Fig. 13) shows that Lac Volant mafic rocks fall in the MORB and back-arc basalt fields. The chemical data therefore suggest that the Lac Volant magma was tholeiitic and was probably emplaced in an intracratonic environment.

In a Th/Yb versus Ta/Yb diagram (Fig. 14), the Lac Volant gabbronorite falls in the calc-alkalic field. It is unlikely that the gabbronorite magma was calc-alkalic, because major element chemistry indicates a tholeiitic affinity. The contradiction may be explained by crustal contamination of the mafic magma. Contamination by crustal material would probably have had an effect on the Th concentration and consequently on the position of the samples in the Th/Yb versus Ta/Yb diagram. In the absence of contamination, discriminant diagrams might show that these rocks originated in a depleted source or an N-type mid-ocean ridge basalt (N-MORB)–type environment. A multi-element plot shows an enrichment in incompatible elements and a negative anomaly in tantalum (Fig. 15). Such a signature could also be the result of crustal contamination.

MODELING THE COMPOSITION OF THE LAC VOLANT MAGMA

Contamination

Contamination of a magma may occur during its ascent through the crust and would be evident if the composition of the incorporated material were very different from that of the magma. Contamination can occur through assimilation of sedimentary and/or felsic rocks or magma mixing. Assimilation of continental crust by a mafic magma would affect the concentrations of elements such as SiO_2, Sr, Ba, Rb, Cs, Th, and the light REE. These elements are strongly enriched in the continental crust, but are present in low concentrations in tholeiitic magmas.

The spectrum of composition of the host for the Lac Volant gabbronorite dike includes gabbronorite, granite, monzonite, and paragneiss. The composition of the Lac Volant dike has been

TABLE 1. MAJOR AND TRACE ELEMENT CONCENTRATIONS IN LAC VOLANT GABBRONORITE

Rock type	Tc-101-a MtGBN	Tc-153-a MtGBN	Tc-184-a MtGBN	Tc-234 GBND	Tc-157-a GBND	Tc-173-A1 GBND	Tc-231-a1 GBND	Tc-289-a GBND	Tc-311-c GBND	Tc-311-b GBND	Tc-263-a GBND
SiO_2	52.00	51.40	49.10	52.60	51.50	50.70	45.60	46.10	47.90	50.40	49.00
TiO_2	1.40	1.70	1.39	1.31	1.03	1.17	2.52	2.50	1.15	1.14	4.45
Al_2O_3	15.30	14.90	15.80	15.60	12.00	16.10	14.60	14.80	16.50	15.80	15.40
FeO	9.79	8.80	7.23	8.83	8.73	7.89	11.30	13.70	7.66	6.85	6.88
Fe_2O_3	1.60	2.52	4.07	1.49	3.10	2.63	3.70	2.20	2.19	2.99	5.05
MnO	0.19	0.20	0.20	0.18	0.20	0.18	0.24	0.24	0.16	0.16	0.13
MgO	6.75	6.49	7.47	6.28	9.49	7.50	6.61	6.58	8.24	7.50	3.92
CaO	9.24	8.72	8.99	8.68	9.27	9.41	9.14	9.09	11.90	9.21	5.92
Na_2O	2.69	2.76	2.71	3.07	2.37	2.47	2.82	2.68	2.01	2.98	4.81
K_2O	0.84	1.21	1.31	0.88	0.59	0.63	1.02	0.70	0.16	1.09	2.32
P_2O_5	0.24	0.38	0.21	0.16	0.18	0.17	0.33	0.33	0.15	0.14	0.61
S	0.11	0.02	0.13	0.15	0.00	0.12	0.13	0.14	0.11	0.09	0.18
LOI	0.00	0.41	1.09	0.37	0.08	0.00	0.99	0.00	1.14	0.95	0.96
Total	100.15	99.51	99.70	99.60	98.54	98.97	99.00	99.06	99.27	99.30	99.63
Mg #	0.55	0.57	0.65	0.56	0.66	0.63	0.51	0.46	0.66	0.66	0.50
Cr (ppm)	130	154	128	69	231	124	76	87	24	166	10
Ni	79	86	105	58	415	145	108	114	74	105	61
Co	51	53	53	52	72	60	67	67	38	58	60
Sc	31	30	30	28	28	29	36	34	6	28	5
V	222	220	218	205	195	232	261	264	238	209	225
Cu	56	54	35	46	198	112	111	133	109	41	106
Zn	122	138	136	129	122	116	154	165	190	129	140
S	1100	200	1300	1500	<100	1200	1300	1400	1100	900	1800
As	0.5	0.9	0.8	0.9	0.3	0.5	0.5	1.8	2.1	0.8	0.8
Ag	1.3	1.0	0.3	0.5	<0.5	0.3	<0.5	0.6	<0.5	1.1	0.4
Se	3.6	3.7	1.9	3.0	1.6	3.2	4.3	4.6	4.8	2.4	4.5
Sb	0.04	0.14	0.13	0.07	0.02		0.19	0.19	0.25	0.08	0.08
Rb	38	20	43	20	10	14	21	16	105	29	32
Cs	0.8	0.4	3.1	1.0	0.6	0.4	1.7	0.8	2.7	1.3	1.2
Ba	369	525	373	379	298	316	256	201	813	267	780
Ta	0.23	0.38	0.14	0.31	0.34	0.27	0.31	0.49	1.75	0.22	1.35
Hf	3.00	4.11	2.76	3.25	2.14	2.76	4.22	4.70	6.44	2.75	4.63
Th	2.35	1.62	2.29	2.90	1.68	1.23	1.31	1.56	2.23	2.19	2.50
U	0.59	0.65	0.68	1.00	0.48	0.37	0.46	0.53	1.89	0.63	0.72
La	16.2	18.9	10.07	14.36	13.53	12.87	9.72	11.50	34.12	10.73	32.55
Ce	35	43.46	26.1	32	29.08	29.29	26.16	30.08	83.07	23.11	73
Nd	28	34.84	22.68	25	22.64	23.22	27.86	31.02	62.24	19.56	54.78
Sm	4.92	6.40	4.63	4.94	4.33	3.97	6.37	6.16	8.69	3.80	7.19
Eu	1.11	1.57	1.35	1.55	1.17	1.36	2.09	2.01	2.71	1.09	2.22
Ho	1.20	1.57	0.99	1.12	0.79	0.64	1.56	2.04	0.43	1.00	0.58
Tb	0.50	0.72	0.61	0.57	0.54	0.57	1.05	1.10	0.51	0.45	0.23
Yb	2.79	3.06	2.87	3.16	2.24	2.41	4.55	4.47	0.99	2.42	0.56
Lu	0.44	0.49	0.43	0.49	0.35	0.34	0.67	0.63	0.15	0.35	0.08

Note: MtGBN—metagabbronorite; GBND—gabbronorite dike; LOI—loss on ignition.

modeled by assuming contamination of the initial magma by paragneiss, based on the analysis of a sample from ~1 km north of the prospect.

The analyzed paragneiss sample is very rich in silica (71%), alumina (13.6%), and incompatible elements. The enrichment of the paragneiss in nickel (117 ppm), copper (176 ppm), and zinc (1.3%) is due to the presence of sulfides (3.4% S). The incompatible element concentrations in the paragneiss are comparable to those of granite, showing high values for thorium, rubidium, and barium (Nabil, 1999). The paragneiss is similar in composition to continental crust (Taylor and McLennan, 1985) and also to average Proterozoic granite (Condie, 1993; Nabil, 1999).

Modeling shows (Fig. 15) that the composition of the Lac Volant gabbronorite dike can be explained by the contamination of an N-MORB-type magma with 15% of country rock paragneiss. The fit is good for most of the oxides, incompatible elements, and REE, but less good for Ba, Sr, and the light REE.

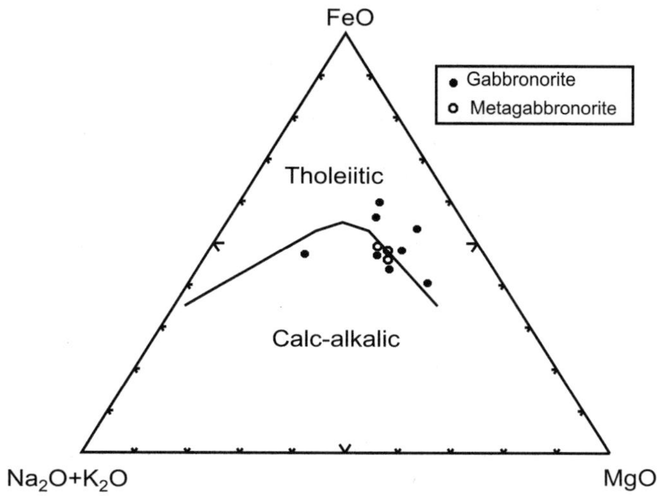

Figure 10. AFM diagram of Irvine and Baragar (1971) showing the tholeiitic affinity of the Lac Volant gabbronorite dike and of its gabbronorite host. The sample rich in Na_2O and K_2O is from a thin, subsidiary dike containing numerous granite enclaves.

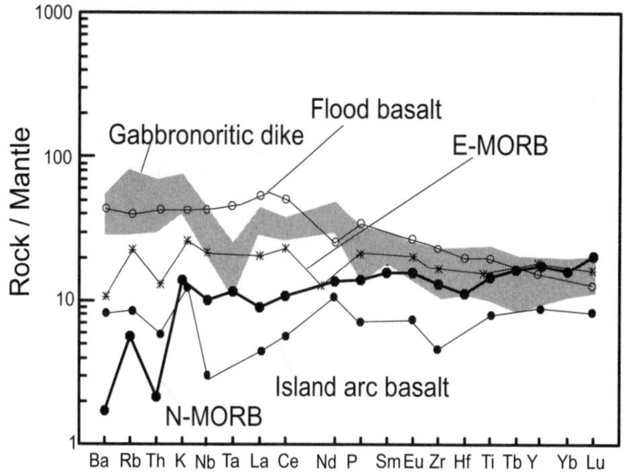

Figure 12. Normalized trace element diagram showing mantle-normalized trace element concentrations in Lac Volant gabbronorite and its gabbronorite host compared to concentrations in other varieties of basalt. The normalization factors are from Thompson (1982), and the value for comparison is taken from Viereck et al. (1989). E-MORB—E-type mid-ocean ridge basalt; N-MORB—N-type mid-ocean ridge basalt.

Fractional Crystallization of Silicates

Crystal fractionation modeling (Nabil, 1999) has shown that the geochemical variations observed in the Lac Volant gabbronorites can be attributed to ~40 to 50% fractional crystallization of clinopyroxene and plagioclase in the proportions 46% plagioclase and 54% clinopyroxene from a parental magma represented by sample TC-96-157. The absence of olivine in the modeling results can be explained by the already evolved nature of the presumed initial magma, represented by the chosen sample. Olivine may have fractionated earlier, from a more primitive magma.

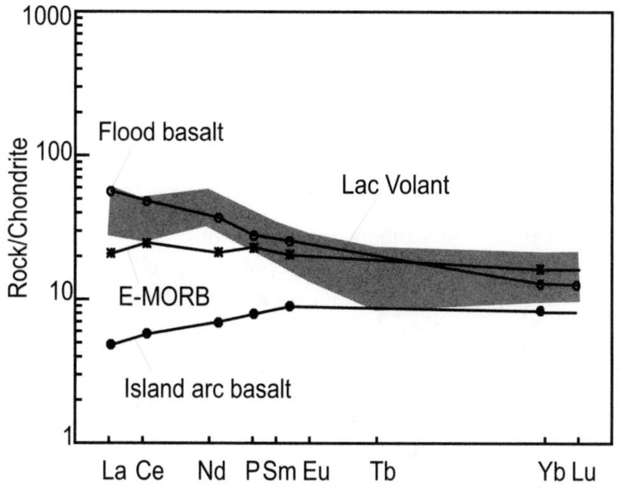

Figure 11. Chondrite-normalized rare earth element concentrations in Lac Volant gabbronorite compared to concentrations in flood basalt, E-type mid-ocean ridge basalt (E-MORB), and island arc basalt (from Wilson, 1989).

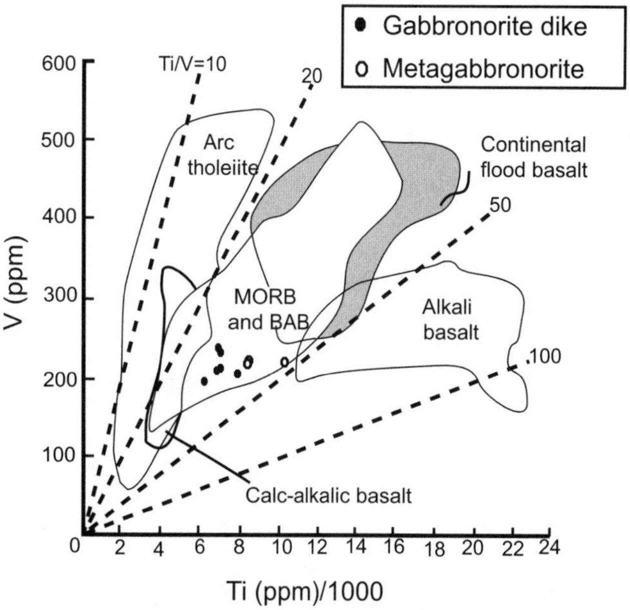

Figure 13. Discriminant diagram (Shervais, 1982) showing the geochemical affinity of Lac Volant mafic rocks. BAB—back-arc basin basalt; MORB—mid-ocean ridge basalt.

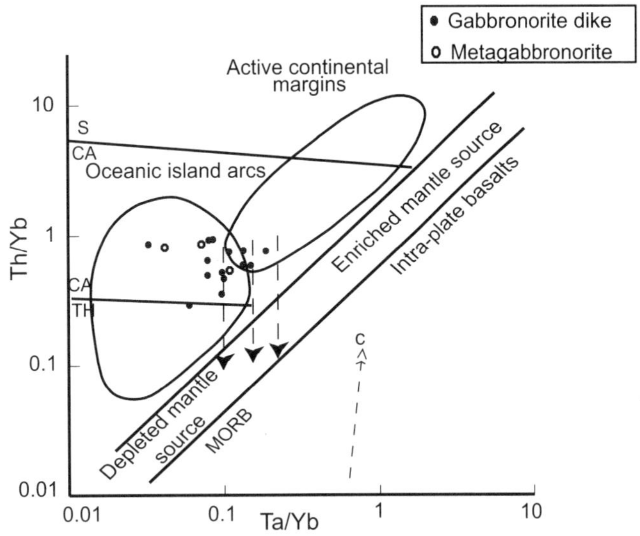

Figure 14. In a Th/Yb versus Ta/Yb diagram of Pearce (1983), Lac Volant gabbronorite samples plot in the calc-alkalic field. S—shoshonitic field; c—contamination; CA—calc-alkalic field; MORB—mid-ocean ridge basalt; TH—tholeiitic field.

Combined Results

The combined results from modeling contamination and fractional crystallization are shown in a Th versus Cr plot in Figure 16. Chrome is an element that is sensitive to fractional crystallization, whereas thorium is sensitive to both fractional crystallization and contamination. According to this diagram, variations in the Th and Cr concentrations in the gabbronorite can be explained by 10% contamination of an N-MORB–type magma with paragneiss, followed by fractionation of plagioclase and pyroxene. The plot also shows the effect of the separation of disseminated and massive sulfides.

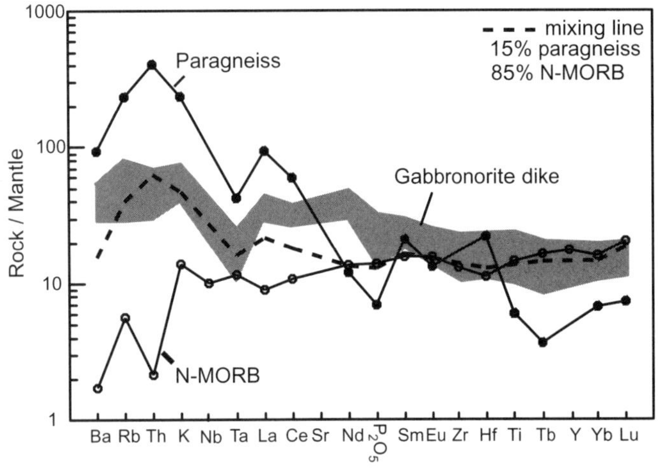

Figure 15. Normalized trace element diagram showing that the composition of Lac Volant gabbronorite can be explained by the contamination of N-type mid-ocean ridge basalt (N-MORB) magma by the addition of 15% paragneiss. The normalization factors are from Viereck et al. (1989).

GEOCHEMISTRY OF THE SULFIDES

The Lac Volant mineralization occurs as massive, matrix, and disseminated sulfides. Table 2 shows the sulfide compositions of mineralized gabbronorite, recalculated on the basis of 100% sulfides using the method of Naldrett (1981).

Sulfur isotope data, Se/S ratios, and the presence of xenoliths have led to the conclusion that many magmatic Ni-Cu deposits, such as those at Noril'sk, Sudbury, and Thompson (Eckstrand et al., 1989); in the Duluth Complex (Ripley and Alawi, 1988; Thériault et al., 1997); at Kambalda (Lesher and Arndt, 1995); and in the Muskox Intrusion (Barnes and Francis, 1995), were formed as a result of the contamination of mafic magmas by sulfur from sulfide-bearing sedimentary rocks or paragneiss. In the following paragraphs, these possibilities are examined with regard to the Lac Volant sulfides.

Selenium (Se) and the S/Se Ratio

Studies of the geochemical behavior of selenium in the natural environment have shown that the chalcophile element selenium is concentrated in volcanic and intrusive rocks (Auclair and Fouquet, 1987). Crustal sedimentary rocks are generally poor in Se compared to rocks derived from the mantle (Eckstrand and Hulbert, 1987). For this reason, the S/Se ratio is used to identify the source of sulfur in magmatic deposits, especially in cases where assimilation of country rocks is suspected to be involved in the genesis of the mineralization (Eckstrand and Cogulu, 1986; Paktunc, 1989).

The Lac Volant sulfides have S/Se ratios of between 9000 and 16,000 (Fig. 17), values that are greater than those of the mantle (~2000 to 5000). These high values are similar to those from contaminated, mineralized norites and sedimentary rocks (argillites) from the Duluth Complex (Thériault et al., 1997). The results shown in Figure 17 can be explained by contamination or by the ratio of silicate magma to sulfide melt (R factor). Modeling results suggest that the R factor for Lac Volant sulfides was between 100 and 200 (Nabil, 1999). In other words, the silicate magma was 100 to 200 times more abundant than the sulfide liquid with which it interacted. By contrast, in rich deposits, the sulfide liquid interacted with a much larger volume of magma (R between 1000 and 10,000). Because elements with high partition coefficients are relatively sensitive to changes in the R factor, a low R factor causes low PGE grades and a high S/Se ratio. In addition, contamination by Se-poor, sulfide-bearing sediments can add sulfur to the magma and as a result increase the S/Se ratio.

Arsenic (As), Antimony (Sb), and Silver (Ag)

Arsenic, antimony, and silver are highly enriched in sedimentary rocks and, consequently, are good indicators of contamination. The concentrations of these elements are high in

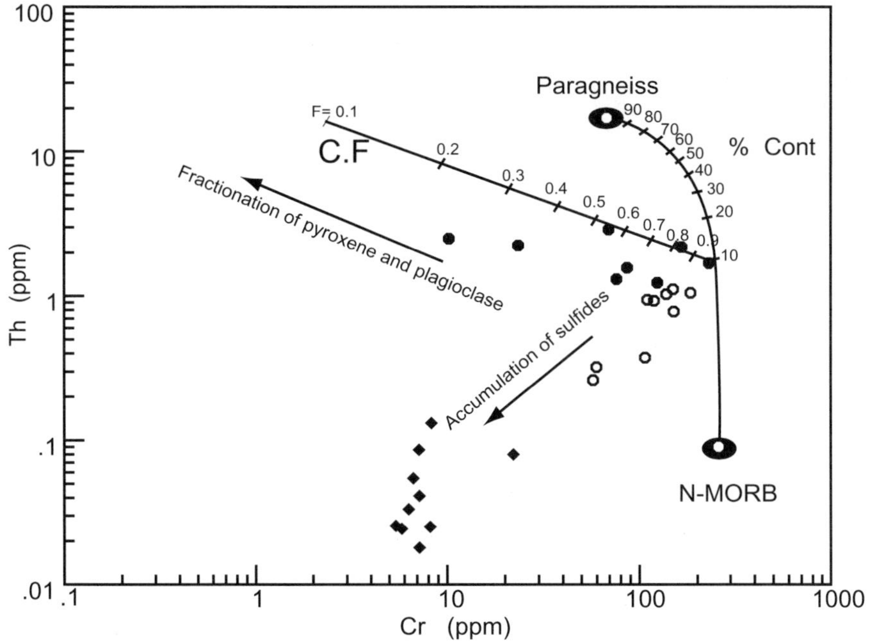

Figure 16. Cr versus Th plot illustrating the effects of the contamination of N-type mid-ocean ridge basalt (N-MORB) magma by para-gneiss and the fractional crystallization of the contaminated magma. CF—fractional crystallization trend; closed circles—gabbronite; cont—contamination; diamonds—massive sulfides; F—fraction of liquid remaining; open circles—disseminated sulfides.

TABLE 2. AVERAGE PGE AND CHALCOPHILE METAL CONCENTRATIONS IN LAC VOLANT SULFIDES

Element	DS	MT	MS
S (%)	40.00	39.78	36.76
Fe (S)	54.33	53.50	50.96
Ni	2.78	2.42	1.74
Cu	1.77	3.31	3.24
Co	0.13	0.13	0.14
As (ppm)	2.91	2.47	0.08
Se	29	32	31
Sb	0.50	0.33	0.04
Ag	14.9	22.0	16.2
Os (ppb)	4	11	2
Ir	1.20	3.43	1.07
Ru	52	30	6
Rh	1.85	3.90	3.10
Pt	74	52	173
Pd	305	487	309
Re	22	22	97
Au	87	65	20

Note: Analytical results have been recalculated to 100% sulfides. MS—massive sulfides (n = 19); DS—disseminated sulfides (n = 6); and MT—matrix sulfides (n = 4).

some magmatic deposits. In the Duluth deposit, for example, arsenic varies from 8 ppm in disseminated sulfides to 300 ppm in massive sulfides (Theriault et al., 1997). At Pechenga, in Russia, arsenic and silver grades reach 300 and 15 ppm, respectively (Barnes et al., 2001). In contrast to these high values, Lac Volant sulfides contain low concentrations of arsenic (1.2 ppm) and antimony (0.2 ppm). These values are similar to those typical of sulfides from the Lorraine mine, Temiscaming, Québec (Bouchaïb, 1992); Lac à Paul (Huss, 2002); and Lac Kenogami (Vaillancourt, 2001). Low concentrations of these elements may be related either to low concentrations in Lac Volant metasediments (As = 20 ppm, Se = 35 ppm, Sb = 3 ppm) or to the loss of these elements during metamorphism.

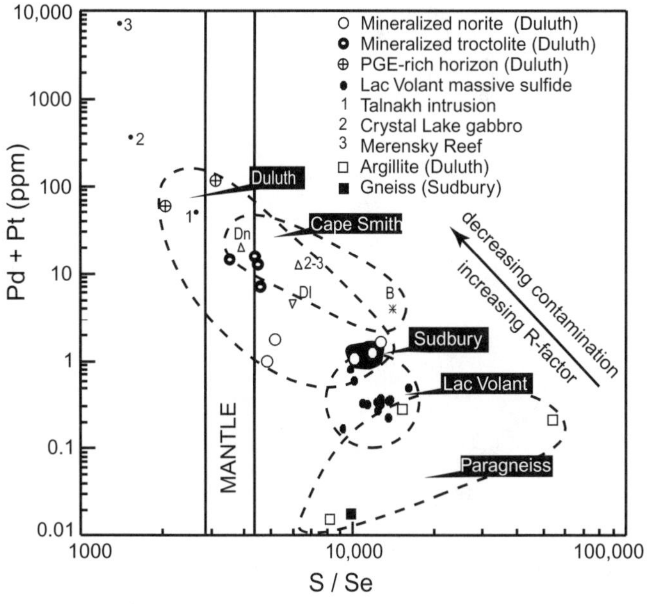

Figure 17. S/Se versus Pt + Pd plot showing Lac Volant mineralized samples compared to other sulfide deposits. The arrows illustrate the effects of increasing the ratio of silicate magma to sulfide melt (R factor) and decreasing the degree of contamination of the magma. 2-3—Delta 2-3 sills; B—Bravo sill; Dl—Deltal sill; Dn—Donaldson.

Fractional Crystallization of Sulfides

Barnes et al. (1988) pointed out that the composition of the parent magma has an effect on the slope of the pattern in a diagram showing mantle-normalized PGE concentrations. A relatively magnesian magma will result in a less fractionated pattern characterized by a relatively low Pd/Ir ratio. The three types of sulfides at Lac Volant have almost identical mantle normalized metal profiles (Fig. 18). The patterns are strongly fractionated (Pd/Ir = 79–746), suggesting that the PGE were derived from an evolved magma. The profiles also indicate that the sulfides are depleted in PGE compared to Ni and Cu. The similarities in the patterns could be the result of homogenization of sulfide compositions during magmatic transport and postmagmatic processes.

It seems likely that fractionation of Lac Volant sulfides could not have been extensive. Their compositions in Fe-Ni-S space are located within the monosulfide solid solution (MSS) field at 1100 °C (Kullerud et al., 1969). Therefore, the temperature interval during which fractionation could have occurred was probably small, thereby reducing the opportunity for the production of a large quantity of evolved sulfide liquid. Nevertheless, it is possible that the irregular, centimeter- to decimeter-size masses of chalcopyrite enclosed within Lac Volant massive sulfides represent blobs of copper-rich, fractionated sulfide liquid that remained trapped in the crystallizing sulfides.

A positive correlation between Pd and Au as a function of Cu in disseminated sulfides may be the result of fractional crystallization, each sulfide bleb acting as a closed system. By contrast, massive sulfides show no correlation between these elements, perhaps because compositions have been modified by open-system, post-solidification processes (Clark, 2003).

The Lac Volant massive sulfides are locally cut by chalcopyrite veins. PGE concentrations were determined in one such vein. A low Pd concentration (46 ppb) compared to Cu (29%) suggests that the vein did not form from a differentiated sulfide liquid. These veins are probably of postmagmatic origin.

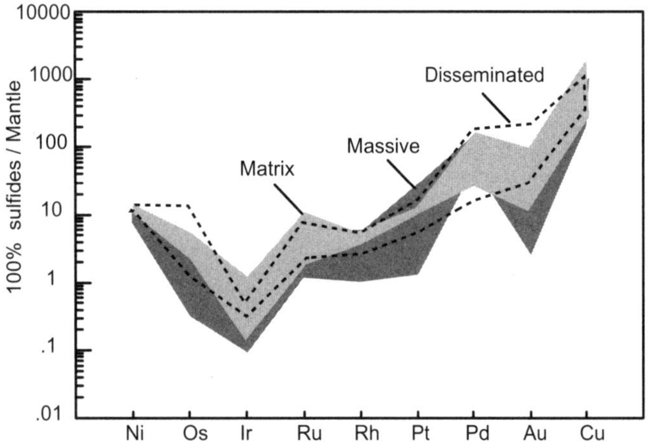

Figure 18. Diagram comparing mantle-normalized base and precious metal concentrations for the three types of Lac Volant sulfides.

Sulfide Segregation Prior to Emplacement

The composition of the sulfides was modeled using the equilibrium fractionation equation developed by Campbell and Naldrett (1979):

$$C_c/C_l = D(R + 1)/(R + D),$$

where C_c is the concentration of the metal in the sulfide, C_l is the concentration of the metal in the silicate magma, D is the partition coefficient of the metal between the sulfide melt and the silicate magma, and R is the ratio of silicate magma to sulfide melt (R factor). The Cu/Pd versus Pd diagram of Barnes et al. (1993) is used to illustrate the results of the modeling (Fig. 19). As shown in the diagram, the proposed parental magma has a high Cu/Pd ratio, which plots outside the expected range for mantle rocks. This indicates that the basaltic magma likely segregated sulfides at depth, becoming depleted in Pd relative to Cu. The metal content of the hypothetical initial magma prior to sulfide segregation can be estimated from the following equilibrium fractionation equation:

$$C_F/C_L = 1/ (1 + X(D - 1)/100),$$

where C_F/C_L (depletion factor) is the concentration of the metal in the fractionated magma divided by the concentration of the metal in the initial magma, D is the partition coefficient of the

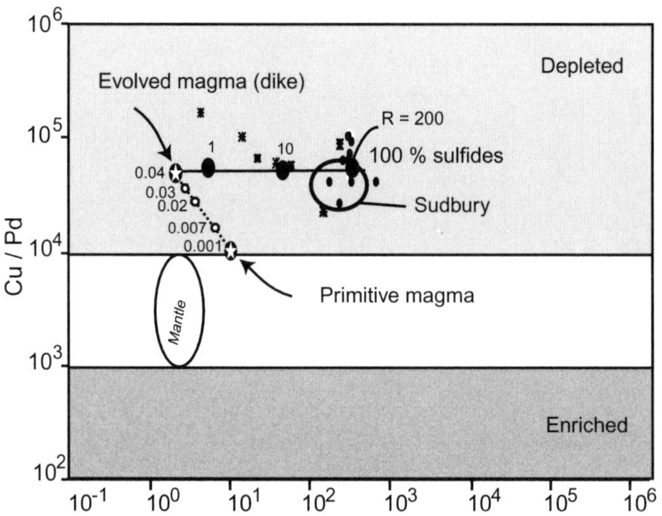

Figure 19. Plot of Cu/Pd versus Pd, showing the composition of Lac Volant sulfides and the results of modeling. The fields of mantle rocks and platinum-group elements (PGE)–dominated deposits are taken from Barnes et al. (1993). Asterisks—disseminated sulfides; closed circles—massive sulfides.

metal between the sulfide melt and the silicate magma, and X is the weight percent of segregated sulfides. Assuming partition coefficients for Cu and Pd of 1000 and 10,000, respectively (Barnes et al., 1993), the calculations show that 0.001–0.04 wt% sulfide may have been removed from the magma at depth, perhaps within an auxiliary magma chamber.

DISCUSSION

The formation of a Ni-Cu-PGE deposit is the result of several fundamental factors (Barnes et al., 1997): (1) the generation of a favorable magma containing sufficient metals and heat; (2) the presence of magma conduits (faults, fractures); (3) the occurrence of assimilation and contamination, resulting in sulfide saturation of the magma; (4) the interaction of the sulfides with a large volume of magma (high R factor); and (5) the efficient accumulation of the sulfides.

Favorability of the Magma

Barnes et al. (1985) showed that a 20 to 25% rate of fusion in the mantle is sufficient to melt all of the sulfides in that part of the mantle, resulting in a metal-enriched magma. In addition, fusion at this rate will result in a hot, MgO-rich magma. The Lac Volant gabbronorite dike was formed from a relatively evolved mafic magma. The magma could not have been in equilibrium with the mantle or represent a primary mantle melt because its Mg # (0.52) is much lower that that of the mantle (greater than 0.66; Wilson, 1989). Petrographic analysis has shown that olivine is not present in the dike. However, it is possible that olivine fractionated at an early stage in the development of the Lac Volant magma. The characteristic breccia structures suggest that the dike was formed from multiple magma injections that came from a magma chamber located at greater depth. The existence of such a chamber is indicated by petrographic data, such as the presence of large phenocrysts of plagioclase (up to 8 cm) and small xenoliths of pyroxenite.

Crustal Magma Conduits

Once a mafic magma is generated in the mantle, rapid ascent is necessary in order to avoid early sulfide separation (Barnes et al., 1988). This is possible in zones of crustal extension (rifts) and in major faults associated with suture zones, e.g., Voisey's Bay (Ryan et al., 1995; Naldrett, 1997). The age of the Lac Volant dike is 1351 ± 6 Ma (U-Pb) (Gobeil et al., 1999), which is comparable to the ages of the anorthosite-mangerite-charnockite-granite Nain Plutonic Suite (1350–1290 Ma; Gower, 1996) and the Rivière-Pentecôte Anorthosite (1354 ± 3 Ma, Martignole et al., 1993; 1365 + 7/–4 Ma, Emslie and Hunt, 1990). In spite of the small quantity of geochronological evidence concerning events between 1400 and 1300 Ma in this part of the Grenville, there is some suggestion that the eastern part of the Grenville was affected at this time by major faults resulting from extensional tectonics, perhaps related to incomplete rifting (Gower, 1996). The age similarities with the previously mentioned igneous events suggest that the Lac Volant magma was related in some way to the igneous activity that generated these intrusions. The Matamec Complex was formed from a large volume of mafic magmas that rose in the crust along a major fault or suture. Several regional-scale faults and fault-parallel mafic dikes have been noted in the Matamec Complex. Mafic magmas, including the Lac Volant magma, probably rose through the crust along such faults. This is suggested by the fact that the Lac Volant dike is fracture-controlled and oriented subparallel to northeast-southwest lineament.

Assimilation, Contamination, and Sulfide Saturation

It is believed that nickel, copper, and PGE were present in the primitive Lac Volant magma in sufficient quantities to produce the observed concentrations of metals in the sulfide phase. However, it is doubtful that the original magma contained enough sulfur to produce the sulfides. Many studies of magmatic nickel deposits have suggested that the nickel sulfide liquids are generated as a result of assimilation by the mafic magma of external S, especially from sedimentary rocks or sulfide-bearing paragneiss (Ripley and Alawi, 1988; Eckstrand et al., 1989; Lesher and Arndt, 1995; Thériault et al., 1997). This hypothesis is based on sulfur isotope data, S/Se ratios, and the presence of xenoliths.

The high values of the S/Se ratio from Lac Volant (9,000 to 16,000) suggest that the magma acquired its sulfur from an external source. On the other hand, sulfur isotope data from Lac Volant (Nabil, 1999) show values of $\delta^{34}S$ between +0.6 and +1.8‰, with an average of +1.3‰. These values are similar to those for mantle-derived sulfur, contrary to data from several magmatic deposits, such as those at Duluth (Thériault et al., 1997), Muskox, and Noril'sk (Barnes et al., 1997), which show large negative values of $\delta^{34}S$. Lac Volant sulfides are poor in elements such as arsenic and antimony, which are good indicators of contamination. The low As and Sb values are probably due either to low concentrations of these elements in the metasediments of the area or to their remobilization during metamorphism. In addition, the gabbronorite has slightly greater mantle-normalized concentrations of light REE and other incompatible elements (e.g., Rb, Th, and Ba) and shows a negative Ta anomaly. This chemical signature is interpreted as the result of assimilation by the magma of lithophile elements from felsic country rocks. Further evidence for assimilation is the local presence in the dike or subsidiary dikes of fragments of quartzite and granite. Electron microprobe analysis of the sulfides has shown that monoclinic pyrrhotite and pyrite are the only iron-rich sulfides, a fact that suggests the addition of sulfur to the mafic magma. Taken together, the geological, petrographic, and geochemical evidence suggests that the Lac Volant magma experienced 10 to 20% contamination by crustal rocks, which contributed sulfur to the magma and affected the dike's geochemical signature (Nabil, 1999).

Sulfide Separation and Interaction of the Sulfide with the Magma

The low PGE concentrations in the Lac Volant magma suggest that, in spite of the presence of faults, magma ascent was slow. This may have been the case for several reasons, such as the great thickness of the crust. The reduction in the speed of magma flow would have hampered transport of the high-density sulfides, and, as a result, they would have accumulated in the conduit. This scenario is supported by geochemical data and modeling results (see earlier discussion).

Modeling has shown (see earlier discussion) that the early separation of 0.04% sulfides from the magma is sufficient to produce the sulfide compositions (for example, the Cu/Pd ratio) observed at Lac Volant. The presence of globules of sulfide at Lac Volant attests to the early accumulation of sulfides. Sulfide globules are considered evidence for immiscibility between sulfide liquid and silicate magma. They are thought to have originated in a mass of accumulated sulfide liquid.

An important parameter involved in the separation of sulfides is the ratio between the masses of silicate magma and sulfide liquid that interacted (the R factor). This factor is responsible for metal variations in magmatic sulfide deposits such as those at Duluth, Muskox, Cape Smith, and Noril'sk (Barnes et al., 1997; Thériault et al., 1997). Modeling sulfide compositions has suggested that the value of the R factor at Lac Volant was low, between 100 and 200 (Nabil, 1999). The disseminated sulfides are slightly richer in nickel than the massive sulfides (Clark, 2003), possibly suggesting a greater degree of interaction with the silicate magma (higher R factor) in the case of the disseminated sulfides. Alternatively, in the manner suggested for Voisey's Bay sulfides (Li and Naldrett, 1999), the metal concentration of disseminated sulfides that separated from an early batch of magma may have been upgraded by reaction with a later pulse of magma that was slightly richer in nickel (Clark, 2003). Low values for the Cu/Zr ratio (Li and Naldrett, 1999) in the gabbronorite dike support this hypothesis. Low Cu/Zr ratios point to chalcophile element depletion of the magma, which can be caused by sulfide separation at depth (Li and Naldrett, 1999).

Fractional Crystallization of Sulfides

Sulfide fractionation may occur within a mass of cooling and crystallizing sulfide liquid. MSS separating early from the sulfide liquid has a greater affinity for Os, Ir, Ru, and Rh relative to Pt, Pd, and Au (Keays and Davidson, 1976; Lightfoot et al., 1984; Barnes et al., 1997). As the MSS crystallizes, the residual liquid becomes poorer in Os, Ir, Ru, and Rh and richer in Pt, Pd, and Au. The residual liquid may escape into the country rocks, forming veins enriched in Pt, Pd, and Au. At the scale of the Lac Volant prospect, no sulfide zone showing varying metal ratios among the PGE and Au has been noted. This observation can be explained by the fact that the bulk composition of the sulfides is situated in the MSS field at 1100 °C (Fig. 4.26 in Nabil, 1999). Liquids with this composition have a restricted temperature range between their liquidus and solidus (Ebel and Naldrett, 1997), suggesting a limited possibility for extensive fractionation.

Nevertheless, small amount of fractional crystallization may explain the observed variations in the Ni/Cu ratio of the Lac Volant sulfides (Clark, 2003). It has been observed that the copper-rich parts of the massive sulfides are composed of irregular to regular veins and pancakelike masses of chalcopyrite surrounded by pyrrhotite. Such masses of chalcopyrite may represent accumulations of differentiated, Cu-rich liquid that have not migrated far from their place of origin. Regular, unzoned distribution of metal (Pt, Pd, Au) in massive sulfides may be explained by the homogenization of the sulfides during transport and postsolidification processes. Closed-system crystallization of the disseminated sulfides may explain the observed correlations among Pd, Au, and Cu.

Sulfide Accumulation

The great variability in the abundance of disseminated sulfides in the dike and breccias suggests preconcentration of sulfides, probably in a magma chamber or other trap located at depth. If the steeply dipping Lac Volant dike had initially been subhorizontal, as suggested by its subconcordant relations with its hosts and its location on the steeply dipping west flank of a large synform, it may be concluded that accumulation of sulfides occurred in the depression in the floor of the dike, now represented by the protuberance in its northwest wall (Fig. 4). Alternatively, if the dike was initially vertical, which is less probable in our opinion, the protuberance might represent a concavity in the wall of the dike in which sulfides accumulated, analogous to the case at Voisey's Bay (Evans-Lamswood et al., 2000). It is also possible that the zigzag in the direction of the dike had an influence on the accumulation of sulfides.

Model for the Origin of the Dike and Its Sulfides

Mineralogical and geochemical evidence points to a magmatic origin for the Lac Volant sulfides. They were formed as a result of sulfide immiscibility in a tholeiitic mafic magma, likely in a magma chamber located at an unknown depth below the dike (Fig. 20, A). The possible existence of the hidden chamber is indicated by petrographic data, such as the phenocrysts of plagioclase measuring several centimeters in length and the small xenoliths of pyroxenite and anorthosite. The brecciated nature of the mineralization and the geochemical data suggest the presence of disseminated, matrix, and probably massive sulfides at depth (Fig. 20, A).

Injection of the magma into the dike was highly dynamic. The magma transported and assimilated fragments of crustal rocks (granite, quartzite) (Fig. 20, B). Crustal assimilation affected the chemical composition of the magma and the sulfides by increasing the concentration of lithophile elements and by increasing the S/Se ratio. Sulfides were entrained by the magma and deposited at favorable sites. The injection of the magma caused fracturing and erosion of the wallrocks of

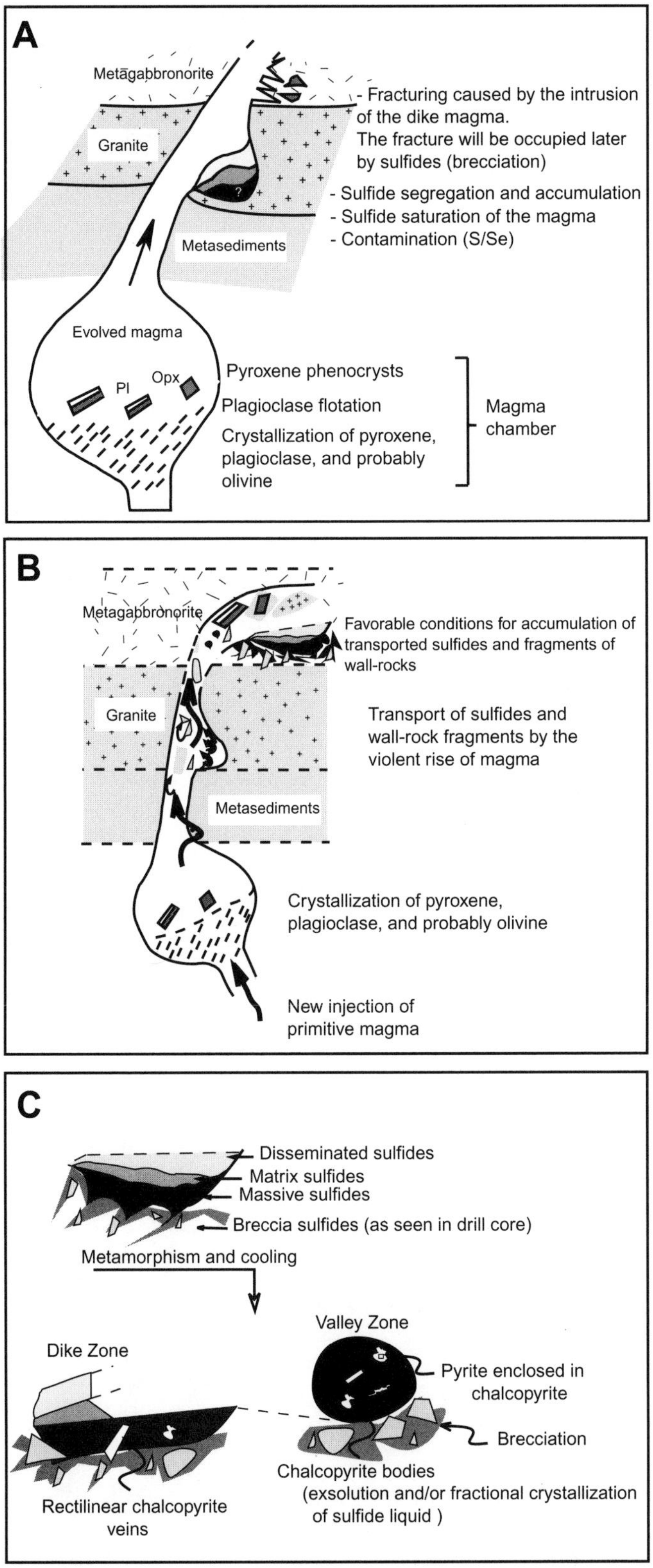

Figure 20. Model for the origin of the mineralization. (A) Magma evolution and sulfide segregation at depth (Pl—plagioclase; Opx—orthopyroxene). (B) Violent injection of new magma, which causes transport of sulfides and fragments of the wallrocks. (C) Sulfide recrystallization caused by metamorphism and deformation.

the dike, bringing about sulfide accumulation in a nearly horizontal trap (Fig. 20, B). Furthermore, the sulfides are mostly located near what appears to be a bend in the dike. This fact suggests that magma flow may have been halted, favoring accumulation. Later metamorphism caused the complete "relatively static" recrystallization of the sulfides, forming large pyrrhotite poikiloblasts. During cooling, textures such as the centimeter-scale pyrite crystals surrounded by chalcopyrite were formed (Fig. 20, C). The effects of deformation on the sulfides are not readily apparent in the field, but it is possible that the apparently isolated body of massive sulfides forming the Valley Zone may be a manifestation of these effects (Fig. 20, C). These sulfides may have been detached from the massive sulfides of the dike during the deformation. Finally, supergene alteration partly transformed the pentlandite into bravoite and violarite.

Relations among the Lac Volant (Grenville), Voisey's Bay (Labrador), and Ertelien and Flåt (Norway) Deposits

The Lac Volant sulfides in some ways resemble Sveconorwegian deposits, such as the Ertelien and Flåt deposits located in southern Norway (Barnes et al., 1988), and the Voisey's Bay deposit (Naldrett and Li, 2000). Similarities with those at Voisey's Bay include the mode of emplacement (dike, proximity of intrusion to regional faults), as well as the age, magma type, petrographic characteristics (sulfide textures, brecciation), and geochemistry (sulfur isotopes, PGE concentrations, and R factor). The Grenville Province extends into Europe as the Sveconorwegian Province in southern Norway and Sweden (Max, 1979; Gower et al., 1990). Located ~40 km northwest of Oslo (Norway), the Ertelien and Flåt nickel sulfide deposits occur within a norite intrusion measuring 600 m by 450 m. These large deposits were in production between 1849 and 1920. The mafic intrusion was emplaced between 1200 and 1370 Ma, probably during the initial rifting related to the Sweconorwegian orogeny (Oftedahl, 1980). These deposits are characterized by low grades of precious metals compared to nickel and copper (Barnes et al., 1988).

CONCLUSIONS

The present petrographic and geochemical study has led to the following conclusions: The Lac Volant mineralization is hosted by an evolved gabbronorite dike. The magma was tholeiitic and similar to N-MORB, and it originated in depleted mantle. The dike was formed in a dynamic system by multiple injections of sterile and sulfide-bearing magmas. Geochemical variations in the dike are due to fractional crystallization of mafic silicates (olivine, orthopyroxene, clinopyroxene) and plagioclase and to contamination by crustal rocks. The dike is chemically similar to the enclosing gabbronorites, suggesting that their respective magmas originated in a similar way.

Petrographic evidence (xenoliths of granite and metasediment) and geochemical evidence (enrichments in Rb, Th, Ba, and light REE; the presence of a Ta anomaly) suggest that the

Lac Volant magma assimilated ~15% paragneiss country rocks. Contamination occurred at depth, before mafic silicates began crystallizing.

Lac Volant sulfides are of magmatic origin. Massive, matrix, and disseminated sulfides have a similar mineralogy. They consist of pyrrhotite (75%), pentlandite (partially transformed to bravoite and violarite), chalcopyrite, and pyrite. The dike and the enclosed sulfides are characterized by a brecciated structure, suggesting a degree of cooling and solidification between magma pulses. Late magma injections appear to have been richer in sulfides than the early pulses.

Metasedimentary xenoliths in the dike and a high S/Se ratio (9,000 to 16,000) suggest that contamination of the magma caused sulfide saturation. The sulfide liquid interacted with only a small volume of silicate magma (R = 100–200).

The various types of sulfides are characterized by low PGE grades compared to Ni and Cu. It appears that the magma lost its PGEs during an early episode of sulfide separation that occurred at depth. Variations in Ni/Cu are possibly due to fractional crystallization of the sulfide melts.

Metamorphism affected sulfide textures through complete recrystallization of the sulfides. Meteoric alteration caused the partial transformation of pentlandite to bravoite and violarite.

The Lac Volant prospect is similar in composition, emplacement style, magma type, and age to certain Norwegian deposits (the Ertelien and Flåt deposits in the Sveconorvegian Province) and to the Voisey's Bay deposit in Labrador. As such it constitutes key evidence that magmatic sulfides can be an important exploration target in the Grenville Province.

ACKNOWLEDGMENTS

This study was financed in part by a Natural Science and Engineering Research Grant (SJB) and by Géologie Québec (TC). Hassan Nabil would like to thank the Université du Québec Fondation for winter stipends and contributions to his travel expenses. We thank André Gobeil and Serge Chevé for their expert help in the field. We also thank André Beaulé and Charlotte Grenier (MRNQ) for the technical drafting of Figures 1 and 2. The authors also thank Marc Constantin and Wolfgang Maier for their constructive reviews and Louise Corriveau for her editorial work, which considerably improved the readability of the manuscript. This is Ministère des Ressources naturelles du Québec publication number 02-5110-01.

REFERENCES CITED

Auclair, G., and Fouquet, Y., 1987, Distribution of selenium in high-temperature hydrothermal sulfide deposits at 13° north, East Pacific Rise: Canadian Mineralogist, v. 25, p. 577–587.

Barnes, S.-J., and Francis, D., 1995, The distribution of platinum-group elements, nickel, copper and gold in the Muskox layered intrusion, Northwest Territories, Canada: Economic Geology, v. 90, p. 135–154.

Barnes, S.-J., Naldrett, A.J., and Gorton, M.P., 1985, The origin of the fractionation of platinum-group elements in terrestrial magmas: Chemical Geology, v. 53, p. 303–323.

Barnes, S.-J., Boyd, R., Korneliussen, A., Nilsson, L.P., Often, M., Pedersen, R.B., and Robins, B., 1988, The use of mantle normalisation in discriminating between the effects of partial melting, crystal fractionation and sulfide segregation on platinum-group elements, gold, nickel and copper: Examples from Norway, *in* Prichard, H.M., et al., eds., Geo-platinum 87: London, Elsevier, p. 113–143.

Barnes, S.-J., Couture, J.-F., Sawyer, E.W., and Bouchaib, C., 1993, Nickel-copper occurrences in the Belleterre-Angliers Belt of Pontiac Subprovince and the use of Cu-Pd ratios in interpreting platinum-group element distributions: Economic Geology, v. 88, p. 1402–1418.

Barnes, S.-J., Zientek, M.L., and Severson, M.J., 1997, Ni, Cu, Au, and platinum element contents of sulfides associated with intraplate magmatism: A synthesis: Canadian Journal of Earth Sciences, v. 34, p. 337–351.

Barnes, S.-J., Melezhik, V.A., and Stanislav, V., 2001, The composition and mode of formation of the Pechenga nickel deposits: Canadian Mineralogist, v. 39, p. 447–471.

Bouchaïb, C., 1992, Distribution du Cu, Ni, Co, ÉGP, Au et Ag dans les sulfures de la Mine Lorraine [M.S. thesis]: Chicoutimi, Université du Québec à Chicoutimi, 73 p.

Campbell, I.H., and Naldrett, A.J., 1979, The influence of silicate:sulfide ratios on the geochemistry of magmatic sulfides: Economic Geology, v. 74, 1503–1505.

Chartrand, F.M., Roy, I., Cayer, A., and Gauthier, M., 2003, The Kwyjibo polymetallic iron oxide deposit, northeastern Grenville Province, Quebec: Canadian Institute of Mining, Metallurgy and Petroleum annual meeting, 4–8 May, Montreal, Program with Abstracts.

Chevé, S., Gobeil, A., Clark, T., Corriveau, L., Perreault, S., Dion, D.-J., and Daigneault, R., 1999, Géologie de la région du lac Manitou (SNRC 22I/14): Québec, Ministère des Ressources naturelles, Québec, RG 99-02, 69 p.

Clark, T., 2003, Métallogénie des métaux usuels et précieux, des éléments radioactifs et des éléments des terres rares, région de la moyenne Côte-Nord, in Synthèse géologique et métallogénique de la partie est de la Province de Grenville: Québec, Ministère des Ressources naturelles, Québec, DV 2002-03, 419 p.

Condie, K.C., 1993, Chemical composition and evolution of the upper continental crust: Contrasting results from surface samples and shales: Chemical Geology, v. 104, p. 1–37.

Ebel, D.S., and Naldrett, A.J., 1997, Crystallisation of sulfide liquids and the interpretation of ore composition: Canadian Journal of Earth Sciences, v. 34, p. 352–365.

Eckstrand, O.R., and Cogulu, E., 1986, Se/S evidence relating to genesis of sulphides in the Crystal Lake Gabbro, Thunder Bay, Ontario: Geological Association of Canada–Mineralogical Association of Canada, Annual Meeting, 19–21 May, Ottawa, Ontario, Program with Abstracts, v. 11, p. 66.

Eckstrand, O.R., and Hulbert, L.J., 1987, Selenium and the source of sulphur in magmatic nickel and platinum deposits: Geological Association of Canada–Mineralogical Association of Canada, Annual Meeting, 25–27 May, Saskatoon, Saskatchewan, Program with Abstracts, v. 12, p. 40.

Eckstrand, O.R., Grinenko, L.N., Krouse, H.R., Paktunc, A.D., Schwann, P.L., and Scoates, R.F.J., 1989, Preliminary data on sulphur isotopes and Se/S ratios, and the source of sulphur in magmatic sulphides from the Fox River Sill, Molson Dykes and Thompson nickel deposits, northern Manitoba: Ottawa, Ontario, Geological Survey of Canada, Current Research, Part C, Paper 89-1C, p. 235–242.

Emslie, R.F., and Hunt, P.A., 1990, Ages and petrogenetic significance of igneous mangerite-charnockite suites associated with massif anorthosites, Grenville Province: Journal of Geology, v. 98, p. 213–231.

Evans-Lamswood, D.M., Butt, D.P., Jackson, R.S., Lee, D.V., Muggridge, M.G., Wheeler, R.I., and Wilton, D.H.C., 2000, Physical controls associated with the distribution of sulfides in the Voisey's Bay Ni-Cu-Co deposit, Labrador: Economic Geology, v. 95, p. 749–769.

Genkin, A.D., Distler, V.V., Lapuntina, I.P., and Flimonova, A.A., 1973, Geochemistry of palladium in copper-nickel ores: Geochemistry International, v. 10, p. 1007–1013.

Gobeil, A., Perreault, S., Clark, T., Chevé, S., Corriveau, L., and Nabil, H., 1996, Cadre géologique et potentiel minéral de la région de Manitou-Nipisso: Québec, Ministère des Ressources naturelles, Québec, DV 97-03, p. 15.

Gobeil, A., Chevé, S., Clark, T., Corriveau, L., Perreault, S., Dion, D.J., and Nabil, H., 1999, Géologie de la région du lac Nipisso (SNRC 22I/13): Québec, Ministère des Ressources naturelles, Québec, RG 98-19, 60 p.

Gower, C.F., 1996, The evolution of the Grenville Province in eastern Labrador, Canada, *in* Brewer, T.S., ed., Precambrian crustal evolution in the North Atlantic Region: Geological Society [London] Special Publication, v. 112, p. 197–218.

Gower, C.F., Ryan, B., and Rivers, T., 1990, Mid-Proterozoic Laurentia-Baltica: An overview of its geological evolution and a summary of the contribution made by this volume, *in* Gower, C.F., et al., eds., Mid-Proterozoic Laurentia-Baltica: St. John's, Newfoundland, Geological Association of Canada Special Paper 38, p. 1–20.

Hogan, H.R., and Grenier, P.A., 1971, Rivière Nipissis-lac Nipisso, comté de Saguenay: Québec, Ministère des Ressources naturelles, Québec, RG-142, 43 p.

Huss, L., 2002, Caractérisation de la minéralisation en Ni-Cu-EGP des indices de la région du Lac à Paul, suite anorthositique de Lac St-Jean [M.S. thesis]: Chicoutimi, Université du Québec à Chicoutimi, 220 p.

Irvine, T.N., and Baragar, W.R.A., 1971, A guide to the chemical classification of the common volcanic rocks: Canadian Journal of Earth Sciences, v. 8, p. 523–548.

Jenkins, J.T., 1957, Région du lac Manitou, Comté de Saguenay: Québec, Ministère des Mines, Québec, RP-349, 8 p.

Keays, R.R., and Davidson, R.M., 1976, Palladium, irridium and gold in the ores and host rocks of nickel sulfide deposits in Western Australia: Economic Geology, v. 71, p. 1214–1228.

Kullerud, G., Yund, R.A., and Moh, G.H., 1969, Phase relations in the Cu-Fe-S, Cu-Ni-S, and Fe-Ni-S systems: Economic Geology Monograph, v. 4, p. 323–343.

Lesher, C.M., and Arndt, N.T., 1995, REE and Nd isotope geochemistry, petrogenesis and volcanic evolution of contaminated komatiites at Kambalda, Western Australia: Lithos, v. 34, p. 127–157.

Li, C., and Naldrett, A.J., 1999, Geology and petrology of the Voisey's Bay intrusion: Reaction of olivine with sulfide and silicate liquids: Lithos, v. 47, p. 1–31.

Lightfoot, P.C., Naldrett, A.J., and Hawkesworth, C.J., 1984, The geology and geochemistry of the Waterfall Gorge section of the Insizwa complex with particular reference to the origin of nickel sulfide deposits: Economic Geology, v. 79, p. 1857–1879.

Martignole, J., Machado, N., and Nantel, S., 1993, Timing of intrusion and deformation of the Rivière-Pentecôte anorthosite (Grenville Province): Journal of Geology, v. 101, p. 652–658.

Max, M.D., 1979, Extent and disposition of Grenville tectonism in the Precambrian continental crust adjacent to the North Atlantic craton: Geology, v. 7, p. 76–78.

Nabil, H., 1999, Caractérisation de la minéralisation en Ni-Cu-Co de l'indice de Lac-Volant, région de Sept-Îles, Québec [M.S. thesis]: Chicoutimi, Université du Québec à Chicoutimi, 189 p.

Naldrett, A.J., 1981, Nickel sulfide deposits: Classification, composition and genesis: Economic Geology, v. 75, p. 628–685.

Naldrett, A.J., 1984, Mineralogy and composition of Sudbury ores: Sudbury, Ontario Geological Survey, Special Volume 1, p. 309–326.

Naldrett, A.J., 1997, Key factors in the genesis of Noril'sk, Sudbury, Jinchuan, Voisey's Bay and other world-class Ni-Cu-PGE deposits: Implications for exploration: Australian Journal of Earth Science, v. 44, p. 283–315.

Naldrett, A.J., and Gasparrini, E.L., 1971, Archean nickel sulphide deposits in Canada: Their classification, geological setting and genesis with some suggestions as to exploration: Sydney, Geological Society of Australia, Special Publication, v. 3, p. 201–226.

Naldrett, A.J., and Li, C., eds., 2003, A special issue on Voisey Bay's Ni-Cu-Co deposit: Economic Geology, v. 95, p. 673–915.

Oftedahl, C., 1980, The geology of Norway: Norges Geologiske Undesokelse Bulletin, v. 54, p. 3–114.

Paktunc, A.D., 1989, Petrology of the St. Stephen intrusion and the genesis of related nickel-copper sulfide deposits: Economic Geology, v. 84, p. 817–840.

Pearce, J.A., 1983, The role of sub-continental lithosphere in magma genesis at destructive plate margins, in Hawkesworth, C.J., and Norry, M.J., eds., Continental basalts and mantle xenoliths: Nantwich, England, Shiva, p. 230–249.

Perreault, S., Clark, T., Gobeil, A., Chevé, S., Dion, D.-J., Corriveau, L., Nabil, H., and Lortie, P., 1996, Le potentiel en Cu-Ni-Co de la région de Sept-Îles: L'indice du lac Volant: Québec, Ministère des Ressources naturelles, Québec, PRO 96-06, 10 p.

Perreault, S., Clark, T., Gobeil, A., Chevé, S., Dion, D.-J., Corriveau, L., Nabil, H., and Lortie, P., 1997, The Cu-Ni-Co potential of the Sept-Îles region: The lac Volant showing: Québec, Ministère des Ressources naturelles, Québec, PRO 97-03, 10 p. (translation of Perreault et al., 1996).

Perry, C., and Roy, I., 1997, Projet Lac Volant 1204, rapport sur les travaux d'exploration: Québec, Ministère des Ressources naturelles, Québec, GM-55933, 150 p.

Raymond, D., Perry, C., and Roy, I., 1997, Le projet de SOQUEM: Une première année d'exploration: Québec, Ministère des Ressources naturelles, Québec, DV 97-03, p. 15.

Ripley, E.M., and Alawi, J.A., 1988, Petrogenesis of pelitic xenoliths at the Babbitt Cu-Ni deposit, Duluth Complex, Minnesota, USA: Lithos, v. 21, p. 143–159.

Rivers, T., Martignole, J., and Davidson, A., 1989, New tectonic divisions of the Grenville Province, southeast Canadian Shield: Tectonics, v. 8, p. 63–84.

Ryan, B., Wardle, R.J., Gower, C.F., and Nunn, G.A.G., 1995, Nickel-copper sulphide mineralisation in Labrador: The Voisey's Bay discovery and its exploration implications: Geological Survey, St. John's, Newfoundland, Department of Natural Resources, Government of Newfoundland and Labrador, Current Research, Report 95-1, p. 177–204.

Saint-Germain, P., 2002, Caractérisation magmatique et géochimique du Complexe de gabbronorite et de monzonite de Matamec, région de Sept-Îles, Grenville oriental (Québec) [M.S. thesis]: Québec, Institut national de la recherche scientifique–Eau, Terre et Envionnement, 192 p.

Saint-Germain, P., and Corriveau, L., 2003, Évolution magmatique et géochimique des gabbronorites et des monzonites du Complexe de Matamec, région de Sept-Îsles, *in* Brisebois, D., and Clark, T., eds., Géologie et ressources minérales de la partie est de la Province de Grenville: Québec, Ministère des Ressources naturelles, Québec, DV 2002-03, 419 p.

Shervais, J.W., 1982, Ti-V plots and petrogenesis of modern and ophiolitic lavas: Earth and Planetary Science Letters, v. 57, p. 101–118.

Taylor, S.R., and McLennan, S.M., 1985, The continental crust: Its composition and evolution: Oxford, England, Blackwell, 312 p.

Thériault, R.D., Barnes, S.-J., and Severson, M.J., 1997, The influence of country rock assimilation and silicate to sulfide ratios (R factors) on the genesis of the Dunka Road Cu-Ni-PGE deposit, Duluth Complex, Minnesota, USA: Canadian Journal of Earth Sciences, v. 34, p. 373–389.

Thompson, R.N., 1982, Magmatism of the British Tertiary volcanic province, Scotland: Journal of Geology, v. 93, p. 603–608.

Vaillancourt, C., 2001, Étude géochimique et économique de la suite mafique et ultramafique de la Baie-à-Cadie au Lac Kénogami, Saguenay-Lac St-Jean, Québec [M.S. thesis]: Chicoutimi, Université du Québec à Chicoutimi, 202 p.

Viereck, L.G., Flower, M.F.J., Hertogen, J., Schmincke, H.U., and Jenner, G.A., 1989, The genesis and significance of N-MORB sub-types: Contributions to Mineralogy and Petrology, v. 102, p. 112–126.

Wilson, M., 1989, Igneous Petrology: London, Chapman and Hall, 466 p.

Manuscript Accepted by the Society August 25, 2003

Printed in the USA

Geological Society of America
Memoir 197
2004

Oxygen-isotope constraints on terrane boundaries and origin of 1.18–1.13 Ga granitoids in the southern Grenville Province

William H. Peck*
Department of Geology, Colgate University, Hamilton, New York 13346, USA
John W. Valley
Department of Geology and Geophysics, University of Wisconsin, Madison, Wisconsin 53706, USA
Louise Corriveau
Natural Resources Canada, Geological Survey of Canada, GSC-Québec, 880 Chemin Ste-Foy, Room 840, Québec City, Québec G1S 2L2, Canada
Anthony Davidson
Geological Survey of Canada, 601 Booth Street, Ottawa, Ontario K1A OE8, Canada
James McLelland
Department of Geology, Colgate University, Hamilton, New York 13346, USA
David A. Farber
Department of Geology and Geophysics, University of Wisconsin, Madison, Wisconsin 53706, USA

ABSTRACT

Granitic rocks related to 1.18 to 1.13 Ga anorthosite-mangerite-charnockite-granite plutonism stitch three terranes in the southwestern Grenville Province (Adirondack Highlands–Morin terrane, Frontenac terrane, Elzevir terrane). Because of the refractory nature of zircon (Zrn), analysis of oxygen-isotope ratios of dated igneous zircon from these rocks allows calculation of $\delta^{18}O$ values of original magmas even if the rocks were subjected to late magmatic assimilation, postmagmatic alteration, or metamorphism. Documented variability in $\delta^{18}O(Zrn)$ for these granitic rocks corresponds to their geographic location. Seven plutons from the central Frontenac terrane (Ontario) have a high average $\delta^{18}O(Zrn) = 11.8 \pm 1.0‰$, which corresponds to $\delta^{18}O$ magma values of 12.4–14.3‰. In contrast, twenty-seven other plutons and dikes of this suite (New York, Ontario, and Québec) average $\delta^{18}O(Zrn) = 8.2 \pm 0.6‰$, with a typical igneous range of 8.6 to 10.3‰ for $\delta^{18}O$ magma values. High $\delta^{18}O$ values in the Frontenac terrane are some of the highest magmatic oxygen-isotope ratios recognized worldwide, but these plutons are not unusual with respect to whole-rock chemistry or radiogenic isotope compositions. Such high $\delta^{18}O$ values can result from mixing between paragneiss ($\delta^{18}O \approx 15‰$) and hydrothermally altered basalts and/or oceanic sediments ($\delta^{18}O \approx 12‰$) in the source region. We propose that high-$\delta^{18}O$, hydrothermally altered basalts and sediments were subducted or underthrust to the base of the Frontenac terrane during closure of an ocean basin between the Frontenac terrane and the Adirondack Highlands at or prior to 1.2 Ga.

Keywords: anorthosite suite, Grenville Province, oxygen isotopes, zircon

*E-mail: wpeck@mail.colgate.edu.

Peck, W.H., Valley, J.W., Corriveau, L., Davidson, A., McLelland, J., and Farber, D.A., 2004, Oxygen-isotope constraints on terrane boundaries and origin of 1.18–1.13 Ga granitoids in the southern Grenville Province, *in* Tollo, R.P., Corriveau, L., McLelland, J., and Bartholomew, M.J., eds., Proterozoic tectonic evolution of the Grenville orogen in North America: Boulder, Colorado, Geological Society of America Memoir 197, p. 163–182. For permission to copy, contact editing@geosociety.org.

INTRODUCTION

In the southwestern Grenville Province different criteria that have been used to identify crustal scale subdivisions include structural style (e.g., Davidson, 1984), ages of igneous suites (e.g., Easton, 1986; Corriveau, 1990; Friedman and Martignole, 1995; McLelland et al., 1996; Corriveau and van Breemen, 2000), age of peak metamorphism (e.g., Mezger et al., 1992, 1993; McLelland et al., 1993), peak metamorphic conditions (e.g., Bohlen et al., 1985; Anovitz and Essene, 1990; Valley et al., 1990; Streepey et al., 1997), and cooling rates (e.g., Mezger et al., 1991; Cosca et al., 1992; Martignole and Reynolds, 1997). Integrated geological and geophysical studies derived from the Lithoprobe project (Carr et al., 2000; Corriveau and Morin, 2000; Martignole et al., 2000) and detailed studies across individual shear zones (e.g., Mezger et al., 1992; van der Pluijm et al., 1994; Busch et al., 1997; Cureton, et al., 1997) also constrain histories of the assembly and cooling of midcrustal blocks.

We use oxygen-isotope compositions of zircons from 1.18 to 1.13 Ga granitic plutons to investigate terrane boundaries in the Allochthonous Monocyclic Belt of the southwestern Grenville Province (Rivers et al., 1989; Fig. 1). Such "mapping" of the lower crust is achieved by isotopic analysis of whole rocks from Phanerozoic plutons (e.g., Solomon and Taylor, 1989), but analysis of zircon is necessary for igneous oxygen-isotope ratios to be determined precisely and accurately in high-grade (commonly polymetamorphic) rocks. Weakly (or para-) magnetic, nonmetamict igneous zircons are systematically recoverable from granitic plutons, and preserve primary igneous oxygen-isotope ratios through conditions ranging from hydrothermal alteration to high-grade metamorphism (e.g., Valley et al., 1994; Gilliam and Valley, 1997; King et al., 1997; 1998; Peck and Valley, 1998; Peck et al., 1999; see Valley, 2003). Their refractory nature allows zircon to "see past" common postmagmatic oxygen-isotope exchanges that affect whole-rock and mineral oxygen-isotope ratios in many orogenic belts. The plutons studied intruded within a similar time span into three major subdivisions of the Allochthonous Monocyclic Belt, spanning >300 km (east to west), and crossing two previously proposed terrane boundaries (Fig. 1). A focus of this study is two anorthosite-mangerite-charnockite-granite (AMCG) suites, namely the 1.16 to 1.13 Ga Morin AMCG suite in the Morin terrane (Québec) (Martignole and Schrijver, 1970; Doig, 1991), and the ca. 1.15 Ga Marcy AMCG suite (Adirondack Highlands) (McLelland and Chiarenzelli, 1990; Clechenko et al., 2002; McLelland et al., 2002). We also examined two suites coeval with AMCG plutonism, namely the 1.17 to 1.16 Ga Chevreuil suite in the Central Metasedimentary Belt (Québec) (Corriveau and van Breemen, 2000) and the 1.18 to 1.15 Ga Frontenac suite in the Central Metasedimentary Belt (Ontario) (Davidson and van Breemen, 2000). Oxygen-isotope ratios of magmas that formed these plutons were determined by analysis of igneous zircon (Valley et al., 1994; this study), allowing recognition of regional domains in the oxygen-isotope ratio of the lower crust (the inferred pluton source region).

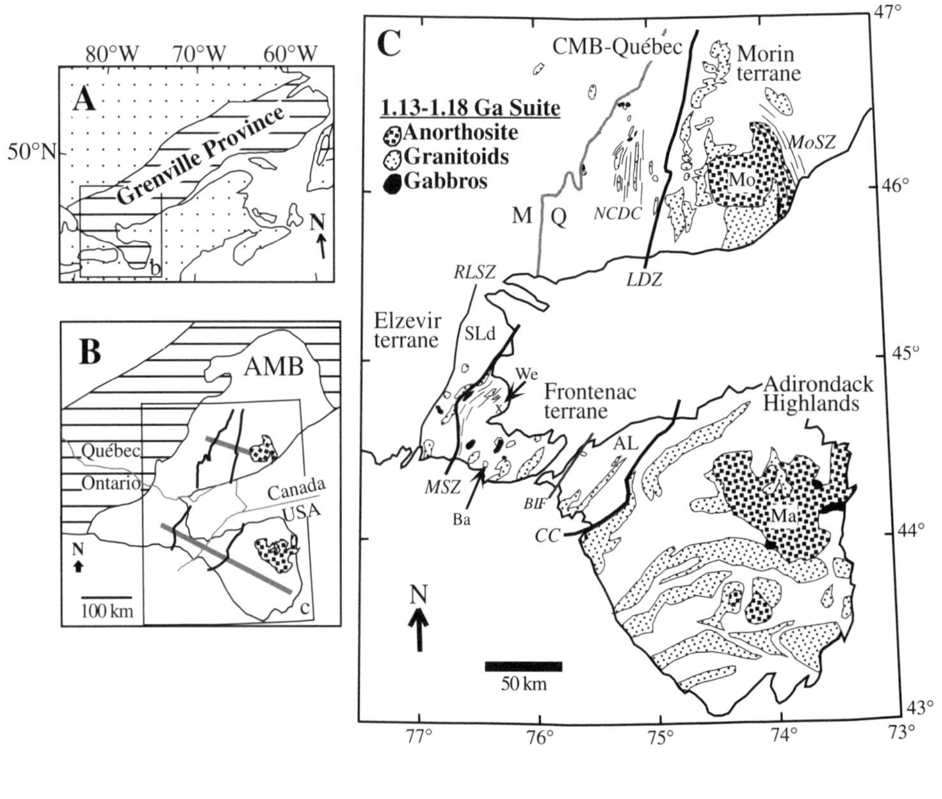

Figure 1. (A) Location of the study area within the Grenville Province. (B) Location map and subdivisions of the Allochthonous Monocyclic Belt (AMB) of the Grenville Province (after Rivers et al., 1989; Davidson, 1995; McLelland et al., 1996; Corriveau et al., 1998). (C) Detail of B. AL—Adirondack Lowlands (Frontenac terrane); Ba—Battersea pluton; BLF—Black Lake fault; CC—Carthage-Colton mylonite zone; CMB-Québec—Central Metasedimentary Belt of Québec; LDZ—Labelle deformation zone; M—marble domain; MSZ—Maberly shear zone; Ma—Marcy anorthosite massif; Mo—Morin anorthosite massif; MoSZ—Morin shear zone; Nominingue-Chénéville deformation zone; Q—quartzite domain of the CMB-Québec; RLSZ—Robertson Lake shear zone; NCDZ—SLd—Sharbot Lake domain of the Elzevir terrane; We—location of the 1.08 Ga Westport pluton. Locations of the oxygen-isotope transects are shown in gray in B.

REGIONAL GEOLOGY

The study area includes the Adirondack Highlands and the Morin terrane (Fig. 1), which are interpreted as 1.3 to 1.25 Ga arc-related rocks accreted to the Laurentian margin ca. 1.2 Ga (1.19 Ga in Wasteneys et al., 1999; 1.22 Ga in Corriveau and van Breemen, 2000). In this model the Central Metasedimentary Belt of Québec and Ontario (subdivided in Ontario into the Elzevir and Frontenac terranes) is the southeastern Laurentian margin at 1.2 Ga (present-day coordinates, Fig. 1). AMCG suite and contemporaneous 1.18 to 1.13 Ga granitic plutons intrude all of these terranes.

Adirondack Highlands

The Adirondack Highlands (Fig. 1) are primarily metaplutonic rocks, dominated by the AMCG suite and minor juvenile metasedimentary rocks ($\varepsilon_{Nd} \sim 1.3$; Daly and McLelland, 1991). The Marcy anorthosite was emplaced at 1.15 ± 0.01 Ga, which is also the age of associated charnockite and mangerite bodies (McLelland and Chiarenzelli, 1990; Clechenko et al., 2002; McLelland et al., 2002). The AMCG suite was overprinted by the granulite-facies Ottawan event at 1.09 to 1.03 Ga (7–8 kilobars, 675–800 °C; Bohlen et al., 1985; McLelland et al., 1996), followed by slow cooling (1–3 °C/m.y.; Mezger et al., 1991). Some rocks near contacts of AMCG plutons preserve low $\delta^{18}O$ values caused by interaction with heated meteoric water during contact metamorphism, indicating that these plutons intruded at relatively shallow depths (<10 km, Valley and O'Neil, 1982; Morrison and Valley, 1988; Clechenko and Valley, 2003). The Adirondack Highlands are bounded on the northwest by the Carthage-Colton mylonite zone (Geraghty et al., 1981).

Frontenac Terrane

The Frontenac terrane of the Central Metasedimentary Belt in Ontario and New York (Fig. 1, C) is made up of metaigneous and metasedimentary rocks, and is bounded on the southeast by the Carthage-Colton mylonite zone (Geraghty et al., 1981). Other lithologic breaks in the Frontenac terrane include the Black Lake fault in New York (Fig. 1, C), which marks the eastern limit of the ca. 1.16 Ga olivine diabase Kingston dike swarm (Davidson, 1995; Carl and deLorraine, 1997). Especially important to this study is a suite of ca. 1.21 Ga calc-alkaline plutons in New York (the Antwerp-Rossie suite), which is interpreted as indicating westward subduction under the eastern Frontenac terrane at this time (Wasteneys et al., 1999).

The Adirondack Lowlands of New York (Fig. 1) are included as part of the Frontenac because of similarities in rock types, magmatic crystallization ages, and timing of metamorphism (Wasteneys et al., 1999). The Adirondack Lowlands experienced amphibolite-facies conditions (640–680 °C, 6–7 kilobars; Bohlen et al., 1985; Kitchen and Valley, 1995), while peak conditions reached the granulite facies in Ontario (Streepey et al., 1997) prior to 1.18–1.16 Ga (McLelland et al., 1993; Mezger et al., 1993; Corfu and Easton, 1997). Metamorphism was broadly coeval with intrusion of monzonite, syenite, granite, and gabbro plutons of the Frontenac suite (van Breemen and Davidson, 1988; Marcantonio et al., 1990; McLelland et al., 1996; Davidson and van Breemen, 2000). In Ontario peak metamorphic conditions predated emplacement of Frontenac suite plutons, which intruded at lower pressures than peak conditions (Anovitz and Essene, 1990; Cosca et al., 1992; Wasteneys, 1994).

The Frontenac terrane is separated from the Sharbot lake domain of the Elzevir terrane on the northwest by the southeast-dipping Maberly shear zone (Fig. 1, C), a locally mylonitic zone of high strain (Corfu and Easton, 1997; Davidson and van Breemen, 2000). Highly strained rocks in the Frontenac hangingwall extend up to15 km southeast of the Maberly shear zone, and are intruded by syntectonically emplaced Frontenac suite plutons.

Sharbot Lake Domain

The Elzevir terrane is divided into several domains (Easton, 1992) that are interpreted as parts of an attenuated (and rifted) marginal basin developed at the edge of pre-Grenvillian Laurentia (e.g., Pehrsson et al., 1996; Smith et al., 1997), or as amalgamated volcanic arc terranes (Carr et al., 2000). The Sharbot Lake domain is the southeasternmost of these divisions (Fig. 1, C; Corfu and Easton, 1997; Davidson and van Breemen, 2000), and is separated from the rest of the Elzevir terrane by the extensional Robertson Lake shear zone (Fig. 3, C).

The metavolcanic and plutonic rocks and associated carbonate metasedimentary rocks of the Sharbot Lake domain are deformed and variably metamorphosed from upper greenschist to granulite facies, reaching 650–700 °C and 8 kilobars near the Maberly shear zone (see Streepey et al., 1997, and references therein). Country rocks are intruded by granitic, monzonitic, and gabbroic rocks of the 1.18–1.15 Ga Frontenac suite (Fig. 1, C; Corfu and Easton, 1997; Davidson and van Breemen, 2000), which are not metamorphosed and are deformed only in the immediate footwall of the Maberly shear zone.

Morin Terrane

The granulite-facies Morin terrane (Fig. 1, C) is dominated by a central anorthosite massif, with voluminous mangerite and monzonite, and minor jotunite and gabbro of the Morin AMCG suite (Martignole and Schrijver, 1970). This magmatism is dated at 1.16 to 1.13 Ga (Emslie and Hunt, 1990; Doig 1991; Friedman and Martignole, 1995; van Breemen and Corriveau, 1995). The Morin terrane is bounded to the west by the Labelle deformation zone, along which high strain occurred between 1.17 and 1.07 Ga (Zhao et al., 1997; Martignole and Friedman, 1998; Corriveau and van Breemen, 2000). Preserved metamorphic conditions are 650–775 °C and 6–7 kilobars for the region (Indares and Martignole, 1990), and were followed by slow cooling (Martignole and Reynolds, 1997). High-grade, deformed metasedimentary rocks occur as inclusions in undeformed Morin AMCG suite rocks (e.g., Martignole and Schrijver, 1970).

Central Metasedimentary Belt (Québec)

The Central Metasedimentary Belt of Québec (Fig. 1, C) is subdivided into a western marble-rich domain (referred to as the marble domain) and an eastern quartzite-rich domain (referred to as the quartzite domain) (Corriveau et al., 1998), interpreted as northern extensions of the Elzevir and Frontenac terranes of Ontario, respectively. Granulite-facies felsic gneiss complexes with mixed volcanic and plutonic protoliths (magmatic age ca. 1.4 Ga; Wodicka et al., this volume) occur within both domains and contain evidence for peak metamorphism at ca. 1.2–1.19 Ga (Corriveau and van Breemen, 2000; ~950 °C and ~10 kilobars, Boggs and Corriveau, this volume). These gneiss complexes were intruded by undeformed commingled dioritic and monzonitic dikes of the 1.17–1.16 Ga Chevreuil suite, while the synkinematic monzonitic, dioritic, and gabbroic sheetlike plutons of this suite were preferentially emplaced into the metasedimentary domains and their deformation zones (Corriveau et al., 1998; Corriveau and Morin, 2000; Corriveau and van Breemen, 2000). In these zones, gneissic host records a metamorphic overprint at 650 °C and 6 kilobars in the western marble domain and 750 °C and 8 kilobars in both the eastern marble and the quartzite domains (Perkins et al., 1982; Indares and Martignole, 1990; Kretz, 1990; Boggs, 1996). These metamorphic conditions resulted from intraplate reactivation at ca. 1.17–1.16 Ga (the ages of monazite, titanite, and syntectonic pegmatite; Friedman and Martignole, 1995; van Breemen and Corriveau, 1995; Corriveau and van Breemen, 2000). Syntectonic Chevreuil suite plutons were emplaced coevally with early members of the Morin AMCG suite to the east (Corriveau and van Breemen, 2000).

AMCG Suite and Contemporaneous Plutonism in the Monocyclic Belt

Because the emplacement style and volume of AMCG suite and contemporaneous plutonic rocks vary across the area, correlation of these rocks between lithotectonic terranes is possible only through precise U-Pb zircon geochronology of rocks that include syenite, monzonite, granite, charnockite, mangerite, and jotunite, and have been the foci of numerous igneous petrology and geochemistry studies (e.g., Buddington, 1939; Heath and Fairbairn, 1969; Martignole and Schrijver, 1970; Barton and Doig, 1977; Ashwal and Wooden, 1983; Shieh, 1985; Wu and Kerrich, 1986; Morrison and Valley, 1988; Lumbers et al., 1990; Marcantonio et al., 1990; McLelland and Whitney, 1990; Valley et al., 1990, 1994; Daly and McLelland, 1991; Doig, 1991; Emslie and Hegner, 1993; McLelland et al., 1993; Owens et al., 1993; Eiler and Valley, 1994; Rockow, 1995; Davidson and van Breemen, 2000; Peck and Valley, 2000). These granitoids have A-type, subalkalic to alkalic chemistries (Irvine and Baragar, 1971), alkali-calcic to calc-alkaline Peacock (1931) indexes, show little iron enrichment in AFM diagrams, are metaluminous (Shand, 1951), have flat heavy rare earth element patterns, and plot as within-plate granites on tectonic discrimination diagrams (e.g., Pearce et al., 1984). These characteristics contrast with those of other A-type granitic rocks that are more subalkalic with respect to $K_2O + Na_2O$ versus SiO_2 (Irvine and Baragar, 1971), and have alkali-calcic or alkalic Peacock indexes (e.g., Anderson, 1983). Neodymium isotope ε_{Nd} values of 1.5 to 2.9, depleted mantle Nd model ages of 1.3 to 1.5 Ga, and initial $^{87}Sr/^{86}Sr$ ratios of 0.704 to 0.705 indicate that these rocks underwent little interaction with Archean crust.

The origin of AMCG suite granitoids is controversial. Many studies indicate that they are not comagmatic with contemporaneous anorthosite, and were likely formed by partial melting of lower crust by anorthosite parent magma (see Ashwal, 1993). This contrasts with interpretations that these granitic rocks are related to anorthosite parent magmas or differentiates (e.g., Frost et al., 2002). In the Central Metasedimentary Belt of Québec, easterly dipping seismic reflectors are interpreted as images of AMCG suite–related monzonite and gabbro sheets that extend from the Moho to the surface along listric high-strain zones (Corriveau and Morin, 2000).

OXYGEN-ISOTOPE METHODS

Minerals and whole-rock powders were analyzed for $\delta^{18}O$ by laser fluorination at the University of Wisconsin, and all data are given in Table 1 or Peck (2000). Oxygen was liberated from mineral separates (Table 1; Appendixes 1–5) by heating with a CO_2 laser in the presence of BrF_5 (see Valley et al., 1995) so that small sample sizes (1–4 mg) can be analyzed with high precision and accuracy, typically better than ±0.1‰. The zircons analyzed are from the same zircon separate used for U-Pb age determination (McLelland, et al., 1988; Emslie and Hunt, 1990; Marcantonio et al., 1990; McLelland and Chiarenzelli, 1990; Chiarenzelli and McLelland, 1991; Corriveau and van Breemen, 2000; Davidson and van Breemen, 2000). Zircons were soaked in cold HF to dissolve or identify radiation damage, and were hand-picked for purity. On the 29 days of mineral analysis, 120 aliquots of garnet standard (UWG-2; Valley et al., 1995) were analyzed. UWG-2 had an average $\delta^{18}O = 5.75 \pm 0.13$‰ (1 s.d. reported, 1 s.e. = 0.01‰) relative to Vienna Standard Mean Ocean Water (VSMOW). This is within the error of the published laboratory average for UWG-2 of 5.74 ± 0.15‰ (n = 1081, 1 s.e. = 0.005‰; Valley et al. 1995). The daily precision of UWG-2 averaged 0.07 ± 0.04‰. Thirty-eight zircon samples were analyzed in duplicate, and reproducibility (half the difference of two analyses) averaged 0.07 ± 0.05‰. The mineral analyses were adjusted by the amount that daily UWG-2 values deviated from 5.8‰, its accepted value (based on $\delta^{18}O$(NBS-28) = 8.59; Valley et al., 1995). This adjustment averaged 0.10‰, and was always <0.30‰.

Whole-rock powders (Appendixes 1–4) were analyzed

using an airlock sample chamber where individual samples were heated and reacted with BrF_5 (Spicuzza et al., 1998). Whole-rock powders are typically reactive with BrF_5 at room temperature, so they were isolated from the lasing chamber with a gate valve, and moved into the lasing chamber only when residual BrF_5 was removed. On the 16 days of whole-rock powder analysis, 63 aliquots of powdered UWG-2 were analyzed; they averaged 5.83 ± 0.15‰ (1s.e. = 0.02‰), and daily precision averaged 0.06 ± 0.02‰. Thirty out of 150 whole-rock powders were duplicated, and reproducibility averaged 0.06 ± 0.05‰. The whole-rock analyses were adjusted by the amount that daily UWG-2 values deviated from 5.8‰, and this adjustment averaged 0.11‰ (maximum adjustment = 0.30‰).

OXYGEN ISOTOPE RESULTS

Whole-rock (WR) $\delta^{18}O$ values from all granitic rocks, except those from the central Frontenac terrane, range from ~7 to 13‰ (Table 1; Appendixes 1–4; Fig. 2). These oxygen-isotope ratios are typical of the high end of variation seen in common granitic rocks (e.g., Taylor and Sheppard, 1986). Granitic rocks in the Frontenac terrane have $\delta^{18}O$(WR) = 9 to 16‰ (Shieh, 1985; Marcantonio et al., 1990; this study), a range much higher than reported igneous values from granitic rocks (see Taylor and Sheppard, 1986). Gabbros from the Central Metasedimentary Belt of Québec range from ~7 to 11‰ (Appendix 2; Fig. 3). The gabbro plutons have $\delta^{18}O$(WR) values comparable to those of gabbros related to AMCG suite plutonism in the Adirondack Highlands (6.8 ± 0.3‰, n = 16; Valley et al., 1994).

There is good correlation between SiO_2 and $\delta^{18}O$(WR) values for granitic rocks of each terrane (Fig. 2). Correlation is best in the Adirondack Highlands, the Morin terrane, and the marble domain (results are not plotted for the Frontenac terrane and the Sharbot Lake domain because of the small number of whole-rock analyses). Eiler and Valley (1994) also showed that individual AMCG plutons in the Adirondack Highlands have a better correlation between SiO_2 and $\delta^{18}O$ than is seen in the complete data set for Adirondack granitic rocks, reflecting

TABLE 1. OXYGEN ISOTOPE ANALYSES OF MAGMATIC ZIRCONS FROM 1.18 TO 1.13 GA GRANITOIDS, SOUTHERN GRENVILLE PROVINCE

Pluton	Sample #	SiO_2 wt%	Zircon sat. *T* (°C)	Zircon mag.	$\delta^{18}O$ zircon	$\delta^{18}O$ zircon	Average zircon	$\delta^{18}O$ cal. WR	$\delta^{18}O$ aver. WR	Age (Ma)
1 Snowy Mountain dome	AM86-7	58.5	730	M-2	8.05	—	8.05	9.37	—	>1130 A
2 Schroon Lake	9.23.85.7	69.2	890	M-2	8.19	—	8.19	9.23	10.23	1125 ± 10 A
3 Hermon granite	DF178	n.d.	~850	NM5	8.34	—	8.34	9.34	—	1149 ± 26 B
4 Edwardsville Syenite	AM87-5	61.5	850	NM-2	11.18	11.03	11.11	12.41	14.57	1164 ± 4 C
5 Battersea	LH86-63	66.1	900	NM0	11.18	11.11	11.15	12.55	13.40	1168 ± 3 D
6 Perth Road	LH87-64	61.0	860	MN0	13.09	12.93	13.01	14.35	13.80	1166 ± 3 D
7 Lyndhurst	LH87-31	68.3	720	MN0	11.11	10.72	10.92	12.97	13.60	1166 ± 3 D
8 South Lake	86DM9c	60.1	870	NM1	11.20	11.06	11.13	12.43	12.48*	1162 ± 3 E
9 Crow Lake	LH87-30	63.6	860	NM0	13.64	13.38	13.51	14.92	15.70	1176 ± 2 D
10 Beales Mills	95DM53	64.2	870	NM3	11.88	11.83	11.86	13.14	14.00	1164 ± 2 F
11 North Crosby	93DM131a	54.4	750	NM0	7.90	7.89	7.90	9.23	9.13	1157 ± 3 F
12 Bennett Bay	93DM19	55.1	630	NM0	7.90	8.03	7.97	9.31	10.01	1162 ± 3 F
13 Silver Lake	93DM20	68.1	840	NM 0.5	6.79	6.81	6.80	7.71	9.25†	1161 +3/–2 F
14 Oso	92DM152	50.0	770	NM0	7.64	—	7.64	8.65	8.58	1154 ± 2 F
15 Grey Valley	96MR43	66.2	860	NM-2	8.49	8.28	8.39	9.44	9.82	n.d.
16 Western Mangerite	96MR21	71.7	870	NM-2	9.41	9.34	9.38	10.53	11.25	n.d.
17 Maskinongé	CQA925A	n.d.	~850	M2	8.36	8.35	8.36	9.36	—	1164 ± 3 H
18 Maskinongé (dike)	CQA925b	n.d.	~850	M2	8.03	8.22	8.13	9.13	—	1168 ± 3 H
19 Mangerite dike	EC84-246	57.1	840	—	9.77	9.60	9.65	10.47	—	1146 +7/–6 I
					9.76	9.64				
					9.69	9.49				
20 Chevreuil	CQA1085	66.1	850	M2	9.31	9.16	9.23	10.31	11.64	1166 ± 2 H
21 Roches	CQA008	62.1	840	M1	7.35	7.51	7.43	8.46	10.34	1166 +8/–4 H
22 Granitic dike	CQA3565B	n.d.	~850	M1	8.80	8.91	8.86	9.86	—	1161 ± 3 H
23 Ecorses	96FN1	57.4	920	NM2	8.26	8.13	8.19	9.04	9.28	n.d.

Note: All oxygen isotope ratios are given in standard per mil (‰) notation relative to Vienna Standard Mean Ocean Water (VSMOW). Zircon mag. indicates magnetic (M) or nonmagnetic (NM) at a particular side tilt (e.g. 2°) of the Frantz Isodynamic Separator. Calculation of zircon saturation temperature follows the method of Watson and Harrison (1983); an average value of ~850 °C is assumed when whole-rock chemistry is not available. Age and geochemistry references: A—Chiarenzelli and McLelland (1991); B—Carl and Sinha (1992); C—McLelland et al. (1993); D—Marcantonio et al. (1990); E—van Breemen and Davidson (1988); F—Davidson and van Breemen (2000); G—Corfu and Easton (1997); H—Corriveau and van Breemen (2000 and personal communication); I—Emslie and Hunt (1990). WR—whole-rock sample; *whole-rock sample 86DM96; †whole-rock sample 94DM29; n.d.—not determined.

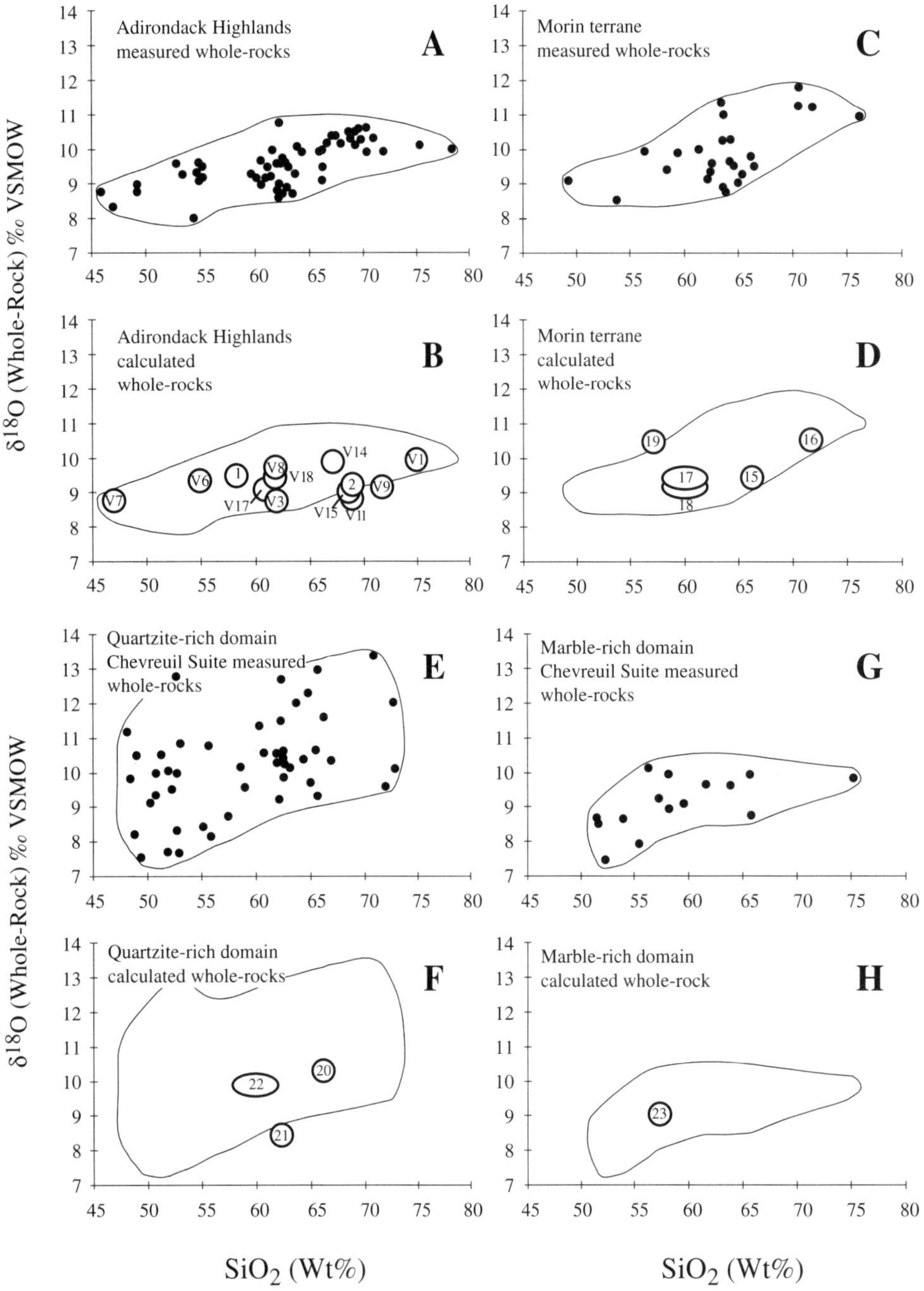

Figure 2. Weight percent SiO_2 versus $\delta^{18}O$ for whole-rock samples (measured or whole-rock values calculated from $\delta^{18}O$ zircon) for 1.13 to 1.18 Ga granitic rocks from the Adirondack Highlands (A and B), Morin terrane (C and D), and Central Metasedimentary Belt of Québec (E–H). Samples with a "V" prefix are from Valley et al. (1994). Adirondack whole-rock $\delta^{18}O$ and SiO_2 data are from Eiler and Valley (1994) and Valley et al. (1994). Calculated whole-rock $\delta^{18}O$ values for samples 17, 18, and 22 (Table 1) are placed to indicate approximate SiO_2 values. Results are not plotted for the Frontenac terrane and the Sharbot Lake domain because of the small number of whole-rock analyses.

pluton-to-pluton variability in assimilants and crystallization histories. This improved correlation is also seen in our data sets, where correlation of $\delta^{18}O$ and SiO_2 for individual plutons is better than for all data taken together from each terrane (not plotted; see Appendixes 1–4). Chevreuil suite granitic rocks of the quartzite domain have the most variation in $\delta^{18}O$(WR), and have high $\delta^{18}O$ values for SiO_2 contents below 60% (Fig. 2, E).

Oxygen-isotope ratios of zircon (Table 1) follow a similar relationship to whole-rock values from terrane to terrane (Fig. 2). For all granitic rocks except those of the Frontenac terrane, $\delta^{18}O$(Zrn) is elevated compared to the values in most igneous rocks, but relatively consistent at 7.4–9.6‰. In the Frontenac terrane, $\delta^{18}O$(Zrn) in granitic rocks ranges from 6.8 to 13.5‰. Zircon from the Lacordaire gabbro has a $\delta^{18}O$ of 6.4‰, which compares well to zircons from two gabbro bodies in the Adirondack Highlands ($\delta^{18}O$ = 6.8 and 6.0‰; Valley et al., 1994). Quartz from the Battersea pluton (Fig. 1, Frontenac terrane) ranges from 13.9 to 14.9‰, and biotite ranges from 9.7 to 10.7‰. In the Perth Road pluton (Fig. 1, Frontenac terrane), $\delta^{18}O$(Qtz) ranges from 14.7 to 17.4‰.

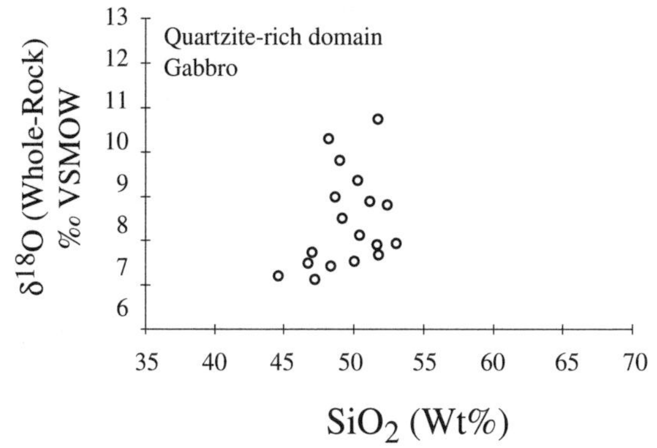

Figure 3. Weight percent SiO_2 versus $\delta^{18}O$ for whole-rock samples of 1.7 Ga gabbros from the quartzite domain of the Central Metasedimentary Belt (Québec).

DISCUSSION AND INTERPRETATIONS

Oxygen-Isotope Ratios of Whole Rocks

Correlation between $\delta^{18}O(WR)$ and SiO_2 (Fig. 2) is consistent with magmatic differentiation that would cause both $\delta^{18}O(WR)$ and SiO_2 to rise as these rocks evolve. This correlation was described in the Adirondack Highlands for granitic rocks from the AMCG suite (Fig. 2, A), and was interpreted as preservation of premetamorphic igneous variation in whole-rock $\delta^{18}O$ values (Eiler and Valley, 1994). In general, we interpret correlation of $\delta^{18}O$ and SiO_2 in other Grenville terranes studied as indicating overall preservation of igneous values in whole rocks as well. Correlation of $\delta^{18}O$ with differentiation indexes is expected from closed-system fractional crystallization, but can explain only a few tenths per mil of the observed variation (e.g., closed-system fractionation at 800 °C would elevate whole-rock $\delta^{18}O$ values by ~0.8‰ for a 20% increase in SiO_2). The observed correlation of SiO_2 and $\delta^{18}O(WR)$ could result from contamination by supracrustal material with high $\delta^{18}O$, or from assimilation fractional crystallization (AFC) processes, which would increase the $\delta^{18}O$ of igneous rocks as they differentiate (e.g., Taylor and Sheppard, 1986). In addition to the consistency with AFC models, good petrologic and isotopic evidence for preservation of igneous $\delta^{18}O$ values and against resetting by exchange with metamorphic fluids is documented in the Adirondack Highlands and in the Frontenac and Morin terranes (e.g., Valley and O'Neil, 1984; Shieh, 1985; Morrison and Valley, 1988; Valley et al., 1990, 1994; Peck and Valley, 2000).

Chevreuil suite granitic rocks of the quartzite domain have the most variation in $\delta^{18}O(WR)$ with respect to SiO_2 content (Fig. 4, A). This could be caused either by higher $\delta^{18}O$ values of potential assimilants or by the tectonic setting of these plutons. These plutons were synkinematically emplaced as sheets into the Nominingue-Chénéville deformation zone (Corriveau and van Breemen, 2000), and high $\delta^{18}O$ values at low SiO_2 contents may reflect extensive interaction with supracrustal materials facilitated by the active tectonic environment of the deformation zone and the large surface area:volume ratios of these plutons. Common field observation of calc-silicate, skarn, and quartzite xenoliths and the absence of biotite-bearing and quartzofeldspathic xenoliths (which are abundant country rocks) point toward efficient assimilation of country rocks that have lower melting temperatures. High and variable $\delta^{18}O(WR)$ for the Chevreuil suite cannot be explained by some country rocks with unusually high $\delta^{18}O$ values; none of the country rocks analyzed in the quartzite domain has an unusual $\delta^{18}O$ value compared to the values of supracrustal rocks elsewhere in the Grenville Province (e.g., Taylor, 1969; Shieh and Schwarcz, 1978; Shieh, 1985; Wu and Kerrich, 1986; Marcantonio et al., 1990; Peck, 2000; Appendixes 1, 2, and 5). This conclusion is supported by oxygen-isotope ratios of vertically layered mafic intrusions associated with the Chevreuil suite in the quartzite domain (Fig. 3). The gabbro plutons show a similar large variation in $\delta^{18}O$ as well as relatively high $\delta^{18}O$ values at low SiO_2 contents, implying similar contamination styles (albeit during different differentiation histories and at different absolute $\delta^{18}O$ values).

Preservation of Igneous Oxygen-Isotope Ratios in Zircons

Zircon is an excellent "reference mineral" from which to estimate oxygen-isotope ratios of metaigneous rocks, inasmuch as low-magnetism, low-radiation-damage zircon retains igneous oxygen-isotope ratios during metamorphism and hydrothermal alteration (Valley et al., 1994; Gilliam and Valley, 1997; King et al., 1997, 1998; Peck and Valley, 1998; Peck et al., 1999; Peck et al., 2003; Valley, 2003). Also, zircon is a high-temperature liquidus phase in many igneous rocks, is refractory, is resistant to recrystallization, and can be dated by U-Pb methods. Inheritance (determined by U-Pb) is uncommon in our zircons, consistent with the alkaline chemistry of many of the plutons, which would promote dissolution of any inherited zircon (Watson and

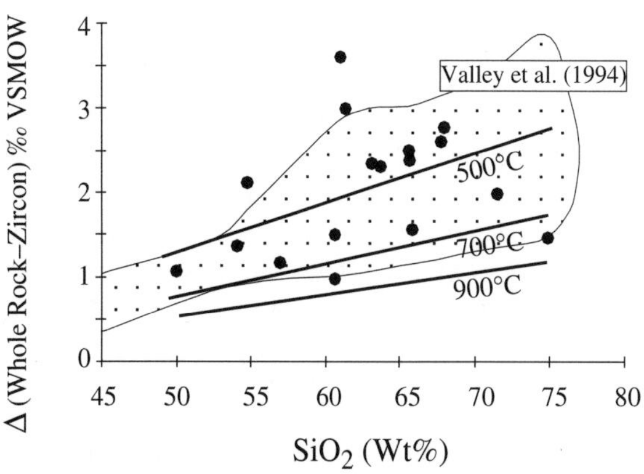

Figure 4. Fractionation between $\delta^{18}O$ (whole rock) and $\delta^{18}O$ (zircon) against SiO_2 for anorthosite-suite granitic rocks. Isotherms are shown, as is the field of data from Valley et al. (1994).

Harrison, 1983); hence zircon is ideal for this regional study of igneous oxygen-isotope geochemistry. Under anhydrous experimental conditions, Watson and Cherniak (1997) determined oxygen diffusion rates that are slow enough to be prohibitive of appreciable intermineral oxygen exchange during geological time (tens of m.y.) at <900 °C. Experiments using free H_2O ("wet" experiments) showed faster oxygen diffusion. When extrapolated to geologic temperatures (e.g., 600 °C), these experimental rates are seven orders of magnitude faster than those in anhydrous experiments. Closure temperatures calculated from "wet" experiments are 500–600 °C for average zircon sizes and slow cooling rates.

Studies of zircon from Adirondack rocks were designed to test the rapid diffusion rates inferred from wet experiments. Thus far, wet experiments do not describe oxygen-isotope systematics of zircons studied from metamorphic rocks (e.g., detrital zircons in quartzite preserve premetamorphic oxygen-isotope values through ~700 °C metamorphism; Valley et al., 1994; Peck and Valley, 1998; Peck et al., 2003). Zircons from orthogneiss do not show the oxygen-isotope zoning caused by slow cooling that wet experiments predict. Given a cooling rate of 2 °C/m.y. after granulite-facies metamorphism, zircons from Adirondack orthogneiss would develop a 0.6 to 0.8‰ fractionation between the smallest and largest sizes analyzed if the wet experiments applied (Peck et al., 1999, 2003). No systematic fractionation is observed between large and small zircon sizes in our samples, indicating that oxygen diffusion rates are significantly slower than that predicted by wet experiments and supporting the interpretation that magmatic oxygen-isotope compositions are preserved in zircon.

Oxygen-Isotope Systematics between Whole Rocks and Zircons

Zircons from our granitic rocks have oxygen-isotope ratios that are broadly consistent with the lower range of measured whole-rock values. Both our data (Table 1) and data from Valley et al. (1994) have variable Δ(WR-Zrn), with some values that are consistent with magmatic temperatures and others that are ~1.0–1.5‰ higher (Fig. 4). High values of $\delta^{18}O$(WR) may indicate assimilation of country rock after zircon crystallization or low-temperature alteration (especially of feldspars) and elevation of $\delta^{18}O$(WR). Fractionations with zircon were calculated after the method of Valley et al. (1994, 2003).

The fractionation between calculated whole-rock $\delta^{18}O$ values (CWRs; Table 1; Fig. 2) and zircon is determined using normative minerals from each rock and published oxygen-isotope fractionation factors. The temperature used for calculating intermineral fractionations is the zircon saturation temperature of Watson and Harrison (1983), which is 750–900 °C for these rocks (Table 1). Calculation of whole-rock $\delta^{18}O$ values from normative minerals versus modal mineralogy (e.g., Valley et al., 1994) does not vary significantly, and the former is a better proxy for $\delta^{18}O$ of the silicate melt (e.g., Matthews et al., 1994). The zircon saturation temperature, an estimate of the temperature at which zircon began to crystallize, is a minimum estimate for zircon crystallization. A range in temperature of 200 °C for zircon crystallization in a granitic melt would raise the $\delta^{18}O$(CWR) only ~0.5‰. Calculated whole-rock $\delta^{18}O$ values fall within the lower envelope of whole-rock data (Fig. 2), which supports the interpretation that zircon retains its igneous $\delta^{18}O$ values during the final stages of assimilation or alteration of these plutons.

High-$\delta^{18}O$ Plutons of the Frontenac Terrane

Oxygen-isotope ratios for whole-rock powders of Frontenac suite granitic rocks from the central Frontenac terrane average ~14‰ (Shieh, 1985; Marcantonio et al., 1990; this study), which is ~4‰ higher than those for other contemporaneous granitic rocks in this study (Figs. 5–7). In both the north-

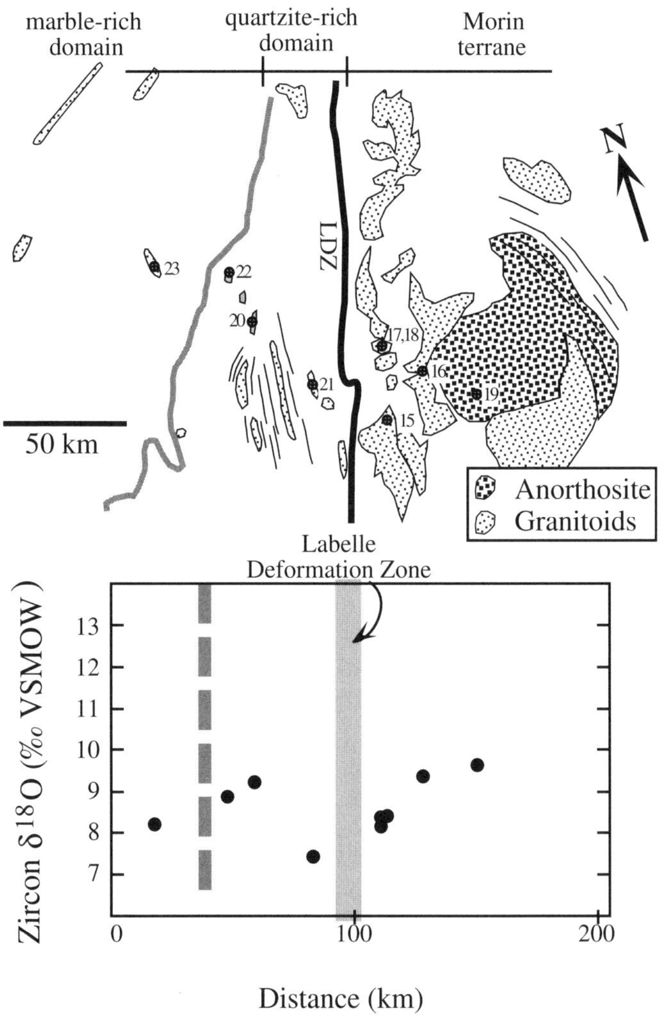

Figure 5. Transect of 1.13–1.18 Ga granitic rocks across the Morin terrane and the Central Metasedimentary Belt (Québec), showing $\delta^{18}O$ (zircon) values as a function of distance perpendicular to terrane boundaries. LDZ—Labelle deformation zone. Numbers refer to plutons in Table 1.

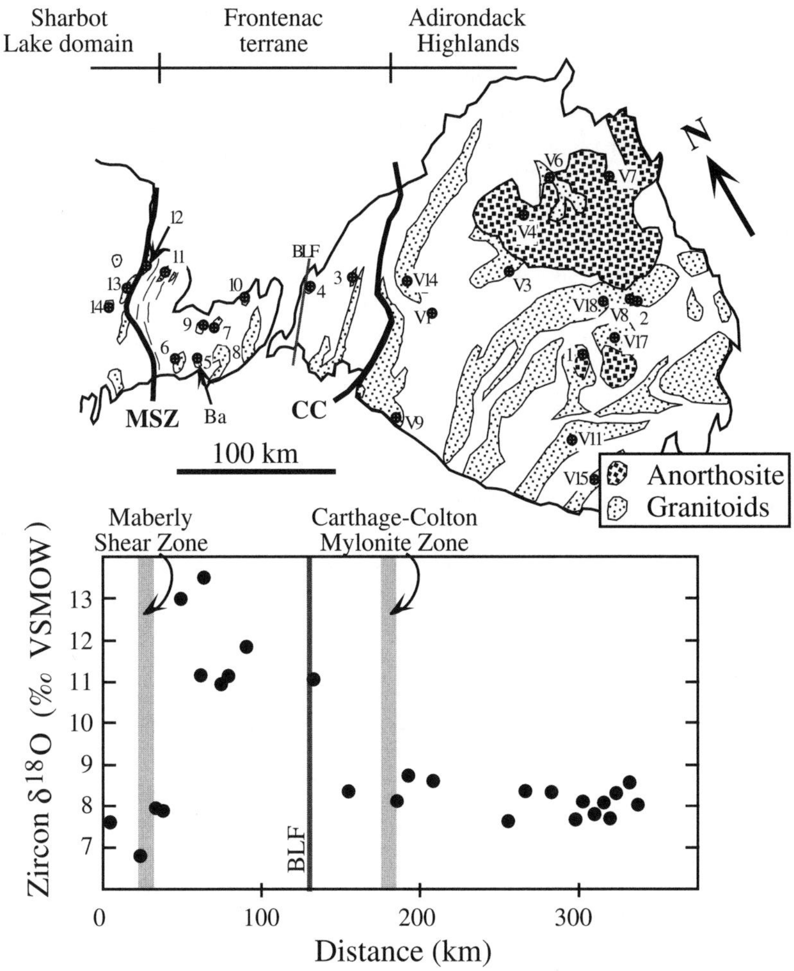

Figure 6. Transect of 1.13–1.18 Ga granitic rocks across the Adirondack Highlands, the Frontenac terrane, and the Sharbot Lake domain (Elzevir terrane), showing $\delta^{18}O$ (zircon) values as a function of distance perpendicular to terrane boundaries. Ba—location of the Battersea pluton; BLF—Black Lake fault; CC—Carthage-Colton mylonite zone; MSZ—Maberly shear zone. Numbers refer to plutons in Table 1.

western Frontenac terrane (Ontario) and the eastern Frontenac terrane (New York), some granitic rocks have lower $\delta^{18}O$ values (i.e., Figs. 3, 6, 11, and 12). The unusually high $\delta^{18}O$ values of some granites from the Frontenac terrane were first recognized by Shieh (1985), who interpreted them as partial melts of metagreywackes or volcanogenic metasedimentary rocks in order to explain both their metaluminous chemistries and their high $\delta^{18}O$. Marcantonio et al. (1990) concluded that the Nd- and Sr-isotope systematics of these plutons could be explained by a mixture of melts derived from depleted mantle and pelitic sedimentary material. They interpreted the unusual oxygen-isotope ratios of these rocks as indicating interaction with high-$\delta^{18}O$, marble-derived CO_2-rich fluids, in effect decoupling radiogenic and stable-isotope systematics. CO_2-rich fluids are necessary for their model because the only reservoirs they considered were melts from depleted mantle, pelitic metasedimentary rocks, and marble. If isotope systems were not decoupled in their calculation, the large amount of marble needed for the high $\delta^{18}O$ values of the plutons would be apparent in lower Si and higher Ca compositions of the granitic rocks (shifting CaO and SiO_2 by ~10 wt%). The strontium-isotope ratio of Grenville marble (Hauer, 1995) is also inconsistent with the composition of these plutons (Marcantonio et al., 1990).

If CO_2-magma interaction was important, the process occurred at depth, and/or the magma was subsequently well mixed. Because CO_2 solubility in granitic magmas is extremely low at crustal pressures, CO_2-magma interaction implies diffusion of CO_2 with high $\delta^{18}O$ values into the magma, and continuous escape of CO_2 in order to shift the magma $\delta^{18}O$. In-filtrating water cannot appreciably change the $\delta^{18}O$ of granitic magmas because of its slow diffusion across the magma–country rock interface (see Taylor and Sheppard, 1986). This applies for CO_2 diffusion as well, but CO_2-magma exchange is even more unlikely given the very low solubility of CO_2 in granitic melts.

The high $\delta^{18}O$ values from the Frontenac terrane require derivation of the granitic rocks from high-$\delta^{18}O$ source materials. These plutons preserve igneous oxygen-isotope fractionations between coexisting minerals (Δ[Qtz–Fsp] = 1.0–1.7‰) and have no correlation between $\delta^{18}O$(WR) and pluton size (Shieh, 1985). Shieh (1985) found no gradients in $\delta^{18}O$(WR) with respect to contacts with high-$\delta^{18}O$ country rock, an obser-

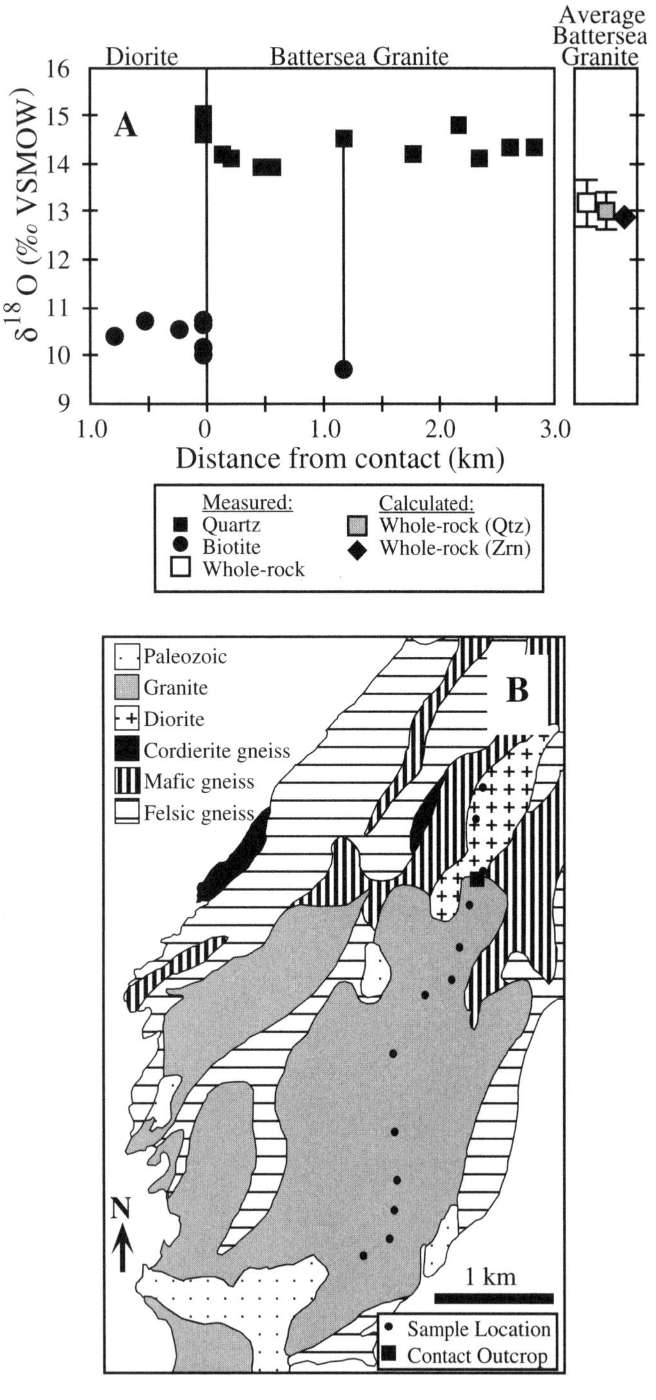

Figure 7. Transect of the Battersea pluton (Frontenac terrane), showing δ^{18}O (quartz) values in granite and δ^{18}O (biotite) in adjacent diorite (A). Geology (B) is from Currie and Ermanovics (1971). The average whole-rock δ^{18}O values calculated in equilibrium with quartz and zircon (shown on right side of figure) agree well with the average measured δ^{18}O (whole rock) (from Shieh, 1985).

vation confirmed by traverses across the plutons. Quartz δ^{18}O values do not vary with location in the Battersea granite (Fig. 7) and in Perth Road plutons (Appendix 5), and homogeneity of δ^{18}O within these plutons supports the absence of interaction with exposed rocks or metamorphic fluids to explain high δ^{18}O values (Shieh, 1985).

Interaction with exposed country rocks cannot explain the high δ^{18}O values in the Frontenac terrane, because younger plutons emplaced in the same country rocks have lower δ^{18}O values (e.g., zircons from the 1.08 Ga Westport pluton [Fig. 1] have δ^{18}O = 8.1‰, compared to the 11.0–13.5‰ zircons from the high-δ^{18}O Frontenac plutons). Granitic rocks of the same age emplaced into similar rocks in adjacent terranes have moderate δ^{18}O(WR) values (8–11‰; e.g., Shieh and Schwarcz, 1978; Wu and Kerrich, 1986; Eiler and Valley, 1994). Given the homogeneity of δ^{18}O(WR) and δ^{18}O(Qtz) on the pluton scale, igneous oxygen-isotope fractionations, and high δ^{18}O(Zrn), we interpret the δ^{18}O values of the Frontenac terrane plutons as inherited from parental materials of the plutons.

The high δ^{18}O(Zrn) of some granitic plutons in the Frontenac terrane (11.81 ± 1.04‰, n = 7) correlates neither with other geochemistry nor with emplacement style. δ^{18}O(Zrn) for the Adirondack Highlands is 8.10 ± 0.36 (n = 13; Valley et al., 1994), and other samples (from Ontario and Québec) have an average of 8.37 ± 0.83‰ (n = 12). This bimodal distribution between the central Frontenac terrane and other terranes reflects neither whole-rock chemistry nor other isotope systems such as Sr and Nd (Heath and Fairbairn, 1969; Barton and Doig, 1977; Ashwal and Wooden, 1983; Marcantonio et al., 1990; Daly and McLelland, 1991; Doig, 1991; Emslie and Hegner, 1993; McLelland et al., 1993). For example, ε_{Nd} values for AMCG granitic rocks (from the Adirondack Highlands) average 2.3 ± 0.7 (n = 5; McLelland et al., 1993), which is indistinguishable from the 2.0 ± 0.5 (n = 5; Marcantonio et al., 1990) of Frontenac terrane samples. The neodymium-isotope compositions of Adirondack granitic rocks are consistent with mixing between melts derived from mantle and paragneiss (Fig. 8). The values for Adirondack samples fall between those for average Frontenac metasedimentary rocks and ca. 1.15 Ga mantle-derived melts, suggesting a 30/70 mixture of the two. Because of the large variability in Frontenac ε_{Nd} values for metasediments, the mixing line does not pass precisely through all measured values. Some Adirondack metasedimentary rocks have higher ε_{Nd} values (+1.3, McLelland et al., 1996) than average Frontenac paragneiss. The high δ^{18}O values of some Frontenac granitic rocks, however, cannot be explained by mixing between 1.15 Ga mantle-derived melts (Fig. 8, M) and metasedimentary rocks.

We propose that the source materials for Frontenac granitic rocks with high δ^{18}O had Mesoproterozoic mantle extraction ages, but that these source materials had a different low temperature history than the source materials for other contemporaneous granitic rocks. Basalt derived from depleted mantle has a

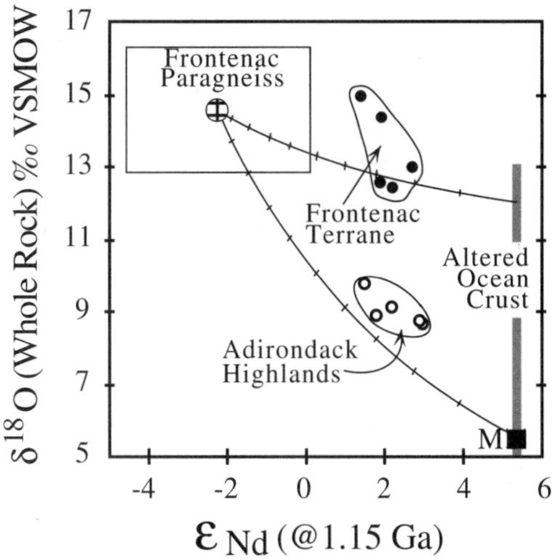

Figure 8. $\delta^{18}O$ (whole rock) versus ε_{Nd} at 1.15 Ga for anorthosite-mangerite-charnockite-granite-related granitic rocks from the Frontenac terrane and Adirondack Highlands, average paragneiss from the Frontenac terrane, melt from unaltered mantle (M—source of MORB, mid-ocean ridge basalt), and melts of ocean crust which underwent low-temperature hydrothermal alteration and/or juvenile oceanic sediments (Shieh, 1985; Muehlenbachs, 1986; Marcantonio et al., 1990; McLelland et al., 1993). Frontenac samples are consistent with mixing between melts of altered ocean crust and paragneiss. The box is ± 1 standard deviation.

$\delta^{18}O$ of ~5.5‰, but hydrothermal alteration on the seafloor is known to change the $\delta^{18}O$ to a higher or lower value, depending on the temperature of alteration (Muehlenbachs, 1986). Seawater alteration at temperatures lower than ~250° will raise $\delta^{18}O$ values, whereas at temperatures higher than ~250°, it will lower $\delta^{18}O$, and basalts altered at intermediate temperatures are essentially unchanged. Strontium-isotope ratios are commonly affected by hydrothermal alteration on the seafloor, but Nd isotopes show only minor shifts (e.g., McCulloch et al., 1980). Melts of metasedimentary rocks and hydrothermally altered basalt (and/or immature, juvenile oceanic sediments) may therefore explain radiogenic and stable isotope results. For example, Sr isotope ratios of Grenville marbles (0.705 to 0.706; Hauer, 1995) do not differ sufficiently from depleted mantle values to cause a large shift in the Sr-isotope ratios of basalt during hydrothermal alteration. The upper mixing line (Fig. 8) is between a hypothetical low-temperature altered basalt-oceanic sediment ($\delta^{18}O$ ~12‰) and Frontenac paragneiss. Frontenac granitic rocks correspond to ~20–30% paragneiss contribution. The Perth Road and Crow Lake plutons have the highest $\delta^{18}O$ values, and may represent mixing between an extremely high-$\delta^{18}O$ component (~15‰) and paragneiss. Hydrothermally altered ocean crust and associated sediments provide a geochemically consistent end-member that accounts for the O-, Sr-, and Nd-isotope compositions of the granitic rocks, does not involve marble, and utilizies little paragneiss so that plutons remain metaluminous. Hydrothermally altered ocean crust has an average oxygen-isotope ratio of ~6‰ (Muehlenbachs, 1986), so the absence of high $\delta^{18}O$ values in adjacent terranes does not imply that hydrothermally altered ocean crust is not present, but implies only that crust under the central Frontenac terrane may have contained more remnants of the upper (high-$\delta^{18}O$) portions of a subducted slab.

TECTONIC IMPLICATIONS

The presence of the AMCG suite and coeval plutons in the Adirondack Highlands, Frontenac terrane, Sharbot Lake domain, Morin terrane, and Central Metasedimentary Belt (Québec) indicates that these crustal blocks were juxtaposed by ca. 1.18 Ga (McLelland et al., 1996; Corriveau et al., 1998; Corriveau and van Breemen, 2000). These plutons have similar trace element patterns, Nd- and Sr-isotope ratios, and depleted mantle Nd model extraction ages that do not correlate with another geochemistry or location. Oxygen-isotope ratios are the only apparent difference among the plutons.

High-$\delta^{18}O$ plutons, as in the central Frontenac terrane, do not exist near the margins of the terrane. The easternmost anomalously high-$\delta^{18}O$ pluton in the Frontenac terrane is the Edwardsville syenite (Fig. 6, #4). To the east, the Hermon granite (Fig. 6, #3) has $\delta^{18}O(Zrn) = 8.3‰$, isotopically identical to the AMCG granitoid suite in the Adirondack Highlands. The westernmost high-$\delta^{18}O$ body is the Perth Road pluton (Fig. 6, #6); the two lower-$\delta^{18}O$ plutons in the western Frontenac terrane (#11 and #12) were synkinematically intruded into the western Frontenac terrane and deformed by the Maberly shear zone (Davidson and van Breemen, 2000). The similarity in oxygen-isotope ratios between these deformed plutons and those in the structurally underlying Sharbot Lake domain may indicate that the normal-$\delta^{18}O$ granitic rocks are derived from the footwall (Sharbot Lake domain) (Fig. 5). An alternate explanation is that the deformation zone at the margin of the Frontenac terrane was a magma conduit (as proposed in Québec by Corriveau et al., 1998, and Corriveau and Morin, 2000), tapping a source region different from that of the undeformed high-$\delta^{18}O$ Frontenac plutons.

The high-$\delta^{18}O$ Edwardsville syenite is on the eastern side of the Black Lake fault, but has a $\delta^{18}O(Zrn)$ identical to plutons in the central Frontenac terrane. This could indicate that the Edwardsville syenite is derived from the same basement as the high-$\delta^{18}O$ plutons in Ontario. Furthermore, basement of the Adirondack Highlands may have extended from the Carthage-Colton mylonite zone under the eastern Frontenac terrane in New York as far as the Hermon granite, which could have been derived from Adirondack Highlands material at depth under the Adirondack Lowlands at ca. 1.15 Ga.

The new oxygen-isotope data are consistent with subduction and storage of ocean crust beneath the Frontenac terrane before 1.18–1.13 Ga. In the tectonic model of Wasteneys et al.

(1999, modified from McLelland et al., 1996), the Frontenac terrane is the trailing margin of the ca. 1.35 to 1.22 Ga Elzevir arc, and west-dipping subduction of ocean crust beneath the Frontenac terrane gave rise to the calc-alkaline ca. 1.21 Ga Antwerp-Rossie suite of the Adirondack Lowlands (Fig. 9). Alternatively, Elzevir subduction from the west could also have been a source of ocean crust in the lower crust of the Frontenac terrane. Collision of the Frontenac terrane with the Adirondack Highlands began after ca. 1.21 Ga Antwerp-Rossie suite plutonism but before generation of the ca. 1.17 Ga Rockport granite and the Hyde School gneiss. Docking and imbrication were accompanied by metamorphism and magmatism. In Québec, collision with the Morin terrane occurred at 1.22 to 1.21 Ga, as manifested by ca. 1.20 to 1.19 Ga metamorphism and anatexis (Corriveau and van Breemen, 2000). The 1.17 to 1.16 Ga Chevreuil suite monzonites of the Central Metasedimentary Belt (Québec) are interpreted as indicating syntectonic intrusion, as are the ca. 1.16 Ga elongate plutons in the Maberly shear zone in the western Frontenac terrane (Corriveau and van Breemen, 2000; Davidson and van Breemen, 2000).

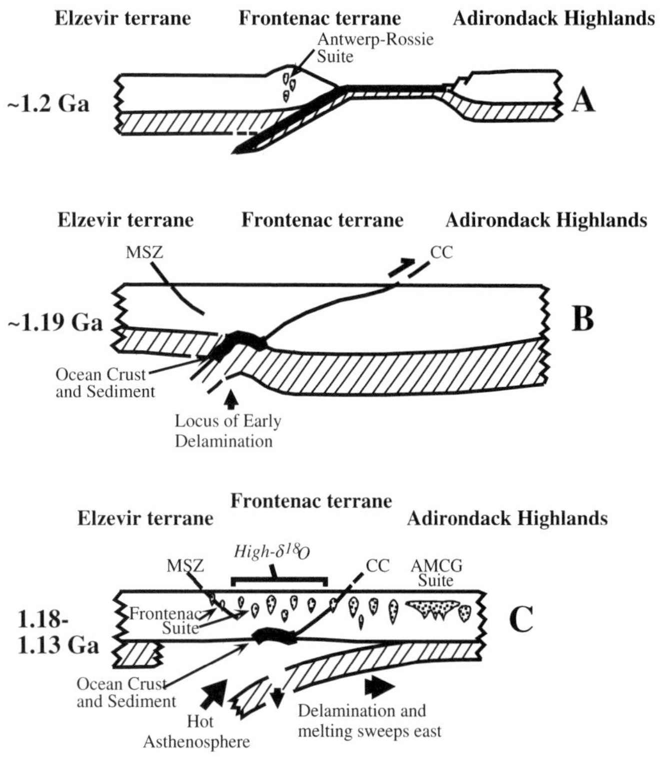

Figure 9. Cartoon (after Wasteneys et al., 1999) of the plate tectonic configuration of subduction and underplating at ca. 1.2 Ga during generation of the calc-alkaline Antwerp-Rossie suite (A), during initial collision of the Frontenac terrane and Adirondack Highlands (B), and during 1.18 to 1.13 Ga anorthosite suite and coeval plutonism (C). Plutonism at 1.18 to 1.13 Ga is interpreted as resulting from postorogenic delamination of lithospheric mantle that progressed to the east, along with decompression melting, ponding of basaltic magmas at the base of the crust, and anatexis of the lower crust to produce granitic magmas (see McLelland et al., 1996). The location of high-$\delta^{18}O$ plutons is consistent with underplating or thrusting of hydrothermally altered ocean crust at the base of the Frontenac terrane before ca. 1.15 Ga. AMCG—anorthosite-mangerite-charnockite-granite; CC—Carthage-Colton mylonite zone; MSZ—Maberly shear zone.

If hydrothermally altered ocean crust was present in the lower crust of the Frontenac terrane during generation of granitic rocks at 1.15 Ga, it was likely either subducted and underplated or underthrust as the orogen closed (Fig. 9). If westward subduction, which caused Antwerp-Rossie plutonism, also subducted hydrothermally altered ocean crust to the base of the Frontenac crust at ca. 1.21 Ga, localization of the Antwerp-Rossie suite in the Frontenac terrane (New York) and high-$\delta^{18}O$ granitic rocks were linked by the geometry of subduction before collision of the Adirondack Highlands–Morin terrane block (Fig. 9). Whatever the direction of subduction, the Central Metasedimentary Belt (Québec) may expose crust that was not underplated because it was farther inboard from the subduction zone than the Frontenac terrane (Ontario). This interpretation is consistent with the lack of juvenile 1.2 Ga Nd model ages in lower crustal xenoliths within the 1.07 Ga Rivard dike in the Central Metasedimentary Belt of Québec (Amelin et al., 1994).

CONCLUSIONS

Zircons preserve the magmatic oxygen-isotope compositions of the high-$\delta^{18}O$ AMCG suite and related granitoid plutons in the Allochthonous Monocyclic Belt of the southern Grenville Province. A traverse from the eastern Adirondacks to the western Frontenac terrane shows pronounced steps of $\delta^{18}O$ in the central Frontenac terrane (Fig. 9). The high-$\delta^{18}O$ plutons of the Frontenac terrane are interpreted as resulting from the melting of a large component of high-$\delta^{18}O$ hydrothermally altered ocean crust and sediments subducted and underplated near the Moho. The lack of gradients in $\delta^{18}O$ on the pluton scale (Fig. 7) and the high $\delta^{18}O$ of zircon indicate that high-$\delta^{18}O$ Frontenac plutons acquired their $\delta^{18}O$ values deep in the crust, and that they were well mixed before crystallization.

The extension of the Frontenac terrane northward along strike in the Central Metasedimentary Belt (Québec) is interpreted to be the quartzite domain, which lacks the distinctive high-$\delta^{18}O$ granitic rocks of Ontario (Fig. 5). A discontinuity along strike in the lower crust of the Frontenac terrane is inferred to reflect differences in location relative to the continental margin in Ontario versus Québec. This discontinuity along strike could also have been caused by the geometry between the Frontenac-quartzite domain block and adjacent terranes, in that subduction and underplating of hydrothermally altered ocean crust and sediments occurred under the Frontenac terrane in Ontario, but not under the extension in Québec.

Basement mapping using stitching plutonic suites can have great utility in delineating exotic crustal blocks, even where there are difficulties in recognizing terrane boundaries. This approach is possible only if the primary, igneous geochemistry of plutonic suites can be rigorously constrained. In this case, analysis of oxygen-isotope ratios in zircon has allowed the inference of large masses of high-$\delta^{18}O$ hydrothermally altered basalts and metasedimentary rocks under part of the Frontenac terrane. The analysis of dated zircons to map the $\delta^{18}O$ values of magmatic source regions has not been applied previously in Precambrian rocks, and will likely assist in delineation of terrane boundaries and discontinuities in the lower crust of other polymetamorphic orogens.

APPENDIX 1. OXYGEN ISOTOPE RATIOS FROM THE MORIN TERRANE, SOUTHERN GRENVILLE PROVINCE

Pluton	Sample number	SiO_2 wt%	$\delta^{18}O$ whole rock	Average whole rock	Other
AMCG suite granitoids					
Grey Valley	CQA877	63.3	11.31, 11.47	11.39	
	CQA878	70.7	11.82	11.82	
	CQA883	70.6	11.31	11.31	
	CQA891	62.7	9.60	9.60	
	CQA928	65.1	9.06	9.06	
	96MR43	66.2	9.82	9.82	Garnet 7.94
					Quartz 11.11, 11.14
Southeastern Mangerite	95MR40	63.7	8.72, 8.87	8.80	
	95MR60	63.3	10.24	10.24	
	95MR62	63.4	10.25	10.25	
	95MR69	64.0	10.30	10.30	
	95MR71	53.6	8.54	8.54	
	95MR72	59.3	9.88	9.88	
	95MR1	58.3	9.52, 9.32	9.42	
	95MR125	56.3	10.00, 9.89	9.95	
Western Mangerite	95MR32	63.6	11.00	11.00	
	95MR35	65.2	9.31, 9.25	9.28	
	95MR81	49.2	9.08	9.08	
	95MR85	76.1	10.93, 11.01	10.97	
	95MR87	66.3	9.55	9.55	
	95MR88	64.4	9.55	9.55	
	96MR21	71.7	11.20, 11.30	11.25	
Janet	CQA1172	64.2	9.61, 9.69	9.65	
	CQA1176	63.6	8.92	8.92	
	CQA1178	62.3	9.37	9.37	
	CQA1181	62.1	9.16	9.16	
Maskinongé	CQA925A				Hornblende 7.75
Mangerite dike	EC84-246	57.1			Quartz 12.95, 12.58
					Garnet 8.78, 8.97, 8.76
					Clinopyroxene 8.52, 8.77
Granitic country rock					
	95MR12		9.90	9.90	
	95MR36		6.65	6.65	
	95MR127		7.26	7.26	
	95MR154		5.69	5.69	
	95MR161		13.68	13.68	
	CQA920	74.5	7.38	7.38	
	CQA921	77.9	7.76, 7.77	7.76	
	98MR7	75.1	8.99, 8.64	8.81	Zircon (NM2) 7.62, 7.43
	98MR6	76.5	8.37	8.37	
	98MR5	77.4	8.26	8.26	

Note: All oxygen isotope ratios are given in standard per mil (‰) notation relative to Vienna Standard Mean Ocean Water (VSMOW). AMCG—anorthosite-mangerite-charnockite-granite. Zircon magnetism indicates magnetic (M) or nonmagnetic (NM) at a particular side tilt (e.g. 2°) of the Frantz Isodynamic Separator. Sample locations are given in Peck (2000).

APPENDIX 2. OXYGEN ISOTOPE RATIOS FROM THE QUARTZITE-RICH DOMAIN OF THE CENTRAL METASEDIMENTARY BELT—QUÉBEC, SOUTHERN GRENVILLE PROVINCE

Pluton	Sample number	SiO_2 wt%	$\delta^{18}O$ whole rock	Average whole rock	Other
Chevreuil suite granitoids					
Armstrong	CQA953	60.7	10.62	10.62	
	CQA1062	70.8	13.40	13.40	
	CQA1067	53.1	10.90	10.90	
	CQA1068	65.6	9.36	9.36	
	CQA1071	52.6	12.77, 12.83	12.80	
	CQA3498a	48.9	8.15, 8.26	8.20	
	CQA3526a	52.2	9.55	9.55	
	CQA3544	62.3	12.71	12.71	
	CQA3547	55.6	10.82	10.82	
	CQA3558	49.0	10.54	10.54	
	CQA4064	52.7	10.02	10.02	
Chreveuil	CQA1085	66.1	11.64	11.64	Hornblende 9.64
	CQA1086d	63.6	12.04	12.04	
	CQA1087	52.6	8.34	8.34	
	CQA1901	60.3	11.23, 11.54	11.38	
	CQA1902	61.9	10.59	10.59	
	CQA3313	51.3	10.52	10.52	
Gagnon	CQA939	64.8	12.31	12.31	
	CQA961	58.9	9.59	9.59	
	CQA964	55.8	8.15	8.15	
	CQA968	72.0	9.61	9.61	
	CQA1372a	66.9	10.39	10.39	
	CQA1951	63.0	10.17	10.17	
	CQA1953	65.6	13.00	13.00	
	CQA1965	72.6	12.05	12.05	
	CQA2206a	48.1	11.20	11.20	
	CQA2200a	62.5	10.28	10.28	
	CQA2647	52.9	7.66, 7.69	7.68	
	CQA2648	64.3	10.40	10.40	
	CQA2651	62.4	10.65	10.65	
	CQA2958	49.4	7.56	7.56	
	CQA3500	51.9	10.08	10.08	
Roches	CQA008	62.1	10.32, 10.36	10.34	Hornblende 7.37
	CQA019	72.8	10.14	10.14	
	CQA023	51.8	7.71	7.71	
	CQA349	65.0	9.77, 9.72	9.74	
	CQA355	62.1	9.20, 9.25	9.22	
	CQA557	55.0	8.45	8.45	
St. Francois	CQA1091a	62.4	10.43	10.43	
	CQA1097b	50.7	9.38	9.38	
	CQA1455a	62.5	9.91	9.91	
Sept Freres	CQA2076a	62.3	11.51	11.51	
Serpent	CQA1022	48.3	9.84	9.84	
	CQA4107	50.8	9.99	9.99	
Sucrerie	CQA134	53.0	7.62	7.62	
	CQA136	58.6	10.10, 10.27	10.18	
	CQA198	57.4	8.75	8.75	
	CQA207	65.4	10.69	10.69	
Unnamed	CQA2605a	50.2	9.16	9.16	
Chevreuil suite gabbros					
Diable	CQA1369a	53.2	9.11	9.11	
	CQA1398a	54.1	9.02	9.02	
	CQA1398b	50.0	8.17	8.17	
	CQA1399a	52.5	8.77	8.77	
	CQA1399b	49.4	8.09	8.09	
	CQA1402a	50.5	9.80	9.80	
	CQA1573a	52.1	8.09	8.09	
	CQA1573b	48.5	7.52	7.52	
Buchesi	CQA1100	49.3	7.24	7.24	
	CQA1105	50.3	7.60	7.60	

APPENDIX 2. *Continued*

Pluton	Sample number	SiO_2 wt%	$\delta^{18}O$ whole rock	Average whole rock	Other
	CQA1806a	49.0	7.84	7.84	
Hydroplane (gab. dike)	CQA2584a	49.1	8.51	8.51	
	CQA2589a	50.4	8.09, 8.16	8.12	
Har-ha-kon	CQA1799	48.6	9.01	9.01	
	CQA1800a	51.1	8.90	8.90	
Lacordaire	CQA1348a	48.1	10.29	10.29	
	CQA1420a	53.0	7.95	7.95	
	CQA1421a	47.2	7.13	7.13	
	CQA1425a	47.0	7.73	7.73	
	CQA1426a	48.4	7.44	7.44	
	CQA1433a	51.7	7.68	7.68	
	CQA2159a	50.1	7.55	7.55	
	CQA2597a	51.7	10.75, 10.74	10.75	
	CQA1427				Zircon (M1) 6.29, 6.59 Hornblende 6.54, 6.47
Country rocks					
Tonalite (BND, 1220 Ma)	CQA3565A				Zircon (M1) 6.36, 6.30 Hornblende 6.45
Pelite	CQA4021a		14.75	14.75	
Pelite	CQA1499		15.85, 15.85	15.85	
Pelite	CQA750	14.52	14.52		
Tonalites (BND)	CQA1659W		10.06, 9.93	9.99	
Tonalites (BND)	CQA4945A2	12.26	12.26		
Tonalites (BND)	CQA4945A2	12.75	12.75		
Tonalites (BND)	CQA5063	9.85	9.85		
Tonalites (BND)	CQA4960	8.77	8.77		
Al-rich facies (BND)	CQA1659H	8.56	8.56		
Al-rich facies (BND)	CQA4957	8.93	8.93		
Al-rich facies (BND)	CQA4961A	8.41	8.41		
Later intrusive					
Diable (1077 Ma)	CQA1369a	53.2	9.11	9.11	Zircon (M2) 7.55, 7.35 Hornblende 7.11

Note: BND—Bondy Gneiss Dome. All oxygen isotope ratios are given in standard per mil (‰) notation relative to Vienna Standard Mean Ocean Water (VSMOW). Zircon magnetism indicates magnetic (M) or nonmagnetic (NM) at a particular side tilt (e.g. 2°) of the Frantz Isodynamic Separator. Sample locations are given in Peck (2000).

APPENDIX 3. OXYGEN ISOTOPE RATIOS FROM THE MARBLE-RICH DOMAIN OF THE CENTRAL METASEDIMENTARY BELT—QUÉBEC, SOUTHERN GRENVILLE PROVINCE

Pluton	Sample number	SiO_2 wt%	$\delta^{18}O$ whole rock	Average whole rock
Chevreuil suite granitoids				
Polonais	CQA1603a	63.8	9.61, 9.73	9.67
Henn	CQA1606	51.6	8.54	8.54
	CQA1607	52.2	7.48	7.48
Baskatong	CQA1161a	61.6	9.69	9.69
	CQA1162	58.1	9.98	9.98
	CQA1167	56.3	10.15	10.15
Ecorses	CQA1116a	58.2	8.97	8.97
	CQA1122	65.9	8.75	8.75
	CQA1126	65.6	9.96, 9.96	9.96
	CQA1127	75.2	9.86	9.86
	CQA1129	51.4	8.70	8.70
	CQA1135a	55.4	7.96	7.96
	CQA1151a	59.5	9.03, 9.15	9.09
	CQA1152	54.0	8.66	8.66
	96FN1	57.4	9.19, 9.36	9.28

Note: All oxygen isotope ratios are given in standard per mil (‰) notation relative to Vienna Standard Mean Ocean Water (VSMOW). Sample locations are given in Peck (2000).

APPENDIX 4. OXYGEN ISOTOPE RATIOS OF WHOLE ROCKS FROM FRONTENAC AND ELZEVIR TERRANES, SOUTHERN GRENVILLE PROVINCE

Lithology/pluton	Sample number	$\delta^{18}O$ whole rock	Average whole rock
Frontenac terrane			
Hermon granite	DF178	14.56	14.56
Edwardsville syenite	AM87-5	13.40	13.40
Battersea	LH86-63	13.80	13.80
Perth Road	LH87-64	13.60	13.60
Lyndhurst	LH87-31	12.36,12.59	12.36
South Lake	86DM9c	15.70	15.70
Crow Lake	LH87-30	13.80,14.19	13.80
Beales Mills	95DM53	7.90,7.89	7.90
Elzevir terrane			
Silver Lake	93DM20	9.25	9.25
Oso	92DM152	8.58	8.58

Note: All oxygen isotope ratios are given in standard per mil (‰) notation relative to Vienna Standard Mean Ocean Water (VSMOW). Sample locations are given in Peck (2000).

APPENDIX 5. OXYGEN ISOTOPE RATIOS OF MINERALS FROM THE BATTERSEA AND PERTH ROAD PLUTONS, FRONTENAC TERRANE, SOUTHERN GRENVILLE PROVINCE

Lithology/pluton	Sample number	$\delta^{18}O$ quartz	Average quartz	$\delta^{18}O$ biotite	Average biotite
Battersea pluton					
Diorite	96BP1	13.91, 13.84	13.88		
Granite	96BP9	14.93, 14.76	14.85		
Granite	96BP10	14.91, 14.94	14.93		
Granite	96BP11	14.21, 14.10	14.16		
Granite	96BP13	14.22, 14.08	14.15		
Granite	96BP14	13.77, 13.94	13.86		
Granite	96BP15	14.45	14.45		
Granite	96BP16	14.29, 14.08	14.19		
Granite	96BP17	14.78	14.78		
Granite	96BP18	14.05, 14.20	14.13		
Granite	96BP19	14.15, 14.43	14.29		
Granite	96BP20	14.31, 14.18	14.25		
Granite	96BP8-A	14.65, 14.64	14.64		
Granite	96BP8-B	14.64	14.64		
Granite	96BP8-C	14.94	14.94		
Granite	96BP8-C			10.63	10.63
Granite	96BP8-D	14.57	14.57		
Diorite	96BP8-E			10.61	10.61
Diorite	96BP8-F			10.67	10.67
Diorite	96BP8-G			10.64	10.64
Diorite	96BP8-H			10.62	10.62
Granite	96BP8-J	13.86	13.86		
Diorite	96BP2			10.67	10.67
Diorite	96BP3			10.42	10.42
Diorite	96BP6			10.12, 10.13	10.12
Diorite	96BP4			10.52	10.52
Granite	96BP15			9.98	9.98
Granite	96BP12			9.69, 9.74	9.71
Diorite	96BP7			10.16, 9.68	9.92
Perth Road pluton					
Syenite	96PR19	16.43, 16.62	16.52		
Granitic gneiss	96PR10	16.91, 16.91	16.91	13.24, 13.63	13.43
Syenite	96PR18	16.14, 16.26	16.20		
Syenite	96PR17	16.82	16.82		
Monzonite	96PR9	14.28, 14.86	14.57		
Granitic gneiss	96PR12	17.00	17.00		

APPENDIX 5. ***Continued***

Lithology/pluton	Sample number	$\delta^{18}O$ quartz	Average quartz	$\delta^{18}O$ biotite	Average biotite
Granitic gneiss	96PR3	17.24	17.24		
Granitic gneiss	96PR4	17.45	17.45		
Granitic gneiss	96PR1	14.80, 15.03	14.92		
Granitic gneiss	96PR2	14.76	14.76		
Monzonite	96PR5	14.87, 15.04	14.96		
Monzonite	96PR8	14.75	14.75		
Monzonite	96PR9	15.19	15.19		
Granitic gneiss	96PR14	17.09	17.09		
Granitic gneiss	96PR15	16.77	16.77		
Granitic gneiss	96PR16	16.49	16.49		
Syenite	96PR20	16.35	16.35		
Syenite	96PR21	15.77, 16.08	15.93		
Syenite	96PR22	16.71	16.71		
Granitic gneiss	96PR13			14.89	14.89
Diorite	96PR6				
	Hornblende = 8.77, 8.87				
	Plagioclase = 11.72, 11.67				
Granitic gneiss	96PR23	17.36, 17.22	17.29		
Granitic gneiss	96PR24	16.89	16.89		
Granitic gneiss	96PR25	17.18	17.18		
Granitic gneiss	96PR26	16.85	16.85		
Granitic gneiss	96PR27	17.71	17.71		
Granitic gneiss	96PR31	16.53	16.53		
Granitic gneiss	96PR32	17.25	17.25		
Granitic gneiss	96PR33	16.65	16.65		
Granitic gneiss	96PR34	16.49	16.49		
Granitic gneiss	96PR35	16.61	16.61		
Monzonite	96PR37	16.89	16.89		

Note: All oxygen isotope ratios are given in standard per mil (‰) notation relative to Vienna Standard Mean Ocean Water (VSMOW). Sample locations are given in Peck (2000).

ACKNOWLEDGMENTS

This work was supported by grants from the National Science Foundation (EAR96-28260), the Department of Energy (93ER14389), and the University of Wisconsin, Department of Geology and Geophysics. Otto van Breemen was instrumental in establishing the geochronology of zircon samples from Ontario and Québec. We thank Mike Spicuzza for assistance in the stable isotope laboratory; Brian Hess for making thin sections; Mike DeAngelis, Chris Frederickson, and Myongsun Kong for their assistance in the field; and James Carl, Don Davis, and Ron Emslie for providing samples. Phil Brown, Guoxiang Chi, Clark Johnson, and Brad Singer offered helpful comments on an early version of the manuscript. Detailed reviews by Margaret Streepey and an anonymous reviewer are gratefully acknowledged. This is Geological Survey of Canada contribution 2002068.

REFERENCES CITED

Amelin, Y., Corriveau, L., and Morin, D., 1994, Constraints on the evolution of Grenvillian lithosphere from Nd-Sr-Pb clinopyroxene and garnet and U-Pb zircon study of pyroxenite and mafic granulitic xenoliths: Reston, Virginia, United States Geological Survey, Circular 1107, p. 5.

Anderson, J.L., 1983, Proterozoic anorogenic granite plutonism of North America, *in* Medaris, L.G., et al., eds., Proterozoic geology: Selected papers from an international Proterozoic symposium: Boulder, Colorado, Geological Society of America Memoir 161, p. 133–154.

Anovitz, L.M., and Essene, E.J., 1990, Thermobarometry and pressure-temperature paths in the Grenville Province of Ontario: Journal of Petrology, v. 31, p. 197–241.

Ashwal, L.D., 1993, Anorthosites. Berlin, Springer-Verlag, 422 p.

Ashwal, L.D., and Wooden, J.L., 1983, Sr and Nd isotope geochronology, geologic history, and origin of the Adirondack Anorthosite: Geochimica et Cosmochimica Acta, v. 47, p. 1875–1885.

Barton, J.M., and Doig, R., 1977, Sr-isotopic studies of the origin of the Morin Anorthosite Complex, Quebec, Canada: Contributions to Mineralogy and Petrology, v. 61, p. 219–230.

Boggs, K., 1996, Retrograde cation exchange in garnets during slow cooling of mid crustal granulites and the P-T-t trajectories from the Mont-Laurier region, Grenville Province, Quebec [M.S. thesis]: Chicoutimi, Université du Québec à Chicoutimi, 333 p.

Bohlen, S.R., Valley, J.W., and Essene, E.J., 1985, Metamorphism in the Adirondacks, 1: Petrology, pressure, and temperature: Journal of Petrology, v. 26, p. 971–992.

Buddington, A.F., 1939, Adirondack igneous rocks and their metamorphism: Boulder, Colorado, Geological Society of America Memoir 7, 354 p.

Busch, J.P., Mezger, K., and van der Pluijm, B.A., 1997, Suturing and extensional reactivation in the Grenville orogen, Canada: Geology, v. 25, p. 507–510.

Carl, J.D., and deLorraine, W.F., 1997, Geochemical and field characteristics of metamorphosed granitic rocks, NW Adirondack Lowlands, New York: Northeastern Geology and Environmental Science, v. 19, p. 276–301.

Carl, J.D., and Sinha, A.K., 1992, Zircon U-Pb age of Popple Hill Gneiss and a Hermon-type granite gneiss, northwest Adirondack Lowlands, New York: Geological Society of America Abstracts with Programs, v. 24, no. 3, p. 11.

Carr, S., Easton, R.M., Jamieson, R.A., and Culshaw, N.G., 2000, Geologic transect across the Grenville orogen of Ontario and New York: Canadian Journal of Earth Sciences, v. 37, p. 193–216.

Chiarenzelli, J., and McLelland, J., 1991, Age and regional relationships of granitoid rocks of the Adirondack Highlands: Journal of Geology, v. 99, p. 571–590.

Clechenko, C.C., and Valley, J.W., 2003, Oscillatory zoning in garnet from the Willsboro Wollastonite Skarn, Adirondack Mts, New York: A record of shallow hydrothermal processes preserved in a granulite facies terrane, Journal of Metamorphic Geology, v. 21, p. 771–784.

Clechenko, C., Valley, J.W., Hamilton, M.A., McLelland, J., and Bickford, M.E., 2002, Direct U-Pb dating of the Marcy Anorthosite, Adirondacks, NY, USA: Twelfth Annual V.M. Goldschmidt Conference: Geochimica et Cosmochimica Acta, v. 66, p. A144.

Corfu, F., and Easton, R.M., 1997, Sharbot Lake terrane and its relationships to Frontenac terrane, Central Metasedimentary Belt, Grenville Province: New insights from U-Pb geochronology: Canadian Journal of Earth Sciences, v. 34, p. 1239–1257.

Corriveau, L., 1990, Proterozoic subduction and terrane amalgamation in the southwestern Grenville Province, Canada: Evidence from ultrapotassic to shoshonitic plutonism: Geology, v. 15, p. 614–617.

Corriveau, L., and Morin, D., 2000, Modeling 3D architecture of western Grenville from surface geology, xenoliths, styles of magma emplacement, and Lithoprobe reflectors: Canadian Journal of Earth Sciences, v. 37, p. 235–251.

Corriveau, L., and van Breemen, O., 2000, Docking of the Central Metasedimentary Belt to Laurentia in geon 12: Evidence for the 1.17–1.16 Ga Chevreuil intrusive suite and host gneisses, Quebec: Canadian Journal of Earth Sciences, v. 37, p. 253–269.

Corriveau, L., Rivard, B., and van Breemen, O., 1998, Rheological controls on Grenvillian intrusive suites: Implications for tectonic analysis: Journal of Structural Geology, v. 20, p. 1191–1204.

Cosca, M.A., Essene, E.J., Kunk, M.J., and Sutter, J.F., 1992, Differential unroofing within the Central Metasedimentary Belt of the Grenville Orogen: Constraints from $^{40}Ar/^{39}Ar$ thermochronology: Contributions to Mineralogy and Petrology, v. 110, p. 211–225.

Cureton, J.S., van der Pluijm, B.A., and Essene, E.J., 1997, Nature of the Elzevir–Mazinaw domain boundary, Grenville Orogen, Ontario: Canadian Journal of Earth Sciences, v. 7, p. 976–991.

Currie, K.L., and Ermanovics, I.F., 1971, Geology of the Loughborough Lake region, Ontario, with a special emphasis on the origin of granitoid rocks: Geological Survey of Canada Bulletin 199, 71 p.

Daly, J.S., and McLelland, J., 1991, Juvenile Middle Proterozoic crust in the Adirondack Highlands, Grenville Province, northeastern North America: Geology, v. 19, p. 119–122.

Davidson, A., 1984, Tectonic boundaries within the Grenville Province of the Canadian Shield: Journal of Geodynamics, v. 1, p. 433–444.

Davidson, A., 1995, A review of the Grenville orogen in its North American type area: AGSO [Australian Geological Survey Organisation] Journal of Australian Geology and Geophysics, v. 16, p. 3–24.

Davidson, A., and van Breemen, O., 2000, The age and extent of the Frontenac plutonic suite in the Central Metasedimentary Belt, Grenville Province, Ontario: Radiogenic Age and Isotope Studies Report 13, Ottawa, Ontario, Geological Survey of Canada Current Research 2000–F.

Doig, R., 1991, U-Pb zircon dates of Morin anorthosite suite rocks, Grenville Province, Québec: Journal of Geology, v. 99, p. 729–738.

Easton, R.M., 1986, Geochronology of the Grenville Province, *in* Moore, J.M., Davidson, A., and Baer, A.J., eds., The Grenville Province: St. John's, Newfoundland, Geological Association of Canada Special Paper 31, p. 127–173.

Easton, R.M., 1992, The Grenville Province and the Proterozoic history of central and southern Ontario, *in* Geology of Ontario: Sudbury, Ontario Geological Survey Special Paper 4, Part 1, p. 715–904.

Eiler, J.M., and Valley, J.W., 1994, Preservation of premetamorphic oxygen isotope ratios in granitic orthogneiss from the Adirondack Mountains, New York, USA: Geochimica et Cosmochimica Acta, v. 58, p. 5525–5535.

Emslie, R.F., and Hegner, E., 1993, Reconnaissance isotopic geochemistry of anorthosite-mangerite-charnockite-granite (AMCG) complexes, Grenville Province, Canada: Chemical Geology, v. 106, p. 279–298.

Emslie, R.F., and Hunt, P.A., 1990, Ages and petrogenetic significance of igneous mangerite-charnockite suites associated with massif anorthosites, Grenville Province: Journal of Geology, v. 98, p; 213–231.

Friedman, R.M., and Martignole, J., 1995, Mesoproterozoic sedimentation, magmatism, and metamorphism in the southern part of the Grenville Province (western Quebec): U-Pb geochronological constraints: Canadian Journal of Earth Sciences, v. 32, p. 2103–2114.

Frost, C.D., Frost, B.R., Bell, J.M., and Chamberlain, K.R., 2002, The relationship between A-type granites and residual magmas from anorthosite: Evidence from the northern Sherman batholith, Laramie Mountains, Wyoming, USA: Precambrian Research, v. 119, p. 45–71.

Geraghty, E.P., Isachsen, Y.W., and Wright, S.F., 1981, Extent and character of the Carthage-Colton Mylonite Zone, northwest Adirondacks, New York: Rockville, Maryland, U.S. Nuclear Regulatory Commission Technical Report NUREG/CR-1865.

Gilliam, C.E., and Valley, J.W., 1997, Low $\delta^{18}O$ magma, Isle of Skye, Scotland: Evidence from zircons: Geochimica et Cosmochimica Acta, v. 61, p. 4975–4981.

Hauer, K.L., 1985, Protoliths, diagenesis, and depositional history of the Upper Marble, Adirondack Lowlands, New York [Ph.D. thesis]: Oxford, Ohio, Miami University, 281 p.

Heath, S.A., and Fairbairn, H.W., 1969, Sr^{87}/Sr^{86} ratios in anorthosites and some associated rocks, *in* Isachsen, Y., ed., Origin of anorthosites and related rocks: Albany, New York State Museum Memoir 18, p. 99–110.

Indares, A., and Martignole, J., 1990, Metamorphic constraints on the tectonic evolution of the allochthonous monocyclic belt of the Grenville Province, western Quebec: Canadian Journal of Earth Sciences, v. 27, p. 371–386.

Irvine, T.N., and Baragar, W.R.A., 1971, A guide to the chemical classification of the common volcanic rocks: Canadian Journal of Earth Sciences, v. 8, p. 523–548.

King, E.M., Barrie, C.T., and Valley, J.W., 1997, Hydrothermal alteration of oxygen isotope ratios in quartz phenocrysts, Kidd Creek Mine, Ontario: Magmatic values are preserved in zircon: Geology, v. 25, p. 1079–1082.

King, E.M., Valley, J.W., Davis, D.W., and Edwards, G.R., 1998, Oxygen isotope ratios of Archean plutonic zircons from granite-greenstone belts of the Superior Province: Indicator of magmatic source: Precambrian Research, v. 92, p. 365–385.

Kitchen, N., and Valley, J.W., 1995, Carbon isotope thermometry in marbles of the Adirondack Mountains, New York: Journal of Metamorphic Geology, v. 13, p. 577–594.

Kretz, R., 1990, Biotite and garnet compositional variation and mineral equilibria in Grenville gneisses of the Otter Lake area, Quebec: Journal of Metamorphic Geology, v. 8, p. 493–506.

Lumbers, S.B., Heaman, L.M., Vertolli, V.M., and Wu, T.-W., 1990, Nature and timing of Middle Proterozoic magmatism in the Central Metasedimentary Belt, Grenville Province, Ontario, *in* Gower, C.F., et al., eds., Mid-Proterozoic Laurentia-Baltica: St. John's, Newfoundland, Geological Association of Canada Special Paper 38, p. 243–276.

Marcantonio, F., McNutt, R.H., Dickin, A.P., and Heaman, L.M., 1990, Isotopic evidence for the crustal evolution of the Frontenac Arch in the Grenville Province of Ontario, Canada: Chemical Geology, v. 83, p. 297–314.

Martignole, J., and Friedman, R., 1998, Geochronological constraints on the last stages of terrane assembly in the central part of the Grenville Province: Precambrian Research, v. 92, p. 145–164.

Martignole, J., and Reynolds, P., 1997, $^{40}Ar/^{39}Ar$ thermochronology along a western Quebec transect of the Grenville Province, Canada: Journal of Metamorphic Geology, v. 15, p. 283–296.

Martignole, J., and Schrijver, K., 1970, Tectonic setting and evolution of the Morin anorthosite, Grenville Province, Quebec: Geological Society of Finland Bulletin, v. 42, p. 165–209.

Martignole, J., Calvert, A.J., Friedman, R., and Reynolds, P., 2000, Crustal evolution along a seismic section across the Grenville Province (western Quebec): Canadian Journal of Earth Sciences, v. 37, p. 291–306.

Matthews, A., Palin, J.M., Epstein, S., and Stolper, E.M., 1994, Experimental study of $^{18}O/^{16}O$ partitioning between crystalline albite, albitic glass, and CO_2 gas: Geochimica et Cosmochimica Acta, v. 58, p. 5255–5266.

McCulloch, M.T., Gregory, R.T., Wasserburg, G.J., and Taylor, H.P., 1980, A neodymium, strontium, and oxygen isotopic study of the Cretaceous Samail Ophiolite and implications for the petrogenesis and seawater-hydrothermal alteration of oceanic crust: Earth and Planetary Sciences Letters, v. 46, p. 201–211.

McLelland, J., and Chiarenzelli, J., 1990, Isotopic constraints on the emplacement age of the Marcy anorthosite massif, Adirondack Mountains, New York: Journal of Geology, v. 98, p. 19–41.

McLelland, J., and Whitney, P., 1990, Anorogenic, bimodal emplacement of anorthositic, charnockitic, and related rocks in the Adirondack Mountains, New York, *in* Stein, H.J., and Hannah, J.L., eds., Ore-bearing granite systems: Petrogenesis and mineralizing processes: Boulder, Colorado, Geological Society of America Special Paper 246, p. 301–315.

McLelland, J., Chiarenzelli, J., Whitney, P., and Isachsen, Y., 1988, U-Pb zircon geochronology of the Adirondack Mountains and implications for their geologic evolution: Geology, v. 16, p. 920–924.

McLelland, J., Daly, J.S., and Chiarenzelli, J., 1993, Sm-Nd and U-Pb isotopic evidence for juvenile crust in the Adirondack Lowlands and implications for the evolution of the Adirondack Mts.: Journal of Geology, v. 101, p. 97–105.

McLelland, J., Daly, J.S., and McLelland, J.M., 1996, The Grenville orogenic cycle (ca. 1350–1000 Ma): An Adirondack perspective: Tectonophysics, v. 265, p. 1–28.

McLelland, J.M., Hamilton, M.A., Bickford, M.E., Clechenko, C., and Valley, J.W., 2002, SHRIMP II Geochronology of the Adirondack AMCG suite, II: Granitoids, gabbros, and crosscutting dikes: Geological Society of America Abstracts with Programs, v. 34, no. 6, p. 271.

Mezger, K., Rawnsley, C.M., Bohlen, S.R., and Hanson, G.N., 1991, U-Pb garnet, sphene, monazite, and rutile ages: Implications for the duration of metamorphism and cooling histories, Adirondack Mts., New York: Journal of Geology, v. 99, p. 415–428.

Mezger, K., van der Pluijm, B.A., Essene, E.J., and Halliday, A.N., 1992, The Carthage-Colton mylonite zone (Adirondack Mountains, New York): The site of a cryptic suture in the Grenville Orogen? Journal of Geology, v. 100, p. 630–638.

Mezger, K., Essene, E.J., van der Pluijm, B.A., and Halliday, A.N., 1993, U-Pb geochronology of the Grenville orogen of Ontario and New York: Constraints on ancient crustal tectonics: Contributions to Mineralogy and Petrology, v. 114, p. 13–26.

Morrison J., and Valley, J.W., 1988, Contamination of the Marcy anorthosite massif, Adirondack Mountains, N.Y.: Petrologic and isotopic evidence: Contributions to Mineralogy and Petrology, v. 98, p. 97–108.

Muehlenbachs, K., 1986, Alteration of the ocean crust and the ^{18}O history of seawater, *in* Valley J.W., Taylor, H.P., and O'Neil, J.R., eds., Stable isotopes in high temperature geological processes: Reviews in Mineralogy, v. 16, p. 425–444.

Owens, B.E., Rockow, M.W., and Dymek, R.F., 1993, Jotunites from the Grenville Province, Quebec: Petrological characteristics and implications for massif anorthosite petrogenesis: Lithos, v. 30, p. 57–80.

Peacock, M.A., 1931, Classifications of igneous rock series: Journal of Geology, v. 39, p. 65–67.

Pearce, J.A., Harris, N.B.W., and Tindle, A.G., 1984, Trace element discrimination diagrams for the tectonic interpretations of granitic rocks: Journal of Petrology, v. 25, p. 956–983.

Peck, W.H., 2000, Oxygen Isotope Studies of Grenville Metamorphism and Magmatism [Ph.D. thesis]: Madison, University of Wisconsin, 320 p.

Peck, W.H., and Valley, J.W., 1998, Oxygen diffusion in zircon: Quartz-zircon disequilibrium in Grenville quartzites: Geological Society of America Abstracts with Programs, v. 30, no. 7, p. 317.

Peck, W.H., and Valley, J.W., 2000, Large crustal input to high $\delta^{18}O$ anorthosite massifs of the southern Grenville Province: New evidence from the Morin Complex, Quebec: Contributions to Mineralogy and Petrology, v. 139, p. 402–417.

Peck, W.H., Valley, J.W., and McLelland, J., 1999, Slow oxygen diffusion in zircon during cooling of Adirondack orthogneiss: Geological Society of America Abstracts with Programs, v. 31, no. 7, p. 103.

Peck, W.H., Valley, J.W., and Graham, C.M., 2003, Slow oxygen diffusion rates in igneous zircons from metamorphic rocks: American Mineralogist, v. 88, p. 1003–1014.

Pehrsson, S., Hanmer, S., and van Breemen, O., 1996, U-Pb geochronology of the Raglan gabbro belt, Central Metasedimentary Belt, Ontario: Implications for an ensialic marginal basin in the Grenville Orogen: Canadian Journal of Earth Sciences, v. 33, p. 691–702.

Perkins, D., Essene, E.J., Marcotty, L.A., 1982, Thermometry and barometry of some amphibolite-granulite facies rocks from the Otter Lake area, southern Quebec: Canadian Journal of Earth Sciences, v. 9, p. 1759–1774.

Rivers, T., Martignole, J., Gower, C.F., and Davidson, A., 1989, New tectonic divisions of the Grenville Province, Southeast Canadian Shield: Tectonics, v. 8, p. 63–84.

Rockow, M.W., 1995, Petrogenesis of the jotunite unit at the Morin Complex, Grenville Province, Quebec: A field, mineralogical and chemical study [Ph.D. thesis]: St. Louis, Missouri, Washington University, 477 p.

Shand, S.J., 1951, Eruptive Rocks: New York, John Wiley and Sons, 360 p.

Shieh, Y., 1985, High-^{18}O granitic plutons from the Frontenac Axis, Grenville Province of Ontario, Canada: Geochimica et Cosmochimica Acta, v. 49, p. 117–123.

Shieh Y., and Schwarcz, H.P., 1978, The oxygen isotope composition of the surface of the Canadian Shield: Canadian Journal of Earth Sciences, v. 15, p. 1773–1782.

Smith, T.E., Holm, P.E., Dennison, N.M., and Harris, M.J., 1997, Crustal assimilation in the Burnt Lake metavolcanics, Grenville Province, southeastern Ontario, and its tectonic significance: Canadian Journal of Earth Sciences, v. 34, p. 1272–1285.

Solomon, G.C., and Taylor, H.P., Jr., 1989, Isotopic evidence for the origin of Mesozoic and Cenozoic granitic plutons in the northern Great Basin: Geology, v. 17, p. 591–594.

Spicuzza, M.J., Valley, J.W., and McConnell, V.S., 1998, Oxygen isotope analysis of whole rock via laser fluorination: An air-lock approach: Geological Society of America Abstracts with Programs, v. 30, no. 7, p. 80.

Streepey, M.M., Essene, E.J., and van der Pluijm, B.A., 1997, A compilation of thermobarometric data from the metasedimentary belt of the Grenville Province, Ontario and New York State: Canadian Mineralogist, v. 35, p. 1237–1247.

Taylor, H.P., 1969, Oxygen isotope studies of anorthosites, with particular reference to the origin of bodies in the Adirondack Mts., NY, *in* Isachsen, Y., ed., Origin of anorthosites and related rocks: Albany, New York State Museum Service Memoir 18, p. 111–134.

Taylor, H.P., and Sheppard, S.M.F., 1986, Igneous Rocks, I: Processes of isotopic fractionation and temperature systematics, *in* Valley, J.W., et al., eds., Stable isotopes in high temperature geological processes: Reviews in Mineralogy, v. 16, p. 227–271.

Valley, J.W., 2003, Oxygen isotopes in zircon, *in* Hanchar, J., and Hoskin, L., eds., Zircon: Reviews in Mineralogy and Geochemistry, v. 53, p. 343–385.

Valley, J.W., and O'Neil, J.R., 1982, Oxygen isotope evidence for shallow intrusion of Adirondack anorthosite: Nature, v. 300, p. 497–500.

Valley, J.W., and O'Neil, J.R., 1984, Fluid heterogeneity during granulite facies metamorphism in the Adirondacks: Stable isotope evidence: Contributions to Mineralogy and Petrology, v. 85, p. 158–173.

Valley, J.W., Bohlen, S.R., Essene, E.J., and Lamb, W., 1990, Metamorphism in the Adirondacks, II: The role of fluids: Journal of Petrology, v. 31, p. 555–596.

Valley, J.W., Chiarenzelli, J.R., and McLelland, J.M., 1994, Oxygen isotope geochemistry of zircon: Earth and Planetary Science Letters, v. 126, p. 187–206.

Valley, J.W., Kitchen, N., Kohn, M.J., Niendorf, C.R., and Spicuzza, M.J., 1995, UWG-2, a garnet standard for oxygen isotope ratios: Strategies for high precision and accuracy with laser heating: Geochimica et Cosmochimica Acta, v. 59, p. 5523–553.

Valley, J.W., Bindeman, I.N., and Peck, W.H., 2003, Empirical calibration of oxygen isotope fractionation in zircon: Geochimica et Cosmochimica Acta, Clayton Volume, v. 67, p. 3257–3266.

van Breemen, O., and Corriveau, L., 1995, Evolution of the Central Metasedimentary Belt in Quebec, Grenville orogen: U-Pb geochronology: International Conference on Tectonics and Metallogeny of Early/Mid Precambrian Orogenic Belt, Precambrian 1995, École Polytechnique de Montrége, Montreal, Program with Abstracts, p. 137.

van Breemen, O., and Davidson, A., 1988, U-Pb zircon ages of granites and syenites in the Central Metasedimentary Belt, Grenville Province, Ontario: Radiogenic Age and Isotope Studies Report 2: Ottawa, Ontario, Geological Survey of Canada Paper 88-2, p. 45–50.

van der Pluijm, B.A., Mezger, K., Cosca, M.A., and Essene, E.J., 1994, Determining the significance of high-grade shear zones by using temperature-time paths, with examples from the Grenville orogen: Geology, v. 22, p. 743–746.

Wasteneys, H.A., 1994, U-Pb geochronology of the Frontenac Terrane, Central Metasedimentary Belt, Grenville Province, Ontario: Geological Association of Canada–Mineralogical Association of Canada, Program with Abstracts, v. 19, p. 18.

Wasteneys, H., McLelland, J., and Lumbers, S., 1999, Precise zircon geochronology in the Adirondack Lowlands and implications for revising plate tectonic models of the Central Metasedimentary Belt and Adirondack Mountains, Grenville Province, Ontario and New York: Canadian Journal of Earth Sciences, v. 36, p. 967–984.

Watson, E.B., and Cherniak, D.J., 1997, Oxygen diffusion in zircon: Earth and Planetary Science Letters, v. 148, p. 527–544.

Watson, B., and Harrison, M., 1983, Zircon saturation revisited: Temperature and composition effects in a variety of crustal magma types: Earth and Planetary Science Letters, v. 64, p. 295–304.

Wu, T., and Kerrich, R., 1986, Combined oxygen isotope-compositional studies of some granitoids from the Grenville Province of Ontario, Canada: Implications for source regions: Canadian Journal of Earth Sciences, v. 23, p. 1412–1432.

Zhao, X., Ji, S., and Martignole, J., 1997, Quartz microstructures and c-axis preferred orientations in high-grade gneisses and mylonites around the Morin anorthosite (Grenville Province): Canadian Journal of Earth Sciences, v. 34, p. 819–832.

Manuscript Accepted by the Society August 25, 2003

Geological Society of America
Memoir 197
2004

L-S shape fabrics in the Mazinaw domain and the issue of northwest-directed thrusting in the Composite Arc Belt, southeastern Ontario

Walfried M. Schwerdtner*
Matthew W. Downey
Sharyn A. Alexander
Department of Geology, University of Toronto, Toronto, Ontario M5S 3B1, Canada

ABSTRACT

Many vertical sections across deeper parts of collisional orogens contain gently dipping thrusts that have the geometric style of shallow-crustal, dip-slip faults. In the southwestern Grenville Province (central and southeastern Ontario), however, structural geologists have found it difficult to document northwest-directed thrusting within large, polydeformed complexes of medium- to high-grade metamorphic rocks. This applies, in particular, to the Mazinaw domain of the eastern Composite Arc Belt, where narrow north- to east-northeast-striking ("longitudinal") zones of high strain have been linked tentatively to ductile thrusting at the sole of Elzevirian (1230–1190 Ma) or Early Ottawan (1085–1015 Ma) fold nappes. During subsequent northwest-southeast shortening, the apparent thrusts behaved as structural detachment horizons or were subhorizontally folded, together with their wallrocks.

Gravity modeling and reflection-seismic profiling have shown that in the present shallow crust most tabular rock bodies and coplanar structural surfaces (including high-strain zones and ductile faults) are horizontal or dip <45°SE. This seems incompatible with the idea that Elzevirian and Early Ottawan thrusts acquired strong curvature in the northwest-southeast vertical plane during upright buckling of fold nappes and other southeast-dipping lithotectonic units.

The longitudinal high-strain zones and their ductilely deformed wallrocks are characterized by steeply inclined to vertical planar fabrics (S) such as geometric systems of strained mineral constituents, pebbles, and volcanic fragments. Linear fabrics (L) coexist with the planar fabrics at many localities, but are commonly weak or seemingly absent. The prevalence of planar fabrics is difficult to explain by shallow-crustal-style ductile thrusting, but these fabrics may have been produced by large-scale strains associated with the formation of midcrustal stretching thrusts.

For a variety of thrust-dip options, we use L < S fabrics to derive the local sense of tangential shear strain within and adjacent to six high-strain zones. The sense of accumulated shear strain proves to be reverse for most realistic options, even if one presumes that the zones rotated during superimposed buckle folding. Mesoscopic Z-folds are common in one high-strain zone, and attest to dextral-reverse shearing and shortening of thrust walls during Mid- to Late Ottawan deformation (1015–980 Ma).

*E-mail: fried@quartz.geology.utoronto.ca.

Schwerdtner, W.M., Downey, M.W., and Alexander, S.A., 2004, L-S shape fabrics in the Mazinaw domain and the issue of northwest-directed thrusting in the Composite Arc Belt, southeastern Ontario, *in* Tollo, R.P., Corriveau, L., McLelland, J., and Bartholomew, M.J., eds., Proterozoic tectonic evolution of the Grenville orogen in North America: Boulder, Colorado, Geological Society of America Memoir 197, p. 183–207. For permission to copy, contact editing@geosociety.org.

The combined results of structural measurement and modeling of shear-strain sense suggest that the high-strain zones and their walls were stretched or shortened progressively in the direction of simultaneous tangential shearing and/or irrotational slip. This means that the longitudinal dislocations of the Mazinaw domain qualify as stretching thrusts that were active in a midcrustal regime of northwest-directed distributed shearing, northwest-southeast pervasive shortening, and associated vertical thickening.

Keywords: lithotectonic boundary, stretching fault, crustal thickening

INTRODUCTION

Northwest-directed thrusting played an important role in the structural evolution of the Grenville orogen in eastern Canada (Davidson et al., 1982; Culshaw et al., 1983; Green et al., 1988; Rivers et al., 1989; Gower et al., 1990; Easton, 1992; Martignole and Calvert, 1996; Rivers, 1997; Rivers and Corrigan, 2000). In the Ontario segment of the Grenville Province, various styles of thrusting have been recognized (for a short summary, see Table 2 *in* Carr et al., 2000). Yet the term *thrust* is commonly qualified by adjectives such as *possible* or *cryptic;* structural evidence for, and clues to the actual style of, faulting remain to be found.

Results of structural mapping, theoretical study, and analogue modeling (Ramsay and Graham, 1970, p. 809–812; Inglis, 1983; Means, 1989) suggest that stretching faults may abound at midcrustal levels of collisional orogens (Means, 1990; Schwerdtner, 1998). The development of such structures leads to distributed tangential shearing and to ductile shortening or elongation of the fault walls during protracted, commonly large, slip (Duebendorfer and Black, 1992; Hajnal et al., 1996; Schwerdtner, 1998; Schwerdtner and Cote, 2001). In the Grenville Province of Ontario, rocks deformed at midcrustal levels are presently exposed throughout the Central Gneiss Belt (Laurentia and Laurentian margin), as well as in the Parry Sound, Bancroft, and Mazinaw domains of the Composite Arc Belt (Figs. 1, 2, and 3).

Twelve lithotectonic divisions, including the Parry Sound domain (Fig. 1) and the Mazinaw domain (the Mazinaw terrane of Easton, 1992), have been identified in the Ontario portion of the Composite Arc Belt (Carr et al., 2000, p. 195). The map pattern of the Mazinaw domain (Figs. 1, 2, and 3), also known as the Mazinaw terrane, is dominated by north- to east-northeast-striking (henceforth called longitudinal) lithotectonic units and kilometer-scale structures such as narrow high-strain zones and nearly upright open to close folds. The high-strain zones are characterized by strongly foliated, locally mylonitic rocks, some of which include highly distorted pebbles, igneous xenoliths, or volcanic fragments. On modern geological maps of the Mazinaw domain and its vicinity, longitudinal zones of high strain have been interpreted as walls of slip planes (ductile-brittle thrust faults) or ductile shear zones situated at the sole of fold nappes (Chappell, 1978; Easton, 1992, p. 843; Ontario Geological Survey, 1992; Corfu and Easton, 1995, p. 962). The boundaries of high-strain zones are gradational, so the horizontal width of the zones is difficult to estimate. Our crude estimates of minimum width vary along and between individual zones and range from tens to hundreds of meters. Most high-strain zones contain a lithotectonic boundary (LTB; Schwerdtner, 1998), such as the stretched border of a granitoid pluton or the unconformable base of the Flinton Group (Moore and Thompson, 1972, 1980).

Two problems create large uncertainties in determining the local horizontal position of LTB surfaces: (1) severe difficulties in identifying the protolith of highly deformed or mylonitic rocks in the field, and (2) the paucity of large outcrops in longitudinal high-strain zones. Elevation differences are generally <50 m along individual high-strain zones, that is, on the same order of magnitude as uncertainties in the local horizontal position of LTB surfaces. This hinders the use of classical projection methods in determining the dip of LTBs or high-strain zones, as well as the construction of vertical sections across the Mazinaw domain. However, detailed gravity modeling furnishes dip values of structural surfaces near the erosion level (Real and Thomas, 1987). For purposes of the following shear sense assessment, we need to know the true dip of LTBs between mapable rock units that border, or fall into, longitudinal high-strain zones.

Lithoprobe Lines 32 and 33 and their reflection-seismic profiles cross the Composite Arc Belt in southeastern Ontario (White et al., 1994, 2000). In the profiles, LTBs and high-strain zones or thrusts may correspond to shallow crustal reflectors, whose average true dip is <25° (cf. Fig. 4). Individual reflectors may be extended as straight lines to the earth's surface, and the approximate dip angles thus obtained are also used in our subsequent modeling of shear-strain conditions.

Linear-planar (L-S) fabrics are commonly steep to vertical within and adjacent to well-exposed segments of longitudinal high-strain zones, and differ markedly in orientation from the slightly to moderately inclined geophysical images of structural surfaces in the present shallow crust of the Mazinaw domain (Fig. 4; Real and Thomas, 1987). The stretching fault concept predicts marked obliquities between the slip plane and principal directions of wallrock strain (Means, 1989; Schwerdtner, 1998), and therefore explains the difference in dip angle between geophysically imaged, structural surfaces ("possible thrusts") and L-S fabrics

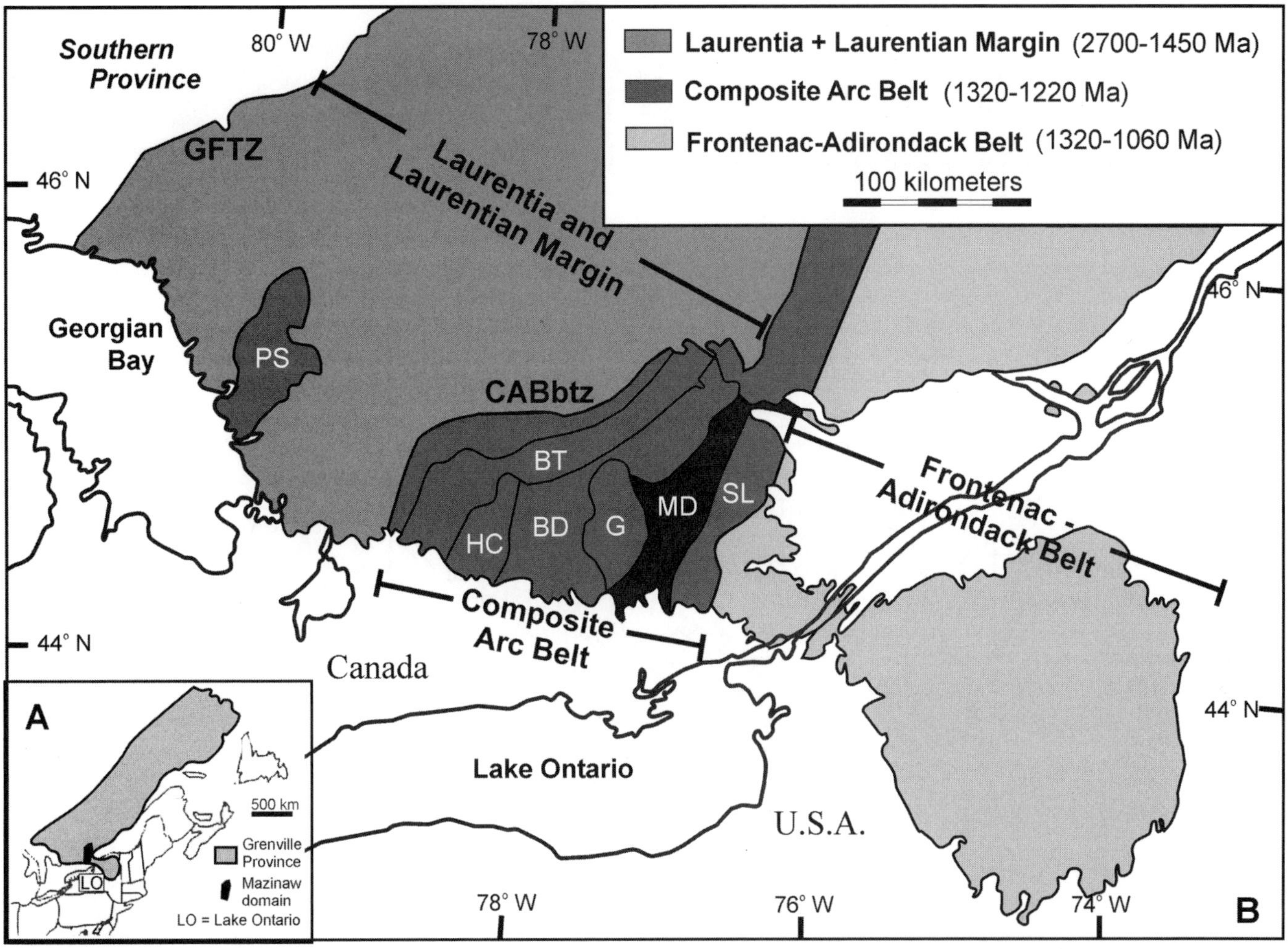

Figure 1. (A) Position of the Mazinaw domain within the Grenville Province, eastern North America. (B) Major divisions of the southwestern Grenville Province, southeastern Ontario, Canada: (1) Pre-Grenvillian Laurentia and its margin, collectively known as the Central Gneiss Belt; (2) the Composite Arc Belt; and (3) the Frontenac-Adirondack Belt. BD—Belmont domain; BT—Bancroft domain; CABbtz—Composite Arc Belt boundary thrust zone; G—Grimsthorpe domain; GFTZ—Grenville Front tectonic zone; HC—Harvey-Cardiff arch; MD—Mazinaw domain; PS—Parry Sound domain, traditionally considered part of the Central Gneiss Belt; SL—Sharbot Lake domain. The Elzevir terrane consists of the Harvey-Cardiff arch, the Grimsthorpe domain, and the Belmont domain. Modified from Carr et al. (2000, Fig. 2b).

within their walls. We test the proposition, formed on the basis of geological maps, geophysical profiles, and other published information, that the high-strain zones and associated thrusts I–VI (Fig. 3) are stretching faults. The test employs a simple graphic technique for deriving the sense of large shear strain (Fig. 5; Schwerdtner, 1998) and relies on L-S fabric data we have acquired recently within and adjacent to six high-strain zones.

BACKGROUND INFORMATION

Geology of the Composite Arc Belt and the Mazinaw Domain

The Composite Arc Belt (Fig. 1) contains repeatedly deformed and metamorphosed volcanic, sedimentary, and plutonic rocks whose protoliths vary in age from ca. 1290 Ma to <1150 Ma (Carr et al., 2000). Effects of at least two major tectonic pulses of the Grenville orogeny have been recognized in the Belt, that is, the 1230–1180 Ma Elzevirian pulse and the 1085–980 Ma Ottawan pulse (Moore and Thompson, 1980; Carr et al., 2000, p. 202–204). In most parts of the belt, rocks have been metamorphosed under amphibolite-facies conditions (Easton, 1992; Carr et al., 2000).

The Mazinaw domain is a 3000-square-kilometer oblong fault block (Fig. 2) that seems to have been created during the Ottawan orogenic pulse (Corfu and Easton, 1995). Uplift and partial erosion of the block have exposed a variety of heterogeneously strained rock units, whose 1020 Ma metamorphism reaches amphibolite facies and is generally at a higher grade than that of similar units observed in the adjacent Grimsthorpe and Sharbot Lake domains (Fig. 1; Easton, 1992). Moreover, rocks presently exposed in these adjacent domains were un-

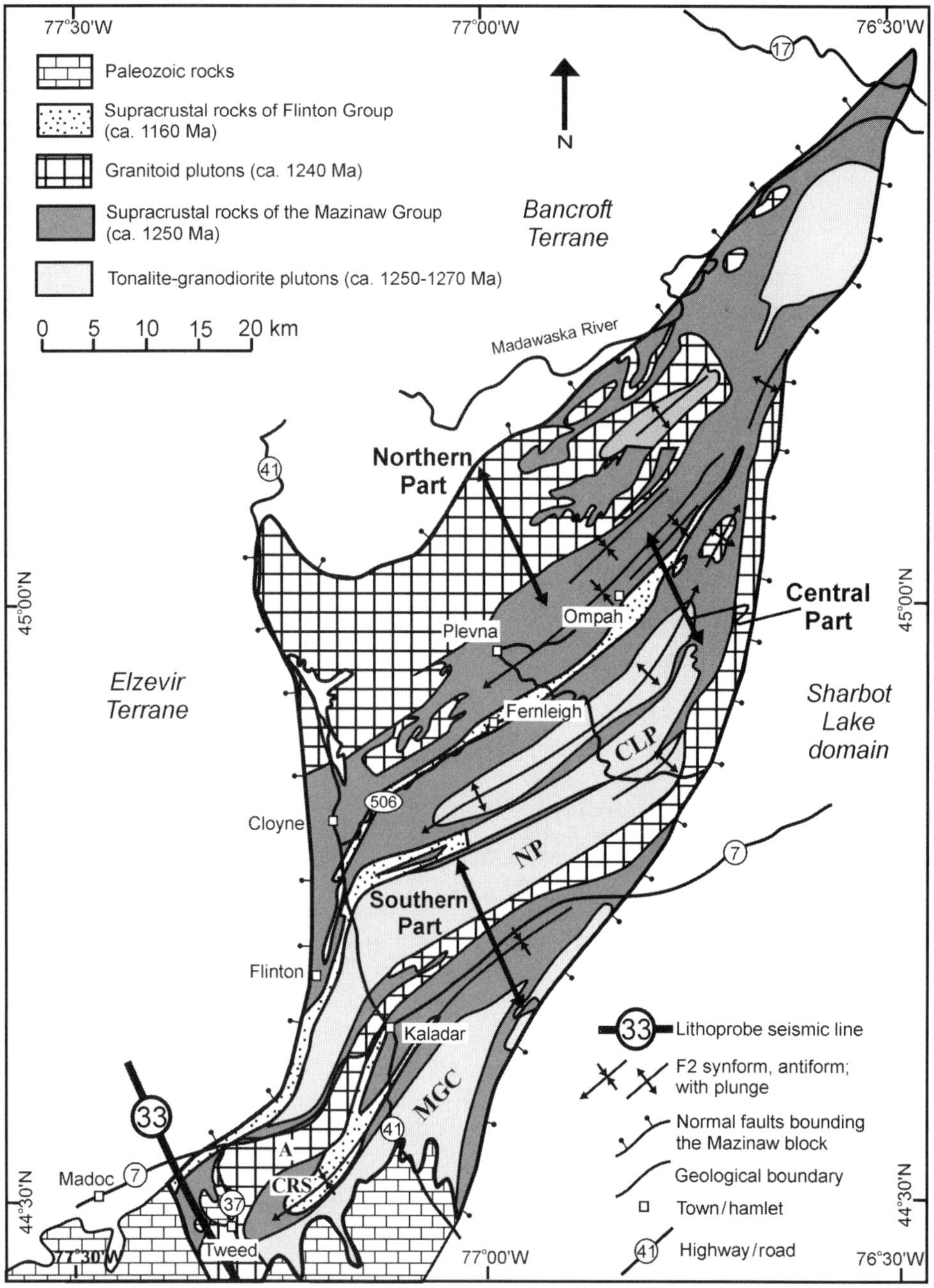

Figure 2. Geological features of the Mazinaw domain of the Composite Arc Belt, Grenville Province. A—Addington pluton; CLP—Cross Lake pluton; CRS—Clare River synform; MGC—Mellon Lake granitoid complex; NP—Northbrook pluton. Modified from Easton (1992) and Corfu and Easton (1995, Fig. 2).

affected by the 1020 Ma metamorphism in the Ottawan orogenic pulse (Corfu and Easton, 1995).

Three kinds of large longitudinal geological features have been delineated in metavolcanic-metasedimentary terrain of the Mazinaw domain (Figs. 2, 3, and 6): distorted granitoid plutons, narrow high-strain zones, and three generations of nearly coaxial folds (F1, F2, F3) whose axial surfaces differ in final shape and orientation (see Ramsay, 1967, p. 533; Passchier et al., 1990, p. 71). North- or east-northeast-striking segments of the longitudinal features are situated in rotated limbs of gentle to open fourth-generation folds (F4) with southeast-trending axial planes (Easton, 1992, p. 849). Together with longitudinal tracts of metavolcanic-metasedimentary rock, the granitoid plutons and F2 structures dominate the geological map pattern (Figs. 2 and 3).

Longitudinal high-strain zones seem to have originated in the overturned limbs of ill-defined F1 nappes (Fig. 6, A), and are represented on recent geological maps by thrust faults (Figs. 2 and 3; Ontario Geological Survey, 1992; Corfu and Easton, 1995). However, northwest-directed thrusting is thought to have commenced before the peak metamorphism and associated ductile deformation at ca. 1020 Ma (Carr et al., 2000). As

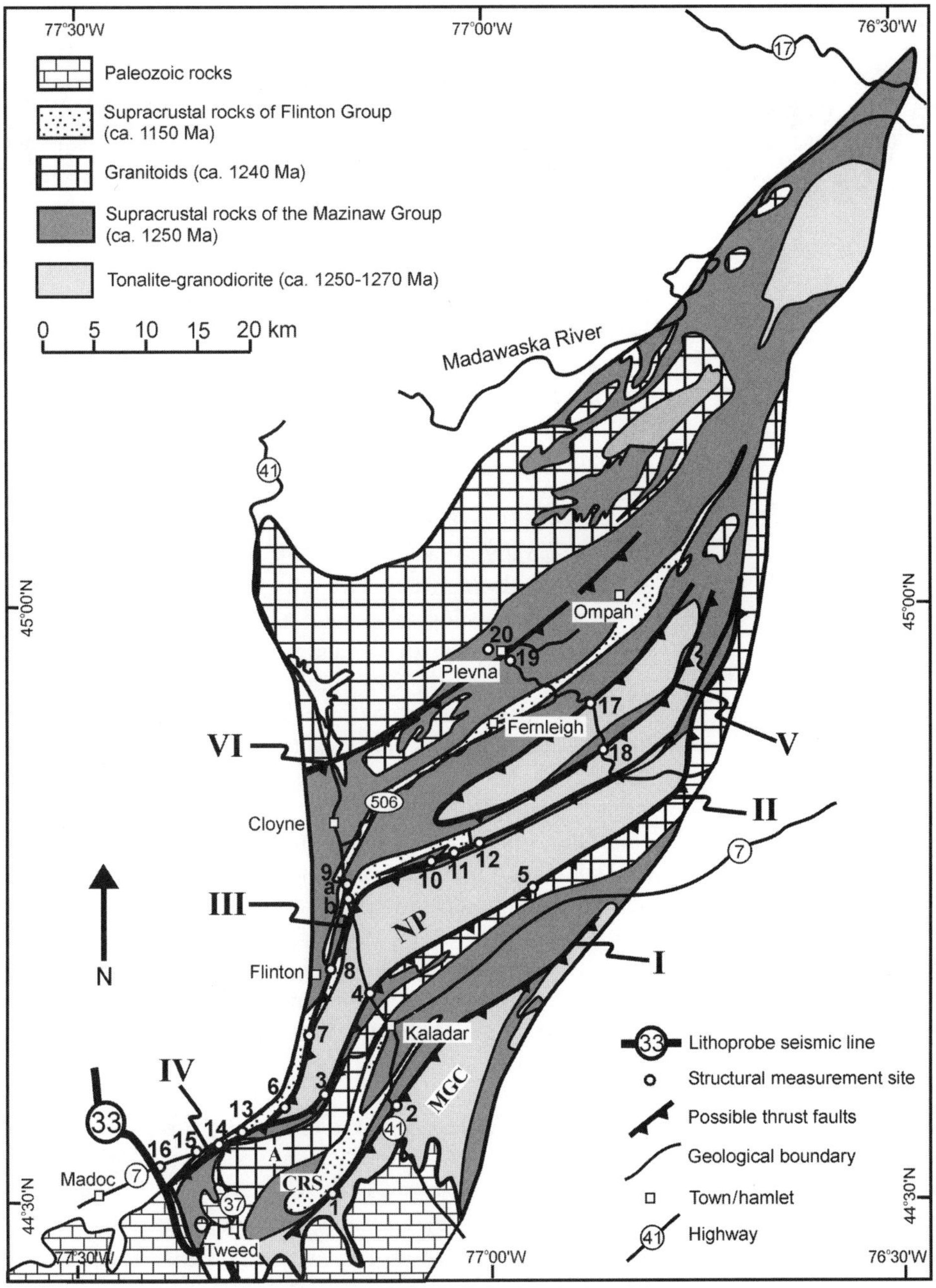

Figure 3. Location map of possible thrust faults (I—VI), structural measurement sites (1—20), and highways or roads within the Mazinaw domain. Thrust labels: I—Mitten thrust; II—Little Skootamatta thrust; III—Henderson Road thrust; IV—southwestern Mooroton shear zone; V—Big Gull thrust; VI—Plevna thrust. Abbreviations: A—Addington pluton; CRS—Clare River synform; MGC—Mellon Lake granitoid complex; NP—Northbrook pluton. Possible thrust faults (I—VI) are as recognized by Easton (1992), Ontario Geological Survey (1992), Corfu and Easton (1995, Fig. 2), and Cureton et al. (1997, p. 976 and 980). Arabic numerals indicate measurement sites (1–20) in or adjacent to high-strain zones (possible thrust faults).

shown on the structural map of the Mazinaw domain (Easton, 1992, p. 843), the trajectories of the present high-strain zones or thrusts locally transect, or depart significantly from, major lithological boundaries. Discordant segments of high-strain zones may be similar to oblique ramps of shallow-crustal faults, and perhaps suggest that the component of bulk translation between the present thrust walls is very large (cf. Ramsay, 1980, p. 97–98).

Macroscopic F2 structures, as exemplified by the Clare River synform (Fig. 2), are upright to slightly inclined and apparently superimposed, in a quasi-coaxial way, on the longitudinal fold nappes and their southeasterly dipping sole thrusts (Fig. 6; Chappell, 1978; Easton and Ford, 1991; Easton, 1992, p. 849). Apart from the questionable southwest closure of thrust V (Fig. 3), however, the longitudinal thrusts show no obvious signs of D2 folding. Moreover, all thrusts of the Mazinaw domain are shown to have southeasterly dips in the Tectonic Assemblages Map (Ontario Geological Survey, 1992). This is explicable by a kinematic model in which D1 thrusts acted as detachment surfaces during D2 folding, and thereby determined the maximum size of F2 structures. In any case, D1 thrust surfaces remained active structural features unless the magnitude

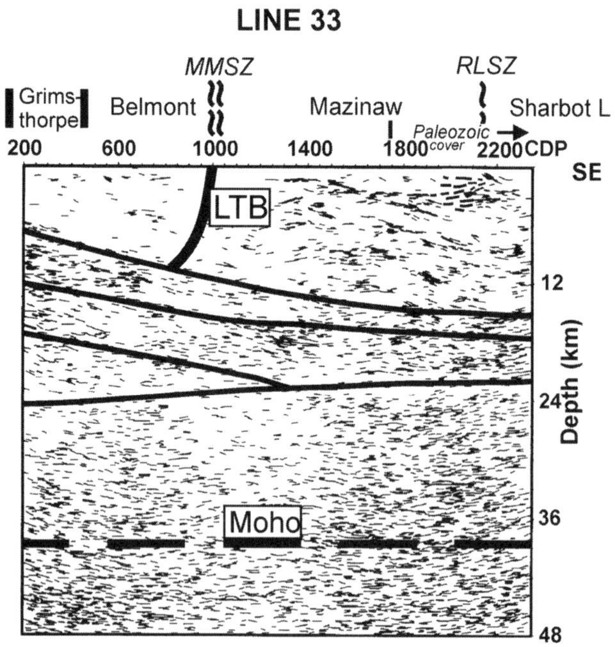

Figure 4. Reflection seismic profile of Lithoprobe Line 33 (cf. Figs. 2 and 3) across the southern Mazinaw domain. RLSZ—Robertson Lake shear zone; MMSZ—Mooroton shear zone. Modified from White et al. (1994, Fig. 8).

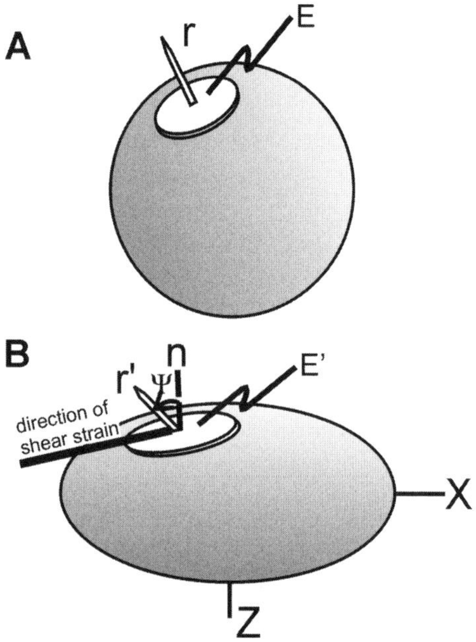

Figure 5. Tangential shear strain at both sides of a material plane (E′) representing a fault or lithotectonic boundary (LTB) segment. E—fault or LTB at the onset of ductile deformation. Undeformed sphere (A) with material elements *E* and *r;* ellipsoid of finite strain with stretched material elements *E*′ and *r*′ (B). The angular departure of *r*′ from the normal (*n*) indicates the sense and magnitude of shear strain. ψ—angle of sinistral shear strain; X, Y—principal directions of finite strain. Modified from Nadai (1950) and Lisle (1998).

of tangential shear stress became very low, at an advanced stage of D2 and during D3 or D4 (see Jaeger, 1962, p. 159–164).

Results of recent geological mapping (Wolff, 1981, 1982; Bright, 1986a, 1986b; Easton, 1988, 1992; Barclay, 1989; Easton and Ford, 1991; Ford, 1992; Easton et al., 1995a, 1995b) and U-Pb geochronology (Connelly et al., 1987; van Breemen and Davidson, 1988; Mezger et al., 1993; Sager-Kinsman and Parrish, 1993; Corfu and Easton, 1995) attest to marked heterogeneity and tectonic complexity in the Mazinaw domain. Thus, the main ductile deformation differs in duration and absolute age between the northern and central parts of the domain (Corfu and Easton, 1995). The starting times of the main deformation also may have differed between the central and southern parts of the Mazinaw domain. However, modern U-Pb geochronology remains to be applied to most rock units in the central and southern parts (cf. Corfu and Easton, 1995; Fig. 2; Ontario Geological Survey, 1992). In particular, no U-Pb zircon date is available for the centrally located Northbrook pluton, which hinders the stratigraphic documentation of thrusts II, III, and V (Fig. 3).

Published geological maps (e.g., Hewitt, 1964) show that a curved panel of metavocanic and metasedimentary rocks separates that Northbrook pluton from its northwestern neighbor, the 1250 Ma Cross Lake pluton (Corfu and Easton, 1995). The panel has a narrow gap, at one locality southeast of Cross Lake (Hewitt, 1964), so the two plutons are ostensibly connected at surface. Thrust III bridges the narrow gap, and completely separates the Northbrook pluton from the Cross Lake pluton (Fig. 3). Unless the plutons are adjacent parts of the same intrusive body, and the dip separation on thrust III is very small (for which there is no compelling evidence at present), the Northbrook and Cross Lake plutons must have originated at different structural levels and later been juxtaposed by thrust III. This casts doubt on the popular notion that the Northbrook and Cross Lake plutons are coeval.

Structural Significance of the Flinton Group

An important characteristic of the Mazinaw domain is the abundant occurrence of deformed conglomerate-rich strata of the Flinton Group (Moore and Thompson, 1980), a sedimentary sequence <1150 Ma old that was deposited unconformably on a complex of arc-related basement rocks ranging in age from 1290 to 1245 Ma (Easton and Ford, 1991; Corfu and Easton, 1995; Carr et al., 2000). Easton (2001) regards the Flinton Group as tripartite in the central and southern Mazinaw domain, where it is represented by the Kaladar and Skootamatta, the Bishop Corners and Beatty, and the Lessard and Bogard formations. All strata of the Flinton Group are thought to have been deposited after the tectonic assembly of the Composite Arc Belt (Easton, 1992; Hanmer et al., 2000). This explains the continuity of metaconglomerate units and associated Flinton Group rocks between the western Mazinaw and the eastern Grimsthorpe domain (Fig. 1), as shown on geological maps by Wilson (1940), Thompson (1972), Moore and Thompson (1980), and Easton (1992, 2001).

Some investigators of the Composite Arc Belt (southeastern Ontario) were skeptical about the Flinton unconformity, and concluded that the boundaries between the metaconglomerate units and adjacent rock masses originated as structural discontinuities or intrusive contacts (Rivers, 1976; Connelly, 1986; Connelly et al., 1987). Regardless of how one interprets the boundaries, there is structural evidence that the rocks of the Mazinaw domain experienced several regional deformations (D1–D4), all of which postdate the deposition of Flinton Group strata (Easton, 1992, p. 849).

Upright to moderately inclined F2-structures in the lithologic boundaries and a prominent first foliation (S1: mineral-grain schistosity, metamorphic lamination, and/or flattened inclusions) typically lack a pervasive, axial-planar foliation (S2) and associated mineral-shape lineations (L2). This points to widespread, passive redeformation (D2) of rocks with L1-S1 fabrics, as we explain more fully later. In the southwest closure of the Clare River synform (Fig. 2), for example, mesoscopic F1 structures in a relict bedding of metasedimentary rocks (Bogart Formation of Easton, 2001) seem to have been slightly deamplified during D2, without acquiring a transverse new foliation (S2). In the same closure, however, highly schistose rocks of the Beatty Formation (Easton, 2001) are strongly crenulated and exhibit an S2 fabric parallel to the axial planes of mesoscopic F2 structures and the Clare River synform. The crenulae (F2) have orthorhombic symmetry and plunge 30°–40°NE. Their geometric style attests to northwest-southeast shortening of the schistosity (S1) in the course of D2, and suggests that the buckling mechanism played a major role in the development of large F2 structures.

Barclay (1989) envisaged a late tectonic pulse of northwest-southeast penetrative shortening that created strong planar fabrics and southwest-plunging folds in the Mitten dike (Bright, 1986a, 1986b) and Flinton Group strata in the southeast limb of the Clare River synform (Easton, 1992, p. 843; Easton, 2001). Barclay disagreed with Chappell's (1978) conclusion that the main foliation in the southern Kaladar area, northeast of Tweed (Fig. 2), was created during D1 and folded on the kilometer scale during D2. However, Barclay's (1989) deformation scenario does not account for the map pattern of the main foliation, in the metavolcanic and metasedimentary rocks of the southwestern Clare River synform (Wilson, 1934, 1940; Currie, 1972), which is explicable by large-scale D2 folding of northeast-southwest–striking, coplanar folia (S1) about a northeasterly plunging axis.

Easton (1992, p. 849) had difficulty in reconciling the map pattern of Flinton Group rocks with published structural scenarios (Moore and Thompson, 1972, 1980; Rivers, 1976; Chappell, 1978; Connelly, 1986). He therefore proposed that the Flinton Group strata were everywhere unaffected by D2, but experienced moderately large strain during D3. However, Flinton Group rocks occur in highly strained walls of several thrusts supposedly created during D1 or early in D2 (Fig. 3). This may be seen as evidence that slip on, and strain concentration at, major lithologic boundaries occurred throughout the deformation history of the rock assemblage in the Mazinaw domain.

Using all geological and geophysical information available, we attempted to draw schematic vertical sections across the southern Mazinaw domain. We relied, in particular, on a new geological map of the Kaladar region (Easton, 2001), which shows the distribution of Flinton Group rocks in southwestern parts of the Clare River synform and the envelope of the Northbrook pluton (Figs. 2 and 3). Unfortunately, we did not succeed in reconciling the map pattern with the results of geophysical profiling, Easton's (2001) updated stratigraphy, and the structural history as summarized earlier. Our lack of success may have had several reasons, but we are most worried about the potential effect of unrecognized premetamorphic faults parallel to apparently stratigraphic contacts in the Clare River synform. The present investigation of longitudinal thrusts, however, does not depend on having sections across the Mazinaw domain.

D1–D2 Thrusting and the Stretching-Fault Concept

According to Means (1989, p. 893), "An unfamiliar class of faults in flowing rock bodies has wall rocks that lengthen or shorten in the slip direction while slip accumulates. . . . Large-scale structures of this kind [stretching faults] may be important tectonic elements below levels in the crust to which faults with rigid walls extend. . . . [Below these levels] the flow associated with faulting is [not] a response to slip, but an independent process driven by the same remote loads that are causing the slip." Stretching faults are expected to develop where LTBs become mechanically incoherent, as exemplified by bedding-parallel segments of sole thrusts in large fold nappes (Ramsay and Huber, 1983, p. 205; Casey and Dietrich, 1997), also called type F thrust sheets in midcrustal shear belts (Hatcher and Hooper, 1992; Hatcher, 2001). The component of net slip (bulk translation) is very large on the sole thrusts of many fold nappes (Ramsay and Huber, 1983, p. 205; Hatcher and Hooper, 1992; Casey and Dietrich, 1997, p. 122), but this need not be true for other stretching faults.

The kinematics of stretching faults were first considered in one or two dimensions (Means, 1990; Duebendorfer and Black, 1992), but have been dealt with more recently in three dimensions (Hajnal et al., 1996; Schwerdtner and Cote, 2001). Half-stretching faults (Means, 1989; Fig. 2 *in* Schwerdtner, 1998) may occur at the boundaries of ductile-extrusion zones situated between rigid blocks (Sanderson and Marchini, 1984; Robin and Cruden, 1994; Jones et al., 1997). In a general three-dimensional setting, however, the progressive strain of stretching faults or LTB walls is expected to be triaxial and markedly heterogeneous. Consequently, the direction of maximum tangential stretching changes with position and time, but is generally oblique to the local slip vector. This results in spatial obliquities between the fault surface and the principal directions of finite strain in the wallrocks. Unlike the walls of translational faults in the brittle upper crust, stretching-fault walls in the middle crust may be characterized, therefore, by discordant shape fabrics such as those of distorted pebbles. The walls of typical stretching thrusts, in particular, are expected to contain L-S fabrics that dip more steeply than the fault surface (Figs. 6 and 7). This casts doubt on the general validity of the time-honored practice of putting the fault surface parallel to the main planar fabric of metamorphic wallrocks exposed at the earth's surface.

In the Mazinaw domain, planar shape fabrics (S) are steep to vertical at most of our structural measurement sites (Fig. 3, Table 1), but, as discussed briefly earlier in this paper, the present style and prevailing attitude of S seem to be a net result of repeated passive deformation during the Elzevirian and Ottawan orogenic pulses. S1 fabrics associated with stretching thrusts of F1 nappes may have been moderately inclined at the end of D1 (Fig. 7, A), and rotated in a passive manner until they became virtually perpendicular to the thrust surface (Fig. 7, B). Very high magnitudes of D2 strain would be required to rotate southeasterly dipping S1 fabrics past the vertical plane (cf. Table 1). The hinge zones of large F2 structures, however, furnish no evidence for such strain magnitudes. This may be attributed to a dominance of the buckling mechanism during D2 folding, at least in the central and southern parts of the Mazinaw domain.

The correct identification of stretching faults in polydeformed metamorphic complexes requires detailed knowledge of the geological history and kinematic path of the wallrocks, which poses a difficult problem for structural geologists (cf. Means, 1989). The problem is particularly serious in the Mazinaw domain, where the structural history and kinematic path of most rock units are poorly known. This, in turn, hinders the kinematic assessment and classification of the longitudinal thrusts

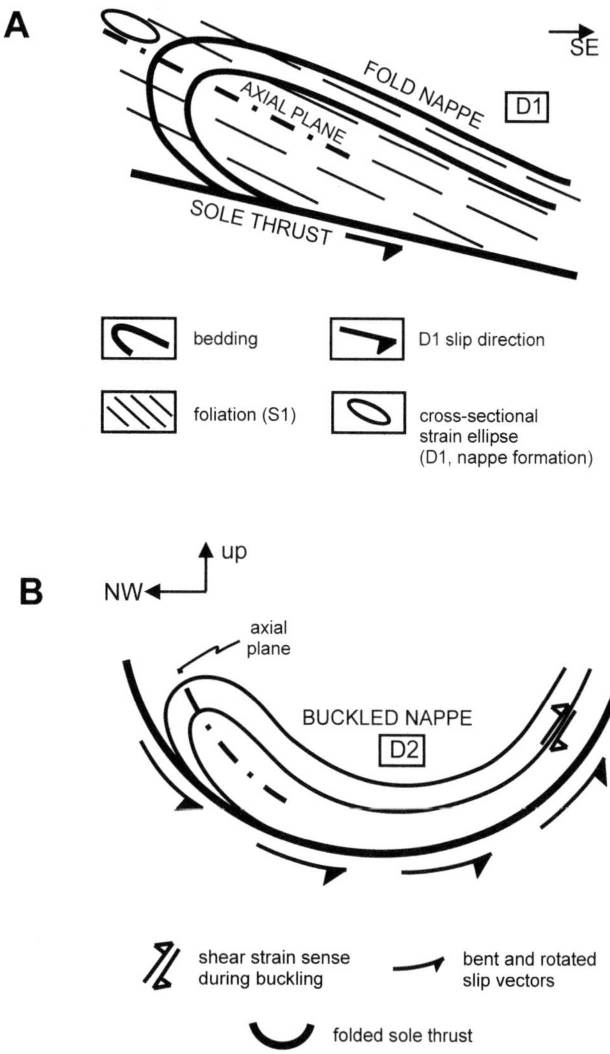

Figure 6. Vertical section across a hypothetical F1 nappe in the Mazinaw domain, at the onset (A) and end (B) of D2 folding. Note the fold-induced change in the apparent sense of D1 slip or tangential shear strain predicted for the southeast limbs of flexural buckle folds, but apparently absent in the actual F2 structures (cf. Fig. 2).

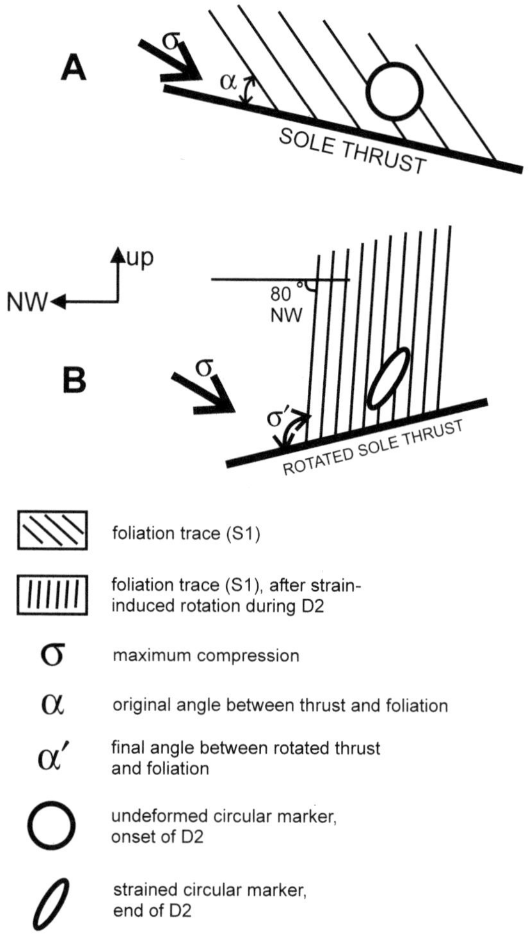

Figure 7. Hypothetical geometric effects of large pervasive strain (D2) not evident in F2 structures of the Mazinaw domain. Traces of S1 and associated sole thrust at the onset (A) and end (B) of subhorizontal D2 shortening expected to precede the buckle folding of parallel strata with low viscosity contrast (cf. Ramsay, 1967, p. 403; Dieterich, 1970). Note the large passive rotation of the S1 trace due to a hypothetical D2 strain (schematic cross section).

TABLE 1. STRUCTURAL FEATURES AND ATTITUDES (IN DEGREES) AT INDIVIDUAL MEASUREMENT SITES

	Type and attitude				
Site	Rock type	Planar feature	Planar attitude	Linear feature	Linear attitude
1	Alaskitic granite (Mitten dike)	Mineral-shape fabric (quartz eyes) and foliation	50/68 NW, 51/72 NW, 48/70 NW, 47/67 NW		
	Mellon Lake granitoids	Mineral lamination and schistosity	50/64 NW, 46/60 NW, 45/72 NW		
2	Biotite-rich and alaskitic granite	Mineral-shape fabrics (mafic clots) and foliation	54/85 NW, 44/87 NW, 44/90, 47/88 NW, 40/88 NW	Mineral-shape lineation (weak)	226/06, 220/06, 232/07, 220/09, 220/10, 226/08, 224/07, 226/09
	Calcitic marble	Mafic laminae	45/86 NW, 46/90, 50/90, 52/74 SE		
	Mellon Lake granitoids	Mafic laminae	40/90, 40/87 NW		
3	Northbrook granitoids	Foliation and schistosity	35/20 SE, 35/25 SE, 15/35 ESE	Mineral lineation	35/01, 209/05
	Hornblende mica schist	Schistosity	16/38 ESE	Mineral lineation	190/15
	Fine-grained amphibolite	Schistosity	30/35 SE, 24/38 SE		
4	Metaconglomerate	Foliation and schistosity	45/40 SE, 33/38 SE, 45/30 SE	Mineral lineation (pebbles)	50/4, 45/4
	Northbrook granitoids	Foliation	56/52 SE, 65/60 SE, 52/47 SE	Mineral lineation (mafic clots)	45/7, 55/8
5	Northbrook granitoids	Mineral-shape fabric and schistosity	56/40 SE		
	Addington Granite with enclaves of amphibolite and mica schist	Mineral lamination	62/48 SE		
	Addington Granite	Mineral foliation	60/58 SE	Weak lineation	62/05, 63/06
6	Metaconglomerate (highly deformed)	Schistosity and foliation	56/80 SE, 50/65 SE	Mineral lineation	231/30
	Northbrook granitoids	Mineral foliation and schistosity	40/55 SE, 45/55 SE, 58/75 SE, 62/75 SE, 50/78 SE	Mineral lineation	232/26
7	Metaconglomerate-Northbrook granitoids	Foliation and schistosity	20/80 ESE, 21/78 ESE		
8	Amphibolite	Mineral schistosity	26/71 ESE, 30/82 ESE, 26/76 ESE	Mineral lineation (mafic clots and grains)	116/72, 80/72
9	Amphibolite	Schistosity	29/85 ESE, 12/83 ESE	Mineral lineation	119/85, 192/78
	Flinton meta conglomerate	Pebble-shape fabric and foliation	34/76 SE, 20/74 ESE, 22/85 ESE, 23/77 ESE	Pebble-shape fabric and mineral lineation	113/77, 195/68
10	Amphibolite and Flinton meta conglomerate	Pebble-shape fabric, foliation, and schistosity	80/70 N, 80/69 N, 82/73 N	Mineral lineation	77/10
11	Amphibolite	Schistosity	79/84 NW, 78/67 NW, 80/78 N	Mineral lineation	268/13, 266/15
12	Northbrook granitoids Amphibolite	Foliation Schistosity	65/85 NW 65/78 NW, 80/78 S, 80/79 S	Mineral lineation	249/06
13	Mica schist	Schistosity	60/70 SE	Mineral lineation	198/64
	Hornblende mica schist	Schistosity	50/80 SE, 45/72 SE	Mineral lineation	205/65
14	Amphibolite	Schistosity	82/70 S	Mineral lineation	199/67
	Marble	Foliation	50/67 SE	Mineral lineation	178/62, 195/60

(continued)

TABLE 1. *Continued*

		Type and attitude			
Site	Rock type	Planar feature	Planar attitude	Linear feature	Linear attitude
	Mica schist	Schistosity	64/77 SE, 70/70 SE	Mineral-grain lineation	217/61, 211/57
13-14	Hornblende mica schist between sites 13 and 14	Z fold enveloping surfaces	37/67 SE, 37/66 SE, 47/61 SE	Z fold hinge line (axes)	188/46, 170/57, 205/30
15	Mica schist	Schistosity	82/85 S, 83/81 S	Mineral lineation	243/68, 213/68
	Amphibolite	Schistosity	92/83 S	Mineral lineation	241/76, 235/77
16	Strained lapilli tuff	Lapilli-shape fabric, schistosity	97/84 S	Lapilli-shape fabric	187/84
	Metaconglomerate	Pebble-shape fabric, foliation	65/72 NW, 82/79 S	Pebble-shape fabric, mineral lineation	335/72, 215/76
17	Amphibolite	Schistosity	56/80 NW, 56/90	Mineral lineation	58/10, 58/07
	Marble	Lamination	60/80 SE		
	Cross Lake granitoids	Lamination	63/80 NW		
18	Cross Lake granitoids	Foliation and shape fabrics (mafic clots, xenoliths)	60/57 SE, 64/64 SE	Shape fabrics and mineral-grain lineation	68/10, 70/22
	Amphibolite	Schistosity	52/76 SE	Mineral-grain lineation	56/32
	Metaconglomerate	Shape fabric	54/75 SE		
19	Laminated marble and associated schist	Lamination	60/78 SE, 62/90	Mineral streaks	238/14
	Granitoid gneiss	Foliation	55/90, 56/85 SE, 58/82 SE	Mineral streaks	240/16
20	Schistose clastic metasediments	Foliation	62/78 NW, 62/82 NW	Weak mineral lineation	237/20, 232/08

and high-strain zones formed during D1–D2. Most important, one cannot rule out the possibility that discordant L-S shape fabrics within, and the associated stretching of, thrust fault walls postdate the main stage of dip slip (cf. Barclay, 1989).

Significance of L-S Shape Fabrics in Ductilely Deformed Metamorphic Rocks

Macroscopically visible shape fabrics with orthorhombic or uniaxial symmetry (Turner and Weiss, 1963, p. 42; Flinn, 1965; Hatcher, 1995, p. 420) are commonly defined by two kinds of preferredly oriented, inequant structural elements: (1) individual mineral grains and (2) polycrystalline inclusions such as strained volcanic fragments or recrystallized porphyroblasts (Turner and Weiss, 1963, p. 97–104; Hobbs et al., 1976, p. 268; Piazolo and Passchier, 2002, p. 26). Geologists recognized many years ago that the shape fabrics of typical metamorphic rocks have a planar component (S; schistosity, gneissic layering, etc.) and a linear component (L; aligned mineral grains, stretched polycrystalline inclusions, etc.). It was also recognized that one component of L-S fabrics is commonly stronger than the other, and that, in any large fold, the relative strength of the components may vary gradually with position. Flinn (1965, 1978) defined the L-S fabric spectrum in terms of an oblateness-prolateness factor (k or K), and saw a close geometric relationship between such fabrics and associated states of finite strain. He focused attention initially on shape fabrics of platy and prismatic metamorphic minerals (micas, amphiboles, etc.), and distinguished between S, S > L, S = L, L > S, and L fabrics (cf. Schwerdtner et al., 1977). If geological objects (polycrystalline inclusions) have no initial preferred orientation, that is, their fabric is isotropic at the onset of a ductile rock deformation, then the oblateness-prolateness factor of a final L-S fabric may indeed correspond to that of an ellipsoid representing the total strain or a late-stage increment of finite magnitude. Accordingly, S and L fabrics correlate, respectively, with oblate and prolate spheroids of finite strain, whereas S > L and L > S correlate, respectively, with oblate and prolate strain ellipsoids. Moreover, the principal axes of an L-S fabric (lineation direction, schistosity normal) are parallel to the local principal directions of a finite strain.

It can be difficult to ascertain, within repeatedly deformed metamorphic rocks, whether their final L-S shape fabrics had isotropic parents (Robin, 1977; Launeau and Robin, 1996). Moreover, stress-induced processes such as pressure solution lead to synkinematic boundary migration or coalescence of closely spaced inclusions, and complicate the simple relationship between fabric axes and principal strain directions (Schwerdtner and Cowan, 1992). Modern structural analysts generally agree, nonetheless, that the lineation of L-S shape fabrics is approximately parallel to the local direction of maximum finite extension and that the schistosity is approximately normal to the local direction of maximum finite shortening (e.g., Dietrich and Durney, 1986; Casey and Dietrich, 1997).

Piazolo and Passchier (2002) made a careful field and laboratory study of a variety of shape lineations in repeatedly deformed greenschist to lower amphibolite-facies metamorphic rocks in the easternmost Variscan Pyrenees of northeast Spain. They found that the final strength of shape lineations within and between ductile shear zones is critically influenced by factors other than finite strain (Piazolo and Passchier, 2002, p. 39). This result seems to cast doubt on the general validity of Flinn's (1965) simple correlation between geometric properties of L-S shape fabrics and those of finite-strain ellipsoids. However, Piazolo and Passchier's (2002) conclusion is based on magnitudes of principal strain (D3) obtained by using equations for straight shear zones with uniform thickness (Ramsay, 1980, p. 92–94), also known as band structures (Cobbold, 1977; Schwerdtner, 1982). By contrast, the Variscan shear zones (D3 structures) studied by Piazolo and Passchier (2002) are variously curved and part of a lozenge-style system that is superimposed on D1-D2 structures. Although the result of the study is of general importance, a proper test of Flinn's L-S fabric principle in heterogeneously strained metamorphic rocks requires measurement of the S component together with the L component.

L-S shape fabrics defined by (1) strained primary inclusions and (2) hornblende grains and other metamorphic minerals abound in the Mazinaw domain. Examples of strained primary inclusions are suboval mafic clots, stretched mafic xenoliths, distorted volcanic breccia fragments, and stretched pebbles in the deformed metaconglomerates of the Flinton Group (Easton, 1992, p. 845). The interpretation of L-S shape fabrics in terms of finite strain is not straightforward if they are derived from preferredly oriented, oblate, or prolate undeformed inclusions. In the Mazinaw domain, this poses a serious problem only in polymictic metaconglomerates and metavolcanic agglomerates, where originally elongate or flat clasts may have had a primary preferred orientation parallel to the depositional layering.

In deformed metabasaltic rocks of the Mazinaw domain, the L-S shape fabric of hornblende grains relates to the net finite strain accumulated under peak metamorphic conditions. By contrast, the shape fabric of passively deformed primary inclusions pertains to the total ductile strain and gives no clue to the deformation path of lithotectonic units. However, modestly strained primary inclusions whose shape fabrics wrap around the hinges of contractional F2 structures reveal that the amount of post-D1 strain can be relatively small.

Tangential Shear Strain at Stretched Lithotectonic Boundaries

Slip vectors on translational faults and other shallow-crustal discontinuities with fixed orientation have played important roles in elucidating modern orogenic belts (e.g., Kearey and Vine, 1990, Chaps. 7 and 9; Moores and Twiss, 1995, p. 289–291). Such vectors are tantamount to particle displacement vectors that are referred to rigid geographic axes and pertain to the total rock deformation or large deformation increments in brittle rocks. Similarly, shear-strain vectors in ductilely deformed rocks pertain to the total strain or large strain increments, but are referred to material surfaces (stretched LTBs and other deformed planar structures) as well as rigid geographic axes (Fig. 1 *in* Cobbold, 1983). The material surfaces may have rotated through large angles in the geographic reference frame while undergoing large finite strain. For example, subhorizontal thrust planes embedded in flowing rock masses (Means, 1989; Schwerdtner, 1998) may have been tilted progressively during ductile shortening and associated thickening of the middle crust. The nature and degree of rotation of deformed material surfaces such as fault planes needs to be assessed in a given region before attempting to make tectonic interpretations of the geographic direction and sense of the total or incremental shear strain.

The sense of a finite shear strain provides no information about flow vorticity (Passchier et al., 1990, p. 17; Hanmer and Passchier, 1991, p. 8–11), originally called "internal rotation" in structural analysis (e.g., Turner and Weiss, 1963). Moreover, "shear-strain sense" must not be equated with "shear sense" as used by students of the kinematic path of severely deformed rocks (e.g., Passchier et al., 1990, p. 55–63; Hanmer and Passchier, 1991). However, the vorticity sense in steady-state deformation, as predicted for a specific material plane in a kinematic hypothesis, must agree with the shear-strain sense on the same plane (cf. Schwerdtner and Cote, 2001).

"Lithotectonic boundary" (LTB) is the family name of geological material surfaces such as strained crustal sutures, stretched unconformities, sheared tops or bottoms of thick beds, and distorted intrusive contacts (Schwerdtner, 1998). LTB surfaces may have been converted into slip planes during regional deformation, while the wallrocks experienced tangential (LTB-parallel) shear strain and were stretched in two dimensions (Schwerdtner, 1998; Schwerdtner and Cote, 2001). Such structures are akin to stretching faults (Means, 1989), and are called "mechanically incoherent LTBs" in the present paper.

The Mazinaw domain is replete with stretched LTBs at which tangential shear strains may be analyzed and used to elucidate the nature of Elzevirian and Ottawan deformations. LTBs other than the base of the <1150 Ma Flinton Group generally originated as intrusive contacts between 1270–1240 Ma plutonic bodies and their metavolcanic or metasedimentary hosts. Some LTBs fall into high-strain zones and, in a crude way, mark the six thrusts shown in Figure 3.

LTB surfaces are soil-covered throughout the central and southern parts of the Mazinaw domain. LTB walls, on the other hand, are commonly exposed in blasted outcrops on Highways 7, 37, and 41, as well as on other paved roads (Figs. 2 and 3). Because of the age difference between Flinton Group strata and their basement complex, one may distinguish, within individual LTB-walls, between the total (Elzevirian plus Ottawan) shear strain of basement rocks and the Ottawan shear strain recorded by Flinton Group rocks (cf. Schwerdtner and Gapais, 1983).

Given the principal ratios of a finite strain and the present or final attitude of the LTB surface, one can determine the direction, sense, and magnitude of tangential shear strain at any locality in an LTB wall (Fig. 5; Nadai, 1950, p. 120–124; Ramsay, 1967, p. 126–129; Lisle, 1998). The graphic derivation of shear-strain sense is simplest where the strain ellipsoid degenerates into an oblate spheroid (pure flattening) or prolate spheroid (perfectly constrictive deformation). No matter what the shape of the local strain ellipsoid, however, the LTB attitude must be specified (known or assumed) before one can analyze the tangential shear strain.

In the present study, we estimated the principal strain ratios only in distorted granitoids whose shape fabrics were demonstrably derived from randomly oriented, undeformed mafic clots. The investigation of tangential shear strain was therefore restricted, at most localities, to finding its direction and sense. This is accomplished by using graphic techniques developed by Lisle (1998) and Schwerdtner (1998). The theoretical basis of these techniques is covered elsewhere (Lisle, 1998; Schwerdtner, 1998; Schwerdtner and Cote, 2001), and is not be repeated in the present paper. Practical aspects of the technique, however, are covered later under the subheading "Wallrocks of Thrust I."

Dip Angles of Large Structural Surfaces

The analysis of tangential shear strain within the walls of large dislocations, LTBs, and other structural surfaces requires that their near-surface dip be known to a first approximation. Throughout the Mazinaw domain, however, the LTB dip is difficult to estimate on the basis of stratigraphic and structural information (as is explained in the introduction). Geophysical profiles may remedy the situation, as detailed in the following paragraphs.

Various geophysical methods have been used to image major planar structures in the shallow crust of the Composite Arc Belt. Among these methods, detailed gravity modeling and reflection seismic profiling have yielded results most easily used for approximating the dip of major structural surfaces at the erosion level. For example, Real and Thomas (1987) have imaged the contacts of rock masses presently exposed in the Mazinaw domain and the vicinity, in different vertical sections, by means of detailed gravity modeling. Five of the six gravity profiles published by these workers transect the Mazinaw domain and thrusts I–VI. By contrast, only thrusts I and IV are crossed by Lithoprobe Line 33 and its reflection seismic profile (White et al., 1994, 2000).

Lithoprobe Line 33 transects the walls of thrust IV east of Madoc (Fig. 3), close to our measurement site 15 at Highway 7, and continues southeastward on dirt roads beyond the northern edge of the Paleozoic cover (White et al., 1994, 2000). The thrust is shown on the Tectonic Assemblages Map (Ontario Geological Survey, 1992), as well as in a recent geological section across the Grenville Province of Ontario (Fig. 4 *in* Carr et al., 2000). As we explained briefly earlier, thrust IV marks the northwestern border of the Mazinaw domain in the Tweed-Madoc area, and falls into a high-strain zone formerly known as the Moira Lake shear zone (Di Brisko, 1989; White et al., 1994) and presently called the Mooroton shear zone (Easton, 1992; Corfu and Easton, 1995; Cureton et al., 1997). Foliated rocks within this shear zone may have been affected by tangential displacements with the sense of normal faults after the 1020 Ma metamorphism and associated ductile deformation (Corfu and Easton, 1995; Carr et al., 2000).

The Mooroton shear zone and thrust IV are spatially associated with a pair of LTBs confining an S-shaped panel of Flinton Group rocks (Fig. 3; Easton, 1992, p. 843). The panel is well exposed in the Cloyne-Flinton area (Wolff, 1981; Easton, 2001), and passes from the Mazinaw domain into the southernmost Elzevir terrane (Figs. 1 and 2). In the reflection seismic profile across the southern Mazinaw domain (Fig. 4), the domain boundary coincides with a seismically transparent listric zone that dips steeply to the north (Fig. 8 *in* White et al., 1994;

Figure 8. Transposed carbonate veinlets (A, B) in highly flattened metaconglomerate (Flinton Group) with relict granitoid pebbles (B), measurement site 15, just north of Highway 7. FGP—flattened granitoid pebble; GTCV—group of transposed carbonate veinlets; QFP—quartz-feldspar vein.

Fig. 2b *in* White et al., 2000). Lithologic contacts and trajectories of the main foliation strike easterly to northeasterly, at thrust IV and within the Mooroton shear zone, and are approximately normal to Line 33.

Northwest and west of Tweed (Figs. 2 and 3), Line 33 passes through a granitoid body in which the main foliation strikes northeasterly to east-northeasterly (Wilson, 1940; Hewitt, 1964). The northeastern contact of the body and the boundaries between adjacent metasedimentary units, however, strike approximately parallel to Line 33. The spatial variation in the strike of planar structures northwest of Tweed makes it difficult to estimate the true dip of major surfaces imaged by southeast-dipping reflectors (Fig. 4).

About 2 km south of Tweed, Line 33 transects thrust I, which lies at the northwestern border of the Mellon Lake granitoid complex, which dates to 1270–1250 Ma (Figs. 2 and 3; Easton, 1992, p. 843; Easton, 2001). This longitudinal thrust can be seen to dip to the southeast on the Tectonic Assemblages Map (Ontario Geological Survey, 1992), and the local shallow crust is replete with southeast-dipping seismic reflectors (see Fig. 4). The poorly exposed borders of, and main foliation within, the southwestern part of the Mellon Lake complex strike northeast to east-northeast near Tweed (Wilson, 1940; Hewitt, 1964). This implies that the Mellon Lake complex, considered by Bright (1986a) and Easton (1992, p. 850) to be the local basement of Mazinaw Group rocks, does not wrap around, and is apparently detached from, the metavolcanic and metasedimentary rocks of the Clare River synform (cf. Chappell, 1978; Barclay, 1989).

This interpretation is supported by detailed gravity modeling (Real and Thomas, 1987, profiles 3 and 4), which images the Mellon Lake complex as a thin sheet that dips gently to the southeast. Thrust I seems to mark the basal contact of this granitoid sheet (cf. Barclay, 1989), which strikes ~45° to the gravity profiles. The sum of the geological and geophysical evidence suggests that the southerly dip of thrust I is larger than the apparent dip of gravity images and seismic reflectors in the shallow crust (Fig. 4).

All other thrusts considered in the present paper are crossed by at least one of the gravity profiles (Real and Thomas, 1987), but none is transected by a reflection seismic line. Figure 3 shows that the distance of our structural measurement sites from Lithoprobe Line 33 is highly varied, with some sites as much as 30 km away. In conjunction with geological maps of the Tweed-Kaladar-Plevna region, however, the reflection seismic profile of Line 33 provides the best general information presently available about the direction and magnitude of LTB dip in the shallow crust. Most reflectors (structural surfaces) are inclined slightly toward the southeast, but horizontal and northwest-dipping reflectors occur locally in the upper part of the profile. The average true dip of seismic reflectors inclined to the southeast is ~21°, an angle obtained on the stereonet after averaging values of easterly apparent dip measured at different localities in the upper profile.

The variation in the dip direction and the apparent dip angle of seismic reflectors (Fig. 4) may be related to the form of structural surfaces defining open to close F2–F3 structures. The shallow crust of the Tweed-Madoc area lacks reflectors with inclinations of 40°–70°, which seems incompatible with supposedly subrecumbent F2-structures (e.g., Fig. 3, thrust V) shown on the Tectonic Assemblages Map (Ontario Geological Survey, 1992). Even open folding of originally southeast-dipping LTBs or thrust surfaces and the main foliation, however, may create a shear sense problem in wall segments that finally attained a westerly dip. More specifically, a rigid-body rotation of easterly dipping thrust walls past the horizontal plane changes the sense of prior tangential shear strain (D1) from reverse to normal (Fig. 6). The tangential shear strain due to D2 buckling (Fig. 6, B), on the other hand, is expected to have the sense of reverse faults or thrusts at all localities in the limbs of nearly upright open to close structures such as the Clare River synform (Fig. 2). Simple reasons for the reverse sense are given in most textbooks of structural geology under the subheading "Flexural Folding" or "Buckle Folding" (e.g., Ramsay, 1967, p. 391–397; Park, 1983, p. 72; Hatcher, 1995, p. 292, 302, 319).

L-S SHAPE FABRICS IN THE MAZINAW DOMAIN

L-S fabrics and other mesoscopic structural features were examined on clean rock outcrops in large granitoid plutons and the southwestern part of the Clare River synform (Fig. 2). However, special emphasis was placed on the study of shape fabrics near the six LTBs or thrust faults shown in Figure 3.

Attitudes of lineation and/or foliation (schistosity, gneissic layering, and other planar fabrics) were obtained at twenty sites in well-exposed wall segments of the thrust faults (Table 1). In addition, L-S shape fabrics were assessed by visual inspection in the field, and oriented rock specimens were sawn for detailed inspection and macroscopic geometric analysis in the laboratory. All macrostructural information thus obtained was used in the subsequent analysis of tangential shear strain, together with realistic dip values for the deformed LTBs or thrusts.

S and S > L fabrics abound in the metasedimentary and metavolcanic wallrocks of the six thrusts investigated, whereas L and L > S fabrics are seen mostly in metaplutonic wallrocks (Table 1). Foliation is steeply inclined to subvertical at most measurement sites, but the lineation direction varies between and within the wall pairs of different thrusts. More specifically, the wallrocks of thrusts I and IV (Fig. 3) have shape lineations that are approximately transverse to the thrust trace (Table 1). By contrast, the lineations at thrusts II, V, and VI are nearly tangential or slightly oblique to the thrust trace. Finally, the attitude of lineations at thrust III varies systematically along the arcuate thrust trace, from nearly tangential to transverse (Table 1). Such variation in lineation attitude and L-S fabric type, between and within wall segments, attests to markedly heterogeneous strain. Moreover, systematic changes in lineation attitude with position are well documented in many collisional orogens (for references and a review of the topic, cf. Connors et al., 2002), and have been explained by a variety of structural or kinematic scenarios. For example, gradual changes in lineation attitude from orogen-subnormal to orogen-parallel in the crystalline-core southern Appalachians are explicable by a marked curvature in the horizontal trajectories of tectonic transport and large-scale flow during the Acadian and Alleghanian orogenies (Reed and Bryant, 1964; Hatcher, 2001). However, no published explanation of lineation attitude changes in collisional orogens, is based on the generalized stretching-fault concept (Schwerdtner, 1998), which permits instantaneous extension or shortening of the wallrocks parallel to the variable direction of local tangential shearing.

Redeformed Rocks with L-S Fabrics

Effects of the Elzevirian pulse and several periods of Ottawan tectonism have been recognized in the Mazinaw block (Easton, 1992), but the L-S shape fabric in strained granitoid bodies such as the Addington and Northbrook plutons lacks conspicuous evidence of repeated ductile deformation. Either the granitoids were strained only once, on mesoscopic scales ranging from 1 cm to 1 m, or their redeformation occurred in a passive manner (as defined, for example, by Hatcher, 1995, p. 314). Such passive redeformation may be caused by melt-dominated flow and thermal softening within the crust (Pavlis, 1996).

The linear shape fabric of modestly strained inclusions is nowhere oblique to the mineral-grain lineation within the inclusions or their rock matrix. For example, in metaplutonic rocks with stretched mafic clots, the direction of clot alignment is everywhere parallel to the hornblende-grain lineation. Pure flattening may have produced a planar grain fabric during D1, and passive reshortening during D2 may have resulted in the final L or L > S fabric. However, such a kinematic scenario is difficult to verify in the absence of actively redeformed ("overprinted") structural features.

Actively Redeformed Fabrics

Strongly foliated rocks that have been transected by narrow shear zones, or subjected to foliation boudinage or flexural folding (crenulation or pervasive buckling) on various mesoscopic scales, occur locally within the metasedimentary and metavolcanic units of the Mazinaw domain. S or Z buckle folds and foliation boudins are widespread in the walls of thrust IV and the Mooroton shear zone, whose hornblende-bearing rocks may have been redeformed under low-grade metamorphic conditions (Easton, 1992, p. 841).

In well-exposed wall segments of thrust III (e.g., on Highway 41, ~8 km south of Cloyne), phyllitic folia wrapping around the pebbles of strained metaconglomerate beds are crenulated (buckled) on the millimeter and centimeter scales. The tiny open buckles are observed throughout the roadcuts of measurement site 9, and reveal that the S > L pebble-shape fabric is a net result of two or more acts of passive deformation. However, the shortening component of the crenulation strain is expected to be <20%, and could not have greatly affected the virtual parallelism between the foliation normal in the rock matrix and the local direction of principal total shortening (cf. Schwerdtner and Cowan, 1992).

Gentle to open mesoscopic buckle folds with subhorizontal hinge lines are commonly imposed on the main foliation of metasedimentary rocks in the Tweed-Kaladar-Fernleigh area, and seem to postdate the coaxial foliation boudinage (to be described later). The late-stage buckling strain is attributable to Late Ottawan crustal thinning of the Mazinaw domain and adjacent divisions of the Composite Arc Belt (Carr et al., 2000).

Vein Systems and Northwest-Directed Thrusting

Closely spaced quartz and carbonate veins occur within and near midcrustal high-strain zones, as noted near thrust IV in the southern Mooroton shear by Easton (1992, p. 841). The veins are most prominent at the base of Flinton Group metaconglomerate and other strained LTBs, and seem to be distorted or even transposed into the strong foliation (Fig. 8). In the southwestern Mooroton shear zone, Easton (1992, p. 841) has identified thin slivers of Flinton Group metasediments, which attest to kilometer-scale slip during regional deformation.

The sum of the evidence suggests that, within or near the high-strain zones of the Mazinaw domain, the basal unconformity and other LTBs were mechanically incoherent, and demonstrates that Ottawan ductile deformation was discontinuous spatially as well as temporally. It seems probable, therefore, that the longitudinal thrusts of the Mazinaw domain are midcrustal stretching faults with unknown amounts of bulk translation between opposite walls.

ANALYSIS OF TANGENTIAL SHEAR STRAIN

On the Tectonic Assemblages Map, all longitudinal thrusts of the Mazinaw domain have easterly to southeasterly dips (Ontario Geological Survey, 1992). This is compatible with the prevailing inclination of seismic reflectors in the shallow crust (Fig. 4; White et al., 1994, 2000), and tantamount to a marked discordance between thrust surfaces and subvertical planar fabrics (Table 1). Easton's (1992, p. 843) geological map, on the other hand, includes folded thrust faults that approximately conform to the northwest- and southeast-dipping limbs of large F2 structures. The uncertainly about the dip direction of thrust surfaces further complicates the following investigation of the shear-strain sense, at least in the northwest limbs of large F2 structures.

Despite modern geophysical imaging of structural surfaces in the shallow crust (Real and Thomas, 1987; White et al., 1994), close estimates of the near-surface dip of LTB or thrust segments are not available at individual measurement sites (Fig. 3). The problem will be addressed by investigating the tangential shear strain for a range of southeasterly dip values. In addition, northwest dip options will be considered at thrusts I and V, as well as at the northwestern border of the southern Mazinaw domain (thrust IV). Except at this border, consideration of thrust dip angles exceeding 60° seems unjustified in view of the geophysical evidence.

Wallrocks of Thrust I

Thrust I is close to the northern edge of the present Paleozoic cover (Fig. 3; Ontario Geological Survey, 1992), and its western terminus remains to be located. The thrust surface or LTB is not exposed in the roadcuts of Highway 41 and of parallel dirt roads northeast of Tweed. Judging from the results of detailed geological mapping (Barclay, 1989) and gravity modeling (Real and Thomas, 1987), the thrust coincides with an incoherent LTB between the Mellon Lake (basement?) complex and a 100- to 150-m-wide panel of strongly laminated marble (not shown in Figs. 2 and 3), biotite-carbonate schist, and associated weakly strained pegmatite (Wilson, 1940; Hewitt, 1964; Wolff, 1981; Easton, 2001). Accordingly, the Mellon Lake complex composes the southeast wall of thrust I. By contrast, the northwest wall is layered, and its boundary and structural relationship with the Clare River synform are uncertain. For purposes of the present study, the northwest wall is regarded as bipartite, and therefore comprises the metasedimentary panel and the Mitten dike (cf. Bright, 1986a; Barclay, 1989; Easton, 2001).

Metaplutonic rocks in the walls of thrust I have a crude mineral lamination, oblate shape fabrics, and concordant quartz-feldspar veinlets at many localities. This points to high levels of accumulated flattening strain (Schwerdtner and Cowan, 1992), and is incompatible with plane deformation (e.g., codirectional pure and simple shear). As is well known to structural geologists and convincingly documented in excellent field examples (e.g., Ramsay and Huber, 1983, p. 205; Hatcher and Hooper, 1992), high wallrock strain generally results in an effective parallelism between the foliation surface and the fault plane. This rule may not apply to all midcrustal dislocations, however, for reasons given earlier. The issue could be settled, in the Mazinaw domain, by drawing objective cross-sections based on stratigraphic and structural observations. The geological data available are insufficient for that task, however, and force us to rely on published geophysical profiles.

In the following graphic investigation of tangential shear strain within the LTB or thrust walls, we utilize variants of simple stereoplots introduced by Schwerdtner (1998, Fig. 4) and explained in the next paragraph. We ignore the problem that the prominent foliation (metamorphic lamination and weak mineral schistosity) in the tonalite-granodiorite of the southeast wall may predate, and possibly have a complex relationship to, the Ottawan thrusting. Pervasive mineral-shape fabrics are best developed in the Mitten dike, whereas most rocks of the metasedimentary panel have a strong flattening foliation (lamination and/or schistosity). At measurement site 1 (Table 1), alaskitic granite of the Mitten dike is pervaded by partly coalescent, polycrystalline quartz eyes (former igneous megacrysts?), whose shape fabric has no conspicuous linear component. Similarly, at measurement site 2, biotite-rich granite with partly coalescent mafic clots of igneous origin is highly strained, but again lacks a conspicuous shape lineation. It is assumed, for purposes of shear-strain derivation, that the mineral fabrics in the Mitten dike and most other parts of the walls are indicative of an axially symmetric flattening strain (represented by oblate spheroids).

Figure 9 depicts the geometric relationship between local normals (Z) of the flattening foliation, and the dip angle of the LTB or thrust plane, and the sense of tangential shear strain at measurement sites 1 and 2. In every stereoplot, the lower hemisphere is divided into fields throughout which the strike or dip components of shear strain have the same sense, for a given dip angle of the LTB or thrust (cf. Fig. 4 *in* Schwerdtner, 1998). The shear-sense fields are bounded by three orthogonal planes: the LTB or thrust surface and perpendicular planes drawn, respectively, through its strike and dip lines (Fig. 9). The wallrocks have been shortened, without tangential shear strain, if the flattening foliation is perpendicular to the LTB or thrust. Small angles between the LTB or thrust plane and Z-directions in highly strained wallrocks point to componental tangential shortening during slip or shear strain with normal or reverse sense (Fig. 10).

In all stereoplots of Figure 9, the Z-directions are close to the line of no strike shear. This implies that the strike component of tangential shear strain is relatively small at measurement sites 1 and 2 and probably throughout the intermediate wall segments of thrust I (Fig. 3). At measurement site 1 (Fig. 9, A and B), the shear-strain sense proves to be reverse for LTB dip angles between 30° and 60°SE, as well as dip angles of <55°NW. (This is revealed by the position of the Z-directions with respect to the LTB or thrust plane and similar to the example of LTB′ depicted schematically in Fig. 10.) The tangential shear strain is very small or vanishes if the great circle representing the LTB or thrust passes through the middle of the Z-direction cluster, that is, the dip of the LTB or thrust is ~20°SE. For dip angles of <15°SE, the tangential shear strain has normal sense, as may be deduced from Figure 9, A. This option is portrayed by the schematic tangent labeled LTB in Figure 10.

At measurement site 2 (Fig. 9, C and D), the tangential shear strain is close to zero if the LTB or thrust plane is horizontal. The position of the Z-direction cluster precludes the possibility of finite shear strain with normal sense. The prevailing reverse sense of the tangential shear strain for northwest dip options (Fig. 9, B and D) rules out a syn-D2, rigid-body rotation of the foliated LTB walls (cf. Fig. 6). In the Clare River synform (as explained earlier in this paper), there is no evidence for pervasive D2 strain capable of causing the very large, hypothetical increase in foliation dip portrayed in Figure 7. This favors structural scenarios in which the steep attitude of actual wallrock fabrics is a primary feature.

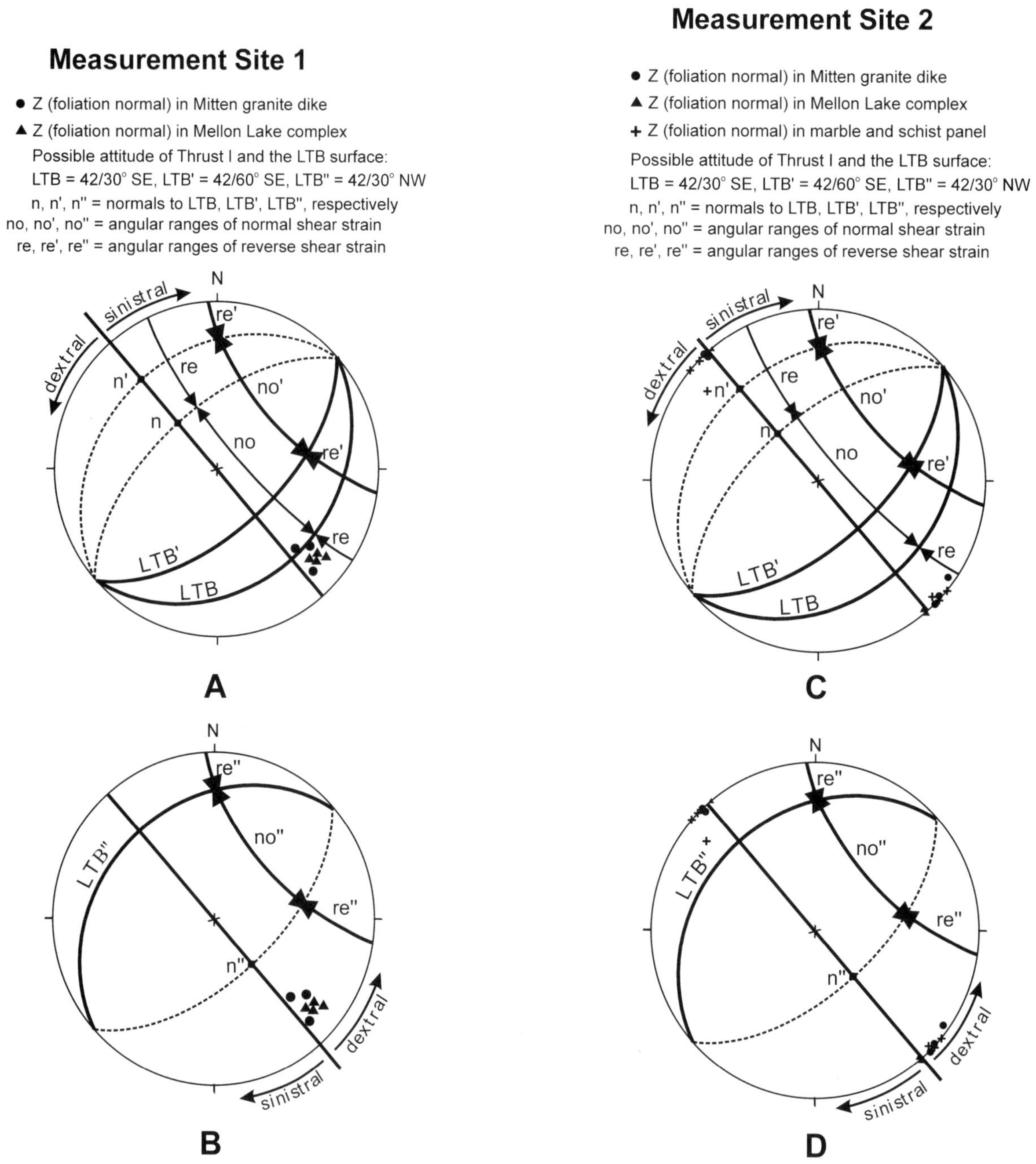

Figure 9. Graphic derivation of shear strain sense in stereographic plots (cf. Schwerdtner, 1998, p. 960) at measurement sites 1 (A, B) and 2 (C, D) in the wallsrocks of thrust I, marked by an incoherent lithotectonic boundary (LTB) (northwest border of the Mellon Lake complex). Structural data used in the plots are listed in Table 1. The shear sense is unaffected by varying the direction and angle of LTB dip (further explanation in the text).

Wallrocks of Thrust II

At measurement sites 3–5, thrust II approximately coincides with the soil-covered southeastern intrusive contact of the 1270–1240 Ma Northbrook pluton (Fig. 3; Wolff, 1981; Easton, 2001). The contact qualifies as a flattened LTB whose trace dips ~40°SE in a northwest-southeast gravity profile (Real and Thomas, 1987, p. 768). Macrostructural relationships at this site were studied in detail by Downey (2002), and are summarized in the following paragraphs.

Deformed Northbrook granodiorite underlies thrust II and its high-strain zone (Fig. 3). The mafic clots of the granodiorite are subhorizontally lineated and seem perfectly prolate in outcrop (Fig. 11). The principal axial ratio of the deformed clots is

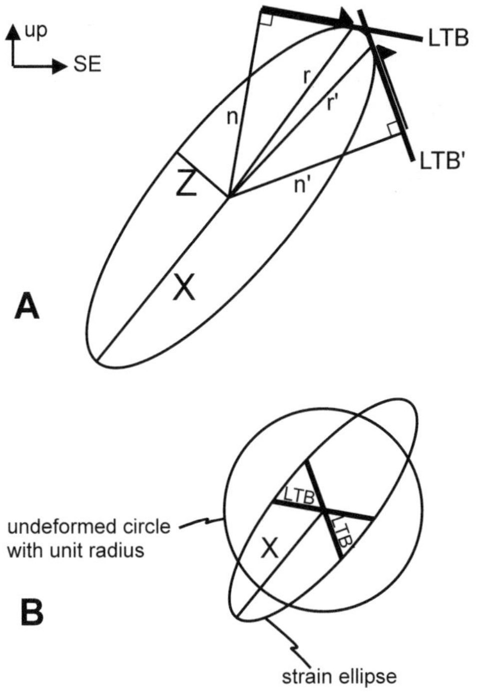

Figure 10. Schematic illustration of determining the sense of tangential shear strain (A), under conditions in which two lithotectonic boundary (LTB) traces are shortened in the shear-strain directions. Note the marked area change in (B) due to the arbitrarily chosen large extension normal to the plane of the cross-sectional strain ellipse. X, Z—principal strain directions; LTB, LTB′—traces of lithotectonic boundaries; *n*, *n*′—normals to LTB and LTB′, respectively; r, r′—oblique radii of strain ellipse. The geometric relationships depicted in (B) apply to metamorphic rocks with a strong mineral foliation and a weak mineral lineation.

6:1 some 50 m north of the high-strain zone on Highway 41 (thrust II).

Metaconglomerate of the Flinton Group and metasedimentary rocks of the basement complex (as interpreted by Wolff, 1981, and Easton, 2001) comprise the high-strain zone and its hangingwall (Fig. 12). All rocks in the high-strain zone and the hangingwall have strong planar fabrics, but mineral-grain lineations and linear shape fabrics are weak or cannot be dis-

Figure 11. Lineated Northbrook granodiorite (cf. Wolff, 1982) without macroscopically visible foliation. From measurement site 4 on Highway 41, ~4.5 km north of Kaladar (Fig. 3).

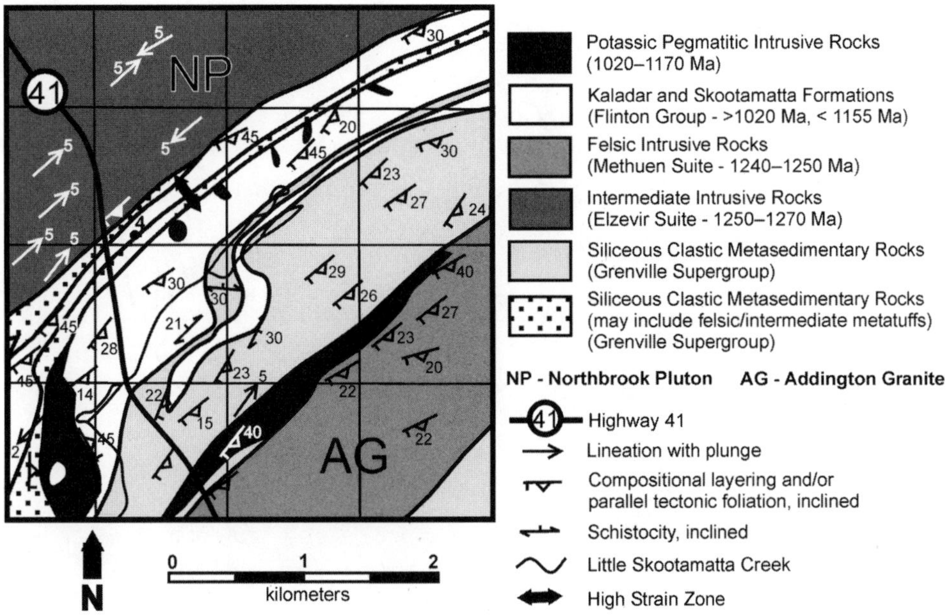

Figure 12. Detailed geological map (Precambrian bedrock) of a small area north of Kaladar (cf. Fig. 2), which includes measurement site 4 (cf. Fig. 3). Modified from Easton (2001), with permission by the Ontario Geological Survey.

cerned in outcrop. The contrast in shape and mineral-grain fabric between the footwall and the hangingwall can be explained by large dip slip on the thrust and/or Elzevirian ductile deformation of the Northbrook pluton.

The internal structure of the hangingwall at Highway 41 and in the vicinity is complicated by an alternation between conglomerate and pebble-free metasedimentary rocks (Fig. 12). This implies to Easton (2001) that the hangingwall north of Kaladar consists of three overturned tight F1 structures (two synclines and an intermediate anticline) in basement rocks, the relict unconformity, and Flinton Group strata.

The LTB normal (*n*) for measurement site 4 falls into the cluster of Z-directions (Fig. 13) if one attempts to derive the shear-strain sense for an LTB dip of 40°SE (Real and Thomas, 1987, p. 768). Similarly, the great circle representing the LTB or thrust passes through the cluster of lineation directions (X). Note that X, Y, and Z have perpendicular tangent planes (not shown in Figs. 5 and 10) and, at least in theory, are directions of no finite shear strain. The geometric relationships depicted in Figure 13 imply, therefore, that the tangential shear strain is effectively zero or at least very small. Alternatively but less likely, the shear-strain magnitude may approach infinity, in which case the shear-strain sense cannot be found using the present technique (see Schwerdtner and Cote, 2001, p. 98).

The average southeast dip of seismic reflectors beneath the southern Mazinaw domain is only ~20°, however. Use of this dip value in Figures 13 and 14 yields a dextral-reverse sense of tangential shear strain at measurement site 4 (Fig. 13). Similar conditions prevail at sites 3 and 5 (Fig. 3; Table 1), but are not depicted in stereoplots.

Within the high-strain zone at measurement site 4, the strong foliation in Mazinaw Group rocks has been thrown into mesoscopic S- or Z-folds whose parallel style is indicative of buckling. The fold-hinge lines parallel a weak mineral lineation with shallow northeasterly plunge (cf. Figure 13, B). The sense of shear strain parallel to the fold-enveloping surfaces is normal, as revealed by the monoclinic geometry of the S- or Z-folds (Fig. 14). Little is known about the strain increment that formed the folds, whose direction of maximum shortening is denoted as Z_i. Judging from the mechanism and the geometric style of folding, Z_i departs from the trace of the enveloping surface in a clockwise direction (Fig. 14). Despite the normal sense of envelope-parallel shear strain, the buckle folds are compatible with northwest-directed thrusting and associated tangential shear strain in the wallrocks.

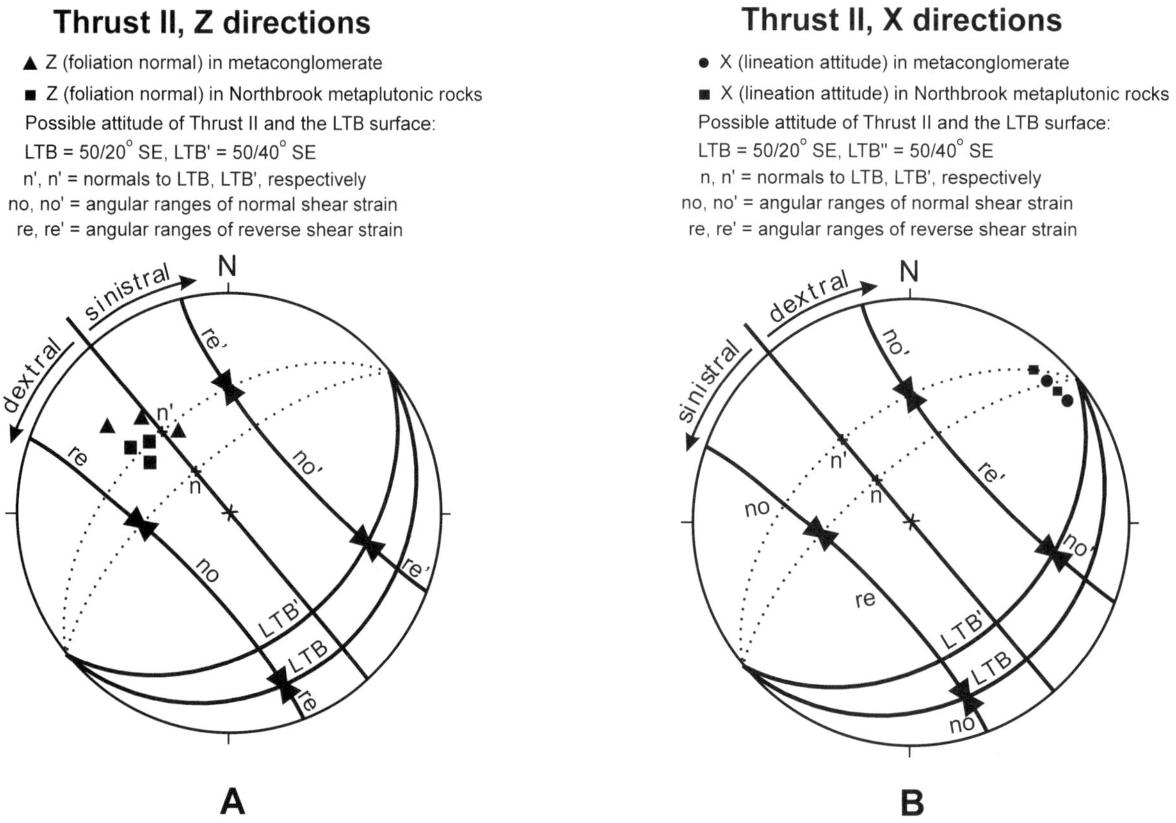

Figure 13. Shear sense derivation by means of foliation normals (A) and lineation directions (B) at measurement site 4, in the walls of thrust II and outcrops on Highway 41 (cf. Fig. 3). Structural data plotted are listed in Table 1.

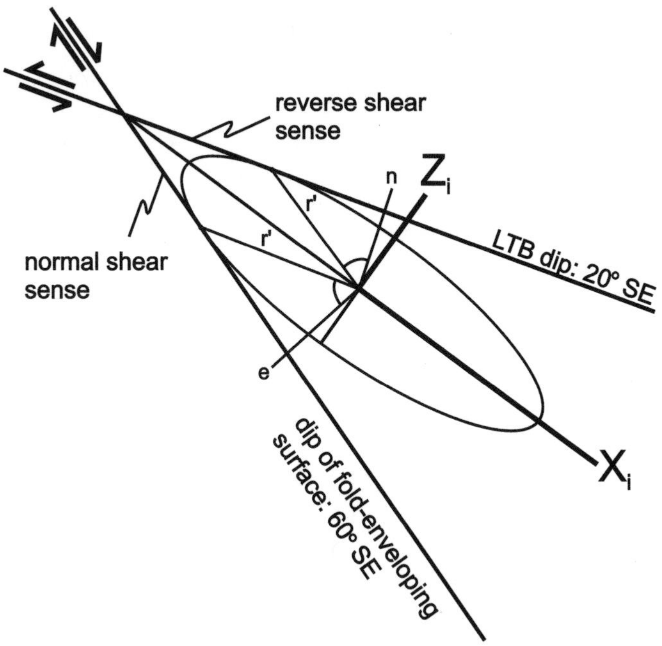

Figure 14. Schematic ellipse of late incremental strain (D2 or D3) in a vertical section across the high-strain zone at measurement site 4. The dip of the lithotectonic boundary (LTB) surface (intrusive contact of the Northbrook pluton) is assumed to coincide with thrust II and dip 20°SE. By contrast, the enveloping surfaces of northeast-plunging Z-folds dip 65°SE. Note that the difference in sense between the dip components of the tangential and fold-envelope-parallel shear strains is compatible with northwest-directed thrusting. X_i, Z_i—principal directions of fold-producing strain; *e*—normal to enveloping surfaces; *n*—LTB normal; *r′*—oblique radii of strain ellipse.

Wallrocks of Thrust III

Shape fabrics within and near the high-strain zone of thrust III are generally oblate and subvertical, and therefore create structural conditions similar to those at thrust I. In contrast to the situation south of the Clare River synform, however, the possibility of northwest- to north-northwest-dipping thrust planes need not be considered at the southwestern thrust III (Fig. 15). This simplifies the assessment of tangential shear strain and furnishes unequivocal evidence of thrust shear in, and transverse tangential shortening of, the wallrocks.

The trajectory of thrust III follows the curved northwest contact of the Northbrook pluton (Fig. 2), and it is convenient to consider its walls as a chain of three segments. Our fabric measurements are confined to the southwest segment (sites 6–9) and the middle segment (sites 10–12). The local trend of the Z-directions departs from the sinistral-dextral boundary line and varies

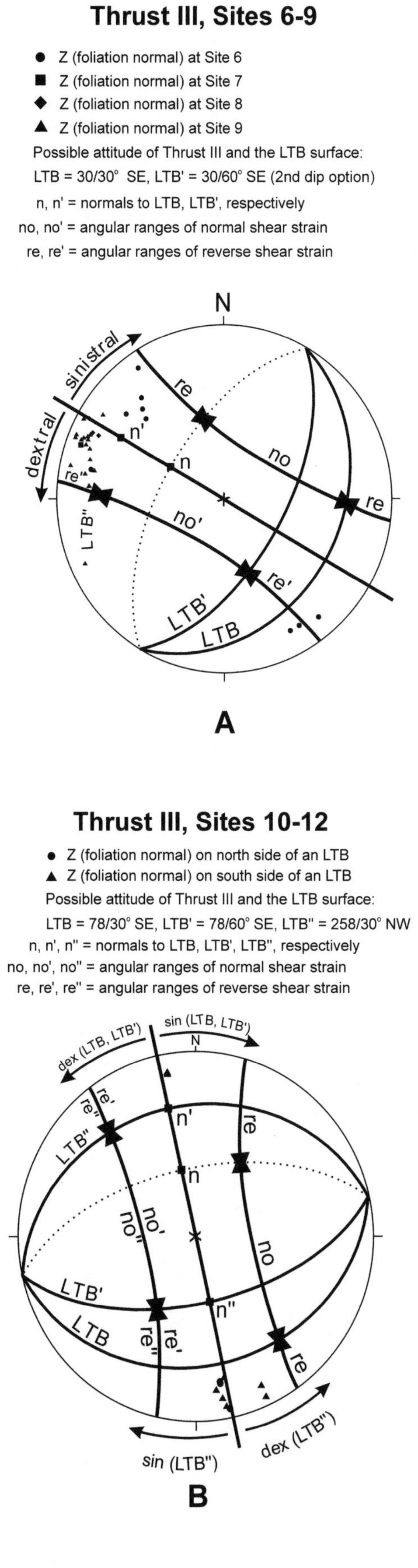

Figure 15. Derivation of the sense of tangential shear strain at thrust III and its lithotectonic boundary (unconformable base of the Flinton Group), at the southwest segment (A) and the middle segment (B) of thrust III.

more widely than in the walls of thrust I. Consequently, the strike-shear component cannot be disregarded in assessing the tangential shear strain in the walls of thrust III. At its southwest segment, most Z-directions plunge to the west-northwest (Fig. 15, A), and point to tangential shear strain with dextral-reverse sense regardless of the southeast dip angle of the LTB or thrust plane. This contrasts with the prevailing south-southeast plunge of Z-directions in the middle segment, where the tangential shear strain vanishes at south-southeast dip values of ~15° (Fig. 15, B). Moreover, the shear-strain sense is dextral-normal if the LTB or thrust plane dips <15°SE.

The northeasternmost part of the Northbrook pluton has been interpreted as an upright antiform (Ontario Geological Survey, 1992), which may warrant a consideration of the north-northwest dip option for the east-northeast-striking middle segment of thrust III and its walls. The option is exemplified by an LTB or thrust plane that dips 30°, and attests to the prevalence of reverse tangential shear strain. As may be seen by detailed inspection of Figure 15, B, the example illustrates conditions that apply to the north-northwest dip range of 5°–65°.

Wallrocks of Thrust IV

The trajectory of thrust IV coincides with the trace of the domain boundary and falls into the southwestern segment of the >100-m-wide Mooroton shear zone (Fig. 3). This high-strain zone contains strongly foliated metavolcanic and metasedimentary rocks whose microstructure, petrology, and geologic history have been elucidated by recent analytical work (Cureton et al., 1997). Amphibolite-grade metamorphic rocks of the zone seem to have acquired their subvertical L-S shape fabrics at midcrustal levels, but the bulk translation due to dip shear between the Mazinaw domain and the Elzevir terrane ceased before 1000 Ma. For an earlier tectonic stage, Cureton et al. (1997) determined from microstructural evidence that the sense of foliation-parallel shearing was reverse (southeastern side up) at the meter scale. Based on this evidence, Cureton et al. (1997, p. 990) suggested that the Mooroton shear zone may have been created by northwest-directed midcrustal thrusting of the Mazinaw domain over the Elzevir terrane. Ductile normal faulting in the shear zone (Corfu and Easton, 1995; Table 2 *in* Carr et al., 2000) seems to have occurred at an advanced stage of the Ottawan orogenic pulse (Easton, 1992, p. 841).

Students of highly strained rocks from kilometer-scale shear zones generally assume that main foliation or mineral schistosity, as seen in hand specimens, is effectively parallel to the shear zone boundaries (cf. Passchier et al., 1990, p. 55–64). At some southwestern localities in the Mooroton shear zone, however, foliation or schistosity has been thrown into southerly-plunging Z-folds. The enveloping surfaces of such mesoscopic Z-folds dip less steeply than foliation surfaces of the long fold limbs that predominate in outcrops with low vertical relief. Moreover, geophysical profiling (Real and Thomas, 1987; White et al., 1994, 2000) does not support the tacit assumption

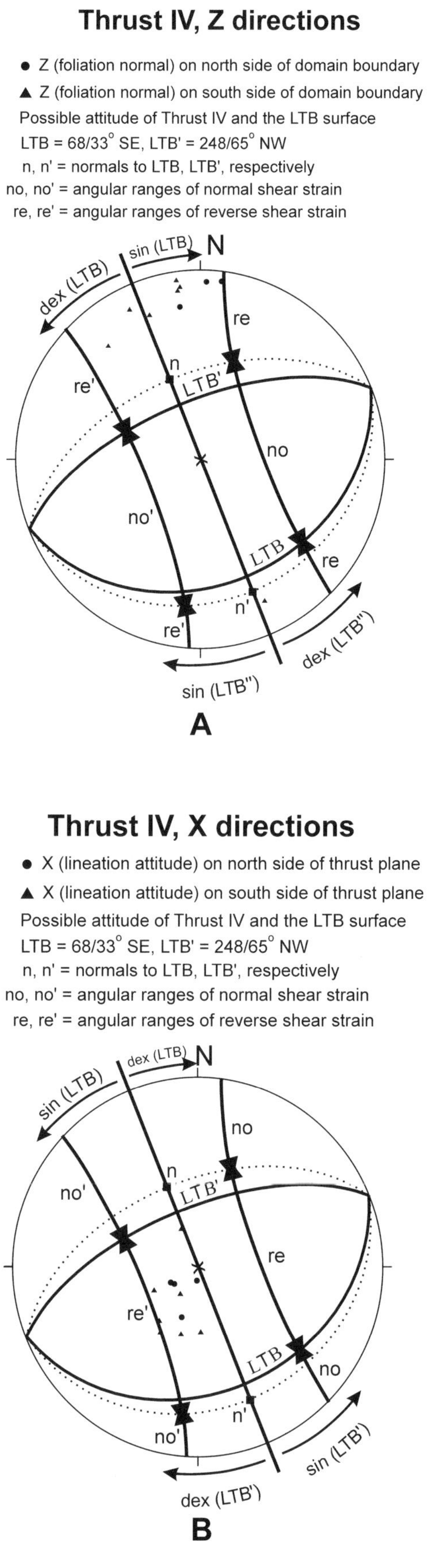

(Cureton et al., 1997) that the southwest segment of the present shear zone dips steeply toward the southeast.

Normals to enveloping surfaces for close to tight mesoscopic S- or Z-folds in the foliation deviate from the local direction of maximum accumulated shortening (Z), even if the fold axis and mineral-shape lineation are parallel to X (the direction of accumulated maximum extension; Schwerdtner and Cowan, 1992). Our shear-sense plot for foliation normals (Fig. 16, A) excludes data from outcrops with trains of mesoscopic Z-folds, but we are uncertain, nonetheless, whether the shear sense based on foliation normals is correct. The foliation problem, however, need not affect the reliability of linear components of shape fabrics as indicators of X, or their use in deriving the sense of tangential shear strain (Fig. 16, B).

In deriving the shear-strain sense in the walls of thrust IV, we made use of two geophysical discontinuities or LTBs that may correspond, respectively, to thrust IV and to the north-northwest–dipping domain boundary. The dip of the discontinuities was obtained, respectively, by gravity modeling (Real and Thomas, 1987) and by reflection-seismic profiling (White et al., 1994). For both structural surfaces, the sense of dip shear proves to be reverse (Fig. 16). The strike shear sense is sinistral for thrust IV (dip: 33°SSE), and dextral for the domain boundary (dip: 68°NNW). However, if the Mooroton shear zone formed at a late stage of Ottawan deformation and was imposed discordantly on thrust IV, it is improper to combine the dip value of 68°NNW with the attitude of L-S shape fabrics developed during Early or Middle Ottawan deformation.

The aforementioned mesoscopic Z-folds are imposed on a foliation that originated under amphibolite-facies metamorphic conditions (Cureton et al., 1997). The fold geometry reveals the sense of a component of late-stage(?) incremental shear strain parallel to the fold-enveloping surfaces (dip: 60°–70°SSE), whereby the shear-strain direction is normal to southerly-plunging hinge lines (Fig. 17). The sense of this shear-strain component is dextral reverse, and points to ductile transpression rather than crustal thinning. Whether normal sense shearing parallel to the north-northwest–dipping domain boundary could be synchronous with the envelope-parallel shear strain is difficult to determine on the basis of the structural information available.

Figure 16. Shear sense derivation by means of foliation normals (A) and lineation directions (B) in the Mooroton shear zone. Structural data (Table 1) were collected at thrust IV and the unconformable base of the Flinton Group (sites 13—15 in Fig. 3), as well as at measurement site 14, near the unconformable base of the Flinton Group (easternmost Elzevir terrane). Lithotectonic boundary (LTB) dips were obtained from geophysical profiles available (Real and Thomas, 1987; White et al., 1994).

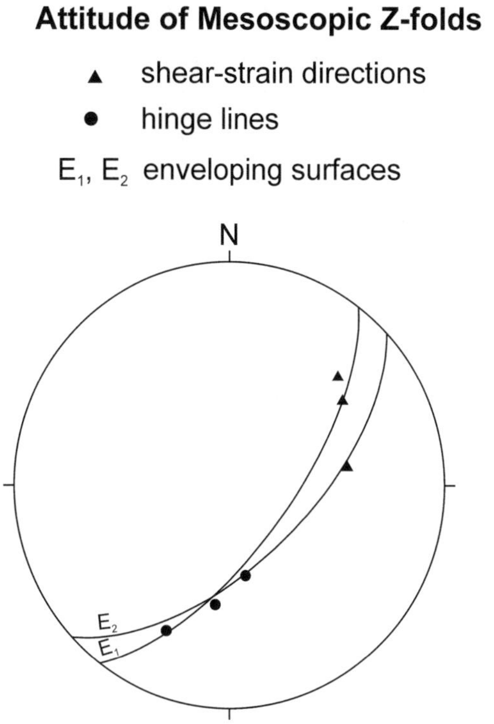

Figure 17. Dextral-reverse sense of a shear-strain component parallel to fold-enveloping surfaces in the Mooroton shear zone or thrust. Structural data from roadcuts between measurement sites 13 and 14 (Alexander's localities 3 and 29). The southerly plunging Z-folds used for the shear-sense derivation may have formed during D3 or D4.

Wallrocks of Thrusts V and VI

The map patterns of the two thrusts are markedly different (Fig. 3), but both are associated with large F2 structures (Fig. 2). This warrants consideration of the northwest dip option (LTB in Figs. 18 and 19) when deriving the sense of tangential shear strain (Fig. 5). The stereoplots in Figures 18 and 19 attest to a dominance of reverse dip shear, but the sense of the strike shear component varies between the two thrusts and their northwest and southeast dip options.

In the walls of thrust V, the strike shear is dextral for the range of southeast-dipping LTBs deemed realistic, and sinistral for the northwest-dipping LTB considered (Figs. 18 and 19). The reverse situation is seen in the walls of thrust VI, whose high-strain zone contains a laminated marble unit similar to that at thrust I.

SUMMARY AND CONCLUSIONS

The 3000 km^2 Mazinaw domain in the Composite Arc Belt, southwestern Grenville Province (southeastern Ontario), contains repeatedly folded, medium- to high-grade metamorphic rocks whose volcanic and sedimentary protoliths range in age from 1290 to <1150 Ma (Corfu and Easton, 1995; Carr et al.,

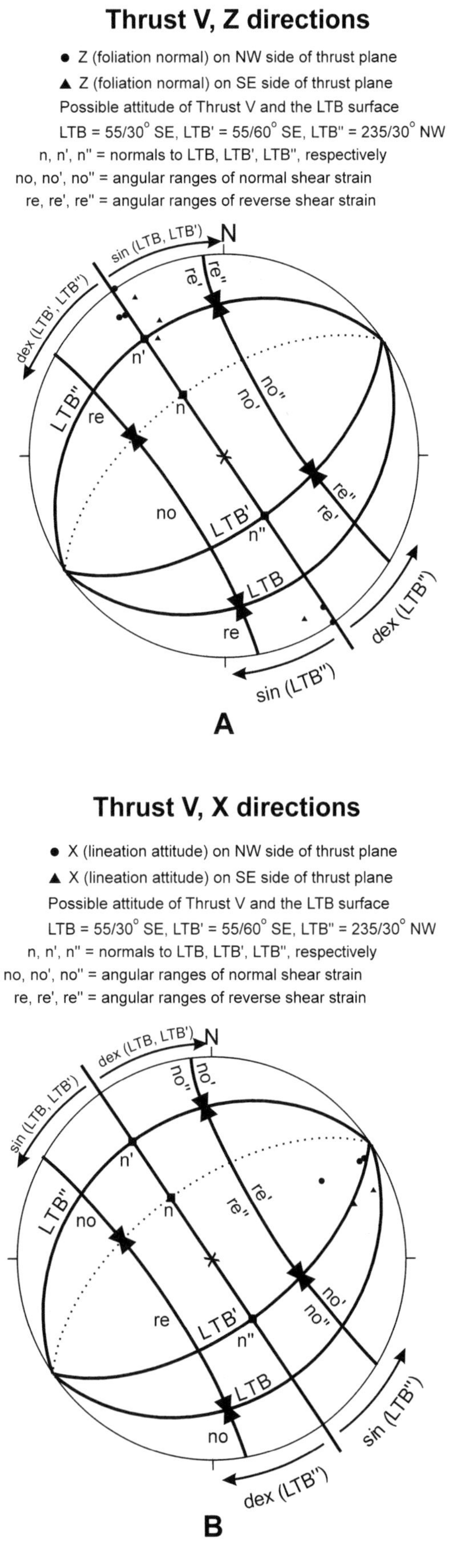

2000). Flinton Group rocks, which include several conglomerate horizons (Moore and Thompson, 1980; Wolff, 1981; Easton, 2001), compose the youngest strained units in the Mazinaw domain. The geological map pattern of the domain is dominated by north- to east-northeast–striking distorted granitoid plutons and large second-generation buckle folds (Hewitt, 1964; Easton, 1992, 2001). Narrow zones of high strain form the borders of, and parallel the northeasterly structural grain within, the Mazinaw domain. In its southern parts, we attempted, but failed, to draw proper or schematic cross-sections compatible with all pieces of geological and geophysical information available. This failure, however, does not preclude the geometric use of linear-planar deformation fabrics in elucidating the structural significance of the high-strain zones.

All high-strain zones studied in this paper have gradational boundaries and horizontal widths of tens to hundreds of meters. The zones are characterized by strained rocks with a strong subvertical flattening foliation and concordant shape fabrics of stretched and/or flattened mineral constituents, distorted pebbles, and volcanic fragments. Most zones and their ductilely deformed walls contain folded or stretched intrusive contacts, as well as the strained unconformable base of the Flinton Group rocks. The shape fabrics of distorted pebbles and other small geological objects correspond to those of metamorphic mineral grains, and are used, in the present paper, to approximate the local principal directions of accumulated strain. Except for the Robertson Lake shear zone at the eastern border of the Mazinaw domain (Busch and van der Pluijm, 1996; Carr et al., 2000; White et al., 2000), all north- to east-northeast–striking high-strain zones have been linked to "possible thrusts" (Easton, 1992; Corfu and Easton, 1995; Cureton et al.,1997), and are represented on the southern sheet of the Tectonic Assemblages Map as east- to south-southeast–dipping faults (Ontario Geological Survey, 1992).

Systems of closely spaced quartz or carbonate veins occur at many localities within, and at the boundaries of, the high-strain zones. The veins are commonly distorted or even transposed into the flattening foliation, and appear to be most prominent at or near contacts of metaconglomerates and metaplutonic units. This suggests that lithotectonic boundaries such as strained segments of the unconformity surface and intrusive contacts were mechanically incoherent in the middle crust, and promoted the synkinematic circulation of igneous and/or metamorphic fluids. No estimate is available as to the magnitude of boundary-parallel slip that apparently accompanied the development of high-strain zones (stretching thrusts). Gravity modeling and reflection seismic profiling have shown that, in the

Figure 18. Shear sense derivation by means of foliation normals (A) and lineation directions (B) in the walls of the folded(?) thrust V and limbs of a large F2 structure. The thrust coincides with lithotectonic boundaries marked by intrusive contacts of strained granitoid plutons (Figs. 2 and 3).

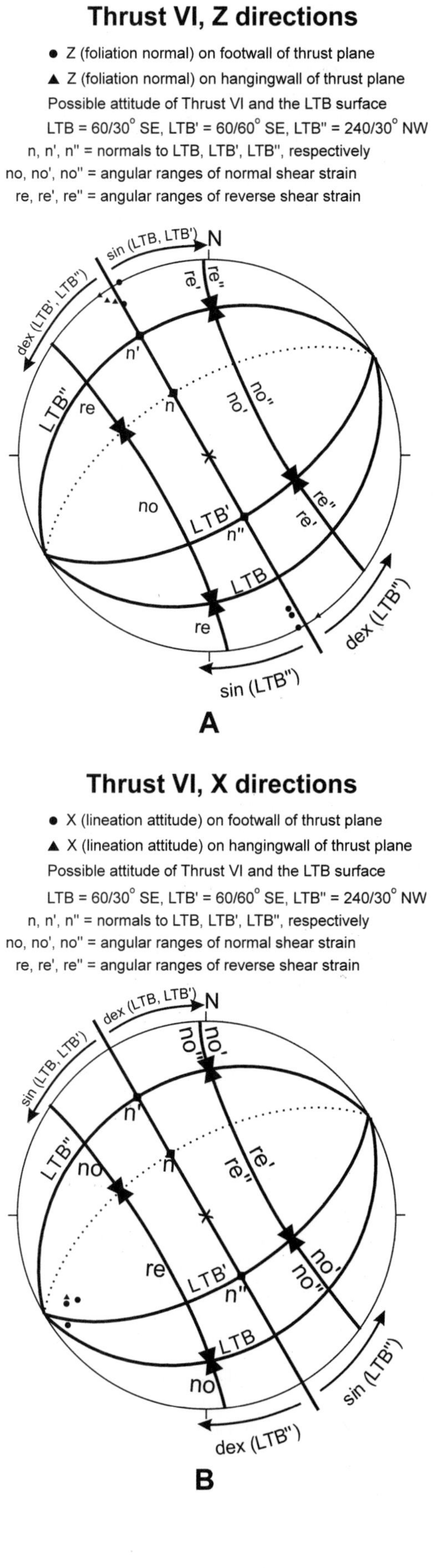

present shallow crust of the Mazinaw domain, most tabular rock bodies and north- to east-northeast–striking structural surfaces (including ductile faults) are subhorizontal or dip <45°SE. For a variety of fault dip options, we determined the sense of tangential shear strain within and adjacent to six high-strain zones. The dip component of accumulated shear strain has the sense of thrusts or reverse faults for most realistic options, even if one presumes that high-strain zones rotated during repeated buckle folding on the kilometer scale. The sense of the strike shear varies with dip option for a given thrust or thrust segment (Figs. 9, 13, 15, 16, 18, and 19). In the high-strain zone of thrust IV, southerly plunging mesoscopic Z-folds attest to a late-stage increment of dextral thrust shear on the fold-enveloping surfaces (Fig. 17). These surfaces are subparallel to thrust IV, but the sense of late incremental strike shear on the surfaces is dextral, and the total strike shear on thrust IV is sinistral (Fig. 16, B).

The results of our field-based study not only elucidate the structural evolution of the Mazinaw domain at a mid-crustal level in the Grenville orogen (southeastern Ontario), but also encourage a broader application of the L-S fabric scheme to strained metamorphic rocks of other collisional orogens. In regard to the Mazinaw domain, our results suggest that the walls of northeast-striking high-strain zones or ductile faults were shortened and/or stretched during midcrustal thickening. This implies that the faults qualify as stretching thrusts on which net slip and associated wallrock shear were directed to the northwest.

ACKNOWLEDGMENTS

This paper is based largely on published geological maps and geophysical profiles, as well as on structural field data collected as part of an ongoing collaborative project on the structural evolution of the Mazinaw domain. We thank our coworkers P.-Y.F. Robin and John Burke for supporting our efforts in the field and the laboratory. The project was financed by the National Sciences and Engineering Research Council Research Grant program, and in-kind support was provided by Pam Sangster, resident geologist at Tweed, Ontario, and her staff. Helpful advice and information were furnished by R.M. Easton of the Ontario Geological Survey in Sudbury. We thank Jim and Elsie Byer for their hospitality and logistic help in 2001–2002. Louise Corriveau, Mike Easton, John Myers, and Phil Simony scrutinized different versions of the manuscript and suggested changes in the line of reasoning and style of presentation.

Figure 19. Shear sense derivation by means of foliation normals (A) and lineation directions (B) in the walls of the folded (?) thrust VI (measurement sites 19, 20). The thrust coincides with a lithotectonic boundary (LTB) at the stretched intrusive contact of a small granitoid body (not shown in Fig. 3).

REFERENCES CITED

Barclay, W.A., 1989, Deformation fabrics along part of the southwest limb of the Clare River synform, Grenville Province, Ontario: Evidence for a late episode of penetrative strain and synkinematic intrusion [M.S. thesis]: Toronto, Ontario, University of Toronto, 216 p.

Bright, T.G., 1986a, Geology of the Mellon Lake area: Sudbury, Ontario Geological Survey, Open File Report 5598, 156 p.

Bright, T.G., 1986b, Petrography of the Precambrian basement rocks in deep drill holes beneath the Phanerozoic cover in southern Ontario: Sudbury, Ontario Geological Survey, Open File Report 5603, 25 p.

Busch, J.P., and van der Pluijm, B.A., 1996, Late-orogenic, plastic to brittle extension along the Robertson Lake shear zone: Implications for the style of deep-crustal extension in the Grenville Orogen, Canada: Precambrian Research, v. 79, p. 41–57.

Carr, S.D., Easton, R.M., Jamieson, R.A., and Culshaw, N.G., 2000, Geologic transect across the Grenville Orogen of Ontario and New York: Canadian Journal of Earth Sciences, v. 37, p. 193–216.

Casey, M., and Dietrich, D., 1997, Overthrust shear in mountain building, *in* Sengupta, S., ed., Evolution of geological structures in micro- to macro-scales: London, Chapman and Hall, p. 119–142.

Chappell, J.F., 1978, The Clare River structure and its tectonic setting [Ph.D. thesis]: Ottawa, Ontario, Carleton University, 184 p.

Cobbold, P.R., 1977, Description and origin of banded deformation structures, I: Regional strain, local perturbations and deformation bands: Canadian Journal of Earth Sciences, v. 14, p. 1721–1731.

Cobbold, P.R., 1983, Kinematic and mechanical discontinuity at a coherent interface: Journal of Structural Geology, v. 5, p. 341–349.

Connelly, J.N., 1986, The emplacement history of the Elzevir batholith and the regional deformation of the supracrustal rocks in the Metasedimentary Belt, Grenville Province, southeastern Ontario [M.S. thesis]: Kingston, Ontario, Queen's University, 152 p.

Connelly, J.N., van Breemen, O., and Hanmer, S.K., 1987, The age of deformation related to emplacement of the Elzevir batholith and its regional implications, Grenville Province, southeastern Ontario: Ottawa, Ontario, Geological Survey of Canada, Paper 87-2, p. 51–54.

Connors, K.A., Ansdell, S.B., and Lucas, S.B., 2002, Development of a transverse to orogen parallel extension lineation in a complex collisional setting, Trans-Hudson Orogen, Manitoba, Canada: Journal of Structural Geology, v. 24, p. 89–106.

Corfu, F., and Easton, R.M., 1995, U-Pb geochronology of the Mazinaw terrane, an imbricate segment of the Central Metasedimentary Belt, Grenville Province, Ontario: Canadian Journal of Earth Sciences, v. 32, p. 954–976.

Culshaw, N.G., Davidson, A., and Nadeau, L., 1983, Structural subdivisions of the Grenville Province in the Parry Sound–Algonquin region, Ontario: Ottawa, Ontario, Geological Survey of Canada, Paper 83-1B, p. 243–252.

Cureton, J.S., van der Pluijm, B., and Essene, E.J., 1997, Nature of the Elzevir-Mazinaw domain boundary, Grenville Orogen, Ontario: Canadian Journal of Earth Sciences, v. 34, p. 976–991.

Currie, A., 1972, Analysis of structure in a part of the Clare River Synform, Ontario [M.S. thesis]: Toronto, Ontario, University of Toronto, 43 p.

Davidson, A., Culshaw, N.G., and Nadeau, L., 1982, A tectono-metamorphic framework for part of the Grenville Province, Parry Sound region, Ontario: Ottawa, Ontario, Geological Survey of Canada, Paper 82-1A, p. 175–190.

Di Brisko, G., 1989, Geology of the Elzevir area, Hastings and Lennox and Addington Counties: Sudbury, Ontario Geological Survey, Miscellaneous Paper 146, p. 169–175.

Dieterich, J.H., 1970, Computer experiments on mechanics of finite-amplitude folds: Canadian Journal of Earth Sciences, v. 7, p. 467–476.

Dietrich, D., and Durney, D.W., 1986, Change in direction of overthrust shear in the Helvetic nappes of western Switzerland: Journal of Structural Geology, v. 8., p. 389–398.

Downey, M.W., 2002, Macro-structural analysis of the Little Skootamatta shear zone, Mazinaw domain, Composite Arc Belt, southeastern Ontario [undergraduate honors thesis]: Toronto, Ontario, University of Toronto, Department of Geology, 53 p.

Duebendorfer, E.M., and Black, R.A., 1992, Kinematic role of transverse structures in continental extension: An example from the Las Vegas Valley shear zone, Nevada: Geology, v. 20, p. 1107–1110.

Easton, R.M., 1988, Precambrian geology of the Darling area, Frontenac and Renfrew counties: Sudbury, Ontario Geological Survey, Open File Report 56, 208 p.

Easton, R.M., 1992, The Grenville Province and the Proterozoic history of central and southern Ontario, *in* Thurston, P.C., et al., eds., Geology of Ontario: Sudbury, Ontario Geological Survey, Special Volume 4, Part 2, p. 715–904.

Easton, R.M., 2001, Precambrian geology, Kaladar area: Sudbury, Ontario Geological Survey, Preliminary Map P. 3441, scale 1:50,000.

Easton, R.M., and Ford, F.D., 1991, Geology of the Mazinaw area (31 C/14): Sudbury, Ontario Geological Survey, Miscellaneous Paper 157, p. 95–106.

Easton, R.M., Ford, F.D., and Cooke, M.A., 1995a, Geology of the Palmerston Lake area (31 F/2SW): Sudbury, Ontario Geological Survey, Preliminary Map P. 3335.

Easton, R.M., Ford, F.D., and Stewart, R., 1995b, Geology of the Mazinaw area (31 C/14NE): Sudbury, Ontario Geological Survey, Preliminary Map P. 3336.

Flinn, D., 1965, On the symmetry principle and the deformation ellipsoid: Geological Magazine, v. 102, p. 36–45.

Flinn, D., 1978, Construction and computation of three-dimensional progressive deformations: Journal of the Geological Society of London, v. 135, p. 291–305.

Ford, F.D., 1992, Geology of the Fernleigh and Ompah "synclines," Palmerston Lake area: Sudbury, Ontario Geological Survey, Miscellaneous Paper 160, p. 41–46.

Gower, C.F., Ryan, A.B., and Rivers, T., 1990, Mid-Proterozoic Laurentia-Baltica: An overview and a summary of the contributions made by this volume: St. John's, Newfoundland, Geological Association of Canada, Special Paper 38, p. 1–20.

Green, A.G., Milkereit, B., Davidson, A., Spencer, C., Hutchinson, D.R., Cannon, W.F., Lee, M.W., Agena, W.F., Behrendt, J.C., and Hinze, W.J., 1988, Crustal structure of the Grenville Front and adjacent terranes: Geology, v. 16, p. 788–792.

Hajnal, Z., Lucas, S.B., White, D.J., Lewry, J.F., Bezdan, S., Stauffer, M.R., and Thomas, M.D., 1996, Seismic reflection images of high-angle faults and linked detachments in the Trans-Hudson orogen: Tectonics, v. 15, p. 427–439.

Hanmer, S., and Passchier, C.W., 1991, Shear-sense indicators: A review: Ottawa, Ontario, Geological Survey of Canada, Paper 90-17.

Hanmer, S., Corrigan, D., Pehrsson, S., and Nadeau, L., 2000, SW Grenville Province, Canada: The case against post–1.4 Ga accretionary tectonics: Tectonophysics, v. 319, p. 33–51.

Hatcher, R.D., 1995, Structural geology (2nd edition): Englewood Cliffs, New Jersey, Prentice-Hall, 525 p.

Hatcher, R.D., 2001, Rheological partitioning during multiple reactivation of the Paleozoic Brevard fault zone, southern Appalachians, USA: Geological Society of London Special Publication 186, p. 257–271.

Hatcher, R.D., and Hooper, R.J., 1992, Evolution of crystalline thrust sheets in the internal parts of mountain chains, *in* McClay, K.R., ed., Thrust tectonics: London, Chapman and Hall, p. 217–234.

Hewitt, D.F., 1964, Madoc Area: Toronto, Ontario Department of Mines, Map 2053.

Hobbs, B.E., Means, W.D., and Williams, P.F., 1976, An outline of structural geology: New York, Wiley and Sons, 571 p.

Inglis, J., 1983, Theoretical strain patterns in ductile zones simultaneously undergoing heterogeneous simple shear and bulk shortening: Journal of Structural Geology, v. 5, p. 369–382.

Jaeger, J.C., 1962, Elasticity, fracture and flow (2nd edition): London, Methuen, 212 p.
Jones, R.R., Holdesworth, R.E., and Bailey, W., 1997, Lateral extrusion in transpression zones: The importance of boundary conditions: Journal of Structural Geology, v. 19, p. 1201–1227.
Kearey, P., and Vine, F.J., 1990, Global tectonics: London, Blackwell Scientific Publications, 302 p.
Launeau, P., and Robin, P.-Y.F., 1996, Fabric analysis using the intercept method: Tectonophysics, v. 267, p. 91–119.
Lisle, R.J., 1998, Simple graphical constructions for the direction of shear: Journal of Structural Geology, v. 20, p. 969–973.
Martignole, J., and Calvert, A.J., 1996, Crustal-scale shortening and extension across the Grenville Province of western Quebec: Tectonics, v. 15, p. 376–386.
Means, W.D., 1989, Stretching faults: Geology, v. 17, p. 893–896.
Means, W.D., 1990, One-dimensional kinematics of stretching faults: Journal of Structural Geology, v. 12, p. 267–272.
Mezger, K., Essene, E.J., van der Pluijm, B., and Halliday, A.N., 1993, U-Pb geochronology of the Grenville Orogen of Ontario and New York: Constraints on ancient crustal tectonics: Contributions to Mineralogy and Petrology, v. 114, p. 13–26.
Moore, J.M., and Thompson, P.H., 1972, The Flinton Group, Grenville Province, eastern Ontario: 24th International Geological Congress, 20–28 August, Montreal, Québec, Proceedings, section 1, p. 221–229.
Moore, J.M., and Thompson, P.H., 1980, The Flinton Group: A late Precambrian metasedimentary succession in the Grenville Province of eastern Ontario: Canadian Journal of Earth Sciences, v. 17, p. 1685–1707.
Moores, E.M., and Twiss, R.J., 1995, Tectonics: New York, W.H. Freeman, 415 p.
Nadai, A., 1950, Theory of flow and fracture of solids: New York, McGraw-Hill, v. 1, 572 p.
Ontario Geological Survey, 1992, Tectonic assemblages map of Ontario, southern sheet, Sudbury, Ontario Geological Survey, Map 2518, scale 1:1,000,000.
Park, R.G., 1983, Foundations of structural geology (1st edition): Glasgow and London, Blackie and Son, 135 p.
Passchier, C.W., Myers, J.S., and Kroner, A., 1990, Field geology of high-grade gneiss terrains: Berlin, Springer-Verlag, 150 p.
Pavlis, T.L., 1996, Fabric development in syntectonic intrusive sheets as a consequence of melt-dominated flow and thermal softening of the crust: Tectonophysics, v. 253, p. 1–31.
Piazolo, S., and Passchier, C.W., 2002, Controls on lineation development in low to medium grade shear zones: A study from the Cap de Creus peninsula, NE Spain: Journal of Structural Geology, v. 24, p. 25–44.
Ramsay, J.G., 1967, Folding and fracturing of rocks: New York, McGraw-Hill, 568 p.
Ramsay, J.G., 1980, Shear zone geometry: A review: Journal of Structural Geology, v. 2, p. 83–99.
Ramsay, J.G., and Graham, R.H., 1970, Strain variations in shear belts: Canadian Journal of Earth Sciences, v. 7, p. 786–813.
Ramsay, J.G., and Huber, M.I., 1983, The techniques of modern structural geology, v. 1: Strain analysis: London, Academic Press, 307 p.
Real, D., and Thomas, M.D., 1987, Gravity models within the southern Central Metasedimentary Belt, Grenville Province, Ontario: Ottawa, Ontario, Geological Survey of Canada, Paper 87-1A, p. 763–772.
Reed, J.C., and Bryant, B., 1964, Evidence for strike-slip faulting along the Brevard zone in North Carolina: Geological Society of America Bulletin, v. 75, p. 1177–1196.
Rivers, T., 1976, Structures and textures of metamorphic rocks, Ompah area, Grenville Province, Ontario [Ph.D. thesis]: Ottawa, Ontario, University of Ottawa, 218 p.
Rivers, T., 1997, Lithotectonic elements of the Grenville Province: Review and tectonic implications: Precambrian Research, v. 86, p. 117–154.
Rivers, T., and Corrigan, D., 2000, Convergent margin on southeastern Laurentia during the Mesoproterozoic: Tectonic implications: Canadian Journal of Earth Sciences, v. 37, p. 359–383.
Rivers, T., Martignole, J., Gower, C.F., and Davidson, A., 1989, New tectonic divisions of the Grenville Province, southeast Canadian Shield: Tectonics, v. 8, p. 63–84.
Robin, P.-Y.F., 1977, Determination of geological strain using randomly oriented strain markers of any shape: Tectonophysics, v. 42, p. T7–T16.
Robin, P.-Y.F., and Cruden, A.R., 1994, Strain and vorticity patterns in ideally ductile transpression zones: Journal of Structural Geology, v. 16, p. 447–466.
Sager-Kinsman, E.A., and Parrish, R.R., 1993, Geochronology of detrital zircons from the Elzevir and Mazinaw terranes, Central Metasedimentary Belt, Grenville Province, Ontario: Canadian Journal of Earth Sciences, v. 30, p. 465–473.
Sanderson, D.J., and Marchini, W.R.D., 1984, Transpression: Journal of Structural Geology, v. 6, p. 449–458.
Schwerdtner, W.M., 1982, Calculation of volume change in ductile band structures: Journal of Structural Geology, v. 4, p. 57–62.
Schwerdtner, W.M., 1998, Graphic derivation of the local sense of shear strain components in stretched walls of lithotectonic boundaries: Journal of Structural Geology, v. 20, p. 957–967.
Schwerdtner, W.M., and Cote, M.L., 2001, Patterns of pervasive shear strain near the boundaries of the La Ronge Domain, inner Trans-Hudson Orogen, western Canadian Shield: Precambrian Research, v. 107, p. 93–116.
Schwerdtner, W.M., and Cowan, E.J., 1992, Theoretical estimates of the angle between the X-Y trace and foliation trajectories in laminated orthogneiss: Journal of Structural Geology, v. 13, p. 11–17.
Schwerdtner, W.M., and Gapais, D., 1983, Calculations of finite incremental deformations in ductile geological materials and structural models: Tectonophysics, v. 93, p. T1–T7.
Schwerdtner, W.M., Bennett, P.J., and Janes, T.W., 1977, Application of L-S fabric scheme to structural mapping and paleostrain analysis: Canadian Journal of Earth Sciences, v. 14, p. 1221–1232.
Thompson, P.H., 1972, Stratigraphy, structure and metamorphism of the Flinton Group in the Bishop Corner–Madoc area, Grenville Province, eastern Ontario [Ph.D. thesis]: Ottawa, Ontario, Carleton University, 268 p.
Turner, F.J., and Weiss, L.E., 1963, Structural analysis of metamorphic tectonites: New York, McGraw-Hill, 545 p.
van Breemen, O., and Davidson, A., 1988, U-Pb zircon ages of granites and syenites in the Central Metasedimentary Belt, Grenville Province, Ontario: Ottawa, Ontario, Geological Survey of Canada, Paper 88-2, p. 45–50.
White, D.J., Easton, R.M., Culshaw, N.G., Milkereit, B., Forsyth, D.A., Carr, S.D., Green, A.G., and Davidson, A., 1994, Seismic images of the Grenville Orogen in Ontario: Canadian Journal of Earth Sciences, v. 31, p. 292–307.
White, D.J., Forsyth, D.A., Asudeh, I., Carr, S.D., Wu, H., Easton, R.M., and Mereu, R.F., 2000, A seismic-based cross-section of the Grenville Orogen in southern Ontario and western Quebec: Canadian Journal of Earth Sciences, v. 37, p. 183–192.
Wilson, M.E., 1934, The Clare River syncline: Transactions of the Royal Society of Canada (Ser. 3), v. 27, sec. 4, p. 7–11.
Wilson, M.E., 1940, Madoc area: Ottawa, Ontario, Geological Survey of Canada, Map 559A, scale 1″ = 1 mile.
Wolff, J.M., 1981, Kaladar: Sudbury, Ontario Geological Survey, Map 2432, scale 1:63,360.
Wolff, J.M., 1982, Geology of the Kaladar area: Sudbury, Ontario Geological Survey, Report 215, 94 p.

Manuscript Accepted by the Society August 25, 2003

Printed in the USA

Geological Society of America
Memoir 197
2004

Early Mesoproterozoic tectonic history of the southwestern Grenville Province, Ontario: Constraints from geochemistry and geochronology of high-grade gneisses

Trond Slagstad*
Nicholas G. Culshaw
Rebecca A. Jamieson
Department of Earth Sciences, Dalhousie University, Halifax, Nova Scotia B3H 3J5, Canada
John W.F. Ketchum
Jack Satterly Geochronology Laboratory, Royal Ontario Museum, 100 Queen's Park, Toronto, Ontario M5S 2C6, Canada

ABSTRACT

The southwestern Grenville Province in Ontario consists of Paleoproterozoic and Mesoproterozoic rocks that formed along the southeastern margin of Laurentia during a significant period of growth of the North American continent. The area investigated consists of high-grade ortho- and paragneisses whose protoliths formed between 1500 and 1350 Ma. Geochemical and geochronological data from the rocks investigated point to a period of arc and back-arc magmatism along the Laurentian margin between ca. 1500 and 1450 Ma. A-type charnockites and granites are temporally and spatially associated with the arc rocks, and have compositions interpreted to indicate derivation from a tholeiitic basaltic underplate and from crustal melting, respectively. This interpretation implies that the arc underwent extension during part of its history. Between ca. 1450 and 1430 Ma, magmatism apparently ceased in the study area and was temporally associated with high-grade metamorphism in the back-arc region, possibly suggesting a change from an extensional to a compressional tectonic regime, although other interpretations are equally likely. Few data exist for the period after ca. 1430 Ma; however, a number of geochronological studies from nearby parts of the Grenville Province suggest that arc magmatism may have moved to a more outboard position on the continental margin. At ca. 1360 Ma, bimodal rhyolitic and basaltic magmatism suggests a return to an extensional tectonic regime. The geochemical data generally support, but suggest refinements to, previously proposed models for the evolution of the southeastern Laurentian margin in the Grenville Province that may be applicable to other parts of the margin.

Keywords: Grenville Province, Central Gneiss Belt, Laurentia, crustal growth

*E-mail: Trond.Slagstad@ngu.no. Current address: Geological Survey of Norway (NGU), Leiv Erikssons vei 39, 7491 Trondheim, Norway.

Slagstad, T., Culshaw, N.G., Jamieson, R.A., and Ketchum, J.W.F., 2004, Early Mesoproterozoic tectonic history of the southwestern Grenville Province, Ontario: Constraints from geochemistry and geochronology of high-grade gneisses, *in* Tollo, R.P., Corriveau, L., McLelland, J., and Bartholomew, M.J., eds., Proterozoic tectonic evolution of the Grenville orogen in North America: Boulder, Colorado, Geological Society of America Memoir 197, p. 209–241. For permission to copy, contact editing@geosociety.org.

INTRODUCTION

Continental growth processes include formation of juvenile magmatic crust at or near the continental margin, accretion of outboard terranes (Bickford, 1988; Condie and Chomiak, 1996), and continental collision (e.g., Yin and Harrison, 2000, and references therein). Where these processes are superimposed, evidence of successive continental growth is likely to be obscured by subsequent accretionary or collisional orogenesis. Unraveling the preorogenic history of strongly deformed, metamorphosed, and dismembered rocks in accretionary and collisional orogenic belts, ancient as well as recent, is a daunting task (see van Staal et al., 1998). In addition to the effects imposed by accretion and continental collision, rocks incorporated in orogenic belts typically span several hundred million years in age, come from a variety of tectonic settings and geographical locations, and may have been involved in several orogenic cycles. This problem, inherent in most large orogenic belts, may be nowhere better illustrated than in the pervasively deformed Central Gneiss Belt of the southwestern Grenville Province in Ontario (Fig. 1).

The Central Gneiss Belt is commonly described as a collage of tectonically stacked lithotectonic units or domains assembled by northwest-directed thrusting during Grenvillian orogenesis (Davidson et al., 1982; Rivers et al., 1989; Nadeau and Hanmer, 1992; Culshaw et al., 1997). Discrete domains, identified based on contrasting lithologies, structural styles, metamorphic histories, and in some cases geophysical properties, are separated by zones of straight gneisses (S tectonites

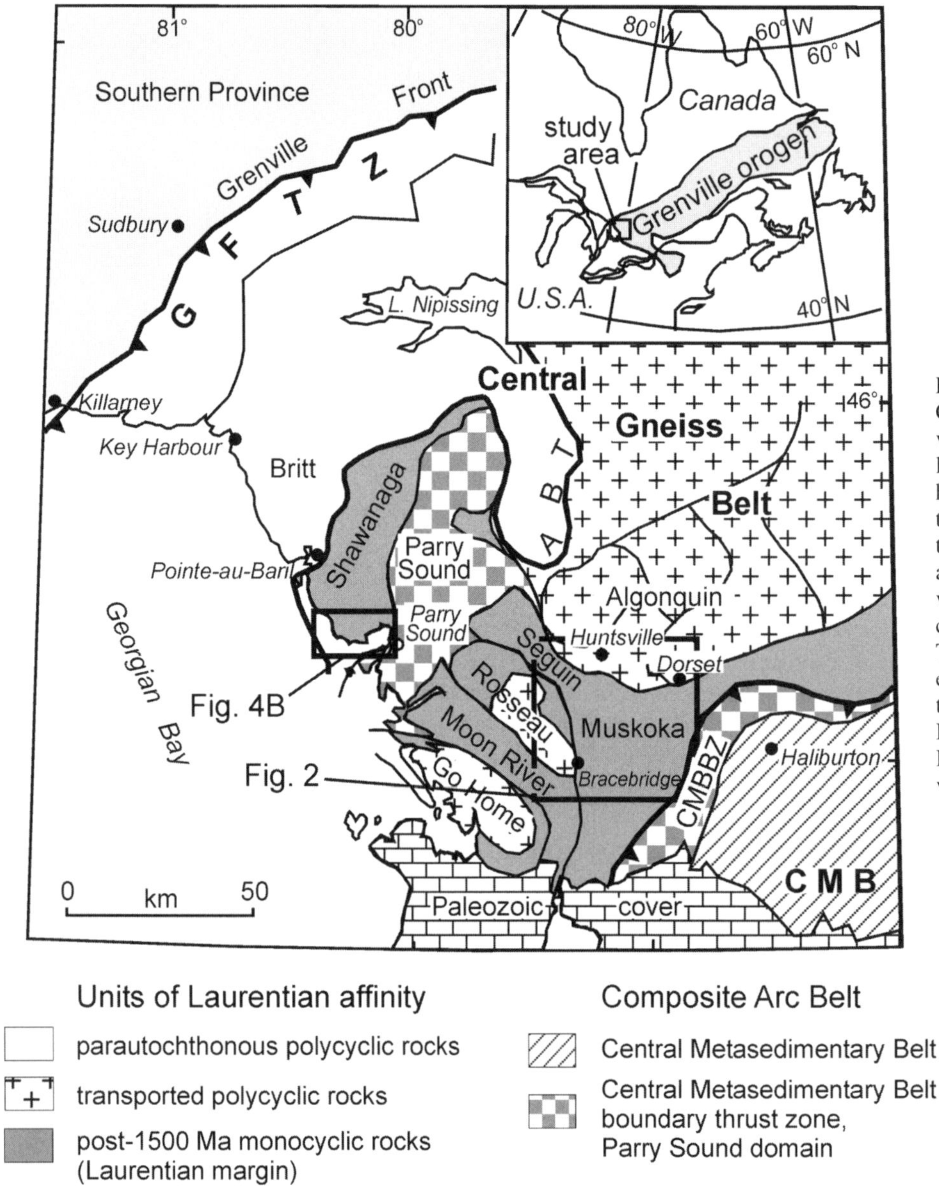

Figure 1. Lithotectonic domains within the Central Gneiss Belt, southwestern Grenville Province, Ontario. The heavy barbed lines indicate major thrust boundaries; the heavy line indicates allochthon boundary thrust, from Ketchum and Davidson (2000); the thin lines represent domain boundaries, after Davidson (1984) and Davidson and van Breemen (1988), with minor modifications by Ketchum and Davidson (2000). The Composite Arc Belt was defined by Carr et al. (2000). ABT—allochthon boundary thrust; CMB—Central Metasedimentary Belt; CMBBZ—Central Metasedimentary Belt boundary thrust zone; GFTZ—Grenville Front tectonic zone.

with a pronounced layering, after Davidson et al., 1982) interpreted to represent ductile shear zones (Davidson et al., 1982; Culshaw et al., 1983, 1997). The Central Gneiss Belt exposes high-grade gneisses with protolith ages ranging from Archean to Late Mesoproterozoic (Ketchum and Davidson, 2000), which include reworked continuations of Archean, Early and Mid-Paleoproterozoic, and Mesoproterozoic orogens (Rivers, 1997, and references therein) representing a variety of probably arc-related tectonic settings, from continental and island arc to intra-arc and back-arc (Culshaw and Dostal, 1997, 2002; Rivers, 1997; Carr et al., 2000; Rivers and Corrigan, 2000). Some units are clearly allochthonous with respect to the surrounding units (Wodicka et al., 1996), and some protoliths were strongly deformed and metamorphosed prior to Grenvillian orogenesis (Corrigan et al., 1994; Ketchum et al., 1994).

Studies of the preorogenic history of rocks in orogenic belts rely heavily on geochronology (e.g., Ketchum et al., 1997; Gebauer, 1999; Söderlund et al., 2002). Geochronological data are necessary in order to identify the sequence of events and correlate rocks that may have formed during the same geological event, and in many cases the robust U-Pb zircon system preserves evidence of several past tectonic events. However, in most cases determining the petrogenesis and tectonic significance of a rock requires information about its composition, in addition to its age. Here we present new geochemical and geochronological data from a variety of rock types from the Central Gneiss Belt in order to constrain their petrogenesis and tectonic setting, and discuss likely tectonic models for the pre-Grenvillian evolution of the area between ca. 1500 and 1350 Ma.

Apart from the detailed studies of Culshaw and Dostal (1997, 2002), proposed tectonic models for the early Mesoproterozoic evolution of the southwestern Grenville Province are typically relatively broad in scope. For example, Rivers and Corrigan (2000) proposed an "Andean"-type model involving continuous subduction of oceanic crust beneath the Laurentian margin, with alternating periods of back-arc extension and compression. Although we agree with their general interpretation, the data presented here suggest differences in detail, and have different implications for the Early Mesoproterozoic geological and magmatic evolution of the North American midcontinent region.

The main contributions of this paper are to (1) demonstrate the robustness of the geochemical system in strongly deformed and metamorphosed rocks, (2) show that, in addition to geochronology, geochemistry is important to consider in studies of preorogenic history, (3) propose a detailed model for the pre-Grenvillian, Early Mesoproterozoic tectonic history of the southwestern Grenville Province, and (4) add to the understanding of how the Laurentian margin evolved during the Mesoproterozoic. Coupled with relatively good exposure, simple logistics, a relatively well-understood Grenvillian history, and large geochronological database, the selected study areas, outlined in Figure 1, are well suited for studies aimed at understanding the pre-Grenvillian history of the Central Gneiss Belt.

REGIONAL GEOLOGY

Wynne-Edwards (1972) described the Grenville Province in Ontario in terms of three major divisions (Fig. 1): the Grenville Front tectonic zone, Central Gneiss Belt, and Central Metasedimentary Belt, the latter two separated by the Central Metasedimentary Belt boundary thrust zone (Hanmer and McEachern, 1992). The Grenville Front tectonic zone is a southeast-dipping crustal-scale thrust zone, active at ca. 1.0 Ga (Haggart et al., 1993; Krogh, 1994), that marks the northwestern limit of Grenvillian deformation and metamorphism. The Central Gneiss Belt structurally overlies the Grenville Front tectonic zone and comprises Late Paleoproterozoic to Early Mesoproterozoic high-grade gneisses that formed along the southeastern margin of the pre-Grenvillian Laurentian craton and were penetratively deformed and tectonically stacked during Grenvillian orogenesis (Culshaw et al., 1997). The Central Metasedimentary Belt structurally overlies the Central Gneiss Belt and mostly comprises post–1.4 Ga arc and back-arc assemblages interpreted to have formed outboard of the southeastern margin of Laurentia (Carr et al., 2000, and references therein). Emplacement of the Central Metasedimentary Belt onto the Central Gneiss Belt took place by thrusting along the Central Metasedimentary Belt boundary thrust zone during Grenvillian orogenesis, although the timing of this event is debated. McEachern and van Breemen (1993) and Corriveau and van Breemen (2000) suggest that collision started ca. 1.2 Ga, whereas Culshaw et al. (1997) and Timmermann et al. (1997), citing the lack of geochronological evidence for metamorphism prior to ca. 1080 Ma in the Central Gneiss Belt in Ontario, argue for continental collision at ca. 1090–1080 Ma. The Parry Sound domain (Fig. 1) is presently located within the Central Gneiss Belt, but is interpreted to represent a thrust-slice originating in the Central Metasedimentary Belt boundary thrust zone or in correlative rocks to the southeast (Wodicka et al., 1996).

The Central Gneiss Belt can be divided into two parts, separated by the allochthon boundary thrust (Fig. 1; Ketchum and Davidson, 2000). Northwest of the allochthon boundary thrust, the Central Gneiss Belt comprises orthogneisses and minor paragneisses whose protoliths formed during two major magmatic episodes at ca. 1750–1600 and 1470–1340 Ma (e.g., Rivers, 1997; Ketchum and Davidson, 2000). Despite extensive reworking during Grenvillian metamorphism and deformation, evidence for multiple phases of pre-Grenvillian tectonism, including significant >1700 Ma and ca. 1450–1350 Ma events, is locally preserved (van Breemen and Davidson, 1988; Tuccillo et al., 1992; Corrigan et al., 1994; Dudás et al., 1994; Ketchum et al., 1994; Jamieson et al., 2001). Krogh et al. (1996) proposed the term "Pinwarian orogeny" for the ca. 1450 Ma event. In contrast, southeast of the allochthon boundary thrust, Culshaw et al. (1997) and Timmermann et al. (1997) found no clear evidence for a pre-Grenvillian metamorphic history in the Shawanaga and Muskoka domains; hence they are "monocyclic" according to the terminology of Rivers et al. (1989). Following Rivers et al.

(1989), we define Grenvillian orogenesis to include tectonic, metamorphic, and plutonic events that took place at or outboard of the Laurentian margin between ca. 1190 and 970 Ma. The distribution of lithological units, structures, and metamorphic grades in the area investigated is largely the result of Grenvillian tectonic processes (e.g., Culshaw et al., 1997; Timmermann et al., 1997; Carr et al., 2000).

The times of protolith formation in the Central Gneiss Belt coincided with major crustal-forming events in the North American midcontinent (Condie, 1986; Bickford, 1988; Van Schmus et al., 1996), and several authors have proposed that events in the Central Gneiss Belt and the adjacent foreland can be correlated with events in the midcontinent region (Davidson, 1986; Easton, 1986; Bickford, 1988; van Breemen and Davidson, 1988; Rivers and Corrigan, 2000; Culshaw and Dostal, 2002). The ca. 1.9–1.6 Ga Penokean, Yavapai-Mazatzal, Killarnean, and Makkovikian orogenies involved formation and accretion of juvenile crust along the southeastern margin of Laurentia (Condie, 1986; Dickin and McNutt, 1990; Davidson et al., 1992; Culshaw et al., 2000). Subsequently, in the intervals 1500–1440 and 1400–1340 Ma, the eastern and southern granite-rhyolite province, respectively, formed along the southeastern margin of Laurentia (Bickford, 1988; Van Schmus et al., 1996). The tectonic setting of the granite-rhyolite provinces is debated. Early models focused on the A-type composition of the rocks and their relative lack of deformation, suggesting an anorogenic setting (e.g., Anderson, 1983; Hoffmann, 1989; Kisvarsanyi and Kisvarsanyi, 1990; Windley, 1993). More recent models, however, emphasize that many of the granites and rhyolites have Nd model ages that are only slightly older than their crystallization ages, suggesting derivation from juvenile continental crust (e.g., Van Schmus et al., 1996; Menuge et al., 2002). The latter authors suggest that the granites and rhyolites most likely formed in an ensialic back-arc setting following accretion of juvenile crust to the Laurentian margin. A problem with this model, however, is that the substrate (i.e., potential source) to the granites and rhyolites is unexposed in the midcontinent region. As pointed out by Culshaw and Dostal (2002), the Central Gneiss Belt may represent the reworked and telescoped continuation of the mid-continental granite-rhyolite provinces; work in this area may therefore contribute to understanding the Mesoproterozoic evolution of the Laurentian margin as a whole.

GEOLOGY OF THE STUDY AREA

The term *domain* is widely used in Grenville literature for segments of crust that, on the basis of lithology, structure, metamorphic grade, geological history, and geophysical signature, are sufficiently distinct to set them apart from adjacent segments (Davidson, 1995). *Domain* is preferred over *terrane* because some tectonic connotations of the latter do not apply in the area; for example, most domains are not bounded by sutures (see Coney et al., 1980), and on palinspastic reconstructions many Grenville domains restore as a single entity (Culshaw et al., 1997).

The rocks investigated were strongly deformed under upper amphibolite- to granulite-facies conditions during Grenvillian orogenesis, and generally form layered gneisses lacking primary igneous or sedimentary structures, although some rocks contain primary megacrystic K-feldspar. For simplicity we generally omit the prefix *meta-* and the generic term *gneiss,* and instead use protolith names as defined by modal compositions. The only exception is the gray gneiss described below, which warrants a distinct grouping because it encompasses a range of protolith compositions with similar geochemical characteristics.

Muskoka and Seguin Domains

Gray Gneiss

The Muskoka domain and the contiguous Seguin subdomain (Fig. 2; hereafter collectively referred to as the Muskoka domain) constitute the uppermost structural level of the Central Gneiss Belt (Culshaw et al., 1983), in the immediate footwall to the Central Metasedimentary Belt boundary thrust zone (Fig. 2). The Muskoka domain is dominated by moderately east- to southeast-dipping, gray, migmatitic orthogneisses that range in composition from gabbro and diorite through granodiorite to volumetrically subordinate granite (Fig. 3). Granodioritic, dioritic, and/or granitic orthogneisses are locally interlayered, imparting the gneisses with a centimeter-scale banded appearance. By analogy with other gneiss terrains (e.g., Passchier et al., 1990), the interlayering is most likely the result of ductile deformation of igneous features. Protolith ages for the gray gneisses range from ca. 1480 to 1430 Ma (Timmermann et al., 1997; Nadeau and van Breemen, 1998; McMullen, 1999), but most ages cluster at ca. 1450 to 1460 Ma. The gneisses were metamorphosed under upper amphibolite- to local granulite-facies conditions during Grenvillian orogenesis (Timmermann et al., 2002), and typically include 30–40 vol% leucosome measured on a series of vertical cross-sections (Slagstad, 2003).

Mafic Enclaves

Elongate mafic enclaves (boudins), ranging in size from a few decimeters up to several meters and concordant with the Grenvillian fabric, are common throughout the Muskoka domain. Nadeau (1990) and Timmermann et al. (2002) interpreted the mafic enclaves to represent disrupted and boudinaged dikes; however, they do not appear to cut any preexisting fabrics or layering, and an origin as xenoliths or cognate inclusions (e.g., Dorais et al., 1990; Kay et al., 1990) cannot be discounted. The composition of the mafic enclaves is unlike that of the Sudbury metadiabase (ca. 1.24 Ga, Dudás et al., 1994), as summarized by Ketchum and Davidson (2000), compatible with the hypothesis that Sudbury metadiabase is restricted to rocks in the footwall of the allochthon boundary thrust (Culshaw et al., 1997; Ketchum and Davidson, 2000).

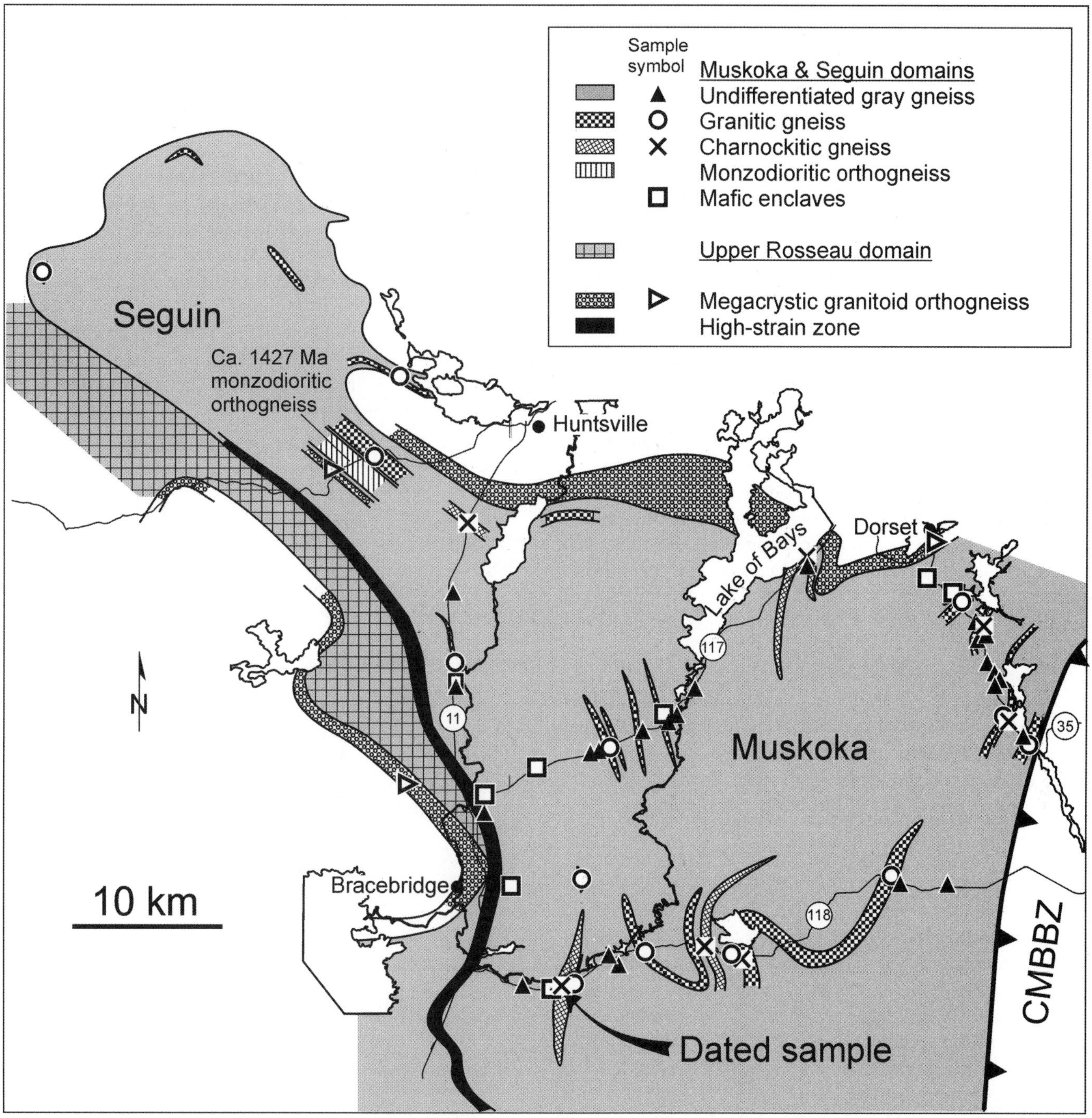

Figure 2. Geological map of the Muskoka domain and the contiguous Seguin subdomain showing dominant lithological units, sample locations, and symbols; major highways, towns, and lakes are included as geographical reference points. UTM (Universal Transverse Mercator) coordinates (zone 17, North American Datum 1983) for dated charnockite sample: 640 050 mE 4980 550 mN. The ca. 1427 Ma age of the monzodioritic orthogneiss is the youngest protolith age obtained from the Muskoka domain (Nadeau and van Breemen, 1998). CMBBZ—Central Metasedimentary Belt boundary thrust zone.

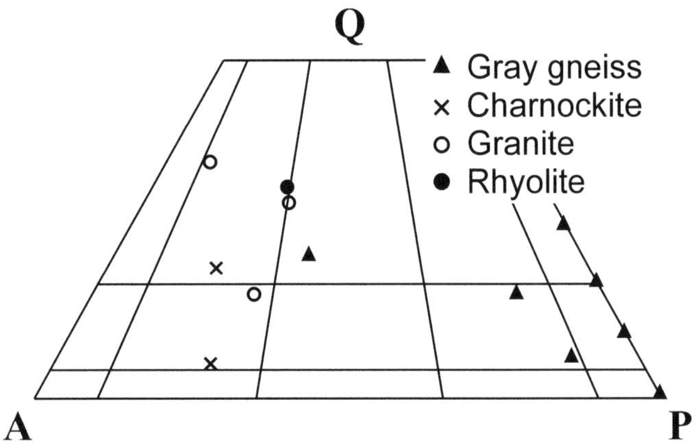

Figure 3. Streckeisen (1973) classification diagram showing representative modal analyses from some of the samples investigated. A—alkali feldspar; P—plagioclase; Q—quartz.

Charnockitic Gneiss

Dark green, two-pyroxene charnockitic gneisses constitute a volumetrically minor part of the Muskoka domain (Fig. 2). They typically form relatively small bodies, up to a few hundred meters across, that may extend along strike for several kilometers, although there is insufficient outcrop to map out individual bodies for more than a few hundred meters. The charnockitic gneisses are recognized in the field based on their green color and textural homogeneity. Leucosome is generally absent, in marked contrast to surrounding rocks. Their modal mineralogy corresponds to two-pyroxene granite to quartz syenite (Fig. 3), but for simplicity they are referred to as charnockite. In one outcrop, the charnockite contains a 0.5 meter–thick mafic sheet that is geochemically indistinguishable from the mafic enclaves. As discussed later, the age of the charnockite is relatively poorly constrained at 1449 +20/–17 Ma, that is, similar to that of the gray gneisses.

Granitic Gneiss

Pink to orange, granitic (granite to quartz syenite, Fig. 3) orthogneiss forms bodies similar in size and distribution to the charnockites (Fig. 2), but do not contain pyroxene. The granites are sparsely migmatitic, with significantly less leucosome than the surrounding gray gneisses. Timmermann et al. (1997) dated one granite body at 1394 ± 13 Ma; however, recent sensitive high-resolution ion microprobe (SHRIMP) data from a geochemically similar granite elsewhere in the Muskoka domain yielded an age of ca. 1470 ± 40 Ma (M. Hamilton, 2002, personal commun.), similar to the gray gneiss and charnockite.

Megacrystic Orthogneiss

Lake of Bays Suite

A unit of K-feldspar megacrystic granitoid orthogneiss, several hundred meters thick and traceable for several tens of kilometers, separates the Upper Rosseau and the Muskoka domains from the underlying domains (Fig. 2). Similar megacrystic orthogneisses do not appear to be present in the Upper Rosseau domain, and they are sparse in the Muskoka and Shawanaga domains. Where dated, the megacrystic orthogneisses have yielded ages similar to those of the gray gneisses (Nadeau and van Breemen, 1998; Timmermann, 1998; McMullen, 1999). In this paper, we refer to these megacrystic orthogneisses as the Lake of Bays suite.

Mann Island Granodiorite and Britt Pluton

Megacrystic orthogneisses are common at lower structural levels of the Central Gneiss Belt in Ontario (Carr et al., 2000). For comparative purposes, we sampled several megacrystic granitoid plutons outside the main study area. In the Britt domain, we sampled the Britt pluton (1456.5 +8.5/–5.9 Ma; van Breemen et al., 1986) and the Mann Island granodiorite (1442 +7/–6 Ma; Corrigan et al., 1994). Both of these plutons intruded older, ca. 1.7 Ga, Laurentian continental crust (Culshaw et al., 1988; Corrigan et al., 1994).

"Marginal Orthogneiss"

In the Shawanaga domain (discussed later), a megacrystic orthogneiss commonly referred to as "marginal orthogneiss" (van Breemen et al., 1986) was sampled. The marginal orthogneiss immediately underlies the Parry Sound shear zone, and has yielded an imprecise age of 1346 +69/–39 Ma (van Breemen et al., 1986).

Shawanaga Domain

The Shawanaga domain is situated in the hangingwall to the Shawanaga shear zone (Figs. 1 and 4) (Culshaw et al., 1994; Ketchum, 1995), which in this area is interpreted to represent an extension of the provincewide allochthon boundary thrust (Ketchum and Davidson, 2000). Culshaw et al. (1988, 1989, 1994, 1997) divided the Shawanaga domain into five units or gneiss associations: the Shawanaga pluton, the Ojibway gneiss association, the Sand Bay gneiss association, the Lighthouse gneiss association, and the "marginal orthogneiss" noted earlier. The Sand Bay gneiss association was the main target of study in the Shawanaga domain, whereas the Ojibway gneiss association and the Shawanaga pluton were investigated only on a reconnaissance level (two samples from each).

Shawanaga Pluton

The Shawanaga pluton (Davidson et al., 1982; Culshaw et al., 1994), dated at ca. 1460 Ma (T.E. Krogh, unpublished data, cited in Culshaw et al., 1994), is a migmatitic, garnet-amphibole-biotite granodiorite, with minor granite and leucogranite, that lies within the Shawanaga shear zone (Ketchum, 1995).

Ojibway Gneiss Association

The Ojibway gneiss association structurally overlies the Shawanaga pluton and extends southeastward to within a few

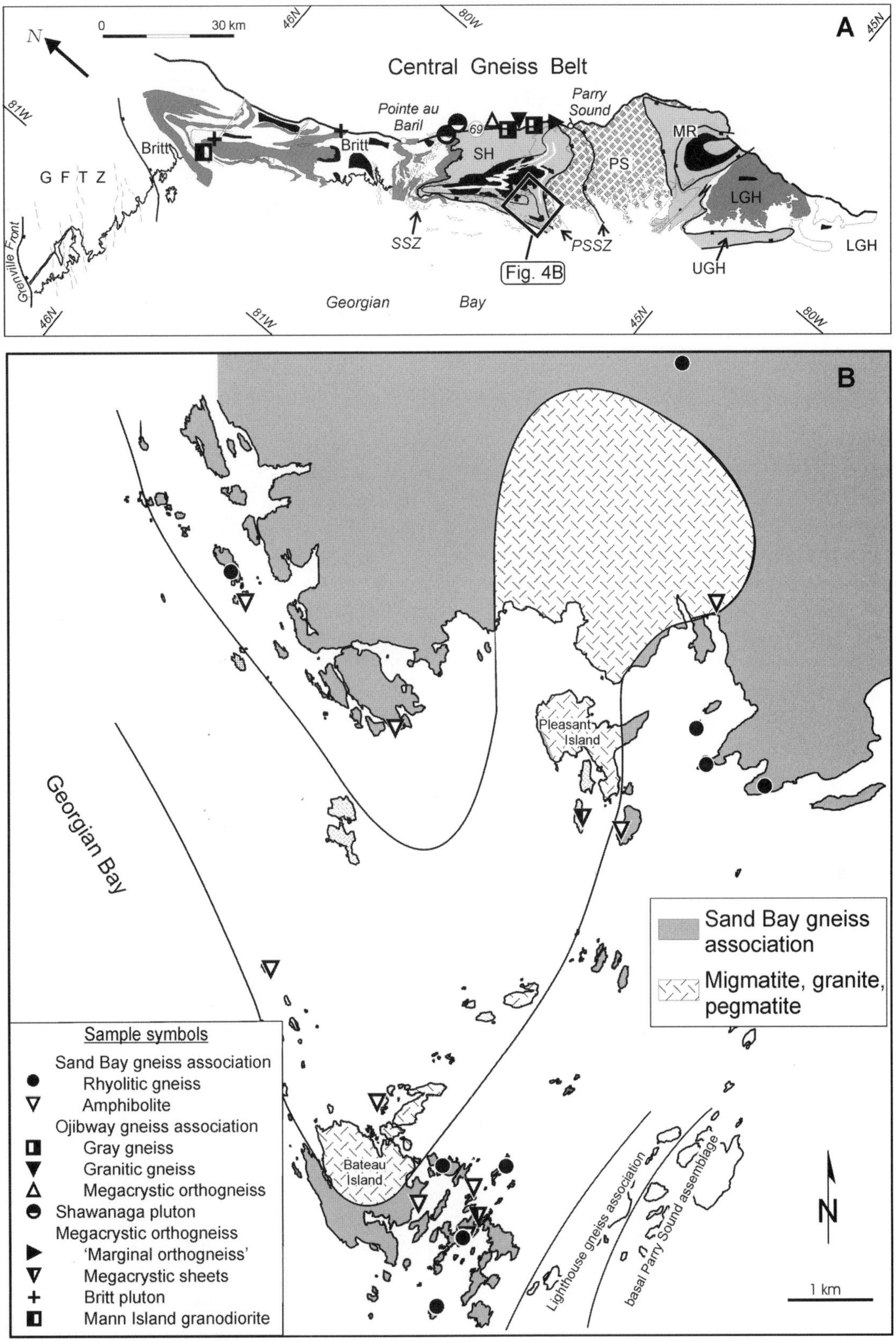

Figure 4. Geological map, sample locations, and symbols of (A) the Central Gneiss Belt along Georgian Bay, from Culshaw and Dostal (2002), and (B) the southwestern part of the Shawanaga domain. Locations of the Lighthouse gneiss association and the basal Parry Sound assemblage after Culshaw and Dostal (2002). The migmatite, granite, pegmatite unit in (B) consists of leucosome-rich migmatites, granites, and heterogeneous pegmatites (Slagstad, 2003). LGH—Lower Go Home domain; MR—Moon River subdomain; PS—Parry Sound domain; PSSZ—Parry Sound shear zone; SH—Shawanaga domain; SSZ—Shawanaga shear zone; UGH—Upper Go Home domain.

kilometers of the Parry Sound domain (Culshaw et al., 1994). Like the Muskoka domain, it is dominated by gray, migmatitic upper amphibolite-facies dioritic to granodioritic orthogneiss (Culshaw et al., 1994), with a protolith age of 1466 ± 11 Ma (T.E. Krogh, unpublished data, cited in Wodicka et al., 2000). The Ojibway gneiss association also contains small granite and megacrystic granitoid bodies similar to those in the Muskoka domain.

Sand Bay Gneiss Association

The Sand Bay gneiss association (Culshaw et al., 1989) comprises a supracrustal assemblage dominated by fine-grained, pink, migmatitic quartzofeldspathic gneisses interpreted to be of volcanic (rhyolitic) origin based on compositional data (Culshaw and Dostal, 1997). Gray quartzofeldspathic gneiss, plagioclase-quartz-biotite schist (Dillon schist, Culshaw et al., 1989), and minor calc-silicate, quartzite, and amphibolite are interlayered with the rhyolites. The contact with the Ojibway gneiss association is isoclinally folded and may be a deformed unconformity or décollement (Culshaw et al., 1994). Detrital zircons from quartzites within the Sand Bay gneiss association yielded ages ranging from >2000 to 1382 Ma (T.E. Krogh, unpublished data, cited in Culshaw and Dostal, 1997), compatible with a Laurentian source. The mineralogical and chemical composition of the Dillon schist suggests derivation from an immature, calcareous, epiclastic sediment (Culshaw and Dostal, 1997), and detrital zircons suggest a contribution from a proximal volcanic source at ca. 1364 Ma (T.E. Krogh, unpublished data, cited in Culshaw and Dostal, 1997). The thin megacrystic granitoid orthogneiss sheets found in a few places in the Sand Bay gneiss association are considered with the marginal orthogneiss in the following discussion.

Amphibolite

Culshaw and Dostal (1997) interpreted the amphibolites in the Sand Bay gneiss association to represent basaltic extrusions, coeval with the rhyolites in the Sand Bay gneiss association. The amphibolites are compositionally similar to the mafic enclaves in the Muskoka domain, and they are referred to collectively as metabasites in this paper.

Lighthouse Gneiss Association

The Lighthouse gneiss association (Culshaw and Dostal, 2002), in the immediate footwall to the Parry Sound shear zone, is the structurally uppermost supracrustal unit of the Shawanaga domain. It includes interlayered pelitic and psammitic gneiss, quartzofeldspathic gneiss, and amphibolite (Wodicka et al., 2000; Culshaw and Dostal, 2002).

Geochemical investigations of the Sand Bay and Lighthouse gneiss associations by Culshaw and Dostal (1997, 2002) suggest the presence of a bimodal suite of continental tholeiitic basalts and rhyolites in the Sand Bay gneiss association, and a progression toward more ocean floor–like (N- to E-type MORB, or mid-ocean ridge basalt) compositions in the lower and upper Lighthouse gneiss association. Culshaw and Dostal (1997, 2002) interpreted this progression to reflect progressive rifting in or behind a continental margin arc; however, it has not yet been established that the Sand Bay and Lighthouse gneiss associations are genetically related. Recent dating of detrital zircons from the Lighthouse gneiss association suggests that the lower unit is related to the Sand Bay gneiss association, whereas the upper unit is distinct (M. Raistrick, 2002, personal commun.).

GEOCHRONOLOGY (CHARNOCKITE)

The charnockites were previously unrecognized and have consequently not been dated. The dated sample was collected in the central part of the Muskoka domain, along Highway 118 (Fig. 2). The sample was crushed in a jaw crusher and reduced to sand-sized particles in a disk mill. Zircons were separated using common separation techniques, including use of a Wilfley water-table, heavy liquids, and magnetic separation. The zircons used for dating were immersed in alcohol and hand-picked with tweezers under a binocular microscope from the nonmagnetic fraction at 1.7A, 10° forward slope, and 0° side tilt on a Frantz isodynamic separator. The zircons selected were abraded using the air abrasion technique of Krogh (1982). A large number of grains were placed in a dissolution capsule with concentrated HF for four hours at 195 °C. Fraction Z6 consists of two rare pristine zircons following this treatment. Isolation of Pb and U by ion exchange column chemistry was performed for fractions Z2–Z5, whereas fractions Z1 and Z6 were loaded directly after dissolution onto a Re filament for mass spectrometry. The isotopic compositions of U and Pb were measured using a single Daly collector with a pulse-counting detector in a solid-source VG354 mass spectrometer at the Royal Ontario Museum in Toronto. The decay constants used are from Jaffey et al. (1971). Uncertainties in text, table, and concordia diagram are given at the 95% confidence level. The discordia line and concordia intercept ages were calculated using the method of Davis (1982).

The zircons are elongate, subhedral to euhedral, and of high quality and resemble primary magmatic zircons (see Poldervaart, 1956). Backscattered electron imaging of mounted and polished representative grains did not reveal any internal growth zoning or core-overgrowth features. The Pb and U isotopic compositions (Table 1) of four colorless, subhedral to euhedral, prismatic grains (Z1–Z4, single-grain analyses), a fraction consisting of three colorless, euhedral grains (Z5), and a fraction consisting of two pristine zircons following HF acid treatment (Z6), were analyzed (Fig. 5). The six analyses are between 1.8 and 3.4% discordant, and a regression through the six points yields an upper intercept age of 1449 +20/–17 Ma and a lower intercept age of 1072 +28/–31 Ma, interpreted to represent the ages of crystallization of the charnockitic magma and Grenvillian metamorphic overprinting, respectively. It is not clear whether the discordance is due to Pb loss during Grenvillian metamorphism or to mixtures of primary and metamorphic

TABLE 1. U-PB ANALYTICAL DATA, CHARNOCKITIC GNEISS

Fraction number and properties[a]	Weight (mg)[b]	U (ppm)	Pb^{rad} (ppm)[c]	Th/U	Total common Pb (pg)[d]	$^{207}Pb^{d}/^{204}Pb$	Corrected atomic ratios[e] $^{206}Pb/^{238}U$	2σ	$^{207}Pb/^{235}U$	2σ	Age (Ma) $^{207}Pb/^{206}Pb$	2σ	Disc.[f] (%)
Z1 subh clr 2:1 pr (1)	0.0036	15	3.5	0.41	0.38	195.3	0.2301	0.0012	2.770	0.018	1367.6	6.9	2.7
Z2 subh clr 2:1 pr (1)	0.0140	17	3.8	0.37	0.74	399.8	0.2233	0.0010	2.646	0.013	1336.5	4.9	3.1
Z3 subh clr 3:1 pr incl (1)	0.0070	13	3.0	0.38	1.14	114.6	0.2203	0.0012	2.582	0.022	1315.7	13.3	2.7
Z4 euh clr 2:1 pr (1)	0.0060	32	7.3	0.34	0.62	390.6	0.2191	0.0007	2.572	0.010	1318.2	4.7	3.4
Z5 euh clr 3:1 pr incl (3)	0.0080	41	8.6	0.39	2.49	155.3	0.2034	0.0005	2.281	0.012	1229.2	8.3	3.2
Z6 subh clr 2:1 pr HF (2)	0.0066	21	5.0	0.39	0.38	490.1	0.2361	0.0008	2.875	0.011	1389.1	3.9	1.8

[a]Abbreviations: Z—zircon; subh—subhedral; euh—euhedral; clr—clear or colorless; pr—prism; incl—fluid and/or opaque inclusions; 2:1, 3:1—length:breadth ratio; number in parentheses indicates number of grains analyzed.
[b]Weighing error ca. ±50%.
[c]Total radiogenic Pb after correction for blank, common Pb, and spike.
[d]Measured.
[e]Ratios corrected for blank and spike Pb, fractionation, and initial common Pb calculated for the age of the sample after Stacey and Kramers (1975).
[f]Percent discordance for the given $^{207}Pb/^{206}Pb$ age.

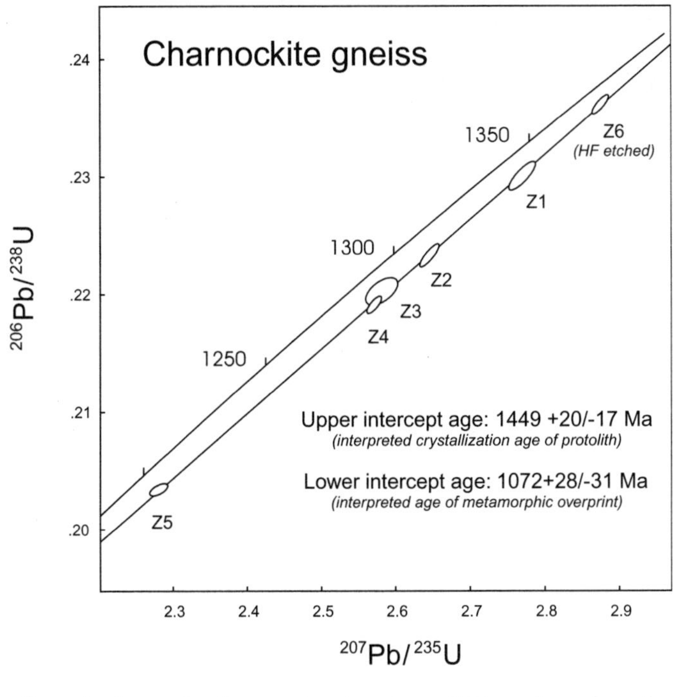

Figure 5. Concordia diagram showing zircon U-Pb analyses from the charnockitic gneiss. The sample location is shown in Figure 2.

(recrystallized?) zircon within individual grains. Both the upper and the lower intercept age are similar to previously reported crystallization and metamorphic ages from the surrounding gray gneisses (e.g., Timmermann et al., 1997).

GEOCHEMISTRY

Analytical and Sampling Procedures

Major and trace elements were determined on fused glass beads and pressed powder pellets, respectively, using standard X-ray fluorescence (XRF) techniques at St. Mary's University, Halifax. Rare earth elements (REE), Hf, and Th were determined at Memorial University, Newfoundland, by inductively coupled plasma–mass spectrometry (ICP-MS) using a Na_2O_2 sintering technique. Dostal et al. (1986) and Longerich et al. (1990) discussed the analytical procedures, uncertainties, and precision of the XRF and ICP-MS analyses, respectively. Analytical results from selected representative samples are presented in Table 2, A–D. A full data set (99 analyses) is presented in Slagstad (2003), and can be obtained in electronic format from the first author on request.

Samples were collected from variably migmatitic outcrops. We have no field, petrographic, or geochemical data to suggest that sparsely migmatitic outcrops lost melt; in particular, the compositions of these rocks are not residual (see Solar and Brown, 2001). The more typical strongly migmatitic outcrops present a greater challenge, because the bulk compositions may have been affected by gain or loss of melt. Samples from these outcrops were selected from mesosomes that lack leucosome on the hand sample scale, although they generally show grain-scale evidence (see Sawyer, 1999) of melting (Timmermann et al., 2002). On the basis of textural evidence, Timmermann et al. (2002) concluded that partial melting in these rocks involved incongruent breakdown of biotite, producing abundant new hornblende. Evidence of this process is lacking from the sampled mesosomes (Slagstad, 2003), suggesting that the mesosomes sampled did not undergo significant melting. We conclude that only small amounts of melt, if any, were added to or segregated from the sampled mesosomes, and that they therefore represent a good approximation to protolith compositions.

Element Mobility

The chemical compositions of highly metamorphosed orthogneisses have been used to determine the petrogenesis and paleotectonic setting of their igneous protoliths, based on the assumption that the concentrations of the elements under consideration remained essentially unchanged during metamorphism and

TABLE 2A. GRAY GNEISSES, MUSKOKA DOMAIN, OJIBWAY GNEISS ASSOCIATION, AND SHAWANAGA PLUTON: GEOCHEMICAL DATA FROM SELECTED REPRESENTATIVE SAMPLES

	Muskoka domain										Ojibway gneiss association		Shawanaga pluton	
	M2707-1	M050711	99.16-Pw	M05077	M130717	M04073-HI	M13079	M25072-1	99.16-Pv	M06073-1	S11064-2	S31052-h	S31057-h	S31057
Easting	668 500	669 380	647 550	671 250	668 800	666 000	668 080	632 700	647 550	634 850	566 580	567 280	557 350	557 350
Northing	5005 580	5002 650	5000 500	4998 830	5003 700	4989 100	5005 150	5002 400	5000 500	4994 050	5033 780	5032 750	5042 730	5042 730
SiO_2 (wt%)	46.72	49.03	53.29	54.72	56.32	58.16	59.69	62.75	69.91	71.29	60.68	63.22	65.38	66.43
TiO_2	2.03	1.56	1.07	1.10	1.06	0.92	0.84	0.77	0.67	0.36	0.83	0.74	0.56	0.53
Al_2O_3	18.31	19.61	17.30	18.86	19.21	18.05	18.43	18.43	15.36	14.62	17.02	16.67	16.55	17.08
Fe_2O_3	12.34	10.73	9.34	7.96	7.55	6.93	5.71	3.96	3.16	2.43	6.17	5.72	4.93	4.06
MnO	0.18	0.13	0.15	0.12	0.12	0.09	0.11	0.07	0.06	0.04	0.11	0.12	0.09	0.06
MgO	4.30	4.42	4.36	3.82	2.91	3.10	1.82	1.20	0.81	0.73	2.04	1.75	1.37	1.09
CaO	7.95	8.63	7.25	7.06	6.17	6.01	4.90	3.06	2.10	1.99	4.65	3.98	3.62	3.29
Na_2O	3.76	4.22	5.05	4.98	4.98	4.58	4.96	4.68	3.94	3.80	4.48	4.71	3.37	3.14
K_2O	2.47	1.61	1.92	1.76	1.97	2.00	2.39	5.11	4.44	4.36	2.92	2.66	3.83	4.96
P_2O_5	1.04	0.53	0.37	0.42	0.38	0.33	0.30	0.20	0.17	0.11	0.26	0.24	0.19	0.15
LOI	0.45	0.45	0.48	0.31	0.17	0.48	0.20	0.40	0.20	0.10	0.50	0.29	0.29	0.29
Total	99.55	100.92	100.58	101.10	100.85	100.65	99.35	100.63	100.82	99.82	99.66	100.10	100.17	101.07
A/CNK	0.79	0.80	0.73	0.82	0.89	0.88	0.94	0.98	1.02	1.00	0.90	0.93	1.02	1.03
A/NK	2.08	2.27	1.67	1.88	1.87	1.87	1.72	1.40	1.37	1.34	1.62	1.57	1.71	1.63
$FeO_t/(FeO_t + MgO)$	0.72	0.69	0.66	0.65	0.70	0.67	0.74	0.75	0.78	0.75	0.73	0.75	0.76	0.77
Ba (ppm)	1469	904	647	970	703	672	790	3631	1266	979	760	630	1138	2039
Rb	50	34	31	34	45	51	85	93	134	142	86	103	107	100
Sr	882	954	636	1094	844	735	655	651	417	430	496	369	370	374
Y	40	35	39	22	27	41	30	27	22	29	35	32	45	20
Zr	578	352	308	313	398	270	410	572	371	192	287	263	259	279
Nb	14	12	8	8	11	8	14	7	11	10	13	13	13	8
Th	1.94	1.54	3.07	1.78	1.67	3.77	10.57	5.65	12.63	14.89	8.42	9.54	12.10	2.33
Pb	b.d.	8	18	31	5	8	14	18	27	20	13	12	15	18
Ga	23	27	21	26	29	23	26	20	18	19	23	21	20	20
Ni	6	7	19	332	9	22	5	b.d.	0	b.d.	7	13	4	4
V	239	216	b.d.	155	142	138	98	88	b.d.	50	1			
13	104	85	83											
Cr	4	b.d.	b.d.	47	16	30	b.d.	b.d.	b.d.	4	9	9	b.d.	7
Hf	12.78	11.36	10.39	8.93	12.15	7.78	13.04	14.23	13.23	5.11	7.96	7.91	7.16	7.78
U	b.d.	b.d.	2	b.d.	b.d.	b.d.	b.d.	1	3	3	2	3	3	3
La	42.37	33.71	40.13	29.30	43.81	46.42	56.42	36.53	47.38	43.52	32.54	33.87	42.74	26.93
Ce	102.98	83.08	91.53	69.22	98.64	105.37	123.96	74.16	102.96	84.11	70.21	69.52	88.59	51.59
Pr	14.25	11.55	11.82	9.72	12.33	14.06	14.70	9.00	11.36	9.57	8.71	8.00	11.38	6.05
Nd	65.51	50.56	47.84	42.03	48.13	57.95	55.22	36.47	41.85	34.73	34.85	29.70	45.62	23.67
Sm	15.55	11.17	10.03	8.85	9.18	11.27	10.48	7.34	7.69	5.89	7.97	6.46	11.72	5.01
Eu	3.95	2.36	2.19	2.16	1.98	2.02	1.81	2.47	1.62	0.96	1.71	1.38	1.51	1.98
Gd	14.12	10.00	8.45	7.37	7.26	9.34	8.32	5.96	5.87	4.70	7.10	5.65	10.69	4.38
Tb	1.92	1.51	1.20	1.04	1.07	1.39	1.21	0.82	0.84	0.70	1.07	0.85	1.69	0.60
Dy	11.07	9.17	7.07	6.06	6.43	8.49	7.04	4.80	4.96	4.30	6.55	5.25	10.47	3.62
Ho	1.94	1.81	1.43	1.21	1.30	1.68	1.40	0.87	1.01	0.87	1.18	0.94	1.83	0.65
Er	5.75	5.56	4.34	3.67	4.00	4.82	4.26	2.75	3.18	2.60	3.68	3.12	5.66	2.05
Tm	0.71	0.73	0.58	0.48	0.52	0.70	0.57	0.38	0.43	0.40	0.51	0.46	0.74	0.29
Yb	4.21	4.45	3.51	2.94	3.31	4.53	3.65	2.49	2.68	2.73	3.32	3.18	4.55	2.01
Lu	0.71	0.67	0.53	0.43	0.50	0.65	0.56	0.44	0.41	0.41	0.58	0.60	0.77	0.37

Note: Coordinates apply to UTM Zone 17 NAD83.

A/CNK–mole $Al_2O_3/(CaO + Na_2O + K_2O)$; A/NK–mole $Al_2O_3/(Na_2O + K_2O)$; b.d.–below detection limit; LOI—loss on ignition.

TABLE 2B. CHARNOCKITE, GRANITE, AND RHYOLITE: GEOCHEMICAL DATA FROM SELECTED REPRESENTATIVE SAMPLES

	Charnockite Muskoka				Granite Muskoka						Granite Ojibway	Rhyolite Sand Bay					
	M100723-2	M080721-1	M300617	M18072	M04074-HI	M07075-1	M150724	M100723-1	M080711	M150712	S11063-2	S30052	S30051	S04066-4	S06064	S02063	S27059
Easting	651 650	656 500	640 050	633 650	662 350	667 100	629 200	651 650	643 300	627 450	566 780	556 580	557 080	559 300	559 500	559 150	556 330
Northing	4984 450	5010 300	4982 550	5012 800	4989 650	5007 600	5022 300	4984 450	4998 150	5017 100	5033 430	5016 230	5017 050	5021 930	5021 550	5025 980	5017 050
SiO_2 (wt%)	57.31	61.86	63.23	67.66	64.33	65.68	69.97	73.41	76.54	76.67	74.62	67.13	75.84	76.75	77.62	78.08	78.78
TiO_2	1.14	0.73	0.77	0.56	0.62	0.45	0.25	0.35	0.21	0.11	0.25	0.76	0.16	0.07	0.08	0.14	0.17
Al_2O_3	16.12	17.08	15.49	14.22	17.97	16.99	14.95	13.44	11.96	12.30	12.84	15.57	12.64	12.57	12.50	11.74	11.59
Fe_2O_3	8.99	6.31	6.53	5.81	3.69	4.00	2.13	3.57	2.74	1.81	1.47	3.24	0.99	1.07	0.79	0.87	1.10
MnO	0.21	0.18	0.18	0.14	0.09	0.10	0.04	0.06	0.03	0.02	0.02	0.06	0.01	0.00	0.00	0.04	0.01
MgO	1.36	0.47	0.52	0.21	0.87	0.39	0.22	0.18	0.06	0.07	0.32	0.94	0.17	0.02	b.d.	0.03	0.01
CaO	2.66	2.92	2.18	1.49	2.56	1.30	1.12	0.83	0.26	0.32	0.80	1.52	0.37	0.81	0.47	0.34	0.37
Na_2O	4.80	4.90	4.95	4.25	4.76	4.48	3.58	3.41	2.23	1.58	2.93	4.38	3.22	2.95	3.47	3.57	3.47
K_2O	5.05	4.87	5.28	5.58	5.43	6.81	6.07	5.76	7.16	7.88	5.65	5.23	5.71	5.24	4.78	4.50	4.18
P_2O_5	0.34	0.22	0.22	0.08	0.18	0.07	0.05	0.07	0.02	0.01	0.03	0.21	0.03	0.01	0.02	0.02	0.01
LOI	0.39	0.00	0.28	0.16	0.47	0.29	0.27	0.00	0.10	0.19	0.17	0.18	0.08	0.13	0.25	0.00	0.08
Total	98.37	99.54	99.63	100.16	100.97	100.55	98.65	101.08	101.31	100.96	99.10	99.23	99.21	99.63	99.98	99.32	99.77
A/CNK	0.89	0.92	0.87	0.90	0.98	0.99	1.03	1.01	1.01	1.05	1.04	1.00	1.04	1.05	1.07	1.03	1.06
A/NK	1.21	1.29	1.12	1.10	1.32	1.16	1.20	1.14	1.05	1.11	1.18	1.21	1.11	1.20	1.15	1.10	1.14
FeO_t/(FeO_t + MgO)	0.86	0.92	0.92	0.96	0.79	0.90	0.90	0.95	0.98	0.96	0.80	0.76	0.84	0.98	n.d.	0.96	0.99
Ba (ppm)	2358	2232	1254	648	2489	720	1077	765	383	325	963	1254	531	362	183	63	41
Rb	43	47	56	66	88	110	130	89	211	108	161	140	201	177	195	284	155
Sr	229	306	185	26	403	68	128	64	72	162	134	239	52	105	11	11	15
Y	140	45	59	120	57	32	29	81	52	84	28	36	11	15	17	34	17
Zr	1916	1364	1526	1249	619	789	287	694	551	370	228	443	109	143	173	225	222
Nb	57	16	27	40	11	11	10	14	23	3	6	19	12	7	17	35	7
Th	3.31	3.82	2.01	4.36	11.41	8.25	15.46	5.71	7.89	3.33	14.76	12.18	13.60	11.73	16.17	26.94	17.22
Pb	3	10	13	10	17	19	24	26	11	26	27	24	16	17	15	14	23
Ga	30	28	29	30	20	19	17	28	23	31	14	22	18	20	22	27	20
Ni	b.d.	b.d.	b.d.	b.d.	5	373	b.d.	b.d.	b.d.	10	b.d.	b.d.	204	b.d.	b.d.	b.d.	b.d.
V	99	67	75	53	72	46	30	40	27	18	33	86	23	15	13	16	13
Cr	b.d.	b.d.	b.d.	b.d.	b.d.	11	b.d.	b.d.	b.d.	b.d.	4	b.d.	b.d.	b.d.	b.d.	b.d.	b.d.
Hf	45.91	29.93	40.08	36.81	13.38	17.49	7.13	14.43	12.41	7.70	6.49	15.18	5.51	5.79	10.67	17.87	14.48
U	3	3	2	4	2	6	4	4	4	3	4	4	3	3	6	8	4
La	136.97	75.10	92.32	177.43	119.76	73.98	21.40	101.32	50.64	54.55	46.10	46.83	34.60	15.03	11.00	50.51	21.19
Ce	302.19	152.22	206.99	398.72	219.86	141.78	73.33	247.33	104.39	122.31	91.95	116.78	53.35	37.46	30.75	72.23	59.66
Pr	40.53	19.27	27.97	51.13	25.52	16.17	5.37	25.60	12.90	13.39	10.57	13.54	4.37	5.02	4.13	5.47	7.78
Nd	166.91	78.48	115.51	211.28	97.92	59.02	22.10	95.73	47.34	53.44	39.90	53.55	11.91	20.53	16.84	13.61	31.58
Sm	33.62	14.36	23.52	46.23	16.98	8.84	5.40	17.99	10.04	13.27	8.02	11.18	1.66	5.41	4.85	2.08	8.36
Eu	5.73	6.80	5.14	4.11	2.88	1.13	1.24	2.32	0.35	0.80	0.93	2.14	0.24	0.20	0.05	0.13	0.05
Gd	30.68	12.06	19.18	38.96	13.54	6.51	5.12	14.80	8.04	12.30	6.36	8.79	1.21	4.63	4.07	1.75	6.81
Tb	4.60	1.80	2.93	5.64	1.96	0.89	0.83	2.52	1.32	2.08	0.82	1.30	0.19	0.73	0.74	0.34	1.04
Dy	28.32	11.00	17.42	33.34	11.69	5.23	5.30	15.88	8.44	13.95	4.49	7.64	1.26	4.33	5.12	2.72	6.10
Ho	5.43	2.22	3.45	5.76	2.28	1.00	1.00	3.17	1.54	2.88	0.70	1.51	0.29	0.82	1.09	0.78	1.17
Er	15.66	6.49	10.46	17.16	6.44	2.97	3.13	9.18	4.67	9.92	1.91	4.71	1.07	2.43	3.78	3.75	3.52
Tm	2.27	0.97	1.37	2.20	0.92	0.46	0.39	1.27	0.71	1.34	0.23	0.64	0.18	0.31	0.55	0.76	0.46
Yb	14.80	6.70	8.79	13.83	6.01	3.30	2.47	7.62	4.77	8.23	1.37	4.07	1.37	1.92	3.49	6.79	2.82
Lu	2.25	1.08	1.40	2.41	0.87	0.57	0.43	1.02	0.75	1.30	0.23	0.63	0.24	0.27	0.51	1.35	0.40

Note: Coordinates apply to UTM Zone 17 NAD83.

A/CNK–mole Al_2O_3/(CaO + Na_2O + K_2O); A/NK–mole Al_2O_3/(Na_2O + K_2O); b.d.–below detection limit; LOI—loss on ignition.

TABLE 2C. MEGACRYSTIC GRANITOID ORTHOGNEISSES: GEOCHEMICAL DATA FROM SELECTED REPRESENTATIVE SAMPLES

	Lake of Bays suite				Ojibway g.a.	Mann Island granodiorite	Britt pluton		"Marginal orthogneiss"	Megacrystic sheets Sand Bay gneiss association	
	M11072	M13071	M15078	2M0706-2-1	S31056-mg	2S28062	2S28063	2S28064	2S30062	S17066	S25061-4
Easting	629 700	665 150	624 800	625 200	561 280	519 080	521 950	—	573 630	556 730	557 980
Northing	4995 950	5011 450	5016 180	5005 600	5038 830	5081 200	5081 400	—	5027 580	5016 500	5020 950
SiO_2 (wt.%)	69.30	60.92	58.67	70.01	60.45	73.02	67.36	62.41	67.27	63.67	71.86
TiO_2	0.40	0.79	0.86	0.40	0.76	0.11	0.99	1.22	0.42	0.61	0.40
Al_2O_3	15.42	17.85	18.64	15.38	17.92	14.82	13.92	14.04	14.94	16.66	14.66
Fe_2O_3	3.17	6.23	6.47	3.28	6.12	1.02	6.24	7.74	3.69	4.78	2.44
MnO	0.05	0.09	0.11	0.07	0.10	0.01	0.15	0.18	0.08	0.08	0.03
MgO	0.83	1.81	1.82	1.00	1.66	0.16	1.11	1.34	1.15	1.54	0.64
CaO	2.79	4.88	5.23	2.45	4.77	1.14	2.69	3.43	2.92	3.76	2.54
Na_2O	3.13	3.40	4.08	4.11	3.79	3.82	3.24	3.57	4.16	4.22	5.14
K_2O	4.44	3.15	3.51	4.53	3.59	5.54	4.94	4.12	3.79	3.59	1.10
P_2O_5	0.10	0.19	0.29	0.12	0.22	0.03	0.32	0.41	0.13	0.19	0.11
LOI	0.19	0.47	0.39	0.20	0.36	0.58	0.49	0.20	0.30	0.56	0.27
Total	99.89	99.78	100.07	101.54	99.73	99.66	100.96	98.46	98.54	99.66	99.18
A/CNK	1.03	1.00	0.93	0.95	0.95	1.03	0.89	0.85	0.92	0.94	1.03
A/NK	1.56	1.99	1.78	1.32	1.78	1.21	1.31	1.36	1.37	1.54	1.53
$FeO_t/(FeO_t + MgO)$	0.77	0.76	0.76	0.75	0.77	0.85	0.83	0.84	0.74	0.74	0.77
Ba (ppm)	973	1573	1835	662	1739	727	1884	2185	536	1169	406
Rb	129	65	84	34	79	81	111	81	118	91	70
Sr	240	405	554	228	505	216	236	287	215	464	311
Y	24	20	33	37	23	54	75	73	43	33	35
Zr	180	342	380	162	385	59	586	607	136	263	205
Nb	11	13	13	11	13	3	23	26	16	10	12
Th	9.17	2.42	6.70	0.87	6.92	3.61	9.08	8.27	8.17	8.92	18.47
Pb	19	11	9	9	13	6	12	8	15	17	20
Ga	20	23	23	23	22	16	19	21	22	21	20
Ni	3	10	b.d.	20	6	5	14	13	5	4	b.d.
V	62	120	111	53	103	23	102	125	56	85	52
Cr	5	14	b.d.	17	12	b.d.	4	5	9	9	b.d.
Hf	5.25	9.07	10.05	9.52	10.20	15.92	12.18	16.87	9.81	7.05	8.37
U	3	1	1	2	1	3	4	3	5	2	3
La	43.01	29.20	37.52	32.91	45.21	74.30	56.50	81.60	47.74	35.57	37.11
Ce	83.79	57.79	81.37	72.11	87.47	151.16	122.28	167.77	90.80	72.72	73.71
Pr	9.57	6.68	10.07	9.36	9.66	18.07	15.01	20.58	10.88	8.73	8.55
Nd	35.56	27.68	40.94	40.45	35.04	71.66	59.09	82.40	41.62	33.64	32.01
Sm	5.96	5.70	9.15	9.88	6.62	15.21	13.06	17.73	7.64	7.28	7.58
Eu	1.31	2.38	2.20	2.16	1.97	3.37	3.37	4.71	1.55	1.48	1.03
Gd	4.73	5.03	8.10	10.64	5.35	12.26	14.03	18.55	6.80	6.25	6.57
Tb	0.69	0.70	1.15	1.52	0.75	1.86	2.18	2.72	1.15	0.92	1.00
Dy	4.06	4.01	6.84	8.99	4.45	11.18	13.96	16.84	7.08	5.74	5.95
Ho	0.80	0.73	1.22	1.37	0.76	1.89	2.33	2.70	1.29	1.02	0.96
Er	2.25	2.21	3.74	3.65	2.38	5.25	6.97	7.74	3.84	3.19	2.88
Tm	0.32	0.28	0.49	0.48	0.32	0.73	1.04	1.10	0.58	0.45	0.40
Yb	2.14	1.83	3.12	2.99	2.09	4.71	6.98	7.17	3.95	2.92	2.63
Lu	0.32	0.34	0.54	0.50	0.39	0.83	1.22	1.25	0.61	0.50	0.48

Note: Coordinates apply to UTM Zone 17 NAD83.

A/CNK—mole $Al_2O_3/(CaO + Na_2O + K_2O)$; A/NK—mole $Al_2O_3/(Na_2O + K_2O)$; b.d.—below detection limit; LOI—loss on ignition.

TABLE 2D. METABASITES: GEOCHEMICAL DATA FROM SELECTED REPRESENTATIVE SAMPLES

	Mafic enclaves, Muskoka domain				Mafic sheet in charnockite	Amphibolite, Sand Bay gneiss association				
	M070713	M10079	M140720	M03083-M	2M1406-3	S13067	S20066-2	S17065	S14063	S29055-A
Easting	664 700	639 480	615 050	632 700	656 500	558 400	555 650	556 650	556 100	554 100
Northing	5009 300	4982 350	5018 150	5002 400	5010 300	5020 800	5017 780	5016 300	5016 630	5023 280
SiO_2 (wt%)	45.10	45.89	46.64	46.64	48.69	46.11	47.24	47.28	48.79	50.99
TiO_2	2.67	1.43	2.44	1.71	1.93	2.22	0.85	2.65	1.57	1.08
Al_2O_3	15.36	16.61	15.47	16.20	13.48	17.68	16.03	17.89	14.92	17.13
Fe_2O_3	16.07	12.87	15.30	12.59	14.88	12.74	12.85	13.47	13.42	11.14
MnO	0.22	0.19	0.22	0.20	0.23	0.16	0.18	0.18	0.22	0.15
MgO	5.99	7.71	5.70	7.33	5.99	6.99	7.99	4.34	5.95	4.68
CaO	8.09	9.07	7.52	8.75	9.42	7.97	9.68	9.42	8.13	8.59
Na_2O	4.15	3.23	2.99	3.49	4.09	3.66	3.21	3.98	4.24	4.16
K_2O	1.02	1.79	2.74	1.96	0.87	1.41	1.65	0.95	2.01	1.44
P_2O_5	0.70	0.17	0.35	0.34	0.26	0.38	0.10	0.44	0.34	0.35
LOI	0.19	0.90	0.60	0.82	0.50	0.84	0.62	0.18	0.44	0.53
Total	99.56	99.86	99.97	100.03	100.33	100.16	100.40	100.77	100.02	100.24
FeO_t/(FeO_t + MgO)	0.71	0.60	0.71	0.61	0.69	0.62	0.59	0.74	0.67	0.68
Mg #	0.42	0.54	0.42	0.54	0.44	0.52	0.55	0.39	0.47	0.45
Ba (ppm)	107	b.d.	533	286	66	77	b.d.	b.d.	150	157
Rb	18	35	64	47	15	26	32	17	87	28
Sr	543	304	253	484	254	496	235	342	305	515
Y	28	21	26	23	29	21	20	29	31	25
Zr	194	75	168	121	131	137	51	168	182	95
Nb	9	3	12	4	6	10	1	7	7	5
Th	0.69	1.05	1.82	0.68	1.36	0.90	0.27	0.00	1.26	1.92
Pb	5	3	5	3	b.d.	6	6	3	14	4
Ga	20	19	22	19	22	19	17	23	20	20
Ni	69	110	59	126	41	110	127	26	56	21
V	339	233	302	245	302	251	178	325	255	204
Cr	54	70	55	120	93	156	102	66	43	44
Hf	4.59	2.03	5.03	2.65	4.45	3.28	1.45	5.88	6.30	1.86
La	18.78	10.00	16.33	12.33	12.43	10.89	2.93	9.83	18.91	15.03
Ce	46.42	22.27	37.61	29.62	28.96	27.04	7.47	27.10	45.41	34.24
Pr	6.95	3.08	5.13	4.24	4.23	3.85	1.17	4.15	6.28	4.85
Nd	33.78	13.95	24.42	19.76	20.12	17.64	5.90	20.55	27.79	21.91
Sm	7.97	3.61	6.74	4.66	5.94	4.76	2.12	6.18	7.06	5.98
Eu	2.76	1.35	2.11	1.72	1.90	1.91	0.82	2.19	1.92	1.66
Gd	7.95	4.06	7.24	4.95	6.21	5.00	2.92	6.85	7.18	5.97
Tb	1.16	0.68	1.10	0.76	1.07	0.74	0.52	1.06	1.12	0.88
Dy	6.99	4.38	6.64	4.74	6.95	4.51	3.67	6.59	7.14	5.34
Ho	1.38	0.88	1.23	0.93	1.27	0.79	0.71	1.31	1.48	0.91
Er	3.84	2.56	3.78	2.62	3.68	2.35	2.32	3.91	4.60	2.74
Tm	0.53	0.37	0.47	0.38	0.52	0.31	0.32	0.50	0.61	0.36
Yb	3.36	2.44	2.93	2.42	3.38	1.91	2.14	3.00	3.84	2.22
Lu	0.49	0.36	0.50	0.35	0.56	0.32	0.37	0.45	0.59	0.39

Note: Coordinates apply to UTM Zone 17 NAD83.
b.d.—below detection limit; LOI—loss on ignition; MG #—mole MgO/(MgO + FeO_t).

deformation. In general, low-field-strength elements (Sr, K, Rb, and Ba) are considered mobile, whereas high-field-strength elements (REE, Y, Th, Zr, Hf, Ti, Nb, and P) and some transition metals (Ni, V, and Cr) are considered immobile (Rollinson, 1993). Element mobility should produce widely scattered patterns in normalized-element plots; however, most elements display no such scatter in the primitive mantle-normalized diagrams presented here (see next section), and the rocks have compositions similar to those of fresh, unmetamorphosed rocks (see following sections). Notable exceptions are Th in the gray gneisses and Ba in the metabasites that display significant variation, possibly because of mobilization. In addition, the distinct differences between suites of rocks in these diagrams indicate only limited element mobility. Another way to assess whether a suite of rocks shows a coherent relationship that can be interpreted in terms of magmatic evolution is to use Pearce element ratios (Pearce, 1968), in which the molar ratios of compatible/incompatible elements are plotted against each other. Pearce element ratio diagrams for the gray gneisses and metabasites, presented in the following sections, indicate that the major element variation within these two suites is similar to that expected for a related suite of magmatic rocks, and the same conclusion can be drawn from Harker variation diagrams.

Despite the potential problems associated with both migmatization and element mobility, the consistency of the geochemical data and their similarity to those for unmetamorphosed magmatic rocks suggest a reasonable approximation to original protolith compositions, allowing for a plausible interpretation of petrogenesis and paleotectonic setting.

Muskoka and Seguin Domains

Gray Gneiss

The gray gneisses range in SiO_2 from 46.7 to 71.3 wt%. They follow a calc-alkaline trend on the AFM ($A = Na_2O + K_2O$, $F = FeO + Fe_2O$, $M = MgO$) diagram of Irvine and Baragar (1971) and have $FeO_t/(FeO_t + MgO)$ ratios <0.80 (Fig. 6, A), typical of calc-alkaline rocks (Frost et al., 2001). In the alumina-saturation diagram, the rocks straddle the boundary between metaluminous and slightly peraluminous compositions (Fig. 6, D), with A/CNK (mole $Al_2O_3/(CaO + Na_2O + K_2O)$) < 1.1, and A/NK (mole $Al_2O_3/(Na_2O + K_2O)$) > 1.2. Harker diagrams (Fig. 7) display relatively well-defined negative trends for most major elements (positive for K_2O) with respect to SiO_2, although there is some scatter in K_2O and Al_2O_3 values and substantial scatter in Na_2O values. The scatter in K_2O and Na_2O may reflect mobilization by secondary processes, although the preservation of a clear positive trend for K_2O suggests limited mobility. Similar scatter in Al_2O_3 was noted by Clynne (1990) in fresh mafic to

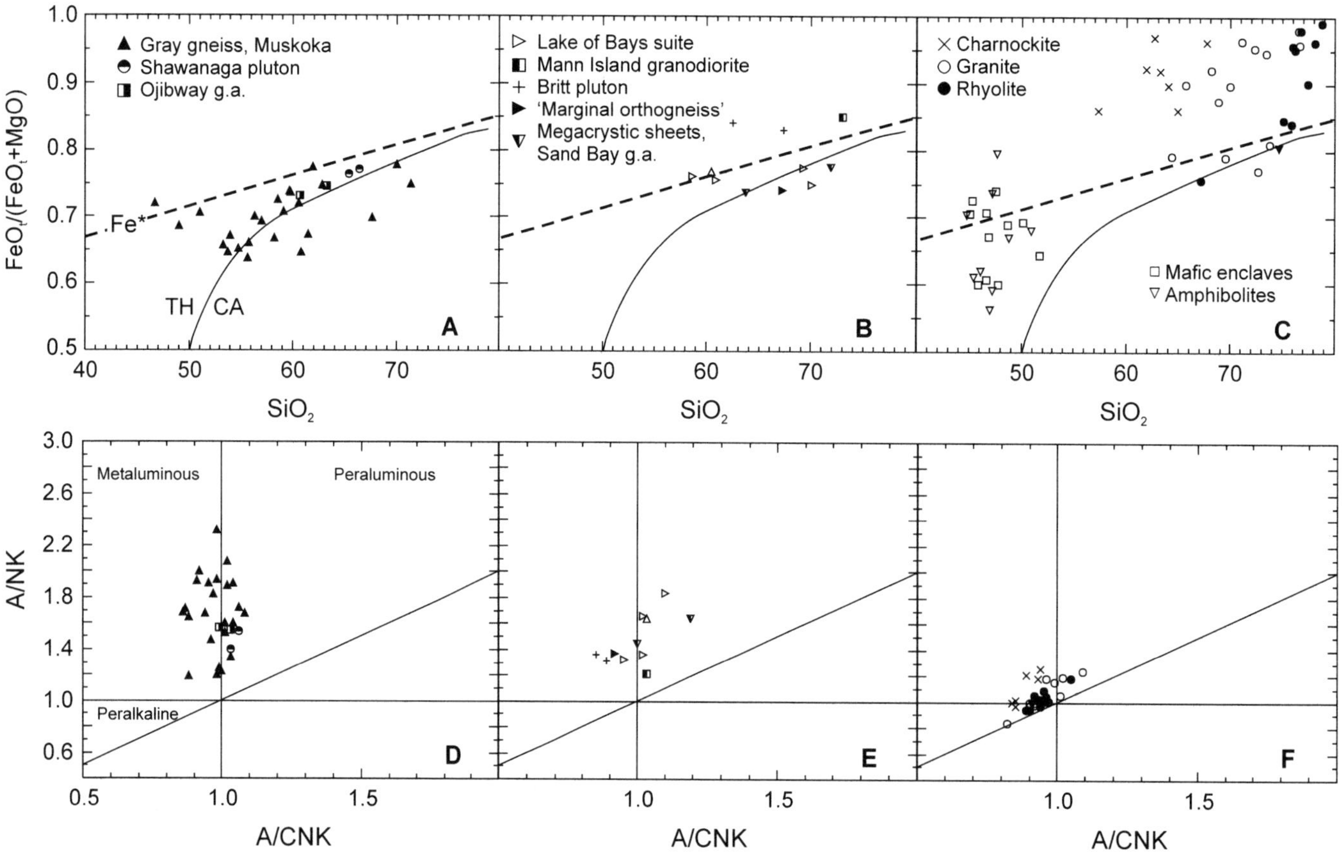

Figure 6. (A–C) Silica contents versus iron enrichment ($FeO_t/(FeO_t+MgO)$) and (D–F) A/CNK (mole $Al_2O_3/(CaO + Na_2O + K_2O)$) versus A/NK (mole $Al_2O_3/(Na_2O + K_2O)$) diagrams. (A, D) gray gneiss, Muskoka domain, Ojibway gneiss association, and Shawanaga pluton; (B, E) Lake of Bays suite, "marginal orthogneiss," megacrystic sheets in Sand Bay gneiss association, Britt pluton, Mann Island granodiorite; (C, F) charnockite, granite, rhyolite, and metabasites. Boundaries after Miyashiro (1974) (A–C) and Maniar and Piccoli (1989) (D–F). The dashed line in (A–C) is the Fe* line of Frost et al. (2001) separating "ferroan" (~tholeiitic) and "magnesian" (~calc-alkaline) rocks at higher and lower ($FeO_t/(FeO_t$ + MgO)), respectively. Most A-type granites are classified as "ferroan" (Frost et al., 2001). g.a.—gneiss association.

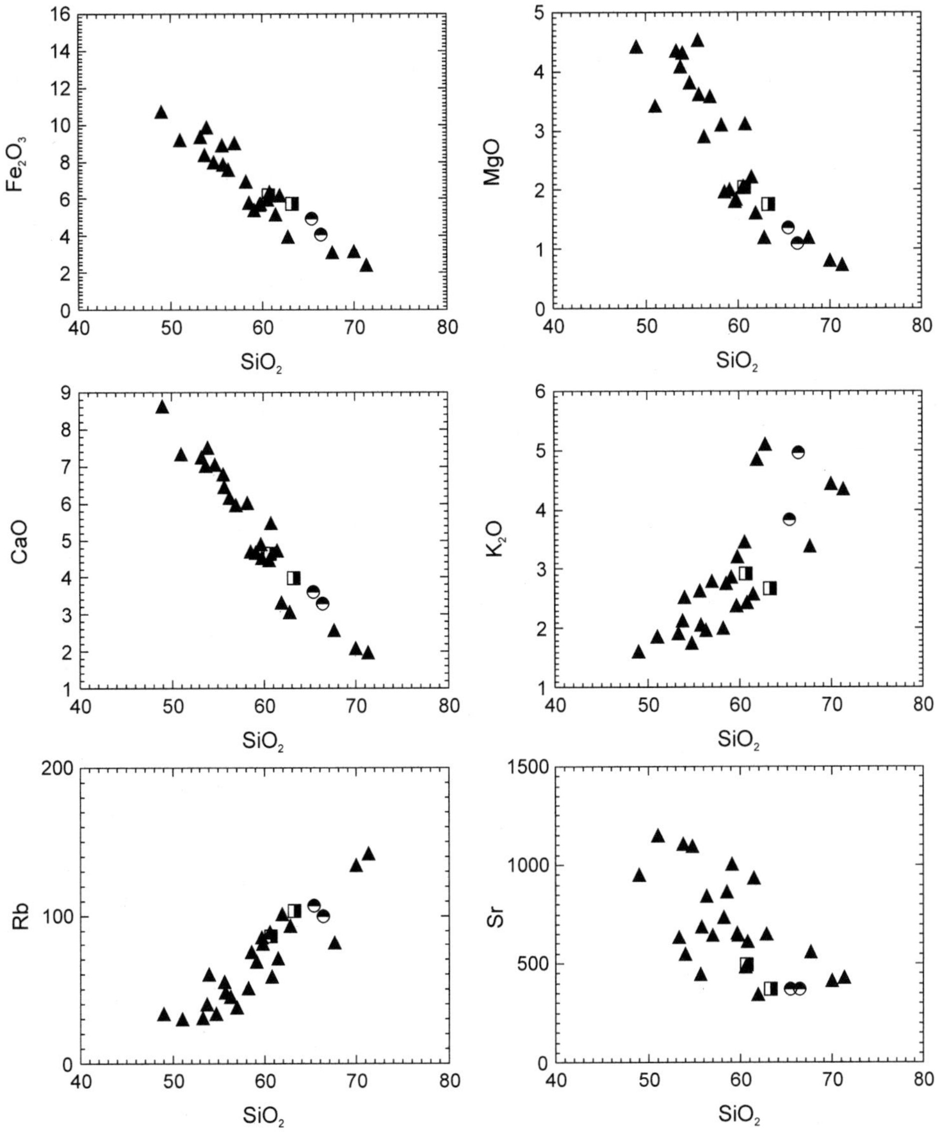

Figure 7. Selected Harker diagrams for the gray gneisses in the Muskoka domain (triangles), the Shawanaga pluton (circles), and the Ojibway gneiss association (squares).

intermediate lavas from the Lassen Volcanic Center in California, and may, thus, be a magmatic feature, or a result of mobility of the reference element (SiO_2 in this case), rather than a result of Al_2O_3 mobility. Figure 8, A and B, shows Pearce element ratio diagrams of Al/K versus (2Ca + Na)/K and Ti/K versus P/K; we chose K as the conserved element in the denominator because it is incompatible in most mafic and intermediate magmas and appears not to be strongly affected by mobilization. The latter assertion is supported by replacing K with Zr, generally assumed to be immobile during high-grade metamorphism, in the denominator, producing similar linear trends (Fig. 8, C and D). The linear trends displayed in these diagrams are consistent with the interpretation that the gray gneisses are part of the same suite of magmatic rocks, and that the geochemical variation within the suite is related to magmatic evolution rather than to element mobility. Trace elements such as Sr and V (compatible in plagioclase and Fe-Ti oxides, respectively) show negative trends with SiO_2, whereas Rb increases with increasing SiO_2.

The REE patterns (Fig. 9, A, inset) are moderately fractionated, with small to moderate negative Eu anomalies ($(Eu/Eu^*)_N = 0.54–0.80$), where $(Eu/Eu^*)_N$ stands for the chondrite-normalized ratio of measured Eu divided by the hypothetical Eu concentration required to produce an REE pattern with no Eu anomaly; the Eu anomalies display a shallow negative trend with increasing SiO_2. Rare earth fractionation ($(La/Yb)_N$) increases from 5.4 to 12.7 and $(La/Sm)_N$ from 1.8 to 4.8 with increasing SiO_2. Light (L)REE are not correlated with SiO_2, whereas heavy (H)REE and particularly middle (M)REE show well-defined negative trends.

The primitive mantle-normalized variation diagram (Fig. 9, A) shows that the rocks have trace element abundances typical

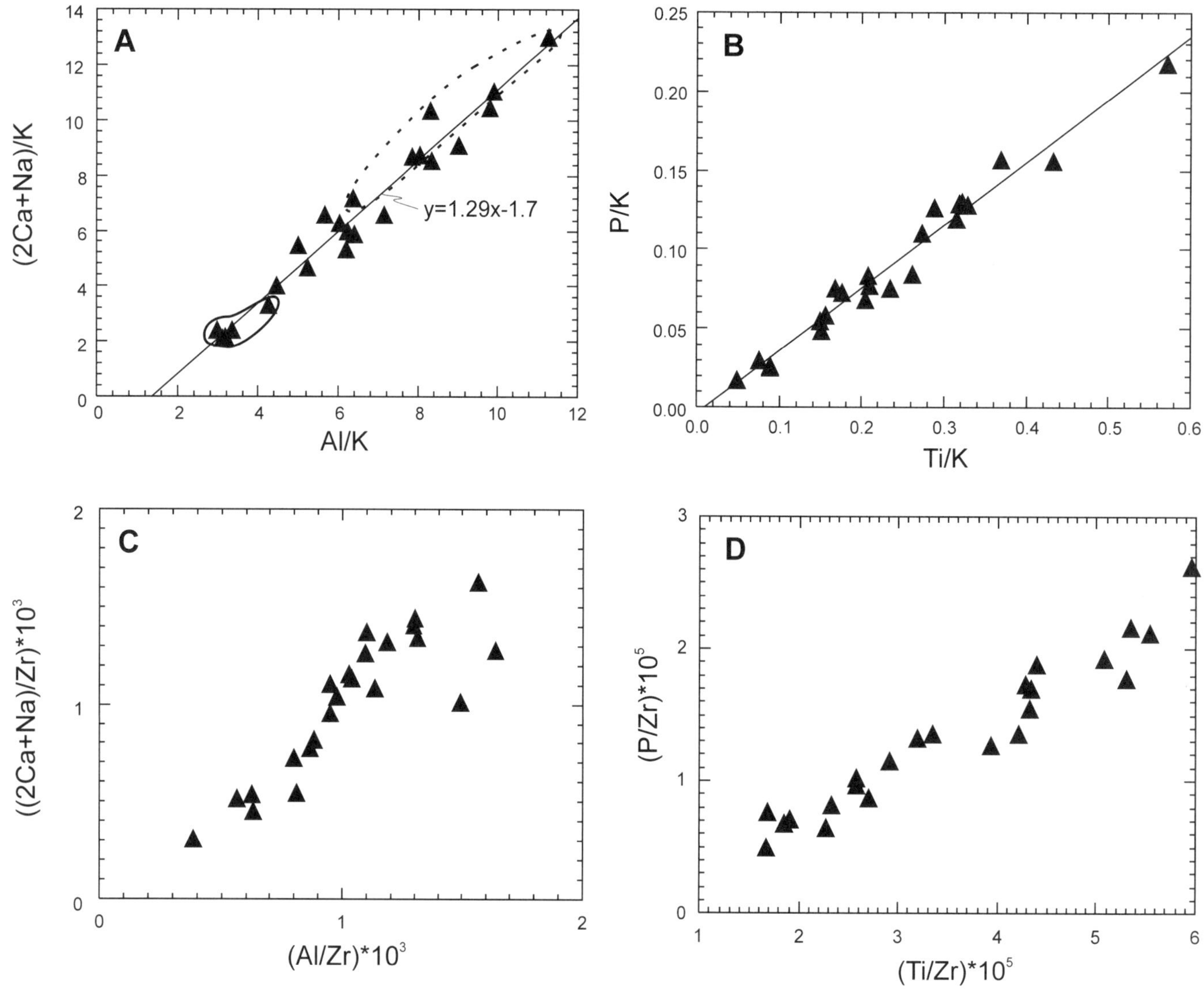

Figure 8. (A) Pearce element ratio diagram of mole Al/K versus (2Ca + Na)/K for the gray gneisses (Muskoka domain), yielding a slope of 1.29, which suggests fractionation of plagioclase and clinopyroxene. Mineral abbreviations after Kretz (1983). The dashed line encloses samples used as the starting composition (<55 wt% SiO_2) for trace element modeling (stage 1, mafic to intermediate gray gneiss); the full line encloses samples representing the most felsic gray gneisses (average SiO_2 = 70 wt%) (stage 2, intermediate to felsic gray gneiss). (B) Pearce element ratio diagram of mole Ti/K versus P/K, suggesting fractionation of apatite and Fe-Ti oxides. (C) and (D) are the same as (A) and (B) but with K replaced by Zr in the denominator.

of magmatic arcs (see Brown et al., 1984), with distinct negative Nb, and small negative P and Ti anomalies. However, Zr values range from 200 to 400 ppm, significantly higher than those of most arc-related rocks. The rocks plot in the magmatic arc field in the Y + Nb versus Rb tectonic discrimination diagram of Pearce et al. (1984) (Fig. 10, A), consistent with their calc-alkaline character and trace element composition.

Charnockites

The charnockites range in SiO_2 from 62 to 68 wt%, and are enriched in alkalis ($Na_2O + K_2O$ = 9.8–11.4) and iron ($FeO_t/(FeO_t + MgO)$ = 0.86–0.97) (Fig. 6, C). Total REE contents are high, and the REE patterns are moderately fractionated ($(La/Yb)_N$ = 4.2–15.8), with a large scatter of Eu anomalies ($(Eu/Eu^*)_N$ = 0.3–1.5) (Fig. 11, A, inset). Though Ba also shows significant scatter and could be construed as mobile, Ba and $(Eu/Eu^*)_N$ display a well-defined linear trend, and the scatter in Ba and Eu anomalies may be better explained by a magmatic, K-feldspar accumulation, process. In the primitive mantle-normalized diagram (Fig. 11, A), the charnockites are depleted in Th and Nb relative to the LREE, and have large negative Sr, P, and Ti anomalies. Except for one sample, the charnockites plot in the field of within-plate granites in tectonic discrimination diagrams (Fig. 10, C).

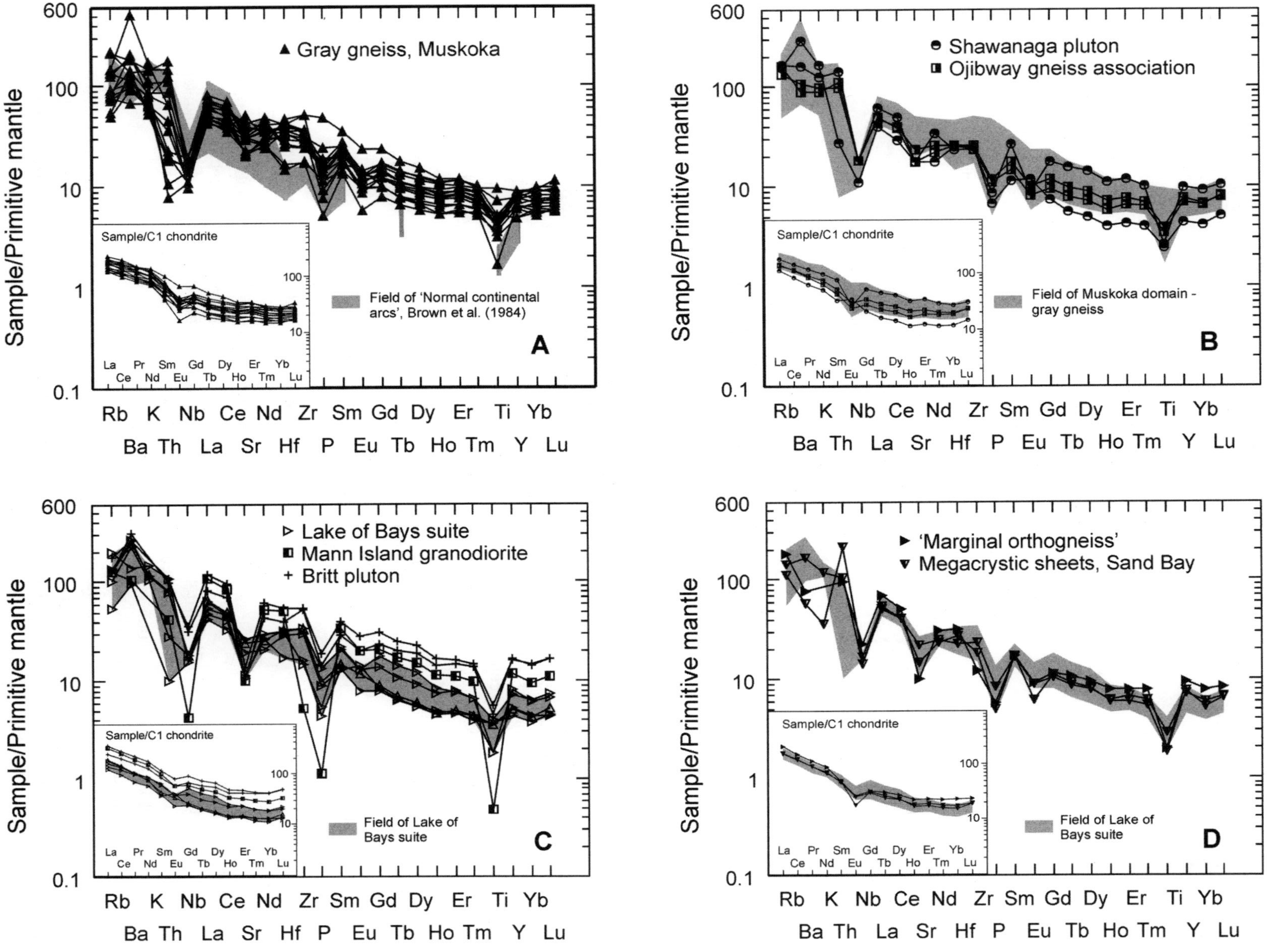

Figure 9. Primitive mantle- and chondrite-normalized trace element variation diagrams for (A) gray gneisses, Muskoka domain; (B) Shawanaga pluton and gray gneiss, Ojibway gneiss association; (C) Lake of Bays suite, Britt pluton, Mann Island granodiorite; (D) "marginal orthogneiss" and megacrystic sheets in the Sand Bay gneiss association. Normalization factors after Sun and McDonough (1989). The gray fields in (A) represent "normal continental arcs" as defined by Brown et al. (1984); note that Brown et al. used normalization factors from Wood (1979), but that this has a negligible effect on the normalized values as compared to the normalization factors of Sun and McDonough (1989). $(Eu/Eu^*)_N = (Eu_N)/((Sm_N + Gd_N)/2)$, where $(Eu/Eu^*)_N$ denotes the chondrite-normalized ratio of measured Eu divided by the hypothetical Eu concentration required to produce an REE pattern with no Eu anomaly.

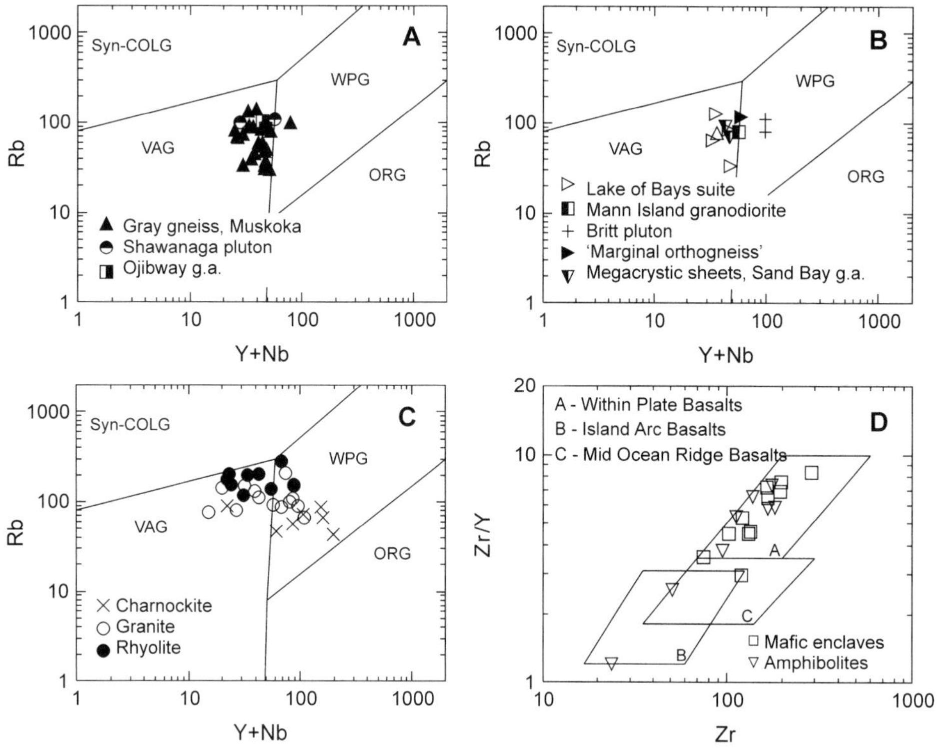

Figure 10. Y + Nb versus Rb tectonic discrimination diagram (Pearce et al., 1984) for (A) gray gneiss in the Muskoka domain, the Shawanaga pluton, and the Ojibway gneiss association; (B) megacrystic granitoid orthogneisses; (C) charnockite, granite, and rhyolite. (D) Zr versus Zr/Y tectonic discrimination diagram (Pearce and Cann, 1973) for the metabasites.

Granites

The granites range in SiO_2 from 64 to 76 wt% and are, like the charnockites, high in alkalis ($Na_2O + K_2O$ = 8.6–10.6) and iron ($FeO_t/(FeO_t + MgO)$ = 0.77–0.98) (Fig. 6, C). They have moderately fractionated REE patterns, with $(La/Yb)_N$ = 9.5–16.4, and large negative Eu anomalies ($(Eu/Eu^*)_N$ = 0.23–0.56) (Fig. 11, B inset). Their total REE contents are high and comparable to those of the charnockites, whereas other incompatible elements (particularly Zr, Y, and Hf) are lower. In the primitive mantle-normalized diagram (Fig. 11, B), the granites are characterized by negative Nb and Eu, and by large negative Sr, P, and Ti anomalies. In tectonic discrimination diagrams, the granites straddle the boundary between within-plate and arc settings (Fig. 10, C).

Megacrystic Orthogneiss

Lake of Bays Suite and "Marginal Orthogneiss"

Samples from the Lake of Bays suite range in SiO_2 from 59 to 70 wt%, and the "marginal orthogneiss" and associated megacrystic sheets in the Sand Bay gneiss association range from 64 to 67 wt%. Despite the apparent age difference between the Lake of Bays suite and the marginal orthogneiss, their major and trace element compositions are indistinguishable from each other and from those of the gray gneisses (Figs. 6, 9, and 10). One sample of a megacrystic orthogneiss from the Ojibway gneiss association is also geochemically similar to samples from the Lake of Bays suite.

Mann Island Granodiorite and Britt Pluton

The Mann Island granodiorite and Britt pluton range in SiO_2 from 62 to 73 wt% and are enriched in iron ($FeO_t/(FeO_t + MgO)$ = 0.83–0.85) relative to the Lake of Bays suite and marginal orthogneiss, which have $FeO_t/(FeO_t + MgO)$ ratios between 0.74 and 0.77, typical of calc-alkaline rocks (Fig. 6, B). The Mann Island granodiorite and Britt pluton are also enriched in incompatible elements relative to the Lake of Bays suite and marginal orthogneiss (Fig. 9, C); they plot in the field of within-plate granites or straddle the boundary between within-plate and arc settings in tectonic discrimination diagrams (Fig. 10, B).

Shawanaga Domain

Shawanaga Pluton and Ojibway Gneiss Association

The four samples of Shawanaga pluton and Ojibway gneiss association contain 61 to 66 wt% SiO_2 and resemble the gray gneisses in terms of major and trace element concentrations (Figs. 6, 7, 9, and 10). The REE patterns are moderately fractionated ($(La/Yb)_N$ = 6.7–9.6), with negative to positive Eu anomalies ($(Eu/Eu^*)_N$ = 0.4–1.3).

Sand Bay Rhyolites

The rhyolites range from 67 to 79 wt% SiO_2; all but one sample have SiO_2 > 75 wt% (Fig. 6, C). The very high SiO_2 could reflect silicification; however, we note that the composition of the rhyolites is comparable to that of little-altered, non-metamorphosed rhyolites elsewhere (see Menuge et al., 2002)

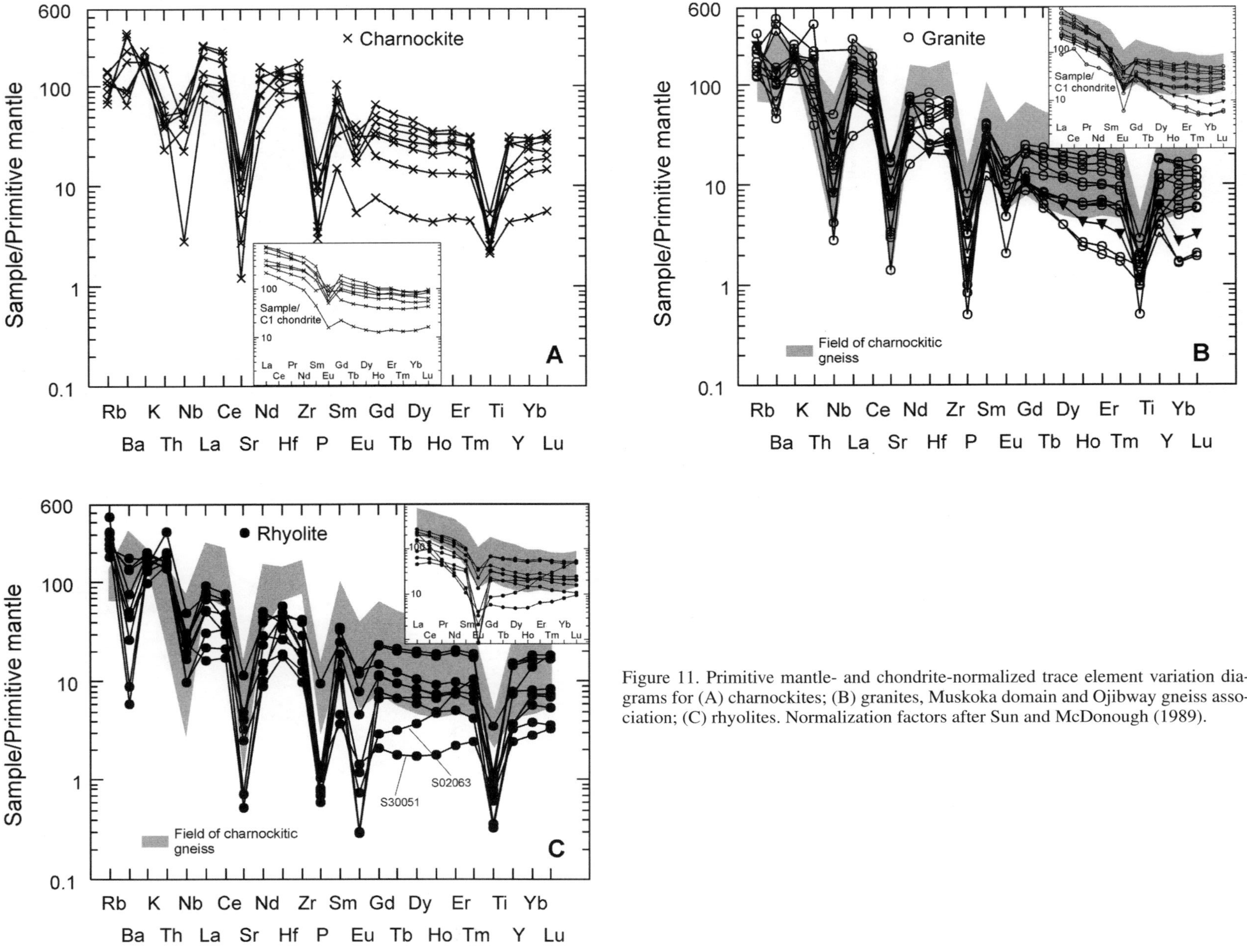

Figure 11. Primitive mantle- and chondrite-normalized trace element variation diagrams for (A) charnockites; (B) granites, Muskoka domain and Ojibway gneiss association; (C) rhyolites. Normalization factors after Sun and McDonough (1989).

(Fig. 12, A). Like the charnockites and granites in the Muskoka domain, the rhyolites are enriched in iron ($FeO_t/(FeO_t + MgO)$ typically >0.85) and alkalis ($Na_2O + K_2O$ = 7.5–9.5). The REE patterns are weakly to moderately fractionated ($(La/Yb)_N$ = 2.3–8.3), with large, negative Eu anomalies ($(Eu/Eu^*)_N$ = 0.02–0.64) (Fig. 11, C, inset). Two of the rhyolite samples (S30051 and S02063) have birdwing-shaped REE patterns with strongly negatively sloping LREE and positively sloping HREE, whereas other element abundances are similar to those of the other rhyolites. The geological significance, if any, of these two samples is uncertain. A possible interpretation is that they are related to severe hydrothermal alteration (Blein et al., this volume), but this possibility has not been investigated in detail here. In the primitive mantle-normalized diagram (Fig. 11, C), the rhyolites are characterized by strong Th enrichment (significantly higher than for the charnockites and granites), moderately negative Nb, and large negative Sr, P, and Ti anomalies. The rocks straddle the boundary between arc and within-plate settings in tectonic discrimination diagrams (Fig. 10, C).

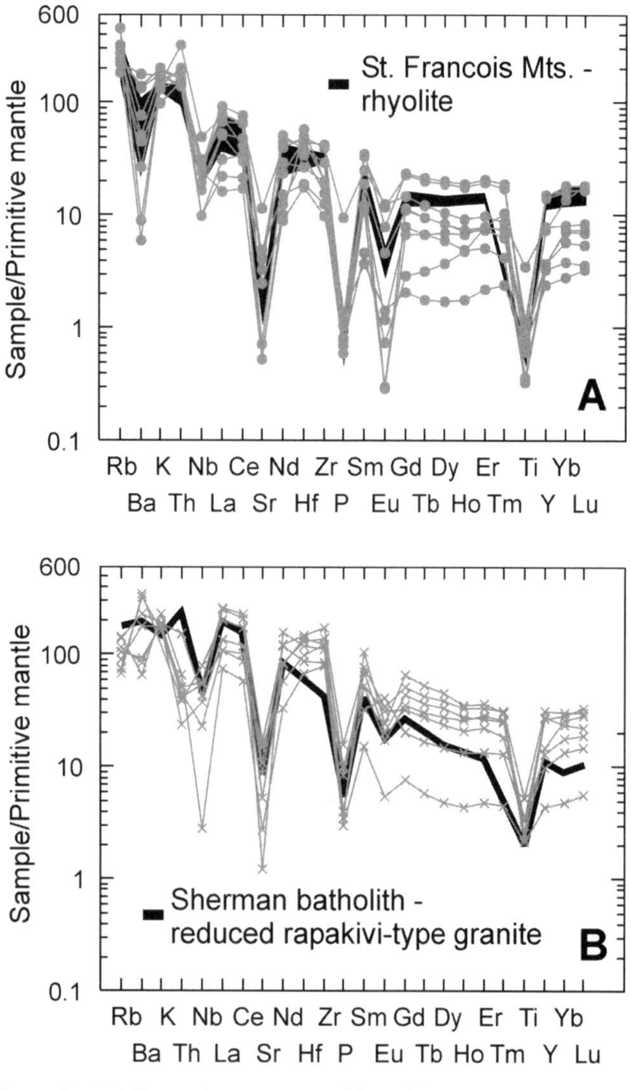

Figure 12. (A) Trace element composition of Sand Bay rhyolite compared to that of rhyolite from the eastern granite-rhyolite province, St. François Mountains (Menuge et al., 2002). (B) Trace element composition of the charnockite compared to that of the Sherman batholith (Frost et al., 1999). Normalization factors after Sun and McDonough (1989).

Metabasites

The mafic enclaves from the Muskoka domain and amphibolites from the Sand Bay gneiss association are geochemically indistinguishable, and they are therefore considered together. However, we emphasize that this does not necessarily mean that they are the same age.

The metabasites range in SiO_2 from 44.8 to 51.0 wt%, with Mg # (mole $MgO/(MgO + FeO_t)$) generally between 0.40 and 0.60, and can be classified as subalkaline basalt to basaltic andesite (Winchester and Floyd, 1977). Ti, Fe_2O_3, and V increase with decreasing Mg #, whereas Ni and Cr decrease (Fig. 13, A). Elements such as Rb, Ba, and Sr show significant scatter (Fig. 14), suggesting secondary alteration. The $FeO_t/(FeO_t + MgO)$ versus SiO_2 (Fig. 6, C) and FeO_t/MgO versus TiO_2 diagrams (Miyashiro, 1974) clearly illustrate the tholeiitic nature of the metabasites. The Si/Ti versus 0.5(Mg + Fe)/Ti Pearce element ratio diagram is consistent with fractionation of phases such as pyroxenes and olivine (see Russell and Nicholls, 1988) (Fig. 13, B).

The metabasites have weakly fractionated REE patterns, with $(La/Yb)_N$ = 2.3–4.9, similar to those of continental flood basalts (Fig. 14, A). Th/La ratios are low, averaging 0.09, arguing against significant crustal contamination; however, evidence of Th mobility in some rocks makes this interpretation highly uncertain. Total REE contents increase and $(Eu/Eu^*)_N$ decreases slightly with decreasing Mg # (Fig. 13, C), whereas $(La/Yb)_N$ remains constant. In the MORB-normalized diagram (Fig. 14), most samples display enrichment in all elements from Th to Sm, and depletion in Y, Yb, and Cr. A negative Nb anomaly is observed for most samples. In tectonic discrimination diagrams the majority of metabasites plot in the field of within-plate basalts (Fig. 10, D). One distinct sample (S20066-2) from the Sand Bay gneiss association has a slightly LREE-depleted pattern, similar to those of ocean floor basalts; however, the geological significance of this sample is uncertain.

PETROGENESIS

Based on geochemical composition, it is convenient to divide the discussion of petrogenesis into three parts: (1) rocks with typical arc geochemical characteristics (gray gneiss), (2) rocks with A-type (defined later) geochemical characteristics (charnockite, granite, and rhyolite), and (3) metabasites (mafic enclaves from the Muskoka domain and amphibolites from the Shawanaga domain). The megacrystic orthogneisses, Shawanaga pluton, and Ojibway gneiss association were investigated only on a reconnaissance level, and a detailed treatment of their petrogenesis is not warranted.

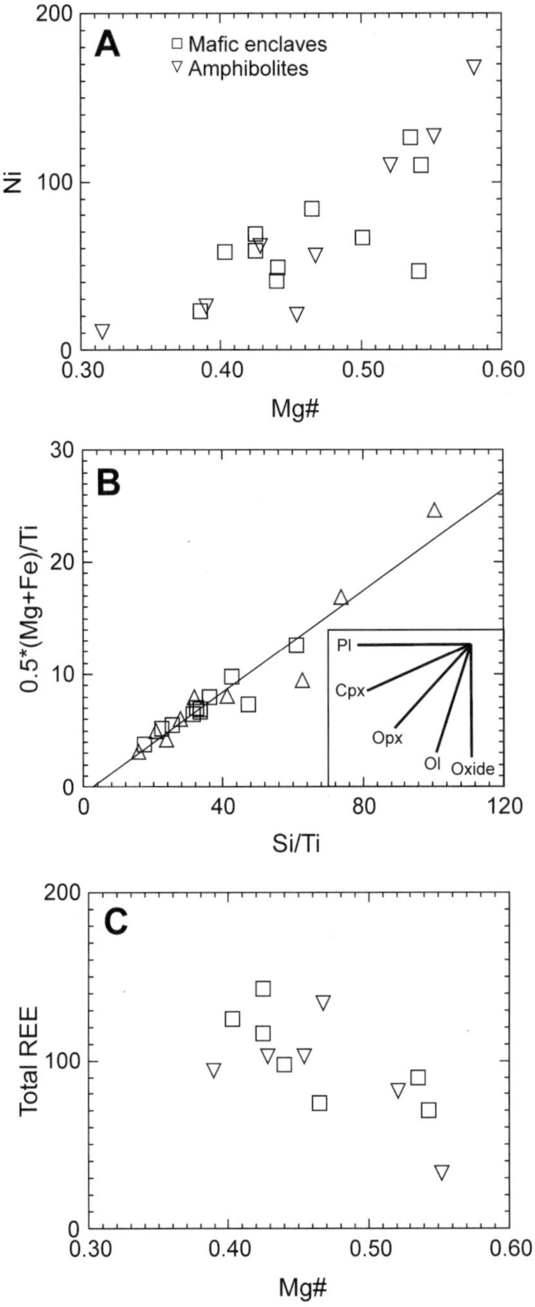

Figure 13. Metabasite variation diagrams. (A) Mg # (mole Mg/(Mg + Fe)) versus Ni. (B) Pearce element ratio diagram of mole Si/Ti versus 0.5(Mg + Fe)/Ti. The inset shows the slopes expected for the fractionation of plagioclase (Pl), pyroxene (Cpx and Opx), olivine (Ol), and Fe-Ti oxides, after Russell and Nicholls (1988). (C) Mg # versus total REE (rare earth element) contents.

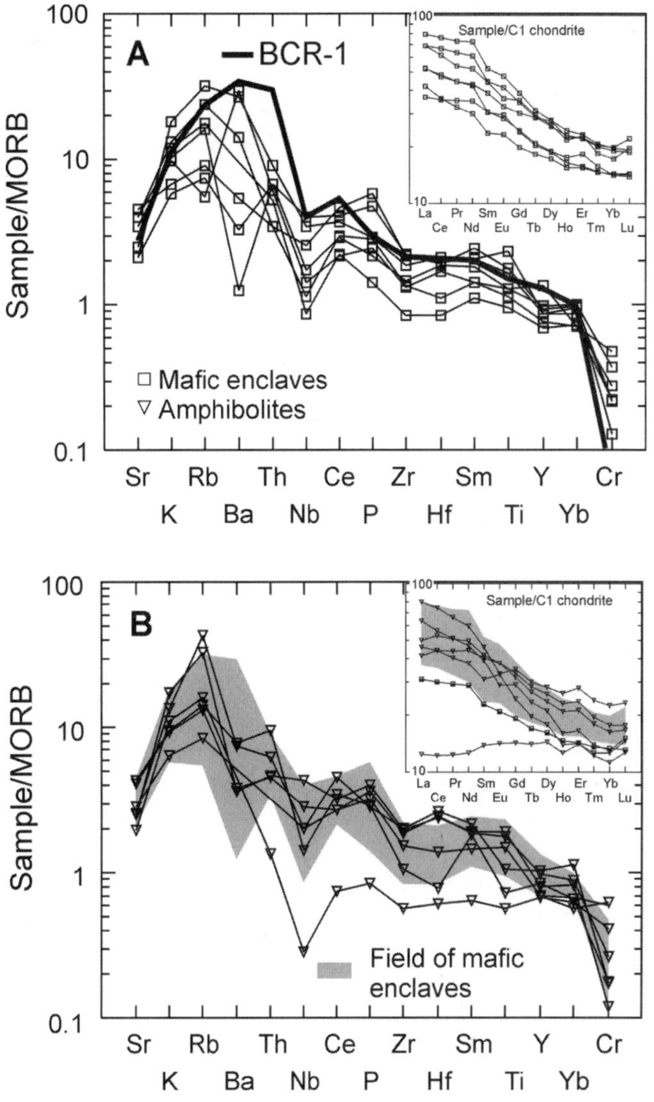

Figure 14. Mid-ocean ridge basalt (MORB)–normalized trace element variation diagram showing (A) mafic enclaves, Muskoka domain; (B) amphibolites, Sand Bay gneiss association. Normalization factors after Pearce (1982). The composition of the Columbia River continental flood basalt (BCR-1; Govindaraju, 1994) is shown in (A) for comparison. The composition of the metabasites is normalized to MORB rather than primitive mantle to facilitate comparison with basalt/amphibolite compositions reported elsewhere, which are typically normalized to MORB.

Gray Gneiss

A positive correlation between MgO and CaO and decreasing Ni and Sr with decreasing MgO are indicative of fractionation of plagioclase and clinopyroxene (see Devine, 1995). Russell and Nicholls (1988) proposed that the molar proportions of plagioclase to clinopyroxene could be determined from the slope in the Al/K versus. (2Ca + Na)/K diagram. The slope for the gray gneisses in this diagram is 1.29 (Fig. 8, A), which corresponds to a (molar) plagioclase/(plagioclase + clinopyroxene) ratio of 3:4. Alternatively, Ti and P can be used as the conserved element (denominator) if Fe-Ti oxides and apatite are absent from the fractionating assemblage, in which case Ti/K versus P/K will define a tight cluster, or ideally a point (Russell and Nicholls, 1988). Plotting Ti/K versus P/K yields a well-defined linear trend for the gray gneisses (Fig. 8, B), suggesting that Ti and P were compatible. We therefore assume that Fe-Ti oxides and apatite, in addition to plagioclase and clinopyroxene, were part of the fractionating assemblage. Increasing $(La/Yb)_N$ and $(La/Sm)_N$ with increasing SiO_2 may indicate fractionation of hornblende (see Arth and Barker, 1976).

Europium anomalies normally become more negative with increasing fractionation if plagioclase is involved, which could be used as an argument against the model proposed earlier. However, Gertisser and Keller (2000) did not observe well-developed negative Eu anomalies in the calc-alkaline Aeolian arc, where it can be shown petrographically that plagioclase was the dominant fractionating phase. They interpreted the lack of negative Eu anomalies to reflect relatively high oxygen fugacities during differentiation.

To test whether fractional crystallization of plagioclase, clinopyroxene, Fe-Ti oxides, hornblende, and apatite can explain the gray gneiss compositions, we modeled changes in REE, Rb, and Sr concentrations assuming Rayleigh fractional crystallization, in which cumulate phases are immediately removed from interaction with the melt after they form (e.g., Hanson, 1978). See Slagstad (2003) for partition coefficients. Rb and Sr were included because they are important petrogenetic indicators and show relatively well-defined trends in bivariate diagrams, suggesting limited mobility. In contrast, Ba shows significant scatter, possibly indicating secondary mobilization, and was not included. However, it is possible that the variation in Ba is a primary magmatic feature as suggested earlier for the charnockites. For example, the only gray gneiss sample with a positive Eu anomaly (M25072–1) is also the sample with the highest Ba contents, but it is unclear whether this sample represents a petrogenetically insignificant outlier or reflects a primary magmatic feature.

Trace element modeling presents a number of uncertainties (see Rollinson, 1993, for a review). One of the most obvious is probably the mineral/melt partition coefficient used in the modeling. Melt composition is the most important single factor controlling partition coefficients. Data availability permitting, we took melt composition (assumed to be similar to rock composition) into account when selecting a partition coefficient; for example, the fractional crystallization modeling for the gray gneisses was done in two stages to account for evolving melt composition. However, the partition coefficients also depend on a number of other factors regarding which we have few constraints, including temperature, pressure, oxygen activity, crystal chemistry, and water content. An additional complexity is that most of the investigated rocks are plutonic; thus, their composition probably represents not melt composition but rather magma composition (melt + entrained cumulate and/or residual minerals). Despite these uncertainties, trace element modeling is useful for assessing, and in some cases refining, petrogenetic models based on major element geochemistry, field observations, and experimental data.

Fractional crystallization modeling of the gray gneisses was performed in two stages to account for changing partition coefficients between mafic or intermediate and felsic compositions. The average composition of the most mafic gray gneisses (all samples with <55 wt% SiO_2; see Fig, 8, A) was used as the starting composition for stage 1, whereas the daughter composition was taken to be intermediate gray gneisses with an average of 59 wt% SiO_2. A good fit between model melt and intermediate gray gneiss was achieved after 30% crystallization (the fractionating assemblage and abundances are given in Fig. 15). The model melt after 30% crystallization in stage 1 was taken to be the starting composition for stage 2 modeling, and the daughter for stage 2 was represented by the felsic gray gneisses (with an average of 70 wt% SiO_2; see Fig. 8, B). A good fit between model melt and felsic gray gneiss was achieved after 30–50% crystallization (see Fig. 15 for details). The total fractionation for stage 1 and 2 was 50–65%.

Both stages were dominated by plagioclase, clinopyroxene, and Fe-Ti oxide (magnetite). Hornblende constitutes a minor part of the model fractionating assemblage, and apatite occurs in trace amounts. Minor zircon was included in stage 2 to account for the depletion in the HREE. The results of the modeling are compatible with the major element data (Pearce element ratios), suggesting that the gray gneisses may have evolved from the dioritic members of the suite by fractional crystallization of plagioclase, clinopyroxene, Fe-Ti oxides, minor hornblende, and trace amounts of apatite. Minor zircon fractionation appears to have affected the most felsic samples.

A-Type Granites (Charnockite, Granite) and Rhyolite

The charnockites, granites, and rhyolites have high $FeO_t/(FeO_t + MgO)$; high alkalis decreasing with SiO_2; high Zr, Y, Hf, REE (except Eu), and Ga/Al; and low MgO, CaO, and Sr, characteristic of A-type granites (Loiselle and Wones, 1979; Whalen et al., 1987; Eby, 1990). A-type granites are typically interpreted to have been emplaced at middle to upper crustal levels in extensional tectonic settings, and are characterized by high magmatic temperatures (Clemens et al., 1986; Creaser et al., 1991) and low fH_2O and fO_2. It is not possible to determine primary fH_2O or fO_2 values for the upper amphibolite- to granulite-facies rocks described here; however, the abundant pyroxene in the charnockites, the scarcity of hydrous minerals in general, and the paucity of leucosome by comparison with the adjacent gray gneisses, suggest that their protoliths were H_2O-deficient. Both the charnockites and the granites form discrete bodies and are petrographically and geochemically distinct from the surrounding gray gneisses, suggesting that their petrological and geochemical features are primary rather than the result of Grenvillian metamorphic processes. Magmatic temperatures can be estimated using the zircon saturation thermometer of Watson and Harrison (1983). As discussed in the section on geochronology, the charnockites contain abundant euhedral zircons that lack distinct inherited cores or metamorphic overgrowths; it is therefore possible that they all crystallized from the charnockite magma, which means that the temperatures cal-

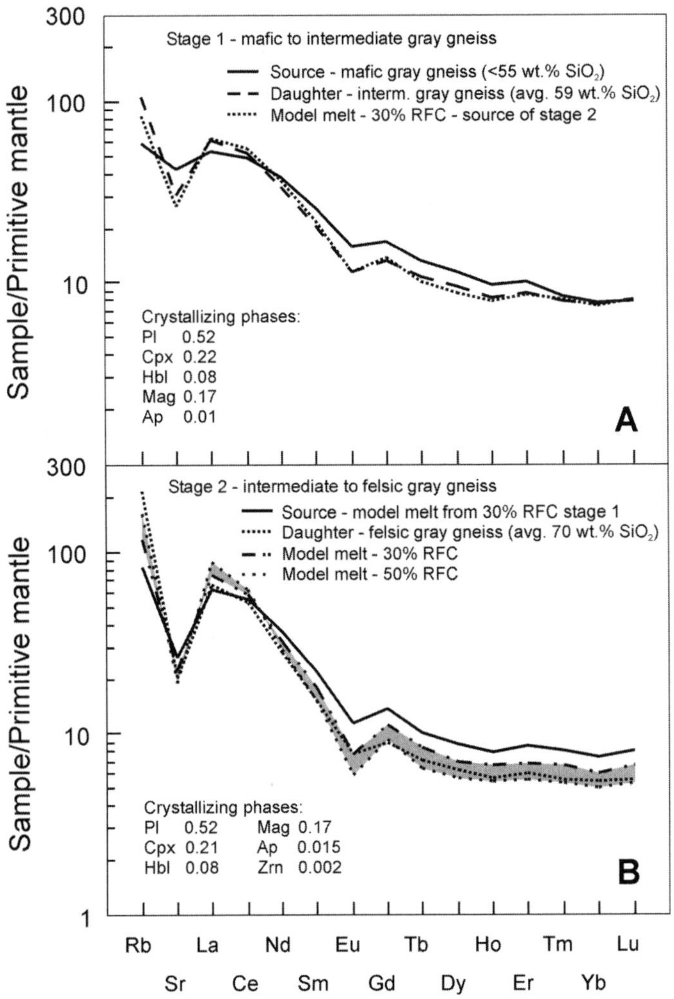

Figure 15. Modeling results for Rayleigh fractional crystallization of (A) a mafic to intermediate magma and (B) an intermediate to felsic magma, simulating the evolution of the gray gneisses. The abundances of crystallizing phases are given as weight proportions. Ap—apatite; Cpx—clinopyroxene; Hbl—hornblende; Mag—magnetite; Pl—plagioclase; RFC—Rayleigh fractional crystallization; Zrn—zircon. Normalization factors after Sun and McDonough (1989).

culated from the Zr content may be interpreted to reflect the liquidus temperature of the magma. Temperatures fall in a narrow range between 931 and 968 °C, consistent with the magmatic temperatures associated with A-type granites (e.g., Creaser and White, 1991). Zircon saturation temperatures for the granites range from 866 to 964 °C, with an average of 904 °C, whereas the rhyolites yield temperatures in the range 755–891 °C, with an average of 819 °C.

The high magmatic temperatures of A-type granites can be explained by intrusion and crystallization of mantle-derived mafic magmas (e.g., Creaser et al., 1991), by high geothermal gradients characteristic of areas undergoing crustal extension (e.g., Sandiford and Powell, 1986), or by a combination of both. Commonly invoked petrogenetic models for A-type granites include partial melting of a residual granulitic source that has previously generated an I-type granite (Collins et al., 1982; Clemens et al., 1986), partial melting of crustal igneous rocks of tonalitic to granodioritic composition (Cullers et al., 1981; Anderson, 1983; Creaser et al., 1991; Patiño Douce, 1997), and extreme differentiation (with or without assimilation) or partial melting of underplated tholeiitic basalts (Turner et al., 1992; Frost et al., 1999). A short review of the various models and a discussion of their applicability to the charnockite, granite, and rhyolite investigated follows.

Partial Melting of a Residual Source

Clemens et al. (1986), Whalen et al. (1987), and Landenberger and Collins (1996) proposed that A-type granites form by partial melting of a granulitic metaigneous source that has previously yielded a hydrous, I-type granitic melt. The model was proposed to explain the low magmatic water activity and high magmatic temperatures of A-type granites, but has been disputed by several authors, based on the chemical composition of A-type granites and experimental data (e.g., Creaser et al., 1991; Cullers et al., 1993; Patiño Douce, 1997). For example, strong iron enrichment appears to be characteristic of A-type granites (e.g., Frost et al., 2001). However, as pointed out by Creaser et al. (1991), partial melts invariably have higher Fe/Mg ratios than the mafic silicates with which they are in equilibrium. Thus, mafic residual silicates such as pyroxene and amphibole should become enriched in Mg after extraction of an I-type melt. Such a residue is unlikely to subsequently yield melts with higher Fe/Mg ratios than those of the I-type melt produced during the first melting episode (Creaser et al., 1991). Following these authors, we consider granulite-facies, residual rocks an unlikely source for the charnockite, granite, and rhyolite.

Partial Melting of a Tonalitic to Granodioritic Source (Gray Gneiss)

Several authors have suggested that granites and rhyolites with A-type characteristics may form by 10–40% partial melting of a tonalitic to granodioritic calc-alkaline source (Anderson and Cullers, 1978; Anderson, 1983; Creaser et al., 1991; Patiño Douce, 1997; Menuge et al., 2002). These granites and rhyolites typically have $SiO_2 > 70$ wt% and $FeO_t < 4$ wt%. In contrast, A-type granites with inferred mantle-derived sources (e.g., Nédélec et al., 1995; Frost et al., 1999) typically have $SiO_2 < 70$ wt% and $FeO_t > 6$ wt%. The granites and rhyolites, but not the charnockites, are compositionally similar to A-type granites inferred to have formed by partial melting of a tonalitic or granodioritic source (Fig. 12, A). To test whether this model is feasible, we used the average gray gneiss (57 wt% SiO_2) as the starting composition for petrogenetic modeling. The modeling assumed equilibrium batch melting, i.e., the melt and residue remained in equilibrium as melting proceeded (e.g., Hanson, 1978); results and details are shown in Figure 16, and partition coefficients are given in Slagstad (2003). A good fit between model melt and observed compositions, except Eu in rhyolite and Sr in rhyolite and granite, was obtained by 10–40% partial melting, leaving a granulitic residue consisting of quartz, plagioclase, pyroxenes, and Fe-Ti oxide. Minor apatite and zircon were included in the residue for the granites, and apatite, zircon, and allanite for the rhyolites. These are common accessory minerals in granitic systems, and partition coefficients of otherwise incompatible elements (e.g., REE) may be extremely high in these phases (Tindle et al., 1988; Wark and Miller, 1993; Bea, 1996). Thus, they may effectively determine the trace element composition of a magma. The results reported earlier must therefore be treated with caution. The model results shown in Figures 16, A and B, are based on partition coefficients appropriate for granitic systems (i.e. SiO_2~ 70 wt.%) (Slagstad, 2003). However, some studies have shown that partition coefficients for Sr and Eu in plagioclase can be significantly higher in high-silica (i.e., SiO_2~ 75 wt%) rhyolites (e.g., Mahood and Hildreth, 1983; Nash and Crecraft, 1985). The rhyolites investigated here can be regarded as high-silica rhyolites, and employing partition coefficients from Nash and Crecraft (1985) yields a good fit between model melt and rhyolite (Fig. 16, C).

In summary, partial melting of an intermediate calc-alkaline source is not feasible for producing the charnockites, but remains a possibility for the granites and appears likely for the rhyolites. A likely source is the abundant gray gneiss in the Muskoka domain and the Ojibway gneiss association, although other similar sources are equally possible. Isotopic data are needed to determine the nature of the source of the granites and rhyolites.

Differentiation from, or Partial Melting of, a Mafic Source

A-type granites, similar in composition to the charnockites, have been described from Madagascar (Nédélec et al., 1995) and Wyoming (Frost et al., 1999). In the Madagascar example, Nédélec et al. (1995) proposed that A-type granites (syenites) formed from a mantle-derived mafic magma by means of fractional crystallization, with hornblende as a major fractionating phase. Frost et al. (1999) proposed partial melting of preexisting tholeiites, or more probably their differentiates, as the most feasible petrogenetic model for the Sherman batholith in Wyoming. Based on their similar composition (Fig. 12, B), we interpret the charnockites we studied to have formed in a similar way to the Sherman batholith. The metabasites have compositions compatible with an origin as underplated basaltic magmas, as we discuss further later. Although the amphibolites in the Sand Bay gneiss association are interpreted to be coeval with the rhyolites; thus, significantly younger than the charnockites, the mafic enclaves

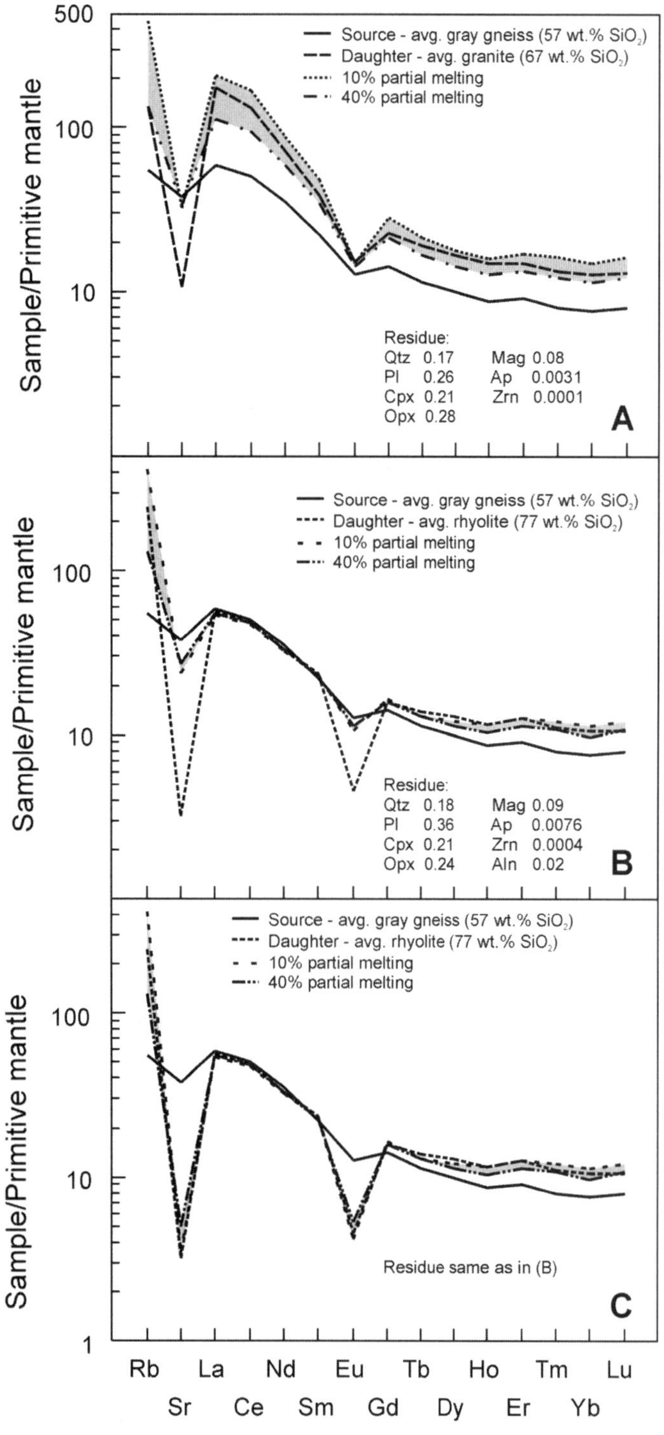

Figure 16. Modeling results for equilibrium partial melting of intermediate gray gneiss. (A) the model melt obtained by 10–40% partial melting is similar to that for the granites for the rare earth elements but enriched in Sr and Rb. (B) the model melt obtained by 10–40% partial melting is similar to that for the rhyolites for most trace elements except for enrichment in Eu and Sr. The results presented in (A) and (B) were obtained using partition coefficients appropriate for felsic systems. (C) identical to (B) except for significantly higher partition coefficients for Eu and Sr in plagioclase, probably more appropriate for high-silica rhyolites (Nash and Crecraft, 1985). The model melt thus obtained is compositionally similar to that for the rhyolites. The abundances of crystallizing phases are given as weight proportions. Aln—allanite; Ap—apatite; Cpx—clinopyroxene; Mag—magnetite; Opx—orthopyroxene; Pl—plagioclase; Qtz—quartz; RFC—Rayleigh fractional crystallization; Zrn—zircon. Normalization factors after Sun and McDonough (1989).

in the Muskoka domain may represent a potential source. This model is tested using trace element modeling; partition coefficients are given in Slagstad (2003).

The modeling yields interesting but inconclusive results (Fig. 17). The charnockites could have formed by means of 80–90% Rayleigh fractional crystallization of a mafic magma, fractionating plagioclase, olivine, clinopyroxene, and minor hornblende (Fig. 17, A). Equilibrium partial melting of a mafic enclave yields a model melt composition that is depleted in REE relative to the average charnockite (Fig. 17, B). Alternatively, the charnockites could have formed via a two-stage process involving fractional crystallization of a mafic magma, followed by partial melting of the fractionated rock, which is similar to the model proposed by Frost et al. (1999) for the Sherman batholith. This alternative was modeled in two stages (Fig. 17, C). Stage 1 involved 40% Rayleigh fractional crystallization of a mafic magma, fractionating the same minerals in similar proportions as shown in Figure 17, A. The model melt composition thus obtained was used as the starting composition for partial melting calculations; the results show that 10–30% partial melting of the fractionated rock, leaving a residue comprising plagioclase, clinopyroxene, and orthopyroxene, produces model melts similar to those for the charnockites. The degree of fractionation during stage 1, arbitrarily set at 40%, determines the REE contents but not the REE pattern of the starting composition for stage 2. Selecting a different value, e.g., 20%, gives similar results but requires lower degrees of partial melting (<10%); thus, the main factors controlling the model results are the fractionating assemblage during stage 1, and the residue during stage 2, not the degree of fractionation. In contrast to the model proposed by Nédélec et al. (1995), hornblende appears to have been relatively insignificant, consistent with the anhydrous nature of these rocks.

Frost et al. (1999) cited the lack of intermediate compositions between mafic rocks and granites as evidence against a fractional crystallization model for the Sherman batholith. Grove and Donnelly-Nolan (1986) suggested that compositional gaps could result from shallow temperature versus composition slopes on liquidus surfaces, allowing large degrees of crystal-

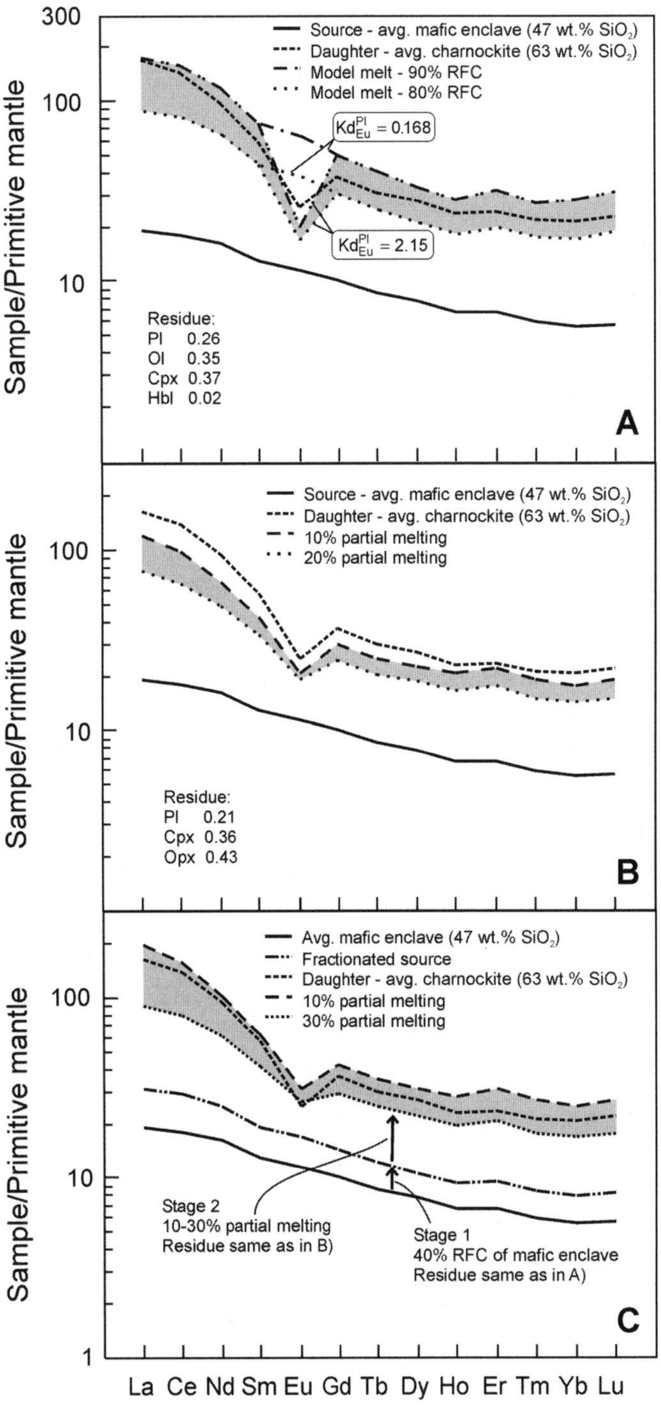

Figure 17. Three possible models for relating the mafic enclaves and the charnockites. (A) Rayleigh fractional crystallization of a basaltic magma (mafic enclave). The fit between model melt and charnockite is good for most elements after 80–90% fractional crystallization, whereas the fit for Eu depends on the partition coefficient for Eu in plagioclase. A partition coefficient of 0.168 (Smith and Humphris, 1998) produces a model melt that is too high in Eu, whereas a partition coefficient of 2.15 (Martin, 1987) produces a model melt depleted in Eu. Both partition coefficients were reported to be appropriate for mafic systems and exemplify some of the inherent problems in trace element modeling. (B) equilibrium partial melting of a basaltic source (mafic enclave). The modeling shows that low degrees of partial melting (10–20%) of a basaltic source produces model melts that are depleted in REE relative to the charnockites. (C) equilibrium partial melting of a fractionated mafic enclave. The source for this model was taken to be the model melt obtained after 40% Rayleigh fractional crystallization of a basaltic magma (mafic enclave), fractionating the same phases and in similar proportions as in (A). Low degrees (10–30%) of partial melting of this fractionated source, leaving a residue similar to (B), yield a model melt similar to the charnockites. Abundances of crystallizing and residual phases given as weight proportions. Cpx—clinopyroxene; Hbl—hornblende; Ol—olivine; Opx—orthopyroxene; Pl—plagioclase; RFC—Rayleigh fractional crystallization; Zrn—zircon. Normalization factors after Sun and McDonough (1989).

lization and compositional change over small temperature intervals. Turner et al. (1992) suggested that the gap between basaltic magmas and A-type granites could be explained by this mechanism. However, the mechanism was originally proposed to explain the commonly observed gap between andesitic and silicic lavas (~60–70 wt% SiO_2) from a single calc-alkaline volcano (e.g., Hildreth, 1981; Bacon and Druitt, 1988), and is not necessarily applicable to A-type granites in general. We therefore conclude that partial melting of underplated tholeiitic basaltic rocks, or their fractionated derivatives, is the most likely petrogenetic model for the charnockites.

Eby (1990) proposed that the source of A-type granites could be identified as having a composition similar to oceanic island basalts or island arc and continental arc basalts, based on Y/Nb ratios <1.2 and >1.2, respectively. With a few exceptions, the charnockites, granites, and rhyolites have Y/Nb ratios >1.2, consistent with derivation from a subduction-influenced source such as the mafic enclaves (see below) or gray gneisses. Isotopic data would help to refine the petrogenetic model(s) for these rocks.

Metabasites

The geochemistry of the metabasites, particularly decreasing Ni and Cr, and increasing total REE contents and constant $(La/Yb)_N$ with decreasing Mg #, is consistent with low-pressure fractionation of phases such as olivine and pyroxene that do not fractionate the REE significantly (e.g., Rollinson, 1993). Shallowly decreasing Al_2O_3, CaO, and $(Eu/Eu^*)_N$ with decreasing Mg # suggests limited plagioclase fractionation. Increasing Ti, Fe, and V, and constant $(La/Yb)_N$ with decreasing Mg #, indicate a lack of Fe-Ti oxides and amphibole in the fractionating assemblage. This result is consistent with the results from the Pearce element ratio diagram (Fig. 13, B).

The metabasites are enriched in incompatible elements relative to MORB (Fig. 14), except Y, Yb, and Cr, typical of within-plate basalts, and have a distinct negative Nb anomaly, characteristic of arc settings. At least three models can account for the low Nb contents: (1) the metabasites are arc-related (Pearce and Peate, 1995), (2) their composition was influenced by crustal assimilation (Davidson et al., 1987), or (3) they were derived from previously subduction-influenced mantle (Barnes et al., 1999). An arc setting cannot be discounted for the mafic enclaves in the Muskoka domain based on composition; in contrast, amphibolites (metamorphosed basalt flows) of similar composition in the supracrustal Sand Bay gneiss association are more likely to have formed in a back-arc than in an arc setting (Culshaw and Dostal, 1997). Assimilation of crustal rocks into mafic magma should be accompanied by fractional crystallization (e.g., DePaolo, 1981). As argued earlier, the metabasites did not fractionate Fe-Ti oxides; thus, Nb should have been incompatible in any fractionating assemblage. We conclude that crustal assimilation cannot explain either the low Nb contents or the low Th/La ratios observed in these rocks. Instead, we suggest that the source of the metabasites was previously subduction-influenced mantle, enriched by fertile asthenospheric mantle invading the mantle wedge as a result of extension and rifting (see Gill, 1984; Hochstaedter et al., 1990).

TECTONIC SIGNIFICANCE

Gray Gneiss

Calc-alkaline rocks are characteristic products of magmatism at convergent plate boundaries, and an arc origin for the gray gneisses in the Muskoka domain is supported by their trace element compositions (McMullen, 1999; this study) and the isotopic data available (Dickin and McNutt, 1990). Based on similarities in composition and age, we suggest that the Shawanaga pluton and the Ojibway gneiss association also formed part of the "Muskoka-Ojibway" arc. The proposed proximity of the Muskoka and the Shawanaga domains prior to Grenvillian orogenesis is consistent with the restored, pre-Grenvillian cross-section of Culshaw et al. (1997). Whether the rocks formed in an evolved island arc, on a rifted continental fragment, or in a continental arc is not easily discerned from the geochemical data. Some constraints are (1) the Nd model ages in the Muskoka domain and the Ojibway gneiss association (Dickin and McNutt, 1990) are similar to or slightly older than crystallization ages, suggesting a setting where the magma interacted with older continental crust, consistent with a continental arc or rifted continental fragment setting; (2) there are no clear indicators of an intervening ocean or back-arc basin between the arc and older Laurentian crust (see Saunders et al., 1979), which appears to support a continental arc setting; and (3) extension-related magmatism at, or inboard of, the Laurentian margin, simultaneous with formation of the Muskoka-Ojibway arc, is indicated by the ca. 1450 Ma Britt pluton and the Mann Island granodiorite (see later) intruding older continental crust in the Central Gneiss Belt and the Grenville foreland, favoring a continental arc setting. We note, however, that one amphibolite from the Sand Bay gneiss association has a composition compatible with an oceanic setting, and that the geological and tectonic significance of the Lighthouse gneiss association in the footwall of the Parry Sound domain is as yet poorly understood. We conclude that the Muskoka-Ojibway arc, with associated ensialic back-arc extension, was located at the southeastern margin of Laurentia at ca. 1480–1430 Ma, although we cannot exclude a more distal position, e.g., on a fragment of rifted continental crust (e.g., Okamura et al., 1998).

Megacrystic Orthogneiss

Lake of Bays Suite

The Lake of Bays suite is interpreted to have formed in a continental magmatic arc, coevally with the gray gneisses in the Muskoka domain and the Ojibway gneiss association. Culshaw et al. (1989) also identified megacrystic orthogneisses of similar age (ca. 1460 Ma, Krogh, 1991) in the Go Home domain (Fig. 1), contiguous with the Muskoka and the Shawanaga domains on restored, pre-Grenvillian cross-sections (Culshaw et al., 1997). The "marginal orthogneiss," including thin sheets of megacrystic orthogneiss in the Sand Bay gneiss association, is compositionally indistinguishable from the Lake of Bays suite. However, the lack of a precise age from the marginal orthogneiss makes it difficult to correlate these units.

Mann Island Granodiorite and Britt Pluton

Despite their similar age, the Mann Island granodiorite and the Britt pluton have within-plate characteristics and intruded pre-1450 Ma continental crust, in contrast to the arc-related signature and lack of older crust typical of the Lake of Bays suite. We consider the most likely tectonic setting for the Mann Island granodiorite and the Britt pluton to be the back-arc region of the postulated ca. 1480–1430 Ma continental arc.

A-Type Granites (Charnockite, Granite) and Rhyolite

Although the charnockite and granite are geochemically distinct from the surrounding gray gneisses, the relatively imprecise geochronological data suggest that they formed at about the same time, and the lack of any evidence for tectonic contacts suggests that they formed in the same general setting. The charnockites have geochemical characteristics indicating derivation from mantle-derived underplated magmas or rocks. Although the geochemical data do not clearly distinguish the process by which they formed, the apparent lack of intermediate precursors to the charnockites favors partial melting rather than fractional crystallization. Lithospheric thinning associated with continued underplating may have provided the heat necessary to melt the previously underplated rocks (e.g., Frost and Frost, 1997). The granites formed at the same time as the charnockites, apparently by partial melting of a crustal source. Although the Sherman batholith (Frost et al., 1999) is about the same age as the charnockite and granite from the Muskoka domain, there is no evidence that its emplacement was spatially or temporally associated with arc magmatism, and Frost et al. (1999) interpreted it to be related to continental rifting. The model proposed for the Muskoka charnockite and granite implies the presence of a tholeiitic basaltic underplate to provide source material, and possibly heat. The mafic enclaves in the Muskoka domain have the appropriate composition to be related to the postulated underplating magmas; however, they may also be younger. The most likely tectonic setting for such an underplate is an extensional tectonic regime, suggesting that, in addition to extension in the back-arc region (the Britt pluton and the Mann Island granodiorite), the arc itself may have undergone extension at ca. 1450 Ma. Unfortunately, the timing of arc-related and A-type magmatism are relatively poorly constrained, making it difficult to propose more detailed tectonic models.

The Sand Bay rhyolites probably formed by partial melting of intermediate calc-alkaline rocks. Lithospheric thinning and associated basaltic underplating, represented by the coeval amphibolites, are considered the most likely heat sources for the inferred crustal melting event. Our data support the conclusions of previous workers (e.g., Culshaw and Dostal, 1997; Rivers and Corrigan, 2000) that the Sand Bay rhyolites formed in an extensional, back-arc setting at ca. 1360 Ma.

Metabasites

Bruhn et al. (1978) described basaltic rocks from southernmost South America with a similar combination of within-plate and arc characteristics and interpreted them to be related to initial stages of back-arc rifting. A significant finding of this study is the complete compositional overlap between mafic enclaves in the Muskoka domain and amphibolites in the Sand Bay gneiss association. The interlayering of rhyolites and amphibolites in the Sand Bay gneiss association (Culshaw and Dostal, 1997) suggests an age of ca. 1360 Ma for the amphibolites. In the absence of radiometric dates from mafic enclaves in the Muskoka domain, the overlap in composition between mafic enclaves and amphibolites could be interpreted in one of two ways: either the mafic enclaves and the amphibolites represent two temporally distinct, but tectonically similar, events, or they formed during the same event at ca. 1360 Ma.

DISCUSSION

The geochemical data presented here, coupled with previously obtained geochronological data, suggest the following tectonic evolution for this part of the southeastern Laurentian margin.

1480–1430 Ma

Arc-related magmatic activity in the Muskoka domain, most likely situated at the Laurentian continental margin, at ca. 1450 Ma is relatively well established based on abundant geochronological data (e.g., Timmermann et al., 1997; Nadeau and van Breemen, 1998; McMullen, 1999), and recent geochemical data (McMullen, 1999; this study). Protolith ages of orthogneisses in the Muskoka domain range from ca. 1480 to 1430 Ma; however, it is worth noting that the younger age limit comes from a "monzodioritic orthogneiss" dated at 1427 +16/–13 Ma (U-Pb zircon) by Nadeau and van Breemen (1998) (Fig. 2), whereas other published protolith ages are >1450 Ma. The monzodioritic orthogneiss contains orthopyroxene and clinopyroxene, unlike the arc-related rocks described herein that lack pyroxene. Thus, the tectonic significance of the monzodioritic orthogneiss and the duration of arc magmatism are currently uncertain. A new finding of this study is that the arc-related rocks may be temporally and spatially related to A-type granites, suggesting intra-arc extension during all or part of this stage of the arc's evolution. In addition to intra-arc extension, the new geochemical data, coupled with published geochronological data, suggest that back-arc magmatism affected the Britt domain at ca. 1450 Ma (van Breemen et al., 1986; Corrigan et al., 1994). Furthermore, Nadeau and van

Breemen (1998) suggested that the Algonquin domain was situated in the back-arc of the Muskoka-Ojibway arc. We conclude that a broad, arc-related extensional regime, e.g., related to slab roll-back, dominated the southeastern Laurentian margin at ca. 1450 Ma. Between ca. 1450 and 1430 Ma, the Britt domain, and possibly the Algonquin domain, underwent granulite-facies metamorphism (Ketchum et al., 1994; Nadeau and van Breemen, 1998; Timmermann, 1998). As noted later, the tectonic setting of the high-grade metamorphism is not clear at present.

1430–1360 Ma

Evidence of geologic activity in the Central Gneiss Belt between ca. 1430 and 1360 Ma is sparse. Carr et al. (2000) and Culshaw et al. (2002) suggested that magmatism may have migrated to a position more outboard of the Laurentian margin at this time. Arc magmatism may be represented by orthogneisses of varied composition in the Parry Sound domain and the Central Metasedimentary Belt boundary thrust zone. For example, magmatism in the Parry Sound domain, which began at ca. 1425 Ma (van Breemen et al., 1986) and continued until ca. 1160 Ma (van Breemen et al., 1986; Wodicka, 1994; Wodicka et al., 1996), produced rocks ranging from granite to anorthosite. The magmatic evolution in the Parry Sound domain has not been studied in detail, but the nearly continuous record of granitoid magmatism with intermittent anorthositic magmatism may point to an evolving arc involving several phases of extension. The relationship, if any, between Laurentia and the Parry Sound domain is unknown, but dating of detrital zircons from a quartzite in the basal Parry Sound assemblage suggests a Laurentian source for some of the clastic sediments (Wodicka et al., 1996).

Corriveau and Morin (2000), Blein et al. (2003), and Wodicka et al. (this volume) interpreted ca. 1.4–1.35 Ga felsic to mafic gneisses, including high-silica rhyolite and metabasalt with a back-arc signature, in the Central Metasedimentary Belt of Québec to represent an island arc built on a thin continental substrate, whereas coeval gneisses farther north have a continental arc signature (Nantel and Pintson, 2002). Hanmer et al. (2000) documented a similar setting farther east. Taken together, the data available from the southwestern and central Grenville Province suggest that the Muskoka-Ojibway arc is part of a long lived arc along the southeastern margin of Laurentia.

<1360 Ma

The Sand Bay gneiss association is widely regarded to have formed at ca. 1360 Ma in a back-arc setting at or near the outermost edge of the Laurentian margin (Culshaw and Dostal, 1997; Rivers and Corrigan, 2000; this study).

Outstanding Problems and Suggestions for Further Work

Although the points listed earlier are compatible with a wide range of data, and are broadly consistent with "continental arc" models proposed previously (e.g., Rivers and Corrigan, 2000), we regard detailed tectonic models for the Late Paleoproterozoic to Early Mesoproterozoic evolution of the Central Gneiss Belt as premature. Instead we point out some outstanding problems and discuss ways to test them. The main unresolved issues raised by this work are as follows:

1. The relationship between arc-related and A-type rocks in the Muskoka domain. The timing of arc and A-type magmatism in the Muskoka domain is only loosely constrained to between 1480 and 1430 Ma. The lack of precise age constraints on the duration of arc magmatism and the temporal relationship between arc and A-type magmatism make models for this time period highly speculative.

2. The significance of granulite-facies metamorphism between 1450 and 1430 Ma. Ketchum et al. (1994) suggested that granulite-facies metamorphism at 1450–1430 Ma may have resulted from accretion of a juvenile arc at this time; however, other interpretations are also possible. For example, Sandiford and Powell (1986) suggested that rocks at lower crustal levels in regions characterized by extension and thinning of the continental crust can undergo granulite-facies metamorphism as a result of the high heat flow associated with such tectonic settings. Dunphy and Ludden (1998) and St-Onge et al. (2000) argued that granulite-facies metamorphism in the Trans-Hudson orogen in northern Québec resulted from heat related to arc plutonism, an interpretation that may also be applicable in the Central Gneiss Belt considering the apparent overlap in magmatic and metamorphic ages. A fourth possible interpretation, essentially a modification of the model proposed by Sandiford and Powell (1986), involves a change in tectonic regime from extension and associated granulite-facies metamorphism in the back-arc region to convergence leading to crustal thickening (see Collins, 2002). In this case, the change from extension to contraction could have been controlled by a number of factors, including the age of the subducting oceanic crust (e.g., old, dense crust favoring slab roll-back and extension; Molnar and Atwater, 1978); subduction of oceanic plateaus or seamounts, inducing shortening (Gutscher et al., 2000); and the rate and angle of convergence (Royden, 1993; Waschbusch and Beaumont, 1996; Pope and Willett, 1998). The 1450–1430 Ma granulite-facies metamorphism in the Britt and Algonquin domains can therefore be interpreted in a number of ways with very different tectonic implications, including accretion or convergence, extension, and arc magmatism, all compatible with the "continental arc" model proposed by Rivers and Corrigan (2000).

3. The relationship between the Parry Sound domain or the Central Metasedimentary Belt boundary thrust zone and the Laurentian margin. The Parry Sound domain and the Central Metasedimentary Belt boundary thrust zone may represent magmatic activity along the Laurentian margin following arc magmatism and high-grade metamorphism between 1480 and 1430 Ma. However, the magmatic evolution and paleogeographic location of the Parry Sound domain and the Central Metasedimentary Belt boundary thrust zone, and their relationship with Laurentia, are relatively unconstrained, which makes interpre-

tations of the evolution of the southwestern Grenville Province between 1430 and 1360 Ma uncertain.

4. The significance of back-arc magmatism in the Sand Bay gneiss association. The Sand Bay gneiss association appears to have formed in a back-arc setting at the Laurentian margin. The identity of the active arc at that time is unknown; possibilities include the Parry Sound domain, the Central Metasedimentary Belt boundary thrust zone, or another, unidentified, arc. In the latter case, it is possible that the arc rifted from the Laurentian margin (e.g., Culshaw and Dostal, 1997), but was not subsequently reaccreted.

This paper and the work of Geringer et al. (1994), Culshaw and Dostal (1997, 2002), and Blein et al. (2003) have shown that geochemical data from high-grade metamorphic rocks can yield significant information about the petrogenesis and tectonic setting of their protoliths that cannot be obtained in any other way. In principle, therefore, answers to several of the problems outlined earlier could be obtained by further geochemical and associated geochronological work:

1. The temporal relationship between arc-related and A-type plutonic rocks in the Muskoka domain could be better constrained with further geochronological work, provided that geochemical data were used to determine the tectonic significance of the dated rocks. The results from integrated geochemical/geochronological studies are likely to improve our understanding of arc evolution significantly. A similar approach should be applied to the Parry Sound domain and the Central Metasedimentary Belt boundary thrust zone in order to address the tectonic significance of the metaigneous rocks in these areas. Furthermore, additional Sm-Nd isotopic data from the rocks described in this paper, and those in the Parry Sound domain and the Central Metasedimentary Belt boundary thrust zone, could yield information about the source to the A-type granites and the substrate upon which the Muskoka-Ojibway arc and the Parry Sound and/or the Central Metasedimentary Belt boundary thrust zone (arcs?) formed. Information about the substrate to the arc-related rocks is important for relating the rocks to the Laurentian margin or to more outboard, possibly oceanic, settings, and similar information could come from dating of detrital or inherited zircons. Evidence for spatial separation between Laurentia and the inferred arcs may come from the identification of accretion-related structures or lithological assemblages (e.g., ophiolites) characteristic of suture zones. In this respect, the findings of ultramafic slivers in the Central Gneiss Belt of Québec may provide keys by which to answer this question (Giguère et al., 1998).

2. Granulite-facies metamorphism in the Britt and Algonquin domains between 1450 and 1430 Ma is compatible with a range of tectonic models, including accretion or convergence, extension, and arc magmatism. The most reliable test for distinguishing between accretion or convergence and extension is to determine the pressure-temperature-time (*P-T-t*) path for the metamorphic event. The predicted *P-T-t* paths for extension-related metamorphism involve heating to granulite-facies conditions at constant or decreasing pressure, followed by cooling at constant or increasing pressure (Sandiford and Powell, 1986), in contrast to *P-T-t* paths from metamorphic belts formed during crustal thickening (England and Thompson, 1984). However, superimposed high-grade Grenvillian metamorphism makes extracting pre-Grenvillian *P-T* information difficult, if not impossible (e.g., Ketchum et al., 1994), and requires further work in areas where Grenvillian tectonic fabrics and metamorphic assemblages are poorly developed (see Boggs and Corriveau, this volume). Determining whether granulite-facies assemblages formed in response to arc magmatism requires better constraints on both the age of granulite-facies metamorphism and the duration of arc magmatism. If arc magmatism ceased prior to granulite-facies metamorphism, any connection between arc magmatism and granulite-facies metamorphism can be discounted.

Our interpretation is generally similar to the model proposed by Rivers and Corrigan (2000), but differs from it in detail. For example, our data indicate that some of their "arc-related" plutons in the Central Gneiss Belt actually formed in a back-arc setting, and we have also identified A-type metaplutonic rocks that may have formed in extensional intra-arc settings (e.g., charnockite and granite in the Muskoka domain). Following Kay et al. (1989) and Menuge et al. (2002), we prefer a back-arc setting for the granite-rhyolite provinces in the midcontinent region, in contrast to the "continental arc" interpretation proposed by Rivers and Corrigan (2000). The substrate to the granite-rhyolite provinces is not exposed; hence, the source regions of the magmas have not been positively identified, although they are inferred to have been previously accreted juvenile arcs (e.g., Kay et al., 1989; Van Schmus et al., 1996; Menuge et al., 2002). Based on the temporal, spatial, and petrogenetic links between arc-related and A-type rocks identified in this study, we suggest that rocks exposed in the Central Gneiss Belt , e.g., in the Muskoka domain, could represent good analogs for the hidden substrate that was the source(s) of A-type silicic rocks in the granite-rhyolite provinces. Our data support a correlation between the Central Gneiss Belt and granite-rhyolite provinces, and we agree with Culshaw and Dostal (2002) that the Central Gneiss Belt represents the tectonically reworked and stacked equivalent of the Proterozoic midcontinental United States. We suggest that further efforts to resolve the Mesoproterozoic evolution of the midcontinent region could benefit from including Central Gneiss Belt rocks in future investigations.

ACKNOWLEDGMENTS

Barrie Clarke reviewed several earlier versions of the paper, resulting in significant improvements. Pat Bogutyn provided able assistance in the field, and together with Kat Eisnor did much of the sample preparation. David Slauenwhite did the XRF analyses, Pam King and Mike Tubrett the ICP-MS analyses. Sandy Grist at Dalhousie University helped with the zircon separation. Gordon Brown made the thin sections. Reviews by Mike Easton and Steve Sheppard, and editorial comments from Louise Corriveau, are greatly appreciated. Scientists and staff at the Geological Survey of Norway are gratefully acknowledged for their support. TS was supported by a Dalhousie University Killam Scholarship and by Ph.D. grant 138552/432 from the Norwegian Research Council (NFR). Field and analytical work was funded by National Sciences and Engineering Research Council (NSERC) research grants to NGC and RAJ.

REFERENCES CITED

Anderson, J.L., 1983, Proterozoic anorogenic granite plutonism of North America: Geological Society of America Memoir 161, p. 133–154.

Anderson, J.L., and Cullers, R.L., 1978, Geochemistry and evolution of the Wolf River Batholith, a late Precambrian rapakivi massif in North Wisconsin, U.S.A.: Precambrian Research, v. 7, p. 287–324.

Arth, J.G., and Barker, F., 1976, Rare earth partitioning between hornblende and dacitic liquid and implications for the genesis of trondhjemitic-tonalitic magmas: Geology, v. 4, p. 534–536.

Bacon, C.R., and Druitt, T.H., 1988, Compositional evolution of the zoned calc-alkaline magma chamber of Mount Mazama, Crater Lake, Oregon: Contributions to Mineralogy and Petrology, v. 98, p. 224–256.

Barnes, C.G., Shannon, W.M., and Kargi, H., 1999, Diverse Mesoproterozoic basaltic magmatism in west Texas: Rocky Mountain Geology, v. 34, p. 263–273.

Bea, F., 1996, Residence of REE, Y, Th, and U in granites and crustal protoliths: Implications for the chemistry of crustal melts: Journal of Petrology, v. 37, p. 521–552.

Bickford, M.E., 1988, The formation of continental crust, Part 1: A review of some principles; Part 2: An application to the Proterozoic evolution of southern North America: Geological Society of America Bulletin, v. 100, p. 1375–1391.

Blein, O., LaFlèche, M.R., and Corriveau, L., 2003, Geochemistry of the granulitic Bondy Gneiss Complex: A 1.4 Ga arc in the Central Metasedimentary Belt, Grenville Province, Canada: Precambrian Research, v. 120, p. 193–218.

Brown, G.C., Thorpe, R.S., and Webb, P.C., 1984, The geochemical characteristics of granitoids in contrasting arcs and comments on magma sources: Journal of the Geological Society of London, v. 141, p. 413–426.

Bruhn, R.L., Stern, C.R., and de Wit, M.J., 1978, Field and geochemical data bearing on the development of a Mesozoic volcano-tectonic rift zone and back-arc basin in southernmost South America: Earth and Planetary Science Letters, v. 41, p. 32–46.

Carr, S.D., Easton, R.M., Jamieson, R.A., and Culshaw, N.G., 2000, Geologic transect across the Grenville orogen of Ontario and New York: Canadian Journal of Earth Sciences, v. 37, p. 193–216.

Clemens, J.D., Holloway, J.R., and White, A.J.R., 1986, Origin of an A-type granite: Experimental constraints: American Mineralogist, v. 71, p. 317–324.

Clynne, M.A., 1990, Stratigraphic, lithologic, and major element geochemical constraints on magmatic evolution at Lassen volcanic center, California: Journal of Geophysical Research, v. 95, p. 19651–19669.

Collins, W.J., 2002, Hot orogens, tectonic switching, and creation of continental crust: Geology, v. 30, p. 535–538.

Collins, W.J., Beams, S.D., White, A.J.R., and Chappell, B.W., 1982, Nature and origin of A-type granites with particular reference to southeastern Australia: Contributions to Mineralogy and Petrology, v. 80, p. 189–200.

Condie, K.C., 1986, Geochemistry and tectonic setting of Early Proterozoic supracrustal rocks in the southwestern United States: Journal of Geology, v. 94, p. 845–864.

Condie, K.C., and Chomiak, B., 1996, Continental accretion: Contrasting Mesozoic and Early Proterozoic tectonic regimes in North America: Tectonophysics, v. 265, p. 101–126.

Coney, P.J., Jones, D.L., and Monger, J.W.H., 1980, Cordilleran suspect terranes: Nature, v. 288, p. 329–333.

Corrigan, D., Culshaw, N.G., and Mortensen, J.K., 1994, Pre-Grenvillian evolution and Grenvillian overprinting of the Parautochthonous Belt in Key Harbour, Ontario: U-Pb and field constraints: Canadian Journal of Earth Sciences, v. 31, p. 583–596.

Corriveau, L., and Morin, D., 2000, Modelling 3D architecture of western Grenville from xenoliths, styles of magma emplacement and Lithoprobe reflectors: Canadian Journal of Earth Sciences, v. 37, p. 235–251.

Corriveau, L., and van Breemen, O., 2000, Docking of the Central Metasedimentary Belt to Laurentia in geon 12: Evidence from the 1.17–1.16 Ga Chevreuil intrusive suite and host gneisses, Québec: Canadian Journal of Earth Sciences, v. 37, p. 253–269.

Creaser, R.A., and White, A.J.R., 1991, Yardea Dacite-Large-volume, high-temperature felsic volcanism from the Middle Proterozoic of South Australia: Geology, v. 19, p. 48–51.

Creaser, R.A., Price, R.C., and Wormald, R.J., 1991, A-type granites revisited: Assessment of a residual-source model: Geology, v. 19, p. 163–166.

Cullers, R.L., Koch, R.J., and Bickford, M.E., 1981, Chemical evolution of magmas in the Proterozoic terrane of the St. Francois Mountains, southeastern Missouri, 2: Trace element data: Journal of Geophysical Research, v. 86, p. 10388–10401.

Cullers, R.L., Stone, J., Anderson, J.L., Sassarini, N., and Bickford, M.E., 1993, Petrogenesis of Mesoproterozoic Oak Creek and West McCoy Gulch plutons, Colorado: An example of cumulate unmixing of mid-crustal, two-mica granite of orogenic affinity: Precambrian Research, v. 62, p. 139–169.

Culshaw, N.G., and Dostal, J., 1997, Sand Bay gneiss association, Grenville Province, Ontario: A Grenvillian rift- (and -drift) assemblage stranded in the Central Gneiss Belt?: Precambrian Research, v. 85, p. 97–113.

Culshaw, N.G., and Dostal, J., 2002, Amphibolites of the Shawanaga domain, Central Gneiss Belt, Grenville Province, Ontario: Tectonic setting and implications for relations between the Central Gneiss Belt and Midcontinental USA: Precambrian Research, v. 113, p. 65–85.

Culshaw, N.G., Davidson, A., and Nadeau, L., 1983, Structural subdivisions of the Grenville Province in the Parry Sound–Algonquin region, Ontario: Québec, Geological Survey of Canada, Current Research, Part B, Paper 83–1B, p. 243–252.

Culshaw, N.G., Corrigan, D., Drage, J., and Wallace, P., 1988, Georgian Bay geological synthesis: Key Harbour to Dillon, Grenville Province of Ontario: Ottawa, Ontario, Geological Survey of Canada, Current Research, Part C, Paper 88–1C, p. 129–133.

Culshaw, N.G., Check, G., Corrigan, D., Drage, J., Gower, R., Haggart, M.J., Wallace, P., and Wodicka, N., 1989, Georgian Bay geological synthesis: Dillon to Twelve Mile Bay, Grenville Province of Ontario: Ottawa, Ontario, Geological Survey of Canada, Current Research, Part C, Paper 89–1C, p. 157–163.

Culshaw, N.G., Ketchum, J.W.F., Wodicka, N., and Wallace, P., 1994, Deep crustal extension following thrusting in the southwestern Grenville Province, Ontario: Canadian Journal of Earth Sciences, v. 31, p. 160–175.

Culshaw, N.G., Jamieson, R.A., Ketchum, J.W.F., Wodicka, N., Corrigan, D., and Reynolds, P.H., 1997, Transect across the northwestern Grenville orogen, Georgian Bay, Ontario: Polystage convergence and extension in the lower orogenic crust: Tectonics, v. 16, p. 966–982.

Culshaw, N.G., Ketchum, J.W.F., and Barr, S., 2000, Structural evolution of the Makkovik Province, Labrador, Canada: Tectonic processes during 200 m.y. at a Paleoproterozoic active margin: Tectonics, v. 19, p. 961–977.

Culshaw, N.G., Jamieson, R.A., Slagstad, T., and Raistrick, M., 2002, The Pre-Grenville magmatic record of the Central Gneiss Belt, Grenville Province, Ontario: A long-lived continental arc: Geological Association of Canada–Mineralogical Association of Canada, Program with Abstracts, v. 27, p. 25.

Davidson, A., 1984, Identification of ductile shear zones in the southwestern Grenville Province of the Canadian Shield, *in* Kröner, A., and Greiling, E., eds., Precambrian tectonics illustrated: Stuttgart, Schweizerbart'sche Verlagsbuchhandlung, p. 263–279.

Davidson, A., 1986, Grenville Front relationships near Killarney, Ontario, *in*

Moore, J.M., et al., eds., The Grenville Province: St. John's, Newfoundland, Geological Association of Canada Special Paper 31, p. 107–117.

Davidson, A., 1995, A review of the Grenville orogen in its North American type area: AGSO [Australian Geological Survey Organisation] Journal of Australian Geology and Geophysics, v. 16, p. 3–24.

Davidson, A., and van Breemen, O., 1988, Baddeleyite-zircon relationships in coronitic metagabbro, Grenville Province, Ontario: Implications for geochronology: Contributions to Mineralogy and Petrology, v. 100, p. 291–299.

Davidson, A., Culshaw, N.G., and Nadeau, L., 1982, A tectono-metamorphic framework for part of the Grenville Province, Parry Sound region, Ontario: Ottawa, Ontario, Geological Survey of Canada, Current Research, Part A, Paper 82–1A, p. 175–190.

Davidson, A., van Breemen, O., and Sullivan, R.W., 1992, Circa 1.75 Ga ages for plutonic rocks from the Southern Province and adjacent Grenville Province: What is the expression of the Penokean orogeny?: Ottawa, Ontario, Geological Survey of Canada, Radiogenic Age and Isotopic Studies, Report 6, Paper 92–2, p. 107–118.

Davidson, J.P., Dungan, M.A., Ferguson, K.M., and Colucci, M.T., 1987, Crust-magma interactions and the evolution of arc magmas: The San Pedro-Pellado volcanic complex, southern Chilean Andes: Geology, v. 15, p. 443–446.

Davis, D., 1982, Optimum linear regression and error estimation applied to U-Pb data: Canadian Journal of Earth Sciences, v. 19, p. 2141–2149.

DePaolo, D.J., 1981, Trace element and isotopic effects of combined wallrock assimilation and fractional crystallization: Earth and Planetary Science Letters, v. 53, p. 189–202.

Devine, J.D., 1995, Petrogenesis of the basalt-andesite-dacite association of Grenada, Lesser Antilles island arc, revisited: Journal of Volcanology and Geothermal Research, v. 69, p. 1–33.

Dickin, A.P., and McNutt, R.H., 1990, Nd model-age mapping of Grenville lithotectonic domains: Mid-Proterozoic crustal evolution in Ontario, *in* Gower, C.F., et al., eds., Mid-Proterozoic Laurentian-Baltica: St. John's, Newfoundland, Geological Association of Canada Special Paper 38, p. 79–94.

Dorais, M.J., Whitney, J.A., and Roden, M.F., 1990, Origin of mafic enclaves in the Dinkey Creek Pluton, central Sierra Nevada Batholith, California: Journal of Petrology, v. 31, p. 853–881.

Dostal, J., Baragar, W.R.A., and Dupuy, C., 1986, Petrogenesis of the Natkusiak continental basalts, Victoria Island, Northwest Territories, Canada: Canadian Journal of Earth Sciences, v. 23, p. 622–632.

Dudás, F.O., Davidson, A., and Bethune, K.M., 1994, Age of the Sudbury diabase dykes and their metamorphism in the Grenville Province, Ontario: Ottawa, Ontario, Geological Survey of Canada, Radiogenic Age and Isotopic Studies, Report 8, Current Research 1994-F, p. 97–106.

Dunphy, J.M., and Ludden, J.N., 1998, Petrological and geochemical characteristics of a Paleoproterozoic magmatic arc (Narsajuaq terrane, Ungava Orogen, Canada) and comparisons to Superior Province granitoids: Precambrian Research, v. 91, p. 109–142.

Easton, R.M., 1986, Geochronology of the Grenville Province, *in* Moore, J.M., et al., eds., The Grenville Province: St. John's, Newfoundland, Geological Association of Canada Special Paper 31, p. 127–173.

Eby, G.N., 1990, The A-type granitoids: A review of their occurrence and chemical characteristics and speculations on their petrogenesis: Lithos, v. 26, p. 115–134.

England, P.C., and Thompson, A.B., 1984, Pressure-temperature-time paths of regional metamorphism, I: Heat transfer during the evolution of regions of thickened continental crust: Journal of Petrology, v. 25, p. 894–928.

Frost, B.R., Barnes, C.G., Collins, W.J., Arculus, R.J., Ellis, D.J., and Frost, C.D., 2001, A geochemical classification for granitic rocks: Journal of Petrology, v. 42, p. 2033–2048.

Frost, C.D., and Frost, B.R., 1997, Reduced rapakivi-type granites: The tholeiite connection: Geology, v. 25, p. 647–650.

Frost, C.D., Frost, B.R., Chamberlain, K.R., and Edwards, B.R., 1999, Petrogenesis of the 1.43 Ga Sherman batholith, SE Wyoming, USA: A reduced, rapakivi-type anorogenic granite: Journal of Petrology, v. 40, p. 1771–1802.

Gebauer, D., 1999, Alpine geochronology of the Central and Western Alps: New constraints for a complex geodynamic evolution: Schweizerische Mineralogische Petrographische Mitteilungen, v. 79, p. 191–208.

Geringer, G.J., Humphreys, H.C., and Scheepers, D.J., 1994, Lithostratigraphy, protolithology, and tectonic setting of the Areachap Group along the eastern margin of the Namaqua Mobile Belt, South Africa: South African Journal of Geology, v. 97, p. 78–100.

Gertisser, R., and Keller, J., 2000, From basalt to dacite: Origin and evolution of the calc-alkaline series of Salina, Aeolian Arc, Italy: Contributions to Mineralogy and Petrology, v. 139, p. 607–626.

Giguère, É., Hébert, E., Sharma, K.N.M., and Cimon, J., 1998, Les roches ultramafiques de la région de Témiscaming et de Fort-Coulonge: Québec, Ministère des Ressources naturelles du Québec, DV-98–05, p. 41.

Gill, J.B., 1984, Sr-Pb-Nd isotopic evidence that both MORB and OIB sources contribute to oceanic island arc magma sources in Fiji: Earth and Planetary Science Letters, v. 68, p. 443–458.

Govindaraju, K., 1994, Compilation of working values and sample description for 383 geostandards: Geostandards Newsletter, v. 18, p. 1–158.

Grove, T.L., and Donnelly-Nolan, J.M., 1986, The evolution of young silicic lavas at Medicine Lake Volcano, California: Implications for the origin of compositional gaps in calc-alkaline series lavas: Contributions to Mineralogy and Petrology, v. 92, p. 281–302.

Gutscher, M.-A., Spakman, W., Bijwaard, H., and Engdahl, E.R., 2000, Geodynamics of flat subduction: Seismicity and tomographic constraints from the Andean margin: Tectonics, v. 19, p. 814–833.

Haggart, M.J., Jamieson, R.A., Reynolds, P.H., Krogh, T.E., Beaumont, C., and Culshaw, N.G., 1993, Last gasp of the Grenville orogeny—Thermochronology of the Grenville Front Tectonic Zone near Killarney, Ontario: Journal of Geology, v. 101, p. 575–589.

Hanmer, S., and McEachern, S.J., 1992, Kinematical and rheological evolution of a crustal-scale ductile thrust zone, Central Metasedimentary Belt, Grenville orogen, Ontario: Canadian Journal of Earth Sciences, v. 29, p. 1779–1790.

Hanmer, S., Corrigan, D., Pehrsson, S., and Nadeau, L., 2000, SW Grenville Province, Canada: The case against post–1.4 Ga accretionary tectonics: Tectonophysics, v. 319, p. 33–51.

Hanson, G.N., 1978, The application of trace elements to the petrogenesis of igneous rocks of granitic composition: Earth and Planetary Science Letters, v. 38, p. 26–43.

Hildreth, W., 1981, Gradients in silicic magma chambers: Implications for lithospheric magmatism: Journal of Geophysical Research, v. 86, p. 10153–10192.

Hochstaedter, A.G., Gill, J.B., and Morris, J.D., 1990, Volcanism in the Sumisu Rift, II: Subduction and non-subduction related components: Earth and Planetary Science Letters, v. 100, p. 195–209.

Hoffman, P.F., 1989, Precambrian geology and tectonic history of North America, *in* Bally, A.W., and Palmer, A.R., eds., The Geology of North America: An overview: Boulder, Colorado, Geological Society of America, Geology of North America, v. A, p. 447–511.

Irvine, T.N., and Baragar, W.R.A., 1971, A guide to the chemical classification of the common volcanic rocks: Canadian Journal of Earth Sciences, v. 8, p. 523–548.

Jaffey, A.H., Flynn, K.F., Glendenin, W.C., Bentley, W.C., and Essling, A.M., 1971, Precision measurements of half-lives and specific activities of ^{235}U and ^{238}U: Physical Reviews C: Nuclear Physics, v. 4, p. 1889–1906.

Jamieson, R.A., Williams, M.L., Jercinovic, M.J., and Timmermann, H., 2001, Chemical and age zoning in monazite from polycyclic paragneiss, Central Gneiss Belt, Grenville Orogen, Ontario: Geological Association of Canada–Mineralogical Association of Canada, Program with Abstracts, v. 26, p. 72.

Kay, S.M., Ramos, V.A., Mpodozis, C., and Sruoga, P., 1989, Late Paleozoic to Jurassic silicic magmatism at the Gondwana margin: Analogy to the Middle Proterozoic in North America?: Geology, v. 17, p. 324–328.

Kay, S.M., Kay, R.B., Citron, G.P., and Perfit, M.R., 1990, Calc-alkaline plutonism in the intra-oceanic Aleutian arc, Alaska, *in* Kay, S.M., and Rapela,

C.W., eds., Plutonism from Antarctica to Alaska: Boulder, Colorado, Geological Society of America Special Paper 241, p. 233–255.

Ketchum, J.W.F., 1995, Extensional shear zones and lithotectonic domains in the southwest Grenville orogen: Structure, metamorphism, and U-Pb geochronology of the Central Gneiss Belt near Pointe-au-Baril, Ontario [Ph.D. thesis]: Halifax, Nova Scotia, Dalhousie University, 341 p.

Ketchum, J.W.F., and Davidson, A., 2000, Crustal architecture and tectonic assembly of the Central Gneiss Belt, southwestern Grenville Province, Canada: A new interpretation: Canadian Journal of Earth Sciences, v. 37, p. 217–234.

Ketchum, J.W.F., Jamieson, R.A., Heaman, L.M., Culshaw, N.G., and Krogh, T.E., 1994, 1.45 Ga granulites in the southwestern Grenville Province: Geologic setting, *P-T* conditions, and U-Pb geochronology: Geology, v. 22, p. 215–218.

Ketchum, J.W.F., Culshaw, N.G., and Dunning, G.R., 1997, U-Pb geochronologic constraints on Paleoproterozoic orogenesis in the northwestern Makkovik Province, Labrador, Canada: Canadian Journal of Earth Sciences, v. 34, p. 1072–1088.

Kisvarsanyi, E.B., and Kisvarsanyi, G., 1990, Alkaline granite ring complexes and metallogeny in the Middle Proterozoic St. Francois mountains, *in* Gower, C.F., et al., eds., Mid-Proterozoic Laurentia-Baltica: St. John's, Newfoundland, Geological Association of Canada Special Paper 38, p. 433–446.

Kretz, R., 1983, Symbols for rock-forming minerals: American Mineralogist, v. 68, p. 277–279.

Krogh, T.E., 1982, Improved accuracy of U-Pb zircon ages by the creation of more concordant systems using an air abrasion technique: Geochimica et Cosmochimica Acta, v. 46, p. 637–649.

Krogh, T.E., 1991, U-Pb zircon geochronology in the western Grenville Province: Paper presented at Lithoprobe Abitibi-Grenville Transect, Workshop III, 25–27 April, University of Toronto.

Krogh, T.E., 1994, Precise U-Pb ages for Grenvillian and pre-Grenvillian thrusting of Proterozoic and Archean metamorphic assemblages in the Grenville Front Tectonic Zone, Canada: Tectonics, v. 13, p. 963–982.

Krogh, T.E., Gower, C.F., and Wardle, R.J., 1996, Pre-Labradorian crust and later Labradorian, Pinwarian and Grenvillian metamorphism in the Mealy Mountains terrane, Grenville Province, eastern Labrador, *in* Gower, C.F., ed., Proterozoic Evolution in the North Atlantic Realm, COPENA ECSOOT-IBTA Conference, 29 July–2 August, Goose Bay, Labrador, Program and Abstracts, p. 106–107.

Landenberger, B., and Collins, W.J., 1996, Derivation of A-type granites from a dehydrated charnockitic lower crust: Evidence from the Chaelundi complex, eastern Australia: Journal of Petrology, v. 37, p. 145–170.

Loiselle, M.C., and Wones, D.R., 1979, Characteristics and origin of anorogenic granites: Geological Society of America, Abstracts with Programs, v. 11, p. 468.

Longerich, H.P., Jenner, G.A., Fryer, B.J., and Jackson, S.E., 1990, Inductively coupled plasma-mass spectrometric analysis of geological samples: A critical evaluation based on case studies: Chemical Geology, v. 83, p. 105–118.

Mahood, G., and Hildreth, W., 1983, Large partition coefficients for trace elements in high-silica rhyolites: Geochimica et Cosmochimica Acta, v. 47, p. 11–30.

Maniar, P.D., and Piccoli, P.M., 1989, Tectonic discrimination of granitoids: Geological Society of America Bulletin, v. 101, p. 635–643.

Martin, H., 1987, Petrogenesis of Archean trondhjemites, tonalites, and granodiorites from eastern Finland: Major and trace element geochemistry: Journal of Petrology, v. 28, p. 921–953.

McEachern, S.J., and van Breemen, O., 1993, Age of deformation within the Central Metasedimentary Belt boundary thrust zone, southwest Grenville orogen: Constraints on the collision of the Mid-Proterozoic Elzevir terrane: Canadian Journal of Earth Sciences, v. 30, p. 1155–1165.

McMullen, S.M., 1999, Tectonic evolution of the Bark Lake area, eastern Central Gneiss Belt, Ontario Grenville: Constraints from geology, geochemistry and U-Pb geochronology [M.S. thesis]: Ottawa, Ontario, Carleton University, 175 p.

Menuge, J.A., Brewer, T.S., and Seeger, C.M., 2002, Petrogenesis of metaluminous A-type rhyolites from the St. Francois Mountains, Missouri, and the Mesoproterozoic evolution of the southern Laurentian margin: Precambrian Research, v. 113, p. 269–291.

Miyashiro, A., 1974, Volcanic rock series in island arcs and active continental margins: American Journal of Science, v. 274, p. 321–355.

Molnar, P., and Atwater, T., 1978, Inter-arc spreading and Cordilleran tectonics as alternates related to the age of the subducted lithosphere: Earth and Planetary Science Letters, v. 41, p. 330–340.

Nadeau, L., 1990, Tectonic, thermal and magmatic evolution of the Central Gneiss Belt, Huntsville region, southwestern Grenville orogen [Ph.D. thesis]: Ottawa, Ontario, Carleton University, 190 p.

Nadeau, L., and Hanmer, S., 1992, Deep-crustal break-back stacking and slow exhumation of the continental footwall beneath a thrusted marginal basin, Grenville orogen, Canada: Tectonophysics, v. 210, p. 215–233.

Nadeau, L., and van Breemen, O., 1998, Plutonic ages and tectonic setting of the Algonquin and Muskoka allochthons, Central Gneiss Belt, Grenville Province, Ontario: Canadian Journal of Earth Sciences, v. 35, p. 1423–1438.

Nantel, S., and Pintson, H., 2002, Géologie de la région du Lac Dieppe (31O/3): Québec, Ministère des Ressources naturelles du Québec, RG 2001–16, 36 p.

Nash, W.P., and Crecraft, H.R., 1985, Partition coefficients for trace elements in silicic magmas: Geochimica et Cosmochimica Acta, v. 49, p. 2309–2322.

Nédélec, A., Stephens, W.E., and Fallick, A.E., 1995, The Panafrican stratoid granites of Madagascar: Alkaline magmatism in a post-collisional extensional setting: Journal of Petrology, v. 36, p. 1367–1391.

Okamura, S., Arculus, R.J., Martynov, Y.A., Kagami, H., Yoshida, T., and Kawano, Y., 1998, Multiple magma sources involved in marginal-sea formation: Pb, Sr, and Nd isotopic evidence from the Japan Sea region: Geology, v. 26, p. 619–622.

Passchier, C.W., Myers, J.S., and Kröner, A., 1990, Field geology of high-grade gneiss terrains: Berlin, Springer-Verlag, 150 p.

Patiño Douce, A.E., 1997, Generation of metaluminous A-type granites by low-pressure melting of calc-alkaline granitoids: Geology, v. 25, p. 743–746.

Pearce, J.A., 1982, Trace element characteristics of lavas from destructive plate boundaries, *in* Thorpe, R.S., ed., Andesites: New York, John Wiley and Sons, p. 525–548.

Pearce, J.A., and Cann, J.R., 1973, Tectonic setting of basic volcanic rocks determined using trace element analyses: Earth and Planetary Science Letters, v. 19, p. 290–300.

Pearce, J.A., and Peate, D.W., 1995, Tectonic implications of the composition of volcanic arc magmas: Annual Review of Earth and Planetary Sciences, v. 23, p. 251–285.

Pearce, J.A., Harris, N.B.W., and Tindle, A.G., 1984, Trace element discrimination diagrams for the tectonic interpretation of granitic rocks: Journal of Petrology, v. 25, p. 956–983.

Pearce, T.H., 1968, A contribution to the theory of variation diagrams: Contributions to Mineralogy and Petrology, v. 19, p. 142–157.

Poldervaart, A., 1956, Zircons in rocks, 2: Igneous rocks: American Journal of Science, v. 254, p. 521–554.

Pope, D.C., and Willett, S.D., 1998, Thermal-mechanical model for crustal thickening in the central Andes driven by ablative subduction: Geology, v. 26, p. 511–514.

Rivers, T., 1997, Lithotectonic elements of the Grenville Province: Review and tectonic implications: Precambrian Research, v. 86, p. 117–154.

Rivers, T., and Corrigan, D., 2000, Convergent margin on southeastern Laurentia during the Mesoproterozoic: Tectonic implications: Canadian Journal of Earth Sciences, v. 37, p. 359–383.

Rivers, T., Martignole, J., Gower, C.F., and Davidson, A., 1989, New tectonic divisions of the Grenville Province, southeast Canadian Shield: Tectonics, v. 8, p. 63–84.

Rollinson, H.R., 1993, Using geochemical data: Evaluation, presentation, interpretation: Essex, England, Longman, 352 p.

Royden, L.H., 1993, The tectonic expression of slab pull at continental convergent boundaries: Tectonics, v. 12, p. 303–325.

Russell, J.K., and Nicholls, J., 1988, Analysis of petrologic hypotheses with Pearce element ratios: Contributions to Mineralogy and Petrology, v. 99, p. 25–35.

Sandiford, M., and Powell, R., 1986, Deep crustal metamorphism during continental extension: Modern and ancient examples: Earth and Planetary Science Letters, v. 79, p. 151–158.

Saunders, A.D., Tarney, J., Stern, C.R., and Dalziel, I.W.D., 1979, Geochemistry of Mesozoic marginal basin floor igneous rocks from southern Chile: Geological Society of America Bulletin, v. 90, p. 237–258.

Sawyer, E.W., 1999, Criteria for the recognition of partial melting: Physics and Chemistry of the Earth, v. 24, p. 269–279.

Slagstad, T., 2003, Muskoka and Shawanaga domains, Central Gneiss Belt, Grenville Province, Ontario: Geochemical and geochronological constraints on pre-Grenvillian and Grenvillian geological evolution [Ph.D. thesis]: Halifax, Nova Scotia, Dalhousie University, 336 p.

Smith, S.E., and Humphris, S.E., 1998, Geochemistry of basaltic rocks from the TAG hydrothermal mound (26°08′N), Mid-Atlantic Ridge, *in* Herzig, P.M., et al., eds., Proceedings of the Ocean Drilling Program: Scientific results, v. 158: College Station, Texas A&M University, Ocean Drilling Program, p. 213–229.

Söderlund, U., Möller, C., Andersson, J., Johansson, L., and Whitehouse, M., 2002, Zircon geochronology in polymetamorphic gneisses in the Sveconorwegian orogen, SW Sweden: Ion microprobe evidence for 1.46–1.42 and 0.98–0.96 Ga reworking: Precambrian Research, v. 113, p. 193–225.

Solar, G.S., and Brown, M., 2001, Petrogenesis of migmatites in Maine, USA: Possible source of peraluminous leucogranite in plutons: Journal of Petrology, v. 42, p. 789–823.

Stacey, J.S., and Kramers, J.D., 1975, Approximation of terrestrial lead isotope evolution by a two-stage model: Earth and Planetary Science Letters, v. 26, p. 207–221.

St-Onge, M.R., Wodicka, N., and Lucas, S.B., 2000, Granulite- and amphibolite-facies metamorphism in a convergent-plate-margin setting: Synthesis of the Quebec-Baffin segment of the Trans-Hudson orogen: Canadian Mineralogist, v. 38, p. 379–398.

Streckeisen, A.L., 1973, Plutonic rocks: Classification and nomenclature recommended by the IUGS subcommission on the systematics of igneous rocks: Geotimes, 18 October, p. 26–30.

Sun, S.-S., and McDonough, W.F., 1989, Chemical and isotopic systematics of oceanic basalts: Implications for mantle composition and processes, *in* Saunders, A.D., and Norry, M.J., eds., Magmatism in the ocean basins: Geological Society of London Special Publication 42, p. 313–345.

Timmermann, H., 1998, Geology, metamorphism, and U-Pb geochronology in the Central Gneiss Belt between Huntsville and Haliburton, southwestern Grenville Province, Ontario [Ph.D. thesis]: Halifax, Nova Scotia, Dalhousie University, 410 p.

Timmermann, H., Parrish, R.R., Jamieson, R.A., and Culshaw, N.G., 1997, Time of metamorphism beneath the Central Metasedimentary Belt boundary thrust zone, Grenville orogen, Ontario: Accretion at 1080 Ma?: Canadian Journal of Earth Sciences, v. 34, p. 1023–1029.

Timmermann, H., Jamieson, R.A., Parrish, R.R., and Culshaw, N.G., 2002, Coeval migmatites and granulites, Muskoka domain, southwestern Grenville Province: Canadian Journal of Earth Sciences, v. 39, p. 239–258.

Tindle, A.G., McGarvie, D.W., and Webb, P.C., 1988, The role of hybridization and crystal fractionation in the evolution of the Cairnsmore of Carsphairn intrusion, southern Uplands of Scotland: Journal of the Geological Society of London, v. 145, p. 11–21.

Tuccillo, M.E., Mezger, K., Essene, E.J., and van der Plujim, B.A., 1992, Thermobarometry, geochronology and the interpretation of *P-T-t* data in the Britt domain, Ontario, Grenville orogen, Canada: Journal of Petrology, v. 33, p. 1225–1259.

Turner, S.P., Foden, J.D., and Morrison, R.S., 1992, Derivation of some A-type magmas by fractionation of basaltic magma: An example from the Padthaway Ridge, south Australia: Lithos, v. 28, p. 151–179.

van Breemen, O., and Davidson, A., 1988, Northeast extension of Proterozoic terranes of mid-continental North America: Geological Society of America Bulletin, v. 100, p. 630–638.

van Breemen, O., Davidson, A., Loveridge, W.D., and Sullivan, R.W., 1986, U-Pb zircon geochronology of Grenville tectonites, granulites and igneous precursors, Parry Sound, Ontario, *in* Moore, J.M., et al., eds., The Grenville Province: St. John's, Newfoundland, Geological Association of Canada Special Paper 31, p. 192–207.

Van Schmus, W.R., Bickford, M.E., and Turek, A., 1996, Proterozoic geology of the east-central Midcontinent basement, *in* van der Pluijm, B.A., and Catacosinos, P.A., eds., Basement and basins of eastern North America: Boulder, Colorado, Geological Society of America Special Paper 308, p. 7–32.

van Staal, C.R., Dewey, J.F., Mac Niocaill, C., and McKerrow, W.S., 1998, The Cambrian-Silurian tectonic evolution of the northern Appalachians and British Caledonides: History of a complex west and southwest Pacific-type segment of Iapetus, in Blundell, D.J., and Scott, A.C., eds., Lyell: The past is the key to the present: Geological Society of London Special Publication 143, p. 199–242.

Wark, D.A., and Miller, C.F., 1993, Accessory mineral behavior during differentiation of a granite suite: Monazite, xenotime and zircon in the Sweetwater Wash pluton, southeastern California, USA: Chemical Geology, v. 110, p. 49–67.

Waschbusch, P., and Beaumont, C., 1996, Effect of a retreating subduction zone on deformation in simple regions of plate convergence: Journal of Geophysical Research, v. 101, p. 28133–28148.

Watson, E.B., and Harrison, T.M., 1983, Zircon saturation revisited: Temperature and composition effects in a variety of crustal magma types: Earth and Planetary Science Letters, v. 64, p. 295–304.

Whalen, J.B., Currie, K.L., and Chappell, B.W., 1987, A-type granites: Geochemical characteristics, discrimination and petrogenesis: Contributions to Mineralogy and Petrology, v. 95, p. 407–419.

Winchester, J.A., and Floyd, P.A., 1977, Geochemical discrimination of different magma series and their differentiation products using immobile elements: Chemical Geology, v. 20, p. 325–343.

Windley, B.F., 1993, Proterozoic anorogenic magmatism and its orogenic connections: Journal of the Geological Society of London, v. 150, p. 39–50.

Wodicka, N., 1994, Middle Proterozoic evolution of the Parry Sound domain, southwestern Grenville orogen, Ontario: Structural, metamorphic, U/Pb, and $^{40}Ar/^{39}Ar$ constraints [Ph.D. thesis]: Halifax, Nova Scotia, Dalhousie University, 357 p.

Wodicka, N., Parrish, R.R., and Jamieson, R.A., 1996, The Parry Sound domain: A far-travelled allochthon? New evidence from U-Pb zircon geochronology: Canadian Journal of Earth Sciences, v. 33, p. 1087–1104.

Wodicka, N., Ketchum, J.W.F., and Jamieson, R.A., 2000, Grenvillian metamorphism of monocyclic rocks, Georgian Bay, Ontario, Canada: Implications for convergence history: Canadian Mineralogist, v. 38, p. 471–510.

Wood, D.A., 1979, A variably-veined sub-oceanic upper mantle: Genetic significance for mid-ocean ridge basalts from geochemical evidence: Geology, v. 7, p. 499–503.

Wynne-Edwards, H.R., 1972, The Grenville Province, *in* Price, R.A., and Douglas, R.J.W., eds., Variations in tectonic styles in Canada: St. John's, Newfoundland, Geological Association of Canada Special Paper 11, p. 263–334.

Yin, A., and Harrison, T.M., 2000, Geologic evolution of the Himalayan-Tibetan orogen: Annual Review of Earth and Planetary Sciences, v. 28, p. 211–280.

Manuscript Accepted by the Society August 25, 2003

Printed in the USA

Geological Society of America
Memoir 197
2004

SHRIMP U-Pb zircon geochronology of the Bondy gneiss complex: Evidence for circa 1.39 Ga arc magmatism and polyphase Grenvillian metamorphism in the Central Metasedimentary Belt, Grenville Province, Québec

Natasha Wodicka*
Geological Survey of Canada, Continental Geoscience Division, 601 Booth Street, Ottawa, Ontario K1A 0E8, Canada
Louise Corriveau
Geological Survey of Canada, Québec Geoscience Centre, P.O. Box 7500, Sainte-Foy, Québec G1V 4C7, Canada
Richard A. Stern**
Geological Survey of Canada, Continental Geoscience Division, 601 Booth Street, Ottawa, Ontario K1A 0E8, Canada

ABSTRACT

We report sensitive high-resolution ion microprobe (SHRIMP) U-Pb zircon ages from high-grade gneisses of the Bondy gneiss complex, a volcano-plutonic arc and back-arc edifice hosting a Cu-Au-Fe oxides hydrothermal system in the Central Metasedimentary Belt, Grenville Province, Québec. Samples of quartzofeldspathic gneiss gave broadly similar results, with zircon cores indicating ages between 1.59 and 1.21 Ga and mantles or whole new zircon crystals giving Grenvillian metamorphic ages. A few analyses indicate minor involvement of Paleoproterozoic and Archean crustal material. Zircon cores are interpreted to have grown during crystallization of the quartzofeldspathic gneiss protoliths, but precise ages could not be determined for any of the samples. The spread of ages is attributed mainly to isotopic disturbance owing to hydrothermal alteration, high-pressure-temperature (*P-T*) metamorphism, and/or recent Pb loss. Only a sample of tonalitic gneiss yielded a well-defined, igneous crystallization age of 1386 ± 10 Ma. Younger zircons from the dated lithologies provide evidence for two episodes of Grenvillian metamorphism: a period of high-*P-T* granulite-facies metamorphism and partial melting at 1.21–1.18 Ga, and a younger, albeit localized, metamorphic overprint at ca. 1.15–1.13 Ga.

The new SHRIMP U-Pb data indicate that the composite volcano-plutonic edifice likely formed, at least in part, at ca. 1.39 Ga. Together with recently published geochemical and Nd data, the new data (1) extend the known distribution of ca. 1.39 Ga arc-related magmatism in the Grenville Province, and (2) suggest that a major portion of the source region to the Bondy gneiss complex was produced by the addition of voluminous juvenile Mesoproterozoic material.

Keywords: Grenville Province, Central Metasedimentary Belt, Bondy gneiss complex, SHRIMP U-Pb geochronology

*E-mail: nwodicka@nrcan.gc.ca.
**Current address: Centre for Microscopy and Microanalysis (M010), The University of Western Australia, 35 Stirling Highway, Crawley, Western Australia 6009, Australia.

Wodicka, N., Corriveau, L., and Stern, R.A., 2004, SHRIMP U-Pb zircon geochronology of the Bondy gneiss complex: Evidence for circa 1.39 Ga arc magmatism and polyphase Grenvillian metamorphism in the Central Metasedimentary Belt, Grenville Province, Québec, *in* Tollo, R.P., Corriveau, L., McLelland, J., and Bartholomew, M.J., eds., Proterozoic tectonic evolution of the Grenville orogen in North America: Boulder, Colorado, Geological Society of America Memoir 197, p. 243–266. For permission to copy, contact editing@geosociety.org.

INTRODUCTION

The Central Metasedimentary Belt in the southwestern Grenville Province (Fig. 1) is generally interpreted as a composite system of post–1300 Ma magmatic arcs and marginal basins (e.g., Windley, 1986; Easton, 1992; Davidson, 1995). The time of accretion of this belt to the southeastern margin of Laurentia, however, is still the subject of much debate (e.g., McEachern and van Breemen, 1993; Timmermann et al., 1997; Wasteneys et al., 1999; Carr et al., 2000; Corriveau and van Breemen, 2000; Hanmer et al., 2000) and may not have been synchronous along the entire strike length of the belt. Besides problems related to the nature and timing of the main collisional event(s), little is known about the basement to supracrustal rocks of the Central Metasedimentary Belt (e.g., Easton 1992; Hildebrand and Easton, 1995), and the question arises as to the nature and age of the crust beneath this belt. The Mont-Laurier area of the Central Metasedimentary Belt is of interest because it contains a series of gneiss complexes that crop out as tectonic domes within two distinctive, north-northeast-trending supracrustal domains (i.e., marble and quartzite domains; Fig. 1, C). The domal structure and distribution of these gneiss complexes suggest that they represent windows of a major lithotectonic domain structurally underlying the supracrustal rocks (Corriveau and Morin, 2000).

A detailed study was undertaken on one of the largest gneiss complexes in the Mont-Laurier area, the Bondy gneiss complex (Figs. 1, C, and 2), to unravel the nature and age of the gneiss protoliths. This area has received considerable attention in recent years, not only because it may have served as basement to supracrustal rocks of the Central Metasedimentary Belt, but also because it is host to a large-scale Cu-Au-Fe oxides hydrothermal system. Although hydrothermal alteration, high-grade metamorphism, anatexis, and polyphase deformation all contributed to the destruction of primary features of protoliths of this gneiss complex, new sensitive high-resolution ion microprobe (SHRIMP) U-Pb zircon age data presented in this paper, in conjunction with geochemical and Nd data (Blein et al., 2003), lend strong support to the possibility that the Bondy gneiss complex represents a ca. 1.39 Ga volcano-plutonic edifice formed in a juvenile arc and back-arc setting. The U-Pb geochronological data also provide evidence for polyphase Grenvillian metamorphism within the gneiss complex. The data allow us to infer a link between the Bondy gneiss complex and Mesoproterozoic juvenile rocks outside of the Central Metasedimentary Belt, including the Central Metasedimentary Belt boundary zone, the Central Granulite Terrane, and allochthonous slices within the Central Gneiss Belt (Fig. 1, B).

GEOLOGICAL SETTING

Regional Framework

The Central Metasedimentary Belt (Wynne-Edwards, 1972), which extends southward from western Québec into Ontario and New York state (Fig. 1, B), consists dominantly of ca. 1300–1250 Ma volcanic, plutonic, volcaniclastic, carbonate, and siliciclastic rocks interpreted to have originated within arcs, rifted arcs, and marginal basins (cf. Easton, 1992; Carr et al., 2000, and references therein). In western Québec, the Central Metasedimentary Belt structurally overlies reworked Archean and Proterozoic rocks of the pre-Grenvillian (following Moore and Thompson [1980], the term *Grenvillian* is used in this paper to describe rocks and events within the range ca. 1250–980 Ma and includes the Elzevirian orogeny) Laurentian margin (Central Gneiss Belt) along a major east-dipping, crustal-scale thrust, i.e., the Cayamant deformation zone, that merges to the north with the granulite-facies Baskatong-Désert dextral shear zone (Fig. 1, C; Sharma et al., 1999; Martignole et al., 2000, and references therein). At depth, the boundary between the two belts is expressed by a major, southeast-dipping, crustal-scale ramp (Baskatong ramp) interpreted to have accommodated strain during and after accretion of the Central Metasedimentary Belt onto the Laurentian margin (Martignole and Calvert, 1996; Corriveau et al., 1998; Martignole et al., 2000). The Cabonga terrane, which comprises Mesoproterozoic upper amphibolite- and granulite-facies metasedimentary and plutonic rocks, is most likely rooted in the Central Metasedimentary Belt (Martignole and Calvert, 1996). To the east, the Central Metasedimentary Belt is separated from the extensive 1165–1135 Ma Morin anorthosite-mangerite-charnockite-granite (AMCG) suite (Emslie and Hunt, 1990; Doig, 1991; Friedman and Martignole, 1995) and the host granulite-facies supracrustal rocks of the Morin terrane by the north-northeast–striking, subvertical Labelle deformation zone (Fig. 1, C; Martignole and Corriveau, 1991; Martignole et al., 2000).

The Central Metasedimentary Belt in western Québec (Fig. 1, C) comprises three discrete lithotectonic domains, including (1) a thin-skinned western marble domain dominated by calcitic and dolomitic marble intercalated with calc-silicate rock, quartzite, and pelitic and quartzofeldspathic gneiss; (2) an eastern quartzite domain composed of mostly quartzite and siliceous gneiss intercalated with metapelite, biotite- or graphite-bearing quartzofeldspathic gneiss, marble, calc-silicate rock, and amphibolite; and (3) a number of quartzofeldspathic and granitic to tonalitic gneiss complexes (e.g., the Bondy gneiss complex; Fig. 1, C) that crop out as elongate structural domes in the core of both supracrustal domains between the Heney and Labelle deformation zones (Corriveau et al., 1998, and references therein; Corriveau and van Breemen, 2000). The marble- and quartzite-rich domains represent the approximate northward continuations of the Elzevir and Frontenac terranes in Ontario, respectively (Fig. 1, B; Corriveau, 1990; Corriveau et al., 1998), whereas the gneiss complexes are interpreted as representing structural windows of a major east-dipping lithotectonic domain beneath the supracrustal domains (Corriveau and Morin, 2000). The lithological transitions between the gneiss complexes and the supracrustal domains are abrupt and are inferred to represent cryptic tectonic boundaries. Recent U-Pb dating indicates that

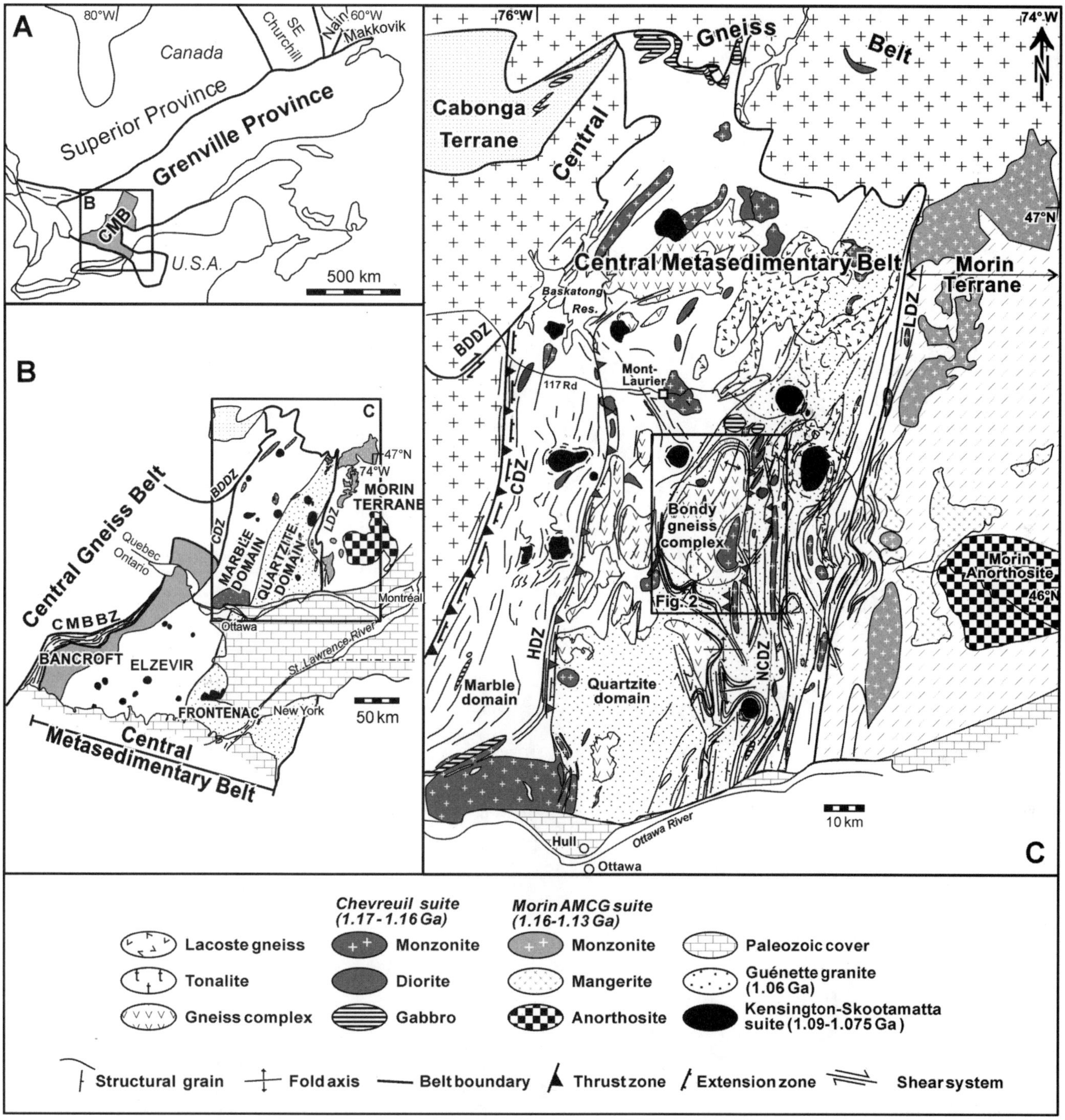

Figure 1. (A) Location of the Central Metasedimentary Belt in the Grenville Province. The box outlines the location of B. (B) Subdivision of terranes and domains in the Central Metasedimentary Belt (after Corriveau et al., 1998). The box outlines the location of C. (C) Generalized geological map of the Central Metasedimentary Belt and adjacent terranes in western Québec (after Davidson, 1998, and Corriveau and van Breemen, 2000). The distribution of gneiss complexes and different plutonic suites in the quartzite (stippled pattern) and marble (unpatterned) domains is emphasized. The box outlines the location of Figure 2. BDDZ—Baskatong-Désert deformation zone; CDZ—Cayamant deformation zone; CMB—Central Metasedimentary Belt; CMBBZ—Central Metasedimentary Belt boundary zone; HDZ—Heney deformation zone; LDZ—Labelle deformation zone; NCDZ—Nominingue-Chénéville deformation zone.

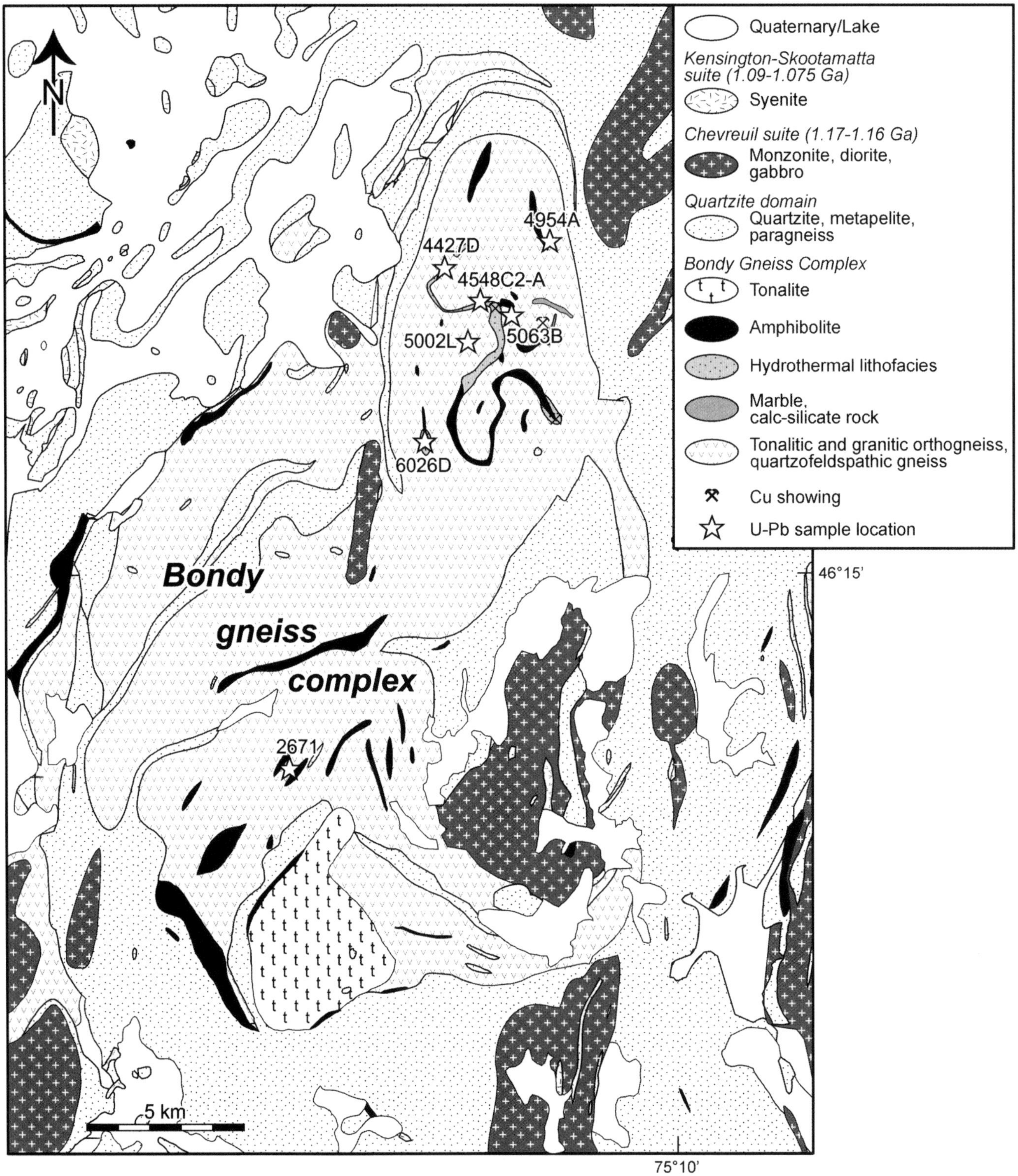

Figure 2. Simplified geological map of the Bondy gneiss complex and its environs (modified from Corriveau and Jourdain, 1992, 2000; Corriveau et al., 1994), showing locations of the SHRIMP U-Pb geochronology samples.

the gneiss complexes represent the oldest component of the Central Metasedimentary Belt, with no known equivalent in neighboring Ontario. In the Lacoste gneiss northeast of Mont-Laurier (Fig. 1, C), a tonalite has a U-Pb zircon crystallization age of 1386 ± 18 Ma (Nantel and Pintson, 2002), similar to a new SHRIMP age reported later for a tonalitic gneiss of the Bondy gneiss complex. The paleotectonic environment for these gneiss complexes is interpreted to have been an arc formed on juvenile crust (Nantel and Pintson, 2002; Blein et al., 2003). Based upon U-Pb ages of detrital zircons, deposition of sediments in the quartzite-rich domain of the Central Metasedimentary Belt took place sometime after ca. 1300 Ma and perhaps as recently as <1209 Ma (Friedman and Martignole, 1995), although the interpretation of the youngest zircon as a detrital grain has recently been questioned (Corriveau and Morin, 2000; Davidson et al., 2002). Nonetheless, the maximum time of deposition of the sediments likely lies in the interval 1246–1275 Ma (Davidson et al., 2002). The detrital zircon ages also attest to contribution from much older sources, possibly as old as Archean (Friedman and Martignole, 1995). Whether or not the older gneiss complexes represent the stratigraphic basement to the marble- and quartzite-rich domains of the Central Metasedimentary Belt remains uncertain (e.g., Wynne-Edwards et al., 1966; Corriveau and Morin, 2000; Harris et al., 2001; Davidson et al., 2002).

Maximum metamorphic conditions preserved across the Central Metasedimentary Belt in western Quebéc range from ~650 °C and 600 MPa along the Cayamant deformation zone to ~750 °C and ~800 MPa in the marble domain, ~950 °C and 1000 MPa in the Bondy gneiss complex, and ~725 °C and ~850 MPa along the eastern boundary of the belt (Fig. 1, C; Kretz, 1980, 1990; Perkins et al., 1982; Indares and Martignole, 1990; Boggs, 1996; Boggs and Corriveau, this volume). U-Pb dating of metamorphic zircon and monazite indicates that peak metamorphism in the Bondy gneiss complex took place between ca. 1.20 and 1.18 Ga (Corriveau and van Breemen, 2000), followed by cooling through ~800–850 °C by ca. 1.18 Ga (Boggs, 1996; Corriveau and van Breemen, 2000). Metamorphic zircon and monazite in supracrustal rocks of the quartzite domain are generally younger, ranging between 1.14 and 1.17 Ga (Friedman and Martignole, 1995; Corriveau and van Breemen, 2000).

The supracrustal rocks and gneiss complexes of the Central Metasedimentary Belt were intruded by three plutonic suites (Fig. 1, C), namely (1) the 1.17–1.16 Ga monzonite-gabbro Chevreuil suite (Corriveau et al., 1998; Corriveau and van Breemen, 2000); (2) the 1.09–1.075 Ga potassic alkaline Kensington-Skootamatta suite, including minette dikes with exotic xenoliths (Corriveau et al., 1990; Corriveau and Gorton, 1993; Morin and Corriveau, 1996); and (3) the 1.06 Ga Guénette granite suite (Friedman and Martignole, 1995). The Chevreuil intrusive suite is broadly coeval with the earliest phases of the Morin AMCG suite in the Morin terrane, the 1.18–1.15 Ga Frontenac plutonic suite (Wasteneys et al., 1999; Davidson and van Breemen, 2000, and references therein) in the Elzevir and Frontenac terranes, and the 1.17 Ga Hyde School Gneiss in the Adirondack Lowlands (Wasteneys et al., 1999). The ca. 1.17–1.16 Ga Chevreuil suite intrusions, which consist of vertically layered gabbro stocks, monzonite-diorite-gabbro sheet intrusions, and a swarm of microdiorite dikes, represent an important chronological field marker for inferring the timing of deformation and metamorphism in the Central Metasedimentary Belt (Corriveau et al., 1998; Corriveau and van Breemen, 2000; Harris et al., 2001). Chevreuil plutons and dikes post-date the regional ca. 1.20–1.18 Ga high-grade metamorphism recorded by gneissic rocks of the Bondy gneiss complex. Throughout this area, the plutons and dikes cut sharply across or include migmatitic paragneisses and felsic gneisses, but are themselves non-deformed or -migmatized. Outside of the rheologically strong gneiss complexes, mafic dikes of the same suite are mildly to strongly deformed and recrystallized at amphibolite facies. A U-Pb zircon crystallization age of 1156 ± 2 Ma (Corriveau and van Breemen, 2000) for a late-stage Chevreuil suite pegmatite in the Nominingue-Chénéville deformation zone indicates that amphibolite-facies deformation had largely waned by this time. Thus, based on the extent of deformation of Chevreuil plutons and dikes and their sheetlike emplacement along the tectonic boundaries of the Central Metasedimentary Belt, Corriveau et al. (1998) argued that the sharp changes in the metamorphic record and the differences in metamorphic ages outlined earlier reflect the imprint of successive tectonometamorphic events strongly partitioned across the mechanically distinct gneiss complexes, supracrustal assemblages, and deformation zones. Corriveau and co-workers (Corriveau et al., 1998; Corriveau and van Breemen, 2000) also argued that docking of the Central Metasedimentary Belt and the Morin terrane to the pre-Grenvillian Laurentian margin predates 1.17 Ga. They suggested that accretion to Laurentia resulted from ca. 1.22–1.21 Ga continent-continent collision, ~20 m.y. prior to high-*PT* regional metamorphism in the gneiss complexes of the Central Metasedimentary Belt. On the basis of emplacement-related fabrics, they suggested that the Chevreuil suite and the coeval Morin AMCG suite intruded during renewed orogenesis. This orogenic pulse (the 1.19–1.14 Ga Shawinigan pulse of Rivers, 1997) is believed to represent a strongly partitioned, compressive, intraplate reactivation event. Strong amphibolite-facies overprinting of Chevreuil dikes and host rocks, particularly along the Cayamant and Nominingue-Chénéville deformation zones (Fig. 1, C; Corriveau et al., 1998a; Harris et al., 2001), could be related to this event.

Bondy Gneiss Complex

The Bondy gneiss complex, one of the largest gneiss complexes of the Central Metasedimentary Belt (Fig. 1, C), is a northeast-southwest–trending dome (30 × 15 km) composed largely of leucocratic tonalitic and granitic orthogneiss with sporadic, meter- to kilometer-thick units of laminated and nonlaminated quartzofeldspathic gneiss, intermediate gneiss, and mafic amphibolite and granulite (Fig. 2). At the southern end of the dome, a relatively homogeneous and leucocratic tonalitic pluton gives way to gneiss along its borders. Owing to deformation, the nature of its contact with surrounding monotonous quartzofeldspathic gneiss is uncertain. Despite granulite-facies metamorphism and variable Rb, U, and/or Th loss, mafic granulite (calc-alkaline basalt), intermediate gneiss (andesite-dacite), and tonalitic gneiss of the Bondy gneiss complex display systematic and highly reproducible major element, trace element, and Nd isotope signatures attributable to volcanic and plutonic rocks formed coevally in a mature island arc setting (Blein et al., 2003). The arc is interpreted to have formed on oceanic or juvenile continental crustal fragments. In contrast, tholeiitic mafic granulite of the Bondy gneiss complex displays geochemical characteristics typical of modern back-arc basalts derived from juvenile and fertile mantle sources. The association of these rocks with laminated quartzofeldspathic gneiss, interpreted as high-silica rhyolite or rhyolitic tuff, suggests the presence of a bimodal volcanic sequence formed during early arc rifting (Blein et al., 2003).

Hydrothermal alteration and subsequent metamorphism reworked the volcano-plutonic rocks of the Bondy gneiss complex, producing a vast array of hydrothermal rocks, including aluminous and magnesian lithofacies, garnetite, and amphibolite (Corriveau et al., 2003; Blein et al., this volume). The main hydrothermal system, host to several Cu-Au showings, defines a kilometer-scale unit folded along a north-south–trending S fold in the northern part of the gneiss complex (Fig. 2). Hydrothermally altered rocks have also been found in smaller satellites outside of the main zone. Geochemical evidence for hydrothermal alteration includes elevated alteration and chlorite-pyrite indices, pervasive and pronounced birdwing-shaped rare earth element (REE) profiles, heavy REE and Zr enrichment, and nonchondritic isovalent trace element ratios (Fu et al., 2003; Blein et al., this volume). Cu-sulfide mineralized zones within the system show a marked increase in zircon content. This contrasts with unaltered host rocks in adjacent areas, where zircon is difficult to find in thin section. On the basis of their very high abundance and unusual beadlike textures (e.g., Rubin et al., 1989; Nesbitt et al., 1999), the zircons in the mineralized zones are interpreted to have grown during hydrothermal alteration of the host rocks (Fu et al., 2003). These hydrothermal zircons have yet to be dated. Although the precise timing of hydrothermal alteration and associated mineralization in the Bondy gneiss complex remains unknown, field observations combined with geochemical data strongly suggest a volcanogenic setting for the cupriferous hydrothermal system (Corriveau et al., 2003; Blein et al., this volume). For example, field evidence for primary lapilli textures in aluminous gneiss associated with polymetallic (Cu-Fe) sulfide and magnetite mineralization indicates that these rocks were derived from a felsic volcanic precursor. These lapillistones are spatially associated with laminated quartzofeldspathic gneiss, which displays centimeter-scale laminae (see later) with striking similarity to flow-banded rhyolite or tuff (e.g., Fig. 5C *in* Rogers, 2001). The sillimanite-garnet-biotite or cordierite-orthopyroxene assemblages in the aluminous gneiss imply severe premetamorphic leaching. In addition, whole-rock geochemistry strongly suggests that the cordierite-orthopyroxene-bearing aluminous gneiss represents the metamorphosed equivalent of chlorite-rich rocks derived from the interaction between felsic volcanic rocks and seafloor hydrothermal fluid (Blein et al., this volume).

Rocks of the Bondy gneiss complex largely escaped syn- to post-Chevreuil suite amphibolite-facies tectonometamorphic overprinting (e.g., Corriveau et al., 1998; Harris et al., 2001). Instead, they preserve high-*P*, high-*T* metamorphic assemblages, including cordierite-orthopyroxene-sillimanite in the hydrothermally leached aluminous gneiss and cordierite-garnet-orthopyroxene in locally derived, postfabric massive leucosomes. Metamorphic conditions attained ~950 °C and 1000 MPa (i.e., ~33 km depth) with isothermal decompression down to ~800 MPa following peak pressure (Boggs, 1996; Boggs and Corriveau, this volume). U-Pb ages obtained for metamorphic zircon indicate that the high-*P-T* regional metamorphism occurred at 1195 ± 7 Ma, with later crystallization of leucosome at 1179 ± 3 Ma (Corriveau and van Breemen, 2000). Metamorphic monazite, interpreted to date the time of cooling through 850–800 °C (Bingen and van Breemen, 1998), yielded ages of between 1185 and 1140 Ma (Boggs, 1996; Corriveau and van Breemen, 2000).

A recent detailed study reveals a complex polydeformation history for the Bondy gneiss complex and overlying supracrustal rocks of the quartzite domain (Harris et al., 2001). Structures developed prior to intrusion of the ca. 1.17–1.16 Ga Chevreuil suite, including an early gneissossity (S_1), isoclinal folds (F_2 and F_3), boudinage and conjugate ductile shear zones (D_4), and upright (F_5) and recumbent (F_6) folds, may have been produced during either thrusting and continued horizontal contraction or regional extension resulting from orogenic collapse or convective thinning of the lithosphere. Peak metamorphic conditions in the Bondy gneiss complex evolved during D_1–D_3, whereas the regional north-northeast–trending antiform responsible for the main outcrop shape of the Bondy gneiss complex (Fig. 1) was produced during F_5 upright folding. By contrast, D_7 structures (conjugate shear zones and north-striking foliations), developed syn- to postintrusion of the Chevreuil suite, imply ESE–WNW contraction and NNE–SSW orogen-parallel extension. A weak, north-striking foliation has been observed only locally within the Bondy gneiss complex, implying that it acted as a rigid buttress during this event. Finally, late open folds (F_8) formed during late Grenvillian SSE–NNW contraction.

Until now, the only age constraint on any of the igneous protoliths in the Bondy gneiss complex has been a thermal ionization mass spectrometry (TIMS) age of 1240 +10/–20 Ma (van Breemen and Corriveau, 1995) for the homogeneous tonalitic pluton in the southern part of the gneiss complex (Fig. 2). This result was interpreted to approximate the time of igneous crystallization. The 1240 +10/–20 Ma age and uncertainties were based upon the lower intercept of three discordant, multigrain fractions and the ^{207}Pb-^{206}Pb age of the youngest of the three fractions. Subsequent cathodoluminescence (CL) imaging and SHRIMP U-Pb geochronology of zircons from the same tonalite sample used in the multigrain study revealed previously unrecognized morphological complexities (O. van Breemen, 2000, personal commun.). Use of CL allowed identification of relict oscillatory igneous zoning in grain interiors overprinted by highly luminescent, locally fracture-controlled domains of recrystallization, as well as ubiquitous metamorphic mantles. The SHRIMP results give a spread of concordant to slightly discordant data points between ca. 1.40 and 1.10 Ga on a concordia plot, and a weighted mean ^{207}Pb-^{206}Pb age of 1.36 ± 0.02 Ga from nine igneous cores is interpreted as a minimum age for the igneous protolith (O. van Breemen, 2000, personal commun.). A more precise SHRIMP age could not be obtained owing to the low U concentrations of igneous cores. In light of these results, the earlier multigrain age of 1240 +10/–20 Ma likely represents mixing of ca. 1.36 Ga protolith zircon and younger recrystallized and mantling zircon. The complicated patterns of recrystallization revealed by CL imagery imply that multigrain or even single-grain dating methods cannot adequately resolve the protolith age of the tonalite. Consequently, the ion microprobe study presented in the following sections was conducted to gain a better understanding of the premetamorphic history of the Bondy gneiss complex.

SAMPLING AND ANALYTICAL METHODS

Samples of representative laminated and nonlaminated quartzofeldspathic gneiss, tonalitic gneiss, and intermediate gneiss were chosen for SHRIMP U-Pb geochronology to determine the nature and age of gneiss protoliths from the Bondy gneiss complex. Sample locations are shown in Figure 2. During the sampling process, special care was taken to avoid anatectic veins in these granulite-facies rocks. Nonetheless, most samples contained a significant number of metamorphic zircons, present either as whole grains or as mantles (see later), that help constrain the timing of high-grade metamorphism in the Bondy gneiss complex.

Zircons were isolated from ~5–30 kg of rock by standard crushing, heavy liquid, and magnetic techniques. The zircons selected were mounted on two separate 2.5 cm–diameter epoxy disks (GSC # IP44 and 57) and polished using diamond compound to reveal the central portions of the grains. The grains were then photographed in plane transmitted and reflected light at high and low magnifications, cleaned, and subsequently evaporatively coated with 6.0 nm of high-purity Au. Prior to SHRIMP analysis, all grains were imaged using a Cambridge Instruments scanning electron microscope equipped with a CL detector in order to identify any internal structures not otherwise evident.

U-Pb isotopic measurements were performed using the SHRIMP II ion microprobe (Stern, 1996) at the J.C. Roddick Ion Microprobe Laboratory, Geological Survey of Canada; the operating and data-processing procedures used are described in further detail by Stern (1997). Selected areas of the zircons were sputtered using a mass-filtered O_2-primary beam operating in Kohler mode. Primary beam currents ranged from ~0.7 to 1.4 nA, generating elliptical pits ~20 μm in diameter. Prior to data collection, the primary beam was rastered across the analysis spot for 3–5 minutes to reduce surface common Pb. Pb/U and Pb/Th ratios were normalized using calibration lines determined from concurrent data for the Kipawa zircon standard ($^{207}Pb/^{206}Pb$ = 0.072265; 993 Ma), grains of which were mounted in each epoxy disk. Correction of the measured isotopic ratios for common Pb was estimated from monitored ^{204}Pb. The decay constants used were those of Steiger and Jäger (1977). In Table 1, corrected ratios and ages are reported with 1σ analytical errors (68% confidence). Weighted mean ages and age intercepts, calculated using the Isoplot/Ex (version 2.2) program of Ludwig (2000), are presented in the text at the 95% confidence level.

Most analyses were divided into different groups based on zircon morphology characteristics, internal structures (core, mantle, zonation), chemistry (U and Th contents, Th/U ratios), and $^{207}Pb/^{206}Pb$ ratios (see Table 1). Analyses obtained for the majority of the samples display a spread of ^{207}Pb-^{206}Pb ages interpreted to indicate, at least in part, variable isotopic disturbance owing to a combination of Mesoproterozoic events together with recent Pb loss. However, where applicable, the weighted means given in this paper were obtained from those analyses with mutually indistinguishable ^{207}Pb-^{206}Pb ages in grains interpreted as belonging to the same zircon population. Finally, it should be noted that several analyses reported in Table 1 are reversely discordant (i.e., the ^{207}Pb-^{206}Pb age is younger than the ^{206}Pb-^{238}U age). Reverse discordance is frequently observed in ion microprobe analyses and is generally attributed to excess radiogenic Pb resulting from redistribution during a geological disturbance (e.g., Williams et al., 1984; Mattinson et al., 1996) or to an analytical artifact of enhanced Pb ion yields owing to differential ionization efficiencies between unknown and standard (McLaren et al., 1994; Wiedenbeck, 1995). In either case, the geochronological information derived from ^{207}Pb-^{206}Pb data remains valid.

RESULTS

Laminated Quartzofeldspathic Gneiss

Laminated quartzofeldspathic gneiss, characterized by pink, leucocratic, continuous, cyclic, and asymmetric centimeter-scale compositional layers, occurs within the northern part of the Bondy gneiss complex. Biotite-rich layers, which typically recur at the same "stratigraphic" position between feldspathic and granitic layers, define the asymmetric layering. This contrasts markedly with biotite-rich melanosomes in migmatites, which are typically developed symmetrically along leucosomes. The cyclic and asymmetric nature of the layering in

TABLE 1. U-PB SHRIMP ANALYTICAL DATA FOR ZIRCONS FROM BONDY GNEISS SAMPLES

Spot name	U (ppm)	Th (ppm)	Th/U	Pb (ppm)	$^{204}Pb/^{206}Pb$	±	f_{206}	$^{208}Pb/^{206}Pb$	±	$^{206}Pb/^{238}U$	±	$^{207}Pb/^{235}U$	±	$^{207}Pb/^{206}Pb$	±	Apparent ages (Ma) $^{206}Pb/^{238}U$	±	$^{207}Pb/^{206}Pb$	±	Conc. (%)
4427D: Laminated quartzofeldspathic gneiss																				
Distinct cores																				
4427D-13.1	97	41	0.42	24	1.42E-4	7.55E-5	0.0025	0.1152	0.0047	0.2440	0.0044	2.980	0.083	0.0886	0.0017	1408	23	1395	37	100.9
4427D-96.2	150	67	0.45	39	7.10E-5	5.73E-5	0.0012	0.1320	0.0033	0.2479	0.0042	3.135	0.084	0.0917	0.0017	1428	21	1461	36	97.7
4427D-107.2	103	46	0.44	24	2.10E-4	8.84E-5	0.0036	0.1310	0.0042	0.2235	0.0039	2.645	0.076	0.0858	0.0018	1300	20	1334	41	97.5
4427D-134.1	44	17	0.39	9	2.30E-4	1.94E-4	0.0040	0.1178	0.0085	0.2023	0.0053	2.418	0.128	0.0867	0.0037	1188	28	1353	85	87.8
Zoned stubby to elongate prisms and subequant crystals																				
4427D-12.1	60	22	0.37	12	3.84E-4	1.56E-4	0.0067	0.1027	0.0074	0.2013	0.0052	2.197	0.104	0.0792	0.0029	1182	28	1176	75	100.5
4427D-105.1	47	14	0.29	10	3.97E-4	1.99E-4	0.0069	0.0992	0.0092	0.2088	0.0068	2.344	0.138	0.0814	0.0037	1223	36	1232	92	99.3
4427D-128.1	146	60	0.41	31	7.70E-5	5.49E-5	0.0013	0.1156	0.0035	0.2036	0.0031	2.225	0.057	0.0793	0.0015	1195	17	1179	38	101.4
4427D-130.1	103	54	0.52	22	1.09E-4	8.47E-5	0.0019	0.1531	0.0045	0.2031	0.0042	2.258	0.078	0.0806	0.0020	1192	22	1212	51	98.3
4427D-131.1	59	29	0.49	13	3.84E-4	1.67E-4	0.0067	0.1402	0.0089	0.2153	0.0059	2.441	0.134	0.0822	0.0036	1257	31	1250	89	100.5
4427D-132.1	221	123	0.56	46	1.12E-4	4.52E-5	0.0019	0.1595	0.0032	0.1951	0.0029	2.124	0.051	0.0780	0.0014	1149	16	1171	35	98.1
4427D-142.1	132	102	0.77	30	7.60E-5	9.02E-5	0.0013	0.2298	0.0048	0.2043	0.0041	2.250	0.071	0.0799	0.0018	1199	22	1194	44	100.4
Metamorphic mantles																				
4427D-13.2	158	30	0.19	29	1.08E-4	4.79E-5	0.0019	0.0584	0.0026	0.1903	0.0032	2.112	0.053	0.0805	0.0013	1123	18	1209	33	92.9
4427D-94.1	784	199	0.25	153	3.30E-6	1.67E-5	0.0006	0.0765	0.0010	0.1957	0.0024	2.171	0.033	0.0804	0.0007	1152	13	1208	17	95.4
4427D-96.1	829	204	0.25	164	1.70E-5	1.33E-5	0.0003	0.0749	0.0010	0.1998	0.0025	2.174	0.031	0.0789	0.0005	1174	13	1170	12	100.4
4427D-107.1	309	71	0.23	64	9.40E-5	3.35E-5	0.0016	0.0683	0.0020	0.2102	0.0033	2.312	0.051	0.0798	0.0011	1230	18	1191	27	103.3
4427D-134.2	241	123	0.51	49	1.14E-4	5.49E-5	0.0020	0.1667	0.0036	0.1884	0.0032	2.020	0.051	0.0778	0.0013	1113	17	1141	33	97.5
5002L: Laminated quartzofeldspathic gneiss																				
Distinct cores																				
5002L-37.2	340	161	0.47	87	2.19E-4	4.71E-5	0.0038	0.1418	0.0027	0.2405	0.0040	2.871	0.067	0.0866	0.0013	1389	21	1351	29	102.8
5002L-70.1	217	81	0.37	52	3.80E-4	9.07E-5	0.0066	0.1052	0.0045	0.2327	0.0047	2.732	0.096	0.0852	0.0023	1349	24	1319	52	102.3
5002L-102.1	396	243	0.61	106	1.81E-4	6.06E-5	0.0031	0.1810	0.0032	0.2443	0.0038	2.946	0.067	0.0875	0.0013	1409	20	1371	28	102.8
5002L-124.1	330	124	0.37	75	1.36E-4	4.10E-5	0.0024	0.1103	0.0025	0.2192	0.0037	2.673	0.060	0.0885	0.0012	1277	19	1393	26	91.7
5002L-144.1	198	85	0.43	49	3.16E-4	7.97E-5	0.0055	0.1327	0.0039	0.2371	0.0037	2.971	0.080	0.0909	0.0018	1372	19	1444	39	95.0
Clear equant to prismatic crystals																				
5002L-43.1	1493	187	0.13	264	3.20E-5	1.53E-5	0.0006	0.0366	0.0011	0.1847	0.0023	2.010	0.035	0.0789	0.0008	1093	13	1170	21	93.4
5002L-65.1	1159	118	0.10	184	5.23E-5	2.19E-5	0.0009	0.0297	0.0011	0.1663	0.0019	1.837	0.026	0.0801	0.0006	992	11	1200	14	82.6
5002L-101.1	2241	268	0.12	457	1.03E-5	4.16E-6	0.0002	0.0354	0.0004	0.2126	0.0023	2.342	0.028	0.0799	0.0003	1243	12	1194	8	104.1
5002L-107.1	424	93	0.22	83	1.16E-4	2.96E-5	0.0020	0.0647	0.0017	0.1987	0.0028	2.139	0.040	0.0781	0.0008	1169	15	1149	20	101.7
5002L-107.2	749	129	0.17	150	5.33E-5	2.19E-5	0.0009	0.0516	0.0012	0.2059	0.0026	2.243	0.035	0.0790	0.0006	1207	14	1172	16	103.0
5002L-112.1	740	88	0.12	116	5.36E-5	1.89E-5	0.0009	0.0356	0.0011	0.1641	0.0023	1.792	0.031	0.0792	0.0007	980	13	1177	17	83.2
5002L-112.2	98	5	0.05	17	3.68E-4	1.80E-4	0.0064	0.0056	0.0068	0.1874	0.0046	1.943	0.104	0.0752	0.0034	1107	25	1074	92	103.1
5002L-118.1	1256	187	0.15	244	7.10E-7	8.32E-6	1E-05	0.0453	0.0007	0.2003	0.0023	2.213	0.031	0.0801	0.0005	1177	13	1200	12	98.1
5002L-135.1	731	96	0.13	130	4.01E-5	1.10E-5	0.0007	0.0392	0.0007	0.1847	0.0025	2.005	0.031	0.0787	0.0005	1093	14	1165	12	93.7
5002L-138.1	679	86	0.13	136	1.30E-4	3.09E5	0.0023	0.0380	0.0014	0.2090	0.0028	2.267	0.041	0.0787	0.0008	1223	15	1164	21	105.1
5002L-141.1	1354	175	0.13	278	4.29E-5	1.02E-5	0.0007	0.0374	0.0007	0.2141	0.0025	2.331	0.032	0.0790	0.0004	1251	13	1171	11	106.8
5002L-147.1	1192	232	0.19	234	2.99E-5	1.39E-5	0.0005	0.0555	0.0008	0.2009	0.0023	2.195	0.030	0.0793	0.0005	1180	12	1179	12	100.1
Metamorphic mantles																				
5002L-37.1	2232	11	0.01	389	3.35E-5	1.02E-05	0.0006	0.0011	0.0004	0.1878	0.0020	2.065	0.026	0.0797	0.0004	1110	11	1190	11	93.2
5002L-68.1	794	95	0.12	152	1.95E-5	1.35E-05	0.0003	0.0377	0.0008	0.1997	0.0025	2.198	0.034	0.0798	0.0006	1174	13	1193	16	98.4
4548C2-A: Laminated quartzofeldspathic gneiss																				
Distinct cores																				
4548C2-17.1	260	66	0.25	54	5.31E-4	1.08E-4	0.0092	0.0789	0.0044	0.2069	0.0038	2.347	0.079	0.0823	0.0021	1212	20	1252	51	96.8
4548C2-34.1	120	36	0.30	27	5.46E-4	1.35E-4	0.0095	0.0844	0.0058	0.2235	0.0048	2.702	0.103	0.0877	0.0025	1300	26	1376	57	94.5
4548C2-54.1	299	130	0.44	65	2.54E-4	9E-5	0.0044	0.1266	0.0042	0.2092	0.0035	2.490	0.070	0.0863	0.0018	1225	19	1345	40	91.0
4548C2-56.1	181	62	0.35	40	8.44E-5	8.51E-5	0.0015	0.1054	0.0041	0.2131	0.0037	2.605	0.072	0.0886	0.0017	1246	20	1396	37	89.2
4548C2-57.1	232	70	0.30	51	1.73E-4	1.17E-4	0.0030	0.0934	0.0060	0.2156	0.0044	2.563	0.094	0.0862	0.0024	1259	23	1343	55	93.7

Clear stubby prism																			
4548C2-39.1	1144	106	0.09	218	8.81E-5	2.30E-5	0.0015	0.0274	0.0011	0.2003	0.0045	2.193	0.055	0.0794	0.0007	1177	24	1183	17	99.5
2671: Quartzofeldspathic gneiss																				
Turbid cores																				
2671-3.1	104	42	0.40	21	3.15E-4	2.12E-4	0.0055	0.1188	0.0090	0.1985	0.0049	2.338	0.128	0.0854	0.0039	1167	26	1325	91	88.1
2671-18.1	111	48	0.44	24	5.27E-4	2.20E-4	0.0091	0.1401	0.0096	0.2027	0.0045	2.274	0.123	0.0814	0.0038	1190	24	1230	94	96.8
2671-23.1	97	68	0.70	22	2.40E-4	1.10E-4	0.0042	0.2112	0.0065	0.2031	0.0045	2.260	0.093	0.0807	0.0026	1192	24	1214	64	98.2
2671-34.1	68	45	0.67	16	2.68E-4	1.83E-4	0.0046	0.1966	0.0089	0.2126	0.0062	2.618	0.133	0.0893	0.0034	1243	33	1411	75	88.1
2671-43.1	74	36	0.49	17	4.39E-4	2.32E-4	0.0076	0.1635	0.0097	0.2163	0.0061	2.790	0.152	0.0935	0.0041	1263	32	1499	84	84.2
2671-43.2	64	23	0.37	16	2.09E-4	2.09E-4	0.0291	0.1176	0.0088	0.2408	0.0057	3.032	0.148	0.0913	0.0036	1391	29	1454	78	95.7
2671-65.1	296	176	0.59	67	8.56E-5	5.01E-5	0.0015	0.1684	0.0035	0.2094	0.0047	2.405	0.061	0.0833	0.0013	1226	20	1277	32	96.0
2671-66.1	49	29	0.59	12	4.67E-4	2.00E-4	0.0081	0.1701	0.0098	0.2222	0.0062	2.732	0.151	0.0892	0.0039	1294	33	1407	87	91.1
2671-71.1	232	82	0.35	47	2.52E-4	6.41E-5	0.0044	0.1009	0.0031	0.2001	0.0033	2.314	0.054	0.0839	0.0013	1176	18	1289	29	91.2
2671-79.1	314	63	0.20	68	6.78E-5	5.47E-5	0.0012	0.0742	0.0028	0.2155	0.0040	2.774	0.069	0.0934	0.0013	1258	21	1496	27	84.1
2671-79.3	517	106	0.20	122	4.64E-5	1.37E-5	0.0008	0.0712	0.0011	0.2353	0.0031	2.949	0.047	0.0909	0.0007	1362	16	1444	15	94.3
2671-79.4	458	138	0.30	115	1.98E-5	2.05E-5	0.0003	0.1023	0.0018	0.2432	0.0032	3.286	0.055	0.0980	0.0008	1403	17	1586	16	88.5
Clear cores of prisms and subequant crystals																				
2671-19.1	294	34	0.12	56	8.18E-5	5.05E-5	0.0014	0.0329	0.0024	0.2012	0.0033	2.199	0.054	0.0793	0.0013	1182	18	1178	33	100.3
2671-21.1	85	6	0.07	16	7.42E-5	1.54E-4	0.0013	0.0225	0.0062	0.1980	0.0054	2.249	0.110	0.0824	0.0031	1165	29	1254	75	92.2
2671-22.1	49	2	0.04	10	4.40E-4	3.35E-4	0.0076	0.0118	0.0128	0.2055	0.0079	2.247	0.197	0.0793	0.0059	1205	43	1180	154	102.1
2671-35.1	127	2	0.02	21	3.07E-4	1.21E-4	0.0053	0.0051	0.0046	0.1801	0.0036	1.992	0.076	0.0802	0.0024	1068	20	1202	61	88.8
2671-36.1	151	2	0.02	26	3.19E-4	8.89E-5	0.0055	0.0032	0.0036	0.1882	0.0043	2.065	0.070	0.0796	0.0018	1112	24	1187	45	93.6
2671-46.1	288	33	0.12	49	1.34E-4	7.56E-5	0.0023	0.0343	0.0030	0.1792	0.0028	1.938	0.049	0.0785	0.0014	1062	16	1158	36	91.7
2671-49.1	104	1	0.01	19	4.92E-4	1.85E-4	0.0085	0.0001	0.0071	0.1927	0.0042	2.129	0.102	0.0801	0.0032	1136	23	1200	81	94.7
2671-53.1	69	6	0.10	12	8.32E-4	2.85E-4	0.0144	0.0318	0.0110	0.1795	0.0041	2.002	0.134	0.0809	0.0049	1064	22	1219	123	87.3
2671-54.1	106	3	0.03	19	5.41E-4	2.72E-4	0.0094	0.0194	0.0106	0.1895	0.0045	2.191	0.142	0.0839	0.0048	1118	24	1290	116	86.7
2671-58.1	100	2	0.02	17	5.19E-4	1.71E-4	0.0090	0.0038	0.0066	0.1796	0.0038	2.016	0.092	0.0814	0.0031	1065	21	1231	77	86.5
2671-63.1	86	7	0.08	17	1.90E-4	1.75E-4	0.0033	0.0243	0.0069	0.2054	0.0047	2.282	0.114	0.0806	0.0034	1204	25	1212	84	99.4
Mantles																				
2671-2.1	454	17	0.04	84	8.30E-5	3.07E-5	0.0014	0.0094	0.0013	0.1986	0.0032	2.130	0.043	0.0778	0.0008	1168	17	1142	21	102.2
2671-34.2	326	18	0.06	54	2.70E-4	9.67E-5	0.0047	0.0151	0.0038	0.1772	0.0032	1.934	0.061	0.0791	0.0019	1052	17	1176	49	89.5
2671-35.2	810	9	0.01	145	2.76E-5	2.89E-5	0.0005	0.0038	0.0012	0.1916	0.0029	2.102	0.046	0.0795	0.0011	1130	16	1186	29	95.3
2671-55.1	289	21	0.07	55	1.51E-4	7.86E-5	0.0026	0.0236	0.0031	0.2006	0.0035	2.134	0.058	0.0772	0.0014	1178	19	1126	38	104.7
2671-71.2	315	20	0.07	57	1.02E-4	5.01E-5	0.0018	0.0204	0.0021	0.1901	0.0032	2.094	0.051	0.0799	0.0013	1122	17	1195	32	93.9
2671-79.2	344	22	0.06	62	7.54E-5	4.17E-5	0.0013	0.0182	0.0017	0.1918	0.0029	2.096	0.043	0.0793	0.0010	1131	16	1179	25	96.0
6026D: Quartzofeldspathic gneiss																				
Distinct cores																				
6026D-48.1	322	124	0.39	76	8.10E-5	2.64E-5	0.0014	0.1082	0.0020	0.2286	0.0032	2.663	0.050	0.0845	0.0009	1327	17	1304	21	101.8
6026D-92.1	261	105	0.40	58	1.10E-4	8.34E-5	0.019	0.1147	0.0038	0.2139	0.0033	2.565	0.0066	0.0870	0.0016	1249	18	1360	37	91.1
Oscillatory zoned prisms																				
6026D-141.2	268	189	0.71	90	2.26E-5	1.92E-5	0.0004	0.2049	0.0021	0.2973	0.0040	4.234	0.069	0.1033	0.0008	1678	20	1684	14	99.6
6026D-143.1	37	35	0.93	23	1.67E-4	1.20E-4	0.0029	0.2626	0.0088	0.4906	0.0134	12.346	0.421	0.1825	0.0031	2573	58	2676	29	96.2
Stubby to elongate prisms																				
6026D-30.1	302	105	0.35	61	1.22E-4	5.06E-5	0.0021	0.0987	0.0032	0.1981	0.0034	2.151	0.054	0.0787	0.0013	1165	18	1166	33	99.9
6026D-32.2	470	175	0.37	99	5.03E-5	1.90E-5	0.0009	0.1096	0.0015	0.2057	0.0025	2.272	0.037	0.0801	0.0007	1206	14	1199	18	100.6
6026D-34.1	205	119	0.58	44	5.58E-5	4.09E-5	0.0010	0.1753	0.0039	0.1988	0.0031	2.219	0.051	0.0809	0.0012	1169	17	1220	30	95.8
6026D-52.1	158	41	0.26	26	5.43E-4	1.02E-4	0.0094	0.0778	0.0048	0.1669	0.0037	1.917	0.068	0.0833	0.0021	995	21	1276	49	78.0
6026D-54.1	160	42	0.27	32	5.46E-5	7.41E-5	0.0010	0.0832	0.0036	0.2027	0.0036	2.271	0.067	0.0813	0.0017	1190	20	1228	42	96.6
6026D-145.1	165	96	0.58	34	2.83E-4	1.09E-4	0.0049	0.1704	0.0057	0.1925	0.0039	2.068	0.076	0.0779	0.0022	1135	21	1145	57	99.2
6026D-147.1	109	30	0.27	22	4.09E-4	1.25E-4	0.0071	0.0742	0.0056	0.2055	0.0041	2.224	0.086	0.0785	0.0024	1205	22	1159	62	103.9
Equant zircons																				
6026D-127.1	195	379	1.9	52	1.15E-4	8.47E-5	0.0020	0.5709	0.0064	0.1856	0.0036	2.017	0.067	0.0788	0.0019	1098	20	1168	49	94.0
6026D-128.1	126	278	2.2	34	2.65E-4	1.24E-4	0.0046	0.6580	0.0112	0.1802	0.0044	1.922	0.083	0.0774	0.0026	1068	24	1131	67	94.4
6026D-131.1	187	355	1.9	49	1.60E-4	6.82E-5	0.0028	0.5685	0.0075	0.1849	0.0030	1.993	0.055	0.0782	0.0016	1094	16	1152	41	95.0
Mantles																				
6026D-32.1	342	64	0.19	67	2.43E-4	6.01E-5	0.0042	0.0587	0.0029	0.2005	0.0036	2.180	0.056	0.0789	0.0013	1178	19	1169	33	100.8
6026D-46.2	166	69	0.42	31	4.02E-4	9.53E-5	0.0070	0.1233	0.0050	0.1802	0.0035	1.919	0.066	0.0772	0.0020	1068	19	1127	53	94.7

(continued)

TABLE 1. *Continued*

Spot name	U (ppm)	Th (ppm)	Th/U	Pb (ppm)	$^{204}Pb/^{206}Pb$	±	f_{206}	$^{208}Pb/^{206}Pb$	±	$^{206}Pb/^{238}U$	±	$^{207}Pb/^{235}U$	±	$^{207}Pb/^{206}Pb$	±	Apparent ages (Ma) $^{206}Pb/^{238}U$	±	$^{207}Pb/^{206}Pb$	±	Conc. (%)
6026D-48.2	96	56	0.58	20	4.80E-4	1.56E-4	0.0083	0.1635	0.0074	0.1961	0.0039	2.083	0.095	0.0770	0.0030	1154	21	1122	79	102.9
4954A: Tonalitic gneiss																				
Distinct core																				
4954A-19.1	131	51	0.39	28	7.80E-4	1.80E-4	0.0135	0.1209	0.0077	0.2045	0.0053	2.506	0.120	0.0889	0.0033	1199	28	1402	73	85.5
Stubby to elongate prisms																				
4954A-1.1	470	131	0.28	89	1.19E-4	3.57E-5	0.0021	0.0815	0.0018	0.1889	0.0030	2.075	0.041	0.0791	0.0008	1116	16	1189	20	93.9
4954A-42.1	403	113	0.28	71	1.38E-4	4.44E-5	0.0024	0.0818	0.0022	0.1765	0.0027	1.986	0.043	0.0816	0.0011	1048	15	1235	27	84.9
4954A-64.1	518	53	0.10	98	4.77E-5	2.39E-5	0.0008	0.0316	0.0011	0.1986	0.0027	2.192	0.036	0.0800	0.0006	1168	14	1198	16	97.5
4954A-68.1	548	468	0.85	148	5.69E-5	1.84E-5	0.0010	0.2387	0.0024	0.2363	0.0033	2.855	0.047	0.0876	0.0006	1368	17	1374	14	99.5
4954A-68.2	613	504	0.82	162	4.01E-5	1.89E-5	0.0007	0.2348	0.0019	0.2327	0.0030	2.803	0.042	0.0874	0.0006	1349	16	1369	12	98.5
4954A-68.3	696	566	0.81	193	4.40E-5	2.63E-5	0.0008	0.2340	0.0020	0.2431	0.0034	2.978	0.049	0.0888	0.0006	1403	17	1401	14	100.2
4954A-107.1	312	213	0.68	78	1.10E-4	5.63E-5	0.0019	0.1964	0.0031	0.2265	0.0036	2.769	0.061	0.0887	0.0012	1316	19	1397	26	94.2
4954A-118.1	849	601	0.71	225	2.41E-5	1.64E-5	0.0004	0.1979	0.0022	0.2396	0.0033	2.899	0.048	0.0878	0.0007	1384	17	1378	15	100.5
4954A-118.2	788	532	0.68	212	2.00E-6	1.30E-5	3E-05	0.1971	0.0013	0.2430	0.0031	2.974	0.043	0.0888	0.0005	1402	16	1399	10	100.2
4954A-128.1	203	109	0.54	50	2.02E-4	8.67E-5	0.0035	0.1493	0.0039	0.2297	0.0036	2.808	0.070	0.0887	0.0015	1333	19	1397	34	95.4
4954A-135.1	603	438	0.73	161	6.19E-5	4.49E-5	0.0011	0.2037	0.0042	0.2393	0.0035	2.928	0.064	0.0888	0.0013	1383	18	1399	28	98.8
4954A-158.1	457	379	0.83	116	1.30E-4	3.70E-5	0.0022	0.2288	0.0023	0.2239	0.0032	2.692	0.048	0.0872	0.0008	1303	17	1365	18	95.4
4954A-160.1	914	147	0.16	174	1.20E-4	2.68E-5	0.0021	0.0521	0.0013	0.1955	0.0028	2.191	0.037	0.0813	0.0006	1151	15	1228	15	93.7
4954A-162.1	346	173	0.50	82	1.31E-4	4.84E-5	0.0023	0.1408	0.0026	0.2242	0.0037	2.713	0.059	0.0878	0.0011	1304	19	1378	25	94.6
4954A-165.1	256	80	0.31	59	3.00E-4	6.90E-5	0.0052	0.0846	0.0032	0.2281	0.0037	2.683	0.074	0.0853	0.0017	1325	20	1322	39	100.2
4954A-169.1	205	118	0.57	51	2.12E-4	7.08E-5	0.0037	0.1594	0.0038	0.2298	0.0042	2.848	0.073	0.0899	0.0014	1334	22	1423	31	93.7
Equant zircons																				
4954A-20.1	343	193	0.56	68	1.68E-4	1.58E-4	0.0096	0.4528	0.0088	0.1852	0.0030	2.069	0.045	0.0811	0.0010	1094	19	1223	24	89.5
4954A-29.1	113	178	1.58	28	5.53E-4	1.32E-4	0.0055	0.2061	0.0071	0.1849	0.0035	2.072	0.084	0.0813	0.0027	1126	20	1228	68	89.1
Metamorphic mantles																				
4954A-29.2	125	92	0.74	26	3.18E-4	4.90E-5	0.0029	0.1629	0.0027	0.1908	0.0037	2.058	0.079	0.0782	0.0024	1095	16	1153	62	97.6
4954A-97.1	64	92	1.45	16	4.30E-4	2.78E-4	0.0075	0.4176	0.0155	0.1936	0.0051	2.168	0.149	0.0812	0.0049	1141	28	1227	123	93.0
4954A-141.1	78	41	0.52	15	9.32E-4	2.24E-4	0.0162	0.1583	0.0096	0.1822	0.0042	2.065	0.113	0.0822	0.0039	1079	23	1250	95	86.3
4954A-165.2	63	88	1.39	15	7.73E-4	4.11E-4	0.0134	0.4101	0.0196	0.1883	0.0050	2.188	0.190	0.0842	0.0067	1113	27	1298	163	85.7
5063B: Intermediate gneiss																				
Core																				
5063B-34.2	596	16	0.03	114	1.82E-5	1.72E-5	0.0003	0.0084	0.0008	0.2036	0.0027	2.250	0.037	0.0802	0.0007	1195	14	1201	17	99.4
Equant, multifaceted zircons																				
5063B-24.1	284	17	0.06	54	1.24E-4	5.07E-5	0.0022	0.0171	0.0023	0.2006	0.0033	2.151	0.053	0.0778	0.0013	1178	18	1141	33	103.3
5063B-41.1	247	2	0.01	44	2.24E-5	4.20E-5	0.0004	0.0038	0.0016	0.1916	0.0027	2.093	0.045	0.0792	0.0011	1130	15	1178	28	95.5
5063B-46.1	2168	179	0.08	427	5.17E-6	4.56E-6	0.0001	0.0240	0.0004	0.2074	0.0023	2.283	0.028	0.0798	0.0004	1215	12	1193	9	101.8
5063B-64.1	209	7	0.03	38	2.90E-4	7.49E-5	0.0050	0.0056	0.0030	0.1936	0.0038	2.111	0.070	0.0791	0.0020	1141	21	1174	50	97.2
Prisms																				
5063B-145.1	207	1	0.01	40	7.62E-5	4.64E-5	0.0013	0.0032	0.0018	0.2078	0.0036	2.297	0.058	0.0802	0.0013	1217	19	1201	32	101.3
5063B-145.2	252	1	0.01	47	5.00E-5	2.80E-5	0.009	0.0031	0.0012	0.1998	0.0029	2.199	0.043	0.0798	0.0009	1174	16	1193	23	98.4
Shell																				
5063B-34.1	762	80	0.11	148	2.76E-5	2.31E-5	0.0005	0.0321	0.0012	0.2037	0.0028	2.176	0.040	0.0775	0.0008	1195	15	1134	21	105.4

Notes: Uncertainties are reported at 1σ and are calculated by numerical propagation of all known sources of error (Stern, 1997). f_{206} refers to mole fraction of total ^{206}Pb that is due to common Pb; data are common-Pb corrected using the 204 method, as described by Stern (1997). Conc. = 100 × ($^{206}Pb/^{238}U$ age)/($^{207}Pb/^{206}Pb$ age).

the laminated quartzofeldspathic gneiss is inferred to be primary (Blein et al., 2003). Based on field observations, as well as major element and trace element geochemistry, these gneissic rocks are best reconciled with a rhyolitic or high-silica rhyolitic tuff protolith. The gneisses display isotopic and geochemical signatures, including comparatively low $\varepsilon Nd_{(T = 1.4\ Ga)}$ values (+4.1 and +5.0), high Zr contents, and low Ti/Zr ratios, characteristic of felsic volcanic rocks related to back-arc rifting (Blein et al., 2003). Their less radiogenic signature, at least compared to that of mafic granulites ($\varepsilon Nd_{(T = 1.4\ Ga)}$ values of +6.2 to +7.0) of the Bondy gneiss complex, may be explained by the assimilation of an older crustal or sedimentary component. Three representative samples of laminated quartzofeldspathic gneiss (4427D, 5002L, and 4548C2-A), all of which come from the northern part of the Bondy gneiss complex in the vicinity of the main hydrothermal system (Fig. 2), were selected for U-Pb geochronology to provide information about the age of both the rhyolitic protolith and high-grade metamorphism. CL images of zircon grains selected from all three samples are shown in Figure 3, and the SHRIMP analytical results are given in Table 1 and plotted in concordia diagrams in Figure 4.

Sample 4427D

Zircons from laminated quartzofeldspathic gneiss sample 4427D are mostly colorless to pale brown or yellow. The dominant morphology comprises subhedral to euhedral, stubby to elongate prisms (2–5:1 aspect ratios) with sharp to slightly rounded edges and terminations. Subequant to rounded crystals are also common. In CL images, grain interiors are typically luminescent and exhibit fine-scale oscillatory to diffuse zoning that is mantled or embayed by darker, U-rich zircon with faint zoning (Fig. 3, A). Approximately 25% of grains contain bright, subrounded cores with faint to prominent concentric growth zoning and relatively thick, weakly luminescent mantles (Fig. 3, B). Most cores show minor fracturing in transmitted light.

Analyses of four subrounded cores yielded Th/U ratios of between 0.39 and 0.45, typical of magmatic zircon (e.g., Maas et al., 1992; Gebauer et al., 1997). Three of the four analyses plot along the concordia curve between ca. 1450 and 1300 Ma (Fig. 4, A); the fourth analysis is strongly discordant with a ^{207}Pb-^{206}Pb age of 1353 Ma (Table 1). Analyses of grain interiors of both oscillatory-zoned prisms and subequant crystals form a separate population with variable Th/U ratios (0.29–0.77) and ^{207}Pb-^{206}Pb ages ranging between 1250 and 1171 Ma. Weakly luminescent mantles generally have higher U contents and lower Th/U ratios (0.19–0.51). Despite these chemical differences, all mantle analyses fall close to the zoned prism and subequant population (Fig. 4, A). The weighted average ^{207}Pb-^{206}Pb age for both the mantles and the oscillatory-zoned grain interiors is 1184 ± 15 (2σ) Ma (MSWD, or mean square of weighted deviates = 0.64), interpreted to date the time of high-grade metamorphism of this laminated quartzofeldspathic gneiss. The significance of the older population of distinct cores in this and all other quartzofeldspathic gneiss samples is discussed later.

Sample 5002L

Zircons recovered from laminated quartzofeldspathic gneiss sample 5002L are subhedral and range from equant to prismatic (2–4:1 aspect ratios). Small, colorless, subrounded to prismatic cores are present in many grains regardless of morphology, and are mantled by relatively thick, yellowish zircon. Oscillatory zoning is not obvious within any of the zircons under CL imaging. About half of the grains are comparatively homogeneous, composed entirely of dark zircon with diffuse coarse zoning and no visible core. Where present, cores are either brightly to mildly luminescent (Fig. 3, C) or nonluminescent and separated from dark, homogeneous mantles by thin luminescent seams (Fig. 3, D).

Spot analyses again indicate two chronologically distinct populations (Fig. 4, B). Analyses of five distinct cores, characterized by moderate U contents (198–396 ppm) and Th/U ratios (0.37–0.61), gave ^{207}Pb-^{206}Pb ages of between 1444 and 1319 Ma, similar to those obtained for the laminated quartzofeldspathic gneiss 4427D (Table 1). Both equant to prismatic crystals and mantles on luminescent grains are distinctly younger than cores, yielding ^{207}Pb-^{206}Pb ages of between 1200 and 1074 Ma. With the exception of analysis 112.2, which comes from the very tip of a dark homogeneous mantle, all analyses from this younger population have high U contents (424–2241 ppm) and comparatively low Th/U ratios (0.01–0.22). Our best estimate of the timing of new zircon growth is 1183 ± 9 (2σ) Ma (MSWD = 1.2), the weighted average age of high-U grain interiors and mantles (n = 13, excluding the youngest, imprecise 112.2 analysis). Linear regression of twelve of the fourteen youngest analyses gives an upper intercept age of 1184 ± 9 Ma (MSWD = 1.2), identical to the weighted mean result.

Sample 4548C2-A

The zircons from this laminated quartzofeldspathic gneiss sample are colorless and euhedral to subhedral. Densely fractured stubby prisms with aspect ratios of less than 3:1 and rounded edges and terminations dominate the population. Approximately 50% of crystals consist of euhedral to rounded cores enveloped by relatively thick mantles. In CL images, cores and grain interiors are typically weakly luminescent, displaying diffuse concentric zoning, whereas the mantles are nonluminescent and largely unzoned (Fig. 3, E and F). As in sample 5002L, brightly luminescent seams, up to 30 μm thick, frequently occur at the boundaries between cores and mantles (Fig. 3, F).

Only six spot analyses were carried out on the clearest, fracture-free areas of zircons from this sample. Five distinct cores, characterized by moderate U contents (120–299 ppm) and similar Th/U ratios (0.25–0.44), yielded moderately discordant data, with ^{207}Pb-^{206}Pb ages ranging between 1396 and 1252 Ma (Table 1; Fig. 4, C). A single stubby prism, with a Th/U ratio of 0.09, yielded a distinctly younger ^{207}Pb-^{206}Pb age of 1183 ± 17 (1σ) Ma. This age is identical, within error, to the metamorphic ages obtained for the other two samples of laminated quartzofeldspathic gneiss.

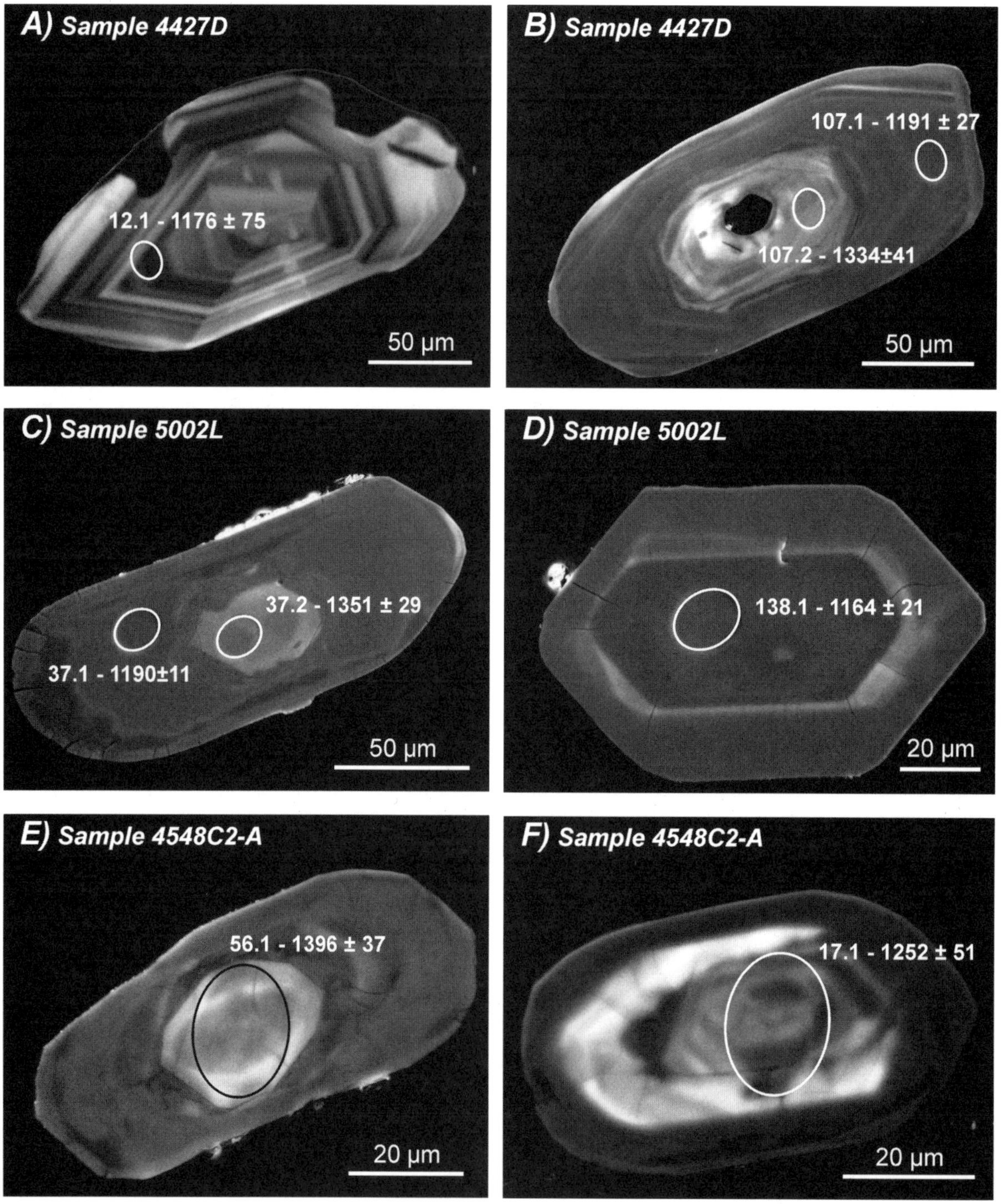

Figure 3. Cathodoluminescence images of selected zircon grains from the laminated quartzofeldspathic gneiss samples illustrating types of recognized internal complexities. The numbers denote spot names and ^{207}Pb-^{206}Pb ages in Ma. SHRIMP analysis pits (white or black ellipses) are ~19–21 µm across. (A) Euhedral, doubly terminated zircon exhibiting oscillatory zoning truncated by high-U embayments, sample 4427D. (B) Elongate zircon with a small subrounded core surrounded by faintly zoned mantle material, sample 4427D. (C) Partly corroded, luminescent core mantled by a thick, dark rim with patchy zoning, sample 5002L. (D) Dark, unzoned grain interior separated from a dark, outer rim by a thin luminescent seam, sample 5002L. (E) Euhedral, luminescent core mantled by a thick, dark rim displaying very faint zoning, sample 4548C2-A. (F) Concentrically zoned, euhedral core separated from an outer, dark rim by a ~8 µm–thick, brightly luminescent seam, sample 4548C2-A.

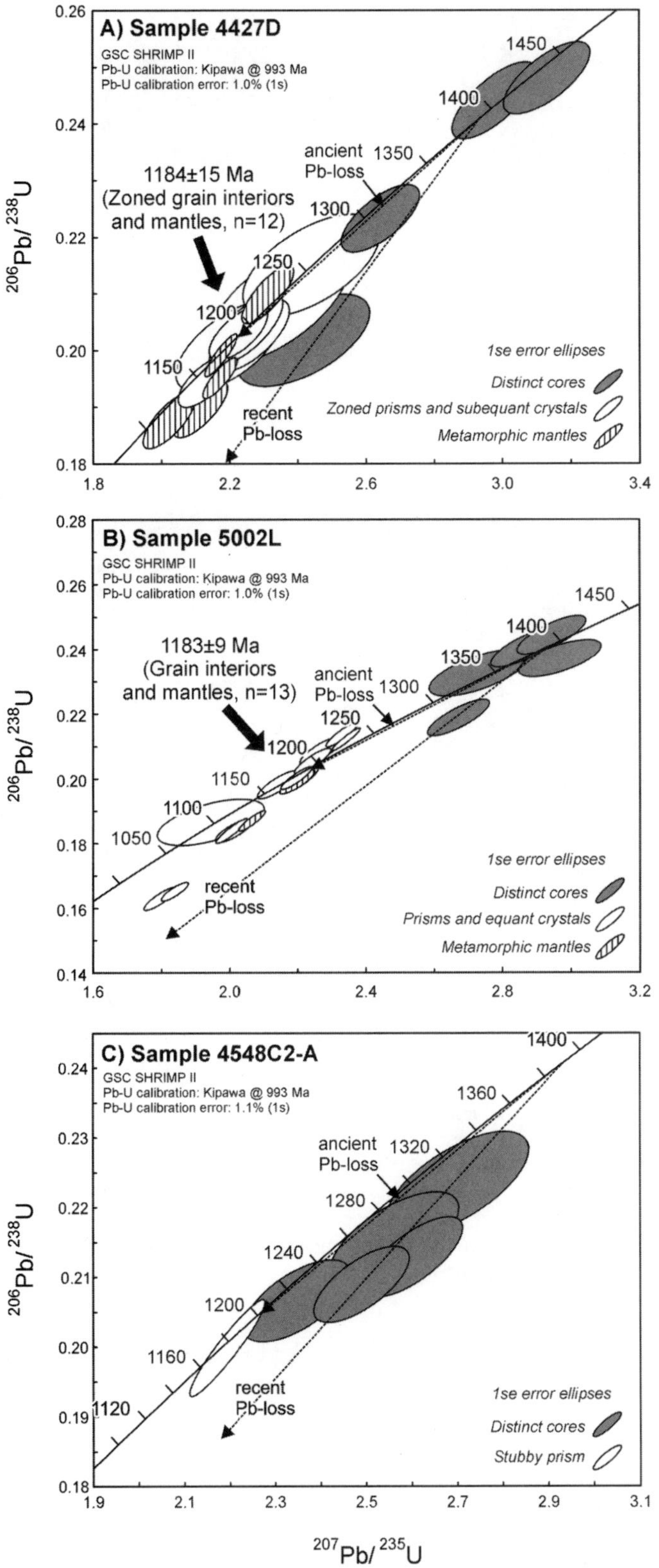

Figure 4. U-Pb concordia diagrams for zircons from laminated quartzofeldspathic samples 4427D (A), 5002L (B), and 4548C2-A (C). Error ellipses are 1σ. Plausible times of Pb loss, assuming a maximum protolith age of ca. 1.39 Ga, are shown by dashed lines for each sample (see text for further details).

Quartzofeldspathic Gneiss

Nonlaminated quartzofeldspathic gneiss of uncertain origin occurs sporadically throughout the Bondy gneiss complex. Typically, hydrothermal alteration and high-grade metamorphism have completely destroyed primary features in these orthopyroxene-biotite-bearing gneisses. Two samples of quartzofeldspathic gneiss (2671 and 6026D), collected from small satellites of hydrothermally altered rocks (Fig. 2), were selected for geochronological work. Despite evidence for hydrothermal alteration (e.g., heavy REE enrichment), geochemical data suggest a tonalitic protolith for sample 2671 (O. Blein, 2002, personal commun.). CL images of zircon grains selected from both samples are shown in Figure 5, and the SHRIMP analytical results are given in Table 1 and plotted in concordia diagrams in Figure 6.

Sample 2671

The colorless, clear to partly turbid zircons from this quartzofeldspathic gneiss sample typically consist of euhedral to subhedral prisms (2–3:1 aspect ratios) and subequant crystals. Rounded crystals are also present. A number of grains contain rounded, turbid or inclusion-free cores mantled by relatively thick zircon. Most turbid cores exhibit diffuse to sharp concentric zoning and are surrounded by more homogeneous, dark mantles under CL imaging (Fig. 5, A). In contrast, clear, inclusion-free cores are typically brightly luminescent and either unzoned or sector zoned (Fig. 5, B). Brightly luminescent seams at the boundaries between cores and mantles are also common in this sample (Fig. 5, A). A few clear grains with no detectable internal zoning are also present.

The turbid cores have low to moderate U contents (49–517 ppm) and variable Th/U ratios (0.20–0.70) (Table 1). The data for these cores scatter below concordia and yielded ^{207}Pb-^{206}Pb ages ranging between 1586 and 1214 Ma (Fig. 6, A). Multiple, closely spaced or partly overlapping analyses were performed within the core regions of grains 43 and 79 in an attempt to provide a better understanding of the age of this quartzofeldspathic gneiss. Three analyses within the core region of grain 79 do not define a linear trend (Fig. 6, A). Subsequent examination of this grain under CL shows that all three analyses partially overlap the brightly luminescent seam at the boundary between core and mantle (Fig. 5, A). This seam appears to be related to corrosion at the original core surface and may have resulted from surface-controlled alteration processes similar to that described by Vavra et al. (1999). Therefore, the variation in ages exhibited by the three analyses likely represents analytical mixing of older zircon and younger, altered zircon affected by one or more Pb-loss events. The two analyses in the core region of grain 43 yielded ^{207}Pb-^{206}Pb ages of 1454 and 1499 Ma. Pooling of these two analyses, together with the two least isotopically disturbed core analyses, 2671-34.1 and -66.1, gives a weighted average ^{207}Pb-^{206}Pb age of 1442 ± 79 (2σ) Ma (MSWD = 0.27). Although this age could be interpreted as the crystallization age of the

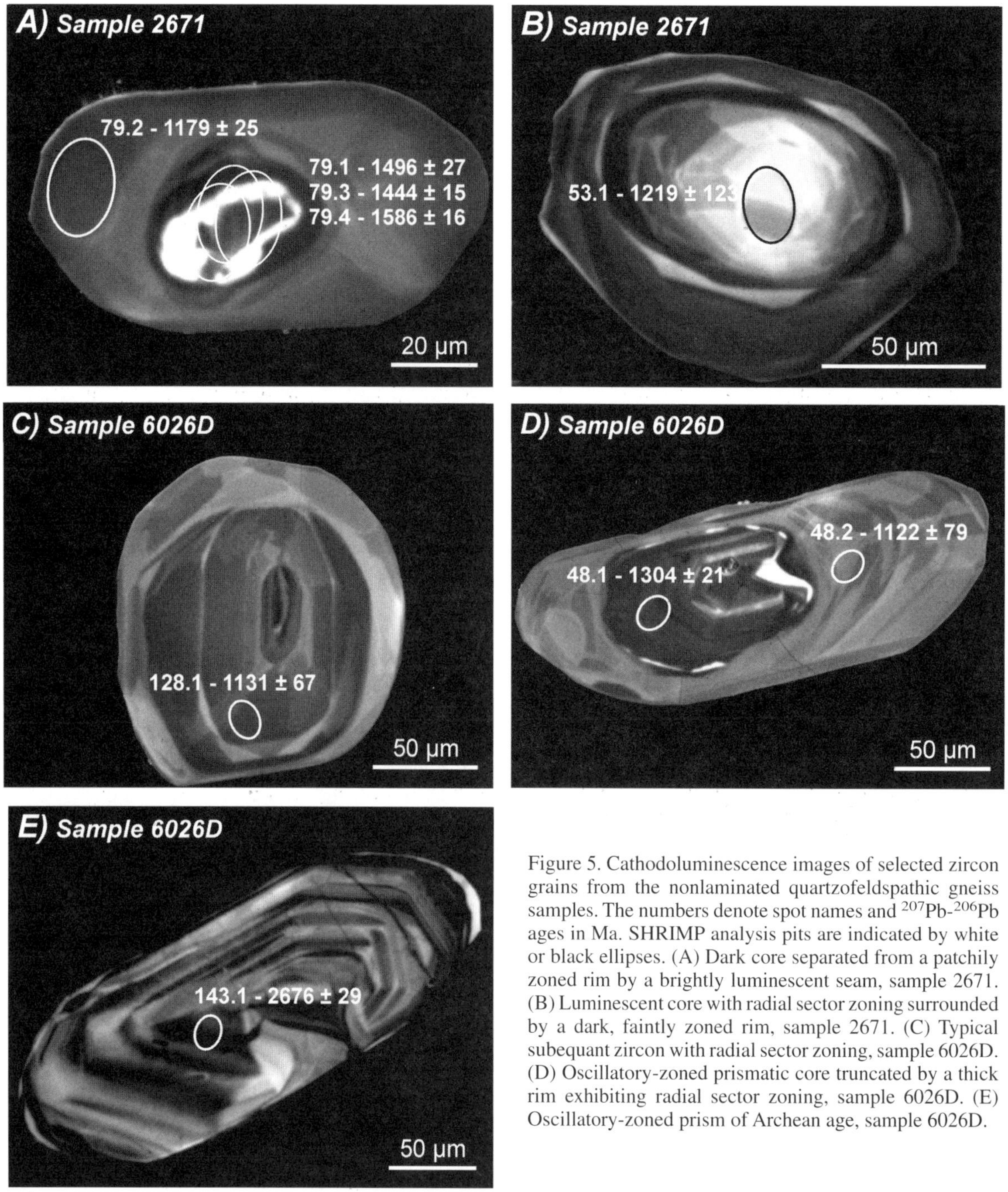

Figure 5. Cathodoluminescence images of selected zircon grains from the nonlaminated quartzofeldspathic gneiss samples. The numbers denote spot names and ^{207}Pb-^{206}Pb ages in Ma. SHRIMP analysis pits are indicated by white or black ellipses. (A) Dark core separated from a patchily zoned rim by a brightly luminescent seam, sample 2671. (B) Luminescent core with radial sector zoning surrounded by a dark, faintly zoned rim, sample 2671. (C) Typical subequant zircon with radial sector zoning, sample 6026D. (D) Oscillatory-zoned prismatic core truncated by a thick rim exhibiting radial sector zoning, sample 6026D. (E) Oscillatory-zoned prism of Archean age, sample 6026D.

tonalitic(?) protolith, this result should be treated with caution given the discordancy of the data and large 1σ errors for all four analyses. Nonetheless, the 1442 ± 79 age falls within error of the age of the tonalitic gneiss sample 4954A (see later). Analysis 2671-79.4 from the core region of grain 79 yielded a distinctly older ^{207}Pb-^{206}Pb age of 1586 Ma, possibly owing to the presence of a xenocrystic component.

The youngest analyses in quartzofeldspathic gneiss sample 2671 comprise clear cores of subequant crystals and stubby prisms as well as mantles on both clear (2671-35.2) and turbid cores (2671-2.1, -34.2, -55.1, -71.2, and -79.2). Clear cores are characterized by low U contents (49–294 ppm) and distinctly low Th/U ratios (0.02–0.12). Mantles on clear and turbid cores have higher U contents (289–810 ppm) but similar Th/U ratios (0.01–0.07). Such low Th/U ratios are typical of zircons grown during high-grade metamorphism (e.g., Williams and Claesson, 1987; Maas et al., 1992). The clear cores have $^{207}Pb/^{206}Pb$ ratios indicating an age of 1189 ± 34 (2σ) Ma (MSWD = 0.28), indis-

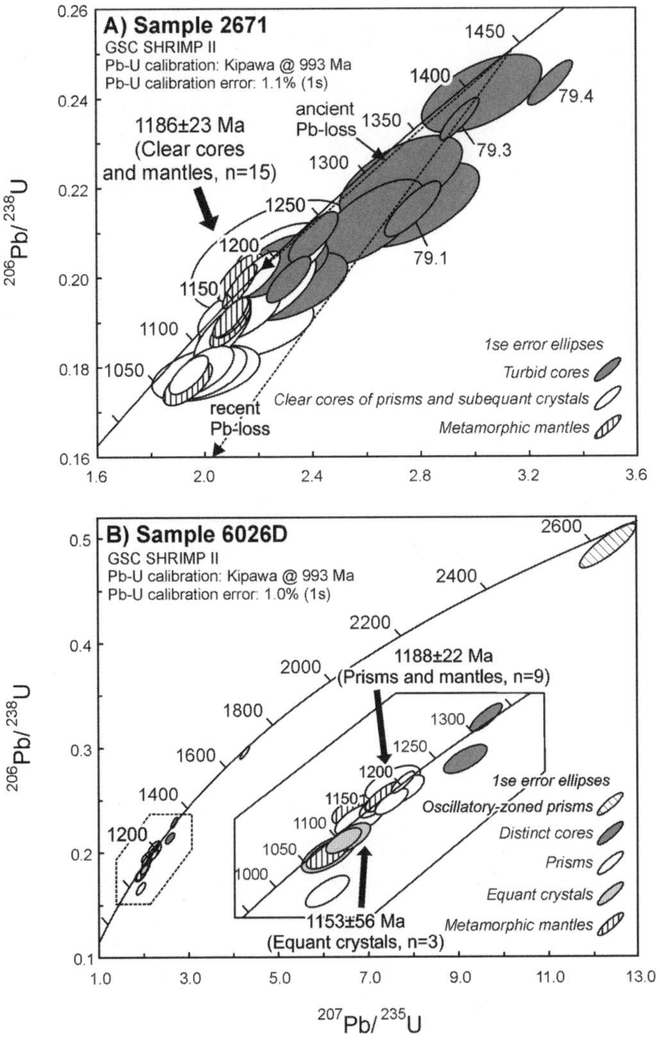

Figure 6. U-Pb concordia diagrams for zircons from nonlaminated quartzofeldspathic samples 2671 (A) and 6026D (B). Error ellipses are 1σ. Plausible times of Pb loss, assuming a protolith age of ca. 1.44 Ga, are shown by dashed lines for sample 2671 (see text for further details). Individual analysis points within the core region of grain 79 (sample 2671) are indicated.

tinguishable from the weighted average age of 1184 ± 30 (2σ) Ma (MSWD = 0.066) for mantles on clear and turbid cores (excluding the two analyses, 2671-2.1 and -55.1, with the lowest $^{207}Pb/^{206}Pb$ ratios). Pooling of these mantle analyses with those of the clear cores (n = 15) indicates an age of 1186 ± 23 (2σ) Ma (MSWD = 0.22). These analyses also fit a regression line with a similar upper intercept age of 1192 ± 48 Ma (MSWD = 0.24). The weighted mean age of 1186 ± 23 Ma is taken as our best estimate of the timing of high-grade metamorphism of this quartzofeldspathic gneiss. This age is also in good agreement with the metamorphic ages obtained for the laminated quartzofeldspathic gneisses.

Sample 6026D

The quartzofeldspathic gneiss sample 6026D has a zircon population dominated by clear, colorless to yellowish, subhedral equant crystals and prisms with 3–6:1 aspect ratios. Fractures oriented parallel to the c-axis are present in most grains, regardless of morphology. In contrast to quartzofeldspathic gneiss 2671, core-overgrowth relationships are uncommon and, in transmitted light, consist of either yellowish cores with colorless shells or colorless, rounded to prismatic cores mantled by relatively thick zircon. In CL images, the majority of grains are luminescent and display coarse, concentric or radial sector zoning (Fig. 5, C). The colorless, rounded to prismatic cores exhibit oscillatory zoning truncated and overgrown by thick, weakly luminescent mantles with radial sector zoning (Fig. 5, D). In contrast, the yellowish cores show coarse concentric zoning mantled by dark, more homogeneous shells. Prismatic crystals with fine-scale oscillatory zoning and relatively thin, luminescent mantles are rare (Fig. 5, E).

Two distinct, colorless cores were analyzed from this sample and gave Th/U ratios of 0.39–0.40 and ^{207}Pb-^{206}Pb ages of 1304 and 1360 Ma, within the range obtained for those from quartzofeldspathic gneiss 2671 (Table 1). However, unlike any other sample of laminated or nonlaminated quartzofeldspathic gneiss, sample 6026D also yielded two significantly older concordant to moderately discordant analyses, 6026D-141.2 and -143.1, with ^{207}Pb-^{206}Pb ages of 1684 and 2676 Ma, respectively (Fig. 6, B). These analyses, characterized by distinctly higher Th/U ratios (0.71–0.93), come from grain interiors of the oscillatory-zoned prisms (Fig. 5, E). On the basis of their distinct zoning patterns, Th/U ratios, and textural context, these grains are interpreted to be xenocrysts.

The younger group of analyses in quartzofeldspathic gneiss 6026D come both from the grain interiors of prismatic and equant crystals and from mantles on prismatic cores (6026D-46.2 and 48.2) and crystals (6026D-32.1) (Table 1; Fig. 6, B). The mantles and grain interiors of prismatic crystals are chemically similar, with moderate to low U and Th contents (<470 and <175 ppm, respectively) and moderate Th/U ratios (0.19–0.58), suggesting that they form part of the same population. Pooling of these analyses gives a weighted mean ^{207}Pb-^{206}Pb age of 1188 ± 22 (2σ) Ma (MSWD = 0.75) (n = 9, excluding the strongly discordant analysis 6026D-52.1). In contrast, the equant zircons are characterized by much higher Th contents (278–355 ppm) and Th/U ratios (1.9–2.2), and their $^{207}Pb/^{206}Pb$ ratios indicate a somewhat younger age of 1153 ± 56 (2σ) Ma (MSWD = 0.10). These results could indicate two discrete thermal pulses in the Bondy gneiss complex at ca. 1.19 and 1.15 Ga (see later).

Tonalitic Gneiss

Granitic to tonalitic gneiss forms an important component of the Bondy gneiss complex. The gneissossity in these rocks is typically defined by discontinuous layers of orthopyroxene and/or biotite. Tonalitic gneiss has a calc-alkaline affinity characterized by enrichment in light REE (Blein et al., 2003). It shows negative Nb, Ta, and Ti anomalies and pronounced positive Zr and Pb anomalies. The tonalitic gneiss has comparatively low $\varepsilon Nd_{(T = 1.4 Ga)}$ values of +4.0–4.1, indicating that continental crustal material was involved in the genesis of these rocks (Blein et al., 2003). One sample of tonalitic gneiss, collected ~3 km northeast of the main hydrothermal system in the northern part of the Bondy gneiss complex (Fig. 2), was chosen for U-Pb geochronology. CL images of the zircon grains selected from this sample are shown in Figure 7, and the SHRIMP analytical results are given in Table 1 and plotted in a concordia diagram in Figure 8.

Sample 4954A

In contrast to the diversified populations of zircons in the laminated and nonlaminated quartzofeldspathic gneisses, the morphology of zircons in tonalitic gneiss 4954A is more uniform, and the amount of crystallization of new zircon as mantles or whole crystals appears relatively limited. Most zircons in this sample occur as highly fractured, stubby to elongate, subhedral prisms with aspect ratios of 2–6:1. Clear, equant, euhedral crystals are less common. Grains with rounded cores are rare. Both prismatic and equant crystals display in CL no growth zoning or only faint growth zoning in their interiors, mantled by mostly thin luminescent zircon (Fig. 7, A). Approximately 10% of grains are characterized by thicker, sector-zoned mantles (Fig. 7, B). Cores, where present, are luminescent and are surrounded by a thick, darker unzoned mantle texturally similar to the interiors of most prismatic grains (Fig. 7, C). Outer, thin luminescent rims are also present on the cored grains.

On a concordia diagram, the spot analyses exhibit a strong bimodal distribution of ages (Fig. 8). The twelve oldest concordant to slightly discordant analyses gave $^{207}Pb\text{-}^{206}Pb$ ages ranging from 1423 to 1322 Ma (Table 1). These analyses were obtained from the grain interiors of ten stubby to elongate prismatic zircons with relatively high U contents (200–850 ppm) and Th/U ratios (0.31–0.85). The single analysis from a distinct core, characterized by a lower U content of 131 ppm and a similar Th/U ratio of 0.39, yielded a similar albeit strongly discordant $^{207}Pb\text{-}^{206}Pb$ age of 1402 Ma. Pooling of all twelve prism analyses gives a weighted mean $^{207}Pb\text{-}^{206}Pb$ age of 1386 ± 10 (2σ) Ma, with a MSWD of 0.87. The consistent morphology of the zircons and the uniform age result support an igneous pro-

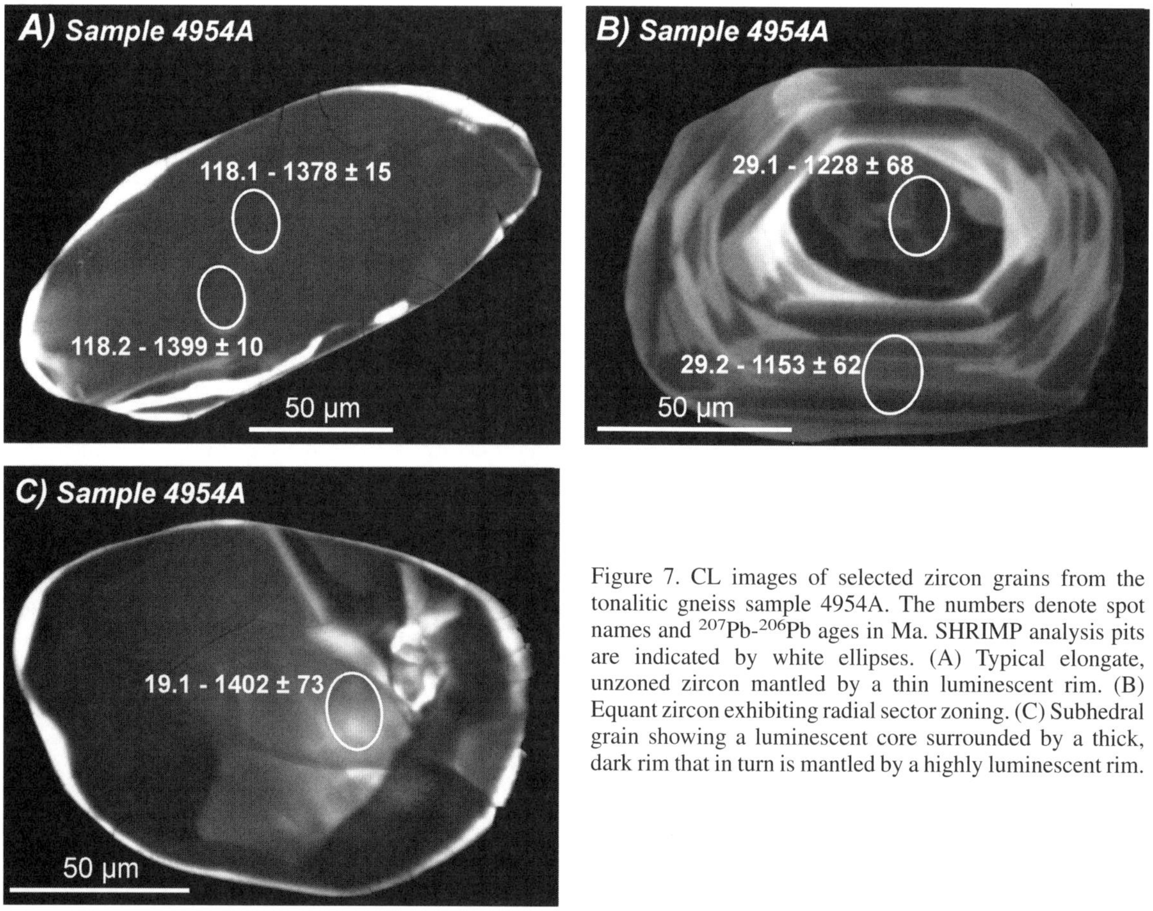

Figure 7. CL images of selected zircon grains from the tonalitic gneiss sample 4954A. The numbers denote spot names and $^{207}Pb\text{-}^{206}Pb$ ages in Ma. SHRIMP analysis pits are indicated by white ellipses. (A) Typical elongate, unzoned zircon mantled by a thin luminescent rim. (B) Equant zircon exhibiting radial sector zoning. (C) Subhedral grain showing a luminescent core surrounded by a thick, dark rim that in turn is mantled by a highly luminescent rim.

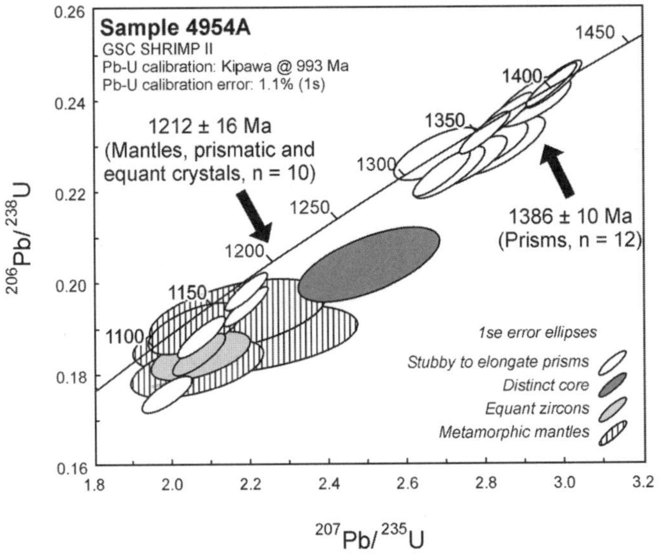

Figure 8. U-Pb concordia diagram for zircons from tonalitic gneiss sample 4954A. Error ellipses are 1σ.

tolith for this gneiss. Thus, we interpret the weighted mean age of 1386 ± 10 Ma as the crystallization age of the tonalitic gneiss precursor.

Four analyses of stubby to elongate prismatic zircons (analyses 4954A-1.1, -42.1, -64.1, and -160.1), although morphologically indistinguishable from those of the older age group, have distinctly lower Th/U ratios (0.10–0.28) and yield ^{207}Pb-^{206}Pb ages similar to those obtained from equant crystals and metamorphic mantles (Table 1; Fig. 8). These analyses may represent disturbed members of the older zircon population through either leaching of radiogenic Pb or recrystallization. Alternatively, they may reflect new zircon growth during high-grade metamorphism and in situ partial melting.

Analyses of equant zircons and thick, sector-zoned mantles in the tonalitic gneiss have relatively low U contents (63–343 ppm) and variable Th/U ratios of 0.52–1.58 (Table 1). Owing to the low U concentrations, the ^{207}Pb-^{206}Pb ages have large 1σ errors (Fig. 8). Nonetheless, these analyses, together with the four youngest analyses of prismatic zircons, have $^{207}Pb/^{206}Pb$ ratios corresponding to an age of 1212 ± 16 (2σ) Ma (MSWD = 0.64). These data fit a regression line with a similar upper intercept age of 1197 ± 28 Ma (MSWD = 0.62). The weighted mean result is interpreted to date the time of high-grade metamorphism of the tonalitic gneiss at 1212 ± 16 Ma.

Intermediate Gneiss

Gneissic rocks of intermediate composition within the Bondy gneiss complex contain plagioclase, K-feldspar, orthopyroxene, garnet, quartz, and biotite, with biotite and orthopyroxene defining the foliation. These rocks display major element and trace element compositions similar to those of modern arc andesites and dacites (Blein et al., 2003). They also have the same Nd isotopic signature as the tonalitic gneisses, implying the involvement of an older crustal component during the petrogenesis of these rocks. One sample of intermediate gneiss, collected from a small satellite of altered rocks immediately east of the main hydrothermal system in the Bondy gneiss complex (Fig. 2), was selected for geochronological work. CL images of zircon grains selected from this sample are shown in Figure 9, and the SHRIMP analytical results are given in Table 1 and plotted in a concordia diagram in Figure 10.

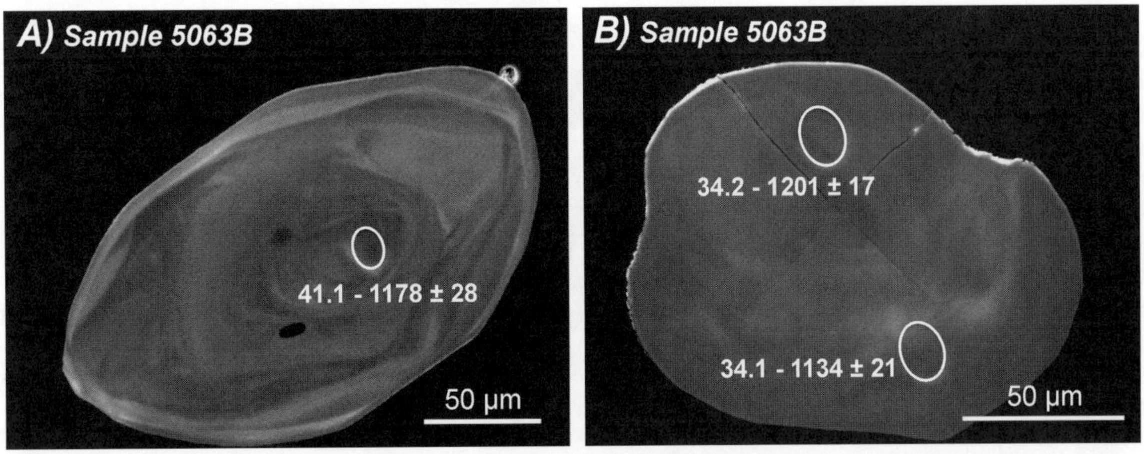

Figure 9. Cathodoluminescence images of selected zircon grains from the intermediate gneiss sample 5063B. The numbers denote spot names and ^{207}Pb-^{206}Pb ages in Ma. SHRIMP analysis pits are indicated by white ellipses. (A) Typical multifaceted zircon (in transmitted light) showing diffuse irregular zoning mantled by a thin dark rim. (B) Structural core with faint zoning mantled by a dark, unzoned shell.

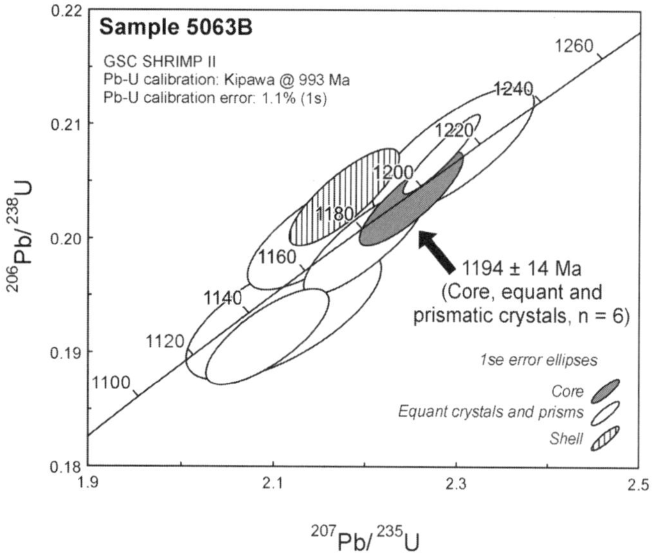

Figure 10. SHRIMP U-Pb concordia diagram for zircons from intermediate gneiss sample 5063B. Error ellipses are 1σ.

Sample 5063B

The intermediate gneiss 5063B yielded abundant clear, colorless to light brown zircons, varying in shape from dominantly equant, multifaceted grains to rounded, elongate prisms. Only a few grains consist of a structural core surrounded by a shell. The equant, multifaceted morphology of the grains is typical of metamorphic zircons (e.g., van Breemen et al., 1986). All morphologies are relatively homogeneous under CL imaging, exhibiting diffuse irregular or radial sector zoning in their interiors (Fig. 9, A). Most grains, including the cored crystals, are surrounded by relatively dark, unzoned mantles (Fig. 9, A and B).

Regardless of the morphology, all zircons analyzed give Grenvillian ^{207}Pb-^{206}Pb ages ranging between 1201 and 1134 Ma (Fig. 10). Six of the eight analyses have moderate U contents (207–596 ppm), low Th contents (1–17 ppm), and extremely low Th/U ratios (<0.06) (Table 1). These analyses come from the grain interiors of colorless, multifaceted equant crystals and prisms, as well as from a structural core. The light brown, multifaceted equant grain 46, although it has a different Th content (179 ppm) and Th/U ratio (0.08) than the colorless grains, yielded a similar ^{207}Pb-^{206}Pb age (1193 Ma; Table 1). Excluding the analysis with the lowest ^{207}Pb/^{206}Pb ratio (5063B-24.1), the remaining six analyses of equant grains, prisms, and core give a weighted mean ^{207}Pb-^{206}Pb age of 1194 ± 14 (2σ) Ma (MSWD = 0.14). We interpret this age as the time of zircon growth during high-grade metamorphism. Linear regression of the six analyses yields an upper intercept age of 1192 ± 18 Ma (MSWD = 0.12), essentially indistinguishable from the weighted mean age. In contrast to these results, the analyzed shell has a distinctly higher U content (762 ppm) and Th/U ratio (0.11), and gave a ^{207}Pb-^{206}Pb age of 1134 ± 21 (1σ) Ma that is statistically younger than the weighted mean age of the main group of analyses. Combined with its distinct chemistry and textural context, this result is interpreted to record an event younger than ca. 1.19 Ga.

DISCUSSION

Gneiss Protolith Formation

Laminated Quartzofeldspathic Gneiss

Overall, the three samples of laminated quartzofeldspathic gneiss gave essentially similar ion microprobe results, taken to suggest that these rocks had a similar history. CL imaging shows that many zircons in the gneiss have complex internal structures, including the presence of distinct cores surrounded by mantles with no zoning or diffuse zoning (Fig. 3). The analyses of mantles, coupled with those of whole zircon grains in all three samples, are quite uniform and indicate an age of ca. 1.184 Ga. Most mantling zircons have low Th/U values (0.01–0.25), a feature typically associated with metamorphic rather than igneous zircon. The general lack of igneous zoning (e.g., oscillatory zoning) in all mantles imaged is also consistent with a metamorphic origin for these zircons. Oscillatory zoning or high Th/U ratios observed in whole grains that give ages identical to those of the mantles could imply that these zircons grew in anatectic melt pods as argued later. On the basis of these observations, we suggest that 1.184 Ga represents not the age of the quartzofeldspathic gneiss protolith, but rather the age of a period of high-grade metamorphism and partial melting as further discussed later. Previously obtained U-Pb ages of 1195 ± 7 Ma and 1179 ± 3 Ma for zircons separated from gneissic rocks of the Bondy gneiss complex and interpreted to have grown under metamorphic and anatectic conditions, respectively (Corriveau and van Breemen, 2000), provide further support for this interpretation.

In contrast to the mantle analyses, most of the core analyses yielded apparent ages ranging from 1.46 to 1.32 Ga. Although it could be argued that the nonuniform ages of these cores reflect a mixed clastic sedimentary source for the quartzofeldspathic gneiss protolith, major element and trace element geochemistry (Blein et al., 2003), coupled with field observations, strongly suggests that these rocks have a volcanogenic (high-silica rhyolites) parentage. In addition, the Th/U ratios of cores are fairly constant for any given sample and range between 0.25–0.61, favoring a common magmatic origin for these zircons. Thus, the distinct zircon cores are interpreted to have grown during crystallization of the rhyolitic protolith, but the precise age(s) of these rocks is difficult to assess because the analyses do not define single concordant groups with similar ^{207}Pb/^{206}Pb ratios, nor do they define statistically meaningful chords. If the interpretation that these rocks formed in a back-arc basin during early arc rifting (Blein et al., 2003) is correct, the U-Pb crystallization age of the arc-related tonalitic gneiss 4954A, that is, ca. 1.39 Ga, would represent a maximum age for the laminated quartzofeldspathic gneiss of the Bondy gneiss complex. The dominance of distinct cores with apparent ages of between 1.32 and 1.40 Ga over grains with ages of greater than 1.40 Ga in all three samples of laminated quartzofeldspathic gneiss is consistent with this interpretation.

Quartzofeldspathic Gneiss

The geochronological data for the nonlaminated quartzofeldspathic gneiss samples 2671 and 6026D show patterns broadly similar to those for the laminated quartzofeldspathic gneiss. With few exceptions, most analyses fall into two broad populations, with protolith zircon cores indicating apparent ages of between 1.59 and 1.21 Ga and metamorphic zircons giving mean ages of 1.19 and 1.15 Ga. Although no precise protolith age could be identified for the quartzofeldspathic gneisses, the SHRIMP ages of turbid cores from sample 2671 are consistent with an early Mesoproterozoic (ca. 1.5–1.4 Ga) origin for the magmatic(?) precursor. In addition, the presence of Paleoproterozoic and Neoarchean zircon xenocrysts or xenocrystic components indicates that old continental crustal material was involved in the genesis of these rocks (see later). The significance of both the complex age spectra of core analyses (e.g., Fig. 6, A) and the younger populations of metamorphic zircons is discussed later.

Tonalitic Gneiss

In contrast to the laminated and nonlaminated quartzofeldspathic gneisses, the tonalitic gneiss 4954A yielded a well-defined crystallization age of 1386 ± 10 Ma. This age falls within error of the SHRIMP U-Pb minimum age of 1.36 ± 0.02 Ga (O. van Breemen, 2000, personal commun.) for the homogeneous tonalitic pluton in the southern part of the Bondy gneiss complex (Fig. 2), suggesting that this pluton may be contemporaneous with the tonalitic gneiss.

The 1386 ± 10 Ma age is taken to represent a reasonable estimate of the age of arc-related magmatism in the Bondy gneiss complex. The tonalitic gneiss from the Bondy gneiss complex is temporally equivalent to the Lacoste gneiss (Fig. 1, C), which has recently been dated at 1386 ± 18 Ma using U-Pb isotope dilution methods (Nantel and Pintson, 2002). In addition, geochemical and Sm-Nd data suggest that rocks of the Lacoste gneiss are mainly juvenile and formed in an arc setting (Nantel and Pintson, 2002). Consequently, age and chemical similarities between the Bondy gneiss complex and Lacoste gneiss support the hypothesis that the gneiss complexes of the Central Metasedimentary Belt in western Québec form a major lithotectonic domain structurally beneath the supracrustal domains (Corriveau and Morin, 2000). Comparison of these data also suggests that the gneiss complexes may all be part of the same arc system that was active during the Mesoproterozoic (see also Corriveau and Morin, 2000).

Isotopic Disturbance

The age scatter observed in core analyses from all samples of laminated and nonlaminated quartzofeldspathic gneiss may be attributable to a number of factors, including minor inheritance and later isotopic disturbance owing to hydrothermal alteration, high-*P-T* metamorphism, and/or more recent Pb loss. The interpretation of the presence of a minor component of inheritance in the nonlaminated quartzofeldspathic gneiss is supported by Nd data for gneissic rocks of the Bondy gneiss complex, which suggest incorporation of an older crustal component (Blein et al., 2003) (see also later). Partial radiogenic Pb loss from igneous zircon during high-*P–T* metamorphism at ca. 1.21–1.18 Ga and/or at recent times is also consistent with the Pb-loss patterns exhibited by the core analyses from several samples (Figs. 4 and 6, A). Disturbance of the zircon U-Pb systematics as a result of hydrothermal alteration is more difficult to evaluate for the following reasons. First, although geochemical data suggest a volcanogenic setting for the cupriferous hydrothermal system in the Bondy gneiss complex (Corriveau et al., 2003; Blein et al., this volume), the precise timing of hydrothermal alteration and associated mineralization could not be resolved from the current SHRIMP data mainly because (1) precise protolith ages could not be determined for any of the host rocks to mineralization, and (2) none of the samples analyzed in this study contain hydrothermal zircons as described by Fu et al. (2003). Second, if hydrothermal alteration occurred sometime between the time of protolith formation and that of high-grade metamorphism, resolution of the timing of such an intermediate Pb-loss episode is hampered by the low angle of discordia arrays (drawn between assumed maximum protolith ages of 1.39 and 1.44 Ga for the laminated and nonlaminated quartzofeldspathic gneisses, respectively, and the time of high-*P-T* metamorphism) with concordia (Figs. 4 and 6, A). However, the presence of highly luminescent seams at boundaries between cores and mantles in several samples does suggest that hydrothermal activity affected the zircon U-Pb systematics. The seams are best developed, although to a variable extent, in samples collected in the vicinity of the main hydrothermal system or from smaller satellites of hydrothermally altered rocks (e.g., samples 4548C2-A, 5002L, 2671, and 6026D). They have not been observed on any of the metamorphic grains, and appear to be absent in the tonalitic gneiss sample 4954A collected several kilometers away from the hydrothermal system (Fig. 2).

In terms of texture and degree of luminescence revealed in CL images, the luminescent seams resemble the CL-bright crystal rims described by Vavra et al. (1999). Thus, following these authors, we attribute the highly luminescent seams described here to surface-controlled alteration (at original core surfaces) resulting from the influx of fluids related to hydrothermal activity. Empirical studies show that hydrothermal fluids containing ingredients such as F, Cl, Na, and Ca can cause severe alteration and Pb loss even from unmetamict zircons (e.g., Sinha et al., 1992). No age could be obtained for the seams, primarily owing to their thinness and low U concentrations. However, their restricted location between protolith cores and metamorphic mantles implies that hydrothermal activity postdated gneiss protolith formation but predated high-grade metamorphism, consistent with field relationships and geochemical data for hydrothermal lithofacies of the Bondy gneiss complex (Blein et al., this volume).

Inheritance

Inherited zircon is suspected in both samples of nonlaminated quartzofeldspathic gneiss and represents an important tool in understanding the tectonic evolution of the Bondy gneiss complex. The oldest ^{207}Pb-^{206}Pb age of ca. 1586 Ma in tonalitic(?) gneiss sample 2671 could be ascribed to some mode of inheritance. This inference is corroborated by comparatively low $\varepsilon Nd_{(T)}$ values (+4.0–4.1) and T_{DM} model ages of 1.57 and 1.58 Ga for tonalitic gneisses of the Bondy gneiss complex, which suggest contamination by a crustal component related to either subducted sediments or arc basement in a juvenile convergent margin setting (Blein et al., 2003). The two ^{207}Pb-^{206}Pb ages of 2676 and 1684 Ma for xenocrysts in the quartzofeldspathic gneiss sample 6026D also demonstrate the presence of at least some Archean and Paleoproterozoic material in the source region. Identification of an inherited component of Pb in the laminated quartzofeldspathic gneiss is more difficult. The oldest ^{207}Pb-^{206}Pb ages of zircon cores, 1444 ± 39 and 1461 ± 36 (1σ) Ma, are not significantly different from the inferred maximum age of 1386 ± 10 (2σ) Ma for these rocks. Although the apparent absence of zircon inheritance in the laminated quartzofeldspathic gneiss does not necessarily rule out some crustal recycling or contamination, it is consistent with back-arc rhyolites elsewhere, which have depleted Nd isotopic compositions similar to those of the upper mantle or lower crustal materials (e.g., Kitami rhyolite of Hokkaido, Japan; Takagi et al., 1999). Thus, taken together, the U-Pb zircon and Nd isotopic data suggest that a major portion of the source region to the Bondy gneiss complex was produced by the addition of voluminous juvenile Mesoproterozoic material and at least some Paleoproterozoic and Archean material, possibly in the form of sedimentary detritus. However, taken alone, the inherited zircon spectra cannot clearly indicate whether the Bondy gneiss complex was built on a thinned continental substrate of Laurentian affinity.

Timing of Metamorphism

The new geochronological data presented earlier allow us to infer at least two episodes of Grenvillian metamorphism in the Bondy gneiss complex. The older group of metamorphic zircons, comprising both mantles and well-formed prismatic or equant crystals from all dated lithologies, yielded generally indistinguishable ages of 1184 ± 15, 1183 ± 9, 1183 ± 17, 1186 ± 23, 1188 ± 22, 1192 ± 18, and 1212 ± 16 Ma. Th/U ratios are fairly constant for any single morphological type of zircon, but can be very different between morphologically different types and/or between different samples (Table 1). The reason for the variability in Th/U chemistry is not entirely clear, and cannot be readily attributed to metamorphic grade (e.g., Vavra et al., 1999). One possible explanation is that during high-grade metamorphism zircons with high Th/U ratios (≥0.2) grew in partial melt segregations, whereas those with low Th/U ratios (≤0.2 to 0.01) grew in crystalline domains via precipitation from U-bearing metamorphic fluids.

The 1.21–1.18 Ga ages reported here correspond well with previously published multigrain, U-Pb zircon ages of 1195 ± 7 and 1179 ± 3 Ma from the Bondy gneiss complex, which have been interpreted to date high-*P-T* granulite-facies metamorphism and in situ partial melting, respectively (Corriveau and van Breemen, 2000). Although the ion microprobe data do not permit the distinction in time between high-grade metamorphism and partial melting, they do indicate that these events were pervasive in the Bondy gneiss complex. Peak metamorphic conditions of ca. 1000 MPa and 950 °C have been estimated for the Bondy gneiss complex (Boggs, 1996; Boggs and Corriveau, this volume). The high pressures may reflect substantial thickening of the crust in response to continental collision (Corriveau and van Breemen, 2000). However, Harris et al. (2001) indicated that the fabric hosting the peak-pressure assemblages could equally well have been developed during ensuing orogenic collapse or rebound of a lithospheric root thickened during terrane assembly. A plausible explanation for the high temperatures includes asthenosphere rise and underplating by voluminous mafic magma (Corriveau and Morin, 2000).

Several lines of evidence suggest that the rocks of the Bondy gneiss complex were affected by a second, albeit localized, metamorphic event. Mineralogical evidence for an overprinting, amphibolite-facies event in the gneiss complex includes the formation of green biotite and sillimanite at the expense of cordierite in aluminous gneiss (Boggs and Corriveau, this volume). The U-Pb data presented earlier provide additional support for the polymetamorphic nature of the Bondy gneiss complex. For example, the intermediate gneiss 5063B contains a 1201 ± 17 (1σ) Ma metamorphic core surrounded by a shell with a distinct chemistry and an apparent age of 1134 ± 21 (1σ) Ma (Table 1; Fig. 9, B). Similarly, in the quartzofeldspathic gneiss 6026D, equant zircons with high Th/U ratios (1.9–2.2) yielded a weighted mean age of 1153 ± 56 (2σ) Ma, whereas metamorphic rims and grain interiors of prismatic crystals have distinctly lower Th/U ratios (0.19–0.58) and gave a mean age of 1188 ± 22 (2σ) Ma. Although the mean age of the different groups of zircons in this sample cannot be shown to be significantly different at the 2σ level, the compositional and textural differences between the two groups provide support for a period of new zircon growth following ca. 1.19 Ga high-*P-T* metamorphism. The SHRIMP results from both samples are not precise enough to constrain the time of new zircon growth, but nevertheless suggest an overprinting event at ca. 1.15–1.13 Ga. The polymetamorphic nature of the Bondy gneiss complex is also supported by the local presence of a weak, north-striking foliation in a ca. 1161 Ma Chevreuil dike (Corriveau and van Breemen, 2000) that cuts across folded Bondy gneisses.

Albeit imprecise, the ca. 1.13 and 1.15 Ga ages inferred for this second metamorphic episode partly overlap the ca. 1.17–1.14 Ga age range (Friedman and Martignole, 1995; Corriveau and van Breemen, 2000) recorded in metamorphic zircon and monazite from supracrustal rocks of the quartzite domain. The ca. 1.39 Ga tonalite from the Lacoste gneiss also yielded a similar metamorphic age of 1152 ± 14 Ma, which Nantel and Pintson (2002) interpreted as being related to emplacement of the Serpent intrusive suite (the age equivalent of the Chevreuil suite). Thus, given these age similarities, it is tempting to correlate the younger metamorphic overprint in the Bondy gneiss complex with the amphibolite-facies overprinting event recorded primarily in Chevreuil dikes and host rocks outside of the rheologically strong gneiss complexes (e.g., Corriveau et al., 1998; Corriveau and van Breemen, 2000). Corriveau and van Breemen (2000) attributed this event to intraplate reactivation during renewed convergence. The lack of isotopic evidence for this younger overprint in the other dated samples is consistent with the interpretation that the Bondy gneiss complex largely behaved as a rigid domain during this event (Corriveau et al., 1998; Harris et al., 2001).

Regional Implications

Trace element and Nd isotopic compositions strongly suggest that the Bondy gneiss complex represents a volcano-plutonic edifice formed in a mature island arc and back-arc setting (Blein et al., 2003). Calc-alkaline mafic granulite, intermediate gneiss (andesite-dacite), and tonalitic gneiss may be related to subduction zone magmatism in a mature island arc or a young continental margin, whereas tholeiitic mafic granulite and laminated quartzofeldspathic gneiss (high-silica rhyolite or rhyolitic tuff) likely formed in a back-arc rift setting. The SHRIMP U-Pb geochronology presented here indicates that the composite volcano-plutonic edifice formed, at least in part, at ca. 1.39 Ga, the age of the tonalitic gneiss precursor. Combined with the Nd data available (Blein et al., 2003), the new geochronological data also suggest that this was a time of major continental-crust formation characterized by a significant amount of juvenile Mesoproterozoic material.

The ca. 1.39 Ga Bondy gneiss complex and the temporally equivalent Lacoste gneiss represent the oldest component of the Central Metasedimentary Belt. The oldest rocks documented so far in neighboring Ontario include 1287–1279 Ma (Heaman et al., 1987; Davis and Bartlett, 1988) tholeiitic volcanic rocks from the Elzevir terrane that likely formed in primitive arcs (Carr et al., 2000, and references therein). In the Adirondack Lowlands, a multigrain U-Pb zircon age of 1416 ± 5 Ma (McLelland et al., 1991) for a leucogranite on Wellesley Island in the St. Lawrence River (Fig. 1) was by far the oldest age obtained in the Central Metasedimentary Belt at the time. However, more recent high-precision, single-grain dating of the Wellesley Island leucogranite indicates that the earlier multigrain age was determined from a mixture of detrital and 1172 Ma igneous zircons (Wasteneys et al., 1999), and that the leucogranite belongs instead to the 1180–1150 Ma Frontenac plutonic suite. Although one could argue that the spread of zircon core ages obtained in the present study reflects a comparable degree of inheritance within the laminated and nonlaminated quartzofeldspathic gneisses, we do not favor such an interpretation for reasons outlined earlier. Furthermore, Nd data reported by Blein et al. (2003) do not support significant inheritance in gneissic rocks of the Bondy gneiss complex.

The 1386 ± 10 Ma age obtained for the arc-related tonalitic gneiss thereby extends the known age range of subduction zone magmatism in the Central Metasedimentary Belt. Elsewhere in the Grenville Province, geochemical and geochronological data signal major juvenile crustal additions to the Laurentian margin, comprising calc-alkaline magmatic arcs and broadly coeval inboard back-arc deposits between ca. 1.5 and 1.3 Ga. However, it remains the subject of much debate whether these arcs originated in an Andean-type continental margin or formed offshore, mainly on oceanic crust, throughout this period (e.g., Corrigan, 1995; Rivers, 1997; Nadeau and van Breemen, 1998; Carr et al., 2000; Hanmer et al., 2000; Rivers and Corrigan, 2000; Culshaw and Dostal, 2002; Slagstad et al., this volume). For example, in the Central Gneiss Belt, the presence of extensive juvenile 1.48–1.40 Ga magmatism testifies to the development of an oceanward-younging ensialic arc along the pre-Grenvillian Laurentian margin (e.g., Culshaw et al., 1997; Nadeau and van Breemen, 1998). Magmatism in the Eastern Granite-Rhyolite Belt of the midcontinental United States coincides in time with that documented in the Central Gneiss Belt (e.g., van Breemen et al., 1986), and a genetic link may exist between the two regions (e.g., Rivers, 1997; Nadeau and van Breemen, 1998; Rivers and Corrigan, 2000; Culshaw and Dostal, 2002; Slagstad et al., this volume). By contrast, a growing body of geological evidence suggests that the younger 1.4–1.33 Ga granitic and tonalitic gneisses of the Parry Sound allochthon within the Central Gneiss Belt may have originated in a magmatic arc initially formed outboard of the pre-Grenvillian Laurentian margin (e.g., Wodicka et al., 1996; Culshaw and Dostal, 2002). The structurally underlying, allochthonous Shawanaga domain contains amphibolite and felsic gneiss interpreted on geochemical grounds to have originated in a back-arc that formed in a Mesoproterozoic active margin, but whether these rocks formed in an Andean-type margin remains uncertain (Culshaw and Dostal, 2002).

To the east, the 1.45 Ga Montauban Group in the Central Granulite Terrane in Québec has been interpreted as a mature island arc or back-arc terrane (e.g., Nadeau et al., 1999) that was subsequently intruded by the 1.4–1.37 Ga La Bostonnais plutonic complex (Hanmer et al., 2000, and references therein). Most workers argue that the La Bostonnais complex formed in an ensialic arc on the southeastern margin of Laurentia (e.g., Corrigan, 1995; Hanmer et al., 2000; Rivers and Corrigan, 2000).

In the Central Metasedimentary Belt boundary zone, ca. 1.37–1.35 Ga calc-alkaline granitic to tonalitic rocks (Lumbers et al., 1990) occur as transported, dismembered plutons. Some workers suggest that these rocks originated in a continental arc on the southeastern margin of Laurentia (e.g., Rivers and Corrigan, 2000), while others argue that the linkage between the Central Metasedimentary Belt boundary zone and Laurentia prior to the onset of the Grenvillian orogeny is still unclear (e.g., Carr et al., 2000). Somewhat younger ca. 1.35–1.30 Ga plutonic rocks of calc-alkaline affinity have been reported from the southeastern Adirondack Mountains and the Green Mountains in Vermont (Ratcliffe et al., 1991; McLelland et al., 1996, and references therein). Nd T_{DM} model ages of 1.45–1.35 Ga and ε_{Nd} values of +3 to +5 for tonalites from these areas indicate derivation from a juvenile source terrane (McLelland et al., 1996, and references therein). Although geochemical data indicate that these rocks formed in a convergent margin setting, they do not clearly discriminate between island arc and Andean-type arc settings.

In summary, the new geochronological results presented here not only extend the known distribution of ca. 1.39 Ga arc-related magmatism in the Grenville Province, but also suggest a link between the Bondy gneiss complex and the ca. 1.45–1.3 Ga juvenile rocks from allochthonous slices within the Central Gneiss Belt and from the Central Metasedimentary Belt boundary zone and the Central Granulite Terrane in Québec and in the Adirondack Highlands. On the basis of geochemical data from the Bondy gneiss complex, Blein et al. (2003) propose a complex setting for the southeastern margin of Laurentia during this period, with either an island arc built on oceanic or thin continental crust alongside a continental margin or a juvenile continental margin within a newly accreted terrane.

The nature and age of the stratigraphic basement to the supracrustal rocks of the Central Metasedimentary Belt in Québec remain problematic. Although there is clear evidence that the Bondy gneiss complex represented the structural basement to overlying supracrustal rocks (e.g., Corriveau and Morin, 2000), it is still unclear whether the gneiss complexes also served as basement for deposition of the Mesoproterozoic sediments. If the latter case, it seems surprising that none of the detrital zircons in a metapelite from the quartzite domain (Friedman and Martignole, 1995) yielded ages of ca. 1.39 Ga. However, a quartzite from the Cabonga terrane, an allochthon rooted in the Central Metasedimentary Belt (Fig. 1, C; Martignole and Calvert, 1996), contains a large population of detrital zircons with ^{207}Pb-^{206}Pb ages ranging from 1350 to 2700 Ma and peaks at 1400 and 1700 Ma (Martignole and Ringuette, 1998), in agreement with protolith and inherited zircon ages in the Bondy gneiss complex. In addition, no zircons younger than 1.3 Ga were found in the quartzite. Similar detrital zircon ages have been reported for a quartzite from the Frontenac terrane (1305–3185 Ma, with a cluster between 1745 and 1890 Ma; Sager-Kinsman and Parrish, 1993) and from the Wellesley Island leucogranite (1475, 1920, and 2540 Ma; Wasteneys et al., 1999). Thus, comparison of these data suggests that the Bondy gneiss complex could have acted as basement to supracrustal rocks of the Central Metasedimentary Belt.

ACKNOWLEDGMENTS

This research was partly funded through a contract with STREX Inc. of the Ste.-Geneviève Group. We thank the Papineau-Labelle Wildlife Reserve for access to the reserve. David Walker and Pat Hunt are thanked for their assistance with cathodoluminescence imaging. Otto van Breemen is thanked for permission to publish data from the homogeneous tonalite sample. We have greatly benefited from discussions with Olivier Blein and Tony Davidson. We would also like to thank John Ketchum and Hardolph Wasteneys for careful and thought-provoking reviews. This paper is Geological Survey of Canada contribution 2002111.

REFERENCES CITED

Bingen, B., and van Breemen, O., 1998, U-Pb monazite ages in amphibolite- to granulite-facies orthogneiss reflect hydrous mineral breakdown reactions: Sveconorwegian Province of SW Norway: Contributions to Mineralogy and Petrology, v. 132, p. 336–353.

Blein, O., LaFlèche, M.R., and Corriveau, L., 2003, Geochemistry of the granulitic Bondy gneiss complex: A 1.4 Ga arc in the Central Metasedimentary Belt, Grenville Province, Canada: Precambrian Research, v. 120, p. 193–217.

Boggs, K., 1996, Retrograde cation exchange in garnets during slow cooling of mid crustal granulites and the *P-T-t* trajectories from the Mont-Laurier region, Grenville Province, Quebec [M.S. thesis]: Chicoutimi, Université du Québec à Chicoutimi, 352 p.

Carr, S.D., Easton, R.M., Jamieson, R.A., and Culshaw, N.G., 2000, Geologic transect across the Grenville orogen of Ontario and New York: Canadian Journal of Earth Sciences, v. 37, p. 193–216.

Corrigan, D., 1995, Mesoproterozoic evolution of the south-central Grenville orogen: Structural, metamorphic and geochronologic constraints from the Mauricie transect [Ph.D. thesis]: Ottawa, Ontario, Carleton University, 285 p.

Corriveau, L., 1990, Proterozoic subduction and terrane amalgamation in the southwestern Grenville Province, Canada: Evidence from ultrapotassic to shoshonitic plutonism: Geology, v. 18, p. 614–617.

Corriveau, L., and Gorton, M.P., 1993, Coexisting K-rich alkaline and shoshonitic magmatism of arc affinities in the Proterozoic: A reassessment of syenitic stocks in the southwestern Grenville Province: Contributions to Mineralogy and Petrology, v. 113, p. 262–279.

Corriveau, L., and Jourdain, V., 1992, Terrane characterization in the Central Metasedimentary Belt of the southern Grenville Orogen, Lac Nominingue map area, Quebec, *in* Current Research, Part C: Ottawa, Ontario, Geological Survey of Canada, Paper 92-1C, p. 81–90.

Corriveau, L., and Jourdain, V., 2000, Lac Nominingue—SNRC 31J06: Québec, Ministère des Ressources naturelles, Map SI-31J06-C3G-00G, scale 1:50,000.

Corriveau, L., and Morin, D., 2000, Modelling 3D architecture of western Grenville from surface geology, xenoliths, styles of magma emplacement, and Lithoprobe reflectors: Canadian Journal of Earth Sciences, v. 37, p. 235–251.

Corriveau, L., and van Breemen, O., 2000, Docking of the Central Metasedimentary Belt to Laurentia in geon 12: Evidence from the 1.17–1.16 Ga Chevreuil intrusive suite and host gneisses, Quebec: Canadian Journal of Earth Sciences, v. 37, p. 253–269.

Corriveau, L., Heaman, L.M., Marcantonio, F., and van Breemen, O., 1990, 1.1 Ga K-rich alkaline plutonism in the SW Grenville Province: U-Pb constraints for the timing of subduction-related magmatism: Contributions to Mineralogy and Petrology, v. 105, p. 473–485.

Corriveau, L., Morin, D., and Madore, L., 1994, Géologie et cibles d'exploration de la partie centre est de la ceinture métasédimentaire du Québec, Province de Grenville, *in* Current Research, Part C: Ottawa, Ontario, Geological Survey of Canada, Paper 1994-C, p. 355–365.

Corriveau, L., Rivard, B., and van Breemen, O., 1998, Rheological controls on Grenvillian intrusive suites: Implications for tectonic analysis: Journal of Structural Geology, v. 20, p. 191–1204.

Corriveau, L., Blein, O., Bonnet, A.-L., Fu, W., Pilote, P., and van Breemen, O., 2003, Field identification of Cu-Au-Fe-oxides bearing hydrothermal systems in undifferentiated gneiss complexes of the Grenville Province: CIM [Canadian Institute of Mining] Montreal 2003 Mining Industry Conference and Exhibition, 4–7 May, Montreal, Québec, Canadian Institute of Mining Technical Paper, CD-ROM.

Culshaw, N.G., and Dostal, J., 2002, Amphibolites of the Shawanaga domain, Central Gneiss Belt, Grenville Province, Ontario: Tectonic setting and implications for relations between the Central Gneiss Belt and Midcontinental USA: Precambrian Research, v. 113, p. 65–85.

Culshaw, N.G., Jamieson, R.A., Ketchum, J.W.F., Wodicka, N., Corrigan, D., and Reynolds, P.H., 1997, Transect across the northwestern Grenville orogen, Georgian Bay, Ontario: Polystage convergence and extension in the lower orogenic crust: Tectonics, v. 16, p. 966–982.

Davidson, A., 1995, A review of the Grenville orogen in its North American type area: AGSO [Australian Geological Survey Organisation] Journal of Australian Geology and Geophysics, v. 16, p. 3–24.

Davidson, A., comp., 1998, Geological map of the Grenville Province, Canada, and adjacent parts of the United States of America: Ottawa, Ontario, Geological Survey of Canada Map 1947A, scale 1:2,000,000.

Davidson, A., and van Breemen, O., 2000, Age and extent of the Frontenac plutonic suite in the Central metasedimentary belt, Grenville Province, southeastern Ontario, *in* Radiogenic Age and Isotopic Studies: Report 13: Ottawa, Ontario, Geological Survey of Canada, Current Research 2000-F4, 15 p.

Davidson, A., Easton, R.M., Corriveau, L., and Martignole, J., 2002, Transect of the southwestern Grenville Province, *in* Geological Association of Canada–Mineralogical Association of Canada, Joint Annual Meeting, May, Saskatoon, Saskatchewan, Field Trip B6 Guidebook, 114 p.

Davis, D.W., and Bartlett, J.R., 1988, Geochronology of the Belmont Lake Metavolcanic Complex and implications for crustal development in the Central Metasedimentary Belt, Grenville Province, Ontario: Canadian Journal of Earth Sciences, v. 25, p. 1751–1759.

Doig, R., 1991, U-Pb zircon dates of Morin anorthosite suite rocks, Grenville Province, Quebec: Journal of Geology, v. 99, p. 729–738.

Easton, R.M., 1992, The Grenville Province and the Proterozoic history of central and southern Ontario, *in* Thurston, P.C., ed., Geology of Ontario: Sudbury, Ontario Geological Survey, Special Volume 4, Part 2, p. 714–904.

Emslie, R.F., and Hunt, P.A., 1990, Ages and petrogenetic significance of igneous mangerite-charnockite suites associated with massif anorthosites, Grenville Province: Journal of Geology, v. 98, p. 213–231.

Friedman, R.M., and Martignole, J., 1995, Mesoproterozoic sedimentation, magmatism, and metamorphism in the southern part of the Grenville Province (western Quebec): U-Pb geochronological constraints: Canadian Journal of Earth Sciences, v. 32, p. 2103–2114.

Fu, W., Corriveau, L., LaFlèche, M.R., and Blein, O., 2003, Birdwing-shaped REE profiles and Nb/Ta, Hf/Sm ratios in the Bondy gneiss complex, Grenville Province, Québec: Sensitive geochemical markers of fossil hydrothermal systems in high-grade metamorphic terrains: CIM [Canadian Institute of Mining] Montreal 2003 Mining Industry Conference and Exhibition, 4–7 May, Montreal, Québec, Canadian Institute of Mining Technical Paper, CD-ROM.

Gebauer, D., Schertl, H.-P., Brix, M., and Schreyer, W., 1997, 35 Ma old ultrahigh-pressure metamorphism and evidence for very rapid exhumation in the Dora Maira Massif, Western Alps: Lithos, v. 41, p. 5–24.

Hanmer, S., Corrigan, D., Pehrsson, S., and Nadeau, L., 2000, SW Grenville Province, Canada: The case against post-1.4 Ga accretionary tectonics: Tectonophysics, v. 319, p. 33–51.

Harris, L.B., Rivard, B., and Corriveau, L., 2001, Structure of the Lac Nominingue–Mont-Laurier region, Central Metasedimentary Belt, Quebec Grenville Province: Canadian Journal of Earth Sciences, v. 38, p. 787–802.

Heaman, L.M., Davis, D.W., Krogh, T.E., and Lumbers, S.B., 1987, Geological evolution of the Central Metasedimentary Belt (CMB), Ontario: A U-Pb perspective: Geological Association of Canada–Mineralogical Association of Canada, Joint Annual Meeting, May, Saskatoon, Saskatchewan, Program with Abstracts, v. 12, p. 54.

Hildebrand, R.S., and Easton, R.M., 1995, An 1161 Ma suture in the Frontenac terrane, Ontario segment of the Grenville orogen: Geology, v. 23, p. 917–920.

Indares, A., and Martignole, J., 1990, Metamorphic constraints on the evolution of the gneisses from the allochthonous monocyclic belt of the Grenville Province, western Québec: Canadian Journal of Earth Sciences, v. 27, p. 371–386.

Kretz, R., 1980, Occurrence, mineral chemistry, and metamorphism of Precambrian carbonate rocks in a portion of the Grenville Province: Journal of Petrology, v. 21, p. 573–620.

Kretz, R., 1990, Biotite and garnet compositional variation and mineral equilibria in Grenville gneisses of the Otter Lake area, Quebec: Journal of Metamorphic Geology, v. 8, p. 493–506.

Ludwig, K.R., 2000, Isoplot/Ex (version 2.2): A geochronological toolkit for Microsoft Excel: Berkeley, California, Berkeley Geochronology Center Special Publication No. 1a, 53 p.

Lumbers, S.B., Heaman, L.M., Vertolli, V.M., and Wu, T.-W., 1990, Nature and timing of Middle Proterozoic magmatism in the Central Metasedimentary Belt, Grenville Province, Ontario, in Gower, C.F., et al., eds., Mid-Proterozoic Laurentia and Baltica: St. John's, Newfoundland, Geological Association of Canada, Special Paper 38, p. 243–276.

Maas, R., Kinny, P.D., Williams, I.S., Froude, D.O., and Compston, W., 1992, The Earth's oldest known crust: A geochronological and geochemical study of 3900–4200 Ma old detrital zircons from Mt. Narryer and Jack Hills, Western Australia: Geochimica et Cosmochimica Acta, v. 56, p. 1281–1300.

Martignole, J., and Calvert, A.J., 1996, Crustal-scale shortening and extension across the Grenville Province of western Quebec: Tectonics, v. 15, p. 376–386.

Martignole, J., and Corriveau, L., 1991, Lithotectonic studies in the Central Metasedimentary Belt of the southern Grenville Province: Lithology and structure of the Saint-Jovite map area, Quebec: Ottawa, Ontario, Geological Survey of Canada, Paper 91-1C, p. 77–87.

Martignole, J., and Ringuette, L., 1998, Dating provenance of metaclastic rocks from allochthonous terranes of the western Quebec portion of the Grenville Province by $^{207}Pb/^{206}Pb$ method with LA(IR)–ICP-MS: Geological Association of Canada–Mineralogical Association of Canada, Joint Annual Meeting, May, Québec, Abstracts Volume, v. 23, p. A-118.

Martignole, J., Calvert, A.J., Friedman, R., and Reynolds, P., 2000, Crustal evolution along a seismic section across the Grenville Province (western Quebec): Canadian Journal of Earth Sciences, v. 37, p. 291–306.

Mattinson, J.M., Graubard, C.M., Parkinson, D.L., and McClelland, W.C., 1996, U-Pb reverse discordance in zircons: The role of fine scale oscillatory zoning and submicron transport of Pb: Geophysical Monographs, v. 95, p. 355–370.

McEachern, S.J., and van Breemen, O., 1993, Age of deformation within the Central Metasedimentary Belt boundary thrust zone, southwest Grenville Orogen: Constraints on the collision of the Mid-Proterozoic Elzevir terrane: Canadian Journal of Earth Sciences, v. 30, p. 1155–1165.

McLaren, A.C., FitzGerald, J.D., and Williams, I.S., 1994, The microstructure of zircon and its influence on the age determination from Pb/U isotopic ratios measured by ion microprobe: Geochimica et Cosmochimica Acta, v. 58, p. 993–105.

McLelland, J., Chiarenzelli, J., and Perham, A., 1991, Age, field, and petrological relationships of the Hyde School Gneiss, Adirondack Lowlands, New York: Criteria for an intrusive igneous origin: Journal of Geology, v. 100, p. 69–90.

McLelland, J., Daly, J.S., and McLelland, J.S., 1996, The Grenville orogenic cycle (ca. 1350–1000 Ma): An Adirondack perspective: Tectonophysics, v. 265, p. 1–28.

Moore, J.M., and Thompson, P.H., 1980, The Flinton Group: A late Precambrian metasedimentary succession in the Grenville Province of eastern Ontario: Canadian Journal of Earth Sciences, v. 17, p. 1685–1707.

Morin, D., and Corriveau, L., 1996, Fragmentation processes and xenolith transport in a Proterozoic minette dyke, SW Grenville Province, Québec: Contributions to Mineralogy and Petrology, v. 125, p. 319–331.

Nadeau, L., and van Breemen, O., 1998, Plutonic ages and tectonic setting of the Algonquin and Muskoka allochthons, Central Gneiss Belt, Grenville Province, Ontario: Canadian Journal of Earth Sciences, v. 35, p. 1423–1438.

Nadeau, L., Brouillette, P., and Hébert, C., 1999, New observations on relict volcanic features in medium-grade gneiss of the Montauban group, Grenville Province, Quebec, in Current Research 1999-E: Ottawa, Ontario, Geological Survey of Canada, p. 149–160.

Nantel, S., and Pintson, H., 2002, Géologie de la région du Lac Dieppe (31O/3): Québec, Ministère des Ressources naturelles du Québec, RG 2001–16, 36 p.

Nesbitt, R.W., Pascual, E., Fanning, C.M., Toscano, M., Sáez, R., and Almodóvar, G.R., 1999, U-Pb dating of stockwork zircons from the eastern Iberian Pyrite Belt: Journal of the Geological Society of London, v. 156, p. 7–10.

Perkins, D., III, Essene, E.J., and Marcotty, L.A., 1982, Thermometry and barometry of some amphibolite-granulite facies rocks from the Otter Lake area, southern Quebec: Canadian Journal of Earth Sciences, v. 19, p. 1759–1774.

Ratcliffe, N.M., Aleinikoff, J.N., Burton, W.C., and Karabinos, P., 1991, Trondhjemitic, 1.35–1.31 Ga gneisses of the Mount Holly Complex of Vermont: Evidence for an Elzevirian event in the Grenville Basement of the United States Appalachians: Canadian Journal of Earth Sciences, v. 28, p. 77–93.

Rivers, T., 1997, Lithotectonic elements of the Grenville Province: Review and tectonic implications: Precambrian Research, v. 86, p. 117–154.

Rivers, T., and Corrigan, D., 2000, Convergent margin on southeastern Laurentia during the Mesoproterozoic: Tectonic implications: Canadian Journal of Earth Sciences, v. 37, p. 359–383.

Rogers, N., 2001, Preliminary report on the stratigraphy and structure of the Bee Lake greenstone belt, Superior Province, northwestern Ontario: Ottawa, Ontario, Geological Survey of Canada, Current Research 2001-C17, 17 p.

Rubin, J.N., Henry, C.D., and Price, J.G., 1989, Hydrothermal zircons and zircon overgrowths, Sierra Blanca Peaks, Texas: American Mineralogist, v. 74, p. 865–869.

Sager-Kinsman, E.A., and Parrish, R.R., 1993, Geochronology of detrital zircons from the Elzevir and Frontenac terranes, Central Metasedimentary Belt, Grenville Province, Ontario: Canadian Journal of Earth Sciences, v. 30, p. 465–473.

Sharma, K.N.M., Singhroy, V.H., Madore, L., Lévesque, J., Hébert, C., and Hinse, M., 1999, Use of radar images in the identification of major regional structures in the Grenville Province, western Quebec: Canadian Journal of Remote Sensing, v. 25, p. 278–290.

Sinha, A.K., Wayne, D.M., and Hewitt, D.A., 1992, The hydrothermal stability of zircons—Preliminary experimental and isotopic studies: Geochimica et Cosmochimica Acta, v. 56, p. 3551–3560.

Steiger, R.H., and Jäger, E., 1977, Subcommission on geochronology: Convention on the use of decay constants in geo- and cosmochronology: Earth Planetary and Science Letters, v. 36, p. 359–362.

Stern, R.A., 1996, The SHRIMP II ion microprobe at the Geological Survey of Canada: Geoscience Canada, v. 23, p. 73–76.

Stern, R.A., 1997, The GSC Sensitive High Resolution Ion Microprobe (SHRIMP): Analytical techniques of zircon U-Th-Pb age determinations and performance evaluation, *in* Radiogenic Age and Isotopic Studies: Report 10: Ottawa, Ontario, Geological Survey of Canada, Current Research 1997-F, p. 1–31.

Tagaki, T., Orihashi, Y., Naito, K., and Watanabe, Y., 1999, Petrology of a mantle-derived rhyolite, Hokkaido, Japan: Chemical Geology, v. 160, p. 425–445.

Timmermann, H., Parrish, R.R., Jamieson, R.A., and Culshaw, N.G., 1997, Time of metamorphism beneath the Central Metasedimentary Belt Boundary Thrust Zone, Grenville Orogen, Ontario: Accretion at 1080 Ma?: Canadian Journal of Earth Sciences, v. 34, p. 1023–1029.

van Breemen, O., and Corriveau, L., 1995, Evolution of the Central Metasedimentary Belt in Quebec, Grenville orogen: U-Pb geochronology: International Conference on Tectonics and Metallogeny of Early/Mid Precambrian Orogenic Belts, Precambrian '95, 27 August–1 September, Université du Québec à Montréal, Program with Abstracts, p. 137.

van Breemen, O., Davidson, A., Loveridge, W.D., and Sullivan, R.W., 1986, U-Pb zircon geochronology of Grenville tectonites, granulites and igneous precursors, Parry Sound, Ontario, *in* Moore, J.M., et al., eds., The Grenville Province: St. John's, Newfoundland, Geological Association of Canada, Special Paper 31, p. 191–207.

Vavra, G., Schmid, R., and Gebauer, D., 1999, Internal morphology, habit and U-Th-Pb microanalysis of amphibolite-to-granulite facies zircons: Geochronology of the Ivrea Zone (Southern Alps): Contributions to Mineralogy and Petrology, v. 134, p. 380–404.

Wasteneys, H., McLelland, J., and Lumbers, S., 1999, Precise zircon geochronology in the Adirondack Lowlands and implications for revising plate-tectonic models of the Central Metasedimentary Belt and Adirondack Mountains, Grenville Province, Ontario and New York: Canadian Journal of Earth Sciences, v. 36, p. 967–984.

Wiedenbeck, M., 1995, An example of reverse discordance during ion microprobe zircon dating: An artifact of enhanced ion yields from a radiogenic labile Pb: Chemical Geology, v. 125, p. 197–218.

Williams, I.S., and Claesson, S., 1987, Isotopic evidence for the Precambrian provenance and Caledonian metamorphism of high grade paragneisses from the Seve Nappes, Scandinavian Caledonides, II: Ion microprobe zircon U-Th-Pb: Contributions to Mineralogy and Petrology, v. 97, p. 205–217.

Williams, I.S., Compston, W., Black, L.P., Ireland, T.R., Foster, J.J., 1984, Unsupported radiogenic Pb in zircon: A cause of anomalously high Pb-Pb, U-Pb and Th-U ages: Contributions to Mineralogy and Petrology, v. 88, p. 322–327.

Windley, B.F., 1986, Comparative tectonics of the western Grenville and the western Himalaya, *in* Moore, J.M., et al., eds., The Grenville Province: St. John's, Newfoundland, Geological Association of Canada, Special Paper 31, p. 341–348.

Wodicka, N., Parrish, R.R., and Jamieson, R.A., 1996, The Parry Sound domain: A far-travelled allochthon? New evidence from U-Pb zircon geochronology: Canadian Journal of Earth Sciences, v. 33, p.1087–1104.

Wynne-Edwards, H.R., 1972, The Grenville Province, *in* Price, R.A., and Douglas, R.J.W., eds., Variations in tectonic styles in Canada: St. John's, Newfoundland, Geological Association of Canada, Special Paper 11, p. 263–334.

Wynne-Edwards, H.R., Gregory, A.F., Hay, P.W., Giovanella, C.A., and Reinhardt, E.W., 1966, Mont-Laurier and Kempt Lake map-areas, Quebec (31 J and 31 O): Ottawa, Ontario, Geological Survey of Canada, Paper 66-32, 32 p.

Manuscript Accepted by the Society August 25, 2003

Printed in the USA

Geological Society of America
Memoir 197
2004

Unraveling growth history of zircon in anatectites from the northeast Adirondack Highlands, New York: Constraints on pressure-temperature-time paths

J. Alcock*
Department of Environmental Sciences, Pennsylvania State University, Abington College, Abington, Pennsylvania 19001, USA
Clark Isachsen
Department of Geoscience, Gould Simpson Bldg., University of Arizona, Tucson, Arizona 85721, USA
Kenneth Livi
Morton K. Blaustein Department of Earth and Planetary Sciences, Johns Hopkins University, Baltimore, Maryland 21218, USA
Peter Muller
Department of Earth Sciences, State University of New York–Oneonta, Oneonta, New York 13820, USA

ABSTRACT

Zircon grains extracted from anatectites in the New Russia gneiss complex have been dated to establish the timing of very high-temperature metamorphism in the northeastern Adirondack Highlands, southwestern Grenville Province. The isotopic and chemical systematics of the zircons studied indicate that (1) zircon grains extracted from anatectic segregations in metagabbro are isotopically homogeneous, implying that only zircon that grew during anatexis is present, and (2) zircons from an anatectic charnockite are complex and preserve evidence of a multistage thermal evolution for the gneiss. This complex history can be deciphered by combining knowledge acquired from backscatter electron images and electron microprobe chemical analyses to guide selection of grain components for isotopic dilution thermal ionization mass spectrometry (ID-TIMS) U-Pb geochronology. The analysis indicates that melt from the metagabbro crystallized at 1054 ± 0.4 (2σ) Ma at $T \geq 915 \pm 50$ °C. Zircon from the charnockitic gneiss yielded ages that date an inherited component at 1143 ± 2 Ma ($^{207}Pb/^{206}Pb$); the emplacement of the protolith at 1112 ± 2 Ma; and the crystallization of the anatectic melt at 1040 ± 1 Ma ($T \geq 825 \pm 50$ °C). Late, retrograde metamorphic zircon yielded a range of ages including a concordant age at 1003 ± 4 Ma.

The method described allows precise dating of anatexis that typically marks the metamorphic peak in granulite-facies terranes and so can be used to constrain model pressure-temperature-time (*P-T-t*) paths. The timing of anatexis and a previously reported age of 1026 Ma for andraditic garnets from skarn adjacent to the Marcy meta-anorthosite establish rapid, nearly isobaric cooling from 915–750 °C in ~30 m.y.

Keywords: Adirondack Highlands, zircon chemistry, zircon geochronology, anatectites, pressure-temperature-time path

---

*E-mail: jea4@psu.edu.

Alcock, J., Isachsen, C., Livi, K., and Muller, P., 2004, Unraveling growth history of zircon in anatectites from the northeast Adirondack Highlands, New York: Constraints on pressure-temperature-time paths, *in* Tollo, R.P., Corriveau, L., McLelland, J., and Bartholomew, M.J., eds., Proterozoic tectonic evolution of the Grenville orogen in North America: Boulder, Colorado, Geological Society of America Memoir 197, p. 267–284. For permission to copy, contact editing @geosociety.org.

INTRODUCTION

Geochronology in high-grade metamorphic terranes is often problematic. Metamorphic reactions can produce mineral growth or dissolution during both the prograde and the retrograde portions of the pressure-temperature-time (*P-T-t*) path. Therefore, a single zircon grain may record a history that includes relict premetamorphic growth, several stages of growth during metamorphism, and open-system behavior during postmetamorphic cooling and uplift. To interpret single-grain ages with confidence may require a detailed study of zoning in each zircon analyzed so that potential mixing of growth stages can be recognized (Schaltegger et al., 1999; Vavra et al., 1999; Whitehouse et al., 1999). Even if a dated grain records only one growth event, accurate interpretation of its age depends on correctly identifying the event that produced that growth (e.g., Bingen et al., 2001).

The problem, while daunting, is not unsolvable. There are two keys to obtaining meaningful ages. One is the ability to identify events within the metamorphic history of the region that are likely to cause the growth of radiogenic minerals. Reactions that result in anatexis and the formation of leucocratic segregations meet this criterion. These reactions are common in granulite-facies terranes, and zircon grown during anatexis should be recognizable as the youngest grains or parts of grains that have a magmatic character. The relative ages of the zones can be determined from their position within complex grains and confirmed by age determination through isotopic analysis.

The second key is that the analytical method used must allow the separation of distinct events. The two most commonly used methods for dating zircon, ion microprobe and isotopic dilution thermal ionization mass spectrometry (ID-TIMS), have strengths and weaknesses in this regard. The ion microprobe allows accurate sampling of small areas of a grain. If a zone of zircon growth associated with anatexis can be identified, it can be probed as a distinct region of the grain. The method's weakness is its imprecision. The 2σ error for a single ion microprobe analysis of a Mesoproterozoic zircon is ~4% (Stern, 1997). Precision can be enhanced to ~1.5% (2σ) by weighted averaging of multiple analyses (Stern, 1997). However, this assumes that all pooled ages have the same true age. In contrast, the ID-TIMS method is quite precise, with a typical 2σ error of less than 0.4% for Mesoproterozoic zircon. However, because whole or partial grains are dated, the potential for obtaining mixed ages that combine zircon grown during different events is high (Wasteneys et al., 1999).

In this paper, we report ages obtained from two anatectites of the New Russia gneiss complex in the northeastern Adirondack Highlands (Fig. 1). These rocks serve as an example of the

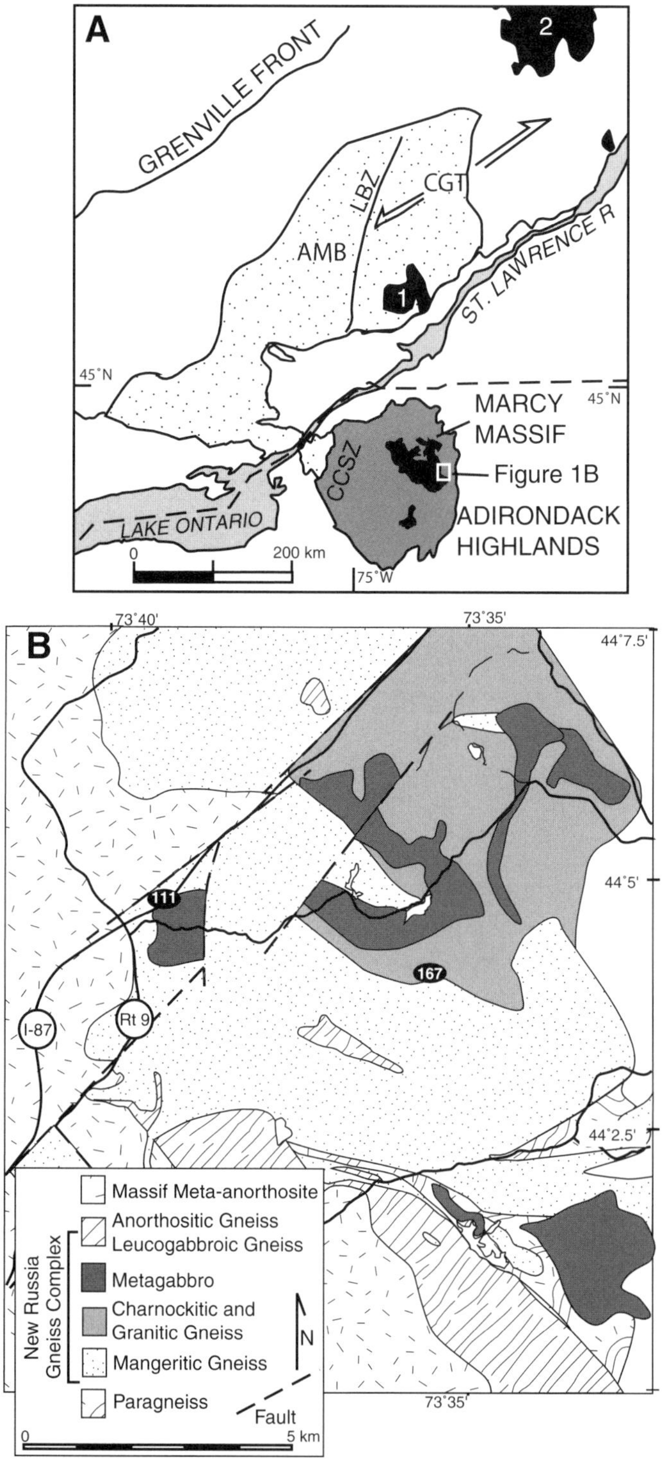

Figure 1. (A) Geologic sketch map showing relationship of the Adirondack Highlands to the allocthonous monocyclic belt (AMB) (Rivers et al., 1989) and the Central Granulite Terrane (CGT) (Wynne-Edwards, 1972). The area shown with a stippled pattern plus the Highlands comprise the AMB. The Labelle shear zone (LBZ) and the Carthage-Colton shear zone (CCSZ) form the western boundary of the Central Granulite Terrane. The Morin and Lac St. Jean anorthosite bodies are identified as 1 and 2, respectively. (B) Bedrock geology within the Witherbee quadrangle, northeastern Adirondack Highlands. Sample locations are indicated by numbers in black ovals. Mangeritic, charnockitic, granitic, gabbroic, and anorthositic gneisses comprise the New Russia gneiss complex. The relationship of anorthositic and leucogabbroic gneiss at the margins of the Marcy meta-anorthosite massif to the New Russia complex remains uncertain.

difficulties associated with U-Pb geochronology in high-grade metamorphic terranes. The region experienced a protracted period (or periods) of granulite-facies metamorphism, and metamorphic minerals (zircon, monazite, garnet, and titanite) yield U-Pb ages of 1160 to 1000 Ma (McLelland and Chiarenzelli, 1990; Mezger et al., 1991). Multiple thermal events that affected the region are also recognized as evidenced by intrusive ages that cluster at 1160–1120, 1100–1080, and 1050–1040 Ma (Silver, 1969; McLelland and Chiarenzelli, 1990; Chiarenzelli and McLelland, 1991; McLelland et al., 1996, 2001b). The study reported here also provides age constraints on potentially closely spaced metamorphic, magmatic, and tectonic events. Analyses by ID-TIMS of single zircon grains from an anatectic segregation in a metagabbroic gneiss yielded concordant ages at ca. 1054 Ma. However, zircon grains extracted from an anatectic, charnockitic gneiss indicated complex U-Pb systematics, necessitating improved spatial resolution. This was achieved by detailed analysis using electron backscatter imaging and electron microprobe point analyses to identify grains or parts of grains dominated by a single period of zircon growth. Chemical analyses proved to be particularly valuable in avoiding misinterpretation of backscatter images. Once identified, zircons that grew in different thermal events were dated using ID-TIMS. This approach resulted in the identification and dating of four distinct stages of zircon growth. We suggest that this method may prove especially valuable for workers attempting to separate closely spaced events and/or for those who do not have access to an ion microprobe.

GEOLOGIC SETTING

The Adirondack Highlands expose an outlier of the Central Granulite Terrane (Wynne-Edwards, 1972) or alternatively the allocthonous monocyclic belt (Rivers et al., 1989) of the Grenville Province of Canada (Fig. 1, A). The Central Granulite Terrane has experienced a complex thermal history. Geochronologic studies from the Morin and Lac St. Jean regions have obtained a range of igneous and metamorphic ages from ca. 1170 to 1000 Ma (Emslie and Hunt, 1990; Doig, 1991; Friedman and Martignole, 1995; Higgins and van Breemen, 1995; Corrigan and van Breemen, 1997; Martignole and Friedman, 1998; Martignole et al., 2000). U-Pb dating of igneous and metamorphic minerals from the Adirondack Highlands has established a similar range of ages. Reported igneous ages of ca. 1160–1120 Ma are common for gabbros, charnockites, mangerites, and granites. Other igneous age groupings are ca. 1100–1080 Ma for a widespread suite of hornblende granites and ca. 1050 Ma for leucogranites of the northwestern and southeastern Highlands (Silver, 1969; Chiarenzelli and McLelland, 1991; McLelland et al., 1996, 2001b). The hornblende granites are gneissic in character with variably developed fabric, but the younger granitoids are described as undeformed (Chiarenzelli and McLelland, 1991; McLelland et al., 1996, 2001b). The apparent syn- to pretectonic nature of the older granitoids and the post-tectonic nature of the younger are interpreted to reflect Ottawan (1090–1050 Ma) deformation in the Adirondacks associated with crustal thickening at a convergent plate margin (Chiarenzelli and McLelland, 1991; McLelland et al., 1996).

U-Pb ages from zircon, garnet, titanite, and monazite that are interpreted to have a metamorphic origin range from ca. 1160 to 1000 Ma, but ages between 1050 and 1000 Ma are more common (McLelland and Chiarenzelli, 1990; Mezger et al., 1991). Samples yielding these later ages include garnets from skarns found adjacent to the Marcy meta-anorthosite (Mezger et al., 1991; Kohn and Valley, 1998). The andraditic garnet from skarn at Cascade Slide yielded a U-Pb age of 1026 ± 4 Ma (Mezger et al., 1991) that is likely to be a growth age (DeWolf et al., 1996) and so fixes a point on the *P-T-t* path at ~750 ± 30 °C, 740 ± 100 MPa (Valley and Essene, 1980). Mezger et al. (1991) infer post-1026 Ma cooling of the Central Highlands by a comparison of ages from different minerals and their closure temperatures with initial cooling at ~4 °C/m.y.

The New Russia Gneiss Complex

In the northeastern Highlands, the regional geology is dominated by granulite-facies, metaigneous rocks. Among these is the New Russia gneiss complex (Fig. 1, B), which lies within an embayment along the eastern margin of the Marcy meta-anorthosite massif (Alcock and Muller, 1999). The complex is a thick, layered assemblage of meta-anorthosite, metagabbro, mangerite, jotunite, charnockite, and granite. All have been metamorphosed under granulite-facies conditions except possibly some granitic gneiss, and all except some internal portions in metagabbro and some small granite masses have a well-developed gneissic fabric. Bodies of the various rocks form sheets and lenses from several meters to several hundred meters in structural thickness. These rocks are chemically and mineralogically similar to other metaigneous rocks of the Highlands and so share rock names; however, the orthogneisses of the New Russia complex are distinctive. Two important features that separate them from similarly named Adirondack rocks are the presence of abundant anatectic segregations that may parallel or crosscut penetrative fabric and the preservation of evidence for very high-temperature metamorphism, as high as 950 ± 50 °C. These characteristics are discussed briefly later. More complete descriptions can be found in Alcock and Muller (1999, 2000, 2001) and Alcock et al. (1999).

The mangeritic, charnockitic, and gabbroic gneisses locally contain numerous anatectic segregations. Enough melting occurred to allow the melt to coalesce and move limited distances through the host to form metatexites in a charnockitic gneiss at location 167 (Fig. 2, A and B). (Metatexites are migmatites in which the leucosomes, the anatectic melt, may disrupt but do not destroy the pre-existing fabric in the host gneiss [Sawyer, 1996, 1998; Milord et al., 2001]). Segregations formed in this way range from thin, felsic bands that parallel foliation to coarse leucosomes that retain a massive igneous texture of interlocking mineral grains and that commonly cut across the gneissic fabric

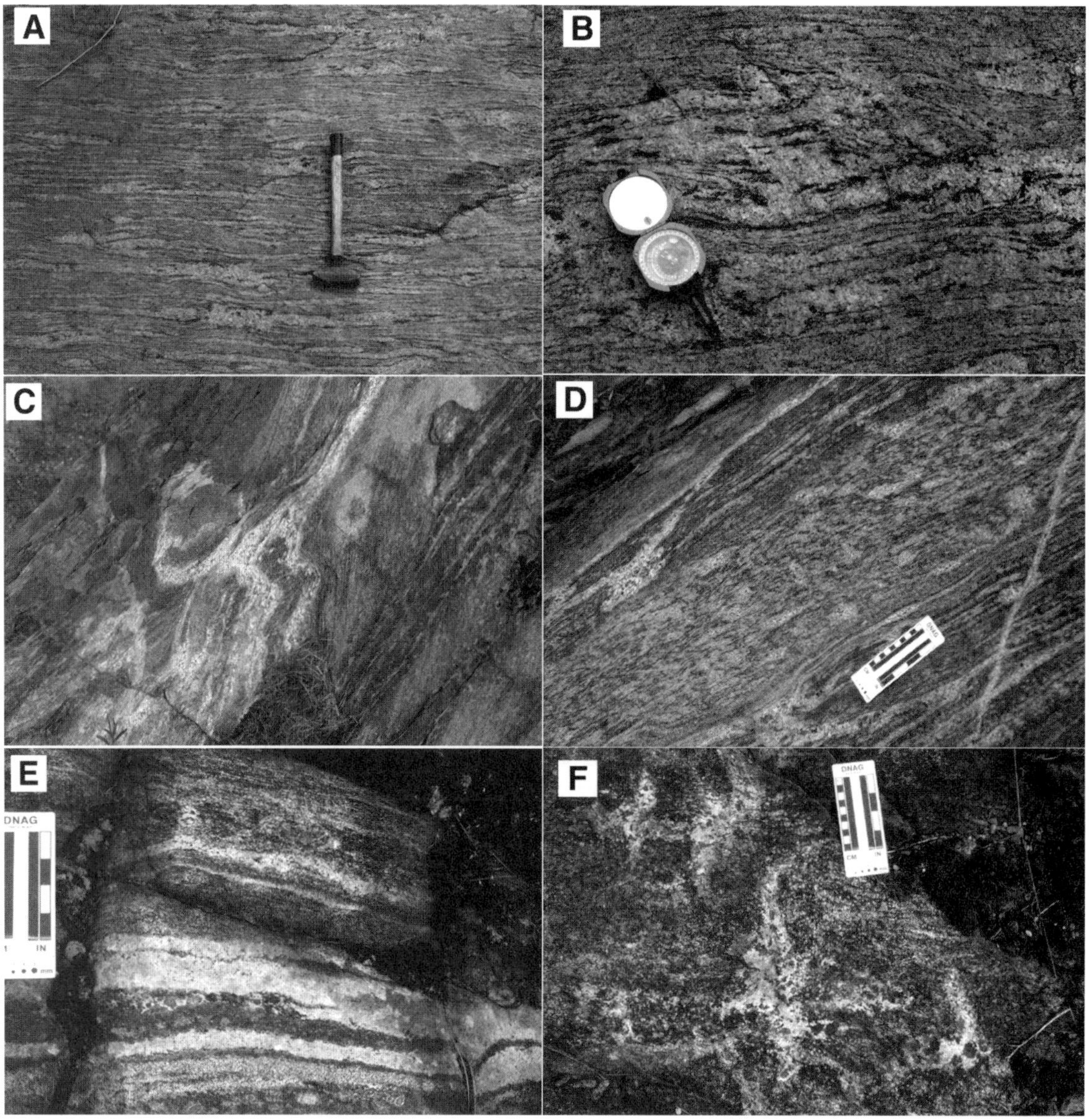

Figure 2. Outcrop photos of New Russia gneiss with anatectic segregations. (A) and (B): Exposure of charnockitic metatexite, location167, with numerous segregations that roughly parallel but disrupt foliation. The hammer and body of the Brunton compass are 45 and 15 cm in length, respectively. (C) and (D): Diatexite in metagabbroic gneiss, location 111 and location 95, which lies in the median strip of Interstate 87 across the lane from location 111. Location 95 is not shown as a separate location in Figure 1 because it would overlie 111 at map scale. The location of the outcrop prevented positioning of scale in C; scale of fabric is similar to that in D. (E): Trondjhemitic sill-like segregations that retain massive igneous texture. (F): Small melt pods occur within garnetiferous restitic gneiss.

of the rock. Anatexis in the metagabbroic gneiss at sample location 111 produced greater separation of melt that created areas of diatexite where the original fabric has been replaced by mobilization of melt and restite to produce a new fabric resulting from material flow (Fig. 2, C and D) (Sawyer, 1996, 1998; Milord et al., 2001) or that produced small sill and dikelike bodies (Fig. 2, E). Leucocratic pods in garnetiferous restite (Fig. 2, F) show textural evidence of back reaction between crystallizing melt and restite minerals that produced symplectites of orthopyroxene (Opx) + plagioclase (Pl) (abbreviations after Kretz, 1983) separating clinopyroxene (Cpx) from matrix Pl and also caused resorbtion of garnet (Grt) (Alcock and Muller, 1999). Deformed segregations in combination with segregations that preserve igneous textures and crosscut regional penetrative fabrics indicate that melting was a syn- to post-tectonic event (Alcock and Muller, 1999, 2001).

The mineralogy of the segregations varies with rock type, from largely Pl (antiperthite) ± Opx in metagabbro to antiperthite

+ Opx + Cpx ± quartz (Qtz) ± mesoperthite in mangeritic gneiss to mesoperthite + Qtz in charnockitic gneiss. This chemical variation is consistent with the segregations' having formed in the host rock by dehydration melting, as each has the appropriate chemistry to have been derived by partial melting of the host (Rushmer, 1991; Burnham, 1992; Springer and Seck, 1997). The melt-forming reactions probably involved the breakdown of pargasitic hornblende (Hbl) with Pl in the metagabbro and of pargasitic Hbl with feldspar + Qtz in mangerite and charnockite. The role of Hbl in the anatectic process is indicated by textural evidence in the metagabbro, where ragged and separated but optically continuous Hbl grains are found included in Grt and Pl in the restite. These Hbl grains often contain large irregular oxides and symplectites of Opx + Pl suggestive of partial replacement of the grain. In addition, areas of the metagabbro that retain primary igneous minerals (Pl + olivine [Ol] + Cpx) and textures and that are without Hbl do not contain leucosomes, indicating that Hbl was a necessary source of fluid for the partial melting reaction. Pargasitic Hbl is also an important phase in the mineral assemblage of both mangerite and charnockite with abundant anatectites. In contrast, felsic charnockitic gneiss with little Hbl to provide fluid to enhance melting and that would, therefore, require temperature to exceed 900 °C to experience significant melting (Whitney 1988) has fewer segregations than nearby metagabbro and mangerite.

Estimates of peak temperatures that affected the New Russia complex are based on the composition of ternary feldspars found in coarse felsic segregations in metagabbro, mangerite, and charnockite. The highest estimates of 950 ± 50 °C from mangerite overlap estimates of 915 ± 50 °C from sample 111A, a trondjheimitic segregation in metagabbro. Estimates from sample 167, a charnockitic gneiss, are lower, ≈825 ± 50 °C (Alcock and Muller, 1999; Alcock et al., 1999). A similar variation in temperature does not emerge if Hbl composition is used to estimate peak conditions (Russ-Nabalek, 1988). Al^{IV} in Hbl from locations sampled as 111A and 167 ranges from 1.77 to 2.0 and from 1.76 to 1.88, respectively, consistent with temperatures >900 °C if the model proposed by Russ-Nabalek (1988) remains valid to pressures near 750 MPa (Alcock and Muller, 1999).

Pressure is estimated to have been 700–1000 MPa during anatexis from garnet growth during dehydration melting in the metagabbro (Rapp and Watson, 1995). Pressures estimated from garnet-pyroxene and garnet-hornblende geobarometers are consistent with the pressure inferred from phase relations (Alcock and Muller, 1999, 2000). Such high pressure and temperature are being documented elsewhere in the Grenville Province, though of different ages (Boggs and Corriveau, this volume).

DESCRIPTION OF SAMPLES

Sample 167 is a charnockitic metatexite gneiss with well-developed fabric and abundant anatectic segregations (Fig. 2, A and B). Mafic-rich layers (restite) contain Opx + Cpx + Hbl + Grt + ilmenite (Ilm) + magnetite (Mag) + minor Qtz + orthoclase (Or). Segregations are almost exclusively mesoperthite + Qtz. Accessory minerals include zircon and apatite. Zircon grains are found in both restite and segregations. In the restite they are elongate to rounded and are most commonly found against hornblende, penetrating feldspar. They also occur as inclusions within hornblende. Some appear complexly zoned with a distinct core region when viewed in plane-polarized light. Grains in melt segregations are typically smaller, prismatic to rectangular in appearance, and do not exhibit optical zonation.

Sample 111A is from a small trondhjemitic dike (≈5 cm by 3 m) that crosscuts the foliation in metagabbroic gneiss. The dike is almost entirely feldspar, mainly sodic antiperthite ± Qtz and accessory zircon. Adjacent to quartz, the feldspar grains have mesoperthite rims with potassium content increasing toward quartz.

One small metagabbroic xenolith in the dike preserves evidence for breakdown of pargasitic Hbl + Grt + feldspar + Cpx in reactions with the dike material to produce Opx, oxides and more sodic feldspar, and new Cpx that forms a nearly continuous rim separating the enclave from the dike material. The similarity of assemblages and textures in the metagabbroic inclusion to those found in the main body of the metagabbro and the structural relationships between anatectic segregations and metagabbro observed in the outcrop lead us to conclude that the dike sampled as 111A is derived from partial melting of the host metagabbro. The trondjhemitic character of the dike is consistent with its originating as melt produced in the metagabbro during dehydration melting of pargasitic Hbl + Pl (Springer and Seck, 1997).

METHODS

Zircon separates were prepared by reducing ~5 kg of rock to less than 35 mesh (0.42 mm) using a jaw crusher and pulverizer followed by standard heavy liquid techniques and magnetic separation. One set of nonmagnetic zircons from each sample was hand-picked and air-abraded following the method of Krogh (1982). A second fraction of nonmagnetic zircons from sample 167 was mounted in epoxy on a glass slide and polished to reveal internal structure. Selected grains mounted in epoxy were then analyzed with the electron microprobe to determine chemistry and for backscatter electron imaging (Hanchar and Rudnick, 1995). Electron microprobe analysis was performed at Johns Hopkins University on the JEOL 8600 Super Probe operating with a 100 nA beam current and 25 keV accelerating potential. Analyses determined concentration data for 28 elements.

Analytical parameters are presented in Table 1. Each analysis required 45 minutes per spot. To avoid longer analytical times a limited number of background positions were analyzed, and each position was then used to calculate background intensities for several rare earth elements (REE). This also helped to avoid potential interference. Because the distance between each background position was large, nonlinear extrapolation correc-

TABLE 1. ZIRCON SEPARATES: ANALYTICAL CONDITIONS AND ERRORS

Element	Na	Mg	Al	Si	P	K	Ca	Ti	Cr
Line	K_a	K_a	K_a	K_a	K_a	K_a	K_a	K_a	K_a
Crystal	TAP	TAP	TAP	TAP	PET	PET	PET	PET	PET
Peak counting time (sec.)	2 × 60	2 × 60	2 × 60	60	2 × 60	3 × 60	3 × 60	3 × 60	4 × 60
Background time (sec.)	120	120	120	120	120	120	120	120	120
Number of iterations	3	3	3	3	3	3	3	3	3
Minimum detection limits (ppm)	25	25	18	43	43	16	15	26	18
Error (2σ) at 0.1 at wt%	±0.042	±0.009	±0.009	±0.008	±0.026	±0.022	±0.017	±0.003	±0.005
Standards	Amelia albite	Zabargad enstatite	Sitkin anorthite	Synthetic zircon	Synthetic YPO_4	Madagascar orthoclase	Sitkin anorthite	Binntal anatase	Chromite

Element	Mn	Fe	Y	Zr	Ba	La	Ce	Pr	Nd
Line	K_a	K_a	L_a	L_a	L_a	L_a	L_a	L_b	L_b
Crystal	LiF	LiF	PET	PET	PET	PET	LiF	LiF	LiF
Peak counting time (sec.)	60	60	60	60	60	60	60	60	60
Bkg time (sec.)	120	120	120	120	120	120	120	120	120
Number of iterations	3	3	3	3	3	3	3	3	3
Minimum detection limits (ppm)	115	127	261	272	95	128	355	379	339
Error (2σ) at 0.1 at wt%	±0.011	±0.006	±0.014	±0.014	±0.009	±0.005	±0.014	±0.015	±0.013
Standards	Rhodonite	Rockport fayalite	Synthetic YPO_4	Synthetic zircon	Benitoite	Synthetic $LaPO_4$	Synthetic $CePO_4$	Synthetic $PrPO_4$	Synthetic $NdPO_4$

Element	Sm	Gd	Dy	Er	Yb	Hf	Ta	Th	U
Line	L_b	L_a	L_b	L_a	L_a	M_a	M_a	M_a	M_b
Crystal	LiF	LiF	LiF	LiF	LiF	TAP	PET	PET	PET
Peak counting time (sec.)	60	60	60	60	60	5 × 60	3 × 60	5 × 60	6 × 60
Bkg Time (sec)	120	120	120	120	120	120	120	120	120
Number of iterations	3	3	3	3	3	3	3	3	3
Minimum detection limits (ppm)	308	235	260	201	189	31	67	46	49
Error (2σ) at 0.1 at wt%	±0.011	±0.008	±0.009	±0.007	±0.006	±0.005	±0.006	±0.006	±0.005
Standards	Synthetic $SmPO_4$	Synthetic $GdPO_4$	Synthetic $DyPO_4$	Synthetic $ErPO_4$	Synthetic $YbPO_4$	Synthetic Hafnon	Synthetic Thorite	Synthetic Thorite	Synthetic UO_2

Note: Error varies with element abundance.

tions were empirically estimated from synthetic zircon and applied to each measured background. Severe overlap between REE peaks was avoided by using the Lβ peaks for Pr, Nd, Sm, and Dy. Other minor overlaps were also corrected. After electron microprobe analysis, grains were removed from epoxy. Some complex grains as identified in the backscatter electron images (Figs. 3 and 4) were dissected with a sharp probe in order to analyze discrete domains within the crystals. In many cases, the domain boundaries coincided with fractures, allowing clean breaks to be made (Fig. 4).

Air-abraded grains and those taken from the grain mount were prepared for ID-TIMS analysis by ultrasonic cleaning in warm 4N HNO_3. Cleaned zircon grains and fragments were dissolved in Teflon microbombs using the method of Parrish (1987) and spiked with ^{205}Pb-^{233}U-^{235}U tracer. Subsequent to anion column separation, the Pb was loaded onto rhenium filaments with silica gel and phosphoric acid. During the course of analyses, procedural blanks ranged from 2 to 5 pg total Pb and <1 to 1 pg U.

Pb isotopic measurements were performed in dynamic multicollector mode using a Micromass Sector 54 solid-source mass spectrometer at the University of Arizona. ^{204}Pb was measured using a Daly electron multiplier, with internal Daly-Faraday gain correction determined by peak switching of ^{205}Pb between the Daly and an off-axis Faraday collector. U isotopic measurements were determined in static multicollector mode. Mass fractionation factors for Pb were determined by replicate analyses of National Bureau of Standards standards SRM 981 and SRM 983 to be 0.06% per atomic mass unit; U fractionation was directly determined. Radiogenic Pb concentrations and U-Pb values were calculated from raw fractionation-corrected mass spectrometer data following Ludwig (1980). Initial Pb compositions were estimated using the two-stage evolutionary model of Stacey and Kramers (1975). Ages were calculated using the decay constants recommended by the International Union of Geological Sciences subcommission on geochronology (Steiger and Jäger, 1977), which are $\lambda^{238}U = 0.155125 * 10^{-9}$ year^{-1} and $\lambda 23^5U = 0.98485 * 10^{-9}$ year^{-1}. Linear regression and calculation of intercepts and uncertainties follow Ludwig (1980). All ages are quoted at the 2σ confidence level.

ELECTRON MICROPROBE RESULTS

Backscatter electron images of the 15 mounted grains from sample 167 that were analyzed are shown in Figure 3. Four dis-

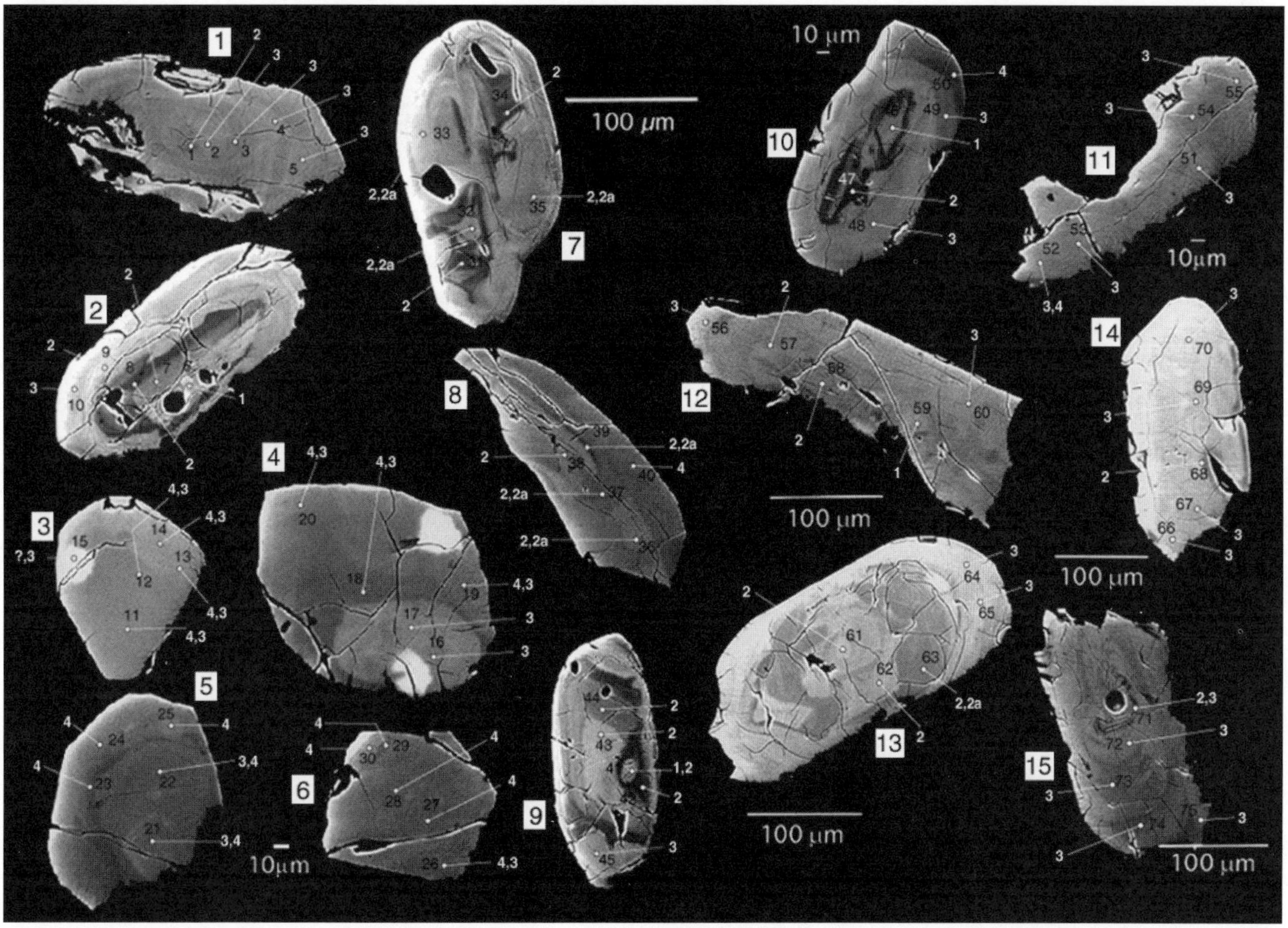

Figure 3. Backscatter electron images of 15 zircon separates representing a cross-section of morphological and optical characteristics. Images and chemical analyses of the grains were used to identify growth stages in the zircon and to choose grains most representative of each growth stage for isotopic dilution thermal ionization mass spectrometry analysis. Grains are identified by black numbers in white rectangles or squares. Locations of chemical analyses are indicated by white dots with black analysis numbers. White numbers connected to dots by lines refer to stages as inferred from appearance. A second number is given if chemical analysis indicates the point belongs to a different growth stage. Scale bar beside grain 7 also provides the scale for grains 1, 2, 3, 4, 5, 8, and 9. Backscatter images obtained using the Johns Hopkins JOEL 8600 Super Probe operating with a 100 nA beam current and 25 keV accelerating potential.

tinct patterns of zonation in the zircon separates were resolvable and interpreted to reflect distinct periods of zircon growth. Small, bright core zones (stage I), as seen in grains 2, 10, and 12 (points 6, 46, and 59), are inferred to be the oldest zone observed. Stage II, the second oldest growth stage on the basis of position within the grains, is characterized by darker core regions with irregular, commonly wispy intergrowths of brighter zircon. Stage II is seen in grains 2, 7, 9, 10, 12, and 13. In grains 9 and 13, zonation in stage II has an oscillatory appearance. Stage III, as observed in grains 1, 14, and 15, is more uniform in appearance but with oscillatory zoning. Stage III also occurs as overgrowths on stage II, for example, in grains 2, 10, 12, and 13. Stage IV, the youngest growth stage, is visible as darker rims and tips on prismatic grains (8, 9, and 10) and as the dark blocky grains 5 and 6. Growth stages II, III, and IV are also recognizable as distinctive grain morphologies. Zircons dominated by stage II growth are typically clear to pinkish, elongate prismatic grains with rounded edges and tips and common inclusions, including occasional brown turbid cores. Brown, turbid prisms with observable concentric zoning typify stage III growth. Stage IV zircons are generally equant to platy grains with irregular, multifaceted shapes and a bronze hue. Grains were also examined by cathodoluminescence. However, the grains were only faintly luminescent and displayed little internal structure.

Quantitative electron microprobe analyses were obtained to test zones identified in backscatter electron images for chemical consistency. Representative analyses are given in Table 2.

Each stage is chemically distinct (Fig. 5). Stage I has high

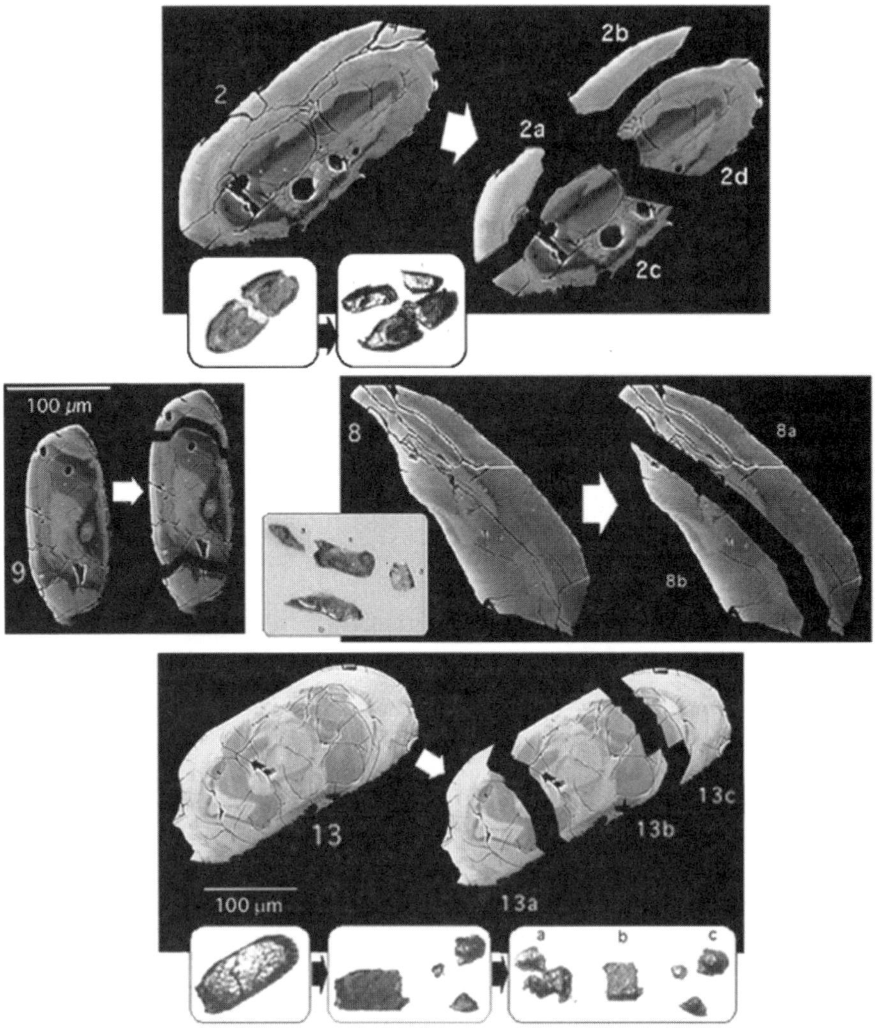

Figure 4. Several zircons that appeared to combine growth stages were broken in attempt to date single growth stages separately. Shown here are whole grain images reproduced from Figure 3, altered images showing graphical representations of the manual fragmentation, and plain light photos of fragments.

HfO_2 content. The majority of analyses of regions identified as stage II follow a distinctive trend, with high Y_2O_3 relative to ThO_2. Seven analyses of stage II areas do not fit this trend and have been separated as stage IIa to reflect the lack of apparent correlation between appearance and chemistry. Stages III and IV have low Y_2O_3/ThO_2 but can be separated by UO_2 content, with stage III having more uranium. Differences in the chemistry of stages that are identified by visual analysis of element-abundance plots are confirmed by statistical tests. Stages II, III, and IV have different compositional means for at least two of the elements Y, Hf, Th, and U at a 97.5% confidence level. Stage I cannot be distinguished by this method because of the small number of analyses within the group. However, a discriminant analysis using Y, Yb, P, Hf, Th, and U concentrations recognizes stage I as distinct. If stage IIa is excluded, discriminant analysis assigns 60 of 68 points to their previously identified group. The "misidentified" points fall in the areas of chemical overlap visible in Figure 5, especially B and C.

TIMS RESULTS

Results of U-Pb analysis of single air-abraded grains from samples 111A and 167 and of polished grains and grain fragments extracted from epoxy after electron microprobe analysis are summarized in Tables 3 and 4 and in Figures 6 and 7. Six grains from sample 111A yield concordant ages averaging 1054 ± 1 Ma. One additional grain is concordant with a $^{207}Pb/^{206}Pb$ age of 1061 ± 2 Ma (Fig. 6).

Air-abraded grains from sample 167 yield a more complex history (Fig. 7). Hand-picked grains included elongate, prismatic grains, presumably of magmatic origin, or equant and multifaceted grains typical of grains grown from solid-state metamorphic reactions (Poldervaart, 1956; Vavra, 1990; Vavra et al., 1999). Grains also exhibited color differences, with prismatic grains ranging from a clear pinkish hue to a brown and turbid appearance. One grain had a visibly darker core region. This grain (viii) yielded a discordant $^{207}Pb/^{206}Pb$ age of 1143 ± 2 Ma,

TABLE 2. REPRESENTATIVE ANALYSES OF ZIRCON SEPARATES

Grain analysis	10.46	12.59	7.31	7.34	9.43	12.58	1.3	3.12	11.54	14.69	15.73	4.19	4.2	6.29	6.3
Growth stage	I	I	II	II	II	II	III	III	III	III	III	IV	IV	IV	IV
Na_2O	0.0245	0.0292	0.0141	0.0245	0.0222	0.0479	0.0506	0.0213	0.0546	0.0276	0.0500	0.0240	0.0210	0.0244	0.0257
MgO	0.0103	0.0009	0.0065	0.0088	0.0108	0.0109	0.0101	0.0102	0.0086	0.0069	0.0103	0.0104	0.0053	0.0125	0.0082
Al_2O_3	b.d.	b.d.	b.d.	b.d.	b.d.	0.0203	b.d.	b.d.	b.d.	b.d.	b.d.	b.d.	b.d.	b.d.	b.d.
SiO_2	32.76	33.05	32.97	32.47	32.52	32.73	32.58	32.69	32.83	32.65	32.11	32.72	32.08	32.65	32.81
P_2O_5	0.0467	0.0729	0.1033	0.1177	0.1150	0.1182	0.0485	0.0545	0.0669	0.0668	0.0569	0.0660	0.0414	0.0563	0.0733
K_2O	0.1212	0.1232	0.1192	0.1223	0.1236	0.1308	0.1192	0.1181	0.1237	0.1176	0.1227	0.1145	0.1094	0.1119	0.1121
CaO	0.0030	b.d.	b.d.	b.d.	b.d.	0.0605	0.0301	0.0016	b.d.	0.0050	0.0284	b.d.	b.d.	b.d.	b.d.
TiO_2	0.3675	0.3772	0.3601	0.3539	0.3688	0.3716	0.3728	0.3772	0.3668	0.3715	0.3640	0.3641	0.3614	0.3572	0.3682
Cr_2O_3	b.d.	b.d.	b.d.	b.d.	b.d.	b.d.	b.d.	b.d.	b.d.	b.d.	b.d.	b.d.	b.d.	b.d.	b.d.
MnO	0.0008	0.0034	0.0013	b.d.	0.0017	b.d.	0.0128	0.0004	0.0075	0.0157	0.0128	b.d.	0.0005	0.0079	b.d.
FeO	b.d.	b.d.	b.d.	b.d.	b.d.	b.d.	b.d.	b.d.	b.d.	b.d.	b.d.	b.d.	b.d.	b.d.	b.d.
Fe_2O_3	0.0123	0.0084	0.0055	0.0117	0.0192	0.0699	0.1889	0.005	0.0395	0.0706	0.2038	0.0035	0.0054	0.0058	0.0018
Y_2O_3	0.0387	0.0490	0.1332	0.1185	0.2531	0.2634	0.0720	0.0627	0.0720	0.0808	0.0683	0.0665	0.0494	0.0406	0.0564
ZrO_2	66.36	64.81	65.94	67.00	66.18	65.45	66.10	65.19	65.57	65.47	66.02	66.12	67.34	66.68	66.14
BaO	b.d.	b.d.	b.d.	0.0038	b.d.	b.d.	b.d.	0.0084	b.d.	b.d.	b.d.	b.d.	b.d.	0.0085	b.d.
La_2O_3	b.d.	0.0076	b.d.	0.0009	b.d.	b.d.	b.d.	b.d.	b.d.	b.d.	0.0001	b.d.	b.d.	b.d.	0.0065
Ce_2O_3	0.0216	0.0087	0.0166	0.0203	0.0215	0.0262	b.d.	0.0174	0.0177	0.0149	0.0092	0.0128	0.0052	0.0133	0.0119
Pr_2O_3	0.0547	0.0523	0.0406	0.0256	0.0425	0.0403	0.0081	0.0323	0.0125	0.0306	0.0447	0.0458	0.0601	0.0396	b.d.
Nd_2O_3	0.0132	0.0068	0.0122	0.0004	0.0320	0.0154	0.0022	0.0123	0.034	0.0126	b.d.	b.d.	0.0070	0.0362	b.d.
Sm_2O_3	b.d.	0.0116	0.0252	0.0286	0.0054	0.0235	0.0121	0.0051	0.0237	0.0119	0.0325	0.0164	0.0079	0.0340	b.d.
Gd_2O_3	b.d.	0.0007	0.0065	0.0046	0.0220	0.0094	0.0025	0.008	b.d.	0.0074	0.0047	0.0038	0.0125	0.0209	0.0033
Dy_2O_3	0.0185	b.d.	0.0338	0.0173	0.0135	0.0337	0.0078	b.d.	0.022	0.0308	0.0196	0.0220	0.0046	b.d.	0.0398
Er_2O_3	0.0216	0.023	0.0335	0.0263	0.0586	0.0487	0.0260	0.0248	0.0223	0.0224	0.0269	0.0321	0.0307	0.0298	0.0307
Yb_2O_3	0.0128	0.0197	0.0400	0.0208	0.0486	0.0599	b.d.	0.0114	0.0155	0.0103	0.0138	0.0243	0.0215	b.d.	0.0179
HfO_2	1.460	1.850	0.979	1.011	1.330	1.260	1.220	1.300	1.230	1.280	1.210	1.240		1.27	1.132
1.165															
Ta_2O_3	0.0730	0.0875	0.0615	0.0663	0.0713	0.0679	0.0746	0.0654	0.068	0.0677	0.0701	0.0648	0.0704	0.0685	0.0620
PbO	0.2013	0.2515	0.0975	0.1116	0.1855	0.1861	0.1472	0.1442	0.1849	0.1780	0.1895	0.1369	0.1392	0.1081	0.1230
ThO_2	0.0098	0.0195	0.0229	0.0171	0.0464	0.0479	0.0096	0.0227	0.0077	0.0383	0.0245	0.0340	0.0220	0.0332	0.0188
UO_2	0.1482	0.1141	0.0363	0.0474	0.1523	0.1772	0.3133	0.1367	0.2664	0.1940	0.2752	0.0572	0.0891	0.0518	0.0744
Total	101.78	100.98	101.06	101.63	101.64	101.27	101.41	100.32	101.04	100.78	100.97	101.18	101.75	101.52	101.15

Note: Analysis number refers to grain and analytical point identified in Figure 3. Stage identity is based on backscatter electron image (Fig. 3). Analytical reports are based on a structural formula with 4 oxygens.

b.d.—concentration below detection limit.

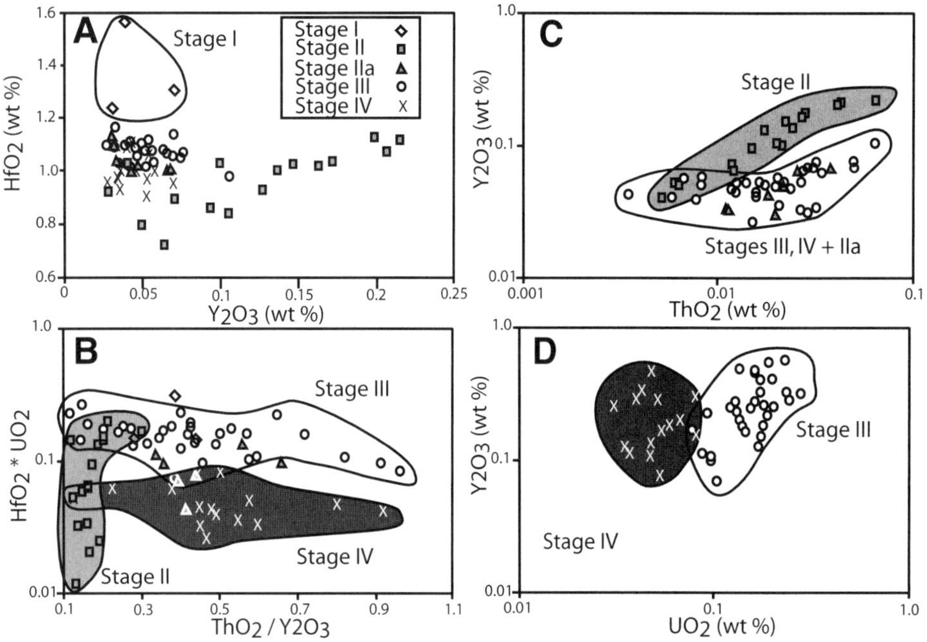

Figure 5. Summary plots of chemical data obtained from zircon separates presented to illustrate differences among groups. Stages identified in the plots are based on chemical analysis (Table 2) and backscatter electron imaging (Fig. 3). The strong correlation between recognized zones in backscatter electron images and zircon chemistry support the inference that the stages reflect zircon growth in different chemical environments. (A) Stage I is recognized by high Hf content. (B) Plot combining weight percents of the four oxides most responsible for between-group differences. Multiplication and division of weight percent is used to present multiple factors and to enhance visual differences among groups. As noted in text, statistical tests support the visual inferences. Stages II, III, and IV can be recognized as distinct groups. Stage IIa, identified in backscatter electron image as stage II, plots outside stage II region and is separated for that reason. (C) Plot of Y_2O_3 against ThO_2 separates stage II from stage III by means of the distinctive trend of high Y relative to Th. Again stage IIa is distinct from stage II. (D) Plot of Y_2O_3 against UO_2 establishes difference between stages III and IV.

inferred to reflect the presence of an inherited component. Other ages from prismatic grains were concordant or nearly concordant at ≈1114 Ma (vi, vii), ≈1090 Ma (i, iii), and ≈1040 to 1050 Ma (ix, x, xii). Equant, multifaceted grains ii and iv yielded ages at ca. 1025 Ma. Most important are the three distinct sets of concordant or nearly concordant ages from prismatic grains because only two were expected, one dating emplacement of the charnockite protolith and the other dating anatexis. It was this inability to resolve the timing of anatexis in the sample that led us to pursue electron microprobe analysis.

Three of the four growth stages identified in backscatter images were successfully isolated for U-Pb analysis. Stage I zircon, correlated with small bright core regions in backscatter electron images, could not be isolated because of its small size. Its inferred presence in grains 2, 10, and 12 is unconfirmed by ID-TIMS analyses, but it cannot be ruled out as a minor component.

Grain 7 and the inner portions of grains 9 and 13 were chosen to represent stage II growth for ID-TIMS analysis. Grain 9 is dominated by stage II growth with the exception of the tips, which appear to be oscillatory-zoned stage III overgrowths. The tips were snapped off for separate U-Pb analysis (Fig. 4), but were too small and/or too low in U and Pb concentration for successful analysis. However, the remaining body of the zircon (9a) was rendered more representative of stage II growth by tip removal and was successfully analyzed, yielding a $^{207}Pb/^{206}Pb$ age of 1104.1 ± 6.9 Ma (–0.34% discordant). Grain 13 was similar to grain 9, with a backscatter electron image showing bright rims, thickest at the tips, overgrowing patchy stage II growth in the interior. It was dissected (Fig. 4) and analyzed successfully as three separate fractions (13a, b, c). The central fragment (13b), which is dominated by stage II, gave a $^{207}Pb/^{206}Pb$ age of 1112.0 ± 1.6 Ma (1.25% discordant). The tips incorporate a mixture of stages II and III. Their $^{207}Pb/^{206}Pb$ ages are 1051.3 ± 1.5 Ma (13a) and 1060.6 ± 1.9 Ma (13c). The backscatter electron image of grain 7 also appears to be dominated by stage II. However, three chemical analyses did not fall within the compositional space of stage II, suggesting that the grain may have a more complex growth history. It yielded a $^{207}Pb/^{206}Pb$ age of 1097.0 ± 5.1 Ma (1.15% discordant). The discordance and the chemical evidence that grain 7 is not pure stage II indicate that the grain experienced postgrowth disturbance. For this reason grain 7 is not considered an accurate measure of stage II growth. However, the ages determined for grains 9a and 13b and single, nonimaged, air-abraded grains vi and vii stand alone as a distinct age grouping at ca. 1100 Ma. The $^{207}Pb/^{206}Pb$ age of 1112 ± 2 Ma for fraction 13b is in close agreement with the ages of grains vi and vii. Together, the three grains produce a combined weighted mean age of 1112 ± 2 Ma, interpreted to date emplacement of the charnockite protolith. If grain 9a is included, the weighted mean age is 1108.1 ± 9 Ma.

Prior to U-Pb analysis, five grains were chosen to represent stage III growth—grains 1, 3, 11, 14, and 15. Grain 3 plots slightly

TABLE 3. U-PB ANALYTICAL DATA

Fractions	Weight (mg)	Concentrations U (ppm)	Pb[a] (ppm)	$^{206}Pb^{b}/^{204}Pb$	$^{208}Pb^{c}/^{206}Pb$	$^{206}Pb^{c}/^{238}U$	% error[d]	$^{207}Pb^{c}/^{235}U$	% error[d]	$^{207}Pb^{c}/^{206}Pb$	% error[d]	Age (Ma) $^{206}Pb/^{238}U$	Age (Ma) $^{207}Pb/^{235}U$	Age (Ma) $^{207}Pb/^{206}Pb$	$^{207}Pb/^{206}Pb$ error (Ma)[d]	Common Pb (pg)	% discordance
Sample 111A, Trondhjemite dike: air-abraded, single whole-grain analyses																	
i (p)	0.0132	510.3	86.1	4819.405	0.025	0.17723	(.3)	1.81899	(.33)	0.07444	(.08)	1051.8	1052.3	1053.4	1.6	12.9	0.16
ii (p)	0.0057	696.8	118.8	3801.449	0.022	0.17965	(.43)	1.85114	(.48)	0.07473	(.20)	1065.1	1063.8	1061.3	3.9	9.2	–0.38
iii (p)	0.0073	609.5	102.8	4258.405	0.021	0.17810	(.38)	1.82825	(.40)	0.07445	(.10)	1056.6	1055.7	1053.6	1.9	9.1	–0.31
iv (p)	0.0012	1289.2	215.4	2309.691	0.018	0.17775	(.28)	1.82493	(.29)	0.07446	(.07)	1054.7	1054.5	1054.1	1.4	5.3	–0.06
vi (p)	0.0230	437.3	73.2	8098.091	0.020	0.17711	(.49)	1.81807	(.51)	0.07445	(.12)	1051.2	1052.0	1053.6	2.4	11.2	0.25
vii (p)	0.0121	508.3	85.1	9634.173	0.022	0.17757	(.19)	1.82289	(.20)	0.07446	(.05)	1053.7	1053.7	1053.8	1.1	4.7	0.02
viii (p)	0.0036	1004.2	168.5	3450.000	0.019	0.17731	(.27)	1.81881	(.70)	0.07440	(.60)	1052.3	1052.3	1052.3	12.0	9.3	0.00
Sample 167, Charnockite gneiss: zircons removed from grain mount																	
1 (III)	0.0017	2629.1	442.4	2836.720	0.051	0.17209	(.14)	1.75398	(.16)	0.07392	(.06)	1023.6	1028.6	1039.3	1.3	14.3	1.63
2c (mix)	0.0005	1054.9	286.2	110.359	0.040	0.17610	(.52)	1.79628	(.71)	0.07398	(.45)	1045.7	1044.1	1040.8	9.0	78.2	–0.50
2d (mix)	0.0005	844.4	138.9	765.453	0.021	0.17478	(.55)	1.79691	(.59)	0.07457	(.21)	1038.4	1044.3	1056.9	4.2	3.6	1.89
3 (III)	0.0030	1381.5	215.4	7052.282	0.022	0.16556	(.26)	1.68170	(.28)	0.07367	(.11)	987.6	1001.6	1032.4	1.4	5.1	5.37
4 (mix)	0.0020	1476.4	225.6	2466.760	0.058	0.15647	(.20)	1.57534	(.23)	0.07302	(.12)	937.1	960.5	1014.5	2.4	7.0	8.19
5 (IV)	0.0013	1067.3	184.4	1548.500	0.068	0.17474	(.22)	1.77779	(.27)	0.07379	(.15)	1038.2	1037.4	1035.7	3.0	4.8	–0.26
6 (IV)	0.0004	485.3	78.8	572.295	0.051	0.16806	(.73)	1.68253	(.76)	0.07261	(.20)	1001.4	1001.9	1003.1	4.0	3.6	0.18
7 (II)	0.0021	1335.8	236.1	2397.898	0.041	0.18339	(.22)	1.92354	(.34)	0.07607	(.25)	1085.4	1089.3	1097.0	5.1	6.6	1.15
8a (mix)	0.0009	3343.1	567.8	2814.247	0.062	0.17395	(.14)	1.77767	(.16)	0.07412	(.06)	1033.8	1037.3	1044.7	1.2	2.8	1.12
8b (mix)	0.0005	3622.8	514.1	2549.119	0.053	0.14651	(.20)	1.48876	(.21)	0.07370	(.06)	881.4	925.8	1033.2	1.2	2.9	15.71
9a (II)	0.0007	1980.2	362.0	313.182	0.047	0.18745	(1.21)	1.97321	(1.25)	0.07634	(.29)	1107.6	1106.4	1104.1	6.9	3.4	–0.34
10 (mix)	0.0024	1895.5	315.9	4082.737	0.034	0.17445	(.13)	1.78971	(.14)	0.07441	(.07)	1036.6	1041.7	1052.6	1.3	7.2	1.65
11 (III)	0.0010	1311.8	221.9	1750.605	0.028	0.17528	(.18)	1.79104	(.22)	0.07411	(.13)	1041.1	1042.2	1044.5	2.6	14.7	0.35
13a (mix)	0.0012	2852.1	480.1	1361.741	0.011	0.17559	(.17)	1.80034	(.19)	0.07436	(.08)	1042.8	1045.6	1051.3	1.5	18.0	0.87
13b (II)	0.0017	2646.6	472.0	2839.182	0.037	0.18590	(.18)	1.96460	(.20)	0.07665	(.08)	1099.1	1103.5	1112.0	1.6	6.0	1.25
13c (mix)	0.0006	2338.9	397.1	1050.290	0.015	0.17470	(.21)	1.79947	(.23)	0.07470	(.09)	1038.0	1045.3	1060.6	1.9	19.4	2.31
14 (III)	0.0068	3926.6	657.9	11721.770	0.064	0.17086	(.08)	1.74121	(.09)	0.07391	(.05)	1016.8	1023.9	1039.0	1.0	14.4	2.31
15 (III)	0.0030	1853.1	314.0	3098.836	0.086	0.16851	(.36)	1.71867	(.39)	0.07397	(.14)	1003.9	1015.5	1040.7	2.8	11.3	3.81
Sample 167, Charnockite gneiss: air-abraded, whole-grain analyses																	
i (p)	0.0011	1067.6	187.8	1730.494	0.055	0.18054	(.39)	1.88681	(.41)	0.07580	(.09)	1069.9	1076.5	1089.7	1.8	5.4	1.96
ii (m)	0.0021	479.9	85.2	1453.486	0.115	0.17357	(.46)	1.75681	(.51)	0.07341	(.20)	1031.8	1029.7	1025.3	4.1	5.1	–0.69
iii (p)	0.0023	485.5	87.5	1631.123	0.063	0.18349	(.39)	1.91893	(.44)	0.07585	(.19)	1086.0	1087.7	1091.1	3.8	5.6	0.50
iv (m)	0.0056	873.2	149.8	5562.383	0.088	0.17127	(.48)	1.73193	(.50)	0.07334	(.11)	1019.1	1020.5	1023.4	2.3	7.0	0.46
vi (p)	0.0022	365.8	68.5	1482.643	0.073	0.18927	(.42)	2.00253	(.45)	0.07673	(.15)	1117.4	1116.4	1114.3	3.1	4.1	–0.31
vii (p)	0.0010	459.4	102.5	226.000	0.051	0.18872	(.83)	1.99565	(1.03)	0.07669	(.57)	1114.5	1114.0	1113.2	1.4	23.7	–0.12
viii (p)	0.0026	2105.5	439.6	597.162	0.072	0.19226	(.14)	2.06418	(.16)	0.07787	(.09)	1133.6	1137.0	1143.5	1.7	106.3	0.94
ix (p)	0.0013	2708.5	464.6	1298.731	0.027	0.17463	(.16)	1.79161	(.18)	0.07441	(.07)	1037.6	1042.4	1052.6	1.5	26.6	1.55
x (p)	0.0016	1442.7	245.7	940.705	0.020	0.17235	(.20)	1.76847	(.22)	0.07442	(.08)	1025.0	1034.0	1052.8	1.5	23.6	2.85
xi (p)	0.0041	2161.3	390.1	2927.506	0.108	0.17474	(.13)	1.78136	(.15)	0.07394	(.06)	1038.2	1038.7	1039.8	1.3	30.4	0.17
xii (p)	0.0012	3425.4	578.1	1237.245	0.026	0.17140	(.18)	1.75569	(.21)	0.07429	(.09)	1019.8	1029.3	1049.4	1.9	32.0	3.05

Note: a—radiogenic Pb; b—measured, not corrected for spike, fractionation, and blank; c—corrected for fractionation, spike, blank, and initial common Pb calculated from Stacey and Kramers (1975); d—errors represent two standard errors of the mean; p—prismatic; m—multifaceted; I, II, III, and IV refer to growth stage of zircon as inferred from backscatter image and chemistry.

TABLE 4. SUMMARY OF ZIRCON ANALYSES

Grain	BSE images (1)	Stages from BSE	Stages from chemistry	$^{207}Pb/^{206}Pb$ age	Grain	BSE images (10)	Stages from BSE	Stages from chemistry	$^{207}Pb/^{206}Pb$ age
1	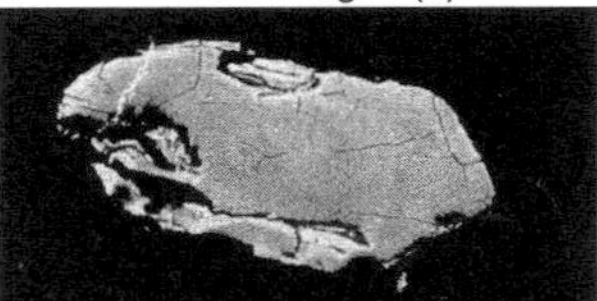	Dominated by stage III. Small stage II core.	Consistent with interpretation of BSE image.	1039 Ma	9	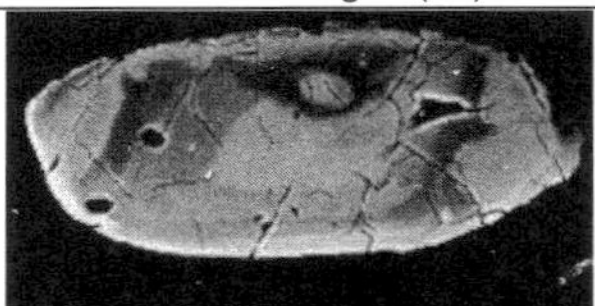	Dominated by stage II with tips of stage III. Small bright area thought to be stage I.	Consistent with BSE image interpretation except no stage I present.	Fragmented grain: 9a—1104 Ma. 9b—Not dated.
2	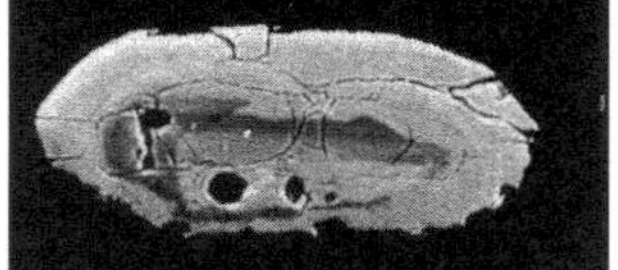	Core zone with stage I and II surrounded by stage III.	Consistent with interpretation of BSE image.	Fragmented grain: 2a and 2b not dated. 2c—1041 Ma. 2d—1057 Ma.	10	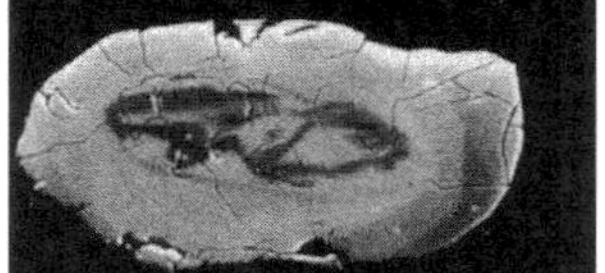	Complex grain with all stages represented.	Consistent with BSE image interpretation.	1052 Ma
3	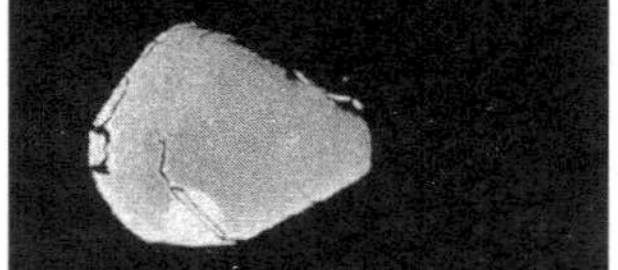	Limited visible zonation inferred to indicate stage IV.	Chemical analyses indicated stage III. Grain was imaged at high angle to long axis.	1032 Ma	11	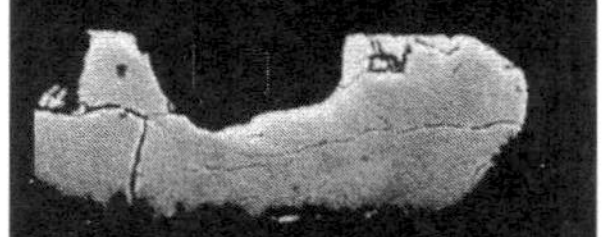	Dominated by stage III.	Consistent with BSE image interpretation. Stage IV found at grain tip.	1044 Ma
4	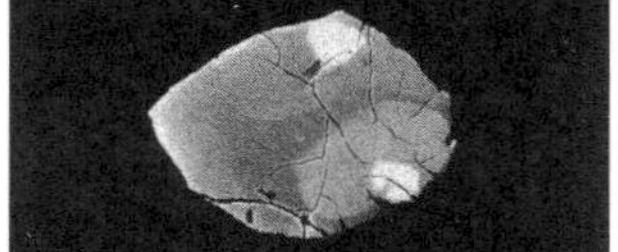	Interpreted to be mix of stages III and IV.	Chemically most like stage III.	Highly discordant grain with Pb/Pb age of 1014 Ma.	12	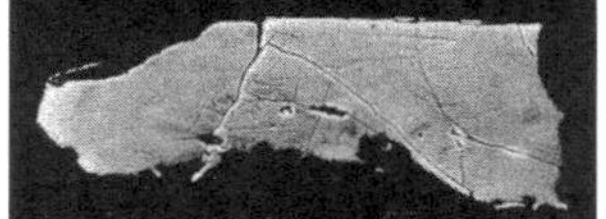	Complex grain with mix of stages I, II, and III.	Consistent with BSE image interpretation.	Not dated.
5	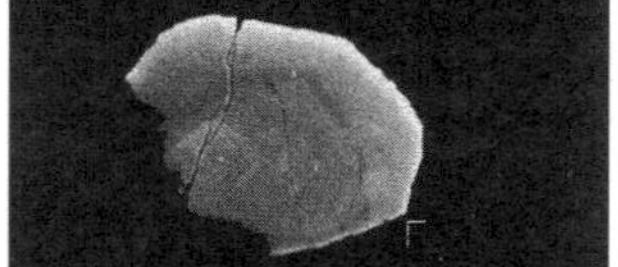	Interpreted to be mix of stages III and IV.	All analyses indicate stage IV.	1036 Ma	13	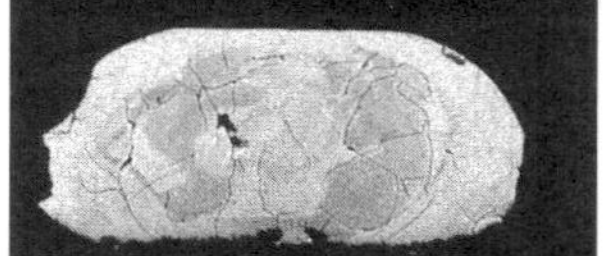	Large internal zone of stage II. Tips are stage III.	Consistent with BSE image interpretation.	Fragmented grain: 13a—1051 Ma. 13b—1112 Ma. 13c—1060 Ma.
6	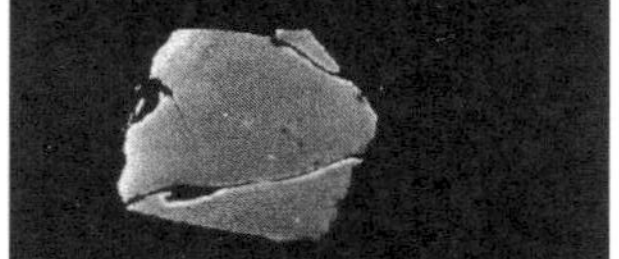	Stage IV	Stage IV	1003 Ma	14	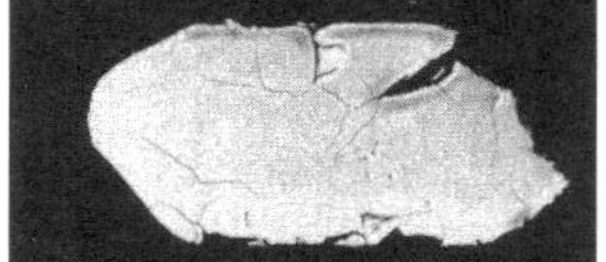	Dominated by stage III. Small area of stage II near inclusion.	Consistent with BSE image interpretation.	1039 Ma
7	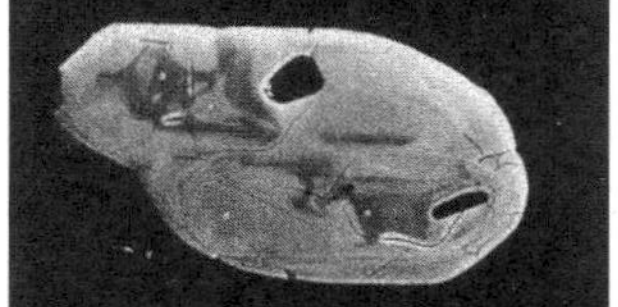	Stage II	3 of 5 analyses do not fall on typical stage II Th/Y trend. Classified as stage IIa.	1097 Ma	15	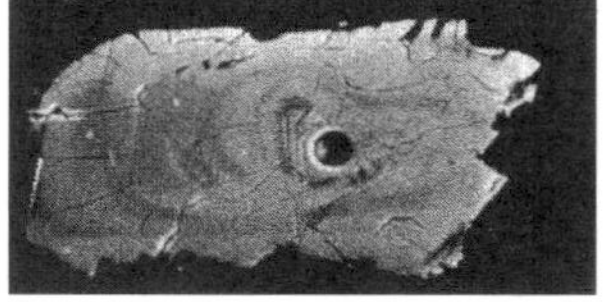	Thought to include small core area with stage II, but dominated by stage III.	Only stage III found in chemical analysis.	1041 Ma
8	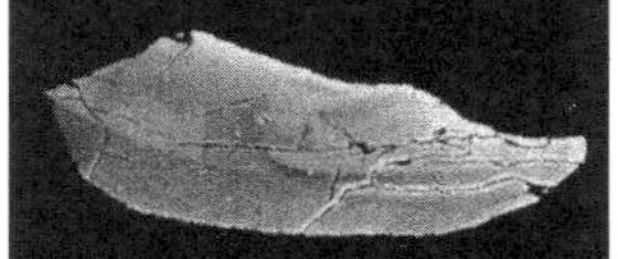	Unusual appearance interpreted to be stage II with considerable stage IV overgrowth.	Chemically a mix of stages including II, IIa, and IV.	Fragmented grain: 8a—1045 Ma. 8b—1033 Ma.					

Note: (1) Backscatter electron (BSE) images have been rescaled to fit cells in table. Refer to Figure 3 for actual sizes.

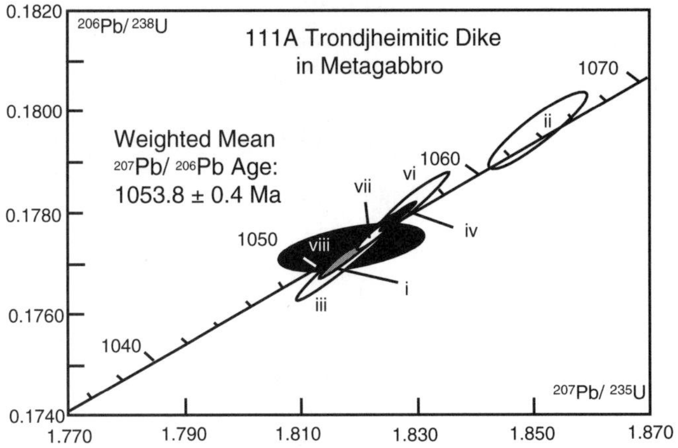

Figure 6. Concordia diagram with data obtained from single air-abraded grains (indicated by lowercase Roman numerals) from sample 111A. Six concordant grains yield an age of 1054 ± 1 Ma, inferred to date anatexis in the host metagabbro. Ellipses show 2σ error.

above and grain 11 slightly below a discordia produced by the other analyses, which yields an age of 1038 ± 3 Ma, with a lower intercept near zero and a mean square of weighted deviates (MSWD) of 0.75. Grain 3, which was not polished or imaged longitudinally, may have contained younger overgrowths concentrated at its tips, yielding a slightly younger $^{207}Pb/^{206}Pb$ age of 1032 ± 1.4 Ma (5.35% discordant). Grain 11, stage III by backscatter electron image and chemistry, gave a $^{207}Pb/^{206}Pb$ age of 1045 ± 2.6 Ma (0.35% discordant). An air-abraded brown prismatic grain (xi) gave an age of 1040 ± 1.3 Ma (0.17% discordant), and its inclusion in the regression produces an age of 1040 ± 1 Ma, with a lower intercept age of 12 ± 91 Ma (MSWD = 0.75). Stage III is interpreted to have grown during anatexis in the charnockite at ca. 1040 Ma.

Two grains were chosen to represent stage IV growth. Grain 6 is the best candidate for pure end-member status. Grain 5 was also chosen to represent stage IV, although it exhibited faint internal zoning. Grain 6 yields a concordant age of 1003 ± 4.0

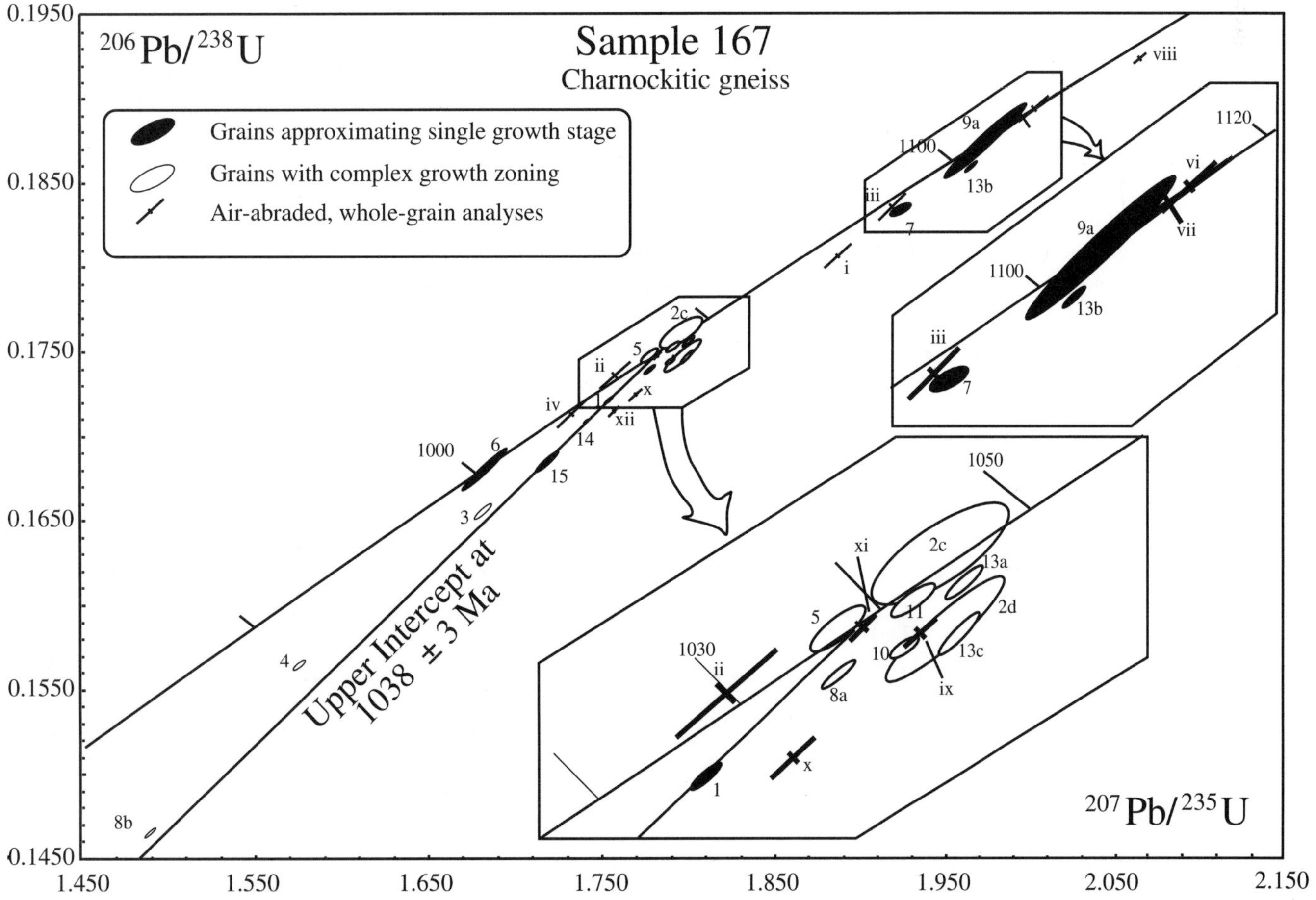

Figure 7. Concordia diagram with data obtained from single-grain analyses of zircons from sample 167 (grains are indicated by one- and two-digit arabic numerals and by lowercase Roman numerals). Included in the plot are data obtained from air-abraded samples that were not imaged using backscatter electron, and also single grains and grain fragments determined to represent specific stages of zircon growth from backscatter electron images and chemical analyses. Ellipses show 2σ error.

Ma (0.18% discordant). Grain 5 yields an age of 1036 ± 3.0 Ma (–0.26% discordant). Two air-abraded grains (ii and iv) with equant, multifaceted morphology yielded ages of ca. 1025 Ma. Taken together these grains provide a poorly constrained age for retrograde metamorphic reactions that produced new growth of zircon either as overgrowths on older grains or as entirely new grains. The concordant 1003 Ma age of grain 6 implies that reactions producing zircon occurred at that time.

Grains 2, 4, 8, 10, 13a, and 13c, which combine aspects of stages I, II, III, and IV, were also analyzed. Grain 4, with its apparent stage IV morphology but stage III chemistry, is highly discordant, with a $^{207}Pb/^{206}Pb$ age of 1014 Ma. Grain fragment 8b, which appears to form the internal part of a prismatic zircon, yields an age intermediate between stage III and stage IV, but is highly discordant and difficult to interpret. The other grains give a family of discordant analyses intermediate in age between stage II and stage III, as would be expected on the basis of chemistry and backscatter image.

DISCUSSION

Analyses of air-abraded single grains from sample 111A yield a consistent result that can be used to date crystallization of the trondhjemitic dike at 1054 ± 1 Ma. We interpret this result to date anatexis in the metagabbro that hosts the dike. We infer a small inherited component for grain ii that yielded the slightly older age of 1061 ± 4 Ma. The simple character of the zircons analyzed supports the inference that the melt is derived from a metagabbro, as it implies a zircon-free or nearly zircon-free source. Olivine metagabbros are unlikely to contain zircon as a significant accessory phase. Therefore, zirconium released to the melt from oxides, Hbl, Grt, and Cpx during anatexis (Poldervaart, 1956; Naslund, 1987; Fraser et al., 1997; Bingen et al., 2001) would react by growing new zircon grains as the melt crystallized. Because relict grains would be rare, there would be little older material to contaminate zircon grown from the melt.

In contrast, zircons extracted from the charnockitic gneiss (sample 167) are complex. As noted earlier, air-abraded single grains yielded five distinct age populations and did not allow an unambiguous determination of the age of anatexis. Backscatter electron imaging and chemical analyses allow recognition of four distinct stages of zircon growth that are identified as zones within individual grains and correlated among grains. Chemical analysis serves as a means to confirm that the stages identified in backscatter electron images represent zircon growth in distinct chemical environments and, by implication, at different times. For example, the shift in Y_2O_3/ThO_2 ratios from stage II to stage III suggests a change in system mineralogy. The presence of garnet in the charnockite during anatexis would have been likely to lower available yttrium (Pyle and Spear, 1999). The higher yttrium concentrations in stage II zircon, therefore, may indicate that the melt that produced the protolith formed from rocks without garnet or by means of reactions that consumed garnet. Lower yttrium concentrations in stages III and IV reflect the presence of garnet in the metamorphic assemblage during and after anatexis. Chemical analysis also serves as a test of the inferred correlation of zoning patterns among grains and allows for recognition of misidentified grains. The good agreement between growth stages identified by backscatter electron imaging and chemistry for 49 of 55 analyses in 11 grains implies that backscatter electron images can serve as a basic discriminant among growth stages. However, the misidentification of four grains underscores the importance of the chemical test when attempting to discern patterns in complexly zoned zircon. The general success of the combined method is also confirmed by the good agreement between expected age groupings based on electron microprobe analysis and the isotopic ages of the individual grains or grain fragments. Table 4 summarizes correlations in zircon zonation recognized in backscatter electron images, electron microprobe analyses, and $^{207}Pb/^{206}Pb$ ages.

Four grains did not exhibit a strong correlation between backscatter electron image and chemistry. Grains 3 and 4, thought to be primarily stage IV zircon, were shown by chemistry to be more like stage III grains. Grains 7 and 8 had six analyses thought to be from stage II that did not fall within the distinctive Y_2O_3/ThO_2 trend typical of that stage. Difficulties with grains 3, 4, and 8 are not surprising, as the backscatter electron image did not provide a clear indication of their growth stage. In addition, grain 3 was viewed at a high angle to its long axis, so the image did not give an accurate representation of zoning. Grain 4 had more internal structure visible in the backscatter electron image than a typical stage IV grain. Its isotopic ratios are highly discordant, suggesting that the grain experienced significant alteration after growth. Grain 8's appearance is unusual and does not conform well to the appearance of a grain of any growth stage; chemically and isotopically it seems to be a mixture of stages II, III, and IV.

Grain 7's mixed chemical signature is more surprising, as its appearance is strongly suggestive of stage II. Chemical characteristics more like those of stage III and IV grains may result from the location of the surface analyzed, which could be near the outer edge of the grain and so include some overgrowth on a partially resorbed stage II grain. Alternatively, Pidgeon et al. (1998) report complex internal zoning in zircon from the Darling batholith similar to that reported here and interpret the patterns to reflect diffusion of elements within the zircon core during high-temperature metamorphism and anatexis. Their model suggests that the loss of oscillatory zoning can be used as a measure of the chemical disturbance of the zircon grain. If they are correct, grain 7 is more disturbed than stage II in grains 9 and 13 and so is less likely to preserve the unusual Y/Th ratio. The older concordant U-Pb ages of samples 9a and 13b, which retain stage II oscillatory zoning, would be consistent with the interpretation of Pidgeon et al. (1998) that oscillatory zonation of core regions can be an important guide to recognizing relatively undisturbed zircon.

Although grains chosen for air abrasion represented the spectrum of observed morphological types, grains analyzed by backscatter electron imaging and electron microprobe analysis offer a distinctly better understanding of the sample's U-Pb systematics. The additional information obtained through electron microprobe analysis allows the precise ages obtained from three of the four growth stages identified to be assigned with confidence to one of three thermal environments that affected the New Russia gneiss complex and that were likely to produce zircon. These are magmatic conditions during intrusion of the gneiss protolith; very high-temperature metamorphism, producing anatexis; and cooling after anatexis, causing retrograde reactions and partial replacement of the peak granulite-facies mineral assemblage. The prismatic nature of stage II grains and the oscillatory growth visible in grains 9 and 13 are consistent with zircon growth from magma (Vavra, 1990; Williams and Claesson, 1987; Pidgeon et al., 1998; Vavra et al., 1999). The ca. 1112 Ma age of stage II, therefore, is inferred to date emplacement of the gneiss protolith. The prismatic habit, oscillatory zoning, and high U content of stage III zircons are also indicative of zircon growth from a melt phase. Because this is the youngest zircon growth from magma, the ca. 1040 Ma age of stage III is interpreted to date anatexis in the charnockitic gneiss. Stage IV grew during retrograde metamorphic reactions as the region cooled from its thermal peak. Rehydration reactions in nearby mangerite gneiss resulted in growth of late Hbl and Grt. The latter occurs as symplectite with Qtz between oxide or Cpx and Pl. Reactions consuming oxides and Cpx could release zirconium and lead to new zircon growth (Poldervaart, 1956; Naslund, 1987). Low U concentration and the platy, multifaceted habit and homogenous internal structure support this interpretation (Vavra, 1990). The 1003 Ma concordant age for grain 6 implies that reactions of this type occurred at that time. It is uncertain whether reactions producing stage IV zircon were limited to a single event at ca. 1000 Ma or occurred over a wider time span during cooling from the metamorphic peak.

The U/Th ratios of the different stages (Fig. 8) do not indicate a clear distinction between zircon growth by magmatic or metamorphic processes. Vavra et al. (1999) report ratios similar to those reported here from zircon overgrowths interpreted to have anatectic origins. They propose that the differences in ratios arise from equilibrated and unequilibrated partitioning of trace elements during zircon growth. Therefore, the Th/U ratio cannot serve as a discriminant for distinguishing the cause of zircon growth, as was reported by Williams and Claesson (1987). Instead, the prismatic or equant habit, the relatively high or low U content, and the contrast between oscillatory zoning and monotonic appearance in the backscatter electron image are a better guide to inferring a magmatic or metamorphic (not anatectic) origin for a particular grain or zone.

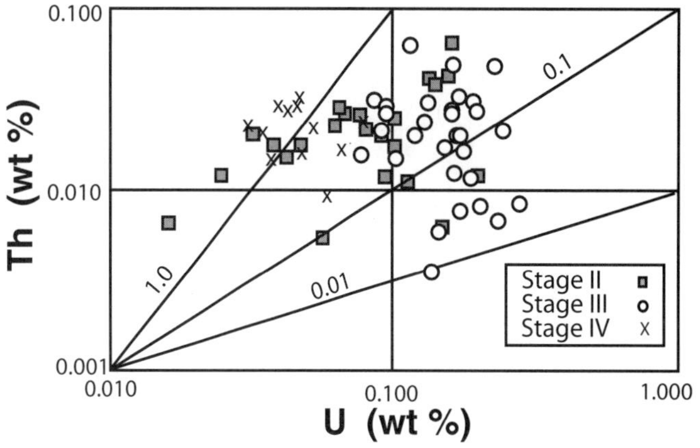

Figure 8. Plot of Th against U from analyses of stages II, III, and IV. Lines labeled 1.0, 0.1, and 0.01 are constant Th/U ratios. Overlap of Th/U among stages prevents its use as a discriminant between metamorphic and igneous zircon as proposed by Williams and Claesson (1987).

Implications for Highlands' Pressure-Temperature-Time Path

The new age data reported here constrain model *P-T-t* paths that attempt to describe Adirondack granulite-facies metamorphism. Figure 9 summarizes the current data. The prograde *P-T* path experienced by the New Russia gneiss remains controversial. The controversy is beyond the scope of this paper, and the new data presented do not impact that early history. For this reason, both models are shown, and readers are referred to the papers of McLelland et al. (2001a, 2001b) and Alcock and Muller (1999, 2000, 2001) if they wish to review the issues of contention.

The cooling path, however, is now better constrained. Crystallization of the trondjhemitic melt (sample 111A) at 915 ± 50 °C (Alcock et al., 1999) occurred at ca. 1054 Ma. Crystallization of the anatectic melt in the charnockite gneiss (sample 167) at 825 ± 50 °C (Alcock and Muller, 1999) occurred at ca. 1040 Ma. Different crystallization temperatures reflect the chemical differences of the two rock types and their anatectic melts. The pressure at peak temperature was between 700 and 1000 MPa, probably closer to 750 MPa (Alcock and Muller, 2000). The 14 m.y. difference in the crystallization ages of segregations in the two rocks may reflect a simple regional cooling from very high-temperature metamorphism. Alternatively, the age difference reflects a more complex metamorphic history that produced a thermal gradient across the New Russia complex (Alcock and Muller, 1999, Alcock et al., 1999), so the peak temperatures experienced by samples 111A and 167 were different and were possibly reached at somewhat different times. In either case, the ages define two points on the complex's *P-T-t* path and imply that the highest temperatures in the complex had cooled by 90 °C in ~14 m.y. or at a rate of ~6.5 °C/m.y.

Reactions forming andradite-rich garnet in skarn rocks adjacent to the Marcy massif meta-anorthosite at 1030 and 1026 Ma (Mezger et al., 1991; Kohn and Valley, 1998) can be used to extend the cooling path to 750 °C and 740 MPa (Valley and Essene, 1980). Cooling was isobaric and continued at a rate in excess of 6 °C/m.y. The isobaric nature of the *P-T-t* path derived

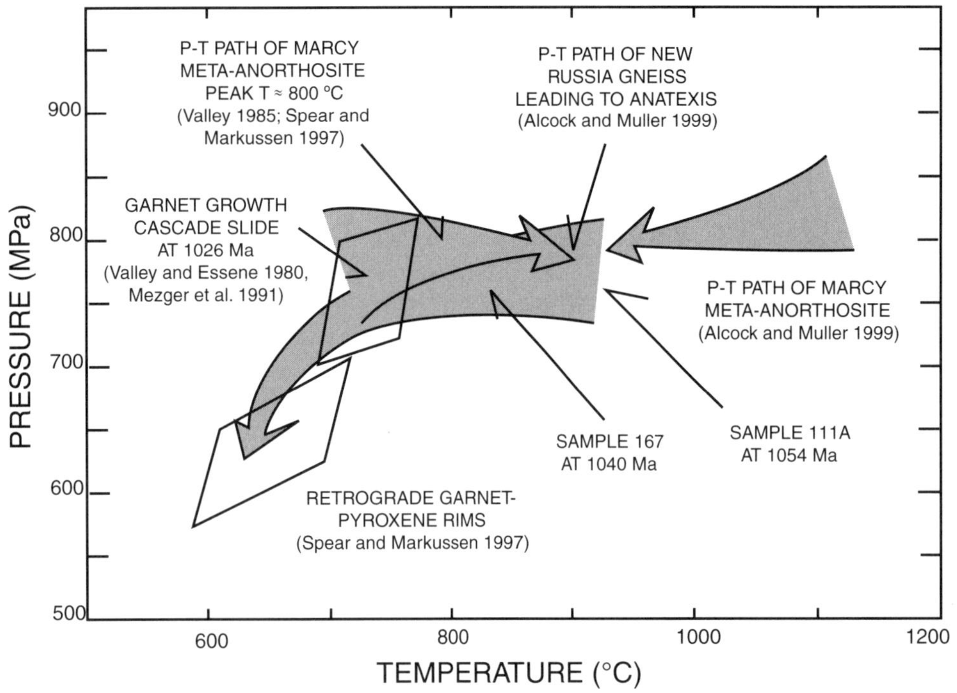

Figure 9. Possible pressure-temperature-time (*P-T-t*) paths proposed for rocks of the Northeastern Adirondack Highlands. Rapid, isobaric cooling from ≥915 °C to ≈750 °C occurred between 1054 and 1026 Ma. The isobaric nature of the cooling path precludes rapid uplift driven by orogenic collapse at that time.

from these data is not consistent with tectonic models that include 1060–1030 Ma uplift of the Highlands driven by orogenic collapse (McLelland et al., 1996, 2001b).

Spear and Markussen (1997) trace continued cooling of the Marcy massif meta-anorthosite to 600 °C, 650 MPa indicating that subsequent cooling occurred in association with limited uplift. Mezger et al. (1991) use ages from hornblende, titanite, rutile, and biotite to infer a cooling path from 750 °C with initial cooling rates of ~4 °C/m.y. Unfortunately, data from radiogenic minerals that might grow or experience closure at lower temperatures and so be used to date cooling to 600 °C and lower are inconsistent and do not define a well-constrained cooling curve for the Highlands (Mezger et al., 1991; Streepey et al., 2000). Mezger et al. (1991) recognize this problem and suggest that local thermal disturbances, probably related to intrusions, are responsible for the complexity of the Highlands' cooling history.

CONCLUSION

Anatexis and related metamorphism can produce new zircon grains that exhibit the characteristics of zircon growth in the presence of a melt phase. Dating zircons of this type requires careful selection of material for isotopic analysis. Anatectites from metagabbro may yield relatively simple zircons and should be considered as potential targets for dating if they are present. The growth history of zircon from anatectites in more felsic rocks is likely to be complex. Imaging techniques, cathodoluminescence or backscatter electron imaging, and electron microprobe chemical analyses to confirm zone identification can be used to recognize grains and parts of grains that grew during the anatectic event. Once identified, grains dominated by zircon crystallized during the anatectic event can be dated with precision using ID-TIMS. The integration of imaging, chemical analyses, and ID-TIMS provides a readily available and relatively inexpensive method by which to obtain high-precision ages that can be correlated to particular thermal events with certainty. The high level of precision can be especially important when attempting to date events that may be separated by relatively short time spans. Application of this method to dating an anatectic charnockite from the New Russia gneiss complex allows the characterization of the growth history of a complex zircon population and the determination of the intrusive and anatectic ages of the gneiss.

The age of the anatectic event that affected the New Russia complex provides a new constraint on model *P-T-t* paths for the northeastern Adirondack Highlands, indicating that rocks like the New Russia gneiss from the vicinity of the Marcy meta-anorthosite experienced relatively rapid, isobaric cooling from very high temperatures after 1050 Ma.

ACKNOWLEDGMENTS

We thank the New York State Geological Survey, Penn State, Abington College, the SUNY Research Foundation, and the Morton K. Blaustein Department of Earth Sciences at Johns Hopkins University for their support of this research. We also recognize the contributions of Yngvar Isachsen to the geosciences, to our understanding of Adirondack geology, and to our lives. His enthusiasm for knowledge and scientific inquiry remains an inspiration. L. Corriveau, L. Solari, and two anonymous reviewers are thanked for their comments on an earlier version of this paper.

REFERENCES CITED

Alcock J., and Muller, P.D., 1999, Very high-temperature, moderate-pressure metamorphism in the New Russia gneiss complex, northeastern Adirondack Highlands, metamorphic aureole to the Marcy anorthosite: Canadian Journal of Earth Sciences, v. 36, p. 1–13.

Alcock J., and Muller, P.D., 2000, Anatexis at 700 to 1000 MPa in the aureole of the Marcy anorthosite, Adirondack Highlands, New York: Contributions to Mineralogy and Petrology, v. 139, p. 643–654.

Alcock J., and Muller, P.D., 2001, Very high-temperature, moderate-pressure metamorphism in the New Russia gneiss complex, northeastern Adirondack Highlands, metamorphic aureole to the Marcy anorthosite: Reply: Canadian Journal of Earth Sciences, v. 38, p. 471–475.

Alcock, J., Myer, K., and Muller, P.D., 1999, Three-dimensional model of heat flow in the aureole of the Marcy anorthosite, Adirondack Highlands, New York: Implications for depth of emplacement: Geological Materials Research, v. 1, no. 4, p. 1–21.

Bingen, B., Austrheim, H., and Whitehouse, M., 2001, Ilmenite as a source for zirconium during high-grade metamorphism? Textural evidence from the Caledonides of Western Norway and implications for zircon geochronology: Journal of Petrology, v. 42, p. 355–375.

Burnham, C.W., 1992, Calculated melt and restite compositions of some Australian granites: Transactions of the Royal Society of Edinburgh: Earth Sciences, v. 83, p. 387–397.

Chiarenzelli, J.R., and McLelland, J.M., 1991, Age and regional relationships of granitoid rocks in the Adirondack Highlands: Journal of Geology, v. 99, p. 571–590.

Corrigan, D., and van Breemen, O., 1997, U-Pb age constraints for the lithotectonic evolution of the Grenville Province along the Mauricie transect, Quebec: Canadian Journal of Earth Sciences, v. 34, p. 299–316.

DeWolf, C.P., Zeissler, C.J., Halliday, A.N., Mezger, K., and Essene, E.J., 1996, The role of inclusions in U-Pb and SM-Nd garnet geochronology: Stepwise dissolution experiments and trace uranium mapping by fission track analysis: Geochimica et Cosmochimica Acta, v. 60, p. 121–134.

Doig, R., 1991, U-Pb zircon dates of Morin anorthosite suite rocks, Grenville Province, Quebec: Journal of Geology, v. 99, p. 729–738.

Emslie, R.F., and Hunt, P.A., 1990, Ages and petrogenetic significance of igneous mangerite-charnockite suites associated with massif anorthosite, Grenville Province: Journal of Geology, v. 98, p. 213–231.

Fraser, G., Ellis, D., and Eggins, S., 1997, Zirconium abundance in granulite-facies minerals, with implications for zircon geochronology in high-grade rocks: Geology, v. 25, p. 607–610.

Friedman, R., and Martignole, J., 1995, Mesoproterozoic sedimentation, magmatism, and metamorphism in the southern part of the Grenville Province (western Quebec): U-Pb geochronological constraints: Canadian Journal of Earth Sciences, v. 32, p. 2103–2114.

Hanchar, J.M., and Rudnick, R.L., 1995, Revealing hidden structures: The application of cathodoluminescence and back-scattered electron imaging to dating zircon from lower crustal xenoliths: Lithos, v. 36, p. 289–303.

Higgins, M.D., and van Breemen, O., 1995, Three generations of anorthosite-mangerite-charnockite-granite magmatism, contact metamorphism, and tectonism in the Saguenay-Lac-Saint-Jean region of the Grenville Province, Canada: Precambrian Research, v. 79, p. 327–346.

Kohn, M.J., and Valley, J.W., 1998, Effects of cation substitutions in garnet and pyroxene on equilibrium oxygen isotopic fractionations: Journal of Metamorphic Geology, v. 16, p. 625–639.

Kretz, R., 1983, Symbols for rock-forming minerals: American Mineralogist, v. 68, p. 277–279.

Krogh, T.E., 1982, Improved accuracy of U-Pb zircon ages by creation of more concordant systems using an air abrasion technique: Geochimica et Cosmochimica Acta, v. 45, p. 637–649.

Ludwig, K.R., 1980, Calculation of uncertainties of U-Pb data: Earth and Planetary Science Letters, v. 46, p. 212–220.

Martignole, J., and Friedman, R., 1998, Geochronological constraints on the last stages of terrane assembly in the central part of the Grenville Province: Precambrian Research, v. 92, p. 145–164.

Martignole, J., Calvert, A.J., Friedman, R., and Reynolds, P., 2000, Crustal evolution along a seismic section across the Grenville Province (western Quebec): Canadian Journal of Earth Sciences, v. 37. p. 291–306.

McLelland, J.M., and Chiarenzelli, J., 1990, Isotopic constraints on emplacement age of anorthositic rocks of the Marcy Massif, Adirondack Mts., New York: Journal of Geology, v. 98, p. 19–41.

McLelland, J.M., Daly, J.S., and McLelland, J.M., 1996, The Grenville Orogenic Cycle (ca. 1350–1000 Ma): An Adirondack perspective: Tectonophysics, v. 265, p. 1–28.

McLelland, J., Hamilton, M., Selleck, B., McLelland, J., Walker, D., and Orrell, S., 2001a, Zircon U-Pb geochronology of the Ottawan Orogeny, Adirondack Highlands, New York: Regional and tectonic implications: Precambrian Research, v. 109, p. 39–72.

McLelland, J.M., Valley, J.W., and Essene, E., 2001b, Very high-temperature, moderate-pressure metamorphism in the New Russia gneiss complex, northeastern Adirondack Highlands, metamorphic aureole to the Marcy anorthosite: Comment: Canadian Journal of Earth Sciences, v. 38, p. 465–470.

Mezger, K., Rawnsley, C.M., Bohlen, S.R., and Hanson, G.N., 1991, U-Pb garnet, sphene, monazite, and rutile ages: Implications for the duration of high-grade metamorphism and cooling histories, Adirondack Mts., New York: Journal of Geology, v. 99, p. 415–428.

Milord, I., Sawyer, E.W., and Brown, M., 2001, Formation of diatexite migmatite and granite magma during anatexis of semipelitic metasedimentary rocks: An example from St. Malo, France: Journal of Petrology, v. 42, p. 487–505.

Naslund, H., 1987, Lamellae of baddeleyite and Fe-Cr spinel in ilmenite from the Basistoppen Sill, east Greenland: Canadian Mineralogist, v. 25, p. 91–96.

Parrish, R.R., 1987, An improved micro-capsule for zircon dissolution in U-Pb geochronology: Isotope Geoscience, v. 66, p. 99–102.

Pidgeon, R.T., Nemchin, A.A., and Hitchen, G.J., 1998, Internal structures of zircons from Archaean granites from the Darling Range batholith: Implications for zircon stability and the interpretation of zircon U-Pb ages: Contributions to Mineralogy and Petrology, v. 132, p. 288–299.

Poldervaart, A., 1956, Zircon in rocks, 2: Igneous rocks: American Journal of Science, v. 254, p. 521–554.

Pyle, J.M., and Spear, F.S., 1999, Yttrium zoning in garnet: Coupling of major and accessory phases during metamorphic reactions: Geological Materials Research, v. 1, no. 6, p. 1–24.

Rapp, R.P., and Watson, E.B., 1995, Dehydration melting of metabasalt at 8–32 kbar: Implications for continental growth and crust-mantle recycling: Journal of Petrology, v. 36, p. 891–931.

Rivers, T., Martignole, J., Gower, C.F., and Davidson, A., 1989, New tectonic divisions of the Grenville Province: Tectonics, v. 8, p. 63–84.

Rushmer, T., 1991, Partial melting of two amphibolites; contrasting experimental results under fluid-absent conditions: Contributions to Mineralogy and Petrology, v. 107, p. 41–59.

Russ-Nabalek, C., 1989, Isochemical contact metamorphism of mafic schist, Laramie Anorthosite Complex, Wyoming: Amphibole compositions and reactions: American Mineralogist, v. 74, p. 530–548.

Sawyer, E.W., 1996, Melt segregation and magma flow in migmatites: Implications for the generation of granite magmas: Transactions of the Royal Society of Edinburgh: Earth Sciences, v. 87, p. 85–94.

Sawyer, E.W., 1998, Formation and evolution of granite magmas during crustal reworking: The significance of diatexites: Journal of Petrology, v. 39, p. 1147–1167.

Schaltegger, U., Fanning, C.M., Günther, D., Maurin, J.C, Schulmann, K., and Gebauer, D., 1999, Growth, annealing and recrystallization of zircon and preservation of monazite in high-grade metamorphism: Conventional and

in-situ U-Pb isotope, cathodoluminescence, and microchemical evidence: Contributions to Mineralogy and Petrology, v. 134, p. 186–201.
Silver, L., 1969, A geochronologic investigation of the Anorthosite Complex, Adirondack Mountains, New York, *in* Isachsen, Y., ed., Origin of anorthosite and related rocks: Albany, New York State Museum Memoir 18, p. 233–252.
Spear, F.S., and Markussen, J.C., 1997, Mineral zoning, P-T-X-M phase relations, and metamorphic evolution of some Adirondack granulites, New York: Journal of Petrology, v. 38, p. 757–783.
Springer, W., and Seck, H.A., 1997, Partial fusion of basic granulites at 5 to 15 kbar: Implications for the origin of TTG magmas: Contributions to Mineralogy and Petrology, v. 127, p. 30–45.
Stacey, J.S., and Kramers, J.D., 1975, Approximation of terrestrial lead isotope evolution by a two-stage model: Earth and Planetary Science Letters, v. 26, p. 207–221.
Steiger, R.H., and Jäger, H., 1977, Subcommission of geochronology: Convention on the use of decay constants in geo- and cosmochronology: Earth and Planetary Science Letters, v. 36, p. 359–362.
Stern, R.A., 1997, The GSC Sensitive High Resolution Ion Microprobe (SHRIMP): Analytical techniques of zircon U-Th-Pb age determinations and performance evaluation, *in* Radiogenic age and isotopic studies: Report 10: Ottawa, Ontario, Geological Survey of Canada, Current Research, 1997-F, p. 1–31.
Streepey, M.M., van der Pluijm, B.A., Essene, E.J., Hall, C.M., and Magloughlin, J.F., 2000, Late Proterozoic (ca. 930 Ma) extension in eastern Laurentia: Geological Society of America Bulletin, v. 112, p. 1522–1530.
Valley, J.W., 1985, Polymetamorphism in the Adirondacks: Wollastonite at contacts of shallowly intruded anorthosite, *in* Tobi, A.C., and Touret, J.L.R., eds., The deep Proterozoic crust in the North Atlantic provinces: Dordrecht, The Netherlands, D. Reidel, p. 217–236.
Valley, J.W., and Essene, E.J., 1980, Åkermanite in the Cascade Slide xenolith and its significance for regional metamorphism in the Adirondacks: Contributions to Mineralogy and Petrology, v. 74, p. 143–152.
Vavra, G., 1990, On the kinematics of zircon growth and its petrogenetic significance: A cathodoluminescence study: Contributions to Mineralogy and Petrology, v. 106, p. 90–99.
Vavra, G., Schmid, R., and Gebauer, D., 1999, Internal morphology, habit and U-Th-Pb microanalysis of amphibolite-to-granulite-facies zircons: Geochronology of the Ivrea Zone (Southern Alps): Contributions to Mineralogy and Petrology, v. 134, p. 380–404.
Wasteneys, H., McLelland, J., and Lumbers, S., 1999, Precise zircon geochronology in the Adirondack Lowlands and implications for revising plate-tectonic models of the Central Metasedimentary Belt and Adirondack Mountains, Grenville Province, Ontario and New York: Canadian Journal of Earth Sciences, v. 36, p. 967–984.
Whitehouse, M.J., Kamber, B.S., and Moorbath, S., 1999, Age significance of U-TH-Pb zircon data from early Archaean rocks of west Greenland—A reassessment based on combined ion-microprobe and imaging studies: Chemical Geology, v. 160, p. 201–224.
Whitney, J.A., 1988, The origin of granite: The role and source of water in the evolution of granitic magmas: Geological Society of America Bulletin, v. 100, p. 1886–1897.
Williams, I.S., and Claesson, S., 1987, Isotopic evidence for the Precambrian provenance and Caledonian metamorphism of high grade paragneisses from the Seve Nappes, Scandinavian Caledonides, 2: Ion microprobe zircon U-Th-Pb: Contributions to Mineralogy and Petrology, v. 97, p. 205–217.
Wynne-Edwards, H.R., 1972, The Grenville Province, *in* Price, R.A., and Douglas, R.J.W., eds., Variations in tectonic styles in Canada: St. John's, Newfoundland, Geological Association of Canada Special Paper 11, p. 263–334.

Manuscript Accepted by the Society August 25, 2003

Printed in the USA

Geological Society of America
Memoir 197
2004

Deformation of the Diana syenite and Carthage-Colton mylonite zone: Implications for timing of Adirondack Lowlands deformation

Graham B. Baird*
Department of Geology and Geophysics, University of Minnesota, 310 Pillsbury Drive SE, Minneapolis, Minnesota 55455-0219, USA
William D. MacDonald
Department of Geological Sciences and Environmental Studies, State University of New York, Vestal Parkway East, Binghamton, New York 13902-6000, USA

ABSTRACT

The Carthage-Colton mylonite zone is a major geothermochronological discontinuity across the northwest Adirondack Mountains of New York, a southern extension of the Grenville Province. A large syenitic gneiss body, the Diana syenite, occurs along most of the southern Carthage-Colton mylonite zone. The present study examined petrofabrics and magnetofabrics of oriented cores and accurately oriented thin-sections to investigate the sources of anisotropy of magnetic susceptibility (AMS) within a central portion of the Diana syenite. Three petrographic foliations, a petrographic lineation, and a magnetic intersection lineation were clearly distinguished. Two of the foliations appear to represent axial planar foliations of the second- and third-phases of regional folding as defined by Wiener (1983). The youngest foliation and the magnetic intersection lineation have not been previously described. This research suggests that folding identified in the Adirondack Lowlands can be traced to at least the southwest margin of the Diana syenite with no obvious discontinuity. Significant implications of this research suggest that: (1) the Adirondack Lowlands deformation likely includes some folding events associated with Ottawan orogen compression, and (2) the kinematics and style of deformation within the Carthage-Colton mylonite zone remain cryptic and cannot conclusively be connected to the fabrics explored in this research.

Keywords: Adirondacks, Carthage-Colton mylonite zone, Diana syenite, Adirondack Lowlands, AMS, anisotropy, magnetic susceptibility, petrofabric, magnetofabric

INTRODUCTION

The Carthage-Colton mylonite zone (Fig. 1), as defined by Geraghty et al. (1981), also referred to by subsequent workers as the Carthage-Colton shear zone, is an ancient crustal shear zone within the Grenville Province. The Carthage-Colton mylonite zone has long been recognized as an important crustal feature associated with the Adirondack Highlands–Lowlands boundary. Recent work (e.g., Streepey et al., 2000, 2001) appears to constrain the style and timing of motion of this large-scale structure. However, specific field-scale structures tied to their (Streepey et al., 2000) proposed terrane-scale motions largely remain unstudied, and consideration of such associations is the logical next step in the analysis of movements along this

*E-mail: bair0042@umn.edu.

Baird, G.B., and MacDonald, W.D., 2004, Deformation of the Diana syenite and Carthage-Colton mylonite zone: Implications for timing of Adirondack Lowlands deformation, *in* Tollo, R.P., Corriveau, L., McLelland, J., and Bartholomew, M.J., eds., Proterozoic tectonic evolution of the Grenville orogen in North America: Boulder, Colorado, Geological Society of America Memoir 197, p. 285–297. For permission to copy, contact editing@geosociety.org.

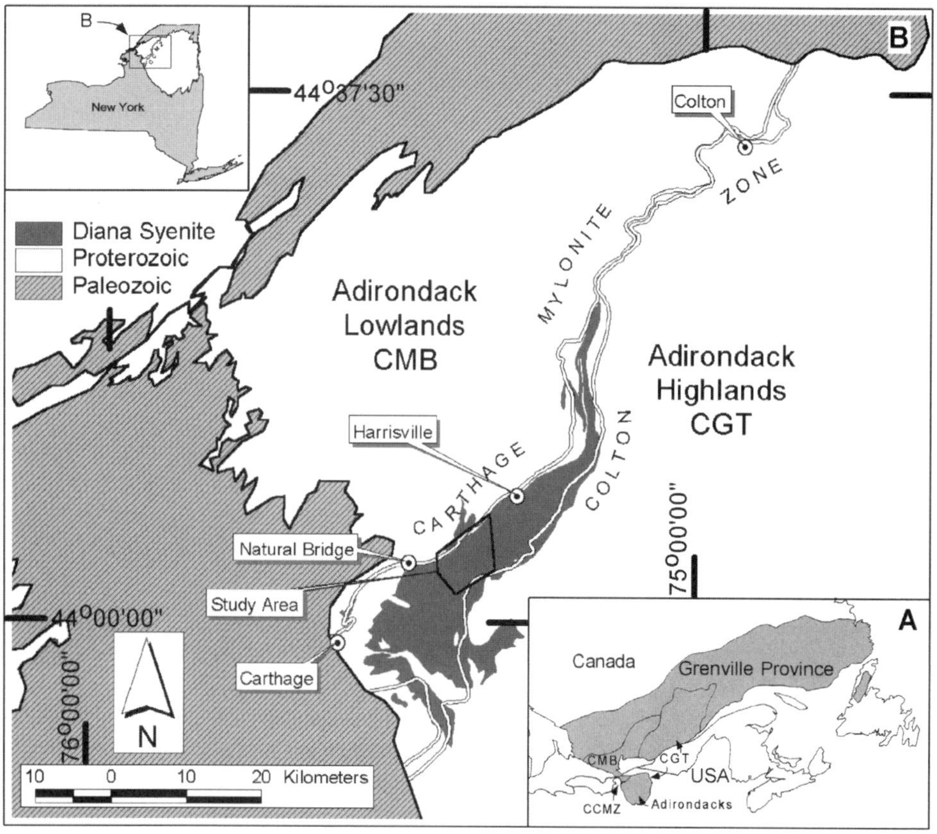

Figure 1. (A) Map of the Grenville Province showing the location of the Carthage-Colton mylonite zone (CCMZ). CMB—Central Metasedimentary Belt; CGT—Central Granulite terrane (after Wynne-Edwards, 1972). (B) Map of the Northwest Adirondacks of New York showing the extent of the Carthage-Colton mylonite zone and the Diana syenite (adapted from Isachsen and Fisher, 1970; Geraghty et al., 1981).

structure. To this end, the present study contributes field and laboratory investigations of rock fabric in the Diana syenite, a dominant rock body along the southern Carthage-Colton mylonite zone, and attempts to relate them to recognized events. The laboratory analyses compare anisotropy of magnetic susceptibility with accurately oriented thin-sections to investigate the relationship between the observed petrofabrics and measured magnetofabrics. This study emphasizes the well-developed mylonitic fabric within the Diana syenite in this part of the Carthage-Colton mylonite zone and its implication for the timing of Adirondack Lowlands deformation.

Regional Geology

The Adirondack Highlands and Lowlands comprise a southern extension of the Proterozoic Grenville Province proper of eastern Canada (Fig. 1). The Adirondack Lowlands exposes a portion of the Frontenac terrane, and is part of the Central Metasedimentary Belt of Wynne-Edwards (1972). The Frontenac terrane connects New York's Adirondack Mountains to the main province in Québec and Ontario. The Lowlands are characterized by metasedimentary rocks (calc-silicates, marble, metapsammites, and metapelites, with scattered orthogneisses) typically metamorphosed to the upper amphibolite facies (McLelland and Isachsen, 1986). Deformation within the Frontenac terrane (including the Lowlands) occurred at ca. 1180–1160 Ma (Corfu and Easton, 1997; Carr et al., 2000). Metaigneous rocks characterize the typically granulite-facies Adirondack Highlands of the Central Granulite terrane of Wynne-Edwards (1972). Local synformal (synclinal?) structures within the Highlands contain paragneisses similar to those of the Lowlands (McLelland and Isachsen, 1986). Highlands deformation includes the 1180–1160 Ma event of the Frontenac terrane, but was also subject to another deformation event at ca. 1060–1000 Ma (Carr et al., 2000).

The Carthage-Colton mylonite zone separates the Highlands and the Lowlands (Fig. 1, B; Geraghty et al., 1981) and is characterized by grain-size reduction accompanied by a strong northwest-dipping foliation. As mapped by Geraghty et al. (1981), the Carthage-Colton mylonite zone varies from 3 m to 5 km wide, from Carthage (southwest) to beyond Colton (northeast), a distance of ~110 km. It has been correlated with the LaBelle shear zone of Québec (e.g., McLelland and Isachsen, 1986; McLelland et al., 1996).

Adirondack tectonic evolution involved three major events (McLelland et al., 1996): (1) an arc-continent collision during the Elzevirian orogeny (ca. 1220–1160 Ma; Wasteneys et al., 1999); (2) an intrusion episode that emplaced the extensive AMCG (anorthosite-mangerite-charnockite-granite) suite (ca. 1180–1130 Ma; McLelland et al., 1996); and (3) the Ottawan orogeny (ca. 1090–1035 Ma; McLelland et al., 2001), thought to be a result of the collision of Laurentia with Amazonia, forming the supercontinent Rodinia (see Weil et al., 1998, and references therein).

Associated with these events, five folding phases can be identified throughout the Adirondacks. The earliest folding event is difficult to identify but produced a strong axial planar foliation and is absent in many Adirondack intrusive rocks. The first and second folding events occurred during peak metamorphic conditions, while the third, fourth, and fifth folding phases formed during retrograde and postmetamorphic conditions, possibly much later than earlier folding (Wiener, 1983; Wiener et al., 1984). The second and third folding phases arc into parallelism with the Carthage-Colton mylonite zone, suggesting a genetic connection (Wiener et al., 1984; McLelland and Isachsen, 1986).

Significance of the Carthage-Colton Mylonite Zone

Recent isotopic information (Mezger et al., 1992; Streepey et al., 2000, 2001) suggests a complicated history for the Carthage-Colton mylonite zone. Mezger et al. (1992), using sphene U-Pb ages, demonstrated a ca. 100 Ma difference between the most recent regional metamorphic ages for the Highlands (ca. 1050–982 Ma) and the Lowlands (ca. 1156–1103 Ma). Sphenes from within a southern portion of the Carthage-Colton mylonite zone show consistent ages of ca. 1098 Ma, which Mezger et al. (1992) interpreted as the Carthage-Colton mylonite zone mylonization event. However, the U-Pb dating of sphenes (1020 Ma) and the ^{40}Ar-^{39}Ar dates of hornblendes (1000 Ma) of mylonite within the Dana Hill metagabbro, in the central Carthage-Colton mylonite zone, suggest a younger mylonitization age (Streepey et al., 2001). The younger mylonite age, in conjunction with field structural interpretations, suggest that early motion along the Carthage-Colton mylonite zone was left lateral strike-slip (Streepey et al., 2001). A large lateral translation and juxtaposition of previously separated terranes could possibly account for the 100 Ma difference in the peak metamorphic ages between the Highlands and Lowlands (Streepey et al., 2001).

Streepey et al. (2000) used the ^{40}Ar-^{39}Ar dates of hornblende and biotite to constrain late Carthage-Colton mylonite zone extension to the interval from ca. 950 to 920 Ma, much later than previous estimates (e.g., ca. 1050 Ma by McLelland et al., 1993). This puts final movement on the Carthage-Colton mylonite zone long after the proposed collapse of the Grenville orogen (e.g., McLelland et al., 1996) and indicates a possible distinct extensional event following the Grenville orogeny but prior to the breakup of Rodinia (Streepey et al., 2000).

Kinematics

Many theories have been proposed to describe the character, timing, and tectonic significance of Carthage-Colton mylonite zone formation. Geraghty et al. (1981) considered the narrower northeast section to be a ductile fold-thrust nappe, and the wider southwestern section a diffuse shear zone forming an open northeast-trending antiform. This interpretation, revised by Isachsen and Geraghty (1986), proposed that the Carthage-Colton mylonite zone is a deep-seated detachment surface similar to those associated with Cordilleran-type core complexes. This suggests that the Carthage-Colton mylonite zone formed in an extensional environment, possibly during the postorogenic collapse of the Grenville orogen. Lumino (1987) reported that the mylonitic fabric is not uniform throughout the Diana syenite, but is more pronounced in ductile deformation zones (DDZs). Analysis of these more strongly deformed areas reveals consistent motion sense—hangingwall down to the northwest on northwest-dipping DDZs. Heyn (1990), in studying tectonites near Russell, northeast of the present study, found ambiguous kinematic indicators in an early high-grade generation of mylonite and attributed their formation to Grenville compression. Heyn (1990) felt the second medium-grade mylonite event was associated with extension. Buddington (1939) suggested that the Lowlands were thrust over the Highlands, while McLelland and Isachsen (1986) report southeast-over-northwest kinematics for the Carthage-Colton mylonite zone. Hall (1984) observed conflicting kinematic indicators but reported that northwest side–down indicators were slightly more common. Most recently, both left-lateral (Streepey et al., 2001) and right-lateral (Johnson et al., this volume) strike-slip motion have been proposed for early Carthage-Colton mylonite zone movement. Both the Streepey and the Johnson interpretations include some normal fault motion following the strike-slip motions. Clearly, there is no consensus for an unequivocal transport model for the Carthage-Colton mylonite zone.

Study Area

The study area covers the central section of the Diana syenite between Natural Bridge and Remington Corners (Figs. 1 and 2). The Diana syenite possesses a fairly uniform lithology, which minimizes structural differences that might otherwise develop due to variable competencies, thus enhancing the usefulness of this location for foliation analysis. The Diana syenite has an intrusive, but locally faulted, northwestern contact with the Lowlands metasediments (e.g., Wiener, 1983).

Two younger granitic gneisses intrude the southeastern Diana syenite (Hargraves, 1969): (1) a coarse-grained hornblende granitic gneiss that appears to be Hawkeye granite (see McLelland et al., 2001) east and southeast of the area depicted in Figure 2, and (2) an aplitic gneiss that fits the description of Lyon Mountain granitic gneiss (see McLelland et al., 2001).

Wiener (1983) states that intrusive rocks to the southeast possess nonmylonitic texture and grade into the mylonitic texture of the Diana syenite. The intrusive northwest contact infolded with Lowlands metasediments, and the intrusion of Highlands granites into the Diana syenite to the southeast suggests that the Diana syenite cannot be considered to have affinities only with either the Highlands or the Lowlands. As mapped

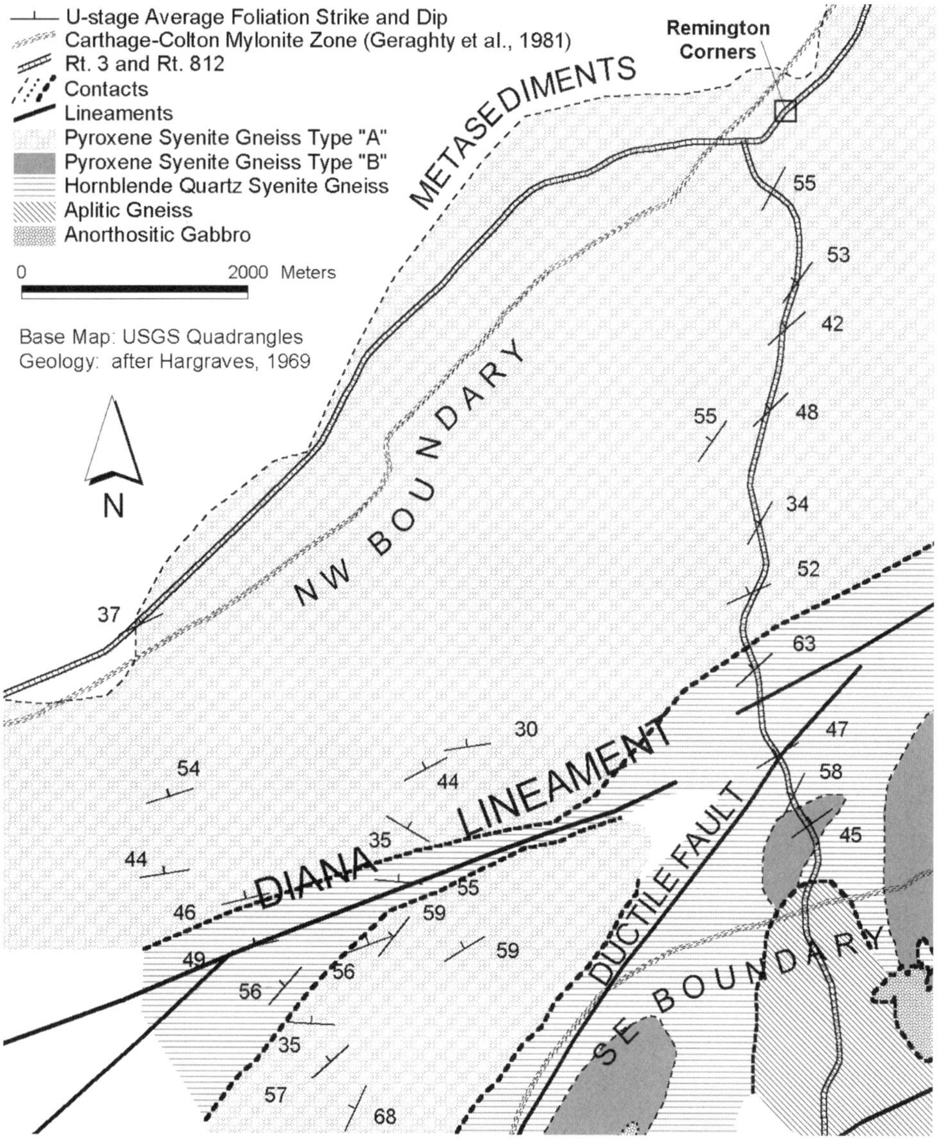

Figure 2. Geologic map of the study area, after Hargraves (1969), with additional foliation data from this study. The numbers indicate dip of foliation.

(Fig. 1, B), portions of the Diana syenite extend beyond both the Highlands and Lowlands margins of the Carthage-Colton mylonite zone.

Ub-Pb zircon dating of the Diana syenite gives ages of 1155 ± 4 Ma (Grant et al., 1986), and it is part of the AMCG suite (McLelland et al., 1996). Hargraves (1969), following the earlier work of Buddington (1939), found two dominant rock types within the Diana syenite: pyroxene syenite, which is prominent in the northwest half of the Diana syenite, and hornblende quartz syenite, predominantly in the southeast. The transition between these two syenite types is spatially associated with the Diana lineament, a zone of brittle deformation 30–300 m wide that extends from Carthage to at least Harrisville. The progression from pyroxene syenite to hornblende quartz syenite, associated with anorthosite to the southwest, is interpreted as remnant intrusive layering (Fig. 2; Hargraves, 1969). Both Buddington (1963) and Hargraves (1969) suggest that the Diana syenite has been folded.

Cartwright et al. (1993), due to the oxygen isotopic signatures of the Diana syenite, conclude that intrusion was shallow (<10 km), similar to the interpretation for the Marcy anorthosite of the Highlands. Subsequent granulite metamorphism was lacking in fluid, and the bulk composition was not changed (Valley et al., 1990).

Wiener (1983), in detailed mapping of the northern margin of the Diana syenite near Harrisville (Fig. 1, B), identified four phases of deformation. The first phase affected only the metasediments to the northwest and was absent in the Diana syenite, implying that syenite intrusion occurred after the first deformation. The second- and third-phase folds, with generally northeast-striking axial planes, isoclinally folded earlier phases. Broad open east-west trending folds represent a fourth phase of folding. The mylonitic texture of the Diana syenite was assigned by Wiener (1983) to the second-phase isoclinal folding because the mylonite foliation is parallel to second-phase axial planes. Wiener (1983) indicated that the third deformation phase folded the mylonite foliation, but folded foliation has not been recognized in the area of this research.

First- and second-phase folding developed under peak metamorphic conditions, with third- and fourth-phase folding forming under retrograde conditions (Wiener, 1983). Mylonitic textures formed axial planar to second-phase folds, at peak conditions according to Wiener (1983), but Lamb (1993) found that mylonitic fabric developed under retrograde conditions around Harrisville, and the fluids accompanying this mylonitization are unique to the Carthage-Colton mylonite zone.

Generalizations regarding Diana Syenite Petrofabric and Minor Structures

The mylonitization associated with the Carthage-Colton mylonite zone has been well documented by several authors (e.g., Geraghty et al., 1981; Wiener, 1983; Lumino, 1987). Mylonitization and fabric development of the Diana syenite varies from weak to very strong (Fig. 3). Trains of opaques and mafic minerals wrapping around alkali-feldspar augens define foliation. Lineation varies from indistinguishable to pervasive and is defined by streaks of recrystallized feldspar aggregates surrounded by opaque and mafic minerals. Lineations within the Carthage-Colton mylonite zone have been reported to be parallel to second-phase fold axes (McLelland and Isachsen, 1986).

Geraghty et al. (1981) describe three relationships between the mylonitic foliation defining the Carthage-Colton mylonite zone and minor folds: (1) mylonitic foliation is axial planar to minor folds, (2) mylonitic foliation is folded by minor folds with a second axial planar mylonitic fabric developed, or (3) mylonitic foliation is folded into an upright open fold without a secondary foliation developed. Geraghty et al. (1981) related minor folds and mylonitic fabric relationships 1 and 2 to Wiener's (1977) second-phase folding without rigorous field investigations.

Three structure types commonly crosscut the northeast-trending fabric: (1) ultramylonite shear zones (Fig. 4, A), (2) brittle faults (such as the Diana lineament; Fig. 2), and (3) ductile faults (Figs. 2 and 4, B). Both Heyn (1990) and Wiener (1983) observed two sets of crosscutting shear zones. Heyn (1990) states that shear zones in the Russell area occurred at peak metamorphic conditions and are folded by the last phase of folding, whereas Cartwright et al. (1993) found that shear zone development occurred at retrograde conditions.

Magnetic Anisotropy and Petrofabric Analysis

Graham (1954) noted that measurement of anisotropy of magnetic susceptibility (AMS) is an underexploited technique for the analysis of petrofabrics. This technique measures the anisotropic induced magnetic response of a sample to an applied magnetic field. The development of analytical instruments for the rapid measurement of AMS has greatly increased the use of magnetic anisotropy as a proxy for fabric analysis of a great variety of rock types (MacDonald and Ellwood, 1987; Tarling and Hrouda, 1993; Kodama, 1995; Borradaile and Henry, 1997).

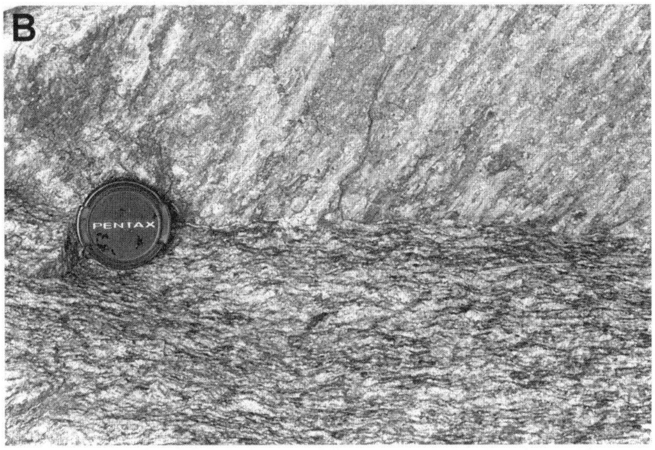

Figure 3. (A) Weakly foliated pyroxene syenite gneiss, type "A." Foliation dips steeply down to the right. Lens cap for scale, 5 cm. (B) Strongly lineated and foliated hornblende quartz syenite gneiss. Prominent lineation is visible in the upper portion of the photo; foliation is subhorizontal in the lower portion of photo. Lens cap for scale, 5 cm.

AMS can be represented by a triaxial ellipsoid with three axes: K_{max} (or K_1), the axis of maximum susceptibility; K_{int} (or K_2), the axis of intermediate susceptibility; and K_{min} (or K_3), the axis of minimum susceptibility (Fig. 5, A). The orientations of these axes are controlled by the type, distribution, shape, and alignment of iron-bearing minerals, especially the oxides and silicates (Tarling and Hrouda, 1993). Magnetite and hornblende are among the most important ferromagnetic and paramagnetic contributors, respectively, to the magnetic susceptibility anisotropy in the Diana syenite.

In most igneous and metamorphic foliated rocks, magnetite distribution follows the foliation and lineation petrofabric (Tarling and Hrouda, 1993). Thus, K_{max} and K_{int} lie in the plane of

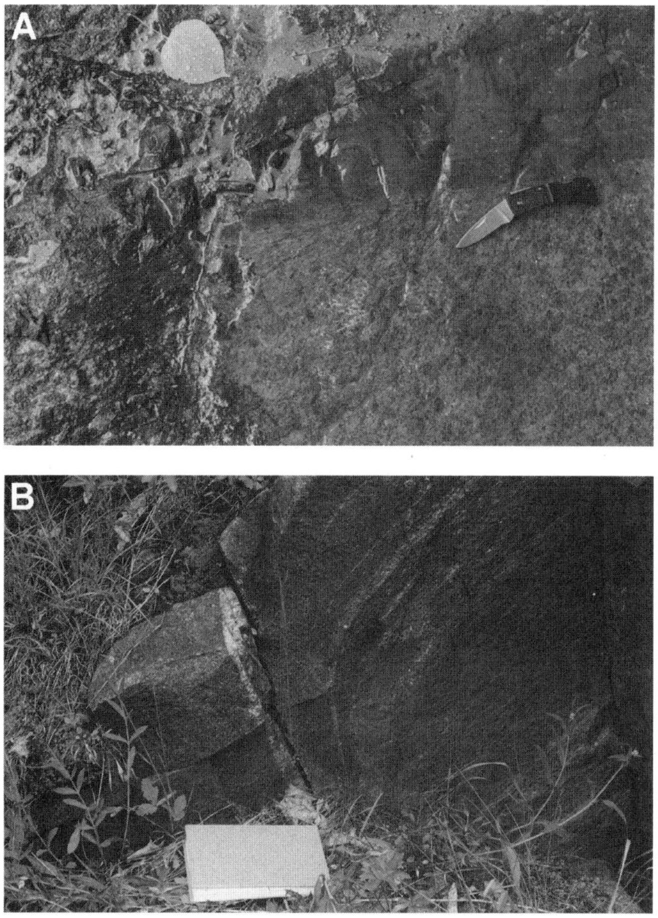

Figure 4. (A) An ultra-mylonite shear zone (above knife) obliquely cuts the prominent foliation. The knife is 12 cm long. The foliation bends tangentially into the zone, indicating dextral relative motion. (B) A ductile fault (070°, 69°SE) as viewed from the southwest showing steep top-to-the-northwest motion. Field book for scale, 21 cm.

the foliation, and K_{min} is normal to the plane of the foliation. In this study, the AMS ellipsoid orientation is referred to as the magnetofabric, by which it is meant that the magnetic "foliation" is represented by the K_{max}-K_{int} plane, or by its perpendicular, the K_{min} axis, and the magnetic "lineation" is represented by K_{max}. Typically, but not always, as will be shown, K_{max} is parallel to the direction of the mineral lineation (Fig. 5, A; e.g., Balsley and Buddington, 1960).

AMS emulates subtle fabrics, making them easier to analyze for petrofabric studies, and AMS may reveal "cryptic" features not easily discernible by visual means. The AMS response can be influenced by subtle petrofabric components, such as magnetite in parallel veinlets in serpentinites (MacDonald and Ellwood, 1988), and magnetite distributed along weakly developed C surfaces in S-C fabrics (Tomezzoli et al., 2003).

Magnetofabric and Petrofabric Comparison

As compared to typical thin-section procedures for fabric analysis, measurement of AMS is very accurate and rapid. Natural variations in rock fabric, and especially superimposed fabrics, commonly cannot be precisely measured in the field, so visible petrofabrics are measured in standard one-inch oriented cores in a Universal stage (U-stage) to improve accuracy (MacDonald and Ellwood, 1988). In this approach, thin-section planes can be more precisely marked for further petrographic analysis and comparison with AMS observations.

METHODS

Approximately six oriented cores were collected at 27 sites across a central section of the Diana syenite (Fig. 2), providing 156 specimens for petrofabric and magnetofabric comparisons. The core surfaces were cleaned on a grinding wheel to remove metal residues that might contribute anomalous AMS effects. Visible foliations were measured in cores mounted in a U-stage jig. Oriented thin-sections were cut parallel and perpendicular to visible petrographic foliations and magnetofoliations. AMS measurements were made with the KLY-2 Kappabridge (Colgate University) and the KLY-3 Kappabridge (SUNY Binghamton) instruments. The U-stage and Kappabridge measurements were corrected to geographic coordinates for evaluation and discussion.

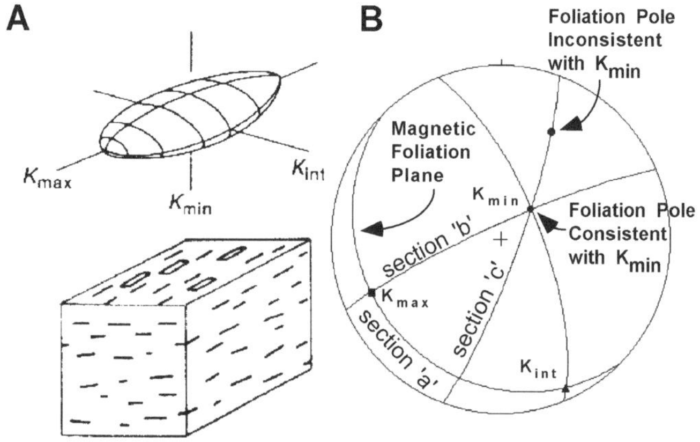

Figure 5. (A) A hypothetical lineated and foliated sample with a corresponding anisotropy of magnetic susceptibility (AMS) ellipsoid. Note that K_{max} (the axis of maximum susceptibility) is parallel to lineation, and K_{min} (the axis of minimum susceptibility) is perpendicular to foliation (after Borradaile, 1988, and Tarling and Hrouda, 1993). (B) Stereonet portraying the relationship between AMS principal axes and inferred petrofabric foliation. Oriented thin-sections were cut along planes a, b, and c, as described in the text.

Oriented Thin-Sections

Twenty oriented thin-sections were cut, each in one of three planes determined by specific criteria. Where the U-stage measured foliation pole is close to the AMS K_{min} axis (within ~10°), the cut was parallel to the K_{max} and K_{int} axes (Fig. 5, B, section "a") in order to reveal the petrofabric explanation for K_{max}. To examine the foliation, the K_{max}-K_{min} section (Fig. 5, B, section "b") was cut. Finally, where the U-stage foliation pole differed significantly from K_{min}, the section was cut along the plane containing K_{min} and the U-stage foliation pole (Fig. 5, B, section "c").

RESULTS

Magnetofabric Analysis

Regionally, the Diana foliation trends northeast and dips northwest, and the magnetofabric generally conforms to this pattern. K_{min} was located in the southeast quadrant (Fig. 6, A); K_{max} and K_{int} lie along a girdle representing a northwest-dipping plane (Fig. 7). However, close inspection of the K_{min} plot shows three clusters supported by density contouring (Fig. 6, B). This raises the possibility that three distinct planes are recorded in the K_{min} axes. The corresponding planes are oriented 196°, 23°NW (M_A—magnetofabric A); 245°, 45°NW (M_B—magnetofabric B); and 219°, 67°NW (M_C—magnetofabric C).

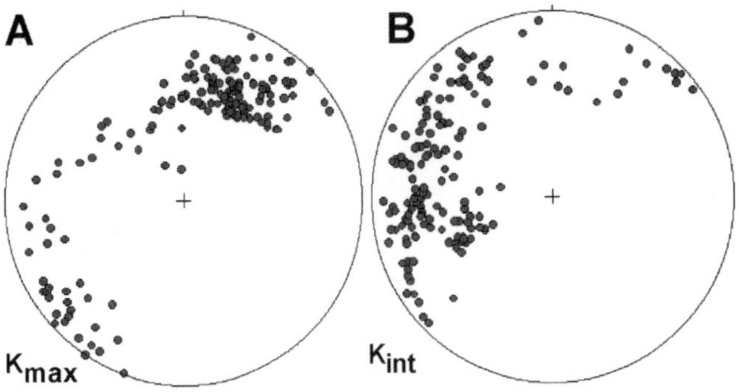

Figure 7. (A) Stereonet of K_{max} (the axis of maximum susceptibility) poles for all specimens. (B) Stereonet of K_{int} (the axis of intermediate susceptibility) poles for all specimens. Note that the K_{max} axes plunge mainly north-northeast and the K_{int} axes plunge mainly west-northwest, corresponding to northwest-dipping magnetofabrics.

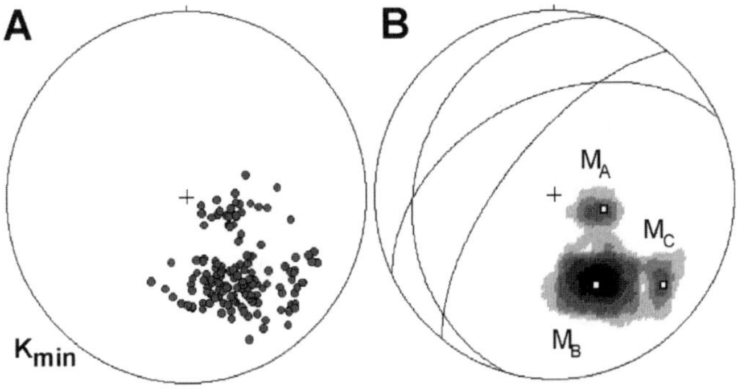

Figure 6. (A) Stereonet of K_{min} (the axis of minimum susceptibility) poles for all specimens. (B) Contour plot of A. Magnetofoliation concentrations are labeled M_A, M_B, and M_C. Open squares are Fisher statistics–determined mean poles, corresponding planes shown as great circles (M_A = 196°, 23°NW; M_B = 245°, 45°NW; M_C = 219°, 67°NW).

Although the scatter in the U-stage measured foliation poles is slightly greater than that of the magnetofabric poles, M_B and M_C agree well with the corresponding U-stage foliation poles (Fig. 8, B, C).

Figure 8, A shows that the M_A magnetic foliation poles deviate consistently from the corresponding U-stage foliation poles. Fisher statistics on the individual populations show that their means are significantly different. The regional distribution of the three magnetofabric domains is shown in Figure 9, where M_B is a widespread domain, M_A is locally but strongly developed along the northeast corner of the study area, and M_C has a patchy development in the southern half of the area, mostly overprinting the M_B domain.

Petrofabric Analysis

Five thin-sections that display M_A were cut along a plane that includes K_{min} and the U-stage measured foliation pole. In

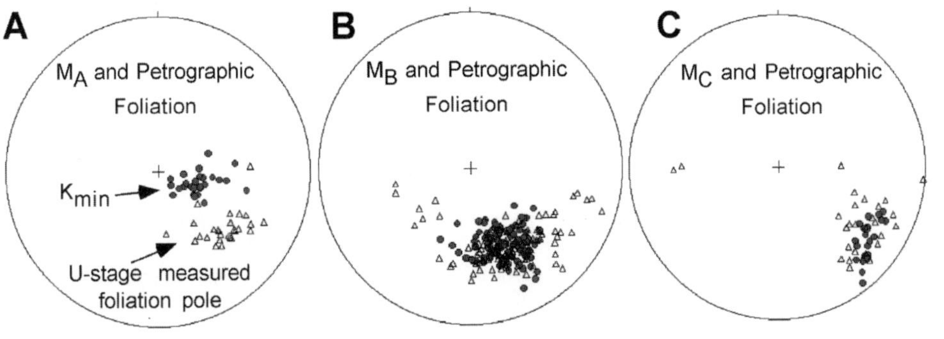

Figure 8. (A) Comparison of the K_{min} (the axis of minimum susceptibility) (circles) of magnetofabric M_A to the U-stage measured foliation poles (triangles) for the same samples. (B) Comparison of the K_{min} of M_B to the U-stage measured foliation pole for the same samples. (C) Comparison of the K_{min} of M_C to the U-stage measured foliation pole for the same samples.

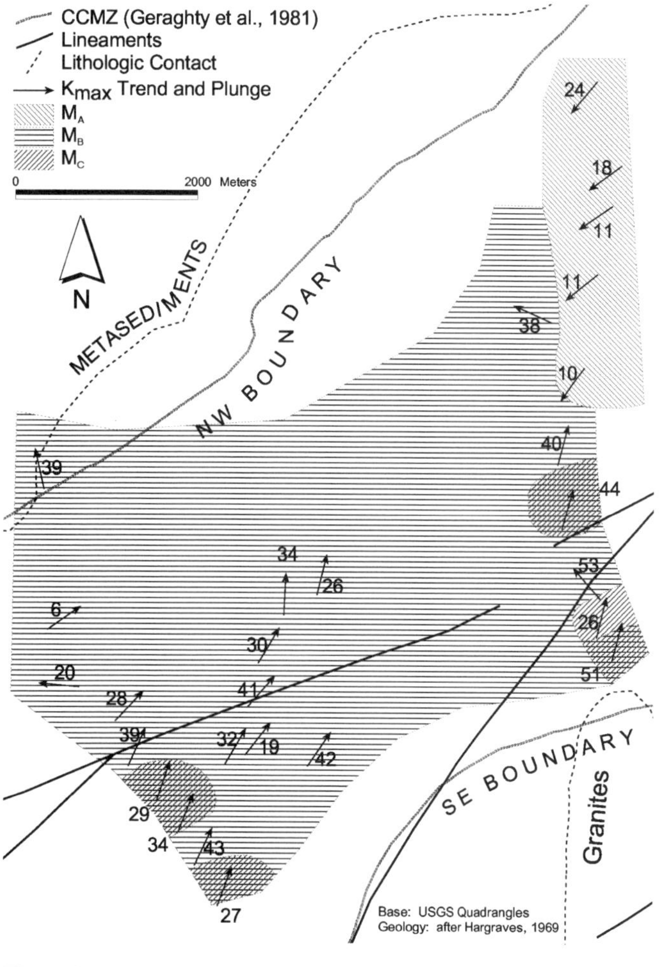

Figure 9. Regional distribution of the domains of magnetofabrics M_A, M_B, and M_C with the corresponding K_{max} (the axis of maximum susceptibility) orientation. The numbers indicate the plunge of K_{max}.

each case two foliations, S_1 and S_2, were clearly displayed, with their apparent relative ages being determined by crosscutting relationships. Typically the more recent foliation (S_2) was dominant (Fig. 10).

The orientations of the two lines of intersection of the two foliations with the plane of the thin-section were measured using a petrographic microscope and enlarged scans of the slide made before the final grinding of the thin-sections (Fig. 10). Eigen analysis was used to fit the planes to the "apparent dips" represented by the foliations of the same relative age in the five "c" thin sections (Fig. 11). The fitted planes have orientations of 221°, 72°NW and 209°, 37°NW for S_1 and S_2, respectively.

The average K_{max} for M_A samples lies at the intersection of the two fitted planes (Fig. 11). Two important conclusions can be drawn from this: (1) the fitted planes have accurately determined the orientations of S_1 and S_2, and (2) the magnetic lineation K_{max} is an intersection lineation analogous to that reported previously by MacDonald and Ellwood (1988). The southwest plunge of this magnetolineation is distinct from other K_{max} orientations, which generally plunge northeast (Fig. 9). This lineation was not detected in hand sample or outcrop.

Where two distinct but equally developed foliations are present, it might be expected that K_{min} would represent their average response. If so, K_{min} should be located in the obtuse angle between the two intersecting foliations. If not equally developed, their joint K_{min} response should lie along the plane perpendicular to their line of intersection. Thus, K_{min} should lie between the poles of petrographic foliations S_1 and S_2, roughly in the middle of the southeast quadrant (Fig. 11). However, K_{min} for these samples is clearly in the "wrong" place, lying slightly north of the S_2 pole (Fig. 8, A). A specific explanation for this apparent anomaly requires further study and should take into account magnetic domain sizes in the oxides as well as the relative contributions from the ferromagnetic and paramagnetic phases of different ages.

Figure 12, A, shows a comparison between the average petrographic lineation of samples possessing the M_B and M_C magnetofabrics and the average K_{max} of those samples. Based on colinearity, here the petrographic lineation appears to strongly dictate the K_{max} direction. The attitude of this petrographic lineation plunges 36° toward 015°, which is also coincident with the intersection of the magnetic foliations of the M_B and M_C domains (Figs. 3, B; 9; and 12, A), suggesting that this petrographic lineation controlling K_{max} may be an intersection lineation. Further, Figure 12, A, shows that M_C and the fitted plane S_1 are nearly parallel. In the absence of the S_2 foliation of domain M_A the resulting magnetofabric is similar to M_C. Overall, kinematic indicators did not provide unequivocal results except for S_2, which consistently showed top-to-the-southeast kinematics (Fig. 10).

Visual inspection of hand specimens of strongly lineated samples and outcrops suggests that not all the mineral lineations are intersection lineations; some may be stretching lineations (Fig. 3, B) or fold-axis ("b"-type) lineations.

S_1 and S_2: S-C Fabric?

The sample displayed in Figure 10 shows a fabric that might be called an S-C fabric, indicating that the S_1 and S_2 foliations developed coevally and not in sequence as suggested by the subscript notation. If this is an S-C fabric, additional observations would support this interpretation: a lineation should be found on the C surface (S_2), and this lineation should be 90° from the inter-

→

Figure 10. (A) Example of an oriented "c" thin-section from the M_A magnetofabric population, with a stereonet showing the direction of cut (S_1 and S_2—foliations; Fol—U-stage measure foliation pole). The thin-section view is roughly down K_{max} (the axis of maximum susceptibility) toward azimuth 234. K_{int}—the axis of intermediate susceptibility; K_{min}—the axis of minimum susceptibility. (B) Foliation interpretation of A showing that S_2 is clearly younger than S_1 and the angle of intersection of the two foliations is 27°. The implied slip direction on S_2 is top to the southeast.

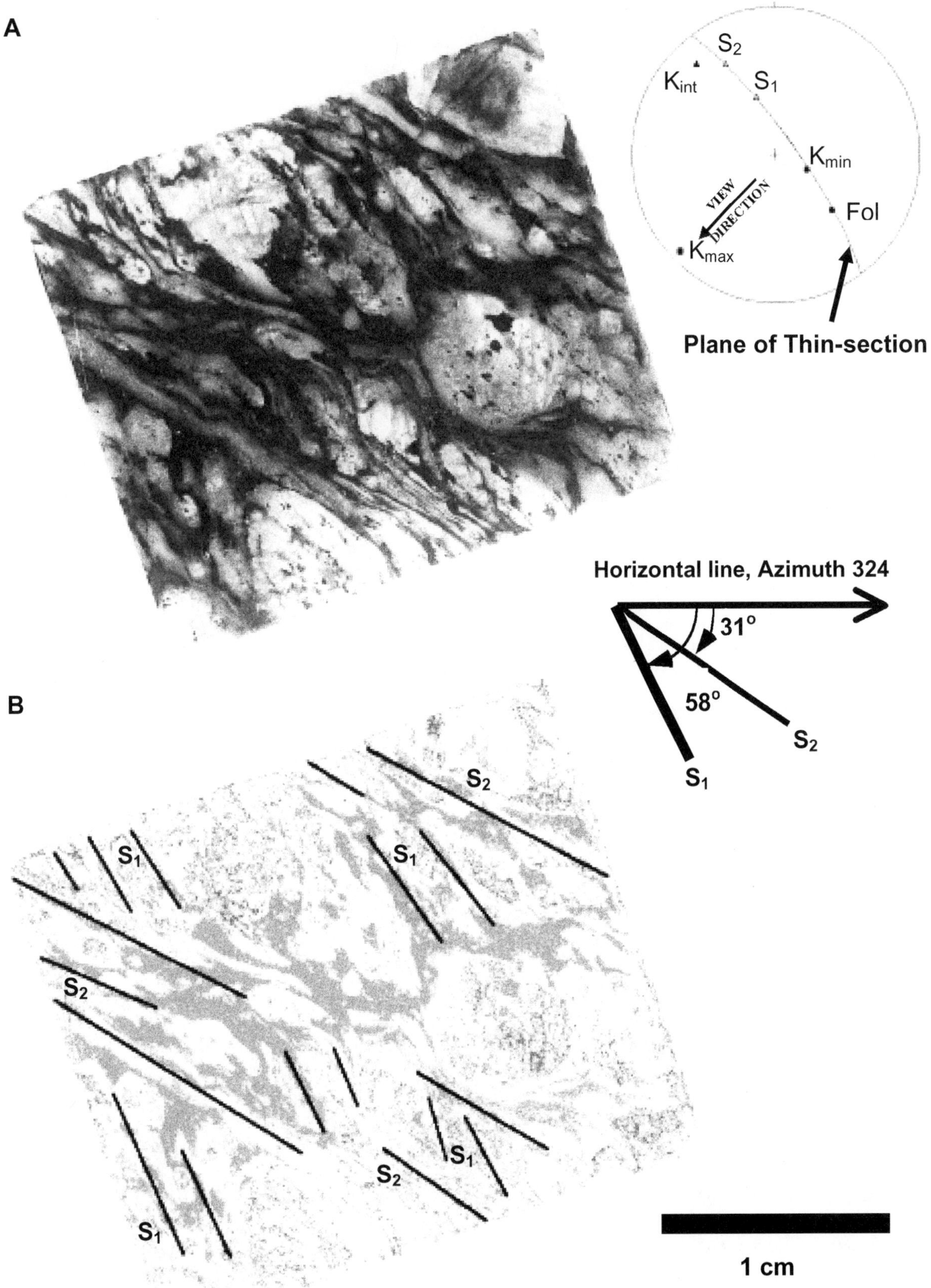
A
S2
Kint
S1
Kmin
VIEW DIRECTION
Fol
Kmax
Plane of Thin-section
Horizontal line, Azimuth 324
31°
58°
S2
S1
B
S2
S1
S1
S2
S1
S2
S1
1 cm

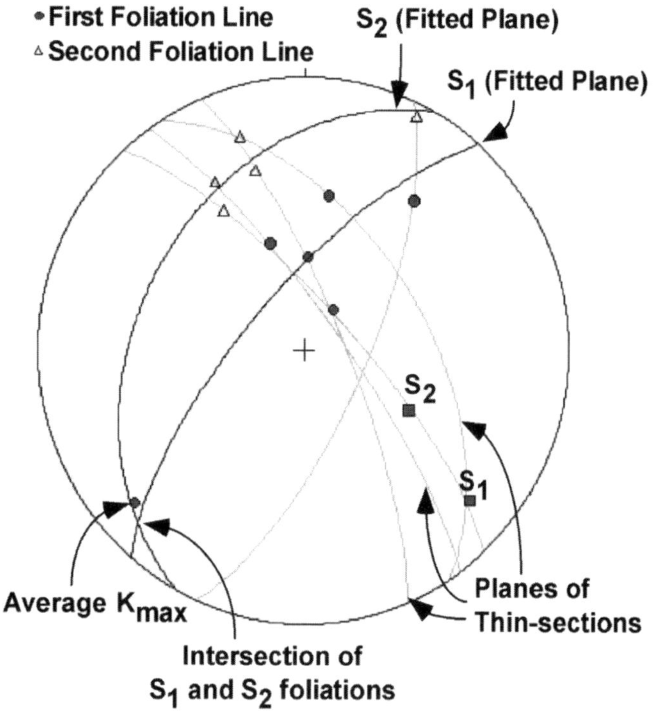

Figure 11. Stereonet showing the method of determining the orientation of foliations S_1 (221°, 72°NW) and S_2 (209°, 37°NW) by means of multiple data from oriented thin-sections of M_A specimens. The squares represent the poles of these fitted planes. The circles and triangles represent the two foliations (S_1 and S_2, respectively) as measured in the oriented "c" thin-sections. In these samples, the magnetolineation K_{max} closely corresponds to the intersection of the fitted S_1 and S_2 petrographic foliations.

section of the S (S_1) and C (S_2) surfaces. Figure 12, A shows that where a lineation is observed it is on the S surface ($M_C = S_1$) without a corresponding C surface. Furthermore, this lineation is ~120° away from the intersection of the S_1 and S_2 surfaces (Fig. 12, A), and our observations suggest that this is could be an intersection lineation and not strictly a transport-parallel lineation. Together this evidence leads us to conclude that the pair of surfaces producing magnetofabric M_A is not an S-C fabric.

DISCUSSION

Our principal observations on magnetofabrics and petrofabrics are summarized in Figure 12, A. We have shown that in the central Diana syenite there are three petrographic foliations with one petrographic lineation (IL_1, Fig. 12, B), which is associated with a K_{max} cluster, and one magnetic intersection lineation (IL_2, Fig. 12, B), which is not associated with any petrographic lineation. The foliations are (1) an east-northeast-striking, moderately northwest-dipping foliation (the petrographic foliation that produced M_B); (2) a northeast-striking, steeply northwest-dipping foliation (magnetofabric M_c and petrofabric S_1); and (3) a northeast-striking, moderately northwest-dipping foliation (petrofabric S_2). A comparison of these petrographic features with the results of Wiener (1983) shows good agreement. Figure 12, C summarizes Wiener's (1983) observations on the third-phase Jenny Creek anticline northeast of Harrisville.

Based on orientation, foliation 1, cited earlier, correlates with the axial plane of second-phase minor folds. Likewise, foliation 2, cited earlier, correlates with Wiener's minor third-phase folds. Furthermore, the major fold axial plane of the third phase of regional deformation is parallel to foliation 2 (Fig. 12, B and C).

Wiener describes the lineation associated with third-phase folds as an intersection lineation or mineral lineation. The fold axis of this major fold (and lineations on average) plunges 55° toward 010° (Wiener, 1983), which is similar to the average K_{max} magnetic lineation (36°, 015°) associated with the petrographic lineation in the study area. Thus, we correlate the lineation IL_1 with Wiener's (1983) third-phase fold axis (Fold Axis, Fig. 12, C).

Not fitting into Wiener's structural framework is our foliation 3 (S_2), cited earlier. This foliation crosscuts what is interpreted as Wiener's (1983) third-phase fold axial planar foliation (Fig. 10), but its orientation is not similar to that of Wiener's fourth-phase folds. Possibly this is a local variation, or a previously undescribed deformation phase. This latest foliation 3 is developed only at the northeast corner of the study area, so resolution of its relation to other structures would call for further work in the area to the east. We found no magnetic expression of any intersection lineation resulting from the intersection of foliation 1 and the latest surface, foliation 3 (Fig. 12, B). Finally, we do not have a satisfactory explanation for the ~40° offset between the M_A magnetofabric and the associated petrographic foliation (Fig. 8, A).

Overall, our results confirm, in part, Wiener's (1983) findings that mylonite foliation was a second-phase axial planar foliation, although the structural foliation pattern is clearly complex. This study has not investigated the numerous discrete postfoliation development structures, such as the brecciated Diana lineament (Fig. 2; see Hargraves, 1969), ultramylonite shear zones (Fig. 4, A; see Wiener, 1983; Lumino, 1987; Heyn, 1990), and a ductile fault zone (Figs. 2 and 4, B; see Geraghty et al., 1981).

Regional Significance

Wiener (1983) showed that the boundary between the metasediments of the Lowlands and the Diana syenite was affected by three folding events. The results of the work presented here are highly suggestive that the second and third folding events of Wiener (1983) can be traced at least to the southeast boundary of the Carthage-Colton mylonite zone south of Harrisville (Fig. 9). It is fairly clear that the fourth-phase folding identified by Wiener (see Wiener et al., 1984) extends across the Carthage-Colton mylonite zone without any interruption

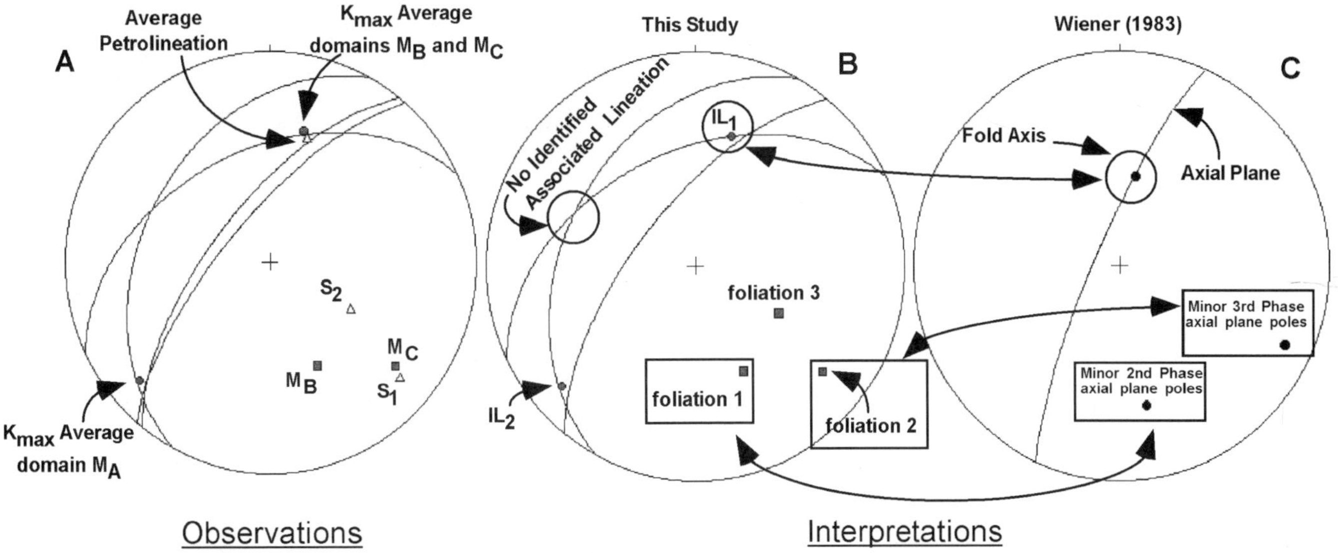

Figure 12. (A) Comparison of observed petrolineation and petrofoliations to anisotropy of magnetic susceptibility measurement results. K_{max}—the axis of maximum susceptibility; M_A, M_B, and M_C—magnetofabrics. (B) The foliation framework derived from this research. IL—intersection lineation; foliation 1 = 245°, 45°NW; foliation 2 = 219°, 67°NW; foliation 3 = 209°, 37°NW. (C) Summary of structural data from the Jenny Creek anticline, a third-phase major fold (modified from Wiener, 1983). The average axial plane as shown is 204°, 80°NW. The average minor fold axial plane poles of third- and second-phase folds are also shown. The average second-phase minor fold axial plane measures 259°, 56°NW. In this fold complex, lineations are parallel to third-phase fold axes and on average plunge 55° toward 010°. Double-headed arrows indicate features that are interpreted to have resulted from the same deformation phase, based on parallelism of structures (see text).

(Wiener et al., 1984). Many workers have suggested the folded nature of the Diana syenite previously (e.g., Buddington, 1963; Hargraves, 1969; Geraghty et al., 1981), but this research ties the observed foliation development to the context of regional folding and shows it is traceable across the Carthage-Colton mylonite zone in this area. This observation is of the highest importance when considering the deformational history of the Adirondack Highlands and Lowlands.

The intrusive age of the Diana syenite is 1155 ± 4 Ma (Grant et al., 1986). Wiener (1983) provides evidence that intrusion is coeval with the second identified regional folding event. This is a consistent pattern for AMCG suite intrusions throughout the Adirondack Highlands as well (Wiener et al., 1984). Therefore, the second phase of folding identified, at least in the Harrisville area, is ca. 1155 Ma; subsequent crosscutting structural surfaces are younger than this, and the first-phase folding event is older than 1155 Ma. This is consistent with the timing constraints of highlands folding provided by Kusky and Loring (2001).

First and second folding phases are thought to be part of continual deformation at peak metamorphic conditions (Wiener, 1983). Retrograde third-phase folding is thought to potentially significantly post-date the second-phase folding (Wiener, 1983). Possibly the mylonite texture that Lamb (1993) showed to be retrogressive in nature is associated with third-phase folding (our foliation 2, correlated to Wiener's third folding phase). Somewhat disturbing is the conclusion of Cartwright et al. (1993) that Diana syenite intrusion was shallow (<10 km). If this occurred at peak metamorphic conditions (Wiener, 1983), temperature gradients at 1155 Ma would have been unusually high, possibly 60°/km, or locally higher. Clearly this needs to be researched further.

Following the tectonic model of McLelland et al. (1996), and the subsequent modifications of Wasteneys et al. (1999) and McLelland et al. (2001), the evidence of this research suggests that first-phase folding must have been a result of Elzevirian orogen compression or must have taken place at least prior to AMCG intrusion. Second-phase folding was possibly Elzevirian as well, because it is thought to be part of continued deformation at peak metamorphic conditions (Wiener, 1983), suggesting that Elzevirian compression continued later than ca. 1160 Ma (Wasteneys et al., 1999). The remaining folding events, third- and fourth-phase folding plus localized fifth-phase folding (Wiener et al., 1984), can be attributed to the Ottawan orogen compression that followed AMCG intrusion.

This suggests that the majority of Lowlands deformation may have been associated with Ottawan compression. Two very important points should be noted: (1) the presence of Ottawan deformation within the Lowlands has not been reported, as Frontenac terrane deformation is thought to have occurred from ca. 1180–1160 Ma (Corfu and Easton, 1997; Carr et al., 2000); and (2) this deformation in the Lowlands occurred without a corresponding thermal event, contrary to deformation of this time in the Highlands (see Mezger et al., 1992). Wasteneys et al. (1999) emphasize continuity across the St. Lawrence River from the Adirondack Lowlands into the Frontenac terrane, and Corfu and Easton (1997) suggest that the Sharbot Lake terrane was continuous with the Frontenac terrane by 1156 Ma. The 1156 ± 2 Ma Maberly stock, which intrudes the Sharbot Lake terrane and is essentially the same age as the Diana syenite, is reported to be undeformed and unmetamorphosed (Corfu and Easton, 1997). This indicates that Ottawan deformation does not extend as far west as the Sharbot Lake terrane. Detailed structural studies comparing the Canadian Frontenac terrane to the Adirondack Lowlands are necessary to confirm this.

Second- and third-phase folds within the Highlands sweep into parallelism with the Carthage-Colton mylonite zone (Wiener et al., 1984; McLelland and Isachsen, 1986) in a manner that suggests at least some component of left-lateral movement for the Carthage-Colton mylonite zone at some time in its history, possibly coeval with these deformation phases or postdating them. The latest foliation found in this research dips to the northwest and has thrust kinematics consistent with the production of a left-lateral map offset. This suggestion of left lateral thrust kinematics for one period of Carthage-Colton mylonite zone motion is tenuous because kinematic indicators within the zone can be equivocal and curved fold axes could be explained in other ways.

Foliations and lineations appear more strongly developed somewhere southwest of the center of the study area, and fabrics are not as nearly as well developed in the northwest or the southeast portions of the Diana syenite. So the zone of maximum offset within the Carthage-Colton mylonite zone, if it is associated with folding in the research area, probably occurs near the middle of the Carthage-Colton mylonite zone, coincident with these stronger fabrics. This conclusion is based on the assumption that the Diana syenite possesses an intrusive northwest contact with the Lowlands metasediments and, in turn, is intruded by other granites of Highlands affinity; therefore, locations accommodating large terrane-scale offset are not likely located at either margin of the Diana syenite. However, major offset may be recorded by other structures entirely.

CONCLUSIONS

The analysis of petrofabric and magnetofabric patterns, supported by U-stage and oriented thin-section study, has revealed multiple generations of foliations in a central cross-section of the Diana syenite within the southwest Carthage-Colton mylonite zone. In field studies, the intersection of two foliations at an acute angle gives the mistaken impression of a single "lenticular" foliation. These were clearly distinguished by measuring AMS axial orientations and petrographic fabric orientations in core specimens oriented in a U-stage. Accurately oriented thin-sections allowed the precise measurement of the orientations of these foliations, the determination of their relative ages, and the sense of relative slip across them.

Though the exact nature of some of the foliations and lineations identified is still not entirely clear, significant conclusions can be drawn: (1) The axial planar foliation patterns observed in the Lowlands in the vicinity of Harrisville can be traced to at least the southeast Carthage-Colton mylonite zone boundary within the Diana syenite. (2) Deformation within the Adirondack Lowlands is not completely limited to Elzevirian deformation; at least two generations of folding have probably occurred as a result of Ottawan compression. The western limit of the Ottawan deformation is not known, but it probably does not extend as far as the western edge of the Frontenac terrane. (3) Carthage-Colton mylonite zone foliation patterns do not suggest a location for a structural discontinuity within this zone in this area. It is unclear whether these fabrics accommodated large offset of the Lowlands with respect to the Highlands.

ACKNOWLEDGMENTS

We thank Art Goldstein (Colgate University) for use of a KLY-2 Kappabridge. The stereographic analyses used the program Estereografica GR, v. 1.0, by E. Cristallini (University of Buenos Aires). GBB thanks Erik Kent for field assistance and advice during nearly every stage of this research and manuscript preparation. We appreciate critical reviews by J.T. Bursnall and an anonymous reviewer. This paper is based on GBB's M.A. thesis, supervised by WDM at the State University of New York at Binghamton.

REFERENCES CITED

Balsley, J.R., and Buddington, A.F., 1960, Magnetic susceptibility anisotropy and fabric of some Adirondack granites and orthogneisses: American Journal of Science, v. 258-A, p. 6–20.

Borradaile, G.J., 1988, Magnetic susceptibility, petrofabrics and strain: Tectonophysics, v. 156, p. 1–20.

Borradaile, G.J., and Henry, B., 1997, Tectonic applications of magnetic susceptibility and its anisotropy: Earth-Science Reviews, v. 42, no. 1–2, p. 49–93.

Buddington, A.F., 1939, Adirondack igneous rocks and their metamorphism: Boulder, Colorado, Geological Society of America Memoir 7, 354 p.

Buddington, A.F., 1963, Isograds and the role of H_2O in metamorphic facies of orthogneisses of the Northwest Adirondack area, New York: Geological Society of America Bulletin, v. 74, p. 1155–1182.

Carr, S.D., Easton, R.M., Jamieson, R.A., and Culshaw, N.G., 2000, Geologic transect across the Grenville orogen of Ontario and New York: Canadian Journal of Earth Science, v. 37, p. 193–216.

Cartwright, I., Valley, J.W., and Hazelwood, A., 1993, Resetting of oxybarometers and oxygen isotope ratios in granulite facies orthogneisses during cooling and shearing, Adirondack Mountains, New York: Contributions to Mineralogy and Petrology, v. 113, p. 208–225.

Corfu, F., and Easton, R.M., 1997, Sharbot Lake terrane and its relationships to Frontenac terrane, Central Metasedimentary Belt, Grenville Province: New insights from U-Pb geochronology: Canadian Journal of Earth Science, v. 34, p. 1239–1257.

Geraghty, E.P., Isachsen, Y.W., and Wright, S.F., 1981, Extent and character of the Carthage-Colton Mylonite Zone, northwest Adirondacks, New York: Washington, D.C., Nuclear Regulatory Commission NUREG/CR-1865, 83 p.

Graham J.W., 1954, Magnetic susceptibility anisotropy, an unexploited petrofabric element: Geological Society of America Bulletin, v. 65, no. 12, p. 1257–1258.

Grant, N., Lepak, R., Maher, T., Hudson, M., and Carl, J., 1986, Geochronological framework for the Grenville rocks of the Adirondack Mountains: Geological Society of America Abstracts with Programs, v. 18, no. 6, p. 620.

Hall, P. C., 1984, Some aspects of deformation fabrics along the Highland/Lowland Boundary, Northwest Adirondacks, New York State [M.S. thesis]: Albany, State University of New York at Albany, 124 p.

Hargraves, R.B., 1969, A contribution to the geology of the Diana Syenite gneiss complex, *in* Isachsen, Y.W., ed., Origin of anorthosite and related rocks: Albany, New York State Museum and Science Service Memoir 18, p. 343–356.

Heyn, T., 1990, Tectonites of the Northwest Adirondack Mountains, New York: Structural and metamorphic evolution [Ph.D. dissertation]: Ithaca, New York, Cornell University, 203 p.

Isachsen, Y.W., and Fisher, D.W., 1970, Geologic map of New York State, Adirondack sheet: Albany, New York State Museum Science Map and Chart Series 15.

Isachsen, Y.W., and Geraghty, E.P., 1986, The Carthage-Colton Mylonite Zone, a major ductile fault in the Grenville Province, *in* Aldrich, M.J., Jr., and Laughlin, A.W., eds., Proceedings of the sixth international conference on basement tectonics, Santa Fe, New Mexico, v. 6, p. 199–200.

Kodama, K.P., 1995, Magnetic fabrics: Review of Geophysics, American Geophysical Union, Washington, DC, Supplemental U.S. National Report to the IUGG [International Union of Geodesy and Geophysics], 1991–1994, p. 129–135.

Kusky, T. M., and Loring, D. P., 2001, Structural and U/Pb chronology of superimposed folds, Adirondack Mountains: Implications for the tectonic evolution of the Grenville Province: Journal of Geodynamics, v. 32, p. 395–418.

Lamb, W.M., 1993, Retrograde deformation within the Carthage-Colton Zone as recorded by fluid inclusions and feldspar compositions: Tectonic implications for the southern Grenville Province: Contributions to Mineralogy and Petrology, v. 114, p. 379–394.

Lumino, K.M., 1987, Deformation within the Diana Complex along the Carthage-Colton Mylonite Zone [M.S. thesis], Rochester, New York, University of Rochester, 104 p.

MacDonald, W.D., and Ellwood, B.B., 1987, Anisotropy of magnetic susceptibility: Sedimentologic, igneous, and structural-tectonic applications: Reviews of Geophysics, v. 25, p. 905–909.

MacDonald, W.D., and Ellwood, B.B., 1988, Magnetic fabric of peridotite with intersecting petrofabric surfaces, Tinaquillo, Venezuela: Physics of the Earth and Planetary Interiors, v. 51, p. 301–312.

McLelland J.M., and Isachsen, Y.W., 1986, Synthesis of geology of the Adirondack Mountains, New York, and their tectonic setting within the southwestern Grenville Province: St. John's, Newfoundland, Geological Association of Canada Special Paper 31, p. 75–94.

McLelland, J.M., Daly, J.S., and Chiarenzelli, J., 1993, Sm-Nd and U-Pb isotopic evidence of juvenile crust in the Adirondack Lowlands and implications for the evolution of the Adirondack Mts.: Journal of Geology, v. 101, p. 97–105.

McLelland J., Daly, J.S., and McLelland, J.M., 1996, The Grenville Orogenic Cycle (ca. 1350–1000): An Adirondack perspective: Tectonophysics, v. 265, p. 1–28.

McLelland J., Hamilton, M., Selleck, B., McLelland, J., Walker, D., and Orrell, S., 2001. Zircon U-Pb geochronology of the Ottawan Orogeny, Adirondack Highlands, New York: Regional and tectonic implications: Precambrian Research, v. 109, p. 39–72.

Mezger, K., van der Pluijm, B.A., Essene, E.J., and Halliday, A.N., 1992, The Carthage-Colton Mylonite Zone (Adirondack Mountains, New York): The site of a cryptic suture in the Grenville Orogen?: Journal of Geology, v. 100, p. 630–638.

Streepey, M.M., van der Pluijm, B.A., Essene, E.J., Hall, C.M., and Magloughlin, J.F., 2000, Late Proterozoic (ca. 930 Ma) extension in eastern Laurentia: Geological Society of America Bulletin, v. 112, p. 1522–1530.

Streepey, M.M., Johnson, E.L., Mezger, K., and van der Pluijm, B.A., 2001, Early history of the Carthage-Colton Shear Zone, Grenville Province, Northwest Adirondacks, New York (U.S.A.): Journal of Geology, v. 109, p. 479–492.

Tarling, D.H., and Hrouda, F., 1993, The magnetic anisotropy of rocks: New York, Chapman & Hall, 217 p.

Tomezzoli, R.N., MacDonald, W.D., and Tickyj, H., 2003, Composite magnetic fabrics and S-C structure in granitic gneiss of Cerro de los Viejos, La Pampa province, Argentina: Journal of Structural Geology, v. 25, no. 2, p. 159–169.

Valley, J.W., Bohlen, S.R., Essene, E.J., and Lamb, W., 1990, Metamorphism in the Adirondacks, II: The role of fluids: Journal of Petrology, v. 31, p. 555–596.

Wasteneys, H., McLelland, J., and Lumbers, S., 1999, Precise zircon geochronology in the Adirondack Lowlands and implications for revising plate-tectonic models of the Central Metasedimentary Belt and Adirondack Mountains, Grenville Province, Ontario and New York: Canadian Journal of Earth Science, v. 36, p. 967–984.

Weil, A.B., Van der Voo, R., Mac Niocaill, C., and Meert, J.G., 1998, The Proterozoic supercontinent Rodinia: Paleomagnetically derived reconstructions for 1100–800 Ma: Earth and Planetary Science Letters, v. 154, p. 13–24.

Wiener, R.W., 1977, Timing of emplacement and deformation of the Diana complex along the Adirondack highlands–northwest lowlands boundary: Geological Society of America Abstracts with Programs, v. 9, no. 3, p. 329–330.

Wiener, R.W., 1983, Adirondack highlands–northwest lowlands "boundary": A multiply folded intrusive contact with fold-associated mylonitization: Geological Society of America Bulletin, v. 94, p. 1081–1108.

Wiener, R.W., McLelland, J.M., Isachsen, Y.W., and Hall, L.M., 1984, Stratigraphy and structural geology of the Adirondack Mountains, New York: Review and synthesis, *in* Bartholomew, M.J., ed., The Grenville Event in the Appalachians and related topics: Boulder, Colorado, Geological Society of America Special Paper 194, p. 1–55.

Wynne-Edwards, H.R., 1972, The Grenville Province, *in* Price, T.A., and Douglas, R.J.W., eds., Variations in tectonic styles in Canada: St. John's, Newfoundland, Geological Association of Canada Special Paper 11, p. 263–334.

Manuscript Accepted by the Society August 25, 2003

Printed in the USA

Geological Society of America
Memoir 197
2004

Slow cooling and apparent tilting of the Adirondack Lowlands, Grenville Province, New York, based on $^{40}Ar/^{39}Ar$ ages

Peter S. Dahl*
Mark E. Pomfrey
Department of Geology, Kent State University, Kent, Ohio 44242, USA
Kenneth A. Foland
Department of Geological Sciences, Ohio State University, Columbus, Ohio 43210, USA

ABSTRACT

$^{40}Ar/^{39}Ar$ incremental-heating measurements of Grenville metamorphic rocks from the Adirondack Lowlands, New York, constrain a ca. 1100–900 Ma history of uplift, exhumation, and cooling in the eastern Frontenac terrane following the ca. 1200–1170 Ma Elzevirian orogeny and the ca. 1080–1070 Ma peak of the subsequent Ottawan orogeny. Four hornblende-biotite pairs yield a time-integrated post-Ottawan cooling rate of 1.6 ± 0.2 °C/m.y. and an inferred exhumation rate of 0.06 ± 0.02 km/m.y. between ca. 1100 Ma (oldest hornblende, northwest Lowlands) and ca. 900 Ma (youngest biotite, southeast Lowlands), based upon closure temperatures of 500 and 300 °C and a 30 ± 10 °C/km geotherm. Moreover, these and ~20 additional cooling ages for hornblende and biotite define parallel northwest-to-southeast younging trends of ~3 m.y./km across ~45 km between the St. Lawrence River and the Carthage-Colton shear zone, which separates the Lowlands and the Adirondack Highlands domains. Two end-member models potentially explain the age trends: (1) ca. 1100–900 Ma cooling of the Lowlands through subhorizontal isotherms, followed by ca. 900–520 Ma regional tilting (~9° ± 3°NW); or (2) ca. 1100–900 Ma cooling of the Lowlands against a warm Highlands footwall, through gently northwest-dipping inclined isotherms that flattened out by ca. 900 Ma. On balance, the regional tilting model is the favored explanation of observed trends in the hornblende and biotite ages. Differential uplift during regional extension most likely tilted the Lowlands, and the ca. 900–520 Ma time frame suggests that breakup of the supercontinent Rodinia was ultimately responsible for these tectonic events and the resultant trends in cooling age.

Keywords: argon geochronology, hornblende, biotite, Adirondack Mountains

INTRODUCTION

The Grenville Province of the Canadian Shield (Fig. 1) is a collage of Precambrian crustal terranes assembled and welded to southeast Laurentia during polyphase orogeny that occurred between ca. 1680 and ca. 850 Ma (Rivers, 1997). This province exhibits continuous exposures of highly deformed and metamorphosed rocks representative of a deeply eroded orogen, thereby also providing an excellent opportunity to study orogenic and postorogenic processes in ancient mountain systems (Cosca et al., 1995; Streepey et al., 2001a). This paper characterizes and interprets the ca. 1100–900 Ma record of these

*E-mail: pdahl@geology.kent.edu.

Dahl, P.S., Pomfrey, M.E., and Foland, K.A., 2004, Slow cooling and apparent tilting of the Adirondack Lowlands, Grenville Province, New York, based on $^{40}Ar/^{39}Ar$ ages, *in* Tollo, R.P., Corriveau, L., McLelland, J., and Bartholomew, M.J., eds., Proterozoic tectonic evolution of the Grenville orogen in North America: Boulder, Colorado, Geological Society of America Memoir 197, p. 299–323. For permission to copy, contact editing@geosociety.org.

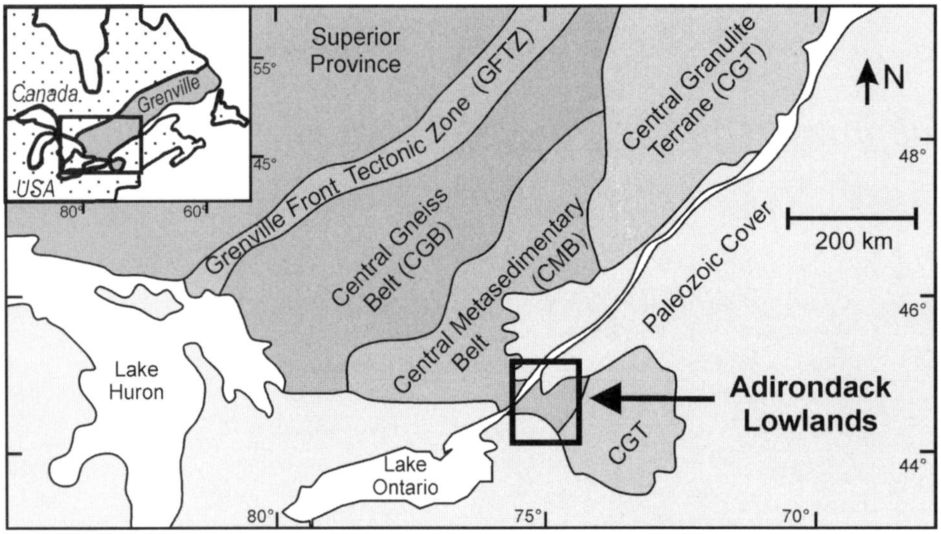

Figure 1. Generalized geologic map showing Mesoproterozoic lithotectonic subdivisions in the western Grenville Province (after Mezger et al., 1991).

processes in the western Adirondack Mountains (New York, USA), a southern outlier of the Grenville Province. The occurrence of Grenville-aged rocks in a belt thousands of kilometers long from Norway to Mexico conveys their global scale and potential relevance in reconstructing the Precambrian assembly and breakup of the supercontinent Rodinia (Ruiz et al., 1988; Cosca et al., 1998; Burrett and Berry, 2000).

The Adirondacks are subdivided into two lithotectonic domains separated by a midcrustal shear zone (Fig. 1). The rugged Highlands domain is part of the Central Granulite terrane (Fig. 1) and is composed of granulite-facies orthogneisses dominated by anorthosite massifs. The relatively flat Lowlands domain comprises the eastern Frontenac terrane of the Central Metasedimentary Belt and is dominated by upper-amphibolite-facies metasedimentary rocks (mainly marbles) and orthogneisses (Figs. 1 and 2). Building upon classic geological studies (e.g., Buddington, 1939, 1963; Engel and Engel, 1958, 1963; deWaard, 1965, 1970), recent workers have shown that metamorphic rocks in the western Adirondacks preserve a complex record of collisional thermotectonism between ca. 1200 and 1040 Ma and exhumation to the midcrust between ca. 1050 and 900 Ma (e.g., Bohlen et al., 1985; Powers and Bohlen, 1985; Bohlen, 1987; McLelland et al., 1988, 1993, 1996, 2001; Valley et al., 1990; Daly and McLelland, 1991; Mezger et al., 1991, 1992, 1993; Kitchen and Valley, 1995; Wasteneys et al., 1999; Streepey et al., 2000, 2001a). Final exhumation of these rocks and cooling to surface temperature occurred between ca. 900 and 520 Ma, as evidenced by nonconformable deposition of the Middle Cambrian Potsdam sandstone (Isachsen and Fisher, 1970).

The boundaries that separate known Grenville terranes from each other are generally southeast-trending ductile faults that were multiply active during the Grenville orogeny (Easton, 1989; Davidson, 1998). The Adirondack Lowlands and Highlands domains are separated by the enigmatic Carthage-Colton shear zone (Fig. 1), a northwest-dipping midcrustal zone of high ductile strain (Isachsen and Geraghty, 1986). The Carthage-Colton shear zone appears to extend into Canada as the Labelle shear zone (Martignole and Reynolds, 1997), and collectively these ductile faults define the Central Metasedimentary Belt–Central Granulite terrane boundary. Conflicting shear-sense indicators convey a complex tectonic evolution for the Carthage-Colton shear zone (e.g., Wasteneys et al., 1999; Streepey et al., 2001a, and references therein), and various models for Central Metasedimentary Belt–Central Granulite terrane juxtaposition have been proposed. Some workers have considered the Carthage-Colton shear zone to be a composite reverse or normal fault that first accommodated Central Metasedimentary Belt–Central Granulite terrane collisional convergence at ca. 1200–1170 Ma (Elzevirian orogeny) and then syn-post-Ottawan collapse of overthickened Adirondack crust to exhume the Highlands footwall at ca. 1050–1040 Ma (Isachsen and Geraghty, 1986; Heyn, 1990; McLelland et al., 1993, 1996, 2001; Corrigan and van Breemen, 1997; Martignole and Reynolds, 1997; Wasteneys et al., 1999). Others have considered the Carthage-Colton shear zone to be a ca. 1050–1040 Ma transpressional suture zone that welded the Central Metasedimentary Belt and the Central Granulite terrane during the Ottawan convergent orogeny, and then accommodated extensional normal faulting at ca. 1040 and ca. 950–930 Ma (e.g., Streepey et al., 2001a).

Diachroneity in the regional geobarometry (i.e., Elzevirian pressures in the Lowlands and Ottawan pressures in the Highlands; McLelland et al., 1996, 2001) precludes the use of metamorphic pressures to constrain juxtaposition history of these domains between 1150 and 520 Ma. Instead, of necessity, such constraints have largely relied upon comparative geochronology of the Lowlands, the Highlands, and the Carthage-Colton shear zone (Mezger et al., 1991, 1992, 1993; Streepey et al., 2000, 2001a).

This paper presents thirty-nine new $^{40}Ar/^{39}Ar$ cooling ages of hornblende and biotite separated from high-grade metamor-

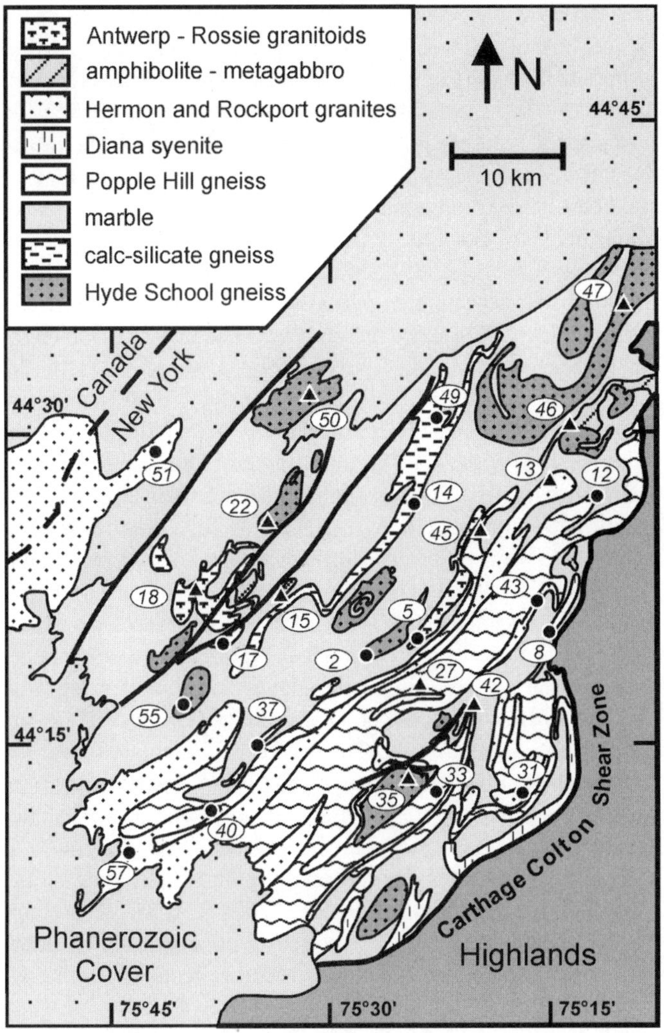

Figure 2. Generalized geologic map of the Mesoproterozoic Adirondack Lowlands domain of the Frontenac terrane. Sample numbers are shown in ovals; prefixes (see Table 1) are omitted. Triangles indicate sample locations from which both hornblende and biotite were analyzed; circles indicate locations from which only biotite was analyzed. Map simplified from Isachsen and Fisher (1970) and Carl and deLorraine (1997).

phic rocks sampled across the Adirondack Lowlands. By extending $^{40}Ar/^{39}Ar$ mineral-age coverage northwestward across the entire Lowlands domain, from the Carthage-Colton shear zone to nearly the St. Lawrence River, these new data provide a syn-post-Ottawan record of differential uplift, exhumation, and cooling in this part of the southern Grenville Province between ca. 1100 and 900 Ma. Moreover, by extending the ~500–300 °C thermochronological record across the Lowlands domain (Fig. 1), the new data serve to narrow the previous ~100 km–wide gap in $^{40}Ar/^{39}Ar$ cooling ages between the eastern Frontenac terrane and the western part of the Central Granulite terrane (Highlands). Merging of $^{40}Ar/^{39}Ar$ data sets (e.g., Cosca et al., 1991, 1992; Streepey et al., 2000, 2001a; this study) offers a more detailed picture of the syn-post-Ottawan history of the Frontenac terrane, the Lowlands domain, and the Lowlands-Highlands boundary zone (Carthage-Colton shear zone) than could emerge from a single data set alone. Thus, this paper summarizes our preliminary reports (Pomfrey et al., 1998a, 1998b; Dahl et al., 2001), which first documented mineral-age patterns in the southern Grenville province and interpreted them in the context of both regional and global tectonics (Rodinia).

GEOLOGIC SETTING AND PREVIOUS GEOCHRONOLOGY

Episodic contractional orogeny culminating in delamination of overthickened crust is considered to have been a major tectonomagmatic theme in the Adirondacks between ca. 1220 and 1050 Ma; these events are thought to have triggered crustal magmatism and extension, with associated crustal buoyancy promoting regional uplift and erosion-controlled exhumation (see McLelland et al., 2001). The Adirondacks have been affected by three major tectonomagmatic events between ca. 1200 and ca. 1000 Ma: (1) the Elzevirian orogeny, which ended by ca. 1170 Ma; (2) the ca. 1175–1130 Ma anorthosite-mangerite-charnockite-granite (AMCG) magmatic event; and, (3) the ca. 1090–1035 Ma Ottawan orogeny (McLelland et al., 1996; Rivers, 1997; Wasteneys et al., 1999). The Elzevirian orogeny was an accretional-collisional event in which outboard volcanic island arcs are thought to have docked with one another and eventually with the southeast margin of Laurentia by ca. 1170 Ma (McLelland et al., 1996; Wasteneys et al., 1999). In the Lowlands, this event ended with syntectonic intrusion of the ca. 1185 Ma Antwerp granite. Crustal thickening caused by Elzevirian terrane accretion culminated in delamination of mantle lithosphere, which triggered widespread extension and AMCG magmatism throughout the Adirondacks (Corrigan and Hanmer, 1997). In the Lowlands, AMCG magmatism was dominated by voluminous granitoid intrusion at ca. 1172 Ma (the Hyde School gneiss and Rockport granite) and ca. 1155–1150 Ma (the Diana syenite and Hermon granite; Fig. 2) into mid- to lower-crustal metasedimentary rocks (Grant et al., 1986; McLelland et al., 1996; Wasteneys and McLelland, 1999). Synchronous mylonitization also occurred locally in the Lowlands (Hudson, 1996).

Regional metamorphism associated with the Elzevirian orogeny culminated during AMCG magmatism, as evidenced by the synchronous (i.e., ca. 1170–1150 Ma) growth of garnet, monazite, and titanite in Lowlands crystalline rocks ($^{207}Pb/^{206}Pb$ ages; Mezger et al., 1991, 1992, 1993). Calcite-graphite carbon-isotope geothermometry (Kitchen and Valley, 1995) of widespread Lowlands marbles indicates that regional-metamorphic temperatures reached 640 ± 30 °C in the central Lowlands and 675 ± 25 °C in the western and eastern Lowlands (Fig. 3). Although these temperature estimates overlap within error, the metamorphic grade is indeed higher in the easternmost Lowlands based upon the north-northeast–trending orthopyroxene isograd (Hoffman, 1982), which provisionally delineates the

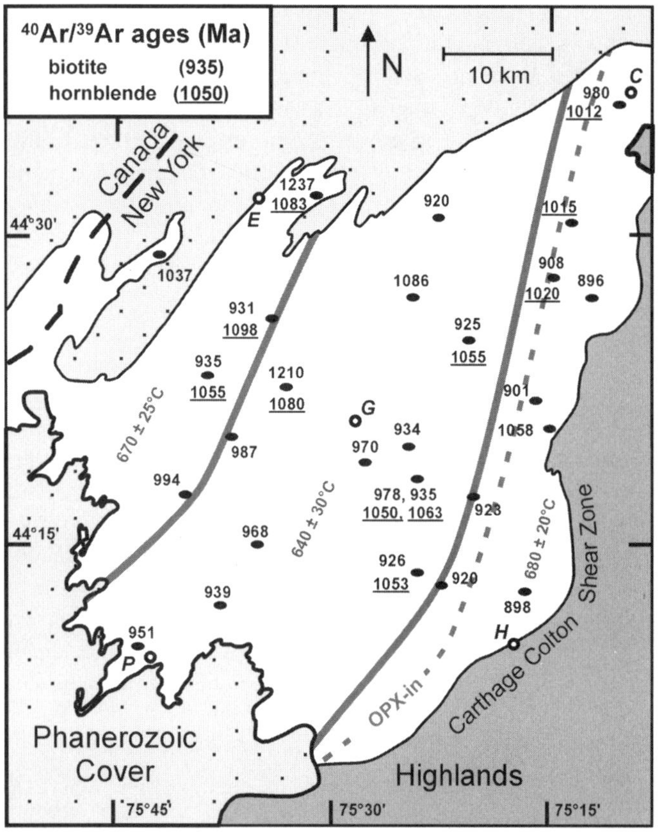

Figure 3. Distributions of $^{40}Ar/^{39}Ar$ ages (this study), regional-metamorphic temperatures (solid lines, after Kitchen and Valley, 1995) and the orthopyroxene isograd (dashed line, after Hoffman, 1982) in the Adirondack Lowlands. Sample locations are labeled with $^{40}Ar/^{39}Ar$ biotite dates (normal font) and/or $^{40}Ar/^{39}Ar$ hornblende dates (underlined). For original sample numbers, see Figure 2 and Table 1. Town abbreviations: C—Colton; E—Edwardsville; G—Gouverneur; H—Harrisville; and P—Philadelphia.

amphibolite-to-granulite-facies transition (Fig. 3). Hornblende geothermometry of amphibole-bearing gneisses just west of this isograd yields metamorphic temperatures ranging from 620 to 680 °C, without significant change along strike (Liogys and Jenkins, 2000). The 1153 Ma intrusion of Diana syenite in the easternmost Lowlands (Fig. 2) locally superimposed ~850 °C contact-metamorphic assemblages on older regional-metamorphic rocks (Powers and Bohlen, 1985). Peak metamorphic pressures recorded in exposed Lowlands rocks are ~6.5 ± 0.5 kilobars (northwest) and ~7.0 ± 0.5 kilobars (southeast) (Bohlen et al., 1985; Powers and Bohlen, 1985), consistent with paleodepths of ~23 ± 2 km and ~25 ± 2 km, respectively. Multiple episodes of folding (predominantly northeast-southwest–trending with northwest-dipping axial planes) occurred in the Lowlands during the Elzevirian-AMCG events. Regional thermobarometric trends (Fig. 3) parallel these regional lithotectonic trends (Fig. 2), suggesting that both sets of northeast-southwest trends originated within the same ca. 1170–1150 Ma time frame, that is, prior to ca. 1100–520 Ma postorogenic cooling.

Between 1103 and 1093 Ma, renewed lithospheric delamination is thought to have triggered another phase of AMCG magmatism in the Highlands, but not in the Lowlands (McLelland et al., 2001), possibly accompanied by rifting and extension along the Carthage-Colton shear zone (Mezger et al., 1992). However, any rifting and extension were abruptly terminated by the Ottawan orogeny, a major collisional event most likely associated with the docking of Amazonia east of the Adirondacks (Hoffman, 1988, 1989) during the assembly of Rodinia. Ottawan thermotectonism pervasively overprinted older, Elzevirian and AMCG-related, assemblages and fabrics in the Highlands (Mezger et al., 1991, 1992, 1993). During this event, the rocks now exposed here were tectonically buried to ~25 km (~7.0–7.5 kilobars) and metamorphosed to granulite-facies conditions (~700–750 °C). The time frame of Ottawan thermotectonism in the Highlands is bracketed between ca. 1090 Ma (deformation of the 1103–1093 Ma Hawkeye granite) and ca. 1047 Ma (the age of the undeformed member of the Lyon Mountain granitoid), with the metamorphic peak in the western Highlands occurring at ca. 1080–1070 Ma (McLelland et al., 2001).

The absence of ca. 1090–1050 Ma U-Pb ages for zircon, monazite, and titanite in the Lowlands suggests that Ottawan metamorphism did not occur here (e.g., Mezger et al., 1991, 1992, 1993; Streepey et al., 2001a). Following this assumption, the fossil thermobarometry (675 °C; 6.0–6.5 kilobars, 21–23 km) inferred from Lowlands mineral subassemblages (calcite-graphite and garnet-pyroxene, respectively) reflects mid- to lower-crustal metamorphic conditions that rocks now at the surface experienced during the ca. 1200–1150 Ma Elzevirian-AMCG tectonomagmatic events. In contrast, the thermobarometry of western Highlands rocks reflects younger conditions indicative of the 1090–1035 Ma Ottawan orogeny (McLelland et al., 1996, 2001), which overprinted 1200–1150 Ma assemblages there (metamorphic pressure unknown).

The assumption that there was no Ottawan metamorphism in the Lowlands is fundamental to previous models for Lowlands-Highlands juxtaposition and thermotectonic evolution of the Carthage-Colton shear zone. Thus, the Lowlands have variably been interpreted as midcrustal cover rocks of a ca. 1050 Ma core complex, which were downdropped to the northwest along the Carthage-Colton shear zone during extensional exhumation of the adjacent Highlands (e.g., Isachsen and Geraghty, 1986; Heyn, 1990; McLelland et al., 1993, 1996, 2001; Corrigan and Hanmer, 1997). Alternatively, Streepey et al. (2001a) have inferred an episode of oblique transpression along the Carthage-Colton shear zone at ca. 1050–1040 Ma, which positioned the Lowlands atop the western Highlands. They further postulate that extensional normal faulting at ca. 1040 and ca. 945–940 Ma then juxtaposed the Lowlands hangingwall into its present position relative to the Highlands footwall, and that the adjacent domains were thermally equilibrated by ca. 920 Ma. Other studies have begun to address the younger ca. 900–520 Ma thermotectonic history of the western Adirondacks region. For example, Heizler and Harrison (1998) have inferred a rifting event in the westernmost Adirondacks at ca. 730 Ma.

SAMPLES AND METHODS

All rock samples were collected from outcrops that exhibited only the prevailing northeast-southwest–trending deformational fabric of presumed Elzevirian age (Fig. 2). Thin sections of all samples were examined by standard petrographic techniques in order to identify any properties of minerals (grain size, color, alteration) that could be related to their inferred ^{40}Ar/^{39}Ar ages or to regional age trends. Wide ranges among all these parameters were observed from sample to sample, as summarized in Table 1, but none that ultimately could be related to the ^{40}Ar/^{39}Ar age or regional trends thereof. Moreover, within any given sample, only one population of hornblende and/or biotite was observed petrographically, suggesting that the dated mineral separates represented a single Elzevirian growth generation in all cases.

Analyses of major and minor elements were performed on all dated minerals using the Cameca SX50 electron microprobe located in the Department of Geological Sciences, Indiana University. The analytical procedures used have been described previously (Pouchou and Pichoir, 1984; Dahl and Dorais, 1996). Representative electron microprobe analyses of biotites and hornblendes are presented in Table 2. Wide ranges

TABLE 1. SUMMARY OF PETROGRAPHIC DATA FOR HORNBLENDES AND BIOTITES, ADIRONDACK LOWLANDS

		Grain size (microns)			Alteration phases of:			
Sample	Formation	Length	Width	Color	Hornblende	Feldspar	Latitude (deg min)	Longitude (deg min)
Hornblende								
AL97-13b	Hermon granite	113 ± 12	40 ± 6	olv	sl Bio	sl sericite	44 27.142	75 15.273
AL97-15a	Amphibolite-metagabbro	1080 ± 130	862 ± 208	olv	mod Bio	sl sericite	44 22.351	75 33.293
AL97-18a	Antwerp granite	97 ± 24	57 ± 10	brn	sl Bio	none	44 23.214	75 38.866
AL97-22b	Hyde School gneiss	169 ± 39	53 ± 15	grn	none	none	44 25.012	75 34.003
AL97-27c	Popple Hill gneiss	99 ± 21	34 ± 7	olv	none	none	44 17.341	75 23.823
AL97-27d	Popple Hill gneiss	96 ± 27	45 ± 12	olv	none	none	44 17.341	75 23.823
AL97-35	Hyde School gneiss	110 ± 16	58 ± 13	brn	mod Bio	none	44 13.581	75 23.110
AL98-42	Hermon granite	70 ± 11	22 ± 6	olv	mod Bio	mod sericite	44 17.052	75 19.922
AL98-45	Antwerp granite	123 ± 41	74 ± 7	grn	none	sl sericite	44 24.878	75 20.338
AL98-46	Amphibolite-metagabbro	93 ± 16	52 ± 12	olv	sl Bio	sl sericite	44 31.289	75 12.522
AL98-47b	Hyde School gneiss	124 ± 22	49 ± 4	olv	none	none	44 35.124	75 10.803
AL98-50	Hyde School gneiss	136 ± 30	58 ± 17	olv	mod Bio	sl sericite	44 31.636	75 31.296
Biotite		Diameter	Thickness		Biotite	Feldspar		
AL97-2a	Antwerp granite	77 ± 44	20 ± 20	red-brn	mod Chl, & Qtz-Fsp	mod sericite	44 18.641	75 27.176
AL97-5b	Antwerp granite	67 ± 26	19 ± 11	brn	none	sl sericite	44 19.076	75 25.409
AL97-8	Diana syenite	74 ± 14	24 ± 12	grn	hvy Chl	mod sericite	44 18.933	75 14.673
AL97-12a	Popple Hill gneiss	137 ± 32	47 ± 28	red	mod Qtz-Fsp	mod sericite	44 26.666	75 11.295
AL97-13a	Hermon granite	50 ± 23	12 ± 4	brn	none	sl sericite	44 27.142	75 15.273
AL97-14a	Calc-silicate gneiss	52 ± 21	10 ± 6	brn	none	none	44 25.840	75 23.820
AL97-15a	Amphibolite	60 ± 12	14 ± 6	brn	sl Qtz-Fsp	sl sericite	44 22.351	75 33.293
AL97-17c	Calc-silicate gneiss	39 ± 14	15 ± 6	red-brn	none	sl sericite	44 19.755	75 38.105
AL97-18a	Antwerp granite	74 ± 22	29 ± 14	red-brn	sl Qtz-Fsp	none	44 23.214	75 38.866
AL97-22b	Hyde School gneiss	69 ± 20	12 ± 5	brn	mod Qtz-Fsp	none	44 25.012	75 34.003
AL97-27c	Popple Hill gneiss	43 ± 20	9 ± 4	brn	none	none	44 17.341	75 23.823
AL97-27d	Popple Hill gneiss	55 ± 18	12 ± 6	brn	none	none	44 17.341	75 23.823
AL97-31	Hermon granite	55 ± 17	30 ± 20	brn	mod Chl	mod sericite	44 11.867	75 16.928
AL97-33	Popple Hill gneiss	52 ± 18	19 ± 6	red-brn	none	sl sericite	44 13.155	75 22.560
AL97-35	Hyde School gneiss	31 ± 13	14 ± 6	red-brn	none	none	44 13.581	75 23.110
AL97-37	Antwerp granite	69 ± 25	21 ± 12	red	mod Chl	sl sericite	44 14.965	75 35.371
AL97-40a	Popple Hill gneiss	58 ± 20	14 ± 5	red-brn	none	hvy sericite	44 11.722	75 37.462
AL98-42	Hermon granite	24 ± 7	6 ± 1	brn	none	mod sericite	44 17.052	75 19.922
AL98-43	Hermon granite	45 ± 18	14 ± 5	brn	none	mod sericite	44 20.967	75 16.334
AL98-45	Antwerp granite	85 ± 18	22 ± 7	red-brn	none	sl sericite	44 24.878	75 20.338
AL98-46	Hyde School gneiss	51 ± 13	11 ± 3	brn	sl Chl	sl sericite	44 31.289	75 12.522
AL98-47b	Hyde School gneiss	65 ± 15	12 ± 5	brn	none	none	44 35.124	75 10.803
AL98-49	Calc-silicate gneiss	37 ± 8	10 ± 2	brn	none	sl sericite	44 30.720	75 21.950
AL98-50	Hyde School gneiss	68 ± 23	20 ± 7	brn	none	sl sericite	44 31.636	75 31.296
AL98-51	Rockport granite	40 ± 8	8 ± 3	brn	none	none	44 27.642	75 42.646
AL98-55	Hyde School gneiss	55 ± 20	14 ± 4	red	sl Qtz-Fsp	hvy sericite	44 17.437	75 40.785
AL98-57	Hermon granite	38 ± 14	11 ± 4	grn-brn	none	mod sericite	44 09.628	75 42.150

Note: Size measurements: biotite (10), hornblende (5). Transmitted light colors: olv—olive; brn—brown; grn—green. Alteration: sl—slight (<10%); mod—moderate (~30%); hvy—heavy (>60%). Minerals: Bio—biotite; Chl—chlorite; Fsp—feldspar; Qtz—quartz. Sample locations in Fig. 2.

TABLE 2. REPRESENTATIVE ELECTRON MICROBE ANALYSES OF HORNBLENDE AND BIOTITE FROM THE ADIRONDACK LOWLANDS, NEW YORK

Sample	13b	15a	18a	22b	27c	27d	35	42	45	46	47b	50	2a	5b	8	12a	13b	14a	15a	17c
Wt%	Hbl	Hbl	Hbl	Hbl	Hbl	Hbl	Hbl	Hbl	Hbl	Hbl	Hbl	Hbl	Bio	Bio	Bio	Bio	Bio	Bio	Bio	Bio
SiO_2	42.72	39.19	45.06	43.39	40.80	40.93	41.42	43.83	42.36	41.91	41.87	39.59	36.04	35.98	39.66	36.75	36.91	37.81	35.34	35.79
Al_2O_3	10.80	13.07	10.32	11.34	13.35	12.70	11.34	9.27	11.67	11.26	12.20	14.14	14.35	13.78	11.69	17.16	14.82	14.88	15.67	15.97
FeO	17.21	18.86	13.75	14.42	19.67	20.02	20.38	15.98	16.36	18.36	16.84	17.57	23.62	23.53	15.26	13.15	17.81	16.28	17.69	18.31
MgO	10.87	9.62	12.37	12.53	7.99	7.67	8.09	11.41	10.60	9.45	10.14	9.06	9.67	9.07	15.87	14.24	13.26	14.85	14.10	11.53
TiO_2	1.21	1.21	1.64	1.26	0.77	1.14	2.14	1.35	1.39	1.20	1.68	1.82	3.42	4.53	1.92	4.56	3.32	2.40	2.57	4.55
MnO	0.56	0.42	0.25	0.37	0.46	0.37	0.21	0.56	0.33	0.43	0.36	0.30	0.18	0.16	0.76	0.03	0.38	0.12	0.11	0.19
Cr_2O_3	0.00	0.00	0.08	0.11	0.00	0.04	0.05	n.d.	n.d.	n.d.	n.d.	n.d.	0.01	0.00	0.01	0.05	0.03	0.03	n.d.	0.07
CaO	11.78	11.48	12.47	11.80	12.43	12.53	11.13	11.38	11.63	11.69	11.68	11.94	0.02	0.01	0.00	0.00	0.01	0.01	0.02	0.00
Na_2O	1.48	2.15	1.45	1.62	1.39	1.32	1.79	1.75	1.34	1.41	1.74	1.51	0.06	0.09	0.05	0.00	0.01	0.00	na	0.00
BaO	0.00	0.00	0.00	0.04	0.00	0.00	0.15	n.d.	n.d.	n.d.	n.d.	n.d.	0.17	0.16	0.15	0.09	0.15	0.02	0.22	0.04
K_2O	1.40	2.10	1.04	1.34	1.10	1.28	1.57	1.31	1.64	1.56	1.41	2.69	9.28	9.46	9.97	9.73	9.44	9.88	9.97	9.85
F	0.53	0.53	0.42	0.35	0.25	0.17	0.56	0.57	0.35	0.31	0.34	0.74	0.98	1.15	2.76	0.85	0.88	0.57	0.83	0.60
Cl	0.15	0.23	0.07	0.05	0.03	0.00	0.12	0.42	0.06	0.39	0.05	0.02	0.15	0.52	0.34	0.14	0.16	0.65	0.09	0.08
O = F	–0.22	–0.23	–0.18	–0.15	–0.11	–0.07	–0.23	–0.24	–0.15	–0.13	–0.14	–0.31	–0.41	–0.49	–1.16	–0.36	–0.37	–0.24	–0.35	–0.25
O = Cl	–0.04	–0.05	–0.02	–0.01	–0.01	0.00	–0.03	–0.10	–0.01	–0.09	–0.01	–0.01	–0.03	–0.12	–0.08	–0.03	–0.04	–0.15	–0.02	–0.02
Total	98.44	98.58	98.71	98.45	98.14	98.10	98.70	97.48	97.57	97.75	98.16	99.07	97.48	97.83	97.20	96.37	96.80	97.10	96.24	96.72
Ions per 23 (Hbl) or 22 (Bio) O:																				
Si	6.449	6.030	6.623	6.440	6.236	6.272	6.343	6.649	6.409	6.417	6.316	6.017	5.544	5.547	5.957	5.415	5.610	5.630	5.372	5.403
Al(IV)	1.551	1.970	1.377	1.560	1.764	1.728	1.657	1.351	1.591	1.583	1.684	1.983	2.456	2.453	2.043	2.585	2.390	2.370	2.628	2.597
Al(VI)	0.370	0.400	0.410	0.424	0.641	0.566	0.391	0.306	0.490	0.449	0.486	0.549	0.144	0.051	0.027	0.395	0.265	0.241	0.179	0.245
Ti	0.137	0.140	0.181	0.140	0.089	0.132	0.247	0.154	0.158	0.139	0.191	0.208	0.395	0.525	0.217	0.505	0.379	0.269	0.294	0.517
Mg	2.446	2.205	2.711	2.771	1.820	1.753	1.846	2.578	2.391	2.155	2.281	2.051	2.216	2.084	3.553	3.128	3.003	3.296	3.194	2.593
Fe	2.173	2.427	1.690	1.790	2.514	2.566	2.611	2.027	2.070	2.351	2.124	2.233	3.039	3.034	1.917	1.621	2.263	2.028	2.248	2.312
Mn	0.071	0.054	0.031	0.046	0.060	0.048	0.027	0.072	0.042	0.056	0.046	0.039	0.024	0.022	0.097	0.003	0.048	0.015	0.014	0.025
Cr	0.000	0.000	0.010	0.012	0.000	0.004	0.006	n.d.	n.d.	n.d.	n.d.	n.d.	0.001	0.000	0.001	0.006	0.004	0.003	n.d.	0.008
Ca	1.905	1.893	1.963	1.877	2.036	2.057	1.827	1.849	1.886	1.918	1.889	1.944	0.003	0.001	0.001	0.001	0.002	0.002	0.004	0.001
Na	0.432	0.642	0.412	0.465	0.413	0.393	0.576	0.514	0.393	0.420	0.510	0.446	0.050	0.046	0.044	0.027	0.045	0.005	0.063	0.013
K	0.270	0.411	0.195	0.254	0.215	0.251	0.307	0.253	0.316	0.304	0.271	0.522	1.820	1.861	1.909	1.829	1.831	1.877	1.933	1.897
Ba	0.000	0.000	0.000	0.003	0.000	0.000	0.009	n.d.	n.d.	n.d.	n.d.	n.d.	0.004	0.006	0.003	0.000	0.001	0.000	n.d.	0.000
Total	15.70	16.05	15.60	15.72	15.63	15.64	15.85	15.75	15.71	15.72	15.78	15.97	15.69	15.63	15.77	15.51	15.84	15.73	15.93	15.61
Mg #	53.0	47.6	61.6	60.8	42.0	40.6	41.4	56.0	53.6	47.8	51.8	47.9	42.2	40.7	65.0	65.9	57.0	61.9	58.7	52.9
F + Cl (wt%)	0.68	0.76	0.49	0.40	0.29	0.17	0.68	0.99	0.41	0.70	0.39	0.76	1.13	1.67	3.09	1.00	1.04	1.22	0.92	0.68

Sample	18a	22b	27c	27d	31	33	35	37	40a	42	43	45	46	47b	49	50	51	55	57
Wt%	Bio	Bio	Bio	Bio	Bio	Bio	Bio	Bio	Bio	Bio	Bio	Bio	Bio	Bio	Bio	Bio	Bio	Bio	Bio
SiO_2	37.27	37.08	36.88	36.13	35.83	35.53	35.86	35.89	34.74	35.91	36.68	36.55	35.54	35.28	35.30	34.66	39.90	38.25	36.17
Al_2O_3	14.42	14.76	16.27	15.37	14.70	14.41	14.01	14.37	16.87	13.51	15.30	14.89	14.86	15.07	15.22	14.88	12.71	13.88	17.51
FeO	16.23	15.97	19.44	20.44	20.28	23.40	21.72	23.17	22.78	17.61	19.59	18.38	19.54	18.95	19.17	18.71	13.06	14.87	19.78
MgO	13.56	14.06	11.64	10.64	11.09	8.66	9.69	8.47	7.53	13.67	10.96	12.74	12.98	12.52	12.02	12.66	17.88	14.36	10.34
TiO_2	4.98	4.82	2.39	3.84	4.23	5.28	4.75	4.65	4.43	4.10	3.65	3.99	2.76	4.05	3.84	4.33	2.21	5.55	2.56
MnO	0.09	0.07	0.22	0.26	0.34	0.10	0.08	0.03	0.20	0.26	0.35	0.22	0.36	0.31	0.23	0.20	0.07	0.08	0.26
Cr_2O_3	0.04	0.07	0.01	0.03	0.01	0.00	0.03	0.03	0.01	n.d.	n.d.	n.d.	n.d.	n.d.	n.d.	n.d.	n.d.	n.d.	n.d.
CaO	0.01	0.01	0.06	0.03	0.02	0.00	0.02	0.01	0.01	0.03	0.01	0.01	0.05	0.01	0.00	0.01	0.00	0.00	0.01
Na_2O	0.05	0.06	0.16	0.29	0.06	0.05	0.45	0.15	0.02	n.d.	n.d.	n.d.	n.d.	n.d.	n.d.	n.d.	n.d.	n.d.	n.d.
BaO	0.05	0.16	0.08	0.08	0.14	0.13	0.14	0.16	0.15	0.12	0.11	0.16	0.21	0.22	0.12	0.16	0.12	0.15	0.14
K_2O	9.82	9.78	9.20	9.48	9.58	9.64	9.36	9.47	9.67	9.87	9.61	9.96	9.65	9.70	9.94	9.87	10.25	9.86	9.99
F	0.81	0.44	0.66	0.28	1.14	0.74	1.06	0.79	0.48	0.96	0.87	0.50	0.58	0.45	0.58	1.14	2.17	0.99	0.55
Cl	0.11	0.07	0.02	0.01	0.09	0.23	0.05	0.96	0.16	0.42	0.13	0.03	0.26	0.03	0.20	0.00	0.05	0.04	0.01
O = F	–0.34	–0.19	–0.28	–0.12	–0.48	–0.31	–0.45	–0.33	–0.20	–0.40	–0.36	–0.21	–0.24	–0.19	–0.24	–0.48	–0.92	–0.42	–0.33
O = Cl	–0.02	–0.02	–0.00	–0.00	–0.02	–0.05	–0.01	–0.22	–0.04	–0.10	–0.03	–0.01	–0.06	–0.01	–0.05	–0.00	–0.01	–0.01	–0.01
Total	97.05	97.14	96.74	96.76	96.99	97.82	96.77	97.59	96.81	95.96	96.87	97.21	96.50	96.39	96.33	96.13	97.50	97.62	96.98
Ions per 23 (Hbl) or 22 (Bio) O:																			
Si	5.549	5.605	5.557	5.494	5.471	5.454	5.530	5.540	5.354	5.500	5.559	5.489	5.434	5.372	5.397	5.323	5.857	5.620	5.456
Al(IV)	2.451	2.395	2.443	2.507	2.529	2.546	2.470	2.460	2.646	2.500	2.441	2.511	2.566	2.628	2.603	2.678	2.143	2.380	2.544
Al(VI)	0.078	0.236	0.447	0.249	0.116	0.061	0.076	0.156	0.419	0.000	0.292	0.126	0.111	0.077	0.140	0.015	0.057	0.024	0.569
Ti	0.558	0.546	0.271	0.439	0.486	0.609	0.552	0.540	0.513	0.473	0.416	0.451	0.318	0.464	0.442	0.500	0.243	0.613	0.291
Mg	3.009	3.171	2.614	2.411	2.523	1.981	2.226	1.949	1.729	3.122	2.476	2.851	2.956	2.841	2.740	2.898	3.912	3.145	2.325
Fe	2.021	2.017	2.450	2.599	2.590	3.003	2.802	2.991	2.936	2.256	2.483	2.309	2.499	2.414	2.451	2.403	1.603	1.828	2.495
Mn	0.011	0.009	0.028	0.034	0.045	0.013	0.010	0.004	0.026	0.033	0.045	0.028	0.046	0.040	0.031	0.026	0.008	0.010	0.030
Cr	0.005	0.008	0.002	0.004	0.001	0.000	0.004	0.003	0.001	n.d.	n.d.	n.d.	n.d.	n.d.	n.d.	n.d.	n.d.	n.d.	n.d.
Ca	0.001	0.002	0.010	0.004	0.003	0.001	0.004	0.001	0.002	0.005	0.002	0.001	0.008	0.001	0.000	0.001	0.000	0.001	0.002
Na	0.014	0.046	0.022	0.023	0.041	0.040	0.041	0.048	0.046	0.037	0.032	0.047	0.063	0.065	0.035	0.048	0.035	0.041	0.040
K	1.864	1.885	1.768	1.839	1.865	1.887	1.842	1.865	1.902	1.928	1.859	1.908	1.883	1.885	1.938	1.933	1.919	1.848	1.922
Ba	0.003	0.004	0.009	0.018	0.003	0.003	0.028	0.009	0.001	n.d.	n.d.	n.d.	n.d.	n.d.	n.d.	n.d.	n.d.	n.d.	n.d.
Total	15.56	15.92	15.62	15.62	15.67	15.60	15.59	15.57	15.57	15.85	15.60	15.72	15.88	15.79	15.78	15.82	15.78	15.51	15.68
Mg #	59.8	61.1	51.6	48.1	49.3	39.8	44.3	39.5	37.1	58.1	49.9	55.3	54.2	54.1	52.8	54.7	70.9	63.2	48.2
F + Cl (wt%)	0.91	0.51	0.67	0.30	1.23	0.98	1.11	1.74	0.64	1.39	1.00	0.53	0.84	0.48	0.78	1.14	2.23	1.04	0.56

Note: Sample prefixes AL97 and AL98 are omitted (see Table 1).
Electron microprobe operating conditions: Beam voltage = 15 kV; beam current = 10 nA.
Total Fe reported as FeO.
n.d.—no data (not analyzed). Mg #—[Mg/(Fe + Mg)] × 100.

in composition characterize both biotite and hornblende among the samples. Values of Mg # ([Mg/(Fe + Mg)] × 100) for biotite range from 37 to 71, and total F + Cl anion contents range from 0.5 to 2.2% by weight. In contrast, the values of Mg # for hornblende range from 39 to 62, and total halogen contents range from 0.2 to 1.0 wt% F + Cl. Moreover, the alkali contents of hornblende vary, with values for Na # ([Na/(K + Na)] × 100) ranging from 43 to 70. Within a given thin section, however, multiple replicate analyses (not shown in Table 2) varied within only ±2% (relative) for major elements, thus supporting the earlier suggestion from petrographic observation that only one growth generation of each mineral was present in each sample.

Incremental step heating was performed on 39 hornblende and biotite mineral separates. Rock samples were crushed, sieved (usually 45–60 mesh), and separated using standard techniques followed by hand-picking. The $^{40}Ar/^{39}Ar$ analyses were performed in the Radiogenic Isotopes Laboratory at Ohio State University using procedures that have been described previously (Foland et al., 1984, 1993), except that some Ar measurements (run numbers with the suffix "M") were made using a new mass spectrometer and associated lines. Aliquots of ~20–30 mg for biotites and 50–60 mg for hornblendes (or a few mg for new equipment) were irradiated in the Ford Nuclear Reactor of the Phoenix Memorial Laboratory at the University of Michigan. The irradiated aliquots were subjected to resistance heating in high-vacuum, low-blank, double-vacuum furnaces to successively higher temperatures, for a period of ~30 minutes at each temperature.

The results of the incremental-heating measurements are summarized in Table 3. Age uncertainties are given at the 1σ level and reflect only internal dispersion that is attributed to analytical uncertainties. The quoted uncertainties do not reflect any provision for uncertainty in the absolute age of the monitor, which is assigned at ±1%. Because this uncertainty is systematic, affecting all ages the same way, it is not included when age uncertainties are quoted in order to emphasize the level of apparent age differences among the plateaus. However, it is appropriate to use the ±1% relative uncertainty when comparing the Ar ages reported here to ages obtained by other methods and workers.

$^{40}Ar/^{39}Ar$ Results

Incremental-heating $^{40}Ar/^{39}Ar$ age results are illustrated in customary age-spectra diagrams in Figures 4 and 5. A full listing of the analytical results (Table 3) is given in the appendix.[1] The age spectra in Figures 4 and 5 range from highly concordant, where variations may be attributed to analytical uncertainties, to highly discordant. The ages interpreted from the step heating are those defined by plateaus. An age plateau is defined by fractions that are continuous and contain a portion of the total $^{39}Ar_K$ released that is not less than 50% and where each fraction of the plateau is internally concordant, overlapping with the plateau age within the 2σ uncertainty for that fraction. The step-heating measurements were also examined in isotope correlation (or "isochron") plots, which are consistent with the plateau and near-plateau dates. However, such plots provide no additional information for the more discordant spectra, given the Precambrian age and highly radiogenic nature of the samples, and thus are not presented or considered in this paper.

More than half of the samples yield plateaus, where the variations in apparent age are within analytical uncertainty, but others do not. Most of these have some age variations greater than expected from analytical errors. Where these apparent age variations are minor in degree, "near-plateaus" are defined. For at least two hornblende separates, discordance is substantial; in these cases an interpreted age is only approximated, as discussed later.

Hornblende

Twelve hornblendes from diverse lithologies sampled across the Lowlands (Fig. 2; Table 1) were analyzed, giving total-gas (or integrated) dates ranging from 956 to 1095 Ma. Six of the hornblendes also yield plateau dates ranging from 1012 ± 3 to 1098 ± 3 Ma (±1σ), two others yield near-plateaus, and three of the other four exhibit discordant but still interpretable spectra (Fig. 4; Table 3). Figure 3 shows the geographic distribution of interpreted ages.

The discordance observed for hornblende spectra appears to arise from two main sources and is mostly or totally due to mineralogical heterogeneities. By far the major discordance is increasing age with progressive Ar release (e.g., see Fig. 4, spectrum 13b). This is interpreted to reflect not partial resetting, but rather intergrowths and inclusions of K-bearing phases, mainly micas and feldspars, as indicated by the concomitant chemical variations. The effects are principally observed in low-age, lower-temperature steps and are evidenced by elevated K/Ca and K/Cl ratios, as seen for all but one hornblende separate. The K-bearing included phases consistently yield younger apparent ages than the host hornblende, which is reasonable for these slowly cooled rocks. Also, these phases occur predominantly as prograde inclusions in the hornblende rather than as retrograde alterations, and as such probably do not date a younger alteration event per se. There is no petrologic or spectral evidence for multiple growth generations of hornblende. A second type of spectral disturbance, generally minor where present, is observed at high temperature (~1030–1070 and ~1120–1150 °C), where the apparent age "dips" to a lower value (samples 18a and 45, respectively; see Fig. 4). These effects may be attributed to K-poor inclusions, generally opaque oxides, and represent artifacts of ^{39}Ar recoil. Basically, recoil of ^{39}Ar during irradiation implants an excess of ^{39}Ar in the inclusion, which, when released due to a high-temperature reaction, produces a low

[1]GSA Data Repository Item 2004060, Supplementary table of argon isotopic data for hornblende and biotite, Adirondack Lowlands, New York, is available on request from Documents Secretary, GSA, P.O. Box 9140, Boulder, CO 80301-9140, USA, editing@geosociety.org, at www.geosociety.org/pubs/ft2004.htm.

TABLE 3. SUMMARY OF $^{40}Ar/^{39}Ar$ INCREMENTAL-HEATING RESULTS FOR HORNBLENDE AND BIOTITE FROM THE ADIRONDACK LOWLANDS, NEW YORK

Sample	Run number	Integrated gas			Plateau		
		K/Ca	K/Cl	Age (Ma)	Temp (°C)	^{39}Ar (%)	Age (Ma)
Hornblende							
AL97-13b	57I29	0.188	11.7	995	1045-1225	56	“1020 ± 4”
AL97-15a	57I18	0.252	20.8	1062	1000-1150	81	1080 ± 3
AL97-18a	58A17M	0.140	26.4	1029			“1055 ± 10
AL97-22b	58A29M	0.147	38.8	1079	960-1000	53	1098 ± 3
AL97-27c	58A23M	0.122	43.3	1033			“1050 ± 10”
AL97-27d	58A251M	0.130	63.2	1050	1150-1400	51	1063 ± 3
AL97-35	57I22M	0.194	50.6	1030	990-1500	66	“1053 ± 4”
AL98-42	60B3M	0.175	3.98	997			
AL98-45	60B6M	0.177	24.2	1033			“1055 ± 10”
AL98-46	60B12	0.227	4.93	1006	930-1020	69	1015 ± 3
AL98-47b	60B14M	0.146	59.3	956	1060-1110	51	1012 ± 3
AL98-50	60B17M	0.240	242	1095	1070-1130	53	1083 ± 3
Biotite							
AL97-2a	57I27	63.2	42.9	966	775-1150	72	970 ± 3
AL97-5b	57I3	43.3	21.3	931	725-1400	94	934 ± 3
AL97-8	57I16	215	28.7	1053	775-1100	93	“1058 ± 3”
AL97-12a	58A4M	31.9	84.5	886	780- 1150	68	“896 ± 3”
AL97-13a	57I5	94.8	51.8	903	730-1150	83	“908 ± 3”
AL97-14a	58A8	38.6	18.9	1046	975-1100	62	1086 ± 3
AL97-15a	60B19M	22.7	122	1201	790-1175	95	1210 ± 3
AL97-17c	58A11M	22.6	139	983	640-1000	70	987 ± 2
AL97-18a	58A16	10.7	112	929	775-1450	88	935 ± 2
AL97-22b	57I7	14.8	129	923	625-1200	97	931 ± 3
AL97-27c	60B25	8.93	329	972	675-1300	91	978 ± 3
AL97-27d	60B27M	4.00	302	932	790-1200	92	“935 ± 3”
AL97-31	57I9	10.1	69.3	895	775-1400	83	“898 ± 3”
AL97-33	57I11M	46.8	37.0	918	730-1120	92	920 ± 2
AL97-35	57I20	9.06	299	918	850-1300	80	926 ± 3
AL97-37	57I14	106	9.03	962	730-1100	84	968 ± 3
AL97-40a	57I25	34.0	66.7	934	725-1100	87	939 ± 2
AL98-42	60A5	22.1	21.0	914	625-1100	95	“923 ± 3”
AL98-43	60A8	55.7	61.8	894	675-1300	87	“901 ± 3”
AL98-45	60A11	26.7	137	923	625-1300	97	925 ± 2
AL98-46	60A16	32.2	37.2	943			
AL98-47b	60A19	18.6	442	972	675-1300	92	980 ± 3
AL98-49	60A22	29.5	48.6	917	675-1100	88	920 ± 2
AL98-50	60A24M	20.0	357	1233	1055-1190	51	1237 ± 3
AL98-51	60A30	103	191	1031	700-1020	82	“1037 ± 3”
AL98-55	60A33	67.9	198	990	675-1300	92	994 ± 3
AL98-57	60A36	163	422	947	625-1300	97	951 ± 3

Notes: Integrated gas—the sum of all the Ar released during incremental analysis. The values for K/Ca ratio, K/Cl ratio, and age are thus for the bulk sample.

Plateau—the age plateau (as defined in the text), if found, or other interpreted age; Temp—the temperature range of increments comprising the plateau; ^{39}Ar—the percentage of the total ^{39}Ar in the fractions of the plateau.

Age—the plateau age. If it is given in quotations, a valid plateau is not developed, but the scatter is not large; here the given age is taken to be the best measure for the sample. For a few hornblendes, discordance is great but an age may be interpreted from the release pattern; for these no %^{39}Ar is given. For samples without entries, no age may be confidently interpreted from the step heating. Age uncertainties are at the 1σ level and do not include uncertainty from the flux monitor, as discussed in the text. See Table 1 and Figure 2 for sample locations.

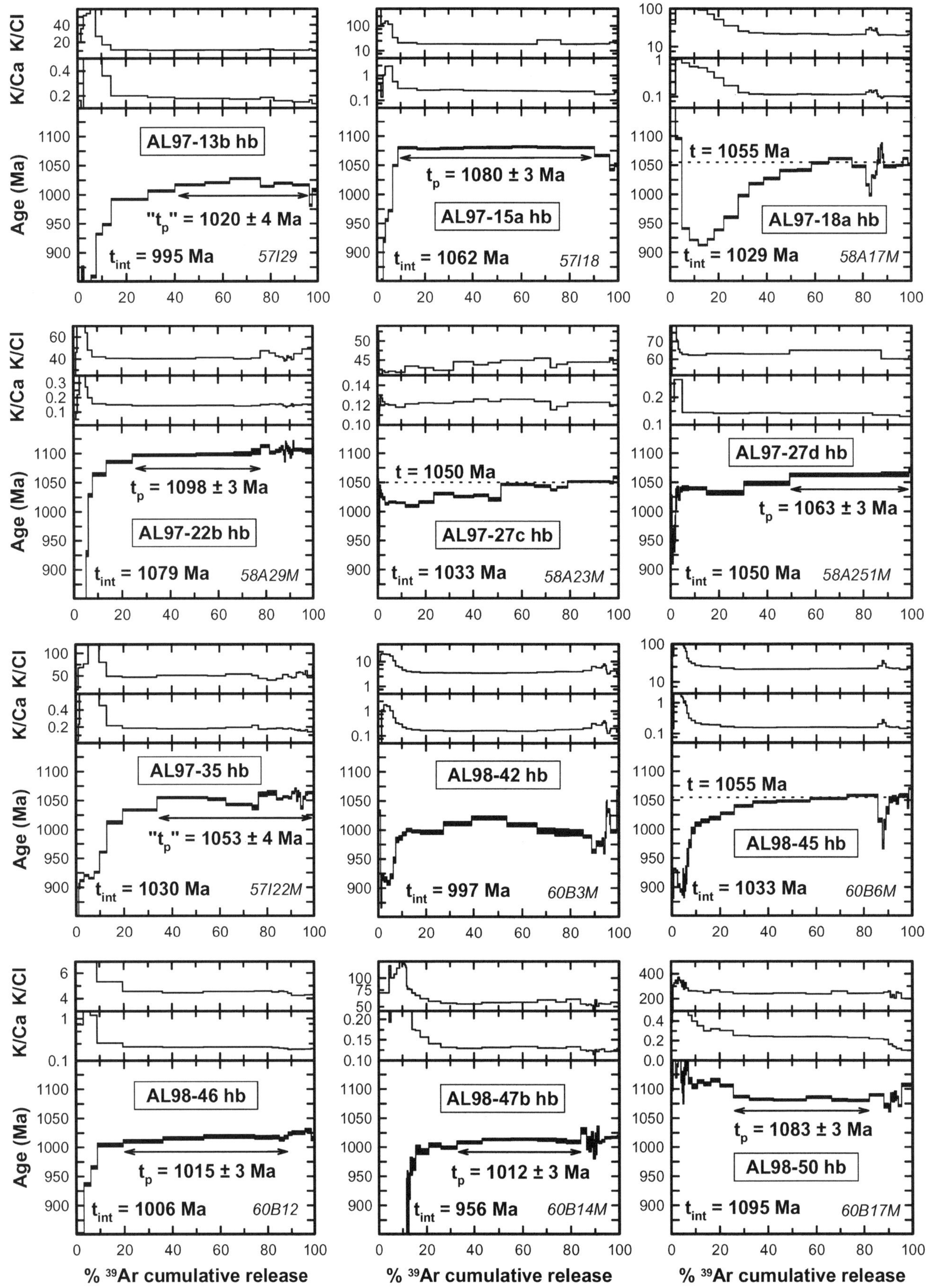
K/Cl
K/Ca
Age (Ma)
AL97-13b hb
"t_p" = 1020 ± 4 Ma
t_int = 995 Ma
57I29
AL97-15a hb
t_p = 1080 ± 3 Ma
t_int = 1062 Ma
57I18
AL97-18a hb
t = 1055 Ma
t_int = 1029 Ma
58A17M
AL97-22b hb
t_p = 1098 ± 3 Ma
t_int = 1079 Ma
58A29M
AL97-27c hb
t = 1050 Ma
t_int = 1033 Ma
58A23M
AL97-27d hb
t_p = 1063 ± 3 Ma
t_int = 1050 Ma
58A251M
AL97-35 hb
"t_p" = 1053 ± 4 Ma
t_int = 1030 Ma
57I22M
AL98-42 hb
t_int = 997 Ma
60B3M
AL98-45 hb
t = 1055 Ma
t_int = 1033 Ma
60B6M
AL98-46 hb
t_p = 1015 ± 3 Ma
t_int = 1006 Ma
60B12
AL98-47b hb
t_p = 1012 ± 3 Ma
t_int = 956 Ma
60B14M
AL98-50 hb
t_p = 1083 ± 3 Ma
t_int = 1095 Ma
60B17M
% ³⁹Ar cumulative release

apparent age. This is evidenced by correlated same-percentage changes in Ca/K, Cl/K, and age. Inspection of the release patterns (Fig. 4) reveals the close correlation of K/Ca and K/Cl variations coupled to variations in age. The spectra must be interpreted carefully with these effects in mind. Most spectra show an increase in age with progressive Ar release, then plateau or level off at higher temperature after the release of Ar from impurities.

Samples 22b, 50, 15a, 27d, 46, and 47b produce well-defined plateaus at 1098 ± 3, 1083 ± 3, 1080 ± 3, 1063 ± 3, 1015 ± 3, and 1012 ± 3 Ma, respectively. Samples 13b and 35 exhibit near-plateaus at 1020 ± 4 and 1053 ± 4 Ma. The restricted discordance of lower apparent age at both low and high temperature is attributed to inclusions and recoil, respectively, as evidenced by related K/Ca and K/Cl variations. Samples 13b, 22b, 27d, 35, 46, and 50 produce plateaus or near-plateaus at relatively higher extraction temperatures where the bulk of variation within the plateaus is considered to be analytical, with perhaps minor recoil effects. All of these spectra except one are convex up. This pattern is correlated with K/Ca and K/Cl variations, indicating the effects of impurity phases even though significant amounts are not apparent in thin section. The exception, hornblende from sample 50, shows older ages for the lower temperature fraction that also shows chemical variation.

Among the remaining hornblende separates, sample 18a (from an amphibolite layer in the Antwerp granite, Fig. 2) is the most discordant (Fig. 4). Low temperature fractions give low ages correlated with high K/Ca and K/Cl ratios; these are attributed to biotite inclusions in the hornblende and/or to minor sericite alteration within the plagioclase impurities. Subsequent fractions progressively increase in age with concomitant decreases in K/Ca, reflecting an increased proportion of Ar from hornblende. The spectrum levels off (with a recoil artifact at high temperature) and gives an apparent age of ca. 1055 Ma, which is the interpreted age, probably a minimum age.

Hornblende 27c (from a large roadcut of Popple Hill gneiss, Fig. 2) yields a spectrum showing progressive increase in apparent age with successive temperature increments. Unavoidable biotite inclusions that degas at lower temperature probably account for this behavior, judging from elevated K/Ca ratios for low temperature increments (Fig. 4). The apparent age levels off at high temperature, at ca. 1050 Ma, which is also the interpreted age. The spectra for both 27c and 27d show a late-fraction drop in apparent age that may be attributed to recoil of ^{39}Ar into inclusions.

Hornblende 45 (from an amphibolite in the Antwerp granite) yields relatively high K/Ca and K/Cl ratios in low temperature increments and younger apparent ages that appear to reflect simultaneous degassing of hornblende and micaceous impurities. The remaining ~70% of the ^{39}Ar levels off but is punctuated by a drop in apparent age that probably reflects ^{39}Ar recoil. An age of ca. 1055 Ma from these fractions is the interpreted age for this hornblende.

Hornblende 42 (from an amphibolite layer in the Hermon granite) produces a broad region of roughly constant age, ca. 1000 Ma, at ~15 to 85% of the released ^{39}Ar. The discordance is substantially different from that of other samples, so the ca. 1000 Ma date is regarded as an approximation only.

Figure 4. ^{40}Ar/^{39}Ar age spectra for hornblende, Adirondack Lowlands. t_p—plateau age (see text for definition); "t_p"—imperfect-plateau age; t_{int}—integrated (i.e., total gas) age. The width of apparent age plotted for each fraction is ±1σ. Uncertainties in plateau ages do not include provisions for systematic uncertainty in the monitor.

Biotite

Twenty-seven biotite separates from diverse lithologies across the Lowlands (Fig. 2; Table 1) yield integrated ^{40}Ar/^{39}Ar dates ranging from 886 to 1233 Ma. The incremental release patterns for biotite are generally concordant, although most samples show significantly younger apparent ages for the lowest temperature fraction(s) (Fig. 5). These fractions also show K/Ca and K/Cl variations indicating that the minor discordance reflects trace impurity phases. Eighteen samples (2a, 5b, 14a, 15a, 17c, 18a, 22b, 27c, 33, 35, 37, 40a, 45, 47b, 49, 50, 55, and 57) yield good plateaus, with plateau dates ranging from 920 ± 2 to 1237 ± 3 Ma (±1σ). Eight of the remaining nine samples (8, 12a, 13a, 27d, 31, 42, 43, and 51) give near-plateaus, and one gives no plateau (46). The reasons for the apparent-age variation observed in most samples remain unclear, pending ultraviolet laser spot dating, although they probably reflect minor mineralogic heterogeneities.

In all cases, the dates from incremental-heating analysis are preferable to total-gas dates, as discussed earlier. Overall these range only from 1012 to 1098 Ma for hornblende, whereas biotite shows a much greater range of 896 to 1237 Ma. However, this ~340 m.y. range for biotite is not compatible with ages reflecting slow cooling (as previously inferred), and thus other factors must be considered. These factors are best considered by examining the regional patterns, which are shown in Figures 3 and 6.

More than two-thirds of the biotite samples fit a regional pattern that is discussed in the following section (see Fig. 6), indicating that most of the plateaus provide ages that are reasonable on independent grounds as well. However, some biotites clearly deviate, and most of these appear to give ages that are too old. Six biotite separates (8, 14a, 15a, 46, 47b, and 50; see Fig. 2) yield anomalously old apparent ages most likely attributable to excess ^{40}Ar. In particular, biotite samples 15a (1210 Ma) and 50 (1237 Ma) are unrealistically old compared to their coexisting hornblendes (1080–1083 Ma; see Table 3 and Fig. 3), and indeed exceed the age of biotite formation per se. These relationships indicate that both of these biotites contain excess ^{40}Ar, yet their spectra fit the definition of a plateau. Biotite 47b yields a plateau age of 980 ± 3 Ma, which is apparently much older than the ca. 895–920 Ma age range represented among the five nearest biotites (see Figs. 3 and 6). Moreover, comparison of the dates for biotite and coexisting hornblende in samples 47b

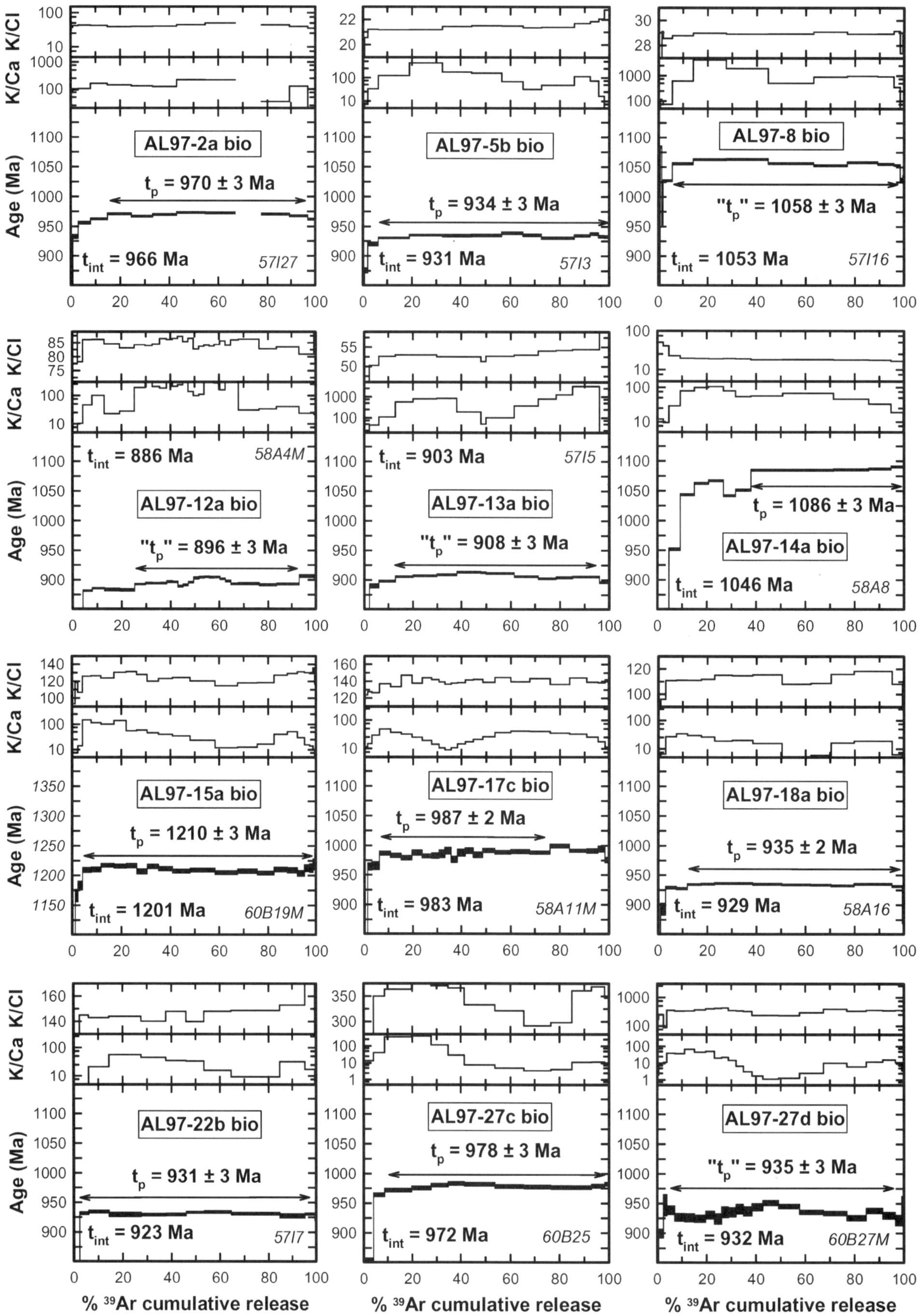

Figure 5 (*on this and following two pages*). $^{40}Ar/^{39}Ar$ age spectra for biotite, Adirondack Lowlands. Conventions as given in caption to Figure 4.

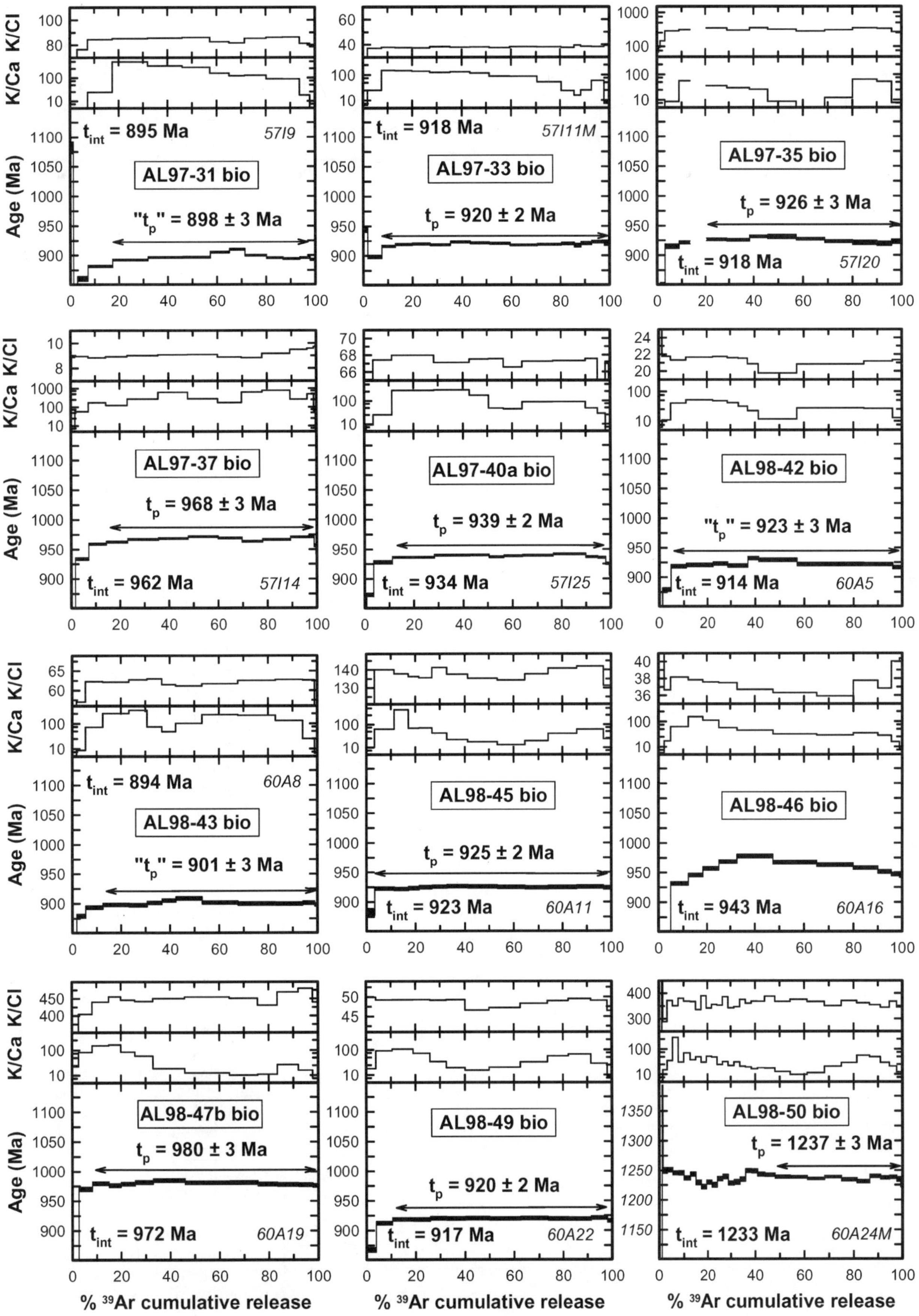

Figure 5 (*continued*).

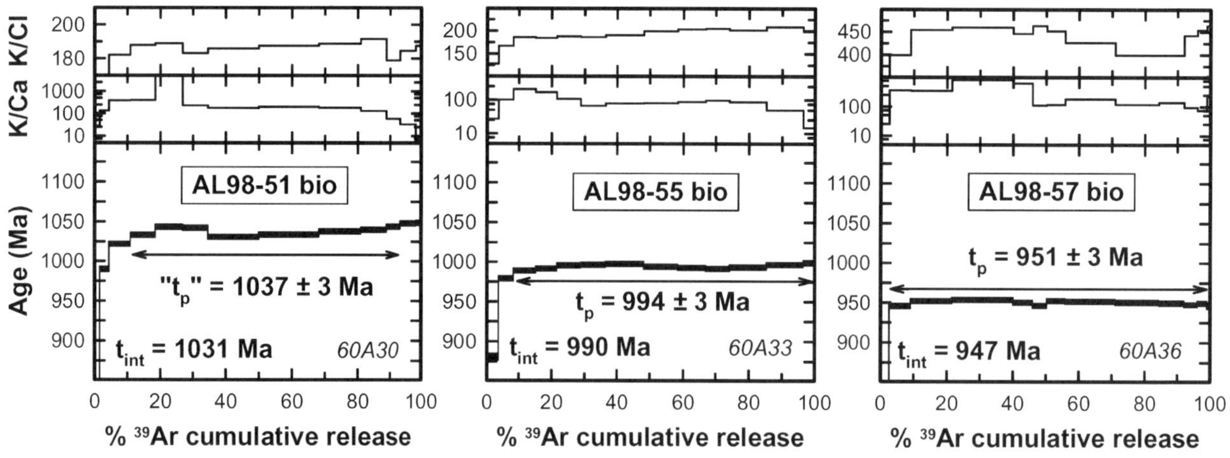

Figure 5 (*continued*).

reveals much smaller apparent-age differences than those exhibited in biotite-hornblende pairs elsewhere in the Lowlands (Figs. 3–6). Taken together, these observations suggest that biotite 47b also contains excess ^{40}Ar, albeit less than biotites 15a and 50. Likewise, biotites 8 (ca. 1058 Ma) and 14a (1086 Ma) are anomalously old compared to hornblendes in nearby localities (Fig. 3), so it is likely that they too have incorporated excess ^{40}Ar.

There is no obvious or compelling explanation as to why these particular biotites contain significant excess ^{40}Ar. This subgroup contains examples of both the most concordant incremental-heating spectrum (e.g., 50) and the most discordant (e.g., 46), showing that internal concordance alone is not a criterion for absence of excess ^{40}Ar, and that a plateau per se is an insufficient albeit necessary criterion for a meaningful age (see, e.g., Foland, 1983). Several of these samples are distinct. For example, sample 8 is a syenite from near the Carthage-Colton shear zone, and its biotite is highly chloritized. Sample 14a contains abundant scapolite with (unaltered) biotite, which is the most chlorine-rich by far. However, there is no common thread connecting these anomalous samples other than apparent excess Ar.

Two other biotites (18a and 22b), located near one another, yield anomalously young dates compared to those nearby (see Figs. 2, 3, and 6). Biotite separates of 18a and 22b yield plateau dates of 935 and 931 Ma, whereas those of 17c, 51, and 55 yield

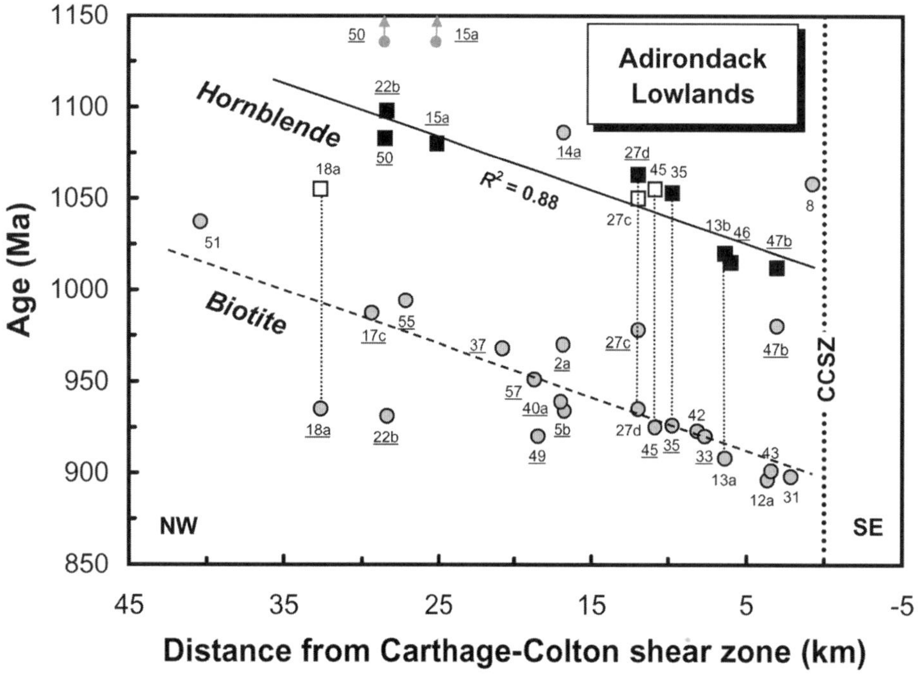

Figure 6. Apparent ^{40}Ar/^{39}Ar ages of hornblendes (squares) and biotites (circles) plotted versus map distance closest (generally orthogonal) to the Carthage-Colton shear zone (CCSZ). Data points are labeled with sample numbers (prefixes omitted); underlined numbers correspond to plateau spectra; filled symbols correspond to plateaus and near-plateaus (point sizes equate to $\pm 1\sigma$ errors in age); and open symbols indicate nonplateaus. Vertical dotted lines indicate pairs of coexisting hornblende and biotite from which individual estimates of cooling rates were inferrred. Plotting off-scale are biotites 15a and 50, which yield >1200 Ma dates indicating excess argon. Linear regression of hornblende data (solid line, exclusive of 18a) suggests that distance from the Carthage-Colton shear zone (CCSZ) accounts for ~90% of the regional age variance; a parallel trend (dotted line) is drawn through favored biotite data. See text for discussion.

plateau or near-plateau dates ranging from 987 to 1037 Ma. There is no unique explanation for these two younger dates, either, and the age spectra are concordant. The 931 Ma date for 22b may reflect the influence of secondary fluids, as evidenced by its close proximity (<10 cm) to a highly altered contact between an amphibolite (a relict mafic dike) and the Hyde School gneiss. Although there is no obvious petrographic evidence, local effects perhaps due to fluid infiltration may have been important for 18a and 22b, which are located just northwest of the Pleasant Lake fault (Fig. 2).

In summary, although there are some significant departures, as outlined earlier, most biotites analyzed (~70%) yield reliable and apparently geologically meaningful results. In contrast, metamorphic biotite ages in adjacent Grenville terranes have been interpreted to be mostly unreliable (Cosca et al., 1991, 1992). In part, this difference may reflect the much larger number of biotite samples included in the present study.

DISCUSSION

Interpretation of $^{40}Ar/^{39}Ar$ Ages

In principle, the 1100–900 Ma $^{40}Ar/^{39}Ar$ dates obtained for Lowlands hornblendes and biotites (Figs. 4 and 5; Table 1) could represent growth, reset, or cooling ages. However, an interpretation of growth ages is ruled out because the last known metamorphism in the Lowlands occurred at 1170–1150 Ma and peaked at 660 ± 50 °C (Fig. 3), whereas the ~500 and ~300 °C temperatures widely accepted for argon closure in hornblende and biotite (McDougall and Harrison, 1999) are much lower. Instead, only interpretations of partially reset (i.e., mixed) ages or of cooling ages are consistent with the relative-age sequence observed in the Lowlands among coexisting amphiboles and micas (Fig. 6; see also Foland et al., 1998) and predicted from crystal-chemical considerations (Dahl, 1996a, 1996b). However, partial resetting would be manifested in hornblende or biotite as multiple growth domains or as compositional heterogeneity, neither of which is observed on the scales of thin section or electron beam (microprobe) in any of the Lowlands samples (Pomfrey et al., 1998a, 1998b). Moreover, as noted earlier, no evidence for Ottawan resetting in the Lowlands has been reported. Thus, in the absence of petrologic, spectral, or tectonic evidence to support it, a scenario of partial resetting of hornblende and biotite is provisionally ruled out as a major factor in controlling the Lowlands $^{40}Ar/^{39}Ar$ dates.

From these considerations, and by process of elimination, it is concluded that the Lowlands $^{40}Ar/^{39}Ar$ dates most likely represent post-Ottawan cooling ages. In fact, the dates themselves are consistent with interpretation as cooling ages. First, the $^{40}Ar/^{39}Ar$ spectra (Figs. 4 and 5) are generally consistent with essentially single hornblende and biotite populations that would characterize simple cooling. Second, with standard closure temperatures of ~500 and ~300 °C for main-trend hornblende and biotite, respectively (Fig. 6), the $^{40}Ar/^{39}Ar$ dates of coexisting mineral pairs reproduce the ~1.5 °C/m.y. cooling rate previously inferred from higher-temperature Pb closure in monazite, titanite, and apatite (Mezger et al., 1991, 1992, 1993). Hence, in the following discussion it is assumed that in samples apparently unaffected by excess ^{40}Ar (Table 3; Fig. 3) the $^{40}Ar/^{39}Ar$ dates largely represent ages of post-Ottawan cooling that reflect regional exhumation.

Age Variations and Regional Trends

The $^{40}Ar/^{39}Ar$ cooling ages of both hornblende and biotite vary systematically across the Lowlands, as shown in Figure 3, and the ages are plotted in Figure 6 as a function of closest (generally orthogonal) map distance from the Carthage-Colton shear zone. Comparison of cooling age with grain size or composition (Tables 1–3; Figs. 3–5) reveals no obvious interrelationships, and no variations in microtexture (e.g., alteration) with distance are evident in thin section (Table 1; Fig. 6). Although some or all of these factors may be important to minor differences among the ages along trend, they do not account for the major ones or for the trends per se. Instead, hornblende and biotite exhibit parallel inverse correlations between age and location (as proxied by distance from northwest to southeast across the Lowlands), which implies a regional thermotectonic basis for the trends.

Of the twelve hornblendes, ten are predominantly pargasitic, hastingsitic, and tschermakitic varieties that define a linear age-distance trend (the other two are edenitic varieties). Linear regression of the hornblende age-distance trend yields an R^2 of 0.88 (Fig. 6), a regional age trajectory of ~2.9 m.y./km across the Lowlands, and a Carthage-Colton shear zone "intercept" of ca. 1010 Ma that coincides with plateau dates reported by Streepey et al. (2001a) for three Carthage-Colton shear zone hornblendes. The high R^2 value suggests that ~90% of the main-trend age variance is somehow related to tectonics (as proxied by distance), whereas the remaining ~10% represented by scatter along the trend is probably related to inherent crystal chemistry and/or superimposed microtexture. Nineteen of the twenty-seven samples of biotite (Fig. 5) adhere to a trend of clearly younger ages that parallels the hornblende trend. This relatively "noisy" trend, shown as a dotted line in Figure 6, gives an apparent Carthage-Colton shear zone "intercept" of ca. 900 Ma. Again, this intercept closely matches the 895–899 Ma plateau dates obtained by Streepey et al. (2000) for two Lowlands biotites sampled closest to the Carthage-Colton shear zone. Since most biotites that do not fit the regional trend fall to the high side, these deviations are apparently due to excess ^{40}Ar, as inferred earlier. Otherwise, the scatter probably reflects a combination of crystal-chemical and microtextural effects on age, but, as with hornblende, these effects cannot explain the inverse age-distance trend per se. The $^{40}Ar/^{39}Ar$ ages of phlogopites sampled from Lowlands marbles (Foland et al., 1998) plot intermediate and parallel to those of hornblende and biotite, thereby providing independent support for the hornblende and biotite age-distance trends inferred in Figure 6.

The apparent parallelism of the age-distance trends for hornblende and biotite (Fig. 6) further implies that retrograde cooling was uniform across the Lowlands during the interval of argon closure of these minerals, as schematically depicted in the temperature-time plot (Fig. 7). Moreover, a scenario of linear slow cooling (versus intervals of no cooling punctuated by episodes of fast cooling) is implied because both the hornblende and the biotite ages vary widely across the Lowlands, whereas episodic fast cooling would have promoted regionally uniform ages among the hornblendes and among the biotites.

The ~115 m.y. age difference inferred between the hornblende and the biotite trends (Fig. 6) yields an estimated cooling rate of ~1.7 °C/m.y., based upon the widely accepted 500 °C and 300 °C closure temperatures, respectively. There are a total of ten hornblende-biotite pairs from individual or adjacent samples. In four pairs both hornblende and biotite conform to the inferred age-distance trends (13b, 27d, 35, 45), with the age differences ranging from 112 to 130 m.y. The nominal ~200 °C difference in closure temperatures applied to each of the four pairs leads to a mean, time-integrated cooling rate of 1.6 ± 0.2 ($\pm 2\sigma$) between ca. 1100 and ca. 900 Ma, which is henceforth reported as 1–2 °C/m.y. Another hornblende-biotite pair (sample 18a) implies a similar cooling rate, although both minerals fall far below their respective main trends (Fig. 6). The other five pairs give anomalous results because biotite variably shows evidence of excess ^{40}Ar (15a, 27c, 47b, 50; see Fig. 6) or of alteration (22b). Based upon these results, it is concluded that regional cooling below 500 °C was slow and uniform across the Lowlands, although cooling through argon closure occurred at progressively later times within the Lowlands—that is, from northwest to southeast in relation to its current position (Figs. 3, 6, and 7).

Lowlands cooling histories for an 1170–900 Ma time frame are proposed in Figure 7, based upon ^{40}Ar/^{39}Ar and U-Pb mineral ages, standard closure temperatures, and mineral-pair thermobarometry. Published U-Pb ages of garnet and monazite for the Lowlands largely reflect the ~1180–1150 Ma interval of peak Elzevirian-AMCG tectonomagmatism, whereas titanites define a younger ca. 1156–1103 Ma range of apparent U-Pb age (Mezger et al., 1991, 1992, 1993). The Pb closure temperatures of all three U-bearing minerals (e.g., Scott and St-Onge, 1995; Dahl, 1997; Villa, 1997) are greater than or equal to peak metamorphic temperatures of 650–700 °C in the Lowlands (Bohlen et al., 1985; Kitchen and Valley, 1995). Thus, whereas garnet and monazite ages date the latest Elzevirian-AMCG regional metamorphism (ca. 1180–1150 Ma; Fig. 7), the younger titanite ages may reflect retrograde growth or postgrowth closure, and so on. The ^{40}Ar/^{39}Ar data for hornblende and biotite (this study) indicate continued regional cooling through ~500 and ~300 °C between ca. 1100 and ca. 900 Ma (Fig. 7, solid-line cooling trajectories). Final cooling of the present surface exposures to 25 °C must have occurred by ca. 520 Ma, based upon the Precambrian-Cambrian nonconformity.

Two sets of thermal histories can be envisioned for the Lowlands between ~750 and ~500 °C, depending upon whether the Lowlands variably experienced delayed post-AMCG cooling (the favored scenario, shown in Fig. 7), or mild Ottawan reheating (not shown), prior to final cooling through 500 °C. Relative uncertainties in the 750–500 °C cooling history of the Lowlands (Fig. 7, dotted lines with question marks) reflect corresponding uncertainty in the 1150–1050 Ma juxtaposition history of the Lowlands and the Highlands. As noted earlier, Ottawan (1090–1045 Ma) metamorphic ages per se have not been reported in the Lowlands (Mezger et al., 1991, 1992, 1993; Streepey et al., 2001a), leading to the proposals that the Lowlands escaped reheating and therefore were separated vertically and/or laterally from the Highlands during Ottawan time.

A model exhumation rate for the Lowlands can be esti-

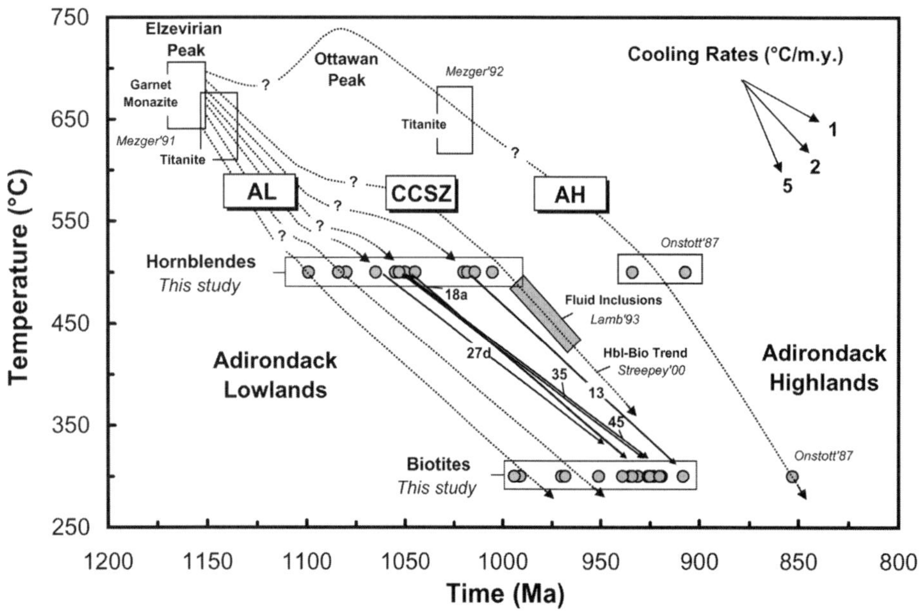

Figure 7. Possible Proterozoic temperature-time cooling trajectories of metamorphic rocks currently exposed in the Adirondack Lowlands (AL), based upon (1) U-Pb mineral ages and peak metamorphic temperatures of Elzevirian orogeny (Bohlen et al., 1985; Mezger et al., 1991, 1992, 1993; Kitchen and Valley, 1995); (2) post-Ottawan ^{40}Ar/^{39}Ar cooling ages of hornblende and biotite (shaded circles), this study; and (3) post-Ottawan hornblende-biotite (Hbl-Bio) cooling trends of Streepey et al. (2000) and Onstott and Peacock (1987) for the Carthage-Colton shear zone (CCSZ) and the western Adirondack Highlands (AH), respectively. Argon closure temperatures of ~500 °C (hornblende) and ~300 °C (biotite) are assumed. See text for discussion.

mated for the 1100–900 Ma hornblende-biotite closure interval. A retrograde geotherm of 30 ± 10 °C/km has been estimated for the southern Grenville province from (1) fluid inclusion studies in the Carthage-Colton shear zone indicating that its retrograde path passed through a pressure of 3–5 kilobars and a temperature of 400–550 °C (Lamb, 1993), and (2) retrograde thermobarometry and aluminosilicate phase relations in the western Central Metasedimentary Belt (Cosca et al., 1991, 1992). The model 1–2 °C/m.y. cooling rate (this study), combined with the 20–40 °C/km geotherm (see earlier), yields a nominal, time-integrated exhumation rate of 0.06 ± 0.02 km/m.y. between ca. 1050 and ca. 900 Ma, with the maximum exhumation rate associated with the lowest (and more likely) geotherm within the estimated range. In comparison, a nominally faster exhumation rate of ~0.13 ± 0.03 km/m.y. is calculated for the adjacent Highlands between ca. 1050 and ca. 900 Ma, assuming the faster cooling (~3–4 °C/m.y.) inferred there during extensional normal faulting along the Carthage-Colton shear zone (Streepey et al., 2000; McLelland et al., 2001). These estimated exhumation rates, coupled with assumptions of a planar Carthage-Colton shear zone (normal fault) dipping ~45°NW between ca. 1050 and ca. 900 Ma, further imply that ~5–7 km of dip-slip displacement (i.e., both vertical and horizontal components of ~4–5 km) occurred during this time frame. Clearly, these first-order calculations are limited by the imprecision of inferred cooling rates, the assumptions noted earlier, and an assumption of the same constant geotherm for both domains during retrograde cooling, for which there is no a priori evidence. Alternatively, presumption of a common exhumation rate (e.g., 0.06 km/m.y.) for both domains implies a model geotherm of ~65 °C/km for the Highlands, which would have been unsustainable for long periods of time and therefore is considered unrealistic.

Geological Explanation of the Regional Age Trends

The parallel $^{40}Ar/^{39}Ar$ age trends younging eastward across the Lowlands (Fig. 6) are attributed to diachronous closure of both the hornblende and the biotite K-Ar systems. These trends are potentially explained by two end-member models or some combination thereof: (1) cooling through subhorizontal isotherms between ca. 1100–900 Ma, followed by gentle westward tilting of the Lowlands between ca. 900–520 Ma; or (2) cooling through gently west-dipping isotherms between ca. 1100–900 Ma, with no tilting. These simple "regional-tilting" and "inclined-isotherm" models, schematically represented in Figures 8 and 9, are necessarily limited by underlying assumptions about the Adirondack geotherm(s) in space and time, as noted earlier. A third potential model, involving progressive resetting of hornblende and biotite cooling ages across the Lowlands, is seemingly precluded by the observation of a consistent ~115 m.y. difference in apparent $^{40}Ar/^{39}Ar$ ages (Figs. 3–6; Table 3). If progressive resetting had occurred from west to east across the Lowlands, a differential kinetic response of hornblende and biotite to resetting would have been manifested as divergence in apparent hornblende-biotite ages, but this is not observed. Accordingly, this model is rejected pending the collection of additional data and not considered further in this paper.

Regional Tilting Model

According to the tilting model (see Figs. 8, A, and 9, A), a stable, long-lived pattern of horizontal isotherms (parallel with topography) was maintained in the midcrustal Lowlands, above and below current levels of exposure, between ca. 1100 Ma (the oldest hornblende plateau age) and at least ca. 920 Ma (the youngest biotite plateau age). Then the Lowlands were gently tilted to the west-northwest during the Neoproterozoic, sometime between ca. 920–900 Ma (final biotite closure) and ca. 520 Ma (nonconformable deposition of Potsdam Sandstone, which remains horizontal). Published U-Pb and $^{40}Ar/^{39}Ar$ ages in rocks from the Carthage-Colton shear zone and environs (Mezger et al., 1992; McLelland et al., 1996, 2001; Streepey et al., 2000, 2001a) suggest that extensional normal faulting occurred along the Carthage-Colton shear zone at ca. 1100, ca. 1050–1040, and/or ca. 945–940 Ma, thereby juxtaposing the Lowlands (hanging-wall) and the Highlands (footwall) in their current structural positions (Fig. 9, B, middle panel). Thus, as originally proposed by Pomfrey et al. (1998a, 1998b), subsequent tilting of the Lowlands and the Carthage-Colton shear zone could have resulted from post-900 Ma differential uplift (i.e., uplift accompanied by counterclockwise rotation) of a previously assembled Lowlands-Highlands crustal block, perhaps initiated by renewed extension (Fig. 9, A, bottom panel). Alternatively, if the Carthage-Colton shear zone is a listric fault, the observed age patterns could be explained by a scenario of post-900 Ma thrusting in which the Lowlands and the Highlands underwent differential counterclockwise rotation during an episode of compression (Fig. 9, A, inset, bottom panel). These potential tilting scenarios are considered further in a later section.

For this model the estimated amount of tilting corresponds to a difference of ~4.3 ± 1.4 km (i.e., ~1.2 ± 0.4 kilobars) in exposed crustal levels across the Lowlands. This estimate is calculated from (1) the regional difference between hornblende and biotite ages (Figs. 3 and 6), (2) the ~1.5 ± 0.5 (i.e., ~1–2) °C/m.y. cooling rate inferred earlier, and (3) the 30 ± 10 °C/km retrograde geotherm suggested for the easternmost Lowlands (Carthage-Colton shear zone; Lamb, 1993; see also Cosca et al., 1991, 1992). The model difference in exposed crustal levels is explained by regional tilting of ~9 ± 3° and erosion following biotite closure; this estimate factors in a slight west-northwest slope (~1°) in the present surface elevation.

The Lowlands age-distance trends are reminiscent of those documented in Phanerozoic terrains, which have also been attributed to postclosure tilting along normal faults (Dallmeyer and Sutter, 1976; Harrison et al., 1989; Haugerud, 1990; Roden-Tice and Wintsch, 2002). However, the Lowlands trend is perhaps unique in that the ages become younger closer to the domain-bounding planar(?) fault (Carthage-Colton shear zone), whereas the ages become older closer to respective listric faults in the cited analogues. This key difference is considered further in a later section.

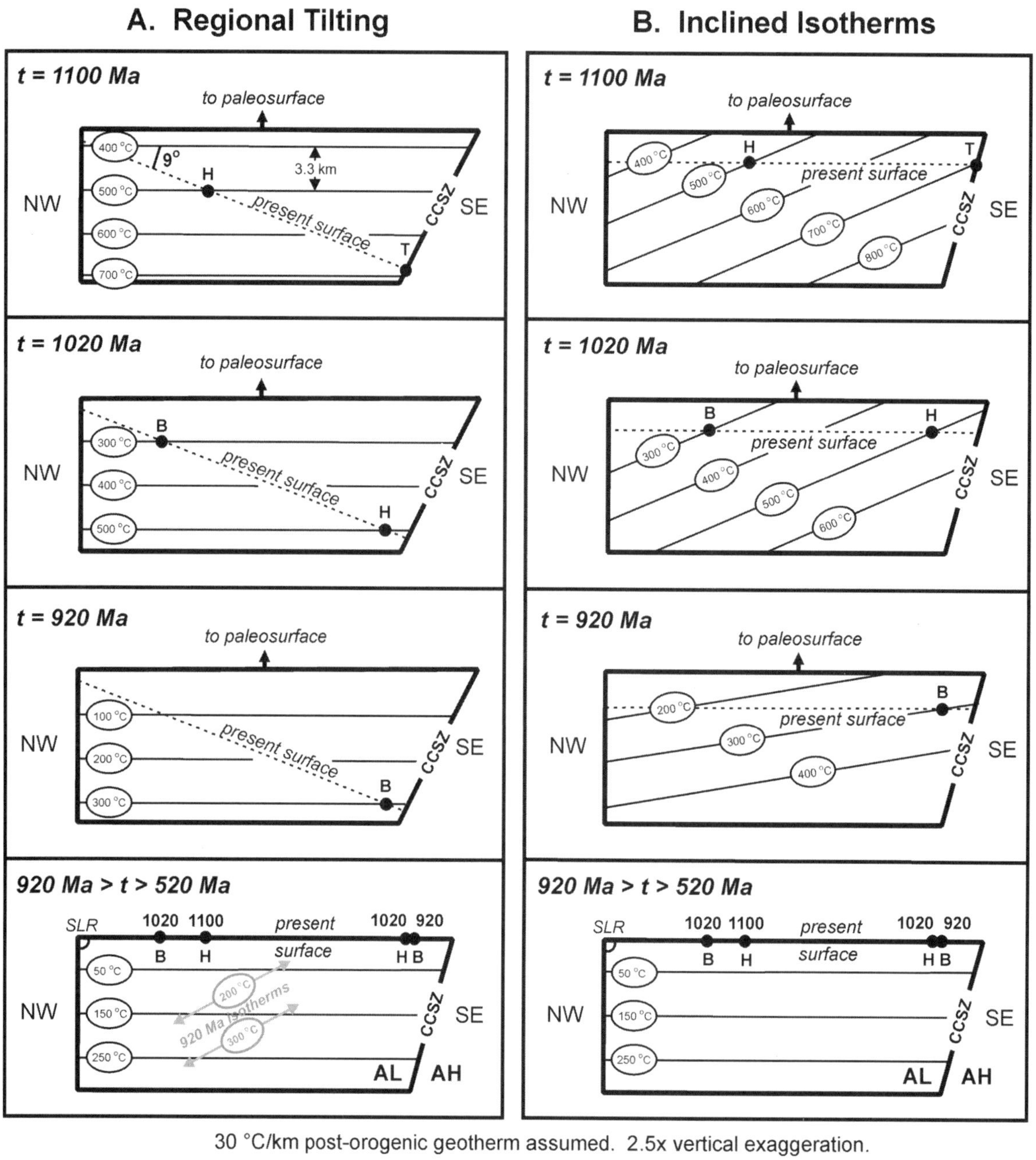

Figure 8. End-member models for the thermal evolution of the Adirondack Lowlands between ca. 1100 Ma and ca. 520 Ma, based upon regional patterns of hornblende and biotite $^{40}Ar/^{39}Ar$ ages (Fig. 6). (A) Regional-tilting model, in which the Lowlands are gently tilted to the west-northwest between ca. 920 and ca. 520 Ma. (B) Inclined-isotherms model, in which inclined isotherms regionally imposed at ca. 1100 Ma gradually decay into horizontal isotherms by ca. 920–520 Ma. Sequential northwest-southeast cross-sections are depicted in both diagrams. "B" and "H" appear corresponding to argon closure of biotites and hornblendes, respectively, at the temperatures and times indicated. CCSZ—Carthage-Colton shear zone (dotted line); AL and AH—Adirondack Lowlands and Highlands; SLR—St. Lawrence River; T—oldest titanite date in the CCSZ (Mezger et al., 1991, 1992, 1993). See text for discussion.

Inclined Isotherm Model

According to this model (see Figs. 8, B, and 9, B), the cooling Lowlands were juxtaposed against a warmer Highlands along the Carthage-Colton shear zone as early as ca. 1100–1040 Ma (i.e., long before the latest extension at ca. 945–940 Ma; Streepey et al., 2001a), which progressively delayed 1100–900 Ma regional cooling from northwest to southeast. A pattern of gently west-northwest–dipping isotherms was thereby maintained throughout the midcrustal Lowlands between ca. 1100 Ma and ca. 900 Ma, while hornblende and biotite cooled slowly through the 500–300 °C closure interval. The isotherms are presumed to have been inclined by ~9 ± 3° initially and to have

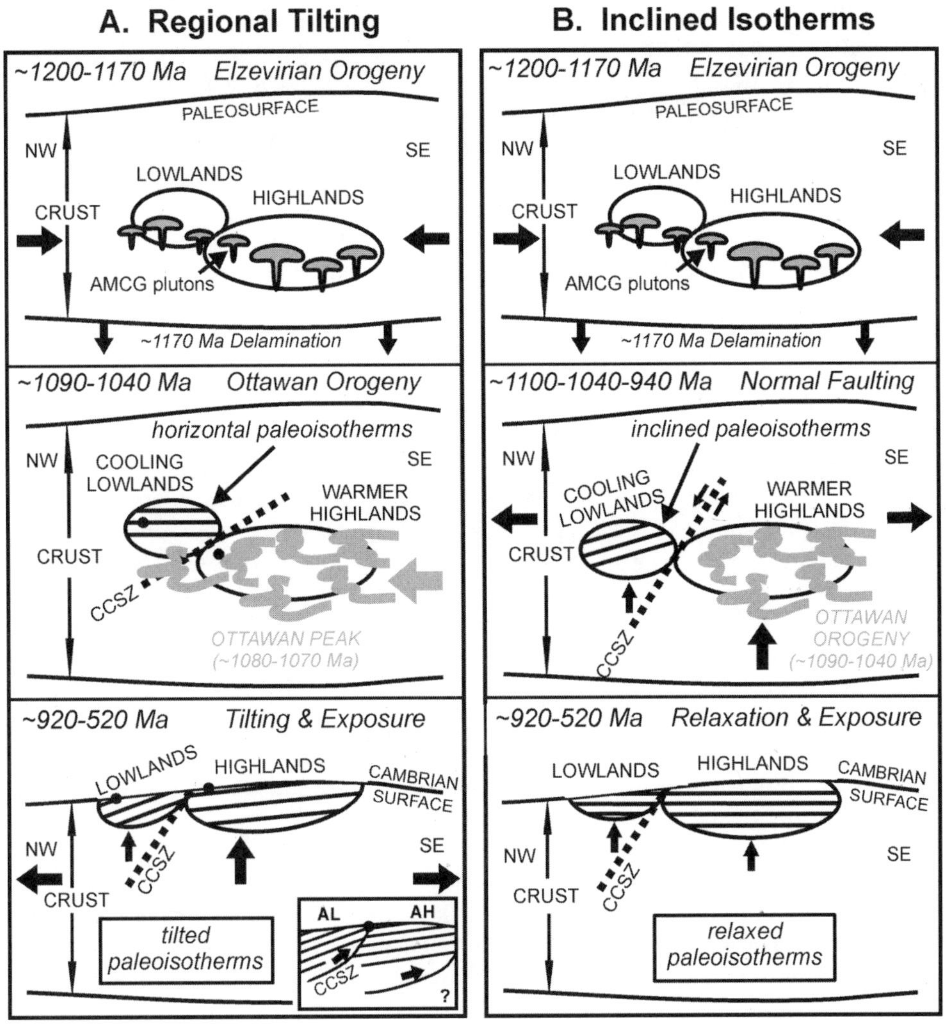

Figure 9. Schematic tectonic diagrams depicting midcrustal thermal evolution of the Lowlands hangingwall and environs, based upon regional patterns of hornblende and biotite $^{40}Ar/^{39}Ar$ ages (Fig. 6). (A) Regional-tilting model, in which the Adirondack Lowlands and Highlands (AL, AH) are differentially tilted to the west-northwest (or rotated counterclockwise; see inset, lower panel) between ca. 920 and 520 Ma. Black dots (middle and bottom panels) schematically indicate relative crustal levels of current exposures in the western Lowlands and western Highlands at ca. 1080–1070 Ma (Ottawan metamorphic peak) and ca. 920–520 Ma (final exposure), respectively. Black dot (inset, bottom panel) depicts the thermal equilibration of the eastern AL and the western AH by ca. 900 Ma and ~300 °C inferred by Streepey et al. (2000, 2001a), although listric faulting along the Carthage-Colton shear zone (CCSZ, dotted line) is conjectural. (B) Inclined-isotherms model, in which inclined isotherms regionally imposed at ca. 1100 Ma gradually decay into horizontal isotherms by ca. 920 Ma. Sequential northwest-southeast cross-sections are depicted in both diagrams, and model relative motions of the Lowlands and Highlands crustal blocks are depicted through time AMCG—anorthosite-mangerite-charnockite-granite suite. See text for discussion.

remained inclined throughout most of the ~200 m.y. interval, gradually relaxing to subhorizontal by ca. 920–900 Ma—that is, when the Lowlands and the western Highlands became thermally equilibrated (Streepey et al., 2000).

In order to maintain a pattern of gently inclined isotherms across the Lowlands for such a long time interval, a necessary (but not sufficient) requirement is a tectonic scenario wherein the Lowlands hangingwall was juxtaposed against a warmer Highlands footwall along the Carthage-Colton shear zone for the entire 1100–900 Ma interval. Indeed, the western Highlands (but not the Lowlands) underwent high-grade metamorphism during the 1090–1050 Ma Ottawan orogeny, both domains cooled slowly thereafter, and the Highlands stayed warmer than the Lowlands until ca. 920 Ma (McLelland et al., 1996; Streepey et al., 2000). Episodic normal faulting along the Carthage-Colton shear zone has been inferred at both ca. 1100 and ca. 1050–1040 Ma (McLelland et al., 1996), and the youngest extensional episode probably ended by ca. 940 Ma (Streepey et al., 2000). These thermotectonic boundary conditions nominally permit cooling across the Lowlands to have been progressively delayed from northwest to southeast (no tilting), but only if the bulk of normal faulting along the Carthage-Colton shear zone occurred relatively early within the 1100–940 Ma time frame, thereby juxtaposing the Lowlands against a warmer Highlands.

Evaluation of the Cooling Models

The two proposed cooling models must be evaluated in terms of independent geological and geochronological constraints, and various permutations of potential cooling and juxtaposition models should point to internally consistent thermotectonic scenarios that account for all observations. Thus, any scenarios for Lowlands-Highlands cooling and prior juxtaposition must account for the following geological boundary conditions, inferred from metamorphic rocks that are now exposed at the same structural level: (1) By ca. 1080–1070 Ma, when prograde Ottawan heating peaked at ~700–750 °C at ~25 km depths in the western Highlands, the western Lowlands were already undergoing retrograde cooling through ~500 °C at shallower crustal depths of ~18 ± 6 km (ca. 1080 Ma hornblende,

Fig. 6; ~30 ± 10 °C/km geotherm assumed; see Fig. 9, A, black dots in middle panel). (2) By ca. 920–900 Ma, the rocks now exposed in the easternmost Lowlands and the westernmost Highlands had cooled through ~300 °C and were thermally equilibrated at a common structural level (Streepey et al., 2000), estimated here at ~10–15 km (i.e., the retrograde geotherm had relaxed to ~30–20 °C/km by ca. 900 Ma; see Fig. 9, A, black dots in bottom panel). (3) Internal consistency between the cooling models (Fig. 8) and the $^{40}Ar/^{39}Ar$ ages (Figs. 4–6) requires that any regional tilting to produce the age-distance trend (Fig. 6) must have occurred after ca. 920–900 Ma (i.e., the youngest Lowlands biotites), whereas any inclined isotherms must have been imposed before ca. 1100–1040 Ma (i.e., the oldest Lowlands hornblendes).

It follows from boundary condition (3) that deciding upon a cooling model, or a combination thereof, partly depends upon knowing the relative magnitude of extensional normal faulting along the Carthage-Colton shear zone through time. However, the fault displacements along the Carthage-Colton shear zone at ca. 1100, ca. 1050–1040, and ca. 945–940 Ma have not been quantified, and, as noted earlier, the extent to which the Lowlands experienced Ottawan thermotectonism remains uncertain. Thus, the previously proposed juxtaposition models are themselves somewhat limited by uncertainty in the relative vertical and lateral positions of present-day Lowlands and Highlands exposures during the 1100–900 Ma interval bracketed by our $^{40}Ar/^{39}Ar$ data (Table 3; Fig. 6). Moreover, the net magnitude of vertical offset across the Carthage-Colton shear zone is obscured by the diachroneity of metamorphic pressures from the Lowlands to the Highlands, as noted earlier. Thus, other geological evidence must be considered in order to evaluate the two cooling models.

Based upon additional evidence, a scenario of post–900 Ma northwestward tilting of the Lowlands is relatively more straightforward to defend than a scenario of ca. 1100–900 Ma inclined isotherms. First, regional tilting is suggested by the fact that the 1172 Ma Rockport granite exhibits crosscutting contacts northwest of the St. Lawrence River, versus only concordant contacts in the Lowlands at nearly the same elevation, suggesting that shallower emplacement levels are exposed to the northwest (Wasteneys et al., 1999). Second, a metamorphic pressure difference (~0–2 kilobars) is observed in surface rocks exposed across the Lowlands; independent geobarometry indicates Elzevirian-AMCG paleopressures of ~6.5 ± 0.5 and ~7.0 ± 0.5 kilobars in the northwest and southeast Lowlands, respectively (Bohlen et al., 1985). Thus, whereas the actual pressures are subject to significant uncertainties, the barometric trend across the Lowlands favors regional tilting over inclined isotherms. Somewhat higher paleopressures of ~7.5–8.0 kilobars recorded in lower-granulite-facies assemblages of the western Highlands (Bohlen et al., 1985) might also be considered consistent with regional tilting, but again, they reflect strictly Ottawan conditions and cannot be used to address differential vertical movement of the Lowlands and the Highlands (cf. Streepey et al., 2000, 2001a). Third, the observed west-to-east progression of Elzevirian upper-amphibolite- to lower-granulite-facies mineral assemblages (see Fig. 3, orthopyroxene isograd) supports a scenario of post-Ottawan regional tilting that also tilted the older, presumably Elzevirian, isograd. Moreover, a proposal of northwestward tilting of the Lowlands is also compatible with the near-constancy of metamorphic temperatures on a northeast-southwest axis along strike (Figs. 2 and 3; Kitchen and Valley, 1995; Liogys and Jenkins, 2000). In addition, the minimum ~3 km of model tilting and the minimum ~20 °C/km estimated geotherm, taken together, imply a minimum ~60 °C difference in peak metamorphic temperature from west to east across the Lowlands, which is consistent with the maximum difference of ~55–90 °C independently recorded in the geothermometry (Fig. 3). However, whereas the age gradient is smooth across the Lowlands, the metamorphic gradient apparently is not (Fig. 3), which may indicate that calcite-graphite temperatures do not always reflect peak temperature throughout the Lowlands. Fourth, both the timing and the angle or direction of Lowlands tilting inferred from the regional age-distance trends (Figs. 6; 8, A; and 9, A) are supported by the results of kinematic/dynamic modeling of the Frontenac terrane recently conducted by Streepey et al. (2001b). They invoked a model of magma-derived upward stresses localized under the western Highlands, coupled with far-field extension (perhaps associated with local rifting at ca. 730 Ma; Heizler and Harrison, 1998), to explain the lengthy time lag between Ottawan orogenic contraction and post-Ottawan extension along the bounding shear zones of the Frontenac terrane (including the Carthage-Colton shear zone; see also Fig. 10). Finally, two-dimensional thermal modeling would be required to determine how far laterally into the Lowlands hangingwall the thermal effects of footwall heating might have been felt and over what time frame. However, it seems unlikely that conductive heat transfer could have been sustained for ~200 m.y. laterally over tens of kilometers. Moreover, most of the normal faulting along the Carthage-Colton shear zone may have occurred at ca. 940 Ma, as inferred later, which, if true, precludes a scenario of a cold Lowlands hangingwall laterally juxtaposed against a warm Highlands footwall during most of the 1100–900 Ma Lowlands cooling interval. These considerations appear to rule out the inclined-isotherms model of progressively delayed cooling from west to east across the Lowlands against a warmer Highlands to the east.

In summary, independent lines of evidence appear to support a scenario of 1100–900 Ma downward cooling of the Lowlands through subhorizontal isotherms followed by NW regional tilting between ca. 900 and ca. 520 Ma, and no known evidence refutes it. In contrast, a scenario of ca. 1100–900 Ma cooling through inclined isotherms sustained from west to east across the Lowlands is difficult to defend, based upon the data available. Thus, the regional tilting model is favored (Figs. 8, A, and 9, A; bottom panels) and the inclined isotherms model is rejected (Figs. 8, B, and 9, B, bottom panels), pending the acquisition of additional information.

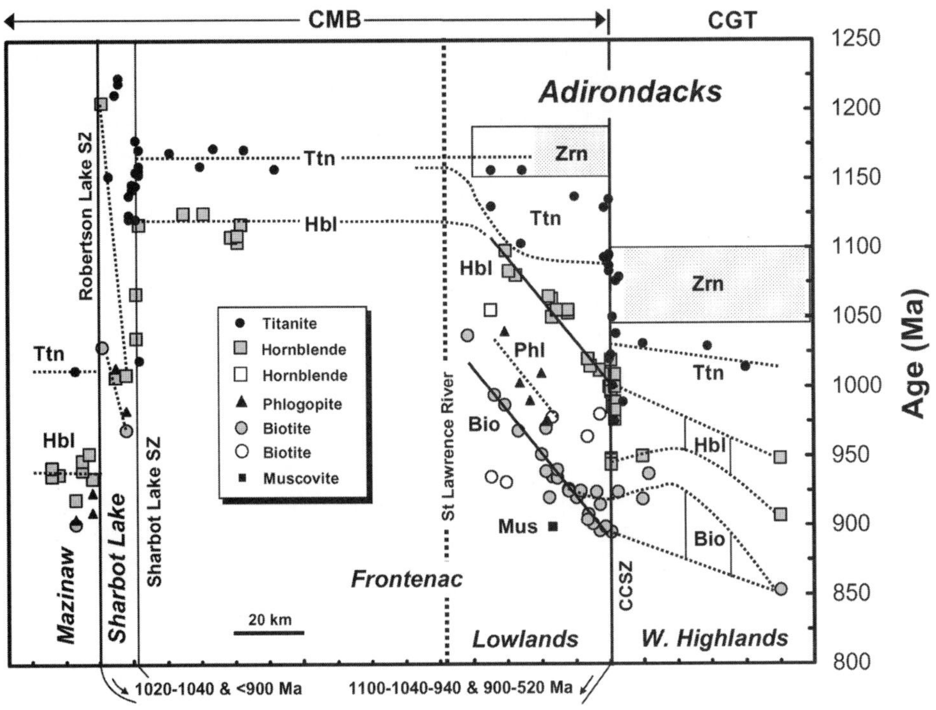

Figure 10. Summary plot of U-Pb and $^{40}Ar/^{39}Ar$ mineral ages (published in the sources listed in this paragraph and in this study, Fig. 6) for the eastern part of the Central Metasedimentary Belt and the western part of the Central Granulite terrane. All distances are referenced to the Carthage-Colton shear zone (CCSZ), as measured normal to regional tectonic strike. U-Pb zircon (Zrn) and titanite (Ttn) ages are from various sources (Mezger et al., 1991, 1992, 1993; McLelland et al., 1996, 2001; Corfu and Easton, 1997; and Streepey et al., 2001a). $^{40}Ar/^{39}Ar$ hornblende (Hbl), biotite (Bio), phlogopite (Phl), and muscovite (Mus) ages are also from various sources (Onstott and Peacock, 1987; Cosca et al., 1991, 1992; Busch and van der Pluijm, 1995; Busch et al., 1996; Foland et al., 1998; Streepey et al., 2000, 2001a; and this study). See text for discussion.

New Constraints on Lowlands-Highlands Juxtapositional History

Two juxtaposition models involving the Carthage-Colton shear zone have been proposed that satisfy the three thermotectonic boundary conditions identified earlier. According to one scenario, the Lowlands were thrust over the Highlands during the Elzevirian orogeny at ca. 1200–1170 Ma (Wasteneys et al., 1999), and late syn-Ottawan collapse at ca. 1050–1040 Ma juxtaposed these domains in their current relative positions (McLelland et al., 2001). In an alternate scenario, these domains were apparently (re)connected during late Ottawan transpression at ca. 1040 Ma, possibly underwent ca. 1040 Ma syn-post-Ottawan collapse, and then experienced yet a later episode of ca. 945–940 Ma extension (Streepey et al., 2000, 2001a, 2001b). Both scenarios permit an intervening episode of ca. 1100 Ma rifting, as inferred by Mezger et al. (1992). Heyn (1990) recognized two main generations of mylonite in the Carthage-Colton shear zone: (1) a high-grade ultramylonite and (2) a younger crosscutting mylonite with medium-grade mineral assemblages and Lowlands-down kinematic indicators. Grant (1993) determined Rb-Sr whole-rock ages of 1175 ± 41 and 941 ± 33 Ma for these two mylonites, respectively, lending support to parts of both tectonic scenarios outlined above, namely, that initial domain suturing occurred at ca. 1200–1170 Ma (reverse faulting) and that final domain juxtaposition occurred at ca. 945–940 Ma (normal faulting). Intervening episodes of rifting, inferred at ca. 1100 and ca. 1040 Ma, and/or of transpression, inferred at ca. 1050 Ma, are not precluded. However, it appears from the inferred exhumation rates that the Highlands were rising faster than the Lowlands between ca. 1080 and ca. 920 Ma, and that the Carthage-Colton shear zone was predominantly an extensional normal fault during this time frame. Thus, following boundary conditions (1) and (2), inferred earlier, the western Lowlands were exhumed from ~18 km to a depth of ~15 km, while the western Highlands were exhumed from ~25 to ~15 km. This implies a net ~7 km of vertical displacement by extensional normal faulting along the Carthage-Colton shear zone at ca. 1100, ca. 1050–1040, and ca. 945–940 Ma prior to final thermal equilibration of the eastern Lowlands and the western Highlands at ca. 920–900 Ma.

The main evidence cited by Streepey et al. (2000) to support extensional normal faulting at ca. 945–940 Ma along the Carthage-Colton shear zone is an ~100 m.y. offset they inferred between two $^{40}Ar/^{39}Ar$ hornblende ages in the east-central Lowlands and one in the western Highlands (1055–1065 Ma vs. 950 Ma). In estimating this offset, they assumed uniform hornblende ages in the two domains. However, the merged hornblende and biotite data sets (Fig. 10) show clear-cut parallel decreases in $^{40}Ar/^{39}Ar$ age from northwest to southeast across the Lowlands, trending to ca. 1000 and ca. 920–900 Ma, respectively, at the Carthage-Colton shear zone, and then flattening out into the Highlands. Thus, it appears that the actual offset in hornblende ages across the Carthage-Colton shear zone is ~50 m.y. or less (cf., no age offset for biotite). Less offset in the hornblende cooling ages translates into less vertical throw along the Carthage-Colton shear zone during ca. 945–940 Ma normal faulting, which is estimated at ~5 km or less (assuming a 20 °C/km geotherm by this time frame and a 2 °C/m.y. cooling rate in the Carthage-Colton shear zone). Thus, only ~2 km or more of the net ~7 km of normal displacement inferred earlier may have occurred between ca. 1100 and ca. 1040 Ma. These first-order approximations may begin to constrain the magnitude of episodic normal faulting along the Carthage-Colton shear zone within the known ca. 1100–940 Ma extensional time frame. If so, most of the normal faulting along the Carthage-Colton shear zone may have occurred during the ca. 945–940 Ma extensional event inferred by Streepey et al. (2001a), which would further preclude the inclined-isotherms cooling model for the Lowlands in terms of boundary condition (3) inferred earlier.

To account for the apparent absence of Ottawan assemblages in the Lowlands, the first tectonic scenario holds that the Lowlands rested sufficiently above and west of the Highlands during the ca. 1080–1070 Ma peak of Ottawan orogeny in the western Highlands. In contrast, the second scenario holds that the two crustal domains were fully separated laterally until the ca. 1040 Ma waning stages of the Ottawan orogeny. However, there is evidence for Ottawan deformation in the Lowlands (Wasteneys et al., 1999), which raises some doubt regarding published claims (Mezger et al., 1991, 1992, 1993; Streepey et al., 2000, 2001a) that there was no Ottawan reheating here, as well as the presumed lateral separation of the Lowlands and Highlands. Moreover, the restricted occurrence of ca. 1156–1103 Ma bulk titanite dates in the easternmost but not the western Frontenac terrane (Fig. 10) raises the possibility that mixed ages in the Lowlands arose from an Ottawan overprint (either prograde or retrograde) of older Elzevirian-AMCG assemblages. Thus, there appears to be no a priori requirement that the Lowlands and the Highlands were laterally separated early during Ottawan time. Removal of this prior constraint allows the possibility that during the Ottawan metamorphic peak (ca. 1080–1070 Ma) the western Lowlands domain of a previously assembled Lowlands-Highlands crustal block resided at shallower structural levels (i.e., 18 ± 6 km and 500 °C; boundary condition 1) atop and west of the structurally deeper western Highlands domain (i.e., ~25 km and 700–750 °C). Therefore, the ca. 1100–850 Ma $^{40}Ar/^{39}Ar$ ages of hornblende and biotite plotted in Figure 7 can be envisioned as reflecting a regime of predominantly post-Ottawan slow cooling of the Lowlands and the Highlands at different crustal levels. Episodic normal faulting along the Carthage-Colton shear zone during cooling then juxtaposed the two domains in their regional pretilting (i.e., pre-920 Ma) configuration.

Between ca. 920 and 520 Ma, the Lowlands domain underwent ~4 km of counterclockwise rotation (i.e., ~9° of northwestward tilting) as the rocks currently exposed were finally exhumed from a depth of ~15 km to the surface. This scenario accounts for the $^{40}Ar/^{39}Ar$ age-distance trends evident in the Lowlands and the Highlands (Figs. 6 and 10), but not for the underlying cause of the regional tilting. Conceptually, the inferred ca. 920–520 Ma rotation (tilting) could have resulted from extension or compression. A model of differential Lowlands uplift accompanied by renewed extension (Figs. 8, A–9, A) was originally proposed by Pomfrey et al. (1998a, 1998b) to explain the $^{40}Ar/^{39}Ar$ age-distance trend (Fig. 6), and a viable mechanism involving magma-derived upward stresses localized under the western Highlands coupled with far-field extension was provided by Streepey et al. (2001a, 2001b). By this model, a presumably planar Carthage-Colton shear zone also underwent counterclockwise rotation along the Highlands, as depicted in Figure 9, A (bottom panel). Alternatively, the inferred ~9° of rotation in the Lowlands (less in the Highlands) and resultant $^{40}Ar/^{39}Ar$ age patterns (Figs. 6 and 10) could have originated by thrusting if the Carthage-Colton shear zone was a listric fault (cf. Roden-Tice and Wintsch, 2002), as also depicted in Figure 9, A (see inset, bottom panel). However, geophysical imaging of Adirondacks crust indicates that the Carthage-Colton shear zone is seismically transparent (Klemperer et al., 1985, cited in Grant, 1993), so there is no way to discern fault-plane geometry at depth, and kinematic indicators in the Carthage-Colton shear zone are not definitive. Thus, selecting the better of the Lowlands-Highlands tilting scenarios requires an alternate approach based upon regional age patterns across the Frontenac and adjacent Grenville terranes (Fig. 10).

A comprehensive mineral-age compilation of the southernmost Grenville province is presented in Figure 10. The regional pattern is one of distinctly *older* mineral ages in the Frontenac (including the Lowlands) terrane of the Central Metasedimentary Belt and relatively *younger* ones in the adjacent Mazinaw and Central Granulite (including Highlands) terranes to the northwest and southeast, respectively. This pattern is observed for all minerals for which regional ages exist, including zircon, monazite, titanite, hornblende, phlogopite, and biotite (Fig. 10); moreover, the same relative-age sequence is observed in all terranes. Also, there is a pronounced asymmetry in the regional age-distance patterns, as highlighted by new $^{40}Ar/^{39}Ar$ age trends in the Lowlands (Figs. 6 and 10) that narrow the previous Highlands-Frontenac age gap, and the major domain-bounding shear zones (the northwest-dipping Carthage-Colton and the southeast-dipping Robertson Lake zones; Fig. 10) separate broad areas of older mineral ages from relatively younger ones. These shear zones also preserve diachronous ca. 945–940 and ca. 900–780 Ma ages of latest extensional normal faulting, respectively ($^{40}Ar/^{39}Ar$ hornblende and feldspar; Streepey et al., 2001a, 2001b). Geobarometric data (Streepey et al., 1997) mirror the regional age trends shown in Figure 10 in that lower pressures are recorded in the Frontenac terrane relative to the Mazinaw terrane and the Highlands domain, although the pressures are not necessarily all synchronous (e.g., from Lowlands to Highlands). Thus, the regional age and (possibly) pressure patterns, combined with inward dips of the major shear zones (Busch and van der Pluijm, 1996; Busch et al., 1996; McLelland et al., 1996), suggest that the Frontenac (including the Lowlands) terrane is a large-scale downdropped crustal block (graben). These features further suggest an extensional rather than a compressional regime between ca. 920 and 520 Ma, which is supported by the inference of ca. 730 Ma reheating in the westernmost Adirondacks that is considered to be rift-related

(related to the breakup of Rodinia; Heizler and Harrison, 1998). Thus, a scenario of late thrusting along a listric Carthage-Colton shear zone appears unlikely, and would seem to require synchronous rifting both to the west and to the east of the Carthage-Colton shear zone, for which there is no evidence. Instead, a combination of ca. 900–800 Ma far-field extension and localized upward stresses centered beneath the western Highlands, as postulated by Streepey et al. (2001b), probably caused the slight northwest tilting of the eastern Frontenac–western Highlands block independently inferred from the $^{40}Ar/^{39}Ar$ age-distance trends documented in this paper.

CONCLUSIONS

A ca. 1200–520 Ma thermotectonic scenario is envisioned for the Adirondack Lowlands based upon new and published data integrated in this paper. According to this scenario, the Adirondack Lowlands were (1) thrust over the Highlands along the Carthage-Colton shear zone (ca. 1200–1170 Ma, Elzevirian orogeny); (2) slowly cooled through subhorizontal isotherms and unroofed (~1–2 °C/m.y.; ~0.04–0.08 km/m.y.) between ca. 1100 and 900 Ma following the Ottawan orogeny, which affected mostly deeper Highlands crust; (3) downdropped and displaced westward by several km along the Carthage-Colton shear zone, against a slightly warmer and thus faster-cooling Highlands footwall (at ca. 1100, ca. 1040, and [mostly] ca. 945–940 Ma); and finally (4) tilted by ~9 ± 3° to the northwest, along with the Carthage-Colton shear zone and the western Highlands, during an episode of renewed uplift and extension (ca. 900–520 Ma). The ca. 1100–1040–940 Ma interval of extensional normal faulting supported in this paper is essentially the metamorphic core complex scenario for the Central Metasedimentary Belt–Central Granulite terrane boundary favored by McLelland et al. (1996, 2001) and Wasteneys et al. (1999), with the Lowlands representing the cover rocks and the Carthage-Colton shear zone representing the detachment fault. If this scenario is correct, then the western Adirondacks represent a perhaps unique midcrustal analog of the more typical shallow-crustal, Cordilleran-type core complexes. The regional tilting scenario documented in this paper also involves the Carthage-Colton shear zone, implying that previously it had a gentler dip of ~30–35° (cf., the present ~40–45° dip). Regional tilting further implies postclosure (i.e., ca. 900–520 Ma) differential uplift of a previously juxtaposed Lowlands-Highlands block, with relatively greater uplift having occurred in the Highlands portion. This scenario of differential uplift accompanied by renewed extension (Figs. 8, A–9, A) was originally proposed by Pomfrey et al. (1998a, 1998b) to explain the age-distance trend (Fig. 6). This scenario has been attributed to far-field extension coupled with magma-related uplift localized under the western Highlands (Streepey et al., 2000, 2001a, 2001b), possibly associated with a ca. 730 Ma rifting event (Heizler and Harrison, 1998).

Finally, the ca. 1090–1040 Ma Ottawan orogeny is thought to have resulted from Mesoproterozoic collision of Amazonia with the Adirondacks during the assembly of supercontinent Rodinia (Hoffman, 1988, 1989), whereas later extension at ca. 945–940 Ma may have heralded the Neoproterozoic breakup of Rodinia that was apparently under way in the western Adirondacks by ca. 730 Ma (Streepey et al., 2000, 2001a; Heizler and Harrison, 1998). Thus, the 1100–900 Ma $^{40}Ar/^{39}Ar$ cooling ages, the Lowlands age-distance trends, and the post-900 Ma tilting of the Lowlands domain—as presented and documented in this paper—can all be interpreted as local thermotectonic manifestations of the assembly and breakup of Rodinia.

ACKNOWLEDGMENTS

We are grateful to Bill deLorraine, Mike Hudson, and Xifan Zhang for support and assistance in the field; to Mike Dorais, Fritz Hubacher, and Jeff Linder for laboratory and computer graphics support; to Phil Whitney and the late Yngvar Isachsen for discussions of Adirondacks geology; and to Jim McLelland and Kelin Whipple for their comments on an earlier manuscript version. We appreciate thoughtful and constructive reviews by Suzanne Baldwin and Robert Wintsch, which led to a much-improved manuscript, and we thank Jim McLelland for his editorial handling of the work. This work was supported by grants from the National Science Foundation (EAR-0106987 and EAR-9614822 to PSD; EAR-0107054, EAR-9701496, EAR-9220344, and EAR-0137546 to KAF).

REFERENCES CITED

Bohlen, S., 1987, Pressure-temperature-time paths and a tectonic model for the evolution of granulites: Journal of Geology, v. 95, p. 617–632.

Bohlen, S., Valley, J., and Essene, E., 1985, Metamorphism in the Adirondacks: Petrology, pressure and temperature: Journal of Petrology, v. 26, p. 971–992.

Buddington, A.F., 1939, Adirondack igneous rocks and their metamorphism: Boulder, Colorado, Geological Society of America Memoir 7, 354 p.

Buddington, A.F., 1963, Metasomatic origin of a large part of the Adirondack phacoliths: A discussion: Geological Society of America Bulletin, v. 74, p. 353.

Burrett, C., and Berry, R., 2000, Proterozoic Australia–Western United States (AUSWUS) fit between Laurentia and Australia: Geology, v. 28, p. 103–106.

Busch, J.P., and van der Pluijm, B.A., 1996, Late orogenic, plastic to brittle extension along the Robertson Lake shear zone: Implications for the style of deep-orogenic extension in the Grenville orogen, Canada: Precambrian Research, v. 77, p. 41–57.

Busch, J.P., van der Pluijm, B.A., Hall, C.M., and Essene, E.J., 1996, Listric normal faulting during postorogenic extension revealed by $^{40}Ar/^{39}Ar$ thermochronology near the Robertson Lake shear zone, Grenville Orogen, Canada: Tectonics, v. 15, p. 387–402.

Carl, J.D., and deLorraine, W.F., 1997, Geochemical and field characteristics of metamorphic granitic rocks, N.W. Adirondack Lowlands, New York: Northeastern Geology and Environmental Sciences, v. 19, p. 276–301.

Corfu, F., and Easton, R.M., 1997, Sharbot Lake terrane and its relationships to Frontenac terrane, Central Metasedimentary Belt, Grenville Province: New insights from U-Pb geochronology: Canadian Journal of Earth Sciences, v. 34, p. 1239–1257.

Corrigan, D., and Hanmer, S., 1997, Anorthosites and related granitoids in the Grenville Orogen: A product of convective thinning of the lithosphere: Geology, v. 25, p. 60–64.

Corrigan, D., and van Breemen, O., 1997, U-Pb age constraints for the lithotectonic evolution of the Grenville Province along the Mauricie transect, Quebec: Canadian Journal of Earth Science, v. 34, p. 299–316.

Cosca, M.A., Sutter, J.F., and Essene, E.J., 1991, Cooling and inferred uplift/erosion history of the Grenville Orogen, Ontario: Constraints from $^{40}Ar/^{39}Ar$ thermochronology: Tectonics, v. 10, p. 959–997.

Cosca, M.A., Essene, E.J., Kunk, M.J., and Sutter, J.F., 1992, Differential unroofing within the Central Metasedimentary Belt of the Grenville Orogen: Constraints from $^{40}Ar/^{39}Ar$ thermochronology: Contributions to Mineralogy and Petrology, v. 110, p. 211–225.

Cosca, M.A., Essene, E.J., Mezger, K., and van der Pluijm, B.A., 1995, Constraints on the duration of tectonic processes: Protracted extension and deep-crustal rotation in the Grenville orogen: Geology, v. 23, p. 361–364.

Cosca, M.A., Mezger, K., and Essene, E.J., 1998, The Baltica-Laurentia connection: Sveconorwegian (Grenvillian) metamorphism, cooling, and unroofing in the Bamble Sector, Norway: Journal of Geology, v. 106, p. 539–552.

Dahl, P.S., 1996a, The crystal-chemical basis for Ar retention in micas: Inferences from interlayer partitioning and implications for geochronology: Contributions to Mineralogy and Petrology, v. 123, p. 22–39.

Dahl, P.S., 1996b, The effects of composition on retentivity of argon and oxygen in hornblende and related amphiboles: A field-tested empirical model: Geochimica et Cosmochimica Acta, v. 60, p. 3687–3700.

Dahl, P.S., 1997, A crystal-chemical basis for Pb retention and fission-track annealing systematics in U-bearing minerals, with implications for geochronology: Earth and Planetary Science Letters, v. 150, p. 277–290.

Dahl, P.S., and Dorais, M.J., 1996, Influence of $F(OH)_{-1}$ substitution on the relative mechanical strength of rock-forming micas: Journal of Geophysical Research, v. 101, no. B5, p. 11519–11524.

Dahl, P.S., Foland, K.A., and Pomfrey, M.E., 2001, $^{40}Ar/^{39}Ar$ thermochronology of hornblende and biotite, Adirondack Lowlands (New York), with implications for evolution of a major shear zone: Geological Society of America Abstracts with Programs, v. 33, no. 6, p. A292.

Dallmeyer, R.D., and Sutter, J.F., 1976, $^{40}Ar/^{39}Ar$ incremental-release ages of biotite and hornblende from variably retrograded basement gneisses of the northeasternmost Reading Prong, New York: Their bearing on Early Paleozoic metamorphic history: American Journal of Science, v. 276, p. 731–747.

Daly, J., and McLelland, J., 1991, Juvenile Middle Proterozoic crust in the Adirondack Highlands, Grenville province, northeastern North America: Geology, v. 19, p. 119–122.

deWaard, D., 1965, The occurrence of garnet in the granulite-facies terrane of the Adirondack Highlands: Journal of Petrology, v. 6, p. 165–191.

deWaard, D., 1970, The anorthosite-charnockite suite of rocks of Roaring Brook Valley in the eastern Adirondacks (Marcy Massif): American Mineralogist, v. 55, p. 2063–2075.

Easton, R.M., 1989, Regional mapping and stratigraphic studies, Grenville province: Ontario Geologic Survey Miscellaneous Papers, v. 146, p. 153–157.

Engel, A., and Engel, C., 1958, Progressive metamorphism and granitization of the major Paragneiss, Northwest Adirondack Mountains, New York: Geological Society of America Bulletin, v. 69, p. 1369–1414.

Engel, A., and Engel, C., 1963, Metasomatic origin of large parts of the Adirondack phacoliths: Geological Society of America Bulletin, v. 74, p. 349–354.

Foland, K.A., 1983, $^{40}Ar/^{39}Ar$ incremental heating plateaus for biotites with excess argon: Isotope Geoscience, v. 1, p. 3–21.

Foland, K.A., Linder, J.S., Lakowski, T.E., and Grant, N.K., 1984, $^{40}Ar/^{39}Ar$ dating of glauconites: Measured ^{39}Ar recoil loss from well-crystallized specimens: Isotope Geoscience, v. 2, p. 241–264.

Foland, K.A., Fleming, T.H., Heimann, A., and Elliot, D.H., 1993, Potassium-argon dating of fine-grained basalts with massive Ar loss: Application of the $^{40}Ar/^{39}Ar$ technique to plagioclase and glass from the Kirkpatrick Basalt, Antarctica: Chemical Geology (Isotope Geoscience Section), v. 107, p. 173–190.

Foland K.A., Dahl, P.S., Pomfrey, M.E., Zhang, X., and Hubacher, F.A., 1998, Contrasting phlogopite and biotite $^{40}Ar/^{39}Ar$ ages from diverse lithologies of the central Adirondack Lowlands, New York: Implications for Ar retention properties of micas: Geological Society of America Abstracts with Programs, v. 30, no. 7, p. A-186.

Grant, N.K., 1993, Annual field trip guide (New York): Oxford, Ohio, Friends of the Grenville, p. 1–63.

Grant, N.K., Lepak, R.J., Maher, T.M., Hudson, M.R., and Carl, J.D., 1986, Geochronological framework for the Grenville rocks in the Adirondack Mountains: Geological Society of America Abstracts with Programs, v. 18, p. 620.

Harrison, T.M., Spear, F.S., and Heizler, M.T., 1989, Geochronologic studies in central New England, II: Post-Acadian hinged and differential uplift: Geology, v. 17, p. 185–189.

Haugerud, R.A., 1990, Comment and Reply on "Geochronologic studies in central New England, II: Post-Acadian hinged and differential uplift": Geology, v. 18, p. 183–185.

Heizler, M.T., and Harrison, T.M., 1998, The thermal history of the New York basement determined from $^{40}Ar/^{39}Ar$ K-feldspar studies: Journal of Geophysical Research, v. 103, p. 795–814.

Heyn, T., 1990, Tectonites of the northwest Adirondack Mountains, New York: Structural and metamorphic evolution [Ph.D. dissertation]: Ithaca, New York, Cornell University.

Hoffman, K.S., 1982, Investigation of the orthopyroxene isograd, northwest Adirondacks [M.S. thesis]: Ann Arbor: University of Michigan.

Hoffman, P.F., 1988, United plates of America, the birth of a craton: Early Proterozoic assembly and growth of Laurentia: Annual Reviews of Earth and Planetary Sciences, v. 16, 543–603.

Hoffman, P.F., 1989, Precambrian geology and tectonic history of North America, *in* Bally, A.W., and Palmer, A.R., eds., The geology of North America—An overview: Boulder, Colorado, Geological Society of America, Geology of North America, v. A, p. 447–512.

Hudson, M.R., 1996, *P-T-X-t* constraints on ductile deformation zones within the Adirondack Lowlands [Ph.D. dissertation]: Oxford, Ohio, Miami University.

Isachsen, Y.W., and Fisher, D.W., 1970, Geologic map of New York: Albany, New York State Museum and Science Service Map and Chart Series, no. 15.

Isachsen, Y.W., and Geraghty, E.P., 1986, The Carthage-Colton Mylonite Zone: A major ductile fault in the Grenville Province: International Basement Tectonics Association Proceedings, v. 6, p. 199–200.

Kitchen, N.E., and Valley, J.W., 1995, Carbon isotope thermometry in marbles of the Adirondack Mountains, New York: Journal of Metamorphic Geology, v. 13, p. 577–594.

Klemperer, S.L., Brown, L.D., Oliver, J.E., Ando, C.J., Czuchra, B.L., and Kaufman, S., 1985, Some results of COCORP seismic reflection profiling in the Grenville-age Adirondack Mountains, New York State: Canadian Journal of Earth Science, v. 22, p. 141–153.

Lamb, W.M., 1993, Retrograde deformation within the Carthage-Colton Zone as recorded by fluid inclusions and feldspar compositions: Tectonic implications for the southern Grenville Province: Contributions to Mineralogy and Petrology, v. 114, p. 379–394.

Liogys, V.A., and Jenkins, D.M., 2000, Hornblende geothermometry of amphibolite layers of the Popple Hill gneiss, north-west Adirondack Lowlands, New York, USA: Journal of Metamorphic Geology, v. 18, p. 513–530.

Martignole, J., and Reynolds, P., 1997, $^{40}Ar/^{39}Ar$ thermochronology along a western Quebec transect of the Grenville province, Canada: Journal of Metamorphic Geology, v. 15, p. 283–296.

McDougall, I., and Harrison, T.M., 1999, Geochronology and thermochronology by the $^{40}Ar/^{39}Ar$ method (2nd ed.): New York, Oxford University Press.

McLelland, J., Chiarenzelli, J., Whitney, P., and Isachsen, Y., 1988, U-Pb geochronology of the Adirondack Mountains and implications for their geologic evolution: Geology, v. 16, p. 920–924.

McLelland, J., Daly, S., and Chiarenzelli, J., 1993, Sm-Nd and U-Pb isotopic evidence of juvenile crust in the Adirondack Lowlands and implications for the evolution of the Adirondack Mts.: Journal of Geology, v. 101, p. 97–105.

McLelland, J., Daly, J.S., and McLelland, J.M., 1996, The Grenville Orogenic Cycle (ca. 1350–1000 Ma): An Adirondack perspective: Tectonophysics, v. 265, p. 1–28.

McLelland, J.A., Hamilton, M., Selleck, B., McLelland, J.M., Walker, D., and Orrell, S., 2001, Zircon U-Pb geochronology of the Ottawan orogeny, Adirondack Highlands, New York: Regional and tectonic implications: Precambrian Research, v. 109, p. 39–72.

Mezger, K., Rawnsley, C.M., Bohlen, S.R., and Hanson, G.N., 1991, U-Pb garnet, sphene, monazite, and rutile ages: Implications for the duration of high-grade metamorphism and cooling histories, Adirondack Mts., New York: Journal of Geology, v. 99, p. 415–428.

Mezger, K., van der Pluijm, B.A., Essene, E.J., and Halliday, A.N., 1992, The Carthage-Colton mylonite zone (Adirondack Mts.): The site of a cryptic suture?: Journal of Geology, v. 100, p. 630–638.

Mezger, K., Essene, E.J., van der Pluijm, B.A., and Halliday, A.N., 1993, U-Pb geochronology of the Grenville Orogen of Ontario and New York: Constraints on ancient crustal tectonics: Contributions to Mineralogy and Petrology, v. 114, p. 13–26.

Onstott, T.C., and Peacock, M.W., 1987, Argon retentivity of hornblendes: A field experiment in a slowly cooled metamorphic terrane: Geochimica et Cosmochimica Acta, v. 51, p. 2891–2903.

Pomfrey, M.E., Dahl, P.S., Foland, K.A., and Hubacher, F.A., 1998a, Argon thermochronology of metamorphic biotite and hornblende from the central Adirondack Lowlands, New York: Implications for the Proterozoic exhumation and cooling history: Geological Society of America Abstracts with Programs, v. 30, no. 2, p. 67.

Pomfrey, M.E., Dahl, P.S., Foland, K.A., and Hubacher, F.A., 1998b, Argon thermochronology of metamorphic biotite and hornblende from the central Adirondack Lowlands, New York, II: Evidence for post–880 Ma, westward tilting of a slowly-cooled Grenville terrane: Geological Society of America Abstracts with Programs, v. 30, no. 7, p. A111.

Pouchou, J.L., and Pichoir, F., 1984, A new model for quantitative X-ray analysis, II: Application to in-depth analysis of heterogeneous samples: Recherche Aerospatiale, v. 5, p. 349–367.

Powers, R.E., and Bohlen, S.R., 1985, The role of synmetamorphic igneous rocks in the metamorphism and partial melting of metasediments, northwest Adirondacks: Contributions to Mineralogy and Petrology, v. 90, p. 401–409.

Rivers, T., 1997, Lithotectonic elements of the Grenville Province: Review and tectonic implications: Precambrian Research, v. 86, p. 117–154.

Roden-Tice, M.K., and Wintsch, R.P., 2002, Early Cretaceous normal faulting in southern New England: Evidence from apatite and zircon fission-track ages: Journal of Geology, v. 110, p. 159–178.

Ruiz, J., Patchett, P.J., and Ortega-Gutierrez, F., 1988, Proterozoic and Phanerozoic basement terranes of Mexico from Nd isotopic studies: Geological Society of America Bulletin, v. 100, p. 274–281.

Scott, D.J., and St-Onge, M.R., 1995, Constraints on Pb closure temperature in titanite based rocks from the Ungava orogen, Canada: Implications for U-Pb geochronology and *P-T-t* path determinations: Geology, v. 23, p. 1123–1126.

Streepey, M.M., Essene, E.J., and van der Pluijm, B.A., 1997, A compilation of thermobarometric data from the metasedimentary belt of the Grenville province, Ontario and New York State: Canadian Mineralogist, v. 35, p. 1237–1247.

Streepey, M.M., van der Pluijm, B.A., Essene, E.J., Hall, C.M., and Magloughlin, J.F., 2000, Late Proterozoic (ca. 930 Ma) extension in eastern Laurentia: Geological Society of America Bulletin, v. 112, p. 1522–1530.

Streepey, M., Johnson, E., Mezger, K., and van der Pluijm, B., 2001a, Early history of the Carthage-Colton shear zone, Grenville Province, northwest Adirondacks, New York (U.S.A.): Journal of Geology, v. 109, p. 479–492.

Streepey, M.M., Lithgow-Bertelloni, C., and van der Pluijm, B.A., 2001b, Uplift and extension along the Proterozoic margin of eastern Laurentia: Geological Society of America Abstracts with Programs, v. 33, no. 6, p. A90.

Valley, J., Bohlen, S.R., Essene, E.J., and Lamb, W., 1990, Metamorphism in the Adirondacks, II: The role of fluids: Journal of Petrology, v. 31, p. 555–596.

Villa, I.M., 1997, Isotopic closure: Terra Nova, v. 10, p. 42–47.

Wasteneys, H., McLelland, J., and Lumbers, S., 1999, Precise zircon geochronology in the Adirondack Lowlands and implications for revising plate-tectonic models of the Central Metasedimentary Belt and Adirondack Mountains, Grenville Province, Ontario and New York: Canadian Journal of Earth Sciences, v. 36, p. 967–984.

MANUSCRIPT ACCEPTED BY THE SOCIETY AUGUST 25, 2003

Printed in the USA

Geological Society of America
Memoir 197
2004

Petrogenesis of prismatine-bearing metapelitic gneisses along the Moose River, west-central Adirondacks, New York

Robert S. Darling*
Department of Geology, State University of New York—College at Cortland, Cortland, New York 13045, USA
Frank P. Florence
Science Division, Jefferson Community College, Watertown, New York 13601, USA
Gregory W. Lester
Department of Earth and Environmental Sciences, Rensselaer Polytechnic Institute, Troy, New York 12180, USA
Philip R. Whitney
New York State Geological Survey, New York State Museum, Albany, New York 12230, USA

ABSTRACT

Metapelitic gneisses exposed along the Moose River in the west-central Adirondack Highlands contain uncommon Mg- and B-rich mineral assemblages. Phases identified in these rocks include cordierite, garnet, sillimanite, spinel, orthopyroxene, tourmaline, and locally prismatine, the B-rich end-member of the kornerupine group. Four distinct mineral assemblages are identified, two of which are interpreted to form by partial melting. Prismatine is interpreted to have formed by the melting of tourmaline + biotite + cordierite + sillimanite. Thermobarometry calculations from net transfer and exchange equilibria record temperatures and pressures of 850 ± 20 °C and 6.6 ± 0.6 kilobars for orthopyroxene + garnet assemblages and 675 ± 50 °C and 5.0 ± 0.6 kilobars for cordierite + garnet + sillimanite + quartz assemblages. Mineral textures and compositions are consistent with formation along a nearly isobaric cooling path similar to that proposed by Spear and Markussen (1997).

Keywords: Adirondacks, prismatine, cordierite orthopyroxene

INTRODUCTION

Kornerupine-group minerals have been described from seven localities in the Grenville Province (Grew, 1996) including one in the Adirondacks (Farrar and Babcock, 1993; Farrar, 1995). Prismatine, the boron-rich end-member of the kornerupine group (Grew et al., 1996), has been positively identified in only three of these locations (Grew, 1996). Furthermore, cordierite + orthopyroxene–bearing metapelites are equally rare, having been described from two locations in the Grenville Province outside the Adirondacks (Reinhardt, 1968; Haase 1981; Perkins et al., 1982; Davidson et al., 1990) and one location in the Adirondack Lowlands (Powers and Bohlen, 1985; Stoddard, 1989). To our knowledge, this is the first report of prismatine in the Adirondacks and the first description of an orthopyroxene + cordierite assemblage in the Adirondack Highlands. Herein we describe, analyze, and interpret prismatine-bearing metapelitic gneisses from along the banks of the Moose River in the west-central Adirondacks of New York, and make inferences regarding the early retrograde metamorphic path.

Rock samples were examined in thin section and minerals were analyzed with Japan Electron Optics Laboratory (JEOL)

*E-mail: darlingr@cortland.edu.

Darling, R.S., Florence, F.P., Lester, G.W., and Whitney, P.R., 2004, Petrogenesis of prismatine-bearing metapelitic gneisses along the Moose River, west-central Adirondacks, New York, *in* Tollo, R.P., Corriveau, L., McLelland, J., and Bartholomew, M.J., eds., Proterozoic tectonic evolution of the Grenville orogen in North America: Boulder, Colorado, Geological Society of America Memoir 197, p. 325–336.

electron microprobes at Cornell University (JXA 8900) and Binghamton University (JXL 8900). Both natural and synthetic minerals were used as oxide standards, and ZAF (atomic number, absorption, fluorescence) correction procedures were employed in all analyses. Prismatine from one sample was analyzed for B and Li using secondary ion mass spectroscopy (SIMS) at the University of New Mexico.

LOCATION AND GEOLOGY

The metapelitic rocks occur in closely associated outcrops along the north and south banks of the Moose River ~5.2 km west of State Highway 28 in the McKeever, New York, U.S. Geological Survey topographic 15-minute quadrangle (Fig. 1). The metapelitic rocks occur within a heterogeneous metasedimentary unit (Fig. 1, BL) that consists principally of quartzite and biotite-quartz-plagioclase gneiss with lesser amounts of calcsilicate rocks, and minor amphibolite, quartzofeldspathic gneiss, and calcite marble. These rocks are interlayered with other metasedimentary undifferentiated rocks (Fig. 1, MU) that contain relatively more calcsilicates and less quartzite than BL, thick tabular bodies of granitic to locally charnockitic gneiss (Fig. 1, CG), and quartz-microcline gneiss with quartzite and calcsilicate granulite layers (Fig. 1, TH). These units occur in a complex, southeast-verging overturned synform bordered on the northwest by a tabular, northwest-dipping body of CG several kilometers thick, and on the southeast by a domical body of batholithic proportions consisting of relatively leucocratic CG. Although the granitic and charnockitic rocks have not been dated, they are lithologically and geochemically similar to felsic rocks of the ca. 1150 Ma anorthosite-mangerite-charnockite-granite (AMCG) suite found throughout much of the Adirondack Highlands (McLelland et al., 2001; Whitney et al., 2002).

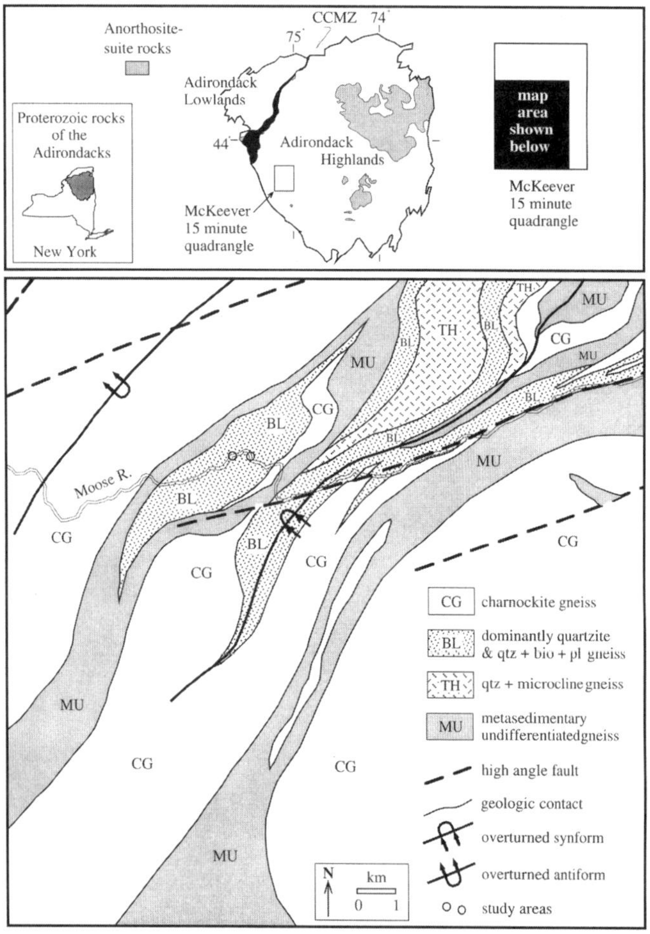

Figure 1. Location map showing study area and local bedrock geology. Bedrock geology, map units, and geologic relations illustrated are taken from Whitney et al. (2002). Bio—biotite; CCMZ—Carthage-Colton mylonite zone; pl—plagioclase; qtz—quartz.

Mineral Assemblages

The mineral assemblages investigated occur in medium- to coarse-grained, variably foliated feldspathic gneisses. There are four distinct assemblages. The first two assemblages, which occur as distinct horizons within well-foliated, medium-grained gneiss, are (1) cordierite (crd) + spinel (spl) + sillimanite (sil) + garnet (grt) + plagioclase (pl) + quartz (qtz) + ilmenite (ilm) + rutile (rut) ± biotite (bio), and (2) cordierite + orthopyroxene (opx) + biotite + perthitic microcline + quartz. The other two assemblages, which occur in coarse-grained lenses parallel to but locally crosscutting the dominant foliation, are (3) prismatine (prs) + microcline + plagioclase + quartz + rutile ± tourmaline (tur) ± biotite ± garnet ± cordierite ± sillimanite, and (4) orthopyroxene + plagioclase + garnet + ilmenite + microcline + quartz ± biotite. These assemblages are described and interpreted in the following sections.

Assemblage 1: Crd + Spl + Sil + Grt + Pl + Qtz + Ilm + Rut ± Bio

Elongate (1.0–2.0 cm) lenses of pale blue cordierite, the cores of which commonly contain fine-grained green spinel and sillimanite (Fig. 2, A), characterize assemblage 1. The sillimanite typically occurs as rims surrounding spinel. These lenses are commonly associated with two forms of garnet: (1) an anhedral, irregular, sillimanite-rich garnet of inferred early origin, and (2) a subhedral to euhedral, equant and generally inclusion-free garnet of inferred late origin. The presence of two generations of garnet is consistent with the common occurrence in the west-central Adirondacks of metapelites with garnet porphyroblasts that have cores rich in sillimanite inclusions and are surrounded by inclusion-free rims (Whitney et al., 2002). The late garnet locally contains fine crystallographically controlled needles of rutile, and often occurs as "buds" surrounding the cordierite + spinel lenses (Fig. 2, B). Ilmenite occurs as elongated grains in all areas except the spinel-rich cordierite cores. In reflected light, ilmenite shows little, if any, hematite exsolution lamellae. Quartz is not present near the cordierite + spinel lenses, but does

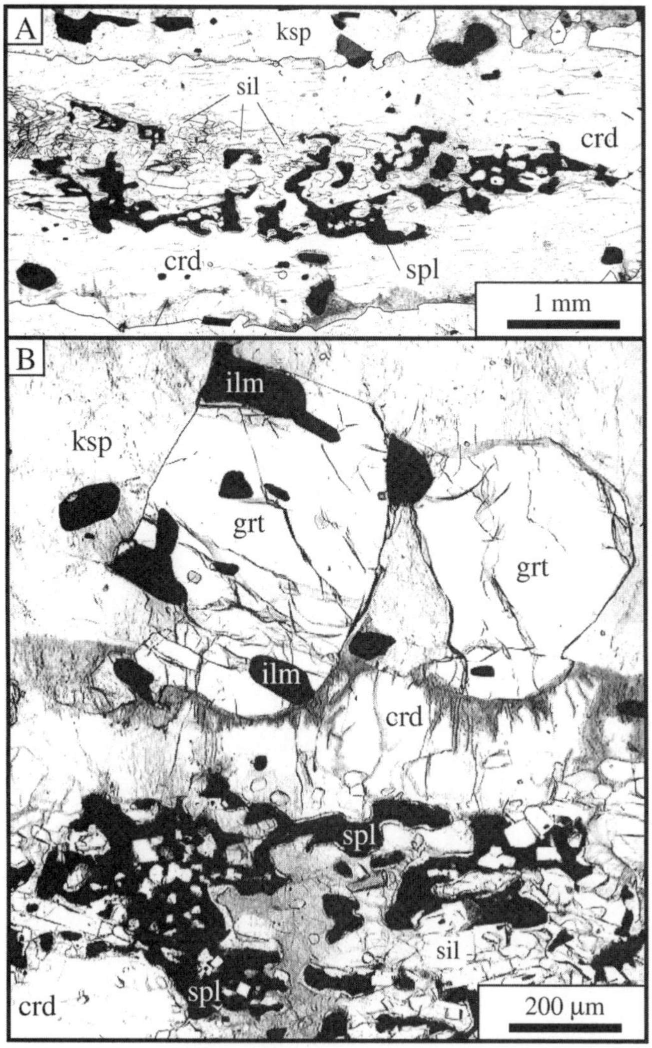

Figure 2. Assemblage 1: (A) Cordierite (crd) + spinel (spl) + sillimanite (sil) lens. (B) Garnet (grt) "buds" growing on cordierite + spinel + sillimanite lens. Ilm—ilmenite.

appear in the surrounding gneisses at distances greater than a few cm. Biotite occurs as subhedral books and is rarely in contact with sillimanite.

Cordierite from this assemblage is Mg-rich, with Mg numbers ranging from 82 to 83 (Table 1). The abundance of molecular H_2O in cordierite is unknown, but all grains measured show high 2V angles, suggesting it is CO_2-rich (Armbruster and Bloss, 1980). No significant compositional differences were noted between early (e) and late (l) garnet, and no compositional zoning was recognized in either variety. Both garnet types have almandine, pyrope, grossular, and spessartine contents ranging from 55.4 to 56.6%, 40.5 to 41.8%, 2.2 to 2.3%, and 0.5 to 0.6%, respectively (Table 1). Biotite is both Ti- and Al-rich, with ~0.33–0.35 and 1.40–1.43 atoms per formula unit, respectively. Also note that biotite has only 0.76–0.77 atoms of K per formula unit (Table 1), suggesting that the twelve-fold site is not completely filled. The Ti and Al abundances are generally consistent with a substitution vector of $2Si^{IV} + Mg^{VI} = 2Al^{IV} + Ti^{VI}$ (Robert, 1976; Dymek 1983), with minor $2Mg^{VI} = \square + Ti^{VI}$ (Dymek, 1983). Both the elevated Ti and Al abundances and the incomplete filling of the twelve-fold site are common in high-grade metamorphic biotites (Dymek, 1983; Guidotti, 1984). The anorthite component of plagioclase ranges from 26.4 to 27.6%. Ilmenite compositions are nearly pure, with Ti formula units greater than 0.98. The ilmenite compositions, coupled with the absence of hematite exsolution lamellae, strongly suggest that low f_{O_2} conditions prevailed during the metamorphism of these rocks. Sillimanite, too, is relatively pure, with Fe^{3+} formula units ranging from 0.012 to 0.013, values also consistent with low f_{O_2} conditions.

The origin of the cordierite + spinel lenses is unknown, but the regularity of their association, and the restriction of spinel to cordierite cores, suggests a pseudomorphic origin. Similar structures occurring in metapelitic rocks on the margin of the Laramide anorthosite complex in Wyoming are interpreted as either staurolite, sapphirine, or sillimanite pseudomorphs, the latter of which may have formed by means of a reaction with biotite, garnet, or a melt phase (Grant and Frost, 1990). The cordierite and spinel compositions in the Moose River rocks (Table 1) suggest that the original bulk composition may have been too Mg-rich to originate from a staurolite precursor. Sapphirine is a more plausible precursor; however, the possibility that the lenses formed by means of a reaction of sillimanite with biotite, Mg-garnet, or a melt phase cannot be ruled out.

The sillimanite rims surrounding spinel and the late garnet buds growing on the cordierite lenses (Fig. 2, B) are interpreted to have formed as a result of the reaction

$$\text{Spinel} + \text{cordierite} + \text{quartz} \rightarrow \text{Garnet} + \text{sillimanite} \quad (1)$$

Because spinel and quartz are never in direct contact, the reaction's progress appears to be controlled by diffusion of silica toward the cordierite lenses and diffusion of Fe and Mg away from spinel (and cordierite). The growth of sillimanite rims around spinel is interpreted to reflect very limited Al diffusion from spinel as compared to Fe and Mg. Reaction 1 is considered further in the discussion section of this paper.

Assemblage 2: Crd + Opx + Bio + Ksp + Qtz

Assemblage 2 is characterized by strongly foliated, biotite-rich, medium-grained gneiss with disseminated orthopyroxene and coarse-grained cordierite knots. In many thin sections, particularly those from near-surface exposures, the orthopyroxene is pervasively altered to a fine-grained, greenish to tan-brown secondary phase of unknown identity. Unaltered orthopyroxene, where present, occurs as anhedral, irregular to equant grains in apparent textural equilibrium with cordierite, biotite, quartz, and K-feldspar (Fig. 3, A). The orthopyroxene cores have Mg numbers ranging from 62 to 63 and weight percent Al_2O_3 ranging from 6.2 to 6.8 (Table 2). Cordierite, as in the previous assem-

TABLE 1. SELECT GARNET, CORDIERITE, PLAGIOCLASE, SILLIMANITE, ILMENITE, BIOTITE, AND SPINEL CORE COMPOSITIONS IN ASSEMBLAGE 1

	MR-98-01-B									MR-98-01-C					
Sample	grt (e)	grt (l)	crd	pl	sil	ilm	ilm	bio	spl	grt (e)	grt (l)	crd	pl	sil	bio
SiO_2	39.81	39.70	49.27	61.22	36.38			37.14	0.10	39.60	39.89	49.95	61.31	36.86	38.03
TiO_2			b.d.	b.d.	0.01	51.32	51.42	6.52	b.d.			b.d.	0.02	b.d.	6.06
Al_2O_3	22.91	22.85	34.13	24.57	63.38	b.d.	0.01	16.90	64.65	23.02	22.97	34.06	24.26	63.00	16.64
FeO	25.81	25.94	4.21	b.d.		47.66	48.46	14.12	26.37	26.28	25.62	4.14	0.05		14.17
Fe_2O_3					0.61									0.63	
MnO	0.26	0.24	0.03	b.d.	0.01	0.39	0.38	0.01	b.d.	0.27	0.28	0.02	b.d.	0.01	b.d.
MgO	10.79	10.63	10.71	b.d.	0.07	0.03	0.03	14.62	10.60	10.54	10.83	11.19	b.d.	0.02	14.64
ZnO									0.69						
CaO	0.79	0.79	b.d.	5.74	0.02			0.01	b.d.	0.83	0.79	0.02	5.50	b.d.	b.d.
Na_2O			0.05	8.06	b.d.			0.10				0.06	8.18	b.d.	0.12
K_2O			b.d.	0.34	b.d.			8.37				0.01	0.41	0.01	8.36
Total	100.37	100.15	98.40	99.93	100.47	99.40	100.30	97.79	102.41	100.54	100.38	99.45	99.73	100.53	98.02
Atoms per formula unit[a]:															
Si	3.007	3.008	4.970	2.720	0.980			2.670	0.003	2.994	3.010	4.985	2.730	0.992	2.722
Ti						0.986	0.981	0.352					0.001		0.326
Al	2.040	2.040	4.058	1.287	2.012			1.432	1.994	2.051	2.043	4.007	1.273	1.998	1.404
Fe	1.631	1.643	0.355		0.012	1.018	1.028	0.849	0.577	1.662	1.617	0.346	0.002	0.013	0.848
Mn	0.016	0.015	0.002			0.008	0.008	0.001		0.017	0.018	0.001			
Mg	1.215	1.201	1.611			0.001	0.001	1.567	0.413	1.188	1.218	1.665		0.001	1.562
Zn									0.013						
Ca	0.064	0.064		0.273				0.001		0.067	0.064	0.002	0.262		
Na			0.009	0.695				0.014				0.011	0.706		0.017
K				0.019				0.767				0.001	0.023		0.762
Cations	7.973	7.972	11.005	4.994	3.008	2.014	2.019	7.652	3.000	7.980	7.969	11.018	4.997	3.003	7.640
Mg #[b]	41.5	41.1	81.9					64.9	41.2	40.5	41.8	82.8			64.8

Note: Core compositions determined by electron microprobe. Grt—garnet; crd—cordierite; pl—plagioclase; sil—sillimanite; ilm—ilmenite; bio—biotite; spl—spinel. Grt (e)—early garnet; grt (l)—late garnet. b.d.—below detection limit.

[a]Formula units for grt, crd, plag, sil, ilm, bio, and spl based on 12, 18, 8, 5, 3, 11, and 4 oxygens, respectively.

[b]Mg # = (Mg/Mg + Fe) × 100 for cordierite and biotite; (Mg/Mg + Fe + Zn) × 100 for spinel; (Mg/Mg + Fe + Ca + Mn) × 100 for garnet.

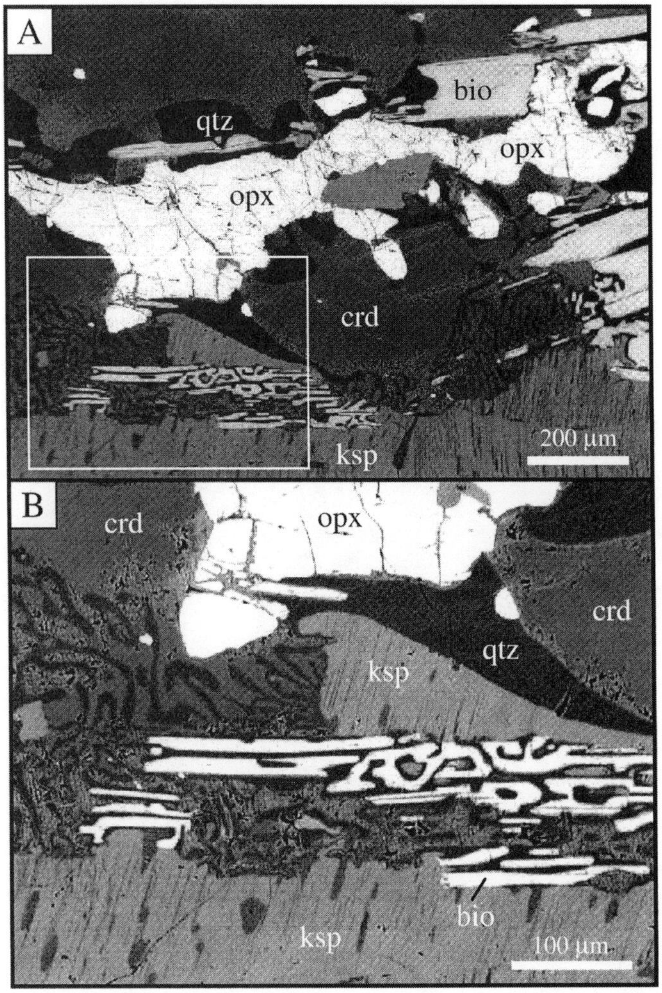

Figure 3. Assemblage 2: (A) Backscattered electron image of orthopyroxene (opx) + cordierite (crd) + biotite (bio) + K-feldspar (ksp) + quartz (qtz) assemblage. Inset area shown in B. (B) Biotite + cordierite + quartz symplectite.

blage, has Mg numbers ranging from 81 to 82. K-feldspar is moderately perthitic, and biotite forms medium-grained, oriented books.

This assemblage also commonly contains biotite + quartz symplectites, locally accompanied by cordierite + quartz symplectites (Fig. 3, B). In both cases, the quartz appears to replace earlier-formed single grains of biotite and cordierite (Fig. 3, B). The biotite + quartz symplectites are interpreted to have formed by means of the retrogressive reaction

$$\text{Orthopyroxene} + \text{cordierite} + \text{K-feldspar} + H_2O \rightarrow \text{Biotite} + \text{quartz} \quad (2)$$

Because biotite is already present in this assemblage, we believe that new biotite growth likely occurred on existing grains of biotite rather than resulting from the nucleation of separate and texturally different grains. The origin of the cordierite + quartz symplectites is less clear. The formation of new biotite obviously occurs at the expense of both cordierite and orthopyroxene, according to reaction 2, but the two reactants may break down differently. We infer that orthopyroxene decomposed only on its grain margins, and that cordierite was chemically replaced by vermicular intergrowths of quartz. At what point along the retrograde path reaction 2 occurred is unknown. If early in the path (supersolidus temperatures), the reaction may have involved a melt phase, whereas a later point on the path (subsolidus temperatures) would favor free H_2O as a reactant.

The results of a major-element traverse across a pair of joined orthopyroxene-cordierite grains are illustrated in Figure 4. The traverse extended more than 100 μm into each mineral, and the spot analysis spacing was 20, 10, 5, and 2 μm, depending the distance to the mineral rim. For convenience, the cation formula units (f.u.) for both cordierite and orthopyroxene were normalized to 18 oxygens. Figure 4 shows three important compositional trends: (1) within 20 μm of the rim, both phases show effects of late retrograde re-equilibration, (2) away from the rim, cordierite shows no evidence of compositional zoning, and (3) from core to rim, orthopyroxene shows a general decrease in Al and an increase in Si, Fe, and Mg. Specifically, within the first eight analyses (100 μm), Al decreases by 0.141 f.u., from 0.881 f.u. in the apparent core to 0.740 f.u. within 20 μm of the rim. Over the same distance, the combined Si + Fe + Mg formula units increase by 0.150 f.u., from 11.125 to 11.275. The similarity of the change in formula units is consistent with Tschermak's substitution (Al^{3+}, Al^{3+}, for (Fe, $Mg)^{2+}$, Si^{4+}) in orthopyroxene. Also note that there is little, if any, change in the orthopyroxene Fe/Mg ratio from core to rim. These compositional trends are considered further in the discussion section of this paper.

Assemblage 3: Prs + Ksp + Pl + Qtz + Rut ± Tur ± Bio ± Grt ± Crd ± Sil

Assemblage 3 is characterized by spectacularly developed, dark green, euhedral prismatine crystals measuring up to 10 cm or more along the c-axis. In surface exposures, prismatine forms loosely disseminated prisms and radiating splays in coarse-grained feldspathic lenses (Fig. 5, A). These lenses range in thickness from 1 cm to several centimeters and are generally parallel to gneissic foliation. Consequently, the prismatine crystals *appear* to have grown only within the plane of the foliation. However, upon closer examination, the prismatine grains are seen to be arranged randomly, but the longest and best-developed crystals formed parallel to the foliation plane. From this we infer that hydrostatic (nondeviatoric) pressure conditions prevailed locally during prismatine formation. It should also be noted that a number of prismatine-bearing feldspathic lenses are located adjacent to fine-grained tourmaline + biotite + plagioclase–rich zones in the hosting gneiss. In these locations, the prismatine-bearing feldspathic lenses texturally invade and appear to form at the expense of the tourmaline-bearing zones (Fig. 5, B). Based on these observations, we interpret the feld-

TABLE 2. SELECT ORTHOPYROXENE AND CORDIERITE CORE COMPOSITIONS FROM ASSEMBLAGE 2

	MR-99-10-A				MR-99-10-C			
Sample	opx	opx	crd	crd	opx	opx	crd	crd
SiO_2	48.94	48.87	49.81	49.42	49.44	49.42	49.64	49.48
TiO_2	0.08	0.09	b.d.	b.d.	0.10	0.12	b.d.	b.d.
Al_2O_3	6.79	6.23	33.70	33.54	6.71	6.64	33.69	33.49
FeO	23.14	23.41	4.49	4.33	22.55	22.82	4.22	4.27
MnO	0.22	0.18	0.03	0.06	0.26	0.24	0.05	0.04
MgO	20.77	21.02	10.78	10.78	21.20	21.36	10.95	10.83
CaO	0.03	0.03	b.d.	b.d.	0.03	0.03	b.d.	b.d.
Na_2O	0.02	b.d.	0.09	0.04	b.d.	b.d.	0.05	0.02
Total	99.99	99.83	98.90	98.17	100.29	100.63	98.60	98.13
Atoms per formula unit[a]:								
Si	0.914	0.916	5.005	5.000	0.918	0.916	4.998	5.005
Ti	0.001	0.001			0.001	0.002		
Al	0.150	0.138	3.991	4.000	0.147	0.145	3.998	3.993
Fe	0.362	0.367	0.377	0.366	0.350	0.354	0.355	0.361
Mn	0.003	0.003	0.003	0.004	0.004	0.004	0.004	0.003
Mg	0.578	0.588	1.615	1.626	0.587	0.590	1.643	1.633
Ca	0.001	0.001			0.001	0.001		
Na	0.001		0.018	0.009			0.010	0.004
Cations	2.010	2.013	11.009	11.005	2.007	2.010	11.008	10.999
Mg #[b]	61.5	61.6	81.1	81.6	62.6	62.5	82.2	81.9

Note: Core compositions determined by electron microprobe. Opx—orthopyroxene; crd—cordierite. b.d.—below detection limit.

[a]Formula units for opx and crd based on 3 and 18 oxygens, respectively.

[b]Mg # = (Mg/Mg + Fe) × 100.

spathic lenses and the prismatine found in them to be of anatectic origin.

Accompanying prismatine in the feldspathic lenses are quartz and fine, disseminated rutile. Coarse-grained garnet and cordierite are locally found as well. Table 3 lists the results of electron microprobe analyses of minerals from two prismatine-bearing samples (MR-99-11 and MR-99-15). Note that two prismatine grains from MR-99-11 were analyzed for B and Li using SIMS at the University of New Mexico. This prismatine contains 0.73 to 0.79 formula units of B and has Mg numbers between 70 and 73 (Table 3). A small amount of Li_2O was also detected. In sample MR-99-11 tourmaline occurs in a relict biotite + plagioclase–rich zone and as inclusions in garnet. Tourmaline does not occur in contact with prismatine. Boron abundance (wt%) in tourmaline was not determined by SIMS analysis, but rather was calculated so that its atoms per formula unit equaled three (Henry and Dutrow, 1996). Tourmaline has an Mg number of 76, and Ca/Ca + Na + K = 0.50 (Table 3). Both are values typical of high-grade metapelites (Henry and Dutrow, 1996). Cordierite and garnet have Mg numbers of 83 and 40, respectively (Table 3). Biotite, as in assemblage 1, is both Ti- and Al-rich, with nearly 0.37 and 1.36 atoms per formula unit, respectively. Similarly, biotite has only 0.8 atoms of K per formula unit (Table 3). As discussed in a previous section, these compositions are common in high-grade metamorphic biotites (Dymek, 1983; Guidotti, 1984). Sillimanite is rare in this assemblage, occurring only as relict anhedral grains surrounded by prismatine.

Based on average mineral compositions in sample MR-99-11 (Table 3), we propose the following prismatine-forming reaction:

$$1.0 \text{ Tourmaline} + 6.6 \text{ sillimanite} + 4.0 \text{ biotite} + 1.0 \text{ cordierite} \rightarrow 4.0 \text{ Prismatine} + 0.61 \text{ rutile} + \text{melt or } (0.91 \text{ plagioclase } (An_{52}) + 3.22 \text{ K-feldspar} + 0.92 \text{ quartz} + H_2O) \quad (3)$$

Reaction 3 is similar to a number of proposed prismatine-forming reactions from other granulite terranes (Grew, 1996), including those in sapphirine-free rocks found in the Reading Prong, New Jersey (Young, 1995), and in Waldheim, Germany (Grew, 1989). In those cases, garnet rather than cordierite was a proposed reactant. The relatively large number of sillimanite moles consumed in reaction 3 and the occurrence of sillimanite only as anhedral grains enveloped by prismatine strongly suggest that the modal abundance of sillimanite may have been the limiting reagent in the formation of prismatine in Moose River rocks.

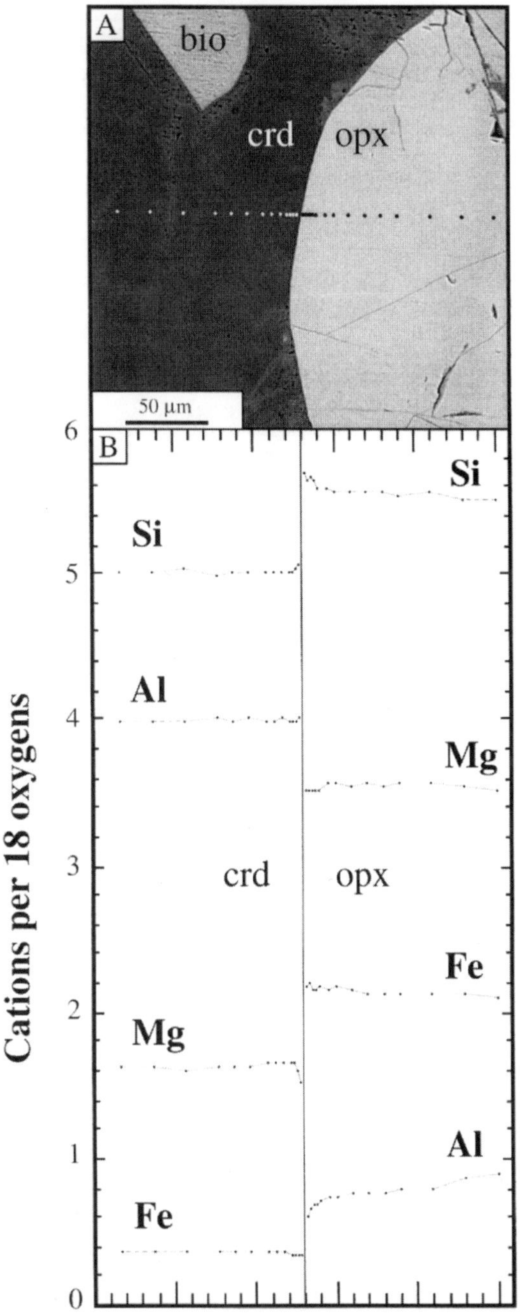

Figure 4. (A) backscattered electron image of adjacent cordierite (crd) + orthopyroxene (opx) grains in assemblage 2 showing location of major element traverse (white dots in crd; black dots in opx). (B). Elemental profiles for Si, Al, Fe, and Mg across cordierite and orthopyroxene grains. Bio—biotite.

Figure 5. Assemblage 3: (A) Prismatine crystals in coarse-grained feldspathic lens, taken parallel to plane of lens. Hammer for scale. (B) Euhedral prismatine in coarse-grained feldspathic lens embaying fine-grained, foliated tourmaline + biotite + plagioclase–rich zones (right center).

Assemblage 4: Opx + Pl + Ksp + Qtz ± Bio ± Grt

Assemblage 4 is characterized by coarse orthopyroxene (>1 cm) in coarse-grained feldspathic lenses locally containing coarse (1–3 mm) biotite books and coarse (4–10 mm) euhedral garnet. The lenses are similar to those hosting prismatine-bearing assemblages in that they are generally parallel to foliation in the surrounding gneisses. Orthopyroxene and plagioclase from this assemblage commonly show a moderately developed to well-developed ophitic texture (Fig. 6). This texture, coupled with the similarity of these lenses to the prismatine-

TABLE 3. SELECT PRISMATINE, TOURMALINE, CORDIERITE, GARNET, BIOTITE, AND SILLIMANITE COMPOSITIONS FROM ASSEMBLAGE 3

	MR-99-11							MR-99-15				
Sample	prs	prs	tur	tur	crd	grt	bio	prs	pl	pl	grt	sil
SiO_2	30.70	30.49	35.30	34.94	49.97	39.48	37.60	30.72	56.44	56.59	39.58	35.82
TiO_2	0.24	0.24	2.01	1.97	b.d.	b.d.	6.84	1.18	b.d.	0.01	b.d.	0.01
Al_2O_3	40.69	41.16	31.28	31.24	33.56	23.13	16.16	41.91	28.12	27.92	22.96	62.94
FeO	9.52	9.78	5.09	5.05	4.12	25.20	12.47	8.05	0.07	0.03	24.82	b.d.
Fe_2O_3	b.d.	b.d.	b.d.	b.d.	b.d.	b.d.	b.d.	b.d.	b.d.	b.d.	b.d.	0.78
MnO	0.04	0.05	b.d.	b.d.	0.03	0.53	0.04	0.05	b.d.	b.d.	0.60	0.02
MgO	14.42	14.55	8.98	8.99	11.38	10.36	16.12	14.90	0.01	0.02	10.89	0.26
CaO	0.04	0.02	2.68	2.67	b.d.	1.66	b.d.	b.d.	9.21	9.11	1.42	b.d.
Na_2O	0.04	0.04	1.38	1.39	0.05	b.d.	0.02	0.04	5.77	5.96	b.d.	b.d.
K_2O	b.d.	b.d.	0.11	0.09	b.d.	b.d.	8.79	b.d.	0.30	0.34	b.d.	b.d.
B_2O^a	3.67	3.40	10.70	10.64								
Li_2O^a	0.16	0.26										
Total	99.52	99.99	97.53	96.98	99.11	100.36	98.04	95.85	99.92	99.98	100.27	99.83
Atoms per formula unit^b:												
Si	3.813	3.780	5.734	5.708	5.004	2.987	2.683		2.532	2.538	2.991	0.972
Ti	0.022	0.022	0.246	0.242			0.367					
Al	5.957	6.015	5.988	6.015	3.961	2.063	1.359		1.487	1.476	2.045	2.013
B	0.787	0.728	3.000	3.000								
Fe	0.989	1.014	0.691	0.690	0.345	1.595	0.744		0.003	0.003	1.569	0.016
Mn	0.004	0.005			0.003	0.034	0.002				0.038	
Mg	2.670	2.689	2.174	2.189	1.699	1.168	1.714		0.001	0.001	1.227	0.011
Ca	0.005	0.003	0.466	0.467		0.135			0.443	0.438	0.115	
Na	0.010	0.010	0.434	0.440	0.01		0.003		0.502	0.518		
K			0.023	0.019			0.799		0.017	0.019		
Li	0.080	0.130										
Cations	14.337	14.396	18.756	18.771	11.021	7.981	7.672		4.984	4.992	7.986	3.013
Mg #^{c}	73.0	70.2	75.9	76.0	83.1	39.8	69.7				41.6	

Note: Core compositions determined by electron microprobe except for B_2O_3 and Li_2O analyses on prismatine. Prs—prismatine; tur—tourmaline; crd—cordierite; grt—garnet; bio—biotite; sil—sillimanite. b.d.—below detection limit.

[a]Determined by SIMS analyses for prismatine; calculated for tourmaline (see text for discussion).

[b]Formula units for prs, tur, crd, grt, bio, pl, and sil based on 21.5, 29, 18, 12, 11, 8, and 5 oxygens, respectively.

[c]Mg # = (Mg/Mg + Fe) × 100 for prismatine, tourmaline, cordierite, and biotite; (Mg/Mg + Fe + Ca + Mn) × 100 for garnet.

bearing feldspathic lenses, strongly suggests that the orthopyroxene-bearing lenses are igneous in origin. Sections of coarse orthopyroxene are locally replaced with biotite + quartz symplectites.

The orthopyroxene and garnet compositions from this assemblage are listed in Table 4. Orthopyroxene and garnet have Mg numbers ranging from 65 to 70 and 44 to 45, respectively. Al abundances in orthopyroxene are similar to those in assemblage 2, ranging from 0.13 to 0.14 atoms per f.u. (based on 3 oxygens). Major-element mapping on the electron microprobe showed no convincing evidence of compositional zoning (for Fe, Mg, and Al) in orthopyroxene or (for Al) in plagioclase from this assemblage. This finding is considered further in the discussion section of this paper.

This assemblage, like the prismatine-bearing assemblage, is interpreted to have formed by means of the partial melting of surrounding lithologies. However, unlike the prismatine-bearing melts, textural evidence for a specific orthopyroxene-bearing melt reaction is lacking, but must have involved bulk rock compositions poorer in Al and B. One possible reaction is

$$\text{Biotite + quartz} \rightarrow \text{Orthopyroxene + K-feldspar + melt} \quad (4)$$

Reaction 4 has been modeled under vapor-absent conditions by Vielzeuf and Holloway (1988) and is discussed in the following section.

THERMOBAROMETRY

Temperature and pressure conditions can be estimated for the low-variance assemblages using petrogenetic grids and net transfer and mineral exchange equilibria. The stability of the

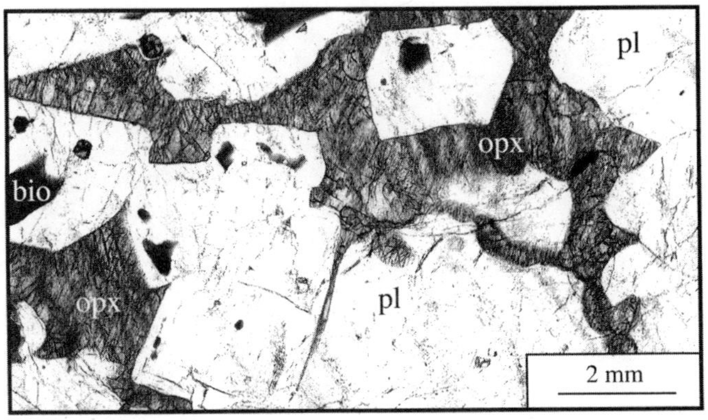

Figure 6. Plane polarized light photomicrograph of ophitic textured plagioclase (pl) + orthopyroxene (opx) in assemblage 4. The orthopyroxene shown is optically continuous. Bio—biotite.

prismatine-bearing assemblages has not been modeled nearly as well as that of B-free systems, but their proximity to and structural association with the other assemblages studied here suggests that the conditions of prismatine formation can be estimated reasonably well.

The petrogenetic grid (Fig. 3) of Spear et al. (1999) shows the following reaction:

Biotite + sillimanite + quartz → Cordierite + garnet + melt (5)

Assemblage 1 contains cordierite + garnet + sillimanite and cordierite + garnet + biotite. Both assemblages are stable on the high-temperature side of reaction 5, which is located at 750 °C at 5 kilobars and 840 °C at 8 kilobars (light-shaded region of Fig. 7). Furthermore, Figure 3 of Spear et al. (1999) shows another reaction relevant to these rocks:

Biotite + garnet + quartz → Cordierite + orthopyroxene + melt (6)

Assemblage 2 contains cordierite + biotite + orthopyroxene. This assemblage is stable on the high-temperature side of reaction 6, which is located at 810 °C at 5 kilobars and 855 °C at 8 kilobars (diagonal-ruled region of Fig. 7). Reactions 5 and 6 intersect at an invariant point at 860 °C and 8.5 kilobars (Fig. 7). At higher temperatures and pressures, orthopyroxene is stable

TABLE 4. SELECT ORTHOPYROXENE AND GARNET CORE COMPOSITIONS FROM ASSEMBLAGE 4

	MR-00-01B		001-B1			
Sample	opx	opx	opx	opx	grt	grt
SiO_2	50.99	50.95	50.75	50.75	39.84	39.56
TiO_2	0.10	0.12	0.13	0.13	0.05	0.01
Al_2O_3	6.41	6.56	6.37	6.16	22.36	22.56
FeO	18.41	18.52	21.30	21.39	25.15	25.06
MnO	0.05	0.06	0.14	0.13	0.40	0.34
MgO	23.76	23.78	21.79	21.83	11.70	12.11
CaO	0.06	0.03	0.07	0.10	0.91	0.89
Na_2O	0.02	b.d.	0.05	0.02	0.01	0.01
K_2O	b.d.	b.d.	b.d.	b.d.	b.d.	b.d.
Total	99.80	100.02	100.60	100.51	100.42	100.54
Atoms per formula unit[a]:						
Si	0.930	0.928	0.931	0.933	3.005	2.980
Ti	0.001	0.001	0.002	0.002	0.003	0.001
Al	0.138	0.138	0.138	0.133	1.988	2.003
Fe	0.281	0.282	0.327	0.329	1.587	1.579
Mn	0.001	0.001	0.002	0.002	0.026	0.022
Mg	0.646	0.646	0.596	0.598	1.315	1.360
Ca	0.001	0.001	0.001	0.002	0.074	0.072
Na	0.001		0.001	0.001	0.001	0.001
K						
Cations	2.000	2.000	1.999	1.999	7.999	8.018
Mg #[b]	69.7	69.6	64.6	64.5	43.8	44.8

Note: Core compositions determined by electron microprobe. Opx—orthopyroxene; grt—garnet. b.d.–below detection limit.

[a]Formula units for opx and grt based on 3 and 12 oxygens, respectively.

[b]Mg # = (Mg/Mg + Fe) × 100 for opx; (Mg/Mg + Fe + Ca + Mn) × 100 for garnet.

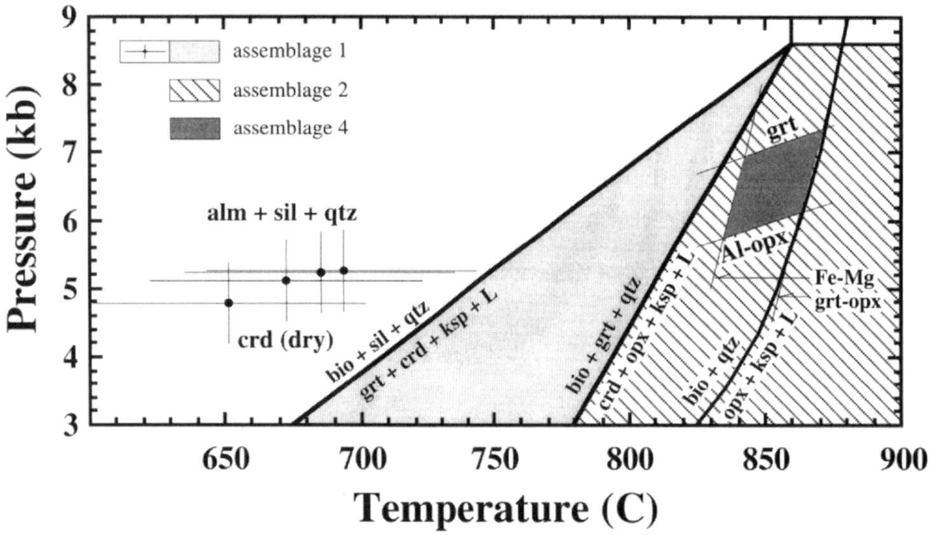

Figure 7. Pressure-temperature diagram showing various univariant reactions and results of exchange and net transfer equilibria calculations on Moose River rocks. Univariant reactions are from Spear et al. (1999) and Vielzeuf and Holloway (1988). See text for discussion. Alm—almandine; bio—biotite; crd—cordierite; grt—garnet; ksp—K-feldspar; opx—orthopyroxene; qtz—quartz; sil—sillimanite.

with sillimanite. Because an orthopyroxene + sillimanite assemblage is not observed in these rocks, the temperature and pressure conditions combined could not have been higher than 860 °C and 8.5 kilobars.

Vielzeuf and Holloway (1988) have modeled the production of orthopyroxene + K-feldspar from biotite + quartz melting (reaction 4) under fluid-absent conditions for a fixed-bulk composition (X_{Mg} = 0.5). Under fluid-absent conditions, reaction 4 is located between 825 °C at 3 kilobars and 879 °C at 9 kilobars. Reaction 4 is shown in Figure 7.

In addition to petrogenetic grid constraints, we have applied a number of net transfer and exchange reaction equilibria to assemblages 1 and 4. First, assemblage 4 comprises garnet + orthopyroxene in apparent textural equilibrium. Using the Fe-Mg exchange model of Harley (1984a) and the garnet + orthopyroxene analyses in Table 4, equilibrium temperatures were calculated and plotted in Figure 7. At 6.5 kilobars, temperatures range from 840 to 867 °C. Univariant curves calculated using Al abundances in orthopyroxene in equilibrium with garnet (Harley, 1984b, as amended in Fitzsimons and Harley, 1994) intersect the Fe-Mg exchange temperatures at conditions of ~850 ± 20 °C and 6.6 ± 0.6 kilobars (heavily shaded region in Figure 7). Note that the inferred temperature and pressure conditions for this assemblage are similar to those for the biotite melting equilibria (reaction 4) of Vielzeuf and Holloway (1988).

Assemblage 1 contains garnet, cordierite, sillimanite, and quartz in apparent textural equilibrium. Equilibrium temperatures and pressures were calculated using Fe-Mg exchange and net transfer equilibria models for the $FeO-Al_2O_3-SiO_2$ and $MgO-Al_2O_3-SiO_2$ systems (Nichols et al., 1992) and the garnet + cordierite analyses from samples MR-98–01-B and MR-98–01-C (Table 1). A dry (water activity or a_{H_2O} = 0) cordierite was assumed in the thermobarometric calculations. The results are illustrated in Figure 7 (closed circles), along with the temperature and pressure uncertainties. Estimates of metamorphic temperature and pressure for assemblage 1 are 675 ± 50 °C and 5.0 ± 0.6 kilobars, respectively.

DISCUSSION

The high-grade conditions of metamorphism experienced by these rocks require that the geothermometry and geobarometry estimates obtained here be considered in light of re-equilibration and reaction progress during retrograde conditions.

The consequences of retrograde reactions are most readily observed in assemblage 2. Measurable zoning of Fe, Mg, and Al in orthopyroxene was observed within ~100 μm of the rim. No compositional variations within individual grains of the other Fe-Mg phases in this assemblage (biotite and cordierite) was observed, except at cordierite grain boundaries where it is in direct contact with orthopyroxene. Fe-Mg variations close to this interface appear to reflect changes in Fe-Mg K_d during retrograde cooling. The penetration distances for zoning apparent in these elements might suggest limited intragranular diffusion. However, Al zoning extends into orthopyroxene more than 100 μm. Al mobility in orthopyroxene was presumably slower than that of Fe and Mg (Dowty, 1980), and preservation of this zoning profile suggests that the compositional trends of the more rapidly diffusing divalent cations may, in fact, be largely homogenized. This would explain the unchanging orthopyroxene Fe/Mg ratio from apparent core to near rim. The extent to which Fe and Mg components diffusing across the orthopyroxene-cordierite interface were accommodated in adjacent grains is not certain, but the mass balance of Al diffusing across the orthopyroxene grain boundary indicates that intergranular exchange operated along the retrograde path in assemblage 2. We point out that widespread symplectite formation in assemblage 2 suggests that relatively abundant aqueous fluid (or melt) was present, at least during the time when this hydration reaction occurred. Furthermore, the relatively rapid intragranular mobility of Fe and Mg in biotite and cordierite at higher temperatures may have facilitated this exchange. This would suggest that the limited Fe-Mg profile for the cordierite and orthopyroxene rims developed only late along the retrograde path.

The absence of Fe, Mg, and Al gradients in orthopyroxene in assemblage 4 contrasts markedly with the compositional trends in assemblage 2. Textural evidence indicates that in assemblage 4 orthopyroxene formed during partial melting, so the lack of observed growth zoning implies that either (1) crystallization took place under relatively uniform conditions, presumably an essentially isothermal melt phase, or (2) growth zoning in orthopyroxene was subsequently homogenized. The absence of recognized retrograde Fe-Mg exchange between orthopyroxene and other Fe-Mg minerals indicates the lack of effective intergranular transport of these components, as suggested by Pattison and Begin (1994) for similar rocks. Intergranular exchange may have shut down due to the loss of a fluid phase, either volatile or melt, during the early cooling of this assemblage. The suggestion that there has been little retrograde modification of the compositions of phases in this assemblage is supported by the consistency of the results from the petrogenetic grid of Spear et al. (1999), Fe-Mg exchange thermometry, Al solubility equilibria, and the fluid-absent biotite melting curve.

Temperatures obtained from assemblage 1 are ~175 °C lower than estimates calculated for assemblage 4. Furthermore, the temperatures obtained from assemblage 1 are inconsistent with the stability of the phases present in assemblage 1 (Fig. 7). The likely reason for the difference in calculated temperatures and known reactions is the difference in closure temperatures at which Fe and Mg are exchanged between specific mineral pairs. The diffusivity of Fe and Mg between garnet and orthopyroxene is known to operate under granulite conditions of metamorphism (e.g., Pattison and Begin, 1994), but Fe and Mg diffusivities in these phases are apparently slower than in cordierite (Thompson, 1976; Lasaga et al., 1977), so orthopyroxene-garnet exchange thermometers record higher temperatures compared with Fe-Mg thermometry between garnet and cordierite. Based on the compositional homogeneity of garnet and the absence of measurable retrograde zoning, we infer that Fe-Mg exchange in garnet ceased at high temperatures, whereas Fe-Mg exchange between cordierite and biotite continued during subsequent cooling, so the temperature estimates reflect closure along the retrograde path experienced by these rocks. The garnet-biotite K_d values preserved are consistent with this interpretation and show the effect of retrograde Fe enrichment in biotite. For example, garnet-biotite thermometry yields unrealistically high calculated temperatures in excess of 945 °C. In a like manner, diffusion of Fe from cordierite sufficient to raise Mg numbers from 77–78 to 81–82 corresponds to a decrease in calculated temperatures of ~125 °C.

Support for the pressure estimate obtained from orthopyroxene-garnet in assemblage 4 is derived from the limited mobility of Al in orthopyroxene, as well as the apparent limited retrograde modification to mineral compositions. The pressure estimate for assemblage 1 is primarily based on cordierite and garnet compositions in the grt + sil + qtz + crd net transfer reaction (Nichols et al., 1992) and is nearly independent of temperature. Therefore, we infer that the Moose River rocks cooled between the established temperature conditions along a nearly isobaric path as illustrated in Figure 8.

Further evidence supporting an isobaric cooling path comes from reaction 1 (from assemblage 1). The petrogenetic grid of Spear et al. (1999) shows reaction 1 as a moderately positively sloping univariant in the FeO-MgO-Al_2O_3-SiO_2 system with cordierite, spinel, and quartz on the high-temperature side of the reaction. Reaction 1 is not shown in Figure 7 because its position is influenced by the distribution of Fe between garnet and spinel (Spear et al., 1999). However, at pressures above 3 kilobars, it plots at temperatures above those seen in reaction 6. Regardless of the distribution of Fe between spinel and garnet (the single spinel analysis in Table 1 shows $X_{Fe,\ spl} = X_{Fe,\ grt}$), reaction 1 has a positive slope, indicating a pressure-temperature path of either decreasing temperature or increasing pressure (or both).

As illustrated in Figure 8, the proposed pressure-temperature path is parallel to, but at a slightly lower pressure (1.2 kb) than, that of Spear and Markussen (1997), with prismatine formation approximately coincident with the high-end conditions.

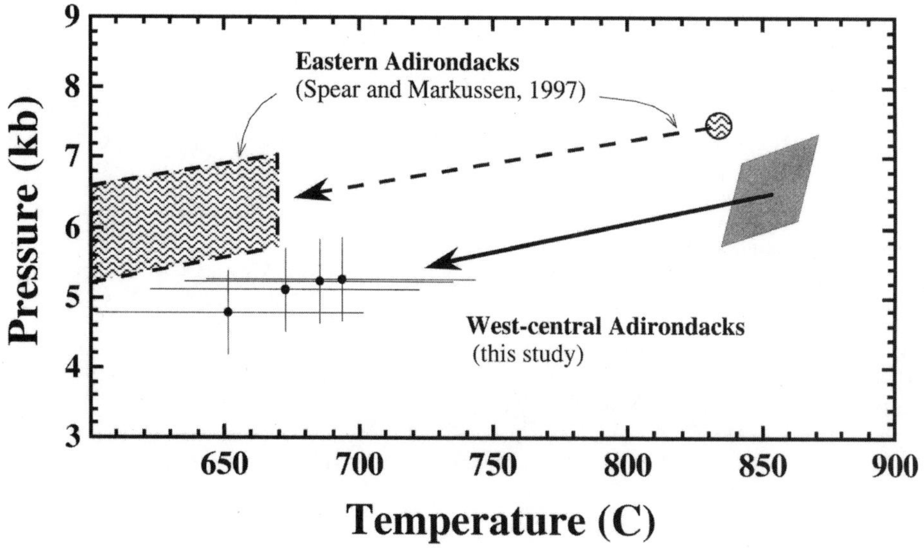

Figure 8. Pressure-temperature (*P-T*) diagram showing *P-T* conditions for Moose River rocks and the proposed *P-T* path. The path of Spear and Markussen (1997) determined for pyroxene-bearing assemblages in the eastern Adirondack Highlands is shown for comparison.

The proposed path is also consistent with studies of Lamb et al. (1991), McLelland et al. (2001), and Darling and Bassett (2002), and supports petrologic arguments of Bohlen (1987).

ACKNOWLEDGMENTS

We are grateful to Ed Grew (University of Maine) and Charles Shearer (University of New Mexico) for performing the SIMS analyses on the prismatine. We are also grateful to Bill Blackburn (Binghamton University) and John Hunt (Cornell University) for their helpful assistance in microprobe analysis. We are indebted to Frank Spear (Rensselaer Polytechnic Institute) and William Peck (Colgate University) for their helpful and constructive reviews of the manuscript.

REFERENCES CITED

Armbruster, T., and Bloss, F.D., 1980, Channel CO_2 in cordierites: Nature, v. 286, p. 140–141.

Bohlen, S.R., 1987, Pressure-temperature paths and tectonic model for the evolution of granulites: Journal of Geology, v. 95, p. 617–632.

Darling, R.S., and Bassett, W.A., 2002, Analysis of natural $H_2O + CO_2 +$ NaCl fluid inclusions in the hydrothermal diamond anvil cell: American Mineralogist, v. 87, p. 69–78.

Davidson, A., Carmichael, D.M., and Pattison, D.R.M., 1990, Metamorphism and geodynamics of the Southwestern Grenville Province, Ontario: International Geological Correlation Program, Project 235-304 International Meeting, Calgary, Alberta, Field Trip #1, 123 p.

Dowty, E., 1980, Crystal-chemical factors affecting the mobility of ions in minerals: American Mineralogist, v. 65, p. 174–182.

Dymek, R.F., 1983, Titanium, aluminum and interlayer cation substitutions in biotite from high grade gneisses, West Greenland: American Mineralogist, v. 68, p. 880–899.

Farrar, S.S., 1995, Mg-Al-B rich facies associated with the Moon Mountain metanorthosite sill, southeastern Adirondacks, NY: Geological Society of America Abstracts with Programs, v. 27, no. 1, p. 42–43.

Farrar, S.S., and Babcock, L.G., 1993, A sapphirine + kornerupine–bearing hornblende spinel periodotite associated with an Adirondack anorthosite sill: Geological Society of America Abstracts with Programs, v. 25, no. 6, p. A265.

Fitzsimons, I.C.W., and Harley, S.L., 1994, The influence of retrograde cation exchange on granulite *P-T* estimates and a convergence technique for the recovery of peak metamorphic conditions: Journal of Petrology, v. 35, p. 543–576.

Grant, J.A., and Frost, B.R., 1990, Contact metamorphism and partial melting of pelitic rocks in the aureole of the Laramide Anorthosite Complex, Morton Pass, Wyoming: American Journal of Science, v. 290, p. 425–472.

Grew, E.S., 1989, A second occurrence of kornerupine in Waldheim, Saxony, German Democratic Republic: Zeitschrift für Geologische Wissenschaften, Berlin, v. 17. p. 67–76.

Grew, E.S., 1996, Borosilicates (exclusive of tourmaline) and boron in rock-forming minerals in metamorphic environments, in Grew, E.S., and Anovitz, L.M., eds., Boron mineralogy, petrology and geochemistry: Washington, D.C., Mineralogical Society of America, Reviews in Mineralogy, v. 33, p. 387–480.

Grew, E.S., Cooper, M.A., and Hawthorne, F.C., 1996, Prismatine: Revalidation for boron-rich compositions in the kornerupine group: Mineralogical Magazine, v. 60, p. 483–491.

Guidotti, C.V., 1984, Micas in metamorphic rocks, *in* Bailey S.W., ed., Micas: Washington, D.C., Mineralogical Society of America, Reviews in Mineralogy, v. 13, p. 357–467.

Haase, J., 1981, A study of grandidierite from an aluminous granulite gneiss near Gananoque, Ontario [B.S. thesis]: Kingston, Ontario, Queen's University, Department of Geological Sciences, 38 p.

Harley, S.L., 1984a, An experimental study of the partitioning of Fe and Mg between garnet and orthopyroxene: Contributions to Mineralogy and Petrology, v. 86, p. 359–373.

Harley, S.L., 1984b, The solubility of alumina in orthopyroxene coexisting with garnet in $FeO\text{-}MgO\text{-}Al_2O_3\text{-}SiO_2$ and $CaO\text{-}FeO\text{-}MgO\text{-}Al_2O_3\text{-}SiO_2$: Journal of Petrology, v. 25 p. 665–695.

Henry, D.J., and Dutrow, B.L., 1996, Metamorphic tourmaline and its petrologic applications, *in* Grew, E.S., and Anovitz, L.M., eds., Boron Mineralogy, Petrology and Geochemistry, Reviews in Mineralogy, v. 33, Mineralogical Society of America, p. 504–557.

Lamb, W.M., Brown, P.E., and Valley, J.W., 1991, Fluid inclusions in Adirondack granulites: Implications for the retrograde P-T path: Contributions to Mineralogy and Petrology, v. 107, p. 472–483.

Lasaga, A.C., Richardson, S.M., and Holland, H.D., 1977, The mathematics of cation diffusion and exchange between silicate minerals during retrograde metamorphism, *in* Saxena, S.X., and Bhattachanji, S., eds., Energetics of Geological Processes, New York, Springer-Verlag, p. 353–388.

McLelland, J., Hamilton, M., Selleck, B., McLelland, J., Walker, D., and Orrell, S., 2001, Zircon U-Pb geochronology of the Ottawan Orogeny, Adirondack Highlands, New York: Regional and tectonic implications: Precambrian Research, v. 109, p. 39–72.

Nichols, G.T., Berry, R.F., and Green, D.H., 1992, Internally consistent gahnitic spinel-cordierite-garnet equilibria in the FMASHZn system: Geothermobarometry and applications: Contributions to Mineralogy and Petrology, v. 111, p. 362–377.

Pattison, D.R.M., and Begin, N.J., 1994, Zoning patterns in orthopyroxene and garnet in granulites: Implications for geothermometry: Journal of Metamorphic Geology, v. 12, p. 387–410.

Perkins, D., Essene, E.J., Marcotty, L.A., 1982, Thermometry and barometry of some amphibolite-granulite facies rocks from the Otter Lake area, southern Quebec: Canadian Journal of Earth Sciences, v. 19, p. 1759–1774.

Powers, R.E., and Bohlen, S.R., 1985, The role of synmetamorphic igneous rocks in the metamorphism and partial melting of metasediments, Northwest Adirondacks: Contributions to Mineralogy and Petrology, v. 90, p. 401–409.

Reinhardt, E.W., 1968, Phase relations in cordierite-bearing gneisses from the Gananoque area, Ontario: Canadian Journal of Earth Sciences, v. 5, p. 455–482.

Robert, J.-L., 1976, Titanium solubility in synthetic phlogopite solid solutions: Chemical Geology, v. 17, p. 213–227.

Spear, F.S., and Markussen, J.C., 1997, Mineral zoning, *P-T-X-M* phase relations, and metamorphic evolution of some Adirondack granulites: Journal of Petrology, v. 38, p. 757–783.

Spear, F.S., Kohn, M.J., and Cheney, J.T., 1999, P-T paths from anatectic pelites: Contributions to Mineralogy and Petrology, v. 134, p. 17–32.

Stoddard, E.F., 1989, Distribution and significance of cordierite in the Northwest Adirondack lowlands: Northeastern Geology, v. 11, p. 40–49.

Thompson, A.B., 1976, Mineral reactions in pelitic rocks, II: Calculation of some *P-T-X* (Fe-Mg) phase relations: American Journal of Science, v. 276, p. 425–454.

Vielzeuf, D., and Holloway, J.R., 1988, Experimental determination of the fluid-absent melting relations in the pelitic system: Contributions to Mineralogy and Petrology, v. 98, p. 257–276.

Whitney, P.R., Fakundiny, R.F., and Isachsen, Y.W., 2002, Bedrock geology of the Fulton Chain-of-Lakes area, west-central Adirondack Mountains, New York: Albany, New York State Museum Map and Chart 44, 123 p. with map.

Young, D.A., 1995, Kornerupine group minerals in Grenville granulite facies paragneiss, Reading Prong, New Jersey: Canadian Mineralogist, v. 33, p. 1255–1262.

Manuscript Accepted by the Society August 25, 2003

Printed in the USA

Geological Society of America
Memoir 197
2004

SHRIMP U-Pb zircon geochronology of the anorthosite-mangerite-charnockite-granite suite, Adirondack Mountains, New York: Ages of emplacement and metamorphism

Michael A. Hamilton*
Geological Survey Canada, 601 Booth Street, Ottawa, Ontario K1A 0E8, Canada
James McLelland*
Department of Geosciences, Skidmore College, Saratoga Springs, New York 12866, USA
Bruce Selleck
Department of Geology, Colgate University, Hamilton, New York 13346, USA

ABSTRACT

Sensitive high-resolution ion microprobe (SHRIMP) analyses of zircons from four charnockitic gneisses and a ferrodiorite dike belonging to the anorthosite-mangerite-charnockite-granite (AMCG) suite of the Adirondack Mountains are consistent with prior multigrain U-Pb zircon determinations that indicated an emplacement age of ca. 1150–1160 Ma for other units of the suite. This result differs from recent assertions that Adirondack anorthosites were emplaced at ca. 1050 Ma. Each of the four charnockites discussed in this paper is intimately associated with adjacent massif anorthosite, with which they commonly exhibit mutually crosscutting and usually gradational relationships. Xenoliths of anorthosite in the charnockitic rocks provide a minimum age for the associated anorthosite. The presence of xenocrysts of blue-gray andesine in the charnockites reflects plucking from a coeval, but not wholly consolidated, anorthositic magma.

Ages determined for the four charnockitic rocks are 1176 ± 9 Ma (Minerva), 1154 ± 17 Ma (Gore Mountain), 1174 ± 25 Ma (Snowy Mountain), and 1164 ± 11 Ma (Diana Complex). Results also indicate an age of 1156 ± 7 Ma for a ferrodiorite dike that crosscuts, and is thought to be comagmatic with, the Oregon Dome anorthosite massif. These five new ages fall within a restricted interval of ca. 1150–1170 Ma that is in close agreement with similar AMCG magmatism in the central and eastern Grenville Province and, as a result, reinforce the proposal that this magmatism took place in response to orogen collapse and delamination following a widespread collisional event at ca. 1200–1160 Ma.

Keywords: anorthosite, Adirondacks, geochronology

*Current address, Hamilton: Jack Satterly Geochronology Laboratory, Department of Geology, University of Toronto, Toronto, Ontario M5S 3B1, Canada. McLelland, corresponding author: jmclelland@citlink.net.

Hamilton, M.A., McLelland, J., and Selleck, B., 2004, SHRIMP U-Pb zircon geochronology of the anorthosite-mangerite-charnockite-granite suite, Adirondack Mountains, New York: Ages of emplacement and metamorphism, *in* Tollo, R.P., Corriveau, L., McLelland, J., and Bartholomew, M.J., eds., Proterozoic tectonic evolution of the Grenville orogen in North America: Boulder, Colorado, Geological Society of America Memoir 197, p. 337–355. For permission to copy, contact editing@geosociety.org.

INTRODUCTION

The Adirondacks, like most of the eastern Grenville Province (Fig. 1), contains large intrusive complexes of andesine anorthosite accompanied by mangerite, charnockite, and granite, together referred to as the AMCG suite (Emslie, 1978). Because of the paucity of zircon in the anorthosites, dating of these complexes has relied on zircon U-Pb ages obtained from the associated granitoids whose field associations, as well as mutually crosscutting relationships, have been utilized to infer contemporaneity. Earlier efforts to separate zircons from Adirondack anorthosite (Silver, 1969; McLelland and Chiarenzelli, 1990) produced fractions greatly dominated by small, round, multifaceted grains of probable metamorphic origin that yielded ages of ca. 1050 Ma using thermal ionization mass spectrometry (TIMS) techniques. Zircons displaying morphology typical of igneous origin were too scarce to allow meaningful multigrain TIMS work. However, McLelland and Chiarenzelli (1990) did succeed in separating enough igneous zircon to obtain one discordant point (6% discordant) that yielded a minimum $^{207}Pb/^{206}Pb$ age of 1113 Ma. In addition, a fraction of baddeleyite from the same sample produced a slightly discordant minimum $^{207}Pb/^{206}Pb$ age of 1087 Ma. These results indicate that the anorthosite was emplaced significantly earlier than the ca. 1050 Ma granulite-facies metamorphism of the Ottawan orogeny, and suggest that it was coeval with the ca. 1155 Ma mangerites, charnockites, and granites of the region.

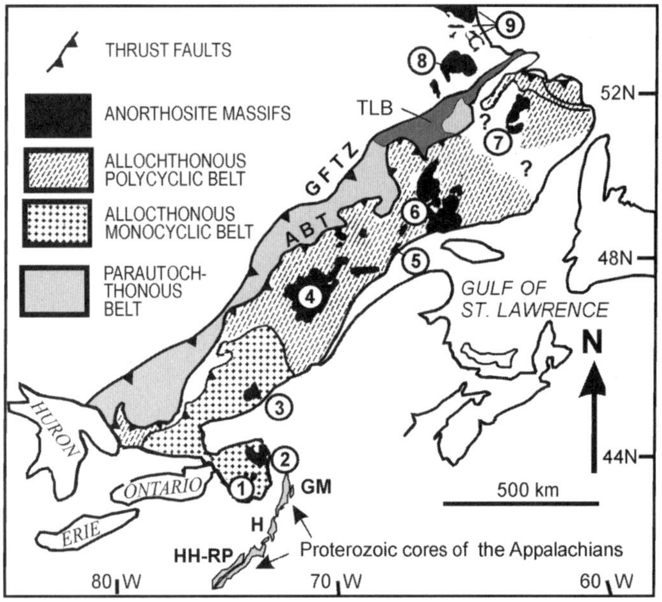

Figure 1. Generalized location map of the Adirondacks as a southwestern extension of the Canadian Grenville Province, whose three major tectonic divisions (Rivers, 1997) are shown. ABT—Allochthon Boundary Thrust; GFTZ—Grenville Front tectonic zone; GM—Green Mountains; H—Housatonic Mountains; HH-RP—Hudson Highlands and Reading prong; TLB (dark gray)—Trans-Labrador batholith. The major anorthosite massifs of the region (with ages) are (1) Oregon Dome (ca. 1150 Ma), (2) Marcy (ca. 1150 Ma), (3) Morin (ca. 1160 Ma), (4) Lac St. Jean (ca. 1150 Ma), (5) Rivière Pentecôte (ca. 1360 Ma), (6) Havre St. Pierre–Atikonak (ca. 1130 Ma), (7) Mealy Mountains (ca. 1650 Ma), (8) Harp Lake (ca. 1450 Ma), (9) Nain Plutonic Suite (ca. 1330–1300 Ma; only southernmost flank shown). Age references: 1–2, Hamilton et al. (2002); 3–7, Emslie and Hunt (1990); 3, Doig (1991); 4, Higgins and van Breemen (1992, 1996) and Hervet et al. (1997); 5, Machado and Martignole (1988) and Martignole et al. (1993); 6, van Breemen and Higgins, (1993); 8, Krogh and Davis (1973); 9, Hamilton et al. (1994, 1998, and references therein).

In recent investigations, anorthosites in the Grenville Province, and elsewhere in eastern Canada, have yielded sufficient zircon to permit direct dating by single-grain or small-fraction TIMS methods (Schärer et al., 1986; Machado and Martignole, 1988; Doig, 1991; Higgins and van Breemen, 1992, 1996; van Breemen and Higgins, 1993; Hamilton et al., 1994, 1998). Similar work is currently under way in the Adirondack Mountains, and early results were recently reported in preliminary form (Clechenko et al., 2002; Hamilton et al., 2002). These results are consistent with those reported in this paper.

In this study we present new data obtained from analysis of individual, small (≤30 μm) growth zones in zircons using the sensitive high-resolution ion microprobe (SHRIMP). Ages have been obtained for four important bodies of charnockitic-mangeritic rock belonging to the AMCG suite, and for a ferrodiorite dike that crosscuts massif anorthosite and is considered to represent a late filter-pressed magma derived from the anorthosite. Similar ferrodiorites are commonly associated with massif anorthosite in the Adirondacks and elsewhere (Emslie, 1978; McLelland et al., 1994). It is argued that these data indicate coeval emplacement of the Adirondack AMCG suite at ca. 1155 Ma.

GEOLOGIC SETTING

The general geology of the Adirondacks is presented in Figure 2, where units are subdivided in terms of lithology and U-Pb zircon age. The 1160–1145 Ma age interval designated for the AMCG suite is based on prior age determinations as well as on the results of this paper. The Marcy massif is the largest of the anorthosite bodies and is identified by name in Figure 2. Also shown in Figure 2 are the Oregon Dome anorthosite massif and the Snowy Mountain anorthosite massif. Mangeritic, charnockitic, and granitic members of the AMCG suite are shown as a single lithologic unit on the map. Previously dated (multigrain) samples are plotted along with those presented in this paper.

All members of the AMCG suite experienced granulite-facies metamorphism and pronounced regional deformation, resulting in large recumbent isoclinal folds and penetrative tectonic fabrics manifested by pencil and ribbon gneisses with L >> S (McLelland, 1984; McLelland et al., 1996). These tectonothermal events are associated with the Ottawan orogeny dated at ca. 1090–1030 Ma (Mezger et al., 1991; McLelland et al., 2001a), and are thought to have resulted from collision of Laurentia with Amazonia (Hoffman, 1991) along a margin that

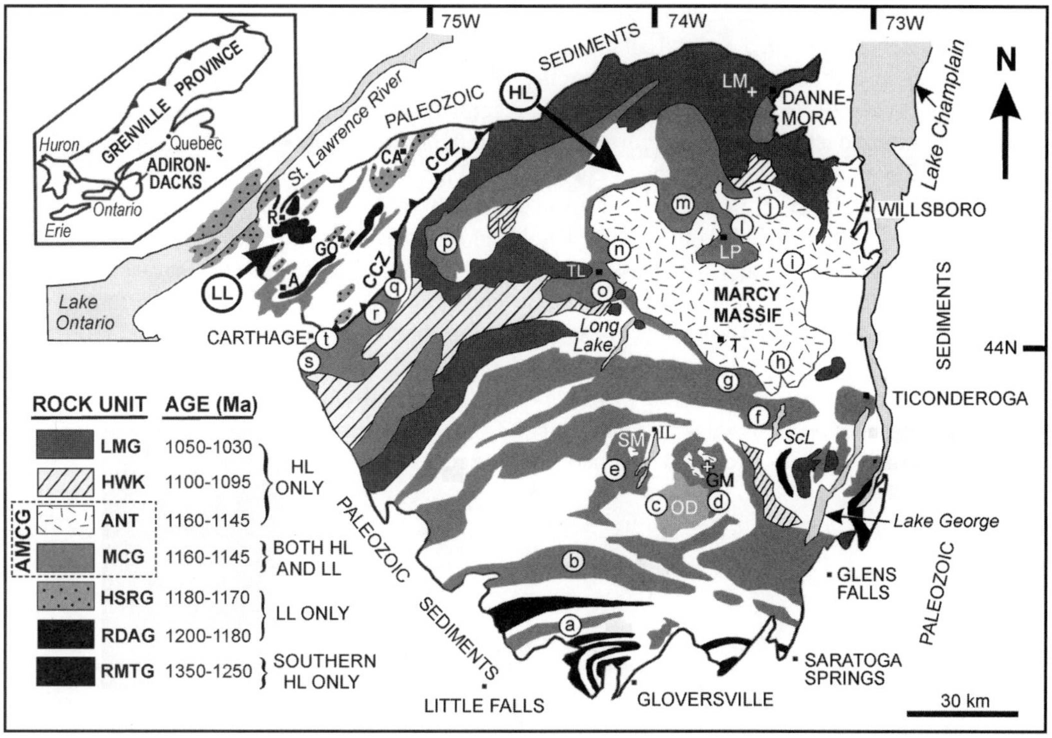

Figure 2. Generalized geological and geochronological map of the Adirondacks. Units designated by patterns and initials consist of igneous rocks dated by U-Pb zircon geochronology, with ages indicated. Units present only in the Highlands (HL) are ANT—anorthosite; HWK—Hawkeye granite; LMG—Lyon Mountain granite; RMTG—Royal Mountain tonalite and granodiorite (southern HL only). Units present in the Lowlands (LL) only are HSRG—Hyde School and Rockport granites (Hyde School also contains tonalite); RDAG—Rossie diorite and Antwerp granodiorite. Granitoid members of the anorthosite-mangerite-charnockite-granite (AMCG) suite (MCG) are present in both the Highlands and the Lowlands. Unpatterned areas consist of metasediments, glacial cover, or undivided units. A—Antwerp; CA—Canton; CCZ—Carthage-Colton mylonite zone; GM—Gore Mountain; GO—Gouvernour; IL—Indian Lake; LM—Lyon Mountain; LP—Lake Placid; OD—Oregon Dome; R—Rossie; ScL—Schroon Lake; SM—Snowy Mountain; T—Tahawus; TL—Tupper Lake. Locations for samples discussed in the text are a—Rooster Hill megacrystic charnockite; b—Piseco leucogranitic ribbon gneiss; c—Oregon Dome ferrodiorite; d—Gore Mountain charnockite; e—Snowy Mountain charnockite; f—Schroon Lake granitic gneiss; g—Minerva charnockite; h—North Hudson metagabbro; i—Woolen Mill gabbro and anorthosite; j—anorthositic pegmatite and clinopyroxene plagioclase dike in the Ausable River at Jay; l—Yard Hill jotunite; m—Bloomingdale mangerite; n—mangeritic dike crosscutting anorthosite northeast of Tupper Lake Village; o—mangerite southeast of Tupper Lake Village; p—rapakivi granite in Stark anticline; q—Oswegatchie leucogranite; r—Diana pyroxene syenite; s—Croghan granitic gneiss; t—Carthage anorthosite.

lay to the southeast of the present-day Appalachians. Effects of the Ottawan orogeny are present throughout the Grenville Province (Rivers, 1997; Rivers and Corrigan, 2000). Ottawan deformation of Adirondack anorthosites is largely confined to the margins of massifs, although local evidence of high strain is widespread in the massif interiors. Deformation at massif margins, and in outlying anorthositic sheets, is similar in style and intensity to that in Adirondack country rocks that fully experienced Ottawan penetrative orogenesis (McLelland and Chiarenzelli, 1990). Interior resistance to strain is the result of the large size of these massifs as well as the highly competent nature of their massive, anhydrous, and plagioclase-rich mineral compositions. This situation is commonly observed within large igneous plutons, especially those rich in plagioclase, e.g., gabbros and anorthosites (McLelland et al., 2001b). Within the plutons, original igneous fabrics and flow foliation are well preserved, although high-temperature, high-pressure garnet + clinopyroxene corona assemblages are ubiquitous and manifest Ottawan thermal effects (McLelland and Whitney, 1977). Recently, Alcock et al. (2001) and Isachsen et al. (2001) asserted that heat from the Marcy massif was responsible for very high-temperature (~800–900 °C) contact metamorphism and anatexis of a charnockite adjacent to the massif that they dated (by means of U-Pb zircon TIMS) at ca. 1040 Ma. They further argued that this age must indirectly date emplacement of the Marcy massif at ca. 1040 Ma. Following this line of reasoning, they concluded that the common presence of minimally deformed igneous fabrics in the massif interior indicates a late- to post-tectonic environment of emplacement for the anorthosite. Such assertions are in direct conflict with observations cited in McLelland and Chiarenzelli (1990), as well as with data set forth in the present paper (see also McLelland et al., 2001a, 2001b).

The compositions of Adirondack AMCG rocks correspond closely with those of AMCG suites elsewhere in the Grenville Province, Labrador, and Scandinavia (Emslie, 1978; Emslie et al., 1994). The anorthosites are composed primarily of calcic andesine and sodic labradorite (An_{46}–An_{52}), and associated granitoids tend to be low in silica, mildly alkaline, and enriched in ferrous iron, thus corresponding to rapakivi-suite rocks as defined by Emslie (1978) as well as to A-type granites (Eby, 1990; McLelland and Whitney, 1990). Representative whole-rock chemical analyses of granitoids discussed in this paper are given in Table 1. Within the Adirondacks, granitoid members of the suite are dominated by mangerite and charnockite, and texturally most orthopyroxene (including inverted pigeonite), as well as subordinate clinopyroxene, consists of large, subhedral grains. Two-pyroxene geothermometry indicates equilibration temperatures of 900–1000 °C and, hence, an igneous origin (Bohlen and Essene, 1977; Nabelek et al., 1987). The lack of primary hornblende reflects the low oxygen fugacity of the original magmas. Elsewhere in the Grenville Province of Canada, Dymek and Owens (1998) described several small anorthosite bodies that contain sodic andesine and labradorite as well as hemo-ilmenite. Emplacement of these relatively oxidized anorthosites has been dated at ca. 1020 Ma, and they are clearly of late to post-Ottawan age. These rocks are distinctly different from the anhydrous andesine massifs of the Adirondacks that are the focus of this paper.

TABLE 1. WHOLE-ROCK CHEMICAL ANALYSES

Oxide (wt%)	AM87-10 Minerva mangerite	AM86-8 Snowy Mt. mangerite	AM87-9 Gore Mt. mangerite	AM87-8 Oregon Dome ferrodiorite	AM86-1 Diana px qz syenite
SiO_2	60.13	59.84	61.05	42.80	65.98
TiO_2	0.90	1.39	0.78	6.04	0.86
Al_2O_3	14.52	16.37	15.98	10.53	15.38
Fe_2O_3*	10.23	6.79	6.70	21.60	4.72
MnO	0.01	0.14	0.05	0.02	0.08
MgO	0.83	1.35	1.64	5.68	0.63
CaO	4.52	5.35	3.63	8.77	2.39
Na_2O	4.83	4.31	3.41	2.17	4.09
K_2O	2.56	3.86	4.76	0.84	5.49
P_2O_5	0.55	0.46	0.42	0.66	0.24
LOI†	0.37	0.38	0.91	0.23	0.25
Total	99.45	100.24	99.33	99.34	100.11
Trace element (ppm)					
Ba	755	1100	680	300	1127
Rb	65	83	160	20	64
Sr	380	410	207	210	523
Y	57	110	43	20	45
Zr	409	1200	270	170	854
Nb	65	25	32	30	12
CIPW§ norms (wt%)					
Q	10.7	8.0	11.0		14.5
Or	15.3	22.2	28.0	4.6	32.5
Ab	40.87	37.3	28.85	18.4	35.0
An	10.4	9.0	14.24	16.5	7.4
Di	7.29	12.5	0.86	19.0	2.5
Hy	7.02	2.6	10.15	16.0	4.5
Mt	4.68	2.6	2.61	6.7	1.74
Il	1.71	2.64	1.48	11.5	1.63
Ap	1.27	0.97	0.97	1.53	0.60
Ol				4.3	
Mol fraction An 100 × (An/An + Ab)	20.3	19.4	33	47.3	17.5

*Total iron expressed as Fe_2O_3.
†LOI—Loss on ignition.
§CIPW—Cross, Iddings, Pirsson, and Washington.

DESCRIPTION OF SAMPLES

Four of the samples analyzed are from sites where zircons were previously dated (Chiarenzelli and McLelland, 1991; McLelland and Chiarenzelli, 1991) by the multigrain TIMS method but yielded highly discordant data. The fifth sample was collected from the Diana Complex (Fig. 2, D) in 1986, but because a multigrain age of ca. 1155 Ma (Grant et al., 1986) had just been reported, its zircons were not analyzed until now. All of the rocks sampled in this study are closely related to specific anorthosite massifs; these relationships are detailed for each sample in the following sections.

Charnockite-Mangerite Relations

All charnockitic-mangeritic samples contain xenocrysts (Fig. 3, A) of andesine plagioclase (An_{43}–An_{50}) identical in composition to plagioclase in adjacent anorthosite massifs and presumably derived from these bodies (Buddington, 1939, 1969; deWaard and Romey, 1969; McLelland and Whitney, 1990). Xenoliths of anorthosite are also present within the charnockitic-mangeritic rocks, but are less common. The xenocrysts are commonly mantled by perthitic feldspar (Fig. 3, A), reflecting reaction of the andesine with the charnockitic magma and the development of reaction textures identical to those described and experimentally produced by Stimac and Wark (1992) and by Wark and Stimac (1992). The abundance of such xenocrysts increases as anorthosite plutons are approached, consistent with the interpretation that these bodies are the source of the xenocrysts and xenoliths in bordering charnockite (deWaard and Romey 1969; Lettney 1969).

On the basis of these observations, we conclude that the xenocryst- and xenolith-bearing charnockite must be younger than, or coeval with, the anorthosite that yielded the inclusions, that is, the age of the charnockite provides a minimum age for emplacement of the anorthosite. In at least two of the samples under consideration, sheets of anorthosite crosscut the charnockite, strengthening the conclusion that they are coeval. This was also the conclusion of Miller (1918), who noted that the margin of the Marcy massif is commonly characterized by a rock that is best described as a mixture between anorthosite and charnockite. He referred to this hybrid as Keene gneiss for the excellent exposures along the Ausable River near the village of Keene. Buddington (1939) and Davis (1969) reached similar conclusions for other portions of the Marcy massif.

An important property of Keene gneiss is the presence of mantled andesine xenocrysts in its charnockitic portions (Fig. 3,

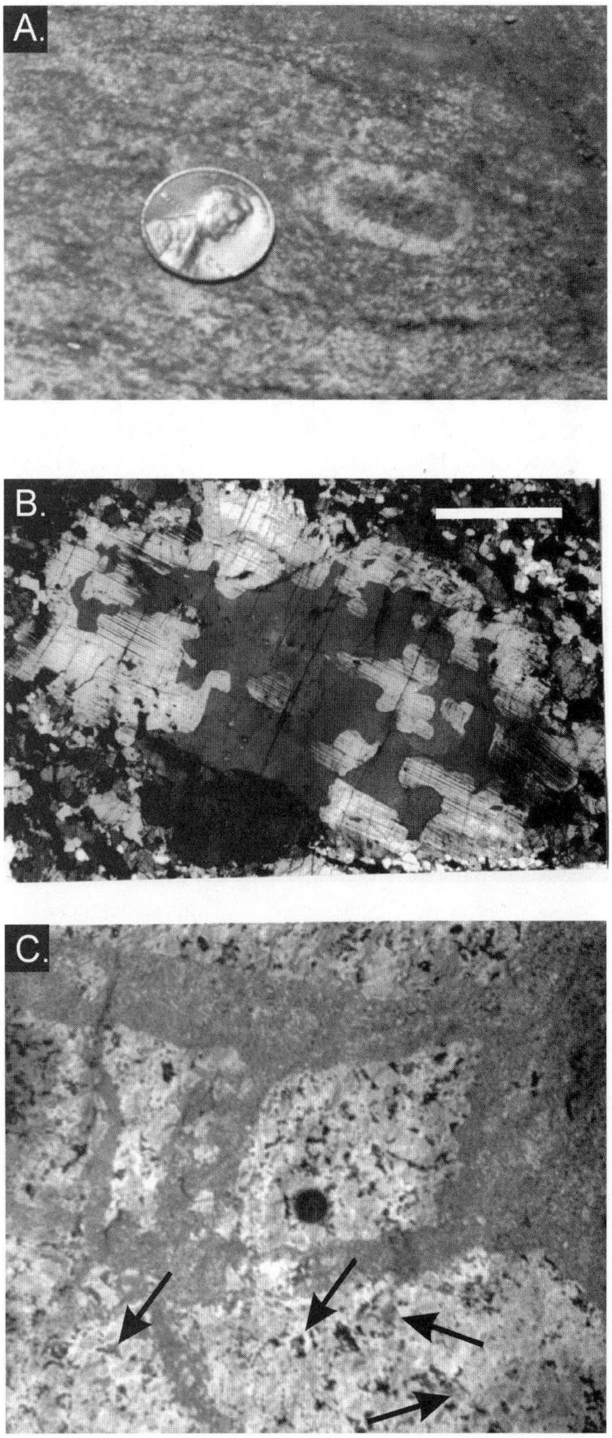

Figure 3. Anorthosite-granitoid relationships in Adirondack anorthosite-mangerite-charnockite-granite (AMCG) rocks. (A) Blue-gray andesine (An_{48}) xenocryst with light-colored reaction rim of perthite, Minerva charnockite (Fig. 2, locality m). Penny, 1.9 cm, for scale. (B) Photomicrograph of alkali feldspar (dark) from charnockite being replaced by plagioclase (An_{40}). The matrix consists of late ferrodioritic differentiate associated with adjacent anorthosite. Sample collected from anorthosite-charnockite transition zone, Cold River, 3 km northeast of the north end of Long Lake. Length of white scale bar is 0.5 cm. (C) Anorthosite (white) intruded by mangeritic Keene gneiss (dark). Black grains in the anorthosite are pyroxene, and gray areas are finer-grained mangerite permeating the anorthosite. Arrows mark several examples. The gray region surrounding the lens cap (5 cm in diameter) has been permeated by mangerite.

A) and plagioclase-mantled perthite in its anorthositic portions (Fig. 3, B). At the contact between anorthosite and xenocryst-bearing charnockite it is common to find the anorthosite in a state of disruption, including permeation by charnockite into interstices between plagioclase grains (Fig. 3, C). Such infiltration suggests that the anorthosite was not yet wholly consolidated when intruded by charnockitic magma. These observations are consistent with those of deWaard and Romey (1969) and of Lettney (1969), who described the presence of charnockite-anorthosite transition rocks surrounding the Snowy Mountain and Oregon Dome–Thirteenth Lake anorthosite massifs. These authors interpreted the transition zones as evidence that the end-member magmas were emplaced coevally. While we agree with this conclusion, we do not agree with the further inference that individual members of the AMCG suite are comagmatic and, in particular, that the anorthosite and granitoids represent differentiation products of a single parental magma (cf. Emslie et al., 1994). This conclusion is the same as that of Buddington (1939, 1969) and Davis (1969), who present a range of observations supporting a coeval but bimodal origin for the Adirondack AMCG suite (see also McLelland and Whitney, 1990). Included in these arguments are (1) the relatively small volume of anorthosite compared to the granitoids and a relative lack of transitional rocks between the two groups, (2) the fact that the transitional rocks that do exist appear to be mechanical mixtures in which feldspars and pyroxenes are commonly out of equilibrium with the enclosing magma, (3) opposing differentiation trends of the granitoids and anorthositic members of the suite, (4) screens of metasediment commonly developed between the anorthosites and granitoids, and (5) the fact that some granitoids preceded emplacement of the anorthosites. With these relationships in mind, we describe the locations and field relationships of the samples collected for SHRIMP U-Pb analysis.

Minerva Charnockite (AM87-10)

This sample was collected from a low roadcut along Route 28N, ~4 km north of the village of Minerva in the eastern Adirondacks (Fig. 2, g). The rock consists entirely of dark, olive-green charnockite containing medium-grained mesoperthite and quartz with minor orthopyroxene and Fe-Ti oxide minerals. A strong linear fabric pervades the rock and dominates a weak foliation developed in response to flattening. Centimeter-scale xenocrysts of blue-gray andesine with perthitic reaction rims occur throughout the outcrop (Fig. 3, A). The abundance of andesine xenocrysts in the charnockite increases northward toward the Marcy anorthosite (Fig. 2). In proximity to the contact, the charnockite contains 15–25% andesine xenocrysts with unreacted cores that are identical to those in the neighboring anorthosite. Compositionally, the rock corresponds to the hybrid Keene gneiss. Eight kilometers east of the Minerva locality, sheets of gabbroic anorthosite intrude the same charnockite and provide evidence supporting the coeval emplacement of these bodies of igneous rock.

Snowy Mountain Charnockite (AM86-8)

The Snowy Mountain charnockite and anorthosite of the central Adirondacks (Fig. 2, SM) were originally studied in detail by deWaard and Romey (1969), who undertook careful mapping of field relationships. They demonstrated that the Snowy Mountain Dome is cored by quasi-circular stocks of andesine anorthosite that grade outward into a more mafic noritic facies that grades, in turn, into olive-green quartz-mesoperthite charnockite that contains andesine xenoliths and xenocrysts. DeWaard and Romey (1969) stressed that the abundance of andesine xenocrysts in charnockite increases as the norite-anorthosite complex is approached, and they were able to contour xenocryst abundance. Near the contact, they identified a charnockite-anorthosite transition zone corresponding to Keene gneiss. The sample of charnockite collected for this study was taken from a roadcut along Route 30, ~5 km north of the Lewey Lake campground and very close to the base of Snowy Mountain near the transition zone (Fig. 2, e). The sample contains numerous andesine xenocrysts.

Gore Mountain Charnockite (AM87-9)

Samples of this unit were collected from outcrops at the rear of the parking lot for the Gore Mountain ski lift in the east central Adirondacks (Fig. 2, GM, d). The rock consists of dark olive-green quartz-mesoperthite charnockite containing xenocrysts of blue-gray andesine exhibiting perthitic reaction rims. The sample locality is in close proximity to the Oregon Dome anorthosite massif to the south and its outlier, the Thirteenth Lake anorthosite, immediately to the north. The xenocrysts were almost certainly derived from these bodies. Lettney (1969) conducted a detailed study of the northwest margin of the Oregon Dome anorthosite and all of the Thirteenth Lake outlier, and emphasized the presence of charnockite-anorthosite transition zones of hybrids corresponding to Keene gneiss at the margins of the anorthosites. He presented textural evidence depicting relationships between units in these zones and argued for coeval emplacement of anorthosite and charnockite in this region.

Oregon Dome Ferrodiorite (AM87-8)

Ferrodiorites of mafic to ultramafic composition occur in small volume as dikes and sheets in all anorthosite massifs of the Adirondacks. Similar rocks are sometimes referred to as apatite-oxide gabbronorites or jotunites (Owens and Dymek, 1992; McLelland et al., 1994). They commonly exhibit crosscutting relationships with the anorthosite and carry xenoliths and xenocrysts of that rock. McLelland et al. (1994) presented evidence that the ferrodioritic and ferrogabbroic bodies represent highly evolved, late-stage interstitial differentiates of the anorthosites, and were filter pressed into their current configuration during late movement of the partially consolidated massifs. A typical ferrodiorite dike crosscutting the Oregon Dome anorthosite along its southern border is shown in Figure 4. The exposure is located on the upper surface of a roadcut on the southwest side of Route 8-30, ~9 km east of Speculator (Fig. 2, OD, c). Contacts between the dike and the anorthosite are sinuous on a small scale and suggest that the two rocks are coeval. A whole-rock chemical analysis of the ferrodiorite is provided in Table 1.

Diana Complex (AM86-1)

This very large granitoid intrusive complex is located along the Carthage-Colton mylonite zone (CCMZ), which separates the Adirondack Highlands from the Adirondack Lowlands (Fig. 2, r). Buddington (1939) and Hargraves (1969) conducted extensive studies of the complex. It consists of a variety of rocks ranging from hornblende granite to pyroxene syenite; a typical whole-rock chemical analysis is given in Table 1. Feldspar is generally mesoperthite; grain size varies from medium to coarse. Although two pyroxenes are commonly present, clinopyroxene occurs as the sole pyroxene in many rocks. Evidence for deformation is present throughout the complex, and is predominantly strong and penetrative, with a northwest direction of tectonic transport. Both ductile and cataclastic deformation, as well as psuedotachylytic textures, are common in proximity to the western contact. Small bodies of andesine anorthosite are present locally near the Diana Complex; the largest of these, the Carthage anorthosite (Fig. 2, t), was interpreted by Buddington (1939) as coeval with the Diana granitoids. The sample chosen for dating was collected from large roadcuts on Route 3 near the western contact of the Diana Complex, ~1 km. south of Harrisville (Fig. 2, r). This sample was originally collected in the early 1980s, but processing was delayed because Grant et al. (1986) had already initiated geochronological studies on rocks from the same outcrops. This earlier investigation yielded a multigrain (zircon, U-Pb) TIMS age of ca. 1155 Ma, although several data points show reversed discordance. Accordingly, it was deemed appropriate to redate this outcrop using modern SHRIMP U-Pb techniques.

Figure 4. Ferrodiorite dike crosscutting Oregon Dome anorthosite (Fig. 2, locality c). The body shown in this photograph is an offshoot of a nearby ~10 m–wide dike. The ferrodiorite contains two generations of plagioclase, one that consists of large andesine (~An_{48}) xenocrysts from the adjacent anorthosite. These are rimmed by more sodic plagioclase (~An_{43}) that corresponds closely to the composition of the second plagioclase, represented by the smaller grains in the dike. The sinuous upper contact of the dike suggests that the anorthosite was not fully consolidated when the dike intruded. The mafic-enriched lower contact of the dike may represent crystal settling, which would be consistent with an original sheetlike form. Length of hammer is 30 cm.

RESULTS OF SHRIMP U-PB ZIRCON GEOCHRONOLOGY

Geochronological results for each of the samples are presented in the following sections. SHRIMP analyses for sample AM86-8 (Snowy Mountain Charnockite) were performed at the SHRIMP II facility at the Australian National University in Canberra, Australia. All other analyses were performed at the SHRIMP II laboratory at the Geological Survey of Canada in Ottawa, Canada. Data are summarized in Table 2. Analytical details are provided in the appendix to this paper. Individual corrected ratios and ages are reported with 1σ analytical errors (68% confidence; Table 2), as are the error ellipses presented in the concordia diagrams. The associated plots of weighted mean $^{207}Pb/^{206}Pb$ ages (insets to concordia diagrams) and calculated mean ages in the text, however, are presented at 95% confidence levels. Statistical treatment and plotting of data were achieved using the Isoplot/Ex program of Ludwig (2001).

Minerva Charnockite (AM87-10)

The Minerva charnockite contains an abundance of large (500–600 µm), prismatic zircons that are represented in the SHRIMP mounts principally by smaller cleavage fragments (Fig. 5). In addition to these, there are abundant small (50–100 µm), round grains, many of which contain irregular cores (Fig. 5, A) or vaguely prismatic cores surrounded by thick, structureless overgrowths (Fig. 5, C). Cathodoluminescence (CL) imaging and SHRIMP analysis confirm that the homogeneous rimming zircon is consistently uranium-rich (400–1650 ppm) relative to the zircon cores (65–280 ppm). Larger zircon fragments exhibit straight internal zoning (Fig. 5, B), or display "feathery" oscillatory growth structures (Fig. 5, D). Both textures are consistent with igneous crystallization; these features occur both with or without thin rims or overgrowths. Both the euhedral to anhedral zircon cores and the large zircon fragments are interpreted as igneous in origin, whereas the round grains and unzoned overgrowths likely result from metamorphic growth. This distinction is corroborated by generally high Th/U for igneous zircon (average = 0.48) compared with lower Th/U (average = 0.21; locally as low as 0.09) for metamorphic zircon in this sample (Table 2).

Results of SHRIMP analyses of zircons from the Minerva charnockite are illustrated in a concordia diagram in Figure 6.

TABLE 2. SHRIMP II U-TH-PB RESULTS FOR ADIRONDACK CHARNOCKITE AND FERRODIORITE SAMPLES

Spot	Loc*	U (ppm)	Th (ppm)	Th/U	$^{206}Pb/^{204}Pb$	$^{207}Pb/^{235}U \pm 1\sigma^{\dagger}$	$^{206}Pb/^{238}U \pm 1\sigma^{\dagger}$	$^{207}Pb/^{206}Pb \pm 1\sigma^{\dagger}$	$^{207}Pb/^{206}Pb$ age $\pm 1\sigma^{\dagger}$	Conc. %§
AM87-10, Minerva charnockite										
10.1	c	279	232	0.861	36153	2.1625 ± 0.0287	0.19689 ± 0.00247	0.07966 ± 0.00024	1188.5 ± 6.1	97.5
2.1	c	140	33	0.243	100000	2.1374 ± 0.0292	0.19474 ± 0.00238	0.07960 ± 0.00038	1187.1 ± 9.4	96.6
17.1	if	132	73	0.575	100000	2.1138 ± 0.0228	0.19347 ± 0.00146	0.07924 ± 0.00054	1178.1 ± 13.6	96.8
5.1	c	155	35	0.231	86430	2.1094 ± 0.0329	0.19331 ± 0.00239	0.07914 ± 0.00064	1175.7 ± 16.1	96.9
7.1	if	275	113	0.426	100000	2.1103 ± 0.0313	0.19404 ± 0.00243	0.07887 ± 0.00052	1169.0 ± 13.1	97.8
4.2	if	212	143	0.699	73964	2.1024 ± 0.0198	0.19351 ± 0.00125	0.07880 ± 0.00048	1167.0 ± 12.1	97.7
4.1	if	216	136	0.648	63052	2.1157 ± 0.0281	0.19549 ± 0.00238	0.07849 ± 0.00031	1159.4 ± 7.9	99.3
26.1	c	76	22	0.295	19444	2.0132 ± 0.0380	0.18739 ± 0.00170	0.07792 ± 0.00120	1144.8 ± 31.0	96.7
3.2	c	65	20	0.325	100000	2.0431 ± 0.0355	0.19030 ± 0.00255	0.07787 ± 0.00074	1143.4 ± 19.0	98.2
12.1	c	84	47	0.580	43085	2.0220 ± 0.0351	0.18977 ± 0.00246	0.07728 ± 0.00078	1128.4 ± 20.2	99.3
1.2	c	77	38	0.505	13116	2.0185 ± 0.0322	0.18945 ± 0.00240	0.07727 ± 0.00064	1128.3 ± 16.6	99.1
9.1	c	77	21	0.287	31656	2.0062 ± 0.0378	0.18901 ± 0.00238	0.07698 ± 0.00097	1120.7 ± 25.2	99.6
8.1	if	978	82	0.087	196078	1.9831 ± 0.0260	0.18702 ± 0.00231	0.07691 ± 0.00024	1118.7 ± 6.3	98.8
15.1	c	780	75	0.100	121212	1.9225 ± 0.0244	0.18442 ± 0.00230	0.07561 ± 0.00011	1084.7 ± 2.8	100.6
16.1	r	987	135	0.141	100000	1.8428 ± 0.0286	0.17834 ± 0.00251	0.07494 ± 0.00037	1067.0 ± 10.0	99.1
2.2	r	1178	216	0.189	114025	1.8204 ± 0.0239	0.17708 ± 0.00220	0.07456 ± 0.00023	1056.5 ± 6.2	99.5
14.1	e	1425	199	0.144	130719	1.8216 ± 0.0223	0.17771 ± 0.00214	0.07434 ± 0.00007	1050.8 ± 2.0	100.3
27.1	e	1563	393	0.260	117647	1.8252 ± 0.0108	0.17812 ± 0.00093	0.07432 ± 0.00017	1050.2 ± 4.5	100.6
24.1	r	537	122	0.234	45188	1.7566 ± 0.0148	0.17161 ± 0.00114	0.07424 ± 0.00033	1048.1 ± 8.9	97.4
12.2	r	562	144	0.264	49310	1.7807 ± 0.0237	0.17407 ± 0.00216	0.07419 ± 0.00026	1046.7 ± 7.1	98.8
10.2	r	706	156	0.228	100000	1.8076 ± 0.0260	0.17672 ± 0.00232	0.07418 ± 0.00033	1046.5 ± 9.1	100.2
23.1	r	693	133	0.199	69881	1.7450 ± 0.0119	0.17088 ± 0.00099	0.07406 ± 0.00022	1043.2 ± 6.1	97.5
13.2	r	420	128	0.314	74683	1.8167 ± 0.0250	0.17797 ± 0.00228	0.07404 ± 0.00026	1042.5 ± 7.2	101.3
1.1	r	572	123	0.222	100000	1.7608 ± 0.0226	0.17285 ± 0.00210	0.07388 ± 0.00021	1038.2 ± 5.9	99.0
11.1	r	809	148	0.189	100000	1.7526 ± 0.0230	0.17230 ± 0.00209	0.07377 ± 0.00028	1035.3 ± 7.6	99.0
16.2	2r	401	42	0.108	100000	1.7175 ± 0.0257	0.17064 ± 0.00223	0.07300 ± 0.00042	1013.9 ± 11.8	100.2
AM86-8, Snowy Mountain charnockite										
16.1	c	53	21	0.388	4988	2.2935 ± 0.1211	0.20001 ± 0.00610	0.08317 ± 0.00326	1273.1 ± 78.6	92.3
9.1	c	62	27	0.431	8027	2.1868 ± 0.1237	0.19410 ± 0.00730	0.08171 ± 0.00308	1238.6 ± 75.7	92.3
11.1	c	160	121	0.756	8525	2.1245 ± 0.1211	0.18875 ± 0.00660	0.08163 ± 0.00331	1236.7 ± 81.7	90.1
5.1	c	210	155	0.739	49480	2.1676 ± 0.0791	0.19611 ± 0.00627	0.08016 ± 0.00112	1200.9 ± 27.8	96.1
1.1	c	152	113	0.745	37	2.1278 ± 0.0621	0.19254 ± 0.00450	0.08015 ± 0.00121	1200.7 ± 30.8	94.5
4.1	c	145	110	0.755	100000	2.1549 ± 0.0769	0.19610 ± 0.00540	0.07974 ± 0.00156	1190.6 ± 39.3	96.9
17.1	c	183	127	0.695	25641	2.0817 ± 0.0533	0.19190 ± 0.00410	0.07867 ± 0.00092	1163.9 ± 23.3	97.2
8.1	c	179	124	0.691	5905	2.1241 ± 0.0806	0.19622 ± 0.00530	0.07851 ± 0.00184	1159.8 ± 47.1	99.6
14.1	c	118	45	0.380	8314	2.1445 ± 0.0836	0.20080 ± 0.00530	0.07746 ± 0.00198	1132.9 ± 51.8	104.1
7.1	c	219	129	0.587	13939	1.8949 ± 0.0704	0.17987 ± 0.00490	0.07641 ± 0.00171	1105.8 ± 45.4	96.4
15.1	r	391	228	0.585	28185	1.7868 ± 0.0460	0.17595 ± 0.00410	0.07365 ± 0.00072	1031.9 ± 20.0	101.2
2.1	r	334	137	0.409	246305	1.7371 ± 0.0473	0.17114 ± 0.00390	0.07361 ± 0.00088	1030.9 ± 24.4	98.8
AM87-9, Gore Mountain charnockite										
11.1	c	70	42	0.615	100000	2.1353 ± 0.0326	0.19592 ± 0.00268	0.07904 ± 0.00042	1173.3 ± 10.5	98.3
16.1	c/if	116	46	0.406	20661	2.1130 ± 0.0273	0.19464 ± 0.00157	0.07874 ± 0.00072	1165.5 ± 18.1	98.4
8.1	c	245	122	0.513	122399	2.0908 ± 0.0294	0.19291 ± 0.00247	0.07861 ± 0.00035	1162.2 ± 8.8	97.8
1.1	c/if	294	62	0.217	100000	2.0719 ± 0.0290	0.19307 ± 0.00247	0.07783 ± 0.00033	1142.6 ± 8.5	99.6
12.2	c	188	83	0.456	25786	2.0932 ± 0.0225	0.19587 ± 0.00147	0.07751 ± 0.00053	1134.3 ± 13.6	101.7
21.1	c/if	83	46	0.572	8568	1.9949 ± 0.0351	0.18805 ± 0.00173	0.07694 ± 0.00107	1119.7 ± 27.9	99.2
13.1	r/m	309	129	0.431	13156	1.8327 ± 0.0268	0.17665 ± 0.00230	0.07525 ± 0.00039	1075.1 ± 10.5	97.5
15.1	r/m	268	140	0.539	21146	1.8552 ± 0.0176	0.17938 ± 0.00125	0.07501 ± 0.00043	1068.7 ± 11.5	99.5

2.1	c	297	179	0.623	147059	1.8699 ± 0.0246	0.18169 ± 0.00223	0.07464 ± 0.00025	1058.9 ± 6.8	101.6
6.1	r/e	366	187	0.528	51387	1.8121 ± 0.0257	0.17616 ± 0.00226	0.07461 ± 0.00035	1057.9 ± 9.4	98.9
19.1	r	92	36	0.402	7087	1.8850 ± 0.0372	0.18328 ± 0.00168	0.07459 ± 0.00122	1057.6 ± 33.3	102.6
14.1	r	304	121	0.412	100000	1.8012 ± 0.0298	0.17568 ± 0.00250	0.07436 ± 0.00051	1051.3 ± 13.9	99.2
5.2	r	172	61	0.367	94607	1.7938 ± 0.0349	0.17541 ± 0.00244	0.07417 ± 0.00089	1046.0 ± 24.3	99.6
3.1	r	170	115	0.700	15669	1.7830 ± 0.0612	0.17443 ± 0.00438	0.07414 ± 0.00152	1045.2 ± 41.9	99.2
20.1	r	273	102	0.385	34783	1.7359 ± 0.0247	0.17005 ± 0.00202	0.07403 ± 0.00048	1042.4 ± 13.2	97.1
11.2	r	200	63	0.326	23507	1.7950 ± 0.0259	0.17597 ± 0.00222	0.07398 ± 0.00042	1041.0 ± 11.4	100.4
12.1	r	372	56	0.156	100000	1.7455 ± 0.0226	0.17126 ± 0.00212	0.07392 ± 0.00019	1039.3 ± 5.3	98.1
7.2	r	288	103	0.370	280112	1.7943 ± 0.0241	0.17613 ± 0.00218	0.07389 ± 0.00029	1038.4 ± 7.9	100.7
5.1	r	400	151	0.390	27747	1.7713 ± 0.0246	0.17452 ± 0.00218	0.07361 ± 0.00035	1030.9 ± 9.5	100.6
AM87-8, Oregon Dome ferrodiorite										
9.2	c	157	44	0.291	100000	2.0592 ± 0.0338	0.18913 ± 0.00275	0.07897 ± 0.00048	1171.3 ± 12.0	95.3
1.3	c	192	91	0.488	119617	2.1157 ± 0.0192	0.19438 ± 0.00121	0.07894 ± 0.00047	1170.7 ± 11.8	97.8
12.1	c	141	43	0.315	19493	2.0747 ± 0.0240	0.19110 ± 0.00147	0.07874 ± 0.00061	1165.6 ± 15.4	96.7
5.1	c	193	59	0.317	26254	2.1011 ± 0.0334	0.19378 ± 0.00264	0.07864 ± 0.00053	1163.0 ± 13.3	98.2
3.1	c	264	140	0.549	92421	2.0776 ± 0.0268	0.19239 ± 0.00233	0.07832 ± 0.00025	1155.0 ± 6.2	98.2
6.2	c	194	58	0.311	31949	2.1106 ± 0.0285	0.19580 ± 0.00245	0.07818 ± 0.00029	1152.0 ± 15.0	100.7
20.1	c	164	47	0.294	184162	2.1261 ± 0.0228	0.19740 ± 0.00147	0.07811 ± 0.00054	1149.8 ± 13.7	101.0
17.1	c	158	63	0.414	25994	2.0530 ± 0.0210	0.19133 ± 0.00128	0.07783 ± 0.00054	1142.4 ± 13.9	98.8
7.1	c	218	82	0.388	88496	2.0558 ± 0.0264	0.19220 ± 0.00110	0.07757 ± 0.00022	1136.0 ± 11.0	100.7
2.1	c	142	42	0.304	25497	2.0380 ± 0.0292	0.19179 ± 0.00246	0.07707 ± 0.00039	1122.9 ± 10.0	100.7
11.1	c	130	40	0.315	38432	1.8937 ± 0.0219	0.18099 ± 0.00129	0.07589 ± 0.00062	1092.1 ± 16.6	98.2
4.1	r	155	58	0.385	15368	1.8410 ± 0.0294	0.17838 ± 0.00221	0.07485 ± 0.00065	1064.5 ± 17.5	99.4
5.2	r	64	25	0.404	6158	1.7486 ± 0.0353	0.16985 ± 0.00223	0.07467 ± 0.00103	1059.6 ± 28.0	95.4
1.1	r	208	112	0.556	31056	1.8495 ± 0.0246	0.18001 ± 0.00221	0.07452 ± 0.00028	1055.6 ± 7.7	101.1
19.1	r	76	32	0.436	19339	1.7531 ± 0.0299	0.17100 ± 0.00147	0.07435 ± 0.00102	1051.1 ± 27.9	96.8
2.2	r	176	88	0.516	23552	1.7784 ± 0.0259	0.17468 ± 0.00218	0.07384 ± 0.00045	1037.1 ± 12.5	100.1
16.1	r	183	88	0.497	11017	1.7838 ± 0.0184	0.17525 ± 0.00109	0.07382 ± 0.00055	1036.6 ± 15.1	100.4
15.1	r	219	98	0.464	16003	1.7509 ± 0.0174	0.17223 ± 0.00111	0.07373 ± 0.00050	1034.1 ± 13.9	99.1
AM86-1, Diana Complex quartz syenite										
8.2	c	188	78	0.429	18932	2.1110 ± 0.0444	0.19214 ± 0.00249	0.07968 ± 0.00120	1189.2 ± 30.0	95.3
2.1	c	133	64	0.501	19069	2.1135 ± 0.0306	0.19250 ± 0.00241	0.07963 ± 0.00047	1187.8 ± 11.7	95.5
7.1	c	117	54	0.475	27878	2.1473 ± 0.0311	0.19600 ± 0.00246	0.07946 ± 0.00046	1183.5 ± 11.5	97.5
9.1	c	134	79	0.612	84602	2.1595 ± 0.0302	0.19736 ± 0.00242	0.07936 ± 0.00043	1181.1 ± 10.6	98.3
18.1	c	66	32	0.504	14641	2.0881 ± 0.0342	0.19260 ± 0.00166	0.07863 ± 0.00101	1163.0 ± 25.8	97.6
13.1	c	436	300	0.710	34495	2.2033 ± 0.0308	0.20362 ± 0.00263	0.07848 ± 0.00031	1159.1 ± 7.9	103.1
6.1	c	219	79	0.374	21317	2.0788 ± 0.0287	0.19216 ± 0.00236	0.07846 ± 0.00039	1158.6 ± 9.9	97.8
5.1	c	203	84	0.425	47371	2.1117 ± 0.0290	0.19586 ± 0.00246	0.07820 ± 0.00033	1151.9 ± 8.3	100.1
8.1	c	187	104	0.578	68306	2.1416 ± 0.0356	0.19885 ± 0.00291	0.07811 ± 0.00049	1149.7 ± 12.4	101.7
10.1	c	226	79	0.363	13734	2.1428 ± 0.0337	0.19925 ± 0.00257	0.07800 ± 0.00059	1146.8 ± 15.0	102.1
4.1	c	132	60	0.470	100000	2.1173 ± 0.0277	0.19839 ± 0.00241	0.07740 ± 0.00027	1131.6 ± 7.1	103.1
14.1	r	68	16	0.241	4606	2.0358 ± 0.0475	0.19136 ± 0.00285	0.07716 ± 0.00125	1125.4 ± 32.6	100.3
1.2	r	84	20	0.243	100000	1.9761 ± 0.0499	0.18597 ± 0.00309	0.07707 ± 0.00131	1122.9 ± 34.4	97.9
11.1	r	59	13	0.226	4847	2.0424 ± 0.0478	0.19242 ± 0.00254	0.07698 ± 0.00137	1120.7 ± 35.8	101.2
15.1	r	177	76	0.447	26399	1.9495 ± 0.0399	0.18371 ± 0.00284	0.07696 ± 0.00089	1120.3 ± 23.3	97.0
1.1	r	162	81	0.520	26137	2.0337 ± 0.0349	0.19173 ± 0.00243	0.07693 ± 0.00078	1119.4 ± 20.2	101.0

Note: Corrections for common Pb made using measured ^{204}Pb. See appendix for analytical details.

* Location of spot analysis or nature of grain analyzed: c—core; r—rim; 2r—second, outermost rim; e—equant and homogeneous; if—fragment of magmatic grain showing simple igneous zoning (igneous fragment); m—probable mixed age (straddles domain boundaries).

†All errors on ratios and ages reported at a 1σ level of uncertainty, absolute, and reflect numerical propagation of all known sources of error.

§Percent concordant = 100 × ($^{206}Pb/^{238}U$ age)/($^{207}Pb/^{206}Pb$ age).

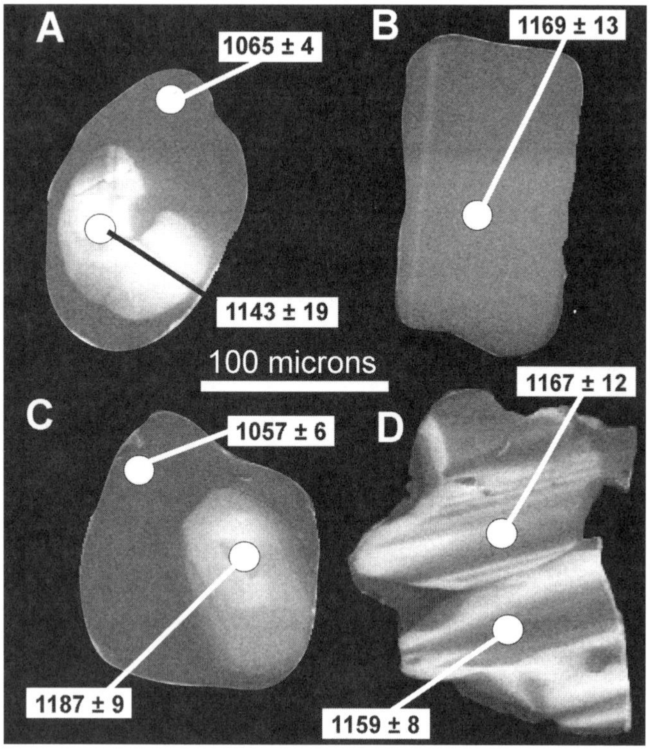

In order to enhance the reliability of the data in Table 2, only those analyses with low degrees of discordance (generally ±4%) were used for sample age assignment and interpretation. The best, clustered SHRIMP ages from unambiguous igneous fragments and cores in this sample define the range ca. 1143–1189 Ma, whereas ages for metamorphic zircon, as defined either by equant, structureless (unzoned) grains or by homogeneous, rounded overgrowths, fall into the range 1035–1067 Ma (Fig. 6; Table 2). Five additional analyses spread along or just below concordia between 1128–1085 Ma. The age of igneous fragments and cores is defined by nine analyses whose grouping produces a well-defined weighted average age of 1176 ± 11 Ma

Figure 5. Cathodoluminescence (CL) images of representative zircons from the Minerva charnockite (AM87-10). (A) Small, round grain containing an irregular core. (B) Large, igneous cleavage fragment showing weak, straight internal zoning. (C) Rounded grain with metamorphic overgrowth cored by older, subhedral igneous grain. (D) Large broken fragment of a "feathery-textured," dominantly igneous grain surrounded by a thin, structureless, dark (high-U) metamorphic overgrowth. In this and subsequent CL images, dots indicate the locations of ion microprobe analyses, with ages (in Ma) given at the ±1σ uncertainty level.

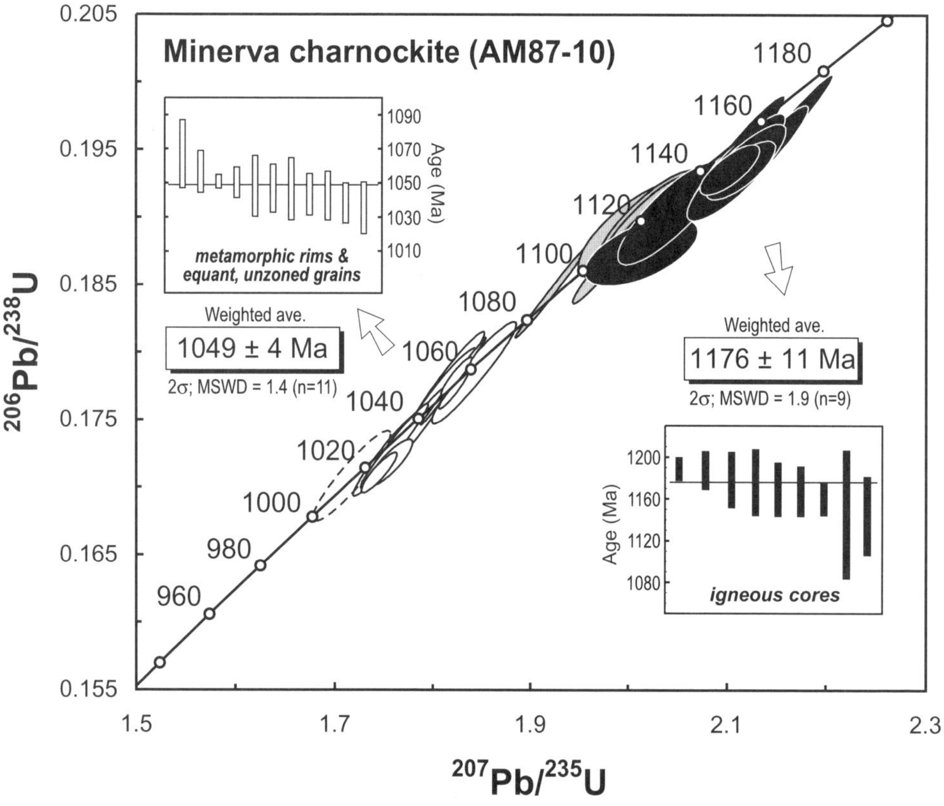

Figure 6. Concordia plot of sensitive high-resolution ion microprobe (SHRIMP) analyses of zircons from the Minerva charnockite (AM87-10) exhibiting bimodal clustering of ages into an igneous group (cores) and a metamorphic group (structureless overgrowths and equant grains). For this plot and succeeding Concordia diagrams, error ellipses from igneous cores used to calculate an intrusion age are shaded black, whereas data used to calculate the age of metamorphic rims are solid white; data not used in calculations are shaded gray (see text for discussion). Dashed ellipse represents a single analysis of a distinct secondary rim (see text for discussion). Error ellipses shown at 1σ level of uncertainty. MSWD—mean square of weighted deviates. Weighted mean plots (insets) show error bars corresponding to $^{207}Pb/^{206}Pb$ ages and 2σ errors, and calculated mean ages at the 95% confidence level.

(95% confidence; mean square of weighted deviates or MSWD = 1.9; Fig. 6, inset). We interpret this age as the time of emplacement of the Minerva charnockite. In contrast, eleven analyses of overgrowths and equant metamorphic zircons define a weighted average age of 1049 ± 4 Ma (MSWD = 1.4; Fig. 6, inset), which falls into the range of regional ca. 1090–1030 Ma Ottawan metamorphism. Intermediate ages not included in either group may be related to intervening events, such as Hawkeye granitic magmatism at ca. 1105–1090 Ma (McLelland et al., 2001a). Alternatively, given the intensity of the Ottawan metamorphic effects as expressed by the development of locally thick rims or overgrowths on older igneous grains, it is possible that the spread of intermediate ages between 1128 and 1085 Ma described earlier represents incomplete Pb loss from ca. 1175 Ma igneous cores during ca. 1050 Ma high-grade metamorphism. A single analysis of a thin outermost rim, which truncates an igneous core but also surrounds a ca. 1067 Ma overgrowth on one zircon grain, yields an age of 1014 ± 12 Ma (1σ; Table 2; Fig. 6, dashed ellipse). This age is interpreted to result from a late, weak thermal event that is recorded locally within the Adirondacks (McLelland et al. 2001a), which occurred between the Ottawan and Rigolet pulses of Grenvillian metamorphism as defined by Rivers (1997). However, the significance of the single age determination as described here requires further substantiation.

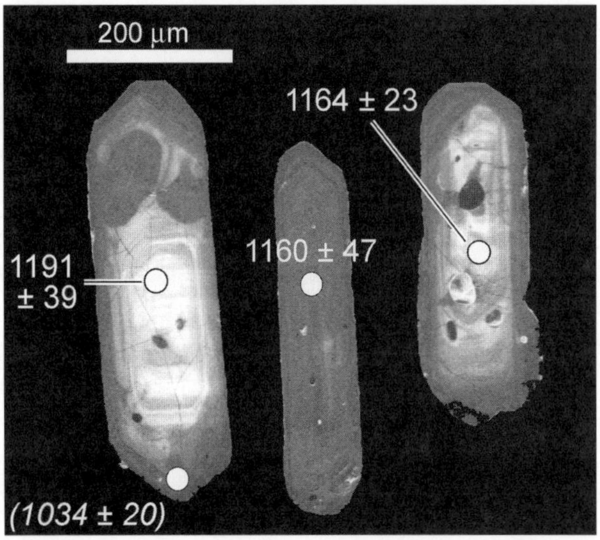

Figure 7. Cathodoluminescence (CL) images of representative zircons from the Snowy Mountain charnockite (AM86-8). Note the generally elongate, prismatic morphologies, and the concentric, oscillatory zoning of broad igneous cores. Thin, partially rounded or irregular rims are dark in CL (higher U), and are interpreted to be metamorphic in origin. Left-hand grain rim was excessively discordant and is not included in Table 2 or Figure 8.

Snowy Mountain Charnockite (AM86-8)

Zircon sizes and morphologies in the Snowy Mountain charnockite are similar to those in the Minerva charnockite. However, elongate, euhedral (100–300 μm), prismatic, and doubly terminated grains are more common, with terminations locally showing partial rounding due to metamorphic overgrowths (Fig. 7). Both the larger and the smaller prismatic grains are interpreted as igneous in origin, and a small number of each group contains cores of older zircon. Small, rounded, discrete metamorphic grains are not abundant in this sample, but thin, partially rounded or irregular rims and overgrowths on older grains are interpreted as having developed during metamorphism (Fig. 7).

Zircons from the Snowy Mountain charnockite proved to be difficult to analyze with the desired precision, generally due to the lower average uranium content of the grains and the specific instrument operating conditions during that analytical session. For purposes of plotting and age determination, we have retained only those analyses showing less than 5% discordance. In Table 2 we also report three very discordant and imprecise analyses that indicate the presence of ca. 1235–1275 Ma cores of pre-AMCG (Elzevirian) origin. The remaining analyses are plotted on a concordia diagram (Fig. 8) and fall into two distinct groups. The oldest seven of these yield a moderately constrained weighted average age of 1174 ± 25 Ma (MSWD = 0.8; average Th/U [n = 7] = 0.70), whereas the younger two analyses give an average age of 1031 ± 30 Ma (average Th/U = 0.50). The older age is interpreted as the age of emplacement of the charnockite. The two younger analyses are represented by overgrowths and are interpreted as metamorphic in origin.

Gore Mountain Charnockite (AM87-9)

Like the other charnockitic rocks discussed in this paper, the Gore Mountain sample contains many large zircons represented by zoned, cleaved, or fractured fragments (Fig. 9, C) up to 250 μm in length and suggestive of appreciably larger original grain dimensions. Smaller 100–200 μm zircons exhibit elongate, prismatic, and doubly terminated morphologies commonly rounded by metamorphic overgrowths (Fig. 9, D). In CL the interiors of these grains show evidence of growth zoning characteristic of crystallization from magma (Fig. 9, D), and the fragments and smaller euhedral cores are both interpreted as igneous in origin. In addition, small equidimensional grains commonly contain cores of older zircon rimmed by metamorphic overgrowths (Fig. 9, A and B).

The zircon population from the Gore Mountain charnockite provided nineteen analyses with less than 3% discordance. The data, plotted on a concordia diagram in Figure 10, exhibit bimodal clustering at 1154 ± 17 Ma and 1041 ± 6 Ma, corresponding to the times of igneous emplacement and regional metamorphism, respectively. Both igneous and metamorphic ages are shown in the representative CL images of Figure 9, and demonstrate the core (igneous) and rim (metamorphic) nature of the two age clusters. Three ellipses shown with gray shading in Figure 10 have intermediate ages between those from cores and

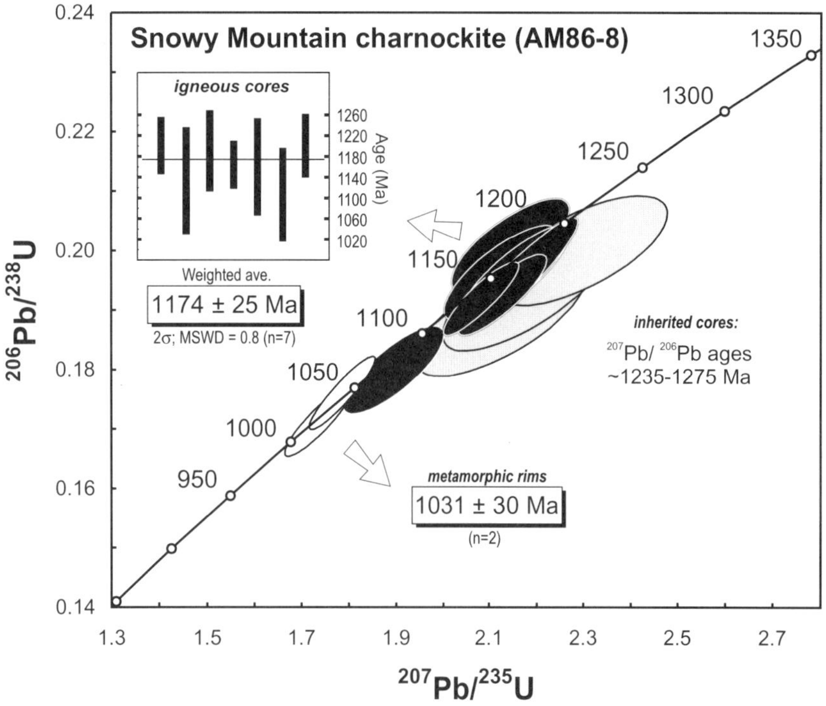

Figure 8. Concordia diagram of SHRIMP analyses of zircons from the Snowy Mountain mangerite (AM86-8) exhibiting bimodal clustering of ages into an igneous group (main population, black ellipses) and a smaller population of metamorphic overgrowths (two youngest analyses, white ellipses). Also shown are three analyses yielding slightly higher $^{207}Pb/^{206}Pb$ ages suggestive of minor inheritance (gray-shaded ellipses). Error ellipse shading, level of uncertainty, and insets as described in Figure 6.

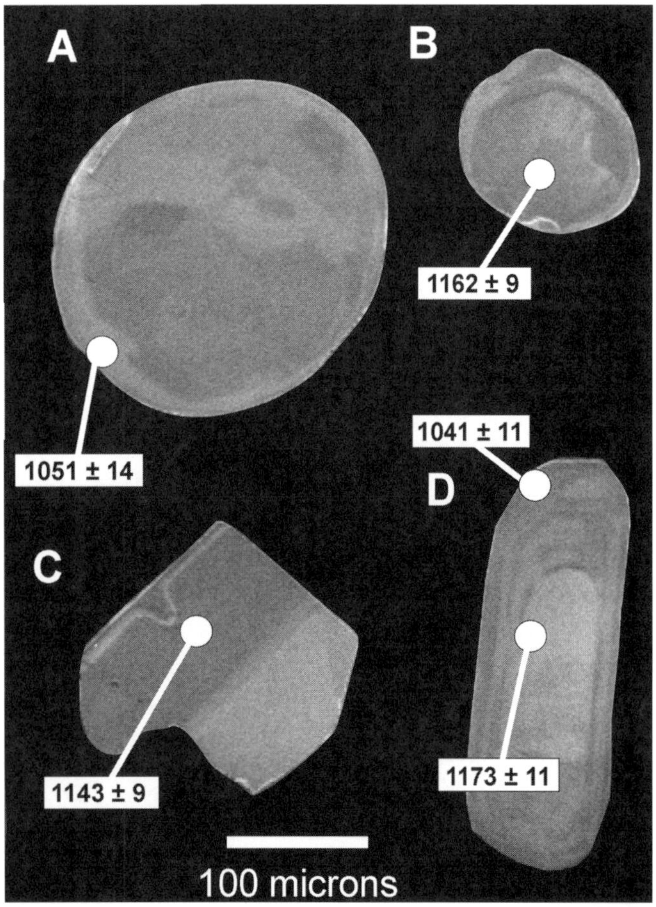

Figure 9. Cathodoluminescence images of Gore Mountain zircons (AM87-9). (A) and (B) Round zircon grains with weakly sector-zoned internal cores rimmed by narrow overgrowths of metamorphic zircon. (C) Large cleavage fragment from a presumed igneous zircon. (D) Euhedral zoned igneous core exhibiting rounding by overgrowth of metamorphic zircon.

those from rims. Two of these analyses (13.1, 15.1) represent spots from thin rims that likely clipped internal, older cores, resulting in mixed ages with no geological significance. The third analysis (2.1) is from a zoned grain core that we interpret to have undergone incomplete lead loss.

Oregon Dome Ferrodiorite (AM87-8)

Zircons from the Oregon Dome ferrodiorite are typical of those found in other Adirondack ferrodiorites. That is, they are relatively small (50–200 μm) grains whose CL images show ubiquitous dark cores overgrown by irregular bright rims (Fig. 11). Some cleavage fragments probably represent original grains exceeding 200 μm, but these larger grains have not yet been found. Some cores that have been fully exposed by grainmount polishing exhibit faint oscillatory zoning (Fig. 11, A), straight or "feathery" zoning, or, rarely, prismatic and euhedrally terminated morphologies (Fig.11, C). However, in most cases, external boundaries to grain cores are smooth or irregular, suggesting resorption and subsequent recrystallization. In almost all cases, these cores are surrounded by comparatively homogeneous

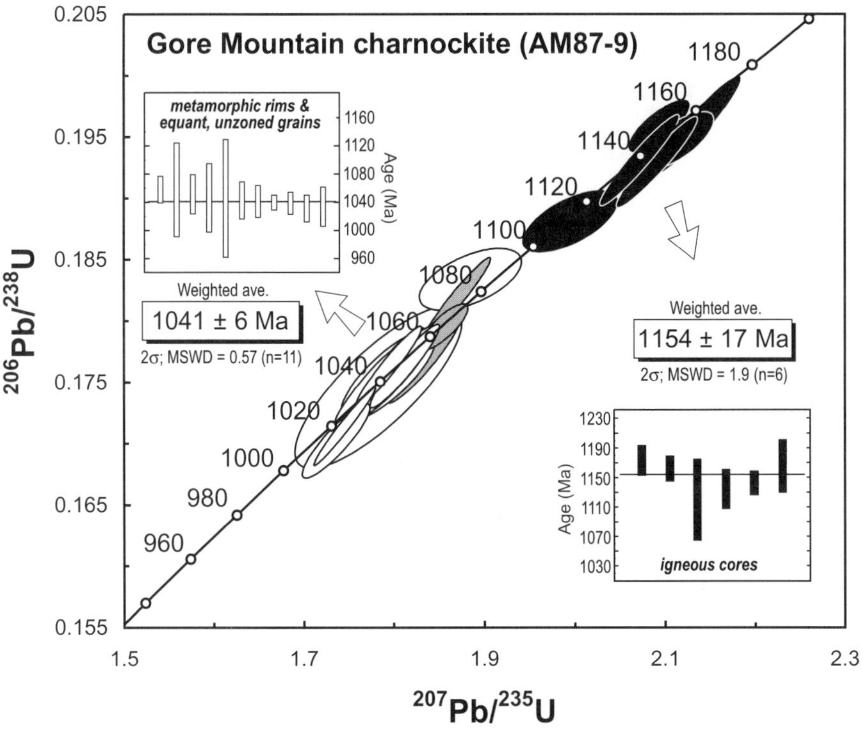

Figure 10. Concordia diagram of SHRIMP analyses of zircons from the Gore Mountain charnockite (AM87-9) exhibiting clustering of ages into a group of cores and a group of metamorphic overgrowths. Error ellipse shading, level of uncertainty, and insets as described in Figure 6. Data not used in age calculation are shaded gray (see text for discussion).

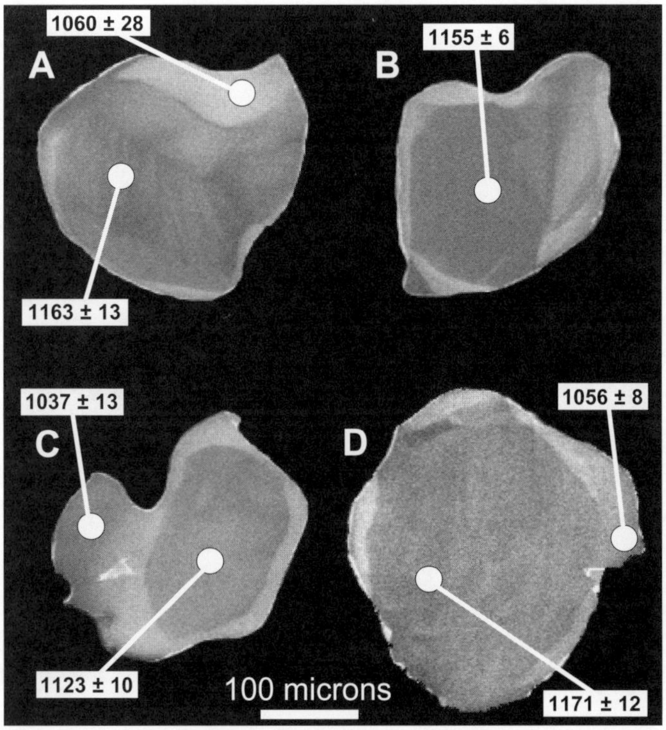

Figure 11. Cathodoluminescence images of zircons from the Oregon Dome ferrodiorite (AM87-8). (A) Broad, irregular igneous core overgrown by metamorphic rim. Although difficult to discern, the core exhibits delicate oscillatory zoning. (B) Small, subhedral igneous zircon with an overgrowth of metamorphic zircon. (C) and (D) Prismatic to rounded igneous zircon cores surrounded by an irregular overgrowth of metamorphic zircon.

(structureless) rims that have slightly lower, but similar, U concentrations. Because of the mafic, igneous nature of the rock, the zircon cores are interpreted to be related to the age of emplacement of the ferrodiorite, whereas the rims reflect subsequent metamorphism.

Results of SHRIMP U-Pb analysis of zircons from the Oregon Dome ferrodiorite are provided in Table 2 and illustrated in a concordia diagram in Figure 12, where the bimodality of the ages is clearly seen. A distinct cluster of igneous core ages range from 1171 to 1136 Ma, and yield a weighted mean age of 1156 ± 7 Ma (95% confidence; MSWD = 1.05). We interpret this age as the time of emplacement of the Oregon Dome ferrodiorite. In contrast, analyses of conspicuous irregular rims have $^{207}Pb/^{206}Pb$ ages that span 1065–1034 Ma, producing a mean age of 1048 ± 10 Ma (95% confidence; MSWD = 0.7; Fig. 12, inset). We interpret this age as best representing the time of superimposed Ottawan metamorphism of this body. Intermediate between these two age clusters are two analyses with $^{207}Pb/^{206}Pb$ ages of 1123 Ma and 1092 Ma (2.1, 11.1; Table 2; Fig. 12, gray shading). These spot analyses come from small igneous domains that are surrounded by thick metamorphic rims, one of which was dated at 1037 Ma (Table 2, analysis 2.2; Fig 11, C). We correspondingly interpret these two ages as reflecting incomplete Pb loss from small zircon cores during high-grade Ottawan metamorphism.

Diana Complex (AM86-1)

The pyroxene quartz syenite of the Diana Complex contains an abundance of elongate, prismatic, and somewhat rounded,

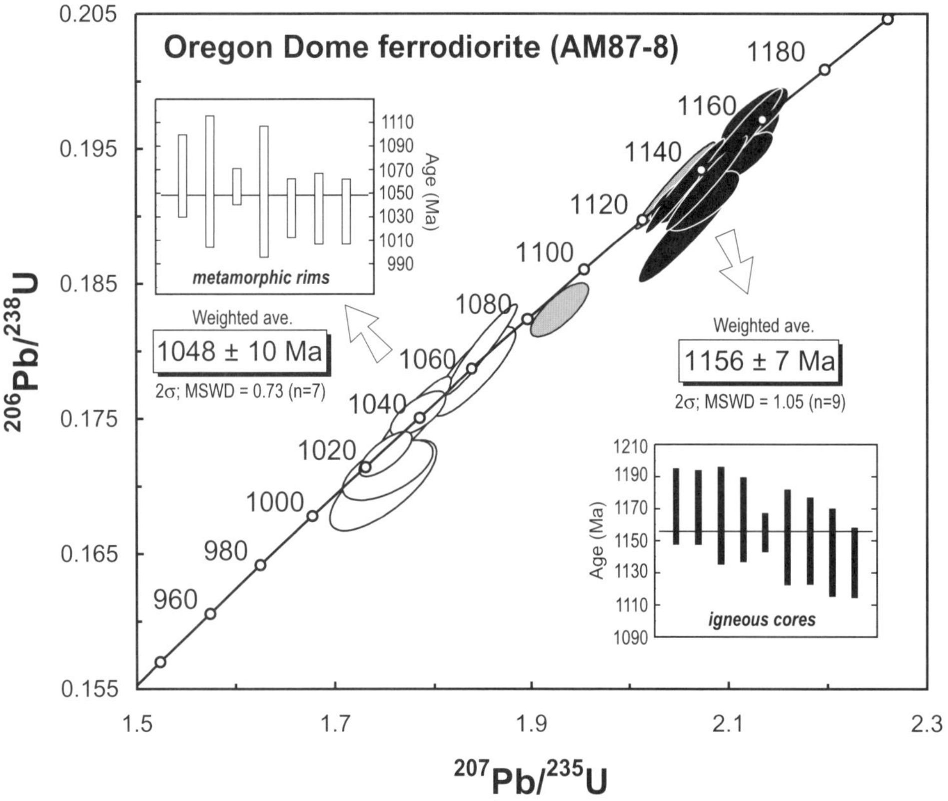

Figure 12. Concordia plot of SHRIMP analyses of zircons for the Oregon Dome ferrodiorite (AM87-8). Note the distinctly bimodal clustering of igneous cores and metamorphic rims. Error ellipse shading, level of uncertainty, and insets as described in Figure 6. Data not used in age calculations are shaded gray (see text for discussion).

doubly terminated zircons that exhibit igneous zoning (Fig. 13). Most grains recovered are between 150 and 200 μm in length, but the fragmental nature of some indicate the existence of much larger grains before sample crushing. With a few exceptions, the grains are interpreted as igneous in origin, although most have subsequently been modified by a younger metamorphism. A number of separated zircons, usually with length:breadth ratios of ~2:1 to 3:1, display elongate cores with igneous zoning (in CL) that is truncated by a relatively structureless thin rim or overgrowth (e.g., Fig. 13, B). None of these rims, however, has yielded a metamorphic age corresponding to the Ottawan orogeny. This is further discussed in the following paragraphs.

A concordia diagram for SHRIMP U-Pb results on zircon from pyroxene quartz syenite of the Diana Complex is presented in Figure 14. Although the data show some complexity, most analyses of unambiguously zoned igneous cores are crowded together on or near concordia; the ten oldest age results are tightly clustered and yield a weighted mean $^{207}Pb/^{206}Pb$ age of 1164 ± 11 Ma (95% confidence; MSWD = 1.8; Fig. 14, inset) This age is interpreted as the time of emplacement of the Diana pyroxene quartz syenite. Spot analyses located entirely within the thin, structureless, truncating rims have $^{207}Pb/^{206}Pb$ ages ranging from ca. 1125 to 1120 Ma (Table 2; Fig. 14, white ellipses). These domains are further distinguished by having lower Th/U (~0.27) than their igneous core counterparts (~0.47). The best estimate of the age of these rims is 1122 ± 29 Ma (Fig. 14, inset; 95% confidence; MSWD = 0.01). Spot analysis 4.1 has a relatively young $^{207}Pb/^{206}Pb$ age (1132 ± 7 Ma; Table 2; Fig. 14, shaded ellipse) but is located in a zoned igneous zircon core from a grain that also has a thin (unanalyzed) bright-CL, low-U core. On textural grounds, we have excluded this analysis from the weighted mean age calculation defined by other young ages that are clearly from metamorphic rims. Inclusion of analysis 4.1 in a mean age calculation with all the other, older, core analyses is also problematic, resulting in a poor statistical fit. We ascribe this behavior to minor Pb loss from this zircon core, possibly during a younger metamorphic event.

We note the spatial proximity of the Diana Complex syenite locality described here to that of the adjacent voluminous ca. 1105–1090 Ma Hawkeye granite (McLelland et al., 2001a). Further, we speculate that the relatively imprecise spread of ages for sporadically developed metamorphic rims may be the result of incomplete Pb loss attending recrystallization of ca. 1164 Ma igneous cores during a ca. 1100 Ma thermal event related to Hawkeye granite plutonism. Our mean age for the metamorphic rims is indistinguishable from the single-grain TIMS age of 1118 ± 2 Ma age reported by Basu and Premo (2001) for rocks from the southeastern flank of the Diana Complex, although it is possible that the TIMS zircon age reflects a mixture of core and overgrowth age domains with or without subsequent Pb loss.

A distinguishing feature of the Diana Complex syenite, in the context of the other samples described here, is the apparent lack of ca. 1040 Ma metamorphic ages (Table 2; Fig. 14) This

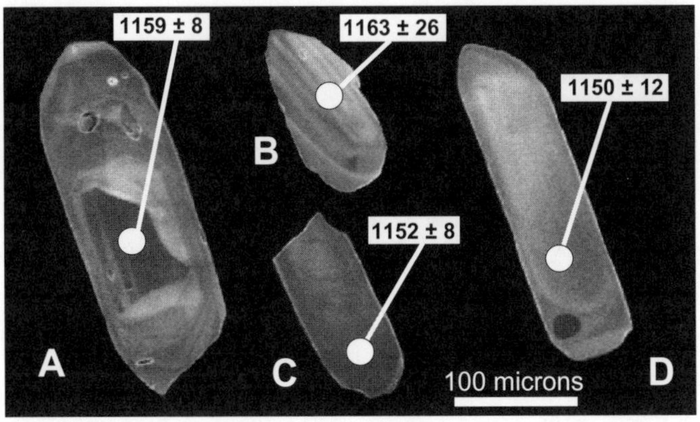

Figure 13. Cathodoluminescence images of zircons from Diana Complex pyroxene quartz syenite (AM86-1).

lack of ca. 1040 Ma overgrowths is intriguing, because the Adirondack Lowlands, located just northwest of the Diana Complex (Fig. 2), are characterized by zircon, titanite, and monazite ages that exceed ca. 1100 Ma (Mezger et al., 1991). In contrast, metamorphic ages in the Highlands, and in the Carthage-Colton Mylonite Zone (Fig. 2, CCMZ), consistently fall into the 1050–1030 Ma range. Several workers (e.g., Mezger et al., 1991; Streepey et al., 2001) interpret this difference as the result of physical separation of the Lowlands and the Highlands until ca. 1050 Ma. The separation is variously attributed to an intervening ocean basin, strike-slip translation, or pre-1050 Ma thrust fault emplacement of the Lowlands on top of the Highlands, followed by juxtaposition during ca. 1050 Ma orogenic collapse. It is beyond the scope of this paper to further address this issue, but the absence of ca. 1040–1050 Ma metamorphic rims on Diana Complex zircons has significant implications for this issue.

DISCUSSION OF RESULTS AND CONCLUSIONS

The Oregon Dome anorthosite can be no younger than the 1156 ± 7 Ma ferrodiorite that crosscuts it, and, based on inferred comagmatic relationships between the two (McLelland et al., 1994), the 1156 Ma age almost certainly applies to both anorthosite and ferrodiorite. For reasons discussed previously, blue-gray andesine xenocrysts in the 1154 ± 17 Ma Gore Mountain charnockite also imply that the adjacent Oregon Dome anorthosite is of the same age. This conclusion is further strengthened by the transitional contact between the Gore Mountain charnockite, the Oregon Dome anorthosite, and the Thirteenth Lake anorthosite, an outlier of the Oregon Dome massif (Lettney, 1969). A similar argument holds for the 1176 ± 11 Ma Minerva charnockite, which, in addition to blue-gray andesine xenocrysts and a transitional contact, shows mutually crosscutting relationships with the adjacent Marcy anorthosite massif (McLelland and Chiarenzelli, 1990). The 1174 ± 25 Ma Snowy Mountain charnockite contains abundant andesine xeno-

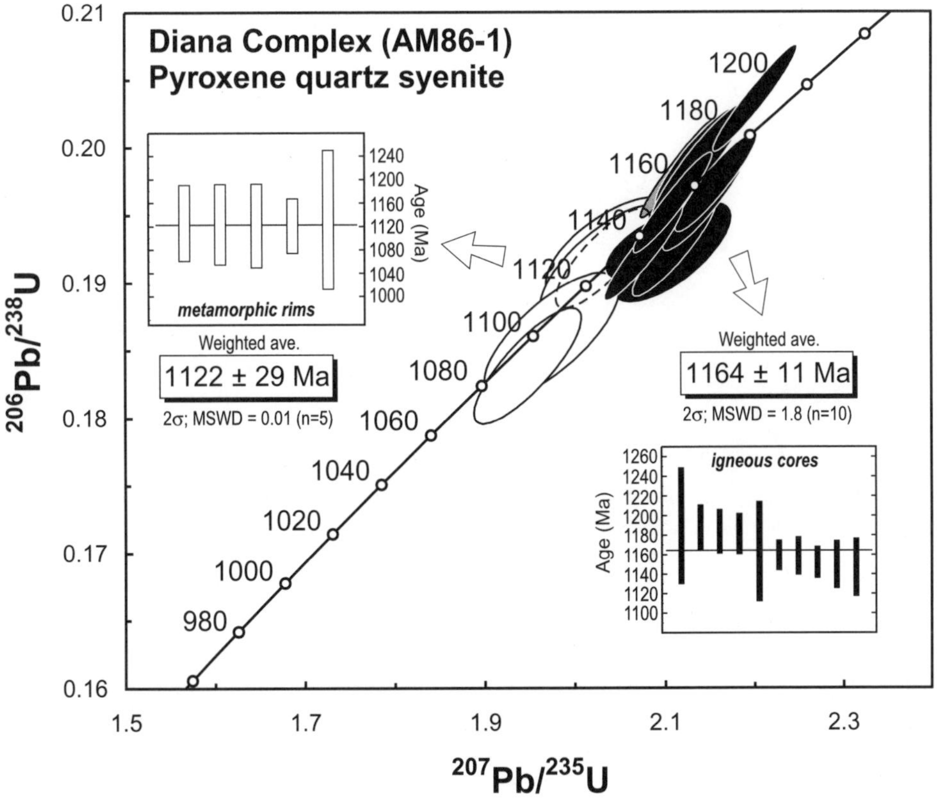

Figure 14. Concordia plot of SHRIMP analyses of zircons from Diana Complex pyroxene quartz syenite (AM86-1). Error ellipse shading and level of uncertainty as described in Fig. 6. Data not used in age calculations is shaded gray (see text for discussion).

crysts plucked from the adjacent Snowy Mountain anorthosite, with which it shows mutually crosscutting relationships (deWaard and Romey 1969). The Diana Complex is compositionally similar to other Adirondack AMCG granitoids and is adjacent to the small Carthage anorthosite massif as well as to several smaller anorthosite bodies on the eastern flank of the complex. The 1164 ± 11 Ma age obtained for the Diana Complex in this study is statistically indistinguishable from the multigrain result of 1155 ± 13 Ma obtained by Chiarenzelli and McLelland (1991) for the southwestern extremity of the Diana Complex just south of the village of Croghan (Fig. 2, C), a consistency indicating that this interval represents the time of emplacement of the entire Complex.

We conclude that anorthosites and granitoids of the Adirondack AMCG suite were emplaced at ca. 1150–1170 Ma (Fig. 15). This age is in close agreement with the 1155 ± 3 Ma (Doig, 1991) and 1160 +7/– 6 Ma (Emslie and Hunt, 1990) ages determined for the Morin complex, as well as with the 1148 ± 4 Ma (Emslie and Hunt, 1990), 1153 ± 7 Ma (Hervet et al., 1997), and 1157–1142 Ma ages (Higgins and van Breemen, 1992, 1996) for the Lac St. Jean anorthosite massif in the central Québec Grenville Province. Although all of these results are U-Pb zircon (± baddeleyite) ages, the data of Emslie and Hunt (1990) are from granitoids associated with anorthosite. In contrast, the data of Doig (1991), Higgins and van Breemen, (1992, 1996), and Hervet et al. (1997) come directly from anorthosite or from related ferrodiorite or troctolite. Given these ages, it is clear that the central portion of the Grenville Province was strongly affected by widespread AMCG magmatism at ca. 1150 Ma. McLelland et al. (1996) and Wasteneys et al. (1999) proposed that this magmatism resulted from delamination following collapse of an overthickened orogen that, in the Adirondack region, was associated with collision of the southeastern margin of Laurentia with the Adirondack Highlands and the Green Mountains during a culminating pulse of the Elzevirian orogeny at ca. 1210–1160 Ma. This event is thought to correlate with the 1190–1160 Ma Shawinigan Pulse to the northeast (Corrigan and van Breemen, 1997), and with the subsequent orogen collapse and delamination-triggered AMCG magmatism in that region as well (Corrigan and Hanmer, 1997). Farther to the northeast, 1126 +7/–6 Ma monzonite adjacent to the Lac Allard lobe of the Havre St. Pierre anorthosite complex (Emslie and Hunt, 1990) appears to be a slightly younger representative of the same magmatic province as that characterized by the Adirondack, Morin, and Lac St. Jean bodies. Clearly the scale of this magmatic event was very large, and crustal additions associated with anorthosite plutonism approximate 20% of areal exposure in this part of the Grenville Province (McLelland, 1989). To the extent that all of these bodies were associated with late orogenic delamination, the still little-understood ca. 1200–1160 Ma orogeny must have been of large dimensions and was attended by substantial removal of overthickened lithosphere.

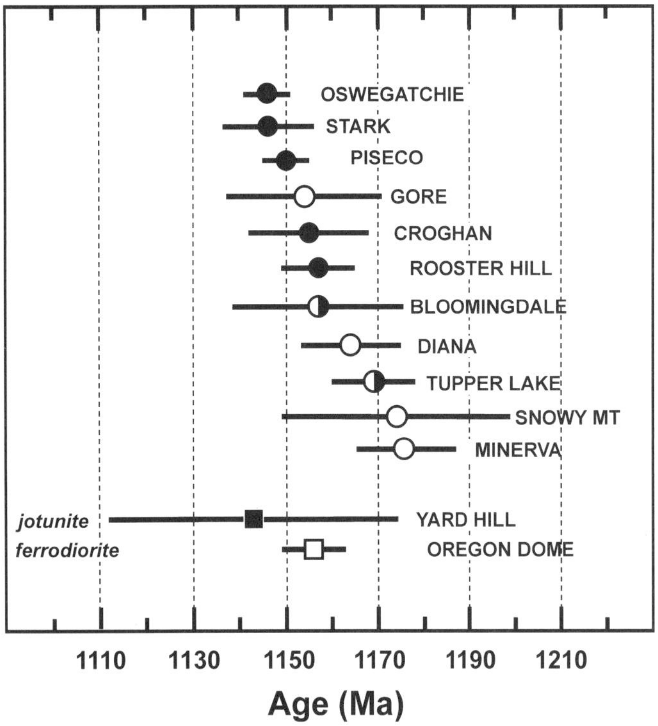

Figure 15. Summary plot of ages of Adirondack anorthosite-mangerite-charnockite-granite samples (circles) and ferrodiorites and jotunite samples (squares). Open symbols represent new sensitive high-resolution ion microprobe (SHRIMP) results of this investigation, half-filled symbols are recent SHRIMP age determinations of McLelland et al. (2002), and filled symbols are published conventional U-Pb ages from previous studies (McLelland and Chiarenzelli, 1990; Chiarenzelli and McLelland, 1991).

Both the ca. 1155 Ma igneous ages and the ca. 1050 Ma metamorphic ages reported in this study have regional significance extending beyond their occurrences in the Adirondacks and the Grenville Province of Canada. Specifically, Tollo et al. (this volume) reported extensive occurrences of 1050 Ma granitic magmatism associated with metamorphism in the Blue Ridge Province of Virginia. Walsh and Aleinikoff (this volume) describe igneous activity of the same age in the Precambrian of western Connecticut. Owens and Tucker (2003), have obtained ca. 1045 Ma ages for the granitic State Farm gneiss in the Goochland terrane of the southeastern Appalachians. Aleinikoff et al. (1996) reported an age of 1045 ± 10 Ma for the Montpelier anorthosite in the Blue Ridge Province of Virginia, and Pettingilll et al. (1984) reported an age of ca. 1045 Ma for the Roseland anorthosite. Aleinikoff et al. (2000) reported monzonitic and granitic gneisses of the Blue Ridge that are dated at ca. 1140–1150 Ma (Group 1 rocks). In addition, the Storm King granite of the Hudson Highlands in southern New York is reported to have a U-Pb (SHRIMP) zircon age of ca. 1174 Ma (J. Aleinikoff and N. Ratcliffe, personal commun.). Therefore, major events recorded in the Adirondacks have important counterparts in the Proterozoic cores of the Appalachian orogen to the south. Both ca. 1050 Ma and ca. 1155 Ma suites are widespread in large portions of the Grenville Province of Canada, and

1050 Ma granitic magmatism is present in the Llano Uplift of Texas (Roback, 1996). These observations reflect the fact that AMCG magmatism and the Ottawan orogeny were of continental scale and of major importance in the evolution of Laurentia. They also imply that most of Precambrian eastern Laurentia was in place by ca. 1155 Ma.

The results presented in this paper as well as the regional results reviewed in the preceding paragraph are in disagreement with the interpretations of Alcock et al. (2001) and Isachsen et al. (2001) cited earlier. Nevertheless, direct dating of zircons from the anorthosite itself supports an AMCG emplacement age of ca. 1155 Ma (Clechenko et al., 2002; Hamilton et al., 2002) consistent with the results and interpretations presented in this paper. Additionally, our new data further substantiate an age of ca. 1150–1160 Ma for emplacement of the regional Adirondack AMCG suite and related complexes to the northeast in the Canadian Grenville Province and to the southeast in the Appalachians.

APPENDIX: ION PROBE (SHRIMP II) SAMPLE PREPARATION AND ANALYTICAL TECHNIQUES

In situ U-Pb analyses were carried out using the sensitive high-resolution ion microprobe (SHRIMP II) facility at the Geological Survey of Canada (GSC), using the analytical and data reduction procedures outlined in detail by Stern (1997). Zircons from each sample were arranged, along with fragments of the GSC laboratory zircon reference standard (BR266, $^{206}Pb/^{238}U$ isotope dilution age = 559 Ma), cast in an epoxy grain mount (GSC mount IP150), and polished sufficiently with diamond compound to reveal the grain centers. The zircon grains were then imaged with a Cambridge Instruments S360 scanning electron microscope equipped with cathodoluminescence (CL) and backscatter detectors in order to identify compositional zoning and fracturing. Scanning electron microscope (SEM) imaging permitted identification of zircon structural growth domains suitable for eventual SHRIMP analysis. Although it was not practical to analyze each zone in every grain, care was taken to analyze all representative domains (e.g., cores, "outer" cores or mantles, and rims) as constrained by the dimension of the SHRIMP spot and the size of the actual zones. In no cases were areas of zircon zones intentionally avoided, except in heavily altered and cracked parts of grains, or regions rich in inclusions that would have been analyzed by the primary beam.

Data were acquired over two separate analytical sessions using a mass-filtered O^- primary beam with a variety of sputtering diameters ranging from ~13 × 16 μm to 25 × 35 μm. Sensitivity for Pb isotopes during both sessions was ~25 cps/ppm/nA (as measured for $^{206}Pb^+$ with an O_2^- primary beam), with mass resolution at 5500 and 5100. Primary O^- beam currents were 4.0–5.0 nA and 14.0–17.0 nA during analysis with the smaller and larger spot sizes, respectively.

Empirical standard calibrations yielded bias corrections for the Pb/U ratios with associated external uncertainties of ±1.0–1.2% (1σ), and these were propagated through to the Pb/U isotopic compositions of the unknowns, together with the counting errors. Correction of the measured isotopic ratios for common Pb was estimated from monitored ^{204}Pb counts.

ACKNOWLEDGMENTS

Much of this work was supported by funds from the Colgate University Research Council made available to JMcL and BS. This support is gratefully acknowledged. We thank Richard Tollo and M.E. Bickford for helpful reviews that substantially improved the quality of this paper.

REFERENCES CITED

Alcock, J., Isachsen, C., and Muller, P., 2001, Conditions and timing of emplacement of the Marcy anorthosite, Northeast Adirondack Highlands, inferred from anatectites in its aureole, Part I: Field relationships and *P-T* estimates: Geological Society of America Abstracts with Programs, v. 33, no. 6, p. 292.

Aleinikoff, J.N., Horton, J.W., Jr., and Walter, M., 1996, Middle Proterozoic age for the Montpelier Anorthosite, Goochland terrane, eastern Piedmont, Virginia: Geological Society of America Bulletin, v. 108, no. 11, p. 1481–1491.

Aleinikoff, J.N., Burton, W.C., Lyttle, P.T., Nelson, A.E., and Southworth, C.S., 2000, U-Pb geochronology of zircon and monazite from Mesoproterozoic granitic gneisses of the northern Blue Ridge, Virginia and Maryland, U.S.A.: Precambrian Research, v. 99, p. 113–146.

Basu, A.R., and Premo, W.R., 2001, U-Pb age of the Diana Complex and Adirondack granulite petrogenesis: Proceedings of the Indian Academy of Sciences, v. 110, p. 385–395.

Bohlen, S.R., and Essene, E.J., 1977, Feldspar and oxide thermometry of granulites in the Adirondack Highlands: Contributions to Mineralogy and Petrology, v. 26, p. 971–992.

Buddington, A.F., 1939, Adirondack igneous rocks and their metamorphism: Boulder, Colorado, Geological Society of America Memoir 7, 354 p.

Buddington, A.F., 1969, Adirondack anorthosite series, *in* Isachsen, W.Y., ed., Origin of anorthosite and related rocks: Albany, New York State Museum Memoir 18, p. 215–231.

Chiarenzelli, J., and McLelland, J., 1991, Age and regional relationships of granitoid rocks of the Adirondack Highlands, New York: Journal of Geology, v. 99, 571–590.

Clechenko, C.C., Valley, J.W., Hamilton, M.A., McLelland, J.M., and Bickford, M.E., 2002, SHRIMP II geochronology of the Adirondack AMCG suite, II: Timing and depth of emplacement of anorthosite in the northeastern Adirondacks: Geological Society of America Abstracts with Programs, v. 34, no. 6, p. 366.

Corrigan, D., and Hanmer, S., 1997, Anorthosites and related granitoids in the Grenville orogen: A product of convective thinning of the lithosphere? Geology, v. 25, p. 61–64.

Corrigan, D., and van Breemen, O., 1997, U-Pb age constraints for the lithotectonic evolution of the Grenville Province along the Mauricie transect, Quebec: Canadian Journal of Earth Science, v. 34, p. 290–316.

Davis, B.T.C., 1969, Anorthositic and quartz syenitic series of the St. Regis quadrangle, New York, *in* Isachsen, W.Y., ed., Origin of anorthosites and related rocks: Albany, New York State Museum and Science Service Memoir 18, p. 281–288.

deWaard, W., and Romey, W.D., 1969, Petrogenetic relationships in the anorthosite-charnockite series of the Snowy Mt. Dome, south-central Adirondacks, New York, *in* Isachsen, W.Y., ed., Origin of anorthosites and related rocks: Albany, New York State Museum and Science Service Memoir 18, p. 307–316.

Doig, R., 1991, U-Pb zircon dates of the Morin anorthosite rocks, Grenville Province, Québec: Journal of Geology, v. 99, 729–738.

Dymek, R., and Owens, B., 1998, A belt of late- to post-tectonic anorthosite massifs, Central Granulite Terrane, Grenville Province, Quebec: Geological Society of America Abstracts with Programs, v. 30, no. 7, p. A-24.

Eby, G.N., 1990, The A-type granitoids: A review of their occurrence and speculations on their petrogenesis: Lithos, v. 26, p. 115–134.

Emslie, R.F., 1978, Anorthosite massifs, rapakivi granites, and late Proterozoic rifting of North America: Precambrian Research v. 7, 61–98.

Emslie, R.F., and Hunt, P.A., 1990, Ages and petrogenetic significance of igneous mangerite-charnockite suites associated with massif anorthosites, Grenville Province: Journal of Geology, v. 98, p. 213–231.

Emslie, R.F., Hamilton, M.A., and Theriault, R.J., 1994, Petrogenesis of a mid-Proterozoic anorthosite-mangerite-charnockite-granite (AMCG) complex: Isotopic and chemical evidence from the Nain plutonic suite: Journal of Geology, v. 102, p. 539–558.

Grant, N.K., Lepak, I., Maher, T.M., Hudson, M.R., and Carl, J.D., 1986, Geochronological framework for the Grenville rocks of the Adirondack Mountains: Geological Society of America Abstracts with Programs, v. 18, p. 620.

Hamilton, M.A., Emslie, R.F., and Roddick, J.C., 1994, Detailed emplacement chronology of magmas of the mid-Proterozoic Nain Plutonic Suite, Labrador: Insights from U-Pb systematics in zircon and baddeleyite: Eighth International Conference on Geochronology, Cosmochronology, and Isotope Geology, 5–11 June, Berkeley, California: Reston, Virginia, United States Geological Survey Circular 1107, p. 124.

Hamilton, M.A., Ryan, A.B., Emslie, R.F., and Ermanovics, I.F., 1998, Identification of Paleoproterozoic anorthositic and monzonitic rocks in the vicinity of the Mesoproterozoic Nain Plutonic Suite, Labrador: U-Pb evidence, *in* Radiogenic age and isotopic studies, Report 11: Ottawa, Ontario, Geological Survey of Canada Current Research 1998-F, p. 23–40.

Hamilton, M.A., McLelland, J.M., Bickford, M.E., Clechenko, C.C., and Valley, J.W., 2002, SHRIMP II geochronology of the Adirondack AMCG suite, I: Emplacement chronology of anorthosites and gabbros: Geological Society of America Abstracts with Programs, v. 34, no. 6, p. 365.

Hargraves, R.B., 1969, A contribution to the geology of the Diana syenite gneiss complex, *in* Isachsen, W.Y., ed., Origin of anorthosites and related rocks: Albany, New York State Museum and Science Service Memoir 18, p. 343–356.

Hervet, M., van Breemen, O., Higgins, M.D., 1997, U-Pb igneous crystallization ages of intrusive rocks near the southeast margin of the Lac St.-Jean anorthosite complex, Grenville Province, Quebec, *in* Radiogenic age and isotope studies, Report 8: Ottawa, Ontario, Geological Survey of Canada, Current Research 1994-F, p. 115–124.

Higgins, M.D., and van Breemen, O., 1992, The age of the Lac-St-Jean anorthosite complex and associated mafic rocks, Grenville Province, Canada: Canadian Journal of Earth Sciences v. 29, p. 1412–1423.

Higgins, M.D., and van Breemen, O., 1996, Three generations of anorthosite-mangerite-charnockite-granite (AMCG) magmatism, contact metamorphism and tectonism in the Saugenay–Lac-St-Jean region of the Grenville Province, Canada: Precambrian Research v. 79, p. 327–346.

Hoffman, P., 1991, Did the breakout of Laurentia turn Gondwana inside out? Science, v. 252, p. 1409–1412.

Isachsen, C., Alcock, J., and Livi, K., 2001, Conditions and timing of emplacement of the Marcy anorthosite, Northeast Adirondack Highlands, inferred from anatectites in its aureole, Part II: U-Pb geochronology: Geological Society of America Abstracts with Programs, v. 33, no. 6, p. 292.

Krogh, T.E., and Davis, G.L., 1973, The significance of inherited zircon on the age and origin of igneous rocks—An investigation of the ages of the Labrador adamellites: Washington, D.C., Carnegie Institution of Washington Yearbook, 72, p. 61–613.

Lettney, C.D., 1969, The anorthosite-norite-charnockite series of the Thirteenth Lake Dome, south-central Adirondacks, New York, *in* Isachsen, W.Y., ed., Origin of anorthosites and related rocks: Albany, New York State Museum and Science Service Memoir 18, p. 329–342.

Ludwig, K.R., 2001, Isoplot/Ex version 2.49—A geochronological toolkit for Microsoft Excel: Berkeley, California, Berkeley Geochronology Center Special Publication No. 1a.

Machado, N., and Martignole, J., 1988, First U-Pb age for magmatic zircons in anorthosites: The case of the Pentecote intrusion in Quebec: Geological Association of Canada Programs with Abstracts v. 13, p. 76.

Martignole, J., Machado, N., and Nantel, S., 1993, Timing of intrusion and deformation of the Rivière Pentecôte anorthosite: Journal of Geology, v. 101, p. 652–658.

McLelland, J., 1984, Origin of ribbon lineation in quartzofeldspathic gneiss in the Adirondacks, *in* Ramsey, J., ed., Linear and planar fabrics of metamorphic rocks: Journal of Structural Geology, v. 6, p. 147–157.

McLelland, J., 1989, Crustal growth associated with anorogenic, mid-Proterozoic anorthosite massifs in northeastern North America, *in* Ashwal, L., ed., Growth of the continental crust: Tectonophysics, v. 161, p. 331–342.

McLelland, J., and Chiarenzelli, J., 1990, Isotopic constraints on the emplacement age of the Marcy massif, Adirondack Mountains, New York: Journal of Geology, v. 98, p. 19–41.

McLelland, J., and Whitney, P.R., 1977, The origin of garnet in the anorthosite-charnockite suite of the Adirondacks: Contributions to Mineralogy and Petrology, v. 60, p. 161–181.

McLelland, J., and Whitney, P.R., 1990, Anorogenic, bimodal emplacement of anorthositic, charnockitic, and related rocks in the Adirondack Mountains, New York, *in* Stein, H., and Hannah, J., eds., Ore-bearing granite systems: Petrogenesis and mineralizing processes: Boulder, Colorado, Geological Society of America Special Paper 246, p. 301–315.

McLelland, J., Ashwal, L., Moore, L., 1994, Chemistry of oxide and apatite rich gabbronorites associated with Proterozoic anorthosite massifs: Examples from the Adirondack Highlands, New York: Contributions to Mineralogy and Petrology, v. 116, p. 225–238.

McLelland, J., Daly, S., and McLelland, J.M., 1996, The Grenville Orogenic Cycle (ca. 1350–1000 Ma): An Adirondack perspective: Tectonophysics, v. 265, p. 1–28.

McLelland, J., Hamilton, M., Selleck, B., McLelland, J.M., Walker, D., Orrell, S., 2001a, Zircon U-Pb geochronology of the Ottawan Orogeny, Adirondack Highlands, New York: Regional and tectonic implications: Precambrian Research, v. 109, p. 39–72.

McLelland, J., Valley, J.W., and Essene, E.J., 2001b, Very high temperature, moderate pressure metamorphism in the New Russia gneiss complex, northeastern Adirondack Highlands, metamorphic aureole to the Marcy anorthosite: Discussion: Canadian Journal of Earth Sciences, v. 38, p. 465–470.

McLelland, J.M., Hamilton, M.A., Bickford, M.E., Clechenko, C.C., Valley, J.W., 2002, SHRIMP II geochronology of the Adirondack AMCG suite: Granitoids, gabbros, and crosscutting dikes: Geological Society of America Abstracts with Programs, v. 34, no. 6, p. 271.

Mezger, K., Rawnsley, C.M., Bohlen, S.R., and Hanson, G.N., 1991, U-Pb garnet, sphene, monazite and rutile ages: Implication of the duration of metamorphism and cooling histories, Adirondack Mountains, New York: Journal of Geology, v. 98, p. 213–231.

Miller, W.J., 1918, Geology of the Lake Placid Quadrangle: Albany, New York State Museum Bulletin 212, 104 p.

Nabelek, P., Lindsley, D.H., and Bohlen, S.R., 1987, Experimental examination of two-pyroxene graphical thermometers using natural pyroxenes with application to metaigneous pyroxenes from the Adirondack Mountains, New York: Contributions to Mineralogy and Petrology, v. 97, p. 66–71.

Owens, B.E., and Dymek, R.F., 1992, Fe-Ti-P-rocks and massif anorthosite: Problems of interpretation illustrated from the Labrieville and St. Urbain plutons, Quebec: Canadian Mineralogist, v. 30, p. 163–190.

Owens, B.E., and Tucker, R.D., 2003, New U-Pb zircon age constraints on the age the State Farm gneiss, Goochland Terrane, Virginia: Geological Society of America Bulletin, v. 115, p. 972–982.

Pettingill, H.S., Sinha, A.K., Tatsumoto, T., 1984, Age and origin of anorthosites, charnockites, and granulites in the Central Virginia Blue Ridge: Contributions to Mineralogy and Petrology, v. 85, p. 279–291.

Rivers, T., 1997, Lithotectonic elements of the Grenville Province: Precambrian Research, v. 86, p. 117–154.

Rivers, T., and Corrigan, D., 2000, Convergent margin on southeastern Laurentia during the Mesoproterozoic: Tectonic implications: Canadian Journal of Earth Sciences, v. 37, p. 359–383.

Roback, R.C., 1996, Characterization and tectonic evolution of a Mesoproterozoic island arc in the southern Grenville Orogen, Llano uplift, central Texas: Tectonophysics, v. 265, p. 29–52.

Schärer, U., Krogh, T.E., and Gower, C.F., 1986, Age and evolution of the Grenville Province in eastern Labrador from U-Pb systematics of accessory minerals: Contributions to Mineralogy and Petrology, v. 94, p. 438–451.

Silver, L., 1969, A geochronologic investigation of the Adirondack Complex, Adirondack Mountains, New York, *in* Isachsen, W.Y., ed., Origin of anorthosites and related rocks: Albany, New York State Museum and Science Service Memoir 18, p. 233–252.

Stern, R.A., 1997, The GSC sensitive high resolution microprobe (SHRIMP): Analytical techniques of zircon U-Th-Pb age determinations and performance evaluation, *in* Radiogenic age and isotopic studies, Report 10: Ottawa, Ontario, Geological Survey of Canada, Current Research 1997-F, p. 1–31.

Stimac, J.A., and Wark, D.A., 1992, Plagioclase mantles on sanidine in silicic lavas, Clear Lake, California: Implications for the origin of rapakivi texture: Geological Society of America Bulletin, v. 104, p. 728–744.

Streepey, M.M., Johnson, E.L., Mezger, K., and van der Pluijm, B.A., 2001, Early history of the Carthage-Colton Shear Zone, Grenville Province, Northwest Adirondacks, New York: Journal of Geology, v. 109, p. 479–472.

van Breemen, O., and Higgins, M.D., 1993, U-Pb zircon age of the southwest lobe of the Havre–St. Pierre anorthosite complex, Grenville Province, Canada: Canadian Journal of Earth Sciences, v. 30, p. 1453–1457.

Wark, D.A., and Stimac, J.A., 1992, Origin of mantled (rapakivi) feldspars: Experimental evidence of a dissolution- and diffusion-controlled mechanism: Contributions to Mineralogy and Petrology, v. 111, p. 345–361.

Wasteneys, H., McLelland, J., and Lumbers, S., 1999, Precise zircon geochronology in the Adirondack Lowlands and implications for revising plate-tectonic models of the Central Metasedimentary Belt and Adirondack Mountains, Grenville Province of Ontario and New York: Canadian Journal of Earth Science, v. 36, p. 967–984.

Watson, E.B., and Harrison, T.M., 1983, Zircon saturation revisited: Temperature and compositional effects in a variety of crustal magma types: Earth and Planetary Science Letters, v. 64, p. 295–304.

Manuscript Accepted by the Society August 25, 2003

Printed in the USA

Geological Society of America
Memoir 197
2004

Right lateral oblique slip movements followed by post-Ottawan (1050–1020 Ma) orogenic collapse along the Carthage-Colton shear zone: Data from the Dana Hill metagabbro body, Adirondack Mountains, New York

Eric L. Johnson*
Department of Geology, Hartwick College, Oneonta, New York 13820, USA
Eric T. Goergen
Department of Geological Sciences, University of Missouri, Columbia, Missouri 65211, USA
Benjamin L. Fruchey
Department of Geology and Geophysics, University of Wyoming, Laramie, Wyoming 82071-3006, USA

ABSTRACT

The Dana Hill metagabbro body is a small (5 × 1.5 km) gabbroic intrusion located along the boundary between the Adirondack Highlands (Central Granulite Terrane) and the Adirondack Lowlands (Central Metasedimentary Belt). The Dana Hill metagabbro body crops out in the belt of highly strained rock known as the Carthage-Colton shear zone. Six generations of deformational and veining events cut the Dana Hill metagabbro body and are used to delineate a deformation or mineralization history for the body. Geothermometric data for multiple early-formed shear zones record temperatures of 770–711 °C ± 50 °C, with early-formed shear zones recording predominately right-lateral oblique-to-strike slip and later events recording oblique-to-dip slip motions. Deformation in the Dana Hill metagabbro body ended with folding and brecciation at greenschist-facies conditions. Early in its deformational history, the Dana Hill metagabbro body records a major episode of high-temperature (>700 °C) fluid infiltration (CO_2-rich) that was marked by scapolite replacement of plagioclase along swarms of parallel hornblende veins. This event was also recorded locally in the Central Metasedimentary Belt terrane, indicating that the two terranes were at or close to the same structural level during this event. Published geochronologic data (Streepey et al., 2001) show that the Dana Hill metagabbro body cooled to 650 °C by 1020 Ma, thereby constraining fluid infiltration and Central Metasedimentary Belt–Central Granulite Terrane juxtaposition to have occurred at or prior to this time. Structural, petrologic, and published geochronologic data support early (Ottawan) oblique-to-strike slip movements along the Carthage-Colton shear zone during the Ottawan orogeny (1090–1030 Ma), followed by a period of orogenic collapse that brought the two terranes to a common structural level prior to 1020 Ma.

Keywords: Grenville, granulite, shear zone

*E-mail: johnsone@hartwick.edu.

Johnson, E.L., Goergen, E.T., and Fruchey, B.L., 2004, Right lateral oblique slip movements followed by post-Ottawan (1050–1020 Ma) orogenic collapse along the Carthage-Colton shear zone: Data from the Dana Hill metagabbro body, Adirondack Mountains, New York, *in* Tollo, R.P., Corriveau, L., McLelland, J., and Bartholomew, M.J., eds., Proterozoic tectonic evolution of the Grenville orogen in North America: Boulder, Colorado, Boulder, Colorado, Geological Society of America Memoir 197, p. 357–378. For permission to copy, contact editing@geosociety.org.

INTRODUCTION

Isolated exposures of rocks belonging to the Grenville Province can be found throughout North America along an intermittent belt extending from northeast Canada south through the Adirondack Mountains, the Green Mountains or Berkshire Highlands, and the Hudson Highlands of New York and New England to the Blue Ridge Mountains of Virginia and the Carolinas and to the Llano Uplift of west-central Texas. Excellent exposures of the Grenville in North America occur in a region extending east of Sudbury, Ontario, where rocks of the Grenville Province are overthrust onto rocks belonging to the Southern Complex and the Superior Province along the Grenville Front tectonic zone (see Fig. 1). Grenville-aged rocks to the south and east of the Grenville Front tectonic zone have been subdivided into a number of smaller terranes separated from one another by shear zones (Davidson, 1986; Rivers et al., 1989; see Fig. 1). Using the terminology of Wynne-Edwards (1972) and Davidson (1986), from west to east these terranes are the Central Granulite Belt, the Central Metasedimentary Belt, and the Central Granulite Terrane (see Fig. 1). An alternative interpretation is provided by Carr et al. (2000), who break the northeast Canadian and U.S. Grenville Province into three belts: the pre-Grenvillian Laurentian margin, the Composite Arc terrane, and the Frontenac-Adirondack Belt. For the purpose of discussion, this paper utilizes the Wynne-Edwards (1972) terminology. The Central Metasedimentary Belt Boundary Zone is a large shear zone that separates the Central Granulite Belt from the Central Metasedimentary Belt (Davidson, 1986). Likewise, the Carthage-Colton shear zone separates the Central Metasedimentary Belt and the Central Granulite Terrane (Geraghty et al., 1981). In Québec, the Central Metasedimentary Belt (Mont-Laurier terrane) and the Central Granulite Terrane (Morin terrane) are separated by the LaBelle shear zone, which is interpreted to represent an extension of the Carthage-Colton shear zone (Indares and Martingole, 1990; Mezger et al., 1992). These terranes and their associated shear zones are interpreted by McLelland et al. (1993b) as representing a series of island arc–continent and continent-continent collisions along the eastern margin of Laurentia. An alternative interpretation of the geology of this region is given by Hanmer et al. (2000), who view the tectonic development of the Grenville of the northeast United States and Canada as an evolving Andean-type margin. In this model, compression and igneous activity are driven by northwest-directed subduction in the interval of 1.4–1.2 Ga punctuated by continent-continent collision at 1.2 Ga.

The Adirondack Mountains in northern New York state provide excellent exposures of highly deformed and variably metamorphosed rocks belonging to the Central Metasedimentary Belt, the Central Granulite Terrane (the Frontenac-Adirondack Belt), and the Carthage-Colton shear zone. Geologic/tectonic relationships between the Central Metasedimentary Belt and the Central Granulite Terrane and the role of the Carthage-Colton shear zone have been a matter of some debate. Geraghty et al. (1981) and Isachsen and Geraghty (1986) interpret the Carthage-Colton shear zone to represent a through-crustal shear zone along which the Central Metasedimentary Belt terrane was thrust over the Central Granulite terrane during the 1.1 billion-year-old Grenville orogeny. Wiener (1983) interprets the Carthage-Colton shear zone to represent the lower limb of a large fold-thrust nappe and does not view the Carthage-Colton shear zone as a boundary between separate (or once-separate) terranes. Isotope data (U/Pb and $^{40}Ar/^{39}Ar$), however, support the assertion that the Carthage-Colton shear zone does indeed separate two terranes with differing thermal histories (Mezger et al., 1991, 1993; McLelland et al., 1993b). Isotopic evidence (cooling dates) clearly show that the Central Granulite Terrane reached granulite-facies conditions during the 1030–1070 Ma

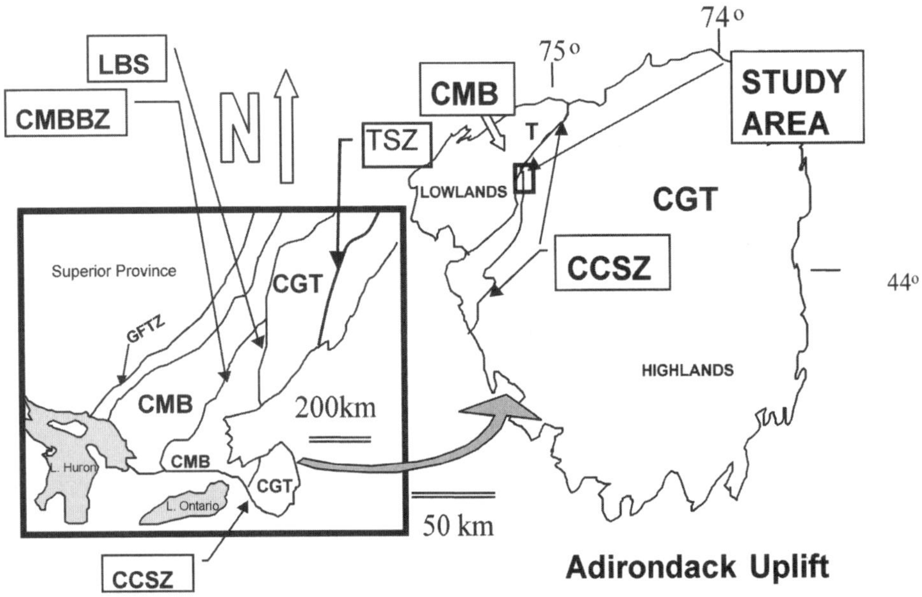

Figure 1. Generalized map of Grenville-aged rocks in the northeastern United States and Canada. CGB—Central Granulite Belt; CGT—Central Granulite terrane; CMB—Central Metasedimentary Belt. The Carthage-Colton shear zone (CCSZ), the Central Metasedimentary Belt Boundary Zone (CMBBZ), the Grenville Front tectonic zone (GFTZ), the LaBelle shear zone (LBS), and the Tawachiche shear zone (TSZ) are also shown. The boxed region is detailed in Figure 2.

Ottawan orogeny (McLelland and Chiarenzelli, 1990). $^{40}Ar/^{39}Ar$ dates in the Central Metasedimentary Belt do not record this event, and U/Pb dates for rutile suggest that maximum temperatures in this terrane during the Ottawan orogeny never exceeded 400 °C (McLelland et al., 1993b; Mezger et al., 1993). The 1070–1030 Ma Ottawan event is believed to represent a continent-continent collision at the close of the Grenville orogenic cycle (McLelland et al., 1993b). It remains problematic, however, that the Frontenac region of the Central Metasedimentary Belt preserves no thermal record of this collision. In addition, the timing of major deformation, including the formation of large fold-thrust nappes in both the Central Granulite Terrane and the Central Metasedimentary Belt remains enigmatic in this model, but the role of the Carthage-Colton shear zone as a major boundary is clear. What is not clear is the pre-Ottawan and syn-Ottawan relationship between the Central Granulite Terrane and the Central Metasedimentary Belt. Both terranes record isotopic evidence for an earlier 1230 Ma mountain-building event (Elzevirian orogeny), followed by a 1150–1130 Ma large-scale thermal event driven by injections of magmas (anorthosite-mangerite-charnockite-granite, or AMCG) into the Central Metasedimentary Belt and the Central Granulite Terrane (McLelland et al., 1993b; Mezger et al., 1993; Martignole and Reynolds, 1997). This led to an unusual history in which the Central Granulite Terrane and the Central Metasedimentary Belt were together prior to and during the 1160 Ma event, separated during the 1050–1070 Ma Ottawan event, and then reunited at some later time. Mezger et al. (1992) identifies the Carthage-Colton shear zone as a major crustal collapse structure that juxtaposed the Central Metasedimentary Belt and Central Granulite Terrane rocks during or after the Ottawan orogeny. Mezger et al. (1992) propose that the Central Metasedimentary Belt and the Central Granulite Terrane were separated by the opening of a small ocean basin (<1150 Ma), allowing these two terranes to follow different pressure-temperature-time paths during the Ottawan orogeny. An alternative view is that the Central Metasedimentary Belt and Central Granulite Terrane rocks were united after the 1160 Ma event, but that during the Ottawan orogeny the Central Metasedimentary Belt rocks were at too high a structural level in the crust to undergo significant thermal heating. In this model, late to syn-Ottawan collapse along the Carthage-Colton shear zone juxtaposed the two terranes at their current structural level (Mezger et al., 1992; McLelland et al., 1993a). Zones of ductile deformation in the Diana Complex, a Central Metasedimentary Belt lithology, record retrograde conditions of 400–550 °C at pressures of 3 to 5 kilobars, which Lamb (1993) attributes to late activity along the Carthage-Colton shear zone during slow uplift and cooling. Martignole and Reynolds (1997) and Zhao et al. (1997) report granulite-grade strike-slip movement along the Labelle and the Morin shear zones in Québec, and Hanmer et al. (2000) report 1.09–1.06 Ga oblique sinestral movements along the Tawachiche shear zone, introducing the possibility that lateral displacements may have been important during the Ottawan phase of the Grenville orogeny. In all of these models, the precise role of the Carthage-Colton shear zone (as well as the Labelle shear zone and the Tawachiche shear zone) is the critical link to understanding the tectonic relationship(s) between the Central Metasedimentary Belt and the Central Granulite Terrane.

This paper presents petrologic and structural data from a complexly deformed metagabbro body (the Dana Hill metagabbro body) that crops out in the Carthage-Colton shear zone (Fig. 2). The Dana Hill metagabbro body was first described by Buddington (1939), but to date the only detailed work on the Dana

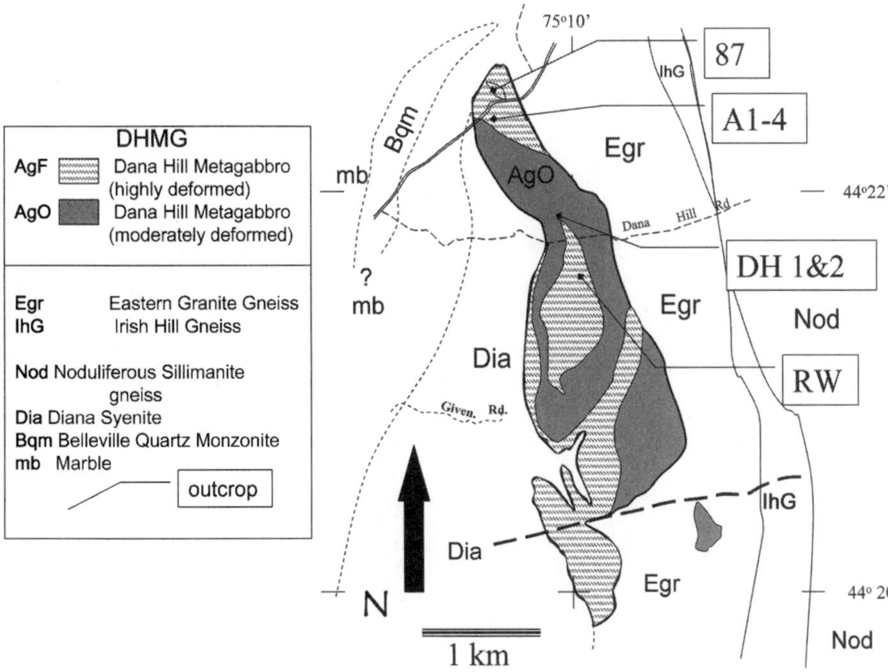

Figure 2. Map of the Dana Hill metagabbro body (DHMG). Outcrop locations are given on the map. After Parodi (1979); Hall (1984); Johnson and Ambers (1985).

Hill metagabbro body consists of M.A. theses by Parodi (1979) and Hall (1984) and a Ph.D. dissertation by Heyn (1990). Hall (1984) mapped the northern portion of the Dana Hill metagabbro body and recognized multiple shearing events within the body. Parodi (1979) mapped the entire Dana Hill metagabbro body and discussed the basic geochemistry of the body. Heyn (1990) did detailed work on the microstructural development of shear zones in the Dana Hill metagabbro body in the context of a larger work on the tectonites of the region. Davis (1981) studied mineral chemistries and deformation in the Diana syenite, directly to the south of the Dana Hill metagabbro body. E.L. Johnson and C. Ambers (1985, personal commun.) remapped the Dana Hill metagabbro body (Fig. 3) and reported on shear-sense indicators both in the Dana Hill metagabbro body and in the surrounding lithologies. Johnson et al. (1994) presented the first detailed work on the petrography of shear zones within the Dana Hill metagabbro body. Based on preliminary geochemical (microprobe) and petrographic analyses, Johnson et al. (1994) recognized granulite-facies, amphibolite-facies, and greenschist-facies deformation zones in the Dana Hill metagabbro body. Streepey et al. (2001) presented U/Pb and $^{40}Ar/^{39}Ar$ cooling ages for the shear zones and veins in the Dana Hill metagabbro body. These workers determined that the Dana Hill metagabbro body cooled through the closure temperature for U/Pb in sphene and the $^{40}Ar/^{39}Ar$ closure temperature for hornblende at 1020 Ma and 1000–973 Ma, respectively. In addition, Streepey et al. (2001) documented some late activity based on 944–947 Ma $^{40}Ar/^{39}Ar$ hornblende ages obtained from two small shear zones in the Dana Hill metagabbro body and attributed these dates to the time of final juxtaposition of the Central Metasedimentary Belt and the Central Granulite Terrane at a common structural level.

This paper documents the style and pressure-temperature conditions of multiple deformational events recorded in the Dana Hill metagabbro body and attempts to use these data, along with geochronologic data from Streepey et al. (2001) to reconstruct the thermal and deformational history of the Carthage-Colton shear zone. This history can then be used to better constrain the relationship between the Central Granulite Terrane and the Central Metasedimentary Belt in this complex region.

METHODS

Outcrop-scale maps of individual exposures within the Dana Hill metagabbro body were prepared to determine the temporal relationships between deformational events. Once a relative age history was determined, various shear zones, veins, and breccia pods were sampled. The strike and dip of the mylonitic foliation and the bearing and plunge of mineral stretching lineations (where present) were carefully recorded and marked on each sample. For samples where structural data could not be determined accurately in the field, oriented drill cores were obtained and structural data were retrieved from the oriented cores. Shear-sense determinations were made in the field and from oriented thin-section examination.

Mineral chemistries were determined using the electron microprobe. Quantitative electron microprobe analyses were performed at the University of Michigan (CAMECA CAMEBAX) and Binghamton University (JEOL 390 superprobe). In all cases, natural mineral and oxide standards were used. An operating voltage of 15 kV with a beam current at 10 na was used. In order to check for the possibility of sodium volitization in scapolite and plagioclase, time-integrated counts were performed in order to identify sodium loss. Corrections were rarely needed, as sodium loss was minimal over the 60-second counting times, but where detected count rates were back-corrected to the zero-time intercept. Oxide minerals were analyzed using mineral oxide standards. A ZAF routine was used to correct raw data. Amphibole mineral formulae were cast based on thirteen cations minus Na, K, and Ca. Fe^{3+} was calculated based on charge balance. Where amphibole formulas were used for geothermometry, the casting routine of Holland and Blundy (1994) was employed. Feldspar analyses were cast based on a total of eight oxygens, and scapolite data were cast based on twelve Al + Si cations. Carbon content was determined by charge balance, and equivalent anorthite content was calculated as X_{EqAn} = (mol Al^{-3})/3 (Rebbert and Rice, 1997).

FIELD AND STRUCTURAL DATA

Figure 2 shows a generalized map of the Dana Hill metagabbro body and surrounding rocks. All contacts between the Dana Hill metagabbro body and surrounding units are highly sheared. Exposures of noduliferous sillimanite geniss (Hall, 1984) close to the southern bounds of the Dana Hill metagabbro body are interesting in that they contain abundant sillimanite in well-defined zones cutting undeformed hornblende granite. McLelland et al. (2002) have described this unusual sillimanite replacement texture from rocks of the late to post-Ottawan (1034 Ma) Lyon Mountain Gneiss. Pods of undeformed pegmatite are common throughout the field area, and a large (3 km diameter) weakly to undeformed granite body is located three kilometers south of the field area (Johnson and Bryan, 2002). Locations for specific outcrops in the Dana Hill metagabbro body are given in Figure 2.

Northern Dana Hill metagabbro body (Outcrops A-4, 87)

A major 30+ m–wide shear zone with a northwest trend (samples CR-6, 7, 11, and 12, H-1A; Fig. 3, A) cuts the northern end of the Dana Hill metagabbro body (outcrops 87 and A-4). Within this zone, fabric development is intense, and the metagabbro shows a well-defined c-s fabric with nearly vertical dips and gently plunging stretching lineations (see Table 1). Both outcrops provide large glacially polished exposures of this shear zone. Between sheared regions were pods of less deformed to undeformed metagabbro (sample CR-9, A-4) that preserve cumulate igneous texture with pronounced amphibole rims on pyroxene cores. Cutting these pods as well as the major

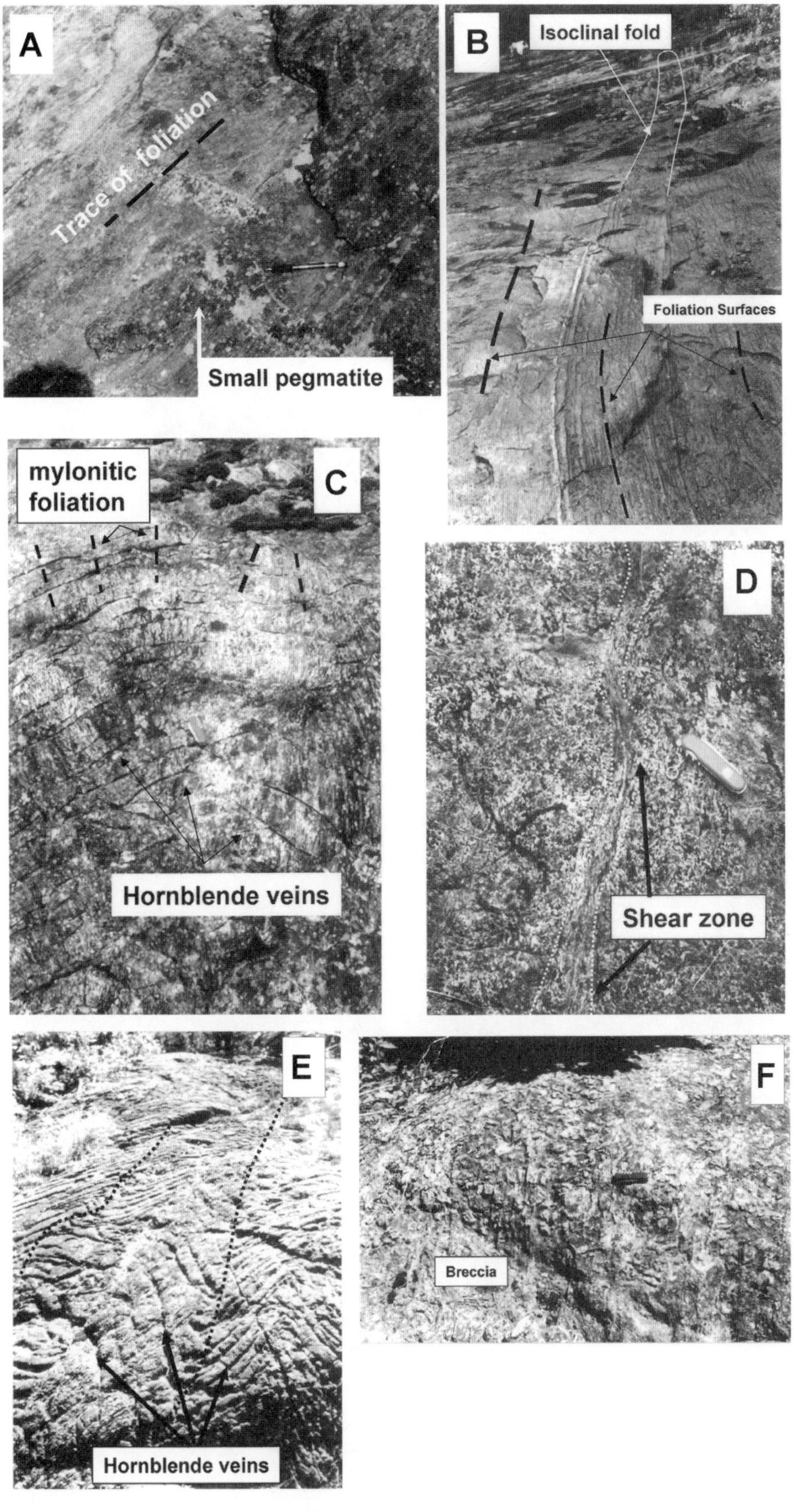

Figure 3. Outcrop photos of the Dana Hill metagabbro body showing the major deformational and veining events present (see Table 1). (A) 30+ m–wide shear zone, outcrop A-4. Foliation dips steeply, with a shallow stretching lineation marked by stringers of sphene and mixtures of clinopyroxene and hornblende. Sample CR-7, A-4. (B) 30 m–wide shear zone from the central outcrop (RW). Note the isoclinal fold nose in the background. Photo taken looking due north from point X in Figure 4. The folded white "veins" consist of plagioclase and are 2 cm wide. Sample RW-H1 and EA-1. Foliation surface is near vertical, with shallowly dipping transport lineation. (C) Parallel amphibole veins in outcrop A-4 (8 cm pocket knife for scale). (D) Centimeter-scale shear zone RW-S1 in outcrop RW (8 cm pocket knife for scale). (E) Folded metagabbro from outcrop 87. The folding is marked by rotated amphibole veins. A preexisting mylonitic foliation fans cross the folds. (F) Pod of brecciated Dana Hill metagabbro in outcrop 87 (8 cm pocket knife for scale). In outcrop, folded metagabbro degrades into these breccia pods (Sample H-6A).

TABLE 1A. PHYSICAL AND CHEMICAL CHARACTERISTICS OF SAMPLES FROM OUTCROPS A-4 AND 87

Outcrop event	Sample number	Orientation	Mineral assemblage	$^{40}Ar/^{39}Ar$ date*	U/Pb date*	T (°C)† (6 kilobars)
Undeformed metagabbro	CR-9, A-4	N.A.	Plagioclase + clinopyroxene + Fe-Ti oxides (amphibole rims on clinopyroxene)	Amphibole rim: 943.8 ± 1.7	N.A.	N.A.
1. 30+ m shear zone	CR-7, CR-12 CR-6, A-4 CR-11, A-4 H1A	Fol: N51W59W Lin: 05 N54W	CR-6, CR-7, CR-11§, and CR-12§: Orthopyroxene + amphibole + titanite + plagioclase H1A: Clinopyroxene + plagioclase + amphibole + titanite	CR-6, A-4: 1005 ± 1.9 Ma	1021–1023 Ma	718–744
2. 3 m shear zone	CR-5, 4, 3, 2	Fol: N47W 76W Lin: 29N55W	CR-5: Amphibole + plagioclase + titanite + quartz	N.A.	N.A.	N.A.
3. Parallel amphibole veins	CR-12	Vein orientation: N09E 67W	CR-12 vein: Amphibole + clinopyroxene + scapolite Matrix: Well-defined halo of scapolite replacing plagioclase close to the vein walls. Vein is 0.3 cm wide.	CR-12, A-4: Matrix: Clinopyroxene mylonite: 996.3 ± 2 Ma. No date from vein.	1021 Ma	N.A.
4. Cm-wide shear zones	CR-3, 87 CR-2, A-2 CR-10, A-4	CR-3: Fol: N44W 74W Lin: 47 N61W CR-2 Fol: N35E 57W Lin: 57 N55W	CR-3, 87: Plagioclase + amphibole + clinopyroxene + Fe-Ti oxides CR-2, A-2: Amphibole + plagioclase + scapolite + titanite + biotite + quartz + Fe-Ti oxides (replacing titanite)	CR-2, A-2: 944 ± 1.9 Ma CR-3, 87: 940.9 ± 3 Ma	N.A.	715
6. Folded mylonitic gabbro	N.A.	N.A.	Folded mylonite with CR-7 assemblage (addition of scapolite near amphibole veins)	N.A.	N.A.	N.A.
Brecciated gabbro	H-6A H-87, 5B	N.A.	Matrix: Albite + chlorite + actinolite + calcite + epidote	N.A.	N.A.	N.A.

*See Streepey et al. (2001).
†Temperature calculated using the plagioclase-amphibole geothermometer of Holland and Blundy (1994).
§Scapolite replaces plagioclase near amphibole veins.
N.A.—not applicable.

TABLE 1B. PHYSICAL AND CHEMICAL CHARACTERISTICS OF SAMPLES FROM OUTCROPS DH-1 AND RW

Outcrop event	Sample number	Orientation	Mineral assemblage	$^{40}Ar/^{39}Ar$ date*	U/Pb date*	T (°C)† (6 kilobars)
1. 30+ m shear zone	RW-H1	Fol: N09W 82 W Lin: 30 346	Amphibole + plagioclase + scapolite	976 ± 2Ma		771
Amphibole-tourmaline vein parallel to RW-H1 foliation	EA-1	Foliated amphibole-tourmaline vein Fol: N30W 90	Amphibole + tourmaline + quartz + titanite	983 ± 2 Ma		
2. 3 m shear zone	RW-S3 (A, B, C)	Fol: N60W 90 Lin:	Plagioclase (granular) + amphibole + scapolite (as corona on amphibole)	1009 ± 3 Ma		
3. Amphibole vein	RW-AVS3	Trend N65E 85S				
4. Cm-wide shear zones	RW-S1 RW-S2 CR-1, DH-1 CR-7, DH-1 CR-6, DH-1	Dextral shears S1: Fol: N01W 87E Lin: N.A. S2: Fol: N09E 70E Lin:N.A. CR-1: Fol: N80W 90 CR-7 Fol: N53W 44NE	RW-S1: Amphibole + scapolite + plagioclase (polygonal) + titanite (minor) RW-S2: CR-6, DH-1: Scapolite + plagioclase + amphibole + biotite + Fe-Ti oxides CR-7, DH-1: Scapolite + amphibole + biotite + plagioclase CR-1, DH-1: Amphibole + scapolite + plagioclase + Fe-Ti oxides	RW-S1: 1000 ± 2 Ma CR-1, DH-1: 989 ± 1 Ma CR-7, DH-1: 985 ± 2 Ma		718
5. Veining event	LHV-1	Late amphibole vein	Amphibole + tourmaline			
6. Brecciated vein	MV-1	Ductile to brittle shear	Highly altered: Amphibole to chlorite			

*See Streepey et al. (2001).
†Temperature calculated using the plagioclase-amphibole geothermometer of Holland and Blundy (1994).
N.A.—not applicable.

shear zone in outcrop A-4 is a smaller shear zone (samples CR-3, 4, and 5, A-4) with intense fabric development. This shear can be traced across the exposure, and, like the larger shear zone that it cuts, has a northwest trend and a gently plunging stretching lineation (see Table 1, A).

Early-formed shear zones are cut by a series of equally spaced (10–20 cm) and parallel amphibole veins 1–3 cm wide (Fig. 3, C). Where the vein intersects preexisting mylonitic fabric foliation, surfaces deflect along the vein wall, yielding dextral shear sense. Parallel amphibole veins are common in both outcrops A-4 and 87. Veins that have not suffered subsequent folding strike north-northeast and dip steeply to the west (see Table 1). A second set of amphibole-tourmaline veins cut the parallel vein set at high angles. These veins tend to be larger in size and randomly spaced. Several small (1 cm to several cm wide) anastamosing shear zones (see Table 1) cut and offset all preexisting structures. All of the veins preserve halos of scapolite about the vein. Late in the deformational history, portions of the Dana Hill metagabbro body were deformed into northwest-trending open to chevron folds with moderate to steeply plunging (30°–64°) hinge lines and steeply dipping axial planes. The folds are defined by rotated amphibole veins (event 3), and, where intersected, the nearly vertical mylonitic fabric (events 1 and 2) fans across the fold noses (see Fig. 3, E). In outcrop 87 (see Fig. 2), folding gradually transitions into zones of brecciation. Chlorite-filled tension gashes are common in the brecciated regions of the outcrop. The sequence of events for these outcrops is given in Table 1.

Central Dana Hill metagabbro body (Outcrops DH and RW)

The open folding observed in the northern outcrops is absent in these exposures; zones of brittle failure, however, are common, usually associated with late faulting. The DH-1 and DH-2 outcrops on Dana Hill Road record multiple generations of centimeter-wide shear zones superimposed on a moderately deformed gabbro. Shear zones anastamoze around lenticular pods of weakly deformed metagabbro, some with well-preserved igneous textures. Shear zones offset one another, preserving at least two discrete generations: right lateral shear set (samples CR-1 and 5, DH-1)and a left lateral shear set (samples CR-6 and 7, DH-1).

Outcrop RW records a more complicated deformation or veining history. A large north-south–trending >30 m–wide shear zone (sample RW-H1) cuts the outcrop. Foliation development in this zone is intense, and dips are nearly vertical. The foliation is axial planar to isoclinal folds, the noses of which are exposed in outcrop RW (see Fig. 3, B). Hinge lines for these folds are vertical, and axial planes trend north-northwest and dip steeply (80°–90°) to the west, with well-developed stretching lineations plunging 30° to azimuth 346°. One large (0.25 m wide) amphibole + tourmaline vein roughly parallels the local mylonitic foliation. The vein margins are deformed, but cut the foliation of the host mylonitic gabbro (sample EA-1). The 30+ m–wide shear zone is clearly cut by a 10 cm–wide left lateral shear zone (sample RW-S3), which is cut by a series of amphibole ± tourmaline veins (Fig. 4, RW-S1A). The amphibole veins are in turn cut by a series of small centimeter-scale right lateral shear zones (RW-S1 and RW-S2; Fig. 3, D) which are cut by a large amphibole-tourmaline vein (see Table 2, B, and Fig. 4, sample LHV). A highly altered amphibole + chlorite cored "shear/fault zone" offsets the boundary of the pod of undeformed gabbro by 1 meter, with a left lateral shear sense (MV-1). Where this zone cuts well-foliated metagabbro, the foliation rotates into the zone. A discrete detachment surface with highly altered vein material is present at the center of the zone (sample MV). This zone cuts all previous shears and veins.

Structural Data

Figure 5, A, is a plot of sixty-five poles to foliations in the Dana Hill metagabbro body. The well-defined maxima at a dip angle of 74° clearly show that the foliation surfaces in the Dana Hill metagabbro body dip steeply to the southwest. In the deformed metagabbro, stretching lineations are pronounced in outcrop and typically are defined by stringers of amphibole or sphene, or both. Foliation and stretching lineation data are plotted in Figure 5, A and B. Note that the stretching lineation data produce a marked maximum with a plunge of 27° to azimuth 321° and a second maximum centered ~49 ° to azimuth 309°.

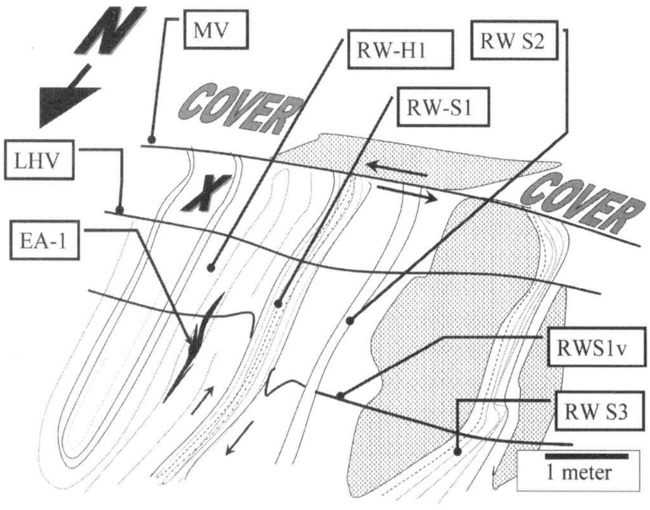

Figure 4. Map of outcrop RW showing all deformation and mineralization events. Heavy black lines represent amphibole or tourmaline veins; stippled region indicates weakly to undeformed gabbro. Arrows indicate offsets, where known. Sample locations are also shown. Note that amphibole vein RW-S1A is cut by shear zones RW-S1 and RW-S2, but cuts shear zone RW-S3, indicating that shear zone RW-S3 is older than either shear zone RW-S1 or shear zone RW-S2 (see text for further discussion).

TABLE 2A. AMPHIBOLE DATA FOR EVENT 1 SHEAR ZONES AND OVERGROWTHS ON IGNEOUS CLINOPYROXENE

Sample	CR-9, A-4	CR-7, A-4	CR-11, A-4	CR-12, A-4	RW-H1 High Cl	RW-H1 Low Cl
SiO_2*	44.21	41.20	39.25	41.26	40.43	39.60
TiO_2	1.51	1.26	1.58	1.01	1.44	1.74
Al_2O_3	11.51	13.91	13.04	12.31	13.27	12.73
Cr_2O_3	0.00	0.00	0.00	N.A.	0.07	0.06
FeO	13.78	15.53	20.04	16.03	15.65	16.91
MnO	0.19	0.06	0.29	0.08	0.10	0.15
MgO	12.90	10.82	8.01	11.03	10.17	9.85
CaO	11.81	10.08	11.50	11.97	12.36	11.82
Na_2O	1.67	1.60	1.70	1.86	1.45	1.32
K_2O	0.93	1.28	1.68	1.43	1.65	2.05
Cl	0.22	0.62	1.25	0.67	1.24	0.66
F	0.43	0.32	0.36	N.A.	0.11	0.48
Total	98.93	96.41	98.7	97.59	97.93	97.36
T Si†	6.40	6.11	6.02	6.20	6.14	6.06
T Al	1.60	1.90	1.98	1.80	1.86	1.94
T Fe^{3+}	0.00	0.00	0.00	0.00	0.00	0.00
T Ti	0.00	0.00	0.00	0.00	0.00	0.00
Σ in T	8.00	8.00	8.00	8.00	8.00	8.00
C Al	0.36	0.54	0.38	.38	0.52	0.36
C Cr	0.00	0.00	0.00	0.00	0.01	0.01
C Fe^{3+}	0.61	1.18	0.62	0.50	0.23	0.50
C Ti	0.16	0.14	0.18	0.12	0.17	0.02
C Mg	2.78	2.39	1.83	2.47	2.30	2.25
C Fe^{2+}	1.06	0.75	1.95	1.51	1.76	1.67
CMn	0.02	0.08	0.04	0.01	0.01	0.02
C Ca	0.00	0.00	0.00	0.00	0.00	0.00
Σ in C	5.00	5.00	5.00	5.00	5.00	5.00
B Mg	0.00	0.00	0.00	0.00	0.00	0.00
B Fe^{2+}	0.00	0.00	0.00	0.00	0.00	0.00
B Mn	0.00	0.00	0.00	0.00	0.00	0.00
B Ca	1.83	1.60	1.89	1.93	2.00	1.94
B Na	0.17	0.40	0.11	.07	0.00	0.06
Σ in B	2.00	2.00	2.00	2.00	2.00	2.00
A Ca	0.00	0.00	0.00	0.00	0.01	0.00
A Na	0.30	.06	0.40	.47	0.43	0.33
A K	0.17	.24	0.33	.27	0.32	0.40
Σ in A	0.47	.30	0.72	.74	0.76	0.73
C Cl	0.05	.16	0.33	.17	0.32	0.17
C F	0.20	.15	0.18	0.00	0.05	0.23
Σ cations	15.47	15.30	15.72	15.74	15.76	15.73
Σ O	23.00	23.00	23.00	23.00	23.00	23.00
An plagioclase	0.49–0.65	0.39–0.44	0.43–0.45	0.44–0.55	0.53–0.49	0.53–0.49

*Major elements expressed in weight percent.
†Elements expressed in moles; T, C, B, and A refer to sites in the amphibole structure.
N.A.—not applicable.

Table 1 shows that that the early-formed shear zones typically have shallowly plunging stretching lineations, whereas the late-formed centimeter-wide shear zones show steeper plunges.

PETROGRAPHIC AND MINERAL CHEMISTRY DATA

Mineral assemblages for each sample studied are presented in Table 1. The petrology of the Dana Hill metagabbro body samples is presented in chronological order (Table 1, events 1–6), as determined from outcrop. Mineral chemistry data are given in Tables 3 and 4 and in Figure 6.

Undeformed Dana Hill Metagabbro

Sample CR-9, A-4 (outcrop A-4) is a cumulate-textured gabbro consisting of large (up to several centimeters) lath-shaped plagioclase crystals with clinopyroxene and Fe-Ti oxide minerals filling the interstices (Fig. 7, A). Plagioclase has a faint

TABLE 2B. REPRESENTATIVE AMPHIBOLE DATA FOR EVENTS 2–6

Sample	CR-5, A-4	Vein CR-12	CR-10, A-4	RW-S3	CR-3, 87	CR-2, A-2	CR-2, A-2	RW-S1	CR-7, DH-1	CR-6, DH-1	LHV-1 vein	H-6A matrix	H-6A clast
SiO_2*	39.24	41.21	42.74	41.06	39.6	44.78	42.09	40.42	42.49	41.86	39.75	53.53	39.99
TiO_2	1.37	1.54	1.63	0.68	1.74	0.82	0.94	1.47	1.45	0.95	1.25	0.02	1.16
Al_2O_3	12.70	13.18	12.68	13.53	12.73	9.84	11.59	13.69	11.99	12.12	12.09	1.15	14.69
Cr_2O_3	0.02	0.00	0.00	N.A.	0.06	0.10	0.11	N.A.	N.A.	0.19	0.06	0.03	0.00
FeO	18.63	16.31	16.61	14.94	16.91	13.48	14.85	14.97	15.03	14.80	18.57	15.29	13.63
MnO	0.18	0.1	0.14	0.15	0.15	0.16	0.11	0.10	0.15	0.19	0.09	0.19	0.12
MgO	8.94	10.33	8.46	10.77	9.85	12.75	11.27	10.22	11.05	11.27	9.00	14.25	11.33
CaO	11.35	11.7	11.53	11.83	11.82	11.65	11.45	11.97	11.67	12.00	11.56	12.53	11.80
Na_2O	1.41	1.79	1.45	1.81	1.32	1.30	1.53	1.35	1.46	1.27	1.74	0.29	2.32
K_2O	1.52	1.2	1.58	1.77	2.05	1.13	1.44	2.06	1.40	1.61	1.36	0.02	1.08
Cl	1.25	0.66	0.99	1.30	0.66	0.73	0.95	1.28	1.16	1.62	2.20	0.01	0.58
F	N.A.	0.21	0.05	N.A.	0.48	0.21	0.12	N.A.	N.A.	0.14	0.00	0.02	0.79
Total	96.32	98.23	97.86	97.56	97.36	96.94	96.46	97.24	97.59	98.02	97.65	97.32	97.48
T Si†	6.05	6.15	6.50	6.19	6.06	6.65	6.37	6.145	6.36	6.30	6.15	7.81	6.01
T Al	1.95	1.845	1.51	1.81	1.94	1.35	1.63	1.86	1.64	1.70	1.85	0.19	1.10
T Fe^{3+}	0.00	0	0.00	0.00	0.00	0.00	0.00	0.00	0.00	0.00	0.00	0.00	0.00
T Ti	0.00	0	0.00	0.00	0.00	0.00	0.00	0.00	0.00	0.00	0.00	0.00	0.00
Σ in T	8.00	8	8.00	8.00	8.00	8.00	8.00	8.00	8.00	8.00	8.00	8.00	8.00
C Al	0.36	0.477	0.77	0.60	0.36	0.38	0.44	0.60	0.47	0.45	0.36	0.01	0.61
C Cr	0.00	0	0.00	0.00	0.01	0.01	0.01	0.00	0.00	0.02	0.01	0.00	0.00
C Fe^{3+}	0.80	0.529	0.00	0.36	0.50	0.47	0.52	0.22	0.41	0.45	0.57	0.17	0.45
C Ti	0.16	0.173	0.19	0.08	0.20	0.09	0.11	0.17	0.16	0.11	0.15	0.00	0.13
C Mg	2.05	2.3	1.92	2.42	2.25	2.82	2.54	2.32	2.46	2.53	2.08	3.10	2.54
C Fe^{2+}	1.60	1.508	2.11	1.53	1.67	1.20	1.37	1.68	1.47	1.41	1.83	1.70	1.27
C Mn	0.02	0.013	0.02	.019	0.02	0.02	0.01	0.01	0.02	0.02	0.01	0.02	0.02
C Ca	0.00	0	0.00	0.00	0.00	0.00	0.00	0.00	0.00	0.00	0.00	0.00	0.00
Σ in C	5.00	5	5.00	5.00	5.00	5.00	5.00	5.00	5.00	5.00	5.00	5.00	5.00
B Mg	0.00	0	0.00	0.00	0.00	0.00	0.00	0.00	0.00	0.00	0.00	0.00	0.00
B Fe^{2+}	0.00	0	0.00	0.00	0.00	0.00	0.00	0.00	0.00	0.00	0.00	0.00	0.00
B Mn	0.00	0	0.00	0.00	0.00	0.00	0.00	0.00	0.00	0.00	0.00	0.00	0.00
B Ca	1.88	1.873	1.88	1.91	1.94	1.86	1.86	1.95	1.87	1.94	1.92	1.96	1.90
B Na	0.13	0.127	0.12	0.09	0.06	0.15	0.14	0.05	0.13	0.07	0.08	0.04	0.10
Σ in B	2.00	2	2.00	2.00	2.00	2.00	2.00	2.00	2.00	2.00	2.00	2.00	2.00
A Ca	0.00	0	0.00	0.00	0.00	0.00	0.00	0.00	0.00	0.00	0.00	0.00	0.00
A Na	0.30	0.391	0.31	0.44	0.33	0.23	0.31	0.35	0.29	0.31	0.44	0.04	0.57
A K	0.30	0.299	0.31	0.34	0.40	0.21	0.28	0.40	0.25	0.31	0.27	0.00	0.21
Σ in A	0.60	0.62	0.61	0.78	0.73	0.45	0.59	0.75	0.56	0.62	0.71	0.05	0.78
C Cl	0.33	0.167	0.26	0.33	0.17	0.18	0.24	0.33	0.29	0.41	0.58	0.00	0.15
C F	0.00	0.099	0.02	N.A.	0.23	0.10	0.06	N.A.	N.A.	0.07	0.00	0.01	0.37
Σ cations	15.60	15.62	15.61	15.78	15.73	15.45	15.86	15.75	15.56	15.62	15.71	15.05	15.78
Σ O	23.00	23.00	23.06	23.00	23.00	23.00	23.00	23.00	23.00	23.00	23.00	23.00	23.00
An plagioclase	0.42–0.49	N.A. (vein)	0.30	0.47	0.32	0.35–0.37	0.35–0.37	0.50–0.52	N.A.	0.50–0.53	N.A.	0.01–0.21	0.43–0.47

*Major elements expressed in weight percent.

†Elements expressed in moles; T, C, B, and A refer to sites in the amphibole structure.

N.A.—not applicable.

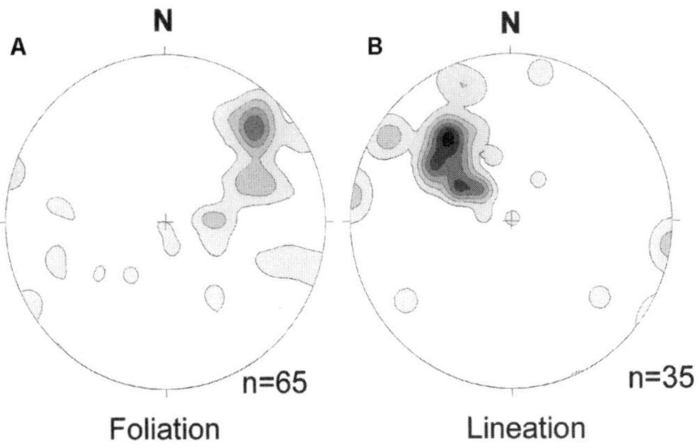

Figure 5. Equal-area lower hemisphere projection of (A) poles to mylonitic foliations and (B) stretching lineations in the Dana Hill metagabbro. Note the shallow plunge on stretching lineation data (see also Table 1).

purple hue in outcrop. In thin section, plagioclase is twinned and clouded by rutile needles. Clinopyroxene crystals are clouded by abundant 5–20 μm opaque solid inclusions. Tschermakitic hornblende along with rounded Fe-Ti oxide minerals forms 50–200 μm rims on clinopyroxene. Plagioclase in contact with these hornblende rims is clear of included rutile, and the boundaries are clean and sharp. Fluorine content of hornblende is high (0.7 wt% average). Fe-Ti Oxide minerals are surrounded by 50–150 μm rims of fibrous amphibole, biotite, and rutile. Abundant fractures cut the sample, and where these fractures cut plagioclase laths they are decorated with rounded epidote and opaque grains replacing plagioclase (Fig. 7, B). In rare instances, scapolite is found replacing plagioclase in these fractures. The cores of many of the large plagioclase laths are altered to mixtures of epidote and white mica. This alteration is associated with and located about small fractures.

Event 1

The large >30 m–wide shear zones (Table 1, event 1; samples RW-H1, CR-6, CR-7, and H-1A) show a limited variability in mineral assemblage. CR-7, A-4 and CR-6, A-4 contain recrystallized clinopyroxene + amphibole + sphene + plagioclase (An_{45-51}), along with accessory minerals (apatite, zircon, ± quartz; see Fig. 7, C). RW-H1 and H1, 87, both lack clinopyroxene in their assemblages, but are otherwise similar. Amphibole compositions for these samples range from ferroan pargasite (RW-H1; CR-7, A-4; and CR-12, A-4) to magnesian hastingsite (CR-11, A-4). Chlorine contents are high for all amphiboles studied, ranging from 2 to 18% hydroxyl site occupation (Table 1; Fig. 6, A). Amphibole and plagioclase chemistries are presented in Tables 3 and 4 and shown in Figure

TABLE 3A. PLAGIOCLASE COMPOSITIONS FOR EVENTS 1 AND 2

Sample	CR-7, A-4	RW-H1	CR-5, A-4*	CR-5, A-4†	CR-5, A-4§	RW-S3
SiO_2**	58.65	55.61	55.05	56.42	55.63	56.17
TiO_2	0.00	0.00	0.00	0.00	0.00	0.00
Al_2O_3	27.53	28.00	28.31	27.80	27.76	27.05
FeO	0.00	0.06	0.17	0.16	0.12	0.12
MnO	0.05	0.00	0.00	0.00	0.00	0.02
MgO	0.00	0.07	0.00	0.00	0.00	0.00
CaO	9.03	9.95	9.86	8.78	9.70	9.61
Na_2O	6.28	5.39	5.41	6.30	5.86	5.93
K_2O	0.20	0.16	0.22	0.18	0.52	0.18
Cl	0.00	0.01	0.00	0.00	0.00	0.01
F	0.00	0.01	0.00	0.00	0.00	0.00
Total	101.70	99.26	99.02	99.64	99.68	99.09
Si††	2.58	2.52	2.50	2.54	2.52	2.55
Al	1.43	1.50	1.52	1.48	1.48	1.45
Ti	0.00	0.00	0.00	0.00	0.00	0.00
Fe^{2+}	0.00	0.00	0.01	0.01	0.00	0.00
Mg	0.00	0.01	0.00	0.00	0.00	0.00
Mn	0.00	0.00	0.00	0.00	0.00	0.00
Ca	0.43	0.48	0.48	0.42	0.47	0.47
Na	0.54	0.47	0.47	0.55	0.513	0.52
K	0.01	0.01	0.01	0.01	0.03	0.01
Σ cations	4.98	4.98	4.98	5.00	5.01	4.99
An	0.44	0.51	0.49	0.43	0.47	0.47

Plagioclase composition from core of zoned crystal.
†Plagioclase composition from rim of zoned plagioclase crystal.
§Composition for unzoned polygonal plagioclase crystal.
*Major elements expressed in weight percent.
††Elements expressed in moles; casting based on 8 total oxygens.

TABLE 3B. PLAGIOCLASE COMPOSITIONS FOR EVENTS 3–6

Sample	CR-10, A-4	CR-3, 87	CR-2, A-2	CR-2, A-2*	RW-S1	CR-6, DH-1	H-6A†	H-6A§
SiO_2**	60.97	60.16	59.30	58.61	55.79	56.96	66.10	56.82
TiO_2	0.00	0.00	0.00	0.00	0.00	0.05	0.24	0.00
Al_2O_3	24.91	24.57	25.04	25.19	27.50	27.36	19.01	27.22
FeO	0.08	0.19	0.07	0.09	0.01	0.15	0.31	0.03
MnO	0.00	0.00	0.00	0.00	0.00	0.00	0.00	0.02
MgO	0.00	0.00	0.00	0.00	0.00	0.00	0.09	0.00
CaO	5.59	6.79	7.82	7.56	10.42	9.72	5.14	9.65
Na_2O	7.68	7.74	7.34	7.42	5.66	5.94	11.07	5.94
K_2O	0.33	0.23	0.19	0.26	0.14	0.18	0.05	0.07
Cl	0.00	0.00	0.00	0.00	0.00	0.00	0.01	0.00
F	0.00	0.00	0.00	0.00	0.00	0.00	0.07	0.04
Total	99.60	99.68	99.76	99.13	99.52	100.36	102.10	99.79
Si††	2.75	2.69	2.66	2.64	2.52	2.55	2.89	2.55
Al	1.33	1.30	1.32	1.34	1.47	1.44	0.98	1.44
Ti	0.00	0.00	0.00	0.00	0.00	0.00	0.01	0.00
Fe^{2+}	0.00	0.01	0.00	0.00	0.00	0.01	0.01	0.00
Mg	0.00	0.00	0.00	0.00	0.00	0.00	0.01	0.00
Mn	0.00	0.00	0.00	0.00	0.00	0.00	0.00	0.00
Ca	0.27	0.33	0.38	0.37	0.51	0.47	0.24	0.47
Na	0.67	0.67	0.64	0.65	0.50	0.52	0.94	0.52
K	0.02	0.01	0.01	0.02	0.01	0.01	0.00	0.00
Σ cations	5.04	5.00	5.00	5.02	4.10	4.99	5.08	4.99
An	0.30	0.32	0.37	0.35	0.50	0.47	0.79	0.47

*Plagioclase grains in contact with amphibole.
†Plagioclase from recrystallized breccia.
§Plagioclase from breccia clasts that preserve granulite-facies mineralogies.
**Major elements expressed in weight percent.
††Elements expressed in moles; casting based on 8 total oxygens.

6. Sample RW-H1 contains scattered amounts of scapolite (mizzonite) partially surrounding some amphibole grains. All samples exhibit a well-annealed polygonal fabric with perfect 120° triple junctions between grains (Fig. 7, C and D). Grain sizes show a narrow range of variation for these samples, averaging in the range of 100–300 μm for polygonal plagioclase (CR-7, CR-11, and CR-12). Recrystallization temperatures for event 1 samples (RW-H1; CR-6 and 7, A-4) measured using the quartz-free geothermometer of Holland and Blundy (1994) range from 744 to 770 °C (for 6 kilobars; see Table 4 and Fig. 8. Plagioclase-amphibole pairs used for geothermometry are in textural equilibrium as determined from petrographic and backscatterd electron imaging.

Event 2

Samples CR-3, 4, and 5 were taken from a meter-wide shear zone that cuts the 30 m–wide shear in outcrop A-4 (event 2; Table 1). This zone contains polygonal plagioclase grains ranging from 150 to 350 μm in diameter (Fig. 7, D). The mineralogy of this sample consists of stringers of polygonal grains of plagioclase (An_{45-49}), Fe-rich hornblende (magnesian hastingsite, FeO_{tot} = 18.6 wt%), sphene with Fe-Ti-oxide cores, and accessory amounts of quartz and apatite. An average of 16.8% of the hydroxl site in amphibole is occupied by chlorine (see Table 1; Fig. 6, A). The high iron content in amphiboles from samples CR-3, 4, and 5 preclude the use of the Holland and Blundy (1994) geothermometer. Some polygonal plagioclase grains are zoned with rim compositions of An_{43} to core compositions of An_{49}. Most polygonal grains, however, show little to no compositional zoning, with an average composition of $An_{\sim 46}$.

Event 2 in outcrop RW (see Table 1) is represented by a 10 cm–wide shear zone. Sample RW-S3 was collected in the center of this shear zone. The sample shows polygonal grains of amphibole with well developed rinds of scapolite. Grain size is quite variable with the finest grains associated with plagioclase ($An_{\sim 47}$). In these regions, average grain size is 20–80 μm. Amphibole is found throughout the slide as clots and stringers. Amphibole chemistry is presented in Table 3, A. Highly embayed clinopyroxene cores are found in some of these clots. Iron-titanium oxide minerals form long (mm) stringers along with highly pleochroic amphibole. A hornblende + tourmaline vein (EA-1) cuts this shear zone and is here assigned to event 2.

Event 3

Figure 9, A, shows a typical hornblende vein cutting event 1 fabric (sample CR-12). The fabric of the host metagabbro is rotated and offset across the 0.3 cm–wide amphibole vein, yielding dextral shear sense that agrees with field observations of fab-

TABLE 4. EQUILIBRIUM OF PLAGIOCLASE-AMPHIBOLE PAIRS USED FOR GEOTHERMOMETRY

CR-7, A-4 Event 1

	Amphibole			Plagioclase		
Oxide	Wt%	Element		Wt%	Element	
SiO_2	41.97	Si	6.25	58.65	Si	2.58
TiO_2	1.26	Ti	0.14	0.00	Ti	0.00
Al_2O_3	13.91	Al	2.44	27.53	Al	1.43
Cr_2O_3	0.00	Cr	0.00	0.00	Cr	0.00
FeO	15.53	Fe	1.40	0.05	Fe	0.00
Fe_2O_3	—	Fe^{3+}	0.53	—	Fe^{3+}	—
MnO	0.06	Mn	0.01	0.00	Mn	0.00
MgO	10.82	Mg	2.40	0.00	Mg	0.00
CaO	10.08	Ca	1.61	9.03	Ca	0.43
Na_2O	1.40	Na	0.40	6.28	Na	0.54
K_2O	1.60	K	0.30	0.20	K	0.01
Cl	0.62	Cl	—	0.00	Cl	0.00
F	0.32	F	—	0.00	F	0.00
					Ab	0.56
Total	96.69			101.70		
Pressure (kilobars)	0		5		10	
Temperature (°C)*	681		734		786	

RH-H1 Event 1

	Amphibole			Plagioclase		
Oxide	Wt%	Element		Wt%	Element	
SiO_2	40.43	Si	6.10	55.88	Si	2.52
TiO_2	1.44	Ti	0.16	0.00	Ti	0.00
Al_2O_3	13.27	Al	2.36	27.76	Al	1.48
Cr_2O_3	0.07	Cr	0.01	0.00	Cr	0.00
FeO	15.56	Fe	1.45	0.04	Fe	0.00
Fe_2O_3	0.00	Fe^{3+}	0.53	—	Fe^{3+}	—
MnO	0.10	Mn	0.01	0.00	Mn	0.00
MgO	10.17	Mg	2.29	0.00	Mg	0.00
CaO	12.36	Ca	1.10	10.28	Ca	0.50
Na_2O	1.45	Na	0.42	5.64	Na	0.49
K_2O	1.65	K	0.32	0.16	K	0.00
Cl	1.24	Cl	—	0.00	Cl	0.01
F	0.11	F	—	0.00	F	0.00
					Ab	0.49
Total	97.60			99.77		
Pressure (kilobars)		0			5	10
Temperature (°C)*		732			764	797

CR-6, DH-1 Event 4

	Amphibole			Plagioclase		
Oxide	Wt%	Element		Wt%	Element	
SiO_2	41.38	Si	6.63	56.06	Si	2.52
TiO_2	1.19	Ti	0.14	0.05	Ti	0.00
Al_2O_3	12.50	Al	2.23	27.59	Al	1.46
Cr_2O_3	0.14	Cr	0.02	0.00	Cr	0.00
FeO	15.14	Fe	1.51	0.052	Fe	0.00
Fe_2O_3	0.00	Fe^{3+}	0.41	—	Fe^{3+}	—
MnO	0.16	Mn	0.16	0.00	Mn	0.00
MgO	10.64	Mg	2.40	0.06	Mg	0.00
CaO	12.01	Ca	1.95	10.07	Ca	0.48
Na_2O	1.16	Na	0.34	5.80	Na	0.51
K_2O	1.72	K	0.33	0.23	K	0.01
Cl	1.71	Cl	—	0.00	Cl	0.00
F	0.15	F	—	0.00	F	0.00
					Ab	0.51
Total	97.90			99.92		
Pressure (kilobars)	0		5		10	
Temperature (°C)*	677		711		745	

CR-2, A-2 Event 4

	Amphibole			Plagioclase		
Oxide	Wt%	Element		Wt%	Element	
SiO_2	42.82	Si	6.41	59.31	Si	2.63
TiO_2	0.85	Ti	0.10	0.00	Ti	0.00
Al_2O_3	11.98	Al	2.11	26.09	Al	1.36
Cr_2O_3	0.00	Cr	0.00	0.00	Cr	0.00
FeO	14.08	Fe	1.36	0.09	Fe	0.01
Fe_2O_3	0.00	Fe^{3+}	0.40	—	Fe^{3+}	—
MnO	0.14	Mn	0.02	0.01	Mn	0.00
MgO	11.72	Mg	2.61	0.00	Mg	0.00
CaO	11.70	Ca	1.88	7.83	Ca	0.37
Na_2O	1.53	Na	0.44	6.91	Na	0.60
K_2O	1.47	K	0.28	0.24	K	0.010
Cl	0.85	Cl	—	0.00	Cl	0.00
F	0.00	F	—	0.00	F	0.00
					Ab	0.61
Total	96.96			100.47		
Pressure (kilobars)		0			5	10
Temperature (°C)		669			706	743

CR-3, 87 Event 4

	Amphibole			Plagioclase		
Oxide	Wt%	Element		Wt%	Element	
SiO_2	41.50	Si	6.28	59.84	Si	2.67
TiO_2	1.70	Ti	0.19	0.00	Ti	0.00
Al_2O_3	12.21	Al	2.18	24.96	Al	1.31
Cr_2O_3	0.00	Cr	0.00	0.00	Cr	0.00
FeO	16.21	Fe	1.76	0.29	Fe	0.01
Fe_2O_3	0.00	Fe^{3+}	0.29	—	Fe^{3+}	—
MnO	0.24	Mn	0.03	0.00	Mn	0.00
MgO	9.99	Mg	2.25	0.00	Mg	0.00
CaO	11.72	Ca	1.90	6.78	Ca	0.32
Na_2O	1.46	Na	0.43	7.89	Na	0.68
K_2O	2.06	K	0.40	0.20	K	0.01
Cl	0.65	Cl	—	0.00	Cl	0.01
F	0.00	F	—	0.00	F	0.00
					Ab	0.67
Total	97.60			100.24		
Pressure (kilobars)	0		5		10	
Temperature (°C)	680		711		743	

* Edenite-richterite rxn: Quartz absent.

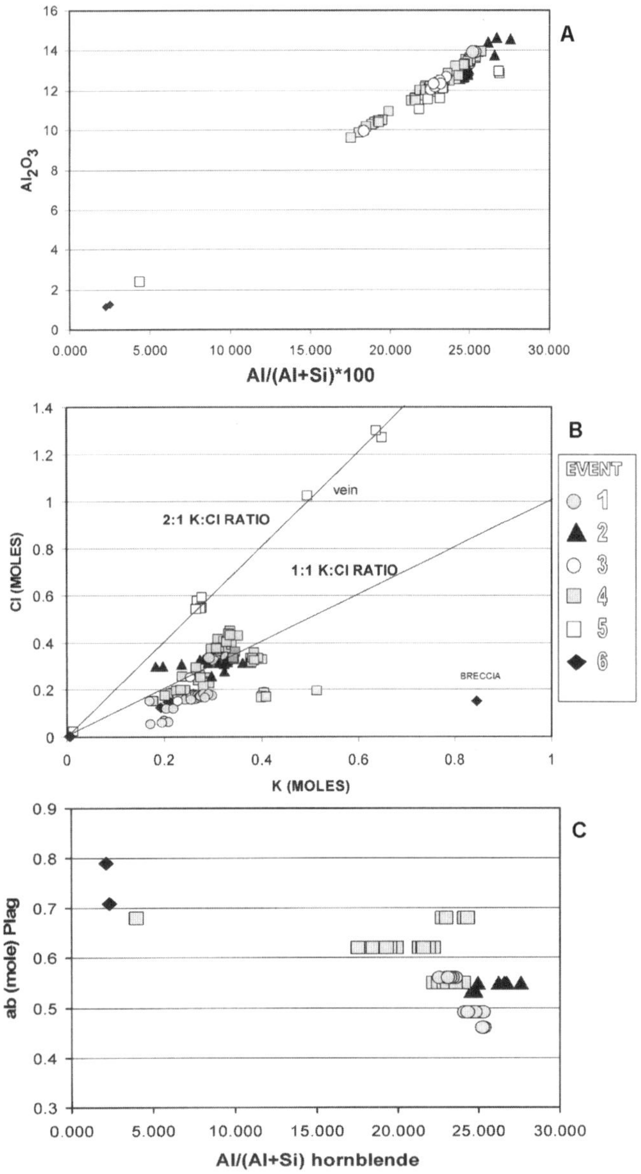

Figure 6. Mineral chemistries from the Dana Hill metagabbro body. (A) Total aluminum (wt% Al_2O_3) versus Al/(Al + Si) for hornblende from the Dana Hill metagabbro body. This type of plot is often used to show variations in grade based on Al content in hornblende (Laird and Albee, 1981). Although the trend of the data is forced, the spread of the data along the trend can be used to infer metamorphic grade. In the case of the Dana Hill metagabbro body samples studied, most plot with moderate to high Al contents. The exception to this trend is event 6, breccia sample H-6A, which shows both high Al hornblende and high actinolite. In thin section, the hornblende is being replaced by actinolite in the breccia matrix. (B) Moles K versus moles Cl. Plot shows a good 1:1 relationship between K and Cl in amphiboles from most samples. Exceptions include event 5, tourmaline + amphibole + scapolite vein LHV-1 (outcrop RW), which plots out a 2:1 Cl to K ratio. The event 6 sample falls well off of the 1:1 line. This sample shows abundant evidence for fluid infiltration during brecciation. (C) Albite content in plagioclase versus total aluminum in hornblende. These quantities are qualitatively tied to metamorphic grade. While there is some separation of event 4 samples from events 1 and 2 samples, all cluster at high total Al values in hornblende.

ric rotations across these veins. The amphibole vein contains hornblende, clinopyroxene, and scapolite. Scapolite (mizzonite) replaces plagioclase in a halo about the margins of the vein, extending 0.3–0.5 cm away from the vein wall (see Fig. 9, A). Amphibole chemistry from within the vein is identical (including chlorine contents) to amphibole chemistry in the host metagabbro. Throughout the sample, amphibole plagioclase (An_{42}), scapolite-amphibole, and scapolite-plagioclase contacts are sharp and straight and often display 120° triple junctions. The results of electron microprobe scans across a vein amphibole to matrix scapolite contact show no systematic chemical changes in either phase at or near the boundary. Compositions for plagioclase-scapolite pairs that are in textural equilibrium are presented in Figure 10. Hornblende veining has been identified in outcrop RW (Fig. 4, RW-S1A). In outcrop, these veins cut event 1 and 2 shear zones, but are clearly cut by event 4 centimeter-scale shear zones. Vein mineralogy consists of hornblende + tourmaline ± sphene ± scapolite.

Event 4

Multiple generations of anastomosing centimeter-wide shear zones that crosscut and offset one another (Fig. 3, E) define event 4 deformation. The zones typically dip steeply, and shear sense, which varies from shear to shear, can be determined by rotation of a preexisting foliation into the zone (see Table 1). The typical assemblage consists of amphibole ± clinopyroxene + scapolite + plagioclase + Fe-Ti oxides. Plagioclase compositions are consistent within individual shear zones, but range from $An_{0.5}$ to $An_{0.32}$ between shear zones. Polygonal scapolite forms distinct halos about amphibole (Fig. 9, B and C). In one sample (CR-2, A-2), a fine rind of plagioclase separates scapolite from amphibole within the halo (see Fig. 9, D). In this sample, plagioclase in contact with amphibole has a composition of $An_{36-38,}$ while plagioclase compositions outside of the scapolite halo range from An_{37} to An_{39}. Biotite is found in a few event 4 samples both as a replacement for amphibole in grains that clearly crosscut the deformation fabric and as blocky grains that follow the foliation in the sample. Sample CR-3, 87, preserves deformation textures (little annealing) and contains the assemblage hornblende + recrystallized clinopyroxene + plagioclase An_{32} + Fe-Ti oxides. Plagioclase-amphibole equilibrium pairs, where present, yield a recrystallization temperature for event 4 shearing in the range of 730–713 °C ± 50 °C for a pressure of 6 kilobars using the quartz-absent plagioclase-amphibole geothermometer of Holland and Blundy (1994) (Fig. 8).

Event 5

Amphibole and amphibole-tourmaline veins (LHV-1 in Fig. 4 and Table 2, A) cut the centimeter-scale shear zones in outcrops A-4 and RW. Tourmaline is common as a veining material in outcrop RW, and hornblende compositions are Cl-rich (see Fig. 6, B). The veins contain mixtures of amphibole,

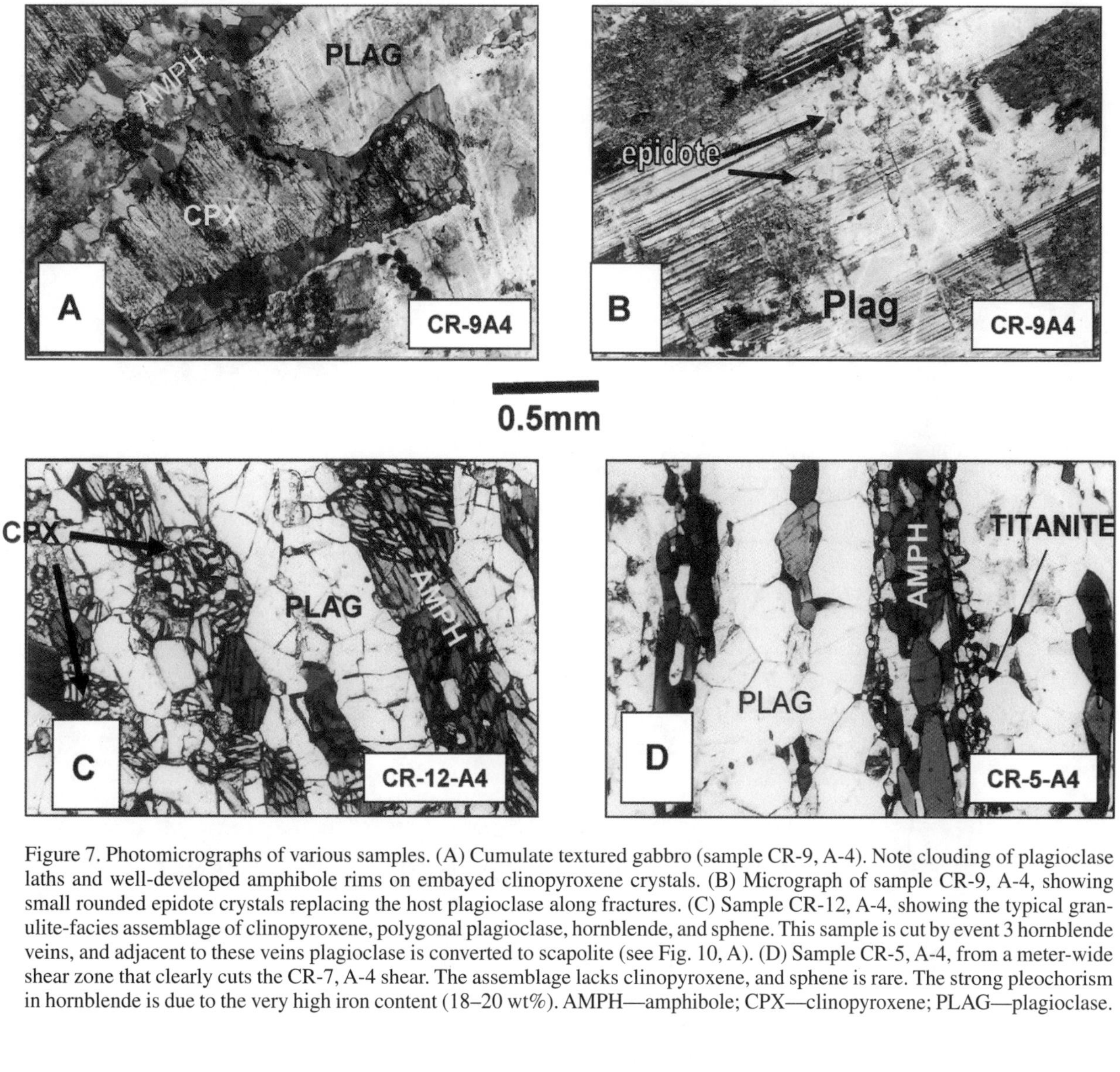

Figure 7. Photomicrographs of various samples. (A) Cumulate textured gabbro (sample CR-9, A-4). Note clouding of plagioclase laths and well-developed amphibole rims on embayed clinopyroxene crystals. (B) Micrograph of sample CR-9, A-4, showing small rounded epidote crystals replacing the host plagioclase along fractures. (C) Sample CR-12, A-4, showing the typical granulite-facies assemblage of clinopyroxene, polygonal plagioclase, hornblende, and sphene. This sample is cut by event 3 hornblende veins, and adjacent to these veins plagioclase is converted to scapolite (see Fig. 10, A). (D) Sample CR-5, A-4, from a meter-wide shear zone that clearly cuts the CR-7, A-4 shear. The assemblage lacks clinopyroxene, and sphene is rare. The strong pleochorism in hornblende is due to the very high iron content (18–20 wt%). AMPH—amphibole; CPX—clinopyroxene; PLAG—plagioclase.

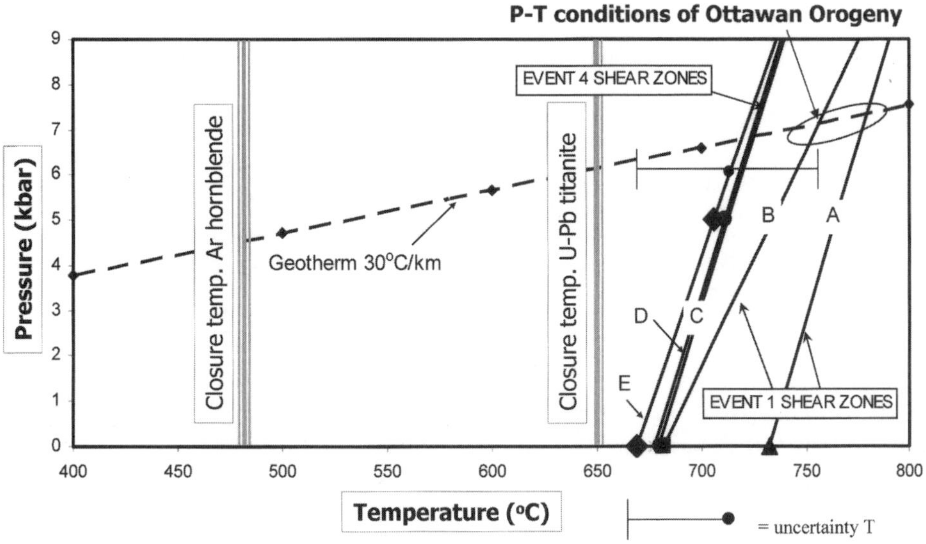

Figure 8. Pressure-temperature (*P-T*) diagram for the Dana Hill metagabbro body samples. Geothermometric data for samples containing amphibole-plagioclase equilibrium pairs (see Table 4) are plotted as lines. The oval shows peak *P-T* conditions for the Ottawan orogeny (from Bohlen et al., 1985). A 30 °C/km geotherm is also shown. Event 1 samples CR-7, A-4 (B) and RW-H (A) record the highest temperatures. Event 4 (cm shear) samples—CR-2, A-2 (E), CR-6, DH-1(D), and CR-3, 87(C)—record lower recrystallization temperatures, but all are above 650 °C.

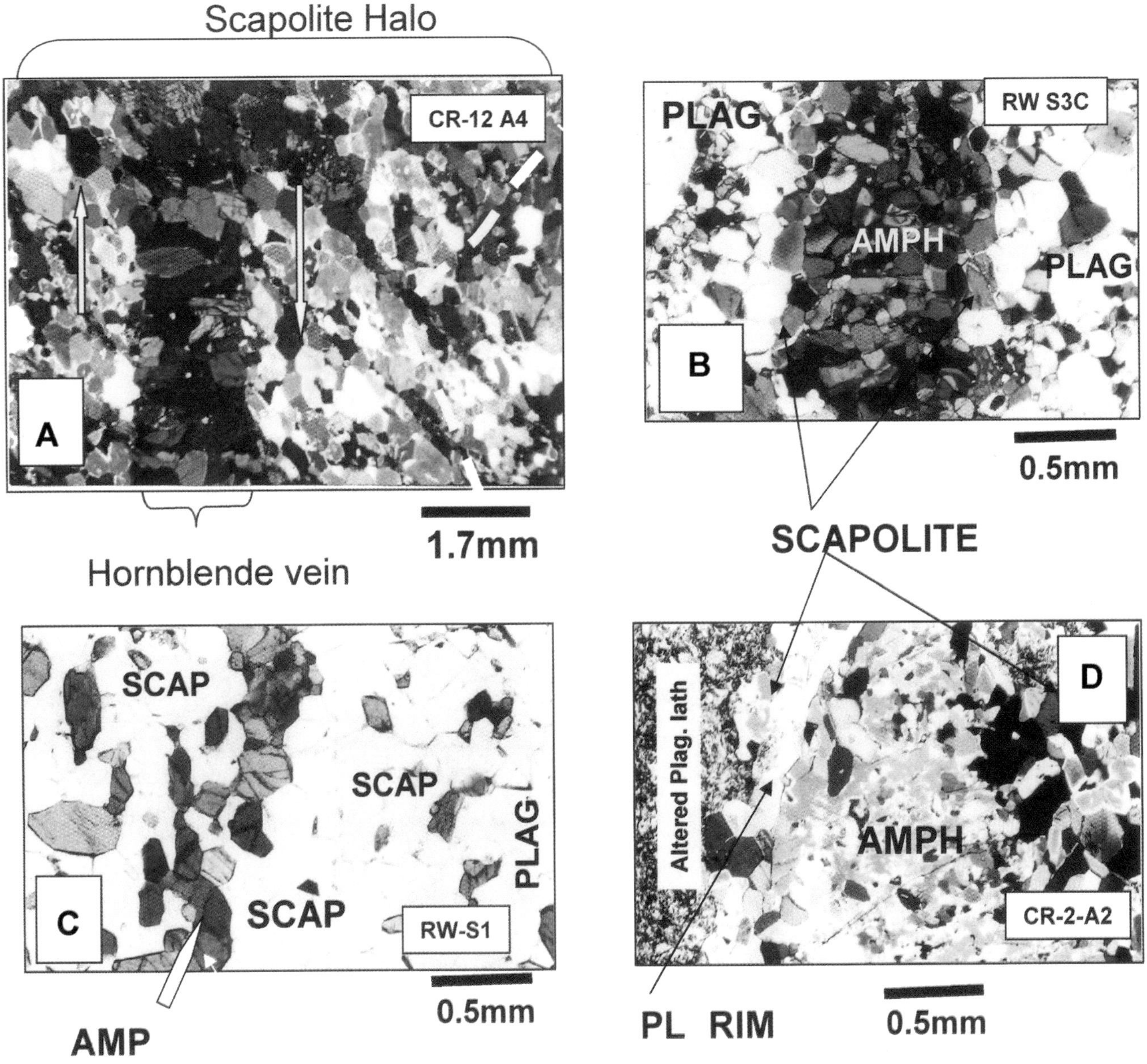

Figure 9 (*on this and following page*). Photomicrographs of event 3–6 samples. (A) Sample CR-12, A-4. This micrograph shows a small hornblende vein cutting a clinopyroxene + amphibole + plagioclase + sphene mylonite. Adjacent to the vein wall, polygonal plagioclase in the host mylonite is converted (pseudomorphed) to polygonal scapolite (scapolite halo). In the outcrop, the mylonitic fabric rotates into and out of these amphibole veins, indicating shearing along the vein. (B) Sample RW-S3C, recrystallized matrix of scapolite + amphibole + plagioclase. Note how the scapolite forms a distinct corona about amphibole. (C) Sample RW-S1, taken from a cm-scale shear zone in outcrop RW. This sample is typical of many of the small-scale shear zones found throughout the Dana Hill metagabbro body. Note the well-developed scapolite coronas about amphibole. (D) Sample CR-2, A-2. This sample was collected from a centimeter-wide shear zone, but, unlike RW-S1, this sample shows the development of a fine rim of plagioclase that separates amphibole from scapolite. Note the different amphibole morphologies (see text for discussion). (E) and (F) Sample H-6A, from a breccia pod in outcrop 87. Note the stringy replacement of amphibole by mixtures of biotite + chlorite + actinolite. F shows same view in crossed polars. See text for discussion. AMPH—amphibole; PLAG—plagioclase; SCAP—scapolite.

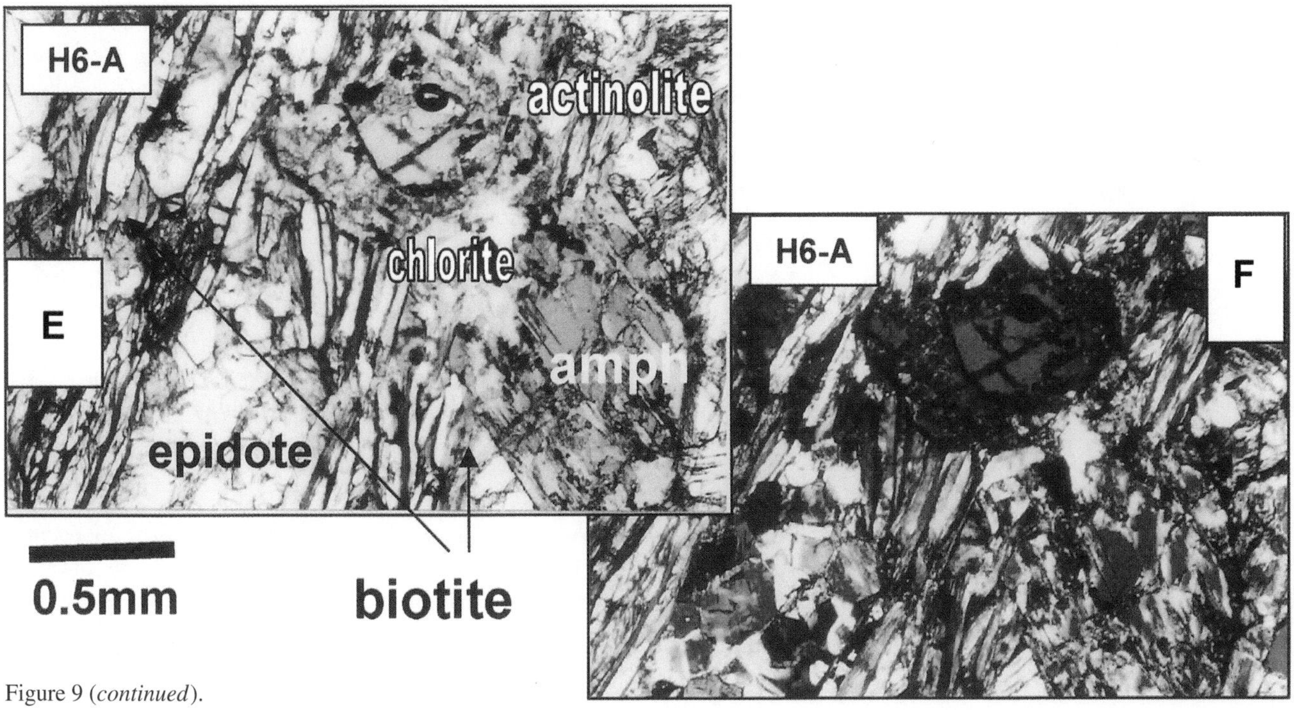

Figure 9 (*continued*).

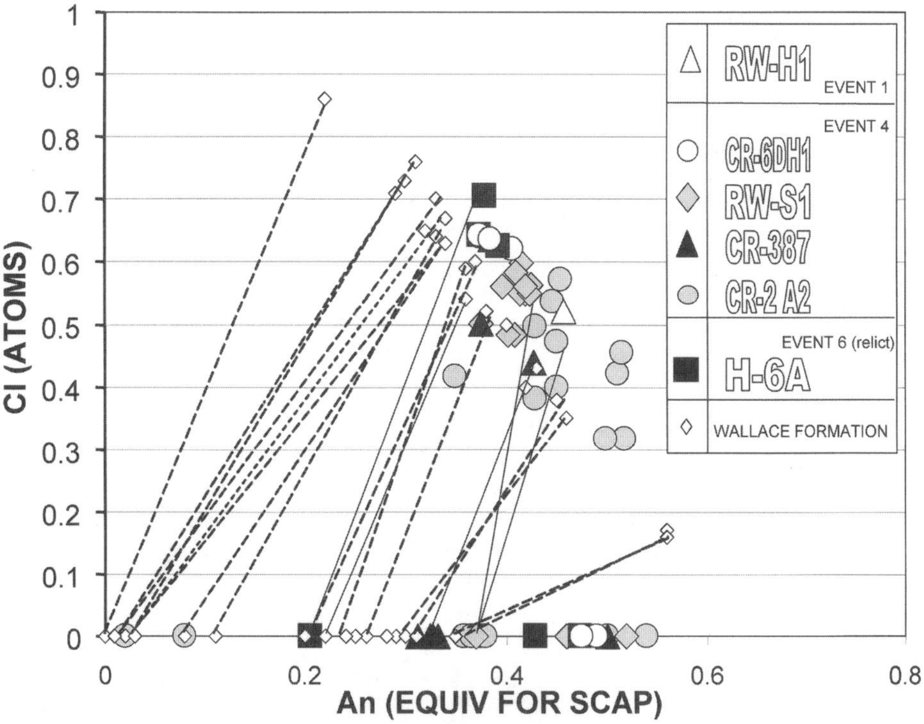

Figure 10. Scapolite-plagioclase chemistries for the Dana Hill metagabbro body as compared to samples from Rebbert and Rice (1997). Tie lines indicate plagioclase-scapolite pairs in textural equilibrium. In all cases, scapolite is enriched in calcium with respect to coexisting plagioclase.

tourmaline, quartz, and scapolite. Grain size is very large (to several centimeters in diameter), with no preferred orientation.

Event 6

Brecciation was accompanied by the growth of actinolite, biotite, and chlorite after hornblende, and the breakdown of scapolite to a mixture of albite, epidote, and calcite (Fig. 9, E and F). Breccia sample H-6A preserves rafts and clots of scapolite-rich mylonitic metagabbro with an invasive matrix of fibrous mats of chlorite, epidote, and actinolite. Hornblende (ferroan pargasite) that has not suffered alteration to actinolite is fluorine-rich (average F = 0.75 wt%; average Cl = 0.57 wt%; Fig. 6, A). The replacement actinolite has negligible amounts of both fluorine and chlorine (total (Cl + F) = 0.1 wt%).

DISCUSSION

Deformation and veining events preserved in the Dana Hill metagabbro body have been carefully mapped at the outcrop level, and the relative timing of each event has been determined. The sequence is presented in Table 1. The earliest deformation event (Table 1) mapped consisted of the creation of a series of very large (30 m width) shear zones cut by a second and smaller (>1 m width) set of shear zones. These early-formed (based on field relationships) shears were cut by northwest-trending amphibole veins. These veins were, in turn, cut by multiple generations of centimeter-scale anastamosing shear zones. Deformation within the Dana Hill metagabbro body concluded with the chevron folding and brecciation.

The overall sequence of events between the northern and the central outcrops of the Dana Hill metagabbro body was strikingly similar, yet direct correlations between events were complex. In the northern outcrops, the appearance of scapolite as a replacement for plagioclase is clearly tied to event 3 vein formation, but in the central outcrop (RW) the appearance of scapolite constrains only postdating of event 1 shearing.

Recrystallization and annealing temperatures calculated using the geothermometer of Holland and Blundy (1994), as well as amphibole and plagioclase chemistries, reveal that events 1 through 4 (see Table 1) took place under a narrow range of temperature conditions (770–711 °C), well within the upper amphibolite to granulite facies. The change in character of the deformation style between event 1 megashears and event 4 anastamosing centimeter-width shears clearly demonstrates that certain conditions attending deformation had changed between these two events. Changes in recrystallization temperatures are insufficient to explain this difference, indicating that some combination of lowered confining pressure (or increase in fluid pressure) and/or decrease in strain rate or strain duration may be responsible. The anastamosing character, variability in shear sense, and transport lineation orientation of event 4 shears show that a more heterogeneous strain field was operating during this time. The emplacement of hundreds of amphibole veins marked a major episode of fluid infiltration into the Dana Hill metagabbro body, separating events 1–2 from event 4. Widespread vein formation and scapolitzation clearly required a sharp rise in fluid pressure (hydrofracturing) or a sharp decline in confining pressure, allowing fluid to effectively infiltrate the Dana Hill metagabbro body. Brecciation and folding (event 6) observed in the Dana Hill metagabbro body occurred at lower greenschist-facies conditions and brought to a close the deformation history of the body.

The nature and timing of fluid infiltration leading to the formation of event 3 hornblende veins is critical to the understanding of the tectonothermal history of the body. The northern outcrops provide a more complete history of deformation and veining, and in these outcrops scapolite growth is clearly tied to the emplacement of a hornblende vein swarm (event 3). However, in the central outcrop RW, scapolite replacement is tied to vein EA, which is assigned to event 2. Scapolite growth in these outcrops is interpreted to have occurred in response to an influx of fluid into the Dana Hill metagabbro body during vein formation. It is important to remember that the veining events are clearly cut by shear zones that record upper-amphibolite to granulite recrystallization temperatures. The strong association of scapolite growth in coronas separating amphibole from plagioclase provides textural evidence for reactions of the type

$$\text{Hornblende (high Cl/K)} + \text{plagioclase} + CO_2 = \text{Hornblende } (Na\uparrow, Al\downarrow: Ca\downarrow, Cl\downarrow) + \text{scapolite} \quad (1)$$

and

$$\text{Hornblende} + \text{plagioclase} + CO_2 + K^+ + Cl^- = \text{Hornblende } (K\uparrow, Cl\uparrow: Al\downarrow, Ca\downarrow) + \text{scapolite} \quad (2)$$

The relationship between potassium and chlorine contents in amphibole (see Fig. 6, B) suggests that both elements may have been added (reaction 2) to the Dana Hill metagabbro body. There is no observed depletion of chlorine in amphibole at scapolite-amphibole boundaries, as would be expected if scapolite production were controlled by reaction 1. Unfortunately, all chemical variability originally present at scapolite-amphibole and plagioclase-amphibole boundaries may have been homogenized during annealing. It is clear that the Dana Hill metagabbro body was infiltrated by CO_2-rich fluids (with or without K and Cl) and that this infiltration event correlates with the emplacement of the parallel hornblende vein swarm in outcrops A-4 and 87. All subsequent shearing or veining events involve scapolite growth at the expense of plagioclase in contact with amphibole. Timing of scapolite formation in the central outcrops is more difficult, because this mineral is found in small amounts in event 1 shear zones, forms well-defined rinds about the deformed and early-formed vein EA-1, and is common in all subsequent shear zones. In this outcrop, scapolite formation is best linked to the formation of vein EA, which may have belonged to events 2 or 3.

Structural data from shear zones (Table 1) show that the associated fabric was steeply dipping, with shallowly plunging transport lineations suggesting a dominant strike-slip motion, at least for the early granulite-facies events. Later shearing events (anastamosing centimeter shears) preserve a wide range of orientations, and although the majority of shear zones have nearly vertical dips, the transport lineations for these later shear zones vary from along-strike to down-dip. Geothermometric data for multiple early-formed shear zones show a narrow range of temperatures spanning 770–711 °C ± 50 °C. Early-formed shears (from outcrop data) preserve granulite-facies mineral assemblages and yield high recrystallization temperatures for a number of discrete deformational events. These data indicate that the early deformation or recrystallization events record predominately strike-oblique slip motions along the Carthage-Colton shear zone that did not radically change the structural level of the Dana Hill metagabbro body. Event 4 shear zones are smaller (centimeter-scale), record granulite to upper amphibolite conditions, and record oblique to dip-slip movements. Deformation of the Dana Hill metagabbro body came to a close with folding and brecciation in greenschist-facies conditions.

Streepey et al. (2001) report U/Pb (sphene) and $^{40}Ar/^{39}Ar$ hornblende cooling ages for the samples used in this study. U/Pb dates from sphene that is well constrained to have grown during shearing yield cooling ages (to 650 °C) of 1020 ± 1.2 Ma (Streepey et al., 2001). Using published cooling rates for the core of the Adirondack massif of 2–4 °C/Ma (Mezger et al., 1991), we estimated that granulite-facies deformation occurred between 1070 and 1033 Ma, which places these events at or near the peak of the Ottawan orogeny (McLelland et al., 2001). U/Pb (sphene) ages for the samples define a well-constrained geochron cooling age of 1020 Ma and a bimodal distribution of $^{40}Ar/^{39}Ar$ (hornblende) cooling ages of 980–1000 Ma and 941–944 Ma. The younger $^{40}Ar/^{39}Ar$ cooling ages come from two event 4 shear zones (CR-3, 87, and CR-2, A-2) that are located close to zones of brecciation in the Dana Hill metagabbro body and may have had their Ar systematics reset. Streepey et al. (2001), however, report flat release spectra for these samples, which argues against late-stage resetting. The 944–947 Ma dates may mark some late recrystallization of amphibole within these shear zones, but it is unlikely that the upper-amphibolite to granulite-facies assemblages within the shear were formed at this time, as amphibole from nearby event 1 shear zones had already cooled to 480 °C by 1005–996 Ma (CR-6, A-4, and CR-12, A-4; see Streepey et al., 2001).

Recrystallization due to deformation and/or annealing within many of these shear zones terminated by 1020 Ma, and by ca. 1000 Ma the majority of the deformed zones passively cooled through the 480 °C closure temperature for argon in hornblende (Streepey et al., 2001). Utilizing the closure temperatures for argon in hornblende (Harrison, 1981) and U/Pb in sphene, we determined that the Dana Hill metagabbro body cooled from peak Ottawan metamorphic conditions (770 °C) at 1070–1030 Ma to 650 °C by 1020 Ma and finally to 480 °C by 1000–980 Ma. Streepey et al. (2001) contend that the late 941–944 Ma $^{40}Ar/^{39}Ar$ dates mark the time when the Central Metasedimentary Belt and the Central Granulite Terrane were brought to a common structural level. Structural, field, geochronologic, and petrologic data indicate that the two terranes (the Lowlands, or Central Metasedimentary Belt, and the Highlands, or Central Granulite Terrane) were juxtaposed at, or very close to, their current positions prior to 1020 Ma. This conclusion is based on the observation that veining event(s) associated with widespread scapolite formation in the Dana Hill metagabbro body are clearly cut by the shear zones that contain upper-amphibolite to granulite-facies assemblages that record recrystallization temperatures exceeding the closure temperature for Ar in hornblende and approach, if not exceed, the closure temperature for U/Pb in sphene (see Fig. 8).

An alternative view is that <1020 Ma (the U/Pb cooling age for sphene), at least two strongly localized heating or deformation events (at 1000–980 Ma and again at 944–947 Ma) drove recrystallization temperatures for events 3 and 4 over 700 °C. It seems unlikely that, if this scenario were responsible for the formation of event 3–4 shear zones and veins, the isotope systematics of the Dana Hill metagabbro body would remain as well behaved as reported by Streepey et al. (2001).

CONCLUSIONS

Based on the structural, petrologic, and geochronologic data available, the tectonothermal history for the Dana Hill metagabbro body and, by extension, the Carthage-Colton shear zone is presented below.

1090–1050 Ma

The Carthage-Colton shear zone acted as a steeply dipping right lateral oblique to strike-slip shear during the Ottawan orogeny. Rocks now exposed at the surface record granulite-facies conditions and high shear strains. Due to the width (30+ m to km) and degree of recrystallization in these early shear zones, right lateral transport may have been, and probably was, on a kilometric scale. The Adirondack Lowlands (Frontenac or Central Metasedimentary Belt) terrane was laterally separated from its current position adjacent to the Adirondack Highlands. During the Ottawan orogeny, right lateral oblique transpressive shearing along the Carthage-Colton shear zone caused the Adrirondack Lowlands (Frontenac/Central Metasedimentary Belt) terrane to override the Highlands terrane (Fig. 11, A).

1050–1020 Ma

The Dana Hill metagabbro body experienced nearly isothermal uplift (Fig. 11, B). In the Dana Hill metagabbro body, initial uplift was marked by fluid infiltration (CO_2-rich), emplacement of amphibole and amphibole-tourmaline veins, and attending scapolite replacement of plagioclase in the Dana Hill meta-

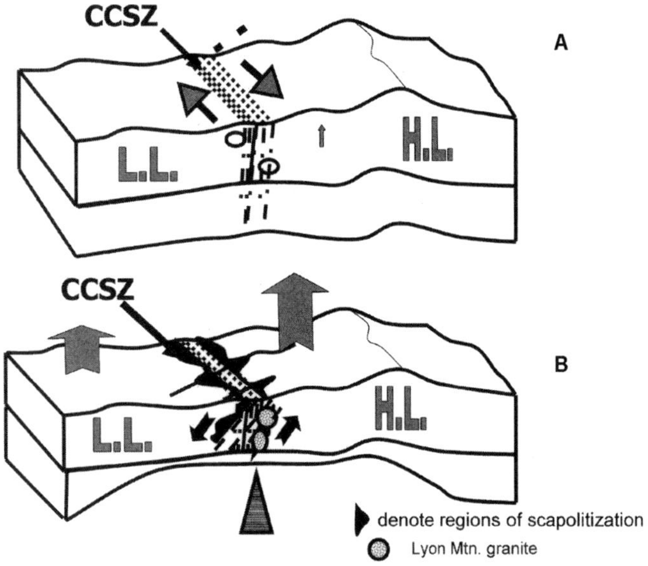

Figure 11. (A) Tectonic model for Ottawan and post-Ottawan movements along the Carthage-Colton shear zone (CCSZ) and (B) relationships between the Central Metasedimentary Belt, or the Adirondack Lowlands (LL), and the Central Granulite terrane, or the Adirondack Highlands (HL). See text for discussion.

gabbro body. The uplift is probably related to thermally driven orogenic collapse along the Carthage-Colton shear zone (McLelland et al., 2001). The nearly isothermal uplift may have been supported by heat input from the rise and emplacement of post-tectonic granite bodies, possibly including the undeformed to weakly deformed (post-tectonic) granites found to the east and south of the Dana Hill metagabbro body (Johnson and Bryan, 2002). All event 1 shears within the Dana Hill metagabbro body lock in a tightly constrained 650 °C cooling age for U/Pb in sphene (Streepey et al., 2001) and later event 4 shears record recrystallization temperatures that exceed the U/Pb closure temperature (see Fig. 8), indicating that shear zone formation had ceased by this time (i.e., no recrystallization ages). The high-temperature (700 °C) chloride metasomatism or scapolitization event reported by Tyler (1980) in rocks adjacent to the Carthage-Colton shear zone in the northern portions of the Adirondack Lowlands agrees with the style and temperatures associated with scapolite formation observed in the Dana Hill metagabbro body, suggesting that these events are related. This observation supports the contention that the Adirondack Highlands (Central Granulite Terrane) and the Adirondack Lowlands (Central Metasedimentary Belt) terranes were juxtaposed at or near to their current position at or prior to 1020 Ma. Evidence for this is provided by both the high recrystallization temperatures (700 °C) for event 1–4 shear zones and the 1000 Ma $^{40}Ar/^{39}Ar$ (hornblende) and cooling ages for shear zones that clearly postdate event 3 vein emplacement and scapolite growth in the Dana Hill metagabbro body. This interpretation agrees well with recent $^{40}Ar/^{39}Ar$ data for hornblende and biotite presented by Dahl et al. (2001), which indicate orogenic collapse along the Carthage-Colton shear zone as early as 1050 Ma. By 1000 Ma, cooling of the Dana Hill metagabbro body brought the temperature below 480 °C (closure for Ar diffusion in hornblende; Harrison, 1981). There is little evidence for shearing in the Dana Hill metagabbro body at this time. Cooling took place between 1020 and 1000 Ma at a rate of 8 °C/Ma, nearly double that proposed by Mezger et al. (1991) for the core of the Adirondacks. This cooling rate most likely represents enhanced cooling of the exhumed Dana Hill metagabbro body to restore thermal equilibrium.

1000–930 Ma

Localized late thermal or shearing events in the Dana Hill metagabbro body set or reset the $^{40}Ar/^{39}Ar$ ages for some Dana Hill metagabbro body shear zones. These deformation events may mark final adjustments between the Central Granulite Terrane and the Central Metasedimentary Belt. The nature of these late events remains an interesting and controversial aspect (see Streepey et al., 2001) requiring further study. Folding and brecciation of the Dana Hill metagabbro body at low greenschist-facies conditions may have occurred during this time.

Our data support the contention that Ottawan phase tectonics along the Carthage-Colton shear zone may have been oblique to strike-slip in nature. Recent work on the Québec (Labelle

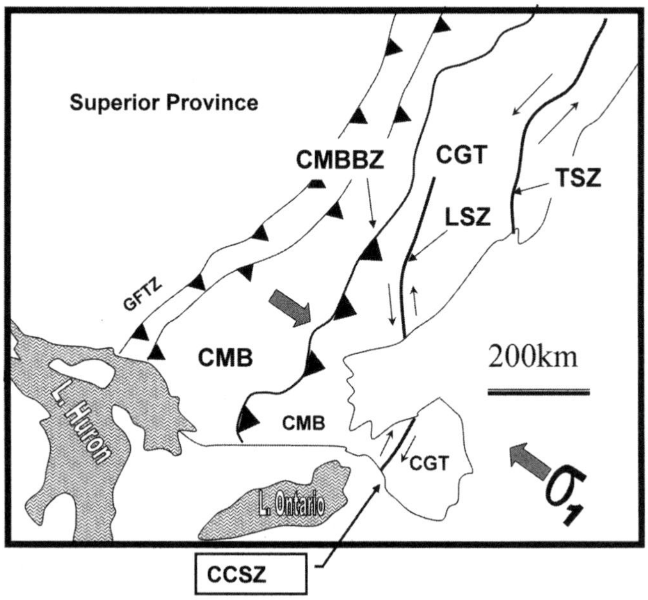

Figure 12. Tectonic map of the Grenville Province during the Ottawan orogeny. Arrows denote shortening direction. This geometry resulted in right lateral oblique shearing along the Carthage-Colton shear zone (CCSZ) while accommodating left-lateral discplacements along the Tawachiche shear zone (TSZ) and the LaBelle shear zone (LBS). CGT—Central Granulite terrane; CMB—Central Metasedimentary Belt; CMBBZ—Central Metasedimentary Belt Boundary Zone; GFTZ—Grenville Front tectonic zone. Adapted from Hanmer et al., 2000.

and Tawachiche shear zones (Martignole and Reynolds, 1997; Zhao et al., 1997; Hanmer et al., 2000) also point to early high-temperature strike slip–oblique slip displacements. Ottawan structures in the Adirondack Highlands indicate strong compression, with east-west trending nappes refolded across northwest-trending hinge lines. Northwest-directed compression across the Central Granulite Terrane (Adirondack Highlands) during the Ottawan orogeny may have been responsible for driving right lateral oblique shearing along the Carthage-Colton shear zone while accommodating left lateral oblique movements along the Tawachiche and Labelle Shear zones (Martignole and Reynolds, 1997; Zhao et al., 1997; Hanmer et al., 2000) to the north in Québec (Fig. 12). In this model, the Carthage-Colton shear zone could represent either an inter- or an intraplate boundary in the greater Grenville Province. Displacements must have been of a great enough magnitude to allow the Central Metasedimentary Belt to escape an Ottawan overprint. Late collapse of the Orogen resulted in reactivation of the Carthage-Colton shear zone as a ductile normal fault bringing rocks of the Central Metasedimentary Belt and Central Granulite Terrane to a common structural level by or prior to 1020 Ma.

ACKNOWLEDGMENTS

The authors thank Dr. David Matty for his critical eye and thoughtful comments and Mr. Ronald White for his "Adirondack" hospitality in providing cold drinks and access to his land. We also thank Mr. William Blackburn for technical advice on all topics of microanalyses, and the Clark family for the use of their Trout Lake cabin. Heartfelt thanks also to Drs. J. Martignole, Michael Easton, and Timothy Horscroft for insightful reviews of an earlier draft of this manuscript and to Drs. M. Streepey and G. Baird for challenging and thoughtful reviews. Finally, ELJ thanks the late Dr. Yngvar Isachsen for being a good friend and for introducing him to the geology of the Adirondack Mountains some fifteen years ago. His gentle presence is greatly missed.

REFERENCES CITED

Bohlen, S.R., Valley, J.R., and Essene, E.J., 1985, Metamorphism in the Adirondacks, I: pressure and temperature: Journal of Petrology, v. 26, p. 971–992.

Buddington, A.F., 1939, Adirondack igneous rocks and their metamorphism: Boulder, Colorado, Geological Society of America Memoir 7, 354 p.

Carr, S.D., Easton, R.M., Jamieson, R.A., and Culshaw, N.G., 2000, Geologic transect across the Grenville orogen of Ontario and New York: Canadian Journal of Earth Science, v. 37, p. 193–213.

Dahl, P., Foland, K., and Pomfrey, M., 2001, $^{40}Ar/^{39}Ar$ thermochronology of hornblende and biotite, Adirondack Lowlands (New York), with implications for evolution of a major shear zone: Geological Society of America Abstracts with Programs, v. 33, no. 6, p. A292.

Davidson, A., 1986, New Interpretations in the southwestern Grenville Province, Ontario, *in* Moore, J.M., et al., eds., The Grenville Province: Ottawa, Ontario, Geological Society of Canada Special Paper 31, p. 61–74.

Davis, M.E., 1981, Petrology and geochemistry of a portion of the Carthage-Colton Mylonite Zone, South Edwards 7.5 minute Quadrangle, NW Adirondacks, New York [M.A. thesis]: State University of New York at Binghamton, 124 p.

Geraghty, E.P., Isachsen, Y., and Wright, S.F., 1981, Extent and character of the Carthage-Colton Mylonite Zone, Northwest Adirondacks, New York: Albany, New York State Geological Survey Report to the U.S. Nuclear Regulatory Commission, 83 p.

Hall, P.C., 1984, Some aspects of deformation fabrics along the highland/lowland boundary, Northeast Adirondacks, New York State [M.S. thesis]: Albany, State University of New York, 86 p.

Hanmer, S., Corrigan, D., Pehrsson, S., and Nadeau, L., 2000, SW Grenville Province, Canada: The case against post-1.4 Ga accretionary tectonics: Tectonophysics, v. 319, p. 33–51.

Harrison, T., 1981 Diffusion of ^{40}Ar in hornblende: Contributions to Mineralogy and Petrology, v. 78, p. 324–331.

Heyn, T., 1990, Tectonites of the northwest Adirondack Mountains, New York: Structural and metamorphic evolution [Ph.D. thesis]: Ithaca, New York, Cornell University, 216 p.

Holland, T., and Blundy, J., 1994, Non-ideal interactions in calcic amphiboles and their bearing on amphibole-plagioclase thermometry: Contributions to Mineralogy and Petrology, v. 116, p. 443–447.

Indares, A., and Martignole, J., 1990, Metamorphic constraints on the tectonic evolution of the allochthonous monocyclic belt of the Grenville Province, western Québec: Canadian Journal of Earth Sciences, v. 27, p. 371–386.

Isachsen, Y., and Geraghty, E.P., 1986, The Carathage-Colton Mylonite Zone: A major ductile fault in the Grenville Province: International Basement Tectonics Association Proceedings, v. 6, p. 199–200.

Johnson, E.L., and Bryan, K., 2002, Sillimanite + quartz replacement of K-feldspar in late to post Ottawan granite bodies in and adjacent to the Carthage Colton Shear Zone, NW Adirondack Mtns, New York State: A case of fluid driven auto-metamorphism?: Geological Society of America Abstracts with Programs, v. 34 , no. 1, p. A6.

Johnson, E.L., Matty, D., and Robinson, C., 1994, *P-T* conditions of multiple deformation/metasomatic events in the Dana Hill Metagabbro, NW Adirondack Mts. New York: Geological Society of America Abstracts with Programs, v. 26, no. 6, p. A26.

Laird, J., and Albee, A.L., 1981, Pressure, temperature, and time indicators in mafic schist: Their application to reconstructing the polymetamorphic history of Vermont: American Journal of Science, v. 281, p. 127–175.

Lamb, W.M., 1993, Retrograde deformation within the Carthage-Colton mylonite zone as recorded by fluid inclusions and feldspar compositions: Tectonic implications for the southern Grenville Province: Contributions to Mineralogy and Petrology, v. 14, p. 379–394.

Martignole, J., and Reynolds, P., 1997, $^{40}Ar/^{39}Ar$ thermochronology along a western Québec transect of the Grenville Province, Canada: Canadian Journal of Earth Science, v. 15, p. 283–296.

McLelland, J.M., and Chiarenzelli, J., 1990, Isotopic constraints on emplacement age of anorthositic rocks of the Marcy massif, Adirondack Mountains, New York: Journal of Geology, v. 98, p. 19–41.

McLelland, J.M., Daly, J.S., and Chiarenzelli, J., 1993a, Sm-Nd and U-Pb isotopic evidence of juvenile crust in the Adirondack Lowlands and implications for the evolution of the Adirondack Mountains: Journal of Geology, v. 101, p. 97–105.

McLelland, J.M., Isachsen, Y., Whitney, P., Chiarenzelli, J., and Hall L., 1993b, Geology of the Adirondack Massif, New York, *in* Rankin, D., ed., Precambrian conterminous, U.S.: Boulder, Colorado, Geological Society of America, Geology of North America, v. C-2, p. 338–353.

McLelland, J.M., McLelland, J., Walker, D., Orrell, S., Hamilton, M., and Selleck, B., 2001, Zircon U-Pb geochronology of the Ottawan Orogeny, Adirondack Highlands, New York: Regional and tectonic implications: Precambrian Research, v. 109, no. 1–2, p. 39–72.

McLelland, J.M., Olson, C., Orrell, S., Goldstein, A, and Cunningham, B., 2002, Structural evolution of a quartz sillimanite vein and nodule complex in a late to post tectonic leucogranite, Western Adirondack Highlands, NY: Journal of Structural Geology, v. 24, no. 6–7, p. 1157–1170.

Mezger, K., Rawnsley, C.M., Bohlen, S.R., and Hanson, G.N., 1991, U-Pb garnet, sphene, monazite, and rutile ages: Implications for the duration of high-grade metamorphism and cooling histories, Adirondack Mts. New York: Journal of Geology, v. 99, p. 415–428.

Mezger, K., van der Pluijm, B.A., Essene, E.J., and Halliday, A.N., 1992, The Carthage Colton Mylonite Zone (Adirondack Mountains, New York): The site of a cryptic suture in the Grenville Orogen? Journal of Geology, v. 100, p. 630–638.

Mezger, K., Essene, E.J, van der Pluijm, B.A., and Halliday, A.N.,1993, U-Pb geochronology of the Grenville Orogen of Ontario and New York: Constraints on ancient crustal tectonics: Contributions to Mineralogy and Petrology, v. 114, p. 13–26.

Parodi, M.R., 1979, Petrology, structure, and geochemistry of the Dana Hill Metagabbro, Russell, New York [M.A. thesis]: State University of New York at Binghamton, 120 p.

Rebbert, C., and Rice, J., 1997, Scapolite-plagioclase exchange: Cl-CO_3 scapolite solution chemistry and implications for peristerite plagioclase: Geochimica et Cosmochimica Acta, v. 61, no. 3, p. 555–567.

Rivers, T., Martignole, J., Grower, C., and Davidson, A., 1989, New tectonic divisions of the Grenville Province: Tectonics, v. 8, p. 63–84.

Streepey, M., Johnson, E.L., Mezger, K., van der Pluijm, B., and Essene, E.J., 2001, Early history of the Carthage-Colton shear zone, Grenville Province, Northwest Adirondacks, New York (U.S.A.): Journal of Geology v. 109, p. 479–492.

Tyler, R.D., 1980, Chloride metasomatism in the southern part of the Pierrepont Quadrangle, Adirondack Mountains, New York [Ph.D. thesis]: State University of New York at Binghamton, 527 p.

Wiener, R.W., 1983, Adirondack Highlands-Lowlands "boundary": A multiply folded intrusive contact with fold-associated mylonitization: Geological Society of America Bulletin, v. 94, p. 1081–1108.

Wynne-Edwards, H.R., 1972, The Grenville Province, *in* Price, R.A., and Douglas, R.J.W., eds., Variations in tectonic styles in Canada: St. John's, Newfoundland, Geological Association of Canada Special Paper 11, p. 263–334.

Zhao, X., Ji, S., and Martignole, J., 1997, Quartz microstructures and c-axis preferred orientations in high-grade gneisses and mylonites around the Morin anorthosite (Grenville Province): Canadian Journal of Earth Sciences, v. 34, p. 819–832.

Manuscript Accepted by the Society August 25, 2003

Printed in the USA

Geological Society of America
Memoir 197
2004

Magmatic-hydrothermal leaching and origin of late to post-tectonic quartz-rich rocks, Adirondack Highlands, New York

Bruce Selleck*
James McLelland
Colgate University, Hamilton, New York 13346, USA
Michael A. Hamilton*
Geological Survey of Canada, 601 Booth Street, Ottawa, Ontario K1A 0E8, Canada

ABSTRACT

Magmatic-hydrothermal processes produced significant metasomatic alteration of country rock, with resultant metal transport and deposition of low-Ti magnetite bodies, during the intrusion of late granitic magmas in the Adirondack Highlands of the Grenville Province. Manifestations of these ore-forming systems occur as sillimanite-bearing quartz-rich rocks within ca. 1040 Ma Lyon Mountain leucogranites in the southwestern Adirondack Highlands. Field and petrographic relationships demonstrate that emplacement of quartz-sillimanite and bull quartz veins into granite was accompanied by leaching of feldspar and other labile minerals from the granite to produce aluminum- and silica-rich residua. These relationships, coupled with fluid inclusion and stable isotope data from quartz in the host granite, in quartz-sillimanite veins, and in massive quartz veins, suggest that crystallization of granite released acidic magmatic fluids that resulted in high-temperature hydrolytic leaching of feldspars. These fluids also transported and redeposited silica and, locally, alumina to form a quartz-rich sillimanite-bearing carapace near the margins of the granite pluton. Later low-temperature hydrothermal processes mobilized silica and altered sillimanite to illite and diaspore. Zircons exhibiting U-Pb ages overlapping those of the host granite are abundant in the leached quartz-rich rock and are interpreted as residual grains incorporated from the granite. The spatial relationship of both high-temperature and low-temperature hydrothermal features relative to the intrusive Lyon Mountain leucogranite strongly suggests that emplacement of granitic magmas drove the hydrothermal system during progressive unroofing and cooling of the Adirondack orogen.

Keywords: Adirondack, hydrothermal granite

*E-mail, Selleck: bselleck@mail.colgate.edu. Current address, Hamilton: Jack Satterly Geochronology Laboratory, Department of Geology, University of Toronto, Toronto, Ontario M5S 3B1, Canada.

Selleck, B., McLelland, J., and Hamilton, M.A., 2004, Magmatic-hydrothermal leaching and origin of late to post-tectonic quartz-rich rocks, Adirondack Highlands, New York, *in* Tollo, R.P., Corriveau, L., McLelland, J., and Bartholomew, M.J., eds., Proterozoic tectonic evolution of the Grenville orogen in North America: Boulder, Colorado, Geological Society of America Memoir 197, p. 379–390. For permission to copy, contact editing@geosociety.org.

INTRODUCTION

High-temperature hydrothermal processes have long been recognized within the Adirondack Highlands of the Grenville Province. In particular, the formation of major low-Ti Kiruna-type Fe-oxide deposits (Alling, 1939; Postel, 1952; Buddington and Leonard, 1962; Leonard and Buddington, 1964; Foose and McLelland, 1995) is associated with leucogranite of the late- to post-tectonic Lyon Mountain granite (Fig. 1). In this paper we describe the development of marginal facies of the Lyon Mountain granite that document quartz and quarz-sillimanite vein emplacement and leaching of host granite that occurred during and after granite intrusion. Using U-Pb zircon geochronology we establish the age relationship between intrusion of the granite and development of its quartz-rich carapace. These distinctive quartz-rich facies represent potential source zones for ore-forming fluids related to the formation of Lyon Mountain granite iron ores, and recognition of these facies in other regions may be important in the exploration and modeling of similar ore-forming systems. Exposures of Lyon Mountain leucogranite near Port Leyden, New York (Fig. 1), display a range of metasomatic or hydrothermal features, including the quartz-sillimanite veins described by McLelland et al. (2002a). The quartz-rich rocks that are the subject of this study occur adjacent to these Lyon Mountain leucogranite exposures.

PREVIOUS STUDIES

The Lyon Mountain granite comprises a distinctive suite of relatively undeformed granites and associated orthogneiss emplaced across wide tracts of the Adirondack Highlands during the waning stages of the ca. 1090–1030 Ma Ottawan orogeny (McLelland et al., 2001). The fluids involved in the Lyon Mountain granite hydrothermal systems may include both magmatic and surface-derived hydrothermal components (McLelland et al., 2002a, 2002b). Fluid inclusion data suggest Na-Cl fluid systems similar to those described by Battles and Barton (1995) and by Barton and Johnson (1996) for Fe-Cu hydrothermal ore deposits in the southwestern United States. In the Lyon Mountain granite, leaching of country rock and early-crystallized igneous rock by circulating fluids resulted in the dissolution of metals and local transport and redeposition of magnetite within host granites and metasedimentary country rock. The leaching of leucogranite by acidic fluids removed cations, especially soda and potash, from feldspars and formed veins of quartz-sillimanite that were disrupted by later magmatic flow (McLelland et al., 2002a, 2002b). The low-Ti magnetite deposits of the Adirondack Highlands are commonly associated with the quartz-albite facies of the Lyon Mountain granite and are interpreted as having resulted from metasomatism by the same Na- and Fe-rich hydrothermal fluids. The Lyon Mountain granite hydrothermal systems operated during and after the intrusion of Lyon Mountain granite plutons at temperatures near 700 °C with fluids derived from the crystallizing magma or from evolved meteoric brines of regional scale. The emplacement age of the Lyon Mountain granite is well constrained by U-Pb zircon dating, and the early stages of hydrothermal activity were coeval with intrusion of Lyon Mountain granites and pegmatite at ca. 1060–1030 Ma (McLelland et al., 2001, 2002a, 2002b).

Postpeak metamorphic infiltration of fluids into the granulite-facies rocks of the Adirondack Highlands has been documented (e.g., Morrison and Valley, 1988; Whitney and Davin, 1987), but the timing of these events is unclear. Some fluid-related alteration is undoubtedly of Paleozoic age, related to flushing of basinal fluids from Paleozoic sediment cover downward, perhaps by means of tectonically driven processes (Whitney and Davin, 1987) in the Taconic or the Acadian event. However, unroofing of the orogen in the Late Proterozoic could have allowed infiltration of surface-derived fluids as soon as pressure-temperature conditions were permissive of brittle behavior and fracture-related fluid flow. Given appropriate regional hydrologic conditions, long-lived hydrothermal systems may have operated in the vicinity of late-stage plutonic complexes like the Lyon Mountain granite. At Lyonsdale, late illite-diaspore-hematite veins document low-temperature hydrothermal alteration in the Lyon Mountain granite. The latest-stage fluids were oxidizing and had relatively high salinity. These low-temperature assemblages are discussed later along with the high-temperature processes cited earlier to provide documentation of the range of field and petrologic features that record magmatic-hydrothermal leaching systems that may be linked to the formation of important ore bodies in the Adirondacks and elsewhere.

GEOLOGIC SETTING OF THE QUARTZ-RICH FACIES

The quartz-rich rocks of this investigation are located near the southwestern margin of the Adirondack Highlands (Fig. 1). The Adirondacks constitute a domical outlier of the Grenville Province that has experienced multiple high-grade metamorphic and intrusive events. The metaigneous and minor metasedimentary rocks that characterize the Adirondack Highlands include early granodioritic and tonalitic rocks (ca. 1350–1300 Ma) that were deformed and metamorphosed at ca. 1170 Ma (Wasteneys et al., 1999). Voluminous magmas of the anorthosite-charnockite-mangerite-granite (AMCG) suite were intruded at ca. 1150 Ma following delamination of the overthickened Elzevirian orogen. After a hiatus of ~60 m.y. the region was intruded by the ca. 1100 Ma Hawkeye granite suite (McLelland et al., 1996). Subsequently, all of the Highlands were subjected to vapor-absent granulite-facies metamorphism and multiple deformations associated with the ca. 1090–1030 Ma Ottawan orogeny (McLelland et al., 2001). Late- to post-tectonic leucogranites of the Lyon Mountain granite (ca. 1060–1030 Ma) are interpreted as related to delamination, rapid uplift, and extensional collapse of the overthickened Grenville orogen (McLelland et al., 2001).

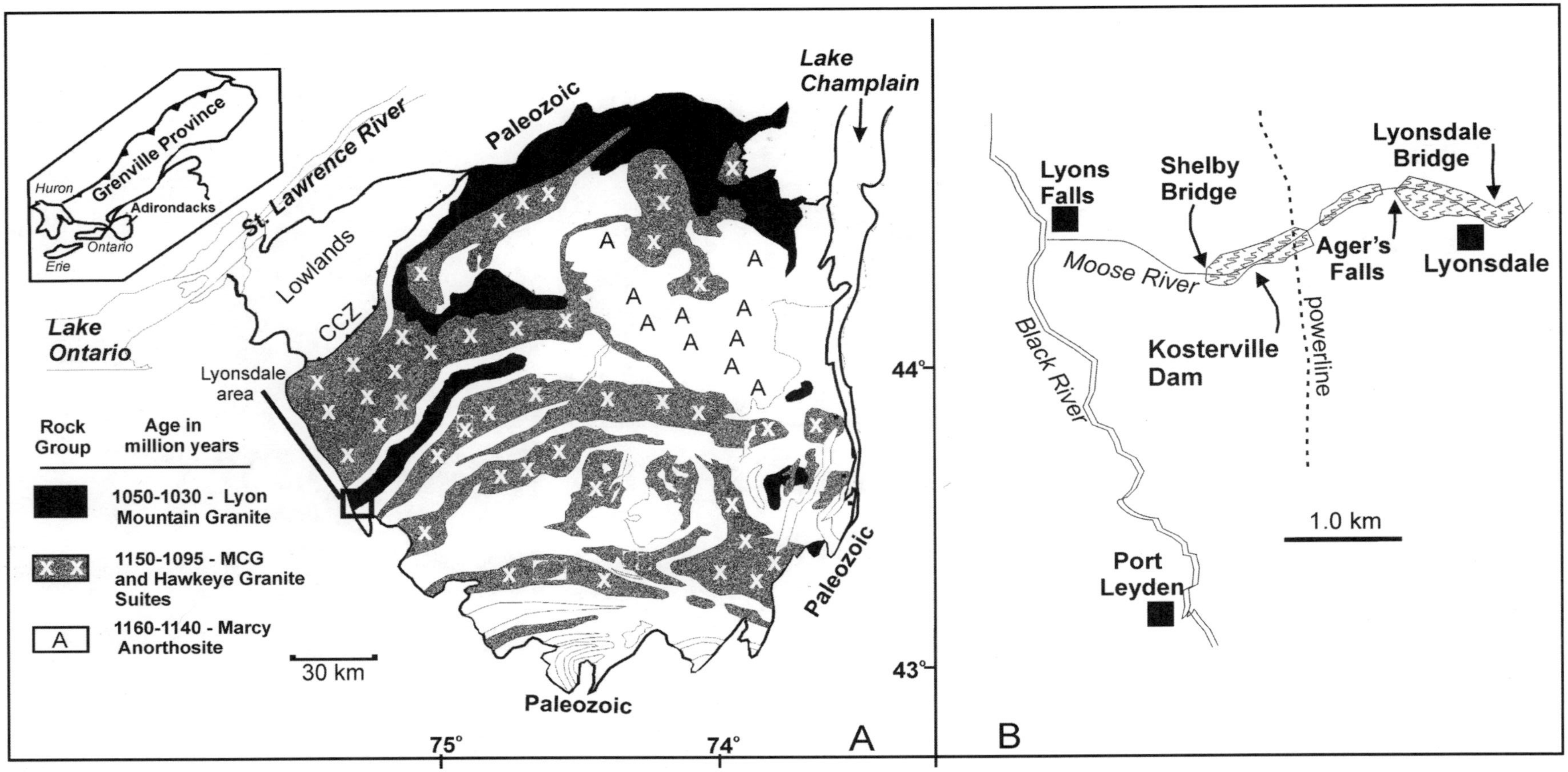

Figure 1. (A) Generalized geological map of the Adirondack Highlands of New York state showing distribution of Lyon Mountain granite and older igneous rocks. CCZ—Carthage-Colton zone separating the Adirondack Highlands from the Adirondack Lowlands. MCG—mangerite-charnockite-granite. Inset shows location of the Adirondacks relative to the Grenville Province. (B) Sketch map of Lyonsdale, New York, and the vicinity showing key localities mentioned in the text.

In the southwestern Adirondack Highlands near Port Leyden, New York, leucogranites of the Lyon Mountain Gneiss (Fig. 1) are found in intrusive contact with older metasedimentary and metaigneous rocks. Dikes of leucogranite crosscut foliated metapelites, and Lyon Mountain granite commonly contains xenoliths of calc-silicate country rock. Large tracts of the leucogranite mass exposed in the bed of the Moose River at Lyonsdale contain quartz-sillimanite segregations. The suite of quartz-rich rocks, exposed on the western margin of the leucogranite mass, are described below.

Quartz-Rich Rock and Related Lithologies near Port Leyden

The quartz-rich facies vary progressively from leucogranite with a variable concentration of narrow veins to vein-rich granite and finally to massive sillimanite-bearing quartz-rich rock containing only a few small, scattered occurrences of leucogranite. The quartz-sillimanite vein–bearing leucogranite facies exposed between Lyonsdale Bridge and Ager's Falls (Fig. 1) consists of sharp-walled, 2–10 cm–thick veins of coarsely crystalline quartz and sillimanite hosted by equigranular granite (Fig. 2, A). These veins contain coarsely crystalline sillimanite as well as minor magnetite, with both concentrated toward the vein centers. The relative timing of the quartz-sillimanite veins is established by the observation that they crosscut country rock calc-silicate schlieren, flow-related layering, and earlier quartz-sillimanite veins within the host leucogranite (McLelland et al., 2002a, 2002b). Within the veins, clusters of tabular sillimanite crystals up to 5 cm in length are aligned parallel to the vein walls. Veins bearing quartz and sillimanite may grade along strike into veins of nearly pure quartz. Sillimanite-bearing veins were formed within the granite by deposition from high-temperature hydrothermal fluids of probable magmatic origin (McLel-

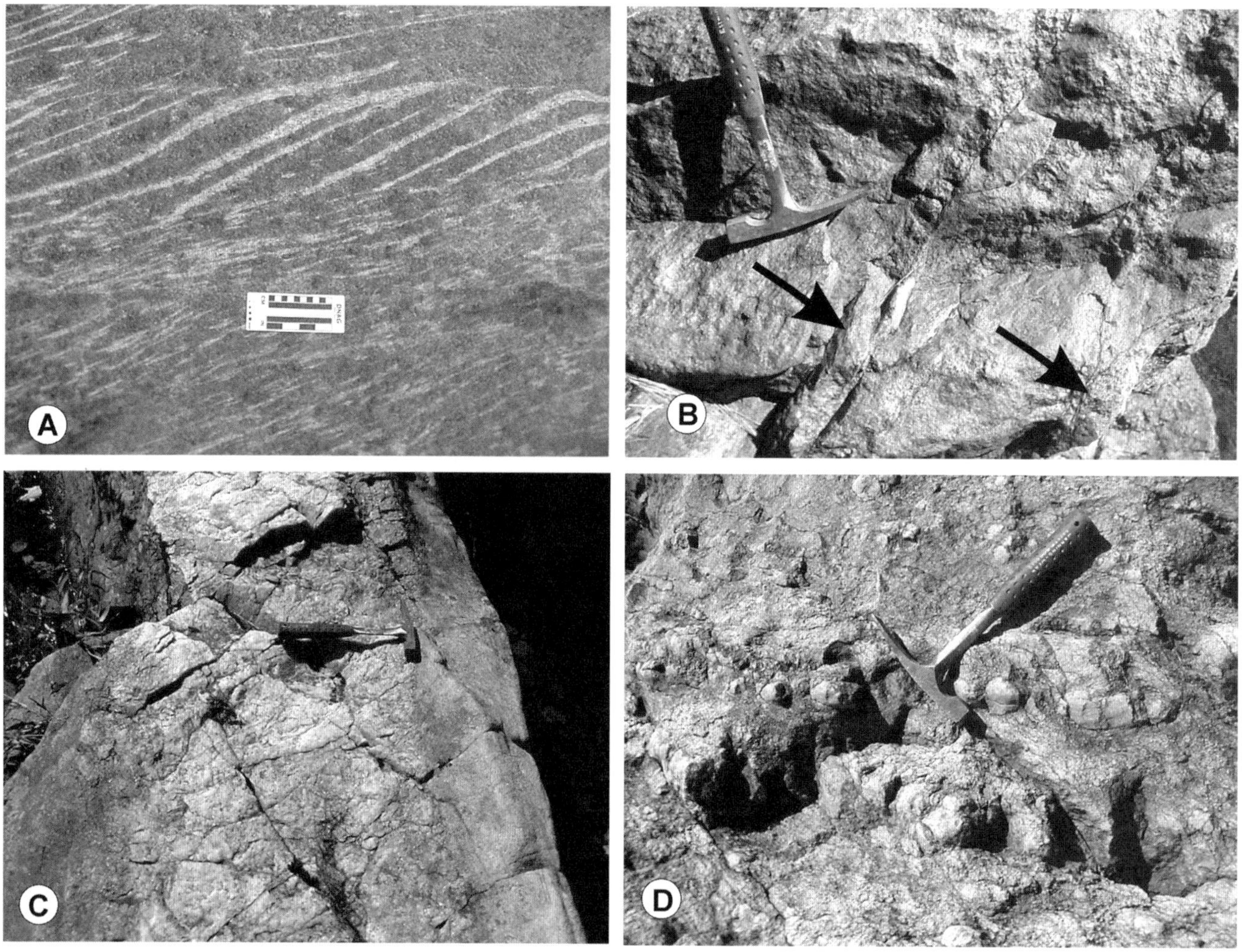

Figure 2. (A) Quartz-sillimanite veins in equigranular leucogranite at Lyonsdale Bridge. Upper part of scale in cm. (B) Centimeter-scale quartz-sillimanite-magnetite veins in weakly foliated quartz-sillimanite bearing nodular granite at Shelby Bridge. Arrows point to quartz-sillimanite-magnetite veins. Hammer is 35 cm in length. (C) Coarsely crystalline "bull quartz" vein in leucogranite of the Lyon Mountain gneiss at the power line locality near Lyonsdale. Hammer head rests on contact. Hammer is 35 cm in length. (D) Nodular quartzite at Shelby Bridge. Lumps resembling sedimentary clasts are boudin of coarsely crystalline quartz in granular, tectonized quartzite. Hammer is 35 cm in length.

land et al., 2002b). These fluids were capable of transporting dissolved silica, and apparently alumina, since the large clusters of sillimanite crystals appear to have grown within the veins via precipitation from fluid and thus represent at least local mobilization of alumina.

Coherent quartz-sillimanite veins are often found within granite that contains isolated, dismembered layers, lenses, knots, and boudin of quartz-sillimanite segregations formed relatively early in Lyon Mountain granite magmatic history. These earlier segregations were then disrupted and deformed by the later flow of the still partly molten magma. If these earlier segregations were formed as sharp-walled veins similar to the latest undisrupted veins, the enclosing granite must have crystallized sufficiently to permit brittle fracture. Alternatively, the highly viscous, partially crystallized leucogranite magma may have fractured due to high local strain rates and elevated fluid pressure (McLelland et al., 2002a).

Thicker decimeter-scale quasi-continuous veins of quartz-sillimanite rock that separate isolated masses of quartz-sillimanite nodule bearing leucogranite occur to the west between Ager's Falls and Kosterville Dam (Fig. 2, B). Magnetite is a persistent accessory in the veins. The sharp contacts between veins of quartz-sillimanite rock and nodular granite suggest that this facies represents successive and continued emplacement of quartz-sillimanite veins into host granite. Exposures of this lithology are commonly striped with centimeter-scale red and green alteration zones related to extensive later low-temperature processes. Wider "bull" quartz veins 50–100 cm thick are also present in these exposures and may grade along strike into pegmatitic leucogranite from nearly pure quartz to typical granite pegmatite over distances of 5–10 m.

Farther to the west at Kosterville Dam (Fig. 1), massive sillimanite-bearing quartz-rich rock, together with minor centimeter-scale quartz-sillimanite veins, occurs as a weakly layered, coarsely crystalline facies with disseminated low-temperature alteration zones. At Shelby Bridge, the later mechanical deformation of nearly pure quartz layers in this facies produced boulder-size, rounded boudins of coarsely crystalline quartz in finer, grain-size reduced granular quartz, forming outcrops that superficially resemble sedimentary conglomerates, but clearly are not (Fig. 2, D). Overall, this east-to-west sequence of quartz-rich rocks is interpreted as representing the marginal zone, or carapace, of a Lyon Mountain granite intrusive body.

Later Alteration Features

Centimeter- to millimeter-scale alteration zones are common in both the quartz-sillimanite veins and the massive quartz-rich facies (hereafter called quartzite) associated with the Lyon Mountain granite in the Lyonsdale area. In outcrop, the alteration zones are characterized by anastamosing networks of veinlets 1–2 mm in thickness, tapering to healed fractures hosting illite and diaspore. These veins have a halo of red and/or green coloration related to minor disseminated hematite and/or chlorite. Hematite spherules are found within healed fractures in quartz. Larger segregations of illite-diaspore are also present, forming irregularly shaped masses of salmon- and green-colored, porcellaneous rock up to 20 cm in overall dimension. Thin sections suggest that the illite-diaspore material forms pseudomorphs after sillimanite in the quartzite (Fig. 3, B and C). In addition, the illite-diaspore rock often contains concentrations of euhedral to subhedral, elongate zircon crystals that strongly resemble zircon grains in typical Lyon Mountain leucogranite (Fig. 3, D).

The illite-diaspore veinlets and segregations within sillimanite-bearing granite and quartz-sillimanite rock are interpreted as low-temperature (<200 °C) features related to a later phase of hydrothermal alteration. This phase included local redistribution of alumina and silica as fluids permeated the fracture network and dissolved and redeposited the low-temperature mineral phases. Dissolution of magnetite also occurred related to the formation of hematite and chlorite. Fluid inclusion data suggest that fluid salinities were high (>20% NaCl equivalent) and oxidizing, based on the common occurrence of minor hematite in the veinlets and as spherules within fluid inclusion trains associated with fractures that continue from veinlets. The low-temperature event may have further concentrated zircon grains as labile minerals were removed, leaving only ultrastable zircon, illite, and diaspore. The zircon-rich illite-diaspore material represents a hydrothermal leachate, resembling the leached, kaolinitic jasperite materials that form residua in the near-surface portions of hydrothermal vents in active volcanic systems (Heald et al., 1987).

Quartz-Calcite Veins near Mckeever

Approximately 12 km east of the Lyonsdale area, exposures of calc-silicate rock and metasedimentary quartzite host decimeter-scale block-structure veins bearing coarse, well-terminated quartz crystals and calcite spar with spherulitic hematite rinds that postdate the quartz and calcite (Darling and Bassett, 2001). Aluminosilicate minerals such as phlogophite and plagioclase in the host rock adjacent to the veins are altered to intergrown chlorite, illite, and diaspore. The relationships of these veins with the alteration features observed in the Lyonsdale exposures are unclear, but suggest that late-stage hydrothermal processes were not limited to the areas of Lyon Mountain leucogranite.

ZIRCON GEOCHRONOLOGY

Single-grain thermal ionization mass spectrometry (TIMS) dating of the nodular leucogranite at Lyonsdale was reported by McLelland et al. (2002a), who obtained an age of 1035 ± 4 Ma that is interpreted as the age of igneous emplacement. In addition, McLelland et al. (2001) obtained a U-Pb zircon TIMS age of 1034 ± 10 Ma for an undeformed pegmatite that crosscuts the leucogranite and quartz-sillimanite veins and the nodules within it, thus constraining granite emplacement and vein or nodule formation to the same brief time interval. These out-

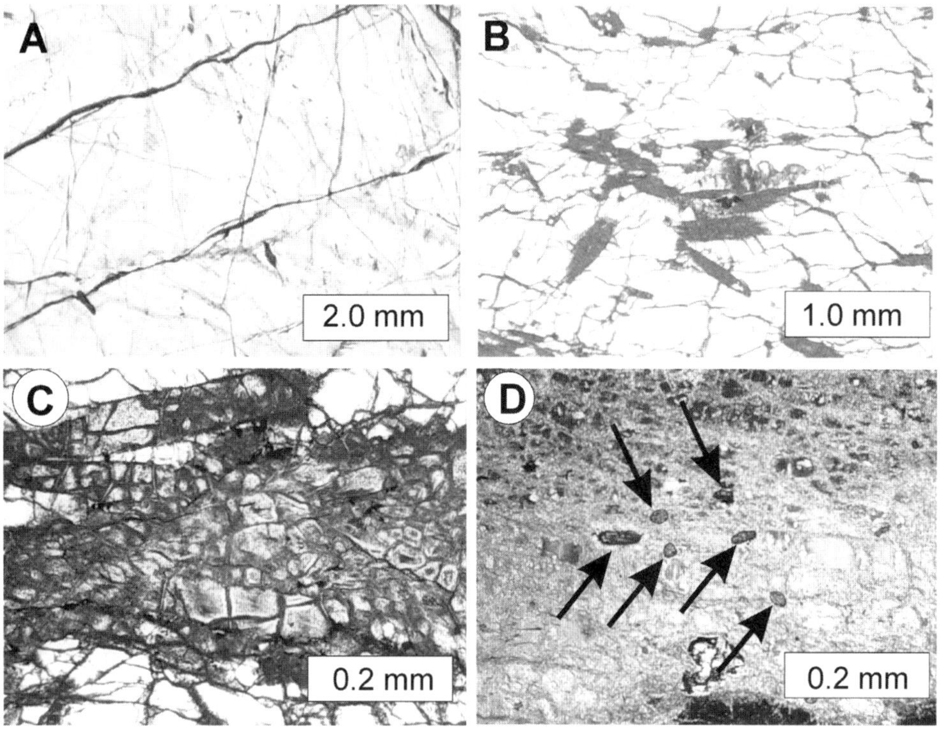

Figure 3. (A) Coarsely crystalline quartz from vein at the Kosterville Dam locality. Note abundant crosscutting sets of fluid inclusions and late fractures with sericite. (B) Illite-diaspore rock from Kosterville Dam. Note apparent pseudomorphs of illite-diaspore after sillimanite crystals. (C) Illite-diaspore segregation in quartzite, Kosterville Dam. (D) Zircons (indicated by arrows) in illite-diaspore segregation at Kosterville Dam. Zircon separates were prepared from this material for sensitive high-resolution ion microprobe (SHRIMP) analyses.

crops are within 0.5 km, along strike, of the quartz-rich rocks discussed here.

Zircons within the diaspore concentrations of the quartz-sillimanite veins and massive quartzite subunits are similar to those in the nearby leucogranite: they are 100–400 microns in length, with length-to-width ratios averaging 3:1, and consist of cores surrounded by relatively thick rims or mantles (Fig. 4). Almost all grains are characterized by a distinctive mottling that has been noted in other Adirondack zircons exposed to large quantities of hydrothermal fluids (McLelland et al., 2001). Due to the presence of inherited cores, the zircons were analyzed in situ by means of ion probe techniques at the Geological Survey of Canada's Sensitive High-Resolution Ion Microprobe (SHRIMP) II Laboratory in Ottawa. Analytical methodology generally followed the protocols described in Hamilton et al. (this volume). Data from these analyses, as well as further instrumental details, are presented in Table 1. Zircons in the diaspore concentrations are texturally characterized by mantles with elevated uranium concentrations ranging from ~1100 to 1800 ppm, and Th/U ratios falling narrowly between 0.1 and 0.2. These characteristics are similar to those described for analogous zircons in the host Lyon Mountain granite and associated pegmatite (McLelland et al., 2001, 2002a). Inherited cores are commonly present in the diaspore zircons and have U concentrations that range from ~500 to 4700 ppm; the most concordant analyses typically range from 500 to 1000 ppm U, while the most discordant cores have ~2380–4700 ppm. As expected, Th/U ratios in the xenocrystic cores span a diverse range, mostly between ~0.3 and 1.4, but extend as low as 0.06 (Table 1).

A concordia plot of all SHRIMP data shows a distinct clustering of mantling zircon near 1050 Ma, together with some strongly discordant points as well as a scattering of older core ages between 1160 and 1230 Ma (Fig. 5, A). A concordia plot of U-Pb data from the zircon mantles is shown in Figure 5, B. A weighted mean calculation of all eight analyses of texturally unambiguous zircon mantles (unshaded ellipses in Fig. 5) yields an age of 1041 ± 8 Ma (mean square of weighted deviates = 0.39). Notably, one broken prismatic and doubly terminated grain (Fig., 4C) gives the same age of ca. 1040–1055 Ma for both core and mantling zircon. Both cores and mantles exhibit faint traces of zoning (Fig. 4, A and B).

Interpretation of the zircon ages from these quartz-rich facies is not straightforward; however, reasonable alternatives can be ascertained. Clearly the sillimanite-bearing veins cannot have a sedimentary origin, since they crosscut the host leucogranite as well as one another. This leaves only the possibility that the veins are of hydrothermal origin. Moreover, they must be younger than the leucogranite (1035 ± 4 Ma) and older than the pegmatite (1034 ± 10 Ma) that crosscuts them (McLelland et al., 2001, 2002a). Note that the age of the leucogranite is within error of the weighted average of the 1041 ± 7 Ma zircons from the illite-diaspore segregations. It is possible that the zircons within the quartz-rich facies are hydrothermal, but rigorous published studies of hydrothermal zircons are scarce, and much of what appears in the literature has been debated (Claoué-Long et al., 1990, 1992; Corfu and Davis, 1991; Kerrich and King, 1993; Nesbitt et al., 1999). In cases where a hydrothermal origin appears certain, the zircons tend to be small and irregular in shape (Rubin et al., 1989). In contrast, the zircons under consideration here exhibit morphologies that are consistent with an

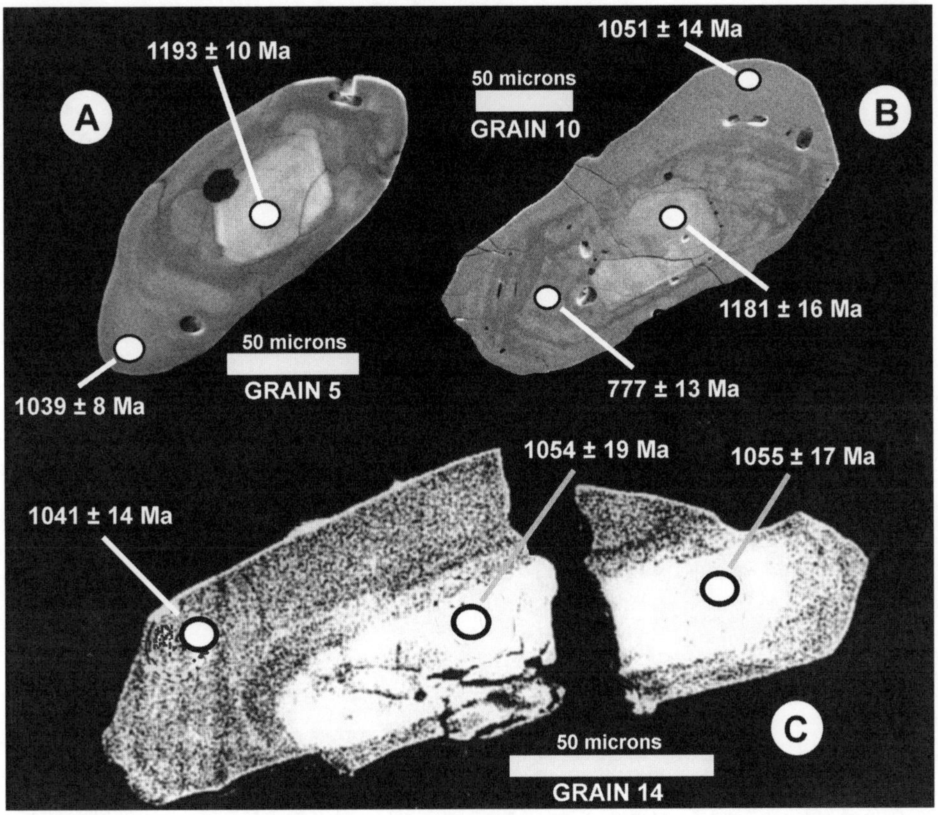

Figure 4. Cathodoluminescence images of zircons from illite-diaspore segregations in quartz-sillimanite vein complex. (A) Grain with inherited core of Elzeverian-age (ca. 1190 Ma) zircon surrounded by mantle of Lyon Mountain granite-age (ca. 1040 Ma) zircon. Very weak zoning in both core and mantle indicates a likely magmatic origin for both. (B) Large grain with inherited anorthosite-mangerite-charnockite-granite or Elzeverian-age core and thick mantle of Lyon Mountain granite age. The anomalous 777 Ma age occurs in a mottled, highly altered portion of the grain. (C) Broken grain with weakly zoned mantle concentric with the euhedral core. Both core and mantle yield Lyon Mountain granite ages.

igneous origin and are identical in morphology, uranium content, and, within error, U-Pb age to zircons within the host leucogranite (McLelland et al., 2002a), and we propose that this was their source. There are two mechanisms by which zircons could have been transferred from the leucogranite to the quartz-rich facies. The first is by plucking of zircons from the granite wallrock of the veins. The second, and preferred, mechanism is by high-temperature (500–650 °C, McLelland et al., 2002b) hydrothermal leaching of alkalis from leucogranite feldspars, leaving behind quartz, sillimanite, and zircon. This mechanism does not rule out some degree of transport for these "lag" deposit constituents, nor does it require all quartz veins in the region to have resulted from alkali leaching. In fact, many of the mutually crosscutting, sharply defined quartz veins are relatively poor in sillimanite and zircon and probably represent redeposited silica. The exceptionally high zircon concentrations in the illite-diaspore rock is, as discussed earlier, the result of an additional phase of low-temperature hydrothermal leaching that removed most of the silica, leaving these unique aluminous and zircon-enriched lag deposits in the massive quartz-rich rock.

OXYGEN ISOTOPES

Oxygen isotope values from the quartz-rich and related rocks in the Lyonsdale area are presented in Table 2. Quartz in host granite and in some quartz-sillimanite nodules shows a range of values from 6.7 to 12.9‰, whereas massive vein quartz and quartz-diaspore rocks have quartz values in the range from 6.5 to 8.3‰. These values suggest that primary igneous quartz within host granite was characteristically isotopically heavier, and that the lighter values associated with quartz-sillimanite nodules and leached quartzites are related to interaction with fluids that circulated within the crystallizing granite and adjacent leached carapace.

McLelland et al. (2002b) concluded that the oxygen isotope composition of water in equilibrium with quartz in the Lyon Mountain granite generally ranged from 9.0 to 13.1‰. Fluids with these values are consistent with either (1) magmatic and/or metamorphic fluids or (2) evolved surface-derived basinal brines. Fluids with somewhat lower $\delta^{18}O$, in the range of 7 to 8‰, were calculated for some isotopically lighter quartz, and were interpreted as magmatic in origin. The data reported here are consistent with high-temperature (500–650 °C) equilibrium between quartz and magmatic fluids derived from the adjacent granite pluton.

Isotopic Results from the Mckeever Quartz-Calcite Veins

Samples of coexisting crystalline quartz and calcite from two block veins at the McKeever locality were analyzed as shown in Table 2. Darling and Bassett (2001) have shown that the McKeever quartz-calcite veins were emplaced at temperatures of 185–232 °C and pressures of 863–1870 bars, based

TABLE 1. SHRIMP II U-TH-PB DATA FOR ZIRCONS IN DIASPORE LENSES IN QUARTZ VEINS

Spot	Location[#]	U (ppm)	Th (ppm)	Th/U	Pb* (ppm)	$^{206}Pb/^{204}Pb$	$^{207}Pb/^{235}U$	$\pm 1\sigma$[§]	$^{206}Pb/^{238}U$	$\pm 1\sigma$[§]	$^{207}Pb/^{206}Pb$	$\pm 1\sigma$[§]	$^{207}Pb/^{206}Pb$ Age (Ma)	$\pm 1\sigma$ (Ma)[§]	% Concentration[†]
1.1	c	1377	422	0.306	295	221729	2.3829	0.0451	0.2123	0.0033	0.0814	0.0007	1231.1	18.0	100.8
2.2	m	1608	159	0.099	282	102145	1.9003	0.0252	0.1848	0.0023	0.0746	0.0003	1057.0	7.1	103.4
3.1	c-a	4717	2960	0.627	426	2905	0.7098	0.0099	0.0852	0.0010	0.0604	0.0004	617.5	14.4	85.4
4.1	c-a	2013	803	0.399	338	29586	1.7276	0.0217	0.1648	0.0019	0.0760	0.0003	1096.3	7.7	89.7
5.1	c	827	690	0.834	186	161290	2.2549	0.0319	0.2048	0.0026	0.0799	0.0004	1193.4	9.9	100.6
5.2	m	1788	261	0.146	302	54113	1.8012	0.0234	0.1768	0.0021	0.0739	0.0003	1038.8	7.9	101.0
6.1	c	916	892	0.974	205	17615	2.1089	0.0328	0.1948	0.0025	0.0785	0.0006	1159.7	14.6	98.9
7.1	c	514	724	1.408	109	28580	2.2148	0.0356	0.2004	0.0026	0.0802	0.0007	1201.3	16.0	98.0
8.1	m	1359	237	0.174	240	46816	1.8637	0.0273	0.1823	0.0023	0.0741	0.0004	1045.3	12.2	103.3
9.1	m	1149	225	0.196	190	100000	1.7267	0.0245	0.1700	0.0022	0.0737	0.0003	1031.9	9.3	98.1
10.1	c	579	326	0.564	115	282486	2.0917	0.0366	0.1912	0.0028	0.0793	0.0006	1180.5	15.5	95.6
10.2	c-a	2664	153	0.057	265	5821	0.9637	0.0132	0.1074	0.0012	0.0651	0.0004	776.9	12.6	84.7
10.3	m	1071	205	0.192	183	154321	1.8031	0.0268	0.1759	0.0022	0.0743	0.0005	1050.8	14.2	99.4
11.1	x	1348	309	0.230	236	26021	1.8772	0.0266	0.1808	0.0023	0.0753	0.0004	1076.2	10.6	99.6
13.1	c	2380	363	0.153	227	14667	0.9259	0.0139	0.1018	0.0012	0.0660	0.0005	806.3	15.8	77.5
13.2	c-a	1426	205	0.144	237	6696	1.7660	0.0245	0.1730	0.0021	0.0740	0.0004	1042.6	11.1	98.7
14.1	m	753	238	0.316	114	100000	1.5777	0.0275	0.1537	0.0021	0.0744	0.0007	1053.5	19.3	87.5
14.2	c	651	235	0.361	100	35778	1.5881	0.0284	0.1546	0.0023	0.0745	0.0006	1054.9	16.7	87.9
14.3	c	1118	179	0.160	195	584795	1.8433	0.0276	0.1807	0.0022	0.0740	0.0005	1041.1	14.1	102.8

Note: Corrections for common Pb made using the measured ^{204}Pb. Data were acquired using a 5–7 nA primary O^- beam with a spot diameter of ~17 × 23 microns. Calibration of Pb/U ratios was referenced to the GSC Kipawa zircon standard (993 Ma).

[#]Locations of spot analyses: c—core; c-a—strongly altered core; m—mantle; x—possible mixed mantle and core.

Pb* — radiogenic Pb.

[§]All errors on ratios and ages reported at 1σ level of uncertainty.

[†]Percent concordant.

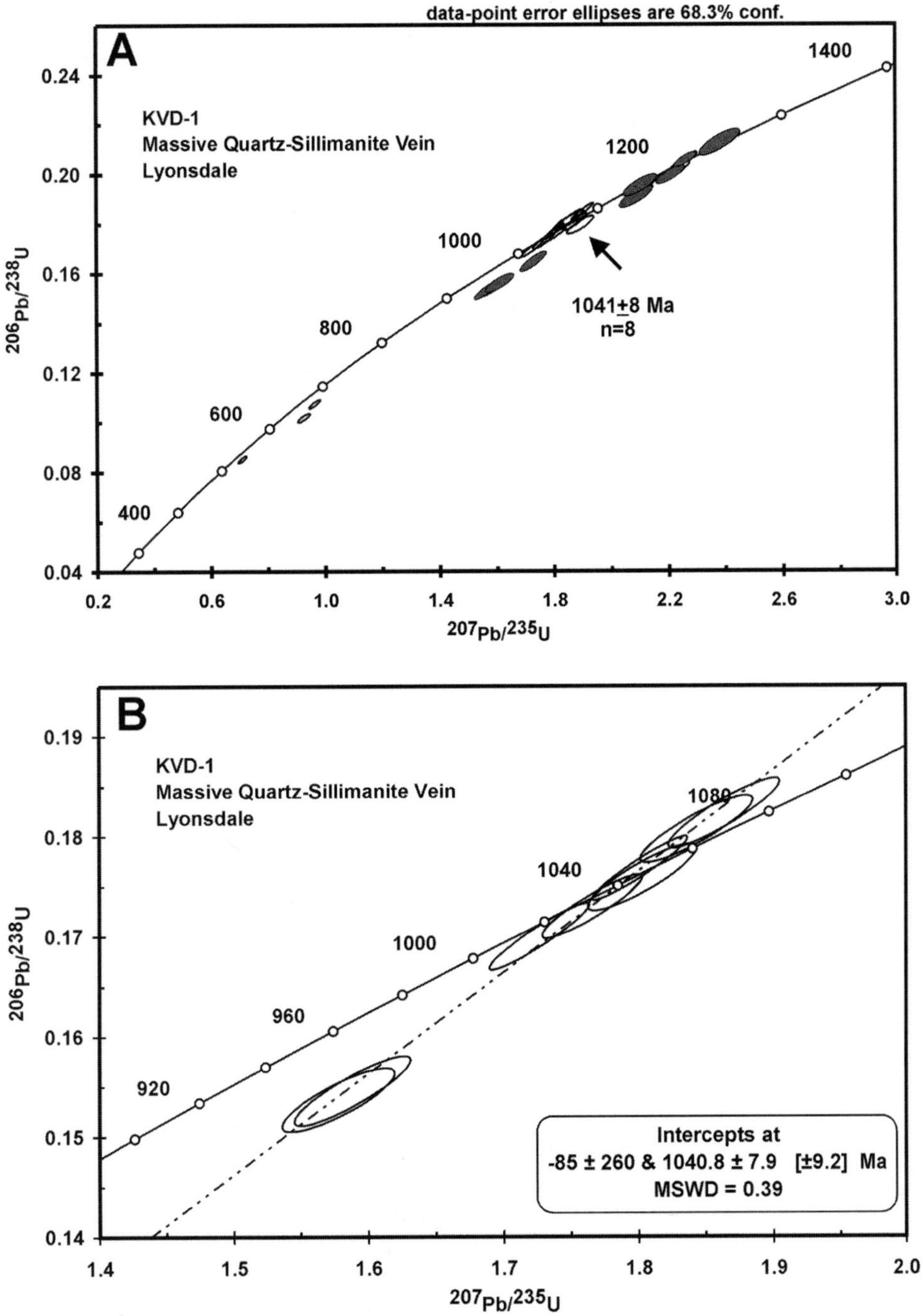

Figure 5. Concordia plots of sensitive high-resolution ion microprobe (SHRIMP) results on zircon from diaspore segregations in quartz-sillimanite vein complex. All error ellipses are shown at the 1σ level of uncertainty. (A) Plot of all U-Pb zircon data from sample KVD-1. Older zircon ages (>1150 Ma) are strictly from grain cores; strongly discordant analyses represent grain cores that are U-rich and pervasively altered. All data from cores are shaded. (B) Expanded view of data for concordant and near-concordant spot analyses from zircon mantles (unshaded). Results for the seven analyses yield a weighted mean $^{207}Pb/^{206}Pb$ age of 1041 ± 7 Ma. Shaded ellipses excluded from age calculation: spots 4.1 (zoned, altered, discordant core) and 11.1 (possibly mixed mantle and slightly overlapped core). MSWD—mean square of weighted deviates.

upon carefully constrained diamond-anvil press studies of CO_2-H_2O inclusions. The results of Darling and Bassett (2001) are consistent with fluid inclusion data from quartz-rich facies in the Lyonsdale area (Table 3).

The $\delta^{18}O$ of water in equilibrium with calcite and quartz in the McKeever locality is relatively light (~–2 to –6‰ SMOW, or standard mean ocean water) compared with those of other Adirondack rock-water systems, including that of the Lyonsdale-area quartzites reported earlier (Fig. 6). For comparative purposes, the McKeever calcite isotope data are plotted with reference to isotopic data from Morrison and Valley's (1988) study of retrograde calcite in Adirondack anorthosite. We suggest that the McKeever veins represent the infiltration of meteoric waters of low $\delta^{18}O$ into rock that had cooled sufficiently from peak metamorphic temperatures to allow brittle fractures to remain open within a zone of hydrothermal circulation driven by nearby Lyon Mountain granitic plutons. This model suggests that other low-temperature hydrothermal alteration features should be common in the vicinity of Lyon Mountain granite rock. Not surprisingly, the low-Ti magnetite deposits found within Lyon

TABLE 2. OXYGEN ISOTOPE ANALYSES OF QUARTZ

Sample number	$\delta^{18}O_{SMOW}$	Description
11-5-96-1	10.11	Lyon Mountain granite at Lyonsdale
11-5-96-2	10.36	Lyon Mountain granite at Lyonsdale
11-5-96-3A	7.99	Lyon Mountain granite at Lyonsdale
11-5-96-3B	7.91	Lyon Mountain granite at Lyonsdale
11-5-96-4	10.00	Lyon Mountain granite at Lyonsdale
11-5-95-5b	7.19	Lyon Mountain granite at Lyonsdale
11-5-96-7	12.95	Calc-silicate skarn at Lyonsdale
11-5-96-8	7.32	Lyon Mountain granite at Lyonsdale
11-5-96-8A	6.70	Lyon Mountain granite at Lyonsdale
AF-a-1	11.45	Quartz-sillimanite nodule at Ager's Falls
AF-a-1a	12.09	Quartz-sillimanite nodule at Ager's Falls
AF-a-2	11.71	Quartz-sillimanite nodule at Ager's Falls
AF-b-1	12.33	Quartz-sillimanite nodule at Ager's Falls
AF-b-2	11.74	Quartz-sillimanite nodule at Ager's Falls
AF-6-12-96a	7.92	Quartz-sillimanite nodule at Ager's Falls
AF-6-12-96b	8.49	Quartz-sillimanite nodule at Ager's Falls
AFb-12-96-4a	8.33	Quartz-sillimanite nodule at Ager's Falls
AFb-12-96-4b	8.19	Quartz-sillimanite nodule at Ager's Falls
HTR-01-2	8.1	"Bull quartz" vein in Lyon Mountain granite at power line locality*
HTR-01-3	8.0	"Bull quartz" vein with minor sillimanite at power line locality*
SB-01-1	8.3	Massive nodular quartzite at Shelby Bridge*
KVD-01-2	7.7	Quartzite hosting illite-diaspore veins at Kosterville Dam*
MRQL-1-Q	6.5	Quartz vein at McKeever*
MRQL-3-Q	7.3	Quartz vein at McKeever*

Note: SMOW—standard mean ocean water.
*Analyses performed at Geochron Labs, Inc. See McLelland et al. (2001) for analytical details on other samples.

Mountain granite elsewhere in the Adirondacks commonly have a low-temperature hydrothermal overprint, including the development of hematite-after-magnetite replacement fabrics in iron ore zones (Buddington, 1965). However, these low-temperature events are not readily distinguished from more widespread alteration of basement rocks throughout the Appalachians that is linked to Paleozoic or younger tectonism (Whitney and Davin, 1987).

TABLE 3. MCKEEVER VEIN ISOTOPIC DATA

Sample number	Mineral	$\delta^{18}O$	$\delta^{13}C$	Temperature	$\delta^{18}O$ water*
MRV-1	Quartz	6.81		185–232	–5.9 to –3.1
MRV-1	Calcite	3.58	–2.80	185–232	–6.3 to –3.9
MRV-2	Quartz	7.33		185–232	–5.4 to –2.6
MRV-2	Calcite	4.05	–2.67	185–232	–5.8 to –2.6

*$\delta^{18}O$ water calculated using the constants of Clayton et al. (1972), assuming an equilibrium range of temperatures based on the fluid inclusion work of Darling and Bassett (2001).

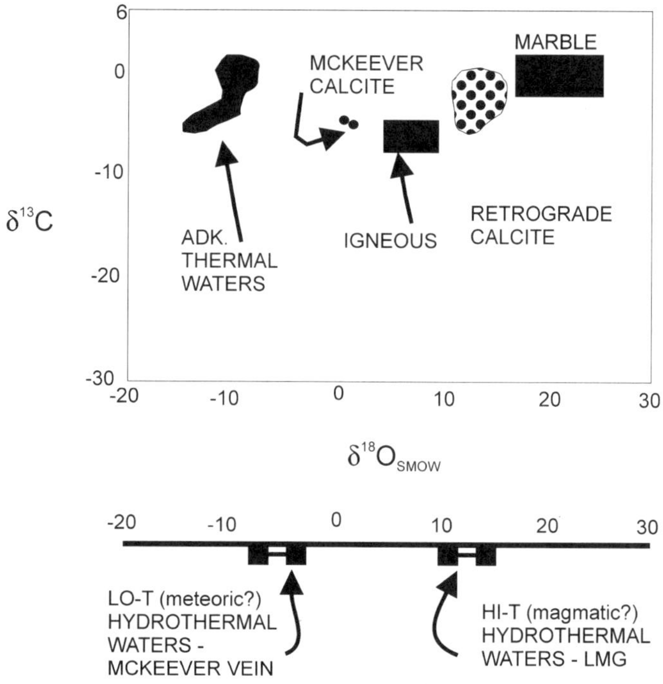

Figure 6. Isotopic data showing range of variation of Adirondack rocks and waters. Marble, retrograde calcite, igneous, and Adirondack (ADK.) thermal waters fields from Morrison and Valley (1988). LMG—Lyon Mountain granite.

FLUID INCLUSIONS

Fluid inclusion studies on Lyon Mountain leucogranites and quartz-sillimanite rock from the Lyonsdale area have previously been reported in McLelland et al. (2002b) and are summarized here. Key results are as follows: (a) The early, high-temperature fluid history of the Lyon Mountain granites is not preserved by primary fluid inclusions. Halite-saturated mixed $H_2O + CO_2$ inclusions with homogenization temperatures in the range from 275 to 300 °C represent fluids that reequilibrated during cooling of the host quartz, perhaps related to the preservation of fractures during uplift and unroofing of the orogen. (b) Lower-temperature (125–200 °C), two-phase, water-only and CO_2-only inclusions were derived from the unmixing of higher-temperature mixed inclusions. (c) Aqueous fluids are Na-Ca-Cl brines at or near halite saturation. It is important that samples of quartz associated with the late, low-temperature alteration illite-diaspore veinlets at Kosterville Dam contain secondary halite-saturated inclusions that also contain spherulitic hematite. This suggests that fluids that invaded the rock at this stage were oxidizing, perhaps due to communication with near-surface hydrologic systems.

DISCUSSION AND CONCLUSIONS

Field relationships and petrographic evidence indicate that quartz-rich rocks associated with Lyon Mountain leucogranite in the southwestern Adirondacks resulted from emplacement of hydrothermal quartz and quartz-sillimanite veins at the margin of a Lyon Mountain granite pluton. This vein emplacement was associated with high-temperature hydrothermal leaching of host granite that removed alkali cations from feldspars and other minerals to produce massive quartz-rich rock resembling typical quartzite along with quartz-sillimanite veins. Illite-diaspore and hematite-chlorite assemblages that resulted from further leaching of quartz and alteration of sillimanite document later low-temperature alteration of the quartz-rich facies. Field relationships in the vicinity of Lyonsdale, New York, suggest that the intruding pluton, comprising the main mass of leucogranite, lay to the east of the leached carapace, which is now represented by the quartz-sillimanite rocks near Kosterville and Shelby Bridge.

Illite-diaspore veins and segregations contain abundant zircons similar to those in the nearby Lyon Mountain leucogranite. While overall morphology and zoning characteristics cannot be used to unequivocally correlate the source of the zircons in the quartz-rich rocks, the close similarity between the quartz-rich facies and granite zircons is striking. The ca. 1034 Ma pegmatite at Lyonsdale Bridge crosscuts quartz-sillimanite veins in leucogranite dated at ca. 1035 Ma, demonstrating that at least some high-temperature vein emplacement occurred nearly synchronously with granite intrusion. Later, low-temperature extraction of quartz and alteration of sillimanite to diaspore and illite further concentrated the granitic zircons by volume reduction of the host rock.

Oxygen isotope data from high-temperature quartz in the Lyon Mountain granite and from quartz-rich facies in the Lyonsdale area indicate that most quartz from unaltered granite is isotopically heavier than quartz in quartz-rich facies. The heavier quartz formed in equilibrium with the granite at temperatures of ~650 °C (McLelland et al., 2002b), whereas the somewhat lighter values reflect interaction with an isotopically lighter fluid of either meteoric or residual magmatic origin. This is consistent with our interpretation of a hydrothermal origin for the quartz-sillimanite veins within the main mass of granite at Lyonsdale and of hydrothermal leaching of the granite to produce the associated carapace of quartz-rich rock. While surface-derived meteoric or basinal brine fluids might provide the appropriate isotopic signature for the quartz in sillimanite-quartz veins, it is unlikely that such fluids would have the acidic, corrosive character documented by the leaching of granite. Therefore, a residual magmatic origin is favored for the hydrothermal fluid in this system.

The isotopic values of quartz and calcite in the ~230 °C McKeever vein are significantly lighter than those of the granite and sillimanite-quartz rocks in the Lyonsdale area, and were in equilibrium with fluids in the range –6 to –2.0‰, strongly suggesting that the fluid was of meteoric origin. Fluid inclusions in the McKeever vein are relatively lower in salinity than fluids in the Lyonsdale rocks (Darling and Bassett, 2001), also supporting involvement of surface-derived waters. Late-stage, high-salinity, hematite-bearing fluid inclusions are associated with the illite-diaspore rock at Kosterville, suggesting further involvement of oxidizing surface-derived brines.

Zircon ages in the quartz-rich facies and the age relationships established by McLelland et al. (2001, 2002a) for the granite and associated quartz-sillimanite rocks at Lyonsdale Bridge constrain the earliest phases of hydrothermal activity as coeval with granite emplacement between ca. 1045 and 1030 Ma. The timing of later stages of hydrothermal activity documented by the McKeever vein and lower-temperature leaching that produced the illite-diaspore assemblages is more difficult to determine. The location of these features in the vicinity of the ca. 1035 Ma Lyon Mountain granite strongly suggests a cogenetic relationship. These later features were generated in brittle fractures and thus provide insight into the unroofing history of the Adirondack orogen (Darling and Bassett, 2001). It is likely that some late-stage hydrothermal features within the Adirondacks are of Paleozoic age and were caused by penetration of saline brines into the basement beneath the younger sedimentary cover. Preliminary monazite age dating in the southern Adirondacks (Storm and Spear, 2002) has shown that Paleozoic events at ca. 500 and 390 Ma were recorded by overgrowths on ca. 1040 Ma metamorphic monazite crystals. Monazite geochronological studies in the McKeever-Lyonsdale area could shed further light on the timing of the low-temperature events and their relationships to granite emplacement and high-temperature hydrothermal activity.

ACKNOWLEDGMENTS

National Science Foundation funding (EAR-9103756) is gratefully acknowledged by JMcL. Both JMcL and BS acknowledge support from the Colgate Research Council. William Willis and Chapin Brackett were of great help during the summers of 1995 and 1996. The authors also appreciate the helpful reviews of the manuscript by Louise Corriveau, Nathalie Marchildon, and one anonymous reviewer.

REFERENCES CITED

Alling, H.L., 1939, Metasomatic origin of the Adirondack magnetite deposits: Economic Geology, v. 34, p. 141–172.

Barton, M.D., and Johnson, D.A., 1996, Evaporite source model for igneous-related Fe-oxide (REE-Cu-Au-U) mineralization: Geology, v. 24, p. 259–262.

Battles, D.A., and Barton, M.D., 1995, Arc-related sodic hydrothermal alteration in the western United States: Geology, v. 23, p. 913–916.

Buddington, A.F., 1965, The Precambrian magnetite deposits of New York and New Jersey: Economic Geology, v. 60, p. 484–511.

Buddington, A.F., and Leonard, B.F., 1962, Regional geology of the St. Lawrence County magnetite district, Northwest Adirondacks, New York: Reston, Virginia, United States Geological Survey Professional Paper 376, 145 p.

Claoué-Long, J.C., King, R.W., and Kerrich, R., 1990, Archaean hydrothermal zircon in the Abitibi greenstone belt: Constraints on the timing of gold mineralization: Earth and Planetary Science Letters, v. 98, p.109–128.

Claoué-Long, J.C., King, R.W., and Kerrich, R., 1992, Reply to comment by F. Corfu and D.W. Davis on "Archaean hydrothermal zircon in the Abitibi greenstone belt: Constraints on the timing of gold mineralization": Earth and Planetary Science Letters, v. 109, p. 601–609.

Clayton, R.N., O'Neil, J.R., and Mayeda, T., 1972, Oxygen isotope exchange between quartz and water: Journal of Geophysical Research, v. 77, p. 3057–3067.

Corfu, F., and Davis, D.W., 1991, Comment on "Archaean hydrothermal zircon in the Abitibi greenstone belt: Constraints on the timing of gold mineralization" by J.C. Claoué-Long, R.W. King, and R. Kerrich: Earth and Planetary Science Letters, v. 104, p. 545–552.

Darling, R.S., and Bassett, W.A., 2001, Analysis of natural $H_2O + CO_2 +$ NaCl fluid inclusions in the hydrothermal diamond anvil cell: American Mineralogist, v. 87, p. 69–78.

Foose, M.P., and McLelland, J.M., 1995, Proterozoic low-Ti iron oxide deposits in New York and New Jersey: Relation to Fe-oxide (Cu-U-Au–rare earth element) deposits and tectonic implications: Geology, v. 23, p. 665–668.

Heald, P., Foley, N.K., and Hayaba, D.O., 1987, Comparative anatomy of volcanic-hosted epithermal deposits: Acid-sulfate and adaluria-sericite types: Economic Geology, v. 82, p. 1–26.

Kerrich, R., and King, R., 1993, Hydrothermal zircon and baddeleyite in Val d'Or Archean mesothermal gold deposits: Characteristics, compositions, and fluid-inclusion properties, with implications for timing of primary gold mineralization: Canadian Journal of Earth Sciences, v. 30, p. 2334–2351.

Leonard, B.F., and Buddington, A.F., 1964, Ore deposits of the St. Lawrence County magnetite district, Northwest Adirondacks, New York: Reston, Virginia, United States Geological Survey Professional Paper 377, 259 p.

McLelland, J., Daly, J.S., and McLelland, J.M., 1996, The Grenville Orogenic Cycle (ca. 1350–1000 Ma): An Adirondack perspective: Tectonophysics, v. 265, p. 1–28.

McLelland, J., Hamilton, M., Selleck, B., McLelland, J., Walker, D., and Orrell, S., 2001, Zircon U-Pb geochronology of the Ottawan Orogeny, Adirondack Highlands, New York: Regional and tectonic implications: Precambrian Research, v. 109, no. 1–2, p. 39–72.

McLelland, J., Goldstein, A., Cunningham, B., Olson, C., and Orrell, S., 2002a, Structural evolution of a quartz-sillimanite vein and nodule complex in a late- to post-tectonic leucogranite, western Adirondack Highlands, New York: Journal of Structural Geology, v. 24, p. 1157–1170.

McLelland, J., Morrison, J., Selleck, B., Cunningham, B., Olson, C., and Schmidt, K., 2002b, Hydrothermal alteration of late- to post-tectonic Lyon Mt. Granitic Gneiss, Adirondack Highlands, New York: Origin of quartz-sillimanite segregations, quartz-albite lithologies, and associated Kiruna-type low-Ti Fe-oxide deposits: Journal of Metamorphic Geology, v. 20, p. 175–190.

Morrison, J., and Valley, J.W., 1988, Post-granulite-facies fluid infiltration in the Adirondack Mountains: Geology, v. 16, p. 513–516.

Nesbitt, R.W., Pascual, E., Fanning, C.M., Toscano, M., Saez, R., and Almodovar, G.R., 1999, U-Pb dating of stockwork zircons from the eastern Iberian Pyrite Belt: Journal of the Geological Society of London, v. 156, p. 7–10.

Postel, A.W., 1952, Geology of the Clinton County magnetite district: Reston, Virginia, United States Geological Survey Professional Paper 237, 88 p.

Rubin, J.N., Henry, C.D., and Price, J.G., 1989, Hydrothermal zircons and zircon overgrowths, Sierra Blanca Peaks, Texas: American Mineralogist, v. 74, p. 865–869.

Storm, L., and Spear, F., 2002, Thermometry, cooling rates and monazite ages of the southern Adirondacks: Geological Society of America Abstracts with Programs, v. 34, no. 1, p. A-6.

Wasteneys, H., McLelland, J., and Lumbers, S., 1999, Precise zircon geochronology in the Adirondack Lowlands and implications for revising plate tectonic models of the Central Metasedimentary Belt and Adirondack Mountains, Grenville Province, Ontario and New York: Canadian Journal of Earth Sciences, v. 36, p. 967–984.

Whitney, P.R., and Davin, M., 1987, Taconic deformation and metasomatism in Proterozoic rocks of the easternmost Adirondacks: Geology, v. 15, p. 500–503.

Manuscript Accepted by the Society August 25, 2003

Printed in the USA

Geological Society of America
Memoir 197
2004

Exhumation of a collisional orogen: A perspective from the North American Grenville Province

Margaret M. Streepey*
Department of Geological Sciences, Florida State University, Tallahassee, Florida 32306-4100, USA
Carolina Lithgow-Bertelloni
Ben A. van der Pluijm
Eric J. Essene
Department of Geological Science, University of Michigan, Ann Arbor, Michigan 48109-1063, USA
Jerry F. Magloughlin
Department of Earth Resources, Colorado State University, Fort Collins, Colorado 80523-1482, USA

ABSTRACT

Combined structural and geochronologic research in the southernmost portion of the contiguous Grenville Province of North America (Ontario and New York State) show protracted periods of extension after the last episode of contraction. The Grenville Province in this area is characterized by synorogenic extension at ca. 1040 Ma, supported by U-Pb data on titanites and ^{40}Ar-^{39}Ar data on hornblendes, followed by regional extension occurring along crustal-scale shear zones between 945 and 780 Ma, as recorded by ^{40}Ar-^{39}Ar analysis of hornblende, biotite, and K-feldspar. By ca. 780 Ma the southern portion of the Grenville Province, from Ontario to the Adirondack Highlands, underwent uplift as a uniform block. Tectonic hypotheses have invoked various driving mechanisms to explain the transition from compression to extension; however, such explanations are thus far geodynamically unconstrained. Numerical models indicate that mechanisms such as gravitational collapse and mantle delamination act over timescales that cannot explain a protracted 300 m.y. extensional history that is contemporaneous with ongoing uplift of the Grenville Province. Rather, the presence of a plume upwelling underneath the Laurentian margin, combined with changes in regional stress directions, permitted the observed uplift and extension in the Grenville Province during this time. The uplift history, while on a slightly different timescale from those of most plume models, is similar to that seen in models of uplift and extension caused by the interaction of a plume with the base of the lithosphere. Some of the protracted extension likely reflects the contribution of far-field effects, possibly caused by tectonic activity in other cratons within the Rodinian supercontinent, effectively changing the stress distributions in the Grenville Province of northeastern North America.

Keywords: Grenville, Rodinia, extension

*E-mail: streepey@quartz.gly.fsu.edu.

Streepey, M.M., Lithgow-Bertelloni, C., van der Pluijm, B.A., Essene, E.J., and Magloughlin, J.F., 2004, Exhumation of a collisional orogen: A perspective from the North American Grenville Province, *in* Tollo, R.P., Corriveau, L., McLelland, J., and Bartholomew, M.J., eds., Proterozoic tectonic evolution of the Grenville orogen in North America: Boulder, Colorado, Geological Society of America Memoir 197, p. 391–410. For permission to copy, contact editing@geosociety.org.

INTRODUCTION

Central to many questions in structural geology and tectonics regarding the evolution of orogens is how crust overthickened by continental collisions is modified and stabilized after an orogenic event. To understand how the crust evolves after orogenesis, it is necessary to study ancient mountain belts, the deep cores of which are exposed at the surface today in high-grade metamorphic terranes. Because results of studies of the temporal evolution of such areas give insight into the time and rates involved in crustal stabilization, these results can be used both to study the general problem of crustal stabilization and to predict the deep behavior of young orogenic belts.

The Grenville Province in northeastern North America is an outstanding, well-studied example of an exposed, deeply eroded, ancient mountain system. The province is affected by a ca. 1.0- to 1.3 billion-year-old set of orogenic events, seen in cratonic blocks worldwide, and culminating in the formation of the supercontinent Rodinia (Hoffman, 1991; Dalziel, 1997). One of the best continuous exposures of Grenville-aged rocks is in northeastern North America between Labrador, Canada, and New York state, where Grenville deformation is thought to have occurred in an arc-continent collision at ca. 1.3–1.2 Ga and a continent-continent collision at ca. 1.1–1.05 Ga (Moore and Thompson, 1980; Easton, 1992; Rivers, 1997; Davidson, 1998; Carr et al., 2000; McLelland et al., 2001). The terrane is characterized by slices of crust that are separated by ductile shear zones in which more of the deformation is concentrated, some of which record normal motion overprinting an earlier contractional history. The colliding craton causing continent-continent collision in this segment of Rodinia is not known, as the proposed collision with Amazonia has recently been questioned by paleomagnetic evidence (Tohver et al., 2002). Early rifting attempts are recorded in some of the blocks of Rodinia (Li et al., 1999; Karlstrom et al., 2000; Dalziel and Soper, 2001; Tack et al., 2001; Timmins et al., 2001). Most major rifting events involving the eastern Laurentian margin (present-day coordinates) appear to have occurred in the late Neoproterozoic. Rifting in this region resulted in the opening of the Iapetus Ocean, which has been dated in the north at ca. 600 Ma (Torsvik et al., 1996; Svenningsen, 2001) and in the south at 570–550 Ma (Torsvik et al., 1996). However, with documented pulses of rifting having occurred in Baltica, Congo, China, and the southwestern United States from ca. 900 Ma to ca. 700 Ma, any extensional activity in the eastern Laurentian block, present-day northeastern North America, during this period may reflect initial stages of Rodinia's breakup (Li et al., 1999; Tanner and Bluck, 1999; Streepey et al., 2000; Dalziel and Soper, 2001; Timmins et al., 2001). Well-exposed Grenville structures in North America provide strong constraints on the nature of extensional activity in the area and also, when compared to extensional activity in other Rodinian blocks that occurred during roughly the same period, on the processes that control the breakup of supercontinents. The driving mechanism(s) for extension in the Laurentian part of the Grenville orogen is the primary focus of this contribution.

Geologic Setting

One of the continuous exposures of rocks that shows Grenville-aged deformation occurs in North America. The eastern edge of the belt abuts the edge of the Appalachian thrust front and is bounded to the west by the Archean Superior Province and other Archean and Proterozoic provinces. Because of the laterally continuous nature of this belt, it offers an excellent opportunity to study lithotectonic relationships in the orogen.

The Grenville Province is composed of lithotectonically distinct blocks representing the autochthonous terrains of the Laurentian craton as well as allochthonous blocks accreted to the Laurentian margin during Grenville orogenesis (Easton, 1992; Rivers, 1997; Davidson, 1998; Hanmer et al., 2000; Fig. 1, inset). These blocks are separated by major crustal-scale shear zones and contain distinct, smaller domains that are also separated by major ductile shear zones (e.g., Davidson, 1984; Easton, 1992). A significant amount of strain recorded by these rocks is concentrated into these zones of deformation, which provide the key to unraveling the tectonic history of the region. In many cases, these shear zones appear to be multiply active, with the latest episode of deformation recording extension, or appearing to record extensional activity, synchronous to Grenville-aged contractional pulses (Mezger et al., 1991b; Culshaw et al., 1994; Busch et al., 1997; Martignole and Reynolds, 1997; Ketchum et al., 1998; Streepey et al., 2001). The current structural expression of the region is of an extensional terrain, and the challenge then lies in determining both the magnitude, timing, and origin of extension as well as the earlier, contractional history of the area.

In this paper, we focus on the eastern Metasedimentary Belt of the Grenville Province and its boundary with the adjacent Granulite Terrane (Fig. 1). This area spans southeastern Ontario and northwestern New York state. The Metasedimentary Belt is one of three major crustal slices that comprise the Grenville Province in this region (Fig. 1). It lies between the Gneiss Belt and the Granulite Terrane and contains variably metamorphosed (greenschist to granulite-facies) metasediments, metagranitoids, and metavolcanic rocks (Easton, 1992).

The Metasedimentary Belt contains several small shear zones that juxtapose lithologically and geochronologically distinct domains. These shear zones within the Metasedimentary Belt dip to the southeast, and the two major boundaries, the Bancroft shear zone and the Robertson Lake shear zone, show late extensional motion. The Carthage-Colton shear zone is located at the eastern edge of the Metasedimentary Belt, and separates it from the Granulite Terrane of the Adirondack Highlands. This shear zone also shows a late extensional history, but dips shallowly to the northwest, creating a grabenlike geometry between the Robertson Lake shear zone and the Carthage-Colton shear zone (Fig. 1).

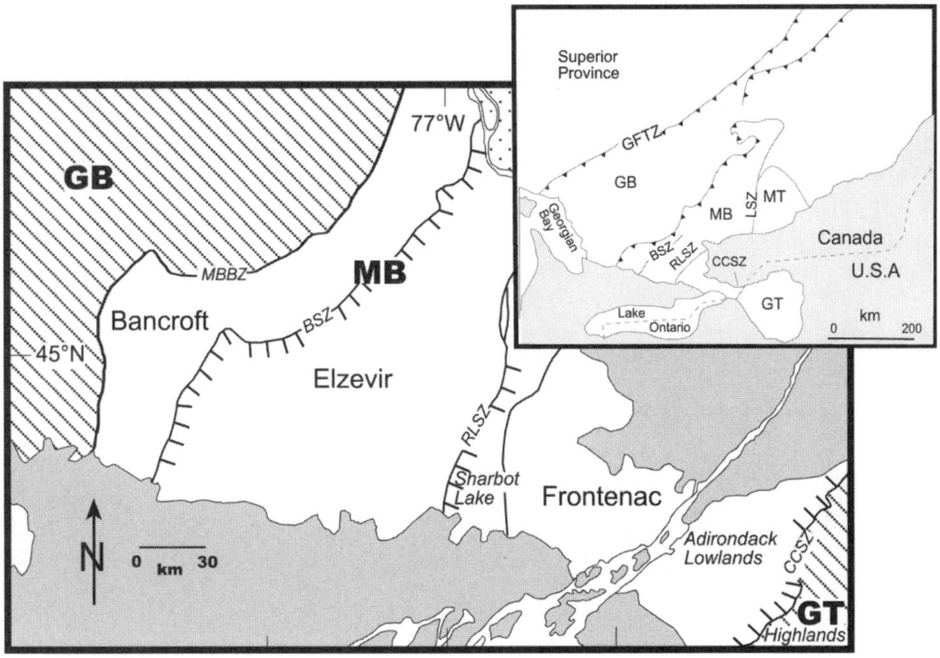

Figure 1. Generalized map of the Metasedimentary Belt (MB) of the Grenville Province (Ontario and New York). The map shows the Metasedimentary Belt in between the Gneiss Belt (GB) and the Granulite Terrane (GT). The Bancroft shear zone (BSZ), Robertson Lake shear zone (RLSZ), and Carthage-Colton shear zone (CCSZ) are shown in their most current expression as normal faults. Other shear zones shown are the Metasedimentary Belt Boundary Zone (MBBZ) and the Sharbot Lake shear zone. The inset map shows the Grenville Province of northeastern North America with the Grenville Front tectonic zone (GFTZ) as it abuts the Archean Superior Province. Other abbreviations: MT—Morin terrane; LSZ—Labelle shear zone. After Streepey et al. (2001).

Studies of the deformation histories of these shear zones require a multidisciplinary approach, with emphasis placed on the field relationships, peak metamorphic pressures and temperatures, and the corresponding geochronologic data that constrain the cooling and exhumation history. Because most of the rocks have experienced more than one phase of deformation and metamorphism, structural relationships in the field can be complex, and field analysis alone is not enough to completely constrain the significance of these boundaries.

This study presents a synthesis of geochronologic information combined with structural analysis and thermobarometric data to describe the kinematics of the uplift or exhumation history of a segment of the Grenville Province in northeastern North America. A summary of ages is given, adding to regional compilations (Cosca et al., 1991, 1992, 1995; Mezger et al., 1991a, 1992, 1993; van der Pluijm et al., 1994). In addition, new ^{40}Ar-^{39}Ar ages from amphiboles in the Adirondack Lowlands and the Adirondack Highlands are presented and further constrain the geologic history of the area.

Whereas the combination of structural, petrologic, and geochronologic information is critical to constructing a kinematic model of the evolution of the region, it does not give a geodynamic picture of the development of the late stages of modification and stabilization of overthickened crust. This information allows us to develop reasonable geologic hypotheses about timing of late, postorogenic extension and the nature of motion between blocks of crust, but it does not explain the physical processes behind the evolution. In addition, geochronologic data are restricted to lithologies and assemblages that contain minerals with the appropriate elements for radiogenic dating. In areas where the appropriate assemblages are not available, the geochronologic results are limited or incomplete and cannot provide a full, detailed cooling history of the rocks.

In order to develop a more geodynamically complete picture of the exhumation history of the Grenville Province, we have developed a two-dimensional numerical model of a slice of crust representing this region. The structures assigned to the model are taken directly from field studies in the region, and the rheologies are assigned based on existing literature (Ranalli, 1995). The numerical models explore possible driving mechanisms for the observed phenomenon of extension in this orogenic belt. From geochronologic and structural information, the timescales involved in the transition from compression to extension have been evaluated and have placed constraints on the amount of displacement across shear zones. Although how this orogenic belt extends following collision is known, why it extends is less evident. It remains uncertain whether extension can be attributed to a single mechanism, such as gravitational collapse, or whether it requires a combination of mechanisms, such as mantle delamination in addition to changes in far-field stresses. Whereas numerical models cannot provide constraints that uniquely solve this problem, they give insights as to whether or not proposed mechanisms can act in a way that fits field observations over the period of time dictated by geochronologic constraints.

GEOCHRONOLOGIC SUMMARY

In studies of ancient metamorphic terranes, motion along ductile shear zones can often be delineated with a combination of ages that yield information on the timing of latest metamorphism and ages that record the cooling or exhumation history of the terrane (e.g., van der Pluijm et al., 1994). In such studies it is critical to constrain the pressure-temperature (*P-T*) conditions of metamorphism in order to determine whether geochronologic ages are ages of cooling from a peak metamorphic event or growth ages. Minerals yield cooling ages if the peak conditions of metamorphism are higher than the closure temperatures of the minerals and growth ages if the minerals can be shown to have grown during metamorphism but at conditions below their closure temperatures. Therefore, in order to best interpret geochronologic data in the eastern portion of the Metasedimentary Belt, it was necessary to initially assess the metamorphic conditions of the terrane.

Figure 2, A and B, shows temperature and pressure maps of the Metasedimentary Belt from Streepey et al. (1997; thermobarometric data from references therein). Metamorphism in the area reached upper-amphibolite to granulite-facies metamorphism. Maximum temperatures in the study area from just west of the Robertson Lake shear zone to just east of the Carthage-Colton shear zone ranged from 600 to 650 °C in and around the Robertson Lake shear zone and increased to the east to 700–750 °C in and around the Carthage-Colton shear zone. Pressures were 600 to 800 MPa over the region.

In order to best interpret radiometric ages from polymetamorphic terranes, it is essential not only to have quantitative

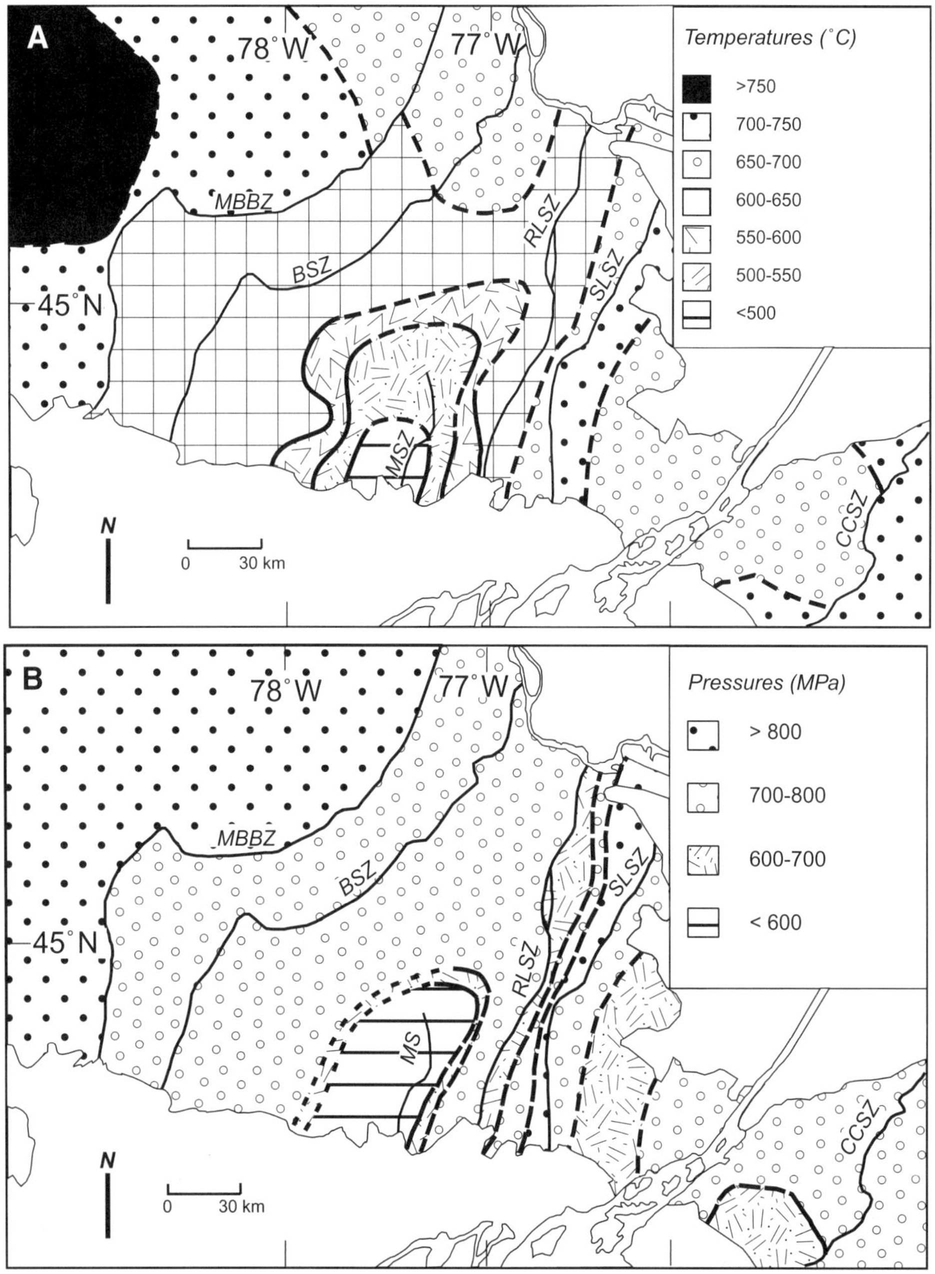

Figure 2. Regional thermobarometric gradients in the Metasedimentary Belt. (A) Contoured temperatures. The lowest temperatures are less than 500 °C in the Elzevir terrane, and the highest temperatures are granulite facies (700 °C and above) in the Gneiss Belt, the Frontenac terrane, and the Adirondack Highlands. (B) Contoured pressures. Pressures generally follow the pattern of temperatures and indicate regional metamorphism in the region. Pressures over most of the area are 700–800 MPa. BSZ—Bancroft shear zone; CCSZ—Carthage-Colton shear zone; MBBZ—Metasedimentary Belt Boundary Zone; RLSZ—Robertson Lake shear zone; SLSZ—Sharbot Lake shear zone.

assessments of *P-T* conditions but also to have accurate closure temperatures for minerals used in analysis and some constraints on the diffusion mechanisms active in isotopic resetting. For this study we consider volume diffusion in grains to be the primary mechanism of isotopic resetting. In addition, published and widely used closure temperatures for the minerals titanite, hornblende, biotite, and K-feldspar are considered appropriate for this study of a slowly cooled, regionally metamorphosed terrane (titanite: 600–700 °C, Mezger et al., 1991a, Scott and St-Onge, 1995; hornblende: 480–500 °C, McDougall and Harrison, 1999; biotite: 300 °C, McDougall and Harrison, 1999; K-feldspar: 150 to 300 °C, Zeitler, 1987, McDougall and Harrison, 1999, Lovera et al., 1991). Because peak temperatures of regional metamorphism are close to or generally exceed the closure temperature of titanite in the U-Pb system, the U-Pb ages of titanite constrain either the timing of latest metamorphism or cooling ages very close to the timing of peak metamorphism in this area. The ^{40}Ar-^{39}Ar ages of hornblende, biotite, and K-feldspar, which have lower closure temperatures than titanite, constrain most of the cooling history of the study area and are considered to have closed to the K-Ar system at some time after peak metamorphism.

The U-Pb ages from zircon, garnet, monazite, and titanite have been determined for the Metasedimentary Belt in numerous studies (Mezger et al., 1993; Corfu and Easton, 1995, 1997; Perhsson et al., 1996; Wasteneys et al., 1999; Corriveau and van Breemen, 2000; McLelland et al., 2001). We provide a brief summary of the metamorphic ages of the Metasedimentary Belt; the reader is referred to reviews by Rivers (1997), Carr et al. (2000), Hanmer et al. (2000), and McLelland et al. (2001) for detailed descriptions of the early metamorphic history of the Metasedimentary Belt. The Metasedimentary Belt yields information on two major periods of "Grenville-aged" orogenesis that represent an arc accretion event from ca. 1300–1190 Ma (the Elzevirian orogeny), culminating in a continent-continent collision at ca. 1080–1020 Ma (the Ottawan orogeny). These events represent two episodes of contraction, possibly separated by extensional events, which are recorded by magmatic activity and emplacement of large anorthosite complexes during these periods (McLelland et al., 1988; Davidson, 1995). The Robertson Lake shear zone and the Carthage-Colton shear zone separate rocks showing a marked discontinuity in metamorphic age, and so are fundamental boundaries separating blocks with distinct metamorphic histories. From the Metasedimentary Belt boundary thrust zone east to the Robertson Lake shear zone (encompassing the Bancroft and Elzevir terranes), metamorphic minerals record geochronological evidence of metamorphism during the latest contractional event during the Ottawan orogeny.

The Frontenac terrane (including the Adirondack Lowlands) from east of the Robertson Lake shear zone to the Carthage-Colton shear zone shows evidence of metamorphism related to the earlier Elzevirian orogeny. It does not, however, appear to record metamorphic ages related to the Ottawan orogeny, indicating that this terrane was either laterally separated from the Elzevir terrane during this period or was at shallow crustal levels (Mezger et al., 1993; Streepey et al., 2000, 2001). However, some investigators have proposed that at least portions of the Adirondack Lowlands may have been deformed during the Ottawan orogeny (Wasteneys et al., 1999), and more detailed isotopic work may be useful in resolving the extent and nature of Ottawan deformation in the Lowlands. East of the Carthage-Colton shear zone, peak metamorphism in the Adirondack Highlands occurred during the Ottawan orogeny, although there is some evidence that this terrane was also deformed during the Elzevirian orogeny (McLelland et al., 1988; McLelland and Chiarenzelli, 1989; Kusky and Lowring, 2001). In this scenario, the Frontenac terrane, bounded by the east-dipping Robertson Lake shear zone and the west-dipping Carthage-Colton shear zone, is a block of crust that has been largely protected from the Ottawan pulse of orogenesis, while the Adirondack Highlands and the Elzevir terrane recorded metamorphic ages during this period.

The cooling history of the terranes immediately adjacent to these two shear zones is discussed in order to evaluate their significance as postorogenic extensional structures. The ages compiled in this study include ages determined from regional U-Pb and ^{40}Ar-^{39}Ar geochronologic data (Busch and van der Pluijm, 1996; Busch et al., 1996b, 1997; Streepey et al., 2000, 2001, 2002, and new results) to provide a complete cooling history for the eastern half of the Metasedimentary Belt. Sample locations and their corresponding ages are given in Table 1. The U-Pb ages of titanite give the timing of early deformation, thought to have some transpressive component along both the east-dipping Robertson Lake shear zone and the west-dipping Carthage-Colton shear zone. The ^{40}Ar-^{39}Ar ages of hornblendes, biotites, and K-feldspars, combined with structural analysis of shear zones, record the timing of later extensional motion along both shear zones. New ^{40}Ar-^{39}Ar ages of hornblendes further detail the cooling history of the crust adjacent to the Carthage-Colton shear zone in northwest New York state.

Robertson Lake Shear Zone

The Robertson Lake shear zone is a multiply active, east-dipping shear zone separating the eastern Elzevir terrane (Mazinaw domain) and the western Frontenac terrane (Sharbot Lake domain) within the Metasedimentary Belt (Fig. 1; Easton, 1988). The latest episode of motion recorded along this zone is down-to-the-east, as shown by shear-sense indicators including S-C, C–C′ fabrics, with these crystal plastic structures crosscut by brittle fabrics delineating an uplift history during deformation (Busch and van der Pluijm, 1996). Its early history records a transpressive event at ca. 1030 Ma, indicating imbrication via sinistral transpression of the Mazinaw and the Sharbot Lake domains (Busch et al., 1997). ^{40}Ar-^{39}Ar cooling ages of hornblendes and micas in the Robertson Lake shear zone show a marked difference across the zone. Combined with the structural information and the nature of the offset, extensional motion had to have occurred to juxtapose the crustal blocks on either side of

TABLE 1. SAMPLE LOCATIONS

Sample	Hornblende age	Biotite age	Domain	Location	
				UTM Coordinates Northing	UTM Coordinates Easting
From Busch et al. (1996)					
RVL118A	1011 Ma	n.d.	Sharbot Lake	4971225	372600
RVL20A	1205 Ma	n.d.	Sharbot Lake	4973050	367025
CDN58	947 Ma	n.d.	Elzevir/Mazinaw	4968970	356050
MTG72	952 Ma	n.d.	Elzevir/Mazinaw	4957060	355300
From Busch and van der Pluijm (1996)					
				Northing	Easting
RVL118B	n.d.	969 Ma	Sharbot Lake	4971225	372600
LVT130B	n.d.	901 Ma	Elzevir/Mazinaw	4990750	361920
From Streepey et al. (2000) and unpublished data					
A102	990 Ma	n.d.	Lowlands	Popple Hill Formation, near Russell.	
A112	1055 Ma	940 Ma	Lowlands	Near Gouverneur, Hwy 58/812, N Jct. of Poplar Hill Rd.	
A114	990 Ma	n.d.	Lowlands	Along Grass River, Hwy 17, W side of road, 1.5 km NW of Russell.	
A117	967 Ma	n.d.	Highlands	0.5 mi. E of Rt. 27 and Brouser's Corner.	
A124	1043 Ma	n.d.	Lowlands	Emoryville Rd., ca. 0.5 mi. NW of power plant, ca 0.25 mi. W of turnoff to Talcville.	
A125	1066 Ma	n.d.	Lowlands	Emoryville Rd., E of Hailsboro, small outcrop north of road, NW of house, 1.8 mi. E of bridge, 2.1 mi. from Island Branch Rd.	
A128	1020 Ma	964 Ma	Lowlands	Trout Lake Rd., N of Edwards, along W side of Trout Lake, S side, W of 2 islands, W side of road. Cut about 40 m long.	
A129	1079 Ma	924 Ma	Lowlands	S. of Trout Lake, Rt. 19 to Edwards, ca. due W of westernmost extreme of Cedar Lake.	
A133	975 Ma	n.d.	Highlands	Crest of White's Hill, White Hill Rd., SE of Parishville about 3 mi. at abandoned lookout. 150′ on small trail to the W.	
A134	972 Ma	n.d.	Highlands	Very low outcrops along both sides of the highway. S side of White Hill, about 1 mi. S of lookout. Outcrop on White Hill Rd., within amphibolite.	
A135	1165 Ma	n.d.	Highlands	Just N of Sterling Rd., near power line, about 150 from junction with Joe Indian Rd.	
A136	943 Ma	937 Ma	Highlands	Small outcrop on Joe Indian Rd. at the crest of a small hill. Located at the N70°W, 60°N symbol on Leonard and Buddington's map.	
A137	1005 Ma	n.d.	Highlands	Rt. 56, about 3 mi. S of Stark. W side of hwy, roadcut of amphibolite migmatites.	
A138	983 Ma	n.d.	Highlands	Small outcrop on E side of Hwy 56, about 1 mi. S of 137, about 4 mi. S of Stark.	
A140	947 Ma	n.d.	Highlands	Hwy 3, just E of Pitcairn. Exposures on embankment N of the road, at crest of hill.	
A142	953 Ma	919 Ma	Highlands	Small outcrop on the E side of Mud Lake Rd. at the junction of Mud Lake and Briggs Rd. (SE corner of intersection).	
A145	935 Ma	n.d.	Highlands	About 2 mi. E of Texas on Texas Rd., 6 mi. from Hwy 812.	
HM95	n.d.	915 Ma	Lowlands	NW of Russell on Rt. 17 in Devil's Elbow, 2 mi. S of Hermon. Sample from E side of outcrop.	
SE596-49	n.d.	895 Ma	Highlands	On Rt. 87, N of Dana Hill Rd., 1.9 mi. S of Whipporwill Corners. Outcrop on NW side of road.	
LB596-31b	n.d.	925 Ma	Lowlands	4.5 mi. from intersection of Hwy 3 and Hwy 812, on 812 S of Balmat. Sample from E side of outcrop.	
LB93	n.d.	904 Ma	Lowlands	Outcrop just N of LB596-31b.	
CN596-56	n.d.	924 Ma	Highlands	0.2 mi. S of junction of Hanson Rd. and Orebed Rd. on Orebed Rd., near Colton. Outcrop on W side of road.	
PP596-60	n.d.	899 Ma	Lowlands	1 mi. SW of Pierrepont on Rt. 2. Outcrop on E side of road.	
From Streepey et al. (2001)					
				Latitude	Longitude
RWS-1	1000 Ma	n.d.	Within CCSZ	North 44 deg., 22 min.	West 75 deg., 10 min.
RWS-3c	999 Ma	n.d.	Within CCSZ	North 44 deg., 22 min.	West 75 deg., 10 min.
EA1	981 Ma	n.d.	Within CCSZ	North 44 deg., 22 min.	West 75 deg., 10 min.
DH98-1	998 Ma	n.d.	Within CCSZ	North 44 deg., 22 min.	West 75 deg., 10 min.
RWH-1	973 Ma	n.d.	Within CCSZ	North 44 deg., 22 min.	West 75 deg., 10 min.
CR7-DH1	989 Ma	n.d.	Within CCSZ	Dana Hill outcrop: On N side of Dana Hill Rd.	
CR1-DH1	980 Ma	n.d.	Within CCSZ	Dana Hill outcrop: On N side of Dana Hill Rd.	
CR3-87	941 Ma	n.d.	Within CCSZ	1.8 km N of intersection between Dana Hill Rd. and Rt. 87, on Rt. 87.	
CR2-A2	948 Ma	n.d.	Within CCSZ	Directly across Rt. 87 from CR3-87.	
DH2-8	1012 Ma	n.d.	Within CCSZ	Directly across Rt. 87 from CR3-87.	
CR6-A4	1005 Ma	n.d.	Within CCSZ	Directly across Rt. 87 from CR3-87.	
CR12-A4	998 Ma	n.d.	Within CCSZ	Directly across Rt. 87 from CR3-87.	
CR9-A4	944 Ma	n.d.	Within CCSZ	Directly across Rt. 87 from CR3-87.	
H87-5b	1019 Ma	n.d.	Within CCSZ	Directly across Rt. 87 from CR3-87.	

n.d.—no data; CCSZ—Carthage-Colton shear zone; UTM—Universal Transverse Mercator.

the shear zone (Busch et al., 1996b; Fig. 3). The timing of the transition from compression to extension can be somewhat constrained by the age of the latest transpressive event in the region at ca. 1030 Ma (Busch et al., 1997), but cannot be directly determined. The termination of extension across the Robertson Lake shear zone cannot be constrained from ^{40}Ar-^{39}Ar analyses of hornblende and biotite, as both show differences of ca. 70–100 Ma across the zone. Results from the analysis of K-feldspar in the region suggest that the entire block was uplifting uniformly by 780 Ma, suggesting termination of extension between 900 and 780 Ma (Streepey et al., 2002).

Ages taken from Busch and van der Pluijm (1996), Busch et al. (1996b, 1997), and Streepey et al. (2002) are compiled in Figure 3. These data constrain the cooling history of the terrane from immediately after orogenesis in the Robertson Lake shear zone area to the time at which the terrane was uplifting as a uniform block. In addition, the geometry of normal fault motion and the amount of displacement along the shear zone are evident (Busch et al., 1996a). The U-Pb ages from titanites document the difference in metamorphic ages between the Mazinaw and the Sharbot Lake domains. As the closure temperature of titanite is ca. 600–700 °C (Mezger et al., 1991a; Scott and St-Onge, 1995), titanite ages give the timing of metamorphism or a cooling age that is very close to the age of metamorphism in the amphibolite to granulite-facies rocks of the Elzevir and the Frontenac terranes. Metamorphism in the Sharbot Lake domain (the hangingwall block of the Robertson Lake shear zone) occurred at ca. 1140 to 1170 Ma (Mezger et al., 1993; Corfu and Easton, 1995; Busch et al., 1997). These ages are generally similar to those across the entire width of the Frontenac terrane (Mezger et al., 1993). However, the youngest metamorphic ages in the Mazinaw domain (the footwall block of the Robertson Lake shear zone) range from ca. 1010 Ma to 1050 Ma (Mezger et al., 1993; Busch et al., 1997), showing a ca. 100-m.y. difference in the timing of the latest metamorphic event across this boundary. Transpressional activity is interpreted to have caused juxtaposition through imbrication of the two terranes at ca. 1030 Ma (Busch et al., 1997).

The ^{40}Ar-^{39}Ar cooling ages of hornblendes, micas, and K-feldspars constrain the post-titanite cooling and unroofing history of the blocks of crust adjacent to the Robertson Lake shear zone and constrain the timing and nature of postorogenic extension along this zone. Hornblende ages for rocks in the vicinity of the Robertson Lake shear zone are shown in Figure 3. The offset in ages shown by the titanite U-Pb geochronology is also shown by the cooling ages of hornblende, with hornblende ages of ca. 1010 Ma in the Sharbot Lake domain (hangingwall), which are at least 60 m.y. older than rocks immediately across the Robertson Lake shear zone in the Mazinaw domain, where hornblende ages are ca. 950 Ma (Busch and van der Pluijm, 1996; Busch et al., 1996a). Biotite, which closes to the K-Ar system at ca. 300 °C, continues to show an offset across the Robertson Lake shear zone, with biotites in the Sharbot Lake domain recording ages of 970 Ma that are at least 70 m.y. older than the 900 Ma biotite ages in the Mazinaw domain (Busch and van der Pluijm, 1996). At the time of biotite closure, rocks were at fairly shallow crustal depths of ~10 to 12 km assuming an average geothermal gradient. Because of the offsets in cooling ages of the rocks that are presently exposed at the surface crustal level, it is clear that extensional activity did not terminate across the Robertson Lake shear zone until sometime after 900 Ma.

The ^{40}Ar-^{39}Ar ages of K-feldspars give some information on the termination of extension along the Robertson Lake shear zone, but do not completely constrain it. Unlike hornblendes or micas, which are considered to have one diffusion domain and therefore a single closure temperature, K-feldspars are thought to have multiple diffusion domains and therefore multiple closure temperatures (Lovera et al., 1991). Analysis of K-feldspars gives, instead of a single age, a temperature-time path for the grain. Thermal modeling of K-feldspar spectra from the Mazinaw and the Sharbot Lake domains show that the two domains were juxtaposed by at least 780 Ma, meaning that the termination of extension across the Robertson Lake shear zone must have occurred between 900 Ma and 780 Ma (Streepey et al., 2002). Thus, postorogenic extension across the Robertson Lake shear zone terminated between 140 and 260 m.y. after the final expression of contractional tectonics in the area at ca. 1040 Ma.

CARTHAGE-COLTON SHEAR ZONE

The Carthage-Colton shear zone separates the eastern Frontenac terrane (Adirondack Lowlands) from the Granulite Terrane (Adirondack Highlands; Fig. 1). It is ~150 km east of the Robertson Lake shear zone and dips to the west, toward the Robertson Lake shear zone. With this geometry, the Frontenac terrane is a grabenlike block bounded by two shear zones dipping toward one another. Upper amphibolite-facies marbles and other metasediments dominate the Adirondack Lowlands lithologies, whereas the Adirondack Highlands are comprised predominantly of granulite-facies metaigneous assemblages. The Carthage-Colton shear zone crops out as a zone of intense deformation between the two terranes, although the exact location of the boundary has been debated (Geraghty et al., 1981).

The Carthage-Colton shear zone had a long history of activity over the duration of the Grenville orogenic cycle. The Adirondack Lowlands and Highlands both appear to have been affected by the ca. 1190 Ma arc-continent collision at the end of the Elzevirian orogeny (Mezger et al., 1991a, 1992; Wasteneys et al., 1999; Kusky and Lowring, 2001). However, only the Adirondack Highlands appear to have been pervasively metamorphosed by the granulite-facies Ottawan orogeny, which has been dated at 1090–1040 Ma (McLelland et al., 1996). The entire Frontenac terrane, from east of the Robertson Lake shear zone to just west of the Carthage-Colton shear zone, appears to have escaped widespread thermal metamorphism and resetting of isotopic ages from this pervasive deformational event, either by being at shallower crustal levels during that period or by being laterally separated.

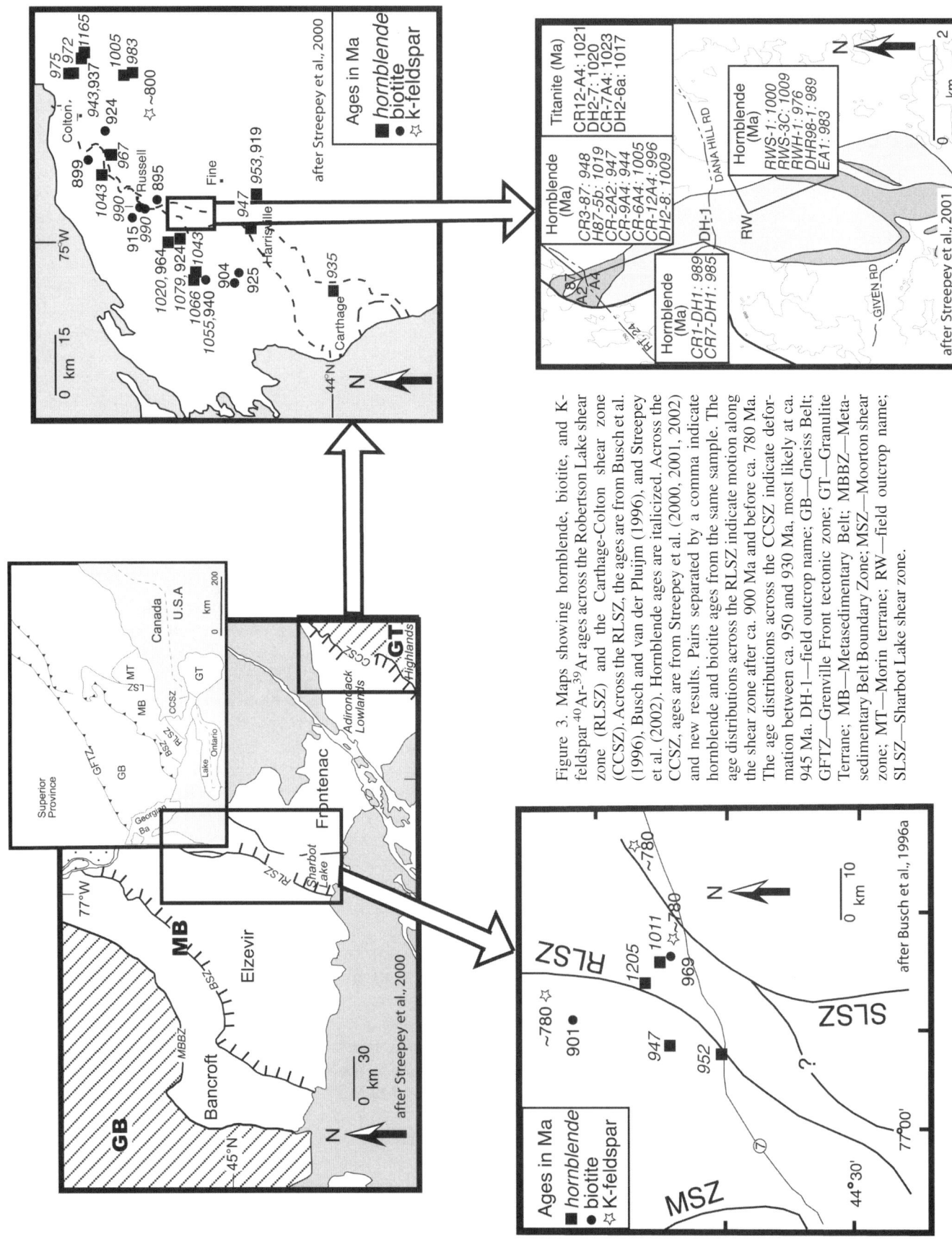

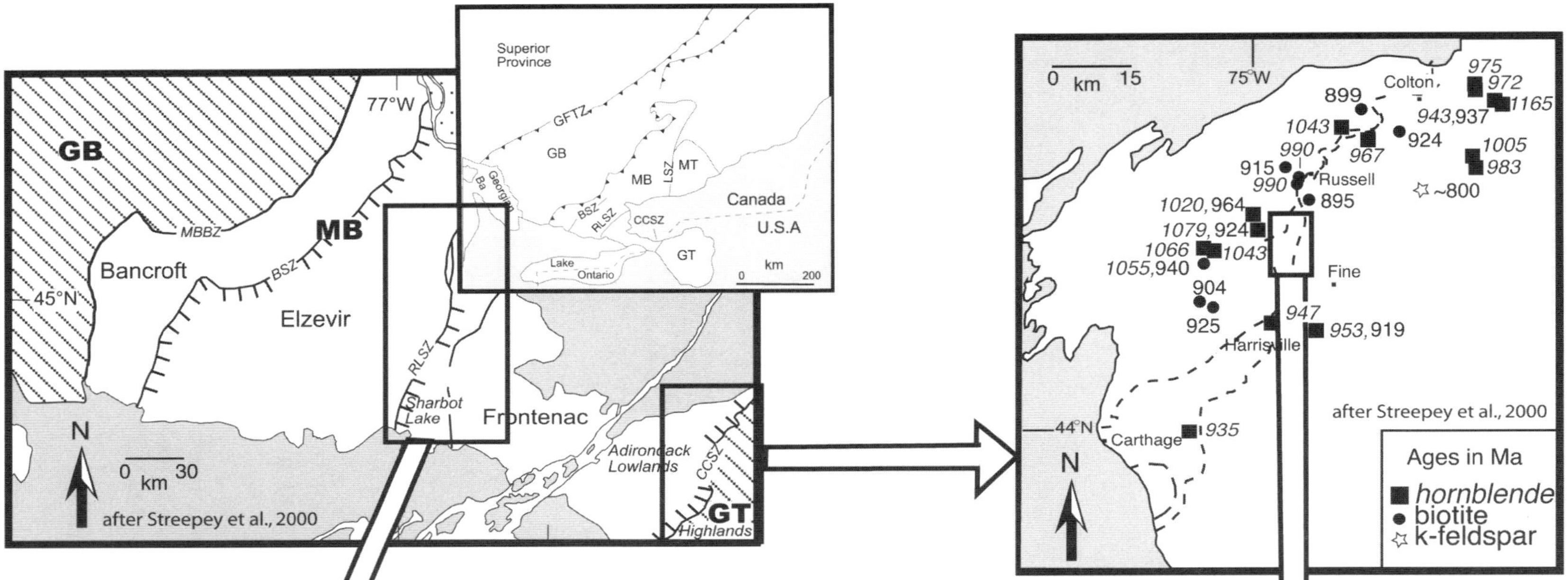

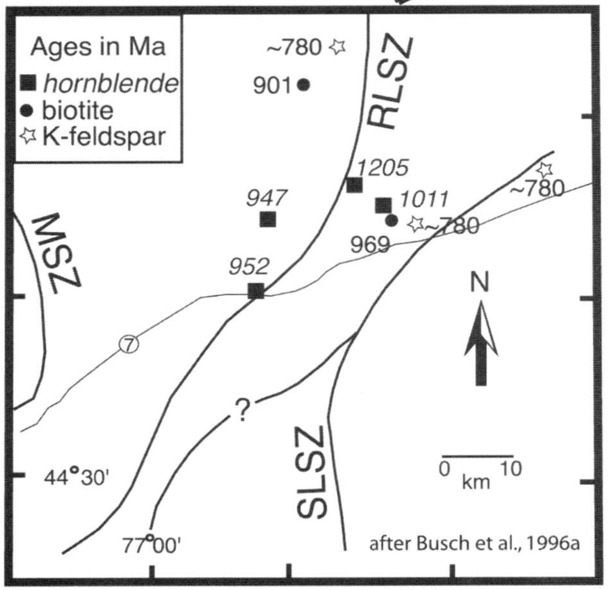

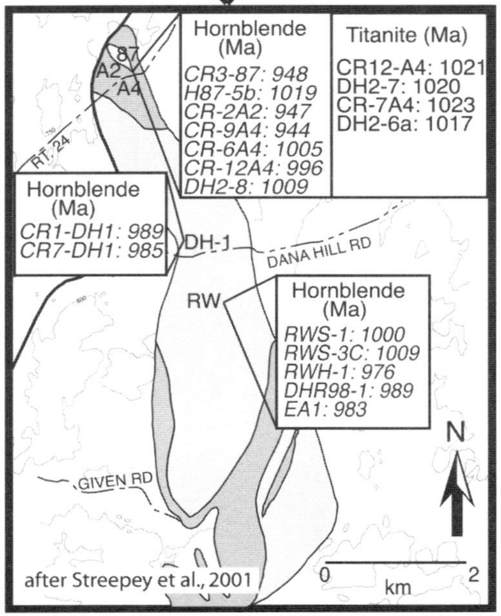

Figure 3. Maps showing hornblende, biotite, and K-feldspar ^{40}Ar-^{39}Ar ages across the Robertson Lake shear zone (RLSZ) and the Carthage-Colton shear zone (CCSZ). Across the RLSZ, the ages are from Busch et al. (1996), Busch and van der Pluijm (1996), and Streepey et al. (2002). Hornblende ages are italicized. Across the CCSZ, ages are from Streepey et al. (2000, 2001, 2002) and new results. Pairs separated by a comma indicate hornblende and biotite ages from the same sample. The age distributions across the RLSZ indicate motion along the shear zone after ca. 900 Ma and before ca. 780 Ma. The age distributions across the CCSZ indicate deformation between ca. 950 and 930 Ma, most likely at ca. 945 Ma. DH-1—field outcrop name; GB—Gneiss Belt; GFTZ—Grenville Front tectonic zone; GT—Granulite Terrane; MB—Metasedimentary Belt; MBBZ—Metasedimentary Belt Boundary Zone; MSZ—Moorton shear zone; MT—Morin terrane; RW—field outcrop name; SLSZ—Sharbot Lake shear zone.

McLelland et al. (1996) proposed that the Carthage-Colton shear zone was the locus for extensional collapse of the Ottawan orogen at ca. 1050 Ma, which resulted in exhumation of the high-grade core of the Adirondack Highlands, presumably while the orogen was still under a contractional stress field and as a direct result of the orogenic event, either through gravitational collapse or through mantle delamination beneath the orogen. Streepey et al. (2001) suggested that the Carthage-Colton shear zone was involved in transpressive deformation similar to that documented across the Robertson Lake shear zone at ca. 1040 Ma. It is clear that the Carthage-Colton shear zone was active during or immediately after the latest episode of Grenville contraction. In addition, cooling ages from hornblendes and biotites show that the Carthage-Colton shear zone was reactivated in an extensional regime similar to that observed along the Robertson Lake shear zone.

We present sixteen new ^{40}Ar-^{39}Ar hornblende analyses from the University of Michigan Radiogenic Isotope Laboratory combined with published ^{40}Ar-^{39}Ar hornblende and biotite ages near and along the Carthage-Colton shear zone (Streepey et al., 2000, 2001). Standard operating procedures for the collection of hornblende and biotite analyses in this laboratory are described in detail in Streepey et al. (2000). Our results are shown in Figure 3, and isotopic data are presented in Table 2.[1] Plateaus were defined as occurring where 50% or more of the total ^{39}Ar was released in three or more consecutive steps and where the ages of the steps overlapped at the 2σ level of error.

The hornblende ages in the Adirondack Lowlands of the Frontenac terrane show that these locations within this slice of crust reached 500 °C at ca. 1036 Ma (Fig. 3; Table 2). Hornblende ages across the Carthage-Colton shear zone are ~947–983 Ma. The offset in ages indicates active movement along the Carthage-Colton shear zone after hornblendes closed to the K-Ar system, or after 950 Ma. The nature of the offset, combined with the regional fabrics, indicates that this motion must have been extensional (Heyn, 1990; Streepey et al., 2000). Some hornblendes in the Dana Hill metagabbro tightly cluster at 945 Ma. Though there is some textural complexity in these samples, with young ages coming from a variety of veins and other textures, these ages fit well within the regional framework of extension (bracketed by regional hornblende and biotite ^{40}Ar-^{39}Ar ages) and indicate that this was likely a time of deformation along the Carthage-Colton shear zone (Streepey et al., 2001). Biotite ^{40}Ar-^{39}Ar ages are ca. 900–930 Ma on both sides of the shear zone, indicating that this block of crust was uniformly uplifting by this time (Streepey et al., 2000). K-feldspar ages are similar to those found in the Robertson Lake shear zone area, further supporting the idea that the entire Metasedimentary Belt was uplifting as a uniform block by ca. 780 Ma (Heizler and Harrison, 1998; Streepey et al., 2002).

NUMERICAL MODELING

Geochronology paired with structural geology shows that extensional motion took place along a large segment of the Metasedimentary Belt well after the Grenville contractional orogeny, during a period that was considered to be relatively quiescent. Application of ^{40}Ar-^{39}Ar and U-Pb geochronology details the kinematics and timing of this transition and the amount and nature of extensional deformation that occurred after it, but gives few constraints on the mechanisms that produced a regional extensional event during this period.

Extension occurred at least 100 m.y. after the latest contractional event in the Grenville Province of New York and southeastern Ontario. Given this timescale, it is difficult to make a causal link between orogenesis and extension. Whereas several attempts have been made to create a Himalayan analog to the Grenville Province (e.g., Windley, 1986), a component of major extension in the latter is clearly postorogenic, so the timescales for the two are different. In Tibet, for example, gravitational potential energy due to the elevated topography of the plateau played a major role in driving collapse (e.g., Shen et al., 2001). However, in the Grenville Province the duration and termination of this extensional event were so much later than compression that gravitational collapse became an ineffective driving mechanism. Earlier synorogenic extensional events (as proposed by McLelland et al., 2001, and references therein) at ca. 1050 Ma were different in nature and almost certainly can be ascribed to processes such as gravitational collapse, mantle delamination, or some combination of these driving mechanisms. The exhumation of the Grenville orogen was not solely the result of erosional and isostatic processes, but was due at least in part to active extension along shear zones, which led to uplift of the region. Geochronologic data have allowed construction of a kinematic model of denudation and unroofing of midcrustal levels of the orogenic belt. When uplift is discussed in the context of this paper, we refer to the exhumation of the core of the Grenville Province and do not constrain a measure of the paleotopographic surface.

A model investigation has been made of the evolution of a block of overthickened crust, with three common driving mechanisms proposed to explain postorogenic extension. We have evaluated the driving forces necessary to generate uplift and extension that match the kinematics of deformation in the Grenville Province as documented through field and laboratory studies. Two critical field observations in this area that must be explained are the continuous regional uplift during the time of extension and the time lag in termination of extensional motion along the Robertson Lake shear zone versus the Carthage-Colton shear zone detailed in the geochronology (see earlier and Fig. 3).

[1]GSA Data Repository item 2004059, Appendix, hornblende spectra, is available on request from Documents Secretary, GSA, P.O. Box 9140, Boulder, CO 80301-9140, USA, editing@geosociety.org, or at www.geosociety.org/pubs/ft2004.htm.

TABLE 2. HORNBLENDE ARGON DATA

Power	^{39}Ar frac	^{39}Ar mol	$^{40}Ar/^{39}Ar$	$^{37}Ar/^{39}Ar$	$^{36}Ar/^{39}Ar$	$^{40}Ar^{*}/^{39}Ar_{K}$	%^{40}Ar atmos	Ca/K	Age (Ma)	1σ error (Ma)
A112-1										
400	0.17977	1.33E-15	82.67166	3.62469	0.01034	79.6151	3.69723	6.65081	1044	2
401	0.12911	9.53E-16	82.13706	3.8956	0.00050	81.9893	0.17990	7.14789	1068	2
402	0.04571	3.37E-16	80.99802	3.88507	–0.00025	81.0706	–0.08961	7.12857	1059	3
420	0.12908	9.52E-16	80.80290	3.97513	–0.00010	80.8318	–0.03576	7.29382	1056	2
440	0.04952	3.65E-16	79.80750	3.82902	–0.00050	79.9564	–0.18657	7.02572	1048	3
460	0.06496	4.79E-16	80.32021	3.94636	–0.00069	80.5247	–0.25459	7.24103	1053	3
480	0.01616	1.19E-16	77.82671	3.64937	–0.00247	78.5576	–0.93913	6.69609	1034	9
560	0.08421	6.21E-16	79.86897	3.63473	0.00025	79.7954	0.09212	6.66923	1046	3
640	0.10076	7.43E-16	80.25100	4.00081	0.00004	80.2387	0.01532	7.34094	1050	3
720	0.03772	2.78E-16	79.24898	3.52251	0.00293	78.3825	1.09337	6.46332	1032	6
800	0.03863	2.85E-16	81.53749	4.13691	0.00093	81.2615	0.33849	7.59066	1061	5
880	0.01055	7.78E-17	80.54386	4.17951	0.00409	79.3344	1.50162	7.66883	1041	14
1000	0.06427	4.74E-16	80.99230	3.91662	0.00074	80.7727	0.27114	7.18646	1056	3
1040	0.02004	1.48E-16	78.72645	3.90584	–0.00173	79.2368	–0.64826	7.16668	1040	8
4800	0.02954	2.18E-16	80.08239	3.91963	–0.00173	80.5944	–0.63935	7.19198	1054	6
J value 9.84667E-03 ± 1.38181E-05										
Total ^{39}K vol = 2.12529E-10 CCNTP/G										
Total gas age = 1051.97 ± 1.39196 Ma										
A114-2										
360	0.00871	6.42E-17	437.13157	4.65082	1.19085	85.2341	80.5015	8.53361	1091	56
400	0.00357	2.63E-17	84.83118	2.64341	0.06967	64.2452	24.2670	4.85029	877	71
440	0.01623	1.20E-16	78.79309	2.49256	0.05687	61.9881	21.3280	4.57350	853	17
480	0.03422	2.53E-16	75.25700	2.55393	0.00739	73.0722	2.90312	4.68611	970	8
520	0.07799	5.75E-16	76.43780	2.69612	0.00273	75.6316	1.05471	4.94701	996	4
560	0.10642	7.85E-16	76.70364	2.71598	0.00102	76.4032	0.39169	4.98345	1004	3
600	0.19840	1.46E-15	76.44511	2.70967	0.00100	76.1492	0.38709	4.97187	1001	2
640	0.06504	4.80E-16	75.72533	2.70733	0.00017	75.6755	0.06580	4.96758	997	4
680	0.05337	3.94E-16	74.36980	2.69303	0.00144	73.9429	0.57402	4.94134	979	4
720	0.14419	1.06E-15	75.23356	2.76182	0.00045	75.1018	0.17513	5.06756	991	3
760	0.00258	1.90E-17	74.30312	2.82691	0.00236	73.6062	0.93794	5.18699	976	57
840	0.20909	1.54E-15	75.25175	2.73198	0.00072	75.0377	0.28445	5.01281	990	2
880	0.04428	3.27E-16	75.13989	2.7655	0.00005	75.1238	0.02142	5.07431	991	6
2000	0.03130	2.31E-16	77.06652	3.00146	0.00076	76.8428	0.29030	5.50727	1008	6
5000	0.00462	3.41E-17	80.45262	3.77874	0.03093	71.3128	11.3605	6.93347	952	40
J value 9.74593E-03 ± 1.69588E-05										
Total ^{39}K vol = 1.50293E-10 CCNTP/G										
Total gas age = 992.403 ± 1.77968 Ma										
A117-1R										
400	0.06297	4.65E-16	113.32355	5.04689	0.04675	99.5093	12.1901	9.26035	1226	3
440	0.13070	9.64E-16	74.12110	4.74351	0.00057	73.9514	0.22895	8.70369	982	3
480	0.17212	1.27E-15	74.65890	4.74913	0.00060	74.4807	0.23869	8.7140	987	2
520	0.10834	7.99E-16	73.44085	4.73651	0.00155	72.9816	0.62534	8.69084	972	3
560	0.08989	6.63E-16	71.93891	4.7711	0.00197	71.3563	0.80987	8.75431	955	4
600	0.05587	4.12E-16	71.29738	4.84565	0.00070	71.0905	0.29017	8.89110	952	5
640	0.04701	3.47E-16	70.75044	4.83448	0.00259	69.9852	1.08161	8.87061	941	4

680	0.02548	1.88E-16	70.30706	4.75113	0.00502	68.8222	2.11197	8.71767	929	4
760	0.08086	5.97E-16	74.37964	4.80444	0.00205	73.7752	0.81265	8.81549	980	3
880	0.04753	3.51E-16	72.89371	4.88858	0.00407	71.6920	1.64858	8.96987	959	6
1000	0.13615	1.00E-15	73.20374	5.31663	0.00025	73.1300	0.10074	9.75528	973	2
2000	0.04257	3.14E-16	73.61771	5.22413	0.00348	72.5902	1.39574	9.58556	968	4
4800	0.00053	3.93E-18	98.69721	5.98711	0.08950	72.2505	26.7958	10.98552	964	390
J value 9.78019E-03 ± 1.65238E-05										
Total ^{39}K vol = 1.44533E-10 CCNTP/G										
Total gas age = 987.869 ± 1.61561 Ma										
A124-1R										
360	0.03019	2.23E-16	125.33636	4.21346	0.04743	111.3220	11.1814	7.73112	1338	5
400	0.17033	1.26E-15	80.50973	3.98635	0.00127	80.1340	0.46669	7.31440	1051	2
401	0.10968	8.09E-16	80.73303	3.89882	0.00148	80.2960	0.54133	7.15380	1053	3
402	0.04250	3.14E-16	81.25107	3.85999	0.00243	80.5319	0.88512	7.08255	1055	5
420	0.01263	9.32E-17	79.31704	3.95845	0.01442	75.0558	5.37241	7.26321	1000	14
460	0.21815	1.61E-15	81.20917	3.91409	0.00086	80.9539	0.31434	7.18182	1060	2
461	0.08099	5.98E-16	81.15436	3.89173	0.00176	80.6357	0.63911	7.14079	1056	3
480	0.10678	7.88E-16	81.52008	3.8357	0.00041	81.3986	0.14902	7.03798	1064	2
520	0.04500	3.32E-16	80.23726	3.91163	0.00240	79.5267	0.88558	7.17730	1045	4
560	0.02183	1.61E-16	79.20467	4.03858	0.00249	78.4681	0.92995	7.41024	1035	8
600	0.00773	5.70E-17	77.39229	4.28706	0.00899	74.7364	3.43173	7.86617	997	22
640	0.01163	8.58E-17	78.55942	4.56397	0.00192	77.9916	0.72279	8.37426	1030	14
720	0.03076	2.27E-16	79.95281	4.34586	–0.00010	79.9817	–0.03614	7.97406	1050	6
800	0.02660	1.96E-16	80.78376	4.17887	0.00348	79.7551	1.27335	7.66765	1048	8
1200	0.05685	4.19E-16	81.00487	4.0396	0.00053	80.8471	0.19477	7.41211	1058	4
4000	0.02837	2.09E-16	81.85268	4.09907	–0.00085	82.1048	–0.30802	7.52123	1071	6
J value 9.87089E-03 ± 1.20594E-05										
Total ^{39}K vol = 2.58136E-10 CCNTP/G										
Total gas age = 1063.43 ± 1.33794 Ma										
A125-1										
400	0.25390	1.87E-15	84.45171	3.90485	0.00647	82.5402	2.26343	7.16486	1056	2
401	0.07938	5.86E-16	84.04833	3.75689	0.00251	83.3075	0.88143	6.89338	1063	4
440	0.13038	9.62E-16	83.99765	3.72575	0.00136	83.5946	0.47983	6.83624	1066	3
480	0.10856	8.01E-16	83.76073	3.70631	0.00069	83.5557	0.24478	6.80057	1065	3
520	0.22332	1.65E-15	84.00499	3.8024	0.00102	83.7027	0.35984	6.97688	1067	2
560	0.08583	6.33E-16	85.40903	3.84262	0.00098	85.1192	0.33935	7.05068	1080	4
600	0.00564	4.16E-17	77.55969	4.87365	0.00266	76.7725	1.01495	8.94248	999	34
680	0.00783	5.77E-17	84.75936	4.43573	–0.01303	88.6097	–4.54267	8.13895	1113	29
820	0.01984	1.46E-16	83.30916	4.05837	–0.00825	85.7468	–2.92602	7.44655	1086	14
1000	0.01869	1.38E-16	83.44706	4.02577	–0.00840	85.9295	–2.97487	7.38673	1088	14
4800	0.06665	4.92E-16	83.66405	4.09005	0.00077	83.4373	0.27103	7.50468	1064	4
J value 9.63302E-03 ± 1.87146E-05										
Total ^{39}K vol= 1.26569E-10 CCNTP/G										
Total gas age = 1065.08 ± 1.92921 Ma										
A128-1										
360	0.02003	1.48E-16	103.37114	0.86044	0.05595	86.8393	15.9927	1.57879	1114	5
380	0.00976	7.20E-17	45.93718	0.55937	0.00118	45.5872	0.76186	1.02636	668	14
420	0.00899	6.63E-17	85.67338	1.20975	0.00381	84.5478	1.3138	2.21972	1092	13

(continued)

TABLE 2. *Continued*

Power	^{39}Ar frac	^{39}Ar mol	^{40}Ar/^{39}Ar	^{37}Ar/^{39}Ar	^{36}Ar/^{39}Ar	^{40}Ar*/^{39}Ar$_K$	%^{40}Ar atmos	Ca/K	Age (Ma)	1σ error (Ma)
460	0.01472	1.09E-16	95.15639	2.10311	0.00229	94.4805	0.71029	3.85892	1186	11
500	0.06145	4.53E-16	83.47260	4.65351	0.00067	83.2755	0.23613	8.53855	1080	2
520	0.16436	1.21E-15	78.39011	4.93948	0.00060	78.2127	0.22632	9.06327	1029	1
530	0.06040	4.46E-16	78.09354	4.64889	0.00062	77.9101	0.23490	8.53007	1026	2
540	0.02521	1.86E-16	71.34198	3.67143	0.00056	71.1764	0.23209	6.73657	957	5
560	0.04059	2.99E-16	75.79147	4.50351	0.00002	75.7861	0.00709	8.26332	1005	3
620	0.10396	7.67E-16	76.91635	4.80667	0.00048	76.7755	0.18312	8.81958	1015	2
660	0.11611	8.57E-16	78.49321	4.95174	0.00055	78.3297	0.20831	9.08576	1031	2
720	0.11533	8.51E-16	79.95906	5.11697	0.00065	79.7680	0.23895	9.38894	1045	2
780	0.08676	6.40E-16	80.74414	4.89144	0.00065	80.5511	0.23908	8.97512	1053	2
1000	0.06995	5.16E-16	80.56078	4.96834	0.00099	80.2677	0.36380	9.11622	1050	2
4000	0.10239	7.55E-16	66.01429	3.44906	0.00134	65.6196	0.59788	6.32855	899	1
J value 9.83709E-03 ± 1.44148E-05										
Total ^{39}K vol = 2.63518E-10 CCNTP/G										
Ttotal gas age = 1022.11 ± 1.28867 Ma										
A129-1										
360	0.02336	1.72E-16	97.70120	6.7107	0.05614	81.1131	16.9784	12.31321	1044	11
500	0.56485	4.17E-15	84.98144	6.63161	0.00053	84.8238	0.18550	12.16809	1079	2
540	0.09031	6.66E-16	84.37694	6.5040	–0.00120	84.7317	–0.42045	11.93394	1079	4
580	0.10879	8.03E-16	85.28425	6.53173	0.00166	84.7932	0.57578	11.98483	1079	3
620	0.02093	1.54E-16	85.02355	6.49323	0.00091	84.7555	0.31527	11.91418	1079	13
1000	0.12513	9.23E-16	82.43000	6.78991	0.00165	81.9417	0.59238	12.45855	1052	3
2000	0.06351	4.69E-16	84.95944	7.47248	0.00029	84.8723	0.10257	13.71097	1080	4
4800	0.00313	2.31E-17	79.77944	6.68513	–0.00745	81.9811	–2.75968	12.26629	1052	88
J value 9.65675E-03 ± 1.79665E-05										
Total ^{39}K Vol= 1.17160E-10 CCNTP/G										
Total gas age = 1075.01 ± 2.10352 Ma										
A133-1										
400	0.10466	7.72E-16	78.38483	3.18793	0.00641	76.4909	2.41619	5.84941	1001	2
440	0.28683	2.12E-15	73.06987	3.12962	0.00103	72.7641	0.41847	5.74242	963	1
441	0.08772	6.47E-16	73.68601	3.14167	0.00127	73.3109	0.50907	5.76453	968	3
460	0.02532	1.87E-16	73.45835	3.18647	0.00229	72.7827	0.91977	5.84673	963	8
500	0.04963	3.66E-16	74.45830	3.22276	0.00102	74.1561	0.40586	5.91332	977	6
540	0.11704	8.64E-16	74.81581	3.20948	0.00134	74.4197	0.52945	5.88895	980	3
580	0.04971	3.67E-16	74.77719	3.13684	0.00234	74.0855	0.9250	5.75567	976	5
620	0.01130	8.34E-17	71.75809	3.13448	0.01051	68.6532	4.32688	5.75134	920	18
700	0.12656	9.34E-16	73.38544	3.21055	0.00100	73.0899	0.40273	5.89092	966	2
780	0.04513	3.33E-16	74.21991	3.26527	0.00039	74.1052	0.15456	5.99132	976	4
880	0.04618	3.41E-16	74.49200	3.20771	0.00037	74.3841	0.14485	5.88571	979	4
1000	0.03645	2.69E-16	73.88380	3.81057	–0.00254	74.6357	–1.01768	6.99187	982	6
1200	0.00925	6.83E-17	74.92031	3.98344	0.00117	74.5737	0.46264	7.30906	981	29
5000	0.00423	3.12E-17	74.50266	3.69954	–0.01149	77.8975	–4.55667	6.78815	1015	33
J value 9.69066E-03 ± 1.73722E-05										
Total ^{39}K vol = 1.89216E-10 CCNTP/G										
Total gas age = 972.995 ± 1.63359 Ma										

A134-1										
360	0.05806	4.28E-16	88.65932	3.11553	0.02434	81.4682	8.11096	5.71657	1061	3
380	0.08990	6.63E-16	74.67086	2.89451	0.00067	74.4739	0.26378	5.31103	991	2
400	0.13506	9.97E-16	72.87802	2.93911	0.00036	72.7716	0.14602	5.39286	973	2
420	0.13788	1.02E-15	72.07033	2.91703	–0.00002	72.0766	–0.00871	5.35235	966	2
430	0.07872	5.81E-16	72.99311	2.94646	0.00062	72.8086	0.25278	5.40635	973	2
440	0.06587	4.86E-16	72.61572	2.93576	0.00075	72.3934	0.30617	5.38672	969	3
460	0.08062	5.95E-16	72.72679	2.93459	0.00137	72.3216	0.55714	5.38457	968	2
480	0.05728	4.23E-16	72.60468	2.93469	0.00191	72.0408	0.77664	5.38475	966	4
500	0.06910	5.10E-16	73.46718	2.96128	0.00186	72.9177	0.74792	5.43354	975	2
520	0.03623	2.67E-16	72.74153	2.89763	0.00251	72.0009	1.01816	5.31675	965	4
560	0.02905	2.14E-16	72.46758	2.99739	0.00192	71.8994	0.78405	5.49980	964	5
600	0.01910	1.41E-16	70.95202	3.05735	0.00210	70.3310	0.87527	5.60982	948	6
680	0.03939	2.91E-16	72.01860	3.25967	0.00347	70.9928	1.42436	5.98105	955	3
760	0.03514	2.59E-16	71.75096	3.20298	0.00185	71.2048	0.76118	5.87703	957	4
840	0.01835	1.35E-16	72.69965	3.60576	0.00535	71.1194	2.17367	6.61607	956	6
4000	0.05026	3.71E-16	72.23752	4.28732	0.00242	71.5225	0.98982	7.86664	960	4
J value 9.82456E-03 ± 1.50837E-05										
Total ^{39}K vol =	3.12587E-10 CCNTP/G									
Total gas age =	974.680 ± 1.34455 Ma									
A135-1										
360	0.00480	3.54E-17	189.06855	3.53975	0.38769	74.5070	60.5926	6.49495	996	29
420	0.03215	2.37E-16	97.21990	2.83047	0.00822	94.7920	2.49733	5.19352	1193	6
460	0.10836	8.00E-16	94.16484	2.76105	0.00110	93.8398	0.34519	5.06615	1184	2
520	0.28549	2.11E-15	92.27292	2.80599	0.00096	91.9885	0.30824	5.14861	1167	1
540	0.17551	1.30E-15	91.59555	2.83532	0.00085	91.3458	0.27267	5.20242	1161	2
560	0.07382	5.45E-16	93.82666	2.84512	0.00231	93.1434	0.72821	5.22040	1178	3
620	0.14952	1.10E-15	91.60731	2.84698	0.00091	91.3373	0.29475	5.22382	1161	2
660	0.06962	5.14E-16	91.92043	2.86217	–0.00076	92.1449	–0.24420	5.25169	1168	3
720	0.03631	2.68E-16	91.53615	2.89577	–0.00059	91.7105	–0.19047	5.31334	1164	6
840	0.04446	3.28E-16	92.45542	2.94274	0.00034	92.3555	0.10808	5.39952	1170	5
4000	0.01996	1.47E-16	92.06961	3.09668	0.00578	90.3630	1.85361	5.68198	1152	9
J value 9.88469E-03 ± 1.11152E-05										
Total ^{39}K vol=	1.54303E-10 CCNTP/G									
Total gas age = 1167.36 ± 1.24451 Ma										
A136-1										
360	0.0290	2.14E-16	107.51473	3.13684	0.05980	89.8435	16.4361	5.75567	1149	8
400	0.08390	6.19E-16	70.25276	3.13401	0.00122	69.8912	0.51466	5.75048	950	4
420	0.16589	1.22E-15	69.97159	3.07525	0.00036	69.8666	0.15005	5.64266	950	2
440	0.2397	1.77E-15	69.01690	3.14081	0.00033	68.9185	0.14258	5.76295	939	1
450	0.07365	5.43E-16	68.46276	3.10302	0.00155	68.0044	0.66950	5.69361	930	5
480	0.04626	3.41E-16	68.16077	3.18853	0.00122	67.8009	0.52798	5.85051	928	5
520	0.08498	6.27E-16	67.39505	3.13191	0.00063	67.2088	0.27636	5.74662	921	2
560	0.02044	1.51E-16	67.85553	3.2227	0.01155	64.4438	5.02793	5.91321	891	20
580	0.01323	9.76E-17	67.22698	3.08304	–0.00313	68.1531	–1.3776	5.65695	931	15
640	0.03406	2.51E-16	70.51416	3.1958	0.00225	69.8502	0.94159	5.86385	949	7
800	0.12837	9.47E-16	68.48829	3.14145	0.00091	68.2200	0.39174	5.76413	932	2
880	0.02252	1.66E-16	67.33619	3.24219	0.00315	66.4059	1.38156	5.94897	913	9

(continued)

TABLE 2. *Continued*

Power	^{39}Ar frac	^{39}Ar mol	$^{40}Ar/^{39}Ar$	$^{37}Ar/^{39}Ar$	$^{36}Ar/^{39}Ar$	$^{40}Ar^{*}/^{39}Ar_K$	%^{40}Ar atmos	Ca/K	Age (Ma)	1σ error (Ma)
4000	0.0580	4.28E-16	69.01521	3.22167	0.00061	68.8346	0.2617	5.91132	939	4
J value 9.91445E-03 ± 1.20341E-05										
Total ^{39}K vol = 2.21028E-10 CCNTP/G										
Total gas age = 943.299 ± 1.37251 Ma										
A137-2										
360	0.07434	5.49E-16	79.54528	2.6555	0.00766	77.2828	2.84427	4.87248	1010	3
380	0.18242	1.35E-15	76.64371	2.56293	0.00054	76.4855	0.20642	4.70262	1002	2
400	0.29765	2.20E-15	77.58981	2.57025	0.00031	77.4988	0.11730	4.71606	1012	6
401	0.02462	1.82E-16	76.83948	2.49381	0.00419	75.6012	1.61151	4.57580	993	7
440	0.15378	1.13E-15	76.76473	2.56553	0.00078	76.5340	0.30057	4.70739	1002	1
460	0.11356	8.38E-16	76.79623	2.55097	0.00117	76.4516	0.44876	4.68068	1002	3
480	0.03633	2.68E-16	77.37406	2.59891	0.00191	76.8109	0.72784	4.76864	1005	3
500	0.01628	1.20E-16	76.30115	2.4661	0.00684	74.2795	2.64957	4.52495	980	9
560	0.02836	2.09E-16	76.61391	2.58639	0.00493	75.1571	1.90149	4.74567	989	6
800	0.05065	3.74E-16	76.83979	2.70719	0.00161	76.3642	0.61893	4.96732	1001	4
4000	0.02203	1.63E-16	77.06461	3.48322	0.00603	75.2818	2.3134	6.39123	990	7
J value 9.70862E-03 ± 1.72082E-05										
Total ^{39}K vol = 2.77185E-10 CCNTP/G										
Total gasage= 1004.41 ± 2.31283 Ma										
A138-1										
360	0.00341	2.51E-17	80.22103	3.86201	0.07427	58.2728	27.3597	7.08626	812	38
400	0.01336	9.86E-17	75.38258	2.98881	0.00238	74.6796	0.93255	5.48406	987	11
420	0.13042	9.62E-16	74.15896	2.96001	0.00045	74.0267	0.17835	5.43121	980	2
440	0.10992	8.11E-16	74.45032	2.96696	0.00062	74.2657	0.24798	5.44396	983	2
460	0.10226	7.55E-16	74.44556	2.98933	−0.00039	74.5612	−0.15534	5.48501	986	2
480	0.08788	6.48E-16	74.58835	3.00933	0.00059	74.4137	0.23416	5.52171	984	3
500	0.11664	8.61E-16	74.50427	2.99731	0.00092	74.2331	0.36397	5.49965	983	2
520	0.07957	5.87E-16	74.44747	2.97613	−0.00003	74.4556	−0.01093	5.46079	985	3
560	0.20707	1.53E-15	74.33688	3.00375	0.00039	74.2214	0.15535	5.51147	982	1
580	0.02001	1.48E-16	73.73513	2.98427	−0.00007	73.7562	−0.02858	5.47572	978	7
640	0.08036	5.93E-16	74.68295	2.98614	0.00071	74.4746	0.27898	5.47916	985	1
4000	0.04910	3.62E-16	74.24016	3.0346	0.00125	73.8720	0.49591	5.56807	979	4
J value 9.75238E-03 ± 1.69034E-05										
Total ^{39}K vol = 2.45945E-10 CCNTP/G										
Total gas age = 982.380 ± 1.49725 Ma										
A140-1										
360	0.04811	3.55E-16	64.05867	1.55391	0.00826	61.6178	3.81037	2.85121	856	4
400	0.15487	1.14E-15	70.53351	2.27729	0.00118	70.1852	0.49382	4.17851	949	1
420	0.07203	5.31E-16	70.73917	2.29034	0.00040	70.6197	0.16889	4.20246	953	2
440	0.12733	9.40E-16	70.67584	2.34235	0.00044	70.5457	0.18414	4.29789	952	3
460	0.05076	3.75E-16	70.53867	2.28231	−0.00001	70.5418	−0.00444	4.18772	952	3
500	0.03876	2.86E-16	70.67347	2.29519	0.00156	70.2130	0.65155	4.21136	949	6

540	0.10871	8.02E-16	72.45699	2.38892	0.00278	71.6342	1.13555	4.38334	964	2
600	0.13072	9.65E-16	71.07979	2.39188	0.00051	70.9285	0.21285	4.38877	956	3
640	0.09072	6.69E-16	71.10540	2.37214	0.00069	70.9027	0.28507	4.35255	956	3
720	0.07920	5.84E-16	70.81559	2.3225	0.00092	70.5424	0.38577	4.26147	952	3
1000	0.06688	4.93E-16	68.26967	2.13221	0.00043	68.1432	0.18525	3.91231	927	3
4000	0.03192	2.35E-16	66.52505	1.88886	0.00066	66.3294	0.29411	3.46580	908	6
J value	9.85806E-03 ± 1.30213E-05									
Total ^{39}K vol =	2.00423E-10 CCNTP/G									
Total gas age =	946.275 ± 1.27061 Ma									
A142-1										
340	0.10645	7.85E-16	86.98428	2.63419	0.03012	78.0844	10.2316	4.83338	1023	2
360	0.00028	2.07E-18	83.87368	−1.95789	0.29163	−2.30401	102.7470	−3.59246	−41	717
440	0.00059	4.38E-18	87.43051	3.62755	−0.09082	114.2680	−30.6958	6.65606	1352	158
460	0.00183	1.35E-17	80.86233	3.13292	0.03457	70.6461	12.6341	5.74848	947	65
500	0.00211	1.55E-17	76.33558	2.88432	0.01508	71.8787	5.83853	5.29233	960	65
600	0.00280	2.07E-17	75.58747	3.20468	0.04974	60.8882	19.4467	5.88015	842	33
660	0.15283	1.13E-15	73.79989	2.59028	0.00183	73.2585	0.73359	4.75281	974	1
680	0.02791	2.06E-16	71.04472	2.62043	0.00324	70.0865	1.34876	4.80813	941	4
700	0.02705	2.00E-16	72.02760	2.60012	0.00257	71.2679	1.05474	4.77086	953	6
720	0.11147	8.22E-16	71.32644	2.56527	0.00112	70.9957	0.46370	4.70692	950	2
740	0.04120	3.04E-16	71.01069	2.5882	0.00227	70.3400	0.94449	4.74899	944	3
800	0.04011	2.96E-16	71.33549	2.55555	0.00076	71.1102	0.31581	4.68908	952	3
880	0.06211	4.58E-16	71.96983	2.61321	0.00125	71.6005	0.51317	4.79488	957	3
920	0.06160	4.55E-16	71.66352	2.55238	0.00195	71.0886	0.80225	4.68327	951	3
1000	0.08337	6.15E-16	70.81080	2.57033	0.00128	70.4329	0.53367	4.71620	945	2
1080	0.17788	1.31E-15	70.92116	2.57601	0.00037	70.8111	0.15518	4.72662	948	1
1160	0.08034	5.93E-16	70.87883	2.58045	0.00103	70.5750	0.42866	4.73477	946	2
4000	0.02009	1.48E-16	70.34298	2.60776	0.00529	68.7793	2.22294	4.78488	927	7
J value	9.76808E-03 ± 1.67233E-05									
Total ^{39}K vol =	2.63802E-10 CCNTP/G									
Total gas age =	959.956 ± 1.43517 Ma									
A145-1										
360	0.00415	3.06E-17	394.93547	2.5978	−0.00141	395.3520	−0.10547	4.76661	2873	27
420	0.03778	2.79E-16	69.46958	2.57393	−0.00355	70.5180	−1.50918	4.72281	955	7
480	0.16300	1.20E-15	68.41997	2.65508	−0.00040	68.5377	−0.17207	4.87171	935	2
490	0.18985	1.40E-15	68.42456	2.65659	−0.00051	68.5766	−0.22220	4.87448	935	1
500	0.01403	1.04E-16	68.09059	2.5561	−0.00679	70.0984	−2.94874	4.69009	951	24
580	0.07373	5.44E-16	68.25978	2.6678	−0.00280	69.0863	−1.21084	4.89505	940	3
640	0.09774	7.21E-16	68.43730	2.63232	0.00040	68.3192	0.17257	4.82994	932	1
680	0.04941	3.65E-16	68.32619	2.64628	0.00041	68.2053	0.17693	4.85556	931	2
720	0.07108	5.24E-16	67.62825	2.60158	0.00048	67.4864	0.20974	4.77354	923	2
920	0.05431	4.01E-16	67.94276	2.63883	0.00075	67.7222	0.32463	4.84189	926	2
1000	0.17403	1.28E-15	68.09987	2.78113	0.00040	67.9816	0.17368	5.10299	929	1
4000	0.07089	5.23E-16	67.82803	2.77444	−0.00023	67.8951	−0.09889	5.09072	928	4
J value	9.90225E-03 ± 1.07859E-05									
Total ^{39}K vol =	3.21601E-10 CCNTP/G									
Total gas age =	947.267 ± 1.11592 Ma									

Model Setup

Postorogenic extension is a feature commonly observed in mountain belts, regardless of the age of the orogen (Dewey, 1988). Physical mechanisms proposed to explain this phenomenon include (a) gravitational collapse; (b) changes in the stress regime, perhaps induced by changes in plate motion that cause either reduced rates of convergence or a transition to extensional stresses; (c) mantle-induced uplift and extension, including the effects of mantle plumes or mantle delamination and slab break-off (e.g., Mareschal, 1994; Burg and Ford, 1997). Gravitational collapse acts on all mountain belts during orogenesis and tends to occur shortly afterward, but the timescales and maximum amount of extension produced by gravitational collapse suggest that it is unlikely to have been the primary driving force behind the observed extension.

Changes in the stress regime acting in the region, referred to in this paper as far-field stresses, are likely to have acted over large areas beyond the core of the orogen and were therefore homogeneous across the relatively narrow region between the two shear zones. In other words, while far-field stresses could have acted over long periods of time and led to the required amount of extension and exhumation, they cannot explain the asymmetric uplift observed in the Grenville Province.

Slab break-off and mantle delamination are often invoked as mechanisms acting in mountain belts (Sacks and Secor, 1990; Platt and England, 1994). Both are suggested as inducing uplift and exhumation of orogens by replacing cold mantle with hotter asthenospheric mantle. However, both require an initial period of subsidence while the slab is in the process of breaking off (and even thereafter), or as the mantle delaminates. This is the required dynamic response of the crust to sinking material. Both mechanisms should also lead initially and locally to compression in the region immediately above the delamination. Our structural, geochronological, and kinematic constraints on the Grenville suggest this was not the case there. Even if the cold and hot mantle had been as much as a thousand degrees apart in temperature, this would have led to only 3% of the difference in density, and hence would have had only an isostatic effect on the order of 2 km, which is less than the observed amount of uplift seen in the Grenville orogen. Furthermore, isostatic uplift acts over short timescales and cannot alone explain the protracted period of exhumation.

A mantle plume will initially generate uplift, via dynamic topography, on the order of 1–2 km (Lithgow-Bertelloni and Silver, 1998; Panasyuk and Hager, 2000). It can also lead to extension, as it induces stresses on the overlying lithosphere as well as adding excess gravitational potential energy due to the added topography. However, the uplift caused by the plume will cease once the head reaches the bottom of the lithosphere.

Given that each individual mechanisms seems unable to explain all geological observations, our study has focused on the impact of a combination of mechanisms on the deformation and uplift of an overthickened crustal block. The block contains shear zones and is subject to different types of stresses acting in the presence of gravity, such as symmetric extensional stresses that stretch the block and vertical stresses that simulate the presence of an upwelling or downwelling mantle. The latter is an approximation of the dynamical effects of mantle convection, particularly of the deformation of the lithosphere subject to such normal stresses. We have not included isostatic effects on topography that would have been derived from the density contrast between the upwelling or downwelling and the normal mantle. We refer to them as bottom heating sources for convenience.

We have solved the equations of motion for an incompressible elastic-plastic material via the finite element method, using the general-purpose finite element code ABAQUS (version 6.2; Hibbit et al., 2000). We considered two-dimensional models of an overthickened block of crust with a geometry similar to that of the Grenville Province in this region at the peak of orogenesis. The dimensions of the block measure 300 km in length, with a depth of 60 km. In this block we site weak zones in positions that are determined by the present-day surface positions of the Robertson Lake shear zone and the Carthage-Colton shear zone (Fig. 4).

We have chosen a simple elastic-plastic rheology, with values determined experimentally or taken from the literature (e.g., Ranalli, 1995). For this study we did not address heterogeneities in rheology as a function of depth (or temperature), but maintained a homogenous crustal material outside the weak zones (the parameters defining crust and weak zones are given in Table 3). Boundary conditions were no-slip in the vertical direction and free-slip in the horizontal direction and at the surface. Rotation in and out of the plane of the page was not allowed. The block was subjected to extensional side loads and vertical tractions that simulated the presence of a plume or a delaminating slab. The time evolution from the latest Grenville compression to the final periods of extension when the crust must have been uplifting as a uniform block was 300 m.y. After a 300-m.y. time evolution, the predicted stresses and displacements were compared to field observations.

Our block of crustal material was subjected to a two-sided extensional load of 100 MPa. The bottom of the block was subjected to a concentrated vertical stress ranging from 10 to 50 MPa, acting upward to simulate a plume or downward to simulate a delaminating mantle. A gravitational body force was applied over the block for the duration of the analysis. The model was then allowed to evolve over a 300-m.y. period.

Results

Models using only far-field stresses simulated by applying a two-sided 100 MPa load to the block show the required amounts of extension but do not explain the temporal differences between zones or the uplift of the block. Models that include only plume stresses explain the uplift of the block but do not generate local stresses around weak zones that are high enough to induce significant amounts of normal faulting. The

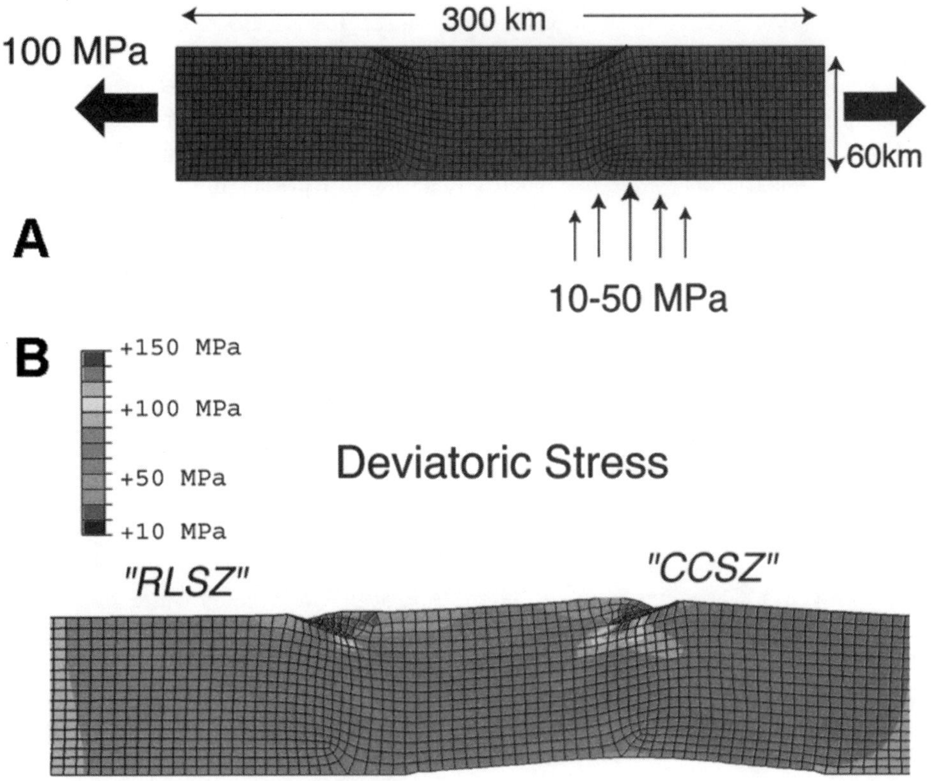

Figure 4. (A) Setup of the finite element model. Dimensions are given, as are boundary conditions and loads during analysis. (B) Contour map of resulting stress distributions. Highest stresses are seen around shear zones, with lowest stresses occurring within the block bounded by shear zones. Model parameters are given in Table 3. CCSZ—Carthage-Colton shear zone; RLSZ—Robertson Lake shear zone.

same behavior is observed if delamination is simulated. However, plume and delamination stresses are different in one important respect. To have generated the asymmetric uplift observed in the region, delamination must have acted locally on the western rather than the eastern edge (by the Robertson Lake shear zone) of the region, since their dynamic effect would have pulled down the crust. Any uplift generated by this mechanism would have required that the mantle previously attached to the crust be replaced by hotter asthenospheric mantle and that the delaminating mantle sink far enough into the mantle to no longer exert stress on the lithosphere. The isostatic uplift from hotter mantle would have acted on timescales comparable to those of postglacial rebound. Models that include only gravitational energy imposed on the block of crust with sited weak zones do not show the appropriate amount of extension. However, the total magnitude of gravitationally driven extension should have changed with the inclusion of paleotopography. The timescales over which it would have acted, however, would have remained the same. We suggest that the most likely physical scenario requires the presence of far-field stresses that induced periodic extension for 250 m.y. and plume stresses that induced the required amount of asymmetric uplift observed in the region. The tilt produced in the Frontenac or Adirondack Lowlands regions matches the tilt inferred by Dahl et al. (2001) from regional hornblende ages.

In this combined model, the results show a 300-m.y. period of uplift and extension, with the greatest uplift occurring underneath the Adirondack region, matching observed geochronologic results that show extensional motion along the Carthage-Colton shear zone due to rapid uplift of the Adirondack Highlands. Plots of contoured deviatoric stresses (Fig. 4, B) show high concentrations of stresses beneath the Carthage-Colton shear zone. The surface around the Robertson Lake shear zone and the Carthage-Colton shear zone regions dropped to very low stresses as the weak zones activated as faults. All strain was confined to the weak zones, indicating that they became the loci of permanent deformation in the region. As expected, shear stresses were highest in the regions immediately adjacent to faults.

TABLE 3. MODEL PARAMETERS

	Crust	Weak zone
Young's modulus (GPa)	80	50
Poisson's ratio	0.25	0.25
Density (kg/m^3)	2650	2650
Yield stress (MPa)	150	15

DISCUSSION

Two-dimensional finite-element models that simulate the Grenville Province at the peak of orogenesis may be used to evaluate the temporal and spatial characteristics of uplift and extension in the region. Postorogenic extension along the Carthage-Colton shear zone occurred at ca. 945 Ma, according to the argon ages of hornblende within the Carthage-Colton shear zone (Streepey et al., 2001). The time of extension across the Robertson Lake shear zone is less precisely known, but is bracketed between 900 Ma and 780 Ma, as documented by ^{40}Ar-^{39}Ar analyses of biotite and K-feldspar in the hangingwall and footwall blocks of the Robertson Lake shear zone. Figure 5 shows the combined kinematic and dynamic model of the postorogenic extensional history of the Metasedimentary Belt. Sometime after 1045 and before 945 Ma, the entire region underwent a transition from a contractional regime to an extensional environment, causing normal motion along the Robertson Lake shear zone and the Carthage-Colton shear zone. Extension terminated along the Carthage-Colton shear zone at ca. 945 Ma, as shown by argon ages of hornblende that are offset across the zone and biotite ages that are not offset along the Carthage-Colton shear zone (Fig. 3), but did not cease along the Robertson Lake shear zone until sometime after 900 Ma. Since extension occurred so much later than contraction, uplift of the crustal blocks that comprise this portion of the Grenville cannot be explained by protracted orogenesis. This differs significantly from the case in southern Asia, which has often been proposed as an analogue to the Grenville Province. In the Himalayan region, India is still underthrusting Asia, leading to continued uplift of the Tibetan plateau even while the plateau is actively extending, probably due to gravitational instabilities or flow in the lower crust (e.g., Shen et al., 2001).

One viable mechanism for generating uplift while also causing active extension is heating from below. This heat source can take several forms, including the formation of a plume, mantle delamination, or slab break-off, all responsible for replacement of crust by hot asthenospheric mantle. In addition, it has long been postulated that the existence of a supercontinent creates a blanket over the mantle (e.g., Gurnis, 1988). With heat produced in the mantle unable to escape via spreading centers, thermal instabilities under the supercontinent can lead to regional uplift and extension.

Results elsewhere in Rodinia indicate that extension is widespread. Bottom-heated sources commonly generate large amounts of magma, creating mafic igneous provinces or massive dike swarms. Whereas dike swarms in the Adirondack region are much younger, on the order of 580 Ma (Geraghty et al., 1979), abundant dike swarms at ca. 780 Ma occur in the western United States (Park et al., 1995). Extensional activity in the Baltica craton has been suggested at ca. 870 Ma (Dalziel and Soper, 2001). In addition, ages of extension have been suggested at ca. 920 Ma along the western edge of the Congo craton (Tack et al., 2001). Mafic dike swarms in the South China block have been dated at ca. 820 Ma. Collectively these results support the idea of widespread extensional activity in the Late Mesoproterozoic or Early Neoproterozoic, before the onset of large-scale rifting of the Rodinian supercontinent along this side of the Laurentian margin. It is possible that other evidence of magmatism in the North American Grenville Province during this period has been obscured by Phanerozoic tectonic activity in the Appalachian orogen.

Finally, bottom heating alone does not result in the pattern of geology seen at the surface today. It would have been necessary to apply a force to the block, approximately equal to a plate-driving force, to generate both the amounts of uplift and the amounts of extensional displacement documented in the field (Fig. 4). Therefore, we surmise that an initial instability in the mantle triggered an adjustment of the plates while Rodinia remained a coherent supercontinent. This adjustment, or change in far-field stress, would have triggered initial stages of breakup of the supercontinent, generating failed rifts or weakening the crust, in advance of later breakup.

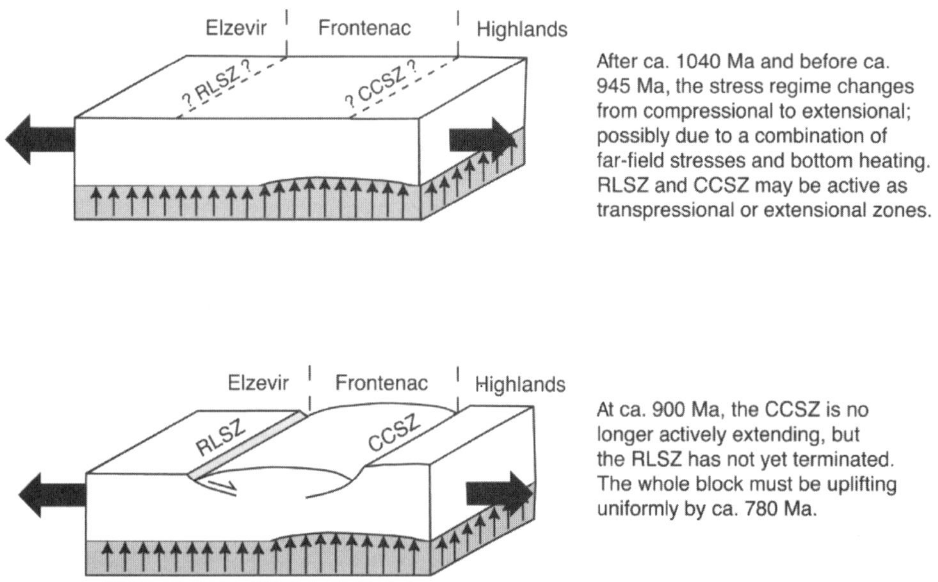
Elzevir
Frontenac
Highlands
? RLSZ ?
? CCSZ ?
After ca. 1040 Ma and before ca. 945 Ma, the stress regime changes from compressional to extensional; possibly due to a combination of far-field stresses and bottom heating. RLSZ and CCSZ may be active as transpressional or extensional zones.
Elzevir
Frontenac
Highlands
RLSZ
CCSZ
At ca. 900 Ma, the CCSZ is no longer actively extending, but the RLSZ has not yet terminated. The whole block must be uplifting uniformly by ca. 780 Ma.

Figure 5. Kinematic and dynamic model of regional extension in the Grenville Province of southeastern Ontario and northwestern New York. Between 1040 and 945 Ma, the entire region shifted from a compressional stress regime to an extensional regime. Extension along the Carthage-Colton shear zone (CCSZ) ended at 945 Ma, while extension across the Robertson Lake shear zone (RLSZ) did not terminate until after 900 Ma.

ACKNOWLEDGMENTS

This research was funded by the David and Lucille Packard Foundation (CLB), National Science Foundation (NSF) grant EAR-9980551 (CLB), NSF grant EAR-9627911 (BvdP), an NSF postdoctoral fellowship (JFM), the Geological Society of America (MMS), and the University of Michigan F. Scott Turner Foundation (MMS). We thank Chris Hall and Marcus Johnson for their generous assistance in the Radiogenic Isotope Geochemistry Laboratory at the University of Michigan. Klaus Mezger and Larry Ruff are thanked for helpful discussions.

REFERENCES CITED

Burg, J.-P., and Ford, M., 1997, Orogeny through time: An overview, *in* Burg, J.-P., and Ford, M., eds., Orogeny through time: Geological Society (London) Special Publication 121, p. 1–17.

Busch, J.P., and van der Pluijm, B.A., 1996, Late orogenic, plastic to brittle extension along the Robertson Lake shear zone: Implications for the style of deep-crustal extension in the Grenville Orogen, Canada: Precambrian Research, v. 77, p. 41–57.

Busch, J.P., Essene, E.J., and van der Pluijm, B.A., 1996a, Evolution of deep-crustal normal faults: Constraints from thermobarometry in the Grenville Orogen, Ontario, Canada: Tectonophysics, v. 265, p. 83–100.

Busch, J.P., van der Pluijm, B.A., Hall, C.M., and Essene, E.J., 1996b, Listric normal faulting during postorogenic extension revealed by $^{40}Ar/^{39}Ar$ thermochronology near the Robertson Lake shear zone, Grenville Orogen, Canada: Tectonics, v. 15, p. 387–402.

Busch, J.P., Mezger, K., and van der Pluijm, B.A., 1997, Suturing and extensional reactivation in the Grenville Orogen, Canada: Geology, v. 25, p. 507–510.

Carr, S.D., Easton, R.M., Jamieson, R.A., and Culshaw, N.G., 2000, Geologic transect across the Grenville orogen of Ontario and New York: Canadian Journal of Earth Sciences, v. 37, p. 193–216.

Corfu, F., and Easton, R.M., 1995, U-Pb geochronology of the Mazinaw Terrane, an imbricate segment of the Central Metasedimentary Belt, Grenville Province, Ontario: Canadian Journal of Earth Sciences, v. 32, p. 959–976.

Corfu, F., and Easton, R.M., 1997, Sharbot Lake terrane and its relationships to Frontenac terrane, Central Metasedimentary Belt, Grenville Province: New insights from U-Pb geochronology: Canadian Journal of Earth Science, v. 34, p. 1239–1257.

Corriveau, L., and van Breemen, O., 2000, Docking of the Central Metasedimentary Belt to Laurentia in Geon 12: Evidence from the 1.17–1.16 Ga Chevreuil Intrusive Suite and host gneisses, Québec: Canadian Journal of Earth Sciences, v. 37, p. 253–269.

Cosca, M.A., Sutter, J.F., and Essene, E.J., 1991, Cooling and inferred uplift/erosion history of the Grenville orogen, Ontario: Constraints from $^{40}Ar/^{39}Ar$ thermochronology: Tectonics, v. 10, p. 959–977.

Cosca, M.A., Essene, E.J., Kunk, M.J., and Sutter, J.F., 1992, Differential unroofing within the Central Metasedimentary Belt of the Grenville Orogen: Constraints from $^{40}Ar/^{39}Ar$ geochronology: Contributions to Mineralogy and Petrology, v. 110, p. 211–225.

Cosca, M.A., Essene, E.J., Mezger, K., and van der Pluijm, B.A., 1995, Constraints on the duration of tectonic processes: Protracted extension and deep-crustal rotation in the Grenville orogen: Geology, v. 23, p. 361–364.

Culshaw, N.G., Ketchum, J.W.F., Wodicka, N., and Wallace, P., 1994, Deep crustal ductile extension following thrusting in the southwestern Grenville Province, Ontario: Canadian Journal of Earth Sciences, v. 31, p. 160–175.

Dahl, P.S., Foland, K.A., and Pomfrey, M.E., 2001, $^{40}Ar/^{39}Ar$ thermochronology of hornblende and biotite, Adirondack Lowlands (New York) with implications for the evolution of a major shear zone: Geological Society of America Abstracts with Programs, v. 33, p. 292.

Dalziel, I.W.D., 1997, Overview: Neoproterozoic-Paleozoic geography and tectonics: Review, hypothesis, environmental speculations: Geological Society of America Bulletin, v. 109, p. 16–42.

Dalziel, I.W.D., and Soper, N.J., 2001, Neoproterozoic extension on the Scottish promontory of Laurentia: Paleogeographic and tectonic implications: Journal of Geology, v. 109, p. 299–317.

Davidson, A., 1984, Tectonic boundaries within the Grenville Province of the Canadian shield: Journal of Geodynamics, v. 1, p. 433–444.

Davidson, A., 1995, A review of the Grenville orogen in its North American type area: Journal of Australian Geology and Geophysics, v. 16, p. 3–24.

Davidson, A., 1998, An overview of Grenville Province geology, Canadian Shield, *in* Lucas, S.B., and St-Onge, M.R., eds., Geology of the Precambrian Superior Province and Precambrian fossils in North America: Boulder, Colorado, Geological Society of America, Geology of North America Series (DNAG), v. 7, p. 207–270.

Dewey, J.F., 1988, The extensional collapse of orogens: Tectonics, v. 7, p. 1123–1139.

Easton, R.M., 1988, The Robertson Lake mylonite zone—A major tectonic boundary in the Central Metasedimentary Belt, eastern Ontario [abs.]: Geological Association of Canada–Mineralogical Association of Canada–Canadian Society of Petroleum Geologists Joint Meeting, 23–25 May, St. John's, Newfoundland, v. 13, p. A34-A35.

Easton, R.M., 1992, The Grenville Province and the Proterozoic history of central and southern Ontario, *in* Thurston, P.C., et al., eds., Geology of Ontario: Sudbury, Ontario Geological Survey Special Volume 4, p. 715–904.

Geraghty, E.P., Isachsen, Y.W., and Wright, S.F., 1979, Dikes of Rand Hill, northeast Adirondacks, as paleostrain indicators: Washington, D.C., Office of Nuclear Regulatory Research, U.S. Nuclear Regulatory Commission, NUREG/CR-0889, 12 p.

Geraghty, E.P., Isachsen, Y.W., and Wright, S.F., 1981, Extent and character of the Carthage-Colton mylonite zone, northwest Adirondacks, New York: Washington, D.C., Office of Nuclear Regulatory Research, U.S. Nuclear Regulatory Commission, NUREG/CR-1865, 105 p.

Gurnis, M., 1988, Large-scale mantle convection and the aggregation and dispersal of supercontinents: Nature, v. 332, p. 695–699.

Hanmer, S., Corrigan, D., Pehrsson, S., and Nadeau, L., 2000, SW Grenville Province, Canada: The case against post-1.4 Ga accretionary tectonics: Tectonophysics, v. 319, p. 33–51.

Heizler, M.T., and Harrison, T.M., 1998, The thermal history of New York basement determined from $^{40}Ar/^{39}Ar$ K-feldspar studies: Journal of Geophysical Research, v. 103, p. 29, 795–29, 814.

Heyn, T., 1990, Tectonites of the northwest Adirondack Mountains, New York: Structural and metamorphic evolution [Ph.D. dissertation]: Ithaca, New York, Cornell University, 203 p.

Hibbit, Karlsson, and Sorensen, Inc., 2000, ABAQUS version 6.2 user's manual: Pawtucket, Rhode Island, Hibbit, Karlsson, and Sorensen, Inc.

Hoffman, P.F., 1991, Did the breakout of Laurentia turn Gondwanaland inside-out? Science, v. 252, p. 1409–1411.

Karlstrom, K.E., Bowring, S.A., Dehler, C.M., Knoll, A.H., Porter, S.M., Sharp, Z., Des Marais, D.J., Weil, A.B., Geissman, J.W., Elrick, M., Timmons, M.J., Keefe, K., and Crossey, L.J., 2000, Chuar group of the Grand Canyon: Record of breakup of Rodinia, associated change in the global carbon cycle, and ecosystem expansion by 740 Ma: Geology, v. 28, p. 619–622.

Ketchum, J.W.F., Heaman, L.M., Krogh, T.E., Culshaw, N.G., and Jamieson, R.A., 1998, Timing and thermal influence of late orogenic extension in the lower crust: A U-Pb geochronological study from the southwest Grenville orogen, Canada: Precambrian Research, v. 89, p. 25–45.

Kusky, T.M., and Lowring, D.P., 2001, Structural and U/Pb chronology of superimposed folds, Adirondack Mountains: Implications for the tectonic evolution of the Grenville Province: Journal of Geodynamics, v. 32, p. 395–418.

Li, Z.X., Li, X.H., Kinny, P.D., and Wang, J., 1999, The breakup of Rodinia: Did it start with a mantle plume beneath South China? Earth and Planetary Science Letters, v. 173, p. 171–181.

Lithgow-Bertelloni, C., and Silver, P.G., 1998, Dynamic topography, plate-driving forces, and the African superswell: Nature, v. 395, p. 269–272.

Lovera, O.M., Richter, F.M, and Harrison, T.M., 1991, Diffusion domains determined by ^{39}Ar released during step heating: Journal of Geophysical Research, v. 96, p. 2057–2069.

Mareschal, J.-C., 1994, Thermal regime and postorogenic extension in collision belts: Tectonophysics, v. 238, p. 471–484.

Martignole, J., and Reynolds, P., 1997, ^{40}Ar/^{39}Ar thermochronology along a western Québec transect of the Grenville Province, Canada: Journal of Metamorphic Geology, v. 15, p. 283–296.

McDougall, I., and Harrison, T.M., 1999, Geochronology and thermochronology by the ^{40}Ar/^{39}Ar method (2nd ed.): Oxford, England, Oxford University Press, 212 p.

McLelland, J., and Chiarenzelli, J., 1989, Age of xenolith-bearing olivine metagabbro, eastern Adirondack Mountains, New York: Journal of Geology, v. 97, p. 373–376.

McLelland, J., Lochhead, A., and Vynhal, C., 1988, Evidence for multiple metamorphic events in the Adirondack Mountains, New York: Journal of Geology, v. 96, p. 279–298.

McLelland, J., Daly, J.S., and McLelland, J.M., 1996, The Grenville orogenic cycle (ca. 1350–1000 Ma): An Adirondack perspective: Tectonophysics, v. 265, p. 1–28.

McLelland, J., Hamilton, M., Selleck, B., McLelland, J., Walker, D., and Orrell, S., 2001, Zircon U-Pb geochronology of the Ottawan Orogeny, Adirondack Highlands, New York: Regional and tectonic implications: Precambrian Research, v. 109, p. 39–72.

Mezger, K., Rawnsley, C.M., Bohlen, S.R., and Hanson, G.N., 1991a, U-Pb garnet, sphene, monazite, and rutile ages: Implications for the duration of high-grade metamorphism and cooling histories, Adirondack Mts., New York: Journal of Geology, v. 99, p. 415–428.

Mezger, K., van der Pluijm, B.A., Essene, E.J., and Halliday, A.N., 1991b, Synorogenic collapse: A perspective from the middle crust: Science, v. 254, p. 695–698.

Mezger, K., van der Pluijm, B.A., Essene, E.J., and Halliday, A.N., 1992, The Carthage-Colton mylonite zone (Adirondack Mountains, New York): The site of a cryptic suture in the Grenville orogen?: Journal of Geology, v. 100, p. 630–638.

Mezger, K., Essene, E.J., van der Pluijm, B.A., and Halliday, A.N., 1993, U-Pb geochronology of the Grenville orogen of Ontario and New York: Constraints on ancient crustal tectonics: Contributions to Mineralogy and Petrology, v. 114, p.13–26.

Moore, J.M., and Thompson, P., 1980, The Flinton group: A late Precambrian metasedimentary sequence in the Grenville Province of eastern Ontario: Canadian Journal of Earth Sciences, v. 17, p. 1685–1707.

Panasyuk, S.V., and Hager, B.H., 2000, Models of isostatic and dynamic topography, geoid anomalies, and their uncertainties: Journal of Geophysical Research, v. 105, p. 28199–28209.

Park, J.K., Buchan, K.L., and Harlan, S.S., 1995, A proposed giant radiating dyke swarm fragmented by the separation of Laurentia and Australia based on paleomagnetism of ca. 780 Ma mafic intrusions in western North America: Earth and Planetary Science Letters, v. 132, p. 129–139.

Pehrsson, S., Hanmer, S., and van Breemen, O., 1996, U-Pb geochronology of the Raglan gabbro belt, Central Metasedimentary Belt, Ontario: Implications for an ensialic marginal basin in the Grenville Orogen: Canadian Journal of Earth Sciences, v. 33, p. 691–702.

Platt, J.P., and England, P.C., 1994, Convective removal of lithosphere beneath mountain belts: Thermal and mechanical consequences: American Journal of Science, v. 294, p. 307–336.

Ranalli, G., 1995, Rheology of the Earth: New York, Chapman and Hall, 413 p.

Rivers, T., 1997, Lithotectonic elements of the Grenville Province: Review and tectonic implications: Precambrian Research, v. 86, p. 117–154.

Sacks, P.E., and Secor, D.T., 1990, Delamination in collisional orogens: Geology, v. 18, p. 999–1002.

Scott, D.J., and St-Onge, M.R., 1995, Constraints on Pb closure temperature in titanite based on rocks from the Ungava Orogen, Canada: Implications for U-Pb geochronology and *P-T-t* path determinations: Geology, v. 23, p. 1123–1126.

Shen, F., Royden, L.H., and Burchfiel, B.C., 2001, Large-scale crustal deformation of the Tibetan plateau: Journal of Geophysical Research, v. 106, p. 6793–6816.

Streepey, M.M., Essene, E.J., and van der Pluijm, B.A., 1997, A compilation of thermobarometric data from the Metasedimentary Belt of the Grenville Province, Ontario and New York State: Canadian Mineralogist, v. 35, p. 1237–1248.

Streepey, M.M., van der Pluijm, B.A., Essene, E.J., Hall, C.M., and Magloughlin, J.F., 2000, Late Proterozoic (ca. 930 Ma) extension in eastern Laurentia: Geological Society of America Bulletin, v. 112, p. 1522–1530.

Streepey, M.M., Johnson, E.L., Mezger, K., and van der Pluijm, B.A., 2001, The early history of the Carthage-Colton Shear Zone, Grenville Province, New York: Journal of Geology, v. 109, p. 479–492.

Streepey, M.M., Hall, C.M., and van der Pluijm, B.A., 2002, ^{40}Ar-^{39}Ar laser analysis of K-feldspar: Constraints on the uplift history of the Grenville Province in Ontario and New York: Journal of Geophysical Research, v. 107 (B11), 2296, doi10.129/2001JB001094.

Svenningsen, O.M., 2001, Onset of seafloor spreading in the Iapetus Ocean at 608 Ma: Precise age of the Sarek dyke swarm, northern Swedish Caledonides: Precambrian Research, v. 110, p. 241–254.

Tack, L., Wingate, M.T.D., Liegeois, J.P., Fernandez-Alonso, M., and Deblond, A., 2001, Early Neoproterozoic magmatism (1000–910 Ma) of the Zadinian and Mayumbian groups (Bas-Congo): Onset of Rodinia rifting at the western edge of the Congo Craton: Precambrian Research, v. 110, p. 307–323.

Tanner, P.W.G., and Bluck, B.J., 1999, Current controversies in the Caledonides: Journal of the Geological Society of London, v. 156, p. 1137–1141.

Timmins, J.M., Karlstrom, K.E., Dehler, C.M., Geissman, J.W., and Heizler, M.T., 2001, Proterozoic multistage (ca. 1.1 and 0.8 Ga) extension recorded in the Grand Canyon supergroup and establishment of northwest- and north-trending tectonic grains in the southwestern United States: Geological Society of America Bulletin, v. 113, p. 163–180.

Tohver, E., van der Pluijm, B.A., Van der Voo, R., Rizotto, G., and Scandolara, J.E., 2002, Paleogeography of the Amazon craton at 1.2 Ga: Early Grenvillian collision with the Llano segment of Laurentia: Earth and Planetary Science Letters, v. 199, p. 185–200.

Torsvik, T.H., Smethurst, M.A., Meert, J.G., Van der Voo, R., McKerrow, W.S., Brasier, M.D., Sturt, B.A., and Walderhaug, H.J., 1996, Continental break-up and collision in the Neoproterozoic and Palaeozoic: A tale of Baltica and Laurentia: Earth Science Reviews, v. 40, p. 229–258.

van der Pluijm, B.A., Mezger, K., Cosca, M.A., and Essene, E.J., 1994, Determining the significance of high-grade shear zones by using temperature-time paths, with examples from the Grenville orogen: Geology, v. 22, p. 743–746.

Wasteneys, H., McLelland, J., and Lumbers, S., 1999, Precise zircon geochronology in the Adirondack Lowlands and implications for revising plate tectonic models of the Central Metasedimentary Belt and Adirondack Mountains, Grenville Province, Ontario and New York: Canadian Journal of Earth Sciences, v. 36, p. 967–984.

Windley, B.F., 1986, Comparative tectonics of the western Grenville and western Himalaya, *in* Price, R.A., and Douglas, R.J.W., eds., Variations in tectonic styles in Canada: St. John's, Newfoundland, Geological Association of Canada Special Publication 11, p. 263–334.

Zeitler, P., 1987, Argon diffusion in partially outgassed alkali feldspar: Insights from ^{40}Ar/^{39}Ar analysis: Chemical Geology, v. 65, p. 167–181.

Manuscript Accepted by the Society August 25, 2003

Printed in the USA

Geological Society of America
Memoir 197
2004

Deciphering multiple Mesoproterozoic and Paleozoic events recorded in zircon and titanite from the Baltimore Gneiss, Maryland: SEM imaging, SHRIMP U-Pb geochronology, and EMP analysis

John N. Aleinikoff*
U.S. Geological Survey, MS 963, Denver Federal Center, Denver, Colorado 80225, USA
J. Wright Horton Jr.
U.S. Geological Survey, MS 926A, 12201 Sunrise Valley Drive, Reston, Virginia 20192, USA
Avery A. Drake Jr.
Scientist Emeritus, U.S. Geological Survey, MS 954, 12201 Sunrise Valley Drive, Reston, Virginia 20192, USA
Robert P. Wintsch
Department of Geological Sciences, Indiana University, Bloomington, Indiana 47405, USA
C. Mark Fanning
Research School of Earth Sciences, Australian National University, Mills Road, Canberra ACT 0200, Australia
Keewook Yi
Indiana University School of Dentistry, Indianapolis, Indiana 46202, USA

ABSTRACT

The Baltimore Gneiss, exposed in antiforms in the eastern Maryland Piedmont, consists of a suite of felsic and mafic gneisses of Mesoproterozoic age. Zircons from the felsic gneisses are complexly zoned, as shown in cathodoluminescence imaging; most zircon grains have multiple overgrowth zones, some of which are adjacent and parallel to elongate cores. Sensitive high-resolution ion microprobe (SHRIMP) analyses of oscillatory-zoned cores indicate that the volcanic protoliths of the felsic gneisses crystallized at ca. 1.25 Ga. These rocks were subsequently affected by at least three Mesoproterozoic growth events, at ca. 1.22, 1.16, and 1.02 Ga. Foliated biotite granite intruded the Baltimore Gneiss metavolcanic sequence at ca. 1075 Ma. The Slaughterhouse Granite (renamed herein) also is Mesoproterozoic, but extremely discordant U-Pb data from high-U, metamict zircons preclude calculating a precise age. The 1.25 Ga rocks of the Baltimore Gneiss are coeval with rocks emplaced in the Grenville Province during the Elzevirian orogeny, and the 1.22 Ga zircon overgrowths are coincident with a later stage of this event. Younger zircon overgrowths formed during the Ottawan phase of the Grenville orogeny.

Backscattered electron imaging of titanites from felsic gneisses and foliated biotite granite reveals that many of the grains contain cores, intermediate mantles, and rims. Electron microprobe traverses across zoned grains show regular variations in composition. SHRIMP ages for titanite from the foliated biotite granite are 374 ± 8, 336 ±

*E-mail: jaleinikoff@usgs.gov.

Aleinikoff, J.N., Horton, J.W., Jr., Drake, A.A., Jr., Wintsch, R.P., Fanning, C.M., and Yi, K., 2004, Deciphering multiple Mesoproterozoic and Paleozoic events recorded in zircon and titanite from the Baltimore Gneiss, Maryland: SEM imaging, SHRIMP U-Pb geochronology, and EMP analysis, *in* Tollo, R.P., Corriveau, L., McLelland, J., and Bartholomew, M.J., eds., Proterozoic tectonic evolution of the Grenville orogen in North America: Boulder, Colorado, Geological Society of America Memoir 197, p. 411–434. For permission to copy, contact editing@geosociety.org.

8, and 301 ± 12 Ma. The ca. 374 Ma age suggests growth of titanite during a thermal event following the Acadian orogeny, whereas the late Paleozoic titanite growth ages may be due to greenschist-facies replacement reactions associated with Alleghanian metamorphism and deformation.

Keywords: Baltimore Gneiss, zircon, titanite, SHRIMP U-Pb geochronology, Grenville

INTRODUCTION

The Baltimore Gneiss, located in the eastern Maryland Piedmont, is exposed in the antiformal cores of refolded recumbent folds (Muller and Chapin, 1984; Fisher, 1989). These antiforms occur in a belt of basement exposures extending from Vermont to Georgia (Fig. 1). Exposures of the Baltimore Gneiss constitute the largest internal massif of Mesoproterozoic (Laurentian?) basement in the central Appalachians (Hatcher, 1984; Drake et al., 1988). The antiquity of the Baltimore Gneiss was first determined by pioneering studies in U-Pb geochronology and corroborated by Rb-Sr dating almost a half century ago (Tilton et al., 1958, 1970; Wetherill et al., 1966, 1968). Subsequent studies in the 1970s confirmed the Mesoproterozoic age of the rocks and documented a multistage history of zircon growth that was unresolvable by conventional dating methods (Grauert et al., 1973; Grauert, 1974).

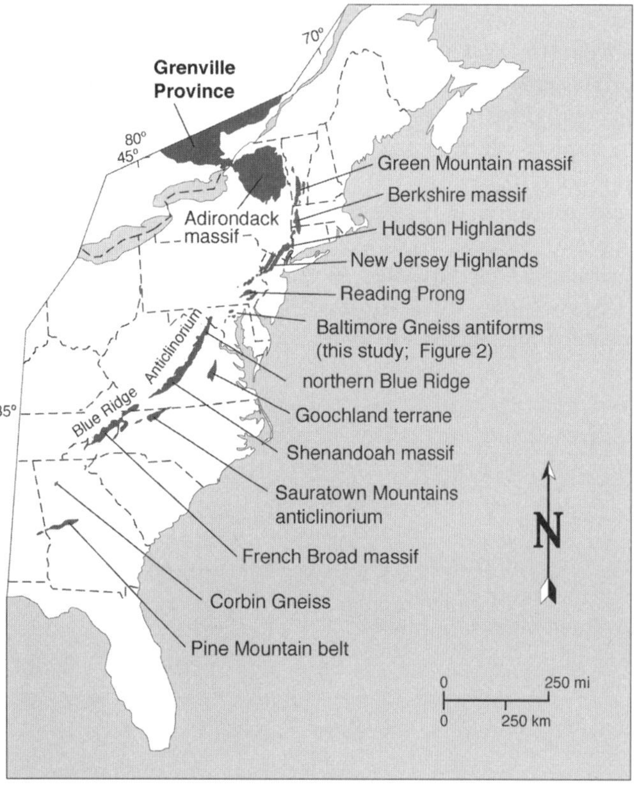

Figure 1. Map showing Mesoproterozoic basement massifs in the eastern United States. Modified from Drake et al. (1988).

In order to decipher the multiple age components of zircon from the Baltimore Gneiss, three samples of felsic gneiss and two samples of granite were collected from within four gneiss antiforms for in situ microanalysis of zircon using the sensitive high-resolution ion microprobe (SHRIMP). In addition, titanite from the granite sample was dated by SHRIMP.

GEOLOGIC SETTING

The Mesoproterozoic Baltimore Gneiss and its Neoproterozoic to early Paleozoic cover sequence, the Glenarm Group, constitute the Baltimore terrane (Williams and Hatcher, 1982, 1983; Horton et al., 1989). The Baltimore Gneiss consists of quartzofeldspathic gneisses of felsic to intermediate composition that contain minor mafic (amphibolite) interlayers (Hopson, 1964; Crowley, 1976). The combination of igneous geochemical compositions and layering of felsic to mafic components containing minor biotite-rich interlayers was interpreted as evidence that the Baltimore Gneiss formed by metamorphism of volcaniclastic and volcanic rocks (Hopson, 1964; Crowley, 1976). These extrusive rocks differ from the predominantly plutonic Mesoproterozoic rocks, some of which are charnockitic, that occur in other parts of the central Appalachians (Drake, 1984; Rankin et al., 1989; Faill, 1997). The unique character of the Baltimore Gneiss has led to hypotheses wherein the Baltimore terrane is more than just an allochthonous slice of Laurentian crust. Previous interpretations propose that the Baltimore terrane is (1) a fragment of Laurentia that rifted away and returned (Rankin, 1975; Thomas, 1977; Muller and Chapin, 1984; Drake et al., 1988, 1989), (2) a Laurentian crustal block displaced by strike-slip faulting (Horton et al., 1989), or (3) part of an exotic Baltimore microcontinent unrelated to Laurentia (Fisher et al., 1979; Faill, 1997).

Exposures of Baltimore Gneiss in the Chattolanee, Clarksville, Mayfield, Phoenix, Texas, Towson, and Woodstock antiforms (Fig. 2) contain the same general suite of rock types. Hopson (1964) described them as banded gneiss, migmatite, veined gneiss, augen gneiss, and granitic gneiss. Crowley (1976) and Crowley et al. (1976) mapped lithologic subdivisions of the Baltimore Gneiss as layered gneiss, augen gneiss, streaked-augen gneiss, and hornblende gneiss, but treated Hopson's (1964, p. 44–45) "granitic gneiss" as a separate unit, named the Slaughterhouse Gneiss.

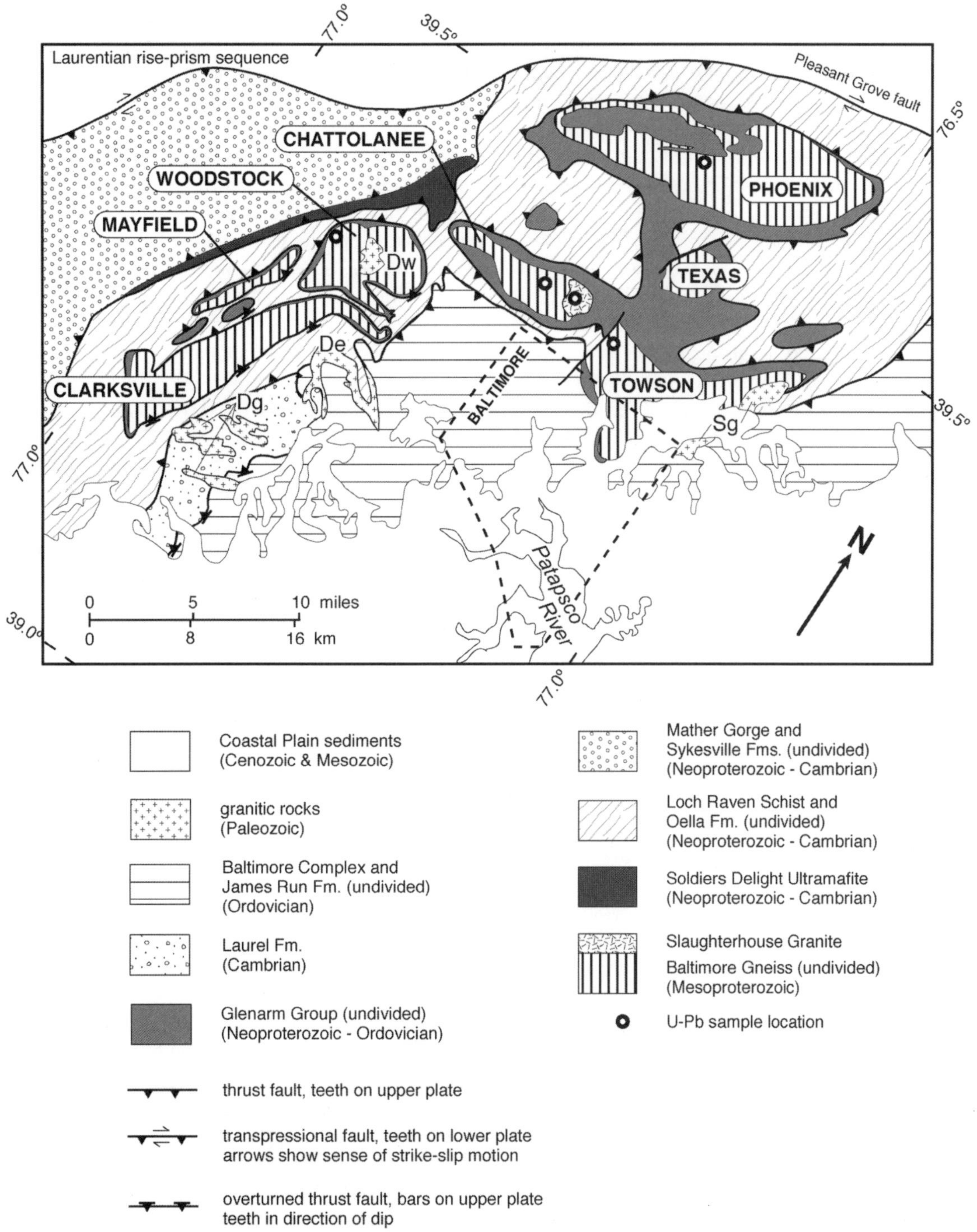

Figure 2. Map showing the general geologic setting of the Baltimore Gneiss in the Maryland Piedmont. It includes exposures in the Phoenix, Texas, Towson, Chattolanee, Woodstock, Mayfield, and Clarksville antiforms. Adapted from Cleaves et al. (1968), Crowley (1976), Crowley et al. (1976), Muller and Chapin (1984), and Drake (1994; 1998a, Fig. 1). Plutonic bodies are De—Ellicott City Granodiorite, Dg—Guilford Granite, Dw—Woodstock Granite, and Sg—Gunpowder Granite.

The Slaughterhouse Gneiss of Crowley (1976) is here renamed the Slaughterhouse Granite, because common use of the term "gneiss" (Jackson, 1997) implies layering uncharacteristic of the unit. The Slaughterhouse Granite contains biotite and muscovite, and is well foliated to almost massive. We concur with Crowley (1976) that it resembles nearby Paleozoic granite except for its restricted occurrence within the Baltimore Gneiss. Crowley's (1976, p. 21) interpretation of the Slaughterhouse as "post-Baltimore Gneiss, pre-Glenarm" is based on mapped plutons "athwart the layering of the Baltimore Gneiss" and truncation by rocks of the Glenarm Group.

Metasedimentary rocks of the Glenarm Group unconformably overlie the Baltimore Gneiss and the Slaughterhouse Granite. Figure 2 presents the geology of our study area and follows Drake's (1994) restricted usage of the Glenarm Group, which includes only the Setters Formation (mainly quartzite and schist) and the overlying Cockeysville Marble. This differs from earlier, broader usage of the Glenarm Series (Knopf and Jonas, 1923; Hopson, 1964; Southwick and Fisher, 1967; Higgins, 1972), Glenarm Supergroup (Crowley, 1976), and Glenarm Group (Drake et al., 1989). Gates et al. (1991, 1999) chose to avoid the Glenarm nomenclature entirely by using individual formation names.

Crowley (1976) suggested that the Cockeysville Marble grades upward into the pelitic Loch Raven Schist, noting layers of marble, quartzite, and conglomerate near the base of the Loch Raven (Rush Brook Member). Fisher (1989) suggested that the Loch Raven may be an offshore equivalent of the Setters Formation. Drake (1994) concluded that the base of the Loch Raven Schist is a thrust fault, now considered by Drake (1998a) and Fleming and Drake (1998) to be above the Laurel Formation.

Along the southeast flank of the Baltimore gneiss domes, all of these rocks have been overthrust by the mafic-ultramafic Baltimore Complex (Higgins, 1977; Morgan, 1977) (also called the Baltimore mafic complex, with a U-Pb age of 489 ± 7 Ma; Sinha et al., 1997), which has, in turn, been overthrust by the predominantly metavolcanic rocks of the James Run Formation (Drake, 1998b). On the northwest flank, successively higher thrust sheets contain the Soldiers Delight Ultramafite (Drake, 1994), the Sykesville Formation (metamorphosed sedimentary mélange), and the Mather Gorge Formation (metagraywacke and schist; Fig. 2).

Previous interpretations that the Baltimore Gneiss cored mantled gneiss domes (Eskola, 1949, p. 470) were questioned by Rodgers (1970, p. 190), who noted map patterns characteristic of refolded recumbent folds. Subsequent investigators concurred with Rodgers (1970) that the Baltimore Gneiss is exposed in the anticlinal cores of polydeformed recumbent folds, commonly termed "anticlines" or "antiforms" for brevity (Crowley, 1976; Fisher et al., 1979; Muller and Chapin, 1984; Fisher, 1989). However, on the basis of a balanced cross-section constructed across the central Appalachians (Elliott et al., 1982), a regional thrust fault is necessary to explain crustal shortening observed to the west in the Blue Ridge, implying that the refolded recumbent fold model is oversimplified (Fisher, 1989).

Petrologic studies of the Baltimore Gneiss in Maryland revealed no evidence for relict granulite-facies mineral assemblages (Olsen, 1999), in contrast to Mesoproterozoic granulites previously described as "Baltimore gneiss" (Wagner and Crawford, 1975) in the Brandywine terrane of southeastern Pennsylvania (Faill, 1997). Mineral assemblages in younger pelitic schists around the Baltimore Gneiss indicate that Paleozoic amphibolite-facies metamorphism was characterized by pressures and temperatures (~6–7 kilobars and 645–685 °C; Olsen, 1999) that decreased outward from the anticlinal cores (Lang, 1990). On the basis of structural interpretations, Gates et al. (1991, 1999) suggested that the arching or doming of recumbent folds may be a result of late Paleozoic transpression.

PREVIOUS GEOCHRONOLOGY

The Baltimore Gneiss has been the subject of numerous geochronologic studies since the mid-1950s. Tilton et al. (1958), in one of the first studies to show different mineral ages from one rock (i.e., a precursor to the concept of cooling ages), established the original antiquity of gneisses from the Towson and Phoenix antiforms by determining $^{207}Pb/^{206}Pb$ ages of zircons of ca. 1120 Ma. They noted that these zircons retain crystal faces, but are somewhat rounded, suggesting modification of igneous zircon by metamorphic or sedimentary processes. Biotite Rb-Sr and K-Ar ages of ca. 300–350 Ma indicated that the gneisses had been metamorphosed in the Paleozoic. Wetherill et al. (1966, 1968) dated many samples from the Maryland Piedmont using the K-Ar and Rb-Sr methods, further establishing the regional geologic context of Carboniferous mineral-cooling ages. They also obtained a Rb-Sr whole-rock isochron age of 1050 ± 100 Ma for gneiss samples from three antiforms. Tilton et al. (1970) reported U-Pb zircon ages from several rocks in the Maryland Piedmont, including a sample of Baltimore Gneiss from the Woodstock antiform that yielded a $^{207}Pb/^{206}Pb$ age of ca. 1250 Ma. They suggested that their previous U-Pb age data from felsic gneisses of the Towson and Phoenix antiforms (ca. 1120 Ma) might represent the resetting of an older age, perhaps represented by the age of the Woodstock antiform sample.

Grauert (1974) provided an interpretation for the different U-Pb zircon ages found by Tilton et al. (1958, 1970). By examining zircons in "cathode luminescence" [sic], he found that most grains from felsic gneiss samples from the Towson and Phoenix antiforms contained multiple growth stages. At least three phases of growth were interpreted to correspond to inherited core material, igneous crystallization, and metamorphic overgrowths. Data from multigrain fractions were discordant, indicating both mixing of age populations and superimposed Pb loss. The isotopic data suggested crystallization ages of ca. 1080–1180 Ma, and disturbance to the isotopic systems in the mid-Paleozoic.

The pioneering and insightful work of Grauert (1974) indicated that multigrain conventional thermal ionization mass spectrometry (TIMS) analyses of zircons from felsic gneisses of the Baltimore Gneiss likely would yield mixed ages. Many of the igneous parts of the grains are partially resorbed and have irregular boundaries. In addition, younger overgrowths locally invaded older parts of the grains along cracks. Thus, modern high-precision TIMS methods, using predigestion abrasion and analysis of single grains or even fragments of grains, could still possibly yield mixed ages. Our solution to this vexing geochronologic problem was to use in situ microanalysis of zircons by SHRIMP (which yields high spatial resolution, but significantly lower precision than TIMS), guided by high-resolution cathodoluminescence (CL) images.

METHODS

Sample Preparation

About 25 kg of rock collected from each outcrop were processed using standard mineral separation techniques, including crushing, pulverizing, and use of a Wilfley Table, a magnetic separator, and heavy liquids. Zircon and titanite separates were concentrated to ~95% purity.

SHRIMP Analyses

Hand-picked zircon and titanite grains ~100–150 μm in length were mounted in epoxy, ground to nearly half-thickness using 1500-grit wet-dry sandpaper, and polished with 6 μm and 1 μm diamond suspension abrasive. Transmitted and reflected light photos were taken of all mounted grains. In addition, CL images of all zircons and backscattered electron images of all titanites were used to reveal internal zoning related to chemical composition.

The analytical procedures for SHRIMP II (Research School of Earth Sciences, Australian National University, Canberra) followed methods described in Williams (1998). The primary oxygen ion beam, operated at ~4–5 nA, excavated an area of ~25–30 μm in diameter to a depth of ~1 μm. The magnet was cycled through the mass stations six times per analysis. Zircon standard AS3 (from 1099 Ma gabbroic anorthosite, Duluth Complex; Paces and Miller, 1993) and titanite standard BLR-1 (1051 Ma metamorphic megacrystal, Ontario; M.D. Schmitz, 2001, personal commun.) were used to calibrate $^{206}Pb/^{238}U$. U and Pb concentrations are accurate to ~10–20%. SHRIMP isotopic data were reduced and plotted using the Squid and Isoplot/Ex programs of Ludwig (2002a, 2002b).

Zircon isotopic data (corrected for common Pb using the measured $^{206}Pb/^{204}Pb$) are shown on $^{238}U/^{206}Pb$– $^{207}Pb/^{206}Pb$ concordia plots (Tera and Wasserburg, 1972) to allow visual assessment of the degree of discordance of each data point. Nearly all analyses have very low common Pb contents (common ^{206}Pb mostly less than 0.2% of total ^{206}Pb; Table 1), so the individual age of each spot analysis is insensitive to the common-Pb correction. Composite ages were determined by calculating the weighted average of $^{207}Pb/^{206}Pb$ ages.

Titanite isotopic data, uncorrected for common Pb, are plotted on a Total Pb (Ludwig, 2002a) Tera-Wasserburg concordia diagram to assess the influence of common Pb, which varied from ~10 to 31% for ^{206}Pb (Table 1). Because of the relatively young ages of all titanites (discussed later), composite ages were determined by two methods: (1) A weighted average of the $^{206}Pb/^{238}U$ ages was calculated using the ^{207}Pb-correction method for each analysis (cf. Stern, 1997; Williams, 1998). The determination of the initial $^{207}Pb/^{206}Pb$ (i.e., common Pb composition) was made by calculating a best-fit line through nine data points (uncorrected for common Pb) with the oldest ages. The y-axis intercept (0.92 ± .20) was used for all titanite analyses, although the uncertainty in the intercept was not propagated through the age calculations. The error in the weighted average composite age incorporates the uncertainty in the age of each spot analysis (related to counting statistics and the 2σ external spot-to-spot error) quadratically combined with the 2σ error of the mean of the standard analyses (Ludwig, 2002b) so that these data can be compared realistically with other data sets. (2) A best-fit regression was calculated through groups of uncorrected data on a Total Pb Tera-Wasserburg plot to determine a lower intercept age and the level of uncertainty.

Electron Microprobe Analyses

Electron microprobe (EMP) analyses of titanite were performed using a JEOL microprobe JXA-8900R at the School of Dentistry, Indiana University, with operating conditions of 15 KV, 40 nA, 2 μm diameter beam size, a 30-second counting time in peak position (except F, which was counted for 20 seconds), and a 10-second counting time in both background positions. Natural standards were used for major elements and synthetic standards for minor elements. The compositions of several titanite grains were analyzed, but the compositions reported later are only for those grains also analyzed by SHRIMP.

RESULTS

Zircon Morphology and Cathodoluminescence Zoning Patterns

Zircons from the three felsic gneiss samples (CV-C-3-1, H-1-96, and Sy-1-96) are similar in appearance. They are light tan to brown and subhedral to euhedral, have length-to-width ratios (l/w) of 2–5, contain numerous inclusions and cracks, and have deformed and partially resorbed crystal faces (Fig. 3, A–C). Most zircons in the foliated biotite granite (sample T-1-96) are colorless to medium brown, euhedral, and have l/w of 2–5. This rock also contains a small population of euhedral, colorless,

TABLE 1. SHRIMP U-TH-PB DATA FOR ZIRCON AND TITANITE FROM FELSIC GNEISSES AND GRANITES, BALTIMORE GNEISS ANTIFORMS, MARYLAND

Sample*	Measured $^{204}Pb/^{206}Pb$	Measured $^{207}Pb/^{206}Pb$	% common ^{206}Pb	U (ppm)	Th/U	$^{206}Pb†/^{238}U$ (Ma)	err³ (Ma)	$^{207}Pb/^{206}Pb$ (Ma)	err§(%)	$^{238}U†/^{206}Pb$	err§ (%)	$^{207}Pb†/^{206}Pb$	err§ (%)
CV-C-3-1, felsic gneiss, Chattolanee antiform													
CVC-1.1 ic	0.000016	.0824	0.03	706	0.64	1263.9	13.9	1249	8	4.620	1.1	.0822	0.4
CVC-2.1 x	0.000028	.0855	0.05	312	0.92	1346.5	15.3	1318	16	4.312	1.2	.0851	0.8
CVC-3.1 m	0.000043	.0811	0.07	764	0.52	1204.2	13.0	1210	9	4.867	1.1	.0805	0.4
CVC-4.1 og	0.000039	.0816	0.07	425	0.22	1249.7	13.9	1221	12	4.681	1.2	.0810	0.6
CVC-5.1 x	0.000055	.0943	0.09	518	0.06	1436.3	15.8	1498	9	3.993	1.1	.0935	0.5
CVC-6.1 og	0.000019	.0787	0.03	427	0.23	1127.4	14.7	1158	13	5.225	1.4	.0784	0.7
CVC-7.1 og	0.000138	.0826	0.23	218	0.21	1247.9	24.0	1213	27	4.690	2.0	.0807	1.4
CVC-8.1 og	0.000149	.0830	0.25	201	0.59	1241.0	18.4	1218	19	4.716	1.5	.0809	1.0
CVC-9.1 ic	0.000096	.0837	0.16	288	0.50	1226.4	15.6	1253	17	4.766	1.3	.0823	0.9
CVC-10.1 og	0.000134	.0821	0.22	285	0.19	1249.8	14.5	1203	18	4.685	1.2	.0802	0.9
CVC-11.1 og	0.000256	.0827	0.43	95	0.46	1228.6	28.7	1174	58	4.775	2.4	.0791	2.9
CVC-12.1 ic	0.000007	.0823	0.01	316	0.61	1249.9	14.4	1251	14	4.673	1.2	.0822	0.7
CVC-13.1 ic	0.000027	.0825	0.04	427	0.38	1230.1	17.0	1247	13	4.752	1.4	.0821	0.7
CVC-14.1 og	N.D.	.0809	N.D.	1154	0.04	1170.4	12.7	1220	12	5.011	1.1	.0809	0.6
CVC-15.1 ic	0.000039	.0828	0.07	467	0.58	1232.6	16.5	1251	11	4.741	1.4	.0822	0.5
CVC-16.1 og	0.000076	.0812	0.13	572	0.20	1180.2	13.2	1200	16	4.973	1.2	.0801	0.8
CVC-17.1 og	0.000885	.0912	1.48	118	0.62	1309.7	23.2	1164	72	4.471	1.8	.0787	3.6
CVC-18.1 x	0.000083	.1064	0.13	346	0.44	1732.8	20.1	1719	10	3.246	1.2	.1052	0.5
CVC-19.1 og	0.000040	.0810	0.07	773	0.27	1180.4	13.9	1209	9	4.970	1.2	.0805	0.4
CVC-19.2 ic	0.000029	.0824	0.05	114	0.47	1244.8	16.9	1245	31	4.695	1.4	.0820	1.6
CVC-20.1 ic	0.000070	.0822	0.12	386	0.22	1259.3	15.1	1227	14	4.643	1.2	.0812	0.7
CVC-21.1 og	0.000256	.0820	0.43	159	0.60	1204.2	15.0	1156	28	4.880	1.3	.0783	1.4
CVC-21.2 og	0.000039	.0793	0.07	926	0.13	1106.5	12.0	1167	9	5.325	1.1	.0788	0.4
H-1-96, felsic gneiss, Phoenix antiform													
H1-96-1.1 ic	0.000064	.0827	0.11	393	0.16	1225.2	17.3	1240	34	4.774	1.5	.0818	1.8
H1-96-1.2 og	0.000231	.0742	0.40	166	0.32	1020.7	13.8	955	35	5.846	1.4	.0709	1.7
H1-96-2.1 ic	0.000007	.0828	0.01	437	0.41	1246.7	18.7	1263	11	4.683	1.6	.0827	0.6
H1-96-3.1 ic	0.000039	.0818	0.06	416	0.16	1208.9	16.1	1229	13	4.844	1.4	.0813	0.7
H1-96-3.2 og	N.D.	.0745	N.D.	125	1.27	1063.6	13.6	1058	22	5.576	1.3	.0746	1.1
H1-96-4.1 m	0.000011	.0810	0.02	1504	0.10	1263.5	13.6	1219	7	4.629	1.1	.0809	0.3
H1-96-5.1 og	0.000105	.0794	0.18	333	0.46	1032.2	11.7	1145	20	5.730	1.2	.0779	1.0
H1-96-6.1 og	0.000013	.0788	0.02	438	0.14	1143.2	19.4	1163	18	5.149	1.8	.0786	0.9
H1-96-7.1 ic	0.000038	.0833	0.06	236	0.11	1252.4	15.3	1263	18	4.661	1.3	.0827	0.9
H1-96-8.1 og	0.000058	.0813	0.10	388	0.47	1233.6	16.0	1209	14	4.747	1.3	.0805	0.7
H1-96-8.2 og	0.000068	.0730	0.12	229	0.24	967.9	11.5	986	25	6.168	1.2	.0720	1.3
H1-96-9.1 og	0.000058	.0812	0.10	315	0.39	1195.5	18.6	1207	18	4.905	1.6	.0804	0.9
H1-96-10.1 ic	0.000074	.0838	0.12	244	0.49	1206.3	14.1	1263	22	4.846	1.2	.0827	1.1
H1-96-11.1 og	0.000035	.0813	0.06	684	0.53	1173.0	14.7	1217	9	5.001	1.3	.0808	0.4
H1-96-12.1 m	0.000239	.0833	0.40	404	0.16	1283.5	17.9	1194	22	4.559	1.4	.0799	1.1
H1-96-13.1 ic	0.000006	.0829	0.01	344	0.30	1318.0	15.4	1264	14	4.420	1.2	.0828	0.7
H1-96-14.1 ic	0.000017	.0825	0.03	459	0.48	1200.8	16.9	1251	11	4.872	1.5	.0822	0.6
H1-96-15.1 og	0.000018	.0796	0.03	2212	0.05	1182.0	13.2	1180	6	4.969	1.2	.0793	0.3
H1-96-15.2 og	0.000013	.0813	0.02	1306	0.62	1275.6	17.1	1225	10	4.582	1.4	.0812	0.5
H1-96-14.2 og	0.000003	.0740	0.01	484	0.14	1026.9	11.6	1040	13	5.788	1.2	.0740	0.6
H1-96-16.1 ic	0.000113	.0847	0.19	253	0.31	1253.2	21.9	1273	17	4.656	1.8	.0831	0.9
H1-96-16.2 og	0.000112	.0777	0.19	1372	0.08	951.4	13.0	1098	12	6.248	1.4	.0761	0.6
H1-96-17.1 m	0.000010	.0800	0.02	1371	0.12	1214.9	16.1	1193	18	4.827	1.4	.0798	0.9
Sy-1-96, felsic gneiss, Woodstock antiform													
Session 1													
SY1-1.1 m	0.000009	.0814	0.02	794	0.66	1233.8	14.4	1229	8	4.742	1.2	.0813	0.4
SY1-2.1 ic	0.000007	.0817	0.01	1241	0.80	1222.6	13.2	1235	6	4.785	1.1	.0816	0.3

SY1-3.1 ic	0.000029	.0822	0.05	820	0.62	1217.8	13.2	1240	10	4.804	1.1	.0818	0.5
SY1-4.1 m	0.000038	.0817	0.06	466	0.56	1231.9	14.2	1224	10	4.751	1.2	.0811	0.5
SY1-5.1 ic	0.000069	.0821	0.12	561	0.58	1195.1	14.6	1225	10	4.902	1.3	.0811	0.5
SY1-6.1 ic	0.000039	.0822	0.07	763	0.61	1224.8	13.4	1237	8	4.776	1.1	.0816	0.4
SY1-7.1 og	0.000029	.0811	0.05	733	0.33	1220.9	13.9	1214	10	4.797	1.2	.0807	0.5
SY1-8.1 ic	0.000018	.0818	0.03	935	0.62	1272.3	13.9	1235	7	4.592	1.1	.0816	0.4
SY1-9.1 og	0.000013	.0790	0.02	638	0.27	1150.1	13.9	1169	14	5.116	1.3	.0789	0.7
SY1-10.1 m	0.000009	.0822	0.02	312	0.49	1245.2	14.4	1248	14	4.692	1.2	.0821	0.7
SY1-9.2 og	0.000031	.0784	0.05	1432	0.04	1095.8	13.7	1147	7	5.385	1.3	.0780	0.4
SY1-11.1 og	0.000046	.0777	0.08	1690	0.04	1072.7	12.7	1121	10	5.511	1.2	.0770	0.5
SY1-12.1 og	N.D.	.0807	0.00	759	0.38	1163.7	13.6	1215	8	5.042	1.2	.0807	0.4
SY1-13.1 ic	0.000017	.0825	0.03	1053	0.67	1263.9	13.6	1252	7	4.620	1.1	.0823	0.3
SY1-14.1 ic	0.000078	.0825	0.13	254	0.45	1170.5	13.5	1232	17	5.008	1.2	.0814	0.9
SY1-15.1 ic	0.000067	.0831	0.11	374	0.47	1235.4	14.7	1249	14	4.731	1.2	.0822	0.7
SY1-16.1 ic	0.000011	.0821	0.02	850	0.63	1241.8	13.4	1245	7	4.707	1.1	.0820	0.4
SY1-17.1 og	0.000032	.0777	0.05	1421	0.02	1023.6	12.1	1128	7	5.784	1.2	.0773	0.3
SY1-17.2 ic	0.000132	.0832	0.22	218	0.40	1162.1	15.8	1230	17	5.046	1.4	.0814	0.9
Session 2													
SY1-1.1 ic	N.D.	.0814	N.D.	340	0.48	1194.4	16.8	1242	14	4.901	1.5	.0818	0.7
SY1-1.2 og	0.000000	.0793	0.00	1406	0.03	1092.9	12.2	1178	7	5.391	1.2	.0793	0.4
SY1-2.1 og	N.D.	.0806	N.D.	388	0.52	1206.0	14.1	1214	12	4.859	1.2	.0807	0.6
SY1-2.2 ic	0.000022	.0823	0.04	565	0.58	1258.5	15.3	1245	10	4.641	1.3	.0820	0.5
SY1-3.1 ic	N.D.	.0827	N.D.	394	0.52	1240.5	13.9	1273	11	4.705	1.2	.0832	0.6
SY1-3.2 og	0.000012	.0777	0.02	1702	0.04	1021.8	11.1	1135	6	5.793	1.1	.0776	0.3
SY1-3.3 m	0.000088	.0687	0.15	139	0.05	619.5	8.9	851	55	9.831	1.5	.0674	2.6
SY1-4.1 og	0.000000	.0790	0.00	1242	0.06	1116.3	12.1	1173	12	5.276	1.1	.0790	0.6
SY1-4.2 og	0.000001	.0812	0.00	776	0.64	1209.7	13.5	1226	8	4.841	1.2	.0812	0.4
SY1-4.3 ic	0.000004	.0823	0.01	683	0.63	1243.5	14.9	1252	9	4.698	1.2	.0823	0.5
SY1-5.1 ic	0.000024	.0821	0.04	1164	0.83	1215.0	13.5	1240	7	4.816	1.2	.0818	0.4
SY1-5.2 og	N.D.	.0807	N.D.	549	0.58	1240.1	13.9	1221	11	4.719	1.2	.0810	0.6
SY1-5.3 og	0.000000	.0727	0.00	717	0.06	956.3	10.6	1005	11	6.240	1.1	.0727	0.5
SY1-6.1 ic	N.D.	.0813	0.00	851	0.70	1219.5	13.4	1230	8	4.799	1.1	.0813	0.4
SY1-6.2 ic	N.D.	.0812	N.D.	267	0.49	1212.6	14.4	1234	15	4.827	1.2	.0815	0.8
SY1-7.1 ic	N.D.	.0819	N.D.	382	0.37	1183.2	14.7	1246	13	4.949	1.3	.0820	0.7
SY1-7.2 og	0.000085	.0807	0.14	761	0.66	1052.5	13.2	1186	10	5.604	1.3	.0796	0.5
SY1-8.1 ic	0.000019	.0819	0.03	269	0.44	1156.6	16.4	1236	15	5.069	1.5	.0816	0.8
SY1-8.2 ic	0.000001	.0815	0.00	796	0.62	1271.0	13.8	1233	7	4.597	1.1	.0815	0.3
SY1-9.1 m	0.000000	.0809	0.00	273	0.49	1216.9	13.7	1220	11	4.812	1.2	.0809	0.6
SY1-9.2 ic	N.D.	.0816	N.D.	839	0.72	1258.7	13.6	1239	7	4.642	1.1	.0817	0.3
T-1-96, foliated biotite granite layer, Towson antiform													
T1-1.1 ic	0.000053	.0761	0.09	541	0.39	1101.3	12.2	1078	12	5.373	1.1	.0754	0.6
T1-2.1 ic	0.000052	.0762	0.09	535	0.42	1104.7	12.7	1082	15	5.355	1.2	.0755	0.7
T1-3.1 ic	0.000058	.0763	0.10	416	0.57	1113.1	12.5	1080	15	5.314	1.2	.0754	0.8
T1-3.2 ic	0.000062	.0754	0.11	962	0.12	1049.6	11.8	1055	11	5.654	1.2	.0745	0.5
T1-4.1 x	0.000122	.0841	0.20	313	0.62	1204.1	13.7	1254	20	4.857	1.2	.0824	1.0
T1-4.2 x	N.D.	.0823	0.00	231	0.18	1167.4	14.1	1254	19	5.016	1.2	.0823	0.9
T1-5.1 x	0.000028	.0803	0.05	609	0.49	1178.4	17.4	1194	15	4.982	1.5	.0799	0.8
T1-6.1 x	0.000210	.0834	0.35	170	0.24	1246.2	24.0	1206	25	4.698	2.0	.0804	1.3
T1-7.1 x	0.000368	.0830	0.62	86	0.67	1280.3	17.3	1142	45	4.582	1.4	.0778	2.3
T1-8.1 x	0.000024	.0817	0.04	308	0.52	1189.5	16.3	1231	13	4.925	1.4	.0814	0.7
T1-9.1 ic	0.000052	.0763	0.09	337	0.53	1122.1	13.0	1084	19	5.268	1.2	.0756	0.9
T1-10.1 x	0.000129	.0836	0.21	345	1.00	1258.6	14.2	1240	15	4.642	1.2	.0818	0.8
T1-11.1 x	N.D.	.0767	N.D.	295	0.47	1110.0	14.1	1130	13	5.317	1.3	.0774	0.7
T1-12.1 x	0.000233	.0867	0.39	409	0.53	1305.0	17.5	1279	16	4.463	1.4	.0834	0.8
T1-12.2 og	0.009487	.2037	16.44	3552	0.55	328.9	3.7	849	252	18.778	1.5	.0674	12.1
T1-13.1 ic	0.000149	.0773	0.25	135	1.39	1088.2	13.6	1074	32	5.441	1.3	.0752	1.6

(continued)

TABLE 1. *Continued*

Sample*	Measured $^{204}Pb/^{206}Pb$	Measured $^{207}Pb/^{206}Pb$	% common ^{206}Pb	U (ppm)	Th/U	$^{206}Pb^{†}/^{238}U$ (Ma)	err[3] (Ma)	$^{207}Pb/^{206}Pb$ (Ma)	err[§](%)	$^{238}U^{†}/^{206}Pb$	err[§] (%)	$^{207}Pb^{†}/^{206}Pb$	err[§] (%)
T1-14.1 x	0.000092	.0833	0.15	114	0.70	1262.7	15.9	1245	23	4.626	1.3	.0820	1.2
T1-14.2 og	0.000137	.0567	0.25	1135	0.01	456.3	5.2	402	25	13.657	1.2	.0547	1.1
T1-15.1 ic	0.000088	.0756	0.15	403	0.20	1036.0	11.6	1049	16	5.733	1.2	.0743	0.8
T1-16.1 og	0.000567	.0810	0.97	388	0.64	627.9	7.3	1012	36	9.634	1.2	.0729	1.8
T1-17.1 ic	0.000019	.0769	0.03	288	0.61	1095.2	12.5	1111	15	5.396	1.2	.0766	0.7
T1-18.1 ic	0.000050	.0764	0.08	249	0.28	1007.0	11.7	1088	32	5.894	1.2	.0757	1.6
T1-1.1ttn-c	0.016429	.2700	24.92	33	0.26	381.0	17.5			12.33	2.6	.2700	1.7
T1-1.2ttn-c	0.015140	.2356	20.94	42	0.25	383.3	17.8			12.90	2.3	.2356	8.0
T1-1.3ttn-i	0.007761	.1635	17.14	90	0.17	324.8	10.1			16.90	2.2	.1635	7.0
T1-1.4ttn-i	0.007689	.1551	15.86	81	0.16	316.2	7.6			17.54	1.8	.1551	2.3
T1-1.5ttn-r	0.014165	.2888	26.65	32	0.24	291.1	15.1			15.75	2.9	.2888	1.8
T1-1.6ttn-i	0.005982	.1333	12.43	105	0.18	333.2	6.8			17.11	1.7	.1333	1.3
T1-1.7ttn-r	0.016983	.3151	29.57	22	0.24	312.3	19.2			14.05	3.1	.3151	4.1
T1-2.1ttn-c	0.012736	.2538	23.10	33	0.26	361.3	15.6			13.34	2.6	.2538	1.5
T1-2.2ttn-c	0.009444	.2012	16.98	50	0.40	379.6	12.3			13.69	2.2	.2012	1.9
T1-2.3ttn-c	0.010190	.1856	15.21	64	0.26	365.8	10.5			14.52	2.0	.1856	1.3
T1-2.4ttn-i	0.005894	.1352	12.70	118	0.17	342.8	6.9			16.58	1.6	.1352	1.2
T1-3.1ttn-c	0.009750	.2098	24.20	59	0.15	365.6	15.0			14.05	2.1	.2098	8.7
T1-3.2ttn-c	0.005161	.1432	10.29	78	0.17	376.7	8.6			14.91	1.9	.1432	1.3
T1-3.3ttn-i	0.005121	.1318	12.14	109	0.13	350.5	8.3			16.28	2.1	.1318	1.2
T1-4.1ttn-c	0.012578	.1991	16.73	53	0.21	379.5	13.4			13.73	2.2	.1991	5.8
T1-4.2ttn-c	0.007965	.1840	15.02	58	0.21	367.2	10.9			14.50	2.1	.1840	2.4
T1-4.3ttn-c	0.008738	.1736	13.79	57	0.20	380.2	10.7			14.19	2.2	.1736	1.4
T1-4.4ttn-i	0.004370	.1196	10.28	122	0.17	340.0	6.3			17.05	1.6	.1196	1.2
T1-4.5ttn-i	0.011757	.2540	31.13	100	0.18	339.6	21.5			14.21	1.8	.2540	13.4
T1-4.6ttn-i	0.010573	.2103	24.32	76	0.19	351.5	39.9			14.62	11.1	.2103	7.1
T1-4.7ttn-r	0.008358	.1682	13.04	99	0.25	304.9	8.1			17.89	1.7	.1682	4.9
T1-4.8ttn-r	0.012369	.2440	21.60	47	0.20	289.3	15.3			16.97	2.6	.2440	9.4
T1-4.9ttn-i	0.007734	.1680	17.80	84	0.16	336.8	10.2			16.17	2.3	.1680	5.0
CV-B-4-1, Slaughterhouse Granite													
CV-B-1.1 og	0.000276	.0779	0.47	1835	0.02	865.2	9.3	1042	21	6.911	1.1	.0740	1.0
CV-B-1.2 x	0.000076	.0803	0.13	1389	0.17	1005.5	11.1	1179	10	5.879	1.1	.0793	0.5
CV-B-2.1 og	0.000021	.0762	0.04	1547	0.01	998.4	11.0	1091	7	5.945	1.1	.0759	0.3
CV-B-2.2 x	0.000351	.0869	0.58	453	0.74	1194.5	13.9	1244	28	4.900	1.2	.0819	1.4
CV-B-3.1 og	0.000135	.0672	0.24	2047	0.03	628.3	6.8	782	14	9.715	1.1	.0652	0.7
CV-B-4.1 og	0.002924	.1235	4.86	2565	0.07	425.0	9.1	1252	258	14.220	1.8	.0823	13.2
CV-B-5.1 og	0.000094	.0775	0.16	1779	0.01	953.1	10.8	1099	8	6.236	1.2	.0762	0.4
CV-B-6.1 og	0.000218	.0751	0.37	2082	0.01	757.1	8.2	985	17	7.953	1.1	.0720	0.8
CV-B-6.2 x	N.D.	.0879	N.D.	213	0.52	1310.5	15.2	1389	17	4.418	1.2	.0883	0.9
CV-B-7.1 og	0.000166	.0679	0.29	2189	0.03	654.7	7.1	791	15	9.309	1.1	.0655	0.7
CV-B-8.1 og	0.009273	.2232	14.95	2625	0.07	353.4	9.8	1523	403	16.935	1.6	.0947	21.4
CV-B-9.1 og	0.006511	.1816	10.59	2766	0.03	300.8	6.5	1445	314	20.035	1.5	.0909	16.5
CV-B-10.1 og	0.000261	.0759	0.45	2026	0.02	787.6	8.5	991	20	7.633	1.1	.0722	1.0
CV-B-11.1 og	0.000096	.0713	0.17	1971	0.01	795.1	8.8	927	10	7.578	1.1	.0700	0.5
CV-B-12.1 og	0.000353	.0797	0.60	1346	0.06	807.9	8.9	1061	23	7.413	1.1	.0747	1.2
CV-B-13.1 og	0.001284	.0893	2.20	1930	0.15	700.0	8.0	954	77	8.633	1.1	.0709	3.8

*Analytical sessions: CV-C-3-1, H-1-96, T-1-96, and Sy-1-96 (session 1) (11/96). CV-B-4-1 and Sy-1-96 (session 2) (4/97). All samples analyzed on SHRIMP II, Research School of Earth Sciences, Australian National University. Abbreviations: ic—igneous core; og—overgrowth; x—xenocryst; m—mixture of two age zones. Analyses with suffix "ttn" are titanite samples; abbreviations following ttn: c—core; i—intermediate zone; r—rim.

†Corrected for common Pb, except titanite analyses from sample T-1-96. $^{206}Pb/^{238}U$ ages corrected for common Pb using the ^{207}Pb-correction method; $^{207}Pb/^{206}Pb$ and $^{238}U/^{206}Pb$ corrected for common Pb using the ^{204}Pb-correction method.

§1σ error.

N.D.—not detected.

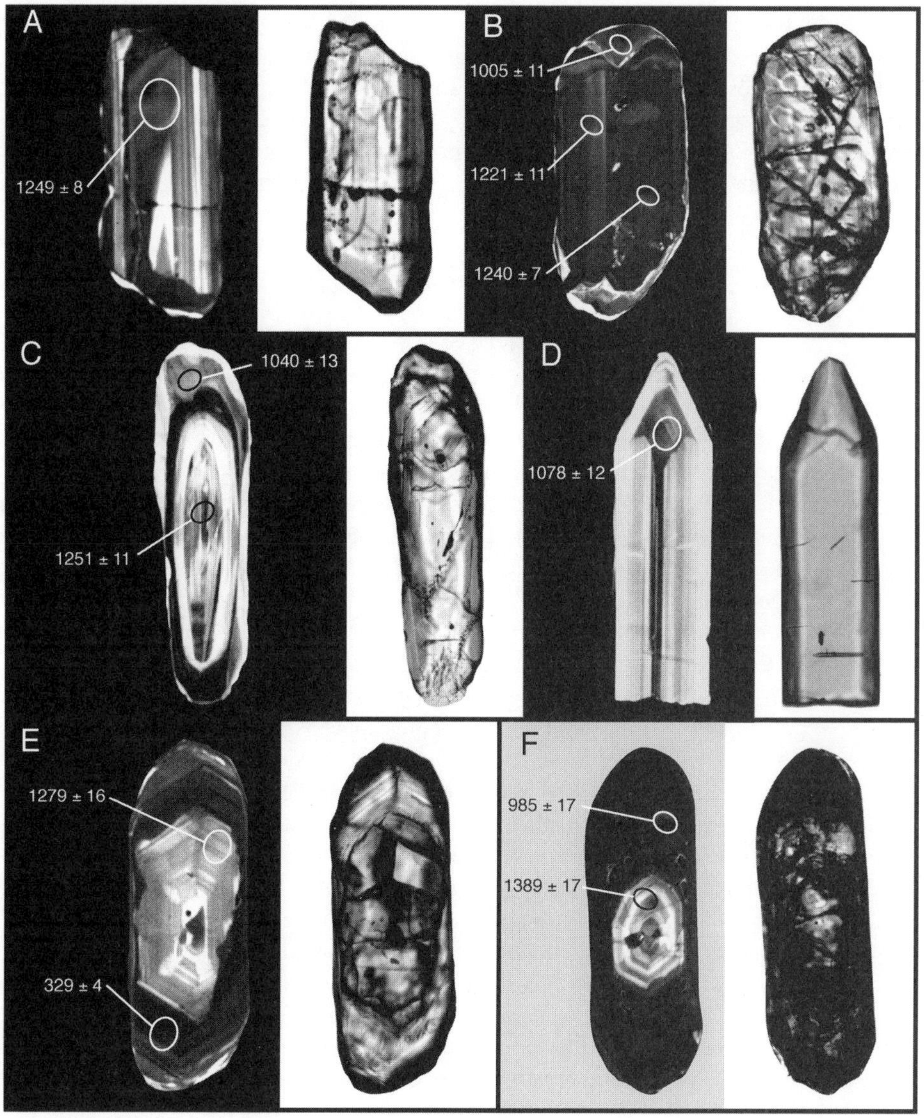

Figure 3. Cathodoluminescence (CL) (on left in each pair) and transmitted light (right) images of representative zircons from Baltimore Gneiss felsic gneiss and granite. Ellipses (~25–30 μm diameter) indicate locations of sensitive high-resolution ion microprobe (SHRIMP) analyses; adjacent dates are $^{207}Pb/^{206}Pb$ ages (±1 σ) in millions of years, except for overgrowth in E, which is a $^{206}Pb/^{238}U$ age. Dark areas in CL images have much higher U than light areas. (A) Sample CV-C-3-1 (Chattolanee antiform felsic gneiss). Oscillatory zoned igneous core and invading metamorphic rim. (B) Sample Sy-1-96 (Woodstock antiform felsic gneiss). Core composed of dark, unzoned igneous zircon (right side) and parallel-growth weakly zoned mantle (left side), both overgrown by metamorphic rim. (C) Sample H-1-96 (Phoenix antiform felsic gneiss). Oscillatory zoned igneous core and four consecutively younger overgrowths. (D) Sample T-1-96 (Towson antiform foliated biotite granite). Oscillatory zoned igneous zircon; no evidence of overgrowth. (E) Sample T-1-96 (Towson antiform foliated biotite granite). Partially resorbed, inherited core overgrown by high-U metamorphic rim of Paleozoic age. (F) Sample CV-B-4-1 (Slaughterhouse Granite). Low-U, inherited core overgrown by very high-U mantle interpreted as igneous. Age is discordant (see Fig. 4).

acicular grains with l/w of 8–10 (Fig. 3, D and E). Zircons from the Slaughterhouse Granite are euhedral and medium to dark brown. Nearly all of the grains have well-preserved, smooth (undeformed) crystal faces (Fig. 3, F). On the basis of zircon morphology alone, it is apparent that zircons from the felsic gneisses have been involved in deformational events that did not affect the zircons in the foliated biotite granite or the Slaughterhouse Granite.

The choice of location of an individual SHRIMP analysis was determined by factors including CL zoning pattern, absence of imperfections such as cracks or inclusions (liquid or solid), and purpose of the analysis (i.e., age of inheritance [source], igneous protolith, and/or metamorphic overgrowth). Because grains are viewed only in reflected light during analysis on the SHRIMP, it was sometimes difficult to locate the primary ion beam exactly on the desired area. For this reason, secondary electron and CL images were taken of all grains after SHRIMP dating to determine whether the excavated analysis location occurred in a single age zone or overlapped two different age zones. Any analysis that was later found to overlap different age zones (determined solely on the basis of CL imaging) was discarded from age calculations; these samples are listed in Table 1 with the suffix *m,* indicating a mixture of ages.

Because the uncertainties of individual $^{207}Pb/^{206}Pb$ ages are typically 10–20 m.y. or more (Table 1), it is possible that data from two different Mesoproterozoic age zones will overlap within limits of their individual errors. To decipher closely spaced geologic events, it is necessary to group analyses on the basis of a priori criteria that are independent of the age analysis. We used CL zoning patterns as the primary guide for grouping data; igneous cores (usually oscillatory-zoned) were grouped separately from intermediate mantles (some oscillatory-zoned, some dark in CL and unzoned) and outermost rims (invariably dark in CL and unzoned). Also eliminated from the core group were ages that are much older than those of typical core analyses (interpreted as inherited xenocrystic material). Igneous crystallization ages were calculated for each sample using isotopic data from the cores. All other ages for each sample were grouped and then statistically deconvoluted using the algorithm of Sambridge and Compston (1994) via the "unmix" routine in Isoplot/Ex (Ludwig, 2002a). In this way, subjective decisions for grouping closely spaced age zones were avoided.

In some grains, core regions are asymmetrically zoned (i.e., as shown in CL images, there are two types of oscillatory zoning, both parallel to the direction of the c-axis, but not concentric), suggesting two distinct periods of growth. In such cases, the older age was included in the igneous core group and the younger age was considered to be due to subsequent overgrowth. A few grains contain interior regions that display oscillatory zoning but yielded concordant ages that are distinctly younger than core ages and as such are grouped with data from mantling regions. Locally, dark (in CL) overgrowths invade cores and occur in the middle of the zircons. The younger material probably formed as a result of partial dissolution along cracks parallel to crystallographic axes or as radiating fractures surrounding cores.

Baltimore Gneiss Felsic Gneisses

Felsic gneiss samples CV-C-3-1 (Chattolanee antiform), H-1-96 (Phoenix antiform), and Sy-1-96 (Woodstock antiform) are interlayered with amphibolite or biotite gneiss. Complete descriptions of their occurrence, location, and mineralogic composition are given in the appendix. Because of the lithologic layering, these rocks have been interpreted to be of metavolcanic origin (Hopson, 1964; Crowley, 1976).

Three cores in zircons from sample CV-C-3-1 (Chattalonee antiform, analyses 2.1, 5.1, and 18.1; Table 1) have $^{207}Pb/^{206}Pb$ ages of 1318, 1498, and 1719 Ma, respectively, interpreted as xenocrystic (inherited) in origin. The weighted average of $^{207}Pb/^{206}Pb$ ages from seven other cores is 1247 ± 9 Ma (Fig. 4, A), interpreted to date the time of crystallization of the protolith of the felsic gneiss. Thirteen additional analyses (one of which is a mixture from two age zones) yielded younger $^{207}Pb/^{206}Pb$ ages, ranging from 1156 to 1220 Ma (Table 1).

Eight analyses of cores from zircons from sample H-1-96 (Phoenix anitform) were grouped to yield a weighted average of $^{207}Pb/^{206}Pb$ ages of 1256 ± 10 Ma (Fig. 4, B), interpreted to date the time of crystallization of the protolith of the felsic gneiss. Fifteen other analyses (three of which were judged to be mixtures from different age zones on the basis of CL imaging of each analytical pit) yielded younger $^{207}Pb/^{206}Pb$ ages, ranging from 986 to 1225 Ma (Table 1).

Zircons from sample Sy-1-96 (Woodstock antiform) were analyzed during two sessions. Isotopic data from ten cores measured during session 1 and eleven cores measured during session 2 yielded weighted averages of $^{207}Pb/^{206}Pb$ ages of 1240 ± 5 and 1241 ± 7 Ma, respectively (Fig. 4, C). The composite age from both sessions is 1240 ± 4 Ma. Younger $^{207}Pb/^{206}Pb$ ages were obtained during session 1 from nine analyses (of which three are mixtures from different age zones on the basis of CL imaging) and during session 2 from ten analyses (of which two are mixtures of from different age zones). The range of younger ages from fourteen analyses from single age zones is 851 to 1226 Ma.

Foliated Biotite Granite

Sample T-1-96, foliated biotite granite, occurs within the layered rocks of the Baltimore Gneiss in the Towson dome. As observed in the outcrop, it appears to have formed as a sill, parallel to the local gneissic layering of the metavolcanic sequence. It is weakly foliated, in contrast to the strongly foliated felsic gneisses (see appendix), and thus we interpret it as an intrusion into the felsic gneisses. Isotopic data from zircons from this sample form two age populations: ten analyses are in the range of ca. 1.18–1.31 Ga, whereas nine analyses yielded a weighted average of $^{207}Pb/^{206}Pb$ ages of 1075 ± 15 Ma (Fig. 4, D). Older ages were obtained from cores of zircons that are rounded, likely partially resorbed, and generally mantled by overgrowths (Fig. 3, E). Many of these cores appear similar to igneous cores in the other felsic gneiss samples. The younger ages are from oscillatory-zoned grains or from oscillatory-zoned overgrowths on rounded cores (Fig. 3, D). The zoning patterns observed in this younger population of zircons (either whole grains or overgrowths) suggest that they formed in a magma. They are quite distinct from the dark (in CL), unzoned overgrowths, interpreted to be metamorphic in origin, on zircons from the felsic gneiss samples. Thus, we conclude that the cores in sample T-1-96 are xenocrystic (inherited), whereas the overgrowths and whole grains that are oscillatory-zoned are considered to have formed during crystallization of the foliated biotite granite protolith at 1075 ± 15 Ma. Three zircons (analyses 12.2, 14.2, and 16.1; Table 1) yielded much younger $^{206}Pb/^{238}U$ ages of 329 ± 4, 456 ± 5, and 628 ± 7 Ma, respectively. These zones may have formed during heating events in the Paleozoic or the Neoproterozoic, as suggested in part by isotopic data from titanite grains (discussed later). Alternatively, these young ages may be due to Pb loss, and thus would be geologically meaningless.

Slaughterhouse Granite

Zircons from the Slaughterhouse Granite consist of oscillatory-zoned cores with very irregular, resorbed boundaries and black (in CL), unzoned overgrowths. The proportion of overgrowth to core ranges from less than 10% to greater than 95%; some grains are composed entirely of dark, unzoned zircon (no core material apparent), whereas other grains are composites of oscillatory-zoned cores and thin rinds of dark (in CL) overgrowths. The majority of zircons from this sample have small relict cores surrounded by volumetrically large overgrowths (Fig. 3, F).

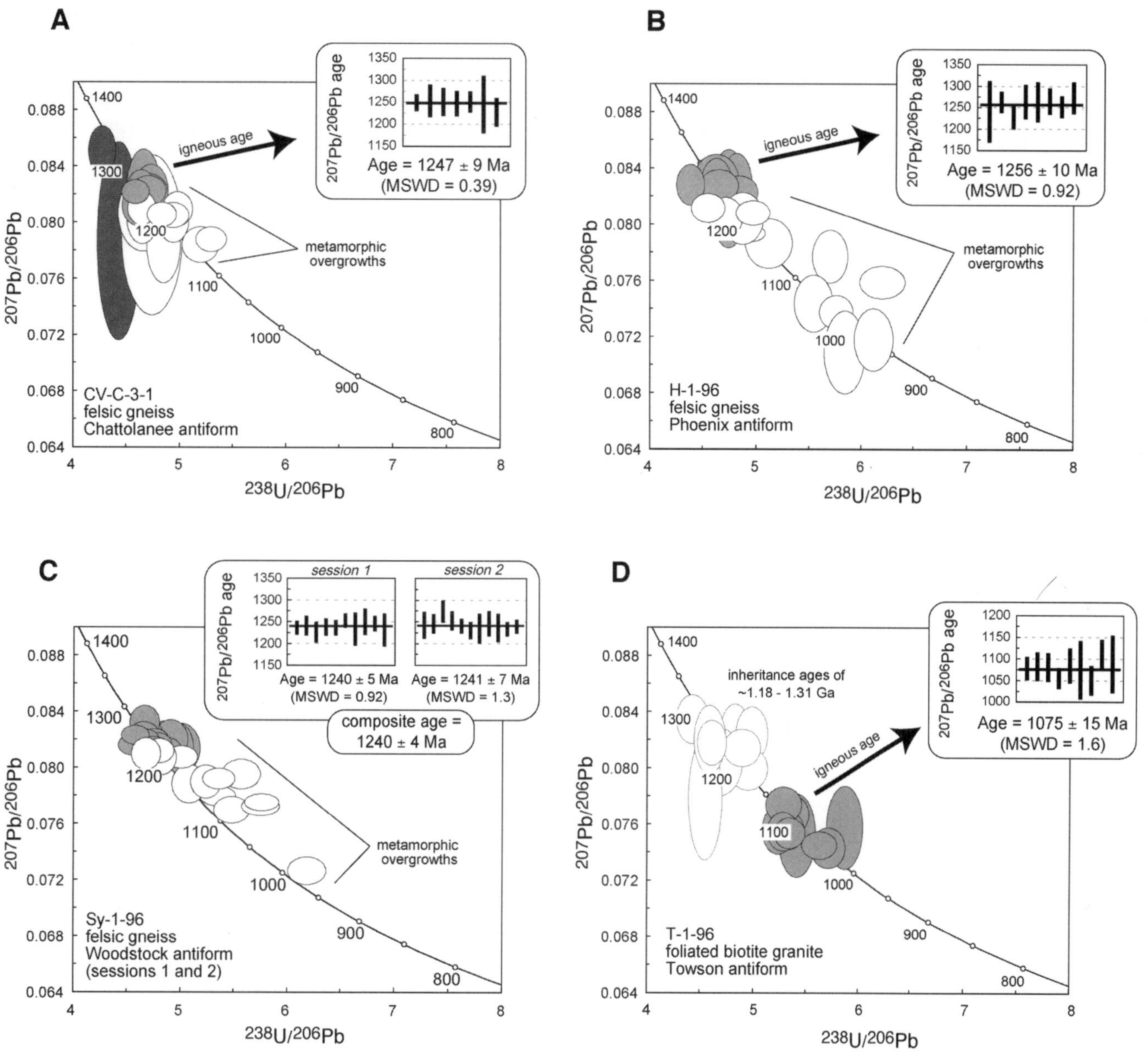

Figure 4. Tera-Wasserburg concordia diagrams and weighted averages plots (insets) for zircons from samples of Baltimore Gneiss. Error ellipses and error bars are plotted as 2σ. Mixed data resulting from the fact that the analysis overlapped two age zones, as indicated by post-analysis CL and secondary electron imaging, are excluded from plots. Medium-shaded error ellipses are used for data interpreted to represent the igneous portions of zircons. Unfilled ellipses are data from metamorphic overgrowths (shown in detail in Fig. 5). Dark-shaded ellipses indicate data obtained from areas interpreted as inherited. MSWD—mean square of weighted deviates. (A) Sample CV-C-3-1 (felsic gneiss, Chattolanee antiform). (B) Sample H-1-96 (felsic gneiss, Phoenix antiform). (C) Sample Sy-1-96 (felsic gneiss, Woodstock antiform). (D) Sample T-1-96 (foliated biotite granite, Towson antiform).

Most SHRIMP analyses of zircons from the Slaughterhouse Granite produced discordant U-Pb age data (Fig. 5; Table 1). Three relatively low-U cores (analyses 1.2, 2.2, and 6.2; Table 1) have $^{207}Pb/^{206}Pb$ ages of 1179 ± 10, 1244 ± 28, and 1389 ± 17 Ma, respectively. The latter two analyses are concordant; the age of 1244 Ma is similar to the crystallization ages of the felsic gneisses into which the Slaughterhouse Granite was intruded, and on this basis the cores are interpreted to be xenocrystic (inherited) in origin. All overgrowths contain high concentrations of U (~1350–2770 ppm; Table 1) and yield discordant age data. $^{207}Pb/^{206}Pb$ ages range from 782 to 1099 Ma; in addition, three analyses that contain high common Pb proportions have Paleozoic $^{206}Pb/^{238}U$ ages (Table 1). The discordant data from the overgrowths form a scattered array and are not

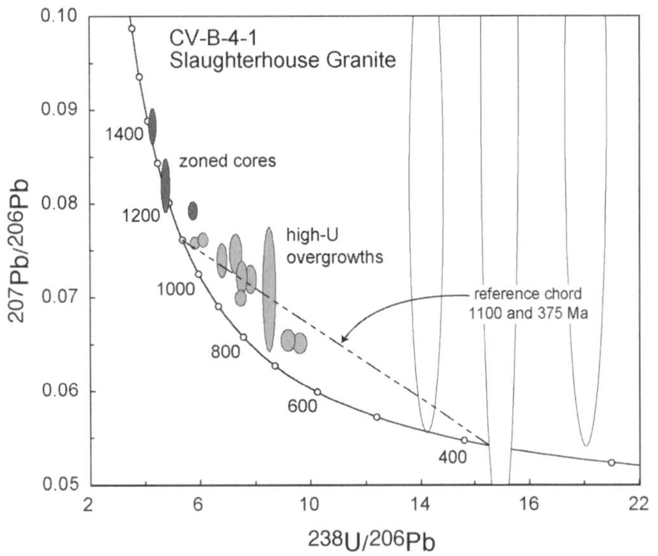

Figure 5. Tera-Wasserburg concordia diagram for zircons from a sample of the Slaughterhouse Granite. All data from igneous mantles are discordant.

amenable to calculating an age (Fig. 5). However, the U-Pb data do provide some age constraints. The Slaughterhouse Granite is younger than ca. 1245 Ma on the basis of its intrusive relation (shown at map scale in Figure 1 but not observable in the field at outcrop scale) to the felsic gneiss (1247 ± 9 Ma) of the Chattolaneee antiform. A zircon core interpreted to be xenocrystic has an age of 1244 ± 28 Ma. The Slaughterhouse Granite is at least as old as ca. 1100 Ma, the oldest $^{207}Pb/^{206}Pb$ age from an overgrowth interpreted as igneous in origin. In addition, the Slaughterhouse Granite lacks the gneissic layering of the older felsic gneisses and resembles dated Paleozoic plutonic rocks (such as the nearby 428 ± 4 Ma Gunpowder Granite; A.K. Sinha, 2002, personal commun.). However, according to map interpretations, the Slaughterhouse Granite is unconformably overlain by sedimentary rocks of the Neoproterozoic-Ordovician Glenarm Group (Crowley, 1976). The U-Pb data plot near a reference chord with intercept ages of 375 and 1100 Ma (Fig. 5). The lower intercept represents the oldest age of titanite from granite gneiss from the nearby Towson dome (see later); the upper intercept is the presumed minimum age of the Slaughterhouse Granite. Three error ellipses plot below this reference chord, implying that a two-stage model of zircon formation and subsequent Pb loss is oversimplified. Although the U-Pb zircon data do not yield an age for the Slaughterhouse Granite, map-scale field relationships and isotopic data suggest that it was emplaced sometime between ca. 1.1 and 1.25 Ga, possibly closer to the younger age because the rock lacks the postintrusion deformational fabrics observed in the 1.25 Ga felsic gneisses.

Zircon Overgrowth Ages

Zircons from the three samples of felsic gneisses have metamorphic overgrowths. Due to overlapping ages and uncertainties, and to the lack of objective evidence needed to determine relative ages, we use the "unmix" routine of Isoplot/Ex (Ludwig, 2002a, based on the algorithm of Sambridge and Compston, 1994) to deconvolute these data sets individually. Six discordant analyses, identified because their 2σ error ellipses plot significantly to the right of the concordia curve (Fig. 6), were excluded from the calculations. The ages derived from

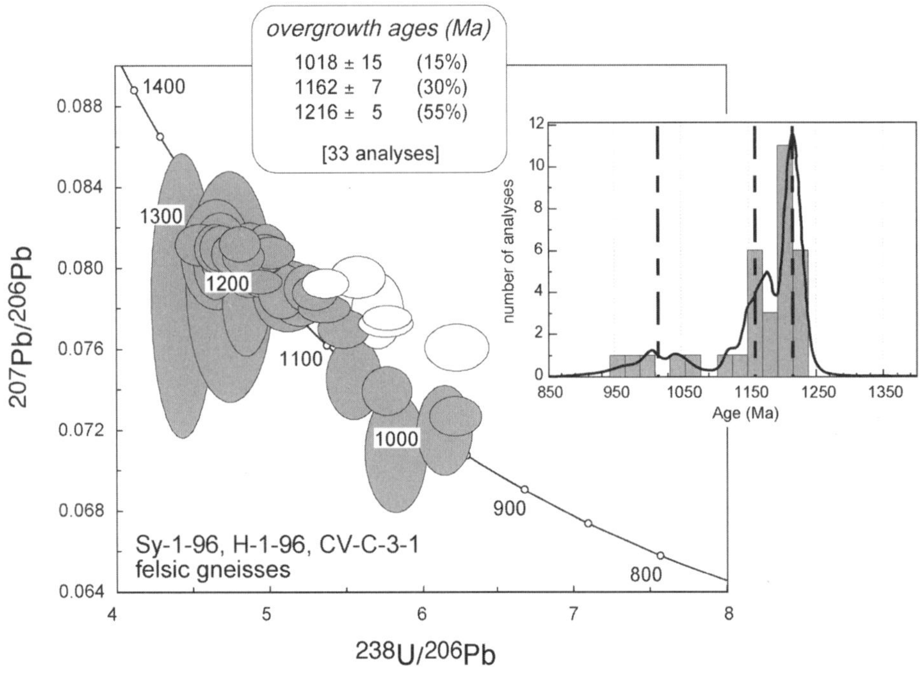

Figure 6. Tera-Wasserburg concordia diagram and relative probability plot (inset) for zircon overgrowths from three samples of felsic gneiss from the Baltimore Gneiss. Dashed lines in inset correspond to calculated overgrowth ages. Discordant data (unfilled ellipses) excluded from calculation. See text for unmixed data from individual samples.

each sample are as follows: (1) CV-C-1 (thirteen analyses): 1211 ± 9 and 1164 ± 15 Ma, (2) H-1-96 (ten analyses): 1198 ± 8 and 1029 ± 20 Ma, and (3) Sy-1-96 (ten analyses): 1219 ± 9, 1147 ± 10, and 1005 ± 22 Ma. These results suggest that the felsic gneisses were affected by postigneous thermal events at ca. 1.21, 1.17, and 1.02 Ga. However, the data sets used to derive these ages are small. An alternative method by which to calculate the times of formation of zircon overgrowths is to combine the three individual sample data sets and then use the "unmix" routine. We believe that it is geologically reasonable to group the individual data sets from the three felsic gneiss samples because the rocks from which the zircons were extracted are (1) similar in composition, (2) areally related, and (3) similar in metamorphic and structural histories. The overgrowth ages derived from the grouped data set are 1216 ± 5, 1162 ± 7, and 1018 ± 15 Ma. Many more analyses are necessary to improve the accuracy and precision of the youngest overgrowth ages.

Titanite

Titanite was obtained from the three felsic gneiss samples and from the foliated biotite granite. We suspected that individual titanite grains might contain multiple age components, as had already been shown for zircon from the Baltimore Gneiss by SHRIMP analysis. Preliminary isotopic analysis indicated that only titanite from the foliated biotite granite contains sufficient U to yield relatively precise U-Pb ages. No other titanites from other samples were analyzed by SHRIMP.

Electron Microprobe Analysis

To evaluate the possible presence of multiple age components, titanites from sample T-1-96 were examined in more detail by backscattered electron (BSE) imaging (Fig. 7) and EMP analysis. The compositions of these titanites are relatively

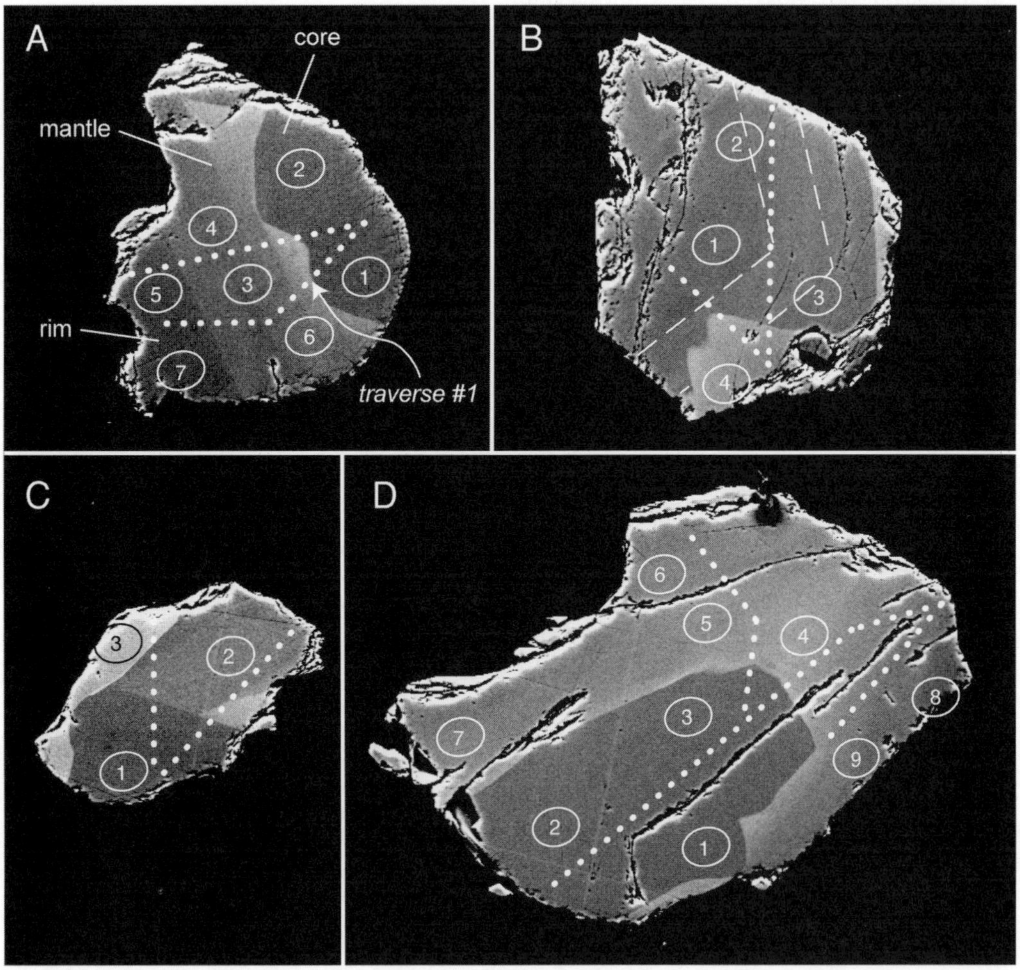

Figure 7. Backscattered electron microprobe (EMP) images of titanite grains from foliated biotite granite sample T-1-96. Ellipses (~25 microns in length) show locations of SHRIMP analyses; numbers correspond to sample numbers in Table 1. Dotted lines indicate locations of EMP traverses. (A) Grain 1. (B) Grain 2; dashed lines indicate internal euhedral zoning pattern. (C) Grain 3. (D) Grain 4.

simple; they have concentrations of CaO, SiO_2, TiO_2, and Al_2O_3 that are typical of metamorphic titanite (cf. Kowallis et al., 1997; Aleinikoff et al., 2002b). Important minor components and their concentrations (in weight percent) include F (~0.7%), Al_2O_3 (2.5–3.5%), total iron as Fe_2O_3 (~0.7%), Y_2O_3 (0.1–1.0%), and Nb_2O_5 (0.2–0.3%; Table 2). The latter four components substitute for TiO_2 in the titanite structure (Deer et al., 1978), with their cation sum equal to almost 1.0 (Fig. 8). Other minor oxides generally present at 0.1 wt% or less include Ce_2O_3, Nd_2O_3, WO_3, and SnO_2 (Table 2). These elements are concentrated in zones visible in BSE imaging (Fig. 7). A traverse across titanite grain 1 (Fig. 7, A) shows that the bright belt in the interior of the grain is caused by a nearly ten-fold increase in the concentration of Y_2O_3, and a smaller increase in Fe_2O_3 (Fig. 9). These and other discontinuities (e.g., in Al_2O_3, rare earth elements, and other minor elements) are compensated for by antithetic changes in the concentrations of CaO and TiO_2 (Fig. 9). Together, these discontinuities define the edge of a mantle overgrowth on a relatively homogeneous core. A third compositional zone is not as conspicuous in BSE imaging, but defines a rim richer in Al_2O_3 and Fe_2O_3, and poorer in Nb_2O_5 and Y_2O_3 (Fig. 9). The significance of these profiles is that, while compositional trends within a zone are smooth (whether flat, decreasing, or rising; note especially Al_2O_3, Fe_2O_3, and Y_2O_3 in Fig. 9), the compositions across zone boundaries are discontinuous. Moreover, euhedral zoning within the core (e.g., Fig. 7, B) is truncated by the mantle, indicating dissolution of the core before precipitation of the mantle. Thus, there are at least three titanite growth events: (1) formation of a core, (2) a dissolution event followed by mantle formation, and (3) overgrowth of the mantle by the rim.

SHRIMP Analysis

To define the ages of these growth events, four grains were analyzed by SHRIMP in twenty-three locations. Using criteria such as BSE imaging, truncation of zoning, internal geometry (i.e., younger zones surround or crosscut older zones), and EMP-defined chemical zoning, three types of analytical locations are designated: (1) cores that are rounded and anhedral, up to 200 μm in diameter, medium to dark gray and oscillatory-zoned in BSE images, and poor in trace elements; (2) mantles that are ~50 μm wide and readily identified by an abruptly bright interface with the core (caused by discontinuous increases in the concentrations of heavy minor elements); and (3) rims that are up to 50 μm wide, are slightly brighter or darker in BSE imaging than the mantle, and show moderate but variable chemical zoning (Fig. 9). The composite weighted average $^{206}Pb/^{238}U$ ages derived from these groups are (1) 374 ± 8 Ma, (2) 336 ± 8 Ma, and (3) 301 ± 12 Ma (Fig. 10, inset). For comparison, deconvolution of the age data using the "unmix" routine in Isoplot/Ex (Ludwig, 2002a) results in ages of 373 ± 8, 340 ± 7, and 309 ± 11 Ma.

In summary, the igneous protoliths of the felsic gneisses of the Baltimore Gneiss crystallized at ca. 1250 Ma. Zircons in

TABLE 2. REPRESENTATIVE ELECTRON MICROPROBE ANALYSES OF TITANITE, SAMPLE T-1-96

Analysis	Core	Mantle			Rim	Max Y
number	75	82	87	103	111	241
SiO_2*	31.40	31.27	31.21	31.39	31.26	30.87
TiO_2	36.61	35.86	36.23	36.95	36.25	34.95
$P2O_5$	0.02	0.03	0.04	0.06	0.00	0.02
Nb_2O_5	0.11	0.24	0.27	0.27	0.19	0.47
Ta_2O_5	0.00	0.22	0.00	0.00	0.00	0.23
Al_2O_3	2.64	2.86	2.71	2.41	2.97	3.00
Fe_2O_3	0.66	0.75	0.71	0.53	0.61	0.80
Y_2O_3	0.08	0.78	0.23	0.06	0.00	1.49
La_2O_3	0.00	0.08	0.00	0.01	0.00	0.00
Ce_2O_3	0.07	0.12	0.15	0.04	0.11	0.11
Pr_2O_3	0.18	0.00	0.05	0.02	0.09	0.00
Nd_2O_3	0.00	0.18	0.07	0.09	0.01	0.12
Sm_2O_3	0.06	0.07	0.07	0.00	0.00	0.08
Dy_2O_3	0.05	0.00	0.00	0.13	0.00	0.23
Er_2O_3	0.06	0.12	0.06	0.03	0.00	0.13
Yb_2O_3	0.00	0.03	0.00	0.02	0.00	0.12
CaO	28.49	27.65	28.45	28.45	28.69	26.89
MgO	0.00	0.01	0.01	0.01	0.01	0.01
MnO	0.08	0.11	0.09	0.09	0.08	0.15
K_2O	0.02	0.01	0.01	0.01	0.01	0.01
ZrO_2	0.00	0.00	0.01	0.01	0.00	0.05
SnO_2	0.01	0.01	0.00	0.02	0.01	0.01
WO_3	0.00	0.00	0.00	0.00	0.00	0.04
F	0.51	0.50	0.64	0.55	0.81	0.37
O = F†	–0.22	–0.21	–0.27	–0.23	–0.34	–0.16
Total	100.84	100.71	100.74	100.92	100.75	100.00
Number of ions based on 5 oxygen atoms						
Si	1.007	1.008	1.001	1.005	0.997	1.009
Ti	0.883	0.869	0.874	0.889	0.870	0.859
P	0.001	0.001	0.001	0.002	0.000	0.000
Nb	0.002	0.004	0.004	0.004	0.003	0.007
Ta	0.000	0.002	0.000	0.000	0.000	0.002
Al	0.100	0.109	0.103	0.091	0.112	0.116
Fe	0.016	0.018	0.017	0.013	0.015	0.020
Y	0.001	0.013	0.004	0.001	0.000	0.026
La	0.000	0.001	0.000	0.000	0.000	0.000
Ce	0.001	0.001	0.002	0.000	0.001	0.001
Pr	0.002	0.000	0.001	0.000	0.001	0.000
Nd	0.000	0.002	0.001	0.001	0.000	0.001
Sm	0.001	0.001	0.001	0.000	0.000	0.001
Dy	0.001	0.000	0.000	0.001	0.000	0.002
Er	0.001	0.001	0.001	0.000	0.000	0.001
Yb	0.000	0.000	0.000	0.000	0.000	0.001
Ca	0.979	0.955	0.977	0.976	0.981	0.942
Mg	0.000	0.001	0.001	0.000	0.000	0.001
Mn	0.002	0.003	0.002	0.002	0.002	0.004
K	0.001	0.001	0.001	0.000	0.000	0.001
Zr	0.000	0.000	0.000	0.000	0.000	0.001
Sn	0.000	0.000	0.000	0.000	0.000	0.000
W	0.000	0.000	0.000	0.000	0.000	0.000
F	0.052	0.051	0.065	0.056	0.081	0.038
Total	3.047	3.040	3.054	3.043	3.064	3.033

*All values in weight percent.
†Oxygen equivalent of fluorine atoms subtracted from total.

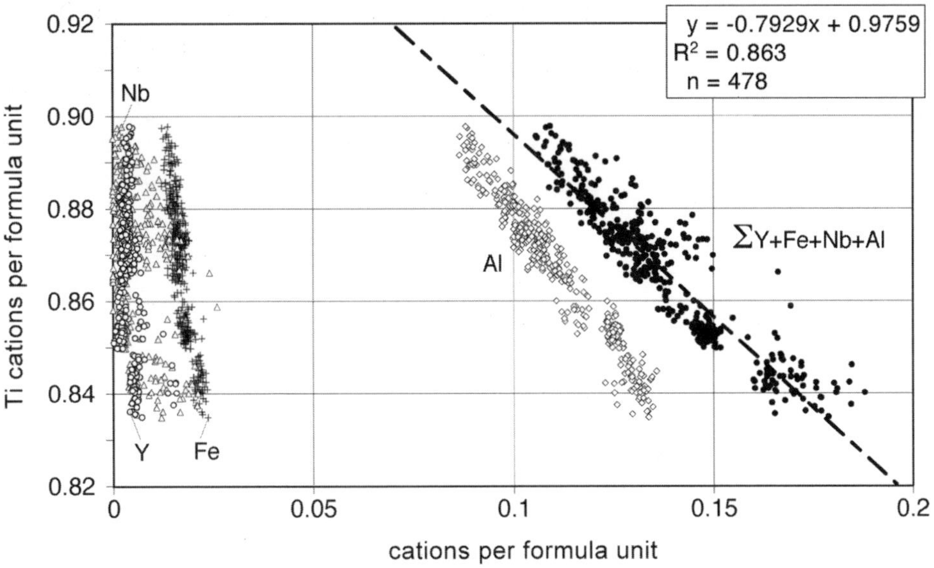

Figure 8. Electron microprobe data from traverses across four titanite grains, showing substitution of Al, plus minor amounts of Y, Fe, and Nb, for Ti.

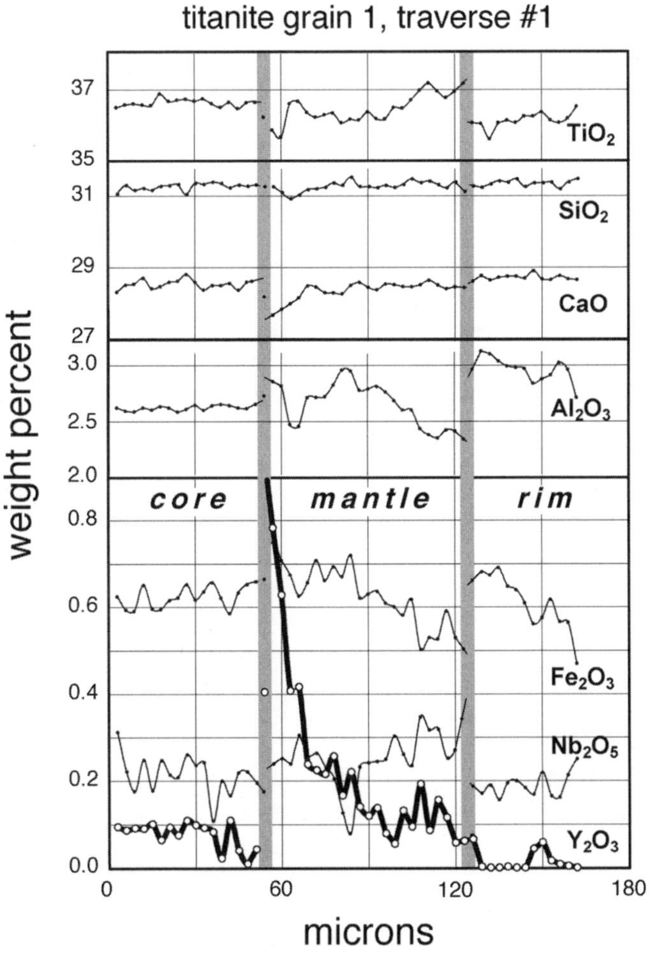

Figure 9. Variations in selected elemental oxides along electron microprobe traverse across titanite grain 1. The larger symbols and heavier line for Y_2O_3 accentuate the profound change in concentration at the core-mantle boundary.

these rocks were affected by subsequent events at ca. 1215, 1170, and 1020 Ma. Biotite granite intruded the felsic gneiss sequence at 1075 ± 15 Ma. The Slaughterhouse Granite also intruded the felsic gneiss sequence during the Mesoproterozoic, but its exact age cannot be determined because of the highly disturbed isotopic systematics. Titanites from biotite granite are only Paleozoic in age. At least three growth events are recorded in zoned titanites, at 374 ± 8, 336 ± 8, and 301 ± 12 Ma. There is no evidence for preservation of Proterozoic ages in the titanites.

DISCUSSION

Origin of Metamorphic Zircon

Zircon overgrowths of Mesoproterozoic age are ubiquitous in the three samples of 1.25 Ga felsic gneiss dated in this study. However, such overgrowths are of only minor significance in zircons from the 1.08 Ga foliated biotite granite and the Mesoproterozoic (but undated) Slaughterhouse Granite. A clue to the cause of overgrowth formation may be found in Th/U data and core morphology from the dated samples. Zircon cores interpreted as igneous in origin are anhedral and have Th/U of 0.1–1.0. The Th/U in zircon overgrowths is 0.02–0.7, most >0.1 (Table 1; Fig. 11, A). This wide range of Th/U in metamorphic zircon is unusual; typically, metamorphic zircons in mafic and felsic rocks have very low Th, and thus, Th/U commonly is less than 0.1 (Rubatto and Gebauer, 2000; Rubatto, 2002). The similarity of Th/U in igneous and metamorphic zircons, coupled with the anhedral shapes of the cores and invasion of the cores by the overgrowths, suggests that igneous zircons were partly dissolved and reprecipitated during the three post–1.25 Ga Proterozoic events. Apparently, the physicochemical conditions necessary for dissolving refractory zircon primarily occurred in the Proterozoic; very few Paleozoic zircon overgrowths were

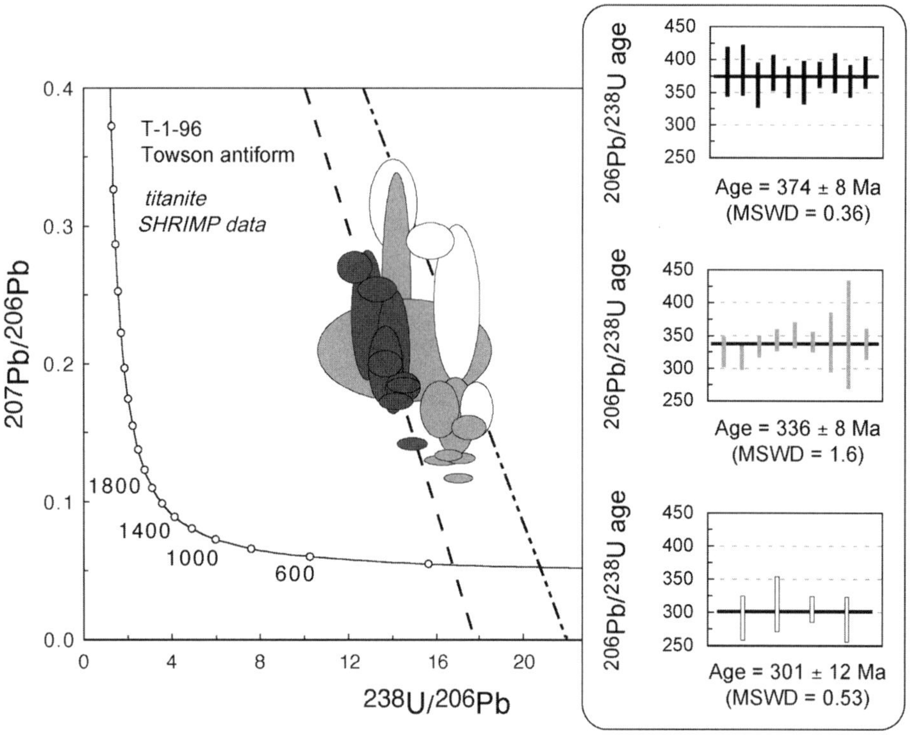

Figure 10. Tera-Wasserburg diagram and weighted averages plots of SHRIMP analyses of titanite from sample T-1-96. Dashed lines represent best-fit regressions through the oldest and youngest data sets, with concordia lower intercept ages of 374 ± 18 and 303 ± 21 Ma, respectively. Identical y-axis intercepts of 0.92 ± 0.20 and 0.94 ± 0.22 reflect the common Pb composition used in calculation of $^{206}Pb/^{238}U$ ages. MSWD—mean square of weighted deviates.

found. Similarly, recent ion microprobe studies of zircons from Grenville and pre-Grenville rocks in the southern Appalachians (Bream et al., 2001; Carrigan et al., 2001) have noted the predominance of Proterozoic overgrowths, and only minor occurrences of Paleozoic (primarily ca. 350 Ma) overgrowths. The relative paucity of metamorphic overgrowths on zircons from the Slaughterhouse Granite and the foliated biotite granite suggests that the primary dissolution events occurred before their emplacement.

Alternatively, the presence or absence of zircon overgrowths may be related to original modal composition. Because of relatively high zircon saturation in mafic magmas (DeLong and Chatelain, 1990), mafic extrusive rocks typically do not contain igneous zircon; however, significant concentrations of Zr (up to several hundred ppm) can be accommodated in clinopyroxene (Ghorbani and Middlemost, 2000). During prograde amphibolite-facies metamorphism, pyroxene is destroyed, liberating Zr. Metamorphic amphibole and garnet both can contain a few tens of ppm Zr (Fraser et al., 1997); excess Zr forms metamorphic zircon (Ayers et al., 2003). In felsic metaigneous rocks, such as the dated samples of Baltimore Gneiss, most Zr originally is contained in igneous zircon, although garnet and hornblende contain trace amounts (10s of ppm) of Zr (Fraser et al., 1997). The Slaughterhouse Granite and the foliated biotite granite probably did not contain garnet or hornblende. Therefore, all Zr may have originally been in zircon, and thus was unavailable to form overgrowths during metamorphism.

Origin of Metamorphic Titanite

The stability of titanite in metamorphic rocks is a function of many factors, including rock composition, pressure and temperature of metamorphism, and oxygen fugacity (e.g., Frost et al., 2000; Castelli and Rubatto, 2002). Numerous studies have noted multiple generations of titanite in metamorphic rocks, typically distinguished by shape and/or color (Corfu, 1996; Verts et al., 1996, and references therein; Essex and Gromet, 2000). Because the closure temperature for thermally activated volume diffusion of Pb in titanite is ~700 °C (Cherniak, 1993; Corfu, 1996; Pidgeon et al., 1996; Zhang and Shärer, 1996), most amphibolite-facies metamorphism (except at the extreme upper temperature limit) would not reset preexisting titanite ages by diffusive Pb loss. Thus, the occurrence of several medium- to high-grade growth events that do not exceed the closure temperature could result in the formation of multiple age populations of titanite.

The EMP and SHRIMP data on titanite from sample T-1-96 illustrate that all of the grains analyzed contain compositional and age zoning. Although the ranges in composition of many elements overlap, some mantle compositions analyzed by EMP show very limited overlap with rims (e.g., Fig. 9). This also is true for Th/U ratios. Cores show a fairly broad range of values (0.15–0.40). However, the Th/U ratios for the mantles (0.14–0.20) consistently are lower that those for the rims (~0.20–0.25; Fig. 11, B).

To allow us to better understand the growth history of titanite, cross-plots of major and minor elements were compiled for

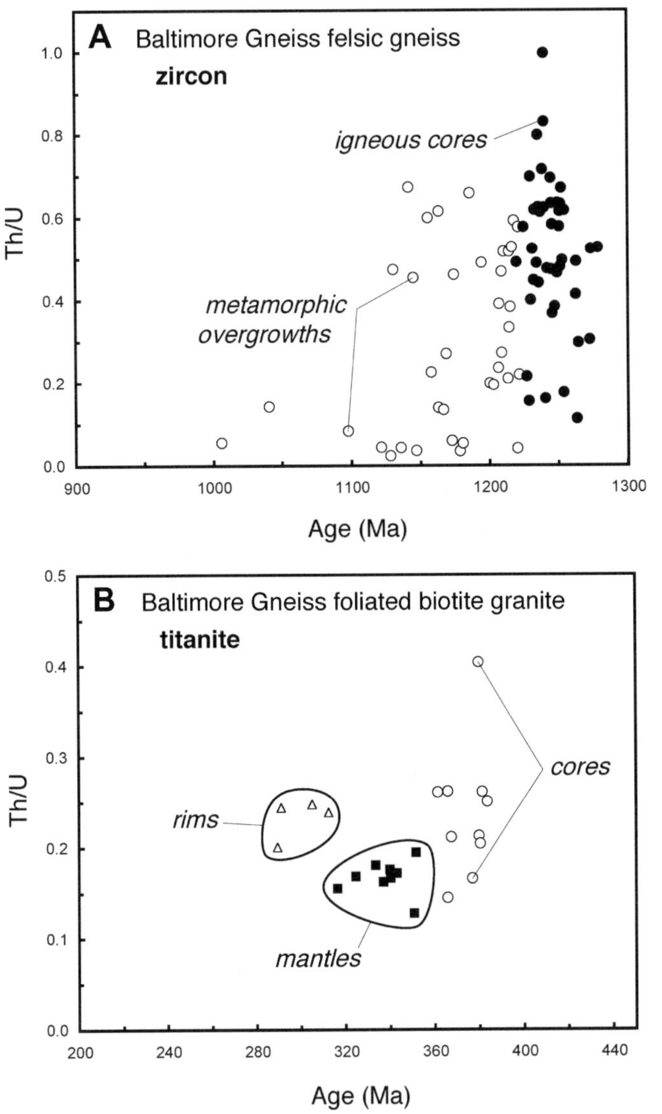

Figure 11. Age versus Th/U for zircon and titanite SHRIMP analyses. (A) Zircon. Filled circles are data from igneous cores; open circles are data from metamorphic overgrowths. Note that although a few overgrowths have very low Th/U typical of metamorphic zircon, most have Th/U similar to igneous cores. (B) Titanite. Open circles are data from the oldest age group (titanite cores, 374 ± 8 Ma), filled squares are data from the intermediate age group (titanite mantles, 336 ± 8 Ma), and open triangles are data from the youngest age group (titanite rims, 301 ± 12 Ma).

each grain. Marked changes in Y and Nb concentrations at zone boundaries are universal. For example, the small oscillations in Al_2O_3 and Fe_2O_3 concentrations in the core of grain 1 define a tight zigzag path (Fig. 12) that reflects a relatively constant activity of Al and Fe^{+3}. Mantle compositions are more variable, with concentrations of Al_2O_3 and Fe_2O_3 exhibiting considerable fluctuation; the outermost mantle has significantly lower Al_2O_3 and Fe_2O_3 values than the core (Fig. 12). The rim compositions define a path that is parallel to the mantle path, but displaced to higher Al concentrations.

The relatively constant core compositions probably reflect growth at or near chemical equilibrium with the local mineral assemblage. The distinctly different compositions in the mantle and the rim (Fig. 12) suggest that these regions formed during major perturbations of this initial equilibrium condition. These perturbations destroyed the ambient solid-fluid equilibrium, leading to dissolution of the core and subsequent precipitation of the mantle and the rim. Core dissolution would have released Y that appears to have fractionated selectively into the precipitating mantle (Fig. 9). Al and Fe also selectively fractionated into the mantle, because their concentrations decay toward the outer edge of the mantle (Figs. 9 and 12). The process that perturbed equilibrium initially increased the activities of Al and Fe^{+3}. Resultant precipitation of mantle and rim eventually reestablished equilibrium conditions with lower concentrations of Al_2O_3 and Fe_2O_3.

The chemical patterns preserved in these titanite compositions independently demonstrate that the zoning evident in backscatter images reflects three discrete growth events. These data substantiate the SHRIMP ages that discriminate the timing of three distinct growth events. The combination of results from the chemical and geochronological methods allows us to conclude that titanite formed by complex geochemical processes over a period of ~75 m.y. in the middle to late Paleozoic.

Coexisting zircons document the occurrence of several Proterozoic heating events. It is likely that titanite originally formed as a metamorphic mineral during the Mesoproterozoic. However, all dated titanites from felsic gneisses of the Baltimore Gneiss have Paleozoic ages; no Proterozoic titanite was found by either TIMS or SHRIMP dating methods. Because the closure temperature of titanite is somewhat higher than the maximum temperatures attained during Paleozoic metamorphism (~645–685 °C; Olsen, 1999), Paleozoic ages of the titanites must result from new growth and/or complete recrystallization, not cooling from a Mesoproterozoic granulite-facies event.

CORRELATION WITH OTHER 1.0–1.3 GA ROCKS IN EASTERN NORTH AMERICA

Proterozoic Tectonic History

The new U-Pb isotopic data presented in this study indicate that the ages of felsic rocks of the Baltimore Gneiss (ca. 1250 Ma) are not typical of other dated Mesoproterozoic rocks in internal and external Appalachian basement massifs of the central and southern Appalachians. For example, granitic rocks of the central and southern Blue Ridge mostly have ages of ca. 1150–1050 Ma (Aleinikoff et al., 2000; Carrigan et al., 2001; Tollo et al., this volume), although Ownby et al. (this volume) report SHRIMP U-Pb ages of ca. 1.20–1.25 for granitic gneisses of the Mars Hill terrane of North Carolina and Tennessee. However, granitic rocks of similar age have been found in central Ver-

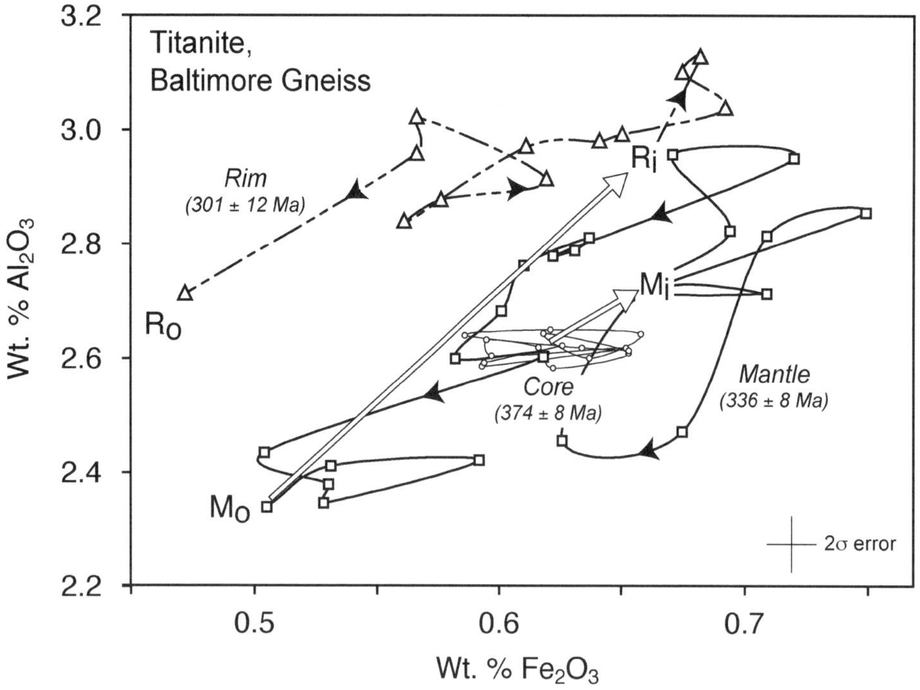

Figure 12. Variation in concentrations of Al_2O_3 and Fe_2O_3 along electron microprobe traverse #1 across titanite grain 1. The core (data points shown as circles) is relatively homogeneous, whereas in the mantle (squares) and rim (triangles) the concentrations of these elemental oxides vary significantly and tend to decrease from the interior to the exterior of each zone along parallel, but separate, trajectories. Traverses were conducted from innermost mantle (M_i) to outer edge of mantle (M_o), and from innermost rim (R_i) to outer edge of grain (R_o), as shown by filled arrowheads. Large unfilled arrows show compositional discontinuities across the core-mantle and mantle-rim boundaries. Uncertainties are ~2% (2σ).

mont (the 1.25 Ga College Hill Suite, Ratcliffe et al., 1991) and the Adirondacks, northern New York (the 1.25 Ga Canada Lake charnockite; McLelland, 2002, personal commun.). Farther north, within the Grenville Province of eastern Canada, the Central Metasedimentary Belt is composed of several tectonically juxtaposed terranes, some of which contain 1.2–1.3 Ga metavolcanic rock sequences and plutonic rock suites (Easton, 1986; Rivers, 1997, and references therein). Basaltic to dacitic rocks of the Tudor Volcanics and the Oak Lake Volcanics have ages of 1286 ± 15 Ma (Silver and Lumbers, 1966), and andesite-dacite and rhyolite of the Belmont Lake Metavolcanic Complex have ages from ca. 1287 ± 11 to 1248 ± 4 Ma (Davis and Bartlett, 1988).

In some cases, comparison of the new SHRIMP U-Pb zircon data from this study with results from previous geochronologic investigations may be inappropriate. Ages for protoliths of felsic gneisses of the Baltimore Gneiss could be determined only by means of in situ microanalysis; all previous dating efforts on these zircons had yielded either mixed ages or metamorphic ages. Thus, attempts to correlate high-grade meta-igneous rocks throughout the Mesoproterozoic basement of eastern North America must consider the possible occurrence of mixed ages because of a combination of inheritance, igneous crystallization, and multiple episodes of metamorphic overgrowth formation in zircon. For example, multigrain fractions of zircon from the Hyde School Gneiss and from leucogranite of Wellesley Island in the Adirondack Lowlands were originally dated by multigrain TIMS as 1230–1285 and 1416 ± 15 Ma, respectively (McLelland et al., 1991). More recently, single-grain TIMS analyses of zircons yielded igneous crystallization ages of ca. 1172 Ma for both rocks (Wasteneys et al., 1999). Thus, efforts to correlate rock units must assess the accuracy and precision of each age.

New SHRIMP U-Pb ages of igneous zircons and metamorphic overgrowths from this study indicate that the Baltimore Gneiss has a geologic history very similar to that of Laurentian crust farther north. The 1.25 Ga felsic rocks of the Baltimore Gneiss are approximately the same age as igneous rocks emplaced in the Grenville Province of eastern Canada during the Elzevirian orogeny, which produced accretionary growth of Laurentia from ca. 1250 Ma to 1190 Ma (Rivers, 1997). Metavolcanic and plutonic rocks of this age also are present in the Adirondacks and in the Green Mountains of Vermont, as discussed earlier.

The overgrowths on igneous zircons from Elzevirian-age rocks of the Baltimore Gneiss indicate subsequent metamorphic events at ca. 1.22, 1.16, and 1.02 Ga. Comparisons to dated rocks in the Grenville Province (Rivers, 1997, and references therein) show that 1.22 Ga zircon overgrowths are coincident with a later stage of the Elzevirian orogeny and that the zircon overgrowths dated at ca. 1.16 and 1.02 Ga, respectively, fit the two oldest of three pulses of crustal shortening that characterize the Ottawan phase of the Grenville orogeny. Furthermore, the age of the 1075 Ma foliated biotie granite from the Baltimore Gneiss in the Towson antiform is well within the span of the Grenville orogeny.

The SHRIMP U-Pb ages of zircon from this study also support earlier interpretations, based on field geology and structural relations, that the Baltimore Gneiss has a different geologic history than other Mesoproterozoic rocks of the central and southern Appalachians, and that it may be exotic with respect to them (Horton et al., 1989). Felsic rocks of the Baltimore Gneiss that contain Elzevirian-age 1.25 Ga igneous zircons are older than granitoid rocks in Grenville massifs of the Reading Prong and the central Appalachian Blue Ridge Province, which typically have ages of ca. 1150–1050 Ma (Aleinikoff et al., 2000). The Elzevirian-age rocks of the Baltimore Gneiss also are older than most well-dated rocks from the internal and external massifs located farther south. Two possible exceptions are the informally named "Forbush orthogneiss" (with a conventional U-Pb upper concordia-intercept age of 1230 ± 6 Ma; McConnell et al., 1988) and the ca. 1.25 Ga granitic gneiss of the Mars Hill terrane (Ownby et al., this volume).

The 1.25 Ga ages of igneous zircon from the Baltimore Gneiss, and Drake's (1990) hypothesis that the Baltimore Gneiss formed in a magmatic arc along the Laurentian margin, are consistent with the age and character of the Elzevirian orogeny in the Grenville Province of eastern Canada (Rivers, 1997). Alternatively, if the Baltimore Gneiss originated as a fragment of Laurentian crust that either rifted away and returned (Thomas, 1977; Muller and Chapin, 1984; Drake et al., 1988) or was displaced by strike-slip faulting (Horton et al., 1989), Elzevirian rocks to the north provide a likely source.

Paleozoic Tectonic History

New SHRIMP U-Pb ages of titanite from the foliated biotite granite intruding the Baltimore Gneiss are consistent with, and augment, existing evidence for the timing of Paleozoic igneous and metamorphic events in the central Appalachian Piedmont. The oldest titanite SHRIMP age from the foliated biotite granite, 374 ± 9 Ma, is similar to a recently determined SHRIMP U-Pb age of 363 ± 3 Ma for zircon from the Guilford Granite (Aleinikoff et al., 2002a; Dg in Fig. 2). The similarity in ages suggests that the core stage of titanite growth in the foliated biotite granite and emplacement of the nearby Guilford Granite occurred simultaneously during a thermal episode following the Devonian Acadian orogeny.

As discussed earlier, the new late Paleozoic U-Pb ages from titanite are considered growth ages and are best interpreted in the context of the regional thermal history, estimated from extant muscovite Rb-Sr and $^{40}Ar/^{39}Ar$ age data in the study area. Muscovite from micaceous quartzite of the Setters Formation in the Towson antiform has a Rb-Sr age (closure temperature of ~500 °C) of ca. 330 ± 14 Ma (uncertainty calculated, using Ludwig, 2002a, assuming a 1σ error of 1% in $^{87}Rb/^{86}Sr$; data from Kohn et al., 1993), whereas $^{40}Ar/^{39}Ar$ ages of muscovite (closure temperature of ~350 °C) in the area are younger: 310 ± 3 Ma (plateau) for muscovite from the Setters Formation in the Phoenix antiform, and 303 ± 3 Ma (from selected incremental heating steps) for muscovite from the Loch Raven Schist on the southeast flank of the Towson antiform (Krol et al., 1999). We suggest that the 330 ± 14 Ma muscovite Rb-Sr age represents regional cooling following a post-Acadian melting event responsible for emplacement of the Guilford Granite at ca. 365 Ma. The 336 ± 8 Ma titanite mantles are identical, within uncertainty, to the Rb-Sr muscovite age and thus grew at a temperature of ~500 °C. During continued cooling, the regional temperature reached ~350 °C (i.e., lower greenschist facies) at ca. 300 Ma, at which time the titanite rims grew. We suggest that these two younger episodes of titanite precipitation occurred when the equilibrium conditions of the original high-grade rocks were disturbed by Alleghanian deformation events. This interpretation is consistent with the local replacement of biotite by titanite observed in thin section (see appendix). Thus, the titanite overgrowths may have formed by means of replacement reactions associated with Alleghanian deformation and tectonic emplacement of the rocks at higher crustal levels. The Alleghanian transpressional arching or doming of recumbent folds proposed by Gates et al. (1999) is a strong candidate for this deformation.

CONCLUSIONS

From our results we have drawn the following conclusions:

1. Zircons from the Baltimore Gneiss have inherited material, igneous cores, and multiple metamorphic overgrowths. The method employed in this study to decipher these stages of zircon growth involved in situ microanalysis using SHRIMP techniques, guided by CL imaging of composition zoning.
2. Volcanic or volcaniclastic protoliths of felsic gneisses of the Baltimore Gneiss, now exposed in antiformal structures, formed at ca. 1.25 Ga. The SHRIMP U-Pb crystallization ages of igneous zircon are (a) Chattolanee antiform felsic gneiss: 1247 ± 9 Ma, (b) Phoenix antiform felsic gneiss: 1256 ± 10 Ma, and (c) Woodstock antiform felsic gneiss: 1240 ± 4 Ma.
3. Granitic rocks emplaced into the metavolcanic sequence include foliated biotite granite (with a SHRIMP U-Pb zircon age of 1075 ± 15 Ma) and the Slaughterhouse Granite (undated due to the high U content of the zircons and the related discordance of U-Pb data). On the basis of field relationships, degree of deformation, and ages of inherited zircon cores and high-U overgrowths, the Slaughterhouse Granite probably has an age between 1.1 and 1.25 Ga.
4. SHRIMP U-Pb data from zircon overgrowths suggest Mesoproterozoic metamorphic events at ca. 1.22, 1.16, and 1.02 Ga.
5. Titanites from the foliated biotite granite are Paleozoic in age. SHRIMP U-Pb ages of 374 ± 8, 336 ± 8, and 301 ± 12 Ma indicate that these rocks were subjected to numerous Paleozoic events, in addition to the Mesoproterozoic events documented by the zircon U-Pb data.
6. The 1.25 Ga rocks of the Baltimore Gneiss are the same age as those emplaced in the Grenville Province during the Elzevirian orogeny, which produced accretionary growth of Laurentia from ca. 1250 to 1190 Ma. Metavolcanic and plutonic rocks of this age are also present in the Adirondacks, the Green Mountains of Vermont, and the Mars Hill terrane of North Carolina

and Tennessee. The 1.22 Ga zircon overgrowths are coincident with a later stage of the Elzevirian orogeny.

7. Younger zircon overgrowths dated at ca. 1.16 and 1.02 Ga, respectively, fit the two oldest of three pulses of crustal shortening that characterize the Ottawan phase of the Grenville orogeny. The 1075 Ma granitic layer from the Baltimore Gneiss in the Towson antiform also was emplaced during the time span of the Ottawan phase.

8. The Baltimore Gneiss has a geologic history that is different from the histories of other Mesoproterozoic rocks in the central and southern Appalachians, and probably is exotic with respect to them.

9. SHRIMP ages of titanite from the Baltimore Gneiss are consistent with existing evidence for the timing of Paleozoic igneous and metamorphic events, and they help to constrain the timing of Alleghanian metamorphism.

APPENDIX: SAMPLE DESCRIPTIONS AND LOCATIONS

The minerals in the rocks named are listed in approximate order of increasing abundance.

CV-C-3-1: Baltimore Gneiss of Chattolanee Antiform

Description: Biotite-quartz-feldspar gneiss, medium-gray (N5) to medium-light-gray (N6), fine- to medium-grained, and flaggy. All rock colors after Goddard et al. (1948).

Location: 39°24′34″N, 76°44′12″W. In a small, north-flowing tributary of Jones Falls, ~350 m south of Hillside Road (about a mile north of Baltimore Beltway I-695).

Notes: Mapped as Baltimore Gneiss "Layered gneiss member" (Crowley et al., 1975; Crowley, 1976). Attitude of gneissic layering and parallel foliation is N85°E, 57°SE; locality is on north limb of anticline, so the structure is overturned.

7.5′ quadrangle: Cockeysville, Maryland (Baltimore County).

Collected: 5/30/96 by Horton and Drake.

Petrography: The biotite-quartz-plagioclase gneiss is fine- to medium-grained and inequigranular, and has feldspar-quartz layers up to 3 mm thick separated by thinner biotite-rich layers up to 1 mm thick. Overall, the gneiss consists of ~36–41% plagioclase, 27–30% quartz, 17–20% biotite, 5–20% potassium feldspar, 0–2% clinozoisite, 1–2% apatite, 1 opaque minerals, and traces (<1%) of muscovite, titanite, and zircon (percentages in all samples were visually estimated primarily from thin sections, with additional information from hand-picked samples). The plagioclase is twinned and untwinned, and locally it contains antiperthitic potassium feldspar that lacks distinct cross-hatched twinning. Quartz is undulose. Feldspar and quartz grains typically are 0.1–3 mm across and generally have straight polygonal boundaries. Minor convex myrmekitic lobes of plagioclase locally embay microcline grains. The biotite-rich layers contain ~50% biotite and as much as 2% muscovite. Oriented biotite flakes up to 5 mm long and 2 mm wide define a strong layer-parallel foliation. Some muscovite flakes (up to 2 mm long) crosscut biotite, and others are intergrown with it. Titanite occurs as foliation-parallel, tabular crystals up to 0.35 mm long and 0.05 mm thick, in and adjacent to biotite. Mineral separates of titanite include distinct populations of pale-yellow anhedral grains and medium-brown, partial wedge-shaped grains. Prismatic zircons are euhedral to subhedral, have aspect ratios of ~4:1, and occur as inclusions in feldspars, quartz, and biotite.

H-1-96: Baltimore Gneiss of Phoenix Antiform

Description: Biotite-quartz-feldspar gneiss, medium-light-gray (N6) to medium-gray (N5), fine- to coarse-grained, inequigranular, gneissose (with alternating light and dark layers, flaser texture, and minor feldspar augen), and variably migmatitic.

Location: 39°30′42″N, 76°40′49″W. In a five-meter-high bluff adjacent to Western Run, near a bend in Western Run Boulevard.

Notes: Mapped as Baltimore Gneiss "Layered gneiss member" (Crowley et al., 1976; Crowley, 1976). Outcrop contains minor interlayered amphibolite (sample H-2-96).

7.5′ quadrangle: Hereford, Maryland (Baltimore County).

Collected: 4/12/96 by Horton, Drake, and Aleinikoff.

Petrography: The biotite-quartz-feldspar gneiss is fine- to medium-grained, inequigranular, and conspicuously layered. Overall, it consists of ~40–60% plagioclase, 20–25% quartz, 20–25% biotite, and 0–12% microcline, with accessory (<1%) epidote minerals, apatite, muscovite, garnet, and zircon. Approximately 80% of the rock is medium- to light-gray gneiss composed largely of plagioclase and lesser amounts of quartz with dark-gray, biotite-rich layers that are mostly 1–2 mm and, less commonly, up to 10 mm thick. The remaining 20% of the gneiss consists of very light-gray to almost white, microcline-rich felsic lenses 3–10 mm thick and up to 5 cm long. Feldspars and quartz grains are typically polygonal, equant, and 0.1 to 2 mm across. Plagioclase is twinned and untwinned, and microcline displays cross-hatched twins. Quartz is undulose, and it occurs as disseminated grains and locally as polycrystalline ribbons up to 7 mm long and 1 mm thick. Biotite-rich layers have ~30–50% biotite intergrown with (in descending order) plagioclase, quartz and microcline, and accessory allanite-epidote and apatite. Biotite flakes mostly are subparallel to the gneissic layering. Coarser biotite (0.5 to 2 mm long) is crosscut at acute angles by finer (.25 to 0.5 mm) biotite (and locally muscovite). Zircons up to 100 microns long have aspect ratios of 3:1 to 4:1, and occur as inclusions in all major minerals. Titanite was not observed in thin section.

Sy-1-96: Baltimore Gneiss of Woodstock Antiform

Description: Hornblende-biotite-plagioclase gneiss, medium-gray (N5), predominantly medium-grained, and inequigranular.

Location: 39°21′07″N, 76°53′47″W. In railroad cut on south side of Patapsco River, 200 m east of road at Marriotsville.

Notes: Amphibolite from the same outcrop has 9% diopside and 4% microcline (Wetherill et al., 1966, dated sample B-65). The next outcrop to the west is quartzite of the Setters Formation.

7.5′ quadrangle: Sykesville, Maryland (Howard County).

Collected: 4/12/96 by Horton, Drake, and Aleinikoff.

Petrography: The hornblende-biotite-plagioclase gneiss is fine- to medium-grained and inequigranular, with compositional layering defined by alternating very light-gray plagioclase-rich layers and thin, dark-gray, hornblende-biotite layers or schistose (biotite-rich) partings. It consists of 43–50% plagioclase, 10–30% brown biotite, 15–20% green hornblende, 7–17% quartz, 1–3% epidote-allanite, 2–3% titanite, and traces of apatite, zircon, and secondary calcite. The thin (0.5–1 mm) dark-gray mafic layers are separated by very light-gray felsic layers 1–2 mm thick. The felsic layers mainly consist of plagioclase (~75%) and lesser amounts of quartz (~25%). Plagioclase commonly is untwinned and weakly zoned. Plagioclase and quartz grains are subequant, range up to 1 mm long and 0.8 mm wide, and have sutured, interlobate boundaries. Some plagioclase-quartz boundaries appear scalloped, with the plagioclase convex and the quartz concave. Mafic layers consist of biotite and hornblende in varied proportions, with minor titanite and epidote. Oriented, tabular biotite crystals, up to 2 mm long and 1 mm thick, define a strong foliation, and hornblende laths up to 2 mm long and 1 mm thick are parallel to the foliation and compositional layering. Allanite-epidote grains are euhedral to subhedral, equant, and invariably associated with biotite. Titanite occurs as 0.1 to 0.3 mm–long, subhedral to anhedral grains in necklacelike tandem arrays associated with biotite, hornblende, and locally epidote, and as smaller, subhedral inclusions in hornblende. Zircon crystals are subhedral, have aspect ratios of 2:1 to 4:1, and occur as randomly oriented inclusions in all of the major minerals.

T-1-96: Foliated Biotite Granite Layer in Baltimore Gneiss of Towson Antiform

Description: Foliated biotite granite, very-light-gray (N8), fine- to medium-grained, and well foliated; occurs as a layer in darker, variably migmatitic, biotite-rich gneiss.

Location: 39°23′55″N, 76°37′27″W. In a creek parallel to Charles Street Avenue, ~15 m upstream from the Chesapeake Street intersection with Charles Street Avenue.

Notes: Within Baltimore Gneiss "Layered gneiss member" (Crowley and Cleaves, 1974; Crowley, 1976). Very close to the locality of sample B-2 of Tilton et al. (1958) and Wetherill et al. (1966), described as dark, fine-grained, biotite-rich gneiss with 20–25% biotite and "veined by quartzofeldspathic material" from a nearby roadcut on Charles Street Avenue between Malvern and Chesapeake Avenues.

7.5′ quadrangle: Towson, Maryland (Baltimore County).

Collected: 4/11/96 by Horton, Drake, and Aleinikoff.

Petrography: The foliated biotite granite layer is fine-grained and essentially equigranular, and has a weak to moderate biotite foliation. Except for sparse, very thin (1–2 mm), leucocratic layers, the granite is nonlayered internally. It consists of 30–37% plagioclase, 25–31% microcline, 28–31% quartz, 5–10% biotite, ≤1% titanite, and traces (<1%) of apatite, clinozoisite, zircon, and an opaque mineral. The foliation is characterized by parallel disseminated biotite flakes and by inequant grains of feldspar and quartz. Plagioclase, microcline, and quartz occur mainly as inequant grains 0.25 to 1 mm across. Plagioclase is unevenly twinned, weakly zoned, and locally myrmekitic adjacent to microcline. Microcline has cross-hatched and patchy twinning. Most quartz is undulose. Biotite flakes are 0.3 to 1 mm thick and widely disseminated. Titanite occurs as 0.05 to 0.1 mm anhedral crystals intergrown with biotite and as inclusions in feldspars. Zircons are euhedral and occur as randomly oriented inclusions in all major minerals.

CV-B-4-1: Slaughterhouse Granite of Chattolanee Antiform

Description: Muscovite-biotite granite gneiss (foliated granite), pinkish-gray (5 YR 8/1) to moderate orange-pink (5 YR 8/4), fine- to medium-grained, inequigranular, strongly foliated, lineated, and only weakly layered.

Location: 39°24′00″N, 76°40′38″W. In Slaughterhouse Branch, on outside (south) side of meander bend, ~850 m west of Rockland and 250 m southeast of Baltimore Beltway I-695. Access from housing development on north side of Old Court Road (#133).

Notes: Slaughterhouse Granite at eastern end of Chattolanee antiform. Previously interpreted as Baltimore Gneiss (Hopson, 1964); distinguished and named Slaughterhouse Gneiss by Crowley (1976), who, based on map relations, interpreted it to be intrusive into Baltimore Gneiss and unconformably overlain by quartzite of the Setters Formation. Foliation attitude is N39°E, 34°SE.

7.5′ quadrangle: Cockeysville, Maryland (Baltimore County).

Collected: 5/30/96 by Horton and Drake.

Petrography: The muscovite-biotite granite gneiss (foliated granite) has a single foliation and lineation defined by elongate, tabular aggregates (up to 10 mm thick) of microcline, plagioclase, and quartz in different proportions; by rods and elongate lenses of polygonal quartz up to 15 mm long, 5 mm wide, and 1 mm thick; and by elongate concentrations of biotite and muscovite up to 40 mm long, 4 mm wide, and 2 mm thick. It has the composition of monzogranite and consists of microcline, plagioclase, and quartz in approximately equal amounts; lesser amounts of biotite (6%) and muscovite (3%); and trace amounts of apatite and zircon. Grains of feldspars and quartz are 0.2 to 2.0 mm across and equant to slightly elongate, and have aspect ratios of 1:1 to 2:1. They have interlocking, mutually embayed to locally curved, polygonal grain boundaries. Plagioclase is twinned and untwinned. Microcline has cross-hatched twinning and is variably perthitic, with irregular blebs of albite. Undulatory extinction in quartz is most conspicuous in the largest grains. Tabular flakes of biotite (0.01 to 0.4 mm thick) and muscovite (0.01 to 0.3 mm thick) have serrated ends. They occur together in foliation-parallel concentrations of interlocking and mutually crosscutting grains, and as individual disseminated crystals generally subparallel to the same foliation. Apatite occurs as randomly oriented, euhedral to subhedral, equant to slightly elongate laths up to 50 microns long, with aspect ratios of 1:1 to 1:3, mainly as inclusions in feldspar and quartz. Zircon occurs as stubby prisms up to 20 microns in length, with aspect ratios of 2:1 to 4:1, observed as randomly oriented inclusions in plagioclase and quartz. No titanite was observed in thin section.

ACKNOWLEDGMENTS

We thank Rebecca Sauer for help with mineral separations. Robert Ayuso, Paul Hackley, Brent Miller, Scott Samson, and Richard Tollo provided excellent technical reviews of the manuscript. Although their comments and criticisms significantly improved the manuscript, the authors take full responsibility for the interpretations and conclusions presented herein.

REFERENCES CITED

Aleinikoff, J.N., Burton, W.C., Lyttle, P.T., Nelson, A.E., and Southworth, C.S., 2000, U-Pb geochronology of zircon and monazite from Middle Proterozoic (1.15–1.05 Ga) granitic gneisses of the northern Blue Ridge, Virginia: Precambrian Research, v. 99, p. 113–146.

Aleinikoff, J.N., Horton, J.W., Jr., Drake, A.A., Jr., and Fanning, C.M., 2002a, SHRIMP and conventional U-Pb ages of Ordovician granites and tonalities in the central Appalachian Piedmont: Implications for Paleozoic tectonic events: American Journal of Science, v. 302, no. 1, p. 50–75.

Aleinikoff, J.N., Wintsch, R.P., Fanning, C.M., and Dorais, M.J., 2002b, U-Pb geochronology of zircon and polygenetic titanite from the Glastonbury Complex, Connecticut, USA: An integrated SEM, EMPA, TIMS, and SHRIMP study: Chemical Geology, v. 188, p. 125–147.

Ayers, J.C., DeLaCruz, K., Miller, C., and Switzer, O., 2003, Experimental study of zircon coarsening in quartzite ± H_2O at 1.0 GPa and 1000 °C, with implications for geochronological studies of high-grade metamorphism: American Mineralogist, v. 88, p. 365–376.

Bream, B.R., Hatcher, R.D., Miller, C.F., Carrigan, C.W., and Fullagar, P.D., 2001, Provenance and geochemistry of Late Proterozoic southern Appalachian crystalline core paragneisses, NC-SC-GA-TN: Geological Society of America Abstracts with Programs, v. 33, no. 6, p. A-29.

Carrigan, C.W., Miller, C.F., Fullagar, P.D., Hatcher, R.D., Bream, B.R., and Coath, C.D., 2001, Age and geochemistry of southern Appalachian basement, NC-SC-GA, with implications for Proterozoic and Paleozoic reconstructions: Geological Society of America Abstracts with Programs, v. 33, no. 6, p. A-29.

Castelli, D., and Rubatto, D., 2002, Stability of Al- and F-rich titanite in metacarbonate: Petrologic and isotopic constraints from a polymetamorphic eclogitic marble of the internal Sesia Zone (Western Alps): Contributions to Mineralogy and Petrology, v. 142, p. 627–639.

Cleaves, E.T., Edwards, J., Jr., and Glaser, J.D., 1968, compilers, Geological map of Maryland: Baltimore, Maryland Geological Survey, scale 1:250,000.

Cherniak, D.J., 1993, Lead diffusion in titanite and preliminary results on the effects of radiation damage on Pb transport: Chemical Geology, v. 110, p. 177–194.

Corfu, F., 1996, Multistage zircon and titanite growth and inheritance in an Archean gneiss complex, Winnipeg River Subprovince, Ontario: Earth and Planetary Science Letters, v. 141, p. 175–186.

Crowley, W.P., 1976, The geology of the crystalline rocks near Baltimore and its bearing on the evolution of the eastern Maryland Piedmont: Baltimore, Maryland Geological Survey Report of Investigations 27, 40 p.

Crowley, W.P., and Cleaves, E.T., 1974, Geologic map of the Towson quadrangle, Maryland: Baltimore, Maryland Geological Survey, scale 1:24,000.

Crowley, W.P., Reinhardt, J., and Cleaves, E.T., 1975, Geologic map of the Cockeysville quadrangle, Maryland: Baltimore, Maryland Geological Survey, scale 1:24,000.

Crowley, W.P., Reinhardt, J., and Cleaves, E.T., 1976, Geologic map of Baltimore County and Baltimore City: Baltimore, Maryland Geological Survey, scale 1:62,500.

Davis, D.W., and Bartlett, J.R., 1988, Geochronology of the Belmont Lake Metavolcanic Complex and implications for crustal development in the Central Metasedimentary Belt, Grenville Province, Ontario: Canadian Journal of Earth Sciences, v. 25, p. 1751–1759.

Deer, W.A., Howie, R.A., and Zussman, J., 1978, Orthosilicates (2nd edition): Cambridge, England, Longman Scientific Technical, v. 1A, 919 p.

DeLong, S.E., and Chatelain, C., 1990, Trace-element constraints on accessory-phase saturation in evolved MORB magma: Earth and Planetary Science Letters, v. 101, p. 206–215.

Drake, A.A., Jr., 1984, The Reading Prong of New Jersey and eastern Pennsylvania: An appraisal of rock relations and chemistry of a major Proterozoic terrane in the Appalachians, *in* Bartholomew, M.J., ed., The Grenville Event and related topics: Boulder, Colorado, Geological Society America Special Paper 194, p. 75–109.

Drake, A.A., Jr., 1990, The regional geologic setting of the Franklin-Sterling Hill district, Sussex County, New Jersey, *in* Character and origin of the Franklin-Sterling Hill ore bodies: Bethlehem, Pennsylvania, Lehigh University, Franklin-Ogdensburg Mineralogical Society Symposium Proceedings Volume, p. 14–31.

Drake, A.A., Jr., 1994, The Soldiers Delight Ultramafite in the Maryland Piedmont, *in* Stratigraphic Notes, 1993: Reston, Virginia, U.S. Geological Survey Bulletin 2076, p. A1–A14.

Drake, A.A., Jr., 1998a, Geologic map of the Kensington quadrangle, Montgomery County, Maryland: Reston, Virginia, U.S. Geological Survey Geological Quadrangle Series Map GQ-1774, scale 1:24,000.

Drake, A.A., Jr., 1998b, Geologic map of the Piedmont in the Savage and Relay quadrangles, Howard, Baltimore, and Arundel Counties, Maryland: Reston, Virginia, U.S. Geological Survey Open File Report 98-757, 30 p.

Drake, A.A., Jr., Hall, L.M., and Nelson, A.E., 1988, Basement and basement-cover relation map of the Appalachian orogen in the United States: Reston, Virginia, U.S. Geological Survey Miscellaneous Investigations Map I-1655, scale 1:1,000,000.

Drake, A.A., Jr., Sinha, A.K., Laird, J., and Guy, R.E., 1989, The Taconic orogen, *in* Hatcher, R.D., Jr., et al., eds., The Appalachian-Ouachita orogen in the United States: Boulder, Colorado, Geological Society of America, The Geology of North America, v. F-2, p. 101–177.

Easton, R.M., 1986, Geochronology of the Grenville province, *in* Moore, J.M., et al., eds., The Grenville Province: Toronto, Ontario, Geological Association of Canada Special Paper 31, p. 127–173.

Elliott, D., Fisher, G.W., and Snelson, S., 1982, A restorable cross section through the Central Appalachians: Geological Society of America Abstracts with Programs, v. 14, no. 7, p. 482.

Eskola, P.E., 1949, The problem of mantled gneiss domes: Quarterly Journal of the Geological Society of London, v. 104, p. 461–476.

Essex, R.M., and Gromet, L.P., 2000, U-Pb dating of prograde and retrograde titanite growth during the Scandian orogeny: Geology, v. 28, no. 5, p. 419–422.

Faill, R.T., 1997, A geologic history of the north-central Appalachians, Part 1: Orogenesis from the Mesoproterozoic through the Taconic orogeny: American Journal of Science, v. 297, p. 551–619.

Fisher, G.W., 1989, Petrology and structure of gneiss anticlines near Baltimore, Maryland, in 28th International Geological Congress, Field Trip Guidebook T204: Washington, D.C., American Geophysical Union, 12 p.

Fisher, G.W., Higgins, M.W., and Zietz, I., 1979, Geological interpretations of aeromagnetic maps of the crystalline rocks in the Appalachians, northern Virginia to New Jersey: Baltimore, Maryland Geological Survey Report of Investigations 32, 43 p.

Fleming, A.T., and Drake, A.A., Jr., 1998, Structure, age, and tectonic setting of a multiply reactivated shear zone in the Piedmont in Washington, D.C., and vicinity: Southeastern Geology, v. 37, no. 3, p. 115–140.

Fraser, G., Ellis, D., and Eggins, S., 1997, Zirconium abundance in granulite-facies minerals, with implications for zircon geochronology in high-grade rocks: Geology, v. 25, p. 607–610.

Frost, B.R., Chamberlain, K.R., and Schumacher, J.C., 2000, Sphene (titanite): Phase relations and role as a geochronometer: Chemical Geology, v. 172, p. 131–148.

Gates, A.E., Muller, P.D., and Valentino, D.W., 1991, Terranes and tectonics of the Maryland and southeast Pennsylvania Piedmont, *in* Schultz, A.P., and Compton-Gooding, E., eds., Geologic evolution of the eastern United States: Martinsville, Virginia Museum of Natural History Guidebook no. 2, p. 1–27.

Gates, A.E., Muller, P.D., and Krol, M.A., 1999, Alleghanian transpressional orogenic float in the Baltimore terrane, central Appalachian Piedmont, *in* Valentino, D.W., and Gates, A.E., eds., The Mid-Atlantic Piedmont: Tectonic missing link of the Appalachians: Boulder, Colorado, Geological Society of America Special Paper 330, p. 127–139.

Ghorbani, M.R., and Middlemost, E.A.K., 2000, Geochemistry of pyroxene inclusions from the Warrumbungle volcano, New South Wales, Australia: American Mineralogist, v. 85, p. 1349–1367.

Goddard, E.N., Traks, P.D., DeFord, R.K., Rove, O.N., Singewald, J.T., and Overbeck, R.M., 1948, Rock-color chart: Washington, D.C., National Research Council, 6 p. [reprinted by Geological Society of America, 1951, 1963, 1970, 1991].

Grauert, B.W., 1974, U-Pb systematics in heterogeneous zircon populations from the Precambrian basement of the Maryland Piedmont: Earth and Planetary Science Letters, v. 23, no. 2, p. 238–248.

Grauert, B., Crawford, M.L., and Wagner, M.E., 1973, U-Pb isotopic analyses of zircons from granulite and amphibolite facies rocks of the West Chester Prong and the Avondale anticline, southeastern Pennsylvania: Washington, D.C., Carnegie Institute of Washington Year Book 72, p. 290–293.

Hatcher, R.D., 1984, Southern and central Appalachian basement massifs, *in* Batholomew, M.J., ed., The Grenville Event in the Appalachians and related topics: Boulder, Colorado, Geological Society of America Special Paper 194, p. 149–153.

Higgins, M.W., 1972, Age, origin, regional relations, and nomenclature of the Glenarm Series, central Appalachian Piedmont: A reinterpretation: Geological Society of America Bulletin, v. 83, p. 989–1026.

Higgins, M.W., 1977, The Baltimore Complex; Maryland and Pennsylvania, *in* Sohl, N.F., and Wright, W.B., eds., Changes in stratigraphic nomenclature by the U.S. Geological Survey, 1976: Reston, Virginia, U.S. Geological Survey Bulletin 1435-A, p. A127-A128.

Hopson, C.A., 1964, The crystalline rocks of Howard and Montgomery Counties, *in* The geology of Howard and Montgomery Counties: Baltimore, Maryland Geological Survey, p. 27–215.

Horton, J.W., Jr., Drake, A.A., Jr., and Rankin, D.W., 1989, Tectonostratigraphic terranes and their Paleozoic boundaries in the central and southern Appalachians, *in* Dallmeyer, R.D., ed., Terranes in the Circum-Atlantic Paleozoic orogens: Boulder, Colorado, Geological Society of America Special Paper 230, p. 213–245.

Jackson, J.A., 1997, Glossary of Geology (4th edition): Alexandria, Virginia, American Geological Institute, 769 p.

Knopf, E.B., and Jonas, A.I., 1923, Stratigraphy of the crystalline schists of Pennsylvania and Maryland: American Journal of Science (5th Ser.), v. 5, p. 40–62.

Kohn, B.P., Wagner, M.E., Lutz, T.M., and Organist, G., 1993, Anomalous Mesozoic thermal regime, central Appalachian Piedmont: Evidence from sphene and zircon fission-track dating: Journal of Geology, v. 101, p. 779–794.

Kowallis, B.J, Christiansen, E.H., and Griffen, D.T., 1997, Compositional variations in titanite: Geological Society of America Abstracts with Programs, v. 29, no. 6, p. 402.

Krol, M.A., Muller, P.D., and Idleman, B.D., 1999, Late Paleozoic deformation within the Pleasant Grove shear zone, Maryland: Results from $^{40}Ar/^{39}Ar$ dating of white mica, *in* Valentino, D.W., and Gates, A.E., eds., The Mid-Atlantic Piedmont: Tectonic missing link of the Appalachians: Boulder, Colorado, Geological Society of America Special Paper 330, p. 93–111.

Lang, H.M., 1990, Regional variation in metamorphic conditions recorded by pelitic schists in the Baltimore area, Maryland: Southeastern Geology, v. 31, no. 1, p. 27–43.

Ludwig, K.R., 2002a. Isoplot/Ex version 3.00: A geochronological toolkit for Mircosoft Excel: Berkeley, California, Berkeley Geochronology Center Special Publication No. 1, 46 p.

Ludwig, K.R., 2002b, Squid, version 1.05: Berkeley, California, Berkeley Geochronology Center Special Publication No. 2, 16 p.

McConnell, K.I., Sinha, A.K., Hatcher, R.D., Jr., and Heyn, T., 1988, Evidence of pre-Grenville crust in the Sauratown Mountains anticlinorium, western North Carolina: Geological Society of America Abstracts with Programs, v. 20, no. 4, p. 279.

McLelland, J., Chiarenzelli, J.R., and Perham, A., 1991, Age, field, and petrological relationships of the Hyde School Gneiss, Adirondack Lowlands, New York: Criteria for an intrusive igneous origin: Journal of Geology, v. 100, p. 97–105.

Morgan, B.A., 1977, The Baltimore Complex, Maryland, Pennsylvania, and Virginia, *in* Coleman, R.G., and Irwin, W.P., eds., North American ophiolites: Portland, Oregon State Department of Geology and Mineral Industries Bulletin 94–95, p. 41–49.

Muller, P.D., and Chapin, D.A., 1984, Tectonic evolution of the Baltimore gneiss anticlines, Maryland, *in* Bartholomew, M.J., ed., The Grenville Event in the Appalachians and related topics: Boulder, Colorado, Geological Society America Special Paper 194, p. 127–148.

Olsen, S.N., 1999, Petrology of the Baltimore Gneiss in the Northeast Towson Dome, Maryland Piedmont, *in* Valentino, D.W., and Gates, A.E., eds., The Mid-Atlantic Piedmont: Tectonic Missing Link of the Appalachians: Boulder, Colorado, Geological Society of America Special Paper 330, p. 113–126.

Paces, J.B., and Miller, J.D., Jr., 1993, Precise U-Pb ages of Duluth Complex and related mafic intrusions, northeastern Minnesota: Geochronological insights to physical, petrogenetic, paleomagnetic, and tectonomagnetic processes associated with the 1.1 Ga Midcontinent Rift System: Journal of Geophysical Research B, Solid Earth and Planets, v. 98, p. 13,997–14,013.

Pidgeon, R.T., Bosch, D., and Bruguier, O., 1996, Inherited zircon and titanite U-Pb systems in an Archean syenite from southwestern Australia: Implications for U-Pb stability of titanite: Earth and Planetary Science Letters, v. 141, p. 187–198.

Rankin, D.W., 1975, The continental margin of eastern North America in the southern Appalachians: The opening and closing of the proto-Atlantic ocean: American Journal of Science, v. 275-A, p. 298–336.

Rankin, D.W., Drake, A.A., Jr., Glover, L., III, Goldsmith, R., Hull, L.M., Murray, D.P., Ratcliffe, N.M., Read, J.F., Secor, D.T., Jr., and Stanley, R.S., 1989, Pre-orogenic terranes, *in* Hatcher, R.D., Jr., et al., eds., The Appalachian-Ouachita orogen in the United States: Boulder, Colorado, Geological Society of America, Geology of North America, v. F-2, p. 7–100.

Ratcliffe, N.M., Aleinikoff, J.N., Burton, W.C., and Karabinos, P., 1991, Trondhjemitic, 1.35–1.31 Ga gneisses of the Mount Holly Complex of Vermont: Evidence for an Elzevirian event in the Grenville basement of the United States Appalachians: Canadian Journal of Earth Sciences, v. 28, p. 77–93.

Rivers, T., 1997, Lithotectonic elements of the Grenville Province: Review and tectonic implications: Precambrian Research, v. 86, p. 117–154.

Rodgers, J., 1970, The tectonics of the Appalachians: New York, Wiley-Interscience, 271 p.

Rubatto, D., 2002, Zircon trace element geochemistry: Partitioning with garnet and the link between U-Pb ages and metamorphism: Chemical Geology, v. 184, n. 1–2, p. 123–138.

Rubatto, D., and Gebauer, D., 2000, Use of cathodoluminescence for U-Pb zircon dating by ion microprobe: Some examples from the Western Alps, *in* Pagel, M., et al., eds., Cathodoluminescence in Geosciences, Berlin, Springer, p. 373–400.

Sambridge, M.S., and Compston, W., 1994, Mixture modeling of multi-component data sets with application to ion-probe zircon ages: Earth and Planetary Science Letters, v. 128, no. 3–4, p. 373–390.

Silver, L.T., and Lumbers, S.B., 1966, Geochronologic studies in the Bancroft-Madoc area of the Grenville Province, Ontario, Canada: Abstracts for 1965: Boulder, Colorado, Geological Society of America Special Paper 87, p. 156.

Sinha, A.K., Hanan, B.B., and Wayne, D.W., 1997, Igneous and metamorphic U-Pb zircon ages from the Baltimore mafic complex, Maryland Piedmont, *in* Sinha, A.K., et al., eds., The nature of magmatism in the Appalachian orogen: Boulder, Colorado, Geological Society of America Memoir 191, p. 275–286.

Southwick, D.L., and Fisher, G.W., 1967, Revision of stratigraphic nomenclature of the Glenarm Series in Maryland: Baltimore, Maryland Geological Survey Report of Investigations 6, 19 p.

Stern, R.A., 1997, The GSC Sensitive High Resolution Ion microprobe (SHRIMP): Analytical techniques of zircon U-Th-Pb age determinations and performance evaluation, *in* Radiogenic Age and Isotopic Studies: Report 10: Ottawa, Ontario, Geological Survey of Canada, Current Research , v. 1997-F, p. 1–31.

Tera, F., and Wasserburg, G.J., 1972, U-Th-Pb systematics in three Apollo 14 basalts and the problem of initial Pb in lunar rocks: Earth and Planetary Science Letters, v. 14, p. 281–304.

Thomas, W.A., 1977, Evolution of Appalachian-Ouachita salients and recesses from reentrants and promontories in the continental margin: American Journal of Science, v. 277, p. 1233–1278.

Tilton, G.R., Wetherill, G.W., Davis, G.L., and Hopson, C.A., 1958, Ages of minerals from the Baltimore gneiss near Baltimore, Maryland: Geological Society of America Bulletin, v. 69, p. 1469–1474.

Tilton, G.R., Doe, B.R., and Hopson, C.A., 1970, Zircon age measurements in the Maryland Piedmont, with special reference to the Baltimore gneiss problems, *in* Fisher, G.W., et al., eds., Studies of Appalachian geology: Central and southern: New York, Wiley-Interscience, p. 429–434.

Verts, L.A., Chamberlain, K.R., and Frost, C.D., 1996, U-Pb titanite dating of metamorphism: The importance of titanite growth in the contact aureole of the Red Mountain pluton, Laramie Mountains, Wyoming: Contributions to Mineralogy and Petrology, v. 125, p. 186–199.

Wagner, M.E., and Crawford, M.L., 1975, Polymetamorphism of the Precambrian Baltimore gneiss in southeastern Pennsylvania: American Journal of Science, v. 275, p. 653–682.

Wasteneys, H., McLelland, J., and Lumbers, S., 1999, Precise zircon geochronology in the Adirondack Lowlands and implications for revising plate-tectonic models of the Central Metasedimentary Belt and Adirondack Mountains, Grenville Province, Ontario and New York: Canadian Journal of Earth Sciences, v. 36, p. 967–984.

Wetherill, G.W., Tilton, G.R., Davis, G.L., Hart, S.R., and Hopson, C.A., 1966, Age measurements in the Maryland Piedmont: Journal of Geophysical Research, v. 71, p. 2139–2155.

Wetherill, G.W., Davis, G.L., and Lee-Hu, C., 1968, Rb-Sr measurements on whole rocks and separated minerals from the Baltimore gneiss, Maryland: Geological Society of America Bulletin, v. 79, p. 757–762.

Williams, H., and Hatcher, R.D., Jr., 1982, Suspect terranes and accretionary history of the Appalachian orogen: Geology, v. 10, p. 530–536.

Williams, H., and Hatcher, R.D., Jr., 1983, Appalachian suspect terranes, *in* Hatcher, R.D., Jr., et al., eds., Contributions to the tectonics and geophysics of mountain chains: Boulder, Colorado, Geological Society of America Memoir 158, p. 33–53.

Williams, I.S., 1998, U-Th-Pb geochronology by ion microprobe, *in* McKibben, M.A., et al., eds., Applications of microanalytical techniques to understanding mineralizing processes: Reviews in Economic Geology, v. 7, p. 1–35.

Zhang, L.-S., and Shärer, U., 1996, Inherited Pb components in magmatic titanite and their consequence for the interpretation of U-Pb ages: Earth and Planetary Science Letters, v. 138, p. 57–65.

Manuscript Accepted by the Society August 25, 2003

Printed in the USA

Geological Society of America
Memoir 197
2004

Geochemical stratigraphy and petrogenesis of the Catoctin volcanic province, central Appalachians

Robert L. Badger*
Department of Geology, State University of New York at Potsdam, Potsdam, New York 13676, USA
A. Krishna Sinha
Department of Geological Sciences, Virginia Tech, Blacksburg, Virginia 24061, USA

ABSTRACT

The Catoctin volcanic province represents a sequence of rift-related tholeiitic magmas erupted during the Late Neoproterozoic opening of the Iapetus Ocean basin. Geochemical data, coupled with stratigraphy for three sections in central Virginia, permit creation of a chemical stratigraphy. The stratigraphy can be divided into chemical subunits based on trends of relatively immobile elements. These trends can be modeled by fractional crystallization of plagioclase, clinopyroxene, and olivine, with few recognizable effects of crustal contamination. Discontinuities between chemical subunits imply replenishment of magma chambers, or contributions from new magma sources with slightly different chemical signatures. Bulk chemical characteristics of the metabasalts can be explained by upper mantle to midcrustal gabbroic fractionation from a mantle source. Chondrite-normalized plots of trace elements indicate that the magmas were derived from a fertile mantle source, in a model similar to those for other flood basalt sequences. Low Nb/Zr and Nb/Y ratios similar to those for magmas generated at the Hawaiian, Iceland, and Reunion hot spots, suggest a plume source. Relatively low Ce/Yb and Sm/Yb ratios imply that Yb was not retained by garnet in the mantle, so the zone of melting must have been above the garnet-spinel transition zone, at a depth of 40–70 km.

Keywords: petrogenesis, metabasalt, trace elements, fractional crystallization, magma-evolution

INTRODUCTION

The Catoctin volcanic province is a sequence of Neoproterozoic mafic lava flows (Reed, 1955; Rankin, 1975; Badger and Sinha, 1988; Badger, 1992, 1999) located in the Blue Ridge Mountains of the central Appalachians. These lava flows were associated with the opening of the Iapetus Ocean basin, and are part of a late Neoproterozoic sequence of mafic volcanic rocks that occur sporadically within the Appalachian chain (Fig. 1), including the Tibbit Hill Volcanics of Vermont and Québec (Coish et al., 1985; Kumarapeli et al., 1989), the Lighthouse Cove Volcanics (Strong and Williams, 1972) and Long Range dikes of Newfoundland (Kamo et al., 1989), the Bakersville dike swarm of North Carolina (Goldberg et al., 1986), the metadiabase dikes of the Hudson Highlands in New York and New Jersey (Ratcliffe, 1987), and the mafic dikes in northern New York (Isachsen et al., 1988; Badger and Coish, 1992; Coish and Sinton, 1992; Badger, 1993b, 1994b).

*E-mail: badgerrl@potsdam.edu.

Badger, R.L., and Sinha, A.K., 2004, Geochemical stratigraphy and petrogenesis of the Catoctin volcanic province, central Appalachians, *in* Tollo, R.P., Corriveau, L., McLelland, J., and Bartholomew, M.J., eds., Proterozoic tectonic evolution of the Grenville orogen in North America: Boulder, Colorado, Geological Society of America Memoir 197, p. 435–458. For permission to copy, contact editing@geosociety.org.

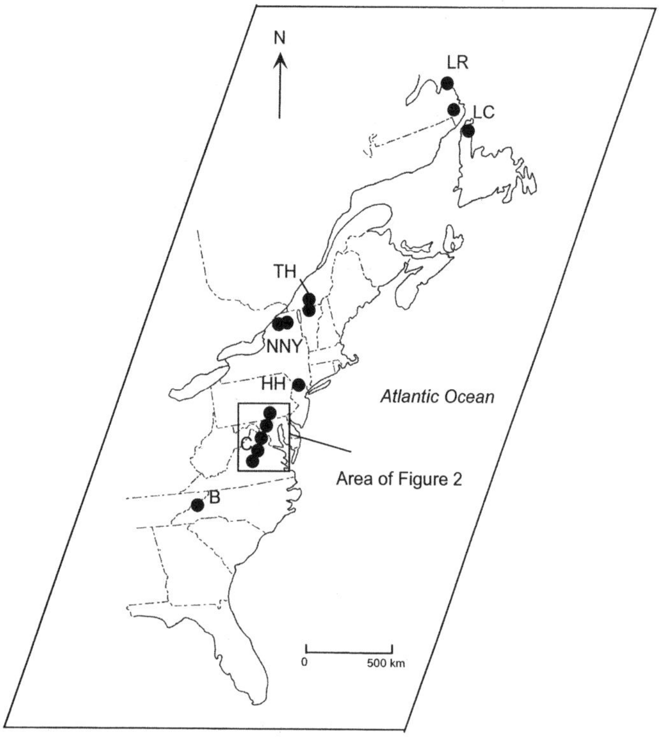

Figure 1. Regional map of eastern North America showing location of Neoproterozoic rift-related magmas discussed in text. LR—Long Range dikes; LC—Lighthouse Cove Volcanics; TH—Tibbit Hill Volcanics; NNY—Northern New York dikes; HH—Hudson Highlands; C—Catoctin volcanic province; B—Bakersville dikes and gabbros.

The regional extent of the Catoctin volcanic province, assuming continuity across the Blue Ridge anticlinorium, would have been at least 11,000 km^2. Therefore, it is the largest flood basalt province in eastern North America. Temporal and spatial correlations with other rift-related magmatic rocks, including the mafic dikes in northern New York and the Long Range dikes of Newfoundland (Puffer, 2002), suggest that the areal extent of Late Neoproterozoic flood basalt volcanism for this province was on the same order of magnitude as that for the better-known and -exposed Columbia River, Deccan Traps, and Paraná flood basalt provinces.

Geochemical analyses of each recognized volcanic flow allows construction of a geochemical stratigraphy, which permits modeling of the petrogenetic evolution of successive flows in terms of crustal contamination, fractional crystallization, and contributions from new magma sources. Analysis of relatively immobile trace elements permits characterization of the mantle source for the magmas, and also permits estimation of the depth of melting.

REGIONAL GEOLOGY

The Blue Ridge Province is cored by the Mesoproterozoic Blue Ridge basement complex (see Sinha and Bartholomew, 1984, for a discussion of rock assemblages). Neoproterozoic to Early Cambrian metaigneous and metasedimentary rock assemblages intruding and overlying the basement complex reflect a long history of rift-related sedimentation, magmatism, crustal extension, and graben formation prior to opening of the Iapetus Ocean basin. The Bakersville dike swarm constitutes the oldest rift-related sequence of mafic rocks, including a sequence of tholeiitic dikes and associated gabbros that intrude the Blue Ridge basement complex in western North Carolina (Fig. 1). The dikes have been dated by Rb-Sr isotopic techniques at 734 ± 26 Ma (Goldberg et al., 1986).

Also intrusive into the Blue Ridge basement complex are a series of Neoproterozoic alkalic plutons with U-Pb ages ranging from ca. 765 to 700 Ma (Su et al., 1994; Fetter and Goldberg, 1995; Tollo and Aleinikoff, 1996; Bailey and Tollo, 1998). Absence of igneous rocks with ages between 700 and 570 Ma has led to the interpretation that Neoproterozoic rifting along the proto-Atlantic margin occurred in two stages. The first phase, while associated with the breakup of the supercontinent Rodinia (Fetter and Goldberg, 1995), was unsuccessful at creating a new plate boundary, whereas the second phase, associated with the Catoctin volcanic province at ca. 570 Ma, succeeded in opening the Iapetus Ocean basin (Badger and Sinha, 1988; Aleinikoff et al., 1995).

Tholeiitic metabasalts of the Catoctin Formation occur on both limbs of the Blue Ridge anticlinorium in central Virginia, and continue northeast into Maryland and southern Pennsylvania, where they wrap around the culmination of the anticlinorium (Fig. 2). Throughout most of the area, the rocks are typically massive to schistose, dark green metabasalt, except in southern Pennsylvania, where metarhyolite predominates (Fauth, 1973). On the east flank of the anticlinorium, some flows have textures indicative of pillow lavas, indicating subaqueous extrusion (Wehr and Glover, 1985). This is consistent with the proposed location of the incipient Iapetus Ocean basin to the east (Wehr and Glover, 1985). On the west limb, columnar jointing indicative of subaerial eruption is commonly preserved.

Stratigraphically beneath the Catoctin Formation on the east limb of the Blue Ridge anticlinorium are several thousand meters of rift-related deep-water clastic and volcanic rocks (Wehr and Glover, 1985) of the Lynchburg, Farquier, and Mechum River Formations. On the west limb of the anticlinorium, the Catoctin Formation rests either unconformably on the Mesoproterozoic basement or disconformably on the Neoproterozoic Swift Run Formation, a thin, discontinuous unit composed of metasandstone and quartz-pebble conglomerate.

Overlying the Catoctin Formation are Latest Neoproterozoic to Early Paleozoic quartzites, phyllites, siltstones, and sandstones of the Chilhowee Formation, marking the transition from rift- to drift-stage continental margin development (Simpson and Eriksson, 1989). The entire Blue Ridge terrane, including the Catoctin volcanic province, has undergone regional low-grade metamorphism. Syn- and postmetamorphic thrusting resulted in telescoping of the Blue Ridge terrane along a series of west-vergent thrust faults (Bartholomew et al., 1981), several of which transect the Catoctin volcanic province (Gathright, 1976). The most recent structural event, the major overthrust of the entire Blue Ridge ter-

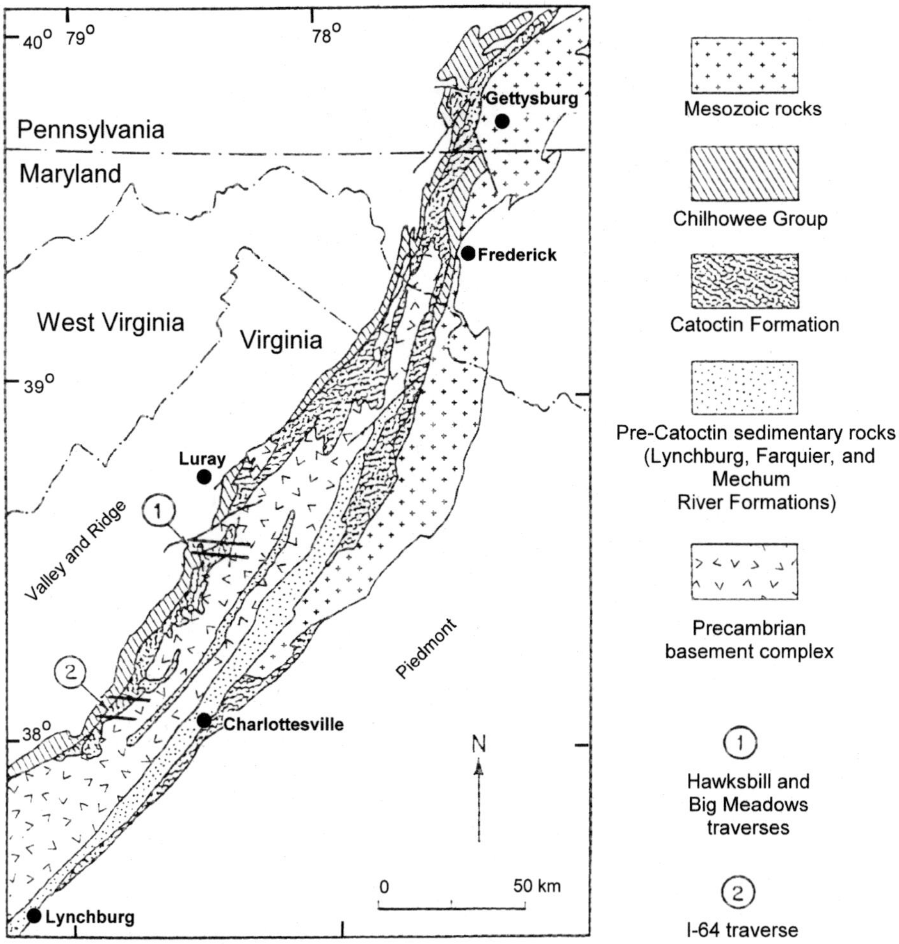

Figure 2. Index map of the central Appalachian mountains showing location of the Catoctin volcanic province. Locations of Hawksbill Mountain, Big Meadows, and I-64 traverses are shown. Adapted from Reed and Morgan (1971).

rane, involves the youngest rocks in the stratigraphic sequence, which are of Mississippian age, suggesting correlation with the regional Permo-Carboniferous Alleghanian event.

The occurrence of a sequence of basaltic rocks several hundred meters thick over an area of several thousand square kilometers suggests rapid extrusion of mantle-derived material and provides the opportunity not only to examine the evolutionary history of the erupted magmas within a stratigraphic framework, but also to characterize the Late Neoproterozoic mantle beneath the central Appalachians.

The area chosen for this study is located on the west flank of the anticlinorium, in an area within and adjacent to Shenandoah National Park, where, despite greenschist-facies metamorphism, a stratigraphic sequence can still be recognized within the volcanic rocks. Individual flows are recognized by columnar jointing, amygdaloidal tops and bottoms, and layers of volcanic breccia or thin metasedimentary units separating flows. The flows are generally mafic in this area, with few felsic units.

ANALYTICAL METHODS

Samples weighing 1–4 kg were processed using standard jaw-crushing techniques. Powders were prepared using a tungsten carbide mill. Pressed-powder pellets and fused-glass discs were prepared from the same powder sample and analyzed by X-ray fluorescence (XRF) on a Phillips model 1450 wavelength-dispersive unit using the techniques of Norrish and Chapell (1977). Pressed-powder pellets were analyzed for Rb, Sr, Ba, Y, and Na; fused-glass discs were analyzed for SiO_2, TiO_2, Al_2O_3, FeO, MgO, MnO, CaO, K_2O, and P_2O_5. Analyses were done at the Virginia Tech Geochemistry Laboratory. Standards for major-element calibrations were made using U.S. Geological Survey rock powders GSP-1, PCC-1, BCR-1, G-2, AGV-1, BHOV-1, QLO-1, and RGM-1. Replicate analyses of the standards yielded the following errors (2σ): SiO_2, ± 1.0 wt%; TiO_2, ± 0.05 wt%; Al_2O_3, ± 0.5 wt%; FeO, ± 0.5 wt%; MgO, ± 1.0 wt%; MnO, ± 0.05 wt%; CaO, ± 0.5 wt%; K_2O, ± 0.05 wt%.

The standards used for Na and Ba were BCR-1, G-2, and GSP-1, with errors (2σ) of ±0.1 wt% for Na and ±3% of the measured value for Ba. The standards used for Rb and Sr were AGV-1, BCR-1, and GSP-1, with errors (2σ) for Sr of ±6 ppm and for Rb of ±2–3 ppm, to a lower limit of ~10 ppm. Rb concentrations for many samples were also measured by isotope dilution on the mass spectrometer.

Zr, Zn, Cu, Ce, and Ni were analyzed by energy-dispersive XRF at the U.S. Geological Survey (USGS) in Reston, Virginia;

Hf, Ta, Th, Sc, Cr, and Co were analyzed by neutron activation, Nb by extraction inductively coupled plasma (ICP), also at the USGS. Relative errors (2σ) for USGS XRF data are ±5%. Average relative errors (2σ) for the Catoctin samples analyzed by neutron activation are as follows: Hf, ±5%; Ta, ±10%; Th, ±15%; Sc, ±5%; Cr, ±8%; Co, ±5%.

STRATIGRAPHY

Within the area of study, the Catoctin Formation consists of dark green, massive metabasalt flows containing irregular pods of light green epidosite (epidote + quartz), interbedded with green to reddish-brown metavolcanic breccia, thin beds of gray to purple phyllite and siltstone, and green to pink arkosic metasandstone. Despite pervasive schistosity, defined by chlorite ± actinolite, individual basaltic flows of 1–50 m thick are recognized by columnar jointing, porphyritic units, vesicular or amygdaloidal margins of flows, and stratigraphic separation by volcanic breccia or metasediments.

Three transects across the Catoctin Formation were carried out on the west limb of the Blue Ridge anticlinoirum in order to characterize the stratigraphy (Fig. 2). One traverse was located along U.S. Interstate 64 (I-64) west of Charlottesville, Virginia. The other two traverses were located within Shenandoah National Park in the area mapped by Reed (1955), southeast of Luray, Virginia. The two Luray sections are ~5 km apart, at Hawksbill Mountain and near Big Meadows campground. They were examined in order to evaluate stratigraphic and chemical continuity along strike. The greatest distance that a single flow was followed continuously was typically ~1 km; however, a distinctive porphyritic flow was traced discontinuously between the two Luray sections and thus provides a useful marker horizon upon which to base stratigraphic and geochemical correlations.

I-64 Traverse

Excellent roadcut exposures along I-64 between Charlottesville and Staunton, Virginia, enable detailed examination of a section of the Catoctin Formation. Younging indicators, including tops and bottoms of volcanic flows, along with graded bedding within Catoctin Formation metasedimentary interbeds and within metasedimentary rocks occurring above and below the formation, indicate that the entire section is overturned. Figure 3 shows the composite stratigraphy for the entire exposed section, and the locations of samples analyzed for whole-rock geochemistry. The cumulative thickness of the exposed section is ~600 m. A significant portion of the section is covered by overburden in three areas. With the covered areas estimated to have a cumulative thickness of 300 to 400 m, it is probable that the total thickness of the formation along the I-64 traverse is ~900 to 1000 m.

The dominant lithology along this transect consists of layers of schistose greenstone that are 1 to 35 m thick. All rocks display a penetrative tectonic cleavage, but there is considerable variation in its spacing and intensity; the centers of flows usually have a less pervasive schistose fabric compared to more intensely deformed margins. The tops of flows have zones of stretched and flattened vesicles 1 to 2 m thick. These vesicles, typically filled with chlorite, are generally 0.25 to 2 cm long,

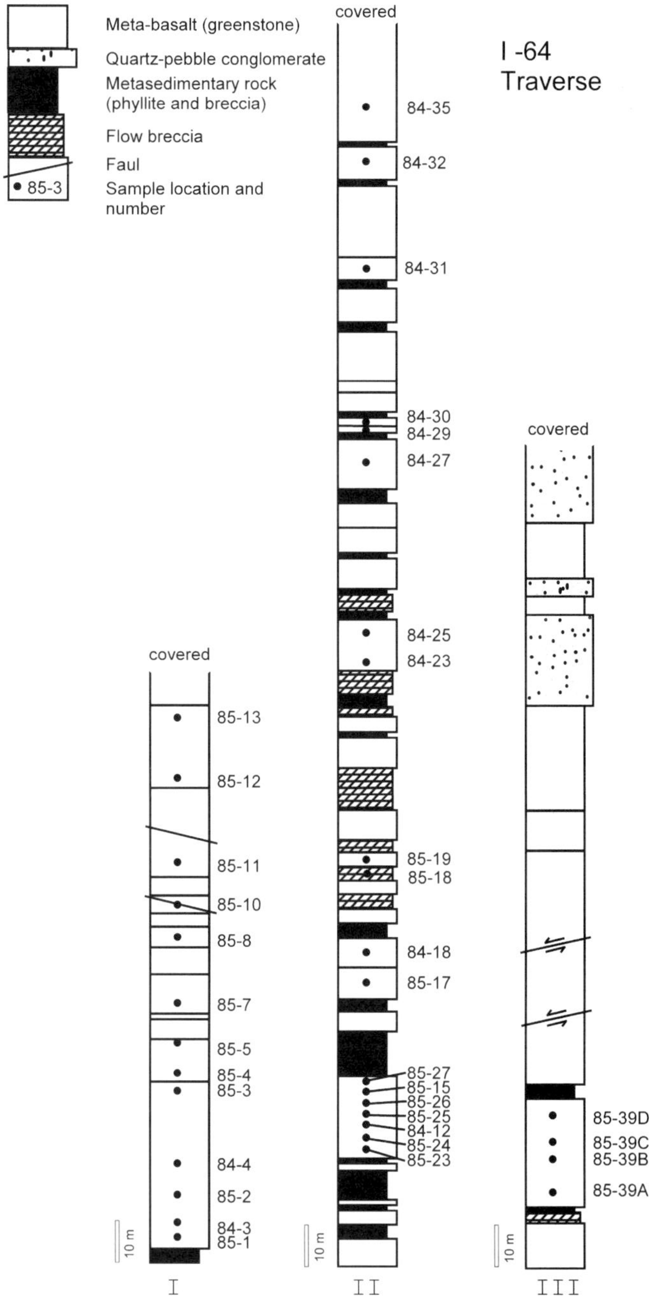

Figure 3. Stratigraphic section for I-64 traverse. The three sections fit on top of one another—section II on top of section I, with a zone covered by overburden separating the two, and section III above section II, also with a zone of overburden separating the two.

with a length-to-width ratio of ~3:1. The bottoms of flows locally also contain zones of vesicles, but these zones are thinner and the vesicles are generally longer, ranging up to several centimeters. The bottoms of flows locally contain "rip-up" fragments of underlying material incorporated during movement of the lavas. Several small faults occur oriented parallel to the tectonic fabric, with offsets ranging from 1 cm to 30 m.

Separating several of the basaltic flows are thin (generally <1 m thick) layers of dark gray to purple, fine-grained phyllite consisting of sericite + magnetite + titanite + epidote + quartz ± chlorite. A few purple phyllite layers contain relict vitroclastic textures, preserving flattened pumice lapilli and flattened vesicles now filled with sericite. These have been interpreted as felsic ash-flow tuff or ignimbrite (Gathright, 1976; Bartholomew, 1977; Gathright et al., 1977). Where vitroclastic textures are not preserved, the purple phyllite is interbedded with metasedimentary rock, suggesting reworking of the felsic tuff by water. Explosive felsic volcanism therefore appears to have been associated with eruption of the Catoctin mafic magmas, as has been documented in metarhyolites of the Catoctin Formation in southern Pennsylvania (Fauth, 1973). These may also correlate with felsic magmatism in the Manhattan Prong of New York (Tollo et al., 2004).

Hawksbill Mountain and Big Meadows Traverses

The stratigraphic sections at Hawksbill Mountain and Big Meadows (Fig. 4) contain nine and ten flows, respectively, and are located ~5 km apart. These sections were mapped in order to evaluate possible stratigraphic and chemical continuity. The stratigraphy is not continuous between the sections, but a distinctive porphyritic flow, the sixth flow from the base at Big Meadows and the seventh from the base at Hawksbill, provides a useful marker bed upon which to base correlations (Fig. 4). At Big Meadows, the absence of the basal flow seen at Hawksbill reflects the pre-Catoctin erosion surface, as shown in Figure 4.

Within the area of the Hawksbill Mountain and Big Meadows traverses, the stratigraphy is upright, as indicated by amygdaloidal zones at the tops of flows and graded bedding in the basal conglomerate beneath the volcanic flows. The total thickness of the volcanic sequence in the two sections measured is 200 to 220 m. The exposed units within these two traverses are from the lower portion of the Catoctin Formation, either overlying metaconglomerate of the Swift Run Formation (Hawksbill) or lying unconformably upon the Blue Ridge basement complex (Big Meadows). At both sites, the upper parts of the formation have been removed by faulting and/or erosion. The predominant rock type consists of 6 to 40 m–thick layers of metabasalt composed of albite + chlorite + epidote + magnetite + titanite ± hematite ± actinolite ± quartz ± relict clinopyroxene. Columnar jointing is common in most flows, with column diameters in the range of 25 to 80 cm. A poorly developed, discontinuous tectonic cleavage defined by chlorite ± actinolite is observed throughout most of the formation. Over the bottom 2 to 3 m of a flow, the foliation is commonly moderately to well developed, but poorly exposed. For a detailed description of the stratigraphic characteristics of each flow, with photographs, see Badger (1992).

CHEMICAL STRATIGRAPHY

Major- and trace-element geochemical data are presented in Tables 1–3. The Mg # was calculated as 100 × molar $Mg^{2+}/(Mg^{2+} + Fe^{2+})$, assuming an original ferric/ferrous ratio of 0.15 (Brooks, 1976). The location of each sample is shown in the stratigraphic columns of Figures 3 and 4. The major-element chemistry is comparable with that of other continental flood basalt provinces. For example, Mg #s range from 28 to 61 for these Catoctin samples, from 31 to 64 for the Columbia River basalts analyzed by Swanson and Wright (1981), and from 36 to 65 for the flows from the Mahabaleshwar section of the Deccan Traps analyzed by Cox and Hawkesworth (1985) and by Sen (1986). SiO_2 values for all three flood basalt provinces are in the 48 to 51 wt% range. TiO_2 is between 2.0 and 3.5 wt% for sixty-two of the seventy Catoctin samples analyzed, similar to the range found for twenty-four of the forty-seven Mahabaleshwar flows analyzed by Najafi et al. (1981), twenty-six of the fifty basalt samples from the Paraná Basin analyzed by Piccirillo et al. (1989), and seventeen of the thirty-one samples of Columbia River basalt analyzed by Swanson and Wright (1981).

Chemical Mobility

The mafic assemblage chlorite, epidote, and albite is clear evidence that the rocks have undergone greenschist-facies metamorphism. Fluid transport of certain cations during metamorphism has long been recognized to result in local compositional changes in metaigneous rocks (Reed and Morgan, 1971). However, careful analysis of such rocks has led to recognition of which cations are least likely to have been mobilized, and thus provide the most reliable indicators of original magmatic compositions (e.g., Pearce and Cann, 1973; Floyd and Winchester, 1975). A method for this type of analysis has been worked out for the Catoctin samples in this study (Badger, 1993a), and is summarized as follows.

Petrographic analysis was used to identify metabasalt flows that retain igneous clinopyroxene and igneous textures. Such features are typically located near the texturally massive center of flows; in contrast, flow margins are characteristically more hydrated and generally do not preserve igneous assemblages or textures. These relations indicate that fluids associated with metamorphism migrated along flow margins where vesicles, interbedded breccias, and sediments provided permeable zones for fluid transport, whereas more massive interiors of flows were subjected to considerably less fluid interaction and therefore less hydration and disruption of the original chemical signature.

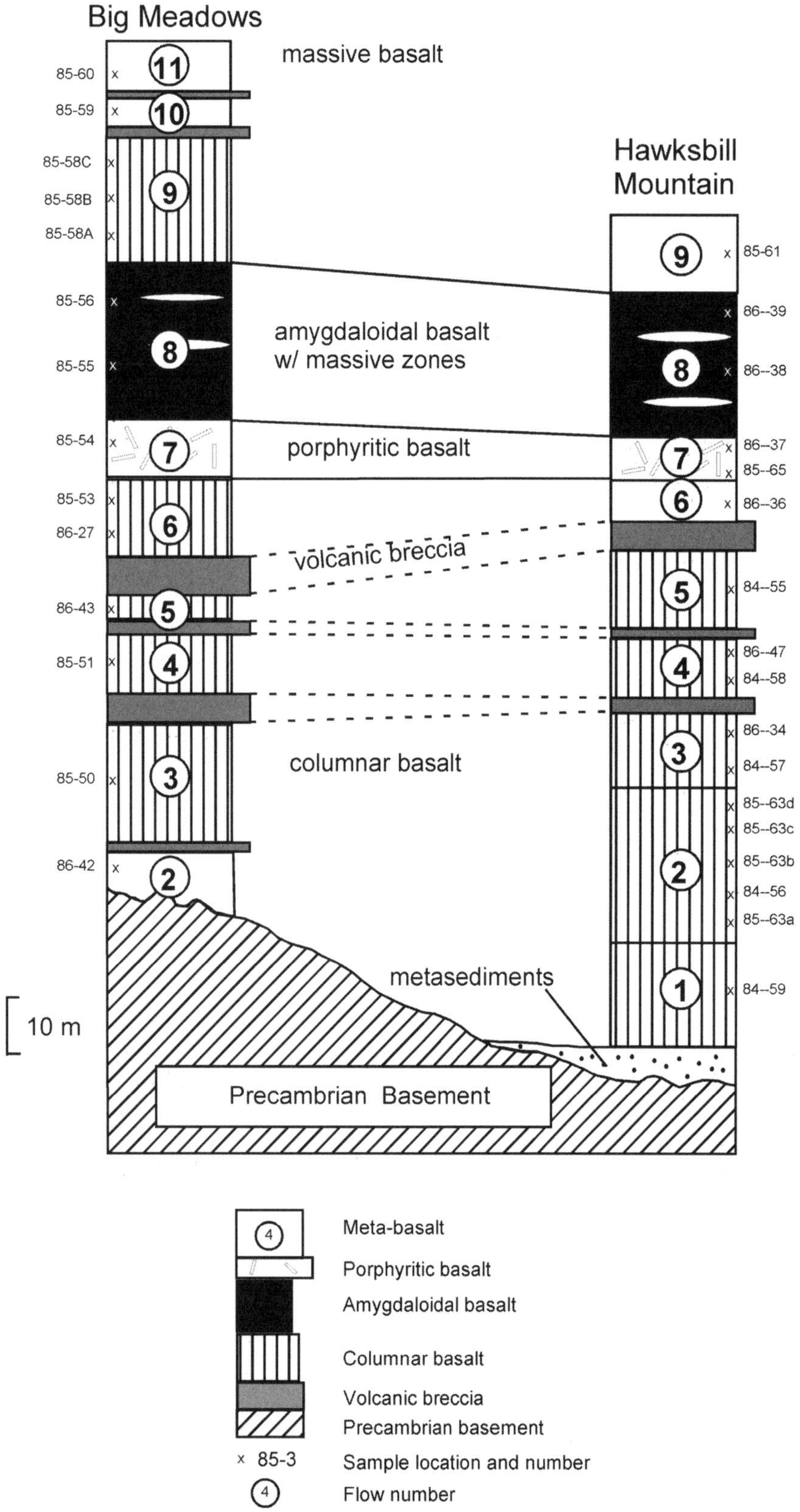

Figure 4. Stratigraphic sections for the Hawksbill Mountain and Big Meadows traverses. Stratigraphic correlation is based on the porphyritic unit and the location of volcanic breccias. Flow missing at the base of Big Meadows traverse is interpreted to reflect the effect of irregular pre-Catoctin erosion.

TABLE 1A. GEOCHEMICAL COMPOSITIONS OF SAMPLES FROM THE I-64 TRAVERSE (MAJOR ELEMENTS AND FOUR TRACE ELEMENTS)

Sample	85-1	84-3	85-2	84-4	85-3	85-4	85-5	85-7	85-8	85-10
Major elements										
SiO_2*	50.14	50.38	49.32	49.33	50.11	46.27	49.18	47.54	44.62	47.15
TiO_2	2.51	2.53	2.60	2.60	2.71	2.77	2.64	2.74	3.19	2.94
Al_2O_3	13.80	13.74	13.84	13.73	13.27	14.97	14.87	14.57	13.78	14.86
FeO†	13.53	13.63	14.43	14.71	14.76	15.30	13.26	14.95	14.67	14.46
MnO	0.24	0.22	0.25	0.24	0.23	0.29	0.24	0.27	0.30	0.24
MgO	5.16	4.67	5.23	5.88	4.63	5.82	5.63	5.84	4.48	3.81
CaO	8.84	8.96	8.60	9.33	8.86	8.16	7.52	8.13	9.44	7.19
Na_2O	3.02	2.51	2.10	2.36	2.42	2.54	2.95	2.81	2.50	2.78
K_2O	0.13	0.44	0.16	0.39	0.55	0.60	0.87	0.02	0.34	1.21
P_2O_5	0.34	0.34	0.35	0.35	0.35	0.31	0.29	0.34	0.42	0.34
LOI	1.96	2.08	3.25	1.78	1.94	2.53	2.40	3.02	6.33	5.50
Total	99.67	99.50	100.13	99.70	98.91	99.56	99.23	99.37	99.03	99.51
Trace elements										
Rb	2.6	11.2	3.4	9.6	14.0	16.0	22.3	3.0	10.7	29.4
Ba	146	490	174	420	217	222	327	39.9	169	434
Sr	354	159	217	220	188	232	187	221	168	78.8
Y	45	46	46	45	49	46	53	50	49	49
Mg #	44.4	41.8	43.2	45.6	39.7	44.4	46.9	45.0	39.0	35.7

Sample	85-11	85-12	85-13	85-15	85-23	85-24	84-12	85-25	85-26	85-27
Major elements										
SiO_2	48.84	48.71	51.43	47.24	49.96	49.49	49.24	50.22	49.19	50.45
TiO_2	2.88	2.48	2.09	2.24	2.28	2.31	2.22	2.30	2.26	2.20
Al_2O_3	14.36	14.34	14.07	14.08	14.82	14.62	14.39	14.70	14.62	14.59
FeO	13.06	13.90	11.92	11.36	12.13	12.57	11.92	13.39	12.19	11.71
MnO	0.25	0.27	0.22	0.19	0.25	0.23	0.25	0.20	0.25	0.20
MgO	2.89	6.25	6.08	4.76	5.66	6.54	5.31	5.84	5.76	5.37
CaO	8.77	8.40	7.61	11.42	7.39	8.32	10.15	7.41	9.08	9.65
Na_2O	3.88	2.29	3.20	2.42	3.11	2.32	1.83	1.73	2.18	2.27
K_2O	0.16	0.02	0.09	0.51	0.10	0.10	0.33	0.24	0.42	0.08
P_2O_5	0.42	0.30	0.28	0.31	0.25	0.27	0.29	0.26	0.26	0.28
LOI	4.99	2.73	2.35	5.35	3.77	2.90	4.26	3.87	3.96	3.57
Total	98.79	99.68	99.33	99.88	99.72	99.67	100.19	100.16	100.17	100.38
Trace elements										
Rb	5.4	1.7	2.8	10.8	3.4	4.3	8.0	5.5	10.3	3.2
Ba	73	59	140	185	48	64	100	105	147	67
Sr	160	170	242	327	238	249	183	168	157	462
Y	56	36	31	30	30	31	31	34	29	35
Mg #	31.7	48.5	51.7	46.8	49.5	52.2	48.3	50.4	49.7	49.0

Sample	85-17	84-18	85-18	85-19	84-23	84-25	84-27	84-29	84-30	84-31
Major elements										
SiO_2	51.55	55.32	42.23	51.34	48.36	46.92	52.90	47.32	50.16	51.09
TiO_2	2.94	2.72	3.67	2.98	3.18	3.25	2.67	2.98	3.08	2.15
Al_2O_3	13.35	13.33	17.87	14.00	14.80	14.20	13.70	14.30	12.99	14.23
FeO	14.20	13.59	16.87	14.11	15.16	14.79	13.24	15.64	14.55	12.45
MnO	0.28	0.20	0.14	0.18	0.27	0.25	0.22	0.30	0.25	0.24
MgO	4.57	2.74	3.10	3.81	4.66	6.11	3.30	5.82	4.89	5.28
CaO	5.76	4.39	5.40	6.25	7.67	6.84	7.41	6.35	7.62	9.82
Na_2O	3.84	4.18	0.90	3.69	2.32	3.38	4.15	3.89	2.74	2.04
K_2O	0.62	1.45	5.89	2.00	1.87	0.71	0.86	0.41	0.92	0.41
P_2O_5	0.41	0.50	0.68	0.49	0.47	0.50	0.41	0.44	0.45	0.30
LOI	1.96	1.37	3.16	1.49	2.40	2.19	0.99	2.15	1.59	2.24
Total	99.45	99.78	99.91	100.34	101.16	99.14	99.45	99.60	99.24	100.26
Trace elements										
Rb	11.5	24.5	111	28.5	41.4	12.1	15.7	10.3	13.1	8.3
Ba	520	1290	1545	1176	707	338	836	167	616	205
Sr	353	227	72.3	376	413	43	517	347	354	232
Y	57	55	35	55	38	48	50	46	46	42
Mg #	40.3	29.7	27.8	36.2	39.2	46.4	34.3	43.8	41.3	47.0

(continued)

TABLE 1A. (*Continued*)

Sample	84-32	84-35	85-39A	85-39B	85-39C	85-39D
Major elements						
SiO_2	49.43	47.65	48.93	49.45	48.67	50.42
TiO_2	2.87	4.03	2.86	2.91	2.91	2.85
Al_2O_3	13.28	13.30	13.53	13.80	13.73	13.34
FeO	14.84	14.73	14.57	13.89	14.85	14.50
MnO	0.29	0.27	0.25	0.21	0.28	0.26
MgO	6.86	4.99	4.66	4.44	4.93	4.99
CaO	5.68	8.36	9.16	9.04	8.27	6.60
Na_2O	2.04	2.37	2.34	2.22	2.08	3.65
K_2O	0.19	0.58	1.29	1.47	1.69	1.39
P_2O_5	0.28	0.79	0.38	0.37	0.37	0.34
LOI	3.44	2.25	2.01	1.74	1.57	1.09
Total	99.20	99.32	99.88	99.54	99.35	99.44
Trace elements						
Rb	4.8	10.4	20.5	19.1	22.1	20.1
Ba	165	670	439	464	721	556
Sr	134	247	237	229	177	193
Y	40	50	36	39	36	38
Mg #	49.2	41.5	40.1	40.1	41.0	41.9

*Major elements expressed in weight percent, trace elements in parts per million.
†Total iron expressed as FeO.

The method used in this study, referred to as the isocon method (Grant, 1986), involves plotting the major-element chemistry of two samples that *should* have the same geochemical signature against one another. For example, consider two samples, X and Y, collected from the same volcanic flow and with identical compositions. If the major oxides, in weight percent, are plotted for sample X on the *x*-axis and for sample Y on the *y*-axis, a straight line with a slope of 1 should pass through each point. Now suppose sample Y has been altered and some oxides have been lost or gained from the system. A line with a slope of 1 (or close to it, depending on total volume loss or gain) will pass through only those geochemical constituents that have not been mobilized, those that have been slightly mobilized will fall near the line, and those that are most mobile will be far from the line.

TABLE 1B. GEOCHEMICAL COMPOSITIONS OF SAMPLES FROM THE I-64 TRAVERSE (TRACE ELEMENTS)

Sample	85-2	85-3	85-4	84-12	84-18	84-27	85-39B	85-39C	85-39D
Th	2.76	2.73	2.82	1.29	2.04	2.29	1.52	1.56	1.47
Zr	226	234	228	159	330	241	200	220	200
Hf	5.85	5.85	6.09	3.80	7.95	6.16	5.29	5.27	5.19
Nb	22	24	22	14	21	17	19	20	19
Ta	1.60	1.65	1.64	1.15	1.59	1.46	1.49	1.42	1.51
Ni	40	35	44	59	31	22	38	41	29
Zn	143	129	141	114	116	107	39	36	38
Cr	35	33	42	100	11	17	49	50	49
Co	51	50	41	47	35	40	52	54	55
Cu	238	136	88	119	N.A.*	38	70	66	53
Sc	38.0	37.8	41.4	33.6	28.2	30.3	33.1	33.5	32.7
La	12	26	21	23	36	16	9	27	26
Ce	50	56	37	37	79	53	42	59	39
Nd	30.4	29.8	30.1	20.3	41.0	31.8	26.9	25.0	24.9
Sm	7.84	7.88	8.05	5.59	10.16	8.50	7.68	7.59	7.47
Eu	2.12	2.15	2.24	1.76	2.71	2.50	2.35	2.32	2.17
Tb	1.27	1.28	1.32	0.89	1.41	1.30	1.21	1.19	1.19
Yb	4.12	3.90	3.96	2.64	4.06	3.90	3.25	3.28	3.15
Lu	0.57	0.56	0.59	0.37	0.56	0.55	0.48	0.46	0.46

*N.A.—Not analyzed.

TABLE 2A. GEOCHEMICAL COMPOSITIONS OF SAMPLES FROM THE BIG MEADOWS TRAVERSE (MAJOR ELEMENTS AND FOUR TRACE ELEMENTS)

Sample	86-42	85-50	85-51	86-43	86-27	85-53	85-54
Major elements							
SiO_2*	54.07	50.09	50.52	50.74	47.79	48.72	49.34
TiO_2	2.49	2.01	1.97	2.22	2.50	2.46	2.23
Al_2O_3	13.48	13.97	14.08	14.09	13.59	13.59	17.47
FeO†	13.33	11.37	11.01	12.86	13.20	13.14	11.22
MnO	0.21	0.20	0.19	0.20	0.19	0.23	0.21
MgO	4.00	6.45	5.98	6.17	5.15	5.81	5.62
CaO	5.59	8.53	8.69	7.25	9.56	9.88	3.58
Na_2O	2.92	3.70	3.80	3.59	4.00	3.42	3.46
K_2O	0.16	0.22	0.18	0.12	0.28	0.10	0.09
P_2O_5	0.35	0.24	0.25	0.29	0.31	0.30	0.26
LOI	1.95	2.23	2.24	2.06	2.34	1.80	3.81
Total	98.55	99.01	98.91	99.59	98.91	99.45	97.29
Trace elements							
Rb	3.6	6.4	5.2	4.0	8.1	3.9	4.7
Ba	361	119	121	88	137	77	66
Sr	152	114	136	100	137	110	69
Y	54	36	40	42	30	39	58
Mg #	38.6	54.3	53.2	50.1	45.0	48.1	51.2
Sample	85-55	85-56	85-58A	85-58B	85-85C	85-59	85-60
Major elements							
SiO_2	51.26	47.92	48.60	53.44	50.38	52.38	52.63
TiO_2	2.47	2.57	2.86	2.80	3.08	2.92	2.99
Al_2O_3	14.54	14.04	13.43	13.12	13.10	13.16	13.40
FeO	12.89	12.62	13.97	12.93	14.01	13.27	13.28
MnO	0.20	0.16	0.21	0.21	0.22	0.22	0.18
MgO	5.00	3.92	5.02	4.24	5.31	4.12	3.69
CaO	6.86	7.85	9.78	6.23	6.77	5.93	5.93
Na_2O	3.12	3.24	1.90	2.74	2.97	2.61	2.64
K_2O	0.08	0.21	0.30	1.58	1.09	2.52	1.97
P_2O_5	0.32	0.31	0.39	0.35	0.37	0.56	0.51
LOI	2.95	6.19	3.45	1.67	1.68	1.45	1.67
Total	99.69	99.03	99.91	99.31	98.98	99.14	98.89
Trace elements							
Rb	4.0	7.3	5.7	25.4	13.0	32.8	20.2
Ba	46	44	180	1054	567	1371	851
Sr	171	61	156	89	249	116	218
Y	38	47	47	45	50	54	62
Mg #	44.9	39.4	43.0	40.1	44.3	39.4	36.8

*Major elements expressed in weight percent, trace elements in parts per million.
†Total iron expressed as FeO.

Using Catoctin Formation samples identified as least altered, based on relict igneous textures and mineral assemblages, plotted against more altered portions of the same flow, it was shown that TiO_2, P_2O_5 and Al_2O_3 were the least mobile major oxides; FeO and MgO showed limited mobility, and CaO and SiO_2 showed higher mobility, whereas K_2O and Na_2O are highly mobile and therefore unreliable for characterization of the magmas. The oxides characterized by least mobility are therefore most useful in establishing the geochemical stratigraphy. These findings are consistent with others showing variable chemical mobility of different oxides within altered basalts (e.g., Gottfried et al., 1983).

Chemical Subunits

Combining geochemical data with the stratigraphy establishes a chemical stratigraphy for the I-64 (Fig. 5), Hawksbill Mountain (Fig. 6), and Big Meadows (Fig. 7) traverses. For the I-64 traverse, recognizing chemical subunits is problematic because portions of the stratigraphic section are covered, and chemical data are not available for all flows. It is worthwhile, however, to show major-element chemical analysis as a function of stratigraphic sequence (Fig. 5) to demonstrate that there are some sharp shifts in geochemical trends that must reflect petrogenetic processes. For the Hawksbill Mountain and Big Mead-

TABLE 2B. GEOCHEMICAL COMPOSITIONS OF SAMPLES FROM THE BIG MEADOWS TRAVERSE (TRACE ELEMENTS)

Sample	85-50	85-53	85-54	85-58A	85-60
Th	1.96	1.41	1.22	2.89	2.45
Zr	164	177	197	242	310
Hf	4.16	4.47	4.66	6.49	7.95
Nb	13	16	12	25	23
Ta	1.24	1.54	0.98	1.90	1.82
Ni	99	80	58	51	27
Zn	36	110	173	137	155
Cr	139	102	147	70	22
Co	54	57	58	51	40
Cu	24	89	N.A.*	123	10
Sc	43.8	43.7	38.8	34.5	31.7
La	21	19	17	20	26
Ce	31	38	36	52	69
Nd	18.9	22.0	20.8	32.2	43.0
Sm	5.44	6.19	6.23	8.61	1.57
Eu	1.38	1.89	1.70	2.49	3.04
Tb	0.95	1.11	1.09	1.30	1.57
Yb	3.11	3.35	2.90	4.02	4.68
Lu	0.42	0.49	0.46	0.56	0.66

*N.A.—Not analyzed.

ows traverses, the chemical stratigraphy is complete for the portion of the traverse exposed, and the stratigraphy can be divided into chemical subunits based on chemical trends and discontinuities between trends. The overall chemical trend reflects increasing TiO_2 and P_2O_5 with decreasing MgO, typical of tholeiitic trends.

The section at Hawksbill Mountain can be divided into two chemical subunits (Fig. 6). Subunit A, containing thirteen samples from the bottom six flows, is marked by increasing TiO_2 (2.0–2.5%), and FeO (10.8–13.7%). Samples from flow 3 (84-57 and 86-34) have the highest CaO content (13–14%) of any flow sampled, reflecting Ca mobility during albitization of surrounding rocks that has been manifested in high-modal epidote, and conversely this flow has very low Na_2O (<1%). Subunit B contains five samples from flows 7 through 9, and reflects increasing TiO_2 (2.0–2.7%), CaO (3.7–7.6%), FeO (12.4–13.6%), and P_2O_5 (0.24–0.32%) and decreasing Al_2O_3 (16.4–13.8%). Subunit A is differentiated from subunit B by the chemical discontinuity between flow 6, the top flow of subunit A, and flow 7, the bottom flow of subunit B. Although the porphyritic nature of flow 7 suggests the possibility that it may have a different genetic affinity, its chemical trends for TiO_2, Al_2O_3, FeO, and P_2O_5 are consistent with the interpretation that it is part of subunit B.

The Big Meadows traverse can also be divided into two major subunits based on TiO_2 and P_2O_5 (Fig. 7). Numbering of the flows is based on the stratigraphic correlation of Figure 4, which reflects the interpretation that flow 1 is missing, due to topographic relief of the pre-Catoctin erosion surface. The bottom sample from Big Meadows (86-42), which would correlate stratigraphically with flow 2 at Hawksbill Mountain, contains 54% SiO_2, and its chemistry does not correlate well with the remainder of the sequence. The high silica content is interpreted to reflect mixing with unconsolidated quartz-rich sediments during extrusion. Because of the possibility of mixing and contamination, this sample is not used for modeling. Subunit A contains four flows (numbers 3 through 6), and reflects increasing TiO_2 (2.0–2.5%), FeO (11.0–13.2%), and P_2O_5 (0.24–0.31%), and decreasing MgO (6.5–5.2%). The second subunit, B, contains flows 7 through 11, and is generally higher in TiO_2, FeO, and P_2O_5, and lower in CaO and MgO, than subunit A. Chemical trends are marked by increasing TiO_2 (2.2–3.1%) and P_2O_5 (0.26–0.56%), and decreasing MgO (5.6–3.7%). The basal flow in subunit B is the porphyritic unit also observed at Hawksbill Mountain. Its petrographic nature, consisting of ~45% modal albitized plagioclase, indicates that it may not have evolved from the same magma chamber from which aphanitic flows were generated. However, chemical trends for TiO_2, MgO, and P_2O_5 are consistent with trends for subunit B. As a result, it is included as the basal member of subunit B. Anomalously high Al_2O_3 and low FeO are due to the abundance of modal feldspar, and low CaO indicates that CaO in plagioclase was lost from the system during albitization.

Chemical trends for subunit A at Big Meadows are similar to chemical trends for subunit A at Hawksbill Mountain, further evidence that the stratigraphic correlation based on location of the porphyritic flow (Fig. 4) is justified. Likewise, chemical trends for subunit B at Big Meadows are similar to chemical trends for subunit B at Hawksbill Mountain, and the chemical discontinuities separating subunits A and B for each traverse are similar. Stratigraphic and geochemical data therefore support correlation between subunits for the two traverses.

The concept of chemical subunits utilized here is similar to that used to distinguish basaltic sills and flows in the Mesozoic Appalachian tholeiite sequence (Froelich and Gottfried, 1985), flow stratigraphy in the Paraná flood basalts (Peate et al., 1990), and basaltic flows in the Columbia River basalt group (Wright et al., 1973; McDougall, 1976; Swanson and Wright, 1981; Goles, 1986).

Trace-element data support the definition of chemical subunits based on major-element compositions. Using combined data from the Hawksbill Mountain and Big Meadows traverses, subunit A has flat Ni and Sc concentrations with variable Mg #, whereas subunit B has increasing Ni and Sc with increasing Mg # (Fig. 8, A and B), probably reflecting pyroxene ± olivine fractionation (as discussed later). A plot of Ni versus Cr (Fig. 8, C) also shows a clear distinction between subunits.

The recognition of chemical subunits implies not that the magmas are a sequence of unrelated flows, but instead that the flows within each subunit may be genetically related. Having ruled out the effects of metamorphic alteration for the elements chosen previously to characterize the chemical trends of the

TABLE 3A. GEOCHEMICAL COMPOSITIONS OF SAMPLES FROM THE HAWKSBILL MOUNTAIN TRAVERSE (MAJOR ELEMENTS AND FOUR TRACE ELEMENTS)

Sample	84-59	85-63A	85-63B	85-63C	85-63D	84-56	84-57	86-34	84-58
Major elements									
SiO_2*	48.55	49.72	49.18	50.56	49.13	49.02	45.11	51.61	46.44
TiO_2	2.01	2.04	2.01	1.93	1.97	1.98	2.21	1.98	2.50
Al_2O_3	14.49	13.90	14.40	13.66	14.86	14.32	14.75	13.93	12.72
FeO†	12.13	11.79	12.08	11.83	10.82	12.18	12.71	12.01	13.16
MnO	0.22	0.23	0.23	0.20	0.21	0.24	0.25	0.22	0.36
MgO	5.28	8.04	7.58	6.23	8.21	5.87	7.32	3.29	9.95
CaO	9.60	7.64	8.24	8.71	8.72	9.03	12.96	13.98	8.77
Na_2O	3.93	2.51	2.63	2.63	2.18	3.85	0.01	0.67	0.01
K_2O	0.33	0.26	0.44	0.92	0.74	0.34	0.01	0.00	0.04
P2O5	0.26	0.25	0.26	0.23	0.24	0.27	0.34	0.31	0.32
LOI	2.87	2.88	2.96	2.96	3.11	2.56	3.69	1.95	4.76
Total	99.67	99.26	100.01	99.86	100.19	99.66	99.36	99.95	99.03
Trace elements									
Rb	9.2	7.4	9.1	17.0	17.8	10.3	2.7	1.1	2.6
Ba	119	143	200	413	350	135	46	19	5
Sr	242	86.1	99.2	88.0	135.9	120	503	809	74
Y	36	38	36	35	33	35	38	37	53
Mg #	47.7	58.8	56.8	52.5	61.4	50.2	54.7	36.5	61.3
Sample	86-47	86-47S	84-55	86-36	85-65	86-37	86-38	86-39	85-61
Major elements									
SiO_2	50.41	51.30	47.79	48.07	48.18	50.42	52.54	51.84	48.73
TiO_2	2.07	2.09	2.26	2.47	2.24	2.04	2.18	2.51	2.68
Al_2O_3	14.30	14.39	14.08	13.74	16.44	15.69	13.60	14.11	13.76
FeO	12.68	12.90	13.69	13.22	12.44	12.49	12.62	12.73	13.62
MnO	0.20	0.19	0.23	0.24	0.24	0.21	0.19	0.22	0.22
MgO	5.04	5.07	6.69	6.24	7.55	5.11	4.66	5.89	5.96
CaO	6.87	6.63	8.66	10.32	3.67	5.67	6.61	5.87	7.56
Na_2O	4.98	4.62	3.46	3.27	3.05	3.69	3.72	3.18	3.34
K_2O	0.09	0.09	0.25	0.07	0.32	0.45	0.87	0.14	0.38
P_2O_5	0.26	0.25	0.26	0.31	0.24	0.24	0.28	0.30	0.32
LOI	1.75	1.77	2.23	1.70	3.67	2.24	1.48	2.45	2.08
Total	98.65	99.30	99.60	99.65	98.04	98.25	98.75	99.24	98.65
Trace elements									
Rb	2.6	2.9	5.9	2.8	8.0	11.9	12.7	4.7	8.8
Ba	94	69	141	71	N.A.*	192	544	88	235
Sr	121	123	153	230	71	179	98	45	115
Y	49	47	41	42	45	42	42	51	42
Mg #	45.5	45.2	50.6	49.7	56.0	46.2	43.6	49.2	47.8

*Major elements expressed in weight percent, trace elements in parts per million.
†Total iron expressed as FeO.
*N.A.—Not analyzed.

magmas, these observed variations can now be evaluated in terms of crustal contamination and fractional crystallization.

PETROGENESIS

Less mobile chemical components can be used to investigate trends within subunits and to analyze discontinuities between them. In this section we investigate the possibility that discontinuities between subunits and variations within each subunit can be the result of crustal contamination. The process of fractional crystallization is also examined to assess whether geochemical trends within subunits can be explained by this process. Next, we select a subset of relatively unaltered samples that we believe most closely approximates the original igneous composition of the magmas, and utilize the resulting data to model evolution from a mantle source. Finally, we speculate on the composition of that mantle source.

Crustal Contamination

Crustal contamination of lava occurs by assimilation of material from the walls of a magma chamber or by assimilation

TABLE 3B. GEOCHEMICAL COMPOSITIONS OF SAMPLES FROM THE HAWKSBILL MOUNTAIN TRAVERSE (TRACE ELEMENTS)

Sample	84-59	85-63A	85-63B	85-63D	86-36	86-37
Th	1.96	2.05	2.05	1.72	1.46	1.09
Zr	159	153	163	153	171	167
Hf	4.16	4.21	4.06	3.98	4.43	4.21
Nb	14	15	16	14	17	12
Ta	1.24	1.21	1.15	1.19	1.46	1.04
Ni	96	90	90	87	86	51
Zn	114	154	141	118	42	142
Cr	139	135	121	135	102	136
Co	54	59	50	52	55	59
Cu	135	143	101	186	55	58
Sc	43.6	43.7	42.1	42.9	43.8	35.2
La	16	23	18	12	13	12
Ce	35	41	25	43	36	28
Nd	18.9	19.9	19.8	19.9	20.9	21.3
Sm	5.44	5.74	5.85	5.71	6.39	5.90
Eu	1.38	1.45	1.68	1.67	1.99	1.84
Tb	0.95	0.99	0.99	0.94	1.11	1.02
Yb	3.11	3.08	3.24	2.95	3.34	3.05
Lu	0.42	0.45	0.46	0.41	0.50	0.45

of displaced material during ascent of a magma through the crust (e.g., Dupuy and Dostal, 1984). Contaminating material may be in the form either of bulk rock that is physically emplaced into the magma, where it disaggregates and melts completely, or of a minimum melt component extracted through partial melting of the country rock by the magma (e.g., Myers et al., 1984).

For the Catoctin magmas, the most likely crustal assimilant is the Blue Ridge basement complex through which the magmas were transported. A representative composition for the Blue Ridge basement complex that can be used for modeling bulk rock assimilation is provided by sample 86-32 (Table 4), a sample of charnockite collected near the Big Meadows and Hawksbill Mountain traverses. An approximation of a minimum melt composition is provided from experiments by Carman et al. (1980) and Carman (1986, personal commun.) performed at 760 °C, 3 kilobars, and H_2O-saturated conditions for a charnockite of the Pedlar Formation (Table 4). Minimum melt composition will vary as a function of P_{Total} and P_{H_2O}, but such variations will not significantly modify the chemical trends predicted by crustal assimilation models. For example, experiments involving partial melting of charnockite at 900 °C and 6.9 ± 0.1 kilobars (Beard et al., 1994, their sample 75-313) produced melts of composition very similar to that of the melt produced by Carman et al. (1980), which is shown in Table 4.

K, Rb, and Ba are the most sensitive indicators of crustal contamination (Dupuy and Dostal, 1984), because their concentrations are significantly greater in the basement complex than in the metabasalts (compare the data in Tables 1, A; 2, A; 3, A; and 4 with that in Pettingill et al., 1984). However, these elements are also those most easily mobilized during hydrothermal alteration, metamorphism, and weathering. In order to address the question of whether chemical discontinuities between subunits could be a result of crustal contamination, and at the same time isolate the effects of hydrothermal alteration, pairs of elements have been chosen such that one element is highly sensitive to crustal contamination (and also to chemical mobility), whereas the other element is less sensitive to con-

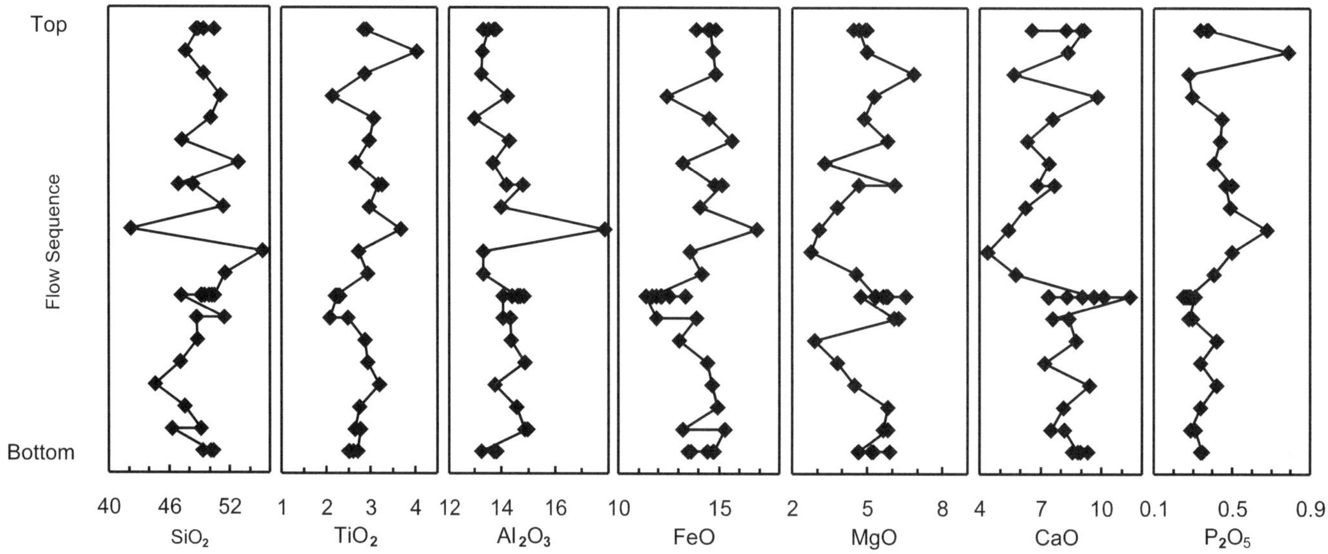

Figure 5. Chemical stratigraphy for the I-64 traverse. Tie lines between successive analyzed flows, disregarding covered sections and unanalyzed flows, demonstrate the variability of magma composition as one moves through the stratigraphic sequence. Local trends such as decreasing MgO or increasing TiO_2 may reflect crystal fractionation, but abrupt changes in trends must reflect other processes.

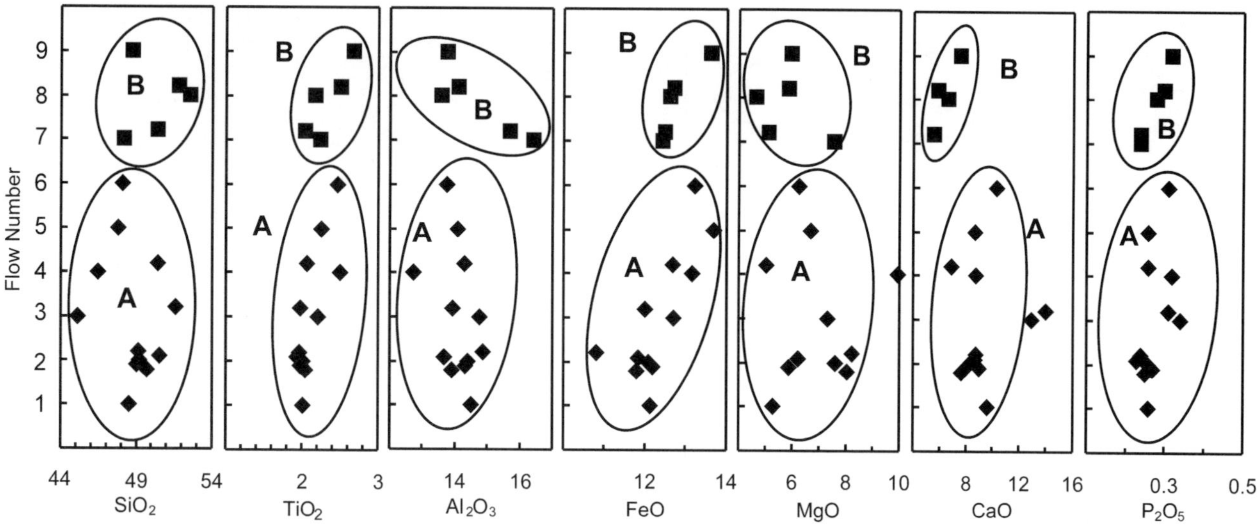

Figure 6. Chemical stratigraphy for the Hawksbill Mountain traverse. Two chemical subunits, A and B, are distinguished by trends of TiO_2, Al_2O_3, FeO, CaO, and P_2O_5.

tamination but is critical to recognition of the chemical discontinuity between subunits.

The chemical discontinuity distinguishing subunits A and B of the Hawksbill Mountain traverse is marked by a decrease in TiO_2, CaO, and P_2O_5, and an increase in Al_2O_3. Plots of TiO_2 versus Rb, Al_2O_3 versus K_2O, and P_2O_5 versus Ba for samples 86-36 (at the top of subunit A) and 86-37 (at the bottom of subunit B), along with chemical trends produced by adding 10, 20, and 30% of either the minimum melt or the bulk crustal component to sample 86-36, demonstrate that the chemical differences between units cannot be explained by crustal contamination (Fig. 9, A–C). Addition of 10–15% of a minimum melt component could account for variation of the more mobile elements, Rb, Ba, and K_2O. However, for the less mobile components, TiO_2 and P_2O_5, 20–30% contamination would be required, and for Al_2O_3 it can be projected that ~100% assimilation (for a 50–50 mixture) would be needed to account for its concentration. Such contaminant quantities are possible, but unlikely, and would have resulted in a SiO_2 content of ~55%, which is clearly not the case. For the contaminated Grande Ronde basalts of the Columbia River basalt group, the crustal component has been calculated to be ~10% (Carlson, 1984). It is therefore unlikely that chemical discontinuities between subunits within the Catoctin Formation are the result of crustal contamination. It is more likely that vari-

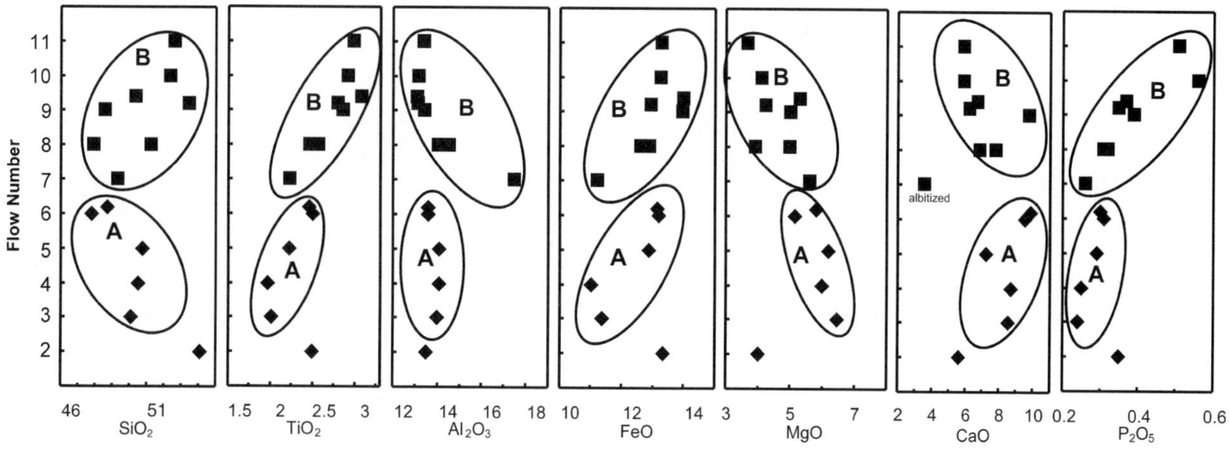

Figure 7. Chemical stratigraphy for the Big Meadows traverse. Chemical subunits A and B are distinguished by trends of TiO_2, Al_2O_3, FeO, CaO, and P_2O_5.

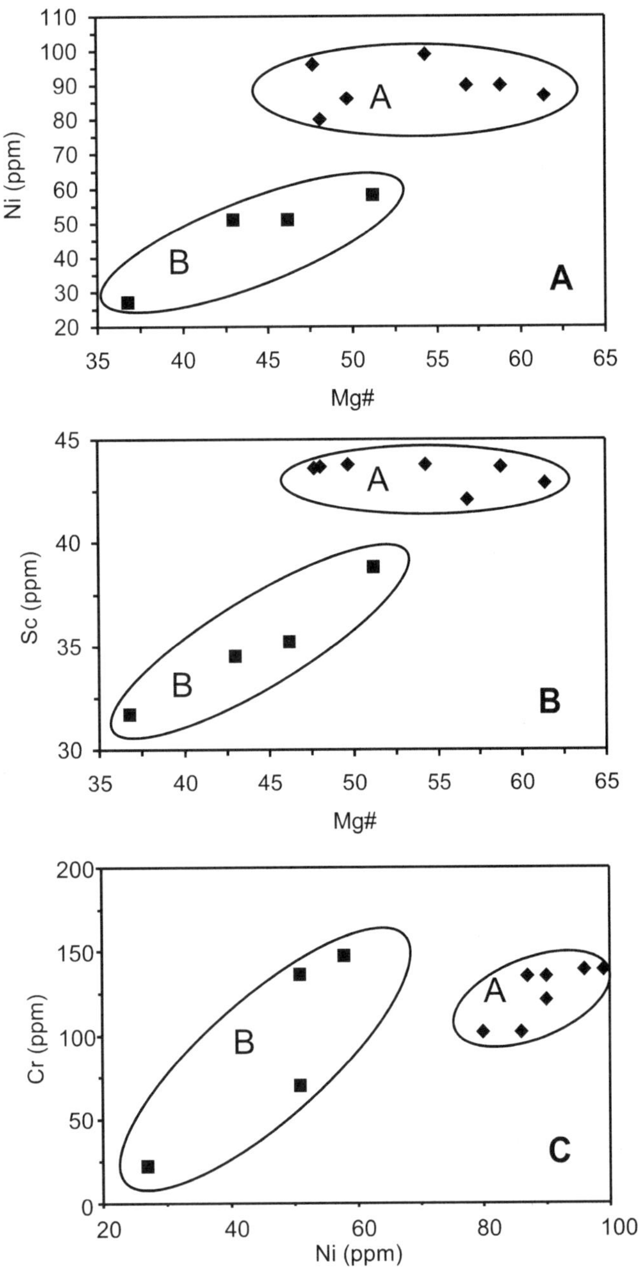

Figure 8. Distinction between chemical subunits A and B using combined trace-element data from the Big Meadows and Hawksbill Mountain traverses. (A) Mg # versus Ni; (B) Mg # versus Sc; (C) Ni versus Cr.

TABLE 4. CRUSTAL MIXING COMPOSITIONS

Sample	Minimum melt*	Basement†
SiO_2	72.24	62.33
TiO_2	0.01	1.22
Al_2O_3	16.13	16.46
FeO	1.35	7.57
MnO	0.04	0.12
MgO	0.03	1.65
CaO	1.38	4.52
Na_2O	3.20	2.83
K_2O	5.62	2.89
P_2O_5	0.00	0.30
Total	100.00	100.00
Rb	99	88
Sr	295	208
Ba	1138§	720

*Minimum melt composition from experiments by Carman et al. (1980) and Carman (1986, personal commun.). Melt composition derived from 48% partial melting of charnockite of the Pedlar Formation at 730 °C, 3 kilobars, and water saturated.

†Blue Ridge basement component is composition of sample 86-32 of charnockite of the Pedlar Formation, normalized to 100%, anhydrous. This composition is similar to that of sample 75-313, used in the melting experiments of Beard et al. (1994).

§Ba estimated for minimum melt using the distribution coefficients of Cox et al. (1979) and the modal mineralogy of the samples before and after partial melting (Carman, 1986, personal commun.) Bulk distribution coefficient for Ba calculated to be 1.58.

ation of K, Rb, and Ba is the result of chemical mobility during hydrothermal alteration, as discussed earlier, and that discontinuities defined by the less mobile chemical components reflect variations in the original composition of the magma. Such discontinuities therefore provide evidence that each chemical subunit reflects a pulse of new magma.

A similar approach can be used to examine variations within individual chemical subunits to determine whether the trends that define each subunit can be explained by increasing amounts of crustal contamination. Subunit B from the Hawksbill Mountain traverse is plotted in Figure 10, along with the effects of adding 10, 20, and 30% each of a bulk crustal or minimum melt component to sample 86-37 from the base of the sequence. Also plotted is the sequence of samples stratigraphically above sample 86-37 in the subunit. Clearly, contamination had little effect on the trend that defines the subunit—increasing TiO_2 and P_2O_5, and decreasing Al_2O_3. Crustal contamination also appears to have had little or no effect on trends of the more alteration-sensitive Rb, K_2O, and Ba concentrations. Contamination would have tended to increase the concentrations of Rb, K_2O, and Ba, but the stratigraphic sequence in subunit B is marked by a decrease in Rb and K_2O, while variations observed in Ba, for one sample, would have required a mix of ~1/3 basalt with 2/3 crustal material. It is far more likely that widely varying concentrations of Ba, K_2O, and Rb are the result of fluid interaction. It is interesting to note that the highest concentrations of Rb, K_2O, and Ba are in the same sample (86-38), while the lowest concentrations of all three are from a sample (86-39) from the same flow. Clearly, these widely varying concentrations do not reflect a mixing process, but must be due to posteruptive chemical mobility during fluid interaction.

Other subunits have been analyzed similarly, and most chemical variations within subunits are not compatible with a contamination model.

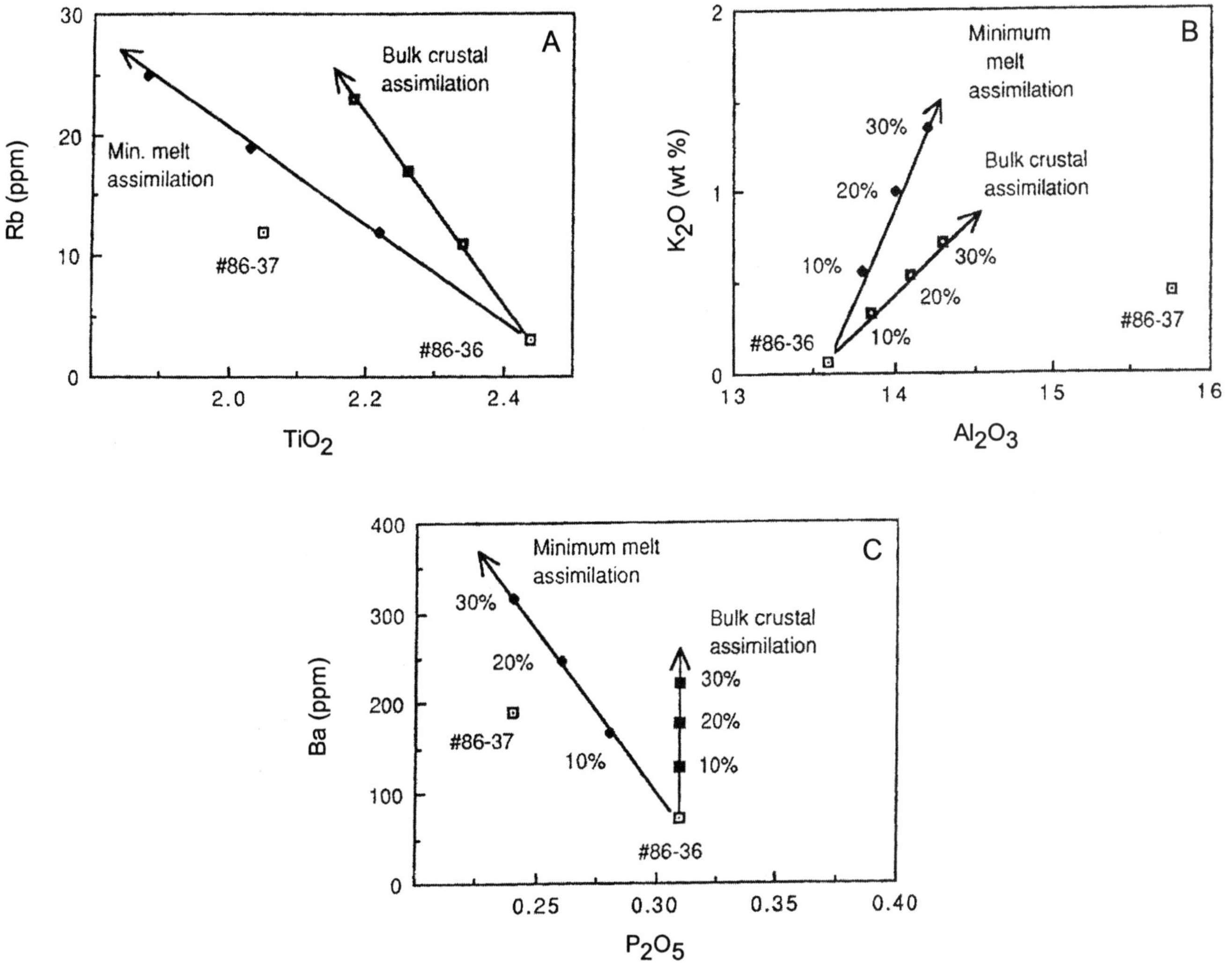

Figure 9. Crustal mixing model for chemical discontinuities between chemical subunits. Chemical effects of adding 10, 20, or 30% of a minimum melt component or a bulk crustal component to Hawksbill Mountain sample 86-36, marking the top of chemical subunit A, in an attempt to derive the basal sample of subunit B, sample 86-37. (A) TiO_2 versus Rb; (B) Al_2O_3 versus K_2O; (C) P_2O_5 versus Ba.

Fractionation Models for Chemical Subunits

The previous analysis demonstrates that both hydrothermal alteration and crustal contamination may be ruled out as probable causes for chemical trends of less mobile components observed in subunits. These trends can, however, be explained by fractionation processes. For modeling crystal fractionation, the mineral phases chosen were those typically present in basalts: clinopyroxene, plagioclase, olivine, ilmenite, and magnetite. The composition chosen for clinopyroxene was $Wo_{40}En_{47}Fs_{13}$, the average composition for cores of large clinopyroxene grains, based on 52 microprobe analyses of Catoctin samples (Badger, 1989; microprobe data available upon request). Plagioclase was chosen as An_{55}, based on microprobe analyses of plagioclase grains from basaltic dikes that are interpreted to have been feeder dikes to Catoctin flows. Olivine composition was chosen as Fo_{70}, a typical value for crustal-level olivine compositions (Roedder and Emslie, 1970). Ilmenite composition was based on two analyses of ilmenite in one of the basaltic dikes. Pure Fe_3O_4 was used for magnetite.

Calculated chemical trends for Mg #, TiO_2, Al_2O_3, FeO, and CaO for flows 8 through 11 in subunit B from the Big Meadows traverse were evaluated for crystal fractionation (Fig. 11, A–D). Flow 7, the porphyritic flow, was not used in order to avoid the weighted chemical effects of its modal mineralogy on fractionation models. For flows with multiple samples, the average composition was used. Arrows show the effect of fractionating 10% plagioclase, 10% clinopyroxene, 10% olivine, 5% magnetite, and 1% ilmenite from flow 8. As can be seen, the chemical trend of each successive flow can be explained by frac-

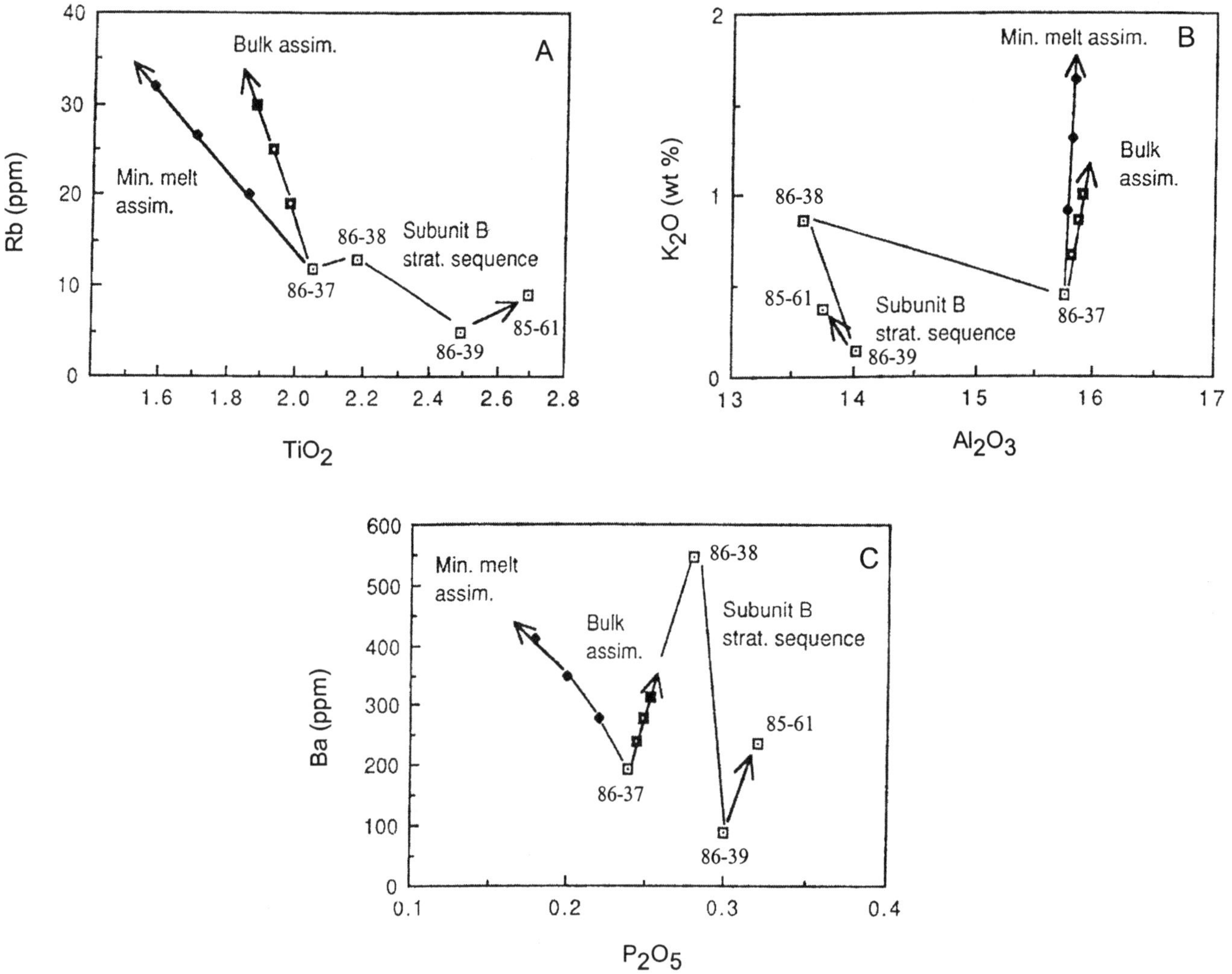

Figure 10. Crustal mixing model for chemical variations and trends within chemical subunits. Addition of 10, 20, and 30% of a minimum melt component or a bulk crustal component to sample 86-37, at the base of subunit B of the Hawksbill Mountain traverse, in an attempt to derive the overlying samples in the stratigraphic sequence of the subunit. (A) TiO_2 versus Rb; (B) Al_2O_3 versus K_2O; (C) P_2O_5 versus Ba.

tionation of up to 10% plagioclase and clinopyroxene, and up to 3% olivine, with possible minor removal of ilmenite and magnetite.

Trace-element data provide further evidence for the involvement of clinopyroxene and olivine as fractionating phases for subunit B of both traverses. As shown earlier (Fig. 8, A and B), there is a strong positive correlation between Sc and Ni with Mg #. Ni, substituting for Mg, is strongly partitioned into both clinopyroxene and olivine (Cox et al., 1979), and Sc, substituting for Ca, is partitioned into clinopyroxene. As a result, decreasing Ni and Sc with Mg # suggests the importance of fractionation of clinopyroxene and olivine.

In order to examine fractionation trends and chemical discontinuities within an entire stratigraphic sequence, pairs of elements or elemental oxides that characterize each subunit within the sequence can be compared. For example, for the Hawksbill Mountain traverse, a continuous trend of decreasing Al_2O_3 with increasing TiO_2 is shown for flows 1 through 6, with minor deviation in flow 3 (Fig. 12, A). A parallel but distinct trend is shown for flows 7 through 9. Likewise, for the Big Meadows traverse, Mg # plotted versus TiO_2 shows an increasing continuum from flows 3 and 4, which are virtually identical in terms of Mg # and TiO_2 concentrations, to flow 5 and flow 6 (Fig. 12, B). There is a marked break between flows 6 and 7, which marks the break between subunits A and B, and then subunit B shows a continuum from flow 7 through flow 11. These trends can be explained by fractionation of mineral phases as discussed earlier, but the compositional breaks that define the subunits must reflect new batches of magma, either from recharge of the existing magma chamber or from extrusion of lava from a new magma chamber.

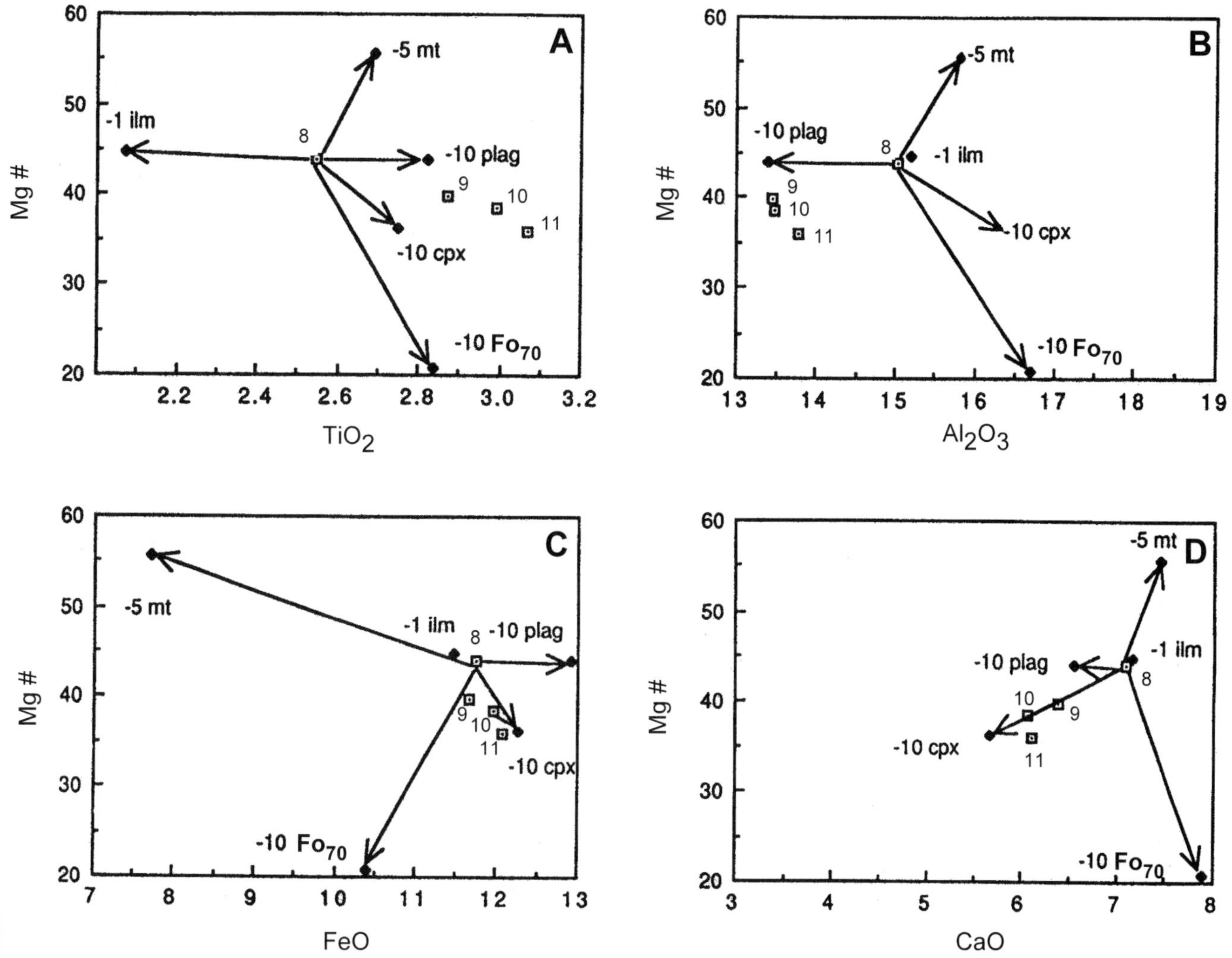

Figure 11. Fractionation model to derive flows 9, 10, and 11 from flow 8 in subunit B, Big Meadows traverse. (A) TiO_2 versus Mg #; (B) Al_2O_3 versus Mg #; (C) FeO versus Mg #; (D) CaO versus Mg #. Lines show fractionation trends for removal of 10% clinopyroxene (Cpx), 10% plagioclase (Pl), 10% olivine (Ol), 5% magnetite (Mt), and 1% ilmenite (Ilm).

Evolution from Source

Having evaluated shallow crustal processes of fractional crystallization and crustal contamination, we next examine evolutionary trends from a mantle source. The composition of the mantle from which continental flood basalts evolve is predominantly evaluated on the basis of trace elements, isotopes, and thermal melting models (see discussions in Carlson, 1991, and Cordery et al., 1997), and this is discussed later in regard to the likely source from which the Catoctin magmas evolved. First, however, we examine major-element trends to determine whether the Catoctin magmas could have been generated by typical fractionating phases from a mantle source, what the major fractionating phases would have been, and roughly what volume of cumulate material would have been removed from the evolving magmas.

Mg #s considerably lower than the values of 70–72 inferred for typical mantle melts (Basaltic Volcanism Study Project, 1981), and Ni concentrations of <100 ppm, considerably lower than typical mantle values (Mysen, 1978), clearly indicate that Catoctin magmas do not represent direct partial melts of mantle material, but must have undergone significant crystal fractionation prior to or during transport to the surface.

Phase equilibria modeling (e.g., Morse, 1980; Stolper, 1980) has shown that mantle sources, despite compositional variations, will yield melts of similar major-element characteristics. Using a calculated partial mantle melt, we were able to model the evolutionary process the Catoctin magmas must have undergone during their ascent through the lithosphere. The composition chosen to represent a primary, partial mantle melt was that of a model picrite, calculated by Cox (1980) from the compositions of mafic flows of flood basalts at Karroo and Paraná. As

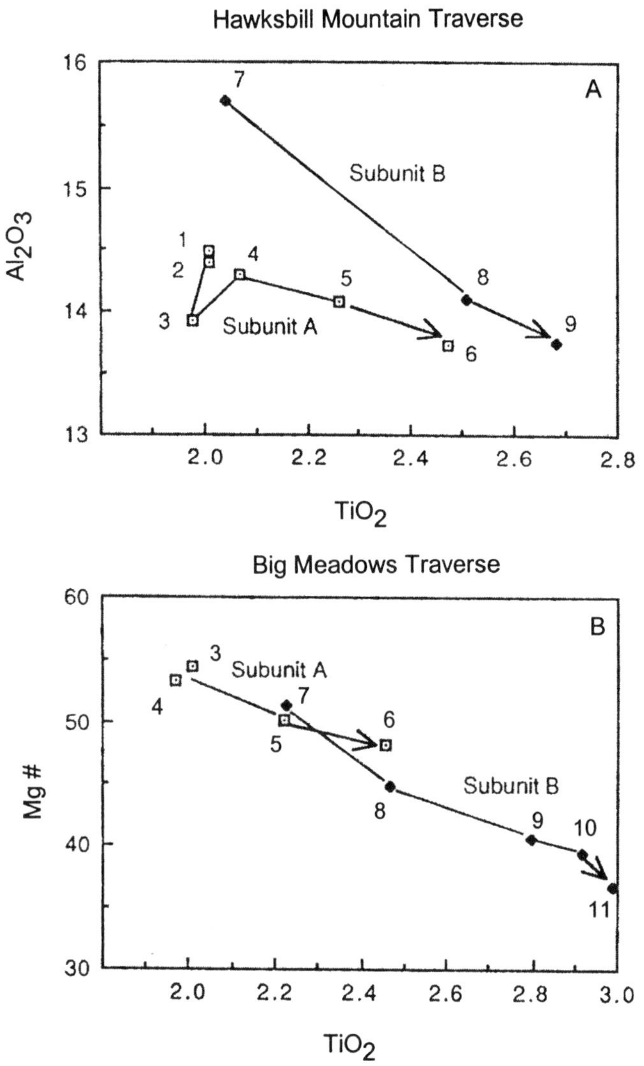

Figure 12. Sequential chemical trends within subunits indicative of fractionation processes. (A) Subunits A and B as distinguished by TiO_2 versus Al_2O_3 at the Hawksbill Mountain traverse. (B) Subunits A and B as distinguished by TiO_2 versus Mg # at the Big Meadows traverse.

representatives of the Catoctin magmas, the ten least altered samples were selected. The selection criteria used were retention of igneous clinopyroxene and igneous textures, and, for all but two, analysis of multiple samples from an individual flow (see discussion in Badger, 1993a).

From the model picrite, the crystal fractionation program of Stormer and Nicholls (1978) was used to remove crystal fractions of olivine, orthopyroxene, clinopyroxene, and plagioclase. For olivine, a composition of Fo_{90} was chosen as representative of typical upper mantle olivine (Roedder and Emslie, 1970; Cox, 1980). An_{70} was chosen as representative of plagioclase compositions at deep crustal or upper mantle depths. Varying pyroxene compositions will have little real effect on their fractionation trends for the oxides chosen in the variation diagrams, so the clinopyroxene composition was the same as that used from fractionation modeling between chemical subunits, $Wo_{40}En_{47}Fs_{13}$. The orthopyroxene composition was from a sample of the Deccan Traps reported by Sen (1986).

Variation diagrams for Mg # versus Al_2O_3, TiO_2, and FeO showing fractionating phases from the model picrite (Fig. 13, A–C) indicate that, if the parental magma had a major-element composition anywhere near the major-element concentration of the model picrite, olivine must have been the dominant mineral on the liquidus. Low Mg #s and low Ni concentrations in Catoctin samples are consistent with significant prior olivine fractionation. Plagioclase must also have been a fractionating phase due to decreasing Al_2O_3 concentrations. The overall major-element geochemistry of the Catoctin magmas can therefore be explained by gabbro fractionation from a partial mantle melt. Crude volume estimates indicate that 40–70% of the original magma was removed. Gabbro fractionation is a process well known to occur in layered mafic intrusions, and such a process is consistent with the interpretation that major-element compositions of the Deccan Traps (Cox and Hawkesworth, 1985; Cox and Dewey, 1986; Sen, 1986) and the Columbia River basalts (Reidel, 1983; Goles, 1986) are controlled by crystal fractionation of olivine or olivine + plagioclase. Large volumes of residual gabbroic cumulates that would have been left behind at the base of the crust, if such a model is correct, could explain the positive gravity anomaly along the Appalachians (see Haworth et al., 1980).

Following the method of Walker et al. (1979), the ten least altered samples selected were plotted in the system plagioclase-silica-olivine (Pl-Si-Ol) as projected from diopside (Fig. 14). All samples plot close to the liquid line of descent (LLD) at 1 atm. Experiments by Stolper (1980) on the effects of pressure on the eutectic composition of the Pl-Si-Ol assemblage show the migration of the eutectic from 1 atm to 10 kilobars along a path close to the trend for the 1 atm LLD. Crystal fractionation of Catoctin magmas, therefore, could have occurred at pressures within the range of 1 atm to ~7 kilobars, or anywhere throughout the upper 20 km of the crust, and not necessarily in a shallow crustal magma chamber.

The major-element geochemical composition of the Catoctin magmas can therefore be explained by gabbro fractionation at upper mantle through midcrustal levels. These magmas were then modified either in moderate- to shallow-level magma chambers or during transport to the surface by crystal fractionation, predominantly of plagioclase, clinopyroxene, and olivine.

Characterization of Source

Trace-element data are commonly used to characterize the source regions from which continental flood basalts evolved (e.g., review by Carlson, 1991). Catoctin trace-element data plotted on commonly used tectonic discrimination diagrams (Fig. 15, A and B) plot in the within-plate tholeiite field, as

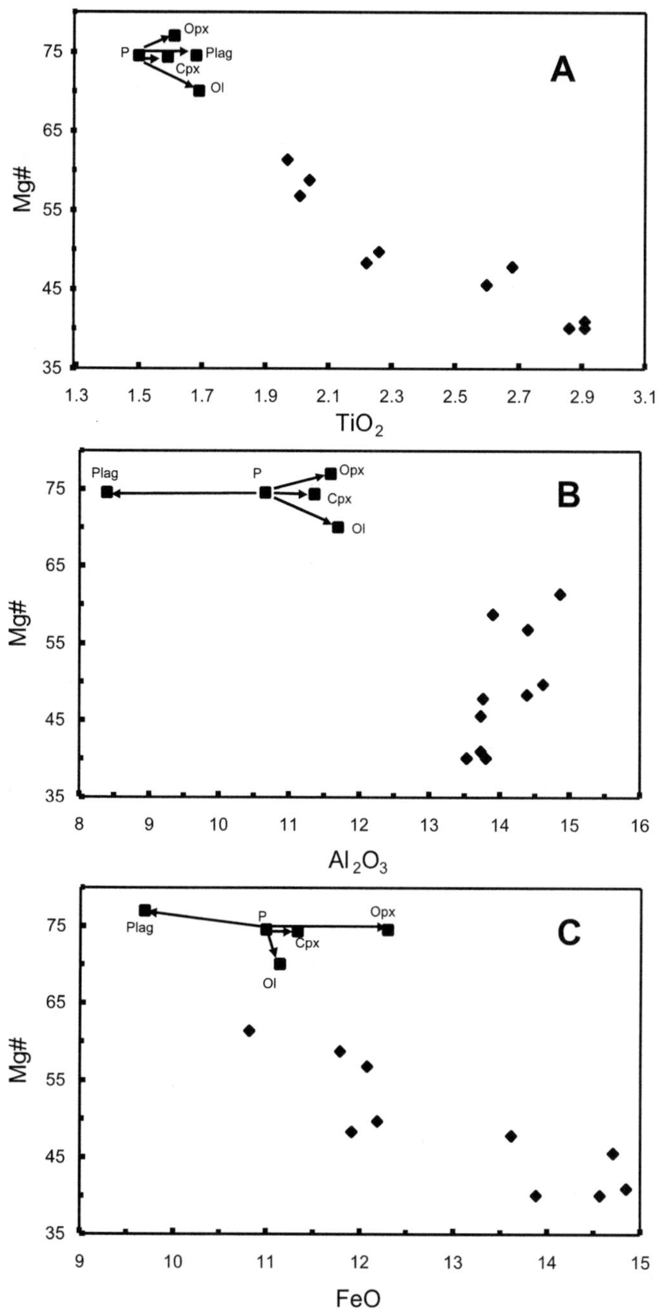

Figure 13. Fractionation diagrams showing trends for removal of 10% olivine (Ol), 10% plagioclase (Pl), 10% orthopyroxene (Opx), and 10% clinopyroxene (Cpx) from a model picrite (P) in order to derive ten least altered samples of Catoctin metabasalts. (A) TiO_2 versus Mg #; (B) Al_2O_3 versus Mg #; (C) FeO versus Mg #.

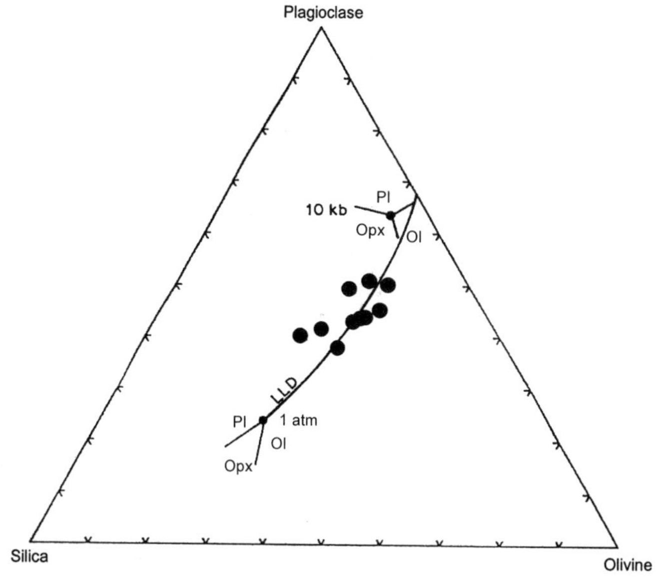

Figure 14. Plagioclase (Pl)–silica (SiO_2)–olivine (Ol) phase diagram. Normative mineralogy of ten selected least altered Catoctin samples projected from diopside onto Pl-SiO_2-Ol plane, with the liquid line of descent (LLD) at 1 atmosphere (after Walker et al., 1979). Eutectic at 10 kb is from Stolper (1980).

would be expected of continental flood basalts. A correlation diagram of Ni versus Mg # (Fig. 16, A) shows that Catoctin magmas were more evolved than normal mid-ocean ridge basalts (N-MORB). Ni versus Cr (Fig. 16, B) data for Catoctin metabasalts plot in the same fields as the Columbia River flood basalts and the Deccan Traps, but are distinctly different from N-MORB and Hawaiian ocean-island basalts. Concentrations of Hf, Ta, and Th for Catoctin magmas overlap fields for Hawaii and Columbia River basalts, but are distinct from those for N-MORB (Fig. 16, C and D).

A Sr-initial ratio of 0.7035 ± 1, obtained using five of the least altered Catoctin samples plus a clinopyroxene mineral separate (Badger and Sinha, 1988), is close to the bulk earth radiogenic Sr evolution trend (Basaltic Volcanism Study Project, 1981). Chondrite-normalized incompatible trace-element patterns (Fig. 17, A) and rare-earth element (REE) trends (Fig. 17, B) show a moderately enriched pattern, considerably different from depleted MORB patterns. Trace-element and Sr isotopic data therefore indicate that the magmas evolved from a fertile magma source, one that had neither been enriched by a subducted slab component nor significantly depleted by previous melt extraction.

Puffer (2002) has compared trace-element concentrations from nine Neoproterozoic rift-related magmas of eastern North America, including the Catoctin Formation, and shown that they all were derived from a mantle source with geochemical similarities to magmas generated by the Hawaiian, Iceland, and Reunion hot spots. Consistent with Puffer's compilation, Catoctin Nb/Y ratios averaging 0.43 and Nb/Zr ratios averaging 0.09 plot in the same field as Hawaiian, Iceland, and Reunion magmas. Puffer (2002) proposed that the eastern North American Neoproterozoic rift-related magmas were plume generated with "considerable input from a subcontinental lithospheric mantle source." An alternative hypothesis is provided by the thermal modeling experiments of Takahashi and Nakajima

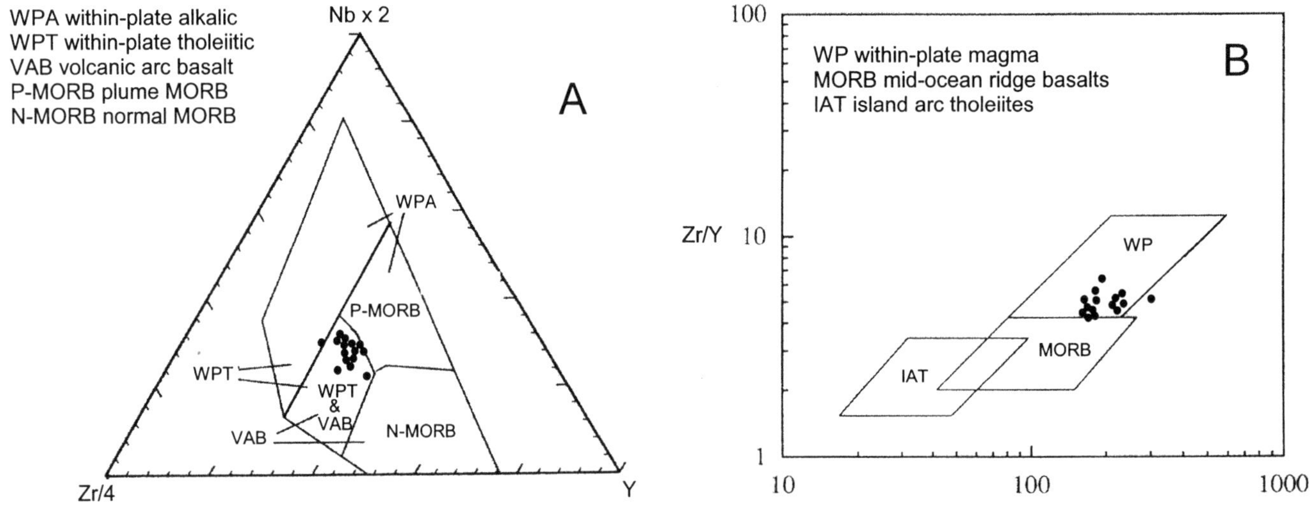

Figure 15. Tectonic discrimination diagrams. (A) Nbx2 versus Zr/4 versus Y after Meschede (1986). (B) Zr/Y versus Zr after Pearce and Norry (1979).

(2002) and Cordery et al. (1997), who proposed that flood basalts are generated by plumes mixed with a significant eclogite component from subducted oceanic slabs at the core/mantle boundary. These rising plumes would have partially melted at lithospheric depths to produce the geochemical signatures typical of flood basalt magmatism. Either model provides a plausible explanation for the geochemical signature of the source from which the Catoctin magmas evolved.

Trace-element data also permit speculation on depth of melting in the source region. Relatively flat heavy rare earth element (HREE) patterns (Fig. 17, B) preclude the presence of garnet in the source, indicating that partial melting must have occurred in spinel-peridotite, above the spinel-garnet transition zone, at depths of less than 80 km (Badger, 1994a). With the data and methods of Ellam (1992) used to interpret depth of melt generation, it was determined that Catoctin Ce/Yb ratios averaging 13.8 and Sm/Yb ratios averaging 2.2 imply that the depth of initial melting for the Catoctin magmas was on the order of 40–70 km (Fig. 18).

An interesting comparison is with Neoproterozoic mafic dikes in the Adirondack Mountains of New York, which have higher average Ce/Yb and Sm/Yb ratios of 26.1 and 2.9,

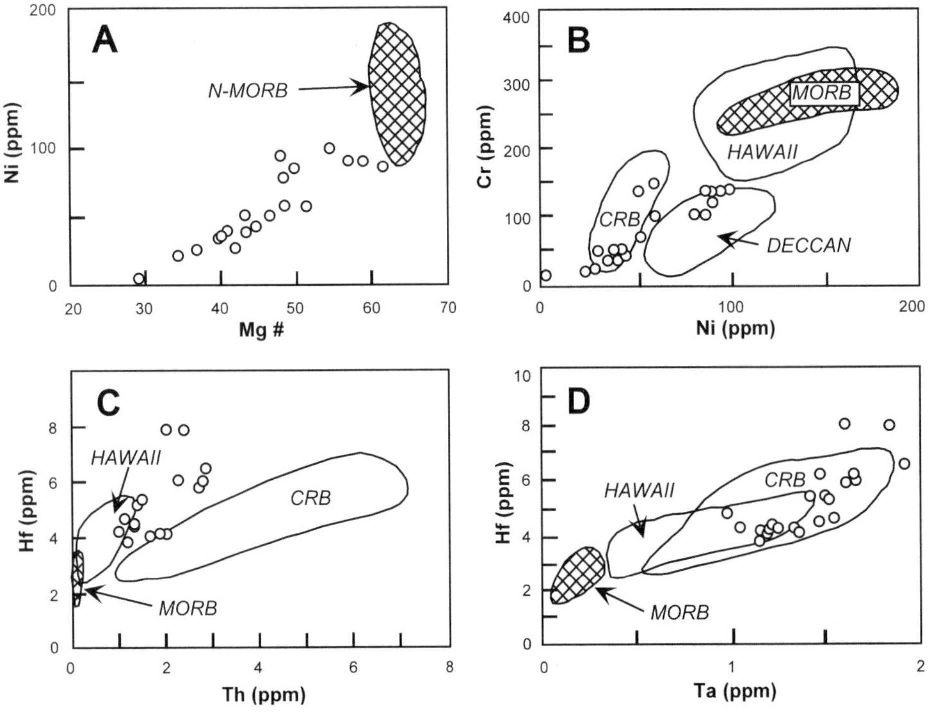

Figure 16. (A) Plot of Mg # versus Ni showing the evolved nature of Catoctin samples as compared to normal mid-ocean ridge basalt (N-MORB; Rhodes et al., 1979). (B) Plot of Ni versus Cr for Catoctin samples, Deccan Traps (Mahoney et al., 1982), Hawaiian tholeiites (Roden et al., 1984), and the Picture Gorge member of the Columbia River basalts (Goles, 1986). (C) Plot of Hf versus Th and (D) plots of Hf versus Ta for Catoctin samples, MORB (Rhodes et al., 1979), Hawaiian tholeiites (Roden et al., 1984; Budahn and Schmitt, 1985), and the Columbia River basalts (CRB; Swanson and Wright, 1981).

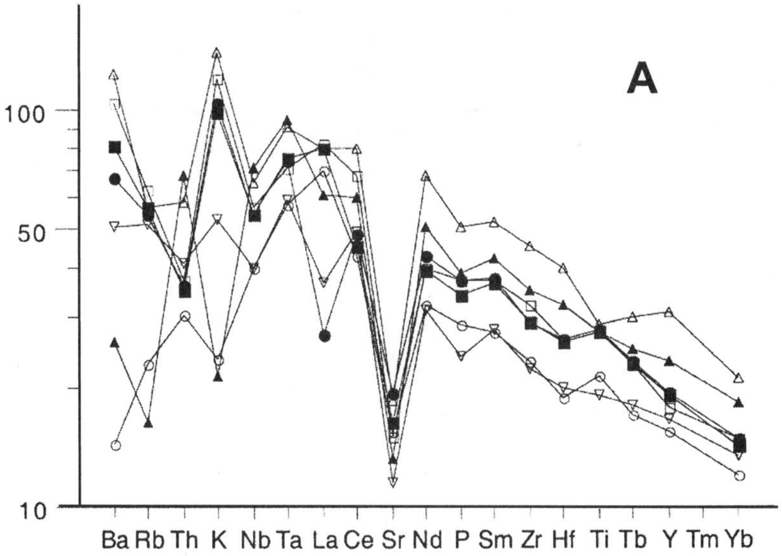

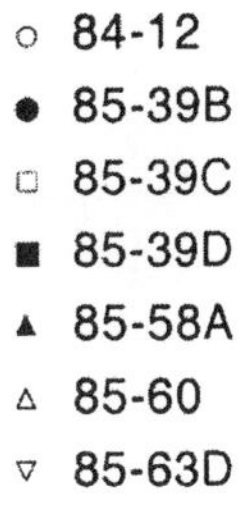

Figure 17. (A) Chondrite-normalized trace-element abundances of Catoctin metabasalts. Sample numbers as shown above. Chondrite values taken from Thompson et al. (1984). Scatter in the large-ion-lithophile elements (Ba, Rb, and K) reflects mobility during greenschist-facies metamorphism. The strong negative Sr anomaly reflects a combination of the fractionation of plagioclase and the loss of Sr from the system during albitization of plagioclase. (B) Chondrite-normalized rare-earth-element abundances of Catoctin metabasalts. Chondrite values taken from Nakamura (1974). The small negative Eu anomaly is indicative of plagioclase fractionation.

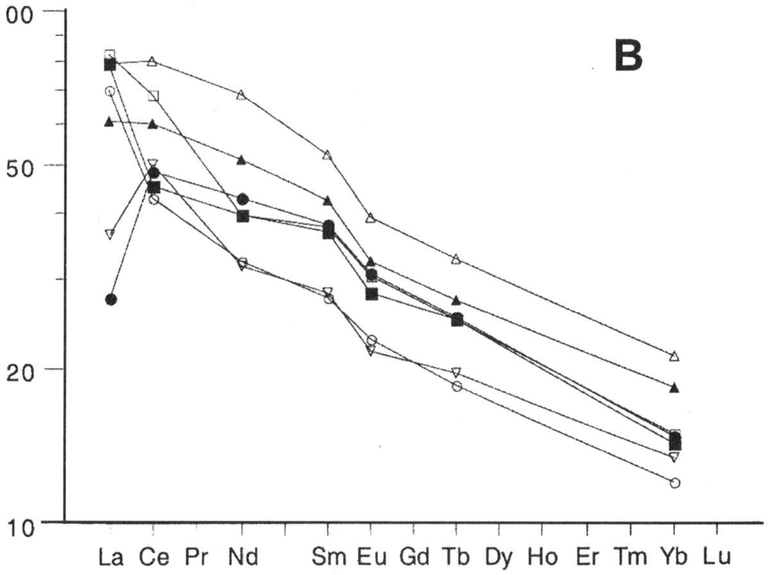

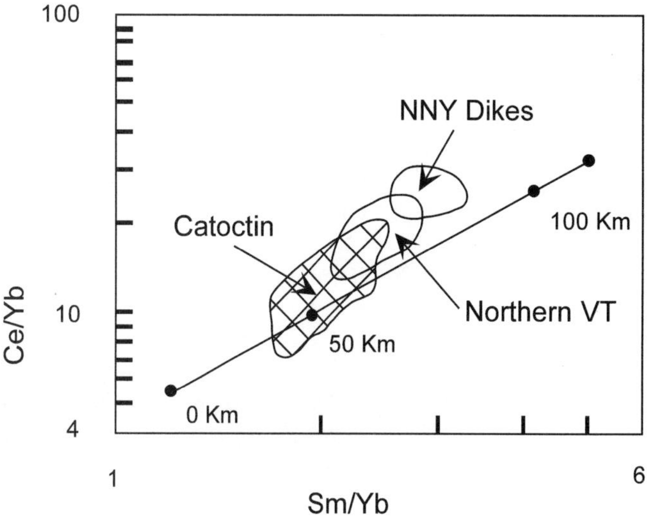

Figure 18. Sm/Yb versus Ce/Yb, after Ellam (1992), inferring depth of melting. Increasing Sm/Yb or Ce/Yb reflect retention of Yb in garnet at deeper levels in the mantle. Northern Vermont (VT) data from Coish et al. (1985). Northern New York (NNY) data from Badger (1994b and previously unpublished data).

respectively (Badger, 1994b). With the data and methods of Ellam (1992), it was determined that the Adirondack data imply a depth of melting of 70–90 km (Fig. 18). The data of Coish et al. (1985) for mafic rocks in northern Vermont, which is closer to the incipient Iapetus rift, imply a depth of melting of 55–75 km. Comparison of these different rift-related suites of rocks may reflect either prerift crustal thickness or synrift crustal thinning. In either case, these limited data show a good correlation between the proximity of the rift zone, the depth of melting, and, by extrapolation, the thickness of the lithosphere.

CONCLUSIONS

Based on major- and trace-element modeling, no evidence was observed for large-scale crustal contamination of the Catoctin magmas by the Blue Ridge basement complex. It is likely that there was some local contamination by unconsolidated sedimentary material (for example, sample 86-42), but this can be readily recognized, and such samples can be excluded from modeling of other petrogenetic processes.

Much of the major-element geochemistry has clearly been disrupted by hydrothermal alteration during metamorphism. Nevertheless, using only the least mobile elements, the petrogenetic evolution of the Catoctin magmas can be interpreted. The model proposed calls for a rising mantle plume that incorporated either an eclogite component from subducted mantle slabs (Cordery et al., 1997; Takahashi and Nakajima, 2002) or subcontinental mantle (Puffer, 2002). Batches of magma were generated by partial melting in the upper mantle above the garnet-spinel transition zone. Each batch of magma must have undergone extensive crystal fractionation of high-Mg olivine and plagioclase at upper mantle to midcrustal depths during transport through the crust. Residence time in the crust was probably short, precluding significant mixing with crustal material. Transport may have been enhanced by fractures generated during crustal extension. Although the magmas were cogenetic, each batch evolved as its own discrete magma system. Crystal fractionation, primarily of plagioclase, clinopyroxene, and olivine during transport, particularly at upper crustal levels, resulted in further chemical differentiation of the lavas and resulted in the development of the chemical subunits that we have recognized.

ACKNOWLEDGMENTS

Encouragement from Dick Tollo is gratefully acknowledged for persuading RLB to retrieve these data from his dissertation, then reanalyze and update it into a presentable form for this publication. Jeff Chiarenzelli kindly helped with illustrations and Janet Bullis with formatting the tables. Critical reviews by Ray Coish, Ron Fodor, Mike Rhodes, and Dick Tollo greatly improved the manuscript.

REFERENCES CITED

Aleinikoff, J.N., Zartman, R.E., Walter, M., Rankin, D.W., Lyttle, P.T., and Burton, W.C., 1995, U-Pb ages of metarhyolites of the Catoctin and Mount Rogers Formations, central and southern Appalachians: Evidence for two pulses of Iapetan rifting: American Journal of Science, v. 295, p. 428–454.

Badger, R.L., 1989, Geochemistry and petrogenesis of the Catoctin Volcanic Province, central Appalachians [Ph.D. thesis]: Blacksburg, Virginia Polytechnic Institute, 337 p.

Badger, R.L., 1992, Stratigraphic characterization and correlation of volcanic flows within the Catoctin Formation, central Appalachians: Southeastern Geology, v. 32, no. 4, p. 175–195.

Badger, R.L., 1993a, Fluid interaction and geochemical mobility in metabasalts: An example from the central Appalachians: Journal of Geology, v. 101, p. 85–95.

Badger, R.L., 1993b, Geochemical affinities of a Late Precambrian basaltic dike, Adirondack lowlands, Northern New York: Geological Society of America Abstracts with Programs, v. 25, no. 2, p. 4.

Badger, R.L., 1994a, Mantle composition and depth of melting during late Proterozoic rifting, central Appalachians: Geological Society of America Abstracts with Programs, v. 26, no. 4, p. 2–3.

Badger, R.L., 1994b, Mantle composition and lithospheric thickness beneath the NW Adirondack region during late Proterozoic Iapetus extension: Geological Society of America Abstracts with Programs, v. 26, no. 3, p. 4.

Badger, R.L., 1999, Geology along Skyline Drive, Shenandoah National Park, Virginia: Helena, Montana, Falcon Press, 100 p.

Badger, R.L., and Coish, R.A., 1992, Highly enriched magmas of the Adirondack lowlands: Correlation with geochemical trends of late Proterozoic rift related magmas of central and western Vermont: Geological Association of Canada/Mineralogic Association of Canada Abstracts, v. 17, p. A5.

Badger, R.L., and Sinha, A.K., 1988, Age and Sr isotopic signature of the CatoctinVolcanic Province: Implications for subcrustal mantle evolution: Geology, v. 16, p. 692–695.

Bailey, C.M., and Tollo, R.P., 1998, Late Neoproterozoic extension-related magma emplacement in the central Appalachians: An example from the Polly Wright Cove Pluton: Journal of Geology, v. 106, p. 347–359.

Bartholomew, M.J., 1977, The geology of the Greenfield and Sherando quadrangles, Virginia: Charlottesville, Virginia Division of Mineral Resources Publication 4, 43 p.

Bartholomew, M.J., Gathright, T.M., and Henika, W.S., 1981, A tectonic model for the Blue Ridge in central Virginia: American Journal of Science, v. 281, p. 1164–1183.

Basaltic Volcanism Study Project, 1981, Basaltic Volcanism on the Terrestrial Planets: New York, Pergamon Press, 1286 p.

Beard, J.S., Lofgren, G.E., Sinha, A.K., and Tollo, R.P., 1994, Partial melting of apatite-bearing charnockite, granulite, and diorite: Melt compositions, restite mineralogy, and petrologic implications: Journal of Geophysical Research, v. 99, p. 21591–21603.

Brooks, C.K., 1976, The Fe_2O_3/FeO ratio of basalt analysis: An appeal for a standardized procedure: Geological Society of Denmark Bulletin, v. 25, p. 119–120.

Budahn, J.R., and Schmitt, R.A., 1985, Petrogenetic modeling of Hawaiian tholeiitic basalts: A geochemical approach: Geochimica et Cosmochimica Acta, v. 49, p. 67–87.

Carlson, R.W., 1984, Isotope constraints on Columbia River flood basalts genesis and the nature of the subcontinental mantle: Geochimica et Cosmochimica Acta, v. 48, p. 2357–2372.

Carlson, R.W., 1991, Physical and chemical evidence on the cause and source characteristics of flood basalt volcanism: Australian Journal of Earth Sciences, v. 38, p. 525–544.

Carman, J.H., Sinha, A.K., and Affholter, K.A., 1980, Equilibrium trends during hydrothermal impulse melting of a charnockitic gneiss: Eos, v. 61, p. 397.

Coish, R.A., and Sinton, C.W., 1992, Geochemistry of mafic dikes in the Adirondack Mountains: Implications for late Proterozoic continental rifting: Contributions to Mineralogy and Petrology, v. 110, p. 500–514.

Coish, R.A., Fleming, F.S., Larsen, M., Poyner, R., and Seibert, J., 1985, Early rift history of the proto-Atlantic Ocean: Geochemical evidence from metavolcanic rocks in Vermont: American Journal of Science, v. 285, p. 351–378.

Cordery, M.J., Davies, G.F., and Campbell, I.H., 1997, Genesis of flood basalts from eclogite-bearing mantle plumes: Journal of Geophysical Research, v. 102, no. B9, p. 20,179–20,197.

Cox, K.G., 1980, A model for flood basalt vulcanism: Journal of Petrology, v. 21, p. 629–650.

Cox, K.G., and Dewey, C.W., 1986, Fractional processes in the Deccan Traps magmas: Comments on the paper by G. Sen—Mineralogy and petrogenesis of the Deccan Traps lava flows around Mahabaleshwar, India: Journal of Petrology, v. 28, p. 235–238.

Cox, K.G., and Hawkesworth, C.J., 1985, Geochemical stratigraphy of the Deccan Traps at Mahabaleshwar, western Ghats, India, with implications for open system magmatic processes: Journal of Petrology, v. 26, p. 355–377.

Cox, K.G., Bell, J.D., and Pankhurst, R.J., 1979, The interpretation of igneous rocks: London, George Allen and Unwin, 450 p.

Dupuy, C., and Dostal, J., 1984, Trace element geochemistry of some continental tholeiites: Earth and Planetary Science Letters, v. 67, p. 61–69.

Ellam, R.M., 1992, Lithospheric thickness as a control on basalt geochemistry: Geology, v. 20, p. 153–156.

Fauth, J.L., 1973, Precambrian welded tuffs, Blue Ridge Province, Maryland and Pennsylvania: Geological Society of America Abstracts with Programs, v. 5, p. 159–160.

Fetter, A.H., and Goldberg, S.A., 1995, Age and geochemical characteristics of bimodal magmatism in the Neoproterozoic Grandfather Mountain rift basin: Journal of Geology, v. 103, p. 313–326.

Floyd, P.A., and Winchester, J.A., 1975, Magma type and tectonic setting discrimination using immobile elements: Earth and Planetary Science Letters, v. 27, p. 211–218.

Froelich, A.J., and Gottfried, D., 1985, Early Jurassic diabase sheets of the eastern United States—A preliminary overview, *in* Robinson, G.R., Jr., and Froelich, A.J., eds., Proceedings of the second U.S. Geological Survey workshop on the early Mesozoic basins of the eastern United States: Reston, Virginia, U.S. Geological Survey Circular 946, p. 79–86.

Gathright, T.M., 1976, Geology of the Shenandoah National Park, Virginia: Charlottesville,Virginia Division of Mineral Resources Bulletin 86, 93 p.

Gathright, T.M., Henika, W.S., and Sullivan, J.L., III, 1977, Geology of the Waynesboro East and Waynesboro West quadrangles, Virginia: Charlottesville, Virginia Division of Mineral Resources Publication 3, 53 p.

Goldberg, S.A., Butler, J.R., and Fullagar, P.D., 1986, The Bakersville dike swarm: Geochronology and petrogenesis of late Proterozoic basaltic magmatism in the southern Appalachian Blue Ridge: American Journal of Science, v. 286, p. 403–430.

Goles, G., 1986, Miocene basalts of the Blue Mountains Province in Oregon, I: Compositional types and their geological settings: Journal of Petrology, v. 27, p. 495–520.

Gottfried, D., Annell, C.S., and Schwarz, L.J., 1983, Geochemistry and tectonic significance of subsurface basalts near Charleston, South Carolina: Clubhouse Crossroads test holes #2 and #3, *in* Studies related to the Charleston, South Carolina, earthquake of 1886—Tectonics and seismicity: Reston, Virginia, U.S. Geological Survey Professional Paper 1313-A, p. A1–A19.

Grant, J.A., 1986, The isocon diagram—A simple solution to Gresens' equation for metasomatic alteration: Economic Geology, v. 81, p. 1976–1982.

Haworth, R.T., Daniels, D.L., Williams, H., and Zeitz, I., 1980, Bouguer gravity map of the Appalachian orogen: St. John's, Memorial University of Newfoundland, Map 3, scale 1:1,000,000.

Isachsen, Y.W., Kelley, W.M., Sinton, C., Coish, R.A., and Heizler, M.J., 1988, Dikes of the Northeast Adirondack region—Introduction to their distribution, orientation, mineralogy, chronology, magnetism, chemistry, and mystery: Plattsburgh, New York State Geological Association Field Trip Guidebook, p. 215–244.

Kamo, S.L., Gowen, C.F., and Krogh, T.E., 1989, Birthdate for the Iapetus Ocean? A precise U-Pb zircon and baddeleyite age for the Long Range dikes, southeast Labrador: Geology, v. 17, no. 7, p. 602–605.

Kumarapeli, P.S., Dunning, G.R., Pintson, H., and Shaver, J., 1989, Geochemistry and U-Pb age of comenditic metafelsites of the Tibbit Hill Formation: Canadian Journal of Earth Sciences, v. 26, p. 1374–1383.

Mahoney, J.J., Macdougall, J.D., Lugmair, G.W., Murali, A.V., Das, M.S., and Gopalan, K., 1982, Origin of the Deccan Trap flows at Mahabaleshwar inferred from Nd and Sr isotopic and chemical evidence: Earth and Planetary Science Letters, v. 72, p. 39–53.

McDougall, I., 1976, Geochemistry and origin of basalt of the Columbia River Group, Oregon and Washington: Geological Society of America Bulletin, v. 87, p. 777–792.

Meschede, M., 1986, A method of discriminating between different types of mid-ocean ridge basalts and continental tholeiites with the Nb-Zr-Y diagram: Chemical Geology, v. 56, p. 207–218.

Morse, S.A., 1980, Basalts and phase diagrams: New York, Springer-Verlag, 493 p.

Myers, J.D., Sinha, A.K., and Marsh, B.D., 1984, Assimilation of crustal material by basaltic magma: Strontium isotopic and trace element data from the Edgecumbe volcanic field, SE Alaska: Journal of Petrology, v. 25, part 1, p. 1–26.

Mysen, B.O., 1978, Experimental determination of nickel partition coefficients between liquid, pargasite, and garnet peridotite minerals and concentration limits of behavior according to Henry's Law at high pressure and temperature: American Journal of Science, v. 278, p. 217–243.

Najafi, S.J., Cox, K.G., and Sukheswala, R.N., 1981, Geology and geochemistry of the basaltic flows (Deccan Traps) of the Mahad-Mahableshwar section, India: Memoire of the Geological Society of India, v. 3, p. 300–315.

Nakamura, N., 1974, Determination of REE, Ba, Mg, Na and K in carbonaceous and ordinary chondrites: Geochimica et Cosmochimica Acta, v. 38, p. 757–775.

Norrish, K., and Chapell, B.W., 1977, X-ray fluorescence spectrometry, *in* Zussman, J., ed., Physical methods in determinative mineralogy (2nd edition): New York, Academic Press, p. 201–273.

Pearce, J.A., and Cann, J.R., 1973, Tectonic setting of basic volcanic rocks determined using trace element analyses: Earth and Planetary Science Letters, v. 19, p. 290–300.

Pearce, J.A., and Norry, M.J., 1979, Petrogenetic implications of Ti, Zr, Y and Nb variations in volcanic rocks: Contributions to Mineralogy and Petrology, v. 69, p. 33–47.

Peate, D.W., Hawkesworth, C.J., Mantovani, M.S.M., and Shukowsky, W., 1990, Mantle plumes and flood basalt stratigraphy in the Paraná, South America: Geology, v. 18, p. 1223–1226.

Pettingill, H.S., Sinha, A.K., and Tatsumoto, M., 1984, Age and origin of anorthosites, charnockites, and granulites in the central Virginia Blue Ridge: Nd and Sr isotopic evidence: Contributions to Mineralogy and Petrology, v. 85, p. 279–291.

Piccirillo, E.M., Civetta, L., Petrini, R., Longinelli, A., Bellieni, G., Comin-Chiaramonti, P., Marques, L.S., and Melfi, A.J., 1989, Regional variations within the Paraná flood basalts (southern Brazil): Evidence for subcontinental mantle heterogeneity and crustal contamination: Chemical Geology, v. 75, p. 103–122.

Puffer, J.H., 2002, A late Neoproterozoic eastern Laurentian superplume: Location, size, chemical composition, and environmental impact: American Journal of Science, v. 302, p. 1–27.

Rankin, D.W., 1975, The continental margin of eastern North America in the southern Appalachians: The opening and closing of the Proto-Atlantic Ocean: American Journal of Science, v. 275-A, p. 298–336.

Ratcliffe, N.M., 1987, High TiO_2 metadiabase dikes of the Hudson Highlands, New York and New Jersey: Possible late Proterozoic rift rocks in the New York recess: American Journal of Science, v. 287, p. 817–850.

Reed, J.C., and Morgan, B.A., 1971, Chemical alteration and spilitization of the Catoctin greenstones, Shenandoah National Park, Virginia: Journal of Geology, v. 79, p. 526–548.

Reed, J.C., Jr., 1955, Catoctin Formation near Luray, Virginia: Geological Society of America Bulletin, v. 66, p. 871–896.

Reidel, S.P., 1983, Stratigraphy and petrogenesis of the Grande Ronde Basalt from the deep canyon country of Washington, Oregon, and Idaho: Geological Society of America Bulletin, v. 94, p. 519–542.

Rhodes, J.M., Dungan, M.A., Blanchard, D.P., and Long, P.E., 1979, Magma mixing at mid-ocean ridges: Evidence from basalts drilled near 22°N on the mid-Atlantic ridge: Tectonophysics, v. 55, p. 35–61.

Roden, M.F., Frey, F.A., and Clague, D.A., 1984, Geochemistry of tholeiitic and alkalic lavas from the Koolau Range, Oahu, Hawaii: Implications for Hawaiian volcanism: Earth and Planetary Science Letters, v. 69, p. 141–158.

Roedder, P.L., and Emslie, R.F., 1970, Olivine-liquid equilibrium: Contributions to Mineralogy and Petrology, v. 29, p. 275–289.

Sen, G., 1986, Mineralogy and petrogenesis of the Deccan Trap lava flows around Mahabaleshwar, India: Journal of Petrology, v. 27, p. 627–663.

Simpson, E.L., and Eriksson, K.A., 1989, Sedimentology of the Unicoi Formation in southern and central Virginia: Evidence for late Proterozoic to early Cambrian rift to passive margin transition: Geological Society of America Bulletin, v. 101, p. 42–54.

Sinha, A.K., and Bartholomew, M.J., 1984, Evolution of the Grenville terrane in the central Virginia Appalachians, *in* Bartholomew, M.J., et al., eds., The Grenville event in the Appalachians and related topics: Boulder, Colorado, Geological Society of America Special Paper 194, p. 175–186.

Stolper, E., 1980, A phase diagram for mid-ocean ridge basalts: Preliminary results and implications for petrogenesis: Contributions to Mineralogy and Petrology, v. 74, p. 13–27.

Stormer, J.C., and Nicholls, J., 1978, XLFRAC: A program for the interactive testing of magmatic differentiation models: Computers and Geosciences, v. 4, p. 143–159.

Strong, D.F., and Williams, H., 1972, Early Paleozoic flood basalts of northwestern Newfoundland: Their petrology and tectonic significance: Geological Association of Canada Proceedings, v. 24, p. 43–54.

Su, Q., Goldberg, S.A., and Fullagar, P.D., 1994, Precise U-Pb zircon ages of Neoproterozoic plutons in the southern Appalachian Blue Ridge and their implications for the initial rifting of Laurentia: Precambrian Research, v. 68, p. 81–95.

Swanson, D.A., and Wright, T.L., 1981, The regional approach to studying the Columbia River Basalt Group: Calcutta, Geological Society of India Memoir 3, p. 58–80.

Takahashi, E., and Nakajima, K., 2002, Melting process in the Hawaiian plume: An experimental study, *in* Hawaiian volcanoes: Deep underwater perspectives: Washington, D.C., American Geophysical Union Geophysical Monograph 128, p. 403–418.

Thompson, R.N., Morrison, M.A., Hendry, G.L., and Parry, S.J., 1984, An assessment of the relative roles of crust and mantle in magma genesis: An elemental approach: Philosophical Transactions of the Royal Society of London (Ser. A), v. 310, p. 549–590.

Tollo, R.P., and Aleinikoff, J.N., 1996, Petrology and U-Pb geochronology of the Robertson River Igneous Suite, Blue Ridge province, Virginia: Evidence for multistage magmatism associated with an early episode of Laurentian rifting: American Journal of Science, v. 296, p. 1045–1090.

Tollo, R.P., Aleinikoff, J.N., Bartholomew, M.J., and Rankin, D.W., 2004, Neoproterozoic A-type granitoids of the central and southern Appalachians: Intraplate magmatism associated with episodic rifting of the Rodinian supercontinent: Precambrian Research, v. 128, p. 3–38.

Walker, D., Shibata, T., and DeLong, S.E., 1979, Abyssal tholeiites from the Oceanographer Fracture Zone: Contributions to Mineralogy and Petrology, v. 70, p. 111–125.

Wehr, F., and Glover, L., 1985, Stratigraphy and tectonics of the Virginia–North Carolina Blue Ridge: Evolution of a late Proterozoic–early Paleozoic hinge zone: Geological Society of America Bulletin, v. 96, p. 285–295.

Wright, T.L., Grolier, M.J., and Swanson, D.A., 1973, Chemical variation related to the stratigraphy of the Columbia River basalt: Geological Society of America Bulletin, v. 84, p. 371–386.

Manuscript Accepted by the Society August 25, 2003

Printed in the USA

Geological Society of America
Memoir 197
2004

Detrital zircon ages and Nd isotopic data from the southern Appalachian crystalline core, Georgia, South Carolina, North Carolina, and Tennessee: New provenance constraints for part of the Laurentian margin

Brendan R. Bream*
Robert D. Hatcher Jr.
Department of Earth and Planetary Sciences, University of Tennessee, Knoxville, Tennessee 37996-1410, USA
Calvin F. Miller
Department of Geology, Vanderbilt University, Nashville, Tennessee 37235, USA
Paul D. Fullagar
Department of Geological Sciences, University of North Carolina, Chapel Hill, North Carolina 27599-3315, USA

ABSTRACT

Sedimentary and metasedimentary rocks within the southern Appalachian Blue Ridge and Inner Piedmont contain a valuable record of Late Proterozoic Laurentian margin evolution following the breakup of Rodinia. Paleogeographic reconstructions and increasing amounts of geochronologic and isotopic data limit the derivation of these paragneisses to the Laurentian and/or west Gondwanan craton(s). Southern Appalachian crystalline core paragneiss samples have ε_{Nd} values between –8.5 and –2.0 at the time of deposition and contain abundant 1.1–1.25 Ga zircon cores with Grenville 1.0–1.1 Ga metamorphic rims. Less abundant detrital zircons are pre-Grenvillian: Middle Proterozoic 1.25–1.6 Ga, Early Proterozoic 1.6–2.1 Ga, and Late Archean 2.7–2.9 Ga. Blue Ridge Grenvillian basement has almost identical ε_{Nd} values and displays the same dominant magmatic core and metamorphic rim zircon ages. Based on our data, nonconformable basement-cover relationships, and crustal ages in eastern North America, we contend that the extensive sedimentary packages in the southern Appalachian Blue Ridge and western Inner Piedmont are derived from Laurentia. ε_{Nd} values from Carolina terrane volcanic, plutonic, and volcaniclastic rocks are isotopically less evolved than southern Appalachian paragneisses and Blue Ridge Grenvillian basement, easily separating this composite terrane from the mostly Laurentian terranes to the west. Neoproterozoic and Ordovician, as well as Grenvillian and pre-Grenvillian, zircons in eastern Inner Piedmont paragneisses indicate that these samples were deposited much later and could have been derived entirely from a Panafrican source or possibly a mixture of Panafrican and recycled Laurentian margin assemblages.

Keywords: detrital zircons, southern Appalachians, provenance, Rodinia, Blue Ridge, Inner Piedmont

*E-mail: bbream@msn.com.

Bream, B.R., Hatcher, R.D., Jr., Miller, C.F., and Fullagar, P.D., 2004, Detrital zircon ages and Nd isotopic data from the southern Appalachian crystalline core, Georgia, South Carolina, North Carolina, and Tennessee: New provenance constraints for part of the Laurentian margin, *in* Tollo, R.P., Corriveau, L., McLelland, J., and Bartholomew, M.J., eds., Proterozoic tectonic evolution of the Grenville orogen in North America: Boulder, Colorado, Geological Society of America Memoir 197, p. 459–475. For permission to copy, contact editing@geosociety.org.

INTRODUCTION

The considerable extent of the Grenvillian orogen that assembled Rodinia is confirmed by the widespread distribution of Grenvillian belts on numerous pre-Grenvillian cratonic margins (e.g., Hoffman, 1991). Although the existence and Late Proterozoic breakup of the Rodinian supercontinent are virtually indisputable, many details concerning Laurentian margin evolution remain controversial and problematic. Sedimentary fill in rift basins along the eastern Laurentian margin and western Iapetus ocean was deposited nonconformably on Grenvillian and likely pre-Grenvillian crust that forms pre-Paleozoic basement of the southern Appalachians. In these sedimentary sequences is a record of the rift-to-drift transition (Wehr and Glover, 1985; Simpson and Sundberg, 1987; Williams and Hiscott, 1987; Simpson and Eriksson, 1989; Thomas, 1991) and later overprinting by Paleozoic orogenies. A complex rift history is verified by reconstructions of the margin (e.g., Rast and Kohles, 1986; Bartholomew, 1992), basement-cover sequence relationships (e.g., Hatcher et al., this volume), and polyphase rifting (e.g., Badger and Sinha, 1988; Aleinikoff et al., 1995; Brewer and Thomas, 2000; Cawood et al., 2001). Deposition of extensive rift fill, in places well over 1000 m thick, along most of the southern Laurentian margin was coeval with a latest Neoproterozoic–Early Cambrian rifting event. Continued rifting, accompanied by deposition of drift facies across the margin, facilitated development of a true passive margin. There is general agreement regarding Laurentian derivation of western Blue Ridge rift-to-drift sedimentary packages. The paucity of internal paragneiss isotopic and geochronologic data, combined with the polydeformed and metamorphosed nature of the crystalline core, however, has resulted in characterization of internal high-grade lithotectonic packages as "disrupted" (Horton et al., 1989) and "problematic" (Goldsmith and Secor, 1993) terranes. New detrital zircon U-Pb ages and Sm-Nd data obtained from Blue Ridge and Inner Piedmont metapsammitic and metapelitic samples place previously unavailable limits on the provenance of southern Laurentian margin deposits.

GEOLOGIC SETTING

Appalachian terranes are exposed along the length of the orogen as continuous and discontinuous elongate curvilinear belts from Alabama to Newfoundland (Williams, 1978), and are separated by major, mostly orogen-scale, faults (Fig. 1). Both suspect and exotic accreted terranes of the southern Appalachians contain sedimentary sequences deposited in a variety of settings on a range of basements. Western Blue Ridge rift-facies sedimentary and volcanic rocks, which were unconformably deposited on rifted Grenvillian basement, include the Ocoee Supergroup and the Mount Rogers, Grandfather Mountain, and Mechum River Formations. Widely distributed eastern Blue Ridge–western Inner Piedmont metapsammite-dominated packages (e.g., Tallulah Falls, Ashe, Sandy Springs, and Emuckfaw Formations) were deposited on small Grenvillian continental fragments and possibly on oceanic crust. Paragneisses of the Hayesville–Soque River and Chunky Gal–Shope Fork thrust sheets, herein referred to as the central Blue Ridge, and Dahlonega gold belt were deposited as immature siliciclastic sediments intruded by, deposited on, or possibly intercalated with Ordovician mafic assemblages of oceanic and/or volcanic arc/back-arc affinities (Spell and Norrell, 1990; Berger et al., 2001; Thomas et al., 2001; Settles et al., 2001). The Sauratown Mountains, Grandfather Mountain, and Pine Mountain windows expose several internal Grenvillian basement granitoids and their associated cover sequences (Fig. 1). Basement-cover relationships and lithologic features similar to those found in the western Blue Ridge are documented in the Pine Mountain (e.g., Schamel et al., 1980; Sears et al., 1981; Kish et al., 1985; Steltenpohl, 1992) and Sauratown Mountains (McConnell, 1988; Walker et al., 1989) windows. Ion microprobe zircon U-Pb geochronologic data favor an Amazonian source for cover-sequence samples from the Pine Mountain window (Steltenpohl et al., this volume).

Several crustal fragments of Laurentian affinity formed during rifting; some were preserved as massifs in the southern Appalachian Blue Ridge (e.g., Toxaway dome), while others were transported and ultimately accreted as exotic terranes onto other cratons (Dalla Salda et al., 1992; Dalziel et al., 1994; Keppie et al., 1996; Thomas and Astini, 1996; Cawood et al., 2001). Based on their structural positions within the orogen, southern Appalachian Grenvillian basement massifs can be described as external massifs that were transported in thrust sheets over the Laurentian margin and internal massifs, which are located beneath major detachments and are exposed in windows (Hatcher, 1984). Western Blue Ridge external basement massifs are distributed within the internal portion of the terrane as a linear outcrop belt along the Blue Ridge Front adjacent to the Valley and Ridge and in the Grandfather Mountain window. Eastern Blue Ridge internal massifs are much smaller, occurring mostly as complexly folded structures (Hatcher et al., this volume).

Rift-related magmatic rocks provide constraints on the timing of Rodinian rifting; unfortunately, western Iapetus geochronologic and paleomagnetic data for these rocks are sparse. Data for the southern Appalachians are limited to Neoproterozoic ages ranging from 730 to 760 Ma for the Bakersville dike swarm (Goldberg et al., 1986; Ownby et al., this volume), Crossnore complex plutons (Su et al., 1994), and Mount Rogers metarhyolite (Aleinikoff et al., 1995). Rift-related magmatic rocks in the adjacent central Appalachians, however, include the Catoctin Formation metarhyolite, dated at 572 ± 5 and 564 ± 9 Ma (Aleinikoff et al., 1995), and the Robertson River suite, dated at 702–735 Ma (Tollo and Aleinikoff, 1996). The two age groups for rift-related magmatism in the Appalachians are interpreted to represent an earlier failed event and a younger successful event that created and opened the Iapetus ocean (Badger and Sinha, 1988; Aleinikoff et al., 1995; Brewer and Thomas, 2000). Cawood et al. (2001) also proposed that the younger event is divisible into two distinct phases: the first, ca. 570 Ma, opened the Iapetus ocean, and the second, 540–535 Ma, evolved into a

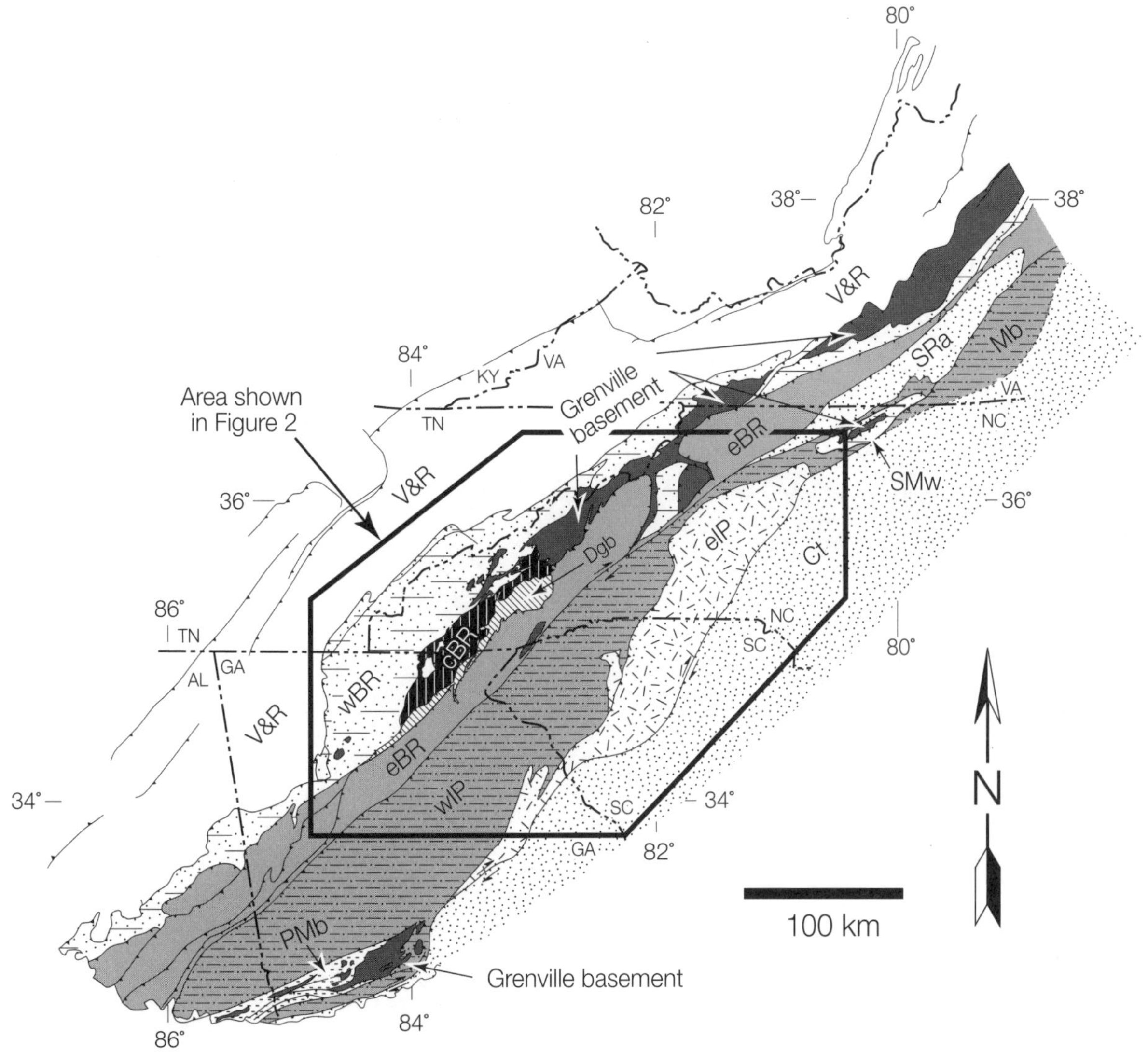

Figure 1. Tectonic map of the southern Appalachians. cBR—central Blue Ridge; Ct—Carolina terrane; Dgb—Dahlonega gold belt; eBR—eastern Blue Ridge; eIP—eastern Inner Piedmont; Mb—Milton belt; PMb—Pine Mountain Block; SMw—Sauratown Mountains window; SRa—Smith River allochthon; V&R—Valley and Ridge; wBR—western Blue Ridge; wIP—western Inner Piedmont. Modified from Hatcher et al. (1990).

true passive margin after fragmentation of much of the southern Laurentian margin, probably again reflecting the rift-to-drift transition (e.g., Wehr and Glover, 1985; Williams and Hiscott, 1987). It was in these complex latest Neoproterozoic to Early Cambrian rift basins, and in nearly coeval to time-transgressive distal slope-rise to ocean floor environments, that paragneisses of the southern Appalachian crystalline core were deposited.

SAMPLES

Samples were collected using several criteria, including quality of mapping and documented tectonic or stratigraphic significance of individual units. Sampling was biased toward metapsammite lithologies that were expected to contain abundant detrital zircons. Fresh unaltered material representative of each unit or lithology was collected. We present new U-Pb detrital zircon ages and Sm-Nd whole-rock isotopic data for 20 metasedimentary samples from the southern Appalachian Blue Ridge and Inner Piedmont. Paleozoic metamorphic grade of these samples ranges from subgreenschist in the westernmost Blue Ridge to middle and upper amphibolite facies in the eastern Blue Ridge and Inner Piedmont (see Figure 2 and Table 1 for sample localities).

Western Blue Ridge Samples

Samples from the western Blue Ridge are from Ocoee Supergroup rift facies. The basal rift facies nonconformably overlie Grenville basement in the western Blue Ridge (King et al., 1958). One sample is from the Snowbird Group, the old-

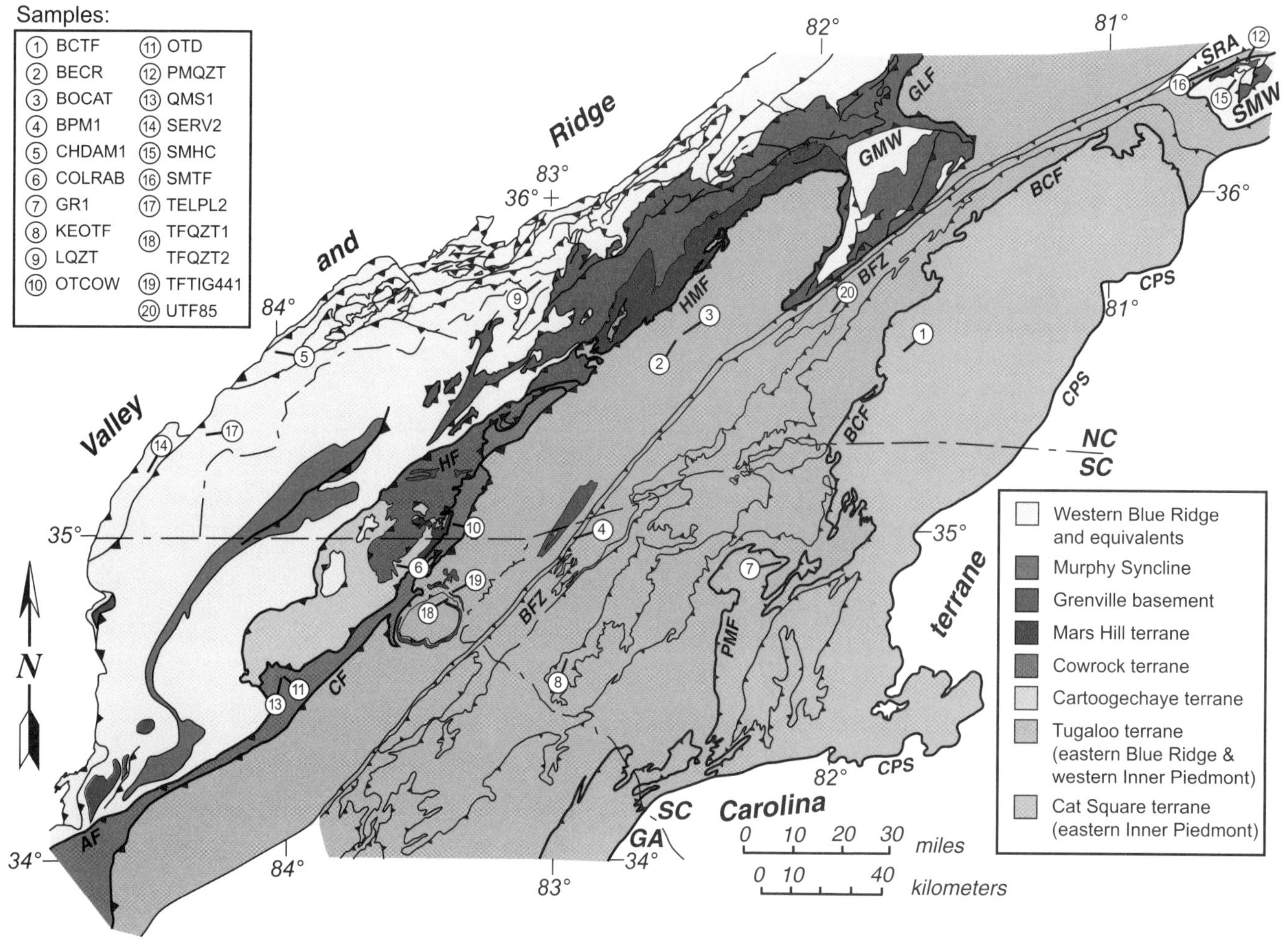

Figure 2. Geologic map of a portion of the southern Appalachians with sample locations. AF—Allatoona fault; BCF—Brindle Creek fault; BFZ—Brevard fault zone; CF—Chattahoochee fault; CPS—Central Piedmont suture; GLF—Gossan Lead fault; GMW—Grandfather Mountain window; HF—Hayesville fault; HMF—Holland Mountain fault; PMF—Paris Mountain fault; SMW—Sauratown Mountains window; SRA—Smith River allochthon. Modified from Hatcher et al. (this volume). Sample abbreviations and labels correspond to those provided in Table 1.

est Ocoee Supergroup unit; one is from the Great Smoky Group; and two are from the Walden Creek Group, the youngest unit in the Ocoee Supergroup. The Longarm Quartzite is predominantly feldspathic sandstone that ranges from <100 to 1,000 m thick in the middle of the Snowbird Group (Montes and Hatcher, 1999). The sample is from the Pigeon River Gorge in North Carolina, collected at the westbound rest area on Interstate 40. Upper Ocoee Supergroup rift samples include a medium- to coarse-grained sandstone sample with small conglomeratic interlayers from the uppermost Dean Formation (Great Smoky Group) on Tennessee 68 southeast of Tellico Plains. Calcareous fine-grained Sandsuck Formation (Walden Creek Group) sandstone was collected in Springtown near Starr Mountain, Tennessee. A matrix-supported polymictic conglomerate from the Sandsuck Formation (Walden Creek Group) was also sampled at Chilhowee Dam on U.S. 129 in southeastern Tennessee. This conglomerate contains large (up to 0.5 m) rounded clasts of mostly milky quartz with subordinate amounts of quartzite, limestone, dolostone, quartz-epidote (from veins), rare granitoid, and black mud (now slate) in a dominantly medium-grained siliciclastic matrix. Attempts were made during sample preparation to isolate matrix material in this sample.

Central Blue Ridge Sample

One sample of central Blue Ridge metasandstone (Coleman River Formation of Hatcher, 1979) was collected on U.S. 76 west of Clayton, Georgia, <1 km east of the Rabun-Towns county line. This unit is the most widely distributed of the Coweeta Group, making up most of the Hayesville–Soque River thrust sheet (Cowrock terrane of Hatcher et al., this volume).

Dahlonega Gold Belt Samples

Three samples were collected from the Otto Formation in the Dahlonega gold belt, a unit consisting of intercalated meta-

TABLE 1. SAMPLES WITH STRATIGRAPHIC ASSOCIATION AND COORDINATES

Sample	Figure 2 location	Formation or unit name	7.5-minute quadrangle	County (state)	Latitude (°N)*	Longitude (°W)*
Western Blue Ridge						
CHDAM1	5	Walden Creek Group (Ocoee Supergroup)	Tallassee	Blount (TN)	35.548	84.050
LQZT	9	Longarm Quartzite, Snowbird Group (Ocoee Supergroup)	Cove Creek Gap	Haywood (NC)	35.702	83.040
SERV2	14	Sandsuck Formation, Walden Creek Group (Ocoee Supergroup)	McFarland	Polk (TN)	35.248	84.444
TELPL2	17	Dean Formation, Great Smoky Group (Ocoee Supergroup)	Tellico Plains	Monroe (TN)	35.336	84.296
Central Blue Ridge						
COLRAB	6	Coleman River Formation	Hightower Bald	Rabun (GA)	34.908	83.612
Dahlonega gold belt						
OTCOW	10	Otto Formation	Prentiss	Macon (NC)	35.060	83.453
OTD	11	Otto Formation	Campbell Mountain	Lumpkin (GA)	34.560	84.074
QMS1	13	Otto Formation	Campbell Mountain	Lumpkin (GA)	34.533	84.018
Eastern Blue Ridge						
BECR	2	Tallulah Falls Formation–Ashe Formation (metagraywacke member)	Dunsmore Mountain	Buncombe (NC)	35.454	82.661
BOCAT	3	Tallulah Falls Formation–Ashe Formation (metagraywacke member)	Asheville	Buncombe (NC)	35.600	82.538
TFQZT1	18	Tallulah Falls Formation (quartzite member)	Tiger	Rabun (GA)	34.817	83.424
TFQZT2	18	Tallulah Falls Formation (quartzite member)	Tallulah Falls	Rabun (GA)	34.736	83.391
TFTIG441	19	Tallulah Falls Formation–Ashe Formation (metagraywacke member)	Tiger	Rabun (GA)	34.838	83.422
Sauratown Mountains window						
SMHC	15	Hogan Creek Formation	Siloam	Surry (NC)	36.271	80.602
PMQZT	12	Pilot Mountain Quartzite	Pinnacle	Surry (NC)	36.344	80.482
Western Inner Piedmont						
BPM1	4	Chauga River Formation (Brevard–Poor Mountain Transitional member)	Salem	Oconee (SC)	34.916	82.914
KEOTF	8	Tallulah Falls Formation–Ashe Formation (aluminous schist member)	Seneca	Oconee (SC)	34.696	82.879
SMTF	16	Tallulah Falls Formation–Ashe Formation (metagraywacke member)	Copeland	Surry (NC)	36.275	80.732
UTF85	20	Tallulah Falls Formation–Ashe Formation (metagraywacke member)	Marion East	McDowell (NC)	35.644	81.987
Eastern Inner Piedmont						
BCTF	1	Metagraywacke biotite-gneiss not presently assigned to a formation	Benn Knob	Rutherford (NC)	35.539	81.702
GR1	7	Metagraywacke biotite-gneiss not presently assigned to a formation	Greer	Spartanburg (SC)	34.917	82.139

*Coordinates are North American Datum 1927–Continental United States (NAD27 CONUS).

pelite and metasandstone (Hatcher, 1988). One sample locality is within Coweeta Hydrologic Laboratory property near Otto, North Carolina, in the footwall of the Shope Fork thrust. Two samples are from the Allatoona thrust sheet in the footwall of the Hayesville–Soque River thrust fault, ~2.5 km west and ~7 km west-northwest of Dahlonega, Georgia, in the Campbell Mountain 7.5-minute quadrangle.

Eastern Blue Ridge and Western Inner Piedmont Samples

Six samples of Tallulah Falls–Ashe Formation were collected from the eastern Blue Ridge and western Inner Piedmont (Tugaloo terrane of Hatcher et al., this volume). The eastern Blue Ridge Tallulah Falls–Ashe Formation was sampled at a roadcut through Beaucatcher Mountain in Asheville, North Carolina; at a site ~200 m north of Bent Creek Gap in the Bent Creek Experimental Forest ~15 km southwest of Asheville, North Carolina, on the Blue Ridge Parkway (the same locality as sample ts-500 of Merschat and Carter, 2002); and on the north flank of the Tallulah Falls dome on U.S. 23–441 ~5 km south-southwest of Clayton, Georgia. The western Inner Piedmont Tallulah Falls–Ashe Formation was sampled in the Forbush thrust sheet (Heyn, 1984) in a railroad cut in the Yadkin River directly southwest of the Sauratown Mountains window, ~13 km east-northeast of Elkin-Jonesville, North Carolina. A second western Inner Piedmont sample was collected from a roadcut 150 m north of Exit 85 on Interstate 40 south of Marion, North Carolina. A third sample was collected on U.S. 76–123 in the Six Mile thrust sheet at Keowee, South Carolina. All of these samples contain middle to upper amphibolite-facies mineral assemblages and are dominantly one-mica (biotite up to 30%), one-feldspar (plagioclase) metagraywackes. Eastern Blue Ridge Tallulah Falls Quartzite was sampled on U.S. 23–441 at Wiley, Georgia; this sample is medium-grained, locally conglomeratic (vein quartz and both plagioclase and K-feldspar clasts) quartzite located within the Tallulah Falls dome, and at this locality contains upright graded beds (Fritz et al., 1989). The locality of the remaining sample, from the Brevard–Poor Mountain transitional member of Hatcher's (1969) Chauga River Formation, is on the west side of South Carolina 11 where it crosses Lake Keowee, in the westernmost Inner Piedmont (Hatcher and Acker, 1984). This sample consists of metasiltstone with mica-fish and thin metagraywacke interlayers.

Sauratown Mountains Window Samples

Both the Hogan Creek Formation and Pilot Mountain Quartzite were sampled in the Sauratown Mountains window. The Hogan Creek Formation consists of metasandstone, pelitic schist, and minor marble in nonconformable contact with Grenville basement orthogneisses of the outer window (Hatcher et al., 1988). The sample is from an abandoned quarry on the Yadkin River ~4 km upstream from Siloam, North Carolina. Pilot Mountain Quartzite overlies the Sauratown Formation, informally named by McConnell (1988) and formally named by Horton and McConnell (1991), and is also in nonconformable contact with Grenville basement. The Pilot Mountain Quartzite is a well-sorted quartz arenite that contains primary sedimentary structures (Walker et al., 1989). The Pilot Mountain Quartzite and Sauratown Formation are restricted to the inner Sauratown Mountains window.

Eastern Inner Piedmont Samples

The eastern Inner Piedmont samples are metagraywackes from mappable lithologic units not currently assigned to a recognized stratigraphy contained within a metapelite-dominated assemblage (Hatcher and Bream, 2002). One sample was collected ~9 km west-northwest of Casar, North Carolina, in Brier Creek, and the other sample was collected ~2.5 km south of Duncan, South Carolina, near the Duncan Correctional Center.

ANALYTICAL METHODS

Some 4–5 kg of each sample were collected for whole-rock Sm-Nd isotopic analyses and detrital zircon separation. Whole-rock powders were prepared from smaller fractions of representative sample material. Samples were cut into thin slabs on a trim saw, rinsed with isopropyl alcohol and deionized water, broken up by hand, and then ground to a fine powder in an alumina ceramic mill. Sm and Nd isotopic data were determined at the Department of Geological Sciences, University of North Carolina at Chapel Hill, with a Micromass VG Sector 54 thermal ionization mass spectrometer using the same analytical technique as outlined in Fullagar et al. (1997).

Standard mineral separation techniques were employed to isolate heavy nonferromagnetic phases. Attempts to split samples into coarse- and fine-grained fractions were made difficult by overall sample hardness; consequently, separates from the coarser-grained western Blue Ridge Sandsuck Formation conglomerate were obtained from a mixture of clast and matrix fragments despite efforts to sample matrix. A minimum of 60–70 zircon grains per sample were hand-picked, mounted in epoxy with zircon standard AS3 or AS57 (1099 Ma; Paces and Miller, 1993) and/or R33 (419 Ma; Stanford University–U.S. Geological Survey Micro Analysis Center), and polished to the approximate center of average-size grains. Zoning and inclusions were identified from cathodoluminescence (CL) images obtained for each grain, and in most cases grains were also photographed with a combination of transmitted and reflected light to augment surface and internal feature identification. Sensitive high-resolution ion microprobe (IMP) data were collected during two sessions at the University of California at Los Angeles (UCLA) using the Cameca IMS-1270 and during four sessions at Stanford University using the SHRIMP-RG (sensitive high-resolution ion microprobe reverse geometry) over two years.

Routine analytical techniques and operating conditions for IMP zircon analyses were followed (e.g., Quidelleur et al., 1997, for the Cameca IMS-1270 technique; Bacon et al., 2000, for the SHRIMP-RG technique) and are briefly summarized here. The primary ion beam produced analytical pits ~1 μm deep and up to ~25 μm by ~30 μm wide (smaller beam sizes, down to ~10 μm, were used for some UCLA data). Before each analysis, the primary beam was used to remove the gold coat and surface contamination on the targeted spot. IMP sessions at UCLA employed fifteen mass peak scans for each spot. The number of scans was reduced to five, with only a slight decrease in precision due in part to a larger primary beam, for the Stanford IMP sessions to increase the total number of analyses per sample. Sample isotopic counts were referenced to counts on a zircon standard (AS3 or AS57; R33 was also used during some Stanford sessions), which was analyzed after approximately every four to six unknown analyses. Most unknown analyses contained minor amounts of common Pb corresponding to modest common Pb corrections. All analyses were corrected for common Pb using a ^{204}Pb correction. Analyzed grains were re-imaged (via backscattered electron and CL imaging) with a Cameca SX-50 electron microprobe at the University of Tennessee–Knoxville to confirm the location and dimensions of IMP analytical spots (Fig. 3).

Analyses with ^{206}Pb*/^{238}U and ^{207}Pb*/^{206}Pb* ages (Pb* denotes radiogenic Pb) that differed by more than 10% were excluded from probability plots in order to increase the likelihood that summed probabilities represent accurate ages; similar studies used comparable values (e.g., Mueller et al., 1992; Cawood and Nemchin, 2001; DeGraff-Surpless et al., 2002). We excluded Paleozoic rim data and constructed summed probability plots of remaining core and non-Paleozoic rim data using ^{206}Pb*/^{238}U ages for analyses <1.6 Ga and ^{207}Pb*/^{206}Pb* ages for analyses >1.6 Ga. For reference, probability curves for all data using ^{207}Pb*/^{206}Pb* ages (normally discordant and ^{207}Pb*/^{206}Pb* ages >800 Ma concordant) and ^{206}Pb*/^{238}U ages (reversely discordant and ^{207}Pb*/^{206}Pb* ages <800 Ma) are provided.

As expected due to the reduced sensitivity of ^{207}Pb*/^{206}Pb* ages for young zircons, most analyses excluded were Late Proterozoic and Paleozoic, although a Middle Proterozoic component with >10% age discrepancies was noted in several samples. The choice of 1.6 Ga was arbitrary (e.g., Cawood and Nemchin, 2001, selected 1.5 Ga), but convenient because of the small number of analyses at or near this age. Statistically, some zircon age ranges within individual samples possibly were not identified (Dodson et al., 1988); therefore our data are better suited for evaluation based on positive rather than negative results (i.e., the presence of an age is more reliable than its absence). Nonetheless, the absence of key zircon ages has significant importance with regard to the dominant source or sources of our samples. Data were pooled into summed probability plots for each tectonic division sampled.

RESULTS

The goal of this research was to address provenance constraints based on new detrital zircon ages and Nd isotopic data for samples from each of the major southern Appalachian tectonic divisions west of the central Piedmont suture and southeast of the Blue Ridge front (Figs. 1 and 2). Isotopic and geochronologic data for metasedimentary rocks must be interpreted together with structural and stratigraphic constraints provided by field observations and detailed mapping. Despite a number of published detrital heavy mineral suite studies (e.g., Carroll et al., 1957; Hadley, 1970), isotopic and geochronologic studies similar to our own in the southern Appalachians are

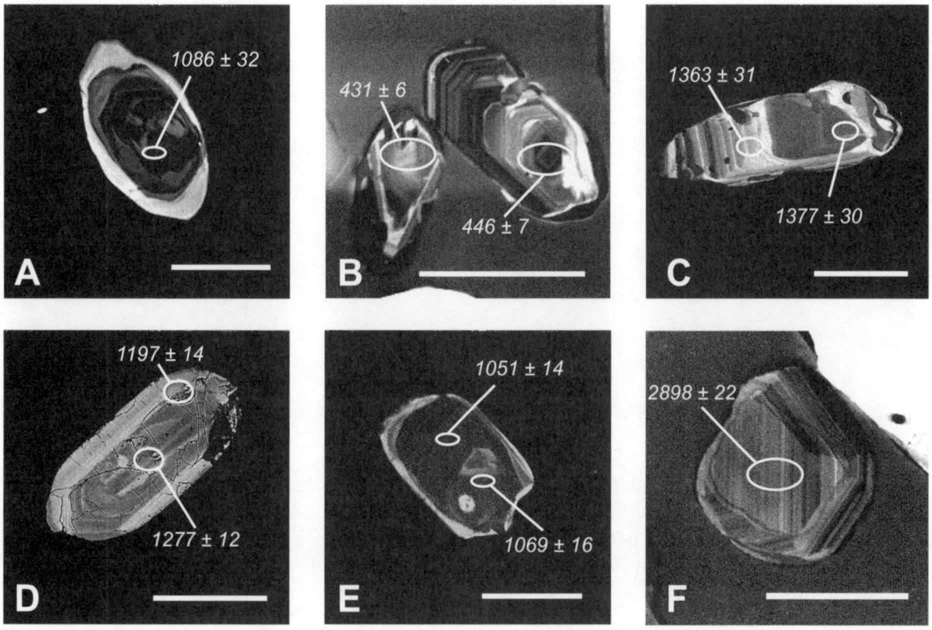

Figure 3. Cathodoluminescence (CL) and backscattered electron (BSE) images of analyzed detrital zircons. Unless otherwise noted, ages are ^{207}Pb*/^{206}Pb* in Ma with 1σ errors. (A) Eastern Blue Ridge Tallulah Falls sample (CL; sample TFTIG441). (B) Eastern Inner Piedmont metagraywacke (CL; sample GR1). (C) Tallulah Falls Formation (CL; sample SMTF). (D) Sauratown Mountains window Hogan Creek Formation (BSE; sample SMHC). (E) Western Inner Piedmont Tallulah Falls Formation (CL; sample UTF85). (F) Dahlonega gold belt Otto Formation (CL; sample QMS1). Note: Spot locations for B and F are approximate. Scale bar is 100 μm.

sparse. Our results necessitate reevaluation of the provenance and crustal affinity of the terranes from which our samples were collected.

Whole-Rock Sm-Nd Data

Modest but increasing Nd isotopic data now exist for crystalline rocks (mainly granitoids) of the Blue Ridge, Inner Piedmont, and Carolina terrane (Fig. 4). Data from metasedimentary rocks of the Blue Ridge and Inner Piedmont (Table 2) yield depleted mantle model (T_{DM}) ages of 1.2–1.9 Ga and ε_{Nd} values from around the time of deposition (~600 Ma) between −2.1 and −8.5, clearly overlapping the Grenville basement data of Carrigan et al. (2003) and Hatcher et al. (this volume). Our data also partially overlap the field defined for samples from the older, more isotopically evolved Mars Hill terrane (Bartholomew and Lewis, 1988, 1992; Carrigan et al., 2003; Ownby et al., this volume). There appears to be little correlation of isotopic maturity with terrane or distance from the Laurentian margin, but notably the oldest T_{DM} ages are from the Sauratown Mountains window and the Dahlonega gold belt.

Based on these data, these rocks could have been derived from a combination of more isotopically evolved (relative to Grenville basement exposed in the southern Appalachians) source(s) and less evolved juvenile crustal source(s), or, more likely, derived directly from rocks of similar age and isotopic maturity to Grenville basement of the Blue Ridge and Sauratown Mountains window. Data from para- and orthogneisses (Wortman et al., 1996; Coler et al., 2000) of the Chopawamsic and Milton terranes also overlap the field for southern Appalachian Grenville basement, whereas Nd data (mostly metaigneous) from the Kings Mountain, Charlotte, Milton, Carolina slate, Raleigh, Kiokee, and eastern slate belts of the Carolina terrane favor generation of Paleozoic magmas from juvenile, isotopically depleted crust with a minor contribution (possibly increasing with time) from older, more evolved crust

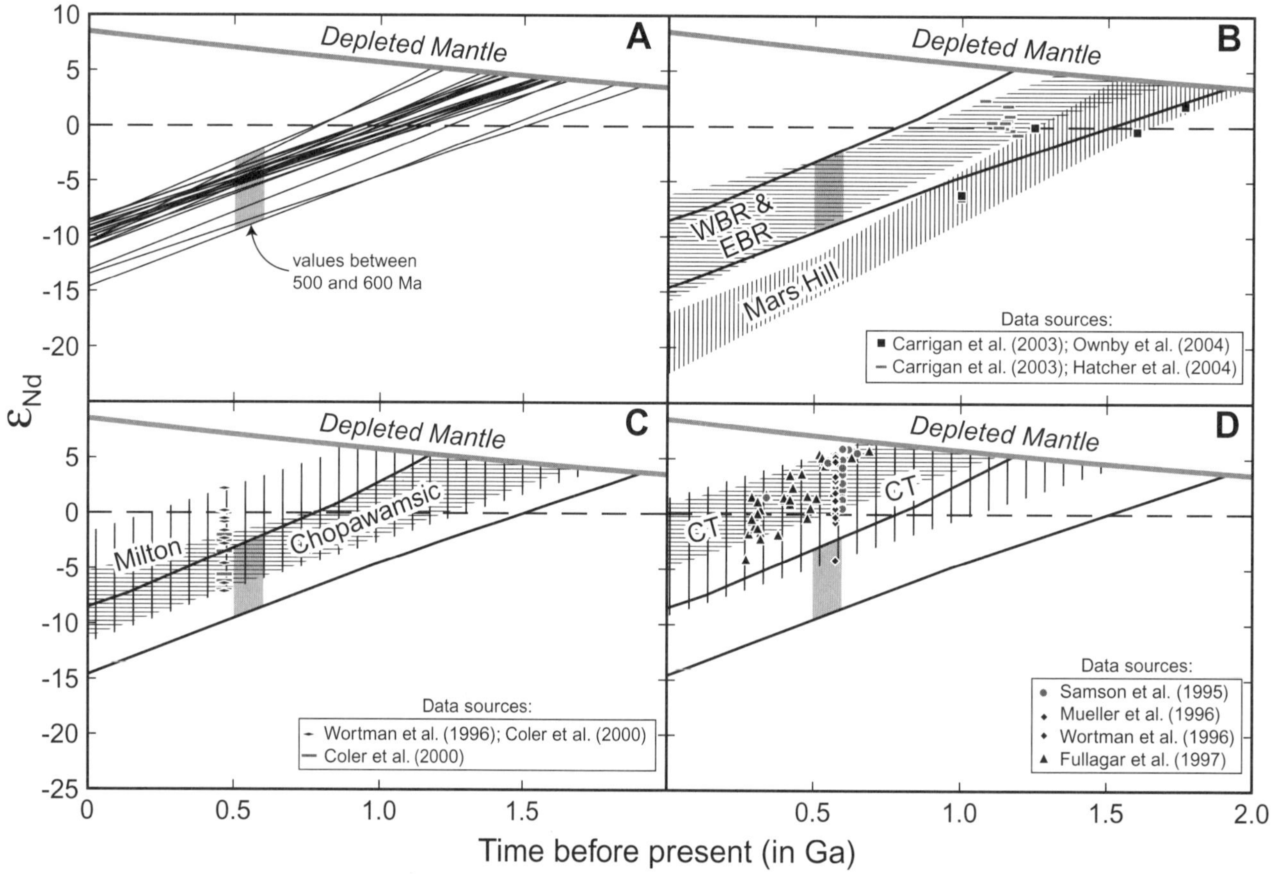

Figure 4. Epsilon Nd evolution over time for metasedimentary samples compared with Blue Ridge basement, Chopawamsic and Milton terranes, and Carolina terrane data. (A) Metasedimentary sample data showing evolution from present to intersection with the depleted mantle. The field outlined is the range of values between 500 and 600 Ma. (B) Metasedimentary data field with data from western and eastern Blue Ridge Grenvillian basement (WBR and EBR; horizontal lines) and Mars Hill terrane (vertical lines). (C) Metasedimentary data field compared with Chopawamsic belt (horizontal lines) and Milton belt (vertical lines) data. (D) Metasedimentary data field compared with Carolina terrane data (CT). Note that the evolution lines in B–D are defined by upper and lower boundaries of data in A, and the symbols represent values at the estimated time of crystallization and/or deposition; data with $^{147}Sm/^{144}Nd > 0.15$ were excluded. Depleted mantle curve from DePaolo (1981).

TABLE 2. SM-ND ISOTOPIC DATA

Sample	Figure 2 location	Sm (ppm)	Nd (ppm)	$^{147}Sm/^{144}Nd$	$^{143}Nd/^{144}Nd$ ($\pm 2\sigma$)†	ε_{Nd} (at present)	ε_{Nd} (at 600 Ma)	T_{DM} (Ga)
Western Blue Ridge								
CHDAM1	5	3.82	21.92	0.10775	0.512093 (04)	−10.6	−3.8	1.37
LQZT	9	3.30	18.71	0.10914	0.512067 (10)	−11.1	−4.4	1.42
SERV2	14	5.66	29.72	0.11783	0.512120 (05)	−10.1	−4.1	1.46
TELPL2	17	7.46	39.57	0.11668	0.512150 (05)	−9.5	−3.4	1.40
Central Blue Ridge								
COLRAB	6	5.41	26.57	0.12594	0.512121 (05)	−10.1	−4.7	1.60
Dahlonega gold belt								
OTCOW	10	8.93	44.75	0.12356	0.512195 (06)	−8.6	−3.0	1.43
OTD	11	7.62	37.48	0.12575	0.512200 (05)	−8.5	−3.1	1.46
QMS1	13	4.30	21.03	0.12660	0.511948 (03)	−13.5	−8.1	1.91
Eastern Blue Ridge								
BECR	2	9.63	47.07	0.12665	0.512146 (06)	−9.6	−4.2	1.57
BOCAT	3	5.14	24.38	0.13050	0.512171 (04)	−9.1	−4.0	1.59
SMTF	16	10.52	52.66	0.12370	0.512091 (07)	−10.7	−5.1	1.61
TFQZT1	18	9.70	55.61	0.10795	0.512184 (03)	−8.9	−2.1	1.24
TFQZT2	18	3.68	19.79	0.11505	0.512100 (07)	−10.5	−4.2	1.45
Sauratown Mountains window								
SMHC	15	4.46	23.55	0.11720	0.511888 (06)	−14.6	−8.5	1.82
PMQZT	12	5.21	27.18	0.11854	0.512064 (06)	−11.2	−5.2	1.56
Western Inner Piedmont								
BPM1	4	8.37	41.23	0.12570	0.512134 (07)	−9.8	−4.4	1.57
KEOTF	8	8.38	39.95	0.12986	0.512185 (05)	−8.8	−3.7	1.56
TFTIG441	19	6.92	32.29	0.13260	0.512152 (05)	−9.5	−4.6	1.67
UTF85	20	11.40	58.35	0.12092	0.512155 (05)	−9.4	−3.6	1.46
Central Inner Piedmont								
BCTF	1	4.80	32.12	0.09238	0.512097 (04)	−10.6	−2.6	1.19
GR1	7	6.73	36.45	0.11430	0.511968 (05)	−13.1	−6.8	1.64

Notes: T_{DM} age calculated according to DePaolo (1981), using $^{147}Sm/^{144}Nd = 0.1967$ and $^{143}Nd/^{144}Nd = 0.512638$ for present-day bulk earth values.

$\varepsilon_{Nd} = [\{^{143}Nd/^{144}Nd]_{Sample} - [^{143}Nd/^{144}Nd]_{Bulk\ earth}\}/\{[^{143}Nd/^{144}Nd]_{Bulk\ earth}\} \times 10^4$. Bulk earth: $^{143}Nd/^{144}Nd = 0.512638$ present day, $^{143}Nd/^{144}Nd = 0.511864$ at 600 Ma.

$^{143}Nd/^{144}Nd$ measured ratios normalized to $^{146}Nd/^{144}Nd = 0.7219$.

†2σ errors are reported as the last two significant digits.

(Fig. 4; Samson et al., 1995; Mueller et al., 1996; Fullagar et al., 1997). Late Proterozoic sources, increasing with distance from the western-eastern Blue Ridge boundary, are noted for Paleozoic granitoids of the southern Appalachian Blue Ridge, Inner Piedmont, and the Carolina terrane (Fullagar et al., 1997); this relationship suggests that magma generation during Paleozoic orogenesis involved larger amounts of juvenile crust away from the Laurentian margin (Fullagar, 2002).

Detrital Zircon Data

Ion microprobe analytical results for this study (Table DR1) are available as GSA Data Repository item 2004022.[1] Summed probabilities for pooled samples (Fig. 5) should be viewed as "relative" abundances because of differences in number and types of samples, number of individual analyses for each division, separation and analytical bias, multiple analyses of grains with unique ages, and consolidation of samples into their respective tectonic terranes. Most detrital grains contain brightly luminescent (in CL) interior zones with oscillatory zoning and relatively unzoned thin darker rims (Fig. 3). Summed probability plots illustrate the abundance of Grenvillian magmatic (1000–1250 Ma) and metamorphic (950–1000 Ma) detrital zircon ages in our samples. Pre-Grenvillian components are also present in all summed probability plots for southern Appalachian crystalline core terranes. Recognized pre-Grenvillian ages are early Middle Proterozoic (1.25–1.6 Ga), Early Proterozoic (1.6–2.1 Ga), and Late Archean (2.7–2.9 Ga). Absence of grains between 2.0 and 2.6 Ga was noted in almost all samples. Except for the eastern Blue Ridge and Sauratown Mountains window, Early Proterozoic zircons occur in each of the terranes. Late Archean zircons were identified in the central Blue Ridge, Dahlonega gold belt, and eastern Inner Piedmont samples.

[1]GSA Data Repository Item 2004022, Table DR1, Ion Microprobe U-Pb Data for Southern Appalachian Detrital Zircons, is available on request from Documents Secretary, GSA, P.O. Box 9140, Boulder, CO 80301-9140, USA, editing@geosociety.org, or at www.geosociety.org/pubs/ft2004.htm.

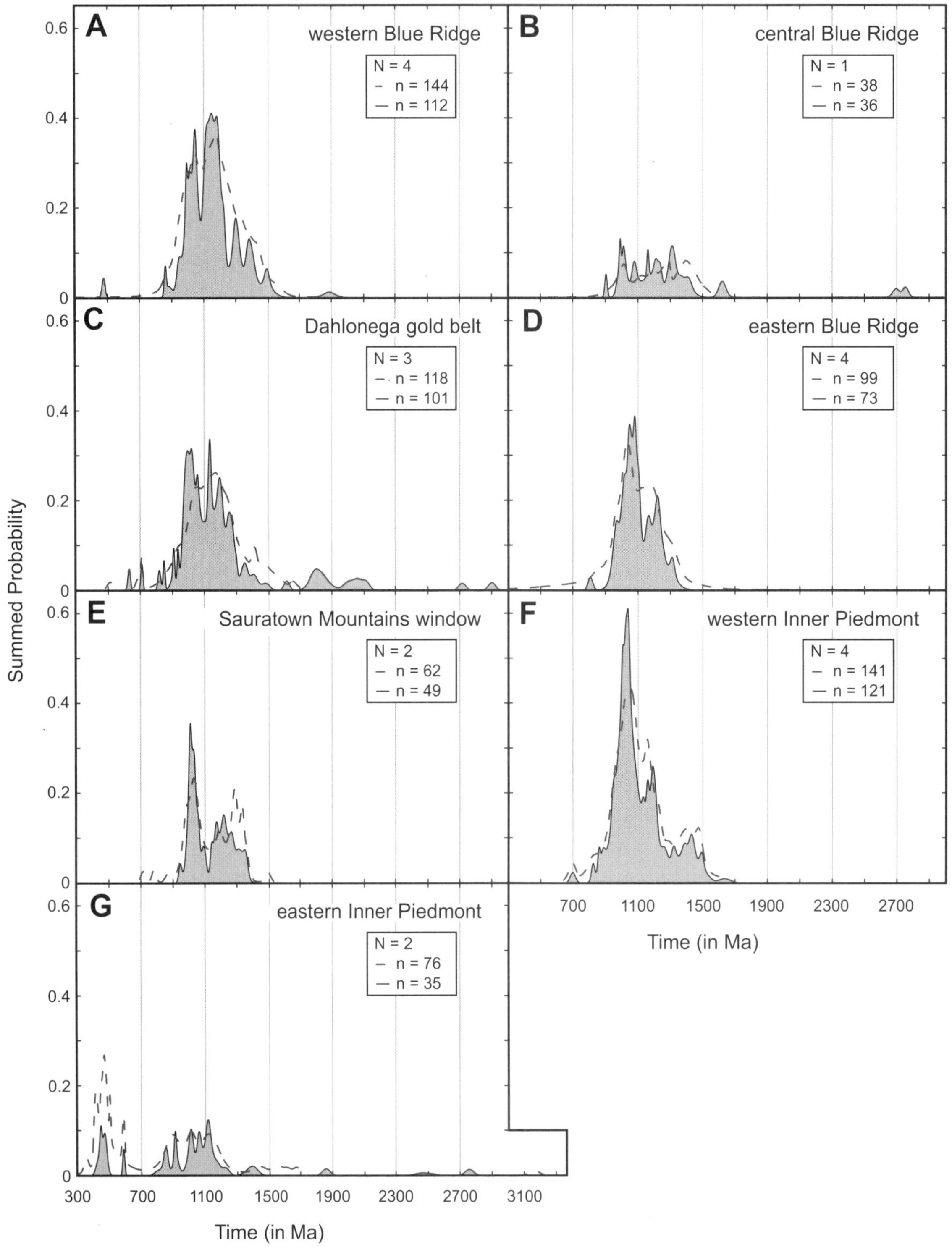

Figure 5. Summed probability plots for detrital zircon data. For reference, probability curves (dashed thicker gray lines) for all the data using $^{207}Pb^{*}/^{206}Pb^{*}$ ages (normally discordant and $^{207}Pb^{*}/^{206}Pb^{*}$ ages greater than 800 Ma concordant) and $^{206}Pb^{*}/^{238}U$ ages (reversely discordant and $^{207}Pb^{*}/^{206}Pb^{*}$ ages less than 800 Ma) are provided. The thinner black lines bounding the shaded areas represent the summed probability of data with $^{206}Pb^{*}/^{238}U$ and $^{207}Pb^{*}/^{206}Pb^{*}$ ages that differ by less than 10%, $^{206}Pb^{*}/^{238}U$ ages for analyses with $^{207}Pb^{*}/^{206}Pb^{*}$ ages <1.6 Ga, and $^{207}Pb^{*}/^{206}Pb^{*}$ ages for analyses with $^{207}Pb^{*}/^{206}Pb^{*}$ ages >1.6 Ga. (A) Western Blue Ridge samples. (B) Central Blue Ridge samples. (C) Dahlonega gold belt samples. (D) Eastern Blue Ridge samples. (E) Sauratown Mountains window samples. (F) Western Inner Piedmont samples. (G) Eastern Inner Piedmont samples. Note that *N* refers to the number of samples from each terrane, *n* refers to the number of analyses plotted, and all plots are at the same scale.

Post-Grenvillian detrital zircons are also present in samples from the Dahlonega gold belt, eastern Blue Ridge, western Inner Piedmont, and eastern Inner Piedmont. These ages are Late Proterozoic and early Paleozoic (eastern Inner Piedmont only) for detrital components.

APPALACHIAN CRUSTAL AFFINITY

Determining crustal affinities for basement and unmetamorphosed, often fossiliferous, sedimentary rocks is more straightforward than clearly identifying crustal affinity for penetratively deformed and metamorphosed sedimentary rocks, such as those found in the southern Appalachian crystalline core. Rifting of the Rodinian supercontinent created numerous crustal fragments and an array of complex depositional basins between the Laurentian and South American cratons (Fig. 6). Evidence exists within the southern Appalachian crystalline core for both exotic (relative to Laurentia) and Laurentian sediment sources. The predominance of 1.0–1.25 Ga detrital zircon ages from the southern Appalachian crystalline core inboard of the Carolina terrane strongly suggests that these basins were all near a Grenvillian source, consistent with the nonconformable nature of most basement-cover contacts in the Blue Ridge and in the Sauratown Mountains window. These Grenvillian ages, however, have limited use when used alone for establishing distinct source(s) because of the nearly ubiquitous presence of Grenvillian belts on the margins of pre-Grenvillian cratons that were near or adjacent to the Laurentian margin during initial rifting stages (e.g., Hoffman, 1991; Rogers, 1996; Dalziel, 1997).

Gondwanan Connections

Early workers positioned the West African craton near or adjacent to the Appalachian segment of Laurentia prior to the breakup of Rodinia (Bird and Dewey, 1970; Hatcher, 1978). More recent work presented convincing evidence that the west Gondwanan cratons of South America (Amazonia and Rio de la Plata) were juxtaposed with the southeastern Laurentian margin (e.g., Dalziel, 1991; Hoffman, 1991; Dalziel, 1997). Exact positions and relative movements of the west Gondwanan Amazonia, Rio de la Plata, Congo, and Kalahari cratons relative to Laurentia are variably portrayed in different paleogeographic reconstructions. In most reconstructions, however, the Amazonia and Rio de la Plata cratons appear as the conjugate margin to Laurentia (e.g., Hoffman, 1991; Dalziel, 1992, 1997; Cawood et al., 2001; Payolla et al., 2002). The Congo craton was also placed opposite the Amazonia craton near the southern Laurentian margin (Hoffman, 1991). Dalziel et al. (2000) suggested that the Kalahari craton may have been close, but still distal, to central Texas at ~1100 Ma. Each of these cratons thus represents a potential source for distal Late Proterozoic to early Paleozoic sediments of the eastern Blue Ridge and Inner Piedmont. The Argentine Precordillera and adjacent Sierra de Pie de Palo is generally accepted as a transported Laurentian fragment (e.g.,

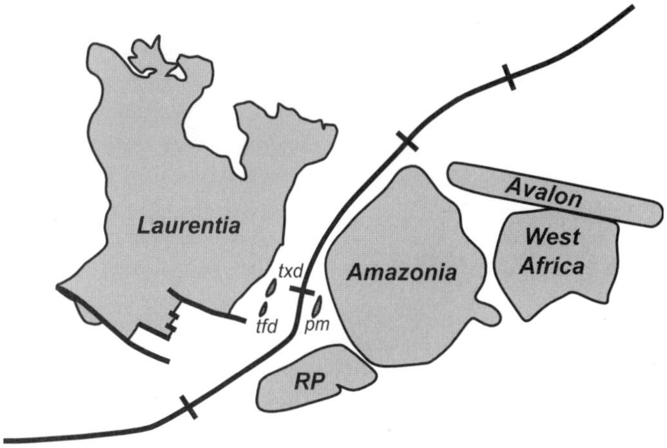

Figure 6. Paleogeography following successful rifting of Rodinia. Modified from Cawood et al. (2001) and Karlstrom et al. (2001). RP—Rio de la Plata craton; txd—Toxaway dome; tfd—Tallulah Falls dome; pm—Pine Mountain block. Compare with Hatcher et al. (this volume), their Figure 14.

Borrello, 1971; Ramos et al., 1986; Astini et al., 1995; Ramos et al., 1998; Vujovich and Kay, 1998; Thomas and Astini, 1999; Casquet et al., 2001); other candidates for transported Laurentian fragments include the South American Chilenia and Cuyania terranes, Oaxacan terrane of Mexico, and possibly the Dashwoods block in the northern Appalachians (e.g., Keppie and Ortega-Gutiérrez, 1999; Keppie and Ramos, 1999; Cawood et al., 2001). Paleozoic transfer of peri-Gondwanan terranes (e.g., the Avalonian-Carolina terrane) onto the Laurentian margin is also widely documented (e.g., Secor et al., 1983; Hatcher, 1989; Keppie and Ramos, 1999). Although the exact provenance of these smaller far-traveled crustal fragments is debatable, their present-day distribution is evidence for complex rifting and subsequent plate interactions at the Laurentian margin.

An exotic origin for internal, eastern Laurentian terranes (e.g., the Jefferson terrane of Horton et al., 1989) is frequently proposed to model subduction and accretionary wedge development at the margin (e.g., Abbott and Raymond, 1984; Rankin et al., 1989). In such models, rocks in the ultramafic and metamorphosed basalt/gabbro-bearing wedge(s) are considered exotic and are separated from the Laurentian margin by a series of Paleozoic faults interpreted to represent the suture and subduction zone locus (e.g., Hatcher, 1978, 1987; Rankin, 1994). Based on correlations with the Late Proterozoic–Cambrian Potomac orogeny in the central Appalachians, Hibbard and Samson (1995) proposed that portions of the southern Appalachians record tectonothermal histories that predate Laurentian passive margin development, possibly having developed in a peri-Gondwanan setting. Correlation of magmatic and tectonic events was also used to link the Laurentian margin with South America (Dalla Salda et al., 1992; Keppie and Ramos, 1999; Hibbard et al., 2003). In particular, comparisons of the Pampean (530–510 Ma), Famatinian (490–470 Ma), and

Achalian (403–382 Ma) magmatic pulses in the southern Sierras Pampeanas (Argentina) revealed similarities to Laurentian margin magmatic activity, thus supporting the possibility that Laurentia and western Gondwana collided ~470 Ma, producing magmas along the margins of both cratons (Stuart-Smith et al., 1999). This latter possibility would then require that subsequent Paleozoic orogenies (late Taconian, Acadian, and Alleghanian) were intracratonic (originally suggested by Bird and Dewey, 1970), unless a rifting event, as yet undocumented, also occurred.

Eastern North American detrital zircon ages and age ranges cited in support of a Gondwanan affinity include (1) combinations of 570 Ma, 1,500–1,550 Ma, and 2,600–2,750 Ma zircons, and absence of 1,000–1,150 Ma zircons from the eastern New England Appalachians (Karabinos and Gromet, 1993); (2) combinations of 515–635 Ma, abundant 1,967–2,282 Ma, and ~2,700 Ma zircons for a Florida basement sandstone (Mueller et al., 1994); (3) the combination of ~550–700 Ma arc magmatism with ~1,000–1,300 Ma and 1,500–1,550 Ma zircons from the Avalon terrane in Nova Scotia (Keppie et al., 1998); (4) the presence of ~550–700 Ma, 2,000–2,200 Ma, and minor components of ~1,000, 1,200, and 1,800 Ma zircons in samples from the late Paleozoic Magdalen basin near the Avalon and Meguma terranes in the Canadian Appalachians (Murphy and Hamilton, 2000); and (5) the presence of Neoproterozoic and 1,100–1,800 Ma with increased contributions of 1,800–2,500 Ma and Archean ages over time for younger Carolina terrane samples (Samson et al., 2001). Based on the presence of 2,200–2,400 Ma and ~1,400 Ma detrital zircons in the Pine Mountain window of Alabama, Heatherington et al. (1997) and Steltenpohl et al. (2001, this volume) suggested that the Hollis Quartzite may have been derived from west Gondwana and that the mixture of Laurentian (or South American Rondonian province) and probable Gondwanan detrital zircons favors deposition of the Hollis Quartzite roughly coeval with separation of the Pine Mountain crustal block from Gondwana. With the exception of the Hollis Quartzite, each of these studies noted the presence of Late Proterozoic to Early Cambrian and Archean detrital zircon components with varying, but minor, Early and Middle Proterozoic detrital zircon components.

Laurentian Connections

A Laurentian connection for the entire metasedimentary package northwest of the Carolina terrane was proposed in Hatcher's (1978) depositional model, wherein the western Blue Ridge and eastern Blue Ridge sedimentary packages were related as equivalent facies within an ocean basin that developed between the Laurentian margin and a rifted fragment of continental basement. In this model, western Inner Piedmont paragneisses were deposited on the eastern margin of the rifted continental basement block. Detailed field work in the eastern Blue Ridge and western Inner Piedmont connected stratigraphic sequences across the Brevard fault zone that separates these divisions (e.g., Heyn, 1984; Hopson and Hatcher, 1988) and identified nonconformable or locally faulted basement-cover contacts around small massifs of the eastern Blue Ridge (Hatcher et al., this volume). These relationships and our new data suggest that eastern Blue Ridge–western Inner Piedmont paragneisses are correlative, and/or perhaps time-transgressive equivalents, to the western Blue Ridge cover sequences that are known to unconformably overlie Grenville basement. Recently published Nd and inherited zircon data demonstrate that Ordovician arc rocks of the Chopawamsic and Milton terranes (Coler et al., 2000) are isotopically more evolved than Gondwanan-affinity arcs (Nance and Murphy, 1994; Samson et al., 1995), and show remarkable similarities to recognized Laurentian arc terranes (e.g., Bronson Hill anticlinorium), Blue Ridge Grenvillian basement, and Blue Ridge–western Inner Piedmont paragneisses (Fig. 3).

Detrital zircons from Newfoundland Humber zone rift-facies, drift-facies, and foreland-basin samples are interpreted to represent material derived from the Superior craton, Early Proterozoic, and Grenvillian orogenic belts, and Neoproterozoic rift-related magmatic rocks (Cawood and Nemchin, 2001). Studies of detrital zircon age components in younger central and northern Appalachian Paleozoic foreland sedimentary sequences also favor Laurentian derivation (Gray and Zeitler, 1997; McLennan et al., 2001). Thus, it is likely that at least a portion of the older rift-and-drift Laurentian margin siliciclastic hinterland (relative to the Paleozoic foreland) source assemblage was originally derived from Laurentia.

DISCUSSION

Although whole-rock Nd isotopic data and detrital zircon ages can provide important information concerning the evolution of metasedimentary rocks, a surprisingly limited amount of these data exist for the southern Appalachians. Three crustal affinities are probable for Blue Ridge and Inner Piedmont metasedimentary samples: (1) Laurentian derivation, (2) west Gondwanan derivation, or (3) a combination of Laurentian and west Gondwanan derivation. Our data (Figs. 4 and 5) taken alone do not preclude any of the aforementioned scenarios, but our data differ from those of similar studies in the southern Appalachians (e.g., Heatherington et al., 1997; Miller et al., 2000; Steltenpohl et al., 2001, this volume) because of the presence of an Early Proterozoic component, albeit minor, for most southern Appalachian divisions. Based on the occurrence of the major age ranges of our data, the source(s) must include Grenvillian rocks and 1,250–1,400 Ma source components for all of the terranes sampled. Additionally, most divisions also contain varying amounts of 1,350–1,500 Ma zircon ages. Archean source(s) are required for our zircon suites in the central Blue Ridge, Dahlonega gold belt, and eastern Inner Piedmont. The Dahlonega gold belt also contains an appreciable detrital zircon component of ~1,600–2,100 Ma, which is not nearly as abundant in the other divisions (limited to single analyses in the western Blue Ridge and eastern Inner Piedmont).

Notably, two grains from the Dahlonega gold belt are 2,000–2,100 Ma, an age range not abundant in the Laurentian craton. Ordovician detrital zircons in the eastern Inner Piedmont samples indicate Late Silurian to Devonian deposition, likely between the proximal Laurentian margin and approaching peri-Gondwanan terrane(s).

Some researchers have contended that the entire package of rocks from the Valley and Ridge and the western Blue Ridge boundary to the southeastern boundary of the Inner Piedmont is Laurentian (e.g., Hatcher, 1978, 1987, 1989); others, however, have proposed that all rocks east of the Allatoona–Hayesville–Holland Mountain–Gossan Lead fault are exotic to North America (e.g., Rankin et al., 1989; Hibbard and Samson, 1995). Our data, as well as data from the Chopawamsic and Milton terranes (Coler et al., 2000), suggest that these terranes were either built on or interacted with Mesoproterozoic crust. Southern Appalachian basement rocks are mostly Grenvillian (Carrigan et al., 2003; Hatcher et al., this volume); studies in the Mars Hill terrane in the western Blue Ridge, however, have identified older whole-rock isotopic components (Monrad and Gulley, 1983; Sinha et al., 1996; Carrigan et al., 2003; Ownby et al., this volume) and pre-Grenvillian zircons with an abundant 1,200–1,300 Ma component and fewer 1,400–1,500 Ma, 1,600–1,900 Ma, and 2,700 Ma components (Fullagar and Gulley, 1999; Carrigan et al., 2003; Ownby et al., this volume). Consequently, Blue Ridge basement rocks or their eroded and/or subsurface equivalents are the best candidates for abundant Grenvillian detrital zircons and lesser amounts of Middle and Early Proterozoic detrital zircons. The combination of 1.3–1.5 Ga and Grenvillian detrital zircons is readily explained by a midcontinent granite-rhyolite terrane source with transport through or along the eroding Grenville orogen.

Our samples are largely dominated by zircons in the ~900–1,400 Ma range and roughly resemble detrital zircon data from Paleozoic sedimentary rocks in the northern Appalachian foreland (Middle Ordovician Austin Glen Member of the Normanskill Formation and Lower Silurian Shawangunk Formation; Gray and Zeitler, 1997; McLennan et al., 2001) and Newfoundland Humber Zone basal sedimentary units (Bradore and Summerside Formations; Cawood and Nemchin, 2001). They contrast markedly with northern Appalachian and Florida basement samples that are largely devoid of Grenvillian detrital zircons (Karabinos and Gromet, 1993; Mueller et al., 1994; Murphy and Hamilton, 2000; Samson et al., 2001). Except for eastern Inner Piedmont zircons, our data essentially lack the Neoproterozoic–Early Cambrian component of Gondwanan terranes and show a reasonable overlap with age distributions constructed for the southern Grenville by Cawood and Nemchin (2001 and references therein). Post-Grenvillian detrital zircon ages could represent sediment input from one or more of the following: pan-African crustal fragments (e.g., the Carolina terrane), Amazonian Brasiliano detritus (Keppie et al., 1998), late Neoproterozoic rift-related igneous rocks (e.g., Crossnore plutons or Catoctin Formation equivalents; Cawood and Nemchin, 2001), or Ordovician magmatic rocks (eastern Inner Piedmont only). Detrital zircon data for the Dahlonega gold belt and the single central Blue Ridge Coleman River Formation sample are distinct due to an Archean component in each and an early Middle Proterozoic age population in the Dahlonega gold belt. Samples from the Dahlonega gold belt represent the best candidates for a mixed Laurentian and Gondwanan affinity; however, this is difficult to explain, because these terranes are structurally beneath the eastern Blue Ridge and western Inner Piedmont, which are most likely derived from Laurentia. One possibility is that sediments of the Dahlonega gold belt and associated Ordovician arcs are younger than surrounding sediments of the Chattahoochee and Hayesville–Soque River thrust sheets and reflect a different sediment-transport direction and/or sources not exposed during initial rifting.

Exotic Pine Mountain block cover sequences place Grenville basement of the block proximal to a west Gondwanan source or sources (Steltenpohl et al., this volume); however, the present position of the block and its unique provenance relative to the now adjacent Inner Piedmont and nearby Blue Ridge may be variably interpreted. Surface geologic data from carbonate horses in the Brevard fault zone (Hatcher et al., 1973), matching Valley and Ridge reflector characteristics with those beneath the Brevard fault zone (Clark et al., 1978), and regional seismic reflection data (Cook et al., 1980; Costain et al., 1989) from the Blue Ridge and Inner Piedmont of Georgia and Carolinas place the crystalline thrust sheets of the eastern Blue Ridge and Inner Piedmont above Valley and Ridge carbonate and clastic rocks. Seismic reflection data from the Pine Mountain block and vicinity (Nelson et al., 1987) provide evidence of Valley and Ridge–type rocks on basement at the northwestern edge of the block. Early accretion of an exotic Pine Mountain block to the Laurentian margin with its Grenville basement and Paleozoic clastic-carbonate cover, followed by development of high- and low-temperature Alleghanian shear zones (Steltenpohl et al., 1992; Student and Sinha, 1992; Steltenpohl and Kunk, 1993; Hooper et al., 1997), and final detachment along the master Appalachian décollement beneath the window is compatible with available data (Hooper and Hatcher, 1988; McBride et al., 2001) and accounts for the present position of the block in the southern Appalachians.

SUMMARY

Laurentian crustal affinity is favored here for paragneisses in the southern Appalachian Blue Ridge and western Inner Piedmont. This view is based on (1) the presence and availability of eastern North America cratonic and basement material with the same ages found in southern Appalachian Blue Ridge and western Inner Piedmont detrital zircons, (2) overlap of Nd isotopic values for southern Appalachian metasedimentary rocks and Grenvillian basement, and (3) a paucity of distinctly Gondwanan detrital zircons. Laurentian provenance requires that (1) marginal basins were isolated from west Gondwanan source(s) from the time of Rodinian rifting at least until the Middle Ordovician, (2) sources include early Middle Proterozoic rocks from the Mid-Continent granite-rhyolite province or the Mars Hill terrane, (3) the Pine Mountain block is either entirely exotic (inclusive of its 1.1 Ga basement) or represents a far-traveled fragment of Laurentia, and (4) paragneisses of the eastern Blue Ridge and western Inner Piedmont are time-transgressive or rift-facies equivalents of western Blue Ridge paragneisses. Gondwanan affinity of some zircons in a few terranes (e.g., Dahlonega gold belt and Cowrock) requires that these sediments were deposited proximal to a Grenville belt in western Gondwana and were sutured to the Laurentian margin during the Taconic and/or Acadian orogenic events. A mixed provenance is plausible for the Dahlonega gold belt but this creates significant tectonic and depositional issues because of its location between and beneath terranes that have connections to Laurentia. Eastern Inner Piedmont samples contain detrital zircon components similar to those in Blue Ridge and western Inner Piedmont samples, but also contain abundant Ordovician and lesser Silurian, Early Devonian, Gondwanan components, thus favoring post-Early Devonian deposition and mixed Laurentian and Gondwanan provenance.

ACKNOWLEDGMENTS

Chris Coath, Mark Harrison, and Kevin McKeegan provided invaluable assistance with the UCLA IMP. Joe Wooden and Anders Meibom provided operational, in addition to interpretive, assistance with the Stanford University IMP. Members of the North Carolina Geological Survey (Leonard Weiner, Carl Merschat, and Mark Carter) and the South Carolina Geological Survey (Butch Maybin) are gratefully acknowledged for accompanying the authors on many of our sampling trips and for numerous conversations in the field regarding this work. National Science Foundation grants EAR-9814800 to RDH and EAR-9814801 to CFM funded this research. Review and comments by Mark Steltenpohl and Stephen Kish improved the focus and quality of the manuscript.

REFERENCES CITED

Abbott, R.N., Jr., and Raymond, L.A., 1984, The Ashe metamorphic suite, northwest North Carolina: Metamorphism and observations on geologic history: American Journal of Science, v. 284, p. 350–375.

Aleinikoff, J.N., Zartman, R.E., Walter, M., Rankin, D.W., Lyttle, P.T., and Burton, W.C., 1995, U-Pb ages of metarhyolites of the Catoctin and Mount Rogers Formations, central and southern Appalachians: Evidence for two phases of Iapetian rifting: American Journal of Science, v. 295, p. 428–454.

Astini, R.A., Benedetto, J.L., and Vaccari, N.E., 1995, The early Paleozoic evolution of the Argentine Precordillera as a Laurentian rifted, drifted and collided terrane: A geodynamic model: Geological Society of America Bulletin, v. 107, p. 253–273.

Bacon, C.R., Persing, H.M., Wooden, J.L., and Ireland, T.R., 2000, Late Pleistocene granodiorite beneath Crater Lake caldera, Oregon, dated by ion microprobe: Geology, v. 28, p. 467–470.

Badger, R.L., and Sinha, A.K., 1988, Age and Sr isotopic signature of the Catoctin volcanic province: Implications for subcrustal mantle evolution: Geology, v. 16, p. 692–695.

Bartholomew, M.J., 1992, Structural characteristics of the Late Proterozoic (post-Grenville) continental margin of the Laurentian craton, *in* Bartholomew, M.J., et al., eds., Basement tectonics, 8: Characterization and comparison of ancient and Mesozoic continental margins—Proceedings of the 8th international conference on basement tectonics: Dordrecht, The Netherlands, Kluwer Academic Publishers, p. 443–467.

Bartholomew, M.J., and Lewis, S.E., 1988, Peregrination of Middle Proterozoic massifs and terranes within the Appalachian orogen, eastern U.S.A., *in* Martinez-Garcia, E., ed., Geology of the Iberian Massif with communications presented at the International Conference on Iberian terranes and their regional correlation: Trabajos de Geologia, University of Oviedo, Spain, v. 17, p. 155–165.

Bartholomew, M.J., and Lewis, S.E., 1992, Appalachian Grenville masifs: Pre-Appalachian translational tectonics, *in* Mason, R., ed., Basement tectonics, 7—Proceedings of the seventh International conference: Dordrecht, The Netherlands, Kluwer Academic Publishers, p. 363–374.

Berger, S., Cochrane, D., Simons, K., Savov, I., Ryan, J.G., and Peterson, V.L., 2001, Insights from rare earth elements into the genesis of the Buck Creek complex, Clay County, NC: Southeastern Geology, v. 40, p. 201–212.

Bird, J.M., and Dewey, J.F., 1970, Lithosphere plate: Continental margin tectonics and the evolution of the Appalachian orogen: Geological Society of America Bulletin, v. 81, p. 1031–1060.

Borrello, A.V., 1971, The Cambrian of South America, *in* Holland, D.H., ed., Cambrian of the New World: London and New York, Wiley-Interscience, p. 385–438.

Brewer, M.C., and Thomas, W.A., 2000, Stratigraphic evidence for Neoproterozoic-Cambrian two-phase rifting of southern Laurentia: Southeastern Geology, v. 39, p. 91–106.

Carrigan, C.W., Miller, C.F., Fullagar, P.D., Bream, B.R., Hatcher, R.D., Jr., and Coath, C.D., 2003, Ion microprobe age and geochemistry of southern Appalachian basement, with implications for Proterozoic and Paleozoic reconstructions: Precambrian Research, v. 120, p. 1–36.

Carroll, D., Neuman, R.B., and Jaffe, H.W., 1957, Heavy minerals in arenaceous beds in parts of the Ocoee Series, Great Smoky Mountains, Tennessee: American Journal of Science, v. 255, p. 175–193.

Casquet, C., Baldo, E., Pankhurst, R.J., Rapela, C.W., Galindo, C., Fanning, C.M., and Saavedra, J., 2001, Involvement of the Argentine Precordillera terrane in the Famatinian mobile belt: U-Pb SHRIMP and metamorphic evidence from the Sierra de Pie de Palo: Geology, v. 29, p. 703–706.

Cawood, P.A., and Nemchin, A.A., 2001, Paleogeographic development of the east Laurentian margin: Constraints from U-Pb dating of detrital zircons in the Newfoundland Appalachians: Geological Society of America Bulletin, v. 113, p. 1234–1246.

Cawood, P.A., McCausland, P.J.A., and Dunning, G.R., 2001, Opening Iapetus: Constraints from the Laurentian margin in Newfoundland: Geological Society of America Bulletin, v. 113, p. 443–453.

Clark, H.B., Costain, J.K., and Glover, L., III, 1978, Structural and seismic reflection studies of the Brevard ductile deformation zone near Rosman, North Carolina: American Journal of Science, v. 278, p. 419–441.

Coler, D.G., Wortman, G.L., Samson, S.D., Hibbard, J.P., and Stern, R., 2000, U-Pb geochronologic, Nd isotopic, and geochemical evidence for the correlation of the Chopawamsic and Milton terranes, Piedmont zone, southern Appalachian orogen: Journal of Geology, v. 108, p. 363–380.

Cook, F.A., Brown, L.D., and Oliver, J.E., 1980, The southern Appalachians and the growth of continents: Scientific American, v. 243, p. 156–168.

Costain, J.K., Hatcher, R.D., Jr., and Çoruh, C., 1989, Appalachian ultradeep core hole (ADCOH) project site investigation, regional seismic lines and geologic interpretation, Plate 8, *in* Bally, A.W., and Palmer, A.R., eds., The geology of North America: An overview, The geology of North America, v. A, and *in* Hatcher, R.D., Jr., et al., eds., The Appalachian–Ouachita orogen in the United States: Boulder, Colorado, Geological Society of America, The Geology of North America, v. F-2.

Dalla Salda, L.H., Cingolani, C., and Varela, R., 1992, Early Paleozoic orogenic belt of the Andes in southwestern South America: Result of Laurentia-Gondwana collision: Geology, v. 20, p. 617–620.

Dalziel, I.W.D., 1991, Pacific margins of Laurentia and East Antarctica–Australia as a conjugate rift pair: Evidence and implications from an Eocambrian supercontinent: Geology, v. 19, p. 598–601.

Dalziel, I.W.D., 1992, On the organization of American plates in the Neoproterozoic and the breakout of Laurentia: GSA Today, v. 2, p. 237–241.

Dalziel, I.W.D., 1997, Neoproterozoic-Paleozoic geography and tectonics: Review, hypothesis, environmental speculation: Geological Society of America Bulletin, v. 109, p. 16–42.

Dalziel, I.W.D., Dalla Salda, L.H., and Gahagan, L.M., 1994, Paleozoic Laurentia-Gondwana interaction and the origin of the Appalachian-Andean mountain system: Geological Society of America Bulletin, v. 106, p. 243–252.

Dalziel, I.W.D., Mosher, S., and Gahagan, L.M., 2000, Laurentia-Kalahari collision and the assembly of Rodinia: Journal of Geology, v. 108, p. 499–513.

DeGraff-Surpless, K., Graham, S.A., Wooden, J.L., and McWilliams, M.O., 2002, Detrital zircon provenance analysis of the Great Valley Group, California: Evolution of an arc-forearc system: Geological Society of America Bulletin, v. 114, p. 1564–1580.

DePaolo, D., 1981, Neodymium isotopes in the Colorado Front Range and crust-mantle evolution in the Proterozoic: Nature, v. 291, p. 193–196.

Dodson, M.H., Compston, W., Williams, I.S., and Wilson, J.F., 1988, A search for ancient detrital zircons in Zimbabwean sediments: Journal of the Geological Society, London, v. 145, p. 977–983.

Fritz, W.J., Hatcher, R.D., Jr., and Hopson, J.L., eds., 1989, Geology of the eastern Blue Ridge of northeast Georgia and the adjacent Carolinas: Georgia Geological Society Guidebooks, v. 9, 199 p.

Fullagar, P.D., 2002, Evidence for Early Mesoproterozoic (and older?) crust in the southern and central Appalachians of North America: Gondwana Research, v. 5, p. 197–203.

Fullagar, P.D., and Gulley, G.L., Jr., 1999, Pre-Grenville uranium-lead zircon age for the Carvers Gap Gneiss in the western Blue Ridge of North Carolina-Tennessee: Geological Society of America Abstracts with Programs, v. 31, no. 3, p. 16.

Fullagar, P.D., Goldberg, S.A., and Butler, J.R., 1997, Nd and Sr isotopic characterization of crystalline rocks from the southern Appalachian Piedmont and Blue Ridge, North and South Carolina, *in* Sinha, A.K., et al., eds., The nature of magmatism in the Appalachian orogen: Boulder, Colorado, Geological Society of America Memoir 191, p. 165–179.

Goldberg, S.A., Butler, J.R., and Fullagar, P.D., 1986, The Bakersville dike swarm: Geochronology and petrogenesis of Late Proterozoic basaltic magmatism in the southern Appalachian Blue Ridge: American Journal of Science, v. 286, p. 403–430.

Goldsmith, R., and Secor, D.T., Jr., 1993, Proterozoic rocks of accreted terranes in the eastern United States, *in* Reed, J.C., Jr., et al., eds., Precambrian: Conterminous United States: Boulder, Colorado, Geological Society of America, The Geology of North America, v. C-2, p. 422–439.

Gray, M.B., and Zeitler, P.K., 1997, Comparison of clastic wedge provenance in the Appalachian foreland using U/Pb ages of detrital zircons: Tectonics, v. 16, p. 151–160.

Hadley, J.B., 1970, The Ocoee Series and its possible correlatives, *in* Fisher, G.W., et al., eds., Studies of Appalachian geology: Central and southern: New York, John Wiley and Sons, p. 247–260.

Hatcher, R.D., Jr., 1969, Stratigraphy, petrology and structure of the low rank belt and part of the Blue Ridge of northwesternmost South Carolina: South Carolina Division of Geology, Geologic Notes, v. 13, p. 105–141.

Hatcher, R.D., Jr., 1978, Tectonics of the western Piedmont and Blue Ridge, southern Appalachians: Review and speculation: American Journal of Science, v. 278, p. 276–304.

Hatcher, R.D., Jr., 1979, The Coweeta Group and Coweeta syncline: Major feature of the North Carolina–Georgia Blue Ridge: Southeastern Geology, v. 21, no. 1, p. 17–29.

Hatcher, R.D., Jr., 1984, Southern and central Appalachian basement massifs, *in* Bartholomew, M.J., ed., The Grenville event in the Appalachians and related topics: Boulder, Colorado, Geological Society of America Special Paper 194, p. 149–153.

Hatcher, R.D., Jr., 1987, Tectonics of the southern and central Appalachian internides: Annual Review of Earth and Planetary Science, v. 15, p. 337–362.

Hatcher, R.D., Jr., 1988, Bedrock geology and regional geologic setting of Coweeta Hydrologic Laboratory in the eastern Blue Ridge, with some discussion of Quaternary deposits and structural controls of topography, *in* Swank, W.T., and Crossley, D.A., Jr., eds., Coweeta symposium, Volume 66: New York, Springer-Verlag, p. 81–92.

Hatcher, R.D., Jr., 1989, Tectonic synthesis of the U.S., Appalachians, Chapter 14, *in* Hatcher, R.D., Jr., et al., eds., The Appalachian-Ouachita orogen in the United States: Boulder, Colorado, Geological Society of America, The Geology of North America, v. F-2, p. 511–535.

Hatcher, R.D., Jr., and Acker, L.L., 1984, Geology of the Salem quadrangle, South Carolina: South Carolina Geological Survey MS-26, 23 p., scale 1:24,000.

Hatcher, R.D., Jr., and Bream, B.R., eds., 2002, Inner Piedmont tectonics focused mostly on detailed studies in the South Mountains and the southern Brushy Mountains, North Carolina: Raleigh, North Carolina Geological Survey, Carolina Geological Society Guidebook, 145 p.

Hatcher, R.D., Jr., Price, V., Jr., and Snipes, D.S., 1973, Analysis of chemical and paleotemperature data from selected carbonate rocks of the southern Appalachians: Southeastern Geology, v. 15, p. 55–70.

Hatcher, R.D., Jr., Hooper, R.J., McConnell, K.I., Heyn, T., and Costello, J.O., 1988, Geometric and time relationships between thrusts in the crystalline southern Appalachians, *in* Mitra, G., and Wojtal, S., eds., Appalachian thrusting: Boulder, Colorado, Geological Society of America Special Paper 222, p. 185–196.

Hatcher, R.D., Jr., Osberg, P.H., Drake, A.A., Jr., Robinson, P., and Thomas, W.A., 1990, Tectonic map of the U.S., Appalachians, *in* Hatcher, R.D., Jr., et al., eds., The Appalachian-Ouachita orogen in the United States: Boulder, Colorado, Geological Society of America, The Geology of North America, v. F-2, scale 1:2,500,000.

Heatherington, A.L., Steltenpohl, M.G., Yokel, L.S., and Mueller, P.A., 1997, Ages of detrital zircons from the Pine Mountain and Inner Piedmont terranes and the Valley and Ridge Province; implications for Neoproterozoic tectonics of the southern Laurentian margin: Geological Society of America Abstracts with Programs, v. 29, no. 6, p. 432.

Heyn, T., 1984, Stratigraphic and structural relationships along the southwestern flank of the Sauratown Mountains anticlinorium [M.S. thesis]: Columbia, University of South Carolina, 192 p.

Hibbard, J.P., and Samson, S.D., 1995, Orogenesis exotic to the Iapetan cycle in the southern Appalachians, *in* Hibbard, J.P., et al., eds., Current perspectives in the Appalachian-Caledonian orogen: Geological Society of Canada Special Paper 41, p. 191–205.

Hibbard, J.P., Tracy, R.J., and Henika, W.S., 2003, Smith River allochthon: A southern Appalachian peri-Gondwanan terrane emplaced directly on Laurentia?: Geology, v. 31, p. 215–218.

Hoffman, P.F., 1991, Did the breakout of Laurentia turn Gondwanaland inside-out?: Science, v. 252, p. 1409–1412.

Hooper, R.J., and Hatcher, R.D., Jr., 1988, Pine Mountain terrane, a complex window in the Georgia and Alabama Piedmont: Evidence from the eastern termination: Geology, v. 16, p. 307–310.

Hooper, R.J., Hatcher, R.D., Jr., Troyer, P.K., Dawson, R.J., and Redmond, C.G., 1997, The character of the Avalon terrane and its boundary with the Piedmont terrane in central Georgia, *in* Glover, L., III, and Gates, A.E., eds., Central and southern Appalachian sutures: Results of the EDGE Project and related studies: Boulder, Colorado, Geological Society of America Special Paper 314, p. 1–14.

Hopson, J.L., and Hatcher, R.D., Jr., 1988, Structural and stratigraphic setting of the Alto allochthon, northeast Georgia: Geological Society of America Bulletin, v. 100, p. 339–350.

Horton, J.W., Jr., and McConnell, K.I., 1991, The western Piedmont, *in* Horton, J.W., Jr., and Zullo, V.A., eds., The Geology of the Carolinas: Knoxville, The University of Tennessee Press, p. 36–58.

Horton, J.W., Jr., Drake, A.A., Jr., and Rankin, D.W., 1989, Tectonostratigraphic terranes and their Paleozoic boundaries in the central and southern Appalachians, *in* Dallmeyer, R.D., ed., Terranes in the Circum-Atlantic Paleozoic orogens: Boulder, Colorado, Geological Society of America Special Paper 230, p. 213–245.

Karabinos, P., and Gromet, P., 1993, Application of single-grain zircon evaporation analyses to detrital grain studies and age discrimination in igneous suites: Geochimica et Cosmochimica Acta, v. 57, p. 4257–4267.

Karlstrom, K.E., Åhäll, K., Harlan, S.S., Williams, M.L., McLelland, J., and Geissman, J.W., 2001, Long-lived (1.8–1.0 Ga) convergent orogen in southern Laurentia, its extensions to Australia and Baltica, and implications for refining Rodinia: Precambrian Research, v. 111, p. 5–30.

Keppie, J.D., and Ortega-Gutiérrez, F., 1999, Middle American Precambrian basement: A missing piece of the reconstructed 1-Ga orogen, *in* Ramos, V.A., and Keppie, J.D., eds., Laurentia-Gondwana connections before Pangea: Boulder, Colorado, Geological Society of America Special Paper 336, p. 199–210.

Keppie, J.D., and Ramos, V.A., 1999, Odyssey of terranes in the Iapetus and Rheic oceans during the Paleozoic, *in* Ramos, V.A., and Keppie, J.D., eds., Laurentia-Gondwana connections before Pangea: Boulder, Colorado, Geological Society of America Special Paper 336, p. 267–276.

Keppie, J.D., Dostal, J., Murphy, J.B., and Nance, R.D., 1996, Terrane transfer between eastern Laurentia and western Gondwana in the early Paleozoic: Constraints on global reconstructions, *in* Nance, R.D., and Thompson, M.D., eds., Avalonian and related peri-Gondwanan terranes of the Circum-North Atlantic: Boulder, Colorado, Geological Society of America Special Paper 301, p. 369–380.

Keppie, J.D., Davis, D.W., and Krogh, T.E., 1998, U-Pb geochronological constraints on Precambrian stratified units in the Avalon composite terrane of Nova Scotia, Canada: Tectonic implications: Canadian Journal of Earth Sciences, v. 35, p. 222–236.

King, P.B., Hadley, J.B., Neuman, R.B., and Hamilton, W.B., 1958, Stratigraphy of Ocoee Series, Great Smoky Mountains, Tennessee and North Carolina: Geological Society of America Bulletin, v. 69, p. 947–966.

Kish, S.A., Hanley, T.B., and Schamel, S., eds., 1985, Geology of the southwestern Piedmont of Georgia: Boulder, Colorado, Geological Society of America Annual Meeting, Field Trip Guide, 47 p.

McBride, J.H., Hatcher, R.D., Jr., and Stephenson, W.J., 2001, Toward reconciliation of seismic reflection and field geologic mapping from the Pine Mountain Belt; preliminary results: Geological Society of America Abstracts with Programs, v. 33, no. 2, p. 74.

McConnell, K.I., 1988, Geology of the Sauratown Mountains anticlinorium: Vienna and Pinnacle 7.5 minute quadrangles, *in* Hatcher, R.D., Jr., ed., Structure of the Sauratown Mountains window, North Carolina: Raleigh, North Carolina Geological Survey, Carolina Geological Society field trip guidebook, p. 51–66.

McLennan, S.M., Bock, B., Compston, W., Hemming, S.R., and McDaniel, D.K., 2001, Detrital zircon geochronology of Taconian and Acadian foreland sedimentary rocks in New England: Journal of Sedimentary Research, v. 71, p. 305–317.

Merschat, C.E., and Carter, M.W., 2002, Bedrock geologic map of the Bent Creek Research and Demonstration Forest, Southern Research Station, USDA Forest Service, including the North Carolina Arboretum and a portion of the Blue Ridge Parkway: Raleigh, North Carolina Geological Survey Geologic Map Series-9, scale 1:12,000.

Miller, C.F., Hatcher, R.D., Jr., Ayers, J.C., Coath, C.D., and Harrison, T.M., 2000, Age and zircon inheritance of eastern Blue Ridge plutons, southwestern North Carolina and northeastern Georgia, with implications for magma history and evolution of the southern Appalachian orogen: American Journal of Science, v. 300, p. 142–172.

Monrad, J.R., and Gulley, G.L., 1983, Age and *P-T* conditions during metamorphism of granulite-facies gneisses, Roan Mountain, NC-TN, *in* Lewis, S.E., ed., Geological investigations in the Blue Ridge of northwestern North Carolina: Boone, North Carolina, Carolina Geological Society Field Trip Guidebook, article IV, p. 41–51.

Montes, C., and Hatcher, R.D., Jr., 1999, Documenting Late Proterozoic rifting in the Ocoee basin, western Blue Ridge, North Carolina: Southeastern Geology, v. 39, p. 37–50.

Mueller, P.A., Wooden, J.L., and Nutman, A.P., 1992, 3.96 Ga zircons from an Archean quartzite, Beartooth Mountains, Montana: Geology, v. 20, p. 327–330.

Mueller, P.A., Heatherinton, A.L., Wooden, J.L., Shuster, R.D., Nutman, A.P., and Williams, I.S., 1994, Precambrian zircons from the Florida basement: A Gondwanan connection: Geology, v. 22, p. 119–122.

Mueller, P.A., Kozuch, M., Heatherington, A.L., Wooden, J.L., Offield, T.W., Koeppen, R.P., Klein, T.L., and Nutman, A.P., 1996, Evidence for Mesoproterozoic basement in the Carolina terrane and speculations on its origin, *in* Nance, R.D., and Thompson, M.D., eds., Avalonian and related peri-Gondwanan terranes of the Circum-North Atlantic: Boulder, Colorado, Geological Society of America Special Paper 304, p. 207–217.

Murphy, J.B., and Hamilton, M.A., 2000, Orogenesis and basin development: U-Pb detrital zircon age constraints on evolution of the late Paleozoic St. Marys basin, central mainland Nova Scotia: Journal of Geology, v. 108, p. 53–71.

Nance, R.D., and Murphy, J.B., 1994, Contrasting basement isotopic signatures and the palinspastic restoration of peripheral orogens: Example from the Neoproterozoic Avalonian-Cadomian belt: Geology, v. 22, p. 617–620.

Nelson, K.D., Arnow, M., Giguere, M., and Schamel, S., 1987, Normal-fault boundary of an Appalachian basement massif: Results of COCORP profiling across the Pine Mountain Belt in western Georgia: Geology, v. 15, p. 832–836.

Paces, J.B., and Miller, J.D., 1993, Precise U-Pb age of Duluth complex and related mafic intrusions, northeastern Minnesota: Geochronological insights into physical, petrogenetic, paleomagnetic, and tectonomagmatic processes associated with the 1.1 Ga Mid-Continent rift system: Journal of Geophysical Research, v. 98, p. 13997–14103.

Payolla, B.L., Bettencourt, J.S., Kozuch, M., Leite, W.B., Fetter, A.H., and Van Schmus, W.R., 2002, Geological evolution of the basement rocks in the east-central part of the Rondônia Tin Province, SW Amazonian craton, Brazil: U-Pb and Sm-Nd isotopic constraints: Precambrian Research, v. 119, p. 141–169.

Quidelleur, X., Grove, M., Lovera, O.M., Harrison, T.M., Yin, A., and Ryerson, F.J., 1997, Thermal evolution and slip history of the Renbu Zedong thrust, southeastern Tibet: Journal of Geophysical Research, v. 102, p. 2659–2679.

Ramos, V.A., Jordan, T.E., Allmendinger, R.W., Mpodozis, C., Kay, S.M., Cortes, M., and Palma, M., 1986, Paleozoic terranes of the central Argentine-Chilean Andes: Tectonics, v. 5, p. 855–880.

Ramos, V.A., Dallmeyer, R.D., and Vujovich, G., 1998, Time constraints on the early Palaeozoic docking of the Precordillera, central Argentina, *in* Pankhurst, R.J., and Rapela, C.W., eds., The proto-Andean margin of Gondwana: London, Geological Society Special Publication 142, p. 143–158.

Rankin, D.W., 1994, Continental margin of the eastern United States: Past and present, *in* Speed, R.C., ed., Phanerozoic evolution of North American continent-ocean transitions: Boulder, Colorado, Geological Society of America, Geology of North America, Continent-Ocean Transects, Accompanying Volume, p. 129–218.

Rankin, D.W., Drake, A.A., Glover, L., III, Goldsmith, R., Hall, L.M., Murray, D.P., Ratcliffe, N.M., Read, J.F., Secor, D.T., Jr., and Stanley, R.S., 1989, Pre-orogenic terranes, in Hatcher, R.D., Jr., et al., eds., The Appalachian-Ouachita orogen in the United States: Boulder, Colorado, Geological Society of America, The Geology of North America, v. F-2, p. 7–100.

Rast, N., and Kohles, K.M., 1986, The origin of the Ocoee Supergroup: American Journal of Science, v. 286, p. 593–616.

Rogers, J.J.W., 1996, A history of the continents in the past three billion years: Journal of Geology, v. 104, p. 91–107.

Samson, S.D., Hibbard, J.P., and Wortman, G.L., 1995, Nd isotopic evidence for juvenile crust in the Carolina terrane, southern Appalachians: Contributions to Mineralogy and Petrology, v. 121, p. 171–184.

Samson, S.D., Secor, D.T., and Hamilton, M.A., 2001, Wandering Carolina: Tracking exotic terranes with detrital zircons: Geological Society of America Abstracts with Programs, v. 33, no. 6, p. 263.

Schamel, S., Hanley, T.B., and Sears, J.W., 1980, Geology of the Pine Mountain window and adjacent terranes in the Piedmont province of Alabama and Georgia: Boulder, Colorado, Geological Society of America, Southeastern Section Meeting Guidebook, 69 p.

Sears, J.W., Cook, R.B., and Brown, D.E., 1981, Tectonic evolution of the western part of the Pine Mountain window and adjacent Inner Piedmont province, *in* Sears, J.W., ed., Contrasts in tectonic style between the Inner Piedmont terrane and the Pine Mountain window: Auburn, Alabama, Alabama Geological Society 18th Annual Field Trip Guidebook, p. 1–13.

Secor, D.T., Samson, S.L., Snoke, A.W., and Palmer, A.R., 1983, Confirmation of the Carolina slate belt as an exotic terrane: Science, v. 221, p. 649–651.

Settles, D.J., Bream, B.R., and Hatcher, R.D., Jr., 2001, The Sally Free mafic complex and associated mafic rocks of the Dahlonega gold belt: Evidence of Ordovician arc volcanism and intrusion northwest of Dahlonega, Georgia: Geological Society of America Abstracts with Programs, v. 33, no. 6, p. 262.

Simpson, E.L., and Eriksson, K.A., 1989, Sedimentology of the Unicoi Formation in southern and central Virginia: Evidence for Late Proterozoic to Early Cambrian rift-to-passive margin transition: Geological Society of America Bulletin, v. 101, p. 42–54.

Simpson, E.L., and Sundberg, F.A., 1987, Early Cambrian age for synrift deposits of the Chilhowee Group of southwestern Virginia: Geology, v. 15, p. 123–126.

Sinha, A.K., Hogan, J.P., and Parks, J., 1996, Lead isotope mapping of crustal reservoirs within the Grenville superterrane I: Central and southern Appalachians: Washington, D.C., American Geophysical Union Geophysical Monograph 95, p. 293–305.

Spell, T.L., and Norrell, G.T., 1990, The Ropes Creek assemblage: Petrology, geochemistry, and tectnoic setting of an ophiolitic thrust sheet in the southern Appalachians: American Journal of Science, v. 290, p. 811–842.

Steltenpohl, M.G., 1992, The Pine Mountain window of Alabama: Basement-cover evolution in the southernmost exposed Appalachians, *in* Bartholomew, M.J., et al., eds., Basement Tectonics, 8: Characterization and comparison of ancient and Mesozoic continental margins—Proceedings of the 8th International Conference on Basement Tectonics, 1988: Dordrecht, The Netherlands, Kluwer Academic Publishers, p. 491–501.

Steltenpohl, M.G., and Kunk, M.J., 1993, $^{40}Ar/^{39}Ar$ thermochronology and Alleghanian development of the southernmost Appalachian Piedmont, Alabama and Southwest Georgia: Geological Society of America Bulletin, v. 105, p. 819–833.

Steltenpohl, M.G., Goldberg, S.A., Hanley, T.B., and Kunk, M.J., 1992, Alleghanian development of the Goat Rock fault zone, southernmost Appalachians: Temporal compatibility with the master decollement: Geology, v. 20, p. 845–848.

Steltenpohl, M.G., Heatherington, A.L., and Mueller, P.A., 2001, Our current understanding of the Grenville event in the southernmost Appalachians, Pine Mountain window, Alabama: Geological Society of America Abstracts with Programs, v. 33, no. 6, p. 29.

Stuart-Smith, P.G., Miró, R., Sims, J.P., Pieters, P.E., Lyons, P., Camacho, A., Skirrow, R.G., and Black, L.P., 1999, Uranium-lead dating of felsic magmatic cycles in the southern Sierras Pampeanas, Argentina: Implications for the tectonic development of the proto-Andean Gondwanan margin, *in* Ramos, V.A., and Keppie, J.D., eds., Laurentia-Gondwana connections before Pangea: Boulder, Colorado, Geological Society of America Special Paper 336, p. 87–112.

Student, J.J., and Sinha, A.K., 1992, Carboniferous U-Pb ages of zircons from the Box Ankle and Ocmulgee faults, central Georgia: Implications for accretionary models: Geological Society of America Abstracts with Programs, v. 24, no. 2, p. 69.

Su, Q., Goldberg, S.A., and Fullagar, P.D., 1994, Precise U-Pb zircon ages of Neoproterozoic plutons in the southern Appalachian Blue Ridge and their implications for the rifting of Laurentia: Precambrian Research, v. 68, p. 81–95.

Thomas, C.W., Miller, C.F., Bream, B.R., and Fullagar, P.D., 2001, Origins of mafic-ultramafic complexes of the eastern Blue Ridge, southern Appalachians: Geochronologic and geochemical constraints: Geological Society of America Abstracts with Programs, v. 33, no. 2, p. 66.

Thomas, W.A., 1991, The Appalachian-Ouachita rifted margin of southeastern North America: Geological Society of America Bulletin, v. 103, p. 415–431.

Thomas, W.A., and Astini, R.A., 1996, The Argentine Precordillera; a traveler from the Ouachita Embayment of North American Laurentia: Science, v. 273, p. 752–757.

Thomas, W.A., and Astini, R.A., 1999, Simple-shear conjugate rift margins of the Argentine Precordillera and the Ouachita embayment of Laurentia: Geological Society of America Bulletin, v. 111, p. 1069–1079.

Tollo, R.P., and Aleinikoff, J.N., 1996, Petrology and U-Pb geochronology of the Robertson River igneous suite, Blue Ridge Province, Virginia; evidence for multistage magmatism associated with an early episode of Laurentian rifting: American Journal of Science, v. 296, p. 1045–1090.

Vujovich, G.I., and Kay, S.M., 1998, A Laurentian? Grenville-age oceanic arc/back-arc terrane in the Sierra de Pie de Palo, Western Sierra Papeanas, Argentina, *in* Pankhurst, R.J., and Rapela, C.W., eds., The proto-Andean margin of Gondwana: London, Geological Society Special Publication 142, p. 159–179.

Walker, D., Driese, S.G., and Hatcher, R.D., Jr., 1989, Paleotectonic significance of the quartzite of the Sauratown Mountains window, North Carolina: Geology, v. 17, p. 913–917.

Wehr, F., and Glover, L., III, 1985, Stratigraphy and tectonics of the Virginia–North Carolina Blue Ridge; evolution of a late Proterozoic-early Paleozoic hinge zone: Geological Society of America Bulletin, v. 96, p. 285–295.

Williams, H., 1978, Tectonic lithofacies map of the Appalachian orogen: St. John's Memorial University of Newfoundland Map no. 1, scale 1:1,000,000.

Williams, H., and Hiscott, R.N., 1987, Definition of the Iapetus rift-drift transition in western Newfoundland: Geology, v. 15, p. 1044–1047.

Wortman, G.L., Samson, S.D., and Hibbard, J.P., 1996, Discrimination of the Milton belt and the Carolina terrane in the southern Appalachians: A Nd isotopic approach: Journal of Geology, v. 104, p. 239–247.

Manuscript Accepted by the Society August 25, 2003

Geological Society of America
Memoir 197
2004

Tectonic evolution of the northern Blue Ridge massif, Virginia and Maryland

William C. Burton*
Scott Southworth
U.S. Geological Survey, MS 926A, National Center, Reston, Virginia 20192, USA

ABSTRACT

Detailed mapping in the Mesoproterozoic northern Blue Ridge massif has delineated ten high-grade metamorphic map units of mostly granitic composition occurring in association with charnockite and paragneiss. U-Pb isotopic dating of zircons from these rocks defines three episodes of protolith intrusion during the interval from 1150 to 1055 Ma. Crosscutting relationships, structural analysis of foliations and lineations, and comparison of deformational fabrics of gneisses with different protolith ages indicate three episodes of Grenvillian deformation under differing stress regimes: D1, post–1140 Ma and pre–1120 Ma, involving regional coaxial compression; D2, post–1055 Ma and pre–1035 (?) Ma, noncoaxial ductile shear; and D3, post–1035 (?) Ma and pre–1030 (?) Ma, late-stage compression. The Short Hill fault, a Paleozoic structure of possibly Grenville-age origin, separates predominantly older gneisses with D1 foliation to the west from predominantly younger gneisses with D2 and D3 structures to the east. Geologic features in the central Blue Ridge massif that are similar to those in the northern Blue Ridge include a central fault zone, the Rockfish Valley fault zone, separating a more charnockitic suite of rocks to the west from a more leucogranitic suite to the east. We propose a model in which the Short Hill fault and the Rockfish Valley fault zone were formerly part of a single discontinuity that was offset by a younger Neoproterozoic normal fault during early rifting of the Laurentian margin. This rift-related normal fault became the locus of intrusion of the 735–702 Ma Robertson River batholith.

Keywords: Blue Ridge, Grenville, Mesoproterozoic, basement, Appalachians, Neoproterozoic

INTRODUCTION

The Mesoproterozoic rocks of the northern Blue Ridge anticlinorium in Virginia and Maryland have in the last fifteen years been the subject of considerable scrutiny in the form of detailed mapping coordinated with geochronologic studies. This project was prompted by a need to better understand bedrock and surficial units and how they relate to the distribution of soil types and groundwater resources. Previously, most of the crystalline rocks in the core of the anticlinorium had not been lithologically subdivided, preventing a better understanding of the Grenville-age history of this part of Laurentia. In this paper we present a model for the tectonic evolution of the Mesoproterozoic rocks and the Grenville-age deformation of the northern Blue Ridge. We build

*E-mail: bburton@usgs.gov.

Burton, W.C., and Southworth, S., 2004, Tectonic evolution of the northern Blue Ridge massif, Virginia and Maryland, *in* Tollo, R.P., Corriveau, L., McLelland, J., and Bartholomew, M.J., eds., Proterozoic tectonic evolution of the Grenville orogen in North America: Boulder, Colorado, Geological Society of America Memoir 197, p. 477–493. For permission to copy, contact editing@geosociety.org.

on the findings summarized in Aleinikoff et al. (2000) and Southworth et al. (2000), with an emphasis on structural relationships and field evidence for multiple episodes of magmatism and deformation. We compare our results with the recent work in the north-central Blue Ridge (Tollo et al., this volume) and discuss possible relationships to rocks in the central Blue Ridge massif.

GEOLOGIC SETTING OF THE NORTHERN BLUE RIDGE

The Mesoproterozoic rocks of the northern Blue Ridge province are located in the core of a north-northeast–trending, west-vergent anticlinorium that measures over 300 km in length (Fig. 1, A). These rocks, consisting of orthogneisses and paragneisses and shown as lithologic unit Yu in Figure 1, B, were intruded and metamorphosed under granulite-facies conditions during the 1.2–1.0 Ga Grenville orogeny (Tilton et al., 1960; Lukert, 1982; Clarke, 1984; Sinha and Bartholomew, 1984; Aleinikoff et al., 2000; Tollo et al., this volume), and intruded by the 735–700 Ma anorogenic Robertson River Igneous Suite (Tollo and Aleinikoff, 1996) (Fig. 1, B, Zrr). In the late Neoproterozoic, the Laurentian margin underwent extension and rifting, locally producing small, isolated half-grabens filled with clastic sediments (Schwab, 1974; Wehr, 1985; Espenshade, 1986; Kline et al., 1991; Fig. 1, B, Zu). As the continent rifted to form the western margin of the Iapetan ocean, sedimentation was followed by dike swarms that fed tholeiitic flood basalts and rhyolitic eruptions that now constitute the ca. 570 Ma Catoctin Formation (Fig. 1, B, Zc; Aleinikoff et al., 1995). Clastic sedimentation along the developing passive margin continued into the Early Cambrian with the deposition of Chilhowee Group quartzose sediments (Fig. 1, B, €c). The Cambrian rocks and underlying late Neoproterozoic volcanic rocks now form the western and eastern limbs of the Blue Ridge anticlinorium. Chilhowee sedimentation was followed by deposition of shelf carbonate and deeper-water shales (Fig. 1, B, O€u) that underlie the Shenandoah Valley and the Valley and Ridge, located west of the Blue Ridge. Orogenesis in the Middle to Late Paleozoic produced the Blue Ridge anticlinorium and a penetrative east-dipping cleavage oriented parallel to the axial plane of the main structure (Cloos, 1947; Mitra and Elliott, 1980; Evans, 1989; Kunk and Burton, 1999), culminating with tectonic transport of Blue Ridge rocks westward along the North Mountain thrust fault (west of the area shown in Fig. 1, B) during the Permo-Carboniferous Alleghanian orogeny (Harris, 1979; Evans, 1989). This was followed by an early Mesozoic rifting event that produced the sedimentary and igneous rocks of the Culpeper and Gettysburg basins (Fig. 1, B, JTRu), located east of the Blue Ridge anticlinorium (Froelich and Olsen, 1985).

As part of this long tectonic history, the Mesoproterozoic rocks of the northern Blue Ridge massif were significantly affected by two post-Grenvillian events. The first event involved intrusion into the massif of the Catoctin feeder dikes, which in this area formed a swarm of north-northeast–trending Neoproterozoic diabase and (rare) rhyolite dikes. Individual dikes average only a few meters in width, and therefore are not shown on the generalized geologic maps in Figures 1 and 2. Nonetheless, they occupy a significant percentage of the total rock volume, ranging from ~20% in the southern part of the study area (Espenshade, 1986) to as much as 60% north of the Potomac River (Burton et al., 1995; Burton and Southworth, 1996). The second event was the Paleozoic deformation that produced the anticlinorium. This deformation occurred under lower greenschist-facies metamorphic conditions and resulted in an overprinting of high-grade Grenvillian fabrics and mineral assemblages by closely spaced, penetrative Paleozoic cleavage and associated retrograde mineralization (Cloos, 1947; Mitra and Elliott, 1980; Boyer and Mitra, 1988). The Paleozoic structural and metamorphic overprint varies in intensity across the massif, and includes localized shear zones, widespread alteration of feldspars, breakdown of biotite and hornblende into chlorite and epidote, and strained and recrystallized quartz grains. The same metamorphic event caused recrystallization of the diabase dikes, producing amphibole + epidote + chlorite + albite mineral assemblages. Although the overprinting effects of Neoproterozoic magmatism and Paleozoic deformation are readily distinguished in outcrop, they obscure the number of Grenvillian features that are available for interpretation in the relatively sparse exposures of the region.

The study area is located at the north end of the Mesoproterozoic basement core of the Blue Ridge anticlinorium and extends from the Loudoun County–Fauquier County (Virginia) line northward to the termination of exposed basement in Maryland (Fig. 2). The Mesoproterozoic rocks are bounded by ridge-forming Late Neoproterozoic to Cambrian cover-sequence rocks (Fig. 2, €Z), which constitute the east (Catoctin Mountain) and west (Blue Ridge–Elk Ridge) limbs of the anticlinorium, as well as a down-faulted inlier (Short Hill Mountain–South Mountain). The Mesoproterozoic map units are the same as those shown on the geologic map of Loudoun County, Virginia (Southworth et al., 2000), except that the nongranitoid rocks (paragneiss, metanorite, and quartzite) are here lumped together for clarity. The orthogneisses are divided into three chronological groups based on U-Pb zircon dating (Aleinikoff et al., 2000), and approximate ages are shown. Mesoproterozoic foliation and lineations are shown with generalized dips and plunges.

The development of the Paleozoic cleavage and metamorphic retrogression becomes generally more pronounced, and the

Figure 1. Geographic and geologic setting of the study area. (A) Location of the study area relative to other areas of Grenville-age rocks in the eastern United States, and to the study area of Bartholomew (1981) and Bartholomew et al. (1991). (B) Generalized geologic map of the northern Blue Ridge anticlinorium showing the study area, adapted from Berg et al. (1980), Cleaves et al. (1968), Davis et al. (2001), and Virginia Division of Mineral Resources (1993). RVFZ—Rockfish Valley fault zone; SHF—Short Hill fault.

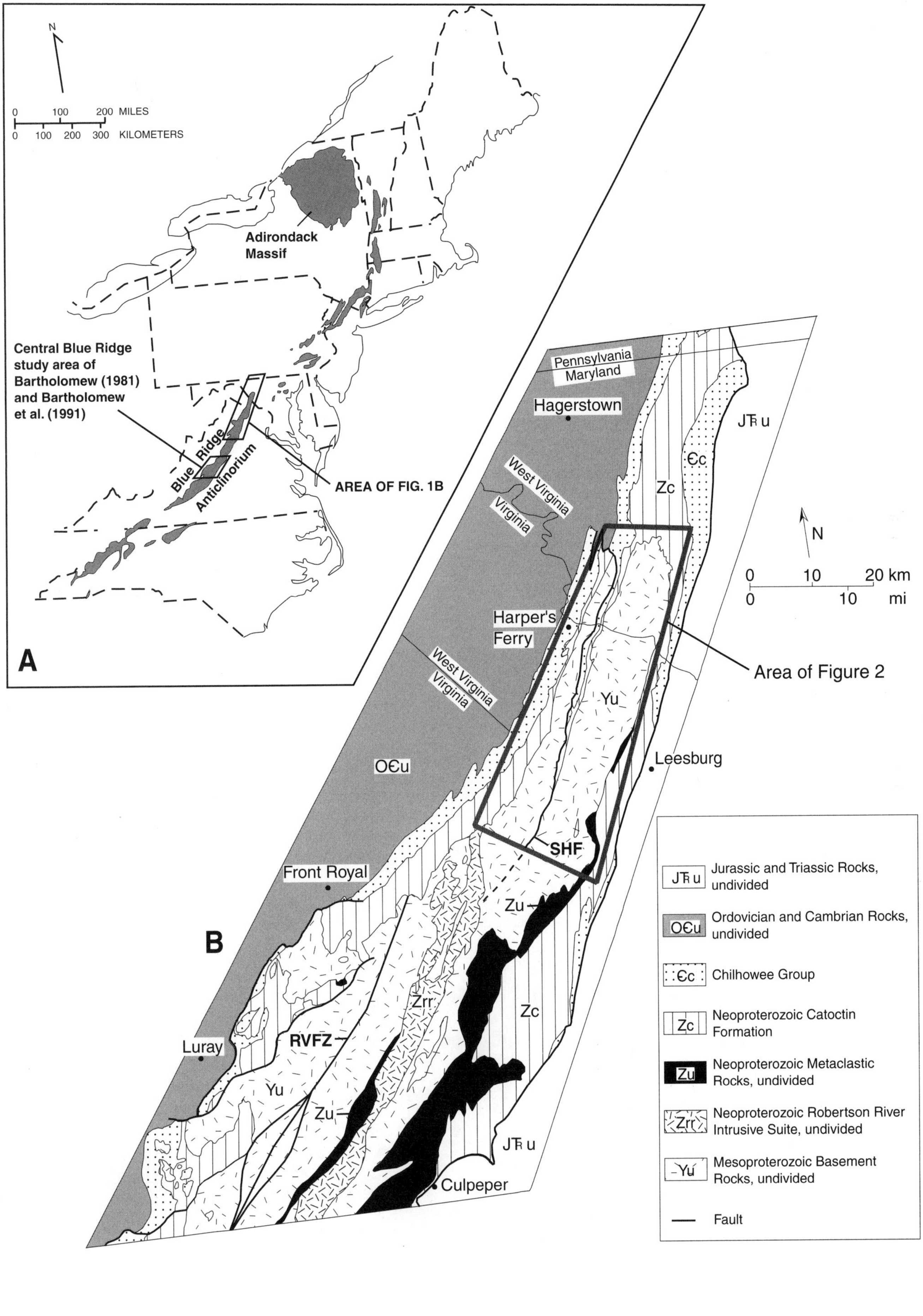

N
0 100 200 MILES
0 100 200 300 KILOMETERS
Adirondack Massif
Central Blue Ridge study area of Bartholomew (1981) and Bartholomew et al. (1991)
Blue Ridge Anticlinorium
AREA OF FIG. 1B
A
Pennsylvania
Maryland
Hagerstown
JTR u
Єc
Zc
West Virginia
Virginia
N
0 10 20 km
0 10 mi
Harper's Ferry
Area of Figure 2
West Virginia
Virginia
Yu
Leesburg
OЄu
SHF
Front Royal
Zu
B
Zrr
Zc
RVFZ
Luray
Yu
Zu
JTR u
Culpeper
JTR u Jurassic and Triassic Rocks, undivided
OЄu Ordovician and Cambrian Rocks, undivided
Єc Chilhowee Group
Zc Neoproterozoic Catoctin Formation
Zu Neoproterozoic Metaclastic Rocks, undivided
Zrr Neoproterozoic Robertson River Intrusive Suite, undivided
Yu Mesoproterozoic Basement Rocks, undivided
Fault

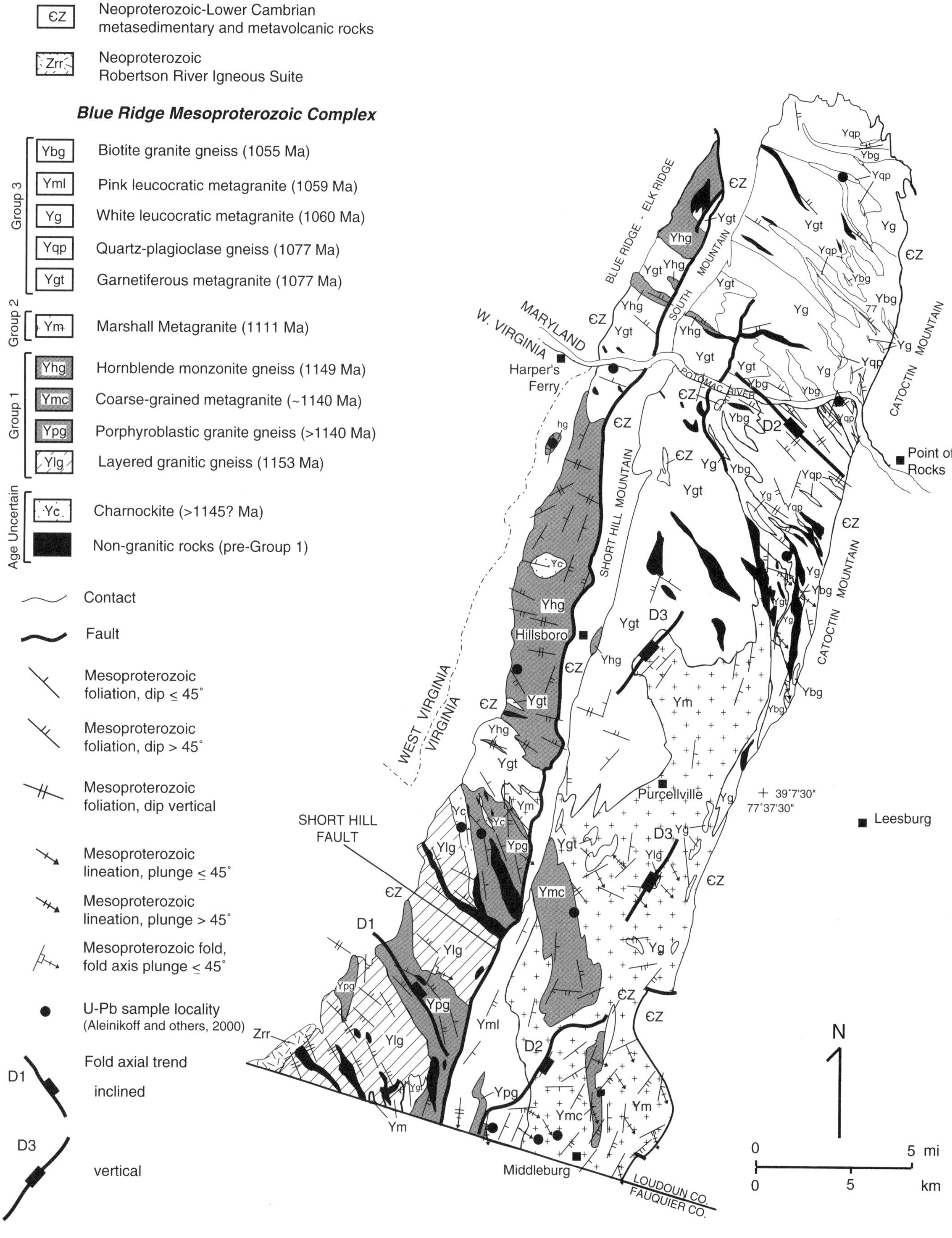
€Z Neoproterozoic-Lower Cambrian metasedimentary and metavolcanic rocks
Zrr Neoproterozoic Robertson River Igneous Suite
Blue Ridge Mesoproterozoic Complex
Group 3
Ybg Biotite granite gneiss (1055 Ma)
Yml Pink leucocratic metagranite (1059 Ma)
Yg White leucocratic metagranite (1060 Ma)
Yqp Quartz-plagioclase gneiss (1077 Ma)
Ygt Garnetiferous metagranite (1077 Ma)
Group 2
Ym Marshall Metagranite (1111 Ma)
Group 1
Yhg Hornblende monzonite gneiss (1149 Ma)
Ymc Coarse-grained metagranite (~1140 Ma)
Ypg Porphyroblastic granite gneiss (>1140 Ma)
Ylg Layered granitic gneiss (1153 Ma)
Age Uncertain
Yc Charnockite (>1145? Ma)
Non-granitic rocks (pre-Group 1)
Contact
Fault
Mesoproterozoic foliation, dip ≤ 45°
Mesoproterozoic foliation, dip > 45°
Mesoproterozoic foliation, dip vertical
Mesoproterozoic lineation, plunge ≤ 45°
Mesoproterozoic lineation, plunge > 45°
Mesoproterozoic fold, fold axis plunge ≤ 45°
U-Pb sample locality (Aleinikoff and others, 2000)
Fold axial trend
D1 inclined
D3 vertical
BLUE RIDGE - ELK RIDGE
SOUTH MOUNTAIN
MARYLAND
W. VIRGINIA
Harper's Ferry
POTOMAC RIVER
CATOCTIN MOUNTAIN
Point of Rocks
SHORT HILL MOUNTAIN
Hillsboro
WEST VIRGINIA
VIRGINIA
SHORT HILL FAULT
Purcellville
39°7'30"
77°37'30"
Leesburg
Middleburg
LOUDOUN CO.
FAUQUIER CO.
N
0 5 mi
0 5 km

task of recognizing Grenvillian features more difficult, from south to north in the field area. There are also zones of concentrated shear and faulting related to Paleozoic deformation, the most significant of which is the Short Hill fault, on the west flank of Short Hill Mountain and South Mountain (Fig. 2). The Short Hill fault defines the west limb of a down-folded syncline of cover-sequence rocks, and follows contacts between Mesoproterozoic units south of the cover rocks (Nickelsen, 1956; Howard, 1991; Southworth and Brezinski, 1996a; Southworth et al., 2000).

LITHOLOGIC CHARACTER OF THE NORTHERN BLUE RIDGE MASSIF

Mesoproterozoic basement rocks of the northern Blue Ridge can be grouped into three categories on the basis of lithology that are also significant with respect to U-Pb geochronology: granitoids of known age, charnockite of uncertain age, and nongranitoid rocks of unknown age (Fig. 2). The nongranitoid rocks are lumped together (Fig. 2, solid black area) because of their relative scarcity in the Blue Ridge and their significance as rocks that predate the widespread granitic intrusive activity. The charnockite bodies are distinguished from the other granitoid rocks because they are generally indicative of high-grade metamorphic conditions, and, as discussed later, their distribution in the Blue Ridge massif has tectonic implications. All of these rocks were subjected to hornblende–granulite-facies metamorphic conditions (Howard, 1991; Southworth et al., 2000).

The nongranitic rocks consist of three dominant lithologies: (1) well-layered garnet + graphite quartzofeldspathic paragneiss; (2) massive hornblende + orthopyroxene + plagioclase metanorite; and (3) quartzite and quartz-rich schist. The garnet + graphite gneiss is rusty-weathering, produces a distinctive light brown soil, and has alternating centimeter-scale garnet-rich and quartzofeldspathic layers; garnets are commonly retrograded to green lensoid clots of fine-grained chlorite and muscovite. Small flakes of graphite are disseminated through the rock, which closely resembles the Border Gneiss of Sinha and Bartholomew (1984) in the central Blue Ridge. The metanorite is gray-weathering, medium- to coarse-grained, and massive to weakly foliated. It contains, in increasing order of abundance, hornblende, orthopyroxene, and plagioclase; a less common dioritic variant contains biotite, hornblende, and plagioclase. Quartzite is white and fine-grained, contains rounded zircons, and has a composite tectonic fabric of both northwest-trending (Grenville-age) and northeast-trending (Paleozoic) elements. Because of the composite fabric, its origin is unclear; however, the rounded zircons, thin lenses of graphite, and mappable pods of garnet-graphite paragneiss suggest a sedimentary origin. In general, the nongranitic rocks occur as isolated, elongate to lensoid bodies within granitic rocks (Fig. 2), which is consistent with an interpretation of their protoliths as screens of sedimentary and volcanic-plutonic country rock within younger plutonic rocks (Burton and Southworth, 1993; Burton et al., 1995).

As was recognized by Jonas (1928) and by Jonas and Stose (1938), the greatest volume by far of Mesoproterozoic rock in the Blue Ridge massif is foliated rock of granitic composition. As a result of our detailed mapping (Southworth et al., 2000), we recognize nine granitic units and one tonalitic unit (Fig. 2). Those units field-termed metagranite typically contain less than 10% modal ferromagnesian minerals (biotite or hornblende) and are less well foliated than those called gneiss, which generally have more than 10% ferromagnesian minerals. With the exception of layered granitic gneiss (Fig. 2, Ylg) the gneisses in the study area lack a strong compositional layering or banding, but all have a Mesoproterozoic tectonic foliation and fit the term *gneiss* (Spry, 1969). The formally named Marshall Metagranite (Jonas, 1928) (Fig. 2, Ym) is weakly to moderately foliated, with 0–5% modal biotite. Taken together, the granitic rocks range in modal composition from syenogranite to granodiorite (Fig. 3, A) and in normative composition from granite to quartz monzonite (Fig. 3, B). Quartz-plagioclase gneiss (Yqp) corresponds modally to tonalite (Fig. 3, A) and falls within the expanded field for normative trondhjemite of Barker (1979; Fig. 3, B). In terms of aluminum saturation, the granitic rocks are peraluminous in composition with the exception of hornblende monzonite gneiss (Yhg), which is metaluminous (Fig. 3, C).

With two possible exceptions, quartz-plagioclase gneiss and layered granitic gneiss, all of the granitoid rocks are considered plutonic in origin. Quartz-plagioclase gneiss (Yqp) may be the by-product of melting of an amphibolitic protolith (Burton and Southworth, 1996), and layered granitic gneiss (Ylg) may be a remobilized felsic metavolcanic rock, based on its fine grain size, well-developed compositional layering, and presence of spherical zircons interpreted as detrital (Aleinikoff et al., 2000).

Charnockite with the mineralogy hornblende + orthopyroxene + microcline + quartz-plagioclase occurs as a few small, isolated bodies in the western part of the northern Blue Ridge massif (Fig. 2, Yc). Modal analyses of two samples of this unit indicate monzogranite and quartz monzonite compositions (Fig. 3, A) equivalent to farsundite and quartz mangerite, respectively, according to Le Maitre et al. (1989). The charnockite is massive to locally strongly foliated, very poorly exposed, and typically found as float boulders that have thick orange-weathering rinds and dark green, fresh surfaces. One mapped body of charnockite is podiform, whereas the others are elongate and appear concordant with surrounding contacts. Due to poor exposure, the nature of the contact relations of the charnockite with adjacent lithologies is unclear. This relative paucity of charnockite contrasts with the geology farther south in the Blue Ridge massif, where charnockite is much more abundant (Bartholomew

Figure 2. Geologic map of the northern Blue Ridge massif, Virginia and Maryland, showing lithologic units, grouped by U-Pb age (Aleinikoff et al., 2000), and Mesoproterozoic foliations and lineations with generalized dips and plunges. Geology adapted from Burton and Southworth (1996) and Southworth et al. (2000).

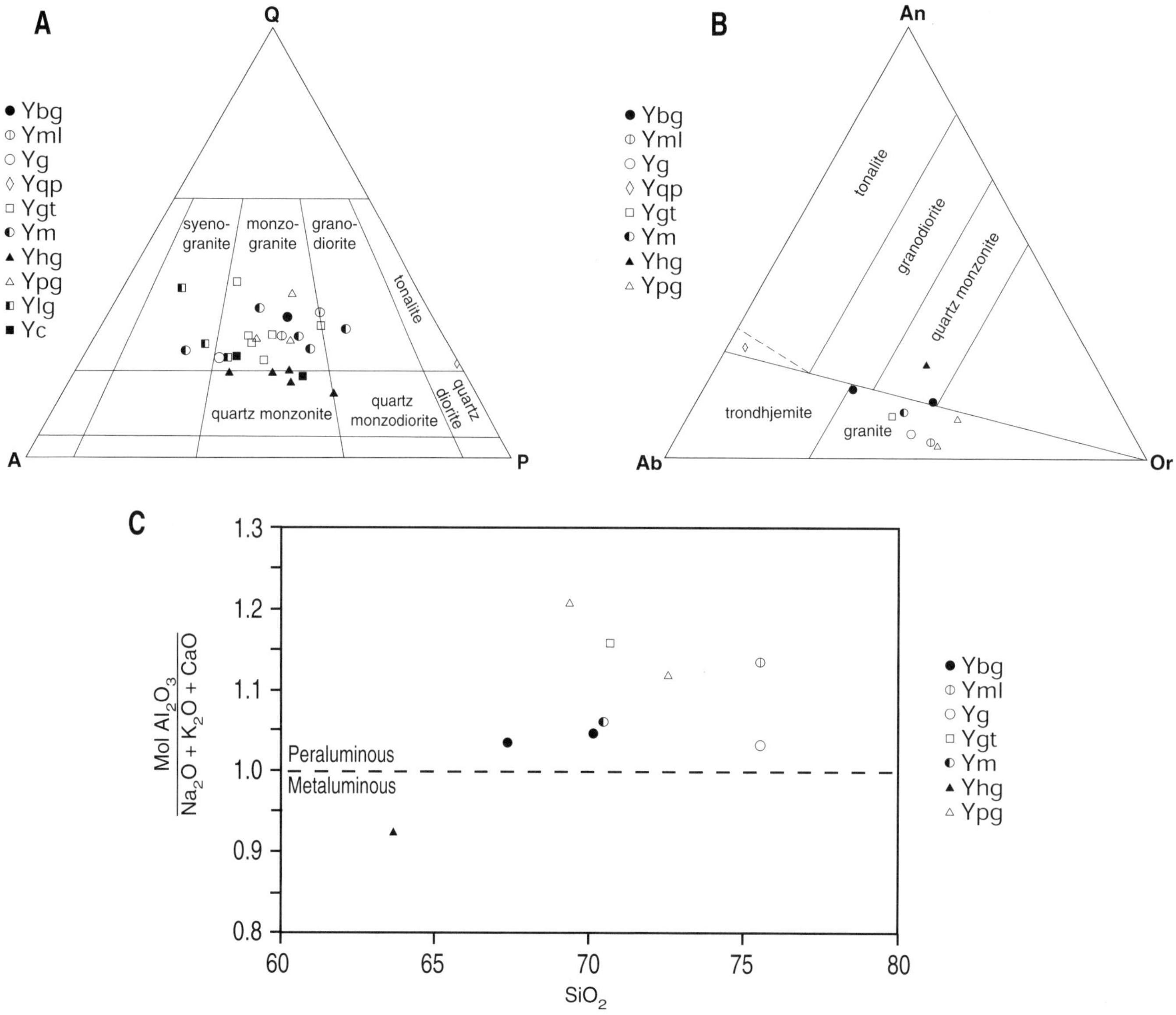

Figure 3. Compositional diagrams for Mesoproterozoic granitic gneisses of the northern Blue Ridge massif. (A) Modal quartz-alkali feldspar-plagioclase plot (Streckeisen, 1976); Q—quartz; A—alkali feldspar; P—plagioclase. (B) Normative feldspar plot (O'Connor, 1965); An—anorthite; Ab—albite; Or—orthoclase. Dashed line shows expanded trondhjemite field of Barker (1979). (C) Plot of SiO_2 (weight percent) versus aluminum saturation index, showing fields for peraluminous and metaluminous granites. Unit symbols include Ybg—biotite granite gneiss; Yml—pink leucocratic metagranite; Yg—white leucocratic metagranite; Yqp—quartz-plagioclase gneiss; Ygt—garnetiferous metagranite; Ym—Marshall Metagranite; Yhg—hornblende monzonite gneiss; Ypg—porphyroblastic granite gneiss; Ylg—layered granitic gneiss; and Yc—charnockite.

and Lewis, 1984; Virginia Division of Mineral Resources, 1993; Tollo et al., this volume).

TEMPORAL AND SPATIAL DISTRIBUTION OF PROTOLITHS

Mesoproterozoic rock units were defined in the field based on composition, appearance, and texture. Our mapping was followed by a comprehensive program of geochronologic analysis in which the eleven map units thought to be metaigneous in origin (Fig. 2) were sampled for zircon and monazite for U-Pb isotopic dating (Aleinikoff et al., 2000). U-Pb ages were obtained from zircon using both thermal ionization mass spectrometry (TIMS) and sensitive high-resolution ion microprobe (SHRIMP) techniques, and the majority of samples produced concordant data or data that yielded reliable concordia intercepts (Table 3 *in* Aleinikoff et al., 2000). The ages, interpreted as times of intrusion of the plutonic protoliths to the gneisses, define three main groups, which we interpret to represent intrusive events during the Grenville orogeny: Group 1, 1150–1140 Ma; Group 2, 1110 Ma; and Group 3, 1075–1055 Ma (Aleinikoff et al., 2000). Groups 1 and 3 are each represented by several map units, whereas Group 2 is represented solely by the Marshall Metagranite (Fig. 2, Ym). Charnockite, the metaigneous rock type with the greatest uncertainty in age, may be as old as or older than Group 1 rocks. The possibly volcanic protolith for layered granitic gneiss (Ylg) is thought to be ca. 1153 Ma (Aleinikoff et al., 2000; sample locality is south of the area shown in Fig. 2). The nongranitic rocks, interpreted as screens of preintrusive country rock, are therefore inferred to be pre–Group 1 in age (Fig. 2).

Aleinikoff et al. (2000) pointed out the similarity of these ages to those of dated magmatic events in the Adirondack Highlands, and McLelland et al. (2001) drew a direct correlation between intrusive events in the two regions in terms of both timing and plutonic composition. Correlative activity includes Group 1 intrusion and emplacement of the anorthosite-mangerite-charnockite-granite (AMCG) plutonic suite in the Adirondacks, Group 2 (Marshall Metagranite) and the Hawkeye granitic suite, and Group 3 and the Lyon Mountain Granite Gneiss suite. McLelland et al. (1996) suggested that AMCG magmatism occurred during delamination of overthickened crust and lithosphere following the Elzevirian orogeny (1350–1185 Ma). McLelland et al. (2001) link the intrusion of the Hawkeye granitic suite to crustal thinning related to formation of the Mid-continent Rift, and intrusion of the Lyon Mountain Granite Gneiss suite to crustal collapse following the Ottawan orogeny.

The distribution of orthogneisses in the Blue Ridge shows a spatial segregation of the different age groups with respect to the Short Hill fault, such that Group 1 rocks occur mostly west of the fault and Group 2 and Group 3 rocks mostly east (Fig. 2). Screens of pre–Group 1 country (nongranitic) rock occur on both sides of the fault within orthogneisses of all three age groups. Exceptions to this orthogneiss spatial pattern include several bodies of Group 1 coarse-grained metagranite (Ymc), porphyroblastic granite gneiss (Ypg), and hornblende monzonite gneiss (Yhg) east of the Short Hill fault, and bodies of Group 2 Marshall Metagranite (Ym) and Group 3 garnetiferous metagranite (Ygt) west of it. These minor exceptions notwithstanding, age groupings of metaigneous rocks in the northernmost Blue Ridge are effectively separated by the Short Hill fault, which mapping indicates also truncates the along-strike outcrop pattern of many lithologic units. The tectonic implications of this are discussed in a later section.

MESOPROTEROZOIC STRUCTURAL FRAMEWORK

Regional Pattern of Foliation, Lineations, and Minor Folds

All basement rocks of the northern Blue Ridge have Grenville-age foliation, including those orthogneisses with the youngest U-Pb ages (1055–1060 Ma; Fig. 2). Foliation in paragneisses is defined by compositional layering that is commonly supplemented by the presence of thin, concordant bodies of aplite and pegmatite. In orthogneisses foliation is expressed by aligned ferromagnesian minerals (biotite and, less commonly, hornblende), planar aggregates of quartz and feldspar, and aplite. Granoblastic textures and triple-junction grain boundaries are ubiquitous in the quartzofeldspathic rocks, consistent with high metamorphic grade and pervasive recrystallization during formation of foliation. The mapped belts of nongranitic rock both east and west of the fault are concordant to foliation in adjacent orthogneiss. All of the foliation is interpreted as tectonic in origin.

In the Group 1 rocks located west of the Short Hill fault, the foliation is typically northwest-striking and vertical to northeast-dipping (Figs. 2 and 4, A). Northwest-striking, steeply dipping foliation is also typical of the Group 3 rocks in the northeast part of the study area (Fig. 2). Widespread northwest-striking Mesoproterozoic foliation has also been mapped in the central Blue Ridge by Bartholomew et al. (1991), in areas where it could be distinguished from the northeast-trending Paleozoic tectonic overprint.

Mesoproterozoic foliation is more variable in orientation, trending both northwest and northeast, in the central and southeast portion of the map area, a region east of the Short Hill fault dominated by Group 2 rocks (Marshall Metagranite) and lesser Group 1 and Group 3 rocks (Figs. 2 and 4, B). In this area, the Mesoproterozoic foliation is locally at high angles to mapped orthogneiss contacts, suggesting development of foliation following intrusion and establishment of initial protolith contacts.

In many areas of Group 2 and Group 3 rocks, particularly in the Marshall Metagranite (Fig. 2, Ym), foliation is accompanied by a lineation that is composed of elongate platy aggregates of biotite and rodded quartz and feldspar (Fig. 5, A and B). The quartzofeldspathic rods are composed of granoblastic grain aggregates that show no internal evidence of crystal lattice strain, indicating recovery under high temperatures (Passchier et al., 1990). Accompanying the lineation in some orthogneiss outcrops are aplite dikes exhibiting ductile minor folds. The hinges of these folds are aligned parallel to the lineation (Fig. 4, B), and their axial surfaces are parallel to foliation (Fig. 5, B and C). The collinear fold hinges and lineations mostly plunge southeast, independent of the strike of the foliation (Figs. 2 and 4, B). The geometry of these structures is similar to that of sheath folds, and the similarity suggests a cogenetic, tectonic origin under conditions of noncoaxial (simple) shear.

Field Evidence for Multiple Episodes of Deformation

Intrusive, crosscutting relations between two distinct types of foliated granitic rock that correlate with recognized map units are exposed in a number of outcrops in the northern Blue Ridge. In all such cases, the intrusive relation agrees with the U-Pb age of the intrusive protolith determined for those rocks. The most

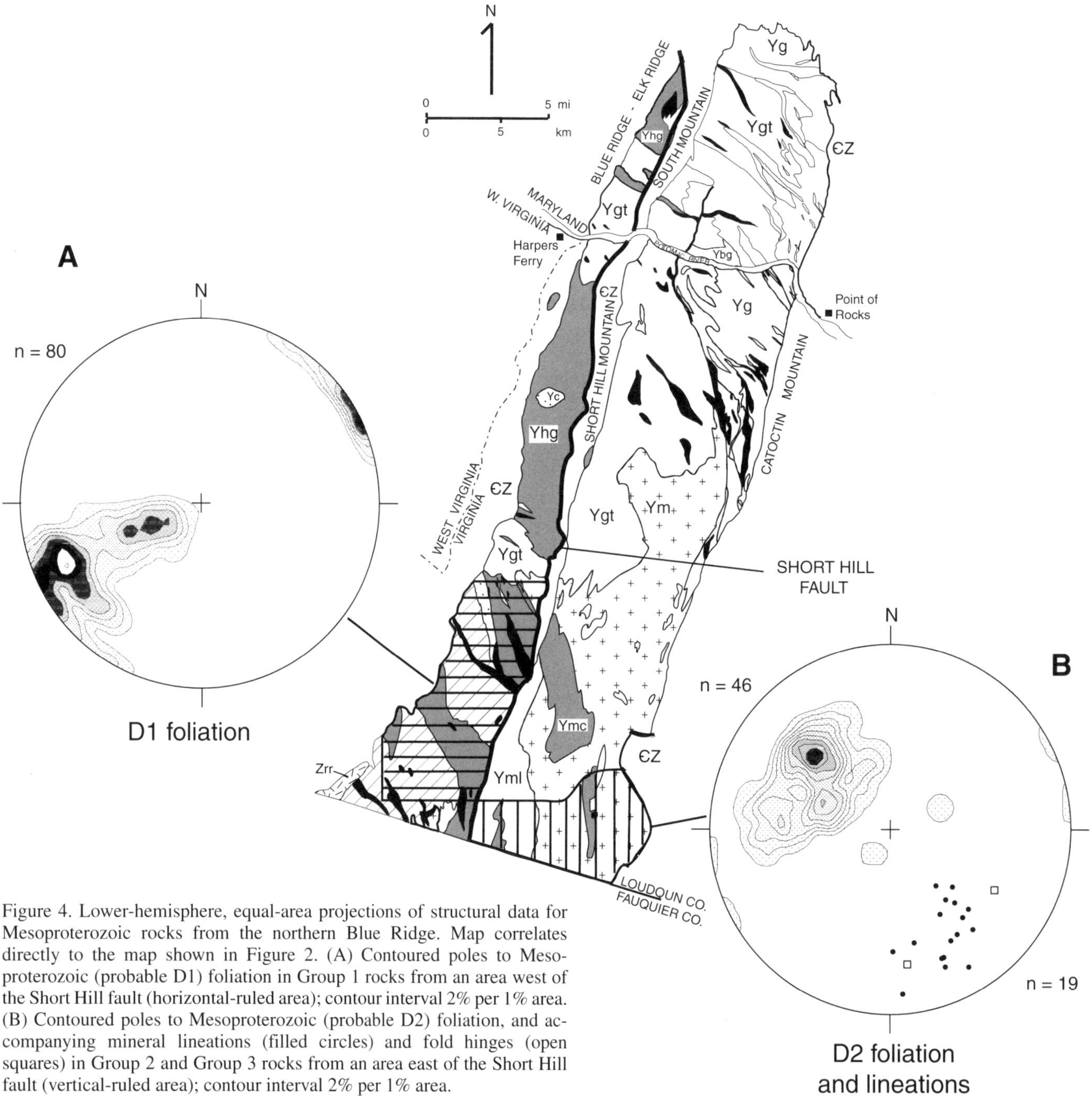

Figure 4. Lower-hemisphere, equal-area projections of structural data for Mesoproterozoic rocks from the northern Blue Ridge. Map correlates directly to the map shown in Figure 2. (A) Contoured poles to Mesoproterozoic (probable D1) foliation in Group 1 rocks from an area west of the Short Hill fault (horizontal-ruled area); contour interval 2% per 1% area. (B) Contoured poles to Mesoproterozoic (probable D2) foliation, and accompanying mineral lineations (filled circles) and fold hinges (open squares) in Group 2 and Group 3 rocks from an area east of the Short Hill fault (vertical-ruled area); contour interval 2% per 1% area.

common example of this is Group 3 garnetiferous metagranite (Fig. 2, Ygt) intruding Group 1 porphyroblastic granite gneiss (Fig. 2, Ypg). At perhaps the best-exposed locality (Fig. 5, F), Ypg with a prominent northwest-trending foliation is intruded by a dike of Ygt with a weak northeasterly foliation. Assuming a tectonic origin for both generations of foliation, this relationship implies that there were at least two episodes of deformation: a D1 producing northwest-striking foliation that is post–Group 1 and pre–Group 3 in age, and a D2 producing northeast-striking foliation that is post–Group 3 in age. In other localities, dikes and sills of garnetiferous metagranite (Ygt) and Marshall Metagranite (Ym) intrude porphyroblastic granite gneiss (Ypg). The existence of two Grenville-age deformational events is inferred even in exposures where a younger foliation cannot be recognized in the crosscutting rock, because the youngest metaplutonic rocks mapped in the study area all locally display a Grenville-age tectonic foliation.

The existence of two regional tectonic events, with distinct deformational styles, is also suggested in map-scale patterns of foliation and lineations in a way that is reminiscent of outcrop

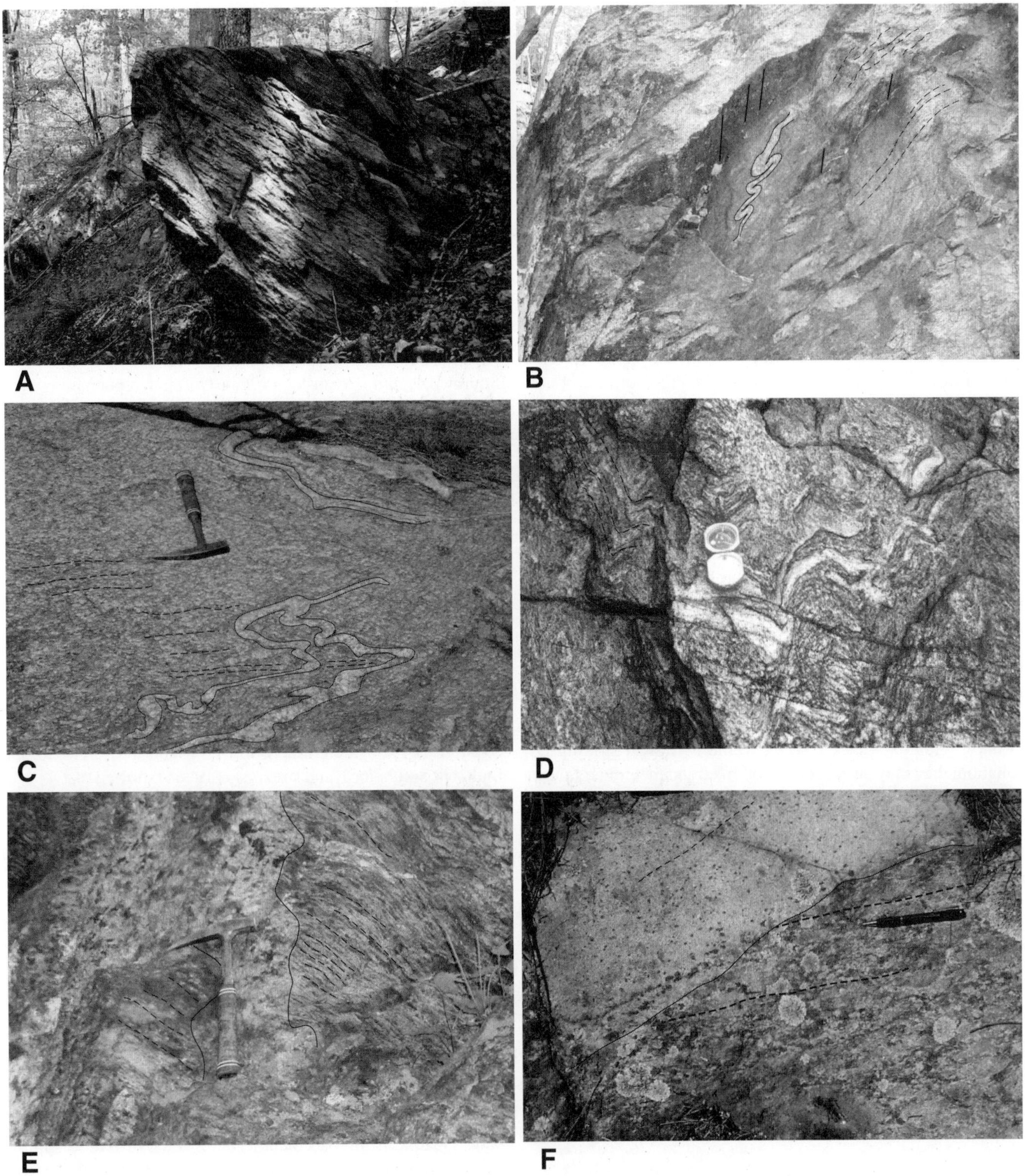

Figure 5. Mesoproterozoic structural fabrics in the northern Blue Ridge. (A) Well-developed D2 foliation (dashed lines) in Group 2 Marshall Metagranite (Fig. 2, Ym), view to the north; more steeply dipping joint set is Paleozoic cleavage. Hammer, for scale, is 33 cm in length. (B) View to the southeast of same outcrop, looking up at overhang, and showing D2 foliation (dashed lines), downdip-plunging mineral lineation (heavy lines), and collinear minor folds in aplite dike (traced). View is 2.5 m across. (C) Recumbent D2 folds in aplite dikes (traced) in Marshall Metagranite, with axial planar D2 foliation (dashed lines); apparent refolded fold in lower aplite dike may be a product of the original intrusive geometry of the dike. Hammer, for scale, is 33 cm in length. (D) 30 m south of 5C outcrop; north-trending D3 upright folds plicate D2 foliation (dashed lines). Open compass, for scale, is 22 cm in length. (E) 30 m southeast of 5A outcrop; undeformed pegmatite cuts D2 foliation (dashed lines). Hammer, for scale, is 33 cm in length. (F) D1 foliation (dashed lines) in Group 1 porphyroblastic granite gneiss (Fig. 2, Ypg), truncated by contact (solid line) with Group 3 garnetiferous metagranite (Fig. 2, Ygt) with D3 foliation (single dashed line). Pen, for scale, is 13.5 cm in length.

patterns. Group 1 intrusive rocks located west of the Short Hill fault, with their predominantly northwesterly structural trends, appear to be truncated by the Group 2 and Group 3 rocks east of the fault, with their variably oriented, northwest- to northeast-striking foliation and coeval down-plunging lineations and fold axes (Figs. 2 and 4). We suggest that the northwesterly trend in the older rocks west of the fault represents the map-scale expression of D1, an episode of northeast-southwest (present orientation) compression or pure shear. In contrast, the variably oriented foliation and the plunging lineations and minor folds in Group 2 and Group 3 rocks east of the fault (Fig. 5, B and C) suggest a D2 event in these younger metaplutonic rocks developed under a different stress regime. We propose a model for the tectonic style of D2, shown in Figure 6, that involves up-from-the-southeast (present orientation) simple shear and simultaneously produced the variable foliations and the plunging folds and lineations, in a fashion similar to that for sheath or tubular folds (Skjernaa, 1989), but at a map scale. The variation of foliation strike and dip over the map area is represented schematically in Figure 6 by a single foliation surface; in reality such a continuous folded surface exists only at outcrop scale, in the form of limbs of minor folds. The granitic protoliths for the Group 2 and Group 3 gneisses were probably still hot and ductile when this deformation occurred. The Group 1 gneisses west of the Short Hill fault had a preexisting high-grade foliation at an orientation unfavorable to the younger stress direction, and would have been less affected. However, the isolated bodies of Group 1 orthogneiss east of the fault that are enveloped by Group 2 and Group 3 granitoids lack a recognizable northwest-striking D1 foliation; they may have been remobilized and their older fabrics destroyed during later intrusion and D2 deformation.

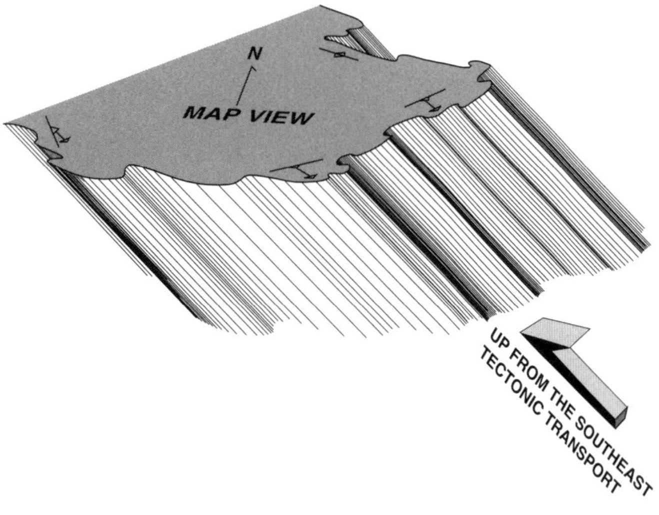

Figure 6. Conceptual diagram illustrating generation of Mesoproterozoic D2 structure through noncoaxial shear in the northern Blue Ridge massif. Hypothetical map-scale folded foliation surface outlines southeast plunging folds and contains collinear mineral lineations. Strike and dip symbols with arrows are map-surface D2 foliations, with arrows representing lineations or axes of minor folds.

Open to tight, upright folds in well-foliated gneisses east of the fault indicate a younger deformation (D3) that is likely associated with late-stage southeast-northwest compression (Fig. 5, D). Where exposed, these folds trend northeast, and are responsible for some map-scale reversals in foliation dip (Fig. 2). Foliation in Group 1 rocks in the western belt that deviates from the D1 northwesterly trend, such as the northeast-striking foliations in the wide belt of hornblende monzonite gneiss (Fig. 2, Yhg), may have been reoriented by D3 folding; alternatively, it could have been the result of Paleozoic deformation that produced a nappelike fold in both basement and cover-sequence rocks nearby that exposes a window of hornblende monzonite gneiss (Yhg) 5 km south of Harper's Ferry (Fig. 2; Nickelsen, 1956; Southworth and Brezinski, 1996a). Unfoliated pegmatite dikes up to a few meters thick that locally intrude strongly foliated gneiss (Fig. 5, E) can be found in all of the Mesoproterozoic orthogneisses, suggesting that at least some pegmatite emplacement postdated the latest plutonic and tectonic activity.

TECTONIC SIGNIFICANCE OF THE SHORT HILL FAULT

The east-dipping Short Hill fault defines the basement-cover contact on the west flank of Short Hill Mountain in Virginia and South Mountain in Maryland (Cloos, 1951; Nickelsen, 1956), and south of there follows the western contact of Group 3 pink leucocratic metagranite (Fig. 2, Yml). It may extend farther, but has not yet been mapped south of the study area. Nickelsen (1956) considered the Short Hill fault a Triassic normal fault, but Southworth and Brezinski (1996b) showed that it probably originated as a synsedimentary normal fault of early Paleozoic age that was reactivated during Late Paleozoic contraction as a reverse or thrust fault. Southworth and Brezinski (1996b) estimate that the net offset along the Short Hill fault in the Paleozoic was ~2 km in a normal sense.

Parker (1968), Bartholomew and Lewis (1984), and Howard (1991) recognized that Mesoproterozoic basement rocks west and east of the Short Hill fault were different and possibly reflected different crustal levels. Parker (1968) suggested that coarse granitoids with two-inch-diameter feldspar phenocrysts and zones of garnets in the rocks west of the fault suggested a higher level of metamorphism and a deeper crustal level relative to rocks east of the fault. Howard (1991) further speculated that the fault potentially marks a significant tectonic boundary between contrasting basement terranes if the differences in the rocks (large garnets in the west and mostly biotite in the east) were not reflective of just protolithic composition.

As discussed earlier, our mapping (Fig. 2) has demonstrated that the Short Hill fault separates contrasting lithologies and tectonic styles. We have mapped coarse-grained and garnet-bearing rocks on both sides of the fault, and so cannot use those criteria as evidence that the rocks record different crustal levels. However, extensive orthopyroxene-bearing granitic rocks, in the form of hypersthene-bearing hornblende monzonite gneiss

(Yhg) and charnockite (Yc), are found only west of the Short Hill fault (Fig. 2). We suggest that the greater abundance of orthopyroxene-bearing granitic rocks west of the fault may reflect a deeper crustal level. In a study of an orthopyroxene-bearing calc-alkaline pluton in Wyoming, Frost et al. (2000) found that magmatic fluid composition played an important role in the crystallization of orthopyroxene, but that crustal depth was also critical, with a difference in depth of ~10 km enough to control the presence or absence of orthopyroxene.

In summary, the evidence suggests that contrasts in lithologies and deformational fabrics west and east of the Short Hill fault may represent differences in crustal levels, but that these differences cannot be explained solely by the estimated 2 km of Paleozoic displacement along the fault (Southworth and Brezinski, 1996b). If the two areas do represent different crustal levels and there is a discontinuity separating them, it must be pre-Paleozoic in origin. One possibility would be a post-D1, pre-D2 Mesoproterozoic extensional fault occupying approximately the same position as the later Short Hill fault. The fault would have served as a conduit for intrusion of Group 2 and Group 3 granitoids, and would have accommodated the shearing during D2 deformation that would otherwise have more strongly affected the western rocks. Another possibility would be a Neoproterozoic extensional fault that was subsequently reactivated, the two episodes of fault movement combined producing the crustal offset. Alternatively, there could be no significant difference in crustal depth west and east of the Short Hill fault, the fault might be no older than Paleozoic, and it might have merely exploited intrusive contacts between Group 2 and Group 3 granitoids and the older (Group 1) rocks.

SUMMARY OF GRENVILLE-AGE INTRUSIVE AND TECTONIC EVENTS

Integrated evidence from field relations and U-Pb isotopic dating of zircon and monazite in the northern Blue Ridge massif suggest a chronology of intrusion and deformation events (Fig. 7). U-Pb ages of zircons indicate that an episode of plutonism (Group 1 event) occurred between ca. 1150 and 1140 Ma (Aleinikoff et al., 2000). Field evidence indicates that a D1 deformational event characterized by northeast-southwest compression (present orientation) occurred following Group 1 intrusion and before Group 2 intrusion. Based on isotopic data available, the Group 2 intrusive episode spanned a short period ca. 1110 Ma and was characterized by emplacement of the Marshall Metagranite. There is no evidence for a deformational event in the interval between Group 2 and Group 3 intrusion events (ca. 1110–1080 Ma). Group 3 intrusive activity started just after 1080 Ma and continued until 1055 Ma (Aleinikoff et al., 2000). A second deformational event, D2, followed Group 3 intrusion, producing variably oriented foliation and plunging folds and lineations, and was followed by D3, a relatively minor event producing local, open to tight outcrop-scale folds and broad map-scale folds.

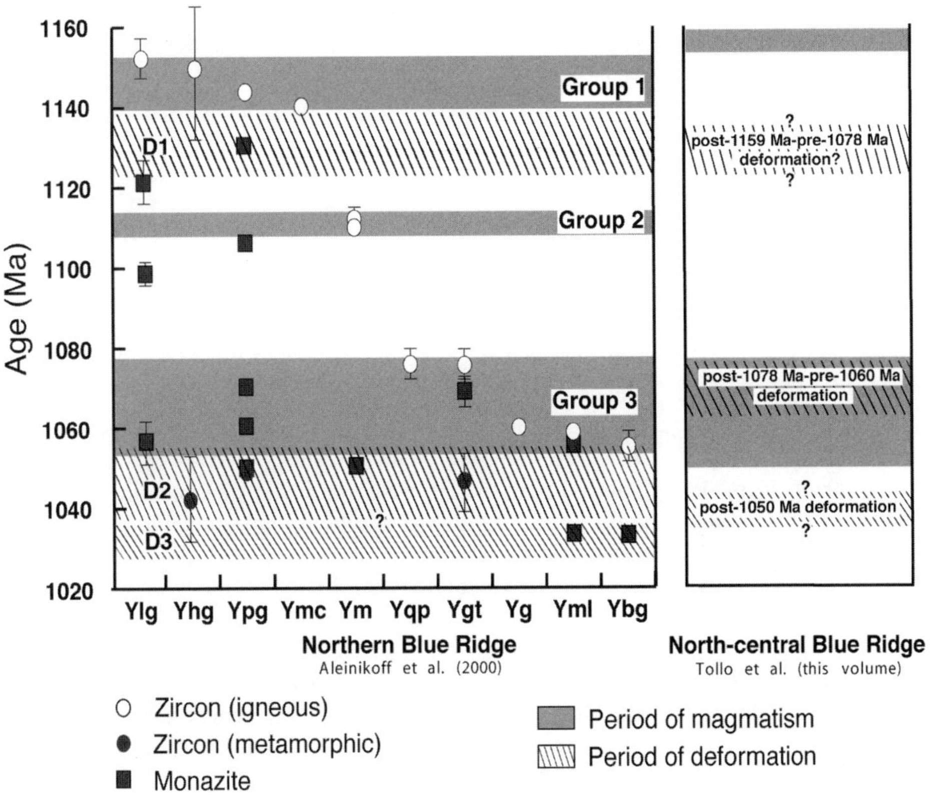

Figure 7. Chronology of intrusion (shaded zones) and deformation (hachured zones) in the northern Blue Ridge massif compared to the north-central Blue Ridge massif, based on zircon and monazite U-Pb data and relative ages of tectonic fabrics. Error bars indicate U-Pb age uncertainties greater than 2 Ma.

Recent field work and U-Pb analysis of zircons and monazites in lithologic units by Tollo et al. (this volume) in the western Blue Ridge massif, ~50 km southwest of our study area, also presents a picture of multiple generations of magmatism and deformation (Fig. 7). These workers obtained a U-Pb zircon age of 1159 ± 14 Ma from a high-silica charnockite, which corresponds to our Group 1 intrusive event; an age of 1078 ± 9 Ma from leucogranite gneiss, which corresponds to our older Group 3 event (Group 3A of Aleinikoff et al., 2000); and ages of 1060 ± 5 Ma and 1050 ± 8 Ma from Old Rag Granite and low-silica charnockite, respectively, which correspond to our younger Group 3 (Group 3B of Aleinikoff et al., 2000) event. They found no evidence for magmatism coincident with our Group 2 episode, the ca. 1110 Ma intrusion of the Marshall metagranite (Fig. 7).

A tectonic foliation is well developed in 1159 Ma high-silica charnockite in the study area of Tollo et al. (this volume), and also in 1078 Ma leucogranite gneiss, both of which are cut by massive to weakly foliated 1060 Ma Old Rag Granite and 1050 Ma low-silica charnockite. The post–1078 Ma, pre–1060 Ma foliation in leucogranite gneiss indicates an episode of deformation slightly older than our D2 (Fig. 7), and signifies the culmination of the Ottawan orogeny (Tollo et al., this volume). The age of the foliation in the 1159 Ma rock is unclear, because foliated, high-silica charnockite and foliated, 1078 Ma leucogranite gneiss are not in contact with each other (Tollo et al., this volume). It is possible that the foliation in the older rock predates that in the younger rock, and perhaps corresponds to the D1 deformational episode of Aleinikoff et al., 2000 (Fig. 7). Weakly developed foliation in the youngest intrusive rocks (Tollo et al., this volume) must be post–1050–1060 Ma in age and probably corresponds to our D2 or D3 deformational event (Fig. 7). Well-developed foliation in biotite gneiss country rock that is intruded by Old Rag Granite and low-silica charnockite must be older than 1060 Ma (Tollo et al., this volume).

U-Pb isotope analyses of monazite (Fig. 7) record the time of cooling through thermal closure at 750 °C (Parrish and Whitehouse, 1999). Two aspects of the monazite data are especially noteworthy: (1) some single samples yield multiple monazite ages, and (2) some of the monazite ages tend to cluster around the time spans for intrusive activity defined by the zircon ages (Aleinikoff et al., 2000) (Fig. 7). The fact that monazite grains with older ages coexist with grains with younger ages indicates that the monazite isotopic systems were not all reset by a single Late Grenville–age thermal event, and that subsequent magmatic and metamorphic activity probably did not produce regional temperatures exceeding 750 °C. The cluster of monazite ages ca. 1030 Ma represents the last high-grade thermal activity in the northern Blue Ridge detected by Aleinikoff et al. (2000), ascribed here to the D3 deformational event (Fig. 7). However, Tollo et al. (this volume) note evidence of younger thermal disturbances, in the form of zircon-rim and monazite ages that range from ca. 1028 to 980 Ma.

COMPARISON WITH THE CENTRAL BLUE RIDGE MASSIF

Until recently, the only detailed mapping-related studies of Grenville-age basement rocks in the Blue Ridge anticlinorium had been conducted by workers in central Virginia, including Bartholomew (1977), Bartholomew et al. (1981), Sinha and Bartholomew (1984), Herz and Force (1984, 1987), Evans (1991), and Bailey and Simpson (1993). Bartholomew et al. (1981) proposed that the central Blue Ridge basement consists of two massifs, the Pedlar massif to the west and the Lovingston massif to the east, separated by the northeast-trending Rockfish Valley fault zone (Fig. 1). The Pedlar massif is dominated by massive charnockite and also contains layered garnet + hypersthene–bearing granulite gneisses that were derived from a volcanic or volcaniclastic protolith during Grenvillian high-grade metamorphism, and subsequently remobilized to produce the charnockite (Hughes et al., 1997). The Lovingston massif contains lesser amounts of charnockite and is composed mostly of gneisses of granitic composition, including the Stage Road Layered Gneiss, which has previously been considered volcaniclastic in origin (Sinha and Bartholomew, 1984). Based on textural and mineralogical differences, including the abundance of Grenville-age garnet in the Pedlar massif and its relative scarcity in the Lovingston, Bartholomew et al. (1981) suggested that the Pedlar massif represents a deeper crustal level and a higher grade of metamorphism than the Lovingston, and that significant translation must have occurred along the Rockfish Valley fault zone to juxtapose the two terranes.

In contrast, Evans (1991) proposed that the mineralogical differences between the two terranes, in particular the predominance of pyroxene in the Pedlar and biotite in the Lovingston, can be largely explained by hydration of the eastern portion of the central Blue Ridge basement during an amphibolite-facies metamorphic event that accompanied post-Grenville, pre-Taconian passive-margin thinning and heating of the crust. The metamorphism preferentially replaced pyroxene-bearing mineral assemblages with biotite and other hydrous phases in the eastern (Lovingston) part of the massif, while the western (Pedlar) portion remained unaltered. This model postulates that during the Grenville orogeny there was no large difference in crustal level between the eastern and the western portions of the central Blue Ridge, and that the Rockfish Valley fault zone did not have the degree of vertical offset previously ascribed to it. The model implies that, in those rocks in which only a Grenville-age fabric is recognized, Grenville-age high-grade mineral assemblages were replaced in situ by younger lower-grade assemblages without the development of a new fabric.

The results of our study provide some new insight into this debate. The northernmost extent of the Rockfish Valley fault zone, as depicted on the geologic map of Virginia (Virginia Division of Mineral Resources, 1993), appears to have an en-echelon relationship to the Short Hill fault to the northeast, with the linear belt of Robertson River intrusions in between (Fig. 1, B). Howard (1991) discussed the general similarities between the Short Hill fault and the Rockfish Valley Fault, and Bartholomew and Lewis (1984, Fig. 6) portrayed the two faults as lateral equivalents. Both the Short Hill fault and the Rockfish Valley fault zone display internal fabrics that indicate only reverse or thrust movement (Bailey and Simpson, 1993; Southworth and Brezinski, 1996b), although the mapped young-on-old-rock relations along the Short Hill fault require an (earlier) extensional component as well (Southworth and Brezinski, 1996b). In terms of lithologic distribution, there is a crude analog in the northern Blue Ridge to the Pedlar and Lovingston terranes and their separation along a fault. All of the recognizable charnockite in the northern Blue Ridge (Fig. 2, Yc) occurs west of the Short Hill fault, as does nearly all of the locally pyroxene-bearing hornblende monzonite gneiss (Fig. 2, Yhg). The hornblende monzonite gneiss displays well-preserved northwest-trending Grenville-age fabric, is locally associated with charnockite, and may be equivalent to the granulite gneisses of the central Blue Ridge. Analogous to the Lovingston terrane in the central Blue Ridge, granitic orthogneisses predominate east of the Short Hill fault, and biotite is the dominant mafic mineral in these gneisses.

However, the biotite-bearing fabrics east of the Short Hill fault are part of syntectonic ductile structures, including tight, plunging folds and mineral lineations, that are consistent with high-grade, Grenville-age metamorphism (Fig. 5, B and C). These fabrics are clearly overprinted by a second-generation planar foliation, trending northeast-southwest and composed of greenschist-facies (biotite-grade) minerals, that is axial-planar to the Blue Ridge anticlinorium and likely Paleozoic in age. There is no evidence for a mica-generating event, intermediate in age between the Grenville-age and Paleozoic events, that is mimetic to the events that produced Grenville-age fabrics; nor is there any petrographic evidence for formerly widespread orthopyroxene-bearing mineral assemblages in the granitoids that were retrograded to biotite-bearing assemblages. $^{40}Ar/^{39}Ar$ cooling ages from biotite and hornblende that define Grenville-age foliation in the northern Blue Ridge basement rocks indicate a Grenville-age (pre–1000 Ma) thermal peak, with no evidence of Neoproterozoic or early Paleozoic resetting of argon systems (Kunk and Burton, 1999). Assuming, therefore, that the analogy between the central and the northern Blue Ridge is valid, Evans's (1991) hypothesis is incorrect, and the mineralogy and fabrics in the eastern portion of the central Blue Ridge that are clearly pre-Paleozoic in age are Grenville in origin. However, analogous to Evans's (1991) suggestion that Paleozoic motion on the Rockfish Valley fault alone cannot explain most of the differences in lithology between the Pedlar and the Lovingston terranes, we propose that lithologic differences and contrasting deformational styles in rocks located on the east and the west sides of the Short Hill fault cannot be explained solely by Paleozoic movement on the fault.

Bailey and Simpson (1993) conducted a structural analysis of tectonic fabrics in basement rocks of the central Blue Ridge that also has implications for the northern Blue Ridge. They found several generations of tectonic fabric and interpreted the following deformational events: (1) Grenville-age compressional deformation that produced high-grade metamorphic foliation; (2) an early phase of post-Grenville extension (northwest-southeast), under at least medium-grade metamorphic conditions, that produced a weak, widespread, low-strain foliation and accompanying lineation; (3) post-Grenville extension (northwest-southeast), under conditions of low metamorphic grade and high, localized strain, that produced shear zones with a normal (southeast-down) sense of motion; and (4) Paleozoic compression (northwest-southeast), under low-grade metamorphic conditions, that produced the Rockfish Valley fault zone and other thrust-related shear zones with a southeast-up sense of motion.

The early, low-strain extensional fabrics contain a ubiquitous southeast-plunging lineation formed from biotite, quartz, and feldspar. Bailey and Simpson (1993) noted the presence of dynamically recrystallized feldspar grains in the lineation fabric as evidence of formation under temperatures of at least 450 °C. Although these fabrics are confined to the area of the Lovingston massif, Bailey and Simpson (1993) did not cite them as evidence for Evans's (1991) model for the formation of Lovingston rocks by hydration and passive-margin thinning. The fabrics bear at least a superficial resemblance to those of our Grenville-age D2 deformation, because both have a biotite-bearing southeast-plunging lineation formed under medium-grade or higher metamorphic conditions, and both clearly postdate an older Grenville-age fabric. However, Bailey and Simpson (1993) reported a southeast-side-down sense of motion on the few kinematic indicators they could find in this fabric, and did not report any folds associated with it, whereas our D2 commonly has plunging folds with hinges that are parallel to the lineations. Despite the differences, we suggest that the early extensional fabric reported by Bailey and Simpson (1993) may in fact be Grenvillian in age. Either it records a compressional event and their few kinematic indicators are erroneous, or our D2 shear-related deformation is in fact extensional rather than compressional, and related to late orogenic crustal collapse.

One of the largest differences between the central and the northern Blue Ridge, according to present geologic interpretation, is the large volume of rock mapped as preintrusive country rock in the central Blue Ridge compared with the northern Blue Ridge. One unit in particular, the Stage Road Layered Gneiss, is widespread in the Lovingston massif and is interpreted as an originally volcaniclastic rock based on pre-Grenvillian Pb^{207}/Pb^{206} ages obtained from rounded zircons that were interpreted to be detrital in origin (Sinha and Bartholomew, 1984). The 1130 Ma age cited by Sinha and Bartholomew (1984) for deposition of the protoliths for both the Stage Road Layered

Gneiss and a granulite gneiss from the Pedlar massif would imply that the central Blue Ridge was at near-surface levels during a time of ductile deformation and intrusion in the northern Blue Ridge (Fig. 7). We present a model below in which the opposite is true: the presently exposed central Blue Ridge represents a deeper crustal level than the northern Blue Ridge. We suggest that a reanalysis of zircons from the Stage Road Layered Gneiss using recently developed analytical techniques, and a reevaluation of its fabric in the field, will permit an intrusive origin for these rocks, as originally interpreted by Bartholomew (1977).

A UNIFYING TECTONIC MODEL FOR THE CENTRAL AND NORTHERN BLUE RIDGE

In comparing the central and the northern portions of the Blue Ridge massif, we have noted some common features. Each area is bisected by a fault zone that may have pre-Paleozoic origins: the Rockfish Valley fault zone in the central Blue Ridge and the Short Hill fault in the northern Blue Ridge. As shown on the Geologic Map of Virginia (Virginia Division of Mineral Resources, 1993), these faults display an en-echelon pattern relative to each other, although both the northern termination of the Rockfish Valley fault zone and the southern termination of the Short Hill fault are presently poorly understood. Each fault separates a comparatively charnockite-rich zone to the west from a charnockite-poor (or absent) biotite granitoid–rich zone to the east. The possible equivalent in the northern Blue Ridge to the Pedlar massif of the central Blue Ridge, the basement rocks west of the Short Hill fault (Figs. 1, B, and 2), has proportionally less charnockite than the Pedlar in the central Blue Ridge and could represent a shallower crustal level, one that is intermediate in depth between the Pedlar and the Lovingston. The distinctive, older, Group 1–dominant rocks of this zone extend southward to the belt of the Neoproterozoic Robertson River batholith (Figs. 1, B, and 2), where they are replaced on the southwest side of the belt by younger (Group 3) granitic rocks, according to Aleinikoff et al. (2000). The Short Hill fault extends at least as far south as the southern end of our study area (Fig. 2), based on existing field work, but Southworth and Brezinski (1996b) suggest that it may extend southwest to the belt of Robertson River rocks (Fig. 1, B).

The Robertson River batholith defines a north-northeast-trending linear belt (Fig. 1, B) that traverses the Blue Ridge anticlinorium from the west limb in the northern Blue Ridge to the east limb in the central Blue Ridge (Robertson River Igneous Suite of Tollo and Lowe, 1994). Plutons forming this multiphase batholith have zircon ages ranging from 735 to 702 Ma (Tollo and Aleinikoff, 1996), and the batholith is thought to have been emplaced during an early phase of rifting of the Laurentian margin because of its highly linear shape, evocative of younger Neoproterozoic rift-related dikes (Bartholomew, 1992), and its A-type granitoid chemistry (Tollo and Aleinikoff, 1996). Bailey and Simpson (1993) noted the geographic proximity of Late Proterozoic alkalic plutons in the central Blue Ridge to extensional shear zones that they considered Late Proterozoic in age because they cut the plutons and are overprinted by Paleozoic thrust-related shear fabrics. They also noted a parallelism between magmatic foliation in the alkalic plutons, the early low-strain extensional fabric in the country rock, and the later high-strain extensional shear zones. Parallelism of fabric in pluton and country rock was also noted by Bailey and Tollo (1998) in their study of the Neoproterozoic alkalic Polly Wright Cove pluton in the central Blue Ridge. Bailey and Simpson (1993) postulated that emplacement of the alkalic plutons occurred during Late Proterozoic stretching of the crust, which was partitioned between low- and high-strain zones, and Bailey and Tollo (1998) similarly concluded that emplacement of the Polly Wright Cove pluton occurred during crustal extension.

Adopting the concept of synextensional plutonism from Bailey and Simpson (1993) and from Bailey and Tollo (1998), and assuming southward extension of the Short Hill fault to the Robertson River batholith, we propose a model in which the northern Blue Ridge massif is a shallower equivalent of the central Blue Ridge that was down-dropped eastward along a normal fault, producing the present-day en-echelon relationship of the two regions (Fig. 8). This fault was also the locus of intrusion by the Robertson River batholith. Its intrusion along an active normal fault would be commensurate with the rift-related environment proposed for it, and would help explain its extremely elongate map pattern, as originally suggested by Bartholomew (1992) and by Bailey and Simpson (1993). Our model explains two major features in the Blue Ridge massif: the elongate nature of the Robertson River batholith, and the spatially offset lithologic and tectonic similarities between the central and the northern Blue Ridge massif (Fig. 8). The model proposes that the Rockfish Valley and Short Hill fault zones were originally part of a single discontinuity that represented the contact between the Pedlar and the Lovingston terranes prior to formation of the crosscutting normal fault associated with the Robertson River batholith. The crosscutting fault down-dropped the northern Blue Ridge to its present position prior to and during intrusion of Robertson River magma, which then was followed by erosion and deposition of a cover sequence of Late Neoproterozoic to Cambrian volcanic and sedimentary rocks.

Testing this proposed model for the central and the northern Blue Ridge will provide a renewed incentive for detailed mapping south of the northern Blue Ridge area we have described in this paper. In particular, we need to investigate the possible southward extension of the Short Hill fault, the nature of the northern termination of the Rockfish Valley fault zone, and the relationship of the two faults to the Robertson River batholith. In addition, the true nature of differences between the Pedlar and the Lovingston terranes, and their possible equivalents in the northern Blue Ridge, need to be examined further in terms of structural and petrologic investigations.

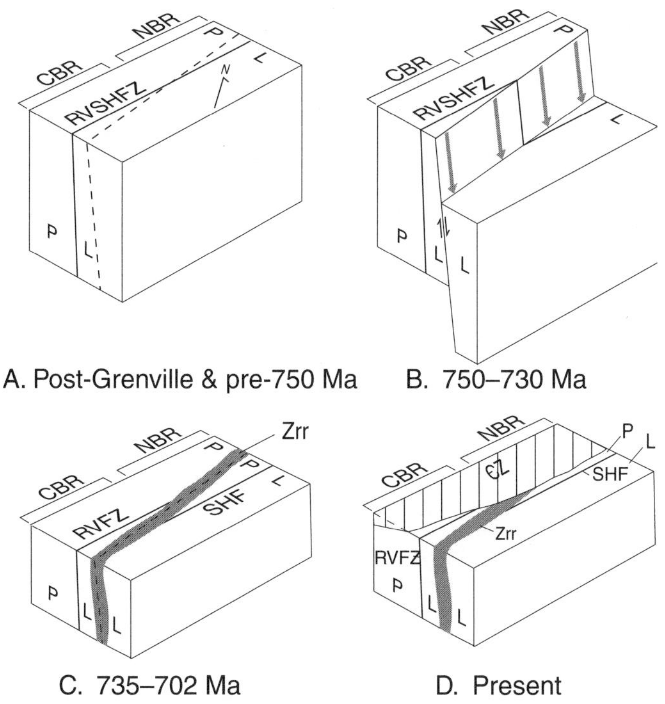

Figure 8. Tectonic model linking the central and the northern Blue Ridge. (A) Post (?)-Grenville and pre–ca. 750 Ma: A single fault zone or discontinuity of uncertain nature, the Rockfish Valley–Short Hill fault zone (RVSHFZ), separates the Pedlar (P) and Lovingston (L) terranes in both the central Blue Ridge (CBR) and the northern Blue Ridge (NBR). Dashed line is trace of later fault. (B) Around 750–730 Ma: The first phase of Neoproterozoic rifting of the Laurentian margin produces a normal fault that offsets the older Rockfish Valley–Short Hill fault zone. (C) 735–702 Ma: The normal fault is the locus of intrusion of the Robertson River batholith (Zrr), separating the older fault into the Rockfish Valley fault zone (RVFZ) segment and the Short Hill fault (SHF) segment. Surface releveled by erosion. (D) Present: Pedlar, Lovingston, and Robertson River rocks partially covered by Cambrian and Late Neoproterozoic cover rocks (€Z). Compare to Figure 1, B.

SUMMARY OF GEOLOGIC EVOLUTION OF THE NORTHERN BLUE RIDGE PROVINCE

In this paper we present and discuss evidence for a sequence of geologic events that have affected the northern Blue Ridge province, summarized as follows. The full extent of these events in the central Blue Ridge is a topic for future investigation.

1. Intrusion of Group 1 granitoids (1150–1140 Ma, Aleinikoff et al., 2000; 1159 Ma, Tollo et al., this volume).
2. D1 deformational event (between 1140 and 1112 Ma) in the northern Blue Ridge (Aleinikoff et al., 2000), and possibly in the north-central Blue Ridge (Tollo et al., this volume), involving coaxial compressional shear.
3. Intrusion of Group 2 Marshall Metagranite (1112 Ma, Aleinikoff et al., 2000).
4. Intrusion of early Group 3 granitoids (1078 Ma, Tollo et al., this volume; 1077 Ma, Aleinikoff et al., 2000).
5. Early D2 deformation (between 1078 and 1060 Ma, Tollo et al., this volume).
6. Intrusion of later Group 3 granitoids (1060–1055 Ma, Aleinikoff et al., 2000; 1060–1050 Ma, Tollo et al., this volume).
7. Development of Grenville-age discontinuity (fault?) between the rocks presently east and west of the Short Hill Fault, and between the Pedlar and the Lovingston terranes in the central Blue Ridge.
8. Later D2 deformation (post–1055 Ma, Aleinikoff et al., 2000; post–1050 Ma, Tollo et al., this volume), involving noncoaxial shear.
9. D3 deformation, involving late Grenville-age compression (post-D2 deformation).
10. Neoproterozoic extension, producing an oblique normal fault and offset between the central and the northern Blue Ridge, intrusion of the Robertson River batholith (Tollo and Aleinikoff, 1996) and other alkalic plutons, development of shear zones in the central Blue Ridge (Bailey and Simpson, 1993), and possibly early movement along the Short Hill fault in the northern Blue Ridge.
11. Early Paleozoic extension along the Short Hill fault (Southworth and Brezinski, 1996b).
12. Late Paleozoic compression, producing thrusting along the Short Hill fault (Southworth and Brezinski, 1996b) and the Rockfish Valley fault zone, followed by folding and development of the axial-planar Blue Ridge–South Mountain cleavage (Mitra and Elliott, 1980; Bailey and Simpson, 1993), and thrusting along the North Mountain thrust fault (Harris, 1979; Evans, 1989).

CONCLUSIONS

In this paper we outline a detailed history of magmatism and deformation for the Mesoproterozoic rocks of the northern Blue Ridge massif. We present evidence for two Grenville-age phases of tectonism that have contrasting deformational styles. We point out differences in Mesoproterozoic lithology, age, and deformational history on opposite sides of the Short Hill fault, and argue that the fault has a long, complex kinematic history extending back to the Grenville orogeny. We point out similarities in lithologic distribution between the northern and the central Blue Ridge, and argue that the Short Hill fault and the Rockfish Valley fault zone played similar tectonic roles in their respective settings. Finally, we present a unifying model for the northern and the central Blue Ridge that explains map-scale features along the length of the massif, including the en-echelon relationship of the Short Hill fault to the Rockfish Valley fault zone, and the linear map pattern of the Robertson River batholith. The model commences with a single fault that was continuous from the central to the northern Blue Ridge and juxtaposed terranes in both regions. During crustal extension in the

Neoproterozoic, the central Blue Ridge was offset from the northern Blue Ridge by movement along a normal fault that trended obliquely across the massif. While still active, this fault became the locus of intrusion by the Robertson River batholith, and the batholith now represents the boundary between the two regions.

By integrating detailed geologic mapping with modern geochronologic dating techniques, we have shown that much can be learned about the Grenville-age tectonic history of the northern Blue Ridge province. We envision application of such an approach along the entire length of the province to investigate longitudinal and latitudinal changes in the orogen, and to gauge the regional extent of the episodes of intrusion and deformation that others and we have documented. In order to test the model we have proposed, we are particularly interested in the possible correlation of central and northern Blue Ridge lithologies and structures, including that of our hornblende monzonite gneiss with the granulite gneisses of the central Blue Ridge; the Marshall Metagranite and younger metaplutonic rocks with rocks of the Lovingston massif; and the Rockfish Valley and Short Hill fault zones with each other and the Robertson River batholith.

ACKNOWLEDGMENTS

This manuscript was considerably improved with the help of R. Tollo, P. Lyttle, G. Petersen, J. Bartholomew, and J. Garihan, and the authors wish to thank them for their careful reviews and suggestions.

REFERENCES CITED

Alenikoff, J.N., Zartman, R.E., Walter, M., Rankin, D.W., Lyttle, P.T., and Burton, W.C., 1995, U-Pb ages of metarhyolites of the Catoctin and Mount Rogers Formations, central and southern Appalachians: Evidence for two pulses of Iapetan rifting: American Journal of Science, v. 295, no. 4, p. 428–454.

Aleinikoff, J.N., Burton, W.C., Lyttle, P.T., Nelson, A.E., and Southworth, C.S., 2000, U-Pb geochronology of zircon and monazite from Mesoproterozoic granitic gneisses of the northern Blue Ridge, Virginia and Maryland, USA: Precambrian Research, v. 99, p. 113–146.

Bailey, C.M., and Simpson, C., 1993, Extensional and contractional deformation in the Blue Ridge Province, Virginia: Geological Society of America Bulletin, v. 105, no. 4, p. 411–422.

Bailey, C.M., and Tollo, R.P., 1998, Late Neoproterozoic extension-related magma emplacement in the central Appalachians: An example from the Polly Wright Cove pluton: Journal of Geology, v. 106, p. 347–359.

Barker, F., 1979, Trondhjemite: Definition, environment, and hypothesis of origin, *in* Barker, F., ed., Trondhjemites, dacites, and related rocks: New York, Elsevier, p. 1–12.

Bartholomew, M.J., 1977, Geology of the Greenfield and Sherando quadrangles, Virginia: Charlottesville, Virginia Division of Mineral Resources Publication 4, 43 p.

Bartholomew, M.J., 1981, Geology of the Roanoke and Stewartsville quadrangle, Virginia: Charlottesville, Virginia Division of Mineral Resources Publication 34, 23 p.

Bartholomew, M.J., 1992, Late Proterozoic continental margin of the Laurentian craton, *in* Bartholomew, M.J., et al., eds., Basement Tectonics 8—Characterization and comparison of ancient and Mesozoic continental margins: Proceedings of the Eighth International Conference on Basement Tectonics held August 1988 in Butte, Montana: Dordrecht, The Netherlands, Kluwer Academic Publishers, p. 443–467.

Bartholomew, M.J., and Lewis, S.E., 1984, Evolution of Grenville massifs in the Blue Ridge geologic province, southern and central Appalachians, *in* Bartholomew, M.J., Force, E.R., Sinha, A.K., and Herz, N., eds., The Grenville event in the Appalachians and related topics: Boulder, Colorado, Geological Society of America Special Paper 194, p. 229–254.

Bartholomew, M.J., Gathright, T.M., II, and Henika, W.S., 1981, A tectonic model for the Blue Ridge in central Virginia: American Journal of Science, v. 281, p. 1164–1183.

Bartholomew, M.J., Lewis, S.E., Hughes, S.S., Badger, R.L., and Sinha, A.K., 1991, Tectonic history of the Blue Ridge basement and its cover, central Virginia, *in* Schultz, A., and Compton-Gooding, E., eds., Geologic evolution of the eastern United States: Martinsville, Virginia Museum of Natural History Guidebook no. 2, p. 57–90.

Berg, T.M., Edmunds, W.E., Geyer, A.R., Glover, A.D., Hoskins, D.M., MacLachlan, D.B., Root, S.E., Sevon, W.D., and Socolow, A.A., 1980, Geologic map of Pennsylvania: Harrisburg, Pennsylvania Department of Environmental Resources, Geological Survey (4th Ser.), scale 1:250,000.

Boyer, S.E., and Mitra, G., 1988, Relations between deformation of crystalline basement and sedimentary cover at the basement/cover transition zone of the Appalachian Blue Ridge Province: Boulder, Colorado, Geological Society of America Special Paper 222, p. 119–136.

Burton, W.C., and Southworth, C.S., 1993, Garnet-graphite paragneiss and other country rocks in granitic Grenvillian basement, Blue Ridge anticlinorium, northern Virginia and Maryland: Geological Society of America Abstracts with Programs, v. 25, no. 4.

Burton, W.C., and Southworth, C.S., 1996, Basement rocks of the Blue Ridge province in Maryland, *in* Brezinski, D.B., and Reger, J.P., eds., Studies in Maryland geology: Baltimore, Maryland Geological Survey Special Publication No. 3, p. 187–203.

Burton, W.C., Froelich, A.J., Pomeroy, J.S., and Lee, K.Y., 1995, Geology of the Waterford quadrangle, VA-MD, and the Virginia portion of the Point of Rocks quadrangle: Reston, Virginia, U.S. Geological Survey Bulletin 2095, 30 p.

Clarke, J.W., 1984, The core of the Blue Ridge anticlinorium in northern Virginia, *in* Bartholomew, M.J., et al., eds. The Grenville event in the Appalachians and related topics: Boulder, Colorado, Geological Society of America Special Paper 194, p. 187–200.

Cleaves, E.T., Edwards, J., Jr., and Glaser, J.D., 1968, Geologic map of Maryland: Baltimore, Maryland Geological Survey, scale 1:250,000.

Cloos, E., 1947, Oolite deformation in the South Mountain fold, Maryland: Geological Society of America Bulletin, v. 56, p. 843–918.

Cloos, E., 1951, Structural geology of Washington County, *in* The physical features of Washington County: Baltimore, Maryland Department of Geology, Mines, and Water Resources, p. 124–163.

Davis, A.M., Southworth, C.S., Reddy, J.F., and Schindler, J.S., 2001, Geologic map database of the Washington DC area featuring data from three 30 × 60–minute quadrangles: Frederick, Washington West, and Fredericksburg: Reston, Virginia, U.S. Geological Survey Open-File Report 01-227, scale 1:100,000.

Espenshade, G.H., 1986, Geology of the Marshall quadrangle, Fauquier County, Virginia: Reston, Virginia, U.S. Geological Survey Bulletin 1560, 60 p., scale 1:24,000.

Evans, M.A., 1989, The structural geometry and evolution of foreland thrust systems, northern Virginia: Geological Society of America Bulletin, v. 101, p. 339–354.

Evans, N.H., 1991, Latest Precambrian to Ordovician metamorphism in the Virginia Blue Ridge: Origin of the contrasting Lovingston and Pedlar basement terranes: American Journal of Science, v. 291, p. 425–452.

Froelich, A.J., and Olsen, P.E., 1985, Newark Supergroup, a revision of the Newark Group in eastern North America, *in* Robinson, G.R., Jr., and Froelich, A.J., eds., Proceedings of the Second U.S. Geological Survey

Workshop on the Early Mesozoic Basins of the Eastern United States, 14–16 November 1984, Reston, Virginia: Reston, Virginia, U.S. Geological Survey Circular 946, p. 1–3.

Frost, B.R., Frost, C.D., Hulsebosch, T.P., and Swapp, S.M., 2000, Origin of the charnockites of the Louis Lake batholith, Wind River Range, Wyoming: Journal of Petrology, v. 41, no. 12, p. 1759–1776.

Harris, L.D., 1979, Similarities between the thick-skinned Blue Ridge anticlinorium and the thin-skinned Powell Valley anticline: Geological Society of America Bulletin, Part 1, v. 90, p. 525–539.

Herz, N., and Force, E.R., 1984, Rock suites in Grenvillian terrane of the Roseland district, Virginia, Part 1: Lithologic relations, *in* Bartholomew, M.J., et al., eds., The Grenville event in the Appalachians and related topics: Boulder, Colorado, Geological Society of America Special Paper 194, p. 187–200.

Herz, N., and Force, E.R., 1987, Geology and mineral deposits of the Roseland district of central Virginia: Reston, Virginia, U.S. Geological Survey Professional Paper 1371, 56 p.

Howard, J.L., 1991, Lithofacies of the Precambrian basement complex in the northernmost Blue Ridge province of Virginia: Southeastern Geology, v. 31, no. 4, p. 191–202.

Hughes, S.S., Lewis, S.E., Bartholomew, M.J., Sinha, A.K., Hudson, T.A., and Herz, N.M., 1997, Chemical diversity and origin of Precambrian charnockitic rocks of the central Pedlar massif, Grenvillian Blue Ridge terrane, Virginia: Precambrian Research, v. 84, p. 37–62.

Jonas, A.I., ed., 1928, Geologic map of Virginia: Charlottesville, Virginia Division of Mineral Resources, scale 1:500,000.

Jonas, A.I., and Stose, G.W., 1938, Geologic map of Frederick County and adjacent parts of Washington and Carroll Counties: Baltimore, Maryland Geological Survey, scale 1:62,500.

Kline, S.W., Lyttle, P.T., and Schindler, J.S., 1991, Late Proterozoic sedimentation and tectonics in northern Virginia, *in* Schultz, A., and Compton-Gooding, E., eds., Geologic evolution of the Eastern United States: Martinsville, Virginia Museum of Natural History Guidebook No. 2, p. 263–294.

Kunk, M.J., and Burton, W.C., 1999, $^{40}Ar/^{39}Ar$ age-spectrum data for amphibole, muscovite, biotite, and K-feldspar samples from metamorphic rocks in the Blue Ridge Anticlinorium, northern Virginia: Reston, Virginia, U.S. Geological Survey Open-File Report 99-552, 111 p.

LeMaitre, R.W., Bateman, P., Dudek, A., Keller, J., Lameyre, J., Le Bas, M.J., Sabine, P.A., Schmid, R., Sorensen, H., Streckeisen, A., Woolley, A.R., and Zanettin, B., 1989, A classification of igneous rocks and glossary of terms: Recommendations of the International Union of Geological Sciences Subcommission of the Systematics of Igneous Rocks: Oxford, England, Blackwell Scientific Publications, 193 p.

Lukert, M.T., 1982, Uranium-lead isotope age of the Old Rag Granite, northern Virginia: American Journal of Science, v. 282, p. 391–398.

McLelland, J., Daly, J.S., and McLelland, J.M., 1996, The Grenville orogenic cycle (ca. 1350–1000 Ma): An Adirondack perspective: Tectonophysics 265, p. 1–28.

McLelland, J., Hamilton, M., Selleck, B., McLelland, J.M., Walker, D., and Orrell, S., 2001, Zircon U-Pb geochronology of the Ottawan Orogeny, Adirondack Highlands, New York: Regional and tectonic implications: Precambrian Research, v. 109, p. 39–72.

Mitra, G., and Elliott, D., 1980, Deformation of basement in the Blue Ridge and the development of the South Mountain cleavage, *in* Wones, D.R., ed., The Caledonides in the USA: Proceedings, International Geological Correlation Program Project 27—Caledonide Orogen, Blacksburg, Virginia, 1979: Blacksburg, Virginia Polytechnic Institute and State University Department of Geological Sciences Memoir 2, p. 307–311.

Nickelsen, R.P., 1956, Geology of the Blue Ridge near Harper's Ferry, West Virginia: Geological Society of America Bulletin, v. 67, p. 230–270.

O'Connor, J.T., 1965, A classification for quartz-rich igneous rocks based on feldspar ratios: Reston, Virginia, U.S. Geological Survey Professional Paper 525-B, p. B79-B84.

Parker, P.E., 1968, Geologic investigation of the Lincoln and Bluemont quadrangles, Virginia: Charlottesville, Virginia Division of Mineral Resources Report of Investigations 14, 21 p.

Parrish, R., and Whitehouse, M., 1999, Constraints on the diffusivity of Pb in monazite, its closure temperature, and its U-Th-Pb systematics in metamorphic terrains, from a TIMS and SIMS study: Journal of Conference Abstracts, v. 4, p. 711.

Passchier, C.W., Myers, J.S., and Kroner, A., 1990, Field geology of high-grade gneiss terrains: Berlin, Springer-Verlag, 150 p.

Schwab, F.L., 1974, Mechum River Formation: Late Precambrian (?) alluvium in the Blue Ridge province of Virginia: Journal of Sedimentary Petrology, v. 44, no. 3, p. 862–871.

Sinha, A.K., and Bartholomew, M.J., 1984, Evolution of the Grenville terrane in the central Virginia Appalachians, *in* Bartholomew, M.J., et al., eds., The Grenville event in the Appalachians and related topics: Boulder, Colorado, Geological Society of America Special Paper 194, p. 175–186.

Skjernaa, L., 1989, Tubular folds and sheath folds: Definitions and conceptual models for their development, with examples from the Grapesvare area, northern Sweden: Journal of Structural Geology, v. 11, no. 6, p. 689–703.

Southworth, C.S., and Brezinski, D.K., 1996a, Geology of the Harper's Ferry quadrangle, Virginia, Maryland, and West Virginia: Reston, Virginia, U.S. Geological Survey Bulletin 2123, 33 p.

Southworth, C.S., and Brezinski, D.K., 1996b, How the Blue Ridge anticlinorium in Virginia becomes the South Mountain anticlinorium in Maryland, *in* Brezinski, D.B., and Reger, J.P., eds., Studies in Maryland geology: Baltimore, Maryland Geological Survey Special Publication No. 3, p. 253–275.

Southworth, C.S., Burton, W.C., Schindler, J.S., and Froelich, A.J., 2000, Digital geologic map of Loudoun County, Virginia: Reston, Virginia, U.S. Geological Survey Open File Report 99-150, scale 1:50,000.

Spry, Alan, 1969, Metamorphic textures: Oxford, England, Pergamon Press, 350 p.

Streckeisen, A.L., 1976, To each plutonic rock its proper name: Earth Science Reviews, v. 12, no. 1., p. 1–33.

Tilton, G.R., Wetherill, G.W., Davis, G.L., and Bass, M.N., 1960, 1000-million-year-old minerals from the eastern United States and Canada: Journal of Geophysical Research, v. 65, p. 4173–4179.

Tollo, R.P., and Aleinikoff, J.N., 1996, Petrology and U-Pb geochronology of the Robertson River Igneous Suite, Blue Ridge province, Virginia: Evidence for multi-stage magmatism associated with an early episode of Laurentian rifting: American Journal of Science, v. 296, p. 1045–1090.

Tollo, R.P., and Lowe, T.K., 1994, Geologic map of the Robertson River Igneous Suite, Blue Ridge province, northern and central Virginia: Reston, Virginia, U.S. Geological Survey Miscellaneous Field Studies Map MF-2229, scale 1:100,000.

Virginia Division of Mineral Resources, 1993, Geologic map of Virginia: Charlottesville, Virginia Division of Mineral Resources, scale 1:500,000.

Wehr, F., 1985, Stratigraphy of the Lynchburg Group and Swift Run Formation, Late Proterozoic (730–750 Ma), central Virginia: Southeastern Geology, v. 25, no. 4, p. 225–239.

Manuscript Accepted by the Society August 25, 2003

Printed in the USA

Geological Society of America
Memoir 197
2004

Mesoproterozoic and Paleoproterozoic crustal growth in the eastern Grenville Province: Nd isotope evidence from the Long Range inlier of the Appalachian orogen

Alan P. Dickin*
School of Geography and Geology, McMaster University, Hamilton, Ontario L8S 4M1, Canada

ABSTRACT

New Nd isotope data are presented for plutonic orthogneisses of the Long Range inlier in western Newfoundland. The inlier represents a fragment of Grenvillian basement located within the Appalachian orogenic belt of eastern Canada. The Nd data suggest that the Long Range is divided into two blocks with different crustal formation ages of ca. 1.75 Ga and 1.55 Ga, interpreted as juvenile arc terranes that were accreted to the Laurentian continent during its Proterozoic evolution. The two blocks are correlated with the Paleoproterozoic and Mesoproterozoic terranes of Labradoria and Quebecia previously recognized in the Grenville Province of the mainland. The younger block in the Long Range is also correlated with the Mesoproterozoic gneisses of the Blair River inlier of Cape Breton Island (Nova Scotia), which has a similar crustal formation age of ca. 1.55 Ga. The two blocks are separated by a boundary that cuts across the Long Range mountains and is correlated with a similar boundary on the mainland, west of Sept Iles. Based upon the sharpness of this boundary, it may be a collisional suture between accreted oceanic arc terranes. Based on published dates for substantial 1.5 Ga plutonism immediately southeast of the Wakeham Group, it is suggested that the terrane boundary probably passes fairly near the north shore of the St. Lawrence River at this point, causing the Labradorian crust in this area to be extensively reworked during the accretion of Quebecia to Labradoria.

Keywords: Nd isotopes, crustal growth, Grenville Province, Appalachian orogen, Newfoundland

INTRODUCTION

The North American shield is a mosaic of crustal terranes of different ages that were accreted together over its geological history. Most of these terranes were generated at convergent plate boundaries, either by subduction-related magmatism or as accretionary prisms derived by erosion of volcanic arcs. Later they were accreted to the cratonic nucleus of North America and often subjected to further magmatic reworking during later crustal accretion events (e.g., see review by Hoffman, 1989).

In order to reconstruct this long history of continental growth it is important to date the time, termed the Crustal Formation Age, when each crustal terrane was first formed in an arc system. Dating this event is often difficult because later geological events tend to overprint the early history of crustal terranes, especially during accretionary orogenic events. This problem is

*E-mail: dickin@mcmaster.ca.

Dickin, A.P., 2004, Mesoproterozoic and Paleoproterozoic crustal growth in the eastern Grenville Province: Nd isotope evidence from the Long Range inlier of the Appalachian orogen, *in* Tollo, R.P., Corriveau, L., McLelland, J., and Bartholomew, M.J., eds., Proterozoic tectonic evolution of the Grenville orogen in North America: Boulder, Colorado, Geological Society of America Memoir 197, p. 495–503. For permission to copy, contact editing@geosociety.org.

particularly severe in the Grenville Province, where intense metamorphism has erased much of the structural evidence that can normally be used to reconstruct a detailed history of geological events. However, because of its resistance to metamorphic resetting, the Sm-Nd isotope method can "see back" through high-grade metamorphic events to record the early history of crustal terranes. Therefore, when supported by igneous crystallization ages from U-Pb dating, Nd isotope analysis provides a powerful technique for determining the crustal formation ages of these terranes.

Based on many previous studies (see review by Dickin, 1995), the depleted mantle model of DePaolo (1981) has been shown to be a reliable indicator of the Nd isotope evolution of the mantle source of orogenic crustal terranes. Therefore, depleted mantle (T_{DM}) ages normally provide a very reliable estimate of the time when a juvenile crustal terrane was first extracted from the mantle source.

A potential source of error in the estimation of crustal formation ages is the occurrence of geological mixing events, such as when an older terrane is partially reworked by younger mantle-derived magmatism (Arndt and Goldstein, 1987). However, if Nd isotope data are supported by other geological evidence, such as igneous crystallization ages based on U-Pb dating, limits can be placed on the magnitude of any mixing processes. For, example, T_{DM} ages typically provide an upper limit for the crustal formation age of a terrane, whereas other techniques, such as U-Pb determination of crystallization ages, provide a minimum age limit for crustal formation. The narrower the bracket between these two ages, the more reliably the formation age of the terrane can be determined.

A good example of the convergence of U-Pb ages and T_{DM} ages is provided by the extension of Archean basement from the Superior Province into the northwestern margin of the Grenville Province. Gneisses of Archean age located southeast of Sudbury yield an average T_{DM} age of 2.73 Ga (Dickin, 1998a), in excellent agreement with the range of U-Pb ages, from 2.70 to 2.75 Ga, in the western Abitibi belt of the Superior Province (e.g., Jackson and Fyon, 1991). Hence, it is concluded that the protoliths of these Grenvillian gneisses originated in juvenile arc terranes of late Archean age, and were rapidly accreted together in the Kenoran orogeny.

MAJOR ACCRETED TERRANES OF THE GRENVILLE BELT

Nd isotope analysis was used by Dickin (2000) to compile a crustal formation age map for most of the exposed Grenville Province, as summarized in Figure 1. This work showed that the province can be interpreted as five large juvenile magmatic arc terranes, along with other areas with a more complex crustal formation history, probably generated in a continental arc setting (e.g., rocks of the Algonquin terrane in Ontario; Nadeau and van Breemen, 1998). The nature of these terranes and the boundaries between them is reviewed briefly next.

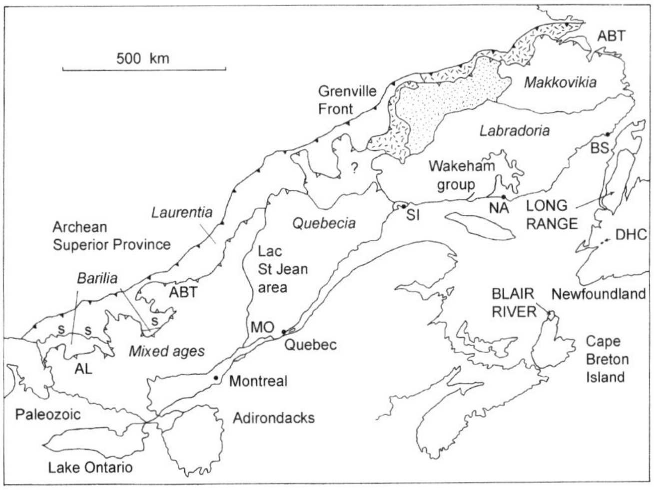

Figure 1. Crustal formation age map of the Grenville Province, after Dickin (2000). Open ticks indicate the location of the Allochthon Boundary Thrust (ABT; Rivers et al., 1989), with the position in western Québec as revised by Dickin and Guo (2001). Coarse stipple—Trans-Labrador batholith; fine stipple—metasedimentary terranes possibly associated with Makkovikia; s—Archean-Proterozoic suture; AL—Algonquin; MO—Montaubon; SI—Sept Iles; NA—Natashquan; BS—Blanc Sablon; DHC—Disappointment Hill complex.

Early geochronological studies showed that the strip of crust inside the northwest margin of the Grenville Province represented the continuation of reworked Archean crust into the Grenville orogenic belt (e.g., see Davidson, 1998, for review). However, the extent of this Archean crust in the Grenville Province was largely unknown until the first Nd isotope mapping work. This allowed a boundary to be drawn between the Laurentian craton and the younger terranes that make up most of the Grenville Province (Dickin and McNutt, 1989). In the French River area near Georgian Bay, this boundary appears to be crosscut by Mesoproterozoic stitching plutons, suggesting that it is a pre-Grenvillian collisional suture (Dickin, 1998b). This "suture" boundary (Fig. 1, s) is exposed between Georgian Bay and the Ontario-Québec border and again northwest of Montreal (Dickin, 2000). In between these two exposures, however, the suture was overridden by Mesoproterozoic rocks of the Allochthonous Polycyclic Belt (Dickin and Guo, 2001), so here the Archean-Proterozoic boundary coincides with the sole thrust of the allochthonous belt, termed the Allochthon Boundary Thrust (ABT) by Rivers et al. (1989).

Nd isotope mapping has revealed the existence of Paleoproterozoic crustal terranes in the Grenville Province of Ontario and Labrador, with formation ages of ca. 1.9 Ga defined by T_{DM} model ages. Minimum ages for these terranes are provided by U-Pb ages of ca. 1.74 Ga in both cases (Scharer and Gower, 1988; Krogh et al., 1992). Also, both of these interpreted terranes (termed Barilia and Makkovikia by Dickin, 2000) were later subjected to intense orogenic or magmatic events that almost completely obscured their original crustal formation history. In the case of eastern Labrador, this magmatic event was caused by the Labradorian orogeny (ca. 1.65 Ga), whose overprinting effects were so strong that many workers (e.g., Scharer, 1991) have not even recognized the existence of an earlier crust-forming event related to the development of the Makkovick Province in the area.

Although Labradorian magmatism caused variable reworking of Makkovikian crust, farther to the south it created a large juvenile arc terrane (termed Labradoria by Dickin, 2000). This has an average crustal formation age (T_{DM} model age) of 1.75 Ga, with a minimum age of 1.65 Ga based on U-Pb dating (Wasteneys et al., 1997). The juvenile Labradorian terrane was itself reworked by a 1.5 Ga magmatic event, termed the Pinwarian orogeny by Tucker and Gower (1994). This event resulted in remelting of older Labradorian-age crust rather than in juvenile crustal formation at 1.5 Ga. However, farther west, a large juvenile arc terrane (Quebecia) is found with an age similar to that of the Pinwarian event, with a maximum age of 1.55 Ga based on Sm-Nd data and a minimum age of 1.45 Ga based on U-Pb dating of the Montauban Group west of Québec (Nadeau and van Breemen, 1994).

It is interesting to note that in the last two cases described the formation age of a juvenile arc terrane (Labradoria and Quebecia) is ~50–100 m.y. older than a magmatic reworking event in the adjacent older terrane. This suggests that these reworking events were actually associated with collisional orogenies that marked the accretion of Labradoria and Quebecia, respectively, to the Laurentian craton. Indeed, the largest magmatic product of the Labradorian orogeny, the Trans-Labrador batholith, exactly corresponds in its extent to the length of the Labradorian terrane (Fig. 1, coarse stipple), suggesting that genesis of the batholith may have been associated with the accretion of this terrane (Dickin, 2000).

In contrast to other areas with a relatively simple pattern of crustal formation ages, an area of complex Nd model age systematics is seen in the southwest Grenville Province (Fig. 1). This area comprises terranes of the Allochthonous Polycyclic Belt and the Allochthonous Monocyclic Belt and was probably an ensialic arc through most of the Mesoproterozoic.

The forgoing discussion has focused on the crustal growth history of the main body of the Grenville Province in Canada. However, important fragments of Grenvillian crust are preserved as basement inliers in the younger Appalachian orogenic belt. The closest of these fragments to the main body of the Grenville Province is the Long Range inlier of western Newfoundland. Therefore, in this paper the crustal formation age structure of this terrane is examined and compared with that of the main body of the province. In addition, a smaller fragment of Grenvillian basement is seen in the Blair River area of Cape Breton Island (Fig. 1). This terrane is discussed in the light of Nd data from the Long Range in the next section.

GEOLOGY OF THE LONG RANGE INLIER

The Long Range inlier consists of a Mesoproterozoic gneiss complex that was extensively intruded by plutons during the Grenville orogeny. The geology has been summarized by Owen (1991), as shown in Figure 2 here. The gneiss complex consists largely of gray gneisses with compositions ranging from quartz diorite through tonalite and granodiorite to granite. The protoliths of these rocks were probably largely plutonic; however, because the rocks are metamorphosed at amphibolite or granulite facies, many original textures have been lost. Therefore, some provenance from volcanic rocks and volcanogenic sediments cannot be excluded (J.V. Owen, personal commun., 2002). Much smaller amounts of pelitic gneiss, quartzite, and mafic gneiss are found in elongate bands up to 1 km across and several kilometers long. The gneiss complex is cut by voluminous plutonic rocks that appear to be dominantly of Grenvillian age, and by a few younger plutons, as well as by the Long Range dike swarm.

The Long Range inlier was recently the subject of a detailed U-Pb dating study (Heaman et al., 2002). This work confirmed the Grenvillian affinity of most of the plutons in the Long Range, including the Lake Michael intrusive suite, the Hooping Harbor satellite pluton, and the Potato Hill pluton (Fig. 2). In contrast, the Taylor Brook gabbro complex was shown to be of Silurian age. Heaman et al. (2002) also dated country rock orthogneisses from three localities (Fig. 2, stars), the first near Western Brook Pond in the southwest and two others from Cat Arm Road in the east. All of these samples gave Grenvillian lower intercept ages, but more variable upper intercept ages.

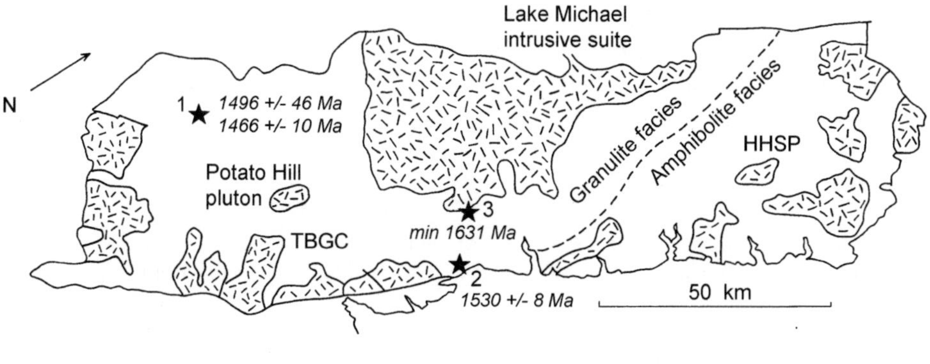

Figure 2. Geological summary map of the Long Range inlier after Owen (1991). Stippled areas represent plutonic rocks intruded into the older gneiss terrane. TBGC—Taylor Brook gabbro complex; HHSP—Hooping Harbor satellite pluton. Stars mark three localities where rocks of the gneisses terrane were dated by U-Pb zircon analysis (Heaman et al., 2002).

In the case of the Western Brook Pond sample and one of the Cat Arm samples, Heaman et al. (2002) obtained Pinwarian upper intercept ages by regressing through the Grenville lower intercept. The Cat Arm sample (Fig. 2, #2) gave a precise age of 1530 ± 8 Ma, whereas the Western Brook Pond data (#1) were somewhat more scattered. A best fit to three discordant points gave an age of 1466 ± 10 Ma, whereas a regression through four discordant points gave an age of 1496 ± 46 Ma (all errors 2σ). For the other Cat Arm sample (#3) the discordant zircons were much more scattered, and no best-fit regression age could be calculated. However, the oldest 207/206 age (1631 Ma, regressed through the origin) represents a minimum protolith age.

The new Pinwarian U-Pb ages are consistent with previously published U-Pb data for the Disappointment Hill gneiss complex south of Grand Lake (Fig. 1, DHC). This is a small Grenvillian fragment which is believed to be closely related to the Long Range inlier, but lies 75 km southwest of the Long Range itself. This complex gives an age of 1498 +9/–8 Ma (Currie et al., 1992), which was interpreted as the age of granulite-facies metamorphism but may also be close to the age of the protolith, interpreted as a granitoid plutonic rock.

A reconnaissance Sm-Nd study of the Long Range was reported by Fryer et al. (1992) as part of a regional investigation of the nature of the crust under Newfoundland. The original data were not reported, but ε Nd values were calculated at 430 Ma for comparison with the initial ratios of Caledonian plutonic rocks in central and eastern Newfoundland. Most of the samples came from Grenville-age (ca. 1020 Ma) plutonic rocks, including the Potato Hill pluton (two samples) and the Lake Michael intrusive suite (five samples). In addition, two samples of the gneiss complex were analyzed from the area between the two plutons. The results were somewhat variable, with ε Nd values of from –5 to –12 at 430 Ma.

The objective of the present study was to map the crustal formation age structure of the Long Range gneiss complex in order to see whether the Nd isotope variations observed by Fryer

TABLE 1. SM-ND DATA FOR GNEISSIC ROCKS FROM THE LONG RANGE, NEWFOUNDLAND

Sample number	Grid ref[@]	Q[#]	P[#]	Nd (ppm)	Sm (ppm)	$^{147}Sm/^{144}Nd$	$^{143}Nd/^{144}Nd$	ε Nd 1.50	T_{DM}[§] (Ga)
Labradoria									
1 LR1	WF 158 376	154	–131	20.75	3.73	.1086	.511645	–2.5	2.03
2 LR3	WF 159 413	171	–114	29.20	5.81	.1203	.511873	–0.2	1.91
3 LR4	WF 136 431	150	–150	20.21	3.35	.1002	.511695	0.2	1.81
4 LR6	WF 091 436	183	–74	26.98	4.39	.0984	.511730	1.2	1.73
5 LR58	VF 982 184	142	17	47.14	9.85	.1263	.512043	1.9	1.74
6 LR60*	VF 966 187	147	–19	136.15	27.35	.1214	.512016	2.3	1.69
7 LR59	VF 956 206	150	0	141.45	30.04	.1284	.512058	1.8	1.76
Quebecia									
8 LR51	WF 019 148	173	–99	29.22	6.29	.1302	.512165	3.6	1.60
9 LR52	WF 015 152	n.d.	n.d.	25.20	5.21	.1249	.512115	3.6	1.59
10 LR54	WF 005 156	148	3	103.59	16.58	.0967	.511876	4.4	1.52
11 LR55*	VF 996 172	137	–59	60.16	10.72	.1078	.512007	4.8	1.49
12 LR56	VF 995 177	n.d.	n.d.	63.95	13.40	.1267	.512141	3.8	1.58
13 LR40	VF 785 240	118	–181	40.18	8.32	.1252	.512139	4.0	1.55
14 LR42	VF 772 248	185	–126	21.07	3.93	.1129	.512058	4.8	1.49
15 LR43*	VF 777 246	162	–112	33.05	7.17	.1311	.512174	3.6	1.60
16 LR44	VF 768 249	137	–21	167.67	29.47	.1062	.511944	3.9	1.56
17 LR45	VF 763 252	213	–21	16.83	3.06	.1100	.511997	4.2	1.53
18 LR37	VF 825 136	153	–25	137.73	30.01	.1317	.512195	3.9	1.57
19 LR39	VF 817 122	135	–36	185.68	42.03	.1368	.512238	3.7	1.59
20 LR14	VF 793 029	78	–44	82.53	12.23	.0895	.511816	4.6	1.50
21 LR11	VE 808 990	84	–232	18.58	3.84	.1251	.512125	3.8	1.58
22 LR20	VE 717 743	199	–135	30.56	4.81	.0952	.511868	4.5	1.51
23 LR21	VE 719 758	48	–219	25.57	5.18	.1224	.512080	3.4	1.60
Plutons									
A LR7*	WF 065 460	148	–175	13.52	2.75	.1232	.512082	3.3	1.62
B LR57*	VF 990 184	220	–88	10.58	1.57	.0900	.511848	5.1	1.47
C LR50	WF 031 143	151	–45	23.81	4.31	.1095	.512121	6.7	1.35
D LR53*	WF 012 153	174	–35	16.97	2.36	.0843	.511834	6.0	1.42

[@]Grid references are given to the nearest 100 m within the 100-km grid squares VE, VF, and WF (Fig. 5).
[#]Quantities Q (quartz) and P (plagioclase) describe the position of samples in the petrochemical Debon and LeFort (1983) diagram (Fig. 3).
[§]T_{DM}—Depleted mantle model age.
*Average of two dissolutions.
n.d.—Not determined.

et al. (1992) had any geographical pattern and whether the Nd isotope signature of the rocks could be related to the arc terranes of Labradoria and Quebecia identified on the mainland. Therefore, large samples of homogeneous orthogneiss were collected where access to the gneiss complex was made possible by forest roads. As far as possible, sampling avoided areas where migmatitic segregation might have disturbed the Sm-Nd system.

The petrochemistry of the gneisses analyzed was evaluated using the grid of Debon and LeFort (1983). Major element analyses performed at Activation Laboratories, Ancaster, Ontario, by inductively coupled plasma-atomic emission spectroscopy were used to calculate the quantities Q (quartz) and P (plagioclase) for the petrochemical grid, which generated an empirical classification of granitoid rocks analogous to that of Streckeisen (1973). These indexes are reported in Table 1 and plotted in Figure 3, where they demonstrate a range of granitoid compositions from dioritic, through tonalitic to granitic gneiss. This distribution closely resembles the model mineral analysis of Owen (1991, p. 8–9), which also showed a trend from quartz-dioritic through tonalitic to granitic gneiss, with occasional more alkaline samples in the quartz-monzonite field. No consistent correlation was observed in the present study between the petrochemistry and isotopic age results.

ANALYTICAL METHODS

Sm-Nd analysis followed our established procedures (e.g., Holmden and Dickin, 1995). After dissolution using HF and HNO_3, samples were split and one aliquot spiked with a mixed ^{150}Nd-^{149}Sm spike. Analysis by this technique yielded Sm/Nd = 0.2280 ± 2 for the reference standard BCR-1. Standard cation and reverse phase column separation methods were used. Nd isotope analyses were performed on a VG isomass 354 mass spectrometer at McMaster University using double filaments and a four-collector peak switching algorithm, and were normalized to a $^{146}Nd/^{144}Nd$ ratio of 0.7219. An average value of 0.51185 ± 2 (2σ population) was determined for the La Jolla standard. Average within-run precision on the samples was ± 0.000012 (2σ). The reproducibility of each model age averages 20 m.y. (2σ), based on empirical experience over several years of analyzing duplicate dissolutions.

RESULTS

Nd isotope data are presented in Table 1 and plotted on a histogram of ε Nd values in Figure 4, A. Epsilon Nd values are calculated at 1.5 Ga, based on the consensus of Pinwarian U-Pb ages discussed earlier from the Long Range and the Disappointment Hill complex. These Nd data are compared with the isotope signatures of the Labradoria and Quebecia terranes of the mainland in Figure 4, B to E, which were also calculated at 1.5 Ga to allow precise comparison. The data in Figure 4, B, come from the Blanc Sablon area, which is the type locality for juvenile Labradorian-age crust (Dickin, 2000). These rocks form a well-defined peak of ε Nd values, largely between +0.5

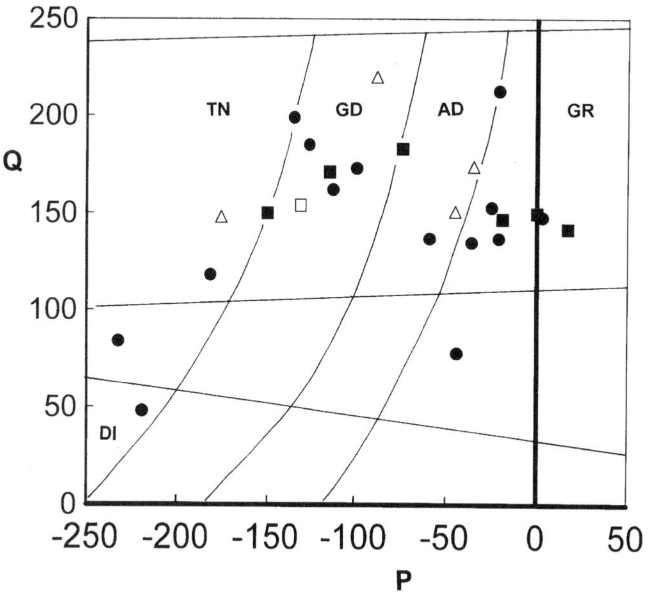

Figure 3. Petrology of samples analyzed, shown on the petrochemical grid of Debon and Le Fort. TN—tonalite; GD—granodiorite; AD—adamellite; GR—granite; DI—diorite. ■—juvenile orthogneisses with a depleted mantle (T_{DM}) age of > 1.7 Ga; ●—juvenile orthogneisses with TDM age of 1.49–1.60 Ga; △—younger plutonic orthogneisses; □—metasediment.

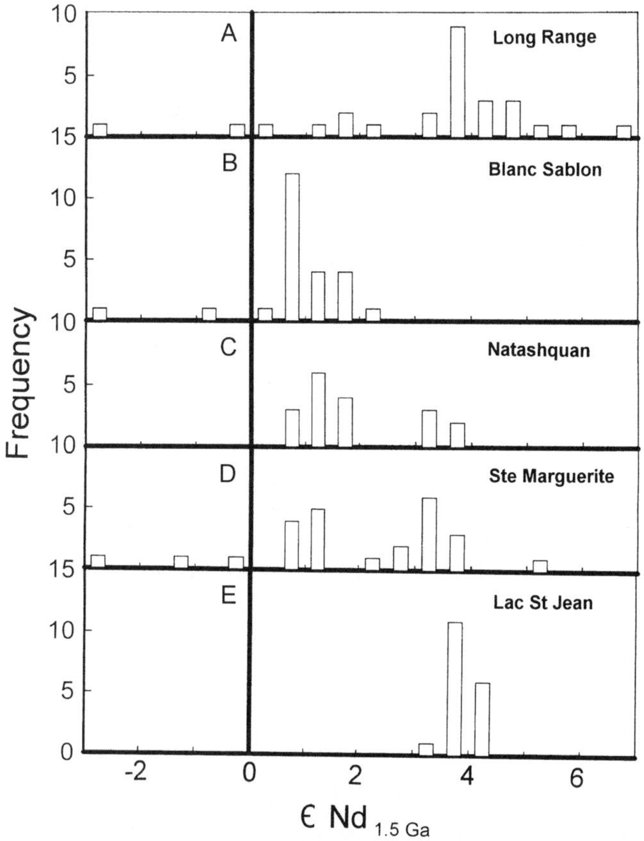

Figure 4. Histogram of ε Nd values at 1.5 Ga. (A) Long Range gneisses; (B) Blanc Sablon; (C) Natashquan; (D) St. Marguerite; (E) Lac St. Jean. Data from Table 1 and Dickin (2000).

and +2 at 1.5 Ga. Southern Labrador is also the type area for the Pinwarian event (Tucker and Gower, 1994). However, the Pinwarian plutons near Blanc Sablon appear to be generated by melting Labradorian crust without a significant juvenile 1.5 Ga component (Dickin, 2000).

The data in Figure 4, C, come from the Natashquan area, immediately southeast of the Wakeham Group (Fig. 1). Labradorian gneisses in this region display a very similar range of ε Nd values to those in Blanc Sablon, but there is also a separate cluster of ε Nd values from +3 to +4, probably representing juvenile 1.5 Ga plutons intruded into the older Labradorian gneisses (Corriveau et al., 2003). Several Pinwarian-age plutons east of this area have also been dated by the U-Pb method (Gower and Krogh, 2002).

Moving farther to the west, in the area of the St. Marguerite River just west of Sept Iles, a terrane boundary was delimited between Labradoria and a juvenile 1.5 Ga terrane named Quebecia (Dickin, 2000). Hence, a bimodal distribution of ε Nd values is again seen (Fig. 4, D). However, the difference between the Natashquan and the St. Marguerite area is that at Natashquan the rocks with younger Nd model ages are scattered randomly among the older gneisses with Labradorian crustal formation ages, whereas at St. Marguerite there is a single boundary separating the two age provinces. Finally, the area around Lac St. Jean represents the core of the Quebecia terrane, with a narrow range of ε Nd values from +3.5 to +4.5 (Fig. 4, E).

Comparison of the four mainland data sets in Figure 4 shows strong similarities, with two distinct peaks from +0.5 to +2.5 and from +3 to +4.5. The Long Range data set is smaller than those for the other areas, but is again resolved into two distinct peaks. Finally, in three of the data sets (Fig. 4, A, B, and D) a few analyses lying to the left of the ε zero line are interpreted as metasediments eroded from older crustal terranes, whereas samples with ε values above +5 are interpreted as younger Mesoproterozoic plutons (since these have Nd model ages under 1.5 Ga).

When the data from Table 1 and Figure 4, A, are plotted on a map of the southern Long Range (Fig. 5), it is possible to draw a line of demarcation between gneisses with crustal formation ages over 1.65 Ga to the north and under 1.65 Ga to the south. This boundary follows the regional foliation directions and lithological boundaries mapped by Owen (1991). The only exception is point A (sample LR7) from Cat Arm road, which is attributed to a younger reworking event in the northern area. Such reworking is consistent with the U-Pb data of Heaman et al. (2002), which showed that the Paleoproterozoic gneisses in this area underwent new zircon growth during the Pinwarian event.

The southern part of the gneiss terrane was sampled over a wide geographical area, with a total north-south distance of 50 km. Sampling in the northern area was less extensive but covered a north-south distance of nearly 30 km. Within both areas, Nd isotope signatures in plutonic orthogneisses vary only slightly outside analytical error. Therefore, the two areas are interpreted as separate crustal blocks with distinct crustal formation ages.

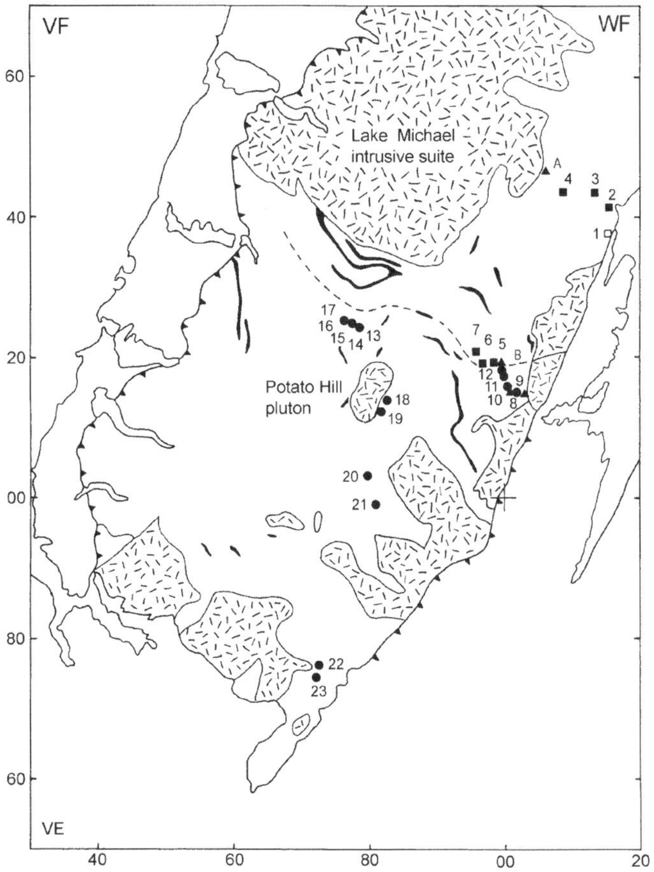

Figure 5. Locations of samples analyzed from the Long Range relative to mapped rock types from Owen (1991). ■—juvenile orthogneisses with depleted mantle (T_{DM}) ages > 1.7 Ga; ●—juvenile orthogneisses with T_{DM} ages of 1.49–1.60 Ga; ▲—younger plutonic orthogneisses; □—metasediment. Dashed line—proposed boundary between blocks correlated with Quebecia and Labradoria, following regional foliation trends and lithological boundaries mapped by Owen (1991); VE, VF, and WF—names of 100-km grid squares.

The average T_{DM} age in the southern area (1.55) is only 50 m.y. older than the consensus of Pinwarian U-Pb ages from the Long Range and the Disappointment Hill complex (Currie et al., 1992; Heaman et al., 2002). This age range is well within the growth time of an arc terrane and provides powerful evidence that the southern block is a juvenile Mesoproterozoic arc terrane. On the other hand, the older crustal formation ages in the northern block are supported by the minimum U-Pb age of 1.63 Ga on Cat Arm road (Fig. 2, locality #3). The average T_{DM} age of 1.75 Ga in these samples is identical to the age of samples from the Blanc Sablon area, and indicates that this is a juvenile Labradorian arc terrane. Hence it is concluded that the crustal formation ages of the northern and southern blocks in the Long Range are identical to the formation ages of the Labradoria and Quebecia terranes on the mainland, and these blocks are therefore interpreted as lateral equivalents of these terranes.

A summary of the Sm-Nd data from the younger and older

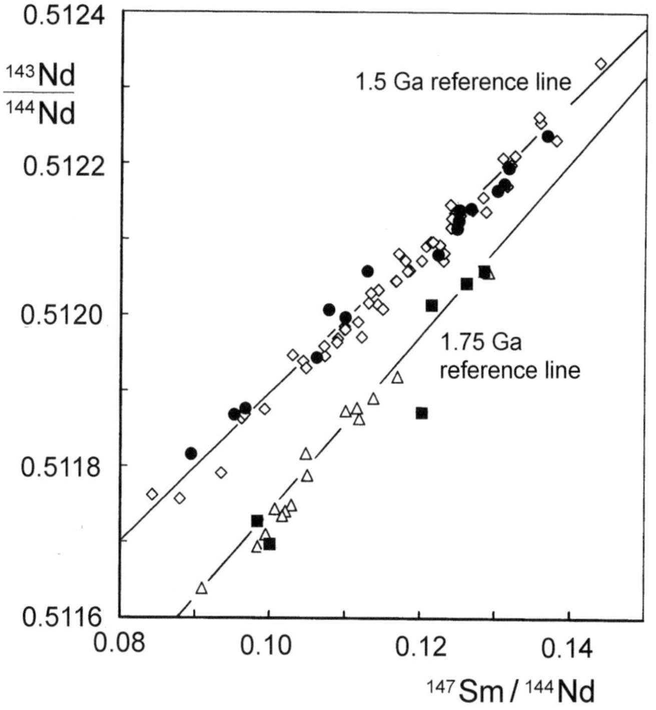

Figure 6. Sm/Nd isochron diagram showing the clustering of Long Range and mainland samples around separate 1.5 Ga and 1.75 reference lines. ●—Southern Long Range; ◇—Quebecia; ■—northern Long Range; △—Labradoria near Blanc Sablon.

blocks in the Long Range is shown in Figure 6, along with a comparison with data suites for the juvenile Quebecia and Labradoria accreted arc terranes on the mainland (Dickin, 2000). It can be seen that the southern Long Range and Québecian samples cluster closely around the 1.5 Ga reference line, consistent with the Pinwarian U-Pb ages from the Long Range. Hence the Sm-Nd isochron age confirms that formation of this crustal block occurred very shortly before 1.5 Ga. On the other hand, Labradorian gneisses from the northern Long Range and the Blanc Sablon area both cluster closely around the 1.75 Ga reference line. Again, this confirms that this is the best estimate for the crustal formation age of this older crustal block.

DISCUSSION

The new data can be compared with the Nd data of Owen et al. (1992) for the Potato Hill pluton on a Nd isotope evolution diagram (Fig. 7). The initial ratios of the monzodiorite phases of the pluton cluster closely, with an ε Nd value close to zero (solid triangles). Two samples of equigranular granite had more scattered Nd signatures, and are represented by open symbols. In the present study, two samples of granitic orthogneiss were collected from an area immediately to the east of the pluton, within the younger block. The evolution lines for these two samples actually pass through the point representing the initial ratio of the monzodiorite, suggesting that this body could have been derived by partial melting of the country rocks. In contrast, the reported Nd signatures of the two samples of country rock gneisses from Fryer et al. (1992) have much lower ε Nd values at 1020 Ma. One of these samples plots close to the evolution line of the three southernmost Labradorian gneisses (Fig. 7), suggesting that the gneissic samples analyzed by Fryer et al. (1992) represent Labradorian gneisses from the older block identified in the present work.

Comparisons can also be made between the new data for the Long Range and Mesoproterozoic gneisses from a Grenvillian terrane from Cape Breton Island named the Blair River inlier. This terrane is cut by Grenvillian- and Caledonian-aged plutons (Miller et al., 1996; Miller and Barr, 2000), but also contains two fragments of gneiss that predate the Grenvillian plutons, the Sailor Brook and Pollett Cove River gneisses. Two Nd isotope analyses have been performed on each of these units (Barr et al., 1998; Miller and Barr, 2000), yielding average T_{DM} model ages of 1.61 and 1.58 Ga, respectively. These fall within the same range as the gneisses from the southern block of the Long Range, suggesting that these gneisses form part of a single terrane extending between the Blair River and the Long Range inliers.

It is further suggested that the age boundary identified in the Long Range mountains should be correlated with a similar boundary identified between Quebecia and Labradoria, west of Sept Iles (Dickin, 2000). An attempt to predict the trend of the boundary in the area under the Gulf of St. Lawrence has been made in Figure 8. This figure shows the Humber zone of the Appalachian orogen, which contains Grenvillian basement that was partially tectonized during the Appalachian orogeny. The

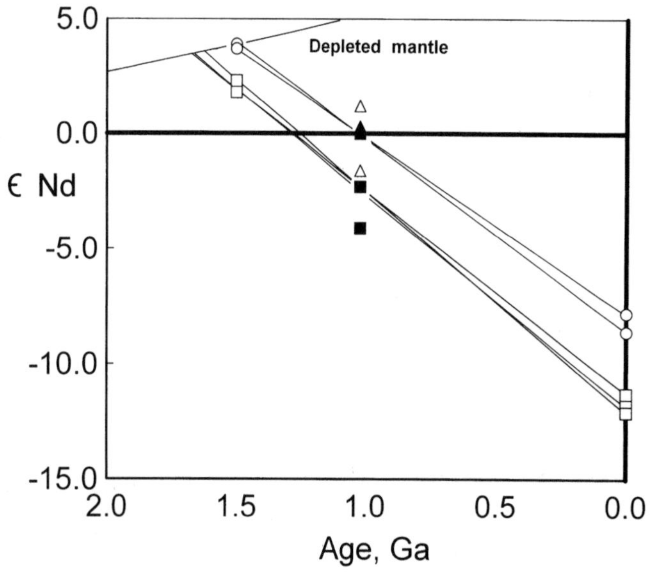

Figure 7. Nd isotope evolution diagram showing the composition of samples from the Potato Hill pluton (▲, △), and two gneisses (■) analyzed by Fryer et al. (1992), relative to evolution lines of gneissic samples from the present study. □ and ○—ε Nd compositions of the northern and southern blocks at the present day and during the Pinwarian event.

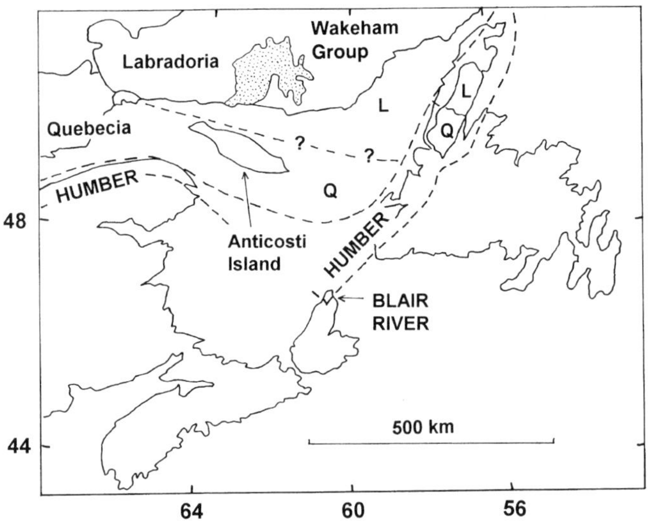

Figure 8. Geological sketch map of the area of the Appalachian tectonic front in eastern Canada showing the possible trend of the Quebecia-Labradoria terrane boundary under the Gulf of St. Lawrence. Q—Quebecia; L—Labradoria.

Humber zone is identified on land and also extrapolated under the Gulf of St. Lawrence using geophysical data (Hall et al., 1998).

The deflection in the Appalachian deformation front south of Anticosti Island makes it difficult to predict a precise trend for the terrane boundary under the gulf. However, based on the large amount of juvenile 1.5 Ga magmatism immediately southeast of the Wakeham terrane, it is suggested that the terrane boundary probably passes fairly close to the coast at this point, allowing the Labradorian crust in this area to be extensively reworked during the accretion of Quebecia to Labradoria. Hence, the boundary is drawn to the north of Anticosti Island. However, the Humber zone is also known to be a zone of strong postorogenic strike-slip faulting (Hall et al., 1998). Therefore, it is likely that the terrane boundary was offset across the Appalachian deformation front, as shown in Figure 8. Further work needs to be done to test and refine the location of the boundary tentatively proposed here.

ACKNOWLEDGMENTS

I thank Catharina Jager and Tuong Nguyen for skilled technical assistance with chemical separations and rock crushing. Louise Corriveau, Victor Owen, and Javier Suarez are thanked for constructive criticism and editorial comments that helped to improve the manuscript. Isotopic analysis at McMaster is partially supported by the National Science and Engineering Research Council.

REFERENCES CITED

Arndt, N.T., and Goldstein, S.L., 1987, Use and abuse of crust-formation ages: Geology v. 15, 893–895.

Barr, S.M., Raeside, R.P., and White, C.E., 1998, Geological correlations between Cape Breton Island and Newfoundland, northern Appalachian orogen: Canadian Journal of Earth Sciences, v. 35, p. 1252–1270.

Corriveau, L., Bonnet, A.-L., van Breemen, O., and Pilote, P., 2003, Tracking the Wakeham Group volcanics and associated Cu-Fe-oxides hydrothermal activity from La Romaine eastward, Eastern Grenville Province, Québec: Québec, Geological Survey of Canada Paper 2003-C12, 11 p.

Currie, K.L., van Breemen, O., Hunt, P.A., and van Berkel, J.T., 1992, Age of high-grade gneisses south of Grand Lake, Newfoundland: Atlantic Geology, v. 28, p. 153–161.

Davidson, A., 1998, An overview of Grenville Province geology, Canadian Shield: Chapter 3, *in* Lucas, S.B., and St-Onge, M.R., coordinators, Geology of the Precambrian Superior and Grenville provinces and Precambrian fossils in North America: Ottawa, Ontario, Geological Survey of Canada, Geology of Canada, no. 7, p. 205–270 (also Boulder, Colorado, Geological Society of America, The Geology of North America, v. C-1).

Debon, F., and LeFort, P., 1983, A chemical-mineralogical classification of common plutonic rocks and associations: Transactions of the Royal Society of Edinburgh: Earth Sciences, v. 73, p. 135–149.

DePaolo, D.J., 1981, Neodymium isotopes in the Colorado Front Range and crust-mantle evolution in the Proterozoic: Nature, v. 291, p. 193–196.

Dickin, A.P., 1995, Radiogenic Isotope Geology, Cambridge University Press.

Dickin, A.P., 1998a, Pb isotope mapping: A case study from the Grenville Province of Ontario: Precambrian Research, v. 91 p. 445–454.

Dickin, A.P., 1998b, Nd isotope mapping of a cryptic continental suture, Grenville Province of Ontario: Precambrian Research, v. 91 p. 433–444.

Dickin, A.P., 2000, Crustal formation in the Grenville Province: Nd-isotope evidence: Canadian Journal of Earth Sciences, v. 37, p. 165–181.

Dickin, A.P., and Guo, A., 2001, The location of the Allochthon Boundary Thrust and the Archean-Proterozoic suture in the Mattawa area of the Grenville Province: Nd isotope evidence: Precambrian Research, v.107, p. 31–43.

Dickin, A.P., and McNutt, R.H., 1989, Nd model age mapping of the southeast margin of the Archean Foreland in the Grenville Province of Ontario: Geology, v. 17, p. 299–302.

Fryer, B.J., Kerr, A., Jenner, G.A., and Longstaffe, F.J., 1992, Probing the crust with plutons: Regional isotopic geochemistry of granitoid intrusions across insular Newfoundland: St. John's, Newfoundland Department of Mines and Energy, Current Research 1992, p. 119–139.

Gower, C.F., and Krogh, T.E., 2002, A U-Pb geochronological review of the Proterozoic history of the eastern Grenville Province: Canadian Journal of Earth Sciences, v. 39, p. 795–829.

Hall, J., Marillier, F., and Dehler, S., 1998, Geophysical studies of the structure of the Appalachian orogen in the Atlantic borderlands of Canada: Canadian Journal of Earth Sciences, v. 35, p. 1205–1221.

Heaman, L.M., Erdmer, P., and Owen, J.V., 2002, U-Pb geochronologic constraints on the crustal evolution of the Long Range Inlier, Newfoundland: Canadian Journal of Earth Sciences, v. 39, p. 845–865.

Hoffman, P., 1989, Precambrian geology and tectonic history of North America, *in* Bally, A.W., and Palmer, A.R., eds., The geology of North America: An overview: Boulder, Colorado, Geological Society of America, The Geology of North America, v. A, p. 447–512.

Holmden, C., and Dickin, A.P., 1995, Paleoproterozoic crustal history of the southwestern Grenville Province: Evidence from Nd isotopic mapping: Canadian Journal of Earth Sciences, v. 32, p. 472–485.

Jackson, S.L., and Fyon, J.A., 1991, The Western Abitibi Subprovince in Ontario, *in* Thurston, P.C., Williams, H.R., Sutcliffe, R.H., and Stott, G.M., eds., Geology of Ontario: Sudbury, Ontario Geological Survey Special Volume 4, part 1, p. 405–484.

Krogh, T.E., Culshaw, N., and Ketchum, J., 1992, Multiple ages of metamorphism and deformation in the Parry Sound–Pointe au Baril area: Vancouver, British Columbia, Lithoprobe (Abitibi-Grenville Workshop IV) Report 33, p. 39.

Miller, B.V., and Barr, S.M., 2000, Petrology and isotopic composition of a Grenvillian basement fragment in the northern Appalachian orogen: Blair

River inlier, Nova Scotia, Canada: Journal of Petrology, v. 41, p. 1777–1804.

Miller, B.V., Dunning, G.R., Barr, S.M., Raeside, R.P., and Jamieson, R.A., 1996, Magmatism and metamorphism in a Grenvillian fragment: U-Pb and $^{40}Ar/^{39}Ar$ ages from the Blair River Complex, northern Cape Breton Island, Nova Scotia, Canada: Geological Society of America Bulletin, v. 108, p. 127–140.

Nadeau, L., and van Breemen, O., 1994, Do the 1.45–1.39 Ga Montauban group and the La Bostonnais complex constitute a Grenvillian accreted terrane? Geological Association of Canada–Mineralogical Association of Canada Programs with Abstracts, v.19, p. A81.

Nadeau, L., and van Breemen, O., 1998, Plutonic ages and tectonic setting of the Algonquin and Muskoka allochthons, Central Gneiss Belt, Grenville Province, Ontario: Canadian Journal of Earth Sciences, v. 35, p. 1423–1438.

Owen, J.V., 1991, Geology of the Long Range inlier, Newfoundland: Ottawa, Ontario, Geological Survey of Canada Bulletin 395.

Owen, J.V., Greenough, J.D., Fryer, B.J., and Longstaffe, F.J., 1992, Petrogenesis of the Potato Hill pluton, Newfoundland: Transpression during the Grenvillian orogenic cycle?: Journal of the Geological Society of London, v. 149, p. 923–935.

Rivers, T., Martignole, J., Gower, C.F., and Davidson, A., 1989, New tectonic divisions of the Grenville Province, southeastern Canadian Shield: Tectonics, v. 8, p. 63–84.

Scharer, U., 1991, Rapid continental crust formation at 1.7 Ga from a reservoir with chondritic isotope signatures, eastern Labrador: Earth and Planetary Science Letters, v. 102, p. 110–133.

Scharer, U., and Gower, C.F., 1988, Crustal evolution in eastern Labrador: Constraints from precise U-Pb ages: Precambrian Research, v. 38, p. 405–421.

Streckeisen, A.L., 1973, Plutonic rocks: Classification and nomenclature recommended by the IUGS [International Union of Geological Sciences] subcommission on the systematics of igneous rocks: Geotimes, 18 October, p. 26–30.

Tucker, R.D., and Gower, C.F., 1994, A U–Pb geochronological framework for the Pinware terrane, Grenville Province, southeast Labrador: Journal of Geology, v. 102, p. 67–78.

Wasteneys, H.A., Kamo, S.L., Moser, D., Krogh, T.E., Gower, C.F., and Owen, J.V., 1997, U-Pb geochronological constraints on the geological evolution of the Pinware terrane and adjacent areas, Grenville Province, southeast Labrador, Canada: Precambrian Research, v. 81, p. 101–128.

Manuscript Accepted by the Society August 25, 2003

Printed in the USA

Geological Society of America
Memoir 197
2004

Geochemistry of the Late Mesoproterozoic Mount Eve granite suite: Implications for Late to post-Ottawan tectonics in the New Jersey–Hudson Highlands

Matthew L. Gorring*
Todd C. Estelle
Department of Earth and Environmental Studies, Montclair State University, Upper Montclair, New Jersey 07043, USA
Richard A. Volkert
New Jersey Geological Survey, Trenton, New Jersey 08625, USA

ABSTRACT

The Mount Eve granite suite is a postorogenic, A-type granitoid suite that consists of several small plutonic bodies occurring in the northwestern New Jersey–Hudson Highlands. Mount Eve granite suite rocks are equigranular, medium- to coarse-grained, quartz monzonite to granite, consisting of quartz, microperthite, and oligoclase, with minor hornblende, biotite, and accessory zircon, apatite, titanite, magnetite, and ilmenomagnetite. Whole-rock analyses indicate that Mount Eve granite is metaluminous to slightly peraluminous (ASI or aluminum saturation index, A/CNK or $Al_2O_3/(CaO + Na_2O + K_2O)$ = 0.62 to 1.12) and has A-type compositional affinity defined by high K_2O/Na_2O (1.4 to 2.8), Ba/Sr (3 to 12), $FeO_t/(FeO_t+MgO)$ (0.77 to 0.87), Ba (400 to 3000 ppm), Zr (200 to 1000 ppm), Y (30 to 130 ppm), Ta (2.5 to 6 ppm), total rare earth elements or REE (300 to 1000 ppm), low MgO (<1 wt%), Cr and Ni (both <5 ppm); and relatively low Sr (200 to 700 ppm). Variably negative Eu anomalies (Eu/Eu* = 0.13 to 0.72, where Eu/Eu* is the chondrite-normalized ratio of measured Eu divided by the hypothetical Eu concentration required to produce REE pattern with no Eu anomaly) and systematic decreases in Sr, Ba, Zr, Hf, Nb, and Ta, with constant total REE content and increasing Ce/Yb and SiO_2 contents, suggest crystallization of feldspars + zircon + titanite ± apatite. Possible modes of origin include dry melting of charnockitic gneisses or Fe-rich mafic to intermediate diorites within the Mesoproterozoic basement. Two possible tectonic mechanisms for generation of Mount Eve granite include (1) residual thermal input from a major lithospheric delamination event during or immediately after peak Ottawan orogenesis (1090–1030 Ma) or (2) broad orogenic relaxation between peak Ottawan and a late (1020–1000 Ma) high-grade, right-lateral transpressional event.

Keywords: granite, A-type, Grenville, New Jersey Highlands, tectonics

*E-mail: gorringm@mail.montclair.edu.

Gorring, M.L., Estelle, T.C., and Volkert, R.A., 2004, Geochemistry of the Late Mesoproterozoic Mount Eve granite suite: Implications for Late to post-Ottawan tectonics in the New Jersey–Hudson Highlands, *in* Tollo, R.P., Corriveau, L., McLelland, J., and Bartholomew, M.J., eds., Proterozoic tectonic evolution of the Grenville orogen in North America: Boulder, Colorado, Geological Society of America Memoir 197, p. 505–523.

INTRODUCTION

Within the Grenville Province of the central and northern Appalachians and the Adirondack Highlands (Fig. 1), there are several examples of weakly to undeformed, late- to postorogenic plutonic rocks that were emplaced sporadically over a period of ~50 to 100 million years and that postdate the major pulse of deformation and metamorphism associated with the Ottawan orogeny (1090–1030 Ma, as defined by Rivers, 1997; McLelland et al., 1996, 2001). These rocks not only place important constraints on the lower age limit for the main pulse of Ottawan orogenic events, but also provide clues to tectonic processes occurring during final stages of the Grenville orogenic cycle. In the Adirondack Highlands, regional lithospheric delamination associated with orogenic collapse of overthickened crust has been proposed to explain large volumes of syn- to Late Ottawan granitoid rocks (Lyon Mountain Gneiss; Whitney and Olmstead, 1988) emplaced during the interval between 1060 and 1045 Ma (McLelland et al., 2001). Younger suites of volumetrically minor and sporadically distributed granitoids were also emplaced in the Adirondacks (McLelland et al., 1996, 2001), the Green Mountains (Ratcliffe et al., 1991), and the New Jersey–Hudson Highlands (Drake et al., 1991; Gates and Krol, 1998; Gates et al., 2001a) during the interval between 1035 and 960 Ma. These younger plutonic suites have received much less attention in the past because of their relatively small volumes and limited areal extent. However, interest in these rocks recently has been renewed due to the important role they play in constraining the

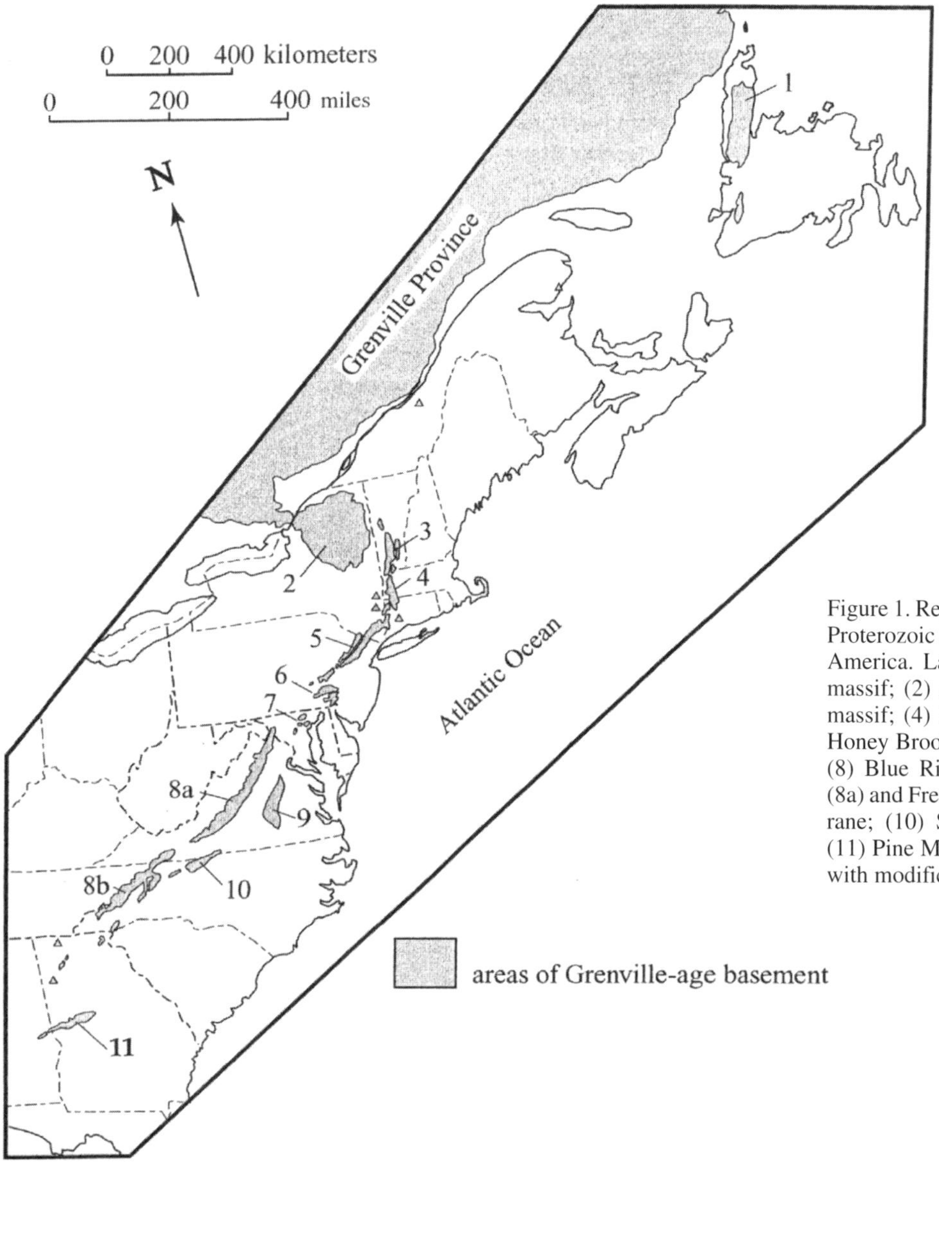

Figure 1. Regional map showing the location of Middle Proterozoic (Grenville-age) basement in eastern North America. Labeled areas include the (1) Long Range massif; (2) Adirondack massif; (3) Green Mountains massif; (4) Berkshire massif; (5) Reading Prong; (6) Honey Brook upland; (7) Baltimore Gneiss antiforms; (8) Blue Ridge Province, including the Shenandoah (8a) and French Broad (8b) massifs; (9) Goochland terrane; (10) Sauratown Mountains anticlinorium; and (11) Pine Mountain belt. Map from Tollo et al. (2004) with modifications after Rankin et al. (1989b).

timing of culminating events of the Ottawan orogeny (e.g., McLelland et al., 2001). Recent recognition of Late to post-Ottawan, high-grade, ductile transpressional shear zones in the central Adirondack Highlands and the New Jersey–Hudson Highlands (Gates, 1995, 1998) has also raised the possibility that this younger plutonic activity could be related to localized crustal extension associated with ductile shearing rather than regional-scale lithospheric delamination.

The Mount Eve granite suite is composed of thirty small (0.2 to 5 km^2), stocklike bodies of undeformed quartz monzonite to granite (Drake et al., 1991) that intrude high-grade Middle Mesoproterozoic basement gneisses of the New Jersey–Hudson Highlands portion of the Reading Prong (Figs. 1 and 2). One of the larger plutons has a U-Pb zircon crystallization age of 1020 ± 4 Ma (Drake et al., 1991), and thus is part of the Late to post-Ottawan plutonic granitoid suite mentioned earlier. Present knowledge of the Mount Eve granite suite comes from geologic mapping by Offield (1967), Drake et al. (1991), and Volkert and Drake (1999). Previously published data for the Mount Eve granite suite include two major-element analyses reported in Drake et al. (1991). This paper presents a more detailed study of the geochemistry of the Mount Eve granite suite in order to elucidate its petrogenesis and implications for Late to post-Ottawan tectonics within this region of the Grenville Province. Based on new data presented here, it has been found that the Mount Eve granite suite has strong A-type chemical affinities and represents partial melting of meta-igneous Mesoproterozoic basement gneisses. The restricted spatial extent, small volumes (~2 to 5 km^3), and temporal association with other volumetrically minor granitoid rocks exposed elsewhere in the New Jersey–Hudson Highlands supports a production mechanism for the Mount Eve granite suite that is related to the onset of a high-grade, dextral transpressional shearing event that occurred during the waning stages (ca. 1020–1000 Ma) of Grenville orogenesis.

REGIONAL GEOLOGIC SETTING

The Mount Eve granite suite is located in northwestern New Jersey and southeastern New York ~60 km northwest of New York City (Fig. 2) in the Mesoproterozoic New Jersey–Hudson Highlands. The New Jersey Highlands, along with the physically contiguous Hudson Highlands of New York state and similar rocks extending into eastern Pennsylvania, are collectively called the Reading Prong, which is one of the largest of several Grenville-age (1300 to 1000 Ma) basement massifs within the core of the Appalachian orogenic belt of eastern North America (Figs. 1 and 2; Rankin, 1975). These basement massifs lie outboard (east) of the main Grenville Province in eastern Canada and record variable amounts of post-Mesoproterozoic metamorphic and deformational overprint (e.g., Rankin et al., 1989a; Gates and Costa, 1998). The New Jersey–Hudson Highlands display evidence of only brittle deformation concentrated along narrow, reactivated Mesoproterozoic shear zones due to Late Paleozoic compression and Mesozoic rifting (Gates, 1995, 1998), and thus have a well-preserved record of Grenville-age metamorphism and deformation.

Rocks of the New Jersey–Hudson Highlands consist of a complex assemblage of paragneiss, orthogneiss, marbles, and intrusive granitoid rocks that were highly deformed and metamorphosed at upper amphibolite- to hornblende-granulite-facies conditions during Grenville orogenesis (Volkert and Drake, 1999). The oldest rocks (of unknown age) in the region are a suite of quartzofeldspathic orthogneiss that include very light-colored, biotite-hornblende-quartz-plagioclase gneiss, charnockitic (orthopyroxene-bearing) quartz diorite, and interlayered amphibolite, collectively called the Losee metamorphic suite (Volkert and Drake, 1999). These rocks are interpreted to represent a continental volcanic-plutonic arc suite of calc-alkaline rocks (Volkert and Drake, 1999). Rocks of the Losee metamorphic suite are lithologically and chemically very similar to tonalitic and charnockitic gneisses found in the southern Adirondacks (McLelland and Chiarenzelli, 1990) and to the Mount Holly Complex in the Green Mountain massif in Vermont (Ratcliffe et al., 1991), which have been dated at ca. 1350 to 1300 Ma. A thick sequence of supracrustal paragneiss, the age of which is not well constrained but is older than 1100 Ma, is interpreted to unconformably overlie the Losee metamorphic suite (Volkert and Drake, 1999). These rocks consist of a heterogeneous package of quartzofeldspathic gneiss, calc-silicate rocks, quartzite, and marble. Although there are few constraints with which to construct a detailed depositional history for the supracrustal sequence, Volkert and Drake (1999) used detailed geochemical data to suggest that these rocks were deposited in an evolving system of continental- to oceanic-arc extensional basins. Amphibolite occurs interlayered with all Late Mesoproterozoic rocks of the New Jersey Highlands (Volkert and Drake, 1999), and the geochemical data available support the interpretation that most amphibolite is of metaigneous origin, representing metamorphosed mafic volcanics with minor occurrences of metamorphosed dikes and plutonic rocks.

Two A-type granitoid suites intrude the Losee metamorphic suite and associated supracrustal paragneisses: the Vernon Supersuite (Volkert and Drake, 1998; Volkert et al., 2000) and the Mount Eve granite suite (Drake et al., 1991). The Vernon Supersuite granitoids underlie ~50% of the New Jersey Highlands and consist of the Byram intrusive suite and the Lake Hopatcong intrusive suite. Conventional multigrain U-Pb geochronology by thermal ionization mass spectrometry (TIMS) on air-abraded zircon fractions extracted from a single sample of the Byram intrusive suite yield upper-intercept ages ranging from 1088 ± 41 Ma to 1122 ± 53 Ma (Drake et al., 1991). Volkert et al. (2000) obtained whole-rock Rb-Sr isochron ages of 1116 ± 41 Ma ($^{87}Sr/^{86}Sr_i$ = 0.70389) for the Byram intrusive suite and 1095 ± 9 Ma ($^{87}Sr/^{86}Sr_i$ = 0.70520) for the Lake Hopatcong intrusive suite based on six individual samples from each suite collected from various localities. The Vernon Supersuite rocks all possess well-developed, high-grade metamorphic

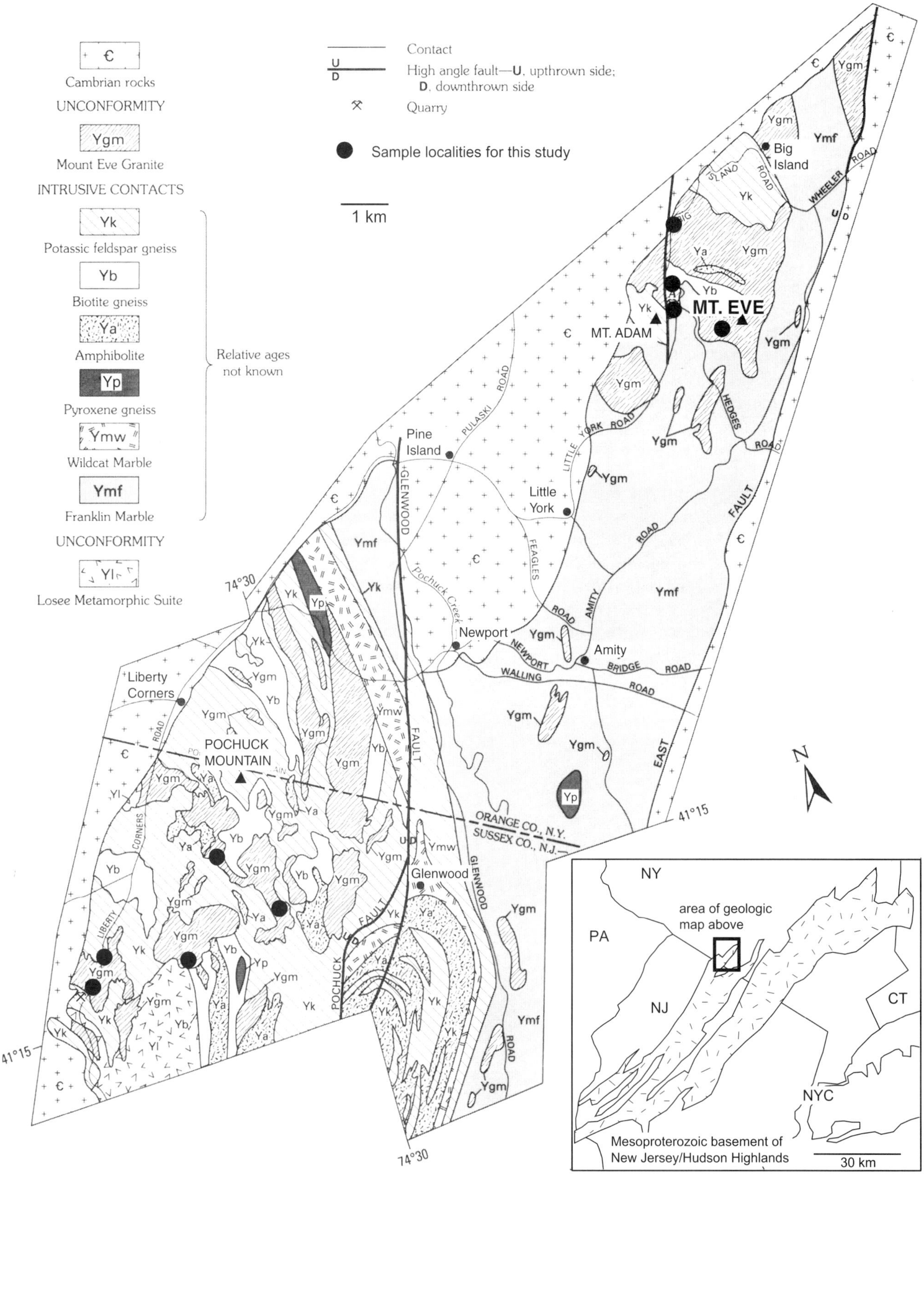

€
Cambrian rocks
UNCONFORMITY
Ygm
Mount Eve Granite
INTRUSIVE CONTACTS
Yk
Potassic feldspar gneiss
Yb
Biotite gneiss
Ya
Amphibolite
Yp
Pyroxene gneiss
Ymw
Wildcat Marble
Ymf
Franklin Marble
Relative ages not known
UNCONFORMITY
Yl
Losee Metamorphic Suite
Contact
U
D
High angle fault—U, upthrown side; D, downthrown side
Quarry
Sample localities for this study
1 km
MT. EVE
MT. ADAM
Big Island
BIG ISLAND ROAD
WHEELER ROAD
HEDGES ROAD
LITTLE YORK ROAD
PULASKI ROAD
Pine Island
Little York
FEAGLES ROAD
AMITY ROAD
Pochuck Creek
Newport
Amity
NEWPORT BRIDGE ROAD
WALLING ROAD
EAST FAULT
GLENWOOD FAULT
GLENWOOD ROAD
POCHUCK FAULT
Liberty Corners
LIBERTY CORNERS ROAD
POCHUCK MOUNTAIN
ORANGE CO., N.Y.
SUSSEX CO., N.J.
Glenwood
74°30
41°15
N
NY
PA
NJ
CT
NYC
area of geologic map above
Mesoproterozoic basement of New Jersey/Hudson Highlands
30 km

fabrics (e.g., gneissic layering) and field relations (e.g., concordant contacts and fold phases and geometries similar to those of surrounding Mesoproterozoic rocks) that indicate these rocks are pre- or syntectonic plutons deformed and metamorphosed during the Ottawan orogeny (Volkert et al., 2000). These granitoids are chemically similar to other plutonic rocks of A-type chemical affinity with ages of 1180–1060 Ma from other parts of the Grenville orogen, including southeastern Canada (Easton, 1986; Lumbers et al., 1990; Davidson 1995), the Adirondacks (Wasteneys et al., 1999; McLelland et al., 2001), the northern Blue Ridge (Tollo and Aleinikoff, 1996; Hughes et al., 1997), and the surrounding Hudson Highlands (Drake et al., 1991; Verrengia and Gorring, 2002). Rocks of Late to post-Ottawan age (<1030 Ma) are volumetrically minor in the New Jersey–Hudson Highlands, but consist of the Mount Eve granite (1020 ± 4 Ma; Drake et al., 1991), the Canada Hill granite (1010 ± 4 Ma; Aleinikoff and Grauch, 1990), the Lake Tiorati diorite (1008 ± 4 Ma; Gates et al., 2001a), and a suite of very late, crosscutting pegmatite dikes (ca. 1000–915 Ma; Gates and Krol, 1998). Similar plutonic activity of Late to post-Ottawan age has been documented elsewhere in the Adirondacks (ca. 1035 Ma, Lyonsdale Bridge pegmatite; 935 Ma, Cathead Mountain leucogranite; McLelland et al., 2001) and in the Green Mountain massif (ca. 960 Ma, Stamford Hill rapakivi granite; Ratcliffe et al., 1991).

THE MOUNT EVE GRANITE SUITE

The Mount Eve granite suite consists of about thirty (0.2 to 5 km^2 each) stocklike bodies of pink to light-gray, massive, medium- to coarse-grained quartz monzonite to granite (International Union of Geological Sciences classification [Streckeisen, 1973], based on modal data from Drake et al., 1991). Based on normative feldspar content calculated from whole-rock major-element analyses reported here, all Mount Eve granite suite samples are classified as granites (Fig. 3). Classification based on normative feldspar was used because detailed point counting of the samples reported in this paper was not undertaken. These intrusive rocks crop out over a over small area (~40 km^2) in the northwestern part of the New Jersey–Hudson Highlands extending ~20 km from near Glenwood, New Jersey, northward to Big Island, New York (Fig. 2). Drake et al. (1991) obtained a U-Pb upper-intercept age of 1020 ± 4 Ma on three air-abraded zircon separates from one sample collected near Mount Eve, New York (Fig. 2), using conventional multigrain TIMS techniques. Except for magmatic flow fabrics documented by Drake et al. (1991) near intrusive contacts, rocks of

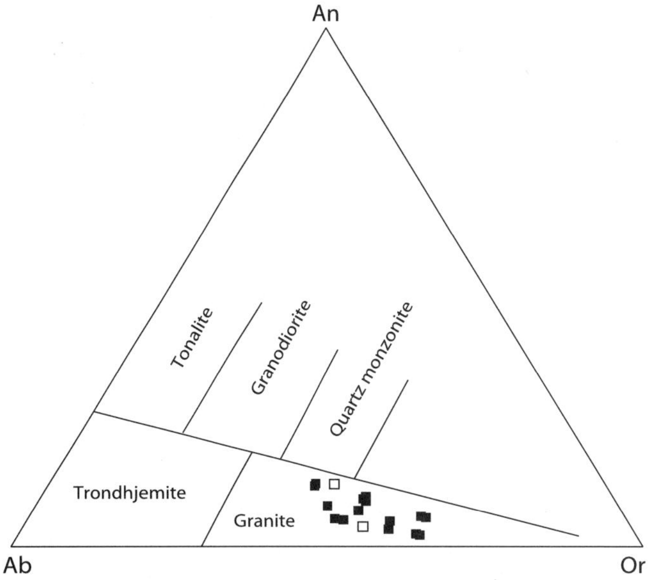

Figure 3. Plot of normative anorthite (An)–albite (Ab)–orthoclase (Or) for the Mount Eve granite suite samples analyzed in this study (filled squares) and two analyses from Drake et al. (1991) (unfilled squares). Diagram after O'Connor (1965).

the Mount Eve granite suite lack a penetrative deformation fabric or evidence for metamorphic recrystallization (Fig. 4). Mapping by Drake et al. (1991) and Volkert and Drake (1999) clearly demonstrates that the granitoid intrusions are discordant to compositional layering, foliation, and folds developed in the surrounding Mesoproterozoic (Grenville-age) country rocks at both outcrop and map scales. Abundant small (<10–15 cm) lenticular xenoliths of amphibolite and quartzofeldspathic gneiss also occur locally in the granitoid intrusions (Figs. C8 and C9 *in* Drake et al., 1991). The U-Pb zircon age and field relations clearly indicate that the Mount Eve granite suite was intruded after the main pulse (or pulses) of high-grade metamorphism and deformation that affected the Mesoproterozoic basement.

The Mount Eve granite suite contains microcline or microcline microperthite, quartz, and oligoclase as the dominant mineral phases. Ferromagnesian minerals typically include both hornblende and biotite. Preliminary electron microprobe data indicate that both the hornblende (hastingsite, according to the classification of Leake et al., 1997) and the biotite are Fe-rich, with Fe/(Fe + Mg) ratios of 0.7 to 0.9, well within the ranges for hornblende and biotite from other Proterozoic A-type granites (Anderson, 1983). The Fe-Ti oxide minerals are magnetite and ilmenomagnetite. Zircon, apatite, allanite, and titanite are common accessory minerals. Titanite has been confirmed by electron microprobe analysis. Primary fluorite has not been observed. The assemblage of titanite + quartz + magnetite and the lack of primary fluorite suggest that parental Mount Eve granite suite magmas were relatively low in magmatic fluorine (<1 wt%; Price et al., 1999) and crystallized under conditions of moderate oxygen fugacity (Wones, 1983).

Figure 2. Location map showing Middle Proterozoic basement of the New Jersey–Hudson Highlands (inset, lower right) and approximate location (heavy lined rectangle in inset, lower right) of geologic map showing the region where the Mount Eve granite suite is exposed. Maps modified from Drake et al. (1991) and Heleneck and Mose (1984), respectively. Abbreviations include NY—New York; PA—Pennsylvania; NJ—New Jersey; CT—Connecticut; NYC—New York City.

Figure 4. Photographs showing the lack of penetrative deformation fabric in the Mount Eve granite suite. Hammer handle in A is 28 cm long; coin (dime) in B is 1.7 cm in diameter.

GEOCHEMISTRY

Fifteen representative samples from four plutons from the type locality at Mount Eve, near Pine Island, New York, and south of Pochuck Mountain, New Jersey, were collected for this study (Fig. 2). Whole-rock major-element compositions indicate that the entire suite is comprised of alkali-rich felsic granitoids with no associated mafic or intermediate rocks (Table 1). The SiO_2 content ranges from 65 to 72 wt%, the Al_2O_3 content ranges from 13.5 to 15.5 wt%, and the total alkali ($Na_2 + K_2O$ content ranges from 8.5 to 10.5 wt% (Table 1; see appendix for analytical methods). Most of the suite is metaluminous (ASI or aluminum saturation index, A/CNK or molar $Al_2O_3/(CaO + Na_2O + K_2O) < 1.0$), although a few samples are mildly peraluminous in composition (A/CNK ~1.1; Fig. 5, A). The Mount Eve granite suite has very high K_2O contents (4.8 to 7.5 wt%) and high K_2O/Na_2O ratios (1.3 to 2.8) that reflect the high modal abundance of K-feldspar in these rocks. For most samples, CaO and MgO contents are generally low (<3 wt% and <0.9 wt%, respectively). Total Fe (as Fe_2O_3) contents and molar $FeO_t/(FeO_t+MgO)$ ratios are high (3 to 5.4 wt% and 0.77 to 0.88, respectively), and most samples plot within the tholeiitic field for granitoid rocks (Fig. 5, B). These major-element characteristics (high K_2O/Na_2O, K_2O/MgO, total Fe, and molar $FeO_t/(FeO_t+MgO)$ and low CaO and MgO) are broadly similar to those of A-type granitoids (Collins et al., 1982; Whalen et al., 1987; Eby, 1992). However, the Mount Eve granite suite has slightly lower $FeO_t/(FeO_t+MgO)$ ratios compared to other Meso- and Neoproterozoic A-type granitoids from the New Jersey Highlands (Vernon Supersuite; Volkert et al., 2000; Fig. 5, B), the northern Blue Ridge (Tollo and Aleinikoff, 1996; Tollo et al., this volume), and the Adirondack massif (Whitney and Olmsted, 1988; McLelland and Whitney, 1990).

Chondrite-normalized rare-earth-element (REE) patterns for the Mount Eve granite suite are characterized by moderate to steep negative slopes (La/Yb = 6 to 40) and generally modest negative Eu anomalies (Eu/Eu* = 0.62 to 0.91, where Eu/Eu* is the chondrite-normalized ratio of measured Eu divided by the hypothetical Eu concentration required to produce REE pattern with no Eu anomaly; Fig. 6). A few samples with noticeably deeper Eu anomalies (Eu/Eu* < 0.4) also have relatively high SiO_2 (~71.5%) and lower Al_2O_3, CaO, Sr, and Ba compared to the rest of the suite (Table 1 and Fig. 6). This suggests a significant role for feldspar in the petrogenesis of these samples through more extensive crystal fractionation of both plagioclase and K-feldspar, as discussed later. Relative to average I-type, calc-alkaline, and peraluminous S-type granitoids, A-types typically have high concentrations of REE, Zr, Nb, Ta, Ga, F, Sn, Zn, and Y and low concentrations of Ba and Sr (Loiselle and Wones, 1979; Whalen et al., 1987; Eby, 1990). The rocks of the Mount Eve granite suite have high concentrations of REE and high-field-strength elements (HFSE) that clearly distinguish them from I- or S-type granites and are similar to those of the older A-type Vernon Supersuite granitoids from the New Jersey Highlands (Fig. 7, A and B) when plotted on the granitoid tectonic discrimination diagrams of Pearce et al. (1984). Most samples of the Mount Eve granite suite plot in the "within-plate granite" field (Fig. 8, A–C), as do many A-type granitoid suites (e.g., Whalen et al., 1987; Eby, 1990). Two samples of Mount Eve granite plot near the boundary of the "volcanic arc" and "ocean ridge" granite fields on the Yb versus Ta diagram (Fig.

TABLE 1. WHOLE-ROCK COMPOSITIONS OF THE MOUNT EVE GRANITE SUITE

Sample	EVE-1	EVE-2	EVE-3	EVE-4	EVE-5	EVE-6	EVE-7	PO-1	PO-2	PO-3a	PO-3b	PO-4	PO-5a	PO-5b	PO-6
Major elements															
SiO_2[#]	67.35	67.14	66.11	65.10	67.24	71.47	65.74	68.02	71.57	68.94	71.35	65.59	69.94	67.00	65.76
TiO_2	0.79	0.70	0.71	0.67	0.66	0.23	0.80	0.51	0.47	0.43	0.49	0.66	0.40	0.58	0.67
Al_2O_3	14.53	14.51	14.19	14.52	14.21	13.51	14.53	14.51	13.60	14.62	14.29	15.28	15.50	14.65	15.28
Fe_2O_3[†]	5.31	4.10	3.09	3.17	3.60	3.13	5.11	3.18	3.18	3.52	3.21	4.72	3.13	4.83	4.72
MnO	0.10	0.08	0.10	0.11	0.08	0.05	0.10	0.02	0.02	0.02	0.03	0.07	0.02	0.05	0.07
MgO	0.74	0.73	0.82	0.73	0.60	0.38	0.89	0.82	0.82	0.61	0.44	0.78	0.54	0.80	0.83
CaO	2.72	3.17	4.61	6.71	2.54	1.52	2.58	1.58	0.86	0.56	1.25	2.78	1.02	2.00	1.72
Na_2O	3.66	3.80	2.60	3.68	3.21	2.88	3.87	2.70	2.76	2.77	3.42	3.47	3.89	3.25	3.48
K_2O	5.21	5.28	7.26	4.76	6.48	7.36	6.10	6.84	5.91	6.94	5.99	6.31	5.49	6.01	6.41
P_2O_5	0.20	0.14	0.06	0.24	0.12	0.02	0.16	0.14	0.12	0.09	0.09	0.15	0.08	0.16	0.15
Total	100.61	99.65	99.55	99.68	98.74	100.56	99.87	98.33	99.30	98.51	100.55	99.81	100.02	99.33	99.09
Trace elements[§]															
La	81	164	39	105	96	242	92	121	115	53	102	76	73	82	91
Ce	256	395	116	257	261	453	256	231	226	127	203	195	137	164	217
Pr	N.A.[##]	N.A.	N.A.	N.A.	38.6	51.6	38.4	27.1	25.3	16.4	24.2	29.0	18.3	20.4	30.6
Nd	138	135	67	107	170	189	170	108	95	68	95	131	73	82	132
Sm	36.9	40.0	16.6	28.7	36.6	33.0	36.6	22.2	17.4	12.2	17.9	28.5	15.0	16.7	28.3
Eu	7.01	7.65	3.65	5.31	7.03	1.39	7.66	4.78	2.18	3.31	4.83	5.23	3.23	3.91	5.76
Gd	N.A.[##]	N.A.	N.A.	N.A.	32.4	31.3	32.5	20.4	16.4	9.7	15.8	25.0	14.2	16.3	25.5
Tb	4.47	4.86	2.12	3.36	4.97	4.01	4.91	3.10	2.27	1.37	2.27	3.82	2.22	2.49	3.98
Dy	N.A.	N.A.	N.A.	N.A.	26.7	18.2	25.0	16.6	11.3	6.5	11.5	20.1	11.3	12.3	19.4
Ho	N.A.	N.A.	N.A.	N.A.	5.16	3.33	4.93	3.07	2.08	1.25	2.19	3.90	2.26	2.48	3.96
Er	N.A.	N.A.	N.A.	N.A.	12.34	7.63	11.34	7.02	4.78	3.35	5.44	9.33	5.55	5.93	9.35
Tm	N.A.	N.A.	N.A.	N.A.	1.75	0.94	1.55	0.93	0.60	0.49	0.79	1.32	0.79	0.79	1.24
Yb	11.1	11.1	6.0	9.2	11.8	6.1	10.5	5.8	3.7	3.4	5.2	8.7	5.4	5.5	8.6
Lu	1.54	1.49	0.79	1.23	1.71	0.87	1.52	0.80	0.51	0.50	0.78	1.29	0.76	0.83	1.25
Sr	410	448	681	660	399	197	441	441	176	409	487	500	307	385	442
Ba	2195	2102	2219	1594	2176	459	2328	2442	985	2250	2246	2160	1468	1825	2160
Cs	0.44	0.44	1.22	1.36	0.85	1.34	0.47	2.95	1.97	1.13	0.95	0.76	1.29	1.03	0.98
Rb	N.A.	N.A.	N.A.	N.A.	170	201	127	189	159	248	167	135	208	164	180
U	2.8	4.7	5.2	8.5	4.3	2.4	2.1	1.2	1.1	4.1	3.8	3.7	3.3	3.8	4.4
Th	3.6	9.7	5.7	14.9	5.9	27.2	3.3	12.0	14.5	8.2	8.2	5.7	10.1	9.5	8.9
Y	122	122	57	92	129	82	119	78	54	30	56	98	57	63	96
Zr	1084	1028	1044	1262	916	238	1163	675	415	639	561	807	490	757	826
Hf	24.4	22.5	23.2	26.2	18.6	6.7	23.3	15.2	10.2	14.6	12.6	18.9	11.0	16.7	18.1
Nb	N.A.	N.A.	N.A.	N.A.	64	24	51	29	20	44	44	38	32	27	40
Ta	3.2	3.3	2.4	2.6	4.0	0.6	3.2	1.4	0.5	2.3	2.0	2.8	2.3	1.3	2.6
Sc	6.5	6.1	5.9	6.4	6.5	4.5	7.3	5.2	10.3	6.2	3.5	7.0	1.8	8.8	7.6
Cr	3	3	2	3	3	3	5	2	4	2	3	5	2	3	4
Ni	14	15	6	12	12	3	15	7	7	6	8	13	6	10	11
Co	4	3	2	3	3	2	4	2	3	2	3	4	2	4	5
V	N.A.	N.A.	N.A.	N.A.	12	4	15	9	22	11	10	22	10	16	19
ASI[††]	0.87	0.82	0.69	0.62	0.84	0.87	0.82	0.99	1.09	1.12	0.99	0.87	1.09	0.95	0.97
K_2O/Na_2O	1.42	1.39	2.79	1.29	2.02	2.55	1.57	2.53	2.14	2.51	1.75	1.82	1.41	1.85	1.84
$FeO_t/(FeO_t + MgO)$[§§]	0.87	0.84	0.77	0.80	0.84	0.88	0.84	0.78	0.78	0.84	0.87	0.84	0.84	0.84	0.84
Eu/Eu*[###]	0.63	0.63	0.72	0.62	0.62	0.13	0.67	0.68	0.39	0.91	0.87	0.59	0.67	0.72	0.65

[#]Major elements expressed in weight percent, trace elements in parts per million.

[†]Total iron expressed as Fe_2O_3.

[§]All trace elements were analyzed by instrumental neutron activation analysis or inductively coupled plasma mass spectrometry, except Sr, Ba, Y, and Zr, which were analyzed by inductively coupled plasma optical emission spectrometry.

[##]N.A.—not analyzed.

[††]ASI—alumina saturation index, A/CNK or molar $Al_2O_3/(CaO + Na_2O + K_2O)$

[§§]$FeO_t/(FeO_t + MgO)$—molar $FeO_t/(FeO_t + MgO)$, where FeO_t is total iron expressed as FeO ($Fe_2O_3 \times 0.9$).

[###]Eu/Eu*—the chondrite-normalized ratio of measured Eu divided by the hypothetical Eu concentration required to produce a rare-earth-element pattern with no Eu anomaly.

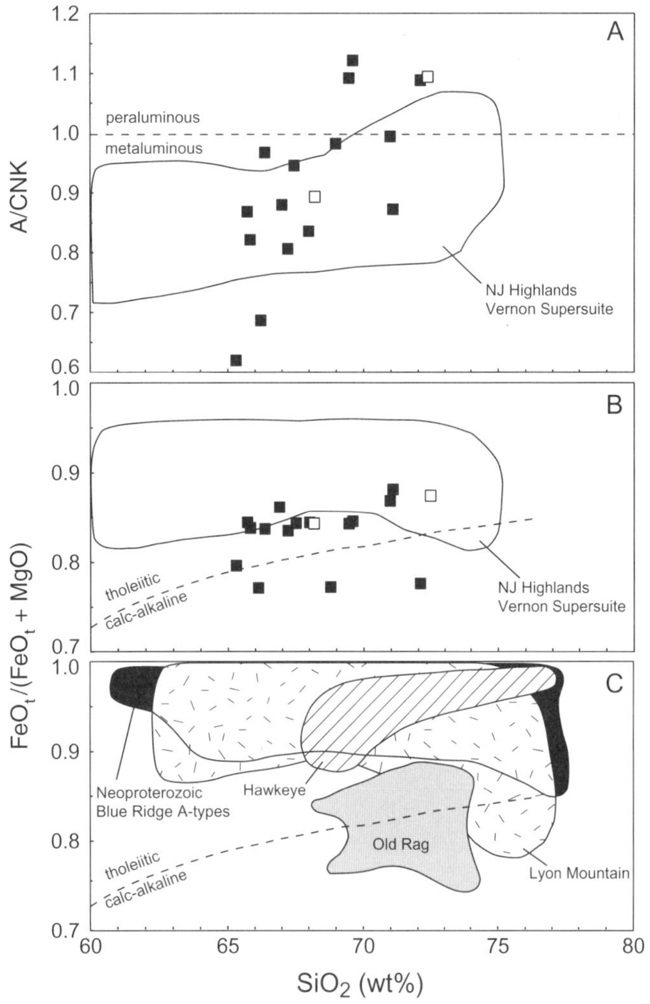

Figure 5. Plots of SiO_2 (wt%) versus (A) ASI or aluminum saturation index, A/CNK or molar $Al_2O_3/(CaO + Na_2O + K_2O)$; (B) $FeO_t/(FeO_t + MgO)$ for the Mount Eve granite suite (filled squares) and for the ca. 1100 Ma Vernon Supersuite (unfilled field; Volkert et al., 2000) of the New Jersey Highlands; and (C) $FeO_t/(FeO_t + MgO)$ for granitoids and granitic gneisses from the Adirondacks, including the 1103–1093 Ma Hawkeye suite (diagonal striped field; J. McLelland, personal commun.) and the 1060–1045 Ma Lyon Mountain Gneiss (open field with small ticks; Whitney and Olmsted, 1988), and also granitoids from the northern Blue Ridge Province, including the 1060 ± 5 Ma Old Rag granite suite (gray shaded field; Tollo et al., this volume) and the Neoproterozoic (765–680 Ma) A-type suite (black shaded field; Tollo and Aleinikoff, 1996; Tollo et al., 2004). The dividing line separating the fields of tholeiitic and calc-alkaline affinity is from Anderson (1983). FeO_t refers to total iron expressed as FeO. Plots B and C modified after Frost et al. (2000).

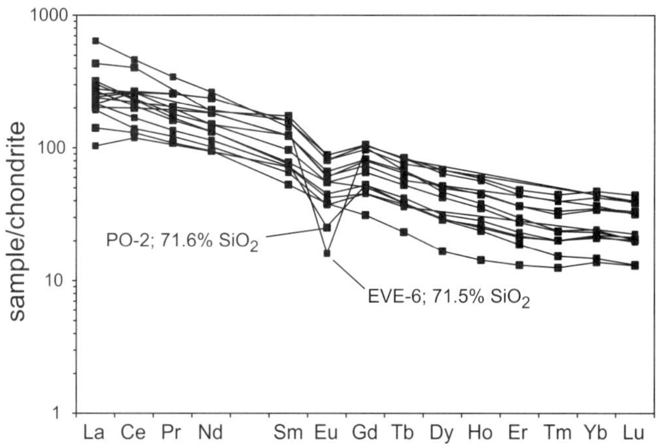

Figure 6. Rare-earth-element (REE) plots of Mount Eve granite suite samples. Normalization factors are based on Leedy chondrite (Masuda et al., 1973) and are La (0.378), Ce (0.978), Pr (0.15), Nd (0.716), Sm (0.23), Eu (0.0866), Gd (0.311), Tb (0.0589), Dy (0.39), Ho (0.087), Er (0.255), Tm (0.039), Yb (0.249), and Lu (0.0387). PO-2 and EVE-6 are high-SiO_2 samples with the lowest Eu/Eu* values (Table 1), where Eu/Eu* is the chondrite-normalized ratio of measured Eu divided by the hypothetical Eu concentration required to produce REE pattern with no Eu anomaly.

8, C). Given the limited volumes and map extent of the Mount Eve granite suite, it is unlikely that these few rocks were generated in a different tectonic setting or involved melting of significantly different crustal source rocks. These two samples have the largest negative Eu anomalies and highest SiO_2 contents (Fig. 6), and thus their relatively low Ta concentrations are attributed to crystal fractionation processes, as discussed later. According to the classifications of Eby (1992), most Mount Eve samples are A_2-type granites (Figs. 9, A and B), indicating their derivation from crustal sources that previously had been generated in a subduction zone or continent-continent collision tectonic environment. This is further supported by the fact that the Mount Eve samples analyzed in this study fall mainly within the compositional range between average continental crust and island-arc basalt (Fig. 10). As noted by previous workers (e.g., Whalen et al., 1987; Eby, 1992; Landenberger and Collins, 1996; Förster et al., 1997), the trace-element discrimination diagrams of Pearce et al. (1984) more accurately reflect the chemistry of crustal source rocks than they identify tectonic settings. With this in mind, the trace-element data for the Mount Eve granite suite indicate derivation from crustal sources that were affected by previous convergent orogenic processes (e.g., the Ottawan orogeny) and were subsequently partially melted during a post-Ottawan crustal heating event.

Trace-element abundances in the Mount Eve granite suite show systematic variation that can be attributed to crystal fractionation of mainly plagioclase and K-feldspar, and accessory zircon and titanite (Figs. 11–13). Strontium and barium concentrations show a progressive decrease with increasing SiO_2 (Fig. 11, A and B), and the two samples with the highest SiO_2 and lowest Sr and Ba also have the lowest Eu/Eu* ratios (Fig. 12, A and B), indicating that feldspars were important crystallizing phases in the Mount Eve suite magmas. In general, the data set more closely parallels fractional crystallization vectors for K-feldspar (Fig. 12, B and C); however, Rb/Ba–Rb/Sr sys-

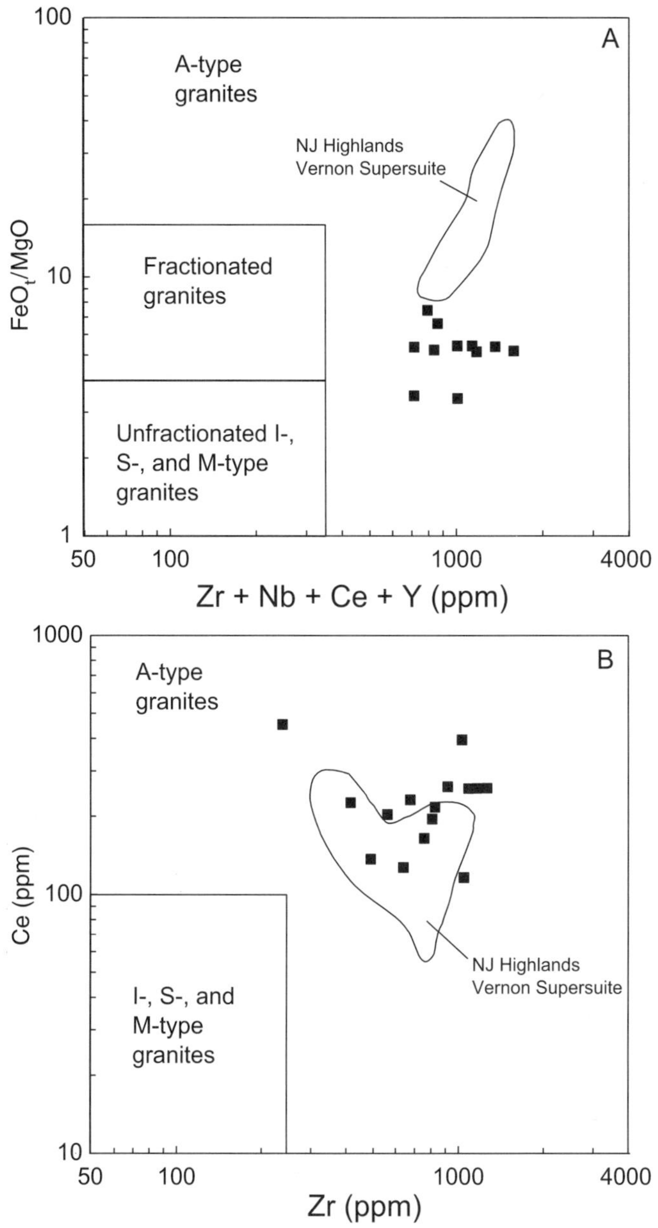

Figure 7. Plots of (A) Zr + Nb + Ce + Y versus FeO_t/MgO and (B) Zr versus Ce for the Mount Eve granite suite (filled squares) compared to the A-typeVernon Supersuite granitoids from the New Jersey Highlands (unfilled field; Volkert et al., 2000). FeO_t refers to total iron expressed as FeO. Diagrams modified after Whalen et al. (1987).

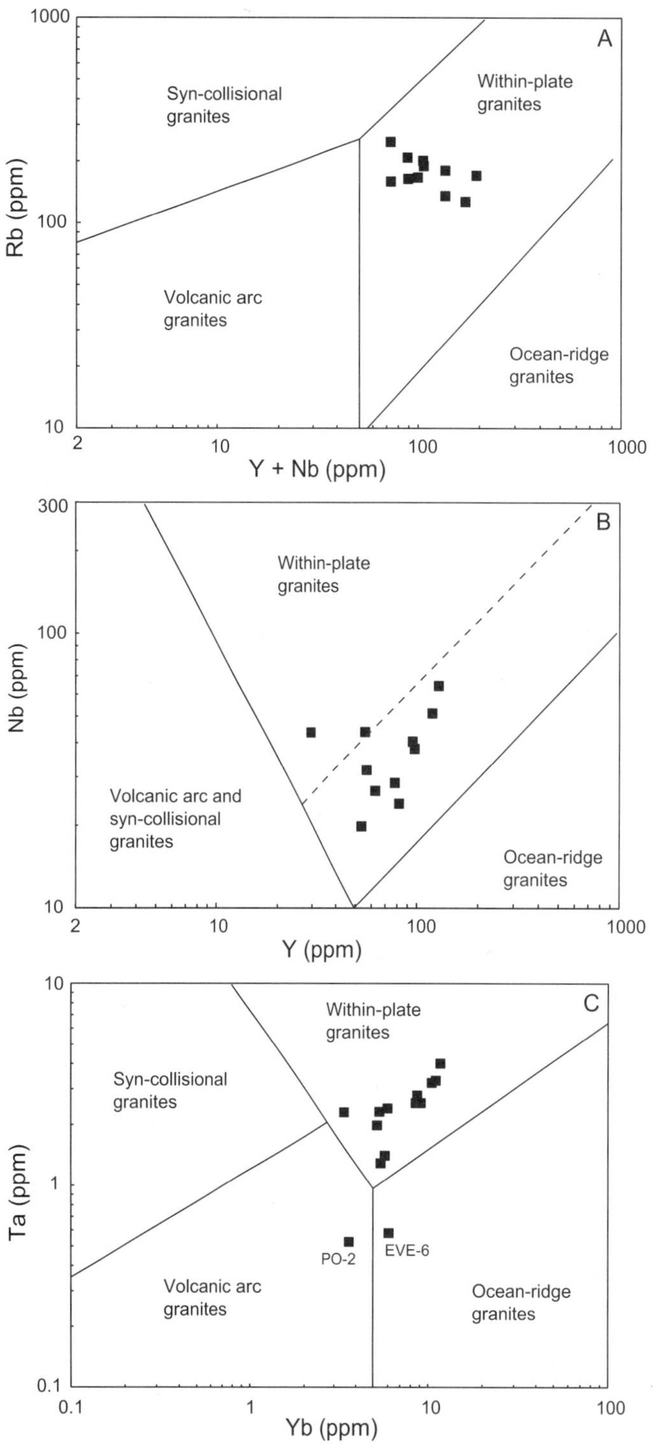

Figure 8. Plots of (A) Y + Nb versus Rb, (B) Y versus Nb, and (C) Yb versus Ta for the Mount Eve granite suite (filled squares) from the study area. Diagrams modified after Pearce et al. (1984). Dashed line in Figure 8, B, is the upper limit of oceanic granites from anomalous ridge segments. PO-2 and EVE-6 are high-SiO_2 samples with the lowest Eu/Eu* values (Table 1), where Eu/Eu* is the chondrite-normalized ratio of measured Eu divided by the hypothetical Eu concentration required to produce REE pattern with no Eu anomaly.

tematics (Fig. 12, D) suggest a mixture of plagioclase and K-feldspar crystallization. Therefore, it is likely that both plagioclase and K-feldspar were important crystallizing phases in the Mount Eve granite suite.

Accessory minerals such zircon, allanite, and titanite are common in granitoid rocks and can exert a strong control on trace-element systematics of granitic liquids due to their large

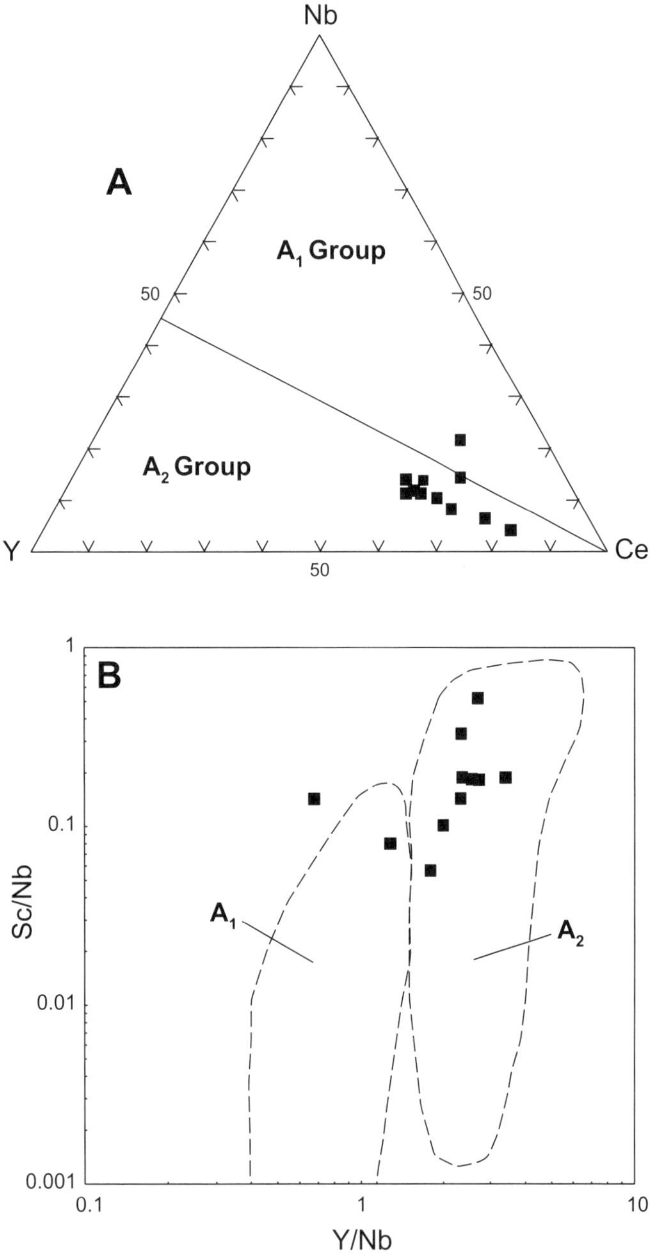

Figure 9. Plots of (A) Nb–Ce–Y and (B) Y/Nb versus Sc/Nb showing the distribution of Mount Eve granite (filled squares) relative to fields (dashed lines in Figure 9, B) for the A_1 and A_2 subtypes of A-type granitoids. Diagrams modified after Eby (1992).

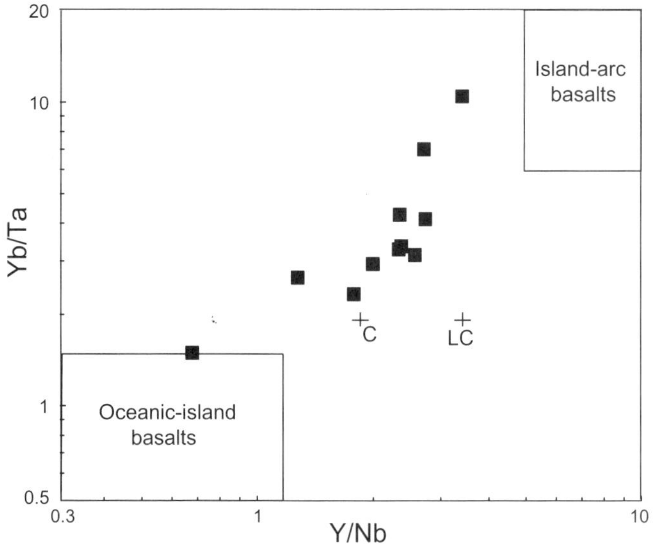

Figure 10. Plot of Y/Nb versus Yb/Ta for Mount Eve granite suite rocks (filled squares). Crosses indicate composition of average continental crust (C) and lower crust (LC), from Taylor and McLennan (1985). Boxes show fields for oceanic-island and island-arc basalts. Diagram modified after Eby (1990).

affect on bulk distribution coefficients for HFSE (Zr, Hf, Nb, and Ta) and REE (Mahood and Hildreth, 1983; Sawka, 1988). In the Mount Eve samples, Zr and Hf both show decreasing trends with increasing SiO_2, indicating that zircon played an important role during crystallization (Fig. 11, C and E). Ta and Nb also show large decreases with increasing SiO_2, which most likely represents the combined effect of fractionation of small amounts of titanite ± Fe-oxides (Fig. 11, D and F). The ubiquitous presence of titanite identified in thin section and by electron microprobe suggests that it was probably a significant phase controlling HFSE variations in the Mount Eve suite. Further evidence for crystal fractionation of both zircon and titanite is displayed by systematic increases in Hf/Ta, Th/Ta, and La/Sm with increasing Ce/Yb (Fig. 13, A–C, respectively). Fractional crystallization vectors in Figure 13, A–C, show that this trace-element variation is consistent with the combined effect of zircon and titanite crystallization using distribution coefficients from Mahood and Hildreth (1983) and Sawka (1988) for these mineral phases. Allanite is essentially precluded from this assemblage because it would have depleted the fractionating granitic liquids in light REE and Th, and thus Ce/Yb would show systematic decreases with Th/Ta and Hf/Ta, which are not observed.

PETROGENETIC MODELS

Several different models for the petrogenesis of A-type granites have been proposed that reflect the wide range of chemical variations observed in these rocks. These models can be grouped into three major categories: (1) those that involve extensive fractional crystallization from basaltic parents (Loiselle and Wones, 1979; Turner et al., 1992), (2) those that involve a combination of mantle and crustal sources via mixing and/or assimilation-fractional crystallization (AFC) processes (e.g., Barker et al., 1975; Foland and Allen, 1991), and (3) those that invoke partial melting of various crustal lithologies, including anhydrous lower crustal granulites (Collins et al., 1982; Whalen

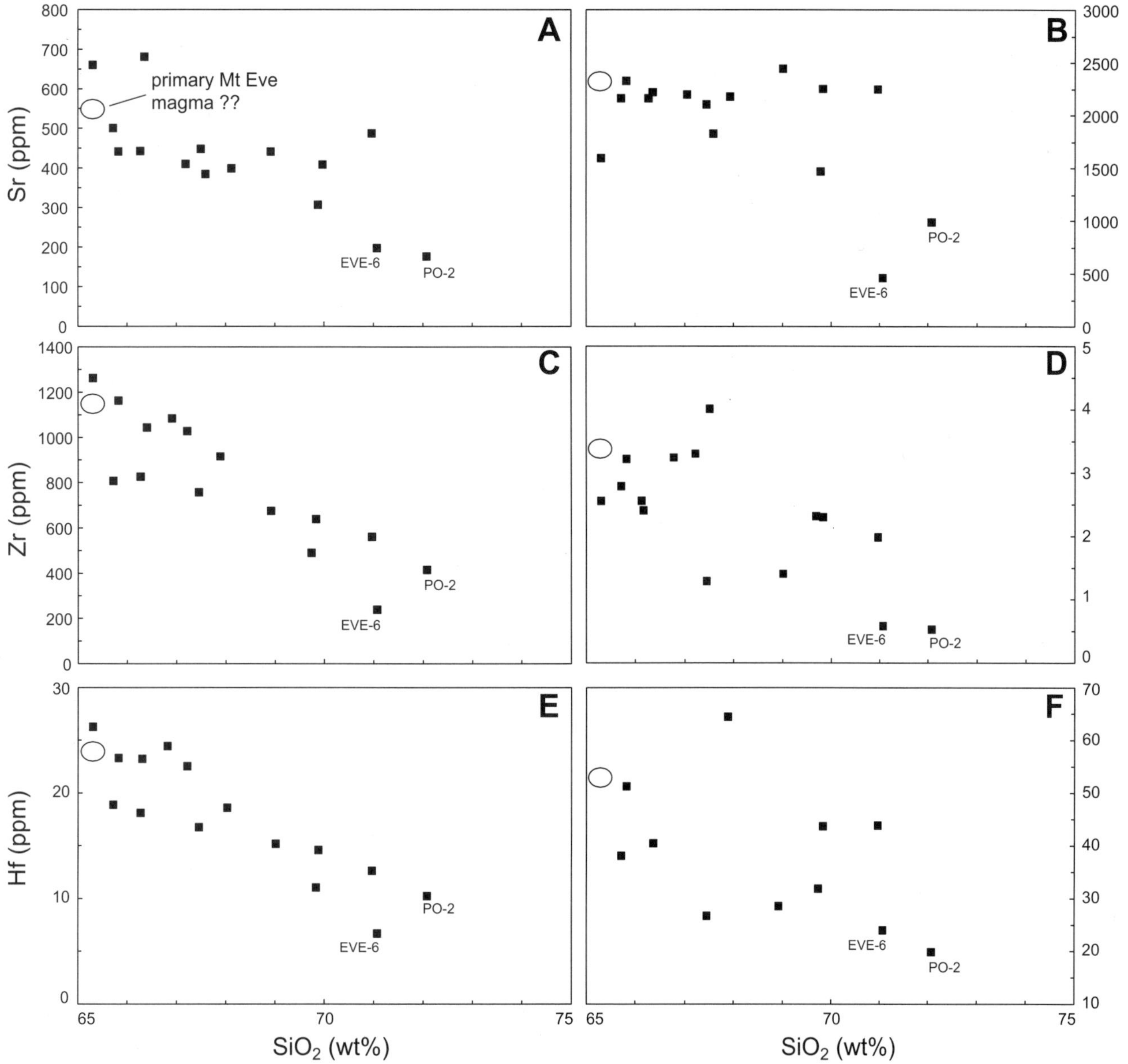

Figure 11. Plots of (A) Sr, (B) Ba, (C) Zr, (D) Ta, (E) Hf, and (F) Nb versus SiO_2 for the Mount Eve granite suite (filled squares). Decreasing Sr and Ba with increasing SiO_2 indicates feldspar crystal fractionation, and decreasing Zr, Hf, Ta, and Nb with increasing SiO_2 indicates zircon + titanite (± Fe-oxide) crystal fractionation. Small, unfilled oval represents the estimated parental magma for the Mount Eve granite suite. PO-2 and EVE-6 are high-SiO_2 samples with the lowest Eu/Eu* values (Table 1), where Eu/Eu* is the chondrite-normalized ratio of measured Eu divided by the hypothetical Eu concentration required to produce REE pattern with no Eu anomaly.

et al., 1987; Landenberger and Collins, 1996), underplated mafic rocks and their differentiates (Frost and Frost, 1997), and intermediate granitoid rocks ranging from tonalite to granodiorite (Anderson, 1983; Sylvester, 1989; Creaser et al., 1991).

Loiselle and Wones (1979) proposed that extensive fractional crystallization of tholeiitic basalts would produce felsic magmas with high FeO/MgO and low oxygen fugacity and water content, characteristic of some A-type granites. This mechanism has been proposed to explain the Red Bluff granitic suite in West Texas (Shannon et al., 1997), the Sybille Intrusion of the Laramie anorthosite complex (Scoates et al., 1996), and the Topsails igneous suite of Newfoundland (Whalen et al., 1996). However, this mechanism is not likely to have formed the Mount Eve granite suite because there is no field evidence for the large volumes of associated mafic rocks and the continuum of compositions that would be expected to have resulted from

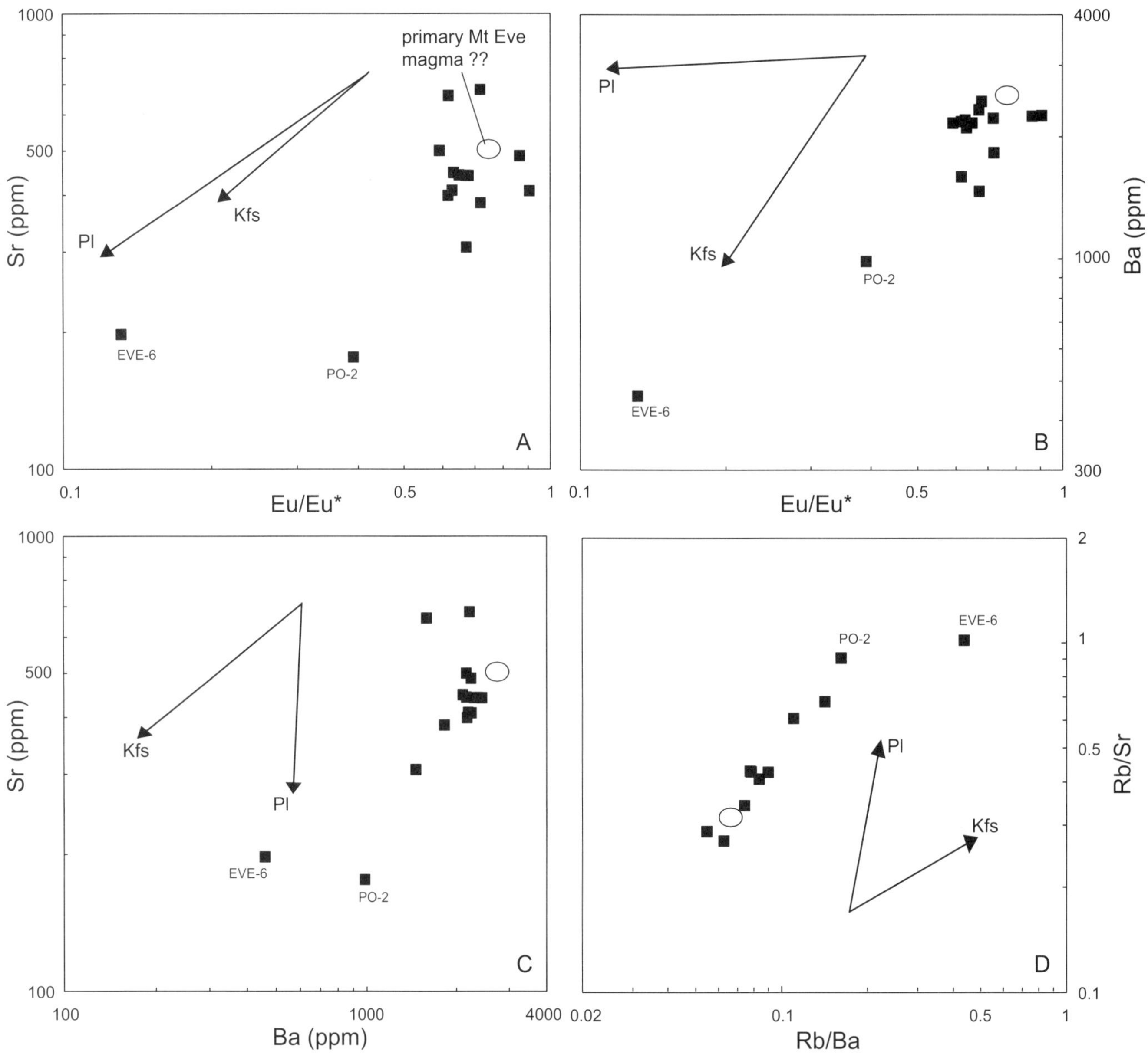

Figure 12. Plots of (A) Sr and (B) Ba versus Eu/Eu*, (C) Sr versus Ba, and (D) Rb/Sr versus Rb/Ba for the Mount Eve granite suite (filled squares). Arrows indicate the direction and magnitude of Rayleigh-type crystal fractionation of plagioclase (Pl; 20%) and K-feldspar (Kfs; 20%) using distribution coefficients from Mahood and Hildreth (1983) and Nash and Crecraft (1985). Small, unfilled oval represents the estimated parental magma for the Mount Eve granite suite. Mineral abbreviations are after Kretz (1983). PO-2 and EVE-6 are high-SiO_2 samples with the lowest Eu/Eu* values (Table 1), where Eu/Eu* is the chondrite-normalized ratio of measured Eu divided by the hypothetical Eu concentration required to produce REE pattern with no Eu anomaly.

such a mechanism. Furthermore, the high concentrations of Sr and modest Eu anomalies in Mount Eve granite suite preclude extensive fractionation of a gabbroic mineral assemblage from a basaltic parent. Although more difficult to completely dismiss, similar arguments can be made against mechanisms involving mixing and/or AFC of mantle-derived basaltic magmas with crust-derived felsic melts (e.g., Barker et al., 1975) to produce the parental magmas of the Mount Eve granite suite. As in the previous model, the expected wide spectrum of spatially associated and demonstrably comagmatic rocks (e.g., gabbro to granite) is not observed. However, this does not entirely preclude mixing and AFC processes. The parental magmas to the Mount Eve granite suite could have been produced by mixing and/or crustal AFC from magmas derived from partial melting

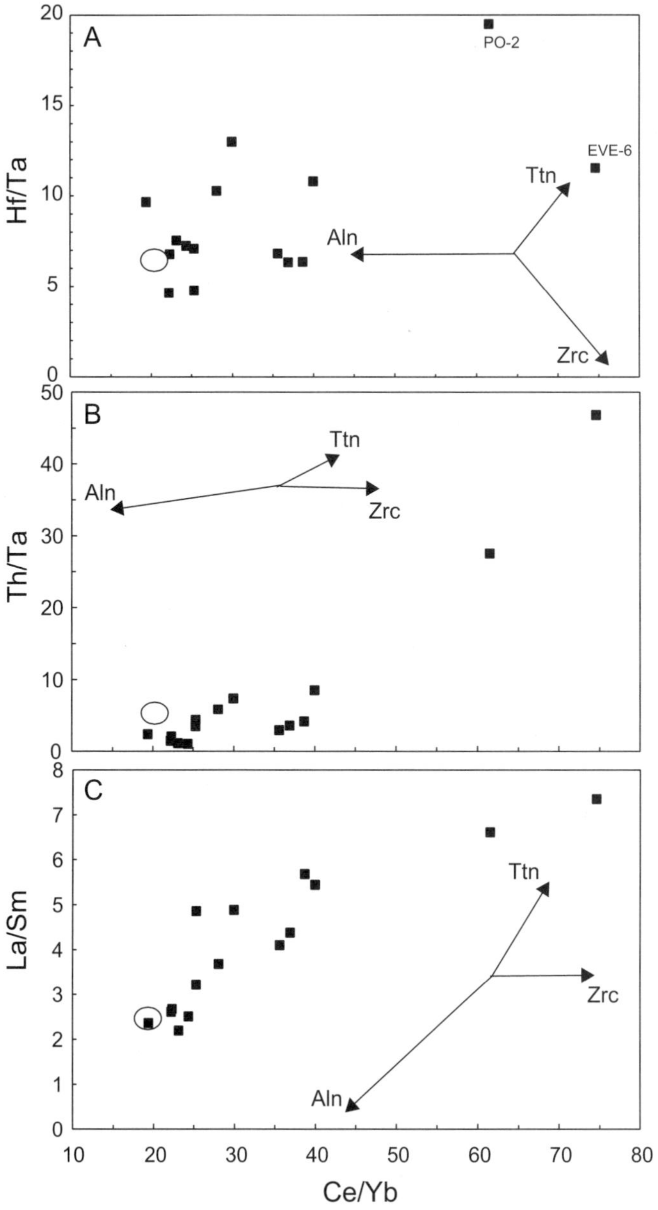

Figure 13. Plots of (A) Hf/Ta, (B) Th/Ta, and (C) La/Sm versus Ce/Yb for the Mount Eve granite suite (filled squares). Arrows indicate the direction and magnitude of Rayleigh-type crystal fractionation of allanite (Aln; 2%), zircon (Zrc; 2%), and titanite (Ttn; 2%) using distribution coefficients from Sawka (1988). Small, unfilled oval represents the estimated parental magma for the Mount Eve granite suite. Mineral abbreviations are after Kretz (1983). PO-2 and EVE-6 are high-SiO_2 samples with the lowest Eu/Eu* values (Table 1), where Eu/Eu* is the chondrite-normalized ratio of measured Eu divided by the hypothetical Eu concentration required to produce REE pattern with no Eu anomaly.

of mafic to intermediate crust (see discussion later). This mechanism is difficult to address without Sr and Nd isotope data on the Mount Eve granite suite, as well as proposed crustal source rocks.

Given the relatively narrow chemical variations and the lack of associated mafic to intermediate rocks, partial melting of crustal rocks is interpreted to have been the dominant petrogenetic process that formed the parental magmas of the Mount Eve granite suite. Supracrustal metasedimentary sequences in the New Jersey–Hudson Highlands are ruled out because they generally lack the appropriate compositions to have produced A-type granites. Dehydration partial melting of micas within aluminosilicate-rich metapelites within the Hudson Highlands has been proposed to have been responsible for forming the highly peraluminous S-type Canada Hill granite (Heleneck and Mose, 1984; Ratcliffe, 1992). In contrast, metaigneous crustal lithologies, commonly cited as possible source rocks for A-type granites (e.g., Collins et al., 1982; Anderson, 1983; Clemens et al., 1986; Whalen et al., 1987; Sylvester, 1989; Creaser et al., 1991; Frost and Frost, 1997), are abundant in the New Jersey–Hudson Highlands (except for tholeiitic ferrodiorites). In particular, rocks of the Losee metamorphic suite (amphibolite, intermediate charnockite, tonalitic gneiss) constitute a large volume of the New Jersey Highlands basement and are considered here as the most representative of possible crustal source rocks for the Mount Eve granite suite. As discussed earlier, geochemical data suggest that the Losee rocks represent a continental-arc sequence, and thus would be appropriate source rocks for A_2-subtype trace elements with the characteristics of those in the Mount Eve granite suite (Fig. 9). In the following sections we explore this hypothesis by comparing the experimental data available for the partial melting of starting materials that are representative of potential source rocks within the Losee metamorphic suite with the least chemically evolved samples of the Mount Eve granite suite (~65–68 wt% SiO_2). Based on the evidence for crystal fractionation within the Mount Eve granite suite, we suggest that the least evolved samples are the best approximation to parental Mount Eve magmas, and therefore best reflect the chemical characteristics of the crustal source(s) from which they were derived.

Partial Melting of Tonalitic to Granodioritic Source Rocks

Partial melting of tonalitic to granodioritic source rocks has been proposed by several workers to explain the origin of A-type granites (Anderson, 1983; Sylvester, 1989; Creaser et al., 1991; Patiño Douce, 1997; Smith et al., 1999). Figure 14, A–D, shows the major-element composition of experimentally produced partial melts of tonalitic to granodioritic sources (e.g., Carroll and Wyllie, 1990; Skjerlie and Johnston, 1993; Patiño Douce, 1997) compared to the major-element compositions of rocks of the Mount Eve granite suite. The majority of the least chemically evolved Mount Eve samples have much lower SiO_2 and higher

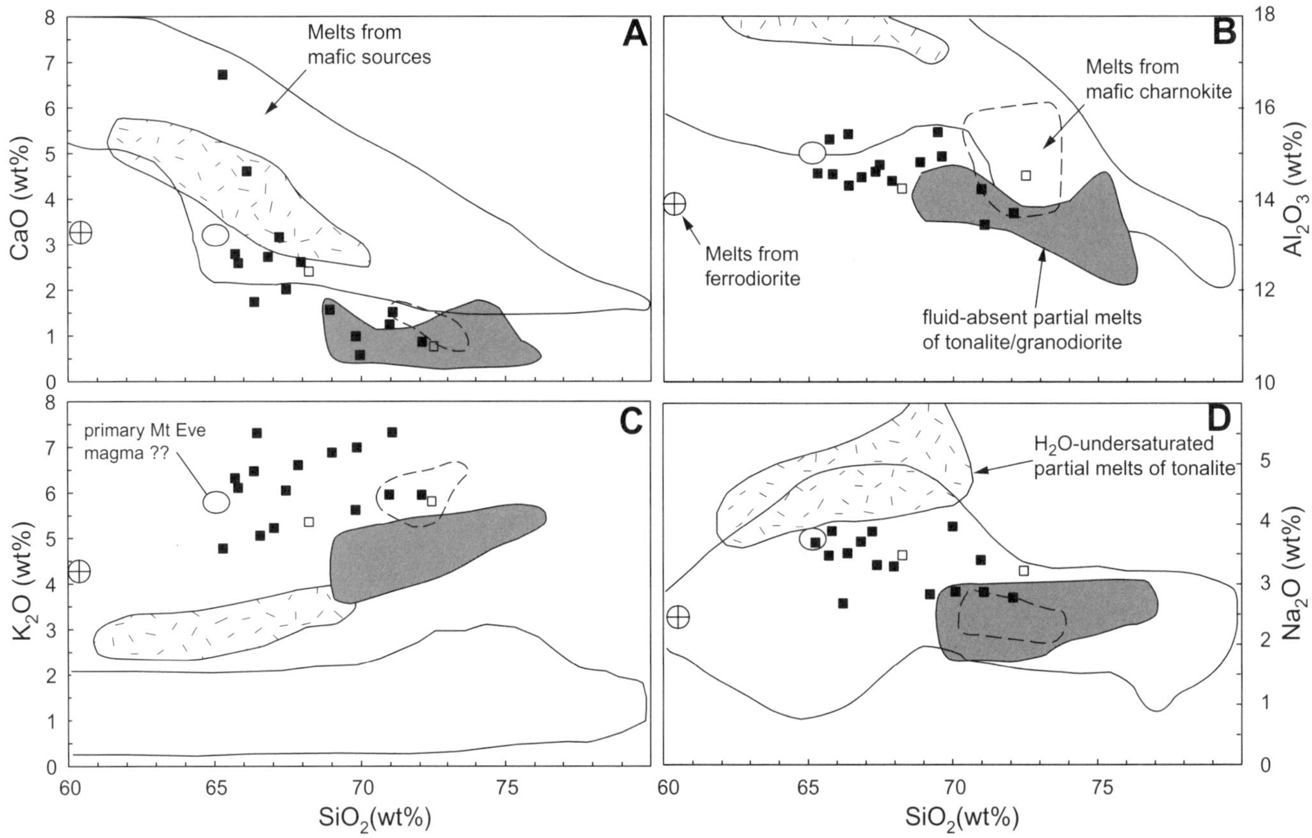

Figure 14. Plots of (A) CaO, (B) Al_2O_3, (C) K_2O, and (D) Na_2O versus SiO_2 for the Mount Eve granite suite samples in this study (filled squares) compared to experimental partial melts of fluid-absent tonalite/granodiorite (gray field; Skjerlie and Johnston, 1993; Patiño Douce, 1997); H_2O-undersaturated tonalite (open field with small ticks; Carroll and Wyllie, 1990); basalts and mafic amphibolites (unfilled field; Beard and Lofgren, 1991; Rushmer, 1991; Rapp and Watson, 1995); Grenville mafic charnockitic (open field with dashed outline; Beard et al., 1994); and ferrodiorite (circle with cross; Scoates et al., 1996). Small, unfilled oval represents the estimated parental magma for the Mount Eve granite suite.

CaO contents than liquids derived from the fluid-absent melting of tonalitic or granodioritic sources. Hydrous partial melting of tonalitic compositions (Carroll and Wyllie, 1990) produces melts with the appropriate SiO_2, CaO, and MgO contents, but the Al_2O_3 and Na_2O contents are much higher and the K_2O content lower than those of the parental Mount Eve Granite magmas. Overall, these comparisons suggest that partial melting of more mafic rocks was most likely involved in the petrogenesis of parental magmas to the Mount Eve Granite suite.

Partial Melting of Mafic Amphibolites

Several workers have discussed the origin of A-type granites by partial melting of mafic source rocks (e.g., Frost and Frost, 1997; Smith et al., 1999). Figure 14, A–D, illustrates that experimentally generated partial melts of basalt and amphibolite can produce intermediate to felsic liquids (Beard and Lofgren, 1991; Rushmer, 1991; Rapp and Watson, 1995) that are similar to the compositions of the least evolved rocks of the Mount Eve granite suite, although their K_2O contents are much lower than those of the Mount Eve granite suite, and to most A-type granites in general. However, most of the partial-melt experiments have been carried out on low- and medium K mafic rocks (<1 wt% K_2O), and Beard and Lofgren (1991) suggest that the alkali content of the melts produced is related to the composition of the starting materials. Chemical data for New Jersey–Hudson Highlands amphibolites indicate a wide variety of basaltic protolith compositions (Volkert and Drake, 1999; Kula and Gorring, 2000; Harclerode et al., 2001), ranging from low-K (<0.4 wt%) tholeiitic basalts to medium-K (1.0 to 1.5 wt%) calc-alkaline basalts. Systematic geochemical investigation of amphibolites within the Losee metamorphic suite has yet to be undertaken, but preliminary data suggest that most are of the medium-K, calc-alkaline type (Kula and Gorring, 2000). Therefore, partial melting of Losee amphibolites could have generated intermediate to felsic A-type liquids with high K_2O and K/Na ratios similar to those of the Mount Eve granite suite. However, as pointed out by Smith et al. (1999), this possibility is difficult to assess because of a lack of experimental data for partial melts of K-rich mafic rocks.

Partial Melting of Charnockitic Gneisses

Other viable lithologies within the Losee metamorphic suite that may have partially melted to produce the Mount Eve granite include a suite of charnockitic intermediate metaigneous rocks ranging in composition from diorite to tonalite (Volkert and Drake, 1999). This suite of rocks is similar in composition to sources involved in petrogenetic models that invoke partial melting of dehydrated charnockitic lower crust to explain the petrogenesis of A-type granites (e.g., Collins et al., 1982; Whalen et al., 1987; Landenberger and Collins, 1996). Tollo and Aleinikoff (1996) suggested partial melting of charnockitic Grenville-age basement in the petrogenesis of parental magmas to metaluminous A_2-type granites of the Robertson River igneous suite (e.g., Laurel Mills granite) in the Blue Ridge Province of Virginia. High-temperature partial melting experiments involving Grenville-age charnockitic rocks (~60–64 wt% SiO_2) conducted by Beard et al. (1994) produced high-SiO_2 (>70 wt%) granitoid melts, and thus these are not appropriate source rocks for parental Mount Eve magmas (Fig. 14, A–D). Volkert et al. (2000) proposed that orthopyroxene-bearing diorite (~55 wt% SiO_2) within the Losee metamorphic suite has the appropriate composition, and upon partial melting could have produced the parental syenitic liquids that formed the widespread and voluminous 1100 Ma A-type Vernon Supersuite magmas. This supports earlier conclusions that more mafic rocks (e.g., amphibolite, diorite) were probably more viable as source rocks for the generation of parental A-type magmas that resemble the least chemically evolved Mount Eve granite suite samples.

A more definitive assessment of the Losee metamorphic suite as the ultimate source of the Mount Eve magmas will necessitate the production of additional isotope and geochemical data on Losee rocks. Nevertheless, we consider the lithologic variability in the Losee metamorphic suite representative of the range of possible crustal sources that could have generated the parental Mount Eve granite magmas. Furthermore, as indicated earlier, it is likely that the parental Mount Eve magmas represent mixtures of partial melts from various source rock compositions in variable proportions. We propose a model similar to that of Frost and Frost (1997), where the metaluminous, relatively oxidized characteristics of the least evolved Mount Eve magmas could be explained as mixtures of intermediate and felsic partial melts derived from the Losee metamorphic suite. These parental magmas would then have crystallized a feldspar-dominated assemblage in the shallow crust to produce the more evolved high-SiO_2 varieties of the Mount Eve granite suite.

IMPLICATIONS FOR GRENVILLE TECTONICS

One of the principal tectonomagmatic events in the Grenville orogen during the late Mesoproterozoic was the Ottawan orogeny (ca. 1090–1030 Ma; e.g., McLelland et al., 1996, 2001). The Ottawan orogeny is thought to have been a Himalayan-style continental collision event with associated crustal thickening, high-grade metamorphism, ductile nappe-style folding in the southeast (e.g., in the Central Granulite Terrane, the Adirondack Highlands, and Appalachian massifs), and brittle northwest-directed thrusting farther west in the orogen (e.g., in the Grenville Front tectonic zone and the Central Metasedimentary Belt). Although the timing of peak Ottawan orogenesis varied spatially, this event severely affected most rocks older than ca. 1060 Ma throughout much of the Grenville orogen (e.g., McLelland, 1996, 2001; Aleinikoff et al., 2000). The age and field relations of the Mount Eve granite suite indicate that its origin is clearly related to post–peak Ottawan tectonic activity, and, as pointed out by Drake et al. (1991), place a lower limit of ca. 1020 Ma on penetrative Ottawan metamorphism and deformation in the New Jersey–Hudson Highlands. Tectonic activity within the Central Granulite terrane and the Adirondack sectors of the Grenville orogen during the interval between 1090 and 1000 Ma was generally characterized by rapid uplift, lithospheric delamination, and extensional collapse along low-angle normal faults within the overthickened core of the Ottawan orogen (McLelland et al., 1996, 2001; Corrigan and Hamner, 1997). Extrapolation of this mechanism to the New Jersey–Hudson Highlands region suggests that lithospheric delamination could have caused deep crustal melting to produce the Mount Eve granite suite. In this model, the A-type characteristics of the Mount Eve granite suite could be explained by partial melting of crustal rocks as hot asthenosphere impinged on the base of the crust subsequent to lithospheric delamination. Similar delamination models have been proposed for the syn- to Late Ottawan, mildy A-type granitoids of the Lyon Mountain Gniess suite (ca. 1060–1045 Ma) in the Adirondacks (McLelland et al., 1996; 2001) and the 1020 to 1010 Ma Labrieville, St. Urbain, and St. Ambroise anorthosite-granite associations in Québec (e.g., Owens et al., 1994; Higgins and van Breemen, 1996). Such major lithospheric delamination in the New Jersey–Hudson Highlands would have been diachronous with syn- to Late Ottawan delamination events in the Adirondacks that produced the Lyon Mountain Gniess granitoid suite. However, a major problem with lithospheric delamination models for the Mount Eve granite suite concerns the small volumes and limited areal extent of the magmatism. Unless delamination was localized, larger volumes and a wider geographic distribution of post-tectonic, undeformed, A-type granitoid magmas would be expected within the New Jersey–Hudson Highlands. Rocks that fit this description have not been documented anywhere in the New Jersey–Hudson Highlands, and localized delamination of a small section of lithosphere beneath the Mount Eve area seems unlikely. These problems notwithstanding, it is possible that residual thermal input from major lithospheric delamination events that occurred in the Adirondacks during or immediately after peak Ottawan orogenesis (ca. 1090–1030 Ma) could have supplied heat to the base of the crust beneath the Mount Eve area. However, such processes do not explain why crustal melting was localized in the Mount Eve area.

An alternative model to lithospheric delamination that might explain the localized nature of the Mount Eve granite suite is crustal heating due to broad orogenic relaxation between peak Ottawan orogenesis and the onset of the post-Ottawan, high-grade, dextral transpressional shearing event that occurred between ca. 1010 and 915 Ma (Gates, 1995, 1998; Fig. 15). This Late Grenville–age transpressional deformation is characterized by a group of vertical, anastomosing shear zones, 0.5 to 2 km wide, that overprint older high-grade fabric elements (Gates, 1998). A post-Ottawan escape-type tectonic event in the central Appalachians resulting from accretion to the north is interpreted to have produced this deformation (Gates et al., 2001b). Localized crustal heating due to upwelling asthenosphere associated with localized extension and/or transtension in the overall dextral transpressional regime could better explain the small volumes and limited areal extent of the Mount Eve granite suite (Fig. 15). This tectonic scenario is also consistent with the "within-plate" and A_2-type trace-element compositions of the Mount Eve granite suite. The strike-slip environment could also explain the temporal association of two other intrusive igneous rock suites of post-Ottawan age within the Hudson Highlands: the Canada Hill granite (1010 ± 6 Ma; Aleinikoff and Grauch, 1990) and the Lake Tiorati diorite (1008 ± 4 Ma; Gates et al., 2001a). Collectively, these rocks, along with the Mount Eve granite suite, consist of small, dispersed plutonic bodies that form a volumetrically minor, chemically diverse group that ranges from A- and S-type granites to calc-alkaline, I-type diorite. A similar transpressional tectonic model has been proposed by Speer et al. (1994) to explain the occurrence of small-volume, chemically diverse plutonic suites of Alleghanian age in the southern Appalachians. This type of model may explain similar small-volume occurrences of Late to post-Ottawan (ca 1030–930 Ma) granitoids elsewhere in the Grenville orogen, particularly in the Adirondacks (e.g., the Lyonsdale Bridge and Cathead Mountain pegmatites; McLelland et al., 2001) and the Green Mountain massif (e.g., the Stamford Hill rapakivi granite; Ratcliffe et al., 1991).

CONCLUSIONS

The Mount Eve granite suite is a relatively homogeneous suite of granitoid rocks that intruded Mesoproterozoic basement of the New Jersey–Hudson Highlands at ca. 1020 Ma during the waning stages of the Ottawan orogeny. The metaluminous to slightly peraluminous granitoids range from quartz monzonite to granite (Drake et al., 1991; Volkert and Drake, 1999) and typically contain Fe-rich hornblende ± biotite and accessory zircon, titanite, apatite, magnetite, and ilmenomagnetite. The Mount Eve granite suite displays A-type chemical characteristics, including high concentrations of Zr, Y, and total REE; high FeO_t/MgO and K_2O/Na_2O; variably negative Eu anomalies; and low MgO, Cr, and Ni. Slight depletion in Nb relative to other HFSE and REE allows further classification of the Mount Eve granite suite into the A_2 chemical subtype (Eby, 1992). Based on experimental and trace-element data, as well as regional geology, the parental Mount Eve granitoids were likely derived from relatively dry and oxidized partial melting of rocks similar to the metaigenous lithologies of the Losee metamorphic suite within the Mesoproterozoic basement. Covariations between key major and trace elements can be explained by crystal fractionation of feldspars and accessory phases (zircon and titanite) from relatively low-SiO_2 parental quartz monzonitic liquids

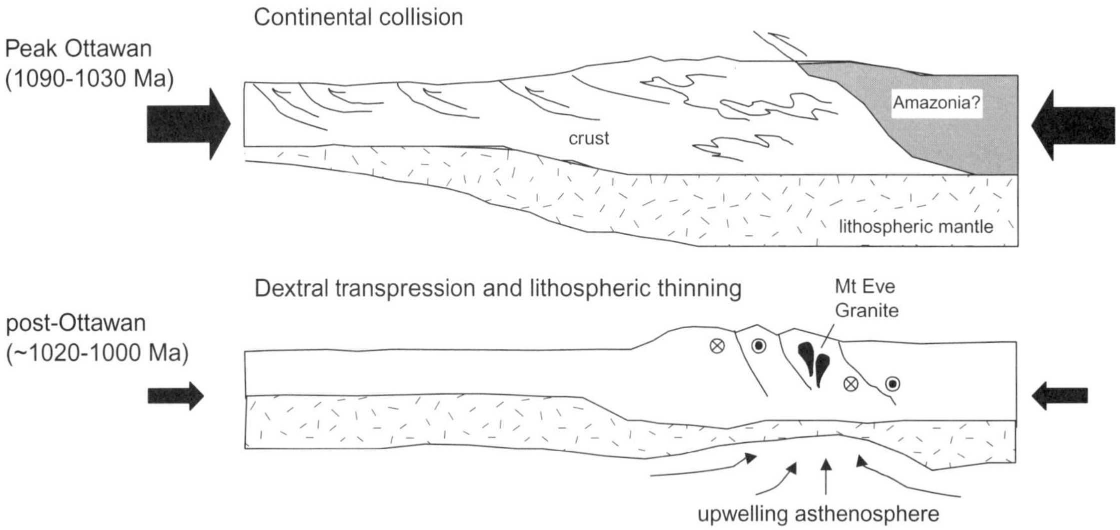

Figure 15. Schematic lithospheric cross-section during the late Middle Proterozoic in the New Jersey Highlands (1100–1000 Ma), modified from McLelland et al. (1996). The Mount Eve granite suite formed from broad lithospheric thinning and the onset of dextral transpressional shearing (e.g., Gates et al., 2001b). Small circles indicate dextral strike-slip motion; those with dots indicate blocks moving toward reader, and those with *x*s represent blocks moving away from reader.

to generate the syenogranites that are higher in SiO_2 and rich in K-feldspar.

The Mount Eve granite suite is part of a group of small-volume, chemically diverse plutonic rocks sporadically emplaced throughout the central and northern Appalachian basement massifs and the Adirondack Highlands between 1030 and 930 Ma. Mechanisms that might have produced the Mount Eve franite suite include (1) residual thermal heating of the base of the crust subsequent to major lithospheric delamination of an overthickened Ottawan orogen, or (2) localized extension and crustal melting due to the onset of Late to post-Ottawan dextral transpressional shearing superimposed on broad orogenic relaxation. The latter mechanism has particular advantages over the former model because it more easily explains the small volumes, limited areal extent, and chemical diversity of these rocks formed late in the Grenville orogenic cycle. Support for a transpressional origin also comes from the similarity of the Mount Eve granite suite in volume, distribution, and lithology to Alleghanian granitoids from the southern Appalachians that are proposed to have formed in a similar strike-slip transpressional tectonic environment.

APPENDIX—ANALYTICAL METHODS

Powders were prepared from samples weighing 10–15 kg by crushing in a hardened steel jaw crusher to a size of ~0.5 cm in diameter. The resulting chips were then thoroughly mixed, and a 50 g split was pulverized in an Al-ceramic shatter box that was used for major- and trace-element analysis. Major- and abundant trace-element (Sr, Ba, Zr, and Y) contents were determined at the Department of Earth and Environmental Studies, Montclair State University, by inductively coupled plasma optical emission spectrometry (ICP-OES) on a JY Ultima C system. Approximately 100 mg of sample were mixed with 400 mg of ultrahigh-purity lithium metatetraborate flux ($^{(R)}$Spex-Certiprep) and fused in high-purity graphite crucibles at 1100 °C for 45 minutes. This technique achieves complete dissolution of the sample, including all accessory phases (e.g., zircon, titanite) based on visual inspection and replicate analyses of U.S. Geological Survey (USGS) standard G-2 and multiple aliquots of EVE-1 (see later). Molten samples were immediately dissolved in 50 ml of 7% HNO_3, and 6.5 ml of this solution was diluted in 50 ml of 2.5% HNO_3 (dilution factor of ~4000x) before ICP-OES analysis.

Experimental water was distilled and subsequently deionized with a Barnstead Nanopure system; the nitric acid used was of trace metal grade and from Fisher Scientific. All analyses were corrected for blank and flux contributions to the analytical signal (<1.5% for all elements). Calibration of instruments was achieved using eight USGS rock standards (BIR-1, DNC-1, BHVO-2, W-2, BCR-2, AGV-2, GSP-2, and G-2). Instrument drift (less than ±2% over the 2- to 3-hour run) was corrected by analyzing a matrix-matched drift monitor that was regularly spaced throughout each analytical run. Analytical precision and accuracy based on six complete dissolutions of USGS standard G-2 was better than 0.7% (2ό standard deviation) for all major elements except Mg and P (1.5% and 2.5%, respectively). Loss on ignition was not determined.

Additional trace-element analyses (of REE, Sc, Cr, Ni, Co, V, Y, Nb, Ta, Hf, U, Th, Rb, and Cs) were performed either by instrumental neutron activation analysis (INAA) at Cornell University on powders or by inductively coupled plasma mass spectrometry (ICP-MS) using a Perkin-Elmer Elan 6000 at Binghamton University. INAA techniques and standards are as given in Kay et al. (1987). INAA precision and accuracy based on replicate analysis of an internal basalt standard (PAL; see Kay et al., 1987) and USGS standard G-2 are 2–5% (2σ) for most elements and ±10% for U, Nd, and Ni. ICP-MS precision and accuracy based on replicate analysis of an internal basalt standard (PAL; see Kay et al., 1987) and USGS standard G-2 are ~5% (2σ) for all elements analyzed. The same flux-dissolved solutions that were analyzed for major elements by ICP-OES were also analyzed by ICP-MS. Complete dissolution of accessory phases was confirmed by replicate analysis of five separate aliquots of EVE-1 that yielded similar precision to that for G-2, as stated above. All ICP-MS analyses were corrected for blank and flux contributions that were <1.5% of the total analytical signal for all elements.

ACKNOWLEDGMENTS

This research was supported by Montclair State University and by the National Science Foundation (DUE-9952667 to the senior author). The authors wish to thank J. Graney (Binghamton University), S.M. Kay (Cornell University), and R. Kay (Cornell University) for analytical assistance with inductively coupled plasma–mass spectrometry and instrumental neutron activation analysis. J. Chiarenzelli, A. Gates, and D. Valentino are thanked for helpful suggestions and lively discussions on the tectonic interpretations of the Mesoproterozoic New Jersey–Hudson Highland rocks. The authors also extend special thanks to J. Hogan and D. Smith for constructive reviews and to R. Tollo for his extreme editorial patience, which greatly improved the manuscript.

REFERENCES CITED

Aleinikoff, J.N., and Grauch, R.I., 1990, U-Pb geochronologic constraints on the origin of a unique monazite-xenotime gneiss, Hudson Highlands, New York: American Journal of Science, v. 290, p. 522–546.

Aleinikoff, J.N., Burton, W.C., Lyttle, P.T., Nelson, A.E., and Southworth, C.S., 2000, U-Pb geochronology of zircon and monazite from Mesoproterozoic granitic gneisses of the northern Blue Ridge, Virginia and Maryland, USA: Precambrian Research, v. 99, p. 113–146.

Anderson, J.L., 1983, Proterozoic anorogenic granite plutonism of North America, *in* Medaris, L.G., et al., eds., Precambrian geology: Selected papers from an international symposium: Boulder, Colorado, Geological Society of America Memoir 161, p. 133–154.

Barker, F., Wones, D.R., Sharp, W.N., and Desborough, G.A., 1975, The Pikes Peak Batholith, Colorado Front Range, and a model for the origin of the gabbro-anorthosite-syenite-potassic granite suite: Precambrian Research, v. 2, p. 97–160.

Beard, J.S., and Lofgren, G.E., 1991, Partial melting of basaltic and andesitic greenstones and amphibolites under dehydration melting and water-saturated conditions at 1, 3, and 6.9 kilobars: Journal of Petrology, v. 32, p. 365–401.

Beard, J.S., Lofgren, G.E., Sinha, A.K., and Tollo, R.P., 1994, Partial melting of apatite-bearing charnockite, granulite, and diorite: Melt compositions, restite mineralogy, and petrologic implications: Journal of Geophysical Research, v. 99, p. 21591–21603.

Carroll, M.R., and Wyllie, P.J., 1990, The system tonalite-H_2O at 15 kbar and the genesis of calc-alkaline magmas: American Mineralogist, v. 75, p. 345–357.

Clemens, J.D., Holloway, J.R., and White, A.J.R., 1986, Origin of A-type granites: Experimental constraints: American Mineralogist, v. 71, p. 317–324.

Collins, W.J., Beams, S.D., White, A.J.R., and Chappell, B.W., 1982, Nature and origin of A-type granites with particular reference to southeastern Australia: Contributions to Mineralogy and Petrology, v. 80, p.189–200.

Corrigan, D., and Hanmer, S., 1997, Anorthosites and related granitoids in the Grenville orogen: A product of convective thinning of the lithosphere?: Geology, v. 25, p. 61–64.

Creaser, R.A., Price, R.C., and Wormald, R.J., 1991, A-type granites revisited: Assessment of a residual-source model: Geology, v. 19, p. 163–166.

Davidson, A., 1995, A review of the Grenville orogen in its North American type area: Journal of Australian Geology and Geophysics, v. 16, p. 3–24.

Drake, A.A., Jr., Aleinikoff, J.N., and Volkert, R.A., 1991, The Mount Eve Granite (Middle Proterozoic) of northern New Jersey and southeastern New York, *in* Drake, A.A., Jr., ed., Contributions to New Jersey geology: Reston, Virginia, U.S. Geological Survey Bulletin, v. 1952, p. C1–C10.

Easton, R.M., 1986, Geochronology of the Grenville Province, *in* Moore, J.M., et al., eds., The Grenville Province: Ottawa, Ontario, Geological Association of Canada Special Paper 31, p. 127–173.

Eby, G.N., 1990, The A-type granitoids: A review of their occurrence and chemical characteristics and speculations on their petrogenesis: Lithos, v. 26, p. 115–134.

Eby, G.N., 1992, Chemical subdivision of the A-type granitoids: Petrogenetic and tectonic implications: Geology, v. 20, p. 641–644.

Foland, K.A., and Allen, J.C., 1991, Magma sources for Mesozoic anorogenic granites of the White Mountains magma series, New England, USA: Contributions to Mineralogy and Petrology, v. 109, p. 195–211.

Förster, H.J., Tischendorf, G., and Trumbull, R.B., 1997, An evaluation of the Rb vs. (Y + Nb) discrimination diagram to infer tectonic setting of silicic igneous rocks: Lithos, v. 40, p. 261–293.

Frost, B.R., Frost, C.D., Hulsbosch, T.P., and Swapp, S.M., 2000, Origin of the charnockites of the Louis Lake batholith, Wind River Range, Wyoming: Journal of Petrology, v. 41, p. 1759–1776.

Frost, C.D., and Frost, R.B., 1997, Reduced rapakivi-type granites: The tholeiitic connection: Geology, v. 25, p. 647–650.

Gates, A.E., 1995, Middle Proterozoic dextral strike-slip event in the central Appalachians: Evidence from the Reservoir fault, New Jersey: Journal of Geodynamics, v. 19, p. 195–212.

Gates, A.E., 1998, Early compression and late dextral transpression within the Grenvillian Event of the Hudson Highlands, NY, USA, *in* Sinha, A.K., ed., Basement tectonics 13: Dordrecht, The Netherlands, Kluwer Academic Publishers, p. 85–98.

Gates, A.E., and Costa, R.E., 1998, Multiple reactivations of rigid basement block margins: Examples in the northern Reading Prong, USA, *in* Gilbert, M.C., and Hogan, J.P., eds., Basement tectonics 12: Central North America and other regions: Dordrecht, The Netherlands, Kluwer Academic Publishers, p. 123–153.

Gates, A.E., and Krol, M.A., 1998, Kinematics and thermochronology of late Grenville escape tectonics from the central Appalachians: Geological Society of America Abstracts with Programs, v. 30, no. 7, p. A-124.

Gates, A.E., Valentino, D., Chiarenzelli, J., and Hamilton, M.A., 2001a, Ages of tectonic events in the Hudson Highlands, NY: Results from SHRIMP analyses: Geological Society of America Abstracts with Programs, v. 33, no. 1, p. A-79.

Gates, A.E., Valentino, D., Chiarenzelli, J., and Hamilton, M.A., 2001b, Deep-seated Himalayan-type syntaxis in the Grenville orogen, NY-NJ-PA: Geological Society of America Abstracts with Programs, v. 33, no. 6, p. A-91.

Harclerode, A., Gorring, M.L., Gates, A.E., and Valentino, D., 2001, Geochemistry and tectonic interpretation of mafic gneisses in the western Hudson Highlands, NY: Geological Society of America Abstracts with Programs, v. 33, no. 1, p. A-10.

Helenek, H.L., and Mose, D.G., 1984, Geology and geochronology of Canada Hill Granite and its bearing on the timing of Grenvillian events in the Hudson Highlands, New York, *in* Bartholomew, M.J., ed., Grenville events and related topics in the Appalachians: Boulder, Colorado, Geological Society of America Special Paper 194, p. 57–73.

Higgins, M.D., and van Breemen, O., 1996, Three generations of anorthosite-mangerite-charnockite-granite (AMCG) magmatism, contact metamorphism and tectonism in the Saguenay–Lac-Saint-Jean region of the Grenville Province, Canada: Precambrian Research, v. 79, p. 327–346.

Hughes, S.S., Lewis, S.E., Bartholomew, M.J., Sinha, A.K., Hudson, T.A., and Herz, N., 1997, Chemical diversity and origin of Precambrian charnockitic rocks of the central Pedlar massif, Grenvillian Blue Ridge Terrane, Virginia: Precambrian Research, v. 84, p. 37–62.

Kay, S.M., Maksaev, V., Moscoso, R., Mpodozis, C., and Nasi, C., 1987, Probing the evolving Andean lithosphere: Mid-late Tertiary magmatism in Chile (29–30°30′S) over the modern zone of subhorizontal subduction: Journal of Geophysical Research, v. 92, p. 6173–6189.

Kretz, R., 1983, Symbols for rock-forming minerals: American Mineralogist, v. 68, p. 277–279.

Kula, J., and Gorring, M.L., 2000, Tectonic setting and protolith for Middle Proterozoic amphibolites within the Losee Gneiss of the New Jersey Highlands: Geological Society of America Abstracts with Programs, v. 32, no. 7, p. A-455.

Landenberger, B., and Collins, W.J., 1996, Derivation of A-type granites from a dehydrated charnockitic lower crust: Evidence for the Chaelundi Complex, Eastern Australia: Journal of Petrology, v. 37, p. 145–170.

Leake, B.E, and 21 others, 1997, Nomenclature of amphiboles: Report of the Subcommittee on Amphiboles of the International Mineralogical Association, Commission on New Minerals and Minerals Names: American Mineralogist, v. 82, p. 1019–1037.

Loiselle, M.C., and Wones, D.R., 1979, Characteristics of anorogenic granites: Geological Society of America Abstracts with Programs, v. 11, p. 468.

Lumbers, S.B., Heaman, L.M., Vertoli, V.M., and Wu, T.-W., 1990, Nature and timing of Middle Proterozoic magmatism in the Central Metasedimentary Belt, Ontario, *in* Gowers, C., et al., eds., Mid-Preoterozoic Laurentia-Baltica: Ottawa, Ontario, Geological Association of Canada Special Paper 38, p. 243–276.

Mahood, G., and Hildreth, W., 1983, Large partition coefficients for trace elements in high-silica rhyolites: Geochimica et Cosmochimica Acta, v. 47, p. 11–30.

Masuda, A., Nakamura, N., and Tanaka, T., 1973, Fine structures of mutually normalized rare-earth patterns of chondrites: Geochimica et Cosmochimica Acta, v. 37, p. 239–248.

McLelland, J., and Chiarenzelli, J., 1990, Geochronological studies in the Adirondack Mts. and the implications of a Middle Proterozoic tonalitic suite, *in* Gowers, C., et al., eds., Mid-Preoterozoic Laurentia-Baltica: Ottawa, Ontario, Geological Association of Canada Special Paper 38, p. 175–194.

McLelland, J., and Whitney, P., 1990, Anorogenic, bimodal emplacement of anorthositic, charnockitic, and related rocks in the Adirondack Mountains, New York, *in* Stein, H.J., and Hannah, J.L., eds., Ore-bearing granite systems, petrogenesis, and mineralizing processes: Boulder, Colorado, Geological Society of America Special Paper 246, p. 301–315.

McLelland, J., Daly, S., and McLelland, J.M., 1996, The Grenville Orogenic Cycle (ca. 1350–1000 Ma): An Adirondack perspective: Tectonophysics, v. 265, p. 1–28.

McLelland, J., Hamilton, M., Selleck, B., McLelland, J., Walker, D., and Orrell, S., 2001, Zircon U-Pb geochronology of the Ottawan Orogeny, Adirondack

Highlands, New York: Regional and tectonic implications: Precambrian Research, v. 109, p. 39–72.
Nash, W.P., and Crecraft, H.P., 1985, Partition coefficients for trace elements in silicic magmas: Geochimica et Comsochimica Acta, v. 49, p. 2309–2322.
O'Connor, J.T., 1965, A classification for quartz-rich igneous rocks based on feldspar ratios: Reston, Virginia, U.S. Geological Survey Professional Paper 525-B, p. B79–B84.
Offield, T.W., 1967, Bedrock geology of the Goshen-Greenwood Lake area, New York: New York State Museum and Science Service Map and Chart Series, v. 9, 78 p.
Owens, B.E., Dymek, R.F., Tucker, R.D., Brannon, J.C., and Podosek, F.A., 1994, Age and radiogenic isotope composition of a late- to post-tectonic anorthosite in the Grenville Province: The Labrieville massif, Quebec: Lithos, v. 31, p. 189–206.
Patiño Douce, A.E., 1997, Generation of metaluminous A-type granites by low-pressure melting of calc-alkaline granitoids: Geology, v. 25, p. 743–746.
Pearce, J.A., Harris, N.B.W., and Tindle, A.G., 1984, Trace element discrimination diagrams for the tectonic interpretation of granitic rocks: Journal of Petrology, v. 25, p. 956–983.
Price, J.D., Hogan, J.P., Gilbert, M.C., London, D., and Morgan, G.B., 1999, Experimental study of titanite-fluorite equilibria in the A-type Mount Scott Granite: Implications for assessing F contents of felsic magma: Geology, v. 27, p. 951–954.
Rankin, D.W., 1975, The continental margin of eastern North America in the southern Appalachians: The opening and closing of the proto-Atlantic Ocean: American Journal of Science, v. 275-A, p. 298–336.
Rankin, D.W., Drake, A.A., Jr., Glover, L., III, Goldsmith, R., Hall, L.M., Murray, D.P., Ratcliffe, N.M., Read, J.F., Secor, D.T., Jr., and Stanley, R.S., 1989a, Pre-orogenic terranes, *in* Hatcher, R.D., Jr., et al., eds., The Appalachian-Ouachita orogen in the United States: Boulder, Colorado, Geological Society of America, The Geology of North America, v. F-2, p. 7–100.
Rankin, D.W., Drake, A.A., Jr., and Ratcliffe, N.M., 1989b, Geologic map of the U.S. Appalachians showing the Laurentian margin and Taconic orogen, *in* Hatcher, R.D., Jr., et al., eds., The Appalachian-Ouachita orogen in the United States: Boulder, Colorado, Geological Society of America, The Geology of North America, v. F-2, plate 2.
Rapp, R.P., and Watson, E.B., 1995, Dehydration melting of metabasalt at 8–32 kbar: Implications for continental growth and crust-mantle recycling: Journal of Petrology, v. 36, p. 891–931.
Ratcliffe, N.M., 1992, Bedrock geology and seismotectonics of the Oscawana Lake quadrangle, New York: Reston, Virginia, U.S. Geological Survey Bulletin 1941-B, 38 p.
Ratcliffe, N.M., Aleinikoff, J.N., Burton, W.C., and Karabinos, P., 1991, Trondhjemitic, 1.35–1.31 Ga gneisses of the Mount Holly Complex of Vermont: Evidence for an Elzevirian event in the Grenville Basement of the United States Appalachians: Canadian Journal of Earth Science, v. 28, p. 77–93.
Rivers, T., 1997, Lithotectonic elements of the Grenville Province: Review and tectonic implications: Precambrian Research, v. 86, p. 117–154.
Rushmer, T., 1991, Partial melting of two amphibolites: Contrasting experimental results under fluid-absent conditions: Contributions to Mineralogy and Petrology, v. 107, p. 41–59.
Sawka, W.N., 1988, REE and trace element variations in accessory minerals and hornblende from the strongly zoned McMurray Meadows Pluton, California: Transactions of the Royal Society of Edinburgh, Earth Sciences, v. 79, p. 157–168.
Scoates, J.S., Frost, C.D., Mitchell, J.N., Lindsley, D.H., and Frost, R.B., 1996, Residual-liquid origin for a monzonitic intrusion in a mid-Proterzoic anorthosite complex: The Sybille intrusion, Laramie anorthosite complex, Wyoming: Geological Society of America Bulletin, v. 108, p. 1357–1371.
Shannon, W.M., Barnes, C.G., and Bickford, M.E., 1997, Grenville magmatism in West Texas: Petrology and geochemistry of the Red Bluff Granitic Suite: Journal of Petrology, v. 38, p. 1279–1305.
Skjerlie, K.P., and Johnston, A.D., 1993, Fluid-absent melting behavior of an F-rich tonalitic gneiss at mid-crustal pressures: Implication for the generation of anorogenic granites: Journal of Petrology, v. 34, p. 785–815.
Smith, D.R., Noblett, J., Wobus, R.A., Unruh, D., Douglass, J., Beane, R., Davis, C., Goldman, S., Kay, G., Gustavson, B., Saltoun, B., and Stewart, J., 1999, Petrology and geochemistry of late-stage intrusions of the A-type, mid-Proterozoic Pikes Peak batholith (Central Colorado, USA): Implications for petrogenetic models: Precambrian Research, v. 98, p. 271–305.
Speer, J.A., McSween, H.Y., and Gates, A.E., 1994, Generation, segregation, ascent, and emplacement of Alleghanian plutons in the southern Appalachians: Journal of Geology, v. 102, p. 249–267.
Streckeisen, A.L., 1973, Plutonic rocks, classification and nomenclature recommended by the IUGS [International Union of Geological Sciences] subcommission on the systematics of igneous rocks: Geotimes, v. 18, p. 26–30.
Sylvester, P.J., 1989, Post-collisional alkaline granites: Journal of Geology, v. 97, p. 261–280.
Taylor, S.R., and McLennan, S.M., 1985, The continental crust: Its composition and evolution: Oxford, England, Blackwell Scientific Publications, 312 p.
Tollo, R.P., and Aleinikoff, J.N., 1996, Petrology and U-Pb geochronology of the Robertson River Igneous Suite, Blue Ridge Province, Virginia: Evidence for multistage magmatism associated with an early episode of Laurentian rifting: American Journal of Science, v. 296, p. 1045–1090.
Tollo, R.P., Aleinikoff, J.N., Bartholomew, M.J., and Rankin, D.W., 2004, Neoproterozoic A-type granitoids of the central and southern Appalachians: Intraplate magmatism associated with episodic rifting of the Rodinian supercontinent: Precambrian Research, v. 128, p. 3–38.
Turner, S.P., Foden, J.D., and Morrison, R.S., 1992, Derivation of some A-type magmas by fractionation of basaltic magma: An example from Padthaway ridge, South Australia: Lithos, v. 28, p. 151–179.
Verrengia, P., and Gorring, M.L., 2002, Geochemistry and tectonic implications of the Storm King Granite, Hudson Highlands, NY: Geological Society of America Abstracts with Programs, v. 34, no. 1, p. A.
Volkert, R.A., and Drake, A.A., Jr., 1998, The Vernon Supersuite: Mesoproterozoic A-type granitoid rocks in the New Jersey Highlands: Northeastern Geology and Environmental Science, v. 20, p. 39–43.
Volkert, R.A., and Drake, A.A., Jr., 1999, Geochemistry and stratigraphic relations of Middle Proterozoic rocks of the New Jersey Highlands, *in* Drake, A.A., Jr., ed., Geologic studies in New Jersey and Eastern Pennsylvania: Reston, Virginia, U.S. Geological Survey Professional Paper 1565-C, 77 p.
Volkert, R.A., Feigenson, M.D., Patino, L.C., Delaney, J.S., and Drake, A.A., Jr., 2000, Sr and Nd isotopic composition, age and petrogenesis of A-type granitoids of the Vernon Supersuite, New Jersey Highlands, USA: Lithos, v. 50, p. 325–347.
Wasteneys, H., McLelland, J., and Lumbers, S., 1999, Precise zircon geochronology in the Adirondack Lowlands and implications for revising plate tectonic models of the Central Metasedimentary Belt and Adirondack Mountains, Grenville Province, Ontario and New York: Canadian Journal of Earth Science, v. 36, p. 967–984.
Whalen, J.B., Currie, K.L., and Chappell, B.W., 1987, A-type granites: Geochemical characteristics, discrimination, and petrogenesis: Contributions to Mineralogy and Petrology, v. 95, p. 407–419.
Whalen, J.B., Jenner, G.A., Longstaffe, F.J., Robert, F., and Gariepy, C., 1996, Geochemical and isotopic (O, Nd, Pb, and Sr) constraints on A-type granite petrogenesis based on the Topsails Igneous Suite, Newfoundland Appalachians: Journal of Petrology, v. 37, p. 1463–1489.
Whitney, P., and Olmsted, J., 1988, Geochemistry and origin of albite gneisses, northeastern Adirondacks, NY: Contributions to Mineralogy and Petrology, v. 99, p. 476–484.
Wones, D.R., 1983, Significance of the assemblage titanite + magnetite + quartz in granitic rocks: American Mineralogist, v. 74, p. 744–749.

MANUSCRIPT ACCEPTED BY THE SOCIETY AUGUST 25, 2003

Printed in the USA

Geological Society of America
Memoir 197
2004

Paleozoic structure of internal basement massifs, southern Appalachian Blue Ridge, incorporating new geochronologic, Nd and Sr isotopic, and geochemical data

Robert D. Hatcher, Jr.*
Brendan R. Bream
Department of Earth and Planetary Sciences, University of Tennessee, 306 Earth and Planetary Sciences Building, Knoxville, Tennessee 37996-1410, USA
Calvin F. Miller
Department of Geology, Vanderbilt University, Nashville, Tennessee 37235, USA
James O. Eckert, Jr.
Department of Geology and Geophysics, Yale University, P.O. Box 208109, New Haven, Connecticut 06520-8109, USA
Paul D. Fullagar
Department of Geological Sciences, CB 3315, Mitchell Hall, University of North Carolina, Chapel Hill, North Carolina 27599-3315, USA
Charles W. Carrigan
Department of Geological Sciences, University of Michigan, Ann Arbor, Michigan 48109-1063, USA

ABSTRACT

A number of Grenvillian basement massifs occur in the southern Appalachian Blue Ridge. The largest are contained in the Blue Ridge anticlinorium, which extends northward from its widest point in western North Carolina to Maryland. The Tallulah Falls dome, Toxaway dome, and Trimont Ridge area contain small internal basement massifs in the eastern and central Blue Ridge of the Carolinas and northeastern Georgia. All are associated with Paleozoic antiformal culminations, but each contains different basement units and contrasting Paleozoic structure.

The Tallulah Falls dome is a broad foliation antiform wherein basement rocks (coarse augen 1158 ± 19 Ma Wiley Gneiss [ion microprobe, $^{207}Pb/^{206}Pb$], medium-grained 1156 ± 23 Ma [$^{207}Pb/^{206}Pb$] and 1126 ± 23 Ma [$^{207}Pb/^{206}Pb$] Sutton Creek Gneiss, and medium-grained to megacrystic 1129 ± 23 Ma Wolf Creek Gneiss [sensitive high resolution ion microprobe, SHRIMP, $^{207}Pb/^{206}Pb$]) form a ring and spiral pattern on the west, south, and southeast sides of the dome. Basement rocks are preserved in the hinges of isoclinal anticlines whose axial surfaces dip off the flanks of the dome. The Wiley Gneiss was intruded by Sutton Creek Gneiss. The Toxaway dome consists predominantly of coarse, banded 1151 ± 17 Ma and coarse augen 1149 ± 32 Ma (SHRIMP $^{206}Pb/^{238}U$) Toxaway Gneiss folded into a northwest-vergent, gently southwest- and northeast-plunging antiform that contains a boomerang structure of Tallulah Falls Formation metasedimentary rocks in the core near the southwest end. The coarse augen gneiss phase constitutes a larger proportion of the Toxaway Gneiss toward the northeast. Field evidence indicates that the augen phase intruded the

*E-mail: bobmap@utk.edu.

Hatcher, R.D., Jr., Bream, B.R., Miller, C.F., Eckert, J.O., Jr., Fullagar, P.D., and Carrigan, C.W., 2004, Paleozoic structure of internal basement massifs, southern Appalachian Blue Ridge, incorporating new geochronologic, Nd and Sr isotopic, and geochemical data, *in* Tollo, R.P., Corriveau, L., McLelland, J., and Bartholomew, M.J., eds., Proterozoic tectonic evolution of the Grenville orogen in North America: Boulder, Colorado, Geological Society of America Memoir 197, p. 525–547. For permission to copy, contact editing@geosociety.org.

banded Toxaway lithology; U/Pb isotopic ages of these lithologies, however, are statistically indistinguishable. The Trimont Ridge massif occurs in an east-west–trending antiform west of Franklin, North Carolina, and consists of felsic gneiss that yielded a 1103 ± 69 Ma SHRIMP $^{207}Pb/^{206}Pb$ age.

An ε_{Nd}-depleted mantle model age of 1.5–1.6 Ga permits derivation of all of these basement rocks (including most from the western Blue Ridge) from eastern granite-rhyolite province crust, except the Mars Hill terrane rocks, which yield 1.8–2.2-Ga model ages. The small Grenvillian internal massifs were probably rifted from Laurentia during the Neoproterozoic, and became islands in the Iapetus ocean that were later swept onto the eastern margin of Laurentia during Ordovician subduction and arc accretion. These massifs were additionally penetratively deformed and metamorphosed during the Taconian and Neoacadian orogenies.

Keywords: Appalachians, Grenvillian internal basement massifs, eastern Blue Ridge, Tallulah Falls dome, Toxaway dome, Trimont Ridge complex

INTRODUCTION

The Appalachians were built on the Neoproterozoic rifted margin of eastern North America through contractional pulses during the Ordovician-Silurian (Taconian), Late Devonian–early Mississippian (Acadian), and late Mississippian-Permian (Alleghanian). A product of most orogenies is new continental crust (Hatcher, 1983, 1999), recycling crust generated during earlier orogenies that is preserved in most orogens as basement massifs or *basement inliers.* In the Appalachians, Caledonides, North American Cordillera, Urals, and many other orogens, the term *basement massif* mostly refers to an isolated block of Precambrian rocks that was formed during an earlier crust-forming tectonothermal event and was incorporated into the younger orogen. In the Alps, however, the term *basement massif* refers to crust that had become part of the European or African cratons during Paleozoic orogenies, and then was recycled during Alpine orogenesis. Similarly, Grenvillian, Taconian, and Acadian (Neoacadian in the southern Appalachians) crust all serve as basement for the Alleghanian event in the Appalachians (Hatcher, 1999). External basement massifs are blocks of older crust that are incorporated into the more external (forelandward) parts of an orogen, whereas internal basement massifs are blocks of older crust that are located in the internal parts of an orogen (Hatcher, 1983). External massifs are more likely to be derived from the nearby craton, but internal massifs can be derived from a variety of locales, not necessarily from the nearby craton, so they can be either proximally derived or parts of exotic terranes.

The Appalachians contain external basement massifs of late Middle Proterozoic Grenvillian (1.2–0.9 Ga) rocks in the western crystalline core that are traceable from the Blue Ridge in Georgia and North Carolina to the Long Range in Newfoundland. Rankin (1976) suggested these external massifs comprise the Appalachian Blue–Green–Long axis. Bartholomew and Lewis (1984) subdivided the Grenvillian basement in western North Carolina and southwestern Virginia into several massifs (Watauga, Globe, Sauras, and Elk River in North Carolina, and Pedlar and Lovingston in Virginia) based on compositional and structural criteria. In addition, a belt of granulite-facies paragneiss containing some orthogneiss was recognized along the eastern flank of the main central Blue Ridge external massif by the compilers of the North Carolina geologic map (Brown et al., 1985), as well as in northeastern Tennessee by Monrad and Gulley (1983) and Gulley (1985). Monrad and Gulley (1983) first suggested this belt contains pre-Grenville (1815 ± 31 Ma Rb/Sr whole rock, 0.7058 $^{87}Sr/^{86}Sr$) rocks represented by the Carvers Gap Gneiss. Bartholomew and Lewis (1984) included the Carvers Gap Gneiss in their Elk River massif, but recognized that this was a probable earlier host assemblage for Grenvillian plutons; later they (Bartholomew and Lewis, 1988) defined this block as the Mars Hill terrane of pre-Grenville crust. Raymond and Johnson (1994) suggested that the Mars Hill could be a Paleozoic accreted terrane. Carrigan et al. (2003) and Ownby et al. (this volume) have reconfirmed the pre-Grenville age of the paragneiss in this terrane, reporting an ion microprobe age of 1.8 Ga from zircon, with most zircons falling in the range of 1.2–1.25 Ga, and ε_{Nd}-depleted mantle model ages of 1.8–2.2 Ga. They also reported that most zircons have 1.0–1.1-Ga metamorphic rims, confirming that the Mars Hill terrane consists of an early Middle Proterozoic crustal block that was recycled into the Grenvillian orogen in late Middle Proterozoic time.

The southernmost external massifs in the northern Georgia Blue Ridge—the Fort Mountain, Corbin, and Salem Church massifs—were studied by McConnell and Costello (1984), and more recently by Li and Tull (1998). The Corbin Gneiss was recently dated (1.106 ± 13 Ga) by Heatherington et al. (1996). In addition, a series of internal massifs of Grenvillian rocks exist to the east of this main axis in the southern Appalachians, exposed in windows deep in the mountain chain, as in the Sauratown Mountains and Pine Mountain windows, and as three small isolated bodies in the central and eastern Blue Ridge in the Carolinas and northeastern Georgia (Fig. 1). Basement and cover in

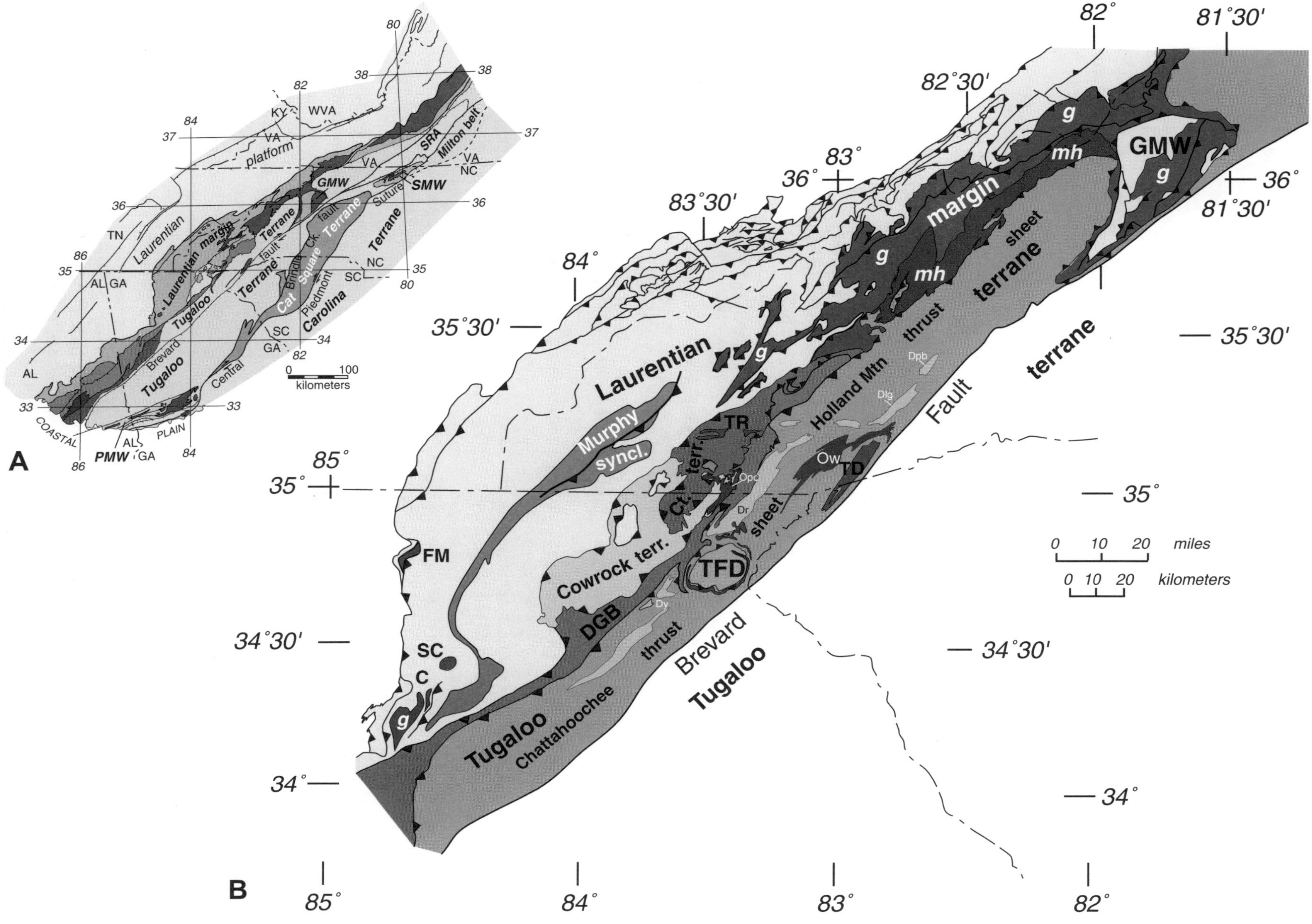

Figure 1. (A) Major tectonic elements of the southern Appalachians. SMW—Sauratown Mountains window. GMW—Grandfather Mountain window. PMW—Pine Mountain window. SRA—Smith River allochthon. (B) Tectonic elements of the southern Appalachian Blue Ridge from northern Georgia to northwestern North Carolina, showing the locations of external and internal Middle Proterozoic basement massifs. g—mostly Grenvillian orthogneisses. Ow—Whiteside Granite. Dy—Yonah granite. Dr—Rabun Granodiorite. Dlg—Looking Glass Granite. Dpb—Pink Beds Granite. opc—Persimmon Creek Gneiss. mh—Mars Hill terrane. DGB—Dahlonega gold belt. GMW—Grandfather Mountain window. FM—Fort Mountain massif. SC—Salem Church massif. C—Corbin massif. TR—Trimont Ridge massif. TFD—Tallulah Falls dome. TD—Toxaway dome. Ct. terr.—Cartoogechaye terrane. Modified from Brown et al. (1985); Hatcher and Hopson (1989); Hatcher et al. (1990); and unpublished maps by M.W. Carter (1998) and D.J. Settles (2001).

the Tallulah Falls and Toxaway domes, and the Trimont Ridge complex, the subjects of this paper, were complexly deformed together during one or more Paleozoic orogenies.

The Sauratown Mountains internal basement massif and window exposes a stack of older and younger thrust sheets revealing both inner and outer windows, possibly arched above a blind Alleghanian duplex (Hatcher et al., 1988; Hatcher, 1989) (Fig. 1). The inner windows expose Grenvillian basement and cover rocks probably of western Blue Ridge basement and rifted-margin rocks (McConnell, 1990; Horton and McConnell, 1991), whereas the outer window is framed by eastern Blue Ridge/western Inner Piedmont rocks of the Ashe/Tallulah Falls Formation and may expose the Hayesville-Gossan Lead fault (Heyn, 1984; Hatcher, 1989). A higher thrust sheet, the Smith River allochthon (Conley and Henika, 1973) of possible Gondwanan affinity (Tracy et al., 2001), flanks the window to the northwest. To the southeast, the Sauratown Mountains window is truncated by the central Piedmont suture, and to the southwest it plunges beneath Inner Piedmont thrust sheets (Fig. 1).

The Pine Mountain internal massif in western Georgia and eastern Alabama contrasts structurally with the Sauratown Mountains window, even though both occupy similar positions immediately northwest of the central Piedmont suture (Hatcher et al., 1988). The Pine Mountain is a complex window surrounded by major dextral (north and south flanks) and thrust (east flank) faults (Hooper and Hatcher, 1988). At the east end of the window, a variety of lithologies and probable Grenvillian structures and textures are preserved inside the window; and a sequence of early Paleozoic(?) cover rocks rests on the basement (Clarke, 1952; Hooper and Hatcher, 1988). The window narrows westward into Alabama and contains several major ductile dextral faults and a late brittle normal fault internally (Sears and Cook, 1984; Steltenpohl, 1988), so that Grenvillian rocks are strongly overprinted by Paleozoic (Alleghanian?) metamorphism and mylonitization (Student and Sinha, 1992).

Our purpose here is to discuss the structure and characterize the basement rocks of three internal basement massifs in the central and eastern Blue Ridge in the Carolinas and northeastern Georgia (Fig. 2). The rocks and Paleozoic structure of each contrast markedly, and basement rocks in each massif responded differently to Paleozoic deformation. The structure of the Tallulah Falls dome basement rocks has not been previously described, except regionally (e.g., Hopson et al., 1989). The Tallulah Falls dome contains three Grenvillian units: the Wiley Gneiss (augen gneiss), the Sutton Creek Gneiss (medium-grained granitoid), and the Wolf Creek Gneiss (medium-grained and megacrystic granitoid). The structure of the Toxaway dome was detailed earlier by Hatcher (1977); this is updated here, incorporating new zircon radiometric age dates, unpublished structural data, and a better understanding of the relationships between the banded and massive augen Toxaway Gneiss lithologies. The Trimont Ridge complex was mapped in detail (Eckert, 1984, 1988), and the metamorphic petrology detailed by Eckert (1988) and Eckert et al. (1989). The complex was speculatively identified as basement by Hatcher and Hopson (1989), now supported by our new data. The dated Trimont Ridge Grenvillian lithology is dominantly medium-grained granitoid, whereas the Trimont Ridge complex is dominated by orthopyroxene-bearing felsic gneiss and amphibolite (containing both orthopyroxene and clinopyroxene). All basement rocks in the three massifs yielded euhedral to subhedral zircons (for ion microprobe geochronology), increasing the likelihood they are orthogneisses (also supported by petrographic, isotopic, and geochemical data, except possibly the Trimont Ridge; Tables 1 and 2).

New geochronologic data presented in this paper (Table 1) were obtained using the sensitive high-resolution ion microprobe with reverse geometry (SHRIMP-RG) located at the U.S. Geological Survey (USGS)–Stanford MicroAnalytical Center at Stanford University, employing established standardized protocols. Our new geochronologic data were obtained using both the USGS–Stanford facility and the Cameca 1270 ion microprobe located at the University of California at Los Angeles. Geochemical data presented here (Table 2) were obtained from X-ray fluorescence techniques and inductively coupled plasma mass spectrometry (ICP-MS) analyses conducted by Activation Laboratories in Ancaster, Ontario. Nd and Sr isotopic data were determined at the University of North Carolina–Chapel Hill. See Bream et al. (this volume) for additional details about analytical techniques.

TALLULAH FALLS AND TOXAWAY DOMES

Introduction and Regional Setting

The Tallulah Falls and Toxaway domes occur in the Chattahoochee thrust sheet in the eastern Blue Ridge of Georgia and South Carolina (Tallulah Falls), and the Carolinas (Toxaway) (Fig. 1). These domes have near circular to elliptical shapes and are foliation domes that interrupt the northeast-southwest trending zone of Tallulah Falls/Ashe Formation rocks that flank both domes immediately northwest of the Brevard fault zone. Foliation and compositional layering (transposed bedding and stromatic migmatite layering) are parallel throughout the central and eastern Blue Ridge and Inner Piedmont. Structural similarities between the two domes end here: they have contrasting structure and distributions of Grenvillian rocks. Moreover, the augen-textured Wiley Gneiss on the Tallulah Falls dome appears similar to the augen phase of the Toxaway Gneiss in the Toxaway dome, but other basement units are texturally and compositionally different.

Immediately northwest and southeast of the Brevard fault zone is the same stratigraphic unit, the Neoproterozoic-Cambrian Tallulah Falls/Ashe Formation. This unit is conformably overlain by the Cambrian(?) Chauga River Formation within and southeast of the Brevard fault zone, then unconformably by the Middle Ordovician Poor Mountain Formation (Hatcher, 2002; Bream, 2003). The Tallulah Falls Formation–Poor Mountain Formation succession is particularly clear in the South

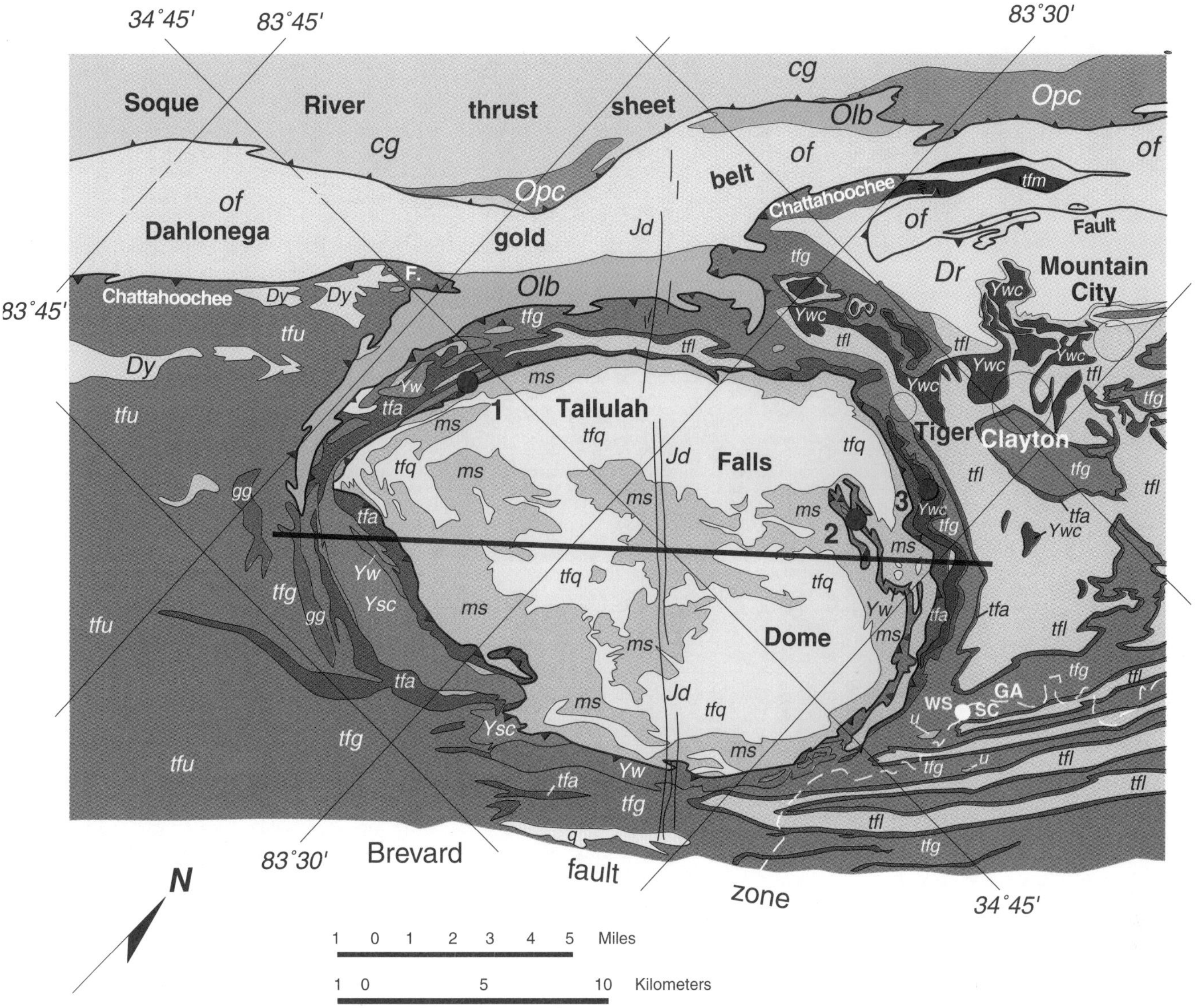

Figure 2. Geologic map of the Tallulah Falls dome and vicinity. Chattahoochee thrust sheet: Ysc—Sutton Creek Gneiss. Yw—Wiley Gneiss. Ywc—Wolf Creek Gneiss. Tallulah Falls Formation members: tfl—Graywacke-schist amphibolite member. tfa—Garnet-aluminous schist member (kyanite- or sillimanite-bearing). tfg—Graywacke-schist member. tfq—Tallulah Falls Quartzite. ms—metasandstone and muscovite-biotite schist beneath the Tallulah Falls Quartzite. tfu—undivided. tfm—migmatite derived from Tallulah Falls Formation metagraywacke. gg—granitoid gneiss, age unknown. q—quartzite and quartzose metasandstone. u—ultramafic rocks. Dr—Rabun granodiorite. Dy—Yonah granitoid. WS—Woodall Shoals (white dot on GA-SC line). Dahlonega gold belt: of—Otto Formation, undivided. Olb—Lake Burton mafic arc complex. Soque River thrust sheet: cg—Coleman River Formation metasandstone and schist. Opc—Persimmon Creek tonalite. Cutting all thrust sheets and the Brevard fault zone: Jd—diabase. Black dots and bold numbers represent the locations of Toxaway Gneiss geochronology and geochemistry samples of Carrigan et al. (2003) and the Wolf Creek Gneiss sample. 1—Sutton Creek Gneiss along Raper Creek (sample SC-RC). 2—Wiley Gneiss, U.S. 23-441 Wiley, Georgia. 3—Wolf Creek Gneiss, Stekoa Creek Road near Stekoa Creek (sample WC). Heavy northeast-southwest line marks the location of the cross section in Figure 3. Modified from Hatcher and Hopson (1989).

Mountains in North Carolina (see Hatcher and Bream, 2002, and articles therein). Both Ordovician (Kennesaw, Whiteside, Henderson, Dysartsville, Caesars Head) and Devonian (Rabun, Yonah, Pink Beds, Looking Glass, Mt. Airy, Stone Mountain–North Carolina) deformed granitoids occur in the eastern Blue Ridge–western Inner Piedmont, indicating deformation of these plutons occurred in the Neoacadian (Late Devonian–Mississippian) event. U/Pb ion microprobe analyses of metamorphic rims of both basement and detrital zircons (Carrigan et al., 2001; Bream, 2003), and U/Pb isotopic ages of monazite determined by conventional dating techniques (Dennis and Wright, 1997) yielded a 360–350-Ma (Neoacadian) age of metamor-

TABLE 1. SHRIMP ANALYTICAL DATA FOR WOLF CREEK GNEISS AND TRIMONT RIDGE GNEISS

Analysis label	Abundances (ppm)		Th/U	$^{204}Pb/^{206}Pb$	f_{206}%	Total				Radiogenic ratio				Age (Ma)				% Conc
	U	Th				$^{238}U/^{206}Pb$	± 1σ	$^{207}Pb/^{206}Pb$	± 1σ	$^{238}U/^{206}Pb$	± 1σ	$^{207}Pb/^{206}Pb$	± 1σ	$^{206}Pb/^{238}U$	± 1σ	$^{207}Pb/^{206}Pb$	± 1σ	
UPST-1 (i)	569	339	0.60	0.000056	0.0009	4.50	0.05	0.0869	0.0017	4.51	0.05	0.0861	0.0021	1292	12	1340	48	96
UPST-2.1	616	267	0.43	0.000277	0.0047	5.39	0.10	0.0846	0.0014	5.42	0.10	0.0807	0.0024	1092	19	1214	59	90
UPST-3.2	125	72	0.58	0.000010	0.0002	5.29	0.14	0.0817	0.0029	5.29	0.14	0.0815	0.0031	1116	27	1234	75	90
UPST-4.1	338	172	0.51	0.000012	0.0002	5.21	0.08	0.0804	0.0014	5.21	0.08	0.0803	0.0014	1133	15	1204	35	94
UPST-5.1	346	214	0.62	0.000113	0.0019	5.86	0.08	0.0760	0.0011	5.87	0.09	0.0744	0.0018	1014	14	1052	49	96
UPST-6.1	80	28	0.35	<0.000001	<0.0001	4.90	0.08	0.0746	0.0042	4.89	0.08	0.0767	0.0043	1200	19	1114	117	108
UPST-7.1	392	164	0.42	0.000014	0.0002	5.53	0.07	0.0772	0.0015	5.53	0.07	0.0770	0.0015	1072	13	1120	39	96
UPST-8.1	304	72	0.24	0.002886	0.0500	6.59	0.12	0.1353	0.0037	6.93	0.15	0.0951	0.0113	869	17	1530	242	57
UPST-9.1	2715	469	0.17	0.000394	0.0067	5.80	0.07	0.0817	0.0005	5.84	0.07	0.0761	0.0013	1018	12	1099	36	93
UPST-10.1	204	120	0.59	0.000040	0.0007	5.72	0.12	0.0779	0.0024	5.72	0.12	0.0773	0.0026	1039	20	1130	67	92
UPST-11.1	236	127	0.54	0.000031	0.0005	5.60	0.11	0.0769	0.0028	5.60	0.11	0.0764	0.0030	1059	19	1106	81	96
UPST-12.1	132	67	0.51	0.000561	0.0094	5.04	0.09	0.0843	0.0037	5.08	0.10	0.0763	0.0050	1157	20	1104	136	105
UPST-13.1	118	37	0.32	0.001483	0.0250	5.16	0.37	0.1079	0.0034	5.29	0.39	0.0871	0.0159	1116	75	1363	399	82
UPST-14.1	287	108	0.38	0.000028	0.0005	5.08	0.11	0.0784	0.0022	5.09	0.11	0.0780	0.0024	1157	24	1146	62	101
UPST-15.1	249	140	0.56	0.000010	0.0002	5.73	0.12	0.0754	0.0026	5.73	0.12	0.0752	0.0026	1037	20	1075	70	97
UPST-16.1	232	51	0.22	0.000169	0.0029	5.80	0.08	0.0750	0.0015	5.82	0.08	0.0726	0.0019	1022	13	1003	53	102
UPST-17.1	3753	1364	0.36	0.000163	0.0028	5.88	0.08	0.0796	0.0009	5.89	0.08	0.0773	0.0011	1011	13	1128	28	90
UPST-18.1	113	89	0.79	0.000688	0.0117	5.61	0.12	0.0768	0.0028	5.67	0.13	0.0668	0.0090	1047	22	833	310	126
UPST-19.1a	326	88	0.27	0.000079	0.0013	5.82	0.10	0.0737	0.0011	5.83	0.10	0.0726	0.0020	1021	16	1003	58	102
UPST-19.1b	1996	1002	0.50	0.000048	0.0008	5.55	0.12	0.0776	0.0007	5.56	0.12	0.0769	0.0008	1067	21	1118	20	95
UPST-20.1	150	72	0.48	0.000132	0.0023	5.92	0.12	0.0721	0.0022	5.93	0.12	0.0702	0.0027	1004	19	935	81	107
UPST-21.1 (i)	1403	290	0.21	0.000023	0.0004	4.43	0.04	0.0871	0.0007	4.44	0.04	0.0868	0.0007	1310	10	1356	15	97
TRG-1.1	189	70	0.37	<0.000001	<0.0001	5.44	0.21	0.0772	0.0036	5.43	0.21	0.0787	0.0048	1089	40	1165	127	94
TRG-2.1	212	104	0.49	0.000279	0.0047	5.58	0.10	0.0773	0.0015	5.60	0.10	0.0733	0.0062	1059	18	1021	182	104
TRG-3.1 (r)	246	67	0.27	0.000053	0.0009	6.97	0.18	0.0754	0.0026	6.98	0.18	0.0747	0.0029	864	21	1060	80	81
TRG-3.2	1208	222	0.18	0.000016	0.0003	7.51	0.15	0.0728	0.0008	7.51	0.15	0.0726	0.0009	806	15	1001	24	80
TRG-4.1	264	59	0.22	0.000030	0.0005	6.63	0.11	0.0774	0.0021	6.63	0.11	0.0770	0.0024	906	13	1120	64	81
TRG-6.1	91	46	0.50	0.009445	0.1582	4.18	0.11	0.2105	0.0115	4.96	0.18	0.0761	0.0247	1184	40	1097	839	108
TRG-7.1	210	57	0.27	0.000010	0.0002	5.85	0.09	0.0782	0.0029	5.86	0.09	0.0780	0.0030	1016	15	1147	77	89
TRG-7.2	72	37	0.51	0.000010	0.0002	7.17	0.14	0.0716	0.0041	7.17	0.14	0.0714	0.0042	841	16	970	124	87
TRG-8.1	163	87	0.53	0.000149	0.0025	5.31	0.12	0.0794	0.0030	5.33	0.12	0.0773	0.0038	1109	23	1129	100	98
TRG-9.1 (r)	146	43	0.29	0.000190	0.0033	5.96	0.14	0.0650	0.0029	5.98	0.14	0.0623	0.0034	997	21	683	120	146
TRG-10.1 (i)	370	98	0.27	<0.000001	<0.0001	4.42	0.07	0.0871	0.0025	4.42	0.07	0.0876	0.0028	1315	20	1374	63	96
TRG-11.1 (r)	223	78	0.35	0.000128	0.0022	6.98	0.19	0.0734	0.0022	6.99	0.19	0.0715	0.0031	862	22	973	91	89

Note: Typical SHRIMP conditions and analytical procedures for zircon analyses were followed (see Bacon et al., 2000). Analyses were referenced to AS57 standard zircon grains; concentrations of U and Th were referenced to SL13 zircon chips. For the radiogenic ratios, common Pb was corrected using measured ^{204}Pb. Five scans on mass peaks for each spot were obtained. Abbreviations: f_{206}%—percentage of ^{206}Pb that is common Pb; (i)—inherited core; % Conc—[($^{206}Pb/^{238}U$ Age)/($^{207}Pb/^{206}Pb$ Age)] × 100, where 100% denotes a concordant analysis; (r)—Rim analyses.

TABLE 2. WHOLE-ROCK MAJOR OXIDE, TRACE ELEMENT, AND ISOTOPIC DATA FOR THE TRIMONT RIDGE AND WOLF CREEK GNEISSES

	Detection limit	TRG2	UPST
SiO_2	0.01 %	68.76	65.53
Al_2O_3	0.01 %	15.06	19.26
Fe_2O_3	0.01 %	6.34	2.69
MnO	0.01 %	0.11	0.03
MgO	0.01 %	1.87	0.67
CaO	0.01 %	1.62	3.34
Na_2O	0.01 %	4.15	6.20
K_2O	0.01 %	1.20	1.37
TiO_2	0.005 %	0.773	0.357
P_2O_5	0.01 %	0.18	0.08
LOI	0.01 %	–0.02	0.37
Total		100.04	99.89
Rb	2 ppm	25	43
Sr	2 ppm	177	1209
Ba	1 ppm	133	274
Y	1 ppm	44	5
Zr	1 ppm	328	136
Hf	0.1 ppm	8.9	3.3
Nb	0.5 ppm	14.4	5.6
Ta	0.1 ppm	1.0	0.3
Zn	1 ppm	52	62
Cu	1 ppm	7	5
V	5 ppm	73	30
Ni	1 ppm	18	2
Cr	0.5 ppm	39.2	4.6
Co	0.1 ppm	12.7	4.4
Sc	0.01 ppm	12.50	4.50
La	0.05 ppm	34.18	3.03
Ce	0.1 ppm	74.5	5.1
Pr	0.02 ppm	8.92	0.65
Nd	0.05 ppm	35.85	2.62
Sm	0.01 ppm	7.59	0.65
Eu	0.005 ppm	1.913	1.083
Gd	0.02 ppm	7.02	0.72
Tb	0.01 ppm	1.26	0.15
Dy	0.02 ppm	8.00	0.93
Ho	0.01 ppm	1.65	0.20
Er	0.01 ppm	4.75	0.65
Tm	0.005 ppm	0.695	0.102
Yb	0.01 ppm	4.63	0.73
Lu	0.002 ppm	0.697	0.131
U	0.05 ppm	2.50	0.74
Th	0.05 ppm	9.80	0.56
$^{87}Rb/^{86}Sr$		0.4062	0.1016
$^{87}Sr/^{86}Sr_{(0)}$		0.71797	0.70688
$^{87}Sr/^{86}Sr_{(i)}$		0.70662	0.70630
$^{147}Sm/^{144}Nd$		0.1292	0.1399
$^{143}Nd/^{144}Nd_{(0)}$		0.51217	0.51222
ε_{Nd}		–9.1	–8.2
$\varepsilon_{Nd(0)}$		0.01	0.4
T_{DM}		1566 Ma	1694 Ma

Note: Detection limit units are percentages for compounds and ppm for the elements. Abbreviations: (i)—initial value; (t)—value at time of crystallization.

phism in the Blue Ridge and Inner Piedmont, further linking peak metamorphism-related penetrative deformation to the Neoacadian. This assemblage of paragneisses, mafic and ultramafic rocks, and granitoids comprises the Tugaloo terrane of Hatcher (2001) (Fig. 1). The eastern boundary of the Tugaloo terrane is the Brindle Creek fault (Giorgis, 1999; Bream et al., 2001) in the eastern Inner Piedmont (see Hatcher, 2002); its western boundary is the Chattahoochee-Holland Mountain fault (Fig. 1). Another attribute of the Tugaloo terrane is the occurrence of mafic and ultramafic rocks of several tectonic affinities and the paucity of Grenvillian basement. The only Grenvillian basement rocks known in the Tugaloo terrane are found in the Tallulah Falls and Toxaway domes (Fig. 1). These attributes and the noncontinental (mostly mid-oceanic-ridge basalt and island arc) geochemical affinities of mafic metavolcanic rocks (Hatcher et al., 1984; Spell and Norell, 1990; Davis, 1993; Hatcher, 2002) suggest the rocks of the Tugaloo terrane were mostly deposited on ocean crust (Hatcher, 1978, 1987, 2002).

Tallulah Falls Dome

Structure and Map Pattern

The Tallulah Falls dome has an almost circular outcrop pattern (Fig. 2) that was clearly outlined in the map by Teague and Furcron (1948). Dominant foliation dips gently in all directions away from a central short axis that plunges gently northeast and southwest. Gentle open late northeast-southwest–trending antiforms and synforms with 2–3-km wavelengths produce the outcrop patterns of rock units in the interior of the dome, but do not affect outcrop patterns of units framing the dome. This pattern provides a clue to the structural complexity of the Tallulah Falls dome. The structural complexity was not completely revealed until the entire dome was mapped in detail (Hopson et al., 1989), although the reconnaissance map by Hatcher (1971) and cross sections by Hatcher (1973) revealed greater complexity than previously had been recognized.

Rock units that crop out in the interior of the dome consist of a two-mica, two-feldspar metasandstone and two-mica schist, overlain by the two-mica, two-feldspar Tallulah Falls Quartzite (Fig. 2). These units are flanked on the northeast, northwest, and partly on the southwest sides of the dome by dominantly one-mica (biotite), one-feldspar (oligoclase) metagraywacke and garnet- and kyanite- or sillimanite-bearing aluminous schist of the Tallulah Falls Formation (Hatcher and Hopson, 1989; Hopson et al., 1989). Correlation of rocks in the interior of the dome with those on the flanks is difficult, because the prominent Tallulah Falls Formation marker units present outside do not appear inside the dome, the Tallulah Falls Quartzite occurs only inside the dome, and the metasandstone inside the dome modally more closely resembles the Otto Formation metasandstone in the Dahlonega gold belt than it does Tallulah Falls metagraywacke (Hatcher, 1980; Hopson et al., 1989) (Fig. 2). If Dahlonega gold belt rocks are inside the dome, the Neoacadian Chattahoochee fault (not an Alleghanian fault, as implied by Livingston and

McKniff, 1967) separates the metasandstone and quartzite from the rocks that frame the Tallulah Falls dome. Ages of detrital zircons, however, suggest that the quartzite and metasandstone do not correlate with Otto Formation metasandstone (Bream et al., this volume), but likely indicate the existence of another pre-Alleghanian fault on this dome. This window interpretation was originally suggested by A.L. Stieve in Hopson et al. (1989) and by Stieve (1989). Livingston and McKniff (1967) first hypothesized that the Tallulah Falls dome is a window analogous to the Grandfather Mountain window, because of its similar location northwest of the Brevard fault zone.

Neoacadian metamorphic isograds surrounding the Tallulah Falls dome reveal additional complexity. Nonmigmatitic kyanite-bearing schist belonging to the Tallulah Falls Formation garnet-aluminous schist member on the immediate flanks of the dome (on all sides) dips beneath migmatitic kyanite, then sillimanite-bearing rocks of the same unit (all labeled *tfa* in Figs. 2 and 3) away from the crest of the dome (Hatcher, 1973). Lack of migmatization in the immediate flanks and interior of the dome suggests the isograds are actually inverted and not related to high-pressure transformation of sillimanite to kyanite or to apparent inversion by faulting.

All Sutton Creek and Wolf Creek Gneiss bodies on the Tallulah Falls dome are consistently enveloped by in-sequence Tallulah Falls Formation, as these bodies comprise the cores of isoclinal anticlines flanked successively by the lower, intermediate, and then the upper members of the Tallulah Falls Formation (Figs. 2 and 3). A smaller body of Wiley Gneiss was mapped by Stieve (1989) within the larger body of Sutton Creek Gneiss on the southwest flank of the dome, where field observations indicate the Sutton Creek Gneiss intruded the Wiley Gneiss. The main body of Wiley Gneiss, with a smaller body of Sutton Creek Gneiss attached at its southwestern end, however, does not have the same enveloping structure by Tallulah Falls Formation as the main Sutton Creek and Wolf Creek bodies, and appears more isolated both stratigraphically and structurally along the east and southeast sides of the dome. This body has an outcrop pattern characterized by a series of interconnected, linear, short-wavelength, 280°-trending folds at its northern end that plunges westward beneath the present erosion surface into the core of the dome. These separate outcrop belts coalesce into a single, larger body of Wiley Gneiss as the outcrop pattern traces a right-handed spiral from the core to the southeast flank of the dome (Fig. 2). Throughout the main outcrop belt of Wiley Gneiss, cover rocks range from Tallulah Falls Formation metagraywacke and truncated garnet-aluminous schist (kyanite-bearing) member rocks in the southeast-to-southwest part of the outcrop belt to the interior metasandstone (that is ultimately overlain by Tallulah Falls Quartzite) in the core of the dome. Stieve (1989) concluded that this body of Wiley Gneiss is a component of a large top-to-the-northwest shear zone that occupies this part of the dome. Sheath folds, with the same shear sense in cover rocks flanking this body, support this conclusion.

The Tallulah Falls Quartzite occurs in the hinge of a reclined isoclinal synform that initially plunges gently westward in the eastern interior part of the dome. The axial surface of this synform has been traced around the outer periphery of the interior of the north, northwest, west, and southwest portions of the

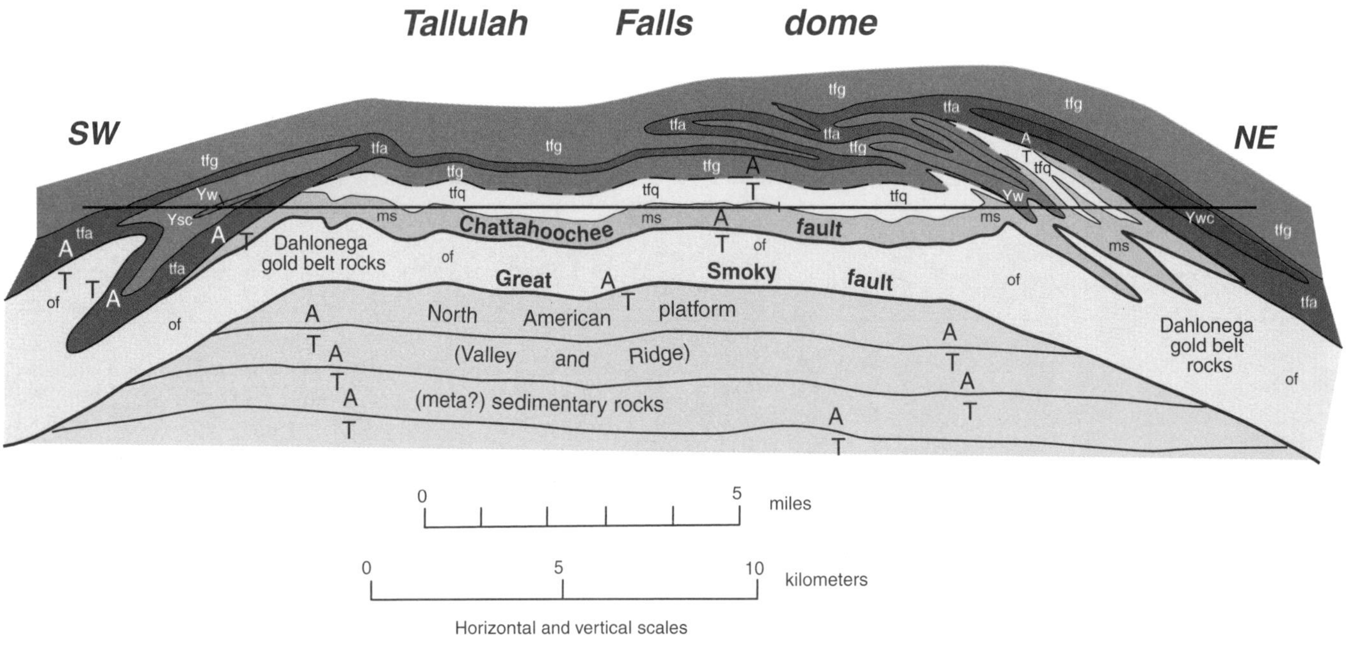

Figure 3. Northeast-southwest cross section through the Tallulah Falls dome, northeastern Georgia. Dashed contact is the projection of the fault inside the Tallulah Falls dome above the present erosion surface. A and T represent fault blocks moving relatively away or toward an observer, respectively. See Figure 2 for explanation of rock unit symbols.

dome; the axial surface systematically changes dip around the dome from northeast (on the northeast flank) to north (on the north flank) to northwest to southwest. A graded quartzite bed exposed on U.S. 441-23 south of Clayton near Wiley, Georgia, on the upright limb (located 200 m north of sample 2, Fig. 2) confirms the rocks here are upright and that the structure is likely to be a syncline.

In summary, the structure of the Tallulah Falls dome appears to be the product of early (Taconian or early Acadian) folding and penetrative deformation; Neoacadian thrusting, polyphase folding, and peak metamorphism; and Alleghanian doming produced by localized duplex formation beneath the Blue Ridge–Piedmont megathrust sheet involving the mechanism described by Hatcher (1991) (Fig. 3). As suggested above, the interior of the dome could be a window, and—if the correlations based on detrital zircon age data from the Tallulah Falls Quartzite, Tallulah Falls Formation metagraywacke, and Dahlonega gold belt metasandstone are correct (Bream et al., this volume)—the fault framing the window cannot be the Chattahoochee fault. Basement rocks on the Tallulah Falls dome are confined to the Chattahoochee thrust sheet and the axial zones of isoclinal anticlines. One basement outcrop belt traces a broken-spiral geometry from inside the dome along the southeast rim southward and southwestward into the outer frame of the window. Separate isoclinal anticlines flanked by nonconformable cover rocks (Fig. 4) partially frame the northeast, southwest, and northwest flanks of the dome. North and northeast of the north flank and outside of the Tallulah Falls dome are dome-and-basin interference folds containing possible Grenvillian Wiley and Wolf Creek Creek Gneisses (Fig. 2). These rocks have not been dated; the southern end of the 374-Ma Rabun pluton (Miller et al., 2000) (*Dr* in Figs. 1 and 2) is also complexly deformed, so it remains unknown if these are additional Grenvillian basement rocks or highly deformed megacrystic and fine-grained Rabun granodiorite.

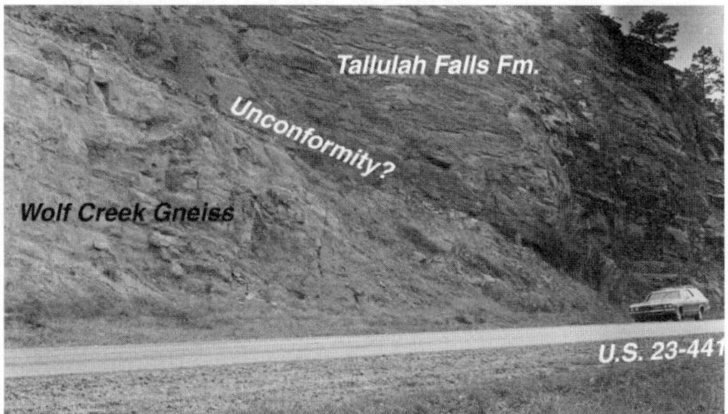

Figure 4. Contact between the Wolf Creek Gneiss (light colored) and Tallulah Falls Formation metagraywacke (dark) on U.S. 23-441 south of Clayton, Georgia, near Big (Stekoa) Creek Falls (~3 km west of sample locality 3, Fig. 2). The sharp contact does not contain evidence that would favor it being a fault; it is probably a nonconformity. The metagraywacke above the contact contains boudins and layers of amphibolite (not visible in photo) and lies sequentially beneath kyanite schist (tfa, Fig. 2), then amphibolite-poor metagraywacke (tfg, Fig. 2), suggesting the sequence is upright. Photograph was taken in 1970 before the cut was seeded.

Complex folding of the southwestern end of the Rabun pluton (*Dr* in Figs. 1 and 2), recognition of several foliations (at least one tectonic), concordant deformational fabrics containing peak metamorphic assemblages outside the pluton, and truncation of the pluton by the Chattahoochee fault (Hatcher and Hopson, 1989; Hatcher et al., 1995) indicate that the dominant structural map patterns in the Tallulah Falls dome (and Toxaway dome discussed below) were formed during the Neoacadian orogeny. Later doming and open folding were probably Alleghanian. Taconian penetrative deformation has not been confirmed in this terrane, but may be contained in undated amphibolite boudins that preserve earlier foliations and folds enclosed in the dominant Neoacadian foliation formed under peak metamorphic conditions. Several of these boudins are present in Tallulah Falls Formation cover rocks around the Tallulah Falls dome, but the most convincing examples of boudins preserving earlier structures are at nearby Woodall Shoals on the Georgia–South Carolina border (see Woodall Shoals field trip stops in Fritz et al., 1989, and Hatcher et al., 1995) (Fig. 2).

Middle Proterozoic Basement: Wiley Gneiss

The Wiley Gneiss was named by Hatcher (1974) for exposures near Wiley, Georgia, in the Tiger, Georgia, 7.5-min quadrangle. It is a coarse-grained feldspar (both oligoclase and microcline)-quartz-biotite-muscovite augen gneiss, with augen composed of microcline and accessory apatite, zircon, clinozoisite, tourmaline, garnet, and opaque minerals (Stieve, 1989) (Fig. 5). Typical Wiley Gneiss is mylonitic and contains a strong penetrative S-C foliation; shear sense determined from a number of exposures throughout the Wiley outcrop belt is consistently oriented top-to-the-northwest. Sheath folds in Tallulah Falls Formation cover rocks likewise contain northwest-oriented sheath axes (Stieve, 1989; Stieve and Hatcher, 1989). These fabrics suggested to Stieve and Hatcher (1989) that the Wiley Gneiss and the immediately associated cover rocks comprise a shear zone that includes the Wiley outcrop belt and adjacent cover. Interestingly, Wiley Gneiss containing its characteristic S-C fabric is clearly intruded by nonmylonitic Sutton Creek Gneiss on the southwest flank of the Tallulah Falls dome truncating the S-C fabric in the Wiley (Stieve, 1989), which suggests the Wiley acquired much of its fabric during the Grenvillian orogeny, shortly after cooling and before intrusion of the Sutton Creek.

The Grenvillian age of the Wiley Gneiss was first suggested by a 1.0-Ga conventional U/Pb isotopic age obtained from zircon by Odom et al. (1973). Stieve and Sinha (1989) reported a Rb/Sr whole-rock age of 1169 ± 19 Ma and a 0.7088 $^{87}Sr/^{86}Sr$ initial ratio for the Wiley Gneiss, confirming its Middle Proterozoic or older heritage. New ε_{Nd} data (Carrigan et al., 2003) yield depleted mantle model ages in the range of 1.3–1.6 Ga,

Figure 5. Wiley Gneiss exposed on U.S. 23-441 at Wiley, south of Clayton, Georgia. Note the large K-spar augen and strong S-C fabric. Knife is 10 cm long.

reconfirming a more primitive source that likely was the eastern granite-rhyolite province. An 1158 ± 19-Ma $^{207}Pb/^{206}Pb$ crystallization age, and a 1056 ± 93-Ma $^{207}Pb/^{206}Pb$ time of metamorphism have recently been determined for the Wiley by Carrigan et al. (2003) using ion microprobe analytical techniques on zircons. The Wiley, however, had to acquire its S-C fabric after 1158 Ma and before 1156 ± 23 Ma (Carrigan et al., 2003; see below), when the Sutton Creek Gneiss was intruded. Aleinikoff et al. (2000) obtained minimum ages as old as 1070 Ma for fabrics in Virginia Blue Ridge Grenvillian basement, but the time required for formation of Wiley Gneiss S-C fabric may indicate that still earlier Grenvillian deformation occurred.

Sutton Creek Gneiss

The Sutton Creek Gneiss was named by Stieve and Sinha (1989) for exposures along Sutton Creek in the Tallulah Falls, Georgia, 7.5-minute quadrangle (Stieve, 1989). The Sutton Creek Gneiss is both texturally and mineralogically more heterogeneous than the Wiley Gneiss, varying from light gray, fine-to-medium–grained uniformly foliated microcline-oligoclase-quartz-biotite granitoid containing relict igneous texture to banded and garnet- and hornblende-bearing varieties (Fig. 6). It contains the accessory minerals zircon, sphene, epidote-clinozoisite, apatite, and opaque minerals. The microcline content of the Sutton Creek Gneiss is consistently greater than that of the Wiley Gneiss (Stieve, 1989).

The Sutton Creek Gneiss was considered a Paleozoic granitoid by Hatcher (1971, 1974), until its map relationships to the Wiley Gneiss and Tallulah Falls Formation were established by Stieve (1989) and it was dated. Stieve and Sinha (1989) reported a Rb/Sr whole-rock age of 1103 ± 14 Ma, with a 0.7063 $^{87}Sr/^{86}Sr$ initial ratio, for the Sutton Creek Gneiss. Carrigan et al. (2003) determined a 1156 ± 23-Ma $^{207}Pb/^{206}Pb$ crystallization age, with a possible 1030 ± 19-Ma $^{207}Pb/^{206}Pb$ age for time of metamorphism of the Sutton Creek. The difference in $^{87}Sr/^{86}Sr$ initial ratios for the Wiley and Sutton Creek Gneisses led Stieve and Sinha (1989) to conclude that the younger Sutton Creek magma was not derived from melting of Wiley Gneiss protolith, and that the Sutton Creek Gneiss was more likely derived from melting of more primitive lower crust. The new ε_{Nd}-depleted mantle model age for the Sutton Creek (Carrigan et al., 2003) of 1.4–1.6 Ga similarly confirms the melting of this magma from a more primitive source.

Wolf Creek Gneiss

The Wolf Creek Gneiss was named by Hatcher (1974) for exposures along Wolf Creek in the Rainy Mountain, Georgia–South Carolina, 7.5-minute quadrangle. It is confined to the east and northeast flanks of the Tallulah Falls dome (Fig. 2). It consists predominantly of light gray, fine-to-medium–grained strongly foliated microcline-oligoclase-quartz-biotite granitoid closely resembling the Sutton Creek Gneiss (Fig. 6) and a megacrystic phase that consists of microcline megacrysts in a groundmass of microcline, plagioclase, quartz, and biotite.

A 1129 ± 23 Ma (MSWD = 0.9, $n = 16$, 2σ error) SHRIMP $^{207}Pb/^{206}Pb$ crystallization age was determined for megacrystic Wolf Creek Gneiss sample UPST (Fig. 7a; Table 1). These data also define a discordia line that intersects the concordia at ~1270 Ma (with a large error), suggesting that the actual time of crystallization may be slightly older than the pooled ^{207}Pb-^{206}Pb ages. At least two of the older (~1350 Ma) grains contain truncated internal zones that we interpret as inherited cores. Overall, the grains have bright (in cathodoluminescence) euhedral cores with thin dark rims and are up to ~400 μm in long dimension. The $^{87}Sr/^{86}Sr$ initial ratio for the Wolf Creek Gneiss is 0.7066 and the ε_{Nd}-depleted mantle model age is 1.2–1.3 Ga (Table 2), indicating the original source of parent material may not be as primitive as those for the Wiley and Sutton Creek Gneisses.

Figure 6. Typical strongly foliated Sutton Creek Gneiss from the main body of Sutton Creek Gneiss (see Fig. 2) in the Clarkesville Northeast 7.5-minute quadrangle. Alternating bands represent more biotite-rich and biotite-poor layers. Photograph from Stieve (1989).

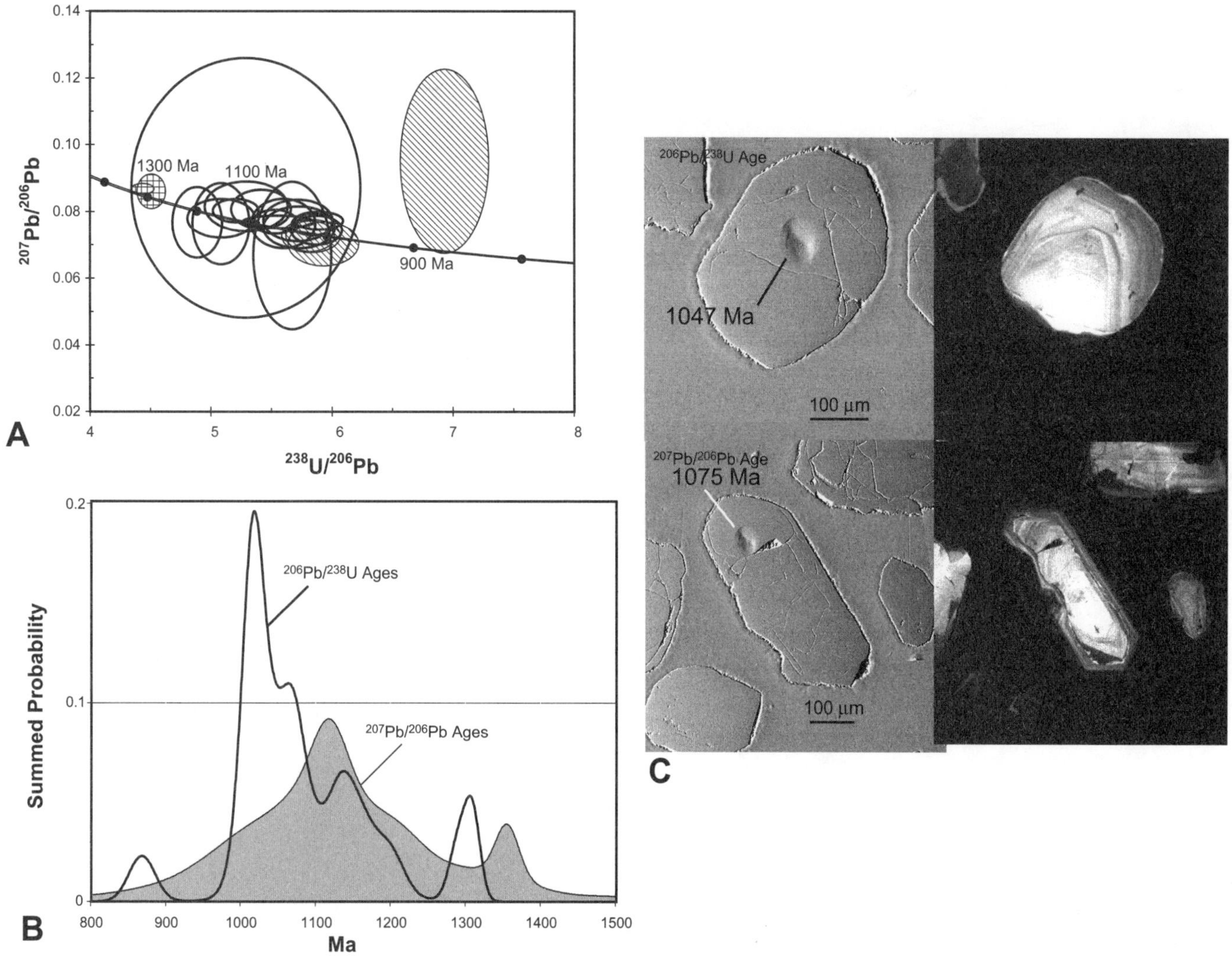

Figure 7. Wolf Creek Gneiss concordia diagram and summed probability plot for SHRIMP RG analytical data. (A) Tera Wasserburg concordia diagram for zircon separates from Wolf Creek Gneiss. Shaded and patterned ellipses were excluded from the pooled age determination (gray denotes Grenville rim, cross hatch denotes inherited core, and sloped lines denote young or discordant analyses). (B) Summed probability plot for Wolf Creek Gneiss ^{206}Pb/^{238}U and ^{207}Pb/^{206}Pb ages. (C) Paired backscattered electron and cathodoluminescence images of typical Wolf Creek Gneiss grains showing the pits left by the primary ion beam and internal fracture and zoning characteristics.

Geochemistry of Basement Rocks

The Wiley and Sutton Creek Gneisses (Carrigan et al., 2003, their Figs. 7, 8, and 9), and all Blue Ridge basement rocks except Wolf Creek Gneiss exhibit similar rare-earth element (REE) distributions, including a negative Eu anomaly. They also mostly plot in the within-plate and volcanic-arc granite fields of standard tectonic discriminant diagrams (Pearce et al., 1984; Carrigan et al., 2003, their Fig. 9) (Fig. 8C). Megacrystic Wolf Creek Gneiss (Table 2), in contrast, consistently plots in the volcanic-arc granites field of standard tectonic discriminant diagrams (Fig. 8C). Although based only on a single analysis, the Wolf Creek Gneiss is chemically distinct from the recognized basement of both the western Blue Ridge and the Mars Hill terrane. Most notably, it has significantly higher Sr content and markedly lower K and Rb values (Fig. 8B). The sample is conspicuously depleted in all REE (except Eu), has a positive Eu anomaly, and is only slightly light REE (LREE)-enriched relative to the other internal basement massif samples (Fig. 8). The strong positive Eu and Sr spikes also suggest plagioclase accumulation in the Wolf Creek sample.

Toxaway Dome

Structure and Map Pattern

The Toxaway dome, like the Tallulah Falls dome, is located in the eastern Blue Ridge Tugaloo terrane immediately northwest of the Brevard fault zone (Figs. 1 and 9). The structure of the Toxaway dome was described by Hatcher (1977); no additional

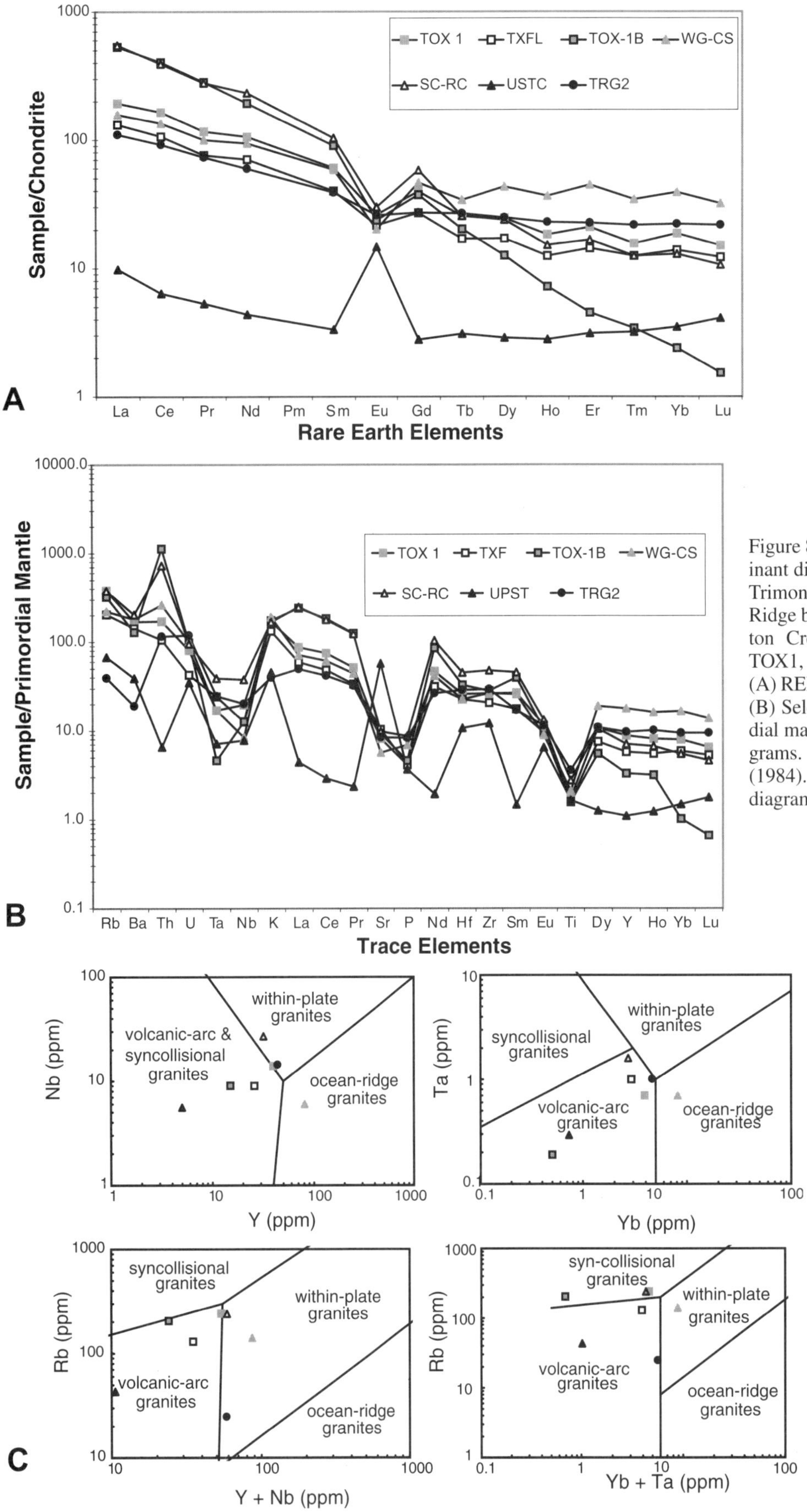

Figure 8. REE, trace element, and tectonic discriminant diagrams for the Wolf Creek Gneiss (UPST), Trimont Ridge Gneiss (TRG2), and eastern Blue Ridge basement data of Carrigan et al. (2003) (Sutton Creek Gneiss—SC-RC; Toxaway Gneiss—TOX1, TXF, TOX-1B; Wiley Gneiss—WG-CS). (A) REE diagram for eastern Blue Ridge basement. (B) Selected trace elements normalized to primordial mantle. (C) Standard tectonic disciminant diagrams. Discriminant diagrams from Pearce et al. (1984). Note that the symbols are the same for each diagram.

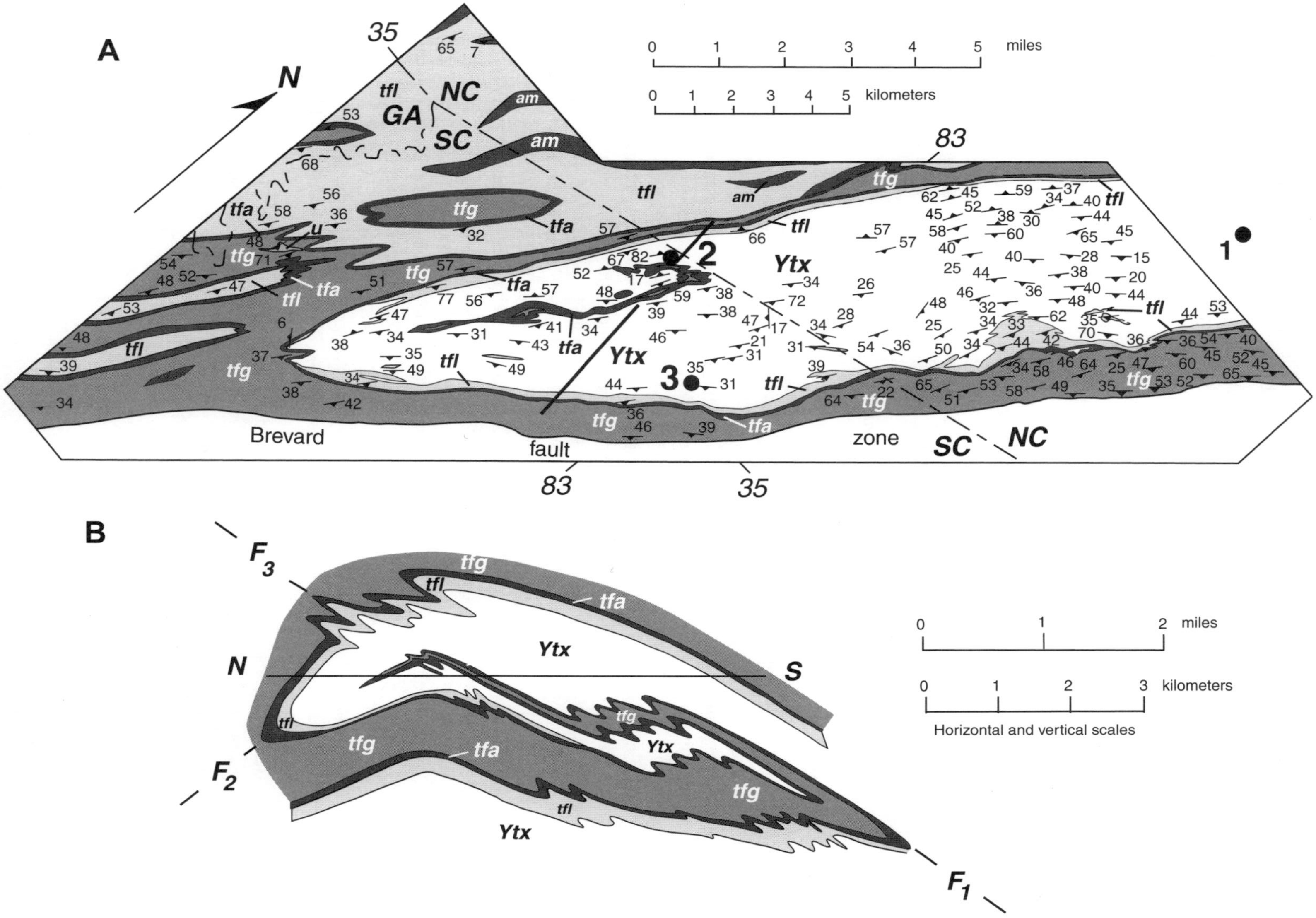

Figure 9. (A) Detailed geologic map southwest end of the Toxaway dome in North and South Carolina. Ytx—Toxaway Gneiss. am–amphibolite. See Figure 2 for explanation of other map symbols. Black dots and bold numbers represent the locations of Toxaway Gneiss geochronology and geochemistry samples of Carrigan et al. (2003). 1—Toxaway Falls (sample TXF). 2—SC Highway 413 overlook (samples TOX 1 and TOX 1B). 3—Bad Creek Powerhouse outlet tunnel (samples BTB3 and GMA2). Dashed contacts in the northwestern part of the map indicate area of less detailed mapping. Modified from Hatcher (1977, his Fig. 2) and Hopson and Hatcher (1988), incorporating data from an unpublished map by M.P. Hartford (1984), and contacts on two small outliers of Tallulah Falls Formation rocks resting on Toxaway Gneiss from Carter et al. (2002). (B) North-south cross section across the Toxaway dome showing the polyphase fold structure required to explain the outcrop pattern of basement and cover rocks. From Hatcher (1977, his Fig. 12).

data have appeared to change the original interpretation (Fig. 9B), so only pertinent details are repeated here. In simple form, the dome consists of an elongated elliptical, northwest-vergent refolded antiformal core of Middle Proterozoic Toxaway Gneiss surrounded by Tallulah Falls Formation; all of the structure is likely the product of Neoacadian deformation. An important clue to the structural complexity of the dome is provided by an isolated boomerang of Tallulah Falls Formation rocks preserved in its core. Roper and Dunn (1970) had earlier mapped this interior belt of Tallulah Falls Formation rocks at the southwest end of the dome to connect with Tallulah Falls Formation rocks on the exterior flanks of the dome. Our subsequent 1:12,000-scale mapping in the relatively inaccessible southwest interior of the dome clearly demonstrated that the Tallulah Falls Formation in the core of the dome does indeed close and is isolated, forming the boomerang structure (Hatcher, 1977) (Fig. 9). The three principal subdivisions of the Tallulah Falls Formation (lower graywacke-schist-amphibolite member, middle garnet-aluminous schist member, and upper graywacke-schist member; Hatcher, 1971) are all represented and are continuously mappable on all of the outside flanks of the dome, with the graywacke-schist-amphibolite member consistently resting on Toxaway Gneiss (except at the southwest end of the dome) (Fig. 9). In the boomerang in the interior and at the southwest exterior end of the dome, the lower member is consistently missing and the garnet-aluminous schist member rests on Toxaway Gneiss. Metamorphic grade of all Tallulah Falls Formation rocks both within and on the flanks of the dome is kyanite, based on consistent occurrence of kyanite in the garnet-aluminous schist member. No indicator minerals have been observed in the Toxaway Gneiss, but most geologists who have studied the Toxaway Gneiss have concluded (based on the greater migmatization of the Toxaway Gneiss than the cover metasedimentary rocks) that it was subjected to Middle Proterozoic upper amphibolite- to granulite-facies metamorphism (McKniff, 1967; Hadley and Nelson, 1971; Hatcher, 1977; Horton, 1982; Carter et al., 2002).

Horton et al. (1989, their Fig. 1) and Rankin et al. (1993, p. 397) suggested the Toxaway Gneiss–Tallulah Falls Formation contact on the Toxaway dome is faulted, based on the logic of separating all rocks of possible Laurentian affinity from their disrupted (and presumed exotic) Jefferson terrane. In making this suggestion, they did not take into account most existing (Hatcher, 1977) and unpublished detailed maps (by R.D. Hatcher and M.P. Hartford) and, more recently, new field data of Carter et al. (2002) from the dome on both sides of the North Carolina–South Carolina border. This contact is rarely exposed and, in the few exposures where the relationships between the two units could be examined, the Toxaway Gneiss below the contact consists of medium-to-coarse-grained, totally annealed, strongly foliated felsic gneiss that is not obviously mylonitic in either hand specimen or thin section. The Tallulah Falls Formation above the contact consists of thoroughly recrystallized and annealed metagraywacke, muscovite-biotite schist, and amphibolite no different from that away from the contact. Mylonitic fabrics are limited to S-C structures in some of the schists that are common throughout the Tallulah Falls Formation in this area and easily related to the high-to-low–temperature (retrogressive) movement history of the Brevard fault zone nearby (Hatcher, 2001). If the contact is faulted, the fault must be very early, follow the same stratigraphic level on all but the southwest flank of the dome, cut upsection in the core of the dome and at the southwest end, and then cut back down again to the northwestern flank. This faulting must also be, despite many years of very detailed geologic mapping, the product of a still earlier unrecognized deformational event. Taking into account the stratigraphic relationships between the two units, the geometry of the contact, and the almost impossible kinematics to produce a faulted contact, a nonconformable relationship at the contact with some local faulting remains the most viable possibility.

Toxaway Gneiss

The name "Toxaway Gneiss" was first suggested by McKniff (1967) for the rocks inside the Toxway dome and later formalized by Hatcher (1977) for exposures at Toxaway Falls on U.S. 64 in the Reid 7.5-minute quadrangle, North Carolina (Fig. 10). It consists of two principal lithologies: a strongly banded orthogneiss, consisting of alternating bands of light-colored, feldspar-quartz and darker-colored feldspar-quartz-biotite gneiss (Fig. 10A); and a microcline-oligoclase-quartz-biotite augen gneiss with K-spar augen, closely resembling the Wiley Gneiss (Fig 10B). Whereas the augen gneiss phase is represented in sporadic surface outcrops (among dominant banded gneiss) and in sizable bodies in the Bad Creek Project underground tunnels (Schaeffer et al., 1979) in the South Carolina portion of the dome, it makes up as much as 50% of the Toxaway Gneiss outcrop area immediately north of the state line in North Carolina (Carter et al., 2002). Field relationships at the Bad Creek powerhouse tunnel suggest that the banded phase was intruded by the augen phase. The augen texture does not appear to be related to an S-C fabric.

The Middle Proterozoic age of the Toxaway Gneiss was first determined by a Rb/Sr whole-rock age of 1197 ± 56 Ma by Fullagar et al. (1979), with a $^{87}Sr/^{86}Sr$ initial ratio of 0.7016 (0.7062, according to Carrigan et al., 2003). The ε_{Nd}-depleted mantle model age for the Toxaway Gneiss samples is 1.3–1.6 Ga (Carrigan et al., 2003), suggesting derivation from the same eastern granite-rhyolite province source as most western Blue Ridge Grenvillian rocks, along with the Wiley and Sutton Creek Gneisses. Carrigan et al. (2003) reported ages of 1151 ± 17 Ma, 1039 ± 11 Ma, and 1149 ± 56 Ma for banded Toxaway Gneiss from the South Carolina Highway 416 (SC 416) overlook just south of the North Carolina state line in the Cashiers 7.5-minute quadrangle, North Carolina–South Carolina–Georgia (their TOX 1 and TOX 1B samples). Another sample of banded Toxaway Gneiss (GMA2 from the Bad Creek powerhouse outlet tunnel in the Reid quadrangle in South Carolina) yielded a con-

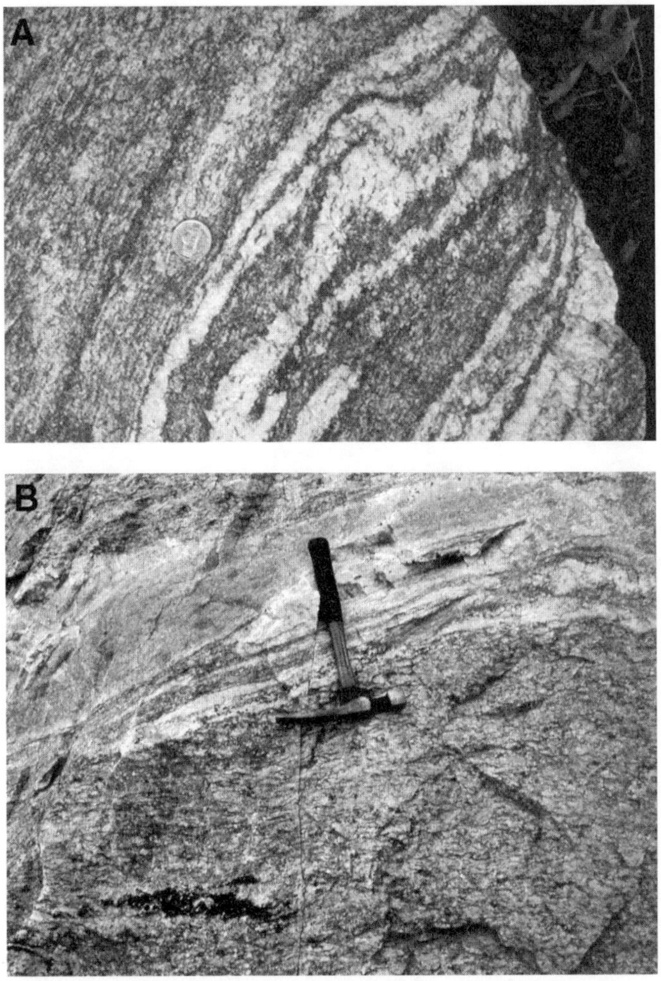

Figure 10. (A) Typical banded Toxaway Gneiss from Upper Whitewater Falls, North Carolina. Light bands are quartz-feldspar dominant; dark bands also contain quartz and feldspars, but biotite is a major component. All structural components of the refolded fold could be the result of Paleozoic folding; alternatively, all could be Grenvillian structures. Involvement of Tallulah Falls Formation cover rocks in complex macroscopic refolding favors the first alternative. (B) Typical augen phase of the Toxaway Gneiss on U.S. 64 at Toxaway Falls, North Carolina.

cordia age of 1157 ± 36 Ma. A sample of banded augen gneiss (BTB3) from the same locality as sample GMA2 yielded a poorly constrained $^{207}Pb/^{206}Pb$ age of 1141 ± 148 Ma. Augen phase Toxaway Gneiss from the Toxaway Falls type locality yielded an ion microprobe concordia age of 1149 ± 32 Ma. Samples from the SC 416 overlook yielded $^{207}Pb/^{206}Pb$ metamorphic ages of 1027 ± 23 to 1039 ± 82 Ma, and the BTB3 sample yielded a 921 ± 45 Ma concordant U-Pb metamorphic age, indicating peak Grenvillian metamorphism occurred at least 100 m.y. after intrusion of the Toxaway Gneiss. Toxaway Gneiss samples plot in the syncollisional and volcanic-arc granite fields of standard tectonic discriminant diagrams using trace element geochemical data (Carrigan et al., 2003) (Fig. 8C).

TRIMONT RIDGE COMPLEX

Structure and Map Pattern

The Trimont Ridge complex is located in the central Blue Ridge Cartoogechaye terrane of Hatcher (2001), where the highest grade of Paleozoic metamorphism is recorded in the southern Appalachians (Figs. 1 and 11). This zone was called the Wayah granulite-facies metamorphic core by Eckert et al. (1989). This Paleozoic event was dated by Moecher and Miller (2000) at Winding Stair Gap on U.S. 64 ~6 km south of the Trimont Ridge massif through a SHRIMP concordia age of 470 ± 7 Ma from metamorphic zircon cores (Fig. 11).

The Trimont Ridge basement massif occurs in a domain of almost east-west trending, steeply dipping structures in the Hayesville–Chunky Gal Mountain–Shope Fork thrust sheet in southwestern North Carolina (Figs. 1 and 11). This domain roughly corresponds to the shape and distribution of Force's (1976) hypersthene zone. The Trimont Ridge basement massif is flanked by biotite gneiss (metagraywacke) and pelitic schist units that were tentatively correlated with the Tallulah Falls Formation and Coweeta Group by Eckert et al. (1989) and Hopson et al. (1989). The Trimont Ridge complex has a mafic component that appears to be intimately related to the dominant felsic (basement) gneiss (Fig. 11). Eckert (1984, his Plate 1) and Eckert et al. (1989, their Fig. 2) suggested the Trimont Ridge massif is overlain by Tallulah Falls Formation and then by Coweeta Group rocks. The characteristic tripartite stratigraphy of the Tallulah Falls Formation in the Tugaloo terrane to the southeast has not been recognized in the Cartoogechaye terrane, and the overlying rocks correlated with the Coweeta Group in the Hayesville-Soque River thrust sheet to the southwest are also dissimilar. Northwest-dipping (inferred southeast-directed) thrust faults appear from the map pattern to be associated with the Trimont Ridge massif. These faults may be kinematically linked to the premetamorphic-peak Hayesville fault. The contact between basement and cover does not appear to be faulted everywhere (Fig. 11).

Trimont Ridge Complex Basement Rocks

The Trimont Ridge metaigneous complex was defined by Eckert and Mohr (1986) as a composite assemblage dominated by low-K, oligoclase-rich garnet and/or orthopyroxene-bearing felsic gneiss, with associated amphibolite (some containing both orthopyroxene and clinopyroxene). Xenoliths of biotite gneiss and amphibolite were noted in the felsic gneiss (Eckert, 1984). Eckert and Mohr (1986) concluded the Trimont Ridge complex formed from an oceanic source, and speculated that the complex is related to the numerous mafic-ultramafic complexes that occur nearby. The Trimont Ridge massif was first speculated to be Middle Proterozoic basement by Hopson et al. (1989) and Hatcher and Hopson (1989), but is only now confirmed here. The layered felsic gneiss samples collected for this study are not

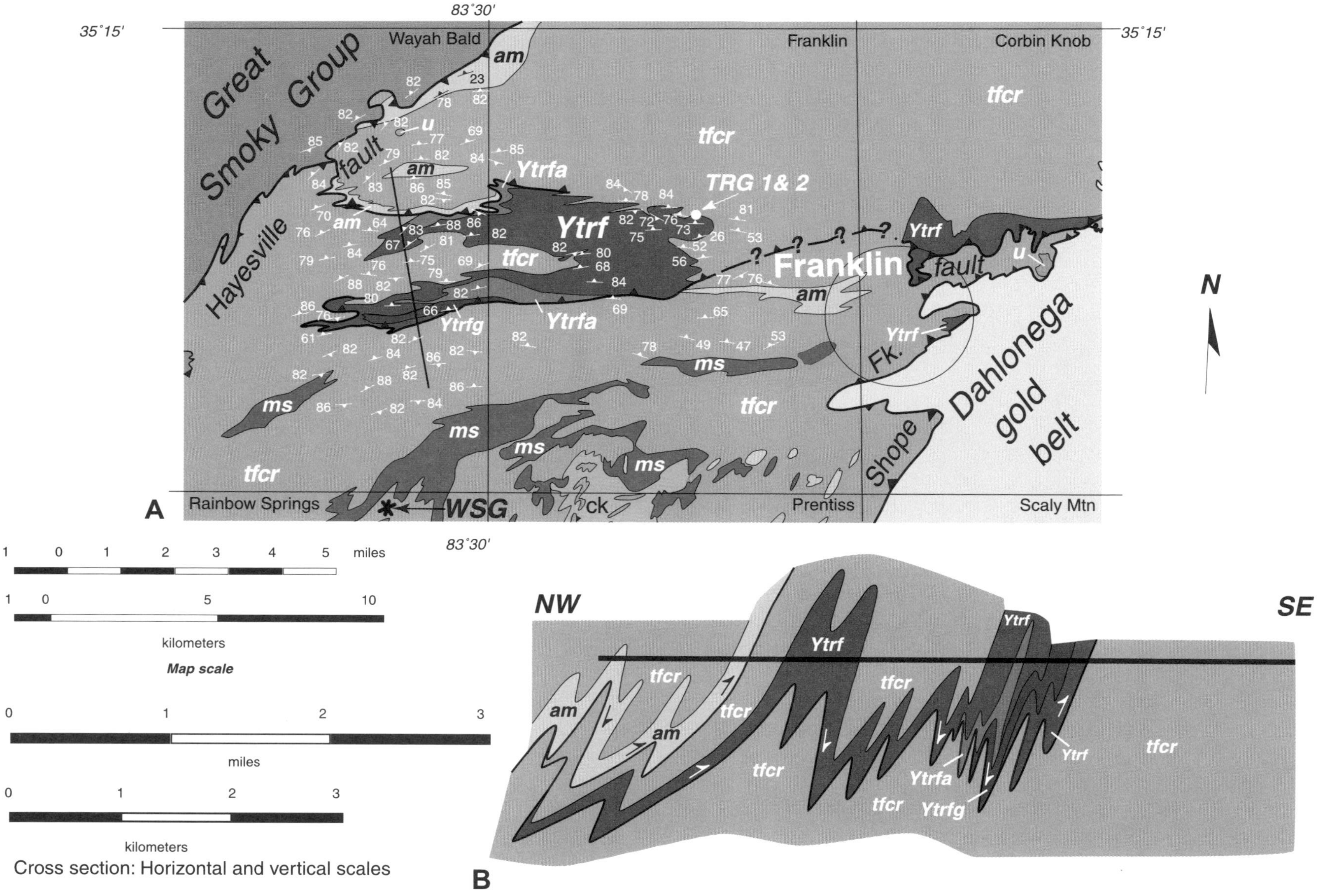

Figure 11. (A) Trimont Ridge complex and vicinity geologic map. Note that the rocks here are mostly steeply dipping, with dominant northwest dip. am—amphibolite interlayered with metasandstone and schist of the tfcr unit. tfcr—Tallulah Falls Formation(?) and Coleman River Formation(?) metasandstone and schist. ms—garnet-rich biotite gneiss with some schist, locally sillimanite-bearing. u—ultramafic rocks. ck—Carroll Knob mafic-ultramafic complex. Ytrf—Trimont Ridge felsic gneiss. Ytrfa—Trimont Ridge felsic gneiss and amphibolite. Ytrfg—Trimont Ridge felsic gneiss and metagabbro. WSG—Winding Stair Gap. Names of 7.5-minute quadrangles are indicated near the northern boundary of each. Location of cross-section line in (B) is indicated by the heavy solid line in the east-central part of the Wayah Bald quadrangle. White dot near the east end of the main Ytrf outcrop belt is the location of geochemical and geochronological samples TRG1 and TRG2. Modified from Hatcher (1980); Eckert (1984); Eckert et al. (1989); J.O. Eckert Jr., unpublished mapping in the Franklin, North Carolina, quadrangle; Hatcher and Hopson (1989); R.D. Hatcher Jr., unpublished mapping in the northern Rainbow Springs and western Wayah Bald quadrangles. (B) Northwest-southeast cross section through the Trimont Ridge massif. See Figure 11A for section location and explanation of symbols. Modified from Eckert (1984, his Plate II).

considered representative of the characteristic assemblages in the complex (Eckert et al., 1989), but are clearly derived from within it (Fig. 11).

Geochronology and Geochemistry

A SHRIMP age of 1103 ± 69 Ma (MSWD = 0.9, *n* = 16, 2σ error $^{207}Pb/^{206}Pb$) age was determined for pooled analyses of one Trimont Ridge felsic gneiss sample. Several grains are interpreted to have Pb loss, and the discordia upper intercept of ~1170 Ma suggests that the time of crystallization could be older. Two Grenvillian rims were analyzed (one shows significant reverse discordance), yielding ages similar to those reported for Grenvillian metamorphism by Carrigan et al. (2003) (Fig. 12). Zircons from the Trimont Ridge Gneiss contain complex zoning patterns and are variably metamict, although this is not directly reflected in the U and Th concentrations of these zones. The $^{87}Sr/^{86}Sr$ initial ratio for the Trimont Ridge basement sample is 0.7063 (Table 2) and the ε_{Nd}-depleted mantle model age is 1.2–1.3 Ga (Fig. 13), similar to that for Wolf Creek Gneiss.

Chemically, the Trimont Ridge sample is distinguished from other basement units by lower K, Rb, and Ba concentrations (Table 2), probably because of the absence of K-spar; a slight positive Eu anomaly; lower LREE concentrations; only slight LREE enrichment; and a flat heavy REE trend (Fig. 8). The sample is also strongly peraluminous and has higher Cr, Ni, and Fe concentrations than expected for a granitoid. Although zircon separates are euhedral and magmatically zoned and xenoliths occur in the felsic gneiss, the limited nature of the data does not preclude a sedimentary protolith for the sample dated. This sample is, however, petrographically and chemically distinct from most of the rocks in the same thrust sheet, so the lack of a typical Grenvillian chemical signature may not eliminate it as basement. Despite these chemical differences, the Nd isotopic values over time are indistinguishable from other basement samples.

DISCUSSION

The three internal basement massifs discussed here have Grenvillian signatures ranging from somewhat to profoundly different isotopically and chemically and exhibit contrasting complex polydeformational Paleozoic histories. The Tallulah Falls dome exhibits the greatest structural complexity, with

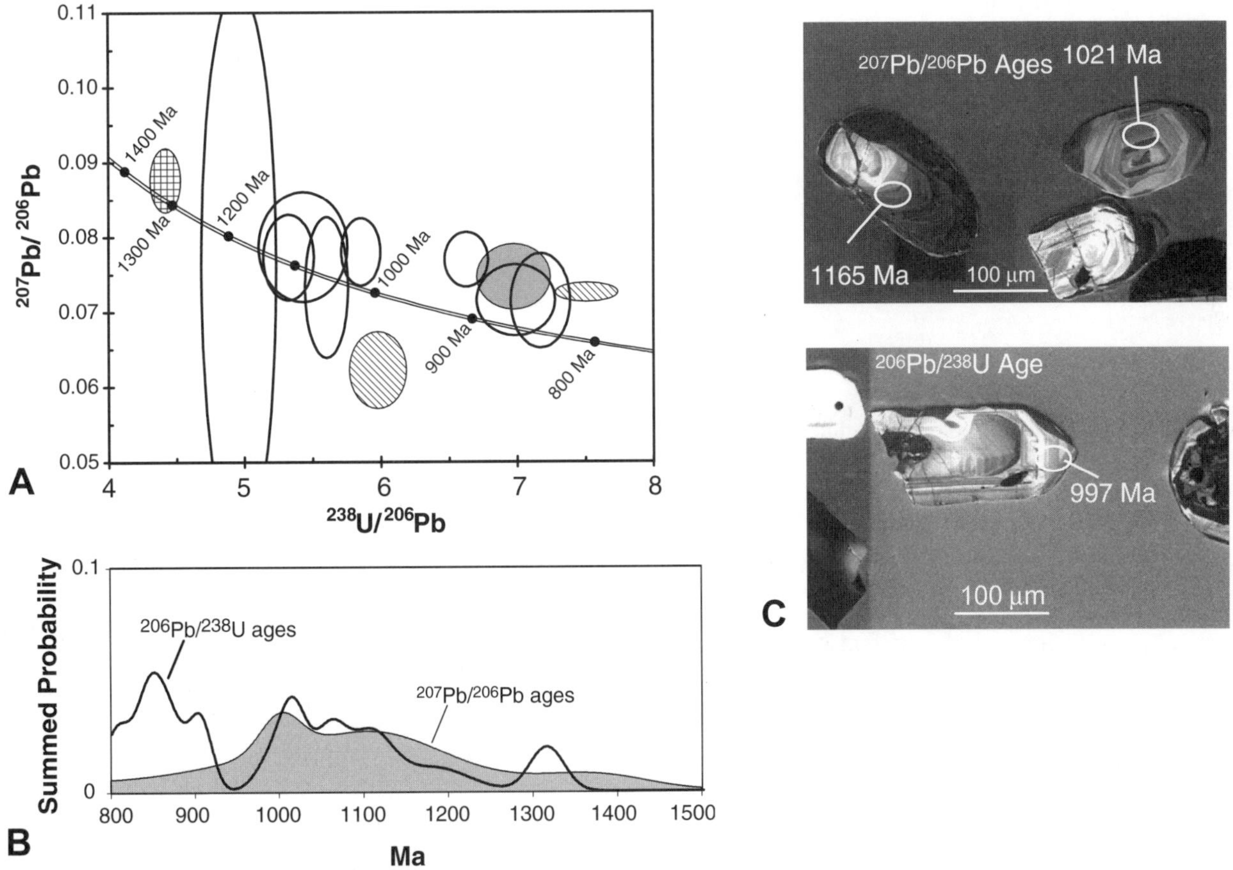

Figure 12. Trimont Ridge Gneiss concordia diagram and summed probability plot for SHRIMP RG analytical data. (A) Tera Wasserburg concordia diagram for zircon separates from the Trimont Ridge Gneiss. Shaded and patterned ellipses were excluded from the pooled age determination (cross hatch denotes inherited core, and sloped lines denote young or discordant analyses). (B) Summed probability plot for Trimont Ridge Gneiss $^{206}Pb/^{238}U$ and $^{207}Pb/^{206}Pb$ ages (shaded field). (C) Cathodoluminescence images of typical Trimont Ridge Gneiss zircons.

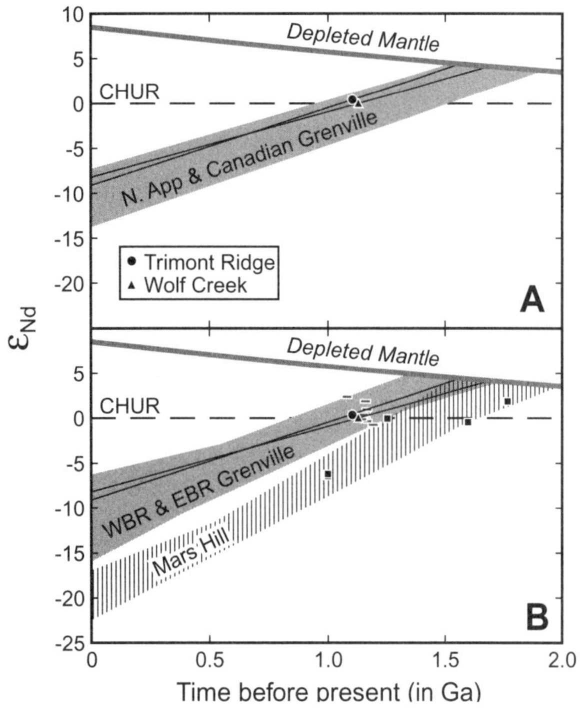

Figure 13. Evolution of ε_{Nd} values with time for the Wolf Creek and Trimont Ridge Gneisses. Symbols are sample values at the estimated time of crystallization (black squares are Mars Hill terrane and gray rectangles are other Blue Ridge basement samples (Fullagar et al., 1997). Note the tight clustering of Blue Ridge Grenville basement data near CHUR at the time of crystallization. (A) Sample data plotted against the composite field for northern Appalachian and Canadian Grenville lithologies (Samson et al., 1995, compiled from Marcantonio et al., 1990; Daly and McLelland, 1991; and Dickin and Higgins, 1992). (B) Sample data plotted against the field defined for the Mars Hill Terrane (Carrigan et al., 2003; Ownby et al., this volume) and western Blue Ridge and Sauratown Mountains window basement samples (Carrigan et al., 2003).

Grenvillian rocks forming the cores of isoclinal antiforms along the flanks of the dome. Basement rocks in isoclinal anticlines on the northeast and southwest flanks of the dome probably remain in nonconformable contact with the Tallulah Falls Formation cover (Figs. 2 and 4). The predominantly Wiley Gneiss basement rocks in the hinge of the antiform that spirals eastward, then southward, and then southwestward from the interior of the dome to the south flank are most certainly faulted at the north end and may be in fault contact with the cover rocks throughout this antiform (Figs. 2 and 3). If so, this body of Grenvillian rocks may have behaved as a more rigid mass that was squeezed upward between layers that behaved much more plastically during deformation. The resulting fault(s) may therefore have small displacement, probably no more than a few km. This hypothesis does not, however, explain how the Wiley (plus Suttton Creek) Gneiss body, localized in the hinge of a superposed antiform, can be traced from the interior of the dome, where it appears to be involved in an isoclinal anticline-syncline pair with basement rocks in the hinge of the synformal anticline and Tallulah Falls Quartzite in the hinge of the syncline. This relationship may only be apparent if the interior of the dome is a window through a major fault. The northern end of the Wiley Gneiss outcrop belt may consist of a small anticlinorium of basement rocks beheaded by the Chattahoochee fault that fortuitously rests today on an isoclinal footwall anticline of metasandstone that resides inside the window. This requires the fault to be locally tightly folded (Fig. 3).

The exposed Toxaway Gneiss body is areally much larger than all of the basement rock bodies on the Tallulah Falls dome combined. The Toxaway dome does not appear to record some of the later Alleghanian(?) deformation recorded by the Tallulah Falls dome, accounting in large part for the differences in outcrop pattern of the two structures. The latest deformation in the Toxaway dome that affected the map pattern was ductile, producing the antiformal culmination and northwestward vergence and overturning of the dome. This deformation probably occurred just after the Neoacadian metamorphic peak. Slight additional tightening could have occurred during the Alleghanian, but this remains unresolved.

The Trimont Ridge basement massif appears to have been involved in Taconian isoclinal folding at near-peak metamorphic conditions. Contacts with the cover rocks are rarely exposed and where they are, it is difficult to make judgments about their nature. Possible truncation of cover rocks suggests the contact may be locally faulted, but a nonconformity is probably present along most of the contact (Fig. 11A).

Geochemical data from most of these basement units do not yield any great surprises, except that the Trimont Ridge sample consistently plots along the boundary between the within-plate granites and volcanic arc granites fields of tectonic discriminant diagrams, despite its low K, whereas the other samples plot in a variety of fields (Fig. 8). The Wolf Creek sample has a positive Eu anomaly and is otherwise chemically different. The ε_{Nd} versus age data for all eastern Blue Ridge samples tightly cluster in the same area, and yield a model age of ~1.5 Ga (Fig. 13). They cluster in a part of the diagram only slightly different from western Blue Ridge basement (Carrigan et al., 2003), but different from the more primitive Mars Hill terrane samples (Carrigan et al., 2003; Ownby et al., this volume). Non–Mars Hill western Blue Ridge samples exhibit greater scatter, probably because of the longer crustal history represented there. Nevertheless, all of the Blue Ridge basement ε_{Nd} versus age data plot within the composite range of ε_{Nd} versus age values of known North American Grenville rocks compiled by Samson et al. (1995) from Marcantonio et al. (1990), Daly and McClelland (1991), and Dickin and Higgins (1992). The 1.5–1.6-Ga model age for the central and eastern Blue Ridge massifs suggests that these rocks could have formed by melting of eastern granite-rhyolite province crust. Although most plutonic rocks of the eastern granite-rhyolite province have ages close to 1.5 Ga, some small plutons in the St. François Mountains, Missouri, have been dated at 1.37 Ga (Bickford and Anderson in Van Schmus et al., 1993). Moreover, Dickin and Higgins (1992) also identified a possible 1.5-Ga crust-forming event within the Grenville province in Canada. Heatherington et al. (1996) reported a 1.6-Ga ε_{Nd}-depleted mantle model age for the 1106 ± 13-Ma (SHRIMP U/Pb) Corbin Gneiss in the western Blue Ridge east of Cartersville, Georgia (Fig. 1). A 1.5-Ga zircon crystallization age component has not been recognized, however, in southern Appalachian basement rocks. The Carvers Gap and Roan Mountain zircons (probably detrital) from the 1.8-Ga Mars Hill terrane (Carrigan et al., 2003; Ownby et al., this volume), as well as detrital zircons from post-Grenville sedimentary and metasedimentary rocks in the southern Appalachians internides, do contain 1.4-, 1.5-, and 1.6-Ga components (Bream, 2003).

The small size of each of the three internal massifs, coupled with the close proximity of the Tallulah Falls and Toxaway massifs in the Tugaloo terrane, suggests that these massifs may be fragments rifted from Laurentia during the later (572–564 Ma; Aleinikoff et al., 1995) rifting stage that successfully opened the Iapetus ocean (Fig. 14). The Trimont Ridge massif could be a similar block, but more isolated than the other two in the Cartoogechaye terrane. Alternatively, but less likely, all three massifs could be rifted fragments of another coevally proximal continent (Gondwana?) that contain the same age and geochemical signature as these massifs and the kinematically more obvious western Blue Ridge Laurentian basement. Either way, the three massifs may be analogous to the small, isolated Precambrian(?) blocks of possible continental crust that Hamilton (1979) identified in Indonesia, far from Asia in Mesozoic-Tertiary arc complexes.

Basement-cover contacts for the three massifs, on detailed geologic maps and where they can be directly examined in the field, appear nonconformable (e.g., Fig. 4) and only locally faulted. If so, these blocks would have been initially surrounded by nascent latest Neoproterozoic-Cambrian (572–564 Ma) oceanic crust that was generated following rifting. Deposition of the overlying Tallulah Falls (Tugaloo terrane) and Tallulah Falls(?) and Coweeta Group(?) (Cartoogechaye terrane) cover rocks would thus have occurred on a combination of oceanic crust and continental fragments, as previously suggested by Hatcher (1978, 1987, 1989). This does not necessarily change the basic definition of the Jefferson disrupted terrane of Horton et al. (1989) and Rankin et al. (1993), except to make the terrane less exotic and more extensive across part of the Inner Piedmont. Detrital zircons from cover rocks in the Tugaloo terrane (from both the eastern Blue Ridge and western Inner Piedmont components) are dominated by a 1.1-Ga (Grenvillian) source terrane internally; western Blue Ridge detrital zircons also indicate a very similar 1.1-Ga source terrane. All detrital zircon suites from these terranes contain the same earlier 1.2-, 1.3-, 1.4-, and 1.5-Ga zircon components (see Bream, 2003; Bream et al., this volume). These factors, coupled with the Grenvillian ages and similarities of geochemical/isotopic signatures of the rocks in the three massifs to known Laurentian Grenvillian rocks, suggest that the Tugaloo and Cartoogechaye terranes contain Laurentian basement and distal Laurentian margin assemblages that were accreted back to Laurentia during the early to middle Paleozoic Taconian and Neoacadian orogenies.

CONCLUSIONS

1. Three internal basement massifs having similar Grenvillian plutonic histories but contrasting Paleozoic structural histories occur in the southern Appalachian central and eastern Blue Ridge. The Tallulah Falls dome has the most complex structure and probably consists of a window in the Tugaloo terrane hanging wall through an earlier fault into the footwall of Tallulah Falls Quartzite and metasandstone. Grenvillian basement rocks occur in the hanging wall in the cores of isoclinal anticlines in both nonconformable and locally faulted contact with the cover rocks. The present structural configuration of the Tallulah Falls dome is a product of deformation during the Acadian and Alleghanian orogenies. Taconian penetrative deformation and folding are probably also present, but are not clearly documentable. The Toxaway dome is largely the product of multiple deformational events of tight to isoclinal folding involving both the basement block and the cover. Some of the earliest deformational phases could be Taconian, but those related to regional peak metamorphic conditions that determined the dominant macroscopic structural patterns are probably Acadian. The Trimont Ridge massif is located in the core of a faulted and isoclinally folded Taconian granulite-facies assemblage in the central Blue Ridge Cartoogechaye terrane.

2. Basement lithologies in the three massifs are all orthogneisses; those on the Toxaway and Tallulah Falls domes were intruded during the same ~1150-Ma event, and that in the Trimont Ridge massif at 1103 ± 69 Ma; all could be considered being intruded during a 1.1–1.16-Ga Grenvillian plutonic event. $^{87}Sr/^{86}Sr$ ratios and ε_{Nd} model ages are similar for most of the orthogneisses, indicating a continental crustal source. The $^{87}Sr/^{86}Sr$ ratio for the Sutton Creek Gneiss is lower than for the others, possibly indicating a more primitive source. These orthogneisses plot in the syncollisional and volcanic granites fields of standard tectonic discriminant diagrams; the Trimont Ridge sample plots consistently on the boundary between the within-plate granite and volcanic-arc granite fields of the same diagrams. Trace element geochemistry of the megacrystic Wolf Creek gneiss contrasts with that of other eastern Blue Ridge basement. Plots of ε_{Nd} versus age for basement rock samples from all three massifs cluster tightly within the composite range of values for known North American Grenville rocks. An ε_{Nd}-depleted mantle model age of 1.5 Ga is consistent with derivation by melting of eastern granite-rhyolite province crust.

3. Taken separately and independently, several of the various geologic map, structural, sedimentological, isotopic, and

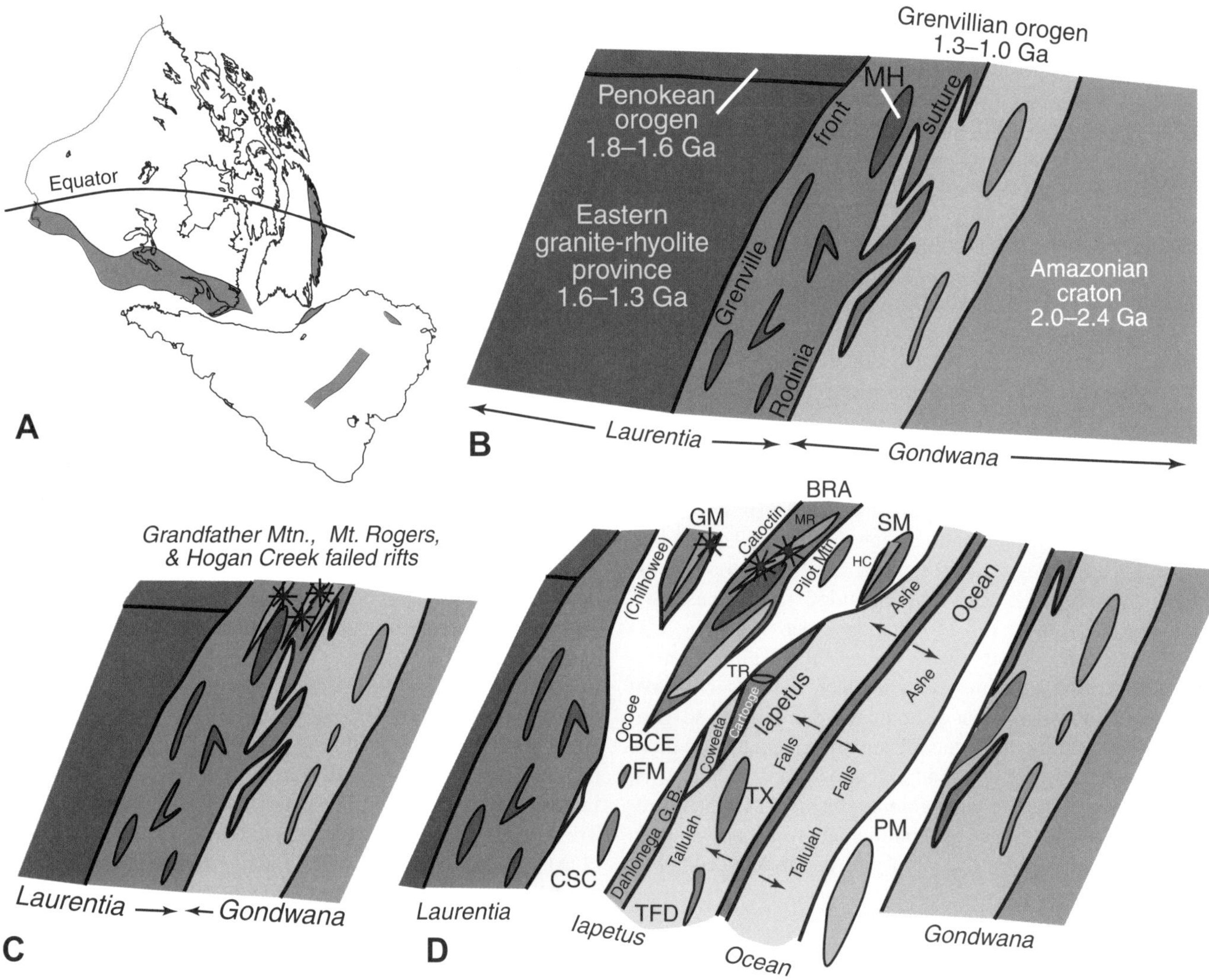

Figure 14. Diagrammatic representation of Laurentian margin evolution from the Grenville orogeny forming Rodinia to Neoproterozoic–early Paleozoic successful rifting and formation of the Iapetus ocean. (A) Possible configuration of Laurentian and Gondwanan cratons at the end of the Proterozoic, showing Grenvillian belts (shaded). (Modified from reconstructions by Hoffman, 1991; Dalziel, 1997; and Cawood et al., 2001.) (B) Representation of likely components of Rodinia and the Grenville rogen along the suture between the Laurentian and Amazonian/Rio de la Plata cratons at the end of the Grenvillian orogenies. Recycled components of the cratons on either side of the suture are shown as outliers and new Grenville crust is shown in lighter shades. MH–Mars Hill terrane (here presumably derived from the Penokean orogen). Mid-continent ages are from Van Schmuus et al. (1993). (C) Failed rifting and bimodal volcanism (asterisks) in "southeastern" Rodinia at 750–720 Ma (Aleinikoff et al., 1995) produced the Grandfather Mountain, Mt. Rogers, and Hogan Creek (Hatcher et al., 1988), and Robertson River assemblage (not shown), farther north in Virginia (Lukert and Banks, 1984) rift facies sedimentary-volcanic and alkali plutonic (e.g., Crossnore) assemblages. (D) Successful rifting and opening of Iapetus 572–564 Ma (Aleinikoff et al., 1995) provided the rifted margin for deposition of the proximal to distal sedimentary sequences off the Laurentian margin. Several of the more distal units (e.g., Tallulah Falls–Ashe) are probably time transgressive and are likely to be much younger than the more proximal deposits. More proximal assemblages were probably deposited on continental or attenuated continental crust, whereas the more distal assemblages were probably deposited on oceanic crust and basement fragments. The configuration of both failed rift deposits shown in (C), and synrift to postrift deposits, and rifted blocks of Grenvillian crust is based on their relative spatial positions in the thrust sheets where they are found today. BCE—Bryson City and Ela domes, and Ravensford anticline; BRA—Blue Ridge anticlinorium; Cartooge—Metasedimentary, metavolcanic, and ultramafic rocks of the Cartoogechaye terrane; CSC—Corbin-Salem Church basement fragment; Dahlonega G.B.—Dahlonega gold belt metasedimentary sequence (could be as young as Ordovician; Thomas, 2001, Bream, 2003); FM—Fort Mountain basement fragment; GM—Grandfather Mountain window basement block and cover metasedimentary and metavolcanic rocks; HC—Hogan Creek Formation failed rift sequence (Sauratown Mountains, SM, outer window); MR—Mount Rogers failed rift sequence and basement rocks; PM—Pine Mountain basement fragment and cover metasedimentary rocks (had to have been located close to Gondwana to receive Gondwanan detrital zircons, but is kinematically difficult to get back to its present position); TFD—Tallulah Falls dome basement fragment(s) and cover metasedimentary rocks; TR—Trimont Ridge basement fragment(s); TX—Toxaway dome fragment. Chilhowee Group is shown in parentheses because it was deposited later, but either directly on basement or on rifted-margin cover (e.g., Hardeman, 1966). (Modified from Hatcher, 1978, his fig. 7, and Bartholomew and Lewis, 1992, their fig. 4.)

geochemical data sets presented here could be used to argue that the three internal basement massifs and their cover in the Tugaloo and Cartoogechaye terranes could have been derived from another continental fragment of Rodinia, such as Gondwana. A combination of all data, however, strongly suggests that the orthogneisses in the three internal basement massifs were all derived as small Laurentian rifted blocks during the latest Neoproterozoic to Early Cambrian rifting event. Cover rocks were deposited during the Early Cambrian to Early Ordovician spreading event as distal offshore Laurentian facies, both on the small basement blocks and adjacent nascent oceanic crust that formed beneath the Tugaloo and Cartoogechaye terranes. These terranes formed part of the collage that was accreted to Laurentia during the Paleozoic Taconian, Neocadian, and Alleghanian crust-forming events.

ACKNOWLEDGMENTS

Support for this research has been provided by National Science Foundation grants GA-1409, GA-20321, EAR-7615564, EAR-8417894, and EAR-9814800 to RDH and EAR-9814801 to CFM. Additional support for work in northeastern Georgia was provided to RDH by the Georgia Geologic Survey, in northwestern South Carolina by the South Carolina Geological Survey, in southwestern North Carolina by the North Carolina Geological Survey, and in the central Toxaway dome by Duke Power Company, Bad Creek Project. The North Carolina Geological Survey provided field support for some of JOE's work in the Trimont Ridge complex. Support for additional field work in the Franklin and Corbin Knob 7.5-minute quadrangles, North Carolina, by JOE was provided by the University of Tennessee Science Alliance Center of Excellence Distinguished Scientist Stipend. Access to the unpublished field data of M.P. Hartford from the Reid quadrangle is appreciated. Access provided by M.W. Carter and C.E. Merschat to two North Carolina Geological Survey unpublished or prepublication maps is also very much appreciated. We gratefully acknowledge J.L. Wooden for access to and assistance with the Stanford University/USGS SHRIMP-RG instrument. The manuscript was improved by critical reviews by J.F. Tull and D.T. Secor Jr. Review comments by L.S. Wiener resulted in additional improvements. We, however, remain culpable for any misrepresentations of fact or interpretation.

REFERENCES CITED

Aleinikoff, J.N., Zartman, R.E., Rankin, D.W., and Burton, W.C., 1995, U-Pb ages of metarhyolites of the Catoctin and Mt. Rogers Formations, central and southern Appalachians: Evidence of two pulses of Iapetan rifting: American Journal of Science, v. 295, p. 428–454.

Aleinikoff, J.N., Burton, W.C., Lyttle, P.T., Nelson, A.E., and Southworth, C.S., 2000, U–Pb geochronology of zircon and monazite from Mesoproterozoic granitic gneisses of the northern Blue Ridge, Virginia and Maryland, USA: Precambrian Research, v. 99, p. 113–146.

Bacon, C.R., Persing, H.M., Wooden, J.L., and Ireland, T.R., 2000, Late Pleistocene granodiorite beneath Crater Lake caldera, Oregon, dated by ion microprobe: Geology, v. 28, p. 467–470.

Bartholomew, M.J., and Lewis, S.E., 1984, Evolution of Grenville massifs in the Blue Ridge province, southern and central Appalachians, *in* Bartholomew, M.J., ed., The Grenville event in the Appalachians and related topics: Geological Society of America Special Paper 194, p. 229–254.

Bartholomew, M.J., and Lewis, S.E., 1988, Peregrination of Middle Proterozoic massifs and terranes within the Appalachian orogen, eastern USA: Trabajos de Geología, Universidad de Oviedo, v. 17, p. 155–165.

Bartholomew, M.J., and Lewis, S.E., 1992, Appalachian Grenville massifs: Pre-Appalachian translational tectonics, *in* Mason, R., ed., Basement tectonics 7: Dordrecht, The Netherlands, Kluwer Academic Publishers, p. 363–374.

Bream, B.R., 2003, Tectonic implications of geochronology and geochemistry of para- and orthogneisses from the southern Appalachian crystalline core [Ph.D. dissertation]: Knoxville, University of Tennessee, 262 p.

Bream, B.R., Hatcher, R.D., Jr., Miller, C.F., and Fullagar, P.D., 2001, Geochemistry and provenance of Inner Piedmont paragneisses, NC and SC: Evidence for an internal terrane boundary?: Geological Society of America Abstracts with Programs, v. 33, no. 2, p. A-65.

Brown, P.M., Burt, E.R., II, Carpenter, P.A., Enos, R.M., Flynt, B.J., Jr., Gallagher, P.E., Horrman, C.W., Merschat, C.E., Wilson, W.F., and Parker, J.M., III, 1985, Geologic map of North Carolina: North Carolina Geological Survey, scale 1:500,000.

Carrigan, C.W., Bream, B.R., Miller, C.F., and Hatcher, R.D., Jr., 2001, Ion microprobe analyses of zircon rims from the eastern Blue Ridge and Inner Piedmont, NC-SC-GA: Implications for the timing of Paleozoic metamorphism in the southern Appalachians: Geological Society of America Abstracts with Programs, v. 33, no. 2, p. A-7.

Carrigan, C.W., Miller, C.F., Fullagar, P.D., Hatcher, R.D., Jr., Bream, B.R., and Coath, C.D., 2003, Ion microprobe age and geochemistry of southern Appalachian basement, with implications for Proterozoic and Paleozoic reconstructions: Precambrian Research, v. 120, p. 1–36.

Carter, M.W., Merschat, C.E., and Wooten, R.M., 2002, Bedrock geologic map of Gorges State Park, Transylvania County, North Carolina: North Carolina Geological Survey, Geologic Map Series 10A, scale 1:12,000.

Cawood, P.A., McCausland, P.J.A., and Dunning, G.R., 2001, Opening Iapetus: Constraints from Laurentian margin in Newfoundland: Geological Society of America Bulletin, v. 113, p. 443–453.

Clarke, J.W., 1952, Geology and mineral resources of the Thomaston quadrangle, Georgia: Georgia Geological Survey Bulletin 59, 99 p., scale 1:62,500.

Conley, J.F., and Henika, W.S., 1973, Geology of the Snow Creek, Martinsville East, Price, and Spray quadrangles, Virginia: Virginia Division of Mineral Resources Report of Investigations 33, 71 p., scale 1:24,000.

Daly, J.S., and McClelland, J.M., 1991, Juvenile Middle Proterozoic crust in the Adirondack Highlands, Grenville province, northeastern North America: Geology, v. 19, p. 119–122.

Dalziel, I.W.D., 1997, Overview: Neoproterozoic-Paleozoic geography and tectonics: Review, hypothesis, and environmental speculation: Geological Society of America Bulletin, v. 109, p. 16–42.

Davis, T.L., 1993, Lithostratigraphy, structure, and metamorphism of a crystalline thrust terrane, western Inner Piedmont, North Carolina [Ph.D. dissertation]: Knoxville, University of Tennessee, 245 p.

Dennis, A.J., and Wright, J.E., 1997, Middle and late Paleozoic monazite U-Pb ages, Inner Piedmont, South Carolina: Geological Society of America Abstracts with Programs, v. 29, no. 12, p. 12.

Dickin, A.P., and Higgins, M.D., 1992, Sm/Nd evidence for a major crust-forming event in the central Grenville province: Geology, v. 20, p. 137–140.

Eckert, J.O., Jr., 1984, Stratigraphy, structure and metamorphism in the east half of the Wayah Bald quadrangle: Key to Paleozoic granulite facies metamorphism in the southern Appalachians [M.S. thesis]: Columbia, University of South Carolina, 411 p.

Eckert, J.O., Jr., 1988, Petrology and tectonic implications of the transition from the staurolite-kyanite zone to the Wayah granulite-facies metamorphic core, southwest North Carolina Blue Ridge: Including quantitative analysis of mineral homogeneity [Ph.D. dissertation]: College Station, Texas A&M University, 337 p.

Eckert, J.O., Jr., and Mohr, D.W., 1986, Trimont Ridge metaigneous complex, southwestern North Carolina: A premetamorphic igneous suite of noncontinental affinity in the Hayesville thrust sheet: Geological Society of America Abstracts with Programs, v. 18, p. 219.

Eckert, J.O., Jr., Hatcher, R.D., Jr., and Mohr, D.W., 1989, The Wayah granulite-facies metamorphic core, southwestern North Carolina: High-grade culmination of Taconic metamorphism in the southern Blue Ridge: Geological Society of America Bulletin, v. 101, p. 1434–1447.

Force, E.R., 1976, Metamorphic source rocks of titanium placer deposits—A geochemical cycle: U.S. Geological Survey Professional Paper 959-B, 14 p.

Fritz, W.J., Hatcher, R.D., Jr., and Hopson, J.L., eds., 1989, Geology of the eastern Blue Ridge of northeast Georgia and the adjacent Carolinas: Georgia Geological Society Guidebooks, v. 9, 199 p.

Fullagar, P.D., Hatcher, R.D., Jr., and Merschat, C.E., 1979, 1200-m.y.-old gneisses in the Blue Ridge province of North and South Carolina: Southeastern Geology, v. 20, p. 69–77.

Fullagar, P.D., Goldberg, S.A., and Butler, J.R., 1997, Nd and Sr isotopic characterization of crystalline rocks from the southern Appalachian Piedmont and Blue Ridge, North and South Carolina, *in* Sinha, A.K., et al., eds., The nature of magmatism in the Appalachian orogen: Geological Society of America Memoir 191, p. 165–179.

Giorgis, S.D., 1999, Inner Piedmont geology of the northwestern South Mountains near Morganton, North Carolina [M.S. thesis]: Knoxville, University of Tennessee, 191 p.

Gulley, G.L., Jr., 1985, A Proterozoic granulite facies terrane on Roane Mountain, western Blue Ridge belt, North Carolina–Tennessee: Geological Society of America Bulletin, v. 96, p. 1428–1439.

Hadley, J.B., and Nelson, A.E., 1971, Geology of the Knoxville 1 by 2 degree quadrangle, North Carolina, Tennessee, and South Carolina: U.S. Geological Survey Miscellaneous Geological Investigations Map I-654, scale 1:250,000.

Hamilton, W.B., 1979, Tectonics of the Indonesian region: U.S. Geological Survey Professional Paper 1078, 345 p.

Hardeman, W.D., 1966, Geologic map of Tennessee: Tennessee Division of Geology, scale 1:250,000.

Hatcher, R.D., Jr., 1971, Geology of Rabun and Habersham Counties, Georgia: A reconnaissance study: Georgia Geological Survey Bulletin 83, 48 p.

Hatcher, R.D., Jr., 1973, Basement versus cover rocks in the Blue Ridge of northeast Georgia, northwestern South Carolina and adjacent North Carolina: American Journal of Science, v. 273, p. 671–685.

Hatcher, R.D., Jr., 1974, Introduction to the tectonic history of northeast Georgia: Georgia Geological Society Guidebook 13-A, 59 p.

Hatcher, R.D., Jr., 1977, Macroscopic polyphase folding illustrated by the Toxaway dome, South Carolina–North Carolina: Geological Society of America Bulletin, v. 88, p. 1678–1688.

Hatcher, R.D., Jr., 1978, Tectonics of the western Piedmont and Blue Ridge: Review and speculation: American Journal of Science, v. 278, p. 276–304.

Hatcher, R.D., Jr., 1980, Geologic map and mineral resources of the Prentiss quadrangle, North Carolina: North Carolina Geological Survey, GM 167-SW, scale 1:24,000.

Hatcher, R.D., Jr., 1983, Basement massifs in the U.S. Appalachians: Geological Journal, v. 18, p. 255–265.

Hatcher, R.D., Jr., 1987, Tectonics of the southern and central Appalachian internides: Annual Review of Earth and Planetary Sciences, v. 15, p. 337–362.

Hatcher, R.D., Jr., 1989, Tectonic synthesis of the U.S. Appalachians, Chapter 14, *in* Hatcher, R.D., Jr., et al., eds., The Appalachian-Ouachita orogen in the United States: Boulder, Colorado, Geological Society of America, The Geology of North America, v. F-2, p. 511–535.

Hatcher, R.D., Jr., 1991, Interactive property of large thrust sheets with footwall rocks—the subthrust interactive duplex hypothesis: A mechanism of dome formation in thrust sheets: Tectonophysics, v. 191, p. 237–242.

Hatcher, R.D., Jr., 1999, Crust-forming processes, *in* Sinha, A.K., ed., Basement tectonics 13: Dordrecht, The Netherlands, Kluwer Academic Publishers, p. 99–118.

Hatcher, R.D., Jr., 2001, Terranes and terrane accretion in the southern Appalachians: An evolved working hypothesis: Geological Society of America Abstracts with Programs, v. 33, no. 2, p. A-65.

Hatcher, R.D., Jr., 2002, Inner Piedmont primer, *in* Hatcher, R.D., Jr., and Bream, B.R., eds., Inner Piedmont tectonics focused mostly on detailed studies in the South Mountains and the southern Brushy Mountains, North Carolina: Carolina Geological Society Guidebook, Raleigh, North Carolina Geological Survey, p. 1–18.

Hatcher, R.D., Jr., and Bream, B.R., eds., 2002, Inner Piedmont tectonics focused mostly on detailed studies in the South Mountains and the southern Brushy Mountains, North Carolina: Carolina Geological Society Guidebook, Raleigh, North Carolina Geological Survey, 145 p.

Hatcher, R.D., Jr., and Hopson, J.L., 1989, Geologic map of the eastern Blue Ridge, northeastern Georgia and the adjacent Carolinas, *in* Fritz, W.J., et al., eds., Geology of the eastern Blue Ridge of northeast Georgia and the adjacent Carolinas: Georgia Geological Society Guidebooks, v. 9, scale 1:100,000.

Hatcher, R.D., Jr., Hooper, R.J., Petty, S.M., and Willis, J.D., 1984, Structure and chemical petrology of three southern Appalachian mafic-ultramafic complexes and their bearing upon the tectonics of emplacement and origin of Appalachian ultramafic bodies: American Journal of Science, v. 284, p. 484–506.

Hatcher, R.D., Jr., Hooper, R.J., McConnell, K.I., Heyn, T., and Costello, J.O., 1988, Geometric and time relationships between thrusts in the crystalline southern Appalachians, *in* Mitra, G., and Wojtal, S., eds., Geometries and mechanisms of thrusting, with special reference to the Appalachians: Geological Society of America Special Paper 222, p. 185–196.

Hatcher, R.D., Jr., Osberg, P.H., Robinson, P., and Thomas, W.A., 1990, Tectonic map of the U.S. Appalachians: Geological Society of America, The Geology of North America, v. F-2, Plate 1, scale 1:2,000,000.

Hatcher, R.D., Jr., Miller, C.F., and Lamb, D.D., 1995, Deformation processes related to emplacement of the Rabun Granite in the eastern Blue Ridge, Georgia and North Carolina, *in* Driese, S.G., ed., Guidebook for field trip excursions, Southeastern Section, Geological Society of America: Knoxville, University of Tennessee, Department of Geological Sciences, Studies in Geology 24, p. 39–56.

Heatherington, A.L., Mueller, P.A., Smith, M.S., and Nutman, A.P., 1996, The Corbin Gneiss: Evidence for Grenvillian magmatism and continental basement in the southernmost Blue Ridge: Southeastern Geology, v. 36, p. 15–25.

Heyn, T., 1984, Stratigraphic and structural relationships along the southwestern flank of the Sauratown Mountains anticlinorium [M.S. thesis]: Columbia, University of South Carolina, 192 p.

Hoffman, P.F., 1991, Did the breakout of Laurentia turn Gondwana inside out?: Science, v. 252, p. 1409–1412.

Hooper, R.J., and Hatcher, R.D., Jr., 1988, The Pine Mountain terrane, a complex window in the Georgia and Alabama Piedmont: Evidence from the eastern termination: Geology, v. 16, p. 307–310.

Hopson, J.L., and Hatcher, R.D., Jr., 1988, Structural and stratigraphic setting of the Alto allochthon, northeast Georgia: Geological Society of America Bulletin, v. 100, p. 339–350.

Hopson, J.L., Hatcher, R.D., Jr., and Stieve, A.L., 1989, Geology of the eastern Blue Ridge, northeastern Georgia and the adjacent Carolinas, *in* Fritz, W.J., et al., eds., Geology of the eastern Blue Ridge of northeast Georgia and the adjacent Carolinas: Georgia Geological Society Guidebooks, v. 9, p. 1–40.

Horton, J.W., Jr., 1982, Geologic map and mineral resources summary of the Rosman quadrangle, North Carolina: North Carolina Geological Survey GM 185-NE, scale 1:24,000.

Horton, J.W., Jr., and McConnell, K.I., 1991, 3. The western Piedmont, *in* Horton, J.W., Jr., and Zullo, V.A., eds., The geology of the Carolinas—Carolina Geological Society 50th Anniversary Volume: Knoxville, The University of Tennessee Press, p. 36–58.

Horton, J.W., Jr., Drake, A.A., Jr., and Rankin, D.W., 1989, Tectonostratigraphic terranes and their Paleozoic boundaries in the central and southern Appalachians, *in* Dallmeyer, R.D., ed., Terranes in the Circum-Atlantic

Paleozoic orogens: Geological Society of America Special Paper 230, p. 213–245.

Li, L., and Tull, J.F., 1998, Cover stratigraphy and structure of the southernmost external basement massifs in the southern Appalachian Blue Ridge: Evidence for two-stage Late Proterozoic rifting: American Journal of Science, v. 298, p. 829–867.

Livingston, J.L., and McKniff, J.M., 1967, Tallulah Falls dome, northeastern Georgia: Another window? [abs]: Geological Society of America Special Paper 115, p. 485–486.

Lukert, M.T., and Banks, P.O., 1984, Geology and age of the Robertson River pluton, *in* Bartholomew, M.J., ed., The Grenville event in the Appalachians and related topics: Geological Society of America Special Paper 194, p. 161–166.

Marcantonio, F., McNutt, R.H., Dickin, A.P., and Heaman, L.M., 1990, Isotopic evidence for the crustal evolution of the Frontenac arc in the Grenville province of Ontario, Canada: Chemical Geology, v. 83, p. 297–314.

McConnell, K.I., 1990, Geology and geochronology of the Sauratown Mountains anticlinorium, northwestern North Carolina [Ph.D. thesis]: Columbia, University of South Carolina, 232 p.

McConnell, K.I., and Costello, J.O., 1984, Basement-cover rock relationships along the western edge of the Blue Ridge thrust sheet in Georgia, *in* Bartholomew, M.J., ed., The Grenville event in the Appalachians and related topics: Geological Society of America Special Paper 194, p. 263–280.

McKniff, J.M., 1967, Geology of the Highlands-Cashiers area, North Carolina [Ph.D. thesis]: Houston, Texas, Rice University, 100 p.

Miller, C.F., Hatcher, R.D., Jr., Ayers, J.C., Coath, C.D., and Harrison, T.M., 2000, Age and zircon inheritance of eastern Blue Ridge plutons, southwestern North Carolina and northeastern Georgia, with implications for magma history and evolution of the southern Appalachian orogen: American Journal of Science, v. 300, p. 142–172.

Moecher, D.P., and Miller, C.F., 2000, Precise age for peak granulite facies metamorphism and melting in the eastern Blue Ridge from SHRIMP U-Pb analysis: Geological Society of America Abstracts with Programs, v. 32, no. 2, p. A-63.

Monrad, J.R., and Gulley, G.L., Jr., 1983, Age and *P-T* conditions during metamorphism of granulite facies gneisses, Roan Mountain, NC-TN, *in* Lewis, S.E., ed., Geological investigations in the Blue Ridge of Northwestern North Carolina: Raleigh, North Carolina, Carolina Geological Society Guidebook, p. 41–51.

Odom, A.L., Kish, S.A., and Leggo, P.J., 1973, Extension of "Grenville basement" to the southern extremity of the Appalachians: Geological Society of America Abstracts with Programs, v. 5, p. 425.

Pearce, J.A., Harris, N.B.W., and Tindle, A.G., 1984, Trace element discrimination diagrams for the tectonic interpretation of granitic rocks: Journal of Petrology, v. 25, p. 956–983.

Rankin, D.W., 1976, Appalachian salients and recesses: Late Precambrian continental breakup and the opening of the Iapetus ocean: Journal of Geophysical Research, v. 81, p. 5605–5619.

Rankin, D.W., Chiarenzelli, J.R., Drake, A.A., Jr., Goldsmith, R., Hall, L.M., Hinze, W.J., Isachsen, Y.W., Lidiak, E.G., McLelland, J., Mosher, S., Ratcliffe, N.M., Secor, D.T., Jr., and Whitney, P.R., 1993, Proterozoic rocks east and southeast of the Grenville front, *in* Reed, J.C., Jr., et al., eds., Precambrian: Conterminous U.S.: Boulder, Colorado, Geological Society of America, Geology of North America, v. C-2, p. 335–461.

Raymond, L.A., and Johnson, P.A., 1994, The Mars Hill terrane; an enigmatic southern Appalachian terrane: Geological Society of America Abstracts with Programs, v. 26, no. 4, p. 59.

Roper, P.J., and Dunn, D.E., 1970, Geology of the Tamassee, Satolah, and Cashiers quadrangles, Oconee County, South Carolina: South Carolina Division of Geology MS-16, scale 1:24,000.

Samson, S.D., Hibbard, J.P., and Wortman, G.L., 1995, Nd isotopic evidence for juvenile crust in the Carolina terrane, southern Appalachians: Contributions to Mineralogy and Petrology, v. 121, p. 171–184.

Schaeffer, M.F., Steffens, R.E., and Hatcher, R.D., Jr., 1979, In situ stress and its relationships to joint formation in the Toxaway Gneiss, northwestern South Carolina: Southeastern Geology, v. 20, p. 129–143.

Sears, J.W., and Cook, R.B., Jr., 1984, An overview of the Grenville basement complex of the Pine Mountain window, Alabama and Georgia, *in* Bartholomew, M.J., ed., The Grenville event in the Appalachians and related topics: Geological Society of America Special Paper 194, p. 281–287.

Spell, T.L., and Norell, G.T., 1990, The Ropes Creek assemblage: Petrology, geochemistry, and tectonic setting of an ophiolitic thrust sheet in the southern Appalachians: American Journal of Science, v. 290, p. 811–842.

Steltenpohl, M.G., 1988, The Pine Mountain window of Alabama: Basement-cover evolution in the southernmost exposed Appalachians, *in* Bartholomew, M.J., et al., eds., Basement tectonics 8: Characterization and comparison of ancient and Mesozoic continental margins: Proceedings of the Eighth International Conference on Basement Tectonics: Dordrecht, The Netherlands, Kluwer Academic Publishers, p. 491–501.

Stieve, A.L., 1989, The structural evolution and metamorphism of the southern portion of the Tallulah Falls dome, northeast Georgia [Ph.D. dissertation]: Columbia, University of South Carolina, 207 p.

Stieve, A.L., and Hatcher, R.D., Jr., 1989, Two sequential shear zones overprinting the Tallulah Falls dome in northeast Georgia, *in* Fritz, W.J., et al., eds., Geology of the eastern Blue Ridge of northeast Georgia and the adjacent Carolinas: Georgia Geological Society Guidebooks, v. 9, p. 111–132.

Stieve, A.L., and Sinha, A.K., 1989, Grenville ages from Rb-Sr whole-rock analysis of two basement gneisses of the Tallulah Falls dome of northeast Georgia, *in* Fritz, W.J., et al., eds., Geology of the eastern Blue Ridge of northeast Georgia and the adjacent Carolinas: Georgia Geological Society Guidebooks, v. 9, p. 57–74.

Student, J.J., and Sinha, A.K., 1992, Carboniferous U-Pb ages of zircons from the Box Ankle and Ocmulgee fault zones, Central Georgia: Implications for accretionary models: Geological Society of America Abstracts with Programs, v. 24, no. 2, p. 69.

Teague, J.H., and Furcron, A.S., 1948, Geology and mineral resources of Rabun and Habersham Counties, Georgia: Georgia Geological Survey, scale 1:125,000.

Thomas, C.W., 2001, Origins of mafic-ultramafic complexes of the eastern Blue Ridge province, southern Appalachians: Geochronological and geochemical constraints [M.S. thesis]: Nashville, Tennessee, Vanderbilt University, 154 p.

Tracy, R.J., Hibbard, J.P., and Henika, W.S., 2001, Implications of a Cambrian (530 Ma) tectonothermal event in the Smith River allochthon, southwestern Virginia: Geological Society of America Abstracts with Programs, v. 33, no. 6, p. A262.

Van Schmus, W.R., and Bickford, M.E., editors, and 23 others, 1993, Transcontinental Proterozoic provinces, *in* Reed, J.C., Jr., Bickford, M.E., Houston, R.S., Link, P.K., Rankin, D.W., Sims, P.K., and Van Schmus, W.R., eds., Precambrian: Conterminous U.S.: Boulder, Colorado, Geological Society of America, Geology of North America, v. C-2, p. 171–334.

Manuscript Accepted by the Society August 25, 2003

Printed in the USA

Geological Society of America
Memoir 197
2004

Geology and geochemistry of granitic and charnockitic rocks in the central Lovingston massif of the Grenvillian Blue Ridge terrane

Scott S. Hughes*
Department of Geosciences, Idaho State University, Pocatello, Idaho 83209, USA
Sharon E. Lewis
Perpetual Metals, 574 West Main Street, Boise, Idaho 83702, USA
Mervin J. Bartholomew
Department of Earth Sciences, University of Memphis, Memphis, Tennessee 38152, USA
A. Krishna Sinha
Department of Geological Sciences, Virginia Tech, Blacksburg, Virginia 24061, USA
Norman Herz
Department of Geology, University of Georgia, Athens, Georgia 30602, USA

ABSTRACT

Basement rocks of the Lovingston massif in the central Blue Ridge anticlinorium include jotunite, anorthosite, charnockitic rocks, and related Fe-Ti oxide- and apatite-rich rocks that display strong similarities to lithologic assemblages elsewhere in the Grenville orogen. The rock units preserve evidence of a protracted history of granulite-facies metamorphism and plutonism from ca. 1.15 to ca. 1.0 Ga, followed by multiple episodes of lower-grade Paleozoic metamorphism. New whole-rock major- and trace-element data indicate systematic chemical trends, incompatible element enrichments relative to the lower crust, and overall uniformity among Lovingston rocks. Quantitative geochemical models and field associations support interpretation of the Grenville-age Archer Mountain Suite and Turkey Mountain Suite plutonic series as being derived from lower crustal sources. Archer Mountain Suite biotite granitoids and leucocratic granitoids and younger Roseland Anorthosite (with the associated Roses Mill pluton) and charnockitic plutons contain a substantial amount of lower crustal chemical component that is equivalent to the average Stage Road Suite. We propose that all of the Grenville-age rocks in the Lovingston massif are related to a broad igneous protolith, represented by the Stage Road Suite, from which remobilized magmatic bodies were derived during orogenesis. Chemical data further indicate that Neoproterozoic intrusions and/or mylonitization along the Rockfish Valley fault had only minor local effects on highly mobile elements near their contacts or localized in shear zones.

Keywords: Lovingston massif, charnockite, anorthosite, geochemistry

*E-mail: hughscot@isu.edu.

Hughes, S.S., Lewis, S.E., Bartholomew, M.J., Sinha, A.K., and Herz, N., 2004, Geology and geochemistry of granitic and charnockitic rocks in the central Lovingston massif of the Grenvillian Blue Ridge terrane, *in* Tollo, R.P., Corriveau, L., McLelland, J., and Bartholomew, M.J., eds., Proterozoic tectonic evolution of the Grenville orogen in North America: Boulder, Colorado, Geological Society of America Memoir 197, p. 549–569. For permission to copy, contact editing@geosociety.org.

INTRODUCTION

Mesoproterozoic crystalline massifs of the central and southern Appalachians were involved in culminating events associated with the ca. 1.0-Ga Grenville orogeny (Bartholomew et al., 1984; Easton, 1986) during development of the supercontinent of Rodinia (Dalziel, 1991, 1997; Moores, 1991; Borg and DePaolo, 1994). Collectively, these allochthonous Appalachian massifs now form the remnants of the eastern margin of the Laurentian portion of Rodinia (Dalziel, 1991; Moores, 1991; Borg and DePaolo, 1994). The massifs occur as basement inliers, which, in central Virginia, form the core of the Blue Ridge anticlinorium that developed as part of the Appalachian orogen. The Grenville-age Blue Ridge terrane (Bartholomew and Lewis, 1988, 1992), comprised in part of the Pedlar and Lovingston massifs of northern and central Virginia, appears to be part of a magmatic terrane that was active before and during the culminating Grenville orogeny (Sinha and Bartholomew, 1984; Bartholomew et al., 1991).

Bartholomew et al. (1991), on the basis of field relations, lithologic similarities, and earlier age determinations (Pettingill et al., 1984; Sinha and Bartholomew, 1984), interpreted the Blue Ridge basement as composed of protoliths and derivative plutons that were emplaced over a protracted period of time. This is consistent with interpretations made elsewhere, mainly, the episodes of magmatism documented for the northern part of the Blue Ridge anticlinorium (Aleinikoff et al., 2000) and for similar episodes in the Adirondack massif (McLelland et al., 1988, 1994). However, many older rocks, such as those involved in earlier events (ca. 1200 Ma or earlier) in the Adirondacks, appear to be absent in the Blue Ridge terrane of Virginia (Bartholomew and Lewis, 1988). Our study, and the data presented in Hughes et al. (1997), provide the first geochemical characterization of the central portions of the Lovingston and Pedlar massifs, respectively, where detailed maps, petrographic descriptions, and older geochronologic data are available for most Grenville-age Blue Ridge rock units (Bartholomew et al., 1981, 1991; Bartholomew and Lewis, 1984; Herz and Force, 1984, 1987; Sinha and Bartholomew, 1984; Owens and Dymek, 1999). Our study focuses on two suites within the Lovingston massif, with its plutonic assemblages of anorthosite, charnockitic rocks, and related Fe-Ti oxide- and apatite-rich rocks that are similar to those occurring elsewhere in the Grenville orogen. Geochemical and lithological relations here may have significant bearing on the tectonic association of the central Blue Ridge terrane with similar rocks located 150 km north in the northern Blue Ridge terrane (Tollo and Aleinikoff, 2001; Burton et al., this volume; Tollo et al., this volume), and possibly with both the Adirondacks (McLelland et al., 1988, 1994, 1996, 2001; Mezger et al., 1993) and inliers in the northern part of the Appalachian orogen (e.g., Miller and Barr, 2000). Palinspastic reconstructions of the allochthonous Grenville-age rocks of the Appalachians place them outboard of the Adirondacks (Bartholomew and Lewis, 1988, 1992; Bartholomew, 1992), thus indicating that the allochthonous Blair River inlier, which is also outboard of the Adirondacks, is perhaps more likely than the inboard Adirondacks to be a tectonic equivalent to the Blue Ridge terrane.

Regional implications of deformational and magmatic episodes during the protracted Grenville orogeny are discussed here to provide additional constraints on the nature of their plutonic protoliths in the Appalachian orogen. Rocks in the study area include compositionally diverse granitoids with charnockitic and noncharnockitic varieties. We address the geochemical, temporal, and spatial relations between later Grenville-age granitoid plutons and surrounding gneissic country rocks, which can be explained by two hypotheses: (1) massif anorthosites, charnockitic plutons, and other granitoid plutons were derived by melting of surrounding Mesoproterozoic layered and foliated gneiss source rocks and subsequent magmatic evolution; or (2) plutons were derived from other sources, possibly mantle-derived primary magmas, which were emplaced in the gneissic country rocks. In either scenario, the Grenville event produced orthopyroxene-bearing plutons that retained igneous textures throughout subsequent retrograde metamorphic events. We also discuss how the data reflect on massif-anorthosite genesis.

GEOLOGIC SETTING

The deep crust of the Mesoproterozoic Virginia Blue Ridge terrane (Fig. 1) is represented by two major massifs (Bartholomew and Lewis, 1992) exposed in the core of the Blue Ridge anticlinorium, a regional fold related to late Paleozoic (Alleghanian) thrusting (Mitra and Elliott, 1980; Mitra and Lukert, 1982; Evans, 1989) in the central and northern Virginia Appalachians. The Pedlar and Lovingston massifs, initially characterized by Bartholomew (1977, 1981) and Bartholomew et al. (1981), are separated along most of their length by the Rockfish Valley fault zone, a Paleozoic fault that dies out northeastward within the core of the Blue Ridge anticlinorium (Bartholomew and Lewis, 1984; Bailey and Simpson, 1993) (Fig. 1). The massifs contain suites of igneous rocks that were emplaced during late stages of Grenville orogenesis into older country rocks, composed of layered and/or foliated gneisses, which formed during earlier Grenvillian orogenic events. The massifs were subsequently subjected to Neoproterozoic and Paleozoic greenschist-facies metamorphism (e.g., Sinha and Bartholomew, 1984; Bartholomew et al., 1991; Evans, 1991; Bartholomew, 1992; Bailey and Simpson, 1993).

According to Bartholomew et al. (1991), the Pedlar and Lovingston massifs are distinguished by (1) their contrasting rock associations; (2) different petrologic, metamorphic, and structural histories; and (3) physiographic aspects that reflect those lithologic assemblages. In the central Blue Ridge terrane, the Pedlar massif is comprised mainly of the Pedlar River Charnockite Suite and older granulite-facies gneisses that form the country rock. Major- and trace-element characteristics of Pedlar massif rocks were interpreted by Hughes et al. (1997) to

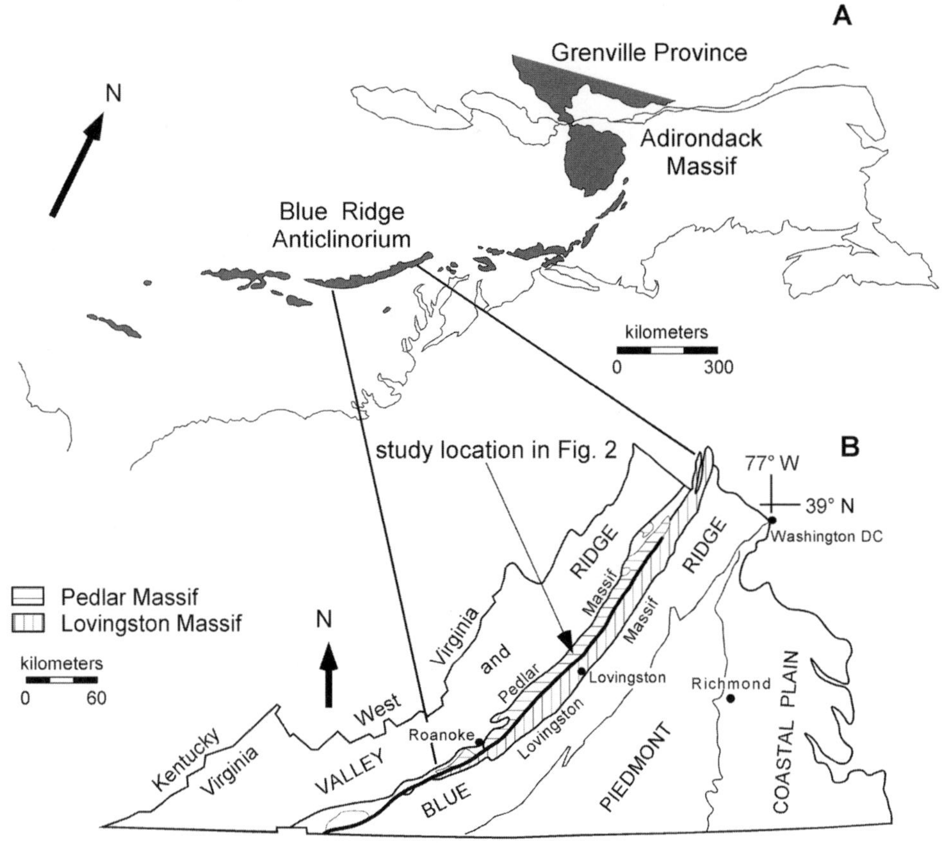

Figure 1. Map illustrating the distribution of major Mesoproterozoic terranes: (A) in eastern North America; (B) the locations of the Blue Ridge anticlinorium in Virginia and inliers of Laurentian terrane in the easternmost part of the Grenville Province. The study area is located in the central part of the Lovingston massif, one of two major tectonic terranes in the Blue Ridge anticlinorium.

reflect complete remobilization by partial melting of pre-existing Proterozoic igneous rocks to form charnockitic magmas that were subsequently emplaced into surrounding gneisses. Field relations in the Lovingston massif broadly support similar processes, although the terrane is lithologically more diverse (Fig. 2). Within this massif lies the Roseland District (Herz and Force, 1987), which is characterized by massif anorthosite, nelsonite, and concentrated zones of rutile and ilmenite mined for titanium into the middle of the twentieth century (Watson and Tabor, 1913; Ross, 1941).

Much of the Lovingston massif is comprised of four compositionally diverse igneous suites of monzogranite, quartz monzonite, granodiorite, and quartz monzodiorite, containing some charnockitic varieties of these granitoid rocks: farsundite, quartz mangerite, charno-enderbite, and quartz jotunite, respectively (nomenclature from Le Maitre et al., 1989). These are the Oventop, Archer Mountain, Turkey Mountain, and Horsepen Mountain suites (Bartholomew and Lewis, 1984). Three general lithologic types are recognized: (1) foliated biotite granitoids; (2) less mafic, leucocratic, coarse-grained granitoids (or granites sensu lato); and (3) charnockitic granitoids. They intruded an older suite of country rocks consisting of layered and foliated gneisses that were subjected to lower granulite-facies (orthopyroxene ± amphibole) metamorphism. In the Lovingston massif, charnockitic plutons (Herz and Force, 1987) and massif anorthosite, with its associated border gneiss, intruded the granitoid plutonic suites. The alkalic Roseland Anorthosite (Herz, 1969) forms the core of 3-km x 15-km elliptical dome and is generally mantled by metasedimentary rocks forming the associated quartzofeldspathic border gneiss (Hillhouse, 1960; Herz and Force, 1984, 1987). Border gneiss of the anorthosite (Fig. 2) is cut by dikes of jotunite and nelsonite.

Grenville-age rocks of the Lovingston massif were intruded by numerous Neoproterozoic A-type granitoids and mafic dikes that are related to Iapetan extension (e.g., Rankin, 1975, 1976; Bartholomew and Lewis, 1984; Bartholomew, 1992; Tollo and Arav, 1992; Tollo and Lowe, 1994; Tollo and Aleinikoff, 1996; Tollo and Hutson, 1996; Tollo et al., 2004). This granitic magmatism is part of the first pulse of a two-stage rifting cycle in the central Appalachians involving pulses dated at 760–702 Ma and ca. 570 Ma by Badger and Sinha (1988) and Aleinikoff et al. (1995). In central Virginia (Fig. 2), the Rockfish River, Polly Wright Cove, and Mobley Mountain plutons are the largest of these A-type granitoids.

K-Ar, U-Pb, and Rb-Sr analyses summarized by Bartholomew et al. (1991) suggest Paleozoic metamorphism occurred at ca. 450–425 Ma. Subsequent retrograde, subgreenschist-facies metamorphism at 303–273 Ma was synchronous with late Paleozoic (ca. 300 Ma) thrusting, and likely contributed to the disturbance of Ordovician K-Ar systematics in the Valley and Ridge Province (Elliot and Aronson, 1987), but its effects in the Blue Ridge are less certain (Bartholomew et al., 1991; Bailey

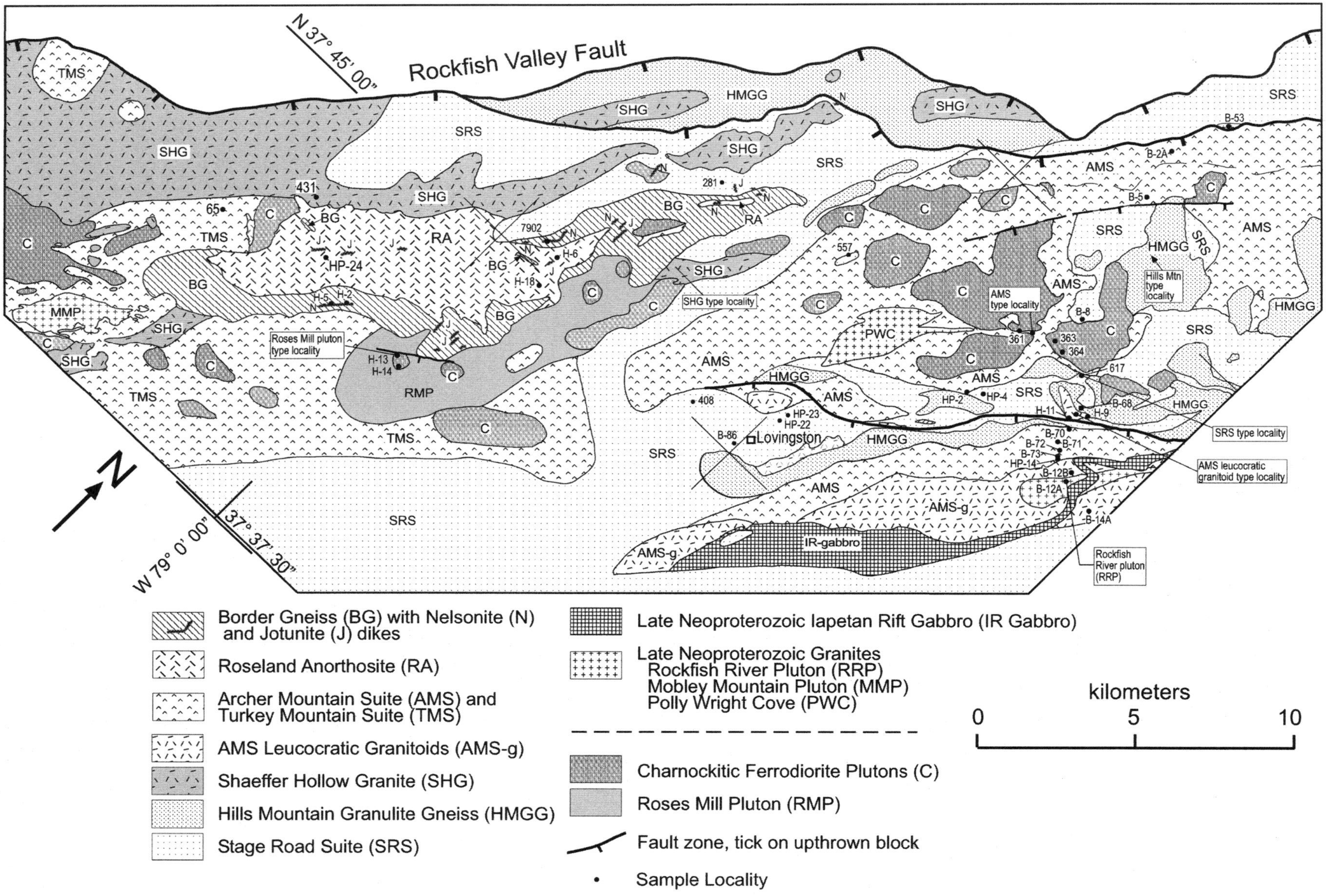

Figure 2. Geologic map of the central segment of the Lovingston massif in the central Blue Ridge anticlinorium, Virginia, showing locations of geochemical samples used in this study. Modified from Moore (1940); Hillhouse (1960); Davis (1974); Bartholomew (1977); Brock (1981); Bailey (1983); Bartholomew and Lewis (1984); Herz and Force (1987); Evans (1991).

and Simpson, 1993). Whereas rocks within the Pedlar massif typically exhibit few effects of these retrograde events, many of the Grenville-age rocks of the Lovingston massif were extensively retrograded (Evans, 1991). Evans (1991) suggested that this retrograde metamorphism was, in part, due to extensive seawater inundation of these crystalline rocks, with high heat flow accompanying emplacement of Neoproterozoic plutons associated with Iapetan rifting. Bailey and Simpson (1993) demonstrated that extensional Neoproterozoic shear zones in the Lovingston massif developed under greenschist- to amphibolite-facies conditions. Along much of the eastern portion of the Lovingston massif, retrograde (upper greenschist-facies) Paleozoic metamorphism obscures the maximum grade of Grenville-age metamorphism, which is assumed to have occurred at upper amphibolite-facies or higher conditions (Bartholomew and Lewis, 1984).

CHRONOLOGY OF GRENVILLE-AGE EVENTS IN THE LOVINGSTON MASSIF

The timing of Grenville events in the central Lovingston massif is based on tentative correlation of these rocks with lithologically similar rocks that were recently dated in the northern Blue Ridge (Aleinikoff et al., 2000; Tollo et al., this volume). Existing geochronology of rocks in our study area was obtained using whole-rock or multiple-grain bulk zircon separates (Sinha and Bartholomew, 1984; Pettingill et al., 1984), which may reflect multiple-age components. Thus, the timing of central Lovingston events can be only broadly correlated to neighboring rock series that were dated by more precise methods. In this regard, we focus more on the comparative geochemical attributes of Grenville-age rocks and their relative ages determined in the field and acknowledge the consistency in absolute ages across the province.

The older layered and foliated gneisses mapped by Bartholomew and Lewis (1984) included the Stage Road Layered Gneiss of Sinha and Bartholomew (1984); in this paper, we change the rank of this unit from a lithodeme to a suite: the Stage Road Suite. Much of the Stage Road Suite at its type locality is comprised of coarse augen gneiss that is interlayered with other older gneisses (Sinha and Bartholomew, 1984) and intruded by the Hills Mountain Granulite Gneiss, a foliated rock that we include in the Stage Road Suite. Pettingill et al. (1984) reported a Rb-Sr whole-rock isochron age for the Stage Road Suite in our study area of 1147 ± 34 Ma. In northern Virginia, the older layered and foliated gneisses mapped by Bartholomew and Lewis (1984) were subdivided isotopically by Aleinikoff et al. (2000) into several lithodemic units (group 1), which yielded zircon and monazite ages ranging from ca. 1153 to 1140 Ma. Thus, we consider the layered and foliated Stage Road Suite to represent a ca. 1150-Ma suite of country rocks of the Lovingston massif. The Stage Road Suite, as thus defined, may be equivalent to a distinct 1160–1140 Ma episode of granitic gneiss and high-silica charnockite formation (Tollo et al., this volume) within the northern end of the Pedlar massif.

Archer Mountain and Turkey Mountain suites of biotite granitoid and leucocratic granitoid plutons intruded the Lovingston massif during the next phase of magmatism, broadly equivalent to the group 3 magmatic event (including the Oventop Suite of Bartholomew and Lewis, 1984) of Aleinikoff et al. (2000). These plutons are synchronous with intrusion of the Pedlar River Charnockite Suite in the Pedlar massif, and with granitic gneiss, leucogranite, and low-silica charnockite (sensu lato) in the northern Blue Ridge. Tollo et al. (this volume) indicate the group 2 pulse was only a minor pulse of activity, represented by the ca. 1111-Ma Marshall Metagranite in the northern Blue Ridge. Lithologic equivalents of the Marshall Metagranite are not identified in central Virginia. The earliest leucocratic granitoid plutons in the Archer Mountain and Turkey Mountain suites appear to be Shaeffer Hollow Granite, which is associated with the Turkey Mountain Suite, and similar unnamed granitic bodies related to the Archer Mountain Suite (Fig. 2), some of which occur as xenoliths within other rocks of the Archer Mountain Suite. Clearly, high precision geochronologic work is needed to resolve the ages of these suites and evaluate the significance of the younger magmatic pulse in the central Blue Ridge.

Small massive charnockitic plutons (*C*, Fig. 2) and the associated large, lithologically complex, Roses Mill pluton intruded the Archer Mountain, Turkey Mountain, and Stage Road suites. The Roses Mill pluton, with concomitant dikes of nelsonite and jotunite, was emplaced after the Roseland Anorthosite. Although the Border Gneiss typically separates the two units (Fig. 2), the Roses Mill pluton contains xenoliths of anorthosite (Herz and Force, 1987), and jotunite dikes intrude the anorthosite. This event is equivalent to or younger than the group 3 episode of Aleinikoff et al. (2000), which Tollo et al. (this volume) consider as the second major episode of magmatism and deformation from 1080 to 1050 Ma. The Roseland Anorthosite, an andesine antiperthite diapir-like body within the massif (Herz and Force, 1984, 1987; Owens and Dymek, 1999), has a Sm-Nd age of 1045 ± 44 Ma, which suggests that it was synchronous with or younger than the Archer Mountain Suite–Turkey Mountain Suite charnockitic plutons. The anorthosite thus likely predated culmination of the Grenville metamorphism of the Archer Mountain Suite, Turkey Mountain Suite, and Roses Mill pluton at ca. 1.0 Ga (Sm-Nd: 1004 ± 36 Ma; Rb-Sr: 1009 ± 26 Ma; Pettingill et al., 1984). The lithologically similar Montpelier Anorthosite in the Goochland terrane of eastern Virginia has similar, but more precise, ages for emplacement (1045 ± 10 Ma) and metamorphism (1011 ± 6 Ma) (Aleinikoff et al., 1996).

GEOCHEMISTRY

Thirty-four samples of Grenville-age lithologic units in the central Lovingston massif were selected for whole-rock geochemical analysis (Table 1). Herz and Force (1987) reported

TABLE 1. WHOLE-ROCK GEOCHEMICAL COMPOSITIONS OF REPRESENTATIVE LOVINGSTON MASSIF SAMPLES

	Map unit and sample number											
	SRS	SRS	SRS	SRS	SRS	SRS	SRS	SRS	SRS	HMGG	AMS-g	AMS-g
Material	617	B-8*	B-86*	281*	408*	557*	HP-4	HP-22	HP-23	B-68	B-14A	B-12B
SiO_2	68.55	64.74	67.34	67.18	65.59	71.54	69.32	61.21	66.16	70.82	64.59	55.13
TiO_2	0.79	0.66	0.96	1.03	0.88	0.51	0.81	1.80	0.97	0.75	0.99	0.57
Al_2O_3	14.30	16.34	14.74	17.29	15.71	15.31	12.84	15.55	14.74	13.69	15.97	17.93
FeO_t†	5.64	5.21	5.70	4.18	4.84	2.57	5.25	5.99	5.57	3.92	5.20	10.43
MnO	0.11	0.10	0.06	0.06	0.07	0.04	0.08	0.09	0.08	0.04	0.08	0.19
MgO	1.04	0.90	1.33	1.08	1.83	0.43	0.37	1.99	1.22	0.49	0.64	2.30
CaO	2.58	3.37	2.05	1.42	2.47	1.02	2.24	3.93	2.64	1.75	4.02	1.83
Na_2O	2.72	3.49	3.04	1.92	3.99	2.13	2.81	4.13	3.29	2.63	3.56	5.06
K_2O	4.06	4.80	4.61	5.47	4.39	6.36	5.98	4.70	5.10	5.61	4.54	6.39
P_2O_5	0.21	0.39	0.17	0.37	0.24	0.09	0.29	0.62	0.24	0.29	0.40	0.17
Original total	97.21	97.93	97.71	100.03	98.03	101.90	98.90	98.53	97.69	97.98	97.47	97.23
Sc	15.0	11.3	18.2	9.7	13.9	5.1	13.3	13.5	15.3	5.7	9.5	9.6
Cr	18.1	5.0	32.0	6.9	24.0	3.7	3.5	44.5	28.7	5.0	3.0	3.0
Co	26.4	6.5	11.1	94.8	36.6	69.2	30.6	29.7	40.0	4.8	4.6	2.9
Rb	104	97	162	85	135	205	138	148	135	314	116	152
Cs	0.21	0.33	0.29	0.09	0.28	0.85	0.06	0.10	0.24	0.25	0.62	0.26
Sr	332	450	160	483	132	112	194	296	222	155	498	154
Ba	1452	1620	1110	2600	930	1210	1891	1554	1660	511	1893	731
La	46.3	62.0	52.0	59.5	40.4	40.2	31.3	78.7	38.8	46.0	90.0	98.0
Ce	87.5	120	97	108	81.4	64.1	63.9	163	80.9	92	168	191
Nd	46.4	73	58	59	36.5	29	31.7	73.7	37.7	62	89	93
Sm	10.1	13.8	10.7	9.77	7.98	5.32	7.99	13.4	8.41	11.6	16.9	17.5
Eu	3.01	4.49	2.04	3.78	2.07	1.61	4.06	2.96	2.01	1.73	4.85	2.65
Tb	1.36	1.79	1.51	1.04	1.08	0.74	1.17	1.39	1.19	1.50	2.21	2.42
Yb	3.26	3.63	3.79	2.07	3.51	1.98	3.23	2.67	3.71	2.44	3.75	5.53
Lu	0.49	0.46	0.58	0.24	0.51	0.22	0.45	0.36	0.52	0.30	0.51	0.81
Zr	420	350	230	350	280	390	1685	507	493	251	1036	1036
Hf	15.4	24.0	15.7	10.4	11.2	15.1	39.9	12.2	12.9	7.0	16.3	25.6
Ta	1.79	0.87	0.92	5.57	2.67	3.86	2.23	2.15	2.91	1.65	1.06	3.44
Th	0.9	0.9	0.8	0.4	0.5	0.7	0.1	3.2	0.9	<0.05	6.0	6.2
U	0.8	2.1	n.a.	1.3	1.0	2.8	0.2	1.2	0.5	<0.1	3.0	3.6
K_2O/TiO_2	5.11	7.23	4.79	5.31	5.00	12.5	7.39	2.62	5.24	7.49	4.60	11.2
A/CNK	1.05	0.96	1.07	1.48	0.99	1.25	0.85	0.82	0.94	1.01	0.88	0.97
$FeO_t/(FeO_t + MgO)$	0.75	0.76	0.71	0.68	0.60	0.77	0.89	0.63	0.72	0.82	0.82	0.72

	AMS-g H-9	AMS-g H-10	AMS-g H-11	SHG 431*	AMS HP-2	AMS B-5	AMS B-70	AMS B-71	AMS B-72	AMS B-73	AMS HP-14	AMS HP-21
SiO_2	75.55	68.04	n.a.	58.74	56.08	65.10	59.70	63.84	66.13	64.93	58.35	61.45
TiO_2	0.21	1.28	n.a.	1.89	2.05	1.13	1.94	1.38	1.31	0.82	2.04	1.84
Al_2O_3	13.70	14.41	n.a.	14.66	15.46	14.05	15.57	14.80	14.38	16.61	14.96	14.83
$FeO_t^{\dagger}$	0.90	4.91	n.a.	8.90	10.44	6.97	7.92	7.35	6.28	4.75	8.67	7.00
MnO	0.01	0.06	n.a.	0.14	0.18	0.10	0.17	0.13	0.10	0.08	0.13	0.10
MgO	0.14	0.91	n.a.	2.36	1.60	0.76	1.49	0.92	0.98	0.79	1.65	1.38
CaO	1.14	3.09	n.a.	5.55	6.79	3.73	5.15	3.91	3.51	3.84	5.70	5.12
Na_2O	2.67	2.82	n.a.	3.63	3.50	3.08	4.70	2.79	3.53	3.39	3.95	3.56
K_2O	5.61	3.85	n.a.	2.88	2.83	4.59	2.39	4.25	3.17	4.45	3.54	3.92
P_2O_5	0.06	0.63	n.a.	1.25	1.09	0.50	0.98	0.63	0.61	0.33	0.99	0.79
Original total	100.71	98.59	—	98.91	97.69	97.55	96.69	96.84	95.41	96.79	97.65	98.42
Sc	1.7	11.2	11.2	19.9	22.7	15.6	16.4	15.0	11.5	10.2	18.3	14.5
Cr	1.7	7.3	8.8	8.9	6.6	3.0	7.0	4.0		2.0	7.1	9.0
Co	37.7	24.0	31.0	39.6	27.4	7.4	11.8	7.8	8.3	4.5	33.1	28.8
Rb	213	183	154	52	64	137	60	117	82	142	73	105
Cs	0.18	0.31	0.05	0.11	0.20	0.54	0.23	0.33	0.24	0.27	0.24	0.32
Sr	279	287	248	580	625	398	545	441	402	459	548	541
Ba	1170	1260	1130	1170	1640	1761	933	1945	1648	1816	1721	1802
La	32.6	77.9	76.2	76.0	93.9	70.0	80.0	78.0	76.0	73.0	86.5	89.2
Ce	36.3	164	168	164	205	177	166	150	141	135	201	197
Nd	8.9	86.4	89.9	99	115	105	102	95	93	86	110	98
Sm	1.26	18.7	18.9	20.3	24.9	17.4	21.1	17.6	16.3	15.3	22.4	20.6
Eu	2.72	3.14	2.47	4.55	5.32	4.89	5.01	5.11	4.38	5.56	4.70	4.24
Tb	0.10	2.01	2.11	2.45	3.10	2.64	2.55	2.14	2.02	1.96	2.73	2.62
Yb	0.33	3.72	4.17	5.38	6.41	4.42	4.32	3.86	3.62	3.91	5.01	5.19
Lu	0.03	0.41	0.48	0.77	0.83	0.64	0.59	0.56	0.57	0.51	0.63	0.67
Zr	125	440	365	410	778	989	915	891	860	732	833	794
Hf	3.9	16.2	13.8	13.8	18.8	27.3	19.6	21.9	18.7	25.3	20.3	17.9
Ta	2.11	5.12	2.82	2.70	2.41	1.19	0.98	0.96	1.06	0.88	2.41	2.52
Th	1.2	1.5	0.5	0.4	0.8	2.5	1.8	2.4	2.9	1.7	2.1	1.9
U	1.1	1.5	1.8	1.5	0.7	3.9	2.0	1.8	1.5	1.9	0.8	0.8
K_2O/TiO_2	26.2	3.00	—	1.52	1.38	4.07	1.23	3.09	2.42	5.41	1.74	2.13
A/CNK	1.09	1.00	—	0.76	0.73	0.84	0.79	0.91	0.92	0.95	0.72	0.76
$FeO_t/(FeO_t + MgO)$	0.79	0.75	—	0.68	0.79	0.84	0.75	0.82	0.78	0.77	0.75	0.74

(continued)

TABLE 1. *Continued*

	Map unit and sample number									
	AMS B-2A	TMS 65	C 363	C 364	C 361	C H-13	C H-14	RA H-6	RA H-18	RA HP-24
SiO_2	61.71	61.57	64.07	63.67	67.78	n.a.	65.14	56.47	58.97	60.76
TiO_2	1.92	1.26	1.01	1.16	0.85	n.a.	0.83	2.01	0.61	0.12
Al_2O_3	11.67	16.25	16.42	15.55	14.14	n.a.	14.79	19.82	18.70	23.60
FeO_t†	12.65	6.50	5.25	6.27	5.22	n.a.	7.71	5.25	5.27	0.34
MnO	0.20	0.09	0.08	0.10	0.08	n.a.	0.12	0.08	0.08	0.01
MgO	2.25	2.15	0.69	0.85	1.18	n.a.	1.33	3.93	3.79	0.08
CaO	2.55	3.18	4.54	4.53	3.29	n.a.	4.35	6.56	6.40	5.12
Na_2O	2.02	2.98	3.47	3.23	2.69	n.a.	2.55	3.75	4.33	6.55
K_2O	4.07	5.25	4.00	4.00	4.53	n.a.	3.04	2.09	1.77	3.36
P_2O_5	0.96	0.78	0.48	0.62	0.23	n.a.	0.14	0.06	0.08	0.05
Original total	97.23	99.08	98.69	98.38	98.98	—	99.50	95.86	99.83	99.57
Sc	19.1	16.3	13.5	13.9	9.9	16.3	18.5	8.4	10.0	0.4
Cr	13.0	7.6	13.5	8.5	21.7	3.9	15.2	124.0	79.6	1.6
Co	13.0	32.6	6.3	7.3	37.0	26.1	44.1	23.6	31.6	20.4
Rb	177	150	110	110	125	68	85	16	14	29
Cs	0.40	0.39	0.50	0.35	0.38	0.29	0.22	0.03	n.a.	0.13
Sr	170	400	545	464	562	571	509	793	856	1154
Ba	833	2000	2423	2040	2410	1340	1510	817	710	1348
La	114	119	77.1	67.8	40.5	65.4	26.8	3.1	2.9	7.6
Ce	244	274	165	145	74.8	139	47.6	4.96	5.06	17.5
Nd	133	140	84	82	35	77	22	1.7	1.8	5.1
Sm	25.6	27.8	18.6	16.7	7.4	16.4	5.90	0.46	0.68	1.41
Eu	3.68	4.64	4.58	4.15	3.49	4.47	3.58	1.32	1.70	2.07
Tb	3.14	3.25	2.20	2.07	0.88	1.89	0.75	0.08	0.14	0.24
Yb	5.46	7.06	5.03	4.57	2.70	4.32	3.37	0.25	0.38	0.93
Lu	0.70	0.89	0.73	0.71	0.34	0.59	0.47	0.02	0.03	0.12
Zr	1059	600	880	890	479	670	390	40	30	182
Hf	27.9	20.8	22.3	16.7	12.3	24.1	13.4	0.8	0.4	4.9
Ta	1.97	3.60	1.09	0.99	2.20	1.84	2.09	0.99	0.92	3.47
Th	2.6	2.2	1.4	0.8	0.8	1.1	0.3	0.0	0.1	4.6
U	6.9	2.7	1.0	0.5	0.6	0.3	1.2	<0.1	0.2	1.1
K_2O/TiO_2	2.12	4.16	3.95	3.46	5.33	—	3.66	1.04	2.91	27.9
A/CNK	0.94	0.99	0.90	0.87	0.92	—	0.96	0.97	0.90	0.99
$FeO_t/(FeO_t + MgO)$	0.76	0.63	0.81	0.80	0.71	—	0.76	0.43	0.44	0.70

Note: Map units are the same as shown in Fig. 2. Analyses are by combined ICP-AES and INAA. For the major oxides, the values are wt% normalized to 100% volatile-free; for trace-elements, the values are ppm. Abbreviations: A/CNK (aluminum saturation index)—Molar $Al_2O_3/(CaO + Na_2O + K_2O)$; n.a.—not analyzed.

*Major oxide analyses provided by a commercial laboratory.

†FeO_t: Total iron expressed as FeO.

major-element analyses, with some trace-element data, for many of the units in the study area. The reconnaissance study by Owens and Dymek (1999) presented additional major- and trace-element data for eight samples in the district. In our study, major-element abundances are considered for overall comparison of units with these previous studies and are used to elucidate possible petrologic associations. Emphasis is placed on rare-earth elements (REE) and other trace elements for discussion of petrologic scenarios.

Analytical Methods

Aliquots obtained from powders used in previous studies (e.g., Sinha and Bartholomew, 1984) were analyzed in the Laboratory for Environmental Geochemistry at Idaho State University. Major-element analyses of many Lovingston massif samples (unpublished) were determined previously, either by a commercial laboratory or by the U.S. Geological Survey laboratory using the "rapid rock" technique. All but seven of the samples reported in Table 1 were re-analyzed for major elements, and all samples were analyzed for trace elements. Major-element data for re-analyzed samples demonstrated no significant interlaboratory bias from earlier results and provided justification for including the seven analyses obtained previously. Major-element concentrations, reported as oxides, and trace elements Sr, Ba, and Zr were determined by inductively-coupled plasma atomic emission spectroscopy (ICP-AES). Sample powders were fused with Li-metaborate flux, and then dissolved in weak nitric acid prior to ICP-AES analysis. Other trace-element data were obtained by instrumental neutron activation analysis (INAA) on separate aliquots, following the procedures described by Hughes et al. (1997). Two major elements, Na as Na_2O and total Fe as FeO, and three trace elements, Ba, Sr, and Zr, were analyzed by INAA as an internal check against ICP-AES values.

Major-element concentrations in many samples indicate low original sums ranging from ~95 to 97 wt%. This discrepancy does not reflect analytical uncertainty, because the precision of each major element was within 1–2% of the value reported; however, it reflects unreported volatile concentrations and Fe oxidation state. Retrograded Lovingston rocks have significant volatile content, mostly sequestered in hydrous minerals, such as biotite, and much of the FeO is expected to be oxidized to Fe_2O_3. Volatiles, such as H_2O, CO_2, F, and Cl, and the Fe oxidation state were not determined, which would require using additional analytical techniques. The lack of this data therefore yields major-element sums lower than 100%. Moreover, previously unpublished rapid rock analyses, where provided, indicated combined H_2O (as H_2O^+ and H_2O^-) and CO_2 values greater than ~1% by weight (typically 2%–3%), and separate FeO (Fe^{2+}) and Fe_2O_3 (Fe^{3+}) values indicating a substantial range in the Fe^{2+}/Fe^{3+} ratio. When all Fe data were adjusted to the same Fe oxidation state and volatiles excluded, the major-element sums of current and previous analyses were essentially identical. Thus, all major elements depicted in diagrams use the volatile-free normalized values to eliminate the effects of volatiles and oxidation state. The low sums reported for some analyses are unlikely to significantly affect geochemical trends or the modeling presented here.

Within the Lovingston massif, the Archer Mountain Suite compositions are divided into representatives of typical lithology (foliated biotite granitoids) and separate groups comprised of five leucocratic granitoid plutons, three charnockitic plutons, and one mylonitic gneiss (B-2A). Comparative units (Fig. 2) are represented by a sample from the Turkey Mountain Suite (quartz monzonite, sample 65) that is similar to the Archer Mountain Suite typical group, and Shaeffer Hollow Granite (quartz monzodiorite, sample 431) that is similar to the Archer Mountain Suite granitic plutons. One comparative unit of the Stage Road Suite is represented by a sample of the Hills Mountain Granulite Gneiss (charnockite, B-68). Samples of less voluminous units that are not directly associated with these two major bodies, but are significant for petrologic evaluation of Grenville-age rocks, include charnockitic bodies in the Roses Mill pluton (H-13 and H-14) and Roseland Anorthosite (H-6, H-18, and HP-24).

Major Elements

Major-element abundances (Table 1) indicate distinct compositional ranges in the Archer Mountain and Stage Road suites (Figs. 3 and 4), as exemplified by SiO_2 concentrations of ~56–68 wt% in the Archer Mountain Suite, exclusive of granitic and charnockitic plutons, and ~61–75 wt% in the Stage Road Suite. Differences between the two suites are apparent in other major-element abundances, especially in FeO (~4.9–9.9 wt% and ~4.2–6.0 wt%, respectively) and CaO (~3.1–6.4 wt% and ~1.0–3.7 wt%, respectively). Charnockitic plutons tend to plot in the region of overlap, where no clear distinction exists between the two suites. Most Lovingston rocks are subaluminous to metaluminous, with A/CNK (molar $Al_2O_3/[CaO + Na_2O + K_2O]$) ratios <1.1. These ratios display a positive correlation with SiO_2 content (Fig. 4), although, not surprisingly, the Roseland Anorthosite samples plot off the trends in Figure 4, with low P_2O_5 (0.05–0.08 wt%) and a large dispersion of other major elements. The overall chemical differences among Lovingston rocks are enhanced when considering the Archer Mountain Suite leucocratic granitoid plutons and charnockitic plutons; however, the Archer Mountain Suite and Stage Road Suite together span nearly the entire compositional range of the massif.

Normative compositions (Fig. 3) indicate a compositional field extending from the farsundite (monzogranite) field through quartz mangerite (quartz monzonite) into quartz jotunite (quartz monzodiorite). Overall, the major-element compositions of rocks in the Lovingston and Pedlar massifs are similar; however, distinctions between the two can be made on the basis of lithologic types, metamorphic grade, and presence of such variants as anorthosite and nelsonite. The Lovingston compositional trend overlaps that of normal Pedlar charnockitic rocks (outlined field in Fig. 3), for which most compositions plot in the farsun-

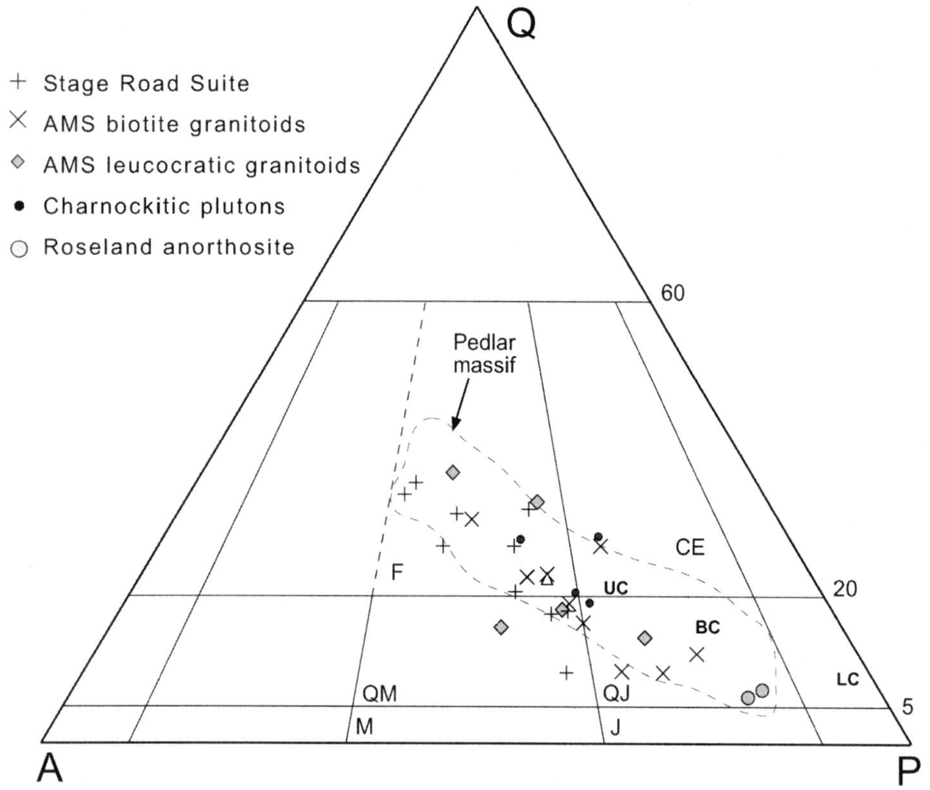

Figure 3. Quartz–alkali feldspar–plagioclase (Q–A–P) ternary diagram, plotted for normative compositions of lithologic units from the Lovingston massif. The compositional field of rocks from the Pedlar massif (Hughes et al., 1997) is shown for comparison. Abbreviations for charnockitic rock names (Streckeisen, 1974) and IUGS granitoids equivalents are: F—farsundite (monzogranite); QM—quartz mangerite (quartz monzonite); M—mangerite (monzonite); CE—charnoenderbite (granodiorite); QJ—quartz jotunite (quartz monzodiorite); J—jotunite (monzodiorite). Average values for upper and lower crust (UC and LC) are from Taylor and McLennan (1995); the average value for bulk crust (BC) is from Rudnick and Fountain (1995).

dite field, although the Pedlar suite (Hughes et al., 1997) has several members closer to the plagioclase apex.

Most Archer Mountain and Stage Road suite samples (including the Archer Mountain Suite leucocratic granitoids, Turkey Mountain Suite sample 65, Shaeffer Hollow Granite, and Archer Mountain and Roses Mill charnockitic plutons) define linear trends on Harker plots (Fig. 4). Most of the compositions are distinctly more "felsic" than the average composition of upper crust in terms of quartz and feldspar components. Relative to the chemical variations portrayed by average lower, bulk, and upper continental crust (LC, BC, and UC, respectively, in Figs. 3 and 4) (Rudnick and Fountain, 1995; Taylor and McLennan, 1995), the Lovingston compositional trends are distinctly lower in MgO and CaO, and higher in TiO_2 and P_2O_5 at equivalent SiO_2 values (Fig. 4). Lovingston rocks also have higher K_2O (Table 1) at equivalent SiO_2 values relative to crustal averages and show a slightly negative Na_2O trend with SiO_2, whereas a slightly positive trend is evident in the reported crustal averages. Despite these differences, the K_2O/TiO_2 and A/CNK ratios follow similar trends, with no significant differences between Lovingston and reported crustal values at equivalent SiO_2 abundances.

Three lithologically unique Lovingston samples were collected to represent possible composition extremes relative to typical rocks in the massif. One of these, the Archer Mountain Suite leucocratic granitoid sample B-12B (56.2 wt% SiO_2), was collected ~1 m from a contact with the 730-Ma Rockfish River pluton (Fig. 2). Comparatively much lower CaO, TiO_2, and P_2O_5 (Fig. 4, C, D, and F, respectively) possibly reflects effects of magmatic differentiation, although these and other chemical differences may be partly related to metasomatic fluids from the Iapetan pluton. Samples more distal to the contact show no appreciable scatter or divergence from chemical trends (Fig. 4, A–H), and thus they do not support regional effects of post-Grenville metasomatism, as proposed by Evans (1991). Another divergent sample, Archer Mountain Suite sample B-2A, was collected from a mylonitized zone near the Rockfish Valley fault (Fig. 2). The composition exhibits significantly higher FeO (Fig. 4, B) and K_2O (5.22 wt%) (Table 1) compared with other samples of the Archer Mountain and Stage Road suites. The mylonite zone apparently did not experience additional chemical disruption, indicated by lack of deviation from the regular trends in other elements (Fig. 4, A and C–H). A third unique sample, H-9, composed mainly of quartz and feldspar, was obtained from an Archer Mountain Suite leucocratic granitoid body located in the northeastern part of the study area (Fig. 2). It has the highest SiO_2 content (75.6 wt%) of the suite, and anchors the high-silica end of Harker plots (Fig. 4, A–H). Although the major-element composition is not extraordinary, the trace element composition of this sample (discussed later in this paper) is significantly different from other Archer Mountain Suite rocks.

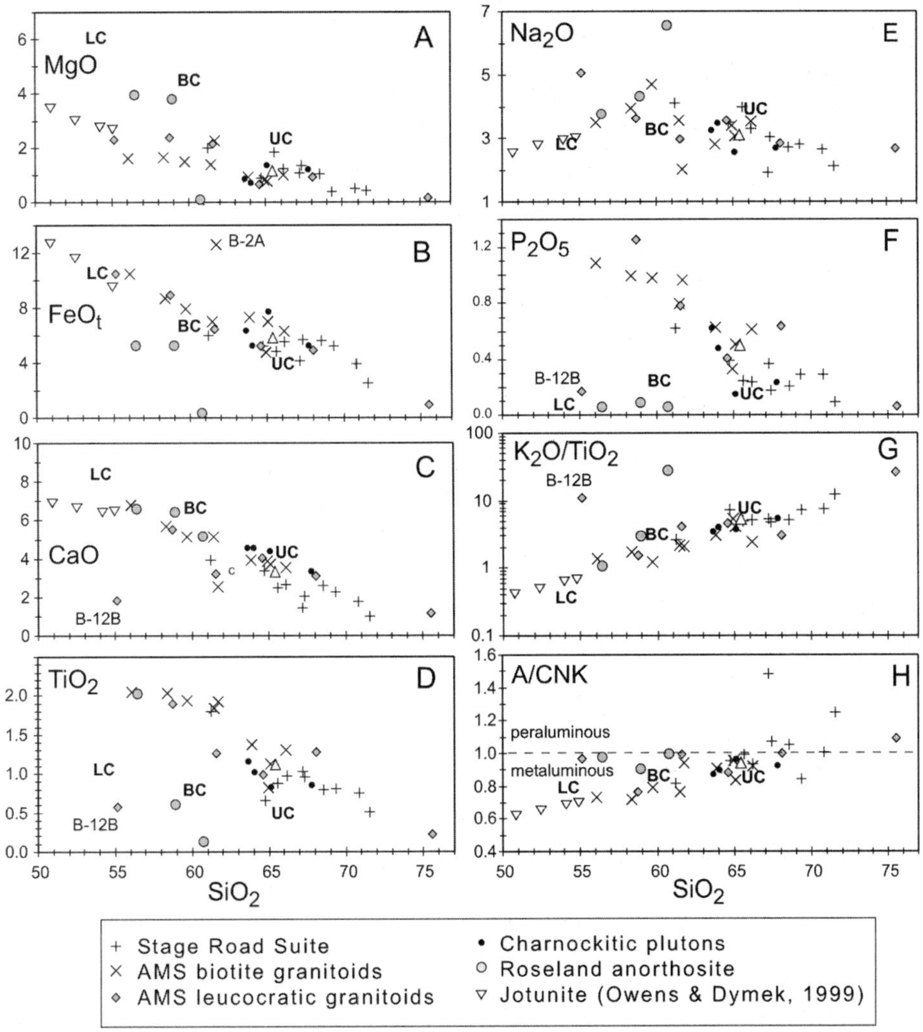

Figure 4. (A–F) Harker diagrams of selected major elements; (G) K_2O/TiO_2 ratios; (H) A/CNK (molar $Al_2O_3/[CaO + Na_2O + K_2O]$) in the Lovingston massif. Plotted analyses include four jotunites from the Lovingston massif reported by Owens and Dymek (1999). Sample numbers refer to Table 1. Average values for upper and lower crust (UC and LC, respectively) are from Taylor and McLennan (1995); the average value for bulk crust (BC) is from Rudnick and Fountain (1995).

Trace Elements

Trace-element abundances (Table 1) indicate compositional differences between the Stage Road and Archer Mountain suites, although considerable overlap exists in some covariation diagrams (Fig. 5). Such incompatible elements as Rb (Fig. 5, A) are weakly correlated with SiO_2 abundances and indicate some separation of chemical fields. SiO_2 concentrations are generally higher in the Stage Road Suite; however, high SiO_2 values of some leucocratic granitoids and charnockitic plutons of the Archer Mountain Suite have this distinction. Notably, the metasomatized and mylonitic samples (B-12B and B-2A, respectively) lie off the main trend. A weak positive correlation exists between Ba and Sr (Fig. 5, B) except for the anorthosite samples, although the "trend" is made up of short segments where the correlation is negative. Better distinction of groups is apparent in the covariation of Th and La (Fig. 5, C); however, in all incompatible element plots, distinct compositional fields are apparent only between the two suites when considering principal lithologies. Compatible elements Cr and Sc are positively correlated, producing separate trends within the Archer Mountain and Stage Road suites. The Stage Road Suite has characteristically higher concentrations of both Cr and Sc and defines a higher slope trend.

Multi-element plots of trace elements normalized to the average lower crust composition (Taylor and McLennan, 1995) demonstrate the overall similarity in relative abundance patterns among principal lithologies (Fig. 6). Overall, patterns in the Stage Road and Archer Mountain suites are similar to those determined for the Pedlar massif (Hughes et al., 1997). Incompatible elements Rb, Ba, and K are variably enriched in the Stage Road Suite (Fig. 6, A) by ~5–50-fold relative to average lower crust. Less enrichment is evident in other incompatible elements, yielding apparent negative anomalies in Cs, Th, Ta, Sr, Ti, and Sc, all having less than 10-fold enrichment factors. The Hills Mountain Granulite Gneiss (B-68) has slightly higher Rb

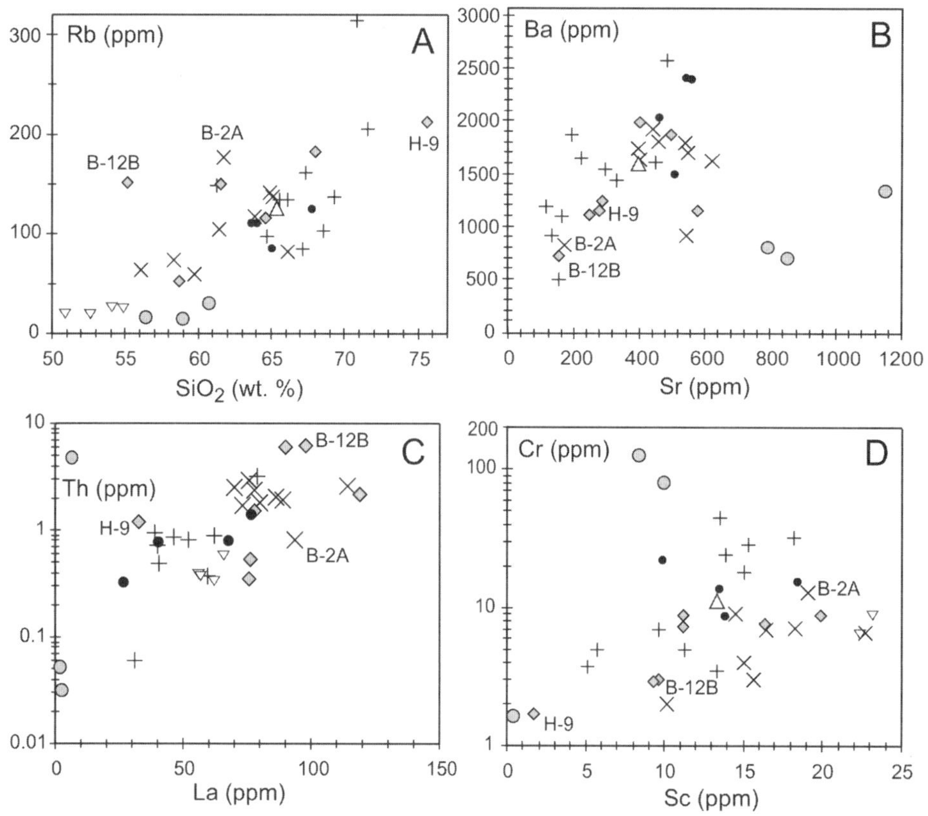

Figure 5. (A) Rb versus SiO_2; (B) Ba versus Sr; (C) Th versus La; and (D) Cr versus Sc in the Lovingston massif. Symbols are the same as in Figure 4.

and lower Ba, Zr, and Hf, but no other significant differences from the primary Stage Road Suite representatives. These minor differences possibly reflect alkali-element mobility during metamorphism, noted by Bartholomew et al. (1981) as evident in irregular-shaped, resorbed potassium feldspar grains. Alternatively, such differences could reflect local effects of possibly inherited monazites, allanites, and zircons, which is evident in the large ranges in Th, Zr, and Hf. Otherwise, no significant difference exists between this unit and remaining members of the Stage Road Suite.

Geochemical patterns of the Archer Mountain Suite and associated rocks (Fig. 6, B and C) show ~4–30-fold enrichments in Rb, Ba, and K and anomalous element behaviors similar to the Stage Road Suite; however, chemical dispersion among Archer Mountain Suite biotite granitoid compositions is less than that of leucocratic granitoids and gneisses of the Stage Road Suite. Comparative units that are geologically related to the Archer Mountain Suite and Stage Road Suite show little divergence from the representative suites. Despite this apparent uniformity, a mildly variant pattern is depicted by the Turkey Mountain suite (Fig. 6, B, sample 65), which represents a separate plutonic event. Even greater variation from the normal pattern is evident in sample B-2A, the mylonitic gneiss, which has a higher Rb/Ba ratio, higher Zr and Hf, and a deeper negative Eu anomaly relative to typical suite compositions. Leucocratic granitoids (Fig. 6, C) display the greatest dispersion relative to the normal suite, especially in incompatible depleted elements, such as Th, and in elements that are depleted (Sr, Ti, Sc) due to compatibility with major phases, or enriched (Zr, Hf) due to partitioning into accessory phases. Relative to other Archer Mountain Suite leucocratic granitoids, the Shaeffer Hollow Granite sample (431, Fig. 6, C) has the lowest Rb and Th (and second-lowest Cs), the highest Sr, Ti, and Sc, and extends the upper limit of incompatible elements Ba, REE, Zr, and Hf, but the overall chemical pattern is not unique.

Chemical differences between Archer Mountain and Stage Road suite compositions are largely attributed to the greater ranges of refractory elements in the Stage Road Suite and Archer Mountain Suite leucocratic granitoids relative to tighter patterns in the normal biotite granitoid Archer Mountain Suite samples. Typical Archer Mountain biotite granitoid compositions (Table 1; Fig. 6, B) have La = 70–94, Sm = 15–25, Hf = 19–27, and Th = 0.8–2.9 ppm; whereas the leucocratic samples (Fig. 6, C without H-9) have La = 76–98, Sm = 17–19, Hf = 14–26, and Th = 0.5–6.2 ppm. By contrast, the Stage Road Suite compositions (Table 1; Fig. 6, A) have La = 31–79, Sm = 5–14, Hf = 10–16, and Th = 0.4–0.9 ppm. As in the case with major oxides, a few divergent compositions are apparent, such as HP-22, a Stage Road Suite rock with 3.2 ppm Th, and HP-4, a Stage Road Suite rock with 0.06 ppm Th and 39.9 ppm Hf (as well as relatively high Zr).

Tectonic discrimination diagrams (Fig. 7) indicate a possible arc component, yet they neither unequivocally support this

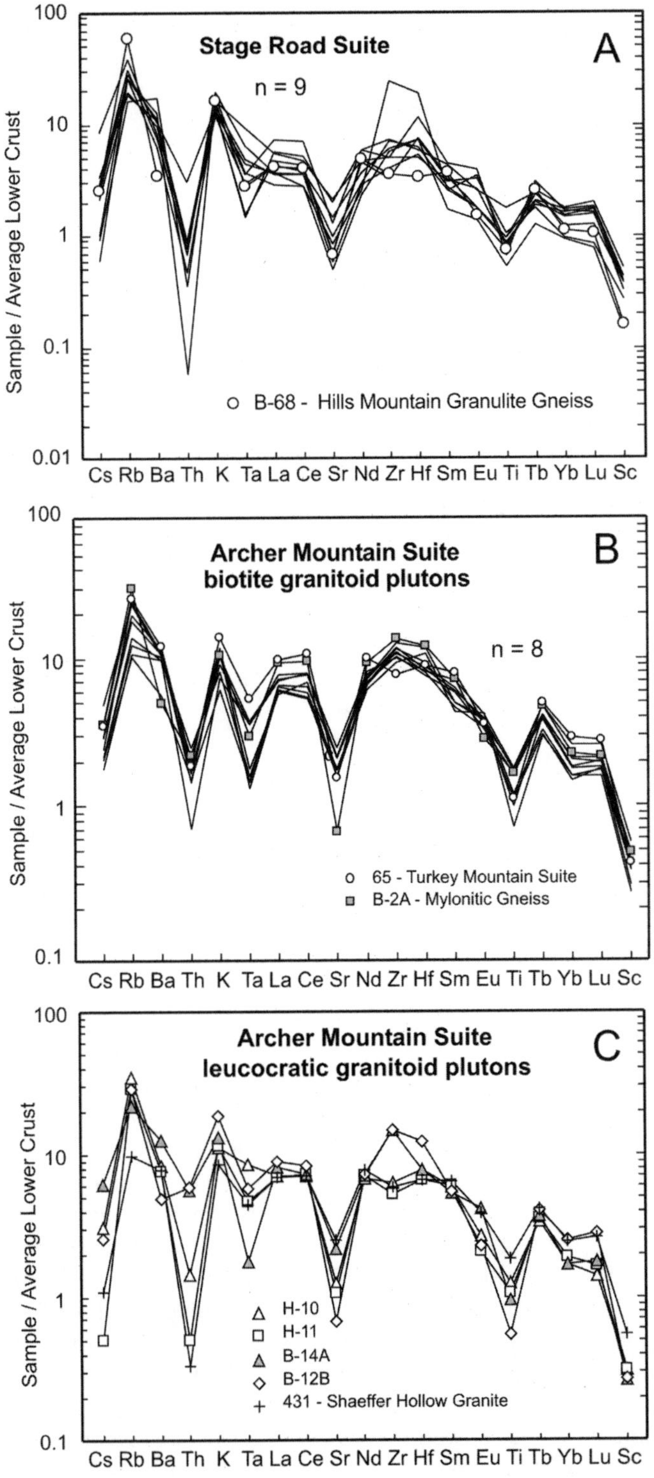

Figure 6. Multiple-element (spider) diagrams of (A) the Stage Road Suite granulite gneisses; (B) Archer Mountain Suite biotite granitoids and related rocks; and (C) Archer Mountain Suite leucocratic granitoids normalized to average lower crust values of Taylor and McLennan (1995).

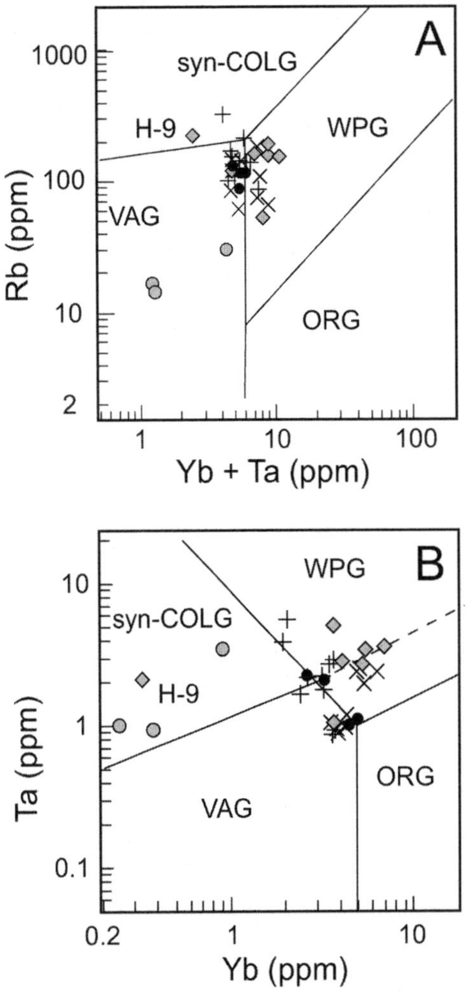

Figure 7. Tectonic discrimination diagrams of (A) Rb versus Yb + Ta; and (B) Ta versus Yb for Lovingston massif compositions, after Pearce et al. (1984). Abbreviations for tectonic fields: ORG—ocean-ridge granites; synCOLG—syn-collisional granites; VAG—volcanic-arc granites; WPG—within-plate granites. Symbols are the same as in Figure 4.

hypothesis nor allow the placement of Lovingston massif samples solely in another tectonic regime. The Rb versus Yb + Ta diagram (Fig. 7, A) shows that most rocks of the Stage Road Suite and the charnockitic plutons plot in the volcanic arc granites (VAG) field; whereas the leucocratic granitoids plot in the within-plate granite (WPG) field, and the biotite granitoids of the Archer Mountain Suite are equally represented in both fields. The Ta versus Yb diagram (Fig. 7, B) indicates that Ta values are sufficiently high to place even fewer samples in the volcanic arc field. Thus, these compositions probably represent, at least in terms of known tectonic discrimination parameters, the influence of multicomponent sources during evolution of the Grenville continental margin.

Rare-Earth Elements

REE patterns (Figs. 8–11), shown normalized to CI chondritic values of Anders and Grevesse (1989), provide detailed constraints on petrologic processes, rock unit associations, and protoliths. Stage Road Suite REE patterns are broadly uniform (Fig. 8), with La_N = 120–320-fold to Lu_N = 8–21-fold. These units are separated into two general groups: one that displays no appreciable Eu anomalies (Fig. 8, A) and the other having both moderately positive and moderately negative Eu anomalies (Fig. 8, B). The Hills Mountain Granulite Gneiss REE pattern, with a distinctly negative Eu anomaly, is indistinguishable from other Stage Road Suite rocks with slight Eu anomalies. Typical Archer Mountain Suite REE patterns (Fig. 9, A) are fairly uniform, with La_N = 300–400-fold to Lu_N = ~20–30-fold and no significant Eu anomalies. This uniformity may imply broadly homogeneous compositional properties and similarities in magma genesis throughout the suite. Although by no means definitive, sample 65, collected in the Turkey Mountain Suite, has ~30% higher REE than the average Archer Mountain Suite biotite granitoids, and it exhibits a moderately negative Eu anomaly (Fig. 9, B).

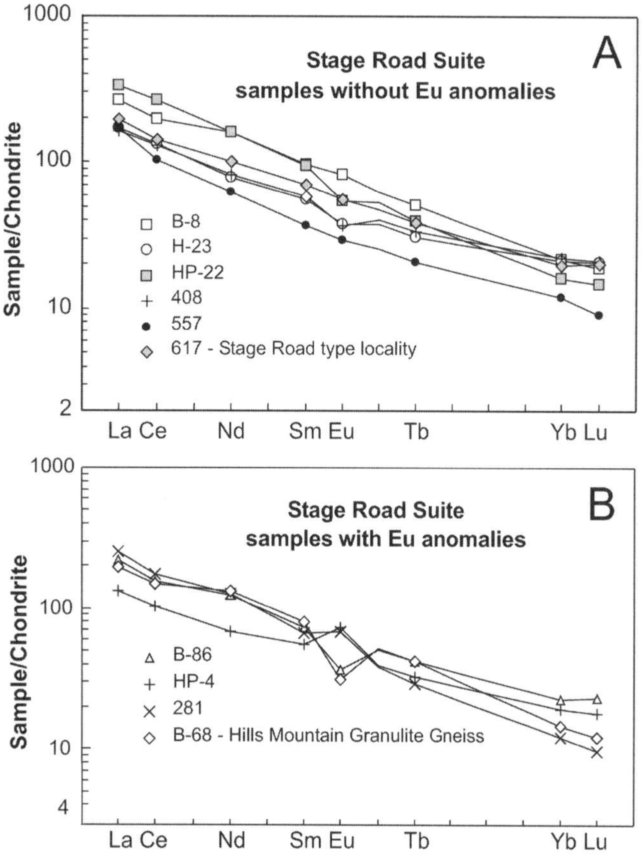

Figure 8. CI chondrite-normalized REE patterns of the Stage Road Suite samples (A) without appreciable Eu anomalies; and (B) samples with Eu anomalies. Chondritic values are from Anders and Grevesse (1989).

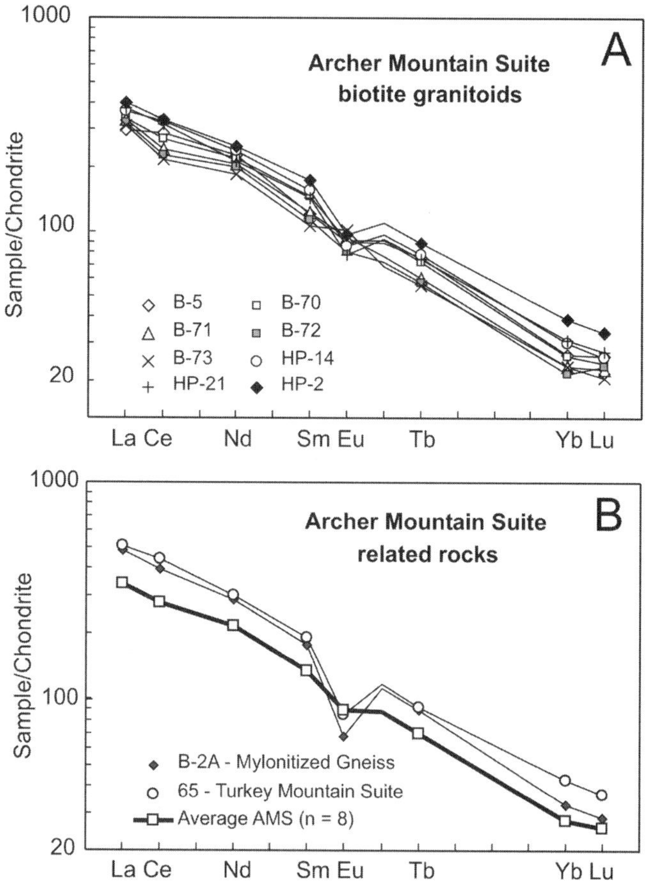

Figure 9. CI chondrite-normalized REE patterns of the Archer Mountain Suite (A) biotite granitoids and (B) related rocks, compared with the average biotite granitoid composition. Chondritic values are from Anders and Grevesse (1989).

The pattern is also nearly equivalent to that of the mylonitized pluton (sample B-2A) in the Archer Mountain Suite, which also exhibits a moderately negative Eu anomaly.

The REE uniformity among many Archer Mountain Suite rocks is further apparent in the leucocratic granitoids and charnockitic plutons (Fig. 10); however, significant differences are represented in several samples. REE patterns of leucocratic granitoids and the Shaeffer Hollow Granite (sample 431) (Fig. 10, A) are indistinguishable from those depicted by the biotite granitoids (Fig. 8), and they display either a slightly negative Eu anomaly or no Eu anomaly. Leucocratic sample H-9 (Fig. 10, A), with a steep low REE pattern and high positive Eu anomaly, exhibits the most divergent composition from this group.

Charnockitic plutons in the Archer Mountain Suite (samples 361, 363, and 364) and Roses Mill pluton (samples H-13 and H-14) and jotunites, including the jotunites of Owens and Dymek (1999), have REE patterns that depict two separate types (Fig. 10, B). The REE patterns of most samples in this group span a narrow range that is nearly equivalent to the average normal Archer Mountain Suite. The second type is depicted by only

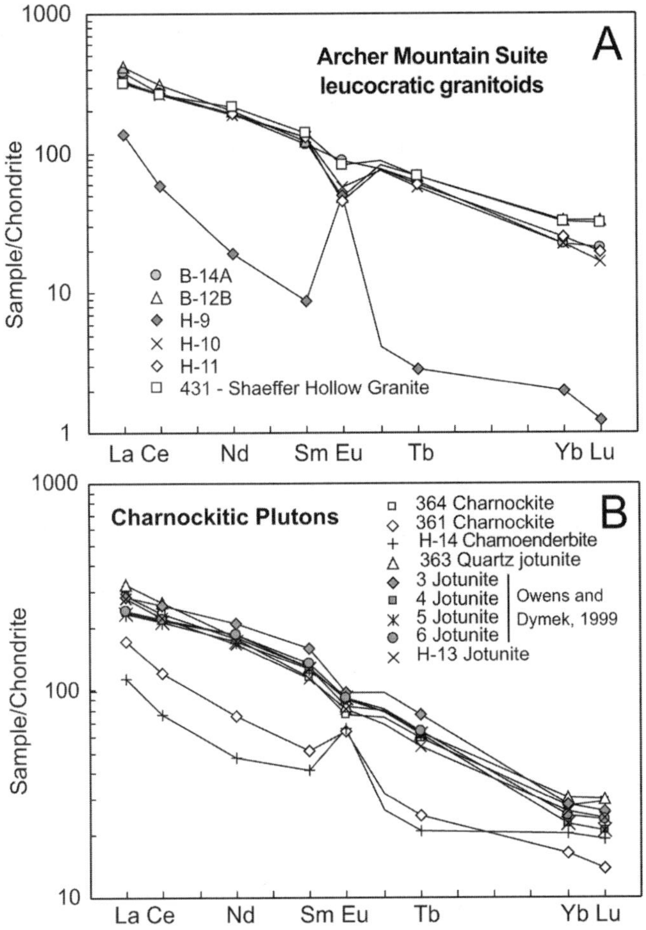

Figure 10. CI chondrite-normalized REE patterns of Archer Mountain Suite (A) leucocratic granitoids and (B) charnockitic plutons in the Lovingston massif. Chondritic values are from Anders and Grevesse (1989).

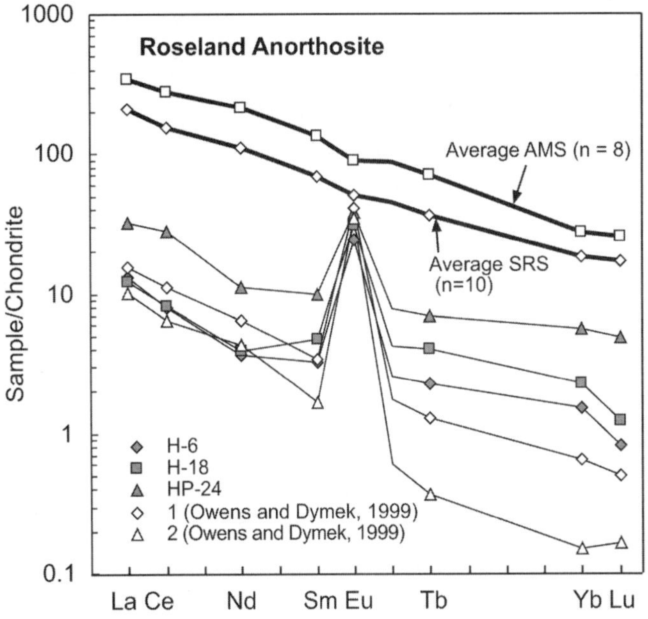

Figure 11. CI chondrite-normalized REE patterns of Roseland Anorthosite samples compared with the average Archer Mountain Suite (AMS) biotite granitoids and average Stage Road Suite (SRS) granulite gneisses. Chondritic values are from Anders and Grevesse (1989).

one charnoenderbite (H-14) and one charnockite (361), which have significantly lower REE patterns and moderately positive Eu anomalies (Fig. 10, B) that complement the REE patterns of other charnockitic rocks.

Positive Eu anomalies characterize three Roseland Anorthosite samples from this study (HP-24, H-18, H-6) and two from Owens and Dymek (1999) (Fig. 11, A). The patterns span a large fractionated range in chondrite-normalized (N) REE abundances from La_N = 10–31 to Lu_N = 0.10–5.0; however, a somewhat narrower range of Eu_N is evident for these samples, reflecting the dominant influence of accumulative plagioclase. The overall anorthosite REE pattern is similar to that of other massif anorthosites. As noted by Owens and Dymek (1999), Eu anomalies in the Roseland Anorthosite are higher than in many other anorthosites, reflecting more strongly reducing conditions that produce greater concentrations of Eu^{2+}, which is partitioned into plagioclase, relative to Eu^{3+}, which is incompatible.

GEOCHEMICAL AND LITHOLOGICAL IMPLICATIONS

Major- and trace-element characteristics of meta-igneous rocks from the Lovingston massif are broadly similar to those of the Pedlar massif (Hughes et al., 1997). More importantly, the geochemical data and rock associations are generally consistent with an extensive early or pre-Grenville magmatic province and possibly reflect a regional geochemical signature throughout the Blue Ridge terrane. In terms of lithologic diversity, the Lovingston massif is comparable with other Grenville terranes, such as the Nain Plutonic Suite (Emslie et al., 1994) and the Blair River Inlier (Miller and Barr, 2000). Both of these suites contain massif anorthosites, as well as various charnockites to jotunites, and Fe-Ti oxide- and apatite-rich rocks.

Harker diagrams of the Lovingston massif (Fig. 4) imply a magmatic trend, and trace-element patterns (Fig. 6) are characterized by depleted Ta, Sr, Ti, and Sc, relative to the lower crust, characteristics evident in both the Lovingston and Pedlar massifs. Overall incompatible trace-element enrichment possibly reflects magma generation from a chemically evolved source, such as lower crust. The associated depletion of Ta, Sr, Ti, and Sc further indicates a protracted history of fractionation prior to Grenville deformation and associated subsequent upper amphibolite- to granulite-facies metamorphism. Depletion of these elements is generally attributed to magmatism at convergent plate margins (Gill, 1981; Pearce et al., 1984); however, moderately negative anomalies in Lovingston samples only marginally support this designation. Major units also exhibit relatively depleted Cs and Th, a common characteristic in the Pedlar massif, although complementary units with anomalously high Th in the Pedlar (Hughes et al., 1997) are not observed in the Lovingston massif.

REE patterns and most trace-element signatures of Grenville-age rocks of the Lovingston massif are essentially uniform and enriched relative to the lower crust. Thus, they likely were derived from fairly uniform lower crust sources, although mantle influence cannot be ruled out entirely. A lower crust source is consistent with the conclusions drawn by Emslie et al. (1994) for the Nain Plutonic Suite, in which trace-element signatures and mineral constraints require granitoid melts to be produced, for the most part, by crustal partial melting. It is also consistent with the petrology and geochemistry of syenitic rocks in the Blair River Inlier (Miller and Barr, 2000) that require either derivation from or contamination by rocks with Nd isotopic signatures equivalent to the gneissic country rocks. The following discussion further outlines these possible associations and the processes relevant to the lithological and chemical attributes of rock units in the Lovingston massif.

Stage Road and Archer Mountain Suites

The Stage Road Suite consists primarily of meter-scale layers of coarse-grained, biotite augen gneiss alternating with thick layers of fine-grained biotite gneiss, biotite schist, and quartzofeldspathic gneiss. The biotite augen gneiss contains 1–5-cm porphyroclasts of feldspar and 1-cm porphyroclasts of quartz. Abundant Paleozoic retrograde (greenschist-facies) minerals (Evans, 1991) preclude determination of Grenville-age mineral assemblages in this unit. However, the Hills Mountain Granulite Gneiss is in intrusive contact with the Stage Road Suite in the northern part of the area (Fig. 2) and contains orthopyroxene assemblages (but no garnet), suggesting that the Stage Road Suite reached lower granulite-facies metamorphic conditions during Grenville metamorphism.

Rocks in the Archer Mountain Suite, including biotite granitoids, leucocratic granitoids, charnockitic plutons, and Turkey Mountain plutons, have chemical abundances that are slightly above (e.g., REE) or fall within the ranges defined by the Stage Road Suite of country rocks. The primary difference between these units is a greater uniformity among Archer Mountain Suite and related rocks compared with surrounding granulitic gneisses. The overall greater chemical variation in the Stage Road Suite relative to the Archer Mountain Suite, supported by extensive field mapping by previous workers, suggests reclassification of the Stage Road Suite as a suite of lithologies, rather than as a singular unit. Also, the overall lower REE patterns in the Stage Road Suite relative to the Archer Mountain Suite may reflect protolithic differences, perhaps including inherited older constituents that were not made more uniform by pervasive magmatism.

Separate trace-element fields of each suite, shown in covariation diagrams (Fig. 5), support consanguinity within each suite. Dispersion within each group follows separate systematic trends, such as the positive relations between Rb versus SiO_2 and Cr versus Sc, which represent significant magmatic differentiation or inherent inhomogeneity. These plots also support consanguinity of the Turkey Mountain Suite and Roses Mill pluton, which are composed of rocks lithologically and chemically similar to the Archer Mountain Suite, but are spatially separated from the Archer Mountain Suite by intervening country rock.

Two hypotheses can explain the observed chemical and lithological patterns noted in trace-element covariation and REE plots in these two igneous suites. One is that the older Stage Road Suite represents a deeply buried volcanic or plutonic series that served as the source for crustal partial melting during later Grenville orogenic events, which also metamorphosed the series to granulite gneiss. The Stage Road Suite was subsequently intruded by plutons that were remobilized from deeper regions in the gneissic sequence. This en masse emplacement hypothesis was favored by Hughes et al. (1997) to explain charnockitic plutons surrounded by chemically similar gneisses in the Pedlar massif. The alternative hypothesis is that these two principal Lovingston units were derived from separate, but perhaps equivalent sources, and they comprised two regional series of diachronous plutonic protoliths produced first during early or pre-Grenville-age magmatism and subsequently during the main Grenville event ca. 100 m.y. later. In this model, current lithological differences must be related to different grades of metamorphism and variable mobility of chemical components.

Roseland Anorthosite and Roses Mill Pluton

Upper granulite-facies mineral assemblages (orthopyroxene + garnet) occur in the vicinity of the Roseland Anorthosite (Bartholomew and Lewis, 1984; Herz and Force, 1984, 1987), where temperatures ranged between 850 °C and 950 °C (Herz and Force, 1987). These units are believed to have been emplaced as a diapiric body into the Lovingston massif prior to or during the ca. 1.0–1.05-Ga peak of the Grenville orogeny (Bartholomew and Lewis, 1984; Herz and Force, 1984, 1987; Sinha and Bartholomew, 1984).

Temperatures of ~850–950 °C in and around the anorthosite (Herz and Force, 1984, 1987) implies equilibration well above the dry solidus (~720–750 °C at 20–25 km) of muscovite granite (Huang and Wyllie, 1973), equivalent to ultrametamorphism (White and Chappell, 1977); hence, anorthosite emplacement may have been accompanied by partial melting. Indeed, dikes of jotunite and nelsonite, which Herz and Force relate to the Roses Mill pluton, intrude the anorthosite, associated border gneiss, and other rock types adjacent to the Roseland Anorthosite. The Roses Mill pluton has a ca. 970-Ma, nearly concordant, zircon age (Herz and Force, 1987). Unlike other charnockitic plutons in this part of the Blue Ridge anticlinorium, the Roses Mill pluton appears to be related to emplacement of the anorthosite during the Grenville-age metamorphic peak.

Geochemical Models

Quantitative geochemical models (Fig. 12) suggest that compositional differences among members of the Archer Mountain and Stage Road suites, including the Roseland Anorthosite and other Lovingston units, are controlled mainly by quartz and feldspar fractionation. They do not predict actual processes, but rather demonstrate the probability of mass transfer related to mineral fractionation or selective sampling of coarse-grained rocks (although large samples were collected to avoid selective sampling). Calculated trends are based on simple mass-balance fractionation of phases in equilibrium with the average Lovingston composition. Vectors for compositional variations that satisfied trends observed on chemical covariation diagrams were used to select a suitable starting composition for each model; however, model-path directions (Fig. 12) would be appropriate to any starting composition. Numbers shown adjacent to modeled curves represent the wt% of mineral added or subtracted. Models are shown for plagioclase (pl) only, equal amounts of plagioclase + quartz (qz), and orthoclase (ksp) only. Partition coefficients (summarized by Rollinson, 1993) were selected to be appropriate for mafic to intermediate compositions, which produced models that satisfied the observed trends far better than did models using relatively high partition coefficients appropriate to more silicic systems.

Chemical models in all cases (Fig. 12, A–F) confirm formation of the Roseland Anorthosite by plagioclase accumulation from a melt represented by a typical State Road Suite composition or some other melt that plots near the average Lovingston composition. Each covariation model depicts a trend that reproduces at least two of the anorthosite compositions by 50–80% plagioclase accumulation. The range in the required amount is due largely to the selection of partition coefficients, and possible effects related to mafic phases or accessory phases that are not shown. Parallel fractionation trends exhibited by the Archer Mountain Suite and related rocks (especially Fig. 12, A, B, and E) would yield compositions similar to the anorthosite, if extended. For example, the Ba versus Sr model (Fig. 12, A)

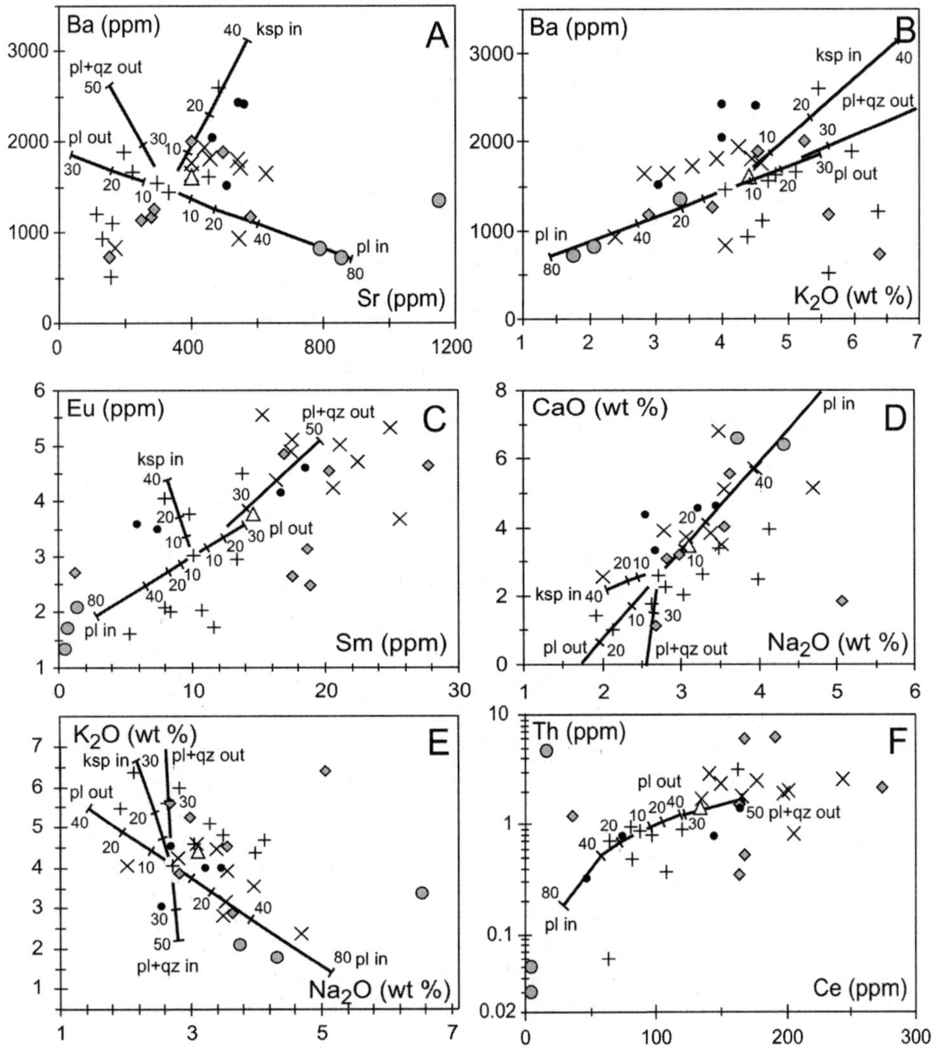

Figure 12. Quantitative models of crystal segregation and accumulation calculated for compositions in equilibrium with the average Lovingston massif composition. Numbers refer to the percent of mass added or subtracted for models of plagioclase (pl) in and out, equal amounts of plagioclase and quartz (qz) out, and K-feldspar (ksp) in. Symbols are the same as in Figure 4.

demonstrates this possibility, which is supported by REE patterns, and by the fact that at least one Archer Mountain Suite leucocratic granitoid (H-9, Fig. 10, A) has an REE pattern nearly identical to the Roseland Anorthosite. The sample was collected in a small granitoid body near H-10 and H-11 (Fig. 2), and the significance of the REE pattern is not clear; however, the large positive Eu anomaly, lower REE relative to other granitoids, and the proximity to other Archer Mountain Suite bodies suggest that it may be a complementary pattern derived during emplacement of the leucocratic granitoids. Overall similarity to the anorthosite REE suggests that it may be derived by similar processes and, as such, offers indirect evidence that these bodies are fractionated residuals from crustal melts, which is consistent with large crustal input documented by Peck and Valley (2000).

Models also demonstrate that plutons in the Archer Mountain Suite are not directly related to simple fractionation of magma represented by the Stage Road Suite (Fig. 12). This is not surprising, because SiO_2 values of rocks in the Stage Road Suite (Fig. 4) are generally higher than those in Archer Mountain suite rocks. Moreover, several models predict possible fractionation scenarios, but there is no consistency in phases required nor in their proportions. However, the models generally indicate scenarios within the Archer Mountain Suite that allow leucocratic and biotite granitoids, charnockitic plutons, and anorthosite to be derived by crystal accumulation or segregation (in some cases, K-feldspar out, not shown, but the opposite of K-feldspar in). Several trends are apparent that show leucocratic granitoid and charnockitic compositions leading away from the cluster of biotite granitoids. These are largely controlled by variable amounts of orthoclase and plagioclase, although no single scenario satisfies all the compositions.

In all models, overall trends of observed compositions roughly follow trends depicted by calculated compositions. Where K-feldspar- and plagioclase-controlled vectors diverge, so do the chemical trends. That Stage Road Suite compositions lie on trends either following or lying parallel to modeled trends that predict Archer Mountain Suite granitoids and related rocks suggests that these two principal components of the massif were derived from similar sources.

Thus, allowing for some dispersion related to local variations in phase proportions, the models (Fig. 12) favor the likelihood that all Lovingston rocks were derived from a uniform source region, probably located in the lower crust. Geochemical similarities require that both these major suites be derived from nearly equivalent sources, at least sources that had similar relative trace element abundances, although their emplacements were separated by a long time. This is consistent with both hypotheses: either rocks deep within the Stage Road Suite sequence were sources for the Archer Mountain Suite, or the Archer Mountain Suite and related magmas were generated from sources separate from, but identical to, those that derived the State Road Suite. Each scenario is subject to further rigorous testing, and neither one can be ruled out entirely; however, several arguments can be made to help determine which process is more likely to have occurred in the Grenville event.

Chemical uniformity among the normal Archer Mountain Suite compositions suggests less overall heterogeneity compared with the Stage Road Suite. Intrusive contacts and other field relations indicate that Archer Mountain Suite biotite and leucocratic granitoids intruded the Stage Road Suite and other gneissic country rocks. Arguably, all of the units initially may have had rather uniform chemical compositions representing a magmatic trend, and the Stage Road Suite experienced greater chemical mobility during metamorphic differentiation; however, there is no reason to suggest that the effects of metamorphic differentiation on juxtaposed rock units having similar compositions would be different during regional Grenville metamorphism. Thus, different degrees of heterogeneity between the two suites probably reflects magmatic processes.

If the Archer Mountain and Stage Road suites are directly related (first hypothesis), then a viable scenario that allows for this connection needs to be outlined. Otherwise, the only conclusion that can be made is that they were derived from separate sources in the lower crust (second hypothesis). Chemical models and covariation diagrams indicate that the Archer Mountain and Stage Road suites do not fall on a continuous magmatic trend. Alternatively, geochemical patterns argue for the possibility that some portions of the Stage Road Suite became mobilized during ultrametamorphism to produce crystal-mush plutons with fairly uniform compositions. Extensive partial melting of relatively heterogeneous Stage Road Suite equivalent rocks would have completely mobilized the mass, including the liquid and residual plagioclase + quartz ± potassium feldspar, capable of accumulation and segregation, or inclusion within emplaced plutonic bodies. These plutons may have experienced crystal segregation and accumulation, consistent with models (Fig. 12), to produce various rock units in the Archer Mountain Suite. Compared with the source, the liquid fraction would have had the higher overall incompatible elements and more uniform REE patterns depicted in the Archer Mountain and Turkey Mountain suites. Residual Stage Road Suite masses would have taken on variable REE signatures, including different heavy REE (Tb to Lu) patterns, dependent on residual mineralogy, and especially REE-compatible residual accessory phases.

Metamorphic Effects on Composition

The chemistry of Grenville-age country rocks may have been affected by emplacement of Neoproterozoic silicic plutons or by subsequent regional metamorphism, as suggested by Evans (1991). Contact metasomatism is evident in the trace-element composition of sample B-12B, an Archer Mountain Suite granite collected within meters of its contact with the 730-Ma Rockfish River pluton (sample B-12A, Fig. 2). The slightly higher K and lower Ba and Sr values in B-12B relative to other Archer Mountain Suite leucocratic granitoids (Figs. 5

and 6, C) are intermediate between the normal Archer Mountain Suite pattern and this silicic pluton. Some of this chemical difference, and especially the significantly lower CaO, TiO_2, and P_2O_5 in B-12B (Fig. 4, C, D, and F, respectively) is probably due to magmatic differentiation. Nonetheless, overall elemental patterns may reflect some interaction between these units near the contact. Notably, exchanges are observed in highly mobile incompatible elements and elements that are compatible in major mineral phases, whereas little or no exchange is apparent for refractory elements, such as the REE. But, whereas minimal assimilative effects are apparent very close to the contact, the absence of similar effects in nearby, slightly more distal samples (B-14A, HP-14, B-73) suggests that such Neoproterozoic metamorphism was significantly more restricted than envisioned by Evans (1991).

A second test of this effect is apparent in mylonitized zones of the massif, regions that might have experienced significant elemental mobility due to pulverization and fluid movement. The mylonitic sample B-2A (Fig. 2) has many chemical similarities to the Turkey Mountain Suite (65), and both are nearly indistinguishable from the rest of the Archer Mountain Suite (Fig. 6, B). Again, this suggests that extensive retrograde metamorphism during the Paleozoic was not induced by vast quantities of sea water infiltrating Grenville-age rocks during Iapetan rifting, as suggested by Evans (1991). The two REE patterns (Fig. 9, B) are easily derived from the normal pattern of the Archer Mountain Suite by simple plagioclase fractionation (± minor phases). We interpret their patterns to represent effects that occurred during magmatic emplacement, with the possibility that late magmatic fluids were filter-pressed and added from adjacent regions.

Alternatively, the variation may be due to granulite-facies metamorphic differentiation, and thus depends on sampling. Significant trace-element variation may be caused by accessory phases, such as zircon, apatite, monazite, and allanite, which strongly fractionate trace elements during solid-state metamorphic differentiation. Such variation will also occur during partial melting, when accessory minerals remain in the residual mass or are selectively melted out of the residue (Bea and Pillar, 1999). However, most of the variation in REE can be explained by magmatic processes, and the effects of accessory minerals during either metamorphic differentiation or low-degree melting would likely yield significant changes in REE and other trace-element patterns.

CONCLUSIONS

Geochemical data for representative samples from the central Lovingston and Pedlar massifs are grossly similar. Post-Grenville, low-grade metamorphism had only minor effects on chemical compositions. Chemical remobilization due to Neoproterozoic intrusion and/or mylonitization along the Rockfish Valley fault was minor and affected only highly mobile elements in samples near their contacts or localized in shear zones.

The Archer Mountain Suite and Turkey Mountain Suite and their associated granitoids and charnockitic plutons are probably related by a common source or protolith that may be represented by the Stage Road Suite. One scenario supported by the geologic and chemical data is that the Stage Road Suite was a deeply buried volcanic or plutonic series and served as the source material for crustal melting during the Grenville orogenesis. The Archer Mountain Suite, Turkey Mountain Suite and related biotite granitoid plutons, and the derivative leucocratic granitoids and charnockite plutons were subsequently intruded into the Stage Road Suite as they became remobilized from deeper crustal regions in the gneissic sequence.

REE patterns and trace-element covariations for the Roseland Anorthosite strongly support magmatic differentiation from crustal melts, although primary mafic melts from deeper regions cannot be ruled out on the basis of the present data. Plagioclase accumulation from an Archer Mountain Suite–like crustal melt is supported by REE patterns in the anorthosite and simple geochemical models of crystal fractionation. Geochemical data support a scenario in which biotite and leucocratic granitoids, anorthosite, and related rocks in the Archer Mountain and Turkey Mountain suites contain a substantial crustal component, most likely represented by the Stage Road Suite.

ACKNOWLEDGMENTS

Chemical analyses for this study were obtained in and supported by the Idaho State University Laboratory for Environmental Geochemistry, Pocatello. Songqiao Chen, analytical technician, performed all the ICP-AES major element analyses that were required to recheck some of the earlier data obtained by various coworkers. Support for INAA neutron irradiations was provided by The Radiation Center at Oregon State University, Corvallis; the U.S. Department of Energy (DOE) Office of New Production Reactors; and Assistant Secretary for Environmental Management under DOE Idaho Operations Office contract DE-AC07-94ID13223. The authors gratefully acknowledge thorough reviews by Jim Beard, Brent Owens, Richard Tollo, and an anonymous reviewer, which significantly improved the quality of the paper and the arguments presented.

REFERENCES CITED

Aleinikoff, J.N., Zartman, R.E., Walter, M., Rankin, D.W., Lyttle, P.T., and Burton, W.C., 1995, U-Pb ages of metarhyolite of the Catoctin and Mount Rogers formations: Evidence for two pulses of Iapetan rifting: American Journal of Science, v. 293, p. 428–454.

Aleinikoff, J.N., Horton, J.W., Jr., and Walter, M., 1996, Middle Proterozoic age for the Montpelier Anorthosite, Goochland terrane, eastern Piedmont, Virginia: Geological Society of America Bulletin, v. 108, p. 1481–1491.

Aleinikoff, J.N., Burton, W.C., Lyttle, P.T., Nelson, A.E., and Southworth, C.S., 2000, U-Pb geochronology of zircon and monazite from Mesoproterozoic granitic gneisses of the northern Blue Ridge, Virginia and Maryland, USA: Precambrian Research, v. 99, p. 113–146.

Anders, E., and Grevesse, N., 1989, Abundances of the elements: Meteoritic and solar: Geochimica et Cosmochimica Acta, v. 53, p. 197–214.

Badger, R.L., and Sinha, A.K., 1988, Age and Sr isotopic signature of the Catoctin volcanic province: Implications for subcrustal mantle evolution: Geology, v. 16, p. 692–695.

Bailey, C.M., and Simpson, C., 1993, Extensional and contractional deformation in the Blue Ridge Province, Virginia: Geological Society of America Bulletin, v. 105, p. 411–422.

Bailey, W.M., 1983, Geology of the northern half of the Horseshoe Mountain quadrangle, Nelson County, Virginia [M.S. thesis]: Athens, Georgia, University of Georgia, 100 p.

Bartholomew, M.J., 1977, Geology of the Greenfield and Sherando quadrangles, Virginia: Virginia Division of Mineral Resources, Publication 1, 43 p.

Bartholomew, M.J., 1981, Geology of the Roanoke and Stewartsville quadrangles, Virginia: Virginia Division of Mineral Resources, Publication 4, 23 p.

Bartholomew, M.J., 1992, Structural characterization of the Late Proterozoic (post-Grenville) continental margin of the Laurentian craton, *in* Bartholomew, M.J., et al., eds., Basement tectonics 8: Characterization and comparison of ancient and Mesozoic continental margins—Proceedings of the 8th International Conference on Basement Tectonics (Butte, Montana, 1988): Dordrecht, The Netherlands, Kluwer Academic Publishers, p. 443–468.

Bartholomew, M.J., and Lewis, S.E., 1984, Evolution of Grenville massifs in the Blue Ridge geologic province, southern and central Appalachians, *in* Bartholomew, M.J., et al., eds., The Grenville event in the Appalachians and related topics: Geological Society of America Special Paper 194, p. 229–254.

Bartholomew, M.J., and Lewis, S.E., 1988, Peregrination of Middle Proterozoic massifs and terranes within the Appalachian orogen, eastern U.S.A: Trabajos de Geología (Spain), v. 17, p. 155–165.

Bartholomew, M.J., and Lewis, S.E., 1992, Appalachian Grenville massifs: Pre-Appalachian translational tectonics, *in* Mason, R., ed., Basement tectonics 7: Dordrecht, The Netherlands, Kluwer Academic Publishers, p. 363–374.

Bartholomew, M.J., Gathright, T.M., II, and Henika, W.S., 1981, A tectonic model for the Blue Ridge in central Virginia: American Journal of Science, v. 281, p. 1164–1183.

Bartholomew, M.J., Force, E.R., Sinha, A.K., and Herz, N., eds., 1984, The Grenville Event in the Appalachians and Related Problems: Geological Society of America Special Paper 194, 287 p.

Bartholomew, M.J., Lewis, S.E., Hughes, S.S., Badger, R.L., and Sinha, A.K., 1991, Tectonic history of the Blue Ridge basement and its cover, central Virginia, *in* Schultz, A., and Compton-Gooding, E., eds., Geologic evolution of the eastern United States, field trip guidebook, NE-SE GSA 1991: Martinsville, Virginia, Virginia Museum of Natural History Guidebook 2, p. 57–90.

Bea, F., and Montero, P., 1999, Behavior of accessory phases and redistribution of Zr, REE, Y, Th, and U during metamorphism and partial melting of metapelites in the lower crust: An example from the Kinzigite Formation of Ivrea-Verbano, NW Italy: Geochimica et Cosmochimica Acta, v. 63, no. 7/8, p. 1133–1153.

Bohlen, S.R., and Mezger, K., 1989, Origin of granulite terranes and the formation of the lowermost continental crust: Science, v. 244, p. 326–329.

Borg, S.C., and DePaolo, D.J., 1994, Laurentia, Australia, and Antarctica as a Late Proterozoic supercontinent: Constraints from isotopic mapping: Geology, v. 22, p. 307–310.

Brock, J.C., 1981, Petrology of the Mobley Mountain granite, Amherst Co., Virginia [M.S. thesis]: Athens, Georgia, University of Georgia, 130 p.

Dalziel, I.W.D., 1991, Pacific margins of Laurentia and East Antarctica-Australia as a conjugate rift pair: Evidence and implications for an Eocambrian supercontinent: Geology, v. 19, p. 598–601.

Dalziel, I.W.D., 1997, Overview: Neoproterozoic-Paleozoic geography and tectonics: Review, hypothesis, environmental speculation: Geological Society of America Bulletin, v. 109, p. 16–42.

Davis, R.G., 1974, Pre-Grenville ages of basement rocks in central Virginia: A model for the interpretation of zircon ages [M.S. thesis]: Blacksburg, Virginia Polytechnic Institute and State University, 46 p.

Easton, R.M., 1986, Geochronology of the Grenville Province, *in* Moore, J.M., et al., eds., The Grenville Province: Geological Association of Canada Special Paper 31, p. 127–173.

Elliot, W.C., and Aronson, J.L., 1987, Alleghanian episode of K-bentonite illitization in the southern Appalachian basin: Geology, v. 15, p. 735–739.

Emslie, R.F., Hamilton, M.A., and Thériault, R.J., 1994, Petrogenesis of a Mid-Proterozoic anorthosite-mangerite-charnockite-granite (AMCG) complex: Isotopic and chemical evidence from the Nain plutonic suite: Journal of Geology, v. 102, p. 539–558.

Evans, M.A., 1989, The structural geometry and evolution of foreland thrust systems, northern Virginia: Geological Society of America Bulletin, v. 101, p. 339–354.

Evans, N.H., 1991, Latest Precambrian to Ordovician metamorphism in the Virginia Blue Ridge: Origin of the contrasting Lovingston and Pedlar basement terranes: American Journal of Science, v. 291, p. 425–452.

Gill, J.B., 1981, Orogenic andesites and plate tectonics: Berlin, Germany, Springer-Verlag, 392 p.

Herz, N., 1969, The Roseland alkalic anorthosite massif, Virginia, *in* Isachsen, Y.W., ed., Origin of anorthosite and related rocks: New York State Museum and Science Service Memoir 18, p. 357–367.

Herz, N., and Force, E.R., 1984, Rock suites in Grenvillian terrane of the Roseland district, Virginia; Part 1. Lithologic relations by Herz and Force; Part 2. Igneous and metamorphic petrology, *in* Bartholomew, M.J., et al., eds., The Grenville Event in the Appalachians and related topics: Geological Society of America Special Paper 194, p. 187–214.

Herz, N., and Force, E.R., 1987, Geology and mineral deposits of the Roseland District of central Virginia: U.S. Geological Survey Professional Paper 1371, 56 p.

Hillhouse, D.N., 1960, Geology of the Piney River–Roseland titanium area, Nelson and Amherst counties, Virginia [Ph.D. dissertation]: Blacksburg, Virginia, Virginia Polytechnic Institute, 129 p.

Huang, W.L., and Wyllie, P.J., 1973, Melting relations of muscovite-granite to 35 kbar as a model for fusion of metamorphosed subducted oceanic sediments: Contributions to Mineralogy and Petrology, v. 42, p. 1–14.

Hughes, S.S., Lewis, S.E., Bartholomew, M.J., Sinha, A.K., Hudson, T.A., and Herz, N., 1997, Chemical diversity and origin of Precambrian charnockitic rocks of the central Pedlar massif, Grenvillian Blue Ridge terrane, Virginia: Precambrian Research, v. 84, p. 37–62.

Le Maitre, R.W., Bateman, P., Dudek, A., Keller, J., Lameyre, J., Le Bas, M.J., Sabine, P.A., Schmid, R., Sorensen, H., Streckeisen, A., Woolley, A.R., and Zanettin, B., eds., 1989, A classification of igneous rocks and glossary of terms: Recommendations of the International Union of Geological Sciences subcommission on the systematics of igneous rocks: Oxford, UK, Blackwell Scientific Publications, 193 p.

McLelland, J., Chiarenzilli, J., Whitney, P., and Isachsen, Y., 1988, U-Pb geochronology of the Adirondack Mountains and implications for their geologic evolution: Geology, v. 16, p. 920–924.

McLelland, J.M., Ashwal, L., and Moore, L., 1994, Composition and Petrogenesis of oxide-, apatite-rich gabbronorites associated with Proterozoic anorthosite massifs: Examples from the Adirondack Mountains, New York: Contributions to Mineralogy and Petrology, v. 116, p. 225–238.

McLelland, J., Daly, J.S., and McLelland, J.M., 1996, The Grenville orogenic cycle (ca. 1350–1000 Ma): An Adirondack perspective: Tectonophysics, v. 265, p. 1–28.

McLelland, J., Hamilton, M., Selleck, B., McLelland, J.M., Walker, D., and Orrell, S., 2001, Zircon U-Pb geochronology of the Ottawan orogeny, Adirondack Highlands, New York: Regional and tectonic implications: Precambrian Research, v. 109, p. 39–72.

Mezger, K., Essene, E.J., van der Pluijm, B.A., and Halliday, A.N., 1993, U-Pb geochronology of the Grenville orogen of Ontario and New York: Constraints on ancient crustal tectonics: Contributions to Mineralogy and Petrology, v. 114, p. 13–26.

Miller, B.V., and Barr, S.M., 2000, Petrology and isotopic composition of a Grenvillian basement fragment in the northern Appalachian orogen: Blair

River inlier, Nova Scotia, Canada: Journal of Petrology, v. 41, no. 12, p. 1777–1804.

Mitra, G., and Elliott, D., 1980, Deformation of basement in the Blue Ridge and the development of the South Mountain cleavage, *in* Wones, D.R., ed., Caledonides in the USA: Blacksburg, Virginia Polytechnic Institute and State University Department of Geological Sciences Memoir 2, p. 307–311.

Mitra, G., and Lukert, M.T., 1982, Geology of the Catoctin–Blue Ridge anticlinorium in northern Virginia, *in* Little, P.T., ed., Central Appalachian geology, NE-SE GSA 1982 field trip guidebook: Washington, D.C., American Geological Institute, p. 83–108.

Moore, C.H., Jr., 1940, Geology and mineral resources of the Amherst quadrangle, Virginia [Ph.D. dissertation]: Ithaca, New York, Cornell University, 92 p.

Moores, E.M., 1991, Southwest U.S.–east Antarctic (SWEAT) connection: A hypothesis: Geology, v. 19, p. 425–428.

Owens, B.E., and Dymek, R.F., 1999, A geochemical reconnaissance of the Roseland Anorthosite complex, Virginia, and comparisons with andesine anorthosites from the Grenville Province, Quebec, *in* Sinha. A.K., ed., Basement tectonics, v. 13: Dordrecht, The Netherlands, Kluwer Academic Publishers, p. 217–232.

Pearce, J.A., Harris, N.B.W., and Tindle, A.G., 1984, Trace element discrimination diagrams for the tectonic interpretation of granitic rocks: London, Geological Society Special Publication, v. 7, p. 14–24.

Peck, W.H., and Valley, J.W., 2000, Large crustal input to high $\delta^{18}O$ anorthosite massifs of the southern Grenville Province: New evidence from the Morin Complex, Quebec: Contributions to Mineralogy and Petrology, v. 139, p. 402–417.

Pettingill, H.S., Sinha, A.K., and Tatsumoto, M., 1984, Age and origin of anorthosites, charnockites, and granulites in the central Virginia Blue Ridge—Nd and Sr isotopic evidence: Contributions to Mineralogy and Petrology, v. 85, p. 279–291.

Rankin, D.W., 1975, The continental margin of eastern North America in the southern Appalachians: The opening and closing of the proto-Atlantic Ocean: American Journal of Science, v. 275-A, p. 298–336.

Rankin, D.W., 1976, Appalachian salients and recesses: Late Precambrian continental breakup and the opening of the Iapetus Ocean: Journal of Geophysical Research, v. 81, p. 5605–5619.

Rollinson, H.R., 1993, Using geochemical data: Evaluation, presentation, interpretation: London, Longman Group UK, 352 p.

Ross, C.S., 1941, Occurrence and origin of the titanium deposits of Nelson and Amherst counties, Virginia: U.S. Geological Survey Professional Paper 198, 59 p.

Rudnick, R.L., and Fountain, D.M., 1995, Nature and composition of the continental crust: A lower crustal perspective: Reviews of Geophysics, v. 33, no. 3, p. 267–309.

Sinha, A.K., and Bartholomew, M.J., 1984, Evolution of the Grenville terrane in the central Virginia Appalachians, *in* Bartholomew, M.J., et al., eds., The Grenville Event in the Appalachians and related topics: Geological Society of America Special Paper 194, p. 175–186.

Streckeisen, A., 1974, How should charnockitic rocks be named? *in* Centenaire de la Société Géologique de Belgique, Géologie des Domaines Cristallins, Liège, p. 349–360.

Taylor, S.R., and McLennan, S.M., 1995, The geochemical evolution of the continental crust: Reviews of Geophysics, v. 33, no. 2, p. 241–265.

Tollo, R.P., and Aleinikoff, J.N., 1996, Petrology and U-Pb geochronology of the Robertson River Igneous Suite, Blue Ridge Province, Virginia—Evidence for multistage magmatism associated with an early episode of Laurentian rifting: American Journal of Science, v. 296, p. 1045–1090.

Tollo, R.P., and Aleinikoff, J.N., 2001, Timing of Grenville-age magmatism and deformation, Blue Ridge province, central Virginia: Geological Society of America Abstracts with Programs, v. 33, no. 2, p. 7.

Tollo, R.P., and Arav, S., 1992, The Robertson River Igneous Suite (Blue Ridge Province, Virginia)—Late Proterozoic anorogenic (A-type) granitoids of unique petrochemical affinity, *in* Bartholomew, M.J., et al., eds., Basement tectonics 8: Characterization and comparison of ancient and Mesozoic continental margins—Proceedings of the 8th International Conference on Basement Tectonics (Butte, Montana, 1988): Dordrecht, The Netherlands, Kluwer Academic Publishers, p. 425–441.

Tollo, R.P., and Hutson, F.E., 1996, 700 Ma rift event in the Blue Ridge province of Virginia: A unique time constraint on pre-Iapetan rifting of Laurentia: Geology, v. 24, p. 59–62.

Tollo, R.P., and Lowe, T.K., 1994, Geologic map of the Robertson River Igneous Suite, Blue Ridge Province, northern and central Virginia: U.S. Geological Survey, Miscellaneous Field Studies Map MF-2229, 1:100,000 scale.

Tollo, R.P., Aleinikoff, J.N., Bartholomew, M.J., and Rankin, D.W., 2004, Neoproterozoic A-type granitoids of the central and southern Appalachians: Intraplate magmatism associated with episodic rifting of the Rodinian supercontinent: Precambrian Research, v. 128, no. 1, p. 3–38.

Watson, T.L., and Tabor, S., 1913, Geology of the titanium and apatite deposits of Virginia: Virginia Division of Mineral Resources Bulletin 3A, 308 p.

White, A.J.R., and Chappell, B.W., 1977, Ultrametamorphism and granitoid genesis: Tectonophysics, v. 43, p. 7–22.

Manuscript Accepted by the Society August 25, 2003

Printed in the USA

Geological Society of America
Memoir 197
2004

Compositional zoning of a Neoproterozoic ash-flow sheet of the Mount Rogers Formation, southwestern Virginia Blue Ridge, and the aborted rifting of Laurentia

Steven W. Novak
Evans East, 104 Windsor Center Drive, Suite 101, East Windsor, New Jersey 08520, USA
Douglas W. Rankin*
U.S. Geological Survey, MS 926A, National Center, Reston, Virginia 20192, USA

ABSTRACT

The 760 Ma Wilburn Rhyolite Member of the Mount Rogers Formation, in southwestern Virginia, is a mineralogically and compositionally zoned welded ash-flow sheet at least 660 m thick. Compositional zoning, preserved despite greenshist-facies metamorphism, developed in the pre-eruptive magma chamber and was inverted during eruption of the ash-flow sheet. Microphenocrysts of aegerine and riebeckite occur at the base of the sheet, riebeckite alone or with biotite at higher levels, and Fe-rich biotite at the top. Alkali feldspar phenocrysts are more potassic, and riebeckite and biotite exhibit decreasing Mg/Fe toward the top of the ash-flow sheet; F content of biotite increases toward the base. The ash-flow tuff is a high-silica rhyolite (SiO_2 = 76.6 wt%); the basal one-sixth of the sheet is interpreted as originally peralkaline, whereas the remainder is metaluminous. Major- and trace-element zoning within the sheet is similar to other well-documented ash-flow sheets: SiO_2, Na_2O, and F increase toward the base of the sheet, whereas Al_2O_3, MgO, CaO, K_2O, and TiO_2 decrease. Concentrations of Be, Rb, Zr, Nb, Sn, Hf, Ta, Th, U, Tb, and Yb increase toward the base; elements more abundant toward the top include Sc, Sr, Ba, La, Ce, Nd, and Eu. These gradients developed in a high-level silicic magma chamber in which peralkaline high-silica rhyolitic magma overlay metaluminous high-silica rhyolite. The 765–740 Ma Crossnore Complex, which includes the Mount Rogers Formation, in the Grenvillian French Broad massif includes A-type granitoids and is interpreted to reflect aborted rifting of Laurentian continental crust. The locus of aborted rifting migrated northeastward to the Shenandoah massif, where A-type magmatism occurred at 735–680 Ma. Continental breakup and opening of the Iapetus Ocean followed Late Neoproterozoic (ca. 572–554 Ma) rifting. Formation of a voluminous high-silica, partly peralkaline magma chamber indicates that the initial pulse of rifting took place in relatively thick continental crust, perhaps explaining why this pulse did not culminate in continental breakup until ~200 m.y. later.

Keywords: ash-flow tuff, peralkaline, rhyolite, rifting, Appalachians

*Corresponding author: dwrankin@usgs.gov.

Novak, S.W., and Rankin, D.W., 2004, Compositional zoning of a Neoproterozoic ash-flow sheet of the Mount Rogers Formation, southwestern Virginia Blue Ridge, and the aborted rifting of Laurentia, *in* Tollo, R.P., Corriveau, L., McLelland, J., and Bartholomew, M.J., eds., Proterozoic tectonic evolution of the Grenville orogen in North America: Boulder, Colorado, Geological Society of America Memoir 197, p. 571–600. For permission to copy, contact editing@geosociety.org.

INTRODUCTION

The Wilburn Rhyolite Member of the Neoproterozoic Mount Rogers Formation in southwestern Virginia is a zoned, welded ash-flow sheet (Novak and Rankin, 1980; Rankin, 1993). It is at least 660 m thick and represents the product of culminating eruptions from the Mt. Rogers volcanic center in the upper part of the Mount Rogers Formation. The great thickness of rhyolitic ash-flow tuff at Mt. Rogers is interpreted to represent the remnants of caldera fill, only part of which is preserved in the thrust sheets that carry the formation. Two sampled stratigraphic sections through the ash-flow sheet indicate chemical and mineralogical zoning that reflects progressive eruption of a compositionally zoned magma chamber. The primary purpose of this paper is to examine the nature and interrelationships of the chemical and mineral changes within the sheet and to use this information to reconstruct the Mt. Rogers magma chamber prior to eruption of the Wilburn Rhyolite Member. In addition, the tectonic setting of the Mount Rogers Formation is reviewed in light of recent work.

LOCAL GEOLOGICAL SETTING

The Mount Rogers Formation, which crops out in southwestern Virginia and adjacent North Carolina and Tennessee on the northwest flank of the allochthonous French Broad massif of the Blue Ridge province (Fig. 1), is composed of ~70% bimodal volcanic rocks and 30% clastic sedimentary rocks (Rankin, 1993). Rocks of the formation have undergone low-grade metamorphism, probably during the Ordovician Taconian orogeny (Laird, 1989; Adams et al., 1995); the prefix "meta-" is assumed to apply to all rocks. Rhyolite makes up 50–60% of the formation and is concentrated in three thick masses interpreted to be the sites of volcanic centers that erupted 500–1000 km^3 of rhyolite. Rocks of the geographically central site (Fig. 2), called the "Mt. Rogers volcanic center," are stratigraphically above basalt (greenstone), which is interlayered with conglomerate, sandstone, pelite, and lesser amounts of rhyolite of the lower part of the formation. In the area of the Mt. Rogers volcanic center, the Mount Rogers Formation is ~3000 m thick.

The Mt. Rogers volcanic center includes metaluminous rhyolite in three stratigraphic members (Rankin, 1993). The basal Buzzard Rock Member consists of low-silica rhyolite lava flows (~300 m thick) containing 5–20% phenocrysts of perthitic alkali feldspar and plagioclase in subequal amounts. The middle Whitetop Rhyolite Member consists of ~750 m of high-silica rhyolite, mostly lava flows, containing up to 10% phenocrysts of quartz and perthitic alkali feldspar occurring in varying proportions. The upper Wilburn Rhyolite Member consists of high-silica welded ash-flow tuff, typically containing 30% quartz and perthitic alkali feldspar (hereafter simply called "perthite") phenocrysts in the main body of the unit. Each succeeding younger rhyolite member locally contains clasts of the underlying units, as well as clasts of granitoid rocks from the underlying Mesoproterozoic crystalline basement. Previous study of the chemical compositions of these rocks indicates that the Wilburn could have been derived from magma typical of the basal Buzzard Rock lava flows through fractionation of ~30% potassium feldspar phenocrysts (Lopez-Escobar, 1972; Rankin et al., 1974). There is no simple relationship between the composition of the Buzzard Rock and the Whitetop, nor between that of the Whitetop and the Wilburn. Field relations indicate that all three members were erupted at the same center.

Aleinikoff et al. (1995) obtained a discordant U-Pb age of 759.4 ± 7.1 Ma on zircons from the Whitetop Rhyolite Member. This age supersedes an older determination of 810 Ma (Rankin et al., 1969) for the rhyolites of the Mt. Rogers volcanic center, but continues the assignment of a Neoproterozoic age to the formation in the time scale of Plumb (1991). The Mount Rogers Formation is part of the Crossnore Complex (formerly the Crossnore plutonic-volcanic group of Rankin, 1970) that includes the Grandfather Mountain Formation and a bimodal suite of plutons and dikes that intrude Mesoproterozoic Grenvillian basement rocks of the French Broad massif of northwestern North Carolina and adjacent Virginia (Rankin, 1993) (Fig. 1). Some of the felsic plutons are peralkaline; most, however, are metaluminous (Rankin, 1975; Tollo et al., 2004). Granitoids of the complex, as well as rhyolite of the Grandfather Mountain Formation, have been dated at ca. 765–740 Ma (U-Pb ages) by Su et al., (1994) and Fetter and Goldberg (1995). Mafic dikes of the Bakersville dike swarm are included in the complex and have a whole-rock Rb-Sr isochron age of 734 ± 26 Ma (Goldberg et al., 1986).

The Mount Rogers Formation nonconformably overlies Mesoproterozoic (~1.1–1.3 Ga) crystalline basement that forms the allochthonous French Broad massif in the Blue Ridge province. At least some of the crystalline rocks of the massif experienced Mesoproterozoic granulite-facies metamorphism. This largely granitoid basement is interpreted to be an outlier within the Appalachian orogen of the Grenville province of the Canadian Shield (Rankin, 1976). Clasts of basement granitoids are common in the conglomerates and locally within all volcanic units of the Mount Rogers Formation. The basement surface on which the Mount Rogers Formation was deposited probably had considerable relief, because nearly every unit in the formation is in local stratigraphic contact with the crystalline basement.

The Mount Rogers Formation is the local base of a cover sequence on Grenvillian basement that, except for some significant hiatuses in the lower part, is mostly continuous through the Lower Ordovician. The first record of orogenic deformation and metamorphism of the cover sequence is the Middle to Late Ordovician Taconian orogeny. The Mount Rogers Formation is paraconformably overlain by the glaciogenic Konnarock Formation (Fig. 2) (Rankin, 1993; Rankin et al., 1994).

These glaciogenic deposits, which are about 1 km thick, were formerly included within the Mount Rogers Formation, but were defined as a separate formation by Rankin (1993), who

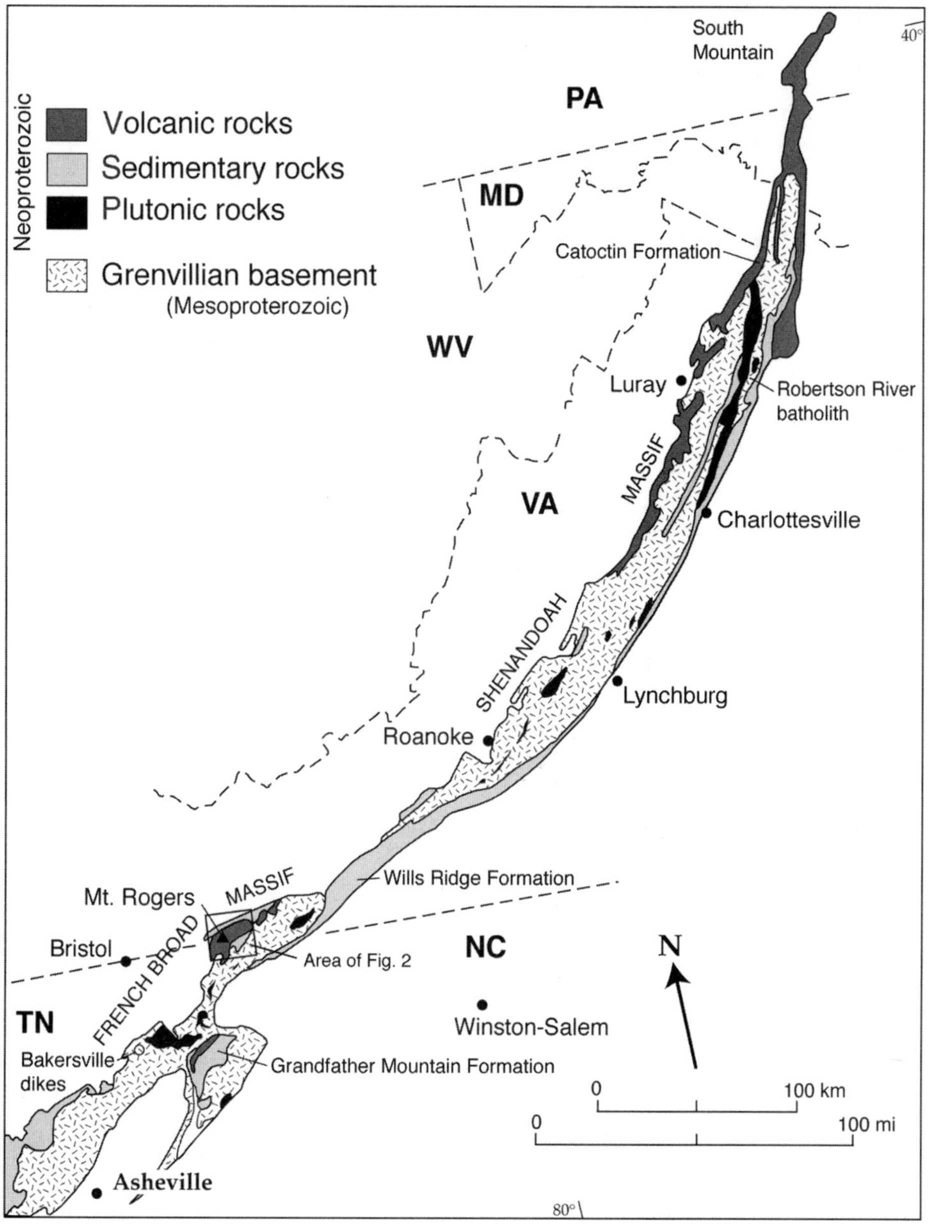

Figure 1. Generalized geologic map of the Blue Ridge province between Asheville, North Carolina, and South Mountain, Pennsylvania, showing Grenvillian massifs, Neoproterozoic rocks, location of the Mt. Rogers area, and the area of Figure 2. Modified from Figure 2 of Tollo et al. (2004).

also discussed the complications of interpreting the Mount Rogers–Konnarock contact. The Konnarock is, in turn, paraconformably overlain by the Late Neoproterozoic and Early Cambrian Unicoi Formation, within which is recorded the rift-to-drift transition of the opening of the Iapetus Ocean. The Tommotian(?) marine trace fossil *Rusophycus* has been described from the local Unicoi Formation ~400 m stratigraphically above the Konnarock Formation (Simpson and Sundberg, 1987). Thin basalt flows occur within the lower part of the Unicoi ~145 m stratigraphically beneath the fossil horizon.

Northwest-directed thrust faulting during the late Paleozoic Alleghanian orogeny complicates interpretations of the geology of the Mt. Rogers area. The Mount Rogers Formation is carried by several slices of the Blue Ridge crystalline thrust sheet and is interpreted to crop out also in the Mountain City window eroded through that thrust sheet (Fig. 2). Most of the Mt. Rogers volcanic center, including outcrops of the Wilburn Rhyolite Member that are the focus of our study, is in the same slice of the Blue Ridge crystalline thrust sheet—the Stone Mountain slice—that carries the bulk of the French Broad massif in southwestern Virginia and adjacent North Carolina. In the Stone Mountain slice, the Wilburn Rhyolite Member is the youngest unit of the Mount Rogers; no stratigraphic top to the formation is preserved in the Stone Mountain slice.

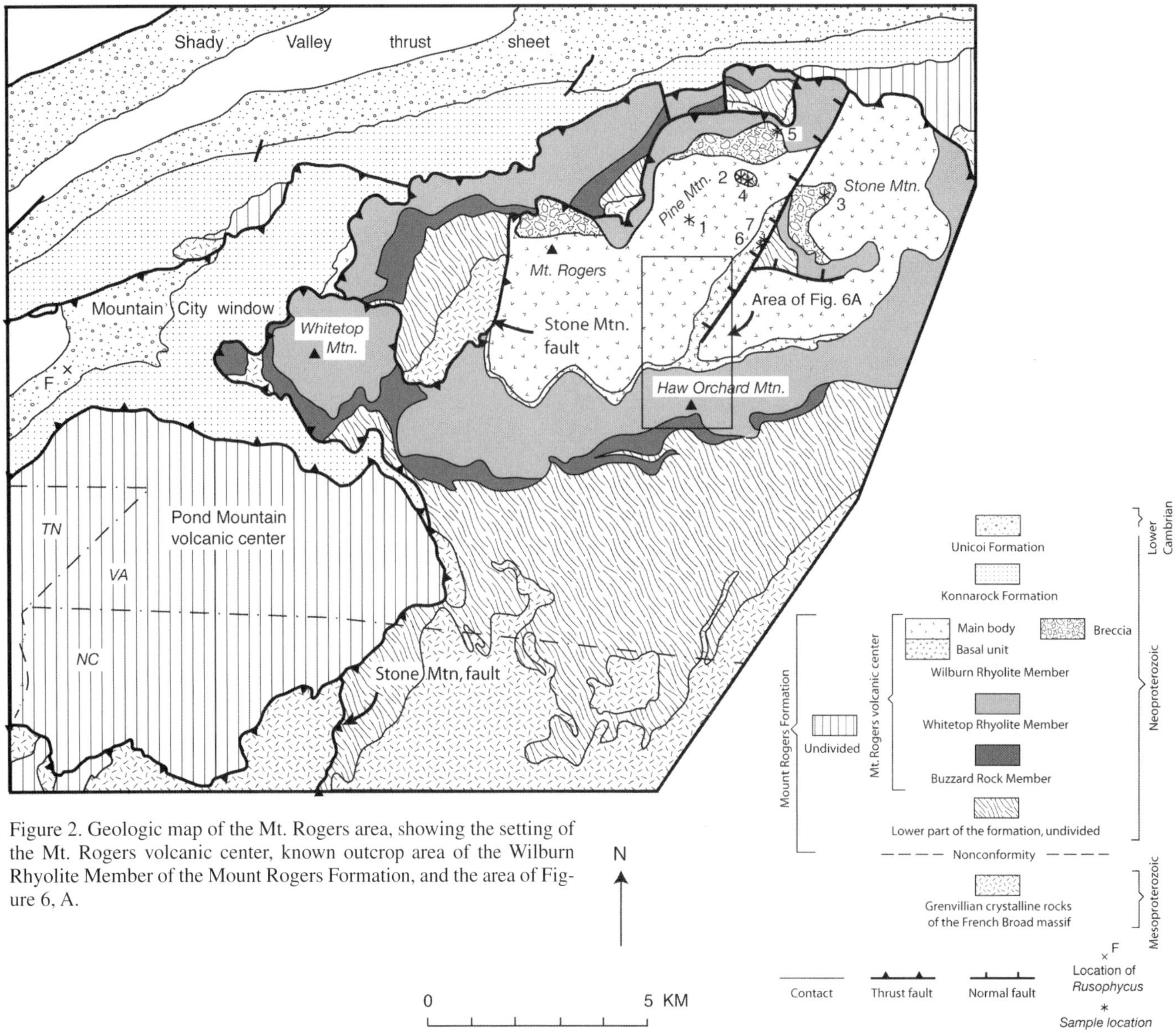

Figure 2. Geologic map of the Mt. Rogers area, showing the setting of the Mt. Rogers volcanic center, known outcrop area of the Wilburn Rhyolite Member of the Mount Rogers Formation, and the area of Figure 6, A.

REGIONAL GEOLOGIC SETTING

Rankin (1976) pointed out that (1) mafic volcanic rocks are common in the Neoproterozoic cover sequence overlying the Grenvillian en echelon external massifs in the Appalachian orogen, (2) intermediate composition volcanic rocks are absent or rare, and (3) felsic volcanic rocks are largely restricted to sites at major bends in Appalachian structural trends. Rankin (1976) suggested that the major bends were inherited from the jagged, rifted margin of Laurentia formed during the opening of the Iapetus Ocean; he argued that the rhyolites were generated at hot spots under triple junctions of intersecting continental rifts. The rhyolites of the Mount Rogers Formation would be one such concentration of felsic volcanic rocks located at a major bend in the trends of the external massifs (called the "Blue-Green-Long axis" by Rankin, 1976). In the terminology of Thomas (1991) this bend, convex toward the craton, would reflect an embayment in the Neoproterozoic Laurentian continental margin.

Several recent developments require revision of these interpretations. It is now recognized that continental rifting in eastern North America occurred in at least two pulses separated by >100 m.y. The first, 765–680 Ma, which generated the Wilburn Rhyolite Member, did not proceed to continental separation; the second pulse, 577–550 Ma, did (Badger and Sinha, 1988; Rankin, 1994; Aleinikoff et al., 1995; Walsh and Aleinikoff, 1999; Cawood et al., 2001; Tollo et al., 2004). The two pulses spatially overlapped between North Carolina and northern Virginia, with older volcanism, represented by the Mount Rogers Formation, dominating in the south and younger volcanism, represented by the Catoctin Formation (Fig. 1), dominating in the north. The thin basalt flows in the Unicoi Formation probably represent the younger pulse of rifting in the Mt. Rogers area. Igneous rocks associated with the younger pulse of rifting occur along the Blue-Green-Long axis at least as far northeast as the Skinner Cove Formation of western Newfoundland (550.5 +3/–2 Ma; Cawood et al., 2001).

Recent work has demonstrated that the older pulse of rifting included a significant episode of Neoproterozoic felsic A-type plutonism and minor (as preserved) volcanism within the allochthonous Shenandoah massif (Fig. 1). U-Pb zircon ages of ten of these bodies, including the Robertson River batholith, range between 735 and 680 Ma (Tollo and Aleinikoff, 1996; Tollo et al., 2004). Tollo et al. (2004) conclude that these granitoid magmas were emplaced as dikelike, vertical sheets at shallow levels in continental crust undergoing active extension. Zircon data for the Blue Ridge province suggest that the age of rift-related magmatism became younger northeastward from ca. 765–740 Ma in the French Broad massif to ca. 735–680 Ma in the Shenandoah massif.

Dated igneous rocks related to the second rifting episode occur along the Blue-Green-Long axis between the northern end of the Blue Ridge in Virginia and western Newfoundland, as well as in the Ottawa graben and north of Sept Îles, Québec, north of the St. Lawrence River (Fig. 7 *in* Cawood et al., 2001). Zircon data indicate ages between 577 and 550 Ma for these rocks. Between North Carolina and southern Québec, no rift-related igneous rocks have yielded ages in the interval 680–577 Ma, suggesting a gap of ~100 m.y. between the two episodes of Neoproterozoic rifting.

Aleinikoff et al. (1995) and Tollo et al. (2004) suggested that the simplistic model of Rankin (1976) of embayments and promontories, reflecting triple junctions with rhyolite concentrated at the triple junctions, also required reconsideration in light of new information. The two major concentrations of rhyolite at Mt. Rogers and South Mountain are nearly 200 m.y. different in age and were not produced during the same rifting event (Aleinikoff et al., 1995). However, the concept of an irregularly rifted Laurentian margin from Alabama (or Texas; see Thomas, 1991) to Newfoundland still seems to be valid. Some of the best evidence for the timing and location of the rift-drift transition comes from the Mt. Rogers area. The presence of large centers of the older felsic volcanism (Mount Rogers Formation) in the same area suggests that the two pulses of rifting were spatially close here, possibly indicating tectonic heredity. Recent delineation of the Robertson River Igneous Suite and related rocks in the Shenandoah massif indicates that it is no longer tenable to suppose that silicic magmatism is concentrated at the triple junctions.

Finally, any reconstruction of the paleogeography of the Blue-Green-Long axis must be approximate at best. Rankin et al. (1991) estimated that the amount of Alleghanian crustal shortening in the vicinity of the Grandfather Mountain window and Mt. Rogers is on the order of 300 km.

DESCRIPTION OF THE WILBURN RHYOLITE MEMBER

The ash-flow origin of the Wilburn Rhyolite Member of the Mount Rogers Formation is clearly demonstrated in outcrops and thin sections. Devitrified and flattened pumice shards are visible in numerous thin sections (Fig. 3, A and B). Crude columnar joints and fiamme (flattened pumice lumps) are present throughout the member, and possible fossil fumaroles are locally present. The columnar joints have been modified by Paleozoic tectonism but are clearly recognizable (Fig. 4, A). Preserved features indicate that the original ash-flow sheet was highly compacted. In the main body of the sheet, fiamme range in length from 1 or 2 cm to a maximum of ~20 cm (or, rarely, 30 cm) and have a flattening ratio of ~10:1 (Fig. 4, B). No systematic or cyclic variation in flattening ratio or size of fiamme was observed through the main body of the sheet. Such a variation, however, could be missed in the field, because fiamme are not visible in many outcrops, presumably because the rocks are recrystallized. Irregularly shaped, calcite-filled vugs, arranged in crudely columnar zones perpendicular to the compaction foliation, are interpreted to be remnants of fossil fumaroles (Fig. 4, C).

In hand specimens, unweathered welded tuff is characterized by an aphanitic, flinty groundmass, typically very dusky purple (5 P 2/2 of the rock color chart, Goddard et al., 1970). The weathered surface is pale purple (5 P 6/2) or pale pink (5 RP 8/2). Textural features, such as fiamme and spherulites, are visible only on weathered surfaces. The main body of the Wilburn Rhyolite Member contains ~30% quartz (typically embayed) and pink perthite phenocrysts. The phenocrysts are mostly 2–3 mm across. Some perthite phenocrysts are as large as 8 mm, and some quartz crystals are as large as 5 mm. Typically, perthite is more abundant than quartz by a ratio of ~3:2; however, in many rocks, the two are present in subequal amounts. The basal few tens of meters (commonly ~30 m) of the Wilburn Rhyolite Member are different from the main body in field appearance, mineral assemblage, and chemical composition. Phenocrysts in this lower, more alkalic part of the sheet are smaller and fewer (typically ~10%) than in the main body. Weathered outcrops in this zone show a mottling of small (1 mm) black spots that are concentrations of microlites of sodic amphibole and opaque oxides. Fiamme tend to be smaller and more compacted. Locally, the fiamme are so flattened that the eutaxitic structure is indistinguishable from flow layering, suggesting postdeposition flow (rheomorphism) in the basal part of the ash-flow sheet. Rheomorphic features are commonly observed in peralkaline ash flows and are caused by the extremely low viscosities of such alkali-rich magmas (Schminke, 1975; Mahood, 1984). In the field, the basal part of the Wilburn is gradational over a few meters into the main body of the ash-flow sheet.

Lithic clasts are common in the basal zone of the Wilburn Rhyolite Member south and southeast of Mt. Rogers (Fig. 2) and also occur locally in the main body of the ash-flow sheet. Lithic fragments in the basal ~10 m of the Wilburn, in places, constitute almost 50% of the rock. Flattened and deformed fiamme are seen between the lithic fragments. The lithic clasts include Whitetop Rhyolite and Buzzard Rock Members, coarsely porphyritic rhyolite, granite, and gneissic granite. Typically Whitetop clasts are the most abundant (up to ~75% of the clasts and as large as 35 cm × 50 cm), followed in abundance by clasts of granitic basement. Clearly, the basal Wilburn ash flows overran a rubbly surface of the Whitetop but also carried fragments of other units, including basement, probably derived from the conduit walls. Zones rich in lithic clasts (Fig. 4, D) higher up in the

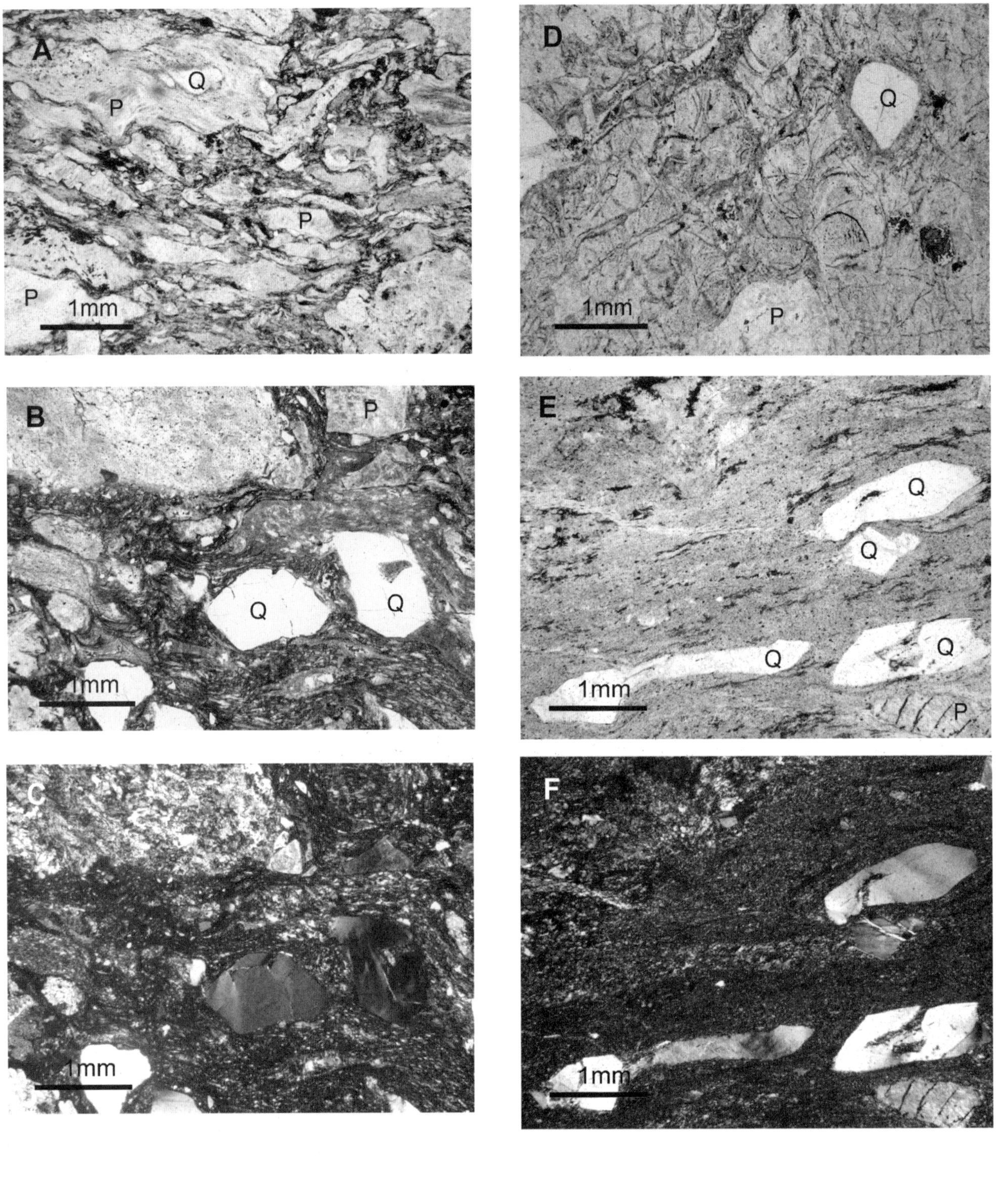

Wilburn probably mark distinct flow units but were not traced in the field.

Distinctive, massive breccia, cropping out over areas as large as 2 km × 0.8 km on the north slopes of Mt. Rogers and Pine Mountain (Fig. 2), suggest the presence of a caldera. The matrix of the breccia is quartz- and perthite-phyric tuff of the Wilburn Rhyolite Member. Neither aegerine nor riebeckite is observed in the matrix. Devitrified shards can be seen in the matrix in some thin sections. Breccia clasts include lava and flow breccia of the Whitetop Rhyolite Member in fragments ranging in diameter from a few millimeters to several meters, as well as welded tuff of the Wilburn, and, rarely, granite. In most outcrops, the Whitetop blocks are the dominant clasts and may constitute 75–80% of the rock (Fig. 4, E). Small areas of ash-flow tuff lacking abundant breccia fragments are present in the area mapped as breccia on Figure 2.

Less densely packed breccia containing clasts of Whitetop Rhyolite in a matrix of ash-flow tuff of the Wilburn crops out in the summit area of Stone Mountain (Fig. 4, F). Clasts commonly constitute ~30% of the outcrops and are as large as 7.5 m × 4.5 m. Fiamme, commonly quite large, are present between the clasts. The ash flow at this locality is interpreted to have overrun a surface littered with blocks of Whitetop Rhyolite. In places, compaction foliation, as defined by the fiamme, warp around the blocks of Whitetop Rhyolite, suggesting that some of the blocks were transported by the ash flow and that the compaction foliation was bent by further compaction after the flow came to rest. The matrix of the ash-flow tuff was at least party vitric, as indicated by the presence of spherulites and perlitic cracking (Fig. 3, D).

Much of the Wilburn Rhyolite Member was penetratively deformed during regional metamorphism. A tectonic foliation is clearly visible in some outcrops, but obscure in others. Regionally, the foliation strikes northeast and dips moderately southeast. Phenocrysts are sheared (Fig. 3, E and F); quartz commonly shows "ribbon extinction." In some samples in which the tectonic foliation is identifiable in thin sections, perthite phenocrysts are tectonically broken (Fig. 3, E and F) and the pieces are separated, indicating extension in the plane of tectonic foliation. The pull-apart fractures are typically filled with quartz, opaque minerals, fluorite, locally calcite, and in the stratigraphically lowest samples, aegerine (prisms as large as 0.1 mm × 0.3 mm) (Fig. 5, A and B). Finally, pressure shadows between phenocrysts, in places, consist of coarser grained intergrowths of quartz, feldspar, and riebeckite, which are oriented parallel to the tectonic foliation. A down-dip mineral lineation, most obvious as elongated perthite phenocrysts, is observed locally. Lithophysae in the Whitetop Rhyolite Member below the base of the Wilburn south of the Massie Gap section (see later in this paper) are flattened in the plane of foliation and elongated down-dip (Fig. 5, C *in* Rankin, 1993). Microlites of opaque minerals and riebeckite are aligned with the tectonic foliation, which is commonly at an angle to the compaction foliation (Fig. 3, E and F). In some outcrops, the tectonic foliation is axial planar to crinkling of the fiamme (Fig. 6, D *in* Rankin, 1993).

SAMPLE SUITE

A suite of twenty-one samples was collected from two sections through the Wilburn Rhyolite Member on Wilburn Ridge, located in the southwestern corner of the Whitetop Mountain 7.5-minute quadrangle (Fig. 6, A; Table 1). The Massie Gap section trends north-northwest on the southern extension of Wilburn Ridge toward Haw Orchard Mountain from an elevation of ~1495 m (4900 ft) to ~1555 m (5100 ft) over a map distance of ~1.0 km. The Québec Branch section trends northwest on the east slope of Wilburn Ridge from an elevation of ~1430 m (4700 ft) near Québec Branch to ~1660 m (5450 ft) near the southern summit of Wilburn Ridge, a map distance of ~1.2 km. Sample elevations are given in feet as well as meters for ease in reference to the contours in feet on the Whitetop Mountain quadrangle.

The Massie Gap section starts at the basal contact of the Wilburn Rhyolite Member with the underlying Whitetop Rhyolite Member and extends stratigraphically upward through the Wilburn as high as compaction foliation (eutaxitic structure) was visible and of a consistent direction of dip. The base of the Wilburn is not exposed in the Québec Branch section, but the section extends from the topographically lowest exposed rock on the northwest side of Québec Branch stratigraphically upward as high as the compaction foliation was visible and indi-

Figure 3. Photomicrographs of the Wilburn Rhyolite Member, Mount Rogers Formation. (A) Welded tuff, showing flattened and devitrified shards, a pumice lump (upper left), and broken phenocrysts of quartz (Q) and perthitic alkali feldspar (P). Dark areas are opaque minerals. The pumice lump contains a quartz and perthite phenocryst. Sample WS4-25, photographed in plane-polarized light (not part of the analyzed suite; location 1 in Fig. 2). (B) Welded tuff, showing flattened and devitrified shards, and phenocrysts of embayed quartz (Q) and broken perthite (P). Object in the upper left consists of extremely fine-grained intergrown quartz and feldspar and is interpreted as crystallized pumice. Shards are outlined by opaque dust. Sample TC1-15, photographed in plane polarized light (not part of the analyzed suite; location 2 in Fig. 2). (C) Same area as in (B), viewed with crossed polarizers. (D) Perlitic cracking in the interior of a fiamme in ash-flow tuff matrix to block breccia. Phenocrysts of quartz (Q), perthite (P), and allanite (A). Most of the thin section shows a pronounced tectonic foliation. Chlorite and sericite are in the perlitic fractures. Sample TC1-10a, photographed in plane-polarized light (not from the analyzed suite; from summit area of Stone Mountain, near location of outcrop shown in Fig. 4, B; location 3 in Fig. 2). (E) Ash-flow tuff, showing recrystallized groundmass and phenocrysts of strongly deformed, embayed quartz (Q) and fractured perthite (P). Fractures in perthite are filled with quartz and opaque dust. Dark patches and streaks are extremely fine-grained (not resolvable in this photomicrograph) clusters of riebeckite and opaque minerals. Streaks of dark minerals define the tectonic foliation, which, in the thin section as a whole, is at an angle of ~25° to flattening foliation defined by axiolites. Sample WS4-45 (S), photographed in plane-polarized light (sample of Québec Branch section, Fig. 6, A). (F). Same area as (E), viewed with crossed polarizers.

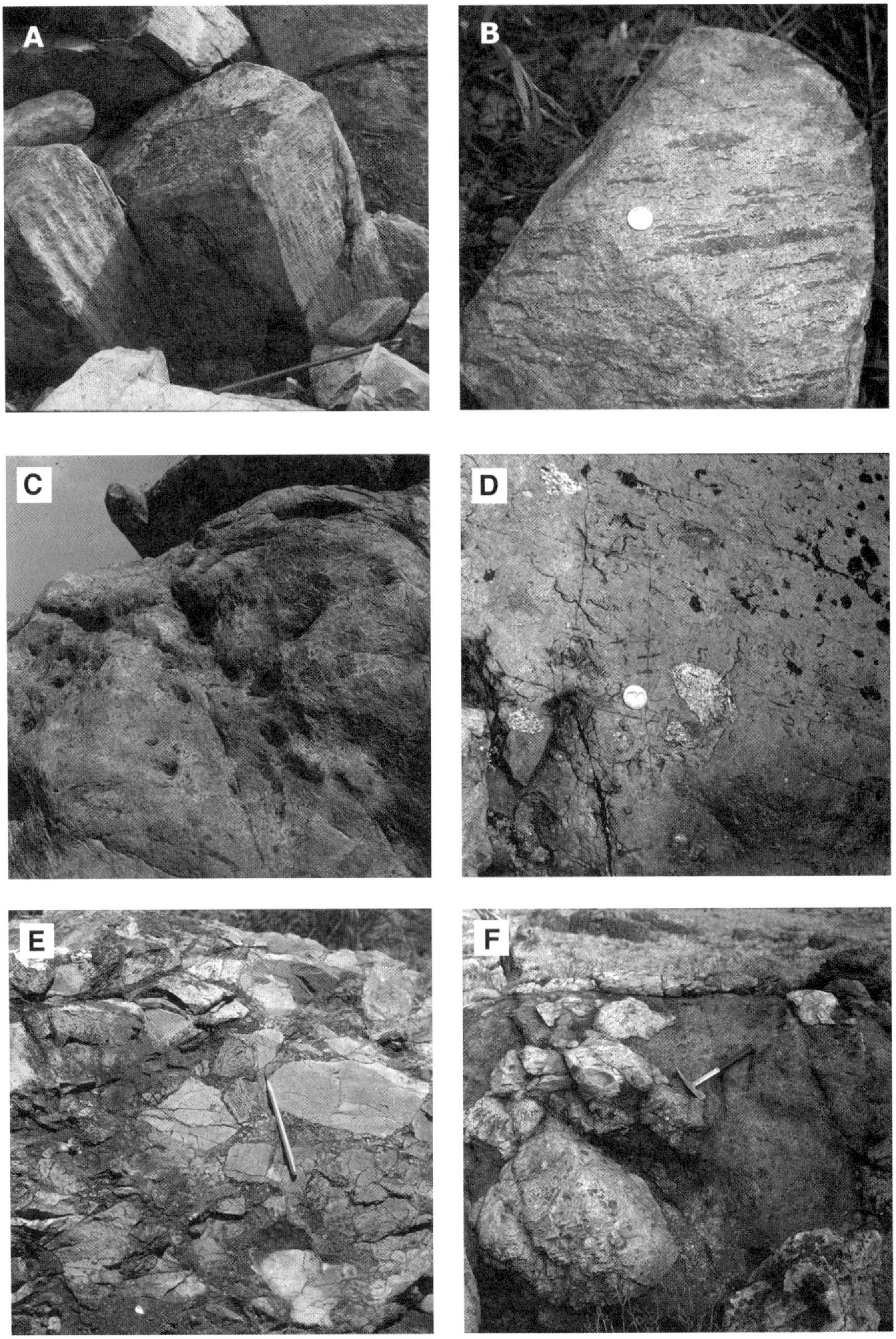
A
B
C
D
E
F

cated a consistent direction of dip. The lowest exposed rock of the Québec Branch section (WS4-43) is phenocryst-poor rhyolite in which flow layering indicative of lava was recognized in the field. This exposure and numerous outcrops to the northeast over a distance of 2.4 km were initially mapped as lava of the Whitetop Rhyolite Member (compare Fig. 2 of this paper with Fig. 3 *in* Rankin, 1993). These rocks, however, contain microlites of aegerine, and sample WS4-43 chemically matches other samples from the basal part of the Wilburn.

Topographic profiles were constructed for each section and combined with the attitude (apparent dip in the line of section) of the compaction foliation near the sample sites to determine the stratigraphic position of each sample in the ash-flow sheet (Fig. 6, B). However, the Massie Gap section is extrapolated well beyond the highest observed compaction foliation. The stratigraphic separation of 660 m between the lowest sample and highest observed northwest dip of the compaction foliation in the line of section is the minimum thickness of the Wilburn Rhyolite Member. If the base of the Wilburn is extrapolated to the topographic low point of the Québec Branch crossing, the minimum thickness of the Wilburn is 735 m. Data from these two sections form the basis for the following discussions regarding compositional zoning of the ash-flow sheet.

The sample suite includes one sample that is not part of the ash-flow sheet of the Wilburn Rhyolite Member. Sample WS4-49 in the Québec Branch section was collected from a roughly circular hypabyssal intrusive body that is a few tens of meters in diameter and composed of phenocryst-poor, high-silica rhyolite with aphanitic groundmass. Phenocrysts of quartz and perthite that are 0.5–1.0 mm in diameter make up ~3% of the rock. Some apparent phenocrysts are polycrystalline aggregates of quartz and perthite. The grain size of the quartz and feldspar groundmass is 0.01–0.05 mm, which is distinctly coarser-grained than that of the ash-flow tuff. Riebeckite and opaque minerals are present in small clusters or as disseminated grains. Tiny aegerine grains occur as inclusions in the perthite phenocrysts. Tectonic foliation is extremely weak. Flow layering in the intrusive body is parallel to the steep contact at the one place the contact was observed. The body contains small xenoliths of coarsely porphyritic rhyolite, granite, and the Wilburn Rhyolite Member. Based on mineralogic and chemical composition, the magma of the intrusion bears a resemblance to the basal unit of the Wilburn.

PETROGRAPHY

The primary phenocrysts in samples from both traverses are quartz, alkali feldspar, and rarely, plagioclase, typically constituting ~30% of the rock except in the basal zone, where they form 10% or less of the rock. The least tectonically deformed quartz phenocrysts are equant or broken and typically embayed (Fig. 5, C and F). Embayments in quartz are recognizable even in strongly sheared phenocrysts (Fig. 3, E and F). The alkali feldspar phenocrysts are blocky subhedral grains, commonly broken prior to tectonism, that now consist of mesoperthite in which the two feldspar phases form coarse patches (Fig. 5, E and F). Alkali feldspar phenocrysts in young rhyolitic ash-flow tuffs are typically cryptoperthitic sanidine (e.g., Boyd, 1961; Scott et al., 1971; Hildreth, 1979; Mahood, 1981). Perthite generally becomes coarser grained with age, longer periods of cooling, or increasing grade of metamorphism (Boyd, 1961; Day and Brown, 1980; Smith, 1988). Day and Brown (1980) describe the textural change in a hypersolvus granite from vein perthite to coarser patch perthite with increasing grade of younger regional metamorphism. Boyd (1961) documented that the cryptoperthite from the Yellowstone Tuff became progressively coarser grained, but remained cryptoperthite, stratigraphically upward in the section and used this as evidence that tuff represented one cooling unit. No change was observed in the texture or coarseness of the patch perthite phenocrysts of the Wilburn Rhyolite from the base to the top of the sampled sections, suggesting that the recrystallization is postemplacement. The patch perthite is interpreted to have developed during low-grade metamorphism. Most feldspar shows cloudy alteration to clay minerals. Embayed perthite is present in several thin sections (for example, Fig. 5, C and D). Quartz and perthite phenocrysts occur both in the groundmass and fiamme (Fig. 5, E). One small broken plagioclase phenocryst was found in sample WS4-34.

Aegerine and riebeckite are the major microphenocrysts in the basal, phenocryst-poor zone of the ash-flow sheet. Extremely fine-grained, poikilitic needles of pale-green aegerine, 0.1–0.45

Figure 4. (A) Welded tuff of the Wilburn Rhyolite Member, showing columnar jointing. Compaction foliation is barely visible on the sides of the columns. Cane in foreground is 90 cm long (from the crest of Wilburn Ridge, near location E, Fig. 6, A). (B) Fiamme, delineating compaction foliation in welded tuff of the Wilburn Rhyolite Member. Coin for scale is 18 mm in diameter. Float block is at an elevation of ~3480 ft in a stream valley ~1 km north of site 5 in Figure 2. (C) Pitted weathering zone at high angle to compaction foliation in welded tuff of the Wilburn Rhyolite Member; interpreted to be a fossil fumarole. The axis of the zone plunges steeply to the right and away from viewer. Knife for scale in the pit near the center of the photograph is 8.3 cm long. Specimen is from the north end of a crag, elevation of ~5050 ft. on the crest of Wilburn Ridge, near location E in Figure 6, A. (D) Zone of scattered lithic clasts in ash-flow tuff of the Wilburn Rhyolite Member. Clasts include granitoid from Grenvillian basement, plagioclase- and perthite-phyric rhyolite, and coarsely porphyritic quartz- and perthite-phyric rhyolite from older parts of Mount Rogers Formation. Coin for scale is 23 mm in diameter. Site is just south of location H in Figure 6, A. (E) Block breccia in the Wilburn Rhyolite Member. Shown are angular blocks of lava of the Whitetop Rhyolite Member, probably derived from the caldera wall, in a matrix of crystal-rich tuff of the Wilburn. Flow layering is visible in some clasts; for example, in the relatively small clast to the left of the pencil (locality 4 in Fig. 2). (F) Light-colored angular blocks of the Whitetop Rhyolite Member, probably derived from the caldera wall, in darker ash-flow tuff of the Wilburn. Hammer is 28.5 cm long (summit area of Stone Mountain, location 3 in Fig. 2).

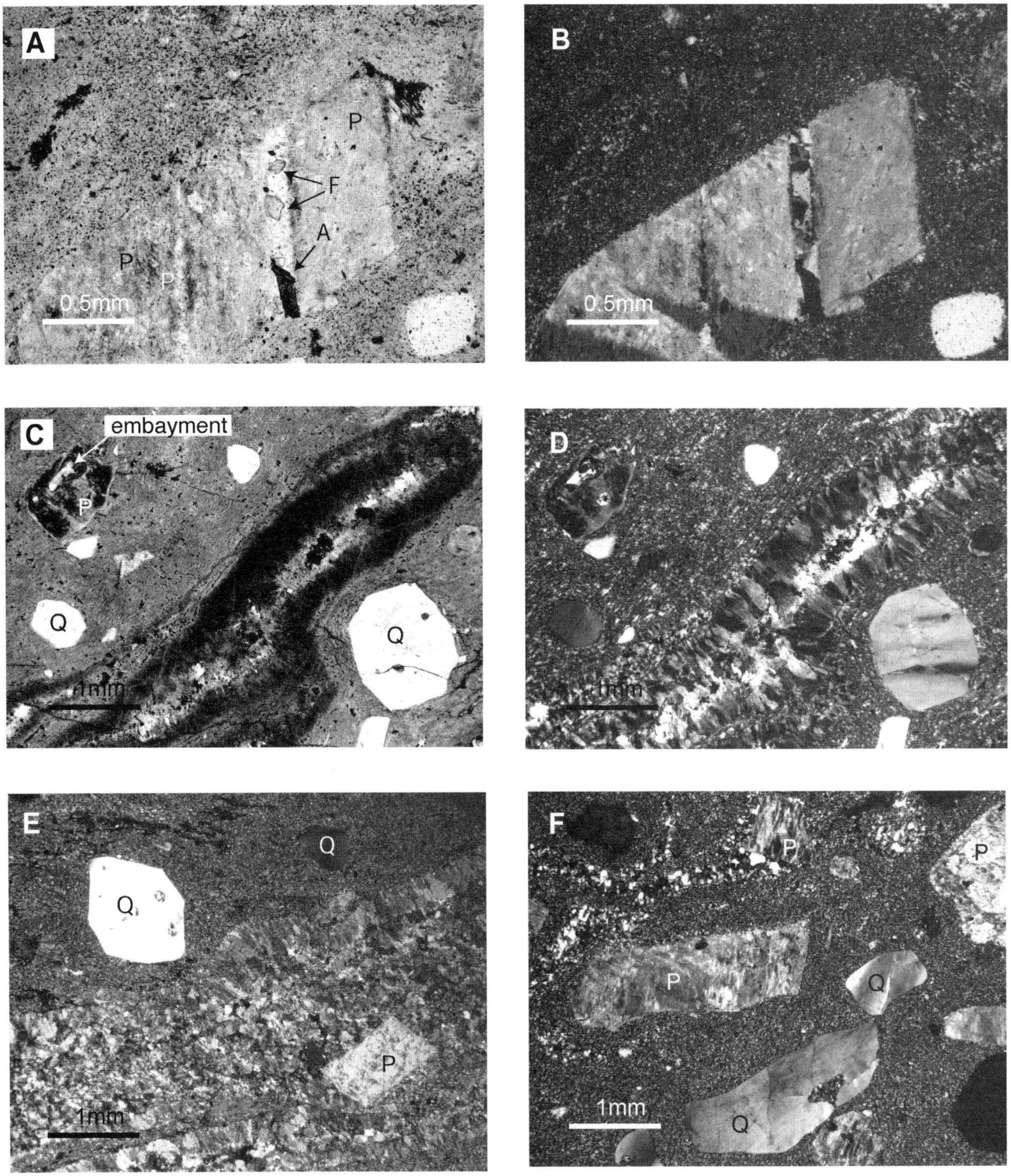
A
P
F
A
P
P
0.5mm
B
0.5mm
C
embayment
P
Q
Q
1mm
D
1mm
E
Q
Q
P
1mm
F
P
P
P
Q
Q
1mm

mm long, form much less than 1% of the rock. These crystals are scattered throughout the groundmass and are commonly intergrown with riebeckite. Aegerine inclusions in the perthite phenocrysts are also common in the basal zone. Microlites of riebeckite, some intergrown with aegerine, are scattered throughout the groundmass. In the basal part of the sheet, riebeckite microlites are more abundant than aegerine and are commonly intergrown with equant opaque minerals. Trains of opaque dust and riebeckite microlites commonly define tectonic foliation, in both the basal and main parts of the ash-flow sheet (Fig. 3, E and 5, E). Riebeckite also occurs as highly poikilitic aggregates consisting of multiple stubby, subhedral prisms that are intergrown and optically continuous (Fig. 7, A and B). Abundant and ubiquitous round quartz and feldspar inclusions give the composite grains a mottled appearance. In the least deformed rocks, riebeckite locally occurs as radiating prisms at the center of spherulites. In one rock, riebeckite occurs as stubby prisms at the periphery of spherulites, oriented radially with respect to the spherulite centers (Fig. 7, C and D). Aegerine and opaque oxides are commonly intergrown with the amphibole. The aggregate grains are as large as 0.2–0.5 mm and relatively abundant in the basal zone, constituting 1–2% of the rock.

Above the basal zone, aegerine is extremely rare and the amount and size of riebeckite aggregates are reduced. In these rocks, the riebeckite aggregates are <0.2 mm in size and make up much less than 1% of the rock. The thickness of this zone of sparser and smaller riebeckite is ~335 m in the Massie Gap section and ~325 m in the Québec Branch section. In the stratigraphically highest parts of the ash-flow sheet, biotite is the only ferromagnesian silicate microphenocryst present and forms irregular 0.2-mm grains displaying a light-green pleochroism. Less commonly, biotite aggregate grains are as large as 1 mm; these are generally altered to chlorite that likely formed as the result of greenschist-facies metamorphism.

Identifiable accessory minerals that individually constitute ≤1% of the Wilburn Rhyolite Member include ilmenite, magnetite, fluorite, and zircon; secondary minerals include minor sericite, chlorite, and calcite. Ilmenite is present throughout the ash-flow sheet as small irregular grains intergrown with riebeckite or, less commonly, as larger euhedral phenocrysts typically rimmed by titanite. Magnetite is ubiquitous as tiny octahedra disseminated through the groundmass or as larger subhedral phenocrysts. Fluorite is relatively common as irregular grains in the groundmass and within cracks in the perthite phenocrysts (Fig. 5, A and B). Zircon is relatively common as clear euhedral grains 0.25 mm in size. Zircon occurs as isolated grains in the groundmass, but is more commonly in aggregates surrounded by magnetite or biotite. Allanite was observed in a few thin sections, including that of sample WS4-35.

The groundmass of much of the ash-flow sheet was initially glassy, as indicated by the presence of flattened relict shards, spherulites, and perlitic texture (Figs. 3, A–D and 7, B–D). Spherulites commonly nucleate around phenocrysts. The groundmass now consists of an extremely fine intergrowth of quartz and feldspar forming an equigranular interlocking fabric with poorly developed grain boundaries. In these greenschist-facies rocks, we have not pursued the distinction between devitrification and crystallization through vapor-phase alteration (Smith, 1960). In some samples, the groundmass is texturally homogeneous except for faint to locally extreme tectonic foliation; no fiamme are preserved. In other samples, what were fiamme are now lens-shaped or elongated areas of coarse and commonly spherulitic or even granophyric quartz and feldspar intergrowths. The grain size in these patches (excluding the spherulites) may reach 0.2 mm, which is much coarser than the surrounding groundmass. The fiamme pseudomorphs typically have axiolitic borders or, for very thin fiamme, the whole feature may be axiolitic (Fig. 5, C–E).

A number of outcrop-scale features shows the same changes from the bottom to the top of each sampled section. In general, lithic inclusions are less common upward; none were observed in the top half of either section. Fiamme are less well preserved in the upper part of the sampled sections and typically are recognized in thin section as intergrowths of quartz and

Figure 5. Photomicrographs of the Wilburn Rhyolite Member. (A) Extensional fracture in perthite (P) phenocryst. The fracture is filled with polycrystalline quartz plus aegerine (A) and fluorite (F). Aligned microlites of riebeckite are at upper right end of the perthite phenocryst. Dark material in groundmass includes riebeckite and opaque minerals. Sample WS4-30A, photographed in plane polarized light (sample of Massie Gap section, locality J in Fig. 6, A). (B) Same sample as in (A), viewed with crossed polarizers. (C) Axiolitic texture in recrystallized welded tuff. Quartz and alkali feldspar fibers are intergrown perpendicular to the borders of the axiolite; spherulites and interstitial quartz and opaque minerals fill the interior. Fiamme are clearly visible in the hand specimen and are the axiolites of the thin section. Phenocrysts are embayed quartz (Q) and embayed perthite (P). Sample TC1-32B, photographed in plane polarized light. Sample is from a float block below outcrop of same material, but the outcrop has less-visible textures. Outcrop is within unit mapped as breccia (not part of analyzed suite; location 5 in Fig. 2). (D) Same sample as in (C), viewed with crossed polarizers. (E) Axiolitic texture in recrystallized welded tuff from the basal part of the Wilburn. Quartz and alkali feldspar fibers are intergrown perpendicular to the borders of the axiolite; spherulites and interstitial quartz, alkali feldspar, riebeckite, and opaque minerals fill the interior. Euhedral, embayed quartz (Q) phenocrysts are in groundmass outside the axiolite. Upper left and bottom right corners of larger quartz phenocrysts are partly recrystallized and drawn out in tails parallel to the tectonic foliation. A perthite phenocryst is inside the axiolite. Dark streaks above larger quartz phenocryst are microlites of riebeckite and opaque minerals defining tectonic foliation. Sample WS4-31A, viewed with crossed polarizers (sample from the Massie Gap section, locality I in Fig. 6, A). (F) Ash-flow tuff with recrystallized groundmass from the upper part of the Massie Gap section, showing phenocrysts of embayed quartz (Q) and perthite (P). Sample WS4-55, viewed with crossed polarizers (locality C in Fig. 6, A).

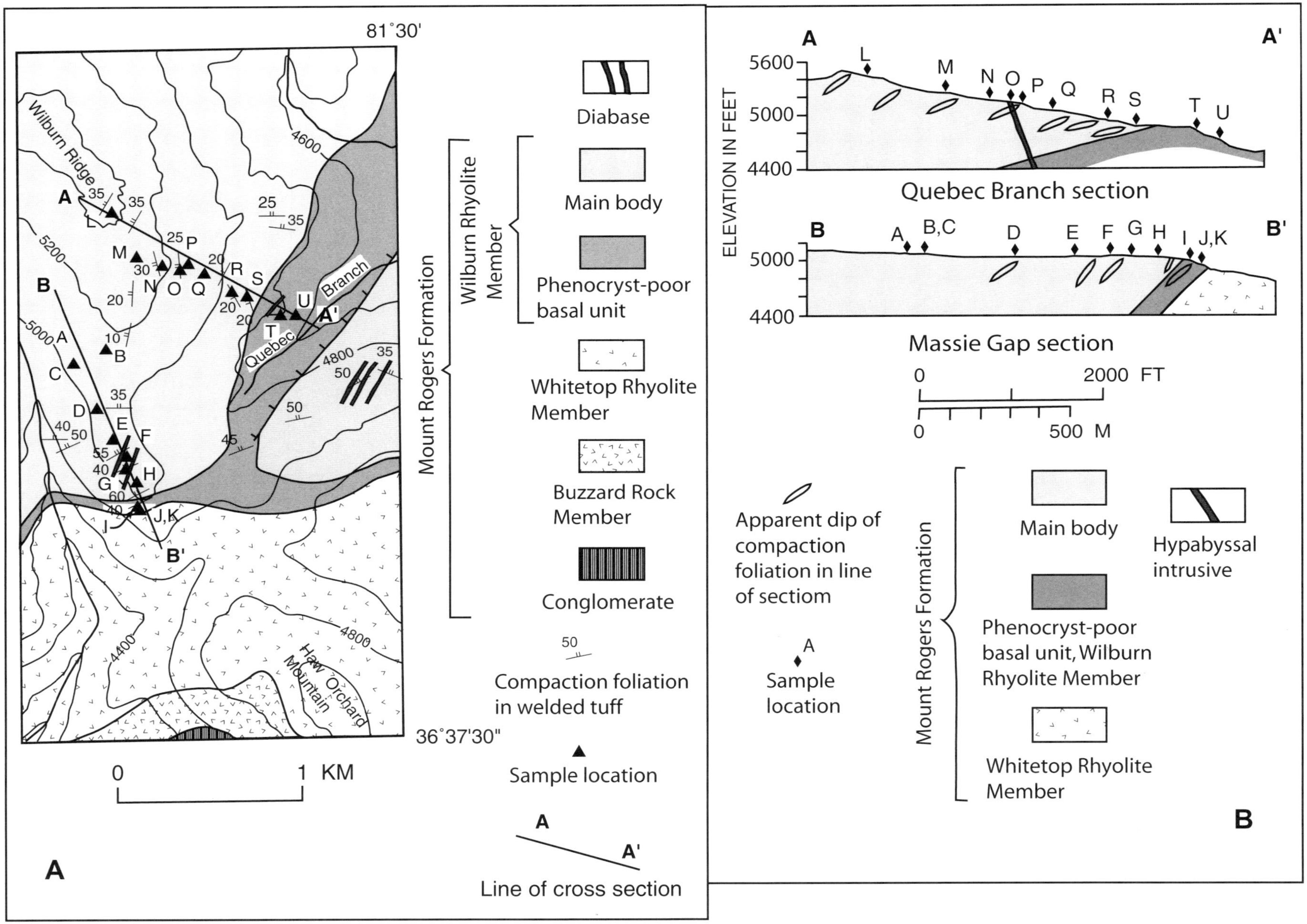

Figure 6. (A) Map showing location of the Massie Gap and Québec Branch sections, sample sites, and lines of profiles AA′ and BB′ of Figure 6, B. Base is from the USGS Whitetop Mountain 7.5 minute quadrangle. The southeast corner of the figure is the southeast corner of the Whitetop Mountain quadrangle. (B) Profiles of sampled sections of the Wilburn Rhyolite Member on Wilburn Ridge. Lines of the profiles are shown in Figure 6, A. Sample locations and attitudes of compaction foliation are projected onto the profile (no vertical exaggeration).

TABLE 1. SAMPLED SECTIONS OF THE WILBURN RHYOLITE MEMBER

Sample number	Symbol	Elevation from map (feet)	Stratigraphic height (m)	Ferromagnesian silicate microlites
Massie Gap Section				
WS4-54	A	5060	600	Biotite*
WS4-53	B	5120	580	Altered (stilpnomelane)
WS4-55	C	5040	580	Biotite*
WS4-36	D	5020	440	Biotite*
WS4-35	E	5050	335	Biotite*
WS4-34	F	5080	250	Biotite*
WS4-33	G	5080	200	Riebeckite*
WS4-32	H	5010	140	Riebeckite*
Top of basal unit as mapped		5000	75	
WS4-31	I	4960	50	Riebeckite Aegerine
WS4-30A	J	4920	18	Riebeckite Aegerine
WS4-30	K	4910	10	Riebeckite* Aegerine
Base of Wilburn Rhyolite Member		4900	0	
Quebec Branch Section				
Crest of ridge: NW dipping compaction foliation			660	
WS4-52	L	5450	620	Biotite
WS4-51	M	5290	435	Biotite
WS4-50	N	5220	360	Biotite
WS4-49 (Hypabyssal Intrusive)	O	5200	330	Riebeckite* Aegerine
WS4-48	P	5160	325	Riebeckite
WS4-47	Q	5100	260	Riebeckite* Aegerine
WS4-46	R	4960	165	Riebeckite
WS4-45	S	4890	130	Riebeckite*
Top of basal unit as mapped		4850	110	
WS4-44	T	4800	50	Riebeckite Aegerine
WS4-43 (Lowest sample in section Base of Wilburn Rhyolite Member unknown)	U	4700	0	Riebeckite* Aegerine
Quebec Branch		4570	–75	

*Minerals analyzed by microprobe.

feldspar that are coarser grained than the adjacent groundmass. The scarcity of observable fiamme in outcrop and the coarser-grained character of the fiamme in thin section in the upper part of the sampled sections suggest a greater degree of recrystallization of the upper part of the ash-flow sheet. Fossil fumaroles are more abundant in the upper parts of the sampled sections, suggesting greater concentration of volatiles higher in the section, possibly resulting from upward migration of volatiles as the sheet cooled following emplacement.

The presence of aegerine and sodic amphibole in the lower parts of the Wilburn Rhyolite clearly demonstrate the peralkaline nature of the rhyolite. These minerals, however, commonly form during pervasive posteruptive vapor-phase alteration of peralkaline tuffs (Wones and Gilbert, 1982) and, thus, they are not true phenocrysts. At least some of the aegerine is primary, however, because inclusions of aegerine are preserved in perthite phenocrysts. The occurrence of linear aggregates of grains paralleling metamorphic foliation indicates redistrib-

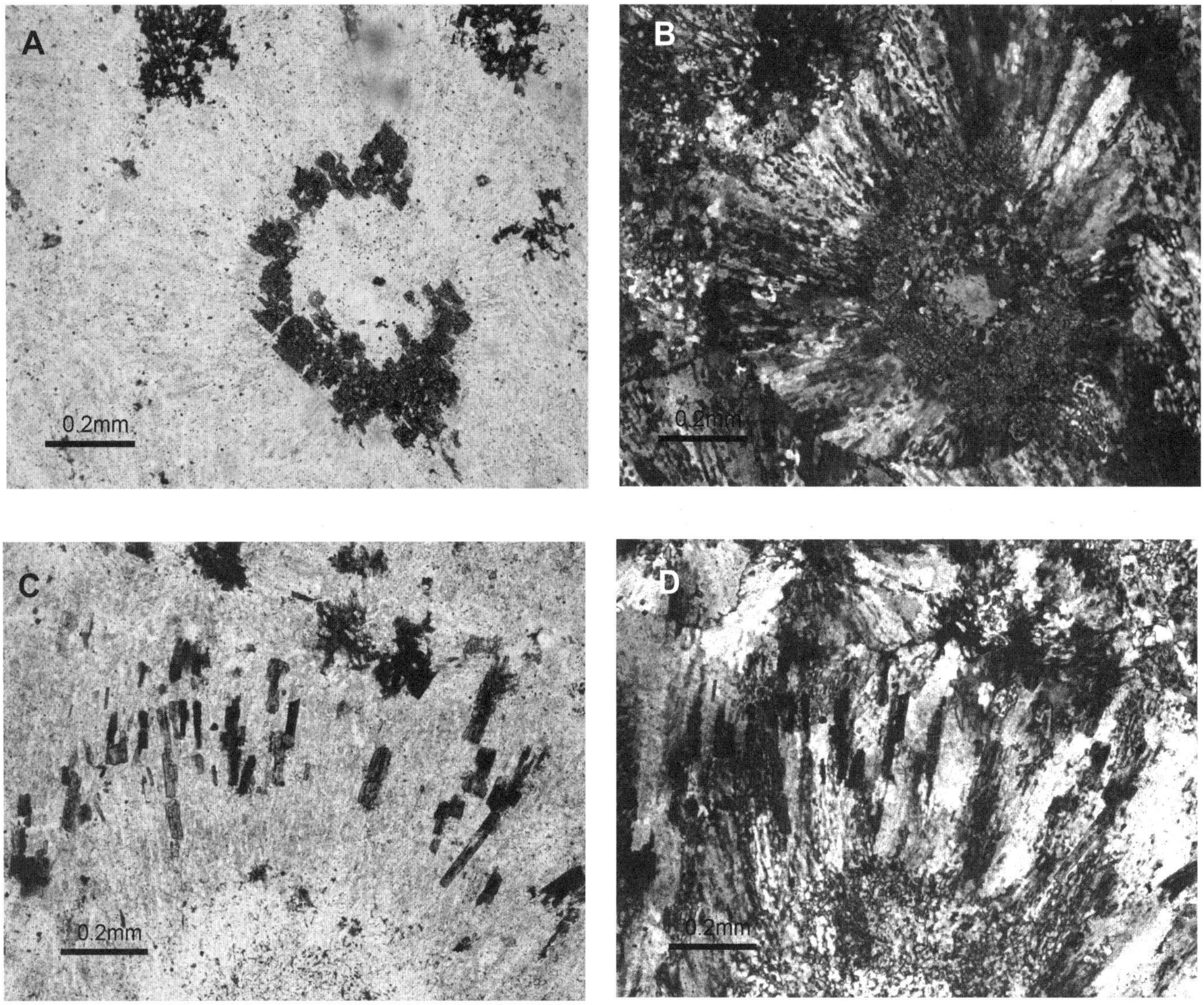

Figure 7. Photomicrographs of the basal unit of the Wilburn Rhyolite Member. (A) Spherulite, showing a concentric zone of polycrystalline, poikilitic, and optically continuous riebeckite between core and radial rim. Riebeckite is oriented parallel to the poorly developed tectonic foliation. Most of the thin section is spherulitic. Tectonic foliation, defined by the riebeckite orientation, is at high angle to the crude compaction foliation. Sample TS1-24, photographed in plane polarized light (not part of the analyzed suite; location 6 in Fig. 2). (B) Same area as in (A), viewed with crossed polarizers. (C) Riebeckite microlites showing radial growth in the outer part of a spherulite. Quartz and alkali feldspar display an incipient granophyric texture in the radial pattern. Sample TS1-23, photographed in plane polarized light (not part of the analyzed suite; location 7 in Fig. 2). (D) Same area as in (C), viewed with crossed polarizers.

ution and recrystallization of mafic minerals during regional metamorphism.

MINERAL CHEMISTRY

Mineral analyses were obtained from carbon-coated thin sections using the automated Applied Research Laboratory model SEMQ microprobe at the U.S. Geological Survey (USGS) in Reston, Virginia. Standard operating conditions for the analyses were: accelerating voltage, 15 kV; sample current, 100 nanoamps; and count time, 20 s or 20,000 counts. A variety of natural silicate and oxide minerals were used as standards. For each spot analysis, the oxides reported were analyzed simultaneously. Three or more spots for each grain were averaged to derive the reported mineral analyses. Mineral formulae were calculated using several different programs. The program of Papike et al. (1974) was used to calculate Fe^{3+} brackets for amphiboles and pyroxenes, as well as site populations for pyroxenes. The program of Goff and Czamanske (1972), modified to accept Zn, was used to calculate amphibole site populations.

Biotite formulae were calculated using a program written by S.D. Ludington (1979, written communication) that was also modified to accept Zn. Feldspar formulae and end members were calculated using a program written by J.C. Rucklidge (1972).

Feldspar Phenocrysts

Feldspar phenocrysts in the welded tuff consist of coarse patchy perthite comprised of albite and microcline. To obtain a bulk composition for these grains, mineral separates were homogenized by heating and then analyzed by microprobe. A series of five of the least altered samples, four from the Massie Gap section and one from the Québec Branch section, were chosen as representative. Bulk rock samples were crushed to 30 mesh and the feldspar separated in a bromoform-acetone mixture. X-ray examination of this material showed a mixture of low albite and microcline (Or_{90}). Both crushed grains and powder were sealed in platinum tubes and heated to 999 ± 4 °C for 7 days to homogenize the feldspar phases. Microprobe analyses of these grains were considered acceptable if compositional variation within a grain was <2%. Each reported analysis represents an average of at least three spot analyses of a grain (Table 2). Bulk compositions range from Or_{46} to Or_{67} and become more potassic at stratigraphically higher levels within the ash-flow sheet. The sodic compositions are similar to phenocrysts reported from other comendites (Novak and Mahood, 1986); the more potassic compositions are similar to phenocrysts commonly found in metaluminous high-silica rhyolites (Hildreth, 1979). Care must be exercised in interpreting feldspar compositions, because postdepositional metamorphism may have induced alkali element exchange and modified the phenocryst compositions (Orville, 1963). However, as with the other compositional changes documented in minerals through the sampled sections, it is difficult to envision a metamorphic process that would induce such large changes in phenocryst composition without more severe alteration of the rocks. We, therefore, feel that the compositional changes reflect the original phenocryst compositions.

TABLE 2. MICROPROBE ANALYSES AND MINERAL FORMULAS FOR HEATED FELDSPAR GRAINS FROM THE WILBURN RHYOLITE MEMBER

	Sample number				
	WS4-30	WS4-33	WS4-35	WS4-36	WS4-51
SiO_2	67.0	66.4	65.4	65.5	66.2
Al_2O_3	18.6	19.9	19.0	19.7	19.1
CaO	0.07	0.09	0.17	0.11	0.04
Na_2O	6.36	6.11	3.71	4.98	3.98
K_2O	7.31	8.09	11.8	10.2	11.2
BaO	0.02	0.02	0.03	0.03	0.03
Total	99.36	100.61	100.11	100.52	100.55
Formula on the basis of 8 oxygen atoms					
Si	3.01	2.97	2.98	2.95	2.99
Al	0.99	1.04	1.02	1.05	1.01
Σ Tet site	4.00	4.01	4.00	4.00	4.00
Ca	0.00	0.00	0.01	0.01	0.00
Na	0.55	0.53	0.33	0.44	0.35
K	0.42	0.46	0.68	0.59	0.64
Σ	0.97	0.99	1.02	1.03	0.99
An	0.3	0.5	0.8	0.5	0.2
Ab	56.8	53.2	32.1	42.5	35.0
Or	42.9	46.3	67.1	57.0	64.8

Note: Oxides are in wt%.

Microphenocrysts

Aegerine

Microprobe analyses of aegerine grains indicate near-end-member compositions (Morimoto et al., 1988) (Table 3). There is little compositional variation among grains within a sample or between samples. The occurrence of aegerine in volcanic rocks is considered one of the hallmarks of peralkaline magmas (Bailey and Schairer, 1966), even though much of the aegerine forms during postdeposition vapor-phase alteration or meta-

TABLE 3. MICROPROBE ANALYSES AND MINERAL FORMULAS OF PYROXENES FROM THE WILBURN RHYOLITE MEMBER

	Sample number	
	WS4-30	WS4-43
SiO_2	53.1	53.0
Al_2O_3	0.91	1.43
Fe_2O_3*	32.4	29.38
MgO	0.14	0.94
CaO	0.69	1.99
Na_2O	13.5	13.1
TiO_2	0.02	0.15
MnO	0.05	0.51
ZnO	0.05	0.13
Total	100.86	100.63
Formula on the basis of 6 oxygen atoms		
Si	2.01	2.01
Al	0.00	0.00
Σ Tet site	2.01	2.01
Al	0.04	0.06
Fe^{3+}	0.92	0.84
Mg	0.01	0.05
Mn	0.00	0.02
Ti	0.00	0.00
Σ M1 site	0.97	0.97
Ca	0.03	0.08
Na	0.99	0.96
Σ M2 site	1.02	1.04

*Calculated from stoichiometry. Oxides in wt%.

morphic recrystallization. Inclusions of aegerine within feldspar phenocrysts at the base of the ash-flow sheet show that peralkaline magma was present; the presence of aegerine in tectonic extension fractures in other feldspar phenocrysts indicates some metamorphic recrystallization.

Riebeckite

Riebeckite occurs as microlites, tiny poikilitic grains, and aggregates in the basal and middle parts of the ash-flow sheet (Fig. 7, A and C). Microprobe analyses of these grains (Table 4) indicate near-end-member riebeckite composition, based on the classification of Leake et al. (1997). Notable characteristics of the riebeckite include the high Fe content, low F content, and presence of Zn. The amount of Fe^{3+} shown is the midpoint value calculated by charge. Mineral formulas were calculated using this midpoint value. Some calculated formulas fulfill ideal stoichiometry well, but many show an excess of Si and deficiency of octahedral cations. These deficiencies may be due to inaccurate SiO_2 values, difficulty in estimating correct Fe^{3+} values, or unanalyzed octahedral cations. The SiO_2 values of the Wilburn amphiboles are similar to those of other analyzed riebeckites (Borley, 1963; Deer et al., 1997). Errors in the estimate of Fe^{3+}

TABLE 4. MICROPROBE ANALYSES AND MINERAL FORMULAS OF AMPHIBOLES FROM THE WILBURN RHYOLITE MEMBER

	A. Massie Gap Section							
	Sample and grain number							
	WS4-30			WS4-32		WS4-33		
	1	2	3	1	2	1	2	3
SiO_2	52.93	51.54	51.33	51.55	50.03	50.84	50.81	52.81
Al_2O_3	0.63	0.70	1.00	0.78	1.22	0.76	0.56	0.18
FeO	32.13	33.21	31.94	33.00	33.31	33.76	35.12	35.41
MgO	2.97	1.70	1.75	1.82	0.27	0.55	0.32	0.00
MnO	0.30	0.51	0.41	0.39	0.73	0.63	0.63	0.43
TiO_2	0.11	0.76	0.77	0.35	1.07	0.79	0.78	0.79
CaO	0.13	0.68	0.46	0.34	0.99	0.48	0.89	0.01
Na_2O	6.93	6.75	6.75	6.43	6.07	6.75	6.06	7.05
K_2O	0.01	0.37	0.62	0.45	1.20	0.66	0.77	0.21
ZnO	0.34	0.30	0.48	0.36	1.00	1.02	1.02	0.22
F	0.29	0.41	0.36	0.24	0.48	N.A.	N.A.	0.08
Total	96.76	96.93	95.89	95.71	96.37	96.25	96.95	97.19
FeO*	18.15	22.73	21.97	20.51	25.52	23.90	24.47	22.71
Fe_2O_3*	15.54	11.65	11.08	13.88	8.66	10.95	11.84	14.11
New total	98.32	98.09	97.00	97.10	97.24	97.34	98.14	98.60
Formula based on 23 oxygen atoms								
Si	8.027	7.993	8.019	8.010	7.968	8.001	7.966	8.122
Al	0.000	0.007	0.000	0.000	0.032	0.000	0.034	0.000
Σ Tet site	8.027	8.000	8.019	8.010	8.000	8.001	8.000	8.122
Al	0.113	0.121	0.184	0.143	0.197	0.141	0.070	0.033
Fe^{3+}	1.774	1.360	1.303	1.623	1.038	1.297	1.397	1.633
Fe^{2+}	2.302	2.948	2.870	2.665	3.399	3.146	3.209	2.921
Mg	0.617	0.393	0.408	0.422	0.064	0.129	0.075	0.000
Mn	0.039	0.056	0.054	0.051	0.056	0.076	0.040	0.056
Ti	0.013	0.089	0.091	0.041	0.128	0.094	0.092	0.091
Zn	0.038	0.034	0.055	0.041	0.118	0.119	0.118	0.025
Σ C site	4.949	5.000	4.967	4.986	5.000	5.000	5.000	4.759
Mn	0.000	0.011	0.000	0.000	0.042	0.008	0.044	0.000
Ca	0.021	0.113	0.077	0.057	0.169	0.081	0.150	0.002
Na	1.979	1.816	1.923	1.937	1.789	1.911	1.807	1.998
Σ B site	2.000	2.000	2.000	1.994	2.000	2.000	2.001	2.000
Na	0.059	0.153	0.122	0.000	0.086	0.149	0.036	0.104
K	0.002	0.073	0.124	0.089	0.244	0.133	0.154	0.041
Σ A site	0.061	0.226	0.246	0.089	0.330	0.282	0.190	0.145
Fe/(Fe+Mg)	0.869	0.916	0.911	0.910	0.986	0.972	0.984	1.000

(continued)

TABLE 4. *Continued*

B. Quebec Branch Section

	WS4-43		WS4-45			WS4-47	WS4-49
	1	2	1	2	3	1	1
SiO_2	51.56	51.91	50.95	50.55	51.02	52.00	51.29
Al_2O_3	0.43	0.64	0.53	0.66	0.54	0.59	0.61
FeO	32.50	28.64	31.75	33.11	32.43	31.86	34.16
MgO	1.17	2.39	0.46	0.15	0.65	0.89	0.10
MnO	1.41	2.04	1.38	0.90	1.04	1.14	0.73
TiO_2	1.55	1.86	1.35	1.36	0.88	0.93	0.64
CaO	0.14	0.23	1.30	1.85	1.59	1.77	0.27
Na_2O	6.28	6.81	5.95	6.07	6.28	5.75	6.67
K_2O	0.18	0.16	0.98	1.00	1.01	1.07	1.10
ZnO	1.30	1.15	1.01	0.44	0.71	0.89	0.75
F	0.10	0.08	0.14	N.A.	0.35	N.A.	N.A.
Total	96.60	95.41	95.79	96.09	96.49	96.89	96.32
FeO*	21.61	18.71	23.49	26.85	24.82	23.27	23.69
Fe_2O_3*	12.10	11.03	9.17	6.95	8.46	9.55	11.64
New Total	97.81	97.00	96.70	96.78	97.34	97.85	97.49
Formula on the basis of 23 oxygen atoms							
Si	8.002	8.016	8.068	8.025	8.069	8.093	8.061
Al	0.000	0.000	0.000	0.000	0.000	0.000	0.000
Σ Tet site	8.002	8.016	8.068	8.025	8.069	8.093	8.061
Al	0.079	0.117	0.099	0.189	0.101	0.108	0.113
Fe^{3+}	1.413	1.282	1.093	0.817	1.007	1.118	1.377
Fe^{2+}	2.805	2.416	3.111	3.578	3.283	3.029	3.114
Mg	0.271	0.550	0.109	0.036	0.153	0.207	0.023
Mn	0.103	0.267	0.185	0.121	0.139	0.150	0.097
Ti	0.181	0.216	0.161	0.162	0.105	0.109	0.076
Zn	0.149	0.131	0.118	0.052	0.083	0.102	0.087
Σ C site	5.000	4.979	4.875	4.955	4.870	4.823	4.886
Mn	0.083	0.000	0.000	0.000	0.000	0.000	0.000
Ca	0.023	0.038	0.221	0.315	0.269	0.295	0.046
Na	1.890	1.962	1.779	1.685	1.731	1.705	1.954
Total Σ B site	1.996	2.000	2.000	2.000	2.000	2.000	2.000
Na	0.000	0.077	0.047	0.183	0.195	0.030	0.078
K	0.036	0.032	0.198	0.162	0.204	0.212	0.221
Σ A site	0.036	0.109	0.245	0.345	0.399	0.243	0.299
Fe/(Fe + Mg)	0.940	0.871	0.975	0.992	0.966	0.952	0.995

Note: Oxides in wt%. N.A.—Not analyzed.
*Calculated values.

are difficult to evaluate, because errors in all of the other elements affect the amount of Fe^{3+} calculated for each analysis. Charge balance should be a requirement of a correct amphibole formula, however, and the estimated Fe^{3+} value best fulfills this. In addition, Fe^{3+}/total Fe ratios are similar to those of riebeckites from similar rocks (Borley, 1963). The possibility of unanalyzed octahedral cations is the most likely cause of low octahedral cation sums. Although an energy-dispersive spectrometer was used to look for all analyzable elements, Li has been reported as a significant constituent of riebeckites by wet chemical analyses (Borley, 1963; Lyons, 1976) and would not be detected by an energy-dispersive unit. If the small amounts of Li reported for other analyzed riebeckites were present in the Wilburn riebeckites, calculated formulae would fall within accepted limits for correct amphibole stoichiometry. The presence of small amounts of Li is thought to be very likely in the Wilburn Rhyolite amphiboles, because these minerals crystallized from a highly differentiated magma.

Chemical analyses show that the amphiboles become more magnesian toward the base of the ash-flow sheet. A charge-coupled substitution results in the amphiboles being more sodic but lower in Fe^{3+} toward the base of the ash-flow sheet (Table 4). These compositional changes mirror those in the bulk rock compositions and argue that, at least for the larger grains analyzed by microprobe, microphenocrysts reflect compositional differences in the magma.

Biotite

Biotite microphenocrysts in the upper parts of the Wilburn ash-flow sheet show major variations in composition including large changes in F content, concomitant changes in Fe/Mg ratios and Al content, and the presence of Zn (Table 5). By assuming charge neutrality, the ideal amount of H_2O expected in the mica can be calculated, and these values are reported in Table 5. The assumption of a full octahedral layer means that any unanalyzed cations will raise the cation fractions in the formula (Ludington, 1974). In the biotites of the Wilburn rhyolite, the low oxide totals, even with calculated H_2O added in, and high Si and K values suggest that some octahedral cation(s) have not been analyzed. As with the amphiboles, small grain size and low abundance makes mineral separation impossible. The possibility of Li being missed is high, because Li is commonly reported

TABLE 5. MICROPROBE ANALYSES AND MINERAL FORMULAS OF BIOTITES FROM THE WILBURN RHYOLITE MEMBER

	Sample and grain number								
	WS4-34		WS4-35		WS4-55			WS4-54	
	1	3	1	WS4-36	2	3	5	1	3
SiO_2	39.00	38.98	38.58	40.18	34.94	37.10	35.33	35.60	35.09
Al_2O_3	11.67	11.54	13.73	12.36	14.13	13.68	13.77	13.22	14.17
TiO_2	0.27	0.46	0.48	0.64	1.23	1.38	0.38	1.28	0.50
FeO	12.92	12.70	18.52	14.62	25.64	22.86	24.54	25.30	25.55
MnO	0.58	0.68	0.52	0.73	0.90	0.68	1.07	0.66	0.43
MgO	16.07	15.91	12.22	14.62	8.00	8.70	8.58	8.72	9.20
CaO	0.06	N.D.	N.D.	N.D.	N.D.	N.D.	N.D.	N.D.	N.D.
Na_2O	N.D.	N.D.	N.D.	N.D.	N.D.	N.D.	N.D.	N.D.	N.D.
K_2O	10.29	10.21	10.35	10.20	7.68	9.70	8.66	8.79	8.39
ZnO	1.58	2.22	0.51	1.14	0.00	1.14	1.40	1.06	1.07
F	3.89	4.18	3.28	2.64	1.24	1.14	1.17	1.21	1.31
Total	96.32	96.89	98.19	97.13	93.76	96.38	94.90	95.84	95.71
H_2O calc	2.78	1.64	1.62	2.02	3.11	3.72	2.15	4.17	3.12
Sum	98.00	98.53	99.81	99.11	96.87	99.06	98.05	99.01	98.83
Si	3.054	3.049	3.012	3.101	2.823	2.959	2.839	2.843	2.778
Al	0.946	0.951	0.988	0.899	1.177	1.041	1.161	1.157	1.222
Σ Z site	4.000	4.000	4.000	4.000	4.000	4.000	4.000	4.000	4.000
Al	0.132	0.113	0.276	0.225	0.168	0.245	0.144	0.088	0.101
Ti	0.016	0.027	0.028	0.037	0.075	0.083	0.023	0.077	0.030
Fe	0.846	0.831	1.209	0.944	1.732	1.525	1.649	1.690	1.692
Mn	0.038	0.045	0.034	0.048	0.062	0.046	0.073	0.045	0.029
Mg	1.876	1.855	1.422	1.682	0.963	1.034	1.028	1.038	1.086
Zn	0.091	0.128	0.029	0.065	0.067	0.083	0.063	0.063	0.062
Σ Y site	3.000	3.000	3.000	3.000	3.000	3.000	3.000	3.000	3.000
Ca	0.005	0.0	0.0	0.0	0.0	0.0	0.0	0.0	0.0
Na	0.0	0.0	0.0	0.0	0.0	0.0	0.0	0.0	0.0
K	1.028	1.019	1.031	1.004	0.792	0.987	0.888	0.896	0.848
Σ X site	1.033	1.019	1.031	1.004	0.792	0.987	0.888	0.896	0.848
F	0.964	1.034	0.810	0.644	0.317	0.288	2.97	0.306	0.328
OH	1.036	0.966	1.190	1.356	1.683	1.712	1.703	1.694	1.672
F/(F + OH)	0.482	0.517	0.405	0.322	0.158	0.144	0.149	0.153	0.164
X_{phlog}	0.63	0.62	0.48	0.57	0.32	0.35	0.34	0.34	0.36
X_{ann}	0.33	0.35	0.33	0.37	0.33	0.39	0.35	0.39	0.31
X_{sid}	0.03	0.03	0.18	0.06	0.35	0.26	0.31	0.27	0.33
Fe/(Fe + Mg)	0.311	0.309	0.460	0.359	0.643	0.596	0.616	0.620	0.609

Note: Oxides in wt%. N.D.—Not detected.
*Calculated from stoichiometry.

in significant amounts in biotite, particularly from phenocrysts in such highly evolved volcanic rocks (Ekstrom, 1972; Ludington, 1974).

Despite problems inherent in calculating precise mineral formulas, it is clear that there are significant changes in biotite composition within the ash-flow sheet. The most notable changes in the biotites of the Wilburn rhyolite are the dramatic Fe/Mg ratio decrease and concomitant F increase toward the base of the ash-flow sheet (Table 5). In addition, the appearance of biotite rather than sodic amphibole in the upper part of the ash-flow sheet suggests a change toward metaluminous compositions.

Fe-Ti Oxides

Opaque oxide minerals are ubiquitous throughout the ash-flow sheet of the Wilburn Rhyolite Member. Microprobe analyses yield low sums of the oxides and low amounts of FeO in apparently homogeneous grains, suggesting that Fe is present mainly as Fe_2O_3. The analyses do suggest that two Fe-Ti oxide phases are present and that these minerals may have been original magnetite and ilmenite solid solutions. Alteration of the oxides precludes their use in oxygen thermometry-barometry. Hence, quantitative temperature estimates of magma conditions are not possible. The analyses also reveal up to several wt% Nb_2O_3 in the ilmenite phase, suggesting that the distribution of ilmenite is significant in controlling the Nb whole-rock content.

GEOCHEMISTRY

Major element analyses were performed at the USGS under Daniel R. Norton, project leader, by Vertie C. Smith, analyst. Contents of trace-elements Co, Cr, Cs, Hf, Rb, Ta, Zn, Sc, and rare-earth elements (REE) were determined by instrumental neutron activation analysis (INAA) at the USGS under Jack Rowe, project leader; P. Foose and L. Schwarz, analysts. U and Th values were determined by delayed neutron analysis at the USGS under H.T. Millard, project leader; analysts H.T. Millard, M. Coughlin, B. Vaughn, M. Schneider, and W. Stang.

Major-Element Chemistry

Major- and trace-element analyses (Tables 6 and 7) indicate that the rocks are high-silica alkali rhyolites. The presence of aegerine indicates that the basal layer of the ash flow must have been peralkaline originally (molecular $[Na_2O + K_2O]/Al_2O_3 > 1$), even though present bulk compositions are not (Bailey and Schairer, 1966). The lower, aegerine-bearing zone of the ash-flow sheet is classified as comendite, because normative femics are <9.2% (MacDonald and Bailey, 1973). A plot of agpaitic index versus aluminum saturation index, however, shows that the present bulk compositions of the suite are subalkaline and primarily metaluminous (Fig. 8). Loss of sodium and volatiles has been demonstrated even for very young recrystallized volcanic rocks (Noble, 1970; Baker and Henage, 1977), so it is not surprising that these low-grade metamorphic rocks are not presently peralkaline. The presence of biotite in the upper part of the sheet suggests that it was metaluminous high-silica rhyolite. Samples from the base of the sheet have the highest agpaitic index. Figure 8 shows that the Wilburn Rhyolite Member compositions overlap the central part of the field of plutonic rocks of the Crossnore Complex (French Broad massif), which are thought to be cogenetic with the Mt. Rogers volcanic center (Rankin, 1975).

Despite low-grade metamorphism, both major- and trace-element geochemical data show regular compositional gradients within the ash-flow sheet (Tables 6 and 7). These analyses show that SiO_2, Na_2O, and F increase toward the base of the sheet, whereas Al_2O_3, MgO, CaO, K_2O, and TiO_2 decrease (Fig. 9). Fluctuations in elemental trends may be related to variations in deposition conditions during the ash-flow eruption, irregular withdrawal from the magma chamber, or local postdepositional alteration. Nevertheless, we propose that it is the overall trend of the changes in the unit that are significant. These trends of relative enrichment and depletion are similar to those found in younger, and better preserved, silicic zoned ash-flow sheets (Smith and Bailey, 1966; Hildreth, 1979, 1981; Smith, 1979; Sawyer and Sargent, 1989). Mineralogical zone boundaries within the ash-flow sheet do not correlate with significant offsets in trends of elemental concentration (Fig. 9).

The stratigraphic thickness of the upper biotite-bearing zone is greater in the Quebec Branch section than it is in the Massie Gap section, but the stratigraphic top of the Wilburn Rhyolite cannot be identified in either section. Other differences in the compositions of the two sections are apparent. Samples at the bases of the two sections are similar, but samples at the top of the Massie Gap section have lower SiO_2 and higher CaO contents than those at the top of the Québec Branch section. These Massie Gap samples have the lowest Thornton and Tuttle (1960) differentiation index as well. Ash-flow tuffs higher in the sections are interpreted to have been erupted sequentially later from deeper in the magma chamber than those from the bottom of the sheet (i.e., the magma chamber was inverted during the eruption). That the ash-flow tuff at the top of the Massie Gap section is less differentiated than that at the top of the Québec Branch section suggests that it came from a deeper level in the magma chamber. It is not known whether stratigraphically higher rocks lie to the northwest of the end of sampling of the Québec Branch section.

Trace-Element Chemistry

Variations in trace-element abundances through the Wilburn Rhyolite Member support the zoned nature of the ash-flow sheet inferred from the major element data (Fig. 10). Elements that increase in abundance toward the base of the sheet include Be, Rb, Zr, Nb, Sn, Hf, Ta, Th, U, Tb, and Yb; elements more abundant toward the top of the sheet include Sc, Sr, Ba, La, Ce, Nd, and Eu. Several elements, such as Li and Cs, show highly variable abundances or much lower abundances than

TABLE 6. GEOCHEMICAL DATA FOR SAMPLES OF THE MASSIE GAP SECTION, WILBURN RHYOLITE MEMBER

	Sample number and symbol										
	WS4-54 A	WS4-53 B	WS4-55 C	WS4-36 D	WS4-35 E	WS4-34 F	WS4-33 G	WS4-32 H	WS4-31 I	WS4-30A J	WS4-30 K
SiO_2	74.21	73.91	73.88	76.07	75.81	76.43	76.74	77.00	76.74	76.44	76.64
Al_2O_3	12.42	12.57	12.83	11.79	12.10	11.76	11.41	11.29	11.57	11.52	11.49
Fe_2O_3	1.31	0.93	1.80	1.75	1.49	1.88	1.18	1.50	1.32	1.39	1.47
FeO	1.00	1.58	1.09	0.45	0.81	0.56	1.08	0.81	0.83	0.99	0.97
MgO	0.16	0.14	0.14	0.09	0.22	0.12	0.07	0.05	0.12	0.03	0.05
CaO	0.64	1.04	0.40	0.20	0.30	0.37	0.44	0.36	0.35	0.32	0.32
Na_2O	3.11	4.36	3.93	2.51	3.28	3.87	3.71	3.73	3.71	3.95	4.01
K_2O	6.14	4.00	5.12	6.66	5.42	4.45	4.79	4.60	4.68	4.46	4.43
TiO_2	0.25	0.25	0.27	0.20	0.21	0.20	0.15	0.15	0.15	0.14	0.14
P_2O_5	0.01	0.02	0.02	0.01	0.01	0.01	0.01	0.01	0.01	0.01	0.01
MnO	0.05	0.05	0.06	0.03	0.03	0.04	0.06	0.05	0.06	0.04	0.04
CO_2	0.31	0.59	0.09	0.00	0.02	0.04	0.05	0.01	0.01	0.02	0.01
H_2O+	0.18	0.41	0.17	0.08	0.01	0.00	0.08	0.09	0.01	0.01	0.05
H_2O-	0.06	0.05	0.05	0.04	0.05	0.04	0.04	0.05	0.07	0.04	0.05
Total	99.85	99.90	99.85	99.88	99.76	99.77	99.81	99.70	99.63	99.36	99.68
F	0.13	0.11	0.11	0.06	0.05	0.11	0.17	0.19	0.22	0.20	0.20
Rb	154	96	129	211	149	135	181	186	214		211
Ba	37	7	22	23	54	21	<2	<2	14		15
Sr	17	21	15	11	16	12	7	8	12		9
La	158	153	171	100	118	110	76	64	70		62
Ce	302	294	325	203	229	209	163	144	153		137
Nd	137	133	146	100	109	103	84	77	87		73
Sm	27.1	29.4	30.7	26.5	23.8	26.7	25.1	25.2	26		22.8
Eu	0.45	0.46	0.51	0.31	0.30	0.3	0.23	0.19	0.23		0.22
Gd	22.2	21.0	23.3	20.5	18.6	22.9	26.4	23.3	28.1		24.2
Tb	4.0	3.8	4.2	4.0	3.4	4.9	5.3	5.1	6.0		5.5
Ho	3.8	3.7	4.5	4.1	3.3	5.2	6.0	5.5	6.7		6.3
Tm	1.5	1.5	1.7	1.6	1.4	2.0	2.3	2.2	2.6		3.0
Yb	10.5	10.1	11.3	10.7	9.9	13.3	14.8	14.9	17.0		17.2
Lu	1.5	1.4	1.6	1.5	1.4	1.8	2.0	2.1	2.3		2.3
Th	17.6	16.4	15.4	18.7	16.3	18.7	24.3	24.8	23.9		27.1
U	2.1	2.2	5.1	3.2	2.5	2.8	3.6	4.3	6.7		4.8
Zr	702	667	739	585	632	619	653	649	693		751
Hf	18.5	17.6	18.4	17.1	17.8	17.8	21.1	22.0	23.7		26.4
Sn	7.5	4.5	4.2	6.1	6.2	6.7	7.6	11	9.5		10
Nb	58	59	60	61	62	65	78	91	93		110
Ta	3.6	3.5	3.4	3.9	3.8	4.2	5.6	6.1	6.4		7.3
W	1.4	1.5	1.2	1.6	1.0	1.1	1.4	1.8	2.3		1.5
Mo	2.9	5.1	2.8	1.1	1.5	1.4	1.6	1.1	1.5		0.8
Co	0.4	0.3	0.4	0.5	0.5	0.5	0.4	0.3	0.5		0.6
Be	4.7	4.8	4.1	3.9	4.0	7.3	7.4	6.3	6.5		8.1
Li	5.4	5.1	4.4	4.9	11	6.7	12	4.9	11		11
Zn	129	113	138	68	60	101	243	212	293		193
Sc	0.72	0.83	0.87	0.32	0.39	0.41	0.08	0.08	0.11		0.17
Cr	<7	3.1	<8	3.4	2.9	1.6	6.4	4.8	7.0		6.3
D I	93.02	94.06	94.41	95.91	95.35	95.31	94.95	95.21	95.56	95.22	94.98
AI	0.948	0.914	0.936	0.962	0.932	0.951	0.990	0.984	0.965	0.983	0.991
ASI	0.961	0.941	1.007	1.008	1.024	0.992	0.944	0.960	0.981	0.968	0.960

Note: Major elements (including F) expressed in wt%; trace elements in ppm. Major element analyses performed by the U.S. Geological Survey under Daniel R. Norton, project leader, Vertie C. Smith, analyst. Measurements of Co, Cr, Cs, Hf, Rb, Ta, Zn, Sc, REE were obtained by INAA at the U.S. Geological Survey under Jack Rowe, project leader; P. Foose and L. Schwarz, analysts. Measurements of U and Th were obtained by delayed neutron analysis at the U.S. Geological Survey under H.T. Millard, project leader; H.T. Millard, M. Coughlin, B. Vaughn, M. Schneider, and W. Stang, analysts. Abbreviations: AI—Agpaitic index (molar Na+K/Al); ASI—aluminum saturation index (Al_2O_3/[$Na_2O + K_2O$ + CaO]); DI—differentiation index of Thornton and Tuttle (1960).

TABLE 7. GEOCHEMICAL DATA FOR SAMPLES FROM THE QUÉBEC BRANCH SECTION, WILBURN RHYOLITE MEMBER

	Sample number and symbol									
	WS4-52	WS4-51	WS4-50	WS4-49	WS4-48	WS4-47	WS4-46	WS4-45	WS4-44	WS4-43
	L	M	N	O	P	Q	R	S	T	U
SiO_2	75.63	75.64	75.67	76.58	77.21	76.99	77.19	77.20	76.59	76.35
Al_2O_3	11.72	11.57	11.81	11.57	11.25	11.26	11.23	11.24	11.59	11.69
Fe_2O_3	2.20	1.79	1.66	1.89	1.80	1.72	1.79	1.87	1.71	1.92
FeO	0.45	0.64	0.88	0.63	0.45	0.53	0.46	0.43	0.67	0.61
MgO	0.11	0.03	0.03	0.02	0.02	0.00	0.03	0.01	0.13	0.05
CaO	0.33	0.64	0.36	0.40	0.34	0.34	0.37	0.33	0.46	0.32
Na_2O	3.26	3.25	2.78	2.79	3.54	3.24	3.56	3.50	4.19	4.34
K_2O	5.37	5.38	6.09	6.08	4.65	5.18	4.66	4.72	4.09	4.04
TiO_2	0.21	0.21	0.21	0.11	0.15	0.15	0.15	0.15	0.14	0.11
P_2O_5	0.01	0.01	0.01	0.01	0.01	0.01	0.01	0.01	0.01	0.01
MnO	0.04	0.04	0.05	0.05	0.03	0.04	0.04	0.04	0.04	0.04
CO_2	0.09	0.07	0.08	0.00	0.03	0.03	0.03	0.01	0.01	0.01
H_2O+	0.09	0.01	0.08	0.05	0.01	0.02	0.06	0.05	0.06	0.07
H_2O-	0.05	0.06	0.06	0.05	0.05	0.04	0.05	0.04	0.04	0.03
Total	99.56	99.34	99.77	100.23	99.54	99.55	99.63	99.60	99.73	99.59
F	0.08	0.10	0.12	0.23	0.13	0.14	0.16	0.14	0.18	0.19
Rb	163		162			192		201	187	
Ba	32		21			22		15	31	
Sr	17		15			16		13	19	
La	112		132			81		79	65	
Ce	224		256			167		157	143	
Nd	107		118			90		88	77	
Sm	28.0		27.3			29.5		24.9	25.7	
Eu	0.32		0.36			0.25		0.23	0.21	
Gd	19.4		23.5			31.5		24.2	26.5	
Tb	4.4		4.6			5.4		4.9	5.8	
Ho	4.2		4.9			8.3		5.1	6.3	
Tm	1.8		1.8			2.7		2.1	2.8	
Yb	11.6		12.2			17.2		13.3	18.2	
Lu	1.6		1.7			2.3		1.9	2.9	
Th	19.4		18.1			40.2		23.8	30.3	
U	2.5		2.2			3.3		3.3	4.4	
Zr	633		646			621		621	758	
Hf	18.3		18.2			21.3		20.9	27.5	
Sn	7.9		5.6			8.4		8.7	10	
Nb	67		60			77		80	100	
Ta	4.3		4.0			5.5		5.8	7.6	
W	1.9		1.2			1.5		2.8	2.2	
Mo	1.0		1.8			0.8		0.8	0.5	
Co	0.5		0.4			0.4		0.3	0.6	
Be	4.6		4.1			7.3		5.8	9.0	
Li	4.9		4.3			6.8		4.2	12	
Zn	131		166			136		146	192	
Sc	0.41		0.43			0.08		0.13	0.20	
Cr	3.7		4.6			4.1		4.5	7.0	
D. I.	93.54	93.65	93.62	94.09	94.69	94.82	94.69	94.67	94.10	94.50
A. I.	0.953	0.965	0.946	0.965	0.965	0.971	0.970	0.966	0.976	0.985
ASI	0.996	0.939	0.999	0.973	0.980	0.975	0.972	0.982	0.954	0.967

Note: Major elements (including F) expressed in wt%; trace elements in ppm. Major element analyses performed by the U.S. Geological Survey under Daniel R. Norton, project leader, Vertie C. Smith, analyst. Measurements of Co, Cr, Cs, Hf, Rb, Ta, Zn, Sc, REE were obtained by INAA at the U.S. Geological Survey under Jack Rowe, project leader; P. Foose and L. Schwarz, analysts. Measurements of U and Th were obtained by delayed neutron analysis at the U.S. Geological Survey under H.T. Millard, project leader; H.T. Millard, M. Coughlin, B. Vaughn, M. Schneider, and W. Stang, analysts. Abbreviations: AI—Agpaitic index (molar Na+K/Al); ASI—aluminum saturation index (Al_2O_3/[Na_2O + K_2O + CaO]); DI—differentiation index of Thornton and Tuttle (1960).

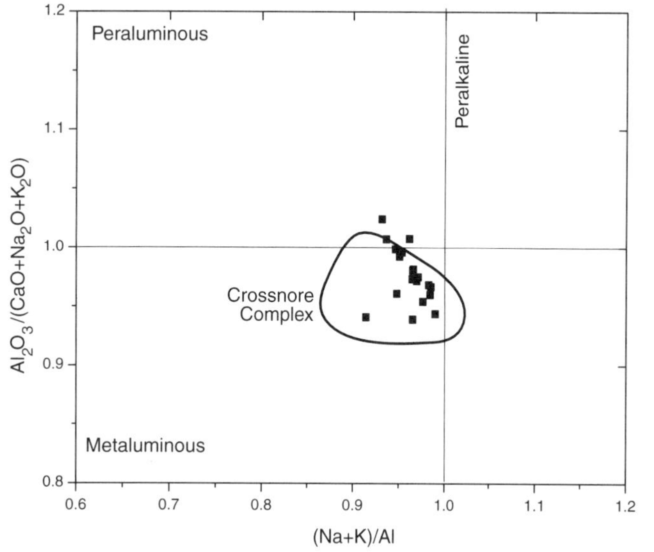

Figure 8. Agpaitic index (molecular [Na + K]/Al) versus aluminum saturation index (molar Al_2O_3/[CaO + Na_2O + K_2O]). Symbols indicate samples of Wilburn rhyolite. The field shows the range of compositions of granitoids of the Crossnore Complex. Data are from Tollo et al. (2004).

would be expected from other trace elements. The content of these relatively light and highly mobile elements is probably strongly affected by alteration (Gottfried et al., 1977). However, other elements that are little affected by alteration, such as Nb, Ta, Hf, and REE, show regular concentration gradients through the ash-flow sheet. As with the major elements, trace elements do not show pronounced offsets in the gradients across mineralogical boundaries in the sheet.

Covariation between Ta and Th for the Wilburn Rhyolite (Fig. 11) defines a linear trend indicating the coherence of the suite and allows comparison with younger, well-documented zoned ash-flow tuffs. Note that samples representing the metaluminous part of the magma system (those with low Th contents) are compositionally continuous with samples representing the peralkaline part. Some of the scatter may be due to alteration or crystal-ash physical separation during eruptive processes. In particular, Th values may be controlled by the abundance of zircon, which, because of its high specific gravity, may be physically fractionated during ash-flow deposition (Hildreth, 1981). The magnitude and slope of the Ta-Th trend is most similar to those of the Tala tuff (Mahood, 1981) and the Kane Wash tuff (Novak and Mahood, 1986) both of which contain both metaluminous and peralkaline compositions.

PETROGENESIS

Field and petrologic characteristics of the Wilburn Rhyolite suggest formation through a series of chronologically closely spaced eruptive events that progressively tapped a zoned magma chamber containing at least 30 km^3 of magma. The presence of at least two zones rich in lithic inclusions in the main body of the sheet suggests more than one flow unit. Hildreth (1981) presented a comprehensive synthesis of factors leading to the development of zoned magma chambers, particularly as shown by volcanic eruptive products. According to this classification, the Wilburn Rhyolite chamber falls within the group of mildly per-

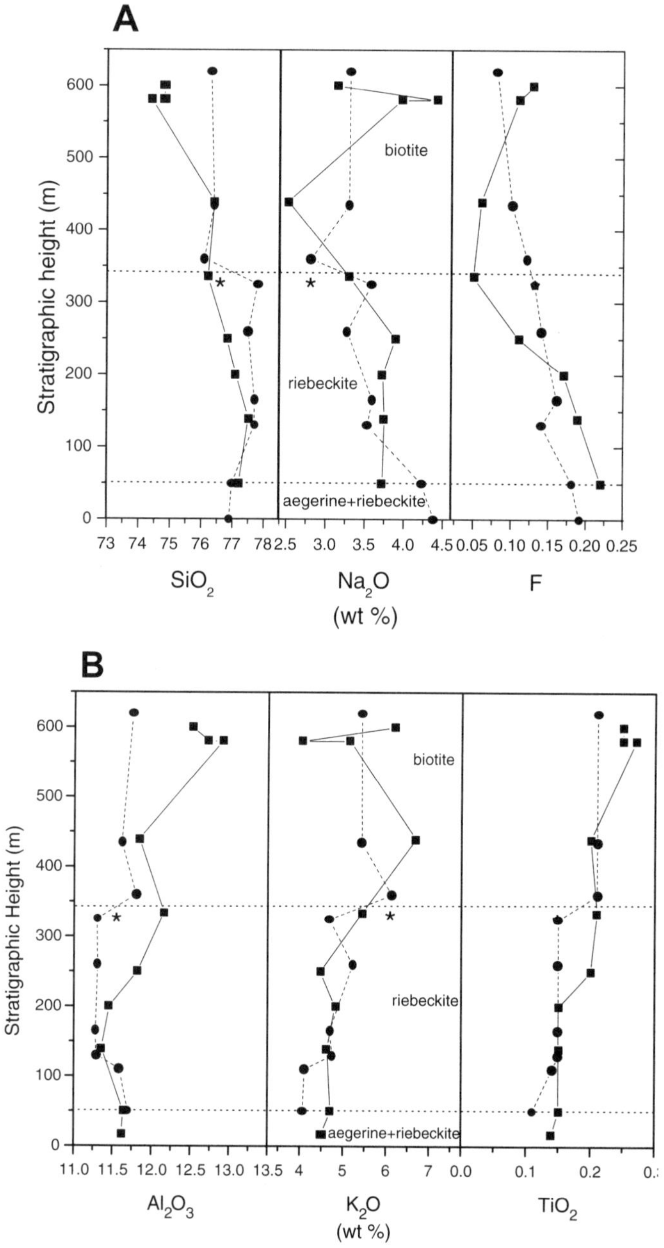

Figure 9. Major element contents (wt%) versus stratigraphic height for the Wilburn Rhyolite Member. (A) Elements enriched toward the base of the ash-flow sheet. (B) Elements enriched toward the top of the ash-flow sheet. Filled squares—Massie Gap section; filled circles—Québec Branch section; asterisk—hypabyssal intrusion. The horizontal dashed lines separate mineralogic zones.

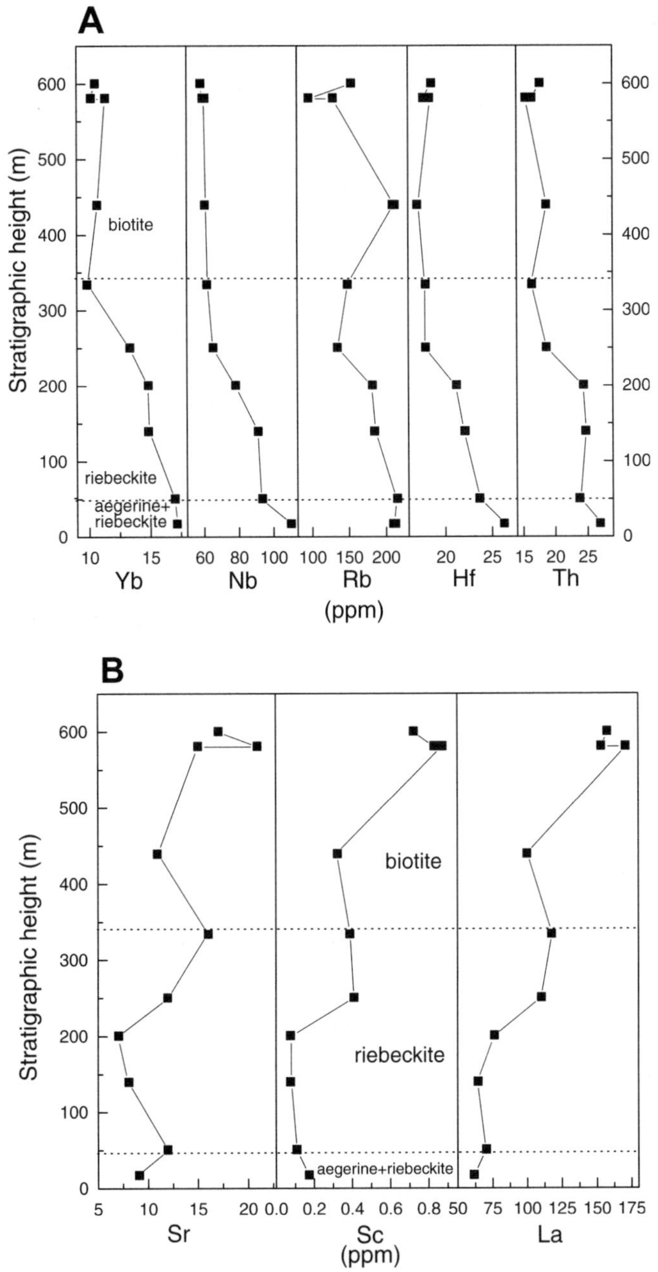

Figure 10. Trace element contents (ppm) versus stratigraphic height for the Massie Gap section of the Wilburn Rhyolite Member. (A) Elements enriched toward the base of the ash-flow sheet. (B) Elements enriched toward the top of the ash-flow sheet. The horizontal lines separate mineralogic zones.

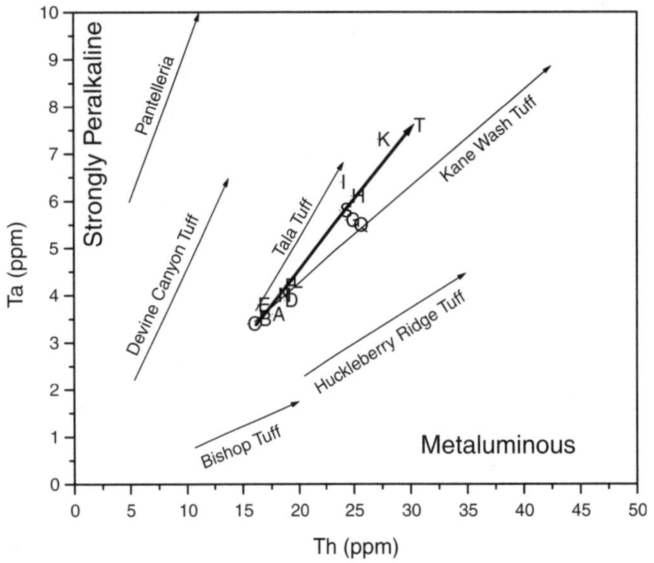

Figure 11. Ta versus Th content for the Wilburn Rhyolite Member. Data for other zoned ash-flow systems are shown for comparison. Heavy arrow indicates enrichment toward the base of the Wilburn Rhyolite ash-flow sheet. Symbols for the Wilburn Rhyolite Member are from Table 1. Data for other systems are from Hildreth (1981) and Novak and Mahood (1986).

alkaline high-silica rhyolites that grade into nonperalkaline roots. Similar compositional zoning characterizes the Tala Tuff (Mahood, 1981) and the Spearhead Member of the Thirsty Canyon Tuff (Noble and Parker, 1974). These authors proposed that such magma chambers might be the result of differentiation of a metaluminous high-silica rhyolite in which the roof zones are enriched in excess Na, Fe, and certain trace elements because of higher volatile fluxes than are found in wholly metaluminous chambers.

The magma that produced the Wilburn Rhyolite shows the closest affinity to A-type granites, although this classification was developed for plutonic rocks. A-type granites are characterized by low CaO, Sc, Cr, Co, Ba, Sr, and Eu contents, high FeO_t/MgO and K_2O/Na_2O ratios, and high REE, Zr, Nb, and Ta (Whalen et al., 1987; Eby, 1990). Trace-element data, such as the high concentration of Ta and Yb (Fig. 12), indicate that the Wilburn Rhyolite compositionally corresponds to "within-plate granites," like most other continental A-type granites (Eby, 1990). Similarly, very high concentrations of other high-field–strength elements, such as Zr and Nb, provide additional evidence for compositional affinity to A-type granites (Tables 6 and 7).

Normative compositions of the Wilburn Rhyolite (Fig. 13), all of which contain >91% (>94% for riebeckite-bearing samples) SiO_2 (Q) +$KAlSi_3O_8$ (Or) + $NaAlSi_3O_8$ (Ab), can be modeled by comparison with the experimental Q-Or-Ab-H_2O system and suggest crystallization at steam saturation pressures of 0.5–2.0 kilobars for the Québec Branch samples and 1.0–3.0 kilobars for the Massie Gap samples. Thus, the Wilburn Rhyolite appears to have erupted from a chamber at shallow crustal depths. Samples from the top of the Massie Gap section plot closer to the Or-Ab sideline than any of the others for either section. The experimental data of Tuttle and Bowen (1958) and Luth (1969) on the Q-Or-Ab system show that higher pressure displaces minimum melt compositions toward the Or-Ab sideline, thus adding further evidence that samples from the top of the Massie Gap sec-

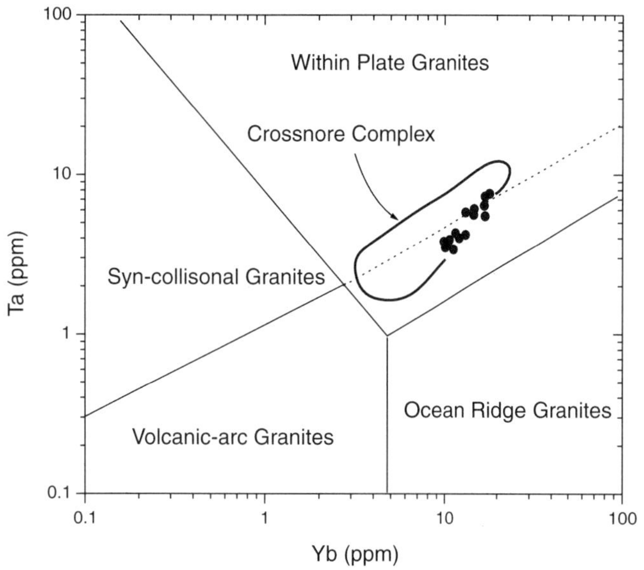

Figure 12. Ta versus Yb content for the Wilburn Rhyolite Member, showing discrimination fields from Pearce et al. (1984). Both Ta and Yb contents increase toward the base of the Wilburn Rhyolite ash-flow sheet. The field shows the range of compositions of granitoids of the Crossnore Complex. Data are from Tollo et al. (2004).

tion represent the deepest levels of the magma chamber. The experimental data of Clemens et al. (1986) show that a mildly peralkaline rhyolite similar to the Wilburn must have a temperature in excess of 800 °C with a total H_2O content <2% to have the phenocryst assemblage present in the rocks.

Phase relationships in the Q-Or-Ab system can also be used to explain the ubiquitous resorption of quartz phenocrysts seen in the ash-flow sheet. Because all magmas were precipitating alkali feldspar solid solution plus quartz, compositions must have lain along the quartz-feldspar cotectic for a given steam saturation pressure. A reduction in lithostatic and (or) steam saturated pressure would shift the cotectic toward the quartz apex (Fig. 13). Thus, a magma that crystallized quartz and alkali feldspar at high pressure would appear to move off the cotectic into the alkali feldspar phase field, where quartz is unstable, resulting in partial resorption. Such a reduction in pressure could be accomplished either by degassing or movement of the magma to higher levels in the crust, or both. Petrographic evidence indicating resorption of both quartz and perthite phenocrysts in the same thin section suggests that the magma chamber underwent more than one cycle of compression and decompression.

The very low Ba and Sr contents and low Eu/Eu of the Wilburn Rhyolite magma implies a significant amount of feldspar fractionation in the parent magma. Bowen (1945) was the first to point out that calcic plagioclase is the first feldspar to crystallize from a silicate melt containing any Ca. Thus, crystallization of plagioclase would deplete Al in the melt relative to Na and K, eventually leading to a peralkaline (molecular $Na_2O + K_2O > Al_2O_3$) composition. This process does not always lead to a peralkaline end product, because phases containing alkalies, such as alkali feldspar, amphibole, or biotite, begin crystallizing. Previous study of rhyolites of the Mt. Rogers volcanic center has shown that the Wilburn Rhyolite magma could have been derived from magma similar to the underlying Buzzard Rock lava flows through crystallization and removal of ~20% feldspar (Lopez-Escobar, 1972). The Buzzard Rock contains phenocrysts of plagioclase and Ba-rich perthite (S.W. Novak, unpublished data). A plot of Ba versus Eu/Eu* shows that the Wilburn rhyolite magmas were much more evolved than the A-type rocks

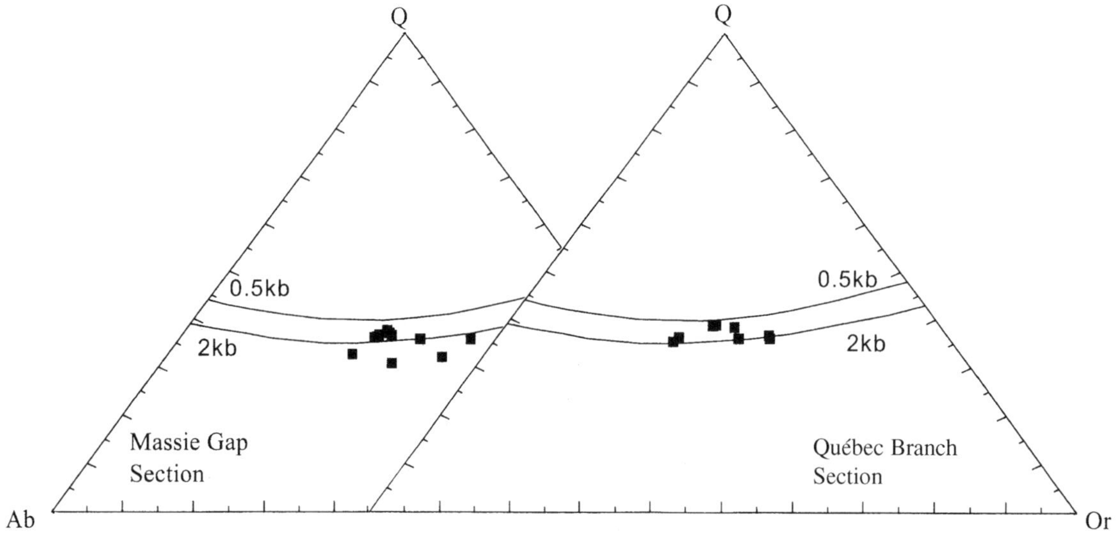

Figure 13. Plots of normative quartz (Q)-orthoclase (Or)-albite (Ab) for sampled sections of the Wilburn Rhyolite Member: (left) Massie Gap section; (right) Québec Branch section. Cotectic data for pH_2O = 0.5 kilobars and 2 kilobars are from Tuttle and Bowen (1958).

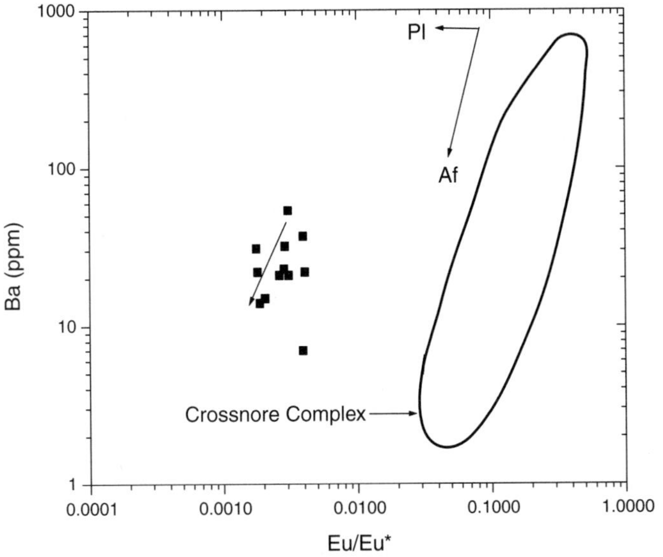

Figure 14. Ba versus Eu/Eu* for the Wilburn Rhyolite Member. Arrow in the symbols indicates the trend toward the base of the ash-flow sheet of the Wilburn Rhyolite Member. Arrows labeled "Pl" and "Af" show the trends calculated for fractionation of 30% plagioclase and alkali feldspar, respectively, as calculated in Eby (1990). The field shows the range of compositions of granitoids of the Crossnore Complex. Data are from Tollo et al. (2004).

studied by Eby (1990), indicating the significance of early plagioclase crystallization in deriving this magma (Fig. 14). An additional feature of this plot is the significant change in Ba with little change in Eu/Eu*, indicating that fractionation in the Wilburn magma chamber must have been dominated by alkali feldspar.

A plot of Sc/Ta versus Yb/Ta has been used to demonstrate the relative importance of pyroxene and zircon fractionation in producing the Robertson River Igneous Suite in the Shenandoah massif (Tollo and Aleinikoff, 1996). This plot (Fig. 15) shows that the Wilburn Rhyolites overlap the compositions of the coeval Crossnore Complex plutonic rocks (Rankin, 1970). Crystallization of pyroxene will preferentially incorporate Sc, leading to a large variation in the Sc/Ta ratio. The nearly vertical trend for the Wilburn Rhyolite indicates the importance of pyroxene fractionation in producing these magmas, with little or no zircon fractionation. The most chemically evolved magmas have the lowest Sc/Ta ratio, resulting from extensive pyroxene fractionation.

Similarly, a plot of Hf/Ta versus Ce/Yb (Fig. 16) has been used to examine the importance of allanite and zircon in the fractional crystallization history of the Robertson River Intrusive Suite (Tollo and Aleinikoff, 1996). Crystallization of allanite, or another light REE (LREE)-rich phase, will cause a decrease in the Ce/Yb ratio, whereas crystallization of zircon causes the

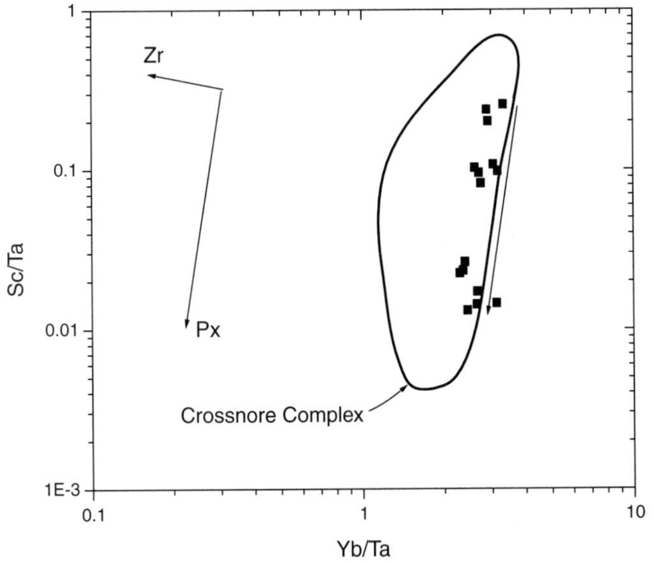

Figure 15. Sc/Ta versus Yb/Ta for the Wilburn Rhyolite Member. Arrow in the symbols indicates the trend toward the base of the ash Wilburn Rhyolite flow sheet. Arrows labeled "Zr" and "Px" show the calculated effects of fractionation of 0.5% zircon or 5% clinopyroxene, respectively, as calculated in Tollo and Aleinikoff (1996). The field shows the range of compositions of granitoids of the Crossnore Complex. Data are from Tollo et al. (2004).

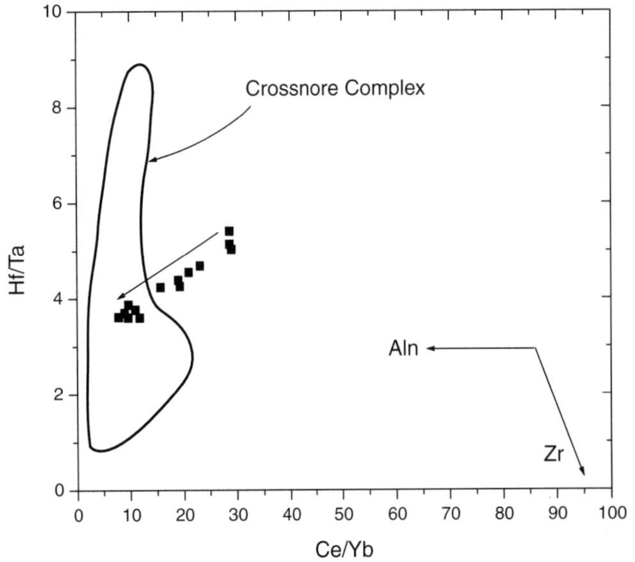

Figure 16. Hf/Ta versus Ce/Yb for the Wilburn Rhyolite Member. Arrow in the symbols indicates the trend toward the base of the Wilburn Rhyolite ash-flow sheet. Arrows labeled "Al" and "Zr" show the calculated effects of fractionation of 0.5% allanite and 0.5% zircon, respectively, as calculated in Tollo and Aleinikoff (1996). The field shows the range of compositions of granitoids of the Crossnore Complex. Data are from Tollo et al. (2004).

Hf/Ta ratio to decrease. Many other peralkaline rhyolites contain chevkinite as a phase that would fractionate LREE (Mahood and Hildreth, 1983; Wolff and Storey, 1984; Novak and Mahood, 1986; Sawyer and Sargent, 1989); the phase is not known to occur in the Wilburn Rhyolite. Figure 16 shows greater variation in the Ce/Yb ratio than in the Hf/Ta ratio for the Wilburn, which is consistent with a predominance of allanite fractionation over zircon. Although the most evolved portion of the Wilburn Rhyolite overlaps the field of the coeval Crossnore Complex, the trend is nearly perpendicular to the trend for the latter, suggesting that zircon was more important in generating the Crossnore Complex magmas. Experimental results have shown that peralkalinity suppresses crystallization of zircon (Watson, 1979). The relatively small amount of zircon fractionation in the Wilburn Rhyolite may be a further indication of the peralkalinity of this magma.

Chondrite-normalized REE plots for the Wilburn sections show progressive changes through the ash-flow sheet. Samples at the base of the sheet have relatively flat patterns with a large Eu anomaly (Fig. 17). With increasing stratigraphic height, LREE are relatively enriched, heavy REE are depleted, and the Eu anomaly lessens. These features indicate the importance of a LREE-fractionating phase, such as allanite, in producing compositional differences in the Wilburn Rhyolite magma chamber. In addition, the large Eu anomaly emphasizes, along with Ba and Sr depletions, the importance of significant feldspar fractionation in the generation of this magma.

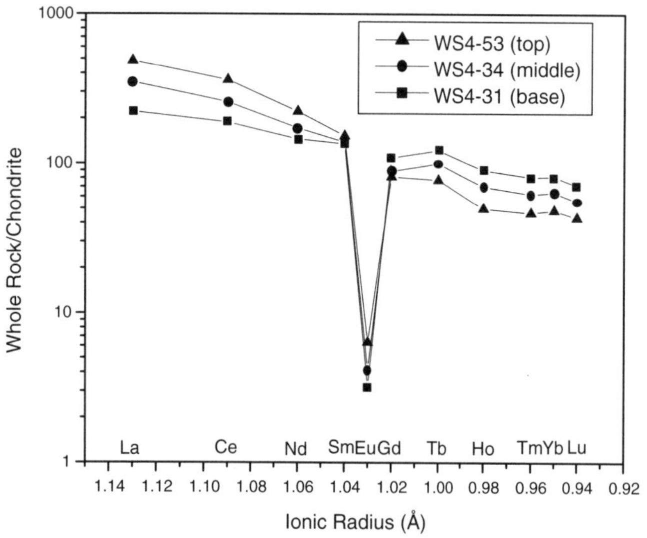

Figure 17. Chondrite-normalized REE contents for selected samples of the Massie Gap section of the Wilburn Rhyolite Member. Sample WS4-31 is from the base of the ash-flow sheet and WS4-53 is from the top. Note the crossover in REE abundances from the top of the sheet to the base.

DEVELOPMENT OF COMPOSITIONAL ZONING

Mineralogical differences in the microphenocryst assemblages suggest a threefold division within the ash-flow sheet of the Wilburn Rhyolite: a basal zone characterized by aegerine plus riebeckite, a middle zone characterized by riebeckite, and an upper zone characterized by biotite. Chemical gradients are continuous within the sheet, however, and do not show consistent deflections or offsets at zone boundaries defined by mineralogy. Thus, it appears that the ash-flow sheet was emplaced by a series of chronologically closely spaced eruptions from a single zoned magma chamber that was biotite bearing (metaluminous) at depth and aegerine bearing (peralkaline) at the top. Concentrations of SiO_2, Na_2O, F, Be, Rb, Zr, Nb, Sn, Hf, Ta, Th, U, Tb, and Yb increase toward the base of the ash-flow sheet (top of the magma chamber), whereas Al_2O_3, MgO, CaO, K_2O, TiO_2, Sc, Sr, Ba, La, Ce, Nd, and Eu increase toward the top of the ash-flow sheet (deeper in the magma chamber). Figure 18 summarizes mineralogical changes and the direction of relative enrichments of elements in the ash-flow sheet.

Although compositionally zoned magma chambers are common in volcanic systems, debate continues about the method of producing the zonation (Smith, 1979; Hildreth, 1979, 1981; Wolff and Storey, 1984; Trial and Spera, 1990; Wolff et al., 1990; De Silva and Wolff, 1995). These authors discussed a number of mechanisms for producing zoning in a magma chamber, including crystal fractionation, roof melting, volatile complexing, and selective extraction of partial melts in the source area. Among these possibilities, the prevailing opinion is that a combination of crystal fractionation and volatile segregation toward the roof of the magma chamber produces the gradients seen in most magma chambers. Isotopic data commonly shows limited wall-rock involvement in silicic systems, and mechanical arguments reject the notion of selective separation of viscous silicic magma from a source melting area (Hildreth, 1981). Formation of a compositionally zoned ash-flow deposit requires a pre-eruptive volatile gradient in the magma chamber (Wolff et al., 1990).

It is clear that the Wilburn Rhyolite magma chamber was zoned prior to eruption. The present state of preservation of the rocks, however, makes it difficult to identify uniquely the process that produced the zoning. Trace-element ratios suggest that fractionation of plagioclase, pyroxene, and a LREE-rich phase, such as allanite or chevkinite, were important in the derivation of the high-silica rhyolite magma of the Wilburn Rhyolite. Indeed, crystal fractionation can generally be invoked to explain the majority of zoning in most such magma chambers. This is thought to occur along the side-walls of the magma chamber, with the differentiated magma rising toward the roof of the system (Wolff et al., 1990). Because of the lifetimes demonstrated for magma chambers of the size of the Wilburn Rhyolite chamber (>50 k.y., Trial and Spera, 1990), it is clear that significant thermal input must be made to the base of the system through basaltic melts (Hildreth, 1981). The relatively

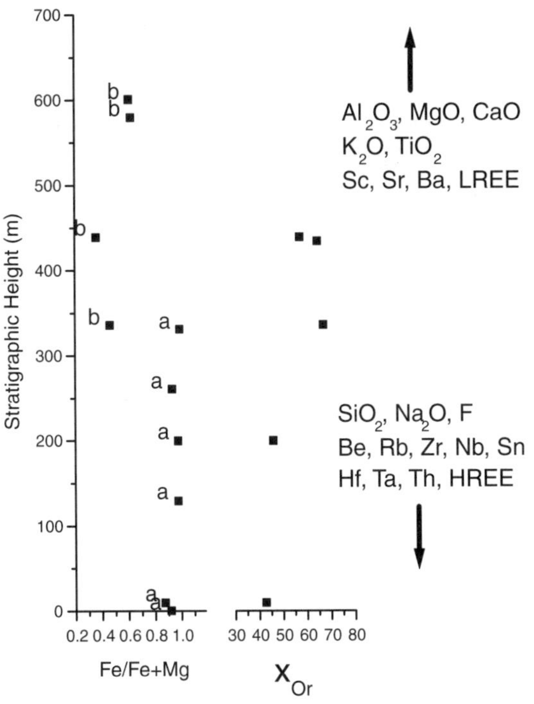

Figure 18. Variations in mineral chemistry and trace-element enrichments as a function of stratigraphic height. Symbols labeled "a" indicate amphibole and those labeled "b" indicate biotite phenocrysts.

high flux of Cl and F from these basal magmas may be important in bringing excesses of alkali and certain trace elements to the roof of the magma chamber (Hildreth 1981; Mahood, 1981). Relatively high Cl contents have been described for the mildly peralkaline parts of several systems (Mahood, 1981; Novak and Mahood, 1986). Support for these ideas comes from a model by Wiebe et al. (1997), which suggests that highly evolved A-type magma may be produced by multiple inputs of basaltic magma to the base of the silicic system. Prolonged mixing and fractional crystallization in the less-evolved base of the magma system could produce the silicic magmas that ascend to form the gradients found in the upper parts of the system.

DISCUSSION AND CONCLUSIONS

The eruptive source of the Wilburn Rhyolite Member has not been positively identified. No fissure source areas have been recognized; no remnant of a volcanic edifice has been recognized. Smith (1960) states that nearly all ash-flow fields with a volume of more than a few cubic miles and a known source region are associated with calderas or other depressions of subsidence. Estimating the volume of the Wilburn ash-flow sheet is hindered by the complex geometry of the thrust sheets in the area. It crops out in several thrust slices, including the Mountain City window and the Shady Valley thrust sheet (included in Mount Rogers Formation, undivided, Fig. 2). The volume of the Wilburn in the Stone Mountain thrust sheet alone is probably on the order of 30 km^3 (>7 mi^3). A caldera source for the Wilburn is likely.

The matrix-poor breccia on the north slope of Mt. Rogers and Pine Mountain (Fig. 2) is interpreted as debris shed from a caldera wall infiltrated with ash from the Wilburn Rhyolite. Absence of sodic pyroxene and amphibole in the matrix of the breccia suggests that the tuff in the matrix does not represent the earliest eruptions of the Wilburn, consistent with caldera collapse in the later stages of the eruptions. These late-stage breccias are in contact (stratigraphic?) with the Whitetop Rhyolite, where it is preserved north of the breccias in the Stone Mountain thrust sheet. This study suggests that the basal Wilburn to the south, stratigraphically overlying the Whitetop, is the product of earlier eruptions. Thus, the Wilburn Rhyolite in the Stone Mountain thrust sheet appears to be in a down-dropped block relative to the Whitetop Rhyolite along the north side of the breccia, suggesting that the Wilburn, here, may be caldera fill. The breccia crops out only at or near the leading edge of the Stone Mountain thrust sheet. The breccia is the best field evidence for a caldera source of the Wilburn. Evidence that the caldera was to the south of the breccia is not compelling, but it is not possible to compare the ash-flow sheet now preserved south of the breccia with the ash-flow sheet that once may have lain north of the breccia. If the caldera lay north of the breccia, it has been removed by erosion.

One of the outstanding features of the Wilburn Rhyolite Member is that, despite posteruptive crystallization and vapor-phase alteration, as well as the effects of low-grade metamorphism and tectonism during the past 760 m.y., subtle mineralogical and compositional gradients are preserved in the ash-flow sheet. The trace-element suite that is relatively enriched in peralkaline rocks has been shown to be elements that have affinity for both F and Cl. Production of peralkaline magmas, therefore, may be influenced by high halogen fluxes from basalt magmas injected below the rhyolitic chamber (Hildreth, 1981). The well-known association of peralkaline volcanic rocks with regions of crustal rifting may indicate that high basalt input beneath the silicic chamber is a necessary ingredient for the development of peralkaline silicic magma.

The association of the Wilburn Rhyolite Member, and associated lavas, with other roughly coeval peralkaline and metaluminous intrusive bodies and dikes of the Crossnore Complex is similar to several rift-related volcanic-plutonic complexes, such as those of the White Mountain batholith of New England (Eby et al., 1992); other provinces summarized by Eby (1990), such as the Nigerian rift, the Arabian Peninsula, and the Topsails Suite of Newfoundland (see also Whalen and Currie, 1990); and the Transbaikal of Russia (Litvinovsky et al., 2002). These provinces are characterized by well-defined, well-separated magma chambers within continental crust and are clearly different from smaller volume, strongly peralkaline volcanic centers of well-developed rifts with thinner continental crust, in which basaltic eruptions predominate, such as the east African rift. Basalt is a significant component of the lower part of the Mount Rogers Formation; mafic dikes constitute a significant volume in the underlying basement rocks (locally as much as 30% over areas as large as 8 km^2; Rankin, 1970), and some mafic dikes cut the Wilburn Rhyolite (Fig. 3). It is not known whether these dikes are part of the ca. 730 Ma Bakersville dike swarm (Crossnore Complex), or belong to the younger Late Neoproterozoic rifting event, or are of both ages. It is common for large rhyolitic calderas to show a paucity of basaltic eruptive products near the caldera, even though basaltic intrusion must ultimately supply the heat needed to generate and sustain such large rhyolitic chambers, because the rhyolitic chamber prevents the more dense basaltic magmas from rising to the surface (Smith, 1979; Hildreth, 1981). Development of the relatively voluminous rhyolitic magma chamber at Mt. Rogers probably indicates that relatively thick continental crust was present in the area prior to its formation, and this may explain why rifting was not complete in this area after eruption of the Wilburn Rhyolite. Relatively voluminous high-silica, mildly-peralkaline magma chambers are known only from the continental interior (Hildreth, 1981) or rifted continental margins (Eby, 1990). Thus, the event that produced the Wilburn Rhyolite magma chamber may have represented the initial pulse in a long-term process of thinning the continental crust that eventually culminated 200 m.y. later in the opening of the Iapetus Ocean.

ACKNOWLEDGMENTS

We thank R.L. Smith for helpful suggestions and interpretation of the trace-element chemistry of the Wilburn Rhyolite. J.S. Huebner and C. Thornber helped with the feldspar homogenization experiment, and J. McGee with the electron microprobe analyses. Thorough reviews by D.B. Stewart, R.P. Tollo, M.C. Gilbert, and S. Seaman resulted in significant improvements in the manuscript.

REFERENCES CITED

Adams, M.G., Stewart, K.G., Trupe, C.H., and Willard, R.A., 1995, Tectonic significance of high-pressure metamorphic rocks and dextral strike-slip faulting along the Taconic suture, *in* Hibbard, J.P., et al., eds., Current perspectives in the Appalachian-Caledonian orogen: Geological Association of Canada, Special Paper 41, p.21–42.

Aleinikoff, J.N., Zartman, R.E., Walter, M., Rankin, D.W., Lyttle, P.T., and Burton, W.C., 1995, U-Pb ages of metarhyolites of the Catoctin and Mount Rogers Formations, central and southern Appalachians: Evidence for two pulses of Iapetan rifting: American Journal of Science, v. 295, p. 428–454.

Badger, R.L., and Sinha, A.K., 1988, Age and Sr isotopic signature of the Catoctin volcanic province: Implications for subcrustal mantle evolution: Geology, v. 16, p. 692–695.

Bailey, D.K., and Schairer, J.F., 1966, The system Na_2O-Al_2O_3-Fe_2O_3-SiO_2 at 1 atmosphere and the petrogenesis of alkaline rocks: Journal of Petrology, v. 7, part 1, p. 114–170.

Baker, B.H., and Henage, L.F., 1977, Compositional changes during crystallization of some peralkaline silicic lavas of the Kenya Rift Valley: Journal of Volcanology and Geothermal Resources, v. 2, p. 17–28.

Borley, G.D., 1963, Amphiboles from the younger granites of Nigeria, part 1, chemical classification: Mineralogical Magazine, v. 33, p. 358–376.

Bowen, N.L., 1945, Phase equilibria bearing on the origin and differentiation of alkaline rocks: American Journal of Science, Daly Volume, v. 243-A, p. 75–89.

Boyd, F.R., 1961, Welded tuffs and flows in the Rhyolite Plateau of Yellowstone Park, Wyoming: Geological Society of America Bulletin, v. 72, p.387–426.

Cawood, P.A., McCausland, P.J.A., and Dunning, G.R., 2001, Opening Iapetus: Constraints from the Laurentian margin in Newfoundland: Geological Society of America Bulletin, v. 113, p. 443–453.

Clemens, J.D., Holloway, J .R., and White, A.J.R., 1986, Origin of an A-type granite: Experimental constraints: American Mineralogist, v. 71, p. 317–324.

Day, H.W., and Brown, V.M., 1980, Evolution of perthite composition and microstructure during progressive metamorphism of hypersolvus granite, Rhode Island, USA: Contributions to Mineralogy and Petrology, v. 72, p. 353–365.

Deer, W.A., Howie, R.A., and Zussman, J., 1997, Rock forming minerals, v. 2B, 2nd edition, Double chain silicates: London, The Geological Society, 764 p.

DeSilva, S.L., and Wolff, J.A., 1995, Zoned magma chambers: The influence of magma chamber geometry on sidewall convective fractionation: Journal of Volcanology and Geothermal Research, v. 65, p. 111–118.

Eby, G.N., 1990, The A-type granitoids: A review of their occurrence and chemical characteristics and speculations on their petrogenesis: Lithos, v. 26, p. 115–134.

Eby, G.N., Krueger, H.W., and Creasy, J.W., 1992, Geology, geochronology, and geochemistry of the White Mountain batholith, New Hampshire, *in* Puffer, J.N., and Ragland, P.C., eds., Eastern North America magmatism: Geological Society of America Special Paper 268, p. 379–398.

Ekstrom, T.K., 1972, Distribution of fluorine among some coexisting minerals: Contributions to Mineralogy and Petrology, v. 34, p. 192–200.

Fetter, A.H., and Goldberg, S.A., 1995, Age and geochemical characteristics of bimodal magmatism in the Neoproterozoic Grandfather Mountain rift basin: Journal of Geology v. 103, p. 313–326.

Goddard, E.N., Trask, P., DeFord, R.K., Rove, O.N., Singewald, J.T., Jr., and Overbeck, R.M., 1970, Rock color chart: Boulder, Colorado, Geological Society of America.

Goff, F.E., and Czamanske, G.K., 1972, Calculation of amphibole structural formulae: Reston, Virginia, U.S. Geological Survey Computer Contribution, 16 p.

Goldberg, S.A., Butler, J.R., and Fullagar, P.D., 1986, The Bakersville dike swarm: Geochronology and petrogenesis of Late Proterozoic basaltic magmatism in the Southern Appalachian Blue Ridge: American Journal of Science, v. 286, p. 403–430.

Gottfried, D., Annell, C.S., and Schwarz, L.J., 1977, Geochemistry of subsurface basalt from the deep corehole (Clubhouse Crossroads Corehole 1) near Charleston, South Carolina—magma type and tectonic implications, *in* Rankin, D.W., ed., Studies related to the Charleston, South Carolina, Earthquake of 1886—A preliminary report: U.S. Geological Survey Professional Paper 1028, p. 91–113.

Hildreth, E.W., 1979, The Bishop Tuff: Evidence for the origin of compositional zonation in silicic magma chambers, *in* Chapin, C.E., and Elston, W.E., eds., Ash-flow tuffs: Geological Society of America Special Paper 180, p. 43–75.

Hildreth, E.W., 1981, Gradients in silicic magma chambers: Implications for lithospheric magmatism: Journal of Geophysical Research, v. 86, p. 10153–10192.

Laird, J., Metamorphism, *in* Hatcher, R.D., Jr., et al., eds., 1989, The Appalachian-Ouachita orogen in the United States: Boulder, Colorado, Geological Society of America, The Geology of North America, v. F-2, p.134–151.

Leake, B.E., Wooley, A.R., Arps, E.S., Birch, W.D., Gilbert, C.M., Grice, J.D., Hawthorne, F.C., Kao, A., Kisch, H.J., Krivovichev, V.G., Linthout, K., Laird, J., Manarino, J.A., Maresch, W.V., Nickel, E.H., Rock, N.M.S., Schumacher, J.C., Smith, D.C., Stephenson, N.C.N., Ungaretti, L., Whittaker, J.W., and Youzhi, G., 1997, Nomenclature of amphiboles: Report of the Mineralogical Association, Committee on New Mineral Names: American Mineralogist, v. 82, p. 1019–1037.

Litvinovsky, B.A., Jahn, B., Znvilevich, A.N., Saunders, A., Poulain, S., Kuzmin, D.V., Reichow, M.K., and Titov, A.V., 2002, Petrogenesis of syenite-granite suites from the Bryansky Complex (Transbaikalia, Russia): Implications for the origin of A-type granitoid magmas: Chemical Geology, v. 189, p. 105–133.

Lopez-Escobar, L., 1972, Appalachian rhyolites: Geochemical data concerning their origin [M.S. thesis]: Cambridge, Massachusetts Institute of Technology, 134 p.

Ludington, S.D., 1974, Application of fluoride-hydroxyl exchange data to natural minerals [Ph.D. dissertation]: Boulder, University of Colorado, 177 p.

Luth, W.C., 1969, The systems $NaAlSi_3O_8$-SiO_2 and $KAlSi_3O_8$-SiO_2 to 20 kb and the relationship between H_2O content, P_{H2O} and P_{total} in granitic magmas: American Journal of Science, v. 267-A, p. 325–341.

Lyons, P.C., 1976, The chemistry of riebeckites of Massachusetts and Rhode Island: Mineralogical Magazine, v. 40, p. 473–479.

MacDonald, R., and Bailey, D.K., 1973, The chemistry of the peralkaline oversaturated obsidians: U.S. Geological Survey Professional Paper 440-N, 37 p.

Mahood, G.A., 1981, Chemical evolution of a late Pleistocene rhyolitic center: Sierra la Primavera, Jalisco, Mexico: Contributions to Mineralogy and Petrology, v. 77, p. 129–149.

Mahood, G.A., 1984, Pyroclastic rocks and calderas associated with strongly peralkaline magmatism: Journal of Geophysical Research, v. 89, p. 8540–8552.

Mahood, G., and Hildreth, W., 1983, Large partition coefficients for trace elements in high-silica rhyolites: Geochimica et Cosmochimica Acta, v. 47, p. 11–30.

Morimoto, N., Fabries, J., Ferguson, A.K., Ginzburg, I.V., Ross, M., Seifert, F.A., Zussman, J., Aoki, K., and Gottardi, G., Nomenclature of pyroxenes: 1988: Mineralogical Magazine, v. 52, p. 535–550.

Noble, D.C., 1970, Loss of sodium from crystallized comendite welded tuffs of the Miocene Grouse Canyon Member of the Belted Range Tuff, Nevada: Geological Society of America Bulletin, v. 81, p. 2677–2688.

Noble, D.C., and Parker, D.F., 1974, Peralkaline silicic volcanic rocks of the western United States: Bulletin Volcanologique, v. 38, p. 803–827.

Novak, S.W., and Mahood, G., 1986, Rise and fall of a basalt-trachyte-rhyolite magma system at the Kane Springs Wash Caldera, Nevada: Contributions to Mineralogy and Petrology, v. 94, p. 352–373.

Novak, S.W., and Rankin, D.W., 1980, Mineralogy and geochemistry of an ash-flow tuff of peralkaline affinity from the Mt. Rogers Formation, Grayson Co., Va.: Geological Society of America Abstracts with Programs, v. 12, p. 203–204.

Orville, P.M., 1963, Alkali ion exchange between vapor and feldspar phases: American Journal of Science, v. 261, p. 201–237.

Papike, J.J., Cameron, K.L., and Baldwin, K., 1974, Amphiboles and pyroxenes: Characterization of other than quadrilateral components and estimates of ferric iron from microprobe data: Geological Society of America Abstracts with Programs, v. 6, p. 1053–1054.

Pearce, J.A., Harris, N.B.W., and Tindle, A.G., 1984, Trace element discrimination diagrams for the tectonic interpretation of granitic rocks: Journal of Petrology, v. 25, p. 956–983.

Plumb, K.A., 1991, New Precambrian time scale: Episodes, v. 14, p. 139–140.

Rankin, D.W., 1970, Stratigraphy and structure of Precambrian rocks in northwestern North Carolina, *in* Fisher, G.W., et al., eds., Studies of Appalachian Geology—Central and southern: New York, Interscience Publishers, p. 227–245.

Rankin, D.W., 1975, The continental margin of eastern North America in the southern Appalachians: The opening and closing of the proto-Atlantic Ocean: American Journal of Science, Tectonics and Mountain Ranges, v. 275-A, p. 298–336.

Rankin, D.W., 1976, Appalachian salients and recesses: Late Precambrian continental breakup and the opening of the Iapetus Ocean: Journal of Geophysical Research, v. 81, p. 5605–5619.

Rankin, D.W., 1993, The volcanogenic Mount Rogers Formation and the overlying glaciogenic Konnarock Formation—Two Late Proterozoic units in southwestern Virginia: U.S. Geological Survey Bulletin 2029, 26 p.

Rankin, D.W., 1994, Continental margin of the eastern United States: Past and present, *in* Speed, R.C., ed., Phanerozoic evolution of North American continent-ocean transitions: Boulder, Colorado, Geological Society of America, Continent-Ocean Transects, Accompanying Volume, p. 129–218.

Rankin, D.W., Stern, T.W., Reed, J.C., Jr., and Newell, M.F., 1969, Zircon ages of felsic volcanic rocks in the upper Precambrian of the Blue Ridge, Appalachian Mountains: Science, v. 166, p. 741–744.

Rankin, D.W., Lopez-Escobar, L., and Frey, F.A., 1974, Rhyolites of the upper Precambrian Mt. Rogers (Virginia) volcanic center: Geochemistry and petrogenesis [abs.]: American Geophysical Union Transactions, v. 55, no. 4, p. 475.

Rankin, D.W., Dillon, W.P., Black, D.F.B., Boyer, S.E., Daniels, D.L., Goldsmith, R., Grow, J.A., Horton, J.W., Jr., Hutchinson, D.R., Klitgord, K.D., McDowell, R.C., Milton, D.J., Owens, J.P., and Phillips, J.D., 1991, E-4, Central Kentucky to Carolina trough: Boulder, Colorado, Geological Society of America, Centennial Continent/Ocean Transect no. 16, 41 p., 2 sheets, scale 1:500,000.

Rankin, D.W., Miller, J.M.G., and Simpson, E.L., 1994, Geology of the Mt. Rogers area, southwestern Virginia Blue Ridge and Unaka belt, *in* Schultz, A., and Henika, W., eds., Fieldguides to Southern Appalachian structure, stratigraphy, and engineering geology: Blacksburg, Virginia Tech Department of Geological Sciences Guidebook No. 10, p. 127–176.

Rucklidge, J.C., 1972, Specifications for Fortran program Superrecal [unpublished report]: Ontario, Toronto University.

Sawyer, D.A., and Sargent, K.A., 1989, Petrologic evolution of divergent peralkaline magmas from the Silent Canyon Caldera Complex, southwestern Nevada volcanic field: Journal of Geophysical Research, v. 94, p. 6021–6040.

Schminke, H.U., 1975, Volcanological aspects of peralkaline silicic welded ash-flow tuffs: Bulletin Volcanologique, v. 38, p. 594–636.

Scott, R.B., Bachinski, S.W., Nesbitt, R.W., and Scott, M.W., 1971, Rate of Al-Si ordering in sanidines from an ignimbrite cooling unit: American Mineralogist, v. 56, p. 1208–1221.

Simpson, E.L., and Sundberg, F.A., 1987, Early Cambrian age for synrift deposits of the Chilhowee Group of south Virginia: Geology, v. 15, p. 123–126.

Smith, R.L., 1960, Zones and zonal variations in welded ash-flows: U.S. Geological Survey Professional Paper 354-F, p. 149–159.

Smith, R.L., 1979, Ash-flow magmatism, *in* Chapin, C.E., and Elston, W.E., eds., Ash-flow tuffs: Geological Society of America Special Paper 180, p. 5–27.

Smith, R.L., and Bailey, R.A., 1966, The Bandelier Tuff: A study of ash-flow eruption cycles from zoned magma chambers: Bulletin Volcanologique, v. 29, p. 83–104.

Su, Q., Goldberg, S.A., and Fullagar, P.D., 1994, Precise U-Pb zircon ages of Neoproterozoic plutons in the southern Appalachian Blue Ridge and their implications for the initial rifting of Laurentia: Precambrian Research, v. 68, p. 81–95.

Thomas, W.A., 1991, The Appalachian-Ouachita rifted margin of southeastern North America: Geological Society of America Bulletin, v. 103, p. 415–431.

Thornton, C.P., and Tuttle, O.F., 1960, Chemistry of igneous rocks. I. Differentiation index: American Journal of Science, v. 258, p. 664–684.

Tollo, R.P., and Aleinikoff, J.N., 1996, Petrology and U-Pb geochronology of the Robertson River Igneous Suite, Blue Ridge Province, Virginia—Evidence for multistage magmatism associated with an early episode of Laurentian rifting: American Journal of Science, v. 296, p. 1045–1090.

Tollo, R.P., Aleinikoff, J.N., Bartholomew, M.J., and Rankin, D.W., 2004, Neoproterozoic A-type granitoids of the Central and Southern Appalachians:

Intraplate magmatism associated with episodic rifting of the Rodinian Supercontinent: Precambrian Research, v. 128, p. 3–38.

Trial, A.F., and Spera, F.J., 1990, Mechanisms for the generation of compositional heterogeneities in magma chambers: Geological Society of America Bulletin, v. 102, p. 353–367.

Tuttle, O.F., and Bowen, N.L., 1958, Origin of granite in the light of experimental studies in the system $NaAlSi_3O_8$-$KAlSi_3O_8$-SiO_2-H_2O: Geological Society of America Memoir 74, 153 p.

Walsh, G.J., and Aleinikoff, J.N., 1999, U-Pb zircon age of metafelsite from the Pinney Hollow Formation: Implications for the development of the Vermont Appalachians: American Journal of Science, v. 299, p. 157–170.

Watson, E.B., 1979, Zircon saturation in felsic liquids: Experimental results and application to trace element geochemistry: Contributions to Mineralogy and Petrology, v. 70, p. 407–419.

Whalen, J.B., Currie, K.L., and Chappell, B.W., 1987, A-type granites: Geochemical characteristics, discrimination and petrogenesis: Contributions to Mineralogy and Petrology, v. 95, p. 407–419.

Whalen, J.B., and Currie, K.L., 1990, The Topsails igneous suite, western Newfoundland; Fractionation and magma mixing in an "orogenic" A-type granite suite, *in* Stein, H.J., and Hannah, J.L., eds., Ore-bearing granite systems; petrogenesis and mineralizing processes: Geological Society of America Memoir 246, p. 287–299.

Wiebe, R.A., Holden, J.B., Coombs, M.L., Wobus, R.A., Schuh, K.J., and Plummer, B.P., 1997, The Cadillac Mountain intrusive complex, Maine: The role of shallow-level magma chamber processes in the generation of A-type granites, *in* Sinha, A.K., et al., eds., The nature of magmatism in the Appalachian orogen: Boulder, Colorado, Geological Society of America Memoir 191, p. 397–418.

Wolff, J.A., and Storey, M., 1984, Zoning in highly alkaline magma bodies: Geological Magazine, v. 121, p. 563–575.

Wolff, J.A., Worner, G., and Blake, S., 1990, Gradients in physical parameters in zoned felsic magma bodies: Implications for evolution and eruptive withdrawal: Journal of Volcanology and Geothermal Research, v. 43, p. 37–55.

Wones, D.R, and Gilbert, M.C., 1982, Amphiboles in the igneous environment, *in* Veblen, D.R., and Ribbe, P.H., eds., Amphiboles: Petrology and experimental phase relations: Mineralogical Society of America Reviews in Mineralogy, v. 9B, p. 355–390.

Manuscript Accepted by the Society August 25, 2003

Printed in the USA

Geological Society of America
Memoir 197
2004

Nd isotopic constraints on the magmatic history of the Goochland terrane, easternmost Grenvillian crust in the southern Appalachians

Brent E. Owens*
Department of Geology, College of William and Mary, Williamsburg, Virginia 23187, USA
Scott D. Samson
Department of Earth Sciences, Syracuse University, Syracuse, New York 13244, USA

ABSTRACT

Nd isotopic compositions for ten samples from the State Farm Gneiss (ca. 1046–1023 Ma), the Montpelier Anorthosite (1045 Ma), and several Neoproterozoic A-type granitoids (ca. 600 Ma) in the Mesoproterozoic Goochland terrane range in initial ε_{Nd} from –0.4 to +1.3. The A-type granitoids reflect Neoproterozoic rifting of the Goochland terrane, and their isotopic compositions are consistent with a substantial contribution from the State Farm Gneiss, or equivalent crust at depth, in their petrogenesis. Protoliths of the State Farm Gneiss and Montpelier Anorthosite were emplaced at approximately the same time, and their Nd isotopic compositions are the same. Based on our results and age, bulk compositional, and isotopic similarities between the State Farm Gneiss and Mesoproterozoic rocks (Pedlar River and Archer Mountain suites) in the Blue Ridge Province, we suggest that the gneiss and anorthosite represent a coeval anorthosite-charnockite suite. The unusually potassic Montpelier Anorthosite is also isotopically similar to, and the same age as, the alkalic Roseland Anorthosite in the Blue Ridge Province. Depleted mantle model-ages for the State Farm Gneiss and Montpelier Anorthosite range from 1.38 to 1.43 Ga. These ages are similar to those of many other Grenvillian crustal blocks (e.g., Adirondacks, Blue Ridge, Llano uplift) along the eastern and southern margins of Laurentia, which show a strong peak from 1.3 to 1.5 Ga. Petrological, geochronological, and geochemical data obtained thus far from the Goochland terrane are consistent with the view that it represents a fragment of Laurentia.

Keywords: Grenville, Virginia, Goochland terrane, Nd isotopes

INTRODUCTION

The Goochland terrane, an isolated block of Mesoproterozoic crust exposed in the Piedmont Province of central Virginia (Fig. 1), was recognized as Grenvillian crust two decades ago (e.g., Glover et al., 1978; Farrar, 1984). Numerous unresolved questions regarding its age, origin, and regional extent include: (1) its status as a native Laurentian or exotic terrane (e.g., Bartholomew and Lewis, 1988, 1992; Keppie et al., 1996); (2) its current extent, particularly to the west and south; (3) the timing of its accretion to Laurentia; (4) the nature and timing of any tectonic interactions with other southern Appalachian terranes

*E-mail: beowen@wm.edu.

Owens, B.E., and Samson, S.D., 2004, Nd isotopic constraints on the magmatic history of the Goochland terrane, easternmost Grenville crust in the southern Appalachians, *in* Tollo, R.P., Corriveau, L., McLelland, J., and Bartholomew, M.J., eds., Proterozoic tectonic evolution of the Grenville orogen in North America: Boulder, Colorado, Geological Society of America Memoir 197, p. 601–608.

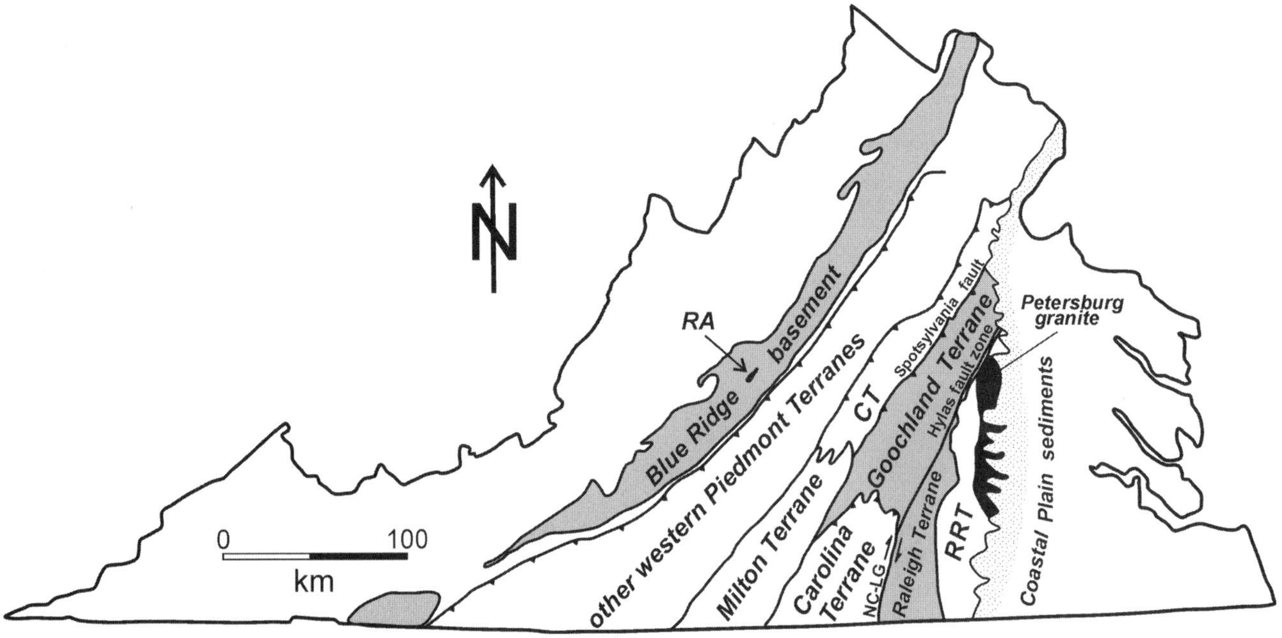

Figure 1. Map of Virginia, showing the location of the Goochland terrane (light shading), Grenvillian basement rocks of the Blue Ridge Province (darker shading), and terranes of the Piedmont Province. The Raleigh terrane, which continues into North Carolina, possibly represents a southern extension of the Goochland terrane. Rocks of the Roanoke Rapids terrane are probably equivalent to those of the Carolina terrane, but both of these are distinct from younger rocks of the Chopawamsic and Milton terranes (Coler et al., 2000). CT—Chopawamsic terrane; NC-LG—Nutbush Creek and Lake Gordon mylonite zones; RA—Roseland Anorthosite; RRT—Roanoke Rapids terrane. Simplified from Horton et al. (1991).

(e.g., the Carolina terrane); and (5) petrogenetic relationships among key units within the terrane. Compositions and precise ages of selected rock types are now available (Aleinikoff et al., 1996; Owens and Tucker, 2003), but isotopic data that could constrain some of the abovementioned unresolved issues have been lacking.

In this paper, we report the first Nd isotopic compositions of rocks from the Goochland terrane, including the State Farm Gneiss, the Montpelier Anorthosite, and several Neoproterozoic granitoids that intruded the State Farm Gneiss. We use these data to (1) evaluate relationships among rocks within the terrane; (2) compare the Mesoproterozoic portions of the Goochland terrane with other known Laurentian massifs, including the Blue Ridge to the west; and (3) at least partially evaluate questions regarding the regional extent of the terrane and its possible relationships to the Chopawamsic and Carolina arc terranes. We anticipate that our data can ultimately help constrain models for the overall origin of the Goochland terrane.

GEOLOGIC SETTING

Background information on the geology of the Goochland terrane, including details regarding deformation and metamorphism is in Farrar (1984), Gates and Glover (1989), Aleinikoff et al. (1996), Farrar and Owens (2001), and Owens and Tucker (2003). Farrar (2001) provided a recent update. The main rock units that compose the terrane include the State Farm Gneiss, Sabot Amphibolite, Maidens Gneiss, and Montpelier Anorthosite (Fig. 2). The State Farm Gneiss is broadly granitic in chemical composition (Owens and Tucker, 2003) and occupies the core of one major and two subsidiary doubly plunging antiformal domes in the eastern part of the terrane (Farrar, 1984). Owens and Tucker (2003) reported U-Pb zircon ages of ca. 1046–1023 Ma for several samples of the gneiss (including three of this study) and interpreted these as crystallization ages. The Sabot Amphibolite (not distinguished in Fig. 2) forms a ~1-km-thick sheet that structurally overlies the main dome cored by State Farm Gneiss. Elsewhere, discontinuous lenses of amphibolite occur within those portions of the Maidens Gneiss near the State Farm Gneiss (Virginia Division of Mineral Resources, 1993). The Maidens Gneiss, which forms most of the rest of the terrane, consists of rocks derived from a variety of different protoliths. The main rock types range from garnet-biotite-quartz-plagioclase gneiss to biotite-quartz-plagioclase-K-feldspar augen gneiss, along with numerous local occurrences of pelitic gneiss, quartzites, amphibolites, and calc-silicate layers (Farrar, 1984). The Montpelier Anorthosite is a small body that lies just northeast of the main dome of State Farm Gneiss (Fig. 2); it is well exposed in a currently active quarry. A U-Pb zircon age of 1045 ± 10 Ma was reported by Aleinikoff et al. (1996), and interpreted as the age of anorthosite crystallization.

Farrar and Givan (1988) and Farrar (1999) originally suggested that a number of granites in the Goochland terrane could be related to Neoproterozoic rifting. This hypothesis was recently

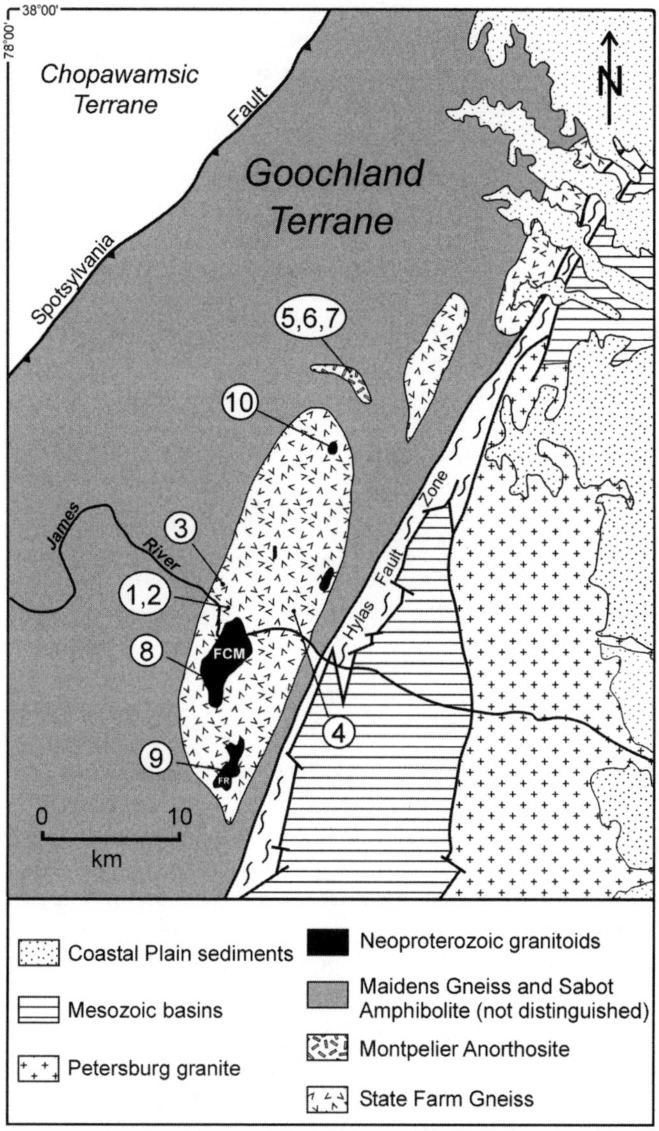

Figure 2. Geologic map of the central portion of the Goochland terrane, showing the sample locations of this study. FCM—Fine Creek Mills granite; FR—Flat Rock granite. Simplified from the geologic map of Virginia (Virginia Division of Mineral Resources, 1993), with some additions from Owens and Tucker (2003).

confirmed by Owens and Tucker (2003), who reported the presence of Neoproterozoic A-type granitoids in the main dome of State Farm Gneiss. These include the Fine Creek Mills and Flat Rock plutons, and several additional smaller bodies (Fig. 2). Owens and Tucker (2003) reported U-Pb zircon crystallization ages for five bodies that range from ca. 654 to ca. 588 Ma. However, all samples show a significant component of Mesoproterozoic inheritance (coupled, in some cases, with probable time-integrated Pb loss), which complicates age interpretations.

The full areal extent of the Goochland terrane is currently a matter of debate, particularly to the south. Farrar (1984) defined the Goochland terrane as the area in the eastern Piedmont Province that contains relict granulite-facies mineral assemblages, although those assemblages are only locally preserved. Clear evidence for this granulite event is documented in central Virginia, but evidence is more equivocal in southern Virginia and northern North Carolina (e.g., Stoddard, 1989; c.f. Farrar, 1998; Farrar and Owens, 2001). Nevertheless, published maps show the terrane extending into northern North Carolina (e.g., Farrar, 1984; Horton et al., 1989, 1991), and Farrar (1985) suggested a correlation with gneisses of the Raleigh metamorphic belt (Raleigh terrane in Fig. 1). A major fault system (extensions of the Nutbush Creek and Lake Gordon mylonite zones in Fig. 1) separates rocks of the Goochland terrane from those of the Raleigh terrane (Horton et al., 1993; Sacks, 1999), but it remains to be determined if this fault system represents a terrane boundary. In addition to this uncertainty to the south, the western limit of the terrane is also poorly constrained in central Virginia. It may correspond to the westernmost extent of the Maidens Gneiss (e.g., Burton and Armstrong, 1997), but earlier maps (e.g., Farrar, 1984; Horton et al., 1991) also include the "Central Piedmont" region (Virginia Division of Mineral Resources, 1993) within the Goochland terrane.

ANALYTICAL TECHNIQUES

Analytical protocols for whole-rock dissolution, chemical separation of Sm and Nd, and mass spectrometry follow those described in Samson et al. (1995). The only exception is that the ^{149}Sm-^{150}Nd tracer "spike" solution was added to samples prior to high temperature hydrofluoric acid dissolution. Additional analytical details are given in Table 1.

SAMPLE DESCRIPTIONS

Owens and Tucker (2003) previously described six samples of this study (Fig. 2), including three examples of State Farm Gneiss (samples 2, 3, and 4) and three Neoproterozoic granitoids (Fine Creek Mills and Flat Rock plutons and a small unnamed pluton; samples 8, 9, and 10, respectively). Whole-rock major- and trace-element compositions of these samples were provided by Owens and Tucker (2003). An additional sample of State Farm Gneiss (sample 1) comes from essentially the same area as sample 2. Three samples of Montpelier Anorthosite (samples 5, 6, and 7) were collected from within the active quarry. All are extremely fresh examples of very coarse-grained, gray anorthosite, which was described by Aleinikoff et al. (1996). The Montpelier Anorthosite is distinctive in containing highly antiperthitic plagioclase, consistent with its unusually potassic bulk composition (~3.5 wt% K_2O; Owens, unpublished results; Farrar and Owens, 2001).

ISOTOPIC RESULTS

Whole-rock Sm-Nd isotopic data for all samples are listed in Table 1, which also includes known or inferred ages for each

TABLE 1. Sm-Nd ISOTOPIC COMPOSITIONS OF GOOCHLAND TERRANE SAMPLES

Sample	Field	Age (Ma)	Sm (ppm)	Nd (ppm)	$^{147}Sm/^{144}Nd$	$^{143}Nd^{\dagger}/^{144}Nd$	$\varepsilon_{Nd}(0)$	$\varepsilon_{Nd}(T)$	T_{DM} (Ga)
State Farm Gneiss									
1	PQ-1	1035a	7.90	44.1	0.1083	0.512085 ± 5	–10.8	+0.9	1.38
2	SF98-1	1046b	21.7	108.4	0.1210	0.512184 ± 3	–8.9	+1.3	1.41
3	SF98-6	1023b	4.28	25.9	0.09978	0.511971 ± 5	–13.0	–0.3	1.43
4	SF98-4	1039b	20.7	105.9	0.1183	0.512147 ± 5	–9.6	+0.9	1.43
Montpelier Anorthosite									
5	MP-017	1045c	1.02	5.54	0.1113	0.512107 ± 4	–10.3	+1.0	1.39
6	MPSS-1	1045c	0.91	4.86	0.1136	0.512114 ± 4	–10.2	+0.9	1.41
7	MP-016	1045c	0.51	2.99	0.1025	0.512027 ± 8	–11.9	+0.7	1.39
Neoproterozoic granitoids									
8	FCM99-1	629b	29.9	146.6	0.1233	0.512360 ± 4	–5.4	+0.5	1.15
9	FR00-1	600b	7.77	24.5	0.1916	0.512642 ± 4	+0.1	+0.5	***
10	SF99-11	600b	37.4	236.9	0.09547	0.512221 ± 5	–8.1	–0.4	1.06

Note: ε_{Nd} (0) indicates present-day epsilon value; present-day bulk Earth values: $^{143}Nd/^{144}Nd$ = 0.512638; $^{147}Sm/^{144}Nd$ = 0.1966. ε_{Nd} (*T*) indicates epsilon value at reported crystallization age. T_{DM} is the depleted mantle model-age, based on the model of DePaolo (1981). Symbols and references for ages: a—Samson (unpublished data); b—Owens and Tucker (2003); c—Aleinikoff et al. (1996); ***—Sample with Sm/Nd too high to provide a meaningful model age.

†Measured ratio, corrected for spike and normalized to $^{146}Nd/^{144}Nd$ = 0.7219. Uncertainties are ±2σ and refer to the least significant digit.

sample based on U-Pb zircon results from previous studies. As a group, the ten samples span a relatively narrow range in initial ε_{Nd} from –0.4 to +1.3 (Fig. 3). Furthermore, samples within each unit (State Farm Gneiss, Montpelier Anorthosite, Neoproterozoic granitoids) typically vary by less than one ε_{Nd} unit and have model ages within 0.1 Ga of one another. Of particular note is the similarity in age, isotopic composition, and depleted mantle model-ages for the State Farm Gneiss and Montpelier Anorthosite samples.

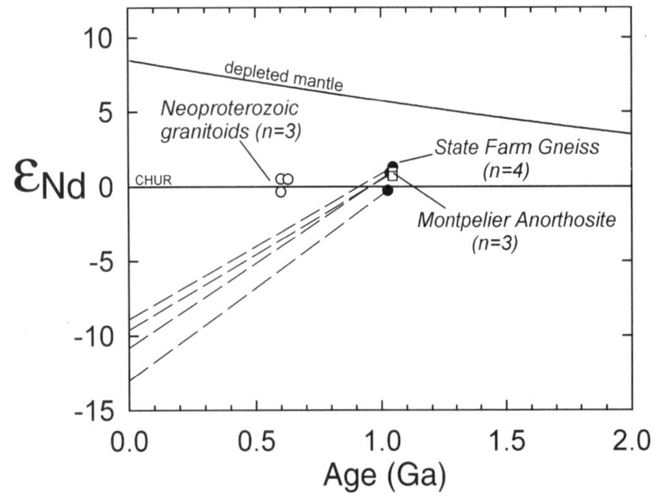

Figure 3. Initial ε_{Nd} versus age for samples from the Goochland terrane. Dashed lines illustrate the change in ε_{Nd} with time for samples of the State Farm Gneiss. CHUR—chondritic uniform reservoir. The depleted mantle curve is from DePaolo (1981).

DISCUSSION

Source Constraints for Neoproterozoic Granitoids

Owens and Tucker (2003) showed that the U-Pb zircon systematics for all Neoproterozoic granitoids in the Goochland terrane are complicated by variable amounts of an inherited component with an average age of ca. 1 Ga. Thus, at depth, these granitoids were derived from or significantly contaminated by Mesoproterozoic crust, which was similar to the State Farm Gneiss. Near-zero ε_{Nd} values displayed by these rocks are also consistent with a significant contribution from evolved crust in their origin, as are their depleted mantle model-ages of 1.15–1.06 Ga (Table 1). However, at ca. 600 Ma, ε_{Nd} values for the State Farm Gneiss range from –5.6 to –3.1 (Fig. 3). These results preclude an origin for the Neoproterozoic granitoids by wholesale anatexis of the State Farm Gneiss or equivalent crust, but they permit significant involvement of such crust in their petrogenesis. Assuming a significant component like the State Farm Gneiss as one of the source materials, the other source component or components, whether in the crust or mantle, must have had a more juvenile Nd isotopic composition than the granitoids themselves. However, the resulting isotopic composition of any mixture of State Farm gneissic material and depleted mantle material would be largely dominated by the isotopic composition of the gneisses, because of their very high Nd content (up to 108 ppm) (Table 1).

State Farm Gneiss and Montpelier Anorthosite: Coeval Charnockitic and Anorthositic Magmatism?

The geochronologic results of Aleinikoff et al. (1996) and Owens and Tucker (2003) indicate that the Montpelier Anorthosite and at least some State Farm Gneiss protoliths were emplaced contemporaneously at ca. 1045 Ma. In addition to this similarity in age, the results of our study show that these units span a similar narrow range in initial ε_{Nd} values (Fig. 3). Such similarities in age and Nd isotopic composition are a characteristic feature of some anorthosite-mangerite-charnockite complexes in the Grenville Province, including the Adirondacks (Ashwal and Wooden, 1983; Daly and McLelland, 1991; Emslie and Hegner, 1993). We suggest that the State Farm and Montpelier bodies represent a similar example of coeval charnockite-anorthosite emplacement, although the State Farm Gneiss lacks orthopyroxene (and is therefore not charnockite). The absence of orthopyroxene could reflect the strong retrograde Paleozoic overprinting that is characteristic of the Goochland terrane (Farrar, 1984).

Further evidence for the mangeritic to charnockitic character of the State Farm Gneiss is provided by the bulk compositional data reported by Owens and Tucker (2003). Specifically, fifteen samples span a range of compositions from quartz monzodiorite to granite (~55–76 wt% SiO_2), which is typical of mangerite-charnockite suites elsewhere. In addition, the State Farm Gneiss shows considerable compositional overlap with other charnockitic suites with respect to minor element oxides, such as TiO_2 and P_2O_5 (e.g., Emslie, 1991). As an example, a plot of wt% SiO_2 versus wt% TiO_2 for samples of the State Farm Gneiss and Mesoproterozoic granitic gneisses from the Pedlar and Lovingston massifs in the Blue Ridge Province to the west (Hughes et al., 1997, this volume) shows significant overlap of data points, confirming the broad compositional similarity of these units (Fig. 4). Furthermore, Blue Ridge and Goochland rocks both range to relatively high levels of TiO_2, as inferred by Farrar (1984), based on the abundance

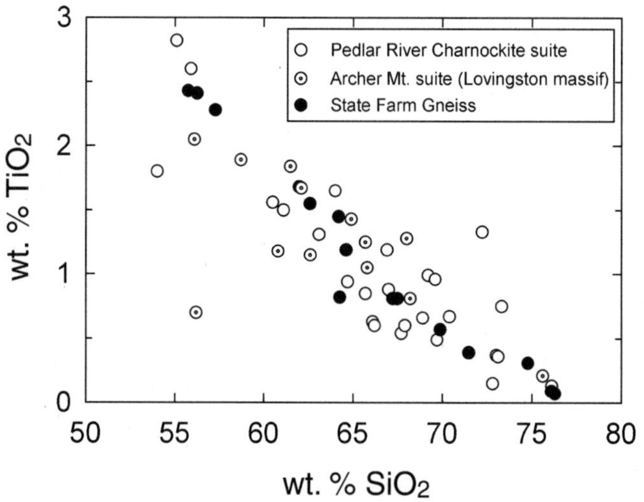

Figure 4. Plot of wt% SiO_2 versus wt% TiO_2 in whole-rock samples of the State Farm Gneiss (Owens and Tucker, 2003) and the Pedlar River and Archer Mountain Suites from the Blue Ridge Province (Hughes et al., 1997,this volume).

of titanite in the State Farm Gneiss. Initial ε_{Nd} values of the State Farm Gneiss also overlap the range of values (~0.8–2.6; 1050 Ma) reported by Pettingill et al. (1984) for the Pedlar River Charnockite Suite.

In addition, our initial ε_{Nd} values (~1.0) for the Montpelier Anorthosite are identical to those (~0.7–1.2) for the Roseland Anorthosite (Pettingill et al., 1984) in the Blue Ridge Province. Furthermore, these anorthosite bodies appear to be the same age (Pettingill et al., 1984; Aleinikoff et al., 1996), and share the distinction of probably being the most alkalic of all terrestrial anorthosites (Herz and Force, 1984, 1987; Owens and Dymek, 1999; Farrar and Owens, 2001).

All of the evidence cited here indicates that anorthositic and possibly charnockitic units in the Goochland terrane and Blue Ridge Province share many geochronological and compositional features. Our new data extend this similarity to Nd isotopic compositions. Thus, from a petrological perspective, a strong link between these closely spaced regions of Grenvillian crust is established. We recognize, however, that the current proximity of these crustal blocks may reflect substantial thrust- and/or translation-related movement of the Goochland terrane since Iapetan rifting (e.g., Bartholomew and Lewis, 1992).

Comparisons with Other Laurentian Massifs

Numerous isolated blocks of Mesoproterozoic (Grenvillian) crust are exposed along the southeastern and southern margins of Laurentia, spanning an area from the Adirondack Mountains in New York to the Franklin Mountains in Texas (and even farther south in the Oaxaca terrane, Mexico). Most of these blocks are of unquestioned Laurentian affinity (i.e., they are not interpreted as exotic terranes), although some are likely rifted fragments that were laterally translated large distances prior to Paleozoic accretion (Bartholomew and Lewis, 1988, 1992). Nd isotopic data are reported for a variety of Mesoproterozoic rocks in many of these blocks, including the Adirondacks, Blue Ridge Province, and Llano Uplift. A histogram of reported depleted mantle model-ages (T_{DM}) for these rocks (Fig. 5) illustrates the dominance of model ages in the range from 1.3 to 1.5 Ga among rocks from all of these blocks, suggesting significant additions of mantle-derived material to the Laurentian crust during this time interval (cf. Patchett and Ruiz, 1989). Calculated model ages for the State Farm Gneiss and Montpelier Anorthosite (1.38–1.43 Ga) (Table 1) also fall within this range, consistent with their derivation from source material that was isotopically similar to sources of other Laurentian terranes.

Relationship to Adjacent Terranes

The central Virginia portion of the Goochland terrane is bounded by faults on its western, eastern, and possibly southern margins. To the west, the Spotsylvania fault (Fig. 1) separates the terrane from the Ordovician Chopawamsic arc terrane (Coler et al., 2000). The Hylas mylonite zone (Gates and Glover, 1989)

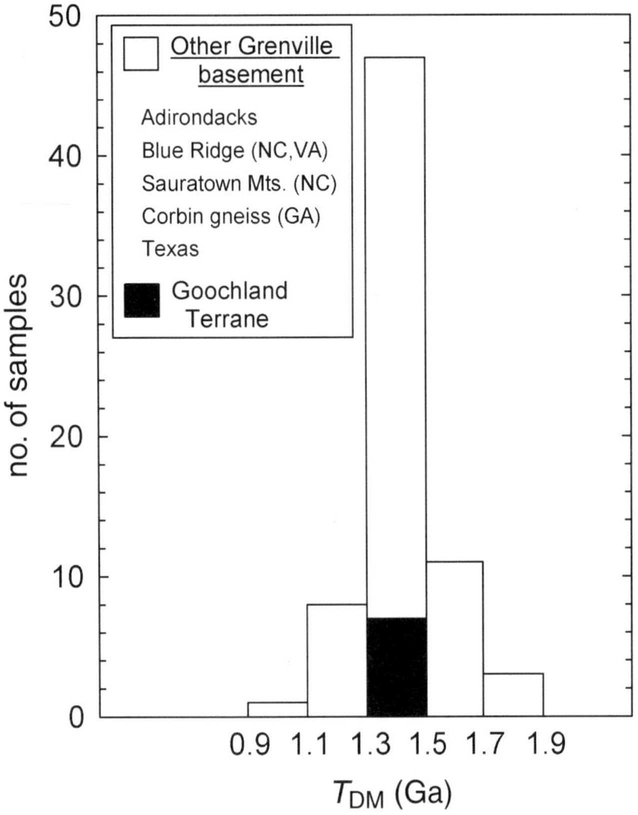

Figure 5. Histogram of reported depleted mantle model-ages (T_{DM}) for miscellaneous Grenvillian basement rocks for an area extending from the Adirondack Mountains of New York to west Texas. Data from this study (State Farm Gneiss and Montpelier Anorthosite) are shown for comparison. Sources of Grenville data: Pettingill et al. (1984); Patchett and Ruiz (1989); Daly and McLelland (1991); McLelland et al. (1993); Heatherington et al. (1996); Fullagar et al. (1997).

occurs to the east, and the current geologic map of Virginia indicates that a major belt of mylonite connects the Hylas zone with the Nutbush Creek and Lake Gordon mylonite zones in southern Virginia (Virginia Division of Mineral Resources, 1993; cf. Horton et al., 1993). This major mylonite zone (Fig. 1) separates the type Goochland terrane area from Raleigh metamorphic belt rocks (or Raleigh terrane) to the south. The Goochland terrane must also be separated by faults from Carolina terrane rocks to the southwest (Fig. 1), but the precise location and nature of this boundary is unclear (Burton and Armstrong, 1997; W.C. Burton, personal commun.). Our isotopic data make it possible to assess, in at least a preliminary way, relationships among the Goochland terrane and the adjacent Chopawamsic, Carolina, and Raleigh terranes.

The Chopawamsic terrane (Horton et al., 1989) in Virginia consists of the Chopawamsic Formation to the west and the more mafic Ta River Metamorphic Suite to the east (Pavlides, 1981). These units contain a variety of felsic and mafic metavolcanic rocks, metasediments, and granitoid plutons; the Chopawamsic terrane was originally interpreted to represent an island arc of possible Cambrian age (Pavlides, 1981). Recent geochronology (Horton et al., 1998; Coler et al., 2000) showed that igneous activity in the Chopawamsic terrane was Ordovician, not Cambrian. Moreover, Coler et al. (2000) reported generally negative initial ε_{Nd} (−7.0 to +0.2) values for a variety of rocks from this terrane (as well as from the Milton terrane to the south) (Fig. 1), indicating a significant contribution from isotopically evolved continental crust to the magmas in the Chopawamsic terrane. Coler et al. (2000) also reported xenocrystic zircons of Mesoproterozoic age (1300–900 Ma) in Chopawamsic metavolcanic rocks, supporting the interpretation that Mesoproterozoic crust was involved in their petrogenesis. Coler et al. (2000) further showed that Grenvillian crust is isotopically appropriate as the evolved component that contributed to Chopawamsic magmas. The similarities in Nd isotopic composition between the Goochland terrane and other Grenville basement raise the possibility that the Chopawamsic terrane arc was built on a Goochland terrane substrate. Although hardly a unique interpretation, it is nonetheless consistent with the current proximity of these terranes.

The Carolina terrane forms the major portion of the Carolina "Zone" of Hibbard and Samson (1995). It consists of numerous deformed and metamorphosed Neoproterozoic to Cambrian magmatic arcs, and Hibbard (2000) argued that all available evidence is consistent with accretion of the zone to Laurentia in the Late Ordovician to Silurian. Samson et al. (1995) reported Nd-isotopic evidence indicating that the older part of the terrane (ca. 650–600-Ma Virgilina sequence) consists of juvenile, mantle-derived crust. In contrast, Mueller et al. (1996) argued that the Nd-isotopic signature of younger rocks (~570–540 Ma Uwharrie-Albemarle sequence) and the presence of xenocrystic Mesoproterozoic zircons reflected involvement of evolved crust in their petrogenesis. If these interpretations are correct, then the younger stages of magmatism in the Carolina terrane were influenced by older (Mesoproterozoic) continental lithosphere. Mueller et al. (1996) suggested two possible sources for contributions from this Mesoproterozoic basement: (1) Grenvillian crust that was rifted from North America, in which case, the Carolina terrane would represent a peri-Laurentian arc; or (2) other Grenvillian crust in a peri-Gondwanan setting (possibly the Sunsas-Rondonian terrane of western Amazonia). Hibbard and Samson (1995) suggested that the Goochland terrane could be a candidate for this Mesoproterozoic basement, an interpretation that is permitted by our isotopic results. Hibbard and Samson (1995) further argued that the Goochland and Carolina terranes were juxtaposed during the ca. 570–545 Virgilina orogeny.

As noted earlier in this paper, the southernmost limit of Mesoproterozoic rocks of the Goochland terrane is a major unresolved problem. Thus, it is not clear whether rocks of the Raleigh terrane (particularly in southern Virginia) are Goochland equivalents (Farrar, 1998) or perhaps higher-grade equivalents of rocks of the Carolina terrane (Horton and Stern, 1994; Coler et al., 1997). Stoddard (1989) found no relict granulite-facies assemblages in the Raleigh Gneiss complex in the Piedmont of eastern North Carolina and suggested that if these rocks are part of the Goochland terrane, then they were completely retrograded by strong Alleghanian overprinting. Coler et al. (1997) evaluated isotopic compositions of Alleghanian granites in the Raleigh terrane and Eastern slate belt and found that three orthogneisses from the Raleigh terrane have T_{DM} ages of 850–630 Ma. In addition, the Nd-isotopic composition of several granites in the Raleigh terrane is inconsistent with their derivation via melting of, or significant contamination by, a Mesoproterozoic source. Therefore, Coler et al. (1997) argued that the extent of Proterozoic crust in this region might be much smaller than previously thought. Our Nd-isotopic results for the State Farm Gneiss confirm that it is isotopically distinct from gneisses of the Raleigh terrane studied by Coler et al. (1997). Furthermore, the Alleghanian granites were not produced by melting of isotopically evolved crust similar to the State Farm Gneiss. For example, at 300 Ma (an average age for the Alleghanian granites), the State Farm Gneiss would have ε_{Nd} values of –6.0 to –9.3, whereas initial ε_{Nd} values for the granites range from –4.4 to +3.0, but cluster at ~0 (Coler et al., 1997). In addition, the granites display relatively low initial $^{87}Sr/^{86}Sr$ values (~0.703–0.708). Thus, these granites lack evidence for significant contamination by older, isotopically evolved crust, such as the State Farm Gneiss. These observations support the conclusions of Coler et al. (1997) that the portion of the Raleigh belt investigated by them is neither underlain by, nor correlative with, the Goochland terrane.

CONCLUSIONS

Neoproterozoic A-type granitoids that intruded the State Farm Gneiss have isotopic compositions consistent with substantial assimilation of that gneiss or equivalent crust at depth. However, the data preclude an origin for the granites solely by anatexis of the State Farm Gneiss. Thus, another more-depleted source (or sources) was also involved in their petrogenesis.

The isotopic data of this study are consistent with the hypothesis that the State Farm Gneiss and Montpelier Anorthosite may represent an original coeval charnockite-anorthosite suite, similar to that proposed by Hughes et al. (this volume) for the Archer Mountain Suite and Roseland Anorthosite in the Virginia Blue Ridge (cf. Herz and Force, 1984). Moreover, the State Farm Gneiss and Montpelier Anorthosite also share age, compositional, and isotopic features with analogous rocks in the Blue Ridge, which may reflect a petrogenetic link between these two blocks of Mesoproterozoic crust. At a minimum, our results unequivocally confirm the overall similarity in Nd-isotopic composition between the Goochland terrane (State Farm Gneiss and Montpelier Anorthosite) and other segments of Mesoproterozoic crust along the eastern and southern margins of Laurentia. The Goochland terrane (at least the State Farm Gneiss) is also a viable candidate for the Mesoproterozoic crust that contributed to magmas of the Chopawamsic terrane, but is less likely to underlie a significant portion of the Raleigh terrane.

Our results provide insights regarding the possibly exotic nature of the Goochland terrane. The hypothesis that the Goochland terrane represents a "transferred terrane" (e.g., Bartholomew and Lewis, 1988, 1992; Keppie et al., 1996) remains viable, but none of the petrological, geochronological, or geochemical data obtained thus far from this terrane suggest that it is truly exotic with respect to Laurentia. More definitive constraints on this ultimate question may come from future geochronological and isotopic studies of the heterogeneous Maidens Gneiss, which forms the major portion of the terrane.

ACKNOWLEDGMENTS

Funding for this work was provided by a grant from the Jeffress Memorial Trust of Virginia (grant number J-468). We thank Stewart Farrar and Jim Hibbard for helpful reviews, and Scott Hughes for providing data for the Archer Mountain Suite.

REFERENCES CITED

Aleinikoff, J.N., Horton, J.W., Jr., and Walter, M., 1996, Middle Proterozoic age for the Montpelier anorthosite, Goochland terrane, eastern Piedmont, Virginia: Geological Society of America Bulletin, v. 108, p. 1481–1491.

Ashwal, L.D., and Wooden, J.L., 1983, Sr and Nd isotope geochronology, geologic history, and origin of the Adirondack Anorthosite: Geochimica et Cosmochimica Acta, v. 47, p. 1875–1885.

Bartholomew, M.J., and Lewis, S.E., 1988, Peregrination of Middle Proterozoic massifs and terranes within the Appalachian orogen, eastern U.S.A.: Trabajos de Geología, University of Oviedo (Spain), v. 17, p. 155–165.

Bartholomew, M.J., and Lewis, S.E., 1992, Appalachian Grenville massifs: Pre-Appalachian translational tectonics, *in* Mason, R., ed., Basement tectonics 7: Dordrecht, The Netherlands, Kluwer Academic Publishers, p. 363–374.

Burton, W.C., and Armstrong, T.R., 1997, Structural and thermobaric history of the western margin of the Goochland terrane in Virginia: Geological Society of America Abstracts with Programs, v. 29, no. 3, p. 7–8.

Coler, D.G., Samson, S.D., and Speer, J.A., 1997, Nd and Sr isotopic constraints on the source of Alleghanian granites in the Raleigh metamorphic belt and Eastern slate belt, southern Appalachians, U.S.A.: Chemical Geology, v. 134, p. 257–275.

Coler, D.G., Wortman, G.L., Samson, S.D., Hibbard, J.P., and Stern, R., 2000, U-Pb geochronologic, Nd isotopic, and geochemical evidence for the correlation of the Chopawamsic and Milton terranes, Piedmont Zone, Southern Appalachian orogen: Journal of Geology, v. 108, p. 363–380.

Daly, J.S., and McLelland, J.M., 1991, Juvenile Middle Proterozoic crust in the Adirondack Highlands, Grenville province, northeastern North America: Geology, v. 19, p. 119–122.

DePaolo, D.J., 1981, Neodymium isotopes in the Colorado Front Range and crust-mantle evolution in the Proterozoic: Nature, v. 291, p. 193–196.

Emslie, R.F., 1991, Granitoids of rapakivi granite-anorthosite and related associations: Precambrian Research, v. 51, p. 173–192.

Emslie, R.F., and Hegner, E., 1993, Reconnaissance isotopic geochemistry of anorthosite-mangerite-charnockite-granite (AMCG) complexes, Grenville Province, Canada: Chemical Geology, v. 106, p. 279–298.

Farrar, S.S., 1984, The Goochland granulite terrane: Remobilized Grenville basement in the eastern Virginia Piedmont, *in* Bartholomew, M.J., ed., The Grenville Event in the Appalachians and related topics: Boulder, Colorado, Geological Society of America Special Paper 194, p. 215–229.

Farrar, S.S., 1985, Tectonic evolution of the easternmost Piedmont, North Carolina: Geological Society of America Bulletin, v. 96, p. 362–380.

Farrar, S.S., 1998, The Goochland terrane, VA-NC, revisited: Geological Society of America Abstracts with Programs, v. 30, no. 4, p. 11.

Farrar, S.S., 1999, Late Proterozoic rifting of Laurentia: Evidence from the Goochland terrane, VA: Geological Society of America Abstracts with Programs, v. 31, no. 3, p. 15.

Farrar, S.S., 2001, The Grenvillian Goochland terrane: Thrust slices of the Late Neoproterozoic Laurentian margin in the southern Appalachians: Geological Society of America Abstracts with Programs, v. 33, no. 6, p. A-28.

Farrar, S.S., and Givan, M.J., 1988, Alkaline-peralkaline (A-type) granites of the eastern Piedmont, North Carolina and Virginia: Geological Society of America Abstracts with Programs, v. 20, p. 263.

Farrar, S.S., and Owens, B.E., 2001, A North-South transect of the Goochland terrane and associated A-type granites—Virginia and North Carolina (field trip guide), *in* Hoffman, C.W., ed., Field Trip Guidebook: Raleigh, North Carolina, Geological Society of America, Southeastern Section, v. 50, p. 75–92.

Fullagar, P.D., Goldberg, S.A., and Butler, J.R., 1997, Nd and Sr isotopic characterization of crystalline rocks from the Southern Appalachian Piedmont and Blue Ridge, North and South Carolina, *in* Sinha, A.K., et al., eds., The nature of magmatism in the Appalachian orogen: Boulder, Colorado, Geological Society of America Memoir 191, p. 165–179.

Gates, A.E., and Glover, L., III, 1989, Alleghanian tectonothermal evolution of the dextral transcurrent Hylas zone, Virginia: Journal of Structural Geology, v. 11, p. 407–419.

Glover, L., III, Mose, D.G., Poland, F.B., Bobyarchick, A.R., and Bourland, W.C., 1978, Grenville basement in the eastern Piedmont of Virginia: Implications for orogenic models: Geological Society of America Abstracts with Programs, v. 10, p. 169.

Heatherington, A.L., Mueller, P.A., Smith, M.S., and Nutman, A.P., 1996, The Corbin Gneiss: Evidence for Grenvillian magmatism and older continental basement in the southernmost Blue Ridge: Southeastern Geology, v. 36, p. 15–25.

Herz, N., and Force, E.R., 1984, Rock suites in Grenvillian terrane of the Roseland district, Virginia, Part 1, Lithologic relations, *in* Bartholomew, M.J., ed., The Grenville Event in the Appalachians and related topics: Boulder, Colorado, Geological Society of America Special Paper 194, p. 187–200.

Herz, N., and Force, E.R., 1987, Geology and mineral deposits of the Roseland District of central Virginia: Reston, Virginia, U.S. Geological Survey Professional Paper 1371, 56 p.

Hibbard, J.P., 2000, Docking Carolina: Mid-Paleozoic accretion in the southern Appalachians: Geology, v. 28, p. 127–130.

Hibbard, J.P., and Samson, S.D., 1995, Orogenesis exotic to the Iapetan cycle in the southern Appalachians, *in* Hibbard, J.P., et al., eds., Current perspectives in the Appalachian-Caledonian orogen: Geological Association of Canada Special Paper 41, p. 191–205.

Horton, J.W., Jr., and Stern, T.W., 1994, Tectonic significance of preliminary uranium-lead ages from the eastern Piedmont of North Carolina: Geological Society of America Abstracts with Programs, v. 26, no. 4, p. 21.

Horton, J.W., Jr., Drake, A.A., Jr., and Rankin, D.W., 1989, Tectonostratigraphic terranes and their Paleozoic boundaries in the central and southern Appalachian, *in* Dallmeyer, R.D., ed., Terranes in the circum-Atlantic Paleozoic orogens: Boulder, Colorado, Geological Society of America Special Paper 230, p. 213–245.

Horton, J.W., Jr., Drake, A.A., Jr., Rankin, D.W., and Dallmeyer, R.D., 1991, Preliminary tectonostratigraphic terrane map of the Central and Southern Appalachians: U.S. Geological Survey, Miscellaneous Investigations Map I-2163, scale 1:2,000,000.

Horton, J.W., Jr., Berquist, C.R., Marr, J.D., Jr., Druhan, R.M., Sacks, P., and Butler, J.R., 1993, The Lake Gordon mylonite zone: A link between the Nutbush Creek and Hylas zones of the eastern Piedmont fault system: Geological Society of America Abstracts with Programs, v. 25, no. 4, p. 23.

Horton, J.W., Jr., Aleinikoff, J.N., Drake, A.A., Jr., and Fanning, C.M., 1998, Significance of middle to late Ordovician volcanic-arc rocks in the central Appalachian Piedmont, Maryland and Virginia: Geological Society of America Abstracts with Programs, v. 30, no. 7, p. A-125.

Hughes, S.S., Lewis, S.E., Bartholomew, M.J., Sinha, A.K., Hudson, T.A., and Herz, N., 1997, Chemical diversity and origin of Precambrian charnockitic rocks of the Pedlar massif, Grenvillian Blue Ridge terrane, Virginia: Precambrian Research, v. 84, p. 37–62.

Keppie, J.D., Dostal, J., Murphy, J.B., and Nance, R.D., 1996, Terrane transfer between eastern Laurentia and western Gondwana in the early Paleozoic: Constraints on global reconstructions, *in* Nance, R.D., and Thompson, M.D., eds., Avalonian and Peri-Gondwanan terranes of the circum–North Atlantic: Boulder, Colorado, Geological Society of America Special Paper 304, p. 369–380.

McLelland, J.M., Daly, J.S., and Chiarenzelli, J., 1993, Sm-Nd and U-Pb isotopic evidence of juvenile crust in the Adirondack lowlands and implications for the evolution of the Adirondack Mts.: Journal of Geology, v. 101, p. 97–105.

Mueller, P.A., Kozuch, M., Heatherington, A.L., Wooden, J.L., Offield, T.W., Koeppen, R.P., Klein, T.L., and Nutman, A.P., 1996, Evidence for Mesoproterozoic basement in the Carolina terrane and speculations on its origin, *in* Nance, R.D., and Thompson, M.D., eds., Avalonian and Peri-Gondwanan terranes of the circum–North Atlantic: Boulder, Colorado, Geological Society of America Special Paper 304, p. 207–217.

Owens, B.E., and Dymek, R.F., 1999, A geochemical reconnaissance of the Roseland anorthosite complex, Virginia, and comparisons with andesine anorthosites from the Grenville Province of Quebec, *in* Sinha, A.K., ed., Basement Tectonics 13: Dordrecht, The Netherlands, Kluwer Academic Publishers, p. 217–232.

Owens, B.E., and Tucker, R.D., 2003, Geochronology of the Mesoproterozoic State Farm Gneiss and associated Neoproterozoic granitoids, Goochland terrane, Virginia: Geological Society of America Bulletin, v. 115, p. 972–982.

Patchett, P.J., and Ruiz, J., 1989, Nd isotopes and the origin of Grenville-age rocks in Texas: Implications for Proterozoic evolution of the United States Mid-continent region: Journal of Geology, v. 97, p. 685–695.

Pavlides, L., 1981, The central Virginia volcanic-plutonic belt: An island arc of Cambrian(?) age: Reston, Virginia, U.S. Geological Survey Professional Paper 1231A, 34 p.

Pettingill, H.S., Sinha, A.K., and Tatsumoto, M., 1984, Age and origin of anorthosites, charnockites, and granulites in the central Virginia Blue Ridge: Nd and Sr isotopic evidence: Contributions to Mineralogy and Petrology, v. 85, p. 279–291.

Sacks, P.E., 1999, Geologic overview of the eastern Appalachian Piedmont along Lake Gaston, North Carolina and Virginia, *in* Sacks, P.E., ed., Geology of the Fall Zone Region along the North Carolina–Virginia State Line: Fieldtrip Guidebook for the 1999 meeting of the Carolina Geological Society, p. 1–15.

Samson, S.D., Hibbard, J.P., and Wortman, G.L., 1995, Nd isotopic evidence for juvenile crust in the Carolina terrane, southern Appalachians: Contributions to Mineralogy and Petrology, v. 121, p. 171–184.

Stoddard, E.F., 1989, The Raleigh Gneiss complex, eastern North Carolina Piedmont: Is it part of the Goochland terrane?: Geological Society of America Abstracts with Programs, v. 21, no. 3, p. 60.

Virginia Division of Mineral Resources, 1993, Geologic Map of Virginia: Virginia Division of Mineral Resources, scale 1:500,000.

Manuscript Accepted by the Society August 25, 2003

Printed in the USA

Geological Society of America
Memoir 197
2004

U-Pb geochronology and geochemistry of a portion of the Mars Hill terrane, North Carolina–Tennessee: Constraints on origin, history, and tectonic assembly

Steven E. Ownby*
Calvin F. Miller*
Peter J. Berquist
Charles W. Carrigan*
Department of Geology, Vanderbilt University, Nashville, Tennessee 37235, USA
Joseph L. Wooden
USGS-SUMAC Ion Probe Lab, Green Building, Stanford University, Stanford, California 94305-2220, USA
Paul D. Fullagar
Department of Geological Sciences CB# 3315, University of North Carolina, Chapel Hill, North Carolina 27599-3315, USA

ABSTRACT

The Mars Hill terrane (MHT), a lithologically diverse belt exposed between Roan Mountain, North Carolina–Tennessee, and Asheville, North Carolina, is distinct in age, metamorphic history, and protoliths from the structurally overlying Eastern Blue Ridge and underlying Western Blue Ridge. MHT lithologies include diverse granitic gneisses, abundant mafic and sparse ultramafic bodies, and mildly to strongly aluminous paragneisses. These lithologies experienced metamorphism in the granulite facies and are intimately interspersed on cm to km scale, reflecting both intrusive and tectonic juxtaposition.

Previous analyses of zircons by high-resolution ion microprobe verified the presence of Paleoproterozoic orthogneiss (1.8 Ga). New data document a major magmatic event at 1.20 Ga. Inherited and detrital zircons ranging in age from 1.3 to 1.9 Ga (plus a single 2.7 Ga core), ubiquitous Sm-Nd depleted mantle model ages ca. 2.0 Ga, and strongly negative ε_{Nd} during Mesoproterozoic time all attest to the pre-Grenville heritage of this crust that was suggested by previous whole-rock Pb and Rb-Sr isotope studies. A single garnet amphibolite yielded a magmatic age of 0.73 Ga, equivalent to the Bakersville dike swarm, which cuts both the MHT and the adjacent Western Blue Ridge. Zircons from this sample display 0.47 Ga metamorphic rims. Zircons from all other samples have well-developed ca. 1.0 Ga metamorphic rims that date granulite-facies metamorphism. Silica contents of analyzed samples range from 45 to 76 wt%, reflecting the extreme diversity observed in the field and the highly variable protoliths.

The MHT contrasts strikingly with basement of the adjacent Eastern and Western Blue Ridge, which comprise relatively homogeneous, 1.1 to 1.2 Ga granitic rocks

*Corresponding author: Miller, calvin.miller@vanderbilt.edu; present address, Carrigan and Ownby: Department of Geological Sciences, University of Michigan, Ann Arbor, Michigan 48109.

Ownby, S.E., Miller, C.F., Berquist, P.J., Carrigan, C.W., Wooden, J.L., and Fullagar, P.D., 2004, U-Pb geochronology and geochemistry of a portion of the Mars Hill terrane, North Carolina–Tennessee: Constraints on origin, history, and tectonic assembly, *in* Tollo, R.P., Corriveau, L., McLelland, J., and Bartholomew, M.J., eds., Proterozoic tectonic evolution of the Grenville orogen in North America: Boulder, Colorado, Geological Society of America Memoir 197, p. 609–632. For permission to copy, contact editing@geosociety.org.

with initial ε_{Nd} values near 0. It appears to have more in common with distant Paleoproterozoic crustal terranes in the Great Lakes region, the southwestern United States, and South America.

Keywords: Appalachians, geochemistry, zircon, SHRIMP, geochronology, Proterozoic, granulite facies, Grenville, Nd isotopes

INTRODUCTION

The Blue Ridge province of the southern Appalachian orogen is divided into western (Western Blue Ridge) and eastern (Eastern Blue Ridge) zones. The Western Blue Ridge is generally thought to be part of Laurentia (native North America), whereas the more structurally and lithologically complex Eastern Blue Ridge is considered to comprise one or more suspect terranes—possibly a rifted and reattached fragment of Laurentia or an exotic terrane(s) (e.g., Hatcher, 1989; Stewart et al., 1997). The Western Blue Ridge includes granitic rocks of Grenville and Neoproterozoic age; Neoproterozoic mafic dikes and intrusions; and overlying, chemically mature metasedimentary rocks that experienced relatively low-grade Paleozoic metamorphism (e.g., Rankin, 1975; Hatcher, 1978; Misra and McSween, 1984; Davis, 1993). The Eastern Blue Ridge comprises an assemblage of less mature, Neoproterozoic–early Paleozoic clastic metasedimentary rocks and mafic to ultramafic bodies of higher metamorphic grade; variably deformed, felsic Paleozoic intrusions; and relatively sparse exposures of Grenville-age granitoid gneisses.

Several workers have noted an assemblage of lithologies at the Eastern Blue Ridge–Western Blue Ridge boundary that, although included by some as part of the Western Blue Ridge, appears to have no counterpart in either area. This assemblage is exposed in a belt that extends at least from the vicinity of Roan Mountain, Tennessee–North Carolina, to northwest of Asheville, North Carolina (Fig. 1) (Merschat, 1977; Gulley, 1982; Bartholomew and Lewis, 1988, 1992; Raymond et al., 1989; Merschat and Wiener, 1990; Johnson, 1994; Raymond and Johnson, 1994; Adams et al., 1995; Stewart et al., 1997; Trupe et al., 2001). It differs from the Eastern Blue Ridge and Western Blue Ridge as follows:

1. The MHT displays widespread evidence for granulite-facies metamorphism (e.g., Merschat, 1977; Gulley, 1985; Adams and Trupe, 1997); granulite-grade rocks are rare in the Eastern and Western Blue Ridge.

2. The MHT contains abundant mafic and some ultramafic rocks interspersed with granitic gneisses on the cm to km scale; the mafic rocks are commonly migmatitic (e.g., Merschat, 1977). Mafic rocks are common in the Eastern Blue Ridge, but only in contact with the metasedimentary sequences, and they are rarely migmatitic; they are rare or absent in the Western Blue Ridge, except for the Neoproterozoic dikes.

3. Field relations and Rb-Sr geochronology suggest that the MHT contains the oldest rocks in the southern Appalachians and lacks Phanerozoic rocks. Metasedimentary rocks that have experienced granulite-facies metamorphism are cut by lower-grade, 730-Ma dikes of the Bakersville swarm (Goldberg et al., 1986), and a whole-rock Rb-Sr isochron has been interpreted to demonstrate a magmatic crystallization age of 1.8 Ga at one locality (Monrad and Gulley, 1983). Whole-rock lead isotope ratios (Sinha et al., 1996) and high initial strontium isotope ratios (Monrad and Gulley, 1983; Fullagar et al., 1979) also suggest antiquity of this terrane. All metasedimentary rocks in the Eastern Blue Ridge and Western Blue Ridge are interpreted to be younger than 730 Ma, and no other igneous rocks from the southeastern United States have reported radiometric ages older than 1.2 Ga.

4. The Mars Hill terrane is the most lithologically diverse basement exposure south of Virginia. Other basement exposures in this region are almost entirely meta-igneous (possibly including some high-grade, feldspathic metasandstone) and generally lack mafic rocks.

Based upon its lithologic distinctiveness, the MHT has been mapped as an important regional unit (or units) and interpreted as a suspect terrane. Merschat (1977) mapped biotite-hornblende migmatite in the vicinity of Mars Hill, North Carolina, and inferred that it was regionally extensive. Gulley (1982, 1985) investigated granulite-facies rocks at Roan Mountain and informally designated metasedimentary lithologies as Cloudland gneiss and mafic and felsic meta-igneous rocks as Carvers Gap gneiss. On the North Carolina state geologic map (Brown

Figure 1. (A) Location of the MHT and its possible extent toward Georgia and Virginia. (B) Localities of samples analyzed for this study and Carrigan et al. (2003). Sample labels all have prefix RM, except for MBCL4 and 5. Inset: Roan Mountain–Bakersville area, where most samples were collected. North Carolina State Highway 261, Tennessee State Highway 143, and Roan Mountain spur road (Forest Service Route 130) are shown for reference. CG—Carvers Gap; MM—Meadlock Mountain; RHB—Roan High Bluff. Only samples for which analytical data are reported are shown. For all samples shown, elemental analyses were done; for circled samples, Rb-Sr-Sm-Nd isotopic analyses; for underlined samples, zircon U-Pb analyses; samples with asterisks were analyzed by Carrigan et al. (2003). Zircon sample CAR 1501 was collected by J.P. Dubé and K.G. Stewart at locality RM2X. Maps modified from Brown et al. (1985) and K.G. Stewart (unpublished data).

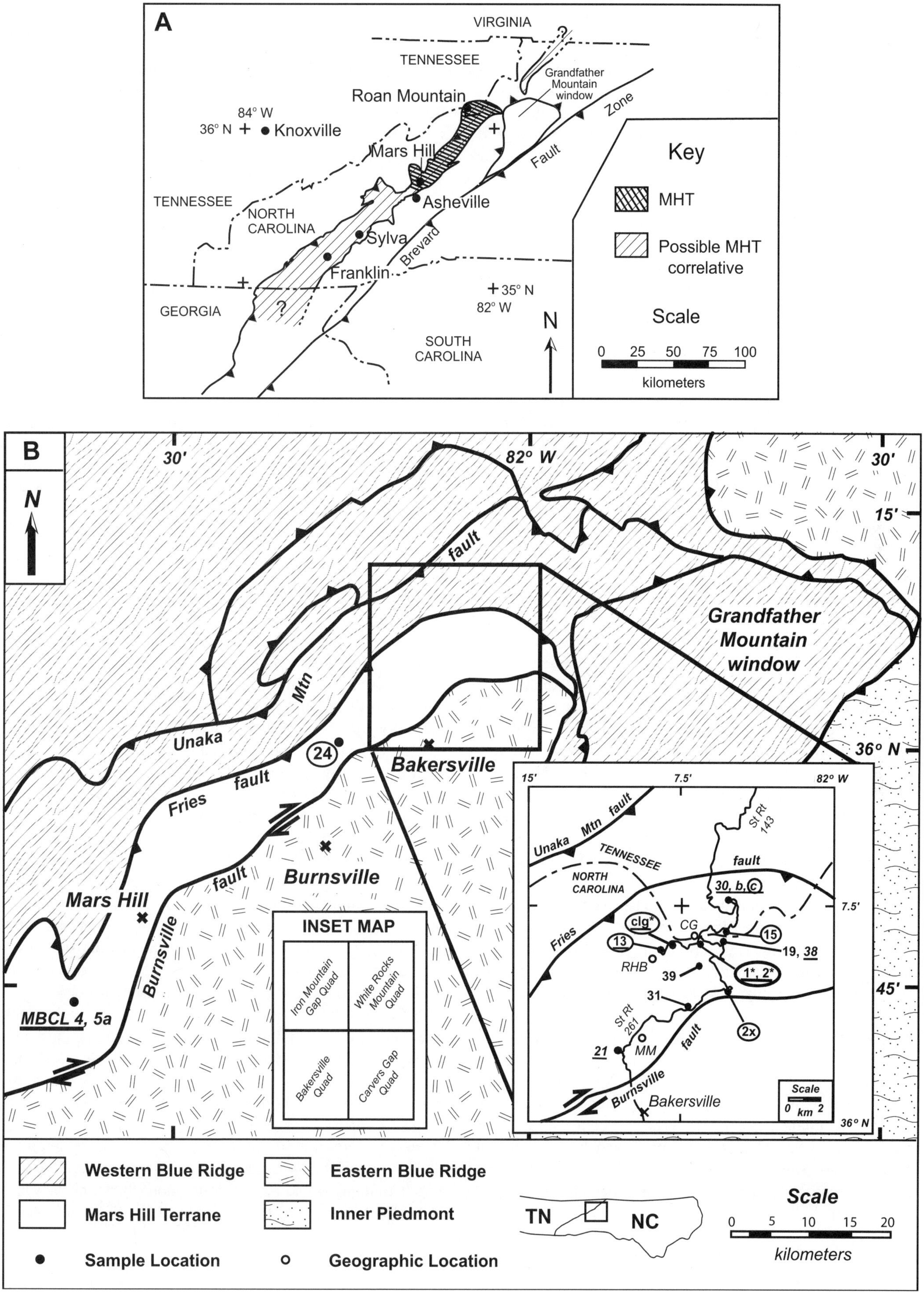

A
VIRGINIA
TENNESSEE
Grandfather Mountain window
Roan Mountain
84° W
36° N
Knoxville
Mars Hill
Zone
Fault
Key
MHT
Possible MHT correlative
TENNESSEE
NORTH CAROLINA
Asheville
Sylva
Brevard
Franklin
GEORGIA
35° N
82° W
N
Scale
0 25 50 75 100
kilometers
SOUTH CAROLINA
B
30'
82° W
30'
15'
N
fault
Grandfather Mountain window
Mtn
Unaka
24
Bakersville
36° N
Fries
fault
fault
Burnsville
Mars Hill
Burnsville
MBCL 4, 5a
INSET MAP
Iron Mountain Gap Quad
White Rocks Mountain Quad
Bakersville Quad
Carvers Gap Quad
15'
7.5'
82° W
Unaka Mtn fault
St Rt 143
TENNESSEE
fault
NORTH CAROLINA
30, b, c
7.5'
clg*
CG
15
Fries
13
19, 38
RHB
39
1*, 2*
45'
31
St Rt 261
2x
MM
21
fault
Burnsville
Bakersville
Scale
0 km 2
36° N
Western Blue Ridge
Eastern Blue Ridge
Mars Hill Terrane
Inner Piedmont
Sample Location
Geographic Location
TN
NC
Scale
0 5 10 15 20
kilometers

et al., 1985), Merschat's biotite-hornblende migmatite unit stretches for 80 km, from Roan Mountain to northwest of Asheville, and similar, possibly related migmatitic biotite gneiss extends another 120 km southwest to the Georgia border (Raymond et al., 1989). Workers from the University of North Carolina at Chapel Hill have mapped the northern portion of this zone as the Pumpkin Patch or Fries thrust sheet and refer to the high-grade rocks as the Pumpkin Patch Metamorphic Suite (Goldberg et al., 1989; Adams et al., 1995; Stewart et al., 1997; Trupe et al., 2001). Bartholomew and Lewis (1988, 1992), Raymond (1987; Raymond et al., 1989), and Brewer and Woodward (1988) identified this general area as a suspect terrane, calling it the Mars Hill terrane, the Cullowhee terrane, and Amphibolitic Basement Complex, respectively. Raymond et al. (1989) and Brewer and Woodward (1988) suggested that the terrane may be a melangelike complex of granitic and mafic material, minutely imbricated as a consequence of ocean basin closure. Raymond et al. (1989) suggested that closure occurred during the late Neoproterozoic–early Paleozoic (post-Bakersville dikes), whereas Brewer and Woodward (1988) interpreted it to have been completed during late Mesoproterozoic–early Neoproterozoic time (pre-Bakersville dikes).

In a reconnaissance study of basement rocks of western North Carolina and northernmost Georgia and South Carolina, Carrigan (2000) and Carrigan et al. (2000, 2001, 2003) noted that Western Blue Ridge and Eastern Blue Ridge basement is uniformly broadly granitic in composition. Zircons from these rocks lack inheritance, yield magmatic ages between 1.08 and 1.19 Ga, and have ca. 1.03 Ga metamorphic rims. An initial ε_{Nd} value near 0 suggests that the crust that these rocks represent was young and possibly juvenile during Grenville time. Only in the MHT did Carrigan identify an older component, with both U-Pb and Sm-Nd data suggesting the presence of Paleoproterozoic crust; this is consistent with the initial Rb-Sr results of Monrad and Gulley (1983) and whole-rock Pb isotope data of Sinha et al. (1996).

The purpose of this paper is to characterize the geochemistry and ages of representative examples of lithologies from the MHT in its better-documented portion between Roan Mountain and Asheville, in order to decipher its geologic history and constrain its relationships with surrounding units. These data will, in turn, contribute to a better understanding of the Proterozoic and Paleozoic assembly of southeastern North America.

METHODS

Samples for both geochronological (5 kg) and geochemical (1 kg) analysis were selected from fresh outcrops to represent both typical lithologies and the diversity of compositions. Sample CAR 1501 had been collected previously by J.P. Dubé and K.G. Stewart. Sample locations are shown in Figure 1, B; precise locations and petrographic descriptions can be found in Ownby (2002).

Zircons were separated using standard procedures, mounted in epoxy, polished, and imaged by cathodoluminescence on the JEOL JSM 5600 scanning electron microscope at Stanford University. Points on the zircon grains ~30–40 μm in diameter were analyzed according to Stanford/U.S. Geological Survey (USGS) Sensitive High-Resolution Ion Microprobe, Reverse Geometry (SHRIMP-RG) Facility procedures (cf. Bacon et al., 2000). Zircon standards R33 (419 Ma) and CZ3 (550 ppm U) were used for U-Pb and U concentration standards, respectively. Standards were provided by the Stanford/USGS facility. Common Pb corrections were based on measured ^{204}Pb and data reduction used SQUID (Version 1.02; Ludwig, 2001).

Powders prepared in an alumina ceramic shatterbox were analyzed for elemental and isotopic analyses. Elemental compositions were determined by Activation Laboratories, Ltd., of Canada, using X-ray fluorescence, inductively coupled plasma mass spectrometry, and instrumental neutron activation analysis. Sm-Nd and Rb-Sr isotopic analyses were performed at the University of North Carolina at Chapel Hill on a VG Micromass Sector 54 multicollector thermal ionization mass spectrometer, following the methods described in Fullagar et al. (1997) and Fullagar and Butler (1979).

PROTOLITH INTERPRETATION: PITFALLS, CRITERIA USED, AND KEY ISSUES

Interpretation of protoliths in the MHT faces formidable obstacles. Primary textures have been obliterated by high-temperature recrystallization, and for the most part, intense ductile deformation has thoroughly modified initial rock-unit geometry and destroyed primary contact relations and textures (see the Petrography and Field Relations section). In this paper, we rely primarily on elemental chemistry for protolith interpretation. Mafic rocks are relatively straightforward: their geochemistry is indistinguishable from common basalts and gabbros and is unlike any common sediments. Likewise, sparse, highly aluminous rocks are clearly metapelites (we have analyzed only a single sample, although Gulley [1982] described numerous samples from the Cloudland gneiss). Other unequivocal protoliths are absent: there are no high-silica rocks (>80 wt% SiO_2), carbonates, or calc-silicates among our samples.

A majority of the rocks of the MHT are feldspar-rich gneisses and granofelses that are intermediate to felsic (55–75 wt% SiO_2) and mildly metaluminous to moderately peraluminous (Fig. 2). Potential protoliths of such rocks include intermediate to felsic igneous rocks and feldspathic sandstones (greywackes, arkoses). There is no entirely reliable way to distinguish igneous rocks from sandstones, because extremely immature sandstones can be identical to their igneous sources. However, there is a strong tendency for clastic sediments to show the imprint of weathering and sorting and therefore to be enriched in quartz and more peraluminous than igneous source rocks. Recognizing that these criteria are not foolproof, we dis-

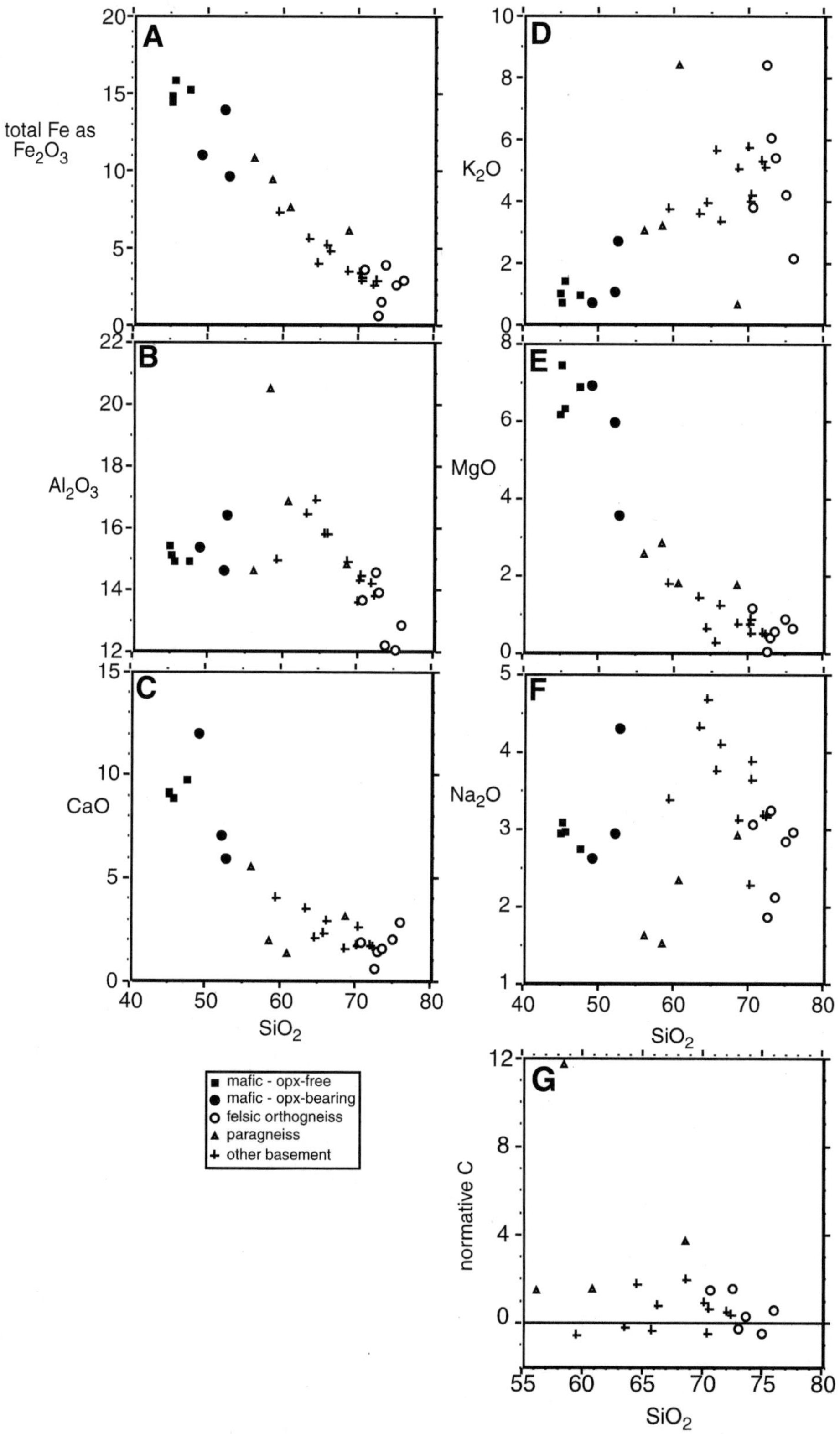

Figure 2. Selected major-element oxides and normative corundum plotted against SiO_2. All concentrations in wt%. Negative normative corundum in panel G is the deficiency in alumina realtive to CaO + Na_2O + K_2O in metaluminous samples (calculated from the same equation as normative corundum). "Other basement" includes Eastern Blue Ridge and Western Blue Ridge basement, from Carrigan (2000) and Carrigan et al. (2003).

tinguish probable sedimentary from igneous protoliths on the following basis: rocks with typical intermediate to felsic igneous SiO_2 and Al_2O_3 concentrations (55–75 and 10–20 wt%, respectively) are considered likely to have igneous protoliths if they do not have unusually high normative corundum (>0 at 55 wt% SiO_2, >2 wt% at 75 wt% SiO_2) or normative quartz (>10 at 55 wt% SiO_2, >45 wt% at 75 wt% SiO_2). Rocks that meet these standards have >50% normative feldspar, also consistent with igneous parentage. A further geochemical test of protoliths, presented in Figure 3, results in the same protolith assignments as the abovementioned criteria for all but one sample. Previous studies in the area have taken similar approaches and concluded that, with the exception of the Cloudland gneiss, igneous protoliths dominate the MHT (Merschat, 1977; Gulley, 1982), although in a review paper, Bartholomew and Lewis (1988) suggested that it is largely metasedimentary.

We emphasize that distinguishing volcanic from plutonic protoliths on the basis of geochemistry is impossible. In some exposures in the MHT, intrusive relationships permit the identification of plutonic rocks, but in most cases the distinction cannot be made.

Zircon age distributions and zoning patterns can also suggest protolith. Excluding metamorphic rims, a single dominant age population is consistent with igneous parentage, although sediment derived from a single-aged source terrane could yield the same result. A modest number of distinctly older zones, especially if they are clearly located within a core, are also consistent with an igneous protolith. Absence of a dominant age population or older ages that are more common than younger ones strongly suggests sedimentary origin. Likewise, fragmented grains may suggest sedimentary transport. Rounded external zircon morphology, however, is an unreliable criterion for sedimentary origin in rocks with histories like these: inherited cores in igneous

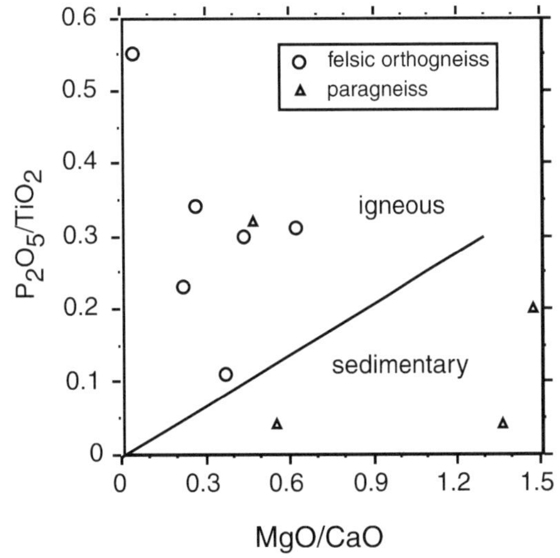

Figure 3. P_2O_5/TiO_2 plotted against MgO/CaO as a discriminator for felsic igneous versus clastic sedimentary protoliths in granulite-facies rocks. After Werner (1987).

rocks typically are rounded by resorption; zircons from very high-grade rocks characteristically have thick overgrowths that impart a subrounded external shape, and partial resorption rounds premetamorphic cores (e.g., Hanchar and Miller, 1993). Zircon data presented in this paper and observed morphologies are mostly consistent with geochemically based protolith assignment; exceptions are discussed in the Geochronology section.

Protolith interpretation is of special interest in two cases: sample RM1, collected along the National Forest spur road 130 to the top of Roan Mountain, and the RM30 exposure (three dated samples) along Tennessee Highway 143, northeast of Roan Mountain. The RM30 exposure and samples are discussed in following sections on field relations, geochemistry, and geochronology. RM1 is a key sample reported in Carrigan et al. (2003) and interpreted as a meta-igneous rock with an age of 1.8 Ga. Only at Roan Mountain has there been a suggestion of rocks of this antiquity in the southeastern United States, and the Carrigan et al. data appear to confirm previous suggestions. If it is in fact metasedimentary, then this age points to an Early Proterozoic source region, but does not verify the existence here of Early Proterozoic rocks. The sample is of massive, unfoliated granofels. The rationale for its interpretation as meta-igneous is as follows:

1. The Carvers Gap gneiss, of which RM1 is part, was interpreted as an igneous complex by Gulley (1982), based primarily on geochemical criteria similar to those discussed above.

2. Previous efforts at dating suggested essentially the same age: Monrad and Gulley (1983) presented a whole-rock Rb-Sr isochron based on samples collected along a 1.5 km road cut traverse that included samples from near site RM1; the isochron appears to be robust and yielded an age of 1.82 Ga, an unlikely result if these were metasedimentary rocks. Furthermore, Fullagar and Gulley (1999) reported a conventional zircon U-Pb upper intercept of 1.84 Ga and lower intercept of 1.08 Ga for a nearby sample.

3. Eight of nine of zircon core analyses of Carrigan et al. (2003) define a discordia with an upper intercept of 1.77 Ga and lower intercept of 1.01 Ga; the only analysis that did not fall on this discordia gave a discordant post-Grenville age and clearly reflected younger lead loss. The well-defined discordia would require that, if this rock were metasedimentary, it had only a single detrital age population and that this population be distinct from detrital populations of all other reported samples from the Southeast (e.g., see Carrigan et al., 2003; Bream et al., this volume, and data in this paper).

4. Internal zoning in zircons from sample RM1 is clearly igneous and, although all grains have metamorphic overgrowths, the magmatic interiors are not truncated, as would be likely in detrital populations (see Fig. 4 *in* Carrigan et al., 2003).

5. The composition of RM1, although fairly high in SiO_2 (75.8 wt%), is within a reasonable igneous range; it is very weakly peraluminous, with only 0.6 wt% normative corundum, and has >50% normative feldspar. We acknowledge that RM1

could be a highly immature meta-arkose, but the preponderance of evidence strongly suggests that it is a meta-igneous rock and supports the notion that there is a 1.8 Ga igneous complex at Roan Mountain.

PETROGRAPHY AND FIELD RELATIONS

The MHT is characterized by a great diversity of lithologies interspersed on all scales. Almost every exposure contains multiple rock types, some with readily interpretable contact relations (dikes, migmatitic leucosomes and melanosomes, pervasive injection zones), but many others with more ambiguous relationships that, at least in some cases, require tectonic juxtaposition. Raymond et al. (1989) described and illustrated exposures in possible correlative rocks to the southwest of Asheville that show similar juxtapositions of lithologies, which they refer to as "block-in-matrix" structures.

Mafic rocks are ubiquitous throughout the MHT. The most readily interpretable are dikes of the Neoproterozoic Bakersville dike swarm (Goldberg et al., 1986; Adams and Trupe, 1997), which crosscut other lithologies and are apparently the youngest rocks of the MHT. Although they commonly preserve relict fine-grained diabasic fabric, the dike rocks are overprinted by amphibolite-facies mineral assemblages (garnet amphibolite). Larger mafic bodies that preserve igneous textures have been interpreted as Bakersville suite intrusions (Goldberg et al., 1986; Adams and Trupe, 1997). The Meadlock Mountain gneiss, a biotite-bearing garnet ± clinopyroxene amphibolite, is unusual in that it forms mappable-scale bodies. Adams et al. (1995) reported that it records peak conditions in the high-*P* portion of the amphibolite facies (13 kb, 725 °C), consistent with the presence of felsic leucosome pods that suggest anatexis during peak metamorphism. Garnet amphibolites lacking orthopyroxene are fairly widespread in the MHT and may correlate with Meadlock Mountain gneiss. Orthopyroxene (opx)-bearing mafic rocks are also abundant, but appear not to form extensive exposures. Gulley (1985) estimated peak granulite-facies conditions for the opx-bearing metabasites and nearby metapelites at Roan Mountain as ~7 kb, 800 °C. The granulite-grade mafic rocks are commonly banded, either with alternating hornblende-rich and opx-rich layers, or with more- and less-felsic layers. In thin section, a garnet-hornblende assemblage rims or replaces opx, indicating a retrograde reaction (lower *T* and/or higher *P*). Ultramafic rocks are present locally (Merschat, 1977; Raymond and Johnson, 1994), but they are exceedingly rare in areas that we sampled (we collected only a single igneous-textured, plagioclase-bearing websterite). We interpret the Bakersville dikes, and probably the larger mafic bodies (Meadlock Mountain gneiss), to be intrusions. It is not evident whether smaller sheets and pods are metamorphosed volcanic rocks, sills, or dismembered dikes and larger intrusions.

Like mafic rocks, felsic gneisses are ubiquitous but variable in field characteristics and composition. None appear to form map-scale plutons. Some are compositionally banded; in others, foliation is defined by weak mafic mineral alignment or by mylonitic fabric; and still others are massive, with prominent, blocky, perthitic K-feldspar. Compositional banding probably reflects both transposed compositional layering in protoliths and deformation-induced metamorphic segregation; protoliths appear to include aluminous sediments, very small intrusions, and probably feldspathic psammites, felsic volcanic rocks, and dismembered larger intrusions. Some are rich in K-feldspar (alkali feldspar granite composition), others are very poor in Kspar (trondhjemitic); most are quartz-rich, but not rich enough to represent quartz-rich sandstone; and some are rich in kyanite and/or sillimanite (Gulley, 1985). Biotite is present in all samples. Opx and clinopyroxene, present in some but not all samples, document the granulite-facies event in the MHT. Where present, the pyroxenes are commonly rimmed by garnet and hornblende, probably reflecting the high-*P* amphibolite-facies event described by Adams et al. (1995) for mafic rocks. Garnet is also commonly present as discrete grains.

Some of the feldspathic banded gneisses are similar to MHT gneisses exposed near Mars Hill that are interpreted as metavolcanics (Merschat and Carter, personal commun.) but, in these intensely deformed and metamorphosed rocks, distinguishing volcanic from intrusive or weakly aluminous sandstone protoliths is difficult (see the Protolith Interpretation and Geochemistry sections for discussion of the distinction between sedimentary and igneous protoliths). Small bodies of massive felsic rock intrude mafic and banded gneisses, indicating that the protoliths were granites. A single plagioclase-biotite-quartz-garnet-scapolite-ilmenite(?) banded gneiss sample (RM13) is of enigmatic origin. Highly aluminous paragneisses with probable shale and aluminous graywacke protoliths are common on Roan Mountain (Gulley, 1985) but rare elsewhere.

The latest events indicated by field and petrographic relations to have affected the MHT include development of local mylonitic shear zones that preserve some unrecovered strain and limited greenschist-facies recrystallization, indicated by minor epidote and fine-grained chlorite, muscovite, and biotite (Gulley, 1985; Adams and Trupe, 1997).

One exposure along Tennessee Highway 143 merits a brief discussion, because it is the site at which three samples (RM30, RM30B, RM30C) were collected that yielded important but somewhat equivocal geochemical and geochronological data. This road-cut exposure lacks the rather chaotic block-in-matrix structure that characterizes much of the MHT, but contact relations among lithologic units are still not straightforward. The road cut is dominated by banded gneiss, mostly gray and intermediate to felsic in appearance (represented by RM30B) and, in part, distinctly more mafic (RM30C). The gray gneiss is mineralogically simple, with feldspars, biotite, quartz, and garnet. The mafic gneiss contains plagioclase, hornblende, clinopyroxene, opx, garnet, biotite, and quartz. Sheets of massive, medium-coarse, felsic granofels up to ~2 m in thickness parallel the foliation of the gneisses. RM30, collected from one of these sheets, comprises abundant feldspars and quartz, with minor biotite, opx, and clinopyroxene. The gneisses and granofels are cut by a fine-grained mafic dike that is probably part of the Bakersville swarm. It is not obvious whether the planar contacts and sheet-like geometry of the three predike lithologies reflect original forms and contacts of the units or tectonic rearrangement and transposition. If the initial geometry is more or less intact, the most reasonable field interpretations are that either (1) all three are part of a depositional sequence—volcanic or sedimentary; or (2) the banded gneisses were intruded by dikes or sills of the granitic granofels protolith (or, conceivably, that the gray banded gneiss was intruded by both the granofels protolith and the mafic gneiss protolith).

GEOCHEMISTRY

Elemental compositions of MHT rocks reflect the lithologic diversity that is evident in the field and constrain possible protoliths (Table 1; Figs. 2–5). Concentrations of SiO_2 in the analyzed samples range from 45 to 76 wt%. The samples with <55 wt% SiO_2 have compositions consistent with mafic magmatic heritage; they have moderately high Al_2O_3, Cr, Ni, and Mg#s (atomic Mg/[Mg + Fe], 0.4–0.6) and are metaluminous and olivine-normative to weakly quartz-normative (Ownby, 2002). In Table 1, Figures 2 and 4–6, and the discussion that follows, these rocks are interpreted to be metamorphosed mafic igneous rocks—diabases, basalts, and/or gabbros—and are subdivided into opx-bearing and opx-free varieties. Compositions of rocks with 70 to 76 wt% SiO_2 suggest felsic igneous protoliths: they have abundant normative feldspar and 27 to 43 wt% normative quartz and are weakly metaluminous to weakly peraluminous. As noted previously, we cannot rule out the possibility that some of these samples are metamorphosed, extremely immature arkoses or volcaniclastic sediments that closely mimic their igneous sources in composition, but the simplest and most plausible interpretation is that most or all are metagranitoids and metarhyolites. Therefore, we refer to them as felsic orthogneisses.

The four analyzed samples with 56 to 69 wt% SiO_2 are all dissimilar to typical igneous rocks. Compared with igneous rocks with similar SiO_2 concentrations, all four are unusually peraluminous, three are unusually high in normative quartz, and three have high Cr and Ni (Table 1; Fig. 2, G) (Ownby, 2002). The peraluminous compositions are a reflection of very low Na_2O and/or CaO concentrations. On a plot of P_2O_5/TiO_2 versus MgO/CaO, three samples plot clearly in the metasedimentary field, whereas all samples interpreted to be felsic orthogneisses plot in the igneous field (Fig. 3) (Werner, 1987). Sample RM31, with extremely high normative corundum and moderate SiO_2, clearly has a pelitic protolith, similar to fairly common lithologies at Roan Mountain described by Gulley (1982, 1985). RM-CLG, a sample of Gulley's Cloudland paragneiss, is also almost certainly metasedimentary, based on its high normative corundum and quartz and concentrations of Ni and Cr; its probable protolith is an impure sandstone, perhaps a quartz-rich greywacke. RM30B has 1.5 wt% normative corundum, very low Na_2O and Sr concentrations, and high Cr and Ni, characteristics that are highly unusual for an igneous rock with 61 wt% SiO_2. We therefore suggest that it is also a metagreywacke (RM30B is discussed further in the following paragraph). RM13 has a highly unusual composition that does not match either typical igneous or sedimentary protoliths; despite having 56 wt% SiO_2, it has low Mg# (0.3), Cr, and Ni, extremely low Na_2O, and is peraluminous. It also is enriched in Sr, Ba, P_2O_5, and TiO_2. This sample may have been derived from an unusual sediment that included both chemically precipitated and insoluble residue components, or it could represent an intensely altered protolith. Because its composition appears to reflect surface or near-surface processes, we group it tentatively with the paragneisses.

Elemental chemistry of the samples from the RM30 road cut on TN Highway 143 suggests that this exposure contains both igneous and sedimentary rocks. RM30B, the gray banded gneiss, is probably metasedimentary, as noted earlier. Mafic banded gneiss RM30C has an igneous composition; it could be a sill or transposed dike, but the simplest interpretation is that it was a basalt flow interbedded with immature sandstones. RM30 has a felsic igneous composition, and therefore its protolith could have been a rhyolitic ash or lava or a dike or sill. Its high Zr concentration (711 ppm) could suggest that it is a metasand-

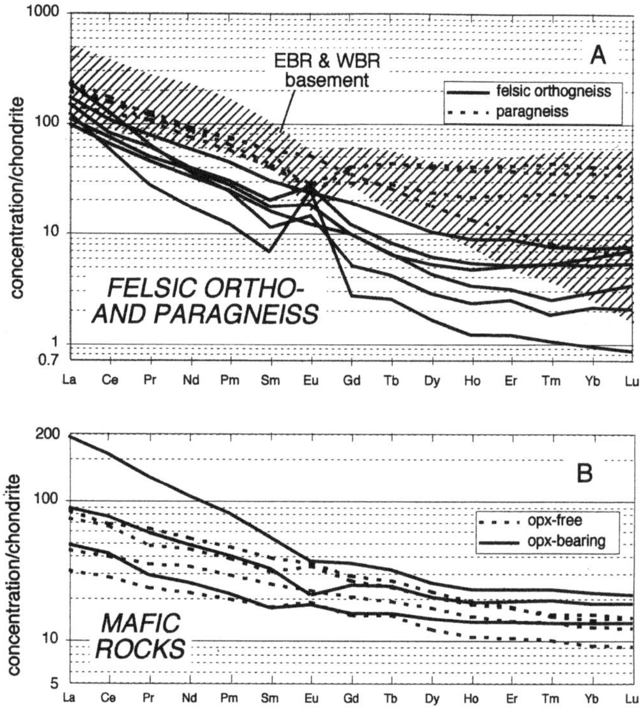

Figure 4. Chondrite-normalized REE abundances (normalization following Boynton, 1984). The field of Eastern Blue Ridge (EBR) and Western Blue Ridge (WBR) basement is from Carrigan (2000) and Carrigan et al. (2003).

TABLE 1. ELEMENTAL DATA

	Sample number																
	Mafic opx-free				Mafic opx-bearing			Felsic orthogneiss						Paragneiss			
	RM2*	RM2X	RM19	RM39	RM24	RM30C	MBCL5A	RM1*	RM15	RM21	RM30	RM38	MBCL4	RM13	RM30B	RM31	RM–CLG*
SiO_2, wt%	47.40	45.44	44.87	45.01	48.97	52.02	52.52	75.79	72.32	74.83	73.40	70.45	72.76	56.01	60.65	58.35	68.48
Al_2O_3	14.89	14.90	15.38	15.10	15.32	14.59	16.38	12.82	14.54	12.03	12.17	13.62	13.90	14.59	16.81	20.50	14.76
Fe_2O_3 (total)	15.12	15.72	14.35	14.77	10.94	13.84	9.52	2.81	0.54	2.56	3.81	3.54	1.47	10.71	7.54	9.39	6.05
MnO	0.20	0.22	0.22	0.21	0.18	0.20	0.15	0.06	0.00	0.04	0.02	0.04	0.02	0.11	0.08	0.17	0.11
MgO	6.85	6.32	6.15	7.42	6.91	5.96	3.56	0.61	0.02	0.87	0.55	1.14	0.37	2.56	1.80	2.81	1.73
CaO	9.69	8.84	9.05	9.14	11.96	7.02	5.91	2.85	0.54	2.02	1.53	1.84	1.40	5.48	1.32	1.91	3.13
Na_2O	2.73	2.95	2.94	3.07	2.61	2.94	4.30	2.96	1.85	2.83	2.11	3.06	3.23	1.61	2.33	1.51	2.91
K_2O	0.91	1.36	0.97	0.70	0.66	1.04	2.69	2.15	8.37	4.16	5.37	3.80	6.05	3.04	8.40	3.16	0.61
TiO_2	2.49	2.94	3.10	2.34	1.00	1.36	2.33	0.27	0.07	0.43	0.64	0.51	0.26	3.63	0.93	1.13	0.80
P_2O_5	0.25	0.41	0.50	0.19	0.10	0.16	0.85	0.06	0.04	0.13	0.07	0.16	0.09	1.16	0.04	0.23	0.03
LOI	–0.09	–0.13	1.45	0.58	0.30	0.30	0.89	0.12	0.94	0.26	0.30	1.59	0.36	0.58	0.12	1.09	0.10
Total	100.44	98.96	98.99	98.52	98.95	99.42	99.10	100.49	99.21	100.14	99.97	99.73	99.91	99.49	100.03	100.25	98.72
norm. C†	–7.64	–6.55	–5.79	–6.90	–11.23	–3.78	–2.34	0.57	1.53	–0.51	0.25	1.49	–0.31	1.44	1.55	11.66	3.68
Rb	14	25	13	7	5	20	71	40	251	121	133	87	157	45	233	90	8
Sr	309	306	272	302	253	168	488	255	330	210	330	583	530	1133	172	423	242
Ba	234	596	291	169	114	185	1175	849	2277	1131	1469	1710	2036	2675	1182	1262	185
Y	30	31	35	21	24	36	46	10	2	6	11	18	4	25	43	71	90
Zr	146	244	225	110	101	148	376	106	37	187	711	353	168	304	314	383	746
Hf	3.8	5.8	5.8	2.8	2.7	3.9	8.4	2.6	0.9	5.2	18.7	9.2	4.0	7.6	8.3	10.3	21.7
Nb	8.3	17.2	21.2	9.9	5.0	8.2	71.4	3.6	3.0	7.7	8.4	6.9	8.1	27.7	15.7	19.1	12.7
Ta	0.92	0.97	1.55	0.68	0.18	0.44	5.89	0.10	0.04	0.26	0.26	0.26	0.41	1.91	0.91	1.21	0.56
Zn	107	126	97	84	70	111	72	44	11	30	51	33	8	132	78	114	75
Cu	63	38	125	92	34	45	39	5	2	26	–1	14	10	10	27	72	1
V	311	303	238	304	266	286	187	34	9	30	36	43	27	168	107	105	69
Ni	62	38	83	67	61	49	10	2	<1	7	3	8	<1	7	23	25	10
Cr	99	67	57	56	197	114	16	5	<0.5	15	3	21	2	11	91	73	113
Co	52.5	55.0	49.8	59.4	42.5	53.5	25.2	5.0	1.4	7.5	6.6	10.2	3.4	24.3	20.5	18.0	11.2
Sc	32.5	31.6	27.0	27.7	44.9	47.7	20.5	6.8	0.3	3.8	3.8	6.4	1.2	10.3	22.3	23.4	19.7
La	13.9	23.0	26.3	9.9	15.1	27.5	89.7	30.4	38.5	34.5	47.2	53.7	71.9	46.4	69.2	60.4	73.7
Ce	32.5	52.9	55.5	23.0	34.4	62.7	174.9	53.1	47.8	61.0	66.5	88.0	101.6	99.8	127.4	125.8	140.4
Pr	4.36	5.88	7.82	2.91	3.59	7.30	18.15	5.56	3.42	6.06	7.41	9.70	7.87	13.23	12.73	15.57	14.36
Nd	20.6	27.4	32.5	13.4	15.7	29.4	65.1	20.0	10.6	22.7	24.1	35.5	21.9	54.3	45.4	56.0	51.4
Sm	5.01	6.17	7.75	3.49	3.42	6.45	10.69	3.16	1.33	3.48	3.93	5.91	2.26	11.16	7.77	11.30	8.41
Eu	1.71	2.51	2.67	1.40	1.36	1.59	2.78	0.91	1.87	1.36	2.01	1.74	1.08	3.81	2.10	1.68	2.08
Gd	5.36	7.03	7.50	4.02	4.14	6.56	9.35	2.57	0.71	2.61	3.21	4.97	1.34	9.07	7.53	10.66	10.57
Tb	0.92	1.16	1.27	0.72	0.75	1.17	1.52	0.31	0.12	0.31	0.39	0.68	0.20	1.21	1.32	2.07	2.08
Dy	5.58	6.65	7.28	3.97	4.73	6.70	8.40	1.70	0.53	1.39	1.98	3.40	0.92	5.83	7.69	12.79	13.38
Ho	1.08	1.33	1.39	0.78	1.00	1.38	1.70	0.34	0.09	0.24	0.39	0.65	0.17	0.97	1.57	2.67	2.96
Er	2.94	3.65	3.73	2.25	2.89	4.07	5.00	1.06	0.26	0.67	1.09	1.88	0.53	2.24	4.75	7.86	8.80
Tm	0.437	0.501	0.486	0.325	0.432	0.627	0.758	0.170	0.034	0.080	0.170	0.245	0.058	0.258	0.735	1.133	1.390
Yb	2.66	3.30	3.02	1.96	2.83	3.91	4.70	1.09	0.20	0.61	1.28	1.57	0.45	1.49	4.65	7.50	8.52
Lu	0.397	0.476	0.476	0.293	0.435	0.597	0.688	0.174	0.027	0.110	0.228	0.246	0.067	0.209	0.712	1.099	1.330
U	0.43	0.26	0.53	0.30	0.32	0.53	1.13	0.10	0.19	0.42	1.03	0.56	0.35	0.14	1.01	2.10	0.43
Th	1.61	2.00	2.03	1.02	3.16	2.73	6.45	0.44	2.55	1.45	3.82	4.35	42.92	0.91	24.85	14.52	4.44

Note: Oxide values are in wt%; trace element values are ppm. Abbreviation: LOI—loss on ignition.

*Data from Carrigan et al. (2003); † normative corundum; negative C values reflect deficiency in Al_2O_3 relative to $CaO + Na_2O + K_2O$ after forming normative apatite.

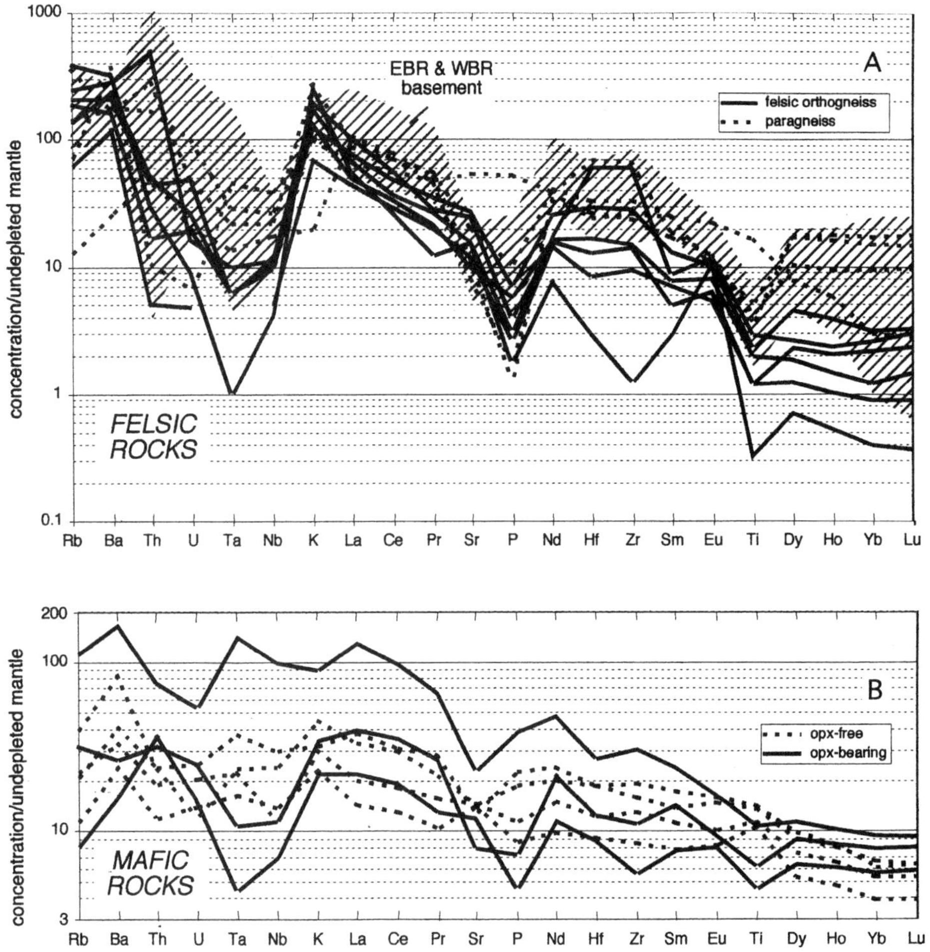

Figure 5. Elemental concentrations normalized to primitive mantle. Normalization values and element sequence are from Sun and McDonough (1989). The field of Eastern Blue Ridge (EBR) and Western Blue Ridge (WBR) basement is from Carrigan (2000) and Carrigan et al. (2003).

stone that was enriched in zircon by sedimentary processes. However, it lacks any other evidence for compositional effects induced by weathering or the mechanical concentration of grains in a sandstone. It has only 0.3 wt% normative corundum and is rich in Sr and not unusually quartz-rich. The high Zr could reflect either origin as a relatively high-*T* rhyolite—the calculated zircon saturation temperature for RM30 is 937 °C—or as a cumulate-rich intrusive rock, consistent with its high Sr and positive Eu anomaly.

Elementally based protolith interpretations suggest that meta-igneous rocks may have been bimodal, with no analyzed samples between 53 and 70 wt% SiO_2. The felsic orthogneisses range from K-poor to highly potassic (2.3 to 8.5 wt% K_2O) and have distinctive rare-earth-element (REE) patterns, with common positive Eu anomalies and low heavy REE (HREE) (Fig. 4, A). Moderate enrichment of incompatible elements with large negative high-field-strength-element (HFSE) anomalies is evident on primitive mantle-normalized spider plots (Fig. 5, A). RM1, the 1.8 Ga felsic sample of Carrigan et al. (2003), is more silicic but much lower in K and Rb than the samples investigated in this study. The paragneisses have no or negative Eu anomalies and much higher HREE contents than do the felsic orthogneisses. Mafic samples, with 45 to 52 wt% SiO_2, are relatively rich in K_2O (0.5 to 1.4, except for MBCL5A, with 2.7 wt%) and other incompatible elements compared with average basalts. The mafic rocks are mildly enriched in light REE (LREE) relative to HREE; MBCL5A is especially LREE rich (Fig. 4, B). All mafic samples have broadly similar incompatible element enrichment patterns. There are, however, subtle but important differences (Figs. 5, B, and 6, and the following discussion).

The U concentrations are for the most part low in analyzed samples (<~1 ppm) and Th/U ratios are high, as is typical of rocks that have undergone granulite-facies metamorphism (e.g., Zartman and Doe, 1981). The mean Th/U ratios of paragneisses (12), felsic orthogneisses (21), and opx-bearing mafic gneisses (6.2) are well above the global average of ~4. In contrast, the average opx-free mafic rock has a ratio of 4.4, possibly because these samples did not experience the highest-grade event.

The MHT samples are compared with analyses of Eastern Blue Ridge and Western Blue Ridge basement (Carrigan, 2000)

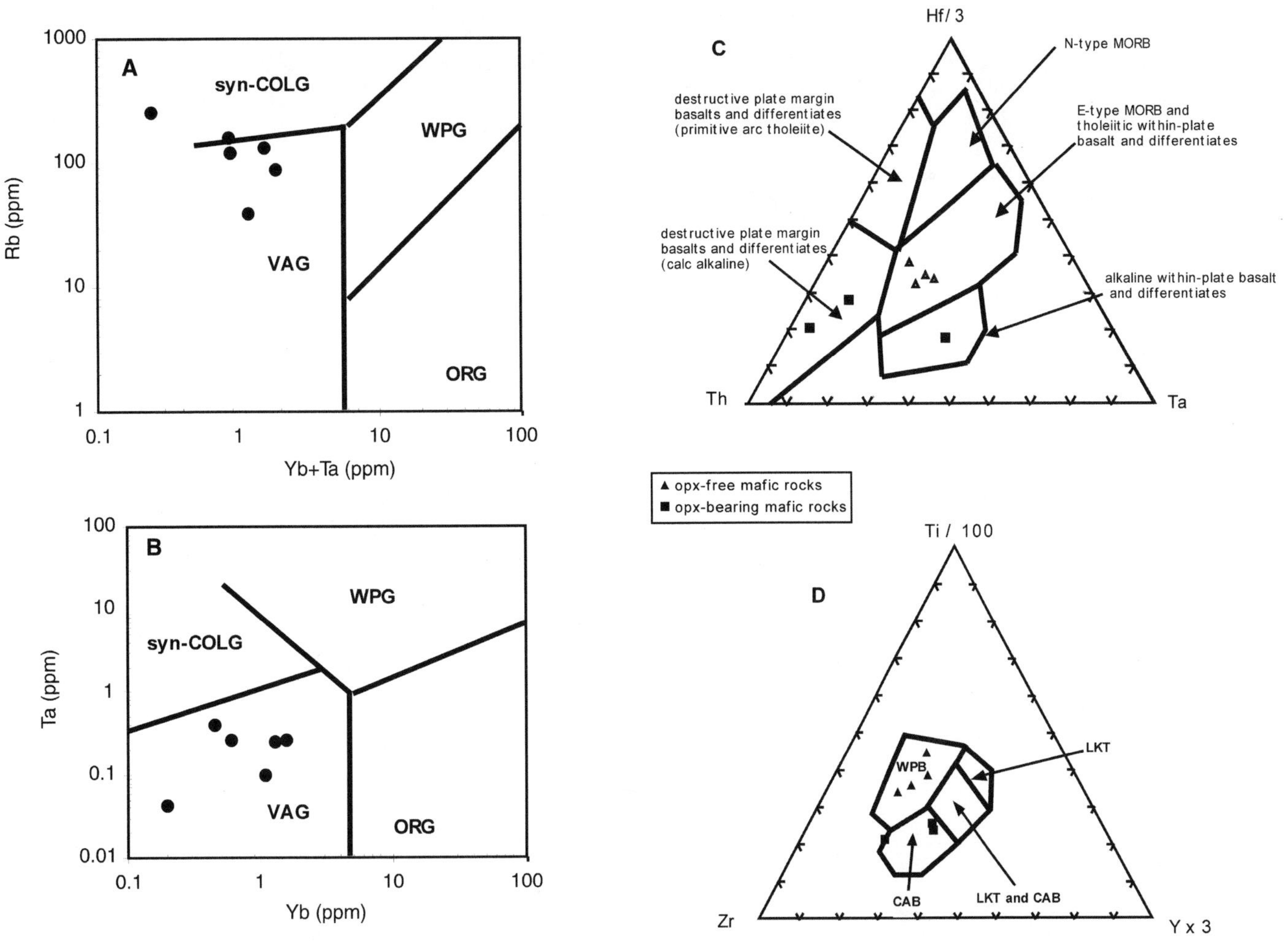

Figure 6. Tectonic discrimination diagrams. Felsic orthogneisses are plotted in 6A and 6B, mafic samples in 6C and 6D. (A) Rb versus Yb + Ta diagram of Pearce et al. (1984). ORG—ocean ridge granite; SYN COLG—syn-collisional granite; VAG—volcanic-arc granite; WPG—within-plate granite. (B) Ta versus Yb diagram of Pearce et al. (1984). Symbols as in panel A. (C) Hf-Th-Ta diagram of Wood (1980). MORB—Mid-oceanic ridge basalt. (D) Ti-Zr-Y diagram of Pearce and Cann (1973). LKT—Low-K tholeiite; WPB—within-plate tholeiite; CAB—calc-alkaline basalt.

in Figures 2, 4, and 5. MHT felsic orthogneisses are distinct from samples of basement from elsewhere in the southern Blue Ridge. The Eastern Blue Ridge and Western Blue Ridge basement gneisses have a wider range of SiO_2, higher REE concentrations (especially HREE), negative Eu anomalies, and somewhat higher incompatible element enrichments.

In standard tectonic discrimination diagrams (Fig. 6), the felsic orthogneisses generally plot together within the fields for arc-related granites, suggesting that they were either generated in an arc setting or derived from arc-generated crust. All four opx-free mafic samples plot as "within-plate basalts." The three opx-bearing mafic rocks plot within the calc-alkaline or arc basalts field on the Ti-Zr-Y diagram of Pearce and Cann (1973); on the Hf-Th-Ta diagram of Wood (1980), two plot as arc basalts, but the incompatible-element-rich rock, MBCL5A, plots alone in the field of alkaline within-plate basalts.

The Sm-Nd isotopic systematics of MHT samples document the antiquity of the crust that they represent (Table 2; Fig. 7). Two felsic orthogneisses, two opx-bearing mafic gneisses, and a paragneiss have Sm-Nd depleted mantle model ages (DePaolo, 1981) of ca. 1.7 to 2.3 Ga, and their calculated ε_{Nd} values during Grenville time were –2 to –7, considerably lower than those of basement granitic gneisses of the Eastern Blue Ridge and Western Blue Ridge at the same time (approximately –1 to +3). Two opx-free mafic samples have much higher ε_{Nd} (calculated as approximately +4 and +5 during Grenville time). Their values at 730 Ma (+1 and +3) are essentially identical to those calculated from whole-rock data for Bakersville dikes (Goldberg et al., 1986) (Fig. 7).

Calculated $^{87}Sr/^{86}Sr$ ratios of the felsic orthogneiss and paragneiss samples during Grenville time (1.0–1.2 Ga) range from 0.706 to 0.714. These high ratios, like the low Nd ratios,

TABLE 2. WHOLE-ROCK ISOTOPIC DATA

	Mafic, opx-free		Mafic, opx-bearing		Felsic gneiss		Paragneiss
	RM2	RM2X	RM24	RM30C	RM1	RM15	RM-CLG
Approximate age (Ga)*	0.73	0.73	1.2	1.2	1.8	1.2	1.2
Rb (ppm)	12.25	22.43	4.95	20.83	39.06	258.37	7.81
Sr (ppm)	310.1	304.9	267.9	147.7	253.6	346.0	249.6
$^{87}Rb/^{86}Sr$†	0.114	0.213	0.054	0.4086	0.446	2.168	0.091
$^{87}Sr/^{86}Sr_{measured}$	0.704732	0.706333	0.707354	0.719332	0.718556	0.742263	0.715261
s.e. $^{87}Sr/^{86}Sr$ (%)	0.0008	0.0010	0.0008	0.0008	0.0007	0.0009	0.0008
$^{87}Sr/^{86}Sr_{initial}$	0.70354	0.70411	0.70643	0.71231	0.70701	0.70500	0.71370
$^{87}Sr/^{86}Sr_{1.15}$‡	0.70286	0.70282	0.70646	0.71260	0.71121	0.70657	0.71376
Nd (ppm)	5.01	6.38	3.68	8.40	2.85	1.48	7.69
Sm (ppm)	19.28	26.99	16.18	37.31	19.00	11.01	48.04
$^{147}Sm/^{144}Nd$†	0.1607	0.1462	0.1407	0.1394	0.0929	0.0831	0.0990
$^{143}Nd/^{144}Nd_{measured}$	0.512625	0.512443	0.512112	0.511915	0.511489	0.511625	0.511637
s.e. $^{143}Nd/^{144}Nd$ (%)	0.0004	0.0009	0.0008	0.0007	0.0012	0.0005	0.0008
$\varepsilon_{Nd,present\ day}$	−0.25	−3.80	−10.26	−14.10	−22.41	−19.76	−19.53
$^{143}Nd/^{144}Nd_{initial}$	0.511856	0.511743	0.511003	0.510817	0.510389	0.510970	0.510857
$\varepsilon_{Nd,initial}$	3.11	0.91	−1.66	−5.31	1.57	−2.31	−4.52
$^{143}Nd/^{144}Nd_{1.15}$‡	0.511412	0.511339	0.511050	0.510863	0.510788	0.510998	0.510890
ε_{Nd1150}	5.06	3.64	−2.02	−5.68	−7.15	−3.04	−5.15
T_{DM} (Ga)	1.20	1.35	1.94	2.32	1.96	1.65	1.82

Note: Data for samples RM2, RM1, and RM-CLG are from Carrigan et al. (2003). Model age T_{DM} is calculated according to DePaolo (1981).
*For calculation of initial ratio.
†Sm/Nd and Rb/Sr uncertainties are estimated to be <1% (2σ).
‡Calculated at 1.15 Ga for comparison.

indicate that they were derived from much older crust. The opx-bearing mafic samples have calculated Grenville ratios of >0.706, whereas the two opx-free mafic samples both have a Grenville ratio of 0.703. Recalculated at 730 Ma, the age of Bakersville dikes (and the age of CAR 1501—see the Geochronology section), both opx-free samples yield ratios of 0.7035 and 0.7041. These ratios are similar to the initial ratio of 0.7044 obtained by Goldberg et al. (1986) from a whole-rock isochron for Bakersville samples. Open-system behavior during granulite-facies metamorphism would tend to lower Rb/Sr ratios and possibly increase $^{87}Sr/^{86}Sr$ ratios through interaction with nearby, more radiogenic Sr reservoirs. Thus, the calculated ratios are probably maxima, but the distinction between opx-bearing and opx-free mafic rocks appears real, as does the similarity of the opx-free samples to Bakersville dikes.

GEOCHRONOLOGY

As noted previously, Monrad and Gulley (1983), Fullagar and Gulley (1999), and Carrigan et al. (2003) have all reported ages of ca. 1.8 Ga for Carvers Gap orthogneiss from Roan Mountain. Fullagar and Gulley (1999) also obtained an upper-intercept age of 1.4 Ga for another Carvers Gap sample. Carrigan et al. (2003) analyzed zircons from four other Roan Mountain samples by ion microprobe, two of Carvers Gap gneiss and two of Cloudland paragneiss. The orthogneiss samples yielded imprecise ages of ca. 1.6 Ga and 1.2 Ga, interpreted as the time of magmatic crystallization. Cores of detrital grains from Cloudland gneiss gave nearly concordant ages ranging from ~1.0 to 1.85 Ga. The youngest ages approach and in some cases are younger than the age of ubiquitous metamorphic rims (ca. 1.03 Ga) and therefore probably reflect Pb loss. However, the abundant concordant ages strongly indicate a range of Mesoproterozoic to Paleoproterozoic detrital ages. Fullagar et al. (1979) reported a whole-rock Rb-Sr age of 1183 ± 65 Ma for granitic gneiss from near Mars Hill.

Sites for SHRIMP U-Pb analysis of zircons from the seven samples investigated for this study were selected and interpreted in part on the basis of zoning evident in cathodoluminescence images. We interpret zoning based on criteria described in Miller et al. (1992, 1998), Hanchar and Miller (1993), and references therein. Almost all grains are strikingly zoned, with bright (less commonly, dark), weakly zoned or unzoned rims that we interpret to be metamorphic and having one or more distinct interior zones. Some grains have zoning that suggests two discrete metamorphic overgrowths, and others have distinct cores with magmatic overgrowths (characterized by euhedral, in some cases oscillatory zoning), all surrounded by a metamorphic rim.

In the following discussion, we briefly describe the zoning (Fig. 8) and U-Pb data (Table 3; Fig. 9) for zircons from each of the samples. In general, metamorphic data are concordant or nearly so, but imprecise owing to low U and Pb concentrations, whereas magmatic and premagmatic data for most samples are discordant. This discordance may in small part reflect beam overlap into two distinct age zones, but the fact that analyses commonly define discordia with young lower intercepts that do

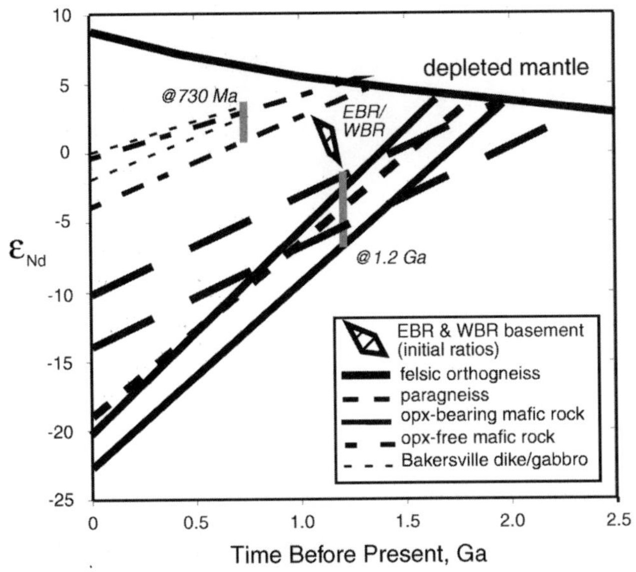

Figure 7. Calculated Nd isotopic evolution of samples analyzed in this study, plus two whole-rock analyses of Bakersville rocks by Goldberg and Dallmeyer (1997). Depleted mantle curve is from DePaolo (1981). Eastern Blue Ridge/Western Blue Ridge basement field from Carrigan (2000) and Carrigan et al. (2003). Gray bars at 730 Ma and 1.2 Ga show ranges of initial values of opx-free mafic rocks + Bakersville dikes and of gabbros and other Mars Hill samples, respectively.

not correspond to ages of any identifiable zones—in several cases, zero-age lower intercepts—indicates that most discordance is a result of Pb loss. The greater discordance of U-rich magmatic zones than of U-poor metamorphic zones is consistent with discordance through Pb loss. There is essentially no correlation between Th/U ratio and zone type (Table 3).

For the most abundant age populations, which we interpret to reflect magmatic crystallization or possibly detrital reworking of a slightly older igneous source, we have estimated age in two ways. First, we have pooled the $^{207}Pb/^{206}Pb$ ages of the more concordant points, and second, we have determined upper intercepts of discordia for those samples for which the data fit well on a regression. For those samples that define a discordia, the upper intercept is invariably within the error of the pooled $^{207}Pb/^{206}Pb$ age. In fact, five samples (all but two) yield ages within error of one another at 1.20 Ga. In Figure 9, both discordia (if defined by the data) and a constant $^{207}Pb/^{206}Pb$ age reference line (from the origin through the approximate magmatic age) are shown. Lower intercepts of the discordia are either near zero or imprecisely defined Paleozoic ages.

The less precise metamorphic ages are estimated in most cases by pooling $^{207}Pb/^{206}Pb$ ages; in some cases, the data can be fit to a discordia. Although the dominant metamorphic age is clearly near 1.0 Ga, there is some evidence that there may be a second, older metamorphic population, but there are too few points to define this age well. Older, apparently detrital and inherited cores are mostly discordant, and we take individual $^{207}Pb/^{206}Pb$ ages as the best estimates for the ages of these zones.

All stated age uncertainties are $\pm 2\sigma$.

RM21—Felsic Orthogneiss

Zircons in RM21 have simple, concentric internal zones, in some cases, euhedral and, rarely, oscillatory (Fig. 8, A). Based on zone morphology alone, it is difficult to distinguish magmatic from inherited portions with certainty, but rare truncated zoned fragments in the centers are the best candidates for inherited cores. Most grains have thick, bright overgrowths that we interpret to be metamorphic; in some cases, these form the rims, but in others, they are surrounded by a thin, darker rim zone.

Five discordant interior points that we interpret to be magmatic fall on a zero-lower-intercept discordia with an upper intercept of 1198 ± 26 Ma (mean square of weighted deviations [MSWD], 0.30) (Fig. 9, A). The four most concordant points yield an identical $^{207}Pb/^{206}Pb$ age (MSWD, 0.40) (Fig. 9, A). Two analyses from cores have discordant $^{207}Pb/^{206}Pb$ ages of 1276 and1538 Ma. The $^{207}Pb/^{206}Pb$ ages of nine points from rims and probable metamorphic interiors average 1026 ± 19 Ma (MSWD, 1.02).

RM38—Felsic, Mylonitic Orthogneiss

Zircons from RM38 (Fig. 8, B) are well formed and prismatic, with euhedral and locally oscillatory zones that we interpret to be magmatic. Dark cores in the magmatic portions are rare. All grains have thin to thick bright, rounded rims that we interpret to be metamorphic, and some have slightly less bright zones inside the bright rims that appear to mark earlier metamorphic growth.

Ten analyses from zones interpreted as magmatic or possibly magmatic define a discordia with an upper intercept of 1200 ± 26 Ma and a lower intercept of 383 ± 110 Ma (MSWD, 1.13) (Fig. 9, B). The five most concordant points have a mean $^{207}Pb/^{206}Pb$ age of 1185 ± 36 Ma (MSWD, 2.2); the slightly younger age and higher MSWD reflect the fact that these points actually lie on the well-defined discordia defined by all ten magmatic points, which has a nonzero lower intercept. One dark unzoned (presumably inherited) core yielded a $^{207}Pb/^{206}Pb$ age of 1511 ± 69 Ma.

MBCL4—Felsic Orthogneiss

Zircons from MBCL4 have very well-defined euhedral, oscillatory magmatic zoning and thin to thick, bright metamorphic rims (Fig. 8, C). No cores are evident. Eleven points from magmatic zones define a discordia with an upper intercept of 1257 ± 26 Ma and a lower intercept of 540 ± 280 (MSWD, 0.35) (Fig. 9, C). The ten most concordant points have a mean

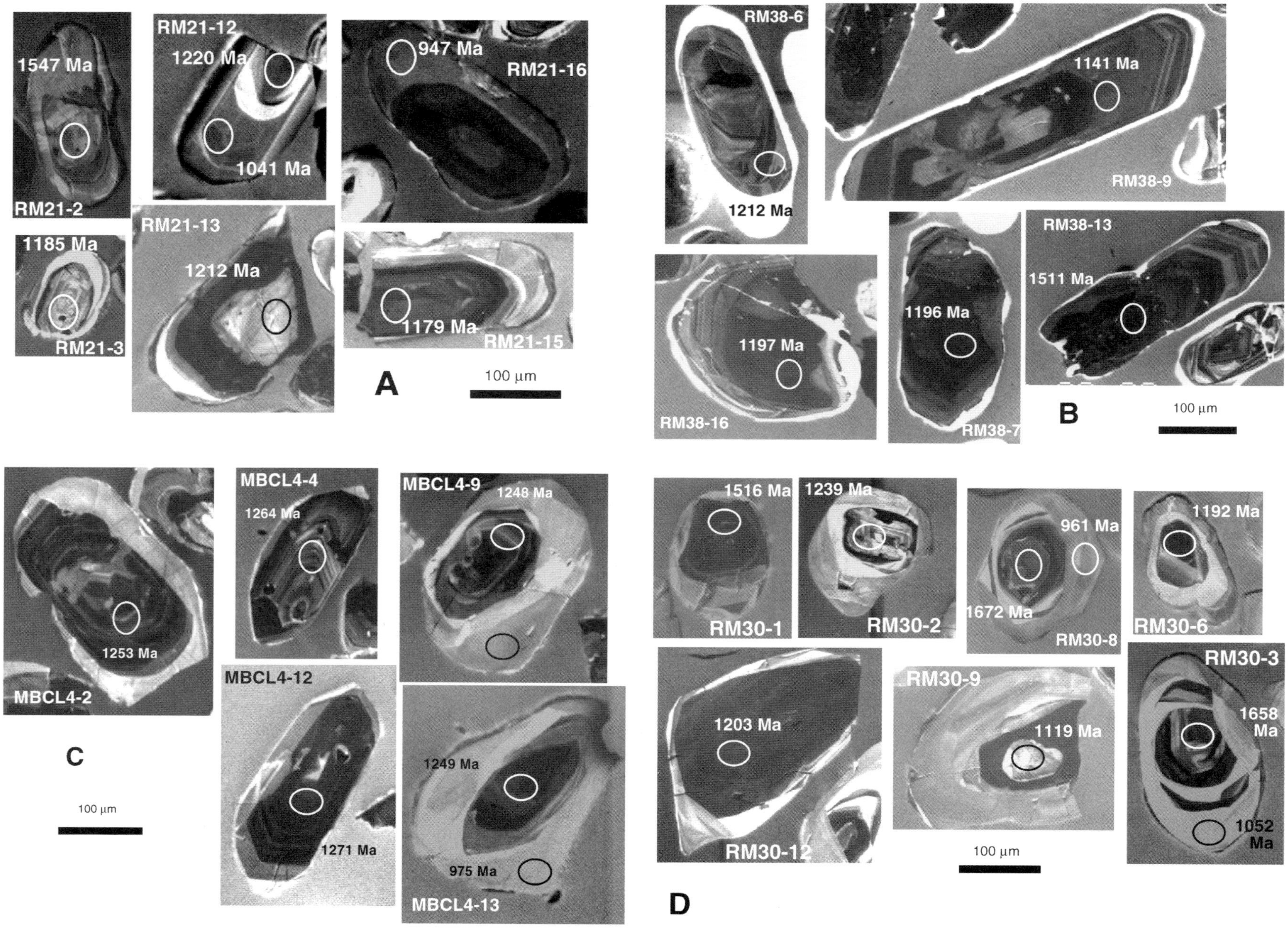
1547 Ma
RM21-2
RM21-12
1220 Ma
1041 Ma
947 Ma
RM21-16
1185 Ma
RM21-3
RM21-13
1212 Ma
1179 Ma
RM21-15
A
100 µm
RM38-6
1212 Ma
1141 Ma
RM38-9
RM38-16
1197 Ma
1196 Ma
RM38-7
RM38-13
1511 Ma
B
100 µm
MBCL4-2
1253 Ma
MBCL4-4
1264 Ma
MBCL4-9
1248 Ma
MBCL4-12
1271 Ma
1249 Ma
975 Ma
MBCL4-13
C
100 µm
1516 Ma
RM30-1
1239 Ma
RM30-2
961 Ma
1672 Ma
RM30-8
1192 Ma
RM30-6
RM30-12
1203 Ma
RM30-9
1119 Ma
RM30-3
1658 Ma
1052 Ma
D
100 µm

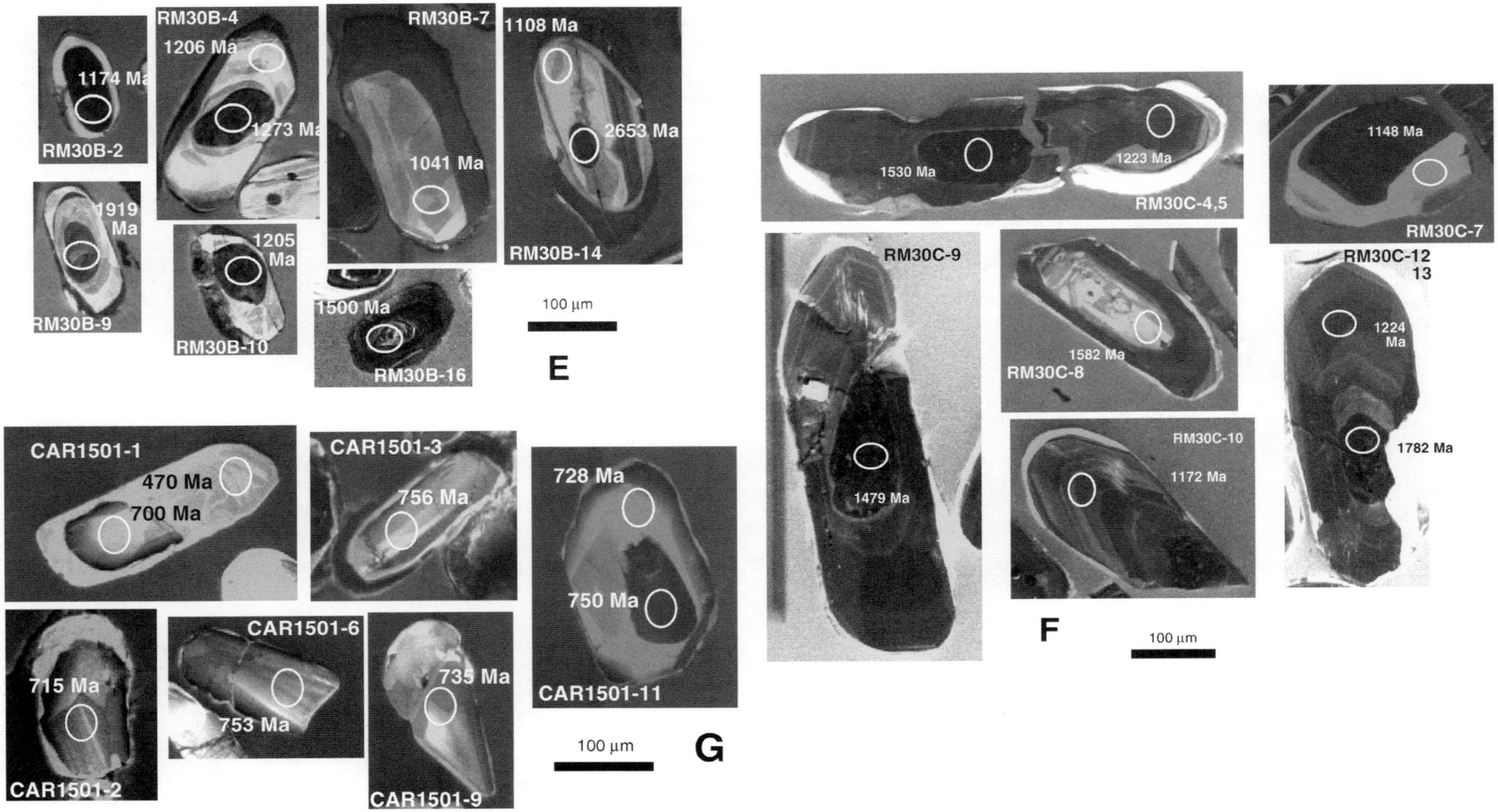

Figure 8. Cathodoluminescence images of representative Mars Hill zircons that were analyzed. Ages are $^{207}Pb/^{206}Pb$, except for CAR 1501 ($^{206}Pb/^{238}U$); ellipses show approximate spot size and location. Labels denote sample and zircon analysis numbers (see Table 3).

TABLE 3. U-PB ZIRCON DATA

Analysis number	Zone type	Common. ^{206}Pb (%)	U (ppm)	Th (ppm)	$^{232}Th/^{238}U$	$^{206}Pb/^{238}U$ age†	2σ error	$^{207}Pb/^{206}Pb$ age§	2σ error	Total $^{238}U/^{206}Pb$	Error (%)	Total $^{207}Pb/^{206}Pb$	Error (%)	$^{207}Pb\#/^{235}U$	Error (%)	$^{206}Pb\#/^{238}U$	Error (%)	Error correlation
RM21-1	DUI	0.02	2180	376	0.18	1054	24	1206	39	5.59	1.17	0.0805	0.98	1.983	1.53	0.179	1.17	0.76
RM21-2	GZC	0.15	188	140	0.77	1316	41	1547	154	4.35	1.55	0.0972	3.93	3.033	4.38	0.229	1.56	0.36
RM21-3	GEC	–0.02	349	123	0.36	1162	31	1185	67	5.06	1.37	0.0794	1.70	2.169	2.18	0.198	1.37	0.63
RM21-4	DZC	0.02	2050	119	0.06	1009	23	1034	31	5.90	1.16	0.0739	0.75	1.723	1.39	0.170	1.16	0.84
RM21-5	BUR	–0.08	87	182	2.17	988	36	1014	159	6.04	1.86	0.0723	3.96	1.669	4.34	0.166	1.86	0.43
RM21-6	BUR	–0.07	88	172	2.02	1068	38	1100	145	5.55	1.81	0.0756	3.67	1.895	4.06	0.180	1.81	0.45
RM21-8	DUC	0.03	3004	278	0.10	929	21	1022	29	6.42	1.18	0.0736	0.70	1.573	1.38	0.156	1.18	0.86
RM21-9	GEC	0.00	2237	334	0.15	1149	26	1276	39	5.10	1.16	0.0832	1.00	2.253	1.53	0.196	1.16	0.76
RM21-10	BUR	–0.07	108	209	2.01	997	34	1016	143	5.98	1.75	0.0725	3.56	1.687	3.94	0.167	1.75	0.44
RM21-11	BUI	–0.10	68	164	2.47	970	38	1103	170	6.13	2.01	0.0755	4.29	1.719	4.69	0.163	2.01	0.43
RM21-12	GZR	0.21	180	345	1.98	1034	30	1041	122	5.73	1.49	0.0758	2.51	1.775	3.37	0.174	1.50	0.44
RM21-12B	GUC	–0.02	299	83	0.29	1031	28	1220	72	5.72	1.40	0.0808	1.83	1.952	2.30	0.175	1.40	0.61
RM21-13	BZC	–0.04	150	133	0.92	971	30	1212	164	6.09	1.54	0.0803	4.18	1.826	4.43	0.164	1.54	0.35
RM21-14	BUR	0.56	66	150	2.33	979	38	765	278	6.12	1.95	0.0693	4.30	1.451	6.89	0.163	1.99	0.29
RM21-15	DEI	–0.01	501	473	0.98	1141	28	1179	54	5.16	1.29	0.0792	1.38	2.121	1.88	0.194	1.29	0.68
RM21-16	GUR	–0.03	176	262	1.54	984	30	947	104	6.07	1.56	0.0704	2.54	1.604	2.98	0.165	1.56	0.53
RM38-1	DUC	0.00	3183	157	0.05	937	21	1134	26	6.34	1.15	0.0775	0.65	1.687	1.32	0.158	1.15	0.87
RM38-6	GEI	0.06	582	195	0.35	1143	28	1212	68	5.13	1.25	0.0811	1.68	2.163	2.14	0.195	1.25	0.58
RM38-7	DUI	0.03	2770	578	0.22	1144	25	1196	25	5.14	1.15	0.0802	0.61	2.146	1.31	0.195	1.15	0.88
RM38-8	GUI	–0.03	307	135	0.46	776	22	1034	126	7.74	1.43	0.0735	3.13	1.314	3.43	0.129	1.43	0.42
RM38-9	GEI	0.09	378	133	0.36	1126	29	1141	73	5.23	1.31	0.0786	1.71	2.048	2.27	0.191	1.31	0.58
RM38-10	DUC	0.05	2357	132	0.06	1048	23	1149	41	5.64	1.15	0.0785	1.02	1.909	1.55	0.177	1.15	0.74
RM38-11	DZC	–0.01	904	143	0.16	925	22	1090	49	6.43	1.21	0.0757	1.23	1.625	1.73	0.155	1.21	0.70
RM38-13	DUC	1.90	2011	292	0.15	1162	28	1511	139	4.88	1.15	0.1105	1.72	2.612	3.86	0.201	1.18	0.31
RM38-15	DUC	–0.01	888	238	0.28	1111	26	1132	45	5.31	1.20	0.0774	1.13	2.010	1.65	0.188	1.20	0.73
RM38-16	DUC	0.00	1240	255	0.21	1164	27	1197	35	5.04	1.19	0.0800	0.89	2.187	1.49	0.198	1.19	0.80
RM38-17	GZI	0.17	351	96	0.28	810	24	1020	109	7.40	1.49	0.0746	2.29	1.362	3.07	0.135	1.49	0.49
MBCL-1	DZC	0.12	729	625	0.89	1012	24	1096	63	5.85	1.22	0.0770	1.24	1.788	2.00	0.171	1.22	0.61
MBCL-2	DEC	–0.01	469	329	0.73	1249	31	1253	56	4.68	1.26	0.0822	1.44	2.426	1.91	0.214	1.26	0.66
MBCL-3	BUR	–0.09	74	246	3.42	955	37	1485	145	6.11	1.93	0.0921	3.87	2.097	4.30	0.164	1.93	0.45
MBCL-3B	DUC	0.08	1136	64	0.06	1079	25	1199	59	5.45	1.20	0.0808	1.38	2.023	1.91	0.183	1.20	0.63
MBCL-4	GEC	0.06	501	980	2.02	1276	41	1264	53	4.57	1.66	0.0833	1.28	2.496	2.14	0.219	1.66	0.78
MBCL-5	DUC	0.00	1561	980	0.65	1275	28	1258	29	4.58	1.16	0.0825	0.74	2.487	1.37	0.219	1.16	0.84
MBCL-6	BZC	0.97	96	114	1.24	1063	37	1203	270	5.49	1.75	0.0884	3.26	1.997	7.10	0.180	1.83	0.26
MBCL-7	DEI	0.37	1063	461	0.45	1129	26	1240	65	5.18	1.19	0.0849	1.10	2.169	2.04	0.192	1.19	0.59
MBCL-7B	GEI	0.06	906	292	0.33	1140	27	1216	60	5.15	1.23	0.0813	1.49	2.162	1.96	0.194	1.23	0.63
MBCL-8	BUR	1.73	63	206	3.39	1012	41	367	389	5.92	2.03	0.0678	4.70	1.233	8.88	0.166	2.06	0.23
MBCL-9	DEC	0.08	374	721	1.99	1215	30	1248	66	4.81	1.29	0.0828	1.58	2.353	2.13	0.208	1.29	0.61
MBCL-9B	BZR	–0.07	92	378	4.23	1012	36	925	155	5.91	1.80	0.0693	3.80	1.633	4.18	0.169	1.80	0.43
MBCL-10	BEC	0.13	360	465	1.34	1210	31	1244	78	4.83	1.31	0.0830	1.71	2.336	2.38	0.207	1.32	0.55
MBCL-11	DZC	0.11	646	671	1.07	1027	25	1029	70	5.78	1.27	0.0744	1.41	1.752	2.15	0.173	1.27	0.59
MBCL-12	DEC	0.00	1301	662	0.53	1257	29	1271	32	4.64	1.18	0.0830	0.82	2.468	1.44	0.216	1.18	0.82
MBCL-13A	DEC	0.26	1529	944	0.64	972	24	1249	50	6.05	1.29	0.0844	0.87	1.866	1.82	0.165	1.29	0.71
MBCL-13B	BUR	0.73	117	275	2.42	1067	34	975	267	5.54	1.65	0.0777	3.10	1.770	6.76	0.179	1.72	0.25
MBCL-15	DZC	0.00	1372	42	0.03	1172	27	1210	33	5.01	1.18	0.0805	0.83	2.217	1.44	0.200	1.18	0.82
MBCL-16	GEI	0.09	361	394	1.13	1172	31	1202	69	5.00	1.35	0.0810	1.63	2.209	2.21	0.200	1.35	0.61
RM30-1	DZC	0.38	1370	741	0.56	1241	29	1516	114	4.63	1.17	0.0977	2.75	2.802	3.25	0.215	1.17	0.36
RM30-2	BZC	–0.02	292	228	0.80	1136	31	1239	101	5.16	1.37	0.0815	2.57	2.183	2.91	0.194	1.37	0.47
RM30-3A	GEC	–0.01	296	301	1.05	1657	45	1658	55	3.41	1.39	0.1017	1.50	4.116	2.04	0.293	1.39	0.68
RM30-3B	BUR	–0.09	80	138	1.78	1068	41	1052	166	5.56	1.95	0.0736	4.16	1.847	4.55	0.180	1.95	0.43
RM30-4	BUR	–0.04	212	49	0.24	995	29	1062	105	5.97	1.50	0.0744	2.62	1.726	3.01	0.167	1.50	0.50
RM30-5	BZC	0.26	754	84	0.11	1199	29	1320	58	4.85	1.26	0.0874	1.42	2.416	1.97	0.206	1.26	0.64
RM30-6	DZC	0.03	1548	108	0.07	1125	26	1192	60	5.23	1.19	0.0800	1.50	2.105	1.92	0.191	1.19	0.62
RM30-7	DZC	0.03	2435	1211	0.51	1273	33	1441	23	4.54	1.35	0.0910	0.59	2.754	1.48	0.220	1.35	0.91
RM30-8	GZC	–0.01	424	520	1.27	1676	42	1672	41	3.37	1.28	0.1026	1.10	4.201	1.69	0.297	1.28	0.76

RM30-8B	BUR	–0.05	164	178	1.12	1064	36	961	123	5.60	1.71	0.0707	3.02	1.753	3.45	0.179	1.71	0.50
RM30-9	BZC	0.14	798	51	0.07	1018	26	1119	57	5.81	1.29	0.0781	1.37	1.822	1.92	0.172	1.29	0.67
RM30-10	DZI	0.00	1571	254	0.17	1194	27	1168	32	4.92	1.17	0.0788	0.80	2.209	1.42	0.203	1.17	0.83
RM30-11	BZC	–0.03	142	122	0.89	1378	41	1383	90	4.20	1.52	0.0878	2.36	2.894	2.80	0.238	1.52	0.54
RM30-12	DUC	0.00	2091	100	0.05	1184	48	1203	28	4.95	2.11	0.0802	0.71	2.233	2.23	0.202	2.11	0.95
RM30B-1	DUC	–0.01	944	15	0.02	1138	28	1199	62	5.16	1.29	0.0800	1.58	2.139	2.04	0.194	1.29	0.63
RM30B-2	DUC	–0.01	966	10	0.01	1163	27	1174	45	5.06	1.20	0.0790	1.14	2.157	1.66	0.198	1.20	0.73
RM30B-3	DUC	0.00	1391	16	0.01	1199	27	1210	51	4.89	1.16	0.0805	1.29	2.271	1.74	0.205	1.16	0.67
RM30B-4A	DUC	–0.01	695	114	0.17	1184	29	1273	58	4.94	1.24	0.0831	1.48	2.322	1.93	0.203	1.24	0.64
RM30B-4B	BZI	–0.07	98	178	1.87	1052	37	1206	138	5.61	1.79	0.0798	3.53	1.979	3.93	0.179	1.79	0.46
RM30B-5	DUI	–0.02	276	205	0.77	1115	30	976	89	5.33	1.39	0.0714	2.18	1.855	2.58	0.188	1.39	0.54
RM30B-6	DUI	0.13	897	109	0.13	1058	25	992	56	5.62	1.24	0.0733	1.26	1.771	1.85	0.178	1.24	0.67
RM30B-7	BZC	–0.07	130	341	2.72	1114	39	1041	147	5.32	1.78	0.0734	3.66	1.918	4.04	0.188	1.78	0.44
RM30B-7B	DUR	–0.01	527	148	0.29	1027	25	996	65	5.80	1.26	0.0723	1.60	1.721	2.03	0.172	1.26	0.62
RM30B-8X	BZC	–0.08	99	207	2.15	961	34	948	157	6.23	1.81	0.0700	3.88	1.566	4.24	0.161	1.81	0.43
RM30B-9	DZC	0.10	438	142	0.34	1365	38	1919	137	4.10	1.30	0.1184	3.76	3.953	4.04	0.244	1.30	0.32
RM30B-10	DUC	–0.01	525	12	0.02	1117	27	1205	60	5.27	1.27	0.0802	1.52	2.103	1.97	0.190	1.27	0.64
RM30B-11	DEC	0.32	640	56	0.09	1201	30	1317	91	4.84	1.26	0.0878	2.04	2.416	2.65	0.206	1.26	0.47
RM30B-12	DUC	0.11	823	33	0.04	1065	27	1118	54	5.55	1.33	0.0778	1.18	1.908	1.89	0.180	1.33	0.70
RM30B-12B	BZR	–0.07	105	201	1.99	1104	49	906	178	5.40	2.26	0.0687	4.37	1.769	4.89	0.185	2.26	0.46
RM30B-13	DUC	0.07	1190	10	0.01	1239	28	1220	44	4.72	1.18	0.0816	0.99	2.363	1.63	0.212	1.19	0.73
RM30B-14	DUC	0.61	672	52	0.08	1581	50	2653	42	3.27	1.29	0.1855	0.88	7.536	1.81	0.304	1.30	0.72
RM30B-14B	BZR	–0.07	116	200	1.78	995	34	1108	140	5.97	1.76	0.0759	3.54	1.769	3.93	0.168	1.76	0.45
RM30B-15	DZC	0.12	258	85	0.34	1367	37	1340	115	4.24	1.39	0.0871	2.85	2.799	3.29	0.236	1.39	0.42
RM30B-8B	DUR	0.08	615	115	0.19	1043	26	922	73	5.72	1.27	0.0704	1.66	1.682	2.18	0.175	1.27	0.58
RM30B-16	DZC	0.06	606	3100	5.28	1562	37	1500	47	3.66	1.22	0.0941	1.22	3.525	1.74	0.273	1.22	0.70
RM30C-1	GEI	–0.01	1204	607	0.52	1132	10	1179	37	5.20	0.43	0.0792	0.93	2.103	1.02	0.192	0.43	0.42
RM30C-2	GEI	–0.01	867	303	0.36	1207	11	1199	42	4.86	0.48	0.0800	1.05	2.272	1.16	0.206	0.48	0.41
RM30C-3	DEC	–0.01	853	661	0.80	1094	11	1123	44	5.40	0.51	0.0770	1.11	1.968	1.22	0.185	0.51	0.42
RM30C-4	DUC	0.42	2288	1399	0.63	1254	11	1530	93	4.57	0.37	0.0987	1.75	2.858	2.51	0.218	0.41	0.16
RM30C-5	GEI	–0.01	634	199	0.32	1121	13	1223	49	5.24	0.58	0.0810	1.26	2.133	1.38	0.191	0.58	0.42
RM30C-6	BUR	–0.02	293	82	0.29	1177	17	1142	72	5.00	0.74	0.0777	1.82	2.146	1.96	0.200	0.74	0.38
RM30C-7	BUR	–0.05	143	38	0.28	1074	26	1148	147	5.50	1.19	0.0776	3.72	1.957	3.88	0.182	1.19	0.31
RM30C-8	BZC	0.14	173	111	0.66	1431	26	1582	80	3.98	0.90	0.0990	1.93	3.380	2.31	0.251	0.91	0.39
RM30C-9	DUC	0.18	2752	185	0.07	1199	8	1479	28	4.81	0.28	0.0941	0.54	2.648	0.78	0.207	0.29	0.37
RM30C-10	GZI	–0.01	602	196	0.34	1163	11	1172	49	5.06	0.50	0.0789	1.23	2.154	1.33	0.198	0.50	0.38
RM30C-11	GZC	–0.01	847	548	0.67	1080	11	1176	44	5.46	0.51	0.0791	1.11	1.999	1.22	0.183	0.51	0.42
RM30C-12	DUC	1.98	1938	787	0.42	1188	35	1782	403	4.69	0.43	0.1261	9.08	3.141	11.06	0.209	0.49	0.04
RM30C-13	DEI	0.03	1051	349	0.34	1180	10	1224	37	4.97	0.44	0.0813	0.93	2.251	1.05	0.201	0.44	0.42
RM30C-14	DEC	–0.01	493	172	0.36	1129	15	1206	57	5.21	0.69	0.0803	1.45	2.128	1.60	0.192	0.69	0.43
RM30C-15	DUC	0.00	2125	211	0.10	1211	9	1432	25	4.79	0.34	0.0903	0.67	2.602	0.75	0.209	0.34	0.46
CAR-1	GEI	–0.14	66	34	0.53	700	33	569	239	8.77	2.39	0.0579	5.60	0.929	5.98	0.114	2.39	0.40
CAR-1B	BUR	–0.32	49	1	0.02	470	43	468	365	13.25	4.67	0.0538	8.64	0.589	9.47	0.076	4.67	0.49
CAR-2	GZC	0.70	93	100	1.10	715	28	572	382	8.51	1.95	0.0648	5.52	0.951	9.00	0.117	2.00	0.22
CAR-3	GEC	1.10	62	58	0.96	756	34	147	652	8.11	2.28	0.0577	6.12	0.824	14.10	0.122	2.39	0.17
CAR-4	DUC	0.16	564	753	1.38	640	17	685	112	9.55	1.32	0.0636	2.08	0.898	2.94	0.105	1.32	0.45
CAR-5	GZC	–0.08	121	66	0.57	738	29	859	169	8.22	2.02	0.0670	4.10	1.136	4.54	0.122	2.02	0.45
CAR-6	GZC	–0.12	83	83	1.04	753	31	797	211	8.06	2.07	0.0647	5.12	1.125	5.45	0.124	2.07	0.38
CAR-7	DZC	–0.03	295	348	1.22	702	21	809	110	8.66	1.53	0.0658	2.64	1.052	3.04	0.115	1.53	0.50
CAR-8A	DZC	–0.02	716	792	1.14	649	16	751	76	9.40	1.27	0.0642	1.80	0.943	2.20	0.106	1.27	0.58
CAR-8B	BZR	–0.45	30	18	0.62	718	50	1008	361	8.43	3.50	0.0691	9.39	1.196	9.56	0.119	3.50	0.37
CAR-9	GZC	–0.16	64	45	0.72	735	33	811	242	8.27	2.30	0.0648	5.91	1.104	6.22	0.121	2.30	0.37
CAR-10	DUC	0.13	681	918	1.39	481	13	445	121	12.91	1.34	0.0569	2.36	0.595	3.04	0.077	1.34	0.44
CAR-11	DZC	–0.05	210	200	0.99	750	25	733	140	8.11	1.72	0.0633	3.33	1.084	3.73	0.123	1.72	0.46
CAR-11B	GZI	–0.22	48	30	0.66	728	40	386	321	8.48	2.80	0.0526	7.39	0.886	7.68	0.118	2.80	0.36

Note: Zone type indicates the nature of zoning at the analytical spot. Note that the abbreviations in the zone type column are concatenated. Abbreviations: B—bright in cathodoluminescence; C—core; D—dark; E—euhedral (~oscillatory) zoned; G—gray; I—interior (not a distinct core) ; R—rim; U—weakly zoned or unzoned; Z—distinctly zoned.
†^{207}Pb-corrected; §^{204}Pb-corrected; #radiogenic.

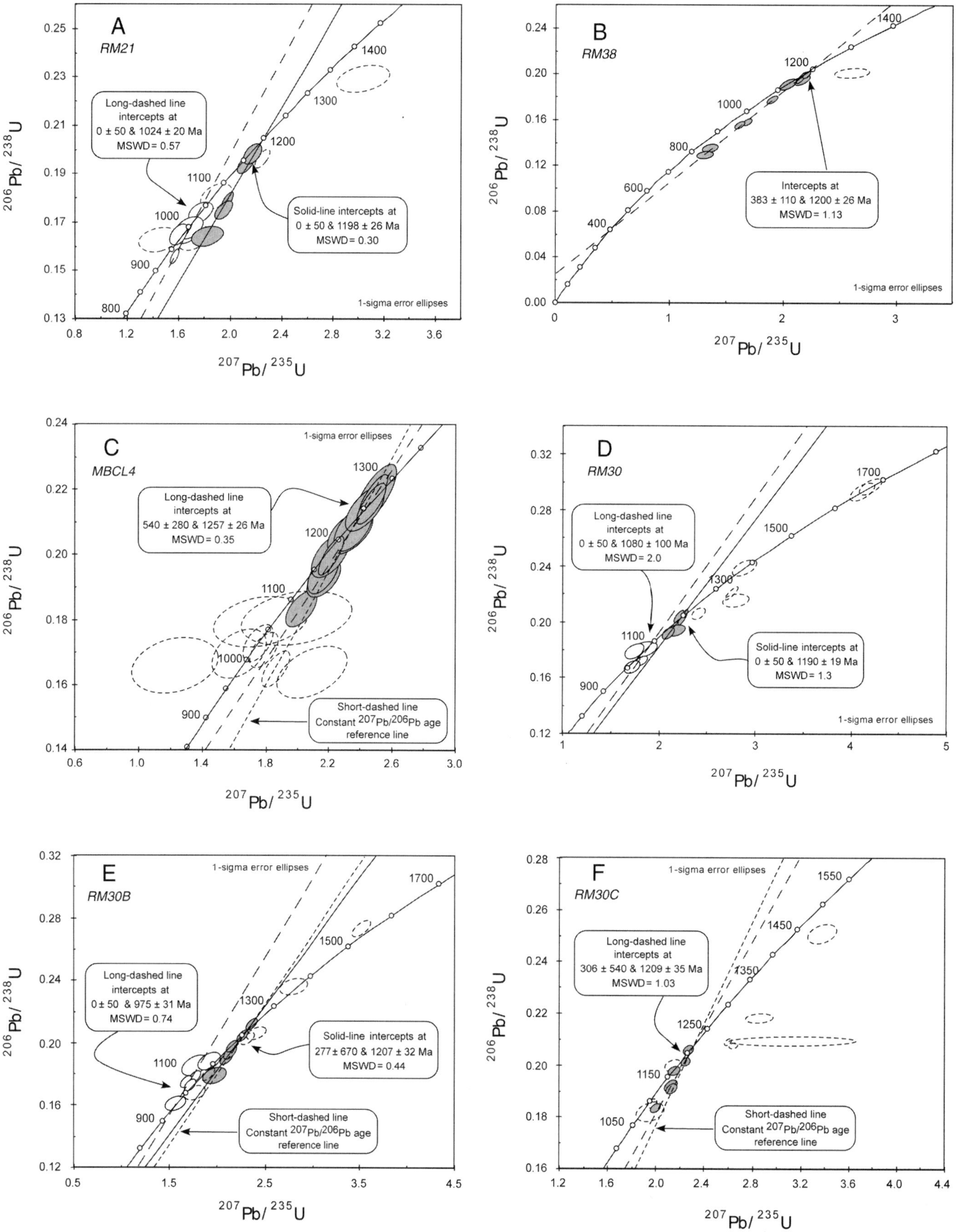
A
RM21
Long-dashed line intercepts at 0 ± 50 & 1024 ± 20 Ma MSWD= 0.57
Solid-line intercepts at 0 ± 50 & 1198 ± 26 Ma MSWD= 0.30
1-sigma error ellipses
B
RM38
Intercepts at 383 ± 110 & 1200 ± 26 Ma MSWD= 1.13
1-sigma error ellipses
C
MBCL4
Long-dashed line intercepts at 540 ± 280 & 1257 ± 26 Ma MSWD= 0.35
Short-dashed line Constant 207Pb/206Pb age reference line
1-sigma error ellipses
D
RM30
Long-dashed line intercepts at 0 ± 50 & 1080 ± 100 Ma MSWD= 2.0
Solid-line intercepts at 0 ± 50 & 1190 ± 19 Ma MSWD= 1.3
1-sigma error ellipses
E
RM30B
Long-dashed line intercepts at 0 ± 50 & 975 ± 31 Ma MSWD= 0.74
Solid-line intercepts at 277 ± 670 & 1207 ± 32 Ma MSWD= 0.44
Short-dashed line Constant 207Pb/206Pb age reference line
1-sigma error ellipses
F
RM30C
Long-dashed line intercepts at 306 ± 540 & 1209 ± 35 Ma MSWD= 1.03
Short-dashed line Constant 207Pb/206Pb age reference line
1-sigma error ellipses
206Pb/238U
207Pb/235U

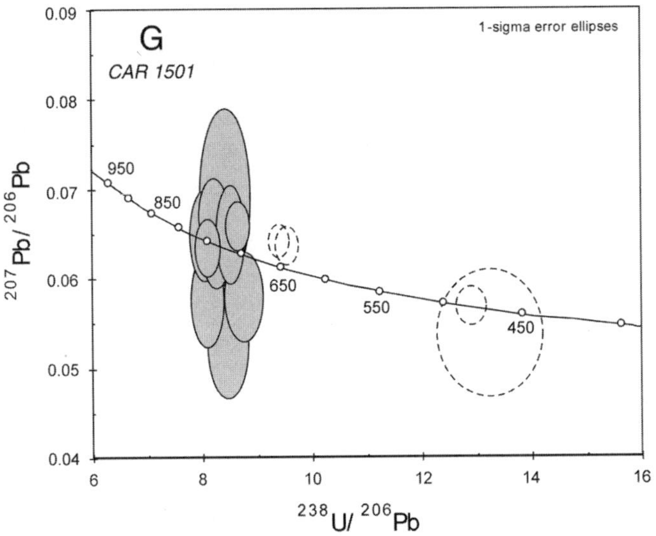

Figure 9 (*on this and previous page*). Concordia plots displaying U-Pb data for analyzed samples. Ellipses represent 1σ uncertainty for individual analyses. Gray ellipses were used to calculate discordia regressions that we interpret to represent ages of magmatic crystallization and Pb loss; open ellipses enclosed by solid lines are regions interpreted as metamorphic, with upper intercepts taken to be the age of metamorphism; and ellipses enclosed by dashed lines are inherited, detrital, or, in some cases, of uncertain origin (possibly a 1.1–1.2 Ga metamorphic event?). Errors in the calculated intercepts are 2σ Solid lines, which extend from the origin through the approximate upper intercept for magmatic points (i.e., they represent constant $^{207}Pb/^{206}Pb$ age), are for reference. (A–F) Conventional concordia plots; (G) Tera-Wasserburg concordia plot for CAR 1501 without common Pb correction.

$^{207}Pb/^{206}Pb$ age of 1245 ± 18 Ma (MSWD, 1.2); as with RM38, the younger age and higher MSWD are consistent with the well-defined discordia with a nonzero lower intercept. Probable metamorphic zones yielded imprecise and, in some cases, strongly discordant or reversely discordant results, but an age of ca. 1.0 Ga is suggested.

RM30 Samples (Highway 143 Road Cut)

Zircons from all three RM30 samples are complexly zoned, with unzoned to weakly zoned rims and, in most cases, multiple distinct interior zones. The best-defined population in all three samples is from interior zones with ages of ca. 1.20 Ga; other interior zones have diverse older ages, suggesting that they represent inherited and/or detrital cores.

RM30—Felsic Orthogneiss(?)

All zircons from RM30 have thin to thick, dark interior zones. Many have some small, concentrically zoned cores inside the dark zones, and all have thin to thick bright rims (Fig. 8, D). Four analyses, three from small euhedral-zoned cores and one from a dark, weakly zoned interior, have $^{207}Pb/^{206}Pb$ ages of 1168 to 1239 Ma and yield a pooled age of 1190 ± 19 Ma (MSWD, 1.3) (Fig. 9, D). Four analyses, three of bright rim zones and a fourth from an interior, give a mean $^{207}Pb/^{206}Pb$ age of 1080 ± 100 Ma. Six zoned cores range in $^{207}Pb/^{206}Pb$ age from 1320 to 1672 Ma, with two concordant at 1658 Ma and 1672 Ma. These cores are presumably either inherited or detrital.

RM30B—Banded Paragneiss(?)

Some zircons from this sample have tiny dark cores, and many have larger concentrically zoned cores (Fig. 8, E) or dark, unzoned to weakly zoned interiors. Almost all of these interiors are surrounded by thick bright, weakly zoned regions that in some cases extend to the rims and in others are surrounded by thin to thick dark outer rims.

Seven analyses that probably represent metamorphic growth (including unzoned dark and bright interiors and rims) have a mean $^{207}Pb/^{206}Pb$ age of 975 ± 31 Ma (MSWD, 0.74) (Fig. 9, E). Six points from weakly to strongly zoned interiors and distinct cores define a discordia with intercepts of 1207 ± 32 Ma and 277 ± 670 Ma (MSWD, 0.44); these points have a mean $^{207}Pb/^{206}Pb$ age of 1211 ± 34 Ma (MSWD, 1.5). Four cores range in $^{207}Pb/^{206}Pb$ age from 1273 to 1500 Ma, another is 1919 Ma, and a sixth is 2653 Ma. If this is indeed a metasedimentary rock, as suggested previously, both the 1200-Ma points and the older ages represent detrital zircon; if the protolith is plutonic or volcanic, 1200 Ma is presumably the crystallization age, and the older ages may be inherited.

RM30C—Mafic, Orthopyroxene-Bearing Banded Orthogneiss

RM30C zircons have euhedral-zoned magmatic interiors, commonly with sector zoning (Fig. 8, F). Well-defined dark or bright cores are fairly common, and most grains have bright, thin metamorphic rims. Seven points from magmatic-like, euhedral-zoned regions define a discordia with intercepts of 1209 ± 35 Ma and 306 ± 540 Ma (MSWD, 1.3) (Fig. 9, F). The same points have a mean $^{207}Pb/^{206}Pb$ age of 1197 ± 24 Ma (MSWD, 1.3). One analysis from a similar area falls off the concordia and yields a younger $^{207}Pb/^{206}Pb$ age of 1123 ± 44 Ma. Two imprecise analyses from rim zones that appear to document metamorphic growth have $^{207}Pb/^{206}Pb$ ages of 1142 ± 72 Ma and 1148 ± 147 Ma. These may suggest an older episode that is evident from metamorphic zones in other samples, or there may have been slight beam overlap with interiors that are older and richer in U. Four inherited (or detrital?) cores have $^{207}Pb/^{206}Pb$ ages between 1432 Ma and 1582 Ma, and another is highly discordant and has an imprecise $^{207}Pb/^{206}Pb$ age of 1782 ± 201 Ma. That these cores are invariably armored by thick magmatic overgrowths suggests that they are inherited.

The zircon data have implications for the interpretation of the RM30 exposure. As discussed previously, field relations suggest that either all three lithologies are part of a depositional sequence, or the protolith of felsic gneiss RM30 was a dike or sill that intruded the banded gneisses. Elemental chemistry is most consistent with RM30 and RM30C being felsic and mafic meta-igneous rocks and RM30B being a metamorphosed feldspathic sandstone. The strikingly similar patterns of zircon zonation and ages are consistent with all three samples being part of a 1.2 Ga volcanic/volcaniclastic sediment sequence that incorporated detrital and xenocrystic zircons.

CAR 1501—Meadlock Mountain Mafic, Orthopyroxene-Free Orthogneiss

Zircons from the sample CAR 1501 have euhedral-zoned interiors and thin to fairly thick bright rims (Fig. 8, G). Twelve data points from the magmatic zones are concordant (or nearly so) and fall between 640 Ma and 756 Ma (Fig. 9, G). When pooled after excluding two young outliers (640 Ma and 649 Ma, possibly reflecting Pb loss), the data yield a weighted mean $^{206}Pb/^{238}U$ age of 728 ± 16 Ma (MSWD, 2.1)—a magmatic age completely different from the remainder of the samples but essentially identical to the 734 ± 26-Ma Rb-Sr age determined by Goldberg et al. (1986) for Bakersville dikes. Two points that apparently represent metamorphic growth are concordant at ca. 475 Ma.

CONCLUSIONS

Age of Mars Hill Crust

Our new data further substantiate the notion that the MHT represents Paleoproterozoic crust. Although we did not date any additional samples that crystallized before 1.3 Ga, ca. 2 Ga Sm-Nd model ages for mafic and felsic samples and abundant inherited and detrital zircons with 1.6–1.9 Ga ages support the ancient heritage suggested by Monrad and Gulley (1983), Sinha et al. (1996), Fullagar and Gulley (1999), and Carrigan et al. (2003). The evidence suggests that the MHT is Paleoproterozoic crust that was intensely reactivated at 1.20 Ga.

Two Generations of Mafic Rocks

The presence in the MHT of the Neoproterozoic Bakersville dike swarm is well established (e.g., Goldberg et al., 1986); however, the age and relationships of other metabasites have been more problematic. The 1.20 Ga age of mafic sample RM30C demonstrates that at least some of the mafic rocks of the MHT are much older than the Bakersville dikes.

Our data suggest that MHT mafic rocks may be roughly divided into two groups, based on the presence or absence of opx. We interpret the presence of opx to reflect granulite-facies metamorphism, probably during the Grenville orogeny; the absence of opx indicates that the rock may have escaped this metamorphism and suggests that its protolith may be post-Grenville. Gulley (1982) and Rainey (1989) demonstrated that in some areas, Bakersville dikes contain opx interpreted to be of metamorphic origin, indicating post-Neoproterozoic granulite-facies metamorphism. Thus, presence or absence of opx is certainly not an entirely reliable discriminator of Neoproterozoic mafic rocks from older samples. However, for the limited number of samples that we have studied, it appears to distinguish populations that are otherwise distinct in petrogenesis and probably in age. The age of RM30C is 1.20 Ga, whereas that of opx-free garnet amphibolite CAR 1501 is 0.73 Ga, identical in age to the Bakersville dikes. The two analyzed opx-free samples have Sr and Nd isotope ratios that match Bakersville dikes (Goldberg et al., 1986; Goldberg and Dallmeyer, 1997); the two analyzed opx-bearing samples have very different ratios that suggest much greater age. The four opx-free samples plot on tectonic discrimination diagrams as within-plate basalts, consistent with Neoproterozoic, rift-related origin, whereas the three opx-bearing samples plot in distinct fields, generally as arc-related basalts. The opx-free samples have normal Th/U ratios, in contrast to the elevated ratios of the opx-bearing samples.

Regardless of the general applicability of the distinction between opx-bearing and opx-free mafic rocks, it is evident that metamorphosed mafic bodies in the MHT reflect both Neoproterozoic, rift-related magmatism and one or more generations of early and possibly pre-Grenville magmatism.

Ages of Magmatism and Metamorphism

Carrigan's recent work (Carrigan, 2000; Carrigan et al., 2003) verified the existence of 1.8 Ga felsic igneous rock at Roan Mountain, as suggested by Monrad and Gulley (1983), and indicated that somewhat younger (ca. 1.6 Ga) Paleproterozoic rock and 1.2 Ga Mesoproterozoic rock were also present. Our new data suggest widespread felsic and mafic magmatism of early Grenville age. By far the dominant magmatism occurred at 1.20 ± 0.01 Ma. A single sample documents a 1.25 Ga magmatic event. The MHT apparently escaped mid- and late Grenville magmatism (post–1.18 Ga), but it was heavily intruded by mafic magma during incipient rifting at 0.73 Ga.

Although generally imprecise, our data for zones interpreted to be metamorphic are consistent with the 1.03 Ga age estimated by Carrigan et al. (2003) for peak metamorphism of basement rocks in the Blue Ridge. There is also a possible suggestion of an earlier episode of metamorphic growth at or prior to ca. 1.1 Ga. Carrigan et al. (2003) found no distinguishable differences in ages of metamorphic rims between Eastern Blue Ridge, Western Blue Ridge, and the MHT. There is also evidence for Ordovician (Taconic) metamorphism from 470 Ma zircon overgrowths from Neoproterozoic sample CAR 1501. There is no other direct evidence for Paleozoic metamorphism in our zircon data, but lower discordia intercepts of Paleozoic age and thin, undated rims on zircons from several samples that were observed under cathodoluminescence may reflect Paleozoic events.

Geologic Evolution

The early stages of the history of the MHT are poorly defined, having been obscured by subsequent events. Available evidence suggests formation of juvenile, probably arc-related crust during the Paleoproterozoic. The Sm-Nd model ages of samples other than Neoproterozoic mafic rocks, the magmatic crystallization age of RM1, and U-Pb ages of detrital and inherited zircons from several samples fall in the range 1.9 ± 0.2 Ga.

A major, possibly bimodal magmatic episode in early Grenville time (1.2 Ga) appears not to have added much new crust, based on Sm-Nd isotopic compositions of samples of this age. The triggering mechanism remains uncertain, as does the relationship between magmatism and sedimentation. Ages of detrital zircon suggest that sedimentation was roughly coeval with magmatism, but neither geochronology nor observed field relations permit us to say whether they were strictly simultaneous (interlayered volcanic and sedimentary strata), sediments were deposited on a slightly older igneous substrate, or sediments were intruded by slightly younger magmas. Field, elemental, and U-Pb data for the outcrop RM30 sample appear to lend credence to synchronous sedimentation and magmatism, but it is only a single exposure and far from conclusive.

Very high-grade metamorphism occurred twice, during late Grenville and Taconic episodes. The conditions attained during these episodes (Adams et al., 1995) and the widespread presence of migmatite, including in Neoproterozoic mafic rocks, suggest that local partial melting accompanied both events. We infer that complex and pervasive deformation observed in the MHT reflects both Grenville and multiple Paleozoic events. It is possible that the small-scale juxtaposition of diverse lithologies may be a consequence of deformation during anatexis. The final ductile deformation, indicated by mylonite zones, probably occurred after peak Paleozoic metamorphism.

Constraints on Relationships to the Eastern and Western Blue Ridge

The Sm-Nd model ages, ages of magmatism, and those of detrital and inherited zircons all indicate that the MHT is a fundamentally older terrane than either the Eastern Blue Ridge or the Western Blue Ridge. It also is clearly different lithologically from the other basement rocks of the southern Blue Ridge; unlike either the Eastern Blue Ridge or Western Blue Ridge, the MHT contains metasedimentary and abundant mafic rocks. It is thus apparent that the MHT is a continental fragment, with an origin distinct from its surroundings. It is plausible that it simply represents an exposed portion of an older lower crust that underlay the more juvenile rocks of the Western Blue Ridge and possibly the Eastern Blue Ridge basement during Grenville time. Fullagar (2002) recognized contributions of older—in part, Paleoproterozoic—crust to the Blue Ridge and adjacent Inner Piedmont. Such contributions may have come from a MHT-like lower crust.

The post–1.1 Ga history of the MHT suggests linkages to the Eastern Blue Ridge and Western Blue Ridge, but the implications of these linkages remain puzzling (cf. Johnson, 1994; Raymond and Johnson, 1994; Adams and Trupe, 1997). All three areas appear to have experienced a profound metamorphic event shortly before 1.0 Ga. Both the Western Blue Ridge and the MHT—but not the Eastern Blue Ridge basement—were intruded by Neoproterozoic mafic dikes, although the MHT seems not to include any of the Neoproterozoic granites that are common in the Western Blue Ridge. Both the MHT and the Eastern Blue Ridge—but not the Western Blue Ridge—underwent similar high-grade Ordovician metamorphism. The data are consistent with, but do not require, the interpretation that all three were in fairly close proximity by late Grenville time. Because of the extent of the global Grenville orogen and of late Grenville metamorphism, the similarity in age of metamorphism may be of limited use as a geographic constraint. The distribution of Bakersville dikes appears to suggest MHT–Western Blue Ridge (but not Eastern Blue Ridge?) proximity during the Neoproterozoic, and high-grade Taconian metamorphism seems to link the Eastern Blue Ridge and MHT, but perhaps not the Western Blue Ridge, during the Ordovician.

Constraints on Relations to Other Ancient Continental Crust

We are unaware of any other exposures of crust of similar antiquity to the MHT in the southeastern United States. The only possible Appalachian correlatives of the MHT that we are aware of are in Virginia and Maryland. The Pedlar and Lovingston massifs in the Virginia Blue Ridge both are highly diverse in terms of lithology (mafic, felsic, and some metasedimentary rocks, wide range of SiO_2) (e.g., Bartholomew and Lewis, 1984; Hughes et al., 1997, 2001). An upper U-Pb concordia intercept of 1.87 Ga for detrital zircon and whole-rock Pb isotope data suggesting a possible Archean source component indicate that the Stage Road layered gneiss, a unit of the Lovingston massif, records input from material that was much older than Grenville (Pettingill et al., 1984; Sinha and Bartholomew, 1984; Sinha et al., 1996). However, the reported Sm-Nd isotopic compositions of analyzed samples of the Stage Road gneiss and other Pedlar and Lovingston lithologies ($\varepsilon_{Nd} > 0$ at 1.1 Ga) (Pettingill et al., 1984) is nowhere near as evolved as that which we have found so far in the MHT. The Goochland terrane, which lies to the east in the Piedmont zone and is of uncertain origin, also is lithologically diverse, but there is no evidence that it includes protoliths much older than Grenville age (no pre-Grenville zircon ages; $\varepsilon_{Nd,\ 1.1\ Ga} \sim 0$, $T_{DM} \sim 1.4$ Ga) (Owens and Samson, 2001). Aleinikoff et al. (this volume) report 1.25 Ga ages for felsic gneisses of the Baltimore Gneiss in the Maryland Piedmont, similar to the dominant MHT magmatic age population. Like the MHT, the Baltimore Gneiss includes both mafic and felsic rocks. However, there appears to be no zircon evidence for Paleoproterozoic precursor crust (no old inheritance), and Pb isotope data reported by Sinha et al. (1996) for Baltimore Gneiss are consistent with other central and southern Appalachian basement but not with the MHT. There are no reported Nd isotopic data for the Baltimore Gneiss.

The nearest known crust of Paleoproterozoic origin is in the Penokean Province near the western Great Lakes. Paleoproterozoic crust is also exposed in the southwestern United States, including parts of the Mojave, Yavapai, and Mazatzal terranes (e.g., Karlstrom et al., 1999). These areas contain diverse lithologies and have compatible estimated crustal formation ages (T_{DM} values of ca. 2 Ga). The MHT is separated from all of these Paleoproterozoic exposures by a vast tract of early Mesoproterozoic rocks (ca. 1.4–1.5 Ga magmatic crystallization ages) that is interpreted to represent juvenile crust (T_{DM} values are only very slightly older) (Van Schmus et al., 1996). If the MHT was once a part of a Paleoproterozoic Laurentian crustal block with the Great Lakes and/or southwestern terranes, and if the midcontinent province is juvenile, then the MHT may be a rifted Laurentian fragment and the midcontinent province an enormous expanse of rift-fill crust. The Pb isotope ratios of southern Appalachian basement have been used as evidence that southern Appalachian crust in general did not originate in its current position with respect to Laurentia, and existing data for the MHT suggest that it is even more distinct from nearby parts of the continent than is true for the rest of the southern Appalachians (Sinha et al., 1996; Sinha and McLelland, 1999; Loewy et al., 2002). An origin adjacent to a distant part of Laurentia might explain this and other discrepancies. Alternatively, the MHT may be an orphan fragment of crust removed from a larger Paleoproterozoic terrane now exposed on another continent, perhaps in West Africa or South America (Rogers, 1996; Tosdal, 1996; Ruiz et al., 1999; Loewy et al., 2002; Pisarevsky et al., 2003).

ACKNOWLEDGMENTS

Kevin Stewart, Carl Merschat, Mark Carter, Mark Adams, and Loren Raymond kindly contributed valuable insights and time in the field with us; we benefited enormously from the work they had done in this region and their generosity in sharing it with us. Brendan Bream, Bob Hatcher, and Jerry Bartholomew all contributed ideas about Appalachian tectonics that contributed to our perspective on Mars Hill problems, and Bream also donated his time and skills in the SHRIMP lab. Richard Tollo, Kevin Stewart, Paul Mueller, John Aleinikoff, and Jerry Bartholomew provided careful reviews of this text that helped us to strengthen and clarify it; the interpretations remain our own, however, and the reviewers should share no blame. The work was supported by NSF grant EAR98-14801 and a Eugene H. Vaughan Jr. research scholarship awarded to SEO by the Vanderbilt Department of Geology.

REFERENCES CITED

Adams, M.G., and Trupe, C.H., 1997, Conditions and timing of metamorphism in the Blue Ridge thrust complex, northwestern North Carolina and western Tennessee, *in* Stewart, K.G., et al., eds., Paleozoic structure, metamorphism, and tectonics of the Blue Ridge of western North Carolina: Carolina Geological Society Field Trip and Annual Meeting, p. 33–47.

Adams, M.G., Stewart, K.G., Trupe, C.H., and Willard, R.A., 1995, Tectonic significance of high-pressure metamorphic rocks and dextral strike-slip faulting along the Taconic suture, *in* Hibbard, J.P., et al., eds., Current perspectives in the Appalachian-Caledonian orogen: Geological Association of Canada, Special Paper 41, p. 21–42.

Bacon, C.R., Persing, H.M., Wooden, J.L., and Ireland, T.R., 2000, Late Pleistocene granodiorite beneath Crater Lake caldera, Oregon, dated by ion microprobe: Geology, v. 28, p. 467–470.

Bartholomew, M.J., and Lewis, S.E., 1984, Evolution of Grenville massifs in the Blue Ridge geologic province, southern and central Appalachians, *in* Bartholomew, M.J., ed., The Grenville Event in the Appalachians and related topics: Boulder, Colorado, Geological Society of America Special Paper 194, p. 229–254.

Bartholomew, M.J., and Lewis, S.E., 1988, Peregrination of middle Proterozoic massifs and terranes within the Appalachian orogen, eastern USA: Trabajos de Geología, v. 17, p. 155–165.

Bartholomew, M.J., and Lewis, S.E., 1992, Appalachian Grenville massifs: Pre-Appalachian translational tectonics, *in* Mason, R., ed., Basement tectonics 7: Dordrecht, The Netherlands, Kluwer Academic Publishers, p. 363–374.

Boynton, W.W., 1984, Cosmochemistry of the rare earth elements, *in* Henderson, P., ed., Rare earth element geochemistry: New York, Elsevier, p. 63–114.

Brewer, R.C., and Woodward, N.B., 1988, The amphibolitic basement complex in the Blue Ridge Province of western North Carolina, Proto-Iapetus?: American Journal of Science, v. 288, p. 953–967.

Brown, P.M., Burt, E.R., II, Carpenter, P.A., Enos, R.M., Flynt, B.J., Jr., Gallagher, P.E., Horrman, C.W., Merschat, C.E., Wilson, W.F., and Parker, J.M., III, 1985, Geologic map of North Carolina: North Carolina Geological Survey, scale 1:500,000.

Carrigan, C.W., 2000, Ion microprobe geochronology of Grenville and older basement in the southern Appalachians [unpublished M.S. thesis]: Nashville, Tennessee, Vanderbilt University.

Carrigan, C.W., Miller, C.F., Hatcher, R.D., Jr., Fullagar, P.D., and Coath, C.D., 2000, Grenville basement of the southern Appalachians, NC-GA—Geochemistry and zircon ion probe geochronology: Geological Society of America Abstracts with Programs, v. 32, no. 2, p. A9.

Carrigan, C.W., Miller, C.F., Fullagar, P.D., Hatcher, R.D., Jr., Bream, B.R., and Coath, C.D., 2001, Age and geochemistry of Southern Appalachian basement, NC-SC-GA, with implications for Proterozoic and Paleozoic reconstructions: Geological Society of America Abstracts with Programs, v. 33, no. 6, p. A29.

Carrigan, C.W., Miller, C.F., Fullagar, P.D., Hatcher, R.D., Jr., Bream, B.R., and Coath, C.D., 2003, Age and geochemistry of Southern Appalachian basement, NC-SC-GA, with implications for Proterozoic and Paleozoic reconstructions: Precambrian Research, v. 120, p. 1–36.

Davis, T.L., 1993, Lithostratigraphy, structure, and metamorphism of a crystalline thrust terrane, western Inner Piedmont, North Carolina [Ph.D. thesis]: Knoxville, University of Tennessee, 245 p.

DePaolo, D.J., 1981, Neodymium isotopes in the Colorado Front Range and crust-mantle evolution in the Proterozoic: Nature, v. 291, p. 193–196.

Fullagar, P.D., 2002, Evidence for early Mesoproterozoic (and older?) crust in the southern and central Appalachians of North America, *in* Rogers, J.J.W., and Santosh, M., eds., Special issue of Gondwana research on "Mesoproterozoic Supercontinent": Gondwana Research, v. 5, p. 197–203.

Fullagar, P.D., and Butler, J.R., 1979, 325–265 m.y.-old granitic plutons in the Piedmont of the southeastern Appalachians: American Journal of Science, v. 279, p. 161–185.

Fullagar, P.D., and Gulley, G.L., Jr., 1999, Pre-Grenville Uranium-Lead zircon age for the Carvers Gap Gneiss in the Western Blue Ridge of North Carolina–Tennessee: Geological Society of America Abstracts with Programs, v. 31, no. 3, p. 16.

Fullagar, P.D., Hatcher, R.D., Jr., and Merschat, C.E., 1979, 1200 m.y. old gneisses in the Blue Ridge province of North Carolina and South Carolina: Southeastern Geology, v. 20, p. 69–77.

Fullagar, P.D., Goldberg, S.A., and Butler, J.R., 1997, Nd and Sr isotopic characterization of crystalline rocks from the southern Appalachian Piedmont and Blue Ridge, North Carolina and South Carolina, *in* Sinha, A.K., et al., eds., The nature of magmatism in the Appalachian orogen: Boulder, Colorado, Geological Society of America Memoir 191, p. 161–185.

Goldberg, S.A., and Dallmeyer, R.D., 1997, Chronology of Paleozoic metamorphism and deformation in the Blue Ridge thrust complex, North Carolina and Tennessee: American Journal of Science, v. 297, p. 488–526.

Goldberg, S.A., Butler, J.R., and Fullagar, P.D., 1986, The Bakersville dike swarm; geochronology and petrogenesis of late Proterozoic basaltic magmatism in the southern Appalachian Blue Ridge: American Journal of Science, v. 286, no. 5, p. 403–430.

Goldberg, S.A., Butler, J.R., Mies, J.W., and Trupe, C.H., 1989, The southern Appalachian orogen in northwestern North Carolina and adjacent states: Washington, D.C., American Geophysical Union IGC Field Trip T365, 55 p.

Gulley, G.L., Jr., 1982, The petrology of granulite facies metamorphic rocks on Roan mountain, Western Blue Ridge, NC-TN [M.S. thesis]: Chapel Hill, University of North Carolina, 163 p.

Gulley, G.L., Jr., 1985, A Proterozoic granulite-facies terrane on Roan Mountain, western Blue Ridge Belt, North Carolina–Tennessee: Geological Society of America Bulletin, v. 96, p. 1428–1439.

Hanchar, J.M., and Miller, C.F., 1993, Zircon zonation patterns and interpretation of crustal histories: Chemical Geology, v. 110, p. 1–13.

Hatcher, R.D., Jr., 1978, Tectonics of the Western Blue Ridge, southern Appalachians: Review and speculation: America Journal of Science, v. 278, p. 276–304.

Hatcher, R.D., Jr., 1989, Tectonic synthesis of the U.S. Appalachians, *in* Hatcher, R.D., Jr., et al., eds., The Appalachian-Ouachita orogen in the United States: Boulder, Colorado, Geological Society of America, The Geology of North America, v. F2, p. 511–535.

Hughes, S.S., Lewis, S.E., Bartholomew, M.J., Sinha, A.K., Hudson, T.A., and Herz, N., 1997, Chemical diversity and origin of Precambrian charnockitic rocks of the central Pedlar Massif, Grenvillian Blue Ridge Terrane, Virginia: Precambrian Research, v. 84, p. 37–46.

Hughes, S.S., Bartholomew, M.J., Lewis, S.E., Sinha, A.K., and Herz, N., 2001, Geochemistry of Precambrian charnockitic rocks of the central Lovingston Massif, Grenvillian Blue Ridge Terrane, Virginia: Geological Society of America, Abstracts with Programs, v. 33, no. 6, p. 28.

Johnson, P.A., 1994, The Mars Hill terrane: an enigmatic southern Appalachian terrane [Senior Honors thesis]: Boone, North Carolina, Appalachian State University.

Karlstrom, K.E., Harlan, S.S., Williams, M.L., McLelland, J., Geissman, J.W., and Åhall, K.-I., 1999, Refining Rodinia: Geologic evidence for the Australia–western U.S. connection in the Proterozoic: GSA Today, v. 10, no. 9, p. 1–7.

Loewy, S.L., Connelly, J.N., and Dalziel, I.W.D., 2002, Pb isotopes as a correlation tool to constrain Rodinia reconstruction: Geological Society of America Abstracts with Programs, v. 34, no. 6, p. 558.

Ludwig, K.R., 2001, SQUID 1.02; a users manual: Berkeley, California, Berkeley Geochronology Center Special Publication no. 2.

Merschat, C.E., 1977, Geologic map and mineral resources summary of the Mars Hill quadrangle, North Carolina: North Carolina Geological Survey, Division of Land Resources, scale 1:24,000.

Merschat, C.E., and Wiener, L.S., 1990, Geology of Grenville-age basement and younger cover rocks in the west-central Blue Ridge, North Carolina: Carolina Geological Society Guidebook, 42 p.

Miller, C.F., Hanchar, J.M., Bennett, V.C., Harrison, T.M., Wark, D.A., and Foster, D.A., 1992, Source region of a batholith: Evidence from lower crustal xenoliths and inherited accessory minerals: Transactions of the Royal Society of Edinburgh: Earth Sciences (Hutton Symposium volume), v. 53, p. 49–62.

Miller, C.F., Hatcher, R.D., Jr., Harrison, T.M., Coath, C., and Gorisch, E.B., 1998, Cryptic crustal events elucidated through zone imaging and ion microprobe studies of zircon, southern Appalachian Blue Ridge, North Carolina–Georgia: Geology, v. 26, p. 419–422.

Misra, K.C., and McSween, H.Y., 1984, Mafic rocks of the southern Appalachians—A review: American Journal of Science, v. 284, p. 294–318.

Monrad, J.R., and Gulley, G.L., Jr., 1983, Age and *P-T* conditions during metamorphism of granulite-facies gneisses, Roan Mountain, NC-TN, *in* Lewis, S.E., ed., Geological Investigations in the Blue Ridge of northwestern North Carolina: Carolina Geological Society Field Trip Guidebook, article 4, p. 1–18.

Owens, B.E., and Samson, S.D., 2001, Nd-isotopic constraints on the magmatic history of the Goochland Terrane, easternmost Grenville crust in the Southern Appalachians: Geological Society of America, Abstracts with Programs, v. 33, no. 6, p. 28.

Ownby, S.E., 2002, Ancient crust of the Mars Hill terrane, North Carolina–Tennessee: Constraints on origin from initial geochemical and geochronological studies [Senior Honors thesis]: Nashville, Tennessee, Vanderbilt University, 107 p.

Pearce, J.A., and Cann, J.R., 1973, Tectonic setting of basic volcanic rocks determined by discriminant analysis using Ti, Zr, and Y: Earth and Planetary Sciences Letters, v. 12, p. 339–349.

Pearce, J.A., Harris, N.B.W., and Tindle, A.G., 1984, Trace element discrimination diagrams for the tectonic interpretation of granitic rocks: Journal of Petrology, v. 25, p. 956–983.

Pettingill, H.S., Sinha, A.K., and Tatsumoto, M., 1984, Age and origin of anorthosites, charnockites, and granulites in the central Virginia Blue Ridge; Nd and Sr isotopic evidence: Contributions to Mineralogy and Petrology, v. 85, p. 279–291.

Pisarevsky, S.A., Wingate, M.T.D., Powell, C.M., Johnson, S., and Evans, D.A.D., 2003, Models of Rodinia assembly and fragmentation, *in* Yoshida, M., and Windley, B.F., eds., Proterozoic east Gondwana: Supercontinent assembly and breakup: Geological Society, London, Special Publication 206, p. 35–55.

Rainey, L., 1989, A study of the late Proterozoic Bakersville Suite; implications for Paleozoic metamorphism in portions of the Blue Ridge thrust complex, western North Carolina and eastern Tennessee [M.S. thesis]: Chapel Hill, University of North Carolina, 161 p.

Rankin, D.W., 1975, The continental margin of eastern North America in the southern Appalachians: The opening and closing of the proto-Atlantic Ocean: American Journal of Science, v. 275A, p. 298–336.

Raymond, L.A., 1987, Terrane amalgamation in the Blue Ridge Belt, southern Appalachian orogen, U.S.A.: Geological Society of America Abstracts with Programs, v. 19, no. 2, p. 125.

Raymond, L.A., and Johnson, P.A., 1994, The Mars Hill terrane; an enigmatic Southern Appalachian terrane: Geological Society of America Abstracts with Programs, v. 26, no. 4, p. 59.

Raymond, L.A., Yurkovich, S.P., and McKinney, M., 1989, Block-in-matrix structures in the North Carolina Blue Ridge Belt and their significance for the tectonic history of the Southern Appalachian orogen, *in* Horton, J.W., Jr., and Rast, N., eds., Melanges and olistostromes of the U.S. Appalachians: Boulder, Colorado, Geological Society of America Special Paper 228, p. 195–215.

Rogers, J.J.W., 1996, A history of continents in the past three billion years: Journal of Geology, v. 104, p. 91–107.

Ruiz, J., Tosdal, R.M., Restrepo, P.A., and Murillo-Muneton, G., 1999, Pb isotope evidence for Colombia–southern Mexico connections in the Proterozoic, *in* Ramos, V.A., and Keppie, J.D., eds., Laurentia-Gondwana connections before Pangea: Boulder, Colorado, Geological Society of America Special Paper 336, p. 183–197.

Sinha, A.K., and Bartholomew, M.J., 1984, Evolution of the Grenville terrane in the central Virginia Appalachians, *in* Bartholomew, M.J., ed., The Grenville Event in the Appalachians and related topics: Boulder, Colorado, Geological Society of America Special Paper 194, p. 175–186.

Sinha, A.K., and McLelland, J.M., 1999, Lead isotope mapping of crustal reservoirs within the Grenville superterrane: II. Adirondack massif, New York, *in* Sinha, A.K., ed., Basement tectonics 13: Dordrecht, The Netherlands, Kluwer Academic Publishers, p. 297–312.

Sinha, A.K., Hogan, J.P., and Parks, J., 1996, Lead isotope mapping of crustal reservoirs within the Grenville Superterrane: I. Central and southern Appalachians: American Geophysical Union Geophysical Monograph 95, p. 293–305.

Stewart, K.G., Adams, M.G., and Trupe, C.H., 1997, Paleozoic structural evolution of the Blue Ridge thrust complex, western North Carolina, *in* Stewart, K.G., et al., eds., Paleozoic structure, metamorphism, and tectonics of the Blue Ridge of western North Carolina: Carolina Geological Society Field Trip and Annual Meeting, p. 21–31.

Sun, S.S., and McDonough, W.F., 1989. Chemical and isotopic systematics of oceanic basalts: Implications for mantle compositions and processes, *in* Saunders, A.D., and Norry, M.J., eds., Magmatism in ocean basins: Geological Society of London Special Publication 42, p. 313–345.

Tosdal, R.M., 1996, The Amazon-Laurentian connection as viewed from the Middle Proterozoic rocks in the central Andes, western Bolivia and northern Chile: Tectonics, v. 15, p. 827–842.

Trupe, C.H., Adams, M.G., and Stewart, K.G., 2001, Diversity of basement rocks along the Eastern-Western Blue Ridge contact in northwestern North Carolina: Geological Society of America Abstracts with Progams, v. 33, no. 6, p. A29.

Van Schmus, W.R., Bickford, M.E., and Turek, A., 1996, Proterozoic geology of the east-central Midcontinent basement, *in* van der Pluijm, B.A., and Catacosinos, P.A., eds., Basement and basins of eastern North America: Boulder, Colorado, Geological Society of America Special Paper 308, p. 7–32.

Werner, C.D., 1987, Saxonian granulites—igneous or lithogeneous? A contribution to the geochemical diagnosis of the original rocks in high-metamorphic complexes, *in* Gerstenberger, H., ed., Contributions to the geology of the Saxonian granulite massif: Sächsisches Granulitegebirge, Zfl-Mitteilungen Nr., v. 133, p. 221–250.

Wood, D.A., 1980, The application of a Th-Ta-Hf diagram to problems of tectonomagmatic classification and to establishing the nature of crustal contamination of basaltic lavas of the British Tertiary volcanic province: Earth and Planetary Sciences Letters, v. 50, p. 11–30.

Zartman, R.E., and Doe, B.R., 1981, Plumbotectonics: The model: Tectonophysics, v. 75, p. 135–162.

Manuscript Accepted by the Society August 25, 2003

Printed in the USA

Geological Society of America
Memoir 197
2004

Pre-Appalachian tectonic evolution of the Pine Mountain window in the southernmost Appalachians, Alabama and Georgia

Mark G. Steltenpohl*
Department of Geology, Auburn University, Auburn, Alabama 36849, USA
Ann Heatherington
Paul Mueller
Department of Geological Sciences, University of Florida, Gainesville, Florida 32611, USA
Joseph L. Wooden
U.S. Geological Survey, 345 Middlefield Road, Menlo Park, California 94025, USA

ABSTRACT

The Pine Mountain window contains the southernmost Grenville basement massif to be found in the Appalachians. Granulite- and upper-amphibolite-facies granitic gneisses that form the basement complex are isotopically dated at 1.1–1.0 Ga. Locally, the gneisses contain rare mafic injections and supracrustal and plutonic xenoliths. The Pine Mountain Group cover sequence nonconformably overlies Grenville basement and is interpreted to correlate with Blue Ridge units as follows: Halawaka/Sparks Schist = Ocoee Supergroup (Late Proterozoic, rift), Hollis Quartzite = Chilhowee Group (Late Proterozoic-Cambrian, rift-to-drift), and Chewacla Marble = Shady Dolomite (Cambro-Ordovician, drift). Facies variations within the sedimentary cover units were cited as evidence for a southward decrease in the extent of the Ocoee rift basins, but new mapping documents the continuity of thick packages of Halawaka (i.e., Ocoee) rocks extending southward beneath the Gulf Coastal Plain. In contrast to upper amphibolite- and granulite-facies metamorphism of the basement during the Grenville event, cover rocks contain staurolite and staurolite-kyanite zone assemblages reflecting Paleozoic Appalachian metamorphism. Sensitive high-resolution ion microprobe (SHRIMP) and conventional single-grain U-Pb datings of detrital zircons from the basal Hollis Quartzite document a distinct population of clear, subrounded zircons of ca. 1.09 Ga, which were most likely derived from underlying Grenville-age gneiss. An older, white/gray population found in the lowermost Hollis is ca. 2.4–2.3 Ga, an age restricted to Gondwanan continents and very limited occurrences in northern Laurentia.

Tectonic reconstructions of Unrug (1997) and others depict southeast Laurentia proximal to the Amazonia and Rio de la Plata cratons during the Neoproterozoic, offering the possibility that they may be the source for 2.4–2.3-Ga zircons in Hollis sediments. Alternatively, the AUSWUS (Australia/Western United States) reconstruction (Karlstrom et al., 2001) places east Antarctica and the Australian Gawler craton, both of which contain abundant 2.4 Ga granites, proximal to the southwestern United States during this time. Depending on the stream systems present during the Neopro-

*E-mail: steltmg@mail.auburn.edu.

Steltenpohl, M.G., Heatherington, A., Mueller, P., and Wooden, J.L., 2004, Pre-Appalachian tectonic evolution of the Pine Mountain window in the southernmost Appalachians, Alabama and Georgia, *in* Tollo, R.P., Corriveau, L., McLelland, J., and Bartholomew, M.J., eds., Proterozoic tectonic evolution of the Grenville orogen in North America: Boulder, Colorado, Geological Society of America Memoir 197, p. 633–646. For permission to copy, contact editing@geosociety.org.

terozoic, zircons from the Gawler may have been transported to the vicinity of the Pine Mountain window. In addition, three clear zircons yield ages of 1.4 Ga, and may have been derived from either the Laurentian Mid-continent granite-rhyolite province or the Rondonian Province of South America. A Chilhowee Group sandstone sample contains a similar mixture of Grenville and Mid-continent/Rondonian-age zircons, but none with ages of 2.4–2.3 Ga.

Keywords: Neoproterozoic, tectonics, geochronology, Appalachians, Rodinia

INTRODUCTION

The Pine Mountain belt is the southernmost internal Grenville basement massif in the Appalachians that retains its primary miogeoclinal cover. As such, it has traditionally figured prominently in studies of how the ancient Laurentian margin was rifted and subsequently contracted during the various phases of the Appalachian orogeny. Comparatively little effort has been made, however, to look at plate evolution prior to the time of Iapetan rifting and to ascertain the Pine Mountain window's prior role in Neoproterozoic plate interactions and the Grenville orogeny. Most workers have concentrated on deciphering structures and metamorphic effects in the stratigraphic cover that nonconformably overlies the Grenville basement. The cover units have a complex polyphase metamorphic/deformational history that has overprinted a record of Neoproterozoic Iapetan rifting, subsequent rift-to-drift transition, and Cambro-Ordovician platform margin creation. Their lithologies and structural/metamorphic history, therefore, provide no information on the earlier history of the Pine Mountain window. Several workers reported "complex" structures and metamorphic features in basement gneisses that probably predate deposition of cover lithologies, but a systematic analysis of these features has not been made. An integrated history for the Pine Mountain window is also essential to resolving the ongoing debate concerning the degree of allochthoneity of the Pine Mountain massif (Hooper and Hatcher, 1988; Steltenpohl et al., 1992; Steltenpohl and Salpas, 1993; West et al., 1995; McBride et al., 2001).

We investigated the pre-Iapetan history of the Pine Mountain window by using U-Pb isotopes to date detrital minerals from the basal metasandstone that mantles the massif, the Hollis Quartzite. For comparison, we also analyzed detrital zircons from the Chilhowee Group of the Alabama Valley and Ridge Province, which clearly is a Cambrian transgressive sandstone deposited along the Laurentian margin. The U-Pb results document, not surprisingly, that Grenville-age (1.2–1.0-Ga) and Mid-continent rhyolite province-age (ca. 1.4 Ga) populations occur within the Chilhowee and Hollis units. The lower part of the Hollis Quartzite, however, contains an additional population of much older, Paleoproterozoic zircons (2.4–2.3 Ga). Because Paleoproterozoic source areas are not known to occur in southern Laurentia, this observation has important implications for Meso- and Neoproterozoic plate interactions between terranes along the southern Laurentian margin, as well as the nature of the Grenville orogeny in this region.

GEOLOGIC BACKGROUND

The Pine Mountain window (Fig. 1) is flanked by wide mylonite zones. The Towaliga fault zone to the northwest juxtaposed the Inner Piedmont terrane, a probable early Paleozoic arc or back-arc complex (Stow et al., 1984; Hatcher, 1987; Horton et al., 1989; Steltenpohl et al., 1990). The Bartletts Ferry fault zone to the southeast separates the Pine Mountain window from the Uchee belt. The Uchee belt is an early Paleozoic arc or back-arc complex, parts of which are interpreted to correspond to the Carolina terrane (Russell, 1978, 1985; Hooper and Hatcher, 1988; Chalokwu and Hanley, 1990; Hanley et al., 1997). These are the southernmost surface exposures of the crystalline Appalachians. Relations between these terranes and those of the Suwannee terrane are concealed by Mesozoic to recent Gulf Coastal Plain sediments to the south. Rocks of the Suwannee terrane are buried beneath these Coastal Plain sediments, only 30–40 km south of exposures in the study area (Fig. 2). The Suwannee terrane is an accreted piece of Gondwanan crust with a late Precambrian to Devonian supracrustal cover sequence (Applin, 1951; King, 1961; Chowns and Williams, 1983; Mueller et al., 1994). Most workers interpret the Pine Mountain massif to have originated near the tip of a large promontory—the Alabama promontory—along the southeastern (modern-day coordinates) margin of the ancient, rifted Laurentian margin (King, 1961; Thomas, 1977), prior to the Appalachian orogeny.

Grenville (ca. 1.2–1.0-Ga) basement massifs in the external parts of the southern Appalachians typically have sedimentary cover units lying nonconformably above them (Hatcher, 1984), but in the internides, only the Pine Mountain window preserves parts of its primary platform cover (Steltenpohl, 1988a). The cover sequence, the Pine Mountain Group (Galpin, 1915; Adams, 1933; Crickmay, 1933, 1952), has a tectonically modified nonconformable contact with the underlying metaplutonic granitic basement complex. In Georgia, the metaplutonic rocks of the complex are subdivided into the Woodland Gneiss, Jeff Davis Granite, Cunningham Granite, or the "charnockite series" (Hewitt and Crickmay, 1937; Clarke, 1952; Schamel et al., 1980). In Alabama, the basement complex is called the "Wacoochee Complex" (Bentley and Neathery, 1970; Raymond

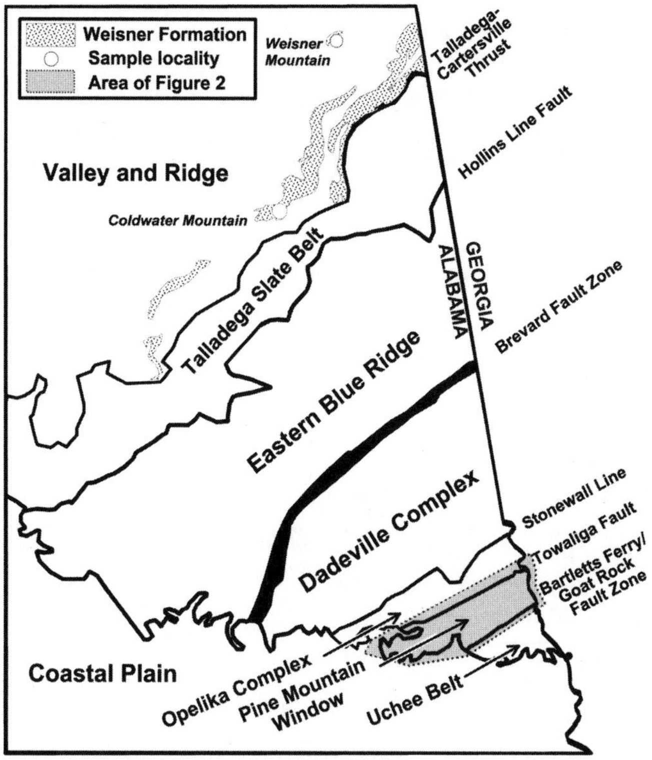

Figure 1. Location map, illustrating pertinent geologic features and the locations of two Weisner Formation samples.

et al., 1988), which comprises variably mylonitized and metamorphosed Whatley Mill Gneiss and associated biotite-feldspar gneisses (Bentley and Neathery, 1970; Sears and Cook, 1984; Raymond et al., 1988; Steltenpohl et al., 1990; Daniell, 1997). Rb-Sr and U-Pb isotopic studies document that the plutonic rocks resulted from the Grenville event, with ages clustering mainly at ca. 1.15 Ga (Odom et al., 1973, 1985; Steltenpohl et al., 1990; Bowring, personal commun., 1995; this study). Xenoliths and preplutonic enclaves are rare within the metaplutonic rocks, and little is known about them (Higgins et al., 1988; Daniell, 1997).

The Pine Mountain Group is the miogeoclinal cover sequence above the metaplutonic complex and comprises, from stratigraphic bottom to top, the Hollis Quartzite, Chewacla Marble, and Manchester Schist (Clarke, 1952; Crickmay, 1952; Bentley and Neathery, 1970; Raymond et al., 1988). No fossils are known from any of the cover units, but they have been interpreted, on a lithologic basis, to correlate with the Lower Cambrian Weisner-Shady-Rome cratonal sequence observed in the foreland (Odom et al., 1973; Sears et al., 1981b; Yokel, 1996; Yokel et al., 1997). Feldspathic metapelitic rocks locally occurring between the Hollis Quartzite and the basement gneiss (Clarke, 1952; Schamel et al., 1980) are called the "Sparks Schist" in Georgia and the "Halawaka Schist" (of the Wachoochee Complex) in Alabama (Clarke, 1952; Bentley and Neathery, 1970; Schamel et al., 1980; Raymond et al., 1988). The Halawaka-Sparks package is interpreted as metamorphosed rift-fill sediments corresponding to the Ocoee Supergroup of the Georgia, Tennessee, and North Carolina Blue Ridge (King et al., 1958; Hadley, 1970; Schamel et al., 1980; Rast and Kohles, 1986).

Middle-amphibolite facies (staurolite-kyanite and/or sillimanite grade) metamorphism recorded in the Pine Mountain cover rocks sharply contrasts with deep-crustal granulite-facies conditions reported for basement units in Georgia (Schamel and Bauer, 1980; Steltenpohl and Moore, 1988; Stieve and Size, 1988), and clearly reflects differences in Appalachian versus Grenvillian evolution. Schamel and Bauer (1980) suggested that the granulite-facies assemblages were retrograded to amphibolite-facies assemblages during prograde metamorphism of the Pine Mountain Group sedimentary cover rocks. This interpretation is supported by reconnaissance *P-T* studies reported in Steltenpohl and Kunk (1993). Calcite-dolomite thermometric estimates for ten Chewacla Marble specimens range from 470 to 525 °C (Steltenpohl and Kunk, 1993). Cover metapelites record rim *P-T* conditions of 534 and 561 °C and 6.2 and 8.0 kilobars. Core-to-rim garnet zonation patterns in one metapelite yield a range of *P-T* conditions from 502 °C and 5 kilobars to 534 °C and 6.2 kilobars, respectively. Calculated rim *P-T* estimates for two basement gneiss samples, however, are 634 °C and 678 °C and 10.1 kilobars and 11.0 kilobars, and one core estimate is 678 °C and 11.0 kilobars.

Timing of staurolite-kyanite/sillimanite–zone metamorphism (Steltenpohl and Moore, 1988) and associated deformation is poorly constrained, due to the lack of fossil evidence and appropriate isotopic dates. Wampler et al. (1970) reported conventional K/Ar mineral-cooling dates ranging from ca. 310 to 270 Ma (Alleghanian). Sears et al. (1981a) argued for an Ordovician (Taconic) peak of metamorphism, based on lithologic correlations and extrapolation of dates reported in other parts of the orogen. Tull (1982) favored a widespread Devonian (Acadian) metamorphic event for the Alabama Piedmont. Steltenpohl and Kunk (1993) reported $^{40}Ar/^{39}Ar$ mineral-cooling dates from the eastern Alabama Piedmont that they interpreted to reflect Alleghanian amphibolite-facies metamorphism and cooling; however, earlier metamorphism is not precluded by these cooling dates.

Structurally, the lithologies of the window are exposed in the eroded core of a multiply deformed, fault-bounded anticlinorium (Fig. 2). Appalachian thrusting resulted in two vertically stacked, basement-cored (F_2) fold nappes—the structurally lower Auburn nappe and the higher Thomaston nappe. Samples collected for our U-Pb isotopic investigation are from the overturned limb of the Auburn nappe (Fig. 2), which is considered the least transported of the two nappes. These fold nappes were subsequently refolded by shallow northeast-plunging, upright (F_3) folds (Sears et al., 1981a; Steltenpohl, 1988a). Structures observed in basement gneisses clearly predate deposition of the cover lithologies (i.e., are pre-Appalachian) (Steltenpohl, 1992), but have not been subjected to a systematic analysis.

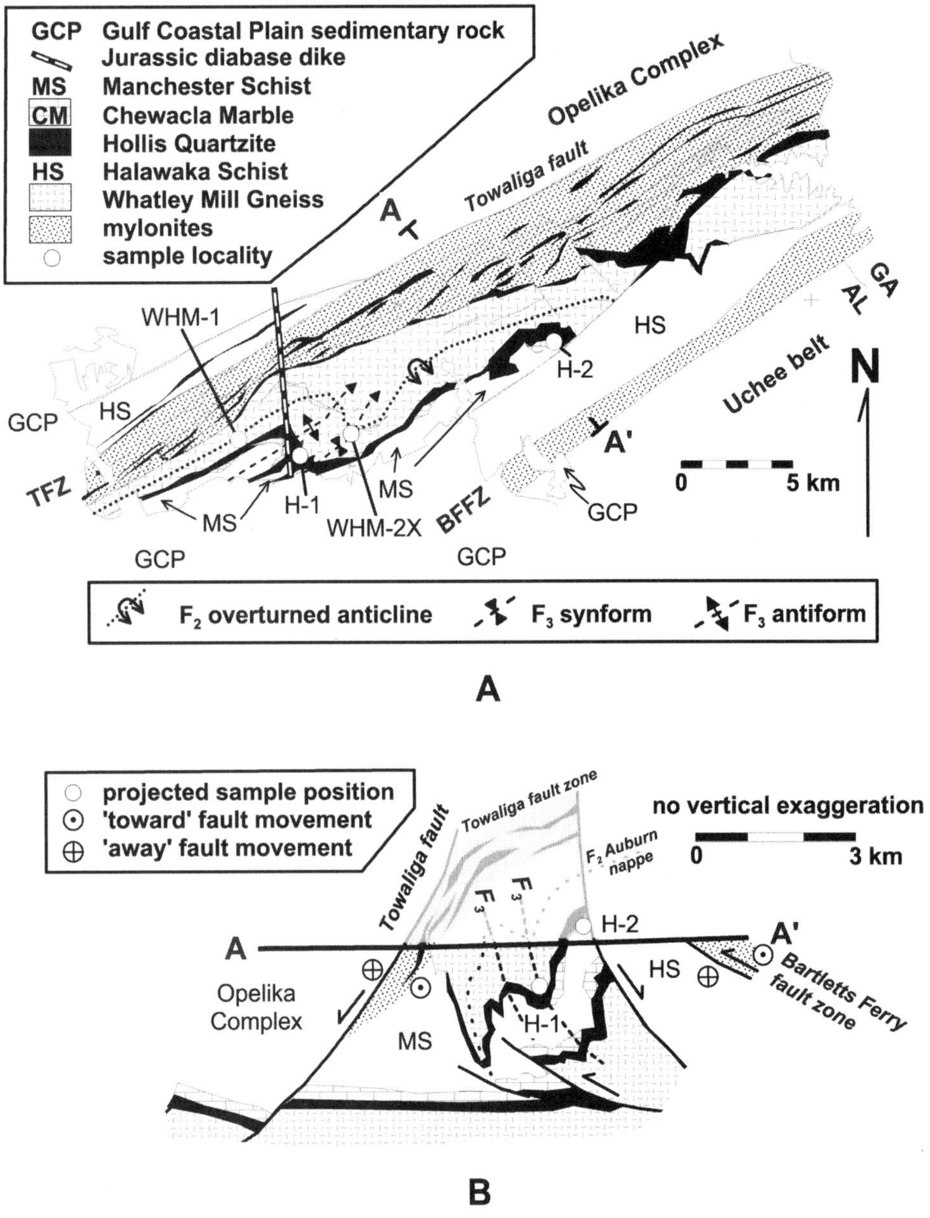

Figure 2. (A) Geologic map of the Pine Mountain window, illustrating sample localities. BFFZ—Bartletts Ferry fault zone; TFZ—Towaliga fault zone. (B) Geologic cross section A–A′ (from panel A) across the Pine Mountain window. The profile shows the structural/stratigraphic relations between Hollis Quartzite samples H-1 and H-2.

The significance of major mylonite zones flanking the Pine Mountain window remains a topic of debate (Hanley and Steltenpohl, 1997). These zones are composite features, with late-stage right-slip structures and fabrics that apparently obscured any earlier history of movement. Latest-stage high-angle cataclastic faults overprint these structures (Schamel et al., 1980; Sears et al., 1981a; Steltenpohl, 1988a,b; Babaei et al., 1991). Clarke (1952) first suggested that the mylonite zones were once connected over the Pine Mountain anticlinorium, such that the entire Alabama-Georgia Piedmont was one continuous thrust sheet. A similar hypothesis was developed by Bentley and Neathery (1970), who argued for emplacement of a "mega nappe" along these fault zones. Schamel et al. (1980) and Sears et al. (1981a) interpreted the mylonite zones to be the folded and faulted segments of a single major thrust, such that the window contained essentially autochthonous/parautochthonous Laurentian basement and cover. These field-derived results supported the Consortium for Continental Reflection Profiling (COCORP) seismic reflection profiles (Cook et al., 1979; Nelson et al., 1985, 1987), which indicated that the window looks through the southern Appalachian master décollement. Hooper (1986) and Hooper and Hatcher (1988), however, argued that the massif is allochthonous. Subsequent isotopic and field/structural data were also compatible with the COCORP interpretation (Steltenpohl et al., 1992; Steltenpohl and Salpas, 1993; West et al., 1995). Recently, McBride et al. (2001), however, reinterpreted the COCORP seismic profile. As a consequence, significant issues regarding the palinspastic position of the décollement remain unresolved.

ANALYTICAL PROCEDURES

Zircons were separated by standard hydraulic and magnetic methods from two samples of the Hollis Quartzite (H-1, Serenity Quarry, near the base of the unit; and H-2, Boral Brick Quarry, from near the top of the unit; see Fig. 2, B), two samples of the Whatley Mill Gneiss protolith (WHM-1, non-xenolith-bearing; and WHM-2X, xenolith-bearing) and two samples of Weisner Formation sandstone (WFCC, Coldwater Mountain; and WFWM, Weisner Mountain). Whole-rock powders of WHM-1 were also prepared for Nd isotopic analysis, which was done by standard methods of thermal ionization mass spectrometry (TIMS) (Heatherington and Mueller, 1991). The U-Pb analyses of the zircons were accomplished by TIMS and by ion probe methods. The analytical results are presented in Table 1.

Zircons analyzed by TIMS were dissolved by standard methods (Krogh, 1973; Parrish et al., 1987; Bowring et al., 1993) and mixed with a ^{205}Pb-^{235}U spike. Laboratory blank values for common Pb were typically <2 pg. Other zircons were analyzed using the sensitive high-resolution ion microprobe, reverse geometry (SHRIMP RG) instrument at the U.S. Geological Survey (USGS)–Stanford ion probe facility and SHRIMP-I at the Australian National University (Compston et al., 1984; Compston and Williams, 1992). The nonmagnetic (6 degree) fraction of the zircon concentrate was mounted in epoxy, polished, and analyzed for U and Pb isotopes.

Ages reported in this paper are of two types. For samples that yield analyses within 2% of concordia, we report error-weighted ages calculated using ISOPLOT (Ludwig, 2002). These ages are based on $^{206}Pb/^{238}U$ and $^{207}Pb/^{206}Pb$ ratios (i.e., the Tera-Wasserburg plot) and do not rely on $^{207}Pb/^{235}U$ ratios. For more discordant samples, the data are evaluated using a traditional concordia diagram and/or $^{207}Pb/^{206}Pb$ ages.

RESULTS

U-Pb Analysis of Zircons from the Cambrian Weisner Formation, Valley and Ridge Province

The Weisner Formation is located in eastern Alabama, in the southernmost Valley and Ridge Province (Fig. 1). It consists of cross-bedded sandstone interbedded with fine-grained siltstone, and is interpreted to reflect a shallow marine deltaic environment. The presence of Skolithos indicates a lower Cambrian depositional age (Bearce, 1997). The U-Pb analyses of zircons from the Weisner Formation allow for comparisons between this native Laurentian unit and the Hollis Quartzite (see later in this paper), which was proposed as a correlative unit (Yokel, 1996; Yokel and Steltenpohl, 1997). The Weisner Formation was sampled at two different localities, Weisner Mountain and Coldwater Mountain (Fig. 1). Zircons from both samples are small, (<75 microns) clear, rounded, colorless, and scarce. Analyses were performed on single and paired grains by conventional methods (TIMS). The $^{207}Pb/^{206}Pb$ ages are 1549 Ma (14% discordant), 1461 Ma (12% discordant), 1394 Ma (9% discordant), 1350 Ma (1% discordant), 1293 Ma (17% discordant), 1283 Ma (4% discordant), and 1206 Ma (63% discordant). This range of ages is consistent with derivation of the zircons from Laurentian sources, such as the Mid-continent granite-rhyolite province (ca. 1.4 Ga), and perhaps the Grenville (ca. 1.2–1.0 Ga), although evidence for the latter is less clear (Hoffman, 1989; Van Schmus et al., 1993). Our results, however, are consistent with what one might expect for this basal Laurentian sandstone.

U-Pb Analysis of Zircons from the Hollis Quartzite, Pine Mountain Window

Two distinct populations of zircon were extracted from a sample of the Hollis Quartzite (H-1) from an outcrop outside of Auburn, Alabama, ~10 m stratigraphically above (i.e., structurally below; see Fig. 2, B) the nonconformity with the basement Whatley Mill Gneiss. The physical appearance of each population is (1) transparent, colorless to amber, and rounded; and (2) opaque, white to gray. All U-Pb data were collected on single grains by either ID-TIMS (University of Florida) or by ion microprobe (USGS–Stanford SHRIMP RG) and are so noted in Table 1. All analyses were performed on single grains.

The $^{207}Pb/^{206}Pb$ ages yielded by the transparent colorless and amber grains that were <10% discordant are listed in Table 1 (Fig. 3, A). Averaging the three most concordant of the younger group of analyses on a Tera-Wasserburg plot yielded an age of 1097 ± 36 Ma (2σ error) (Fig. 3, B). This age and the first six ages mentioned directly above correspond to the Grenville event and are similar to the age of the nearby Whatley Mill Gneiss protolith (see below) and zircon ages reported for the Corbin Gneiss of Georgia (Odom et al., 1973; Heatherington et al., 1996). The 1410 and 1449-Ma ages correspond to the Mid-continent granite-rhyolite province (e.g., Van Schmus et al., 1993) and overlap with the range of Mesoproterozoic ages recorded in the Weisner zircons. Our analyses might indicate that the "Grenville" Hollis zircons are somewhat younger than those from the Weisner Formation (c.f., ca. 1000–1200 Ma for the former and 1206–1293 Ma for the latter), although the small number of grains analyzed makes it uncertain whether this reflects real differences in the ages of the Grenville rocks in the provenance.

The white-to-gray grains extracted from the Hollis Quartzite were very discordant (most ≥50%) with Paleoproterozoic $^{207}Pb/^{206}Pb$ ages (Table 1). The milky white-to-gray opacity of the older zircons is interpreted to result from metamictization and alteration, which likely contributed to the significant Pb loss and discordance. Zircons analyzed by TIMS yielded $^{207}Pb/^{206}Pb$ ages from 1833 to 2325 Ma. SHRIMP data for other zircons

TABLE 1. U-PB DATA FOR ZIRCON ANALYSES

Sample	$^{207}Pb^*/^{235}U$	$^{207}Pb^*/^{235}U$ Percentage error (1σ)	$^{206}Pb^*/^{238}U$	$^{206}Pb^*/^{238}U$ Percentage error (1σ)	$^{206}Pb^*/^{238}U$ Age (Ma)	$^{206}Pb^*/^{238}U$ Absolute error (1σ)	$^{207}Pb^*/^{206}Pb^*$ Age (Ma)	$^{207}Pb^*/^{206}Pb^*$ Absolute error (1σ)	Percentage discordance	Air-abraded?
Whatley Mill Gneiss										
WHM-1: SHRIMP data, grain.spot										
WHM-1 1.1	1.93	2.6	0.183	2.1	1085	21	1102	30	2	NA
WHM-1 2.1	1.97	2.7	0.185	2.1	1095	21	1122	32	2	NA
WHM-1 2.2	1.98	3.3	0.182	2.2	1079	22	1168	49	8	NA
WHM-1 3.1	1.74	2.4	0.17	2.2	1010	21	1051	21	4	NA
WHM-1 4.1	1.83	2.7	0.178	2.1	1054	21	1059	33	0	NA
WHM-1 5.1	1.91	3	0.182	2.2	1078	22	1097	40	2	NA
WHM-1 6.1	1.74	3	0.173	2.2	1031	21	1008	41	–2	NA
WHM-1 7.1	1.73	2.6	0.17	2.1	1015	20	1032	32	2	NA
WHM-1 8.1	1.84	3	0.182	2.3	1077	22	1029	40	–5	NA
WHM-1 9.1	1.84	3	0.175	2.2	1040	21	1096	40	5	NA
WHM-1 10.1	1.87	2.6	0.183	2.1	1083	21	1050	29	–3	NA
WHM-1 11.1	1.83	2.7	0.179	2.1	1062	21	1049	34	–1	NA
WHM-1 12.1	1.68	2.3	0.166	2	993	19	1018	22	3	NA
WHM-1 13.1	1.79	2.3	0.172	2	1026	19	1079	21	5	NA
WHM-1 14.1	2.31	2.5	0.209	2.1	1226	23	1197	27	–2	NA
WHM-2X: SHRIMP data, grain.spot										
WHM-2X 1.1	1.86	2.6	0.179	1.9	1060	19	1083	34	2	NA
WHM-2X 2.1	1.93	2.6	0.191	1.9	1125	20	1030	36	–9	NA
WHM-2X 3.1	1.87	2.3	0.175	1.9	1037	18	1144	26	9	NA
WHM-2X 4.1	1.82	2.9	0.185	2	1094	20	964	44	–13	NA
WHM-2X 5.1	1.86	2.3	0.174	1.9	1036	18	1134	25	9	NA
WHM-2X-6.1	1.95	2.1	0.183	1.8	1083	18	1129	19	4	NA
WHM-2X 7.1	1.9	2.4	0.184	1.9	1089	19	1062	28	–3	NA
WHM-2X 8.1	1.67	3	0.169	1.9	1005	18	982	46	–2	NA
WHM-2X 9.1	1.85	2.3	0.174	1.9	1037	18	1116	27	7	NA
WHM-2X 10.1	1.75	2.5	0.172	1.9	1024	18	1032	34	1	NA
WHM-2X 11.1	1.7	2.2	0.17	1.8	1010	17	1007	22	0	NA
WHM-2X 12.1	1.75	5.2	0.169	2	1005	19	1071	97	6	NA
WHM-2X 13.1	1.72	2	0.17	1.8	1014	17	1027	16	1	NA
WHM-2X 14.1	1.76	2	0.171	1.8	1016	17	1057	19	4	NA
WHM-2X 15.1	1.95	2.4	0.183	1.9	1085	19	1124	28	3	NA
WHM-2X 16.1	1.79	2.1	0.17	1.8	1012	17	1104	19	8	NA
Hollis Quartzite										
H-1: opaque grains, SHRIMP data grain.spot										
HW-1.1	1.82	9.4	0.098	4.3	605	12	2152	71	72	NA
HW-2.1	4.349	4.7	0.214	4.4	1248	25	2320	11	46	NA

HW-3.1	1.992	6.4	0.107	4.3	658	13	2157	38	69	NA
HW-4.1	7.005	6.8	0.334	5.4	1858	44	2370	30	22	NA
HW-5.1	4.016	7.6	0.215	5.4	1255	31	2171	42	42	NA
HW-6.1	3.674	4.6	0.187	4.4	1103	22	2261	8	51	NA
HW-7.1	2.934	9.4	0.238	5.3	1378	33	1410	70	2	NA
H-1: clear grains, SHRIMP data, grain.spot										
HC-1.1	2.909	16.1	0.237	11.2	1370	69	1406	102	3	NA
HC-1.2	1.372	10.5	0.129	5.3	785	20	877	31	11	NA
HC-2.1	2.004	44.4	0.185	10.9	1094	55	1162	481	6	NA
HC-3.1	2.922	6.8	0.233	5.6	1348	34	1449	30	7	NA
HC-4.1	1.817	32.8	0.184	12.7	1088	64	977	324	−11	NA
HC-5.1	2.163	27.1	0.188	9	1113	46	1276	260	13	NA
HC-6.1	1.96	6.8	0.184	5	1086	25	1133	41	4	NA
H-1: opaque grains, TIMS data										
H-1 wh A	3.656	0.3	0.179	0.3	1061.2	3.6	2325	1.2	54	Yes
H-1 wh B	1.897	0.2	0.1	0.2	611.7	0.9	2205.3	1	72	No
H-1 wh C	4.086	0.2	0.21	0.2	1227.9	2.9	2242.4	0.7	45	Yes
H-1 wh E	1.311	0.9	0.085	0.2	525.1	1.2	1833.1	14.5	71	Yes
H-1 wh F	1.115	0.3	0.065	0.1	407.5	0.6	2014.2	3.9	80	Yes
H-1 gr A	2.306	0.2	0.135	0.2	813.6	1.9	2019.4	1.4	60	Yes
H-1 gr B	2.652	0.1	0.156	0.1	934.5	1.2	2004.4	0.6	53	Yes
H-1: clear grains, TIMS data										
H-1 clr A	1.665	0.4	0.163	0.1	971	1.4	1048.8	7	7	Yes
H-1 clr B	1.651	0.7	0.164	0.3	979.7	2.7	1012.5	11	3	Yes
H-1 clr D	2.285	0.3	0.194	2	1144.9	2.3	1321.7	5	13	Yes
H-1 clr E	1.563	2.2	0.157	0.3	940	2.8	991.6	42.5	5	Yes
H-1 clr G	1.006	2	0.105	0.8	646.5	5.3	903	34.5	28	Yes
H-1 clr H	1.839	0.3	0.176	0.2	1043.3	2.5	1093.5	3.3	5	Yes
H-1 clr N	1.465	0.4	0.134	0.4	811.2	2.9	1178.2	3.3	31	No
H-1 clr O	1.372	0.5	0.143	0.4	861.7	3.5	915.7	7	6	No
H-1 clr R	1.816	1.3	0.161	1.3	964.4	12.2	1236.7	2.2	22	No
H-2, TIMS data										
H-2 m2-1	0.894	0.4	0.097	0.2	597	0.5	831.3	8	28	No
H-2 m2-2	1.776	0.3	0.162	0.3	965.5	1.3	1189.9	1.9	19	No
H-2 m2-3	1.651	0.3	0.154	0.1	921.3	0.6	1145	4.2	20	No
H-2 m2-4	0.615	0.3	0.071	0.2	443.7	0.4	695.5	5.5	36	No
H-2 m2-5	1.448	0.2	0.144	0.1	864.5	0.6	1019.4	1.7	15	No
H-2 m3-1	0.96	0.2	0.098	0.1	601.5	0.4	963	3.8	38	No
H-2 m4-1	1.671	0.2	0.16	0.1	956	0.6	1090.1	2	12	No

Note: * refers to radiogenic Pb. Abbreviations: NA—not applicable; SHRIMP—sensitive high-resolution ion microprobe; TIMS—thermal ionization mass spectrometry.

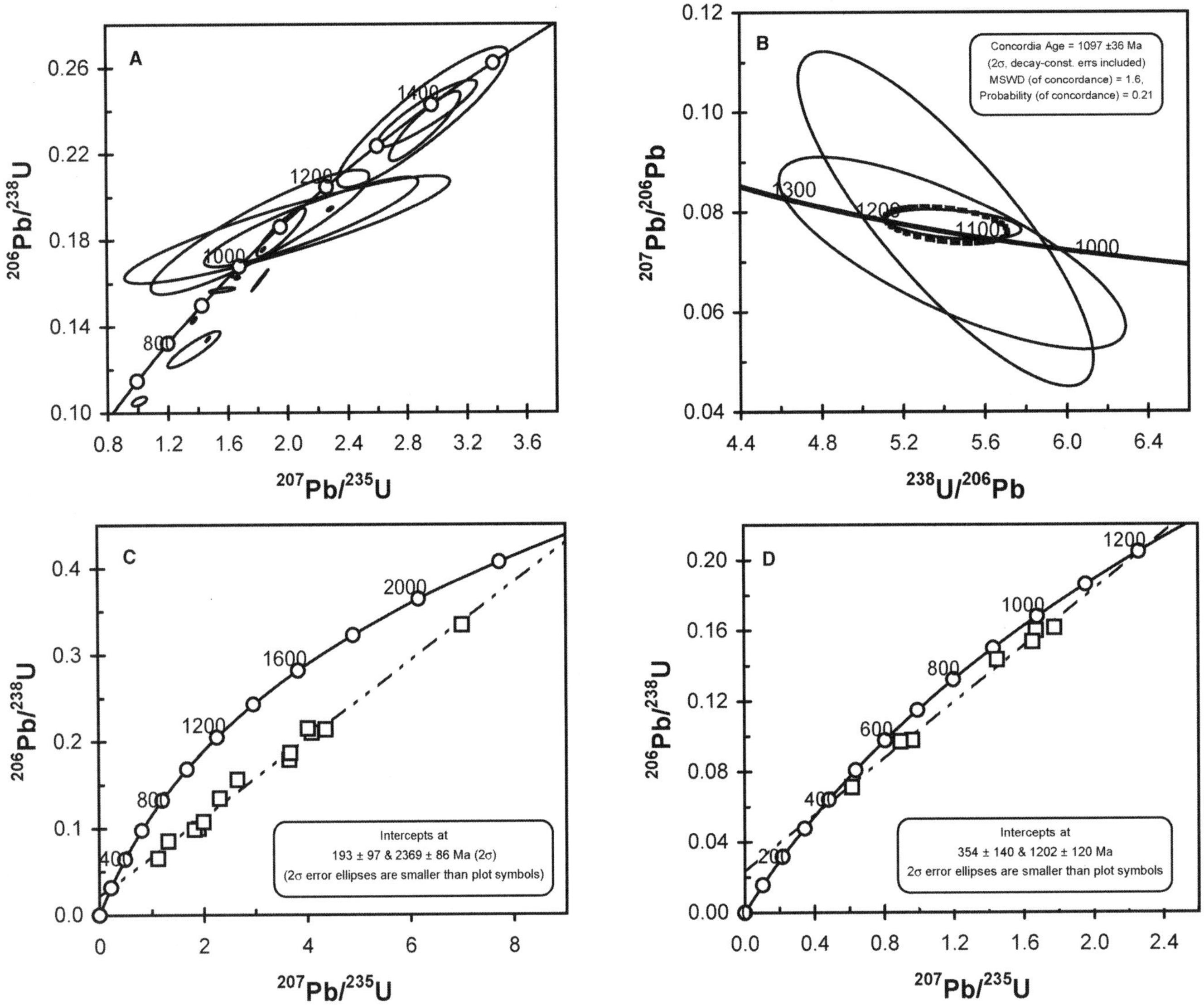

Figure 3. Conventional and Tera-Wasserburg concordia plots for Hollis Quartzite samples. Ages on the concordia curves are Ma. (A) Conventional concordia plot of U-Pb data for clear and amber zircons extracted from sample H-1. Error envelopes are 2σ, with decay constant errors included. (B) Tera-Wasserburg concordia plot, showing U-Pb data for the three most concordant ca. 1.1-Ga zircons extracted from sample H-1. Dashed ellipse represents the average age of the three samples, 1097 ± 36 Ma. Error envelopes are 2σ. (C) Conventional concordia plot of U-Pb data for white and gray opaque zircons extracted from sample H-1. The 2σ error ellipses were too small to display at this scale; therefore, data points are represented as squares. (D) Conventional concordia plot of U-Pb data for zircons extracted from sample H-2. The 2σ error ellipses were too small to display at this scale; therefore, data points are represented as squares. Intercept errors are 2σ.

from these grains yielded ages from 2152 to 2370 Ma. Plotting both TIMS and SHRIMP data yields a linear array with an upper intercept of 2369 ± 86 Ma (2σ error) (Fig. 3, C), approximately equivalent to the $^{207}Pb/^{206}Pb$ age of the least discordant individual grain. The large error from this array might be attributable to multiple original crystallization ages for this suite of zircons, mixed-age domains created by inadvertent analysis of overgrowths, and/or past episodes of Pb loss. The discordant nature of the grains makes it difficult to precisely identify crystallization age(s), but it is highly likely that one or more sources with ages of ca. 2.4–2.3 Ga contributed detritus to the Hollis. The scarcity of lithologic units of this age in North America suggests that a non-Laurentian source is more probable (discussed later).

The lower intercept age of ca. 193 Ma for H-1 corresponds to the time of Mesozoic rifting and diabase injection in the region, although a large error exists on the intercept. That the Auburn dike (ca. 200 Ma) (Hames et al., 2000) passes within 800 m of this sample location (Fig. 2) suggests that thermal adjustments and/or fluid activity related to its intrusion may have caused partial resetting of the U-Pb zircon systematics and may have overprinted evidence of any earlier resetting caused by Paleozoic events.

A second sample of Hollis Quartzite (H-2), taken from near the top of the section, ~18 m below the contact with the overlying Chewacla marble, was also examined. The gray-to-white opaque zircon population present in H-1 was completely absent from H-2. All grains were transparent, colorless to amber, and rounded. The SHRIMP U-Pb analysis produced only Grenvillian and younger $^{207}Pb/^{206}Pb$ ages: 1190 ± 4 Ma, 1145 ± 8 Ma, 1090 ± 4 Ma, 1019 ± 3 Ma, 963 ± 8 Ma, and 831 ± 16 Ma (Fig. 3, D). In general, these ages are consistent with crystallization events during the Grenville orogeny and, perhaps, Pb loss during the Paleozoic. In addition, the ca. 350-Ma lower intercept is consistent with evidence for post–Early Mississippian metamorphism in the nearby Talladega slate belt (Fig. 2) (Gastaldo et al., 1993; Steltenpohl et al., 2001; Tull, 2002) and a ca. 337-Ma $^{40}Ar/^{39}Ar$ date suggested for hornblende from the Pine Mountain window (Steltenpohl and Kunk, 1993).

Because of the potential significance of the Paleoproterozoic zircons from the Hollis Quartzite, we also investigated zircons from metaplutonic basement rocks, to explore whether the 2.4–2.3-Ga grains may have been derived from these Grenville gneisses, (e.g., inherited by the plutonic protoliths of the gneisses). Some portions of these metaplutonic rocks contain rare, hornblende-biotite gneiss enclaves, which are relatively finer grained and have a more mafic composition than the encapsulating granodioritic Whatley Mill Gneiss (Daniell, 1997). Zircons from two samples were separated and examined using the USGS–Stanford ion microprobe.

Sample WHM-1 is a non-xenolith-bearing, coarse-grained porphyroblastic biotite-granodiorite gneiss (Sears and Cook, 1984). All grains were transparent, colorless to amber, and rounded; virtually identical in appearance to the grains of Grenville and Mid-continent granite-rhyolite–province age found in each of the sedimentary samples. This rock lacked the distinctive white and gray Paleoproterozoic zircon grains found in the Hollis Quartzite. Of the fifteen U-Pb analyses performed on fourteen grains, all results were ≤8% discordant. Eight analyses were <2% discordant, and yielded $^{207}Pb/^{206}Pb$ ages from 1008 to 1197 Ma (all errors are 2σ). Although some of the ages overlap within 2σ error, the 1197-Ma grain is older than the others within 1σ error (Fig. 4, A). The seven most concordant analyses of the younger group yielded an error-weighted average of 1063 ± 14 Ma (2σ) (Fig. 4, B). We suggest that crystallization of the protolith to the Whatley Mill Gneiss took place at 1063 ± 14 Ma, and interpret the 1197-Ma grain as inheritance from an older event. The depleted mantle Sm-Nd model age (DePaolo, 1981) of the xenolith-free sample of Whatley Mill Gneiss is 1.39 Ga,

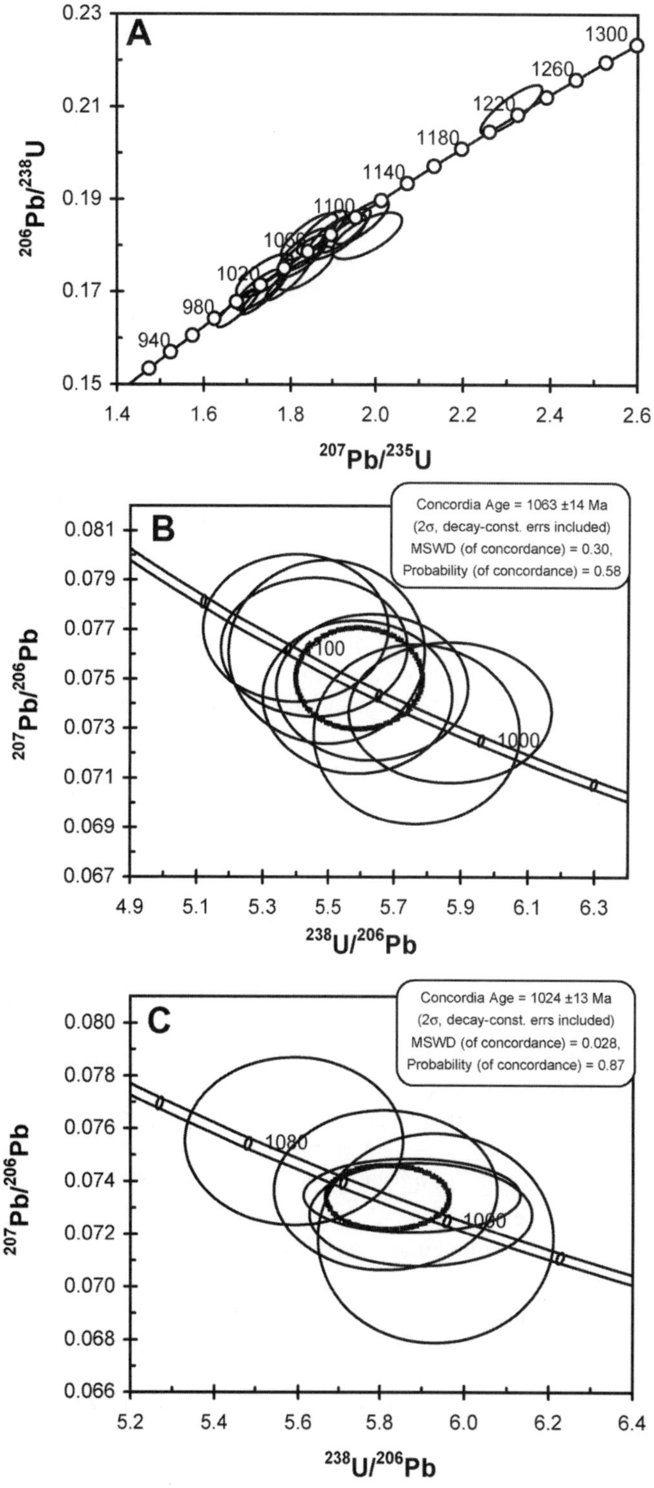

Figure 4. Conventional and Tera-Wasserburg concordia plots for Whatley Mill Gneiss samples. Error envelopes are 2σ; ages on concordia curves are Ma. (A) Conventional concordia plot of U-Pb data for white and gray opaque zircons extracted from sample WHM-1. (B) Tera-Wasserburg concordia plot of U-Pb data for the seven most concordant ca. 1.1-Ga zircons extracted from sample WHM-1. Dotted ellipse represents the average age of the seven samples, 1063 ± 14 Ma (2σ). (C) Tera-Wasserburg concordia plot of U-Pb data for the five most concordant zircons extracted from sample WHM-2X. Dotted ellipse represents the average age of the seven samples, 1024 ± 13 Ma.

based on present-day $^{143}Nd/^{144}Nd = 0.51204$ and $^{147}Sm/^{149}Nd = 0.1033$. The model age of the Whatley Mill Gneiss is consistent with its derivation from, and/or interaction with, older crust.

A sample of Whatley Mill Gneiss containing densely intermixed enclaves of intermediate composition was also examined (WHM-2X). Based on field and petrographic observations, Daniell (1997) interpreted such enclaves to contain the oldest lithologies associated with the Whatley Mill Gneiss. Sixteen zircons were examined, and five of these were ≤2% discordant, with $^{207}Pb/^{206}Pb$ ages of 982 ± 92 Ma, 1007 ± 44 Ma, 1027 ± 32 Ma, and 1083 ± 68 Ma. A weighted average age (Tera-Wasserburg) for five of these grains is 1024 ± 13 Ma (2σ error), slightly younger than the age derived for Whatley Mill gneiss sample WHM-1 (Fig. 4, C). Previously, Odom et al. (1985) determined ages of 1078 and 1165 Ma, respectively, for the Woodland Gneiss and Cunningham Granite basement orthogneisses from the Pine Mountain window in Georgia. These ages are in general agreement with the ages reported here for the Whatley Mill Gneiss in Alabama and suggest a general geochronologic similarity of the Pine Mountain basement over a wide geographic area.

In summary, the data demonstrate that at least three different sources, with Grenvillian, Mid-continent, and Paleoproterozoic ages, contributed sediment to the Hollis Quartzite protolith. The Grenvillian source may be local (e.g., the Whatley Mill Gneiss protolith). The ages of the Whatley Mill Gneiss (1063 ± 14 Ma and 1024 ± 14 Ma) and the youngest Hollis Quartzite zircons (1097 ± 36 Ma) are nearly equivalent within error. However, more distant sources cannot be ruled out. The 1.4-Ga age zircons may also have come from a Laurentian source, such as the Mid-continent granite-rhyolite province, but the source of Paleoproterozoic-age grains is likely to be non-Laurentian.

DISCUSSION AND CONCLUSIONS

The occurrence of a distinct population of 2.4–2.3 Ga zircons within the Hollis Quartzite has important implications for paleogeographic reconstructions for the time period of Hollis deposition. No 2.4–2.3 Ga terranes that could provide a source for these zircons in the Hollis are known in southern Laurentia. The closest potential sources for zircons of this age in North America are small, isolated localities in Canada or western Montana (e.g., Hoffman, 1989; Bickford et al., 1994; Mueller et al., 1996; Riller et al., 1999). It is unlikely, however, that rivers draining these areas could transport the 2.4–2.3 Ga zircons without mixing in zircons of different ages that coexist in those localities, as well as others from all points in between.

Alternatively, 2.4–2.3 Ga source regions are abundant in Gondwanan cratons. For example, some tectonic reconstructions for the Late Proterozoic (700–500 Ma) place eastern Laurentia adjacent to the Amazonian and Rio de la Plata cratons (Unrug, 1997; Weil et al., 1998) (Fig. 5). If the depositional age of the Hollis is latest Neoproterozoic to Cambrian, as was proposed based on lithologic correlations with units in the foreland (Yokel, 1996; Yokel and Steltenpohl, 1997; Tull and Steltenpohl, 2001), then Paleoproterozoic lithologies within these Gondwanan cratons (Cahen and Snelling, 1984; Bernasconi, 1987; Teixeira et al., 1989) may have served as a source of sediment. This is somewhat problematic, however, because the Amazonian orogenic event is reported as marginally younger (2.2 Ga) than the ages of zircons extracted from the Hollis Quartzite (Bernasconi, 1987). An alternative source for these zircons is suggested by the AUSWUS (Australia/Western United States) reconstruction of Karlstrom et al. (2001), which places the Gawler craton of Western Australia along the Proterozoic margin of the southwestern United States until ca. 600 Ma. Given a favorable stream system or longshore drift, 2.4 Ga zircons derived from granites and gneisses of the Gawler craton, Australia (Daly and Fanning, 1990), or East Antarctica (Oliver et al., 1983; Carson et al., 2002) may have traveled to the vicinity of the window, whether it was located approximately south (present-day orientation) of its modern position or somewhat to the west.

One of the difficulties of identifying a cratonal provenance for the Hollis is that only the lowermost sample of the Hollis contains the 2.4–2.3 Ga population (in addition to Grenville-age grains), and that the stratigraphically higher sample contains only Grenville-age zircons. This observation makes it difficult to argue for derivation of the (lower) Hollis from any of the aforementioned cratons (including western Laurentia) via large river systems, longshore transport, or the like, because of their demonstrated chronologic complexity (e.g., panafrican to Archean in many Gondwanan cratons). Although the absence of panafrican (ca. 650–550 Ma) grains in the Hollis could be attributed to a pre-panafrican depositional age, the lack of Archean and other Paleoproterozoic ages is more likely to reflect deposition in a basin with a restricted provenance (Tull and Steltenpohl, 2001). In this regard, note that a depositional age as old as ca. 1.0 Ga is permitted, but not required, by the detrital zircon ages.

Two populations of Weisner Formation zircons (i.e., 1.4–1.2 Ga) are similar in morphology and age to the two younger populations found in the Hollis Quartzite, although a distinct 1.1-Ga component is lacking. In addition, a 1.6-Ga component is present that was not observed in the Hollis Quartzite. Laurentian sources may be invoked as the source of these zircons. For example, 1.2 Ga corresponds to the early stages of the Grenville event; 1.4–1.2 Ga to the Mid-continent granite-rhyolite province (e.g., Van Schmus et al., 1993); and 1.6 Ga to plutons of the Central Plains orogen (e.g., Sims, 1990; Van Schmus et al., 1993). The presence of the 1.6-Ga component and the absence of a 1.1 Ga component may be explained by a more craton-ward location of the Weisner sandstone with respect to the Hollis Quartzite, and does not by itself rule out a correlation between the two units. More importantly, however, is the lack of

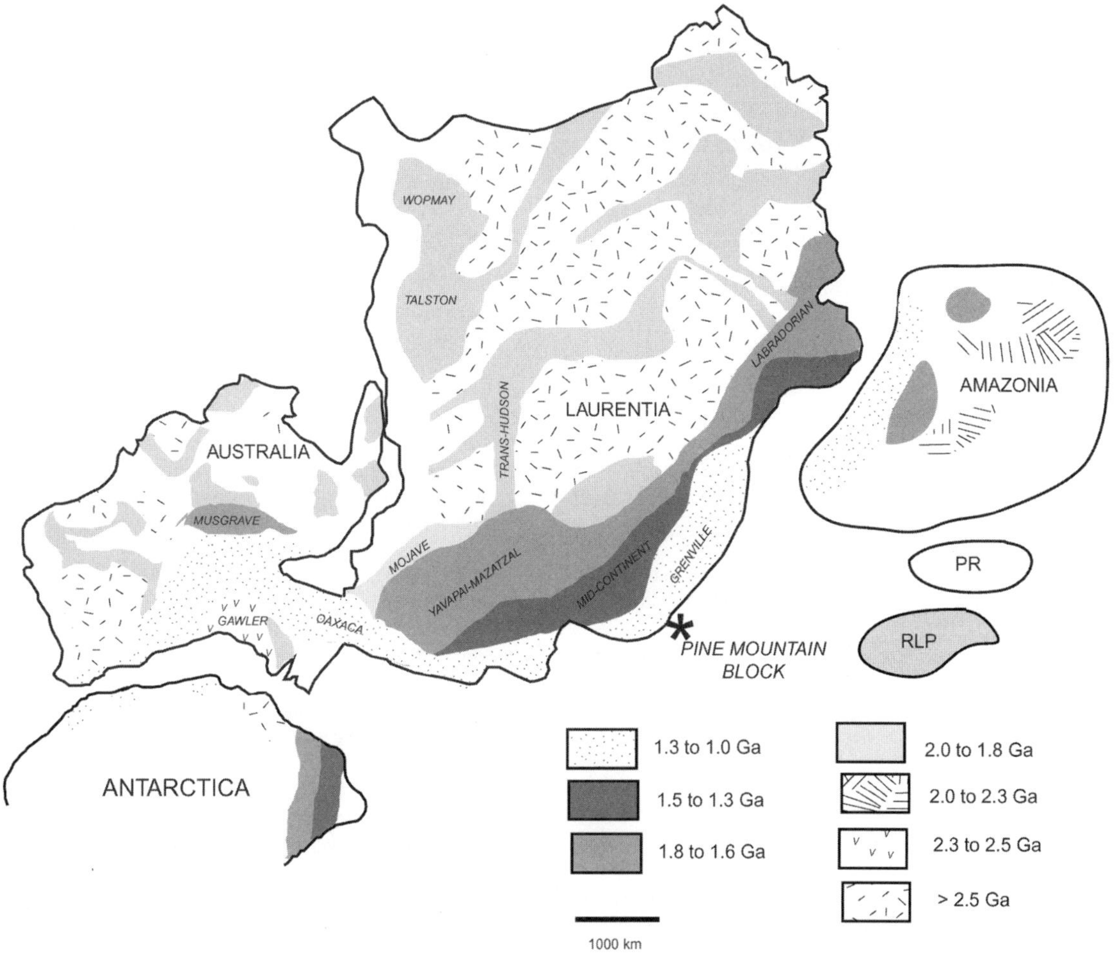

Figure 5. Laurentia and surrounding continents at 1.3–1.0 Ga, according to the AUSWUS reconstruction (Karlstrom et al., 1999, 2001; Brookfield, 1993), with locations of crustal provinces >1 Ga. PR—Parana block; RLP—Rio de la Plata craton. Modified from Karlstrom et al. (2001) and Unrug (1997), with additional data from Hoffman (1989, 1991), Nance and Murphy (1996), Condie and Rosen (1994), Restropo-Pace et al. (1997), Bernasconi (1987), Brito Neves and Cordani (1991), Teixeira et al. (1989), and Parker (1993).

the older, >2 Ga population found in the lower Hollis. Although the sedimentological dynamics of the ancient Laurentian margin are not well constrained, a simple first-order interpretation of the combined geologic and isotopic data is that the Weisner sediment was derived from Laurentian sources.

In summary, the presence of apparent 2.4–2.3 Ga zircons and Grenville-age zircons in the lower part of the Hollis Quartzite suggests that the Pine Mountain massif may have had a history distinct from other Appalachian Grenville massifs. In particular, it appears the Pine Mountain block may have received detritus from both a Laurentian and a non-Laurentian (probably Gondwanan) source that contained at least some 2.4–2.3 Ga rocks This suggests at least two somewhat distinct possibilities:

1. The Hollis was deposited during the rift phase of the separation of a Gondwanan conjugate to the Grenville margin of the southern Appalachians, such as the Amazonian craton or another Gondwanan craton that may have faced the Pine Mountain block across proto-Iapetus.

2. The Paleoproterozoic detritus was derived from 2.4–2.3 Ga units of East Antarctica or South Australia (i.e., the AUSWUS plate configuration is correct), and the Pine Mountain block may have been located significantly west of other Grenville massifs during the Neoproterozoic. If this was the case, the Pine Mountain block may be more closely related to the Grenville of Oaxaca or Texas than to the Grenville of eastern Laurentia.

In either case, the Pine Mountain block may have originated as a (Laurentian?) Grenvillian continental fragment between the ancient Laurentian and Gondwanan margins that was thrust-emplaced during an Appalachian or older event (Thomas, 1977; Hooper and Hatcher, 1988).

ACKNOWLEDGMENTS

Acknowledgment is made of the donors of The Petroleum Research Fund, administered by the American Chemical Society, for support of this research (ACS-PRF 23762-GB2 to MS); the National Science Foundation (EAR96-14342 to PM); Joseph Wooden and the staff of the USGS–Stanford Microanalytical Center; Allen Nutman for assistance with SHRIMP-I analyses; and the USGS Educational Mapping Program (MS).

REFERENCES CITED

Adams, G.I., 1933, General geology of the crystalline rocks of Alabama: Journal of Geology, v. 41, p. 159–173.

Applin, P.L., 1951, Preliminary report on buried, pre-Mesozoic rocks in Florida and adjacent areas: Reston, Virginia, U.S. Geological Survey Circular 91, 28 p.

Babaei, H.A., Hadizadeh, J., Babaei, A., and Ghazi, A.M., 1991, Timing and temperature of cataclastic deformation along segments of the Towaliga fault zone, western Georgia, U.S.A.: Journal of Structural Geology, v. 13, p. 579–586.

Bearce, D.W., 1997, Lower Cambrian rocks of the Talladega slate belt in eastern Alabama, *in* Yokel, L.S., et al., eds., Comparison of the Pine Mountain Block basement-cover sequence with the Lower Cambrian clastic-carbonate sequence of the Talladega slate belt: Tuscaloosa, Alabama, Alabama Geological Society, Guidebook for the 46th Annual Southeastern Section of the Geological Society of America, 93 p.

Bentley, R.D., and Neathery, T.L., eds., 1970, Geology of the Brevard fault zone and related rocks of the Inner Piedmont of Alabama: Tuscaloosa, Alabama, Alabama Geological Society, 8th Annual Field Trip Guidebook, 119 p.

Bernasconi, A., 1987, The major Precambrian terranes of eastern South America: A study of their regional and geochronological evolution: Precambrian Research, v. 37, p. 107–124.

Bickford, M., Collerson, K., and Lewery, J., 1994, Crustal history of the Rae and Hearne provinces, southwestern Canadian Shield, Saskatchewan: Constraints from geochronologic and isotopic data: Precambrian Research, v. 68, p. 1–21.

Bowring, S.A., Grotzinger, J.P., Isachsen, C.E., Knoll, A.H., Pelechaty, S.M., and Kolosov, P., 1993, Calibrating rates of Early Cambrian evolution: Science, v. 261, p. 1293–1298.

Brito Neves, B.B., and Cordani, U.G., 1991, Tectonic evolution of South America during the Late Proterozoic: Precambrian Research, v. 53, p. 23–40.

Brookfield, M.E., 1993, Neoproterozoic Laurentia-Australia fit: Geology, v. 21, p. 683–686.

Cahen, L., and Snelling, N., 1984, The geochronology and evolution of Africa: Oxford, England, Clarendon, 512 p.

Carson, C.J., Ague, J.J., Grove, M., Coath, C.D., and Harrison, T.M., 2002, U-Pb isotopic behavior of zircon during upper amphibolite facies fluid infiltration in the Napier Complex, east Antarctica: Earth and Planetary Science Letters, v. 199, p. 287–310.

Chalokwu, C.I., and Hanley, T.B., 1990, Geochemistry, petrogenesis, and tectonic setting of amphibolites from the southernmost exposure of the Appalachian Piedmont: Journal of Geology, v. 98, p. 725–738.

Chowns, T.M., and Williams, C.T., 1983, Pre-Cretaceous rocks beneath the Georgia Coastal Plain—Regional implications, *in* Gohn, G.S., ed., Studies related to the Charleston, South Carolina, earthquake of 1886—Tectonics and seismicity: Reston, Virginia, U.S. Geological Survey Professional Paper 1313-L, p. L1–L42.

Clarke, J.W., 1952, Geology and mineral resources of the Thomaston quadrangle, Georgia: Atlanta, Georgia, Georgia Geological Survey Bulletin 59, 99 p.

Compston, W., and Williams, I., 1992, Ion probe ages for the British Ordovician and Silurian stratotypes, *in* B. Webby and J. Laurie, eds., Global perspectives on Ordovician geology: Rotterdam, The Netherlands, Bolkema, p. 59–67.

Compston, W., Williams, I.S., and Meyer, C., 1984, U-Pb geochronology of zircons from lunar breccia 73217 using a sensitive high mass-resolution ion microprobe, in Proceedings, Lunar and Planetary Science Conference, 14th: Journal of Geophysical Research, v. 89, p. B525–B534.

Condie, K.C., and Rosen, O.M., 1994, Laurentia-Siberia connection revisited: Geology, v. 22, p. 168–170.

Cook, R.A., Albaugh, D.S., Brown, L.D., Kaufman, S., Oliver, J.E., and Hatcher, R.D., Jr., 1979, Thin-skinned tectonics in the crystalline southern Appalachians: COCORP seismic-reflection profiling of the Blue Ridge and Piedmont: Geology, v. 7, p. 563–568.

Crickmay, G.W., 1933, The occurrence of mylonites in the crystalline rocks of Georgia: American Journal of Science, v. 26, p. 161–177.

Crickmay, G.W., 1952, Geology of the crystalline rocks of Georgia: Atlanta, Georgia, Georgia Geological Survey Bulletin 46, p. 32–36.

Daly, S.J., and Fanning, C.M., 1990, Archaean geology of the Gawler Craton, South Australia, *in* Glover, J.E., and Ho, S.E., compilers, 3rd International Archaean Symposium, Perth, 1990, extended abstracts: Perth, Western Australia, Geoconferences, p. 91–92.

Daniell, Neil, 1997, Geochemical evidence for separate protoliths for the Whatley Mill Gneiss, Pine Mountain massif, Alabama [M.S. thesis]: Auburn, Alabama, Auburn University, 89 p.

DePaolo, D., 1981, Neodymium isotopes in the Colorado Front Range and crust-mantle evolution in the Proterozoic: Nature, v. 291, p. 193–196.

Galpin, S.L., 1915, A preliminary report on the feldspar and mica deposits of Georgia: Atlanta, Georgia, Geological Survey of Georgia Bulletin 30, 190 p.

Gastaldo, R.A., Guthrie, G.M., and Steltenpohl, M.G., 1993, Mississippian fossils from southern Appalachian metamorphic rocks and their implications for late Paleozoic tectonic evolution: Science, v. 262, p. 732–734.

Hadley, J.B., 1970, The Ocoee Series and its possible correlatives, *in* Fisher, G.W., et al., eds., Studies in Appalachian Geology: Central and Southern: New York, Wiley Interscience, p. 247–259.

Hames, W.E., Renne, P.R., and Ruppel, C., 2000, New evidence for geologically-instantaneous emplacement of earliest Jurassic Central Atlantic Magmatic Province basalts on the North American margin: Geology, v. 28, no. 9, p. 859–862.

Hanley, T., and Steltenpohl, M.G., 1997, Mylonites and other fault-related rocks of the Pine Mountain and Uchee belts of eastern Alabama and western Georgia: Tuscaloosa, Alabama, Alabama Geological Society, Field Trip Guidebook for the 46th Annual Southeastern Section of the Geological Society of America, 78 p.

Hanley, T.B., Chalokwu, C.I., and Steltenpohl, M.G., 1997, Constraints on the location of the Carolina/Avalon terrane boundary in the southernmost exposed Appalachians, western Georgia and eastern Alabama, *in* Glover, L., III, and Gates, A.E., eds., Central and southern Appalachian sutures: Results of the EDGE project and related studies: Boulder, Colorado, Geological Society of America Special Paper 314, p. 15–24.

Hatcher, R.D., Jr., 1984, Southern and central Appalachian basement massifs, *in* Bartholomew, M.J., ed., The Grenville Event in the Appalachians and related topics: Boulder, Colorado, Geological Society of America Special Paper 194, 294 p.

Hatcher, R.D., Jr., 1987, Tectonics of the southern and central Appalachian internides: Annual Reviews of Earth and Planetary Sciences, v. 15, p. 337–362.

Heatherington, A., and Mueller, P., 1991, Geochemical evidence for Triassic rifting in southwestern Florida: Tectonophysics, v. 188, p. 291–302.

Heatherington, A.L., Mueller, P.A., and Nutman, A.P., 1996, Neoproterozoic magmatism in the Suwannee terrane: Implications for terrane correlation, *in* Nance, R.D., and Thompson, M., eds., Avalonian and related peri-Gondwanan terranes of the circum-North Atlantic: Boulder, Colorado, Geological Society of America Special Paper 304, p. 257–268.

Hewitt, D.F., and Crickmay, G.W., 1937, The Warm Springs of Georgia, their geologic relations and origin: Reston, Virginia, U.S. Geological Survey Water Supply Paper 819, 37 p.

Higgins, M.W., Atkins, R.L., Crawford, T.J., Crawford, R.F., Brooks, R., and Cook, R.B., 1988, The structure, stratigraphy, tectonostratigraphy and evolution of the southernmost part of the Appalachian orogen: Reston, Virginia, U.S. Geological Survey Professional Paper 1475, 173 p.

Hoffman, P.F., 1989, Precambrian geology and tectonic history of North America, *in* Bally, A.W., and Palmer, A.R., eds., The geology of North America, an overview: Boulder, Colorado, Geological Society of America, Geology of North America: v. A, p. 447–512.

Hoffman, P.F., 1991, Did the breakout of Laurentia turn Gondwanaland inside-out?: Science, v. 252, p. 1409–1412.

Hooper, R.J., 1986, Geologic studies at the east end of the Pine Mountain window and adjacent Piedmont, central Georgia, v. I and II [Ph.D. thesis]: Columbia, University of South Carolina, 374 p.

Hooper, R.J., and Hatcher, R.D., Jr., 1988, Pine Mountain terrane, a complex window in the Georgia and Alabama Piedmont: Evidence from the eastern termination: Geology, v. 16, p. 307–310.

Horton, J.W., Jr., Avery, A.D., Jr., and Rankin, D.W., 1989, Tectonostratigraphic terranes and their Paleozoic boundaries in the central and southern Appalachians: Boulder, Colorado, Geological Society of America Special Paper 230, p. 213–245.

Karlstrom, K.E., Williams, M.L., McLelland, J., Geissman, J.W., and Ahall, K.-I., 1999, Refining Rodinia: Geologic evidence for the Australia-Western U.S. connection in the Proterozoic: GSA Today, v. 9, no. 10, p. 1–7.

Karlstrom, K.E., Ahall, K.-I., Harlan, S.S., Williams, M.L., McLelland, J., and Geissman, J.W., 2001, Long-lived (1.8–1.0 Ga) convergent orogen in southern Laurentia, its extensions to Australia and Baltica, and implications for refining Rodinia: Precambrian Research, v. 111, p. 5–30.

King, P.B., 1961, The subsurface Ouachita structural belt east of the Ouachita Mountains, *in* Flawn, P.T., et al., eds., The Ouachita system: Austin, Texas, University of Texas Publication 6120, p. 83–98, 347–361.

King, P.B., Hadley, J.B., Neuman, R.B., and Hamilton, W., 1958, Stratigraphy of the Ocoee Series, Great Smoky Mountains, Tennessee and North Carolina: Geological Society of America Bulletin, v. 69, p. 947–966.

Krogh, T.E., 1973, A low contamination method for hydrothermal decomposition of zircon and extraction of U and Pb for isotopic age determinations: Geochimica et Cosmochimica Acta, v. 37, p. 485–494.

Ludwig, K.R., 2002, Isoplot/Ex version 2.49, A geochronological toolkit for Microsoft Excel: Berkeley, California, Berkeley Geochronology Center Special Publication 1a.

McBride, J.H., Hatcher, R.D., Jr., and Stephenson, W.J., 2001, Toward a reconciliation of seismic reflection and field geologic mapping from the Pine Mountain belt: Preliminary results: Geological Society of America Abstracts with Programs, v. 33, no. 2, p. A-74.

Mueller, P.A., Heatherington, A.L., Wooden, J.L., Shuster, R.D., Nutman, A.P., and Williams, I.S., 1994, Precambrian zircons from the Florida basement: A Gondwanan connection: Geology, v. 22, p. 119–122.

Mueller, P., Heatherington, A., D'Arcy, K., Wooden, J., and Nutman, A., 1996, Contrasts between Sm-Nd and U-Pb zircon systematics in the Tobacco Root batholith, Montana: Implications for the determination of crustal age provinces; Tectonophysics, v. 265, p. 169–179.

Nance, R.D., and Murphy, J.B., 1996, Basement isotopic signatures and Neoproterozoic paleogeography of Avalonian-Cadomian and related terranes in the circum-North Atlantic, *in* Nance, R.D., and Thompson, M.D., eds., Avalonian and related peri-Gondwanan terranes of the circum–North Atlantic: Boulder, Colorado, Geological Society of America Special Paper 304, p. 333–346.

Nelson, K.D., Arnow, J.A., Giguere, M., and Schamel, S., 1987, Normal-fault boundary of an Appalachian basement massif?: Results of COCORP profiling across the Pine Mountain belt in western Georgia: Geology, v. 15, p. 832–836.

Nelson, K.D., Arnow, J.A., McBride, J.H., Willemin, J.H., Huang, J., Zheng, L., Oliver, J.E., Brown, L.D., and Kaufman, S., 1985, New COCORP profiling in the southeastern United States. Part I: Late Paleozoic suture and Mesozoic rift basin: Geology, v. 13, p. 714–718.

Odom, A.L., Kish, S.A., and Leggo, P.J., 1973, Extension of "Grenville Basement" to the southern extremity of the Appalachians: U-Pb ages of zircons: Geological Society of America Abstracts with Programs, v. 10, p. 196.

Odom, A.L., Kish, S.A., and Russell, C.W., 1985, U-Pb and Rb-Sr geochronology of the basement rocks in the Pine Mountain Belt of southwest Georgia, *in* Kish, S.A., et al., eds., Geology of the Southwestern Piedmont of Georgia: Boulder, Colorado, Geological Society of America Field Trip Guidebook, p. 12–14.

Oliver, R.L., Cooper, J.A., and Truelove, A.J., 1983, Petrology and zircon geochronology of Herring Island and Commonwealth Bay and evidence for Gondwana reconstruction, *in* Oliver, R.L., et al., eds., Antarctic earth science, 4th International Symposium on Antarctic Earth Sciences, Adelaide, Australia, 1982: Cambridge, England, Cambridge University Press, p. 5–10.

Parker, A.J., 1993, Geological framework, *in* Drexel, J.F., et al., eds., The geology of South Australia, v. 1, The Precambrian: Eastwood, Australia, South Australia Geological Survey Bulletin 54, p. 9–24.

Parrish, R.R., Roddick, J.C., Loveridge, W.D., and Sullivan, R.W., 1987, Uranium-lead analytical techniques at the Geochronology laboratory, Geological Survey of Canada, in Radiogenic age and isotopic studies, report 1: Québec, Geological Survey of Canada Paper 87–2, p. 3–7.

Rast, N., and Kohles, K.M., 1986, The origin of the Ocoee Supergroup: American Journal of Science, v. 286, p. 593–616.

Raymond, D.E., Osborne, W.E., Copeland, C.W., and Neathery, T.L., 1988, Alabama stratigraphy: Tuscaloosa, Alabama, Alabama Geological Survey Circular 140, 97 p.

Restrepo-Pace, P.A., Ruiz, J., Gehrels, G., and Cosca, M., 1997, Geochronology and Nd isotopic data of Grenville-age rocks in the Columbian Andes: New constraints for Late Proterozoic–Early Paleozoic paleocontinental reconstructions of the Americas: Earth and Planetary Science Letters, v. 150, p. 427–441.

Riller, U., Schwerdtner, W., Halls, A., and Card, K., 1999, Transpressive tectonism in the eastern Penokean orogen, Canada: Consequences for Proterozoic crustal kinematics and continental fragmentation: Precambrian Research, v. 93, p. 51–70.

Russell, G.S., 1978, U-Pb, Rb-Sr, and K-Ar isotopic studies bearing on the development of the southernmost Appalachian orogen, Alabama [Ph.D. thesis]: Tallahassee, Florida State University, 197 p.

Russell, G.S., 1985, Reconnaissance geochronological investigations in the Phenix City Gneiss and Bartletts Ferry mylonite zone, *in* Kish, S.A., et al., eds., Geology of the southwestern Piedmont of Georgia: Tallahassee, Florida, Florida State University, Geological Society of America, Southeastern Section, Field Trip Guidebook, p. 9–11.

Schamel, S., and Bauer, D.T., 1980, Remobilized Grenville basement in the Pine Mountain window, *in* Wones, D.R., ed., Proceedings of The Caledonides in the U.S.A.: Blacksburg, Virginia, Department of Geosciences, Virginia Polytechnic Institute and State University, Memoir 2, p. 313–316.

Schamel, S., Hanley, T.B., and Sears, J.W., 1980, Geology of the Pine Mountain window and adjacent terranes in the Piedmont Province of Alabama and Georgia: Tuscaloosa, Alabama, Alabama Geological Society, Geological Society of America, Southeastern Section, Field Trip Guidebook, 69 p.

Sears, J.W., and Cook, R.B., 1984, An overview of the Grenville basement complex of the Pine Mountain window, Alabama and Georgia, *in* Bartholomew, M.J., The Grenville Event in the Appalachians and related topics: Boulder, Colorado, Geological Society of America Special Paper 194, p. 281–287.

Sears, J.W., Cook, R.B., and Brown, D.E., 1981a, Tectonic evolution of the western part of the Pine Mountain window and adjacent Inner Piedmont province, *in* Sears, J.W., ed., Contrasts in tectonic style between the Inner Piedmont terrane and the Pine Mountain window: Tuscaloosa, Alabama, Alabama Geological Society, 18th Annual Field Trip Guidebook, 61 p.

Sears, J.W., Cook, R.B., Gilbert, O.E., Jr., Carrington, T.J., and Schamel, S., 1981b, Stratigraphy and structure of the Pine Mountain window, Georgia and Alabama, *in* Wigley, P.B., ed., Latest thinking on the stratigraphy of selected areas in Georgia: Atlanta, Georgia, Georgia Geological Survey Information Circular 54A, p. 41–54.

Sims, P.K., 1990, Precambrian basement map of the northern Mid-continent,

U.S.A.: Reston, Virginia, U.S. Geological Survey Miscellaneous Investigations Map I-1853-A, scale 1:1,000,000.

Steltenpohl, M.G., 1988a, Geology of the Pine Mountain imbricate zone, Lee County, Alabama: Tuscaloosa, Alabama, Alabama Geological Survey, Circular 136, 15 p.

Steltenpohl, M.G., 1988b, Kinematics of the Towaliga, Bartletts Ferry, and Goat Rock fault zones, Alabama: The late Paleozoic dextral shear system in the southernmost Appalachians: Geology, v. 16, p. 852–855.

Steltenpohl, M.G., 1992, The Pine Mountain window of Alabama: Basement-cover evolution in the southernmost exposed Appalachians, *in* Bartholomew, M.J., eds., Basement tectonics 8: Characterization of Ancient and Mesozoic Continental Margins—Proceedings of the 8th International Conference on Basement Tectonics, Butte, Montana, 1988: Dordrecht, The Netherlands, Kluwer Academic Publishers, p. 491–501.

Steltenpohl, M.G., and Kunk, M.J., 1993, $^{40}Ar/^{39}Ar$ thermochronology and Alleghanian development of the southernmost Appalachian Piedmont, Alabama and southwest Georgia: Geological Society of America Bulletin, v. 105, p. 819–833.

Steltenpohl, M.G., and Moore, W.B., 1988, Metamorphism in the Alabama Piedmont: Tuscaloosa, Alabama, Alabama Geological Survey Circular 138, 27 p.

Steltenpohl, M.G., and Salpas, P.A., 1993, Geology of the southernmost exposed Appalachian Piedmont rocks along the Alabama fall line: Boulder, Colorado, Geological Society of America Field Trip Guidebook III, 204 p.

Steltenpohl, M.G., Drummond, M.S., and Goldberg, S.A., 1990, Petrogenesis and structural evolution of ductile deformation zones in Pine Mountain window basement gneisses, Lee County, Alabama: Geological Survey of Alabama Bulletin 137, 39 p.

Steltenpohl, M.G., Goldberg, S.A., Hanley, T.B., and Kunk, M.J., 1992, Evidence for Alleghanian development of the Goat Rock fault zone, Alabama and southwest Georgia: Temporal compatibility with the master décollement: Geology, v. 20, p. 845–848.

Steltenpohl, M.G., Hames, W.E., Kunk, M.J., and Mies, J.W., 2001, $^{40}Ar/^{39}Ar$ data from the western Blue Ridge and their implications for time of metamorphism: Geological Society of America Abstracts with Programs, v. 33, no. 2, p. 6.

Stieve, A.L., and Size, W.B., 1988, Corona structures: Key to the recognition of Grenville basement within the Pine Mountain window, Georgia: Southeastern Geology, v. 28, p. 225–236.

Stow, S.H., Neilson, M.J., and Neathery, T.L., 1984, Petrography, geochemistry and tectonic significance of the amphibolites of the Alabama Piedmont: American Journal of Science, v. 284, p. 416–436.

Teixeira, W., Tassinari, C.C.G., Cordani, U.G., and Kawashita, K., 1989, A review of the geochronology of the Amazonian Craton: Tectonic implications: Precambrian Research, v. 42, p. 213–227.

Thomas, W.A., 1977, Evolution of the Appalachian-Ouachita salients and recesses from reentrants and promontories in the continental margin: American Journal of Science, v. 277, p. 1233–1278.

Tull, J.F., 1982, Stratigraphic framework of the Talladega slate belt, Alabama Appalachians, in Bearce, D.N., et al., eds., Tectonic studies in the Talladega and Carolina slate belts, Southern Appalachian Orogen: Boulder, Colorado, Geological Society of America Special Paper 191, 165 p.

Tull, J.F., 2002, Southeastern margin of the middle Paleozoic shelf, southwesternmost Appalachians: Regional stability bracketed by Acadian and Alleghanian tectonism: Geological Society of America Bulletin, v. 114, p. 643–655.

Tull, J.F., and Steltenpohl, M.G., 2001, Speculations about the potential palinspastic relationships between the Talladega and Pine Mountain belts, *in* Steltenpohl, M.G., and Tull, J.F., eds., Comparison of the Pine Mountain basement-cover sequence with the Lower Cambrian to Lower Ordovician clastic-carbonate sequence of the Talladega slate belt: Alabama Geological Society field trip guidebook, p. 58–66.

Unrug, R., 1997, Rodinia to Gondwana: The geodynamic map of Gondwana supercontinent assembly: GSA Today, v. 7, no. 1, p. 1–6.

Van Schmus, W.R., Bickford, M.E., Sims, P.K., Anderson, R.R., Shearer, C.K., and Treves, S.B., 1993, Proterozoic geology of the western mid-continent basement, *in* Reed, J.S., Jr., et al., eds., Precambrian: Conterminous U.S.: Boulder, Colorado, Geological Society of America, Geology of North America, v. C-2, p. 239–258.

Wampler, J.M., Neathery, T.L., and Bentley, R.D., 1970, Age relations in the Alabama Piedmont, *in* Bentley, R.D., and Neathery, T.L., eds., Geology of the Brevard fault zone and related rocks of the Inner Piedmont of Alabama: Alabama Geological Society 8th Annual Field Trip Guidebook, p. 81–90.

Weil, A.B., Van der Voo, R., Mac Niocaill, C., and Meert, J.G., 1998, The Proterozoic supercontinent Rodinia: Paleomagnetically derived reconstructions for 1100 to 800 Ma: Earth and Planetary Science Letters, v. 154, p. 13–24.

West, T.E., Jr., Secor, D.T., Jr., Pray, J.R., Boland, I.B., and Maher, H.D., Jr., 1995, New field evidence for an exposure of the Appalachian décollement at the east end of the Pine Mountain terrane, Georgia: Geology, v. 23, p. 621–624.

Yokel, L.S., 1996, Geology of the Chewacla Marble and associated units, Lee County, Alabama [M.S. thesis]: Auburn, Alabama, Auburn University, 106 p.

Yokel, L.S., and Steltenpohl, M.G., 1997, Laurentian cover lithologies of the Pine Mountain massif, Alabama, and potential correlations with the Valley and Ridge and Western Blue Ridge units, *in* Bearce, D., ed., Comparison of the Pine Mountain block basement-cover sequence with the Lower Cambrian clastic-carbonate sequence in the Talladega slate belt: Alabama Geological Society Field Trip Guidebook, p. 1–26.

Yokel, L.S., Tull, J.F., Steltenpohl, M.G., Johnson, L.W., and Bearce, D., 1997, Comparison of the Pine Mountain block basement-cover sequence with the Lower Cambrian clastic-carbonate sequence in the Talladega slate belt: Alabama Geological Society Field Trip Guidebook, 93 p.

Manuscript Accepted by the Society August 25, 2003

Printed in the USA

Geological Society of America
Memoir 197
2004

Petrologic and geochronologic evolution of the Grenville orogen, northern Blue Ridge Province, Virginia

Richard P. Tollo*
Department of Earth and Environmental Sciences, George Washington University, 2029 G Street, NW, Washington, D.C. 20052, USA
John N. Aleinikoff
U.S. Geological Survey, MS 963 Federal Center, Denver, Colorado 80225, USA
Elizabeth A. Borduas
Department of Earth and Environmental Sciences, George Washington University, 2029 G Street, NW, Washington, D.C. 20052, USA
Paul C. Hackley
U.S. Geological Survey, 12201 Sunrise Valley Drive, Reston, Virginia 20192, USA
C. Mark Fanning
Research School of Earth Sciences, Australian National University, Mills Road, Canberra ACT 0200, Australia

ABSTRACT

Basement rocks in the northern Virginia Blue Ridge include petrologically diverse granitoids and granitic gneisses that collectively record over 100 m.y. of Grenville orogenic history. New U-Pb sensitive high-resolution ion microprobe (SHRIMP) isotopic analyses of zircon indicate igneous crystallization ages of 1159 ± 14 Ma (high-silica charnockite), 1078 ± 9 Ma (leucogranite gneiss), 1060 ± 5 Ma (Old Rag magmatic series), and 1050 ± 8 Ma (low-silica charnockite). These ages, together with SHRIMP and thermal ionization mass spectrometry (TIMS) ages from previous studies, define three intervals of Grenville-age magmatic activity: ca. 1160–1140 Ma (Magmatic Interval I), ca. 1112 Ma (Magmatic Interval II), and ca. 1080–1050 Ma (Magmatic Interval III). Field relations and ages of crosscutting igneous units indicate that a high-grade deformation event, likely associated with Ottawan orogenesis, occurred between 1078 and 1050 Ma. All rocks display tholeiitic affinity and trace-element concentrations indicative of derivation from heterogenous sources. The low-silica charnockite exhibits A-type geochemical affinity; however, all other meta-igneous rocks are compositionally transitional between A-types and fractionated I-types. Similar ages of magmatism in the Blue Ridge and Adirondacks indicate that meta-igneous rocks in both massifs define age clusters that both predate and postdate the main pulse of local Ottawan orogenesis. Late- to postorogenic A-type magmatism is represented by the 1050 Ma low-silica charnockite in the Blue Ridge and the 1060–1045 Ma Lyon Mountain granitic gneiss in the Adirondacks. Zircons from Blue Ridge granitoids emplaced during Magmatic Interval III preserve evidence of thermal effects associated with waning stages of Ottawan orogenesis at ca. 1020 Ma and 980 Ma.

Keywords: charnockite, Grenville, Blue Ridge, A-type granite, geochronology

*E-mail: rtollo@gwu.edu.

Tollo, R.P., Aleinikoff, J.N., Borduas, E.A., Hackley, P.C., and Fanning, C.M., 2004, Petrologic and geochronologic evolution of the Grenville orogen, northern Blue Ridge province, Virginia, *in* Tollo, R.P., Corriveau, L., McLelland, J., and Bartholomew, M.J., eds., Proterozoic tectonic evolution of the Grenville orogen in North America: Boulder, Colorado, Geological Society of America Memoir 197, p. 647–677. For permission to copy, contact editing@geosociety.org.

INTRODUCTION

Studies of basement rocks occurring in cratonic blocks and adjacent to continental margins provide the most direct means of elucidating the nature and extent of plate tectonic processes that influenced the local geodynamic evolution of such regions in the distant past. In the Appalachian orogen of eastern North America, Precambrian basement occurs as a series of discontinuous inliers that collectively preserve evidence of the cratonic foundation on which the mountain belt was developed and the Precambrian activity that affected the margin during an earlier plate tectonic cycle. The Blue Ridge in Virginia includes the largest area of exposed Laurentian crust in the Appalachians and, together with other inliers, preserves important evidence bearing on geologic processes involved in Mesoproterozoic amalgamation of Rodinia. Basement rocks in this region are dominantly meta-igneous in origin, with crystallization ages corresponding to the 1.3–1.0 Ga interval of the Grenville orogenic cycle (McLelland et al., 1996). This paper presents results of an integrated field and geochemical study of basement rocks occurring in the Blue Ridge of northern Virginia. The discussion focuses on the timing of Grenville-age events, possible implications of compositional constraints on likely magmatic sources, and the tectonic role of granitoids and related rocks in the geodynamic evolution of the Grenville orogen. These rocks provide considerable insight into the nature and extent of local Grenville-age orogenesis and serve as a basis for correlation of events recorded by basement in neighboring inliers. The local understanding and regional correlation of crustal evolution derived from such studies provide critical benchmarks for models of cratonic configurations and plate interactions that accompanied global-scale Grenville orogenesis.

GEOLOGICAL EVOLUTION OF THE BLUE RIDGE

The Blue Ridge Province is one of a series of inliers that expose Laurentian basement within thrust sheets of the Appalachians (Rankin et al., 1989a) (Fig. 1). In the central Appalachians, the Blue Ridge Province consists of two massifs: the Shenandoah, located mostly in Virginia, and the French Broad, which extends across western North Carolina to southwestern Virginia (Fig. 1). These massifs are separated by a narrow, fault-bounded sheet of Neoproterozoic clastic rocks in southwestern Virginia (Rankin et al., 1989a,b). The Blue Ridge in Virginia is an allochthonous fault-bounded composite thrust sheet that was transported northwestward during Permian-Carboniferous Alleghanian orogenesis (Hatcher, 1989). The geologic history of the Blue Ridge includes >1.0 b.y. of tectonic activity and has been shaped by processes responsible for creation and destruction of both the Mesoproterozoic Grenville and Paleozoic Appalachian orogens.

Mesoproterozoic igneous and metamorphic rocks constituting the Blue Ridge Basement Complex in Virginia (Virginia Division of Mineral Resources, 1993) form the core of the Blue Ridge anticlinorium, a regional, northeast-trending and plunging Alleghanian structure that is overturned toward the northwest (Fig. 2). Basement consists largely of massive to weakly foliated plutonic rocks of granitoid composition that intruded lithologically diverse, variably foliated gneisses recrystallized at upper amphibolite- to granulite-facies conditions (Bartholomew and Lewis, 1984; Sinha and Bartholomew, 1984). Following an extended interval of nearly 250 m.y., during which deep crustal portions of the Grenville orogen were locally uplifted and exposed at the surface, the area underwent two episodes of Neoproterozoic regional extension. The initial episode (765–700 Ma) was accompanied by intrusion of numerous A-type granitoids and local eruption of compositionally similar volcanic rocks, but did not result in continental fragmentation (Tollo and Aleinikoff, 1996); the later episode (570 Ma) produced abundant mafic volcanism and ultimately resulted in creation of Iapetus (Badger and Sinha, 1988; Aleinikoff et al., 1995). The predominantly

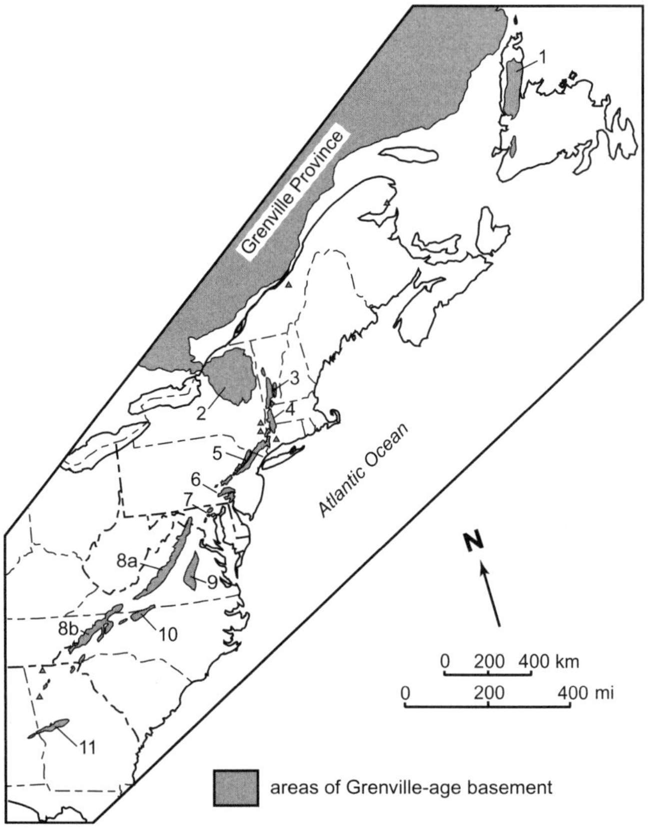

Figure 1. Locations of major occurrences of Grenville-age basement in eastern North America. Labeled areas: 1—Long Range massif; 2—Adirondack massif; 3—Green Mountains massif; 4—Berkshire massif; 5—Reading prong–New Jersey Highlands–Hudson Highlands; 6—Honey Brook upland; 7—Baltimore Gneiss antiforms; 8—Blue Ridge Province, including Shenandoah (8a) and French Broad (8b) massifs; 9—Goochland terrane; 10—Sauratown Mountains anticlinorium; 11—Pine Mountain belt. Modified after Rankin et al. (1989b).

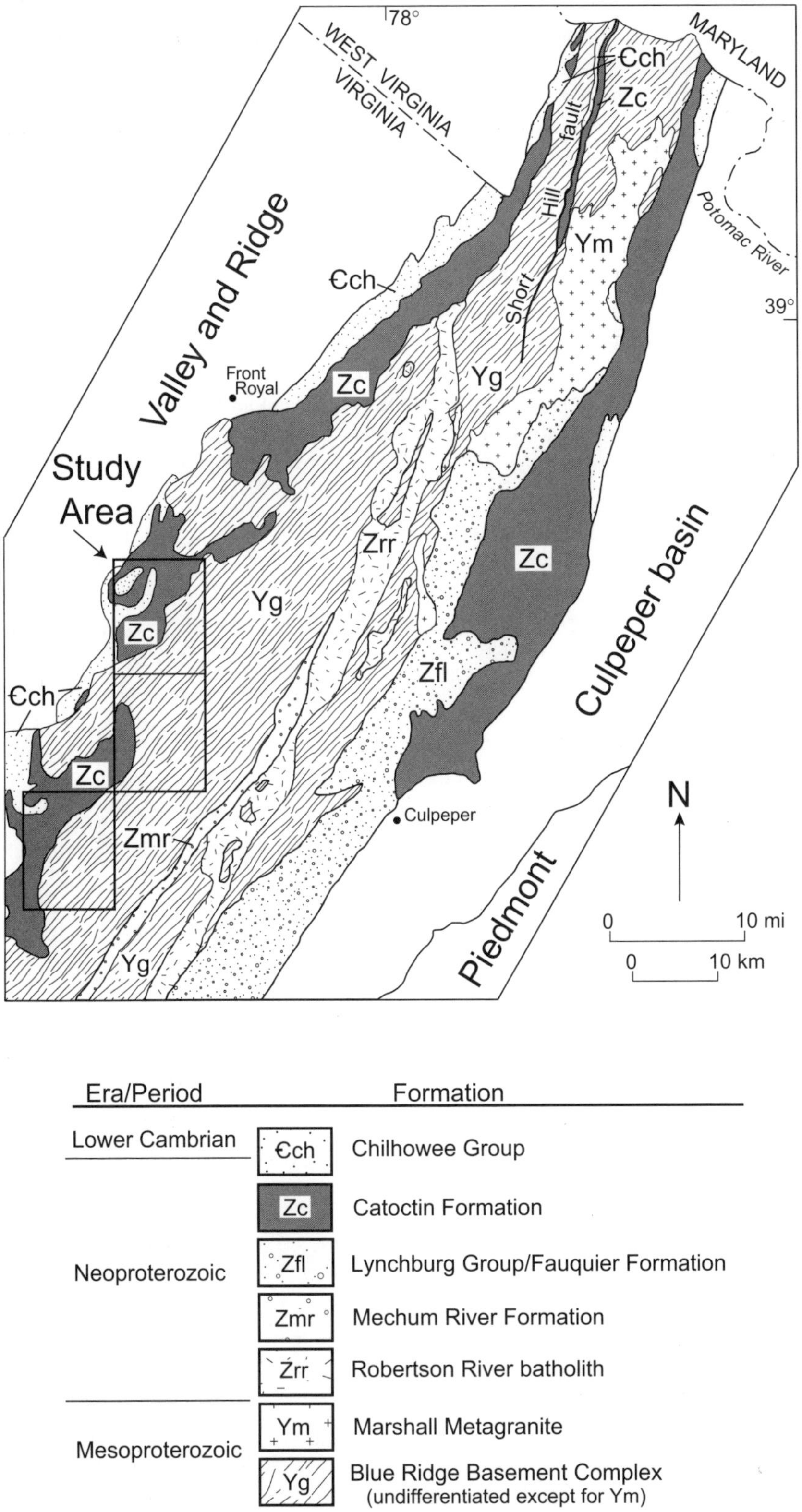

Figure 2. Generalized geologic map of the Blue Ridge anticlinorium in Virginia, showing location of quadrangles included in Figure 3. In the Grenville-age basement, only the Marshall Metagranite has been distinguished. Modified after Virginia Division of Mineral Resources (1993).

basaltic rocks of the Neoproterozoic Catoctin Formation and associated extension-related sedimentary strata produced during the later episode form a cover sequence nonconformably overlying the high-grade Blue Ridge basement and collectively produce a regional outcrop pattern that defines the anticlinorium (Fig. 2). These rocks are, in turn, overlain by Neoproterozoic to Cambrian clastic sedimentary strata of the Chilhowee Group deposited on the developing passive margin of Laurentia and marking the rift-to-drift transition in regional sedimentation patterns (Simpson and Eriksson, 1989). Repeated amalgamation of island arcs and other composite terranes, followed by final closure of Iapetus, led to a series of Paleozoic orogenies that culminated with Permo-Carboniferous Alleghanian activity (Hatcher, 1989). These events affected the Blue Ridge by producing retrograde greenschist-facies mineral assemblages (Evans, 1991), resetting some isotopic systems (Aleinikoff et al., 1995), and developing tectonic cleavage in mineralogically suitable rocks throughout the anticlinorium (Mitra and Elliott, 1979).

PREVIOUS STUDIES OF BLUE RIDGE BASEMENT ROCKS

Basement rocks within the Shenandoah massif include mostly meta-igneous plutonic rocks of variable composition, including anorthosite, and pervasively deformed gneisses that range from mafic to felsic in composition (Bartholomew and Lewis, 1984; Rankin et al., 1989a). Studies of basement rocks involving detailed field mapping of wide areas have been undertaken in central and northern portions of the Shenandoah massif (Bartholomew and Lewis, 1984; Sinha and Bartholomew, 1984; Aleinikoff et al., 2000 and references included therein; this study). Within each area, investigators have recognized an association of older, generally pervasively deformed, gneissic rocks that constitute country rock intruded by one or more generations of typically massive to weakly deformed plutons.

In the central Shenandoah massif, south of the field area of this study shown in Figure 3, two sequences of basement litholo-

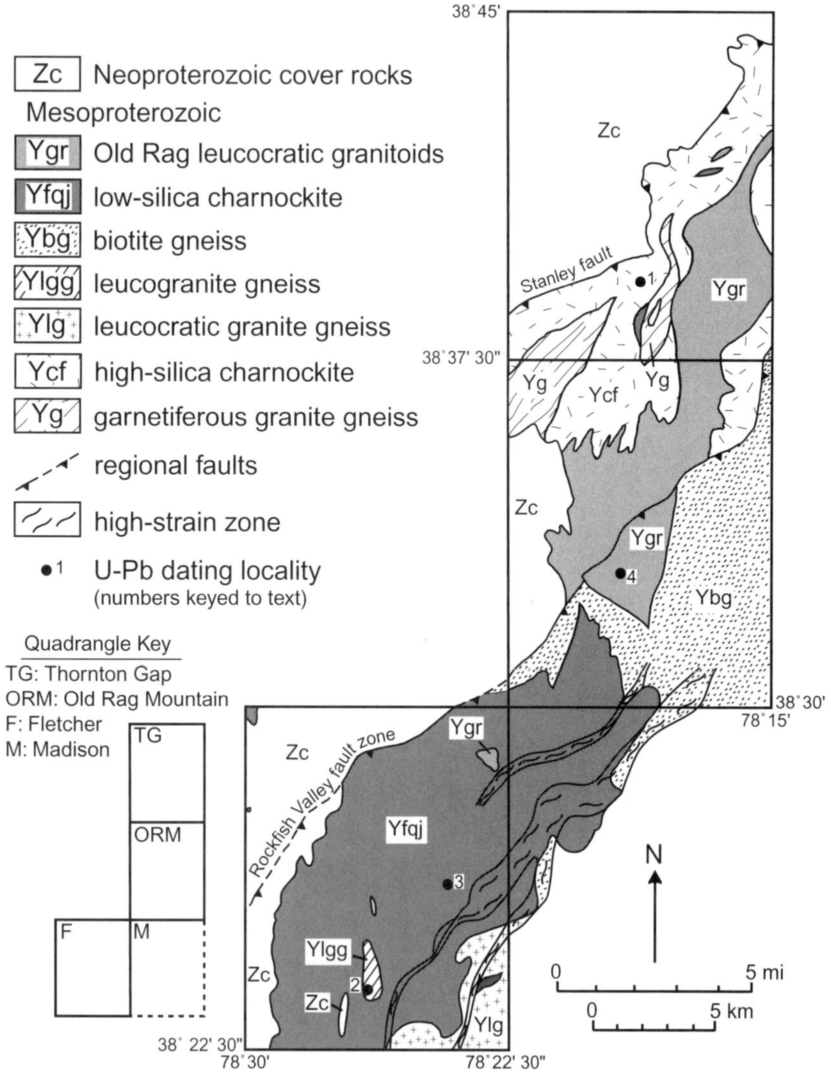

Figure 3. Geologic map of basement rocks in the study area (quadrangles with solid borders). Geologic relations for adjoining Madison quadrangle from Bailey et al. (2003).

gies were recognized by Bartholomew and Lewis (1984) east and west of the Rockfish Valley fault (Fig. 3), a narrow, northeast-trending zone of ductile deformation that separates lithologically distinct blocks of Mesoproterozoic crustal rocks within southern and central sections of the massif (Bartholomew et al., 1981; Sinha and Bartholomew, 1984). East of the fault, where basement recrystallized at upper amphibolite- to lower granulite-facies conditions during Grenville orogenesis, the generally biotite-bearing Stage Road Layered Gneiss constitutes the most areally extensive Mesoproterozoic country-rock type. These rocks were intruded by plutons of the Archer Mountain Suite composed primarily of biotite quartz monzonite and younger, massive charnockite (sensu lato). The alkalic Roseland Anorthosite Complex, an association of anorthosite, gabbroic sheets, and nelsonite bodies, also occurs in this area and intrudes the lithologically diverse Border Gneiss, which separates the complex from surrounding rocks (Sinha and Bartholomew, 1984). West of the fault in this area, where basement rocks display mineralogical evidence of crystallization at higher temperature and pressure granulite-facies conditions, Mesoproterozoic country rocks are dominated by the orthopyroxene (opx) ± garnet-bearing Lady Slipper and Nellysford granulite gneisses (Bartholomew and Lewis, 1984). Mapping indicates that these deformed rocks were intruded by abundant opx-bearing plutons of the Pedlar River Charnockite Suite and related bodies. Although field mapping has demonstrated the relative ages and sequence of events characterizing these rocks, modern isotopic analyses employing sensitive high-resolution ion microprobe (SHRIMP) techniques on zircons and other appropriate U-bearing minerals have not been undertaken and, as a result, a precise chronology of events is not yet available. Bartholomew et al. (1981) proposed that Mesoproterozoic basement in central Virginia defines two crustal blocks that were juxtaposed during middle Paleozoic thrusting along the Rockfish Valley fault zone. According to these authors, the crustal blocks are distinguished by contrasting rock associations that reflect differences in petrologic and structural evolution. However, mineral-chemical studies by Evans (1991) support an alternative hypothesis, suggesting that lithologic differences characterizing basement in this area resulted from differing responses to amphibolite-facies metamorphism that likely occurred during pre-Middle Ordovician burial. In this model, basement rocks throughout the entire central Shenandoah massif experienced similar granulite-facies conditions; however, early Paleozoic retrograde reactions were more advanced east of the Rockfish Valley fault zone by a combination of (1) enhanced fluid flow resulting from an inferred greater degree of crustal extention and fracturing, (2) possibly higher heat flow, and (3) the presence of a thick prism of overlying late Precambrian sediments that overlay the eastern block of Mesoproterozoic basement.

Basement rocks extending over a large area of the anticlinorium core in northern Virginia and Maryland were mapped by the U.S. Geological Survey (McDowell and Milton, 1992; Burton et al., 1995; Southworth, 1994, 1995; Burton and Southworth, 1996; Southworth and Brezinski, 1996). These studies indicate that the area is underlain by a diverse array of pervasively deformed gneisses and variably deformed granitoids, including charnockite (sensu lato). Ages obtained by U-Pb isotopic analyses of zircon and monazite, utilizing both isotope-dilution thermal ionization mass spectrometry (ID-TIMS) and SHRIMP techniques, indicate that the rocks define three age clusters (Aleinikoff et al., 2000). Oldest rocks (ca. 1153–1140 Ma; Group 1) include generally strongly foliated granitic to monzonitic gneisses and less strongly foliated granite and charnockite that occur mostly as inliers within younger plutons. Weakly to moderately foliated bioitite granitoid assigned to the Marshall Metagranite (Espanshade, 1986) is widespread in the northern Blue Ridge anticlinorium (Fig. 2) and has been dated by ID-TIMS techniques (two samples) at ca. 1111 Ma (Group 2) (Aleinikoff et al., 2000). No rocks intermediate in age between Group 1 and the Marshall Metagranite are recognized in the area. A distinctly younger series of rocks includes compositionally and texturally diverse granitoids and gneisses. A subset of these rocks includes massive to strongly foliated quartz-plagioclase gneiss and massive to weakly foliated garnet-bearing metagranite, both of which crystallized at ca. 1077 Ma (Group 3A) (Aleinikoff et al., 2000). Slightly younger rocks in the group include gneiss, metagranite, and leucocratic granitoids with crystallization ages ranging from ca. 1060 to 1055 Ma (Group 3B). Northwest-trending (D1) foliation that developed in most of the Group 1 rocks is cut by massive to weakly foliated garnetiferous granite (Group 3A), suggesting that D1 deformation occurred in the time interval 1145–1077 Ma. A second period of deformation (D2), recorded in the rocks of groups 2 and 3, has a maximum age of 1055 Ma and was followed by D3 folding. Differences in the geochronologic sequence of events recorded by rocks located east and west of the northeast-trending Short Hill fault (Fig. 2) in this area suggest that significant displacement occurred along the fault during Grenville-age orogenesis.

BASEMENT ROCKS OF THE STUDY AREA

The study area is located in the core of the Blue Ridge anticlinorium in north-central Virginia (Fig. 2). The area was chosen because of the diversity of rock types exposed in the region and because previous 1:100,000-scale mapping showed that coarse-grained leucocratic granitoids of Mesoproterozic age intruded coarse-grained charnockitic rocks, a field relationship that was consistent with the relatively imprecise isotopic ages available at the time (Tilton et al., 1960; Gathright, 1976; Lukert, 1982).

Many of the rocks described in this paper bear primary opx and are interpreted as igneous in origin, thus meeting the requirements of LeMaitre et al. (1989) for use of the special nomenclature applied to charnockitic rocks. Such terms include "charnockite" (sensu strictu, opx-bearing syenogranite), "farsundite" (opx-bearing monzogranite), "opdalite" (opx-bearing granodiorite), "mangerite" (opx-bearing monzonite), and "jotunite" (opx-bearing monzodiorite). However, many of these terms are used sporadically in the geologic literature and the general term "charnockite" (sensu lato) is sometimes applied to opx-bearing rocks of widely varying bulk composition (e.g., Kilpatrick and Ellis, 1992). In this paper, the more widely used general nomenclature (e.g., "opx monzogranite") is employed to describe the relevant rocks. Special terms suggested by LeMaitre are given in parentheses following first usage. The term "charnockitic" is used to refer to all opx-bearing rocks of inferred igneous origin.

Field Relations

Basement rocks investigated in this study occur in three 7.5-minute quadrangles located in the north-central Blue Ridge (Fig. 2). The study area includes a diverse assemblage of generally meta-igneous basement lithologies and several significant ductile fault zones, including a northeast-trending segment of the northern Rockfish Valley fault zone that bifurcates the area (Fig. 3). Results from geologic mapping, petrologic studies, and U-Pb isotopic analyses of zircon indicate that basement rocks in this region define a polyphase sequence of deformation and igneous emplacement spanning nearly 100 m.y. The oldest rocks located west of the Rockfish Valley fault zone include strongly foliated, medium- to coarse-grained opx + garnet monzogranite (farsundite) gneiss (Yg in Fig. 3) that occurs only as elongate inliers within younger plutonic rocks. The inliers are interpreted as large screens included within younger granitic plutons that intruded and isolated blocks of the monzogranite gneiss. Mapping indicates that the inliers are enclosed within moderately to strongly foliated, coarse- to very coarse–grained opx ± amphibole syenogranite and monzogranite (charnockite and farsundite; hereafter referred to as "high-silica charnockite"; Ycf), which forms a large plutonic body that only occurs west of the Rockfish Valley fault zone. The high-silica charnockite pluton is, in turn, cut by dikes and sheets of massive to moderately foliated, fine- to medium-grained, biotite ± garnet ± opx granite and leucogranite (Ygr) that occur on both map- and outcrop-scale. These leucocratic rocks can be physically traced and correlated compositonally to the similarly leucocratic, typically very coarse-grained Old Rag Granite that underlies Old Rag Mountain, located 2 km east of the Rockfish Valley fault zone (U-Pb dating locality number 4 in Fig. 3). Hackley (1999) collectively designated these leucocratic granitoids as the Old Rag magmatic series.

Strongly foliated, medium- to coarse-grained, biotite granite and granodiorite gneiss (collectively termed "biotite gneiss," Ybg) constitutes probable Mesoproterozic country rock located east of the Rockfish Valley fault zone (Fig. 3). In contrast to basement rocks exposed west of the Rockfish Valley fault zone, the pervasively deformed biotite gneiss does not contain evidence of original primary opx (Hackley, 1999). Field relations suggest that biotite gneiss is intruded by the Old Rag magmatic series and by rocks forming a pluton composed of massive to weakly foliated, coarse- to very coarse–grained opx ± amphibole monzogranite and quartz monzodiorite (farsundite and quartz jotunite; hereafter referred to as "low-silica charnockite"; Yfqj). This typically dark-green, plutonic rock is similar in appearance to the high-silica charnockite located west of the Rockfish Valley fault zone, but can be distinguished in the field by the characteristically abundant opx and lower concentrations of modal quartz. Field mapping indicates that the low-silica charnockite locally includes meter-scale segregations of leucocratic granitoid and pegmatite that typically contain abundant amphibole and magnetite. A second gneissic unit occurs as map-scale inliers and meter-scale xenoliths within low-silica charnockite east of the Rockfish Valley fault zone. This distinctive, coarse- to very coarse-grained biotite + muscovite leucogranite gneiss (Ylgg) is pervasively deformed and displays a pronounced foliation defined by ovoid coarse-grained feldspar porphyroclasts that, in individual xenoliths, is truncated by surrounding low-silica charnockite (Fig. 4). The relatively undeformed low-silica charnockite is locally intruded by dikes of massive to moderately foliated, medium- to coarse-grained biotite ± garnet ± opx granite and leucogranite (Ygr in Fig. 3) that is similar in appearance and composition to leucocratic granitoids recognized west of the fault zone. Field relations thus indicate that these leucocratic granitoids constitute the youngest Mesoproterozoic rocks presently recognized on both sides of the fault zone.

Mineralogy

The coarse grain size, locally pervasive weathering, porphyroblastic and phenocrystic texture, and strongly anisotropic fabric that characterize many of the lithologic units investigated in this study resulted in wide variations in modal data that were not indicative of the more limited mineralogic heterogeneity observed in many units in the field. Moreover, the wide variation in modal data was inconsistent with the more limited range of bulk compositions indicated for most rocks by major-element chemical analyses and calculated normative compositions. Lithologic nomenclature therefore was determined by using normative data obtained from the least weathered rocks and applying the chemical equivalents proposed by Streckeisen and Le Maitre (1979) for the standard International Union of Geological Sciences (IUGS) quartz-alkali feldspar-plagioclase ternary diagram (Fig. 5). The granitic normative compositions, typical hypidiomorphic textures, characteristic feldspar porphyroblasts, cross-cutting field relations, and local development of pegmatitic segregations suggest that most of the rocks examined in this study are igneous in origin. Moreover, normative compositions for all granitic units plot to the high-pressure, low-pH_2O side of the ternary minimum in the system quartz-albite-orthoclase (not shown), supporting the likelihood of an igneous origin. Perthitic alkali feldspar (dominantly microcline) containing locally abundant patch exsolution and plagioclase containing closely spaced, lenticular exsolution lamellae occur in all rocks, with alkali

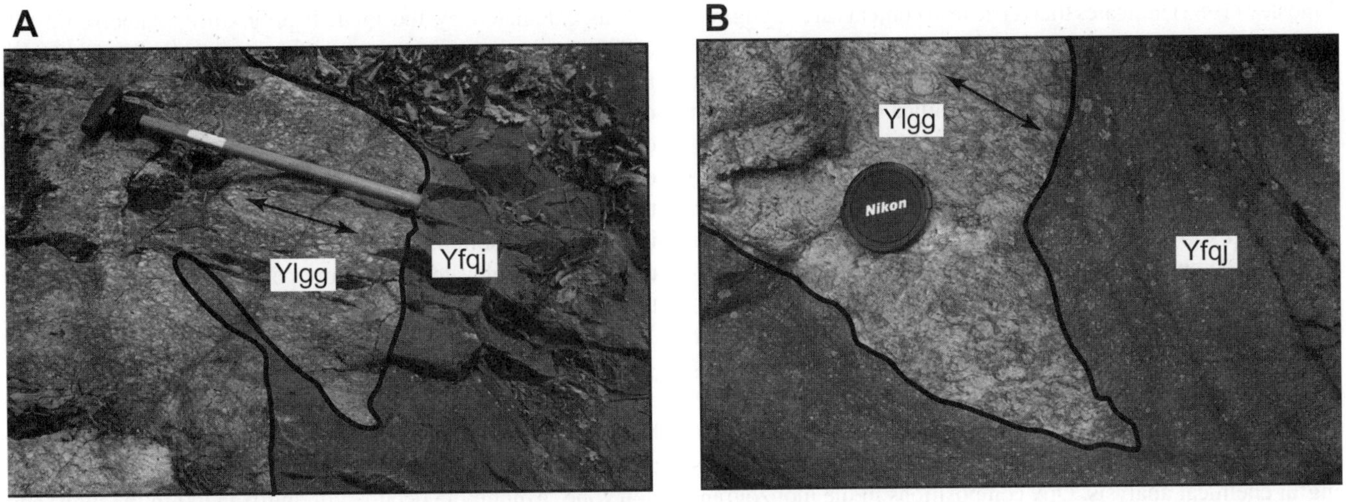

Figure 4. (A) Fragmented xenolith of leucogranite gneiss (Ylgg) intruded by low-silica charnockite (Yfqj). Hammer handle is 82 cm long. (B) foliation defined by aligned feldspar porphyroclasts within leucogranite gneiss xenolith truncated at contact with nonfoliated low-silica charnockite. Camera lens cap is 6.5 cm in diameter. Arrows in each panel are oriented parallel to foliation.

feldspar dominant in rocks of felsic bulk composition and plagioclase dominant in more mafic varieties. In most rocks, the feldspars form a textural network of euhedral to subhedral grains with quartz and ferromagnesian minerals occurring within interstices. Prominent porphyroblastic feldspars are characteristic of most rocks, especially the charnockites (Ycf and Yfqj in Fig. 5) and leucogranite gneiss (Ylgg).

Ferromagnesian minerals, including garnet, biotite, and opx, constitute <4 modal % of the granite and leucogranite (Ygr in Fig. 5) of the Old Rag magmatic series (Hackley, 1999). Biotite is the dominant ferromagnesian mineral in the granite and granodiorite gneiss (Ybg) and an accessory in other units. However, in most cases, biotite displays textural evidence of varying degrees of recrystallization, probably resulting from Paleozoic retrograde metamorphism. Opx occurs with garnet in the monzogranite gneiss, and with amphibole and rare clinopyroxene in the low- and high-silica charnockites. Microprobe analyses were used to further characterize the amphiboles and pyroxenes, because these ferromagnesian silicate minerals serve to distinguish the largely granitic rocks and preserve evidence bearing on the igneous origins of the protoliths.

Pyroxenes

All pyroxenes analyzed in this study display strong Quad (Ca-Fe-Mg) compositional features, according to the criteria of Morimoto (1988). Application of the charge-balance criteria of

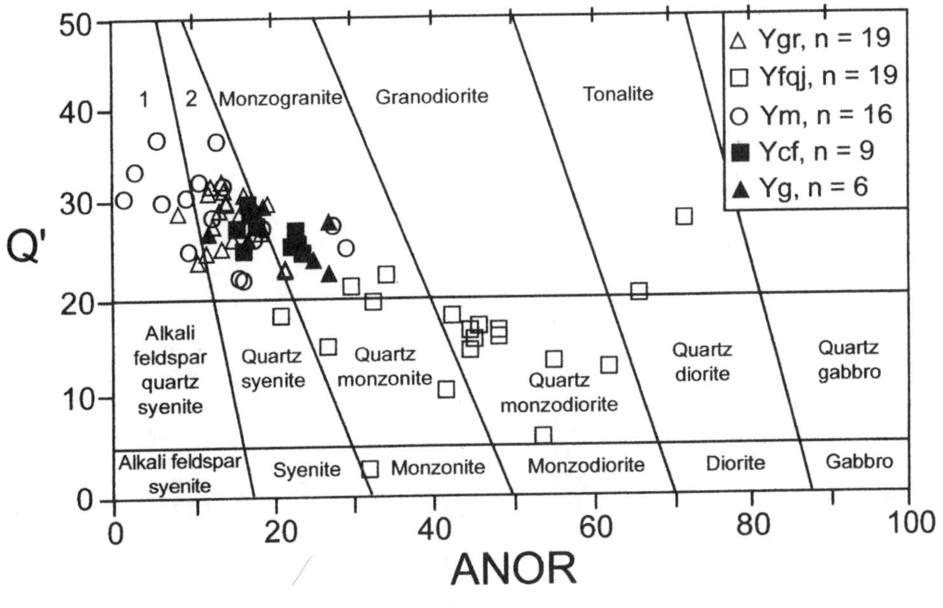

Figure 5. Normative (anorthite/[orthoclase + anorthite]) × 100 versus normative (quartz/[quartz + orthoclase + albite + anorthite]) × 100 (ANOR versus Q′). Normative compositions calculated with $Fe^{2+}/Fe_{total} = 0.9$. 1—Alkali feldspar granite; 2—syenogranite; Ycf—high-silica charnockite; Yfqj—low-silica charnockite; Yg—garnet monzogranite gneiss; Ygr—granite and leucogranite of the Old Rag magmatic series; Ym—Marshall Metagranite. After Streckeisen and Le Maitre (1979).

Lindsley (1983) indicates that ferric iron contents are negligible. Pyroxene compositions show very little variation both within individual grains and between grains in single thin sections. Clinopyroxenes in the high- and low-silica charnockites differ primarily in their Fe/Mg ratios, with pyroxenes in the high-silica charnockite markedly more Fe-rich (average $Wo_{43}En_{22}Fs_{35}$) than those occurring in the low-silica charnockite (average $Wo_{42}En_{28}Fs_{30}$) (Table 1). Optically resolvable exsolution is rare. Opx compositions in the charnockitic rocks are characterized by consistently low values of wollastonite component. Pyroxenes in the high-silica charnockite are significantly more Fe-rich (average $Wo_2En_{27}Fs_{71}$) than those occurring in fine-grained charnockitic dikes (average $Wo_2En_{37}Fs_{61}$) associated with the mineralogically similar, low-silica charnockite pluton. Opx compositions in the latter are typically too altered for meaningful chemical analysis. Opx compositions in the monzogranite gneiss are compositionally similar ($Wo_1En_{28}Fs_{71}$) to those in the high-silica charnockite. Optically resolvable exsolution in opx is rare in all rocks.

Compositions of coexisting pyroxenes in high-silica charnockite (Ycf) indicate equilibration temperatures of <500 °C, as calculated by the QUILF program of Andersen et al. (1993) for pressures consistent with observed granulite-facies mineral assemblages. Because such temperatures fall hundreds of degrees below likely crystallization temperatures for igneous charnockites (Kilpatrick and Ellis, 1992) and because the pyroxenes exhibit evidence of thorough compositional homogenization, the rocks are inferred to have undergone an extended interval of subsolidus re-equilibration at crustal temperatures that remained elevated but below peak granulite-facies conditions.

Amphiboles

Amphiboles occurring in the charnockites are typically brown and closely associated with opx. Where in contact, opx and amphibole are locally separated by a narrow optical transition zone. Amphibole occurs with pyroxene(s) in the high- and low-silica charnockites, fine-grained charnockitic dikes that crosscut the low-silica charnockite, and in leucogranitic pegmatites that occur locally within the latter. Compositions of the amphiboles correspond to hornblende (sensu lato) according to the criteria of Deer et al. (1992) and to the calcic group defini-

TABLE 1. COMPOSITIONS OF PYROXENES IN CHARNOCKITIC ROCKS

	Sample number and map unit									
	96-1 Yg Opx	96-1 Yg Opx	96-10C Ycf Opx	96-10C Ycf Cpx	96-10C Ycf Opx	96-10C Ycf Cpx	97-46C Yfqj Cpx	97-46C Yfqj Cpx	97-43C Yfqj dike Opx	97-43C Yfqj dike Opx
SiO_2	47.69	47.71	48.01	49.99	48.83	50.01	50.36	50.12	49.66	49.67
TiO_2	0.11	0.08	0.09	0.17	0.10	0.19	0.24	0.21	0.12	0.10
Al_2O_3	1.78	1.54	0.97	1.14	0.67	1.37	1.91	1.82	0.81	0.75
Cr_2O_3	0.02	0.01	0.02	0.01	0.01	0.03	0.00	0.00	0.01	0.01
FeO	40.37	40.26	40.41	20.35	40.22	20.43	16.74	17.40	35.54	35.76
MnO	0.44	0.45	0.66	0.32	0.71	0.32	0.42	0.44	0.82	0.80
NiO	0.01	0.00	0.01	0.01	0.01	0.00	0.01	0.01	0.02	0.00
MgO	9.60	9.42	8.54	7.15	8.71	7.22	9.18	9.34	12.36	12.45
CaO	0.36	0.34	0.98	20.36	0.91	20.42	20.99	20.58	0.77	0.82
Na_2O	0.02	0.01	0.04	0.34	0.02	0.37	0.46	0.42	0.02	0.02
K_2O	0.02	0.02	0.08	0.03	0.04	0.05	0.03	0.03	0.02	0.02
Total	100.41	99.86	99.80	99.86	100.24	100.42	100.34	100.39	100.15	100.39
Cations calculated on the basis of 6 oxygen atoms										
Si	1.939	1.950	1.972	1.969	1.989	1.959	1.944	1.939	1.978	1.975
Ti	0.004	0.003	0.003	0.005	0.003	0.005	0.007	0.006	0.004	0.003
Al	0.085	0.074	0.047	0.053	0.032	0.063	0.087	0.083	0.038	0.035
Cr	0.000	0.000	0.000	0.000	0.000	0.000	0.000	0.000	0.000	0.000
Fe	1.373	1.376	1.388	0.670	1.370	0.669	0.540	0.563	1.184	1.190
Ni	0.015	0.016	0.023	0.011	0.024	0.011	0.014	0.014	0.028	0.027
Mn	0.000	0.000	0.000	0.000	0.000	0.000	0.000	0.000	0.001	0.000
Mg	0.581	0.574	0.523	0.420	0.529	0.422	0.528	0.538	0.734	0.738
Ca	0.016	0.015	0.043	0.859	0.040	0.857	0.868	0.853	0.033	0.035
Na	0.002	0.001	0.003	0.026	0.001	0.028	0.034	0.032	0.001	0.002
K	0.001	0.001	0.004	0.002	0.002	0.002	0.001	0.001	0.001	0.001
Total	4.016	4.010	4.006	4.015	3.990	4.016	4.023	4.029	4.002	4.006
Wo	0.8	0.8	2.2	44.1	2.0	44.0	44.8	43.6	1.7	1.8
En	29.5	29.2	26.8	21.5	27.3	21.6	27.3	27.5	37.6	37.6
Fs	69.7	70.0	71.0	34.4	70.7	34.4	27.9	28.8	60.7	60.6
Number of analyses	3	3	3	5	3	4	4	5	3	3

tion of Leake et al. (1997). Data from electron microprobe traverses of individual grains indicate little evidence of compositional zoning. Analyses of multiple grains in single samples indicate restricted compositional variation. Calculated ferric iron contents, which were estimated using the International Mineralogical Association–recommended procedure of adjusting the sum Si + Al + Cr + Ti + Fe + Mg + Mn to 13 and redistributing total iron according to charge-balance considerations (Leake et al., 1997; see also discussion in Rock and Leake, 1984), suggest low Fe^{3+}/Fe^{2+} ratios. Amphiboles in the high-silica charnockite have markedly lower $Mg/(Mg + Fe_{total})$ than do amphiboles in the low-silica charnockite (Table 2). Amphiboles in fine-grained charnockitic dikes that crosscut the low-silica charnockite are more highly magnesian and lower in Al than amphiboles in the host rock, whereas amphiboles from pegmatitic segregations occurring in the low-silica charnockite are more Fe-rich than those in the host rock.

Amphiboles in the Blue Ridge rocks display compositional similarities to amphiboles observed in charnockites from the calc-alkaline Louis Lake batholith in the Wind River Range of Wyoming, which were intepreted by Frost et al. (2000) to represent primary phases that crystallized from igneous melts. The similarity in the compositions of amphiboles occurring in the low-silica charnockite and in pegmatites that were probably derived by localized, in situ fractionation suggests that these amphiboles were likewise magmatic in origin, crystallizing as primary magmatic phases exhibiting slight Fe-enrichment in the more compositionally evolved pegmatites. The more magnesian compositions of amphiboles in the low-silica charnockite dikes probably reflect crystallization from relatively unevolved melts that subsequently intruded the main low-silica charnockite pluton. Amphiboles of likely igneous origin occur in charnockitic rocks of tholeiitic affinity from other locations, including (1) mangerites from the Lofoten-Vesterålen area of northern Norway (Malm and Ormaasen, 1978); (2) amphibole charnockite and differentiated pegmatites from the Bjerkreim-Sokndal layered intrusion in southwestern Norway (Duchesne and Wilmart, 1997); and (3) quartz gabbros, quartz monzogabbros, and quartz monzonites of the Bunger Hills in East Antarctica (Sheraton et al., 1992), suggesting that the occurrence of igneous amphiboles in charnockitic rocks is not uncommon.

Geochronology

The U-Pb isotopic ages obtained by SHRIMP techniques are reported for four basement units within the study area. Sam-

TABLE 2. COMPOSITIONS OF AMPHIBOLES IN CHARNOCKITIC UNITS AND ASSOCIATED PEGMATITE

	Sample number and map unit									
	96-4C Ycf	96-4C Ycf	96-10C Ycf	96-10C Ycf	97-46C Yfqj	97-46C Yfqj	97-43C Yfqj dike	97-43C Yfqj dike	97-47C Yfqj peg	97-47C Yfqj peg
SiO_2	41.95	41.73	42.01	42.09	42.62	42.16	43.74	43.97	42.20	41.99
TiO_2	2.10	2.10	1.59	1.59	1.82	2.02	1.65	1.51	2.28	2.22
Al_2O_3	9.81	9.94	10.62	10.51	10.46	11.06	10.00	9.86	10.43	10.54
Cr_2O_3	0.01	0.01	0.04	0.03	0.01	0.00	0.02	0.03	0.00	0.01
FeO	22.88	22.07	21.59	21.23	18.87	18.20	16.51	16.16	19.07	19.01
MnO	0.16	0.15	0.14	0.13	0.21	0.19	0.15	0.18	0.25	0.22
MgO	7.11	7.00	7.54	7.70	9.32	9.21	11.41	11.49	9.22	9.03
CaO	10.69	10.92	11.07	11.17	11.23	11.10	11.02	11.42	11.02	11.02
Na_2O	2.03	1.85	1.92	2.09	1.74	1.66	1.88	1.83	1.93	1.78
K_2O	1.79	1.79	1.72	1.53	2.05	2.13	1.90	2.03	2.00	2.08
F	1.39	1.28	0.93	1.07	1.26	1.23	1.78	1.49	1.29	1.38
Cl	0.08	0.08	0.13	0.13	0.30	0.31	0.05	0.05	0.33	0.35
O = F, Cl	–0.60	–0.55	–0.42	–0.48	–0.48	–0.48	–0.48	–0.48	–0.48	–0.48
Total	99.39	98.38	98.88	98.78	99.43	98.80	99.63	99.55	99.54	99.16
Cations calculated on the basis of 23 oxygen atoms										
Si	6.482	6.490	6.464	6.475	6.475	6.426	6.556	6.576	6.422	6.422
Ti	0.244	0.245	0.184	0.184	0.208	0.232	0.186	0.170	0.261	0.255
Al	1.786	1.823	1.927	1.906	1.873	1.987	1.768	1.739	1.871	1.900
Cr	0.001	0.002	0.005	0.004	0.001	0.000	0.002	0.004	0.000	0.001
Fe	2.957	2.871	2.779	2.731	2.398	2.320	2.069	2.021	2.427	2.431
Mn	0.021	0.020	0.018	0.017	0.028	0.024	0.019	0.023	0.032	0.029
Mg	1.638	1.624	1.730	1.765	2.112	2.093	2.550	2.563	2.092	2.060
Ca	1.770	1.820	1.825	1.842	1.828	1.813	1.769	1.830	1.797	1.806
Na	0.609	0.559	0.573	0.625	0.512	0.491	0.545	0.529	0.568	0.528
K	0.354	0.355	0.337	0.299	0.397	0.415	0.364	0.387	0.389	0.406
Total	15.862	15.809	15.842	15.848	15.832	15.801	15.828	15.842	15.859	15.838
Mg/(Fe + Mg)	0.36	0.36	0.38	0.39	0.47	0.47	0.55	0.56	0.46	0.46
Number of analyses	5	5	3	2	3	5	2	5	4	6

ples weighing ~50 kg were collected from outcrops selected on the basis of field, petrographic, and geochemical criteria. Zircon and monazite were extracted using standard mineral separation techniques, including crushing, pulverizing, Wilfley Table, magnetic separator, and heavy liquids. Individual hand-picked grains were mounted in epoxy, ground nearly to half-thickness using 1500-grit wet-dry sandpaper, and polished with 6-μm and 1-μm diamond suspension abrasive. All grains were photographed in transmitted and reflected light; zircon was imaged in cathodoluminescence (CL) and monazite was imaged in scanning electron microscope (SEM) backscatter. Isotopic analyses were done on the SHRIMP II at the Research School of Earth Sciences at Australian National University in Canberra, following the methods described in Compston et al. (1984) and Williams and Claesson (1987). The primary oxygen ion beam, operated at 4–5 nA, excavated a pit ~25 μm in diameter and 1 μm deep. The magnet was cycled through the mass stations six times per analysis. In every fourth analysis, elemental fractionation was corrected by analyzing a reference zircon of known age, either AS3 (1099-Ma gabbroic anorthosite from the Duluth Complex [Paces and Miller, 1993]) or R33 (419-Ma quartz diorite from the Braintree Complex, Vermont [R. Mundil, written commun., 1999; S. Kamo, written commun., 2001]). Raw data were reduced and plotted using the Squid and Isoplot/Ex programs of Ludwig (2001a,b). Ages were determined by calculating the weighted average of $^{207}Pb/^{206}Pb$ or $^{206}Pb/^{238}U$ ages (see later in this paper). Concentrations (Table 3) are believed to be accurate to ~10–20%. All errors for ages are calculated at the 95% confidence limit.

Data are reported for four lithologic units, including high-silica charnockite (Ycf), leucogranite gneiss (Ylgg), low-silica charnockite (Yfqj), and Old Rag Granite. The latter constitutes part of the Old Rag magmatic series which includes consanguineous leucocratic granites and leucogranites collectively labeled as "Ygr" in Figure 3. The high-silica charnockite and Old Rag Granite were dated previously by conventional methods involving multigrain samples (Tilton et al., 1960; Lukert, 1982). However, because zircons in nearly all of the rocks exhibit evidence of multiple age domains, we decided to re-analyze these rocks, along with the two other samples, utilizing the high spatial resolution of SHRIMP to distinguish multiple ages within single grains.

High-silica charnockite (sample SNP96–10)

A sample of high-silica charnockite was collected for isotopic analysis from the south portal of Mary's Rock tunnel in Shenandoah National Park (location 1 in Fig. 3). The roadcut is composed of coarse-grained, foliated charnockite that contains prominent feldspar megacrysts. Zircons from this rock define two populations based on external morphology: (1) euhedral, elongate prisms, typically exhibiting dipyramidal terminations and length-to-width ratios (l/w) of ~5–6; and (2) nearly equant grains (Fig. 6, A). Examination of the prismatic zircons in CL reveals distinctly zoned core regions surrounded by narrow, unzoned rims. The concentric pattern of oscillatory compositional zoning that is characteristic of most cores is interpreted to be the result of magmatic crystallization. The equant zircons contain patchwork zoning and are interpreted as products of metamorphic recrystallization. Results from thirteen analyses of cores within the prismatic grains yield a weighted average of the $^{207}Pb/^{206}Pb$ ages of 1159 ± 14 Ma, which we interpret as the age of igneous crystallization (Fig. 7, A). Analyses of equant grains and overgrowths on prismatic grains yield a composite age of 1052 ± 14 Ma, which probably corresponds to a period of thermal disturbance in the region. The $^{206}Pb/^{238}U$ ages for cores and overgrowths are 1147 ± 16 and 1055 ± 13 Ma, respectively, identical within uncertainty to the $^{207}Pb/^{206}Pb$ ages.

Leucogranite gneiss (sample SNP99–90)

Leucogranite gneiss was collected from outcrops recently exposed by catastrophic flooding of the Conway River north of Fletcher (location 2 in Fig. 3). Foliated, coarse-grained leucogranite gneiss, containing abundant feldspar porphyroblasts and displaying evidence of pervasive ductile deformation occurs as meter- to decimeter-scale xenoliths enclosed within massive to weakly foliated, low-silica charnockite. Foliation within the leucogranite xenoliths is truncated at the margins of individual blocks, where they are in contact with charnockite. Zircons from this rock are typically subhedral and prismatic with l/w ratios of 3–5 (Fig. 6, B). CL imaging indicates that grains typically include broad cores, characterized by concentric, oscillatory zoning surrounded by distinct, unzoned rims. Thirteen analyses of cores yield a weighted average of the $^{207}Pb/^{206}Pb$ ages of 1078 ± 9 Ma, which is interpreted as the crystallization age of the igneous protolith (Fig. 7, B). The weighted average of the $^{206}Pb/^{238}U$ ages is 1073 ± 21 Ma, in agreement with, but less precise than, the age calculated from the $^{207}Pb/^{206}Pb$ ages. Analyses of a limited number of overgrowths suggest two periods of metamorphism at 1028 ± 10 and 997 ± 19 Ma (Fig. 7, B).

Low-silica charnockite (sample SNP99–93)

Low-silica charnockite was collected for isotopic analysis from newly exposed outcrops (location 3 in Fig. 3), produced as a result of a series of landslides that occurred in Madison County in 1995. The rock at this location is slightly foliated, dark gray-green, coarse-grained charnockite that is correlated petrographically and geochemically with the massive charnockite that encloses xenoliths of leucogranite gneiss at the locality from which the latter unit was collected for dating (see the preceding paragraph). Zircons from the low-silica charnockite occur as euhedral to subhedral prismatic grains, characterized by l/w ratios of 4–5 (Fig. 6, C). CL imaging indicates the presence of cores displaying concentric, oscillatory zoning that is interpreted as reflecting compositional variation resulting from igneous crystallization. Unzoned rims around zoned cores typically

TABLE 3. SHRIMP U-TH-PB DATA FOR DATED SAMPLES

Sample number	Measured $^{204}Pb/^{206}Pb$	Measured $^{207}Pb/^{206}Pb$	Common ^{206}Pb (%)	U (ppm)	Th/U	$^{206}Pb/^{238}U$ (Ma)†	Error§ (Ma)	$^{207}Pb/^{206}Pb$ (Ma)	Error§ (Ma)	$^{238}U/^{206}Pb$†	Error§ (%)	$^{207}Pb/^{206}Pb$†	Error§ (%)
SNP96-10 high-silica charnockite, Mary's Rock tunnel, Shenandoah National Park, 38°39′10″N, 78°18′43″W													
SNP10-1.1c	0.001046	0.0868	1.39	126	0.44	1069.6	23.0	983	220	5.561	2.3	0.0719	10.8
SNP10-1.2c	0.000044	0.0787	0.07	742	0.29	1148.7	12.8	1148	10	5.127	1.2	0.0780	0.5
SNP10-2.1c	0.000367	0.0872	1.17	96	0.52	1129.4	19.5	1246	54	5.193	1.8	0.0820	2.7
SNP10-3.1m	0.000386	0.0818	0.87	151	0.39	1056.4	15.1	1103	72	5.604	1.5	0.0763	3.6
SNP10-4.1c	0.000300	0.0831	0.72	275	0.41	1121.3	15.2	1168	31	5.252	1.4	0.0789	1.6
SNP10-5.1c	0.000197	0.0802	0.19	416	0.38	1160.7	15.4	1131	32	5.076	1.4	0.0774	1.6
SNP10-6.1c	0.000183	0.0812	0.14	442	0.59	1197.2	15.3	1162	18	4.908	1.3	0.0786	0.9
SNP10-7.1c	0.000104	0.0789	0.06	681	0.53	1156.4	16.9	1132	22	5.096	1.5	0.0774	1.1
SNP10-8.1m	0.000180	0.0774	0.03	592	0.31	1123.3	19.3	1063	27	5.268	1.8	0.0748	1.3
SNP10-8.2c	0.000072	0.0806	0.31	1393	0.28	1148.6	17.9	1188	11	5.118	1.6	0.0796	0.6
SNP10-9.1m	0.000416	0.0782	0.48	308	0.40	1048.0	12.4	995	49	5.678	1.2	0.0723	2.4
SNP10-9.2m	0.000306	0.0795	0.50	360	0.36	1077.5	21.1	1072	39	5.498	2.0	0.0751	2.0
SNP10-10.1m	0.000578	0.0811	0.64	278	0.39	1088.8	17.5	1011	80	5.453	1.7	0.0729	4.0
SNP10-10.2c	0.000094	0.0791	0.30	1093	0.02	1109.5	13.2	1142	17	5.317	1.2	0.0778	0.9
SNP10-11.1m	0.000133	0.0755	0.31	220	0.46	1010.5	11.9	1030	25	5.887	1.2	0.0736	1.2
SNP10-12.1m	0.000140	0.0751	0.11	313	0.38	1046.3	13.2	1016	26	5.682	1.3	0.0731	1.3
SNP10-13.1m	0.000078	0.0749	n.d.	371	1.08	1069.0	14.3	1035	19	5.552	1.4	0.0738	1.0
SNP10-14.1c	0.000191	0.0803	0.23	296	0.66	1157.9	16.9	1137	24	5.087	1.5	0.0776	1.2
SNP10-15.1c	0.000144	0.0803	0.18	201	0.53	1166.4	13.9	1154	20	5.045	1.2	0.0783	1.0
SNP10-16.1c	0.000214	0.0815	0.48	131	0.53	1132.4	14.3	1159	37	5.201	1.3	0.0785	1.9
SNP10-17.1m	0.000160	0.0775	n.d.	280	0.36	1169.2	13.9	1074	32	5.051	1.2	0.0752	1.6
SNP10-18.1m	0.000285	0.0777	0.28	502	0.17	1077.5	23.7	1032	25	5.508	2.3	0.0736	1.2
SNP10-19.1c	0.000169	0.0818	0.34	283	0.56	1169.2	14.1	1182	27	5.026	1.2	0.0794	1.4
SNP10-20.1m	0.000290	0.0806	0.47	222	0.50	1111.7	13.7	1108	34	5.314	1.3	0.0765	1.7
SNP10-21.1m	0.000025	0.0757	0.07	723	1.08	1072.1	17.9	1078	21	5.525	1.7	0.0754	1.0
SNP10-22.1m	0.000196	0.0764	0.32	418	0.85	1033.2	15.1	1031	29	5.753	1.5	0.0736	1.4
SNP10-23.1m	0.000067	0.0761	0.29	671	0.81	1033.3	12.0	1074	12	5.742	1.2	0.0752	0.6
SNP99-90 leucogranite gneiss, Conway River, 38°23′54″N, 78°26′16″W													
SNP-90-1.1c	0.000040	0.0752	0.07	161	0.48	1065.0	12.2	1059	17	5.568	1.2	0.0746	0.8
SNP-90-2.1c	0.000121	0.0756	0.21	144	0.47	1046.9	26.9	1039	24	5.673	2.7	0.0739	1.2
SNP-90-3.1c	0.000000	0.0750	0.00	140	0.28	1019.3	11.9	1069	17	5.825	1.2	0.0750	0.9
SNP-90-3.2r	0.000286	0.0780	0.49	784	0.25	1039.3	11.6	1039	29	5.716	1.2	0.0739	1.4
SNP-90-4.1r	0.000043	0.0728	0.07	396	0.38	1021.3	10.8	993	12	5.832	1.1	0.0722	0.6
SNP-90-5.1c	0.000067	0.0767	0.11	217	0.43	1118.0	46.3	1089	27	5.287	4.3	0.0758	1.4
SNP-90-6.1c	0.000015	0.0745	0.03	173	0.45	1073.0	16.3	1050	16	5.527	1.6	0.0743	0.8
SNP-90-6.2r	0.000003	0.0739	0.00	518	0.28	1016.3	10.7	1037	10	5.851	1.1	0.0738	0.5
SNP-90-7.1r	0.000042	0.0734	0.07	215	0.29	974.7	10.8	1009	20	6.117	1.2	0.0728	1.0
SNP-90-8.1r	0.000053	0.0731	0.09	192	0.24	991.3	11.1	996	25	6.015	1.2	0.0723	1.2
SNP-90-8.2r	0.000631	0.0831	1.07	1828	0.12	984.0	10.9	1046	40	6.048	1.1	0.0742	2.0
SNP-90-9.1c	0.000004	0.0761	0.01	190	0.44	1076.0	12.0	1095	14	5.500	1.2	0.0760	0.7

(continued)

TABLE 3. *Continued*

Sample number	Measured $^{204}Pb/^{206}Pb$	Measured $^{207}Pb/^{206}Pb$	Common ^{206}Pb (%)	U (ppm)	Th/U	$^{206}Pb/^{238}U$ (Ma)†	Error§ (Ma)	$^{207}Pb/^{206}Pb$ (Ma)	Error§ (Ma)	$^{238}U/^{206}Pb$†	Error§ (%)	$^{207}Pb/^{206}Pb$†	Error§ (%)
SNP-90-10.1c	n.d.	0.0772	n.d.	334	0.41	1144.6	31.3	1127	19	5.151	2.8	0.0772	0.9
SNP-90-11.1c	n.d.	0.0760	n.d.	234	0.41	1145.4	39.0	1095	21	5.155	3.5	0.0760	1.1
SNP-90-12.1r	0.000024	0.0734	0.04	549	0.35	993.2	13.3	1014	9	5.998	1.4	0.0730	0.5
SNP-90-13.1c	0.000006	0.0751	0.01	229	0.52	1086.7	11.9	1069	14	5.451	1.1	0.0750	0.7
SNP-90-13.2r	0.000003	0.0737	0.00	567	0.46	1045.6	11.2	1031	9	5.683	1.1	0.0736	0.4
SNP-90-14.1c	0.000005	0.0755	0.01	163	0.55	1030.9	11.8	1080	17	5.754	1.2	0.0754	0.9
SNP-90-15.1c	0.000001	0.0759	0.00	241	0.44	1092.4	16.4	1091	13	5.416	1.6	0.0758	0.7
SNP-90-16.1c	0.000007	0.0759	0.01	217	0.52	1116.7	13.0	1089	15	5.294	1.2	0.0758	0.7
SNP-90-17.1c	0.000043	0.0752	0.07	293	0.47	1098.2	11.9	1057	14	5.394	1.1	0.0746	0.7
SNP99-93 low-silica charnockite, Kinsey Run, 38°26′18″N, 78°24′3″W													
SNP-93-1.1c	0.000102	0.0764	0.17	108	0.92	1035.9	14.8	1067	31	5.729	1.5	0.0749	1.5
SNP-93-2.1c	0.000038	0.0752	0.06	265	0.36	1051.1	11.4	1059	13	5.645	1.1	0.0746	0.6
SNP-93-2.2r	0.000065	0.0739	0.11	252	0.34	1077.1	11.8	1014	18	5.514	1.1	0.0730	0.9
SNP-93-3.1c	0.000073	0.0749	0.12	69	0.64	1027.4	18.3	1038	27	5.785	1.8	0.0739	1.4
SNP-93-3.2r	0.000034	0.0740	0.06	265	0.35	1070.2	12.7	1029	17	5.548	1.2	0.0736	0.8
SNP-93-4.1c	0.000028	0.0762	0.05	242	0.78	1075.6	11.9	1088	15	5.504	1.1	0.0758	0.7
SNP-93-5.1r	0.000072	0.0743	0.12	274	0.36	1077.7	11.7	1022	14	5.509	1.1	0.0733	0.7
SNP-93-6.1c	0.000004	0.0740	0.01	793	0.23	1086.2	11.5	1040	8	5.460	1.1	0.0739	0.4
SNP-93-6.2c	0.000031	0.0751	0.05	255	0.37	1072.8	13.8	1059	17	5.526	1.3	0.0747	0.9
SNP-93-7.1c	0.000007	0.0746	0.01	292	0.52	1049.4	12.2	1056	13	5.655	1.2	0.0745	0.6
SNP-93-7.2r	0.000022	0.0745	0.04	323	0.36	1060.7	11.4	1046	14	5.595	1.1	0.0742	0.7
SNP-93-8.1c	0.000000	0.0767	0.00	185	0.75	1082.8	12.2	1113	14	5.460	1.2	0.0767	0.7
SNP-93-9.1c	0.000029	0.0745	0.05	125	0.89	1065.8	12.7	1044	23	5.568	1.2	0.0741	1.1
SNP-93-9.2r	0.000017	0.0747	0.03	229	0.35	1061.6	11.7	1055	15	5.588	1.1	0.0745	0.8
SNP-93-10.1r	0.000010	0.0748	0.02	296	0.34	1076.1	12.0	1060	12	5.508	1.2	0.0747	0.6
SNP-93-11.1c	0.000024	0.0735	0.04	386	0.34	1038.1	12.1	1017	15	5.728	1.2	0.0731	0.8
SNP-93-11.2c	0.000005	0.0746	0.01	1092	0.35	1054.8	10.8	1055	6	5.625	1.1	0.0745	0.3
SNP-93-12.1c	0.000111	0.0748	0.19	167	0.65	1135.2	13.0	1020	32	5.220	1.2	0.0732	1.6
SNP-93-12.2r	0.000029	0.0734	0.05	239	0.36	1064.0	12.2	1013	16	5.585	1.2	0.0729	0.8
SNP-93-13.1c	0.000073	0.0755	0.12	98	0.40	1091.3	17.2	1053	33	5.430	1.6	0.0744	1.7
SNP-93-14.1c	0.000097	0.0743	0.17	167	0.45	1070.2	13.7	1011	24	5.552	1.3	0.0729	1.2
SNP-93-15.1c	0.000113	0.0742	0.19	92	0.74	1107.3	14.8	1003	28	5.361	1.4	0.0726	1.4
SNP-93-15.2r	0.000048	0.0736	0.08	245	0.36	1070.4	11.7	1011	16	5.551	1.1	0.0729	0.8
OR97-35 Old Rag Granite, Old Rag Mountain, Shenandoah National Park, 38°33′6″N, 78°18′59″W													
OR97-1.1r	0.000196	0.0758	0.29	2423	0.03	1023.6	19.2	1014	21	5.813	1.9	0.0730	1.0
OR97-1.2x	0.000088	0.0793	n.d.	293	0.49	1185.6	23.0	1148	22	4.962	2.0	0.0780	1.1

OR97-2.1c	0.000014	0.0752	n.d.	1547	0.43	1081.4	20.2	1070	7	5.478	1.9	0.0750	0.4
OR97-3.1r	0.000014	0.0722	n.d.	3190	0.05	1032.8	19.5	986	9	5.767	2.0	0.0720	0.4
OR97-3.2x	0.000024	0.0807	0.11	394	0.21	1191.3	23.5	1205	29	4.923	2.0	0.0803	1.5
OR97-3.3x	n.d.	0.0854	n.d.	147	0.80	1341.1	36.9	1358	31	4.320	2.9	0.0869	1.6
OR97-4.1c	0.000011	0.0753	n.d.	5337	0.58	1165.2	21.7	1073	5	5.069	1.9	0.0752	0.3
OR97-5.1r	0.000081	0.0727	n.d.	4880	0.04	1035.0	19.3	974	8	5.757	1.9	0.0716	0.4
OR97-5.2r	0.000895	0.0842	1.51	4509	0.28	972.8	18.6	968	57	6.140	2.0	0.0714	2.8
OR97-6.1c	0.000002	0.0745	n.d.	9133	0.25	1172.7	21.8	1054	3	5.039	1.9	0.0745	0.1
OR97-7.1c	n.d.	0.0748	n.d.	971	0.32	1085.6	21.5	1077	11	5.454	2.1	0.0753	0.5
OR97-8.1x	0.000010	0.0762	n.d.	1373	0.51	1126.0	21.1	1096	9	5.247	1.9	0.0760	0.5
OR97-9.1c	n.d.	0.0749	n.d.	1520	0.33	1113.5	20.9	1070	13	5.314	1.9	0.0750	0.6
OR97-10.1r	0.000016	0.0728	n.d.	2539	0.16	1124.0	22.0	1002	18	5.278	2.0	0.0726	0.9
OR97-11.1r	n.d.	0.0732	n.d.	1138	0.18	1022.7	19.3	1021	9	5.817	2.0	0.0733	0.4
OR97-11.2c	0.000008	0.0750	n.d.	1825	0.45	1086.9	20.7	1064	7	5.450	2.0	0.0748	0.3
OR97-12.1c	0.000005	0.0746	n.d.	1912	0.42	1146.1	21.5	1057	7	5.160	2.0	0.0746	0.3
OR97-13.1c	0.000026	0.0747	n.d.	1482	0.32	1079.1	20.4	1051	12	5.494	2.0	0.0744	0.6
OR97-14.1c	0.000012	0.0742	n.d.	1938	0.20	1092.3	20.8	1042	22	5.428	2.0	0.0740	1.1
OR97-15.1c	0.000022	0.0748	n.d.	1251	0.60	1078.2	20.4	1054	9	5.499	2.0	0.0745	0.4
OR97-16.1r	n.d.	0.0760	n.d.	1876	0.18	1112.3	22.1	1095	11	5.314	2.1	0.0760	0.6
OR97-16.2r	0.000015	0.0728	n.d.	4900	0.03	1069.7	20.1	1004	10	5.556	1.9	0.0726	0.5
OR97-17.1r	0.000074	0.0738	n.d.	10256	0.22	1155.9	21.5	1007	6	5.126	1.9	0.0728	0.3
OR97-18.1r	0.000049	0.0744	n.d.	7240	0.24	1083.3	20.4	1034	5	5.477	2.0	0.0737	0.3
OR97-19.1c	0.000039	0.0757	0.31	719	0.42	1016.3	19.3	1073	13	5.841	2.0	0.0752	0.6
OR97-20.1c	n.d.	0.0747	n.d.	1170	0.64	1091.2	20.5	1066	9	5.428	1.9	0.0749	0.5
OR97-21.1c	0.000011	0.0747	n.d.	1233	0.61	1083.4	20.6	1056	9	5.471	2.0	0.0745	0.4
OR97mon-1.1r	0.000049	0.0741	n.d.	13124	10.65	1147.3	23.1	1026	7	5.162	2.1	0.0734	0.3
OR97mon-1.2r	0.000033	0.0738	n.d.	12433	15.16	1055.1	21.2	1022	7	5.632	2.1	0.0733	0.3
OR97mon-2.1c	0.000033	0.0753	0.17	13595	10.07	1038.5	26.2	1064	6	5.714	2.6	0.0748	0.3
OR97mon-2.2c	0.000034	0.0752	0.19	13545	11.46	1030.5	20.7	1060	16	5.761	2.1	0.0747	0.8
OR97mon-3.1c	0.000022	0.0746	n.d.	17146	7.70	1063.2	22.1	1049	8	5.580	2.2	0.0743	0.4
OR97mon-3.2c	0.000025	0.0747	0.29	22568	7.09	992.3	26.1	1050	18	5.994	2.7	0.0743	0.9
OR97mon-4.1c	0.000083	0.0760	0.21	19188	7.46	1047.4	22.2	1064	10	5.664	2.2	0.0748	0.5
OR97mon-4.2r	0.000020	0.0736	0.01	22498	7.18	1028.9	20.7	1023	6	5.780	2.1	0.0733	0.3
OR97mon-5.1r	0.000025	0.0712	n.d.	30717	7.20	1023.8	24.4	953	15	5.827	2.5	0.0709	0.7
OR97mon-5.2r	0.000021	0.0742	0.18	21853	7.56	1006.5	20.5	1040	12	5.910	2.1	0.0739	0.6
OR97mon-6.1r	0.000038	0.0732	0.06	16564	7.46	1006.5	21.0	1006	12	5.918	2.2	0.0727	0.6
OR97mon-6.2r	0.000024	0.0741	0.03	24660	7.53	1037.5	20.8	1035	5	5.727	2.1	0.0738	0.2

Note: Analytical sessions: SNP96-10, November 1997; OR97-35, April 1998; SNP99-90 and SNP99-93, November 2000. All samples were analyzed on SHRIMP II, Research School of Earth Sciences, Australian National University, Canberra. Analyses with suffix "mon" are monazite samples. Abbreviations: c—core; m—metamorphic; n.d.—not detected; r—rim; x—xenocryst.

†Corrected for common Pb. $^{206}Pb/^{238}U$ ages corrected for common Pb using the ^{207}Pb-correction method; $^{207}Pb/^{206}Pb$ ages and $^{207}Pb/^{206}Pb$ and $^{238}U/^{206}Pb$ corrected for common Pb using the ^{204}Pb-correction method.

§1σ error.

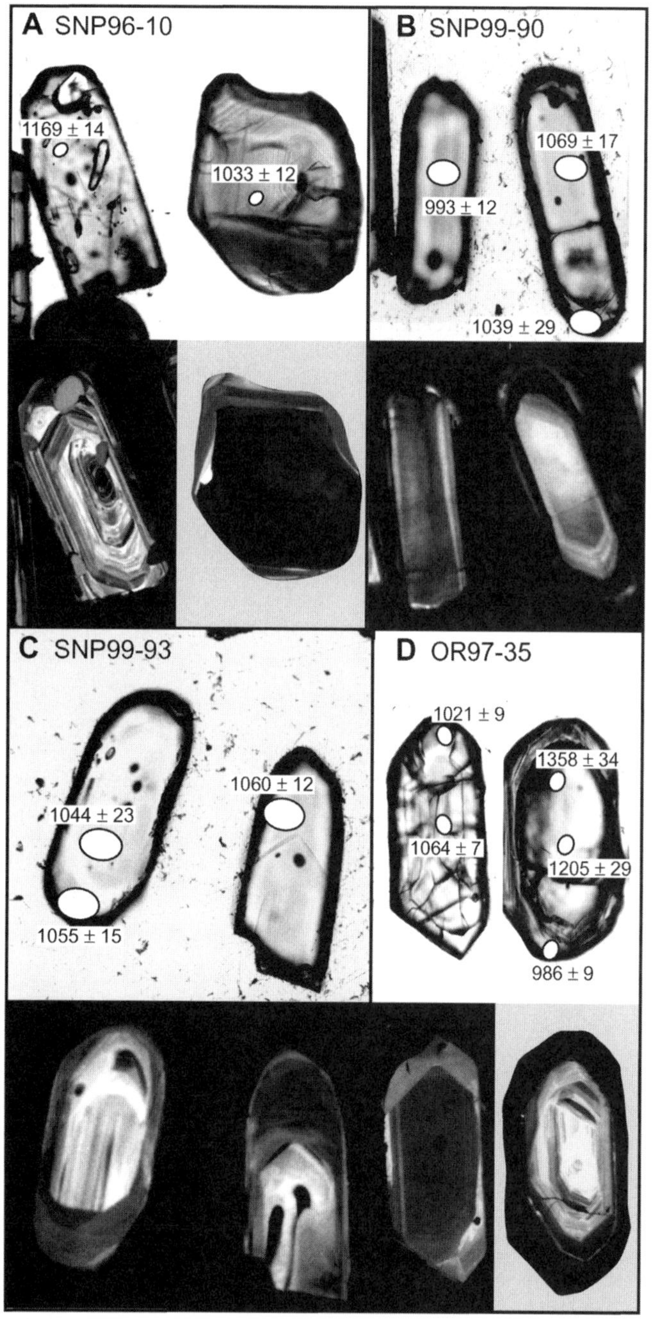

Figure 6. Images of zircons from various samples. White ellipses indicate locations of SHRIMP analyses and are ~25–30 microns in diameter and 0.5–1 micron deep. (A) Transmitted-light microscope image of high-silica charnockite (sample SNP96–10). (B) Transmitted-light microscope image of leucogranite gneiss (sample SNP99–90). (C) CL image of low-silica charnockite (sample SNP99–93). (D) CL image of Old Rag Granite (sample OR97–35).

contain higher U concentrations than is found in the cores. Eighteen analyses of cores yield a weighted average of the $^{207}Pb/^{206}Pb$ ages of 1050 ± 8 Ma, which is interpreted as the time of igneous crystallization of the charnockite (Fig. 7, C). The weighted average age using the $^{206}Pb/^{238}U$ ages is 1065 ± 12 Ma, within uncertainty of, but less precise than, the age calculated from the $^{207}Pb/^{206}Pb$ ages. Five analyses of the overgrowths indicate a subsequent episode of thermal flux, related to metamorphism at 1018 ± 14 Ma.

Old Rag Granite (sample OR97–35)

Coarse-grained, leucocratic Old Rag Granite was collected from near the summit of Old Rag Mountain (location 4 in Fig. 3) for isotopic analysis. The rock at this location is unfoliated and occurs within a mappable pluton of dominantly coarse-grained, leucocratic granite exposed east of the Rockfish Valley fault zone (Gathright, 1976; Hackley, 1999). Zircons from this rock are generally dark brown and include euhedral to subhedral prismatic grains with l/w ratios of 3–6 (Fig. 6, D). CL imaging indicates that nearly all of the zircons have oscillatory, zoned cores and dark overgrowths. Information from CL zoning patterns, combined with geochronology, result in the following age groups:

1. Oscillatory, zoned cores have a wide range of U contents (720–9130 ppm; Table 3) and Th/U ratios of 0.20–0.64, typical for zircons of igneous origin. As shown on Figure 7, D, several of these analyses are reversely discordant (i.e., $^{207}Pb/^{206}Pb$ age < $^{206}Pb/^{238}U$ age, plotting to the left of the concordia curve). Although this unusual age relationship could be due to U loss or Pb gain, it is unlikely that redistribution of U or Pb is responsible for the ages, given that the SHRIMP analyses were carried out on pristine, undisturbed areas of the zircons. A more likely cause, documented by Williams and Hergt (2000), is that, in high-U (greater than ~2500 ppm) zircons, the $^{206}Pb/^{238}U$ ages are too old, due to a machine-induced bias related to metamictization of the grains. For these types of zircons, we use the $^{207}Pb/^{206}Pb$ ages (which are not affected by the high-U bias) to calculate the age. Thirteen analyses of the oscillatory, zoned cores from zircons from the Old Rag Granite yield a weighted average of the $^{207}Pb/^{206}Pb$ ages of 1060 ± 5 Ma (Fig. 7, D), which we interpret as the time of emplacement of the Old Rag Granite and, by extension, the Old Rag magmatic series.
2. Unzoned, dark (in CL) areas overgrowing igneous cores have very high U contents (1138–10,256 ppm). Ratios of Th/U of 0.03–0.24 are lower than in the igneous cores, due to much lower Th contents, typical of metamorphic origin. Ages of overgrowths indicate two episodes of postmagmatic crystallization, occurring at 1019 ± 15 and 979 ± 11 Ma. These ages are similar to ages of overgrowths in zircons from the leucogranite gneiss and likewise are interpreted to result from metamorphic thermal events.
3. Five analyses of oscillatory zoned cores yield ages that are older than the cores interpreted to have formed during emplacement of the Old Rag magma at ~1060 Ma (Table 3). Two

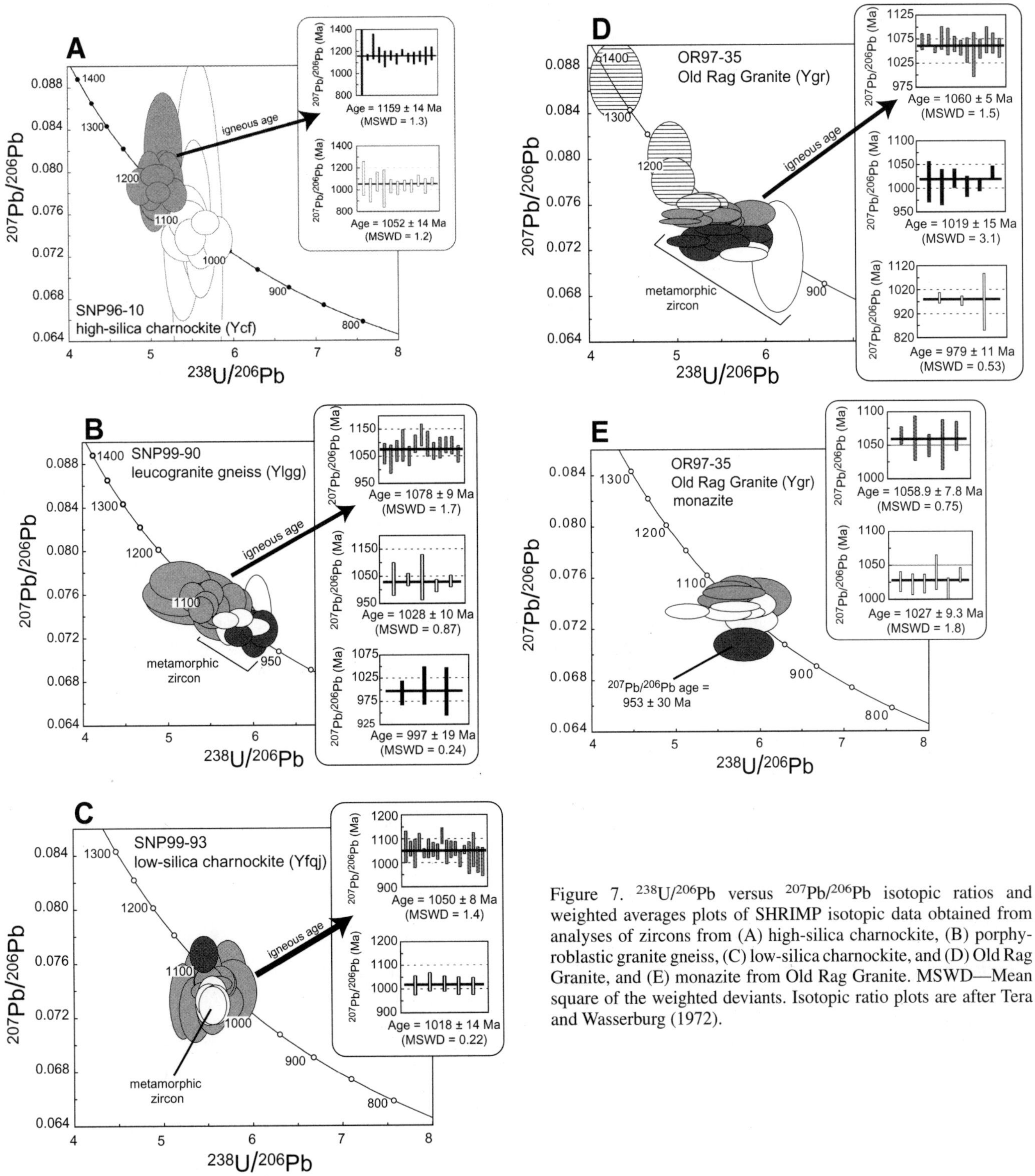

Figure 7. $^{238}U/^{206}Pb$ versus $^{207}Pb/^{206}Pb$ isotopic ratios and weighted averages plots of SHRIMP isotopic data obtained from analyses of zircons from (A) high-silica charnockite, (B) porphyroblastic granite gneiss, (C) low-silica charnockite, and (D) Old Rag Granite, and (E) monazite from Old Rag Granite. MSWD—Mean square of the weighted deviants. Isotopic ratio plots are after Tera and Wasserburg (1972).

analyses are ~1095 Ma and have U contents of 1373 and 1876 ppm. Three other analyses are 1148, 1205, and 1368 Ma and have distinctly lower U contents of 293, 394, and 147 ppm, respectively. All of these samples are interpreted as xenocrystic in origin.

Monazite from the Old Rag Granite was also dated by SHRIMP methods. Compositional zoning (displayed by back-scattered electrons imaging) suggested that many of the grains are composed of multiple-age components, similar to zircons from the same sample. A limited number of analyses from five monazite grains yielded results that agree with the zircon SHRIMP age data (Fig. 7, E). Five analyses yield an age of 1059 ± 8 Ma, identical within uncertainty to the zircon age of 1060 ± 5 Ma, which is interpreted as the time of emplacement of the Old Rag magma. Six other monazite analyses yield an age of 1027 ± 9 Ma, and one grain is 953 ± 30 Ma, both within uncertainty of the zircon overgrowth ages of 1019 ± 15 and 979 ± 11 Ma, respectively.

The occurrence of dikes of Old Rag-type leucocratic granitoids cutting low-silica charnockite at several localities in the study area indicates that, at least locally, rocks of the Old Rag magmatic series are younger than the low-silica charnockite. These field relationships and the new isotopic ages therefore suggest that these penecontemporaneous intrusives were likely emplaced during intervals that overlapped in both space and time.

In summary, igneous rocks were emplaced at 1159 ± 14 Ma (high-silica charnockite), 1078 ± 9 Ma (leucogranite gneiss), 1050 ± 8 Ma (low-silica charnockite), and 1060 ± 5 (Old Rag Granite). Overgrowths on magmatic zircons in the high-silica charnockite formed at ca. 1055 Ma, probably in response to thermal perturbations related to emplacement of the low-silica charnockite and Old Rag Granite. These younger granitoids were subsequently affected by Grenville thermal events at ca. 1020 and 980 Ma. The Old Rag Granite also contains xenocrystic zircons, some of which are several hundred m.y. older than igneous zircons that formed during emplacement of the granite. Thus, the previous multigrain TIMS age of 1115 Ma for the Old Rag (Lukert, 1982) is likely a composite age of several components, including cores (both igneous and inherited) and at least two ages of overgrowths, and is now superseded by the new SHRIMP data.

Geochemistry

The new isotopic ages presented in this paper indicate that granitoids in the study area correspond temporally to the Group 1 (1155–1140 Ma) and Group 3 (1077–1055 Ma) magmatic intervals defined by Aleinikoff et al. (2000) in the northern Blue Ridge. The following discussion will concentrate on the compositional systematics and petrogenesis of the most thoroughly sampled, demonstrably meta-igneous rocks in the study area. The Group 1 temporal equivalents discussed here include the monzogranite gneiss (Yg) and high-silica charnockite (Ycf); Group 3 equivalents include leucogranites (Ygr) of the Old Rag magmatic series and the low-silica charnockite (Yfqj). Additional data for the Marshall Metagranite, dated at 1112 ± 3 and 1111 ± 2 Ma (Group 2) by Aleinikoff et al. (2000), are also presented. These data were obtained from samples collected by the late W.G. Leo and made available by the U.S. Geological Survey. The Marshall Metagranite is one of the most thoroughly sampled basement units in the northern Blue Ridge province and provides important insight into the chemical evolution of magmatic rocks associated with the 1100-Ma episode of regional Grenville-age orogenesis.

Compositional Characteristics

Most elemental concentrations characterizing rocks examined in this investigation are interpreted to reflect igneous processes based on the following criteria: (1) well-defined major-element trends that are similar to those characterizing other charnockitic suites of inferred igneous origin (McLelland and Whitney, 1990; Kilpatrick and Ellis, 1992) (see Figs. 8 and 9); (2) trace-element concentrations and ratios that are similar to compositional data for granitic rocks of likely comparable crustal origin from other orogens (Pearce et al., 1984; Eby, 1992) (see Figs. 10–14); and (3) close correspondence in trace-element characteristics to compositions of younger, undeformed and either unrecrystallized or mildly recrystallized granitoids from other localities (Collins et al., 1982; Whalen and Currie, 1990) (see Fig. 12). The typically mild textural and mineralogical recrystallization and limited geochemical re-equilibration observed in the Blue Ridge rocks likely results from the inherently low water content of the granitoids and the observed tendency for fluid movement associated with Paleozoic retrograde processes to be channeled through faults, lithologic contacts, and pre-existing fractures (Wilson and Tollo, 2001). Malm and Ormaasen (1978), Sheraton et al. (1992), Zhou et al. (1995), and Duchesne and Wilmart (1997) have observed similar evidence of magmatic processes preserved in charnockitic rocks of Precambrian age and igneous origin from other orogenic belts.

Meta-igneous rocks of the study area and the Marshall Metagranite define two broad compositional groups, together spanning a wide range in composition (Fig. 8). The 1050 ± 8 Ma low-silica charnockite includes nearly all rocks with SiO_2 contents <66 wt%. In contrast, nearly all samples from the other basement units are characterized by SiO_2 contents >68 wt%. Considered in terms of temporal groupings, Group 1 (monzogranite gneiss and high-silica charnockite) and Group 2 (Marshall Metagranite) rocks are exclusively compositionally evolved. Group 3 rocks (low-silica charnockite and Old Rag magmatic series) define a broad range in composition spanning nearly the entire spectrum of analyzed samples. The low-silica charnockite exhibits considerable variation in composition but is separated from the coeval leucogranites of the Old Rag magmatic series by a compositional gap extending from 65 to 68% SiO_2. Kilpatrick and Ellis (1992) demonstrated that steep declines in TiO_2 and P_2O_5 with increasing silica content, trends here defined mainly by samples of the low-silica charnockite

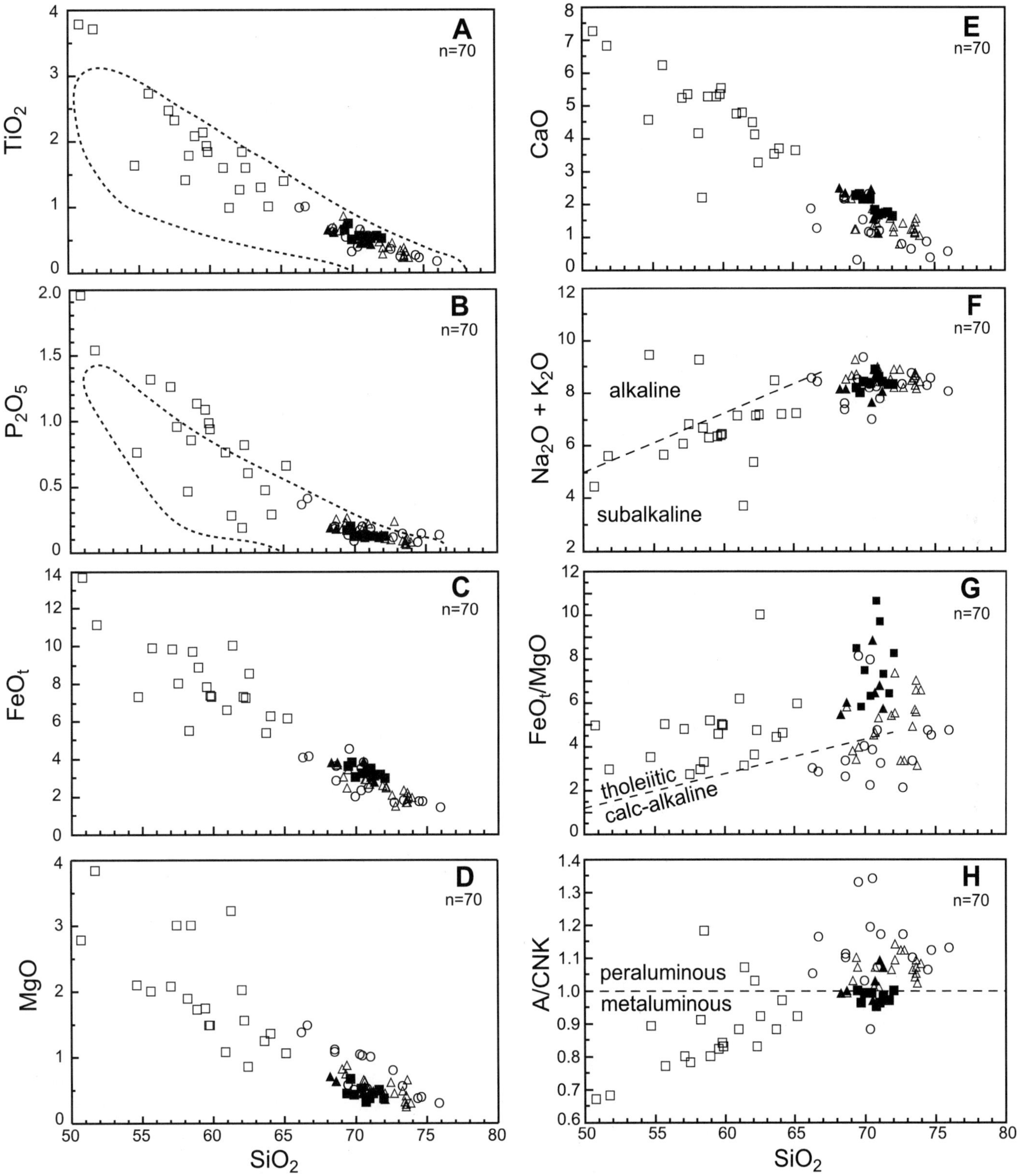

Figure 8. SiO_2 versus (A) TiO_2, (B) P_2O_5, (C) FeO_t, (D) MgO, (E) CaO, (F) $Na_2O + K_2O$, (G) FeO_t/MgO, and (H) aluminum saturation index (A/CNK = molar $Al_2O_3/[CaO + Na_2O + K_2O]$) for meta-igneous rocks of the study area and the Marshall Metagranite. All data are expressed in wt%. Filled triangle—garnet monzogranite gneiss; filled square—high-silica charnockite; open circle—Marshall Metagranite; open square—low-silica charnockite; open triangle—granite and leucogranite of the Old Rag magmatic series. FeO_t refers to total iron expressed as FeO. Dashed line in A and B encloses compositional field of igneous charnockites from Kilpatrick and Ellis (1992). Dashed line separating tholeiitic and calc-alkaline fields in G is from Miyashiro (1974). Dashed line separating alkaline and subalkaline fields in H is from Irvine and Baragar (1971).

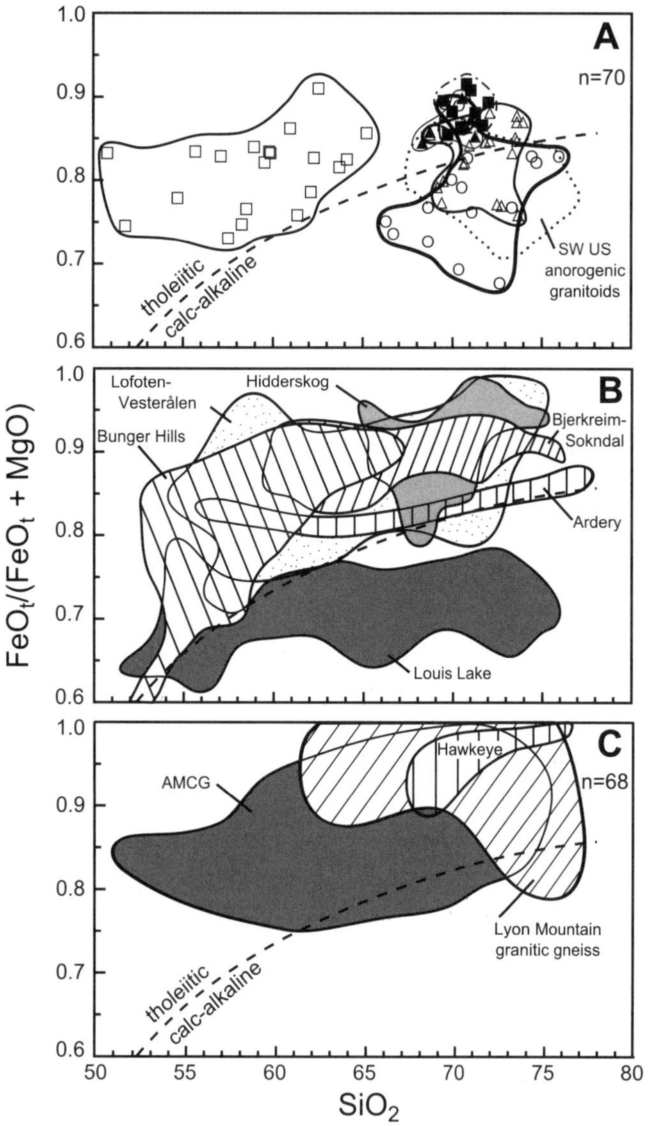

Figure 9. SiO_2 versus $FeO_t/(FeO_t + MgO)$ for rock of various regions. (A) Meta-igneous rocks of this study. Dotted field encloses compositions of peraluminous anorogenic granitoids of 1.33–1.44 Ga age from the southwestern United States (SW US) (Anderson, 1983). (B) Charnockitic suites from other locations, including the Hidderskog massif (Zhou et al., 1995) and Bjerkreim-Sokndal layered intrusion from southwest Norway (Duchesne and Wilmart, 1997), Lofoten-Vesterålen terrane of northern Norway (Malm and Ormaasen, 1978), Bunger Hills intrusions of East Antarctica (Sheraton et al., 1992), Ardery charnockitic intrusions of the Antarctic Windmill Islands (Kilpatrick and Ellis, 1992), and Louis Lake batholith from Wyoming (Frost et al., 2000). (C) Granitoids and granitic gneisses from the Adirondacks, including the 1155–1130-Ma AMCG suite, the 1103–1093 Ma Hawkeye suite, and the 1060–1045 Ma Lyon Mountain granitic gneiss (McLelland and Whitney, 1990; additional data courtesy of J. McLelland). Filled triangle—garnet monzogranite gneiss; filled square—high-silica charnockite; open circle—Marshall Metagranite; open square—low-silica charnockite; open triangle—granite and leucogranite of the Old Rag magmatic series. Dividing line separating fields of tholeiitic and calc-alkaline affinity is from Anderson (1983). Modified after Frost et al. (2000).

(Fig. 8, A and B), constitute compositional features characteristic of opx-bearing granitoids of igneous origin. In contrast, opx-bearing granitoids of metamorphic origin typically display less pronounced decreases in TiO_2 and P_2O_5 (Kilpatrick and Ellis, 1992). Characterized by steep slopes similar to those defined by opx-bearing rocks of inferred igneous origin by Kilpatrick and Ellis (1992), the rocks examined in this study are likewise considered to have magmatic protoliths. The biotite-bearing Marshall Metagranite and all of the garnet-bearing rocks, including the Group 1 monzogranite gneiss and Group 3 leucogranites, are moderately peraluminous with corundum-normative compositions (Fig. 8, H). The high-silica charnockite (Group 1) defines a relatively tight cluster of transitional to mildly peraluminous compositions, whereas the low-silica charnockite (Group 3) displays considerable variation, extending from strongly to mildly metaluminous compositions.

All of the rocks analyzed in this study are subalkaline (Fig. 8, F) and most exhibit Fe-enrichment characteristic of tholeiitic affinity (Figs. 8, G, and 9, A). Similar Fe-enrichment and tholeiitic affinity is characteristic of many other charnockitic suites, including the Hidderskog massif (Zhou et al., 1995) and Bjerkreim-Sokndal layered intrusion from southwest Norway (Duchesne and Wilmart, 1997), Lofoten-Vesterålen terrane of northern Norway (Malm and Ormaasen, 1978), Bunger Hills intrusions of East Antarctica (Sheraton et al., 1992), and Ardery charnockitic intrusions of the Antarctic Windmill Islands (Kilpatrick and Ellis, 1992) (Fig. 9, B). Granitoids and granitic gneisses from the Adirondacks—including the 1160–1130 Ma anorthosite-mangerite-charnockite-granite (AMCG) suite, 1103–1093 Ma Hawkeye suite, and 1060–1045 Ma Lyon Mountain granitic gneiss—exhibit similar tholeiitic characteristics (Fig. 9, C) (McLelland and Whitney, 1990; McLelland et al., 1996, 2001). Such charnockitic suites stand in contrast to the calc-alkaline Louis Lake batholith from the Wind River range in Wyoming, which is characterized by lower $FeO_t/(FeO_t + MgO)$ ratios, especially at moderate to high silica content (Fig. 9, B). The relatively low $FeO_t/(FeO_t + MgO)$ ratios characteristic of the Marshall Metagranite, which overlap data from the high-SiO_2 range of the Louis Lake batholith, illustrate the difficulty inherent in distinguishing tholeiitic versus calc-alkaline affinity in granitoids rocks characterized by high silica content and low FeO_t + MgO concentrations. Indeed, Anderson (1983) noted that compositions of geochemically comparable, mildly peraluminous granitoids of tholeiitic affinity and 1.44–1.31 Ga age from Arizona and New Mexico similarly indicated erroneous calc-alkaline affinity (Fig. 10, G, *in* Anderson, 1983) that was not substantiated by mineralogic, geochemical, or tectonic characteristics. In the absence of other corroborative evidence, as discussed below, the low $FeO_t/(FeO_t + MgO)$ ratios of the Marshall Metagranite are likewise interpreted to reflect the effects of source composition and conditions of magma emplacement rather than calc-alkaline affinity.

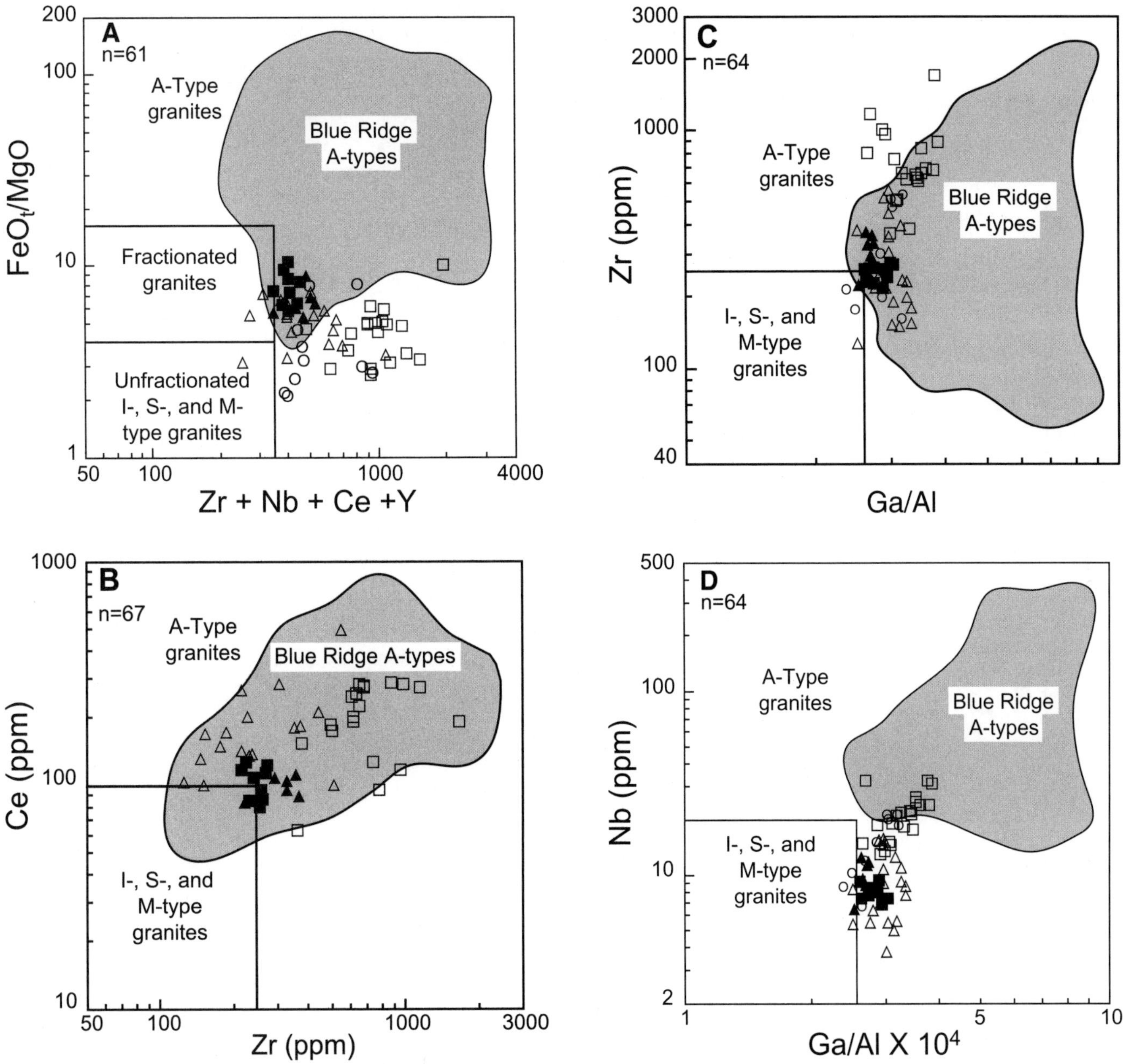

Figure 10. Comparison of meta-igneous rocks from the study area and Neoproterozoic A-type granitoids from the central Appalachians. (A) Zr + Nb + Ce + Y versus FeO$_t$/MgO; (B) Zr versus Ce; (C) Ga/Al × 10^4 versus Zr; (D) Ga/Al × 10^4 versus Nb. Filled triangle—garnet monzogranite gneiss; filled square—high-silica charnockite; open circle—Marshall Metagranite; open square—low-silica charnockite; open triangle—granite and leucogranite of the Old Rag magmatic series; shaded field—150 analyses from Tollo and Aleinikoff (1996) and Tollo et al. (2004). Modified after Whalen et al. (1987).

Granitoid Typology

The relatively high iron contents (Figs. 8, G and 10, A), enriched concentrations of some high-field-strength elements (Fig. 10), and elevated Ga/Al ratios (Fig. 10, C and D) that characterize the Blue Ridge meta-igneous rocks suggest compositional similarities to A-type granitoids. Indeed, A-type compositional affinities have been noted for other Precambrian charnockite-granite suites, including those from the Adirondacks (McLelland and Whitney, 1990) and southwestern Norway, where examples include the Hidderskog meta-charnockite intrusion (Zhou et al., 1995) and felsic charnockites of the Bjerkreim-Sokndal layered intrusion (Duchesne and Wilmart,

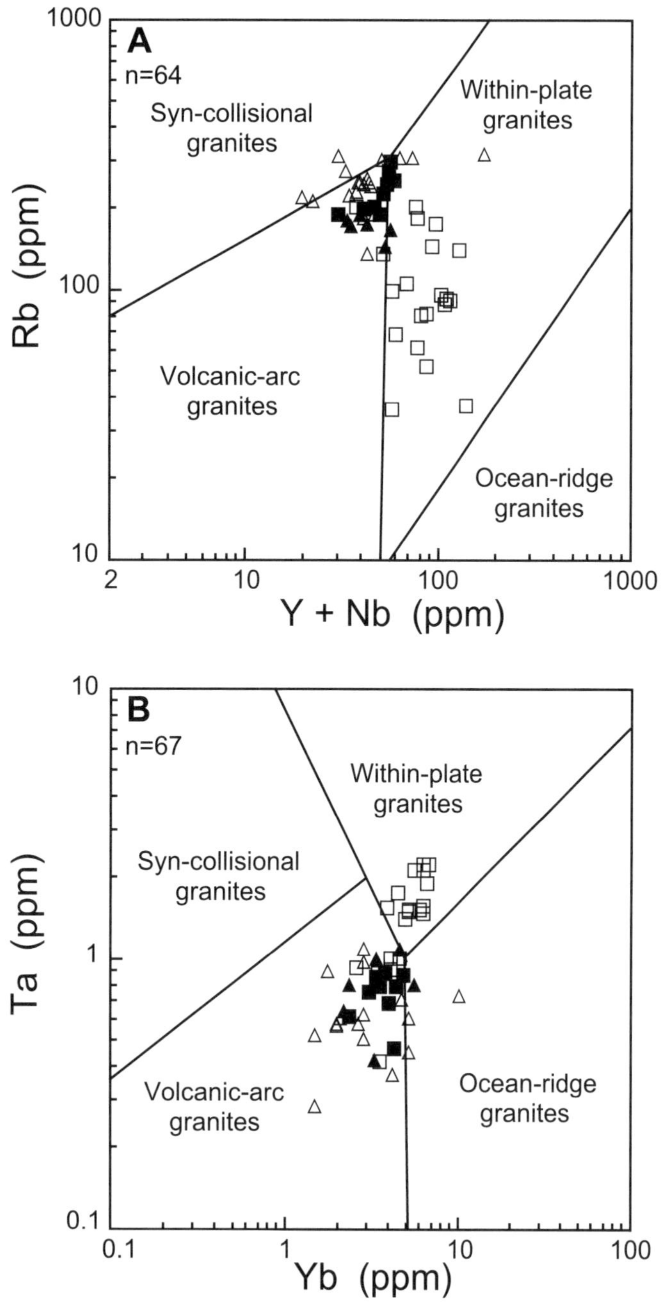

Figure 11. (A) Y + Nb versus Rb and (B) Yb versus Ta for meta-igneous rocks from the study area. Filled triangle—garnet monzogranite gneiss; filled square—high-silica charnockite; open circle—Marshall Metagranite; open square—low-silica charnockite; open triangle—granite and leucogranite of the Old Rag magmatic series. Modified after Pearce et al. (1984).

1997). All of the Blue Ridge rocks display the typical A-type features of high Zr + Ce contents and elevated Ga/Al ratios, relative to other granite types (Fig. 10, B–D). However, most of these rocks do not exhibit the very high FeO_t/MgO ratios (Fig. 10, A), high Nb contents (Fig. 10, D), low CaO (at $SiO_2 < 65$ wt%) contents (Table 4), low Sc concentrations (Table 4), and alkaline affinity (Fig. 8, F) typical of most A-type granitoids (Whalen et al., 1987; Eby, 1990). Moreover, unlike the majority of A-types, most Blue Ridge granitoids have trace-element compositions that plot across field boundaries in the geochemical discrimination diagrams of Pearce et al. (1984) (Fig. 11, A and B). Collectively, most of the Blue Ridge rocks display compositional features that are transitional between A-type and fractionated I-type granitoids. Only the low-silica charnockite consistently exhibits A-type characteristics. Other studies have illustrated the difficulty inherent in distiguishing granite typology in similar rocks, such as the Ardery Charnockitic intrusions from Antarctica (Creaser and White, 1991; Kilpatrick and Ellis, 1992). Landenberger and Collins (1996) demonstrated that, in the absence of diagnostic mineral assemblages, distiguishing between closely associated I-types and A-types is most efficiently accomplished at low-silica content, where I-types have characteristically higher MgO, CaO, and Sr concentrations and lower amounts of FeO, Ba, Zr, and Zn. Correspondingly, typological classification of the Blue Ridge rocks is hindered by the lack of silica-poor members within each intrusive unit other than the low-silica charnockite. Thus, the data presented in this study suggest A-type affinity for the low-silica charnockite but are equivocal for all other units, which are here provisionally classified as geochemically transitional between A-types and fractionated I-types.

Tectonic Setting and Source Characteristics

Previous studies have demonstrated that geochemical discrimination diagrams, such as Figure 11, A and B, more accurately and reliably reflect the compositions of magmatic sources than the type of tectonic environments (Pearce et al., 1984; Sylvester, 1989; Förster et al., 1997). Correspondingly, the Blue Ridge granitoids that plot in the volcanic-arc compositional field in Figure 11, A and B, are not considered to have formed within a subduction-related tectonic regime because they (1) do not display the marked decoupling of large-ion lithophile and high-field-strength elements that is characteristic of most subduction-related granitic magmatic systems (Pearce et al., 1984) (Fig. 12, A–C); (2) exhibit concentrations of high-field-strength elements similar to ocean-ridge granitoids, a characteristic of within-plate magmatic systems (Pearce et al., 1984); (3) have ages that indicate no direct temporal association with regional orogenesis; and (4) are not associated with voluminous rocks of granodioritic and dioritic composition, as is typical of many modern magmatic arcs (Condie, 1997).

Most of the Blue Ridge rocks plot in a region of the Pearce et al. (1984) diagrams (Fig. 11, A and B) that is characteristic of

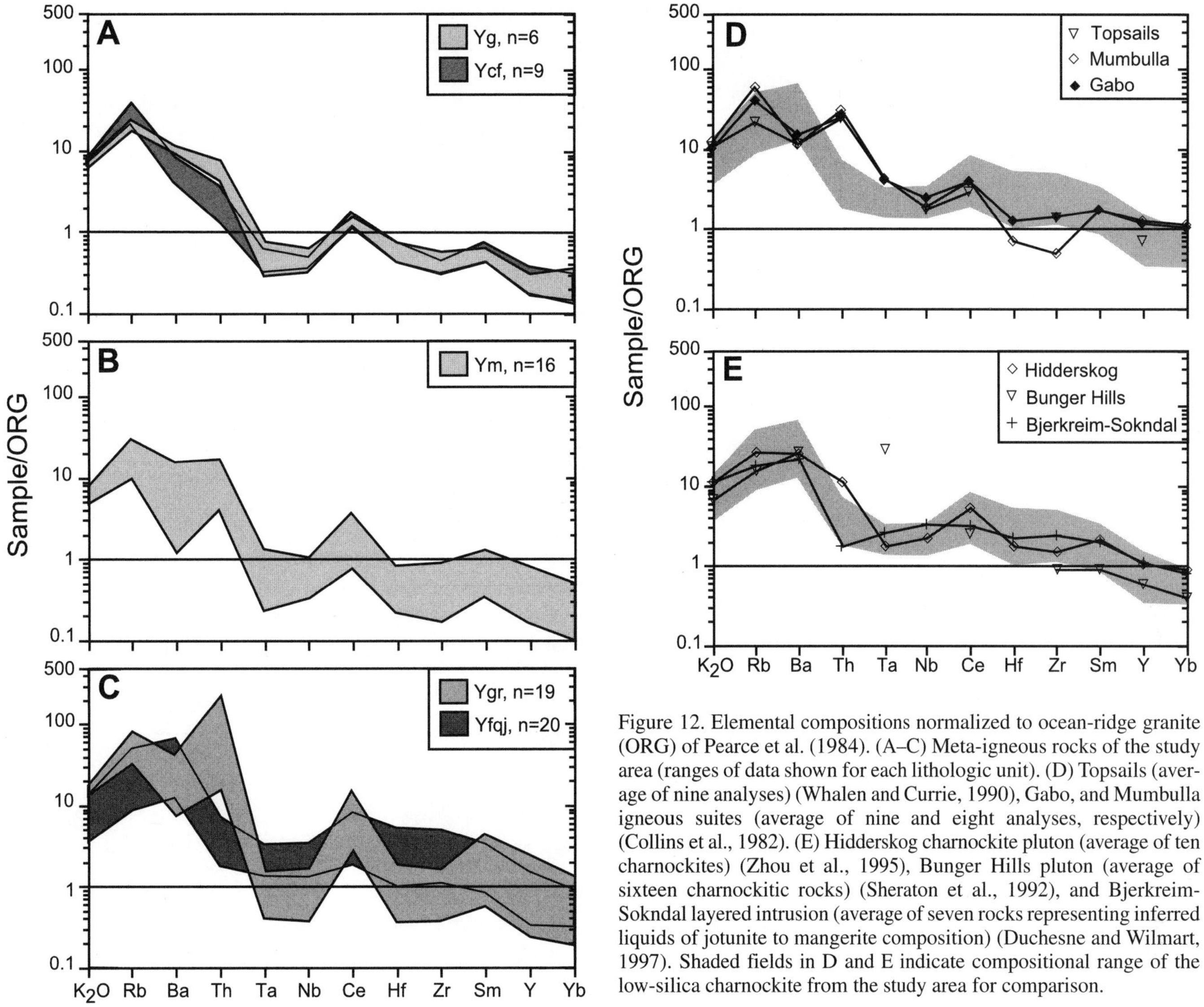

Figure 12. Elemental compositions normalized to ocean-ridge granite (ORG) of Pearce et al. (1984). (A–C) Meta-igneous rocks of the study area (ranges of data shown for each lithologic unit). (D) Topsails (average of nine analyses) (Whalen and Currie, 1990), Gabo, and Mumbulla igneous suites (average of nine and eight analyses, respectively) (Collins et al., 1982). (E) Hidderskog charnockite pluton (average of ten charnockites) (Zhou et al., 1995), Bunger Hills pluton (average of sixteen charnockitic rocks) (Sheraton et al., 1992), and Bjerkreim-Sokndal layered intrusion (average of seven rocks representing inferred liquids of jotunite to mangerite composition) (Duchesne and Wilmart, 1997). Shaded fields in D and E indicate compositional range of the low-silica charnockite from the study area for comparison.

granitoids emplaced in broadly defined postorogenic geologic settings (Sylvester, 1989; Förster et al., 1997). The relatively flat patterns defined by high-field-strength elements Hf through Yb for all Blue Ridge rocks (Fig. 12, A–C) indicate the influence of a crust-dominated source similar to that of the "within-plate" granitoids of Pearce et al. (1984). The Blue Ridge granitoids also display a marked trough for Ta and Nb in the ocean-ridge granite (ORG)-normalized diagrams. Similar troughs characterize most of the subduction-related "volcanic-arc" granites and arc-continent "collision" granites of Pearce et al. (1984) and thus appear to be a source-related characteristic resulting from involvement of arc crust in magma genesis. As noted by previous studies (Pearce et al., 1984; Maniar and Piccoli, 1989; Sylvester, 1989; Eby, 1990, 1992; Förster et al., 1997), granites linked to postorogenic processes and/or within-plate tectonic environments include both anorogenic and postorogenic types. Such suites commonly include A-type granitoids (Eby, 1992). The compositional signatures of such granitoids vary considerably and, especially for postorogenic types, typically reflect derivation from mixed sources (Sylvester, 1989; Eby, 1990, 1992; Förster et al., 1997). Eby (1992) used geochemical characteristics and inferred geologic settings to propose a twofold subdivision of A-type granitoids, including (1) characteristically peralkaline types that are generally associated with anorogenic tectonic environments (e.g., hot spots, plumes, and continental rift zones) and (2) typically metaluminous, postorogenic types emplaced at variable intervals following regional orogenesis. Differences in inferred sources for the two groups are manifested in trace-element compositions and ratios that indicate that anorogenic types (A_1) are most likely derived through differentiation

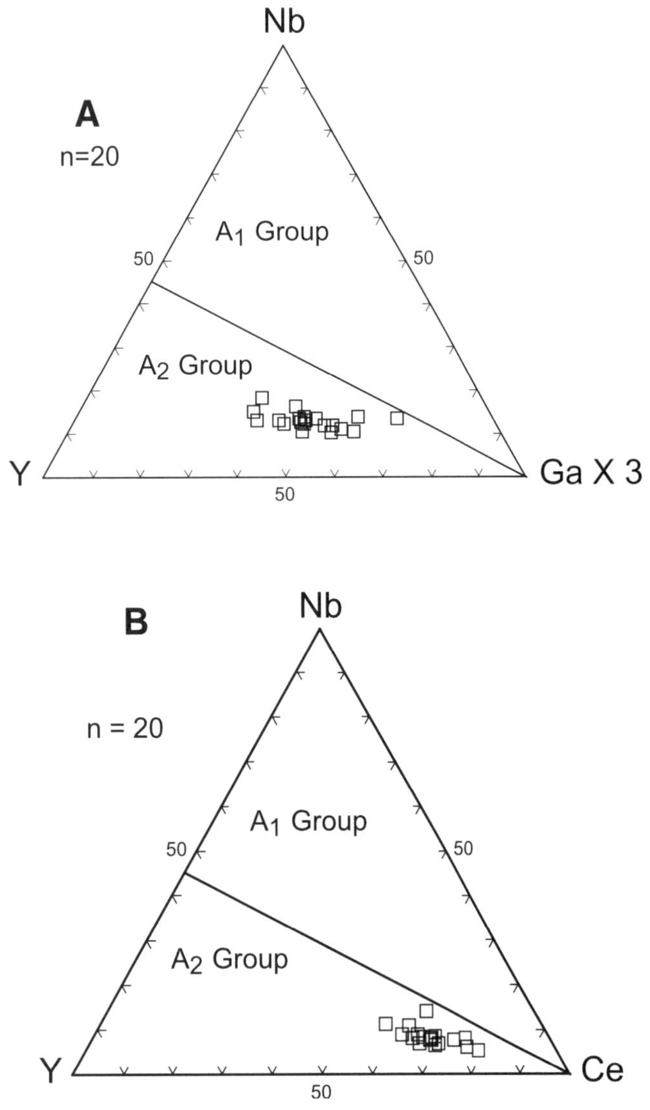

Figure 13. Ternary plots of (A) Nb-Y-3xGa and (B) Nb-Y-Ce for the low-silica charnockite from the study area, indicating compositional affinity to the A_2 group of A-type granitoids of Eby (1992). After Eby (1992).

of magmas produced from sources similar in composition to ocean-island basalts, whereas postorogenic (A_2) types are likely derived through melting of continental crust or underplated crust that has experienced a previous cycle of orogenic magmatism (Eby, 1992). Trace-element compositions and ratios presented in this study indicate that the low-silica charnockite corresponds geochemically to the A_2 group of Eby (1992) (Fig. 13) and that precursor magmas for all Blue Ridge rocks were likely derived from heterogeneous sources bracketed in composition by average continental crust and island-arc basalt (Fig. 14).

Although broadly correlative to Eby's A_2 group, the Blue Ridge rocks are characterized by concentrations of source-sensitive trace elements that are especially similar to those of the undeformed Paleozoic Topsails igneous suite from Newfoundland and the Gabo and Mumbulla suites from Australia (Fig. 12, D), each of which was included in Eby's original compilation of A_2 examples. The Late Ordovician to Silurian Topsails igneous suite of western Newfoundland (Whalen and Currie, 1990) consists of a diverse group of crustally derived, bimodal igneous rocks formed in an extensional regime developed in proximity to an active orogenic margin that was undergoing a protracted period of collisional tectonic events. The voluminous suite consists mainly of peraluminous to peralkaline, subsolvus to hypersolvus granites derived from partial melting of depleted continental crust (Whalen and Currie, 1990). The Late Devonian Gabo and Mumbulla suites of southeastern Australia (Collins et al., 1982) consist of metaluminous granitoids emplaced into the Lachlan Fold Belt following a period of extensive Early Devonian I-type granitic magmatism (Collins et al., 1982). These Australian A-type granitoids were derived through partial melting of lower crustal granulitic rocks that previously experienced an episode of melt production (Collins et al., 1982). The close correspondence in both concentrations and patterns of source-sensitive trace elements between the Topsails, Gabo, and Mumbulla suites (Fig. 12, D) and the Blue Ridge rocks suggests

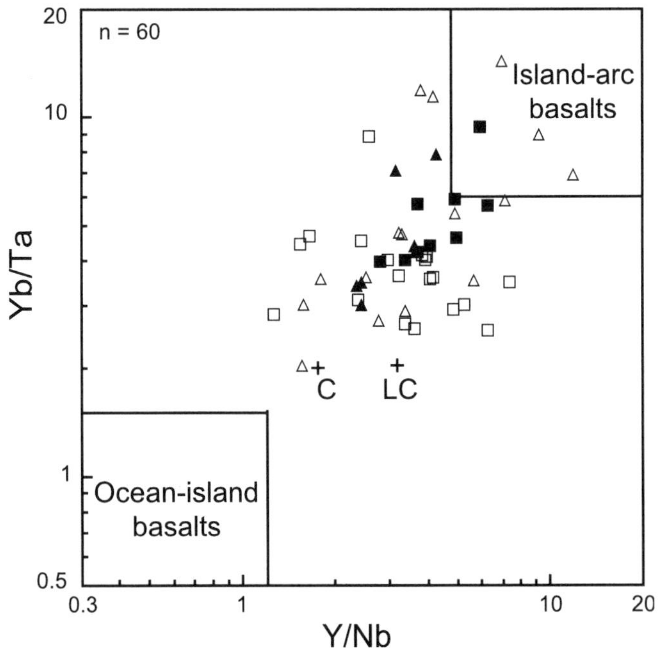

Figure 14. Y/Nb versus Yb/Ta for meta-igneous rocks of the study area. Crosses indicate compositions of average continental crust (C) and lower crust (LC) from Taylor and McLennan (1985). Boxes show fields of ocean-island basalts and Nb- and Ta-depleted island-arc basalts. Filled triangle—garnet monzogranite gneiss; filled square—high-silica charnockite; open circle—Marshall Metagranite; open square—low-silica charnockite; open triangle—granite and leucogranite of the Old Rag magmatic series. Modified after Eby (1990).

that the latter were likely derived through similar processes involving lower crustal sources.

As discussed earlier, most Blue Ridge granitoids display geochemical characterisitics that are transitional between A-types and fractionated I-types, with only the low-silica charnockite consistently displaying A-type geochemical affinity. The occurrence of granitoids with such transitional features underscores the important, first-order control of source composition and the subsequent effects of fractionation on the characteristic geochemical signature of various types of granitoid rocks. However, the existence of such granitoids also raises questions regarding the efficacy of attempts to recognize distinct typological categories in some parts of the granite spectrum.

The occurrence of A-type and relatively evolved I-type granitoids that are closely related in space and time is not unusual in orogenic belts, and, as demonstrated in examples from eastern Australia and the study area, the A-type granitoids are typically younger than associated I-types (Landenberger and Collins, 1996; this study). The nature of petrologic mechanisms leading to the development of A-type magmas in particular has been the subject of considerable debate in recent decades and has involved discussion of mechanisms ranging from fractionation of mantle-derived basaltic magma to melting of various crustal sources (Collins et al., 1982; Eby, 1990, 1992; Creaser et al., 1991; Turner et al., 1992). The low degree of fractionation of rare-earth elements (average La/Yb = 22) and absence of significant negative Eu anomaly (average Eu/Eu* = 0.6, where Eu/Eu* is a measure of the magnitude of the Eu anomaly equal to the chondrite-normalized Eu concentration divided by the value interpolated from neighboring elements) of the most chemically primitive rocks from the low-silica charnockite unit indicate that an origin through extreme fractionation of a basaltic parent, involving significant removal of plagioclase, is unlikely. Based on similar compositional criteria, such a mechanism also is untenable for the other, more geochemically evolved, transitional rocks. Trace-element compositions and source-sensitive ratios of the Blue Ridge transitional rocks indicate a derivation through partial melting of crustal sources (Fig. 14) and suggest that the compositions of such sources and the conditions under which melting occurred were primary factors controlling the geochemical characterisitics of the resulting magmas. Creaser et al. (1991) discussed the effect of source characteristics on production of granitic melts of I-type and A-type composition, suggesting that metaluminous A-types are derived through partial melting of lower crustal rocks of tonalitic to granodioritic composition, whereas I-types are more commonly generated from more chemically primitive, typically basaltic to tonalitic, sources. In contrast, Landenberger and Collins (1996) demonstrated that coexisting I-type and A-type granitoids of the subduction-related Chaelundi Complex in eastern Australia were generated by water-undersaturated partial melting of compositionally similar, mafic to intermediate lower crustal sources, but that the A-type parent magmas resulted from lower water activity conditions during partial melting. The low SiO_2 content of the least chemically evolved rocks of the low-silica charnockite from this study suggests derivation from similar, relatively unevolved and possibly mafic, sources; however, the restricted compositional range of the other rocks provides little evidence bearing on the extent of petrologic evolution characterizing sources of the transitional rocks. Nevertheless, examples such as the Chaelundi Complex and rocks of the study area illustrate that spatially and temporally related granitoids of both I-type and A-type affinity may constitute a locally significant magmatic component produced during the posttectonic stages of orogen evolution.

The close compositional affinity between tholeiitic charnockites of igneous origin and largely metaluminous A-type granitoids has been noted in several recent studies (Kilpatrick and Ellis, 1992; Landenberger and Collins, 1996; Frost et al., 2000). Kilpatrick and Ellis (1992) emphasized the typically higher $Mg/(Mg + Fe^{2+})$ ratios of igneous charnockites relative to A-types in defining a charnockitic magma type ("C-type") that they proposed to be distinct from petrologic lineages characterizing A-type (more Fe-rich) and unfractionated I-type (more Mg-rich) granitoids. While recognizing compositional differences that likely result primarily from variations in melt production and differentiation, Landenberger and Collins (1996) identified similarities in trace-element characteristics that suggest a common origin for igneous charnockites and metaluminous A-types. They proposed that both magma types may be produced by partial melting of a dehydrated, K-rich, mafic to intermediate source with differences in melt composition reflecting variations in the anhydrous mineral assemblage. However, Frost et al. (2000) discounted the existence of a separate C-type granite lineage, proposing instead that igneous charnockites can be broadly divided into tholeiitic and calc-alkaline groups on the basis of $FeO_t/(FeO_t + MgO)$ ratios (Fig. 9). According to this classification, the Fe-rich tholeiitic charnockites typically display A-type characteristics and are especially similar compositionally to metaluminous varieties of the latter. The Blue Ridge rocks examined in this study provide an additional example, demonstrating the close petrologic linkage between igneous charnockites and metaluminous A-type granites, and offer further evidence that compositional distinction between A-types and fractionated I-types is difficult when the latter granitoids are characterized by relatively evolved compositions (Eby, 1990; Champion and Chappell, 1992).

Other Mesoproterozoic charnockitic rocks of tholeiitic affinity, including the Hidderskog, Bjerkreim-Sokndal, and Bunger Hills examples cited earlier, display trace-element characteristics that are similar to the Blue Ridge rocks (Fig. 12, E) and, because each of these Precambrian intrusions was emplaced in a nonorogenic setting, provide useful analogs for the tectonic role of the Blue Ridge lithologies. The 1160-Ma Hidderskog charnockite pluton from southwest Norway was emplaced in a largely anorogenic setting prior to the main Sveconorwegian

TABLE 4. GEOCHEMICAL COMPOSITION OF REPRESENTATIVE SAMPLES

	Sample number and map unit							
	Yg SNP96-1	Yg SNP96-6	Yg SNP96-8	Ycf SNP96-4	Ycf SNP96-10(z)	Ycf SNP98-63	Ym RT-005A	Ym RT-181
SiO_2	68.37	71.09	70.78	72.10	69.78	70.82	68.69	71.16
TiO_2	0.64	0.43	0.43	0.52	0.74	0.48	0.60	0.54
Al_2O_3	14.93	14.32	13.97	13.22	13.64	13.77	15.83	14.17
$Fe_2O_3^{\dagger}$	4.22	3.38	3.85	3.28	4.23	3.65	3.14	3.50
MnO	0.06	0.07	0.08	0.05	0.07	0.03	0.03	0.05
MgO	0.70	0.45	0.54	0.36	0.66	0.31	1.07	0.99
CaO	2.45	1.11	1.54	1.62	2.29	1.81	2.15	1.18
Na_2O	3.19	2.92	2.82	2.32	2.38	2.77	3.78	2.74
K_2O	4.94	5.86	5.72	5.99	5.63	6.12	3.79	5.02
P_2O_5	0.18	0.12	0.11	0.12	0.19	0.12	0.19	0.17
Total	99.66	99.74	99.82	99.59	99.60	99.87	99.27	99.53
Rb	170.3	174.6	165.4	295.6	199.3	225.5	78.1	222
Ba	986	1210	1090	458	668	755	1370	730
Sr	175	106	113	67	128	107	560	120
Pb	30	27	16	33	28	31	11	16
Th	12.0	7.3	8.8	16.9	2.3	5.8	12.2	23.9
U	n.d.	1.2	1.0	2.2	0.7	0.8	0.78	2.2
Zr	330	371	360	276	260	225	290	290
Hf	10.70	12.90	12.20	13.10	11.20	8.9	7.36	8.49
Nb	9.6	12.5	11.4	7.4	9.3	8.5	6.6	11.7
Ta	0.64	1.00	1.08	0.86	0.85	0.78	0.59	0.89
Ni	7	3	3	3	3	4	14	27
Zn	81	65	66	45	63	55	63	66
Cr	34	27	21	26	28	31	8.4	18
Ga	21	20	20	21	21	20	22	20
Sc	9.08	10.90	9.15	6.29	8.90	8.48	4.91	10.3
V	36	14	12	16	25	17	33	32
La	58.0	53.9	66.2	72.1	52.8	67.0	56.5	61.5
Ce	105.0	89.5	111.0	122.0	94.3	127	109	124
Nd	47.4	42.6	57.8	54.4	46.6	59	47	58.1
Sm	10.00	9.04	11.40	12.10	10.30	11.73	9.6	13.8
Eu	1.82	1.97	1.90	1.28	2.09	1.57	1.74	1.46
Tb	1.25	1.26	1.64	1.84	1.50	1.65	0.914	1.81
Yb	2.2	3.4	4.7	4.9	3.4	3.6	1.7	3.9
Lu	0.23	0.39	0.57	0.54	0.42	0.61	0.21	0.495
Y	26.2	30.4	44.6	48.9	38.2	44.4	22.8	44.9
$Ga/Al \times 10^4$	2.66	2.64	2.71	3.00	2.91	2.74	2.63	2.67
$FeO_t/(FeO_t + MgO)$	0.84	0.87	0.87	0.89	0.85	0.91	0.73	0.76
AI	0.71	0.78	0.78	0.78	0.73	0.81	0.65	0.70
ASI	0.99	1.09	1.03	1.00	0.96	0.95	1.11	1.17

Note: Major elements expressed in wt%; trace elements in ppm. Elements in italics were analyzed by neutron activation; others by X-ray fluorescence. AI (agpaitic index) = molar (Na +K)/Al; ASI (aluminum saturation index) = molar $Al_2O_3/(CaO + Na_2O + K_2O)$; (z) zircon analyzed (this study);

orogeny (1040–980 Ma) (Zhou et al., 1995). Chemical data indicate a compositionally complex source that might have included a small component of subduction-related materials. The 913 Ma Bjerkreim-Sokndal layered intrusion from southwestern Norway was emplaced during a brief episode of magmatic activity that postdated regional deformation by ~60 m.y. and was derived, at least in part, through differentiation of parental magma of jotunitic composition (Duchesne and Wilmart, 1997). Data for inferred jotunitic and mangeritic liquid compositions are plotted in Figure 12, E. The 1170–1160 Ma Bunger Hills plutons from Antarctica were emplaced after regional granulite-facies metamorphism, which peaked at ca. 1190 Ma; precursor magmas were derived through melting of heterogeneous sources (Sheraton et al., 1992). Data for charnockitic rocks from the Bunger Hills intrusions are plotted in Figure 12, E. The lack of strong decoupling between large-ion lithophile and high-field-strength elements and the lack of depletion (relative to ocean-ridge granites) of the Hf through Yb segment of these patterns are features of these Precambrian charnockitic systems that are similar to the "within-plate" granitoids of Pearce et al. (1984)

Sample number and map unit									
Ym RT-244-1	Ylgg SNP90-90(z)	Ybg OR97-25	Ybg OR97-39	Ygr OR96-4	Ygr OR97-35 (z)	Ygr OR97-54	Yfqj SNP97-39	Yfqj SNP99-89	Yfqj SNP99-93(z)
74.51	75.35	61.84	68.17	72.16	72.18	73.79	65.25	50.86	59.60
0.26	0.04	1.52	0.91	0.36	0.28	0.30	1.39	3.77	2.12
13.09	13.74	15.78	15.27	14.31	14.44	13.63	14.59	13.02	14.88
1.91	0.47	7.01	4.02	2.68	2.76	2.21	6.82	15.19	8.70
0.05	0.01	0.12	0.06	0.05	0.06	0.05	0.11	0.21	0.12
0.36	0.02	1.22	0.83	0.44	0.34	0.64	1.04	2.77	1.73
0.85	0.36	4.70	1.57	1.27	1.15	1.53	3.60	7.26	5.24
3.17	2.59	2.50	2.96	2.48	2.85	3.23	2.82	2.55	2.93
5.13	7.35	4.31	5.81	5.71	6.04	4.89	4.39	1.88	3.42
0.07	0.02	0.67	0.30	0.12	0.10	0.10	0.65	2.02	1.08
99.40	99.95	99.66	99.88	99.58	100.19	100.37	100.65	99.54	99.81
203	177.4	106.1	195.5	183.2	305.1	209.4	172.9	36.8	94.1
400	1302	1874	816	745	425	489	801	1018	1673
99	111	524	184	166	80	210	293	570	583
22	32	21	46	45	65	31	21	15	24
21.4	9.05	1.9	93	82.4	75.1	40.3	2.9	2.4	3.9
3.7	0.04	0.7	4.3	3.5	9.5	n.d.	n.d.	0.6	1.8
210	31	476	547	218	230	127	689	681	642
6.74	1.53	20.6	19.8	7.66	8.98	5.01	21.78	20.2	20.6
9.1	0.5	20	15.7	5.5	10.9	5.4	24	32.5	22.4
0.54	0.08	1.32	1.10	n.d.	0.72	0.52	2.11	2.1	1.55
8	2	3	4	4	2	4	3	4	3
59	5	155	118	60	70	36	149	310	194
4	20	22	23	29	25	40	26	19	26
20	19	25	24	21	25	18	28	26	27
4.74	0.64	15.04	6.42	5.86	6.27	4.03	16.70	33.20	18.18
9	2	65	34	19	8	15	33	148	76
107.0	64.3	78.8	220.6	128.8	112.0	52.8	119.1	117.3	118.6
125	93.1	174.3	477.0	265.0	202.0	75	276.4	272	256
100	25	87	183	82.0	80.6	33.0	155.0	144	124
24.6	2.78	18.27	32.03	19.69	20.80	5.20	31.22	26.7	22.6
1.8	1.27	4.09	1.76	1.54	0.91	0.97	3.71	5.14	4.52
3.28	0.18	2.54	2.06	2.86	3.61	0.50	3.10	1.46	3.2
8.23	0.54	4.5	1.86	5.0	10.2	1.5	5.6	6.3	6.3
1.04	0.08	0.61	0.28	0.62	1.10	0.19	0.63	0.98	0.84
96.3	2.8	55.2	28.0	36.4	62.6	17.1	74.1	106.5	82.3
2.89	2.98	2.99	2.97	2.77	3.27	3.35	3.63	3.77	3.53
0.83	0.89	0.84	0.81	0.85	0.88	0.87	0.85	0.83	0.84
0.82	0.85	0.56	0.73	0.72	0.78	0.78	0.64	0.48	0.59
1.06	1.05	0.91	1.09	1.14	1.09	1.08	0.92	0.67	0.80

n.d.—not detected.

†Total iron expressed as Fe_2O_3.

and resemble the trace-element characteristics of the Blue Ridge rocks. This similarity suggests that such trace-element features are a distinguishing characteristic of Precambrian granitoids developed within, or in physical proximity to, orogens, but produced during nonorogenic periods.

The nearly coeval Old Rag magmatic suite and low-silica charnockite, which together constitute the final magmatic episode in the study area, define a continuous trend on Harker variation diagrams and span a range in silica content of nearly 25 wt% (Fig. 8, A–E). Nevertheless, these rocks display contrasts in geochemical compositions that indicate separate petrologic lineages. The generally higher-silica rocks of the Old Rag magmatic series exhibit more pronounced, negative Eu anomalies (Fig. 15) and, at similar silica values, comparable Sr concentrations (Table 4), features that are interpreted to indicate greater fractionation of plagioclase relative to the low-silica charnockite. A comagmatic relationship is precluded, however, by nearly identical FeO/MgO ratios (Fig. 8, G) and similar $Na_2O + K_2O$ values (Fig. 8, F). Moreover, these units show differences in source-related characteristics, as indicated by differences in

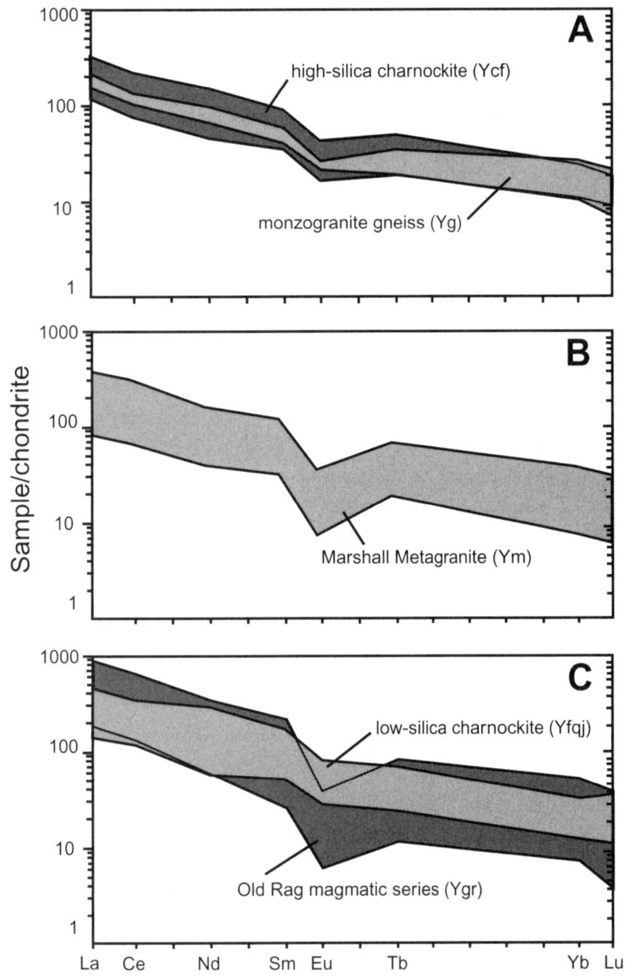

Figure 15. Rare-earth-element compositions of meta-igneous rocks of the study area, normalized to chondritic abundances of Anders and Ebihara (1982). Fields indicate ranges in composition of lithologic units. (A) Monzogranite gneiss and high-silica charnockite. (B) Marshall Metagranite. (C) Low-silica charnockite and the Old Rag magmatic series.

Y/Nb and Yb/Ta ratios (Figs. 11, B, and 14). The similar overall slope, but less pronounced Ta-Nb trough (Fig. 12, C), defined by the low-silica charnockite, suggests derivation from comparably complex "within-plate" sources that may have involved less island-arc component.

Summary

The Blue Ridge rocks preserve evidence of dominantly tholeiitic magmatism produced through melting of compositionally heterogeneous crustal sources that are geochemically similar to those which produced the "within-plate" granitoids of Pearce et al. (1984). In addition, most Blue Ridge units exhibit Nb and Ta depletion that likely reflects involvement of a source component of calc-alkaline compositional affinity. All rocks display some geochemical features of the postorogenic A_2 group of A-type granitoids of Eby (1992); however, only the low-silica charnockite can be classified as an A_2 type with confidence. All other rocks appear to be geochemically transitional between A-types and fractionated I-types (Fig. 10), exhibiting compositional characteristics that underscore the difficulties involved in distinguishing between granitoid types in such tectonic environments. Similarities in trace-element compositions between the Blue Ridge meta-igneous rocks and Precambrian charnockites of A-type affinity from other localities indicate derivation from similar within-plate crustal sources and suggest that such compositional characteristics are a distinguishing feature of Precambrian granitoids emplaced in orogenic belts during nonorogenic episodes. Most of the Blue Ridge rocks examined in this study are compositionally evolved, with SiO_2 contents ≥66 wt%; however, possible parental magmatic protoliths of low-silica composition are relatively rare in the northern Blue Ridge. The contemporaneous low-silica charnockite and leucogranites of the Old Rag magmatic series have similar $Na_2O + K_2O$ values and FeO_t/MgO ratios (Fig. 8, F and G, respectively), and different source-sensitive, trace-element characteristics (Fig. 12, C), and thus are unlikely to constitute a magmatic lineage.

Tectonic Significance

The new geochronologic data presented in this study both augment and expand on the threefold temporal model of Aleinikoff et al. (2000), which defined periods of magmatism at ca. 1153–1140 Ma (Group 1), 1112–1111 Ma (Group 2), and ca. 1077–1055 Ma (Group 3) (Fig. 16). To emphasize the temporal nature of the magmatic periods and avoid confusion with petrologic and geochemical groupings of rocks, we introduce the term "magmatic interval" to refer to the time spans defined by emplacement ages of rocks in each of the three groups of Aleinikoff et al. (2000). In the study area, Magmatic Interval I includes the high-silica charnockite (Ycf) and the geochemically similar monzogranite gneiss (Yg), which indicates that magmatism occurred at 1159 Ma and possibly earlier. These rocks correspond to the Group 1 intrusives of Aleinikoff et al. (2000). Magmatic Interval II includes the 1112–1111 Ma Marshall Metagranite, which has no correlative unit in the study area. In the study area, the leucogranite gneiss (Ylgg, 1078 ± 9 Ma) corresponds temporally to Group 3A rocks of Aleinikoff et al. (2000), whereas both the low-silica charnockite and Old Rag magmatic series (1050 ± 8 and 1060 ± 5 Ma, respectively) correlate in age to the Group 3B intrusives recognized in the northernmost Blue Ridge by these authors.

Field and geochronologic data from the study area bracket a period of deformation of probable regional significance. The 1078 ± 9 Ma leucogranite gneiss occurs as xenoliths and inliers within the low-silica charnockite (Fig. 3). The characteristic ductile fabric, defined by aligned feldspar porphyroclasts in the leucogranite gneiss, is locally truncated at contacts with typically nonfoliated low-silica charnockite (Fig. 4), indicating that ductile deformation occurred between 1078 and 1050 Ma. Because the low-silica charnockite is locally intruded by similarly nonfoliated leucogranite of the Old Rag magmatic series

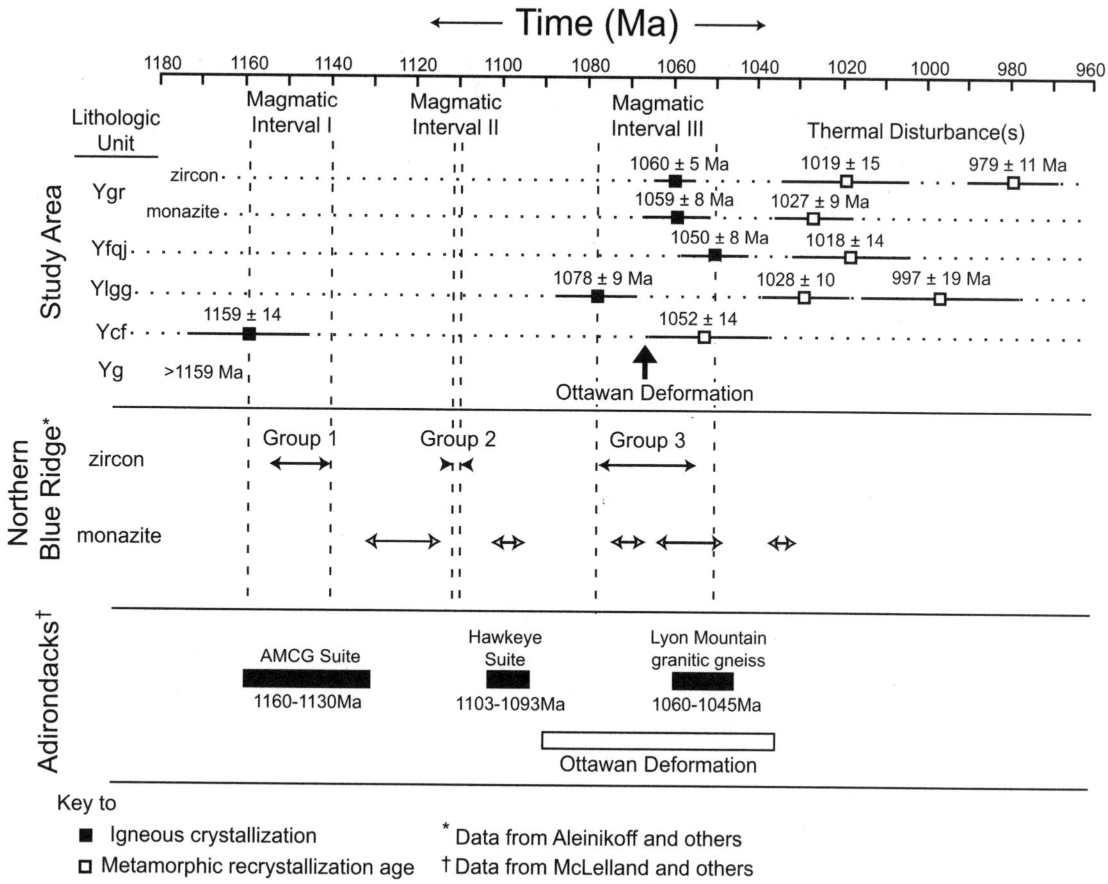

Figure 16. Timeline showing chronology of Grenville-age tectonomagmatic events in the study area, northernmost Virginia Blue Ridge (Aleinikoff et al., 2000), and Adirondacks (McLelland et al., 2001). New isotopic data are shown for the study area; ranges of data are shown for other areas.

that was emplaced at 1060 ± 5 Ma, the minimum age for the deformation may be ~10 m.y. older. The age of this Blue Ridge deformation falls within the period of principal tectonomagmatic activity associated with the Ottawan orogeny in the Adirondacks, which has been constrained to the interval 1090–1035 Ma (McLelland et al., 2001). As a result, ductile fabrics and gneissosity developed in the monzogranite gneiss, high-silica charnockite, and leucogranite gneiss are interpreted to be a likely result of Ottawan orogenesis in the northern Blue Ridge. Aleinikoff et al. (2000) recognized a northwest-striking foliation (D1) developed in their Group 1 rocks and noted that the foliation is cut discordantly by garnetiferous metagranite of age 1077 ± 4 Ma. These observations suggest an earlier period of deformation that occurred in the northernmost Blue Ridge in the interval 1145–1077 Ma.

Geochronologic data from Aleinkoff et al. (2000) and this study indicate that Blue Ridge rocks preserve evidence of two intervals of lithologically diverse magmatism occurring at 1159 to ca. 1140 Ma and from ca. 1078 to 1050 Ma. These intervals correspond temporally to episodes of AMCG magmatism recognized elsewhere in the Grenville province (Rivers, 1997). Nonorogenic (within-plate) magmatism is a dominant feature of the Grenville orogen during the interval 1170–1040 Ma and, in most cases, intrusive activity is associated with regional extension following periods of orogenesis (Rivers, 1997). An origin related to crustal extension is possible for the Blue Ridge granitoids; however, direct evidence of extensional activity is presently lacking in the northern Blue Ridge. Evidence of deformation related to Ottawan orogenesis is preserved in the 1078 ± 9 Ma leucogranite gneiss, and it is likely that the compositionally bimodal magmatism that occurred at 1060–1050 Ma (Magmatic Interval III) developed in response to lithospheric readjustments following orogeny. At present, there is no recognized deformational event that may have preceded magmatism developed during Magmatic Interval I (1159 to ca. 1140 Ma) in the Blue Ridge; however, orogenic pulses of appropriate age have been documented elsewhere in the Grenville orogen (Rivers, 1997).

The 1112–1111 Ma Marshall Metagranite defines the short-lived Magmatic Interval II, which preceded local Ottawan orogenesis by at least 30 m.y. Granitic rocks of similar age have been noted in the Adirondacks, where McLelland et al. (2001) interpreted the 1103–1093 Ma Hawkeye granite suite to be a result of far-field thermal effects related to development of the Mid-continent rift. This model may also be applicable to the Blue Ridge, which was located, even in its palinspastically restored position (Bartholomew and Lewis, 1992), directly along strike of the Mid-Michigan rift. The latter rift forms an eastern extension of the Mid-continent rift system (Jachens et al., 1993). The Marshall Metagranite corresponds in age to the earliest major phase of extension and bimodal magmatism within the rift system, which occurred at 1109–1087 Ma (Van Schmus et al., 1993).

Intervals of magmatic activity defined by crystallization ages of igneous protoliths in the northern Blue Ridge show a remarkable similarity in timing to episodes of igneous intrusion in the Adirondacks (McLelland et al., 1996, 2001). Nevertheless, post-1160-Ma igneous rocks of the Blue Ridge and Adirondacks display significant compositional differences that bear directly on petrogenesis. Nearly all granitic rocks of the Adirondacks, including the 1155–1130 Ma AMCG suite, the 1103–1093 Ma Hawkeye granitic suite, and the 1060–1045 Ma Lyon Mountain granitic gneiss, are characterized by high Nb and Y content (McLelland and Whitney, 1990; additional data courtesy of J. McLelland), suggestive of A-type affinity and derivation from within-plate crustal sources (Pearce et al., 1984; Eby, 1990). In contrast, most Blue Ridge granitoids examined in this study have lower Nb and Y content and trace-element signatures that are transitional between A-types and fractionated I-type granites. Moreover, greater ranges in source-sensitive trace elements suggest that the Blue Ridge rocks were derived from more compositionally diverse sources. Such differences in composition of igneous rocks produced during nearly simultaneous episodes of magmatic activity are likely a consequence of the physical separation of the massifs and resulting differences in source characteristics. Nevertheless, similarities in source-sensitive trace-element characteristics between the 1050 ± 8 Ma low-silica charnockite (Yfqj) of the study area and the 1060–1045 Ma Lyon Mountain granitic gneiss of the Adirondacks, both of which show more consistent within-plate geochemical signatures, indicate that both areas experienced simultaneous episodes of postorogenic, A-type magmatism following Ottawan orogenesis.

CONCLUSIONS

Grenville-age basement of the northern Blue Ridge includes a lithologically complex suite of granitic gneisses, granitoids, and charnockites similar to rocks documented in the central Blue Ridge and in other high-grade Precambrian terranes. In the northern Blue Ridge, the gneisses tend to be older than associated granitoids and typically occur as inliers within younger plutonic rocks; charnockites span the entire age range.

The mineralogic compositions of most rocks indicate crystallization from relatively dry magmas at granulite-facies conditions. Amphiboles of probable igneous origin occur in both high- and low-silica charnockites, and are present in pegmatites that are locally derived through fractionation of the latter.

The U-Pb isotopic analyses of zircons define two intervals of magmatic activity that correspond to Groups 1 and 3 of Aleinikoff et al. (2000). Magmatic activity in the northern Blue Ridge occurs at ca. 1160–1140 Ma (Magmatic Interval I), ca. 1112 Ma (Magmatic Interval II), and ca. 1080–1050 Ma (Magmatic Interval III). Field relations and ages of crosscutting igneous units allow recognition of a deformational event that occurred between ca. 1078 and 1050 Ma. This episode of deformation may represent local effects of regional Ottawan orogenesis. Field relations and geochronologic data indicate that most of the rocks in the study area were emplaced during nonorogenic episodes that both predated and postdated the 1078–1050 Ma deformational event.

Basement rocks of the study area are characterized by a broad range in major-element geochemical composition. Most rocks exhibit tholeiitic affinity and trace-element evidence of derivation from heterogeneous sources that probably experienced one or more cycles of continent-continent collision or island-arc magmatism. All rocks except the A-type low-silica charnockite are compositionally transitional between A-types and fractionated I-type granitoids. Silica-rich leucogranites and low-silica charnockite emplaced during Magmatic Interval III exhibit compositional evidence indicating that the rocks are not part of a single magmatic lineage.

Intervals of magmatic activity defined by rocks of the northern Blue Ridge correspond closely to episodes of magmatism documented in the Adirondacks. Intrusive rocks associated with Magmatic Intervals I and III in both regions are interpreted to have been produced in nonorogenic tectonic settings. Rocks emplaced during Magmatic Interval II correlate temporally to magmatic activity that occurred during early stages in the development of the Mid-continent rift system. Similarities in trace-element compositions of the 1050 ± 8 Ma low-silica charnockite in the Blue Ridge and the 1060–1045 Ma Lyon Mountain granitic gneiss in the Adirondacks indicate that both areas experienced nearly coeval postorogenic A-type magmatic activity following Ottawan orogenesis.

Similarities in compositional features between the Blue Ridge plutonic rocks and other Precambrian charnockite-granite suites, including the Hidderskog, Bunger Hills, and Bjerkreim-Sokndal examples discussed in this paper, indicate that granitoids of tholeiitic affinity with within-plate source characteristics constitute a common lithologic component of many orogenic belts worldwide. Such granitoid suites appear to be developed within, or in close proximity to, orogens, but are produced during nonorogenic periods.

ACKNOWLEDGMENTS

This research was supported by the Educational Geologic Mapping Program administered by the U.S. Geological Survey, and by contracts from the Virginia Department of Mines, Minerals, and Energy. Our appreciation is extended to G.N. Eby and S.J. Seaman for their thorough reviews and helpful comments regarding the original manuscript. The authors thank C. Murphy and E. du Bray of the U.S. Geological Survey for locating and making available sample powders of the Marshall Metagranite and other basement rocks of the northern Blue Ridge. J.M. Rhodes, P. Dawson, and M. Vollinger of the Ronald B. Gilmore Laboratory at the University of Massachusetts in Amherst are also gratefully acknowledged for their efforts in making possible timely trace-element analyses for some samples. W. Hodges helped with mineral separations. R.P.T. extends his sincere gratitude to B. Morgan of the U.S. Geological Survey for assistance in making some of the U-Pb isotopic analyses possible. The participation of E.A.B. in this project was supported in part by an award from the Mineralogical Society of the District of Columbia. The prodigious efforts of undergraduate students A. Antignano, C. Claflin, S. Nyman, and E. Pogue in support of this project are also gratefully acknowledged.

REFERENCES CITED

Aleinikoff, J.N., Zartman, R.E., Walter, M., Rankin, D.W., Lyttle, P.T., and Burton, W.C., 1995, U-Pb ages of metarhyolites of the Catoctin and Mount Rogers Formations, central and southern Appalachians: Evidence for two pulses of Iapetan rifting: American Journal of Science, v. 295, p. 428–454.

Aleinikoff, J.N., Burton, W.C., Lyttle, P.T., Nelson, A.E., and Southworth, C.S., 2000, U-Pb geochronology of zircon and monazite from Mesoproterozoic granitic gneisses of the northern Blue Ridge, Virginia and Maryland, USA: Precambrian Research, v. 99, p. 113–146.

Anders, E., and Ebihara, M., 1982, Solar system abundances of the elements: Geochimica et Cosmochimica Acta, v. 46, p. 2363–2380.

Andersen, D.J., Lindsley, D.H., and Davidson, P.M., 1993, QUILF: A Pascal program to assess equilibria among Fe-Mg-Mn-Ti oxides, pyroxenes, olivine, and quartz: Computers and Geosciences, v. 19, no. 9, p. 1333–1350.

Anderson, J.L., 1983, Proterozoic anorogenic granitic plutonism of North America, *in* Medaris, L., et al., eds., Proterozoic geology: Selected papers from an international symposium: Boulder, Colorado, Geological Society of America Memoir 161, p. 133–154.

Badger, R.L., and Sinha, A.K., 1988, Age and Sr isotopic signature of the Catoctin volcanic province: Implications for subcrustal mantle evolution: Geology, v. 16, p. 692–695.

Bailey, C.M., Berquist, P.J., Mager, S.M., Knight, B.D., Shotwell, N.L., and Gilmer, A.K., 2003, Bedrock geology of the Madison Quadrangle, Virginia: Charlottesville, Virginia, Virginia Department of Mineral Resources Publication 157.

Bartholomew, M.J., and Lewis, S.E., 1984, Evolution of Grenville massifs in the Blue Ridge geologic province, southern and central Appalachians, *in* Bartholomew, M.J., et al., eds., The Grenville event in the Appalachians and related topics: Boulder, Colorado, Geological Society of America Special Paper 194, p. 229–254.

Bartholomew, M.J., and Lewis, S.E., 1992, Appalachian Grenville massifs: Pre-Appalachian translational tectonics, *in* Mason, R., ed., Basement tectonics 7: Dordrecht, The Netherlands, Kluwer Academic Publishers, v. 7, p. 363–374.

Bartholomew, M.J., Gathright, T.M., II, and Henika, W.S., 1981, A tectonic model for the Blue Ridge in central Virginia: American Journal of Science, v. 281, p. 1164–1183.

Burton, W.C., and Southworth, C.S., 1996, Basement rocks of the Blue Ridge province in Maryland, *in* Brezinski, D.K., and Reger, J.P., eds., Studies in Maryland geology: Baltimore, Maryland, Maryland Geological Survey Special Publication 3, p. 187–203.

Burton, W.C., Froelich, A.J., Pomeroy, J.S., and Lee, K.Y., 1995, Geology of the Waterford quadrangle, Virginia and Maryland, and the Virginia part of the Point of Rocks quadrangle: U.S. Geological Survey Bulletin 2095, 30 p., scale 1:24,000.

Champion, D.C., and Chappell, B.W., 1992, Petrogenesis of felsic I-type granites: An example from northern Queensland, *in* Brown, P.E., and Chappell, B.W., eds., The Second Hutton Symposium on the origin of granites and related rocks: Boulder, Colorado, Geological Society of America Special Paper 272, p. 115–126.

Collins, W.J., Beams, S.D., White, A.J.R., and Chappell, B.W., 1982, Nature and origin of A-type granites with particular reference to southeastern Australia: Contributions to Mineralogy and Petrology, v. 80, p. 189–200.

Compston, W., Williams, I.S., and Meyer, C., 1984, U-Pb geochronology of zircons from lunar breccia 73217 using a sensitive high-resolution ion-microprobe: Proceedings of the 14th Lunar Science Conference: Journal of Geophysical Research B, v. 98, p. 525–534.

Condie, K.C., 1997, Plate tectonics and crustal evolution, 4th edition: Boston, Butterworth-Heinemann, 282 p.

Creaser, R.A., and White, A.J.R., 1991, Yardea Dacite—Large-volume, high-temperature felsic volcanism from the Middle Proterozoic of South Australia: Geology, v. 19, p. 48–51.

Creaser, R.A., Price, R.C., and Wormald, R.J., 1991, A-type granites revisited: Assessment of the residual-source model: Geology, v. 19, p. 163–166.

Deer, W.A., Howie, R.A., and Zussman, J., 1992, An introduction to the rock-forming minerals: Essex, England, Addison Wesley Longman, 696 p.

Duchesne, J.C., and Wilmart, E., 1997, Igneous charnockites and related rocks from the Bjerkreim-Sokndal layered intrusion (southwest Norway): A jotunite (hypersthene monzodiorite)-derived A-type granitoid suite: Journal of Petrology, v. 38, p. 337–369.

Eby, G.N., 1990, The A-type granitoids: A review of their occurrence and chemical characteristics and speculations on their petrogenesis: Lithos, v. 26, p. 115–134.

Eby, G.N., 1992, Chemical subdivision of the A-type granitoids: Petrogenetic and tectonic implications: Geology, v. 20, p. 641–644.

Espanshade, G.H., 1986, Geology of the Marshall quadrangle: U.S. Geological Survey Bulletin 1560, 60 p., scale 1:24,000.

Evans, N., 1991, Latest Precambrian to Ordovician metamorphism in the Virginia Blue Ridge: Origin of the contrasting Lovingston and Pedlar basement massifs: American Journal of Science, v. 291, p. 425–452.

Förster, H.J., Tischendorf, G., and Trumbull, R.B., 1997, An evaluation of the Rb vs. (Y + Nb) discrimination diagram to infer tectonic setting of silicic igneous rocks: Lithos, v. 40, p. 261–293.

Frost, B.R., Frost, C.D., Hulsebosch, T.P., and Swapp, S.M., 2000, Origin of the charnockites of the Louis Lake batholith, Wind River Range, Wyoming: Journal of Petrology, v. 41, p. 1759–1776.

Gathright, T.M., II, 1976, Geology of the Shenandoah National Park in Virginia: Charlottesville, Virginia Division of Mineral Resources Bulletin 86, 93 p.

Hackley, P.H., 1999, Petrology, geochemistry, field relations of the Old Rag Granite and associated charnockitic rocks, Old Rag Mountain 7.5-minute quadrangle, Madison and Rappahannock Counties, Virginia [M.S. thesis]: Washington, D.C., George Washington University, 244 p.

Hatcher, R.H., Jr., 1989, Tectonic synthesis of the U.S., Appalachians, *in* Hatcher, R.D., Jr., et al., eds., The Appalachian-Ouachita orogen in the United States: Boulder, Colorado, Geological Society of America, The Geology of North America, v. F-2, p. 511–535.

Irvine, T.N., and Baragar, W.R.A., 1971, A guide to the chemical classification of the common volcanic rocks: Canadian Journal of Earth Sciences, v. 8, p. 523–548.

Jachens, R.C., Simpson, R.W., Blakely, R.J., and Saltus, R.W., 1993, Isostatic residual gravity map of part of the conterminous United States and some adjacent parts of Canada, *in* Reed, J.C., Jr., et al., eds., Precambrian: Conterminous U.S.: Boulder, Colorado, Geological Society of America, The Geology of North America, v. C-2, plate 4.

Kilpatrick, J.A., and Ellis, D.J., 1992, C-type magmas: Igneous charnockites and their extrusive equivalents, *in* Brown, P.E., and Chappell, B.W., eds., The

Second Hutton Symposium on the origin of granites and related rocks: Boulder, Colorado, Geological Society of America Special Paper 272, p. 155–164.

Landenberger, B., and Collins, W.J., 1996, Derivation of A-type granites from a dehydrated charnockitic lower crust: Evidence from the Chaelundi Complex, eastern Australia: Journal of Petrology, v. 37, p. 145–170.

Leake, B.E., Woolley, A.R., Arps, C.E.S., Birch, W.D., Gilbert, M.C., Grice, J.D., Hawthorne, F.C., Kato, A., Kisch, H.J., Krivovichev, V.G., Linthout, K., Laird, J., Mandarino, J.A., Maresch, W.V., Nickel, E.H., Rock, N.M.S., Schumacher, J.C., Smith, D.C., Stephenson, N.C.N., Ungaretti, L., Whittaker, E.J.W., and Youzhi, G., 1997, Nomenclature of amphiboles: Report of the Subcommittee on Amphiboles of the International Mineralogical Association, Commission on New Minerals and Mineral Names: American Mineralogist, v. 82, p. 1019–1037.

Le Maitre, R.W., Bateman, P., Dudek, A., Keller, J., Lameyre, J., Le Bas, M.J., Sabine, P.A., Schmid, R., Sørensen, H., Streckeisen, A., Woolley, A.R., and Zanettin, B., 1989, A classification of igneous rocks and glossary of terms: Recommendations of the International Union of Geological Sciences Subcommision on the Systematics of Igneous Rocks: Oxford, Blackwell Scientific, 193 p.

Lindsley, D.H., 1983, Pyroxene thermometry: American Mineralogist, v. 68, p. 477–493.

Ludwig, K.R., 2001a, Isoplot/Ex, rev. 249, A geochronological toolkit for Microsoft Excel: Berkeley, California, Berkeley Geochronology Center Special Publication 1a, 58 p.

Ludwig, K.R., 2001b, Squid 1.02, A user's manual: Berkeley, California, Berkeley Geochronology Center Special Publication 2, 22 p.

Lukert, M.T., 1982, Uranium-lead isotope age of the Old Rag Granite, northern Virginia: American Journal of Science, v. 282, p. 391–398.

Malm, O.A., and Ormaasen, D.E., 1978, Mangerite-charnockite intrusives in the Lofoten-Vesterålen area, North Norway: Petrography, chemistry, and petrology: Norges Geologiske Undersokelse, v. 338, p. 38–114.

Maniar, P.D., and Piccoli, P.M., 1989, Tectonic discrimination of granitoids: Geological Society of America Bulletin, v. 101, p. 635–643.

McDowell, R.C., and Milton, D.J., 1992, Geologic map of the Round Hill quadrangle, Clarke and Loudoun Counties, Virginia, and Jefferson County, West Virginia: Reston, Virginia, U.S. Geological Survey Geologic Quadrangle GQ-1702, scale 1:24,000.

McLelland, J., and Whitney, P., 1990, Anorogenic, bimodal emplacement of anorthositic, charnockitic, and related rocks in the Adirondack Mountains, New York, *in* Stein, H.J., and Hannah, J.L., eds., Ore-bearing granite systems; petrogenesis and mineralizing processes: Boulder, Colorado, Geological Society of America Special Paper 246, p. 301–315.

McLelland, J., Daly, J.S., and McLelland, J.M., 1996, The Grenville orogenic cycle (ca. 1350–1000 Ma): An Adirondack perspective: Tectonophysics, v. 256, p. 1–28.

McLelland, J., Hamilton, M., Selleck, B., McLelland, J., Walker, D., and Orrell, S., 2001, Zircon U-Pb geochronology of the Ottawan orogeny, Adirondack Highlands, New York: Regional and tectonic implications: Precambrian Research, v. 109, p. 39–72.

Mitra, G., and Elliott, D., 1979, Deformation of basement in the Blue Ridge and development of the South Mountain cleavage, *in* Wones, D.R., ed., Proceedings, Caledonides in the USA, IGCP Project 27: Blacksburg, Virginia, Virginia Polytechnic and State University Memoir 2, p. 307–312.

Miyashiro, A., 1974, Volcanic rock series in island arcs and active continental margins: American Journal of Science, v. 274, p. 321–355.

Morimoto, N., 1988, Nomenclature of pyroxenes: Mineralogical Magazine, v. 52, p. 535–550.

Paces, J.B., and Miller, J.D., Jr., 1993, Precise U-Pb ages of Duluth Complex and related mafic intrusions, northeastern Minnesota; geochronological insights to physical, petrogenetic, paleomagnetic, and tectonomagmatic processes associated with the 1.1 Ga Midcontinental rift system: Journal of Geophysical Research B, Solid Earth and Planets, v. 98, p. 13,997–14,013.

Pearce, J.A., Harris, N.B.W., and Tindle, A.G., 1984, Trace element discrimination diagrams for the tectonic interpretation of granitic rocks: Journal of Petrology, v. 25, p. 956–983.

Rankin, D.W., Drake, A.A., Jr., Glover, L., III, Goldsmith, R., Hall, L.M., Murray, D.P., Ratcliffe, N.M., Read, J.F., Secor, D.T., Jr., and Stanley, R.S., 1989a, Pre-orogenic terranes, *in* Hatcher, R.D., Jr., et al., eds., The Appalachian-Ouachita orogen in the United States: Boulder, Colorado, Geological Society of America, The Geology of North America, v. F-2, p. 7–100.

Rankin, D.W., Drake, A.A., Jr., and Ratcliffe, N.M., 1989b, Geologic map of the U.S. Appalachians showing the Laurentian margin and Taconic orogen, *in* Hatcher, R.D., Jr., et al., eds., The Appalachian-Ouachita orogen in the United States: Boulder, Colorado, Geological Society of America, The Geology of North America, v. F-2, plate 2.

Rivers, T., 1997, Lithotectonic elements of the Grenville Province: Review and tectonic implications: Precambrian Research, v. 86, p. 117–154.

Rock, N.M.S., and Leake, B.E., 1984, The International Mineralogical Association amphibole nomenclature scheme: Computerization and its consequences: Mineralogical Magazine, v. 48, p. 211–227.

Sheraton, J.W., Black, L.P., and Tindle, A.G., 1992, Petrogenesis of plutonic rocks in a Proterozoic granulite-facies terrane—The Bunger Hills, East Antarctica: Chemical Geology, v. 97, p. 163–198.

Simpson, E.L., and Eriksson, K.A., 1989, Sedimentology of the Unicoi Formation in southern and central Virginia: Evidence for late Proterozoic to Early Cambrian rift-to-drift passive margin transition: Geological Society of America Bulletin, v. 101, p. 42–54.

Sinha, A.K., and Bartholomew, M.J., 1984, Evolution of the Grenville terrane in the central Virginia Appalachians, *in* Bartholomew, M.J., et al., eds., The Grenville event in the Appalachians and related topics: Boulder, Colorado, Geological Society of America Special Paper 194, p. 175–186.

Southworth, S., 1994, Geologic map of the Bluemont quadrangle, Loudoun and Clarke Counties, Virginia: Reston, Virginia, U.S. Geological Survey Geologic Quadrangle GQ-1739, scale 1:24,000.

Southworth, S., 1995, Geologic map of the Purcellville quadrangle, Loudoun County, Virginia: Reston, Virginia, U.S. Geological Survey Geologic Quadrangle GQ-1755, scale 1:24,000.

Southworth, S., and Brezinski, D.K., 1996, Geology of the Harpers Ferry quadrangle, Virginia, Maryland, and West Virginia: U.S. Geological Survey Bulletin 2133, 33 p., scale 1:24,000.

Streckeisen, A.J., and Le Maitre, R.W., 1979, A chemical approximation to the modal QAPF classification of the igneous rocks: Neues Jahrbuch für Mineralogie, Abhandlungen, v. 136, p. 169–206.

Sylvester, P.J., 1989, Post-collisional alkaline granites: Journal of Geology, v. 97, p. 261–280.

Taylor, S.R., and McLennan, S.M., 1985, The continental crust: Its composition and evolution: Oxford, Blackwell Scientific, 312 p.

Tera, F., and Wasserburg, G.J., 1972, U-Th-Pb systematics in three Apollo 14 basalts and the problem of initial Pb in lunar rocks: Earth and Planetary Science Letters, v. 14, p. 281–304.

Tilton, G.R., Wetherill, G.W., David, G.L., and Bass, N.M., 1960, 1,000 million-year-old minerals from eastern United States and Canada: Journal of Geophysical Research, v. 65, p. 4173–4179.

Tollo, R.P., and Aleinikoff, J.N., 1996, Petrology and U-Pb geochronology of the Robertson River Igneous Suite, Blue Ridge Province, Virginia: Evidence for multistage magmatism associated with an early episode of Laurentian rifting: American Journal of Science, v. 296, p. 1045–1090.

Tollo, R.P., Aleinikoff, J.N., Bartholomew, M.J., and Rankin, D.W., 2004, Neoproterozoic A-type granitoids of the central and southern Appalachians: Intraplate magmatism associated with episodic rifting of the Rodinian supercontinent: Precambrian Research, v. 128, no. 1, p. 3–38.

Turner, S.P., Foden, J.D., and Morrison, R.S., 1992, Derivation of some A-type magmas by fractionation of basaltic magma: An example from the Pathway Ridge, South Australia: Lithos, v. 28, p. 151–179.

Van Schmus, W.R., and Bickford, M.E., eds. 1993, Transcontinental Proterozoic provinces, *in* Reed, J.C., Jr., et al., eds., Precambrian: Conterminous U.S.:

Boulder, Colorado, Geological Society of America, The Geology of North America, v. C-2, p. 171–334.

Virginia Division of Mineral Resources, 1993, Geologic Map of Virginia: Charlottesville, Virginia, Virginia Division of Mineral Resources, scale 1:500,000.

Whalen, J.B., and Currie, K.L., 1990, The Topsails igneous suite, western Newfoundland; fractionation and magma mixing in an "orogenic" A-type granite suite, *in* Stein, H.J., and Hannah, J.L., eds., Ore-bearing granite systems; Petrogenesis and mineralizing processes: Boulder, Colorado, Geological Society of America Special Paper 246, p. 287–299.

Whalen, J.B., Currie, K.L., and Chappell, B.W., 1987, A-type granites: Geochemical characteristics, discrimination and petrogenesis: Contributions to Mineralogy and Petrology, v. 95, p. 407–419.

Williams, I.S., and Claesson, S., 1987, Isotopic evidence for Precambrian provenance and Caledonian metamorphism of high-grade paragneisses from the Seve Nappes, Scandinavian Caledonides, II, ion microprobe zircon U-Th-Pb: Contributions to Mineralogy and Petrology, v. 97, p. 205–217.

Williams, I.S., and Hergt, J.M., 2000, U-Pb dating of Tasmanian dolerites: A cautionary tale of SHRIMP analysis of high-U zircon, *in* Woodhead, J.D., et al., eds., Beyond 2000: New Frontiers in Isotope Geoscience, Abstracts and Proceedings: Mulgrave, Australia, Eastern Press, p. 185–188.

Wilson, E.W., and Tollo, R.P., 2001, Geochemical distinction and tectonic significance of Mesozoic and Late Neoproterozoic dikes, Blue Ridge province, Virginia: Geological Society of America Abstracts with Programs, v. 33, no. 2, p. A70.

Zhou, X.Q., Bingen, B., Demaiffe, D., Liegeois, J.P., Hertogen, J., Weis, D., and Michot, J., 1995, The 1160 Ma Hidderskog meta-charnockite: Implications of this A-type pluton for the Sveconorwegian belt of Vest Agder (SW Norway): Lithos, v. 36, p. 51–66.

Manuscript Accepted by the Society August 25, 2003

Printed in the USA

Geological Society of America
Memoir 197
2004

Deciphering the Grenville of the southern Appalachians through evaluation of the post-Grenville tectonic history in northwestern North Carolina

Charles H. Trupe*
Department of Geology and Geography, P.O. Box 8149, Georgia Southern University, Statesboro, Georgia 30460-8149, USA
Kevin G. Stewart
Department of Geological Sciences, CB# 3315, University of North Carolina, Chapel Hill, North Carolina 27599-3315, USA
Mark G. Adams
Department of Geology, Appalachian State University, Boone, North Carolina 28608 and Unimin Corporation, Harris Mining Company Rd., Spruce Pine, North Carolina 28777, USA
John P. Foudy*
Department of Geological Sciences, CB# 3315, University of North Carolina, Chapel Hill, North Carolina 27599-3315, USA

ABSTRACT

The western Blue Ridge Province of the southern Appalachians contains a rich record of the Mesoproterozoic Grenville orogeny, but subsequent Paleozoic metamorphic events have variably overprinted Grenville rocks, and Paleozoic thrusting has telescoped Grenville rock units. Effects of Paleozoic orogenesis must be unraveled to decipher the Grenville record. The Grenville rocks in the Blue Ridge of northwestern North Carolina and eastern Tennessee reside in a stack of Alleghanian thrust sheets that lie above the Grandfather Mountain and Mountain City windows. The composite Fries thrust sheet is the structurally highest unit in the stack and contains rocks of the eastern Blue Ridge juxtaposed against Grenville basement rocks (Pumpkin Patch Metamorphic Suite) along the Devonian Burnsville fault, which is the only identifiable Acadian structure in the thrust stack. West of the Grandfather Mountain window, the Sams Gap–Pigeonroost thrust splays off the Fries fault. Only the Fries and Sams Gap–Pigeonroost sheets appear to have been affected by Ordovician Taconic metamorphism. Below the Fries and Sams Gap–Pigeonroost sheets, Grenville basement rocks in the Fork Ridge and Linville Falls–Stone Mountain thrust sheets display widespread greenschist-facies metamorphism and deformation associated with Alleghanian thrusting. The lowest basement sheet is the Little Pond Mountain thrust sheet, which experienced only Late Paleozoic chlorite-grade metamorphism. Palinspastic restoration of the Grenville rocks to their pre-Paleozoic relative positions places rocks of the Pumpkin Patch Metamorphic Suite outboard of western Blue Ridge Grenvillian rocks.

Keywords: Appalachians, Grenville, thrust faults, Blue Ridge, palinspastic

*E-mail, corresponding author, Trupe: chtrupe@georgiasouthern.edu; e-mail and present address, Foudy: John.P.Foudy@ExxonMobil.com, ExxonMobil, 233 Benmar Street, GP3-452, Houston, Texas 77060.

Trupe, C.H., Stewart, K.G., Adams, M.G., and Foudy, J.P., 2004, Deciphering the Grenville of the southern Appalachians through evaluation of the post-Grenville tectonic history in northwestern North Carolina, *in* Tollo, R.P., Corriveau, L., McLelland, J., and Bartholomew, M.J., eds., Proterozoic tectonic evolution of the Grenville orogen in North America: Boulder, Colorado, Geological Society of America Memoir 197, p. 679–695. For permission to copy, contact editing@geosociety.org.

INTRODUCTION

The Blue Ridge of the southern Appalachians in western North Carolina and eastern Tennessee (Fig. 1) is a complex of thrust sheets that consist predominantly of Grenville-age high-grade metamorphic rocks. These Grenville rocks form the Laurentian basement upon which Paleozoic cover sequences were deposited. They were subsequently subjected to Iapetan and multiple Paleozoic orogenic events; their present geometry is largely due to Paleozoic structural processes. Post-Grenville fabrics, metamorphic assemblages, and structures obscure relations of the Grenville rocks.

Arthur Keith (1903, 1905, 1907) first recognized many of the Precambrian rocks in the western Blue Ridge, and subsequent studies have relied on Keith's mapping and nomenclature (e.g., Rodgers, 1953; King and Ferguson, 1960; Bryant and Reed, 1970; Brown, 1985). Historically, these Grenville rocks were mapped as units of undifferentiated basement (e.g., Rodgers, 1953; King and Ferguson, 1960), which ignores the internal structure of the crystalline thrust complex. Thus, to understand the nature of the Grenville orogenic belt in this region, both the detailed internal geology and the structural architecture of the thrust complex must be understood. Bartholomew and Lewis (1984) were among the first to attempt this by dividing the Grenville basement rocks into massifs with internal lithologic units and bounding faults. Butler and co-workers (e.g., Butler, 1973; Butler et al., 1987; Goldberg et al., 1986a,b, 1989, 1992) noted the differences in Paleozoic metamorphic and deformational histories among tectonic units of the Blue Ridge thrust complex in the area surrounding the Grandfather Mountain window.

This study focuses on rocks west of the Grandfather Mountain window along the North Carolina–Tennessee border, which tend to show less Paleozoic retrogressive metamorphism than the rocks nearer the window, where recent 1:24,000-scale mapping by the North Carolina Geological Survey and us is the basis for a more complete model for the structural geometry of the Grenville rocks. In this paper, we provide a new tectonic map and a restoration of the Grenville belts to their pre-Paleozoic positions. This reconstruction provides an up-to-date, high-resolution picture of the arrangement of the Grenville belts and

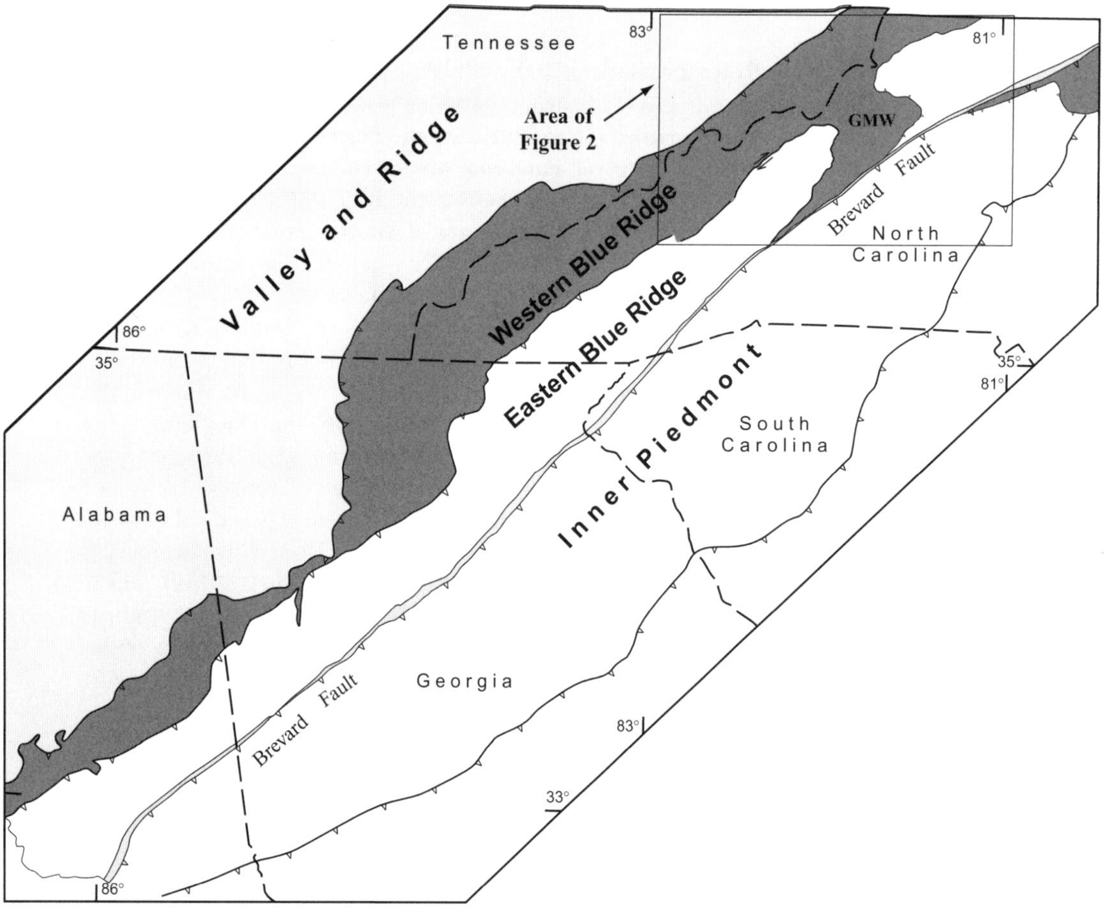

Figure 1. Generalized geology of the southern Appalachians. The outlined area shows the location of Figure 2 and is the focus of this study. GMW—Grandfather Mountain window. Modified from Rankin et al. (1989).

can serve as a framework for subsequent studies of Grenville rocks in the southern Appalachians.

GENERAL GEOLOGY OF THE BLUE RIDGE

In the vicinity of the Grandfather Mountain window (Fig. 2, A), the Blue Ridge thrust complex consists of a series of crystalline thrust sheets bounded by shear zones up to 1 km thick (Goldberg et al., 1986b, 1989; Butler et al., 1987). In this paper, we include all thrust sheets that contain basement rocks between the Brevard fault zone and the Iron Mountain fault in the complex, including the rocks in the Grandfather Mountain window (Fig. 2, A). We focus our discussion here on the thrust sheets outside the window, which are our tectonic subdivisions of the Blue Ridge. In contrast to earlier studies (e.g., Goldberg et al., 1989, 1992), we have used the convention of naming thrust sheets after the basal bounding fault.

Blue Ridge thrust sheets display differences in *P-T* and deformation histories and generally show a decrease in Paleozoic regional metamorphic grade from the hinterland toward the foreland (Goldberg et al., 1989, 1992; Stewart et al., 1997). The thrust complex thus preserves an inverted Paleozoic metamorphic gradient resulting from combined structural and metamorphic processes.

The structurally highest (Alleghanian) Fries thrust sheet contains both eastern Blue Ridge (Ashe and Alligator Back Metamorphic Suites) (Rankin, 1970; Abbott and Raymond, 1984; Raymond et al., 1989) and basement rocks, separated by the Burnsville-Gossan Lead fault, a Devonian dextral strike-slip fault (Adams et al., 1995a; Trupe et al., 2003). The eastern Blue Ridge rocks consist of amphibolite-facies metavolcanic and metasedimentary rocks, Paleozoic granitic intrusive rocks, and ultramafic bodies. Adjacent Grenville rocks northwest of the Burnsville fault make up a heterolithic suite of amphibolite- to granulite-facies gneisses that are intruded by abundant plutons and mafic dikes of the Neoproterozoic Bakersville Suite (Fig. 2, B) (734-Ma Rb-Sr whole-rock age) (Goldberg et al., 1986a). Bakersville intrusive rocks contain Paleozoic amphibolite- to granulite-facies mineral assemblages.

Grenville basement rocks and Bakersville Suite mafic rocks are also present in the thrust sheets structurally below the Fries sheet (Fig. 2, A and B). The Fork Ridge thrust sheet (Fig. 2, A) contains basement gneisses with a pervasive Paleozoic greenschist-facies overprint. The Linville Falls–Stone Mountain thrust sheet contains basement intruded by abundant Neoproterozoic Beech-Crossnore Suite peralkaline granitic plutons (Fig. 2, B) and is also characterized by Paleozoic greenschist-facies metamorphism. The Sams Gap–Pigeonroost thrust sheet (Fig. 2, A) lies between the Fries and Fork Ridge thrust sheets and contains rocks with Paleozoic amphibolite facies metamorphism. The lowermost Little Pond Mountain thrust sheet contains Cambrian Chilhowee Group sedimentary rocks nonconformably over Grenville basement rocks that exhibit a very low-grade Paleozoic metamorphic overprint (Goldberg et al., 1989).

THRUST SHEETS OF THE BLUE RIDGE THRUST COMPLEX

We define the thrust sheets of the Blue Ridge on the basis of rock type, metamorphic grade, and the bounding faults. The thrust sheets were emplaced during the late Paleozoic (325–265 Ma) Alleghanian orogeny. Taconic (480–450 Ma) or Acadian faults (410–360-Ma) within the basement thrust sheets northwest of the Burnsville fault are unknown, but may exist. Although Rodgers (1953) proposed that the Devil Fork fault was the northeastern extension of the Taconic Greenbrier fault, our recent work indicates it is a postmetamorphic Alleghanian fault.

Fries Thrust Sheet

The Fries thrust sheet is a composite sheet with high-grade Grenville basement rocks of the western Blue Ridge in its lower portion, overlain by eastern Blue Ridge rocks (Ashe Metamorphic Suite) southeast of the Burnsville fault (Fig. 2, A). It is bounded by the Fries fault on the northwest and the Brevard fault zone on the southeast. Basement rocks in the Fries thrust sheet include granitic gneiss, layered biotite-hornblende gneiss, amphibolite, calc-silicate rocks, and mica gneiss and schist. These rocks are intensely folded and locally migmatitic; Paleozoic metamorphic grade is upper-amphibolite to granulite facies (Fig. 2, B).

The Ashe Metamorphic Suite in the upper part of the Fries thrust sheet contains amphibolite-facies metasedimentary rocks, including mica schist and gneiss, amphibolite, calc-silicate rocks, dunite, and altered ultramafic rocks, all intruded by granitoid rocks of the Silurian-Devonian Spruce Pine Plutonic Suite (Brobst, 1962; Rankin et al., 1973; Kish, 1983; McSween et al., 1991). Eclogite occurs locally at the base of the Ashe Metamorphic Suite (Willard and Adams, 1994; Adams et al., 1995a; Abbott and Raymond, 1997; Abbott and Greenwood, 2000). The Ashe Metamorphic Suite has been interpreted as a combination of metamorphosed oceanic crust and sediments formed east of the Laurentian margin, possibly as an accretionary mélange containing fragments of dismembered ophiolite (Hatcher, 1978; Abbott and Raymond, 1984; Hatcher et al., 1984; Horton et al., 1989; Raymond et al., 1989; Misra and Conte, 1991; Adams et al., 1995a).

Fries Fault

The Fries fault is a thick, greenschist-grade shear zone that separates high-grade basement rocks in the hanging wall from greenschist-facies basement of the Fork Ridge thrust sheet and amphibolite-facies basement of the Sams Gap–Pigeonroost thrust sheet in the footwall (Fig. 2, A). Microstructures and mineral assemblages in Fries mylonites indicate top-to-northwest shear sense under greenschist-facies conditions (Mies, 1991; Trupe, 1997a,b; Hibbard et al., 2001). Timing of thrusting is interpreted as late Paleozoic (Alleghanian), based on the retrogression of regional metamorphic assemblages, shearing of

A

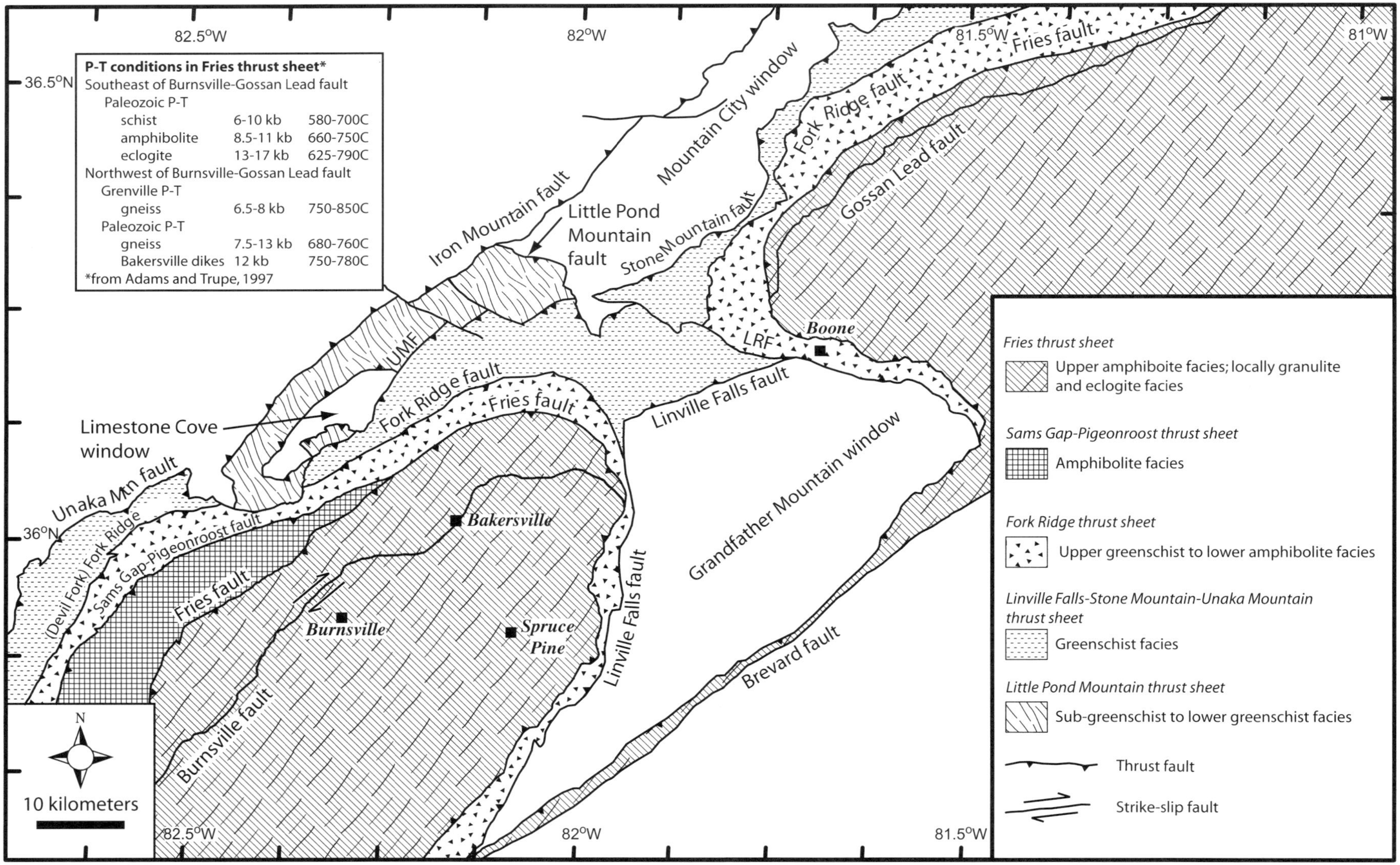

82.5°W
82°W
81.5°W
81°W
36.5°N
36°N
P-T conditions in Fries thrust sheet*
Southeast of Burnsville-Gossan Lead fault
Paleozoic P-T
schist 6-10 kb 580-700C
amphibolite 8.5-11 kb 660-750C
eclogite 13-17 kb 625-790C
Northwest of Burnsville-Gossan Lead fault
Grenville P-T
gneiss 6.5-8 kb 750-850C
Paleozoic P-T
gneiss 7.5-13 kb 680-760C
Bakersville dikes 12 kb 750-780C
*from Adams and Trupe, 1997
Iron Mountain fault
Little Pond Mountain fault
Mountain City window
Stone Mountain fault
Fork Ridge fault
Fries fault
Gossan Lead fault
Boone
LRF
UMF
Linville Falls fault
Grandfather Mountain window
Limestone Cove window
Unaka Mtn fault
(Devil Fork) Fork Ridge
Sams Gap-Pigeonroost fault
Bakersville
Burnsville
Spruce Pine
Burnsville fault
Brevard fault
N
10 kilometers
82.5°W
82°W
81.5°W
Fries thrust sheet
Upper amphiboite facies; locally granulite and eclogite facies
Sams Gap-Pigeonroost thrust sheet
Amphibolite facies
Fork Ridge thrust sheet
Upper greenschist to lower amphibolite facies
Linville Falls-Stone Mountain-Unaka Mountain thrust sheet
Greenschist facies
Little Pond Mountain thrust sheet
Sub-greenschist to lower greenschist facies
Thrust fault
Strike-slip fault

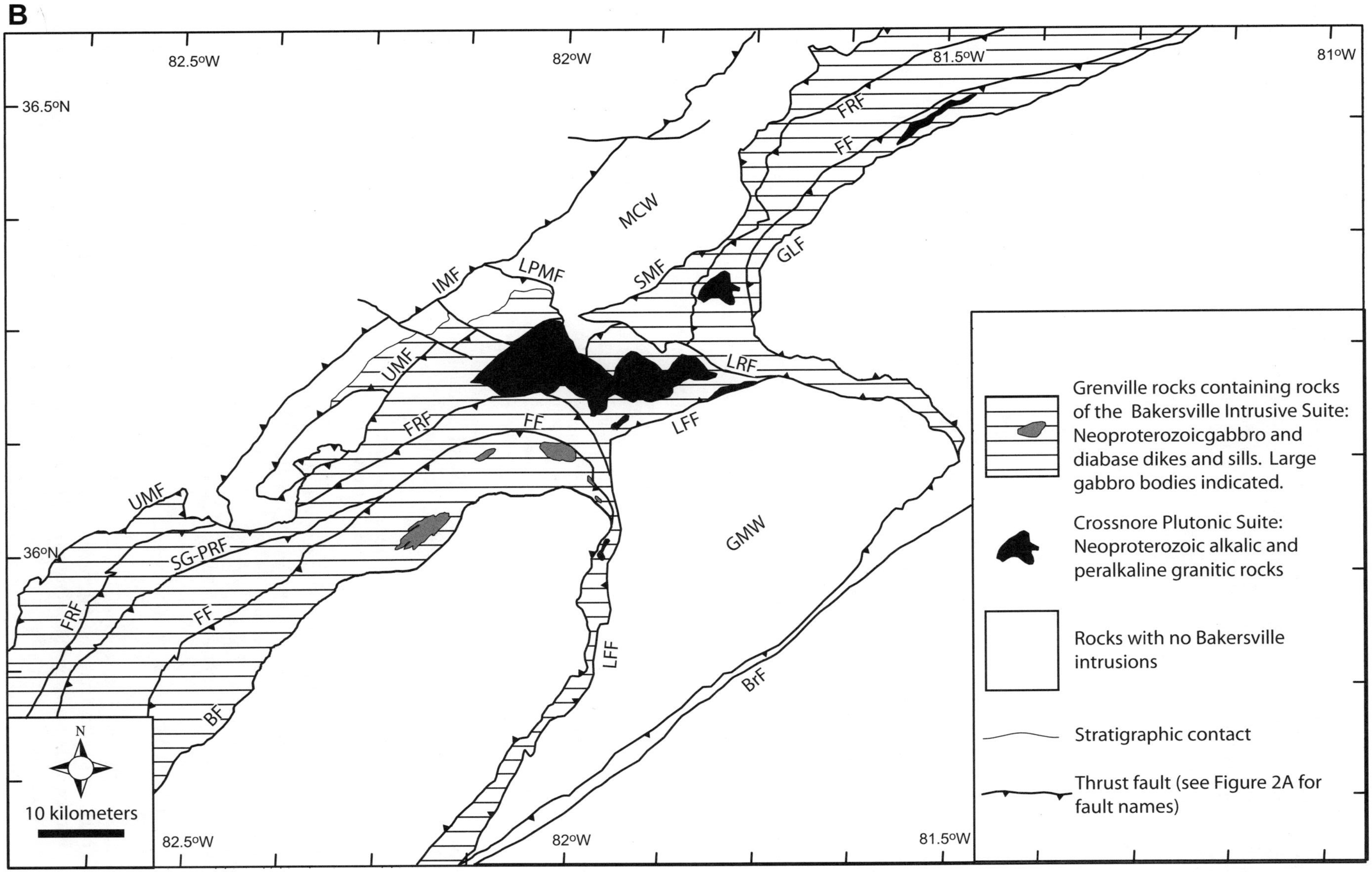

Figure 2. (A) Tectonic map of the Blue Ridge in northwestern North Carolina and eastern Tennessee, showing major Alleghanian thrust sheets and Paleozoic metamorphic grade. (B) Map of the Blue Ridge thrust sheet showing the distribution of major Neoproterozoic intrusive rocks. BF—Burnsville fault; BrF—Brevard fault; FF—Fries fault; FRF—Fork Ridge fault; GLF—Gossan Lead fault; GMW—Grandfather Mountain window; IMF—Iron Mountain fault; LFF—Linville Falls fault; LPMF—Little Pond Mountain fault; LRF—Long Ridge fault; MCW—Mountain City window; SG-PRF—Sams Gap–Pigeonroost fault; SMF—Stone Mountain fault; UMF—Unaka Mountain fault. Modified from Keith (1903, 1904, 1905, 1907); Rodgers (1953); Stose and Stose (1957); Shekarchi (1959); King and Ferguson (1960); Bryant and Reed (1970); Rankin et al. (1972); Merschat (1977, 1993); Bartholomew and Gryta (1980); Bartholomew and Lewis (1984); Brown (1985); Rankin et al. (1989); Adams (1990, 1995); Mies (1990); Adams et al., 1995a; Burton (1996).

Spruce Pine pegmatites, and correlation with faults that have yielded Alleghanian ages (Abbott and Raymond, 1984; Goldberg et al., 1989, 1992; McSween et al., 1989; Mies, 1990). Goldberg and Dallmeyer (1997) reported ages of ca. 325 Ma for shearing in the Fries fault northwest of Bakersville (Fig. 2, A) ($^{40}Ar/^{39}Ar$ muscovite ages, 323 ± 1 to 336 ± 1 Ma; Rb-Sr muscovite ages of 324 ± 3 to 326 ± 6 Ma).

Burnsville Fault

We have mapped the Ashe-basement contact (west of the Grandfather Mountain window) as a dextral strike-slip fault, the Burnsville fault, which formed under relatively high-grade (amphibolite-facies) conditions (Adams et al., 1995a; Trupe et al., 2003). The Burnsville fault is interpreted as a Devonian structure that was subsequently carried northwestward within Alleghanian thrust sheets (Adams et al., 1995a; Stewart et al., 1997; Trupe et al., 2003).

The Burnsville fault is a mixed-rock shear zone containing slices of both Ashe and basement rocks. The shear zone is characterized by steeply dipping northeast-striking foliation and mineral-stretching lineations that trend northeast-southwest (Raymond and Johnson, 1992; Adams et al., 1995a; Burton, 1996; Trupe et al., 2003). A recent U-Pb zircon age for sheared pegmatite within the Burnsville fault is 377 Ma, indicating that the Burnsville fault is an Acadian structure with its latest motion no older than Devonian (Trupe et al., 2003). The $^{40}Ar/^{39}Ar$ cooling ages from the Fries thrust sheet (Goldberg and Dallmeyer, 1997) show that the Burnsville fault must be older than ca. 360 Ma.

Sams Gap–Pigeonroost Thrust Sheet and Fault

The Sams Gap–Pigeonroost thrust sheet is bounded below by the Sams Gap–Pigeonroost fault and above by the Fries fault (Fig. 2, A). Rocks of the Sams Gap–Pigeonroost thrust sheet were mapped as Late to Middle Proterozoic biotite granitic gneiss (Merschat, 1977; Brown, 1985). Lithologies within the Sams Gap–Pigeon Roost thrust sheet are primarily layered granitic gneiss, biotite gneiss, biotite-hornblende gneiss, and numerous dikes and metagabbro bodies of the Bakersville Intrusive Suite (Merschat, 1977).

The Pigeonroost fault was originally defined by Carter et al. (1998) as an Alleghanian fault west of the Grandfather Mountain window that places Grenville granulite-facies gneisses upon biotite-granitic gneisses. Bartholomew and Lewis (1984) extended the Fork Ridge fault from north of the Grandfather Mountain window to the area west of Burnsville, North Carolina (Fig. 2, A), and joined the fault with the Devil Fork fault (Fig. 2, A) (see also Brown, 1985). Part of the trace of their Fork Ridge fault coincides with the Sams Gap–Pigeonroost fault.

Our recent work indicates that the Sams Gap–Pigeonroost fault joins the Fries fault northwest of Bakersville, North Carolina (Fig. 2, A). This interpretation makes the Sams Gap–Pigeonroost fault a diverging splay off the Fries fault. The Fork Ridge fault appears to be the continuation of the Devil Fork fault (see later in this paper) and therefore, the Sams Gap–Pigeonroost fault must lie southeast of the Devil Fork–Fork Ridge fault.

The Sams Gap–Pigeonroost fault is characterized by a kilometer-thick mylonite zone that is particularly well exposed in the Sams Gap, North Carolina–Tennessee quadrangle along highway U.S. 23. Kinematic indicators from the mylonites show consistent top-to-northwest shear sense. Quartz in the mylonites typically occurs as monocrystalline ribbons with undulose extinction, subgrain development, and deformation bands. Feldspar is fractured and shows partial alteration to sericite. Plagioclase is commonly sausseritized. These features are typical of Alleghanian mylonites in the Blue Ridge and are consistent with deformation at lower greenschist-facies conditions (Passchier and Trouw, 1992).

Fork Ridge Thrust Sheet and Fault

The Fork Ridge thrust sheet occurs beneath the Fries sheet and above the Linville Falls thrust sheet. This sheet forms the hanging wall of the Linville Falls fault along the west side of the Grandfather Mountain window (southeast of the branch line between the Linville Falls and Fork Ridge faults) and along the northeast side of the window, east of the Long Ridge fault (Fig. 2, A). The Fork Ridge sheet contains layered gneiss and granitic gneiss and Neoproterozoic Beech-Crossnore and Bakersville intrusive rocks.

The Fork Ridge fault (Bartholomew and Gryta, 1980; Bartholomew and Lewis, 1984; Brown, 1985) is a thrust fault that places gray, well-layered gneiss (Cranberry Mine Gneiss) upon more massive, greenish-pink gneiss (Watauga River Gneiss). The fault is characterized by a zone of greenschist-grade mylonites exhibiting top-to-northwest shear sense (Adams, 1990).

Near the Grandfather Mountain window, essentially no change in Paleozoic metamorphic grade occurs across the fault (Adams, 1990). However, farther to the southwest, metamorphosed Bakersville dikes exhibit different assemblages in the hanging wall and footwall of the fault. In the hanging wall of the Fork Ridge fault, relict igneous pyroxenes within Bakersville gabbro commonly contain thin rims of blue-green metamorphic hornblende, as well as grains of matrix hornblende. Abundant metamorphic biotite and epidote occur, along with lesser amounts of chlorite and actinolite. In the footwall of the Fork Ridge fault, Bakersville intrusives contain little to no hornblende, but they do contain abundant chlorite, actinolite, and lesser amounts of biotite.

Bartholomew and Lewis (1984) postulated that the Fork Ridge fault is correlative with the Devil Fork fault (Rodgers, 1953) to the southwest. Our mapping supports this interpretation. North of the Grandfather Mountain window, the sinistral Long Ridge fault cuts the Fork Ridge fault and has displaced the latter's trace by at least 6 km (Adams, 1990).

Linville Falls Thrust Sheet and Fault

The Linville Falls thrust sheet overlies the Grandfather Mountain window along its northern margin and is bounded on the southeast by the Linville Falls fault and on the northwest by the Stone Mountain–Unaka Mountain fault system (Fig. 2, A). Mesoproterozoic gneisses, Neoproterozoic granitic plutons, and minor Neoproterozoic Bakersville mafic dikes dominate the lithology of the thrust sheet. The Mesoproterozoic gneisses (primarily Watauga River granodiorite) (Bartholomew and Lewis, 1984) include rocks that were metamorphosed or crystallized during the Grenville orogeny between ca. 1177 and 950 Ma (Fullagar and Bartholomew, 1983). The Neoproterozoic granitic plutons include those of the Crossnore-Beech intrusive suite (Fig. 2, B) (741 Ma U-Pb zircon age) (Su et al., 1994).

The Linville Falls fault is named for an exposure on the western side of the Grandfather Mountain at Linville Falls (Bryant and Reed, 1970). Bryant and Reed (1970) described the fault as a thin mylonite zone, separating Cranberry Gneiss in the hanging wall from Chilhowee Group metasandstone of the Tablerock thrust sheet in the footwall. Our recent detailed mapping shows the fault zone to be a thick shear zone (≤1 km thick) along the entire length of the Linville Falls fault (Adams, 1990; Trupe, 1997a). The shear zone consists of mylonite and ultramylonite derived from basement gneisses and contains tectonic slices of exotic rocks (Trupe, 1997b). Along the western side of the Grandfather Mountain window, sheared Linville Falls metagranite and Grenville basement gneiss in the Fork Ridge thrust sheet form the hanging wall of the fault (Trupe, 1997a).

Mineral assemblages and deformation mechanisms in Linville Falls mylonites indicate that shearing occurred under greenschist-facies conditions (e.g., Bryant and Reed, 1970; Wojtal and Mitra, 1988; Trupe, 1989, 1997a,b; Newman and Mitra, 1993). Mineral-stretching lineations trend northwest-southeast and kinematic indicators are consistent with top-to-northwest shear sense (Trupe, 1989, 1997a,b; Adams, 1990). Late cataclasis overprints mylonitic fabrics in the Linville Falls fault (Trupe, 1989, 1997a,b; Adams, 1990, 1995; Adams and Su, 1996). Cataclasites are also present in the Unaka Mountain and Stone Mountain fault systems (King and Ferguson, 1960; Bryant and Reed, 1970; Diegel and Wojtal, 1985; O'Hara, 1992; Adams and Su, 1996).

In the Linville Falls thrust sheet, the Long Ridge fault connects the Linville Falls fault along the northern margin of the Grandfather Mountain window with the Stone Mountain fault (Fig. 2, A) (Adams, 1990). The Long Ridge fault is a northwest-striking sinistral strike-slip shear zone that truncates the Fork Ridge fault (Adams, 1990). Mylonitic rocks in the shear zone exhibit greenschist-facies assemblages and deformation features and are overprinted by cataclastic deformation (Adams, 1990; O'Hara, 1992; Adams and Su, 1996). Adams (1990) interpreted the Long Ridge fault as a low-angle tear fault in the Linville Falls thrust sheet. Adams and Su (1996) reported an age of ca. 300 Ma for mylonitic rocks from the Long Ridge shear zone.

Little Pond Mountain Thrust Sheet and Fault

The Little Pond Mountain thrust sheet (Fig. 2, A) is bounded on the southeast by the Unaka Mountain fault, on the northeast by the Little Pond Mountain fault (Rodgers, 1953), and is overridden on the northwest by the Iron Mountain thrust sheet. King and Ferguson (1960: 14) mapped Grenville rocks of the Little Pond Mountain thrust sheet as "Granitic rocks south of Pond and Little Pond Mountains" and "Plutonic complex of Pardee Point and Buck Ridge." Goldberg et al. (1986b) named this sheet the "Pardee Point thrust sheet." Diegel (1988) remapped the Little Pond Mountain fault and called rocks in the hanging wall the "Little Pond Mountain sheet."

The Little Pond Mountain thrust sheet contains amphibolite-facies Mesoproterozoic granitic gneiss and layered gneiss, and Neoproterozoic Bakersville intrusions, overlain nonconformably by lower Paleozoic sedimentary rocks (Goldberg et al., 1992). These rocks were affected only slightly by Paleozoic metamorphism and preserve Grenville fabrics and assemblages, as well as the original Bakersville Suite igneous textures (Goldberg et al., 1989; Butler, 1991).

Rodgers (1953) originally interpreted the Little Pond Mountain fault as a southwest-dipping thrust fault. King and Ferguson (1960) suggested, instead, that the fault was a transcurrent fault with right-lateral strike-slip displacement. Rodgers (1970) later stated that the Little Pond Mountain fault combines both thrust and dextral wrench slip. Diegel (1988) considered the fault to be a low-angle thrust fault, part of the Doe Ridges roof thrust of the Mountain City window that places basement and Cambrian Chilhowee Group rocks on Cambrian Rome Formation rocks (Diegel, 1988, p.139). The Little Pond Mountain thrust is characterized by a thick zone of carbonate-evaporite breccia, postulated to be the result of Alleghanian décollement thrusting at the base of the Rome Formation under brittle conditions of deformation (Diegel, 1988).

PALEOZOIC METAMORPHISM IN THE BLUE RIDGE THRUST COMPLEX

Relatively few published quantitative *P-T* estimates exist for rocks in the study area. Most quantitative estimates have focused on Ashe Metamorphic Suite rocks in the eastern Blue Ridge (see later in this paper). However, Neoproterozoic Bakersville intrusive rocks are common in most of the Grenvillian basement rocks (Fig. 2, B) and are useful indicators of Paleozoic metamorphism in the Blue Ridge thrust complex.

Fries Thrust Sheet

Ashe Metamorphic Suite

Rocks of the Ashe Metamorphic Suite preserve evidence of several metamorphic events. Eclogite-facies rocks occur locally near the base of the suite (Willard and Adams, 1994; Abbott and

Raymond, 1997). Metasedimentary rocks of the Ashe Metamorphic Suite contain amphibolite-facies assemblages (garnet + staurolite + kyanite ± sillimanite) (e.g., Abbott and Raymond, 1984; Adams and Trupe, 1997). Peak *P-T* estimates in these rocks are ~6–10 kilobars and 580–750 °C (Abbott and Raymond, 1984; McSween et al., 1989; Willard and Adams, 1994; Adams et al., 1995a; Adams and Trupe, 1997; Hewitt, 1999). The Ashe Metamorphic Suite rocks were subsequently affected by shearing in the Burnsville fault that deformed, but did not significantly retrograde, the amphibolite-facies assemblages (Adams et al., 1995a). To the northeast, near the Grandfather Mountain window, rocks of the Ashe Metamorphic Suite were overprinted during greenschist-facies retrogression associated with northwest-directed thrusting (Adams et al., 1995a; Trupe, 1997a,b).

Published isotopic ages suggest that Ashe Metamorphic Suite rocks experienced both Taconic and Acadian orogenesis (see summaries by Butler [1991] and Goldberg and Dallmeyer [1997]). A recent U-Pb zircon age (ca. 460 Ma) for eclogite-facies metamorphism (Miller et al., 2000) is consistent with early Taconic metamorphism in the Ashe Metamorphic Suite. The Sm-Nd, Rb-Sr, and $^{40}Ar/^{39}Ar$ ages reported by Goldberg and Dallmeyer (1997) suggest Taconic metamorphism followed by Devonian cooling.

The time of formation of the pervasive metamorphic foliation in the Ashe Metamorphic Suite is constrained by crosscutting Silurian-Devonian Spruce Pine intrusive rocks. Pegmatite ages are 430–377 Ma (e.g., Kish, 1983; Butler, 1991; Trupe et al., 2003). The pegmatites are concordant and discordant to the main metamorphic foliation. Butler (1973) suggested that the pegmatites intruded the rocks of the Ashe Metamorphic Suite shortly after the thermal peak of metamorphism, based on their occurrence almost entirely in the kyanite and higher metamorphic zones and on the absence of contact metamorphic aureoles. Thus, field relationships of the Spruce Pine rocks constrain the time of formation of pervasive metamorphic foliation in the Ashe Metamorphic Suite to older than the 430–377 Ma ages of the pegmatites.

Grenville rocks in the Fries thrust sheet

Grenville gneisses in the Fries thrust sheet contain amphibolite- to granulite-facies assemblages (garnet + hornblende + plagioclase ± biotite ± orthopyroxene) and are locally migmatitic, indicating temperatures sufficiently high for anatexis (Gulley, 1985; Rainey, 1989; Butler, 1991; Adams et al., 1995a). These gneisses contain refolded folds, indicating multiple deformation, which are crosscut by Neoproterozoic Bakersville intrusive rocks. Bakersville mafic rocks within the Fries thrust sheet were subsequently metamorphosed under amphibolite- to granulite-facies conditions (Goldberg et al., 1989, 1992; Rainey, 1989; Adams et al., 1995b).

Paleozoic radiometric ages within these Grenville gneisses are similar to those of the Ashe Metamorphic Suite. Goldberg and Dallmeyer (1997) reported a ca. 460 Ma age (Sm-Nd mineral isochron) from these gneisses, interpreted as the age of Taconic metamorphism. Hornblende cooling ages (^{40}Ar-^{39}Ar) are ca. 400–370 Ma, suggesting Acadian cooling (Goldberg and Dallmeyer, 1997).

Sams Gap–Pigeonroost Thrust Sheet

Locally, the rocks of the Sams Gap–Pigeonroost thrust sheet preserve up to three metamorphic events. Grenville metamorphism ranges from amphibolite facies to local granulite facies (Merschat, 1977). Grenville fabrics are cut by Bakersville dikes, which, in turn, exhibit varying degrees of an amphibolite-facies metamorphic overprint (Merschat, 1977; Kuchenbuch, 1979). Although no radiometric ages are known from the Sams Gap–Pigeonroost thrust sheet, the amphibolite-facies event in this thrust sheet is likely to be correlated with the amphibolite-facies event in Grenville gneisses of the Fries thrust sheet (Goldberg and Dallmeyer, 1997). Mylonitic rocks within the thrust-sheet bounding shear zones typically preserve a greenschist-facies overprint, which is presumably Alleghanian (Late Paleozoic).

Fork Ridge and Linville Falls Thrust Sheets

Grenville gneisses of the Fork Ridge and Linville Falls thrust sheets contain layers of garnet amphibolite, suggesting early amphibolite-facies metamorphic conditions. Migmatites are ubiquitous, indicating thermal conditions at least high enough to induce partial melting (Bartholomew and Lewis, 1984; Adams, 1990). Virtually all of the rocks in the Fork Ridge sheet preserve a greenschist- to lower amphibolite-facies overprint (Bryant and Reed, 1970; Trupe, 1989; Adams, 1990; Adams and Su, 1996).

Timing of metamorphism is constrained by crosscutting relations and radiometric isotope dates. Neoproterozoic intrusive rocks crosscut gneissic layering in Grenville rocks. These intrusive rocks do not show evidence of the same high-grade metamorphism that affected the Grenville rocks, but are deformed by greenschist-facies shearing (Fig. 3). Thus, the amphibolite-facies metamorphism that affected Grenville rocks must be older than the 741–734 Ma rocks of the Crossnore-Beech and Bakersville intrusive suites. Additionally, Dallmeyer (1975), Fullagar and Bartholomew (1983), and Goldberg and Dallmeyer (1997) interpreted isotopic dates to represent metamorphic ages for basement rocks in the Linville Falls sheet at ca. 1.0 Ga.

Greenschist-facies metamorphism in the Fork Ridge and Linville Falls thrust sheets overprints igneous assemblages in Neoproterozoic intrusive rocks and is well developed near Alleghanian shear zones (Adams and Trupe, 1991, 1997). Isotopic ages of Alleghanian shear zones, such as the Linville Falls and Long Ridge faults, are ca. 300 Ma (Van Camp and Fullagar, 1982; Schedl et al., 1992; Adams and Su, 1996). The Long Ridge fault truncates the Fork Ridge fault, indicating that the Fork Ridge fault is older than 300 Ma (Adams, 1990; Adams and Su, 1996).

Figure 3. Foliated Bakersville dike crosscutting Grenville-age compositional layering in basement gneiss, Fork Ridge thrust sheet. Alleghanian shearing has overprinted both gneissic layering and original dike contact. Both rocks have greenschist-facies metamorphic minerals. Lens cap is 5.5 cm in diameter.

Little Pond Mountain Thrust Sheet

Grenville gneisses of the Little Pond Mountain thrust sheet contain garnet amphibolite within the gneissic layering, and migmatization is prominent throughout these gneisses. These observations indicate that metamorphic conditions reached at least amphibolite facies and that temperature conditions were sufficient to induce anatexis. Thin-section observations indicate that the amphibolite-facies minerals in the Grenville rocks of the Little Pond Mountain sheet do not show significant by greenschist-facies retrogression (Butler, 1991). Additionally, Butler (1991) stated that Cambrian Chilhowee Group rocks in the thrust sheet experienced anchimetamorphic conditions, resulting from burial during Alleghanian thrusting, and were probably no higher than chlorite grade. Thus, the Little Pond Mountain thrust sheet was subjected to only very low-grade metamorphism during the Alleghanian orogeny.

Bakersville mafic intrusions crosscut gneissic layering in the Grenville rocks of the Little Pond Mountain thrust sheet, indicating that the metamorphism and formation of the layering predates the intrusion of the 734 Ma Bakersville rocks. Virtually all the documented radiometric ages from minerals from the gneisses are Precambrian (Long et al., 1959; Dallmeyer, 1975; McCrary, 1997), suggesting that the rocks of the Little Pond Mountain sheet were not affected by Paleozoic metamorphism, or that temperatures were not high enough to reset the isotope systematics in those minerals.

INTERPRETATION OF THE PALEOZOIC STRUCTURAL AND METAMORPHIC HISTORY OF THE BLUE RIDGE THRUST COMPLEX

We have combined detailed field studies with existing mapping and data to formulate a new model for Paleozoic juxtaposition of Grenville terranes in the study area. To fully understand the Grenville rocks, it is essential to understand the Paleozoic structural and thermal overprint (Fig. 2, A). Understanding this overprint permits us to postulate where the Grenville rocks were prior to the Paleozoic orogenies and to suggest possible Grenville structures within the basement rocks.

Our interpretation of the structural geology and timing of faulting differs from most previous interpretations in several ways:

1. In our study area, no faults northwest of the Devonian Burnsville fault are known to be older than Alleghanian. Radiometric ages and greenschist-facies assemblages in shear zone rocks support this interpretation.
2. In our model, the Sams Gap–Pigeonroost fault branches from the Fries fault northwest of Bakersville, North Carolina (Fig. 2, A). We suggest that movement on the Fries fault transferred to the Sams Gap–Pigeonroost fault in the southwestern part of the study area (see later in this paper).
3. We have interpreted the Fork Ridge fault as a splay off the Linville Falls fault, with a branch line near the northwest corner of the Grandfather Mountain window. This is in contrast to Bartholomew and Lewis' (1984) interpretation that the Fork Ridge fault represents the reemergence of the Linville Falls fault to the northwest.
4. We interpret the contact between the Ashe Metamorphic Suite and Grenville rocks as a Devonian strike-slip fault—the Burnsville fault—that was carried passively in the Fries thrust sheet during the Alleghanian. Previous workers interpret the eastern-western Blue Ridge contact in this area as a thrust fault of either Taconic (Hayesville fault) or Alleghanian (Fries fault) age (e.g., Rankin et al., 1972; Abbott and Raymond, 1984; Hatcher, 1989; Mies, 1990).

Paleozoic Kinematic History of the Blue Ridge Thrust Complex

Ordovician collision of the Piedmont terrane and Laurentian crust above an east-dipping subduction zone resulted in the Taconic orogeny (Hatcher, 1989) and caused amphibolite- to granulite-facies metamorphism in the Ashe Metamorphic Suite and the Laurentian margin. Initial juxtaposition of the eastern and western Blue Ridge rocks occurred during this event.

After collision of the Piedmont terrane with Laurentia, at least part of the Taconic suture evolved into a dextral strike-slip fault (the Burnsville fault) (Adams et al., 1995a). The Burnsville fault was active under amphibolite-facies conditions; earlier regional metamorphic assemblages were deformed during this shearing. Following regional uplift and cooling during the Devonian and Mississippian (Goldberg and Dallmeyer, 1997), Alleghanian thrusting progressed from the southeast toward the northwest (Fig. 4). The Fries thrust cuts through Laurentian crust that records early Paleozoic amphibolite- to granulite-facies metamorphism (Fig. 2, A). Northwest transport

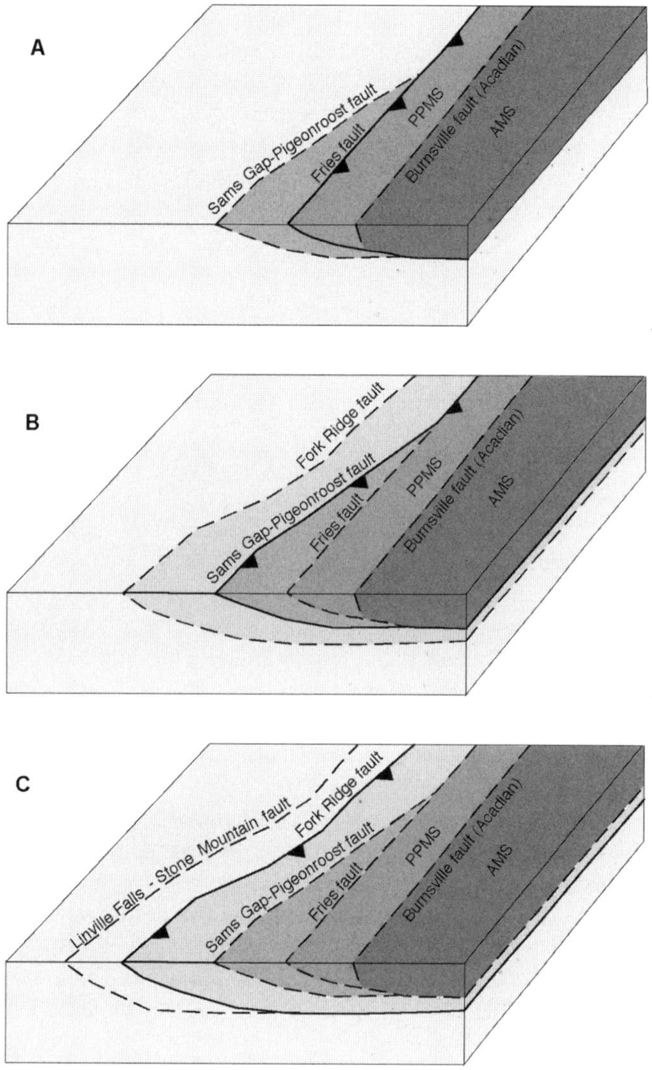

Figure 4. Schematic models showing sequence of thrust development for major thrust sheets in the Blue Ridge thrust complex. Active faults in each diagram are drawn with a solid line and thrust teeth; inactive faults are drawn with a dashed line. (A) Fries thrust carries composite sheet containing Pumpkin Patch Metamorphic Suite (PPMS) and Ashe Metamorphic Suite (AMS), which are separated by the Acadian Burnsville fault. Sams Gap–Pigeonroost fault forms as a diverging splay off the Fries. (B) Motion on the Fries is transferred to the Sams Gap–Pigeonroost fault south of the branch line. Fork Ridge fault breaks underneath and forward of the Sams Gap–Pigeonroost–Fries thrust. (C) Linville Falls–Stone Mountain fault breaks forward as a splay off the Fork Ridge fault, then overrides the Little Pond Mountain thrust sheet (not shown). Deep duplexing underneath the thrust stack arched the overlying rocks to create the structural culminations now exposed in the Grandfather Mountain and Mountain City windows.

of the Fries thrust sheet was probably accompanied by movement of the Piedmont thrust sheet and decapitated the Taconic suture, transporting it to the northwest (Rankin et al., 1991; Stewart et al., 1997). In the southwestern part of the study area, a branch line formed between the Fries and Sams Gap–Pigeonroost faults, and movement on the Fries was transferred to the Sams Gap–Pigeonroost fault (Fig. 4). This resulted in thrusting of granulite- and amphibolite-facies rocks of the Sams Gap–Pigeonroost thrust sheet over lower-grade rocks of the Fork Ridge thrust sheet.

In our model, the pervasive Alleghanian greenschist metamorphism in the Fork Ridge and Linville Falls thrust sheets was induced by emplacement of the Fries thrust sheet. Basement rocks juxtaposed by these faults were not involved in high-grade Taconic metamorphism, because these rocks were too far west of the Taconic metamorphic front. The Fork Ridge fault branches from the Linville Falls fault near the Grandfather Mountain window and juxtaposes Grenville rock units of similar Paleozoic metamorphic grade (Cranberry Mine and Watauga River gneisses).

Emplacement of the Linville Falls thrust sheet produced low-grade Alleghanian metamorphism in the Little Pond Mountain thrust sheet (Butler, 1991). During the late Alleghanian, the Blue Ridge thrust sheets were folded by duplexing of basement and cover rocks beneath the Grandfather Mountain and Mountain City windows (e.g., Boyer and Elliott, 1982; Rankin et al., 1991; Schedl et al., 1992). The Little Pond Mountain fault formed at this time (Diegel, 1988).

MAJOR GRENVILLE ROCK UNITS OF THE BLUE RIDGE

Watauga River Gneiss

The Watauga River Gneiss is present within the Linville Falls thrust sheet along the northern margin of the Grandfather Mountain window (Fig. 5) (Bartholomew and Gryta, 1980; Bartholomew and Lewis, 1984; Adams, 1990). It is a distinctive K-feldspar–rich gneiss, first mapped as a separate igneous protolith by Bartholomew and Gryta (1980) in the Sherwood, North Carolina, quadrangle. Bartholomew and Lewis (1984) proposed it as a formal lithodemic unit (within the Watauga River massif). Bartholomew and Gryta (1980) described the Watauga River Gneiss at its type locality as a granodiorite gneiss with a massive gneiss facies and a mylonitic facies. The massive gneiss facies is a pinkish-green to greenish-gray, medium- to coarse-grained porphyroclastic gneiss composed of microcline, plagioclase, perthite, quartz, and biotite in a matrix of fine-grained quartz, plagioclase, chlorite, biotite, muscovite, epidote, magnetite, sphene, and apatite. The gneiss is variably foliated, locally migmatitic, and cut by small pegmatites and aplite dikes. Bartholomew and Gryta (1980) attributed mylonitization of the Watuaga River Gneiss to Paleozoic thrusting along the Fork Ridge fault. Its distinctive pinkish-green color is useful in mapping tectonic boundaries in the western Blue Ridge.

Similar K-feldspar–rich gneisses were previously mapped as Cranberry Granite and Max Patch Granite (Keith 1903, 1904, 1907), unakite gneiss (Shekarchi, 1959), and migmatite of the Watauga River area (Hamilton in King and Ferguson, 1960).

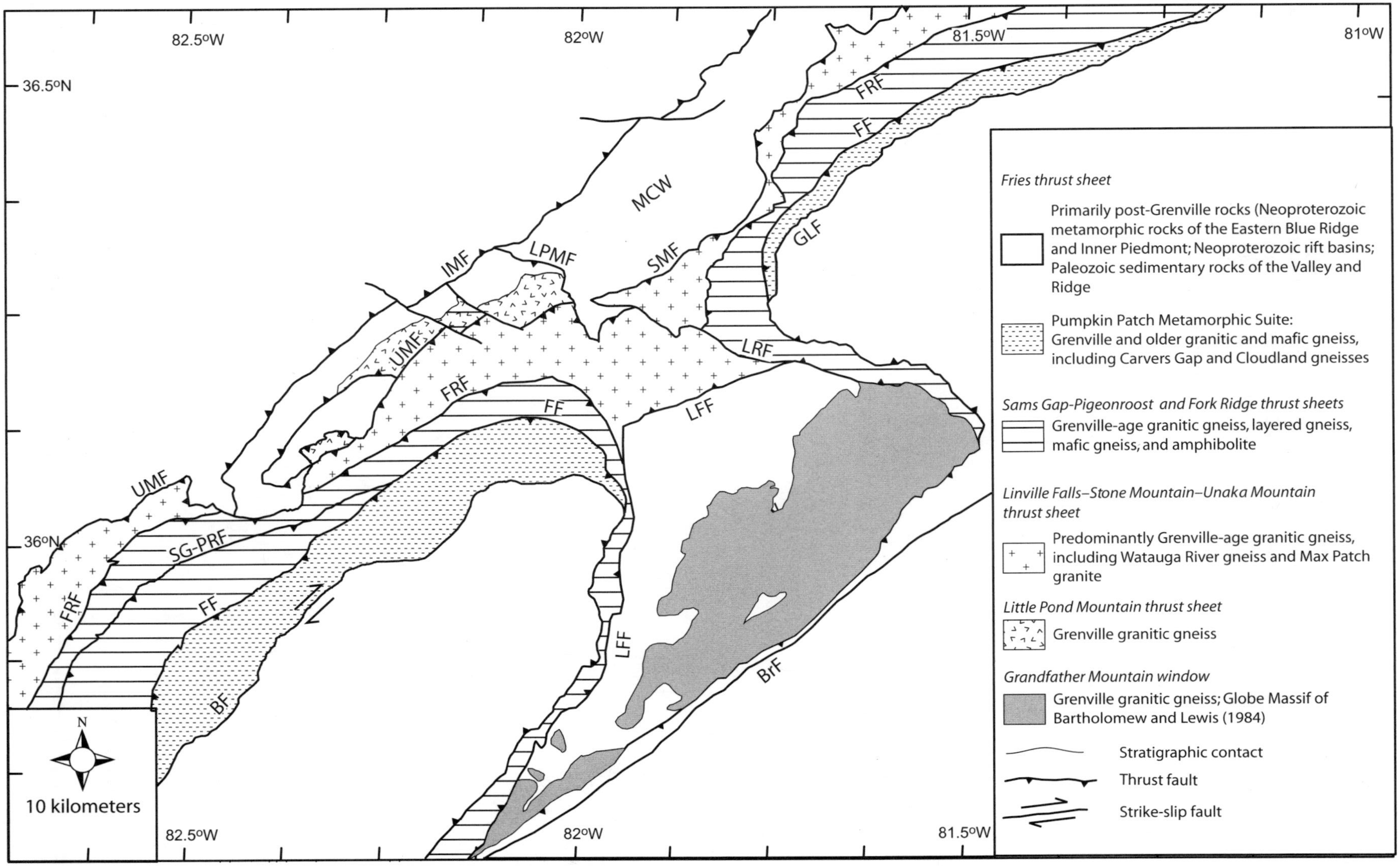

Figure 5. The Blue Ridge thrust complex, showing distribution of Grenville rocks. Abbreviations and sources as in Figure 2.

This unit was included as part of the Cranberry Gneiss by Bryant and Reed (1970) and is in the granodioritic gneiss unit on the Geologic Map of North Carolina (Brown, 1985). Fullagar and Bartholomew (1983) reported a Rb-Sr whole-rock age of ca. 1175 Ma for the Watauga River Gneiss, which they interpreted as the crystallization age.

Cranberry Granite and Cranberry Mine Gneiss

The Cranberry Granite and Cranberry Mine gneisses comprise a mixture of granitic gneiss and layered biotite gneiss and schist that occurs mainly in the hanging wall of the Fork Ridge fault (Fig. 5). The rocks are typically light to dark gray, equigranular to porphyroclastic, fine- to coarse-grained, well foliated to massive, and variably sheared.

These rocks were mapped as part of the Cranberry Granite (Keith, 1903) and Cranberry Gneiss (Bryant and Reed, 1970) on early maps of the western Blue Ridge. Keith (1903) first described the Cranberry Granite as a Precambrian plutonic complex encompassing a major portion of the Blue Ridge thrust sheet; he also included the older layered gneisses in this unit (Bartholomew and Lewis, 1984). Bryant and Reed (1970) renamed Keith's unit the "Cranberry Gneiss," because of the pervasive metamorphic features in the rock. They mapped essentially all basement rocks north of the Grandfather Mountain window as Cranberry Gneiss, but recognized that it was a heterogeneous unit. Bartholomew and co-workers (Bartholomew and Gryta, 1980; Bartholomew et al., 1983) delineated distinct units within the basement, retaining Cranberry Gneiss for the one clearly igneous unit originally described by Keith (1907). Bartholomew and Lewis (1984) renamed the older layered gneiss, which is intruded by the Cranberry Gneiss, as the "Cranberry Mine Layered Gneiss," after the exposures at the Cranberry mine. Fullagar and Bartholomew (1983) obtained a Rb-Sr whole-rock age of ca. 1020 Ma for the Cranberry Mine Layered Gneiss and interpreted this as a Grenville metamorphic age.

We suggest a redefinition of the names "Cranberry Gneiss" and "Cranberry Mine Layered Gneiss" as follows. Keith (1903) originally described the granitic Cranberry Gneiss as an igneous rock. Bartholomew and Lewis (1984) stated that the Cranberry Gneiss is intrusive into the Cranberry Mine Layered Gneiss, and noted its igneous texture and composition. Thus, we propose that the name "Cranberry Gneiss" be changed to "Cranberry Granite," in keeping with the igneous protolith of the unit. Additionally, Adams (1990) noted that the older layered gneiss was not always layered, as the name "Cranberry Mine Layered Gneiss" implies. We therefore suggest that the Cranberry Mine Layered Gneiss be redefined as the Cranberry Mine Gneiss.

Pumpkin Patch Metamorphic Suite

The Pumpkin Patch Metamorphic Suite is a distinctive Grenville suite of layered granitic gneisses and paragneisses present in the Fries thrust sheet between the Fries and Burnsville faults (Fig. 5). These rocks make up the amphibolitic basement complex of Brewer and Woodward (1988) and are included in the migmatitic biotite-hornblende gneiss unit on the Geologic Map of North Carolina (Brown, 1985). Lithologies include biotite gneiss and schist, biotite-hornblende gneiss, amphibolite, granitic gneiss, altered ultramafic rock, and garnet-kyanite gneiss (see Appendix). The rocks in this suite are intimately interlayered, and the suite is characterized by well-defined compositional layering and tight to isoclinal, contorted folds. The predominant Grenville metamorphic grade is amphibolite facies, although granulite-facies assemblages occur locally (Merschat, 1977; Gulley, 1985; Adams and Trupe, 1997).

We propose the name "Pumpkin Patch Metamorphic Suite" as a formal lithodemic unit (North American Commission on Stratigraphic Nomenclature, 1983) for these rocks (see Appendix). Although the suite contains both paragneisses and orthogneisses, including pre-Grenville igneous rocks (Berquist et al., 2003; Ownby et al., this volume) (see Appendix), the common characteristics of all rocks of the suite are high-grade Grenville metamorphism and deformation.

DISCUSSION

The Grenville rocks in our study area have experienced both Taconic and Alleghanian deformation and metamorphism. The pattern and grade of metamorphism for each event, however, are distinct and recognizable. The grade of Taconic metamorphism is highest in the Fries thrust sheet (amphibolite to granulite), decreases slightly in the underlying Sams Gap–Pigeonroost thrust sheet (amphibolite), and is absent from all the remaining lower thrust sheets. Although the Fries and Sams Gap sheets show evidence of high-grade metamorphism, preserved Grenville fabrics and relict igneous textures in Bakersville rocks demonstrate that penetrative Taconic deformation was not widespread.

Alleghanian greenschist-facies metamorphism is strongest in the Linville Falls–Stone Mountain and Fork Ridge thrust sheets. The Little Pond Mountain thrust sheet shows very low-grade Alleghanian metamorphism. Alleghanian greenschist-facies metamorphism in the Sams Gap–Pigeonroost and Fries thrust sheets is restricted to retrograde metamorphism associated with the bounding shear zones.

The foregoing observations suggest that basement in the Little Pond Mountain thrust sheet represents typical Grenville basement in the study area. In the Linville Falls–Stone Mountain, Fork Ridge, and Fries thrust sheets north of the Grandfather Mountain window, multiple Alleghanian fault zones occur close together. This region of amalgamated shear zones tends to obscure the Grenville fabrics in the rocks, making this a poor area to study Grenville features. Grenville features are best preserved west of the Grandfather Mountain window, along the west side of Figure 2, A.

The Pumpkin Patch Metamorphic Suite within the Fries thrust sheet has been interpreted as a post-Grenville, Precambrian mélange (Brewer and Woodward, 1988) and a Taconic mélange (Raymond et al., 1989). Because Bakersville intrusive rocks crosscut fabrics within the Pumpkin Patch Metamorphic Suite, we suggest that if these rocks represent a mélange, it is a Grenville mélange. This interpretation is consistent with new geochronology and geochemistry from the Pumpkin Patch Metamorphic Suite (Ownby et al., this volume). If this is the case, it represents the only known Grenville mélange in the southern Appalachians, and presumably marks a Grenville tectonic boundary.

The break-forward sequence of Alleghanian thrusting depicted in Figure 4 is consistent with the pattern of Taconic metamorphism present in the Blue Ridge thrust sheets. Only the Sams Gap–Pigeonroost and Fries thrust sheets show evidence of Taconic metamorphism. When restored to their pre-Alleghanian position, they are the farthest east and therefore closest to the Taconic metamorphic core. Grenville rocks of the Little Pond Mountain, Linville Falls, and Fork Ridge thrust sheets lie well west of this Taconic metamorphic front. The Linville Falls and Fork Ridge sheets overlie basement rocks within the Grandfather Mountain window (Globe massif of Bartholomew and Lewis, 1984), as well as overlying the Little Pond Mountain thrust sheet rocks, and must have been situated east of these rocks prior to Alleghanian thrusting. A palinspastic map of the Grenville terranes is shown in Figure 6. Displacement on the Alleghanian thrusts is not well constrained, although minimum transport distances can be obtained from the map patterns of the thrust sheets (i.e., thrust sheets must have been transported at least as far as their map width).

The restoration shows the distribution of Grenville tectonic and lithologic provinces following the main phase of Mesoproterozoic orogenesis. The rocks of the Fries thrust sheet are part of the Grenville Mars Hill terrane (Bartholomew and Lewis, 1988) and occupied a position on the outer edge of the Grenville orogen, a position consistent with the palinspastic reconstruction presented by Bartholomew and Lewis (1988, 1992). Raymond et al. (1989) included these rocks in the Cullowhee terrane and suggested that they represent a metamorphosed Paleozoic mélange. Brewer and Woodward (1988) interpreted these same rocks as a Precambrian mélange associated with closure of a post-Grenville ocean basin. The nature of the Pumpkin Patch Metamorphic Suite is consistent with an origin as an accretionary complex, and our reconstruction is consistent with the idea that the Mars Hill terrane represents material accreted to the Laurentian margin during the Grenville orogeny.

The restored positions of the Elk River and Wautaga massifs (Fig. 6) are similar to those proposed by Bartholomew and

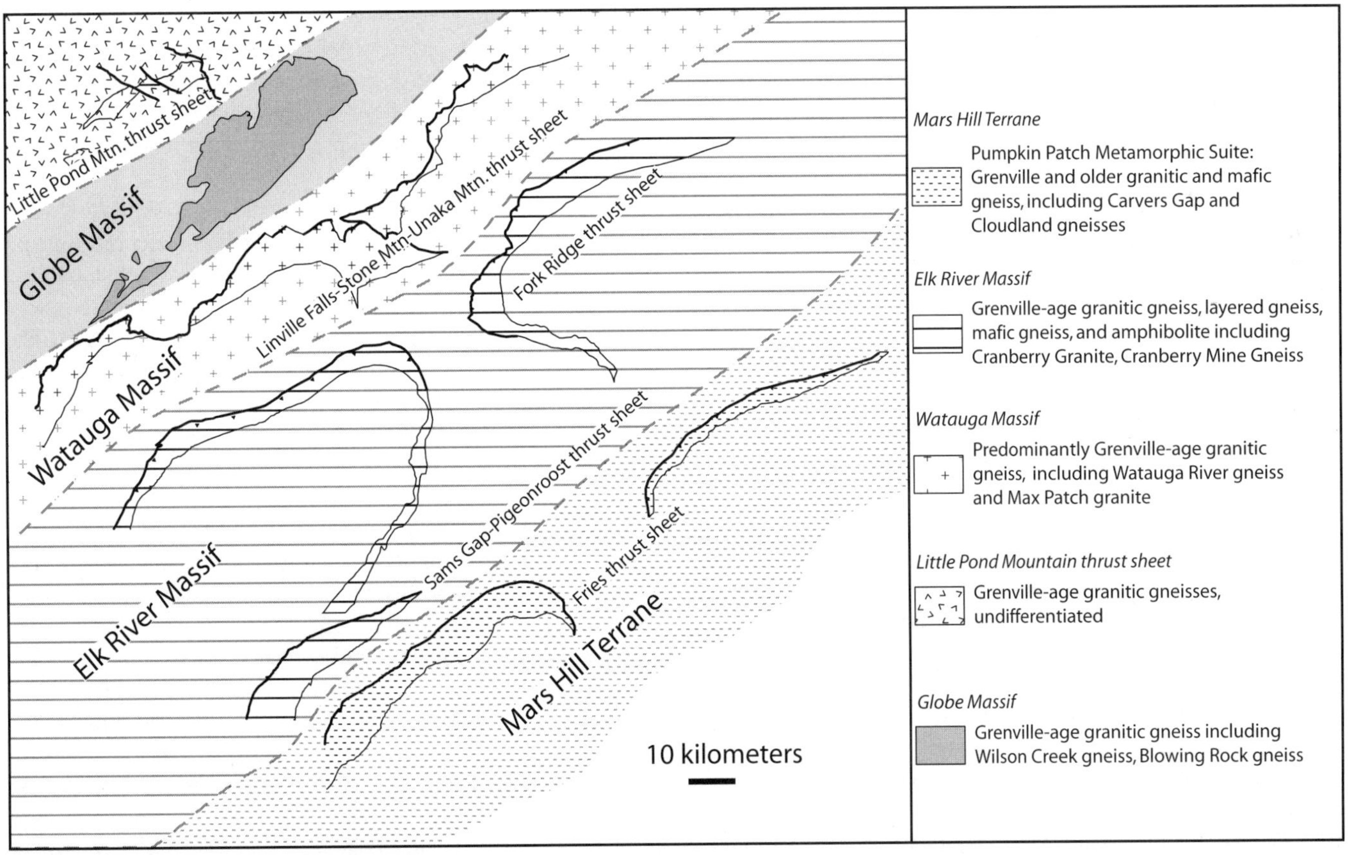

Figure 6. Restoration of Grenville massifs and terranes of the Blue Ridge to relative pre-Paleozoic positions, with outlines of present-day thrust sheets from Figure 2.

Lewis (1988, 1992), although we have included rocks in the Watauga massif that they put in the Elk River massif. Our mapping indicates that Watauga River Gneiss and Max Patch granite are present within our mapped Linville Falls–Stone Mountain–Unaka Mountain thrust sheet. Zircon from the Max Patch granite yielded an age of ca. 1 Ga (Officer et al., 2003), which may indicate that the granitic igneous rocks in this terrane represent a late-Grenville phase of felsic magmatism that was concentrated in this belt.

Our palinspastic reconstruction is also consistent with the pattern of Neopoterozoic rift-related rocks that are present in the Blue Ridge. Neoproterozoic mafic dikes intruded all of the Grenville belts. Rocks of the Bakersville Intrusive Suite have been mapped in all of the belts, except within the rocks of the Grandfather Mountain window. Figure 6, however, shows that the Neoproterozoic dikes and mafic volcanic rocks inside the Grandfather Mountain window are likely part of the same rift-related igneous activity that produced the Bakersville rocks.

CONCLUSIONS

During the Paleozoic, Grenville rocks in the Blue Ridge of northwestern North Carolina and eastern Tennessee were primarily affected by Alleghanian deformation and greenschist-facies metamorphism. The only Grenville rocks affected by Taconic orogenesis are those of the Pumpkin Patch Metamorphic Suite within the Fries thrust sheet and the rocks of the Sams Gap–Pigeonroost thrust sheet. Rocks of this suite may well represent a Grenville accretionary complex, but not a Paleozoic complex.

Effects of Paleozoic tectonism must first be deciphered to understand the Grenville orogeny in this part of the Blue Ridge. Near the Grandfather Mountain window, Alleghanian shearing is intense and widespread throughout Grenville basement rocks. West of the window, Grenville features are more easily discerned. This area therefore presents an excellent opportunity for the study of pre-Paleozoic rocks and structures.

APPENDIX. DESCRIPTION AND TYPE LOCALITY OF THE PUMPKIN PATCH METAMORPHIC SUITE

The Pumpkin Patch Metamorphic Suite is the name applied to the Grenville metamorphic rocks within the Fries thrust sheet, which lie structurally below the Ashe Metamorphic Suite. The type locality is designated as the outcrop along a roadcut of Jacks Creek Road (S.R. 1336) in the Huntdale, North Carolina 7.5-minute quadrangle, ~150 m west of the intersection with S.R. 1338 (36°00′20″N, 82°16′42″W). The name is derived from the Pumpkin Patch thrust sheet of Goldberg et al. (1986b) and Butler et al. (1987). Previously named lithodemes included in the Pumpkin Patch Metamorphic Suite are the Carvers Gap Granulite Gneiss and Cloudland Granulite Gneiss (Gulley, 1985).

The Pumpkin Patch Metamorphic Suite comprises a variety of metamorphic rock types, including biotite gneiss, biotite schist, biotite-hornblende gneiss, amphibolite, granitic gneiss (including compositions ranging from granite to granodiorite), two-pyroxene garnet granulite, altered ultramafic rock, and garnet-kyanite gneiss. These rock types are complexly interlayered, and the composite suite is characterized by well-defined compositional layering and tight to isoclinal, contorted folds.

The predominant metamorphic grade of the Pumpkin Patch Metamorphic Suite is amphibolite facies. Locally, some rocks contain granulite facies assemblages (e.g., garnet + orthopyroxene + clinopyroxene) (Merschat, 1977; Gulley, 1985; Adams and Trupe, 1997) and some yield granulite-facies conditions from cation exchange geothermobarometers (e.g., Carvers Gap and Cloudland gneisses) (Gulley, 1985). Near Late Paleozoic shear zones, the rocks exhibit a retrograde greenschist-facies overprint of the higher-grade assemblages and fabrics.

The Grenville metamorphic age of the suite is confirmed by crosscutting relationships and isotopic ages. Neoproterozoic (734 Ma) (Goldberg et al., 1986a) Bakersville Suite mafic dikes and gabbro bodies intrude the Pumpkin Patch Metamorphic Suite, crosscutting layering and folds in the Grenville gneisses. Gneisses within the suite have yielded ages ranging from 1800 to 800 Ma (Fullagar et al., 1979; Monrad and Gulley, 1983; Fullagar et al., 1999; Miller et al., 1999). The older ages reflect a pre-Grenville crustal component (Fullagar, 2002), whereas ages of ca. 1200 Ma and 800 Ma are interpreted as Grenville metamorphism and post-Grenville cooling, respectively (Fullagar et al., 1979; Monrad and Gulley, 1983). The U-Pb zircon ion-probe ages range from 1.0 to 2.7 Ga for detrital cores in metasedimentary rocks, 1.3 to 1.8 Ga for cores in meta-igneous rocks, and 1.0 to 1.1 Ga for metamorphic rims (Berquist et al., 2003; Ownby et al., this volume).

Along its southeastern boundary, the Pumpkin Patch Metamorphic Suite is separated from the Ashe Metamorphic Suite by the Burnsville fault. Along it northwestern boundary, the thrust sheet containing the Pumpkin Patch Metamorphic Suite is separated from the Fork Ridge thrust sheet by the Fries fault. The outcrop belt of the suite thus ranges from a few kilometers to ~12 km wide.

ACKNOWLEDGMENTS

Financial support for this research was provided by many sources, including the National Science Foundation (EAR-9316033), University Research Council of the University of North Carolina–Chapel Hill, Sigma Xi, the University of North Carolina Department of Geological Sciences, and Georgia Southern University. Partial support provided by the U.S. Geological Survey, National Cooperative Geologic Mapping Program. Many colleagues and former students have contributed to the ideas expressed in this paper. These include Richard N. Abbott, Mervin J. Bartholomew, Forrest H. Burton, J. Robert Butler, Jean-Pierre Dubé, Steven A. Goldberg, Lauren K. Hewitt, James P. Hibbard, Benjamin S. Johnson, Laura D. Mallard, Jonathan W. Mies, Brent V. Miller, Calvin F. Miller, Loren

A. Raymond, Issac D. Standard, and Cheryl Waters Tormey. We thank Carl E. Merschat and Mark W. Carter of the North Carolina Geological Survey–Asheville for providing unpublished geologic maps. We thank Mark Evans and Virginia Peterson for thoughtful reviews of this manuscript.

REFERENCES CITED

Abbott, R.N., Jr., and Greenwood, J.P., 2000, Retrograde metamorphism of eclogite in the eastern Blue Ridge Province of the southern Appalachian Mountains; implications regarding the origin of the Ashe Metamorphic Suite: Geological Society of America Abstracts with Programs, v. 32, no. 2, p. 1.

Abbott, R.N., Jr., and Raymond, L.A., 1984, The Ashe Metamorphic Suite, northwest North Carolina: Metamorphism and observations on geologic history: American Journal of Science, v. 284, p. 350–375.

Abbott, R.N., Jr., and Raymond, L.A., 1997, Petrology of pelitic and mafic rocks in the Ashe and Alligator Back Metamorphic Suites northeast of the Grandfather Mountain window, *in* Stewart, K.G., et al., eds., Paleozoic structure, metamorphism, and tectonics of the Blue Ridge of western North Carolina: Chapel Hill, North Carolina, Carolina Geological Society 1997 Field Trip Guidebook, p. 87–101.

Adams, M.G., 1990, The geology of the Valle Crucis area, northwestern North Carolina [M.S. thesis]: Chapel Hill, University of North Carolina, 95 p.

Adams, M.G., 1995, The tectonothermal evolution of part of the Blue Ridge thrust complex, northwestern North Carolina [Ph.D. thesis]: Chapel Hill, University of North Carolina, 193 p.

Adams, M.G., 2000, The nature of the eastern-western Blue Ridge contact across the Grandfather Mountain window: Geological Society of America Abstracts with Programs, v. 32, no. 2, p. 1.

Adams, M.G., and Su, Q., 1996, The nature and timing of deformation in the Beech Mountain thrust sheet between the Grandfather Mountain and Mountain City windows in the Blue Ridge of northwestern North Carolina: Journal of Geology, v. 104, p. 197–213.

Adams, M.G., and Trupe, C.H., 1991, Significance of shear zones in the Beech Mountain thrust sheet, northwestern North Carolina and northeastern Tennessee: Geological Society of America Abstracts with Programs, v. 23, no. 1, p. 2.

Adams, M.G., and Trupe, C.H., 1997, Conditions and timing of metamorphism in the Blue Ridge thrust complex, northwestern North Carolina and eastern Tennessee, *in* Stewart, K.G., et al., eds., Paleozoic structure, metamorphism, and tectonics of the Blue Ridge of western North Carolina: Chapel Hill, North Carolina, Carolina Geological Society 1997 Field Trip Guidebook, p. 33–48.

Adams, M.G., Stewart, K.G., Trupe, C.H., and Willard, R.A., 1995a, Tectonic significance of high-pressure metamorphic rocks and dextral strike-slip faulting in the southern Appalachians, *in* Hibbard, J., et al., eds., Current perspectives in the Appalachian-Caledonian orogen: St. John's, Newfoundland, Geological Association of Canada Special Paper 41, p. 21–42.

Adams, M.G., Trupe, C.H., Goldberg, S.A., Stewart, K.G., and Butler, J.R., 1995b, Pressure-temperature history of high-grade metamorphic rocks along the eastern-western Blue Ridge boundary, northwestern North Carolina: Geological Society of America Abstracts with Programs, v. 27, no. 2, p. 33.

Bartholomew, M.J., and Gryta, J.J., 1980, Geologic map and mineral resources report of the Sherwood quadrangle, North Carolina–Tennessee: Raleigh, North Carolina, North Carolina Division of Natural Resources and Community Development, GM 214-SE and MRS, 8 p.

Bartholomew, M.J., and Lewis, S.E., 1984, Evolution of Grenville massifs in the Blue Ridge geologic province, central and southern Appalachians, *in* Bartholomew, M.J., editor, The Grenville Event in the Appalachians and related topics: Boulder, Colorado, Geological Society of America Special Paper 194, p. 229–254.

Bartholomew, M.J., and Lewis, S.E., 1988, Peregrination of middle Proterozoic massifs and terranes within the Appalachian Orogen, eastern U.S.: Trabajos de Geología, Universidad de Oveida, v. 17, p. 155–165.

Bartholomew, M.J., and Lewis, S.E., 1992, Appalachian Grenville massifs: Pre-Appalachian translational tectonics, *in* Mason, R., ed., Basement tectonics 7: Dordrecht, The Netherlands, Kluwer Academic, v. 7, p. 363–374.

Bartholomew, M.J., Lewis, S.E., Wilson, J.R., and Gryta, J.J., 1983, Deformational history of the region between Grandfather Mountain and Mountain City Windows, *in* Lewis, S.E., ed., Geologic investigations in the Blue Ridge of northwestern North Carolina: North Carolina Geological Survey and Carolina Geological Society Guidebook: Raleigh, North Carolina, North Carolina Geological Survey, Article I, 30 p.

Berquist, P.J., Miller, C., Wooden, J., Fullagar, P., Ownby, S., and Carrigan, C., 2003, The Mars Hill terrane: Extent, age, and origin of the oldest rocks in the southeastern USA: Geological Society of America Abstracts with Programs, v. 35, no. 1, p. 19.

Boyer, S.E., and Elliott, D., 1982, Thrust systems: American Association of Petroleum Geologists Bulletin, v. 66, p. 1196–1230.

Brewer, R.C., and Woodward, N.B., 1988, The amphibolitic basement complex in the Blue Ridge Province of western North Carolina, proto-Iapetus?: American Journal of Science, v. 288, p. 953–967.

Brobst, D.A., 1962, Geology of the Spruce Pine district, Avery, Mitchell, and Yancey Counties, North Carolina: Reston, Virginia, U.S. Geological Survey Bulletin, 1122-A, 26 p.

Brown, P.M., compiler, 1985, Geologic map of North Carolina: Raleigh, North Carolina, North Carolina Geological Survey, scale 1:500,000.

Bryant, B., and Reed, J.C., Jr., 1970, Geology of the Grandfather Mountain window and vicinity, North Carolina and Tennessee: Reston, Virginia, U.S. Geological Survey Professional Paper 615, 190 p.

Burton, F.H., 1996, Kinematic study of the Taconic suture, west-central North Carolina [M.S. thesis]: Chapel Hill, University of North Carolina, 114 p.

Butler, J.R., 1973, Paleozoic deformation and metamorphism in part of the Blue Ridge thrust sheet, North Carolina: American Journal of Science, v. 273-A, p. 72–88.

Butler, J.R., 1991, Metamorphism, *in* Horton, J.W., Jr., and Zullo, V.A., eds., The geology of the Carolinas: Carolina Geological Society fiftieth anniversary volume: Knoxville, University of Tennessee Press and Carolina Geological Society, p. 127–141.

Butler, J.R., Goldberg, S.A., and Mies, J.W., 1987, Tectonics of the Blue Ridge west of the Grandfather Mountain window, North Carolina and Tennessee: Geological Society of America Abstracts with Programs, v. 19, no. 2, p. 77.

Carter, M.W., Merschat, C.E., Williams, S.T., and Wiener, L.S., 1998, West-central Blue Ridge tectonic synthesis revisited, western North Carolina: Geological Society of America Abstracts with Programs, v. 30, no. 4, p. 6.

Dallmeyer, R.D., 1975, $^{40}Ar/^{39}Ar$ ages of biotite and hornblende from a progressively metamorphosed basement terrane: Their bearing on the interpretation of release spectra: Geochimica et Cosmochimica Acta, v. 39, p. 1655–1669.

Diegel, F.A., 1988, The Rome Formation décollement in the Mountain City window, Tennessee; A case for involvement of evaporites in the genesis of Max Meadows–type breccias, *in* Mitra, G. and Wojtal, S., eds., Geometries and mechanisms of thrusting with special reference to the Appalachians: Boulder, Colorado, Geological Society of America Special Paper 222, p. 137–164.

Diegel, F.A., and Wojtal, S., 1985, Structural transect in SW Virginia and NE Tennessee, *in* Woodward, N.B., ed., Field trips in the Southern Appalachians: Knoxville, University of Tennessee, Department of Geological Sciences, Studies in Geology 9, p. 70–143.

Fullagar, P.D., 2002, Evidence for early Mesoproterozoic (and older?) crust in the southern and central Appalachians of North America: Gondwana Research, v. 5, p. 197–203.

Fullagar, P.D., and Bartholomew, M.J., 1983, Rubidium-strontium ages of the Watauga River, Cranberry, and Crossing Knob gneisses, northwestern North Carolina, *in* Lewis, S.E., ed., Geologic investigations in the Blue

Ridge of northwestern North Carolina: Chapel Hill, North Carolina, Carolina Geological Society Guidebook, North Carolina Geological Survey, Article II, 29 p.

Fullagar, P.D., Hatcher, R.D., Jr., and Merschat, C.E., 1979, 1200-m.y.-old gneisses in the Blue Ridge Province of North and South Carolina: Southeastern Geology, v. 20, p. 69–77.

Fullagar, P.D., Su, Q., and Gulley, G.L., Jr., 1999, Pre-Grenville uranium-lead zircon age for the Carvers Gap Gneiss in the western Blue Ridge of North Carolina–Tennessee: Geological Society of America Abstracts with Programs, v. 31, no. 3, p. 16.

Goldberg, S.A., and Dallmeyer, R.D., 1997, Chronology of Paleozoic metamorphism and deformation in the Blue Ridge thrust complex, North Carolina and Tennessee: American Journal of Science, v. 297, p. 488–526.

Goldberg, S.A., Butler, J.R., and Fullagar, P.D., 1986a, The Bakersville dike swarm: Geochronology and petrogenesis of Late Proterozoic basaltic magmatism in the southern Appalachian Blue Ridge: American Journal of Science, v. 286, p. 403–430.

Goldberg, S.A., Butler, J.R., and Mies, J.W., 1986b, Subdivision of the Blue Ridge thrust complex, western North Carolina and adjacent Tennessee: Geological Society of America Abstracts with Programs, v. 18, no. 6, p. 616–617.

Goldberg, S.A., Butler, J.R., Mies, J.W., and Trupe, C.H., 1989, The southern Appalachian orogen in northwestern North Carolina and adjacent states: Washington, D.C., American Geophysical Union, International Geological Congress Field Trip T365 Guidebook, 55 p.

Goldberg, S.A., Butler, J.R., Trupe, C.H., and Adams, M.G., 1992, The Blue Ridge thrust complex northwest of the Grandfather Mountain window, North Carolina and Tennessee, *in* Dennison, J.M., and Stewart, K.G., eds., Geologic field guides to North Carolina and vicinity: Chapel Hill, University of North Carolina, Department of Geology, Geologic Guidebook no. 1, p. 213–233.

Gulley, G.L., Jr., 1985, A Proterozoic granulite-facies terrane on Roan Mountain, western Blue Ridge belt, North Carolina–Tennessee: Geological Society of America Bulletin, v. 96, p. 1428–1439.

Hatcher, R.D., Jr., 1978, Tectonics of the western Piedmont and Blue Ridge, southern Appalachians: Review and speculation: American Journal of Science, v. 278, p. 276–304.

Hatcher, R.D., Jr., 1989, Tectonic synthesis of the U.S. Appalachians, *in* Hatcher, R.D., Jr., et al., eds., The Appalachian-Ouachita orogen in the United States: Boulder, Colorado, Geological Society of America, The Geology of North America, v. F-2, p. 511–535.

Hatcher, R.D., Jr., Hooper, R.J., Petty, S.M., and Willis, J.D., 1984, Structure and chemical petrology of three southern Appalachian mafic-ultramafic complexes and their bearing upon the tectonics of emplacement and origin of Appalachian ultramafic bodies: American Journal of Science, v. 284, p. 484–506.

Hewitt, L.K., 1999, The thermobarometric history along a transect of the Ashe Metamorphic Suite, southwest of the Grandfather Mountain window, northwestern North Carolina [M.S. thesis]: Chapel Hill, University of North Carolina, 162 p.

Hibbard, J.P., Stewart, K.G., and Henika, W.S., 2001, Framing the Piedmont Zone in North Carolina and southern Virginia, *in* Hoffman, C.W., ed., Fiftieth annual meeting, Geological Society of America, Southeastern Section, Field Trip Guidebook: Raleigh, North Carolina, Geological Society of America, Southeastern Section, p. 1–26.

Horton, J.W., Drake, A.A., Jr., and Rankin, D.W., 1989, Tectonostratigraphic terranes and their boundaries in the central and southern Appalachians, *in* Dallmeyer, R.D., ed., Terranes in the circum-Atlantic Paleozoic orogens: Boulder, Colorado, Geological Society of America Special Paper 230, p. 213–245.

Keith, A., 1903, Description of the Cranberry quadrangle, North Carolina–Tennessee: Reston, Virginia, U.S. Geological Survey Geologic Atlas, folio 90.

Keith, A., 1904, Description of the Asheville quadrangle, North Carolina–Tennessee: Reston, Virginia, U.S. Geological Survey Geologic Atlas, folio 116.

Keith, A., 1905, Description of the Mount Mitchell quadrangle, North Carolina–Tennessee: Reston, Virginia, U.S. Geological Survey Geological Atlas, folio 124.

Keith, A., 1907, Description of the Roan Mountain quadrangle, North Carolina–Tennessee: Reston, Virginia, U.S. Geological Survey Geological Atlas, folio 151.

King, P.B., and Ferguson, H.W., 1960, Geology of northeasternmost Tennessee, with description of the basement rocks by Warren Hamilton: Reston, Virginia, U.S. Geological Survey Professional Paper 311, 160 p.

Kish, S.A., 1983, A geochronological study of deformation and metamorphism in the Blue Ridge and Piedmont of the Carolinas [Ph.D. dissertation]: Chapel Hill, University of North Carolina, 220 p.

Kuchenbuch, P.A., 1979, Petrology of some metagabbro bodies in Mars Hill quadrangle, North Carolina: Geological Society of America Abstracts with Programs, v.11, no. 4, p. 186.

Long, L.E., Kulp, J.L., and Eckelmann, F.D., 1959, Chronology of major metamorphic events in the southeastern United States: American Journal of Science, v. 257, p. 585–603.

McCrary, C.R., III, 1997, Pressure-temperature-time history of Grenville basement gneisses in the Pardee Point thrust sheet near Hampton, Tennessee: Geological Society of America Abstracts with Programs, v. 29, no. 3, p. 57.

McSween, H.Y., Jr., Abbott, R.N., and Raymond, L.A., 1989, Metamorphic conditions in the Ashe Metamorphic Suite, North Carolina Blue Ridge: Geology, v. 17, p. 1140–1143.

McSween, H.Y., Jr., Speer, J.A., and Fullagar, P.D., 1991, Plutonic rocks, *in* Horton, J.W., Jr., and Zullo, V.A., eds., The geology of the Carolinas: Carolina Geological Society fiftieth anniversary volume: Knoxville, University of Tennessee Press and Carolina Geological Society, p. 109–127.

Merschat, C.E., 1977, Geologic map and mineral resources summary of the Mars Hill quadrangle, North Carolina: Raleigh, North Carolina, North Carolina Geological and Mineral Resources Section Map GM191-SE, scale 1:24,000, 15 p.

Merschat, C.E., 1993, Geologic map and mineral resources summary of the Barnardsville quadrangle, North Carolina: North Carolina Department of Natural Resources and Community Development, Division of Land Resources, Geological Survey Section, GM 200-SW and MRS 200-SW, 21 p.

Mies, J.W., 1990, Structural and petrologic studies of mylonite at the Grenville basement–Ashe Formation boundary, Grayson County, Virginia to Mitchell County, North Carolina [Ph.D. dissertation]: Chapel Hill, University of North Carolina, 330 p.

Mies, J.W., 1991, Planar dispersion of folds in ductile shear zones and kinematic interpretation of fold hinge girdles: Journal of Structural Geology, v. 13, p. 281–297.

Miller, B.V, Stewart, K.G., Miller, C.F., and Thomas, C.W., 2000, U-Pb ages from the Bakersville, North Carolina eclogite: Taconian eclogite metamorphism followed by Acadian and Alleghanian cooling: Geological Society of America Abstracts with Programs, v. 32, no. 2, p. 62.

Miller, C.F., Vinson, S.B., Fullagar, P.D., Hatcher, R.D., Jr., and Coath, C.D., 1999, Ages of Precambrian zircon from the Blue Ridge and Inner Piedmont, North Carolina–South Carolina–Georgia—An ion microprobe study: Geological Society of America Abstracts with Programs, v. 31, no. 3, p. 30.

Misra, K.C., and Conte, J.A., 1991, Amphibolites of the Ashe and Alligator Back Formations, North Carolina: Samples of Late Proterozoic–Early Paleozoic oceanic crust: Geological Society of America Bulletin, v. 103, p. 737–750.

Monrad, J.R., and Gulley, G.L., Jr., 1983, Age and *P-T* conditions during metamorphism of granulite-facies gneisses, Roan Mountain, NC-TN, *in* Lewis, S.E., ed., Geologic investigations in the Blue Ridge of northwestern North Carolina: Carolina Geological Society Field Trip Guidebook: Raleigh, North Carolina, North Carolina Geological Survey, Article IV, 29 p.

Newman, J., and Mitra, G., 1993, Lateral variations in mylonite zone thickness as influenced by fluid-rock interactions, Linville Falls fault, North Carolina: Journal of Structural Geology, v. 15, p. 849–863.

North American Commission on Stratigraphic Nomenclature (NACSN), 1983,

North American Stratigraphic Code: American Association of Petroleum Geologists Bulletin, v. 67, p. 841–875.

Officer, N.D., Berquist, P.J., Miller, C.F., Fullagar, P., Wooden, J., and Carrigan, C.W., 2003, Western Blue Ridge basement of northeast TN and northwestern NC: Age, geochemistry, and possible relationships to Proterozoic rocks of the southeastern USA: Geological Society of America Abstracts with Programs, v. 35, no. 1, p. 19.

O'Hara, K., 1992, Major- and trace-element constraints on the petrogenesis of a fault-related pseudotachylyte, western Blue Ridge province, North Carolina: Tectonophysics, v. 204, p. 279–288.

Passchier, C.W., and Trouw, R.A.J., 1992, Microtectonics: Berlin, Springer-Verlag, 289 p.

Rainey, L., 1989, A study of the Late Proterozoic Bakersville Suite: Implications for Paleozoic metamorphism in portions of the Blue Ridge thrust complex, western North Carolina and eastern Tennessee [M.S. thesis]: Chapel Hill, University of North Carolina, 161 p.

Rankin, D.W., 1970, Stratigraphy and structure of Precambrian rocks in northwestern North Carolina, *in* Fisher, G.W., et al., eds., Studies of Appalachian geology—Central and southern: New York, Wiley Interscience, p. 227–245.

Rankin, D.W., Espenshade, G.H., and Neuman, R.B., 1972, Geologic map of the west half of the Winston-Salem quadrangle, North Carolina, Virginia, and Tennessee: Reston, Virginia, U.S. Geological Survey Map I-709-A, Scale 1:250,000.

Rankin, D.W., Espenshade, G.H., and Shaw, K.W., 1973, Stratigraphy and structure of the metamorphic belt in northwestern North Carolina and southwestern Virginia: A study from the Blue Ridge across the Brevard fault zone to the Sauratown Mountains anticlinorium: American Journal of Science, v. 273-A, p. 1–40.

Rankin, D.W., Drake, A.A., Jr., Glover, L., III, Goldsmith, R., Hall, L.M., Murray, D.P., Ratcliffe, N.M., Read, J.F., Secor, D.T., Jr., and Stanley, R.S., 1989, Pre-orogenic terranes, *in* Hatcher, R.D., Jr., et al., eds., The Appalachian-Ouachita orogen in the United States: Boulder, Colorado, Geological Society of America, Geology of North America, v. F-2, p. 7–100.

Rankin, D.W., Dillon, W.P., Black, D.F.B., Boyer, S.E., Daniels, D.L., Goldsmith, R., Grow, J.A., Horton, J.W., Jr., Hutchinson, D.R., Klitgord, K.D., McKowell, R.C., Miltop, D.J., Owens, J.P., and Phillips, J.D., with contributions by Bayer, K.C., Butler, J.R., Elliott, D.W., and Milici, R.C., 1991, Centennial Continent/Ocean Transect #16, E-4, Central Kentucky to the Carolina Trough: Boulder, Colorado, Geological Society of America, 41 p.

Raymond, L.A., and Johnson, P.A., 1992, The Ashe Metamorphic Suite–Cullowhee terrane contact in the Flat Creek area, Blue Ridge Belt, North Carolina: A re-activated fault: Geological Society of America Abstracts with Programs, v. 24, no. 2, p. 59–60.

Raymond, L.A., Yurkovich, S.P., and McKinney, M., 1989, Block-in-matrix structures in the North Carolina Blue Ridge belt and their significance for the tectonic history of the southern Appalachian orogen, *in* Horton, J.W., Jr., and Rast, N., eds., Mélanges and olistostromes of the U.S. Appalachians: Boulder, Colorado, Geological Society of America Special Paper 228, p. 195–215.

Rodgers, J., 1953, Geologic map of east Tennessee with explanatory text: Nashville, Tennessee, Tennessee Division of Geology, Bulletin 58, 168 p.

Rodgers, J., 1970, The tectonics of the Appalachians: New York, Wiley Interscience, 271 p.

Schedl, A., McCabe, C., Montanez, I., Fullagar, P.D., and Valley, J., 1992, Alleghanian regional diagenesis: A response to the migration of modified metamorphic fluids derived from beneath the Blue Ridge–Piedmont thrust sheet: Journal of Geology, v. 100, p. 339–352.

Shekarchi, E., 1959, The geology of the Flag Pond quadrangle, Tennessee–North Carolina [Ph.D. dissertation]: Knoxville, University of Tennessee, 140 p.

Stewart, K.G., Adams, M.G., and Trupe, C.H., 1997, Paleozoic structural evolution of the Blue Ridge thrust complex, western North Carolina, *in,* Stewart, K.G., et al., eds., Paleozoic structure, metamorphism, and tectonics of the Blue Ridge of western North Carolina: Chapel Hill, North Carolina, Carolina Geological Society 1997 Field Trip Guidebook, p. 21–32.

Stose, A.J., and Stose, G.W., 1957, Geology and mineral resources of the Gossan Lead district and adjacent areas in Virginia: Charlottesville, Virginia Division of Mineral Resources Bulletin 72, 233 p.

Su, Q., Goldberg, S.A., and Fullagar, P.D., 1994, Precise U-Pb zircon ages of Neoproterozoic plutons in the southern Appalachian Blue Ridge and their implications for the initial rifting of Laurentia: Precambrian Research, v. 68, p. 81–95.

Trupe, C.H., 1989, Microstructural analysis of mylonites from the Blue Ridge thrust complex, northwestern North Carolina [M.S. thesis]: Chapel Hill, University of North Carolina, 101 p.

Trupe, C.H., 1997a, Deformation and metamorphism in part of the Blue Ridge thrust complex, northwestern North Carolina [Ph.D. dissertation]: Chapel Hill, University of North Carolina, 176 p.

Trupe, C.H., 1997b, Structural relationships in the Linville Falls shear zone, Blue Ridge thrust complex, northwestern North Carolina, *in* Stewart, K.G., et al., eds., Paleozoic structure, metamorphism, and tectonics of the Blue Ridge of western North Carolina: Chapel Hill, North Carolina, Carolina Geological Society 1997 Field Trip Guidebook, p. 49–66.

Trupe, C.H., Stewart, K.G., Adams, M.G., Waters, C.L., Miller, B.V., and Hewitt, L.K., 2003, The Burnsville fault: Evidence for the timing and kinematics of Acadian dextral transform tectonics in the southern Appalachians: Geological Society of America Bulletin, v. 115, p. 1365–1376.

Van Camp, S.G., and Fullagar, P.D., 1982, Rb-Sr whole-rock ages of mylonites from the Blue Ridge and Brevard zone of North Carolina: Geological Society of America Abstracts with Programs, v. 14, no. 1–2, p. 92–93.

Willard, R.A., and Adams, M.G., 1994, Newly discovered eclogite in the southern Appalachian orogen, northwestern North Carolina: Earth and Planetary Science Letters, v. 123, p. 61–70.

Wojtal, S., and Mitra, G., 1988, Nature of deformation in some fault rocks from Appalachian thrusts, *in* Mitra, G., and Wojtal, S., eds., Geometries and mechanism of thrusting, with special reference to the Appalachians: Boulder, Colorado, Geological Society of America Special Paper 222, p. 17–33.

Manuscript Accepted by the Society August 25, 2003

Geological Society of America
Memoir 197
2004

Mesoproterozoic rocks of the New Jersey Highlands, north-central Appalachians: Petrogenesis and tectonic history

Richard A. Volkert*
New Jersey Geological Survey, P.O. Box 427, Trenton, New Jersey 08625, USA

ABSTRACT

The New Jersey Highlands preserve a diverse assemblage of Mesoproterozoic rocks, whose geologic evolution correlates in part to other Grenville terranes in eastern North America, mainly north of the Blue Ridge. Voluminous undated—but postulated to be >1200 Ma—calc-alkaline metaplutonic and metavolcanic rocks of the Losee Metamorphic Suite were formed in a continental-margin magmatic arc, and metamorphosed supracrustal rocks >1174 Ma were formed along eastern Laurentia in a marginal back-arc basin. They include terrigenous to shallow marine metasandstones, a shelf sequence of sand-dominated metaclastic rocks, stromatolitic marble, locally pillowed mafic volcanics, and deep-water volcanogenic metagraywacke. Subduction and calc-alkaline magmatism had ceased by 1176 Ma, but timing of accretion of the arc complex to Laurentia is uncertain.

Between 1160 and 1130 Ma, supracrustal rocks were intruded by thin sheets and dikes of meta-anorthosite and megacrystic amphibolites likely coeval with anorthosite, mangerite, charnockite, granite (AMCG) magmatism in the Adirondack Highlands. Meta-anorthosite and megacrystic amphibolites in New Jersey were intruded by A-type granite of the 1110 ± 25 Ma Byram and 1095 ± 9 Ma Lake Hopatcong Intrusive Suites. Emplacement of these suites was followed by upper amphibolite- to granulite-facies metamorphism at 1090–1030 Ma, during the Ottawan orogeny. Sheets of late synorogenic microperthite alaskite were emplaced during the latter part of Ottawan orogenesis. Postorogenic 1020 ± 4 Ma Mount Eve Granite and undeformed 1029 ± 1 Ma trondhjemite date the close of the Ottawan orogeny in the New Jersey Highlands. Undeformed discordant pegmatites and small granite bodies were emplaced between 1004 and 989 Ma.

Keywords: Mesoproterozoic, Grenville, New Jersey Highlands, geochemistry, tectonics

INTRODUCTION

Mesoproterozoic rocks of the New Jersey Highlands have been a subject of study for more than a century, largely due to past economic interest in the >300 magnetite mines that are distributed throughout the region and in the the marble-hosted Zn-Mn-Fe deposits at Franklin and Sterling Hill. Work by geologists from the New Jersey Zinc Company (Hague et al., 1956) and the U.S. Geological Survey (Hotz, 1952; Sims, 1958; Offield, 1967; Drake, 1969; Baker and Buddington, 1970) led to lithologic refinement of the four-unit subdivision of the Mesoproterozoic rocks developed by early workers (e.g., Spencer et al., 1908).

From 1984 until 1990, reconnaissance-scale mapping of the Highlands was undertaken jointly by the New Jersey Geologi-

*E-mail: rich.volkert@dep.state.nj.us.

Volkert, R.A., 2004, Mesoproterozoic rocks of the New Jersey Highlands, North-Central Appalachians: Petrogenesis and tectonic history, *in* Tollo, R.P., Corriveau, L., McLelland, J., and Bartholomew, M.J., eds., Proterozoic tectonic evolution of the Grenville orogen in North America: Boulder, Colorado, Geological Society of America Memoir 197, p. 697–728. For permission to copy, contact editing@geosociety.org.

cal Survey and the U.S. Geological Survey for the new bedrock geologic map of New Jersey (Drake et al., 1996) and from 1991 until the present, the New Jersey Geological Survey has conducted detailed 1:24,000-scale mapping. This work has built significantly on previous studies and regional syntheses involving the Highlands (e.g., Rankin et al., 1993) and, for the first time, has permitted a comprehensive comparison of the Mesoproterozoic rocks of the western and eastern Highlands that are separated by Paleozoic cover rocks. Equally important, it has permitted a comparison of the Mesoproterozoic rocks of the New Jersey Highlands and other eastern North American Grenville terranes. The latter has proven critical in understanding the geologic evolution of Mesoproterozoic rocks in the north-central Appalachians in the context of the broader Grenville orogenic cycle. Much has been accomplished in deciphering the tectonic setting of the formation, relative stratigraphy, and geologic evolution of the Mesoproterozoic rocks of the Highlands through integrated mapping and geochemical studies. However, limitations imposed by the paucity of geochronology, the results of which are summarized in Table 1, have prevented further refinement of the ages of the major rock suites. The importance of high-precision geochronology in supporting detailed studies of the Highlands is recognized, but U-Pb age dating of rock suites remains an unattainable goal for the New Jersey Geological Survey.

This paper discusses the lithologic, stratigraphic, geochemical, and metamorphic framework of Mesoproterozoic rocks in New Jersey, based on previous work and recent mapping, and presents a tectonic synthesis that is consistent with the available data. The terminology proposed by Moore and Thompson (1980) for the succession of orogenic events constituting the Grenville orogenic cycle, which includes the Elzevirian orogeny at ca. 1300–1200 Ma and the Ottawan orogeny at ca. 1100–1000 Ma, is followed in this paper.

GEOLOGIC SETTING

The New Jersey Highlands, along with the contiguous Hudson Highlands in southern New York and Reading Prong in eastern Pennsylvania, constitutes one of the largest Grenville terranes extending along eastern North America (Fig. 1). Most of the 1000 km^2 area of the New Jersey Highlands is underlain by rocks of Mesoproterozoic age that are separated into western and eastern domains (Fig. 2) by downfaulted unmetamorphosed Paleozoic cover rocks of the Green Pond Mountain region. Basement of the two domains is linked beneath the cover. Mesoproterozoic rocks in the western Highlands are nonconformably overlain by a locally preserved sequence of Neoproterozoic siliciclastic and felsic volcanic rocks (Chestnut Hill Formation) (Drake, 1984; Volkert, 2000). Throughout the Highlands, Mesoproterozoic rocks are intruded by a northeast-trending swarm of Neoproterozoic alkalic to tholeiitic diabase dikes related to the breakup of Rodinia (Volkert and Puffer, 1995). The Early Cambrian Hardyston Formation rests nonconformably on Mesoproterozoic basement and on the Neoproterozoic rocks, where present.

Along the eastern border of the Highlands, Mesoproterozoic rocks are in fault contact with Late Triassic to Early Jurassic terrestrial fluvial and lacustrine sedimentary rocks and Early Jurassic tholeiitic basalts, and locally with sedimentary rocks of Cambrian and Ordovician age (Fig. 2). Along the western border, Mesoproterozoic rocks are nonconformably overlain by, or in fault contact with, a passive margin clastic and carbonate shelf sequence of Cambrian and Ordovician age.

METAMORPHIC CONDITIONS

Mesoproterozoic rocks of the New Jersey Highlands were metamorphosed at upper amphibolite- to granulite-facies con-

TABLE 1. SUMMARY OF ISOTOPIC AGES OF THE NEW JERSEY HIGHLANDS

Rock unit	Lithology	Isotopic age (Ma)	Technique	Reference
Losee Metamorphic Suite				
	Anatectic metatrondhjemite	1029 ± 1	U-Pb TIMS (zircon)	J.N. Aleinikoff (personal commun., 2003)
	Anatectic metatrondhjemite	1030 ± 12	U-Pb SHRIMP (zircon)	J.N. Aleinikoff (personal commun., 2003)
Byram Intrusive Suite				
	Hornblende granite	1116 ± 41	Rb-Sr whole-rock	Volkert et al. (2000b)
	Hornblende granite	1110 ± 25	Rb-Sr whole-rock	R.A. Volkert (this study)
Lake Hopatcong Intrusive Suite				
	Pyroxene granite	1095 ± 9	Rb-Sr whole-rock	Volkert et al. (2000b)
	Pyroxene granite	1097 ± 18	Rb-Sr whole-rock	R.A. Volkert (this study)
Mount Eve Granite				
	Hornblende-biotite granite	1020 ± 4	U-Pb TIMS (zircon)	Drake et al. (1991b)
	Granite pegmatite	1004 ± 3	U-Pb TIMS (zircon)	R.E. Zartman (personal commun., 2003)
Pegmatites				
	Granite pegmatite	998–989	U-Pb TIMS (zircon, titanite)	R.D. Tucker (personal commun., 2003)

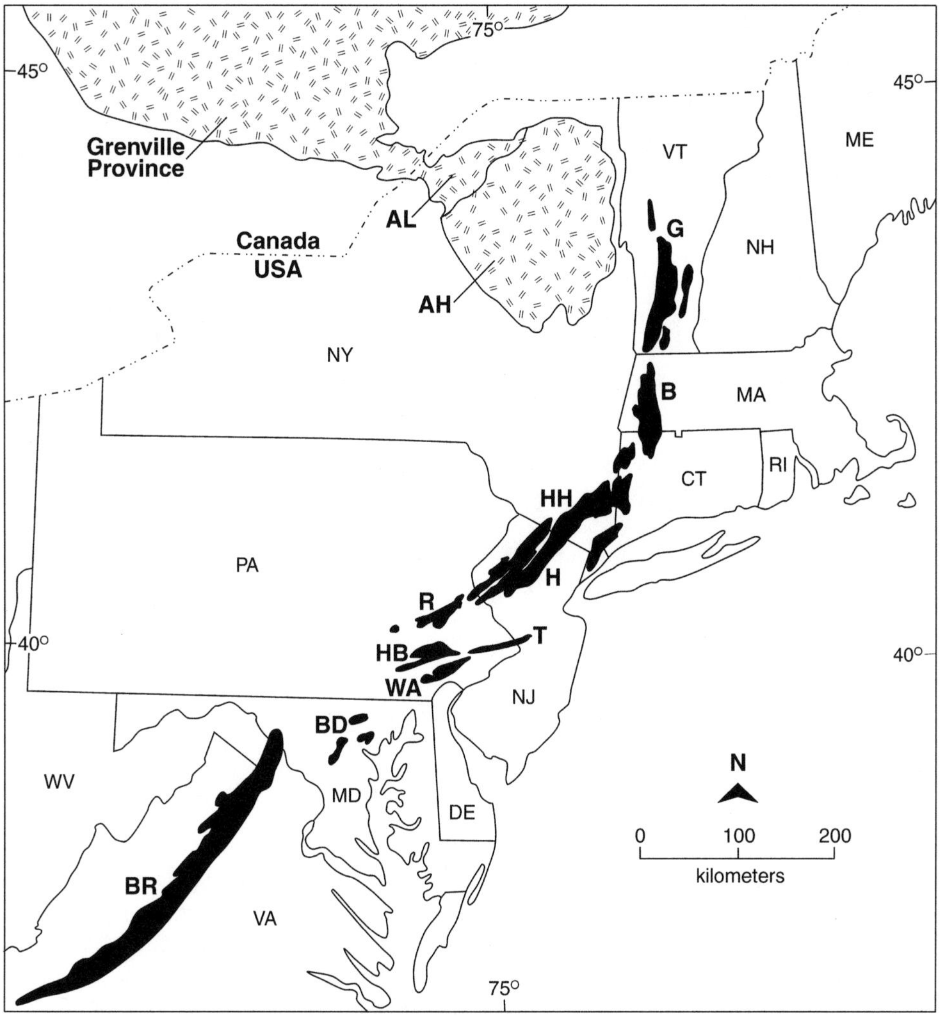

Figure 1. Distribution of Grenville-age rocks in the central and northern Appalachians (solid bodies), the Adirondack Highlands (AH) and Lowlands (AL), and the Grenville Province of southeastern Canada. B—Berkshire Mountains; BD—Baltimore Gneiss domes; BR—Blue Ridge; G—Green Mountains; H—New Jersey Highlands; HB—Honey Brook Upland; HH—Hudson Highlands; R—Reading Prong; T—Trenton Prong; WA—West Chester and Avondale Massifs. Modified from Rankin et al. (1993).

ditions during the Ottawan orogeny. Timing of this 1090–1030 Ma metamorphism is reasonably well constrained by field relationships and available geochronologic data. A maximum age is provided by a Rb-Sr whole-rock isochron of 1095 ± 9 Ma (Volkert et al., 2000b) from penetratively foliated granite of the Lake Hopatcong Intrusive Suite, which was emplaced immediately prior to the Ottawan event. A minimum age is provided by a U-Pb zircon age of 1020 ± 4 Ma (Drake et al., 1991a) from undeformed postorogenic Mount Eve Granite and U-Pb zircon ages of 1029 ± 1 Ma and 1030 ± 12 Ma (J.N. Aleinikoff, unpublished data) (Table 1) from undeformed anatectic trondhjemite mobilized from older, well-foliated metatonalite of the Losee Metamorphic Suite. These ages are discussed in more detail elsewhere in the paper. Metamorphic mineral separates of Grenville-age basement from beneath the New Jersey Coastal Plain define a less precise Rb-Sr isochron age of 1025 ± 36 Ma (Sheridan et al., 1999) that also falls within this range.

Available pressure and temperature estimates were compiled for the western and eastern Highlands and, whereas most of the *P-T* data are from the former, the results appear to be applicable to both areas. Prograde mineral assemblages of paragneisses are the same, and orthopyroxene is present in mafic and intermediate orthogneisses in both areas, indicating similar peak *P-T* conditions representative of granulite facies. Granulite-facies mineral assemblages include orthopyroxene + clinopyroxene + plagioclase + hornblende ± garnet in mafic rocks; clinopyroxene + orthopyroxene + plagioclase + quartz in intermediate rocks, garnet + biotite + sillimanite + plagioclase + K-feldspar + quartz in quartzofeldspathic gneiss; and clinopyroxene + plagioclase + quartz ± scapolite ± calcite in calcsilicate rocks.

In the western Highlands, Carvalho and Sclar (1988) obtained a minimum temperature of 760 °C for metamorphism, based on franklinite-gahnite intergrowths in zinc ore hosted by the Franklin Marble in the Franklin quadrangle (Fig. 2). Johnson (1990) used the garnet-plagioclase-sillimanite-quartz geobarometer to calculate pressures of 410–490 MPa and the garnet-rutile-ilmenite-plagioclase-quartz geobarometer to cal-

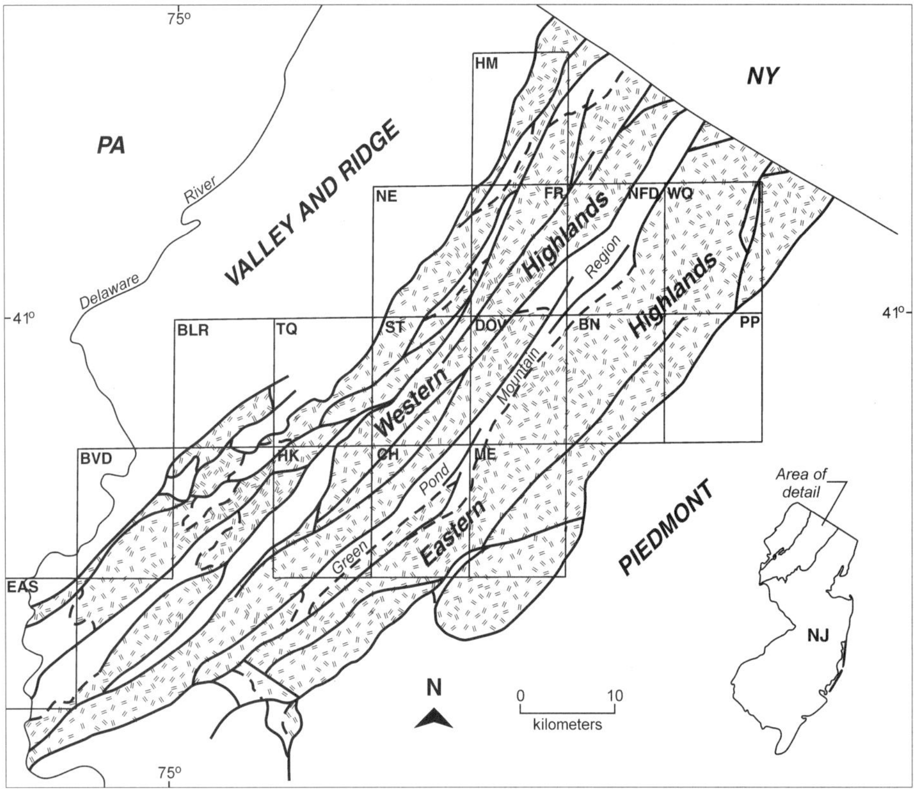

Figure 2. Simplified geologic map of northern New Jersey, showing Mesoproterozoic rocks within fault-bounded blocks of the western and eastern Highlands (cross-hatched), Paleozoic rocks of the Valley and Ridge, and Mesozoic rocks of the Piedmont. Unpatterned areas within the Highlands are underlain by Paleozoic rocks. Solid lines are faults; dashed lines are unconformities. Abbreviations of 7.5-minute quadrangles are: BLR—Blairstown; BN—Boonton; BVD—Belvidere; CH—Chester; DOV—Dover; EAS—Easton; FR—Franklin; HK—Hackettstown; HM—Hamburg; ME—Mendham; NE—Newton East; NFD—Newfoundland; PP—Pompton Plains; ST—Stanhope; TQ—Tranquility; WQ—Wanaque. Adapted from Drake et al. (1996).

culate a single pressure of 710 MPa for quartzofeldspathic gneiss in the same area. Based on the compositions of Fe-Ti oxides from mineral assemblages in magnetite ore in the Franklin quadrangle, Puffer et al. (1993a) obtained a temperature of 730 °C for the quartzofeldspathic gneiss host and 705–750 °C for coexisting magnetite-ilmenite pairs in the ore. Volkert and Gorring (2001) calculated temperatures of 760–785 °C and pressures of 490–540 MPa from amphibolites in the Easton, Blairstown, and Tranquility quadrangles using Al-in-hornblende geobarometry and hornblende-plagioclase geothermometry.

In the eastern Highlands in New Jersey and New York, Dallmeyer (1974) obtained temperatures of 700–750 °C and pressures of 400–550 MPa from equilibrium mineral assemblages in outcrops of rusty, graphitic biotite-quartz-feldspar gneiss. Young (1995) used the garnet-biotite geothermometer to calculate temperatures of 670–740 °C and the garnet-rutile-ilmenite-sillimanite-quartz geobarometer to calculate pressures of 620–800 MPa from mineral assemblages in the same gneiss from the Dover quadrangle. Based on graphite crystallinity, Volkert et al. (2000a) calculated a less precise minimum temperature of 685 °C from outcrops of the same gneiss in the eastern Highlands. In the Dover quadrangle, Young and Cuthbertson (1994) calculated a load pressure of 600–650 MPa during crystallization of monzonitic rocks of the Lake Hopatcong Suite using the pigeonite geobarometer. Except for the 800 MPa pressure estimate of Young (1995) and the 710 MPa estimate of Johnson (1990), the data from both the western and eastern Highlands are in reasonably good agreement and yield overlapping temperatures of 670–780 °C and pressures of 410 to ~620 MPa during regional metamorphism associated with the Ottawan orogeny.

The *P-T* data provide little evidence for a metamorphic discontinuity between the western and eastern Highlands that is related to the Ottawan event. The significance of the two higher-pressure estimates is currently not well understood. They may preserve evidence for a deeper level of burial during the Ottawan event, or they may record an older, pre-Ottawan high-pressure metamorphism. More work is needed to peer through the overprinting effects of Ottawan metamorphism to detect *P-T* differences between the various structural blocks related to polymetamorphism and constrain the age of these tectonothermal events.

Mesoproterozoic rocks in the New Jersey Highlands were locally retrograded to greenschist facies during post-Ottawan tectonic events associated with Taconian and Alleghanian compressional orogenesis and Mesozoic extension; these effects are most common along fault zones (e.g., Hull et al., 1986).

Rates of postpeak metamorphic cooling in the north-central Appalachians following Ottawan orogenesis are provided by hornblende $^{40}Ar/^{39}Ar$ ages of 880 Ma, obtained from the Honey Brook Upland in eastern Pennsylvania (Crawford and Hoersch, 1984), 919 ± 5 Ma from the eastern New Jersey Highlands (R.A. Volkert, this study), 870 ± 20 to 916 ± 20 Ma from the western

New Jersey Highlands, and 890 ± 20 to 939 ± 20 Ma from the western New York Hudson Highlands (Dallmeyer et al., 1975). These data suggest that the region remained above the ~500 °C blocking temperature for Ar retention in amphibole during regional uplift and unroofing until ca. 900 Ma.

BASEMENT ROCKS

Calc-alkaline metaplutonic and metavolcanic rocks of the Losee Metamorphic Suite are inferred to be the oldest rocks exposed in the New Jersey Highlands (Drake, 1984; Volkert and Drake, 1999), where they underlie ~30% of the region (Fig. 3) and form a basement assemblage to the other Mesoproterozoic rocks. They are as yet undated, but are postulated to have an age of >1200 Ma, based on mineralogical and geochemical similarity to dated calc-alkaline rocks in the Adirondack Highlands (McLelland and Chiarenzelli, 1990), Green Mountains (Ratcliffe et al., 1991), and the Central Metasedimentary Belt in southeastern Canada and the contiguous Adirondack Lowlands (e.g., Corfu and Easton, 1995; Wasteneys et al., 1999). The Losee Metamorphic Suite is divisible into (1) leucocratic rocks, which have the chemical composition of dacite, tonalite, and trondhjemite and (2) charnockitic (orthopyroxene-bearing) rocks, which have the chemical composition of andesite, dacite, minor rhyolite, diorite, quartz diorite, and tonalite. Associated mafic rocks include basalt and basaltic andesite of calc-alkaline and tholeiitic composition. Coeval intrusive potassic rocks are conspicuously absent in the suite.

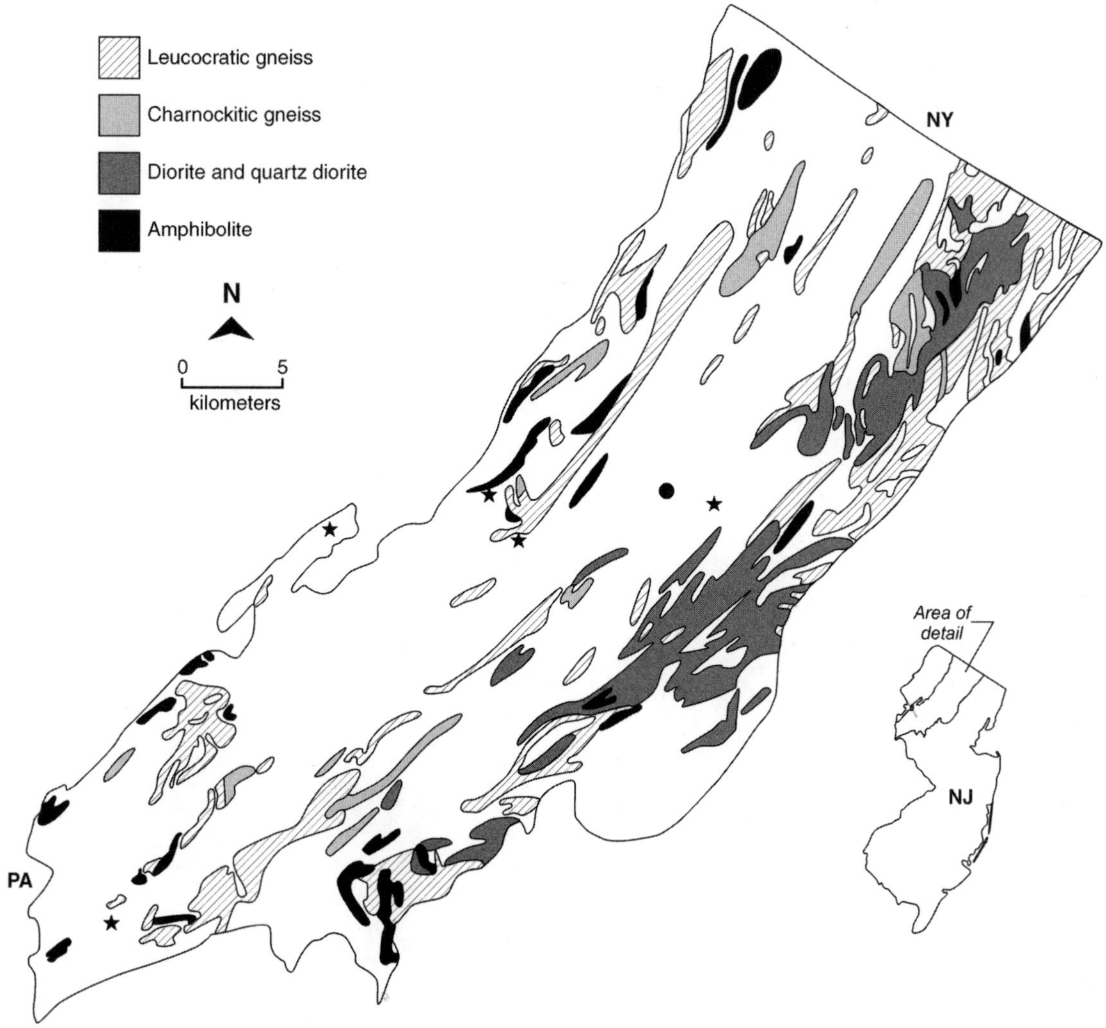

Figure 3. Distribution of leucocratic and charnockitic rocks of the Losee Metamorphic Suite and amphibolite in the New Jersey Highlands. Also shown are locations of megacrystic amphibolites (stars) and meta-anorthosite (filled circle). Amphibolites associated with the Losee Metamorphic Suite and supracrustal amphibolites are undifferentiated in this figure. Modified from Volkert and Drake (1999).

Leucocratic Rocks

Leucocratic rocks in the Losee Suite are voluminous and occur in the western and eastern Highlands in nearly equal abundance (Fig. 3). These rocks characteristically weather white, are light greenish-gray on fresh surfaces, and range texturally from medium-grained, well-layered gneiss (Fig. 4, A), to well-foliated gneiss (Fig. 4, B), to massive, medium- to coarse-grained, undeformed variants (Fig. 4, B and C). Foliated gneiss is the most abundant type and occurs in equal abundance in both the western and eastern Highlands, where it underlies virtually all of the area of leucocratic rocks (Fig. 3). Massive, poorly foliated to undeformed rocks collectively underlie <1% of the remaining area. All of the leucocratic rocks are composed essentially of quartz and plagioclase (oligoclase to andesine), with variable amounts of amphibole, biotite, clinopyroxene, and Fe-Ti oxides. Garnet and rutile are locally present.

Representative geochemical compositions of leucocratic rocks (Table 2) have high Na_2O/K_2O ratios of ~4 and thus, fall exclusively in the fields of tonalite and trondhjemite on a normative feldspar diagram (Puffer and Volkert, 1991). Leucocratic rocks are calc-alkaline and have an alkali-lime index of 61 (Fig. 5), comparable with indices of 60–65 for Adirondack tonalites (McLelland and Chiarenzelli, 1990) and 62 for tonalites in the Green Mountains (Ratcliffe et al., 1991). Leucocratic Losee rocks form a well-defined calc-alkaline trend on a plot of log $(CaO/[Na_2O+K_2O])$ (Fig. 5) and an AFM diagram (Fig. 6). Note that the metatonalites display no discernible textural, mineralogical, or geochemical differences between the western and eastern Highlands. Leucocratic rocks typically have >14 wt% Al_2O_3 (Table 2) and light rare-earth-element (REE) enrichment (La/Yb ~23, Yb ~0.72 ppm). Arth (1979) proposed using Al_2O_3 and Yb abundances to discriminate between formation in a continental arc (Al_2O_3 >14.5 wt%, Yb <1.5 ppm) and oceanic arc environment (Al_2O_3 <14.5 wt%, Yb >1.5 ppm). By these criteria, Losee rocks have a continental arc signature that is quite similar to the high-Al_2O_3 and low-Yb calc-alkaline rocks of the Green Mountains described by Ratcliffe et al. (1991).

Based on the geochemical data and field relationships, some undeformed, massive-textured leucocratic rocks in the Highlands were interpreted by Puffer and Volkert (1991) to be trondhjemitic melts produced by partial melting of older

Figure 4. Outcrop photographs, showing principal textural variants of leucocratic rocks of the Losee Metamorphic Suite. (A) Well-layered and foliated quartz-plagioclase metadacite gneiss, Wanaque quadrangle. (B) Contact between foliated quartz-plagioclase metatonalite gneiss containing thin amphibolite layers (right side of photograph) and undeformed trondhjemite (left side), generated through melting of the former, Franklin quadrangle. (C) Metatonalite gneiss (left side of photograph) and amphibolite intruded by undeformed trondhjemite (right side), Franklin quadrangle. Note the thin trondhjemite vein (above and left of hammer) that cuts both amphibolite and metatonalite gneiss. Hammer is 33 cm long.

TABLE 2. REPRESENTATIVE ANALYSES OF LEUCOCRATIC ROCKS OF THE LOSEE METAMORPHIC SUITE

	Sample number						
	Metatonalite					Trondhjemite	
	7	36	638	28	471	1	380
SiO_2	66.60	67.50	71.40	64.40	70.40	73.20	77.90
TiO_2	0.30	0.23	0.17	0.49	0.42	0.15	0.10
Al_2O_3	19.00	18.20	16.40	16.00	15.20	15.30	12.10
Fe_2O_3	1.01	0.64	0.57	1.82	1.10	0.37	0.17
FeO	0.20	1.00	0.80	2.50	1.30	0.56	0.30
MnO	0.01	0.03	0.06	0.08	0.03	0.07	0.01
MgO	0.18	0.88	0.45	2.02	1.37	0.29	0.20
CaO	3.44	3.92	4.05	4.39	2.21	2.80	0.77
Na_2O	6.56	5.62	5.12	4.68	5.45	4.70	4.95
K_2O	1.43	1.71	0.83	1.10	1.98	1.40	2.77
P_2O_5	0.03	0.10	0.06	0.15	0.06	0.04	0.02
LOI	0.25	0.47	0.31	0.93	0.62	0.43	0.47
Total	99.01	100.30	100.22	98.56	100.14	99.31	99.76
Ba	220	450	450	220	390	n.d.	420
Cr	n.d.	20	20	40	31	n.d.	10
Nb	10	20	10	<10	n.d.	n.d.	10
Rb	20	80	30	20	50	n.d.	40
Sr	130	750	790	610	450	n.d.	120
Y	10	<10	10	<10	n.d.	n.d.	20
Zr	180	70	40	80	80	n.d.	80
La	n.d.	15	4.00	10	14	5.00	n.d.
Ce	n.d.	33	10	26	29	8.00	n.d.
Nd	n.d.	6.00	5.00	10	12	4.00	n.d.
Sm	n.d.	1.20	0.80	2.10	2.30	0.50	n.d.
Eu	n.d.	0.40	0.20	0.60	0.80	0.30	n.d.
Tb	n.d.	0.10	0.20	0.20	0.10	0.10	n.d.
Yb	n.d.	0.30	0.70	0.60	0.20	0.30	n.d.
Lu	n.d.	0.12	0.10	0.10	0.10	0.11	n.d.

Note: Units for chemical compositions are wt%; those for elements are ppm. Analyses are from Volkert and Drake (1999); Puffer and Volkert (1991). n.d.—Not determined; LOI—loss on ignition.

metadacite and metatonalite. Evidence supporting this interpretation includes the REE data and the depletion in undeformed rocks of Al_2O_3, TiO_2, FeO, and MgO and enrichment of SiO_2, compared with well-foliated metatonalite (Table 2). Partial melting of the latter has formed a mafic biotite-amphibole–rich residue with high FeO_t, MgO, and heavy REE (Puffer and Volkert, 1991). This residue is a common feature in both the western and eastern Highlands, where melting of metatonalite has occurred. The absence of metamorphic foliation in anatectic trondhjemite that is unrelated to differences in mineralogy or grain size between it and well-foliated precursor metatonalite and a gradational contact between both rock types (Fig. 4, B) further supports an anatectic origin. In the Franklin quadrangle, undeformed trondhjemite that grades into well-foliated metatonalitic gneiss has yielded a U-Pb thermal ionization mass spectrometry (TIMS) age of 1029 ± 1 Ma and a sensitive high-resolution ion microprobe (SHRIMP) age of 1030 ± 12 Ma (Table 1). An age of 1030 Ma is interpreted here to record local anatexis of the host metatonalite, supporting the conclusions of Puffer and Volkert (1991) and providing an important temporal constraint on the close of the Ottawan orogeny in the Highlands.

Puffer and Volkert (1991) proposed that the tonalite, trondhjemite, and dacite protoliths originated in a continental-margin arc environment that was dominated by calc-alkaline magmatism, generated through partial melting of a basaltic source. Preliminary isotopic work (see Volkert et al. [2000b] for analytical methods) on well-foliated leucocratic gneiss has yielded whole-rock initial $^{87}Sr/^{86}Sr$ ratios of 0.702149–0.704770 and initial $^{143}Nd/^{144}Nd$ ratios of 0.512013–0.512123 (Table 3), calculated for an estimated age of 1250 Ma, which is consistent with an origin from a mafic source. Daly and McLelland (1991) report similar initial $^{143}Nd/^{144}Nd$ ratios of 0.511956–0.512119 for 1.35–1.3 Ga tonalities from the Adirondack Highlands.

Charnockitic Rocks

Calc-alkaline, plagioclase-rich rocks that contain modal hypersthene occur throughout the Highlands (Fig. 3). Based on

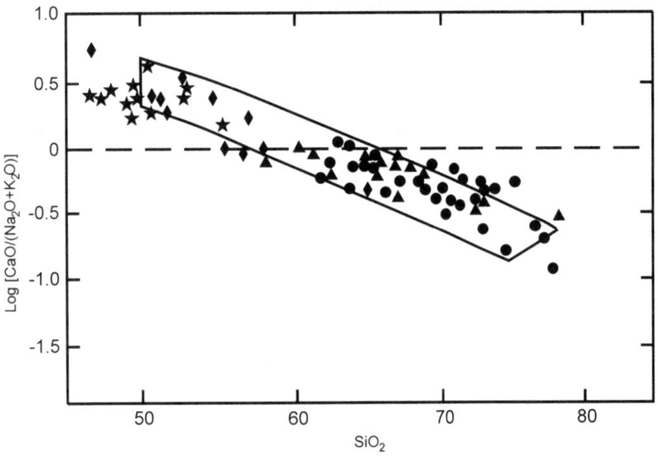

Figure 5. Log (CaO/[Na_2O + K_2O]) versus SiO_2 plot of leucocratic and charnockitic rocks and amphibolites of the Losee Suite. The enclosed field for calc-alkaline collisional rocks is modified from Brown (1982). Circles—leucocratic rocks; diamonds—charnockitic metaplutonic rocks; stars—amphibolite; triangles—charnockitic metavolcanic rocks. Data are from Puffer and Volkert (1991) and Volkert and Drake (1999); additional amphibolite data from Maxey (1971) and Kula and Gorring (2000).

field relationships and geochemical similarity to leucocratic rocks, these rocks were interpreted by Volkert (1993) to be part of the Losee Suite. Layered charnockitic and leucocratic gneisses commonly exhibit gradational contacts and, along with amphibolite, are intercalated, strongly suggesting a cogenetic origin. The presence of highly sodic charnockitic rocks within the Losee does not in itself provide unequivocal proof for a pre-1200-Ma age assignment, because charnockites are a recognized component of the 1160–1130 Ma AMCG complexes (McLelland et al., 2001) of the Adirondack Highlands. However, the Adirondack AMCG charnockites have characteristic A-type geochemical compositions (McLelland and Whitney, 1990) that are quite distinct from the calc-alkaline characteristics of charnockitic rocks of the Losee Suite.

Two distinct types of charnockitic rocks are recognized in the New Jersey Highlands. One type consists of massive, rela-

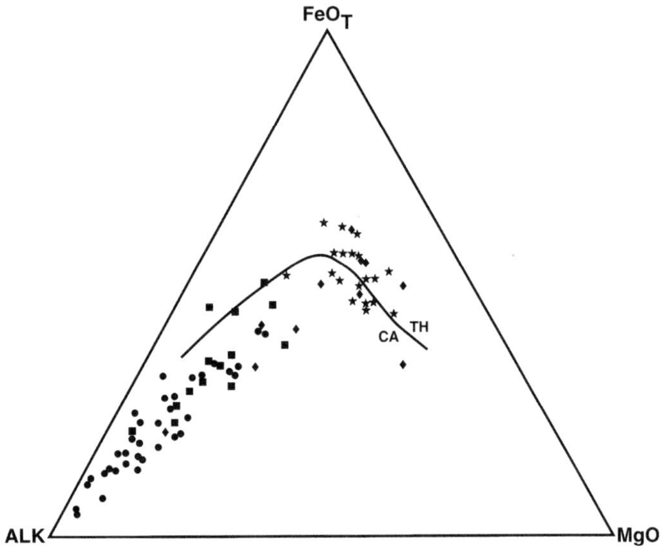

Figure 6. AFM plot of leucocratic and charnockitic rocks and amphibolites of the Losee Suite. ALK = K_2O + Na_2O; FeO_t is the total iron expressed as FeO. Line shows boundary between tholeiitic (TH) and calc-alkaline (CA) fields (from Irvine and Baragar, 1971). Circles—leucocratic rocks; diamonds—charnockitic metaplutonic rocks; squares—charnockitic metavolcanic rocks; stars—amphibolite. Data are from Puffer and Volkert (1991), Volkert and Drake (1999), and additional amphibolite data from Maxey (1971) and Kula and Gorring (2000).

TABLE 3. SM-ND ISOTOPE DATA

Suite	Lithology	Sm (ppm)	Nd (ppm)	$^{147}Sm/^{144}Nd$	$^{143}Nd/^{144}Nd$	$\varepsilon_{Nd}(t)$
Losee						
	Diorite	5.87	25.83	0.1387	0.512249	1.68
	Diorite	5.93	25.18	0.1437	0.512410	4.00
	Charnockitic gneiss	3.08	20.37	0.9220	0.511959	3.44
	Charnockitic gneiss	11.40	63.79	0.1090	0.512156	4.60
	Anatectic trondhjemite	0.20	1.30	0.7620	0.512060	7.96
	Leucocratic gneiss	1.18	6.24	0.1281	0.512121	0.87
	Leucocratic gneiss	0.36	2.39	0.9760	0.512013	3.64
	Leucocratic gneiss	1.12	6.07	0.1169	0.512123	2.69
Byram						
	Monzogranite	14.5	73.8	0.1198	0.512140	1.21
	Syenogranite	14.6	62.2	0.1238	0.512180	1.56
	Quartz monzonite	17.6	83.0	0.1276	0.512270	2.74
Lake Hopatcong						
	Syenogranite	6.7	28.9	0.1420	0.512360	2.24
	Syenogranite	18.9	60.5	0.1908	0.512740	2.95
	Monzogranite	15.3	68.3	0.1364	0.512330	2.38

Note: Byram and Lake Hopatcong data from Volkert et al. (2000b). ε_{Nd} calculated for an age of 1250 Ma for the Losee Suite.

tively homogeneous medium-grained, foliated metaplutonic rocks (Fig. 7, A) that predominate in the eastern Highlands, where they form large intrusive bodies. These rocks are mainly calc-alkaline and, less commonly, tholeiitic (Fig. 6). Although unzoned, charnockitic bodies exhibit regional variations in chemical composition that range from diorite and quartz diorite (norite) to hypersthene tonalite (enderbite). The other type of charnockitic rock consists of medium-grained, well-layered, and foliated quartz-rich gneisses (Fig. 7, B) that are quite widespread. These rocks are calc-alkaline in composition (Fig. 6) and are predominantly metadacite and meta-andesite, with very minor amounts of metarhyolite (Volkert and Drake, 1999). Charnockitic metadacites are more abundant and widely distributed in the Highlands than are andesitic rocks, although both types occur throughout the region. Charnockitic metaplutonic and metavolcanic rocks weather light gray or tan, are greenish-gray to brownish-gray on fresh surfaces, and display greasy lusters. These rocks are composed of plagioclase (oligoclase to andesine), clinopyroxene, amphibole, hypersthene, and variable amounts of quartz, biotite, and Fe-Ti oxides.

Representative geochemical compositions (Table 4) indicate that charnockitic meta-andesite has somewhat lower SiO_2 and higher TiO_2, FeO, CaO, P_2O_5, and Cr than do charnockitic metadacite or leucocratic rocks of the Losee Suite. Metaplutonic rocks are metaluminous and, compared with charnockitic metavolcanics, have more mafic compositions that are enriched in Al_2O_3, FeO, MgO, CaO, and Cr. Except for some tholeiitic mafic rocks, the Losee Suite follows an overlapping and continuous trend on a diagram of CaO-Al_2O_3-(FeO_t + MgO) (Volkert and Drake, 1999), and typically defines continuous calc-alkaline trends (Figs. 5 and 6) from Fe-enrichment in most diorite and amphibolites to progressive Fe-depletion in metavolcanic charnockitic and leucocratic rocks, accompanied by a systematic decrease in Al_2O_3, MgO, and CaO with increasing SiO_2. Based on the available geochemical data, leucocratic and charnockitic metadacites and meta-andesites may be the volcanic equivalents of tonalite and diorite, respectively. Preliminary whole-rock isotope analyses indicate initial $^{87}Sr/^{86}Sr$ ratios of 0.70246–0.70490 for charnockitic plutonic rocks and 0.70380–0.70496 for layered charnockitic gneiss, similar to the ratios obtained from leucocratic rocks. Both types of charnockitic rocks have similar whole-rock initial $^{143}Nd/^{144}Nd$ ratios of 0.511959–0.512410 and moderately high initial ε_{Nd} (Table 3), calculated for an estimated age of 1250 Ma.

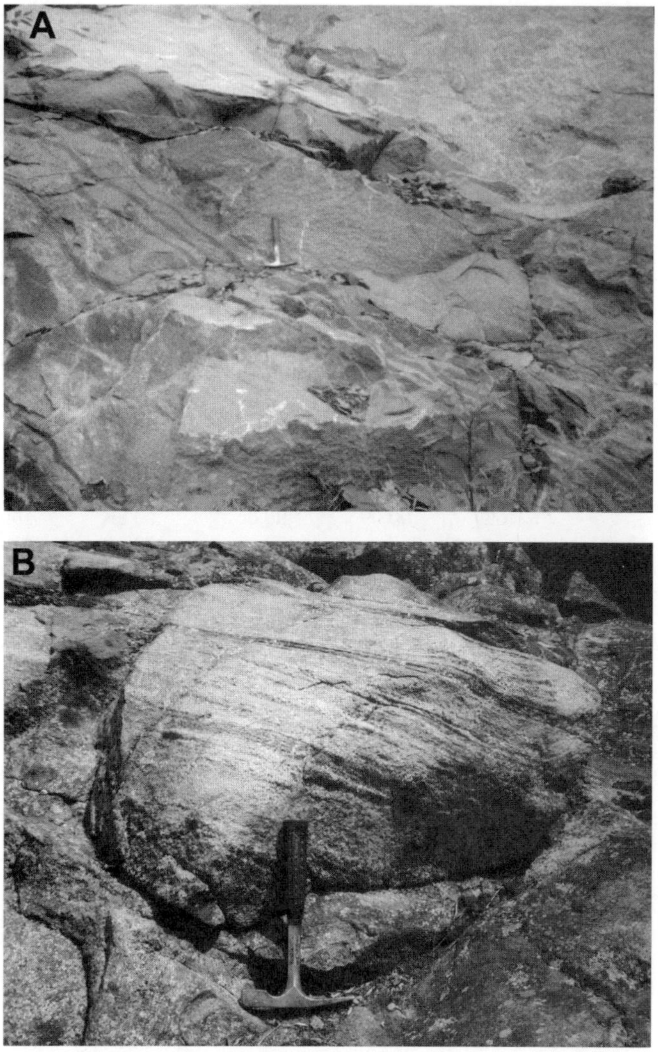

Figure 7. Outcrop photographs of charnockitic rocks of the Losee Metamorphic Suite. (A) Massive-textured metaplutonic rock mapped as diorite, Boonton quadrangle. Note thin mafic interlayers. (B) Layered metavolcanic rock mapped as hypersthene-quartz-plagioclase gneiss, Newfoundland quadrangle. Hammer is 33 cm long.

Amphibolites

Medium-grained, well-foliated amphibolites that contain hornblende + plagioclase + augite ± hypersthene ± biotite are commonly intercalated with the leucocratic and charnockitic variants of the Losee Suite (Fig. 4, C), although they are volumetrically a relatively minor component. Geochemical analyses of major oxides are available for Losee amphibolites (Maxey, 1971; Puffer et al., 1993b; Volkert and Drake, 1999) but trace-element data are sparse and restricted to samples from the Franklin quadrangle. However, the combined data permit the following observations. Losee amphibolites are basalts to basaltic andesites with 49–56 wt% SiO_2. They range from tholeiitic to calc-alkaline (Fig. 6) and have mg# values of 40–60 (mg# is the molar ratio Mg/[Mg + Fe^{2+}]). Compared with normal mid-oceanic–ridge basalt (N-MORB), they are depleted in some high-field-strength elements (HFSE), such as Hf (~1.52 ppm) and TiO_2 (~0.97 wt%), as well as Cr (~87 ppm) and Ni (~52 ppm), and are enriched in light REE (La ~11 ppm; La/Yb_N ~2.94) and Th (~0.24 ppm) (Kula and Gorring, 2000). They display negative Ta-Nb anomalies on primitive mantle-normalized multi-element diagrams (not shown) and Nb depletion relative to La (Nb/La ~0.37) that, along with the low HFSE, are suggestive

TABLE 4. REPRESENTATIVE ANALYSES OF CHARNOCKITIC ROCKS OF THE LOSEE METAMORPHIC SUITE

	Sample number							
	Metavolcanic gneiss				Metaplutonic rocks			
	423	138	31	37	561	172	381	80
SiO_2	60.30	67.20	68.70	66.00	52.40	55.40	51.70	56.80
TiO_2	1.83	0.54	0.45	0.62	0.26	1.45	1.58	0.62
Al_2O_3	15.10	15.90	15.60	15.70	20.60	17.30	15.30	19.20
Fe_2O_3	3.71	1.25	0.77	3.15	1.93	1.55	2.00	1.85
FeO	4.70	2.40	2.60	3.40	3.10	6.30	8.80	3.60
MnO	0.12	0.09	0.06	0.11	0.09	0.13	0.17	0.12
MgO	2.09	1.32	1.17	0.82	6.43	2.89	6.48	3.01
CaO	5.68	4.07	3.98	3.75	12.50	7.15	8.60	6.52
Na_2O	4.07	4.17	5.27	5.02	2.50	5.57	3.60	5.60
K_2O	1.49	1.94	1.16	1.34	0.30	1.42	0.70	1.24
P_2O_5	0.61	0.16	0.12	0.21	0.06	0.47	0.26	0.22
LOI	0.39	0.62	0.08	0.39	0.23	0.31	0.31	0.47
Total	100.09	99.66	99.96	100.51	100.41	99.94	99.50	99.25
Ba	860	870	160	420	90	920	270	170
Cr	17	10	20	<10	230	37	270	40
Nb	20	20	10	10	<10	40	20	20
Rb	30	40	30	70	10	30	20	20
Sr	570	590	330	320	830	940	400	680
Y	70	<10	10	20	10	20	20	30
Zr	270	190	120	250	<10	220	90	100

Note: Units for chemical compositions are wt%; those for elements are ppm. Metavolcanic rocks mapped as hypersthene-quartz-plagioclase gneiss; metaplutonic rocks mapped as diorite. Analyses are from Volkert and Drake (1999). LOI—loss on ignition.

of an arc-related origin. On tectonomagmatic diagrams of TiO_2-MnO-P_2O_5 and Hf/3-Th-Ta (not shown) and mg# versus TiO_2 (Fig. 8), Losee amphibolites plot mainly in the field of island-arc tholeiites or destructive margin basalts. Pearce (1983) proposed using ratios of Th/Yb and Ta/Yb of basalts to screen for the effects of crustal contamination and subduction-zone enrichment, because these processes enrich Th but not Ta and because Yb is unaffected by these enrichment processes, eliminating variations due to partial melting or fractionational crystallization. Losee amphibolites lack a vertical trend of increasing Th/Yb and plot within a mantle array transitional between normal and enriched MORB (Fig. 9). Uniformly low Th/La (~0.027) of Losee amphibolites similarly implies limited crustal (or sediment) contamination or enrichment by subduction-zone fluids.

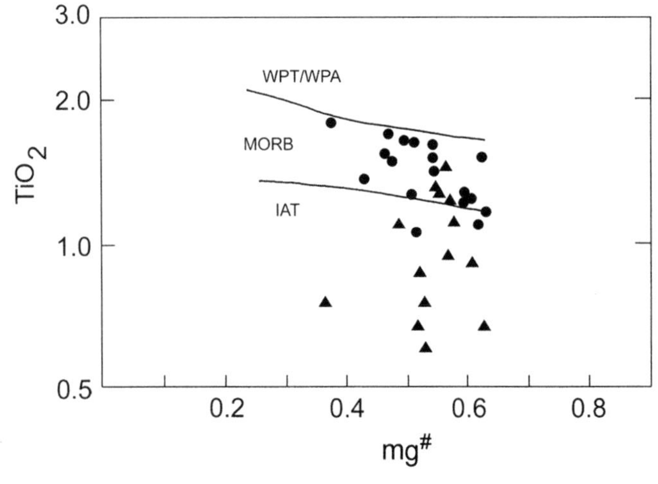

Figure 8. Log TiO_2 versus mg# (Basaltic Volcanism Study Project, 1981) for supracrustal amphibolites (circles) and Losee suite amphibolites (triangles) from the New Jersey Highlands. Fields: IAT—island-arc tholeiite; MORB—midocean-ridge basalt; WPT/WPA—within-plate tholeiitic/alkalic basalt. Data are from Maxey (1971), Volkert and Drake (1999), and Kula and Gorring (2000).

On an AFM diagram (Fig. 6), some Losee amphibolites and diorites plot along a different tholeiitic trend. Unfortunately, trace-element data are not available for these rocks. They may represent mafic and intermediate magmas from a different source, or temporal variations in magmatism, as younger arcs are characterized predominantly by tholeiitic rocks that progress toward calc-alkaline compositions as the arc matures (e.g., Wilson, 1989).

SUPRACRUSTAL ROCKS

Age

The age of the supracrustal succession in the New Jersey Highlands remains problematic. The Losee Metamorphic Suite was interpreted by Volkert (1993) to be overlain by the supracrustal rocks, based on the author's recognition of a nonconformity in outcrop in the Wanaque quadrangle in the eastern Highlands (Fig. 10). This contact is characterized by a slight angular discordance between layered charnockitic gneiss and amphibo-

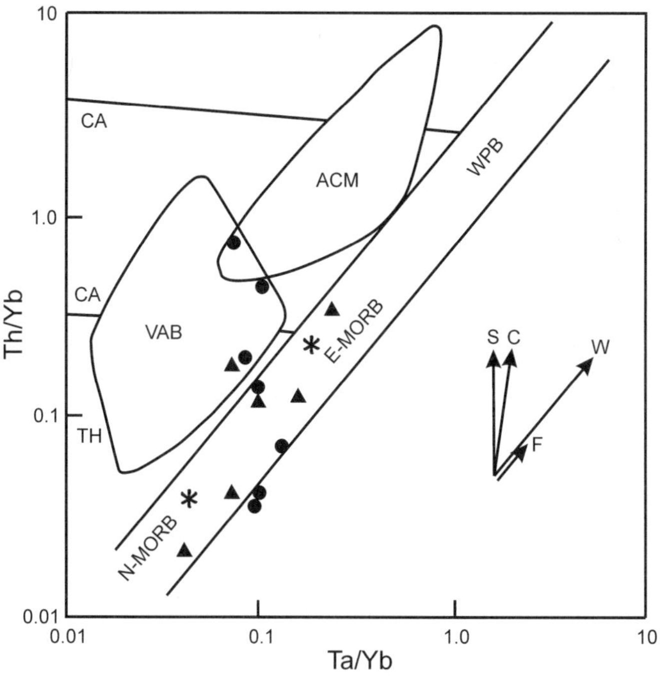

Figure 9. Th/Yb versus Ta/Yb for supracrustal amphibolites (circles) and Losee amphibolites (triangles). TH and CA denote tholeiitic and calc-alkaline boundaries, respectively. Arrows show vectors for subduction zone enrichment (S), crustal contamination (C), within-plate enrichment (W), and fractional crystallization (F). Fields: ACM—active continental margins; VAB—basalts from volcanic arcs; WPB—intraplate settings. Average composition of N-MORB and E-MORB from Sun and McDonough (1989). Diagram modified from Pearce (1983). Data are from Kula and Gorring (2000).

lite and the overlying metaquartzite and quartzofeldspathic gneiss (meta-arkose), which is interpreted to represent the basal part of the succession. Elsewhere in the Highlands, the supracrustal rocks have been isoclinally folded with the Losee Suite, and contacts between them are now concordant. Whether this represents the original concordance, or is a consequence of tectonic transposition during Ottawan orogenesis has not been determined. A minimum age for the supracrustal rocks of 1174 ± 15 Ma is provided by granite that intrudes paragneisses in the Hudson Highlands (Ratcliffe and Aleinikoff, 2001a), which maintain continuity along the strike to the south with similar paragneisses in New Jersey. Gates et al. (2001a) obtained U-Pb SHRIMP detrital zircon ages of 2000–1200 Ma from paragneiss in the Hudson Highlands. If the age of the youngest detrital zircon provides a maximum age, then the supracrustal rocks may have been deposited between 1200 and 1174 Ma.

North of the western Highlands, Offield (1967) interpreted rocks in the Hudson Highlands similar to the Losee Suite to constitute basement to the local supracrustal sequence and suggested the possibility of a nonconformity between them. Metasedimentary successions similar to those of the New Jersey Highlands occur in the Green Mountains (Ratcliffe et al., 1991) and Adirondack Highlands (Wiener et al., 1984). McLelland et al. (1996) noted strong similarities between the metasediments in the eastern Adirondacks and Green Mountains and proposed their continuity between the two areas. Ratcliffe et al. (1991) interpreted paragneisses in the Green Mountains to nonconformably overlie 1.35–1.31 Ga calc-alkaline rocks of the Mount Holly Complex, fixing an age of <1.31 Ga for their deposition. They also proposed that paragneisses in the Berkshire Mountains (Washington Gneiss) are likely a southern continuation of the same succession in the Green Mountains. Metasedimentary rocks of the Adirondack Highlands (Lake George Group) rest nonconformably on charnockitic and granitic gneisses of the Piseco Group (Wiener et al., 1984). They include possibly >1300 Ma paragneisses, metaquartzite, and marble that represent deposition in a shelf-type environment (McLelland et al., 1996). It is particularly noteworthy that graphite deposits in New Jersey and the Adirondack Highlands are hosted by the same lithologies. The entire stratigraphy of the Spring Pond Formation of the Lake George Group, particularly the Dixon Schist that hosts the deposits (Alling, 1917), is remarkably similar to that of metasediments hosting graphite deposits in New Jersey (Volkert et al., 2000a).

Widespread and abundant supracrustal rocks of the Adirondack Lowlands are intruded by the ca. 1207 Ma calc-alkaline Rossie-Antwerp Suite, fixing a minimum age for their deposition (Wasteneys et al., 1999). Lowlands metasediments are lithologically similar to the New Jersey succession. They include various paragneisses, metaquartzite, and two marble layers (the upper one hosting the Balmat Zn deposit), which may have been deposited in either a shallow-water marine shelf environment or in an intracratonic basin (deLorraine, 2001). Despite the absence of volcanics in the Lowlands succession, comparisons between the supracrustal rocks in New Jersey and the Adirondacks are appealing. The western parts of both regions contain the most extensive successions, in which thick marble layers

Figure 10. Possible angular nonconformity between feldspathic metaquartzite (above) and layered charnockitic gneiss and amphibolite (below) in the Wanaque quadrangle. Pencil is 13.8 cm long.

host large Zn deposits, and the eastern parts of both regions contain thin marble layers, more extensive metaclastic gneisses, and numerous graphite deposits.

Stratigraphy and Lithology

Collectively, supracrustal rocks in the New Jersey Highlands are widespread and underlie ~20% of the region (Fig. 11). These rocks predominate in the western Highlands, where they attain an estimated maximum thickness of ~2100 m (Fig. 12), and are much thinner in the east, although thickness estimates in this area are difficult to constrain. Included in the supracrustal succession are quartzofeldspathic gneisses, metaquartzite, calc-silicate rocks, marble, and amphibolite. Ultramafic rocks are absent, and mafic volcanics compose <5% of the succession. Inferred stratigraphic relations, constructed principally for the western Highlands, are shown in Figure 12.

Metasiliciclastic rocks

Quartzofeldspathic gneisses occur in equal abundance in both areas of the Highlands. These rocks include a range of gneisses whose protoliths were predominantly arkose (K-feldspar gneiss) and graywacke (biotite-quartz-feldspar gneiss) (Volkert and Drake, 1999). Representative geochemical compositions are given in Table 5. Their heterogeneous textures, layered association with metaquartzites, and geochemical characteristics (especially the siliceous compositions and variable Zr/Nb and Zr/Y ratios) support a sedimentary origin. Quartzofeldspathic gneisses are typically pinkish-white to pinkish-gray, locally rusty, and range in texture from well layered to massive. These rocks are composed primarily of quartz, K-feldspar, and oligoclase. Ferromagnesian minerals, where present, include biotite, amphibole, and/or clinopyroxene. Garnet, sillimanite, zircon, apatite, and Fe-Ti oxides occur as common accessory phases. Graphite and sulfides are confined to rusty

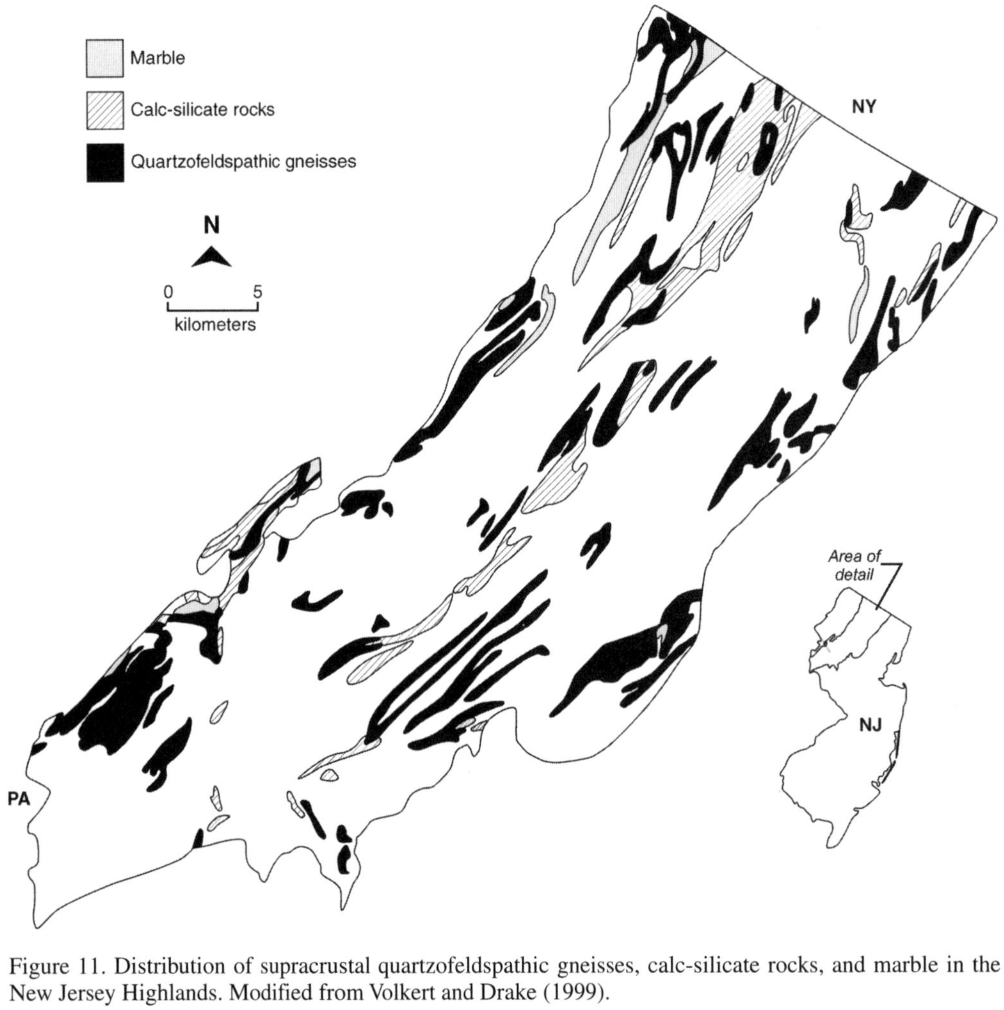

Figure 11. Distribution of supracrustal quartzofeldspathic gneisses, calc-silicate rocks, and marble in the New Jersey Highlands. Modified from Volkert and Drake (1999).

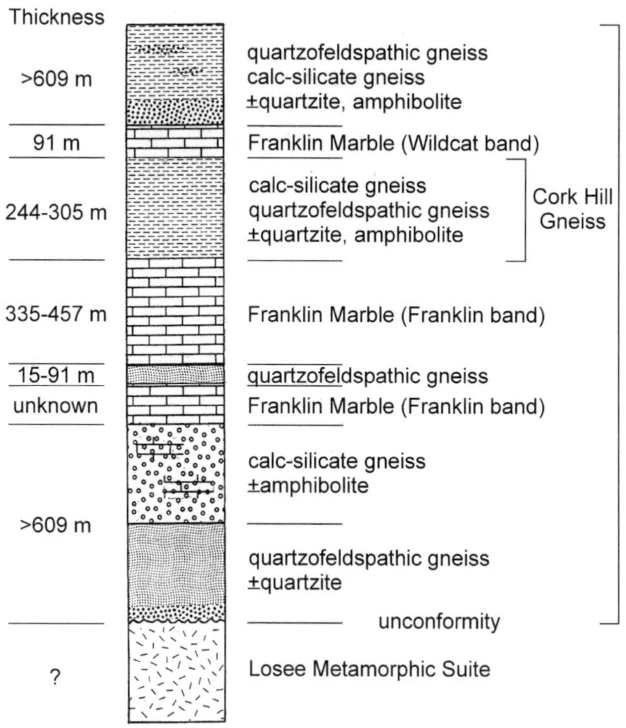

Figure 12. Generalized stratigraphic column of supracrustal rocks compiled mainly for the western New Jersey Highlands. Thickness of column blocks and symbols are not to scale. Modified from Volkert 2001. The column is a compilation of work by Hague et al. (1956), Baker and Buddington (1970), and Volkert and Drake (1999).

biotite-quartz-feldspar gneiss. Some K-feldspar gneiss in the Pompton Plains and Dover quadrangles very locally preserves relict graded beds, thin layers of pebble conglomerate with deformed and flattened pebbles, and trough crosslaminated beds, despite the effects of high-grade metamorphism and recrystallization.

In the Chester and Hackettstown quadrangles, K-feldspar gneiss locally contains a paleoplacer deposit of monazite (Volkert, 1993; Volkert and Drake, 1999) that forms a persistent unit as much as a few hundred meters thick that is continuous along strike for several kilometers. Markewicz (1965) identified a high content of monazite in the gneiss (≤15%) and sparse amounts of magnetite, ilmenite, rutile, and zircon. These appear to represent a relict suite of heavy minerals and imply an origin of either a fluvial deposit or beach sand (Volkert, 1993).

Metaquartzite is sparsely exposed throughout the Highlands, where it occurs as thin lenses and layers 0.5–0.8 m thick interlayered with quartzofeldspathic gneisses, calc-silicate gneiss, and marble. The unit lacks textural evidence suggestive of an origin by metamorphic segregation or anatexis and can locally be traced along strike for distances of >1 km. Representative geochemical compositions are given in Table 6. Metaquartzite is characteristically light gray, vitreous, medium-grained, and well layered to massive. Where associated with quartzofeldspathic gneisses, it commonly contains biotite, garnet, graphite, and locally abundant tourmaline; where associated with calc-silicate gneiss, it contains mainly diopside. Protoliths for most metaquartzite are interpreted as pure to impure quartz-

TABLE 5. REPRESENTATIVE ANALYSES OF SUPRACRUSTAL QUARTZOFELDSPATHIC GNEISSES

	Sample number								
	K-feldspar gneiss				Biotite-quartz-feldspar gneiss				
	16	530	81	155	65	130	192	311	333
SiO_2	74.50	77.50	72.00	73.00	66.60	54.90	77.50	67.00	75.30
TiO_2	0.26	0.27	0.28	0.36	1.16	1.13	0.19	0.67	0.49
Al_2O_3	11.90	10.50	13.80	12.50	14.80	23.40	11.80	13.70	11.30
Fe_2O_3	2.78	1.72	2.00	2.56	2.47	2.45	0.79	3.93	2.40
FeO	0.90	0.40	1.00	1.90	4.10	2.80	0.80	2.20	1.60
MnO	0.03	0.02	0.03	0.04	0.09	0.03	0.01	0.01	0.08
MgO	0.12	0.45	0.37	1.04	1.57	2.11	0.51	1.80	1.14
CaO	0.27	0.21	1.05	1.27	3.24	0.51	1.32	0.66	1.44
Na_2O	3.02	1.58	2.90	1.58	2.83	2.48	3.25	3.76	2.59
K_2O	5.79	5.53	5.32	5.18	2.46	7.39	3.17	5.43	2.98
P_2O_5	0.03	0.07	0.06	0.07	0.49	0.08	0.03	0.10	0.10
LOI	0.39	0.85	1.16	0.77	0.31	1.93	0.70	0.47	0.47
Total	99.93	99.10	99.97	100.27	100.12	99.21	100.07	99.73	99.89
Ba	690	980	630	640	450	1000	920	610	880
Cr	23	70	21	20	30	90	10	70	110
Nb	40	50	20	30	20	30	<10	30	30
Rb	120	190	240	130	120	230	110	180	120
Sr	30	60	110	70	180	120	260	30	150
Y	80	150	70	90	50	120	<10	30	40
Zr	780	190	290	390	310	300	150	170	250

Note: Units for chemical compositions are wt%; those for elements are ppm. Analyses are from Volkert and Drake (1999). LOI—loss on ignition.

TABLE 6. REPRESENTATIVE ANALYSES OF METAQUARTZITE

	Sample number					
	5	7	9	12	1	79
SiO_2	90.74	89.08	89.79	92.01	86.38	88.60
TiO_2	0.01	0.24	0.04	0.45	0.12	0.34
Al_2O_3	3.46	3.86	2.84	1.12	2.81	5.20
Fe_2O_3*	1.14	1.06	2.93	3.62	0.50	1.49
MnO	0.00	0.06	0.01	0.01	0.01	0.03
MgO	0.13	1.00	0.21	0.21	0.07	0.55
CaO	0.26	0.24	0.74	0.30	0.30	0.18
Na_2O	0.79	0.21	0.48	0.39	1.05	0.61
K_2O	3.36	3.37	2.69	0.83	2.21	1.78
P_2O_5	0.00	0.00	0.00	0.00	0.00	0.03
LOI	0.14	0.21	0.32	0.38	0.50	0.85
Total	100.03	99.33	100.05	99.32	93.95†	99.57
Ba	n.d.	n.d.	n.d.	n.d.	n.d.	520
Cr	n.d.	n.d.	n.d.	n.d.	n.d.	30
Nb	n.d.	n.d.	n.d.	n.d.	n.d.	<10
Rb	n.d.	n.d.	n.d.	n.d.	n.d.	60
Sr	n.d.	n.d.	n.d.	n.d.	n.d.	90
Y	n.d.	n.d.	n.d.	n.d.	n.d.	30
Zr	n.d.	n.d.	n.d.	n.d.	n.d.	190

Note: Units for chemical compositions are wt%; those for elements are ppm. Analyses from Volkert (1997). Abbreviations: n.d.—Not determined; LOI—loss on ignition.
*Total Fe as Fe_2O_3.
†Sample contains 6.58 wt% sulfur.

ose sandstone, although some thin discontinuous lenses and layers in marble may have originally been chert.

Calc-silicate rocks consist of pyroxene-bearing gneisses that encompass a range of compositions, whose protoliths likely were calcareous sandstone, quartzose and argillaceous carbonate rocks, cherty dolomite, and carbonate-bearing volcaniclastic sandstone (Volkert and Drake, 1999). Representative geochemical compositions are given in Table 7. Calc-silicate rocks are typically light gray, green, or greenish gray, locally rusty, and range texturally from well layered to massive. Ferromagnesian minerals consist of diopside ± amphibole and biotite. Quartz, oligoclase, K-feldspar, carbonate, epidote, titanite, scapolite, sulfides, and Fe-Ti oxides occur in varying proportions in different rock types.

Marble

Marble is a volumetrically minor rock, occurring most abundantly in the northwestern Highlands (Fig. 11), but it is important in stratigraphic and sedimentological reconstructions and as a host for Zn and Fe deposits. New Jersey Zinc Company geologists (Hague et al., 1956) subdivided marble in the northwest Highlands into the 335–457-m-thick lower Franklin band, which hosts the Zn deposits at Franklin and Sterling Hill, and the ~91-m-thick upper Wildcat band (Fig. 12). These two bands are separated by a 244–305-m-thick sequence of quartzofeldspathic and calc-silicate gneisses and amphibolite, known informally as the Cork Hill Gneiss (Hague et al., 1956). Despite the mineralogical similarity of the clastic rocks between the marble bands, chemical compositions are sufficiently different (Volkert and Drake, 1999) to preclude an interpretation as simple repetition due to folding (i.e., lower marble, clastics, upper marble, clastics). Marble in the northwest Highlands is white to light gray, fine to coarse crystalline, well layered to massive, and mostly calcitic and locally dolomitic. Common accessory minerals include graphite, phlogopite, and variable amounts of chondrodite, calcic amphibole, and clinopyroxene.

Marble in the southwestern Highlands (Fig. 11) is spatially associated with the same lithologies as that in the northwestern Highlands. It occurs in thick lenses that are calcitic toward the north, in the Blairstown and Belvidere quadrangles, and predominantly dolomitic to the south, in the Easton quadrangle (Fig. 2). The northern quadrangles contain phlogopite and graphite ± tourmaline, and are locally associated with tourmaline-bearing metaquartzite and borosilicate-bearing gneiss. Marble in the Easton quadrangle contains abundant talc and serpentine and is associated with tremolite-rich rock (Peck, 1904; Drake, 1969). The Mg-rich composition and locally abundant disseminated tourmaline suggest that marble in the southwest Highlands may, at least in part, be an evaporite deposit (Volkert, 2001). A similar origin was proposed by Kearns (1977) for the Franklin band in southern New York.

Marble in the eastern Highlands (Fig. 11) occurs as thin disconnected layers and small bodies, typically <30 m thick. They are primarily calcitic and contain accessory calcic clinopyrox-

TABLE 7. REPRESENTATIVE ANALYSES OF CALC-SILICATE GNEISSES

	Sample number								
	Type A			Type B1			Type B2		
	284	242	260	63	121	384	1045	58	85
SiO_2	52.80	56.60	59.20	65.40	67.50	67.60	67.90	75.80	77.90
TiO_2	0.87	0.99	0.44	0.65	0.73	0.64	0.16	0.57	0.46
Al_2O_3	11.80	10.80	11.70	14.80	14.00	14.70	16.80	10.30	8.48
Fe_2O_3	2.30	3.44	3.36	2.13	2.58	0.96	0.45	1.17	1.15
FeO	7.50	5.00	3.90	2.20	1.30	1.90	0.70	1.20	0.40
MnO	0.19	0.11	0.06	0.12	0.06	0.15	0.03	0.04	0.04
MgO	3.61	5.09	3.45	1.42	1.56	1.40	0.74	1.44	0.87
CaO	18.30	10.80	9.05	5.06	3.87	4.83	3.10	2.54	3.76
Na_2O	1.90	5.39	6.63	6.42	6.20	6.92	5.00	2.34	0.63
K_2O	0.79	0.52	0.70	0.90	0.78	0.31	3.97	2.64	3.65
P_2O_5	0.14	0.63	0.64	0.16	0.16	0.19	0.07	0.14	0.08
LOI	0.23	0.47	0.62	0.85	0.62	0.62	0.16	1.54	0.62
Total	100.43	99.84	99.75	100.11	99.36	100.22	99.08	99.72	98.04
Ba	40	130	40	130	290	150	670	380	560
Cr	50	100	40	<10	40	110	20	45	30
Nb	20	30	30	20	20	20	20	30	30
Rb	30	20	20	20	30	20	100	100	130
Sr	380	70	70	260	180	230	170	180	150
Y	<10	80	80	90	40	90	<10	160	40
Zr	100	220	110	230	240	250	10	240	150
Fe_2O_3 + MgO	13.41	13.53	10.71	5.75	5.44	4.26	1.89	3.81	2.46
Al_2O_3/SiO_2	0.22	0.19	0.20	0.23	0.21	0.22	0.25	0.14	0.11
$Al_2O_3/(CaO + Na_2O)$	0.58	0.67	0.75	1.29	1.39	1.25	2.07	2.11	1.93

Note: Units for chemical compositions are wt%; those for elements are ppm. Analyses from Volkert and Drake (1999). LOI—loss on ignition.

ene, phlogopite, secondary antigorite, and trace amounts of graphite. At many locations in the eastern Highlands, marble occurs as two discrete bands. The lower band is commonly underlain by calc-silicate gneiss ± metaquartzite and amphibolite and overlain by calc-silicate gneiss, rusty quartzofeldspathic gneiss, amphibolite, and calc-silicate gneiss. An upper marble band is overlain by calc-silicate gneiss, rusty quartzofeldspathic gneiss, and amphibolite. The sequence resembles, in more compressed form, the stratigraphy and lithologic associations of the Franklin band, the Cork Hill Gneiss, and the Wildcat band in the western Highlands (Fig. 12).

Marble in the western Highlands locally preserves features consistent with an origin as a marine deposit. Stromatolites have been recognized in the lower (Franklin) marble band in the Newton East quadrangle (Volkert, 2001); ~9 km to the southwest in the Stanhope quadrangle, the upper (Wildcat) marble band contains a relict reeflike structure (Fig. 13, A) recently discovered by the author. The inferred top of the reef structure is irregular and shows evidence of growth into the enclosing marble. The base is built on a thin layer of carbonate pebble conglomerate that may represent reef talus. The reef structure has a relatively massive appearance, with faint internal layering and arcuate laminations along the upper surface that resemble organic structures (Fig. 13, B), possibly stromatolites, although they are not well preserved.

Amphibolites

Amphibolites in the supracrustal succession are medium-grained, well-foliated gneisses composed of hornblende + plagioclase ± augite ± biotite ± Fe-Ti oxides. They occur at several different stratigraphic intervals (Fig. 12) in both areas of the Highlands. Some amphibolites, interlayered with calc-silicate gneiss of known sedimentary parentage, commonly weather rusty, are sulfidic, and may represent metamorphosed shale. However, most amphibolites have a geochemical affinity to metabasalt, indicating that protoliths were most likely mafic volcanics (Maxey, 1971; Puffer et al., 1993b; Volkert and Drake, 1999). This type is most abundant in the northwestern Highlands, where it stratigraphically overlies the Wildcat Marble band and is locally pillowed (Hague et al., 1956) (Fig. 14).

The relatively few amphibolites that have been geochemically analyzed are interlayered with calc-silicate (metagraywacke) gneisses in the eastern Highlands or directly overlie the Franklin (Fig. 15) and Wildcat Marble bands in the western Highlands. Representative geochemical compositions of both types (Table 8) have ≤52 wt% SiO_2, mg# of 46–63, and differ from amphibolites in the Losee Suite in lacking compositions corresponding to basaltic andesites. They also differ from Losee amphibolites in having higher TiO_2 (~1.42 wt%), Hf (~2.86 ppm), Cr (~253 ppm), Th (~0.47 ppm), light REE (La ~27 ppm, La/Yb_N ~4.41), and, in some samples, Ni (~85 ppm). Although

Figure 13. (A) Massive-textured reef structure enclosed in the Wildcat band of the Franklin Marble, Stanhope quadrangle. Arrow gives location of the close-up in panel B. (B) Upper part of the reef, showing arcuate laminations and structures (beneath pencil) that may be organic in origin. Pencil is 13.8 cm long.

Figure 14. Relict pillow structures in amphibolite from the Hamburg quadrangle. Arrow points to coin 1.9 cm in diameter.

these values are typical of MORB, ratios of Nb/La, La/Ta, Th/Nb, Th/Ta, Th/Yb, Th/Hf, and Hf/Ta divide the supracrustal amphibolites into two groups. Those from above the marble bands and some from the Wanaque quadrangle are tholeiitic and MORB-like (Table 8, samples 1–4, 6, and 7). Others from the Wanaque and Greenwood Lake quadrangles are calc-alkaline and more like island-arc basalts, or have characteristics transitional between MORB and arc basalts (Table 8, samples 5 and 8). Supracrustal amphibolites are similarly divided between the fields of MORB and island-arc or destructive-margin basalts on tectonomagmatic diagrams of TiO_2-MnO-P_2O_5 and Hf/3-Th-Ta (not shown), and mg# versus TiO_2 (Fig. 8). They all display negative Ta and Nb anomalies on a primitive mantle-normalized trace-element diagram (Fig. 16). Both types plot along a different trend than that of Losee amphibolites in Figure 9 (they are displaced toward higher Th), suggesting that the magmas have been modified by a subduction-zone component. Thus, supracrustal amphibolites display geochemical characteristics of both MORB and volcanic-arc basalts. Basalts displaying mixed characteristics of both types are typical of back-arc basin settings, in

Figure 15. Conformable contact between supracrustal amphibolite (top) and stratigraphically underlying Franklin Marble band, Newton East quadrangle. Hammer is 33 cm long.

TABLE 8. REPRESENTATIVE ANALYSES OF SUPRACRUSTAL AMPHIBOLITES

	Sample number							
	Western Highlands				Eastern Highlands			
	1	2	3	4	5	6	7	8
SiO_2	47.44	49.16	47.14	50.61	48.34	51.13	51.36	52.03
TiO_2	1.27	1.55	1.26	1.02	1.42	1.63	1.59	1.27
$Al2O_3$	14.43	16.05	17.61	16.52	16.94	15.27	14.93	15.39
Fe_2O_3*	14.15	10.19	12.08	10.13	n.d.	n.d.	n.d.	n.d.
FeO†	n.d.	n.d.	n.d.	n.d.	10.86	10.18	12.04	8.67
MnO	0.36	0.17	0.21	0.15	0.24	0.06	0.12	0.07
MgO	7.50	8.24	5.75	6.28	7.37	6.85	7.03	7.35
CaO	10.33	11.17	10.10	8.80	10.27	9.60	8.48	10.05
Na_2O	3.07	3.17	4.23	5.21	3.20	3.58	3.13	3.87
K_2O	0.89	0.29	1.37	1.10	0.69	1.63	1.22	1.19
P_2O_5	0.56	n.d.	0.25	0.19	0.67	0.07	0.09	0.11
Total	100.00	100.00	100.00	100.00	100.00	100.00	100.00	100.00
Ba	230	195	701	356	240	196	183	42
Sr	530	487	406	327	425	151	133	159
Th	n.d.	0.24	n.d.	n.d.	0.45	0.16	0.15	1.37
Hf	n.d.	3.50	n.d.	n.d.	1.86	3.03	3.25	2.64
Nb	20	8.20	n.d.	n.d.	3.40	6.80	6.40	5.60
Ta	n.d.	0.48	n.d.	n.d.	0.20	0.40	0.38	0.33
Y	30	n.d.	26	21	n.d.	n.d.	n.d.	n.d.
Zr	90	85	106	109	n.d.	n.d.	n.d.	n.d.
Sc	n.d.	27	33	30	36	41	37	41
Cr	120	231	380	285	255	227	272	448
Ni	n.d.	91	98	96	135	24	33	35
Co	n.d.	28	n.d.	n.d.	50	40	46	38
La	n.d.	15.10	n.d.	n.d.	17.60	20.50	25.30	18.90
Ce	n.d.	31.60	n.d.	n.d.	41.50	51.00	60.00	41.90
Nd	n.d.	18.50	n.d.	n.d.	28.30	31.00	31.50	21.30
Sm	n.d.	4.48	n.d.	n.d.	6.25	7.27	6.30	4.90
Eu	n.d.	1.30	n.d.	n.d.	2.00	1.50	1.71	1.24
Tb	n.d.	0.95	n.d.	n.d.	0.93	1.34	1.13	0.93
Yb	n.d.	3.39	n.d.	n.d.	2.40	4.12	3.93	3.00
Lu	n.d.	0.55	n.d.	n.d.	0.37	0.63	0.61	0.45

Note: Analyses normalized to 100% anhydrous. Units for chemical compositions are wt%; those for elements are ppm. Samples 1 and 2 overlie Wildcat Marble; samples 3 and 4 overlie Franklin Marble; samples 5–8 are from Kula and Gorring (2000) interlayered with calc-silicate metagraywacke. Abbreviation: n.d.—Not determined.
*Total Fe as Fe_2O_3.
†Total Fe as FeO.

which modification of the mantle source by subduction-zone fluids occurs during an early stage of spreading and more typical MORB is produced during a later stage (Tarney et al., 1977; Wilson, 1989). Based on these geochemical characteristics, supracrustal amphibolites are tentatively interpreted to have formed in a back-arc basin undergoing extension.

Depositional Environment

Attempts to interpret the supracrustal rocks are hindered by the high grade of metamorphism, structural complications, and large volumes of younger intrusive rocks that mask the stratigraphic relationships. Nevertheless, Volkert and Drake (1999) used an extensive geochemical database, in combination with detailed mapping and previous work by others (e.g., Hague et al., 1956; Baker and Buddington, 1970), to develop a tentative stratigraphy and interpretation of the depositional environment. Elements that are relatively immobile during high-grade metamorphism of sandstones (see, e.g., Roser and Nathan, 1997), such as TiO_2, Al_2O_3, Cr, Nb, and Zr, were used as a basis for interpretations. Because all the rocks analyzed were metasandstones, geochemical effects associated with grain-size variation were avoided. Bhatia and Crook (1986) demonstrated that TiO_2, Nb, and Zr are particularly sensitive discriminators of the tectonic setting of deposition and provenance of such rocks, with TiO_2 and Cr serving as a proxy for the mafic component and Nb and Zr for the felsic component. Using ratios of these elements from magmatic rocks in New Jersey of mafic, intermediate, and felsic composition, a set of reference values was compiled that provided a useful tool for making gross interpretations of

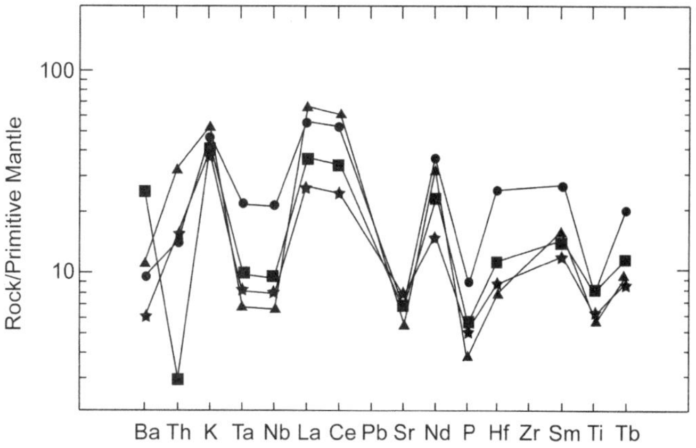

Figure 16. Primitive mantle-normalized trace-element diagram of supracrustal amphibolites from the New Jersey Highlands. Compiled from the geochemical data of Kula and Gorring (2000). Normalization values from Sun and McDonough (1989).

source-sediment composition. The Ti/Zr ratio was found to be a particularly good discriminant, with amphibolites averaging 75, charnockitic Losee rocks 32, leucocratic Losee rocks 19, and potassic granites of the Byram and Lake Hopatcong Intrusive Suites averaging identical values of 4.3. Discriminant function analysis was also used, because of its utility in characterizing provenance signatures of clastic rocks (Bhatia and Crook, 1986; Roser and Korsch, 1988). Despite the effects of some major-element mobility during regional metamorphism, the distribution of supracrustal rocks on the resultant plot (Fig. 17) is consistent with the immobile-element ratios, supporting its use here in provenance interpretation.

The earliest recognized part of the metasiliciclastic sequence in the New Jersey Highlands consists of arkose and lithic sandstones (K-feldspar gneisses) (Table 5, samples 16, 530, 81, and 155), feldspathic and quartzose sandstones (metaquartzite) (Table 6), and graywacke (biotite-quartz-feldspar gneiss) (Table 5, samples 130, 192, 311, and 333; Table 7, type B2) of terrigenous to shallow-marine affinity. These lithologies occur in abundance in the western and eastern Highlands. Meta-arkose has Ti/Zr of 6.0, Cr/Nb of 1.2, and Zr concentrations (375 ppm) that are comparable to the potassic granites (4.3 and 0.88, respectively). Metagraywackes that overlie meta-arkose have similar Ti/Zr of 5–14 and Cr/Nb of 1.0. Metaclastic rocks in this part of the succession have lithologic and geochemical characteristics typical of sediments that were receiving detritus from an uplifted source of predominantly felsic composition (Fig. 17). Interlayered metaquartzites have SiO_2/Al_2O_3 ratios of ~35, indicative of relatively mature sediment formed from the recycling of older quartz-rich basement rocks (Bhatia, 1983).

The terrigenous sequence is transitional to a marine sequence of calcareous graywacke and siliceous carbonate rocks (calc-silicate gneisses), quartzose sandstone (metaquartzite), limestone and dolomitic carbonate rocks (marble), and locally carbonaceous graywacke (biotite-quartz-feldspar gneiss) that are interlayered and have gradational contacts. Marbles clearly reflect sedimentation in a marine environment, as suggested by the stable isotope data and occurrence of stromatolites and a reef structure (Fig. 13) in the western Highlands. The latter features constrain the environment to shallow marine, possibly subtidal to intertidal. In this paper, marble in the eastern Highlands is also interpreted as a marine deposit, based on the similar lithologic associations and stable isotope data that overlaps the values for marble in the west ($\delta^{13}C_{carbonate}$ = –1.5 to 1.0 per mil) (Johnson et al., 1990; Meredith et al., 2003; R.A. Volkert, this study). However, the local occurrence in eastern Highlands marbles of a nontectonic carbonate breccia may indicate gravitational slumping of the carbonate shelf and a somewhat deeper-water, mid- to outer-shelf environment of deposition. This is consistent with the spatial association of these marbles with metagraywacke that was deposited in a deeper-water shelf environment.

The metasiliciclastic rocks in the marine succession occur in equal abundance throughout the New Jersey Highlands. Protoliths must have contained a large amount of sand and relatively little mud, because the rocks are quite siliceous and pelites are

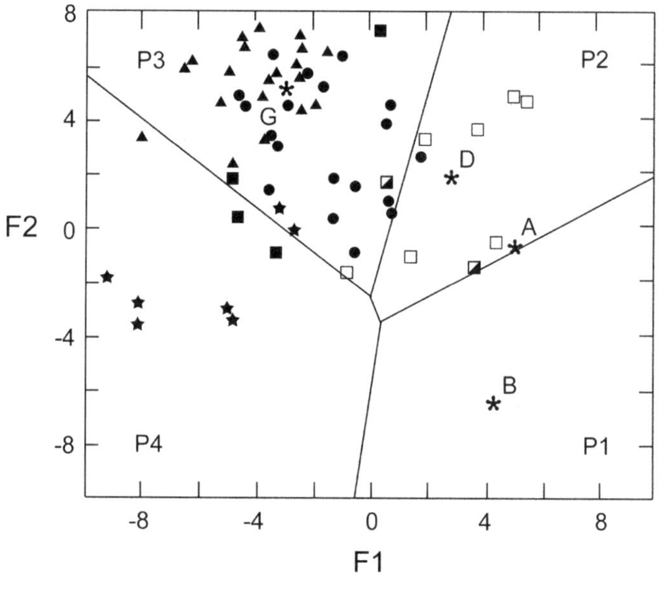

Figure 17. Supracrustal rocks of the Highlands on the tectonic discrimination diagram of Roser and Korsch (1988). F1 versus F2 calculated from discriminant function analysis using Al_2O_3, TiO_2, Fe_2O_{3t}, MgO, CaO, Na_2O, and K_2O. Fields: P1—Mafic volcanic arc; P2—intermediate volcanic arc; P3—active continental margin; P4—recycled orogen. G, D, A, and B are average compositions of granite, dacite, andesite, and basalt, respectively, from Le Maitre (1976). Circles—feldspathic metagraywacke; squares—calcareous metagraywacke (half-filled—type A; open—type B1; filled—type B2; see text for explanation); stars—metaquartzite; triangles—meta-arkose. Data are from Volkert and Drake (1999).

rare in the Highlands. Metaquartzites in the marine succession have SiO_2/Al_2O_3 of ~11, reflecting an origin from a less mature sediment source. Geochemical compositions indicate that most of the marine metaclastics in the Highlands above the marble are metamorphosed graywacke and mixed metavolcaniclastic rocks that consist of feldspathic (Table 5, sample 65) and calcareous types (Table 7, types A and B1). Both types of metagraywacke have Ti/Zr ratios of 15–22 and Cr/Nb of 3.2 that are comparable to calc-alkaline rocks of the Losee Suite (Ti/Zr = 19–32, Cr/Nb = 2–2.7) and to sandstones formed in an active continental margin (Ti/Zr = 15.3, Cr/Nb = 2.4) or continental island-arc setting (Ti/Zr = 19.7, Cr/Nb = 6) (Bhatia and Crook, 1986). Thus, the marine metaclastic rocks that occur stratigraphically above the marble reflect the input of calc-alkaline arc-related volcanogenic detritus of dacitic and andesitic composition (Fig. 17). Geochemical differences between the metagraywackes in the terrigenous and marine successions likely represent proximity to different sediment sources. Those reflecting a continental source may have formed in a shallow shelflike environment along the cratonic margin side of the basin, whereas those reflecting an arclike source may have formed more distally and closer to the arc. Some of the latter metagraywackes are sulfidic and graphitic, suggesting that the sediments were formed under locally anoxic bottom-water conditions.

It is clear that synchronous sedimentation and volcanism were occurring in this basin environment in the New Jersey Highlands and that the greater volume of mafic volcanics in the northwestern Highlands implies closer proximity to a volcanic center. Interestingly, the largest volumes of supracrustal amphibolites are within a few km of the marble-hosted Zn-Mn-Fe deposits at Franklin and Sterling Hill that are interpreted as having a seafloor hydrothermal origin (e.g., Johnson et al., 1990; Johnson, 2001 and references therein). Seafloor vents possibly associated with this volcanism present an attractive hypothesis for the source of the metals in the Zn-Mn-Fe deposits in the Franklin quadrangle.

Intercalated mafic metavolcanics of MORB and arclike affinity are inconsistent with the formation of the supracrustal succession along a passive margin. Episodic rifting, required to account for the occurrence of mafic volcanics at different stratigraphic intervals, suggests that deposition occurred in a basin that was undergoing extension. Protoliths of the metasediments resemble those from an interval of pre- to syntectonic sedimentation, from early rift-related facies, and the development of a shelf sequence, to the deposition of volcanogenic sediments and silting of the carbonate shelf in response to closure of the basin. Volkert and Drake (1999) interpreted the basin setting to be intracratonic and to have formed along the Laurentian margin. However, the lithologic associations and mixed MORB and arclike geochemistry of intercalated mafic metavolcanics are probably more compatible with a back-arc basin setting, which is still interpreted to have occurred along the eastern Laurentian margin. Gunderson (1986) proposed a similar basin environment for the supracrustal rocks in the Hudson Highlands in New York and northern New Jersey Highlands.

META-ANORTHOSITE AND RELATED ROCKS

Rocks of anorthositic composition are sparsely exposed, but are nonetheless present in the New Jersey Highlands. Young (1969) recognized plagioclase-rich gneiss in a conformable layer 55 m thick in the Dover quadrangle (Fig. 3), which he interpreted as a meta-anorthosite sill, based on geochemical data (Table 9, sample 3). This gneiss is gray, medium grained, and well foliated. It is composed predominantly of plagioclase (An_{44-48}) and varying amounts of biotite, amphibole, clinopyroxene, orthopyroxene, apatite, and magnetite (Young and Icenhower, 1989). A thin mafic layer composed of plagioclase (An_{44-60}), amphibole, clinopyroxene, orthopyroxene, apatite, and magnetite that occurs at the base of the unit was interpreted by Young (1969) to be gabbroic anorthosite, based on geochemical data (Table 9, sample 4). Recent mapping in the Dover quadrangle by the author has identified two additional thin layers of probable meta-anorthosite, located northeast of those recognized by Young.

Megacrystic amphibolites have been identified by the author in the Easton, Blairstown, Tranquility, Stanhope, and Dover quadrangles (Fig. 3), where they form conformable layers 0.5–2 m thick that appear to be dikes. These amphibolites contain plagioclase megacrysts (An_{29-44}) ≤13 cm long (Fig. 18) in a medium-grained, well-foliated groundmass of amphibole (magnesiohastingsite), plagioclase (An_{18-38}), and local biotite, augite, apatite, and Fe-Ti oxides. The chemical compositions (Table 9, samples 1 and 2) are consistent with an alkalic to tholeiitic basalt protolith that formed through partial melting of an enriched, subduction-modified, garnet-free lithospheric mantle source, implying magma generation at a depth of <65 km (Volkert and Gorring, 2001). These amphibolites are very likely genetically related to the meta-anorthosite gneiss in the Dover quadrangle; both rock types are part of a study in progress.

Megacrystic amphibolites appear to be analogous to the mafic rocks associated with anorthosite massifs in the Adirondack Highlands, described by Olsen (1992). Both the New Jersey and Adirondack mafic rocks are tholeiitic (Y/Nb ratios of 4.7 and 5.3, respectively) olivine-normative basalts. They have similar geochemical characteristics, including high TiO_2 (2.5–3.5 wt%), Al_2O_3 (14–19 wt%), FeO_t (10–16 wt%), and comparable trace and REE abundances (Table 9), forming a distinct Al-Fe mafic magma type characteristic of anorthosite associations (Olsen and Morse, 1990). Megacrystic amphibolites are also lithologically and geochemically similar to Mesoproterozoic mafic anorthosite (hornblende gabbro) dikes associated with anorthosite in the Honey Brook Upland in eastern Pennsylvania, as described by Crawford and Hoersch (1984) (Table 9, sample 5). Based on field relationships (discussed later in this paper), geochemical similarities between the New Jersey and Adiron-

TABLE 9. REPRESENTATIVE ANALYSES OF MEGACRYSTIC AMPHIBOLITES AND POSSIBLE RELATED ROCKS

	Sample number					
	1 (n = 4)	2 (n = 6)	3 (n = 5)	4 (n = 2)	5 (n = 4)	6 (n = 20)
SiO_2	46.87	47.40	51.02	41.35	46.90	48.27
TiO_2	1.74	2.46	0.68	4.25	1.98	2.66
Al_2O_3	17.67	15.29	25.78	13.45	17.94	15.84
Fe_2O_3*	12.86	13.97	4.12	18.08	13.59	n.d.
FeO†	n.d.	n.d.	n.d.	n.d.	n.d.	13.37
MnO	0.17	0.19	0.10	0.26	0.13	0.20
MgO	5.61	5.15	1.53	7.04	3.31	5.89
CaO	8.42	8.11	8.68	10.50	9.22	8.75
Na_2O	3.33	3.45	4.31	1.39	3.75	3.29
K_2O	1.82	1.88	1.98	1.22	1.12	1.29
P_2O_5	0.22	0.75	0.09	1.11	0.46	0.45
LOI	1.18	0.93	1.15	0.95	1.79	n.d.
Total	99.89	99.58	99.44	99.60	100.19	100.01
Sr	385	363	299	208	n.d.	277
Rb	56	63	102	29	n.d.	32
Ba	436	449	484	495	n.d.	286
Sc	24	27	10	50	n.d.	30
Zr	124	226	120	335	n.d.	204
La	8.27	39.30	11.87	40.66	n.d.	24.50
Ce	20.47	80.57	26.06	85.09	n.d.	64.20
Nd	15.07	44.07	11.01	55.83	n.d.	38.02
Sm	4.46	10.27	2.14	3.61	n.d.	9.22
Eu	1.55	2.48	1.45	2.71	n.d.	2.66
Tb	0.79	1.44	0.46	2.97	n.d.	1.56
Yb	2.66	4.22	0.93	6.46	n.d.	5.12
Lu	0.40	0.63	0.12	0.87	n.d.	0.80

Note: Units for chemical compositions are wt%; those for elements are ppm. Samples 1 and 2, type A and type B megacrystic amphibolites, respectively (Volkert and Gorring, 2001); sample 3, meta-anorthosite (Young and Icenhower, 1989); sample 4, gabbroic anorthosite (Young and Icenhower, 1989); sample 5, mafic anorthosite from Honey Brook Upland (Crawford and Hoersch, 1984); sample 6, amphibolites, Adirondack Highlands (Olsen, 1992). Abbreviations: n.d.—Not determined; LOI—loss on ignition.

*Total Fe as Fe_2O_3.

†Total Fe as FeO.

dack amphibolites, and the spatial association of both types of amphibolites with meta-anorthosite, megacrystic mafic rocks and meta-anorthosite in the New Jersey Highlands are interpreted here to be coeval with AMCG magmatism in the Adirondacks, dated by McLelland et al. (2001) at 1160–1130 Ma.

GRANITES

Two suites of voluminous A-type granites (Byram and Lake Hopatcong) and comparatively small and localized occurrences of postorogenic granites and pegmatites intrude rocks of the Losee Suite and the supracrustal succession in the New Jersey Highlands. The Byram Intrusive Suite (Drake et al., 1991b) and the Lake Hopatcong Intrusive Suite (Drake and Volkert, 1991) together constitute the Vernon Supersuite (Volkert and Drake, 1998). These rocks are both widespread and abundant, underlying ~50% of the Highlands (Fig. 19). Postorogenic rocks include the Mount Eve Granite (Fig. 19), thin sheets of weakly foliated, mildly discordant microperthite alaskite, widespread undeformed pegmatites, and locally occurring, thin, discordant undeformed granite bodies.

Vernon Supersuite

Rocks of the Vernon Supersuite form linear belts with continuous strike lengths of ≤80 km. Within these belts, rocks of both suites display similar isoclinal fold phases and geometries. Contacts between the suites are gradational and conformable, lack evidence for crosscutting relationships, and display indications of magma commingling along their margins. Contacts between these suites and other Mesoproterozoic rocks are largely concordant and all share the same penetrative metamorphic foliation and isoclinal fold structures resulting from Ottawan orogenesis. However, at both map and outcrop scale, the granites locally display weakly to moderately discordant contacts against supracrustal rocks—especially against rocks of

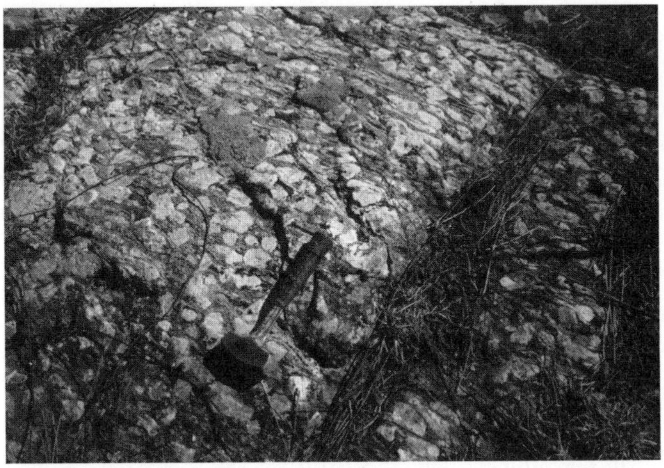

Figure 18. Megacrystic amphibolite from the Blairstown quadrangle. Hammer is 28 cm long.

the Losee Metamorphic Suite. In the Mendham quadrangle, Lake Hopatcong granite displays a mildly discordant contact against diorite. In the Wanaque and Boonton quadrangles, Byram granite has discordant contacts against diorite and contains xenoliths of leucocratic Losee gneiss and diorite up to several tens of meters long near the margins of the granite sheets.

Samples of Byram granite provide whole-rock Rb-Sr isochron ages of 1116 ± 41 Ma and 1110 ± 25 Ma; Lake Hopatcong samples yield granite whole-rock Rb-Sr ages of 1095 ± 9 Ma and 1097 ± 18 Ma (Volkert et al., 2000b) (Table 1). Disturbance of the Rb-Sr systematics was interpreted by Volkert et al. (2000b) to be minimal, despite the potential effects of isotopic homogenization related to high-grade metamorphism, for the following reasons. The absence of scatter of the points on the Rb-Sr isochrons, the coherent trends for Ba versus Rb/Ba and Sr versus Rb/Sr, and the slightly different initial Sr isotope ratios of both suites argues against homogenization. Moreover, it is

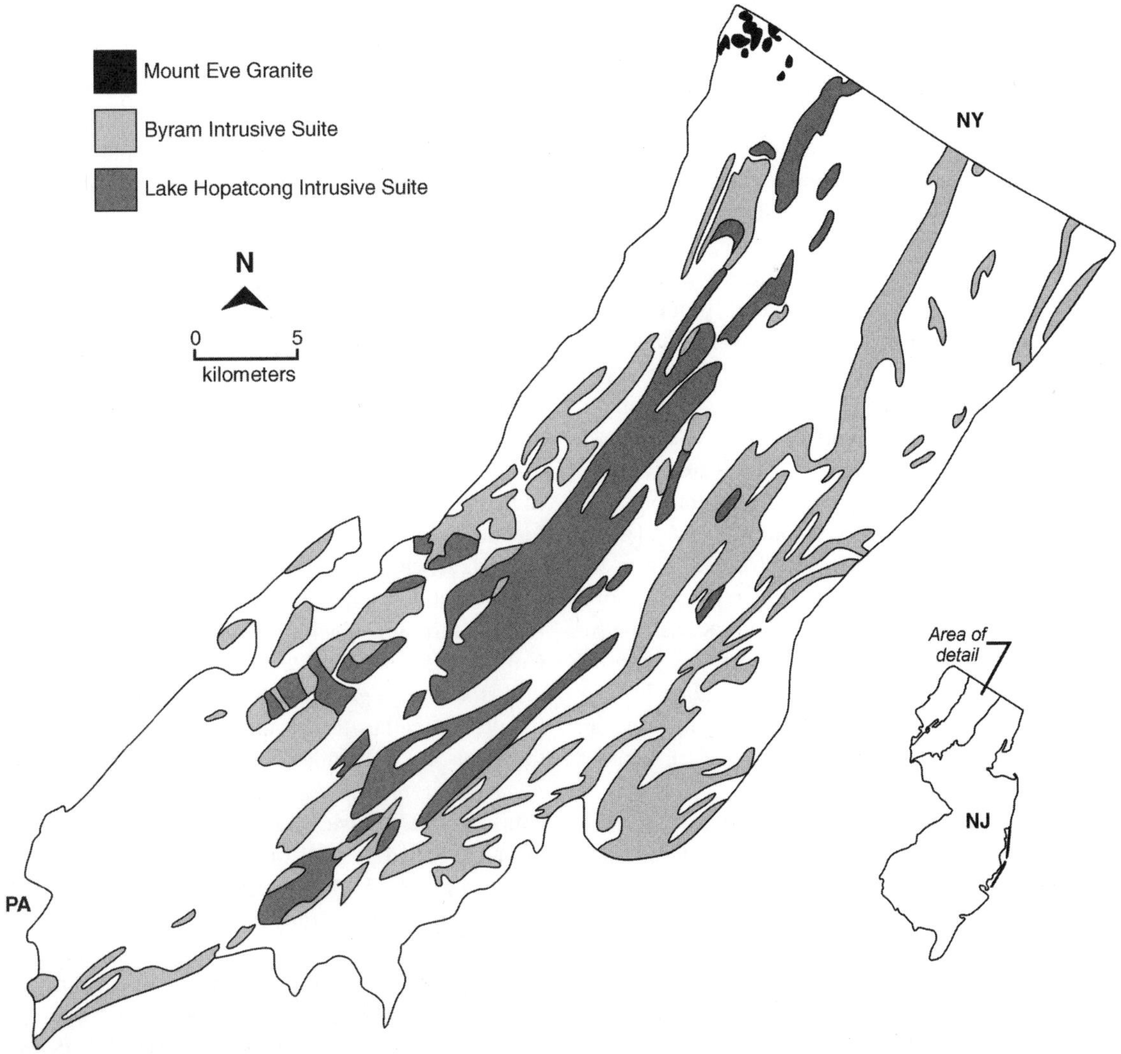

Figure 19. Distribution of granitic rocks of the Byram and Lake Hopatcong Intrusive suites (Vernon Supersuite) and the Mount Eve Granite in the New Jersey Highlands. Modified from Volkert and Drake (1999).

unlikely that samples collected kilometers apart experienced the same Rb depletion and ^{87}Sr enrichment required to give approximately equivalent ages for both suites. Field relationships offer additional support for the Rb-Sr ages, because these suites intrude meta-anorthosite and megacrystic amphibolites that are likely coeval with AMCG-age rocks in the Adirondack Highlands and Canadian Grenville Province dated at 1160–1130 Ma (Corrigan and Hanmer, 1997; McLelland et al., 2001). Other evidence for an age of ca. 1100 Ma for the Byram and Lake Hopatcong suites is provided by overgrowths on zircons from Hudson Highlands granites that have an age of 1114 ± 15 Ma (Ratcliffe and Aleinikoff, 2001b), and mildly discordant $^{207}Pb/^{206}Pb$ ages of 1110–1074 Ma from five fractions of abraded zircons (Drake et al., 1991b). These 1114 to ca. 1100 Ma ages are older than the onset of Ottawan orogenesis, but they overlap the age of the Byram and Lake Hopatcong granites and provide evidence for a thermal event at this time that may be associated with emplacement of these suites. Samples of Byram and Lake Hopatcong granite from the New Jersey Highlands have not been dated using U-Pb techniques. This is an important step in confirming an age of ca. 1100 Ma for these suites, and in determining whether the Lake Hopatcong Suite is, in fact, slightly younger.

Rocks of the Byram Intrusive Suite form a continuous differentiation series that includes monzonite, quartz monzonite, granite, alaskite, and pegmatite (Volkert, 1995). Pegmatites consist of two populations, a pre-orogenic group that is foliated and shares conformable contacts and fold structures with adjacent lithologies, and a more abundant postorogenic group that is undeformed and discordant. All Byram rocks contain hastingsite ± biotite (annite) as ferromagnesian phases. The Fe/(Fe + Mg) molar ratios of these minerals are high (0.84–0.95 for amphiboles and 0.80–0.99 for biotites) and overlap the bulk-rock Fe/(Fe + Mg) ratios (Table 10). These high values are comparable with the ratios of 0.72–0.95 reported for amphiboles and biotites from other Proterozoic A-type granites (e.g., Anderson, 1983; Smith et al., 1997). Biotite typically formed late in the crystallization sequence and coexists with or replaces amphibole in Byram rocks. Feldspars include microcline microperthite and oligoclase, although some mesoperthite, mantled by microcline and oligoclase, is preserved in many rocks. The Fe-Ti oxides are magnetite and ilmenite. Zircon and apatite are common accessory phases; fluorite and various REE-bearing minerals are locally present.

Lake Hopatcong rocks also form a continuous differentiation series that includes monzonite, quartz monzonite, granite, and alaskite (Volkert, 1995). All Lake Hopatcong rocks contain hedenbergite. Fayalitic olivine and sparse ferrosilite and Fe-pigeonite locally coexist with hedenbergite (Young and Cuthbertson, 1994). Hastingsite mantles or replaces clinopyroxene and occurs in separate grains in some rocks. Hedenbergite is Fe-rich and has Fe/(Fe + Mg) molar ratios of 0.88–0.93 that overlap the bulk-rock ratio (Table 11). Feldspars are primarily mesoperthite and oligoclase, and less commonly microcline microperthite or microantiperthite. Fe-Ti oxides are ilmenite and magnetite. Pyrite, zircon, titanite, and apatite are ubiquitous accessory phases.

Byram and Lake Hopatcong rocks are metaluminous to weakly peraluminous (A/CNK = 0.7–1.1) and have very similar concentrations of major and most trace elements (Tables 10 and 11). Both suites have near continuous ranges of SiO_2 (from 58 to 75 wt%) and distinctive chemical compositions, characterized by high total alkalies ($K_2O + Na_2O$ of 7–11 wt%), Fe, F, Cl, Nb, Zr, Y, Ga, light REE, and low Mg, Ca, and Sr. They form an overlapping and continuous differentiation trend from monzonite through granite on Harker variation diagrams and on an AFM diagram (Volkert et al., 2000b). Both suites plot exclusively within the A-type granite field on major- and trace-element discriminant diagrams (Volkert, 1995) and have similar Ga/Al × 10^4 ratios of 4–6 (Fig. 20) that are characteristic of A-type granites (Whalen et al., 1987). Both suites plot in the field of within-plate granites on tectonomagmatic discriminant diagrams (Fig. 21).

Byram and Lake Hopatcong rocks formed through partial melting of a common upper mantle or juvenile lower crustal parent, possibly dioritic in composition (Volkert, 1995). The resulting syenitic liquids underwent fractionation to more siliceous compositions, with Byram and Lake Hopatcong magmas evolving independently. Crystallization occurred over a near identical range of T (≥ 850–550 °C) and relatively low fo_2 (10^{-16} to 10^{-23}) that falls on or slightly below the quartz-fayalite-magnetite buffer curve (Volkert et al., 2000b). Frost and Frost (1997) proposed that low fo_2 values reflect an origin from a similarly low-fo_2 parent. Such low fo_2 values, together with the typically extreme Fe enrichment, suggest a tholeiitic (basalt or ferrodiorite) affinity for magmas with these characteristics. These characteristics are shared by both Byram and Lake Hopatcong rocks. Taken together with initial ε_{Nd} values of 1.21–2.95 (Table 3) and high concentrations of large-ion lithophile and HFSE in granites of both suites (Volkert et al., 2000b), these characteristics are consistent with derivation from a tholeiitic parent. Amphibole was probably stabilized as an early crystallizing phase in Byram magma through substitution of F^- into the OH^- structure (e.g., Collins et al., 1982).

Hornblende granite that is mineralogically similar to Byram granite, as well as the Mount Eve Granite, has been mapped in the Hudson Highlands (e.g., Hotz, 1952; Offield, 1967) and much of it probably correlates to the Storm King Granite (Berkey, 1907; Lowe, 1950). The U-Pb zircon SHRIMP dating of Storm King Granite from the Hudson Highlands indicates an age of 1174 ± 15 Ma (Ratcliffe and Aleinikoff, 2001a). Other hornblende granites from the Hudson Highlands of uncertain affinity have yielded a U-Pb SHRIMP age of 1176 ± 15 Ma from zircon cores (J.N. Aleinikoff, unpublished data).

Granitic gneisses from the northern Virginia Blue Ridge have provided U-Pb zircon (SHRIMP and TIMS) ages of 1160–1140 Ma (Aleinikoff et al., 2000; Tollo et al., this volume), bracketing New Jersey by AMCG-age magmatism. Whether potassic

TABLE 10. REPRESENTATIVE ANALYSES OF THE BYRAM INTRUSIVE SUITE

	Granite					Quartz monzonite and monzonite		
Sample	613	263	10	200	66	38	147	70
SiO_2	74.80	70.90	72.90	70.70	71.60	66.30	61.60	58.10
TiO_2	0.23	0.51	0.35	0.55	0.24	0.46	0.66	0.86
Al_2O_3	13.00	12.30	12.50	13.50	12.60	16.00	17.10	18.50
Fe_2O_3	0.40	2.40	1.70	2.09	1.62	1.32	3.07	2.20
FeO	1.10	3.50	2.00	2.70	2.10	2.70	3.70	4.30
MnO	0.03	0.07	0.05	0.06	0.04	0.05	0.17	0.12
MgO	0.25	0.28	0.30	0.39	0.15	0.36	0.52	1.01
CaO	1.34	1.93	1.32	1.57	1.56	2.38	2.22	4.62
Na_2O	2.97	3.31	2.76	3.44	3.03	5.20	4.50	5.00
K_2O	5.52	4.38	5.79	5.06	4.91	4.79	6.59	4.33
P_2O_5	0.04	0.11	0.08	0.10	0.02	0.08	0.12	0.32
Cl	0.12	0.15	0.10	n.d.	0.08	n.d.	n.d.	n.d.
F	0.07	0.11	0.04	n.d.	0.08	n.d.	n.d.	n.d.
LOI	0.23	0.20	0.01	0.08	0.77	0.08	0.06	0.47
Total	99.91	99.89	99.76	100.24	98.80	99.72	100.31	99.83
Ba	800	1100	910	810	470	1300	1100	2000
Cr	10	<10	50	10	10	20	<10	12
Nb	<10	60	40	50	20	40	20	20
Rb	190	130	150	140	160	70	90	60
Sr	100	130	140	80	70	230	150	510
Y	50	100	110	90	300	60	60	40
Zr	270	1300	450	880	390	910	1500	760
Ga	30	32	30	n.d.	32	n.d.	n.d.	n.d.
La	69.30	n.d.	n.d.	66	n.d.	44.90	48.60	n.d.
Ce	140	n.d.	n.d.	132	n.d.	102	117	n.d.
Nd	66	n.d.	n.d.	63	n.d.	60	54	n.d.
Sm	13.10	n.d.	n.d.	12.60	n.d.	12.20	10.70	n.d.
Eu	2.72	n.d.	n.d.	3.29	n.d.	3.91	3.84	n.d.
Tb	2.00	n.d.	n.d.	1.90	n.d.	1.80	1.90	n.d.
Yb	6.31	n.d.	n.d.	6.27	n.d.	3.53	6.95	n.d.
Lu	0.92	n.d.	n.d.	0.92	n.d.	0.59	1.05	n.d.
A/CNK	0.98	0.90	0.94	0.96	0.96	0.89	0.92	0.87
K_2O/Na_2O	1.86	1.32	2.10	1.47	1.49	0.92	1.46	0.87
$Ga/Al \times 10^4$	4.36	4.92	4.54	n.d.	4.80	n.d.	n.d.	n.d.
Fe/(Fe + Mg)	0.81	0.95	0.92	0.92	0.96	0.91	0.92	0.85

Note: Units for chemical compositions are wt%; those for elements are ppm. A/CNK = molar $Al_2O_3/(CaO + Na_2O + K_2O)$; Fe/(Fe + Mg) = molar $FeO_{total}/(FeO_{total} + MgO)$. Analyses from Volkert (1995). Abbreviations: n.d.—Not determined; LOI—loss on ignition.

granites coeval with the Storm King Granite in the Hudson Highlands occur in New Jersey is uncertain; its recognition, if present, requires U-Pb dating of hornblende granite from both areas of the New Jersey Highlands.

Mount Eve Granite

The postorogenic Mount Eve Granite occurs in the western Highlands, in northernmost New Jersey and adjacent New York (Fig. 19), where it comprises nearly thirty relatively small intrusions. Mount Eve Granite lacks a penetrative metamorphic foliation, but displays what I interpret to be a weak magmatic fabric. Intrusive bodies exhibit discordant contacts with adjacent Mesoproterozoic rocks, contain sparse xenoliths of local country rock, and produced thermal aureoles containing local wollastonite where intrusive into the Franklin Marble (Volkert, 1993). The Mount Eve Granite has yielded a U-Pb TIMS age of 1020 ± 4 Ma (Drake et al., 1991a) from concordant and mildly discordant zircon fractions, and a U-Pb TIMS age of 1004 ± 3 Ma from associated pegmatites (R.E. Zartman, unpublished data). The field relationships and geochronology support an interpretation of postorogenic emplacement and provide an important temporal constraint on the close of the Ottawan orogeny in the north-central Appalachians.

Mount Eve Granite is pinkish-gray, medium to coarse grained, and is composed of quartz, microcline microperthite, oligoclase, amphibole, biotite, and magnetite. The A-type Mount Eve Granite is metaluminous to weakly peraluminous, has major- and trace-element abundances that overlap granites of the Byram Suite, and contains similar calcic amphibole and biotite with high molar Fe/(Fe + Mg) ratios (Gorring et al., this volume). Recent studies suggest that the Mount Eve Granite

TABLE 11. REPRESENTATIVE ANALYSES OF THE LAKE HOPATCONG INTRUSIVE SUITE

	Sample number							
	Granite					Quartz monzonite and monzonite		
	200	506	65	537	2	127	42	8
SiO_2	70.80	71.70	75.00	68.90	71.40	64.40	61.30	59.00
TiO_2	0.49	0.42	0.28	0.37	0.14	1.03	0.99	0.91
Al_2O_3	14.30	12.20	11.30	14.10	14.60	12.80	16.10	17.90
Fe_2O_3	2.04	4.03	0.82	3.49	0.32	4.52	5.10	4.02
FeO	1.30	1.70	1.10	1.30	1.58	3.40	3.10	2.90
MnO	0.05	0.04	0.04	0.06	0.09	0.13	0.13	0.19
MgO	0.36	0.13	0.15	0.26	0.36	0.97	0.77	0.79
CaO	1.39	1.01	1.50	1.27	1.29	2.74	2.40	3.37
Na_2O	5.56	3.19	2.79	4.65	4.63	3.84	5.74	5.56
K_2O	3.53	5.21	5.62	4.90	5.49	4.46	4.37	4.84
P_2O_5	0.08	0.04	0.02	0.06	0.08	0.31	0.18	0.25
Cl	0.03	0.01	0.03	n.d.	n.d.	0.09	n.d.	n.d.
F	0.02	0.02	0.01	n.d.	n.d.	0.01	n.d.	n.d.
LOI	0.23	0.01	1.08	0.16	0.31	0.20	0.06	0.08
Total	100.13	99.68	99.70	99.52	100.29	98.90	100.24	99.81
Ba	1300	1100	1000	1000	800	950	1300	2400
Cr	20	10	30	<10	20	70	10	16
Nb	20	30	70	30	<10	20	20	20
Rb	50	120	160	90	120	40	50	70
Sr	150	350	80	120	80	150	180	420
Y	30	40	140	100	50	110	60	20
Zr	750	1100	710	840	710	1400	1500	1100
Ga	31	35	36	n.d.	n.d.	33	n.d.	n.d.
La	21	61.20	n.d.	65.60	27.80	n.d.	36	n.d.
Ce	49	170	n.d.	144	56	n.d.	81	n.d.
Nd	27	100	n.d.	75	42	n.d.	42	n.d.
Sm	6.30	18.70	n.d.	15.70	6.80	n.d.	9.00	n.d.
Eu	2.96	3.99	n.d.	2.36	3.70	n.d.	5.90	n.d.
Tb	1.10	2.80	n.d.	3.00	1.40	n.d.	1.40	n.d.
Yb	4.09	7.76	n.d.	9.06	3.00	n.d.	4.75	n.d.
Lu	0.68	1.15	n.d.	1.32	0.61	n.d.	0.73	n.d.
A/CNK	0.92	0.96	0.84	0.92	0.92	0.79	0.87	0.87
K_2O/Na_2O	0.63	1.63	2.01	1.05	1.19	1.16	0.76	0.87
$Ga/Al \times 10^4$	4.10	5.42	6.02	n.d.	n.d.	4.87	n.d.	n.d.
Fe/(Fe + Mg)	0.89	0.97	0.88	0.94	0.83	0.88	0.90	0.89

Note: Units for chemical compositions are wt%; those for elements are ppm. A/CNK = molar $Al_2O_3/(CaO + Na_2O + K_2O)$; Fe/(Fe + Mg) = molar $FeO_{total}/(FeO_{total} + MgO)$. Analyses from Volkert (1995). Abbreviations: n.d.—Not determined; LOI—loss on ignition.

may have formed through dry melting of charnockite or an Fe-rich mafic to intermediate source under conditions of lithospheric delamination, or during a right-lateral transpressional shearing event following the peak of the Ottawan orogeny (Gorring et al., this volume).

Other Intrusive Rocks

Undated postorogenic granites are present locally throughout the New Jersey Highlands. Mildly to strongly discordant contacts and lack of a penetrative fabric imply a late syn- to postorogenic age. These intrusions are quite felsic and probably represent crustal melts related to postcollisional crustal thickening and orogenic collapse. The most abundant type is weakly foliated microperthite alaskite, which forms thin sheets and screens that have concordant to mildly discordant contacts. They are medium- to coarse-grained rocks composed of quartz, microcline microperthite, and oligoclase. These leucocratic granites lack the concordant contacts and penetrative foliation of alaskitic variants of the Byram and Lake Hopatcong suites, consistent with an interpreted younger age than rock of those suites. Field relationships imply an age bracketed by the 1095-Ma Lake Hopatcong granite and undeformed 1030-Ma trondhjemite. Leucogranites of similar age include the late- to postorogenic 1060–1045 Ma Lyon Mountain Granite in the Adirondack Highlands (McLelland et al., 2001), the 1080–1040 Ma granite of the Tyringham Gneiss in the Berkshire Mountains (Ratcliffe and Aleinikoff, 2001a), and the 1077–1055 Ma granites in the Virginia Blue Ridge (Aleinikoff et al., 2000).

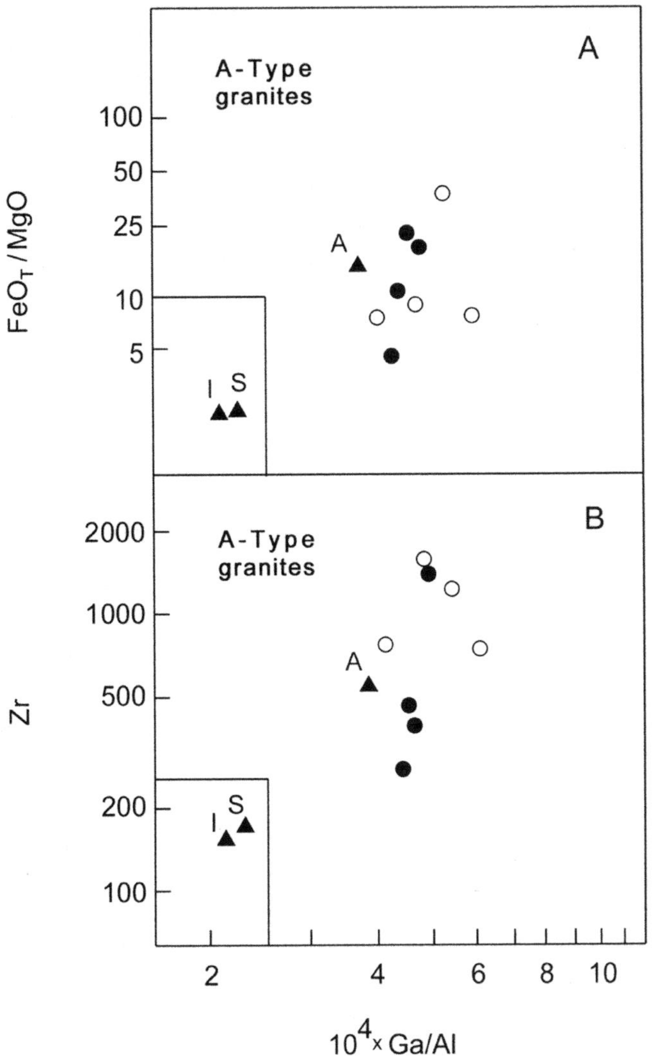

Figure 20. Plot of Byram (solid circles) and Lake Hopatcong (open circles) granites (after Volkert, 1995) on diagrams of Ga/Al × 10^4 versus (A) FeO_t + MgO and (B) Zr from Whalen et al. (1987). A, I, and S are average composition of A, I, and S-type granites, respectively.

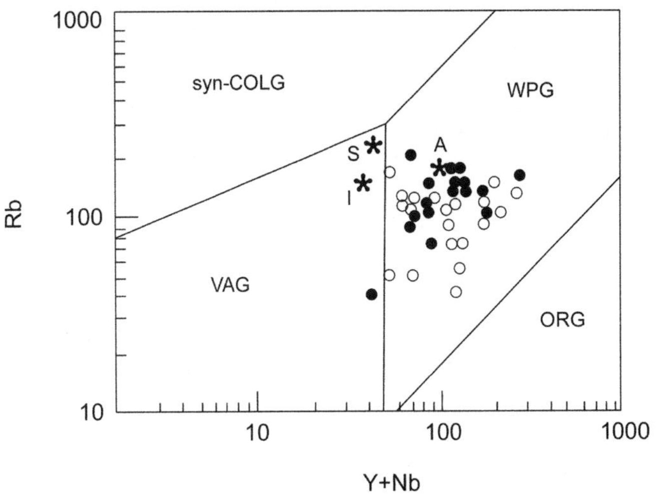

Figure 21. Plot of Byram (solid circles) and Lake Hopatcong (open circles) granites (after Volkert et al., 2000b) on the Rb versus Y + Nb tectonic discriminant diagram of Pearce et al. (1984). Fields: ORG—ocean ridge granite; syn COLG—syn-collision; VAG—volcanic arc; WPG—within-plate. A, I, and S are average composition of A, I, and S-type granites, respectively. Average composition of A-, I-, and S-type granites from Whalen et al. (1987).

In the New Jersey Highlands, other occurrences of undated (but clearly postorogenic) granitic rocks consist of the Pompton Pink Granite (informal usage of Lewis, 1909) in the Pompton Plains quadrangle and undeformed pegmatites composed of graphic granite in the Newton East and Franklin quadrangles, neither of which is extensive enough to be shown in Figure 19. The Pompton Granite, which was used as dimension stone for the Smithsonian Institution in Washington, D.C. (Lewis, 1909), is a pinkish-white, coarse- to very coarse-grained, undeformed rock composed essentially of quartz, microcline, and microperthite. It occurs as small irregular-shaped bodies that have discordant contacts with, and contain xenoliths of, foliated country rock. Undeformed pegmatites that lack discernible fabric and are composed of graphic granite occur in a series of small irregular-shaped bodies as much as tens of meters thick that intrude the Franklin Marble. Contacts with the marble are discordant and locally enveloped by thin zones of undeformed coarse-grained calc-silicate rock. Along their margins, the graphic granite pegmatites contain small xenoliths of marble and trace amounts of graphite likely mobilized from the marble. Ubiquitous graphic intergrowths and the structural relationships suggest that emplacement occurred at a relatively high level in the crust.

The youngest postorogenic rocks are widespread granite pegmatites that have strongly discordant contacts. Dated pegmatites have U-Pb (titanite, zircon) ages of 998 and 989 Ma (R.D. Tucker, unpublished data) (Table 1). Similar undeformed pegmatites and intrusive rocks in the Hudson Highlands were emplaced between 1010 ± 6 Ma (Canada Hill Granite) and 965 ± 10 Ma (Grauch and Aleinikoff, 1985; Aleinikoff and Grauch, 1990). Whether these ages constitute a continuum or are discrete populations cannot be resolved through the limited sampling completed to date. Nevertheless, these postorogenic rocks present clear evidence for thermal activity continuing well past the peak of granulite-facies metamorphism.

GEOLOGIC EVOLUTION OF THE HIGHLANDS

The geologic evolution of Mesoproterozoic rocks in the New Jersey Highlands spanned an interval of at least 250 Ma. From oldest to youngest, the major tectonomagmatic events include:

1. Calc-alkaline magmatism and volcanism at >1200 Ma in a continental-margin arc located along a convergent margin;

2. Deposition of terrigenous to marine sediments and coeval mafic volcanics in a back-arc basin;

3. Termination of subduction and calc-alkaline magmatism by 1176 Ma, and possible accretion of the arc complex to Laurentia by this time as well;

4. Emplacement of meta-anorthosite sills and megacrystic amphibolite dikes between 1160 and 1130 Ma;

5. Emplacement of voluminous A-type granites at ca. 1100 Ma;

6. Crustal thickening, northwest-directed thrusting, granulite-facies metamorphism, and anatexis at 1090–1030 Ma during Ottawan orogenesis; and

7. Emplacement of late synorogenic granites sometime before 1030 Ma and postorogenic granites and pegmatites at 1004–989 Ma.

These events are described more fully in the following sections.

Continental-Margin Arc Magmatism

The undated Losee Suite is interpreted as being part of the widespread continental-margin arc magmatism that existed outboard of the Laurentian margin prior to 1200 Ma, based on the following evidence. The lithologic, geochemical, and isotopic similarity of the calc-alkaline rocks of the New Jersey Highlands to other Grenville-age arc-related occurrences to the north is striking and suggests a comparable origin in a magmatic-arc environment. Calc-alkaline rocks were formed in arc environments outboard of eastern Laurentia throughout the Mesoproterozoic until ca. 1190–1170 Ma, at which time calc-alkaline subduction-related magmatism had terminated (e.g., Rivers, 1997; Wasteneys et al., 1999; Hanmer et al., 2000). Hence, it is unlikely that the large amount of calc-alkaline metaplutonic and metavolcanic rocks in the New Jersey Highlands were generated after this time, and these rocks clearly represent a substantial contribution of isotopically juvenile material to the crust in the north-central Appalachians at ca. 1200 Ma.

Plutonic and volcanic rocks of the Losee Metamorphic Suite were formed in a magmatic-arc environment as shown by the range of rock types, calc-alkaline chemical characteristics, and the arc-related affinity of interlayered mafic rocks. A continental-margin arc setting is favored, based on the evidence discussed earlier in this paper. Moreover, thin crust typical of oceanic arcs produces large volumes of tholeiitic mafic rocks (Wilson, 1989) that are vastly subordinate in the Highlands to rocks of intermediate to felsic composition. Thus, the Losee Suite manifests geochemical characteristics of a mature arc. However, production of Losee arc magma beneath relatively thick crust is unlikely, based on the geochemical and preliminary isotope data, as well as the conspicuous absence of associated plutonic rocks of quartz monzonitic or granodioritic composition.

The widespread occurrence of dacitic and tonalitic (leucocratic) rocks of similar geochemical and isotopic composition throughout the Highlands suggests that the various crustal blocks comprising the western and eastern regions may be related and constitute parts of the same magmatic-arc complex. In both areas, the close spatial association of leucocratic and charnockitic rocks, similarity of geochemical and Sr and Nd isotope data between both rock types, and ubiquitous association of mafic rocks provide further support for this interpretation. However, the confinement of diorite intrusions to a linear belt in the eastern Highlands (Fig. 3) complicates a single-arc interpretation for the entire calc-alkaline assemblage. These intrusions may identify this block as representing a somewhat deeper level of the arc, or they may be a component of a separate arc.

Most tectonic models for the Elzevirian orogeny propose westward migration of the arc and back-arc complexes and collision with eastern Laurentia, culminating in the Elzevirian orogeny (e.g., McLelland et al., 1996; Corrigan and Hanmer, 1997; Rivers and Corrigan, 2000). In the Grenville Province, accretion of the Central Metasedimentary Belt to Laurentia had occurred by 1190 Ma (Rivers, 1997 and references therein); this was followed by accretion of the Adirondack Highlands–Green Mountain block to Laurentia by 1172 Ma (Wasteneys et al., 1999). This collisional orogen was characterized by cessation of subduction and calc-alkaline magmatism, northwest-directed thrusting, crustal thickening, and metamorphism (e.g., Corrigan and Hanmer, 1997; Rivers, 1997). A tectonic boundary representing a suture from this event has not been recognized in the New Jersey Highlands. However, evidence presented in this paper and elsewhere (Puffer and Volkert, 1991; Volkert and Drake, 1999) firmly establishes the presence of a calc-alkaline magmatic arc in the New Jersey Highlands segment of the north-central Appalachians outboard of the Laurentian margin prior to 1200 Ma. Furthermore, the evidence requires the accretion of this arc and associated back-arc basin to Laurentia. The timing of this event is not well constrained, but may have occurred prior to 1176 Ma, at which time calc-alkaline rocks correlative with the Losee Suite and supracrustal rocks in the Hudson Highlands were intruded by AMCG-age granites. If correct, the timing would overlap that proposed for accretion of the Adirondack Highlands–Green Mountain block to Laurentia, and would imply that a collisional event coeval with the Elzevirian orogeny had affected the Laurentian margin at this time in the north-central Appalachians.

Circa 1200 Ma Sedimentation

Inferences regarding the age of the supracrustal rocks in the New Jersey Highlands may be made, based on field relationships, geochemistry, and temporal constraints from the Hudson Highlands. The U-Pb SHRIMP detrital zircon ages of 2000–1200 Ma were obtained from rusty biotite-quartz-feldspar paragneiss in the Hudson Highlands (Gates et al., 2001a), which maintains continuity along strike with similar gneiss in New Jersey. The youngest detrital zircon may provide a maximum age of 1200 Ma for deposition of the succession. A minimum age of 1174 Ma is provided by granite in the Hudson Highlands (Ratcliffe and Aleinikoff, 2001a) that intrudes supracrustal rocks. If correct, then the age of supracrustal rocks in the New Jersey Highlands would be older than successions in southeastern Canada, such as the Flinton and St. Boniface Groups dated at 1180 to ca. 1100 Ma (Corrigan and Hanmer, 1997 and references therein), and would overlap the age of Grenville Supergroup successions in the Central Metasedimentary Belt of southeastern Canada and the Adirondacks, dated at ca. 1300 to 1183 Ma (Wasteneys et al., 1999; Rivers and Corrigan, 2000 and references therein). These successions are very similar to the sequence of supracrustal rocks in the New Jersey Highlands; they consist of metamorphosed siliciclastic rocks, marble, and, except for the Adiirondack Lowlands, volcanic and volcaniclastic rocks. Most of these successions are interpreted to have formed in a back-arc basin setting (Smith and Holm, 1990; summary in Rivers and Corrigan, 2000). Associated mafic metavolcanic rocks in Canada also display geochemical characteristics consistent with a back-arc basin setting that Smith et al. (2001) interpreted to have formed on rifted and attenuated continental crust near the southeastern margin of Laurentia, with the Adirondack Highlands and Green Mountains representing the associated arc to the east.

The lithologic and geochemical characteristics of the metasediments in New Jersey, along with newly acquired geochemical data for intercalated mafic metavolcanic rocks, are most consistent with formation of this succession in a back-arc basin setting. However, this interpretation requires the supracrustal rocks to overlap the age of the Losee Suite, as the back-arc presumably would have been destroyed during accretion of the arc to Laurentia. Three possible scenarios that may explain the apparent contradiction are:

1. The Losee Suite and supracrustal rocks are not significantly different in age, and the contact between them represents a volcanic edifice slightly older than deposition of the supracrustals, as suggested by Ratcliffe et al. (1991) for the apparent nonconformable relationships in the Mount Holly Complex;
2. Supracrustal rocks in the New Jersey Highlands have more than one age and environment of formation; and
3. The supracrustal rocks are the same age, but formed as a passive margin sequence along the margin of a terrane trailing to the east that did not participate in the collisional orogen involving the arc.

The first scenario is possible because the field relationships suggest, but do not prove, that the succession rests nonconformably above the Losee Suite. In terms of the second scenario, the presence in the western and eastern Highlands of rocks representative of both the terrestrial and marine successions (and their similar lithologic and geochemical characteristics in both areas) suggests continuity of the deposits and the absence of a tectonic boundary between them. Thus, a single-basin interpretation is favored for supracrustal rocks in both areas of the Highlands. The third scenario can probably be ruled out as well, because, as stated earlier, the intercalated mafic metavolcanics are not consistent with formation of the supracrustal rocks in a passive margin environment.

Reinterpretation of the supracrustal succession as a back-arc basin fill does not alter the fact that sedimentation in the basin was initiated by rifting and led to the deposition of terrigenous clastics. The arkosic nature of the basal metasediments and their association with mature quartzose sandstones would appear to preclude rocks of the highly sodic Losee Suite as a sediment source. Following terrigenous sedimentation, the basin subsided below sea level, and a shallow marine shelf sequence was deposited on the continental side of the basin. The dramatic thinning of marble in the eastern Highlands and preponderance there of volcanogenic metagraywacke suggests a transition from a shelf environment to the west (present coordinates) to a deeper-water environment to the east that was more proximal to an arclike sediment source.

Magmatism Between 1200 and 1100 Ma

Following the 1190–1170-Ma collisional Elzevirian orogeny, overthickened crust underwent convective thinning or delamination, with replacement of delaminated lithospheric mantle by hot asthenosphere (e.g., Corrigan and Hanmer, 1997). As a result, mantle melts accumulated at the base of the crust, where they underwent varying amounts of fractionation to produce intrusions of gabbroic to anorthositic composition. Heat from the mantle melts facilitated crustal melting, which led to the generation of the potassic and charnockitic components of AMCG complexes of the Adirondack Highlands and the Grenville Province of southeastern Canada (e.g., Whitney, 1983; McLelland et al., 1996; Corrigan and Hanmer, 1997). AMCG complexes display characteristic "anorogenic" within-plate geochemical features (Windley, 1989), which contrast with the calc-alkaline characteristics of older arc-related suites. Based on the geochemistry of Vernon Supersuite granites and the Mount Eve Granite, the same "anorogenic" within-plate characteristics must have persisted continuously until after the Ottawan orogeny.

Megacrystic amphibolites and meta-anorthosite in the New Jersey Highlands formed following the termination of calc-alkaline magmatism in the Losee arc. They represent an interval of AMCG-age magmatism in the Highlands that is otherwise lacking or has yet to be recognized. Field relationships of meta-anorthosite suggest emplacement as sheets into the supracrustal rocks (Young, 1969; Young and Icenhower, 1989), whereas the occurrence of megacrystic amphibolites in an east-west array (Fig. 3) is more consistent with their emplacement as dikes. It is noteworthy that meta-anorthosite sills in the Adirondack Highlands, which are petrogenetically related to massif anorthosites, were also emplaced into supracrustal rocks (Whitney, 1983; Wiener et al., 1984).

AMCG magmatism in the Adirondack Highlands falls in the interval 1160–1130 Ma (McLelland et al., 2001), although a younger (1090–1050 Ma) pulse is recognized in the Canadian Grenville Province (Corrigan and Hanmer, 1997). Meta-anorthosite and megacrystic amphibolites in New Jersey were emplaced into supracrustal rocks, implying an age of <1174 Ma and they are intruded by Byram and Lake Hopatcong rocks that provide a minimum age of >1100 Ma. These bounding ages overlap the older pulse of AMCG magmatism and provide support for the meta-anorthosite and megacrystic amphibolites in New Jersey being coeval with 1160–1130 Ma AMCG complexes elsewhere in the Grenville.

Megacrystic amphibolites are tentatively interpreted to have formed roughly contemporaneously with meta-anorthosite in the Highlands through plagioclase separation from a common mafic magma that ponded at the base of the crust. Premature tapping of this magma by extensional, crustal-scale fractures may have interrupted the extensive fractionation and plagioclase separation necessary to form more voluminous anorthosite intrusions in the Highlands. Instead, this process resulted in the emplacement of meager amounts of anorthosite as thin sills and some amphibolites as thin dikes containing abundant megacrystic plagioclase. Alternatively, the amount of lithospheric delamination and resultant mafic magma underplating of the crust beneath the Highlands at this time may have been small. This would be consistent with the apparent absence of potassic AMCG-type granitoids in the Highlands, because the relatively small amount of ponded mantle would have been insufficient to melt the lower crust.

Voluminous A-type granites of the Byram and Lake Hopatcong Intrusive Suites were generated through partial melting of a common mantle-derived or juvenile lower-crustal source, based on the geochemical and isotopic data. Emplacement of these suites occurred prior to regional compression associated with the Ottawan orogeny, as suggested by the penetrative metamorphic foliation in both suites and the structural relationships with enclosing country rocks discussed previously. These suites were generated during a minor pulse of A-type magmatism at ca. 1100 Ma and, if their interpreted age is correct, the Byram and Lake Hopatcong Suites are among only a few other occurrences of 1100-Ma granites that are recognized in the North American Grenville orogen outside the area of the Mid-continent rift. These include the A-type 1103–1093 Ma Hawkeye granite suite in the Adirondack Highlands (McLelland et al., 2001), 1120 Ma Red Bluff granitic suite in Texas (Smith et al., 1997), and transitional A- to I-type 1112–1111 Ma Marshall Metagranite in the Blue Ridge of Virginia and Maryland (Aleinikoff et al., 2000; Tollo et al., this volume). Byram and Lake Hopatcong granites, as well as the other occurrences mentioned, are regarded as having been emplaced into an extensional tectonic setting. Development of the Mid-continent rift and associated bimodal magmatism at 1109–1094 Ma (Cannon, 1994) overlaps the age of each of these occurrences. McLelland et al. (2001) proposed that lithospheric thinning and crustal melting in the Adirondacks led to the production of Hawkeye granite magmas as a result of far-field effects from rifting of the Mid-continent. A similar interpretation is proposed for the Marshall Metagranite in the Blue Ridge of northern Virginia (Tollo et al., this volume). Rifting and magmatism of the Mid-continent coincide quite closely with the interpreted ages of the Byram and Lake Hopatcong Suites, and crustal extension within the Highlands during their emplacement may have also had a causal relationship to rifting in the Mid-continent.

Ottawan Orogenesis and Postorogenic Magmatism

The tectonic setting at ca. 1100 Ma along eastern Laurentia prior to the Ottawan orogeny was compressional and involved continent-continent collision, crustal thickening, and northwest-directed thrusting (e.g., Windley, 1989). The thermal peak of Ottawan orogenesis in the Highlands occurred between 1090 and 1030 Ma. A maximum age of 1090 Ma for this event is provided by penetratively foliated granite of the Lake Hopatcong Intrusive Suite dated at 1095 ± 9 Ma. The poorly foliated Mount Eve Granite was emplaced at 1020 ± 4 Ma, following the waning stages of the Ottawan orogeny, and metadacite and tonalite underwent local melting at 1029 ± 1 Ma to produce small intrusions of undeformed trondhjemite. These events provide a constraint for a minimum age of 1030 Ma for Ottawan orogenesis. Mesoproterozoic rocks were metamorphosed at upper amphibolite- to hornblende-granulite-facies conditions during the Ottawan event, with *P-T* data from the western and eastern Highlands preserving overlapping temperatures of 670–780 °C and pressures of 410 to ~620 MPa. The pervasive northeast-trending, southeast-dipping metamorphic fabric in the Mesoproterozoic rocks was acquired as a result of northwest-directed compression during this tectonothermal event, although it is unknown whether any of this fabric is relict from an earlier pre-Ottawan event.

Sheets of weakly foliated microperthite alaskite were emplaced as late synorogenic intrusions that are bracketed by the 1095 Ma Lake Hopatcong granites and the undeformed 1030 Ma anatectic trondhjemites. Undeformed pegmatites and other postorogenic intrusive rocks were emplaced between 1004 and 989 Ma and provide evidence for continued thermal activity following the metamorphic peak. This magmatism may be related to an episode of right-lateral transpressional shear produced by tectonic escape following the Ottowan orogeny, as proposed by Gates et al. (2001b), or it may represent crustal melting related to orogenic collapse following Ottawan collisional crustal thickening.

CONCLUSIONS

Mesoproterozoic rocks of the New Jersey Highlands were metamorphosed at 1090–1030 Ma during a tectonothermal event coeval with the Ottawan orogeny. Peak metamorphic conditions during this event reached granulite facies. Overlapping *T* of 670–780 °C and *P* of 410 to ~620 MPa are recorded from the western and eastern Highlands. Evidence for a high-pressure event that occurred at 710–800 MPa is limited to two samples and is presently not well understood. These higher *P* estimates may indicate a deeper level of burial during the Ottawan event, or they may be relict from an older pre-Ottawan event.

The undated (but postulated >1200 Ma) Losee Metamorphic Suite forms a calc-alkaline basement assemblage of metamorphosed plutonic (diorite, quartz diorite, and tonalite) and volcanic equivalents (andesite, dacite, and minor rhyolite) and associated tholeiitic and calc-alkaline metabasalt and basaltic andesite that were generated in a continental-margin magmatic arc outboard of the Laurentian margin. Subduction and calc-alkaline magmatism had ceased regionally by 1176 Ma. Timing of the accretion of the arc complex to the eastern margin of Laurentia is uncertain but may also have occurred by this time.

Supracrustal rocks in the New Jersey Highlands consist of metamorphosed terrigenous arkose (quartzofeldspathic gneiss) and quartzose sandstone (metaquartzite), a shallow marine shelf sequence of sand-dominated clastic rocks that reflect the input of continentally-derived detritus, and stromatolitic marble intercalated with locally pillowed mafic metavolcanics of MORB and volcanic arc affinity. These are overlain by a sequence of volcanogenic metagraywacke, reflecting the input of arc-derived detritus. The entire succession is consistent with formation in a back-arc basin developed on thin continental crust inboard of the Losee volcanic arc. The age of this succession is not well constrained but is likely >1174 Ma.

Meta-anorthosite sheets and megacrystic amphibolite dikes appear to be geochemically and temporally equivalent to 1160–1130-Ma AMCG complexes in the Adirondack Highlands. Other AMCG correlatives—specifically, potassic granites—have not been recognized in the New Jersey Highlands but do occur in the Hudson Highlands of southern New York, where they have been dated at 1176–1174 Ma.

Voluminous A-type granites of the Byram and Lake Hopatcong Intrusive suites were generated through partial melting of a common mantle-derived or juvenile lower crustal source as a possible result of extension related to rifting in the Mid-continent. Emplacement at ca. 1100 Ma shortly preceded granulite-facies metamorphism during the Ottawan orogeny. The Byram and Lake Hopatcong suites may constitute a distinct pulse of A-type magmatism at 1100 Ma that has counterparts in the Adirondack Highlands (Hawkeye Granite Suite), Blue Ridge (Marshall Metagranite), and Grenville of Texas (Red Bluff Granitic Suite).

A maximum age for the Ottawan orogeny in the Highlands is constrained by penetratively foliated rocks of the Lake Hopatcong Intrusive Suite dated at 1095 ± 9 Ma. A minimum age is provided by the 1020 ± 4 Ma postorogenic Mount Eve Granite and undeformed 1029 ± 1 Ma anatectic trondhjemite. Small volumes of weakly foliated granite were emplaced during and following the waning stages of Ottawan orogenesis, and undeformed coarse-grained granites and discordant pegmatites were emplaced between 1004 and 989 Ma, well past the metamorphic thermal peak.

Based on the commonality of metamorphic *P-T* conditions and lithologies of the western and eastern Highlands, faults partitioning the region into structural blocks lack evidence for an origin as tectonic boundaries separating disparate terranes. This continuity between both areas of the Highlands suggests a single —although tectonically shortened—terrane that experienced a common geologic evolution.

The New Jersey Highlands correlates in part to other Grenvillian terranes in eastern North America, mainly north of the Blue Ridge, suggesting that similar magmatic, tectonic, sedimentary, and metamorphic processes extended to the north-central Appalachians.

ACKNOWLEDGMENTS

Support for regional geologic mapping that provided the basis for much of the work presented here was obtained through the COGEOMAP and STATEMAP programs administered by the U.S. Geological Survey. I thank the organizers of this volume for their invitation to contribute, and the editorial assistance and support of Richard Tollo throughout the preparation of this paper is gratefully acknowledged. I am also very grateful to the following individuals: Mark Feigenson for permission to use preliminary Sr and Nd isotope data from the Losee Metamorphic Suite; John Aleinikoff, Robert Tucker, Robert Zartman, and Paulus Moore for permission to use unpublished U-Pb ages from the New Jersey Highlands; Matthew Gorring for permission to use unpublished amphibolite geochemical data; and Donald Monteverde for assistance in confirming the reef structure in the Franklin Marble. William Peck, John Puffer, Richard Tollo, and an anonymous reviewer provided very helpful comments that significantly improved the paper. William Graff and Brian Volkert generously assisted with the cartography.

REFERENCES CITED

Aleinikoff, J.N., and Grauch, R.I., 1990, U-Pb geochronologic constraints on the origin of a unique monazite-xenotime gneiss, Hudson Highlands, New York: American Journal of Science, v. 290, p. 522–546.

Aleinikoff, J.N., Burton, W.C., Lyttle, P.T., Nelson, A.E., and Southworth, C.S., 2000, U-Pb geochronology of zircon and monazite from Mesoproterozoic granitic gneisses of the northern Blue Ridge, Virginia and Maryland, USA: Precambrian Research, v. 99, p. 113–146.

Alling, H.L., 1917, The Adirondack graphite deposits: Albany, New York, New York State Museum Bulletin no. 199, 150 p.

Anderson, J.L., 1983, Proterozoic anorogenic granite plutonism of North America, *in* Medaris, L.G., Jr., et al., eds., Proterozoic Geology: Selected papers from an international symposium: Boulder, Colorado, Geological Society of America Memoir 161, p. 133–154.

Arth, J.G., 1979, Some trace elements in trondhjemite—Their implications to magma genesis and paleotectonic setting, *in* Barker, F., ed., Trondhjemites, dacites, and related rocks: Amsterdam, Elsevier, p. 123–132.

Baker, D.R., and Buddington, A.F., 1970, Geology and magnetite deposits of the Franklin quadrangle and part of the Hamburg quadrangle, New Jersey: Reston, Virginia, U.S. Geological Survey Professional Paper 638, 73 p.

Basaltic Volcanism Study Project, 1981, Basaltic volcanism on the terrestrial planets: New York, Pergamon, 1286 p.

Berkey, C.P., 1907, Structural and stratigraphic features of the basal gneisses of the Highlands: Albany, New York, New York State Museum Bulletin 107, p. 361–387.

Bhatia, M.R., 1983, Plate tectonics and geochemical composition of sandstones: Journal of Geology, v. 91, p. 611–627.

Bhatia, M.R., and Crook, K.A.W., 1986, Trace element characteristics of graywackes and tectonic setting discrimination of sedimentary basins: Contributions to Mineralogy and Petrology, v. 92, p. 181–193.

Brown, G.C., 1982, Calc-alkaline intrusive rocks: Their diversity, evolution, and relation to volcanic arcs, *in* Thorpe, R.S., ed., Andesites—Orogenic andesites and related rocks: New York, John Wiley & Sons, p. 437–461.

Cannon, W.F., 1994, Closing of the Midcontinent rift—A far-field effect of Grenvillian compression: Geology, v. 22, p. 155–158.

Carvalho, A.V., III, and Sclar, C.B., 1988, Experimental determination of the $ZnFe_2O_4$-$ZnAl_2O_4$ miscibility gap with application to franklinite-gahnite exsolution intergrowths from the Sterling Hill zinc deposit, New Jersey: Economic Geology, v. 83, p. 1447–1452.

Collins, W.J., Beam, S.D., White, A.J.R., and Chappell, B.W., 1982, Nature and origin of A-type granites with particular reference to southeastern Australia: Contributions to Mineralogy and Petrology, v. 80, p. 189–200.

Corfu, F., and Easton, M., 1995, U-Pb geochronology of the Mazinaw terrane, an imbricate segment of the Central Metasedimentary Belt, Grenville Province, Ontario: Canadian Journal of Earth Sciences, v. 32, p. 959–976.

Corrigan, D., and Hanmer, S., 1997, Anorthosites and related granitoids in the Grenville orogen: A product of convective thinning of the lithosphere?: Geology, v. 25, p. 61–64.

Crawford, W.A., and Hoersch, A.L., 1984, The geology of the Honey Brook Upland, southeastern Pennsylvania, *in* Bartholomew, M.J., ed., The Grenville Event in the Appalachians and related topics: Boulder, Colorado, Geological Society of America Special Paper 194, p. 111–125.

Dallmeyer, R.D., 1974, Metamorphic history of the northeastern Reading Prong, New York and northern New Jersey: Journal of Petrology, v. 15, p. 325–359.

Dallmeyer, R.D., Sutter, J.F., and Baker, D.J., 1975, Incremental $^{40}Ar/^{39}Ar$ ages of biotite and hornblende from the northeastern Peading Prong: Their bearing on late Proterozoic thermal and tectonic history: Geological Society of America Bulletin, v. 86, p. 1435–1443.

Daly, J.S., and McLelland, J.M., 1991, Juvenile Middle Proterozoic crust in the Adirondack Highlands, Grenville Province, northeastern North America: Geology, v. 19, p. 119–122.

deLorraine, W.F., 2001, Metamorphism, polydeformation, and extensive remobilization of the Balmat zinc ore bodies, northwest Adirondacks, New York, *in* Slack, J.F., ed., Proterozoic iron and zinc deposits of the Adirondack Mountains of New York and the New Jersey Highlands, Part 1: Littleton, Colorado, Society of Economic Geologists Guidebook Series, v. 35, p. 25–54.

Drake, A.A., Jr., 1969, Precambrian and lower Paleozoic geology of the Delaware Valley, New Jersey–Pennsylvania, *in* Subitsky, S., ed., Geology of selected areas in New Jersey and eastern Pennsylvania and guidebook of excursions: New Brunswick, New Jersey, Rutgers University Press, p. 51–131.

Drake, A.A., Jr., 1984, The Reading Prong of New Jersey and eastern Pennsylvania: An appraisal of rock relations and chemistry of a major Proterozoic terrane in the Appalachians, *in* Bartholomew, M.J., ed., The Grenville event in the Appalachians and related topics: Boulder, Colorado, Geological Society of America Special Paper 194, p. 75–109.

Drake, A.A., Jr., and Volkert, R.A., 1991, The Lake Hopatcong Intrusive Suite (Middle Proterozoic) of the New Jersey Highlands, *in* Drake, A.A., Jr., ed., Contributions to New Jersey geology: Reston, Virginia, U.S. Geological Survey Bulletin 1952, p. A1–A9.

Drake, A.A., Jr., Aleinikoff, J.N., and Volkert, R.A., 1991a, The Mount Eve Granite (Middle Proterozoic) of northern New Jersey and southeastern New York, *in* Drake, A.A., Jr., ed., Contributions to New Jersey geology: Reston, Virginia, U.S. Geological Survey Bulletin 1952, p. C1–C10.

Drake, A.A., Jr., Aleinikoff, J.N., and Volkert, R.A., 1991b, The Byram Intrusive Suite of the Reading Prong—Age and tectonic environment, *in* Drake, A.A., Jr., ed., Contributions to New Jersey geology: Reston, Virginia, U.S. Geological Survey Bulletin 1952, p. D1–D14.

Drake, A.A., Jr., Volkert, R.A., Monteverde, D.H., Herman, G.C., Houghton, H.F., Parker, R.A., and Dalton, R.F., 1996, Bedrock geologic map of New Jersey: U.S. Geological Survey Miscellaneous Investigations Series Map I-2540-A, scale 1:100,000.

Frost, C.D., and Frost, B.R., 1997, Reduced rapakivi-type granites: The tholeiite connection: Geology, v. 25, p. 647–650.

Gates, A.E., Valentino, D.W., Chiarenzelli, J.R., Gorring, M.L., and Hamilton, M.A., 2001a, Bedrock geology, geochemistry and geochronology of the Grenville Province in the western Hudson Highlands, New York, *in* Gates A.E., ed., New York State Geological Association Guidebook, p. 176–206.

Gates, A.E., Valentino, D.W., Chiarenzelli, J.R., and Hamilton, M.A., 2001b, Deep-seated Himalayan-type syntaxis in the Grenville orogen, NY-NJ-PA: Geological Society of America Abstracts with Programs, v. 33, no. 6, p. A-91.

Grauch, R.I., and Aleinikoff, J.N., 1985, Multiple thermal events in the Grenvillian orogenic cycle: Geochronologic evidence from the northern Reading Prong, New York–New Jersey: Geological Society of America Abstracts with Programs, v. 17, no. 7, p. 596.

Gunderson, L.C., 1986, Geology and geochemistry of the Precambrian rocks of the Reading Prong, New York and New Jersey—Implications for the genesis of iron-uranium-rare earth deposits: Reston, Virginia, U.S. Geological Survey Circular 974, USGS Research on Energy Resources—1986 Program and Abstracts, p. 19.

Hague, J.M., Baum, J.L., Herrman, L.A., and Pickering, R.J., 1956, Geology and structure of the Franklin-Sterling area, New Jersey: Geological Society of America Bulletin, v. 67, p. 435–474.

Hanmer, S., Corrigan, D., Pehrsson, S., and Nadeau, L., 2000, SW Grenville Province, Canada: The case against post-1.4 Ga accretionary tectonics: Tectonophysics, v. 319, p. 33–51.

Hotz, P.E., 1952, Magnetite deposits of the Sterling Lake, N.Y.–Ringwood, N.J., area: Reston, Virginia, U.S. Geological Survey Bulletin 982-F, p. 153–244.

Hull, J., Koto, R., and Bizub, R., 1986, Deformation zones in the Highlands of New Jersey, *in* Husch, J.M., and Goldstein, F.R., eds., Geology of the New Jersey Highlands and radon in New Jersey: Field Guide and Proceedings of the third annual meeting of the Geological Association of New Jersey, p. 19–66.

Irvine, T.N., and Baragar, W.R.A., 1971, A guide to the chemical classification of the common volcanic rocks: Canadian Journal of Earth Sciences, v. 8, p. 523–543.

Johnson, C.A., 1990, Petrologic and stable isotopic studies of the metamorphosed zinc-iron-manganese deposit at Sterling Hill, New Jersey [Ph.D. dissertation]: New Haven, Connecticut, Yale University, 108 p.

Johnson, C.A., 2001, Geochemical constraints on the origin of the Sterling Hill and Franklin zinc deposits, and the Furnace magnetite bed, northwestern New Jersey, *in* Slack, J.F., ed., Proterozoic iron and zinc deposits of the Adirondack Mountains of New York and the New Jersey Highlands, Part 1: Littleton, Colorado, Society of Economic Geologists Guidebook Series, v. 35, p. 89–97.

Johnson, C.A., Rye, D.M., and Skinner, B.J., 1990, Petrology and stable isotope geochemistry of the metamorphosed zinc-iron-manganese deposit at Sterling Hill, New Jersey: Economic Geology, v. 85, p. 1133–1161.

Kearns, L.E., 1977, The mineralogy of the Franklin Marble, Orange County, New York [Ph.D. dissertation]: Newark, University of Delaware, 211 p.

Kula, J., and Gorring, M.L., 2000, Tectonic setting and protolith for Middle Proterozoic amphibolites within the Losee Gneiss of the New Jersey Highlands: Geological Society of America Abstracts with Programs, v. 32, no. 7, p. A-455.

Le Maitre, R.W., 1976, The chemical variability of some common igneous rocks: Journal of Petrology, v. 17, p. 589–637.

Lewis, J.V., 1909, Building stones of New Jersey: New Jersey Geological Survey annual report of the state geologist for the year 1908: Trenton, MacCrellish and Quigley, p. 55–124.

Lowe, K.E., 1950, Storm King Granite at Bear Mountain, N.Y.: Geological Society of America Bulletin, v. 61, p. 137–190.

Markewicz, F.J., 1965, Chester monazite belt [unpublished report on file in the office of the New Jersey Geological Survey]: Trenton, New Jersey Geological Survey, 6 p.

Maxey, L.R., 1971, Metamorphism and origin of Precambrian amphibolites of the New Jersey Highlands [Ph.D. dissertation]: New Brunswick, New Jersey, Rutgers University, 156 p.

McLelland, J., and Whitney, P., 1990, Anorogenic bimodal emplacement of anorthositic, charnockitic, and related rocks in the Adirondack Mountains, New York, *in* Stein, H.J., and Hannah, J.L., eds., Ore-bearing granite systems: Petrogenesis and mineralizing processes: Boulder, Colorado, Geological Society of America Special Paper 246, p. 301–315.

McLelland, J., Daly, J.S., and McLelland, J.M., 1996, The Grenville orogenic cycle (ca. 1350–1000 Ma): An Adirondack perspective: Tectonophysics, v. 265, p. 1–28.

McLelland, J., Hamilton, M., Selleck, B., McLelland, J., Walker, D., and Orrell, S., 2001, Zircon U-Pb geochronology of the Ottawan orogeny, Adirondack Highlands, New York: Regional and tectonic implications: Precambrian Research, v. 109, p. 39–72.

McLelland, J.M., and Chiarenzelli, J.R., 1990, Geochronological studies in the Adirondack Mountains and the implications of a Middle Proterozoic tonalitic suite, *in* Gower, C.F., et al., eds., Mid-Proterozoic Laurentia-Baltica: St. John's, Newfoundland, Geological Association of Canada Special Paper 38, p. 175–194.

Meredith, M.T., Doverspike, B.A., and Peck, W.H., 2003, Stable isotope geochemistry of the Franklin Marble (Grenville Province), New Jersey: Geological Society of America Abstracts with Programs, v. 35, no. 3, p. 96.

Moore, J.M., Jr., and Thompson, P.H., 1980, The Flinton Group: A late Precambrian metasedimentary succession in the Grenville Province of eastern Ontario: Canadian Journal of Earth Sciences, v. 17, p. 1685–1707.

Offield, T.W., 1967, Bedrock geology of the Goshen–Greenwood Lake area, N.Y.: Albany, New York, New York State Museum and Science Service, Map and Chart Series no. 9, 78 p.

Olsen, K.E., 1992, The petrology and geochemistry of mafic igneous rocks in the anorthosite-bearing Adirondack Highlands, New York: Journal of Petrology, v. 33, p. 471–502.

Olsen, K.E., and Morse, S.A., 1990, Regional Al-Fe mafic magmas associated with anorthosite-bearing terranes: Nature, v. 344, p. 760–762.

Pearce, J.A., 1983, Role of the subcontinental lithosphere in magma genesis at active continental margins, *in* Hawkesworth, C.J., and Norry, M.J., eds., Continental basalts and mantle xenoliths: Nantwich, England, Shiva, p. 230–249.

Pearce, J.A., Harris, N.B.W., and Tindle, A.G., 1984, Trace element diagrams for the tectonic interpretation of granitic rocks: Journal of Petrology, v. 25, p. 956–983.

Peck, F.B., 1904, The talc deposits of Phillipsburg, N.J., and Easton, Pa.: Trenton, MacCrellish and Quigley, New Jersey Geological Survey, annual report of the state geologist, p. 161–185.

Puffer, J.H., and Volkert, R.A., 1991, Generation of trondhjemite from partial melting of dacite under granulite facies conditions: An example from the New Jersey Highlands, USA: Precambrian Research, v. 51, p. 115–125.

Puffer, J.H., Pamganamamula, R.V., and Davin, M.T., 1993a, Precambrian iron deposits of the New Jersey Highlands, *in* Puffer, J.H., ed., Geologic traverse across the Precambrian rocks of the New Jersey Highlands: Field guide and proceedings of the tenth annual meeting of the Geological Association of New Jersey, p. 56–95.

Puffer, J.H., Osian, E., Gilchrist, S., Connolly, M., Forrest, C., Mullarkey, D., and Ratti, J., 1993b, Amphibolites and pyroxenites of the New Jersey Highlands, *in* Puffer, J.H., ed., Geologic traverse across the Precambrian rocks of the New Jersey Highlands: Field Guide and proceedings of the tenth annual meeting of the Geological Association of New Jersey, p. 96–116.

Rankin, D.W., Chiarenzelli, J.R., Drake, A.A., Jr., Goldsmith, R., Hall, L.M., Hinze, W.J., Isachsen, Y.W., Lidiak, E.G., McLelland, J., Mosher, S., Ratcliffe, N.M., Secor, D.T., Jr., and Whitney, P.R., 1993, Proterozoic rocks east and southeast of the Grenville front, *in* Reed, J.C., Jr., et al., eds., Precambrian: Conterminous U.S.: Boulder, Colorado, Geological Society of America, The Geology of North America, v. C-2, p. 335–461.

Ratcliffe, N.M., and Aleinikoff, J.N., 2001a, New insights into the tectonic history of the Mesoproterozoic basement of the northern Appalachians: One coherent terrane?: Geological Society of America Abstracts with Programs, v. 33, no. 6, p. A-90.

Ratcliffe, N.M., and Aleinikoff, J.N., 2001b, Pre-Ottawan deformation and transpressional faulting in the Hudson Highlands of New York based on SHRIMP zircon ages of the Storm King granite and the syntectonic Canopus pluton: Geological Society of America Abstracts with Programs, v. 30, no. 1, p. A5.

Ratcliffe, N.M., Aleinikoff, J.N., Burton, W.C., and Karabinos, P., 1991, Trondhjemitic 1.35–1.31 Ga gneisses of the Mount Holly Complex of Vermont: Evidence for an Elzevirian event in the Grenville basement of the United States Appalachians: Canadian Journal of Earth Sciences, v. 28, p. 77–93.

Rivers, T., 1997, Lithotectonic elements of the Grenville Province: Review and tectonic implications: Precambrian Research, v. 86, p. 117–154.

Rivers, T., and Corrigan, D., 2000, Convergent margin on southeast Laurentia during the Mesoproterozoic: Tectonic implications: Canadian Journal of Earth Sciences, v. 37, p. 359–383.

Roser, B.P., and Korsch, R.J., 1988, Provenance signatures of sandstone-mudstone suites determined using discriminant function analysis of major element data: Chemical Geology, v. 67, p. 119–139.

Roser, B.P., and Nathan, S., 1997, An evaluation of elemental mobility during metamorphism of a turbidite sequence (Greenland Group, New Zealand): Geological Magazine, v. 134, p. 219–234.

Sheridan, R.E., Maguire, T.J., Feigenson, M.D., Patino, L.C., and Volkert, R.A., 1999, Grenville age of basement rocks in Cape May NJ well: New evidence for Laurentian crust in U.S. Atlantic Coastal Plain basement Chesapeake terrane: Journal of Geodynamics, v. 27, p. 623–633.

Sims, P.K., 1958, Geology and magnetite deposits of the Dover district, Morris County, New Jersey: Reston, Virginia, U.S. Geological Survey Professional Paper 287, 162 p.

Smith, D.R., Barnes, C., Shannon, W., Roback, R., and James, E., 1997, Petrogenesis of Mid-Proterozoic granitic magmas: Examples from central and west Texas: Precambrian Research, v. 85, p. 53–79.

Smith, T.E., and Holm, P.E., 1990, The geochemistry and tectonic significance of pre-metamorphic minor intrusions of the Central Metasedimentary Belt, Grenville Province, Canada: Precambrian Research, v. 48, p. 341–360.

Smith, T.E., Harris, M.J., Huang, C.H., and Holm, P.E., 2001, The geochemical nature of the Sharbot Lake domain, Central Metasedimentary Belt, Ontario: Canadian Journal of Earth Sciences, v. 38, p. 1037–1057.

Spencer, A.C., Kummel, H.B., Wolff, J.E., Salisbury, R.D., and Palache, C., 1908, Franklin Furnace, New Jersey: Reston, Virginia, U.S. Geological Survey Geologic Atlas, Folio 161, 27 p.

Sun, S.S., and McDonough, W.F., 1989, Chemical and isotopic systematics of oceanic basalts: Implications for mantle composition and processes, *in* Saunders, A.D., and Norry, M.J., eds., Magmatism in the ocean basins: London, Geological Society of London Special Publication 42, p. 313–345.

Tarney, J., Saunders, A.D., and Weaver, S.D., 1977, Geochemistry of volcanic rocks from the island arcs and marginal basins of the Scotia Arc region, *in* Talwani, M., and Pitman, W.C., III., eds., Island arcs, deep sea trenches and back arc basins: Washington, D.C., American Geophysical Union, p. 367–377.

Volkert, R.A., 1993, Geology of the Middle Proterozoic rocks of the New Jersey Highlands, *in* Puffer, J.H., ed., Geologic traverse across the Precambrian rocks of the New Jersey Highlands: Field guide and proceedings of the tenth annual meeting of the Geological Association of New Jersey, p. 23–55.

Volkert, R.A., 1995, The Byram and Lake Hopatcong Intrusive Suites: Geochemistry and petrogenetic relationship of A-type granites from the New Jersey Highlands: Northeastern Geology and Envrionmental Sciences, v. 17, p. 247–258.

Volkert, R.A., 1997, Occurrence and origin of graphite deposits in Middle Proterozoic rocks of the New Jersey Highlands, *in* Benimoff, A.I., and Puffer, J.H., eds., The economic geology of northern New Jersey: Field guide and proceedings of the fourteenth annual meeting of the Geological Association of New Jersey, p. 1–19.

Volkert, R.A., 2000, Basaltic magmatism and synrift sedimentation associated with Iapetan rifting in the New Jersey Highlands: Geological Society of America Abstracts with Programs, v. 32, no. 1, p. A-81.

Volkert, R.A., 2001, Geologic setting of Proterozoic iron, zinc, and graphite deposits, New Jersey Highlands, *in* Slack, J.F., ed., Proterozoic iron and zinc deposits of the Adirondack Mountains of New York and the New Jersey Highlands, Part 1: Littleton, Colorado, Society of Economic Geologists Guidebook Series, v. 35, p. 59–73.

Volkert, R.A., and Drake, A.A., Jr., 1998, The Vernon Supersuite: Mesoproterozoic A-type granitoid rocks in the New Jersey Highlands: Northeastern Geology and Environmental Sciences, v. 20, p. 39–43.

Volkert, R.A., and Drake, A.A., Jr., 1999, Geochemistry and stratigraphic relations of Middle Proterozoic rocks of the New Jersey Highlands: Reston, Virginia, U.S. Geological Survey Professional Paper 1565-C, 77 p.

Volkert, R.A., and Gorring, M.L., 2001, Mesoproterozoic megacrystic amphibolites, New Jersey Highlands: Petrogenesis and tectonic implications: Geological Society of America Abstracts with Programs, v. 33, no. 6, p. A-90.

Volkert, R.A., and Puffer, J.H., 1995, Late Proterozoic diabase dikes of the New Jersey Highlands—A remnant of Iapetan rifting in the north-central Appalachians: Reston, Virginia, U.S. Geological Survey Professional Paper 1565-A, 22 p.

Volkert, R.A., Johnson, C.A., and Tamashausky, A.V., 2000a, Mesoproterozoic graphite deposits, New Jersey Highlands: Geologic and stable isotopic evidence for possible algal origins: Canadian Journal of Earth Sciences, v. 37, p. 1665–1675.

Volkert, R.A., Feigenson, M.D., Patino, L.C., Delaney, J.S., and Drake, A.A., Jr., 2000b, Sr and Nd isotopic compositions, age and petrogenesis of A-type granitoids of the Vernon Supersuite, New Jersey Highlands, USA: Lithos, v. 50, p. 325–347.

Wasteneys, H., McLelland, J., and Lumbers, S., 1999, Precise zircon geochronology in the Adirondack Lowlands and implications for revising plate-tectonic models of the Central Metasedimentary Belt and Adirondack Mountains, Grenville Province, Ontario and New York: Canadian Journal of Earth Sciences, v. 36, p. 967–984.

Whalen, J.B., Currie, K.L., and Chappell, B.W., 1987, A-type granites: Geochemical characteristics, discrimination and petrogenesis: Contributions to Mineralogy and Petrology, v. 95, p. 407–419.

Whitney, P.R., 1983, A three-stage model for the tectonic history of the Adirondack region, New York: Northeastern Geology, v. 5, p. 61–72.

Wiener, R.W., McLelland, J.M., Isachesen, Y.W., and Hall, L.M., 1984, Stratigraphy and structural geology of the Adirondack Mountains, New York: Review and synthesis, *in* Bartholomew, M.J., ed., The Grenville event in the Appalachians and related topics: Boulder, Colorado, Geological Society of America Special Paper 194, p. 1–55.

Wilson, M., 1989, Igneous petrogenesis: Boston, Unwin Hyman, 466 p.

Windley, B.F., 1989, Anorogenic magmatism and the Grenvillian orogeny: Canadian Journal of Earth Sciences, v. 26, p. 479–489.

Young, D.A., 1969, Petrology and structure of the west-central New Jersey Highlands [Ph.D. dissertation]: Providence, Rhode Island, Brown University, 194 p.

Young, D.A., 1995, Kornerupine-group minerals in Grenville granulite-facies paragneiss, Reading Prong, New Jersey: The Canadian Mineralogist, v. 33, p. 1255–1262.

Young, D.A., and Cuthbertson, J., 1994, A new ferrosilite and Fe-pigeonite occurrence in the Reading Prong, New Jersey, USA: Lithos, v. 31, p. 163–176.

Young, D.A., and Icenhower, J.P., 1989, A metamorphosed anorthosite sill in the New Jersey Highlands: Northeastern Geology, v. 11, p. 56–64.

Manuscript Accepted by the Society August 25, 2003

Printed in the USA

Geological Society of America
Memoir 197
2004

U-Pb geochronology and evolution of Mesoproterozoic basement rocks, western Connecticut

Gregory J. Walsh*
U.S. Geological Survey, P.O. Box 628, 87 State Street, Room 324, Montpelier, Vermont 05601, USA
John N. Aleinikoff
U.S. Geological Survey, Box 25046, MS 963, Denver, Colorado 80225, USA
C. Mark Fanning
Research School of Earth Sciences, The Australian National University, Mills Road, Canberra ACT 0200, Australia

ABSTRACT

Geologic mapping and U-Pb geochronology by ion microprobe on zircon, titanite, and monazite in the New Milford quadrangle, western Connecticut indicate Mesoproterozoic events at ca. 1.3, 1.05, and 0.99 Ga in the Laurentian basement rocks. Pink granite gneiss (1311 ± 7 Ma) intruded a paragneiss sequence during the early stages of the Elzevirian orogeny. During the Ottawan orogeny, syn-tectonic anatexis produced a belt of stromatic migmatite at 1057 ± 10 Ma. Ottawan igneous activity included syn-tectonic intrusion of abundant sills of biotite granite gneiss, dated at 1050 ± 14 and 1048 ± 11 Ma, and intrusion of the Danbury augen granite at 1045 ± 8 Ma. Overgrowths on igneous zircon and metamorphic zircon in hornblende gneiss indicate that terminal Grenville metamorphism occurred at ca. 993 ± 8 Ma.

Late Ordovician syn-tectonic events included intrusion of a leucogranite dike into the Brookfield Gneiss at 453 ± 6 Ma and intrusion of the Candlewood Granite at 443 ± 7 Ma. A monazite age from the Candlewood Granite of 445 ± 9 Ma agrees with the zircon age. A second phase of migmatization in the basement rocks is associated with the injection of numerous granitic leucosomes at 444 ± 6 Ma along the margin of the Candlewood Granite. Titanite ages from 431 to 406 Ma indicate several high-grade heating events from the Silurian to the Early Devonian. The lack of Grenville-age titanite in the basement suggests that post-Taconian heating was sufficient to completely reset old titanite in the massif.

Keywords: geochronology, zircon, titanite, monazite, Elzevirian, Ottawan, Grenville, Taconian, Candlewood Granite, Danbury augen granite, New Milford massif, Hudson Highlands

*E-mail: gwalsh@usgs.gov.

Walsh, G.J., Aleinikoff, J.N., and Fanning, C.M., 2004, U-Pb geochronology and evolution of Mesoproterozoic basement rocks, western Connecticut, *in* Tollo, R.P., Corriveau, L., McLelland, J., and Bartholomew, M.J., eds., Proterozoic tectonic evolution of the Grenville orogen in North America: Boulder, Colorado, Geological Society of America Memoir 197, p. 729–753. For permission to copy, contact editing@geosociety.org.

INTRODUCTION

Mesoproterozoic Laurentian basement rocks form discontinuous massifs in western Connecticut, southeastern New York, and northern New Jersey (Fig. 1). The Hudson Highlands, the contiguous New Jersey Highlands, and many smaller unnamed massifs form the Reading Prong, a ~250-km-long belt of Laurentian basement rocks that extends from Reading, Pennsylvania, to Lake Waramaug, Connecticut (Drake, 1984; Rodgers, 1985; Rankin et al., 1989). In Connecticut, the easternmost limit of these massifs is situated immediately west of Cameron's Line (Rodgers, 1985). Cameron's Line marks the eastern limit of the autochthonous Cambrian-Ordovician Iapetan carbonate shelf sequence in western New England and corresponds to a major Ordovician fault (Rodgers et al., 1959; Rodgers, 1971, 1985; Hatch and Stanley, 1973; Hall, 1980) that is interpreted as the root zone for the Taconic allochthons (Stanley and Ratcliffe, 1985). All the Laurentian basement rocks of the northern Appalachians crop out west of Cameron's Line, except the rocks of the Chester and Athens domes, which are exposed in a tectonic window in southeastern Vermont (Ratcliffe and Armstrong, 1995; Ratcliffe et al., 1997; Ratcliffe, 2000a,b). In Connecticut, the Laurentian basement occurs mainly in three large massifs—the Hudson Highlands, the Housatonic Highlands, and the Berkshire massif. These three massifs and a number of smaller unnamed massifs are collectively referred to as the "Highlands massifs" on the Connecticut State map (Rodgers, 1985). In this paper, we use the term "Hudson Highlands" to describe only the Hudson Highlands massif proper. Other massifs in the 7.5-min. New Milford, Connecticut, quadrangle include the New Milford, Sherman, and Morrissey Brook massifs (Fig. 2).

The basement rocks in western Connecticut have been interpreted as Precambrian orthogneisses and paragneisses (Clarke, 1958; Hall et al., 1975; Dana, 1977; Jackson, 1980; Rodgers, 1985). Some authors report that these rocks are more similar to rocks of the Berkshire massif and Housatonic Highlands than they are to rocks of the western Hudson Highlands and southern and central Reading Prong (Harwood and Zietz, 1974; Hall et al., 1975; Drake, 1984; Rankin et al., 1989).

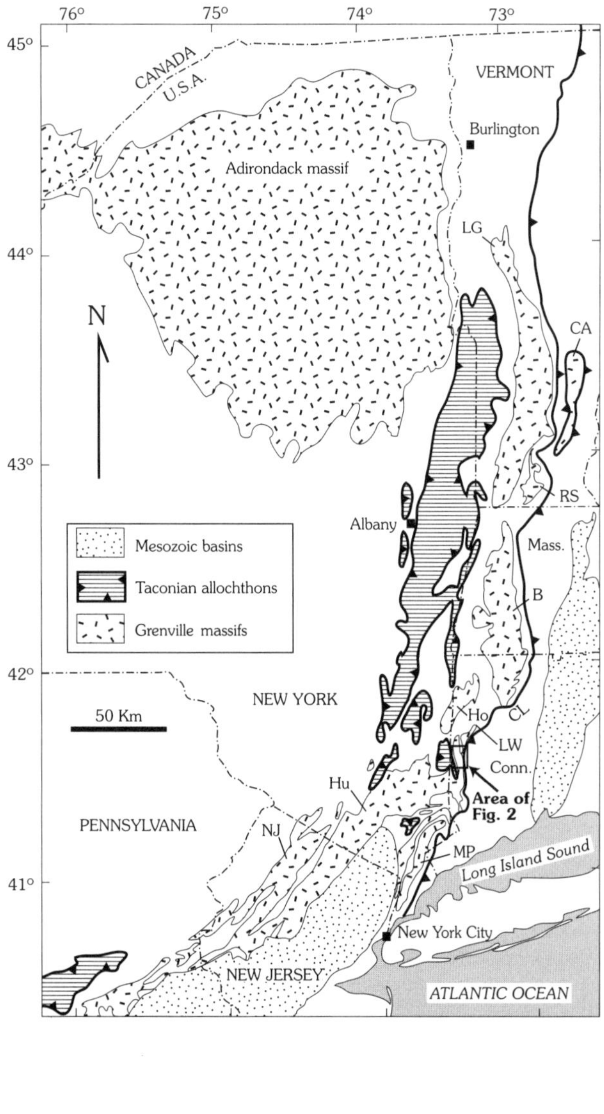

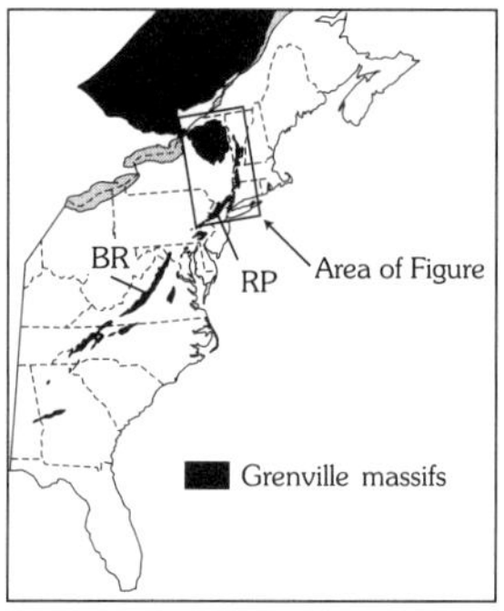

Figure 1. Simplified geologic map of the northeastern United States, showing the distribution of Precambrian basement massifs and the location of the New Milford quadrangle in western Connecticut (area of Figure 2). Massifs: B—Berkshire massif; BR—Blue Ridge; CA—Chester and Athens domes; HO—Housatonic Highlands; HU—Hudson Highlands; LG—Lincoln Mountain–Green Mountain massif; MP—Manhattan Prong; NJ—New Jersey Highlands; RP—Reading Prong; RS—Rayponda and Sadawga domes. Features discussed in the text: CL—Cameron's Line; LW—Lake Waramaug.

Ratcliffe (1980, 1992), Ratcliffe et al. (1985), and Ratcliffe and Burton (1990) indicate that the eastern and western Hudson Highlands have similar paragneiss sequences but different structural configurations across the Canopus fault. The boundary between the eastern and western Hudson Highlands is the Canopus fault (Ratcliffe, 1980, 1992).

Previous geochronologic studies on the rocks of the eastern Hudson Highlands are limited to whole-rock and mineral K/Ar and Rb-Sr ages that loosely constrain the oldest ages of the rocks to a "Grenville" event between ca. 1.3 and 1.1 Ga (Prucha et al., 1968; Hall et al., 1975; Helenek and Mose, 1976, 1978). In this paper, we use the term "Grenville orogenic cycle" of Moore and Thompson (1980) to describe a period of multiple orogenic events between ca. 1.8 and 0.98 Ga (Rivers, 1997; Karlstrom et al., 2001), which are related to continent-continent collisions that culminated at ca. 0.98 Ga (Rivers, 1997) and marked the final assembly of Rodinia (Dalziel, 1997). In addition to the Proterozoic igneous and metamorphic events, all of the basement rocks in the New Milford quadrangle of western Connecticut are overprinted by early Paleozoic metamorphic and igneous events (Long and Kulp, 1962; Clark and Kulp, 1968; Dallmeyer et al., 1975; Hall et al., 1975; Dallmeyer and Sutter, 1976; Helenek and Mose, 1976; Sutter et al., 1985; Hames et al., 1991; Sevigny and Hanson, 1995). We present new data on the geochronology and tectonic evolution of the area using U-Pb zircon, titanite, and monazite geochronology determined by sensitive high resolution ion microprobe (SHRIMP) from Precambrian gneisses and early Paleozoic intrusive rocks of the New Milford quadrangle.

GEOLOGIC SETTING

Previous mapping in the basement rocks of the New Milford area includes work to the north by Dana (1977) in the Lake Waramaug area and by Jackson (1980) in the Kent quadrangle. To the south, Clarke (1958) mapped the Danbury quadrangle. Rodgers (1985) compiled unpublished manuscript maps of the New Milford quadrangle by K.G. Caldwell and G.V. Carroll for the State Bedrock Geological Map. The map of Ratcliffe and Burton (1990) represents the closest recent mapping in the eastern Hudson Highlands, two quadrangles to the west in the Poughquag quadrangle, New York.

In the New Milford area, the Mesoproterozoic basement consists of upper amphibolite-facies gneisses (Hall et al., 1975) unconformably overlain by amphibolite-facies metasedimentary and metavolcanic rocks (Hall et al., 1975; Rodgers, 1985) (Fig. 2). The basement gneisses are unconformably overlain by the Neoproterozoic to Cambrian Dalton Formation and Cambrian to Ordovician Stockbridge Formation (Rodgers, 1985). The Neoproterozoic-Ordovician cover sequence is best preserved in the north-central part of the New Milford quadrangle. On the west side of the New Milford massif (Fig. 2), the Ordovician Walloomsac Formation unconformably overlies both the Dalton Formation and the Precambrian gneisses. Here, the Stockbridge Formation was eroded away prior to the deposition of the Walloomsac Formation along a Middle Ordovician unconformity during the Taconic orogeny (Zen, 1967; Rodgers, 1971, 1985; Jacobi, 1981). Late Middle to early Late Ordovician fossils constrain the age of the Walloomsac Formation in New York (Potter, 1972; Ratcliffe, 1974; Finney, 1986; Ratcliffe et al., 1999). Allochthonous rocks of the Neoproterozoic to Cambrian Manhattan Schist are tectonically above the basement rocks and the autochthonous cover sequence (Fig. 2). The source, or root zone, of the allochthonous rocks is represented by Cameron's Line, a tectonic contact that separates basement and autochthonous platform sequence rocks to the west from ocean-rise sequence rocks with fragments of oceanic crust to the east (Rodgers, 1985; Stanley and Ratcliffe, 1985). The rocks east of Cameron's Line include metasedimentary rocks of the Cambrian to Ordovician Rowe and Ratlum Mountain Schists and the Ordovician Brookfield Gneiss (Fig. 2) (Rodgers, 1985). If the allochthonous rocks in the central and western part of the New Milford quadrangle truly do root from Cameron's Line, and the Walloomsac here is time-correlative with the dated rocks in New York, then the maximum age of movement along Cameron's Line is given by the Blackriveran age (ca. 454 Ma) of the Walloomsac Formation (Ratcliffe et al., 1999), which lies tectonically beneath the allochthonous Manhattan Schist. The minimum age of movement along Cameron's Line is less well constrained. Sevigny and Hanson (1995) provide a composite intrusive age of 453 ± 3 Ma from three plutons of the Brookfield Gneiss east of Cameron's Line. The Brookfield Gneiss is synkinematic with respect to the D2 structures, and the dominant mylonitic foliation along Cameron's Line and in the rocks on both sides of the fault is the D2 fabric, suggesting that the fault was active during the D2 event (ca. 453 Ma). Thus, the paleontologic and geochronologic data suggest that transport of the allochthons in this area occurred after ca. 454 Ma, and may have been completed by ca. 453 Ma, the time of the Brookfield emplacement and development of early D2 fabric. Sevigny and Hanson (1995) also report a postkinematic titanite age of 443 ± 3 Ma from the Brookfield Gneiss and a postkinematic pegmatite age at 445 ± 1.5 Ma. They report younger concordant ages, however, of 435 ± 3 and 438 ± 2 Ma for xenotime from granite with blastomylonitic fabric west of Cameron's Line, suggesting that movement along the fault may have continued into the Silurian. Currently, the Cameron's Line fault and the dominant foliation in the surrounding rocks are overturned and dip steeply to the west, due to subsequent Acadian folding (Rogers, 1985; Panish, 1992; Sevigny and Hanson, 1995).

Basement Gneisses

New mapping in the New Milford quadrangle by Walsh (2003) subdivides the basement rocks into six gneiss units (Fig. 2): (1) layered biotite gneiss (Ybg), (2) biotite granitic gneiss (Ybgg), (3) hornblende gneiss and amphibolite, (4) migmatite gneiss (OYmig), (5) pink granite gneiss (Ygg), and (6) Danbury augen granite of Clarke (1958) (Yag).

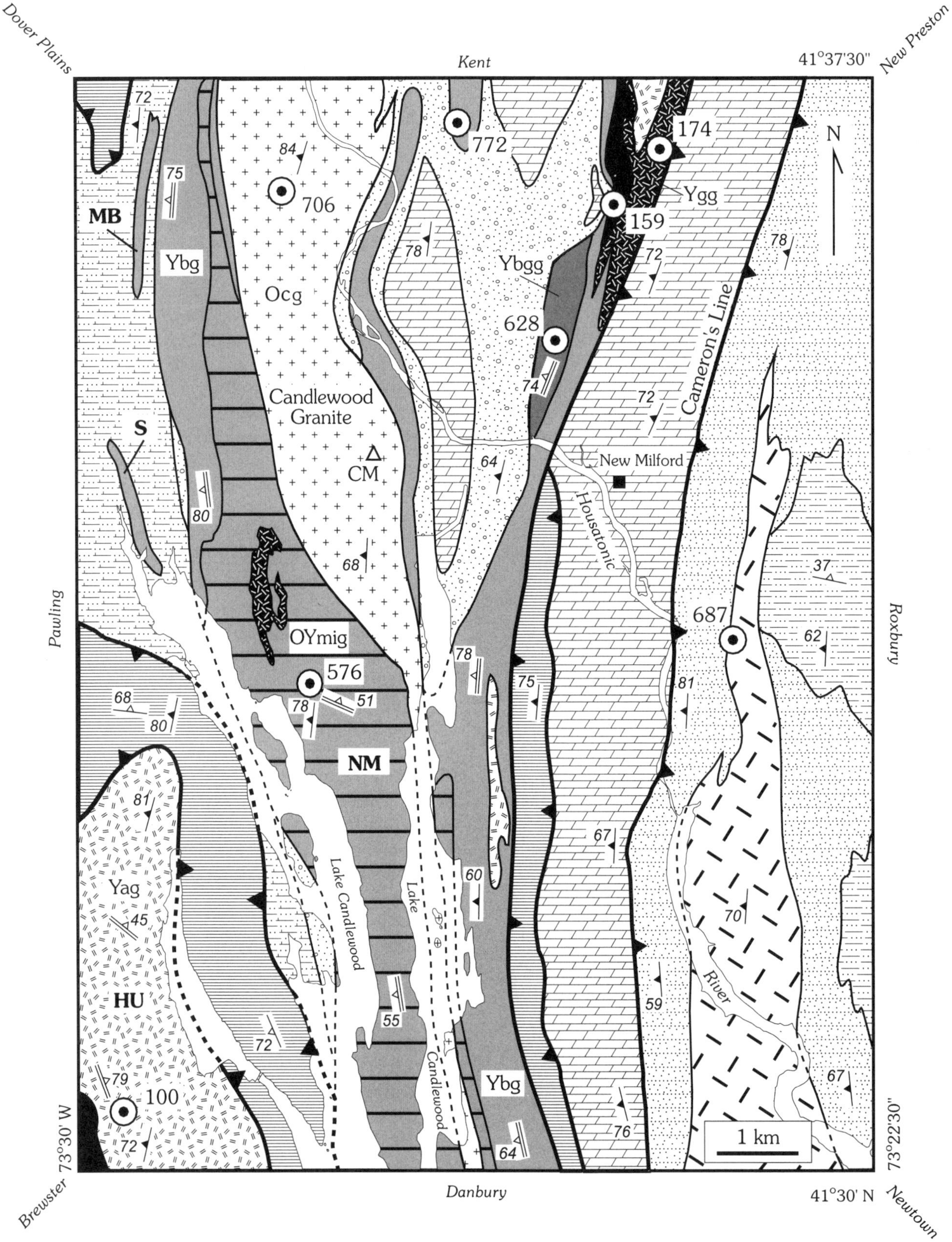
Dover Plains
Kent
41°37'30"
New Preston
Pawling
Roxbury
73°30' W
73°22'30"
Brewster
Danbury
41°30' N
Newtown
N
MB
S
HU
NM
Ybg
Ocg
Candlewood Granite
CM
OYmig
Ybgg
Ygg
Yag
Cameron's Line
New Milford
Housatonic
River
Lake Candlewood
Lake Candlewood
772
706
174
159
628
576
687
100
1 km

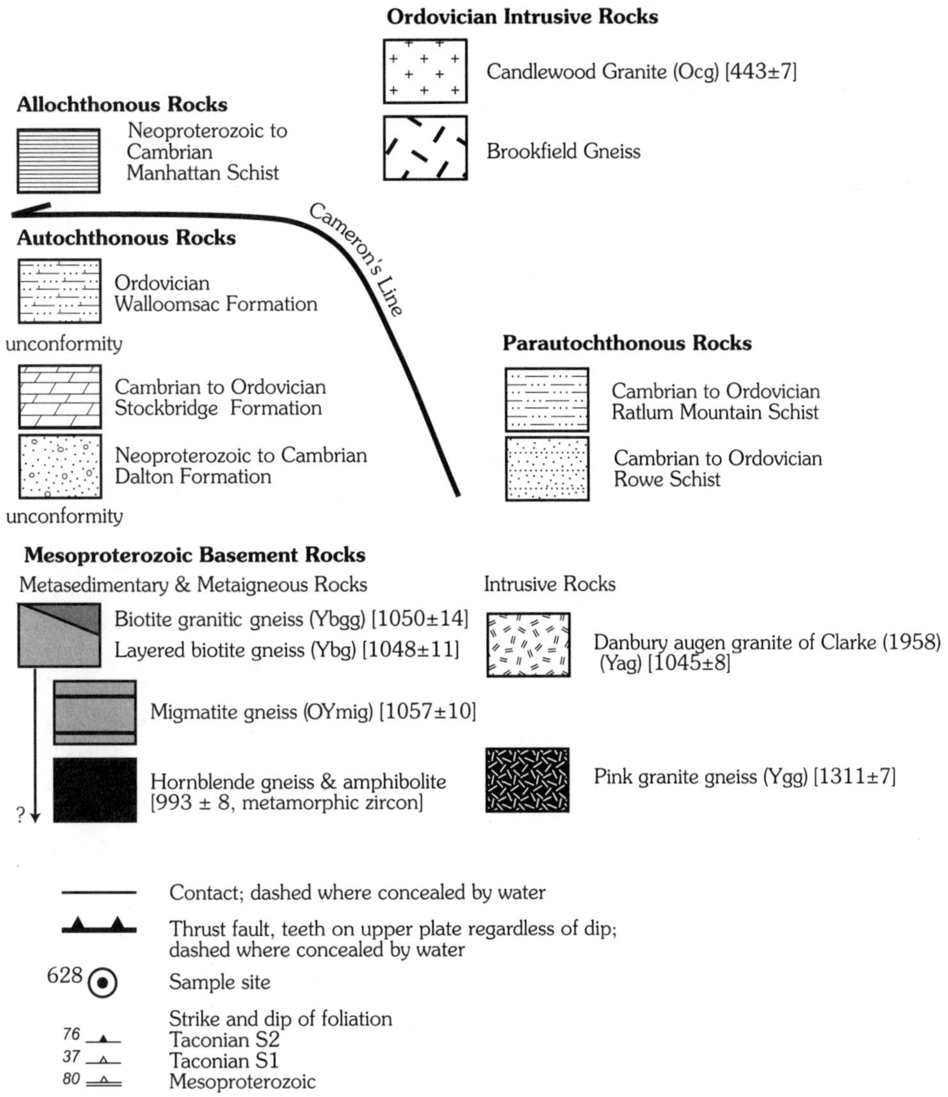

Figure 2 (*on this and previous page*). Generalized geologic map of the 7.5-min. New Milford quadrangle, Connecticut, after new mapping by Walsh (2003) showing sample sites discussed in the text. The U-Pb zircon ages are shown in brackets []. CM—Candlewood Mountain; HU—Hudson Highlands; MB—Morrissey Brook inlier; NM—New Milford massif; S—Sherman inlier.

All of the rocks contain a gneissosity that is truncated by the unconformity at the base of the autochthonous Neoproterozoic to Ordovician cover sequence. This gneissosity is, therefore, Precambrian in age. The gneisses are also deformed during at least two early Paleozoic deformation events, the second of which produced the dominant foliation (S2) in the cover rocks. Along the east side of the New Milford massif, the dominant foliation in the gneisses is re-oriented into subparallelism with the S2 foliation along a syn-D2 fault zone (Fig. 2). Elsewhere in the quadrangle, the Precambrian gneissosity is locally at a high-angle to the younger Paleozoic fabric. The regional trend of the New Milford massif is generally north-south, parallel to the Precambrian gneissosity. To the west, Ratcliffe and Burton (1990) and Ratcliffe (1992) recognize three generations of Precambrian structures (YF_1, YF_2, and YF_3). Their first generation of folds (YF_1) is conjectural and is identified by stratigraphic repetitions, intersection lineations, and an older locally preserved gneissosity. Their second generation (YF_2) is associated with the dominant gneissic fabric and hornblende-granulite–facies metamorphism. Their third generation (YF_3) is related to right-lateral shear zones. In the New Milford area, the most conspicuous gneissosity and folds are second-generation (our

YD2). Early repetitions of paragneisses are not recognized in the New Milford area, although locally, two generations of gneissosity can be observed at the same outcrop (our YD1 and YD2) (e.g., see Fig. 4). Where Precambrian folds are preserved in outcrops in the New Milford quadrangle, only one generation of folds is observed (YD2). The scarcity of observations of the older, poorly preserved gneissosity (YD1) makes it difficult to assess the regional trend of this cryptic fabric. No evidence of Precambrian right-lateral shearing was observed in the New Milford area.

Layered Biotite Gneiss (Ybg)

Layered biotite gneiss (Ybg) is the most widespread rock of the basement gneisses (Fig. 2). It consists of well-layered, light gray biotite-microcline-quartz-plagioclase gneiss, biotite-quartz-plagioclase gneiss, amphibolite, quartz-rich biotite gneiss, and rare vitreous quartzite and calc-silicate gneiss. In this and subsequent descriptions, minerals are listed in order of increasing abundance. The unit is interpreted as a paragneiss with abundant layer-parallel sills of granitic gneiss. Light gray leucocratic biotite-microcline-quartz-plagioclase granitic gneiss occurs everywhere within the Ybg unit as layers ranging from centimeters to meters thick (Fig 3). Sample NM772 (see Fig. 2) is from a 30-cm-thick layer of this leucocratic gneiss, and is bounded by 20-cm- and 10-cm-thick layers of amphibolite. The lithologic layering of the leucocratic gneiss is generally subparallel to the dominant foliation in the rock, but locally, lenses of the leucocratic gneiss converge and appear to crosscut interlayered amphibolite and paragneiss. Crosscutting relations are rarely exposed, and relative age relations of the leucocratic gneiss layers within the Ybg are often equivocal and, therefore, some of these leucocratic layers may represent metasedimentary or metavolcanic rocks, or lit-par-lit intrusive or metasomatic layers.

Figure 3. Map view photograph of the layered biotite gneiss unit (Ybg), showing layers of light gray leucocratic biotite-quartz-microcline-plagioclase gneiss alternating with layers of dark amphibolite. North is to the right. Hammer is 33 cm long.

Biotite Granitic Gneiss (Ybgg)

The biotite granitic gneiss is a well-foliated, moderately layered, gray biotite-microcline-quartz-plagioclase gneiss with a generally uniform texture and composition. It occurs as a single separate map unit in the New Milford massif, northwest of downtown New Milford (Fig. 2). Contacts with the adjacent Ybg unit are not exposed, but Ybgg is interpreted as an intrusive body into Ybg, as it resembles the thin layers of leucocratic biotite gneiss found within the Ybg, and it does not possess the compositional heterogeneity associated with Ybg. Sample NM628 (see Fig. 2) is from the biotite granitic gneiss unit (Ybgg).

Hornblende Gneiss and Amphibolite

The hornblende gneiss and amphibolite unit is poorly to well-foliated, fine to medium grained epidote-biotite-quartz-hornblende-plagioclase gneiss to coarse-grained hornblende-plagioclase gneiss. Two belts are mapped separately in the New Milford quadrangle, one in the northern part of the map and a second in the southwestern corner of the area. Sample NM159 comes from the northern belt of hornblende gneiss and amphibolite (Fig. 2) and is a well-foliated amphibolite gneiss with accessory titanite and opaque minerals. Layers of amphibolite and hornblende gneiss, too small to map separately, occur throughout the layered biotite gneiss (Ybg) and migmatite gneiss (OYmig). Xenoliths of mafic rocks also occur within the pink granite gneiss (Ygg) and Danbury augen granite (Yag).

Migmatite Gneiss (OYmig)

The migmatite gneiss (OYmig) occurs along the west side of the New Milford massif in a north-south trending belt that spans the length of the quadrangle (Fig. 2). The map unit consists of stromatic migmatite containing very well-layered, gray biotite-microcline-quartz-plagioclase gneiss, biotite-quartz-plagioclase gneiss, and amphibolite with abundant layer-parallel white to pink leucosomes of granitic pegmatite and granitic gneiss. The host rock for the migmatite gneiss (OYmig) is the layered biotite gneiss unit (Ybg). The contact between OYmig and Ybg is gradational over a distance of tens of meters and is defined based on the greater abundance of leucosomes within OYmig. Locally, the migmatite is dominated by leucosomes to the extent that the original character of the Ybg host rock is almost completely replaced (Fig. 4). The migmatite gneiss contains folds that predate the dominant Paleozoic folds (F2) and all structures in the cover rocks, suggesting that the older folds are Mesoproterozoic (YD2). In addition to abundant layer-parallel leucosomes, the migmatite gneiss contains abundant dikes and veins of light gray to pink granitic gneiss, aplite, and pink migmatitic granite gneiss that postdate the Mesoproterozoic folding and gneissosity (Fig. 5). The younger intrusions are deformed and contain the dominant fabric (F2 folds and S2 foliation) found in the cover sequence rocks and Ordovician granites. These field relations suggest that

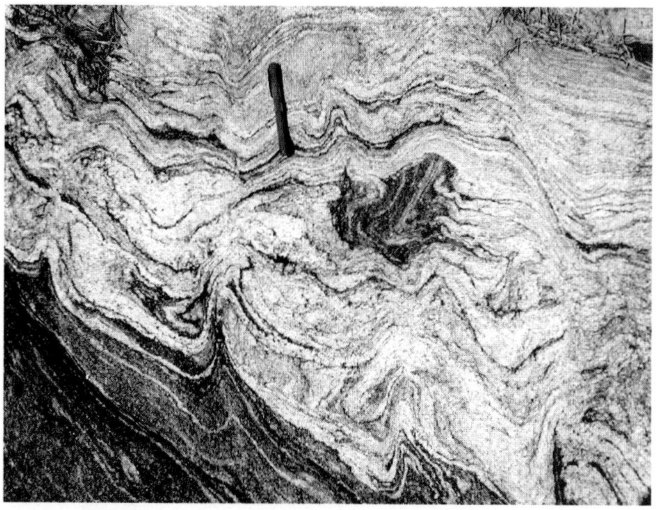

Figure 4. Map view photograph of the migmatite gneiss (OYmig). The dominant foliation is the Precambrian gneissosity; which contains folded, rootless, tight to isoclinal folds (YD2) of older Precambrian gneissosity (YD1), both of which are deformed by upright Taconian (Paleozoic) F2 folds (parallel to pen). Sample NM576B comes from the leucocratic material. The darker rock in the photograph is amphibolite and biotite-rich gneiss that is typical of rocks in the layered biotite gneiss unit (Ybg). The top of the photo is north. Pen is 15 cm long.

the migmatite gneiss has a composite history of Mesoproterozoic migmatization by anatexis, followed by early Paleozoic (Ordovician) migmatization by injection. Sample NM576A is from the relatively younger pink granitic gneiss component (Fig. 5), whereas sample NM576B is from the relatively older layered migmatite gneiss component (Fig. 4) in the same outcrop.

Pink Granite Gneiss (Ygg)

The pink granite gneiss is a well-foliated, tan weathering, light pink biotite-quartz-plagioclase-microcline gneiss that has a generally uniform texture and modal composition of a granite. It occurs as two separate map units in the New Milford massif, one in the west-central part of the massif and a second in the northern part, north of downtown New Milford (Fig. 2). Contacts with the adjacent rock units are not exposed, but Ygg contains xenoliths of layered biotite gneiss and amphibolite and is interpreted as an intrusive body. Sample NM174 is from the pink granite gneiss (Ygg). Foliation trends in the pink granite gneiss are not distinctly different from the migmatite gneiss (OYmig) and layered biotite gneiss units (Ybg).

Danbury Augen Granite (Yag)

Clarke (1958) named the Danbury augen granite for exposures on the west side of Lake Candlewood in the towns of Danbury and New Fairfield, Connecticut. Rice and Gregory (1906) originally used the term "Danbury granodiorite" to describe the same rock, but on their map, they showed the unit as far more extensive than do Clarke (1958) and Rodgers (1985). The rock is a coarse-grained, weakly to moderately well-foliated hornblende-biotite-plagioclase-microcline granite gneiss with deformed phenocrysts of microcline ≤4 cm across and 10 cm long (Fig. 6). Locally, the microcline augen form a pronounced lineation that plunges steeply down the dip of the foliation or steeply to the north and northwest (Clarke, 1958). Clarke (1958) attributed the lineation to movement during emplacement. The Danbury augen granite does not possess the older (YD1) gneissosity seen rarely in the other basement gneisses, but xenoliths within the augen granite do contain the older fabric, suggesting

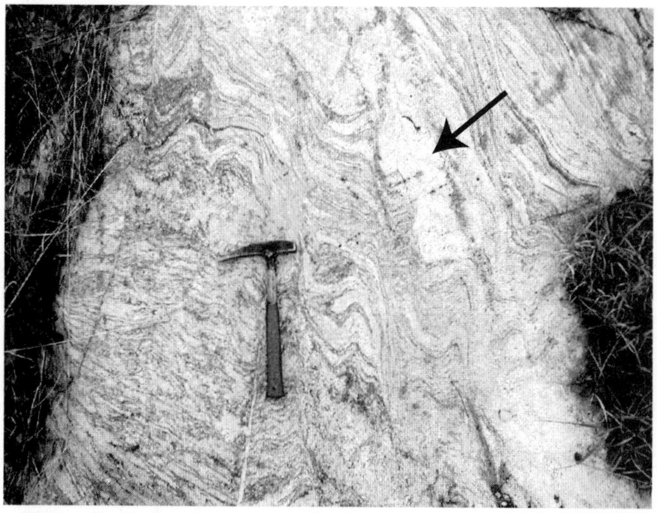

Figure 5. Map view photograph of the migmatite gneiss (OYmig). Light colored granitic leucosome in the upper central part of the image (arrow) cross cuts the Precambrian gneissosity and the older gray stromatic migmatite, but is deformed by the Taconian F2 folds (parallel to hammer handle). Sample NM576A comes from this younger granitic material. Hammer is 33 cm long.

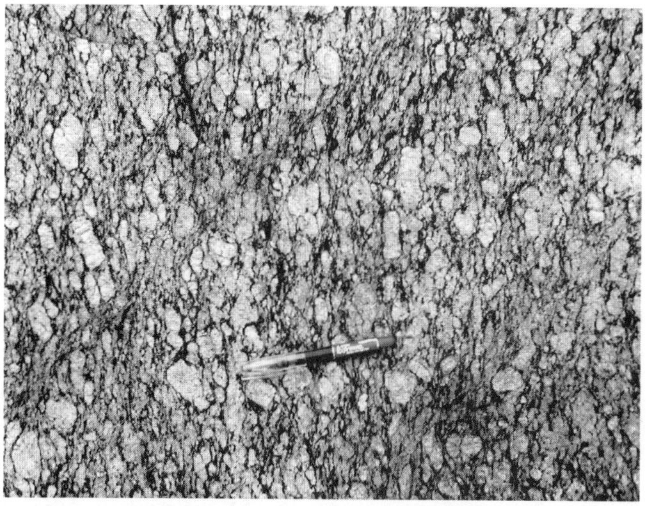

Figure 6. Map view photograph of the Danbury augen granite (Yag), showing the large microcline phenocrysts. The Precambrian gneissosity trends across the photo from top to bottom. The foliation that cuts across the gneissosity from lower left to upper right is Taconian S2. The top of the photo is north-northwest. Pen is 15 cm long.

that the augen granite entirely postdates the earlier deformation event. The Danbury augen granite is variably deformed by the Mesoproterozoic foliation (YD2). Contacts with the adjacent rock units are not exposed, but the augen granite contains xenoliths of amphibolite, hornblende gneiss, layered biotite gneiss, and pink granite gneiss. The augen granite occurs in three separate belts in the New Milford quadrangle. The largest body is a pluton exposed in the southwestern part of the map, west of Lake Candlewood (Fig. 2), which is coextensive with Clarke's (1958) type locality to the south. Two smaller plutons intrude the gneisses of the New Milford massif along its eastern margin, and one of these extends into the adjacent Kent quadrangle to the north (Jackson, 1980). Unlike the larger pluton in the southwestern part of the quadrangle, the two smaller plutons to the northeast are highly deformed and mylonitized along a fault zone that is parallel to the dominant S2 foliation. This fault zone forms the eastern border of the New Milford massif. Sample NM100 is from the larger, less deformed pluton in the southwestern part of the map (Fig. 2).

Ordovician Intrusive Rocks

Two Ordovician plutons are recognized in the New Milford quadrangle: (1) Candlewood Granite (Ocg) and (2) the Brookfield Gneiss.

Candlewood Granite (Ocg)

The name "Candlewood Granite" is redefined here to include the large granite pluton west of Cameron's Line that extends from the Kent quadrangle, through the west-central part of the New Milford quadrangle, and into the Danbury quadrangle to the south (Fig. 2). Gregory and Robinson (1907) originally used the term "Thomaston Granite" to define numerous granitic bodies in western Connecticut. In the Kent quadrangle, Jackson (1980) used the term "Candlewood Lake Pluton" specifically to describe the large intrusive body that passes through Lake Candlewood. Since that time, others have used both the terms "Candlewood Lake Pluton" and "Candlewood Lake granite" to refer to this intrusive body (Mose and Wenner, 1980; Amenta et al., 1982; Mose and Nagel, 1982; Amenta and Mose, 1985). Although the granite does pass through Lake Candlewood (not "Candlewood Lake"), it is poorly exposed along the lake. New mapping in the New Milford quadrangle by Walsh (2003) shows that the type locality of the granite is more properly located at Candlewood Mountain (Fig. 2), which is almost entirely underlain by the granite. For this reason, we believe the abbreviated term "Candlewood Granite" is more appropriate.

The Candlewood Granite is a light gray-, tan-, or white-weathering, medium gray, weakly to moderately foliated, fine- to medium-grained, muscovite-biotite-quartz-plagioclase-microcline granite that contains accessory garnet and tourmaline. The granite is locally porphyritic with phenocrysts (as large as 1 cm) of microcline (Fig. 7, A). The Candlewood Granite contains xenoliths of basement gneiss and autochthonous cover

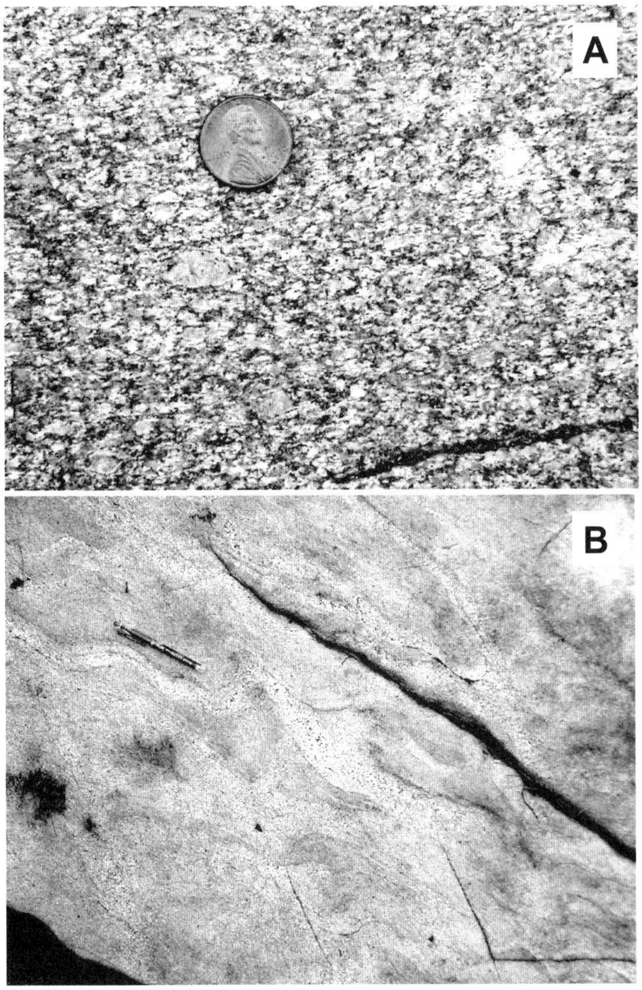

Figure 7. The Candlewood Granite. (A) Map view of the phenocrystic (microcline) granite at the sample site (NM706). Penny is 19 mm in diameter. (B) Cross-sectional photograph of the Candlewood Granite on Candlewood Mountain, showing rarely observed complex igneous flow foliation (parallel to the pen) deformed by the west-dipping (to the right) F2 folds. Left is to the east, view is to the south. Pen is 15 cm long.

sequence rocks, including the Dalton and Stockbridge Formations. Many smaller, unnamed granite dikes occur throughout the map area, both east and west of Cameron's Line. Igneous flow foliation is locally present in the Candlewood Granite and is generally deformed by F2 folds (Fig. 7, B). Most outcrops of the Candlewood Granite lack the flow foliation, and the granite is texturally more homogeneous in these areas. The most conspicuous planar fabric in Candlewood Granite and the smaller intrusive bodies is the S2 foliation. The granite bodies are considered syn-tectonic (D2) intrusions, because they cut F2 folds in the cover rocks, contain F2 folds of flow foliation, contain the S2 foliation, and generally intrude along the axial surface of F2 folds, subparallel to the regional trend of the S2 foliation. Sample NM706 comes from a porphyritic phase in the center of the

Candlewood Granite at its widest spot in the quadrangle (Figs. 2 and 7, A).

Brookfield Gneiss

The Brookfield Gneiss (Rodgers, 1985) or Brookfield plutonic series (Clarke, 1958) crops out as a large pluton that extends southward into the Danbury quadrangle east of Cameron's Line (Fig. 2). Clarke (1958) and Sperandio (1974) recognize several phases in the Brookfield Gneiss that range from granite to diorite. The Brookfield Gneiss intrudes previously deformed parautochthonous Cambrian to Ordovician metasedimentary and metavolcanic rocks of the Rowe and Ratlum Mountain Schists (Fig. 2). The new mapping in the New Milford quadrangle (Walsh, 2003) and previous work by Sevigny and Hanson (1995) indicate that the Brookfield postdates isoclinal folds in the parautochthonous rocks. Sevigny and Hanson (1995) do not state the number of isoclinal fold generations that are truncated by the Brookfield Gneiss, but the new mapping in the New Milford quadrangle (Walsh, 2003) indicates that the pluton cuts two generations of isoclinal folds, both of which are presumed to be of Taconian origin (F1 and F2). Similar to the Candlewood Granite, the Brookfield Gneiss appears to be a syn-tectonic pluton that intruded during the development of the second-generation folds in the host rocks. The U-Pb zircon age of the Brookfield Gneiss is 453 ± 3 Ma (Sevigny and Hanson, 1995). Fine-grained, well-foliated aplite dikes intrude the Brookfield Gneiss, one of which was sampled (NM687). The aplite dikes are also considered syn-D2, because they intrude the Brookfield Gneiss parallel to the S2 schistosity, yet contain the S2 foliation as the most conspicuous planar fabric.

GEOCHRONOLOGY

Because Proterozoic rocks of the New Milford massif have been deformed by several Proterozoic and Paleozoic events, we considered it likely that minerals to be dated from these rocks (i.e., zircon, titanite, and monazite) would contain multiple age components. Previous U-Pb geochronologic studies of Grenville rocks, such as in the western Hudson Highlands (Ratcliffe and Aleinikoff, 2001), Baltimore Gneiss domes (Aleinikoff et al., this volume) and Shenandoah National Park (Tollo et al., this volume), have shown that zircons from these rocks contain complexly intergrown cores and overgrowths. Analysis of such complicated grains by conventional thermal ionization mass spectrometry (TIMS), even of single grains or parts of single grains, would probably yield meaningless age data that are the result of mixing two or more age components. To properly determine the ages preserved within complex zircon, we used SHRIMP, an analytical instrument with sufficiently high spatial resolution to distinguish different age components.

Methods

Zircon, titanite, and monazite were extracted from rock samples collected at outcrops in the New Milford quadrangle using standard mineral separation techniques, including crushing, pulverizing, Wilfley Table, magnetic separator, and heavy liquids. Individual grains of each mineral were handpicked, mounted in epoxy, ground to nearly half-thickness using 1500-grit wet-dry sandpaper, and polished using 6 μm and 1 μm diamond suspension. All grains were photographed in reflected and transmitted light. Compositional zoning of titanite and monazite is shown by backscatter imaging, whereas zircon was imaged in cathodoluminescence (CL).

As suspected, the internal zoning displayed by CL imaging indicates that most of the zircons from rocks of the New Milford quadrangle contain cores and overgrowths similar to Grenville-age zircon from adjacent areas. Zircon and titanite were analyzed on SHRIMP II, at the Research School of Earth Sciences, Australian National University, Canberra; monazite was analyzed on the U.S. Geological Survey–Stanford SHRIMP-reverse geometry (RG). The analytical procedures for SHRIMP U-Pb dating follow methods described in Compston et al. (1984) and Williams and Claesson (1987). The primary oxygen ion beam, operated at ~4–5 nA for SHRIMP II and ~10 nA for SHRIMP-RG, excavated an area of ~25–30 μm in diameter to a depth of ~1 μm. The spectrometer magnet was cycled through the mass stations six times per analysis. Elemental fractionation was corrected by analyzing mineral standards of known age every fourth analysis:

1. Zircon standard R33 (419 Ma from quartz diorite, Braintree Complex, Vermont) (R. Mundil, 1999, personal commun.; S.L. Kamo, 2001, personal commun.);
2. Titanite standard BLR-1 (1051-Ma metamorphic megacrystal, Ontario) (M.D. Schmitz, 2001, personal commun.); and
3. Monazite standard TM-1 (1768 Ma from Thompson Mine, Manitoba) (I. Williams, 1999, personal commun.).

Calculated concentrations of Pb and U are believed to be accurate to ~10%–20%.

The U-Pb isotopic data from zircon, titanite, and monazite were reduced and plotted using the Squid and Isoplot/Ex programs of Ludwig (2002a,b). Common Pb in zircon and monazite is corrected using model Pb isotopic compositions from Stacey and Kramers (1975). Corrected data for zircon are plotted on $^{238}U/^{206}Pb$- $^{207}Pb/^{206}Pb$ concordia diagrams (Tera and Wasserburg, 1972) to visually assess the degree of discordancy of each data point (Fig. 8). Composite ages are determined by calculating the weighted average of either $^{207}Pb/^{206}Pb$ ages (for Proterozoic rocks, except sample NM772; see below) or $^{206}Pb/^{238}U$ ages (for Paleozoic rocks, except sample NM687; see below). Nearly all analyses of zircon and monazite have very low common Pb contents (common ^{206}Pb is mostly <0.2% of total ^{206}Pb; Table 1) so that the individual age of each spot analysis is quite insensitive to the common Pb correction. Titanite contains much higher proportions of common Pb (3%–52% common ^{206}Pb). For titanite, initial common Pb ratios calculated using a Total Pb plot (Ludwig, 2002a) are used to determine individual

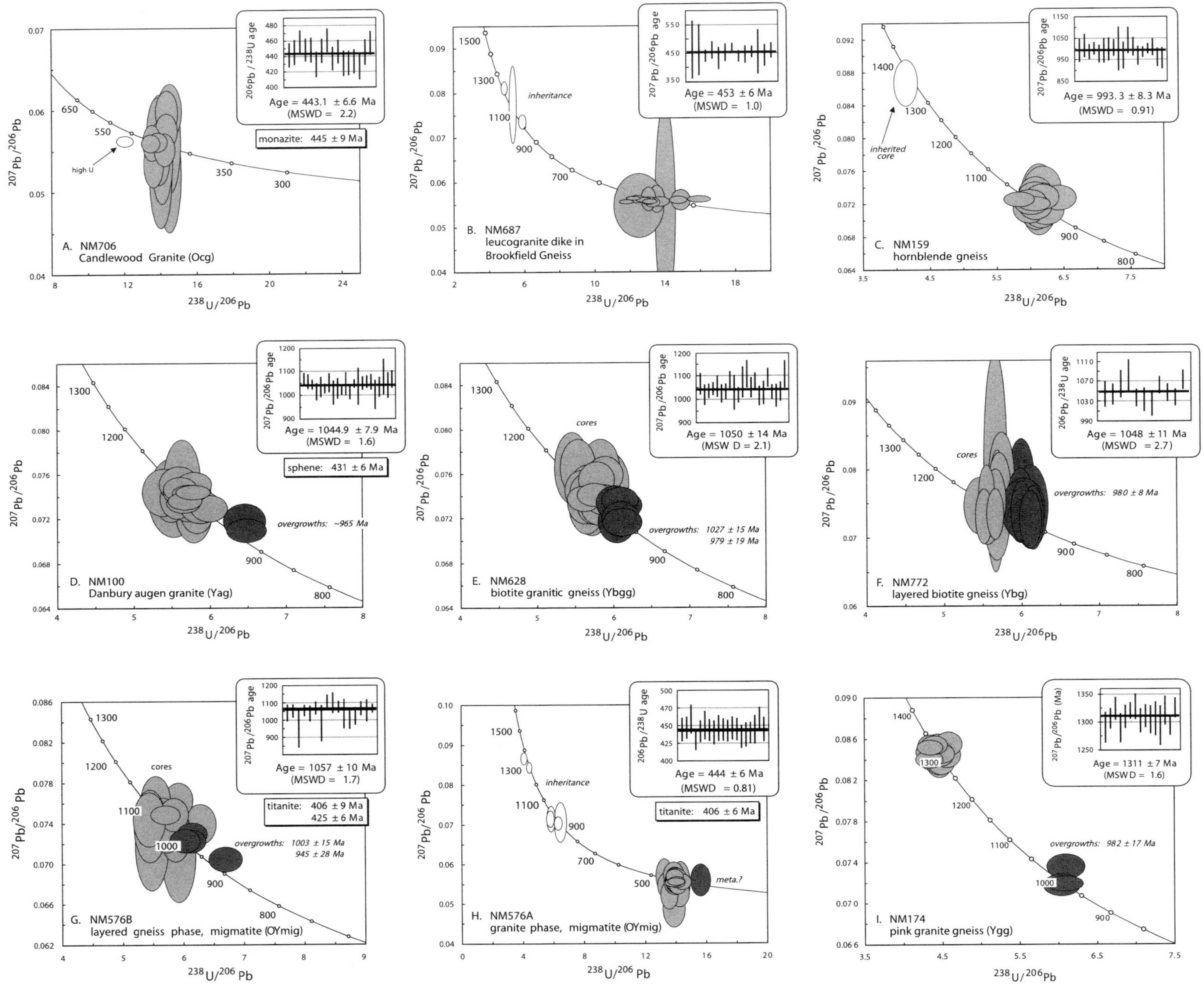

Figure 8. Tera-Wasserburg (1972) plots and weighted average plots of U-Pb zircon data from the New Milford quadrangle. See text for discussion of data points used for regression and weighted average ages. Concordia plotted with 2σ error ellipses.

TABLE 1. SHRIMP U-Th-Pb DATA FOR ZIRCON, TITANITE, AND MONAZITE FROM GNEISSES AND GRANITIC ROCKS, NEW MILFORD QUADRANGLE, CONNECTICUT

Sample	Measured $^{204}Pb/^{206}Pb$	Measured $^{207}Pb/^{206}Pb$	Common ^{206}Pb (%)	U (ppm)	Th/U	$^{206}Pb/^{238}U$ (Ma)*	Error (Ma)	$^{207}Pb/^{206}Pb$ (Ma)	Error (Ma)	$^{238}U/^{206}Pb^*$	Error (%)	$^{207}Pb/^{206}Pb^*$	Error (%)
NM706 Candlewood Granite†													
706-1.1	0.000397	.0601	0.53	291	0.35	442	7			14.13	1.7	.0543	4.2
706-2.1	0.000506	.0592	0.42	258	0.32	445	8			14.06	1.8	.0518	4.8
706-3.1	0.000097	.0574	0.16	1412	0.18	459	7			13.56	1.6	.0560	1.6
706-4.1	0.000089	.0574	0.19	1273	0.32	448	7			13.89	1.6	.0561	0.8
706-5.1	0.000081	.0574	—	7462	0.24	514	8			12.07	1.5	.0562	0.5
706-5.2	0.000218	.0601	0.52	297	0.18	448	8			13.87	1.7	.0569	3.3
706-6.1	0.000339	.0610	0.69	200	0.32	429	8			14.51	1.8	.0560	4.5
706-7.1	0.000258	.0588	0.36	426	0.36	447	7			13.94	1.7	.0550	2.6
706-8.1	0.000071	.0570	0.10	1626	0.45	461	7			13.50	1.6	.0560	0.8
706-9.1	0.000120	.0569	0.16	652	0.30	437	7			14.27	1.6	.0552	1.5
706-10.1	0.000286	.0581	0.28	639	0.23	446	7			14.00	1.6	.0539	2.3
706-11.1	0.000696	.0623	0.84	208	0.38	431	8			14.54	1.8	.0521	5.4
706-12.1	0.000036	.0602	0.59	280	0.34	431	7			14.39	1.7	.0597	1.8
706-13.1	0.000170	.0599	0.54	410	0.48	433	7			14.36	1.7	.0574	1.5
706-14.1	0.000092	.0612	0.72	235	0.27	426	8			14.55	1.8	.0598	2.3
706-15.1	0.000242	.0603	0.56	237	0.19	445	8			13.98	1.8	.0568	3.6
706-16.1	0.000454	.0601	0.49	1032	0.30	457	7			13.66	1.6	.0534	2.2
m706-1.1	0.000050	.0562	0.07	1016	25.76	438	6			14.21	1.4	.0555	0.8
m706-2.1	0.000058	.0564	0.10	1859	15.12	433	6			14.38	1.4	.0555	0.7
m706-3.1	0.000178	.0565	0.01	709	42.29	468	6			13.33	1.4	.0539	1.5
m706-4.1	0.000032	.0559	—	3843	15.48	460	6			13.54	1.3	.0554	0.5
m706-5.1	0.000043	.0558	0.01	2343	17.37	439	6			14.19	1.3	.0552	0.6
m706-6.1	0.000073	.0560	0.05	1677	17.53	439	6			14.21	1.5	.0550	0.7
m706-7.1	0.000073	.0560	—	1343	22.18	457	6			13.64	1.3	.0549	0.8
m706-8.1	0.000043	.0560	0.03	2195	21.37	444	6			14.02	1.3	.0554	0.6
m706-9.1	0.000058	.0563	0.00	2155	18.33	463	6			13.45	1.3	.0554	0.6
m706-10.1	0.000047	.0564	0.07	1921	16.30	445	6			14.01	1.3	.0557	0.6
m706-11.1	0.000082	.0568	0.14	989	26.03	440	6			14.15	1.5	.0556	0.9
m706-12.1	0.000030	.0561	0.09	2049	11.82	427	6			14.61	1.4	.0557	0.6
m706-13.1	0.000089	.0572	0.26	1408	32.44	415	6			15.04	1.4	.0559	1.0
m706-14.1	0.000058	.0571	0.19	1339	20.30	436	6			14.27	1.5	.0563	0.7
m706-15.1	0.000039	.0557	0.00	2402	13.06	438	6			14.25	1.4	.0551	0.5
NM687 leucogranite dike that intrudes the Brookfield Gneiss													
687-1.1	0.000325	.0610	0.60	1543	0.85	459.6	6.3			13.53	1.4	.0563	2.2
687-1.2	—	.0810	0.13	385	0.35	1195.6	15.9			4.90	1.4	.0810	0.7
687-2.1	0.000610	.0652	1.05	6044	0.19	475.8	6.3			13.06	1.4	.0563	1.9
687-3.1	0.001321	.0746	2.20	4601	0.27	444.5	6.9			14.03	1.7	.0543	17.5
687-4.1	0.000037	.0562	—	4723	0.22	473.3	6.2			13.14	1.3	.0557	0.4
687-5.1	0.000055	.0570	—	12232	0.28	500.4	7.3			12.40	1.5	.0562	0.6
687-5.2	—	.0745	0.09	245	0.23	1005.9	13.6			5.92	1.4	.0735	0.9
687-6.1	0.000057	.0563	—	4839	0.20	468.8	6.2			13.27	1.4	.0555	0.8
687-6.2	—	.0773	0.14	94	0.13	1097.8	17.9			5.38	1.6	.0773	4.6
687-7.1	0.000004	.0560	0.10	8063	0.15	422.2	5.5			14.76	1.3	.0560	0.6
687-8.1	0.000033	.0579	—	5689	0.19	508.6	6.8			12.20	1.4	.0564	0.3
687-9.1	0.000026	.0561	—	4761	0.19	525.5	7.2			11.81	1.4	.0557	0.4
687-10.1	0.000898	.0689	1.33	4640	0.17	496.6	22.7			12.52	4.7	.0548	5.3

(continued)

TABLE 1. *Continued*

Sample	Measured $^{204}Pb/^{206}Pb$	Measured $^{207}Pb/^{206}Pb$	Common ^{206}Pb (%)	U (ppm)	Th/U	$^{206}Pb/^{238}U$ (Ma)*	Error (Ma)	$^{207}Pb/^{206}Pb$ (Ma)	Error (Ma)	$^{238}U/^{206}Pb$*	Error (%)	$^{207}Pb/^{206}Pb$*	Error (%)
687-11.1	0.000060	.0577	0.01	6174	0.21	475.5	6.3			13.08	1.4	.0558	0.6
687-12.1	0.000015	.0563	—	4719	0.20	486.2	6.9			12.78	1.4	.0561	0.4
687-13.1	0.000178	.0597	0.44	3341	0.20	417.5	6.1			14.93	1.5	.0561	1.7
687-14.1	0.000076	.0578	0.10	6031	0.23	453.0	6.8			13.74	1.5	.0557	0.8
687-15.1	0.000007	.0563	0.22	6627	0.22	390.9	6.5			15.96	1.7	.0562	0.5
NM159 hornblende gneiss													
159-1.1	0.000123	.0740	0.21	277	0.28	964	15	993	26	6.192	1.7	.0723	1.3
159-2.1	0.000032	.0735	0.05	199	0.17	978	16	1015	25	6.093	1.7	.0730	1.2
159-3.1	0.000069	.0730	0.12	426	0.20	989	16	987	16	6.033	1.7	.0720	0.8
159-3.2	0.000043	.0732	0.07	787	0.26	992	15	1002	12	6.010	1.6	.0726	0.6
159-4.1	0.000105	.0732	0.18	488	0.22	980	15	978	19	6.094	1.6	.0717	0.9
159-5.1	0.000161	.0745	0.27	286	0.30	984	16	993	27	6.059	1.7	.0722	1.3
159-5.2	0.000100	.0739	0.17	243	0.20	976	17	999	25	6.109	1.8	.0725	1.2
159-6.1	0.000177	.0892	0.29	162	0.56	1383	23	1355	25	4.184	1.7	.0867	1.3
159-6.2	0.000042	.0733	0.07	366	0.22	927	15	1006	17	6.443	1.6	.0727	0.9
159-7.1	0.000300	.0768	0.51	96	0.17	971	17	1001	49	6.146	1.9	.0725	2.4
159-8.1	0.000204	.0744	0.35	281	0.17	977	16	970	30	6.111	1.7	.0714	1.5
159-9.1	0.000089	.0752	0.15	203	0.16	961	17	1040	29	6.198	1.8	.0740	1.4
159-10.1	0.000076	.0740	0.13	316	0.20	980	16	1012	20	6.080	1.6	.0729	1.0
159-11.1	0.000121	.0731	0.21	336	0.21	994	16	970	20	6.006	1.6	.0714	1.0
159-12.1	0.000000	.0720	0.00	921	0.13	981	15	986	10	6.083	1.6	.0720	0.5
159-13.1	0.000024	.0729	0.04	945	0.19	1018	15	1001	10	5.849	1.6	.0726	0.5
159-14.1	0.000087	.0740	0.15	382	0.23	963	15	1008	17	6.193	1.6	.0728	0.9
159-15.1	0.000090	.0724	0.15	427	0.16	956	16	961	19	6.252	1.7	.0711	0.9
159-15.2	0.000130	.0728	0.22	269	0.20	968	15	957	23	6.175	1.7	.0710	1.1
NM100 Danbury augen granite													
100-1.1	0.000046	.0758	0.08	576	0.27	1021	16	1072	12	5.814	1.6	.0751	0.6
100-2.1	0.000081	.0759	0.14	462	0.32	1025	17	1062	14	5.793	1.7	.0748	0.7
100-3.1	0.000034	.0748	0.06	985	0.27	1032	16	1049	9	5.754	1.6	.0743	0.4
100-3.2	0.000028	.0735	0.05	655	0.28	1009	18	1016	17	5.898	1.8	.0731	0.8
100-4.1	0.000127	.0756	0.22	503	0.35	1061	16	1037	21	5.596	1.6	.0738	1.0
100-4.2	0.000011	.0738	0.02	894	0.25	1042	16	1031	9	5.700	1.6	.0736	0.4
100-5.1	0.000084	.0757	0.14	290	0.51	1041	17	1056	19	5.704	1.7	.0745	0.9
100-6.1	0.000148	.0750	0.25	392	0.31	1018	16	1012	25	5.848	1.6	.0729	1.2
100-7.1	0.000081	.0748	0.14	494	0.32	1036	17	1031	21	5.737	1.7	.0736	1.0
100-8.1	0.000043	.0740	0.07	860	0.29	1047	16	1025	10	5.678	1.6	.0734	0.5
100-9.1	0.000046	.0725	0.08	982	0.13	924	16	980	18	6.470	1.7	.0718	0.9
100-9.2	0.000193	.0771	0.33	281	0.95	1094	21	1052	24	5.415	2.0	.0744	1.2
100-10.1	0.000074	.0740	0.13	848	0.26	991	15	1012	13	6.011	1.6	.0729	0.7
100-10.2	0.000019	.0747	0.03	1349	0.21	1024	15	1053	7	5.804	1.6	.0744	0.4
100-11.1	0.000049	.0747	0.08	614	0.18	1069	25	1040	40	5.551	2.4	.0740	2.0
100-12.2	0.000035	.0714	0.06	750	0.16	922	14	956	12	6.494	1.6	.0709	0.6
100-13.1	0.000072	.0755	0.12	565	0.34	1052	17	1055	13	5.640	1.6	.0745	0.7
100-14.1	0.000036	.0752	0.06	617	0.51	1052	16	1060	12	5.638	1.6	.0747	0.6
100-15.1	0.000015	.0749	0.02	765	0.15	1026	16	1059	16	5.790	1.6	.0747	0.8
100-15.2	0.000203	.0755	0.35	178	0.37	1025	17	1003	31	5.809	1.7	.0726	1.5
100-16.1	0.000115	.0755	0.20	311	0.61	1022	16	1037	21	5.817	1.6	.0738	1.0
100-17.1	0.000233	.0789	0.39	166	0.69	1046	17	1084	38	5.665	1.7	.0756	1.9
100-18.1	0.000219	.0773	0.37	151	0.54	1075	18	1045	27	5.516	1.7	.0741	1.3

100-19.1	0.000099	.0765	0.17	317	0.49	1054	19	1072	17	5.623	1.8	.0751	0.9
t100-1.1 c	0.007094	.1643	11.75	67	2.64	441	9			12.45	2.0	.1643	2.4
t100-2.1 c	0.005828	.1300	8.04	85	2.04	438	7			13.08	1.6	.1300	3.0
t100-2.2 r	0.012632	.2130	18.33	32	2.10	401	12			12.71	2.5	.2130	3.2
t100-3.1 c	0.009124	.1938	14.97	41	5.51	426	11			12.45	2.4	.1938	2.3
t100-4.1 c	0.004453	.1311	8.17	80	1.54	434	8			13.19	1.7	.1311	1.4
t100-4.2 r	0.015766	.3093	29.49	17	0.43	400	27			11.01	6.3	.3093	2.1
t100-5.1 c	0.015550	.2564	21.73	30	0.28	435	14			11.22	2.8	.2564	2.9
t100-5.2 i	0.004159	.1163	6.59	101	1.62	430	9			13.55	2.2	.1163	2.4
t100-5.3 r	0.015274	.2539	22.97	23	0.95	445	17			10.78	3.3	.2539	2.0
t100-6.1 r?	0.021565	.3222	30.91	16	1.72	437	24			9.86	4.5	.3222	3.8
t100-6.2 c	0.006096	.1422	9.39	73	1.99	427	8			13.22	1.8	.1422	1.5
t100-7.1 c	0.006350	.1514	10.37	62	3.98	434	9			12.88	1.9	.1514	1.8
t100-7.2 r	0.006348	.1556	11.63	57	2.59	425	9			12.97	2.0	.1556	1.5
t100-8.1 r?	0.013080	.2488	22.43	26	4.16	424	22			11.42	5.0	.2488	1.9
t100-8.2 c?	0.017315	.2973	28.03	21	2.60	433	18			10.36	3.5	.2973	2.1
NM628 biotite granite gneiss													
628-1.1	0.000074	.0760	0.13	285	0.50	1042	16	1066	21	5.695	1.6	.0749	1.0
628-2.1	0.000094	.0746	0.16	379	0.69	1029	16	1020	20	5.781	1.6	.0732	1.0
628-2.1	0.000062	.0747	0.11	760	0.50	1052	16	1036	13	5.648	1.6	.0738	0.6
628-3.2	0.000014	.0719	0.02	432	0.28	973	16	976	14	6.136	1.7	.0717	0.7
628-4.1	0.000051	.0749	0.09	666	0.44	1029	16	1046	12	5.775	1.6	.0742	0.6
628-5.1	0.000078	.0756	0.13	228	0.50	1056	17	1056	20	5.619	1.7	.0745	1.0
628-5.2	0.000029	.0738	0.05	307	0.30	986	16	1025	17	6.040	1.6	.0734	0.9
628-6.1	0.000080	.0748	0.14	355	0.48	1010	16	1033	16	5.893	1.6	.0737	0.8
628-7.1	0.000085	.0767	0.14	342	0.51	1022	16	1082	18	5.806	1.6	.0755	0.9
628-8.1	0.000352	.0786	0.60	122	0.64	1035	18	1030	35	5.742	1.8	.0736	1.7
628-8.2	0.000088	.0744	0.15	710	0.18	969	15	1019	14	6.151	1.6	.0732	0.7
628-9.1	0.000151	.0774	0.26	116	0.82	1002	20	1074	31	5.927	2.0	.0752	1.6
628-10.1	0.000104	.0780	0.18	153	0.45	1012	17	1109	28	5.857	1.7	.0765	1.4
628-11.1	0.000119	.0761	0.20	274	0.55	1044	17	1051	19	5.688	1.6	.0744	1.0
628-12.1	0.000120	.0783	0.20	153	0.52	1078	18	1110	34	5.486	1.7	.0766	1.7
628-13.1	0.000115	.0747	0.20	320	0.52	1020	16	1015	19	5.832	1.6	.0731	0.9
628-14.1	0.000123	.0753	0.21	329	0.53	1051	17	1029	22	5.651	1.6	.0735	1.1
628-15.1	0.000050	.0766	0.08	293	0.46	1004	16	1093	18	5.911	1.6	.0759	0.9
628-15.2	0.000044	.0743	0.08	514	0.29	975	15	1031	13	6.111	1.6	.0736	0.7
628-16.1	0.000220	.0763	0.37	251	0.49	1047	17	1018	28	5.680	1.7	.0731	1.4
628-17.1	0.000141	.0751	0.24	380	0.48	997	16	1016	22	5.975	1.7	.0731	1.1
628-18.1	0.000152	.0739	0.26	387	0.33	984	15	979	22	6.064	1.6	.0717	1.1
628-19.1	0.000107	.0752	0.18	300	0.24	1010	16	1034	18	5.888	1.6	.0737	0.9
628-20.1	0.000099	.0734	0.17	331	0.31	976	15	985	18	6.114	1.6	.0720	0.9
628-21.1	0.000039	.0771	0.07	225	0.67	1063	17	1110	27	5.566	1.7	.0766	1.4
NM772 layered biotite gneiss													
772-3.1	—	.0724	—	338	0.41	1044.2	12.0	996	64	5.70	1.2	.0724	3.2
772-4.1	—	.0729	—	528	0.41	967.7	8.4	1013	53	6.16	0.9	.0729	2.6
772-5.1	—	.0727	—	619	0.38	973.4	9.3	1004	48	6.13	1.0	.0727	2.4
772-6.1	—	.0754	—	539	0.25	1043.7	9.7	1079	50	5.68	0.9	.0754	2.5
772-7.1	—	.0735	—	518	0.80	1064.5	12.9	1027	50	5.58	1.2	.0735	2.5
772-8.1	—	.0747	—	392	0.37	991.1	12.0	1061	60	6.00	1.2	.0747	3.0
772-9.1	—	.0747	—	897	1.64	1081.6	15.7	1059	46	5.48	1.5	.0747	2.3
772-10.1	0.001268	.0990	2.11	920	0.28	1036.6	8.0	1225	159	5.68	1.1	.0811	8.1
772-11.1	—	.0766	—	316	0.57	1033.1	11.0	1112	62	5.73	1.1	.0766	3.1
772-12.1	—	.0791	—	265	1.75	1025.7	12.0	1174	66	5.76	1.2	.0791	3.3
772-13.1	—	.0739	—	556	0.32	990.8	8.8	1039	49	6.01	0.9	.0739	2.4

(continued)

TABLE 1. ***Continued***

Sample	Measured $^{204}Pb/^{206}Pb$	Measured $^{207}Pb/^{206}Pb$	Common ^{206}Pb (%)	U (ppm)	Th/U	$^{206}Pb/^{238}U$ (Ma)*	Error (Ma)	$^{207}Pb/^{206}Pb$ (Ma)	Error (Ma)	$^{238}U/^{206}Pb$*	Error (%)	$^{207}Pb/^{206}Pb$*	Error (%)
772-14.1	—	.0741	—	535	0.47	986.0	18.0	1045	51	6.04	1.9	.0741	2.5
772-15.1	—	.0745	—	357	0.39	971.1	14.0	1055	67	6.13	1.5	.0745	3.3
772-16.1	—	.0766	—	661	0.56	1062.4	8.0	1110	43	5.57	0.8	.0766	2.2
772-17.1	—	.0734	—	750	0.61	1046.9	8.0	1026	42	5.68	0.8	.0734	2.1
772-18.1	—	.0767	—	738	2.34	1037.4	7.7	1112	65	5.71	0.7	.0767	3.3
772-19.1	—	.0731	—	790	0.09	1073.1	8.9	1017	45	5.53	0.8	.0731	2.2
772-19.1	—	.0780	—	278	1.10	985.6	11.4	1146	67	6.01	1.2	.0780	3.4
NM576A migmatite (pink granite)													
576A-1.1	0.000150	.0587	0.36	176	0.69	444	8			14.00	1.7	.0565	2.2
576A-1.2	0.000042	.0563	0.05	830	0.02	447	7			13.93	1.6	.0557	0.8
576A-2.1	0.000187	.0577	0.19	183	0.81	459	9			13.57	2.1	.0550	2.5
576A-1.3	0.000141	.0579	0.31	689	0.03	430	7			14.48	1.6	.0559	1.4
576A-2.2	0.000087	.0571	0.17	719	0.03	442	7			14.09	1.6	.0558	1.0
576A-3.1	0.000465	.0560	0.01	154	0.39	450	9			13.93	2.1	.0492	3.8
576A-4.1	0.000084	.0727	—	210	0.23	1013	16			5.88	1.7	.0715	1.2
576A-4.2	0.000490	.0636	0.97	855	0.14	444	7			14.03	1.6	.0565	2.1
576A-5.1	0.000025	.0868	—	243	0.22	1392	26			4.16	1.9	.0864	0.7
576A-5.2	0.000256	.0744	0.56	63	0.12	920	22			6.51	2.5	.0707	2.8
576A-5.3	0.000990	.0718	1.91	1223	0.06	467	7			13.30	1.6	.0574	3.0
576A-6.1	0.000153	.0738	0.13	136	0.61	1006	18			5.93	1.9	.0716	1.8
576A-6.2	0.000167	.0590	0.40	894	0.11	440	8			14.14	1.8	.0565	1.9
576A-7.1	0.000000	.0577	0.23	190	0.38	446	8			13.91	1.9	.0577	2.2
576A-7.2	0.000328	.0610	0.64	463	0.07	444	7			14.02	1.7	.0562	2.2
576A-8.1	0.000043	.0848	0.08	301	0.85	1296	20			4.49	1.6	.0842	0.7
576A-8.2	0.000000	.0705	0.01	392	0.25	939	15			6.38	1.7	.0705	0.9
576A-8.3	0.000027	.0563	0.07	1419	0.10	443	7			14.07	1.6	.0559	0.7
576A-9.1	0.000195	.0592	0.41	286	1.12	446	7			13.94	1.7	.0563	2.3
576A-10.1	0.000039	.0579	0.26	522	0.32	445	7			13.97	1.6	.0573	1.2
576A-11.1	0.000236	.0591	0.43	205	0.56	434	7			14.35	1.8	.0556	3.9
576A-12.1	0.000447	.0598	0.51	130	0.61	437	8			14.30	1.8	.0532	2.7
576A-13.1	0.000245	.0586	0.36	961	0.09	440	7			14.18	1.6	.0550	1.6
576A-13.2	0.000201	.0587	0.36	243	0.82	444	9			14.01	2.1	.0558	2.2
576A-14.1	0.000283	.0567	0.07	246	1.02	459	8			13.61	1.7	.0526	2.8
576A-14.2	0.000929	.0699	1.88	846	0.38	398	7			15.65	1.7	.0563	2.9
576A-15.1	0.000312	.0586	0.35	147	0.55	445	8			14.03	1.8	.0541	3.8
t576A-1.1	0.004990	.1282	9.04	76	1.91	419	11			13.53	2.6	.1282	1.8
t576A-2.1	0.008020	.1563	12.54	40	1.38	416	10			13.13	2.5	.1563	1.8
t576A-3.1	0.007176	.1593	12.93	37	1.64	408	11			13.34	2.6	.1593	2.0
t576A-4.1	0.010252	.1856	16.17	32	3.93	412	12			12.71	2.9	.1856	2.0
t576A-5.1	0.010021	.1633	13.48	35	2.66	384	10			14.11	2.5	.1633	1.9
t576A-6.1	0.014365	.2577	25.10	21	11.43	408	16			11.45	3.8	.2577	2.1
t576A-7.1	0.004329	.1163	7.63	97	1.06	399	8			14.47	2.0	.1163	1.6
t576A-8.1	0.006741	.1491	11.70	50	1.35	397	10			13.90	2.5	.1491	1.9
t576A-9.1	0.008804	.1517	12.01	42	2.59	401	11			13.70	2.8	.1517	1.8
t576A-10.1	0.012176	.3971	42.66	9	7.73	347	18			10.37	4.8	.3971	2.5
t576A-10.2	0.007315	.1475	11.45	72	1.74	411	10			13.45	2.5	.1475	2.0
t576A-11.1	0.010993	.2405	22.95	32	8.41	423	13			11.35	3.1	.2405	2.2
t576A-11.2	0.007588	.1756	15.11	36	0.59	367	10			14.51	2.7	.1756	2.0
t576A-12.1	0.004191	.1159	7.55	87	2.76	408	8			14.17	2.1	.1159	1.6
t576A-12.2	0.006976	.1812	15.60	39	1.96	425	15			12.38	3.5	.1812	2.7

t576A-13.1	0.004584	.1257	8.77	70	1.90	405	9			14.05	2.1	.1257	1.6
t576A-13.2	0.002839	.1248	8.69	121	3.66	397	8			14.39	2.1	.1248	2.0
t576A-14.1	0.008430	.1670	13.95	149	1.09	381	11			14.14	2.7	.1670	4.6
t576A-14.2	0.006887	.1167	7.71	312	2.15	384	13			15.02	3.4	.1167	3.2
t576A-15.1	0.005764	.1471	11.45	49	1.06	394	10			14.05	2.5	.1471	2.1
t576A-16.1	0.005760	.1302	9.22	60	3.08	433	10			13.07	2.3	.1302	1.7
t576A-16.2	0.005673	.1502	11.84	50	1.13	395	11			13.96	2.7	.1502	2.2
t576A-17.1	0.015936	.3598	37.76	12	12.86	403	23			9.64	5.2	.3598	3.0
t576A-18.1	0.003239	.1083	6.60	98	1.24	412	8			14.13	2.0	.1083	2.3
t576A-18.2	0.002689	.1024	5.85	69	0.65	420	9			13.99	2.1	.1024	1.6
t576A-19.1	0.006484	.1772	15.11	63	0.94	422	15			12.56	3.6	.1772	2.4
t576A-20.1	0.006861	.1316	9.53	55	1.50	397	12			14.23	3.1	.1316	1.9
t576A-20.2	0.014066	.2254	21.36	55	1.00	343	10			14.40	2.9	.2254	2.1
t576A-2.2	0.013845	.2645	25.88	21	6.55	437	19			10.57	4.0	.2645	3.8
NM576B migmatite (gray layered gneiss)													
576B-1.1	0.000090	.0753	0.15	220	0.71	945	15	1041	22	6.308	1.7	.0740	1.1
576B-1.2	—	.0743	—	263	0.13	1023	16	1051	16	5.806	1.7	.0744	0.8
576B-2.1	0.000314	.0753	0.54	71	0.71	1003	18	951	52	5.956	1.9	.0708	2.6
576B-2.2	0.000028	.0749	0.05	550	0.06	1038	17	1055	13	5.720	1.7	.0745	0.7
576B-3.1	0.000066	.0748	0.11	358	0.19	1044	16	1037	20	5.689	1.6	.0738	1.0
576B-3.2	0.000064	.0759	0.11	244	0.87	1087	18	1069	16	5.449	1.8	.0750	0.8
576B-4.1	0.000243	.0753	0.42	80	1.21	1078	21	981	50	5.517	2.0	.0718	2.4
576B-4.2	—	.0762	—	551	0.12	997	15	1104	18	5.954	1.6	.0764	0.9
576B-5.1	0.000099	.0775	0.17	84	0.86	1061	18	1099	27	5.580	1.8	.0761	1.4
576B-5.2	0.000000	.0730	0.00	392	0.15	966	15	1015	13	6.173	1.6	.0730	0.6
576B-6.1	0.000046	.0759	0.08	404	0.17	1042	16	1076	14	5.690	1.6	.0753	0.7
576B-6.2	0.000211	.0768	0.36	109	0.92	1081	21	1037	40	5.488	2.0	.0738	2.0
576B-7.1	0.000173	.0753	0.30	158	0.96	1010	17	1009	28	5.899	1.7	.0728	1.4
576B-7.2	0.000071	.0743	0.12	355	0.28	1033	17	1022	23	5.757	1.7	.0733	1.1
576B-8.1	0.000045	.0756	0.08	263	0.45	1011	16	1066	18	5.876	1.7	.0749	0.9
576B-9.1	0.000103	.0759	0.17	115	1.00	1083	21	1054	29	5.476	2.0	.0745	1.5
576B-10.1	0.000024	.0753	0.04	731	0.08	1034	17	1066	10	5.741	1.7	.0749	0.5
576B-11.1	0.000031	.0728	0.05	443	0.23	971	16	996	13	6.146	1.7	.0723	0.6
576B-12.1	0.000025	.0727	0.04	775	0.14	985	16	996	13	6.055	1.7	.0724	0.6
576B-13.1	0.000048	.0712	0.08	474	0.15	892	14	945	14	6.724	1.7	.0706	0.7
t576B-1.1 br-r	0.003771	.0975	4.91	149	0.78	420	6			14.12	1.4	.0975	1.4
t576B-1.2 br-c	0.002115	.0875	3.71	196	0.77	433	5			13.86	1.3	.0875	1.6
t576B-2.1 br-c	0.002861	.0963	4.74	149	1.05	430	6			13.82	1.4	.0963	3.2
t576B-3.1 br-r	0.006715	.1351	9.40	72	1.42	401	8			14.13	1.9	.1351	3.6
t576B-3.2 br-c	0.002344	.0898	3.98	176	0.78	430	5			13.92	1.2	.0898	2.2
t576B-4.1 br-r	0.004485	.1219	7.85	86	1.49	403	8			14.28	1.9	.1219	3.8
t576B-4.2 br-c	0.002513	.0935	4.45	158	1.01	414	6			14.41	1.4	.0935	2.0
t576B-5.1 br-c	0.003112	.1017	5.38	133	1.01	421	6			14.03	1.5	.1017	2.9
t576B-6.1 br-c	0.004409	.1281	8.45	98	1.05	420	8			13.61	1.8	.1281	3.7
t576B-6.2 br-r	0.010083	.1952	16.42	39	1.09	401	12			13.02	2.8	.1952	3.5
t576B-7.1 br-c	0.002735	.0939	4.49	208	0.78	419	6			14.22	1.3	.0939	3.4
t576B-8.1 br-r?	0.005905	.1687	13.31	58	3.37	404	11			13.39	2.5	.1687	3.6
t576B-8.2 br-r?	0.007194	.1627	12.59	60	2.18	413	12			13.23	2.8	.1627	3.9
t576B-9.1 cl-c	0.017343	.3645	36.37	17	0.91	319	42			12.56	12.9	.3645	3.0
t576B-10.1 cl-c	0.007437	.1638	12.76	53	1.30	401	11			13.60	2.4	.1638	4.5
t576B-10.2 cl-r	0.015499	.2781	26.15	19	1.02	383	18			12.07	4.1	.2781	2.8
t576B-7.2 br-r	0.006787	.1655	12.91	51	1.72	416	10			13.07	2.3	.1655	2.1
t576B-11.1 br c	0.003299	.1050	5.73	126	1.14	435	7			13.50	1.5	.1050	2.2

(continued)

TABLE 1. *Continued*

Sample	Measured $^{204}Pb/^{206}Pb$	Measured $^{207}Pb/^{206}Pb$	Common ^{206}Pb (%)	U (ppm)	Th/U	$^{206}Pb/^{238}U$ (Ma)*	Error (Ma)	$^{207}Pb/^{206}Pb$ (Ma)	Error (Ma)	$^{238}U/^{206}Pb$*	Error (%)	$^{207}Pb/^{206}Pb$*	Error (%)
t576B-12.1 br-c	0.006033	.1380	9.74	76	1.80	397	9			14.22	2.1	.1380	2.6
t576B-12.2 br-r	0.012848	.2549	23.36	26	0.80	418	24			11.45	5.4	.2549	4.0
t576B-13.1 cl-c	0.021162	.3708	36.86	12	1.03	449	30			8.75	5.6	.3708	2.9
t576B-14.1 cl-c	0.026319	.4134	42.00	10	0.89	367	28			9.90	5.7	.4134	3.4
t576B-15.1 cl-c	0.020937	.3206	31.04	17	2.03	424	26			10.15	4.8	.3206	5.3
t576B-16.1 cl-c	0.024173	.3830	38.38	12	1.79	407	37			9.46	7.9	.3830	4.2
t576B-17.1 cl-c	0.029059	.5027	52.41	5	5.64	379	43			7.86	7.8	.5027	4.7
t576B-18.1 br-r	0.005525	.1206	7.70	114	1.78	405	8			14.22	1.8	.1206	5.3
t576B-18.2 br-r	0.007027	.1384	9.75	82	1.09	412	9			13.67	2.0	.1384	4.3
NM174 pink granite gneiss													
174-1.1	0.000104	.0854	0.17	657	0.31	1303	20	1291	13	4.466	1.6	.0839	0.7
174-2.1	0.000045	.0852	0.07	1083	0.41	1294	20	1307	8	4.495	1.6	.0846	0.4
174-3.1	0.000055	.0862	0.09	843	0.35	1331	20	1325	10	4.363	1.6	.0854	0.5
174-3.2	0.000095	.0749	0.16	1419	0.02	978	16	1030	14	6.09	1.7	.0736	0.7
174-4.1	0.000068	.0848	0.11	594	0.38	1294	21	1290	12	4.499	1.6	.0839	0.6
174-5.1	0.000036	.0853	0.06	688	0.34	1299	20	1310	10	4.478	1.6	.0848	0.5
174-6.1	0.000015	.0854	0.02	1356	0.53	1334	20	1321	7	4.353	1.6	.0852	0.3
174-7.1	0.000017	.0858	0.03	618	0.30	1267	19	1328	11	4.590	1.6	.0856	0.6
174-8.1	0.000036	.0850	0.06	628	0.39	1300	21	1303	10	4.476	1.6	.0844	0.5
174-9.1	0.000025	.0851	0.04	676	0.40	1338	21	1311	9	4.343	1.6	.0848	0.5
174-10.1	0.000058	.0853	0.10	581	0.35	1273	20	1305	12	4.571	1.6	.0845	0.6
174-11.1	0.000041	.0852	0.07	732	0.39	1333	20	1306	15	4.360	1.6	.0846	0.7
174-12.1	0.000049	.0848	0.08	394	0.35	1301	20	1294	17	4.474	1.6	.0841	0.9
174-13.1	0.000052	.0860	0.08	537	0.29	1306	20	1321	12	4.449	1.6	.0852	0.6
174-14.1	0.000028	.0845	0.05	1046	0.40	1338	21	1294	8	4.345	1.7	.0841	0.4
174-15.1	0.000000	.0855	0.00	933	0.37	1317	20	1326	8	4.409	1.6	.0855	0.4
174-16.1	0.000062	.0726	0.11	608	0.12	988	15	979	12	6.04	1.6	.0717	0.6
174-17.1	0.000033	.0724	0.06	1211	0.04	981	15	983	9	6.08	1.6	.0719	0.4

Note: Analytical sessions: Zircon: NM100, NM159, NM174, NM567A, NM576B, NM628, NM706 (July 2001, SHRIMP II, Research School of Earth Sciences, Australian National University, Canberra), NM687 (September 2001, USGS-Stanford SHRIMP-RG), NM772 (March 2002, SHRIMP-RG). Titanite: NM100, NM576A, NM576B (July 2001, SHRIMP II); monazite: NM706 (November 2001, SHRIMP-RG). Analyses with prefix t are titanite samples; analyses with prefix m are monazite. Errors are 1σ. Abbreviations: c—core; i—intermediate zone; r—rim. A — indicates not detected; blank entires indicate age not calculated.

*Corrected for common Pb, except titanite analyses. $^{206}Pb/^{238}U$ ages corrected for common Pb using the ^{207}Pb-correction method; $^{207}Pb/^{206}Pb$ and $^{238}U/^{206}Pb$ corrected for common Pb using the ^{204}Pb-correction method.

†Latitude and Longitude of samples (NAD27 datum):
Candlewood Granite (NM706): 41°36′39.4″N, 73°27′57.8″W.
Leucogranite dike (NM687): 41°33′36.1″N, 73°23′49.8″W.
Hornblende gneiss (NM159): 41°36′31.4″N, 73°24′55.1″W.
Danbury augen granite (NM100): 41°30′27.5″N, 73°29′41.8″W.
Biotite granite gneiss (NM628): 41°35′30.6″N, 73°25′29.4″W.
Layered biotite gneiss (NM 772): 41°37′09.2″N, 73°26′23.6″W.
Migmatite: pink granite (NM576A) and gray layered gneiss (NM576B): 41°33′20.7″N, 73°27′52.4″W.
Pink granite gneiss (NM174): 41°37′00.2″N, 73°24′25.3″W.

$^{206}Pb/^{238}U$ ages. Final composite titanite ages are determined using the weighted average calculation.

Weighted average calculations of $^{206}Pb/^{238}U$ ages incorporate both the 2σ external spot-to-spot error of the standard and the 2σ error of the mean of the standard (Ludwig, 2002a). All errors for weighted average ages are reported at the 95% confidence limit.

Results

Ages of samples are discussed below in order of youngest to oldest.

Candlewood Granite

Zircons from the Candlewood Granite (sample NM706) are prismatic and euhedral; most are cracked. Length-to-width ratios (l/w) are 2–6. The CL images show that most grains contain primarily fine concentric, oscillatory zoning, usually overgrown by a thin dark (in CL) rim (Fig. 9, A). Some grains contain rounded, partially resorbed cores, suggestive of inherited (xenocrystic) material. The U-Pb isotopic data are straightforward and form a coherent group of sixteen analyses with a weighted average age of 443.1 ± 6.6 Ma (Fig. 8; Table 1). One analysis (706-5.1; Table 1) of a dark overgrowth has an older $^{206}Pb/^{238}U$ age of 514 Ma, probably due to very high U content, which causes an instrumental bias in Pb/U (Williams and Hergt, 2000). The $^{207}Pb/^{206}Pb$ age of this analysis (461 ± 10 Ma, 1σ; Table 1) is unaffected by the high U. The core of this zircon (analysis 706-5.2) is 448 ± 8 Ma and is included in the age calculation. Monazite from the Candlewood Granite shows very subtle compositional zoning in backscatter electron imaging (BSE), suggestive of an igneous origin, and yields an age of 445 ± 9 Ma, within the uncertainty of the zircon age.

Leucogranite Aplite Dike

Zircons from a leucogranite aplite dike (sample NM687) that intrudes the Brookfield Gneiss are prismatic (l/w = 3–5), euhedral, and cracked. Many grains contain obvious (in transmitted light) rounded cores and thick overgrowths. In CL, the cores are light to white and have oscillatory zoning, whereas the overgrowths are black and weakly zoned to unzoned (Fig. 9, B). Analyses of three cores give $^{207}Pb/^{206}Pb$ ages of 1222–1028 Ma (Table 1), interpreted as inherited xenocrystic material. Overgrowths contain very high U content (1543–12,232 ppm; Table 1). The $^{206}Pb/^{238}U$ ages of such high-U zircon frequently are too old (Williams and Hergt, 2000), as shown by the horizontal spread of data on Figure 8, necessitating use of the $^{207}Pb/^{206}Pb$ ages for the age calculation. The weighted average of the $^{207}Pb/^{206}Pb$ ages is 453 ± 6 Ma, interpreted as the time of emplacement of the leucogranite aplite dike.

Hornblende Gneiss

Elongate (l/w = 3–4) and equant (l/w = 1) zircons occur in hornblende gneiss (sample NM159). In both morphologies, the CL zoning is patchwork, typical of metamorphic origin (Fig. 9, C). Some grains have rounded cores, some of which have oscillatory zoning and some of which are white in CL (i.e., very low U). We presume that the protolith of the hornblende gneiss probably had a basaltic composition and did not contain zircon; Zr was probably sequestered in primary pyroxene. During the prograde portion of high-grade metamorphism, pyroxene reacted to form amphibole, releasing Zr to form metamorphic zircon. If early high-grade metamorphism reached pyroxene granulite facies, subsequent metamorphism and hydration of pyroxene at hornblende granulite facies may have produced metamorphic zircon. One rounded core (analysis 159-6.1; Table 1) has a $^{207}Pb/^{206}Pb$ age of 1355 Ma; all other analyses, from both cores

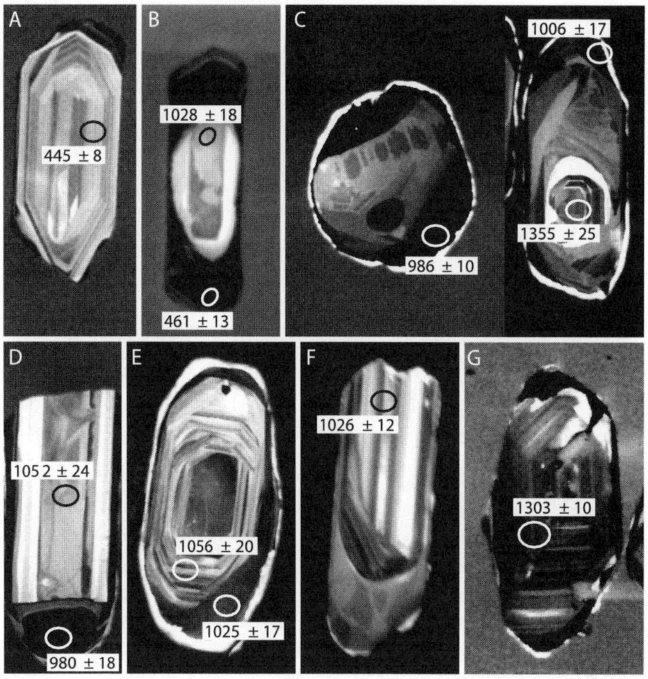

Figure 9. CL images of selected zircons from samples of igneous and meta-igneous rocks, New Milford quadrangle, Connecticut. Locations of SHRIMP analyses shown by ellipses (~25–30 μm in length). Proterozoic dates are $^{207}Pb/^{206}Pb$ ages; Paleozoic dates are $^{206}Pb/^{238}Pb$ ages. Uncertainties shown with all ages are 1σ. (A) Candlewood Granite (sample NM706). Concentric oscillatory zoning of igneous zircon with very thin dark overgrowth. (B) Leucogranite dike that intrudes Brookfield Gneiss (sample NM687). Rounded, inherited cores with dark mantle are interpreted as igneous. (C) Hornblende gneiss (sample NM159). Equant metamorphic zircon with patchwork CL zoning; elongate metamorphic zircon with inherited core. (D) Danbury augen granite (sample NM100). Oscillatory zoned igneous core and dark, unzoned metamorphic overgrowth. (E) Biotite granitic gneiss (sample NM628). Oscillatory-zoned igneous core and unzoned metamorphic overgrowth. (F) Layered biotite gneiss (sample NM772). Oscillatory-zoned igneous core and patchwork-zoned, presumably metamorphic, overgrowth. (G) Pink granite gneiss (sample NM174). Oscillatory-zoned core that is partially embayed and rimmed by dark, presumably metamorphic, overgrowth.

and rims yield a weighted average of the $^{207}Pb/^{206}Pb$ age of 993.3 ± 8.3 (Fig. 8). We interpret this age to be the time at which the protolith of the hornblende gneiss was metamorphosed to amphibolite grade (i.e., pyroxene of igneous or metamorphic origin reacted to form amphibole). The rounded zircon cores suggest that either the protolith of the hornblende gneiss was a volcanic rock or that the cores are related to an older (ca. 1355 Ga) high-grade metamorphism.

Danbury Augen Granite

Zircons from the Danbury augen granite (sample NM100) are elongate, (l/w = 3–4) and euhedral; a few are cracked. In CL, these zircons have fine concentric oscillatory zoning preserved throughout most of the grains (Fig. 9, D). A few zircons have very small, dark (in CL) overgrowths. The weighted average of the $^{207}Pb/^{206}Pb$ ages for the oscillatory-zoned cores is 1044.9 ± 7.9 Ma (Fig. 8), interpreted as the time of emplacement of the gneiss protolith. Analyses of two overgrowths suggest an age of ca. 965 Ma. Titanite from this sample is 431 ± 6 Ma.

Biotite Gneiss

To better understand the ages of the different biotite gneiss units (Ybg and Ybgg in Fig. 2), two samples were dated. Sample NM628 was collected from an ~1 m thick layer in the eastern belt of biotite granitic gneiss (Ybgg). This ~20-m-wide outcrop is composed almost entirely of layers of biotite granitic gneiss. The homogeneous character of the rock in this outcrop suggests that the protolith of this rock may have been intrusive, although an extrusive origin cannot be ruled out. Sample NM772 was collected from the central belt of layered biotite gneiss (Ybg), from an outcrop of thin (~0.3–0.5-m), interlayered felsic (biotite granite) gneiss and mafic (amphibolite) rocks. This lithologic variability indicates that the sampled felsic gneiss is either a sill or a layered metavolcanic rock.

Elongate (l/w = 2–4) and equant (l/w = 1) zircons occur in biotite granitic gneiss sample NM628 (Ybgg). Elongate zircons are subhedral to euhedral and show fine concentric oscillatory zoning, indicative of an igneous origin. Many of these grains have thin, dark (in CL) overgrowths (Fig. 9, E). Equant zircons contain cores that are fragments of oscillatory-zoned igneous zircon, plus thin to broad, dark (in CL) overgrowths of probable metamorphic origin. The weighted average of the $^{207}Pb/^{206}Pb$ ages for the oscillatory-zoned cores (including one from an equant grain) is 1050 ± 14 Ma (Fig. 8), interpreted as the time of crystallization of the protolith of the biotite granitic gneiss (Ybgg). Overgrowth ages, measured on rims of both elongate and equant zircons, are 1027 ± 15 and 979 ± 19 Ma.

Zircons from the biotite granite gneiss sample NM772 (in the layered Ybg unit) are elongate (l/w = 3–4) and show two types of CL zoning. Some grains contain large oscillatory-zoned cores with thin dark (in CL) or patchwork overgrowths; other grains contain small, rounded oscillatory zoned cores and broad patchwork mantles (Fig. 9, F). Due to the relatively poor performance of the instrument (uncharacteristically low sensitivity) during analysis of sample NM772, individual $^{206}Pb/^{238}U$ ages have significantly lower uncertainty than the $^{207}Pb/^{206}Pb$ ages. Thus, the age of zircon from this sample is calculated using the weighted average of the $^{206}Pb/^{238}U$ ages. Oscillatory-zoned cores yield a crystallization age of 1048 ± 11 Ma, within the uncertainty of the age (1050 ± 14 Ma) of the other biotite granitic gneiss sample (NM628); mantles and overgrowths are 980 ± 8 Ma (Fig. 8). Although field evidence suggests that some of the components of the layered biotite gneiss unit (Ybg) are the oldest rocks in the New Milford quadrangle, the zircon U-Pb data indicate that the two sampled layers (NM628 and NM772) are significantly younger than the pink granite gneiss (Ygg) (see below). Possible causes for this apparent discrepancy between field and isotopic data are addressed in the discussion section.

Migmatite Gneiss

Migmatite gneiss (OYmig) is a major unit that occurs in a wide north-south belt through the center of the New Milford quadrangle. To better understand the timing of migmatization, we collected two phases of the migmatite: (1) leucocratic gray layered gneiss (NM576B), and (2) crosscutting pink granite (NM576A).

Elongate (l/w = 4–6) and equant (l/w = 1) zircons occur in the leucocratic gray, layered gneiss (sample NM576B). The CL imaging shows that most of the elongate zircons consist of light colored, oscillatory-zoned cores with broken terminations, overgrown by darker gray, concentric oscillatory-zoned mantles. Most grains also contain a very thin, black (in CL) outer rim that is too small to be analyzed (Fig. 10, A). The $^{207}Pb/^{206}Pb$ ages on cores and mantles are the same, within uncertainty, resulting in a weighted average age of 1057 ± 10 Ma (Fig. 8). This age is interpreted as the time of migmatization and emplacement of leucosomes during Mesoproterozoic anatexis. This age is the same, within uncertainty, as the two ages from the biotite gneiss units Ybg and Ybgg (1048 ± 11 and 1050 ±14 Ma). Complex relations in the migmatite, combined with similar ages for the gray layered gneiss (NM576B) and the two biotite gneiss samples (NM628 and NM772) allow an interpretation in which the sampled gray layered gneiss was equivalent material to the biotite gneiss samples, with migmatization occuring later, perhaps during subsequent metamorphism (ca. 1.0 Ga). Equant grains generally contain a small, irregularly shaped, oscillatory-zoned core and a broad black (in CL) overgrowth. Analyses of three dark overgrowths in equant grains, plus one overgrowth on an elongate grain, suggest metamorphic ages of 1003 ± 9 and ca. 950 Ma.

Titanite from the gray, layered gneiss (NM576B) was dated to provide age constraints for the timing of migmatization and/or subsequent metamorphism. Medium to dark brown titanites from the gray, layered gneiss are relatively high in U (98–208 ppm). These grains are overgrown by thin, light brown to colorless overgrowths (darker in BSE) that have lower U concentrations (5–114 ppm); whole colorless grains also occur. The

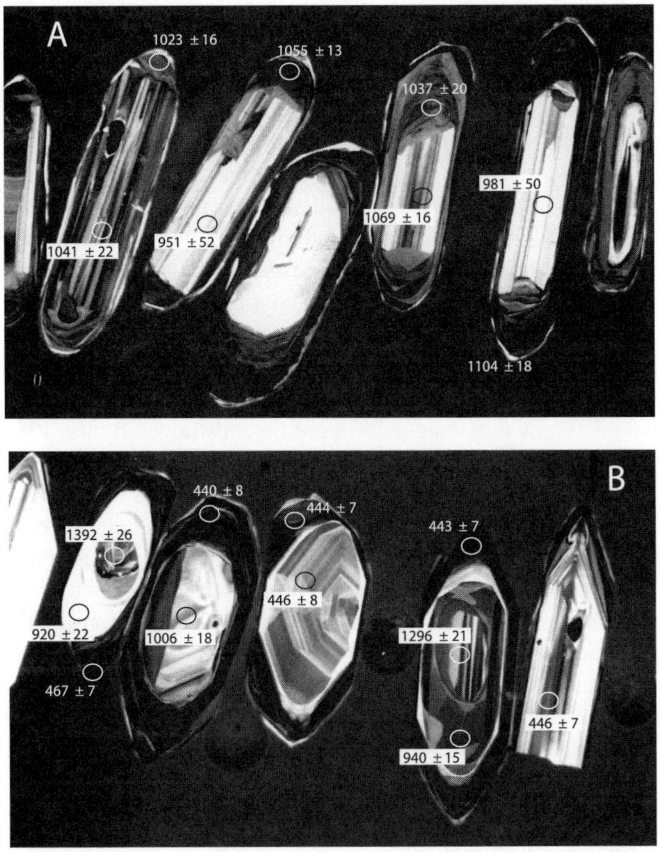

Figure 10. CL images of zircon from migmatite samples. (A) Layered gray gneiss phase (sample NM576B). Light, oscillatory-zoned cores, overgrown mostly on the pyramidal tips by darker zoned material. In all examples shown here, zoned pyramidal overgrowths are the same age (within error) of the lighter cores. Some grains have thin black outermost overgrowths that are mostly too small to be dated. (B) Pink granite phase (sample NM576A). Most of the zircons contain multiple, rounded, inherited cores of Proterozoic age, overgrown by Ordovician igneous zircon.

weighted average of the $^{206}Pb/^{238}U$ ages for the brown titanite cores is 425 ± 6 Ma. Whole colorless grains and overgrowths yield a weighted average age of 406 ± 9 Ma.

Zircons from the crosscutting pink granite gneiss phase of the migmatite (sample NM576A) are elongate (l/w = 3–4) and contain obvious cores and overgrowths in transmitted light. In CL, the zircons are shown to be exceedingly complex; many grains contain at least four distinct growth zones (Fig. 10, B). Most grains consist of rounded cores and rounded mantles that have Proterozoic ages (Table 1); these are interpreted as inherited (xenocrystic) material. Outermost overgrowths that form the external crystal morphology (pyramidal terminations and prism faces) are dark in CL and have oscillatory-zoning. A few prismatic grains lack xenocrystic cores and are composed entirely of concentric oscillatory-zoned material. The weighted average of the $^{206}Pb/^{238}U$ ages of the rims and oscillatory-zoned grains is 444 ± 6 Ma, interpreted as the age of crystallization of the pink granite and, by extension, the time of Paleozoic migmatization by injection. The age of this granite is identical to the age of the Candlewood Granite, and the two rocks are quite similar; the Candlewood Granite is generally gray and locally light pink, whereas the granitic phase of the migmatite is a distinct pink color. Titanite from the pink granite contains very subtle patchy zoning, only observable at maximum contrast in BSE. These grains contain relatively low U (8–291 ppm, mostly <100 ppm) and are pale yellow to colorless. The weighted average of $^{206}Pb/^{238}U$ ages from all but three analyses is 405 ± 6 Ma. Three analyses are slightly younger and may represent later events; alternatively, the younger ages might be due to a small amount of Pb-loss.

Pink granite gneiss

Zircons from the pink granite gneiss (sample NM174, Ygg) are euhedral (l/w = 2–3), cracked, and contain many inclusions. Most grains contain fine concentric oscillatory zoning; a few grains also have dark (in CL) incomplete thin overgrowths (Fig. 9, G). The weighted average of the igneous core ages is 1311 ± 7 Ma (Fig. 8), interpreted as the time of emplacement of the granite protolith. Three overgrowths yield a weighted average age of 982 ± 17 Ma. Because the pink granite gneiss (Ygg) contains xenoliths of some of the rock types in the biotite gneiss unit (amphibolite and quartzofeldspathic gneiss of Ybg), it is clear that some parts of the Ybg unit shown in Figure 2 must be older than the 1311 Ma of the pink granite gneiss (Ygg). Where dated, the biotite granite gneiss component of Ybg (NM772) is significantly younger (1048 Ma) than the pink granite gneiss. We suggest, therefore, that Ybg is a composite map unit, consisting of younger biotite granite gneiss layers (ca. 1.05 Ga) and older paragneiss. The paragneiss was intruded by pink granite gneiss at 1311 Ma, and the age of the host paragneiss is as yet unknown. Alternatively, it is possible, but highly unlikely, that the dated zircons from the pink granite gneiss are entirely xenocrystic and that the pink granite gneiss is younger than the biotite gneiss units (Ybg and Ybgg). Given the relatively pristine character of the zircons, the complete lack of ca. 1.05 Ga age data, and the tight grouping of ages, we conclude that the pink granite gneiss really is ca. 1.3 Ga.

In summary, our U-Pb geochronology indicates that the protolith of the pink granite gneiss intruded at ca. 1.31 Ga. This event occurred ~260 m.y. before a significant period of intrusion and migmatization at ca. 1.05 Ga. Ages of zircon overgrowths suggest that the basement rocks later experienced a significant metamorphic event at ca. 0.99 Ga. Ordovician granitic rocks were emplaced at ca. 453–443 Ma. In the Candlewood Granite, a single monazite age of 445 ± 9 Ma agrees, within uncertainty, with the zircon age of 443 ± 7 Ma. A similar age of 444 ± 6 Ma from within the basement gneisses indicates a phase of Ordovician migmatization by injection along the margin of the Candlewood Granite. The U-Pb ages of titanite in three samples indicate two episodes of Paleozoic metamorphism, at ca. 430–425 and 405 Ma.

DISCUSSION

Our U-Pb SHRIMP data from rocks in the New Milford quadrangle provide new insights on the evolution of the Mesoproterozoic basement. The data show the occurrence of three Mesoproterozoic thermal events (c. 1.3 Ga, 1.05 Ga, and 0.99 Ga), overprinted by a Late Ordovician igneous event (c. 443–453 Ma) with thermal pulses recorded by the growth of titanite into the Early Devonian (431–406). These data are summarized in Figure 11.

1.3 Ga Rocks

The oldest date from the area comes from the pink granitic gneiss (Ygg; ca. 1311 Ma). This age represents the oldest reliably dated rock in Connecticut. The previous oldest ages from nearby basement massifs are limited and less well-constrained Rb-Sr whole-rock ages (Prucha et al., 1968; Hall et al., 1975; Helenek and Mose, 1976). Most older ages from the area are questionable, however, and in this paper, we refer only to more reliable SHRIMP or single-crystal zircon TIMS data. Although limited dates were available to early workers, both Clarke (1958) and Hall et al. (1975) considered these granite gneisses to be among the oldest granitic rocks in the eastern Hudson Highlands. Dana (1977) and Jackson (1980) interpreted the pink granite gneiss unit as the youngest basement gneiss and stated that it cut all other rocks in the basement. Considering that previously unrecognized Ordovician pink granitic migmatite gneiss also occurs in the basement of the New Milford area (Sample NM576A), it is possible that some of these younger granitic rocks have been mapped as Precambrian rocks. Our findings suggest that some of the units mapped by Rodgers (1985) as Precambrian pink granitic gneiss on the Connecticut State geologic map need to be re-evaluated.

Although unequivocal crosscutting relations between the pink granite gneiss (Ygg) and the other gneisses were not observed in the New Milford quadrangle, the presence of layered biotite gneiss and amphibolite, interpreted as screens and xenoliths within the Ygg, suggests that the granite gneiss intruded a sequence of paragneiss. The 1311 Ma age of the Ygg unit implies that only part of the unit mapped as layered biotite gneiss (Ybg) must be older than 1311 Ma, as indicated by the 1048 Ma age from the sampled layer in Ybg (NM772). The minimum age of the paragneiss might be constrained by the 1311 Ma age of the pink granite gneiss, if our interpretation that the Ygg unit is intrusive into a paragneiss sequence is correct. The single 1355 Ma age from an inherited core in the amphibolite gneiss could possibly provide an age constraint on the older paragneiss sequence but, without further data, the age of the host paragneiss is speculative. An alternative interpretation for the 1311 Ma age could allow the Ygg unit to be the oldest rock in the quadrangle, and all other rocks in Ybg could be metavolcanic and metasedimentary rocks deposited unconformably on the pink granite gneiss. This interpretation is not favored, however, as there is no evidence of unconformable relationships (e.g., truncated older fabrics in the pink granite gneiss) or of the presence of basal conglomerates.

Recent U-Pb zircon ages from the southern Adirondacks (McLelland and Chiarenzelli, 1990) and the Mount Holly Complex in the Green Mountain massif (Ratcliffe et al., 1991) rep-

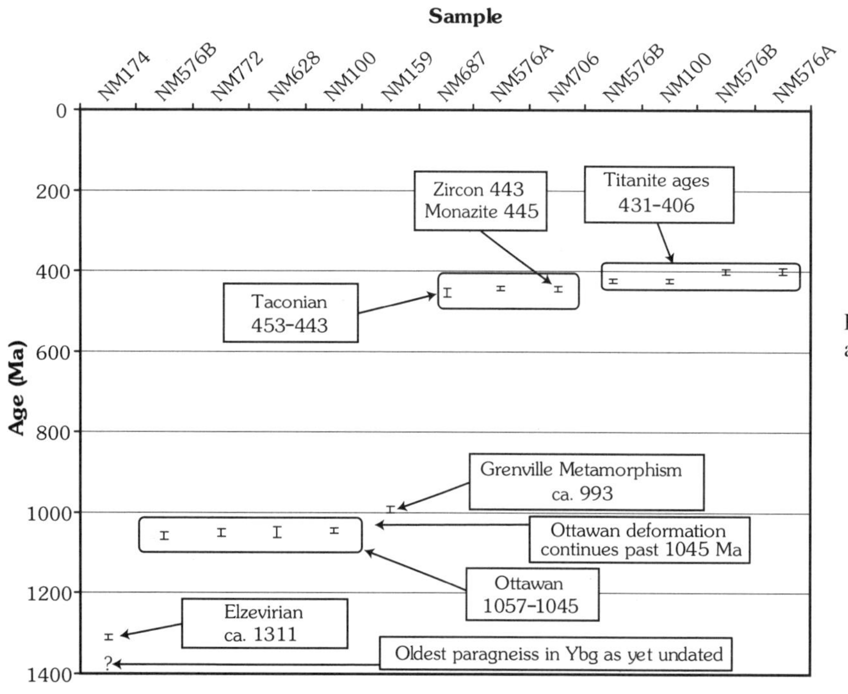

Figure 11. Summary of zircon, titanite, and monazite ages from the New Milford quadrangle.

resent the closest rocks with similar ages to the pink granite gneiss. McLelland and Chiarenzelli (1990) report TIMS ages of 1336–1302 Ma from tonalitic gneiss in the Adirondacks. Metavolcanics and metaintrusives of the Mount Holly Complex include dacitic, trondhjemitic, and tonalitic rocks dated between 1357 and 1308 Ma (Ratcliffe et al., 1991). Ratcliffe et al. (1991) were the first to demonstrate a correlation between rocks of the Green Mountain massif and the Adirondacks, and our new data support a correlation of this age group with the rocks of the New Milford massif, extending the belt of ca. 1.3 Ga rocks considerably to the south. No rocks of similar age have been reported from the Reading Prong. Ratcliffe et al. (1991) note that similar rocks are absent from the Housatonic Highlands and the Berkshire massif, but that trondhjemitic gneisses similar to the rocks in the Mount Holly Complex are present along the Hudson River in the Hudson Highlands. Aleinikoff et al. (2000) note the absence of rocks of similar age in the northern Blue Ridge.

The ca. 1.3 Ga event in the Adirondacks and Green Mountains has been attributed to calc-alkaline arc magmatism along the margin of Laurentia during the earliest stages of the Elzevirian orogeny (McLelland et al., 1988; McLelland and Chiarenzelli, 1990; Ratcliffe et al., 1991). The Elzevirian orogeny spans the period from ca. 1300 to 1160 Ma in the Central Metasedimentary Belt of Southern Ontario (Wasteneys et al., 1999).

1.05 Ga Rocks

Geochronologic data from four samples of gneiss from the New Milford quadrangle indicate a significant thermal event from ca. 1057 to 1045 Ma. There are no reliable ages from the eastern Hudson Highlands or western Connecticut within this range. Gates et al. (2001a) report similar ages of ca. 1057–1019 Ma on metamorphic overgrowths on zircon in the western Hudson Highlands. Regionally, these ages correlate well with the timing of the Ottawan orogeny in the Adirondacks (McLelland et al., 2001). There, the Ottawan orogeny spans the period of 1090–1035 Ma and is associated with Himalayan style continent-continent collision (Rivers, 1997; McLelland et al., 2001). In the Adirondacks, rocks of the Lyon Mountain Granite Gneiss suite range in age from 1060 to 1045 Ma (Chiarenzelli and McLelland, 1991; McLelland et al., 2001), closely matching the ages from the New Milford quadrangle. McLelland et al. (2001) state that the Lyon Mountain Granite Gneiss suite possesses deformational fabrics that imply a late to post-tectonic timing for the suite.

In the New Milford area, the oldest (ca. 1057 Ma) phase of igneous activity during this period is associated with metamorphism and migmatization (OYmig) of the paragneisses of the layered biotite gneiss sequence (Ybg). The migmatization is considered syn-tectonic with the dominant phase of Mesoproterozoic deformation in the area as it cuts, and yet is deformed by, the dominant folding event seen in the basement rocks (Fig. 4). In the New Milford area, the dominant deformation is a second-generation event (YD2). In the western Hudson Highlands, the dominant deformational event is also second-generation, but has been dated between 1174 and 1144 Ma, and migmatization is associated with a third-generation event during intrusion of the Canada Hill Granite at ca. 1010 Ma (Aleinikoff and Grauch, 1990; Ratcliffe, 1992; Ratcliffe and Aleinikoff, 2001). Following migmatization, igneous activity is associated with intrusion of abundant sills of biotite granitic gneiss parallel to the dominant gneissosity. The age of the sills corresponds to the ages of the sampled layers of Ybg (NM772) at 1048 ± 11 Ma and Ybgg (NM628) at 1050 ± 14 Ma. The Danbury augen granite intruded the other gneisses at 1045 ± 8 Ma. The Danbury augen granite may be considered a late tectonic intrusion, because it cuts the layering and dominant folds in the other gneisses, yet contains the weakly to moderately developed dominant Mesoproterozoic gneissosity, implying that Ottawan tectonics in the area ended after 1045 Ma.

Mesoproterozoic rocks in the Poughquag quadrangle, New York, are quite similar to rocks in the New Milford quadrangle, but no ages are available for those rocks (Ratcliffe and Burton, 1990). In the Poughquag quadrangle, Berkey and Rice (1921) and Ratcliffe and Burton (1990) point out complex relationships associated with the Reservoir Gneiss, including intrusive granite gneiss and migmatite that postdate layered biotite-quartz-plagioclase gneiss and amphibolite paragneisses. Ratcliffe and Burton (1990) state, however, that not all of the Reservoir Gneiss is intrusive, because contact relationships between intrusive granite gneiss and other layered felsic gneisses are not always clear; thus, the Reservoir Gneiss contains many of the complexities observed in the New Milford area.

Late to post-tectonic Ottawan granites recognized in the New Jersey Highlands and western Hudson Highlands are not yet recognized in the eastern Hudson Highlands. For example, the Mount Eve Granite, dated at 1020 ± 4 Ma (Drake et al., 1991), is a relatively undeformed postorogenic granite that constrains Ottawan deformation in the western New Jersey Highlands (Estelle et al., 2001). The younger Canada Hill Granite in the western Hudson Highlands, dated at ca. 1010 Ma (Aleinikoff et al., 1982; Aleinikoff and Grauch, 1990), is interpreted as a syn-tectonic Ottawan pluton (Ratcliffe, 1992; Ratcliffe and Aleinikoff, 2001), suggesting that Ottawan deformation and regional metamorphism in the New Jersey and Hudson highlands gets younger to the east. Prior to our findings, Ratcliffe and Aleinikoff (2001) suggested that the Canada Hill Granite and the synchronous deformation at ca. 1020 Ma might be the only Ottawan event in the Hudson Highlands and that much of the deformation and igneous activity occurred prior to 1144 Ma. Gorring et al. (2001) report a SHRIMP zircon age of 1008 ± 4 Ma from a diorite pluton in the southwestern Hudson Highlands and interpret the diorite as a post-Ottawan intrusion.

The new ca. 1.05 Ga ages from the New Milford area represent the first Ottawan ages from the eastern Hudson Highlands and western Connecticut massifs. Igneous rocks with similar ages have not yet been reported from the western Hudson Highlands or elsewhere in the Reading Prong. Conversely, rocks from the western Hudson Highlands—for example, the Storm King Granite (1174 ± 8 Ma) and the Canopus Pluton (1144 ± 13 Ma) of Ratcliffe and Aleinikoff (2001)—record a period of igneous activity during the Shawinigan pulse of Rivers (1997) from ca. 1.19 to 1.14 Ga that is not yet recognized in the eastern Hudson Highlands or Connecticut. We speculate that the absence of this age group in the eastern Hudson Highlands may be a function of limited high-precision geochronology, and not the absence of the rocks.

Possible age-correlative rocks to the south include 1060–1055-Ma granitic gneisses: Group 3B, Northern Blue Ridge, Virginia (Aleinikoff et al., 2000) and the Old Rag Granite, central Virginia, (Tollo et al., this volume). In summary, the new ages from the New Milford area confirm the presence of significant dynamothermal events during the Ottawan orogeny, as defined by McLelland et al. (2001).

0.99 Ga Metamorphism

Ages of zircon overgrowths and metamorphic zircons indicate a significant metamorphic event at ca. 0.99 Ga. Mesoproterozoic rocks of the Hudson Highlands and western Connecticut massifs experienced granulite-facies and amphibolite-facies metamorphism (Hall et al., 1975; Murray, 1976) that was subsequently overprinted by amphibolite-facies metamorphism during the Ordovician and Devonian (Dallmeyer et al., 1975; Dallmeyer and Sutter, 1976; Dana, 1977; Jackson, 1980; Sutter et al., 1985; Hames et al., 1991). Hall et al. (1975) document lower granulite-facies metamorphism in the western Hudson Highlands of eastern New York and upper amphibolite-facies metamorphism in the eastern Hudson Highlands of New York and Connecticut. In the New Milford quadrangle, the basement rocks experienced subsequent Paleozoic metamorphism that reached upper amphibolite facies, indicated by sillimanite–K-feldspar assemblages in the cover rocks. Both the basement and cover rocks west of Cameron's Line experienced subsequent retrogression of sillimanite–K-feldspar assemblages to sillimanite (fibrolite)-muscovite assemblages during the Devonian (Hames et al., 1991).

Ages of Mesoproterozoic metamorphism in the Hudson Highlands range from 1.05 to 0.91 Ga. Aleinikoff et al. (1982) report ca. 1000 Ma metamorphic zircons from hornblende gneiss in the Peekskill, New York, quadrangle in the western Hudson Highlands. Aleinikoff and Grauch (1990) report a metamorphic age of ca. 985 Ma from a monazite-xenotime rock adjacent to the Canada Hill Granite and report some monazite ages as young as 915 Ma. The rocks near the Canada Hill Granite are subsequently cut by postmetamorphic pegmatites at ca. 965 Ma (Grauch and Aleinikoff, 1985). Elsewhere in the Reading Prong, Gates et al. (2001b) document Ottawan granulite-facies metamorphism at ca. 1050 Ma from metamorphic zircons that are overprinted by right-lateral fault zones dated between ca. 1008 and 915 Ma by Ar/Ar thermochronology. Gates et al. (2001a) report ages of 1010 and 1007 Ma for metamorphic zircons from the western Hudson Highlands. Our new data show that a Grenville metamorphic event in the New Milford massif occurred as late as 993 Ma, in general agreement with other metamorphic ages from the Hudson Highlands.

Early Paleozoic Intrusive Rocks

Our new data demonstrate a Late Ordovician age event for the Candlewood Granite (443 ± 7 Ma) and a leucogranite aplite dike (453 ± 6 Ma) that cuts the Brookfield Gneiss. West of Cameron's Line, a second phase of migmatization in the basement rocks is associated with the injection of numerous granitic leucosomes at 444 ± 6 Ma along the margin of the Candlewood Granite. This phase of Ordovician migmatization was not previously recognized in western Connecticut basement massifs. These findings could explain why some previous workers believed that some pink granitic gneisses were the youngest Precambrian intrusive rocks in the basement (Dana, 1977; Jackson, 1980).

Our age of the Candlewood Granite refines less well-constrained Rb-Sr whole-rock ages of the granite that range from 442 to 426 Ma (Mose and Nagel, 1982; Amenta and Mose, 1985). Mose and Nagel (1982) sampled the Candlewood Granite in the Danbury, New Milford, and Kent quadrangles and reported ages that got older from south to north (426, 435, and 442 Ma, respectively). Amenta and Mose (1985) also report a Rb-Sr whole-rock age of 440 ± 4 Ma for the Sunset Hill granite east of Cameron's Line, and note the similarities between the two granites across Cameron's Line. Recent work by Sevigny and Hanson (1995) indicates an age of 453 ± 3 Ma for the Brookfield Gneiss by single-crystal TIMS U-Pb analysis. This age is the same, within uncertainty, as the age of our leucogranite aplite dike (453 ± 6 Ma, NM687) that cuts the Brookfield Gneiss. Sevigny and Hanson (1995) also report U-Pb zircon TIMS ages of 446–438 Ma for the Newtown Gneiss, a composite pluton with compositions that range from quartz diorite to granite. The Candlewood Granite clearly cuts Mesoproterozoic basement rocks, and some of the unanalyzed zircons show suspected xenocrystic cores. The Brookfield and Newtown gneisses intrude the parautochthonous rocks east of Cameron's Line, and their zircons also show Mesoproterozoic inheritance either from basement rocks directly or sediments derived from the Laurentian crust (Sevigny and Hanson, 1995).

Early Paleozoic Titanite Ages

No titanite of Grenville age was found in Proterozoic basement rocks in the New Milford quadrangle. Instead, titanite formed in the gneisses at ca. 430–425 Ma and ca. 405 Ma. Sevigny and Hanson (1995) obtained ages of ca. 450–420 Ma from titanite from the Ordovician Brookfield Gneiss and Newtown Gneiss just to the east of Cameron's Line. Their older ages probably represent time of cooling of igneous titanite shortly after emplacement of the plutons; the younger ages coincide with our older titanite ages from basement rocks and suggest a period of metamorphic growth of titanite in the Silurian. In contrast to our titanite data, Sevigny and Hanson (1995) found no evidence for an additional Devonian (Acadian) growth episode.

We suggest that the Paleozoic ages of titanite in Proterozoic gneisses in the New Milford quadrangle represent growth events, not times of cooling, because the closure temperature of the U-Pb system in titanite (~700 °C) (Cherniak, 1993; Corfu, 1996; Pidgeon et al., 1996; Zhang and Shärer, 1996) is greater than the regional metamorphic temperature. Ordovician metamorphism, synchronous with emplacement of the Candlewood Granite and Brookfield Gneiss at ca. 450–445 Ma, attained sillimanite–K-feldspar grade (Dana, 1977; Jackson, 1980) that was retrogressed to sillimanite-muscovite (Hames et al., 1991). Jackson (1980) estimates maximum *P-T* conditions of ~5 kilobars and 650 °C during sillimanite–K-feldspar grade metamorphism in the adjacent Kent quadrangle. Thus, the regional metamorphic temperature in the Ordovician was less than the titanite closure temperature, implying that the titanite ages must record growth episodes. More data are necessary to determine if the Silurian and Devonian growth events are related to heating pulses or deformational events.

CONCLUSIONS

New mapping and U-Pb geochronology by SHRIMP on zircon, titanite, and monazite in the New Milford quadrangle, Connecticut indicate the following:

1. The oldest dated rock in the New Milford massif is a pink granite gneiss (1311 ± 7 Ma). Outcrops interpreted as xenoliths in the pink granite gneiss suggest that this rock intrudes a paragneiss sequence of biotite-quartz-plagioclase gneiss, amphibolite, and rare calc-silicate rock and quartzite, although unequivocal crosscutting relations are not observed. The paragneiss sequence is still undated. The closest potentially correlative intrusive rocks of similar age occur in the Green Mountains (Ratcliffe et al., 1991) and Adirondacks (McLelland and Chiarenzelli, 1990), and are attributed to the early stages of the Elzevirian orogeny.

2. A significant dynamothermal event during the Ottawan orogeny affected all basement rocks in the quadrangle. Syn-tectonic anatexis produced a belt of stromatic migmatite in the central part of the quadrangle at 1057 ± 10 Ma. Igneous activity included syn-tectonic intrusion of abundant sills of biotite granitic gneiss between 1050 ± 14 and 1048 ± 11 Ma, followed by the Danbury augen granite at 1045 ± 8 Ma. The closest potentially correlative rocks of similar age occur in the Adirondacks (Chiarenzelli and McLelland, 1991; McLelland et al., 2001) and Northern Blue Ridge (Aleinikoff et al., 2000), although slightly younger syn- to late-tectonic rocks do occur in the western Hudson Highlands.

3. Overgrowths on igneous zircon from several samples and new metamorphic zircon in hornblende gneiss indicate that terminal Grenville metamorphism occurred at 993 ± 8 Ma.

4. Late Ordovician syn-tectonic (Paleozoic D2) activity included intrusion of a leucogranite aplite dike in the Brookfield Gneiss at 453 ± 6 Ma east of Cameron's Line and intrusion of the Candlewood Granite at 443 ± 7 Ma west of Cameron's Line. A monazite age from the Candlewood Granite of 445 ± 9 Ma corroborates the zircon age. West of Cameron's Line, a second phase of migmatization is associated with the injection of numerous granitic leucosomes at 444 ± 6 Ma into the basement and cover rocks along the margin of the Candlewood Granite.

5. Titanite in Proterozoic gneisses formed at ca. 430–425 and ca. 405 Ma. No Proterozoic titanite is preserved in the gneisses. Titanite ages represent growth episodes, not cooling ages, suggesting that heating and/or tectonic/deformational events occurred in the Silurian and Devonian.

ACKNOWLEDGMENTS

We thank Joe Wooden of the U.S. Geological Survey for help with SHRIMP-RG analyses. Rhonda Driscoll and Ezra Yacob of the U.S. Geological Survey in Denver, Colorado, helped with mineral separations. Thanks go to William Burton, Craig Dietsch, and Nicholas Ratcliffe for critical reviews of this paper. The manuscript also benefited from helpful comments by James McLelland.

REFERENCES CITED

Aleinikoff, J.N., and Grauch, R.I., 1990, U-Pb geochronologic constraints on the origin of a unique monazite-xenotime gneiss, Hudson Highlands, New York: American Journal of Science, v. 290, no. 5, p. 522–546.

Aleinikoff, J.N., Grauch, R.I., Simmons, K.R., and Nutt, C.J., 1982, Chronology of metamorphic rocks associated with uranium occurrences, Hudson Highlands, New York–New Jersey: Geological Society of America Abstracts with Programs, v. 14, no. 1-2, p. 1.

Aleinikoff, J.N., Burton, W.C., Lyttle, P.T., Nelson, A.E., and Southworth, C.S., 2000, U-Pb geochronology of zircon and monazite from Mesoproterozoic granitic gneisses of the northern Blue Ridge, Virginia and Maryland, USA: Precambrian Research, v. 99, p. 113–146.

Amenta, R.V., and Mose, D.G., 1985, Tectonic implications of Rb-Sr ages of granitic plutons near Cameron's Line in western Connecticut: Northeastern Geology, v. 7, no. 1, p. 11–19.

Amenta, R.V., Mose, D.G., Nagel, S., and Tunsoy, A., 1982, Rb-Sr ages and tectonic implications of granitic plutons adjacent to Cameron's Line in western Connecticut: Geological Society of America Abstracts with Programs, v. 14, no. 1-2, p. 1.

Berkey, C.P., and Rice, M., 1921, Geology of the West Point quadrangle, New York: New York State Museum and Science Service Bulletin nos. 225–226, 152 p.

Cherniak, D.J., 1993, Lead diffusion in titanite and preliminary results on the effects of radiation damage on Pb transport: Chemical Geology, v. 110, p. 177–194.

Chiarenzelli, J.R., and McLelland, J.M., 1991, Age and regional relationships of granitoid rocks of the Adirondack Highlands: Journal of Geology, v. 99, no. 4, p. 571–590.

Clark, G.S., and Kulp, J.L., 1968, Isotopic age study of metamorphism and intrusion in western Connecticut and southeastern New York: American Journal of Science, v. 266, no. 10, p. 865–894.

Clarke, J.W., 1958, The bedrock geology of the Danbury quadrangle, Connecticut: Connecticut Geological and Natural History Survey, Quadrangle Report No. 7, scale 1:24,000, 47 p.

Compston, W., Williams, I.S., and Meyer, C., 1984, U-Pb geochronology of zircons from lunar breccia 73217 using a sensitive high-resolution ion-microprobe: Proceedings, 14th Lunar Science Conference: Journal of Geophysical Research, v. 98B, p. 525–534.

Corfu, F., 1996, Multistage zircon and titanite growth and inheritance in an Archean gneiss complex, Winnipeg River Subprovince, Ontario: Earth and Planetary Science Letters, v. 141, p. 175–186.

Dallmeyer, R.D., and Sutter, J.F., 1976, $^{40}Ar/^{39}Ar$ incremental-release ages of biotite and hornblende from variably retrograded basement gneisses of the northeasternmost Reading Prong, New York; their bearing on early Paleozoic metamorphic history: American Journal of Science, v. 276, no. 6, p. 731–747.

Dallmeyer, R.D., Sutter, J.F., and Baker, D.J., 1975, Incremental $^{40}Ar/^{39}Ar$ ages of biotite and hornblende from the northeastern Reading Prong; their bearing on late Proterozoic thermal and tectonic history: Geological Society of America Bulletin, v. 86, p. 1435–1443.

Dalziel, I.W.D., 1997, Overview: Neoproterozoic-Paleozoic geography and tectonics: Review, hypothesis, environmental speculations: Geological Society of America Bulletin, v. 109, p. 16–42.

Dana, R.H., Jr., 1977, Stratigraphy and structural geology of the Lake Waramaug area, western Connecticut [M.S. thesis]: Amherst, University of Massachusetts, scale 1:24,000, 108 p.

Drake, A.A., Jr., 1984, The Reading Prong of New Jersey and Eastern Pennsylvania: An appraisal of rock relations and chemistry of a major Proterozoic terrane in the Appalachians, *in* Bartholomew, M.J., ed., The Grenville Event in the Appalachians and related topics: Boulder, Colorado, Geological Society of America Special Paper 194, p. 75–109.

Drake, A.A., Jr., Aleinikoff, J.N., and Volkert, R.A., 1991, The Mount Eve Granite (middle Proterozoic) of northern New Jersey and southeastern New York: U.S. Geological Survey Bulletin B-1952, p. C1–C10.

Estelle, T.C., Gorring, M.L., and Volkert, R.A., 2001, Geochemistry of the late Mesoproterozoic Mount Eve Granite: Implications for late- to post-Ottawan tectonics in the NJ/Hudson Highlands: Geological Society of America Abstracts with Programs, v. 33, no. 6, p. 91.

Finney, S.C., 1986, Graptolite biofacies and correlation of eustatic, subsidence, and tectonic events in the Middle to Upper Ordovician of North America: Palaios, v. 1, no. 5, p. 435–461.

Gates, A.F., Valentino, D.W., Chiarenzelli, J.R., and Hamilton, M.A., 2001a, Ages of tectonic events in the Hudson Highlands, NY: Results from SHRIMP analyses: Geological Society of America Abstracts with Programs, v. 33, no. 1, p. 79.

Gates, A.F., Valentino, D.W., Chiarenzelli, J.R., and Hamilton, M.A., 2001b, Deep-seated Himalayan-type syntaxis in the Grenville orogen, NY-NJ-PA: Geological Society of America Abstracts with Programs, v. 33, no. 6, p. 91.

Gorring, M.L., Gates, A., Valentino, D., and Chiarenzelli, J., 2001, Magmatic history and geochemistry of meta-igneous rocks from the SW Hudson Highlands, NY: Tectonic implications and regional correlations: Geological Society of America Abstracts with Programs, v. 33, no. 6, p. 91.

Grauch, R.I., and Aleinikoff, J.N., 1985, Multiple thermal events in the Grenvillian orogenic cycle—Geochronologic evidence from the northern Reading Prong, New York–New Jersey: Geological Society of America Abstracts with Programs, v. 17, no. 7, p. 596.

Gregory, H.E., and Robinson, H.H., 1907, Preliminary geological map of Connecticut: Hartford, Connecticut, Connecticut Geological and Natural History Survey, Bulletin no. 7, 39 p.

Hall, L.M., 1980, Basement-cover relations in western Connecticut and southeastern New York, *in* Wones, D.R., ed., The Caledonides in the USA: Blacksburg, Virginia, Virginia Polytechnic Institute Department of Geological Sciences Memoir, no. 2, p. 299–306.

Hall, L.M., Helenek, H.L., Jackson, R.A., Caldwell, K.G., Mose, D., and Murray, D.P., 1975, Some basement rocks from Bear Mountain to the Housatonic Highlands, *in* Ratcliffe, N.M., ed., Guidebook for field trips in western Massachusetts, northern Connecticut and adjacent areas of New York, New England Intercollegiate Geological Conference, 67th Annual Meeting: New York, City College of New York, p. 1–29.

Hames, W.E., Tracy, R.J., Ratcliffe, N.M., and Sutter, J.F., 1991, Petrologic, structural, and geochronologic characteristics of the Acadian metamorphic overprint on the Taconide zone in part of southwestern New England: American Journal of Science, v. 291, no. 9, p. 887–913.

Harwood, D.S., and Zietz, I., 1974, Configuration of Precambrian rocks in southeastern New York and adjacent New England from aeromagnetic data: Geological Society of America Bulletin, v. 85, p. 181–188.

Hatch, N.L., Jr., and Stanley, R.S., 1973, Some suggested stratigraphic relations in part of southwestern New England: U.S. Geological Survey Bulletin B-1380, 83 p.

Helenek, H.L., and Mose, D., 1976, Structure, petrology and geochronology of the Precambrian rocks in the central Hudson Highlands, *in* Johnsen, J.H., ed., Guidebook to field excursions, meeting no. 48: Poughkeepsie, New York State Geological Association, p. B1–B27.

Helenek, H.L., and Mose, D., 1978, Geology and geochronology of Precambrian rocks in the Lake Carmel region, Hudson Highlands, New York: Geological Society of America Abstracts with Programs, v. 10, no. 2, p. 47.

Jackson, R.A., 1980, Autochthon and allochthon of the Kent quadrangle, western Connecticut [Ph.D. dissertation]: Amherst, University of Massachusetts, scale 1:24,000, 146 p.

Jacobi, R.D., 1981, Peripheral bulge; a causal mechanism for the Lower/Middle Ordovician unconformity along the western margin of the Northern Appalachians: Earth and Planetary Science Letters, v. 56, p. 245–251.

Karlstrom, K.E., Ahall, K.-I., Harlan, S.S., Williams, M.L., McLelland, J., and Geissman, J.W., 2001, Long-lived (1.8–1.0 Ga) convergent orogen in southern Laurentia, its extensions to Australia and Baltica, and implications for refining Rodinia: Precambrian Research, v. 111, p. 5–30.

Long, L.E, and Kulp, J.L., 1962, Isotopic age study of the metamorphic history of the Manhattan and Reading prongs: Geological Society of America Bulletin, v. 73, p. 969–995.

Ludwig, K.R., 2002a, Squid, version 1.05: Berkeley, California, Berkeley Geochronology Center Special Publication no. 2, 16 p.

Ludwig, K.R., 2002b, Isoplot/Ex version 3.00, a geochronological toolkit for Microsoft Excel: Berkeley, Berkeley Geochronology Center Special Publication no. 2, 46 p.

McLelland, J., and Chiarenzelli, J., 1990, Isotopic constraints on the emplacement age of the Marcy massif, Adirondack Mountains, New York: Journal of Geology, v. 98, p. 19–41.

McLelland, J., Hamilton, M., Selleck, B., McLelland, J., Walker, D., and Orrell, S., 2001, Zircon U-Pb geochronology of the Ottawan orogeny, Adirondack Highlands, New York: Regional and tectonic implications: Precambrian Research, v. 109, no. 1–2, p. 39–72.

McLelland, J.M., Chiarenzelli, J., Whitney, P., and Isachsen, Y., 1988, U-Pb zircon geochronology of the Adirondack Mountains and implications for their geologic evolution: Geology, v. 16, no. 10, p. 920–924.

Moore, J.M., Jr., and Thompson, P.H., 1980, The Flinton Group: A late Precambrian metasedimentary succession in the Grenville Province of eastern Ontario: Canadian Journal of Earth Sciences, v. 17, no. 12, p. 1685–1707.

Mose, D.G., and Nagel, M.S., 1982, Chronology of metamorphism in western Connecticut: Rb-Sr ages, *in* Joesten, R., and Quarrier, S.S., eds., Guidebook for fieldtrips in Connecticut and south central Massachusetts, New England Intercollegiate Geological Conference, 74th annual meeting: Hartford, Connecticut, Connecticut Geological and Natural History Survey Guidebook no. 5, p. 247–262.

Mose, D., and Wenner, D., 1980, Parallel Rb-Sr whole-rock isochrons and $^{18}O/^{16}O$ data from selected Appalachian plutons: Geological Society of America Abstracts with Programs, v. 12, no. 4, p. 202.

Murray, D.P., 1976, Chemical equilibrium in epidote-bearing calc-silicates and basic gneisses, Reading Prong, New York [Ph.D. dissertation]: Providence, Rhode Island, Brown University, 256 p.

Panish, P.T., 1992, The Mt. Prospect region of western Connecticut: Mafic plutonism in Iapetus strata and thrust emplacement onto the North American margin, *in* Robinson, P., and Brady, J.B., eds., Guidebook for field trips in the Connecticut Valley region of Massachusetts and adjacent areas, New England Intercollegiate Geological Conference, 84th Annual Meeting: Amherst, University of Massachusetts, p. 398–423.

Pidgeon, R.T., Bosch, D., and Bruguier, O., 1996, Inherited zircon and titanite

U-Pb systems in an Archean syenite from southwestern Australia: Implications for U-Pb stability of titanite: Earth and Planetary Science Letters, v. 141, p. 187–198.

Potter, D.B., 1972, Stratigraphy and structure of the Hoosic Falls area, New York–Vermont, east-central Taconics: Albany, New York, New York State Museum, Map and Chart Series, no. 19, 71 p.

Prucha, J.J., Scotford, D.M., and Sneider, R.M., 1968, Bedrock geology of parts of Putnam and Westchester counties, New York, and Fairfield County, Connecticut: Albany, New York, New York State Museum, Map and Chart Series, no. 11, 26 p.

Rankin, D.W., Drake, A.A., Jr., Glover, L., III, Goldsmith, R., Hall, L.M., Murray, D.P., Ratcliffe, N.M., Read, J.F., Secor, D.T., Jr., and Stanley, R.S., 1989, Pre-orogenic terranes, *in* Hatcher, R.D., Jr., et al., eds., The Appalachian-Ouachita orogen in the United States: Boulder, Colorado, Geological Society of America, The geology of North America, v. F-2, p. 7–100.

Ratcliffe, N.M., 1974, Bedrock geologic map of the State Line quadrangle, Columbia County, New York, and Berkshire County, Massachusetts: U.S. Geological Survey Geologic Quadrangle Map GQ-1142, scale 1:24,000.

Ratcliffe, N.M., 1980, Brittle faults (Ramapo Fault) and phyllonitic ductile shear zones in the basement rocks of the Ramapo seismic zones, New York and New Jersey, and their relationship to current seismicity, *in* Manspeizer, W., ed., Field studies of New Jersey geology and guide to field trips, New York State Geological Association, 52nd annual meeting: Albany, New York, New York State Geological Survey, p. 278–311.

Ratcliffe, N.M., 1992, Bedrock geology and seismotectonics of the Oscawana Lake quadrangle, New York: U.S. Geological Survey Bulletin 1941, scale 1:24,000.

Ratcliffe, N.M., 2000a, Bedrock geologic map of the Cavendish quadrangle, Windsor County, Vermont: U.S. Geological Survey Geologic Quadrangle Series Map GQ-1773, scale 1:24,000.

Ratcliffe, N.M., 2000b, Bedrock geologic map of the Chester quadrangle, Windsor County, Vermont: U.S. Geological Survey Geologic Investigations Series Map I-2598, scale 1:24,000.

Ratcliffe, N.M., and Aleinikoff, J.N., 2001, Pre-Ottawan deformation and transpressional faulting in the Hudson Highlands of New York based on SHRIMP zircon ages of the Storm King Granite and the syntectonic Canopus pluton: Geological Society of America Abstracts with Programs, v. 33, no. 1, p. 5.

Ratcliffe, N.M., and Armstrong, T.R., 1995, Preliminary bedrock geologic map of the Saxtons River 7.5 × 15 minute quadrangle, Windham and Windsor Counties, Vermont: U.S. Geological Survey Open-File Report 95-482, scale 1:24,000.

Ratcliffe, N.M., and Burton, W.C., 1990, Bedrock geologic map of the Poughquag quadrangle, New York: U.S. Geological Survey Geologic Quadrangle Map GQ-1662, scale 1:24,000.

Ratcliffe, N.M., Buden, R.V., and Burton, W.C., 1985, Ordovician ductile deformation zones in the Hudson Highlands and their relationship to metamorphic zonation in the cover rocks of Dutchess County, New York, *in* Tracy, R.J., ed., Guidebook for fieldtrips in Connecticut and adjacent areas of New York and Rhode Island, New England Intercollegiate Geological Conference, 77th Annual Meeting: Hartford, Connecticut, Connecticut Geological and Natural History Survey Guidebook no. 6, p. 25–60.

Ratcliffe, N.M., Aleinikoff, J.N., Burton, W.C., and Karabinos, P.A., 1991, Trondhjemitic 1.35–1.31 Ga gneisses of the Mount Holly Complex, of Vermont—Evidence for an Elzevirian event in the Grenville basement of the United States Appalachians: Canadian Journal of Earth Sciences, v. 28, no. 1, p. 77–93.

Ratcliffe, N.M., Armstrong, T.R., and Aleinikoff, J.N., 1997, Stratigraphy, geochronology, and tectonic evolution of the basement and cover rocks of the Chester and Athens domes, *in* Grover, T.W., et al., eds., Guidebook to field trips in Vermont and adjacent New Hampshire and New York, New England Intercollegiate Geological Conference, 89th Annual Meeting: Castleton, Vermont, Castleton State College, trip B6, p. 1–55.

Ratcliffe, N.M., Harris, A.G., and Walsh, G.J., 1999, Tectonic and regional metamorphic implications of the discovery of Middle Ordovician conodonts in cover rocks east of the Green Mountain Massif, Vermont: Canadian Journal of Earth Sciences, v. 36, no. 3, p. 371–382.

Rice, W.N., and Gregory, H.E., 1906, Manual of the geology of Connecticut: Hartford, Connecticut, Connecticut Geological and Natural History Survey Bulletin, no. 6, 273 p.

Rivers, T., 1997, Lithotectonic elements of the Grenville Province; review and tectonic implications: Precambrian Research, v. 8, p. 17–154.

Rodgers, J., 1971, The Taconic orogeny: Geological Society of America Bulletin, v. 82, p. 1141–1177.

Rodgers, J., 1985, Geological map of Connecticut: Connecticut Geological and Natural History Survey, scale 1:250,000.

Rodgers, J., Gates, R.M., and Rosenfeld, J.L., 1959, Explanatory text for preliminary geological map of Connecticut: Hartford, Connecticut, Connecticut Geological and Natural History Survey Bulletin, no. 84, 64 p.

Sevigny, J.H., and Hanson, G.N., 1995, Late-Taconian and pre-Acadian history of the New England Appalachians of southwestern Connecticut: Geological Society of America Bulletin, v. 107, p. 487–498.

Sperandio, R.J., 1974, The geology of the Brookfield diorite, New Milford quadrangle, Connecticut [M.S. thesis]: Washington, D.C., George Washington University, 91 p.

Stacey, J.S., and Kramers, J.D., 1975, Approximation of terrestrial lead isotope evolution by a two-stage model: Earth and Planetary Science Letters, v. 26, p. 207–226.

Stanley, R.S., and Ratcliffe, N.M., 1985, Tectonic synthesis of the Taconian orogeny in western New England: Geological Society of America Bulletin, v. 96, p. 1227–1250.

Sutter, J.F., Ratcliffe, N.M., and Mukasa, S.B., 1985, $^{40}Ar/^{39}Ar$ and K-Ar data bearing on the metamorphic and tectonic history of western New England: Geological Society of America Bulletin, v. 96, p. 123–136.

Tera, F., and Wasserburg, G.J., 1972, U-Th-Pb systematics in three Apollo 14 basalts and the problem of initial Pb in lunar rocks: Earth and Planetary Science Letters, v. 14, p. 281–304.

Walsh, G.J., 2003, Bedrock geologic map of the New Milford quadrangle, Litchfield and Fairfield Counties, Connecticut: U.S. Geological Survey Open-File Report 03-487, scale 1:24,000.

Wasteneys, H., McLelland, J., and Lumbers, S., 1999, Precise zircon geochronology in the Adirondack Lowlands and implications for revising plate-tectonic models of the Central Metasedimentary Belt and Adirondack Mountains, Grenville Province, Ontario and New York: Canadian Journal of Earth Sciences, v. 36, no. 6, p. 967–984.

Williams, I.S., and Claesson, S., 1987, Isotopic evidence for the Precambrian provenance and Caledonian metamorphism of high-grade paragneisses from the Seve Nappes, Scandinavian Caledonides, II. Ion microprobe zircon U-Th-Pb: Contributions to Mineralogy and Petrology, v. 97, p. 205–217.

Williams, I.S., and Hergt, J.M., 2000, U-Pb dating of Tasmanian dolerites: A cautionary tale of SHRIMP analysis of high-U zircon, *in* Woodhead, J.D., et al., eds., Beyond 2000: New frontiers in isotope geoscience, Lorne, Australia, abstracts and proceedings: Melbourne, Australia, University of Melbourne, p. 185–188.

Zen, E-an, 1967, Time and space relationships of the Taconic allochthon and autochthon: Boulder, Colorado, Geological Society of America Special Paper 97, 107 p.

Zhang, L.-S., and Shärer, U., 1996, Inherited Pb components in magmatic titanite and their consequence for the interpretation of U-Pb ages: Earth and Planetary Science Letters, v. 138, p. 57–65.

MANUSCRIPT ACCEPTED BY THE SOCIETY AUGUST 25, 2003

Printed in the USA

Geological Society of America
Memoir 197
2004

U-Pb geochronology and Pb isotopic compositions of leached feldspars: Constraints on the origin and evolution of Grenville rocks from eastern and southern Mexico

Kenneth L. Cameron*
Robert Lopez*
Department of Earth Sciences, University of California, Santa Cruz, California 95064, USA
Fernando Ortega-Gutiérrez
Luigi A. Solari
J. Duncan Keppie
Carlos Schulze
Departamento de Geología, Instituto de Geología,
Universidad Nacional Autónoma de México, 04510 México, D.F.

ABSTRACT

Mesoproterozoic igneous protoliths that were metamorphosed to granulite facies during the Grenville orogeny are exposed south of the Ouachita suture in eastern and southern Mexico, and clasts of Mesoproterozoic granitoids occur in a Paleozoic conglomerate in northeastern Mexico. These rocks can be divided into two groups, based on crystallization ages. The older group has arc and backarc geochemical signatures and was emplaced between ca. 1235 Ma and ca. 1115 Ma. The younger group, an anorthosite-mangerite-charnockite-granite (AMCG) suite, is not well dated, but present data suggest that it was emplaced between ca. 1035 Ma and ca. 1010 Ma. The Mexican terrane was probably a trailing margin from ca. 1115 Ma until the Grenville orogeny at 990 ± 5 Ma that produced widespread granulite-facies metamorphism. There is no structural or metamorphic evidence in Mexico for a major, widespread orogenic event before the emplacement of the AMCG rocks; thus, the model relating the origin of AMCG suites to delamination or convective thinning of subcontinental lithosphere following collision and crustal thickening may not apply to the Mexican Grenville terrane. The Pb isotopic compositions of acid-leached potassium feldspars define a tight linear array on a $^{207}Pb/^{204}Pb$ versus $^{206}Pb/^{204}Pb$ diagram. This array is interpreted to reflect mixing between two end-member reservoirs. The radiogenic component was presumably continental in origin (perhaps subducted Paleoproterozoic sediments), whereas the nonradiogenic end-member was most likely ca. 1200 Ma mantle. The Pb isotopic compositions of feldspars from eastern and southern Mexico are distinct from those of feldspars from Laurentia as represented by the Grenville Province and Grenville granites of Texas.

Keywords: Mexico, Grenville orogeny, common Pb, U-Pb geochronology

*E-mail, corresponding author: Cameron, rocks@ucsc.edu; present address: Lopez, Department of Geology and Oceanography, West Valley Community College, Saratoga, California 95070, USA.

Cameron, K.L., Lopez, R., Ortega-Gutiérrez, F., Solari, L.A., Keppie, J.D., and Schulze, C., 2004, U-Pb geochronology and Pb isotopic compositions of leached feldspars: Constraints on the origin and evolution of Grenville rocks from eastern and southern Mexico, *in* Tollo, R.P., Corriveau, L., McLelland, J., and Bartholomew, M.J., eds., Proterozoic tectonic evolution of the Grenville orogen in North America: Boulder, Colorado, Geological Society of America Memoir 197, p. 755–769. For permission to copy, contact editing@geosociety.org.

INTRODUCTION

The crystalline basement of much of Mexico and Central America is believed to be Grenville rocks (Ortega-Gutiérrez et al., 1995; Manton, 1996). Phanerozoic sedimentary and volcanic rocks blanket most of that area, where Precambrian basement, consisting of Grenville granulites, is exposed in eastern and southern Mexico (Fig. 1). In addition to these basement exposures of granulite, Mesoproterozoic granitoids have been described recently from Coahuila, in northern Mexico, where they occur as cobbles and boulders in Paleozoic flysch (Lopez et al., 2001). Ortega-Gutiérrez et al. (1995) proposed that the Grenville basement of Mexico south of the Ouachita suture is a middle Proterozoic microcontinent, referred to as "Oaxaquia" ("wä hä′ ke a"), that was accreted to the southern margin of Laurentia in the late Paleozoic. Oaxaquia is inferred to extend from the Ouachita suture south for ~2000 km into Central America. It probably underlies an area of ~1 × 10^6 km^2, and thus, is a major segment of the worldwide Grenville orogen.

We have three objectives in this paper. The first is to present the results of U-Pb dating of the Grenville Novillo Gneiss exposed in northeastern Mexico (Fig. 1). These results are fairly typical of those from elsewhere in Mexico, and they illustrate some of the problems and challenges encountered in attempting to date the granulites exposed there. The second objective is to summarize the U-Pb zircon dates from the Mexican basement rocks south of the Ouachita suture. These dates place important constraints on the tectonic evolution of a major segment of the worldwide Grenville orogen. The third objective is to present Pb isotopic data from acid-leached feldspars from the Mexican rocks. These results hold promise for elucidating the provenance of Oaxaquia.

There are two small exposures of Mesoproterozoic rocks north of the Ouachita suture in Mexico. These are Laurentian basement (Blount, 1993; Mosher, 1998), and are excluded from discussion in this paper.

ROCK TYPES OF THE NOVILLO GNEISS

The Novillo Gneiss Complex, exposed in northeastern Mexico (Fig. 1), is fairly small, with a total outcrop area of only 35 km^2, and the best exposures of orthogneisses are found in the

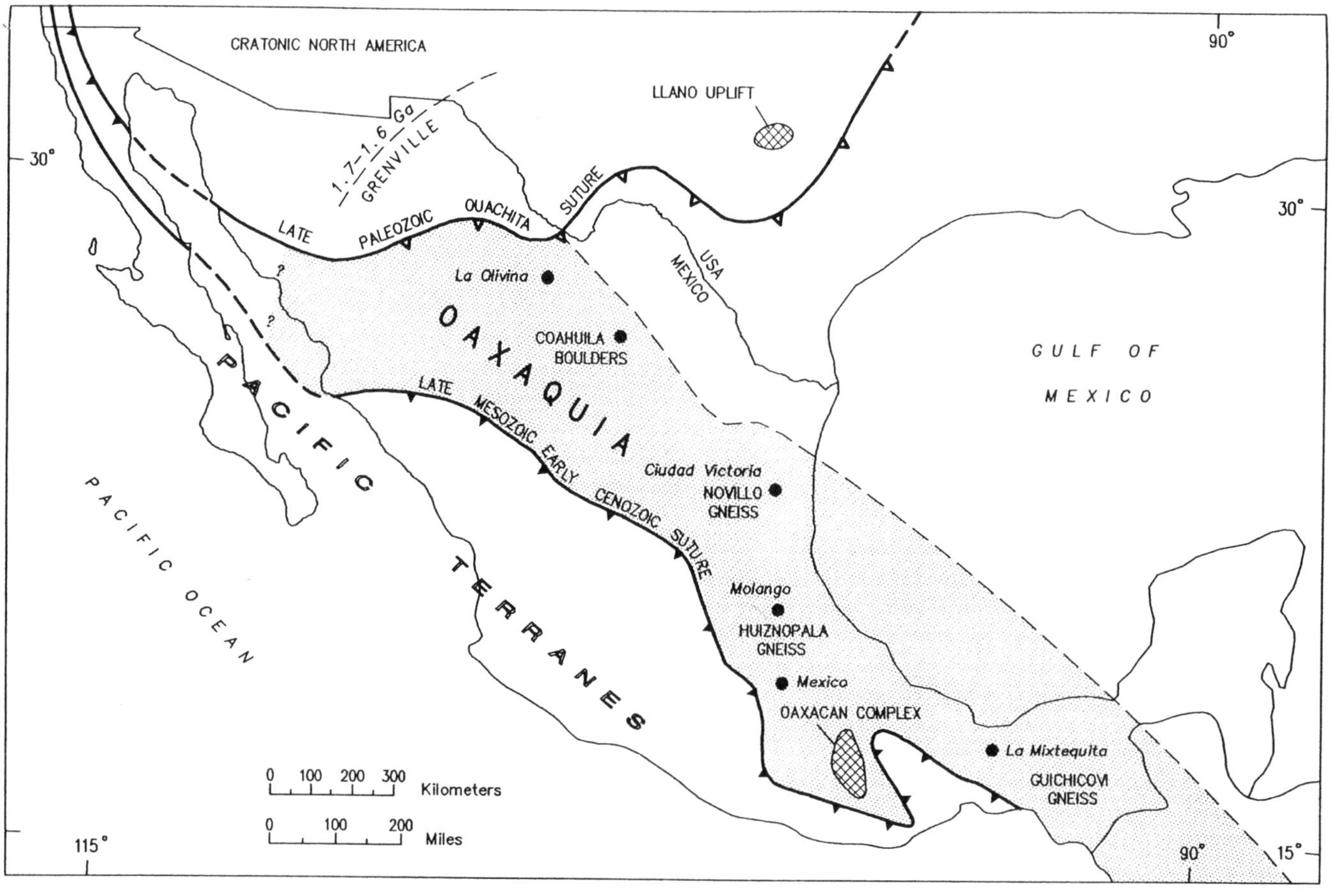

Figure 1. Location map. The four exposures of Grenville granulites in eastern and southern Mexico are the Novillo Gneiss, Huiznopala Gneiss, Oaxacan Complex, and Guichicovi Gneiss. Pluma Hidalgo, mentioned in the text, is in the southern Oaxacan Complex. Cobbles and boulders of Grenville granitoids are found in northern Mexico in the state of Coahuila. The Ouachita suture that runs from Texas southwestward into Mexico was formed by the late Paleozoic collision of Laurentia and Gondwana.

creek bed of Novillo Canyon. The complex was mapped in some detail along streams by Ramirez-Ramirez (1992) at a scale of 1:50,000.

Field relations and U-Pb geochronology results presented later in this paper show that the orthogneisses of Novillo Canyon can be divided into an older group of gabbroic and potassic granulites and a younger anorthosite-mangerite-charnockite-granite (AMCG) suite. All these rocks have been overprinted by granulite-facies metamorphism.

The older gabbroic granulites (Fig. 2) contain the primary assemblage plagioclase + clinopyroxene + orthopyroxene + garnet. The older potassic granulites are intermediate to silicic in composition, and are characterized by high K_2O (Table 1, N-5 and N-10) and abundant quartz and microcline. Most specimens contain fresh garnet, and one sample has fresh clinopyroxene. These rocks may have been charnockites, but the pyroxenes in most specimens are completely altered, and no fresh hypersthene was found. One of the analyzed potassic granulites plots in the within-plate granitoid field on the Nb-Y discriminant diagram, and the other falls within the volcanic-arc granitoid field (Fig. 3).

Granites of the AMCG suite are very leucocratic and occur primarily as centimeter- to meter-scale dikes that intrude the older rocks, most conspicuously the gabbroic granulites (Fig. 2). The three analyzed granites all plot within the volcanic-arc granitoid field (Fig. 3). However, these granites may have been crustal melts; consequently, where they plot on the Nb-Y discriminant diagram may have no tectonic significance.

The ferrogabbroic granulites occur as meter-scale dikes that intrude the anorthosites and all other units mentioned above.

Figure 2. An outcrop of the Novillo Gneiss in Novillo Canyon. Light dikes of AMCG granite intruded dark gabbroic granulite parallel to foliation in the older rock.

These dikes are similar in composition to rocks from AMCG suites of Canada and the Adirondacks, variously referred to as ferrograbbros, ferrodiorites, and low-Al gabbros, but they have even higher TiO_2 and P_2O_5 concentrations (Tables 1 and 2).

The following is a summary of events recognized by field relations in the Novillo Gneiss: (1) emplacement of the older gabbroic and potassic suites; (2) folding; (3) emplacement of the anorthosites and granites of the AMCG suite; (4) emplacement of the ferrogabbroic dikes that cut across the granitic dikes; (5) polyphase deformation [foliation (S1)], which produced two

TABLE 1. ANALYSES OF THE NOVILLO GNEISS

	Sample number												
	Older gabbroic granulites			Potassic granulites		Granites			Ferrogabboric granulite dikes				
	N-32	N-16	N-31	N-10	N-5	N-19	N-8	N-6	N-14	N-33	N-18	N-20	N-13
SiO_2	48.76	51.91	53.33	57.41	67.36	72.17	74.47	75.46	47.96	50.21	51.59	51.82	52.40
TiO_2	2.12	1.28	0.44	2.28	0.89	0.07	0.06	0.10	6.20	5.57	5.27	5.67	5.09
Al_2O_3	13.87	17.27	19.05	15.85	14.97	14.96	13.90	13.38	12.70	13.38	12.90	14.02	13.00
FeO*	16.45	13.15	8.41	10.32	5.30	0.60	0.48	0.93	12.92	13.90	12.59	10.85	11.01
MnO	0.27	0.20	0.16	0.09	0.08	0.01	0.01	0.01	0.15	0.22	0.16	0.10	0.11
MgO	5.87	5.33	4.97	1.99	0.58	0.30	0.21	0.16	3.24	3.47	3.45	3.50	2.20
CaO	9.41	7.03	9.04	3.97	2.43	0.71	1.00	1.65	8.95	8.48	7.37	8.06	13.57
Na_2O	2.56	2.70	3.74	3.16	2.97	3.01	3.24	2.91	2.68	2.02	2.41	2.95	0.14
K_2O	0.37	0.72	0.72	4.00	5.14	8.00	6.58	5.24	1.72	0.15	1.82	0.16	0.03
P_2O_5	0.32	0.42	0.13	0.92	0.27	0.16	0.05	0.15	3.50	2.62	2.44	2.87	2.45
Fe*/(Fe* + Mg)	0.61	0.58	0.49	0.74	0.84	0.53	0.56	0.76	0.69	0.69	0.67	0.63	0.74
Nb	7	11	4	35	10	3	3	2	30	30	25	27	26
Zr	66	85	56	796	1140	30	56	221	663	549	429	596	481
Y	35	25	19	74	33	9	3	5	57	50	45	40	44
Sr	122	157	330	418	297	164	150	195	763	519	711	603	2173
Rb	14	16	13	60	95	171	112	88	19	7	26	10	5

Note: Major elements recalculated to 100% anhydrous. Units for chemical compositions are wt%; those for elements are ppm. Total Fe reported as FeO*.

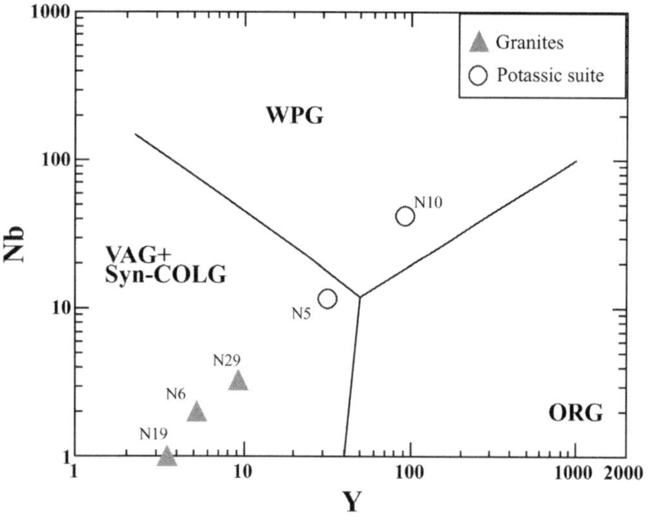

Figure 3. The Nb-Y tectonic discriminant diagram for granitic rocks. The numbers associated with each symbol are sample numbers. ORG—orogenic granites; Syn-COLG—syn-collisional granites; VAG—volcanic-arc granites; WPG—within-plate granites. From Pearce et al. (1984).

sets of folds [F2 and F3] that deform the AMCG suite; and (6) granulite metamorphism coeval with deformation.

GEOCHRONOLOGY OF THE NOVILLO GNEISS

Older Rocks

Sample N-10 is an augen gneiss from the potassic suite. Zircons from N-10 are typically multifaceted and range in morphology from roughly spherical to ovate to elongate. Cathodoluminescence images of typical grains from N-10 are shown in Figures 4 and 5, A and B. These grains show igneous zoning without obvious metamorphic overgrowths. Seven fractions were dated, and all except one (Table 3, M) were single grains. The spherical to ovate grains yielded the oldest ages. Fractions O1 and O2 were fragments of one grain that broke up during abrasion and were dated separately. Both fragments are concordant at 1165 ± 2 and 1167 ± 4 Ma, respectively. The oldest Pb-Pb age is 1174 ± 1 Ma, and a regression of all the data has an upper intercept of 1175 ± 16 Ma (Fig. 4, A), which we interpret as the crystallization age of the sample. The lower intercept is consistent with the age of granulite-facies metamorphism, discussed below.

Sample N-5 is another potassic granulite, and the U-Pb results for it are very similar to those from N-10. Two fractions are only 0.3% discordant at ca. 1165 Ma, but there is one older fraction with a Pb-Pb age of 1181 ± 4 Ma. Regression of all the data yields an upper intercept of 1184 ± 30 Ma, with a mean square of weighted deviates (MSWD) of 4.3 (Fig. 4, B). Omitting fraction J (Fig. 4) produces an upper intercept of 1173 ± 21 Ma with a MSWD of 2.6 (not shown). The grain shown in Figure 5, C, appears to have an igneous core with a wide, homogenous metamorphic rim, whereas that shown in Figure 5, D, exhibits only igneous structures.

The older gabbroic granulites have low Zr concentrations, and only a few zircon grains were recovered from N-32, the single sample from this group that was processed. Regression of the three youngest fractions produces an upper intercept of ca. 1000 Ma (with a huge error), which is near the time of granulite metamorphism. Thus, these fragments are interpreted as metamorphic zircons that have been variably affected by more recent Pb loss (Fig. 4, C). The remaining fraction, F2, is ~2% discordant, with a Pb-Pb age of 1151 ± 2 Ma. A regression of this frac-

TABLE 2. COMPARISION OF AMCG GABBROIC ROCKS

	Adirondacks	Quebec	Labrador	Labrador	Mexico
Unit	Marcy	Morin	Harp Lake	Harp Lake	Novillo
Rock name	Ferrogabbro	Ferrogabbro	Low-Al gabbro	Ferrodiorite	Ferrogabbro
Number of samples in average	5	4	17	17	5
SiO_2	41.35	47.63	48.77	49.93	50.80
TiO_2	4.84	3.11	2.60	3.03	5.56
Al_2O_3	13.00	16.11	14.66	14.41	13.20
FeO*	21.58	14.42	14.81	15.73	12.25
MnO	0.31	0.22	0.22	0.23	0.15
MgO	4.58	3.42	5.96	3.40	3.17
CaO	9.74	10.11	9.47	8.02	9.29
Na_2O	2.16	3.05	2.68	3.30	2.04
K_2O	0.72	0.87	0.56	1.20	0.77
P_2O_5	1.72	1.04	0.26	0.74	2.77
Fe*/(Fe* + Mg)	0.73	0.70	0.58	0.72	0.68

Note: Major elements recalculated to 100% anhydrous. Units for chemical compositions are wt%.Total Fe reported as FeO*. Data are from Ashwal (1993) and Table 1.

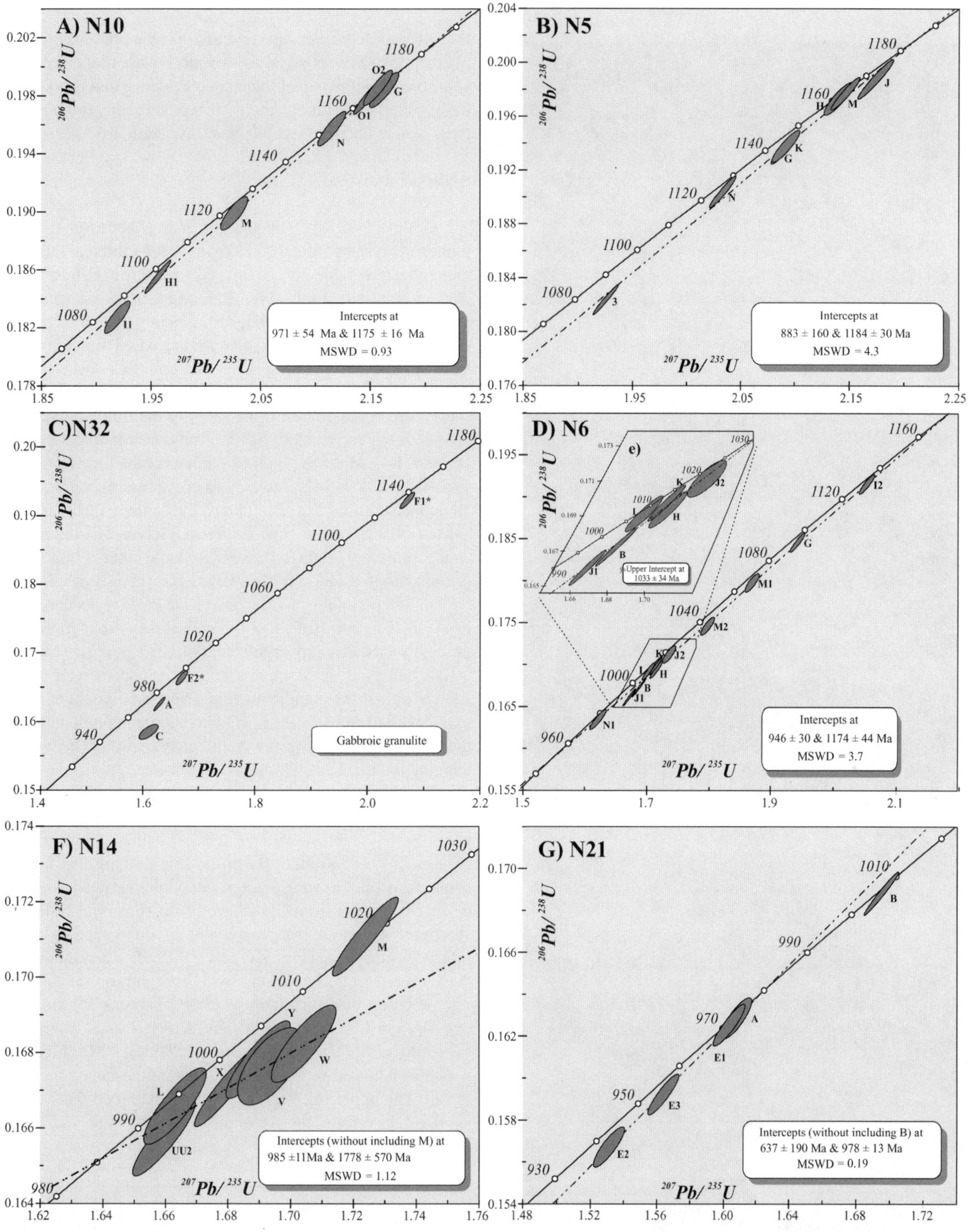

Figure 4. Concordia diagrams with U-Pb analyses for zircons from the Novillo Gneiss. Ellipses indicate 2σ errors. Intercepts were calculated using the program of Ludwig (1998). See Table 3 and text for a description of the samples. (A) Sample N-10; (B) sample N-5; (C) sample N-32; (D) sample N-6; (E) inset in panel D, showing a detail from that panel; (F) sample N-14; (G) sample N-21.

tion with a fixed lower intercept of 1000 Ma produces an upper intercept of 1247 ± 48 Ma. All we can say about the age of this gabbroic granulite is that there is no evidence that it is substantially older than other dated Mesoproterozoic orthogneisses from eastern and southern Mexico (discussed in a later section).

AMCG Suite

Granite N-6 was selected for dating, because it had the highest Zr concentration—221 ppm—of the three granite samples collected (Table 2). The imaged grains from N-6 have complex internal structures (Fig. 5, E and F); we interpret most grains to be dominated by igneous rather than metamorphic components. The morphologies of the dated fractions range from spherical to stubby prisms, but there was no clear relationship between age and morphology. The results for this sample are rather complex, but we offer the following interpretations. A regression of all fractions produces an upper intercept of 1174 ± 44 Ma (Fig. 4, D). We interpret the four oldest fractions (I2, G, M1, M2; Table 3) to contain an inherited zircon component similar in age to that of the "older rocks" discussed above (i.e., ca. 1.2 Ga). The youngest fraction, N1, has a low U concentration (97 ppm) compared with the others (Table 3) and is concordant at 982 ± 6 Ma. We interpret it as a metamorphic zircon. The remaining six fractions have Pb-Pb ages that cluster between 1028 and 1012 Ma (Table 3). If they are regressed as a group, they yield an upper intercept of 1033 ± 34 Ma (Fig. 4, D, inset labeled E), which we interpret as the approximate crystallization age of the granitic magma. Even these younger zircons may contain some of the ca. 1.2-Ga inherited component; thus, the crystallization age of the granite may be somewhat younger than the intercept age.

The ferrogabbroic granulite dikes as a group have rather high Zr concentrations (compared, e.g., with the older gabbroic granulites) (Table 2). We selected N-14, the sample with the highest Zr concentration, for processing, hoping that it would contain igneous zircon. However, cathodoluminescence revealed that most zircons from N-14 have broad rims with little or no structure, suggesting that most grains are dominated by a metamorphic component (Fig. 5, G and H). The U-Pb systematics of the analyzed zircons and their low U concentrations support the interpretation that all fractions (except perhaps M) are dominated by a metamorphic component. A regression of all fractions except M yields a lower intercept of 985 ± 11 Ma (Fig. 4, F), which is believed to be the age of granulite-facies metamorphism. The upper intercept is poorly constrained but is within approximate error of the age of the older suite of rocks. Fraction

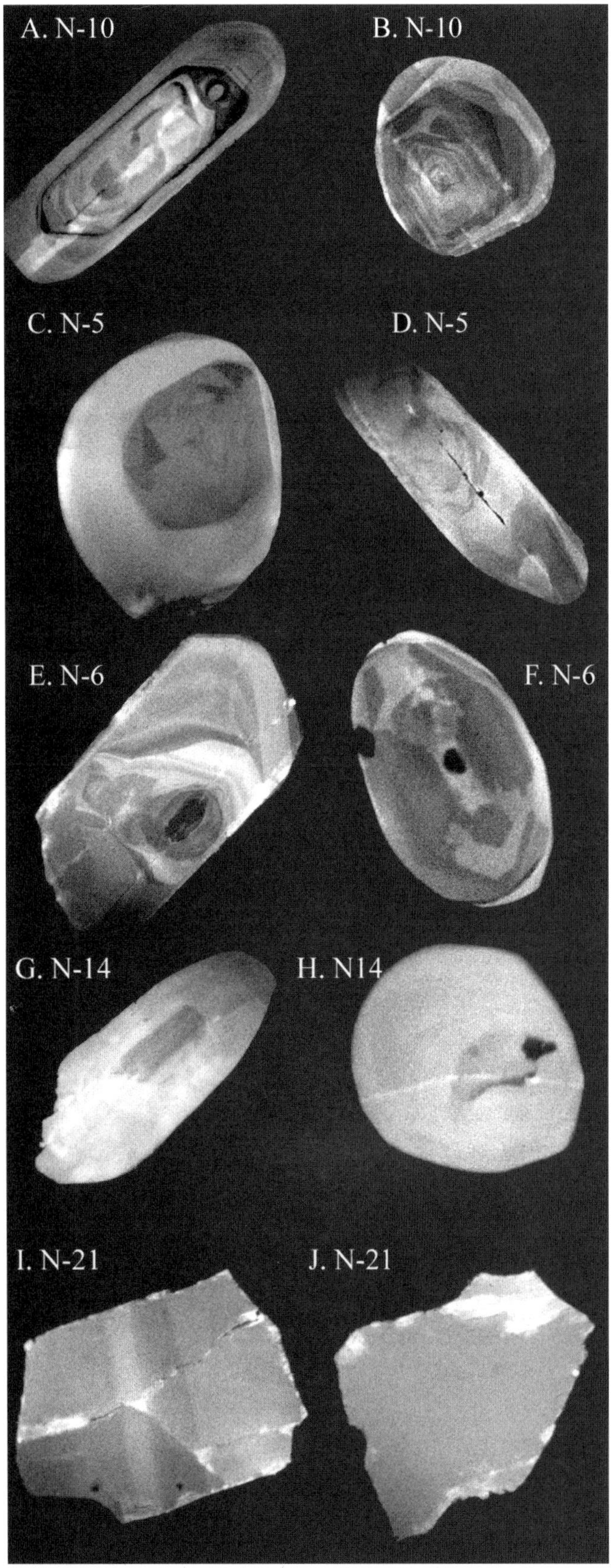

Figure 5. Cathodoluminescence images of zircons from the Novillo Gneiss. See Table 3 and text for a description of the samples. (A, B) Sample N-10; (C, D) sample N-5; (E, F) sample N-6; (G, H) sample N-14; (I, J) sample N-21.

TABLE 3. U-PB GEOCHRONOLOGICAL DATA FOR THE NOVILLO GNEISS

Fraction	Description	Weight (μg)	U (ppm)	Total Pb (ppm)	Common Pb (pg)	$^{206}Pb/^{204}Pb$ Measured	$^{207}Pb^*/^{206}Pb^*$ Atomic ratio	$^{207}Pb^*/^{206}Pb^*$ Error (%)	$^{208}Pb^*/^{206}Pb^*$ Atomic ratio	$^{206}Pb^*/^{238}U$ Atomic ratio	$^{206}Pb^*/^{238}U$ Error (%)	$^{207}Pb^*/^{235}U$ Atomic ratio	$^{207}Pb^*/^{235}U$ Error (%)	Age (Ma) $^{206}Pb^*/^{238}U$	Age (Ma) $^{207}Pb^*/^{235}U$	Age (Ma) $^{207}Pb^*/^{206}Pb^*$
N-5, potassic granulite																
G	1:1.5 (1)	78	138	28	31	17270	0.07831	0.04	0.11576	0.19359	0.10	2.0903	0.10	1141	1146	1155 ± 1
H	ovate, abr (2)	150	143	29	33	7820	0.07867	0.04	0.12574	0.19725	0.07	2.1397	0.08	1161	1162	1164 ± 1
3	1:4 (3)	94	89	17	5	10494	0.07652	0.06	0.13414	0.18235	0.27	1.9239	0.28	1080	1089	1109 ± 1
J	ovate, abr (1)	23	55	12	13	1288	0.07935	0.17	0.13514	0.19869	0.60	2.1738	0.62	1168	1173	1181 ± 4
K	ovate, abr (1)	60	150	31	37	2921	0.07831	0.05	0.12397	0.19365	0.12	2.0910	0.13	1141	1146	1155 ± 1
M	sph, abr (1)	54	151	31	26	3923	0.07875	0.05	0.11803	0.19764	0.16	2.1460	0.16	1163	1164	1166 ± 1
N	sph, abr (2)	74	106	21	6	16664	0.07744	0.04	0.12029	0.19034	0.13	2.0323	0.13	1123	1126	1133 ± 1
N-6, granite																
G	1:1.5, abr (1)	59	327	68	14	15640	0.07635	0.03	0.22284	0.18448	0.09	1.9420	0.09	1091	1096	1104 ± 1
H	1:3, abr (1)	20	222	47	33	1444	0.07337	0.08	0.34232	0.16934	0.21	1.7130	0.22	1008	1013	1024 ± 2
B	1:2 (1)	20	338	63	19	3644	0.07308	0.07	0.21680	0.16721	0.54	1.6848	0.55	997	1003	1016 ± 2
I2	1:2, abr (1)	38	408	87	18	10127	0.07775	0.04	0.21301	0.19143	0.08	2.0522	0.09	1129	1133	1141 ± 1
J1	sph, mf (10)	76	203	38	16	10078	0.07291	0.04	0.22849	0.16603	0.20	1.6690	0.21	990	997	1011 ± 1
J2	sph, mf (4)	32	214	41	9	7919	0.07351	0.93	0.22704	0.17113	0.16	1.7344	0.19	1018	1021	1028 ± 2
K	1:2, abr (1)	36	309	59	12	9861	0.07317	0.05	0.23297	0.16973	0.20	1.7124	0.21	1011	1013	1019 ± 1
L	1:2, abr (1)	21	294	58	13	5117	0.07292	0.07	0.28552	0.16911	0.19	1.7001	0.20	1007	1009	1012 ± 2
M1	1:2, abr (1)	33	282	60	35	2922	0.07543	0.09	0.28457	0.17966	0.14	1.8685	0.17	1065	1070	1080 ± 2
M2	1:2, abr (1)	22	305	59	6	10925	0.07462	0.09	0.21847	0.17451	0.14	1.7954	0.17	1037	1044	1058 ± 2
N1	1:2, abr (1)	16	97	19	16	1001	0.07186	0.28	0.27042	0.16342	0.61	1.6191	0.67	976	978	982 ± 6
N-10 augen gneiss																
I1	1:3, abr (1)	41	289	55	22	6206	0.07613	0.05	0.12122	0.18270	0.09	1.9177	0.10	1082	1087	1099 ± 1
H1	sph, mf, abr (1)	40	388	79	33	5638	0.07644	0.04	0.19480	0.18549	0.16	1.9548	0.16	1097	1100	1107 ± 1
G	sph, mf, abr (1)	59	258	54	25	7411	0.07906	0.04	0.12897	0.19837	0.07	2.1624	0.08	1167	1169	1174 ± 1
M	1:3, abr (3)	45	235	47	11	11187	0.07737	0.08	0.13602	0.18986	0.11	2.0254	0.14	1121	1124	1131 ± 2
N	sph, mf, abr (1)	25	316	64	11	8634	0.07837	0.07	0.12055	0.19569	0.13	2.1145	0.15	1152	1154	1156 ± 1
O1	ovate, abr (1)	30	210	43	8	10192	0.07871	0.07	0.11987	0.19791	0.24	2.1478	0.25	1164	1164	1165 ± 2
O2	fragment of above	13	201	43	13	945	0.07880	0.22	0.11504	0.19852	0.47	2.1569	0.52	1167	1167	1167 ± 4
N-14, ferrogabbroic granulite																
L	1:2, abr (1)	82	33	6	8	3784	0.07239	0.19	0.10821	0.16656	0.32	1.6625	0.37	993	994	997 ± 4
M	eq to ov, abr (11)	62	72	13	12	4036	0.07308	0.09	0.11014	0.17107	0.24	1.7238	0.25	1018	1017	1016 ± 2
UU2	sph, mf (6)	123	24	4	20	1529	0.07260	0.17	0.08643	0.16573	0.35	1.6590	0.39	989	993	1003 ± 3
V	flatish (6)	121	38	7	18	2692	0.07326	0.36	0.10302	0.16762	0.29	1.6931	0.46	999	1006	1021 ± 7
W	1:2 (9)	117	50	9	49	1241	0.07346	0.16	0.10843	0.16822	0.24	1.7039	0.29	1002	1010	1027 ± 3
X	sph, mf (6)	142	32	5	13	3517	0.07290	0.10	0.10006	0.16705	0.28	1.6791	0.30	996	1001	1011 ± 2
Y	ovate, mf (6)	127	42	7	11	4879	0.07301	0.17	0.11151	0.16780	0.32	1.6893	0.37	1000	1004	1014 ± 4
N-21, anorthosite pegmatite																
A	frag, abr (1)	97	104	18	33	5482	0.07160	0.23	0.14461	0.16281	0.34	1.6074	0.41	972	973	975 ± 5
B	frag, abr (1)	133	79	14	75	4531	0.07283	0.06	0.15332	0.16878	0.30	1.6948	0.31	1005	1007	1010 ± 1
E3	frag, abr (2)	26	99	16	10	2591	0.07122	0.18	0.12888	0.15919	0.40	1.5633	0.43	952	956	964 ± 4
E1	frag, abr (1)	52	82	14	8	5758	0.07154	0.06	0.11970	0.16250	0.12	1.6030	0.13	971	971	973 ± 2
E2	frag, abr (1)	33	84	14	7	3710	0.07083	0.16	0.13231	0.15668	0.28	1.5303	0.33	938	943	953 ± 4
N-32, mafic granulite																
C	frag, dk, abr (1)	11	239	41	23	1130	0.07362	0.62	0.15543	0.15840	0.58	1.6079	0.86	948	973	1031 ± 13
A	sph, abr (1)	24	299	55	56	1304	0.07266	0.07	0.20550	0.16245	0.23	1.6275	0.24	970	981	1004 ± 2
F1	frag (1)	29	352	71	10	11927	0.07817	0.09	0.13504	0.19214	0.16	2.0708	0.18	1133	1139	1151 ± 2
F2	frag (3)	22	960	164	28	7769	0.07274	0.08	0.11307	0.16630	0.11	1.6680	0.13	992	996	1007 ± 2

Note: Number in parenthesis is number of grains; aspect ratio of elongate grains given. Abbreviations: abr—abraded; dk—dark; eq—equant; frag—angular fragment; mf—multifaceted; sph—spherical; ov—ovate.

*Ratios corrected for fractionation, 10-pg lab blank, and initial common Pb is from feldspars; ages and errors calculated using Pbdat (Ludwig, 1991).

M is concordant at 1016 ± 2 Ma. This may be the crystallization age of the dike, but we doubt it for the following reasons. First, the U concentration of fraction M is only slightly higher than that of other zircons from the rocks (Table 3). Second, fraction M consisted of eleven grains indistinguishable in morphology from most other grains from the sample. It seems unlikely that this one multigrain fraction would contain only igneous zircons. Third, if the rock contained fairly abundant igneous zircons with an age of ca. 1016 Ma, then we would expect to find analyzed fractions that contained both igneous and metamorphic components lying on a chord between ca. 1016 Ma and ca. 985 Ma. In practice, that chord would be indistinguishable from concordia, and there are no clear examples of such fractions. Our preferred interpretation of fraction M is that it contained both an inherited igneous and a metamorphic component. However, because fraction M lies well off the discordia defined by the other fractions from this rock (Fig. 4, F), we interpret the inherited component in M to be significantly younger than the inherited component present in the other fractions. It may have been a xenocrystic zircon from some older AMCG-suite wall rock, perhaps with an age similar to the granites (e.g., ca. 1035 Ma). In any case, the Pb-Pb age of fraction M would be the minimum age of the inherited component.

A sample of quartz-bearing anorthositic pegmatite (N-21) was collected that contained millimeter-size zircons that were visible in the hand-specimen. The zircon fragments from this pegmatite appear to be metamorphic, because they are typically homogenous and structureless under cathodoluminescence, except for broad bands in some (Fig. 5, I and J). Regression of the five analyzed fractions yields the rather useless upper intercept age of 1006 ± 54 Ma (not shown). Regression of the four youngest fractions alone results in an upper intercept of 978 ± 13 Ma (Fig. 4, G). The remaining fraction, B, is ~0.5% discordant with a Pb-Pb age of 1010 ± 1 Ma and is interpreted to contain an inherited igneous component.

Previous Geochronologic Studies

The only previous U-Pb date from the Novillo Gneiss was published in an abstract by Silver et al. (1994). They stated that zircons from a garnetiferous charnockite yielded a concordant age of 1018 ± 3 Ma. They did not discuss the significance of the age, but we offer two possible interpretations. They may have collected a charnockite from the AMCG suite containing zircons that preserved the igneous crystallization age. Alternatively, their sample may have been from the older potassic suite, and the 1018-Ma date may represent a mixture of igneous and metamorphic zircons. Only two other geochronologic studies of the Novillo Gneiss have been published in the past 35 years. Patchett and Ruiz (1997) determined Sm-Nd garnet–whole-rock cooling ages of 887 ± 3 Ma and 904 ± 4 Ma. Denison et al. (1971) reported two hornblende K-Ar ages from granitic gneisses of 871 ± 18 Ma and 910 ± 18 Ma, and a K-Ar age of white mica from a marble of 920 Ma.

REVIEW OF U-PB GEOCHRONOLOGY OF BASEMENT ROCKS FROM EASTERN AND SOUTHERN MEXICO

A chart of all available igneous crystallization ages from eastern and southern Mexico is shown in Figure 6. The granitic cobbles and boulders from Coahuila have igneous textures, although some have experienced low-grade metamorphism. All remaining samples are orthogneisses metamorphosed to granulite facies during the Grenville orogeny.

There is only scattered evidence for crust older than 1300 Ma south of the Ouchita suture (Fig. 1). Three of the Coahuila samples, all banded granitoids (Lopez et al., 2001), contain inherited zircons that yield rather poorly constrained upper intercept ages between ca. 2000 Ma and ca. 1800 Ma (Fig. 6). In the areas of exposed granulites, protolith ages older than 1300 Ma have only been found in the Oaxacan complex. The paleosome of a migmatite in the northern part of the complex has an upper intercept age of 1400 ± 58 Ma (Solari et al., 2003), and two orthogneisses from Pluma Hildalgo, in the southern part of complex, have upper intercept ages of 1384 ± 60 Ma and 1311 ± 41 Ma.

The remaining Mesoproterozoic granitoids and orthogneisses can be divided into two groups, probably separated in age by ~100 m.y. The older group generally has arc and backarc geochemical signatures (Lawlor et al., 1999; Keppie et al., 2001, 2003). This phase of activity was under way by 1230 Ma and lasted until at least 1117 Ma. Rocks of this arc-backarc group are found in each study area, and there is no obvious variation in ages related to geographical position.

Anorthosites and related rocks (i.e., the AMCG suite) are found in each basement exposure, but this suite has proven difficult to date. The most convincing age was obtained from a garnet-orthopyroxene mafic gneiss (sample 6498; Fig. 6) associated with an anorthosite body in the northern Oaxacan complex. Zircons from this sample are prismatic, with aspect rations of 4:1; under cathodoluminescence, they show typical igneous growth zoning with narrow metamorphic rims (Keppie et al., 2003). Results from four single abraded crystals yield an upper intercept age of 1012 ± 12 Ma, which is inferred to be the time of intrusion. As previously discussed, granite sample N-6 from the Novillo Gneiss is interpreted to have a crystallization age of 1033 ± 34 Ma. Three anorthositic samples—N-21, this study; PJL6, Lawlor et al. (1999); and 6398, Keppie et al. (2003)—have been dated, but the results are not shown on Figure 6, because the zircons are dominantly metamorphic.

Solari et al. (2003) referred to the Grenville orogeny in Mexico as the "Zapotecan tectonothermal event." Geochronologic results relevant to that event are shown in Figure 7. Four samples were dated by intercepts (indicated on Figure 7); all other ages are based on concordant results. The end of the Zapotecan event in the Oaxacan complex is well constrained by ages of post-tectonic pegmatites (Fig. 7). These pegmatites are samples 66D98 (978 ± 3 Ma) from northern Oaxaca (Solari et al.,

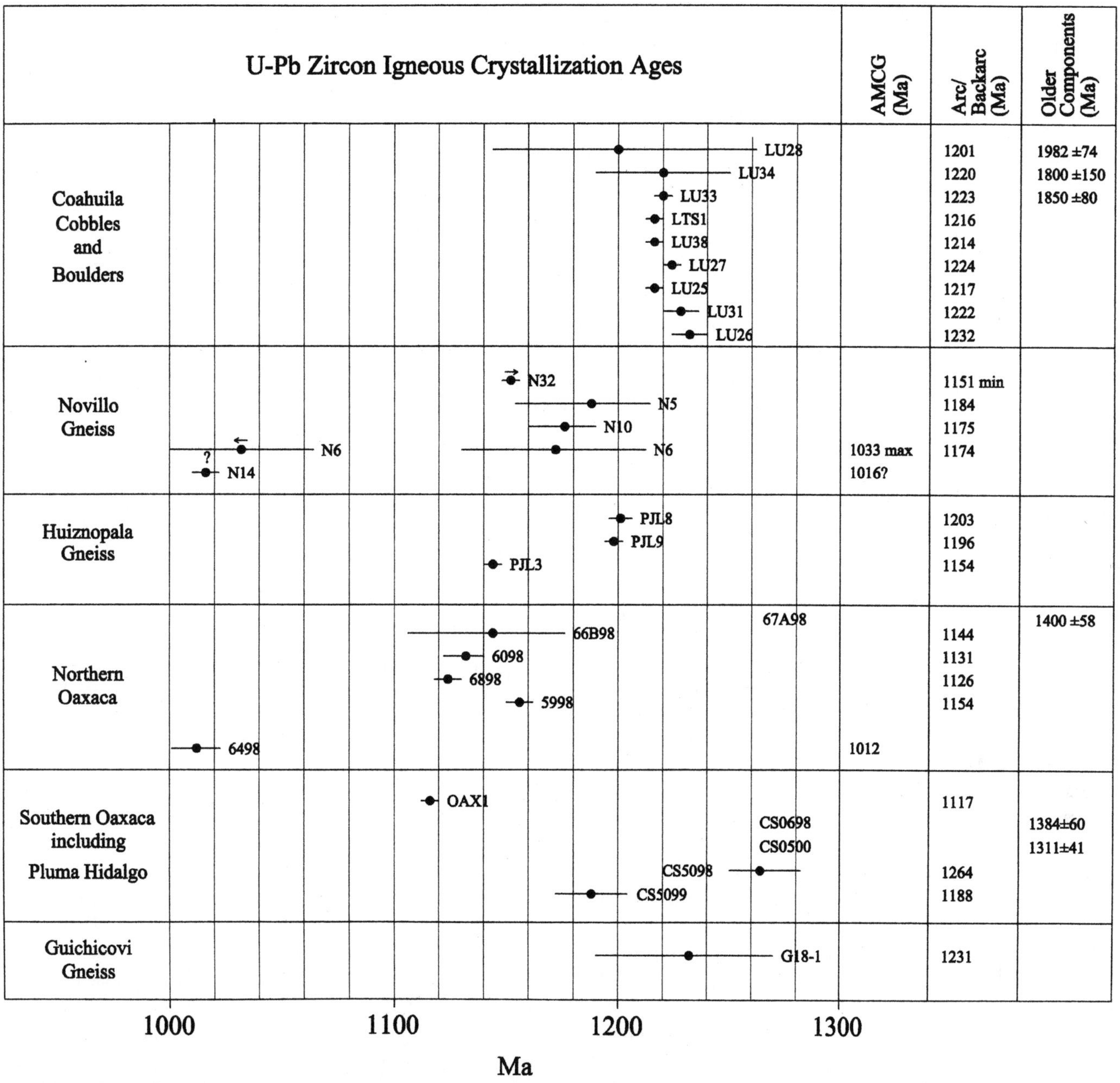

Figure 6. U-Pb zircon igneous crystallization ages from Grenville rocks from eastern and southern Mexico. Sample numbers are shown for each point. Horizontal line through the point is the reported uncertainty on the age in m.y. Left-pointing arrow over the dot indicates a maximum age; right-pointing arrow a minimum age; see text for discussion of sample N-14. The older component ages from Coahuila are upper intercept ages of inherited components in the top three arc-backarc granitoids. The older component ages from Oaxacan are interpreted to be protolith crystallization ages. All dating except that from Guichicovi Gneiss was done at the University of California, Santa Cruz. Data are from: Coahuila boulders, Lopez et al. (2001); Novillo Gneiss, this study; Huiznopala Gneiss, Lawlor et al. (1999); northern Oaxaca, Keppie et al. (2003), Solari et al. (2003); southern Oaxaca, Keppie et al. (2001); Pluma Hidalgo, Schulze, Cameron, and Lopez (unpublished data); Guichicovi Gneiss, Weber and Kohler (2001).

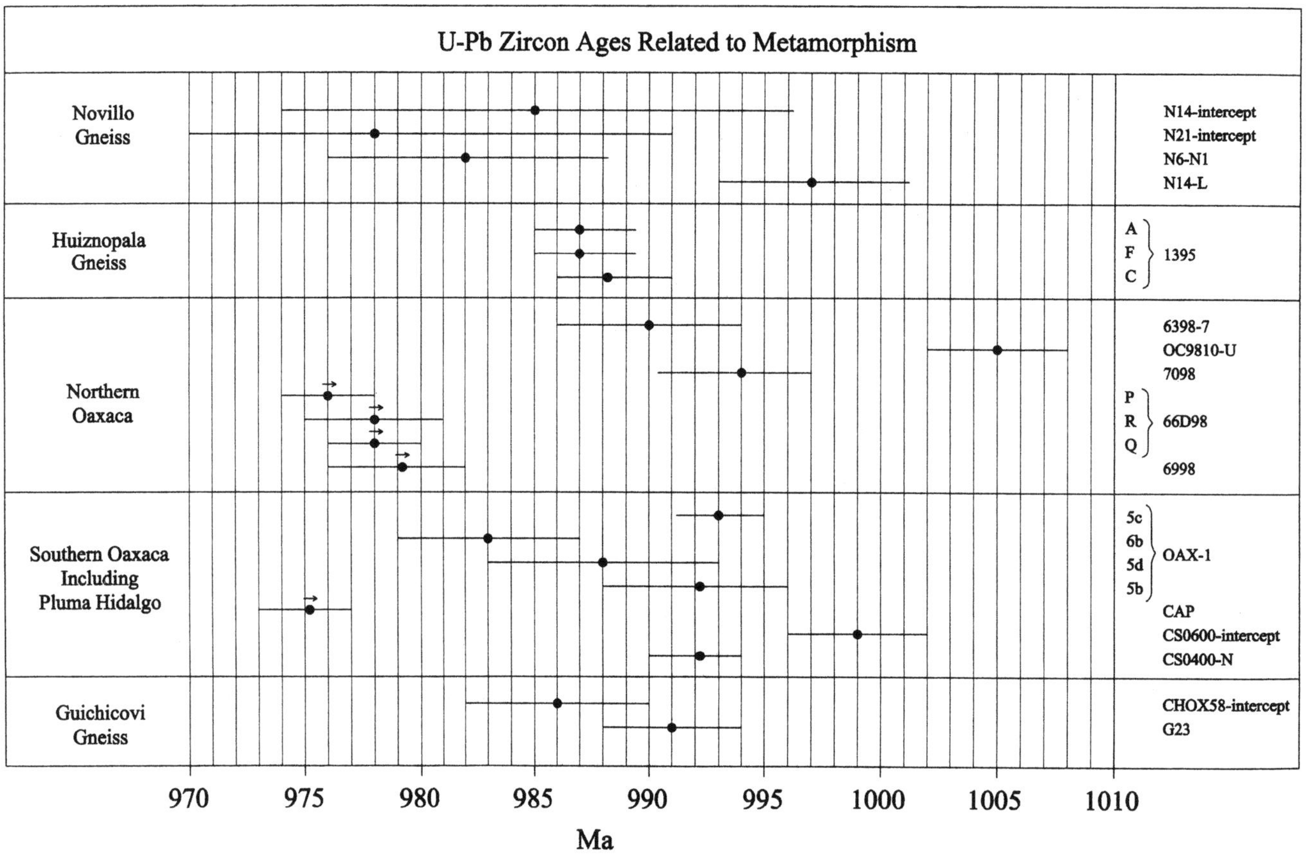

Figure 7. U-Pb zircon ages related to metamorphism. Samples numbers are shown in the right-hand column, and ages from intercepts are identified. Other ages are from concordant results, generally from single grains. Where there is more than one age from a sample, the fraction number is given. The horizontal line through the point is the reported uncertainty on the age in m.y. Right-pointing arrows above are ages from post-tectonic pegmatites. Data sources are the same as Figure 6, except for sample CHOX58 (Guichicovi), which is from Ruiz et al. (1999).

2003) and CAP (975 ± 2 Ma) from Pluma Hidalgo, in the southern part of the Oaxaca complex (Schulze, 2003, personal commun.). Solari et al. (2003) described structural aspects of a third pegmatite, 6998 (979 ± 3 Ma), that suggest that it also postdates deformation. Two-thirds of the concordant dates older than these pegmatites lie between 995 and 985 Ma; in fact, all metamorphic ages, except the two oldest, are within error of 990 ± 5 Ma. The two oldest fractions have ages of 999 ± 3 Ma and 1004 ± 3 Ma, but these may contain a small component of inherited igneous zircon. Like the igneous ages, there is no obvious variation in metamorphic ages related to geographical position.

PB ISOTOPIC COMPOSITIONS OF LEACHED FELDSPARS

Results of Pb isotopic analyses of hydrofluoric acid–leached feldspars from all four exposures of Grenville granulites in Mexico and from the Coahuila granitoid boulders are presented in Table 4. Potassium feldspars generally have higher concentrations of Pb than plagioclase feldspars, and consequently, their Pb isotopic compositions are less sensitive than plagioclase to contamination by trace amounts of radiogenic phases. The boundary of the field shown in Figure 8 was drawn based on the twenty-three analyses of potassium feldspar reported in Table 4. Compositions of twelve of the seventeen plagioclase and plagioclase + potassium feldspar mixtures (Table 4) also lie in that field, and they too are plotted in Figure 8. The remaining five Mexican samples have Pb isotopic compositions more radiogenic than the potassium feldspar field (Fig. 9; Table 4); we suspect that their isotopic compositions may be affected by trace inclusions of radiogenic minerals. The feldspar results define a remarkably linear array on Figure 8. The compositions of most of the feldspars, including multiple samples from all localities, cluster near the center of the field. There is no clear variation related to geographical position, although six of the seven most radiogenic feldspars shown in Figure 8 are from Coahuila and Guichicovi.

The linear array depicted in Figure 8 could have originated in at least two ways: (1) formation of a isotopically homogenous crust at ca. 1200 Ma, followed by development of a secondary isochron between 1200 Ma and 1000 Ma, when the feldspars last equilibrated with the host rock during the granulite metamor-

TABLE 4. PB ISOTOPIC COMPOSITIONS OF FELDSPARS

Sample	$^{206}Pb/^{204}Pb$	$^{207}Pb/^{204}Pb$	$^{208}Pb/^{204}Pb$	Type
Coahuila boulders				
LTS1	17.256	15.500	36.752	K-spar
LU27	17.150	15.470	36.327	K-spar
LU38	17.221	15.486	36.569	Plag + K-spar
LU25	17.199	15.483	36.481	K-spar
LU26	17.631	15.550	37.310	K-spar
LU31	17.204	15.478	36.697	K-spar
LU37	17.440	15.530	37.130	K-spar
LU28	17.285	15.527	36.935	K-spar
LU29	17.424	15.527	36.557	Plag + K-spar
LU33	17.326	15.509	36.758	K-spar
LU34*	18.979	15.645	39.256	Plag
Novillo Gneiss				
4183	17.268	15.486	36.770	Plag
4177*	17.861	15.554	37.447	Plag
4180	17.190	15.490	36.636	K-spar
4178	17.556	15.533	37.425	Plag + K-spar
N-5	17.093	15.457	36.501	K-spar
N-6	17.171	15.484	37.123	K-spar
N-10	17.281	15.474	36.636	K-spar
N-14	17.179	15.485	36.624	Plag
N-32	17.550	15.502	36.954	Plag
Huiznopala Gneiss				
PJL3	17.361	15.498	36.755	K-spar
PJL6	17.090	15.465	36.411	Plag
PJL8	17.251	15.494	36.420	Plag + K-spar
PJL9	16.933	15.442	36.266	K-spar
PJL10	17.306	15.485	36.309	Plag
Oaxacan Complex				
Oax001	16.945	15.471	36.403	K-spar
CON215	17.221	15.501	36.602	Plag
6298	17.236	15.502	36.590	Plag
6898	17.294	15.506	36.571	K-spar
6998	17.151	15.483	36.608	K-spar
6098	17.031	15.474	36.403	K-spar
66B98	17.338	15.501	36.778	K-spar
66B98B	17.233	15.515	36.813	K-spar
CS0500C	17.034	15.451	36.545	Plag + K-spar
Guichicovi Gneiss				
G-23	17.304	15.518	36.657	K-spar
G-11*	19.166	15.687	38.122	Plag
G-22*	17.458	15.522	36.699	K-spar
G04-3	17.888	15.574	39.318	Plag + K-spar
G09-6	17.256	15.504	36.677	K-spar
MX05a*	19.043	15.653	38.427	Plag

Note: Data for Coahuila boulders are from Lopez (1997); data for Huiznopala Gneiss are from Lawlor et al. (1999).

*Samples that plot outside the field of Mexican potassium feldspars on Figure 9.

phism; or (2) mixing between different isotopic reservoirs. The first is unlikely, because feldspars from the granitoids and granulites plot along the same trend. This strongly suggests that little radiogenic Pb grew in the protoliths of the granulites between the times of igneous crystallization and recrystallization of the feldspars during the Grenville-age metamorphism. Furthermore, the slope of the array (0.12) is too steep to be modeled by possibility (1). Modeling requires a protolith crystallization age of ca. 2000 Ma, which is inconsistent with U-Pb zircon results discussed in the previous section.

We favor possibility (2), which requires two end-member reservoirs that were each homogeneous with respect to Pb isotopic compositions over long distances (e.g., ~1200 km between the Guichicovi Complex and Coahuila) (Fig. 1). The radiogenic end-member was presumably continental in origin, and may have been added to the magmas by source-region contamination (i.e., subducted sediments) or by crustal assimilation (e.g., Hogan and Sinha, 1991). The three samples of Coahuila banded granitoids, which contain a component of ca. 1900 Ma crust, plot near the upper end of the array, and the radiogenic end-member may have been about that age. The nonradiogenic end-member of the mixing array could have been either mantle-derived magma or pre-Grenville lower crust.

Based on whole-rock Pb isotopic data, Ruiz et al. (1999) concluded that a major suture divides the basement of northern and southern Mexico and that northern Mexico (i.e., Coahuila boulders, Novillo Gneiss, Huiznopala Gneiss) is composed of "offset pieces of the Grenville Province." The clustering near the center of the ellipse of multiple feldspar samples from all Mexican localities fails to provide support for the existence of such a major suture.

COMPARISON OF THE PB ISOTOPIC COMPOSITIONS OF FELDSPARS FROM MEXICO WITH THOSE FROM THE GRENVILLE PROVINCE AND TEXAS

In several recent reconstructions of the supercontinent of Rodinia, some or all of the Mesoproterozoic rocks of eastern and southern Mexico have been shown as continuous with, or displaced from, the Laurentian basement of Texas (e.g., Karlstrom et al., 1999; Ruiz et al., 1999; Burrett and Berry, 2000). In contrast, as mentioned in the introduction to this paper, Ortega-Gutiérrez et al. (1995) proposed that those Mexican rocks are exposures of a single middle Proterozoic microcontinent, Oaxaquia (Fig. 1), which was accreted to the southern margin of Laurentia during the late Paleozoic Alleghanian-Ouachita orogeny. Keppie et al. (2001) advocated a position adjacent to Amazonia for Oaxaquia at ca. 1 Ga, and Lopez et al. (2001) argued that the basement of Oaxaquia has Gondwanan affinities, based on the Panafrican age of a cobble from Coahuila.

Pb isotopic compositions may shed light on the provenance of Oaxaquia. This approach has been used in a number of recent

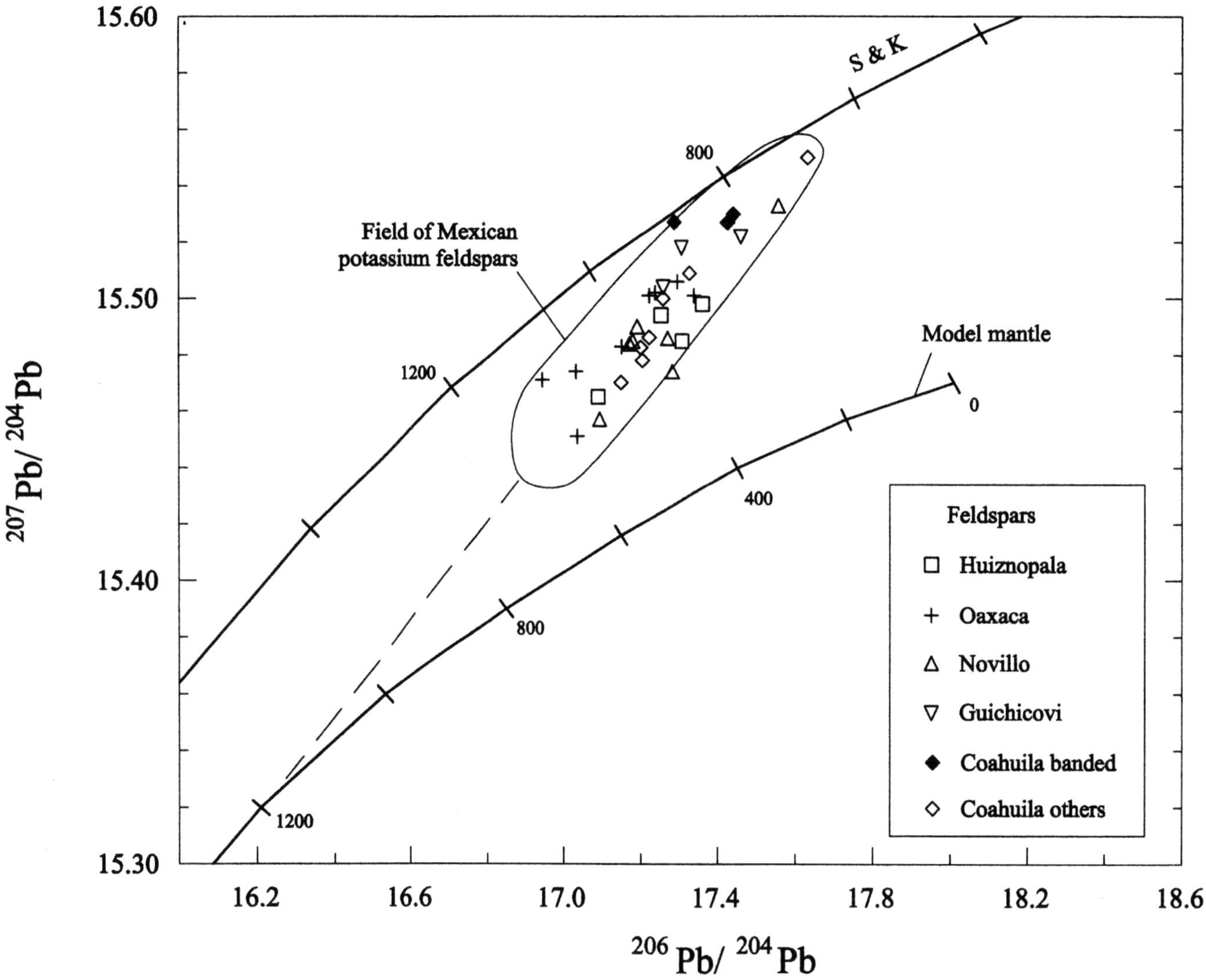

Figure 8. $^{207}Pb/^{204}Pb$ versus $^{206}Pb/^{204}Pb$, showing the compositions of feldspars from Mexico. Boundaries of the encircled field are based entirely on the compositions of potassium feldspars; however, the compositions of plagioclase feldspars that lie in the field are also plotted. Data are from Table 4. Model growth curves for average crust (S&K; Stacey and Kramers, 1975) and mantle (Zartman and Haines, 1988) are shown.

studies of Grenville terranes (e.g., Keppie and Ortega-Gutiérrez, 1999; Kay et al., 1996; Sinha et al., 1996; Tosdal, 1996; Lawlor et al., 1999; Ruiz et al., 1999). All these studies have relied primarily on data from whole rocks. The whole-rock Pb isotopic compositions are very scattered, probably due in part to differences in sample preparation (some rock powders were acid leached and others were not), to including data from both ortho- and paragneisses, and to the growth of radiogenic Pb over the past b.y. in samples with quite variable and often high U/Pb and Th/Pb ratios. Many of these problems would be eliminated if Pb isotopic comparisons were limited to data from acid-leached feldspars (ideally potassium feldspars) from orthogneisses.

Unfortunately, there are few Pb isotopic data sets from feldspars from Grenville rocks to compare with those from Mexico; however, two are relevant, one from the Grenville Province and the other from Texas. DeWolf and Mezger (1994) discussed the results of their feldspar analyses from the Grenville Province in terms of two groups, the northern and southern Grenville arrays (NGA, SGA, respectively; Fig. 9). The NGA represents samples from a belt ~200 km wide, adjacent to the Grenville front and near the Archean Superior craton. The isotopic composition of Pb in these feldspars was interpreted to reflect mixing between juvenile mantle-derived magma and Archean upper and lower crust. Sample localities for the SGA extend from the boundary with the NGA southeastward for ~500 km, including the Adirondack Highlands. DeWolf and Mezger interpreted this area to be underlain by juvenile crust produced by mantle-derived magmas at ca. 1.5 Ga. The SGA field is elongate and was modeled as a secondary isochron that evolved from 1.5 to 1.1 Ga, when Pb exchange in the feldspars closed shortly after

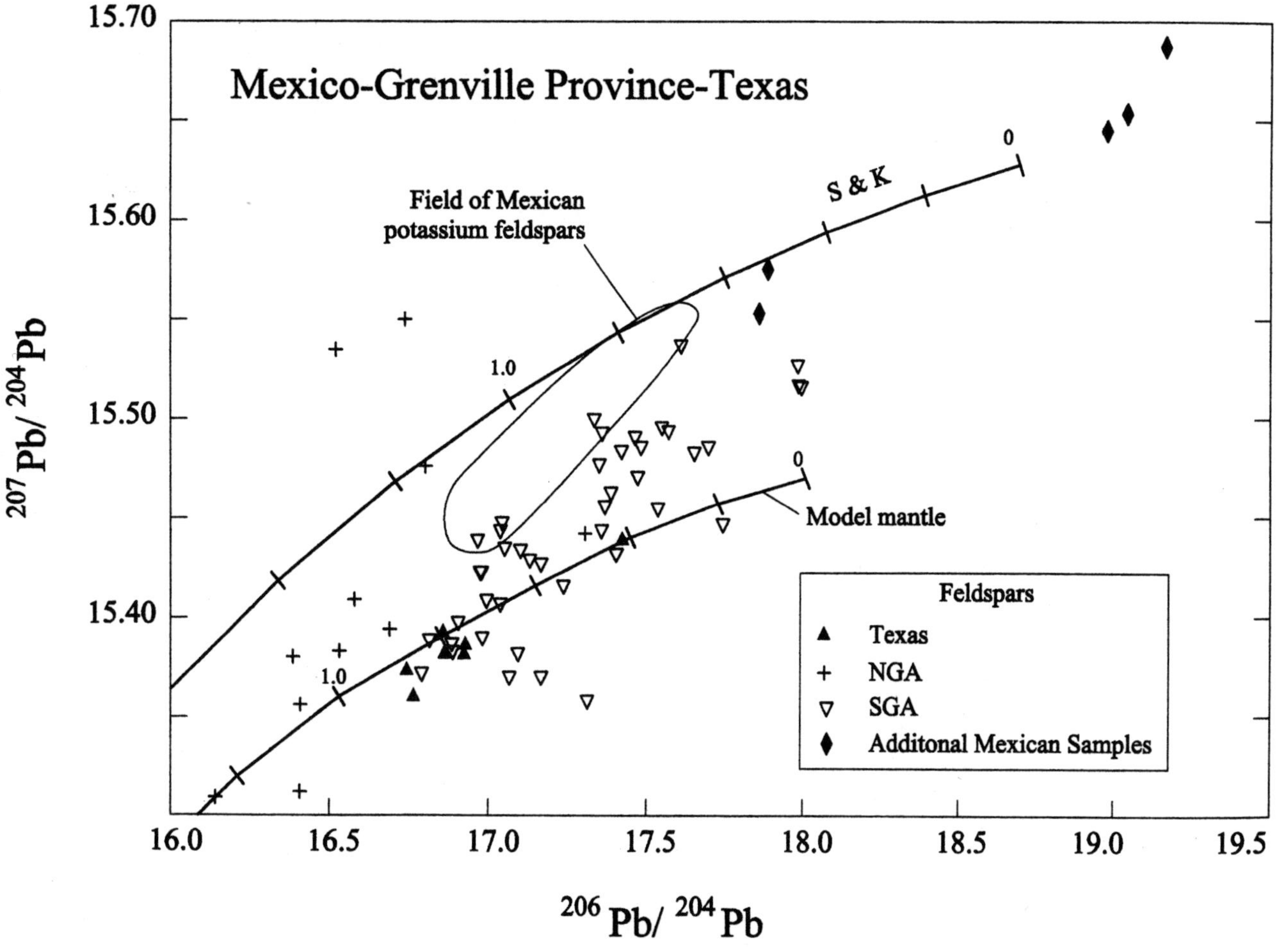

Figure 9. $^{207}Pb/^{204}Pb$ versus $^{206}Pb/^{204}Pb$, comparing Pb isotopic compositions of feldspars from Mexico with those from Texas and the Grenville Province. The field of Mexican potassium feldspars and the model growth curves are the same as shown in Figure 8. Northern Grenville array (NGA) and southern Grenville array (SGA) are from DeWolf and Mezger (1994). Texas data are from Smith et al. (1997).

Grenville-age metamorphism. There is slight overlap between the Mexican feldspars and the upper portion of the SGA field (Fig. 9), but most Mexican feldspars, both potassium and plagioclase, are distinct in Pb isotopic compositions from those of the Grenville Province.

Smith et al. (1997) report analyses of eight feldspar separates from ca. 1.1 Ga "anorogenic" granites from central and west Texas (five and three analyses, respectively). The granites are separated by ~600 km, and they intrude basements of different ages and characteristics. West Texas is underlain by the 1.5–1.3 Ga granite-rhyolite province, and the ca. 1.1 Ga granites are interpreted as "having a direct derivation from mantle sources via extended fractional crystallization of basaltic parental magmas, with minor crustal assimilation" (Smith et al., 1997: 60). In contrast, the granites from central Texas are believed to be anatectic melts from slightly older, juvenile crustal sources (Smith et al., 1997). With one exception, the Pb isotopic compositions of the feldspars from the Texas granites cluster tightly, and all points lie far below the Mexican samples on Figure 9. (Note that all eight points are plotted, but only seven are resolved in the figure, due to the overlap of data.) Although the Texas dataset is small, there is no evidence from those Pb isotopic compositions to support the hypothesis that the basement of eastern and southern Mexico is an extension of the Laurentian basement of Texas.

DISCUSSION AND CONCLUSIONS

The Mesoproterozoic granitoids and orthogneisses of eastern and southern Mexico can be divided into two groups. The older group generally has arc and backarc geochemical signatures and most members of this group were emplaced between ca. 1235 and ca. 1115 Ma. The younger group consists of anorthosites and related rocks. This AMCG suite is not well dated, but present data suggest that it was emplaced between ca. 1035 and ca. 1010 Ma.

The Grenville orogeny in Mexico, known locally as the Zapotecan tectonothermal event (Solari et al., 2003), produced widespread granulite-facies metamorphism. Most metamorphic zircons grew at 990 ± 5 Ma, and post-tectonic pegmatites were emplaced at 978 ± 3 Ma. The Zapotecan event was coeval with the Rigolet phase of the Grenville orogeny in the Grenville Province (Rivers, 1997).

Oaxaquia was most likely a trailing margin, but possibly was a transform margin from ca. 1115 Ma until the Zapotecan/Rigolet event, because there is no evidence of arc magmatism in this time period. Likewise, arc magmatism was absent from the Grenville Province at this time (Rivers, 1997). This argues against Oaxaquia (or at least the exposed portions of Oaxaquia) being the "enigmatic southeastern continental mass" (Davidson, 1995: 3) that collided with the eastern margin of Laurentia during the Grenville orogeny.

Two periods of AMCG magmatism have been recognized in the Grenville Province: an older one associated with the Elzevirian orogeny and a younger one with the Ottawan orogeny (e.g., McLelland et al., 1996; Corrigan and Hanmer, 1997). In each period, AMCG magmatism overlapped with and continued well beyond the peak orogenic event. The older AMCG suite was deformed and metamorphosed during the Ottawan orogeny, but the younger suite is generally undeformed and not affected by regional metamorphism. Both AMCG suites in the Grenville Province are interpreted to have formed in response to delamination or convective thinning of subcontinent lithosphere, following collision and crustal thickening (e.g., McLelland et al., 1996; Corrigan and Hanmer, 1997; Rivers, 1997). This model may not apply to Mexico. Although the Mexican AMCG rocks are not well dated, there is no structural or metamorphic evidence for a major collisional event older than the suite. Only one major, widespread orogeny, the Zapotecan/Rigolet event at 990 ± 5 Ma, has been recognized in the Mexican Grenville terrane, and the AMCG suite was deformed and metamorphosed at that time.

The conclusion of Ortega-Gutiérrez et al. (1995) that eastern and southern Mexico is underlain by a single, coherent, Middle Proterozoic microcontinent, Oaxaquia, is strongly supported by: (1) the tight clustering and minimal geographic variation of potassium feldspar Pb isotopic compositions, (2) the similarity in age of the older arc-backarc suite throughout the area, and (3) the identical (within error) age of granulite metamorphism throughout the region.

There is no evidence from the Pb isotopic compositions of feldspars to support the hypothesis that the Grenville rocks of eastern and southern Mexico are an extension of the Laurentian basement of Texas.

ACKNOWLEDGMENTS

This research was supported by National Science Foundation grant EAR 9909459 to KLC, a UC Mexico–United States of America–Consejo Nacional de Ciencias y Technológia (CONACyT) grant to KLC and FO-G, CONACyT grants (0225P-T9506 and 25705-T), Programa de Apoyo a Proyectos de Investigación e Innovación (PAPIIT)–Universidad Nacional Autónoma de México (UNAM) grants (IN116999 and IN107999) to JDK and FO-G, and a UNAM-PAEP student grant to LAS. We thank Bodo Weber for providing feldspar samples from the Guichicovi complex and for reviewing the manuscript. Barbra Martiny provided the Pb isotopic composition of feldspar sample CON215.

REFERENCES CITED

Ashwal, L.D., 1993, Anorthosites. New York, Springer-Verlag, 422 p.

Blount, J.G., 1993, The geochemistry, petrogenesis, and geochronology of the Precambrian meta-igneous rocks of Sierra Del Cuervo and Cerro El Carrizalillo, Chihuahua, Mexico [Ph.D. thesis]: Austin, University of Texas, 242 p.

Burrett, C., and Berry, R., 2000, Proterozoic Australia–western United States (AUSWUS) fit between Laurentia and Australia: Geology, v. 28, p.103–106.

Corrigan, D., and Hanmer, S., 1997, Anorthosites and related granitoids in the Grenville orogen: A product of convective thinning of the lithosphere?: Geology, v. 25, p.61–64.

Davidson, A., 1995, A review of the Grenville orogen in its North American type area: Journal of Australian Geology and Geophysics, v. 16, p. 3–24.

Denison, R.E., Burke, W.H., Jr., Hetherington, E.A., Jr., and Otto, J.B., 1971, Basement rock framework of part of Texas, southern New Mexico and northern Mexico, *in* Seewald, K. and Sundeen, D. eds., The geological frame of the Chihuahua Tectonic Belt: Midland, West Texas Geological Society Publication 71-59, p. 3–14.

DeWolf, C.P., and Mezger, K., 1994, Lead isotopic analyses of leached feldspars: Constraints on the early crustal history of the Grenville orogen: Geochimica et Cosmochimica Acta, v. 58, p. 5537–5550.

Hogan, J.P, and Sinha, A.K., 1991, The effect of accessory minerals on the redistribution of lead isotopes during crustal anatexis, a model: Geochimica et Cosmochimica Acta, v. 55, p. 335–348.

Karlstrom, K.E., Harlan, S.S., Williams, M.L., McLelland, J., Geissman, J.W., and Ahall, K.I., 1999, Refining Rodinia: Geologic evidence for the Australia–western U.S. connection in the Proterozoic: GSA Today, v. 9, no. 10, p. 1–7.

Kay, S.M., Orrell, S., and Abbruzzi, J.M., 1996, Zircon and whole rock Nd-Pb isotopic evidence for a Grenville age and a Laurentian origin for the basement of the Precordillera in Argentina: Journal of Geology, v. 104, p. 637–648.

Keppie, J.D., and Ortega-Gutiérrez, F., 1999, Middle American Precambrian basement: A missing piece of the reconstructed 1-Ga orogen, *in* Ramos, V., and Keppie, J.D., eds., Laurentia-Gondwana connections before Pangea: Boulder, Colorado, Geological Society of America Special Paper 336, p. 199–210.

Keppie, J.D., Dostal, J., Ortega-Gutiérrez, F., and Lopez, R., 2001, A Grenvillian arc on the margin of Amazonia: Evidence from the southern Oaxacan Complex, southern Mexico: Precambrian Research, v. 112, no. 3–4, p. 165–181.

Keppie, J.D., Dostal, J., Cameron, K.L., Solari, L.A., Ortega-Gutiérrez, F., and Lopez, R., 2003, Geochronology and geochemistry of Grenvillian igneous suites in the northern Oaxacan Complex, southern Mexico: Tectonic implications. Precambrian Research, v. 120, p. 365–389.

Lawlor, P.J., Ortega-Gutiérrez, F., Cameron, K.L., Ochoa-Camarillo, H., Lopez, R.L, and Sampson, D.S, 1999, U-Pb geochronology, geochemistry and provenance of the Grenvillian Huiznopala Gneiss of eastern Mexico: Precambrian Research, v. 94, p. 73–99.

Lopez, R.L., 1997. The pre-Jurassic geotectonic evolution of the Coahuila terrane, northwestern Mexico: Grenville basement, a late Paleozoic arc, Triassic plutonism, and the events south of the Ouachita suture [Ph.D. thesis]: Santa Cruz, University of California, 147 p.

Lopez, R.L., Cameron, K.L., and Jones, N.W., 2001, U-Pb zircon evidence for a Grenvillian arc reworking ~1.85 Ga crust and for Pan African magmatism in Coahuila, northeastern Mexico: Precambrian Research, v. 107, p. 195–214.

Ludwig, K.R., 1991, PBDAT: A computer program for processing Pb-U-Th isotope data, version 1.20: Reston, Virginia, U.S. Geological Survey Open-File Report 88-542, 40 p.

Ludwig, K.R., 1998, Isoplot/Ex version 1.00b: Berkeley, California, Berkeley Geochronology Center Special Publication no. 1, 43 pages.

Manton, W.I., 1996, The Grenville of Honduras: Geological Society of America Abstracts with Programs, v. 28, no. 7, p. A-493.

McLelland, J., Daly, J.S., and McLelland, J.M., 1996, The Grenville orogenic cycle (ca. 1350–1000 Ma): An Adirondack perspective: Tectonophysics, v. 265, p. 1–28.

Mosher, S., 1998, Tectonic evolution of the southern Laurentian Grenville orogenic belt: Geological Society of America Bulletin, v. 110, p. 1357–1375.

Ortega-Gutiérrez, F., Ruiz, J., and Centeno-Garcia, E., 1995, Oaxaquia, a Proterozoic microcontinent accreted to North America during the late Paleozoic: Geology, v. 23, p. 1127–1130.

Patchett P.J., and Ruiz, J., 1987, Nd isotopic ages of crust formation and metamorphism in the Precambrian of eastern and southern Mexico: Contributions to Mineralogy and Petrology, v. 96, p. 523–528.

Pearce, J.A., Harris, N.B.W., and Tindle, A.G., 1984, Trace element discrimination diagrams for the tectonic interpretation of granitic rocks: Journal of Petrology, v. 25, p. 956–983.

Ramirez-Ramirez, C., 1992, Pre-Mesozoic geology of the Huizachal-Peregrina anticlinorium, Ciudad Victoria, Tamaulipas, and adjacent parts of eastern Mexico [Ph.D. thesis]: Austin, University of Texas, 317 p.

Rivers, T., 1997, Lithotectonic elements of the Grenville Province: Review and tectonic implications: Precambrian Research, v. 86, p. 117–154.

Ruiz, J., Tosdal, R.M., Restrepo, P.A., and Murillo-Muneton, G., 1999, Pb isotope evidence for a Colombia–southern Mexico connection in the Proterozoic, *in* Ramos, V., and Keppie, J.D., eds., Laurentia-Gondwana connections before Pangea: Boulder, Colorado, Geological Society of America Special Paper 336, p. 183–197.

Silver, L.T., Anderson, A.T., and Ortega-Gutiérrez, F., 1994, The "thousand million year" orogeny in eastern and southern Mexico: Geological Society of America Abstracts with Programs, v. 26, no. 7, p. A-48.

Sinha, A.K., Hogan, J.P., and Parks, J., 1996, Lead isotope mapping of crustal reservoirs within the Grenville Superterrane: I. Central and southern Appalachians, *in* Earth processes: Reading the isotopic code: American Geophysical Union Monograph 95, p. 293–305.

Smith, D.R., Barnes, C.G., Shannon, W., Roback, R.C., and James, E., 1997, Petrogenesis of Mid-Proterozoic granitic magmas: Examples from central and west Texas: Precambrian Research, v. 85, p. 53–79.

Solari, L.A., Keppie, J.D., Ortega-Gutiérrez, F., Cameron, K.L., Lopez, R., and Hames, W.E., 2003, 990 Ma and 1,100 Ma Grenvillian tectonothermal events in the northern Oaxacan Complex, southern Mexico: Roots of an orogen: Tectonophysics, v. 365, p. 257–282.

Stacey J.S., and Kramers, J.D., 1975, Approximation of terrestrial lead isotope evolution by a two-stage model: Earth and Planetary Science Letters, v. 26, p. 207–221.

Tosdal, R.M., 1996, The Amazon-Laurentian connection as viewed from the Middle Paleozoic rocks in the central Andes, western Bolivia and northern Chile: Tectonics, v. 15, p. 827–842.

Weber, B., and Kohler, H., 2001, Sm-Nd, Rb-Sr, and U-Pb geochronology of a Grenville fragment in southern Mexico: Origin and geologic history of the Guichicovi complex: Precambrian Research, v. 96, p. 245–262.

Zartman, R.E., and Haines, S.M., 1988, The plumbotectonic model for Pb isotopic systematics among major terrestrial reservoirs—A case for bi-directional transport: Geochimica et Cosmochimica Acta, v. 52, p. 1327–1339.

Manuscript Accepted by the Society August 25, 2003

Printed in the USA

Geological Society of America
Memoir 197
2004

U-Pb and $^{40}Ar/^{39}Ar$ constraints on the cooling history of the northern Oaxacan Complex, southern Mexico: Tectonic implications

J. Duncan Keppie*
Luigi A. Solari
Fernando Ortega-Gutiérrez
Instituto de Geología, Universidad Nacional Autónoma de México, 04510 México D.F., México
Amabel Ortega-Rivera
Centro de Geociencias, Campus Juriquilla, Universidad Nacional Autónoma de México, Apdo. Postal 1-742, Centro Querétaro, Qro. 76001, México
James K.W. Lee
Department of Geology, Queens University, Kingston, Ontario K7L 3NG, Canada
Robert Lopez
Geology Department, West Valley College, Saratoga, California 95070, USA
Willis E. Hames
Department of Geology, Auburn University, Auburn, Alabama, 36830, USA

ABSTRACT

To better define the cooling history of the northern Oaxacan Complex, titanite and phlogopite from metasedimentary calc-silicate and biotite from a pegmatite were collected. All these rocks were involved in the granulite-facies Zapotecan orogeny between ca. 1004 and 978 ± 3 Ma, inferred to result from underthrusting the Oaxacan Complex beneath an arc or continent. Fragments of 2 × 5 cm^2 titanite crystals yielded a concordant U-Pb age of 968 ± 9 Ma, whereas $^{40}Ar/^{39}Ar$ analyses of phlogopite and biotite gave ages of 945 ± 10 Ma and 856 ± 10 Ma, respectively. These ages are inferred to date cooling through 660–700 °C, 450 °C, and 300–350 °C, respectively. When combined with published ages (Sm-Nd garnet, $^{40}Ar/^{39}Ar$ hornblende, Rb-Sr biotite and whole rock, and K-Ar biotite and K-feldspar) the data define a two-stage cooling curve: (1) 8 °C/m.y. between 978 and 945 Ma, cooling through 450 °C by which time the rocks had risen through a depth of 15 km; and (2) 2 °C/m.y., which, by extrapolation, brought the rocks to the surface between 710 and 760 Ma. The first stage of exhumation is interpreted in terms of tectonic switching between steep and flat slab subduction, a result of interactions of a ridge, a plume, or an oceanic plateau with the trench. The second stage may be related to thermal relaxation of the lithosphere, ending with the breakup of Rodinia, which brought the rocks to the surface.

Keywords: Oaxaca, Mexico, geochronology, tectonics, Rodinia

*E-mail: duncan@servidor.unam.mx.

Keppie, J.D., Solari, L.A., Ortega-Gutiérrez, F., Ortega-Rivera, A., Lee, J.K.W., Lopez, R., and Hames, W.E., 2004, U-Pb and $^{40}Ar/^{39}Ar$ constraints on the cooling history of the northern Oaxacan Complex, southern Mexico: Tectonic implications, *in* Tollo, R.P., Corriveau, L., McLelland, J., and Bartholomew, M.J., eds., Proterozoic tectonic evolution of the Grenville orogen in North America: Boulder, Colorado, Geological Society of America Memoir 197, p. 771–781. For permission to copy, contact editing@geosociety.org.

INTRODUCTION

In referring to granulites, Jamieson et al. (1998, p. 49) stated: "These high grade rocks are unlikely to be exhumed until a subsequent tectonic episode, if ever." This statement is based on geophysical modeling of subduction and collision that shows decoupling at a depth of 30 km of the middle-upper orogenic crust from a 5–10 km thick, hot (≥700 °C), weak lower crust, the latter characterized by flat-lying syn-metamorphic foliations. In this context, it is pertinent to determine the exhumation history of ca. 1 Ga granulites (Oaxacan Complex) developed at 700–825 °C and 7.2–8.2 kilobars (Mora et al., 1986) that are unconformably overlain by the lowest Ordovician (Tremadocian) sedimentary rocks in southern Mexico. In this region, the only significant tectonic events between the ca. 1 Ga Zapotecan orogeny (Solari et al., 2003) and deposition of the Tremadocian rocks (Robison and Pantoja-Alor, 1968) may be associated with the breakup of Rodinia at ca. 755 Ma and the opening of Iapetus at ca. 600–550 Ma (Dalziel, 1997). However, a biotite–whole-rock Rb-Sr date of 876 ± 9 Ma from the northern Oaxacan Complex (Patchett and Ruiz, 1987) indicates that cooling below 300 °C had occurred before the onset of either of these extensional events. This paper reports new U-Pb and $^{40}Ar/^{39}Ar$ mineral ages, which broadly confirm previous geochronology, and discusses various exhumation models for the granulite-facies Oaxacan Complex following the Zapotecan orogeny. This discussion may be relevant to the Grenville Province, because the cooling history of the two regions is remarkably similar (Keppie and Ortega-Gutiérrez, 1999).

GEOLOGICAL SETTING

Rocks ca. 1 Ga in age are exposed in several inliers in Mexico and are inferred to form a basement terrane, which is called "Oaxaquia" (Keppie and Ortega-Gutiérrez, 1999), beneath the backbone of Mexico that probably extended beneath the Chortis block in Honduras, an area of 1,000,000 km^2 (Fig. 1). The pre–980 Ma history of ca. 1 Ga rocks of Mexico has been the subject of several recent papers: Keppie et al. (2003) and Solari et al. (2003) on the northern Oaxacan Complex; Keppie et al. (2001) on the southern Oaxacan Complex; Weber and Köhler (1999) on the Guichicovi Gneiss; Lawlor et al. (1999) on the Huiznopala Gneiss; and Cameron et al. (this volume) on the Novillo Gneiss. A summary of the ca. 1 Ga rocks of Mexico and Honduras was published by Keppie and Ortega-Gutiérrez (1999). Thus, only a brief review of the pre–980 Ma geological history of the northern Oaxacan Complex is presented here.

The largest inlier of ca. 1 Ga rocks exposes the Oaxacan Complex, which consists of paragneisses and within-plate, rift-related orthogneisses (anorthosite, charnockite, metagabbro, and metagranite) intruded at ≥1140 Ma and 1012 ± 12 Ma (U-Pb zircon ages) (Keppie et al., 2003). The recent discovery of inverted pigeonite of igneous origin in the 1012 ± 12 Ma suite indicates a shallow level of intrusion. These rocks were involved in two polyphase tectonothermal events: the ca. 1100 Ma Olmecan event and the 1004 ± 3- to 979 ± 3-Ma Zapotecan event (based on concordant U-Pb zircon ages) (Solari et al., 2003). The Olmecan event produced migmatites during polyphase deformation and is synchronous with intrusion of ca. 1117 Ma rift-related granite (Keppie et al., 2001). This led Solari et al. (2003) to suggest that the Olmecan event was extensional. The Zapotecan event involved polyphase deformation under granulite-facies metamorphic conditions that Solari et al. (2003) infer to be related to underthrusting of the northern Oaxacan Complex beneath an arc or a continent, the heat for metamorphism being supplied by intrusion of the immediately preceding rift-related igneous suite. Peak metamorphic conditions were estimated at 700–825 °C at 7.2–8.2 kilobars under restricted P_{H_2O} conditions (Mora et al., 1986). *P-T* terminal conditions during retrograde metamorphism were estimated at 480 °C and 5 kilobars (Keppie and Ortega-Gutiérrez, 1999), based on the presence of coronas of actinolite-biotite-epidote around clinopyroxene (Bloomfield and Ortega-Gutiérrez, 1975). At 917 ± 6 Ma, the northern Oaxacan Complex was intruded by an arc-related granitoid pluton at a depth between 6 and 15 km, based on minimum melt compositions and some melt textures (Ortega-Obregon et al., 2003). During the Mesozoic, the rocks were deformed by northnorthwest-trending, open, upright folds (Solari et al., 2004).

The rocks in the northern Oaxacan Complex are subdivided into two gently north-dipping thrust slices: a lower, 1012 ± 12 Ma anorthosite-gabbro-charnockite-granite that intrudes a ≥1340 Ma igneous body migmatized at ca. 1106 Ma, and an upper paragneiss unit intruded by within-plate ≥1140 Ma charnockite and metasyenite (Fig. 1) (Keppie et al., 2003; Solari et al., 2003). Two granulite-facies calc-silicate samples in the upper thrust slice were collected from bodies that cut across the foliation in the Oaxacan Complex, indicating that thermal metamorphism outlasted the deformation. They are inferred to have a sedimentary protolith remobilized during the granulite-facies metamorphism. The calc-silicate rock consists of scapolite, feldspar, pyroxene, and titanite. The third sample was collected from a quartzofeldspathic pegmatite intruded during the latter stages of the Zapotecan orogeny under upper amphibolite-facies conditions (Solari et al., 2003). This pegmatite yielded a concordant U-Pb age of 978 ± 3 Ma (Solari et al., 2003).

Previous U-Pb geochronology on single zircon grains suggests that the granulite-facies metamorphism lasted from 1004 ± 3 to 978 ± 3 Ma (Keppie et al., 2003; Solari et al., 2003). Sm-Nd isotopic analyses of garnets from charnockite and paragneiss of the Oaxacan Complex have yielded ages of 938 ± 4 Ma and 963 ± 3 Ma, respectively (Patchett and Ruiz, 1987). The latter age is inferred to date cooling through ≥600 °C, as the garnet composition is similar to that used by Humphries and Cliff (1982). $^{40}Ar/^{39}Ar$ laser fusion analyses of hornblende indicate cooling through 500 °C by 977 ± 12 Ma (Solari et al., 2003). A Rb-Sr biotite–whole-rock age of 876 ± 9 Ma was interpreted to date cooling through 300 °C (Patchett and Ruiz, 1987). These

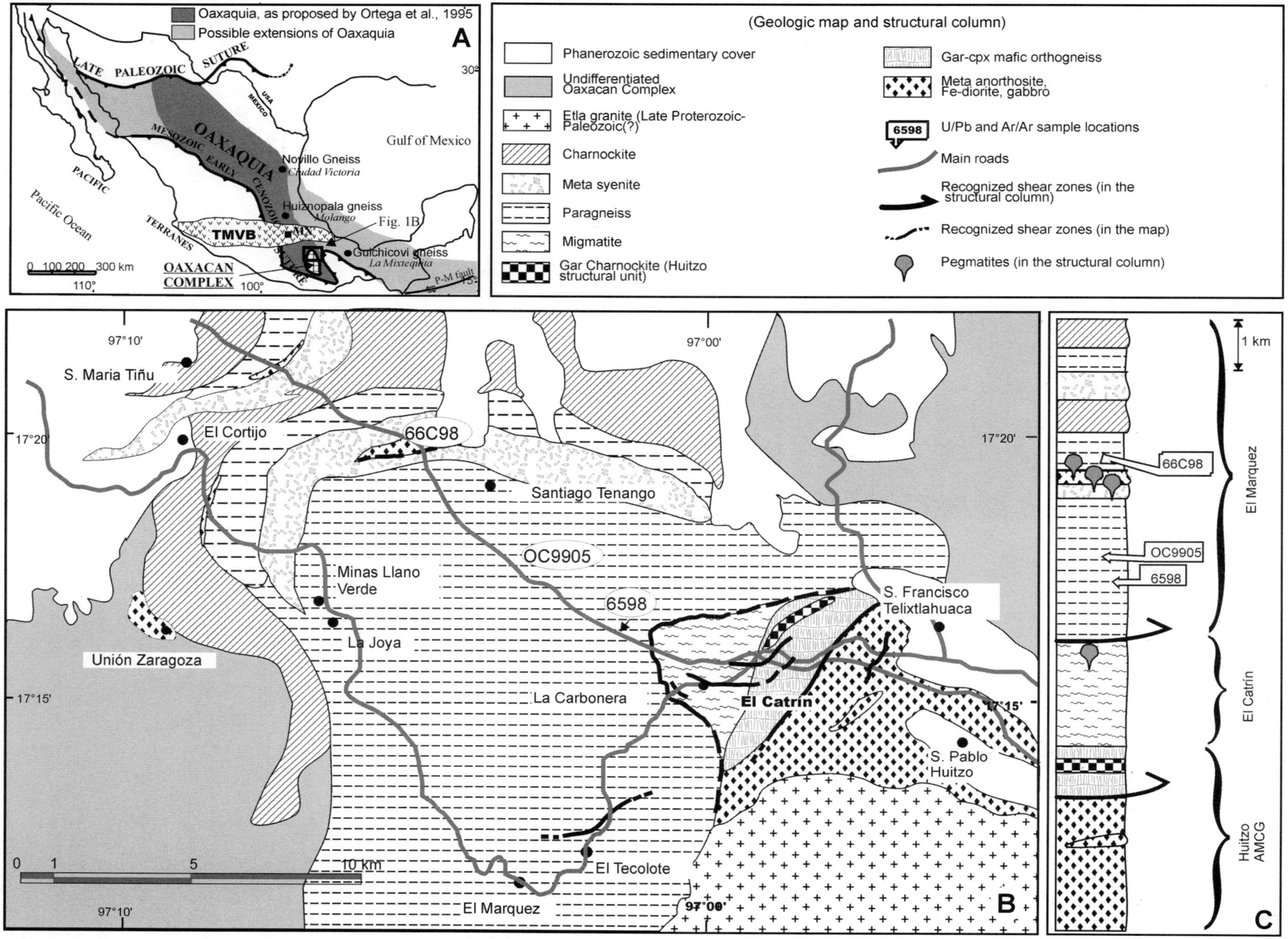

Figure 1. Maps and section, showing the locations of dated samples. (A) The location of the Oaxacan Complex in Mexico and the extent of Oaxaquia. TMVB—Trans-Mexican Volcanic Belt. (B) Geological map of the northern Oaxacan Complex. (C) Structural sequence in the northern Oaxacan Complex. Numbers in boxes are sample numbers. MSWD—Mean squares weighted deviation. Modified after Solari et al. (2003).

complement the earlier dating of biotite by Rb-Sr and K-Ar methods that yielded ages of 900 ± 35 Ma and 930–862 ± 30 Ma, respectively, whereas microcline gave a Rb-Sr age of 797 ± 35 Ma (ages recalculated using modern decay constants after Fries et al. [1962]; Fries and Rincon-Orta [1965]). Exhumation and burial is recorded by deposition of basal Ordovician (Tremadocian) sedimentary rocks containing shallow-water fauna nonconformably on the Oaxacan granulites (taxonomy of Robison and Pantoja-Alor [1968]. revised by Shergold [1975]) (Cox and Fortey, 1990).

GEOCHRONOLOGY

Two fragments of large titanite crystals (1–2 cm across and 5 cm in length) separated from a crosscutting calc-silicate dyke at km 207 on Federal Highway 190 were analyzed for U-Pb isotopes using the procedures described in Lopez et al. (2001). The petrography and geochemistry of this dyke (authors' unpublished data) is consistent with remobilization of metasedimentary rocks under granulite facies (Ortega-Gutiérrez, 1984). They yielded concordant data with an age of 968 ± 9 and 968 ± 16 Ma (2σ errors) (Table 1, sample 6598; Fig. 2, A). Estimates of the closure temperature of titanite vary with such factors as size and cooling rate, and Frost et al. (2000) have calculated a closure temperature of 660 °C for grains with a radius of 0.1 mm cooling at 10 °C/Ma. However, Verts et al. (1996) report that titanite ages were reset at temperatures >700 °C in the contact aureole of the Laramie Anorthosite Complex. Given the large size of the titanite crystals analyzed in this study, the 968 ± 9 Ma age is inferred to represent cooling through a temperature in the range of 660–700 °C, but probably closer to 700 °C.

Small subhedral phlogopite crystals (1 mm in diameter) were separated from a crosscutting calc-silicate body with a sedimentary protolith metamorphosed under granulite-facies (Dostal et al., 2004) at km 204 on Federal Highway 190. Ten $^{40}Ar/^{39}Ar$ laser fusion analyses of ten individual crystals using the methodology of Hames and Bowring (1994) yielded a mean age of 945 ± 10 Ma (2σ) for the air-corrected data (Table 2; Fig. 2, B). Note that the phlogopite analyses are characterized by radiogenic yields generally >98% ($^{36}Ar/^{40}Ar$ ratios generally $<1 \times 10^{-4}$) and the data are well correlated in terms of the $^{40}Ar/^{39}Ar$ ratio and thus, age. One analysis seems distinct from the cluster of the remaining nine, and is interpreted to contain ~3% extraneous argon of atmospheric composition. As the data show an indication of extraneous, nonatmospheric argon contamination, the mean age of 945 ± 10 Ma (2σ standard error) is interpreted as the best estimate of the time of cooling through the argon retention temperature for phlogopite. In view of the crystal size and using the diffusion data of Giletti (1974), we calculate a temperature of 450 ± 25 °C for a cooling rate of 8.8 °C/m.y., a significantly higher temperature than is typical of biotite. Because the diameters of these phlogopite crystals were relatively small and restricted in range, we expect that the crystals would have reached isotopic closure at similar times and with limited development of diffusion gradients; this prediction is

TABLE 1. U-PB ISOTOPIC ANALYSIS OF TITANITE FROM CALCSILICATE SAMPLE 6598

Fraction	Weight (mg)	U (ppm)	Total Pb (ppm)	Common Pb (pg)	$^{206}Pb/^{204}Pb$	$^{206}Pb/^{204}Pb$ Error (%)	$^{207}Pb^*/^{206}Pb^*$	$^{207}Pb^*/^{206}Pb^*$ Error (%)	$^{208}Pb^*/^{206}Pb^*$	$^{208}Pb^*/^{206}Pb^*$ Error (%)	$^{206}Pb^*/^{238}U$	$^{207}Pb^*/^{235}U$	$^{207}Pb^*/^{206}Pb^*$	$^{206}Pb^*/^{238}U$	$^{207}Pb^*/^{235}U$	$^{207}Pb^*/^{206}Pb^*$	Disconformable (%)
					Raw data†						Atomic ratio††			Age (Ma)†††			
1. Titanite, abraded	1.291	8	2	170	425	1.4	0.1050	0.024	1.0446	0.2	0.16197	1.5939	0.07137	968	968	968 ± 16	0.0
2. Titanite, abraded	0.531	32	10	410	430	0.8	0.1046	0.08	1.0592	0.08	0.16214	1.5961	0.0714	969	969	969 ± 9	0.0

Note: Pb* denotes radiogenic Pb. Titanite sample dissolution and ion exchange chemistry modified after Krogh (1973) and Mattinson (1987) in Parrish (1987)-type microcapsules. Titanite chemistry with 1N HBr. Fraction 1 used a mixed $^{208}Pb/^{235}U$ spike, whereas fraction 2 used a mixed $^{205}Pb/^{235}U$ spike. Isotopic data were measured on a VG 54-30 sector multicollector mass spectrometer with a pulse-counting Daly detector at the University of California, Santa Cruz.

†Observed isotopic ratios are corrected for mass fractionation of 1‰ for both ^{208}Pb- and ^{205}Pb-spiked fractions. Fractions spiked with the mixed $^{235}U/^{205}Pb$ tracer are also corrected for spike and blank relative contributions.

††Decay constants used: $^{238}U = 1.55125 \times 10^{-10}$; $^{235}U = 9.48485 \times 10^{-10}$; $^{238}U/^{235}U = 137.88$. Estimated uncertainties of the U/Pb ratio are ± 0.4, based on replicate analyses of a single zircon standard fraction (see Lopez et al., 2001).

†††$^{207}Pb^*/^{206}Pb^*$ age uncertainties are 2σ and from the data reduction program PBDAT of Ludwig (1991). Total processing Pb-blank amount varied between 2 pg and 30 pg, generally averaging <10 pg. Initial Pb compositions are from isotopic analysis of feldspar separates.

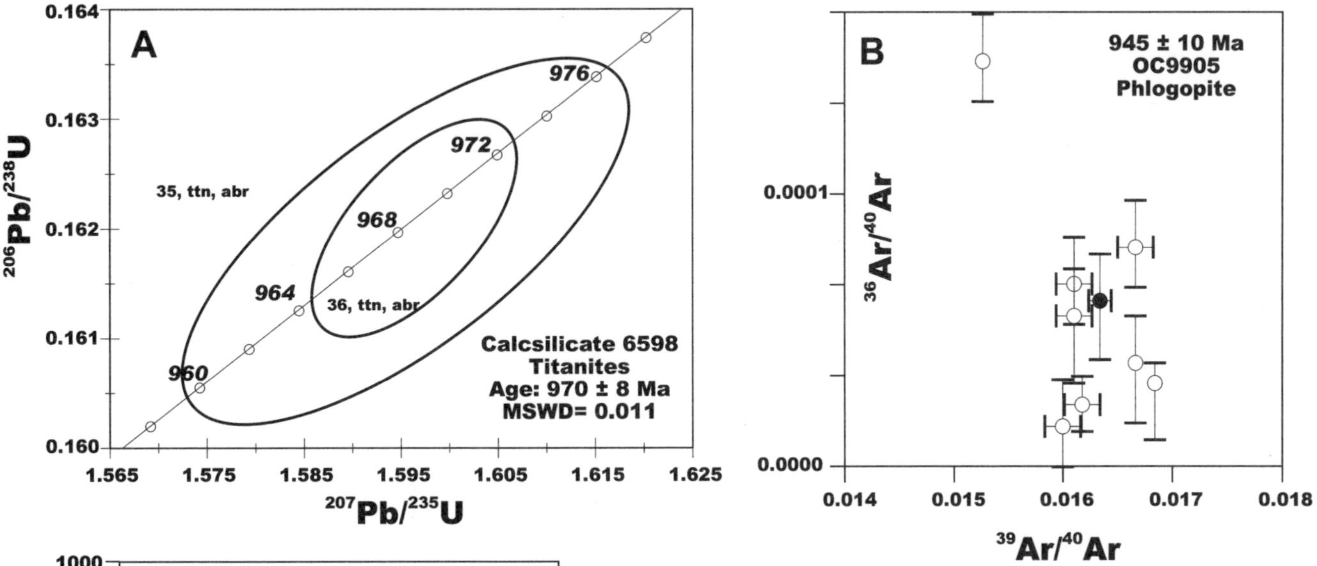

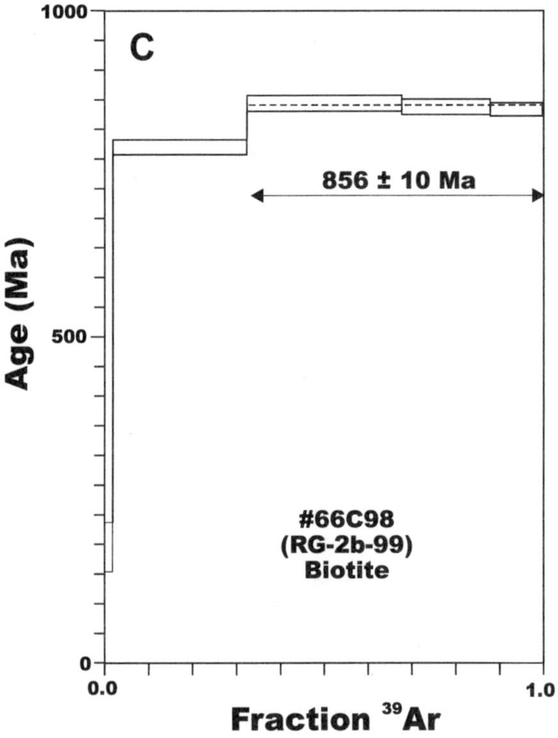

Figure 2. Age data. (A) U-Pb concordia plot for data from titanite in calc-silicate sample 6598. (B) $^{36}Ar/^{40}Ar$ versus $^{39}Ar/^{40}Ar$ laser fusion data from phlogopite in calc-silicate sample OC9905. The black symbol represents two coincident points. (C) Calculated age versus fraction ^{39}Ar released in laser step-heating experiments from biotite in pegmatite sample 66C98.

verified by the relatively consistent $^{40}Ar/^{39}Ar$ ratios and ages calculated.

Biotite (1–2 mm in size) separated from a pegmatite (sample 66C98) was analyzed at the Queen's University $^{40}Ar/^{39}Ar$ Geochronology Laboratory in Kingston, Ontario, Canada, using the $^{40}Ar/^{39}Ar$ laser step-heating procedure outlined in Clark et al. (1998). This pegmatite had previously yielded a concordant U-Pb zircon age of 978 ± 3 Ma and is inferred to have been intruded during upper amphibolite-facies conditions (Solari et al., 2003). The biotite yielded discordant $^{40}Ar/^{39}Ar$ data, with a plateau age of 856 ± 10 Ma (2σ) from the three high-temperature steps that released 67% of the gas (Fig. 2, C; Table 2). This suggests that the ca. 856-Ma ages provide minimum ages for cooling through a blocking temperature of 300–350 °C (recalculated using 2 °C/m.y. and 1–2 mm crystal size) (Harrison et al., 1985). The considerable range in age determined for the biotite is consistent with the relatively large size of the crystals and is also suggestive of a slower cooling rate through biotite closure temperatures (~300–350 °C) rather than through the closure of the phlogopite. The two low-temperature steps decrease from the plateau to an age of 178 ± 38 Ma, a pattern typical of spectra that have suffered partial outgassing. Although the uncertainties are large, the ca. 178 Ma increment may reflect a younger thermal event, such as the approximately synchronous

TABLE 2. ARGON ISOTOPIC ANALYSES OF BIOTITE (66C98, RG2-99, AOR-L-303) AND PHLOGOPITE (OC9905)

J value: 0.007583 ± 0.000070 Mass: 1.0 mg Volume ^{39}K: 27.21 × 10^{-10} cm^3 NTP Integrated age: 827.04 ± 8.72 Ma

Plateau age: 855.96 ± 9.67 Ma (67.4% of ^{39}Ar, steps marked by <)

Power	^{36}Ar/^{40}Ar	^{39}Ar/^{40}Ar	r	Ca/K	%40Atm	%^{39}Ar	^{40}Ar*/^{39}K	Age (Ma)
1	0.002661± 0.000147	0.015661 ± 0.000529	0.513	0.186	78.54	1.69	13.682 ± 3.027	178.09 ± 37.52
2	0.000209 ± 0.000042	0.012896 ± 0.000164	0.041	0.045	6.15	30.9	72.776 ± 1.358	792.82 ± 11.97
<3.50	0.000059 ± 0.000039	0.012213 ± 0.000152	0.011	0.042	1.72	35.5	80.469 ± 1.371	859.39 ± 11.65
<4.50	0.000099 ± 0.000046	0.012174 ± 0.000128	0.017	0.044	2.9	19.86	79.762 ± 1.402	853.37 ± 11.95
<5.00	0.000079 ± 0.000036	0.012308 ± 0.000142	0.025	0.045	2.29	12.05	79.382 ± 1.264	850.13 ± 10.79

Power	^{40}Ar	^{39}Ar	^{38}Ar	^{37}Ar	^{36}Ar	Blank ^{40}Ar	Atmospheric ^{40}Ar/^{36}Ar
1	22.369 ± 0.687	0.362 ± 0.005	0.247 ± 0.008	0.011 ± 0.003	0.063 ± 0.000	0.03	288.024
2	497.493 ± 4.823	6.459 ± 0.053	4.281 ± 0.078	0.038 ± 0.016	0.109 ± 0.000	0.021	288.024
<3.50	603.493 ± 4.196	7.419 ± 0.076	4.720 ± 0.096	0.040 ± 0.019	0.039 ± 0.000	0.024	288.024
<4.50	338.707 ± 2.273	4.154 ± 0.033	2.722 ± 0.076	0.025 ± 0.014	0.037 ± 0.000	0.023	288.024
<5.00	203.348 ± 1.729	2.525 ± 0.019	1.657 ± 0.044	0.016 ± 0.009	0.019 ± 0.000	0.025	288.024

OC9905 Phlogopite
J value: 0.01132 ± 0.00006 Integrated age: 945 ± 10 Ma
Initial ^{40}Ar/^{36}Ar: 736+339

Analysis	^{36}Ar/^{40}Ar	^{37}Ar/^{40}Ar	^{38}Ar/^{40}Ar	^{39}Ar/^{40}Ar	K/Ca	K/Cl	Ca/Cl	%40Rad
51	0.0000923 ± 0.0000123	0.0001 ± 0.0001	0.0000192 ± 0.000000496	0.0155 ± 0.0000827	64.4	86	1.3	97.3
52	0.00000809 ± 0.000013	0 ± 0.0001	0.0000394 ± 0.00000147	0.016 ± 0.0000912	172.7	43.2	0.3	99.8
53	0.0000422 ± 0.0000111	0.0001 ± 0.00004	0.0000451 ± 0.00000108	0.0162 ± 0.000171	139.2	38.3	0.3	98.8
54	0.0000485 ± 0.00000694	0.0002 ± 0.00004	0.0000372 ± 0.000000934	0.0166 ± 0.000129	40.8	47.7	1.2	98.6
55	0.0000353 ± 0.00000831	0.0001 ± 0.00004	0.0000313 ± 0.000000528	0.0164 ± 0.000146	70.3	55.6	0.8	99
56	0.0000186 ± 0.00000773	0.0001 ± 0.00004	0.0000644 ± 0.00000247	0.0167 ± 0.0000823	159.1	27.6	0.2	99.4
57	0.0000348 ± 0.00000875	0.0001 ± 0.00005	0.0000579 ± 0.00000264	0.0164 ± 0.000117	85.7	30.1	0.4	99
58	0.0000133 ± 0.00000591	0.0024 ± 0.00007	0.0000353 ± 0.00000138	0.0161 ± 0.000242	3.5	48.5	13.8	99.6
59	0.000034 ± 0.0000163	0.0001 ± 0.00005	0.0000356 ± 0.00000155	0.0162 ± 0.000187	66.4	48.5	0.7	99
60	0.0000256 ± 0.0000111	0.0001 ± 0.00004	0.0000528 ± 0.000000964	0.0166 ± 0.000148	88	33.5	0.4	99.2

Note: Radiogenic elements are denoted by *. NTP—Normal temperature and pressure (25 °C and 1 atm). All Ar isotope values are volumes in units of 10^{-10} cm^3 NTP.

tectonic unroofing of the Oaxacan Complex recorded in the overlying Paleozoic strata (Centeno-Garcia and Keppie, 1999) and the Jurassic tectonothermal event recorded in the adjacent Acatlan Complex (Powell et al., 1999).

INTERPRETATION AND CONCLUSIONS

Combining these new ages with published data allows the cooling history of the northern Oaxacan Complex to be quite precisely defined (Fig. 3). It may be divided into two stages: an early stage of relatively rapid cooling between ca. 978 and 945 Ma, followed by a slower cooling stage. The early stage shows that, following the Zapotecan granulite-facies metamorphism, the northern Oaxacan Complex cooled from 700 °C through 450 °C at a rate of 8 °C/m.y. Note that the previously published hornblende age falls slightly off the mean curve, which suggests that either there is excess argon (not supported by the data) (Solari et al., 2003) or that the closure temperature is too low. Given the large size of the hornblende crystals (up to 3–4 mm) and their high Fe-Ti composition, it would appear that a closure temperature at least 100–200 °C higher might be in order. Using published estimates for peak *P-T* conditions of 7.2–8.2 kilobars (26.7–30.4 km) at 700–825 °C for the northern Oaxacan Complex (Mora et al., 1986) yields a geothermal gradient of 23–31 °C/km at the end of the Zapotecan orogeny. This straddles the medium- to lower-pressure metamorphic gradients. Using previous *P-T* estimates of 480 °C and 5 kilobars during retrograde metamorphism (Keppie and Ortega-Gutiérrez, 1999), one may estimate $\delta P/\delta T$ (the change in pressure/change in temperature) rate to be 6–15 bars/°C. Thus, by 945 Ma, when the rocks had cooled to 450 °C, the northern Oaxacan Complex had risen above a depth of 15 km, consistent with intrusion of the 917 ± 6-Ma Etla pluton at a depth of 6–15 km (Ortega-Obregon et al., 2003). This represents >11–16 km of exhumation during the first stage of cooling, at a mean rate of 0.33–0.49 mm/yr. This is higher than the maximum erosion rate of 0.25 mm/yr (Harley, 1992), which suggests that the northern Oaxacan Complex probably underwent an additional component of tectonic erosion. To date, all greenschist- to amphibolite-facies shear zones in the northern Oaxacan Complex have yielded Paleozoic ages (Solari et al., 2004). Cooling rates in other Proterozoic granulites are generally between 1 and 5 °C/m.y. (Harley, 1992): the 10 °C/m.y. recorded only in the Imataca Complex of Venezuela and the Pielavesi block in Finland was attributed to overthrusting and a local magmatic heat source, respectively. In contrast, the northern Oaxacan Complex is inferred to have been thrust beneath an arc or continent during the Zapotecan orogeny (Solari et al., 2003), which is consistent with the shallow level of intrusion of the 1012 ± 12 Ma anorthosite-mangerite-charnockite-granite (AMCG) suite.

However, after 945 Ma, the 450–150 °C cooling rate is 2 °C/m.y. If this rate continued until the rocks reached the surface, they would have been exposed between 710 and 760 Ma, considerably earlier than the age of the nonconformably overlying Tremadocian rocks (489 ± 1 to 478 ± 4 Ma) (Okulitch, 2002).

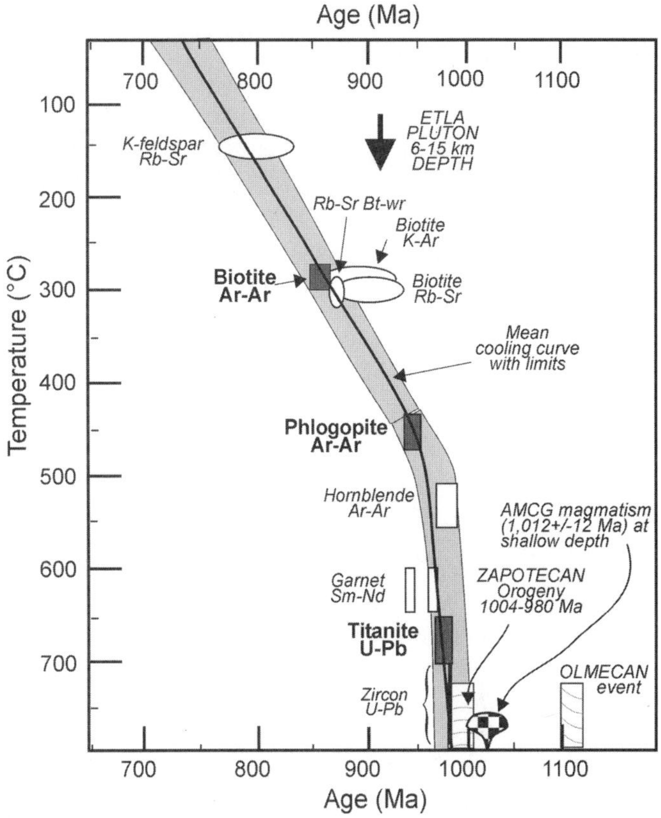

Figure 3. Cooling ages for the northern Oaxacan Complex plotted on a graph of temperature versus age. Ages presented in this paper are **boldface;** other data are from published sources (see Fries et al., 1962; Fries and Rincon-Orta, 1965; Patchett and Ruis, 1987; Keppie et al., 2003; Solari et al., 2003). wr—Whole rock.

Field observation and thermomechanical modeling of subduction-collision orogens suggest that granulite-facies rocks remain at depth until brought up by a subsequent tectonic event, such as orogenesis or rifting (Windley, 1993; Jamieson et al., 1998). The only potential tectonic events between the Zapotecan orogeny and the Tremadocian are the breakup of Rodinia at ca. 755–700 Ma and the birth of Iapetus at ca. 600–550 Ma (Hoffman, 1991; Dalziel, 1997). Clearly, these events are younger than the early cooling stage, and the first 200 m.y. of the second cooling stage. Therefore, some other mechanism for exhumation of the northern Oaxacan Complex must be sought. Solari et al. (2003) infer that the Zapotecan orogeny resulted from underthrusting the Oaxacan Complex beneath an arc or a continent. This tectonothermal event closely followed or overlapped intrusion of the 1012 ± 12 Ma rift-related AMCG suite (Keppie et al., 2003). The heat required to produce the Zapotecan granulite-facies metamorphism has been attributed to the intrusion of this AMCG suite (Solari et al., 2003). Although this process is consistent with *P-T-t* conditions in the northern Oaxacan Complex, it does not explain its exhumation without a subsequent tectonic event.

Several geodynamic models have been proposed that could explain exhumation of the granulitic rocks during the same orogenic cycle (Fig. 4):

1. Delamination or convective thinning of the orogenic root (Houseman et al., 1981; England and Houseman, 1988; Platt and England, 1993);
2. Forceful extrusional exhumation in a transpressive orogen (Thompson et al., 1997);
3. Overriding of a plume (Gibson and Stevens, 1998; Murphy et al., 1998);
4. Collision of a midoceanic ridge with the trench; thus, producing slab windows (Brown, 1998); and
5. Tectonic switching, involving alternating extension and compression produced by switching between steep and flat subduction (Collins, 2002).

The delamination/convective thinning model (Model 1) was proposed for the Grenville Province by Corrigan and Hanmer (1997) to explain the apparent alternation of crustal shortening and extension every 30 m.y. that produced crustal thickening followed by intrusion of intraplate AMCG suites and denudation, respectively. It is evident that extension followed shortening, because the 1090–1050 Ma AMCG suite has not undergone regional metamorphism and is virtually undeformed (Corrigan and Hanmer, 1997). In constrast, in the Oaxacan Complex, extension precedes shortening. Thus, intrusion of the 1012 ± 12 Ma AMCG suite immediately predates and overlaps the early stages of the Zapotecan orogeny (1004–979 ± 3 Ma) (Keppie et al., 2003; Solari et al., 2003). Furthermore, there is a difference of 90 m.y. between the 1106 ± 6 Ma Olmecan (?)extensional tectonothermal event and intrusion of the AMCG suite. Thus, no crustal thickening appears to have preceded intrusion of the AMCG suites in the northern Oaxacan Complex, and so this mechanism appears to be at odds with the geological record in the northern Oaxacan Complex.

The forceful extrusional exhumation model of Thompson et al. (1997) (Model 2) assumes that a tabular vertical zone set in a transpressional orogen is subjected to simultaneous flattening and side-parallel shearing, which induces upward and horizontal extrusion. The exhumation rate is related to the convergence rate between the colliding plates and the angle between the plate boundary and the displacement vector. Although this model is capable of deforming and rapidly exhuming granulitic rocks during one cycle, one would expect the extrusion zone to display steeply dipping fabrics. Such a feature is absent in the northern Oaxacan Complex, which consists of gently north-dipping thrust slices that were not refolded into upright folds until the Mesozoic (Solari et al., 2004).

Overriding of a plume causes flattening of the subducting slab and exhumation over the associated swell, which may reach the dimensions of ≤400 × 2400 km^2 (Murphy et al., 1998) (Model 3). This flattening causes a sequence of events: (1) inboard migration of the magmatic arc; (2) a period of magmatic

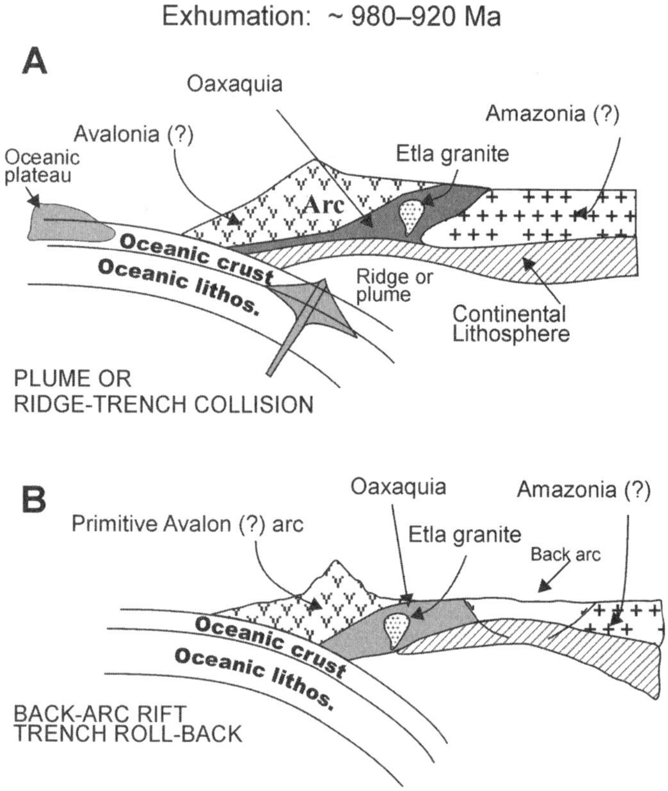

Figure 4. Candidate models for the rapid 978–945 Ma exhumation of granulitic rocks. (A) Over-riding of a plume, a midoceanic ridge, and/or an oceanic plateau. (B) Backarc rifting.

quiescence, as the plume burns through the overlying lower lithosphere; (3) voluminous intracrustal magmatism, which is mainly felsic, as the plume impinges on the lower crust; (4) inboard migration of the deformation front; and (5) rapid exhumation and cooling (0.5–1.5 km/m.y. and 25 °C/m.y.) (Murphy et al., 1998; Keppie and Krogh, 1999). Migration of the plume head causes these effects to be diachronous.

Collision of a midoceanic ridge with the trench produces various types of triple junctions, of which TTF-TTR and FFT-RTF evolve slab windows (Cronin, 1992), where F = transform fault, R = ridge, and T = trench (Model 4). Such collisions may also produce flat-slab subduction and were proposed for the switch from arc to rift-related plutonism above the slab-window, low- to medium-pressure metamorphism (e.g., the Abukuma *P-T* path passes through 700 °C at 11 kilobars, 750 °C at 4 kilobars), associated polyphase deformation, and rapid exhumation in the forearc region of Japan (Brown, 1998). These effects are also diachronous, due to migration of the triple point. The relative rates of plate motions and the relative orientations of plate boundaries determine the sequence of events recorded in adjacent regions (Fig. 4). Thus, in some situations, convergence follows extension (e.g., subduction resuming after ridge-trench collision); in others, convergence precedes extension (subduction followed by ridge-trench collision).

The tectonic switching mechanism proposed by Collins (2002) (Model 5) for granulite-facies terrains involves the alternation of two regimes:

1. Steeply dipping subduction, involving trench rollback and extension of the arc that is split by a backarc, under which the asthenosphere rises to <40 km, inducing adiabatic decompression melting and granulite-facies metamorphism; and
2. Flat-slab subduction, due to the arrival of an oceanic plateau at the trench, causing contractional deformation that was focused in the thermally softened, arc-backarc region, which cools quickly because it is now isolated from the asthenosphere.

The size of the oceanic plateaus limits the size of the resulting orogen to <500 × 100 km^2, and the orogenic event is short-lived (5–15 m.y.) and diachronous both across and along the orogenic belt. Note that flat-slab subduction is also associated with (1) overriding a plume (Model 3), and indeed, many oceanic plateaux are the products of plumes; (2) ridge-trench collision (Model 4), because young oceanic lithosphere is more buoyant; and (3) high convergence rates.

Thus, models 1 and 2 appear to be eliminated for the northern Oaxacan Complex on geological grounds. Models 3 and 4 involve flat-slab subduction associated with a plume or a ridge, which, although consistent with the general absence of associated arc magmatism, is not in accord with the absence of plume and slab-window magmatism. Diachronism is predicted by both models 3 and 4; however, the age of granulite-facies metamorphism from all the ca. 1 Ga inliers in Mexico is the same within error (Cameron et al., this volume). However, the limited amount of isotopic data from other ca. 1 Ga inliers spread over an area of 200 × 1000 km^2 precludes comparison of their subsequent cooling histories at present.

With the present database, the tectonic switching model (Model 5) appears to be the most consistent with the geological record of the northern Oaxacan Complex, which involved a sequence of extension-contraction-extension events reflected as: (1) intrusion of rift-related plutons at 1012 ± 12 Ma at a shallow depth; (2) the contractional Zapotecan orogenic event at 1004 ± 3 to 978 ± 3 Ma at a depth of 26–31 km; (3) retrograde metamorphism at a depth of ≤15 km and arc-related magmatism at 917 ± 6 Ma at a depth of 6–15 km. Such a sequence is consistent with both the depth changes from shallow to deep to shallow and changes in the angle of subduction: steep-shallow-steep. The absence of arc magmatism until ca. 917 Ma is explicable if the northern Oaxacan Complex was located in the backarc region during the emplacement of the rift-related ca. 1012 Ma magmatism, and the Zapotecan orogeny was associ-

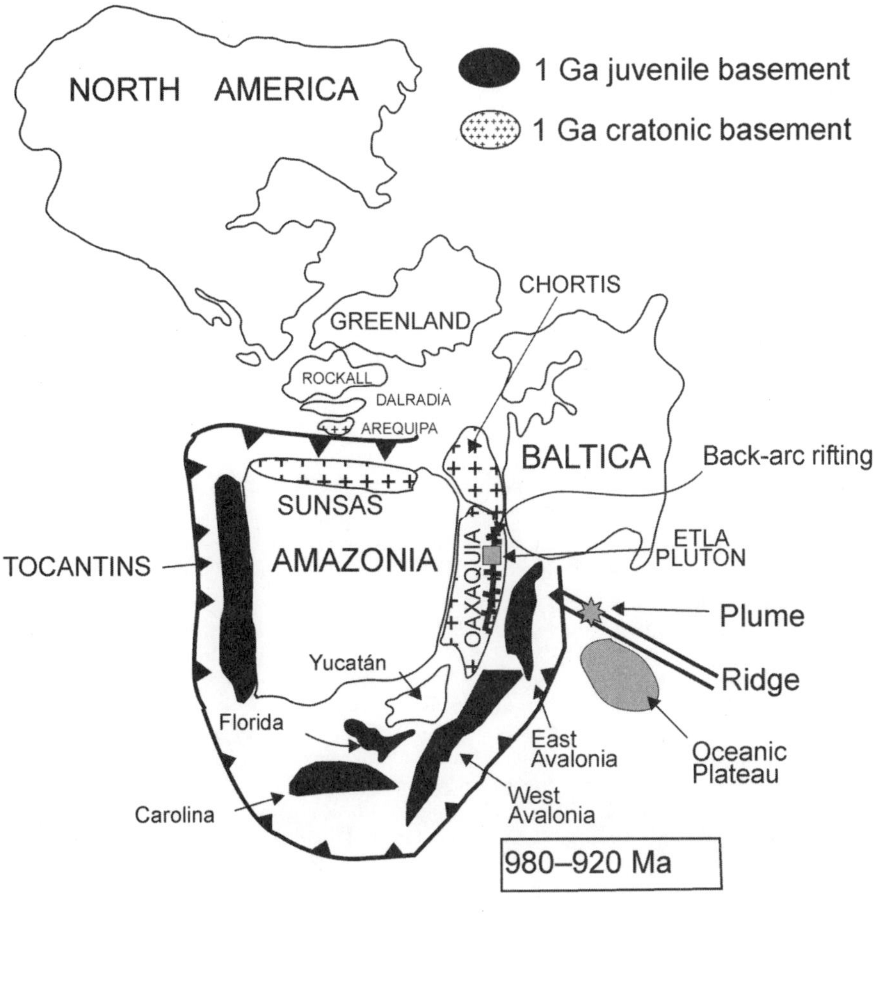

Figure 5. Reconstruction of Rodinia, showing the inferred location of Oaxaquia and the collision of a midoceanic ridge, a plume, and/or an oceanic plateau with the trench, followed by backarc rifting. Modified after Keppie et al. (2001).

ated with flat slab subduction. The limited amount of arc magmatism at ca. 917 Ma suggests that the northern Oaxacan Complex lay on the periphery of the arc adjacent to the rifting backarc. As flat-slab subduction can also be produced by both overriding a plume and ridge-trench collision (models 3 and 4), these mechanisms cannot be ruled out without additional data.

Models 3–5 also require subduction of an ocean on one side of Oaxaquia. This is consistent with recently published reconstructions. Thus, Keppie and Ortega-Gutiérrez (1995) and Keppie et al. (2001) proposed that Oaxaquia lay along the periphery of Rodinia, adjacent to Amazonia, at ca. 1 Ga, with a subducting ocean on one side (Fig. 5). Approximately contemporaneous primitive arc magmatism around Amazonia is indicated in at least two places:

1. The 940–630 Ma Tocantins province of central Brazil (Goiás juvenile magmatic arc) was thrust onto the Amazon craton during development of the Paraguay belt, a Late Neoproterozoic (ca. 630 Ma) belt (Pimentel et al., 2000); and
2. The 1.2–1.0 Ga juvenile arc of Avalonia and Carolinia, which was placed adjacent to Oaxaquia on recent reconstructions (Murphy et al., 1999, 2000; Keppie et al., 2001).

The alternating flat-steep–subduction model proposed for the ca. 1012–917 Ma history of the northern Oaxacan Complex may have implications for similar features in the Grenville Province. If such a mechanism is applied to convergent and extensional events in the Grenville Province, a different geometry and/or relative rates of plate motions are required to produce the opposite sequence to that observed in the Oaxacan Complex of southern Mexico. These models may be tested by searching for such things as diachronism along the belt.

The slower cooling rate recorded by the northern Oaxacan Complex between 945 and 760–710 Ma is similar to that in most granulite terrains (1–5 °C/m.y.) (Harley, 1992), and may result from cessation of an active margin, leading to slow thermal relaxation and exhumation of the region and terminating in rifting associated with the breakup of Rodinia, which brought the northern Oaxacan Complex to the surface. Fission track ages may better define the lowest temperature increment.

ACKNOWLEDGMENTS

Funding for various aspects of this project was provided by Consejo Nacional de Ciencias y Technológia grants (0255P-T9506 and 25705-T), Programa de Apoyo a Proyectos de Investigación e Innovación grants (IN116999 and IN10799) to JDK, FO-G and AO-R, National Science and Engineering Research Council grant to JKWL, a Mexico–United States of America grant to KLC and FO-G, and an Internal Research grant from the Instituto de Geologia, Universidad Nacional Autónoma de México to JDK and AO-R. We thank Ken Cameron and Pete Holden (University of California, Santa Cruz) for assistance with the U-Pb isotopic analyses, and Carlos Ortega for sample preparation and for drafting the figures. We also thank J.P. Hogan, R. Moraes, and an anonymous reviewer for their constructive comments on the manuscript.

REFERENCES CITED

Bloomfield, K., and Ortega-Gutiérrez, F., 1975, Notas sobre la petrologia del Complejo Oaxaqueño: Instituto de Geología, Universidad Nacional Autonoma de Mexico Boletin, v. 95, p. 23–48 (in Spanish).

Brown, M., 1998, Ridge-trench interactions and high-*T*–low-*P* metamorphism, with particular reference to the Cretaceous evolution of the Japanese Islands, *in* Treloar, P.J., and O'Brien, P.J., eds., What drives metamorphism and metamorphic reactions?: London, Geological Society Special Publication 138, p. 137–168.

Centeno-Garcia, E., and Keppie, J.D., 1999, Latest Paleozoic–Early Mesozoic structures in the central Oaxaca terrane of southern Mexico: Deformation near a triple junction: Tectonophysics, v. 301, p. 231–242.

Clark, A.H., Archibald, D.A., Lee, A.W., Farrar, E., and Hodgson, C.J., 1998, Laser probe $^{40}Ar/^{39}Ar$ ages of early- and late-stage alteration assemblages, Rosario porphyry copper-molybdenum deposit, Collahuasi District, I Region, Chile: Economic Geology, v. 93, p. 326–337.

Collins, W.J., 2002, Hot orogens, tectonic switching, and creation of continental crust: Geology, v. 30, p. 535–538.

Corrigan, D., and Hanmer, S., 1997, Anorthosites and related granitoids in the Grenville orogen: A product of convective thinning of the lithosphere?: Geology, v. 25, p. 60–64.

Cox, L.R.M., and Fortey, R.A., 1990, Biogeography of Ordovician and Silurian faunas, *in* McKerrow, W.S., and Scotese, C.R., eds., Paleozoic paleogeography and biogeography: London, Geological Society Memoir 12, p. 97–104.

Cronin, V.S., 1992, Types and kinematic stability of triple junctions: Tectonophysics, v. 207, p. 287–301.

Dalziel, I., 1997, Overview: Neoproterozoic-Paleozoic geography and tectonics: Review, hypotheses and environmental speculations: Geological Society of America Bulletin, v. 109, p. 16–42.

Dostal, J., Keppie, J.D., Macdonald, H., and Ortega-Gutiérrez, F., 2004, Sedimentary origin of calcareous intrusions in the 1 Ga Oaxacan Complex, Southern Mexico: tectonic implications: International Geology Reviews (in press).

England, P.C., and Houseman, G.A., 1988, The mechanics of the Tibetan Plateau, *in* Shackleton, R.M., Dewey, J.F., and Windley, B.F., eds., Tectonic evolution of the Himalayas and Tibet: Philosophical Transactions of the Royal Society of London, v. A326, p. 301–319.

Fries, C., Jr., and Rincon-Orta, C., 1965, Nuevas aportaciones geocronológicas y técnicas empleadas en el Laboratorio de Geocronometría: Universidad Nacional Autónoma de México, Instituto de Geología Boletin, v. 73, p. 57–133 (in Spanish).

Fries, C., Jr., Schmitter, E., Damon, P.E., Livingston, D.E., and Erikson, R., 1962, Edad de las rocas metamórficas en las cañones de La Peregrina y de Caballeros, parte centro-occidental de Tamaulipas: Universidad Nacional Autónoma de México, Instituto de Geología Boletin, v. 64, p. 55–69 (in Spanish).

Frost, B.R., Chamberlain, K.R., and Schumacher, J.C., 2000, Sphene (titanite): Phase relations and role as a geochronometer: Chemical Geology, v. 172, p. 131–148.

Gibson, R.L., and Stevens, G., 1998, Regional metamorphism due to anorogenic intracratonic magmatism, *in* Treloar, P.J., and O'Brien, P.J., eds., What drives metamorphism and metamorphic reactions?: London, Geological Society Special Publication 138, p. 121–135.

Giletti, B., 1974, Studies in argon diffusion I: Argon in phlogopite mica, *in* Hoffman, A.W., et al., eds., Geochemical transport and kinetics: Washington, D.C., Carnegie Institute of Washington Publication 634, p. 107–115.

Hames, W.E., and Bowring, S.A., 1994, An empirical evaluation of argon diffusion geometry in muscovite: Earth and Planetary Science Letters, v. 124, p. 1612–1617.

Harley, S.L., 1992, Proterozoic granulite terrains, *in* Condie, K.C., ed., Proterozoic crustal evolution: Developments in Precambrian Geology, v. 10, p. 301–360.

Harrison, T.M., Duncan, I., and McDougall, I., 1985, Diffusion of ^{40}Ar in biotite: Temperature, pressure, and compositional effects: Geochimica et Cosmochimica Acta, v. 49, p. 2461–2468.

Hoffman, P.F., 1991, Did the breakout of Laurentia turn Gondwanaland inside out?: Science, v. 252, p. 1409–1412.

Houseman, G., McKenzie, D., and Molnar, P., 1981, Convective instability of a thermal boundary layer and its relevance to the thermal evolution of continental convergent belts: Journal of Geophysical Research, v. 86, p. 6115–6132.

Humphries, F.J., and Cliff, R.A., 1982, Sm-Nd dating and cooling history of Scourian granulites, Sutherland, NW Scotland: Nature, v. 295, p. 515–517.

Jamieson, R.A., Beaumont, C., Fullsack, P., and Lee, B., 1998, Barrovian regional metamorphism: Where's the heat?, *in* Treloar, P.J., and O'Brien, P.J., eds., What drives metamorphism and metamorphic reactions?: London, Geological Society Special Publication 138, p. 23–52.

Keppie, J.D., and Krogh, T.E., 1999, U-Pb geochronology of Devonian granites in the Meguma Terrane of Nova Scotia, Canada: Evidence for hotspot melting of a Neoproterozoic source: Journal of Geology, v. 107, p. 555–568.

Keppie, J.D., and Ortega-Gutiérrez, F., 1995, Provenance of Mexican terranes: Isotopic constraints: International Geology Reviews, v. 37, p. 813–824.

Keppie, J.D., and Ortega-Gutiérrez, F., 1999, Middle American Precambrian basement: A missing piece of the reconstructed 1-Ga orogen, *in* Ramos, V.A., and Keppie, J.D., eds., Laurentia-Gondwana connections before Pangea: Boulder, Colorado, Geological Society of America Special Paper 336, p. 199–210.

Keppie, J.D., Dostal, J., Ortega-Gutiérrez, F., and Lopez, R., 2001, A Grenvillian arc on the margin of Amazonia: Evidence from the southern Oaxacan Complex, southern Mexico: Precambrian Research, v. 112, no. 3–4, p. 165–181.

Keppie, J.D., Dostal, J., Cameron, K.L., Solari, L.A., Ortega-Gutiérrez, F., and Lopez, R., 2003, Geochronology and geochemistry of Grenvillian igneous suites in the northern Oaxacan Complex, southern Mexico: Tectonic implications: Tectonophysics, v. 120, p. 365–389.

Krogh, T.E., 1973, A low-contamination method for hydrothermal decomposition of zircon and extraction of U and Pb for isotopic age determinations: Geochimica et Cosmochimica Acta, v. 37, p. 485–494.

Lawlor, P.J., Ortega-Gutiérrez, F., Cameron, K.L., Ochoa-Camarillo, H., Lopez, R., and Sampson, D.E., 1999, U-Pb geochronology, geochemistry, and provenance of the Grenvillian Huiznopala Gneiss of Eastern Mexico: Precambrian Research, v. 94, p. 73–99.

Lopez, R.L., Cameron, K.L., and Jones, N.W., 2001, Evidence for Paleoproterozoic, Grenvillian, and Pan-Africa age crust beneath northeastern Mexico: Precambrian Research, v. 107, p. 195–214.

Ludwig, K.R., 1991, PbDat: A computer program for processing Pb-U-Th isotope data, version 1.24: Reston, Virginia, U.S. Geological Survey Open-File Report 88-542.

Mattinson, J.M., 1987, U-Pb ages of zircons: A basic examination of error propagation: Chemical Geology, v. 66, p. 151–162.

Mora, C.I., Valley, J.W., and Ortega-Gutiérrez, F., 1986, The temperature and pressure conditions of Grenville-age granulite-facies metamorphism of the Oaxacan Complex, southern México: Universidad Nacional Autónoma de México, Instituto de Geologia Revista, v. 5, p. 222–242.

Murphy, J.B., Oppliger, G.L., Brimhall, G.H., Jr., and Hynes, A., 1998, Plume-modified orogeny: An example from the western United States: Geology, v. 26, p. 731–734.

Murphy, J.B., Keppie, J.D., Dostal, J., and Nance, R.D., 1999, Neoproterozoic–Early Paleozoic evolution of Avalonia, *in* Ramos, V.A., and Keppie, J.D., eds., Laurentia-Gondwana connections before Pangea: Boulder, Colorado, Geological Society of America Special Paper 336, p. 253–266.

Murphy, J.B., Strachan, R.A., Nance, R.D., Parker, K.D., and Fowler, M.B., 2000, Proto-Avalonia: A 1.2–1.0 Ga tectonothermal event and constraints for the evolution of Rodinia: Geology, v. 28, p. 1071–1074.

Okulitch, A.V., 2002, Geological time chart: Québec, Geological Survey of Canada, 1 p.

Ortega-Gutiérrez, F., 1984, Evidence of Precambrian evaporites in the Oaxacan granulite complex of southern México: Precambrian Research, v. 23, p. 377–393.

Ortega-Obregon, C., Keppie, J.D., Solari, L.A., Ortega-Gutiérrez, F., Dostal, J., Lopez, R., Ortega-Rivera, A., and Lee, J.K.W., 2003, Geochronology and geochemistry of the 917 Ma, calc-alkaline Etla granitoid pluton (Oaxaca, southern Mexico): Evidence of post-Grenvillian subduction along the northern margin of Amazonia: International Geology Review, v. 45, p. 596–622.

Parrish, R.R., 1987, An improved micro-capsule for zircon dissolution in U-Pb geochronology: Chemical Geology, v. 66, p. 99–102.

Patchett, P.J., and Ruiz, J., 1987, Nd isotopic ages of crust formation and metamorphism in the Precambrian of eastern and southern Mexico: Contributions to Mineralogy and Petrology, v. 96, p. 523–528.

Pimentel, M.M., Fuck, R.A., Jost, H., Ferreira Filho, C.F., and de Araújo, S.M., 2000, The basement of the Brasilia fold belt and the Goiás magmatic arc, *in* Cordani, U.G., et al., eds., Tectonic evolution of South America: Rio de Janeiro, Brazil, 31st International Geological Congress, p. 195–230.

Platt, J.P., and England, P.C., 1993, Convective removal of lithosphere beneath mountain belts: Thermal and mechanical consequences: American Journal of Science, v. 293, p. 307–336.

Powell, J.T., Nance, R.D., Keppie, J.D., and Ortega-Gutiérrez, F., 1999, Tectonic significance of the Magdalena Migmatite, Acatlan Complex, southern Mexico: Geological Society of America Abstracts with Programs, v. 31, no. 7, p. A-294.

Robison, R., and Pantoja-Alor, J., 1968, Tremadocian trilobites from Nochixtlan region, Oaxaca, Mexico: Journal of Paleontology, v. 42, p. 767–800.

Shergold, J.H., 1975, Late Cambrian and Early Ordovician trilobites from the Burke River Structural Belt, western Queensland, Australia: Canberra, Department of Mineral and Energy, Bureau of Mineral Resources, Geology and Geophysics, Bulletin 153, 221 p.

Solari, L.A., Keppie, J.D., Ortega-Gutiérrez, F., Cameron, K.L., Lopez, R., and Hames, W.E., 2003, 990 Ma and 1,100 Ma Grenvillian tectonothermal events in the northern Oaxacan Complex, southern Mexico: Roots of an orogen: Tectonophysics, v. 365, p. 257–282.

Solari, L.A., Keppie, J.D., Ortega-Gutiérrez, F., Ortega-Rivera, A., Hames, W.E., and Lee, J.K.W., 2004, Phanerozoic structures in the Grenvillian northern Oaxacan Complex, southern Mexico: The result of thick-skinned tectonics: International Geology Reviews (in press).

Thompson, A.B., Schulmann, K., and Jezek, J., 1997, Thermal evolution and exhumation in obliquely convergent (transpressive) orogens: Tectonophysics, v. 280, p. 171–184.

Verts, L.A., Chamberlain, K.R., and Frost, C.D., 1996, U-Pb sphene dating of metamorphism: The importance of sphene growth in the contact aureole of the Red Mountain pluton, Laramie Mountains, Wyoming: Contributions to Mineralogy and Petrology, v. 125, p. 186–199.

Weber, B., and Köhler, H., 1999, Sm/Nd, Rb/Sr, and U-Pb geochronology of a Grenville terrane in southern Mexico: Origin and geologic history of the Guichicovi complex: Precambrian Research, v. 96, p. 245–262.

Windley, B.F., 1993, Uniformitarianism today: Plate tectonics is the key to the past: Journal of the Geological Society of London, v. 150, p. 7–19.

Manuscript Accepted by the Society August 25, 2003

Printed in the USA

Geological Society of America
Memoir 197
2004

Tectonic evolution of the eastern Llano uplift, central Texas: A record of Grenville orogenesis along the southern Laurentian margin

Sharon Mosher*
April M. Hoh
Jostin A. Zumbro*
Department of Geological Sciences, Jackson School of Geosciences, University Station C1100, University of Texas at Austin, Austin, Texas 78712, USA
Joseph F. Reese
Department of Geosciences, Edinboro University of Pennsylvania, Edinboro, Pennsylvania 16444, USA

ABSTRACT

Mesoproterozoic metamorphic rocks exposed in the eastern Llano uplift, central Texas, were involved in a ca. 1150–1116 Ma Grenville-age orogenic event along the southern margin of Laurentia. Collision of the exotic Coal Creek arc and a southerly continental block with Laurentia tectonically telescoped and stacked three distinct lithotectonic domains. A major ductile shear zone forms the contact between the Coal Creek ensimatic arc terrane, the southwesternmost domain, and rocks with Laurentian affinities to the north. Directly north of the arc boundary, supracrustal rocks of the Packsaddle domain represent basinal sedimentary and volcanic rocks deposited along the southern Laurentian margin. Farther north, granitic gneisses of the Valley Spring domain, which consist of both plutonic and supracrustal rocks, represent a Laurentian continental-margin arc emplaced beneath the Packsaddle domain along a shear zone.

Recent mapping in the northeastern uplift shows that the Valley Spring domain records a polyphase deformational history equivalent to that observed in the Packsaddle domain of the southeastern uplift. Early deformation occurred under uppermost amphibolite-facies metamorphic conditions and was accompanied by partial melting and formation of foliation-parallel leucosomes, consistent with deformation deeper in the orogenic pile. Mylonitic rocks in the shear zone separating these two domains show thrusting to the northeast, similar to that shown by the shear zone separating the Coal Creek arc from the Packsaddle domain. The Valley Spring domain igneous complex adjacent to the shear zone is highly attenuated into thin sheets in this zone.

The Grenville event in the eastern uplift is characterized by polyphase ductile deformation synchronous with upper amphibolite-facies dynamothermal metamorphism. Deformation progressed from northeast-directed ductile thrusting and folding

*E-mail, corresponding author: mosher@mail.utexas.edu; present address, Zumbro: 3 Hutton Center Drive, Suite 200, Santa Ana, California 92707, USA.

Mosher, S., Hoh, A.M., Zumbro, J.A., and Reese, J.F., 2004, Tectonic evolution of the eastern Llano Uplift, central Texas: A record of Grenville orogenesis along the southern Laurentian margin, *in* Tollo, R.P., Corriveau, L., McLelland, J., and Bartholomew, M.J., eds., Proterozoic tectonic evolution of the Grenville orogen in North America: Boulder, Colorado, Geological Society of America Memoir 197, p. 783–798. For permission to copy, contact editing@geosociety.org.

(D_1, D_2), which accommodated collision-related crustal thickening and contraction, to polyphase regional-scale folding (D_3, D_4, D_5), which accommodated continued collision-related north- to northeast-directed contraction. The kinematics of the deformation in the eastern uplift are consistent with the northeastward collision of an exotic arc terrane and a southern continental block with the generally east-trending Laurentian margin. No evidence of transcurrent motion along the margin has been observed.

Keywords: Grenville, Llano Uplift, Rodinia, collision, Mesoproterozoic

INTRODUCTION

The amalgamation of major continents at the end of the Mesoproterozoic, leading to the formation of the supercontinent Rodinia, resulted in many Grenville-age orogenic belts with generally similar orogenic histories (e.g., Dalziel et al., 2000). Laurentia, which plays a central position in Rodinian reconstructions, has a Grenville-age orogenic belt along both its east-trending southern margin and its northnortheast-trending eastern margin (Fig. 1, inset) (Mosher, 1998). The timing and style of orogenesis along the two margins are similar, leading to the question of whether (1) a single continent-continent collision resulted in a continuous Grenville orogenic belt along the entire Laurentian margin or (2) two separate continents collided with Laurentia nearly contemporaneously, producing orogenic belts along the orthogonal margins. The answer to this question is critical for reconstructing the Rodinian supercontinent and lies in the kinematics of the deformation along the two margins.

The eastern margin of Laurentia records continent-continent collision at ca. 1.1–1.0 Ga, which telescoped previously formed or accreted island arcs, allochthonous terranes, and older basement with their transport generally orthogonal to the Laurentian margin (Moore and Thompson, 1980; Davidson, 1986; Easton, 1986; Rivers et al., 1989; Corriveau et al., 1990; Culotta et al., 1990; Gower et al., 1990; Indares, 1993; Connelly et al., 1995; Gower, 1996). The latest movement on the frontal thrusts is dated at 995–980 Ma (Krogh, 1994).

Along the southern margin of Laurentia, the high-grade collisional core of the Grenville-age orogen is exposed in the Llano uplift of central Texas, as much as 300 km behind the deformation front, and the frontal region is exposed in west Texas, where metamorphic rocks were thrust over foreland Mesoproterozoic rocks (Fig. 1, inset) (Mosher, 1998). Within the southeastern Llano uplift, an accreted exotic-arc terrane was thrust northeastward over basinal supracrustal rocks that record polyphase deformation and thrusting along ductile shear zones, with tectonic transport to the northeast (Mosher, 1998). In west Texas, metamorphic rocks record northwestward transport associated with oblique, dextral transpression (Grimes 1999; Grimes and Mosher, 2003). These rocks decrease in grade northward from mid-amphibolite to lower-greenschist facies, where they are thrust over sedimentary and volcanic rocks in the foreland (Grimes, 1999); the underlying narrow fold-and-thrust belt also formed in a transpressive setting, although the shear sense is unknown (Soegaard and Callahan, 1994). Mosher (1998) called on the collision of an indentor continent from the south to explain the generally northward transport that diverges from northwest in west Texas to northeast in central Texas. Syn-metamorphic deformation in the orogen core occurred ca. 1150–1116 Ma, whereas deformation in west Texas was later, with the latest motion on frontal thrusts ca. 1030–981 Ma (Mosher, 1998; Grimes, 1999).

The nearly orthogonal orientation of the Laurentian margins and interpreted convergence directions imply collision of separate continental blocks (Mosher, 1998; Grimes, 1999; Dalziel et al., 2000; Tohver et al., 2002; Grimes and Mosher, 2003). Collision of a single eastern continental block, as implied by Bickford et al. (2000) and Karlstrom et al. (2001), would require bulk transcurrent motion and associated deformation along the entire southern margin. In the Llano uplift, evidence from the southeastern uplift (Carter, 1989; Nelis et al., 1989; Reese, 1995; Reese and Mosher, 2004) was used to support collision of a continental block from the south or southwest. The possibility remains, however, that deformation in this area could be solely related to accretion of the arc terrane, and the rest of the uplift could record transcurrent-related deformation. In this paper, we present the results of two structural studies in the northeastern uplift that demonstrate that the entire eastern uplift has a similar tectonic and kinematic history and that deformation is not partitioned between convergent and transcurrent motion.

Lithotectonic Domains

The Mesoproterozoic metamorphic rocks of the Llano uplift are divided into three regionally extensive lithotectonic domains, which are from north to south in the eastern uplift: the Valley Spring domain, the Packsaddle domain, and the Coal Creek domain (Fig. 1) (Mosher, 1996, 1998; Roback, 1996a). The Valley Spring domain consists of plutonic and supracrustal rocks interpreted to represent a continental-margin arc, terrigenous clastic rocks, and some older continental basement (Mosher, 1998; Zumbro, 1999; Reese et al., 2000). It consists predominantly of quartzofeldspathic gneisses, with minor inter-

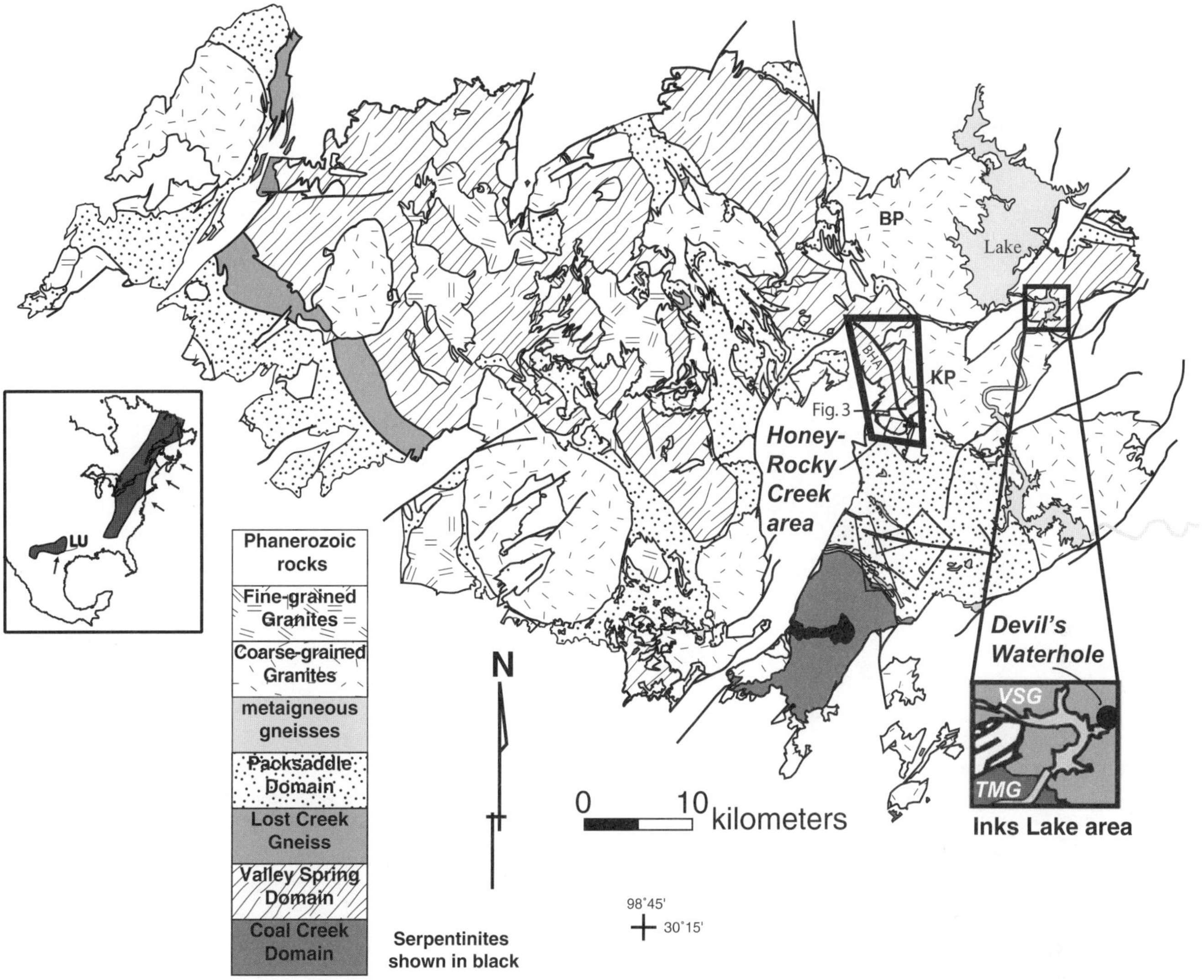

Figure 1. Geologic map of the Llano uplift. Location of study areas in the northeastern uplift are outlined by boxes (thick lines); inset of Inks Lake area shows location of Devil's Waterhole. Location of Babyhead Anticline (BHA) and Figure 3 are shown in Honey Creek–Rocky Creek area. Location of previous studies in Packsaddle domain are shown by boxes (thin lines). Inset (left) shows the east- and north-northeast–trending Grenville-age orogens along the southern and eastern margins of Laurentia, respectively. The interpreted direction of convergence is shown. BP—Lake Buchanan pluton; KP—Kingsland pluton; LU—Llano uplift. Map is modified from Barnes (1981) and Reed (1999).

calated marble, mafic and pelitic schists, and amphibolite (Barnes, 1981) with igneous crystallization ages ranging from 1288 ± 2 to 1232 ± 4 Ma and one anomalous age of 1366 ± 3 Ma (Walker, 1992; Reese et al., 2000; Connelly, unpublished data; Roback, unpublished data). The structurally uppermost Valley Spring domain is an intrusive complex juxtaposed with the Packsaddle domain along a ductile shear zone; the structurally lower parts contain sedimentary and perhaps volcanic rocks intermixed with plutonic rocks (Zumbro, 1999). The structurally overlying Packsaddle domain contains a package of supracrustal rocks that are interpreted as shallow shelf/slope sediments, with interbedded volcanic and volcanoclastic deposits deposited on the flank of the continental-margin volcanic arc (Garrison, 1981; Mosher, 1993). Volcanic rocks range in age from 1274 ± 2 to 1244 ± 5 Ma and are intruded by minor 1238–1256 Ma granitic sills and dikes (Roback, 1996a; Reese et al., 2000). The Coal Creek domain represents an exotic ensimatic arc terrane comprised of a 1326 ± 2 to 1275^{+2}_{-1} Ma mafic to intermediate plutonic complex. The Coal Creek domain has Pb and Nd isotopic signatures, as well as petrographic and geochemical characteristics, that are distinctly different from the rocks of the Packsaddle and Valley Spring domains; the Coal Creek domain records an earlier deformational and metamorphic history (Roback, 1996a; Whitefield, 1997; Mosher, 1998). This domain is separated from the rest of the uplift by a 2–3-km-wide ductile shear zone that generally strikes northwest (Carter, 1989; Roback, 1996a; Whitefield, 1997; Mosher, 1998).

Deformation, Metamorphism and Late Plutonism

Mesoproterozoic metamorphic rocks of the uplift are everywhere polydeformed synchronous with an early dynamothermal, high- to moderate-pressure upper-amphibolite- to granulite- (transitional- to eclogite-) facies metamorphism (Wilkerson et al., 1988; Carlson and Schwarze, 1997; Carlson, 1998; Mosher, 1998). Multiple phases of deformation producing five generations of folds with associated penetrative to moderately well-developed metamorphic fabrics are documented in many parts of the uplift, along with ductile shear zones in the southeastern uplift (Denny and Gobel, 1986; Carter, 1989; Gillis, 1989; Nelis et al., 1989; Mosher, 1993, 1998; Reese, 1995; Roback, 1996a; Roback et al., 1999; Hoh, 2000; Hunt, 2000; Reese and Mosher, 2004). Pressures and temperatures for the earliest metamorphism range from 15 to 21 kilobars and 750 °C in the western uplift (Carlson, 1998; Anderson, 2001) to 6 to 8 kilobars and ~585 °C in the northwestern uplift (Carlson and Reese, 1994; Carlson and Schwarze, 1997).

A second low-pressure, mid-amphibolite–facies metamorphism (Wilkerson et al., 1988; Carlson and Schwarze, 1997; Carlson, 1998) resulted in reheating and rehydration of the rocks during intrusion of abundant late syntectonic to post-tectonic plutons (Town Mountain Granites and related finer grained intrusions) (Bebout and Carlson, 1986; Reed, 1999; Rougvie et al., 1999). This metamorphism was synchronous with late-stage deformation and also statically overprinted dynamic textures in many areas (Reed, 1999). This late metamorphic event had temperatures ranging from 475 to 625 °C and pressures of 2.75 ± 0.75 kilobars (Carlson and Schwarze, 1997).

The timing of metamorphism and synchronous deformation across the uplift is constrained between ca. 1147 and 1115 Ma by U-Pb dating of metamorphic minerals (zircons, titanite, rutile, and monazite) (Roback, 1996b; Roback and Carlson, 1996; Mosher, 1998) that record the highest temperatures but not the highest pressures. Dating of protoliths requires the pervasive deformation to postdate 1239–1232 Ma (Walker, 1992). Voluminous late granites (1126–1070 Ma) intrude all domains (Walker, 1992; Reed, 1999; Roback et al., 1999), indicating juxtaposition by this time (Mosher, 1998). Of these late granites, early syntectonic plutons (1126–1118 Ma) (Reed, 1999; Roback et al., 1999) are locally folded by late regional kilometer-scale (F_5) folds or are deformed by local shear zones, but they crosscut and postdate all earlier structures (Reed, 1999). Crosscutting undeformed pegmatites (1116 ± 3 Ma) and dikes (1098^{+3}_{-2} Ma) locally put an upper time limit on deformation, and circular ganitic plutons (1091–1070 Ma) appear to be post-tectonic (Mosher, 1998), although deformation of the country rock and margins of some plutons indicates at least local shortening during intrusion (Reed, 1999).

STRUCTURE AND METAMORPHISM

The deformational and metamorphic history of the Packsaddle domain in the southeastern uplift is well documented (Carter, 1989; Nelis et al., 1989; Reese, 1995; Mosher, 1993, 1998; Reese and Mosher, 2004) (Fig. 1). In that domain, well-layered supracrustal rocks are multiply folded, metamorphosed, and juxtaposed with the well-studied Coal Creek domain along a major shear zone that generally strikes northwest (Carter, 1989; Roback, 1996a; Whitefield, 1997). The Valley Spring domain in the northeastern uplift consists predominately of granitic gneisses. Two detailed studies (Zumbro, 1999; Hoh, 2000) document that the Valley Spring domain in the eastern uplift had undergone the same deformational and metamorphic history as the Packsaddle domain of the southeastern uplift. In the next sections, we summarize the Packsaddle domain deformational and metamorphic history and then present new data for the Valley Spring domain adjacent to the Packsaddle domain (Honey Creek–Rocky Creek area) and for an area farther northeast on the opposite side of the post-tectonic Kingsland pluton (Inks Lake area) (Fig. 1).

Packsaddle Domain of the Southeastern Uplift

In the southeastern uplift, supracrustal and minor plutonic rocks in the Packsaddle domain were affected by five phases of ductile deformation (Carter, 1989; Nelis et. al., 1989; Mosher, 1993, 1998; Reese, 1995; Reese and Mosher, 2004). D_2 structures are the most penetrative and pervasive; S_2 forms the dominant foliation and is axial planar to abundant F_2 isoclinal folds. S_2 is generally defined by metamorphic differentiation, but is mylonitic in shear zones at domain boundaries and locally within the Packsaddle domain. An earlier metamorphic foliation (S_1) that is locally well developed and axial planar to F_1 folds is folded about F_2 axes. Rare F_1 and F_2 folds form mushroom-shape interference patterns. S_1 and S_2 and lithologic contacts are generally parallel and only distinguishable in F_2 fold hinges.

Three other metamorphic fabrics and associated fold generations are readily recognized in the field. S_3, S_4, and S_5 are generally crenulation cleavages with metamorphic differentiation in the limbs and are distinguished by their superposition and axial planar relationship to associated fold generation. S_4 strikes at a high angle to S_3 and S_5 (Fig. 2, A, C, and E). Associated fold generations are identified based on the folding of pre-existing fabrics and fold generations. F_3 folds are northeast-verging, southeast-plunging, tight, chevron to semi-chevron folds (Fig. 2, A). F_4 folds are north- to north-northeast–verging, chevron, tight to closed asymmetric folds characterized by an east-striking axial plane (Fig. 2, C). F_4 folds are variable in orientation and fall on a southwest-dipping girdle. F_5 folds are northeast-verging, generally southeast-plunging, tight to open folds that are correlated with late regional-scale open folds affecting the uplift (Fig. 2, E). Minor extensional structures, including extensional crenulation cleavages, were reported.

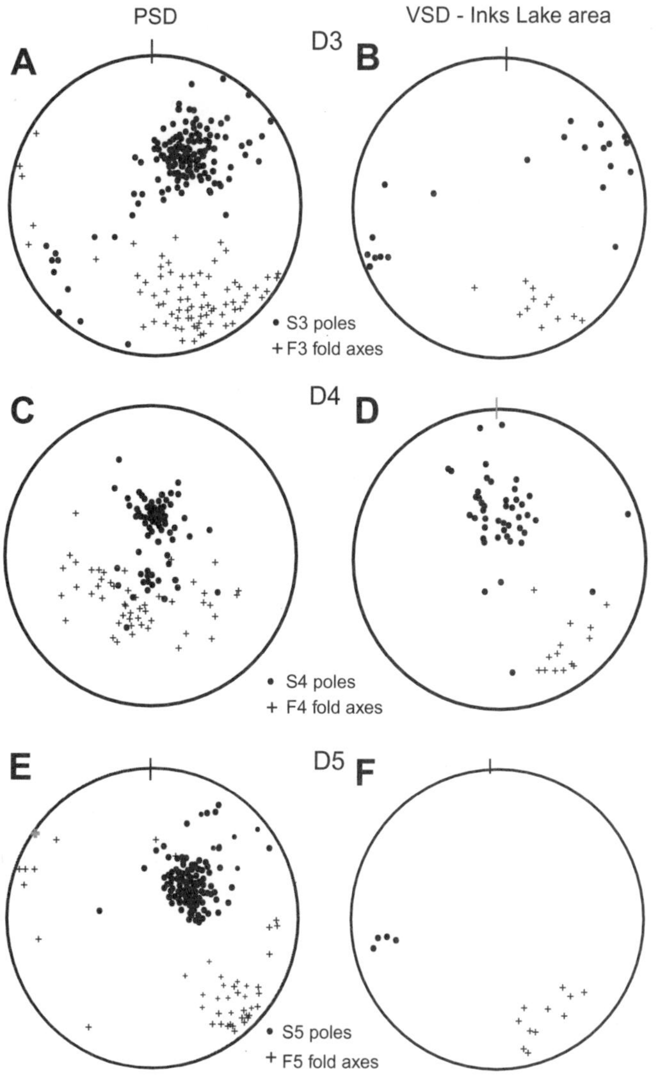

Figure 2. Orientations of structural elements D_3, D_4, and D_5. Packsaddle domain is shown in (A), (C), and (E); Valley Spring domain in (B), (D), and (F). See text for discussion. Equal-area, lower-hemisphere stereonets. PSD—Packsaddle domain; VSD—Valley Spring domain.

P-T conditions early in the deformational history (syn- to post-D_2, pre-D_5) in the Packsaddle domain (Fig. 1) are estimated at 700 °C and 7–8 kilobars, based on the reaction of staurolite + quartz → garnet + sillimanite and the H content of the staurolites (Carlson and Reese, 1994). This estimate is supported by dynamic rotational recrystallization of feldspar porphyroclasts, especially in mylonites, which indicates upper amphibolite facies or higher (Fitz Gerald and Stunitz, 1993a,b; Pryer, 1993; Passchier and Trouw, 1996). (Note that we use the terminology of Poirier [1985] throughout the paper; rotational recrystallization is equivalent to the subgrain rotation-recrystallization of Passchier and Trouw [1996].) Metamorphic conditions were roughly similar throughout deformation. Micas and amphiboles of similar compositions are aligned parallel to all foliations. Metamorphic segregation of biotite into limbs of crenulation cleavages for all fabrics is nearly ubiquitous. Evidence of a subsequent low-pressure, high-temperature thermal overprint associated with intrusion of late syn- to post-tectonic plutons is locally present; both static and dynamic textures are documented, depending on location (Carter, 1989; Nelis et al., 1989; Reese, 1995; Reed, 1999).

Valley Spring Domain Adjacent to the Packsaddle Domain (Honey Creek–Rocky Creek Area)

The Packsaddle domain of the southeastern uplift is in structural contact with the Valley Spring domain along a major shear zone (Reese, 1995; Reese and Mosher, 2004). The uppermost part of the Valley Spring domain consists of a plutonic complex that overlies a heterogeneous package of orthogneisses and paragneisses (Zumbro, 1999) (Fig. 3). Rocks adjacent to the boundary are mylonitic, whereas the plutonic complex immediately beneath the shear zone was attenuated into thin sheets with extensive boudinage. Mylonites also occur throughout the upper plutonic complex.

The dominant structures are a penetrative S_2 foliation, a pervasive intersection lineation, and large-scale upright to overturned F_5 folds that deform the entire Valley Spring domain and the Valley Spring/Packsaddle domain contact and together form the regional Babyhead antiform (Fig. 1; Stenzel, 1935). These folds have a curved map pattern, and their orientation is apparently controlled by the shape of the adjacent Kingsland pluton (Town Mountain Granite; Figure 1). Two other penetrative foliations (S_1 and S_3) and a third, locally developed foliation (S_4), plus three locally preserved fold generations (F_2, F_3 and F_4) are observed, which are directly correlative to structures in the Packsaddle domain. Boudinage and veining are widespread.

Shallow (10–15°) plunges of F_5 folds, deforming the Valley Spring–Packsaddle domain contact exposed in Honey Creek, coupled with relatively flat topography, result in extensive exposures of the upper plutonic complex that is immediately adjacent to the contact over much of the Honey Creek–Rocky Creek area. Thus, structures in this area are dominated by those related to juxtaposition of the Valley Spring and Packsaddle domains. Almost continuous exposure of these units allows determination of structural and primary relationships between units, including intrusive contacts (Fig. 3). Only in the northern parts of this area do structurally lower units crop out in the core of the regional Babyhead antiform. These sparse exposures consist of thinly interlayered, laterally discontinuous pelitic rocks, arkoses, and gneisses, as well as abundant aplite, granite, and pegmatite intrusions (presumably related to Town Mountain granites). These units best express multiple generations of folds and foliations. One outcrop adjacent to the Kingsland pluton contains multiple generations of leucosomes, refolded folds, and a rock composition similar to that exposed in the Inks Lake area, indicating a similar history for these structurally lower, more heterogeneous units.

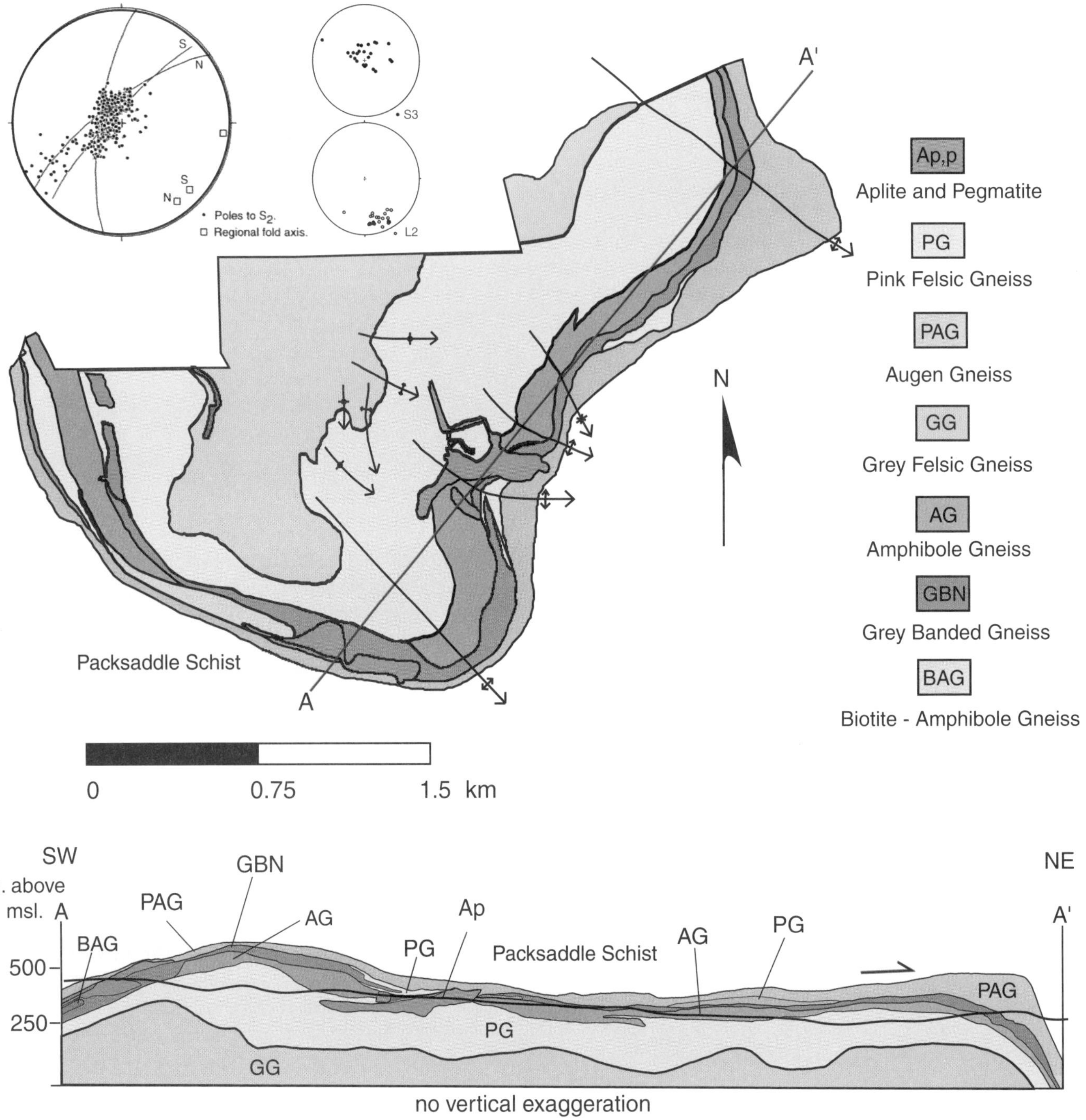

Figure 3. Geologic map and cross section of the uppermost Valley Spring Gneiss plutonic complex along Honey Creek, Cap Mountain Quadrangle, Llano County, Texas. Note the attenuation of the plutonic complex and the intrusive relationships. Folds on map shown are F_4 (east-plunging) and F_5 (southeast-plunging). Map is of the southern portion of Honey Creek–Rocky Creek area (Fig. 1). For cross section, the units above the black topographic profile line were projected down the plunge into the line of section. Stereonets show orientations of poles to S_2 and S_3, and L2 lineations in the map area shown. Note that for S_2, the three girdles and regional fold axes correspond to those for the northern (N) and southern (S), large F_5 folds and the interpreted regional east-trending F_4 fold. msl—Mean sea level. Map and cross section are modified from Zumbro (1999).

Early Structures

S_1, S_2, and commonly, S_0 form a composite foliation and compositional layering that generally cannot be separated. It is defined by millimeter- to centimeter-scale metamorphic segregations and layering of mafic and felsic minerals, alternating bands of different feldspar compositions and/or contents, diffuse bands that are rich in iron oxides, quartz-rich layers, or mineralogical layers that are now composed of secondary alteration products, such as epidote, biotite, sericite, and, rarely, muscovite. Only the presence of syn-metamorphic deformation clearly distinguishes S_1 and S_2 from S_0, which is the compositional layering that is considered to be the original layering.

S_1 is defined by aligned, elongated, rotationally recrystallized microcline, quartz and plagioclase, and, where present, biotite and hornblende. S_1 can only be distinguished from S_2 in F_2 fold hinges, where it is folded by F_2 and S_2 parallels these axial surfaces. Although this relationship is rarely observed in the field, it is pervasive in thin section, particularly in the gneisses. Coarser layers contain a well-defined metamorphic foliation that forms abundant folds, with a pronounced S_2 axial planar foliation. These folds are tight to isoclinal and ~2 mm in wavelength. Rare porphyroblasts of feldspar contain S_1, defined as aligned inclusions of amphibole and epidote, at a high angle to S_2 in the matrix.

S_2 is a pervasive, well-developed, locally mylonitic, gneissic foliation, the dominant foliation of the Valley Spring domain. S_2 is axial planar to isoclinal folds of S_0/S_1 and is folded by all later fold generations. S_2 is defined by aligned, elongated, rotationally recrystallized microcline, quartz and plagioclase, and, if present, hornblende, biotite, and epidote. Most F_2 folds are observed in thin section, although in one unit, F_2 folds abundant thin pegmatites.

The mylonitic fabric within the shear zones in the Valley Spring domain and at the Valley Spring–Packsaddle domain boundary parallels S_2 in all other units (Fig. 4). Some mylonites contain asymmetric porphyroclasts (0.5–3. cm in diameter) with recrystallized sigma tails indicating a top-to-northeast or, rarely, top-to-north sense of shear. Augen are generally microcline and contain tails of rotationally recrystallized microcline and quartz. Rarely, some augen are plagioclase with recrystallized plagioclase tails. Asymmetric augen are well developed in thin sections cut perpendicular to a southeast-plunging lineation and are very poorly developed to nonexistent in those sections cut parallel to the lineation. Samples with well-developed lineations contain two or three intersecting foliations. Many samples that contain asymmetric augen do not contain a well-developed lineation. Thus, the lineation is an intersection lineation, not a stretching lineation.

Extreme attenuation and extensive chocolate-tablet boudinage of the plutonic complex that forms the uppermost Valley Spring domain occurred as the result of shortening perpendicular to S_2 and extension parallel to it. These intrusive bodies were attenuated into thin sheets, making them appear to be a layered supracrustal sequence (Figs. 3 and 4); however, intrusive relationships are readily apparent in the field. In addition, highly attenuated gneisses are interlayered with garnet clinopyroxenites, most likely representing mafic sills, and foliated pegmatites that together form a series of nested chocolate-tablet boudins. Extension within the shallowly dipping S_2 was roughly north-south and east-west. Boudinage affecting other units show north-south extension as well, along with northeast-southwest and northwest-southeast extension, which, together, suggests that shortening perpendicular to S_2 was the most important factor in the formation of these structures.

Figure 4. Highly attenuated plutonic complex and mylonites at the Valley Spring–Packsaddle domain boundary. Pronounced fabric is a composite S_1/S_2 foliation.

The garnet-clinopyroxenite boudins range from large lenticular bodies to almost perfectly round spheres of garnet clinopyroxenite up to 1 m in diameter that are completely wrapped by gneissic S_1/S_2 foliation (Fig. 5). The boudin shapes indicate a very high competency contrast with surrounding felsic gneisses. Individual boudins are separated by half a meter to tens of meters and, in some cases, appear completely isolated (Fig. 5). In many cases, these layers were attenuated on either end into 5–6-cm-wide, highly foliated schistose sheets, which either pinch out completely or are connected to another large lenticular body along strike. This foliation is defined by aligned amphiboles that formed during retrogression of the garnet-clinopyroxene mineral assemblage. The margins of some boudins also contain a similarly defined foliation, which is parallel to that of enveloping gneisses and pegmatites. In some cases, amphibolitic layers pinch and swell, suggesting nearly total replacement of the garnet clinopyroxenite prior to or synchronous with boudinage. The nested foliated pegmatites contain isoclinally folded S_1 with S_2 axial planar (i.e., F_2 folds). This, plus wrapping of boudins by S_2, indicates boudinage occurred post-S_2, but the relationship to S_3 is unknown. Given extensive shortening perpendicular to S_2, it seems likely that deformation continued after S_2, with the same stress system causing both attenuation and boudinage. It is unlikely that the

Figure 5. Isolated garnet clinopyroxenite boudin in felsic gneiss, in uppermost Valley Spring domain. This boudin still retains much of its round shape, but has thin tails of foliated and retrogressed material in the neck regions. Note the wrapping of the boudin by S_2.

boudinage was related to either F_4/S_4 or F_5, which are open structures with axial planes at high angles to S_2. Retrogression and associated foliation formation must have been synchronous with some boudinage, however, because it is localized in boudin necks, and only where layers are retrogressed do they thin or pinch out. Thus, boudinage most likely started when the rocks were competent, producing rounded shapes; as deformation progressed, an influx of fluids caused retrogression, allowing further boudinage and flattening. Two stages of boudinage cannot be ruled out, however.

S_3 is a pervasive foliation visible in almost all thin sections (>300) but is only locally observed in the field. S_3 is a shallowly dipping foliation (Fig. 3) that overprints S_1/S_2 at a low angle (~30–35°). Although the low angle between S_2 and S_3 is suggestive of an S-C shear-related fabric, S_3 clearly crosscuts S_2 and is not deflected into it. In addition, the angular relationship between S_2 and S_3 changes, suggesting the presence of large-scale F_3 folds, although hinges are not visible in the field. In thin section, S_3 is defined by elongated, rotationally recrystallized microcline, quartz, plagioclase, and, if present, hornblende and biotite. Rare recumbent F_3 folds in schistose amphibolite gneisses and biotite amphibole schists are symmetric, tight to isoclinal folds of S_2. Locally, S_3 is a crenulation cleavage with new biotite and amphibole that grew parallel to axial planes. Generally, S_3 is best expressed in outcrops rich in amphibole, biotite, or muscovite or in highly weathered quartzofeldspathic gneisses, where the fabric produced a plane for weathering.

Lineations on S_2 are ubiquitous features in all units across the area. These lineations trend generally southeast with fairly shallow plunges (L2) (Fig. 3). The lineation is a composite of $S_0/S_1/S_3/S_3$ intersections. Sections cut parallel to the lineation usually only show one foliation, whereas those cut perpendicular to the lineation show two or three. In addition, most F_2 folds are observed in sections cut perpendicular to the lineation and not in those cut parallel to the lineation. Rare F_3 folds are also southeast plunging. As noted earlier, this lineation is nearly orthogonal to the shear direction shown by recrystallized augen tails. Nearly parallel orientations of S_0, S_1, S_2, and the slightly oblique S_3 cause the very pronounced intersection lineation in many units.

These early structures are correlative with those in the Packsaddle domain of the southeastern uplift. The contact between the Valley Spring and Packsaddle domains is everywhere a ductile shear zone with mylonitic fabric parallel to S_2 and asymmetric augen showing top-to-northeast movement in both the Valley Spring and Packsaddle domains. In outcrops of the contact between domains, the S_2 foliation is concordant in both biotite amphibole schist (Packsaddle domain) and augen gneiss (Valley Spring domain) and is similarly defined. S_3 is also expressed in both and is nearly identical in orientation. In thin section, S_3 in the Packsaddle domain is defined by aligned hornblende, biotite, and elongate quartz and feldspar. In the Valley Spring domain, it is defined by elongate quartz and microcline and fine-grained biotite.

Late Structures

Some later stage folds and foliations in the Valley Spring domain adjacent to the Packsaddle domain are not as directly correlative, but are interpreted to correspond to F_4 and S_4 in the Packsaddle domain. Amphibole and biotite-amphibole schists and gneisses are deformed by a west-northwest–striking, upright crenulation cleavage (wavelength of ~0.5–1 cm), with shallowly plunging axes and rare biotites parallel to the axial planes. This cleavage crenulates S_2 and is not in an orientation consistent with either F_3 or F_5 axial planes, but has a similar orientation and style to S_4 found in the adjacent Packsaddle domain, where it is axial planar to F_4 folds that fold F_3/S_3. Map-scale folds similar to F_4 in the Packsaddle domain are observed as well. An open, antiform-synform pair with shallow east- to east-southeast plunges is observed folding the contact with the Packsaddle domain (Fig. 3) and similar, smaller-scale (~1.5 m wavelength) folds with moderate west plunges affect felsic gneisses farther north.

Late-stage F_5 folds have large wavelengths (up to ~0.5 km), fold S_0–S_3 and the Valley Spring/Packsaddle domain contact (Fig. 3), and reorient S_4. No S_5 was observed. F_5 are generally very gentle to open, asymmetric folds with steep to overturned eastern limbs. These folds together comprise the regional Babyhead antiform; they are the dominant folds in cross section and cause the map pattern (Fig. 3). F_5 plunge shallowly southeast near the Packsaddle domain contact, but are south- to south-southeast-plunging farther north and near the contact with the Kingsland pluton of the Town Mountain granites (Figs. 1 and 3). The F_5 fold orientation diverges up to 50° but follows the curved margin of the pluton. Some smaller, shallowly plunging, north- to north-northeast–trending folds adjacent to the pluton have granite intruded along the axial planes. These F_5 folds, which also affect the Packsaddle domain near the contact, are correlated with late regional F_5 folds throughout the uplift, including those in the Packsaddle domain in the southeastern uplift.

Minor late-stage boudinage is characterized by brittle quartz-filled fractures (some en echelon) and little ductile flow of surrounding rock. It clearly occurred under different conditions than the extensive boudinage described earlier; however, the timing relative to F_4 and F_5 is indeterminate. In the northern part of the area, boudins are asymmetrical with down-to-southwest normal movement.

At the Kingsland pluton margin, local migmatization, brecciation, boudinage, and folding are observed, and granites, pegmatites, and aplites crosscut and lie parallel to gneissic foliation. Locally, chaotic, swirled migmatites contain segregations of mafic and felsic material resulting from partial melting. In one agmatic migmatite, numerous blocky-ended, veined blocks and boudins of garnet clinopyroxenite and foliated biotite gneiss are surrounded by foliated pink felsic leucosomes. Granite commonly fills the necks of boudins. Folds within the migmatites generally trend approximately north-south, parallel to the pluton margin. Apparently, intrusion of Kingsland pluton caused partial melting in the Valley Spring domain country rock synchronous with late-stage folding and boudinage.

The Valley Spring domain is intruded by several large (map-scale) aplite bodies that crosscut the dominant S_2 fabric and contain misoriented xenoliths of well-foliated country rock. Some aplites show one or two foliations at their margins, but foliation decreases in intensity or disappears in their interiors. No foliations are oriented appropriately to be S_4 or S_5; locally, one appears to be correlative to S_3 in the country rock, but elsewhere, an aplite clearly cuts S_3. Foliations are defined by elongated quartz, rotationally recrystallized microcline, and aligned biotite. One aplite is boudinaged and S_2 in the country rock wraps the boudins. Although the age of these aplites is unknown, they are interpreted as related to late-stage (Town Mountain) syntectonic granites. Several attempts at dating them were unsuccessful.

Timing of the late deformation is roughly constrained. Curvature of regional F_5 folds around the Kingsland pluton plus relationships observed between folds and the granite at the pluton margin strongly suggest that F_5 folding was just prior to, or essentially synchronous with, intrusion of the granite. Aplites that cut the dominant S_2 foliation but show foliations at the margins also suggest some deformation postdated late intrusions.

Metamorphism

Because of the quartzofeldspathic nature of most of the units, metamorphic conditions are best determined from recrystallization mechanisms. In thin section, microcline is moderately to extremely elongate parallel to S_1, S_2 and S_3, in part, as a result of grain boundary migration and grain growth. Microcline augen and some larger microcline in the groundmass, however, have extensive rotational recrystallization on margins perpendicular to the shortening direction, resulting in elongated relict grains. The groundmass consists of rotationally recrystallized microcline and plagioclase, and some augen also have recrystallized tails. Quartz forms elongated ribbon-like grains parallel to S_1, S_2, and S_3 and irregular grains with amoeboid grain boundaries that locally include micas or amphibole aligned parallel to S_2. These quartz textures indicate rapid grain-boundary migration and grain growth. Both quartz and feldspar textures—particularly the degree of rotational recrystallization and recrystallized grain size of the feldspars—and the evidence for rapid grain-boundary migration in the quartz and feldspar are indicative of upper amphibolite facies or higher (White and Mawer, 1986; Passchier et al., 1990; Tullis, 1990; Fitz Gerald and Stunitz, 1993a,b; Pryer, 1993; Passchier and Trouw, 1996).

Garnet clinopyroxenites provide the best mineral-assemblage estimate of metamorphic conditions. Although *P-T* work has not been done on these samples, the textures described below are indicative of retrogressed, moderately high-pressure and temperature mafic assemblages, similar to those described elsewhere in the uplift (Wilkerson et al., 1988; Carlson and Schwarze, 1997; Carlson, 1998). Garnets up to 2 mm in diameter are surrounded by radiating symplectitic coronas of amphibole, magnetite, oligoclase, and clinopyroxene. The groundmass also contains abundant fine-grained symplectic intergrowths, where earlier pyroxene (with recognizable relict shapes) reacted to form amphibole + secondary clinopyroxene + oligoclase. Rutile rims magnetite, and in turn, is rimmed by titanite. Magnetite is abundant in symplectitic coronas and as larger crystals in the groundmass. Near margins of garnet clinopyroxenite boudins, garnet is totally replaced and foliation is defined by amphibole. In the interior of the largest boudin (5 m in diameter), a compositional layering oblique to the surrounding foliation is defined by elongate clumps of garnet and parallel symplectitic clusters of relict pyroxene-rich layers; these relict tabular to lath-shaped crystals are randomly oriented. Evidently, the moderately high-pressure and temperature mineral assemblage formed under nondeviatoric stress conditions but overgrew a compositional layering, either an original igneous layering or an early metamorphic segregation. Thus, the timing of this earlier metamorphism cannot be tied to the deformational history. However, the boudinage occurred, at least in part, during retrogression; thus, some post-S_2 shortening and boudinage was at amphibolite facies. In the quartzofelspathic gneisses, plagioclase are oligioclase in composition and amphiboles are generally hornblende, consistent with amphibolite-facies metamorphism.

A late-stage, lower-pressure and temperature metamorphism and accompanying hydration overprints most rocks, with fluids most likely coming from the adjacent Kingsland pluton. Actinolite is ubiquitous and, in some cases, appears as overgrowths on hornblende. Biotite replaces amphibole and chlorite replaces both biotite and amphibole; muscovite and clinozoisite are observed replacing plagioclase. Nearly complete replacement and mimicking of higher-temperature metamorphic minerals by retrograde minerals makes it difficult to constrain conditions of deformation. Granoblastic textures representing static metamorphism are observed adjacent to the pluton.

Inks Lake Area

The Inks Lake area (Devil's Waterhole of Inks Lake State Park) (Fig. 1) is northeast of the previously described area, on the opposite side of the Kingsland pluton, and consists of units similar to those exposed in the structurally lower core of the Babyhead F_5 antiform. Here, quartzofeldspathic gneisses, an amphibole gneiss, pelitic gneisses and schists, and a 1253^{+3}_{-2} Ma foliated sill (Roback, unpublished data, *in* Mosher, 1998) are multiply deformed and intruded by an undeformed granite and granite dikes related to Town Mountain Granites. Nearly continuous exposure shows crosscutting foliations, leucosomes, and superposed fold generations. Detailed structural mapping (1:600 scale) identified five S-surfaces and four associated generations of folds (Plates 1 and 2 *in* Hoh, 2000). Dominant structural features are a metamorphic segregation, defining the dominant foliation (S_1/S_2), and southeast-plunging isoclinal folds (F_3 and F_5). Lithologic layering is parallel to S_1/S_2 and is folded by a large-scale F_5 antiform, which causes lithologic units to repeat across the fold, including the foliated sill. Four of the S-surfaces (S_1/S_2, S_3, and S_4) are penetrative throughout the field area. S_5 is found only locally in F_5 fold hinges.

The five foliations are distinguished by their relationships to fold generations and crosscutting relationships. All are high-*T* metamorphic fabrics, but their degree of development and/or definition differs. Because most lithologies are quartz and feldspar rich and are dynamically recrystallized, foliations are not well expressed in thin section. All foliations, however, are distinct from the original compositional layering, S_0, within and separating major identifiable lithologies. Differentiation of fold generations was made on the basis of which S-surface was folded or was axial planar, fold superposition (Fig. 6), and fold geometry and style. Orientation of fold axes is less useful, because most fold generations are nearly coaxial.

Figure 6. Multiply folded, leucosome-rich outcrop at Inks Lake. Prominent foliation (S_2) and a leucosome generation is folded by isoclinal F_3 folds with S_3 axial planar. Axial planes of these F_3 folds are refolded by open F_4 folds; an axial planar S_4 dips shallowly to the left.

At least three generations of leucosomes are recognized (Fig 6). All generations of leucosomes appear to be the result of in situ melting of host rock, rather than intrusions of felsic material into these rocks. Generally, the adjacent rock is more micaceous and the boundary with the leucosome is gradational. Leucosomes are no wider than 1–2 cm, with spacing generally >5 cm. All lithologies except unfoliated granites show evidence of partial melting. Leucosomes are composed primarily of microcline ± plagioclase ± quartz ± biotite ± chlorite ± muscovite, with variable amounts of each mineral in the different leucosomes. Chlorite and muscovite are observed in association with leucosomes in some thin sections, most likely resulting from the breakdown of biotite. Leucosomes tend to be microcline dominant, even when the host rock has more plagioclase than microcline. Leucosomes form pre-F_2/S_2, syn-F_2/S_2, post-F_2/S_2–pre-F_3, and syn-F_3, with slight variations in the nature of these leucosomes between generations.

Early Structures

S_1 is parallel to S_2 in much of the field area and is only distinguished from the latter in hinges of F_2 folds, where S_1 is folded and S_2 is axial planar. No F_1 folds were recognized in the field. Where S_1 is identified, it is defined by alignment of micas or amphiboles and/or metamorphic segregation of micas or amphiboles versus quartz and feldspars into compositional layers. In quartzofeldspathic units, where biotite is sparse, identification is difficult. Quartz and feldspar are elongate and aligned parallel to S_1. Both quartz and feldspar are affected by rotational recrystallization and grain-boundary migration and grain growth. Sillimanite is parallel to S_1 in pelitic rocks. A generation of light pink, wispy leucosomes is parallel to S_1. These leucosomes are generally foliated by S_2, highly boudined, and folded by F_2 folds; feldspar and quartz are rotationally recrystallized.

S_2 is a pervasive metamorphic segregation in most lithologies. It strikes predominantly northwest and dips to the southwest and northeast, defining a large F_5 antiform. F_2 folds are southeast-plunging isoclinal folds that fold S_1 and have the dominant S_2 parallel to axial planes.

S_2 is defined primarily by layers of felsic versus mafic minerals, formed by metamorphic segregation (Fig. 6). These mafic minerals include biotite, amphibole, opaques, and clinopyroxene, all of which are aligned parallel to S_2. These felsic minerals include elongate quartz, microcline, and plagioclase, plus muscovite and sillimanite in more pelitic rocks. Quartz and feldspar are elongated and dynamically recrystallized by rotational recrystallization and grain-boundary migration and grain growth. A generation of pink leucosomes is parallel to S_2. These leucosomes are boudinaged and folded by later fold generations. Syn-S_2 leucosomes are deformed but not foliated; feldspar and quartz were rotationally recrystallized.

F_2 folds are isoclinal, fold a metamorphic foliation (S_1) and an associated generation of leucosomes, have S_2 axial planar, and are most abundant in thin section, where they are identified by remnant hinges defined by biotite or elongated quartz and feldspar parallel to S_1 foliation. F_2 folds range in size from ~3 cm to 1.5 m in amplitude. Because the S_0/S_1 fabric is much weaker than the dominant S_2 fabric, F_2 folds have a faint, wispy appearance relative to later fold generations.

S_3 is subparallel to S_1/S_2 throughout much of the field area but is clearly distinguished from S_1/S_2 in hinges of isoclinal F_3 folds, where S_3 cuts across folded S_1/S_2 fabric (Fig. 6). In general it strikes northwest and is steeply dipping (Fig. 2, B). It is a penetrative, but generally poorly developed, foliation that is much less pronounced than S_1/S_2 and S_4. F_3 folds are the most pervasive fold generation in the field and are very distinct, southeast-plunging (Fig. 2, B) isoclinal folds of S_1/S_2 with an axial planar S_3 foliation (Fig. 6). An L3 intersection lineation between S_1/S_2 and S_3 parallel to F_3 is well developed.

S_3 is defined by alignment of biotite, opaques, and elongate quartz and feldspar, and/or a metamorphic segregation of felsic versus mafic minerals. Sillimanite in pelitic rocks also parallels S_3. Locally, large elongate to amoeboid quartz overgrows the aligned biotite that defines S_3, indicating that rapid grain-boundary migration occurred after S_3. A generation of pink leucosomes is parallel to S_3. These leucosomes are highly boudined and folded by subsequent generations of folds.

F_3 folds are distinguished in the field from F_2 folds by folding of the pronounced S_2 fabric (Fig. 6). In general, they have a slightly more open style and are less wispy in appearance. Leucosomes, S_1, and S_2 are folded by F_3 folds, which are cut by S_4 and S_5. F_3 folds occur at all scales—from thin-section to outcrop. F_3 folds are difficult to identify in thin section, however, even when field samples of distinct F_3 fold hinges are sectioned. Growth and dynamic recrystallization of quartz and feldspar during and after F_3 folding greatly modifies S_2, obscuring fold hinges. Where F_3 folds are identified, they have elongate quartz in the limbs (S_2) and larger granoblastic, dynamically recrystallized quartz and feldspar in the hinges; thus, temperatures were high during F_3/S_3.

Late Structures

S_4 is a penetrative, shallowly south-dipping foliation (Fig. 6) that strikes east-west at a high angle to all other foliations (Fig. 2, D) and clearly cuts all previous fabrics and fold generations. F_4 are open folds that clearly fold F_3 (Fig. 6). Although they also plunge southeast, they have a more easterly trend than all other fold generations (Fig 2, D). An L4 intersection lineation between $S_1/S_2/S_3$ and S_4 parallels F_4.

S_4 is a poorly to well-developed foliation in thin section, defined by aligned biotite, amphibole, pyroxene, and elongated quartz and feldspar. Quartz and feldspar are dynamically recrystallized by rotational recrystallization and grain-boundary migration. No leucosome generation parallels S_4 or is observed accompanying F_4/S_4.

F_4 folds are most easily recognized where they fold F_3 axial surfaces, because of the high angle between their axial planes (Fig. 6). F_4 folds in thin section are open folds that fold aligned biotites, amphiboles, quartz, feldspar, sillimanite, muscovite, and F_3 folds.

Between F_3 and F_5, the rocks were extended to the northwest parallel to S_2, forming boudins (~0.5 m). These boudins affect S_2 and F_3 structures. Boudin necks are filled with quartz, indicating that melt was not available to fill interstices at the time of their formation. Boudins were then folded by F_5 folds.

S_5 is a locally well-developed foliation that is axial planar to the F_5 folds and is best expressed in the western portion of the area, in the hinge of a large F_5 antiform. S_5 strikes north-northwest, and dips steeply eastward (Fig 2, F). F_5 folds are outcrop to regional in scale, southeast plunging, upright to overturned to the southwest (Fig 2, F), tight to open folds. An L5 intersection lineation between $S_1/S_2/S_3/S_4$ and S_5 parallels F_5.

S_5 is defined by alignment of micas and amphiboles, and can be quite penetrative, where present. Some rotationally recrystallized, elongated quartz (but no feldspar) parallels S_5 foliation. No leucosomes parallel S_5 or are recognized accompanying F_5/S_5.

F_5 fold hinges contain a crenulation cleavage and a series of parasitic folds that, along with their more open nature, makes F_5 folds distinct from F_3 folds. Superposed relations with F_3 are also observed. The scale of these folds ranges on the order of hundreds of meters to about a half meter.

Undeformed to weakly deformed crosscutting granite dikes cut all structures in the field area. These granites are most likely related to Town Mountain granites, represented by the nearby Kingsland and Lake Buchanan plutons (Fig. 1). Dikes that crosscut F_5 axial planes are undeformed, but those parallel to F_5 axial planes are slightly boudinaged and contain fabric in boudin necks.

Timing of Deformation

Careful structural examination of the 1253^{+3}_{-2} Ma foliated sill (Roback, unpublished data, *in* Mosher, 1998) and surrounding host rock shows that both the sill and the amphibole gneiss contain all structural generations observed in the host amphibolite (i.e., the sill contains S_1–S_4 and is folded F_2–F_5) (Hoh, 2000). Later slip occurred along parts of the sill boundary, and quartz and feldspar veins are locally present, causing an apparent S_1/S_2 discordance locally. Elsewhere, however, the dominant composite S_1/S_2 fabric clearly crosscuts the sill boundary or is parallel, and thus, is common to both the sill and country rock. Thus, similar to the rest of the uplift, deformation postdates crystallization of Valley Spring domain plutonic units. Amphibolite gneiss intruded by the sill yields a U-Pb date of 1129 ± 3 Ma for metamorphic zircons (Hoh, 2000; Connelly, unpublished data). Because these rocks are clearly intruded by the 1253^{+3}_{-2} Ma foliated sill and the zircons are metamorphic, the determined age is interpreted to reflect the time of the dynamothermal metamorphism and most likely the breakdown of clinopyroxene to hornblende.

Relationship of structures from the Inks Lake area to the southeastern uplift Packsaddle domain

Five generations of structures documented for the Inks Lake area are similar in style, orientation, superposed relationship, and general deformation conditions to those previously studied in the southeastern uplift Packsaddle domain. Nearly identical development of structures and the same superposed relationships argue strongly for correlation. In addition, the orientation of structures is directly comparable. Most F_2 in both areas are southeast trending; they maintain this general orientation despite later folding, because subsequent F_3 and F_5 folds are generally coaxial. Only F_4 in both areas diverge from a generally southeast trend. A comparison of S_3–S_5 and F_3–F_5 data from the Packsaddle domain (Carter, 1989; Nelis et al., 1989; Reese, 1995) and data from the Inks Lake area is shown in Figure 2 and discussed here.

S_3 data from the Packsaddle domain (Fig. 2, A) and this study (Fig. 2, B) plot on a northeast-trending girdle, showing the effect of F_5 folding, with a majority of poles plotting in the northeastern quadrant of the stereonet. (Note that S_3 in the Honey Creek area also falls on the same girdle: see Fig. 3.) F_3 fold axes plot predominantly in the southeastern quadrant of the stereonets for all areas (Fig. 2, A and B); later folding modified their orientation somewhat, causing the spread observed in the data. The plunge in the Packsaddle domain ranges from moderate (Nelis et al., 1989) to shallow (Reese, 1995), and Inks Lake data overlap data from both studies.

S_4 data are predominantly east-west striking, south-dipping foliations in all study areas (Fig 2, C and D). S_4 data have a significant component of scatter associated with them, most likely as a result of F_5 folding, but fall on a north-south–trending girdle, reflecting their original east-west strike. All F_4 data plot on a west-northwest–trending, south-dipping girdle on the stereonets (Fig. 2, C and D), indicating changes in F_4 from a southwest-plunge to a southeast-plunge, most likely as a result of F_5 folding.

S_5 data for the Packsaddle domain show S_5 as a northwest-striking, moderately southwest-dipping fabric (Fig. 2, E). In the Inks Lake areas, S_5 and F_5 axial planes have similar strikes (north-northwest) but are more upright to steeply northeast-dipping (Fig. 2, F). Changes in F_5 axial planes from southwest-dipping (Packsaddle and Valley Spring domain near Packsaddle domain boundary) (Fig. 2, E) to upright and steeply northeast-dipping (Inks Lake area) (Fig. 2, F) may reflect the distance from the colliding Coal Creek arc terrane. F_5 fold axes are predominantly southeast-trending in all areas (Fig. 2, E and F). South- to southwest-trending late folds observed in the Honey Creek–Rocky Creek area follow the margin of the Kingsland pluton, suggesting a genetic relationship.

In summary, structures seen in the Inks Lake area are correlative in style, geometry, sequence, orientation, and general deformation conditions to those in the Packsaddle domain of the southeastern uplift. Continuity of these structures across the Valley Spring–Packsaddle domain boundary in the Honey Creek–Rocky Creek area further substantiates the correlation.

Metamorphism

The rocks are migmatites that record a polymetamorphic history, with an early dynamothermal metamorphism followed by a later reheating event. Dynamothermal metamorphism resulted in metamorphic mineral growth and leucosome formation that are temporally linked to the first three phases of deformation (F_1/S_1–F_3/S_3). Metamorphic mineral growth, with an absence of leucosomes, is temporally linked to F_4/S_4–F_5/S_5. Temperature and pressure conditions during deformation are constrained by dynamic recrystallization mechanisms, mineral reactions, and the presence of leucosomes, as discussed later. The later reheating event is accompanied by static recrystallization textures in quartz and feldspar. This static recrystallization is seen predominantly in rock units adjacent to large (Town Mountain–related) granite intrusions.

Quartz and feldspar in all lithologies were dynamically recrystallized. Quartz shows major grain-boundary migration, which caused elongation parallel to fabrics S_1–S_5 and minor subsequent rotational recrystallization. Quartz textures are indicative of rapid grain-boundary migration, which caused quartz to overgrow and contain micas parallel to S_1–S_3. Such migration occurs in quartz at amphibolite-facies conditions or higher (White and Mawer, 1986; Passchier et al., 1990; Tullis, 1990; Fitz Gerald and Stunitz, 1993a, b; Pryer, 1993). Feldspar shows rotational recrystallization, which results in elongate grains parallel to fabrics S_1–S_3. Feldspar undergoes rotational recrystallization only during upper amphibolite-facies or higher metamorphism (Fitz Gerald and Stunitz, 1993a,b; Pryer, 1993; Passchier and Trouw, 1996). Potassium feldspar is still thermodynamically stable during D_4 and D_5, but shows a mantling of very fine-grained new grains, indicating grain-boundary migration recrystallization rather than rotational recrystallization, which suggests lower temperature deformation at this time (e.g., Fitz Gerald and Stunitz, 1993a, b).

The amphibole gneiss has the major metamorphic mineral assemblage of quartz + microcline + plagioclase + biotite ± amphibole ± opaques ± clinopyroxene. Biotite is parallel to all five fabrics (S_1–S_5); thus, it was forming throughout the deformation. Clinopyroxene and amphibole are aligned parallel to S_1/S_2 in addition to biotite; however, the three minerals did not form synchronously. Reaction textures show clinopyroxene breaking down to form amphibole, which, in turn, was replaced by biotite. The breakdown reaction of pyroxene to amphibole may be accompanied by the growth of metamorphic zircons. The aligned clinopyroxene grew during S_1/S_2, but because biotite and amphibole mimic clinopyroxene orientation, the timing of their growth is uncertain.

Pelitic gneiss has a metamorphic mineral assemblage of quartz + microcline + plagioclase + biotite + sillimanite ± cordierite ± muscovite. Reactions involving sillimanite in the pelitic gneiss place constraints on *P-T* conditions during deformation. Sillimanite is parallel to S_1–S_3 fabrics and thus was stable throughout formation of F_1/S_1–F_3/S_3. Hence, deformation occurred in the sillimanite portion of *P-T* space, with temperatures greater than ~500 °C and pressures < 10 kilobars (Fig. 7). The presence of sillimanite and rotationally recrystallized microcline both parallel to S_1–S_3 further constrains *P-T* conditions of syn-S_1–S_3 metamorphism. Sillimanite and potassium feldspar are both stable, which means that deformation prior to F_4/S_4 (F_1/S_1–F_3/S_3) occurred in *P-T* space above the second sillimanite isograd; therefore, at the uppermost amphibolite facies (Yardley, 1989; Spear, 1993) (Fig. 7). Textural evidence sug-

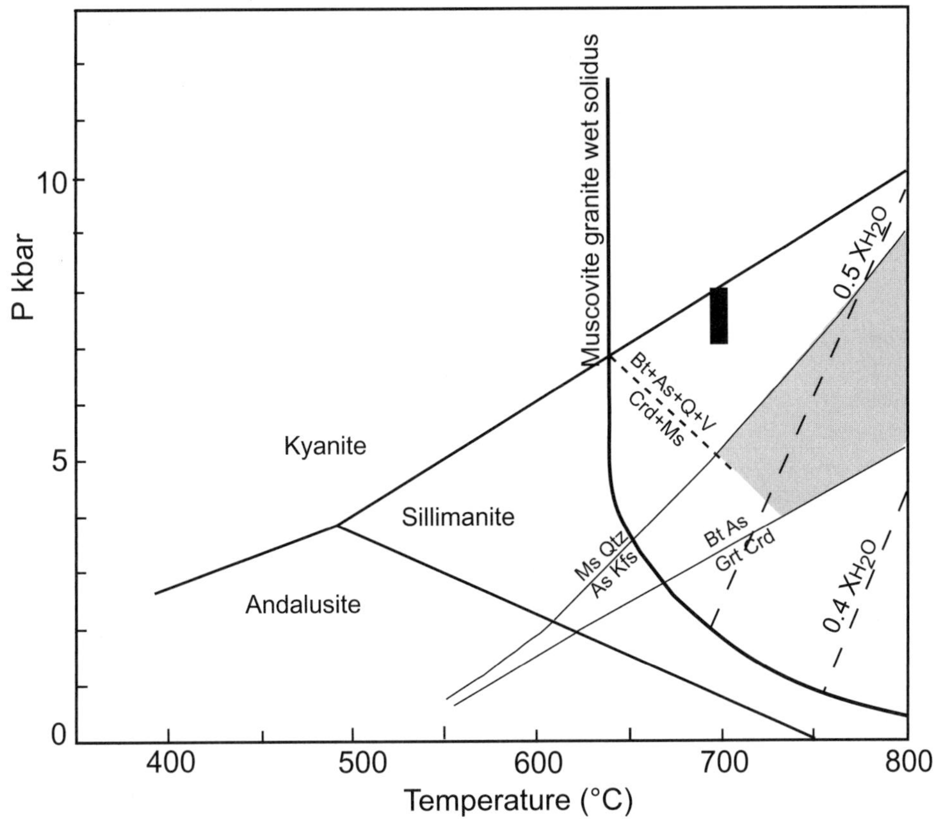

Figure 7. *P-T* grid, showing selected reactions to illustrate the approximate *P-T* conditions for the rocks at Inks Lake during F_2/S_2 (shaded region). The *P-T* estimate for the Packsaddle domain in the southeastern uplift (black box) is shown for reference. The dashed line represents a reaction based on experimental studies by Seifert (1970) and Bird and Fawcett (1973). The rest of the grid is from Spear (1993). The muscovite granite wet solidus is shown to illustrate the approximate temperatures for melting, with 1.0, 0.5, and 0.4 XH_2O in melt shown.

gests this reaction did not go to completion, however, during the formation of S_1. Although sillimanite is aligned parallel to a foliation that is folded by F_2 folds—requiring sillimanite to have formed synchronous with formation of the S_1 fabric—biotite, muscovite, cordierite, and potassium feldspar are also parallel to the S_1 fabric, indicating a disequilibrium assemblage. Cordierite is not stable after F_1/S_1 in pelitic compositions, as cordierite and muscovite are replaced by S_2-parallel biotite and sillimanite. Therefore, during S_2 formation, the reaction cordierite + muscovite → biotite + sillimanite occurred (Fig. 7). Experimental studies of the cordierite breakdown reaction (Seifert, 1970; Bird and Fawcett, 1973) indicate pressures >5 kilobars and temperatures greater than ~660 °C for F_2/S_2 metamorphic conditions (Fig. 7). Using the two reactions (cordierite + muscovite → sillimanite + phlogopite and muscovite + quartz → sillimanite + microcline) constrains deformation conditions for S_2/F_2 to pressures <10 kbar and temperatures greater than ~660 °C (shaded area, Fig. 7).

After F_3/S_3, *P-T* conditions decreased. F_3 folds of sillimanite aligned parallel to S_2 are completely enclosed within single biotite and muscovite crystals. Biotite and muscovite replace sillimanite after F_3/S_3, which is indicative of a drop in temperature. Muscovite grains that contain F_3 folds of sillimanite-defined S_2 have quartz inclusions, suggesting that post-F_3, the second sillimanite isograd is crossed in the opposite direction, with sillimanite + microcline → muscovite + quartz. The decrease in temperature precludes further leucosome development after F_3.

Leucosomes are composed predominantly of quartz, microcline and plagioclase, which require temperatures greater than ~650 °C to melt under wet conditions. In most pelitic compositions, migmatization does not occur until *P-T* conditions are near the second sillimanite isograd (Yardley, 1989), where muscovite begins to break down, liberating water. The amount of water liberated is just enough to induce melting at these *P-T* conditions. The melt forms parallel to foliations, because water is released by dehydration and instantly forms melt under the same deformation conditions as the associated fabrics. Formation of leucosomes constrains the metamorphic conditions to uppermost amphibolite facies or higher during F_1/S_1–F_3/S_3, consistent with dynamic constraints and mineral assemblages.

Some evidence of static recrystallization is present through much of the field area, but evidence of earlier dynamic recrystallization is not obliterated. Static textures are most common in the western part of the field area, adjacent to a large granitic (Town Mountain) dike, suggesting it is associated with its emplacement.

In summary, the metamorphic history of these rocks can be well constrained by formation of key metamorphic minerals, partial melting, and dynamic recrystallization mechanisms. Dynamic recrystallization mechanisms of quartz and feldspar indicate that F_1/S_1–F_3/S_3 occurred in upper amphibolite-facies conditions or higher. The presence of sillimanite growing in equilibrium with potassium feldspar require *P-T* conditions higher than the second sillimanite isograd, requiring uppermost amphibolite-facies metamorphic conditions. Leucosome generation during S_1–S_3 supports upper amphibolite-facies *P-T* conditions, as partial melting occurred during breakdown and dehydration of muscovite at the second sillimanite isograd. Peak conditions occurred during F_2/S_2 and are constrained to temperatures >660 °C and pressures <10 kilobars. After D_3, sillimanite was replaced by muscovite, and no leucosomes were generated, suggesting that metamorphic conditions dropped but remained at amphibolite facies through formation of F_5/S_5.

DISCUSSION AND CONCLUSIONS

The deformation history of the Valley Spring domain is equivalent to that of the Packsaddle domain and shows thrusting and multiple phases of folding, with a northeastward direction of tectonic transport. The boundary between the two domains is a ductile shear zone, where the Packsaddle domain was thrust northeastward over the Valley Spring domain. S_2 is mylonitic in both the Packsaddle and Valley Spring domains at the contact and is concordant with S_2 in the underlying Packsaddle domain and overlying Valley Spring domain. S_3 also can be traced across the Valley Spring–Packsaddle domain boundary and dips southwest. For both domains, an earlier S_1 is folded by F_2 folds with S_2 axial planar. The shear zone is folded by F_5 and apparently F_4 folds that show northeast and northward transport, respectively.

The uppermost Valley Spring domain was attenuated as a result of extreme shortening perpendicular to S_2, at least some of which is post-S_2. Together with the shear zone, this indicates that the Valley Spring and Packsaddle domain boundary represents a high-strain zone associated with their juxtaposition. This shortening differs from extreme transposition and shearing observed over a 2–3 km zone in the Packsaddle domain at the Coal Creek–Packsaddle domain boundary and suggests that, although the Valley Spring–Packsaddle domain boundary involves thrusting, the weight of overlying rocks contributed significantly to the strain, reflecting the deeper position in the orogenic pile. Although prograde upper amphibolite-facies conditions existed during S_1/S_2 formation, at least some clinopyroxenite boudinage occurred during amphibolite-facies regression, suggesting decreasing temperatures before and/or during S_3/F_3.

Structurally beneath this zone, thin, well-layered units within the Valley Spring domain show the same history as the Packsaddle domain in the southeastern uplift. Structures are similar in style, orientation, superposed relationships, and general deformation conditions to those in the Packsaddle domain. The main differences are the presence of extensive leucosomes associated with S_1–S_3, the clear evidence for uppermost amphibolite-facies metamorphism, and the pervasiveness of F_3 compared with its relative scarcity in the Packsaddle domain. Comparison of the *P-T* estimate for the central Packsaddle domain with that for the Inks Lake area (Fig. 7) strongly suggests higher temperature conditions for the Inks Lake area during D_1–D_3, which is supported by the lack of partial melting within the Packsaddle domain. Higher temperature is consistent with these units being the structurally deepest exposed levels of the orogenic pile in the eastern uplift. If so, then pressures would also be expected to be at least as high as those in the Packsaddle domain. The prominence of F_3/S_3 and associated partial melting suggests that temperatures may have remained somewhat higher for a longer span of time in the deformation history at depth. Although F_4/S_4 and F_5/S_5 apparently formed at lower temperatures than did earlier structures, they were still at amphibolite-facies conditions. Direction of transport for early structures (through F_3/S_3) at Inks Lake is not as easy to determine as it is for structures farther south, because of their position within a large F_5 antiform. F_4/ S_4 show northward transport, but F_5 are either upright or slightly overturned to the southwest. As mentioned previously, this may reflect the increasing distance from the colliding Coal Creek arc terrane, or possibly the influence of the Kingsland pluton, depending on the timing of intrusion relative to F_5 folding.

Timing of metamorphism and synchronous deformation is dated at ca. 1129 Ma for the Inks Lake area, which falls in the range of metamorphic dates from across the uplift (Roback, 1996b), including the Packsaddle domain in the southeastern uplift. Late-stage deformation of some granitic bodies (aplites, pegmatites; Town Mountain Granite related) in both the Honey Creek–Rocky Creek area and Inks Lake areas is consistent with the presence of late (1126–1118 Ma) syntectonic granites elsewhere in the uplift (Reed, 1999). F_5 fold axes also appear to follow the margin of the Kingland pluton, further suggesting a genetic relationship between the latest folding and plutonism. Either emplacement caused the deformation, or more likely—given the predominance of F_5 folds across the uplift and folding of some plutons by F_5 folds—the folding and intrusion were essentially contemporaneous (Reed, 1999). Not all late plutons and associated dikes intruded at the same time, however; granitic dikes at Inks Lake crosscut F_5 axial planes and are undeformed. Late syntectonic granites across the uplift were emplaced at low pressure but high temperature. Therefore, D_5 most likely occurred at shallower depths, but the kinematics remained the same as earlier (D_1–D_3) deformation. Only F_4 shows a more northward direction of transport.

Thus, the Valley Spring domain of the northeastern uplift shows a tectonic history that is consistent with the structure of the Packsaddle domain of the southeastern uplift, indicating northeastward structural stacking and transport. No evidence of transcurrent motion is observed. Therefore, strain was not partitioned into transcurrent motion in the northeastern uplift away from the exotic-arc (Coal Creek) terrane. Instead, northeastward transport of the Coal Creek arc caused structural imbrication of the three domains plus thrusting and multiple phases of folding within both the basinal supracrustal rocks of the Packsaddle domain and the continental marginal arc rocks and terrigenous sedimentary rocks of the Valley Spring domain. Temperature and the degree of flattening strain increased downward in the orogenic pile, with uppermost amphibolite-facies metamorphism and partial melting at the deepest levels exposed. The deformation and metamorphic history is most consistent with northeastward collision of the exotic arc along an east-trending Laurentian margin.

Mosher (1998) previously proposed that a southern continental block collided as well as the Coal Creek arc terrane. The presence of a colliding southern continental block is not required to produce the conditions observed in the eastern uplift, although it would help explain the pronounced, composite early fabrics and the intensity of the strain. In the western uplift, however, no remnant of the arc terrane is observed, and the very high pressures recorded (15–21 kbars; Wilkerson et al., 1988; Carlson and Schwarze, 1997; Carlson, 1998; Anderson 2001), coupled with the deformation, necessitate collision of a continental block with Laurentia.

Deformation kinematics in west Texas (Grimes, 1999; Grimes and Mosher, 2003) record northwestward transport, whereas the eastern Llano uplift records northeastward transport, supporting the idea of a collision of an indentor with Laurentia, as proposed by Mosher (1998). Deformation in west Texas could also represent escape tectonics at the edge of the colliding continent (Grimes and Mosher, 2003), similar to the mechanism proposed by Gates (1995) in the Appalachians, with deformation subsequent to the main collision.

Timing of the deformation varies across the orogen as well. Deformation at the outermost extent of the orogen in west Texas is younger (1057–978 Ma) (Grimes, 1999) than that within the ~300-km-inboard orogen core in the Llano uplift (ca. 1150–1116 Ma), consistent with deformation propagating onto the continent during collision. Between 1124 and 1070 Ma, the Llano uplift was intruded by voluminous granitic plutons; the earliest are syn-tectonic, and the latest show deformation of the margins and surrounding country rock during intrusion. The latter is compatible with continued shortening in the orogen core, although deformation could be the result of forcible emplacement. Uplift and cooling of rocks in the Llano uplift occurred between 1100 and 1030 Ma (Rougvie et al., 1999), overlapping and predating deformation along the margin. Thus, the tectonic history is similar to that observed in the Himalayas, where early collisional tectonism is followed by uplift and cooling synchronous with continued shortening along the orogen margin.

Our results from the eastern Llano uplift indicate that normal convergence occurred along the southern margin of Laurentia during a Grenville-age orogeny. The timing and style of Grenville orogenesis along the east-trending southern and north-northeast–trending eastern margins of Laurentia are very similar, but the normal convergence along both margins requires collision of two separate continental blocks from different directions at different times.

ACKNOWLEDGMENTS

This work was funded by National Science Foundation Grant EAR-9706692 to SM, James N. Connelly, and William D. Carlson. The authors also thank the Geology Foundation of the University of Texas, Austin, for further support.

REFERENCES CITED

Anderson, S.D., 2001, High-pressure metamorphism in the western Llano uplift recorded by garnet-clinopyroxenites in Mason County, Texas [M.A. thesis]: Austin, University of Texas, 221 p.

Barnes, V.E., 1981, Llano sheet, geologic atlas of Texas: Austin, University of Texas, Austin, U.S. Bureau of Economic Geology, 1 sheet.

Bebout, G.E., and Carlson, W.D., 1986, Fluid evolution and transport during metamorphism: Evidence from the Llano uplift, Texas: Contributions to Mineralogy and Petrology, v. 92, p. 518–529.

Bickford, M.E., Soegaard, K., Nielsen, K.C., and McLelland, J.M., 2000, Geology and geochronology of Grenville-age rocks in the Van Horn and Franklin Mountains area, West Texas; implications for the tectonic evolution of Laurentia during the Grenville: Geological Society of America Bulletin, v. 112, p. 1134–1148.

Bird, G.W., and Fawcett, J.J., 1973, Stability relations of Mg-chlorite-muscovite and quartz between 5 and 10 kb water pressure: Journal of Petrology, v. 14, p. 415–428.

Carlson, W.D., 1998, Petrologic constraints on the tectonic evolution of the Llano uplift, *in* Hogan, J.P., and Gilbert, M.C., eds., Basement tectonics 12: Proceedings of the 12th International Conference on Basement Tectonics: Dordrecht, The Netherlands, Kluwer Academic Press, p. 3–28.

Carlson, W.D., and Reese, J.F., 1994, Nearly pure iron staurolite in the Llano uplift and its petrologic significance: American Mineralogist, v. 79, p. 154–160.

Carlson, W.D., and Schwarze, E., 1997, Petrological significance of prograde homogenization of growth zoning in garnet: An example from the Llano uplift: Journal of Metamorphic Geology, v. 15, p. 631–644.

Carter, K., 1989, Grenville orogenic affinities in the Red Mountain area, Llano uplift, Texas: Canadian Journal of Earth Sciences, v. 26, p. 1124–1135.

Connelly, J.N., Rivers, T., and James, D.T., 1995, Thermotectonic evolution of the Grenville Province of western Labrador: Tectonics, v. 124, p. 202–217.

Corriveau, L., Heaman, L.M., Marcantonio, F., and van Breemen, O., 1990, 1.1 Ga K-rich alkaline plutonism in the southwest Grenville Province: Contributions to Mineralogy and Petrology, v. 105, p. 473–485.

Culotta, R.C., Pratt, T., and Oliver, J., 1990, A tale of two sutures: COCORP'S deep seismic surveys of the Grenville Province in the eastern U.S., mid-continent: Geology, v. 18, p. 646–649.

Dalziel, I.W.D., Mosher, S., and Gahagan, L.M., 2000, Laurentia-Kalahari collision and the assembly of Rodinia: Journal of Geology, v. 108, p. 499–513.

Davidson, A., 1986, New interpretation of the southwestern Grenville Province, *in* Moore, J.M., Jr., et al., eds., The Grenville Province: St. John's, Newfoundland, Geological Association of Canada Special Paper 31, p. 61–74.

Denny, J.H., and Gobel, V.W., 1986, Structural-geologic framework and emplacement of late Proterozoic Lone Grove and Northern Granite Mountain plutons, northeastern Llano uplift, central Texas, *in* The Geological Society of America annual meeting guidebook, field trip 13: Boulder, Colorado, Geological Society of America, p. 21–39.

Easton, R.M., 1986, Geochronology of the Grenville Province, *in* Moore, J.M., Jr., et al., eds., The Grenville Province: St. John's, Newfoundland, Geological Association of Canada Special Paper, v. 31, p. 127–173.

Fitz Gerald, J.D., and Stunitz, H., 1993a, Deformation of granitoids at low metamorphic grade. I: Reactions and grain size reduction: Tectonophysics, v. 221, p. 269–297.

Fitz Gerald, J.D., and Stunitz, H., 1993b, Deformation of granitoids at low metamorphic grade. II: Granular flow in albite-rich mylonites: Tectonophysics, v. 221, p. 299–324.

Garrison, J.R., 1981, Coal Creek serpentinite, Llano uplift, Texas: A fragment of an incomplete Precambrian ophiolite: Geology, v. 9, p. 225–230.

Gates, A.E., 1995, Middle Proterozoic dextral strike-slip event in the central Appalachians: Evidence from the Reservoir Fault, NJ: Journal of Geodynamics, v. 19, p. 195–212.

Gillis, G.M., 1989. Polyphase deformation of the Middle Proterozoic Coal Creek Serpentinite, Llano uplift, Texas [M.A. thesis]: Austin, University of Texas, 103 p.

Gower, C.F., 1996, The evolution of the Grenville Province in eastern Labrador, Canada, *in* Brewer, T.S., ed., Precambrian crustal evolution in the North Atlantic region: Boulder, Colorado, Geological Society of America Special Publication 112, p. 197–218.

Gower, C.F., Ryan, B., and Rivers, T., 1990, Mid-Proterozoic Laurentia-Baltica: An overview of its geological evolution and a summary of the contributions made by this volume, *in* Gower, C.F., et al., eds., Mid-Proterozoic Laurentia-Baltica: St. John's, Newfoundland, Geological Association of Canada Special Paper 38, p. 1–20.

Grimes, S.W., 1999, The Grenville orogeny in West Texas; structure, kinematics, metamorphism and depositional environment of the Carrizo Mountain Group [Ph.D. dissertation]: Austin, University of Texas, 372 p.

Grimes, S.W., and Mosher, S., 2003, Structure of the Carrizo Mountain group, southeastern Carrizo Mountains, west Texas: A transpressional zone of the Grenville orogen: Tectonics, v. 22, no. 1, 1003, doi 10.1029/200ITC001316.

Hoh, A.M., 2000, Deformational history of the Valley Spring domain in the northeastern Llano uplift, Devil's Waterhole, Inks Lake State Park, Burnet County, Texas [M.A. thesis]: Austin, University of Texas, 139 p.

Hunt, B.B., 2000, Mesoproterozoic structural evolution and lithologic investigation of the western Llano uplift, Mason County, central Texas [M.A. thesis]: Austin, University of Texas, 138 p.

Indares, A., 1993, Eclogized gabbros from the eastern Grenville Province: Textures, metamorphic context and implications: Canadian Journal of Earth Science, v. 30, p. 159–173.

Karlstrom, K.E., Ahall, K.-I., Harlan, S.S., Williams, M.L., MacLelland, J., and Geissman, J.W., 2001, Long-lived (1.8–1.0 Ga) convergent orogen in southern Laurentia, its extensions to Australia and Baltica, and implications for refining Rodinia: Precambrian Research, v. 111, p. 5–30.

Krogh, T.E., 1984, Precise U-Pb ages for Grenvillian and pre-Grenvillian thrusting of Proterozoic and Archean metamorphic assemblages in the Grenville front tectonic zone, Canada: Tectonics, v. 13, p. 963–982.

Mosher, S., 1993, Western extensions of Grenville age rocks: Texas, *in* Reed, J.C., Jr., et al., eds., Precambrian: Conterminous U.S.: Boulder, Colorado, Geological Society of America, Geology of North America, v. C-2, p. 365–378.

Mosher, S., 1996, Geology of the eastern Llano uplift, *in* Mosher, S., ed., Guide to the Precambrian geology of the eastern Llano uplift: Geological Society of America, 30th Annual south-central section meeting guidebook: Austin, Department of Geological Sciences, University of Texas, p. 3–7.

Mosher, S., 1998, Tectonic evolution of the southern Laurentian Grenville orogenic belt: Geological Society of America Bulletin, v. 110, p. 1357–1375.

Moore, J.M., Jr., and Thompson, P.H., 1980, The Flinton group: A late Precambrian metasedimentary succession in the Grenville Province of eastern Ontario: Canadian Journal of Earth Science, v. 17, p. 1685–1707.

Nelis, M.K., Mosher, S., and Carlson, W.D., 1989, Grenville-age orogeny in the Llano uplift of central Texas: Deformation and metamorphism of the Rough Ridge formation: Geological Society of America Bulletin, v. 101, p. 876–883.

Passchier, C.W., and Trouw, R.A.J., 1996, Microtectonics: Heidelberg, Springer-Verlag, 289 p.

Passchier, C.W., Myers, J.S., and Kröner, A., 1990, Field geology of high-grade gneiss terrains: Heidelberg, Springer-Verlag, 150 p.

Poirier, J.-P., 1985, Creep of crystals: Cambridge, U.K., Cambridge University Press, 260 p.

Pryer, L.L., 1993, Microstructures in feldspars from a major crustal thrust zone: The Grenville front, Ontario, Canada: Journal of Structural Geology, v. 15, p. 21–36.

Reed, R.M., 1999, Emplacement and deformation of late syn-orogenic, Grenville-age granites in the Llano uplift, central Texas [Ph.D. dissertation]: Austin, University of Texas, 271 p.

Reese, J.F., 1995, Structural evolution and geochronology of the southeastern Llano uplift, central Texas [Ph.D. dissertation]: Austin, University of Texas, 172 p.

Reese, J.F., and Mosher, S., 2004, Kinematic constraints on Rodinia reconstructions from the core of the Texas Grenville orogen: Journal of Geology, v. 112, p. 185–205.

Reese, J.F., Mosher, S., Connelly, J.N., and Roback, R.C., 2000, Mesoproterozoic chronostratigraphy of the southeastern Llano uplift, central Texas: Geological Society of America Bulletin v. 112, p. 278–291.

Rivers, T., Martignole, J., Gower, C.F., and Davidson, A., 1989, New tectonic divisions of the Grenville Province, southeast Canadian shield: Tectonics, v. 8, p. 63–84.

Roback, R.C., 1996a, Characterization and tectonic evolution of a Mesoproterozoic island arc in the southern Grenville orogen, Llano uplift, central Texas: Tectonophysics, v. 265, p. 29–52.

Roback, R.C., 1996b, Mesoproterozoic polymetamorphism and magmatism in the Llano uplift, central Texas: Geological Society of America Abstracts with Programs, v. 28, no. 7, p. 377.

Roback, R.C., and Carlson, W.D., 1996, Constraining the timing of high-*P* metamorphism in the Llano uplift through geochronology of eclogitic rocks: Geological Society of America Abstracts with Programs, v. 28, no. 1, p. 60–61.

Roback, R.C., Hunt, B, B., and Helper, M.A., 1999, Mesoproterozoic tectonic evolution of the western Llano uplift, central Texas; the story in an outcrop, *in* Frost, C.D., ed., Proterozoic magmatism of the Rocky Mountains and environs (Part I): Rocky Mountain Geology, v. 34, no. 2, p. 275–287.

Rougvie, J.R., Carlson, W.D., Copeland, P., and Connelly, J.N., 1999, Late thermal evolution of Proterozoic rocks in the northeastern Llano uplift, central Texas: Precambrian Research, v. 94, p. 49–72.

Seifert, F. 1970, Low-temperature compatibility relations of cordierite in haplopelites of the system K_2O-MgO-Al_2O_3-SiO_2-H_2O: Journal of Petrology, v. 11, p. 73–99.

Soegaard, K., and Callahan, D.M., 1994, Late Middle Proterozoic Hazel formation near Van Horn, West Texas: Evidence for transpressive deformation in Grenvillian basement: Geological Society of America Bulletin, v. 106, p. 413–423.

Spear, F.S., 1993, Metamorphic phase equilibria and pressure-temperature-time paths: Washington, D.C., Mineralogical Society of America, 799 p.

Stenzel, H.B., 1935, Pre-Cambrian structural conditions in the Llano region, *in* Sellards, E.H., and Baker, C.L., eds., The geology of Texas, vol. II: Structural and economic geology: University of Texas Bulletin, no. 3401, p. 74–79.

Tohver, E., van der Pluijm, B.A., van der Voo, R., Rizzotto, G., and Scandolara, J.E., 2002, Paleogeography of the Amazon craton at 1.2 Ga: Early Grenvillian collision with the Llano segment of Laurentia: Earth and Planetary Science Letters, v. 199, p. 185–200.

Tullis, J., 1990, Experimental studies of deformation mechanisms and microstructures in quartzo-feldspathic rocks, *in* Barber, D.J., and Meredith, P.G., eds., Deformation Processes in minerals, ceramics and rocks: London, Unwin Hyman, p. 190–227.

Walker, N., 1992, Middle Proterozoic geologic evolution of Llano uplift, Texas: Evidence from U-Pb zircon geochronometry: Geological Society of America Bulletin, v. 104, p. 494–504.

White and Mawer, 1986, Extreme ductility of feldspar from a mylonite, Parry Sound, Canada: Journal of Structural Geology, v. 8, p. 133–144.

Whitefield, C., 1997, A Sm-Nd isotopic study of the Coal creek domain and Sandy Creek shear zone [M.A. thesis]: Austin, University of Texas, 159 p.

Wilkerson, A., Carlson, W.D., and Smith, D., 1988, High-pressure metamorphism during the Llano orogeny inferred from Proterozoic eclogitic remnants: Geology, v. 16, p. 391–394.

Yardley, B.W.D., 1989, An introduction to metamorphic petrology: New York, John Wiley and Sons, 248 p.

Zumbro, J., 1999, A structural, petrologic, and geochemical investigation of the Valley Spring Gneiss of the southeastern Llano uplift, central Texas [M.A. thesis]: Austin, University of Texas, 428 p.

Manuscript Accepted by the Society August 25, 2003

Printed in the USA

Index

C

H

N

O

P

Q

R

S

T

U

V